8th
Edition

Calculus and Analytic Geometry

George B. Thomas, Jr.
Massachusetts Institute of Technology

Ross L. Finney
U.S. Naval Postgraduate School

Addison-Wesley Publishing Company

Reading, Massachusetts   Menlo Park, California   New York
Don Mills, Ontario   Wokingham, England   Amsterdam   Bonn
Sydney   Singapore   Tokyo   Madrid   San Juan   Milan   Paris

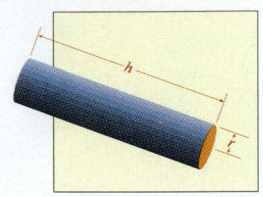 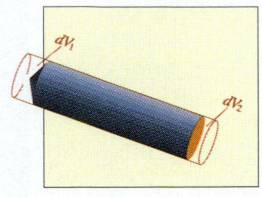 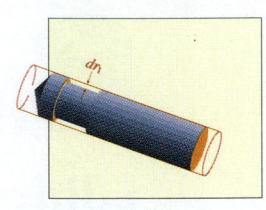 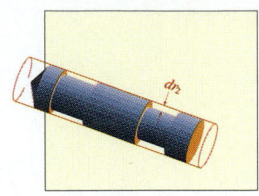

About the cover: Calculus is the mathematics of motion and change. We can use calculus to find out how rapidly the volume of a metal machine part changes as we cut a slot in it on a lathe. Or we can calculate the part's mass from a knowledge of its shape and density. Calculus can also tell us how precisely we must grind the metal to control its mass to one one-thousandth of a gram.

Sponsoring Editor:	Jerome Grant
Development Editor:	David M. Chelton
Managing Editor:	Karen Guardino
Production Supervisor:	Jack Casteel
Copy Editor:	Barbara Flanagan
Proofreader:	Joyce Grandy
Text Design:	Martha Podren
Art Consultant:	Joseph Vetere
Art Coordinator:	Connie Hulse
Electronic Illustration:	Tech-Graphics
Production Editorial Services:	Barbara Pendergast
Manufacturing Supervisor:	Roy Logan
Cover Design:	Marshall Henrichs

Library of Congress Cataloging-in-Publication Data

Thomas, George Brinton, 1914–
    Calculus and analytic geometry / by George B. Thomas, Jr. and Ross
    L. Finney — 8th ed.
      p. cm.
    Includes index.
    ISBN 0-201-52929-7 (set) –   ISBN 0-201-53286-7 (pt. 1).
    ISBN 0-201-53287-5 (pt. 2) –   ISBN 0-201-50900-8 (special edition)
    1. Calculus.   2. Geometry, Analytic.   I. Finney, Ross L. II. Title
QA303, T42   1992   91–8848
515', 15—cc20   CIP

Reprinted with corrections, May 1992.

2 3 4 5 6 7 8 9 10 DO 95949392

Contents

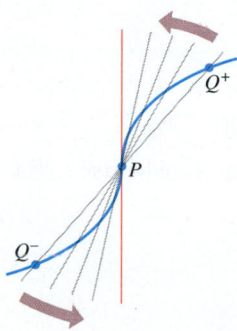

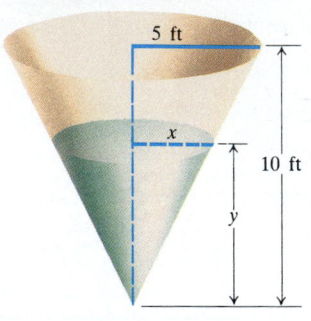
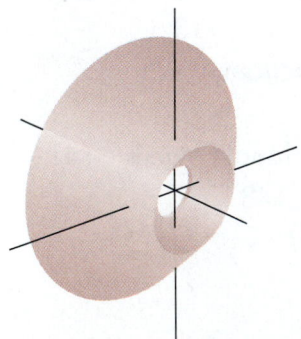

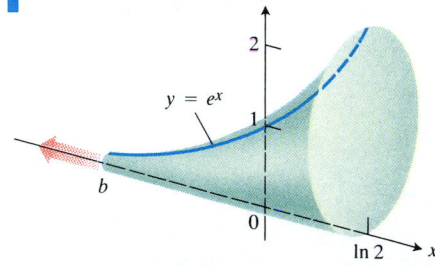

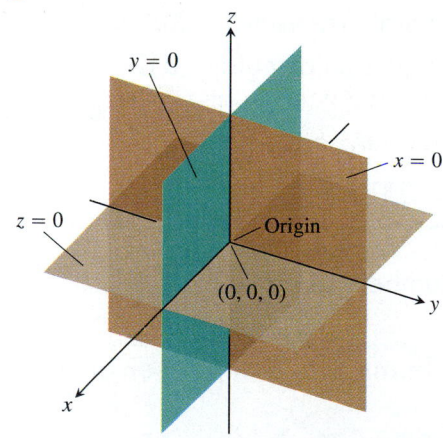

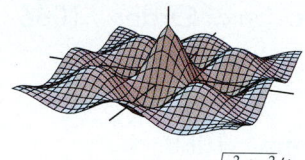
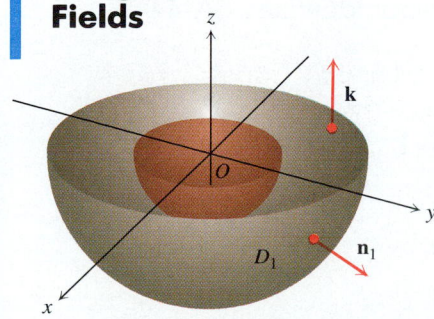

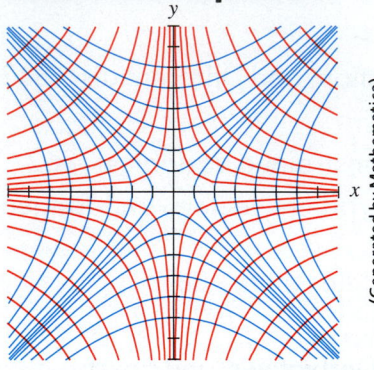

Preface

This eighth edition of *Calculus and Analytic Geometry* is intended for the standard three-semester or four-quarter calculus sequence in the freshman and sophomore years of college. The prerequisites are the usual studies of algebra and trigonometry, which are reviewed at the beginning of Chapter 1 and in the Appendices at the end of the book.

Our goals in this edition have been to shorten the text without omitting significant topics; to continue our emphasis on visualization as an aid to understanding calculus; and to offer more options for integrating technology into the mainstream course. We believe the result is a more exciting and more motivating book, while being easier for instructors to adapt to their particular teaching situations. Throughout the revision, we have maintained the mathematical level of previous editions, presenting calculus as a practical tool students will use in their later academic and professional careers. Many applications in the worked examples and in the exercises have been updated, but continue to illustrate the core mathematics students need to learn.

Content Changes for the Eighth Edition

Chapter 1 has a shorter review of precalculus topics, allowing readers to get to the derivative a little sooner. Section 1.3 shows that hand-held calculators can do more than just compute numerical answers; they can also be used to explore patterns of mathematical expressions. The idea of target values in Section 1.4 helps set up the epsilon-delta definition of limit in Section 1.7. Circles and parabolas are discussed in Chapter 1 because they occur often in illustrations of calculus concepts in the first semester.

Chapter 2 begins with the formal definition of derivative and then moves quickly to show how derivatives are calculated easily with rules based on the limit theorems presented in Chapter 1. Section 2.7 on Newton's Method contains a brief (optional) discussion of the fascinating area of chaos theory, illustrating the surprising mathematics that can sometimes lie behind simple equations.

The Mean Value Theorem has been moved nearer the beginning of **Chapter 3** to give the theoretical justification before rather than after the applications of derivatives. The result is a shorter and tighter presentation of this material. At the same time, the chapter starts with related rates of change to give students earlier practice in applying derivatives to realistic situations.

Integration begins with the indefinite integral in **Chapter 4,** including a discussion of initial-value problems and mathematical modeling. The idea of using integration in important mathematical models is woven throughout **Chapter 5.** We have tried to emphasize the recurring pattern of formulating an appropriate Riemann sum, finding its limit, and comparing the limit with previously known results, so students will be better able to deal with kinds of problems they have not seen before.

Chapter 6 on transcendental functions includes material on hyperbolic functions, which has been reduced from the coverage in the previous edition. Section 6.6 on rates of growth has several examples on measuring the effectiveness of algorithms, with applications to computer science.

The presentation of techniques of integration in **Chapter 7** has been condensed without omitting any significant topics. Again, emphasis is on teaching general problem-solving methods applicable to situations not explicitly covered in the text, so that students will be prepared to solve problems in new contexts.

Sequences and series have been condensed into one chapter, **Chapter 8,** again without omitting major topics. (Several minor discussions have been deleted.) We have tried to emphasize the importance of Taylor and Maclaurin series from the start of the chapter, pointing to them as the motivating reason for studying some of the chapter's other topics.

The presentation of analytic geometry in **Chapter 9** has been combined with parametric equations, plane curves, and polar coordinates. While these topics are important in calculus, their coverage has been reduced to allow room for other topics now demanding place in the crowded mainstream calculus syllabus.

Chapters 10 and 11 on vectors and vector-valued functions remain largely unchanged from the previous edition. Drawing lessons to help students visualize and draw in three dimensions are incorporated into Chapter 11.

Functions of two or more variables are discussed in **Chapter 12,** in an organization meant to parallel the presentation of the single-variable material. Applications of partial differential equations to physics and engineering appear in several exercise sets.

Chapters 13 and 14 on multiple integrals and integration in vector fields are much the same as in the seventh edition. However, **Chapter 15** on differential equations has been cut back to serve more as an introduction to a topic that is usually taught as a separate course.

Features of the Text

Mathematical level: We have taken care to maintain the level of rigor of previous editions while striving for a more informal and accessible writing style. We try to explain things carefully without either belaboring the obvious or jumping ahead of the reader's understanding.

Art and design: Our introduction of four-color figures in the seventh edition was well received by our adopters. This edition is now completely four-color. As before, color is used pedagogically to highlight the most important parts of figures and help students visualize in three dimensions. We have increased the number of figures over the seventh edition to appeal to students' geometric intuition in explaining basic principles. Many complicated graphs and surfaces have been rendered in *Mathematica* ® to ensure their accuracy. Overall, this edition has more figures and more exercises that require students to draw and interpret figures.

Applications: It has been a hallmark of this book through the years that we illustrate applications of calculus with real data based on situations students are likely to encounter later on. Most of our examples and exercises are directed toward science and engineering, such as calculating numerically the cross-section area of a solar-powered car (p. 306).

Worked examples: We have explicitly shown more of the steps involved in the worked examples, as well as the reasoning that leads from one step to the next. In addition, we have made a special point of correlating examples with the stated problem-solving strategies, so students see how these strategies work in practice.

Exercise sets: We have added new exercises in this edition at several levels of difficulty, from drill questions to challenging problems. Many of the new exercises require a calculator or computer grapher for their solution, and are so labeled. Our goal has not been to take a group of exercises and simply graph them, but to use the power of computer graphing to explore new ideas and new relations. We continue to include sets of **Review Questions** that ask students to think about the concepts presented without trying to calculate numerical answers. And we still end each chapter with a collection of **Miscellaneous Exercises** that extend the student's grasp of the material in new ways.

Integration of technology: Students do not need a calculator or computer grapher to use this book. However, for those who do have access to technological aids and would like to use them, we have included several technological features in this edition. We have already mentioned exercises for calculators and computer graphers. In addition, several optional computer programs, written in BASIC, have been added. Many sections of the text also include a list of programs from the **Calculus Explorer** that may be used as computational aids both for working the exercises in that section and for general exploration. Additional information about software for calculus is presented in the Supplements portion of this Preface.

Enhancements to learning: In the seventh edition we introduced a number of strategy boxes to summarize particular problem-solving methods, such as those used in solving related rates of change problems, optimization problems, and so on. We have expanded the number of these boxes to include more types of problems.

Several **Drawing Lessons** were also introduced in the seventh edition, aimed at helping students draw lines, planes, and curved surfaces in three dimensions. We have added lessons for drawing curves in two dimensions as well.

One further pedagogical feature of the seventh edition was the occasional use of flowcharts to summarize methods of attacking extended calculus problems, such as the problem of determining series convergence. This feature, too, has been expanded, to include flowcharts on such topics as l'Hôpital's rule and indefinite integration.

Supplements for the Instructor

Complete Solutions Manual: This two-volume supplement contains the worked-out solutions for all exercises in the text.

Complete Answer Book: Contains the answers to all exercises in the text.

OmniTest: Based on the learning objectives of the text, this computerized testing system allows the instructor to generate tests or quizzes easily. As an algorithm-driven system, OmniTest easily creates multiple versions of the same test by automatically inserting random numbers into model problems. While the numbers are random, they are constrained to produce reasonable answers. Questions may be selected in any combination of open-ended, multiple-choice, and true-false formats. Instructors may assign an instructor code and level of difficulty to each model problem, and may print tests with variable spacing. OmniTest is available for MacIntosh, and IBM PC/compatibles.

Printed Test Bank: Three versions of tests for each chapter in the text are included in multiple-choice, open-ended, and true-false formats. The Printed Test Bank also contains printed answer keys and student worksheets for each test.

Transparency Masters: Includes a selection of key definitions, theorems, proofs, formulas, tables, and figures. These may be copied onto transparency acetates for overhead projection.

Supplements for the Student

Student Study Guide: By Maurice Weir, Naval Postgraduate School. Organized to correspond with the text, this workbook in a semi-programmed format increases student proficiency.

Student Solutions Manual: By Thomas Cochran and Michael Schneider, of Belleville Area College. This manual is designed for the student and contains carefully worked-out solutions to all of the odd-numbered exercises in the text.

Options for Integrating Technology

Analyzer*:   This program is a tool for exploring functions in calculus and many other disciplines. It can graph a function of a single variable and overlay graphs of other functions. It can differentiate, integrate, or iterate a function. It can find roots, maxima and minima, and inflection points, as well as vertical asymptotes. In addition, Analyzer* can compose functions, graph polar and parametric equations, make families of curves, and make animated sequences with changing parameters. It exploits the unique flexibility of the MacIntosh wherever possible, allowing input to be either numeric (from the keyboard) or graphic (with a mouse). Analyzer* runs on MacIntosh II, Plus, and SE computers.

The Calculus Explorer: Consisting of 27 programs ranging from functions to vector fields, this software enables the instructor and student to use the computer as an "electronic chalkboard." The Explorer is highly interactive and allows for manipulation of variables and equations to provide graphical visualization of mathematical relationships that are not intuitively obvious. The Explorer provides user-friendly operation through an easy-to-use menu-driven system, extensive on-line documentation, superior graphics capability, and fast operation. An accompanying manual includes sections covering each program, with appropriate examples and exercises. Available for IBM PC/compatibles.

InSight: A calculus demonstration software program that enhances understanding of calculus concepts graphically. The program consists of ten simulation scenarios. Each scenario presents an application and takes the user through the solution visually. The format is interactive. Available for IBM PC/compatibles.

The Student Edition of DERIVE®:   A streamlined version of the professional program, the Student Edition of DERIVE is a powerful, yet exceptionally easy-to-use computer algebra system for numerical, symbolic, and graphical computation. Its menu-driven interface and on-line help make the software user-friendly and easy to learn. The accompanying manual is pedagogically oriented and introduces users to the capabilities of the program step by step. The Student Edition of DERIVE® frees students from performing long mathematical calculations by hand, thereby enhancing their learning experience by allowing them to spend more time in mathematical exploration. Available for IBM PC/compatibles.

The Student Edition of MathCAD®: A powerful free-form scratchpad, this Student Edition has all the problem-solving capabilities of the professional version of MathCAD®. Available for the IBM PC/compatibles.

Master Grapher and 3D Grapher: A powerful, interactive graphing utility for functions, polar equations, parametric equations, and other functions in two and three variables. Prepared by Franklin Demana and Bert Waits of Ohio State University. Available for MacIntosh, Apple, and IBM PC/compatibles.

Mathematica® Laboratories for Calculus I Using Mathematica: By Margaret Höft, The University of Michigan–Dearborn. An inexpensive collection of *Mathematica* lab experiments consisting of material usually covered in the first term of the calculus sequence.

Math Explorations Series: Each manual provides problems and explorations in calculus. Intended for self-paced and laboratory settings, these books are an excellent complement to Thomas/Finney.

Exploring Calculus with a Graphing Calculator, Second Edition, by Charlene E. Beckmann and Ted Sundstrom of Grand Valley State University.

Exploring Calculus with Mathematica®, by James K. Finch and Millianne Lehmann of the University of San Francisco.

Exploring Calculus with Derive®, by David C. Arney of the United States Military Academy at West Point.

Exploring Calculus with the IBM® PC Version 2.0, by John B. Fraleigh and Lewis I. Pakula of the University of Rhode Island.

Acknowledgments

We would like to thank and acknowledge the contributions of our revision planning survey participants. Their comments and suggestions helped to shape our ideas for this Eighth Edition.

Linda Allen, Texas Tech University

William Arlinghaus, Lawrence Technological University

Lewis D. Blake, Duke University

Phyllis Boutilier, Michigan Technological University

Major James Boutner, U.S. Air Force Academy

Wayne Britt, Louisiana State University

David H. Carlson, San Diego State University

Misium Castroconde, University of California–Irvine

Robert Connelly, Cornell University

Robin Gottlieb, Harvard University

Hiroshi Gunji, University of Wisconsin–Madison

Leon M. Hale, University of Missouri–Rolla

Jennifer Johnson, University of Utah

Jeuel LaTorre, Clemson University

John Lawlor, University of Vermont

John C. Mainhuber, University of Maine

Francis J. Narcowick, Texas A & M University

Charles Okonkwo, Arizona State University

Cathryn Olsen, State University of New York–Buffalo

Sanford Segal, University of Rochester

Betty Travis, University of Texas–San Antonio

Constantine Tsatsos, New Jersey Institute of Technology

Paul Tseng, University of Washington

Jaak Vilms, Colorado State University

Bertram Walsh, Rutgers University

Dr. Richard Wheeden, Rutgers University

Many valuable contributions to this Eighth Edition were made by people who reviewed the manuscript as it developed through its various stages.

Martin Bartelt, Christopher Newport College

Therlene Boyett, Valencia Community College

Fred Brauer, University of Wisconsin–Madison

David Collingwood, University of Washington

S. A. R. Disney, University of New South Wales

John Erdman, Portland State University

David Furuto, Brigham Young University–Hawaii Campus

Stuart Goldenberg, California Polytechnic State University–San Luis Obispo

Ralph Grimaldi, Rose-Hulman Institute of Technology

Bernard Harris, Northern Illinois University

Arnold Insel, Illinois State University

David Johnson, Lehigh University

Cecilia Knoll, Florida Institute of Technology

J. J. Koliha, University of Melbourne

Elton Lacey, Texas A & M University

James Lang, Valencia Community College

Melvin D. Lax, California State University–Long Beach

Millianne Lehmann, University of San Francisco

Stanley M. Luckawecki, Clemson University

Arthur Moore, Orange Coast College

Daniel Moran, Michigan State University

James Nicholson, Clemson University

James Osterburg, University of Cincinnati

F. J. Papp, University of Michigan–Dearborn

Thomas W. Rishel, Cornell University

J. Rod Smart, University of Wisconsin–Madison

Kirby C. Smith, Texas A & M University

Joseph Stephen, Northern Illinois University

Monty J. Strauss, Texas Tech University

Sally Thomas, Orange Coast College

Henry Zatzkis, New Jersey Institute of Technology

We would particularly like to express our gratitude to Richard A. Askey, University of Wisconsin–Madison, and Richard W. Hamming, U.S. Naval Postgraduate School, for their continuing and thoughtful advice, and to thank Curtis F. Gerald, California Polytechnic State University, Emeritus, for developing the computer programs in this edition.

We want to express our special appreciation for the generous advice and help given to us by Thomas Cochran and Michael Schneider, Department of Mathematics, Belleville Area College, as they developed the text's answer section and solutions manuals.

We owe a special thanks to Charles Slavin of the University of Maine at Orono and James Lang of Valencia Community College for proofreading the book in galley pages.

We would also like to thank Laura R. Finney for keyboarding the manuscript and for proofreading the book in manuscript, galleys, and pages.

We also wish to express our thanks to the many other contributors whose names we have not been able to mention.

Any errors that appear are the responsibility of the authors. We will appreciate having these brought to our attention.

State College, PA
Monterey, CA

G.B.T., Jr.
R.L.F.

Prologue: What Is Calculus?

Calculus is the mathematics of motion and change. Where there is motion or growth, where variable forces are at work producing acceleration, calculus is the right mathematics to apply. This was true in the beginnings of the subject, and it is true today.

Calculus was first created to meet the mathematical needs of the scientists of the seventeenth century. Differential calculus dealt with the problem of calculating rates of change. It enabled people to define slopes of curves, to calculate the velocities and accelerations of moving bodies, to find the firing angle that gave a cannon its greatest range, and to predict the times when planets would be closest together or farthest apart. Integral calculus dealt with the problem of determining a function from information about its rate of change. It enabled people to calculate the future location of a body from its present position and a knowledge of the forces acting on it, to find the areas of irregular regions in the plane, to measure the lengths of curves, and to locate the centers of mass of arbitrary solids.

Before the mathematical developments that culminated in the great discoveries of Sir Isaac Newton (1642–1727) and Baron Gottfried Wilhelm Leibniz (1646–1716), it took the astronomer Johannes Kepler (1571–1630) twenty years of concentration, record-keeping, and arithmetic to discover the three laws of planetary motion that now bear his name:

1. Each planet travels in an ellipse that has one focus at the sun (Fig. P.1).

2. The radius vector from the sun to a planet sweeps out equal areas in equal intervals of time.

3. The squares of the periods of revolution of the planets about the sun are proportional to the cubes of their orbits' semimajor axes. If T is the length of a planet's year and a is the semimajor axis of its orbit, then the ratio T^2/a^3 has the same constant value for all planets in the solar system.

With calculus, deriving Kepler's laws from Newton's laws of motion is but an afternoon's work. Kepler described how the solar system worked—Newton and Leibniz, with their calculus, explained why.

Today, calculus and its extensions in mathematical analysis are far reaching indeed, and the physicists, mathematicians, and astronomers who first invented the subject would surely be amazed and delighted, as we hope you will be, to see what a profusion of problems it solves and what a wide range of fields now use it in the mathematical models that bring understanding about the universe and the world around us.

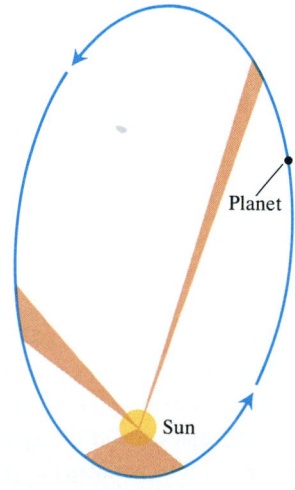

P.1 A planet moving about its sun. The shaded regions have equal areas. According to Kepler's second law, the planet takes the same amount of time to traverse the curved outer boundary of each region. The planet therefore moves faster near the sun than it does farther away.

(a)

(b)

P.2   Calculus helped us predict that moons would travel in elliptical orbits about their planets; it also helped us to launch cameras and telescopes to observe the planets of our solar system. This photograph of Jupiter (a), taken on February 13, 1979, by the Voyager I space probe, shows two of its moons, Io (left) and Europa. Photograph (b) shows the Astro-1 group of telescopes on board the space shuttle Columbia in December 1990. One of the goals of this mission was to investigate the magnetic fields of Jupiter. We describe the effects of magnetic fields on moving electrical charges with calculus.

Economists use calculus to forecast global trends. Oceanographers use calculus to formulate theories about ocean currents and meteorologists use it to describe the flow of air in the upper atmosphere. Biologists use calculus to forecast population size and to describe the way predators like foxes interact with their prey. Medical researchers use calculus to design ultrasound and x-ray equipment for scanning the internal organs of the body. Space scientists use calculus to design rockets and explore distant planets. Psychologists use calculus to understand optical illusions in visual perception. Physicists use calculus to design inertial navigation systems and to study the nature of time and the universe. Hydraulic engineers use calculus to find safe closure patterns for valves in pipelines. Electrical engineers use it to design stroboscopic flash equipment and to solve the differential equations that describe current flow in computers. Sports equipment manufacturers use calculus to design tennis rackets and baseball bats. Stock market analysts use calculus to predict prices and assess interest rate risk. Physiologists use calculus to describe electrical impulses in neurons in the human nervous system. Drug companies use calculus to determine profitable inventory levels and timber companies use it to decide the most profitable time to harvest trees. The list is practically endless, for almost every professional field today uses calculus in some way.

"The calculus was the first achievement of modern mathematics," wrote John von Neumann (1903–1957), one of the great mathematicians of the present century, "and it is difficult to overestimate its importance. I think it defines more unequivocally than anything else the inception of modern mathematics; and the system of mathematical analysis, which is its logical development, still constitutes the greatest technical advance in exact thinking."*

World of Mathematics, Vol. 4 (New York: Simon and Schuster, 1960), "The Mathematician," by John von Neumann, pp. 2053-2063.

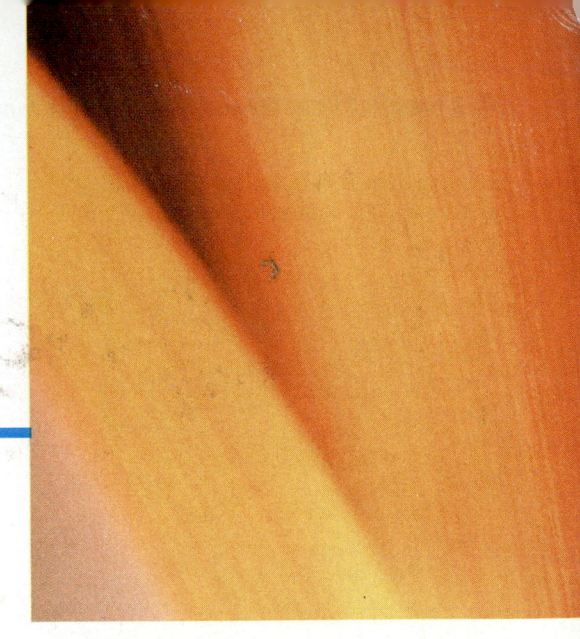

THE RATE OF CHANGE OF A FUNCTION

Overview   This chapter reviews the most important things you need to know to start learning calculus. It also discusses some of the relations between calculus and computation and introduces one of the chief ideas of calculus, the derivative, as a way to describe rates of change and slopes of curves.

Calculus is built on the notion of limit, and derivatives are just one of the many kinds of limits calculus deals with. The rules for calculating limits are straightforward; most of the limits we need can be found with a combination of direct substitution and algebra. Proving that the calculation rules always work, however, is a more subtle affair that requires a formal definition of limit.

One of the important uses of limits in calculus is to test functions for continuity. Continuous functions are widely used in science because they serve to model an enormous range of natural behavior. In Section 1.10, we shall see what makes continuous functions special and we shall work mainly with continuous functions thereafter.

1.1 Cartesian Coordinates and Equations for Lines

We set the stage for calculus by assigning numerical coordinate pairs to all points in the plane. The coordinates make it possible to describe lines and other important curves with coordinate equations.

We begin with two number lines that cross at their zero points at right angles. Each line represents the real numbers, which are the numbers that can be represented by decimals. Figure 1.1 shows the usual way of drawing the lines, with one line horizontal and the other vertical. The horizontal line is called the **x-axis** and the vertical line the **y-axis.** The point at which the lines cross is the **origin.**

On the x-axis, the positive number a lies a units to the right of the origin, and the negative number $-a$ lies a units to the left of the origin. On the y-axis, the positive number b lies b units above the origin while the negative number $-b$ lies b units below the origin.

The coordinates defined here are often called **Cartesian** coordinates, in honor of their chief inventor, René Descartes (1596–1650).

1.1   In Cartesian coordinates, the scaling on each axis is symmetric about the origin.

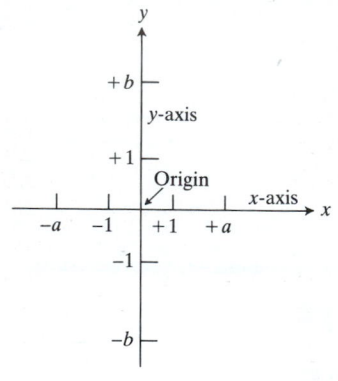

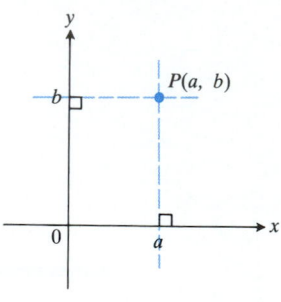

1.2   The ordered pair (a, b) corresponds to the point where the perpendicular to the x-axis at a crosses the perpendicular to the y-axis at b.

With the axes in place, we assign a pair (a, b) of real numbers to each point P in the plane. The number a is the number at the foot of the perpendicular from P to the x-axis. The number b is the number at the foot of the perpendicular from P to the y-axis (Fig. 1.2). The notation (a, b) is read "$a\, b$."

The number a from the x-axis is the **x-coordinate** of P. The number b from the y-axis is the **y-coordinate** of P. The pair (a, b) is the **coordinate pair** of the point P. It is an **ordered pair,** with the x-coordinate first and y-coordinate second. To show that P has the coordinate pair (a, b), we sometimes write the P and (a, b) together: $P(a, b)$.

The construction that assigns an ordered pair of real numbers to each point in the plane can be reversed to assign a point in the plane to each ordered pair of real numbers. The point assigned to the pair (a, b) is the point where the perpendicular to the x-axis at a crosses the perpendicular to the y-axis at b. Thus, the assignment of coordinates is a one-to-one correspondence between the points of the plane and the set of all ordered pairs of real numbers.

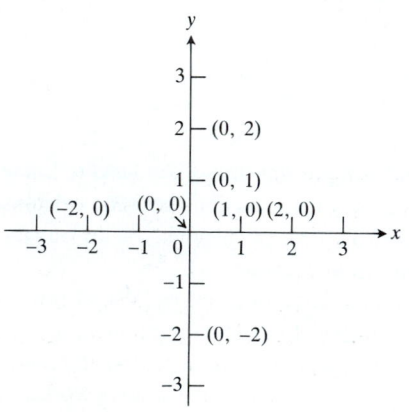

1.3   Points on the axes can be labeled in two ways.

The points on the coordinate axes now have two kinds of numerical labels: single numbers from the axes and paired numbers from the plane (Fig. 1.3). Every point on the x-axis has a zero y-coordinate, and every point on the y-axis has a zero x-coordinate. The origin is the point $(0, 0)$.

Directions and Quadrants

Motion from left to right along the x-axis is said to be motion in the **positive x-direction.** Motion from right to left is in the **negative x-direction.** Along the y-axis, the positive direction is up, and the negative direction is down. The coordinates of each point in the plane give the point's **directed distances** from the coordinate axes. The x-coordinate is the directed distance from the y-axis; the y-coordinate is the directed distance from the x-axis.

The origin divides the x-axis into the **positive x-axis** to the right of the origin and the **negative x-axis** to the left of the origin. Similarly, the origin divides the y-axis into the **positive y-axis** and the **negative y-axis.** The axes divide the plane into four regions called **quadrants,** numbered I, II, III, and IV (Fig. 1.4).

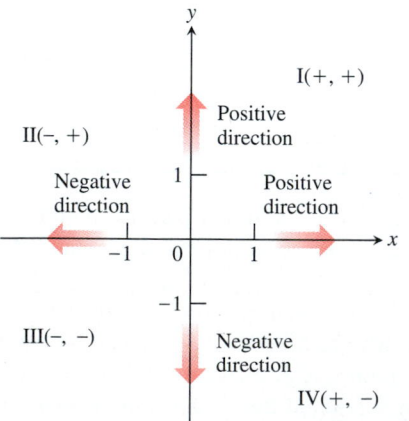

1.4   Directions along the axes: The values of x and y increase in the positive direction and decrease in the negative direction. Roman numerals label the quadrants.

A Word About Scales

When we plot data in the coordinate plane or graph formulas whose variables have different units of measure, we do not need to use the same scale on the two axes. If we plot time vs. thrust for a rocket motor, for example, there is no reason to place the mark that shows 1 sec on the time axis the same distance from the origin as the mark that shows 1 lb on the thrust axis.

When we graph functions whose variables do not represent physical measurements and when we draw figures in the coordinate plane to study their geometry and trigonometry, we try to make the scales on the axes identical. A vertical unit of distance then looks the same as a horizontal unit. As on a surveyor's map or a scale drawing, line segments that are supposed to have the same length will look as if they do and angles that are supposed to be congruent will look congruent.

Computer displays and calculator displays are another matter. The vertical and horizontal scales on machine-generated graphs usually differ, and there are corresponding distortions in distances, slopes, and angles. Circles may look like ellipses, rectangles may look like squares, right angles may appear to be acute or obtuse, and so on. Circumstances like these require us to take extra care in interpreting what we see.

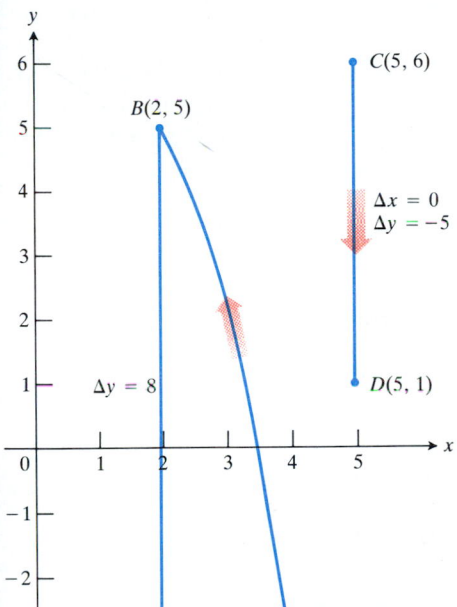

1.5 From A to B, $\Delta x = -2$ and $\Delta y = 8$. From C to D, $\Delta x = 0$ and $\Delta y = -5$.

Increments

One of the many reasons calculus has proved to be so useful over the years is that it is the mathematics that makes the connection between rates of change and slopes of smooth curves. Explaining this connection is one of the goals of the present chapter. The basic plan is first to define what we mean by the slope of a line and then to define the slope of a curve at each point on the curve to be the limit of slopes of selected secant lines through that point. Just how that is done will become clear as the chapter goes on. Our first step is to find a practical way to calculate the slopes of lines. We find it in the notion of increment.

When a particle in the plane moves from one point to another, the net changes or increments in its coordinates are found by subtracting the coordinates of the point where the particle starts from the coordinates of the point where the particle stops.

Example 1   From $A(4, -3)$ to $B(2, 5)$, the increments in the x- and y-coordinates (Fig. 1.5) are

$$\Delta x = 2 - 4 = -2, \qquad \Delta y = 5 - (-3) = 8.$$

The symbols Δx and Δy in Example 1 are read "delta x" and "delta y." The letter Δ is a Greek capital d, for "difference." Neither Δx nor Δy denotes multiplication; Δx is not "delta times x" nor is Δy "delta times y."

DEFINITION

> **Increments** are net changes. When a particle moves from (x_1, y_1) to (x_2, y_2), the increments in its coordinates are
>
> $$\Delta x = x_2 - x_1 \qquad \text{and} \qquad \Delta y = y_2 - y_1. \qquad (1)$$

Example 2   From $C(5, 6)$ to $D(5,1)$ the increments (Fig. 1.5) are

$$\Delta x = 5 - 5 = 0, \qquad \Delta y = 1 - 6 = -5.$$

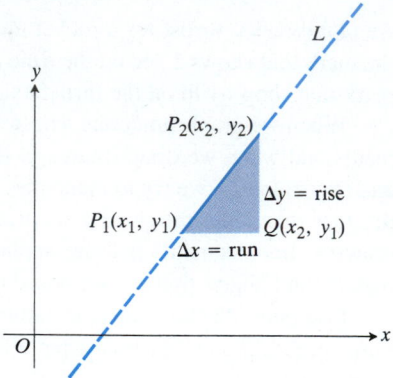

1.6   The slope of the line is
$$m = \frac{\Delta y}{\Delta x} = \frac{\text{rise}}{\text{run}}.$$

Slopes of Nonvertical Lines

All lines except vertical lines have slopes. We calculate slopes from changes in coordinates. Once we see how this is done, we shall also see why vertical lines are an exception.

To begin, let L be a nonvertical line in the plane. Let $P_1(x_1, y_1)$ and $P_2(x_2, y_2)$ be two points on L (Fig. 1.6). We call $\Delta y = y_2 - y_1$ the **rise** from P_1 to P_2 and $\Delta x = x_2 - x_1$ the **run** from P_1 to P_2. Since L is not vertical, $\Delta x \neq 0$ and we may define the **slope** of L to be $\Delta y / \Delta x$, the amount of rise per unit of run. It is conventional to denote the slope by the letter m.

DEFINITION

> The **slope** of a nonvertical line is
> $$m = \frac{\text{rise}}{\text{run}} = \frac{\Delta y}{\Delta x} = \frac{y_2 - y_1}{x_2 - x_1}. \qquad (2)$$

Suppose that instead of choosing the points P_1 and P_2 to calculate the slope in Eq. (2), we choose a different pair of points P'_1 and P'_2 on L and calculate

$$m' = \frac{y'_2 - y'_1}{x'_2 - x'_1} = \frac{\Delta y'}{\Delta x'}. \qquad (3)$$

Will we get the same value for the slope? In other words, will m' equal m? The answer is yes because m and m' are the ratios of corresponding sides of similar triangles (Fig. 1.7). The slope of a line depends only on how steeply the line rises or falls and not on the points we use to calculate it.

1.7   Because triangles $P_1 Q P_2$ and $P'_1 Q' P'_2$ are similar,
$$m' = \frac{\Delta y'}{\Delta x'} = \frac{\Delta y}{\Delta x} = m.$$

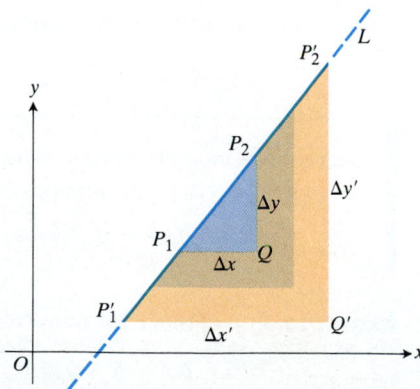

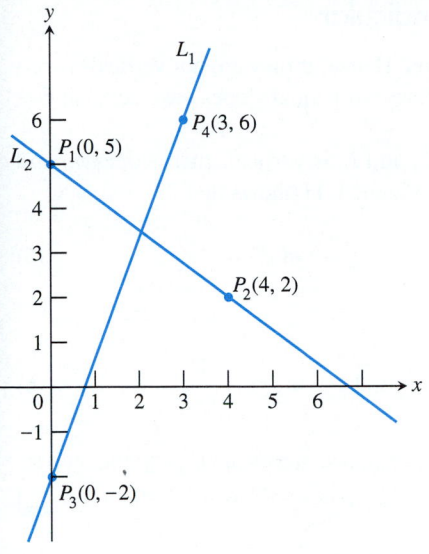

A line that goes uphill as x increases has a positive slope. A line that goes downhill as x increases has a negative slope. (See Fig. 1.8.) A horizontal line has slope zero. The points on it all have the same y-coordinate, so $\Delta y = 0$.

The formula $m = \Delta y / \Delta x$ does not apply to vertical lines because Δx is zero along a vertical line. We express this by saying that vertical lines have no slope or that the slope of a vertical line is undefined.

Angles of Inclination

The **angle of inclination** of a line that crosses the x-axis is the smallest angle we get when we measure counterclockwise from the x-axis around the point of intersection (Fig. 1.9). The angle of inclination of a horizontal line is taken to be $0°$. Thus, an angle of inclination may have any measure from $0°$ up to but not including $180°$.

The slope of a nonvertical line is the tangent of the line's angle of inclination. Figure 1.10 shows why this is true. If m denotes the slope and ϕ the angle, then

$$m = \tan \phi. \tag{4}$$

1.8   The slope of L_1 is
$$m = \frac{\Delta y}{\Delta x} = \frac{6 - (-2)}{3 - 0} = \frac{8}{3}.$$

This means that $3\Delta y = 8\Delta x$ for every change of position on the line.

The slope of L_2 is
$$m = \frac{\Delta y}{\Delta x} = \frac{2 - 5}{4 - 0} = \frac{-3}{4}.$$

This means that y decreases 3 units every time x increases 4 units.

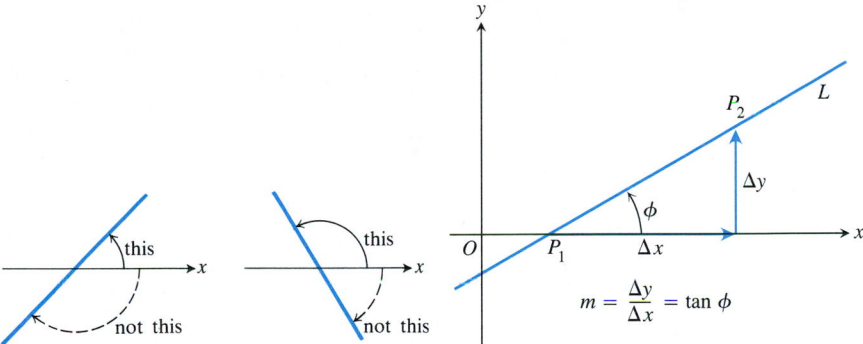

1.9   Angles of inclination are measured counterclockwise from the x-axis.

1.10   The slope of a nonvertical line is the tangent of its angle of inclination.

Railroads and Highways

Civil engineers calculate the slope of a roadbed by calculating the ratio of the distance it rises or falls to the distance it runs horizontally. They call this ratio the **grade** of the roadbed, usually written as a percentage. Along the coast, railroad grades are usually less than 2%. In the mountains, they may go as high as 4%. Highway grades are usually less than 5%.

In analytic geometry we calculate slopes the same way, but we usually do not express them as percentages.

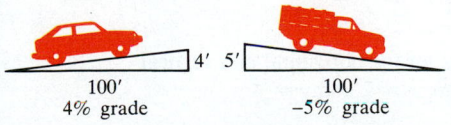

100'
4% grade

100'
−5% grade

Example 3   The slopes of lines become increasingly large as their angles of inclination approach $90°$ from either side. A few keystrokes on a calculator will show you that, after rounding,

ϕ approaching 90° from below	ϕ approaching 90° from above
$\tan 89.9° = 573$	$\tan 90.1° = -573$
$\tan 89.99 = 5730$	$\tan 90.01 = -5730$
$\tan 89.999 = 57300$	$\tan 90.001 = -57300$
$\tan 89.9999 = 57300\ 0$	$\tan 90.0001 = -57300\ 0$

We say that the slope of a line "becomes infinite" as its angle of inclination approaches $90°$. However, vertical lines themselves have no slope.

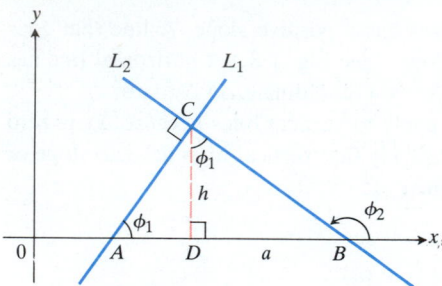

1.11 ΔADC is similar to ΔCDB. Hence ϕ_1 is also the upper angle in ΔCDB. From the sides of ΔCDB, we read $\tan \phi_1 = a/h$.

Lines That Are Parallel or Perpendicular

Parallel lines have equal angles of inclination. Hence, if they are not vertical, parallel lines have the same slope. Conversely, lines with equal slopes have equal angles of inclination and are therefore parallel.

If neither of two perpendicular lines L_1 and L_2 is vertical, their slopes m_1 and m_2 are related by the equation $m_1 m_2 = -1$. Figure 1.11 shows that

$$m_1 = \tan \phi_1 = \frac{a}{h}, \qquad \text{while} \qquad m_2 = \tan \phi_2 = -\frac{h}{a}. \qquad (5)$$

Hence,

$$m_1 m_2 = \left(\frac{a}{h}\right)\left(-\frac{h}{a}\right) = -1 \qquad \text{and} \qquad m_2 = -\frac{1}{m_1}. \qquad (6)$$

Example 4   The slope of a line perpendicular to a line of slope 3/4 is $-4/3$.

Equations for Lines

DEFINITION

An **equation for a line** is an equation that is satisfied by the coordinates of every point on the line but is not satisfied by the coordinates of points that lie elsewhere.

Horizontal and Vertical Lines

The standard equations for the horizontal and vertical lines through a point (a, b) are simply $y = b$ and $x = a$ (Fig. 1.12). A point (x, y) lies on the horizontal line through (a, b) if and only if $y = b$. It lies on the vertical line through (a, b) if and only if $x = a$.

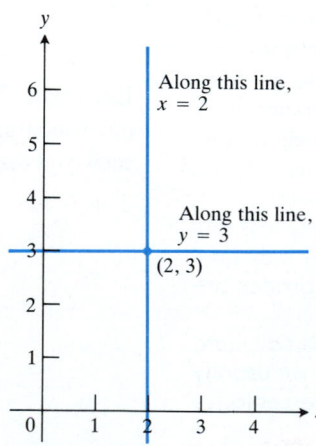

1.12   The standard equations for the horizontal and vertical lines through the point $(2, 3)$ are $x = 2$ and $y = 3$.

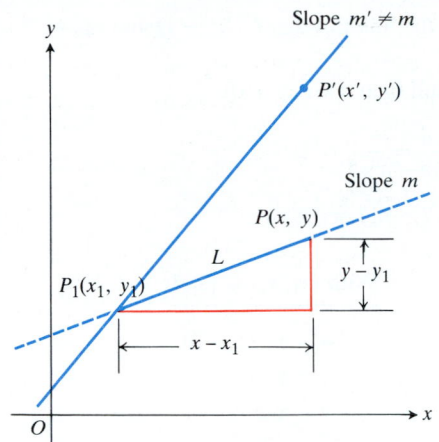

1.13 If L is the line through $P_1(x_1, y_1)$ whose slope is m, then other points $P(x, y)$ lie on this line if and only if slope $PP_1 = m$. This fact gives us the point–slope equation for L.

Point–Slope Equations

To write an equation for a nonvertical line L, it is enough to know its slope m and the coordinates of a point $P_1(x_1, y_1)$ on it. If $P(x, y)$ is any other point on L (Fig. 1.13), then $x \neq x_1$ and we can write the slope of L as

$$\frac{y - y_1}{x - x_1}. \tag{7}$$

We can then set this expression equal to m to get

$$\frac{y - y_1}{x - x_1} = m, \tag{8}$$

or

$$y - y_1 = m(x - x_1). \tag{9}$$

Equation (9) is an equation for L, as we can check right away. Every point (x, y) on L satisfies the equation—even the point (x_1, y_1). What about points not on L? If $P'(x', y')$ is a point not on L (Fig. 1.13), then the slope m' of $P'P_1$ is different from m, and the coordinates x' and y' of P' do not satisfy Eq. (9).

DEFINITION

The equation

$$y - y_1 = m(x - x_1) \tag{10}$$

is the **point–slope equation** of the line that passes through the point (x_1, y_1) with slope m.

Equation (10) is the form we usually use to write equations for lines.

Example 5   Write an equation for the line that passes through the point $(2, 3)$ with slope $-3/2$.

Solution

$$y - y_1 = m(x - x_1) \qquad \left(\begin{array}{l}\text{Start with the general} \\ \text{point–slope equation, Eq. (10).}\end{array}\right)$$

$$y - 3 = -\frac{3}{2}(x - 2)$$

$$y = -\frac{3}{2}x + 6.$$

This is an equation for the line (Fig. 1.14).

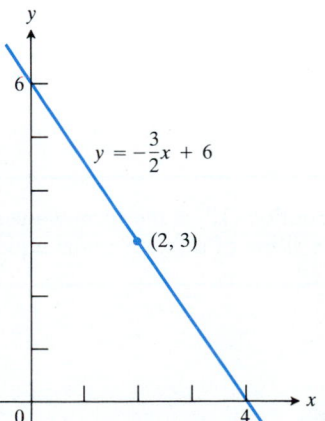

1.14   The line in Example 5.

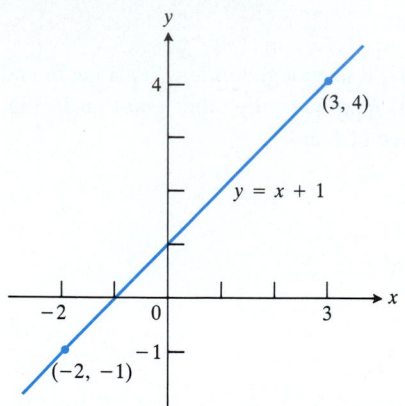

1.15   The line in Example 6.

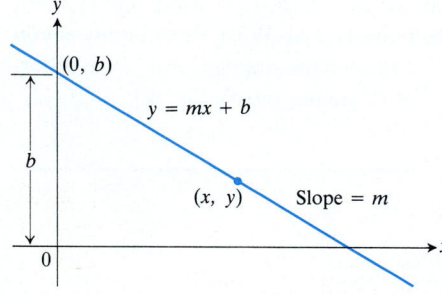

1.16   The line with slope m and y-intercept b. Its standard equation is $y = mx + b$.

DEFINITION

Equation (13) is the form we use to find the slope of a line from its equation.

Example 6   Write an equation for the line through $(-2, -1)$ and $(3, 4)$.

Solution   We first calculate the slope and then use Eq. (10):

$$m = \frac{-1-4}{-2-3} = \frac{-5}{-5} = 1.$$

The (x_1, y_1) in Eq. (10) can be either $(-2, -1)$ or $(3, 4)$:

With $(x_1, y_1) = (-2, -1)$	With $(x_1, y_1) = (3, 4)$
$y - (-1) = 1 \cdot (x - (-2))$	$y - 4 = 1 \cdot (x - 3)$
$y + 1 = x + 2$	$y - 4 = x - 3$
$y = x + 1$	$y = x + 1$

same result

Either way, $y = x + 1$ is an equation for the line (Fig. 1.15).

Slope–Intercept Equations

Figure 1.16 shows a line with slope m and y-intercept b. If we take $(x_1, y_1) = (0, b)$ in the point–slope equation for the line, we find that

$$y - b = m(x - 0). \tag{11}$$

When rearranged, this becomes

$$y = mx + b. \tag{12}$$

The equation

$$y = mx + b \tag{13}$$

is the **slope–intercept equation** of the line with slope m and y-intercept b.

Example 7   The slope–intercept equation of the line with slope 2 and y-intercept 5 is

$$y = 2x + 5.$$

Example 8   Find the slope and y-intercept of the line $8x + 5y = 20$.

Solution   Solve the equation for y to put the equation in slope–intercept form. Then read the slope and y-intercept from the equation:

$$8x + 5y = 20$$
$$5y = -8x + 20$$
$$y = -\frac{8}{5}x + 4.$$

The slope is $m = -8/5$. The y-intercept is $b = 4$.

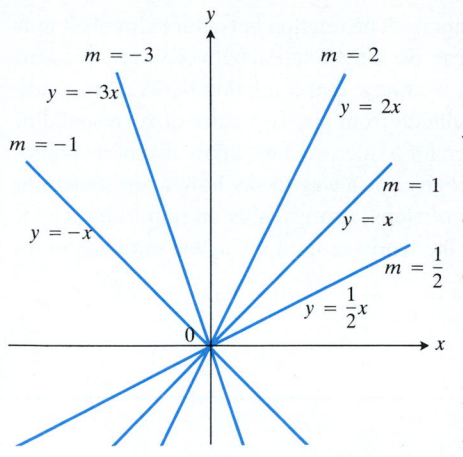

1.17   The line $y = mx$ passes through the origin with slope m.

Example 9   *Lines Through the Origin.*   When a nonvertical line passes through the origin, its y-intercept is $b = 0$ and the equation $y = mx + b$ simplifies to $y = mx$. The line $y = (1/2)x$ passes through the origin with slope $1/2$. The line $y = x$ passes through the origin with slope 1. The line $y = -x$ is the line through the origin with slope -1 (Fig. 1.17).

The General Linear Equation

The equation

$$Ax + By = C \qquad (A \text{ and } B \text{ not both zero}) \qquad (14)$$

is called the **general linear equation** because its graph is always a line and because every line has an equation in this form. We shall not take the time to prove this, but notice that all the equations in this section can be arranged in this form. A few of them are in this form already.

Slope

1. The slope of the nonvertical line through $P_1(x_1, y_1)$ and $P_2(x_2, y_2)$, $x_1 \neq x_2$, is

$$m = \frac{\text{rise}}{\text{run}} = \frac{y_2 - y_1}{x_2 - x_1} = \frac{\Delta y}{\Delta x}.$$

2. $m = \tan \phi$ (ϕ is the angle of inclination).

3. Vertical lines have no slope.

4. Horizontal lines have slope zero.

5. For lines that are neither horizontal nor vertical, it is handy to remember:
   a) they are parallel $\Leftrightarrow m_2 = m_1$;

   b) they are perpendicular $\Leftrightarrow m_2 = -1/m_1$.

   (The symbol $\Leftrightarrow$ is read "if and only if.")

Equations for Lines

$x = a$	Vertical line through (a, b)
$y = b$	Horizontal line through (a, b)
$y = mx + b$	Slope–intercept equation
$y - y_1 = m(x - x_1)$	Point–slope equation
$Ax + By = C$	General linear equation (A and B not both zero)

Applications—The Importance of Lines and Slope

Light travels along lines, as do bodies falling from rest in a planet's gravitational field or coasting under their own momentum (like a hockey puck gliding across the ice). We often study such motions by writing equations for the lines. Many important variables are related by linear equations. The relation between Fahrenheit temperature and Celsius temperature is linear. So is the relation between water pressure and depth in an ocean or a lake. Once we know that a relation between two variables is linear, we can find the exact relation from any two pairs of corresponding values the same way we find an equation for a line when we know two of its points.

Slope itself is important because it gives us a way to say how steep something is (roadbeds, roofs, stairs). The notion of slope also gives us an important way to measure how rapidly things change in the world around us, as we shall see in the chapters ahead.

EXERCISES 1.1

In Exercises 1–4, a particle moves from A to B. Find the net changes Δx and Δy in the particle's coordinates.

1. $A(-3, 2)$, $B(-1, -2)$
2. $A(-1, -2)$, $B(-3, 2)$
3. $A(-3.2, -2)$, $B(-8.1, -2)$
4. $A(\sqrt{2}, 4)$, $B(0, 1.5)$

Plot the points A and B in Exercises 5–8. Then find the slope (if any) of the line they determine. Also find the common slope (if any) of the lines perpendicular to line AB.

5. $A(-1, 2)$, $B(-2, -1)$ 6. $A(2, -1)$, $B(-2, 1)$
7. $A(2, 3)$, $B(-1, 3)$ 8. $A(1, 2)$, $B(1, -3)$

In Exercises 9–12, find an equation for (a) the vertical line and (b) the horizontal line through the given point.

9. $(-1, 4/3)$ 10. $(\sqrt{2}, -1.3)$
11. $(0, -\sqrt{2})$ 12. $(-\pi, 0)$

In Exercises 13–16, write an equation for the line through P with slope m.

13. $P(-1, 1)$, $m = 1$ 14. $P(-2, 2)$, $m = 1/2$
15. $P(0, b)$, $m = 2$ 16. $P(a, 0)$, $m = -2$

In Exercises 17–20, write an equation for the line through the two points.

17. $(1, 1)$, $(2, 1)$ 18. $(-2, 0)$, $(-2, -2)$
19. $(0, 0)$, $(2, 3)$ 20. $(-2, 1)$, $(2, -2)$

In Exercises 21–24, write an equation for the line with slope m and y-intercept b.

21. $m = 1$, $b = \sqrt{2}$ 22. $m = -1/2$, $b = -3$
23. $m = -5$, $b = 2.5$ 24. $m = 1/3$, $b = -1$

Quick graphing.   Graph the lines in Exercises 25–28 by taking the following steps:

1. Find the x-intercept by setting y equal to 0 in the equation.

2. Find the y-intercept by setting x equal to 0.

3. Plot the intercepts and draw the line.

25. $3x + 4y = 12$
26. $x + 2y = -4$
27. $\sqrt{2}x - \sqrt{3}y = \sqrt{6}$
28. $1.5x - y = -3$

In Exercises 29–34, find equations for the lines through P that are (a) parallel and (b) perpendicular to the given line.

29. $P(2, 1)$, $y = x + 2$
30. $P(0, 0)$, $3x - y = 5$
31. $P(1, 2)$, $x + 2y = 3$
32. $P(-2, 2)$, $2x + y = 4$
33. $P(-2, 4)$, $x = 5$
34. $P(-3, -2)$, $y = 3$

Increments and Motion

35. A particle starts at $A(-2, 3)$ and its coordinates change by increments $\Delta x = 5$, $\Delta y = -6$. Find its new position.

36. A particle starts at $A(6, 0)$ and its coordinates change by increments $\Delta x = -6$, $\Delta y = 0$. Find its new position.

37. The coordinates of a particle change by $\Delta x = 5$ and $\Delta y = 6$ as it moves from $A(x, y)$ to $B(3, -3)$. Find x and y.

38. A particle started at $A(1, 0)$, circled the origin once counterclockwise, and returned to $A(1, 0)$. What were the net changes in its coordinates?

Applications

39. *Insulation.*   By measuring slopes in Fig. 1.18, estimate the temperature change in degrees per inch for (a) the gypsum wallboard; (b) the fiberglass insulation; (c) the wood sheathing.

40. *Insulation.*   According to Fig. 1.18, which of the materials in Exercise 39 is the best insulator? the poorest? Explain.

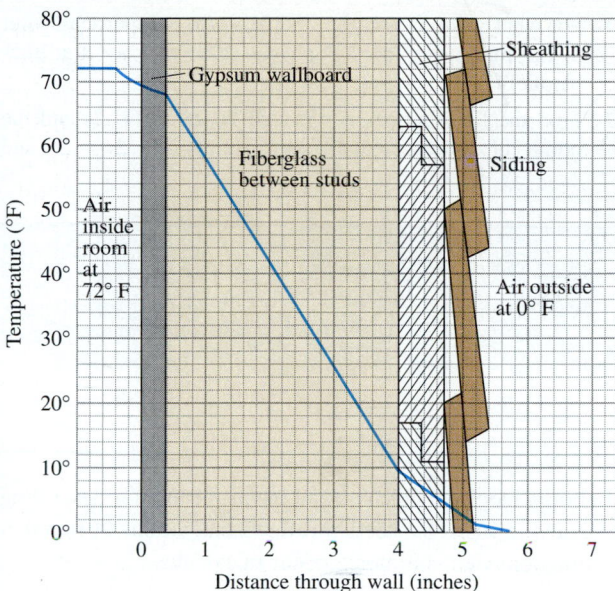

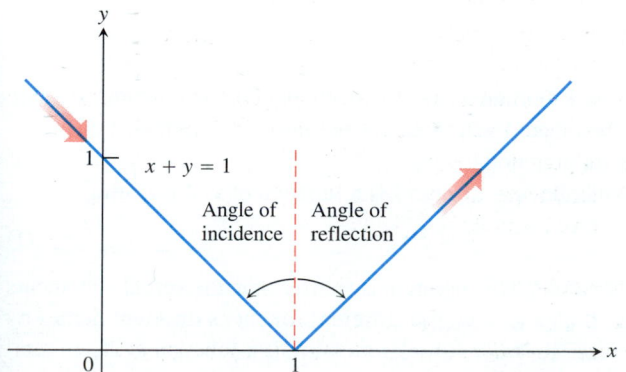

1.18 The temperature changes in the wall in Exercises 39 and 40 (*Source: Differentiation*, by W. U. Walton et al., Project CALC, Education Development Center, Inc. Newton, Mass. [1975], p. 25.)

41. *Pressure under water.* The pressure p experienced by a diver under water is related to the diver's depth d by an equation of the form $p = kd + 1$ (k a constant). At the surface, the pressure is 1 atmosphere. The pressure at 100 meters is about 10.94 atmospheres. Find the pressure at 50 meters.

42. *Reflected light.* A ray of light comes in along the line $x + y = 1$ from the second quadrant and reflects off the x-axis (Fig. 1.19). The angle of incidence is equal to the angle of reflection. Write an equation for the line along which the departing light travels.

1.19 The path of the light ray in Exercise 42. Angles of incidence and reflection are measured from the perpendicular.

43. *Fahrenheit vs. Celsius.* If we plot Fahrenheit temperature readings against Celsius temperature readings in a coordinate plane, the points we plot always lie along a straight line. The line passes through the point (0, 32) because $F = 32$ when $C = 0$. It also passes through the point (100, 212) because $F = 212$ when $C = 100$.
  a) Sketch the line in the CF-plane.
  b) Find a formula that expresses F in terms of C.
  c) Is there a temperature at which a Fahrenheit thermometer gives the same numerical reading as a Celsius thermometer? If so, what is it?

44. CALCULATOR *The Mt. Washington Cog Railway.* The steepest part of the Mt. Washington Cog Railway in New Hampshire has a phenomenal 37.1% grade. Along this part of the track, the passengers in the front of the car are 14 ft above those in the rear. About how far apart are the front and rear rows of seats, as measured along the floor of the car?

Geometry

45. A rectangle with sides parallel to the axes has vertices at (3, −2) and (−4, −7).
  a) Find the coordinates of the other two vertices.
  b) Find the area of the rectangle.

46. The rectangle in Fig. 1.20 has sides parallel to the axes. It is three times as long as it is wide. Its perimeter is 56 units. Find the coordinates of the vertices A, B, and C.

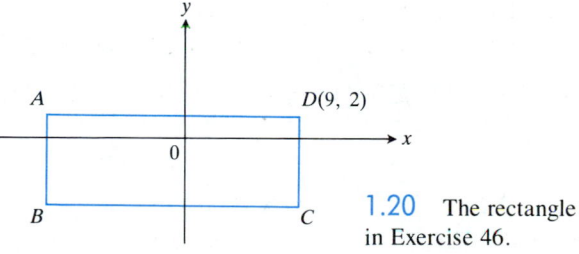

1.20 The rectangle in Exercise 46.

47. The line through the points (1, 1) and (2, 0) cuts the y-axis at the point (0, b). Find b by using similar triangles.

48. A 90° rotation counterclockwise about the origin takes (2, 0) to (0, 2) and (0, 3) to (−3, 0), as shown in Fig. 1.21.

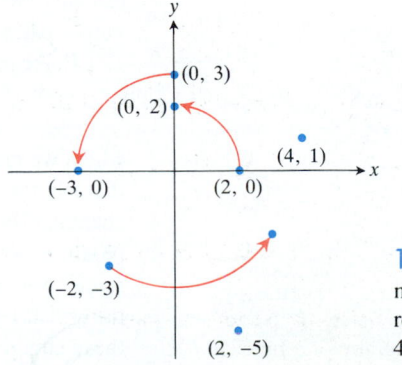

1.21 The points moved by the 90° rotation in Exercise 48.

Where does the rotation take each of the following points?

a) (4, 1)   b) (−2, −3)   c) (2, −5)

d) (x, 0)   e) (0, y)   f) (x, y)

g) What point is taken to (10, 3)?

49. Three different parallelograms have vertices at (−1, 1), (2, 0), and (2, 3). Sketch them and give the coordinates of the missing vertices.

50. CALCULATOR How large a slope can you calculate with your calculator? To find out, continue the list of tangent values in Example 3. The best we could do was tan (89.999 99999 99) = 57295 77951 31 and tan (90.000 00000 01) = −57295 77951 31.

51. For what value of k is the line $2x + ky = 3$ perpendicular to the line $x + y = 1$? For what value of k are the lines parallel?

52. Find the line that passes through the point (1, 2) and the point of intersection of the lines $x + 2y = 3$ and $2x − 3y = −1$.

EXPLORER PROGRAM

Name a Function   Offers practice in matching formulas to standard graphs

---

1.2 Functions and Their Graphs

Functions are the main building blocks of calculus and the major tool for describing the real world in mathematical terms. This section reviews the notions of function and graph and describes functions that commonly occur in calculus.

Functions

As you know, the pressure in the boiler of a power plant depends on the steam temperature, and the area of a circle depends on its radius. In each case, the value of one variable quantity, call it y, depends on the value of another quantity, say x. Since the value of y is completely determined by the value of x, we say that y is a function of x.

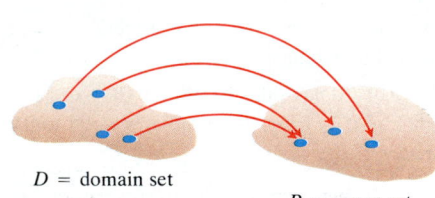

$D = $ domain set

$R = $ range set

1.22 A function from set D to set R assigns an element of R to each element in D.

In mathematics, any rule that assigns to each element in one set some element from another set is called a **function**. The sets may be sets of numbers, sets of number pairs, sets of points, or sets of objects of any kind. The sets do not have to be the same. All the function has to do is assign some element from the second set to each element in the first set (Fig. 1.22). Thus a function is like a machine that assigns an output to every allowable input. The **inputs** x make up the **domain** of the function. The **outputs** $f(x)$ make up the function's **range** (Fig. 1.23).

DEFINITION

A **function** from a set D to a set R is a rule that assigns a single element of R to each element in D.

The word *single* in the definition means that each input in the function's domain has only one output in the range. Each input appears just once in the list of input–output pairs defined by the function.

Euler invented a symbolic way to say "y is a function of x" by writing

$$y = f(x), \tag{1}$$

which we read "y equals f of x." This notation is shorter than the verbal statements that say the same thing. It also lets us give different functions different names by changing the letters we use. To say that boiler pressure is a function of steam temperature, we can write $p = f(t)$. To say that the area of a circle is a function of its radius, we can write $A = g(r)$. (Here we use a g because we just used f for something else.) We have to know what the variables p, t, A, and r mean, of course, for these equations to make sense.

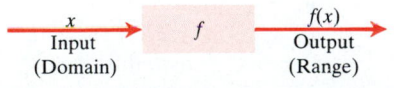

x → Input (Domain) f $f(x)$ → Output (Range)

1.23   A flow diagram for a function f.

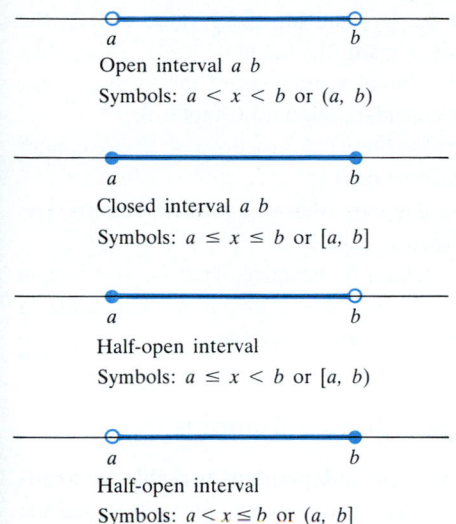

1.24   The domains and ranges of many functions are finite intervals like these.

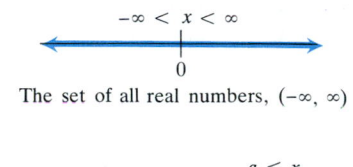

The set of all real numbers, $(-\infty, \infty)$

The set of numbers greater than a, (a, ∞)

The set of numbers greater than or equal to a, $[a, \infty)$

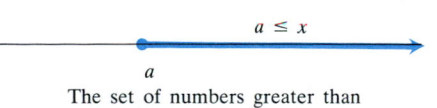

The set of numbers less than b, $(-\infty, b)$

The set of numbers less than or equal to b, $(-\infty, b]$

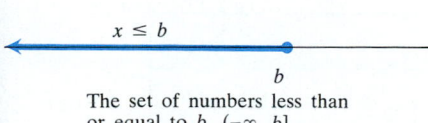

1.25   Rays on the number line and the line itself are called *infinite intervals*. The symbol ∞ (infinity) in the notation is used merely for convenience; it is not to be taken as a suggestion that there is a number ∞.

Real-Valued Functions of a Real Variable

In most of our work, functions have domains and ranges that are sets of real numbers. Such functions are called **real-valued functions of a real variable** and are usually defined by formulas or equations.

In addition to giving a function a useful name, the notation $y = f(x)$ gives a way to denote specific values of a function. For instance, the value of f at $x = 5$ is written $f(5)$("f of 5").

Example 1   The function A defined by the rule $A(r) = \pi r^2$ gives the area of a circle as a function of its radius r. The area of a circle of radius 2 is $A(2) = \pi(2)^2 = 4\pi$.

In the context of geometry, the domain of the function $A = \pi r^2$ is the set of all possible radii—in this case the set of all positive real numbers. The range is also the set of positive real numbers.

Example 2   *The Function $y = x^2$.*   The formula $y = x^2$ defines the number y to be the square of the number x. If $x = 5$, then $y = 5^2 = 25$.

The domain is the set of allowable x-values—in this case the set of all real numbers. The range, which consists of the resulting y-values, is the set of nonnegative real numbers.

Many Domains and Ranges Are Intervals

The domains and ranges of many functions are intervals of real numbers (Fig. 1.24). The set of all real numbers that lie *strictly between* two fixed numbers a and b is an **open interval**. The interval is "open" at each end because it contains neither of its endpoints. Intervals that contain both endpoints are **closed**. Intervals that contain one endpoint but not both are **half-open**. The domains and ranges of functions can also be **infinite intervals** (Fig. 1.25).

The endpoints of an interval make up the interval's **boundary** and are called **boundary points**. The remaining points make up the interval's **interior** and are called **interior points**. Closed intervals contain their boundary points; open intervals do not. Every point of an open interval is an interior point of the interval.

Example 3   In each of the following functions, the domain is taken to be the largest set of real x-values for which the formula gives real y-values. We shall say more about this convention in a moment.

Function	Domain (x)	Range (y)
$y = x^2$	$(-\infty, \infty)$	$[0, \infty)$
$y = \sqrt{1 - x^2}$	$[-1, 1]$	$[0, 1]$
$y = \dfrac{1}{x}$	$(-\infty, 0) \cup (0, \infty)$	$(-\infty, 0) \cup (0, \infty)$
$y = \sqrt{x}$	$[0, \infty)$	$[0, \infty)$
$y = \sqrt{4 - x}$	$(-\infty, 4]$	$[0, \infty)$

The formula $y = x^2$ gives a real y-value for any real number x.

Leonhard Euler (1707–1783)

Leonhard Euler, the dominant mathematical figure of his century and the most prolific mathematician who ever lived, was also an astronomer, physicist, botanist, and chemist and an expert in Oriental languages. He was the first scientist to give the function concept the prominence in his work that it has in mathematics today. Euler's collected books and papers fill 70 volumes. His introductory algebra text, written originally in German (Euler was Swiss), is still read in English translation.

The formula $y = \sqrt{1 - x^2}$ gives a real y-value for every value of x in the closed interval from -1 to 1. Beyond this domain, the quantity $1 - x^2$ is negative and its square root is not a real number. (Complex numbers of the form $a + bi$, where $i = \sqrt{-1}$, are excluded from our consideration until Chapter 8.)

The formula $y = 1/x$ gives a real y-value for every x except $x = 0$. We cannot divide 1 (or any other number, for that matter) by 0.

The formula $y = \sqrt{x}$ gives a real y-value only when x is positive or zero. The number $y = \sqrt{x}$ is not a real number when x is negative.

In $y = \sqrt{4 - x}$, the quantity $4 - x$ cannot be negative. That is, $4 - x$ must be greater than or equal to 0. In symbols, $0 \le 4 - x$ or $x \le 4$. The formula $y = \sqrt{4 - x}$ gives a real y-value for any x less than or equal to 4.

Variables, Zero, and a Convention About Domains

The variable x in a function $y = f(x)$ is called the **independent variable,** or **argument,** of the function. The variable y, whose value depends on x, is called the **dependent variable.**

We must keep two restrictions in mind when we define functions. First, we *never divide by 0*. When we see $y = 1/x$, we must think "$x \ne 0$." Zero is not in the domain of the function. When we see $y = 1/(x - 2)$, we must think "$x \ne 2$."

The second restriction is that we shall deal exclusively with real-valued functions (except for a very short while later in the book). We therefore have to restrict our domains when we have square roots or other even roots. If $y = \sqrt{1 - x^2}$, we should think "x^2 must not be greater than 1. The domain must not extend beyond the interval $-1 \le x \le 1$."

We observe a convention about the domains of functions defined by formulas. If the domain is not stated explicitly, then the domain is automatically the largest set of x-values for which the formula gives real y-values. If we wish to exclude values from this domain, we must say so. We write "$y = x^2, x \ge 0$" to exclude negative values from the domain of the function $y = x^2$. We write "$y = x^2$" (unrestricted) if we want the domain to be the entire set of real numbers.

Graphs and Graphing

The points (x, y) in the plane whose coordinates are the input–output pairs of a function $y = f(x)$ make up the function's **graph.** The graph of the function $y = x + 2$, for example, is the line $y = x + 2$. It is the set of points (x, y) in which y equals $x + 2$.

Example 4   Graph the function $y = x^2$ over the interval $-2 \le x \le 2$.

These are the steps we take to graph a function.

Solution   STEP 1: Make a table of xy-pairs that satisfy the function rule, in this case the equation $y = x^2$.

x	$y = x^2$
-2	4
-1	1
0	0
1	1
2	4

STEP 2: Plot the points (x, y) whose coordinates appear in the table.

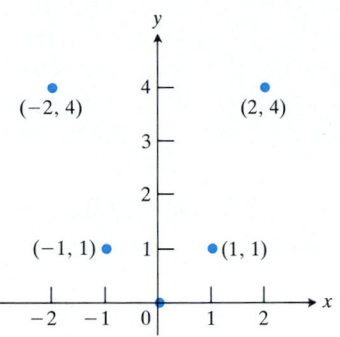

STEP 3: Draw a smooth curve through the plotted points. Label the curve with its equation.

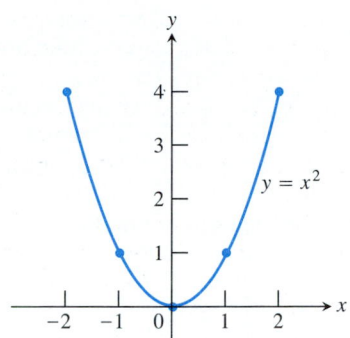

How do we know that the graph of $y = x^2$ doesn't look like one of *these* curves?

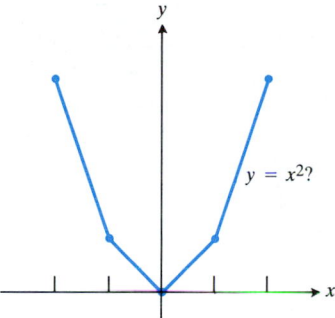

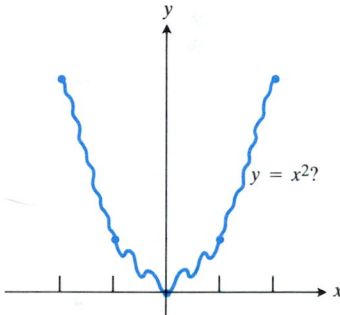

To find out, we might plot more points. But how would we then connect *them*? The basic question would still remain: How do we know for sure what the graph does between the points we plot?

The answer lies in calculus, as we shall see in Chapter 3. There we shall learn to use a marvelous mathematical tool called a *derivative* to find a curve's exact shape between plotted points. Meanwhile, we shall have to settle for plotting individual points and connecting them as best we can.

Example 5   Graph the function $y = \sqrt{4 - x}$.

Solution   STEP 1: Determine the function's domain. In this case, the domain is the set of values for which $4 - x \geq 0$, or $x \leq 4$.

STEP 2: Make a table.

x	$\sqrt{4 - x}$
4.0	0
3.75	0.5
2.0	$\sqrt{2} \approx 1.4$
0	2
-2	$\sqrt{6} \approx 2.4$

STEP 3: Plot the points and draw the curve.

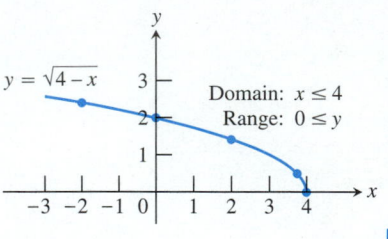

1.26   Useful graphs of functions.

Figures 1.26 and 1.27 show the graphs of functions frequently used in calculus. For a review of the trigonometric functions, see Appendix 2.

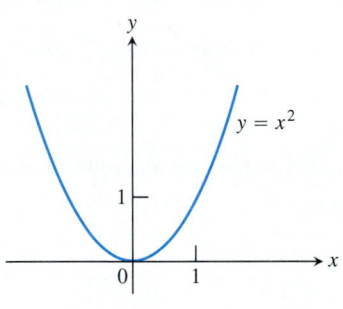

Domain: $-\infty < x < \infty$
Range: $0 < y < \infty$

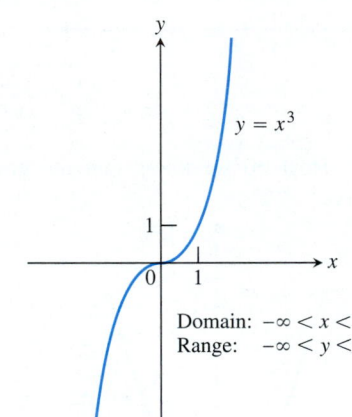

Domain: $-\infty < x < \infty$
Range: $-\infty < y < \infty$

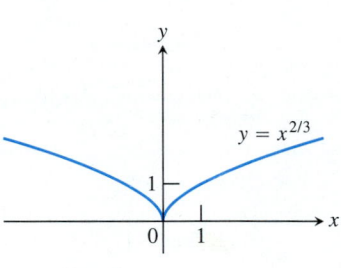

Domain: $-\infty < x < \infty$
Range: $0 < y < \infty$

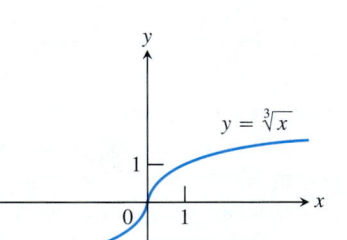

Domain: $0 \le x < \infty$
Range: $0 \le y < \infty$

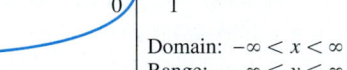

Domain: $-\infty < x < \infty$
Range: $-\infty < y < \infty$

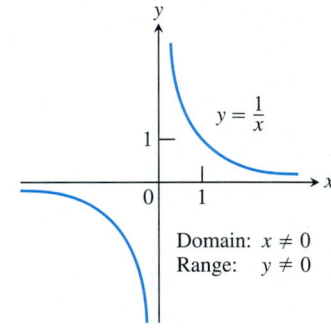

Domain: $x \ne 0$
Range: $y \ne 0$

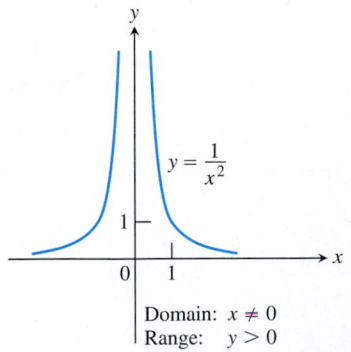

Domain: $x \ne 0$
Range: $y > 0$

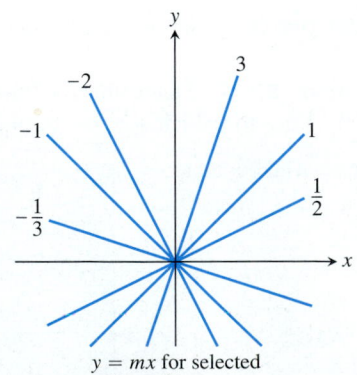

$y = mx$ for selected values of m
Domain: $-\infty < x < \infty$
Range: $-\infty < y < \infty$

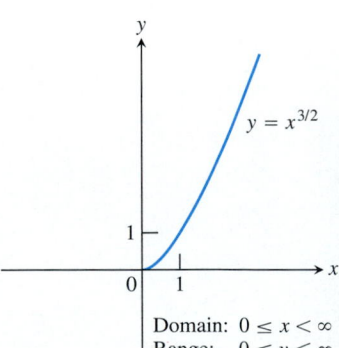

Domain: $0 \le x < \infty$
Range: $0 \le y < \infty$

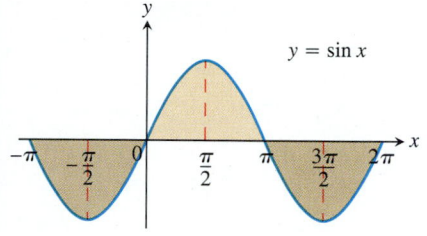

$y = \sin x$

Domain: $-\infty < x < \infty$
Range: $-1 \le y \le 1$

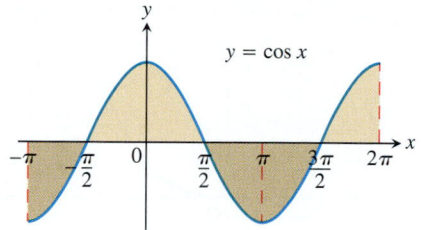

$y = \cos x$

Domain: $-\infty < x < \infty$
Range: $-1 \le y \le 1$

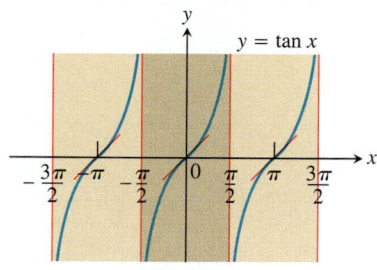

$y = \tan x$

Domain: All real numbers except odd
integer multiples of $\pi/2$
Range: $-\infty < y < \infty$

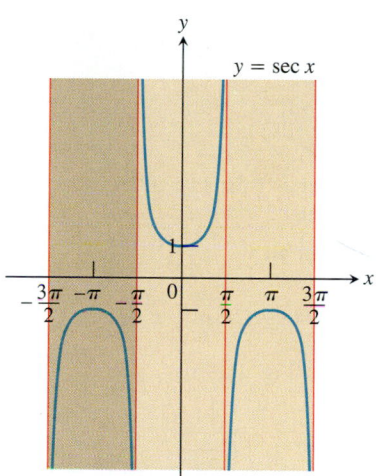

$y = \sec x$

Domain: $x \ne \pm\dfrac{\pi}{2}, \pm\dfrac{3\pi}{2}, \ldots$
Range: $y \le -1$ and $y \ge 1$

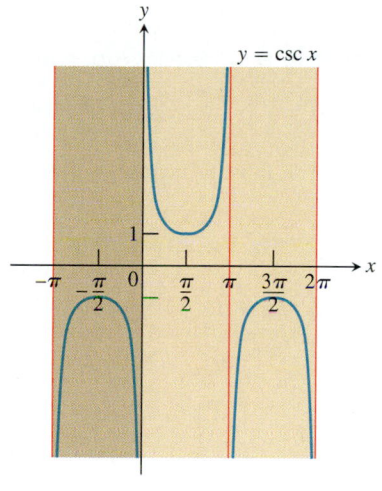

$y = \csc x$

Domain: $x \ne 0, \pm\pi, \pm 2\pi, \ldots$
Range: $y \le -1$ and $y \ge 1$

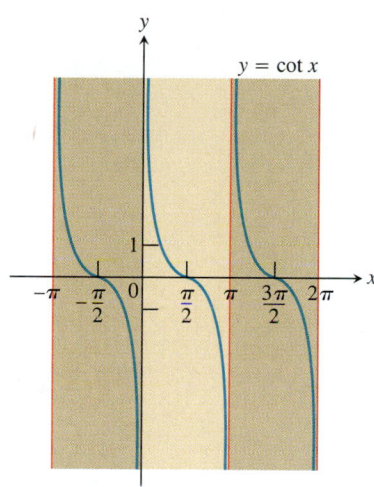

$y = \cot x$

Domain: $x \ne 0, \pm\pi, \pm 2\pi, \ldots$
Range: $-\infty < y < \infty$

1.27   The graphs of the six basic trigonometric functions as functions of radian measure.

Even Functions and Odd Functions—Symmetry

DEFINITIONS

A function $y = f(x)$ is an **even** function of x if $f(-x) = f(x)$ for every x in the function's domain. It is an **odd** function of x if $f(-x) = -f(x)$ for every x in the function's domain.

The names *even* and *odd* come from powers of x. If y equals an even power of x, as in $y = x^2$ or $y = x^4$, it is an even function of x (because $(-x)^2 = x^2$ and $(-x)^4 = x^4$). If y equals an odd power of x, as in $y = x$ or $y = x^3$, it is an odd function of x (because $(-x)^1 = -x$ and $(-x)^3 = -x^3$).

Saying that a function $y = f(x)$ is even is equivalent to saying that its graph is symmetric about the y-axis. Since $f(-x) = f(x)$, the point (x, y) lies on the curve if and only if the point $(-x, y)$ lies on the curve (Fig. 1.28a).

Saying that a function $y = f(x)$ is odd is equivalent to saying that its graph is symmetric with respect to the origin. Since $f(-x) = -f(x)$, the point (x, y) lies on the curve if and only if the point $(-x, -y)$ lies on the curve (Fig. 1.28b).

The graphs of polynomials in even powers of x are symmetric about the y-axis. The graphs of polynomials in odd powers of x are symmetric about the origin.

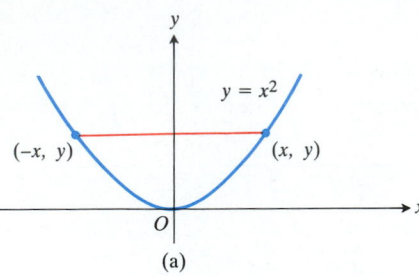

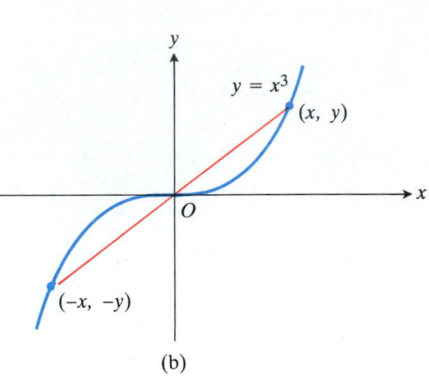

1.28 (a) The graph of an even function is symmetric about the y-axis. (b) The graph of an odd function is symmetric about the origin.

Example 6   *Even, Odd, and Neither*

$f(x) = x^2$	Even function: $(-x)^2 = x^2$ for all x Symmetry about y-axis
$f(x) = x^2 + 1$	Even function: $(-x)^2 + 1 = x^2 + 1$ for all x Symmetry about the y-axis (Fig. 1.29a)
$f(x) = x$	Odd function: $(-x) = -(x)$ for all x Symmetry about the origin
$f(x) = x + 1$	Not odd: $f(-x) = -x + 1$, but $-f(x) = -x - 1$. The two are not equal. Not even: $(-x) + 1 \neq x + 1$. (Fig. 1.29b)

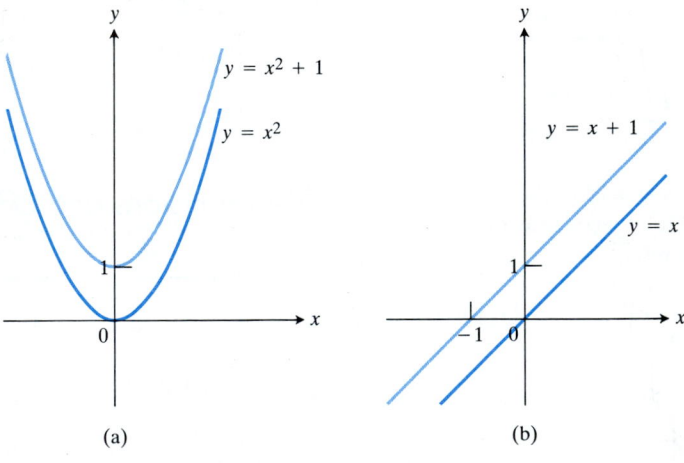

1.29 (a) When we add 1 to $y = x^2$, the resulting function is still even and its graph is still symmetric about the y-axis. (b) When we add 1 to $y = x$, the resulting function is no longer odd. The symmetry about the origin is lost.

Example 7   *The Basic Trigonometric Functions*

Even	*Odd*
$\cos(-x) = \cos x$ for all x	$\sin(-x) = -\sin x$ for all x
$\sec(-x) = \sec x$ for all x	$\tan(-x) = -\tan x$ for all x
	$\csc(-x) = -\csc x$ for all x
	$\cot(-x) = -\cot x$ for all x

Notice the symmetries in the graphs in Fig. 1.27.

Integer-Valued Functions

The greatest integer less than or equal to a number x is called the *greatest integer in* x. Because each real number x corresponds to only one greatest integer, the greatest integer in x is a function of x. The symbol for it is $\lfloor x \rfloor$, which is read "the greatest integer in x." Other common notations are $[x]$ and $[\![x]\!]$, read the same way. Figure 1.30 shows the graph.

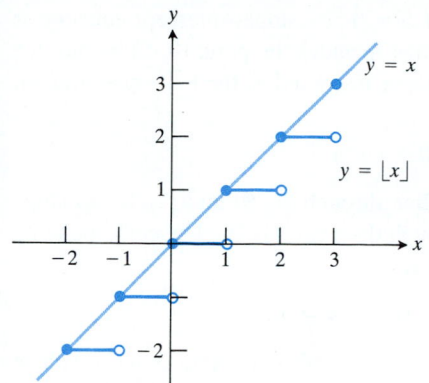

1.30 The graph of $y = \lfloor x \rfloor$ and its relation to the line $y = x$. As the figure shows, $\lfloor x \rfloor$ is less than or equal to x, so it provides an integer floor for x.

Example 8   *Values of the Greatest Integer Function* $y = \lfloor x \rfloor$

Positive $\lfloor 1.9 \rfloor = 1, \ \lfloor 2.0 \rfloor = 2, \ \lfloor 2.4 \rfloor = 2$

Zero $\lfloor 0.5 \rfloor = 0, \ \lfloor 0 \rfloor = 0$

Negative $\lfloor -1.2 \rfloor = -2, \ \lfloor -0.5 \rfloor = -1$

Notice that if x is negative, $\lfloor x \rfloor$ may have a larger absolute value than x does. ⌋

The notation $y = \lfloor x \rfloor$ comes from computer science, where it is used to denote the result of rounding x down to the nearest integer. You can think of it as the integer floor for x, and the notation is chosen to suggest just that. The companion notation

$$\lceil x \rceil \qquad \text{integer ceiling for } x$$

denotes the result of rounding x up to the nearest integer. It gives the *least integer greater than or equal to x* (Fig. 1.31).

Notice, for later reference, that the integer floor and ceiling functions exhibit points called *discontinuities,* where the functions jump from one value to the next without taking on the intermediate values. They jump like this at every integer value of x.

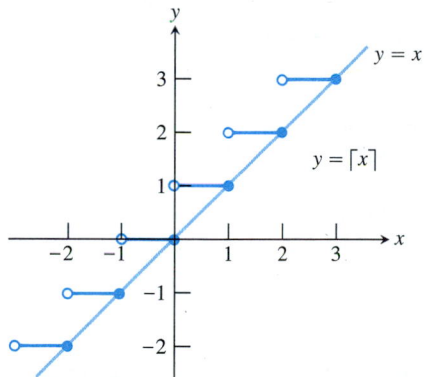

1.31 The graph of $y = \lceil x \rceil$ and its relation to the line $y = x$. As the figure shows, $\lceil x \rceil$ is greater than or equal to x, so it provides an integer ceiling for x.

Functions Defined in Pieces

While some functions are defined by single formulas, others are defined by applying different formulas to different parts of their domains.

Example 9   The values of the function

$$y = f(x) = \begin{cases} -x & \text{for } x < 0, \\ x^2 & \text{for } 0 \le x \le 1, \\ 1 & \text{for } x > 1, \end{cases}$$

are given by the formula $y = -x$ when $x < 0$, by the formula $y = x^2$ when $0 \le x \le 1$, and by the formula $y = 1$ when $x > 1$. The function is *just one function,* however, whose domain is the entire set of real numbers (Fig. 1.32). ⌋

Example 10   Suppose that the graph of a function $y = f(x)$ consists of the line segments shown in Fig. 1.33. Write a formula for f.

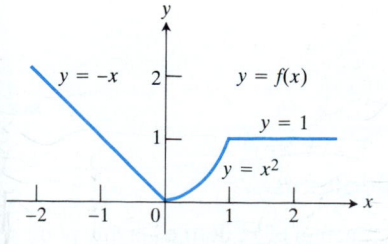

1.32 To graph the function $y = f(x)$ shown here, we apply different formulas to different parts of its domain.

Solution   We find formulas for the segments from $(0, 0)$ to $(1, 1)$ and from $(1, 0)$ to $(2, 1)$ and piece them together in the manner of Example 9.

Segment from (0, 0) to (1, 1). The line through $(0, 0)$ and $(1, 1)$ has slope

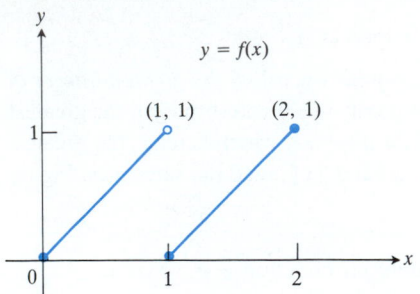

1.33   The graph of the function $y = f(x)$ (Example 10) shown here consists of two line segments. The segment on the left contains the endpoint at the origin (shown by a heavy dot) but does not contain the endpoint (1, 1). The segment on the right contains both endpoints.

$m = (1 - 0)/(1 - 0) = 1$ and y-intercept $b = 0$. Its slope–intercept equation is $y = x$. The segment from (0, 0) to (1, 1) that includes the point (0, 0) but not the point (1, 1) is the graph of the function $y = x$ restricted to the half-open interval $0 \leq x < 1$, namely,

$$y = x, \qquad 0 \leq x < 1.$$

Segment from (1, 0) to (2, 1). The line through (1, 0) and (2, 1) has slope $m = (1 - 0)/(2 - 1) = 1$ and passes through the point (1, 0). The corresponding point–slope equation for the line is therefore

$$y - 0 = 1(x - 1), \qquad \text{or} \qquad y = x - 1.$$

The segment from (1, 0) to (2, 1) that includes both endpoints is the graph of $y = x - 1$ restricted to the closed interval $1 \leq x \leq 2$, namely,

$$y = x - 1, \qquad 1 \leq x \leq 2.$$

Formula for the function $y = f(x)$ shown in Fig. 1.33. We obtain a formula for f on the interval $0 \leq x \leq 2$ by combining the formulas we obtained for the two segments of its graph:

$$f(x) = \begin{cases} x & \text{for } 0 \leq x < 1, \\ x - 1 & \text{for } 1 \leq x \leq 2. \end{cases}$$

Example 11   The domain of the "step" function $y = g(x)$ graphed in Fig. 1.34 is the closed interval $0 \leq x \leq 3$. Find a formula for $g(x)$.

Solution   The graph consists of three horizontal line segments. The left segment is the half-open interval $0 \leq x < 1$ on the x-axis, which we may think of as a portion of the line $y = 0$:

$$y = 0, \qquad 0 \leq x < 1.$$

The second segment is the portion of the line $y = 1$ that lies over the closed interval $1 \leq x \leq 2$:

$$y = 1, \qquad 1 \leq x \leq 2.$$

The third segment is the half-open interval $2 < x \leq 3$ on the line $y = 0$:

$$y = 0, \qquad 2 < x \leq 3.$$

The values of g are therefore given by the three-piece formula

$$g(x) = \begin{cases} 0 & \text{for } 0 \leq x < 1, \\ 1 & \text{for } 1 \leq x \leq 2, \\ 0 & \text{for } 2 < x \leq 3. \end{cases}$$

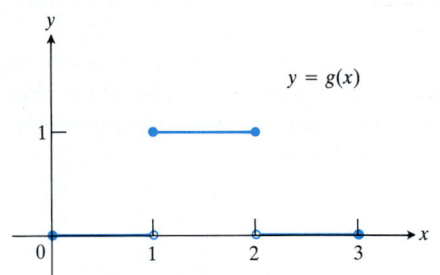

1.34   Functions like the one graphed here are called *step functions*. Example 11 shows how to write a formula for g.

Sums, Differences, Products, and Quotients

The sum $f + g$ of two functions of x is itself a function of x, defined at any point x that lies in both domains. The same holds for the differences $f - g$ and $g - f$, the product $f \cdot g$, and the quotients f/g and g/f, as long as we exclude any points that require division by zero.

Example 12   The functions f and g defined by the formulas

$$f(x) = \sqrt{x} \quad \text{and} \quad g(x) = \sqrt{1 - x}$$

can combine in the following ways.

Function	Formula	Domain
f	$f(x) = \sqrt{x}$	$0 \leq x$
g	$g(x) = \sqrt{1 - x}$	$x \leq 1$
$f + g$	$(f + g)(x) = f(x) + g(x) = \sqrt{x} + \sqrt{1 - x}$	$0 \leq x \leq 1$ (Intersection of domains of f and g)
$f - g$	$(f - g)(x) = f(x) - g(x) = \sqrt{x} - \sqrt{1 - x}$	$0 \leq x \leq 1$
$g - f$	$(g - f)(x) = g(x) - f(x) = \sqrt{1 - x} - \sqrt{x}$	$0 \leq x \leq 1$
$f \cdot g$	$(f \cdot g)(x) = f(x)g(x) = \sqrt{x(1 - x)}$	$0 \leq x \leq 1$
f/g	$\dfrac{f}{g}(x) = \dfrac{f(x)}{g(x)} = \sqrt{\dfrac{x}{1 - x}}$	$0 \leq x < 1$ ($x = 1$ excluded)
g/f	$\dfrac{g}{f}(x) = \dfrac{g(x)}{f(x)} = \sqrt{\dfrac{1 - x}{x}}$	$0 < x \leq 1$ ($x = 0$ excluded)

Composite Functions

Suppose that the outputs of a function g can be used as inputs of a function f. We can then hook g and f together to form a new function whose inputs are the inputs of g and whose outputs are the numbers $f(g(x))$, as in Fig. 1.35. We say that the function $f(g(x))$ (pronounced "f of g of x") is the **composite** of g and f. It is made by **composing** g and f in the order first g, then f. The usual "stand-alone" notation for this composite is $f \circ g$, which is read "f of g." Thus, the value of $f \circ g$ at x is $(f \circ g)(x) = f(g(x))$.

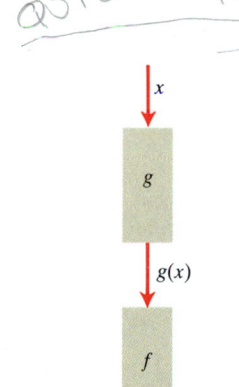

1.35   Two functions can be composed when the range of the first lies in the domain of the second.

Example 13   Find a formula for $f(g(x))$ if $g(x) = x^2$ and $f(x) = x - 7$. Then find the value of $f(g(2))$.

Solution   To find $f(g(x))$, we replace x in the formula for $f(x)$ by the expression given for $g(x)$:

$$f(x) = x - 7$$

$$f(g(x)) = g(x) - 7 = x^2 - 7.$$

We then find the value of $f(g(2))$ by substituting 2 for x:

$$f(g(2)) = (2)^2 - 7 = 4 - 7 = -3.$$

Changing the order in which functions are composed usually changes the result.

Example 14   Find a formula for $g(f(x))$ if $g(x) = x^2$ and $f(x) = x - 7$. Then find $g(f(2))$.

Solution   To find $g(f(x))$, we replace x in the formula for $g(x)$ by the expression for $f(x)$:

$$g(x) = x^2$$

$$g(f(x)) = (f(x))^2 = (x - 7)^2.$$

To find $g(f(2))$, we then substitute 2 for x:

$$g(f(2)) = (2 - 7)^2 = (-5)^2 = 25.$$

The parentheses in the notation for composite functions tell which function comes first:

The notation $f(g(x))$ says "first g, then f." To calculate $f(g(2))$, calculate $g(2)$ and then apply f.

The notation $g(f(x))$ says "first f, then g." To calculate $g(f(2))$, calculate $f(2)$ and then apply g.

❋ A Useful One-Line Function Table Generator

The following BASIC program will generate a table of values for any function you can key into your computer. We illustrate the program with the commands for evaluating $y = f(x) = x^3 - 4x$ at six points evenly spaced across the interval [0, 2].

First, call up BASIC and enter the function's formula with the line

```
10 DEF FNF(X) = X ^ 3 − 4 * X
```

Then enter the command RUN. Finally, enter the line

```
FOR X = 0 TO 2 STEP .4:   PRINT X, FNF(X): NEXT
```

The screen will list the xy pairs for $x = 0, 0.4, 0.8, 1.2, 1.6$, and 2. You can refine the list by decreasing the step size.

EXERCISES 1.2

In Exercises 1–8, find the domain and range of each function.

1. $y = 1 + x^2$

2. $y = 1 - \sqrt{x}$

3. $y = \sqrt{-x}$

4. $y = \sqrt{-|x|}$

5. $y = |\tan x|$

6. $y = |\sec x|$

7. $y = \lfloor \sin x \rfloor$

8. $y = \lfloor 2x \rfloor$

Say whether the functions in Exercises 9–20 are even, odd, or neither. Try to answer without writing anything down (except the answer).

9. $y = x$

10. $y = x^2$

11. $y = x^3$

12. $y = x^4$

13. $y = x + 1$

14. $y = x + x^2$

15. $y = x^2 + 1$

16. $y = x + x^3$

17. $y = \dfrac{1}{x^2 - 1}$

18. $y = \dfrac{1}{x + 1}$

19. $y = \dfrac{x}{x^2 - 1}$

20. $y = \dfrac{x^2}{x^2 - 1}$

Find the domains and ranges of the functions in Exercises 21–28. Then graph the functions. What symmetries, if any, do the graphs have?

21. $y = 2x^2$

22. $y = -x^2$

23. $y = x^2 - 9$

24. $y = 4 - x^2$

25. $y = -x^3$

26. $y = -\sqrt[3]{x}$

27. $y = -\dfrac{1}{x}$

28. $y = -\dfrac{1}{x^2}$

The formulas in Exercises 29–32 define s as a function of t. Graph the function in the ts-plane (t-axis horizontal, s-axis vertical). What symmetries do the graphs have?

29. $s = 2 \sin t$ **30.** $s = \sin 2t$

31. $s = -2 \cos t$ **32.** $s = \cos(t/2)$

33. The formula $w = 1/\sqrt{s}$ defines w as a function of s.
   a) Can s be negative?
   b) Can s be 0?
   c) What is the domain of the function?

34. The formula $z = \sqrt{(1/r) - 1}$ defines z as a function of r.
   a) Can r be negative?
   b) Can r be 0?
   c) Can r be greater than 1?
   d) What is the domain of the function?

35. Which of the graphs in Fig. 1.36 could be the graph of
   a) $y = x^2 - 1$? Why?
   b) $y = (x - 1)^2$? Why?

36. Which of the graphs in Fig. 1.36 could *not* be the graph of $y = 4x^2$? Why?

(i)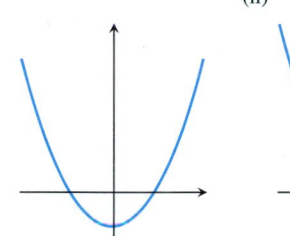

(ii)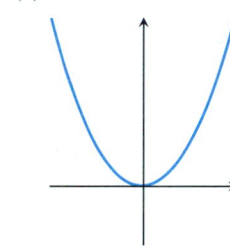

(iii)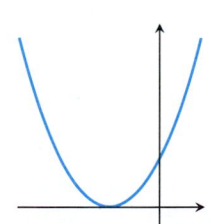

(iv)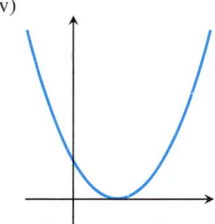

1.36   The graphs for Exercises 35 and 36.

37. For what values of x does (a) $\lfloor x \rfloor = 0$? (b) $\lceil x \rceil = 0$?

38. Does $\lfloor x \rfloor$ ever equal $\lceil x \rceil$? Explain.

39. Graph each function over the given interval.
   a) $y = x - \lfloor x \rfloor$, $\quad -3 \leq x \leq 3$
   b) $y = \lfloor x \rfloor - \lceil x \rceil$, $\quad -3 \leq x \leq 3$

40. *Integer parts of decimals.* When x is positive or zero, $\lfloor x \rfloor$ is the integer part of the decimal representation of x. What is the corresponding description of $\lceil x \rceil$ when x is negative or zero?

41. Make a table of values with $x = 0, 1$, and 2, and graph the function
$$y = \begin{cases} x, & 0 \leq x \leq 1, \\ 2 - x, & 1 < x \leq 2. \end{cases}$$

42. Make a table of values with $x = 0, 1$, and 2, and graph the function
$$y = \begin{cases} 1 - x, & 0 \leq x \leq 1, \\ 2 - x, & 1 < x \leq 2. \end{cases}$$

Graph the functions in Exercises 43–46.

43. $y = \begin{cases} 3 - x, & x \leq 1, \\ 2x, & 1 < x \end{cases}$ **44.** $y = \begin{cases} 1/x, & x < 0, \\ x, & 0 \leq x \end{cases}$

45. $y = \begin{cases} 1, & x < 5, \\ 0, & 5 \leq x \end{cases}$ **46.** $y = \begin{cases} 1, & x < 0, \\ \sqrt{x}, & x \geq 0 \end{cases}$

47. Find formulas for the functions graphed in Fig. 1.37.

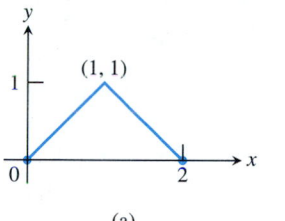

(a)

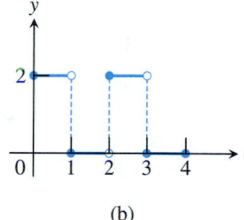

(b)

1.37   The graphs for Exercise 47.

48. Find formulas for the functions graphed in Fig. 1.38.

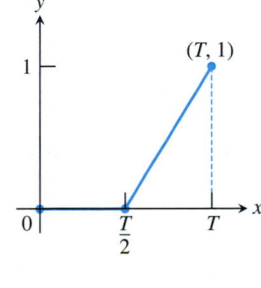

(a)

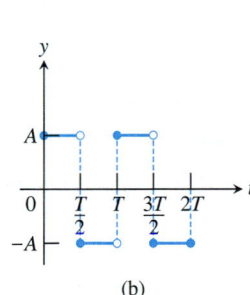

(b)

1.38   The graphs for Exercise 48.

In Exercises 49 and 50 find the domains of f and g and the corresponding domains of $f + g$, $f - g$, $f \cdot g$, f/g, and g/f.

49. $f(x) = x$, $\quad g(x) = \sqrt{x - 1}$

50. $f(x) = \sqrt{x + 1}$, $\quad g(x) = \sqrt{x - 1}$

51. If $f(x) = x + 5$ and $g(x) = x^2 - 3$, find the following:
   a) $f(g(0))$ b) $g(f(0))$
   c) $f(g(x))$ d) $g(f(x))$
   e) $f(f(-5))$ f) $g(g(2))$
   g) $f(f(x))$ h) $g(g(x))$

52. If $f(x) = x + 1$ and $g(x) = x - 1$, find the following.
a) $f(g(0))$ b) $g(f(0))$
c) $f(g(1))$ d) $g(f(1))$
e) $f(g(x))$ f) $g(f(x))$

53. Copy and complete the following table.

$g(x)$	$f(x)$	$f \circ g(x)$
a) $x - 7$	$\sqrt{x}$	
b) $x + 2$	$3x$	
c)	$\sqrt{x - 5}$	$\sqrt{x^2 - 5}$
d) $\dfrac{x}{x - 1}$	$\dfrac{x}{x - 1}$	
e)	$1 + \dfrac{1}{x}$	x
f) $\dfrac{1}{x}$		x

54. *A magic trick.* You may have heard of a magic trick that goes like this: Take any number. Add 5. Double the result. Subtract 6. Divide by 2. Subtract 2. Now tell me your answer, and I'll tell you what you started with.

Pick a number and try it.

You can see what is going on if you let x be your original number and follow the steps to make a formula $f(x)$ for the number you end up with.

Even vs. Odd

55. a) Show that the product $y = \sin x \cos x$ is an odd function of x.
b) Show that the product of an even function and an odd

function is always odd (on their common domain).

56. a) Show that the function $y = \sin^2 x$ is an even function of x (even though the sine itself is odd).
b) Show that the square of an odd function is always even.
c) Show that the product of any two odd functions is even (on their common domain).

Computer Grapher or Graphing Calculator

57. *(Continuation of Example 12)* Graph the functions $f(x) = \sqrt{x}$ and $g(x) = \sqrt{1 - x}$ together with (a) their sum, (b) their product, (c) their two differences, (d) their two quotients.

58. a) Graph $\sec x$ and $\cos x$ together for $-10 \leq x \leq 10$. Use different colors if available. At what values of x do the graphs intersect? What is the value of $\cos x$ when $\sec x$ is undefined?
b) Repeat part (a) with $\csc x$ and $\sin x$ in place of $\sec x$ and $\cos x$.

59. Graph $\tan x$ and $\cot x$ together for $-7 \leq x \leq 7$. Use different colors if available. At what points do the graphs cross? What is the value of $\tan x$ when $\cot x$ is undefined? What is the value of $\cot x$ when $\tan x$ is undefined?

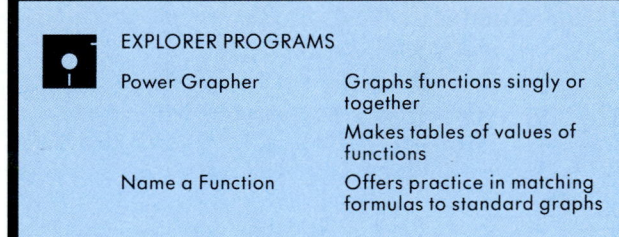

EXPLORER PROGRAMS	
Power Grapher	Graphs functions singly or together
	Makes tables of values of functions
Name a Function	Offers practice in matching formulas to standard graphs

1.3 Calculus and Computation

In this section, we look at some of the important functions on a scientific calculator, including e^x, $\ln x$, and some of the trigonometric functions. We also discuss how to use a calculator to explore various properties of functions, such as the possibility that two functions are equal or differ by some constant value. We then encounter a mysterious pattern (that will later be explained by calculus) and explore the way Fermat wanted to use slopes of lines to define slopes of curves.

Calculators are basically hand-held computers with small displays. With larger machines, we can see graphs in more detail and enter long programs for special-purpose computations. We can run large-scale simulations to predict changes in the economy or to model how galaxies interact when they collide. We shall say more about modeling in Chapter 4.

Evaluating Functions

Calculators enable us to evaluate most functions in calculus to six or more decimal places with only a few key presses. Calculators work either with algebraic notation or with reverse polish notation. Here are two typical calculations.

Example 1   *Algebraic Notation.*   To find the sine of 2 radians on a Casio fx-7000G, set the calculator in radian mode and press $\boxed{\text{SIN}}\,\boxed{2}\,\boxed{\text{EXE}}$.

Key	Operation	Display
$\boxed{\text{SIN}}$	Choose the function.	SIN
$\boxed{2}$	Enter the input value.	SIN 2
$\boxed{\text{EXE}}$	Execute the calculation.	0.9092 97426 8

On another algebraic machine, the appropriate key sequence might be $\boxed{\text{SIN}}\,\boxed{2}\,\boxed{=}$ or simply $\boxed{2}\,\boxed{\text{SIN}}$. There can be quite a bit of variation.

Example 2   *Reverse Polish Notation.*   To find the sine of 2 radians on an HP 28S, set the calculator in radian mode and press $\boxed{2}\,\boxed{\text{ENTER}}\,\boxed{\text{SIN}}$.

Key	Operation	Display
$\boxed{2}$	Put 2 in the command line.	2
$\boxed{\text{ENTER}}$	Enter it in stack level 1.	2
$\boxed{\text{SIN}}$	Find the sine of the number in stack level 1.	0.9092 97426 826

Questions of Accuracy

Although machines can calculate only the first few digits of an infinite decimal like

$$\sin 2 = 0.9092 \ 97426 \ 826 \ldots ,$$

they give estimates to more decimal places than you usually find in printed tables. They are also easier to carry and faster to use than tables.

However, when your inputs require the calculator to do arithmetic with numbers that are very small or very large, the errors associated with rounding and truncation (clipping without rounding) may produce a sizable error in the final result. Sometimes we can avoid round-off and truncation errors by restructuring the calculation. For example, suppose we want to evaluate the quotient

$$Q = \frac{\dfrac{1}{x + \Delta x} - \dfrac{1}{x}}{\Delta x} \tag{1}$$

when $x = 0.7$ and $\Delta x = 10^{-8}$. (We shall see why we might want to do this when we get to Section 1.6.) A direct evaluation gives

$$Q = \frac{1.4285 \ 71408 \ 8 - 1.4285 \ 71429}{10^{-8}}$$

$$= -\frac{0.0000 \ 00021}{10^{-8}} = -2.1. \tag{2}$$

The numbers originally in the numerator of Q are precise to ten digits, but their difference is precise to only two digits. We get a bad cancellation and a relatively imprecise answer.

We get a better result if we combine the fractions in the numerator of Q before we calculate, to get

$$Q = \frac{1}{\Delta x}\left(\frac{x - (x + \Delta x)}{(x + \Delta x)x}\right) = \frac{1}{\Delta x}\left(\frac{-\Delta x}{(x + \Delta x)x}\right) = -\frac{1}{(x + \Delta x)x}. \tag{3}$$

Now evaluation gives

$$Q = -\frac{1}{(0.7 + 10^{-8})(0.7)} = -2.0408\ 16298, \tag{4}$$

which is correct to 10 digits. If we just subtract, as we did in Eq. (2), we lose 8 digits and cannot hope to be anywhere near right.

At other times, the errors are an unavoidable consequence of how the calculator works, although some calculators work better than others. When we calculated the tangent of 89.99999° on three popular models, we got three different answers:

Calculator A: tan 89.99999° = 57295 45

Calculator B: tan 89.99999° = 57295 77

Calculator C: tan 89.99999° = 57295 80

Only Calculator B returned an answer that was correct in all seven places.

In this book, we shall not ask for calculations that push machines over the edge this way, unless the purpose of the exercise is to explore a calculator's limitations or to reveal the risk of attempting a particular kind of computation.

Exponential and Logarithmic Functions

In addition to having keys for trigonometric functions and functions like x^2, $\sqrt{x}$, and $1/x$, scientific calculators have keys for

e^x (the exponential function e to the x),

$\ln x$ (the natural logarithm of x),

$\log x$ (the base-10 logarithm of x).

The number e is about 2.7 1828 1828, and the function e^x plays a key role in describing growth, as does the natural logarithm function. The graphs of $y = e^x$ and $y = \ln x$ are shown in Figs. 1.39 and 1.40. Since $\log x = (\ln x)/\ln 10$, as we shall explain later in the book, the graph of $\log x$ is just a scaled-down version of the graph of $\ln x$. We shall study these and a number of other so-called transcendental functions in Chapter 6.

As you look at Figs. 1.39 and 1.40, notice the difference in the rates at which e^x and $\ln x$ grow as x increases. The function e^x grows rapidly — exponentially, in fact (which is where the adverb comes from). As we shall see, e^x eventually outgrows any power of x, even $x^{1,000,000}$, as x increases. In contrast, the natural logarithm $\ln x$ eventually grows more slowly as x increases than any fractional power of x, even $x^{1/1,000,000}$. As we shall see in Chapter 6, the functions e^x and $\ln x$ provide the standards by which we measure the growth rates of other functions.

Solving Equations

Calculus provides the mathematical justification for many of the things we do with a calculator, including the ways we solve equations.

The solutions of an equation $f(x) = 0$ are the x-coordinates of the points where

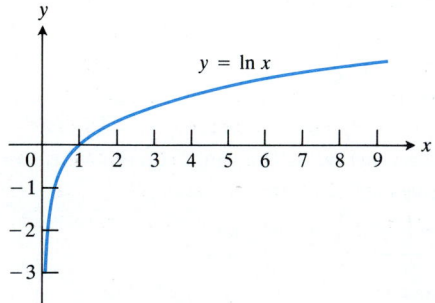

1.39   The graph of $y = e^x$.
Domain: $-\infty < x < \infty$
Range: $y > 0$

1.40   The graph of $y = \ln x$.
Domain: $x > 0$
Range: $-\infty < y < \infty$

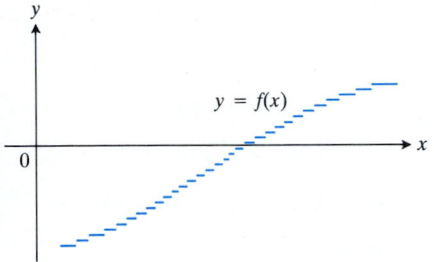

1.41   The graph of f steps across the axis without touching it. Hence, f changes from negative to positive without becoming zero in between. The equation $f(x) = 0$ has no solution.

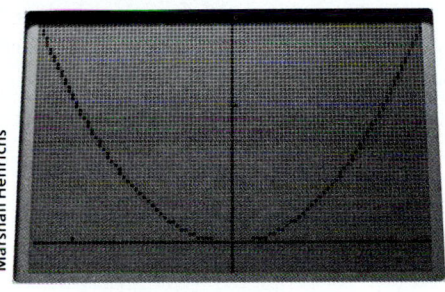

Marshall Henrichs

1.42   On a graphing calculator, the graph of $y = x^2 + 0.0001$ appears to pass through the origin even though y is never zero.

the graph of the equation $y = f(x)$ touches or crosses the x-axis. On a graphing calculator we can try to find these points by displaying the graph of f and reading the coordinates of the points of contact.

At least, that's the idea. But is there really a solution there? Does a curve that crosses the axis really touch the axis as it crosses? Perhaps it steps across the axis instead, without touching it, like the miniature step function in Fig. 1.41.

Also, does a curve that appears to be tangent to the axis really touch it? The function $y = x^2 + 0.0001$ is never zero, but this fact is certainly not revealed by the calculator screen in Fig. 1.42.

We answer these and many other questions about graphs by applying the theorems of calculus, as you will see in the next few chapters. Calculus will tell us when there is a solution (and when there isn't) and how many solutions we may find. It will also furnish us with a first-rate method for calculating solutions to as many decimal places as we please.

Comparing Function Values

In the 1640s, Henry Bond, a British teacher of surveying and navigation, discovered a close agreement between the tables of values of two functions used in navigation. In 1645, in Norwood's *Epitome of Navigation,* he published his conjecture that the two functions were equal. The conjecture became widely known, and in 1666 the mathematician Nicolas Mercator (no relation to the mapmaker) offered a prize to anyone who could settle it.

It was finally settled by James Gregory in 1668, but with a geometric proof so long that even the genius Edmund Halley (of comet fame) found it nearly too tiresome to read. The first intelligible proof involved calculus and was published by Newton's university mathematics teacher Isaac Barrow in 1669. One of the functions in question was

$$f(x) = \ln\left|\tan\left(\frac{x}{2} + \frac{\pi}{4}\right)\right|.$$

Barrow accomplished his proof by showing that the other function was

$$g(x) = \ln|\sec x + \tan x|,$$

which can be transformed into f with trigonometric identities.

Bond was led to conjecture the equality of the two functions by comparing tables of their values. Today, we can make such comparisons at will with a calculator. If we suspect that two functions f and g are equal or that they differ, say, by some constant value, we can test the possibility by calculating $f(x) - g(x)$ for a number of values of x. If the difference varies from one value of x to another, our suspicion is unfounded. But if the calculated differences are equal, or very nearly so, we may be on to something. Of course, knowing that $f(x) - g(x) = C$ for a few values of x does not prove equality for all values of x. But it makes the possibility worth investigating and gives a target value for C. You will find some nice examples in Exercises 7–12.

Successive Square Roots

Calculators can also reveal interesting behavior on the part of an individual function. If we start with a positive number like $x = 2$ and press the square root key

Pierre de Fermat (1601–1665)

Fermat, a skilled linguist and one of the seventeenth century's greatest mathematicians, refused to publish his work and rarely wrote completed descriptions even for his personal use. His famous unproved "last theorem" (that $a^n = b^n + c^n$ has no positive integer solutions for a, b, and c if $n > 2$) is known only from a note he jotted in the margin of a book. His name slipped into relative obscurity until the late 1800s, and it was only from a four-volume edition of his works published at the turn of the century that the true importance of his many achievements became clear.

Besides the work in physics and number theory for which he is best known, he found the areas under curves as limits of sums of rectangle areas (as we do today) and developed a method for finding the centroids of shapes bounded by curves in the plane. The standard formula for the first derivative of a polynomial function, the formulas for calculating arc length and finding the area of a surface of revolution, and the second derivative test for extreme values of functions can all be found in his papers. We shall see what these are as the text continues.

repeatedly, here is what we find:

$$x_0 = 2$$

$$x_1 = \sqrt{x_0} = 1.4142\ 13562$$
$$x_2 = \sqrt{x_1} = 1.1892\ 07115$$
$$x_3 = \sqrt{x_2} = 1.0905\ 07733$$
$$x_4 = \sqrt{x_3} = 1.0442\ 73782$$
$$x_5 = \sqrt{x_4} = 1.0218\ 97149$$
$$x_6 = \sqrt{x_5} = 1.0108\ 89286$$

$$x_7 = \sqrt{x_6} = 1.0054\ 29901$$
$$x_8 = \sqrt{x_7} = 1.0027\ 11275$$
$$x_9 = \sqrt{x_8} = 1.0013\ 5472$$
$$x_{10} = \sqrt{x_9} = 1.0006\ 77131$$
$$x_{11} = \sqrt{x_{10}} = 1.0003\ 38508$$
$$x_{12} = \sqrt{x_{11}} = 1.0001\ 6924$$
$$x_{13} = \sqrt{x_{12}} = 1.0000\ 84616$$

We could continue this further, but the two patterns we want you to see are already emerging. One is that the numbers seem to be approaching the value 1. Indeed they must, as we shall see in Chapter 8. The other pattern, less obvious perhaps but more intriguing, is that each keypress divides the decimal part of the number nearly in half. This behavior, too, is explained by calculus, as we shall see in Chapter 2.

These two phenomena have nothing particular to do with the starting number 2. Taking successive square roots of any number greater than 1 brings you ever closer to 1, eventually dividing the decimal part roughly in half with each keypress.

Is This Idea Any Good?

One of the ways Fermat tried to define the slope of a curve at a point P was to run a secant line through P and a nearby point Q and then watch the slope of the secant line as Q approached P (see Fig. 1.43). Every time he did this, the secant slopes seemed to approach a particular value that depended only on the location of P, so he wanted to call this value the slope of the curve at P. Was this a good idea?

To find out, we might try it at a particular point on a particular curve, say at the point $P\,(0, 0)$ on the curve $y = \sin x$. In the notation of Fig. 1.44, the point Q is the point $(\Delta x, \sin \Delta x)$. The slope of the secant line is therefore

$$m_{\text{sec}} = \frac{\sin \Delta x - 0}{\Delta x - 0} = \frac{\sin \Delta x}{\Delta x}. \tag{5}$$

What happens to the value of this ratio as Q approaches P along the curve?

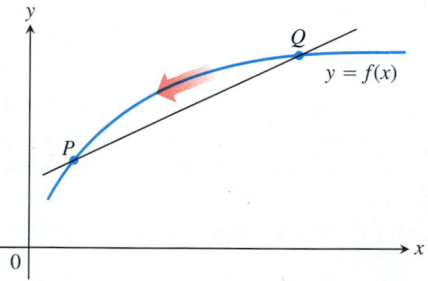

1.43 Fermat wanted to find out whether the slope of the secant shown here would approach a fixed value as Q approached P. If it did, he would call this value the slope of the curve at P.

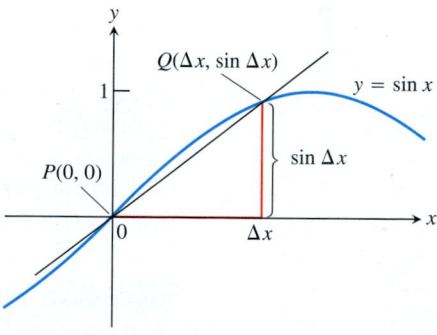

1.44 The slope of secant PQ is rise/run $= (\sin\Delta x)/\Delta x$.

To see what to expect, we might try it with a calculator, starting with $\Delta x = 1$ and dividing by 10 each time to get $\Delta x = 0.1, 0.01$, and so on. If we do this with the calculator in radian mode, we get the following values.

Δx (radians)	$(\sin \Delta x)/\Delta x$
1.0	0.8414 70984 8
0.1	0.9983 34166
0.01	0.9999 98333
0.001	0.9999 998
0.0001	0.9999 99
0.0000 1	1.0
0.0000 01	1
0.0000 001	1
0.0000 0001	1
0.0000 00001	1
0.0000 00000 1	1
0.0000 00000 01	0

$\left(\begin{array}{l} \text{The calculator thinks that} \\ \sin(0.00001) = 0.00001. \end{array} \right)$

$\left(\begin{array}{l} \Delta x \text{ is now so small that} \\ \text{the calculator thinks} \\ \sin(\Delta x) = 0. \end{array} \right)$

What are we to make of these numbers? It looks as if the values of $(\sin \Delta x)/\Delta x$ are steadily approaching 1. At least they do so until we reach the limit of our calculator's ability to distinguish $\sin \Delta x$ from Δx. Then the numbers remain at 1 until we reach the limit of the calculator's ability to distinguish $\sin \Delta x$ from 0.

If the calculator could work with arbitrarily many decimal places, would the ratios approach 1 forever? Would it be a good idea to define the slope of the sine curve at $x = 0$ to be 1? The answers, obtained from calculus, as you will soon see, are both yes.

EXERCISES 1.3

1. a) For how large a value of x can your calculator evaluate e^x?

b) Calculate e^x for $x = -1, -10, -100$, and -1000. How far to the left of zero can your calculator evaluate e^x without getting 0 or an error message?

2. a) For how large a value of x can your calculator evaluate $\ln x$? How far above the x-axis is the graph of $y = \ln x$ at this point?

b) The function $y = \ln x$ is defined for all positive values of x. What is the smallest value of x for which your calculator can find $\ln x$? How far below the x-axis is the graph of $y = \ln x$ at this point? How close is the graph to the y-axis there?

3. GRAPHING CALCULATOR   Graph $y = \sin x$ and estimate the value x^* where the graph first crosses the positive x-axis. To 9 decimal places, $\pi = 3.1415\ 92654$.

4. GRAPHING CALCULATOR   Graph the function
$$y = 10^{-6}(x - 1000)^3.$$

5. Evaluate the composite function
$$y = e^{(\ln x)}$$
for various values of x. What do you find? (This will be explained in Chapter 6.)

6. Evaluate the composite
$$y = \ln(e^x)$$
for various values of x. What do you find? (This will be explained in Chapter 6.)

In Exercises 7–12, test the functions to indicate whether they could differ by a constant value on some domain of x-values. If they appear to do so, see if you can prove that they really do and find an appropriate domain for x.

7. $\dfrac{x}{x+1}$ and $\dfrac{-1}{x+1}$

8. $\dfrac{x^2+3}{x^2+1}$ and $\dfrac{2}{x^2+1}$

9. $\tan x \sin 2x$ and $-2\cos^2 x$

10. $2\cos^2 x$ and $\cos 2x$ 11. $\ln 2x$ and $\ln x$

12. x and $\sqrt{x^2}$

13. *Successive square roots*

a) Enter the number 3 into your calculator. Then press the square root key repeatedly, pausing between keypresses to read the display. With each keypress you will find the decimal part of the display approximately halved.

b) Repeat part (a) with other numbers greater than 1.

14. *Continuation of Exercise 13.*   One way to describe the halving of the decimal parts of the square roots in Exercise 13 is to say that the number-line distance between 1 and the square root is approximately halved each time. What happens if you start with a positive number that is less than 1 instead of greater than 1? Do successive square roots approach 1 the same way? Try it with $x = 0.5$.

15. *Successive tenth roots.*   If you have a $\boxed{\sqrt{}}$ key or some other key that will enable you to calculate tenth roots, calculate successive tenth roots of 2, pausing to view each new display. What pattern do you see? (This will be explained in Chapter 2.)

16. *Continuation of Exercise 15.*   Repeat Exercise 15, starting with 0.5 instead of 2. What pattern do you see now?

17. *A fast estimate of $\pi/2$.*   Set your calculator in radian mode and enter the number $x_0 = 1$. Then calculate the successive numbers $x_1, x_2, \ldots$ in the following list.

$$x_0 = 1$$
$$x_1 = 1 + \cos 1$$
$$x_2 = x_1 + \cos x_1$$
$$x_3 = x_2 + \cos x_2$$
$$x_4 = x_3 + \cos x_3$$
$$\vdots$$
$$x_n = x_{n-1} + \cos x_{n-1} \quad \left(\begin{array}{c} \text{Formula for} \\ \text{generating the} \\ \text{sequence} \end{array} \right)$$

The numbers you calculate will soon begin to repeat at the value $x^* = 1.5707\ 96327$, which is $\pi/2$ to nine decimal places. Figure 1.45 explains why.

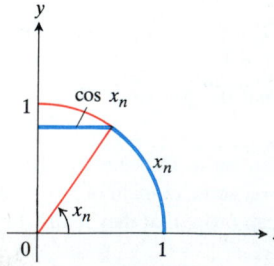

1.45   The length $\pi/2$ of the circular arc is approximated by $x_n + \cos x_n$ (Exercise 17).

18. *Radians vs. degrees.*   What happens if you use degrees instead of radians to calculate the slope of the curve $y = \sin x$ at $x = 0$? To find out, set your calculator in degree mode and make a table of values for

$$m_{\text{sec}} = \frac{\sin \Delta x}{\Delta x}.$$

Use $\Delta x = 0.1, 0.01, 0.001$, and so on, as far as your calculator can go. What do you find? Now multiply each value of m_{sec} in your table by the number $180/\pi$. What do you find?

19. *The function x^x (for calculators with an $\boxed{x^y}$ key).*   The rules of exponents tell us that $a^0 = 1$ if a is any number different from zero. They also tell us that $0^n = 0$ if n is any positive number.

If we tried to extend these rules to include the case 0^0, we would get conflicting results. The first rule would say $0^0 = 1$ while the second would say $0^0 = 0$.

We are not dealing with a question of right or wrong here. Neither rule applies as it stands, so there is no contradiction. We could, in fact, define 0^0 to have any value we wanted as long as we could sell other people on our idea.

What value would you like 0^0 to have? Here are some calculations that might help you decide: Calculate x^x for $x = 0.1, 0.01, 0.001$, and so on, as far as your calculator can go. Write down the value you get each time. What pattern do you see?

We shall say more about the function $y = x^x$ in Chapter 6.

20. *A reason you might want 0^0 to be something other than 0 or 1 (for calculators with an $\boxed{x^y}$ key).*   As the number x increases through positive values, the numbers $1/x$ and $1/(\ln x)$ approach zero. What happens to the number

$$f(x) = \left[\frac{1}{x} \right]^{1/(\ln x)}$$

as x increases?

To find out, evaluate f for $x = 10, 100, 1000$, and so on, as far as your calculator can reasonably go. What pattern do you see? (Chapter 6 will explain what is going on.)

21. *The slope of $y = \ln x$ at $x = 1$.*   In this exercise, we use Fermat's idea to find a numerical candidate for the slope of the curve $y = \ln x$ at the point $P(1, 0)$ where the graph crosses the x-axis (Fig. 1.46).

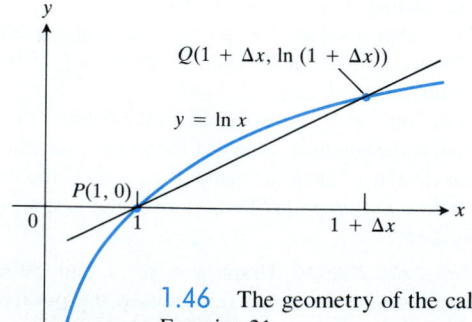

1.46   The geometry of the calculations in Exercise 21.

Make a table of values of

$$m_{\text{sec}} = \frac{\ln(1 + \Delta x) - 0}{(1 + \Delta x) - 1} = \frac{\ln(1 + \Delta x)}{\Delta x}$$

for $\Delta x = 0.1, 0.01, 0.001$, and so on, as far as your calculator can go. As far as you can tell, what would be a good value for the curve's slope at $x = 1$?

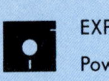

EXPLORER PROGRAMS

Power Grapher   Graphs calculator functions

Makes tables of values of functions and differences of functions

1.4 Absolute Value (Magnitude) and Target Values

As you may already know, the absolute value of a number is its numerical size or magnitude, and the function that sends each number to its magnitude is called the absolute value function. In calculus, we use absolute values to measure distances on the number line, to create formulas that define intervals, and to control function values. This section gives the details.

The Absolute Value Function

The **absolute value** or **magnitude** of a number x, denoted by $|x|$ (read "the absolute value of x"), is defined by the formula

$$|x| = \begin{cases} x \text{ if } x \geq 0 \\ -x \text{ if } x < 0. \end{cases} \tag{1}$$

The vertical lines in the symbol $|x|$ are called *absolute value bars*.

Example 1

$$|3| = 3, \qquad |0| = 0, \qquad |-5| = 5$$

The function $y = |x|$ is called the *absolute value function*. Its graph lies along the line $y = x$ when $x \geq 0$ and along the line $y = -x$ when $x < 0$ (Fig. 1.47).

Another way to find the absolute value of a number is to square the number and then take the positive square root:

$$|x| = \sqrt{x^2}. \tag{2}$$

1.47 The absolute value function.

Example 2

$$|2| = \sqrt{2^2} = \sqrt{4} = 2 \qquad \text{and} \qquad |-2| = \sqrt{(-2)^2} = \sqrt{4} = 2$$

To solve an equation that contains absolute values, we write equivalent equations without absolute values and then solve as usual.

Example 3   Solve the equation $|2x - 3| = 7$.

Solution   The equation says that $2x - 3 = \pm 7$, so there are two possibilities:

$$2x - 3 = 7 \qquad 2x - 3 = -7 \qquad \binom{\text{Equivalent equations without}}{\text{absolute values}}$$

$$2x = 10 \qquad 2x = -4 \qquad \text{(Solve as usual.)}$$

$$x = 5 \qquad x = -2$$

The equation $|2x - 3| = 7$ has two solutions: $x = 5$ and $x = -2$.

When we do arithmetic with absolute values, we always use the following rules.

Arithmetic with Absolute Values

1. $|-a| = |a|$ A number and its negative have the same absolute value.

2. $|ab| = |a||b|$ The absolute value of a product is the product of the absolute values.

3. $\left|\dfrac{a}{b}\right| = \dfrac{|a|}{|b|}$ The absolute value of a quotient is the quotient of the absolute values.

Example 4

$$|-\sin x| = |\sin x|$$

$$|-2(x + 5)| = |-2|\,|x + 5| = 2|x + 5|$$

$$\left|\frac{3}{x}\right| = \frac{|3|}{|x|} = \frac{3}{|x|}$$

The absolute value of a sum of two numbers is never larger than the sum of their absolute values. When we put this in symbols, we get the important triangle inequality.

The Triangle Inequality

$$|a + b| \le |a| + |b| \qquad \text{for all numbers } a \text{ and } b \tag{3}$$

Example 5

The number $|a + b|$ is less than $|a| + |b|$ if a and b have different signs. In all other cases, $|a + b|$ equals $|a| + |b|$.

$$|-3 + 5| = |2| = 2 < |-3| + |5| = 8$$

$$|3 + 5| = |8| = 8 = |3| + |5|$$

$$|-3 - 5| = |-8| = 8 = |-3| + |-5|$$

Notice that absolute value bars in expressions such as $|-3 + 5|$ also work like parentheses: We do the arithmetic inside *before* we take the absolute value.

Absolute Values and Distance

The numbers $|a - b|$ and $|b - a|$ are always equal because

$$|a - b| = |(-1)(b - a)| = |-1|\,|b - a| = |b - a|. \tag{4}$$

They give the distance between the points a and b on the number line (Fig. 1.48).

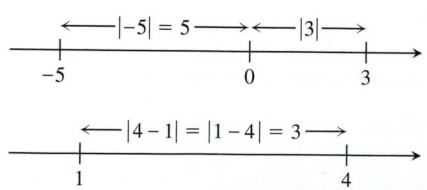

1.48   Absolute values give distances between points on the number line.

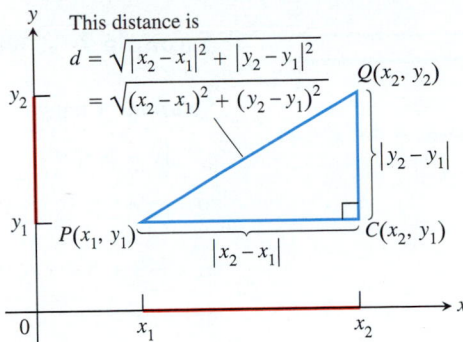

1.49 To calculate the distance between $P(x_1, y_1)$ and $Q(x_2, y_2)$, apply the Pythagorean theorem to triangle PCQ.

In the plane, the distance between points is calculated with a formula that comes from the Pythagorean theorem (Fig. 1.49).

Distance Formula for Points in the Plane

The distance between $P(x_1, y_1)$ and $Q(x_2, y_2)$ is

$$d = \sqrt{(x_2 - x_1)^2 + (y_2 - y_1)^2}. \tag{5}$$

Example 6   The distance between $P(-1, 2)$ and $Q(3, 4)$ is

$$\sqrt{(3 - (-1))^2 + (4 - 2)^2} = \sqrt{(4)^2 + (2)^2} = \sqrt{20} = \sqrt{4 \cdot 5} = 2\sqrt{5}.$$

Absolute Values and Intervals

The connection between absolute values and distance gives a new way to write formulas for intervals.

The inequality $|a| < 5$ says that the distance from a to the origin is less than 5. This is the same as saying that a lies between -5 and 5 on the number line. In symbols,

$$|a| < 5 \quad \Leftrightarrow \quad -5 < a < 5. \tag{6}$$

The set of numbers a with $|a| < 5$ is the open interval from -5 to 5 (Fig. 1.50).

The general rule is this:

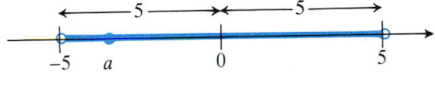

1.50 $|a| < 5$ means $-5 < a < 5$.

$\Rightarrow$ means "implies" (and is read that way).
$\Leftrightarrow$ means "if and only if."

Relation Between Intervals and Absolute Values

If D is any positive number, then

$$|a| < D \quad \Leftrightarrow \quad -D < a < D, \tag{7}$$

$$|a| \leq D \quad \Leftrightarrow \quad -D \leq a \leq D. \tag{8}$$

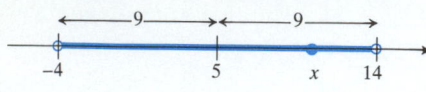

1.51 $|x - 5| < 9$ means $-4 < x < 14$.

Example 7   What values of x satisfy the inequality $|x - 5| < 9$?

Solution   Change

$$|x - 5| < 9$$

to $-9 < x - 5 < 9$ (Eq. (7) with $a = x - 5$ and $D = 9$)

to $-9 + 5 < x < 9 + 5$ $\left(\begin{array}{l}\text{Adding a number to both sides of an}\\ \text{inequality gives an equivalent inequality.}\\ \text{Adding 5 here isolates the } x.\end{array}\right)$

or $-4 < x < 14.$

The steps we just took are reversible, so the values of x that satisfy the inequality $|x - 5| < 9$ are the numbers in the interval $-4 < x < 14$ (Fig. 1.51). Here 14 is the upper bound on x and -4 is the lower bound.

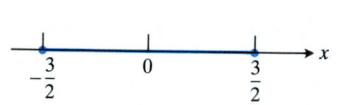

1.52   The midpoint of a finite interval is the average of the endpoint values (Example 8).

Example 8   Describe the interval $-3 < x < 5$ with an absolute value inequality of the form $|x - x_0| < D$.

Solution   We average the endpoint values to find the interval's midpoint:

$$\text{midpoint } x_0 = \frac{-3 + 5}{2} = \frac{2}{2} = 1.$$

The midpoint lies 4 units away from each endpoint (Fig. 1.52). The interval therefore consists of the points that lie within 4 units of the midpoint, or the points x with

$$|x - 1| < 4.$$

Example 9   What values of x satisfy the inequality $\left|\dfrac{2x}{3}\right| \leq 1$?

Solution   Change

$$\left|\frac{2x}{3}\right| \leq 1$$

to $-1 \leq \dfrac{2x}{3} \leq 1$ (Eq. (8) with $a = 2x/3$ and D = 1)

to $-3 \leq 2x \leq 3$ $\left(\begin{array}{l}\text{Multiplying both sides of an inequality by a}\\ \text{positive number gives an equivalent inequality.}\end{array}\right)$

to $-\dfrac{3}{2} \leq x \leq \dfrac{3}{2}.$ $\left(\begin{array}{l}\text{Dividing both sides of an inequality by a}\\ \text{positive number gives an equivalent inequality.}\end{array}\right)$

1.53   The inequality $|2x/3| \leq 1$ holds on the interval $-3/2 \leq x \leq 3/2$.

The original inequality holds for x in the closed interval from $-3/2$ to $3/2$ (Fig. 1.53).

Example 10   What values of x satisfy the inequality

$$\left|5 - \frac{2}{x}\right| < 1?$$

Solution   Change

$$\left|5 - \frac{2}{x}\right| < 1$$

to $-1 < 5 - \dfrac{2}{x} < 1$ (Eq. (7) with $a = 5 - 2/x$ and $D = 1$)

to $-6 < -\dfrac{2}{x} < -4$ $\left(\begin{array}{l}\text{Subtracting a number, in this case 5, from}\\ \text{both sides of an inequality gives an}\\ \text{equivalent inequality.}\end{array}\right)$

to $4 < \dfrac{2}{x} < 6$ $\left(\begin{array}{l}\text{Multiplying both sides of an equality by}\\ -1 \text{ reverses the inequality.}\end{array}\right)$

to $2 < \dfrac{1}{x} < 3$ (Divide by 2.)

to $\dfrac{1}{3} < x < \dfrac{1}{2}.$ $\left(\begin{array}{l}\text{Take reciprocals. When the numbers involved have}\\ \text{the same sign, taking reciprocals reverses an}\\ \text{inequality.}\end{array}\right)$

The original inequality holds if and only if x lies between $1/3$ and $1/2$.

Keeping Function Outputs near Target Values

We sometimes want the outputs of a function $y = f(x)$ to lie near a particular target value y_0. This need can come about in different ways. A gas station attendant, asked for $5.00 worth of gas, will try to pump the gas to the nearest cent. A mechanic grinding a 3.385-inch cylinder bore will not let the bore exceed this value by more than 0.002 in. A pharmacist making ointments will measure the ingredients to the nearest milligram.

So the question becomes: How accurate do our machines and instruments have to be to keep the outputs within useful bounds? When we express this question with mathematical symbols, we ask: How closely must we control x to keep $y = f(x)$ within an acceptable interval about some particular target value y_0? The following examples show how to answer this question.

Example 11   *Controlling a Linear Function.*   How close to $x_0 = 4$ must we hold x to be sure that $y = 2x - 1$ lies within 2 units of $y_0 = 7$?

Solution   We are asked: For what values of x is $|y - 7| < 2$? To find the answer, we first express $|y - 7|$ in terms of x:

$$|y - 7| = |(2x - 1) - 7| = |2x - 8|.$$

The question then becomes: What values of x satisfy the inequality $|2x - 8| < 2$? To find out, we solve the inequality:

$$|2x - 8| < 2$$
$$-2 < 2x - 8 < 2$$
$$6 < 2x < 10$$
$$3 < x < 5.$$

To keep y within 2 units of $y_0 = 7$, we must keep x within 1 unit of $x_0 = 4$ (Fig. 1.54).

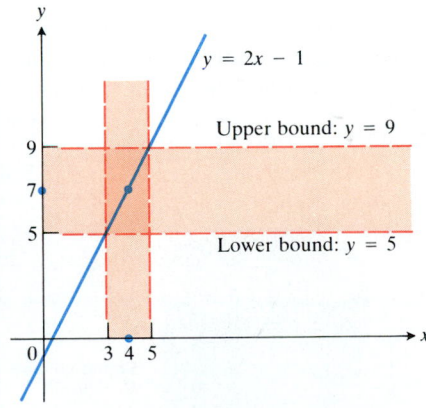

1.54   Keeping x between 3 and 5 will keep $y = 2x - 1$ between $y = 5$ and $y = 9$.

Example 12   *Controlling the Area of a Circle.*   In what interval about $r_0 = 10$ must we hold r to be sure that $A = \pi r^2$ lies within π square units of $A_0 = 100\pi$?

Solution   We want to know the values of r for which $|A - A_0| < \pi$. To find them, we solve the inequality:

$$|A - A_0| < \pi$$

$$|\pi r^2 - 100\pi| < \pi$$

$$-\pi < \pi r^2 - 100\pi < \pi$$

$$-1 < r^2 - 100 < 1$$

$$99 < r^2 < 101$$

$$\sqrt{99} < r < \sqrt{101}.$$ $\left(\begin{array}{l}\text{For } a, \text{ } b, \text{ and } c \text{ nonnegative,} \\ a < b < c \Leftrightarrow \sqrt{a} < \sqrt{b} < \sqrt{c}.\end{array}\right)$

The interval of possible radii is the open interval from $r = \sqrt{99}$ to $r = \sqrt{101}$ (Fig. 1.55).

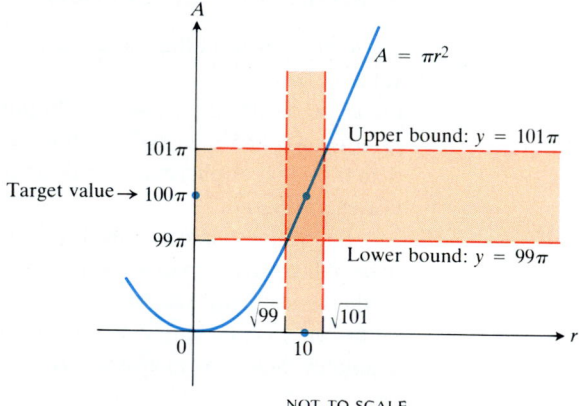

1.55   Keeping r between $\sqrt{99}$ and $\sqrt{101}$ will keep πr^2 between 99π and 101π.

NOT TO SCALE

Example 13   *Why the Stripes on a 1-Liter Kitchen Measuring Cup Are About a Millimeter Wide.*   The interior of a typical 1-L measuring cup is a right circular cylinder of radius 6 cm (Fig. 1.56). The volume of water we put in the cup is therefore a function of the level h to which the cup is filled, the formula being

$$V = \pi 6^2 h = 36\pi h.$$

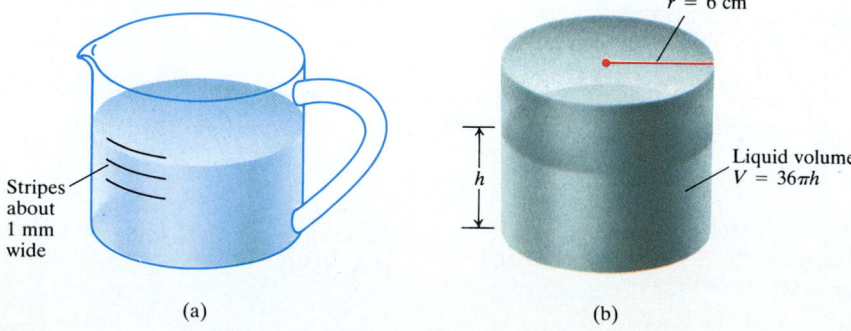

1.56   A typical 1-liter measuring cup (a), modeled in (b) as a right circular cylinder of radius $r = 6$ cm. To get a liter of water to the nearest 1%, how accurately must we measure h? See Example 13.

Stripes about 1 mm wide

(a)

(b)

How closely do we have to measure h to measure out one liter of water (1000 cm³) with an error of no more than 1% (10 cm³)?

Solution   In terms of V and h, we want to know in what interval to hold values of h to make V satisfy the inequality

$$|V - 1000| = |36\pi h - 1000| \leq 10.$$

To find out, we solve the inequality:

$$|36\pi h - 1000| \leq 10$$

$$-10 \leq 36\pi h - 1000 \leq 10$$

$$990 \leq 36\pi h \leq 1010$$

$$\frac{990}{36\pi} \leq h \leq \frac{1010}{36\pi}$$

$$8.8 \leq h \leq 8.9. \qquad \text{(Values found with a calculator)}$$

rounded up,   rounded down,
to be safe   to be safe

The interval in which we should hold h is about 0.1 centimeter wide (1 millimeter). With stripes 1 millimeter wide, we can therefore expect to measure a liter of water with an accuracy of 1%, which is more than enough accuracy for cooking.

Properties of Inequalities Used in This Section

1. $a < b \quad \Rightarrow \quad a + c < b + c$

2. $a < b \quad \Rightarrow \quad a - c < b - c$

3. $a < b$ and $c > 0 \quad \Rightarrow \quad ac < bc$

4. $a < b$ and $c > 0 \quad \Rightarrow \quad \dfrac{a}{c} < \dfrac{b}{c}$

5. $a < b \quad \Rightarrow \quad -b < -a$

6. $a < b \quad \Rightarrow \quad \dfrac{1}{b} < \dfrac{1}{a}$ $\left(\begin{array}{l}\text{Requires } a \text{ and } b \text{ both positive}\\ \text{or both negative}\end{array}\right)$

These hold for $\leq$ as well as for $<$.

EXERCISES 1.4

In Exercises 1–6, find the distances between the given points.

1. $(1, 0)$ and $(0, 1)$

2. $(2, 4)$ and $(-1, 0)$

3. $(2\sqrt{3}, 4)$ and $(-\sqrt{3}, 1)$

4. $(2, 1)$ and $(1, -1/3)$

5. (a, b) and $(0, 0)$

6. $(0, y)$ and $(x, 0)$

Find the absolute values in Exercises 7–10.

7. $|-3|$

8. $|2 - 7|$

9. $|-2 + 7|$

10. $|1.1 - 5.2|$

11. If $2 < x < 6$, which of the following statements about x are true and which are false?

a) $0 < x < 4$

b) $0 < x - 2 < 4$

c) $1 < \dfrac{x}{2} < 3$

d) $\dfrac{1}{6} < \dfrac{1}{x} < \dfrac{1}{2}$

e) $1 < \dfrac{6}{x} < 3$

f) $|x - 4| < 2$

g) $+6 < -x < 2$

h) $+6 < -x < +2$

12. If $-1 < y - 5 < 1$, which of the following statements about y are true and which are false?

a) $4 < y < 6$

b) $|y - 5| < 1$

c) $y > 4$

d) $y < 6$

e) $0 < y - 4 < 2$

f) $2 < \dfrac{y}{2} < 3$

g) $\dfrac{1}{6} < \dfrac{1}{y} < \dfrac{1}{4}$

h) $-6 < y < -4$

Solve the equations in Exercises 13–18.

13. $|x| = 2$

14. $|x - 3| = 7$

15. $|2t + 5| = 4$

16. $|1 - t| = 1$

17. $|8 - 3s| = 9$

18. $\left|\dfrac{s}{2} - 1\right| = 1$

In Exercises 19–24, match each absolute value inequality with the interval it determines.

Absolute value inequality   **Interval**

19. $|x + 3| < 1$

a) $-2 < x < 1$

20. $\left|\dfrac{x}{2}\right| < 1$

b) $-1 < x < 3$

21. $|1 - x| < 2$

c) $-2 < x < 2$

22. $|2x - 5| \le 1$

d) $-4 < x < 4$

23. $\left|\dfrac{x - 1}{2}\right| < 1$

e) $-4 < x < -2$

f) $2 \le x \le 3$

24. $\left|\dfrac{2x + 1}{3}\right| < 1$

g) $-2 \le x \le 2$

The inequalities in Exercises 25–32 define intervals. Describe each interval with inequalities that do not involve absolute values.

25. $|y| < 2$

26. $|y| \le 2$

27. $|r - 1| \le 2$

28. $|r + 2| < 1$

29. $|3s - 7| < 2$

30. $|2s + 5| < 1$

31. $\left|\dfrac{t}{2} - 1\right| \le 1$

32. $\left|2 - \dfrac{t}{2}\right| < \dfrac{1}{2}$

Describe the intervals in Exercises 33–36 with absolute value inequalities of the form $|x - x_0| < D$. It may help to draw a picture of the interval first.

33. $3 < x < 9$

34. $-3 < x < 9$

35. $-5 < x < 3$

36. $-7 < x < -1$

Each of Exercises 37–42 gives a function $y = f(x)$, a number E, and a target value y_0. In what interval must we hold x in each case to be sure that $y = f(x)$ lies within E units of y_0?

37. $y = x^2$, $E = 0.1$, $y_0 = 100$

38. $y = x^2 - 5$, $E = 1$, $y_0 = 11$

39. $y = \sqrt{x - 7}$, $E = 0.1$, $y_0 = 4$

40. $y = \sqrt{19 - x}$, $E = 1$, $y_0 = 3$

41. $y = 120/x$, $E = 1$, $y_0 = 5$

42. $y = 1/(4x)$, $E = 1/2$, $y_0 = 1$

Each of Exercises 43–46 gives a function $y = f(x)$, a number E, a point x_0, and a target value y_0. In what interval about x_0 must we hold x in each case to be sure that $y = f(x)$ lies within E units of y_0? Describe the interval with an absolute value inequality of the form $|x - x_0| < D$.

43. $y = x + 1$, $E = 0.5$, $x_0 = 3$, $y_0 = 4$

44. $y = 2x - 1$, $E = 1$, $x_0 = -2$, $y_0 = -5$

45. $y = -(x/2) + 1$, $E = 1/2$, $x_0 = 6$, $y_0 = -2$

46. $y = -2x - 2$, $E = 0.2$, $x_0 = -3$, $y_0 = 4$

47. CALCULATOR *Grinding engine cylinders.* Before contracting to grind engine cylinders to a cross-section area of 9 in^2, you need to know how much deviation from the ideal cylinder diameter of $x_0 = 3.385$ in. you can allow and still have the area come within 0.01 in^2 of the required 9 in^2. To find out, you let $A = \pi(x/2)^2$ and look for the interval in which you must hold x to make $|A - 9| \le 0.01$. What interval do you find?

48. *Manufacturing electrical resistors.* Ohm's law for electrical circuits, like the one shown in Fig. 1.57, states that $V = RI$. In this equation, V is a constant voltage, I is the current in amperes, and R is the resistance in ohms. Your firm has been asked to supply the resistors for a circuit in which V will be 120 volts and I is to be 5 ± 0.1 amperes. In what interval does R have to lie for I to be within 0.1 ampere of the target value $I_0 = 5$?

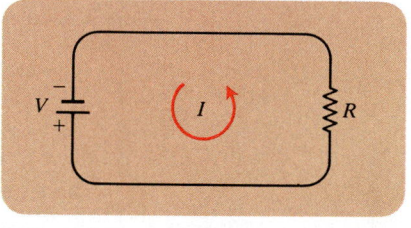

1.57 The circuit in Exercise 48.

49. Do not fall into the trap $|-a| = a$. The equation does not hold for all values of a.

a) Find a value of a for which $|-a| \ne a$.

b) For what values of a does the equation $|-a| = a$ hold?

50. For what values of x does $|1 - x|$ equal $1 - x$? For what values of x does it equal $x - 1$?

51. Compare the domains and ranges of the functions $y = \sqrt{x^2}$ and $y = (\sqrt{x})^2$.

52. Find $f(x)$ if $g(x) = \sqrt{x}$ and $(g \circ f)(x) = |x|$.

53. Find $g(x)$ if $f(x) = x^2 + 2x + 1$ and $(g \circ f)(x) = |x + 1|$.

54. Find functions $f(x)$ and $g(x)$ whose composites satisfy the two equations

$$(g \circ f)(x) = |\sin x| \quad \text{and} \quad (f \circ g)(x) = (\sin \sqrt{x})^2.$$

1.5 Shifts, Circles, and Parabolas

In this section, we show how to change an equation to move its graph up or down or to the right or left in the coordinate plane. We practice mostly with circles and parabolas (because they make useful examples in calculus), but the methods we use apply to other curves as well. We shall have more to say about parabolas and circles when we discuss conic sections in Chapter 9.

How to Shift a Graph

To shift the graph of a function $y = f(x)$ straight up, add a positive constant to the right-hand side of the formula $y = f(x)$.

Example 1   Adding 1 to the right-hand side of the formula $y = x^2$ to get $y = x^2 + 1$ shifts the graph up 1 unit (Fig. 1.58).

To shift the graph of a function $y = f(x)$ straight down, subtract a positive constant from the right-hand side of the formula $y = f(x)$.

Example 2   Subtracting 2 from the right-hand side of the formula $y = x^2$ to get $y = x^2 - 2$ shifts the graph down 2 units (Fig. 1.58).

To shift the graph of $y = f(x)$ to the left, add a positive constant to x.

Example 3   Adding 3 to x in $y = x^2$ to get $y = (x + 3)^2$ shifts the graph 3 units to the left (Fig. 1.59).

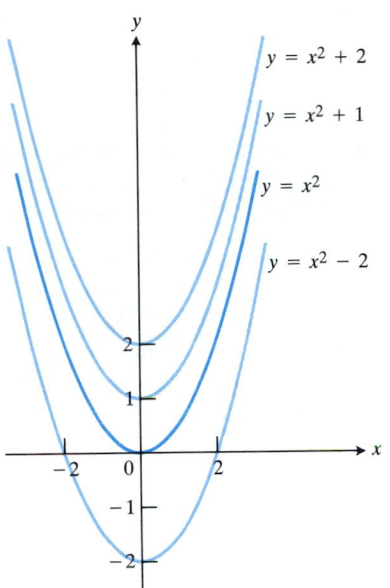

$y = x^2 + 2$
$y = x^2 + 1$
$y = x^2$
$y = x^2 - 2$

1.58   To shift the graph of $f(x) = x^2$ up (or down), we add positive constants to (or subtract them from) the formula for f.

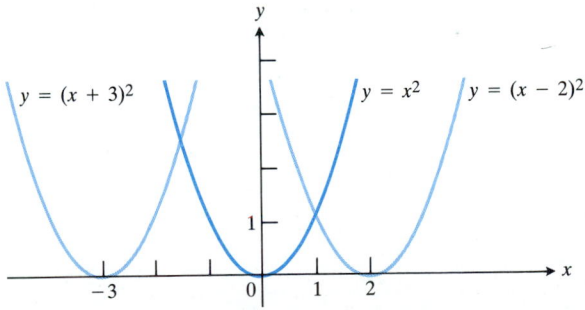

$y = (x + 3)^2$ $y = x^2$ $y = (x - 2)^2$

1.59   To shift the graph of $y = x^2$ to the left, we add a positive constant to x. To shift the graph to the right, we subtract a positive constant from x.

To shift the graph of $y = f(x)$ to the right, subtract a positive constant from x.

Example 4   Subtracting 2 from x in $y = x^2$ to get $y = (x - 2)^2$ shifts the graph 2 units to the right (Fig. 1.59).

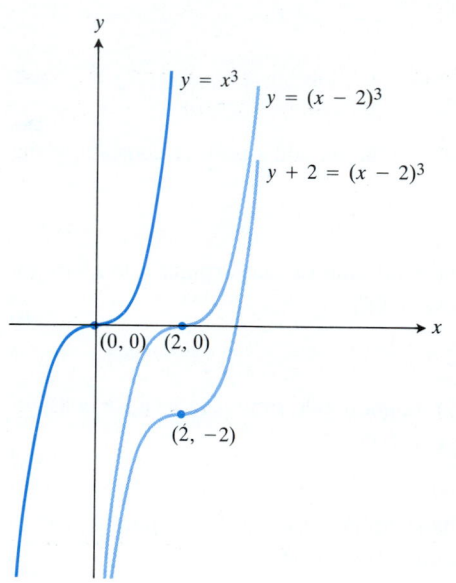

$y = x^3$

$y = (x - 2)^3$

$y + 2 = (x - 2)^3$

$(0, 0)$ $(2, 0)$

$(2, -2)$

1.60 The graph of $y = x^3$ shifted 2 units right and then 2 units down.

Shift Formulas $(c > 0)$

Vertical shifts

$$y = f(x) + c \quad \text{or} \quad y - c = f(x). \qquad \text{Shifts the graph of } f \text{ up } c \text{ units}$$

$$y = f(x) - c \quad \text{or} \quad y + c = f(x) \qquad \text{Shifts the graph of } f \text{ down } c \text{ units}$$

Horizontal shifts

$$y = f(x + c) \qquad \text{Shifts the graph of } f \text{ left } c \text{ units}$$

$$y = f(x - c) \qquad \text{Shifts the graph of } f \text{ right } c \text{ units}$$

The formula for a vertical shift upward is often written as $y - c = f(x)$ instead of $y = f(x) + c$. But there is another way to think of it. For $y - c$ to match a given x, the value of y has to be c units larger than before. The graph of y vs. x therefore must be c units higher.

Example 5   At each value of x, the graph of $y - 1 = x^2$ is 1 unit higher than the graph of $y = x^2$. See Fig. 1.58, where the graph is labeled $y = x^2 + 1$.

The formula for a vertical downward shift of c units is often written as $y + c = f(x)$.

Example 6   The graph of $y + 2 = (x - 2)^3$ lies 2 units below the graph of $y = x^3$ and 2 units to the right (Fig. 1.60).

Equations for Circles in the Plane

DEFINITIONS

> A **circle** is the set of points in a plane whose distance from a given fixed point in the plane is a constant. The fixed point is the **center** of the circle. The constant distance is the **radius** of the circle.

To write an equation for the circle of radius a centered at the point $C\,(h, k)$, we let $P\,(x, y)$ denote a typical point on the circle (Fig. 1.61). The statement that CP equals a then becomes

$$\sqrt{(x - h)^2 + (y - k)^2} = a$$

or
$$(x - h)^2 + (y - k)^2 = a^2. \tag{1}$$

If $CP = a$, then Eq. (1) holds. If Eq. (1) holds, then $CP = a$. Equation (1) is therefore an equation for the circle.

1.61 The standard equation for the circle shown here is $(x - h)^2 + (y - k)^2 = a^2$.

> **The Standard Equation for the Circle of Radius a Centered at the Point (h, k)**
>
> $$(x - h)^2 + (y - k)^2 = a^2 \tag{2}$$

Example 7   The standard equation for the circle of radius 2 centered at the point (3, 4) is

$$(x - 3)^2 + (y - 4)^2 = (2)^2$$

or

$$(x - 3)^2 + (y - 4)^2 = 4.$$

There is no need to square out the x and y terms in this equation. In fact, it is better not to do so. The present form reveals the circle's center and radius.

Example 8   Find the center and radius of the circle

$$(x - 1)^2 + (y + 5)^2 = 3.$$

Solution   Comparing

$$(x - h)^2 + (y - k)^2 = a^2 \qquad \text{with} \qquad (x - 1)^2 + (y + 5)^2 = 3$$

shows that $h = 1$, $k = -5$, and $a = \sqrt{3}$. The center is the point $(h, k) = (1, -5)$. The radius is $a = \sqrt{3}$.

For circles centered at the origin, h and k are 0 and Eq. (2) simplifies to $x^2 + y^2 = a^2$.

The Standard Equation for the Circle of Radius a Centered at the Origin

$$x^2 + y^2 = a^2 \tag{3}$$

The circle of radius 1 unit centered at the origin is called **the unit circle.**

Notice that the circle $(x - h)^2 + (y - k)^2 = a^2$ is the same as the circle $x^2 + y^2 = a^2$ with its center shifted from the origin to the point (h, k). The shift formulas we have been using for graphs of functions apply to equations of any kind. Shifts to the right and up are accomplished by subtracting positive values of h and k. Shifts to the left and down are accomplished by subtracting negative values of h and k.

Example 9   If the circle $x^2 + y^2 = 25$ is shifted two units to the left and three units up, its new equation is

$$(x - (-2))^2 + (y - 3)^2 = 25 \qquad \text{or} \qquad (x + 2)^2 + (y - 3)^2 = 25.$$

As Eq. (2) says it should be, this is the equation of the circle of radius 5 centered at $(h, k) = (-2, 3)$.

The points that lie inside the circle $(x - h)^2 + (y - k)^2 = a^2$ are the points less than a units from (h, k). They satisfy the inequality

$$(x - h)^2 + (y - k)^2 < a^2 \tag{4}$$

and make up the region we call the **interior** of the circle (Fig. 1.62).

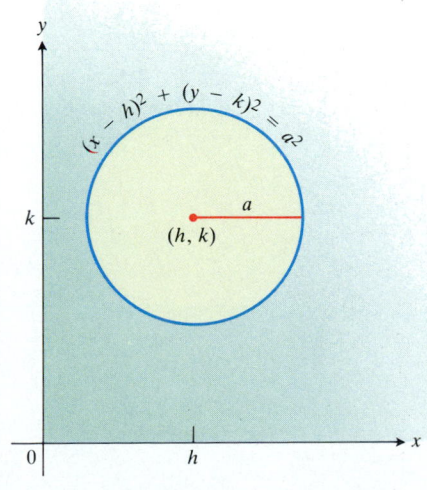

1.62   The interior and exterior of the circle $(x - h)^2 + (y - k)^2 = a^2$.

The circle's **exterior** consists of the points that lie more than a units from (h, k). These points satisfy the inequality

$$(x - h)^2 + (y - k)^2 > a^2. \tag{5}$$

Example 10

Inequality	Region
$x^2 + y^2 < 1$	Interior of the unit circle
$x^2 + y^2 \leq 1$	Unit circle plus its interior
$x^2 + y^2 > 1$	Exterior of the unit circle
$x^2 + y^2 \geq 1$	Unit circle plus its exterior

DRAWING LESSON

Tip for Drawing Circles: Circle first, axes later

1.
The basic shape:

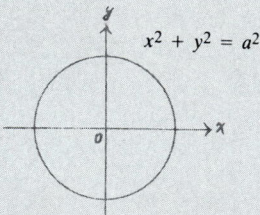

2.
The basic shape with axes in different positions:

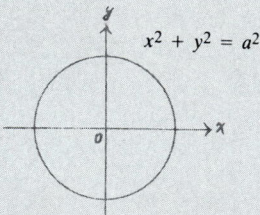

$x^2 + y^2 = a^2$

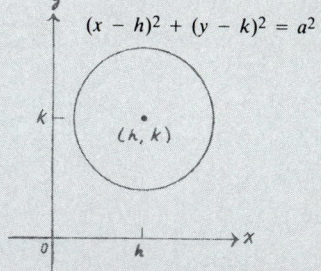

$(x - h)^2 + (y - k)^2 = a^2$

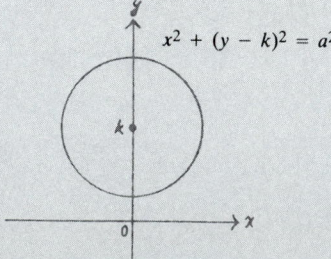

$x^2 + (y - k)^2 = a^2$

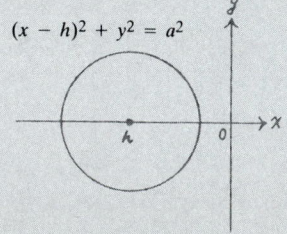

$(x - h)^2 + y^2 = a^2$

Equations for Parabolas

DEFINITIONS

> A **parabola** is the set of points in a plane that are equidistant from a given fixed point and fixed line in the plane. The fixed point is the parabola's **focus.** The fixed line is the parabola's **directrix.**

If the focus F were to lie on the directrix L, the parabola would be a line through F perpendicular to L. We regard this as a degenerate case and assume from now on that F never lies on L.

As you may know, parabolas have relatively simple equations when their foci (read "FOE-sigh") and directrices ("di-REC-tri-sees") straddle the coordinate axes. We begin with parabolas that open upward.

Parabolas That Open Upward   Suppose the focus lies at $F(0, p)$ on the positive y-axis and that the directrix is the line $y = -p$ (Fig. 1.63). In the notation of the figure, a point $P(x, y)$ lies on the parabola if and only if

$$PF = PQ. \tag{6}$$

From the distance formula,

$$PF = \sqrt{(x - 0)^2 + (y - p)^2} = \sqrt{x^2 + (y - p)^2}, \tag{7}$$

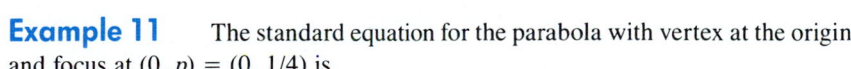

$$PQ = \sqrt{(x - x)^2 + (y - (-p))^2} = \sqrt{(y + p)^2}. \tag{8}$$

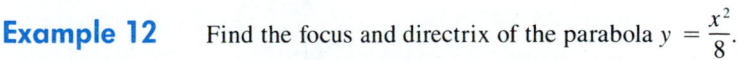

When we equate these expressions, square, and simplify, we get

$$y = \frac{x^2}{4p}. \qquad \text{(Standard equation)} \tag{9}$$

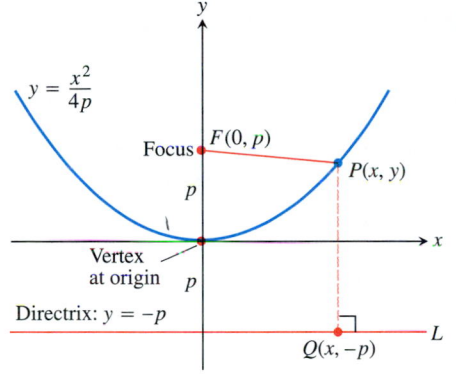

1.63   The parabola $y = x^2/4p$. Notice that the vertex lies halfway between the directrix and focus.

This equation reveals the parabola's symmetry about the y-axis. We call the y-axis the **axis** of the parabola (short for "axis of symmetry").

The point where a parabola crosses its axis, midway between the focus and directrix, is called the **vertex** of the parabola. The vertex of the parabola $y = x^2/4p$ lies at the origin. The number p is the **focal length** of the parabola, and $4p$ is the **width of the parabola at the focus** (Exercise 45).

Example 11   The standard equation for the parabola with vertex at the origin and focus at $(0, p) = (0, 1/4)$ is

$$y = \frac{x^2}{4(1/4)}, \qquad \text{or} \qquad y = x^2. \qquad \text{(Eq. (9) with } p = 1/4)$$

Example 12   Find the focus and directrix of the parabola $y = \dfrac{x^2}{8}$.

Solution

STEP 1: Find the value of p in the standard equation:

$$y = \frac{x^2}{8} \quad \text{is} \quad y = \frac{x^2}{4p} \quad \text{with} \quad p = 2.$$

STEP 2: Find the focus and directrix for this value of p:

$$\text{Focus:} \quad (0, p) = (0, 2) \qquad \text{Directrix:} \quad y = -p = -2.$$

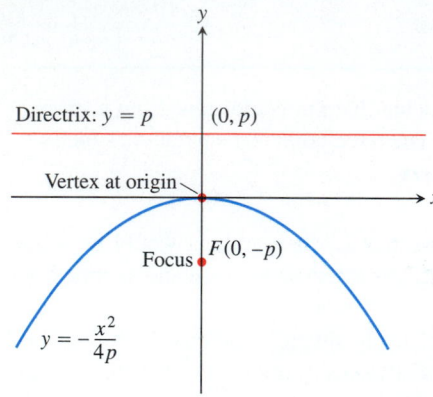

1.64   The parabola $y = -x^2/4p$.

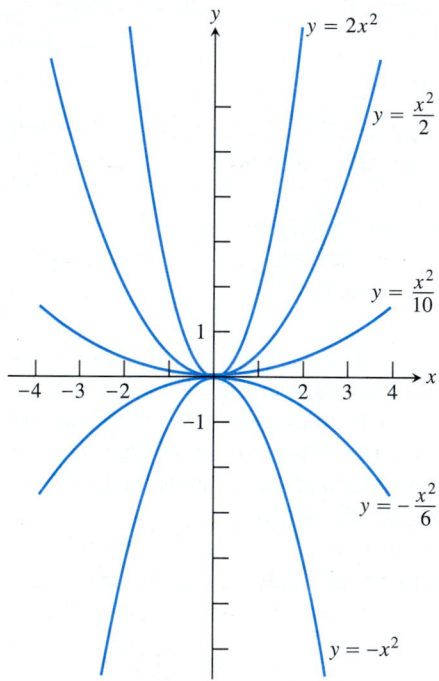

1.65   The parabola $y = ax^2$ opens upward if a is positive and downward if a is negative.

1.66   Typical examples of the parabola $y = ax^2 + bx + c$. In Chapter 3, we shall learn a quick way to locate the vertices of parabolas like these, using calculus.

Parabolas That Open Downward   If the parabola opens downward with its focus at $F(0, -p)$ and its directrix the line $y = p$ (Fig. 1.64), Eq. (9) becomes

$$y = -\frac{x^2}{4p}. \tag{10}$$

From Eqs. (9) and (10), we can see that any equation — such as $y = 3x^2$ or $y = -5x^2$ — that has the form

$$y = ax^2 \tag{11}$$

is the equation of a parabola that is symmetric about the y-axis with vertex at the origin. The parabola opens upward if a is positive and downward if a is negative (Fig. 1.65). The number a also serves as a scaling factor. The parabola widens as a approaches zero and narrows as a becomes numerically large.

The Parabola $y = ax^2 + bx + c$ To shift the graph of the parabola $y = ax^2$ horizontally, we rewrite its equation as

$$y = a(x - h)^2. \tag{12}$$

This shifts the parabola to the right if h is positive and to the left if h is negative.

To shift the parabola vertically as well, we change the equation to

$$y - k = a(x - h)^2. \tag{13}$$

This shifts the parabola up if k is positive and down if k is negative.

Normally, there would be no point in squaring out the right-hand side of Eq. (13). In this case, however, we can learn something from doing so because the resulting equation, when rearranged, takes the form

$$y = ax^2 + bx + c. \tag{14}$$

This tells us that the graph of every equation of the form $y = ax^2 + bx + c$ is a shifted version of the parabola $y = ax^2$. Why? Because the steps that take us from Eq. (13) to (14) can be reversed to take us from (14) back to (13).

The Graph of $y = ax^2 + bx + c$

The graph of $y = ax^2 + bx + c$ is a parabola whose axis is parallel to the y-axis. The parabola opens upward if $a > 0$ and downward if $a < 0$. (See Fig. 1.66 for typical examples.)

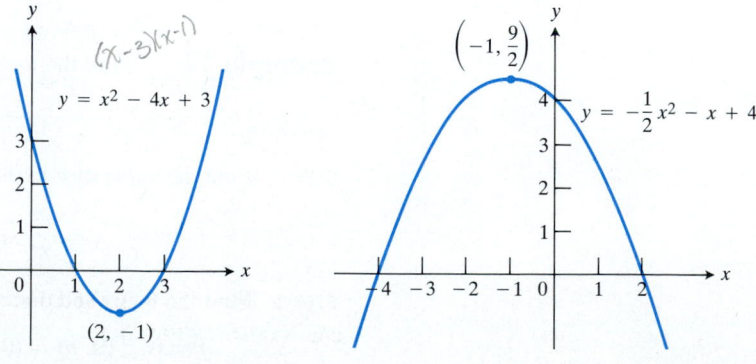

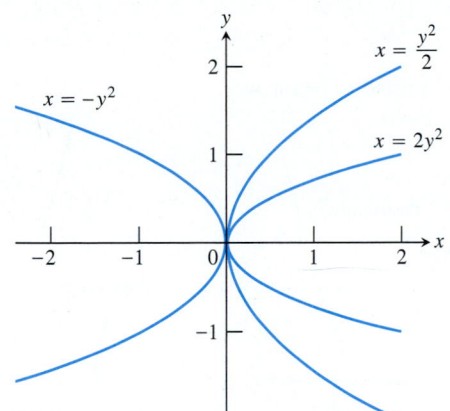

1.67   The parabola $x = ay^2$ opens to the right if $a > 0$ and to the left if $a < 0$.

Parabolas That Open to the Right or Left   If we interchange x and y in the formula $y = ax^2$, we obtain the equation

$$x = ay^2. \tag{15}$$

With the roles of x and y now reversed, the graph is a parabola whose axis is the x-axis. The vertex still lies at the origin. The parabola opens to the right if $a > 0$ and to the left if $a < 0$ (Fig. 1.67).

Example 13   The formula $x = y^2$ gives x as a function of y but does *not* give y as a function of x. If we solve for y, we find that

$$y = \pm\sqrt{x}.$$

For each positive value of x we get *two* values of y instead of the required single value.

When taken separately, however, the formulas

$$y = \sqrt{x} \quad \text{and} \quad y = -\sqrt{x}$$

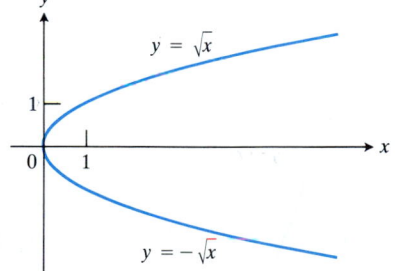

1.68   The graphs of the functions $y = \sqrt{x}$ and $y = -\sqrt{x}$ join at the origin to make the graph of the equation $x = y^2$.

do define functions of x. Each formula gives exactly one value of y for each possible value of x. The graph of $y = \sqrt{x}$ is the upper half of the parabola $x = y^2$. The graph of $y = -\sqrt{x}$ is the lower half. The two graphs meet at the origin (Fig. 1.68).

The Reflective Properties of Parabolas

The chief application of parabolas involves their use as reflectors of light and radio waves. Rays originating at a parabola's focus are reflected out of the parabola parallel to the parabola's axis (in Fig. 1.67, the x-axis). Similarly, rays coming in parallel to the axis are reflected toward the focus. We shall see why this is so when we get into the calculus.

This property is used in parabolic mirrors and telescopes, in automobile headlamps, in spotlights of all kinds, in radar and microwave antennas, and in solar collectors. Parabolas are also used in bridge construction, wind tunnel photography, and submarine tracking.

EXERCISES 1.5

1. Figure 1.69 shows the graph of $y = x^2$ shifted to two new positions. Write equations for the new graphs.

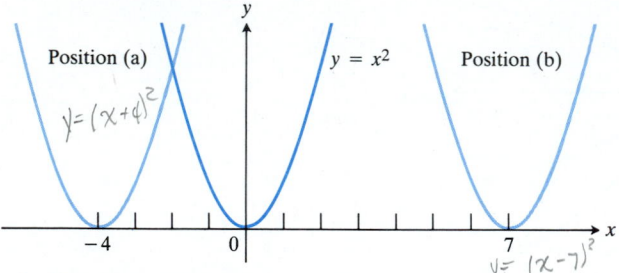

1.69 The parabolas in Exercise 1.

2. Figure 1.70 shows the graph of $y = x^2$ shifted to two new positions. Write equations for the new graphs.

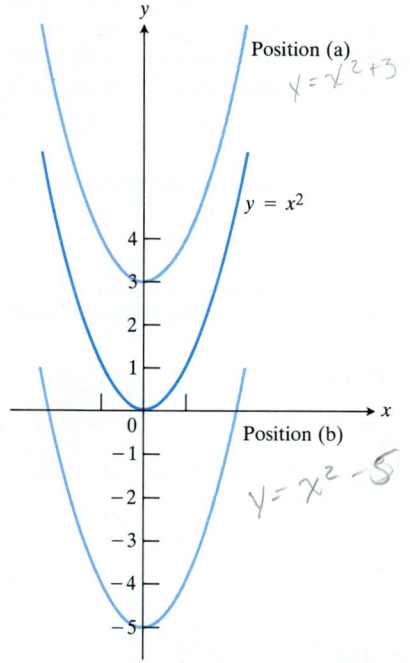

1.70 The parabolas in Exercise 2.

3. Match the equations listed in (a)–(d) to the graphs in Fig. 1.71.
a) $y + 4 = (x - 1)^2$
b) $y - 2 = (x - 2)^2$
c) $y - 2 = (x + 2)^2$
d) $y + 2 = (x + 3)^2$

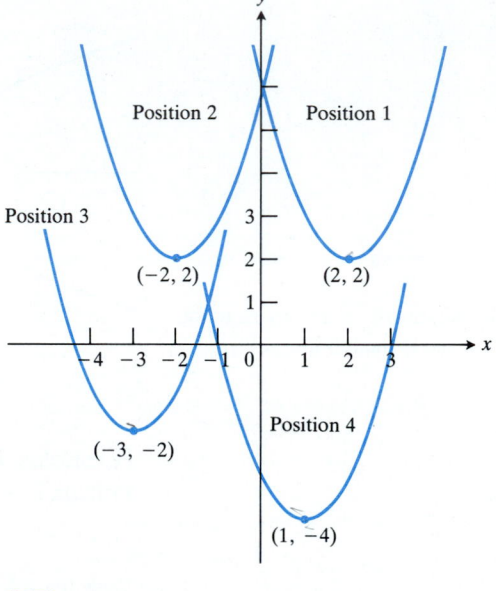

1.71 The parabolas in Exercise 3.

4. Figure 1.72 shows the graph of $y = x^2$ shifted to four new positions. Write an equation for each new graph.

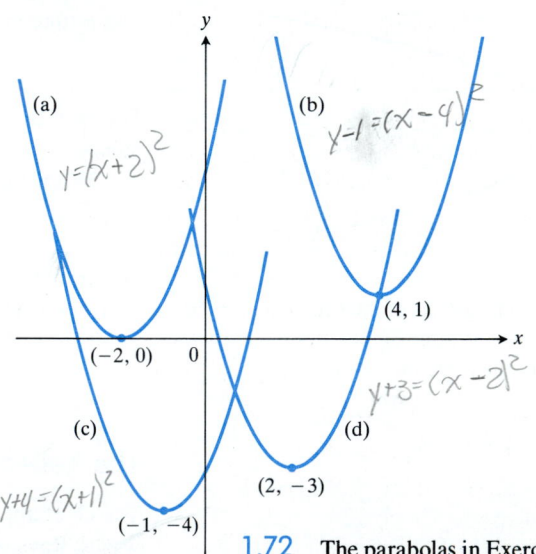

1.72 The parabolas in Exercise 4.

In Exercises 5–8, find an equation for the circle with the given center $C(h, k)$ and radius a. Then draw the circle in the coordinate plane.

5. $C(0, 2), \quad a = 2$
6. $C(-2, 0), \quad a = 3$
7. $C(3, -4), \quad a = 5$
8. $C(1, 1), \quad a = \sqrt{2}$

Write equations for the circles in Exercises 9–12.

9.

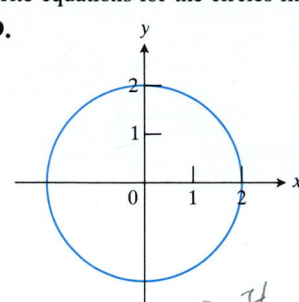

$(y-1)^2(x-1)^2 = 1$

10.

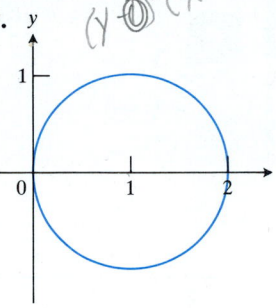

$y = x^2$

$x^2 + y^2 = 4$

11.

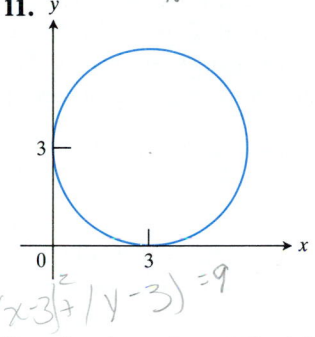

$(x-3)^2 / (y-3) = 9$

12.

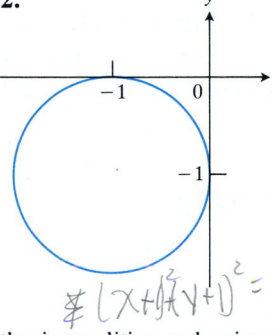

$\neq (x+1)(y+1)^2 = 1$

Describe the regions defined by the inequalities and pairs of inequalities in Exercises 13 and 14.

13. a) $x^2 + y^2 > 1$ EXT CIRCLE
   b) $x^2 + y^2 < 4$ INT
   c) the inequalities in (a) and (b) together.

14. a) $x^2 + y^2 \geq 1$ EXT CIRCI + CIRCLE
   b) $x^2 + y^2 \leq 4$ INT CI.II IL II
   c) the inequalities in (a) and (b) together.

15. Write an inequality that describes the points that lie inside the circle with center $C(-2, -1)$ and radius $a = \sqrt{6}$.
$(x+2)^2 + (y+1)^2 < 6$

16. Write an inequality that describes the points that lie outside the circle with center $C(-4, 2)$ and radius $a = 4$.
$(x+4)^2 + (y-2) \geq 16$

In Exercises 17–20, write an equation for the parabola with the given focus and directrix.

17. Focus: $(0, 4)$ Directrix: $y = -4$

18. Focus: $(0, 1/4)$ Directrix: $y = -1/4$

19. Focus: $(0, -3)$ Directrix: $y = 3$

20. Focus: $(0, -1/2)$ Directrix: $y = 1/2$

In Exercises 21–24, find the parabola's focus and directrix. Then make a sketch showing the parabola, focus, and directrix together.

21. $y = x^2/2$ **22.** $y = -x^2$

23. $y = -x^2/4$ **24.** $y = x^2/16$

Exercises 25–34 tell how many units and in what directions the graphs of the given equations are to be shifted. Give an equation

for the shifted graph in each case.

25. $y = x^2$ Down 3, left 2

26. $x^2 + y^2 = 5$ Up 3, left 4

27. $y + 1 = (x + 1)^2$ Up 1, right 1 $y = x^2$

28. $y = x^3$ Right 1

29. $y = \sqrt{x}$ Left 4

30. $y = 2x - 7$ Up 7

31. $y - 5 = \dfrac{1}{2}(x + 1)$ Down 5, right 1

32. $x = y^2$ Left 1

33. $y = \sqrt{-x}$ Right 9

34. $y = 1/x$ Up 1, right 1

Sketch the curves in Exercises 35–40.

35. $y = \sqrt{x - 1}$ **36.** $y = -\sqrt{x + 4}$

37. $y = 1 + \sin x$ **38.** $y = 1 - \cos x$

39. $y = \cos(x - (\pi/2))$ **40.** $y = \sin(x + (\pi/2))$

41. Match the following functions to the graphs in Fig. 1.73.
   a) $y = |x - 3|$
   b) $y = -|x - 3|$
   c) $y = |x + 3|$
   d) $y = |x| - 3$

i)

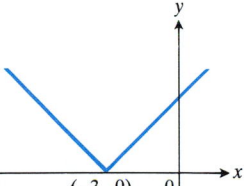

ii)

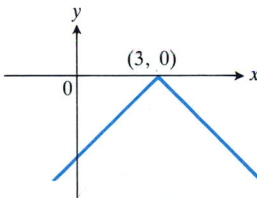

iii)

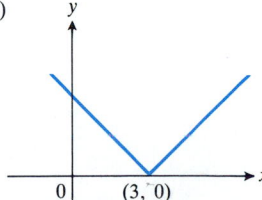

iv)

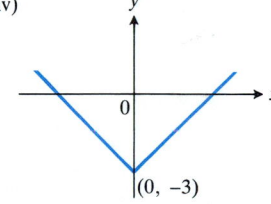

1.73   The graphs for Exercise 41.

42. Graph the following functions.
   a) $y = |x - 1| - 1$
   b) $y = -|x + 1| - 1$

43. The line $y = mx$ is shifted horizontally and vertically to make it pass through the point (x_0, y_0). What is the line's new point–slope equation?

44. The line $y = mx$ is shifted vertically to make it pass through the point $(0, b)$. What is the line's new slope–intercept equation?

45. To verify the statement that $4p$ is the width of the parabola $y = x^2/4p$ at the focus $(p > 0)$, show that the line $y = p$ cuts the parabola in points that are $4p$ units apart.

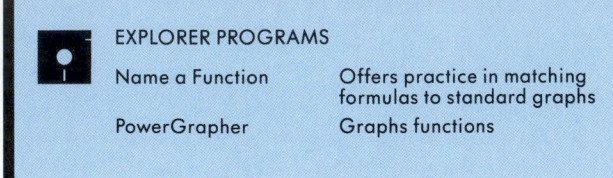

EXPLORER PROGRAMS

| Name a Function | Offers practice in matching formulas to standard graphs |
| PowerGrapher | Graphs functions |

1.6 Slopes, Tangent Lines, and Derivatives

We now get our first view of the role calculus plays in describing change. We begin with the average rate at which a quantity changes over a period of time and end with a way to describe the rate at which that same quantity is changing at any instant. We accomplish the transition by identifying average rates of change with slopes of secant lines.

Average Rates of Change

We encounter average rates of change in such forms as average speeds (distance traveled divided by elapsed time, say, in miles per hour), growth rates of populations (in percent per year), and average monthly rainfall (in inches per month). The **average rate of change** in a quantity over a period of time is the amount of change divided by the time it takes.

Experimental biologists often want to know the rates at which populations grow under controlled laboratory conditions. Figure 1.74 shows data from a fruit fly–growing experiment, the setting for our first example.

Example 1   *The Average Growth Rate of a Laboratory Population.*

graph in Fig. 1.74 shows how the number of fruit flies *(Drosophila)* grew in a controlled 50-day experiment. The graph was made by counting flies at regular intervals, plotting a point for each count, and drawing a smooth curve through the plotted points.

There were 150 flies on day 23 and 340 flies on day 45. This gave an increase of $340 - 150 = 190$ flies in $45 - 23 = 22$ days. The average rate of change in the

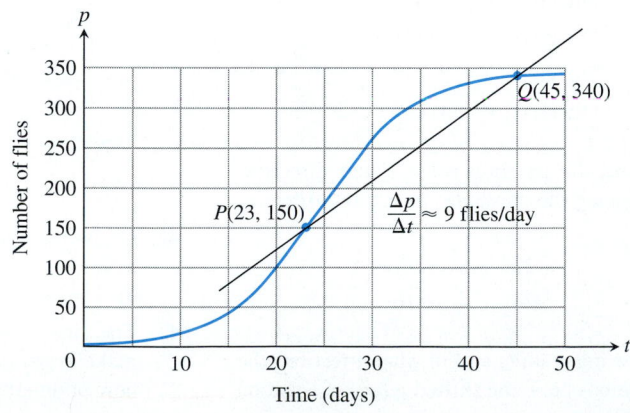

1.74   Growth of a fruit fly population in a controlled experiment. (*Source: Elements of Mathematical Biology* by A. J. Lotka, 1956, Dover, New York, p. 69.)

population from day 23 to day 45 was therefore

$$\text{Average rate of change:} \qquad \frac{\Delta p}{\Delta t} = \frac{340 - 150}{45 - 23} = \frac{190}{22} \approx 9 \text{ flies/day.} \qquad (1)$$

The average rate of change in Eq. (1) is also the slope of the secant line through the two points

$$P(23, 150) \qquad \text{and} \qquad Q(45, 340)$$

on the population curve. (A line through two points on a curve is called a **secant** to the curve.) We can calculate the slope of the secant PQ from the coordinates of P and Q:

$$\text{Secant slope:} \qquad \frac{\Delta p}{\Delta t} = \frac{340 - 150}{45 - 23} = \frac{190}{22} \approx 9 \text{ flies/day.} \qquad (2)$$

By comparing Eqs. (1) and (2) we can see that the average rate of change in (1) is the same number as the slope in (2), units and all. We can always think of an average rate of change as the slope of a secant line.

In addition to knowing the average rate at which the population grew from day 23 to day 45, we may also want to know how fast the population was growing on day 23 itself. To find out, we can watch the slope of the secant PQ change as we back Q along the curve toward P. The results for four positions of Q are shown in Fig. 1.75.

In terms of geometry, what we see as Q approaches P along the curve is this: The secant PQ approaches the tangent line AB that we drew by eye at P. This means that within the limitations of our drawing, the slopes of the secants approach the slope of the tangent, which we calculate from the coordinates of A and B to be

$$\frac{350 - 0}{35 - 15} = 17 \text{ flies/day.}$$

In terms of population change, what we see as Q approaches P is this: The average growth rates for increasingly smaller time intervals approach the slope of the tangent to the curve at P (17 flies per day). The slope of the tangent line is therefore the number we take as the rate at which the fly population was changing on day $t = 23$.

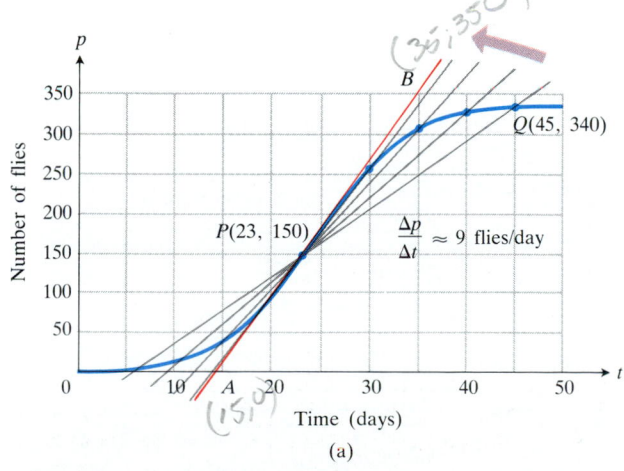

1.75 (a) Four secants to the fruit fly graph of Fig. 1.74 through the point $P(23, 150)$. (b) The slopes of the four secants.

Q	Slope of $PQ = \Delta p/\Delta t$ (flies/day)
(45, 340)	$(340 - 150)/(45 - 23) \approx 8.6$
(40, 330)	$(330 - 150)/(40 - 23) \approx 10.6$
(35, 310)	$(310 - 150)/(35 - 23) \approx 13.3$
(30, 265)	$(265 - 150)/(30 - 23) \approx 16.4$

(b)

How Do You Find a Tangent to a Curve?

This was the dominant mathematical question of the early seventeenth century and it is hard to overestimate how badly the scientists of the day wanted to know the answer. In optics, the tangent determined the angle at which a ray of light entered a curved lens. In mechanics, the tangent determined the direction of a body's motion at every point along its path. In geometry, the tangents to two curves at a point of intersection determined the angle at which the curves intersected. Descartes went so far as to say that the problem of finding a tangent to a curve was "the most useful and most general problem not only that I know but even that I have any desire to know."

Defining Slopes and Tangent Lines

The moral of the fruit fly story seems to be that we should define the rate at which the value of the function $y = f(x)$ is changing with respect to x at any particular value $x = x_1$ to be the slope of the tangent to the curve $y = f(x)$ at $x = x_1$. But how are we to define the tangent line at an arbitrary point P on the curve and deduce its slope from the formula $y = f(x)$?

The answer that Fermat finally found in 1629 proved to be one of that century's major contributions to calculus. We still use his method of defining tangents to produce formulas for slopes of curves and rates of change. It goes like this:

1. We start with what we *can* calculate, namely the slope of a secant through P and a point Q nearby on the curve.

2. We find the limiting value of the secant slope (if it exists) as Q approaches P along the curve.

3. We take this number to be the slope of the curve at P and define the tangent to the curve at P to be the line through P with this slope.

Example 2   Find the slope of the parabola $y = x^2$ at the point $P(2, 4)$. Write an equation for the tangent to the parabola at this point.

Solution   We begin with a secant line that passes through $P(2, 4)$ and a point $Q(2 + h, (2 + h)^2)$ nearby on the curve (Fig. 1.76). We then write an expression for the slope of the secant PQ and watch what happens to the slope as Q approaches P along the curve.

The slope of secant PQ is

$$\text{Secant slope} = \frac{\Delta y}{\Delta x} = \frac{(2 + h)^2 - 2^2}{h} = \frac{h^2 + 4h + 4 - 4}{h}$$

$$= \frac{h^2 + 4h}{h} = h + 4. \tag{3}$$

As Q approaches P along the curve, h approaches 0 and the secant slope, $h + 4$, approaches $0 + 4 = 4$. We describe this behavior by saying that the limit of $h + 4$ as h approaches 0 is 4, and we write

$$\lim_{h \to 0} (h + 4) = 4. \qquad \text{("The limit as } h \text{ approaches 0 of } (h + 4) \text{ is 4.")}$$

We take 4 to be the slope of the parabola at P.

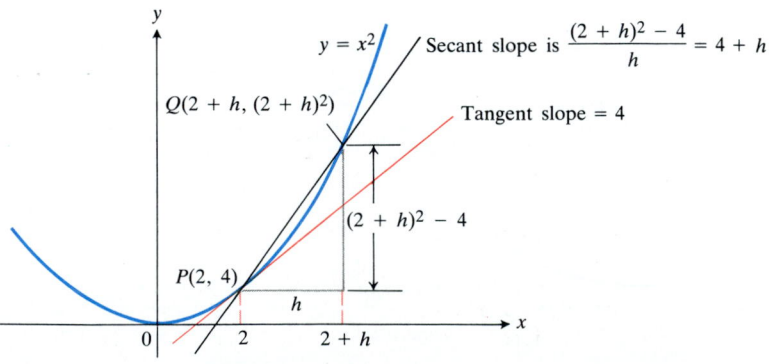

1.76   The slope of the secant PQ approaches 4 as Q approaches P along the curve.

The tangent to the parabola at P is the line through P with slope 4:

Point: (2, 4)

Slope: $m = 4$

Equation: $y - 4 = 4(x - 2)$
$$y = 4x - 8 + 4$$
$$y = 4x - 4$$

The mathematics we just used to find the slope of the parabola $y = x^2$ at the point $P(2, 4)$ will find the slope of the parabola at any other point, too. Here's how it works.

Example 3   Find the slope of the parabola $y = x^2$ at any point on the curve.

Solution   Let $P(x_1, x_1^2)$ be the point. In the notation of Fig. 1.77, the slope of the secant line through P and any nearby point $Q(x_1 + h, (x_1 + h)^2)$ is

$$\text{Secant slope} = \frac{\Delta y}{\Delta x} = \frac{(x_1 + h)^2 - x_1^2}{h}$$

$$= \frac{x_1^2 + 2x_1h + h^2 - x_1^2}{h}$$

$$= \frac{2x_1h + h^2}{h} = 2x_1 + h.$$

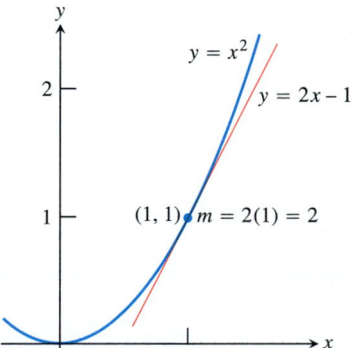

1.77   The slope of the secant PQ is $[(x_1 + h)^2 - x_1^2]/h = 2x_1 + h$.

The limit of the secant slope as Q approaches P along the curve is

$$\lim_{Q \to P} (\text{secant slope}) = \lim_{h \to 0} (2x_1 + h) = 2x_1 + 0 = 2x_1.$$

Since x_1 can be any value of x, we can omit the subscript 1. At any point (x, y) on the parabola, the slope is

$$m = 2x.$$

When $x = 2$, for example, the slope is 4, as in Example 2.

The next example shows how to use the slope formula from Example 3 to find equations for tangent lines.

Example 4   Find the point–slope equation for the tangent to the parabola $y = x^2$ at the point (1, 1) (Fig. 1.78).

Solution   We use the slope formula $m = 2x$ from Example 3.

Point: (1, 1)

Slope: $m = 2x = 2(1) = 2$

Equation: $y - 1 = 2(x - 1)$
$$y - 1 = 2x - 2$$
$$y = 2x - 1$$

1.78   The slope of the tangent at a point (x, y) on the parabola $y = x^2$ is $m = 2x$.

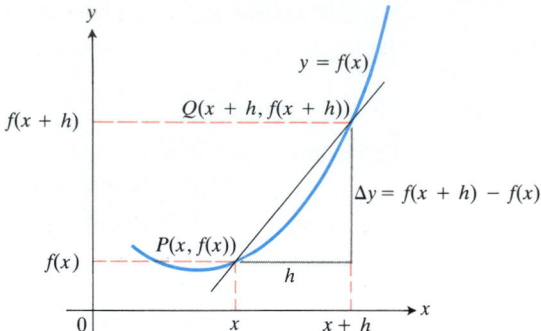

1.79   The slope of the line PQ is
$$\frac{f(x + h) - f(x)}{h}.$$

The Derivative of a Function

The function $m = 2x$ that gives the slope of the parabola $y = x^2$ at x is called the derivative of the function $y = x^2$.

To find the derivative of an arbitrary function $y = f(x)$ (when the function has one—we'll come back to that), we simply repeat for f the steps we took in Examples 2 and 3 for x^2. We start with an arbitrary point $P(x, f(x))$ on the graph of f, as in Fig. 1.79. The slope of the secant line through P and a nearby point $Q(x + h, f(x + h))$ is then

$$\text{Secant slope} = \frac{\Delta y}{\Delta x} = \frac{f(x + h) - f(x)}{h}. \tag{4}$$

The slope of the graph at P is the limit of the secant slope as Q approaches P along the graph. We find this limit from the formula for f by calculating the limit

$$\lim_{h \to 0} \frac{f(x + h) - f(x)}{h}. \tag{5}$$

The fraction $(f(x + h) - f(x))/h$ is the **difference quotient** for f at x.

The limit in Eq. (5) is itself a function of x. We denote it by f' ("f prime") and call it the **derivative** of f. Its domain is a subset of the domain of f. For most of the functions in this book, f' will be defined at all or all but a few of the points where f is defined. We shall state the definition of derivative formally in Chapter 2, after we have studied limits in detail.

If the limit in (5) exists, we say that f **has a derivative (is differentiable) at** x. If f has a derivative at every point of its domain, we call f **differentiable.**

When $f'(x)$ exists, it is called the **slope** of the curve $y = f(x)$ at x. The line through the point $(x, f(x))$ with slope $f'(x)$ is then called the **tangent** to the curve at x.

The most common notations for the derivative of a function $y = f(x)$, besides $f'(x)$, are

y'	("y prime")	(Nice and brief)
$\dfrac{dy}{dx}$	("$dy\ dx$")	$\left(\begin{array}{l}\text{Names the variables and}\\\text{has a "}d\text{" for derivative}\end{array}\right)$
$\dfrac{df}{dx}$	("$df\ dx$")	(Emphasizes the function's name)
$D_x(f)$	("Dx of f")	$\left(\begin{array}{l}\text{Emphasizes the idea that taking the}\\\text{derivative is an operation performed on }f\end{array}\right)$
$\dfrac{d}{dx}(f)$	("$d\ dx$ of f")	(Ditto)

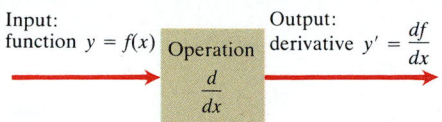

1.80 Flow diagram for the operation of taking a derivative with respect to x.

We also read dy/dx as "the derivative of y with respect to x" and df/dx as "the derivative of f with respect to x." See Fig. 1.80.

The Slopes of Lines

We now have two definitions for the slope of a line $y = mx + b$: the number m and, at each point, the derivative of the function $f(x) = mx + b$. Whenever we bring in a new definition, it is a good idea to be sure that the new and old definitions agree on objects to which they both apply. We do this in the next example.

Example 5   Show that the derivative of the function $f(x) = mx + b$ is the slope of the line $y = mx + b$.

The algebraic steps we use to calculate $f'(x)$ directly from the definition are always the same:

1. Write out $f(x)$ and $f(x + h)$.
2. Subtract $f(x)$ from $f(x + h)$.
3. Divide by h.
4. Take the limit as $h \to 0$.

Solution   The idea is to show that the derivative of $f(x) = mx + b$ has the constant value m. To see that this is so, we calculate the limit in (5) with $f(x) = mx + b$. The calculation takes four steps.

STEP 1:   Write out $f(x)$ and $f(x + h)$:

$$f(x) = mx + b$$

$$f(x + h) = m(x + h) + b = mx + mh + b.$$

STEP 2:   Subtract $f(x)$ from $f(x + h)$:

$$f(x + h) - f(x) = mh.$$

STEP 3:   Divide by h:

$$\frac{f(x + h) - f(x)}{h} = \frac{mh}{h} = m.$$

STEP 4:   Take the limit as $h \to 0$:

$$\lim_{h \to 0} \frac{f(x + h) - f(x)}{h} = \lim_{h \to 0} (m) = m.$$

The derivative of f does indeed have the constant value m.

Typical Derivative Calculations

Example 6   Find dy/dx if $y = 1/x$.

Solution   We take $f(x) = 1/x$ and $f(x + h) = 1/(x + h)$ and then form the quotient

$$\frac{f(x + h) - f(x)}{h} = \frac{\dfrac{1}{x + h} - \dfrac{1}{x}}{h}$$

$$= \frac{1}{h} \cdot \frac{x - (x + h)}{x(x + h)} \tag{6}$$

$$= \frac{1}{h} \cdot \frac{-h}{x(x + h)} = \frac{-1}{x(x + h)}.$$

We then take the limit as $h \to 0$:

$$\frac{dy}{dx} = \lim_{h \to 0} \frac{f(x + h) - f(x)}{h} = \lim_{h \to 0} \frac{-1}{x(x + h)} = \frac{-1}{x(x + 0)} = -\frac{1}{x^2}.$$

Example 7   Find $f'(x)$ if $f(x) = x^2 - 2x$.

Solution   We subtract $f(x) = x^2 - 2x$ from

$$f(x + h) = (x + h)^2 - 2(x + h) = x^2 + 2hx + h^2 - 2x - 2h$$

and form the difference quotient

$$\frac{f(x + h) - f(x)}{h} = \frac{(x^2 + 2hx + h^2 - 2x - 2h) - (x^2 - 2x)}{h}$$

$$= \frac{h^2 + 2hx - 2h}{h}$$

$$= h + 2x - 2.$$

We then take the limit as $h \to 0$:

$$f'(x) = \lim_{h \to 0} \frac{f(x + h) - f(x)}{h}$$

$$= \lim_{h \to 0} (h + 2x - 2) = 0 + 2x - 2 = 2x - 2.$$

Example 8   Show that the derivative of $y = \sqrt{x}$ is $\dfrac{dy}{dx} = \dfrac{1}{2\sqrt{x}}$ for $x > 0$.

Solution   We use Eq. (5) with $f(x + h) = \sqrt{x + h}$ and $f(x) = \sqrt{x}$ to form the quotient

$$\frac{f(x + h) - f(x)}{h} = \frac{\sqrt{x + h} - \sqrt{x}}{h}. \tag{7}$$

Unfortunately, this will involve division by 0 if we try to calculate the limit by replacing h with 0. We therefore look for an equivalent expression in which this difficulty does not arise. If we rationalize the numerator in Eq. (7), we find

$$\frac{\sqrt{x + h} - \sqrt{x}}{h} = \frac{\sqrt{x + h} - \sqrt{x}}{h} \cdot \frac{\sqrt{x + h} + \sqrt{x}}{\sqrt{x + h} + \sqrt{x}}$$

$$= \frac{(x + h) - x}{h(\sqrt{x + h} + \sqrt{x})} = \frac{1}{\sqrt{x + h} + \sqrt{x}}. \tag{8}$$

Now as h approaches 0, the denominator in the final form approaches $\sqrt{x} + \sqrt{x} = 2\sqrt{x}$, which is positive because $x > 0$. Therefore,

$$\frac{dy}{dx} = \lim_{h \to 0} \frac{\sqrt{x + h} - \sqrt{x}}{h} = \lim_{h \to 0} \frac{1}{\sqrt{x + h} + \sqrt{x}}$$

$$= \frac{1}{\sqrt{x + 0} + \sqrt{x}} = \frac{1}{2\sqrt{x}}. \tag{9}$$

Example 9   Find an equation for the tangent to the curve $y = \sqrt{x}$ at $x = 4$.

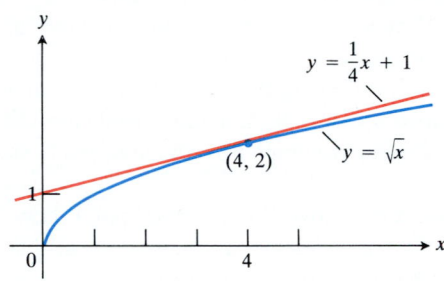

1.81 The curve $y = \sqrt{x}$ and its tangent at (4, 2). The slope of the tangent is found by evaluating dy/dx at $x = 4$.

Solution The slope at $x = 4$ is the value of the function's derivative there. Equation (9) gives $dy/dx = 1/(2\sqrt{x})$, so the slope is

$$\frac{1}{2\sqrt{x}}\Big|_{x=4} = \frac{1}{2\sqrt{4}} = \frac{1}{4}. \qquad \left(\begin{array}{l}\text{The symbols }\big|_{x=4}\text{ are read}\\\text{``at }x\text{ equals 4.''}\end{array}\right)$$

The tangent is the line through the point $P(4,\sqrt{4}) = (4, 2)$ with slope 1/4:

Point: (4, 2)

Slope: $\dfrac{1}{4}$

Equation: $y - 2 = \dfrac{1}{4}(x - 4)$

$$y = \frac{1}{4}x - 1 + 2 = \frac{1}{4}x + 1.$$

The tangent is the line $y = (1/4)x + 1$ (Fig. 1.81).

When Does a Function *Not* Have a Derivative at a Point?

As we know, a function has a derivative at a point x_0 if the slope of the secant line through $P(x_0, f(x_0))$ and a nearby point Q on the graph approaches a limit as Q approaches P. Whenever the secant fails to take up a limiting position as Q approaches P, the derivative does not exist. Typically, a function whose graph appears otherwise smooth will fail to have a derivative at a point where the graph has a corner, jump, or vertical tangent (Fig. 1.82).

1.82 As the pictures here suggest, a function has no derivative at a point if the graph there has a corner or jump where the secant lines come to more than one limiting position (points a and b). Nor can it have a derivative at a point where the secant lines become vertical as Q approaches P. At such a point (c) or (d), the curve has a vertical tangent but no slope.

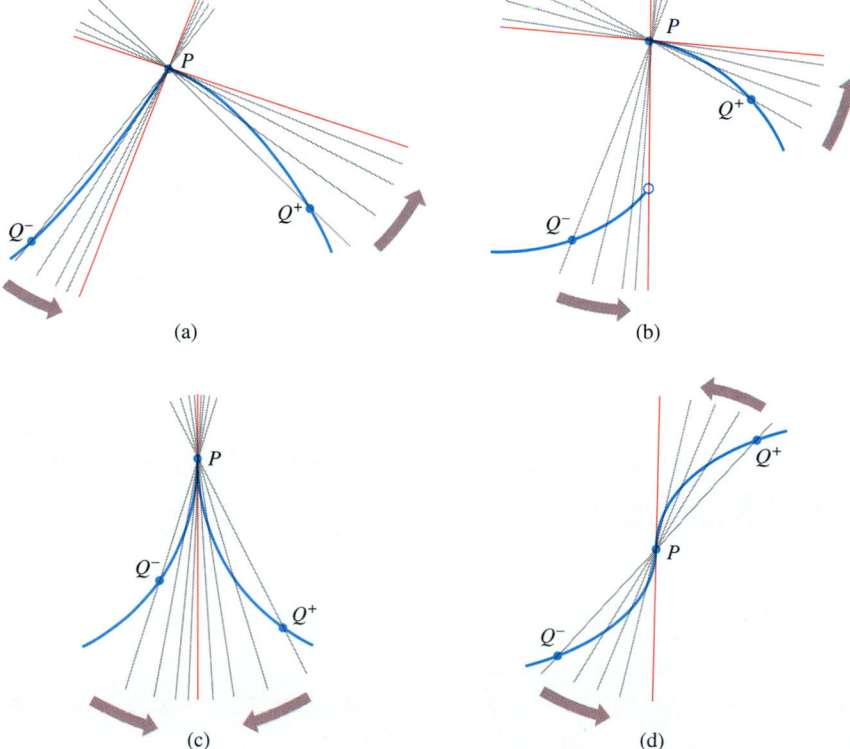

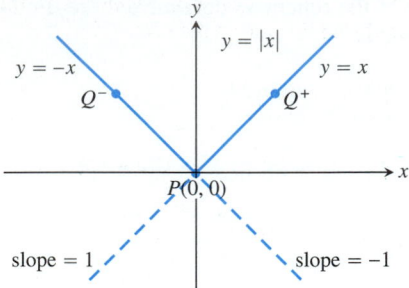

1.83   The graph of the absolute value function has only two secants through the corner point $P(0, 0)$: the line $y = x$ and the line $y = -x$.

Example 10   The absolute value function $f(x) = |x|$ has no derivative at $x = 0$ but has a derivative at every other point.

The only secants to the graph of f at the origin $P(0, 0)$ are the lines $y = x$ and $y = -x$ (Fig. 1.83). For f to have a derivative at $x = 0$, these secants would have to approach a common line as Q approaches P along the graph from the right and from the left, but they do not.

The absolute value function does have a derivative at every other point, however. For $x > 0$, the values of f are those of the function $y = x$, whose derivative at each point is 1 (Example 5). For $x < 0$, the values of f are those of the function $y = -x$, whose derivative at each point is -1. Thus,

$$\frac{df}{dx} = \begin{cases} 1, & x > 0, \\ -1, & x < 0. \end{cases}$$

Which Functions Are Differentiable?

Most of the functions we have seen so far are differentiable. Polynomial functions are differentiable, as are logarithmic, exponential, and trigonometric functions. Composites of differentiable functions are differentiable, and so are sums, differences, products, powers, and quotients of differentiable functions (where defined). We shall explain all this as the book continues.

Where We Go from Here

Derivatives arise in nearly every applied field today. Economists use derivatives to predict costs and analyze profitability. Physicists use them to describe motions of all kinds and to predict the effects of forces. Chemists use derivatives to describe diffusion and to measure the rates at which chemical reactions take place. Electrical engineers use derivatives in the equations that describe currents in electric circuits. And, of course, derivatives are important in mathematics.

To understand the contribution of derivatives to all these fields, we need to be able to calculate derivatives with more facility than we have right now. Most derivatives can be calculated quickly with rules that enable us to avoid the kind of time-consuming application of Fermat's difference quotient we have encountered so far. We still need the difference quotient, but to derive the rules, not to apply them. To derive the rules, we need to know more about limits. So we shall set derivatives aside for the moment and turn our attention to limits. Then, in Chapter 2, we shall derive rules for rapid differentiation and begin the developments that account for the importance of calculus.

EXERCISES 1.6

In Exercises 1–6, use the four algebraic steps on page 53 to find the derivative of the function $y = f(x)$ at the given point. Then write an equation for the tangent to the curve at the point and graph the function and tangent together.

1. $y = x^2 + 1$, $(2, 5)$ **2.** $y = -x^2$, $(1, -1)$

3. $y = 4 - x^2$, $(-1, 3)$

4. $y = (x - 1)^2 + 1$, $(1, 1)$

5. $y = \sqrt{x}$, $(1, 1)$

6. $y = 1/x^2$, $(-1, 1)$

In Exercises 7–16, use the four algebraic steps on page 53 to find the derivative dy/dx of the function $y = f(x)$. Then find the slope of the curve $y = f(x)$ at the given value of x.

7. $y = x^2, \quad x = 1$

8. $y = 2x^2 - 5, \quad x = 2$

9. $y = \dfrac{2}{x}, \quad x = -2$

10. $y = \dfrac{x}{x + 1}, \quad x = 0$

11. $y = x + \dfrac{9}{x}, \quad x = -3$

12. $y = 1 + \sqrt{x}, \quad x = 1$

13. $y = \sqrt{2x}, \quad x = \dfrac{1}{2}$

14. $y = \dfrac{1}{\sqrt{x}}, \quad x = 4$

15. $y = \dfrac{1}{\sqrt{2x + 3}}, \quad x = -1$

16. $y = x^3, \quad x = -2$

When we change the symbols for the dependent and independent variables of a function from x and y to other letters, we change the derivative's notation to match. Except for this change, the steps we use to calculate the derivative remain the same. Find the values of the derivatives in Exercises 17–22 at the given points.

17. $\dfrac{dr}{ds}$ if $r = \dfrac{s^2}{2}, \quad s = \sqrt{2}$

18. $\dfrac{du}{dv}$ if $u = v^2 - v, \quad v = -\dfrac{1}{2}$

19. $\dfrac{ds}{dt}$ if $s = \dfrac{1}{t + 1}, \quad t = 2$

20. $\dfrac{dv}{dt}$ if $v = t - \dfrac{1}{t}, \quad t = -1$

21. $\dfrac{dp}{dq}$ if $p = \sqrt{q + 1}, \quad q = 3$

22. $\dfrac{dz}{dw}$ if $z = \sqrt{2w + 3}, \quad w = \dfrac{1}{2}$

23. The graph of the function $y = f(x)$ in Fig. 1.84 is made of line segments joined end to end.
a) At which points of the open interval $(-4, 6)$ is the derivative of f not defined? (Ignore the endpoints $x = -4$ and $x = 6$ for the moment. We discuss derivatives at endpoints in the next section.)

b) Graph the derivative of f. Call the vertical axis the y'-axis. The graph should show a step function.

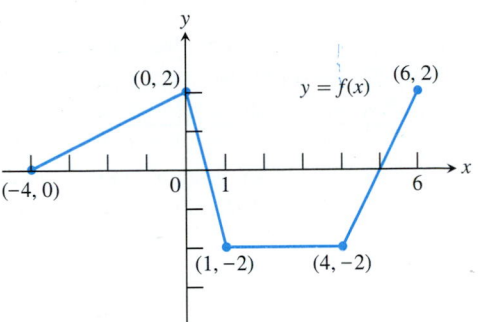

1.84   The graph for Exercise 23.

24. *Recovering a function from its derivative*
a) Use the following information to graph the function f over the closed interval $[-2, 5]$.
   i) The graph of f is made of closed line segments joined end to end.
   ii) The graph starts at the point $(-2, 3)$.
   iii) The derivative of f is the step function in Fig. 1.85.
b) Repeat part (a) assuming that the graph starts at $(-2, 0)$ instead of $(-2, 3)$.

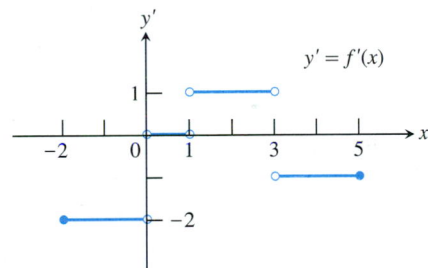

1.85   The derivative graph for Exercise 24.

25. Graph the derivative of $f(x) = |x|$ (Example 10). Then graph the function $y = |x|/x, \quad x \neq 0$. What can you conclude?

Computer Grapher or Graphing Calculator

26. Graph $y = 2x$ for $-4 \leq x \leq 4$. Then, on the same screen, graph

$$y = \frac{(x + h)^2 - x^2}{h}$$

for $h = 2, 1, 0.5,$ and 0.1. Use different colors if available. Then try $h = -2, -1, -0.5,$ and -0.1. What happens as h approaches 0?

27. Graph $y = 1/(2\sqrt{x})$ in a window that has $-1 \leq x \leq 10$. Then, on the same screen, graph

$$y = \frac{\sqrt{x + h} - \sqrt{x}}{h}$$

for $h = 4, 2, 1$, and 0.5. Use different colors if available. Then try $h = -4, -2, -1$, and -0.5. What happens as h approaches 0?

28. Graph $y = -1/x^2$ in a window that has $-2 \le x \le 2$. Then, on the same screen, graph

$$y = \frac{1/(x + h) - 1/x}{h} = -\frac{1}{x(x + h)}$$

for $h = 1, 0.5, 0.2, 0.1$. Use different colors if available. Then try $h = -1, -0.5, -0.2, -0.1$. What happens as h approaches 0?

EXPLORER PROGRAMS

Derivatives — Automatically graphs the derivative of any function you key in

Secant Lines — Draws secant lines on command and displays their slopes as they change from one position to the next; also draws tangents when they exist

---

1.7 Limits of Function Values

We know we need to calculate limits to find derivatives. But just what *are* limits and what is the best way to calculate them? In this section, we find out.

Our plan is first to develop a technical definition of limit and then to establish the basic calculation rules that will enable us to find limits quickly, without a lot of fuss. We can then return to derivatives in Chapter 2 and use these rules to calculate derivatives.

As it turns out, the mathematics we need to make the notion of limit precise enough to be useful is the same mathematics we used in Section 1.4 to study target values of functions. There, we had a function $y = f(x)$, a target value y_0, and an upper bound E on the amount of error we could allow the output value y to have. We wanted to know how close we had to keep x to a particular value x_0 to make y lie within E units of y_0. In symbols, we were asking for a value of D that would make the inequality $|x - x_0| < D$ imply the inequality $|y - y_0| < E$. The number D described the amount by which x could differ from x_0 and still give y-values that approximated y_0 with an error less than E.

In the limit discussions that follow, we shall use the traditional Greek letters δ (delta) and ϵ (epsilon) in place of the English letters D and E. These are the letters that Cauchy and Weierstrass used in their pioneering work on continuity in the nineteenth century. In their arguments, δ meant "différence" (French for *difference*) and ϵ meant "erreur" (French for *error*).

The Definition of Limit

Suppose we are watching the values of a function $f(x)$ as x approaches x_0 (without taking on the value x_0 itself). What do we have to know about the values of f to say that they have a particular number L as their limit? What observable pattern in their behavior would guarantee their eventual approach to L?

Certainly we want to be able to say that $f(x)$ stays within one-tenth of a unit of L as soon as x stays within a certain radius r_1 of x_0, as shown in Fig. 1.86. But that in itself is not enough, because as x continues on its course toward x_0, what is to prevent $f(x)$ from jittering about within the interval from $L - 1/10$ to $L + 1/10$ without tending toward L?

We need to say also that as x continues toward x_0, the number $f(x)$ has to get still closer to L. We might say this by requiring $f(x)$ to lie within $1/100$ of a unit of L for all values of x within some smaller radius r_2 of x_0. But this is not enough either. What if $f(x)$ then skips about within the interval from $L - 1/100$ to $L + 1/100$, without heading toward L?

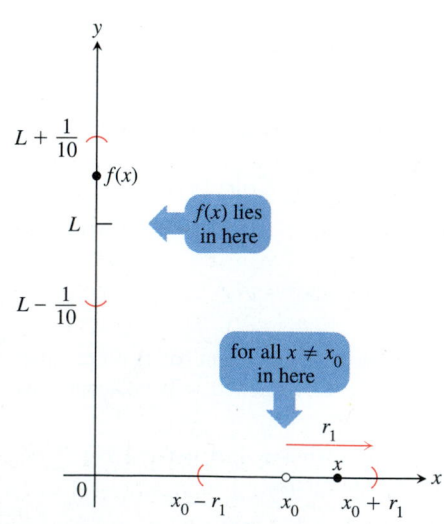

1.86 A preliminary stage in the development of the definition of limit.

We had better require that $f(x)$ lie within $1/1000$ of a unit of L after a while. That is, for all values of x within some still smaller radius r_3 of x_0, all the values of $y = f(x)$ should lie in the interval

$$L - \frac{1}{1000} < y < L + \frac{1}{1000}.$$

This still does not guarantee that $f(x)$ will now move toward L as x approaches x_0. Even if $f(x)$ has not skipped about before, it might start now. We need more.

We need to require that for *every* interval about L, no matter how small, we can find an interval of numbers about x_0 whose f-values all lie within that interval about L. In other words, given any positive radius ϵ about L, there should exist some positive radius δ about x_0 such that for all x within δ units of x_0 (except x_0 itself) the values $y = f(x)$ lie within ϵ units of L (Fig. 1.87). If f satisfies these requirements, we will say that

$$\lim_{x \to x_0} f(x) = L.$$

Here, at last, is a mathematical way to say "the closer x gets to x_0, the closer $y = f(x)$ must get to L."

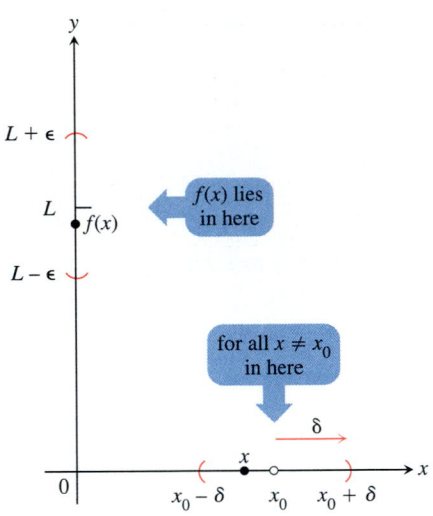

1.87 The relation of δ and ϵ in the definition of limit.

DEFINITION

> The **limit** of $f(x)$ as x approaches x_0 is the number L if the following criterion holds:
>
> Given any radius $\epsilon > 0$ about L there exists a radius $\delta > 0$ about x_0 such that for all x,
>
> $$0 < |x - x_0| < \delta \quad \text{implies} \quad |f(x) - L| < \epsilon. \tag{1}$$

To return to the notions of error and difference, we might think of machining something like a generator shaft to a close tolerance. We try for diameter L, but since nothing is perfect we must be satisfied to get the diameter $f(x)$ somewhere between $L - \epsilon$ and $L + \epsilon$. The δ is the measure of how accurate our control setting for x must be to guarantee this degree of accuracy in the diameter of the shaft.

Examples: Testing the Definition

Whenever someone proposes a new definition, it is a good idea to see if it gives results that are consistent with past experience. For instance, our experience tells us that as x approaches 1, the number $5x - 3$ approaches $5 - 3 = 2$. If our new definition were to lead to some other result, we would want to throw the definition out and look for another one. The following three examples are included in part to show that the definition in Eq. (1) gives the kinds of results we want.

Example 1   *Testing the Definition.*   Show that

$$\lim_{x \to 1} (5x - 3) = 2.$$

Solution   In the definition of limit, we set $x_0 = 1$, $f(x) = 5x - 3$, and $L = 2$. To show that $\lim_{x \to 1} (5x - 3) = 2$, we need to show that for any number $\epsilon > 0$ there exists a number $\delta > 0$ such that for all x,

$$0 < |x - 1| < \delta \quad \Rightarrow \quad |(5x - 3) - 2| < \epsilon. \tag{2}$$

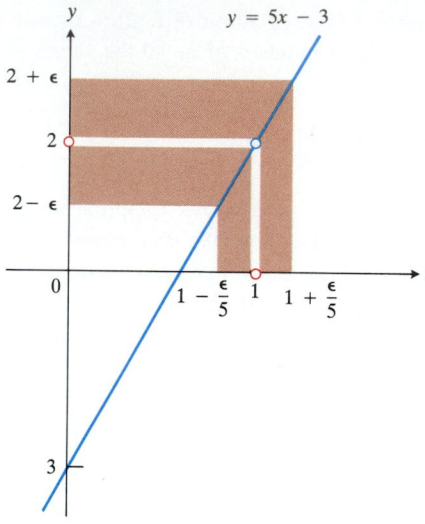

1.88 For the function $f(x) = 5x - 3$, we find that $|x - 1| < \epsilon/5$ will guarantee $|f(x) - 2| < \epsilon$ (Example 1).

The open circle on the graph does not mean that the function has no value at $x = 1$ but is meant to emphasize that the existence of $\lim_{x \to 1} f(x)$ does not depend on the value of f at $x = 1$.

To find a suitable value for δ, we solve the ϵ-inequality:

$$|(5x - 3) - 2| < \epsilon$$
$$|5x - 5| < \epsilon$$
$$5|x - 1| < \epsilon$$
$$|x - 1| < \epsilon/5.$$

The last line tells us that the original ϵ-inequality, and hence the implication in statement (2), will hold if we choose $\delta = \epsilon/5$ (Fig. 1.88).

The value $\delta = \epsilon/5$ is not the only value that will make the implication in (2) hold. Any smaller positive δ will do as well. The definition does not ask for a "best" δ, just one that will work.

Example 2 $\lim_{x \to x_0} x = x_0$. Show that for any number x_0,

$$\lim_{x \to x_0} (x) = x_0.$$

Solution In the definition of limit, we set $f(x) = x$ and $L = x_0$. To show that $\lim_{x \to x_0} (x) = x_0$, we must show that for any $\epsilon > 0$ there exists a $\delta > 0$ such that for all x,

$$0 < |x - x_0| < \delta \quad \Rightarrow \quad |x - x_0| < \epsilon.$$

The implication will hold if δ is ϵ itself or any smaller positive number (Fig. 1.89).

Example 3 *Limits of Constant Functions.* Let $f(x) = k$ be the function whose outputs have the constant value k. Show that for any number x_0,

$$\lim_{x \to x_0} f(x) = k.$$

Solution In the definition of limit, we set $f(x) = k$ and $L = k$. To show that $\lim_{x \to x_0} f(x) = k$, we must show that for any $\epsilon > 0$ there exists a $\delta > 0$ such that for all x,

$$0 < |x - x_0| < \delta \quad \Rightarrow \quad |k - k| < \epsilon.$$

This implication will hold for any positive δ because $|k - k| = 0$ is less than every positive ϵ for all x (Fig. 1.90).

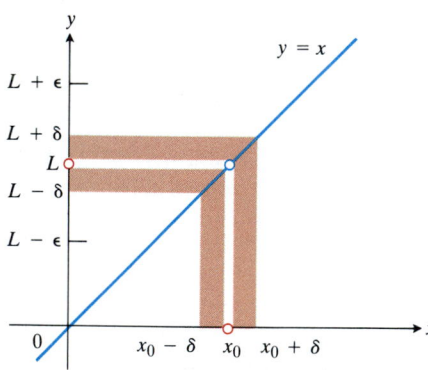

1.89 For the function $f(x) = x$, we find that $|x - x_0| < \delta$ will guarantee $|f(x) - x_0| < \epsilon$ whenever $\delta \le \epsilon$ (Example 2.)

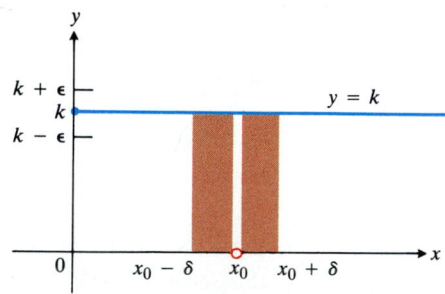

1.90 For the function $f(x) = k$, we find $|f(x) - k| < \epsilon$ for any positive δ (Example 3).

Finding Deltas for Given Epsilons

Here are four numerical examples.

Example 4   Find the largest $\delta > 0$ such that if $0 < |x + 2| < \delta$, then $|3x + 6| < 0.03$.

Solution   We rewrite the inequality $|3x + 6| < 0.03$ to see what it says about $|x + 2|$:

$$|3x + 6| < 0.03$$
$$3|x + 2| < 0.03$$
$$|x + 2| < 0.01$$

We now see that $|3x + 6| < 0.03$ if and only if $|x + 2| < 0.01$. The largest δ is 0.01. If $|x + 2| < 0.01$, then $|3x + 6| < 0.03$. If $|x + 2| \geq 0.01$, then $|3x + 6| \geq 0.03$.

The interval of x-values for which $|f(x) - L| < \epsilon$ is not always symmetric about x_0. When it is not, we take δ to be the distance from x_0 to the nearer endpoint.

Example 5   Find the largest interval symmetric about $x_0 = 5$ lying in the interval $2 < x < 10$.

Solution   We draw a picture to show $x_0 = 5$ in the interval (Fig. 1.91). The endpoint nearer $x_0 = 5$ is the point $x = 2$, which is 3 units away. We can take δ as large as 3 (but no larger) and have the symmetric interval

$$5 - \delta < x < 5 + \delta$$

lie in the interval $2 < x < 10$. The interval we seek is the interval $2 < x < 8$.

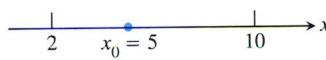

1.91   The interval in Example 5.

Example 6   Find the largest interval about $x_0 = 5$ for which

$$|\sqrt{x - 1} - 2| < 1$$

(Fig. 1.92). Then find a value of δ for which

$$|x - 5| < \delta \quad \Rightarrow \quad |\sqrt{x - 1} - 2| < 1.$$

Solution   We rewrite the inequality $|\sqrt{x - 1} - 2| < 1$ in the form $a < x < b$:

$$|\sqrt{x - 1} - 2| < 1$$
$$-1 < \sqrt{x - 1} - 2 < 1$$
$$1 < \sqrt{x - 1} < 3$$
$$1 < x - 1 < 9$$
$$2 < x < 10.$$

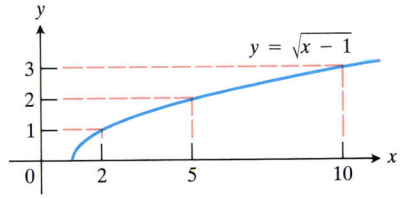

1.92   The graph for Example 6.

The original inequality holds for all x in the interval $2 < x < 10$.

As in Example 5, we may take δ to be 3, the distance from $x_0 = 5$ to the nearer endpoint $a = 2$. If δ is 3 or any smaller positive number, the inequality $|x - 5| < \delta$ will automatically place x between 2 and 10 to make $|\sqrt{x - 1} - 2| < 1$.

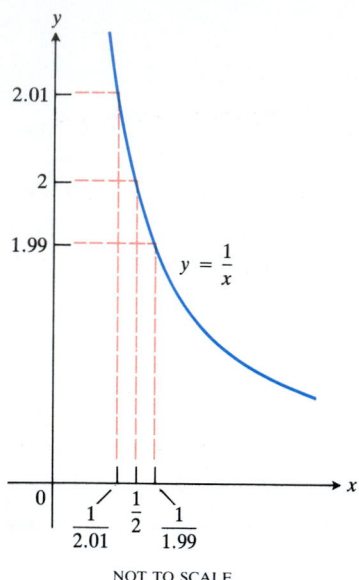

1.93   The graph for Example 7.

Example 7   Find the largest interval centered at $x_0 = 1/2$ for which

$$\left|\frac{1}{x} - 2\right| < 0.01$$

(Fig. 1.93). Then find a value of δ for which

$$\left|x - \frac{1}{2}\right| < \delta \quad \Rightarrow \quad \left|\frac{1}{x} - 2\right| < 0.01.$$

Solution   We rewrite the inequality $|(1/x) - 2| < 0.01$ in the form $a < x < b$:

$$\left|\frac{1}{x} - 2\right| < 0.01$$

$$-0.01 < \frac{1}{x} - 2 < 0.01$$

$$1.99 < \frac{1}{x} < 2.01$$

$$\frac{1}{2.01} < x < \frac{1}{1.99}. \qquad \text{(Take reciprocals and reverse the inequality.)}$$

The original inequality holds for all x in the open interval from $1/2.01$ to $1/1.99$.

Again, we can take δ to be the distance from x_0 to the nearer endpoint, in this case the distance from $x_0 = 1/2$ to the nearer of the points $1/2.01$ and $1/1.99$. To find out which is nearer, we calculate the distances and compare:

$$\text{Distance from } \frac{1}{2} \text{ to } \frac{1}{2.01}: \qquad \frac{1}{2} - \frac{1}{2.01} = \frac{2.01 - 2}{4.02} = \frac{0.01}{4.02} = \frac{1}{402}.$$

$$\text{Distance from } \frac{1}{2} \text{ to } \frac{1}{1.99}: \qquad \frac{1}{1.99} - \frac{1}{2} = \frac{2 - 1.99}{3.98} = \frac{0.01}{3.98} = \frac{1}{398}.$$

The smaller of these is $1/402$. If δ has this value or any smaller positive value, the inequality $|x - (1/2)| < \delta$ will automatically place x between $1/2.01$ and $1/1.99$ to make $|(1/x) - 2| < 0.01$.   ◼

Calculation Rules

The following theorems free us from having to appeal to the ϵ-δ definition every time we need to find a limit.

THEOREM 1

Limit Rules

The following rules hold if $\lim_{x \to c} f_1(x) = L_1$ and $\lim_{x \to c} f_2(x) = L_2$.

1. *Sum Rule:* $\lim [f_1(x) + f_2(x)] = L_1 + L_2$

2. *Difference Rule:* $\lim [f_1(x) - f_2(x)] = L_1 - L_2$

3. *Product Rule:* $\lim f_1(x) \cdot f_2(x) = L_1 \cdot L_2$

4. *Constant Multiple Rule:* $\lim k \cdot f_2(x) = k \cdot L_2$ (any number k)

5. *Quotient Rule:* $\lim \dfrac{f_1(x)}{f_2(x)} = \dfrac{L_1}{L_2}$ if $L_2 \neq 0$.

The limits are all taken as $x \to c$, and L_1 and L_2 are real numbers.

In words, the formulas in Theorem 1 say:

1. The limit of the sum of two functions is the sum of their limits.
2. The limit of the difference of two functions is the difference of their limits.
3. The limit of a product of two functions is the product of their limits.
4. The limit of a constant times a function is the constant times the limit of the function.
5. The limit of a quotient of two functions is the quotient of their limits, provided the denominator does not tend to zero.

We have included a formal proof of Theorem 1 in Appendix 4. Informally, we can paraphrase the theorem in terms that make it highly reasonable: When x is close to c, $f_1(x)$ is close to L_1 and $f_2(x)$ is close to L_2. Then we naturally think that $f_1(x) + f_2(x)$ is close to $L_1 + L_2$; $f_1(x) - f_2(x)$ is close to $L_1 - L_2$; $f_1(x)f_2(x)$ is close to L_1L_2; $kf_2(x)$ is close to kL_2; and $f_1(x)/f_2(x)$ is close to L_1/L_2 if L_2 is not zero.

Here is an example of what Theorem 1 can do for us.

Example 8   We know from Examples 2 and 3 that $\lim_{x \to c} x = c$ and $\lim_{x \to c} k = k$. The various parts of Theorem 1 now let us combine these results to calculate other limits.

a) $\lim_{x \to c} x^2 = \lim_{x \to c} x \cdot x = c \cdot c = c^2$ (Product)

b) $\lim_{x \to c} (x^2 + 5) = \lim_{x \to c} x^2 + \lim_{x \to c} 5$ (Sum)

$= c^2 + 5$ (from (a))

c) $\lim_{x \to c} 4x^2 = 4 \lim_{x \to c} x^2$ (Constant multiple)

$= 4c^2$ (from (a))

d) $\lim_{x \to c} (4x^2 - 3) = \lim_{x \to c} 4x^2 - \lim_{x \to c} 3$ (Difference)

$= 4c^2 - 3$ (from (c))

e) $\lim_{x \to c} x^3 = \lim_{x \to c} x^2 \cdot x = c^2 \cdot c = c^3$ (Product and (a))

f) $\lim_{x \to c} (x^3 + 4x^2 - 3) = \lim_{x \to c} x^3 + \lim_{x \to c} (4x^2 - 3)$ (Sum)

$= c^3 + 4c^2 - 3$ (from (e) and (d))

g) $\lim_{x \to c} \dfrac{x^3 + 4x^2 - 3}{x^2 + 5} = \dfrac{\lim_{x \to c} (x^3 + 4x^2 - 3)}{\lim_{x \to c} (x^2 + 5)}$ (Quotient)

$= \dfrac{c^3 + 4c^2 - 3}{c^2 + 5}.$ (from (f) and (b))

Example 8 shows the remarkable strength of Theorem 1. From the two simple observations that $\lim_{x \to c} x = c$ and $\lim_{x \to c} k = k$ we can immediately work our way to limits of all polynomials and most **rational functions** (ratios of polynomials). As in part (f) of Example 8, the limit of any polynomial $f(x)$ as x approaches c is $f(c)$, the number we get when we substitute $x = c$. As in part (g) of Example 8, the limit of the ratio $f(x)/g(x)$ of two polynomials is $f(c)/g(c)$, provided $g(c)$ is different from 0.

THEOREM 2

Limits of Polynomials Can Be Found by Substitution

If $f(x) = a_n x^n + a_{n-1} x^{n-1} + \cdots + a_0$ is any polynomial function, then

$$\lim_{x \to c} f(x) = f(c) = a_n c^n + a_{n-1} c^{n-1} + \cdots + a_0. \qquad (3)$$

THEOREM 3

The Limit of a Rational Function Can Be Found by Substitution When the Denominator Is Different from Zero

If $f(x)$ and $g(x)$ are polynomials, then

$$\lim_{x \to c} \frac{f(x)}{g(x)} = \frac{f(c)}{g(c)} \qquad \text{(provided } g(c) \neq 0\text{).} \qquad (4)$$

Example 9

a) $\lim_{x \to 3} x^2(2 - x) = \lim_{x \to 3} (2x^2 - x^3) = 2(3)^2 - (3)^3 = 18 - 27 = -9.$

b) Same limit, found another way:

$$\lim_{x \to 3} x^2(2 - x) = \lim_{x \to 3} x^2 \cdot \lim_{x \to 3} (2 - x) = (3)^2 \cdot (2 - 3) = 9 \cdot (-1) = -9.$$

c) $\lim_{x \to 2} \dfrac{x^2 + 2x + 4}{x + 2} = \dfrac{(2)^2 + 2(2) + 4}{2 + 2} = \dfrac{12}{4} = 3.$

In Example 9c, we can use Eq. (4) to find the limit of $f(x)/g(x)$ because the value of the denominator, $g(x) = x + 2$, is different from 0 when $x = 2$. In the next example, the denominator is 0 when $x = 2$, so we cannot apply Eq. (4) directly. We have to rewrite the fraction $f(x)/g(x)$ first.

Example 10

$\lim_{x \to 2} \dfrac{x^3 - 8}{x^2 - 4}$

(Substitution will not give the limit because $x^2 - 4 = 0$ when $x = 2$.)

$= \lim_{x \to 2} \dfrac{(x - 2)(x^2 + 2x + 4)}{(x - 2)(x + 2)}$

(We factor the numerator and denominator and see that $(x - 2)$ is a common factor.)

$= \lim_{x \to 2} \dfrac{x^2 + 2x + 4}{x + 2}$

(Canceling $(x - 2)$, we have an equivalent form whose limit we can now find by substitution.)

$= 3$

(as in Example 9c)

Example 10 illustrates an important point about limits: The limit of a function $f(x)$ as x approaches c *never* depends on what happens at $x = c$. The limit, if it exists at all, is entirely determined by the values f has when $x \neq c$. In Example 10, the quotient $f(x) = (x^3 - 8)/(x^2 - 4)$ is not even defined at $x = 2$. Yet it has a limit as x approaches 2 and that limit is 3.

In the next example, the function is defined at every value of x, but the function's limit as x approaches 2 is not the same as the function's value at $x = 2$.

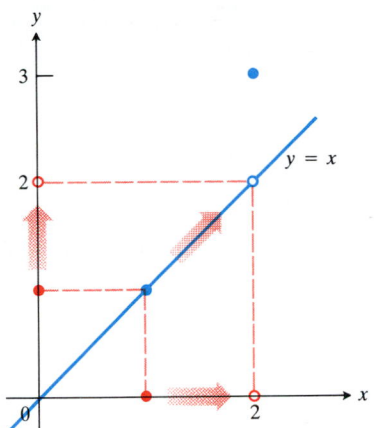

1.94   The graph of

$$f(x) = \begin{cases} x, & x \neq 2 \\ 3, & x = 2. \end{cases}$$

Notice that $f(x) \rightarrow 2$ as $x \rightarrow 2$ even though $f(2)$ itself is 3 (Example 11).

Example 11   If

$$f(x) = \begin{cases} x, & x \neq 2 \\ 3, & x = 2, \end{cases}$$

then

$$\lim_{x \to 2} f(x) = \lim_{x \to 2} (x) = 2, \quad \text{while} \quad f(2) = 3.$$

As always, the limit is determined by the function's approach behavior, not by what happens at $x = 2$ (Fig. 1.94).

Examples 10 and 11 show how special polynomial functions really are. If $f(x)$ is a polynomial function, then $\lim_{x \to c} f(x)$ is always $f(c)$ itself. Functions with this remarkable property are extremely useful in mathematics, as we shall see in Section 1.10.

Right-hand Limits and Left-hand Limits

Sometimes the values of a function $f(x)$ tend to different limits as x approaches a number c from different sides. When this happens, we call the limit of f as x approaches c from the right the **right-hand limit** of f at c, and the limit as x approaches c from the left the **left-hand limit** of f at c.

The notation for the right-hand limit is

$$\lim_{x \to c^+} f(x) \qquad \text{(``the limit of f as x approaches c from the right'')}.$$

The (+) says that x approaches c through values above c on the number line.

The notation for the left-hand limit is

$$\lim_{x \to c^-} f(x) \qquad \text{(``the limit of f as x approaches c from the left'')}.$$

The (−) says that x approaches c through values below c on the number line.

As you might expect, right-hand limits and left-hand limits obey all the limit rules of Theorems 1–3. The right-hand limit of the sum of two functions is the sum of their right-hand limits, and so on.

One-sided vs. Two-sided Limits

We sometimes call $\lim_{x \to c} f(x)$ the **two-sided limit** of f at c to distinguish it from the **one-sided** right-hand and left-hand limits of f at c. If the two one-sided limits of f exist at c and are equal, their common value is the two-sided limit of f at c. Conversely, if the two-sided limit of f at c exists, the two one-sided limits exist and have the same value as the two-sided limit.

THEOREM 4

A function $f(x)$ has a limit as x approaches c if and only if the right-hand and left-hand limits at c exist and are equal. In symbols,

$$\lim_{x \to c} f(x) = L \quad \Leftrightarrow \quad \lim_{x \to c^+} f(x) = L \quad \text{and} \quad \lim_{x \to c^-} f(x) = L. \qquad (5)$$

Theorem 4 will be proved formally at the end of the section, but the next examples show what is going on.

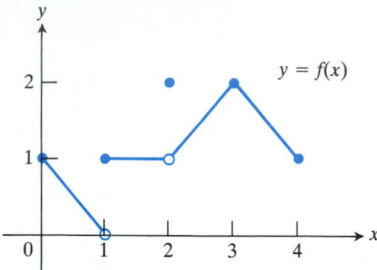

1.95 Example 12 discusses the limit properties of the function $y = f(x)$ graphed here.

Example 12   All the following statements about the function $y = f(x)$ graphed in Fig. 1.95 are true.

At $x = 0$: $\displaystyle\lim_{x \to 0^+} f(x) = 1$

At $x = 1$: $\displaystyle\lim_{x \to 1^-} f(x) = 0$ even though $f(1) = 1$,

$\displaystyle\lim_{x \to 1^+} f(x) = 1$,

$f(x)$ has no limit as $x \to 1$ (The right- and left-hand limits at 1 are not equal.)

At $x = 2$: $\displaystyle\lim_{x \to 2^-} f(x) = 1$,

$\displaystyle\lim_{x \to 2^+} f(x) = 1$,

$\displaystyle\lim_{x \to 2} f(x) = 1$ even though $f(2) = 2$

At $x = 3$: $\displaystyle\lim_{x \to 3^-} f(x) = \lim_{x \to 3^+} f(x) = \lim_{x \to 3} f(x) = f(3) = 2$

At $x = 4$: $\displaystyle\lim_{x \to 4^-} f(x) = 1$.

At every other point c between 0 and 4, $f(x)$ has a limit as $x \to c$.

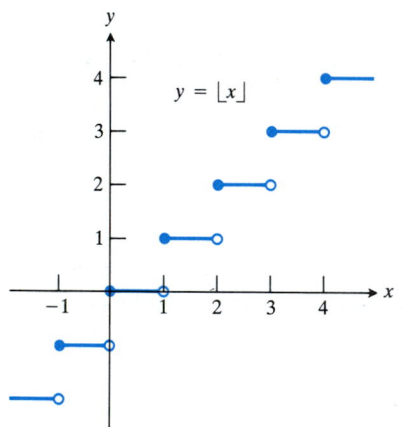

1.96 At each integer, the greatest integer function $y = \lfloor x \rfloor$ has different right-hand and left-hand limits (Example 13).

Example 13   The greatest integer function $f(x) = \lfloor x \rfloor$ has different right-hand and left-hand limits at each integer (Fig. 1.96). The limit as x approaches an integer n from the right is n, while the limit as x approaches n from the left is $n - 1$. The two-sided limit, $\lim_{x \to n} \lfloor x \rfloor$, fails to exist at every integer.

In the examples we have seen so far, the functions that failed to have limits at various points did so because the right-hand and left-hand limits at the points were not equal. The function in the next example fails to have a limit because neither the right-hand limit nor the left-hand limit exists at all.

Example 14   Show that the function $y = \sin(1/x)$ has no limit as x approaches 0 from either side (Fig. 1.97).

Solution   As x approaches 0, its reciprocal, $1/x$, grows without bound and the values of $\sin(1/x)$ cycle repeatedly from -1 to 1. There is no single number L that the function's values all get close to as x approaches 0. This is true even if we restrict x to positive values or to negative values. The function has neither a right-hand limit nor a left-hand limit as x approaches 0.

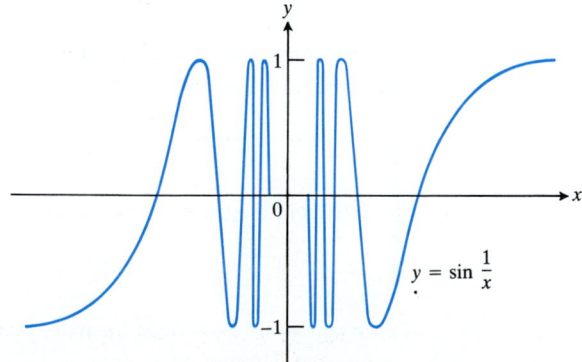

1.97   The function $y = \sin(1/x)$ has neither a right-hand nor a left-hand limit as x approaches 0 (Example 14).

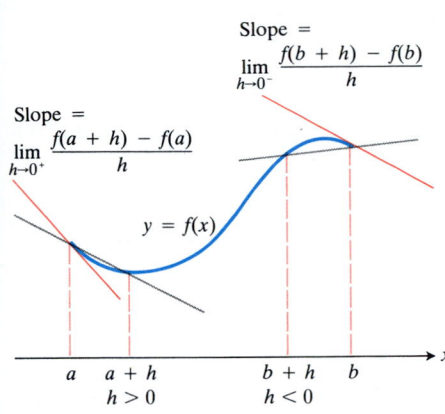

1.98 Derivatives at endpoints are one-sided limits.

Differentiable on a Closed Interval: One-sided Derivatives

A function $y = f(x)$ is **differentiable on a closed interval** $[a, b]$ if it has a derivative at every interior point and if the limits

$$\lim_{h \to 0^+} \frac{f(a + h) - f(a)}{h} \qquad \text{(Right-hand derivative at } a\text{)}$$

$$\lim_{h \to 0^-} \frac{f(b + h) - f(b)}{h} \qquad \text{(Left-hand derivative at } b\text{)}$$

exist at the endpoints a and b. In the right-hand derivative, h is positive and $a + h$ approaches a from the right. In the left-hand derivative, h is negative and $b + h$ approaches b from the left (Fig. 1.98).

Right-hand and left-hand derivatives may be defined at any point in a function's domain. Theorem 4 tells us that a function has a (two-sided) derivative at a point if and only if the function's right-hand and left-hand derivatives are defined and are equal at that point.

Example 15   The function $y = \begin{cases} x^2, & x \le 0 \\ 2x, & x > 0 \end{cases}$

(Fig. 1.99) has no derivative at $x = 0$ because the right-hand and left-hand derivatives are different there. The slope of the parabola on the left is $2(0) = 0$ (Section 1.6, Example 3). The slope of the line on the right is 2. ∎

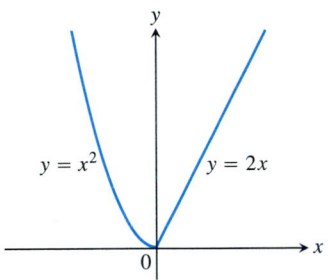

1.99   The function graphed here has different right-hand and left-hand derivatives at the origin (Example 15).

How Limit Theorems Are Proved

Although we shall not ask you to prove limit theorems yourself, we want to present a typical proof, if only to support our claim that having a precise definition of limit now makes it possible for us to prove the limit theorems on which calculus depends. Our example will be the proof of the Sum Rule for limits (the first part of Theorem 1). You can find a proof of the rest of Theorem 1 in Appendix 4.

Theorem 1, Part 1:   If $\lim_{x \to x_0} f_1(x) = L_1$ and $\lim_{x \to x_0} f_2(x) = L_2$, then $\lim_{x \to x_0} (f_1(x) + f_2(x)) = L_1 + L_2$.

Proof   To show that $\lim_{x \to x_0} (f_1(x) + f_2(x)) = L_1 + L_2$, we must show that for any $\epsilon > 0$ there exists a $\delta > 0$ such that for all x

$$0 < |x - x_0| < \delta \quad \Rightarrow \quad |f_1(x) + f_2(x) - (L_1 + L_2)| < \epsilon. \qquad (6)$$

Suppose, then, that ϵ is a positive number. The number $\epsilon/2$ is positive too, and because $\lim_{x \to x_0} f_1(x) = L_1$ we know that there is a $\delta_1 > 0$ such that for all x,

$$0 < |x - x_0| < \delta_1 \quad \Rightarrow \quad |f_1(x) - L_1| < \frac{\epsilon}{2}. \qquad (7)$$

Because $\lim_{x \to x_0} f_2(x) = L_2$, there is also a $\delta_2 > 0$ such that for all x,

$$0 < |x - x_0| < \delta_2 \quad \Rightarrow \quad |f_2(x) - L_2| < \frac{\epsilon}{2}. \qquad (8)$$

Now, either δ_1 equals δ_2 or it doesn't. If δ_1 equals δ_2, the implications in (7) and (8) both hold true for their common value δ. Taken together, (7) and (8) then say that, for all x, $0 < |x - x_0| < \delta$ implies

$$|f_1(x) + f_2(x) - (L_1 + L_2)| = |(f_1(x) - L_1) + (f_2(x) - L_2)|$$

$$\leq |(f_1(x) - L_1)| + |f_2(x) - L_2| \qquad \text{(Triangle inequality)}$$

$$< \frac{\epsilon}{2} + \frac{\epsilon}{2}$$

$$< \epsilon.$$

If $\delta_1 \neq \delta_2$, let δ be the smaller of δ_1 and δ_2. The implications in (7) and (8) then both hold for all x such that $0 < |x - x_0| < \delta$. As before,

$$|f_1(x) + f_2(x) - (L_1 + L_2)| < \epsilon.$$

Either way, we know that given any $\epsilon > 0$ there exists a $\delta > 0$ such that for all x,

$$0 < |x - x_0| < \delta \quad \Rightarrow \quad |f_1(x) + f_2(x) - (L_1 + L_2)| < \epsilon.$$

According to the ϵ-δ definition of limit, then,

$$\lim_{x \to x_0} (f(x_1) + f(x_2)) = L_1 + L_2.$$

The Relation Between One-sided and Two-sided Limits

The formal definitions of right-hand and left-hand limits go like this:

DEFINITIONS

Right-hand Limit: $\lim\limits_{x \to x_0+} f(x) = L$

The limit of $f(x)$ as x approaches x_0 from the right is the number L if the following criterion holds (Fig. 1.100):

Given any radius $\epsilon > 0$ about L there exists a radius $\delta > 0$ to the right of x_0 such that for all x,

$$x_0 < x < x_0 + \delta \quad \Rightarrow \quad |f(x) - L| < \epsilon. \tag{9}$$

Left-hand Limit: $\lim\limits_{x \to x_0-} f(x) = L$

The limit of $f(x)$ as x approaches x_0 from the left is the number L if the following criterion holds (Fig. 1.101):

Given any radius $\epsilon > 0$ about L there exists a radius $\delta > 0$ to the left of x_0 such that for all x,

$$x_0 - \delta < x < x_0 \quad \Rightarrow \quad |f(x) - L| < \epsilon. \tag{10}$$

By comparing Eqs. (9) and (10) with Eq. (1), we can see the relation between the one-sided limits just defined and the two-sided limit defined earlier. If we subtract x_0 from the δ-inequalities in Eqs. (9) and (10), they become

$$0 < x - x_0 < \delta \quad \Rightarrow \quad |f(x) - L| < \epsilon \tag{11}$$

and

$$-\delta < x - x_0 < 0 \quad \Rightarrow \quad |f(x) - L| < \epsilon. \tag{12}$$

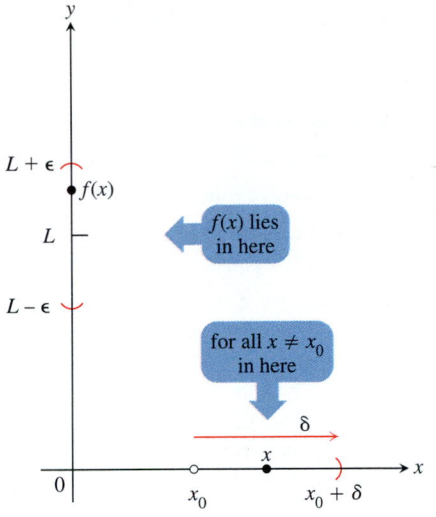

1.100 Diagram for the definition of right-hand limit.

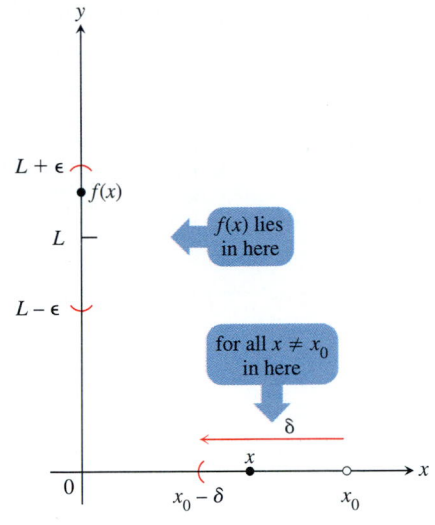

1.101 Diagram for the definition of left-hand limit.

Together, Eqs. (11) and (12) say the same thing as

$$0 < |x - x_0| < \delta \quad \Rightarrow \quad |f(x) - L| < \epsilon, \qquad (13)$$

which is Eq. (1) in the definition of limit. In other words, $f(x)$ has limit L at x_0 if and only if the right-hand and left-hand limits of f at x_0 exist and equal L.

EXERCISES 1.7

Limit Calculations

Find the limits in Exercises 1–10.

1. $\lim_{x \to 1} (3x - 1)$

2. $\lim_{x \to 1/3} (3x - 1)$

3. $\lim_{x \to 5} x^2$

4. $\lim_{x \to 2} x(2 - x)$

5. $\lim_{x \to 0} 5(2x - 1)$

6. $\lim_{x \to -1} 3x^2(2x - 1)$

7. $\lim_{x \to 2} 5(2x - 1)(x + 1)$

8. $\lim_{x \to 6} 8(x - 5)(x - 7)$

9. $\lim_{x \to 1} (x^3 + 3x^2 - 2x - 17)$

10. $\lim_{x \to -2} (x^3 - 2x^2 + 4x + 8)$

Find the limits in Exercises 11–20.

11. $\lim_{x \to 2} \dfrac{x + 3}{x + 6}$

12. $\lim_{x \to 5} \dfrac{4}{x - 7}$

13. $\lim_{x \to -1} \dfrac{x + 3}{x^2 + 3x + 1}$

14. $\lim_{x \to -5} \dfrac{x^2}{5 - x}$

15. $\lim_{y \to 2} \dfrac{y^2 + 5y + 6}{y + 2}$

16. $\lim_{y \to 0} \dfrac{4 - y}{3 + y^3}$

17. $\lim_{y \to -3} \dfrac{y^2 + 4y + 3}{y^2 - 3}$

18. $\lim_{y \to 3} \dfrac{1}{y^4 - 9y^2 + y}$

19. $\lim_{x \to -1} \dfrac{x^3 - 5x + 7}{-x^3 + x^2 - x + 1}$

20. $\lim_{t \to 2} \dfrac{3t^3 - 10t - 3}{t^4 - 14}$

Find the limits in Exercises 21–30.

21. $\lim_{x \to -5} \dfrac{x^2 + 3x - 10}{x + 5}$

22. $\lim_{x \to 2} \dfrac{x^2 - 7x + 10}{x - 2}$

23. $\lim_{x \to 1} \dfrac{x - 1}{x^2 - 1}$

24. $\lim_{x \to 1} \dfrac{x^2 + x - 2}{x^2 - 1}$

25. $\lim_{x \to -3} \dfrac{x + 3}{x^2 + 4x + 3}$

26. $\lim_{x \to -2} \dfrac{x^2 + x - 2}{x^2 - 4}$

27. $\lim_{x \to 5} \dfrac{x - 5}{x^2 - 25}$

28. $\lim_{x \to -5} \dfrac{x + 5}{x^2 - 25}$

29. $\lim_{x \to 2} \dfrac{2x - 4}{x^3 - 2x^2}$

30. $\lim_{x \to 0} \dfrac{5x^3 + 8x^2}{3x^4 - 16x^2}$

31. Let $f(x) = \begin{cases} 3 - x, & x < 2 \\ \dfrac{x}{2} + 1, & x > 2. \end{cases}$

a) Find $\lim_{x \to 2^+} f(x)$
and $\lim_{x \to 2^-} f(x)$.

b) Does $\lim_{x \to 2} f(x)$ exist?
If so, what is it?
If not, why not?

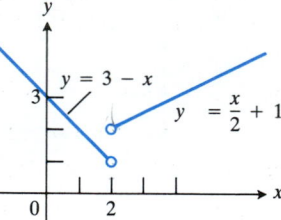

32. Let $f(x) = \begin{cases} 3 - x, & x < 2 \\ 2, & x = 2 \\ \dfrac{x}{2}, & x > 2. \end{cases}$

a) Find $\lim_{x \to 2^+} f(x)$
and $\lim_{x \to 2^-} f(x)$.

b) Does $\lim_{x \to 2} f(x)$ exist?
If so, what is it?
If not, why not?

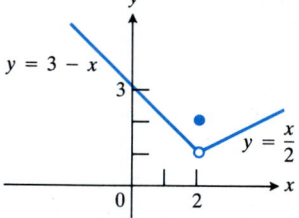

33. Which of the following statements are true of the function $y = f(x)$ graphed here?

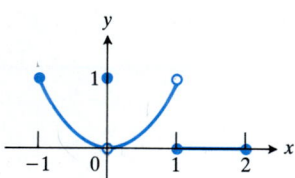

a) $\lim_{x \to -1^+} f(x) = 1$

b) $\lim_{x \to 0^-} f(x) = 0$

c) $\lim_{x \to 0^-} f(x) = 1$

d) $\lim_{x \to 0^-} f(x) = \lim_{x \to 0^+} f(x)$

e) $\lim_{x \to 0} f(x)$ exists.

f) $\lim_{x \to 0} f(x) = 0$

g) $\lim_{x \to 0} f(x) = 1$

h) $\lim_{x \to 1} f(x) = 1$

i) $\lim_{x \to 1} f(x) = 0$

j) $\lim_{x \to 2^-} f(x) = 2$

34. Which of the following statements are true of the function graphed here?

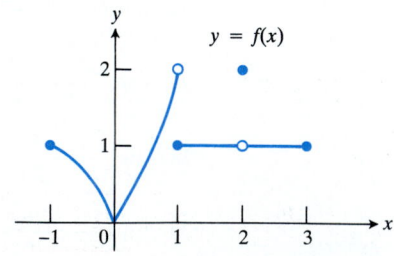

a) $\lim_{x \to -1^+} f(x) = 1$

b) $\lim_{x \to 2} f(x)$ does not exist.

c) $\lim_{x \to 2} f(x) = 2$

d) $\lim_{x \to 1^-} f(x) = 2$

e) $\lim_{x \to 1^+} f(x) = 1$

f) $\lim_{x \to 1} f(x)$ does not exist.

g) $\lim_{x \to 0^+} f(x) = \lim_{x \to 0^-} f(x)$

h) $\lim_{x \to c} f(x)$ exists at every c in $(-1, 1)$.

i) $\lim_{x \to c} f(x)$ exists at every c in $(1, 3)$.

35. a) Graph $f(x) = \begin{cases} x^3, & x \neq 1 \\ 0, & x = 1. \end{cases}$

b) Find $\lim_{x \to 1^-} f(x)$ and $\lim_{x \to 1^+} f(x)$.

c) Does $\lim_{x \to 1} f(x)$ exist? If so, what is it? If not, why not?

36. a) Graph $f(x) = \begin{cases} 1 - x^2, & x \neq 1 \\ 2, & x = 1. \end{cases}$

b) Find $\lim_{x \to 1^+} f(x)$ and $\lim_{x \to 1^-} f(x)$.

c) Does $\lim_{x \to 1} f(x)$ exist? If so, what is it? If not, why not?

Graph the two functions in Exercises 37 and 38. Then answer these questions:

a) At what points c in the domain of f does $\lim_{x \to c} f(x)$ exist?

b) At what points does only the left-hand limit exist?

c) At what points does only the right-hand limit exist?

37. $f(x) = \begin{cases} \sqrt{1 - x^2} & \text{if } 0 \leq x < 1 \\ 1 & \text{if } 1 \leq x < 2 \\ 2 & \text{if } x = 2 \end{cases}$

38. $f(x) = \begin{cases} x & \text{if } -1 \leq x < 0 \text{ or } 0 < x \leq 1 \\ 1 & \text{if } x = 0 \\ 0 & \text{if } x < -1 \text{ or } x > 1 \end{cases}$

39. Let $f(x) = \begin{cases} 0, & x \leq 0 \\ \sin \dfrac{1}{x}, & x > 0. \end{cases}$

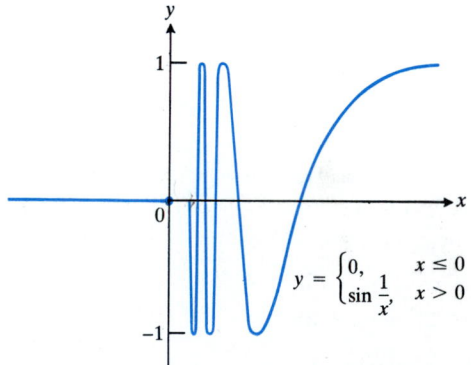

a) Does $\lim_{x \to 0^+} f(x)$ exist? If so, what is it?

b) Does $\lim_{x \to 0^-} f(x)$ exist? If so, what is it?

c) Does $\lim_{x \to 0} f(x)$ exist? If so, what is it? If not, why not?

40. Let $f(x) = \begin{cases} 0, & x = 0 \\ x \sin \dfrac{1}{x}, & x \neq 0. \end{cases}$

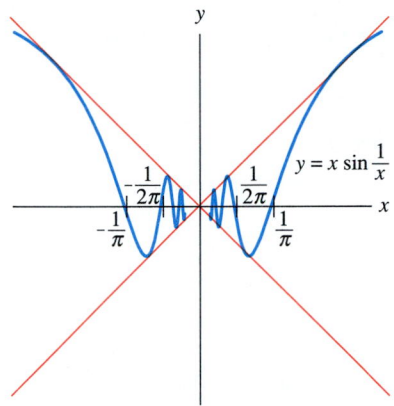

(Generated by Mathematica)

a) Does $\lim_{x \to 0^+} f(x)$ exist? If so, what is it?
b) Does $\lim_{x \to 0^-} f(x)$ exist? If so, what is it?
c) Does $\lim_{x \to 0} f(x)$ exist? If so, what is it? If not, why not?

41. Find
a) $\lim_{x \to 0^+} \lfloor x \rfloor$;
b) $\lim_{x \to 0.5} \lfloor x \rfloor$;
c) $\lim_{x \to 2^-} \lfloor x \rfloor$.

42. Find
a) $\lim_{x \to 0^+} \dfrac{x}{|x|}$;

b) $\lim_{x \to 0^-} \dfrac{x}{|x|}$.

43. Suppose $\lim_{x \to 4} f(x) = 0$ and $\lim_{x \to 4} g(x) = -3$. Find
a) $\lim_{x \to 4} (g(x) + 3)$;
b) $\lim_{x \to 4} x f(x)$;
c) $\lim_{x \to 4} g^2(x)$;
d) $\lim_{x \to 4} \dfrac{g(x)}{f(x) - 1}$.

44. Suppose $\lim_{x \to b} f(x) = 7$ and $\lim_{x \to b} g(x) = -3$. Find
a) $\lim_{x \to b} (f(x) + g(x))$;
b) $\lim_{x \to b} f(x) \cdot g(x)$;
c) $\lim_{x \to b} 4g(x)$;
d) $\lim_{x \to b} f(x)/g(x)$.

Use right-hand and left-hand derivatives to show that the functions graphed in Exercises 45 and 46 are not differentiable at the indicated point P.

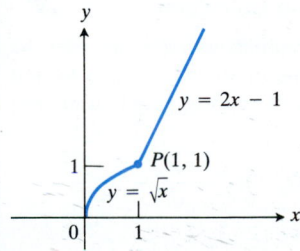

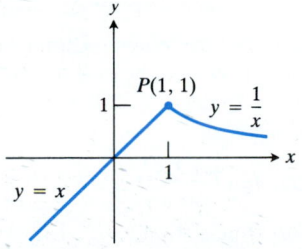

The Formal Definition of Limit

In Exercises 47–54, sketch the interval (a, b) on the x-axis with the point x_0 inside. Then find the largest value of $\delta > 0$ such that $|x - x_0| < \delta$ implies $a < x < b$.

47. $a = 1, \quad b = 7, \quad x_0 = 5$

48. $a = 1, \quad b = 7, \quad x_0 = 2$

49. $a = -7/2, \quad b = -1/2, \quad x_0 = -3$

50. $a = -7/2, \quad b = -1/2, \quad x_0 = -3/2$

51. $a = -5, \quad b = 3, \quad x_0 = 1$

52. $a = -5, \quad b = 3, \quad x_0 = -2$

53. $a = 4/9, \quad b = 4/7, \quad x_0 = 1/2$

54. $a = 2.7591, \quad b = 3.2391, \quad x_0 = 3$

Use the graphs in Exercises 55–60 to find a $\delta > 0$ such that for all x,

$$0 < |x - x_0| < \delta \quad \Rightarrow \quad |f(x) - L| < \epsilon.$$

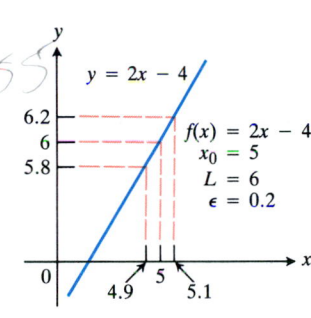

NOT TO SCALE

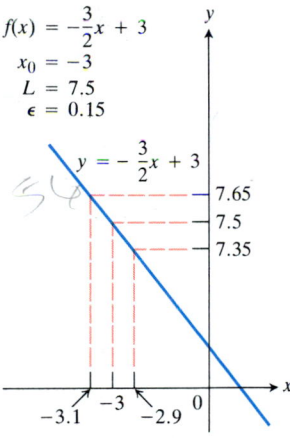

NOT TO SCALE

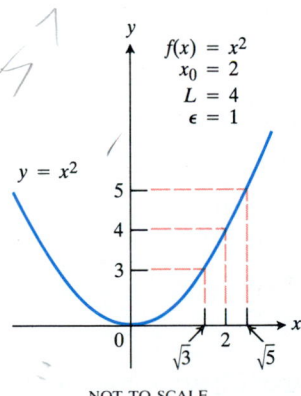

NOT TO SCALE

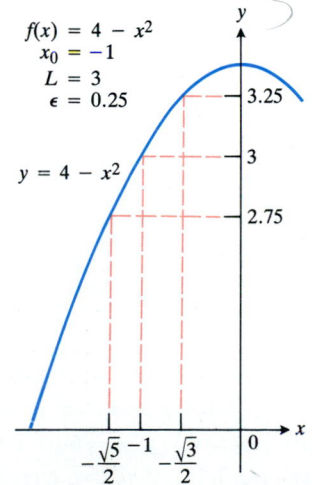

59.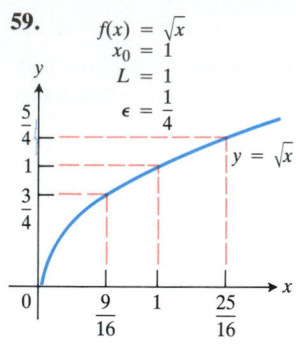

$f(x) = \sqrt{x}$
$x_0 = 1$
$L = 1$
$\epsilon = \dfrac{1}{4}$

60.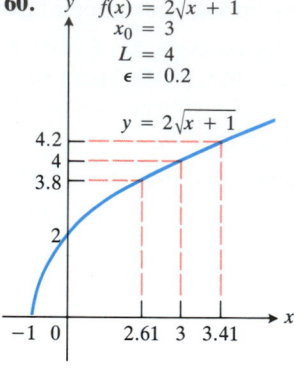

$f(x) = 2\sqrt{x+1}$
$x_0 = 3$
$L = 4$
$\epsilon = 0.2$

NOT TO SCALE

For each function $y = f(x)$ and number $\epsilon > 0$ in Exercises 61–68, find the set of x-values for which $|f(x) - 4| < \epsilon$.

61. $f(x) = x + 1$, $\epsilon = 0.01$

62. $f(x) = 2x - 2$, $\epsilon = 0.02$

63. $f(x) = x^2 - 5$, $\epsilon = 0.05$

64. $f(x) = 4 - x^2$, $\epsilon = 0.04$

65. $f(x) = \sqrt{19 - x}$, $\epsilon = 0.03$

66. $f(x) = \sqrt{x + 1}$, $\epsilon = 0.1$

67. $f(x) = 1/x$, $\epsilon = 0.1$

68. $f(x) = \dfrac{1}{x - 2}$, $\epsilon = 0.5$

Each of Exercises 69–78 gives a function $f(x)$, a point x_0, and a positive number ϵ. Find $L = \lim_{x \to x_0} f(x)$. Then find a number $\delta > 0$ such that for all x,

$$0 < |x - x_0| < \delta \quad \Rightarrow \quad |f(x) - L| < \epsilon.$$

69. $f(x) = 2x + 3$, $x_0 = 1$, $\epsilon = 0.01$

70. $f(x) = 3 - 2x$, $x_0 = 3$, $\epsilon = 0.02$

71. $f(x) = 4x - 2$, $x_0 = 1/2$, $\epsilon = 0.02$

72. $f(x) = -3x - 2$, $x_0 = -1$, $\epsilon = 0.03$

73. $f(x) = \dfrac{x^2 - 4}{x - 2}$, $x_0 = 2$, $\epsilon = 0.05$

74. $f(x) = \dfrac{x^2 + 6x + 5}{x + 5}$, $x_0 = -5$, $\epsilon = 0.05$

75. $f(x) = \sqrt{x - 7}$, $x_0 = 11$, $\epsilon = 0.01$

76. $f(x) = \sqrt{1 - 5x}$, $x_0 = -3$, $\epsilon = 0.5$

77. $f(x) = 4/x$, $x_0 = 2$, $\epsilon = 0.4$

78. $f(x) = 4/x$, $x_0 = 1/2$, $\epsilon = 0.04$

In Exercises 79 and 80, find the largest $\delta > 0$ such that for all x,

$$0 < |x - 4| < \delta \quad \Rightarrow \quad |f(x) - 5| < \epsilon.$$

79. $f(x) = 9 - x$, $\epsilon = 0.01, 0.001, 0.0001$, arbitrary $\epsilon > 0$

80. $f(x) = |3x - 7|$, $\epsilon = 0.003, 0.0003$, arbitrary $\epsilon > 0$

81. Given $\epsilon > 0$, find an interval $I = (5, 5 + \delta)$, $\delta > 0$, such that if x lies in I then $\sqrt{x - 5} < \epsilon$. What limit is being verified?

82. Given $\epsilon > 0$, find an interval $I = (4 - \delta, 4)$, $\delta > 0$, such that if x lies in I then $\sqrt{4 - x} < \epsilon$. What limit is being verified?

Calculator Exercises

83. a) CALCULATOR   Estimate the value of

$$\lim_{x \to 1} \frac{x^2 - 1}{x - 1}$$

by taking $x = 1.1, 1.01, 1.001$, and so on as far as your calculator can go.

b) Find the limit's exact value.

84. a) CALCULATOR   Estimate the value of

$$\lim_{x \to 3} \frac{x^2 - 2x - 3}{x^2 - 4x + 3}$$

by taking $x = 3.1, 3.01, 3.001$, and so on as far as your calculator can go.

b) Find the limit's exact value.

85. CALCULATOR   By taking $x = 1.1, 1.01, 1.001$, and so on as far as your calculator can go, estimate the value of

$$\lim_{x \to 1} \frac{\ln(x^2)}{\ln x}.$$

(Chapter 6 will explain what is going on here.)

86. CALCULATOR   It is sometimes easy to guess the value of a limit once the limit is known to exist.

a) Ignoring the question of whether the following limit exists (it does and is finite), use a calculator to guess its value:

$$\lim_{h \to 0} \frac{\sqrt{4 + h} - 2}{h}$$

First take $h = 0.1, 0.01, 0.001, \ldots$, continuing until you are ready to guess the right-hand limit. Then test your guess by using $h = -0.1, -0.01. \ldots$

b) Relate the limit in (a) to a derivative.

87. CALCULATOR   To estimate the value of the derivative of $f(x) = \sqrt{9 - x^2}$ at $x = 0$, write out the appropriate difference quotient and proceed as in Exercise 86(a).

Computer Grapher or Graphing Calculator

Estimate the following limits by graphing each function near the origin. We will be able to verify the limits in Exercises 88 and 89 in Section 1.9. The limit in Exercise 90 will have to wait until Chapter 6.

88. $\lim\limits_{x \to 0} \dfrac{\sin 3x}{x}$

89. $\lim\limits_{x \to 0} \dfrac{1 - \cos x}{x^2}$

90. $\lim\limits_{x \to 0^+} x^{1/x}$

1.8 Limits Involving Infinity

In this section, we describe what it means for the values of a function to approach infinity and what it means for a function $f(x)$ to have a limit as x approaches infinity. Although there is no real number infinity, the word *infinity* is useful for describing how a function $y = f(x)$ behaves when the magnitude of either x or y increases without bound.

Limits as $x \to \infty$ or $x \to -\infty$

The function

$$f(x) = \frac{1}{x}$$

is defined for all real numbers except $x = 0$. As Fig. 1.102 suggests,

a) $1/x$ is small and positive when x is large and positive;

b) $1/x$ is large and positive when x is small and positive;

c) $1/x$ is large and negative when x is small and negative;

d) $1/x$ is small and negative when x is large and negative.

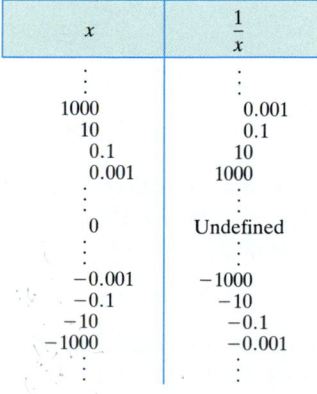

x	$\dfrac{1}{x}$
⋮	⋮
1000	0.001
10	0.1
0.1	10
0.001	1000
⋮	⋮
0	Undefined
⋮	⋮
−0.001	−1000
−0.1	−10
−10	−0.1
−1000	−0.001
⋮	⋮

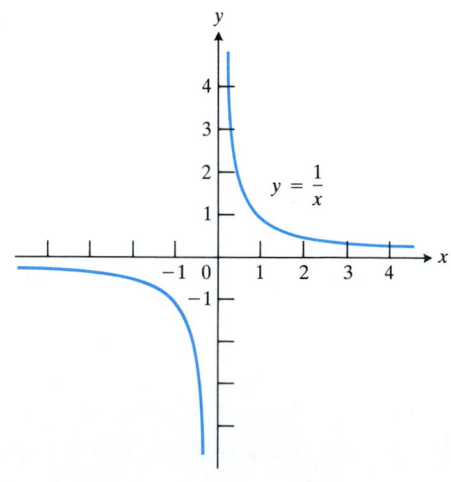

1.102   The graph of $y = 1/x$.

We summarize these facts by saying:

a) As x tends to ∞, $1/x$ approaches 0.

b) As x approaches 0 from the right, $1/x$ tends to ∞.

c) As x approaches 0 from the left, $1/x$ tends to $-\infty$.

d) As x tends to $-\infty$, $1/x$ approaches 0.

The symbol ∞, **infinity,** does not represent any real number. We cannot use ∞ in arithmetic in the usual way, but it is convenient to be able to say things like "the limit of $1/x$ as x approaches infinity is 0."

DEFINITIONS

$$\lim_{x \to \infty} f(x) = L \qquad \text{and} \qquad \lim_{x \to -\infty} f(x) = L$$

1. The **limit** of the function $f(x)$ **as x approaches infinity** is the number L if the following criterion holds:
   Given any number $\epsilon > 0$, there exists a number M such that for all x,

$$x > M \quad \text{implies} \quad |f(x) - L| < \epsilon. \tag{1}$$

2. The **limit** of $f(x)$ **as x approaches negative infinity** is the number L if the following criterion holds:
   Given any number $\epsilon > 0$, there exists a number N such that for all x,

$$x < N \quad \text{implies} \quad |f(x) - L| < \epsilon. \tag{2}$$

Calculation Rules for Functions with Finite Limits as $x \to \pm\infty$

Our strategy is again the one that worked so well in Section 1.7. We find the limits of two "basic" functions as $x \to \infty$ and $x \to -\infty$ and then, to find everything else, we use a theorem about limits of algebraic combinations. In Section 1.7, the basic functions were the constant function $y = k$ and the identity function $y = x$. Here, the basic functions are $y = k$ and the reciprocal $y = 1/x$.

Example 1   Show that if f is the constant function whose outputs have the constant value $f(x) = k$, then

$$\lim_{x \to \infty} f(x) = \lim_{x \to \infty} (k) = k$$

and

$$\lim_{x \to -\infty} f(x) = \lim_{x \to -\infty} (k) = k.$$

Solution   In the definition of limit, we set $f(x) = k$ and $L = k$. To show that $\lim_{x \to \pm\infty} f(x) = k$, we must show that for any $\epsilon > 0$ there exist numbers M and N such that for all x,

$$x > M \quad \text{or} \quad x < N \quad \Rightarrow \quad |k - k| < \epsilon.$$

This implication will hold for any M and N because $|k - k| = 0$ is less than every positive ϵ for all x.

Example 2   Show that $\lim_{x \to \infty} (1/x) = 0$ and $\lim_{x \to -\infty} (1/x) = 0$.

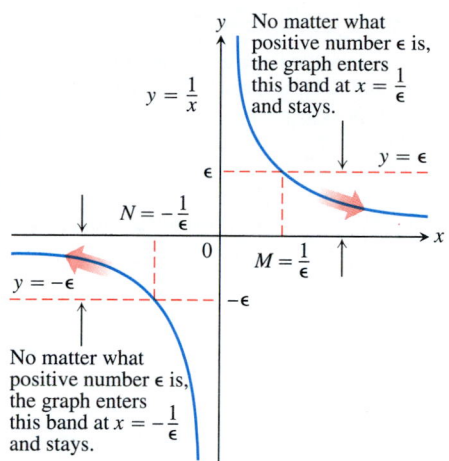

$y = \frac{1}{x}$

No matter what positive number ϵ is, the graph enters this band at $x = \frac{1}{\epsilon}$ and stays.

$y = \epsilon$

ϵ

$N = -\frac{1}{\epsilon}$

$M = \frac{1}{\epsilon}$

$y = -\epsilon$

$-\epsilon$

No matter what positive number ϵ is, the graph enters this band at $x = -\frac{1}{\epsilon}$ and stays.

1.103   The geometry behind the argument in Example 2.

Solution   In the definition of limit, we set $f(x) = 1/x$ and $L = 0$. To show that $\lim_{x \to \pm\infty} (1/x) = 0$, we must show that for any $\epsilon > 0$ there exist numbers M and N such that for all x,

$$x > M \quad \text{or} \quad x < N \quad \Rightarrow \quad \left| \frac{1}{x} - 0 \right| = \left| \frac{1}{x} \right| < \epsilon.$$

The inequality on the right is equivalent, in turn, to

$$1 < \epsilon |x|, \qquad \frac{1}{\epsilon} < |x|, \qquad \text{and} \qquad |x| > \frac{1}{\epsilon}.$$

The latter will hold for $x > M$ and $x < N$ if we choose M to be $1/\epsilon$ and N to be $-1/\epsilon$ (Fig. 1.103).

We now have enough specific information to calculate the limits of a wide variety of rational functions as $x \to \pm\infty$ by using the limit rules in the following theorem.

THEOREM 5

Properties of Finite Limits as $x \to \pm\infty$

The following rules hold if

$$\lim_{x \to \infty} f_1(x) = L_1 \qquad \text{and} \qquad \lim_{x \to \infty} f_2(x) = L_2$$

and L_1 and L_2 are (finite) real numbers.

1. *Sum Rule:* $\lim_{x \to \infty} [f_1(x) + f_2(x)] = L_1 + L_2$

2. *Difference Rule:* $\lim_{x \to \infty} [f_1(x) - f_2(x)] = L_1 - L_2$

3. *Product Rule:* $\lim_{x \to \infty} f_1(x) \cdot f_2(x) = L_1 \cdot L_2$

4. *Constant Multiple Rule:* $\lim_{x \to \infty} k \cdot f_1(x) = k \cdot L_1$ (any number k)

5. *Quotient Rule:* $\lim_{x \to \infty} \frac{f_1(x)}{f_2(x)} = \frac{L_1}{L_2}$ if $L_2 \neq 0$

These properties hold for $x \to -\infty$ as well as $x \to \infty$.

These properties are just like the properties we stated in Section 1.7 for limits as $x \to c$, and we use them the same way.

Example 3

a) $\lim_{x \to \infty} \left[5 + \frac{1}{x} \right] = \lim_{x \to \infty} 5 + \lim_{x \to \infty} \frac{1}{x} = 5 + 0 = 5$ (Sum Rule and known values)

b) $\lim_{x \to -\infty} \frac{4}{x^2} = \lim_{x \to -\infty} 4 \cdot \lim_{x \to -\infty} \frac{1}{x} \cdot \lim_{x \to -\infty} \frac{1}{x} = 4 \cdot 0 \cdot 0 = 0$ $\left(\begin{array}{l} \text{Product Rule and} \\ \text{known values} \end{array} \right)$

Limits of Rational Functions as $x \to \pm\infty$

To find the limit of a rational function as $x \to \pm\infty$ (when the limit exists), we divide the numerator and denominator by the highest power of x in the denominator. What happens then depends on the degrees of the polynomials involved.

The *degree* of $a_n x^n + a_{n-1}x^{n-1} + \cdots + a_1 x + a_0$ is n, the largest exponent. The leading coefficient is a_n.

Example 4   *Numerator and Denominator of the Same Degree*

a) $\displaystyle\lim_{x\to\infty} \frac{-x}{7x + 4} = \lim_{x\to\infty} \frac{-1}{7 + (4/x)}$ $\left(\begin{array}{l}\text{Divide numerator and denominator}\\\text{by highest power of } x \text{ in denominator,}\\\text{in this case } x\,.\end{array}\right)$

$$= \frac{-1}{7 + 0} = -\frac{1}{7}$$

b) $\displaystyle\lim_{x\to-\infty} \frac{2x^2 - x + 3}{3x^2 + 5} = \lim_{x\to-\infty} \frac{2 - (1/x) + (3/x^2)}{3 + (5/x^2)}$ $\left(\begin{array}{l}\text{Divide numerator and}\\\text{denominator by } x^2.\end{array}\right)$

$$= \frac{2 - 0 + 0}{3 + 0} = \frac{2}{3}$$

Example 5   *Degree of Numerator Less Than Degree of Denominator*

$$\lim_{x\to\infty} \frac{5x + 2}{2x^3 - 1} = \lim_{x\to\infty} \frac{(5/x^2) + (2/x^3)}{2 - (1/x^3)}$$ $\left(\begin{array}{l}\text{Divide numerator and}\\\text{denominator by } x^3.\end{array}\right)$

$$= \frac{0 + 0}{2 - 0} = 0.$$

Summary for Rational Functions

$\displaystyle\lim_{x\to\pm\infty} \frac{f(x)}{g(x)}$ is

a) *zero if* deg(f) < deg(g),

b) *the ratio of the leading coefficients if* deg(f) = deg(g),

c) *infinite if* deg(f) > deg(g).

See Exercise 47.

The rule for finding limits of rational functions as $x \to \pm\infty$ is this: If the numerator and denominator have the same degree, the limit is the ratio of the leading coefficients. If the degree of the numerator is less than the degree of the denominator, the limit is zero. If the degree of the numerator is greater than the degree of the denominator, the limit is infinite, as we shall see in a moment.

Lim $f(x) = \infty$ or lim $f(x) = -\infty$

As suggested by the behavior of $1/x$ as $x \to 0$, we sometimes want to say such things as

a) $\displaystyle\lim_{x\to c} f(x) = \infty$,   b) $\displaystyle\lim_{x\to c^+} f(x) = \infty$,   c) $\displaystyle\lim_{x\to c^-} f(x) = \infty$,   (3)

d) $\displaystyle\lim_{x\to\infty} f(x) = \infty$,   e) $\displaystyle\lim_{x\to-\infty} f(x) = \infty$.

In every instance, we mean that the value of $f(x)$ eventually exceeds any positive number B. That is, given any positive real number B no matter how large, the values of f satisfy the condition

$$f(x) > B \tag{4}$$

if x lies in some restricted set, usually depending on B:

In (a) the set has the form $0 < |x - c| < \delta$.

In (b) the set is an interval $c < x < c + \delta$ to the right of c.

In (c) the set is an interval $c - \delta < x < c$ to the left of c.

In (d) the set is an infinite interval $M < x < \infty$.

In (e) the set is an infinite interval $-\infty < x < N$.

By replacing the condition $f(x) > B$ in (4) by the condition

$$f(x) < -B\,, \tag{5}$$

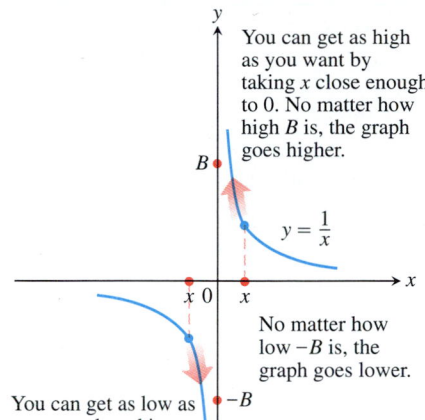

You can get as high as you want by taking x close enough to 0. No matter how high B is, the graph goes higher.

B

$y = \dfrac{1}{x}$

No matter how low $-B$ is, the graph goes lower.

You can get as low as you want by taking x close enough to 0.

$-B$

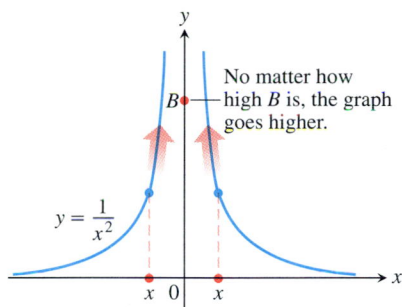

No matter how high B is, the graph goes higher.

$y = \dfrac{1}{x^2}$

1.104   The graphs of the functions in Example 6.

where $-B$ is a negative real number, we can similarly define the statements

$$\lim_{x \to c} f(x) = -\infty, \qquad \lim_{x \to c^+} f(x) = -\infty, \qquad \lim_{x \to c^-} f(x) = -\infty, \tag{6}$$

$$\lim_{x \to \infty} f(x) = -\infty, \qquad \lim_{x \to -\infty} f(x) = -\infty.$$

Example 6   (Fig. 1.104)

a) $\displaystyle \lim_{x \to 0^+} \frac{1}{x} = \infty$

b) $\displaystyle \lim_{x \to 0^-} \frac{1}{x} = -\infty$

c) $\displaystyle \lim_{x \to 0^+} \frac{1}{x^2} = \infty$

d) $\displaystyle \lim_{x \to 0^-} \frac{1}{x^2} = \infty$

Example 7   (Fig. 1.105)

$$\lim_{x \to 1^+} \frac{1}{x - 1} = \infty \qquad \lim_{x \to 1^-} \frac{1}{x - 1} = -\infty$$

The idea here is to think about the number $x - 1$. As x approaches 1 from above, $x - 1$ approaches 0 from above. The reciprocal $1/(x - 1)$ stays positive and increases beyond all bounds.

On the other hand, if x approaches 1 from below, $x - 1$ remains negative and approaches 0 from below. Its reciprocal $1/(x - 1)$ is negative as well and approaches $-\infty$.

Example 8   *Degree of Numerator Greater Than Degree of Denominator*

a) $\displaystyle \lim_{x \to \infty} \frac{2x^2 - 3}{7x + 4} = \lim_{x \to \infty} \frac{2x - (3/x)}{7 + (4/x)}$ $\left(\begin{array}{l}\text{Divide by the highest power of}\\ x \text{ in the denominator.}\end{array}\right)$

$\qquad\qquad = \infty$ $\left(\begin{array}{l}\text{The numerator approaches } \infty\\ \text{and the denominator approaches}\\ 7, \text{ so their ratio approaches } \infty.\end{array}\right)$

b) $\displaystyle \lim_{x \to -\infty} \frac{2x^2 - 3}{7x + 4} = \lim_{x \to -\infty} \frac{2x - (3/x)}{7 + (4/x)}$ $\left(\begin{array}{l}\text{Divide by the highest power of}\\ x \text{ in the denominator.}\end{array}\right)$

$\qquad\qquad = -\infty$ $\left(\begin{array}{l}\text{The numerator approaches } -\infty\\ \text{and the denominator approaches}\\ 7, \text{ so their ratio approaches } -\infty.\end{array}\right)$

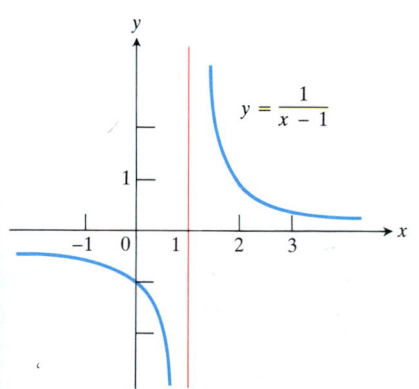

1.105   Near $x = 1$, the function $y = 1/(x - 1)$ behaves the way the function $y = 1/x$ behaves near $x = 0$. Its graph is the graph of $y = 1/x$ shifted one unit to the right.

The procedure in Example 8 is typical. To find the limit as $x \to \pm\infty$ of a rational function in which the degree of the numerator exceeds the degree of the denominator, divide by the highest power of x in the denominator. The limit of the new denominator will then be finite, and the limit of the new numerator will be infinite. The limit of the ratio can be either $+\infty$ or $-\infty$, depending on the signs assumed by the numerator and denominator as x becomes numerically large.

EXERCISES 1.8

Find the limits of the functions defined by the expressions in Exercises 1–14 (a) as $x \to \infty$ and (b) as $x \to -\infty$.

1. $\dfrac{2x + 3}{5x + 7}$

2. $\dfrac{2x^3 + 7}{x^3 - x^2 + x + 7}$

3. $\dfrac{x + 1}{x^2 + 3}$

4. $\dfrac{3x + 7}{x^2 - 2}$

5. $\dfrac{1 - 12x^2}{4x^2 + 12}$

6. $\dfrac{2x^2 + 3}{-x^2 + x}$

7. $\dfrac{3x^2 - 6x}{4x - 8}$

8. $\dfrac{x^4}{x^3 + 1}$

9. $\dfrac{1}{x^3 - 4x + 1}$

10. $\dfrac{10x^5 + x^4 + 31}{x^6}$

11. $\dfrac{7x^3}{x^3 - 3x^2 + 6x}$

12. $\dfrac{9x^4 + x}{2x^4 + 5x^2 - x + 6}$

13. $\dfrac{-2x^3 - 2x + 3}{3x^3 + 3x^2 - 5x}$

14. $\dfrac{-x^4}{x^4 - 7x^3 + 7x^2 + 9}$

Use the Product Rule to find the limits of the functions defined by the expressions in Exercises 15–18 (a) as $x \to \infty$ and (b) as $x \to -\infty$.

15. $\left(\dfrac{-x}{x + 1}\right)\left(\dfrac{x^2}{5 + x^2}\right)$

16. $\left(\dfrac{2}{x} + 1\right)\left(\dfrac{5x^2 - 1}{x^2}\right)$

17. $\left(\dfrac{1 - x^2}{1 + 2x^2}\right)\left(\dfrac{8x^2 + 7x}{4x^2}\right)$

18. $\left(\dfrac{x - 3}{x^2 - 5x + 4}\right)\left(\dfrac{x}{x - 1}\right)$

Find the limits in Exercises 19–28.

19. $\lim\limits_{x \to 0^+} \dfrac{1}{3x}$

20. $\lim\limits_{x \to 0^+} \dfrac{5}{2x}$

21. $\lim\limits_{x \to 2^+} \dfrac{1}{x - 2}$

22. $\lim\limits_{x \to 2^-} \dfrac{1}{x - 2}$

23. $\lim\limits_{x \to 2^+} \dfrac{x}{x - 2}$

24. $\lim\limits_{x \to 2^-} \dfrac{x}{x - 2}$

25. $\lim\limits_{t \to -3^+} \dfrac{1}{t + 3}$

26. $\lim\limits_{t \to -3^-} \dfrac{1}{t + 3}$

27. $\lim\limits_{t \to -3^+} \dfrac{t}{t + 3}$

28. $\lim\limits_{t \to -3^-} \dfrac{t}{t + 3}$

Find the limits in Exercises 29–36.

29. $\lim\limits_{t \to \infty} \dfrac{t^2 - 2t + 3}{2t^2 + 5t - 3}$

30. $\lim\limits_{r \to -\infty} \dfrac{3r}{1 - 2r^2}$

31. $\lim\limits_{s \to \infty} \left(\dfrac{1}{s^4} + \dfrac{1}{s^2} + 1\right)$

32. $\lim\limits_{y \to -\infty} \dfrac{9y^3 - 7y}{y^3 + 1}$

33. $\lim\limits_{z \to \infty} \dfrac{z^3 - 7z + 10}{2z^2 - 7}$

34. $\lim\limits_{x \to \infty} \dfrac{8x^{23} - 7x^2 + 5}{2x^{23} + x^{22}}$

35. $\lim\limits_{t \to \infty} \left(\dfrac{t^2 + 5}{3t^2 - 2} - \dfrac{5t - 7}{6t + 8}\right)$

36. $\lim\limits_{y \to -\infty} \left(\dfrac{y + 2}{y + 3} - \dfrac{y^3 + 2}{y^3 + 3}\right)$

37. Find $\lim \dfrac{1}{x^2 - 4}$ as

   a) $x \to 2^+$ b) $x \to 2^-$

   c) $x \to -2^+$ d) $x \to -2^-$.

38. Find $\lim \dfrac{x}{x^2 - 1}$ as

   a) $x \to 1^+$ b) $x \to 1^-$

   c) $x \to -1^+$ d) $x \to -1^-$.

39. Find $\lim \dfrac{x^2 - 1}{2x + 4}$ as

   a) $x \to -2^+$ b) $x \to -2^-$

   c) $x \to \infty$ d) $x \to -\infty$.

40. Find $\lim \left(x^2 + \dfrac{4}{x}\right)$ as

   a) $x \to 0^+$ b) $x \to 0^-$

   c) $x \to \infty$ d) $x \to -\infty$.

Find the limits in Exercises 41–44.

41. $\lim\limits_{x \to 0^+} \dfrac{\lfloor x \rfloor}{x}$

42. $\lim\limits_{x \to 0^-} \dfrac{\lfloor x \rfloor}{x}$

43. $\lim\limits_{x \to \infty} \dfrac{|x|}{|x| + 1}$

44. $\lim\limits_{x \to -\infty} \dfrac{x}{|x|}$

45. Let $f(x) = \begin{cases} \dfrac{1}{x}, & x < 0 \\ -1, & x \geq 0. \end{cases}$

   Find $\lim f(x)$ as $x \to -\infty, 0^-, 0^+$, and ∞.

46. Let $f(x) = \begin{cases} \dfrac{x - 2}{x - 1}, & x \leq 0 \\ \dfrac{1}{x^2}, & x > 0. \end{cases}$

   Find $\lim f(x)$ as $x \to -\infty, 0^-, 0^+$, and ∞.

47. Let $f(x) = a_n x^n + a_{n-1} x^{n-1} + \cdots + a_1 x + a_0$ be a polynomial of degree n and $g(x) = b_m x^m + b_{m-1} x^{m-1} + \cdots + b_1 x + b_0$ a polynomial of degree m. Show that $\lim_{x \to \infty} f(x)/g(x)$ is a_n/b_m if $m = n$, 0 if $m > n$, and infinite if $m < n$. (*Hint:* Divide the numerator and denominator of the fraction by x^m. What happens to x^n/x^m as $x \to \infty$ if $m = n$? If $m > n$? If $m < n$?)

48. CALCULATOR *Limits of ratios of logarithms*

   a) By taking $x = 10, 100, 1000$, and so on as far as your calculator can go, estimate the value of

$$\lim_{x \to \infty} \frac{\ln(x + 1)}{\ln x}.$$

   b) Does the 1 in $\ln(x + 1)$ really matter? Suppose you have 999 there instead. What do you get for the value of

$$\lim_{x \to \infty} \frac{\ln(x + 999)}{\ln x}?$$

c) Estimate the value of

$$\lim_{x \to \infty} \frac{\ln x^2}{\ln x}.$$

d) Estimate the value of

$$\lim_{x \to \infty} \frac{\ln x}{\log x}.$$

All the behavior you see here will be explained in Chapter 6.

Computer Grapher or Graphing Calculator

Estimate the following limits by graphing each function for large values of x. (When we get information this way we still have to verify that the limit is what we think, but the graph gives us a target value to work with.)

49. $\displaystyle\lim_{x \to \infty} \frac{2x^2 + 5x + 34}{5x^2 - 734}$

50. $\displaystyle\lim_{x \to \infty} \left(1 + \frac{1}{x}\right)^x$

(We shall learn more about this limit in Chapter 6.)

51. $\displaystyle\lim_{x \to \infty} \sqrt{x + 54} - \sqrt{x}$

EXPLORER PROGRAMS

Limit Problems	Provides practice in determining when limits exist and in finding them when they do
PowerGrapher	Graphs all the functions in this section

1.9 The Sandwich Theorem and $(\sin \theta)/\theta$

One of the most useful facts in calculus is that $\lim_{\theta \to 0} (\sin \theta)/\theta = 1$ when θ is measured in radians. As we shall see in Chapter 2, this beautiful and simple result is the key to measuring the rates at which all trigonometric functions of θ change their values as θ changes.

This limit is not the kind we can evaluate by substituting $\theta = 0$, so we have to find it in a more subtle way. What we do is sandwich the fraction $(\sin \theta)/\theta$ between the number 1 and a fraction that is known to approach 1 as θ approaches 0. This tells us that $(\sin \theta)/\theta$ approaches 1 as well.

The theorem we rely on for this argument is the Sandwich Theorem.

THEOREM 6

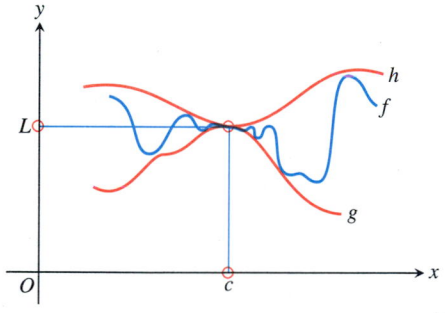

1.106 If the graph of f is sandwiched between the graphs of functions g and h, and if g and h have limit L as x approaches c, then f must have limit L as well.

The Sandwich Theorem

Suppose that

$$g(x) \le f(x) \le h(x)$$

for all $x \ne c$ in some interval about c and that

$$\lim_{x \to c} g(x) = \lim_{x \to c} h(x) = L.$$

Then

$$\lim_{x \to c} f(x) = L$$

(Fig. 1.106).

The theorem holds for one-sided as well as two-sided limits. The idea is that if the values of f are sandwiched between the values of two functions that approach L, then the values of f approach L too. We have included a proof in Appendix 4.

The following example shows how the Sandwich Theorem is typically used to calculate limits.

Example 1   Show that

$$\lim_{\theta \to 0} \sin \theta = 0 \qquad \text{and} \qquad \lim_{\theta \to 0} \cos \theta = 1.$$

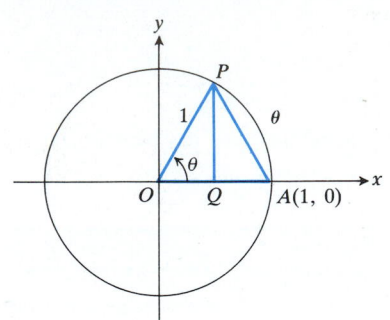

1.107   From the geometry of this figure, drawn for $\theta > 0$, we get the inequality

$$\sin^2\theta + (1 - \cos \theta)^2 < \theta^2.$$

Establishing this inequality is the chief step in showing that $\sin \theta \to 0$ and $\cos \theta \to 1$ as $\theta \to 0$ (Example 1).

Solution   To calculate these limits, we picture θ as the radian measure of an angle in standard position (Fig. 1.107). The circle in the figure is a unit circle, so $|\theta|$ also equals the length of the circular arc AP. The length of the line segment AP is therefore less than $|\theta|$.

Triangle APQ is a right triangle with legs of length

$$QP = |\sin \theta|, \qquad AQ = 1 - \cos \theta.$$

From the Pythagorean theorem and the fact that $AP < |\theta|$, we get

$$\sin^2\theta + (1 - \cos \theta)^2 = (AP)^2 < \theta^2. \tag{1}$$

The terms on the left side of Eq. (1) are both positive, so each is smaller than their sum and hence is less than θ^2:

$$\sin^2\theta < \theta^2 \qquad \text{and} \qquad (1 - \cos \theta)^2 < \theta^2. \tag{2}$$

By taking square roots we can see this is equivalent to saying that

$$|\sin \theta| < |\theta| \qquad \text{and} \qquad |1 - \cos \theta| < |\theta| \tag{3}$$

or

$$-|\theta| < \sin \theta < |\theta| \qquad \text{and} \qquad -|\theta| < 1 - \cos \theta < |\theta|. \tag{4}$$

Now let θ approach 0. Since $-|\theta|$ and $|\theta|$ both approach 0, we may apply the Sandwich Theorem to the inequalities in (4) and conclude that $\sin \theta$ and $1 - \cos \theta$ approach 0 as θ approaches 0. Hence,

$$\lim_{\theta \to 0} \sin \theta = 0 \qquad \text{and} \qquad \lim_{\theta \to 0} \cos \theta = 1.$$

Example 2   Building on the results in Example 1, we have

$$\lim_{x \to 0} \tan x = \lim_{x \to 0} \frac{\sin x}{\cos x} = \frac{\lim\limits_{x \to 0} \sin x}{\lim\limits_{x \to 0} \cos x} = \frac{0}{1} = 0.$$

We now extend the results in Example 1 to show that $\lim_{\theta \to 0} (\sin \theta)/\theta = 1$ when θ is measured in radians. Figure 1.108 shows the graph of $(\sin \theta)/\theta$ and Table 1.1 shows values of the function for θ near 0. During the proof we draw on a formula from geometry that says that the area cut from a unit circle by a central angle of θ radians is $\theta/2$. Figure 1.109 shows where this formula comes from.

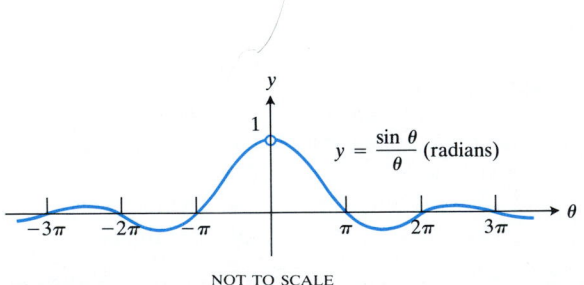

NOT TO SCALE

1.108   The graph of $f(\theta) = (\sin \theta)/\theta$ in the Cartesian θy-plane has a "hole" at $\theta = 0$ where f is not defined. To fill the hole and get a continuous curve, we would extend the definition of f by setting $f(0)$ equal to 1.

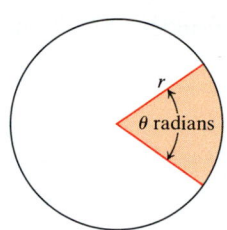

Sector area $= \dfrac{\theta}{2\pi} \cdot$ Circle area

$= \dfrac{\theta}{2\pi} \cdot \pi r^2$

$= \dfrac{1}{2} r^2 \theta$ (Usual form)

$= \dfrac{\theta}{2}$ (If $r = 1$)

1.109   The formula for the area of a sector of a unit circle is $A = \theta/2$.

TABLE 1.1

θ (radians)	sin θ	(sin θ)/θ
0.1	0.0998 33416 65	0.9983 34166 5
0.01	0.0099 99833 334	0.9999 83333 4
0.001	0.0009 99999 8333	0.9999 99833 3
0.0001	0.0000 99999 99983	0.9999 99998 3

Example 3   Show that if θ is measured in radians, then

$$\lim_{\theta \to 0} \frac{\sin \theta}{\theta} = 1. \tag{5}$$

Solution   Our plan is to show that the right-hand and left-hand limits are both 1. We will then know that the two-sided limit is 1 as well.

To show that the right-hand limit is 1, we begin with values of θ that are positive and less than $\pi/2$ (Fig. 1.110). We compare the areas of $\triangle OAP$, sector OAP, and $\triangle OAT$ and note that

$$\text{Area } \triangle OAP < \text{Area sector } OAP < \text{Area } \triangle OAT. \tag{6}$$

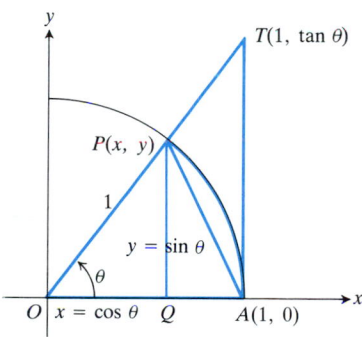

1.110   Area $\triangle OAP <$ area sector $OAP <$ area $\triangle OAT$.

Equation (8) is where the radian measurement comes in: The area of sector OAP is $\theta/2$ only if θ is measured in radians.

We can express these areas in terms of θ as follows:

$$\text{Area } \triangle OAP = \frac{1}{2} \text{ base} \times \text{height} = \frac{1}{2}(1)(\sin \theta) = \frac{1}{2} \sin \theta, \tag{7}$$

$$\text{Area sector } OAP = \frac{1}{2} r^2 \theta = \frac{1}{2}(1)^2 \theta = \frac{\theta}{2}, \tag{8}$$

$$\text{Area } \triangle OAT = \frac{1}{2} \text{ base} \times \text{height} = \frac{1}{2}(1)(\tan \theta) = \frac{1}{2} \tan \theta, \tag{9}$$

so

$$\frac{1}{2} \sin \theta < \frac{1}{2} \theta < \frac{1}{2} \tan \theta. \tag{10}$$

The inequality in (10) will go the same way if we divide all three terms by the positive number (1/2)sin θ:

$$1 < \frac{\theta}{\sin \theta} < \frac{1}{\cos \theta}. \tag{11}$$

We next take reciprocals in (11), which reverses the inequalities:

$$\cos \theta < \frac{\sin \theta}{\theta} < 1. \tag{12}$$

Because $\cos \theta$ approaches 1 as θ approaches 0, the Sandwich Theorem tells us that

$$\lim_{\theta \to 0^+} \frac{\sin \theta}{\theta} = 1. \tag{13}$$

The limit in Eq. (13) is a right-hand limit because we have been dealing with values of θ between 0 and $\pi/2$, but we obtain the same limit for $(\sin \theta)/\theta$ as θ approaches 0 from the left. For if α is positive and $\theta = -\alpha$, then

$$\frac{\sin \theta}{\theta} = \frac{\sin(-\alpha)}{-\alpha} = \frac{-\sin(\alpha)}{-\alpha} = \frac{\sin \alpha}{\alpha}. \tag{14}$$

Therefore,

$$\lim_{\theta \to 0^-} \frac{\sin \theta}{\theta} = \lim_{\alpha \to 0^+} \frac{\sin \alpha}{\alpha} = 1. \tag{15}$$

Together, Eqs. (13) and (15) imply that $\lim_{\theta \to 0} (\sin \theta)/\theta = 1$.

Knowing the limit of $(\sin \theta)/\theta$ helps us to calculate related limits.

Example 4

$$\lim_{x \to 0} \frac{\sin 3x}{x} \qquad \left(\begin{matrix} \text{Eq. (5) does not apply.} \\ \text{Substitute } \theta = 3x. \end{matrix} \right)$$

$$= \lim_{\theta \to 0} \frac{\sin \theta}{\theta/3} \qquad (x = \theta/3 \text{ and } \theta \to 0 \text{ as } x \to 0)$$

$$= \lim_{\theta \to 0} 3 \frac{\sin \theta}{\theta} \qquad (\text{Eq. (5) applies now.})$$

$$= 3 \lim_{\theta \to 0} \frac{\sin \theta}{\theta} = 3 \cdot 1 = 3$$

Example 5

$$\lim_{x \to 0} \frac{\tan x}{x} = \lim_{x \to 0} \frac{\sin x}{x} \frac{1}{\cos x} = \lim_{x \to 0} \frac{\sin x}{x} \cdot \lim_{x \to 0} \frac{1}{\cos x} = 1 \cdot 1 = 1$$

Limits as $x \to \pm\infty$

The Sandwich Theorem also holds for limits as $x \to \pm\infty$.

Example 6   Show that $\lim_{x \to \infty} \frac{\sin x}{x} = 0$.

Solution   Since $-1 \le \sin x \le 1$, the inequality

$$-\frac{1}{x} \le \frac{\sin x}{x} \le \frac{1}{x}$$

holds for all positive values of x. Both $-1/x$ and $1/x$ approach 0 as $x \to \infty$, so $(\sin x)/x$ approaches 0 as well. Figure 1.108 (which graphs the quotient as a function of θ instead of x) shows how rapidly the amplitudes of the oscillations decrease as the curve moves out from the origin.

Applications

The function $(\sin x)/x$ plays a key role in many scientific fields, and its occurrence in calculus is not an isolated event. It arises in such diverse fields as quantum physics (where it appears in solutions of the wave equation) and electrical engineering (in signal analysis and signal filter design) as well as in the mathematical fields of differential equations and probability theory.

EXERCISES 1.9

Find the limits in Exercises 1–10.

1. $\lim\limits_{x \to 0} \dfrac{1 + \sin x}{1 + \cos x}$

2. $\lim\limits_{x \to 0} \dfrac{x^2 + 1}{1 - \sin x}$

3. $\lim\limits_{x \to 0} \dfrac{\sin x}{2x^2 - x}$

4. $\lim\limits_{x \to 0} \dfrac{x + \sin x}{x}$

5. $\lim\limits_{\theta \to 0^+} \dfrac{\theta}{\sin \theta}$

6. $\lim\limits_{h \to 0^-} \dfrac{h}{\tan h}$

7. $\lim\limits_{x \to 0} \dfrac{\sin 2x}{x}$

8. $\lim\limits_{x \to 0} \dfrac{x}{\sin 3x}$

9. $\lim\limits_{x \to 0} \dfrac{\tan 2x}{2x}$

10. $\lim\limits_{x \to 0} \dfrac{\tan 2x}{x}$

Find the limits in Exercises 11–18.

11. $\lim\limits_{x \to \infty} \dfrac{\sin 2x}{x}$

12. $\lim\limits_{x \to -\infty} \dfrac{\cos x}{x}$

13. $\lim\limits_{t \to \infty} \left[2 + \dfrac{\sin t}{t} \right]$

14. $\lim\limits_{x \to \infty} \dfrac{x + \sin x}{x + \cos x}$

15. $\lim\limits_{x \to \infty} x \sin \dfrac{1}{x}$

(*Hint*: Set $\theta = 1/x$ and let $\theta \to 0^+$.)

16. $\lim\limits_{u \to \infty} \left[1 + \cos \dfrac{1}{u} \right]$

17. $\lim\limits_{x \to -\infty} \dfrac{\cos (1/x)}{1 + (1/x)}$

(*Hint*: Set $\theta = 1/x$ and let $\theta \to 0^-$.)

18. $\lim\limits_{x \to -\infty} \left[1 + \dfrac{2}{x} \right] \left[\cos \dfrac{1}{x} \right]$

19. Does $\lim\limits_{x \to 0} (\sin x)/|x|$ exist? If so, what is it? If not, why not?

20. *Sandwich Theorem.* As we saw in Example 3, we can sometimes use the Sandwich Theorem to calculate the limit

of a fraction whose numerator and denominator both approach zero. Another example is the fraction

$$f(x) = \frac{x \sin x}{2 - 2 \cos x},$$

which satisfies the inequality

$$1 - \frac{x^2}{6} < \frac{x \sin x}{2 - 2 \cos x} < 1$$

when x is an angle in radians close to 0 (Fig. 1.111). Use this inequality to find $\lim\limits_{x \to 0} f(x)$. Inequalities like this come from infinite series (Chapter 8).

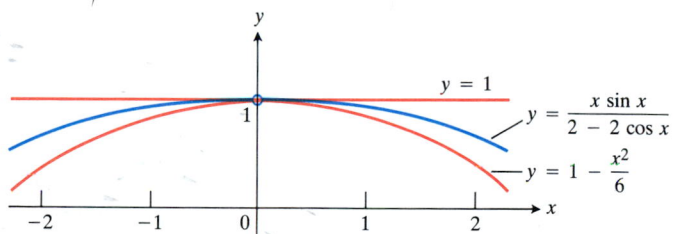

1.111   For all $x \neq 0$, the graph of $f(x) = (x \sin x)/(2 - 2 \cos x)$ lies between the line $y = 1$ and the parabola $y = 1 - x^2/6$ (Exercise 20).

21. a) CALCULATOR   Estimate the value of

$$\lim_{x \to 0} \frac{1 - \cos x}{x^2}$$

by taking $x = 0.1, 0.01, 0.001$, and so on as far as your calculator can go.

b) *Sandwich Theorem.*   Use the inequality

$$\frac{1}{2} - \frac{x^2}{24} < \frac{1 - \cos x}{x^2} < \frac{1}{2}$$

(Fig. 1.112) to find the exact value of the limit in (a). Inequalities like this come from infinite series (Chapter 8).

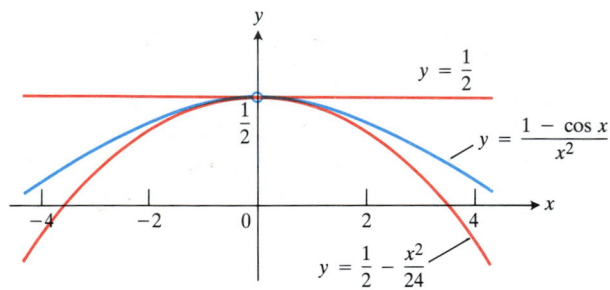

1.112   For all $x \neq 0$, the graph of $f(x) = (1 - \cos x)/x^2$ lies between the line $y = 1/2$ and the parabola $y = (1/2) - x^2/24$ (Exercise 21).

c) Find the limit in (a) by multiplying numerator and denominator by $1 + \cos x$ and using trigonometry to produce an equivalent expression whose limit you now know.

22. The area formula $A = (1/2)r^2\theta$ derived in Fig. 1.109 for radian measure has to be changed if the angle is measured in degrees. What should the new formula be?

As we mentioned, the Sandwich Theorem also holds for limits as $x \to \pm\infty$. Use the theorem to find the limits in Exercises 23 and 24.

23. Find $\lim\limits_{x \to \infty} f(x)$ and $\lim\limits_{x \to -\infty} f(x)$ if

$$\frac{2x^2}{x^2 + 1} < f(x) < \frac{2x^2 + 5}{x^2}.$$

24. The greatest integer function. Find $\lim\limits_{x \to \infty} \lfloor x \rfloor / x$ and $\lim\limits_{x \to -\infty} \lfloor x \rfloor / x$ given that

$$\frac{x - 1}{x} < \frac{\lfloor x \rfloor}{x} \leq 1 \qquad (x \neq 0).$$

1.10 Continuous Functions

When we plot function values generated in the laboratory or collected in the field, we often connect the plotted points with an unbroken curve to show what the function's values are likely to have been at the times we did not measure. In doing so, we are assuming that we are working with a continuous function, a function whose outputs vary continuously with the inputs and do not jump from one value to another without taking on all the values in between.

Continuous functions are the functions we normally use in the equations that describe numerical relations in the world around us. They are the functions we use to find a cannon's maximum range or a planet's closest approach to the sun. They are also the functions we use to describe how a body moves through space or how the speed of a chemical reaction changes with time. In fact, so many observable physical processes proceed continuously that throughout the eighteenth and nineteenth centuries it rarely occurred to anyone to look for any other kind of behavior. It came as quite a surprise when the physicists of the 1920s discovered that the vibrating atoms in a hydrogen molecule can oscillate only at discrete energy levels, that light comes in particles, and that, when heated, atoms emit light in discrete frequencies and not in continuous spectra.

As a result of these and other discoveries, and because of the heavy use of discrete functions in computer science and statistics, the issue of continuity has become one of practical as well as theoretical importance. As scientists, we need to know when continuity is called for, what it is, and how to test for it.

The Definition of Continuity

A function $y = f(x)$ that can be graphed over each interval of its domain with one continuous motion of the pen is an example of a **continuous function.** The height of the graph over the interval varies continuously with x. At each interior point of the function's domain, like the point c in Fig. 1.113, the function value $f(c)$ is the limit of the function values on either side. At the left-hand endpoint a, the function

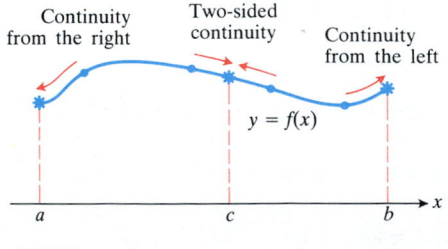

1.113   Continuity at points a, b, and c.

value $f(a)$ is the limit of function values from the right. The function value $f(b)$ is the limit of function values from the left.

DEFINITIONS

A function $y = f(x)$ is

1. **continuous at an interior point** c of its domain if $\lim_{x \to c} f(x) = f(c)$;
2. **continuous at a left endpoint** a of its domain if $\lim_{x \to a^+} f(x) = f(a)$;
3. **continuous at a right endpoint** b of its domain if $\lim_{x \to b^-} f(x) = f(b)$;
4. **continuous** if it is continuous at each point of its domain;
5. **discontinuous at a point** c if it is not continuous at c (which makes c a **point of discontinuity of** f).

How to Test for Continuity at a Point

To test for continuity at a point, we apply the following test.

The Continuity Test

The function $y = f(x)$ is continuous at $x = c$ if and only if *all three* of the following statements are true.

1. $f(c)$ exists (c lies in the domain of f).
2. $\lim_{x \to c} f(x)$ exists (f has a limit as $x \to c$).
3. $\lim_{x \to c} f(x) = f(c)$ (the limit equals the function value).

The limit in the continuity test is to be two-sided if c is an interior point of the domain of f; it is to be the appropriate one-sided limit if c is an endpoint of the domain.

Example 1   When applied to the function $y = f(x)$ in Fig. 1.114 at the points $x = 0, 1, 2, 3$, and 4, the continuity test gives the following results.

a) f is continuous at $x = 0$ because
   i) $f(0)$ exists   (it equals 1),
   ii) $\lim_{x \to 0^+} f(x) = 1$ (f has a limit as $x \to 0^+$),
   iii) $\lim_{x \to 0^+} f(x) = f(0)$ (the limit equals the function value).

b) f is discontinuous at $x = 1$ because $\lim_{x \to 1} f(x)$ does not exist. The function fails part 2 of the test. (The right-hand and left-hand limits exist at $x = 1$, but they are not equal.)

c) f is discontinuous at $x = 2$ because $\lim_{x \to 2} f(x) \neq f(2)$. The function fails part 3 of the test.

d) f is continuous at $x = 3$ because
   i) $f(3)$ exists   (it equals 2),
   ii) $\lim_{x \to 3} f(x) = 2$ (f has a limit as $x \to 3$),
   iii) $\lim_{x \to 3} f(x) = f(3)$ (the limit equals the function value).

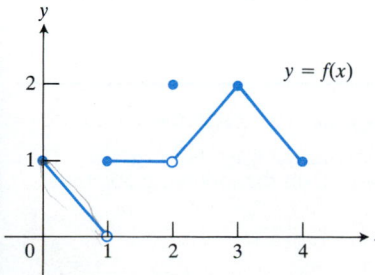

1.114   This function, defined on the closed interval [0, 4], is continuous at $x = 0, 3$, and 4 and discontinuous at $x = 1$ and 2.

e) f is continuous at $x = 4$ because
 i) $f(4)$ exists   (it equals 1),
 ii) $\lim_{x \to 4^-} f(x) = 1$ (f has a limit as $x \to 4^-$),
 iii) $\lim_{x \to 4^-} f(x) = f(4)$ (the limit equals the function value).

Example 2   The function $y = 1/x$ is continuous at every value of x except $x = 0$. The function is not defined at $x = 0$ and therefore fails part 1 of the continuity test at $x = 0$.

Example 3   *Discontinuities of the Greatest Integer Function.*   At every integer, the greatest integer function $y = \lfloor x \rfloor$ fails to have a limit and so fails part 2 of the test (Fig. 1.115).

Example 4   *Continuity of the Sine and Cosine.*   We know that the sine and cosine are continuous at $x = 0$ because

$$\lim_{x \to 0} \sin x = 0 = \sin 0 \qquad \text{and} \qquad \lim_{x \to 0} \cos x = 1 = \cos 0$$

(limit results from Section 1.9). In Chapter 2, we shall see that the sine and cosine are continuous at every other point as well.

Example 5   *Continuity of Polynomial Functions and Rational Functions.*   We saw in Section 1.7 that $\lim_{x \to c} f(x) = f(c)$ for any polynomial function and that if $g(x)$ is also a polynomial function, then $\lim_{x \to c} f(x)/g(x) = f(c)/g(c)$ at every point where the quotient is defined. For instance,

$$f(x) = x^4 + 20 \qquad \text{and} \qquad g(x) = 5x(x - 2)$$

are continuous at every value of x and

$$\frac{f(x)}{g(x)} = \frac{x^4 + 20}{5x(x - 2)}$$

is continuous at every value of x except $x = 0$ and $x = 2$.

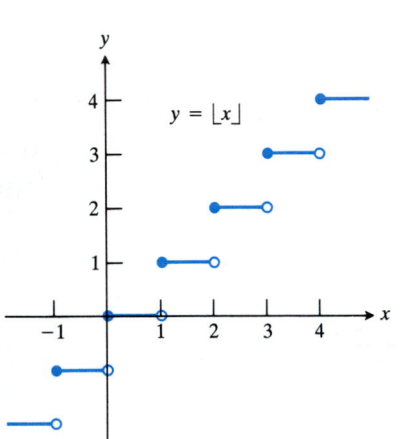

1.115   The greatest integer function $y = \lfloor x \rfloor$ is discontinuous at every integer (but continuous at every other point).

Algebraic Combinations of Continuous Functions Are Continuous

As you may have guessed, algebraic combinations of continuous functions are continuous at every point at which they are defined.

THEOREM 7

Algebraic Properties of Continuous Functions

If the functions f and g are continuous at $x = c$, then the following combinations are continuous at $x = c$:

1. Sums: $f + g$
2. Differences: $f - g$
3. Products: $f \cdot g$
4. Constant multiples: $k \cdot g$ (any number k)
5. Quotients: f/g (provided $g(c) \neq 0$)

Proof   Theorem 7 is a special case of the Limit Rule Theorem in Section 1.7. If the latter were restated for the functions f and g, it would say that if $\lim_{x \to c} f(x) = f(c)$ and $\lim_{x \to c} g(x) = g(c)$, then

1. $\lim_{x \to c} [f(x) + g(x)] = f(c) + g(c)$;
2. $\lim_{x \to c} [f(x) - g(x)] = f(c) - g(c)$;
3. $\lim_{x \to c} f(x)g(x) = f(c)g(c)$;
4. $\lim_{x \to c} kg(x) = kg(c)$ (any number k);
5. $\lim_{x \to c} \dfrac{f(x)}{g(x)} = \dfrac{f(c)}{g(c)}$ (provided $g(c) \neq 0$).

In each case, the limit and the appropriate function value exist and are equal. Similar arguments with right-hand and left-hand limits establish the theorem for continuity at endpoints.

Example 6   *A Continuous Extension.*   Is it possible to define $f(2)$ in a way that extends

$$f(x) = \frac{x^2 + x - 6}{x^2 - 4}$$

to be continuous at $x = 2$? If so, what value should $f(2)$ have?

Solution   For f to be continuous at $x = 2$, $f(2)$ must equal $\lim_{x \to 2} f(x)$. Does f have a limit at $x = 2$ and, if so, what is it? To answer this question, we try to factor the numerator and denominator of the expression for $f(x)$ to see if we can rewrite it to avoid division by zero when $x = 2$. We find

$$f(x) = \frac{x^2 + x - 6}{x^2 - 4}$$

$$= \frac{(x - 2)(x + 3)}{(x - 2)(x + 2)} = \frac{x + 3}{x + 2}.$$

Therefore,

$$\lim_{x \to 2} f(x) = \lim_{x \to 2} \frac{x + 3}{x + 2}$$

$$= \frac{2 + 3}{2 + 2} = \frac{5}{4},$$

and defining $f(2) = 5/4$ will make

$$f(2) = \lim_{x \to 2} f(x).$$

The extended function

$$f(x) = \begin{cases} \dfrac{x^2 + x - 6}{x^2 - 4} & \text{if } x \neq 2, \\[2ex] \dfrac{5}{4} & \text{if } x = 2, \end{cases} \tag{1}$$

is continuous at $x = 2$ because $\lim_{x \to 2} f(x)$ exists and equals $f(2)$ (Fig. 1.116).

The function in Eq. (1) is called the **continuous extension** of the original function to the point $x = 2$.

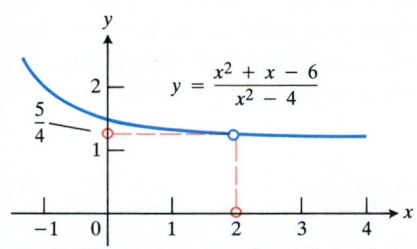

1.116   The graph of $f(x) = \dfrac{x^2 + x - 6}{x^2 - 4}$.

Differentiable Functions Are Continuous

A function is continuous at every point at which it has a derivative.

THEOREM 8	If f has a derivative at $x = c$, then f is continuous at $x = c$.

Proof   Our task is to show that $\lim_{x \to c} f(x) = f(c)$ or, equivalently, that

$$\lim_{x \to c} [f(x) - f(c)] = 0. \tag{2}$$

To this end, we let $P(c, f(c))$ be a point on the graph of f and let $Q(x, f(x))$ be a point nearby (Fig. 1.117). The slope of the secant PQ is

$$\text{Secant slope} = \frac{f(x) - f(c)}{x - c}.$$

By definition, the derivative of f at c is the limiting value of this slope as Q approaches P along the curve, which means in this case the limit as $x \to c$:

$$f'(c) = \lim_{x \to c} \frac{f(x) - f(c)}{x - c}. \tag{3}$$

Why should the mere existence of this limit imply that $[f(x) - f(c)] \to 0$ as $x \to c$? Because, with the denominator $x - c$ going to zero, the quotient can have a finite limit only if the numerator goes to zero at the same time. Indeed, this is exactly what we find if we apply the Limit Product Rule from Section 1.7:

$$\lim_{x \to c} [f(x) - f(c)] = \lim_{x \to c} \left[(x - c) \frac{f(x) - f(c)}{x - c} \right]$$

$$= \lim_{x \to c} (x - c) \cdot \lim_{x \to c} \frac{f(x) - f(c)}{x - c} \quad \begin{pmatrix} \text{Limit Product} \\ \text{Rule} \end{pmatrix}$$

$$= 0 \cdot f'(c) = 0. \qquad \text{(Known values)}$$

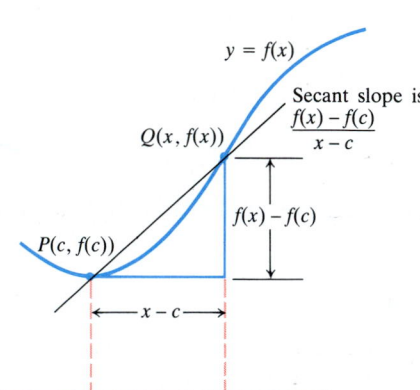

1.117   Figure for the proof that a function is continuous at every point at which it has a derivative.

Example 7   The function $f(x) = \sqrt{x}$ is continuous. It is differentiable if $x > 0$ by Example 8, Section 1.6, and continuous at $x = 0$ because $\lim_{x \to 0^+} \sqrt{x} = 0$ (Exercise 39).

Continuous Functions Need Not Be Differentiable

As you know, the absolute value function is continuous at $x = 0$ but fails to have a derivative there. Using the same idea, we can construct a sawtooth graph to define a continuous function that fails to have a derivative at infinitely many points (Fig. 1.118). But can a continuous function fail to have a derivative at *every* point?

The answer, surprisingly, is yes, as Karl Weierstrass proved in 1872. One of his formulas (there are many like it) was

$$f(x) = \sum_{n=0}^{\infty} \left[\frac{2}{3}\right]^n \cos(9^n \pi x), \tag{4}$$

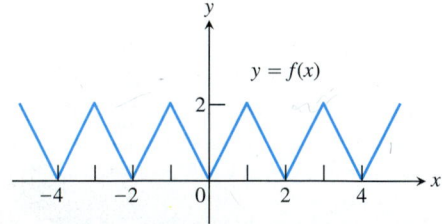

1.118   The continuous sawtooth function graphed here fails to have a derivative at each integer value of x.

a formula that expresses f as an infinite sum of cosines with increasingly short periods. By adding wiggles to wiggles infinitely many times, so to speak, the formula produces a graph that is too rough to have a tangent anywhere.

**Karl Theodor Wilhelm Weierstrass
(1815–1897)**

In his (successful) attempt to put mathematical analysis on a sound logical foundation, German mathematician Karl Weierstrass developed the now accepted δ-ϵ definitions of limit and continuity. In his Berlin lectures in the 1860s, he also proved the Max-Min Theorem for continuous functions, a theorem that earlier mathematicians had assumed without proof.

In the first half of the nineteenth century, most mathematicians believed and many texts "proved" that continuous functions were differentiable. Their idea of function was limited to functions defined by algebraic formulas. They did not have Gustav Dirichlet's modern definition based on sets and probably would not have allowed Weierstrass's formula (Eq. 4) as one that defined a function.

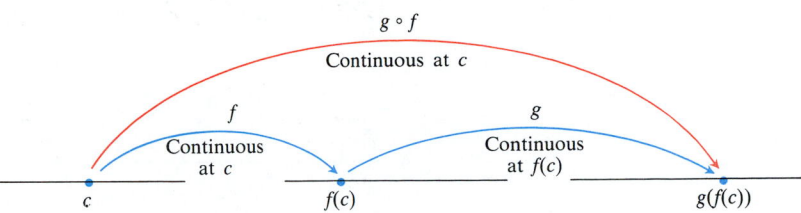

1.119 Composites of continuous functions are continuous.

Continuous curves that fail to have a tangent anywhere play an important role in chaos theory, in part because there is no way to measure their lengths. We shall see what length has to do with derivatives in Section 5.4.

Composites of Continuous Functions Are Continuous

All composites of continuous functions are continuous. This means that composites like

$$y = \sin \sqrt{x} \qquad \text{and} \qquad y = |\cos x|$$

are continuous at every point at which they are defined. The idea is that if $f(x)$ is continuous at $x = c$, and $g(x)$ is continuous at $x = f(c)$, then $g \circ f$ is continuous at $x = c$ (Fig. 1.119).

THEOREM 9

> If f is continuous at c, and g is continuous at $f(c)$, then the composite $g \circ f$ is continuous at c.

You will find an outline of the proof in Appendix 4, Exercise 6.

If a composite function $g \circ f$ is continuous at a point $x = c$, its limit as $x \to c$ is $g(f(c))$.

Example 8

a) $\displaystyle\lim_{x \to 1^+} \sin \sqrt{x - 1} = \sin \sqrt{1 - 1} = \sin 0 = 0$

b) $\displaystyle\lim_{x \to 0} |1 + \cos x| = |1 + \cos 0| = |1 + 1| = 2$

Properties of Continuous Functions

We study continuous functions because they are useful in mathematics and its applications. It turns out that every continuous function is some other function's derivative, as we shall see in Chapter 4. The ability to recover a function from information about its derivative is one of the great powers given to us by calculus. Thus, given a formula $v(t)$ for the velocity of a moving body as a continuous function of time, we shall be able, with the calculus of Chapters 2, 3, and 4, to produce a formula $s(t)$ that tells how far the body has traveled from its starting point at any instant.

In addition, a function that is continuous at every point of a closed interval

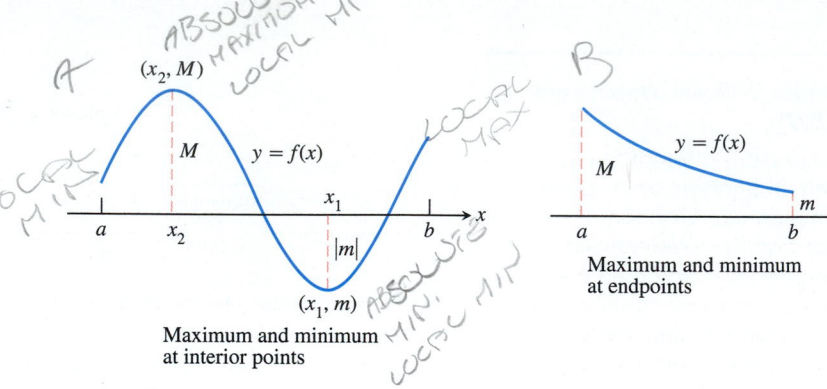

1.120   Typical arrangements of a continuous function's absolute maxima and minima on a closed interval.

Maximum and minimum at interior points

Maximum and minimum at endpoints

Maximum at interior point, minimum at endpoint

Minimum at interior point, maximum at endpoint

$[a, b]$ has an absolute maximum value and an absolute minimum value on this interval. We always look for these values when we graph a function, and we shall see the role they play in problem solving (Chapter 3) and in the development of the integral calculus (Chapters 4 and 5).

Finally, a function f that is continuous at every point of a closed interval $[a, b]$ assumes every value between $f(a)$ and $f(b)$. We shall see some consequences of this in a moment.

The proofs of these properties require a detailed knowledge of the real number system and we shall not give them here.

THEOREM 10

The Max-Min Theorem for Continuous Functions

If f is continuous at every point of a closed interval $[a, b]$, then f takes on both an absolute maximum value M and an absolute minimum value m somewhere in that interval. That is, for some numbers x_1 and x_2 in $[a, b]$ we have $f(x_1) = m, f(x_2) = M$, and $m \leq f(x) \leq M$ at every other point x of the interval (Fig. 1.120).

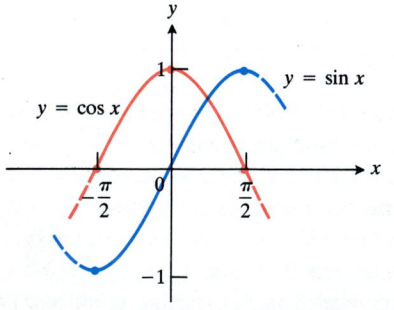

1.121   The sine and cosine on the interval $[-\pi/2, \pi/2]$ (Example 9).

Example 9   On $[-\pi/2, \pi/2]$, the cosine takes on a maximum value of 1 (once) and a minimum value of 0 (twice). The sine takes on a maximum value of 1 and a minimum value of -1 (Fig. 1.121).

As Figs. 1.122 and 1.123 show, the requirements that the interval be closed and that f be continuous are key ingredients of Theorem 10. Without them, the conclusion of the theorem need not hold.

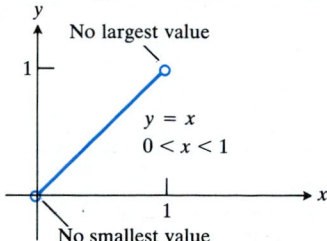

1.122 On an open interval, a continuous function need not have either a maximum or a minimum value. The function $f(x) = x$ has neither a largest nor a smallest value on $(0, 1)$.

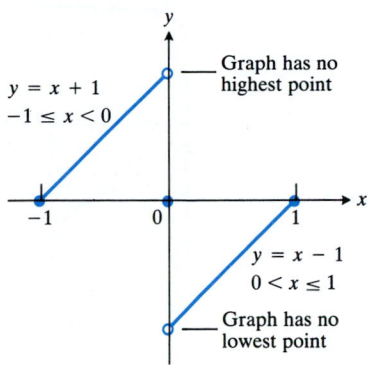

1.123 Even a single point of discontinuity can keep a function from having either a maximum or a minimum value on a closed interval. The function

$$y = \begin{cases} x + 1, & -1 \le x < 0 \\ 0, & x = 0 \\ x - 1, & 0 < x \le 1 \end{cases}$$

is continuous at every point of $[-1, 1]$ except $x = 0$, yet its graph over $[-1, 1]$ has neither a highest nor a lowest point.

THEOREM 11

The Intermediate Value Theorem for Continuous Functions

A function $y = f(x)$ that is continuous on a closed interval $[a, b]$ takes on every value between $f(a)$ and $f(b)$ (Fig. 1.124).

A Consequence for Graphing: Connectivity Suppose we want to graph a function $y = f(x)$ that is continuous throughout some interval I on the x-axis. Theorem 11 tells us that the graph of f over I will never move from one y-value to another without taking on the y-values in between. The graph of f over I will be **connected**: It will consist of a single, unbroken curve, like the graph of $y = \sin x$. The graph of f will not have jumps like the graph of the greatest integer function or separate branches like the graph of $y = \tan x$.

The Consequence for Root Finding We call a solution of the equation $f(x) = 0$ a **root** or **zero** of the function f. The Intermediate Value Theorem tells us that if f is continuous, then any interval on which f changes sign must contain a zero of the function. In other words, there has to be a zero of f between any point where the graph of f lies above the x-axis and any point where the graph lies below.

This observation is the basis of the way we solve equations of the form $f(x) = 0$ with a graphing calculator or computer grapher (when f is continuous). The solutions are the x-intercepts of the graph of f. We graph the function $y = f(x)$ over a large interval to see roughly where its zeros are. Then we zoom in on the intersection points one at a time to estimate their coordinates to as many decimal places as we need (or as the machine will allow). Figure 1.125 shows a typical sequence of steps in the graphical solution of the equation $x^3 - x - 1 = 0$.

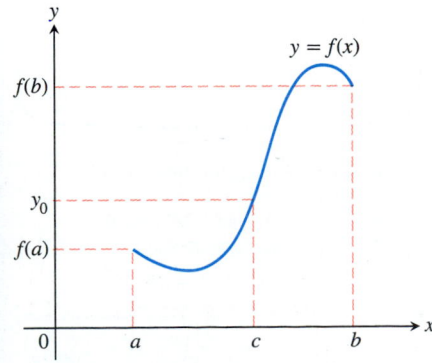

1.124 The Intermediate Value Theorem says that if f is continuous on $[a, b]$ and if y_0 is a value between $f(a)$ and $f(b)$, then $y_0 = f(c)$ for some c in $[a, b]$.

Graphical procedures for solving equations and finding zeros of functions, while instructive, are relatively slow. We usually get faster results from numerical methods, as you will see in Section 2.7.

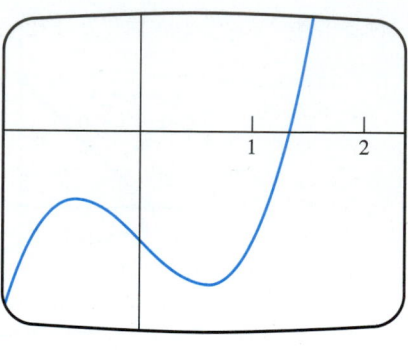

First we make a graph with a relatively large scale. It reveals a root (zero) between $x = 1$ and $x = 2$.

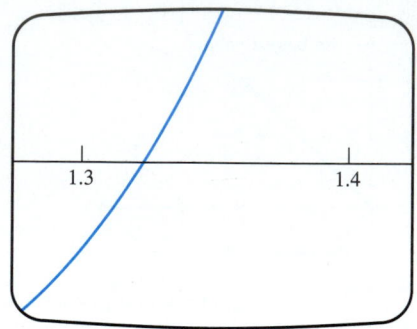

We change the viewing window to $1 \le x \le 2, -1 \le y \le 1$. We now see that the root lies between 1.3 and 1.4.

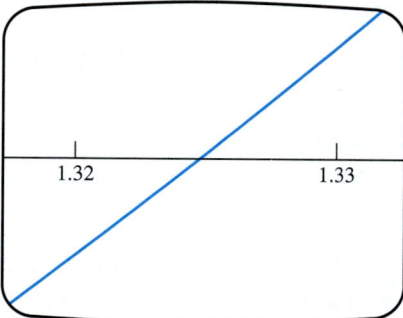

We change the window to $1.3 \le x \le 1.35, -0.1 \le y \le 0.1$. This shows that the root lies in the interval [1.32, 1.33].

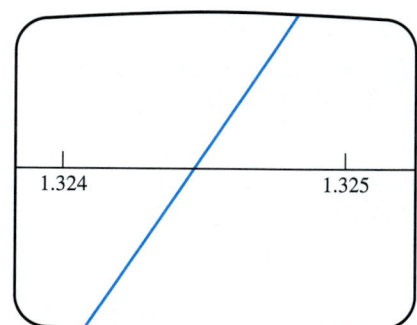

We change the window to $1.32 \le x \le 1.33, -0.01 \le y \le 0.01$. This shows that the root x lies between 1.324 and 1.325.

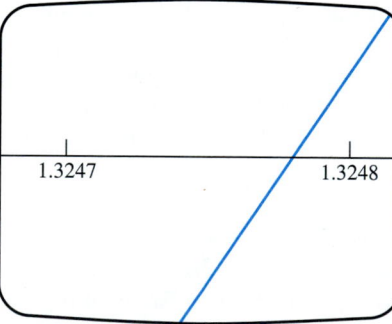

We change the window to $1.324 \le x \le 1.325, -0.001 \le y \le 0.001$. The root is now seen to lie between 1.3247 and 1.3248.

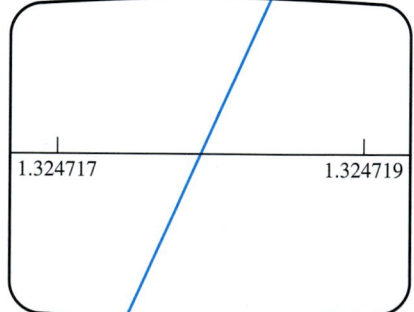

After two more enlargements, we arrive at a screen that shows the root to lie between 1.324717 and 1.324719. To five decimal places, the root is $x = 1.32472$.

1.125   A graphical solution of the equation $x^3 - x - 1 = 0$. We graph the function $f(x) = x^3 - x - 1$ and, with successive screen enlargements, estimate the coordinates of the point where the graph crosses the x-axis.

The Intermediate Value Property of Derivatives

It comes in handy now and then to know that derivatives have the intermediate value property: If f has a derivative at every point of a closed interval $[a, b]$, then f' assumes every value between $f'(a)$ and $f'(b)$. We shall refer to the property briefly in Chapter 3 but shall make no attempt to prove it. There are proofs in more advanced texts.

This property of derivatives allows a partial answer to the question When is a function defined on an interval the derivative of some other function throughout that interval? The (partial) answer is Only when it has the intermediate value property. Step functions are not derivatives of other functions, for example.

The question of when a function is a derivative is one of the central questions in all calculus, and Newton and Leibniz's answer to this question revolutionized the world of mathematics. In Chapter 4 we shall see what that answer was.

EXERCISES 1.10

Exercises 1–6 are about the function graphed in Fig. 1.126, whose formula is

$$f(x) = \begin{cases} x^2 - 1, & -1 \le x < 0 \\ 2x, & 0 \le x < 1 \\ 1, & x = 1 \\ -2x + 4, & 1 < x < 2 \\ 0, & 2 < x \le 3. \end{cases}$$

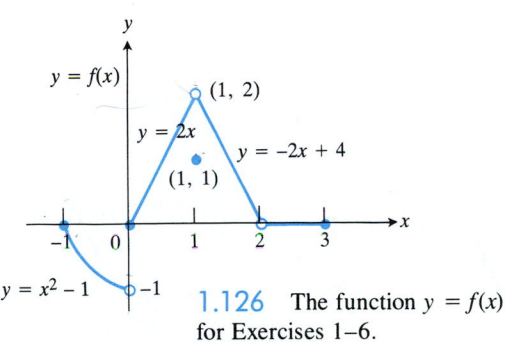

1.126 The function $y = f(x)$ for Exercises 1–6.

1. a) Does $f(-1)$ exist?
  b) Does $\lim_{x \to -1^+} f(x)$ exist?
  c) Does $\lim_{x \to -1^+} f(x) = f(-1)$?
  d) Is f continuous at $x = -1$?

2. a) Does $f(1)$ exist?
  b) Does $\lim_{x \to 1} f(x)$ exist?
  c) Does $\lim_{x \to 1} f(x) = f(1)$?
  d) Is f continuous at $x = 1$?

3. a) Is f defined at $x = 2$? (Look at the definition of f.)
  b) Is f continuous at $x = 2$?

4. At what values of x is f continuous?

5. a) What is the value of $\lim_{x \to 2} f(x)$?
  b) What value should be assigned to $f(2)$ to make f continuous at $x = 2$?

6. To what new value should $f(1)$ be changed to make f continuous at $x = 1$?

At which points are the functions in the following exercises in Section 1.7 continuous?

7. Exercise 31   **8.** Exercise 32

9. Exercise 33   **10.** Exercise 34

Find the points (if any) at which the functions in Exercises 11–20 are *not* continuous.

11. $y = \dfrac{1}{x - 2}$ **12.** $y = \dfrac{1}{(x + 2)^2}$

13. $y = \dfrac{x}{x + 1}$ **14.** $y = \dfrac{x + 1}{x^2 - 4x + 3}$

15. $y = |x - 1|$ **16.** $y = \dfrac{x + 3}{x^2 - 3x - 10}$

17. $y = \dfrac{x^3 - 1}{x^2 - 1}$ **18.** $y = \dfrac{1}{x^2 + 1}$

19. $y = \dfrac{\cos x}{x}$ **20.** $y = \dfrac{|x|}{x}$

21. The function $f(x)$ is defined by $f(x) = (x^2 - 1)/(x - 1)$ when $x \ne 1$ and by $f(1) = 2$. Is f continuous at $x = 1$? Explain.

22. Define $g(3)$ in a way that extends $g(x) = (x^2 - 9)/(x - 3)$ to be continuous at $x = 3$.

23. Define $h(2)$ in a way that extends $h(x) = (x^2 + 3x - 10)/(x - 2)$ to be continuous at $x = 2$.

24. Define $f(1)$ in a way that extends $f(x) = (x^3 - 1)/(x^2 - 1)$ to be continuous at $x = 1$.

25. Define $g(4)$ in a way that extends $g(x) = (x^2 - 16)/(x^2 - 3x - 4)$ to be continuous at $x = 4$.

26. How should $f(2)$ be redefined in Fig. 1.114 to make the function continuous at $x = 2$?

27. What value should be assigned to a to make the function

$$f(x) = \begin{cases} x^2 - 1, & x < 3 \\ 2ax, & x \geq 3 \end{cases}$$

continuous at $x = 3$?

28. What value should be assigned to b to make the function

$$g(x) = \begin{cases} x^3, & x < 1/2 \\ bx^2, & x \geq 1/2 \end{cases}$$

continuous at $x = 1/2$?

29. Find

a) $\lim_{x \to 0} \dfrac{\sqrt{1 + \cos x}}{2}$

b) $\lim_{x \to 0} \cos\left[1 - \dfrac{\sin x}{x}\right]$.

30. Find

a) $\lim_{x \to 0} \sqrt{\tan x + 3 \sec x}$

b) $\lim_{x \to 0} \sin\left[\dfrac{\pi}{2} \cos(\tan x)\right]$.

31. At what values of x does the function in Fig. 1.114 take on its maximum value? Does the function take on a minimum value? Explain.

32. At what values (if any) does the function in Fig. 1.126 take on a maximum value? a minimum value? Explain.

33. Does the function $y = x^2$ have a maximum value on the open interval $-1 < x < 1$? a minimum value? Explain.

34. On the closed interval $0 \leq x \leq 1$ the greatest integer function $y = \lfloor x \rfloor$ takes on a minimum value $m = 0$ and a maximum value $M = 1$. It does so even though it is discontinuous at $x = 1$. Does this violate the Max-Min Theorem? Why?

35. Show that a continuous function that is never zero has a fixed sign.

36. Show that the function

$$f(x) = \begin{cases} 0, & -1 \leq x < 0 \\ 1, & 0 \leq x \leq 1 \end{cases}$$

is not the derivative of any function on the interval $-1 \leq x \leq 1$. (*Hint:* Does f have the intermediate value property? What does its graph look like?)

37. Show that the greatest integer function $y = \lfloor x \rfloor$ is not the derivative of any function on the interval $(-\infty, \infty)$.

38. a) Use the ϵ-δ definition of limit to prove that $\lim_{x \to 0} |x| = 0$.

b) Prove that $y = |x|$ is continuous at every value of x.

39. Use the ϵ-δ definition of right-hand limit (end of Section 1.7) to prove that $\lim_{x \to 0^+} \sqrt{x} = 0$.

Graphing Calculator or Computer Grapher

40. Explain why the following five statements ask for the same information.

a) Find the roots of $f(x) = x^3 - 3x - 1$.

b) Find the x-coordinates of the points where the curve $y = x^3$ crosses the line $y = 3x + 1$.

c) Find all the values of x for which $x^3 - 3x = 1$.

d) Find the x-coordinates of the points where the cubic curve $y = x^3 - 3x$ crosses the line $y = 1$.

e) Solve the equation $x^3 - 3x - 1 = 0$.

Use a graphing calculator or computer grapher to solve the equations in Exercises 41–45.

41. $x^3 - 3x - 1 = 0$

42. $2x^3 - 2x^2 - 2x + 1 = 0$

43. $\cos x = x$

44. $x^x = 2$

45. $\sqrt{x} + \sqrt{1 + x} = 4$

EXPLORER PROGRAMS

Continuity	Provides practice with the three-step definition of continuity at a point
Bisections, Secants, Newton	Graphs any function you key in and approximates its roots by any of four numerical methods you select

---

REVIEW QUESTIONS

1. How can you write the equation for a line if you know the coordinates of two points on the line? The line's slope and the coordinates of one point on the line? The line's slope and y-intercept? Give examples.

2. What are the standard equations for lines perpendicular to the coordinate axes?

3. How are the slopes of mutually perpendicular lines related? Give examples.

4. When a line is not vertical, what is the relation between its slope and its angle of inclination?

5. What is a function? Give examples. How do you graph a real-valued function of a real variable?

6. Name some typical functions and draw their graphs.

7. What is an even function? An odd function? What symmetries do the graphs of such functions have? Give examples. Give an example of a function that is neither odd nor even.

8. When is it possible to compose one function with another? Give examples of composites and their values at various points. Does the order in which functions are composed ever matter?

9. Define the function $y = |x|$. Give examples of numbers and their absolute values. How are $|-a|$, $|ab|$, $|a/b|$, and $|a + b|$ related to $|a|$ and $|b|$?

10. How are absolute values used to describe intervals of real numbers?

11. Show by example how absolute values are used to control function values.

12. How can you write an equation for a circle in the xy-plane if you know its radius a and the coordinates (h, k) of its center? Give examples.

13. What inequality is satisfied by the coordinates of the points that lie inside the circle of radius a centered at (h, k)? What inequality is satisfied by the coordinates of the points that lie outside this circle?

14. The graph of a function $y = f(x)$ in the xy-plane is shifted 5 units to the left and then 3 units up. Write an equation for the new graph.

15. What is a parabola? What are typical equations for parabolas? What reflective property do parabolas have?

16. What is a derivative? A right-hand derivative? A left-hand derivative? How are they related? Give examples.

17. What significance do derivatives have?

18. You have been asked to calculate the limit of a function $f(x)$ as x approaches a finite number c. What theorems are available for calculating the limit? Give examples to show how the theorems are used.

19. What is the relation between one-sided and two-sided limits? How is this relation sometimes used to calculate a limit or to prove that a limit does not exist? Give examples.

20. What are the formal definitions of (two-sided) limit, right-hand limit, and left-hand limit? In particular, what does it mean to say that $\lim_{x \to 2} f(x) = 5$, $\lim_{x \to 0^-} g(x) = 3$, and $\lim_{x \to -1^+} h(x) = 7$?

21. What is the procedure for finding the limit of a rational function as $x \to \pm\infty$? When is the limit 0? Infinite?

22. CALCULATOR You used a calculator to estimate $\lim_{\theta \to 0} (\sin \theta)/\theta$ by evaluating the quotient at $\theta = 0.1$, 0.01, 0.001, and so on. You found, to your surprise, that instead of approaching 1 the quotients approached 0.01745 32925 $\approx \pi/180$. What did you do wrong?

23. How is a function's differentiability at a point related to its continuity there (if at all)?

24. What test can you use to find out whether a function $y = f(x)$ is continuous at a point $x = c$? Give examples of functions that are continuous at $x = 0$ and of functions that are discontinuous at $x = 0$. They do not have to be examples from the book; you may make up your own.

25. What can be said about the continuity of polynomial functions and rational functions? About composites of continuous functions?

26. What are the important theorems about continuous functions? Can functions that are not continuous be expected to have the properties guaranteed by these theorems? Give examples.

27. What does continuity have to do with solving equations?

MISCELLANEOUS EXERCISES

1. A particle in the plane moved from $A(-2, 5)$ to the y-axis in such a way that Δy equaled 3 Δx. What were the particle's new coordinates?

2. *Kepler's second law for masses moving in a straight line.* An object's center of mass moves at a constant velocity v along a straight line past the origin. Figure 1.127 shows the coordinate system and the line of motion. The dots show positions that are 1 second apart. Why are the areas A_1, A_2,

1.127   If an object's center of mass moves in a straight line past the origin at a constant velocity v, the line joining it to the origin sweeps out equal areas $A_1, A_2, \ldots, A_5$ in equal times (Exercise 2). The axes are scaled in kilometers. Time, denoted by t, is measured in seconds.

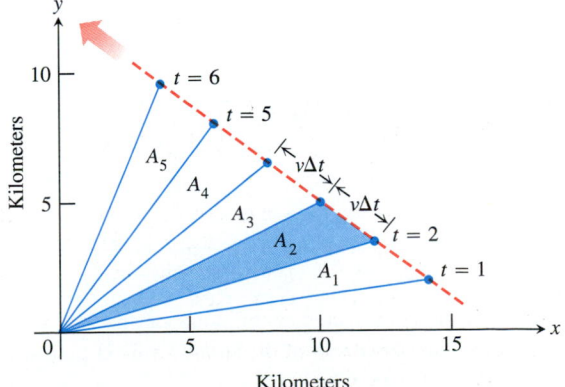

..., A_5 in the figure all equal? (*Hint:* Drop an altitude from the origin to the line of motion.) As in Kepler's second law (see the Prologue), the line that joins the object's center of mass to the origin sweeps out equal areas in equal times.

3. a) Plot the points $A(8, 1)$, $B(2, 10)$, $C(-4, 6)$, $D(2, -3)$, and $E(14/3, 6)$.
    b) Find the slopes of the lines, AB, BC, CD, DA, CE, and BD.
    c) Do any four of the five points A, B, C, D, and E form a parallelogram?
    d) Are any three of the five points collinear? How do you know?
    e) Which of the lines determined by the five points pass through the origin?

4. *Distance from a point to a line*
    a) Find an equation for the line through $P(1, -3)$ perpendicular to the line $L: 2y - 3x = 4$.
    b) Find the coordinates of the point Q in which the two lines meet.
    c) Find the distance from P to L. (It's the distance from P to Q.)

5. *Midpoints.* Find the coordinates of the midpoint of the line segment joining the points $P_1(x_1, y_1)$ to $P_2(x_2, y_2)$.

6. If A, B, C, and C' are constants and not both of A and B are zero, show that
    a) the lines $Ax + By = C$ and $Ax + By = C'$ either coincide or are parallel and
    b) the lines $Ax + By = C$ and $Bx - Ay = C'$ are perpendicular.
    (*Hint:* Divide your argument into two cases: $A \neq 0$ and $B \neq 0$.)

7. Express the area A and the circumference C of a circle as functions of the circle's radius r. Then express A as a function of C.

8. a) Show that the function $f(x) = x - (1/x)$ is odd.
    b) Show that $f(1/x) = -f(x)$.

Find "two-line" formulas for the functions graphed in Exercises 9 and 10.

9.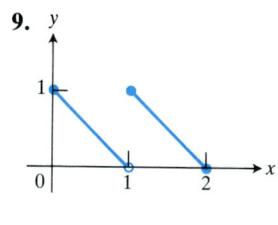

10.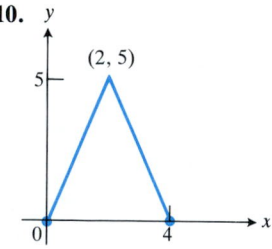

11. Do the points $A(6, 4)$, $B(4, -3)$, and $C(-2, 3)$ form an isosceles triangle? A right triangle? How do you know?

12. Find the coordinates of the point on the line $y = 3x + 1$ that is equidistant from $(0, 0)$ and $(-3, 4)$.

13. Graph the following equations.

    a) $y = \dfrac{|x - 2|}{x - 2}$ b) $y = \dfrac{x - |x|}{2}$ c) $y = \dfrac{x + |x|}{2}$

14. Graph the following equations.

    a) $y = |\cos x|$
    b) $y = \dfrac{\cos x + |\cos x|}{2}$
    c) $y = \dfrac{\cos x - |\cos x|}{2}$

15. a) Graph the equation $|x| + |y| = 1$.
    b) Graph the equation $y + |y| = x + |x|$.

16. *Max(a, b) and min(a, b).* Show that the expression
    $$\max(a, b) = \frac{a + b}{2} + \frac{|a - b|}{2}$$
    equals a if $a \geq b$ and equals b if $b \geq a$. In other words, max(a, b) gives the larger of the two numbers a and b. Find a similar expression for min(a, b), the smaller of a and b.

17. *Controlling the flow from a draining tank.* Torricelli's law says that if you drain a tank like the one in Fig. 1.128, the rate y at which the water runs out is a constant times the square root of the water's depth. As the tank drains, x decreases and so does y, but y decreases less rapidly than x. The value of the constant depends on the size of the exit valve.

    Suppose that for the tank in question, $y = \sqrt{x}/2$. You are trying to maintain a constant exit rate of $y_0 = 1$ ft³/min by refilling the tank with a hose from time to time. How deep must you keep the water to hold the rate to within 0.2 ft³/min of $y_0 = 1$? Within 0.1 ft³/min of $y_0 = 1$?

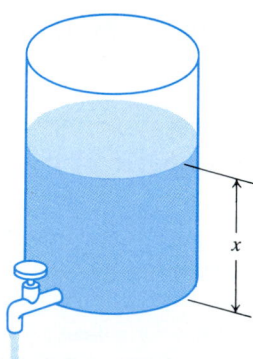

Exit rate y ft³/min

1.128   The tank in Exercise 17.

In other words, in what interval must you keep x to hold y within 0.2 (or 0.1) unit of $y_0 = 1$?

Remark: What if we want to know how long it will take the tank to drain if we do not refill it? We cannot answer such a question with the usual equation time = amount/rate because the rate changes as the tank drains. We could always open the valve, sit down with a watch, and wait; but with a large tank or a reservoir, that might take hours or even days. With calculus, we will be able to find the answer in just a minute or two, as you will see if you do Exercise 27 in Section 4.2.

18. *Dimension changes in equipment.*   As you probably know, most metals expand when heated and contract when cooled, and people sometimes have to take this into account in their work. Boston and Maine Railroad crews try to lay track at temperatures as close to 65°F as they can, so the track won't expand too much in the summer or shrink too much in the winter. Surveyors have to correct their measurements for temperature when they use steel measuring tapes.

The dimensions of a piece of laboratory equipment are often so critical that the machine shop in which it is made has to be held at the same temperature as the laboratory where the part is to be installed. And once the piece is installed, the laboratory must continue to be held at that temperature.

A typical aluminum bar that is 10 cm wide at 70°F will be

$$y = 10 + (t - 70) \times 10^{-4}$$

centimeters wide at a nearby temperature t. As t rises above 70, the bar's width increases; as t falls below 70, the bar's width decreases.

Suppose you have a bar like this made for a gravity wave detector you are building. You need the width of the bar to stay within 0.0005 cm of the ideal 10 cm. How close to 70°F must you maintain the temperature of your laboratory to achieve this? In other words, how close to $t_0 = 70$ must you keep t to be sure that y lies within 0.0005 of $y_0 = 10$?

19. Show that $\lim (|0 + h| - |0|)/h$ is 1 if $h \to 0^+$ and -1 if $h \to 0^-$ and hence that $f(x) = |x|$ has no derivative at $x = 0$.

20. Suppose that the domain of the real-valued function $f(x)$ is $(-\infty, \infty)$, that $f(0) \neq 0$, and that $f(x + h) = f(x) \cdot f(h)$ for all values of x and h.
a) Show that $f(0) = 1$. (*Hint:* Let $h = x = 0$.)
b) Show that if f has a derivative at zero, then f has a derivative at every real number x and that $f'(x) = f(x) \cdot f'(0)$.

We will encounter a function with these properties in Chapter 6.

21. Suppose that $f(x)$ and $g(x)$ are defined for all x and that $\lim_{x \to c} f(x) = -7$ and $\lim_{x \to c} g(x) = 0$. Find the limit as $x \to c$ of the following functions.
a) $3f(x)$
b) $(f(x))^2$
c) $f(x) \cdot g(x)$
d) $\dfrac{f(x)}{g(x) - 7}$
e) $\cos(g(x))$
f) $|f(x)|$

22. Suppose that $f(x)$ and $g(x)$ are defined for all x and that $\lim_{x \to 0} f(x) = 1/2$ and $\lim_{x \to 0} g(x) = \sqrt{2}$. Find the limits as $x \to 0$ of the following functions.
a) $-g(x)$
b) $g(x) \cdot f(x)$
c) $f(x) + g(x)$
d) $1/f(x)$
e) $x + f(x)$
f) $\dfrac{f(x) \cdot \sin x}{x}$

23. Graph the function

$$f(x) = \begin{cases} 1, & x \leq -1 \\ -x, & -1 < x < 0 \\ 1, & x = 0 \\ -x, & 0 < x < 1 \\ 1, & x \geq 1 \end{cases}$$

and answer the following questions.
a) Find the right-hand and left-hand limits of f at -1, 0, and 1.
b) Does f have a limit as x approaches -1? 0? 1? If so, what is it? If not, why not?
c) At which of the points $x = -1$, 0, 1, if any, is f continuous?

24. Repeat the questions in Exercise 23 for the function

$$f(x) = \begin{cases} 0, & x \leq -1 \\ |2x|, & -1 < x < 1 \\ 0, & x = 1 \\ 1, & x > 1. \end{cases}$$

25. Use the inequality $0 \leq |\sqrt{x} \sin(1/x)| \leq \sqrt{x}$ to find $\lim_{x \to 0^+} \sqrt{x} \sin(1/x)$. See Fig. 1.129.

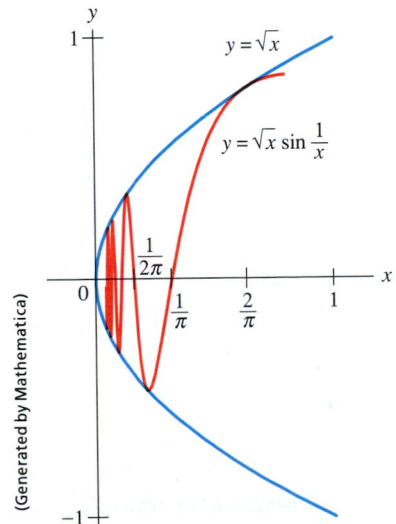

(Generated by Mathematica)

1.129   The graph of $y = \sqrt{x} \sin(1/x)$ (Exercise 25).

26. Use the inequality $0 \le |x^2 \sin (1/x)| \le x^2$ to find $\lim_{x \to 0}$ $x^2 \sin(1/x)$. See Fig. 1.130.

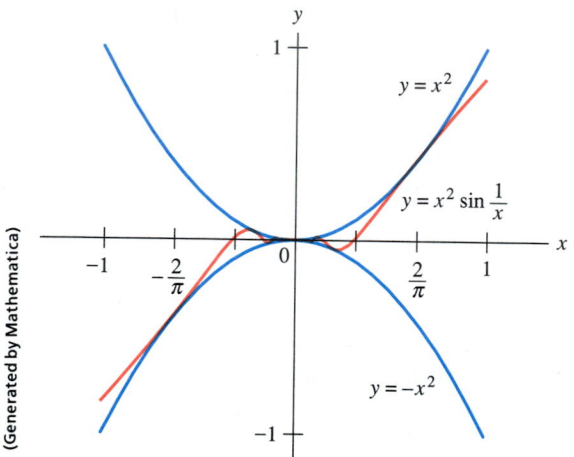

1.130   The graph of $y = x^2 \sin(1/x)$ (Exercise 26).

27. a) Graph the function

$$f(x) = \begin{cases} x \sin(1/x), & x \ne 0 \\ 0, & x = 0. \end{cases}$$

The graph lies in a bow-shaped region of the plane bounded by the lines $y = x$ and $y = -x$.

b) Show that f is continuous at $x = 0$. (*Hint:* First show that $|x \sin(1/x)| \le |x|$ for all x. Then decide how small $|x - 0|$ needs to be to make $|x \sin(1/x) - 0|$ less than ϵ.)

28. Suppose f is a real-valued function whose values are bounded from above by a constant M, i.e., whose values satisfy the inequality $f(x) \le M$. Show that if $\lim_{x \to c} f(x) = L$, then $L \le M$.

(*Hint:* Instead of proving directly that $L \le M$, prove instead that $L > M$ is false. Assume $L > M$, observe that $(L - M)/2$ is positive, take $\epsilon = (L - M)/2$ in the definition of limit, and arrive at a contradiction.)

29. *Wrong descriptions of limit.*   Show by example that the following statements are wrong.

a) The number L is the limit of $f(x)$ as x approaches x_0 if $f(x)$ gets closer to L as x approaches x_0.

b) The number L is the limit of $f(x)$ as x approaches x_0 if, given any $\epsilon > 0$, there is a value of x for which $|f(x) - L| < \epsilon$.

30. Graph the function

$$f(x) = \begin{cases} 4 - 2x, & x < 1 \\ 6x - 4, & x \ge 1. \end{cases}$$

Then, given $\epsilon > 0$, find the largest δ for which $f(x)$ lies between $y = 2 - \epsilon$ and $y = 2 + \epsilon$ for x in the interval $I = (1 - \delta, 1 + \delta)$.

31. Let $f(x) = |x - 5|/(x - 5)$. Find the set of x-values for which $1 - \epsilon < f(x) < 1 + \epsilon$, for $\epsilon = 4, 2, 1$, and $1/2$. (*Hint:* First graph f for $x > 5$ and $x < 5$.)

32. Suppose $0 < \epsilon < 4$. Find the largest $\delta > 0$ with the property that $f(x) = x^2$ lies between $y = 4 - \epsilon$ and $y = 4 + \epsilon$ for all x in the interval $I = (2 - \delta, 2 + \delta)$. What happens to δ as ϵ decreases toward 0? Graph δ as a function of ϵ.

Each of Exercises 33–40 gives a function $f(x)$, a point x_0, and a positive number ϵ. Find $L = \lim_{x \to x_0} f(x)$. Then find a number $\delta > 0$ such that for all x,

$$0 < |x - x_0| < \delta \quad \Rightarrow \quad |f(x) - L| < \epsilon.$$

33. $f(x) = 5x - 10$, $\quad x_0 = 3$, $\quad \epsilon = 0.05$

34. $f(x) = 5x - 10$, $\quad x_0 = 2$, $\quad \epsilon = 0.05$

35. $f(x) = 5x - 10$, $\quad x_0 = 1$, $\quad \epsilon = 0.05$

36. $f(x) = 5x - 10$, $\quad x_0 = 0$, $\quad \epsilon = 0.05$

37. $f(x) = \sqrt{x - 5}$, $\quad x_0 = 9$, $\quad \epsilon = 1$

38. $f(x) = \sqrt{2x - 3}$, $\quad x_0 = 2$, $\quad \epsilon = 1/2$

39. $f(x) = 2/x$, $\quad x_0 = 2$, $\quad \epsilon = 0.1$

40. $f(x) = 1/(4x)$, $\quad x_0 = 1/4$, $\quad \epsilon = 1/20$

Find the limits in Exercises 41–48.

41. $\lim_{x \to \infty} \dfrac{x + \sin x}{2x + 5}$

42. $\lim_{x \to 0} \dfrac{x}{\tan 3x}$

43. $\lim_{x \to \infty} \dfrac{x \sin x}{x + \sin x}$

44. $\lim_{x \to 2} \dfrac{x^2 - 4}{x^3 - 8}$

45. $\lim_{x \to a} \dfrac{x^2 - a^2}{x - a}$

46. $\lim_{x \to 1} \dfrac{1 - \sqrt{x}}{1 - x}$

47. $\lim_{h \to 0} \dfrac{(x + h)^2 - x^2}{h}$

48. $\lim_{h \to 0} \dfrac{\sqrt{x + h} - \sqrt{x}}{h}$

49. *Polynomials of odd degree.*   Show that every polynomial function $f(x) = a_n x^n + a_{n-1} x^{n-1} + \cdots + a_1 x + a_0$ has at least one real zero if n is odd.

50. *The function $f(x) = \sin(1/x)$ has no continuous extension to $x = 0$.*   Show that there is no way to define $f(0)$ to make the extended function continuous. (*Hint:* Which steps in the test for continuity will always fail for the extended function?)

51. The function $y = 1/x$ does not take on either a maximum or a minimum on the interval $0 < x < 1$ even though the function is continuous on this interval. Does this contradict the Max-Min Theorem for continuous functions? Why?

52. What are the maximum and minimum values of the function $y = |x|$ on the interval $-1 \le x < 1$? Notice that the interval is not closed. Is this consistent with the Max-Min Theorem for continuous functions? Why?

53. True, or false? If $y = f(x)$ is continuous, with $f(1) = 0$ and $f(2) = 3$, then f takes on the value 2.5 at some point between $x = 1$ and $x = 2$. Explain.

54. Show that there is at least one value of x for which $x + \cos x = 0$.

55. *A surprising result.* Let f be defined on an open interval (a, b) and suppose that there is a point c in (a, b) at which f is both positive and continuous. Show that f is positive over an entire subinterval $(c - \delta, c + \delta)$ of (a, b). (*Hint*: Take $\epsilon = f(c)/2$ in the definition of continuity at c.)

Notice that even though f is defined throughout (a, b) it is not required to be continuous at any point except c. Continuous and positive at a single point are enough to make f positive throughout an interval.

56. *Continuity is a pointwise phenomenon.* It is possible for a function defined on the entire real line to be continuous at just one point. For example, let

$$f(x) = \begin{cases} x^2 & \text{if } x \text{ is rational} \\ 0 & \text{if } x \text{ is irrational.} \end{cases}$$

a) Show that f is continuous at $x = 0$.

b) Use the fact that every nonempty interval of real numbers contains both rational numbers and irrational numbers to show that f is discontinuous at every nonzero value of x.

57. *Properties of inequalities.* If a and b are any two real numbers, we say a is less than b and write $a < b$ if (and only if) $b - a$ is positive. If $a < b$ we also say that b is greater than a ($b > a$). Prove the following properties of inequalities.

a) If $a < b$, then $a + c < b + c$ and $a - c < b - c$ for any real number c.

b) If $a < b$ and $c < d$, then $a + c < b + d$. Is it also true that $a - c < b - d$? If so, prove it; if not, give a counterexample.

c) If a and b are both positive (or both negative) and $a < b$, then $1/b < 1/a$.

d) If $a < 0 < b$, then $1/a < 0 < 1/b$.

e) If $a < b$ and $c > 0$, then $ac < bc$.

f) If $a < b$ and $c < 0$, then $bc < ac$.

58. *Properties of absolute values.*

a) Prove that $|a| < |b|$ if and only if $a^2 < b^2$.

b) Prove that $|a + b| \leq |a| + |b|$.

c) Prove that $|a - b| \geq ||a| - |b||$.

d) Prove by mathematical induction that

$$|a_1 + a_2 + \cdots + a_n| \leq |a_1| + |a_2| + \cdots + |a_n|.$$

(Mathematical induction is reviewed in Appendix 5.)

e) Using the result from (d), prove that

$$|a_1 + a_2 + \cdots + a_n| \geq |a_1| - |a_2| - \cdots - |a_n|.$$

CHAPTER

2

DERIVATIVES

Overview   In Chapter 1 we saw that the slope of a curve at a point is defined as the limit of secant slopes. This limit, called a derivative, enabled the scientists of the seventeenth century and their successors to formulate precise and workable definitions of tangent and instantaneous rate of change.

So far, however, the definition of derivative has proved to be workable only in the sense that, with enough time, we can work it out. But that was before we studied limits. We now know enough to be able to calculate derivatives rapidly, and that is the main goal of this chapter—to learn how to calculate derivatives rapidly.

The notion of derivative is one of the most important ideas in calculus. Any subject that uses calculus, and most subjects do, has applications of derivatives. We shall discuss applications mainly in the next chapter. However, to give some sense of the importance of derivatives, we have included in this chapter some frequently used applications to estimation, root finding, and the study of motion along a straight line.

2.1 Differentiation Rules

The process of calculating a derivative is called **differentiation.** This section describes how to differentiate functions without having to apply the definition of derivative each time. This will give us the facility we need to calculate the velocities, accelerations, and other important rates of change we shall encounter in the next section.

The Definition of Derivative

We define the derivative of a function as a limit in the following way.

DEFINITIONS

The **derivative** of a function f is the function f' whose value at x is the number

$$f'(x) = \lim_{h \to 0} \frac{f(x+h) - f(x)}{h}. \tag{1}$$

The fraction on the right-hand side of this equation is the **difference quotient** for f at x. It is sometimes called **Fermat's difference quotient.**

If the limit in Eq. (1) exists, we say that f **has a derivative (is differentiable) at x.** If f has a derivative at every point of its domain, we call f **differentiable.** If f is differentiable, we call its graph a **differentiable curve.**

We now apply the formal definition of derivative to produce efficient rules for differentiation and show by example how these rules are used.

Integer Powers, Multiples, Sums, and Differences

The first rule of differentiation is that the derivative of every constant function is zero.

RULE 1

Derivative of a Constant

If c is a constant, then $\quad \dfrac{d}{dx}(c) = 0.$

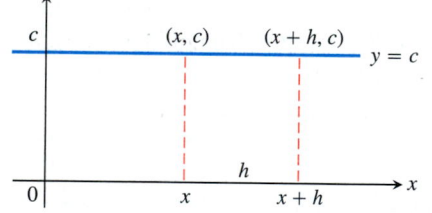

2.1   The rule $(d/dx)(c) = 0$ is another way to say that the values of constant functions never change and that the slope of a horizontal line is zero at every point.

Proof of Rule 1   We apply the definition of derivative to $f(x) = c$, the function whose outputs have the constant value c (Fig. 2.1). At every value of x, we find that

$$\lim_{h \to 0} \frac{f(x+h) - f(x)}{h} = \lim_{h \to 0} \frac{c - c}{h} = \lim_{h \to 0} 0 = 0.$$

The next rule is about derivatives of positive integer powers of x.

RULE 2

Power Rule for Positive Integer Powers of x

If n is a positive integer, then

$$\frac{d}{dx}(x^n) = nx^{n-1}. \tag{2}$$

To apply the power rule, we subtract 1 from the original exponent (n) and multiply the result by n.

Example 1

$$\frac{d}{dx}(x) = \frac{d}{dx}(x^1) = 1 \cdot x^0 = 1, \qquad \frac{d}{dx}(x^2) = 2x^1 = 2x,$$

$$\frac{d}{dx}(x^3) = 3x^2, \qquad \frac{d}{dx}(x^4) = 4x^3$$

Proof of Rule 2   We set $f(x) = x^n$ and find the limit as h approaches zero of

$$\frac{f(x + h) - f(x)}{h} = \frac{(x + h)^n - x^n}{h}. \qquad (3)$$

Since n is a positive integer, we can apply the algebra formula

$$a^n - b^n = (a - b)(a^{n-1} + a^{n-2}b + \cdots + ab^{n-2} + b^{n-1}),$$

with $(a - b) = h$, $a = x + h$, and $b = x$, to replace the expression $(x + h)^n - x^n$ in Eq. (3) by a form that is divisible by h. The resulting division tells us that

$$\frac{f(x + h) - f(x)}{h} = \frac{(x + h)^n - x^n}{h}$$

$$= \frac{(h)[(x + h)^{n-1} + (x + h)^{n-2}x + \cdots + (x + h)x^{n-2} + x^{n-1}]}{h}$$

$$= \underbrace{(x + h)^{n-1} + (x + h)^{n-2}x + \cdots + (x + h)x^{n-2} + x^{n-1}}_{n \text{ terms, each with limit } x^{n-1} \text{ as } h \to 0}. \qquad (4)$$

Hence

$$\frac{d}{dx}(x^n) = \lim_{h \to 0} \frac{f(x + h) - f(x)}{h} = nx^{n-1}.$$

The next rule says that when a differentiable function is multiplied by a constant, its derivative is multiplied by the same constant.

RULE 3

The Constant Multiple Rule

If u is a differentiable function of x and c is a constant, then

$$\frac{d}{dx}(cu) = c\frac{du}{dx}. \qquad (5)$$

In particular, if n is a positive integer, then

$$\frac{d}{dx}(cx^n) = cnx^{n-1}. \qquad (6)$$

Example 2   The derivative formula

$$\frac{d}{dx}(3x^2) = 3 \cdot 2x = 6x$$

says that if we deform the graph of $y = x^2$ by multiplying each y-coordinate by 3, then we multiply the slope at each point by 3 (Fig. 2.2).

Example 3   *A Useful Special Case.*   The derivative of the negative of a differentiable function is the negative of the function's derivative. Rule 3 with $c = -1$ gives

$$\frac{d}{dx}(-u) = \frac{d}{dx}(-1 \cdot u) = -1 \cdot \frac{d}{dx}(u) = -\frac{du}{dx}.$$

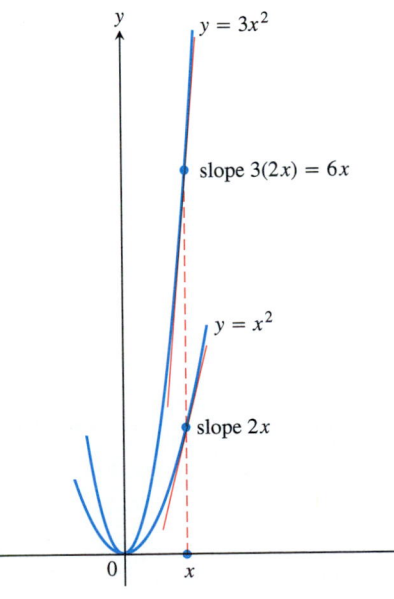

y

$y = 3x^2$

slope $3(2x) = 6x$

$y = x^2$

slope $2x$

0 x x

2.2   The graphs of $y = x^2$ and $y = 3x^2$. Multiplying the y-coordinates by 3 multiplies the slopes by 3.

Proof of Rule 3

$$\frac{d}{dx}\,cu = \lim_{h\to 0}\frac{cu(x+h)-cu(x)}{h} \qquad \left(\begin{array}{l}\text{Derivative definition} \\ \text{with } f(x)=cu(x)\end{array}\right)$$

$$= c\,\lim_{h\to 0}\frac{u(x+h)-u(x)}{h} \qquad \text{(Limit property)}$$

$$= c\,\frac{du}{dx} \qquad\qquad\qquad (u \text{ is differentiable.})$$

The next rule says that the derivative of the sum or difference of two differentiable functions is the sum or difference of their derivatives.

RULE 4

The Sum and Difference Rule

If u and v are differentiable functions of x, then their sum and difference are differentiable at every point where u and v are both differentiable. At such points,

a) $\dfrac{d}{dx}(u+v)=\dfrac{du}{dx}+\dfrac{dv}{dx},$ b) $\dfrac{d}{dx}(u-v)=\dfrac{du}{dx}-\dfrac{dv}{dx}.$

Similar equations hold for more than two differentiable functions, as long as the number of functions involved is finite.

Example 4

a) $y=x^4+12x$

$$\frac{dy}{dx}=\frac{d}{dx}(x^4)+\frac{d}{dx}(12x)$$

$$=4x^3+12$$

b) $y=\dfrac{7x^2}{3}-5$

$$\frac{dy}{dx}=\frac{d}{dx}\left(\frac{7}{3}x^2\right)-\frac{d}{dx}(5)$$

$$=\frac{7}{3}\cdot 2x-0=\frac{14}{3}x$$

c) $y=x^3+3x^2-5x+1$

$$\frac{dy}{dx}=\frac{d}{dx}(x^3)+\frac{d}{dx}(3x^2)-\frac{d}{dx}(5x)+\frac{d}{dx}(1)$$

$$=3x^2+3\cdot 2x-5+0$$

$$=3x^2+6x-5$$

Notice that we can differentiate any polynomial term by term, the way we differentiated the polynomials in Example 4.

Proof of Rule 4 Rule 4(b) is a special case of Rule 4(a); we do not need to prove both.

$$\frac{d}{dx}(u - v) = \frac{d}{dx}[u + (-v)]$$

$$= \frac{du}{dx} + \frac{d}{dx}(-v) \qquad \binom{\text{Rule 4(a) with the function } v \text{ replaced}}{\text{by the function } -v}$$

$$= \frac{du}{dx} - \frac{dv}{dx}$$

We may always think of the difference of two functions as the sum of a function and another's negative. Any general statement we make about a sum of functions applies equally well to differences and combinations of sums and differences.

We now use the two steps of mathematical induction (Appendix 5) to prove that Rule 4 holds for an arbitrary finite sum $u_1 + u_2 + \cdots + u_n$ of n differentiable functions.

Step 1 is to prove the rule for $n = 2$. Calling the functions u and v (to avoid subscripts), we apply the derivative definition with $f(x) = u(x) + v(x)$:

$$\frac{d}{dx}[u(x) + v(x)] = \lim_{h \to 0} \frac{[u(x + h) + v(x + h)] - [u(x) + v(x)]}{h}$$

$$= \lim_{h \to 0} \left[\frac{u(x + h) - u(x)}{h} + \frac{v(x + h) - v(x)}{h} \right] \qquad (7)$$

$$= \lim_{h \to 0} \frac{u(x + h) - u(x)}{h} + \lim_{h \to 0} \frac{v(x + h) - v(x)}{h}$$

$$= \frac{du}{dx} + \frac{dv}{dx}.$$

Step 2 is to show that if the rule holds for any positive integer $n = k \geq 2$, it holds for $n = k + 1$ as well. So suppose that

$$\frac{d}{dx}(u_1 + u_2 + \cdots + u_k) = \frac{du_1}{dx} + \frac{du_2}{dx} + \cdots + \frac{du_k}{dx}. \qquad (8)$$

Then

$$\frac{d}{dx}(\underbrace{u_1 + u_2 + \cdots + u_k}_{\substack{\text{Call the function} \\ \text{defined by this sum } u.}} + \underbrace{u_{k+1}}_{\substack{\text{Call this} \\ \text{function } v.}})$$

$$= \frac{d}{dx}(u_1 + u_2 + \cdots + u_k) + \frac{du_{k+1}}{dx} \qquad \left(\text{Rule 4 for } \frac{d}{dx}(u + v)\right)$$

$$= \frac{du_1}{dx} + \frac{du_2}{dx} + \cdots + \frac{du_k}{dx} + \frac{du_{k+1}}{dx}. \qquad \text{(Eq. (8))}$$

With these steps verified, the mathematical induction principle now guarantees the rule for every integer $n \geq 2$.   ∎

As we saw in Section 1.6, the *slope* of a differentiable curve $y = f(x)$ at a point $P(x, f(x))$ on the curve is $f'(x)$. The line through P with this slope is the *tangent* to the curve at P. Of particular importance, as we shall see in Chapter 3, are any points where the tangent is horizontal. In the next example, we differentiate a fourth-degree polynomial to find these points.

Example 5   Does the curve $y = x^4 - 2x^2 + 2$ have any horizontal tangents? If so, where?

Solution   The horizontal tangents, if any, occur where the slope dy/dx is zero. To find these points, we

1. Calculate dy/dx: $\dfrac{dy}{dx} = \dfrac{d}{dx}(x^4 - 2x^2 + 2) = 4x^3 - 4x$.

2. Solve the equation $\dfrac{dy}{dx} = 0$ for x: $4x^3 - 4x = 0$

$$4x(x^2 - 1) = 0$$

$$x = 0, 1, -1.$$

The curve in question has horizontal tangents at $x = 0, 1$, and -1. The corresponding points on the curve (found from the equation $y = x^4 - 2x^2 + 2$) are $(0, 2)$, $(1, 1)$, and $(-1, 1)$. See Fig. 2.3.

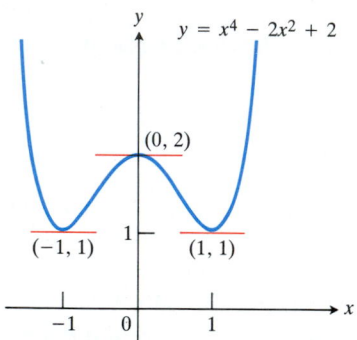

$y = x^4 - 2x^2 + 2$

$(0, 2)$

$(-1, 1)$ $(1, 1)$

2.3   The curve $y = x^4 - 2x^2 + 2$ and its horizontal tangents. The tangents were located by setting dy/dx equal to zero and solving for x, as in Example 5.

Second and Higher Order Derivatives

The derivative $y' = dy/dx$ is the **first derivative** of y with respect to x. The first derivative may also be a differentiable function of x. If so, its derivative

$$y'' = \frac{dy'}{dx} = \frac{d}{dx}\left(\frac{dy}{dx}\right) = \frac{d^2y}{dx^2}$$

is called the **second derivative** of y with respect to x. If y'' is differentiable, its derivative

$$y''' = \frac{dy''}{dx}$$

is the **third derivative** of y with respect to x. The names continue as you imagine they would, with

$$y^{(n)} = \frac{d}{dx}\, y^{(n-1)}$$

denoting the **nth derivative** of y with respect to x.

How to Read the Symbols for Higher Order Derivatives

y'	(y prime)
y''	(y double prime)
$\dfrac{d^2y}{dx^2}$	(d squared y dx squared)
y'''	(y triple prime)
$y^{(n)}$	(y super n)

Example 6   The first four derivatives of $y = x^3 - 3x^2 + 2$ are

First derivative: $y' = 3x^2 - 6x$,

Second derivative: $y'' = 6x - 6$,

Third derivative: $y''' = 6$,

Fourth derivative: $y^{(4)} = 0$.

The function has derivatives of all orders, the fifth and later derivatives all being zero.

Products

While the derivative of the sum of two functions is the sum of their derivatives, and the derivative of the difference of two functions is the difference of their derivatives, the derivative of the product of two functions is not the product of their derivatives. For instance,

$$\frac{d}{dx}(x \cdot x) = \frac{d}{dx}(x^2) = 2x,$$

while

$$\frac{d}{dx}(x) \cdot \frac{d}{dx}(x) = 1 \cdot 1 = 1.$$

The derivative of a product is the sum of *two* products, as we now explain.

RULE 5

The Product Rule

The product of two differentiable functions u and v is differentiable, and

$$\frac{d}{dx}(uv) = u\frac{dv}{dx} + v\frac{du}{dx}. \tag{9}$$

In *prime notation*, Eq. (9) becomes

$$(uv)' = uv' + vu' \tag{10}$$

As with the Sum and Difference Rule, the Product Rule is understood to hold only at values of x where u and v both have derivatives. At such a value of x, the derivative of the product uv is u times the derivative of v plus v times the derivative of u.

Proof of Rule 5

$$\frac{d}{dx}(uv) = \lim_{h \to 0} \frac{u(x+h)v(x+h) - u(x)v(x)}{h}.$$

To change this fraction into an equivalent one that contains difference quotients for the derivatives of u and v, we subtract and add $u(x+h)v(x)$ in the numerator. Then,

$$\frac{d}{dx}(uv) = \lim_{h \to 0} \frac{u(x+h)v(x+h) - u(x+h)v(x) + u(x+h)v(x) - u(x)v(x)}{h}$$

$$= \lim_{h \to 0}\left[u(x+h)\frac{v(x+h) - v(x)}{h} + v(x)\frac{u(x+h) - u(x)}{h}\right] \tag{11}$$

$$= \lim_{h \to 0} u(x+h) \cdot \lim_{h \to 0}\frac{v(x+h) - v(x)}{h} + v(x) \cdot \lim_{h \to 0}\frac{u(x+h) - u(x)}{h}.$$

As h approaches zero, $u(x+h)$ approaches $u(x)$ because u, being differentiable at x, is continuous at x. The two fractions approach the values of du/dx at x and dv/dx at x. In short,

$$\frac{d}{dx}(uv) = u\frac{dv}{dx} + v\frac{du}{dx}.$$

Picturing the Product Rule

In the special case that u and v are positive, increasing functions of x and the derivatives in Eq. (9) can be calculated by letting h approach zero from above, the Product Rule has a nice geometric interpretation. For we can represent the quantities involved in the following way.

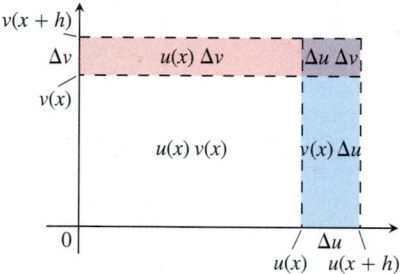

The total shaded area is

$$u(x + h)\, v(x + h) - u(x)\, v(x)$$

$$= u(x + h)\Delta v + v(x + h)\,\Delta u - \Delta u\,\Delta v.$$

Dividing through by h gives

$$\frac{u(x + h)\, v(x + h) - u(x)\, v(x)}{h}$$

$$= u(x + h) \cdot \frac{\Delta v}{h} + v(x + h)\,\frac{\Delta u}{h}$$

$$- \Delta u\,\frac{\Delta v}{h}.$$

As $h \to 0^+$,

$$\Delta u\,\frac{\Delta v}{h} \to (0)\frac{dv}{dx} = 0,$$

leaving

$$\frac{d}{dx}(uv) = u\,\frac{dv}{dx} + v\,\frac{du}{dx}.$$

Example 7 Find the derivative of $y = (x^2 + 1)(x^3 + 3)$.

Solution From the Product Rule with

$$u = x^2 + 1, \qquad v = x^3 + 3,$$

we find

$$\frac{d}{dx}[(x^2 + 1)(x^3 + 3)] = (x^2 + 1)(3x^2) + (x^3 + 3)(2x)$$

$$= 3x^4 + 3x^2 + 2x^4 + 6x$$

$$= 5x^4 + 3x^2 + 6x.$$

Example 7 can be done as well (perhaps better) by multiplying out the original expression for y and differentiating the resulting polynomial. We do that now as a check:

$$y = (x^2 + 1)(x^3 + 3) = x^5 + x^3 + 3x^2 + 3,$$

$$\frac{dy}{dx} = 5x^4 + 3x^2 + 6x,$$

in agreement with our first calculation.

There are times, however, when the Product Rule *must* be used. We might need to differentiate a product like $x \sin x$. Or we might have only numerical values to work with, as in the next example.

Example 8 Let $y = uv$ be the product of the functions u and v. Find $y'(2)$ if

$$u(2) = 3, \qquad u'(2) = -4, \qquad v(2) = 1, \qquad \text{and} \qquad v'(2) = 2.$$

Solution From the Product Rule, in the form

$$y' = (uv)' = uv' + vu',$$

we have

$$y'(2) = u(2)v'(2) + v(2)u'(2) = (3)(2) + (1)(-4) = 6 - 4 = 2.$$

Quotients

Just as the derivative of the product of two differentiable functions is not the product of their derivatives, the derivative of the quotient of two functions is not the quotient of their derivatives. What happens instead is this:

RULE 6

The Quotient Rule

At a point where $v \neq 0$, the quotient $y = u/v$ of two differentiable functions is differentiable and

$$\frac{d}{dx}\left(\frac{u}{v}\right) = \frac{v\dfrac{du}{dx} - u\dfrac{dv}{dx}}{v^2}. \tag{12}$$

As with the earlier combination rules, the Quotient Rule holds only at values of x at which u and v both have derivatives.

Proof of Rule 6

$$\frac{d}{dx}\left(\frac{u}{v}\right) = \lim_{h\to 0} \frac{\dfrac{u(x+h)}{v(x+h)} - \dfrac{u(x)}{v(x)}}{h}$$

$$= \lim_{h\to 0} \frac{v(x)u(x+h) - u(x)v(x+h)}{hv(x+h)v(x)}.$$

To change the last fraction into an equivalent one that contains the difference quotients for the derivatives of u and v, we subtract and add $v(x)u(x)$ in the numerator. This allows us to continue with

$$\frac{d}{dx}\left(\frac{u}{v}\right) = \lim_{h\to 0} \frac{v(x)u(x+h) - v(x)u(x) + v(x)u(x) - u(x)v(x+h)}{hv(x+h)v(x)}$$

$$= \lim_{h\to 0} \frac{v(x)\dfrac{u(x+h)-u(x)}{h} - u(x)\dfrac{v(x+h)-v(x)}{h}}{v(x+h)v(x)}.$$

Taking the limit in the numerator and denominator now gives the Quotient Rule.

Example 9   Find the derivative of $y = \dfrac{x^2 - 1}{x^2 + 1}$.

Solution   We apply the Quotient Rule with $u = x^2 - 1$ and $v = x^2 + 1$:

$$\frac{dy}{dx} = \frac{(x^2 + 1)\cdot 2x - (x^2 - 1)\cdot 2x}{(x^2 + 1)^2}$$

$$= \frac{2x^3 + 2x - 2x^3 + 2x}{(x^2 + 1)^2}$$

$$= \frac{4x}{(x^2 + 1)^2}.$$

Negative Integer Powers of x

The rule for differentiating negative integer powers of x is the same as the rule for differentiating positive integer powers of x.

RULE 7

Power Rule for Negative Integer Powers of x

If n is a negative integer and $x \neq 0$, then

$$\frac{d}{dx}(x^n) = nx^{n-1}. \tag{13}$$

Proof of Rule 7   The proof uses the Quotient Rule in a clever way. If n is a negative integer, then $n = -m$ where m is a positive integer. Hence, $x^n = x^{-m} = 1/x^m$ and

$$\frac{d}{dx}(x^n) = \frac{d}{dx}\left(\frac{1}{x^m}\right)$$

$$= \frac{x^m \cdot \dfrac{d}{dx}(1) - 1 \cdot \dfrac{d}{dx}(x^m)}{(x^m)^2} \qquad \left(\begin{array}{l}\text{Quotient Rule}\\ \text{with } u = 1 \text{ and } v = x^m\end{array}\right)$$

$$= \frac{0 - mx^{m-1}}{x^{2m}} \qquad \left(\begin{array}{l}\text{Since } m > 0,\\ \dfrac{d}{dx}(x^m) = mx^{m-1}\end{array}\right)$$

$$= -mx^{-m-1}$$

$$= nx^{n-1}. \qquad (\text{Since } -m = n)$$

Example 10

$$\frac{d}{dx}\left(\frac{1}{x}\right) = \frac{d}{dx}(x^{-1}) = (-1)x^{-2} = -\frac{1}{x^2}$$

$$\frac{d}{dx}\left(\frac{4}{x^3}\right) = 4\frac{d}{dx}(x^{-3}) = 4(-3)x^{-4} = -\frac{12}{x^4}$$

Example 11   Find an equation for the tangent to the curve

$$y = x + \frac{2}{x}$$

at the point $(1, 3)$ (Fig. 2.4).

Solution   The slope of the curve is

$$\frac{dy}{dx} = \frac{d}{dx}(x) + 2\frac{d}{dx}\left(\frac{1}{x}\right) = 1 + 2\left(-\frac{1}{x^2}\right) = 1 - \frac{2}{x^2}.$$

The slope at $x = 1$ is

$$\left.\frac{dy}{dx}\right|_{x=1} = \left[1 - \frac{2}{x^2}\right]_{x=1} = 1 - 2 = -1.$$

The line through $(1, 3)$ with slope $m = -1$ is

$$y - 3 = (-1)(x - 1) \qquad (\text{Point–slope equation})$$

$$y = -x + 1 + 3$$

$$y = -x + 4.$$

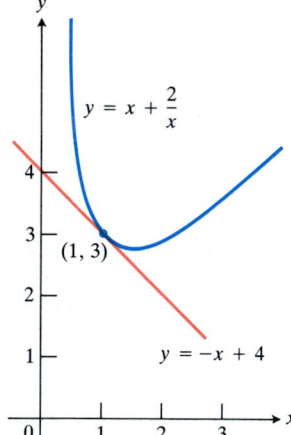

$y = x + \dfrac{2}{x}$

$(1, 3)$

$y = -x + 4$

2.4   We find the slope of the tangent to the curve $y = x + (2/x)$ at the point $(1, 3)$ by evaluating y' at $x = 1$. (The curve also has a third-quadrant portion not shown here. We shall see how to graph functions like this in Chapter 3.)

Choosing Which Rules to Use

The choice of which rules to use in solving a differentiation problem can make a difference in how much work you have to do. Here is an example.

Example 12   Do not use the Quotient Rule to find the derivative of

$$y = \frac{(x-1)(x^2-2x)}{x^4}.$$

Instead, expand the numerator and divide by x^4:

$$y = \frac{(x-1)(x^2-2x)}{x^4} = \frac{x^3 - 3x^2 + 2x}{x^4} = x^{-1} - 3x^{-2} + 2x^{-3}.$$

Then use the Sum and Power Rules:

$$\frac{dy}{dx} = -x^{-2} - 3(-2)x^{-3} + 2(-3)x^{-4} = -\frac{1}{x^2} + \frac{6}{x^3} - \frac{6}{x^4}.$$

EXERCISES 2.1

In Exercises 1–4, find dy/dx and d^2y/dx^2 when y equals the given expression. Try to answer without writing down the original polynomial.

1. $-10x^2$

2. $1 - x + x^2 - x^3$

3. $\dfrac{x^4}{4} - \dfrac{x^3}{3} + \dfrac{x^2}{2} - x + 3$

$x^3 - x^2 + x - 1$

4. $2x^4 - 4x^2 - 8$

Find all nonzero derivatives of the functions in Exercises 5–8.

5. $y = x^2 - 1$

6. $y = \dfrac{x^3}{3} + \dfrac{x^2}{2} - 5$

7. $y = \dfrac{x^4}{2} - \dfrac{3}{2}x^2 - x$

8. $y = \dfrac{x^5}{120}$

In Exercises 9–12, find dy/dx when y equals the given expression. Find each derivative in two ways: (a) by applying the Product Rule and (b) by multiplying the factors to produce a sum of simpler terms to differentiate. In these examples, (b) is faster, but it is not always possible to multiply this way.

9. $(x + 1)(3 - x^2)$

10. $x^2\left(x + 5 + \dfrac{1}{x}\right)$

11. $(x - 1)(x^2 + x + 1)$

12. $\left(x + \dfrac{1}{x}\right)\left(x - \dfrac{1}{x}\right)$

In Exercises 13–20, find dy/dx when y equals the given expression.

13. $\dfrac{2x + 5}{3x - 2}$

14. $\dfrac{2x + 1}{x^2 - 1}$

15. $\dfrac{x^2 - 4}{x + 0.5}$

16. $\dfrac{x^2 - 1}{x^2 + x - 2}$

17. $(1 - x)(1 + x^2)^{-1}$

18. $(2x - 7)^{-1}(x + 5)$

19. $\dfrac{1}{(x^2 - 1)(x^2 + x + 1)}$

20. $\dfrac{(x + 1)(x + 2)}{(x - 1)(x - 2)}$

In Exercises 21–30, find $y' = dy/dx$ and $y'' = d^2y/dx^2$ when y equals the given expression.

21. $\dfrac{3}{x^2}$

22. $-\dfrac{1}{x}$

23. $\dfrac{5}{3x}$

24. $-\dfrac{3}{x^7}$

25. $x + 1 + \dfrac{1}{x}$

26. $\dfrac{12}{x} - \dfrac{4}{x^3} + \dfrac{1}{x^4}$

27. $\dfrac{x^3 + 7}{x}$

28. $\dfrac{x^2 + 5x - 1}{x^2}$

29. $\dfrac{(x - 1)(x^2 + x + 1)}{x^3}$

30. $\dfrac{(x^2 + x)(x^2 - x + 1)}{x^4}$

The Product Rule can be extended by mathematical induction to show that any finite product $y = u_1 u_2 \cdots u_n$ of differentiable functions is differentiable on their common domain and that

$$\frac{dy}{dx} = \frac{du_1}{dx} u_2 \cdots u_n + u_1 \frac{du_2}{dx} \cdots u_n + \cdots \qquad (14)$$

$$+ u_1 u_2 \cdots u_{n-1} \frac{du_n}{dx}.$$

Use Eq. (14) to find the derivatives of the products in Exercises 31–36. Then check your answer by multiplying the factors to produce a polynomial to differentiate.

31. $y = x(x - 1)(x + 1)$

32. $y = (x - 1)(x + 1)(x^2 + 1)$

33. $y = (1 - x)(x + 1)(3 - x^2)$

34. $y = x^2(x - 1)(x^2 + x + 1)$

35. $y = (x - 1)(x + 1)(x - 2)(x + 2)$

36. $y = (x - 1)(x + 2)(x - 3)(x + 4)$

37. Find ds/dt if

a) $s = \dfrac{t}{1 + t^2}$

b) $s = t^2(t + 1)(1 - t^2)$.

38. Find $dr/d\theta$ if

a) $r = \dfrac{2 + \theta^3}{2 - \theta^3}$

b) $r = \left(\dfrac{1}{3} - \dfrac{\theta^2}{2}\right)\left(3 + 2\theta^{-2}\right)$.

39. Find dw/dz if

a) $w = \dfrac{z^2 + 1}{z^2 + 2z + 1}$

b) $w = (z + z^2)(z^{-1} + z^{-2})$.

40. Find dp/dq if

a) $p = \dfrac{1}{(q + 1)^2 + (q - 1)^2}$

b) $p = \left(\dfrac{q^2 + q}{q^3}\right)\left(\dfrac{q^4 - q^3}{q^7}\right)$.

41. Suppose u and v are functions of x that are differentiable at $x = 0$ and that

$$u(0) = 5, \quad u'(0) = -3, \quad v(0) = -1, \quad v'(0) = 2.$$

Find the values of the following derivatives at $x = 0$.

a) $\dfrac{d}{dx}(uv)$ b) $\dfrac{d}{dx}\left(\dfrac{u}{v}\right)$ c) $\dfrac{d}{dx}\left(\dfrac{v}{u}\right)$ d) $\dfrac{d}{dx}(7v - 2u)$

42. Suppose u and v are differentiable functions of x and that

$$u(1) = 2, \quad u'(1) = 0, \quad v(1) = 5, \quad v'(1) = 0.$$

Find the values of the following derivatives at $x = 1$.

a) $\dfrac{d}{dx}(uv)$ b) $\dfrac{d}{dx}\left(\dfrac{u}{v}\right)$ c) $\dfrac{d}{dx}\left(\dfrac{v}{u}\right)$ d) $\dfrac{d}{dx}(7v - 2u)$

43. Find an equation for the line that passes through the point $(2, 3)$ on the curve $y = x^3 - 3x + 1$ perpendicular to the curve's tangent there.

44. Find the tangents to the curve $y = x^3 + x$ at the points where the slope is 4. What is the smallest slope on the curve? At what value of x does the curve have this slope?

45. Find the points on the curve $y = 2x^3 - 3x^2 - 12x + 20$ where the tangent is parallel to the x-axis.

46. Find the x- and y-intercepts of the line that is tangent to the curve $y = x^3$ at the point $(-2, -8)$.

47. Find the tangents to
Newton's Serpentine

$$y = \dfrac{4x}{x^2 + 1}$$

at the origin and the point $(1, 2)$.

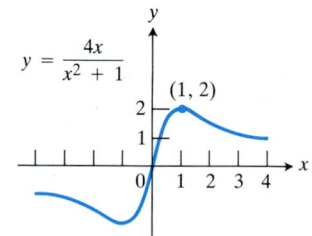

$$y = \dfrac{4x}{x^2 + 1}$$

48. Find the tangent to the
Witch of Agnesi

$$y = \dfrac{8}{x^2 + 4}$$

at the point $(2, 1)$.

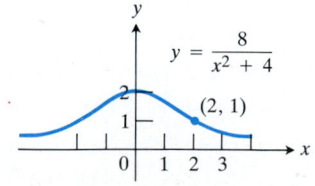

$$y = \dfrac{8}{x^2 + 4}$$

There is a nice story about this curve in the historical note on Agnesi in Chapter 9.

49. The curve $y = ax^2 + bx + c$ passes through the point $(1, 2)$ and is tangent to the line $y = x$ at the origin. Find a, b, and c.

50. The curves $y = x^2 + ax + b$ and $y = cx - x^2$ have a common tangent line at the point $(1, 0)$. Find a, b, and c.

51. *Pressure and volume.*   If the gas in a closed container is maintained at a constant temperature T, the pressure P is related to the volume V by a formula of the form

$$P = \dfrac{nRT}{V - nb} - \dfrac{an^2}{V^2},$$

in which a, b, n, and R are constants. Find dP/dV and d^2P/dV^2.

52. *The body's reaction to medicine.*   The reaction of the body to a dose of medicine can sometimes be represented by an equation of the form

$$R = M^2\left(\dfrac{C}{2} - \dfrac{M}{3}\right),$$

where C is a positive constant and M is the amount of medicine absorbed in the blood. If the reaction is a change in blood pressure, R is measured in millimeters of mercury. If the reaction is a change in temperature, R is measured in degrees, and so on.

Find dR/dM. This derivative, as a function of M, is called the sensitivity of the body to the medicine. In Chapter 3 we shall see how to find the amount of medicine to which the body is most sensitive. (*Source: Some Mathematical Models in Biology,* Revised Edition, R. M. Thrall, J. A. Mortimer, K. R. Rebman, R. F. Baum, eds., December 1967, PB-202 364, p. 221; distributed by N.T.I.S., U.S. Department of Commerce.)

53. Show that the Constant Multiple Rule,

$$\dfrac{d}{dx}(cu) = c\,\dfrac{du}{dx},$$

is a special case of the Product Rule.

54. Show that the Power Rule

$$\dfrac{d}{dx}(x^n) = nx^{n-1}$$

for positive integer powers of x is a special case of the generalized product rule (Eq. 14) preceding Exercises 31–36.

2.2 Velocity and Other Rates of Change

In this section, we see how derivatives provide the mathematics we need to understand the way things change in the world around us. With derivatives, we can describe the rates at which water reservoirs empty, populations change, rocks fall, the economy changes, and an athlete's blood sugar varies with exercise. We begin with free fall, the kind of fall that takes place in a vacuum near the surface of the earth.

Free Fall

Near the surface of the earth, all bodies fall with the same constant acceleration. The distance a body falls after it is released from rest is a constant multiple of the square of the time elapsed. At least, that is what happens when the body falls in a vacuum, where there is no air to slow it down. The square-of-time rule also holds for dense, heavy objects like rocks, ball bearings, and steel tools during the first few seconds of their fall through air, before their velocities build up to where air resistance begins to matter. When air resistance is absent or insignificant and the only force acting on a falling body is the force of gravity, we call the way the body falls *free fall*.

The equation we write to say that the distance an object falls from rest is proportional to the square of the time elapsed is

$$s = \frac{1}{2} g t^2. \tag{1}$$

In this equation, s is distance, t is time, and g, as we shall see in a moment, is the constant acceleration given to an object by the force of gravity.

Example 1   The value of g in the equation $s = (1/2)gt^2$ depends on the units used to measure t and s. With t in seconds (the usual unit),

$$g = 32 \text{ ft/sec}^2 \qquad s = \frac{1}{2}(32)t^2 = 16t^2 \qquad (s \text{ in feet}),$$

$$g = 9.80 \text{ m/sec}^2 \qquad s = \frac{1}{2}(9.80)t^2 = 4.9t^2 \qquad (s \text{ in meters}).$$

The abbreviation ft/sec^2 is read "feet per second squared" or "feet per second per second," and m/sec^2 is read "meters per second squared."

Figure 2.5 shows the free fall of a heavy ball bearing released from rest at time $t = 0$. During the first 2 sec, the ball falls

$$s(2) = 16(2)^2 = 16 \cdot 4 = 64 \text{ ft.}$$

Example 2   How long did it take the ball bearing in Fig. 2.5 to fall the first 14.7 m?

Solution   The free-fall equation for s in meters and t in seconds is

$$s = 4.9t^2$$

(from Example 1). To find the time it took the ball bearing to cover the first 14.7 m,

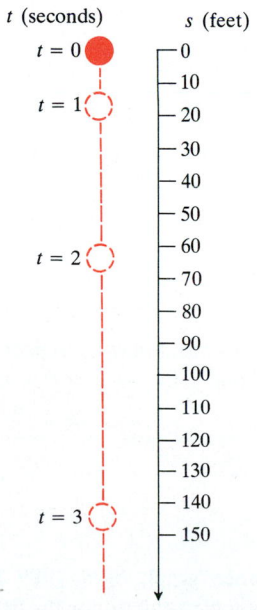

2.5   Distance fallen by a ball bearing released from rest at $t = 0$ sec.

we substitute $s = 14.7$ and solve for t:

$$14.7 = 4.9t^2$$

$$t^2 = \frac{14.7}{4.9} = 3$$

$$t = \sqrt{3}. \qquad \left(\begin{array}{l}\text{Elapsed time increases from } t = 0 \text{ so}\\ \text{we ignore the negative root.}\end{array}\right)$$

It took the ball $t = \sqrt{3}$ or about 1.732 sec to fall the first 14.7 m.

Velocity

Suppose we have a body moving along a coordinate line and we know that its position at time t is $s = f(t)$. As the body moves along, it has a velocity at each particular instant and we want to find out what that velocity is. The information we seek must somehow be contained in the formula $s = f(t)$, but how do we find it?

We reason like this: In the interval from any time t to the slightly later time $t + \Delta t$, the body moves from position $s = f(t)$ to position

$$s + \Delta s = f(t + \Delta t) \tag{2}$$

Position at time t . . .   and at time $t + \Delta t$

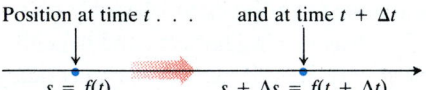

2.6 A model of the positions of a body moving along a coordinate line at time t and shortly later at time $t + \Delta t$.

(Fig. 2.6). The body's net change in position, or displacement, for this short time interval is

$$\Delta s = f(t + \Delta t) - f(t). \tag{3}$$

The body's average velocity for the time interval is Δs divided by Δt.

DEFINITIONS

The **displacement** of a body moving along a line from position $s = f(t)$ to position $s = f(t + \Delta t)$ is $\Delta s = f(t + \Delta t) - f(t)$. The body's **average velocity** for the time interval from t to $t + \Delta t$ is

$$v_{av} = \frac{\text{displacement}}{\text{travel time}} = \frac{\Delta s}{\Delta t} = \frac{f(t + \Delta t) - f(t)}{\Delta t}. \tag{4}$$

To find the body's velocity at the exact instant t, we take the limit of the average velocity over the interval from t to $t + \Delta t$ as the interval gets shorter and shorter and Δt shrinks to zero. Here is where the derivative comes in. For, as we now know, this limit is the derivative of f with respect to t.

DEFINITION

Instantaneous velocity (velocity) is the derivative of position with respect to time. If the position function of a body moving along a line is $s = f(t)$, the body's velocity at time t is

$$v(t) = \frac{ds}{dt} = \lim_{\Delta t \to 0} \frac{f(t + \Delta t) - f(t)}{\Delta t}. \tag{5}$$

Example 3   Figure 2.7 shows a time-to-distance graph of a 1989 Ford Thunderbird SC. The slope of the secant PQ is the average velocity for the 10-sec interval from $t = 5$ to $t = 15$ sec, in this case 40 m/sec or 144 km/h. The slope of the tangent at P is the speedometer reading at $t = 5$ sec, about 20 m/sec or 72

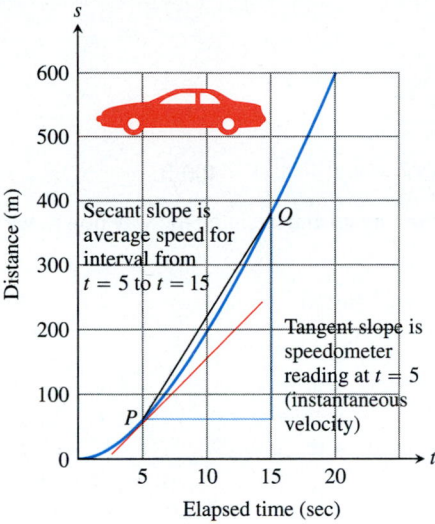

2.7 The time-to-distance data for Example 3.

km/h. The car's top velocity is 235 km/h (146 mph). (*Source: Car and Driver, March 1989.*)

Example 4   *Velocity During Free Fall*

Free-fall equation from Example 1	Corresponding velocity equation	Velocity units
$s = \frac{1}{2} g t^2$	$v = \frac{ds}{dt} = gt$	
$s = 16t^2$	$v = 32t$	ft/sec
$s = 4.9t^2$	$v = 9.8t$	m/sec

The velocity of the falling ball bearing t sec after release is $v = 32t$ ft/sec (Fig. 2.5).

At $t = 2$: $v = 32(2) = 64$ ft/sec,

At $t = 3$: $v = 32(3) = 96$ ft/sec.

Example 5   A dynamite blast blows a heavy rock straight up with a launch velocity of 160 ft/sec (about 109 mph) (Fig. 2.8a). It reaches a height of $s = 160t - 16t^2$ ft after t sec.

a) How high does the rock go?

b) How fast is the rock traveling when it is 256 ft above the ground on the way up? On the way down?

Solution   (a) To find how high the rock goes, we find the value of s when the rock's velocity is zero. The velocity is

$$v = \frac{ds}{dt} = \frac{d}{dt}(160t - 16t^2)$$

$$= 160 - 32t \text{ ft/sec.}$$

2.8 The graph of the rock's height s as a function of time (Example 5). The curve in (b) is not the path of the rock itself.

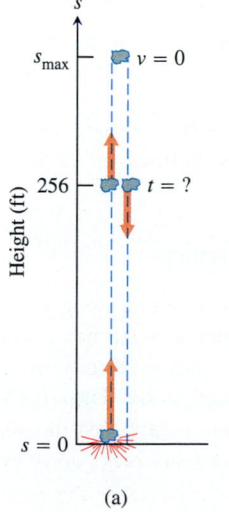

(a)

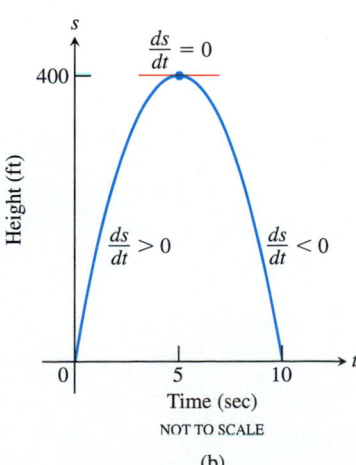

(b)

The velocity is zero when

$$160 - 32t = 0, \qquad \text{or} \qquad t = 5 \text{ sec.}$$

The rock's height at $t = 5$ sec is

$$s_{max} = s(5) = 160(5) - 16(5)^2 = 800 - 400 = 400 \text{ ft.}$$

b) To find the rock's velocity at 256 ft on the way up and again on the way down, we find the two values of t for which

$$s(t) = 160t - 16t^2 = 256. \tag{6}$$

To solve Eq. (6) we write

$$16t^2 - 160t + 256 = 0$$

$$16(t^2 - 10t + 16) = 0$$

$$(t - 2)(t - 8) = 0$$

$$t = 2 \text{ sec}, \quad t = 8 \text{ sec.}$$

The rock is 256 ft above the ground 2 sec after the explosion and again 8 sec after the explosion. The rock's velocities at these times are

$$v(2) = 160 - 32(2) = 160 - 64 = 96 \text{ ft/sec,}$$

$$v(8) = 160 - 32(8) = 160 - 256 = -96 \text{ ft/sec.}$$

Why is the downward velocity negative? It has to do with how we set up the coordinate system. Since s measures height from the ground up, changes in s are positive as the rock rises and negative as the rock falls (Fig. 2.8b).

Speed

If we drive to a friend's house and back at 30 mph, say, the speedometer will show 30 on the way over but it will not show -30 on the way back. The speedometer always shows speed, and speed is the absolute value of velocity. Speed measures the rate of forward progress regardless of direction.

DEFINITION

Speed is the absolute value of velocity.

Example 6   When the rock in Example 5 passed the 256-ft mark, its forward speed was 96 ft/sec on the way up and $|-96| = 96$ ft/sec again on the way down.

Acceleration

In studies of motion along a coordinate line, we usually assume that the body's position function $s = f(t)$ has a second derivative as well as a first. The first derivative gives the body's velocity as a function of time; the second derivative gives the body's acceleration. Thus, the velocity is how fast the position is changing, and the acceleration is how fast the velocity is changing. The acceleration measures how quickly the body picks up or loses speed.

DEFINITION

Acceleration is the derivative of velocity with respect to time. If a body's position at time t is $s = f(t)$, then the body's acceleration at time t is

$$a = \frac{dv}{dt} = \frac{d^2s}{dt^2}. \tag{7}$$

Example 7   The acceleration of the rock in Example 5 is

$$a = \frac{dv}{dt} = \frac{d}{dt}(160 - 32t) = 0 - 32 = -32 \text{ ft/sec}^2.$$

The minus sign confirms that the acceleration is downward, in the negative s direction. Whether the rock is going up or down, it is subject to the same constant downward pull of gravity.

Other Rates of Change

Although it is natural to think of rates of change in terms of motion and time, there is no need to be so restrictive. We can define the average rate of change of any function over any interval of its domain as the change in the function divided by the length of the interval. We can then go on to define the instantaneous rate of change as the limit of average change as the length of the interval goes to zero.

DEFINITIONS

The **average rate of change** of a function $f(x)$ over the interval from x to $x + h$ is

$$\text{average rate of change} = \frac{f(x + h) - f(x)}{h}.$$

The (**instantaneous**) **rate of change** of f at x is the derivative

$$f'(x) = \lim_{h \to 0} \frac{f(x + h) - f(x)}{h},$$

provided the limit exists.

It is conventional to use the word *instantaneous* even when x does not represent time.

Why Peas Wrinkle

British geneticists have recently discovered that the wrinkling trait comes from an extra piece of DNA that prevents the gene that directs starch synthesis from functioning properly. With the plant's starch conversion impaired, sucrose and water build up in the young seeds. As the seeds mature, they lose much of this water, and the shrinkage leaves them wrinkled.

Example 8   *Sensitivity to Change.*   The Austrian monk Gregor Johan Mendel (1822–1888), working with garden peas and other plants, provided the first scientific explanation of hybridization. His careful records showed that if p (a number between 0 and 1) is the frequency of the gene for smooth skin in peas (dominant) and $(1 - p)$ is the frequency of the gene for wrinkled skin in peas, then the proportion of smooth-skinned peas in the population at large is

$$y = 2p(1 - p) + p^2 = 2p - p^2.$$

The graph of y versus p in Fig. 2.9(a) suggests that the value of y is more sensitive to a change in p when p is small than when p is large. Indeed, this is borne out by the derivative graph in Fig. 2.9(b), which shows that dy/dp is close to 2 when p is near 0 and close to 0 when p is near 1.

2.9   (a) The frequency $y = 2p - p^2$ of smooth skin in garden peas as a function of the prevalence p of the dominant smooth-skin gene. (b) The graph of dy/dp shows, among other things, how much y will change in response to a small change in p: a lot, if p is near 0, but not much if p is near 1.

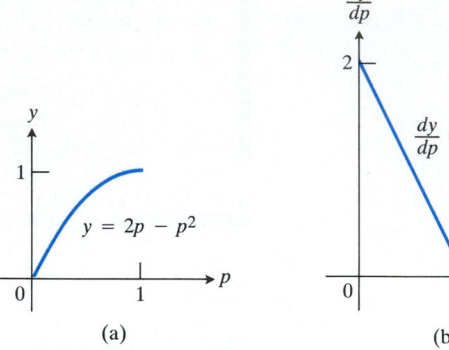
(a)   (b)

We will be able to say more about how sensitive functions are to changes in their variables when we get to Section 2.6.

Derivatives in Economics

Economists often call the derivative of a function the **marginal value** of the function.

Example 9   *Marginal Cost.*   Suppose it costs a company $c(x)$ dollars to produce x tons of steel in a week. It costs more to produce $x + h$ tons per week, and the cost difference, divided by h, is the average increase in cost per ton per week:

$$\frac{c(x + h) - c(x)}{h} = \text{average increase in cost.} \tag{8}$$

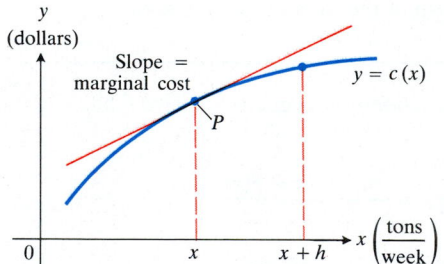

The limit of the ratio as $h \to 0$ is the marginal cost when x tons of steel are produced:

$$c'(x) = \lim_{h \to 0} \frac{c(x + h) - c(x)}{h} = \text{marginal cost.} \tag{9}$$

2.10   Weekly steel production: $c(x)$ is the cost of producing x tons per week. The cost of producing an additional h tons is $c(x + h) - c(x)$.

How are we to interpret this derivative? First of all, it is the slope of the graph of c at the point marked P in Fig. 2.10. But there is more.

Figure 2.11 shows an enlarged view of the curve and its tangent at P. We can see that if the company, currently producing x tons per week, increases production by one ton, then the incremental cost $\Delta c = c(x + 1) - c(x)$ of producing that one ton is approximately $c'(x)$. That is,

$$\Delta c \approx c'(x) \qquad \text{when} \qquad \Delta x = 1. \tag{10}$$

2.11   As weekly steel production increases from x to $x + 1$ tons, the cost curve rises by the amount Δc. The tangent line rises by the amount slope · run = $c'(x) \cdot 1 = c'(x)$. Since $\Delta c/\Delta x \approx c'(x)$, we have $\Delta c \approx c'(x)$ when $\Delta x = 1$.

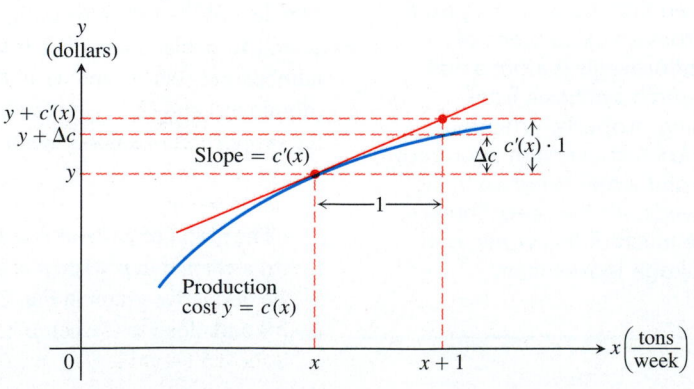

Herein lies the economic importance of marginal cost: It estimates the cost of producing one unit beyond the present production level.

Example 10   *Marginal Cost (continued).*   Suppose it costs

$$c(x) = x^3 - 6x^2 + 15x$$

dollars to produce x radiators and your shop is currently producing 10 radiators a day. About how much extra will it cost to produce one more radiator a day?

Solution   The cost of producing one more radiator a day when 10 are produced is about $c'(10)$. Since

$$c'(x) = \frac{d}{dx}(x^3 - 6x^2 + 15x)$$

$$= 3x^2 - 12x + 15,$$

$$c'(10) = 3(100) - 12(10) + 15$$

$$= 195.$$

The additional cost will be about \$195.

Example 11   *Marginal Revenue.*   If

$$r(x) = x^3 - 3x^2 + 12x$$

gives the dollar revenue from selling x thousand candy bars, the marginal revenue when x thousand are sold is

$$r'(x) = \frac{d}{dx}(x^3 - 3x^2 + 12x)$$

$$= 3x^2 - 6x + 12.$$

As with marginal cost, the marginal revenue function estimates the increase in revenue that will result from selling one additional unit. If you currently sell 10 thousand candy bars a week, you can expect your revenue to increase by about

$$r'(10) = 3(100) - 6(10) + 12 = \$252$$

if you increase sales to 11 thousand bars a week.

Estimating f' from a Graph of f

When we record data in the laboratory or in the field, we usually imagine that we are recording the values of a function $y = f(x)$. We might be recording the pressure in a gas as a function of volume or the size of a population as a function of time. To see what the function looks like, we usually plot the data points and fit them with a curve. Even if we have no formula for the function from which to calculate the derivative $y' = f'(x)$, we can still graph f': We estimate the slopes on the graph of f and plot these slopes. The following examples show how this is done and what we can learn from the graph of f'.

Example 12   On April 23, 1988, the human-powered airplane *Daedalus* flew a record-breaking 119 km from Crete to the island of Santorini in the Aegean Sea,

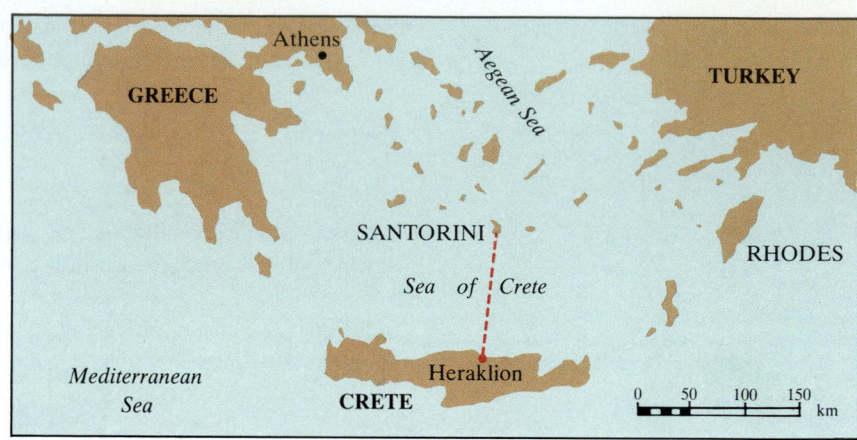

Map 2.1   *Daedalus's* flight path on April 23, 1988.

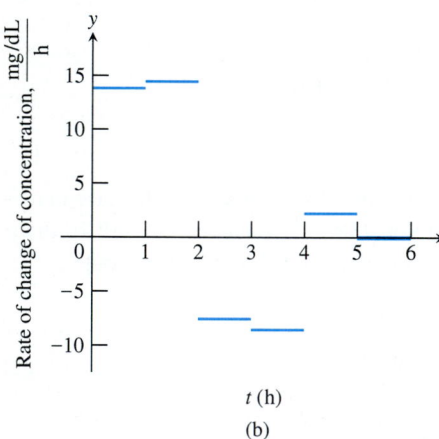

(a)

(b)

2.12   (a) The sugar concentration in the blood of a *Daedalus* pilot during a 6-h preflight endurance test. (b) The derivative of the pilot's blood-sugar concentration shows how rapidly the concentration rose and fell during various portions of the test. (*Source: The Daedalus Project: Physiological Problems and Solutions* by Ethan R. Nadel and Steven R. Bussolari, *American Scientist*, Vol. 76, No. 4, July–August 1988, p. 358.)

southeast of mainland Greece. During the 6-h endurance tests before the flight, researchers monitored the prospective pilots' blood-sugar concentrations. The concentration graph for one of the athlete-pilots is shown in Fig. 2.12(a), where the concentration in milligrams/deciliter is plotted against time in hours.

The graph is made of line segments connecting data points. The constant slope of each segment gives an estimate of the derivative of the concentration between measurements. We calculated the slopes from the coordinate grid and plotted the derivative as a step function in Fig. 2.12(b). To make the plot for the first hour, for instance, we observed that the concentration increased from about 79 mg/dL to 93 mg/dL. The net increase was $\Delta y = 93 - 79 = 14$ mg/dL. Dividing this by $\Delta t = 1$ h gave the rate of change:

$$\frac{\Delta y}{\Delta t} = \frac{14}{1} = 14 \text{ mg/dL per h.}$$

When we have so many data that the graph we get by connecting the data points resembles a smooth curve, we may wish to plot the derivative as a smooth curve. The next example shows how this is done.

Example 13   Graph the derivative of the function $y = f(x)$ in Fig. 2.13(a).

Solution   We draw a pair of axes, marking the horizontal axis in x-units and the vertical axis in y'-units (Fig. 2.13b). Next we sketch tangents to the graph of f at frequent intervals and use their slopes to estimate the values of $y' = f'(x)$ at these points. We plot the corresponding (x, y') pairs and connect them with a smooth curve.

From the graph of $y' = f'(x)$ we see at a glance

1. where f's rate of change is positive, negative, or zero;

2. the rough size of the growth rate at any x and its size in relation to the size of $f(x)$;

3. where the rate of change itself is increasing or decreasing.

2.13 We made the graph of $y' = f'(x)$ in (b) by plotting slopes from the graph of $y = f(x)$ in (a). The vertical coordinate of B' is the slope at B, and so on. The graph of $y' = f'(x)$ is a visual record of how the slope of f changes with x.

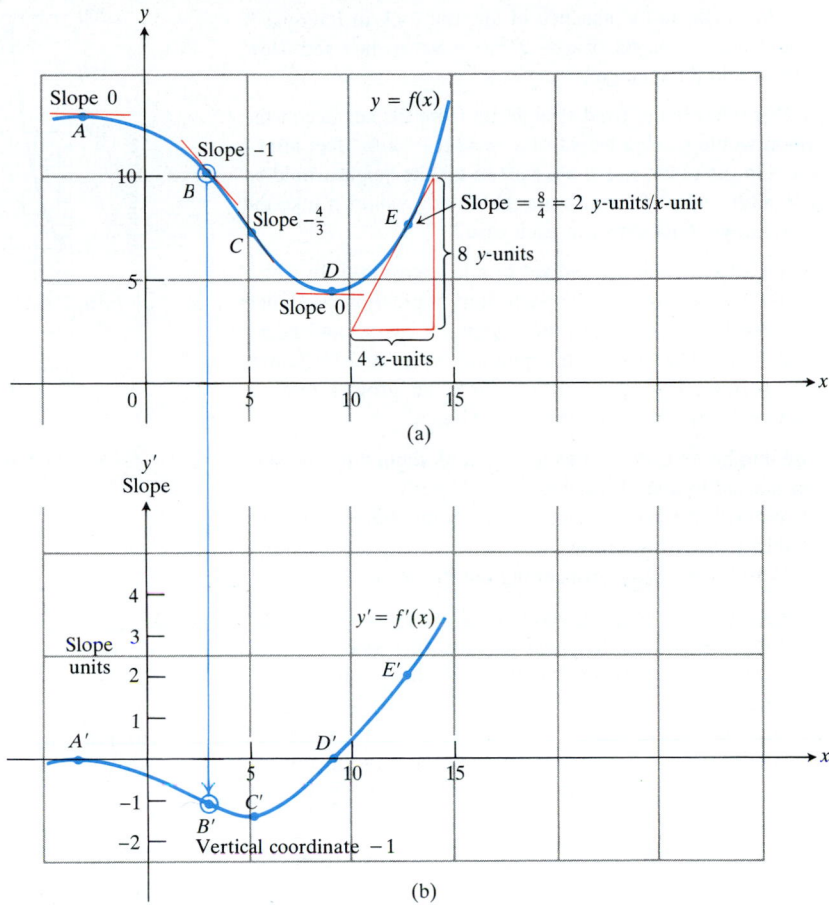

EXERCISES 2.2

Exercises 1–6 give the position $s = f(t)$ of a body moving along a coordinate line during a time interval $a \le t \le b$, with s in meters and t in seconds.

  a) Find the body's velocity and acceleration at the beginning and end of the interval.

  b) Find the body's displacement and average velocity for the interval.

1. $s = 0.8t^2$, $0 \le t \le 10$ (free fall on the moon)

2. $s = 1.86t^2$, $0 \le t \le 0.5$ (free fall on Mars)

3. $s = t^3 - 3t^2 + 2t$, $0 \le t \le 2$

4. $s = t^3 - 3t^2 + 5$, $0 \le t \le 2$

5. $s = \dfrac{125}{t} - \dfrac{625}{t^3}$, $1 \le t \le 5$

6. $s = \dfrac{25}{t + 5}$, $-4 \le t \le 0$

7. The equations for free fall at the surfaces of Mars and Jupiter (s in meters, t in seconds) are $s = 1.86t^2$ for Mars, $s = 11.44t^2$ for Jupiter. How long will it take a rock falling from rest to reach a velocity of 16.6 m/sec (about 59.8 km/h) on each planet?

8. A rock thrown vertically upward from the surface of the moon at a velocity of 24 m/sec (about 86.4 km/h) reaches a height of $s = 24t - 0.8t^2$ m in t sec.

  a) Find the rock's velocity and acceleration. (The acceleration in this case is the acceleration of gravity on the moon.)

  b) How long did it take the rock to reach its highest point?

  c) How high did the rock go?

  d) How long did it take the rock to reach half its maximum height?

  e) How long was the rock aloft?

9. On the earth, in the absence of air, the rock in Exercise 8 would reach a height of $s = 24t - 4.9t^2$ m in t sec. How high would the rock go?

10. A 45-caliber bullet fired straight up from the surface of the moon would reach a height of $s = 832t - 2.6t^2$ feet after t sec. On the earth, in the absence of air, its height would be $s = 832t - 16t^2$ ft after t sec. How long would it take the bullet to get back down in each case?

11. When a bactericide was added to a nutrient broth in which bacteria were growing, the bacterium population continued to grow for a while but then stopped growing and began to decline. The size of the population at time t (hours) was $b(t) = 10^6 + 10^4 t - 10^3 t^2$. Find the growth rates at (a) $t = 0$; (b) $t = 5$; and (c) $t = 10$ hours.

12. The number of gallons of water in a tank t min after the tank has started to drain is $Q(t) = 200(30 - t)^2$.
   a) What is the average rate at which the water flows out during the first 10 min?
   b) How fast is the water running out at the end of 10 min?

13. *Marginal cost.*   Suppose that the dollar cost of producing x washing machines is $c(x) = 2000 + 100x - 0.1x^2$.
   a) Find the marginal cost when 100 washing machines are produced.
   b) Show that the marginal cost when 100 washing machines are produced is approximately the cost of producing one more washing machine after the first 100 have been made, by calculating the latter cost directly.

14. *Marginal revenue.*   Suppose the dollar revenue from selling x custom-made office desks is

$$r(x) = 2000\left(1 - \frac{1}{x + 1}\right).$$

   a) Find the marginal revenue when x desks are produced.
   b) Use the function $r'(x)$ to estimate the increase in revenue that will result from increasing production from 5 desks a week to 6 desks a week.
   c) Find the limit of $r'(x)$ as $x \to \infty$. How would you interpret this number?

15. The position of a body at time t sec is $s = t^3 - 6t^2 + 9t$ m. Find the body's acceleration each time the velocity is zero.

16. The velocity of a body at time t sec is $v = 2t^3 - 9t^2 + 12t - 5$ m/sec. Find the body's speed each time the acceleration is zero.

17. When a model rocket is launched, the propellant burns for a few seconds, accelerating the rocket upward. After burnout, the rocket coasts upward for a while and then begins to fall. A small explosive charge pops out a parachute shortly after the rocket starts down. The parachute slows the rocket to keep it from breaking when it lands.

   Figure 2.14 shows velocity data from the flight of a model rocket. Use the data to answer the following.

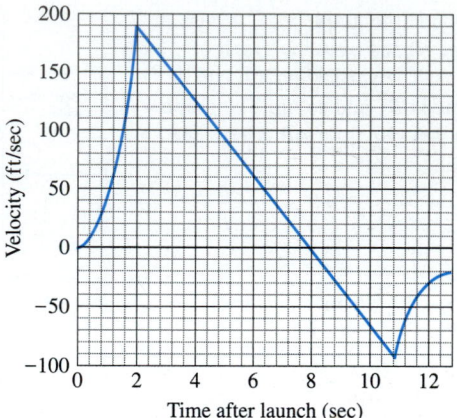

2.14   The velocity of the model rocket in Exercise 17.

   a) How fast was the rocket climbing when the engine stopped?
   b) For how many seconds did the engine burn?
   c) When did the rocket reach its highest point? What was its velocity then?

18. *Pisa by parachute (continuation of Exercise 17).*   On August 5, 1988, Mike McCarthy of London jumped from the Tower of Pisa and then opened his parachute in what he said was a world record low-level parachute jump of 179 ft. Make a rough sketch to show the shape of the graph of his downward velocity during the jump.

19. Figure 2.15 shows a multiflash photograph of two balls falling from rest. The rulers in the figure are marked in centimeters. Use the equation $s = 490t^2$ to answer the following questions.
   a) How long did it take the balls to fall the first 160 cm? What was their average velocity for the period?
   b) How fast were the balls falling when they reached the 160-cm mark? What was their acceleration then?
   c) About how fast was the light flashing (flashes per second)?

20. The graph in Fig. 2.16 shows the position s of a truck traveling on a highway. The truck starts at $t = 0$ and returns 15 hours later at $t = 15$.
   a) Use the technique described in Example 13 to graph the truck's velocity $v = ds/dt$ for $0 \le t \le 15$. Then repeat the process, with the velocity curve, to graph the truck's acceleration dv/dt.
   b) Suppose $s = 15t^2 - t^3$. Graph ds/dt and d^2s/dt^2 and compare your graphs with those in (a).

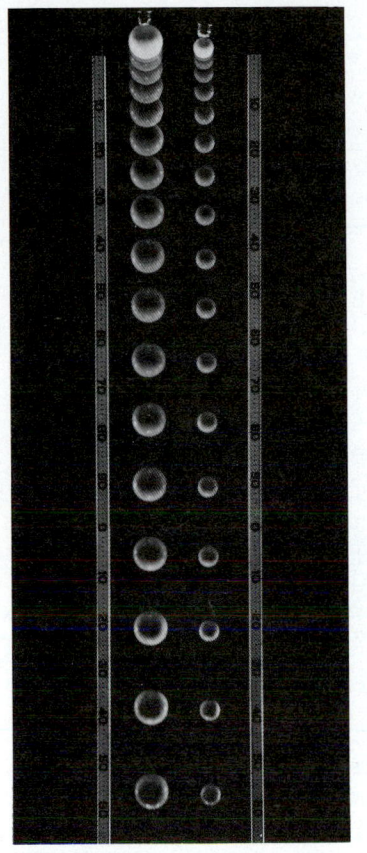

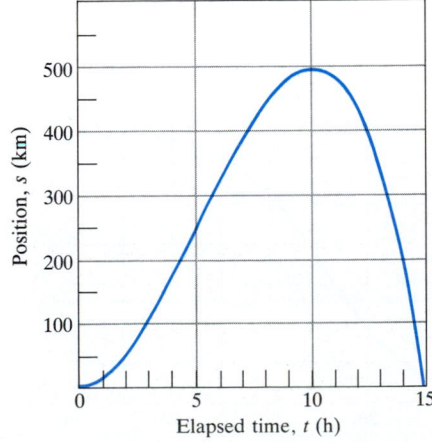

2.15 Two balls falling from rest (Exercise 19).

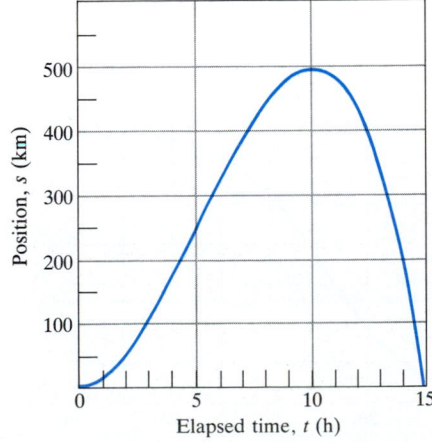

2.16 The time vs. position graph for the truck in Exercise 20.

Exercises 21 and 22 are about the graphs in Fig. 2.17. The graphs in part (a) show the numbers of rabbits and foxes in a small Arctic population. They are plotted as functions of time for 200 days. The number of rabbits increases at first, as the rabbits reproduce. But the foxes prey on the rabbits and, as the number of foxes increases, the rabbit population levels off and then drops. Figure 2.17(b) shows the graph of the derivative of the rabbit population. We made it by plotting slopes, as in Example 13.

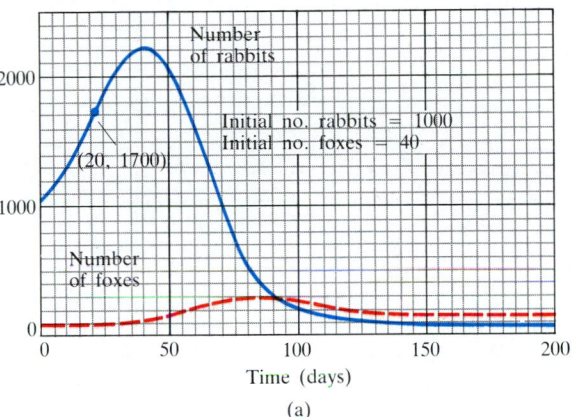

(a)

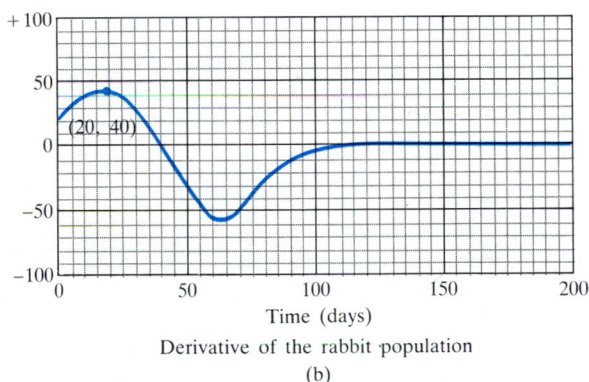

Time (days)

Derivative of the rabbit population

(b)

2.17 Rabbits and foxes in the Arctic predator-prey food chain in Exercises 21 and 22. (*Source: Differentiation*, by W. U. Walton et al., Project CALC, Education Development Center, Inc., Newton, Mass., 1975, p. 86.)

21. a) What is the value of the derivative of the rabbit population in Fig. 2.17 when the number of rabbits is largest? Smallest?
   b) What is the size of the rabbit population in Fig. 2.17 when its derivative is largest? Smallest?

22. In what units should the slopes of the rabbit and fox population curves be measured?

Match the graphs of the functions in Exercises 23–26 with the graphs of the derivatives in Fig. 2.18.

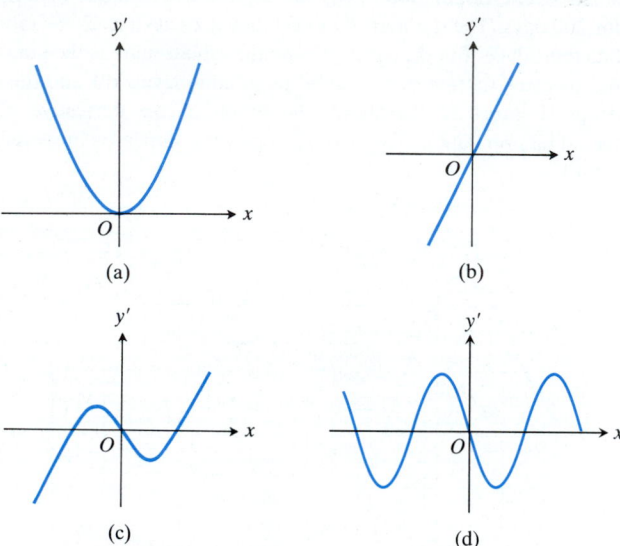

(a)   (b)

(c)   (d)

2.18   The derivative graphs for Exercises 23–26.

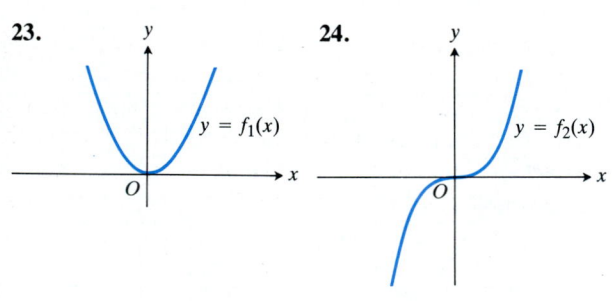

23. $y = f_1(x)$

24. $y = f_2(x)$

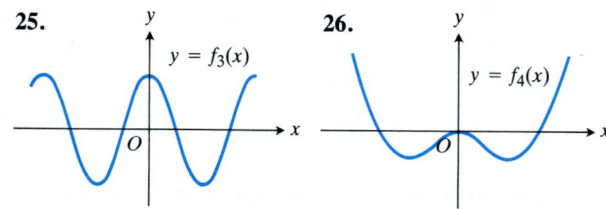

25. $y = f_3(x)$

26. $y = f_4(x)$

27. Figure 2.19 shows the graph of a function $y = f(x)$ defined on the interval $[-3, 7]$. The graph is made of line segments joined end to end.
a) Graph the function's derivative.
b) At what points of $[-3, 7]$ is the derivative not defined?

28. *Growth in the economy.*   The graph in Fig. 2.20 shows the average annual percentage change $y = f(t)$ in the U.S. gross national product (GNP) for the years 1983–1988. Graph dy/dt (where defined). (*Source: Statistical Abstracts of the United States, 110th Edition,* U.S. Department of Commerce, p. 427.)

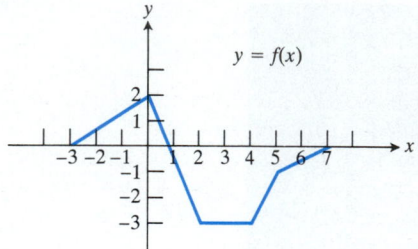

2.19   The graph for Exercise 27.

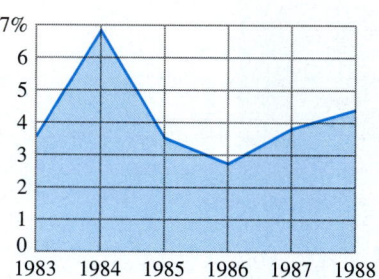

2.20   The graph for Exercise 28.

29. *Fruit flies (Example 1, Section 1.6, continued).*   Populations starting out in closed environments grow slowly at first, when there are relatively few members, then more rapidly as the number of reproducing individuals increases and resources are still abundant, then slowly again as the population reaches the carrying capacity of the environment.
a) Use the graphical technique of Example 13 to graph the derivative of the fruit fly population introduced in Section 1.6. The graph of the population is reproduced here as Fig. 2.21. What units should be used on the horizontal and vertical axes for the derivative's graph?

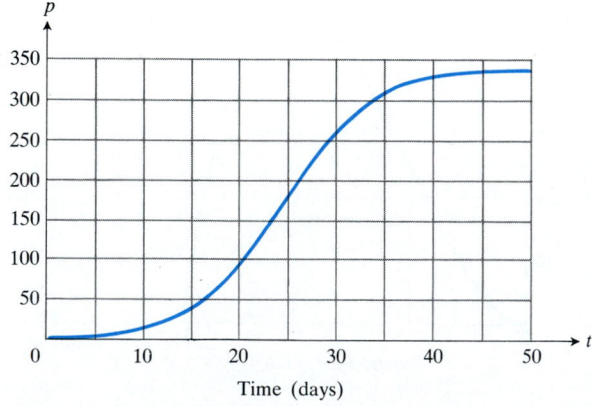

Time (days)

2.21   The graph for Exercise 29.

b) During what days does the population seem to be increasing fastest? Slowest?

30. CALCULATOR *Volcanic lava fountains.* Although the November 1959 Kilauea Iki eruption on the island of Hawaii began with a line of fountains along the wall of the crater, activity was later confined to a single vent in the crater's floor, which at one point shot lava 1900 ft straight into the air (a world record). What was the lava's exit velocity in feet per second? In miles per hour? (*Hint:* If v_0 is the exit velocity of a particle of lava, its height t sec later will be $s = v_0 t - 16t^2$ ft. Begin by finding the time at which $ds/dt = 0$. Neglect air resistance.)

Graphing Calculator or Computer Grapher

Graph the functions in Exercises 31–34 together with their first and second derivatives.

31. $s = 200t - 16t^2$, $t \geq 0$ (A heavy object fired straight up from the earth's surface at 200 ft/sec)

32. $s = 0.8t^2$ (Free fall on the moon, in meters and seconds)

33. $r = 2000\left(1 - \dfrac{1}{x+1}\right)$, $x \geq 0$ (The revenue function from Exercise 14)

34. $s = t^3 - 6t^2 + 9t$ (The position of the body in Exercise 15)

EXPLORER PROGRAMS

Derivatives	Keys in any function $y = f(x)$. The program will graph f and f' together, f and f'' together, or f' and f'' together.
PowerGrapher	Graphs functions and their derivatives in different colors in a common display.

2.3 Derivatives of Trigonometric Functions

Trigonometric functions are important because so many of the phenomena we want information about are periodic (electromagnetic fields, heart rhythms, tides, weather). A surprising and beautiful theorem from advanced calculus says that every periodic function we are likely to use in mathematical modeling can be written as an algebraic combination of sines and cosines, so the derivatives of sines and cosines play a key role in describing important changes. This section shows how to differentiate the six basic trigonometric functions.

The Derivative of the Sine

We begin by showing that

$$\lim_{h \to 0} \frac{\cos h - 1}{h} = 0. \tag{1}$$

We divide the identity

$$\frac{1 - \cos 2\theta}{2} = \sin^2 \theta \qquad \text{(Appendix 2, Eq. (22))}$$

by θ to get

$$\frac{1 - \cos 2\theta}{2\theta} = \frac{\sin \theta}{\theta} \cdot \sin \theta.$$

Letting θ approach zero then gives

$$\lim_{\theta \to 0} \frac{1 - \cos 2\theta}{2\theta} = \lim_{\theta \to 0} \frac{\sin \theta}{\theta} \cdot \lim_{\theta \to 0} \sin \theta = 1 \cdot 0 = 0.$$

Replacing 2θ by h gives Eq. (1).

By definition, the derivative of $y = \sin x$ is the limit

$$\frac{dy}{dx} = \lim_{h \to 0} \frac{\sin(x + h) - \sin x}{h}. \tag{2}$$

To calculate this limit, we combine Eq. (1) with two other results:

1. $\sin(x + h) = \sin x \cos h + \cos x \sin h,$ (Appendix 2, Eq. (16))

2. $\lim\limits_{h \to 0} \dfrac{\sin h}{h} = 1.$ (Section 1.9)

Then, taking all limits as $h \to 0$, we have

$$\frac{dy}{dx} = \lim \frac{\sin(x + h) - \sin x}{h}$$

$$= \lim \frac{\sin x \cos h + \cos x \sin h - \sin x}{h}$$

$$= \lim \frac{\sin x (\cos h - 1) + \cos x \sin h}{h} \tag{3}$$

$$= \lim \sin x \cdot \lim \frac{\cos h - 1}{h} + \lim \cos x \cdot \lim \frac{\sin h}{h}$$

$$= \sin x \cdot 0 + \cos x \cdot 1$$

$$= \cos x.$$

In short,

$$\frac{d}{dx} \sin x = \cos x. \tag{4}$$

Example 1

a) $y = x^2 - \sin x$ $\dfrac{dy}{dx} = 2x - \dfrac{d}{dx}(\sin x)$ (Difference Rule)

    $\qquad\qquad\qquad\qquad = 2x - \cos x$

b) $y = x^2 \sin x$ $\dfrac{dy}{dx} = x^2 \dfrac{d}{dx}(\sin x) + 2x \sin x$ (Product Rule)

    $\qquad\qquad\qquad\qquad = x^2 \cos x + 2x \sin x$

c) $y = \dfrac{\sin x}{x}$ $\dfrac{dy}{dx} = \dfrac{x \cdot \dfrac{d}{dx}(\sin x) - \sin x \cdot 1}{x^2}$ (Quotient Rule)

    $\qquad\qquad\qquad\qquad = \dfrac{x \cos x - \sin x}{x^2}$

In case you are wondering why calculus uses radian measure when the rest of the world seems to use degrees, the answer is provided by the argument that the

derivative of the sine is the cosine. The derivative of $\sin x$ is $\cos x$ *only* if x is measured in radians. The argument requires that when h is a small increment in x,

$$\lim_{h \to 0} (\sin h)/h = 1.$$

This is true only for radian measure, as we discussed in Section 1.9.

The Derivative of the Cosine

To calculate the derivative of the cosine, we use the equation

$$\cos(x + h) = \cos x \cos h - \sin x \sin h. \qquad \text{(Appendix 2, Eq. (15))}$$

Then, with all limits taken as $h \to 0$, we have

$$\frac{dy}{dx} = \lim \frac{\cos(x + h) - \cos x}{h} \qquad \text{(Derivative definition)}$$

$$= \lim \frac{\cos x \cos h - \sin x \sin h - \cos x}{h}$$

$$= \lim \frac{\cos x(\cos h - 1) - \sin x \sin h}{h}$$

$$= \cos x \cdot \lim \frac{\cos h - 1}{h} - \sin x \cdot \lim \frac{\sin h}{h}$$

$$= \cos x \cdot 0 - \sin x \cdot 1 \qquad \text{(Eq. (1))}$$

$$= -\sin x.$$

In short,

$$\frac{d}{dx} \cos x = -\sin x. \tag{5}$$

Notice the minus sign. The derivative of the sine is the cosine, but the derivative of the cosine is *minus* the sine (Fig. 2.22).

2.22 The sine and cosine plotted together. The cosine gives the sine's slope; the sine gives the negative of the cosine's slope.

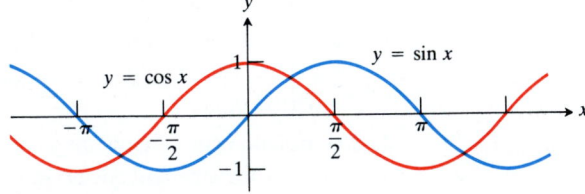

Example 2

a) $y = 5x + \cos x$

$$\frac{dy}{dx} = \frac{d}{dx}(5x) + \frac{d}{dx}(\cos x) \qquad \text{(Sum Rule)}$$

$$= 5 - \sin x$$

b) $y = \sin x \cos x$

$$\frac{dy}{dx} = \sin x \frac{d}{dx}(\cos x) + \cos x \frac{d}{dx}(\sin x) \qquad \text{(Product Rule)}$$
$$= \sin x(-\sin x) + \cos x(\cos x)$$
$$= \cos^2 x - \sin^2 x.$$

c) $y = \dfrac{\cos x}{1 - \sin x}$

$$\frac{dy}{dx} = \frac{(1 - \sin x)\dfrac{d}{dx}(\cos x) - \cos x \dfrac{d}{dx}(1 - \sin x)}{(1 - \sin x)^2} \qquad \text{(Quotient Rule)}$$
$$= \frac{(1 - \sin x)(-\sin x) - \cos x(0 - \cos x)}{(1 - \sin x)^2}$$
$$= \frac{1 - \sin x}{(1 - \sin x)^2} \qquad (\sin^2 x + \cos^2 x = 1)$$
$$= \frac{1}{1 - \sin x}$$

Simple Harmonic Motion

The motion of a body bobbing up and down on the end of a spring is a *simple harmonic motion*. The next example describes a case in which there are no opposing forces like air friction to slow the motion down.

Example 3   A body hanging from a spring (Fig. 2.23) is stretched 5 units beyond its rest position and released at time $t = 0$ to bob up and down. Its position at any later time t is

$$s = 5 \cos t.$$

What are its velocity and acceleration at time t?

Solution   We have

Position: $\quad s = 5 \cos t$

Velocity: $\quad \dfrac{ds}{dt} = \dfrac{d}{dt}(5 \cos t) = 5\dfrac{d}{dt}(\cos t) = -5 \sin t$

Acceleration: $\quad \dfrac{dv}{dt} = \dfrac{d}{dt}(-5 \sin t) = -5\dfrac{d}{dt}(\sin t) = -5 \cos t.$

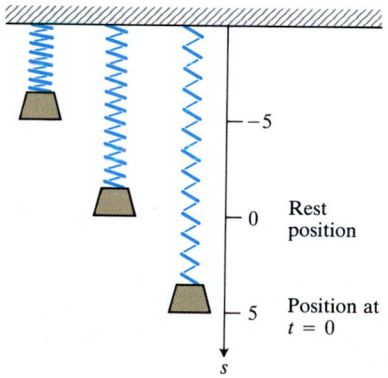

2.23   The body in Example 3.

Here is what we can learn from these equations:

1. As time passes, the body moves up and down between $s = 5$ and $s = -5$ on the s-axis. The amplitude of the motion is 5. The period of the motion is 2π, the period of $\cos t$.

2. The function $\sin t$ attains its greatest magnitude (1) when $\cos t = 0$, as the graphs of the sine and cosine show (Fig. 2.22). Hence, the body's speed, $|v| = 5|\sin t|$, is greatest every time $\cos t = 0$, i.e., every time the body passes its rest position.

    The body's speed is zero when $\sin t = 0$. This occurs when $\cos t = \pm 1$, at the endpoints of the interval of motion.

3. The acceleration, $a = -5 \cos t$, is zero only at the rest position, where $\cos t = 0$. When the body is anywhere else, the spring is either pulling on it or pushing on it. The acceleration is greatest in magnitude at the points farthest from the origin, where $\cos t = \pm 1$.

The Derivatives of the Other Basic Functions

Because $\sin x$ and $\cos x$ are differentiable functions of x, the functions

$$\tan x = \frac{\sin x}{\cos x} \qquad \sec x = \frac{1}{\cos x}$$

$$\cot x = \frac{\cos x}{\sin x} \qquad \csc x = \frac{1}{\sin x}$$

are differentiable at every value of x at which they are defined. Their derivatives, calculated from the Quotient Rule, are given by the following formulas.

$$\frac{d}{dx}\tan x = \sec^2 x \qquad (6) \qquad\qquad \frac{d}{dx}\sec x = \sec x \tan x \qquad (7)$$

$$\frac{d}{dx}\cot x = -\csc^2 x \quad (8) \qquad\qquad \frac{d}{dx}\csc x = -\csc x \cot x \qquad (9)$$

Notice the minus signs in the equations for the cotangent and cosecant.

To show how a typical calculation goes, we derive Eq. (6). The other derivations are left as exercises.

Example 4   Find dy/dx if $y = \tan x$.

Solution

$$\frac{d}{dx}\tan x = \frac{d}{dx}\left(\frac{\sin x}{\cos x}\right) = \frac{\cos x \dfrac{d}{dx}(\sin x) - \sin x \dfrac{d}{dx}(\cos x)}{\cos^2 x}$$

$$= \frac{\cos x \cos x - \sin x(-\sin x)}{\cos^2 x} = \frac{\cos^2 x + \sin^2 x}{\cos^2 x}$$

$$= \frac{1}{\cos^2 x} = \sec^2 x$$

Example 5   Find y'' if $y = \sec x$.

Solution

$$y = \sec x$$

$$y' = \sec x \tan x \qquad\qquad\qquad \text{(Eq. (7))}$$

$$y'' = \frac{d}{dx}(\sec x \tan x)$$

$$= \sec x \frac{d}{dx}(\tan x) + \tan x \frac{d}{dx}(\sec x)$$

$$= \sec x(\sec^2 x) + \tan x(\sec x \tan x)$$

$$= \sec^3 x + \sec x \tan^2 x$$

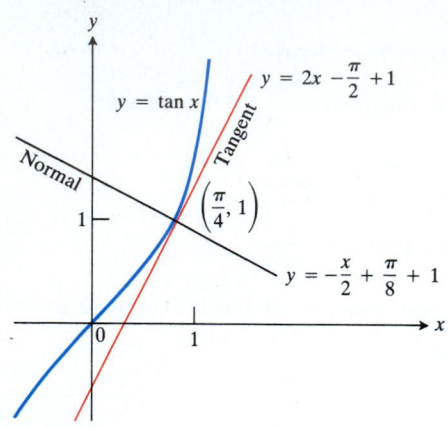

2.24 Example 7 shows how to find equations for the tangent and normal lines shown here.

The Word **Normal**

When analytic geometry was developed in the seventeenth century, European scientists still wrote about their work and ideas in Latin, the one language that all educated Europeans could read and understand. The word *normalis*, which scholars used for "perpendicular" in Latin, became *normal* when they discussed geometry in English.

Example 6

a) $\dfrac{d}{dx}(3x + \cot x) = 3 + \dfrac{d}{dx}(\cot x) = 3 - \csc^2 x$

b) $\dfrac{d}{dx}\left(\dfrac{2}{\sin x}\right) = \dfrac{d}{dx}(2\csc x) = 2\dfrac{d}{dx}(\csc x)$

$$= 2(-\csc x \cot x) = -2\csc x \cot x$$

Example 7
Find the lines that are tangent and normal (perpendicular) to the curve $y = \tan x$ at the point $(\pi/4, 1)$ (Fig. 2.24).

Solution The slope of the curve at $(\pi/4, 1)$ is the value of $dy/dx = \sec^2 x$ at $x = \pi/4$:

$$\left.\frac{dy}{dx}\right|_{x=\pi/4} = \sec^2\left(\frac{\pi}{4}\right) = \left(\sqrt{2}\right)^2 = 2.$$

The tangent is the line

$$y - 1 = 2\left(x - \frac{\pi}{4}\right), \quad \text{or} \quad y = 2x - \frac{\pi}{2} + 1.$$

The normal has slope $m = -1/2$, so its point–slope equation is

$$y - 1 = -\frac{1}{2}\left(x - \frac{\pi}{4}\right), \quad \text{or} \quad y = -\frac{x}{2} + \frac{\pi}{8} + 1.$$

Continuity

Since the six basic trigonometric functions are differentiable, they are also continuous (by Theorem 8, Section 1.10). For each one, $\lim_{x\to a} f(x) = f(a)$ whenever $f(a)$ is defined. This means that we can calculate the limits of most trigonometric functions as $x \to a$ by evaluating them at $x = a$.

EXERCISES 2.3

In Exercises 1–12, find dy/dx when y equals the given expression. Simplify your answers.

1. $1 + x - \cos x$

2. $\dfrac{1}{x} + 5\sin x$

3. $\csc x - 5x + 7$

4. $x \sec x$

5. $x^2 \cot x$

6. $3x + x\tan x$

7. $\sin x \sec x$

8. $\tan x \cot x$

9. $\dfrac{4}{\cos x}$

10. $\dfrac{\cos x}{x}$

11. $\dfrac{x}{1 + \cos x}$

12. $\dfrac{\cot x}{1 + \cot x}$

In Exercises 13–16, find ds/dt.

13. $s = 2\sin t - \tan t$

14. $s = t^2 - \sec t$

15. $s = 2t + \cot t$

16. $s = t\csc t$

In Exercises 17–20, find $dr/d\theta$. Simplify your answers.

17. $r = 4 - \theta^2 \sin \theta$

18. $r = \theta \sin \theta + \cos \theta$

19. $r = \sec \theta \csc \theta$

20. $r = \cos \theta(1 + \sec \theta)$

In Exercises 21–24, find dp/dq. Simplify your answers.

21. $p = 5 + \dfrac{1}{\tan q}$

22. $p = \dfrac{2}{\csc q} - \dfrac{1}{\sec q}$

23. $p = \dfrac{\sin q + \cos q}{\cos q}$ **24.** $p = \dfrac{\cos q}{1 + \sin q}$

25. Find y'' if $y = \csc x$.

26. Find $y^{(4)} = d^4y/dx^4$ if
a) $y = \sin x$,
b) $y = \cos x$.

In Exercises 27–30, find equations for the lines that are tangent and normal to the curve $y = f(x)$ at the given point.

27. $y = \sin x$, $(0, 0)$

28. $y = \tan x$, $(0, 0)$

29. $y = \cos x$, $(\pi, -1)$

30. $y = 1 + \cos x$, $(\pi/2, 1)$

31. Show that the graphs of $y = \sec x$ and $y = \cos x$ have horizontal tangents at $x = 0$.

32. Show that the graphs of $y = \tan x$ and $y = \cot x$ never have horizontal tangents.

Do the graphs of the functions in Exercises 33–36 have any horizontal tangents in the interval $0 \le x \le 2\pi$? If so, where? If not, why not?

33. $y = x + \sin x$

34. $y = 2x + \sin x$

35. $y = x + \cos x$

36. $y = x + 2 \cos x$

Find the limits in Exercises 37–40.

37. $\lim\limits_{x \to 2} \sin \left(\dfrac{1}{x} - \dfrac{1}{2} \right)$

38. $\lim\limits_{x \to \pi} \sec(1 + \cos x)$

39. $\lim\limits_{x \to 0} (\sec x + \tan x)$

40. $\lim\limits_{x \to 0} \cos \left(\dfrac{\cos^2 x - 1}{x^2} \right)$

41. Is there a value of b that makes

$$f(x) = \begin{cases} x + b, & x < 0 \\ \cos x, & x \ge 0 \end{cases}$$

continuous at $x = 0$? If so, what is it? If not, why not?

42. *Simple harmonic motion.*   The equations in (a) and (b) give the position $s = f(t)$ of a body moving along a coordinate line. Find each body's velocity, speed, and acceleration at time $t = \pi/4$.
a) $s = 2 - 2 \sin t$
b) $s = \sin t + \cos t$

43. Find equations for the lines that are tangent and normal to the curve $y = \sqrt{2} \cos x$ at the point $(\pi/4, 1)$.

44. Find the points on the curve $y = \tan x$, $-\pi/2 < x < \pi/2$, where the tangent is parallel to the line $y = 2x$.

In Exercises 45 and 46, find an equation for (a) the horizontal tangent, (b) the tangent to the curve at the indicated point P.

45.

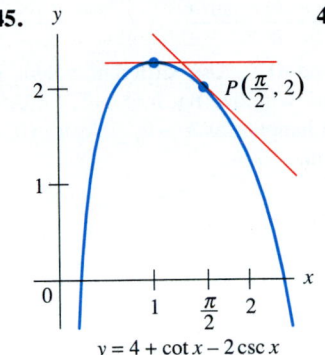

$y = 4 + \cot x - 2 \csc x$

(Generated by Mathematica)

46.

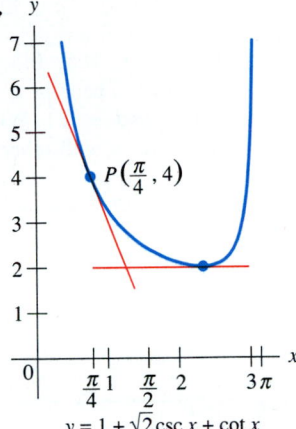

$y = 1 + \sqrt{2} \csc x + \cot x$

47. CALCULATOR   Although $\lim\limits_{h \to 0} (1 - \cos h)/h = 0$, it turns out that

$$\lim\limits_{h \to 0} \frac{1 - \cos h}{h^2} \ne 0.$$

Try finding out what the limit is by taking $h = 1, 0.1, 0.01$, and so on, as far as your calculator can go. (L'Hôpital's rule in Section 3.6 will provide a quick way to confirm your answer.)

48. CALCULATOR   *Radians vs. degrees.*   What happens if you use degrees instead of radians to calculate

$$\lim\limits_{h \to 0} \frac{\sin h}{h}?$$

To find out, set your calculator in degree mode and make a table of values of $(\sin h)/h$ for $h = 0.1, 0.01, 0.001$, and so on, as far as your calculator can go. Now multiply each entry in the table by $180/\pi$. What do you find? What would the derivative of $\sin x$ be if x were measured in degrees instead of radians? (Look at Eq. 3 to find out.)

49. a) Derive Eq. (7) by writing $\sec x$ as $1/\cos x$ and differentiating with respect to x.
b) Derive Eq. (9) by writing $\csc x$ as $1/\sin x$ and differentiating with respect to x.

50. Derive Eq. (8) by writing $\cot x$ as $(\cos x)/(\sin x)$ and differentiating with respect to x.

Computer Grapher or Graphing Calculator

51. Graph $y = \tan x$ and its derivative together over the interval $-\pi/2 < x < \pi/2$.

52. Graph $y = \cot x$ and its derivative together over the interval $0 < x < \pi$.

53. Graph $y = \cos x$ for $-\pi \le x \le 2\pi$. Then, on the same screen, graph

$$y = \frac{\sin(x + h) - \sin x}{h}$$

for $h = 1$, 0.5, 0.3, and 0.1. Use different colors if available. Then, in a new window, try $h = -1$, -0.5, -0.3, and -0.1. What happens as $h \to 0^+$? As $h \to 0^-$? Experiment with other values of h.

EXPLORER PROGRAMS

Derivatives	Graphs the first and second derivatives of any function you key in
PowerGrapher	Graphs functions and their derivatives in a common display

2.4 The Chain Rule

We now know how to differentiate $\sin x$ and $x^2 - 4$, but how do we differentiate a composite like $\sin(x^2 - 4)$? The answer is, with the Chain Rule, which says that the derivative of the composite of two differentiable functions is the product of their derivatives evaluated at appropriate points. The Chain Rule is probably the most widely used differentiation rule in mathematics. This section describes the rule and how to use it. We begin with some examples.

Example 1   The function $y = 6x - 10 = 2(3x - 5)$ is the composite of the functions $y = 2u$ and $u = 3x - 5$. How are the derivatives of these three functions related?

Solution   We have

$$\frac{dy}{dx} = 6, \qquad \frac{dy}{du} = 2, \qquad \frac{du}{dx} = 3.$$

Since $6 = 2 \cdot 3$,

$$\frac{dy}{dx} = \frac{dy}{du}\frac{du}{dx}.$$

Example 2   The function $y = 9x^2 + 6x + 1 = (3x + 1)^2$ is the composite of $y = u^2$ and $u = 3x + 1$. The derivatives involved are

$$\frac{dy}{dx} = \frac{d}{dx}(9x^2 + 6x + 1) = 18x + 6 = 6(3x + 1) = 6u,$$

$$\frac{dy}{du} = \frac{d}{du}(u^2) = 2u,$$

$$\frac{du}{dx} = \frac{d}{dx}(3x + 1) = 3.$$

Once again,

$$\frac{dy}{dx} = \frac{dy}{du}\frac{du}{dx}.$$

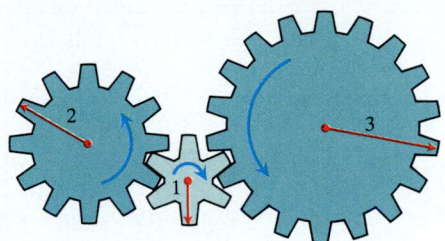

C: *y* turns   B: *u* turns   A: *x* turns

2.25 When gear A takes *x* turns, gear B takes *u* turns and gear C takes *y* turns. By comparing circumferences or counting teeth, we see that $dy/du = 1/2$ and $du/dx = 3$. What is dy/dx?

Example 3   In the gear train in Fig. 2.25, the ratios of the radii of gears A, B, and C are 3:1:2. If gear A turns *x* times, then gear B turns $u = 3x$ times and gear C turns $y = u/2 = (3/2)x$ times. In terms of derivatives,

$$\frac{dy}{du} = \frac{1}{2} \qquad \text{(C turns at one-half B's rate.)}$$

$$\frac{du}{dx} = 3 \qquad \text{(B turns at 3 times A's rate.)}$$

In this example, too, we can calculate dy/dx by multiplying dy/du by du/dx:

$$\frac{dy}{dx} = \frac{3}{2} = \frac{1}{2} \cdot 3 = \frac{dy}{du}\frac{du}{dx}.$$

The Chain Rule

The preceding examples all work because the derivative of a composite $f \circ g$ of two differentiable functions is the product of their derivatives evaluated at appropriate points. This is the observation we state formally as the Chain Rule. As in Section 1.2, the notation $f \circ g$ ("f of g") denotes the composite of the functions f and g, with f following g. The value of $f \circ g$ at a point x is $(f \circ g)(x) = f(g(x))$.

The Chain Rule (First Form)

Suppose that $f \circ g$ is the composite of the differentiable functions $y = f(u)$ and $u = g(x)$. Then $f \circ g$ is a differentiable function of x whose derivative at each value of x is

$$(f \circ g)'_{\text{at } x} = f'_{\text{at } u = g(x)} \cdot g'_{\text{at } x}. \tag{1}$$

In short,

$$(f \circ g)'(x) = f'(g(x)) \cdot g'(x). \tag{2}$$

See Fig. 2.26.

Equations (1) and (2) name the function involved as well as the dependent and independent variables. Once we know what the functions are, as we usually do in any particular example, we can get by with writing the Chain Rule a shorter way.

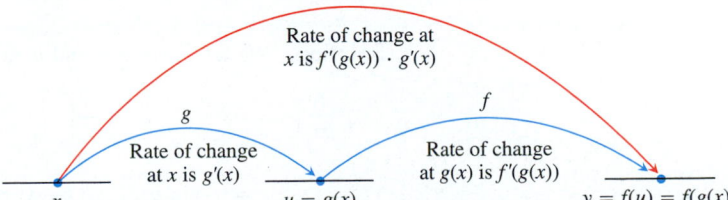

2.26 Rates of change multiply: The derivative of f of g at x is the derivative of f at the point $g(x)$ times the derivative of g at x.

> **Chain Rule (Shorter Form)**
>
> If y is a differentiable function of u, and u is a differentiable function of x, then y is a differentiable function of x and
>
> $$\frac{dy}{dx}\bigg|_{x} = \frac{dy}{du}\bigg|_{u(x)} \cdot \frac{du}{dx}\bigg|_{x}. \tag{3}$$

Equation (3) still tells how each derivative is to be evaluated. When we do not need to be told that, we can get along with an even shorter form.

> **Chain Rule (Shortest Form)**
>
> If y is a differentiable function of u, and u is a differentiable function of x, then
>
> $$\frac{dy}{dx} = \frac{dy}{du}\frac{du}{dx}. \tag{4}$$

You might think it would be a relatively easy matter to prove the Chain Rule by starting with the derivative definition the way we started the proofs of the Product and Quotient Rules. Unfortunately, this is the way the *hard* proof starts. The (relatively) easy proof begins with an equation in Section 2.6. We shall prove the Chain Rule there.

Like different instruments in a doctor's bag, each form of the Chain Rule makes some task a little easier. We shall use them all in the examples that follow. But remember—they all express the same one rule: The derivative of a composite of differentiable functions is the product of their derivatives.

Example 4   Suppose $f \circ g$ is a composite of the differentiable functions $y = f(u)$ and $u = g(x)$, that $g'(2) = 4$, and that $f'(g(2)) = -5$. Find $(f \circ g)'$ at $x = 2$.

Solution   Equation (2) gives

$$(f \circ g)'(2) = f'(g(2)) \cdot g'(2) = (-5) \cdot 4 = -20.$$

Example 5   Find dy/dx at $x = 0$ if $y = \cos\left(\frac{\pi}{2} - 3x\right)$.

Solution   We let $y = \cos u$ and $u = (\pi/2) - 3x$ and apply Eq. (3):

$$\frac{dy}{dx}\bigg|_{x=0} = \frac{dy}{du}\bigg|_{u=\pi/2} \cdot \frac{du}{dx}\bigg|_{x=0} \qquad \left(u = \frac{\pi}{2} \text{ when } x = 0\right)$$

$$= -\sin u\big|_{u=\pi/2} \cdot (-3) = 3\sin\frac{\pi}{2} = 3 \cdot 1 = 3.$$

Example 6   Find dy/dx if $y = \sin(x^2 - 4)$.

Solution   We take $y = \sin u$ and $u = x^2 - 4$ and apply Eq. (4):

$$\frac{dy}{dx} = \frac{dy}{du}\frac{du}{dx} \qquad \text{(Eq. (4))}$$

$$= \cos u \cdot 2x$$

$$= \cos(x^2 - 4) \cdot 2x$$

$$= 2x \cos(x^2 - 4).$$

Integer Powers of Differentiable Functions

The Chain Rule enables us to differentiate powers like $y = \sin^5 x$ and $y = (2x + 1)^{-3}$ because these powers are composites:

$$y = \sin^5 x \quad \text{is} \quad u^5 \quad \text{with} \quad u = \sin x;$$

$$y = (2x + 1)^{-3} \quad \text{is} \quad u^{-3} \quad \text{with} \quad u = 2x + 1.$$

If u is any differentiable function of x and $y = u^n$, then the Chain Rule in the form

$$\frac{dy}{dx} = \frac{dy}{du}\frac{du}{dx}$$

gives

$$\frac{dy}{dx} = \frac{d}{du}(u^n) \cdot \frac{du}{dx}$$

$$= nu^{n-1}\frac{du}{dx}. \qquad \left(\begin{array}{l}\text{Differentiating } u^n \text{ with respect to } u \text{ itself}\\ \text{gives } nu^{n-1}.\end{array}\right)$$

Integer Powers of a Differentiable Function

If u^n is an integer power of a differentiable function $u(x)$, then u^n is differentiable and

$$\frac{d}{dx}u^n = nu^{n-1}\frac{du}{dx}. \qquad (5)$$

Example 7

a) $\dfrac{d}{dx}\sin^5 x = 5\sin^4 x\dfrac{d}{dx}(\sin x)$ 　　　　$\left(\begin{array}{l}\text{Eq. (5) with}\\ u = \sin x,\ n = 5\end{array}\right)$

　　　　　　$= 5\sin^4 x \cos x$

b) $\dfrac{d}{dx}(2x + 1)^{-3} = -3(2x + 1)^{-4}\dfrac{d}{dx}(2x + 1)$ 　　$\left(\begin{array}{l}\text{Eq. (5) with}\\ u = 2x + 1,\ n = -3\end{array}\right)$

　　　　　　　　$= -3(2x + 1)^{-4}(2)$

　　　　　　　　$= -6(2x + 1)^{-4}$

The "Inside-Outside" Rule

It sometimes helps to think about the Chain Rule the following way. If $y = f(g(x))$, Eq. (2) tells us that

$$\frac{dy}{dx} = f'(g(x)) \cdot g'(x). \tag{6}$$

In words, Eq. (6) says: To find dy/dx, differentiate the "outside" function f and leave the "inside" $g(x)$ alone; then multiply by the derivative of the inside.

Example 8

$$\underset{\text{inside}}{\frac{d}{dx} \sin \underset{}{(x^2 + x)}} = \underset{\substack{\text{inside} \qquad \text{derivative} \\ \text{left alone} \; \text{of the inside}}}{\cos(x^2 + x) \cdot (2x + 1)}$$

outside · derivative of the outside

Repeated Use

We sometimes have to use the Chain Rule two or more times to get the job done. Here is an example.

Example 9

a) $\dfrac{d}{dx} \cos^2 3x = 2 \cos 3x \cdot \dfrac{d}{dx} (\cos 3x)$ (Power (Chain) Rule)

$\qquad = 2 \cos 3x (-\sin 3x) \dfrac{d}{dx} (3x)$ (Chain Rule again)

$\qquad = 2 \cos 3x (-\sin 3x)(3)$

$\qquad = -6 \cos 3x \sin 3x$

b) $\dfrac{d}{dx} \sin(1 + \tan 2x) = \cos(1 + \tan 2x) \cdot \dfrac{d}{dx} (1 + \tan 2x)$ (Chain Rule)

$\qquad = \cos(1 + \tan 2x) \cdot \sec^2 2x \cdot \dfrac{d}{dx} (2x)$ (Chain Rule)

$\qquad = 2 \cos(1 + \tan 2x) \sec^2 2x$

Derivative Formulas That Include the Chain Rule

Many of the derivative formulas you will encounter in your scientific work already include the Chain Rule.

If f is a differentiable function of u, and u is a differentiable function of x, then substituting $y = f(u)$ in the Chain Rule formula

$$\frac{dy}{dx} = \frac{dy}{du} \frac{du}{dx}$$

leads to the formula

$$\frac{d}{dx} f(u) = f'(u) \frac{du}{dx}. \tag{7}$$

When we spell this out for the functions whose derivatives we have studied so far, we get the formulas in Table 2.1.

TABLE 2.1
Derivative formulas that include the Chain Rule

$$\frac{d}{dx} u^n = nu^{n-1} \frac{du}{dx} \quad (n \text{ an integer})$$

$$\frac{d}{dx} \sin u = \cos u \frac{du}{dx}$$

$$\frac{d}{dx} \cos u = -\sin u \frac{du}{dx}$$

$$\frac{d}{dx} \tan u = \sec^2 u \frac{du}{dx}$$

$$\frac{d}{dx} \cot u = -\csc^2 u \frac{du}{dx}$$

$$\frac{d}{dx} \sec u = \sec u \tan u \frac{du}{dx}$$

$$\frac{d}{dx} \csc u = -\csc u \cot u \frac{du}{dx}$$

Example 10

a) $\frac{d}{dx} \sin(-x) = \cos(-x) \frac{d}{dx} (-x)$ $\qquad \left(\frac{d}{dx} \sin u = \cos u \frac{du}{dx} \text{ with } u = -x \right)$

$\qquad\qquad\quad = -\cos(-x)$

$\qquad\qquad\quad = -\cos x \qquad\qquad$ (The cosine is even.)

b) $\frac{d}{dx} \tan \left(\frac{1}{x} \right) = \sec^2 \frac{1}{x} \cdot \frac{d}{dx} \left(\frac{1}{x} \right)$ $\qquad \left(\frac{d}{dx} \tan u = \sec^2 u \frac{du}{dx} \text{ with } u = \frac{1}{x} \right)$

$\qquad\qquad\quad = \sec^2 \frac{1}{x} \cdot \left(-\frac{1}{x^2} \right)$

$\qquad\qquad\quad = -\frac{1}{x^2} \sec^2 \frac{1}{x}$

c) $\frac{d}{dx} \cos(\cos x) = -\sin(\cos x) \cdot \frac{d}{dx} (\cos x)$ $\qquad \left(\begin{array}{l} \frac{d}{dx} \cos u = -\sin u \frac{du}{dx} \\ \text{with } u = \cos x \end{array} \right)$

$\qquad\qquad\quad = -\sin(\cos x) \cdot (-\sin x)$

$\qquad\qquad\quad = \sin x \sin(\cos x)$

✳ Melting Ice Cubes

In mathematics, we tend to use letters like f, g, x, y, and u for functions and variables. However, applied fields use letters like V, for volume, and s, for side, that come from the names of the things being modeled. The letters in the Chain Rule then change too, as in the next example.

Example 11   *The Melting Ice Cube.*   How long will it take an ice cube to melt?

Solution   As with all applications to science, we start with a mathematical model. Let us assume that the cube has side length s, so that its volume is $V = s^3$, and that V and s are differentiable functions of time t. Suppose also that the cube's volume decreases at a rate that is proportional to its surface area. This assumption seems reasonable enough when we think that the melting takes place at the surface: Changing the amount of surface changes the amount of ice exposed to melt. In mathematical terms,

$$\frac{dV}{dt} = -k(6s^2).$$

The minus sign is there because the volume is decreasing. We assume that the positive proportionality factor k is constant. (It probably depends on many things, however, such as the relative humidity of the surrounding air, the air temperature, and the incidence or absence of sunlight, to name only a few.)

Finally, we need at least one more piece of information: How long will it take a specific percentage of the ice cube to melt? We have nothing to guide us unless we make one or more observations, but now let us assume a particular set of conditions in which the cube lost 1/4 of its volume during the first hour. (You could use letters instead of particular numbers: say $n\%$ in r hours. Then your answer would be in terms of n and r.)

Mathematically, we now have the following problem.

Given: $V = s^3$ and $\dfrac{dV}{dt} = -k(6s^2)$

$V = V_0$ when $t = 0$

$V = (3/4)V_0$ when $t = 1$ h

Find: The value of t when $V = 0$

We apply the Chain Rule to differentiate $V = s^3$ with respect to t:

$$\frac{dV}{dt} = 3s^2 \frac{ds}{dt}.$$

We set this equal to the given rate, $-k(6s^2)$, to get

$$3s^2 \frac{ds}{dt} = -6ks^2,$$

$$\frac{ds}{dt} = -2k.$$

The side length is *decreasing* at the constant rate of $2k$ units per hour. Thus, if the initial length of the cube's side is s_0, the length of its side one hour later is $s_1 = s_0 - 2k$. This equation tells us that

$$2k = s_0 - s_1.$$

The melting time is the value of t that makes $2kt = s_0$. Hence,

$$t_{\text{melt}} = \frac{s_0}{2k} = \frac{s_0}{s_0 - s_1} = \frac{1}{1 - (s_1/s_0)}.$$

But

$$\frac{s_1}{s_0} = \frac{\left(\frac{3}{4}V_0\right)^{1/3}}{(V_0)^{1/3}} = \left(\frac{3}{4}\right)^{1/3} = 0.91. \qquad \text{(Calculator, rounded)}$$

Therefore,

$$t_{\text{melt}} = \frac{1}{1 - 0.91} = 11 \text{ h}. \qquad \text{(Again)}$$

If 1/4 of the cube melts in 1 h, it will take about 10 h more for the rest of it to melt.

If we were natural scientists interested in testing the assumptions on which our mathematical model is based, our next step would be to run a number of experiments and compare their outcomes with the model's predictions. One practical application might lie in analyzing the proposal to tow large icebergs from polar waters to offshore locations near southern California, where the melting ice could provide fresh water. As a first approximation, we might imagine the iceberg to be a large cube or rectangular solid, or perhaps a pyramid. We shall say more about mathematical modeling in Section 4.2.

EXERCISES 2.4

In Exercises 1–28, find dy/dx when y equals the given expression.

1. $\sin(x + 1)$

2. $\sec(\pi x/2)$

3. $x \cos 5x$

4. $\dfrac{\cos 2x}{1 + \sin 2x}$

5. $\sin^3 x$

6. $\sin(x^2)$

7. $x^2 \tan(1/x)$

8. $\sin 2x \cos 3x$

9. $x + \sec(x^2 + 1)$

10. $\csc(x^2 + 7x)$

11. $\cos(\sin x)$

12. $\sec(\tan x)$

13. $(x - 2)^4$

14. $(4 - 3x)^9$

15. $(2 \sin x + 5)^{-5}$

16. $(1 + \cot x)^{-6}$

17. $\left(1 - \dfrac{x}{7}\right)^7$

18. $\left(\dfrac{x}{2} - 1\right)^{-10}$

19. $\left(1 + x - \dfrac{1}{x}\right)^{-4}$

20. $\left(\dfrac{x}{5} + \dfrac{1}{5x}\right)^5$

21. $(2x - 5)^4(x + 1)^8$

22. $(x + 1)^{-2}(x - 5)^{-1}$

23. $(2x^3 + 3x^2 + 6x + 6)^6$

24. $(\sin x - x \cos x)^{-4}$

25. $-(\sec x + \tan x)^{-1}$

26. $(\csc x + \cot x)^{-1}$

27. $\sin\left(\dfrac{x - 2}{x + 3}\right)$

28. $\left(\dfrac{\sin x}{1 + \cos x}\right)^2$

In Exercises 29–32, find ds/dt.

29. $s = (2t + 1)^{-4}$

30. $s = \left(\dfrac{t}{t + 1}\right)^6$

31. $s = \dfrac{4}{3\pi} \sin 3t + \dfrac{4}{5\pi} \sin 5t$

32. $s = (\pi - 4t) \cos(\pi - 4t)$

In Exercises 33–36, find $dr/d\theta$.

33. $r = \tan(2 - \theta)$

34. $r = \sec 2\theta \tan 2\theta$

35. $r = \dfrac{\cot 3\theta}{2 + \cot 3\theta}$

36. $r = \sin(\theta + \sin \theta)$

In Exercises 37–42, find dy/dx when y equals the given expression. You will need to use the Chain Rule two or three times in each case.

37. $\sin^2(3x - 2)$

38. $\sec^2 5x$

39. $(1 + \cos 2x)^2$

40. $(1 - \tan(x/2))^{-2}$

41. $\sin(\cos(2x - 5))$

42. $(1 + \cos^2 7x)^3$

Find y'' in Exercises 43–46.

43. $y = \tan x$

44. $y = 9 \tan(x/3)$

45. $y = \cot x$

46. $y = \cot(3x - 1)$

In Exercises 47–52, find the value of $(f \circ g)'$ at the given value of x.

47. $f(u) = u^5 + 1, \quad u = g(x) = \sqrt{x}, \quad x = 1$

48. $f(u) = 1 - \dfrac{1}{u}, \quad u = g(x) = \dfrac{1}{1 - x}, \quad x = -1$

49. $f(u) = \cot \dfrac{\pi u}{10}, \quad u = g(x) = 5\sqrt{x}, \quad x = 1$

50. $f(u) = u + \dfrac{1}{\cos^2 u}, \quad u = g(x) = \pi x, \quad x = 1/4$

51. $f(u) = \dfrac{2u}{u^2 + 1}, \quad u = g(x) = 10x^2 + x + 1, \quad x = 0$

52. $f(u) = \left(\dfrac{u - 1}{u + 1}\right)^2, \quad u = g(x) = \dfrac{1}{x^2} - 1, \quad x = -1$

What happens if you can write a function as a composite in different ways? Do you get the same derivative each time? The Chain Rule says you should. Try it with the functions in Exercises 53–56.

53. Find dy/dx if $y = \cos(6x + 2)$ by writing y as a composite with
  a) $y = \cos u$ and $u = 6x + 2$;
  b) $y = \cos 2u$ and $u = 3x + 1$.

54. Find dy/dx if $y = \sin(x^2 + 1)$ by writing y as a composite with
  a) $y = \sin(u + 1)$ and $u = x^2$;
  b) $y = \sin u$ and $u = x^2 + 1$.

55. Find dy/dx if $y = x$ by writing y as the composite of
  a) $y = (u/5) + 7$ and $u = 5x - 35$;
  b) $y = 1 + (1/u)$ and $u = 1/(x - 1)$.

56. Find dy/dx if $y = \sin(\sin(2x))$ by writing y as the composite of
  a) $y = \sin u$ and $u = \sin 2x$;
  b) $y = \sin(\sin u)$ and $u = 2x$.

57. Find ds/dt when $\theta = 3\pi/2$ if $s = \cos \theta$ and $d\theta/dt = 5$.

58. Find dy/dt when $x = 1$ if $y = x^2 + 7x - 5$ and $dx/dt = 1/3$.

59. What is the largest value the slope of the curve $y = \sin(x/2)$ can ever have?

60. Write an equation for the tangent to the curve $y = \sin mx$ at the origin.

61. Find the lines that are tangent and normal to the curve $y = 2 \tan(\pi x/4)$ at $x = 1$.

62. *Orthogonal curves.* Two curves are said to cross at right angles if their tangents are perpendicular at the crossing point. The technical word for "crossing at right angles" is *orthogonal*. Show that the curves $y = \sin 2x$ and $y = -\sin(x/2)$ are orthogonal at the origin (Fig. 2.27).

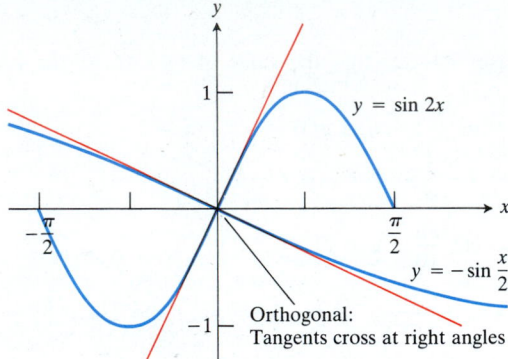

2.27 The curves in Exercise 62.

63. Suppose that functions f and g and their derivatives with respect to x have the following values at $x = 2$ and $x = 3$.

x	$f(x)$	$g(x)$	$f'(x)$	$g'(x)$
2	8	2	1/3	−3
3	3	−4	2π	5

Find the derivatives with respect to x of the following combinations at the given value of x.
a) $2f(x)$ at $x = 2$
b) $f(x) + g(x)$ at $x = 3$
c) $f(x) \cdot g(x)$ at $x = 3$
d) $f(x)/g(x)$ at $x = 2$
e) $f(g(x))$ at $x = 2$
f) $\sqrt{f(x)}$ at $x = 2$
g) $1/g^2(x)$ at $x = 3$
h) $\sqrt{f^2(x) + g^2(x)}$ at $x = 2$

64. Suppose that the functions f and g and their derivatives with respect to x have the following values at $x = 0$ and $x = 1$.

x	$f(x)$	$g(x)$	$f'(x)$	$g'(x)$
0	1	1	5	1/3
1	3	−4	−1/3	−8/3

Find the derivatives with respect to x of the following combinations at the given value of x.
a) $5f(x) - g(x)$, $x = 1$
b) $f(x)\,g^3(x)$, $x = 0$
c) $\dfrac{f(x)}{g(x) + 1}$, $x = 1$
d) $f(g(x))$, $x = 0$
e) $g(f(x))$, $x = 0$
f) $(x^{11} + f(x))^{-2}$, $x = 1$
g) $f(x + g(x))$, $x = 0$

65. *Running machinery too fast.* Suppose that a piston is moving straight up and down and that its position at time t sec is

$$s = A\cos(2\pi bt),$$

with A and b positive. The value of A is the amplitude of the motion, and b is the frequency (number of times the piston moves up and down each second). What effect does doubling the frequency have on the piston's velocity and acceleration? (Once you find out, you will know why machinery breaks when you run it too fast.)

66. *Temperatures in Fairbanks, Alaska.* The graph in Fig. 2.28 shows the average Fahrenheit temperature in Fairbanks, Alaska, during a typical 365-day year. The equation that gives the temperature on day x is

$$y = 37\sin\left[\frac{2\pi}{365}(x - 101)\right] + 25.$$

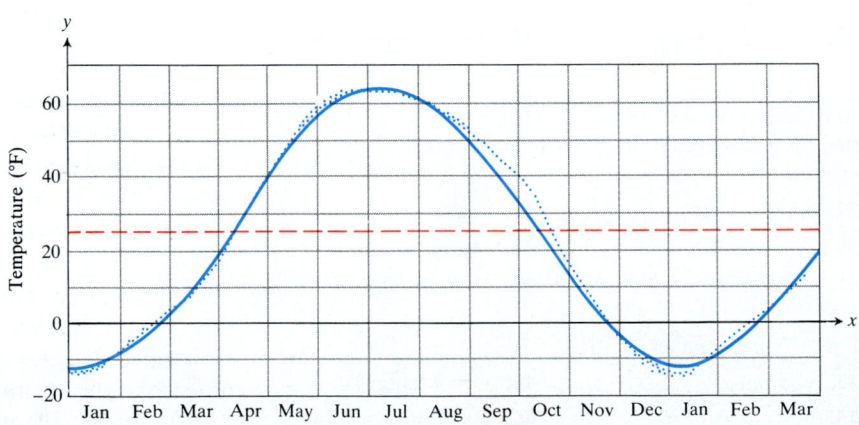

2.28 Normal mean air temperatures at Fairbanks, Alaska, plotted as data points. The approximating sine function is

$$f(x) = 37\sin\left[\frac{2\pi}{365}(x - 101)\right] + 25.$$

a) On what day is the temperature increasing the fastest?

b) CALCULATOR About how many degrees per day is the temperature increasing when it is increasing at its fastest?

67. A particle moves along the x-axis with velocity $dx/dt = f(x)$. Show that the particle's acceleration is $f(x)f'(x)$.

68. Suppose that $f(x) = x^2$ and $g(x) = |x|$. Then the composites
$$(f \circ g)(x) = |x|^2 = x^2 \quad \text{and} \quad (g \circ f)(x) = |x^2| = x^2$$
are both differentiable at $x = 0$ even though g itself is not differentiable at $x = 0$. Does this contradict the Chain Rule? Explain.

Computer Grapher or Graphing Calculator

69. *The derivative of* $\sin(x^2)$

a) Graph $\sin(x^2)$ and its derivative $2x \cos(x^2)$ together for $-2 \le x \le 4.8$. Notice how the derivative portrays the steepening of the graph of $\sin(x^2)$ as x moves away from the origin.

b) Graph $2x \cos(x^2)$ for $-2 \le x \le 4.8$. Then, on the same screen, graph
$$y = \frac{\sin[(x + h)^2] - \sin(x^2)}{h}$$
for $h = 0.5$ and 0.1. Use different colors, if available. Experiment with other values of h, including negative values. What do you see happening as $h \to 0$? Explain.

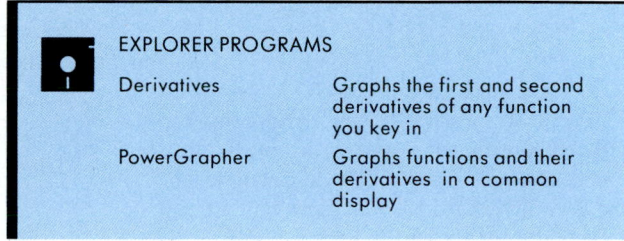

EXPLORER PROGRAMS

Derivatives	Graphs the first and second derivatives of any function you key in
PowerGrapher	Graphs functions and their derivatives in a common display

2.5 Implicit Differentiation and Fractional Powers

An equation like $y^5 + \sin xy = 0$ defines y as a differentiable function of x but does not let us solve for y in terms of x. However, we can still find dy/dx with a technique called *implicit differentiation*. This section describes the technique and uses it to extend the Power Rule for differentiation to include fractional exponents.

Implicit Differentiation

The graph of an equation $F(x, y) = 0$ is not the graph of a function of x if some of the vertical lines that cross it intersect it more than once. For example, the numbers y_1, y_2, and y_3 in Fig. 2.29 all correspond to the same x-value, $x = x_0$. However, various parts of a curve $F(x, y) = 0$ may well be the graphs of functions of x. The arc AB in Fig. 2.29 is the graph of a function of x, and so are arcs BC and CD.

Sometimes we can find explicit formulas for the functions defined by an equation $F(x, y) = 0$, but usually we cannot. The equation $F(x, y) = 0$ has defined the functions *implicitly* but not *explicitly*.

When may we expect the functions defined by an equation $F(x, y) = 0$ in a region of the xy-plane to be differentiable? The answer, from a theorem in advanced mathematics, is when F is continuous (in a sense to be described in Chapter 12) and the first derivatives of F with respect to each variable, with the other variable held fixed, are continuous throughout the region. Derivatives like these are important, and we shall have more to say about them in Chapter 12.

Example 1 Find dy/dx if $y^2 = x$.

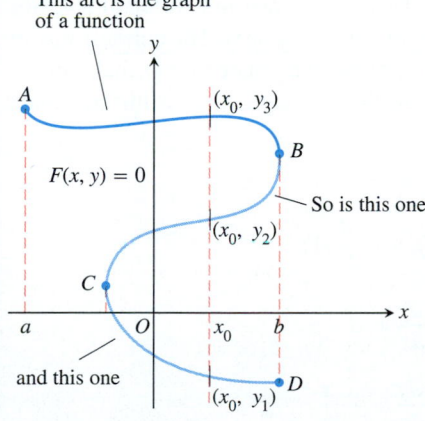

This arc is the graph of a function

2.29 As a whole, the curve $F(x, y) = 0$ is not the graph of a function of x. Some of the vertical lines that cross it intersect it more than once. However, the curve can be divided into separate arcs that *are* the graphs of functions of x.

Solution The equation $y^2 = x$ defines two differentiable functions of x, namely $y = \sqrt{x}$ and $y = -\sqrt{x}$ (Fig. 2.30). We know how to find the derivative of each of these for $x > 0$, from Example 8 in Section 1.6. But suppose we knew only that the equation $y^2 = x$ defined y as one or more differentiable functions of x for $x > 0$ without knowing exactly what these functions were. Could we still find dy/dx?

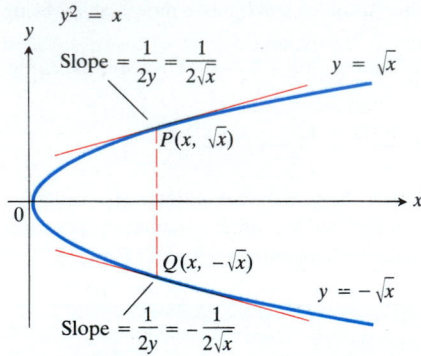

2.30   The equation $y^2 - x = 0$, or $y^2 = x$, as it is usually written, defines two differentiable functions of x on the interval $x \geq 0$. Example 1 shows how to calculate the derivatives of these functions without solving the equation $y^2 = x$ for y.

The answer in this case is yes. To find dy/dx we simply differentiate both sides of the equation $y^2 = x$ with respect to x, treating y as a differentiable implicit function of x. When we do this, we get

$$y^2 = x$$

$$2y\,\frac{dy}{dx} = 1 \qquad \left(\begin{array}{l}\text{The Chain Rule gives} \\ (d/dx)y^2 = 2y\,(dy/dx).\end{array}\right)$$

$$\frac{dy}{dx} = \frac{1}{2y}.$$

How does this compare with what happens when we solve $y^2 = x$ for y first and then differentiate?

With $y = \sqrt{x}$	With $y = -\sqrt{x}$
$\dfrac{dy}{dx} = \dfrac{1}{2\sqrt{x}}$	$\dfrac{dy}{dx} = -\dfrac{1}{2\sqrt{x}}$
$= \dfrac{1}{2y}$	$= \dfrac{1}{2(-\sqrt{x})}$
	$= \dfrac{1}{2y}$

In both cases, the derivative is given by the formula we obtained without solving for y, so the two methods agree.

Example 2   The graph of $F(x, y) = x^2 + y^2 - 1 = 0$ is the circle $x^2 + y^2 = 1$. Taken as a whole, the circle is not the graph of any single function of x (Fig. 2.31). Each x in the interval $-1 < x < 1$ gives two values of y, namely $y = \sqrt{1 - x^2}$ and $y = -\sqrt{1 - x^2}$, instead of the required single value.

The upper and lower semicircles are the graphs of the functions $f(x) = \sqrt{1 - x^2}$ and $g(x) = -\sqrt{1 - x^2}$. These functions are differentiable for $|x| < 1$ because they are composites of differentiable functions. The quickest way to find their derivatives, however, is not to differentiate the square root formulas but to differentiate both sides of the original equation, treating y as a differentiable implicit function of x:

$$x^2 + y^2 = 1$$

$$\frac{d}{dx}(x^2) + \frac{d}{dx}(y^2) = \frac{d}{dx}(1)$$

$$2x + 2y\,\frac{dy}{dx} = 0$$

$$\frac{dy}{dx} = -\frac{x}{y}$$

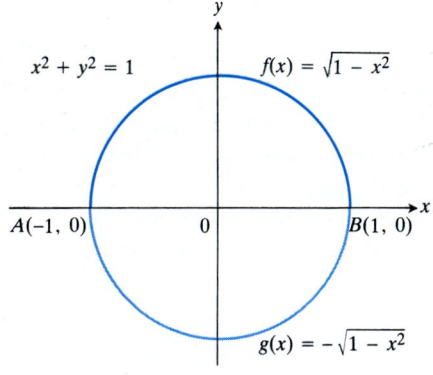

2.31   The graph of the equation $F(x, y) = x^2 + y^2 - 1 = 0$ is the complete circle $x^2 + y^2 = 1$. The upper semicircle AB is the graph of the function $f(x) = \sqrt{1 - x^2}$. The lower semicircle AB is the graph of $g(x) = -\sqrt{1 - x^2}$.

This formula for dy/dx is simpler than either of the formulas we would get by differentiating f and g and holds for all points on the curve above or below the x-axis. The formula is also easy to evaluate at any such point. At $(\sqrt{2}/2, \sqrt{2}/2)$, for instance,

$$\frac{dy}{dx} = -\frac{\sqrt{2}/2}{\sqrt{2}/2} = -1.$$

Implicit Differentiation Takes Four Steps

1. Differentiate both sides of the equation with respect to x.
2. Collect the terms with dy/dx on one side of the equation.
3. Factor out dy/dx.
4. Solve for dy/dx by dividing.

To calculate the derivatives of other implicitly defined functions we simply proceed as in Examples 1 and 2: We treat y as a differentiable implicit function of x and apply the already familiar rules of differentiation to differentiate both sides of the defining equation. This procedure is called **implicit differentiation.**

Example 3   Find dy/dx if $2y = x^2 + \sin y$.

Solution

$$2y = x^2 + \sin y$$

$$\frac{d}{dx}(2y) = \frac{d}{dx}(x^2) + \frac{d}{dx}(\sin y) \qquad \left(\begin{array}{l}\text{Differentiate both}\\\text{sides with respect to } x.\end{array}\right)$$

$$2\frac{dy}{dx} = 2x + \cos y \frac{dy}{dx}$$

$$2\frac{dy}{dx} - \cos y \frac{dy}{dx} = 2x$$

$$(2 - \cos y)\frac{dy}{dx} = 2x$$

$$\frac{dy}{dx} = \frac{2x}{2 - \cos y}$$

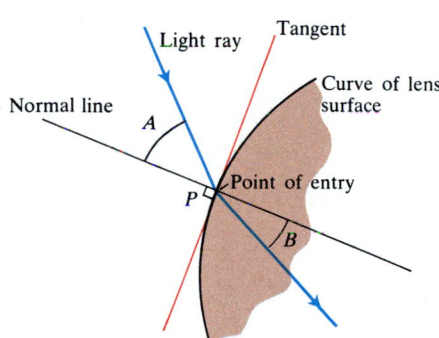

2.32   The profile or cutaway view of a lens, showing the bending (refraction) of a light ray as it passes through the lens surface.

Lenses, Tangents, and Normal Lines

In the law that describes how light changes direction as it enters a lens, the important angles are the angles the light makes with the line perpendicular to the surface of the lens at the point of entry (angles A and B in Fig. 2.32). This line is called the *normal to the surface* at the point of entry. In a profile view of a lens like the one in Fig. 2.32, the normal is the line perpendicular to the tangent to the profile curve at the point of entry.

The profiles of lenses are often described by quadratic curves. When they are, we can use implicit differentiation to find the tangents and normals.

Example 4   Find the tangent and normal to the curve $x^2 - xy + y^2 = 7$ at the point $(-1, 2)$. (See Fig. 2.33.)

Solution   We first use implicit differentiation to find dy/dx:

$$x^2 - xy + y^2 = 7$$

$$\frac{d}{dx}(x^2) - \frac{d}{dx}(xy) + \frac{d}{dx}(y^2) = \frac{d}{dx}(7) \qquad \left(\begin{array}{l}\text{Differentiate both sides}\\\text{with respect to } x \ldots\end{array}\right.$$

$$2x - \left(x\frac{dy}{dx} + y\frac{dx}{dx}\right) + 2y\frac{dy}{dx} = 0 \qquad \left.\begin{array}{l}\ldots \text{treating } xy \text{ as a product}\\\text{and } y^2 \text{ as a power.}\end{array}\right)$$

$$(2y - x)\frac{dy}{dx} = y - 2x \qquad \text{(Collect terms.)}$$

$$\frac{dy}{dx} = \frac{y - 2x}{2y - x} \qquad \text{(Solve as usual.)}$$

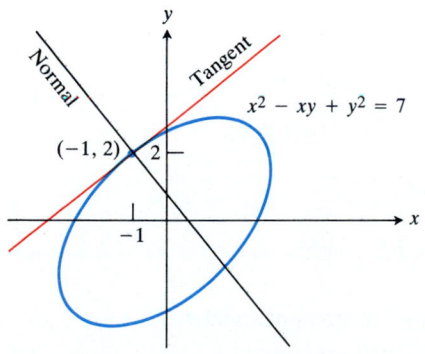

2.33   The graph of $x^2 - xy + y^2 = 7$ is an ellipse. The slope of the curve at the point $(-1, 2)$ is $(dy/dx)_{(-1, 2)} = 4/5$ (Example 4).

Helga von Koch's Snowflake Curve (1904)

Start with an equilateral triangle, calling it Curve 1. On the middle third of each side, build an equilateral triangle pointing outward. Then erase the interiors of the old middle thirds. Call the expanded curve Curve 2. Now put equilateral triangles, again pointing outward, on the middle thirds of the sides of Curve 2. Erase the interiors of the old middle thirds to make Curve 3. Repeat the process, as shown, to define an infinite sequence of plane curves. The limit curve of the sequence is Koch's snowflake curve.

The snowflake curve is too rough to have a tangent at any point. In other words, the equation $F(x, y) = 0$ defining the curve does not define y as a differentiable function of x or x as a differentiable function of y at any point. We shall encounter the snowflake curve again when we study length in Section 5.4.

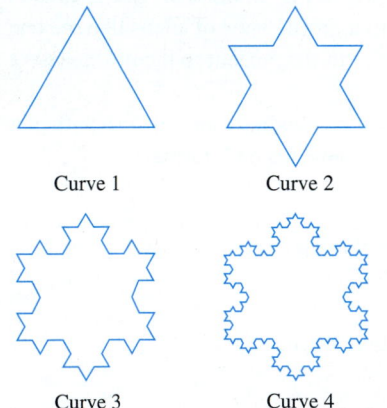

Curve 1   Curve 2

Curve 3   Curve 4

We then evaluate the derivative at $x = -1$, $y = 2$, to obtain

$$\frac{dy}{dx}\bigg|_{(-1, 2)} = \frac{y - 2x}{2y - x}\bigg|_{(-1, 2)}$$

$$= \frac{2 - 2(-1)}{2(2) - (-1)} = \frac{4}{5}.$$

The tangent to the curve at $(-1, 2)$ is

$$y - 2 = \frac{4}{5}(x - (-1))$$

$$y = \frac{4}{5}x + \frac{14}{5}.$$

The normal to the curve at $(-1, 2)$ is

$$y - 2 = -\frac{5}{4}(x + 1)$$

$$y = -\frac{5}{4}x + \frac{3}{4}.$$

Using Implicit Differentiation to Find Derivatives of Higher Order

Implicit differentiation can also produce derivatives of higher order. Here is an example.

Example 5   Find d^2y/dx^2 if $2x^3 - 3y^2 = 7$.

Solution   To start, we differentiate both sides of the equation with respect to x to find $y' = dy/dx$:

$$2x^3 - 3y^2 = 7$$

$$\frac{d}{dx}(2x^3) - \frac{d}{dx}(3y^2) = \frac{d}{dx}(7)$$

$$6x^2 - 6yy' = 0 \tag{1}$$

$$x^2 - yy' = 0$$

$$y' = \frac{x^2}{y} \qquad (\text{if } y \neq 0).$$

We now apply the Quotient Rule to find y'':

$$y'' = \frac{d}{dx}\left(\frac{x^2}{y}\right) = \frac{2xy - x^2y'}{y^2} = \frac{2x}{y} - \frac{x^2}{y^2}y'. \tag{2}$$

Finally, we substitute $y' = x^2/y$ to express y'' in terms of x and y:

$$y'' = \frac{2x}{y} - \frac{x^2}{y^2}\left(\frac{x^2}{y}\right) = \frac{2x}{y} - \frac{x^4}{y^3}. \tag{3}$$

The second derivative is not defined at $y = 0$ but is given by Eq. (3) when $y \neq 0$.

Fractional Powers of Differentiable Functions

We know that the Power Rule

$$\frac{d}{dx} u^n = nu^{n-1} \frac{du}{dx} \tag{4}$$

holds when n is an integer. Our goal now is to show that it holds when n is a fraction. We will then be able to differentiate functions like

$$y = x^{4/3} \quad \text{and} \quad y = (\cos x)^{-1/5}$$

that were beyond our reach before.

Power Rule for Fractional Exponents

If n is any rational number, then

$$\frac{d}{dx} x^n = nx^{n-1}, \tag{5}$$

provided $x \neq 0$ if $n - 1 < 0$ (i.e., if $n < 1$).

If n is a rational number and u is a differentiable function of x, then u^n is a differentiable function of x and

$$\frac{d}{dx} u^n = nu^{n-1} \frac{du}{dx}, \tag{6}$$

provided $u \neq 0$ if $n < 1$.

The restrictions $x \neq 0$ if $n < 1$ and $u \neq 0$ if $n < 1$ are there to protect against inadvertent attempts to divide by zero. There is nothing mysterious about these restrictions. They come up quite naturally in practice, as the next example shows.

Example 6

a) $\dfrac{d}{dx}(x^{1/2}) = \dfrac{1}{2}x^{-1/2} = \dfrac{1}{2\sqrt{x}}$ $\left(\text{Eq. (5) with } n = \dfrac{1}{2}\right)$

function defined for $x \geq 0$ derivative defined only for $x > 0$

b) $\dfrac{d}{dx}(x^{1/5}) = \dfrac{1}{5}x^{-4/5}$ $\left(\text{Eq. (5) with } n = \dfrac{1}{5}\right)$

function defined for all x derivative not defined at $x = 0$

c) $\dfrac{d}{dx}(1 - x^2)^{1/2} = \dfrac{1}{2}(1 - x^2)^{-1/2}(-2x)$ $\left(\begin{array}{l}\text{Eq. (6) with } u = 1 - x^2 \\ \text{and } n = \dfrac{1}{2}\end{array}\right)$

function defined for $-1 \leq x \leq 1$

$$= \frac{-x}{(1 - x^2)^{1/2}}$$

derivative defined only for $-1 < x < 1$

The derivatives of the functions $x^{4/3}$ and $(\cos x)^{-1/5}$ are defined wherever the functions themselves are defined, as we see in the next example.

Example 7

a) $\dfrac{d}{dx} x^{4/3} = \dfrac{4}{3} x^{1/3}$

b) $\dfrac{d}{dx} (\cos x)^{-1/5} = -\dfrac{1}{5} (\cos x)^{-6/5} \dfrac{d}{dx} (\cos x)$

$\qquad\qquad\qquad\quad = -\dfrac{1}{5} (\cos x)^{-6/5} (-\sin x)$

$\qquad\qquad\qquad\quad = \dfrac{1}{5} \sin x (\cos x)^{-6/5}$

Proof of the Power Rule for Fractional Exponents We prove Eq. (5) first and then apply the Chain Rule to get Eq. (6).

To prove Eq. (5), let p and q be integers with $q > 0$ and suppose that $y = x^{p/q}$. Then

$$y^q = x^p.$$

This equation is an algebraic combination of powers of x and y, so the advanced theorem we mentioned at the beginning of the section assures us that y is a differentiable function of x. Since p and q are integers (for which we already have the Power Rule), we can differentiate both sides of the equation with respect to x and obtain

$$qy^{q-1} \frac{dy}{dx} = px^{p-1}.$$

Hence, if $y \neq 0$,

$$\frac{dy}{dx} = \frac{p}{q} \frac{x^{p-1}}{y^{q-1}} = \frac{p}{q} \frac{x^{p-1}}{(x^{p/q})^{q-1}} = \frac{p}{q} \frac{x^{p-1}}{x^{p-p/q}} = \frac{p}{q} x^{(p/q)-1}.$$

This proves Eq. (5).

To prove Eq. (6), we let $y = u^{p/q}$ and apply the Chain Rule in the form

$$\frac{dy}{dx} = \frac{dy}{du} \frac{du}{dx}.$$

From Eq. (5), $(d/du)u^{p/q} = (p/q)u^{(p/q)-1}$. Hence

$$\frac{dy}{dx} = \frac{p}{q} u^{(p/q)-1} \frac{du}{dx}$$

and we're done.

EXERCISES 2.5

Find dy/dx in Exercises 1–18.

1. $y = x^{9/4}$

2. $y = x^{-3/5}$

3. $y = \sqrt[3]{x}$

4. $y = \sqrt[4]{x}$

5. $y = (2x + 5)^{-1/2}$

6. $y = (1 - 6x)^{2/3}$

7. $y = x\sqrt{x^2 + 1}$

8. $y = \dfrac{x}{\sqrt{x^2 + 1}}$

9. $x^2y + xy^2 = 6$

10. $x^3 + y^3 = 18xy$

11. $2xy + y^2 = x + y$

12. $x^3 - xy + y^3 = 1$

13. $x^2y^2 = x^2 + y^2$

14. $(3x + 7)^2 = 2y^3$

15. $x^3 + 3x^2y + 3xy^2 + y^3 = 0$

16. $x^2 = \dfrac{x - y}{x + y}$

17. $y = \sqrt{1 - \sqrt{x}}$

18. $y = 3(2x^{-1/2} + 1)^{-1/3}$

Find dy/dx in Exercises 19–26.

19. $y = \sqrt{1 + \cos 2x}$

20. $y = \sqrt{\sec 2x}$

21. $y = 3(\csc x)^{3/2}$

22. $y = [\sin(x + 5)]^{5/4}$

23. $x = \tan y$

24. $x = \sin y$

25. $x + \tan(xy) = 0$

26. $x + \sin y = xy$

In Exercises 27–30, use implicit differentiation to find dy/dx and then d^2y/dx^2.

27. $x^2 + y^2 = 1$

28. $x^{2/3} + y^{2/3} = 1$

29. $y^2 = x^2 + 2x$

30. $y^2 + 2y = 2x + 1$

31. If $x^3 + y^3 = 16$, find the value of d^2y/dx^2 at the point $(2, 2)$.

32. If $xy + y^2 = 1$, find the value of d^2y/dx^2 at the point $(0, -1)$.

In Exercises 33–36, find the lines that are (a) tangent and (b) normal to the curve at the given point.

33. $x^2 + xy - y^2 = 1$, $(2, 3)$

34. $x^2 + y^2 = 25$, $(3, -4)$

35. $x^2y^2 = 9$, $(-1, 3)$

36. $y^2 - 2x - 4y - 1 = 0$, $(-2, 1)$

37. Find the two points where the curve $x^2 + xy + y^2 = 7$ crosses the x-axis, and show that the tangents to the curve at these points are parallel. What is the common slope of these tangents?

38. Find points on the curve $x^2 + xy + y^2 = 7$ (a) where the tangent is parallel to the x-axis and (b) where the tangent is parallel to the y-axis. (In the latter case, dy/dx is not defined, but dx/dy is. What value does dx/dy have at these points?)

39. Assume that the equation $2xy + \pi \sin y = 2\pi$ defines y as a differentiable function of x. Find dy/dx when $x = 1$ and $y = \pi/2$.

40. Find an equation for the tangent to the curve $x \sin 2y = y \cos 2x$ at the point $(\pi/4, \pi/2)$.

41. *The eight curve.* Find the slopes of the figure 8–shaped curve $y^4 = y^2 - x^2$ at the two points shown in Fig. 2.34.

42. *The cissoid of Diocles (c. 200 B.C.).* Find equations for the tangent and normal to the cissoid of Diocles $y^2(2 - x) = x^3$ (Fig. 2.35) at the point $(1, 1)$.

43. Which of the following could be true if $f''(x) = x^{-1/3}$?

a) $f(x) = \dfrac{3}{2}x^{2/3} - 3$

b) $f(x) = \dfrac{9}{10}x^{5/3} - 7$

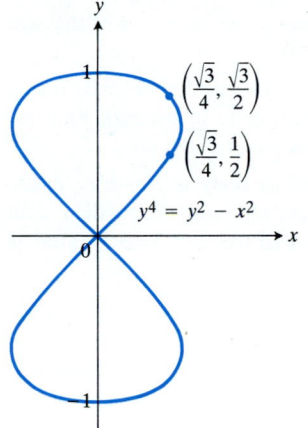

2.34   The curve in Exercise 41.

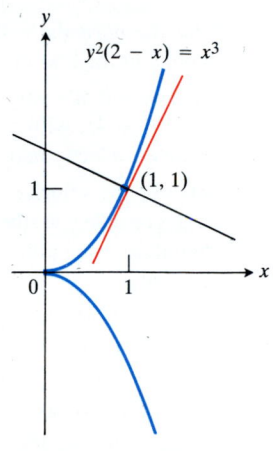

2.35   The cissoid of Diocles (Exercise 42).

c) $f'''(x) = -\dfrac{1}{3}x^{-4/3}$

d) $f'(x) = \dfrac{3}{2}x^{2/3} + 6$

44. The line that is normal to the curve $x^2 + 2xy - 3y^2 = 0$ at $(1, 1)$ intersects the curve at what other point?

45. Show that if it is possible to draw three normals from the point $(a, 0)$ to the parabola $x = y^2$ (Fig. 2.36), then a must be greater than $1/2$. One of the normals is the x-axis. For what value of a are the other two normals perpendicular?

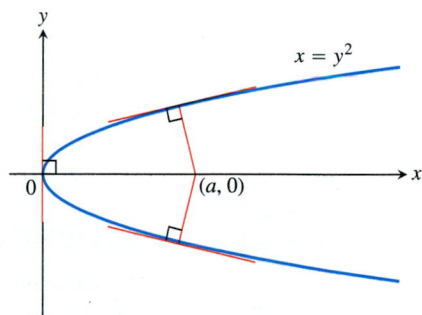

2.36   The parabola in Exercise 45.

46. *Orthogonal curves.* Two curves are *orthogonal* at a point of intersection if their tangents there cross at right angles. Show that the curves $2x^2 + 3y^2 = 5$ and $y^2 = x^3$ (Fig. 2.37) are orthogonal at $(1, 1)$ and $(1, -1)$.

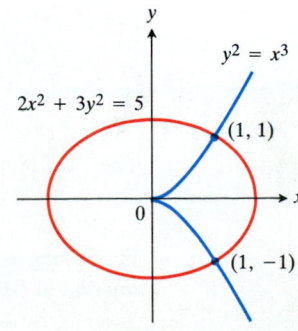

2.37   The curves in Exercise 46.

47. Find the normals to the curve $xy + 2x - y = 0$ that are parallel to the line $2x + y = 0$.

48. The position of a particle moving along a coordinate line is $s = \sqrt{1 + 4t}$, with s in meters and t in seconds. Find the particle's velocity and acceleration at $t = 6$ sec.

49. Suppose the velocity of a falling body is $v = k\sqrt{s}$ m/sec (k a constant) at the instant the body has fallen s m from its starting point. Show that the body's acceleration is constant.

50. The velocity of a heavy meteorite entering the earth's atmosphere is inversely proportional to $\sqrt{s}$ when it is s km from the earth's center. Show that the meteorite's acceleration is inversely proportional to s^2.

51. *Temperature and the period of a pendulum.*   For oscillations of small amplitude (short swings), we may safely model the relationship between the period T and the length L of a simple pendulum with the equation

$$T = 2\pi\sqrt{\frac{L}{g}},$$

where g is the constant acceleration of gravity at the pendulum's location. If we measure g in centimeters per second squared, we measure L in centimeters and T in seconds. If the pendulum is made of metal, its length will vary with temperature, either increasing or decreasing at a rate that is roughly proportional to L. In symbols, with u being temperature and k the proportionality constant,

$$\frac{dL}{du} = kL.$$

Assuming this to be the case, show that the rate at which the period changes with respect to temperature is $kT/2$.

2.6 Linearization and Differentials

Sometimes we can approximate complicated functions with simpler ones that give the accuracy we want for specific applications without being so hard to work with. It is important to know how to do this, and in this section we study the simplest of the useful approximations. For reasons that will be clear in a moment, the approximation is called a *linearization*. It is based on the ideas of derivative and tangent line.

We also introduce a new symbol, dx, for an increment in a variable x. This symbol is called the differential of x. In the physical sciences, it is used more frequently than Δx. In mathematics, differentials are used to estimate changes in function values, as we shall see toward the end of this section.

Linearizations Are Linear Replacement Formulas

As you can see in Fig. 2.38, the tangent to a curve $y = f(x)$ lies close to the curve near the point of tangency. For a brief interval to either side, the y-values along the tangent line give good approximations to the y-values on the curve. Therefore, to simplify the expression for the function near this point, we propose to replace the formula for f over this interval by the formula for its tangent line.

In the notation of Fig. 2.39, the tangent passes through the point $(a, f(a))$ with slope $f'(a)$, so its point–slope equation is

$$y - f(a) = f'(a)(x - a),$$

or

$$y = f(a) + f'(a)(x - a). \tag{1}$$

Thus, the tangent line is the graph of the function

$$L(x) = f(a) + f'(a)(x - a). \tag{2}$$

For as long as the line remains close to the graph of f, $L(x)$ will give a good approximation to $f(x)$.

2.38 The more we magnify the graph of a function near a point where the function is differentiable, the flatter the graph becomes and the more it resembles its tangent. You can see this happening in the four views of the curve $y = x^2$ shown here. As the magnification increases in the vicinity of the point (1, 1), the curve flattens and comes more and more to resemble its tangent line $y = 2x - 1$. Indeed, in the fourth frame, our computer no longer shows any difference between the two.

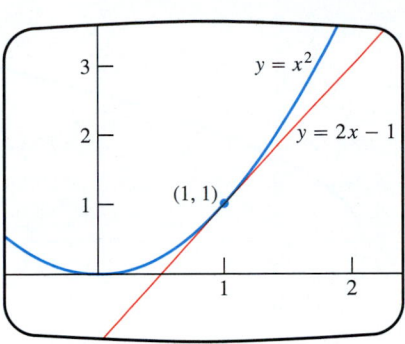

$y = x^2$ and its tangent $y = 2x - 1$ at (1, 1).

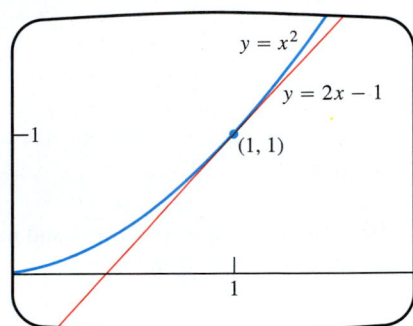

Tangent and curve very close near (1, 1).

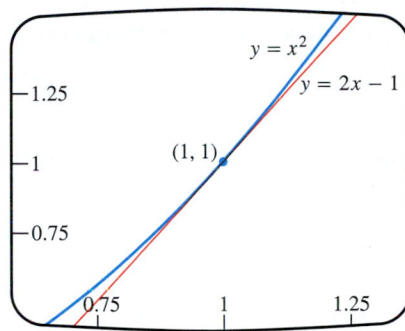

Tangent and curve very close throughout entire x-interval shown.

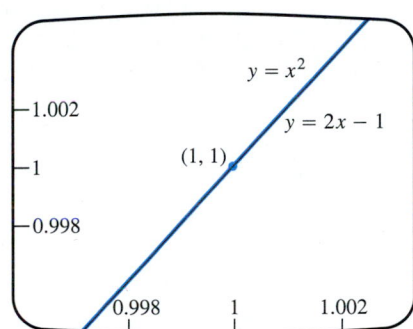

Tangent and curve closer still. Computer screen cannot distinguish tangent from curve on this x-interval.

DEFINITIONS

Linearization and Standard Linear Approximation

If $y = f(x)$ is differentiable at $x = a$, then

$$L(x) = f(a) + f'(a)(x - a) \tag{3}$$

is the **linearization** of f at a. The approximation

$$f(x) \approx L(x)$$

is the **standard linear approximation** of f at a. The point a is the **center** of the approximation.

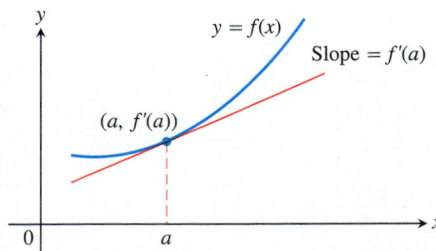

2.39 The equation of the tangent line is $y = f(a) + f'(a)(x - a)$.

Example 1   Find the linearization of $f(x) = \sqrt{1 + x}$ at $x = 0$.

Solution   We evaluate Eq. (3) for $f(x) = \sqrt{1 + x}$ and $a = 0$.
The derivative of f is

$$f'(x) = \frac{1}{2}(1 + x)^{-1/2} = \frac{1}{2\sqrt{1 + x}}.$$

Its value at $x = 0$ is 1/2. We substitute this along with $a = 0$ and $f(0) = 1$ into Eq. (3):

$$L(x) = f(a) + f'(a)(x - a) = 1 + \frac{1}{2}(x - 0) = 1 + \frac{x}{2}.$$

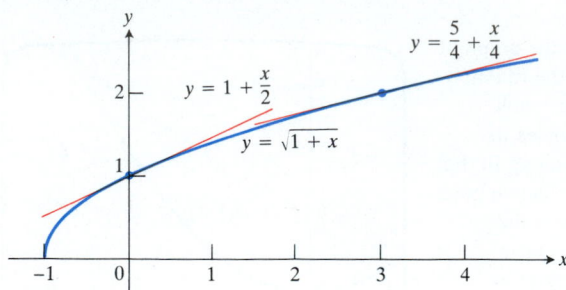

2.40 The graph of $y = \sqrt{1 + x}$ and its linearizations at $x = 0$ and $x = 3$.

The linearization of $\sqrt{1 + x}$ at $x = 0$ is $L(x) = 1 + \dfrac{x}{2}$. See Fig. 2.40.

In case you are wondering how close the approximation

$$\sqrt{1 + x} \approx 1 + \frac{x}{2}$$

really is, we can try a few values with a calculator:

$$\sqrt{1.2} \approx 1 + \frac{0.2}{2} = 1.10, \qquad \text{(Accurate to 2 decimals)}$$

$$\sqrt{1.05} \approx 1 + \frac{0.05}{2} = 1.025, \qquad \text{(Accurate to 3 decimals)}$$

$$\sqrt{1.005} \approx 1 + \frac{0.005}{2} = 1.0025 \ 0. \qquad \text{(Accurate to 5 decimals)}$$

The approximation becomes more accurate as we move toward the center, $x = 0$, and less accurate as we move away. As Fig. 2.40 suggests, the approximation will probably be too crude to be useful if we move out as far, say, as $x = 3$. To approximate $\sqrt{1 + x}$ near $x = 3$, we had best find its linearization at $x = 3$.

Example 2   Find the linearization of $f(x) = \sqrt{1 + x}$ at $x = 3$.

Solution   We evaluate Eq. (3) for $f(x) = \sqrt{1 + x}$, $f'(x) = 1/(2\sqrt{1 + x})$, and $a = 3$. With

$$f(3) = 2, \qquad f'(3) = \frac{1}{2\sqrt{1 + 3}} = \frac{1}{4}.$$

Eq. (3) gives

$$L(x) = 2 + \frac{1}{4}(x - 3) = 2 + \frac{x}{4} - \frac{3}{4} = \frac{5}{4} + \frac{x}{4}.$$

Thus, near $x = 3$,

$$\sqrt{1 + x} \approx \frac{5}{4} + \frac{x}{4}.$$

At $x = 3.2$, this linearization gives

$$\sqrt{1 + x} = \sqrt{1 + 3.2} \approx \frac{5}{4} + \frac{3.2}{4} = 1.250 + 0.800 = 2.050,$$

which differs from $\sqrt{4.2} = 2.04939$ by less than one thousandth. The linearization from Example 1 gives

$$\sqrt{1 + x} = \sqrt{1 + 3.2} \approx 1 + \frac{3.2}{2} = 1 + 1.6 = 2.6,$$

a result that is off by more than 25%. The linearization at $x = 3$ is clearly the one to use for values of $\sqrt{1 + x}$ near 3.

Do not be misled by the calculations here into thinking that whatever we do with a linearization is better done with a calculator. In practice, we would never use a linearization to find the value of a particular square root. That is not what linearizations are for. The utility of the linearizations in Examples 1 and 2 lies in their ability to replace the complicated formula $\sqrt{1 + x}$ by a simpler formula. If we have to work with $\sqrt{1 + x}$ for values of x close to 0 and can tolerate the small amount of error involved, we can safely work with $1 + (x/2)$ instead. Of course, we then need to know just how much error there really is. We shall look at that in a moment but will not have the full answer until Chapter 3.

Common Linear Approximations for x Numerically Small

$$\sin x \approx x$$

$$\cos x \approx 1$$

$$\tan x \approx x$$

(See Exercises 13 and 14.)

Example 3   The most important linearization for roots and powers is

$$(1 + x)^k \approx 1 + kx \qquad (x \approx 0; \text{ any number } k) \tag{4}$$

(Exercise 22). This approximation, good for values of x near zero, has broad application.

Approximation (x numerically small)	Source: Eq. (4) with . . .
$\sqrt{1 + x} \approx 1 + \dfrac{x}{2}$	($k = 1/2$)
$\dfrac{1}{1 - x} = (1 - x)^{-1} \approx 1 + (-1)(-x) = 1 + x$	$\left(\begin{array}{l}k = -1; \ -x \text{ in} \\ \text{place of } x\end{array}\right)$
$\sqrt[3]{1 + 5x^4} = (1 + 5x^4)^{1/3} \approx 1 + \dfrac{1}{3}(5x^4) = 1 + \dfrac{5}{3}x^4$	$\left(\begin{array}{l}k = 1/3; \ 5x^4 \\ \text{in place of } x\end{array}\right)$
$\dfrac{1}{\sqrt{1 - x^2}} = (1 - x^2)^{-1/2} \approx 1 + \left(-\dfrac{1}{2}\right)\left(-x^2\right) = 1 + \dfrac{x^2}{2}$	$\left(\begin{array}{l}k = -1/2; \ -x^2 \\ \text{in place of } x\end{array}\right)$

Example 4   Find the linearization of $f(x) = \tan x$ at $x = 0$.

Solution   We use the equation

$$L(x) = f(a) + f'(a)(x - a)$$

with $f(x) = \tan x$ and $a = 0$. Since

$$f(0) = \tan(0) = 0, \qquad f'(0) = \sec^2(0) = 1,$$

we have $L(x) = 0 + 1(x - 0) = x$. Near $x = 0$ (Fig. 2.41),

$$\tan x \approx x.$$

Example 5   Find the linearization of $f(x) = \cos x$ at $x = \pi/2$.

Solution   We use the equation

$$L(x) = f(a) + f'(a)(x - a)$$

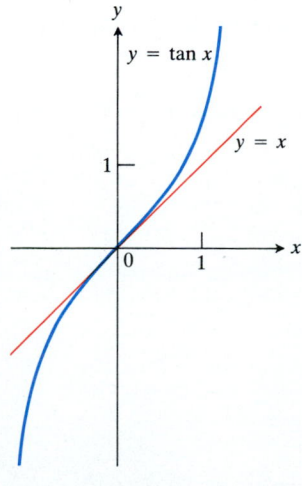

2.41   Near $x = 0$, $\tan x \approx x$.

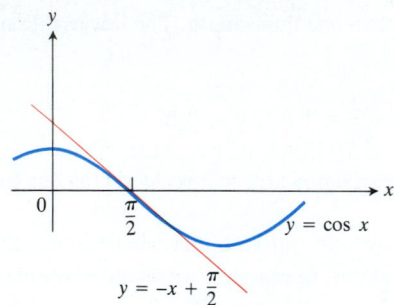

2.42 The graph of $y = \cos x$ and its linearization at $x = \pi/2$. Near $x = \pi/2$, $\cos x \approx -x + (\pi/2)$.

with $f(x) = \cos x$ and $a = \pi/2$. Since

$$f(\pi/2) = \cos(\pi/2) = 0, \qquad f'(\pi/2) = -\sin(\pi/2) = -1,$$

the linearization is

$$L(x) = 0 - 1\left(x - \frac{\pi}{2}\right) = -x + \frac{\pi}{2}.$$

See Fig. 2.42.

Estimating Change with Differentials

Suppose we know the value of a differentiable function $f(x)$ at a particular point x_0 and want to predict how much this value will change if we move nearby to the point $x_0 + h$. If h is small, f and its linearization L at x_0 will change by nearly the same amount. Since the values of L are always simple to calculate, calculating the change in L gives a practical way to estimate the change in f.

In the notation of Fig. 2.43, the change in f is

$$\Delta f = f(x_0 + h) - f(x_0).$$

The corresponding change in L is

$$\begin{aligned}
\Delta L &= L(x_0 + h) - L(x_0) \\
&= f(x_0) + f'(x_0)[(x_0 + h) - x_0] - f(x_0) \qquad (5) \\
&= f'(x_0)h.
\end{aligned}$$

The formula for Δf is usually as hard to work with as the formula for f. The formula for ΔL, however, is always simple to work with. As you can see, the change in L is just a constant times h.

The change $\Delta L = f'(x_0)h$ is usually described with the more suggestive notation

$$df = f'(x_0)\,dx, \qquad (6)$$

in which df denotes the change in the linearization of f that results from the change dx in x. We call dx the **differential** of x, and df the corresponding **differential** of f.

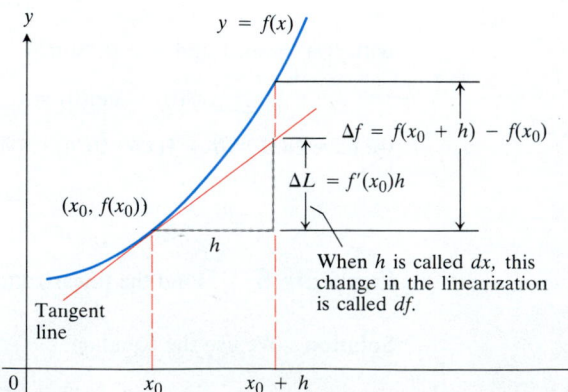

2.43 If h is small, the change in the linearization of f is nearly the same as the change in f.

If $y = f(x)$ and we divide both sides of the equation $dy = f'(x)\, dx$ by dx, we obtain the familiar equation

$$\frac{df}{dx} = f'(x).$$

This equation now says that we may regard the derivative df/dx as a quotient of differentials. In many calculations, it is convenient to be able to think this way. For example, in writing the Chain Rule as

$$\frac{dy}{dx} = \frac{dy}{du}\frac{du}{dx},$$

we can think of the derivatives on the right as quotients in which the du's cancel to produce the fraction on the left. This gives a quick check on whether we remembered the rule correctly.

Example 6   The radius of a circle increases from an initial value of $r_0 = 10$ by an amount $dr = 0.1$ (Fig. 2.44). Estimate the corresponding increase in the circle's area $A = \pi r^2$ by calculating dA. Compare dA with the true change ΔA.

Solution   To calculate dA, we apply Eq. (6) to the function $A = \pi r^2$:

$$dA = A'(r_0)dr = 2\pi r_0\, dr.$$

We then substitute the values $r_0 = 10$ and $dr = 0.1$:

$$dA = 2\pi(10)(0.1) = 2\pi.$$

The estimated change is 2π square units.

A direct calculation of ΔA gives

$$\Delta A = \pi(10.1)^2 - \pi(10)^2 = (102.01 - 100)\pi = \underbrace{2\pi}_{dA} + \underbrace{0.01\pi}_{\text{error}}.$$

The error in the estimate dA is 0.01π square unit. As a percentage of the circle's original area, the error is quite small, as we can see from the following calculation:

$$\frac{\text{error}}{\text{original area}} = \frac{0.01\pi}{100\pi} = 0.01\%.$$

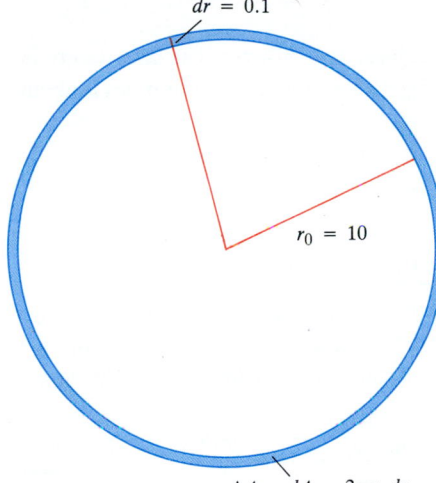

dr = 0.1

$r_0 = 10$

$\Delta A \approx dA = 2\pi r_0\, dr$

2.44   When dr is small compared with r_0, as it is when $dr = 0.1$ and $r_0 = 10$, the differential $dA = 2\pi r_0\, dr$ gives a good estimate of ΔA. (See Example 6.)

Absolute, Relative, and Percentage Change

What is the difference again between Δf and df? The increment Δf is the change in f; the differential df is the change in the linearization of f. Unlike Δf, the differential df is always simple to calculate, and it gives a good estimate of Δf when the change in x is small.

As we move from x_0 to a nearby point, we can describe the corresponding change in the value of f in three ways:

	True	Estimate
Absolute change	Δf	df
Relative change	$\dfrac{\Delta f}{f(x_0)}$	$\dfrac{df}{f(x_0)}$
Percentage change	$\dfrac{\Delta f}{f(x_0)} \times 100$	$\dfrac{df}{f(x_0)} \times 100$

Example 7   Estimate the percentage change that will occur in the area of a circle if its radius increases from $r_0 = 10$ units to 10.1 units.

Solution   From the preceding table, we have

$$\text{estimated percentage change} = \frac{dA}{A(r_0)} \times 100.$$

With $dA = 2\pi$ (from Example 6) and $A(r_0) = 100\pi$, the formula gives

$$\frac{dA}{A(r_0)} \times 100 = \frac{2\pi}{100\pi} \times 100 = 2\%.$$

Example 8   *The Earth's Surface Area.*   Suppose the earth were a perfect sphere and we determined its radius to be 3959 ± 0.1 miles. What effect would the tolerance of ± 0.1 have on our estimate of the earth's surface area?

Solution   The surface area of a sphere of radius r is $S = 4\pi r^2$. The uncertainty in the calculation of S that arises from measuring r with a tolerance of dr miles is about

$$dS = \left(\frac{dS}{dr}\right) dr = 8\pi r \, dr.$$

With $r = 3959$ and $dr = 0.1$,

$$dS = 8\pi(3959)(0.1) = 9950 \text{ mi}^2$$

to the nearest square mile, which is about the area of the state of Maryland (Fig. 2.45). In absolute terms this might seem like a large error. However, 9950 mi^2 is a relatively small error when compared with the calculated surface area of the earth:

$$\frac{dS}{\text{calculated } S} = \frac{9950}{4\pi(3959)^2} \approx \frac{9950}{196{,}961{,}284} \approx 0.005\%.$$

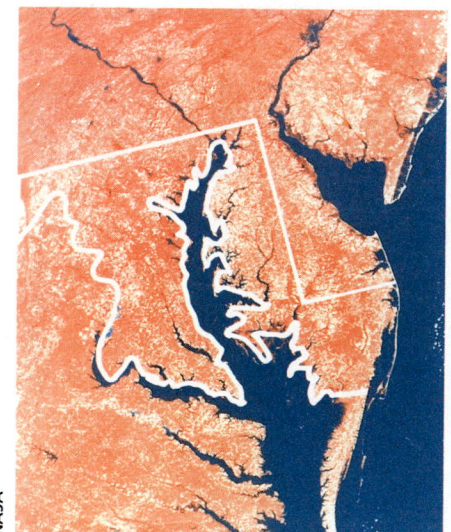

2.45   If we underestimated the radius of the earth by 528 ft during a calculation of the earth's surface area, we would leave out an area the size of the state of Maryland. (See Example 8.)

Example 9   About how accurately should we measure the radius r of a sphere to calculate the surface area $S = 4\pi r^2$ within 1% of its true value?

Solution   We want any inaccuracy in our measurement to be small enough to make the corresponding increment ΔS in the surface area satisfy the inequality

$$|\Delta S| \leq \frac{1}{100} S = \frac{4\pi r^2}{100}. \tag{7}$$

We replace ΔS in this inequality with

$$dS = \left(\frac{dS}{dr}\right) dr = 8\pi r \, dr.$$

This gives

$$|8\pi r \, dr| \leq \frac{4\pi r^2}{100}, \qquad \text{or} \qquad |dr| \leq \frac{1}{8\pi r} \cdot \frac{4\pi r^2}{100} = \frac{1}{2} \frac{r}{100}.$$

We should measure the radius with an error dr that is no more than 0.5% of the true value.

Sensitivity to Change

The equation $df = f'(x)\, dx$ tells how sensitive the output of f is to a change in input at different values of x. The larger the value of f' at x, the greater is the effect of a given change dx.

Example 10   You want to calculate the height of a bridge by timing how long it takes a heavy stone you drop to splash into the water below and then using the equation $s = 16t^2$. How sensitive will your calculation be to a 0.1-sec error in measuring the time?

Solution   The size of ds in the equation

$$ds = 32t\, dt$$

depends on how big t is. If $t = 2$ sec, the error caused by $dt = 0.1$ is only

$$ds = 32(2)(0.1) = 6.4 \text{ ft.}$$

Three seconds later, at $t = 5$ sec, the error caused by the same dt is

$$ds = 32(5)(0.1) = 16 \text{ ft.}$$

The Error in the Approximation $\Delta f \approx f'(a)\, \Delta x$

How well does the quantity $f'(a)\, \Delta x$ estimate the true increment $\Delta f = f(a + \Delta x) - f(a)$? We measure the error by subtracting one from the other:

$$\text{Approximation error} = \Delta f - f'(a)\, \Delta x$$

$$= f(a + \Delta x) - f(a) - f'(a)\, \Delta x$$

$$= \underbrace{\left(\frac{f(a + \Delta x) - f(a)}{\Delta x} - f'(a) \right)}_{\text{Call this part } \epsilon .} \Delta x, \qquad (8)$$

$$= \epsilon \cdot \Delta x .$$

As $\Delta x \to 0$, the difference quotient

$$\frac{f(a + \Delta x) - f(a)}{\Delta x}$$

approaches $f'(a)$ (remember the definition of $f'(a)$), so the quantity in parentheses in Eq. (8) becomes a very small number (which is why we called it ϵ). In fact,

$$\epsilon \to 0 \quad \text{as} \quad \Delta x \to 0.$$

Thus, when Δx is small, the approximation error $\epsilon \Delta x$ is smaller still.

$$\underbrace{\Delta f}_{\substack{\text{true} \\ \text{change}}} = \underbrace{f'(a)\, \Delta x}_{\substack{\text{estimated} \\ \text{change}}} + \underbrace{\epsilon \Delta x}_{\text{error}} \qquad (9)$$

While we do not know exactly how small the error is and will not be able to make much progress on this front until Section 3.7, there is something worth noting here, namely the *form* taken by the equation.

If $y = f(x)$ is differentiable at $x = a$, and x changes from a to $a + \Delta x$, the change Δy in f is given by an equation of the form

$$\Delta y = f'(a)\,\Delta x + \epsilon\,\Delta x, \qquad (10)$$

in which $\epsilon \to 0$ as $\Delta x \to 0$.

Surprising as it may seem, just knowing the form of Eq. (10) enables us to prove the Chain Rule, as we shall see in a moment.

Derivatives in Differential Notation

Every formula like

$$\frac{d(u + v)}{dx} = \frac{du}{dx} + \frac{dv}{dx}$$

has a corresponding differential formula like

$$d(u + v) = du + dv$$

that comes from multiplying both sides by dx (Table 2.2).

Example 11

a) $d(3x^2 - 6) = 6x\,dx$

b) $d(\cos 3x) = -\sin 3x\,d(3x) = -3\sin 3x\,dx$

c) $d\,\dfrac{x}{x + 1} = \dfrac{(x + 1)dx - x\,d(x + 1)}{(x + 1)^2} = \dfrac{x\,dx + dx - x\,dx}{(x + 1)^2} = \dfrac{dx}{(x + 1)^2}$

A differential on one side of an equation always calls for a differential on the other side of the equation. We never have $dy = 3x^2$ but, instead, $dy = 3x^2 dx$.

A Proof of the Chain Rule

You may recall our saying in Section 2.4 that the proof we wanted to give of the Chain Rule began with an equation in Section 2.6, the present section. The equation we were referring to is Eq. (10), and the proof goes like this:

Our goal is to show that if $f(u)$ is a differentiable function of u, and $u = g(x)$ is a differentiable function of x, then the composite $y = f(g(x))$ is a differentiable function of x. More precisely, if g is differentiable at x_0 and f is differentiable at $g(x_0)$, then the composite is differentiable at x_0 and

$$\left.\frac{dy}{dx}\right|_{x = x_0} = f'(g(x_0)) \cdot g'(x_0). \qquad (11)$$

Suppose that Δx is an increment in x and that Δu and Δy are the corresponding increments in u and y. As you can see in Fig. 2.46,

$$\left.\frac{dy}{dx}\right|_{x = x_0} = \lim_{\Delta x \to 0} \frac{\Delta y}{\Delta x},$$

so our goal is to show that this limit is $f'(g(x_0)) \cdot g'(x_0)$.

TABLE 2.2
Formulas for differentials

$$dc = 0$$
$$d(cu) = c\,du$$
$$d(u + v) = du + dv$$
$$d(uv) = u\,dv + v\,du$$
$$d\left(\frac{u}{v}\right) = \frac{v\,du - u\,dv}{v^2}$$
$$d(u^n) = nu^{n-1}du$$
$$d(\sin u) = \cos u\,du$$
$$d(\cos u) = -\sin u\,du$$
$$d(\tan u) = \sec^2 u\,du$$
$$d(\cot u) = -\csc^2 u\,du$$
$$d(\sec u) = \sec u \tan u\,du$$
$$d(\csc u) = -\csc u \cot u\,du$$

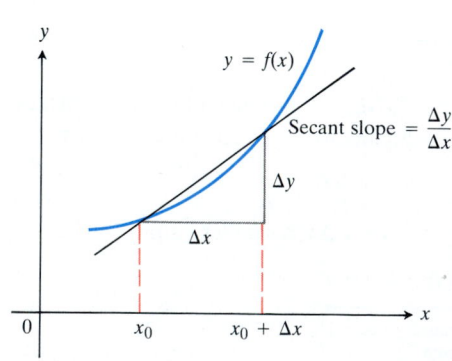

2.46 The graph of y as a function of x. The derivative of y with respect to x at $x = x_0$ is $\lim_{\Delta x \to 0} \Delta y / \Delta x$.

By virtue of Eq. (10),

$$\Delta u = g'(x_0)\,\Delta x + \epsilon_1\,\Delta x$$
$$= [g'(x_0) + \epsilon_1]\,\Delta x, \tag{12}$$

where $\epsilon_1 \to 0$ as $\Delta x \to 0$. Similarly,

$$\Delta y = f'(u_0)\,\Delta u + \epsilon_2\,\Delta u = [f'(u_0) + \epsilon_2]\,\Delta u, \tag{13}$$

where $\epsilon_2 \to 0$ as $\Delta u \to 0$. Combining the equation of Δu with the one for Δy gives

$$\Delta y = [f'(u_0) + \epsilon_2]\,[g'(x_0) + \epsilon_1]\,\Delta x, \tag{14}$$

so

$$\frac{\Delta y}{\Delta x} = f'(u_0)g'(x_0) + \epsilon_2\,g'(x_0) + f'(u_0)\epsilon_1 + \epsilon_2\epsilon_1.$$

Since ϵ_1 and ϵ_2 go to zero as Δx goes to zero, the three terms on the right vanish in the limit and

$$\lim_{\Delta x \to 0} \frac{\Delta y}{\Delta x} = f'(u_0)g'(x_0) = f'(g(x_0)) \cdot g'(x_0).$$

This concludes the proof.

* The Conversion of Mass to Energy

Here is an example of how the approximation

$$\frac{1}{\sqrt{1 - x^2}} \approx 1 + \frac{1}{2}x^2 \tag{15}$$

from Example 3 is used in an applied problem.

Newton's second law,

$$F = \frac{d}{dt}(mv) = m\frac{dv}{dt} = ma,$$

is stated with the assumption that mass is constant, but we know this is not strictly true because the mass of a body increases with velocity. In Einstein's corrected formula, mass has the value

$$m = \frac{m_0}{\sqrt{1 - v^2/c^2}}, \tag{16}$$

where the "rest mass" m_0 represents the mass of a body that is not moving and c is the speed of light, which is about 300,000 km/sec. When v is very small compared with c, v^2/c^2 is close to zero and it is safe to use the approximation

$$\frac{1}{\sqrt{1 - v^2/c^2}} \approx 1 + \frac{1}{2}\left(\frac{v^2}{c^2}\right)$$

(Eq. 15 with $x = v/c$) to write

$$m = \frac{m_0}{\sqrt{1 - v^2/c^2}} \approx m_0\left[1 + \frac{1}{2}\left(\frac{v^2}{c^2}\right)\right] = m_0 + \frac{1}{2}m_0v^2\left(\frac{1}{c^2}\right),$$

or

$$m \approx m_0 + \frac{1}{2} m_0 v^2 \left(\frac{1}{c^2} \right). \tag{17}$$

Equation (17) expresses the increase in mass that results from the added velocity v.

In Newtonian physics, $(1/2)m_0 v^2$ is the kinetic energy (KE) of the body, and if we rewrite Eq. (17) in the form

$$(m - m_0)c^2 \approx \frac{1}{2} m_0 v^2,$$

we see that

$$(m - m_0)c^2 \approx \frac{1}{2} m_0 v^2 = \frac{1}{2} m_0 v^2 - \frac{1}{2} m_0 (0)^2 = \Delta(\text{KE}),$$

or

$$(\Delta m)c^2 \approx \Delta(\text{KE}). \tag{18}$$

In other words, the change in kinetic energy $\Delta(\text{KE})$ in going from velocity 0 to velocity v is approximately equal to $(\Delta m)c^2$.

With c equal to 300,000 m/sec, Eq. (18) becomes

$$\Delta(\text{KE}) \approx 90{,}000{,}000{,}000 \, \Delta m \text{ joules,}$$

and we see that a small change in mass can create a large change in energy. The energy released by a 20-kiloton atomic bomb, for instance, is the result of converting only one gram of mass to energy. The products of the explosion weigh only one gram less than the material exploded. A penny weighs about three grams.

EXERCISES 2.6

In Exercises 1–6, find the linearization $L(x)$ of $f(x)$ at $x = a$.

1. $f(x) = x^4$ at $x = 1$

2. $f(x) = x^{-1}$ at $x = 2$

3. $f(x) = x^3 - x$ at $x = 1$

4. $f(x) = x^3 - 2x + 3$ at $x = 2$

5. $f(x) = \sqrt{x}$ at $x = 4$

6. $f(x) = \sqrt{x^2 + 9}$ at $x = -4$

You want linearizations that will replace the functions in Exercises 7–12 over intervals that include the given points x_0. To make your subsequent work as simple as possible, you want to center each linearization not at x_0 but at a nearby integer $x = a$ at which the given function and its derivative are easy to evaluate. What linearization do you use in each case?

7. $f(x) = x^2 + 2x$, $x_0 = 0.1$

8. $f(x) = x^{-1}$, $x_0 = 0.6$

9. $f(x) = 2x^2 + 4x - 3$, $x_0 = -0.9$

10. $f(x) = 1 + x$, $x_0 = 8.1$

11. $f(x) = \sqrt[3]{x}$, $x_0 = 8.5$

12. $f(x) = \dfrac{x}{x + 1}$, $x_0 = 1.3$

In Exercises 13–18, find the linearization $L(x)$ of the given function at $x = a$. Then graph f and L together near $x = a$.

13. $f(x) = \sin x$ at $x = 0$

14. $f(x) = \cos x$ at $x = 0$

15. $f(x) = \sin x$ at $x = \pi$

16. $f(x) = \cos x$ at $x = -\pi/2$

17. $f(x) = \tan x$ at $x = \pi/4$

18. $f(x) = \sec x$ at $x = \pi/4$

19. Use the formula $(1 + x)^k \approx 1 + kx$ to find linear approximations of the following functions for values of x near zero.

a) $(1 + x)^2$

b) $\dfrac{1}{(1 + x)^5}$

c) $\dfrac{2}{1 - x}$

d) $(1 - x)^6$

e) $3(1 + x)^{1/3}$

f) $\dfrac{1}{\sqrt{1 + x}}$

20. *Faster than a calculator.*   Use the approximation $(1 + x)^k \approx 1 + kx$ to estimate

a) $(1.002)^{100}$

b) $\sqrt[3]{1.009}$.

21. Find the linearization of $f(x) = \sqrt{x + 1} + \sin x$ at $x = 0$. How is it related to the individual linearizations for $\sqrt{x + 1}$ and $\sin x$ at $x = 0$?

22. We know from the Power Rule that the equation

$$\frac{d}{dx}(1 + x)^k = k(1 + x)^{k - 1}$$

holds for every rational number k. In Chapter 6, we shall show that it holds for every irrational number as well. Assuming this result for now, verify Eq. (4) by showing that the linearization of $f(x) = (1 + x)^k$ at $x = 0$ is $L(x) = 1 + kx$ for any number k.

23. CALCULATOR   In Section 1.3 we promised to explain what happened when you took successive square roots of 2 by entering 2 in your calculator and pressing the square root key repeatedly. As you may recall, or as you will find if you try it now, each keypress divides the decimal part of the display approximately in half. The explanation comes from the fact that the linearization of $\sqrt{1 + x}$ at $x = 0$ is $1 + (x/2)$. The x is the decimal part of the display $(1.x)$, and each new square root is about $1 + (x/2)$.

  If you have not done so already, enter 2 into your calculator and press $\boxed{\sqrt{}}$ repeatedly to see what happens.

24. CALCULATOR   If you have not already done so, turn to Section 1.3 and do Exercise 14 there. If you have a key that will enable you to calculate tenth roots, do Exercise 15 as well.

In Exercises 25–30, each function changes value when x changes from x_0 to $x_0 + dx$, as shown schematically in Fig. 2.47. Find

a) the change $\Delta f = f(x_0 + dx) - f(x_0)$;

b) the value of the estimate of $df = f'(x_0)\,dx$;

c) the approximation error $|\Delta f - df|$.

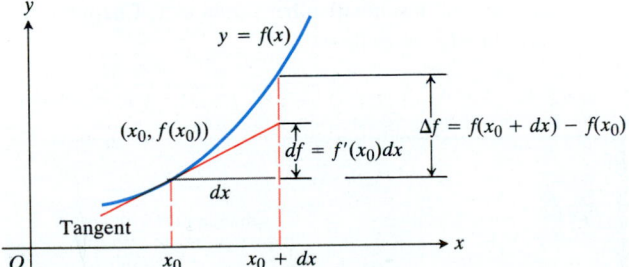

2.47   Schematic diagram for the changes in Exercises 25–30.

25. $f(x) = x^2 + 2x$, $x_0 = 0$, $dx = 0.1$

26. $f(x) = 2x^2 + 4x - 3$, $x_0 = -1$, $dx = 0.1$

27. $f(x) = x^3 - x$, $x_0 = 1$, $dx = 0.1$

28. $f(x) = x^4$, $x_0 = 1$, $dx = 0.1$

29. $f(x) = x^{-1}$, $x_0 = 0.5$, $dx = 0.1$

30. $f(x) = x^3 - 2x + 3$, $x_0 = 2$, $dx = 0.1$

In Exercises 31–36, write a differential formula that estimates the given change in volume or surface area.

31. The change in the volume $V = (4/3)\pi r^3$ of a sphere when the radius changes from r_0 to $r_0 + dr$

32. The change in the surface area $S = 4\pi r^2$ of a sphere when the radius changes from r_0 to $r_0 + dr$

33. The change in the volume $V = x^3$ of a cube when the edge lengths change from x_0 to $x_0 + dx$

34. The change in the surface area $S = 6x^2$ of a cube when the edge lengths change from x_0 to $x_0 + dx$

35. The change in the volume $V = \pi r^2 h$ of a right circular cylinder when the radius changes from r_0 to $r_0 + dr$ and the height does not change

36. The change in the lateral surface area $S = 2\pi rh$ of a right circular cylinder when the height changes from h_0 to $h_0 + dh$ and the radius does not change

37. The radius of a circle is increased from 2.00 to 2.02 m. (a) Estimate the resulting change in area. (b) Express the estimate in (a) as a percentage of the circle's original area.

38. The diameter of a tree was 10 in. During the following year, the circumference grew 2 in. About how much did the tree's diameter grow? The tree's cross-section area?

39. The edge of a cube is measured as 10 cm with an error of 1%. The cube's volume is to be calculated from this measurement. Estimate the percentage error in the volume calculation.

40. About how accurately should you measure the side of a square to be sure of calculating the area within 2% of its true value?

41. The diameter of a sphere is measured as 100 ± 1 cm, and the volume is calculated from this measurement. Estimate the percentage error in the volume calculation.

42. Estimate the allowable percentage error in measuring the diameter d of a sphere if the volume is to be calculated correctly to within 3%.

43. The height and radius of a right circular cylinder are equal, so the cylinder's volume is $V = \pi h^3$. The volume is to be calculated from a measurement of h and must be calculated with an error of no more than 1% of the true value. Find approximately the greatest error that can be tolerated in the measurement of h, expressed as a percentage of h.

44. a) About how accurately must the interior diameter of a

10-m-high cylindrical storage tank be measured to calculate the tank's volume to within 1% of its true value?

b) About how accurately must the tank's exterior diameter be measured to calculate the amount of paint it will take to paint the side of the tank within 5% of the true amount?

45. A manufacturer contracts to mint coins for the federal government. How much variation dr in the radius of the coins can be tolerated if the coins are to weigh within 1/1000 of their ideal weight? Assume that the thickness does not vary.

46. *Continuation of Example 10.* Show that a 5% error in measuring t will cause about a 10% error in calculating s from the equation $s = 16t^2$.

47. *Unclogging arteries.* In the late 1830s, the French physiologist Jean Poiseuille ("pwa · zoy") discovered the formula we use today to predict how much the radius of a partially clogged artery has to be expanded to restore normal flow. His formula, $V = kr^4$, says that the volume V of a fluid flowing through a small pipe or tube in a unit of time at a fixed pressure is a constant times the fourth power of the tube's radius r. How will a 10% increase in r affect V?

48. *The effect of flight maneuvers on the heart.* The amount of work done by the heart's main pumping chamber, the left ventricle, is given by the equation

$$W = PV + \frac{Vpv^2}{2g},$$

where W is the work per unit time (power), P is the average blood pressure, V is the volume of blood pumped out during the unit of time, p is the density of the blood, v is the average velocity of the exiting blood, and g is the acceleration of gravity.

When P, V, p, and v remain constant, W becomes a function of g, and the equation takes the simplified form

$$W = a + \frac{b}{g} \qquad (a, b \text{ constant}) \qquad (19)$$

As a member of NASA's medical team, you want to know how sensitive W is to apparent changes in g caused by flight maneuvers, and this depends on the initial value of g. As part of your investigation, you decide to compare the effect on W of a given change dg on the moon, where $g = 5.2$ ft/sec², with the effect the same change dg would have on Earth, where $g = 32$ ft/sec². You use Eq. (19) to find the ratio of dW_{moon} to dW_{Earth}. What do you find?

49. *Sketching the change in a cube's volume.* The volume $V = x^3$ of a cube with edges of length x increases by an amount ΔV when x increases by an amount Δx. Show with a sketch how to represent ΔV geometrically as the sum of the volumes of

a) three slabs of dimensions x by x by Δx;

b) three bars of dimensions x by Δx by Δx;

c) one cube of dimensions Δx by Δx by Δx.

The differential formula $dV = 3x^2\,dx$ estimates the change in V with the three slabs.

50. *Measuring the acceleration of gravity.* When the length L of a clock pendulum is held constant by controlling its temperature, the pendulum's period T depends on the acceleration of gravity g. The period will therefore vary slightly as the clock is moved from place to place on the earth's surface, depending on the change in g. By keeping track of ΔT, we can estimate the variation in g from the equation $T = 2\pi(L/g)^{1/2}$ that relates T, g, and L.

a) With L held constant and g as the independent variable, calculate dT and use it to answer (b) and (c).

b) If g increases, will T increase or decrease? Will a pendulum clock speed up or slow down? Explain.

c) A clock with a 100-cm pendulum is moved from a location where $g = 980$ cm/sec² to a new location. This increases the period by $dT = 0.001$ sec. Find dg and estimate the value of g at the new location.

Derivatives in Differential Form

Find dy in Exercises 51–60.

51. $y = x^3 - 3x$

52. $y = x\sqrt{1 - x^2}$

53. $y = 2x/(1 + x)$

54. $y = (\sin x)/(1 + \cos x)$

55. $y + xy = x$

56. $xy^2 + x^2y = 4$

57. $y = 4\tan^2(x/2)$

58. $y = \sin^3(x/3)$

59. $y = x^2 - \cos(x^2)$

60. $y = 2\cot\sqrt{x}$

Computer Grapher or Graphing Calculator

61. *Reading derivatives from graphs.* The idea that differentiable curves flatten out when magnified can be used to estimate the values of the derivatives of functions at particular points. We magnify the curve until the portion we see looks like a straight line through the point in question, and then we use the screen's coordinate grid to read the slope of the curve as the slope of the line it resembles.

a) To see how the process works, try it first with the function $y = x^2$ at $x = 1$. The slope you read should be 2.

b) Then try it with the curve $y = e^x$ at $x = 1$, $x = 0$, and $x = -1$. In each case, compare your estimate of the derivative with the value of e^x at the point. What pattern do you see? Test it with other values of x. Chapter 6 will explain what is going on.

EXPLORER PROGRAM

PowerGrapher   Graphs functions and their linearizations together by constructing tangents to curves at points of your choice

2.7 Newton's Method

When exact formulas are not available for solving an equation $f(x) = 0$, we can turn to numerical techniques from calculus for approximating the solution we seek. One of these techniques is Newton's method or, as it is more accurately called, the Newton-Raphson method. It is based on the idea of using a tangent line to replace the graph of $y = f(x)$ near the points where f is zero. Once again, linearization is the key to solving a practical problem.

If you have access to a computer, you can use the program at the end of this section to do the arithmetic. If not, you can still see how the technique works, and you can work the beginning exercises at the end of the section by hand.

The Procedure for Newton's Method

1. Guess a first approximation to a root of the equation $f(x) = 0$. A graph of $y = f(x)$ will help.

2. Use the first approximation to get a second, the second to get a third, and so on. To go from the nth approximation x_n to the next approximation x_{n+1}, use the formula

$$x_{n+1} = x_n - \frac{f(x_n)}{f'(x_n)}, \tag{1}$$

where $f'(x_n)$ is the derivative of f at x_n.

We first show how the method works and then go to the theory behind it.

In our first example we find decimal approximations to $\sqrt{2}$ by estimating the positive root of the equation $f(x) = x^2 - 2 = 0$.

Example 1

Find the positive root of the equation

$$f(x) = x^2 - 2 = 0.$$

Solution With $f(x) = x^2 - 2$ and $f'(x) = 2x$, Eq. (1) becomes

$$x_{n+1} = x_n - \frac{x_n^2 - 2}{2x_n}. \tag{2}$$

To use our calculator efficiently, we rewrite Eq. (2) in a form that uses fewer arithmetic operations:

$$x_{n+1} = x_n - \frac{x_n^2 - 2}{2x_n} = \frac{x_n^2 + 2}{2x_n}$$

$$= \frac{x_n}{2} + \frac{1}{x_n}. \tag{3}$$

The equation

$$x_{n+1} = \frac{x_n}{2} + \frac{1}{x_n} \tag{4}$$

enables us to go from each approximation to the next by the following steps.

Algorithm, Recursion, and Iteration

It is customary to call a specified sequence of computational steps like the one in Newton's method an *algorithm*. When an algorithm proceeds by repeating a given set of steps over and over, using the answer from the previous step as the input for the next, the algorithm is called *recursive* and each repetition is called an *iteration*. Newton's method is one of the really fast recursive techniques for finding roots.

	Operation	Result	Example
$\boxed{x}$	Enter x.	x	1.5
$\boxed{\frac{1}{x}}$ $\boxed{\text{STO}}$	Store the reciprocal.	$\frac{1}{x}$	0.66667
$\boxed{\frac{1}{x}}$	Take the reciprocal of the display again.	x	1.5
$\boxed{\div}$ $\boxed{2}$	Divide by 2.	$\frac{x}{2}$	0.75
$\boxed{+}$ $\boxed{\text{RCL}}$ $\boxed{=}$	Add memory to display.	$\frac{x}{2} + \frac{1}{x}$	1.41667

With the starting value $x_0 = 1$, we get the results in the first column of the following table. (To five decimal places, $\sqrt{2} = 1.41421$.)

	Error	Number of correct figures
$x_0 = 1$	-0.41421	1
$x_1 = 1.5$	$+0.08579$	1
$x_2 = 1.41667$	0.00245	3
$x_3 = 1.41422$	0.00001	5

Newton's method is the method used by most calculators to calculate roots because it converges so fast (more about this later). If the arithmetic in the table in Example 1 had been carried to 13 decimal places rather than 5, then going one step further would have given $\sqrt{2}$ correctly to more than 10 decimal places.

Example 2   Find the x-coordinate of the point where the curve $y = x^3 - x$ crosses the horizontal line $y = 1$.

Solution   The curve crosses the line when $x^3 - x = 1$ or $x^3 - x - 1 = 0$. When does $f(x) = x^3 - x - 1$ equal zero? The graph of f (Fig. 2.48) shows a single root, located between $x = 1$ and $x = 2$. We apply Newton's method to f with the starting value $x_0 = 1$. The results are displayed in Table 2.3 and Fig. 2.49.

TABLE 2.3
The result of applying Newton's method to $f(x) = x^3 - x - 1$ with $x_0 = 1$

n	x_n	$f(x_n)$	$f'(x_n)$	$x_{n+1} = x_n - \dfrac{f(x_n)}{f'(x_n)}$
0	1	-1	2	1.5
1	1.5	0.875	5.75	1.3478 26087
2	1.3478 26087	0.1006 82174	4.4499 05482	1.3252 00399
3	1.3252 00399	0.0020 58363	4.2684 68293	1.3247 18174
4	1.3247 18174	0.0000 00925	4.2646 34722	1.3247 17957
5	1.3247 17957	-5×10^{-10}	4.2646 32997	1.3247 17957

At $n = 5$ we come to the result $x_5 = x_4 = 1.3247\ 17957$. When $x_{n+1} = x_n$, Eq. (1) shows that $f(x_n) = 0$. We have found a solution of $f(x) = 0$ to nine decimals.

2.48   The graph of $f(x) = x^3 - x - 1$ crosses the x-axis just once, at a point between $x = 1$ and $x = 2$.

$f(x) = x^3 - x - 1$

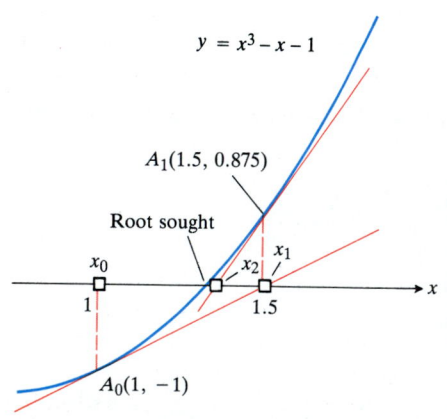

2.49   The first three x-values in Table 2.3.

$y = x^3 - x - 1$

$A_1(1.5, 0.875)$

Root sought

$A_0(1, -1)$

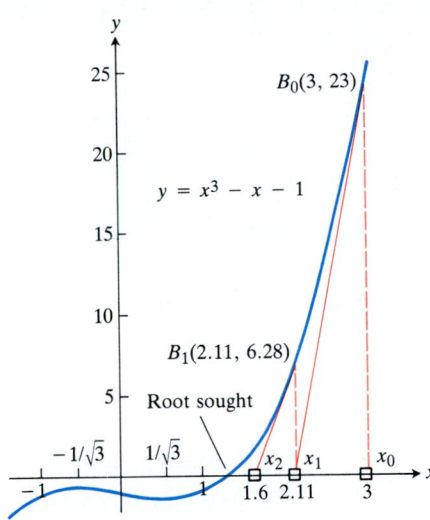

2.50 Any starting value x_0 to the right of $x = 1/\sqrt{3}$ will lead to the root.

In Fig. 2.50, we have indicated that the process in Example 2 might have started at the point $B_0(3, 23)$ on the curve, with $x_0 = 3$. Point B_0 is quite far from the x-axis, but the tangent at B_0 crosses the x-axis at about $(2.11, 0)$, so x_1 is still an improvement over x_0. If we use Eq. (1) repeatedly as before, with $f(x) = x^3 - x - 1$ and $f'(x) = 3x^2 - 1$, we confirm the nine-place solution $x_6 = x_5 = 1.3247\ 17957$ in six steps.

The curve in Fig. 2.50 has a high turning point at $x = -1/\sqrt{3}$ and a low turning point at $x = +1/\sqrt{3}$. We would not expect good results from Newton's method if we were to start with x_0 between these points, but we can start any place to the right of $x = 1/\sqrt{3}$ and get the answer. It would not be very clever to do so, but we could even begin far to the right of B_0, for example with $x_0 = 10$. It takes a bit longer, but the process still converges to the same answer as before.

The Theory Behind the Method

We use the tangent to approximate the graph of $y = f(x)$ near the point $P(x_n, y_n)$, where $y_n = f(x_n)$ is small, and we let x_{n+1} be the value of x where that tangent line crosses the x-axis. (We assume that the slope $f'(x_n)$ of the tangent is not zero.) The equation of the tangent is

$$y - y_n = f'(x_n)(x - x_n). \tag{5}$$

We put $y_n = f(x_n)$ and $y = 0$ into Eq. (5) and solve for x to get

$$x_{n+1} = x_n - \frac{f(x_n)}{f'(x_n)}.$$

See Fig. 2.51.

From another point of view, every time we replace the graph of f by one of its tangent lines we are replacing the function by one of its linearizations L. We then solve $L(x) = 0$ to estimate the solution of $f(x) = 0$.

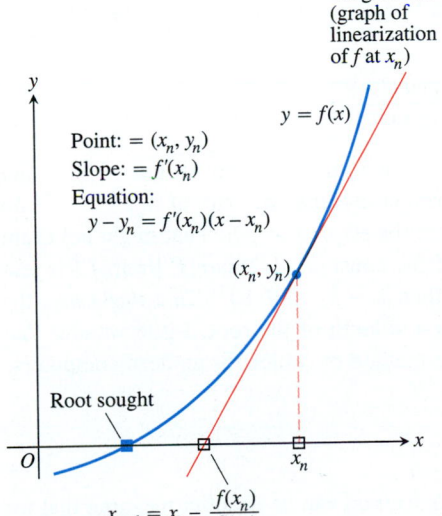

2.51 The geometry of the successive steps of Newton's method. From x_n we go up to the curve and follow the tangent line down to find x_{n+1}.

Strengths and Limitations of Newton's Method

Newton's method does not work if $f'(x_n) = 0$. In that case, choose a new starting point. Of course, $f(x) = 0$ and $f'(x) = 0$ may have a common root. To detect whether this is so, we could first find the solutions of $f'(x) = 0$ and check the value of $f(x)$ at such places. Or we could graph f and f' together on a computer to look for places where the graphs might cross the x-axis together.

Newton's method does not always converge. For instance, if

$$f(x) = \begin{cases} \sqrt{x - r} & \text{for } x \geq r, \\ -\sqrt{r - x} & \text{for } x < r, \end{cases} \tag{6}$$

the graph will be like the one in Fig. 2.52. If we begin with $x_0 = r - h$, we get $x_1 = r + h$, and successive approximations go back and forth between these two values. No amount of iteration brings us closer to the root r than our first guess.

If Newton's method does converge, it converges to a root of $f(x)$. However, the method may converge to a root different from the one we expect if the starting value is not close enough to the root we seek. Figure 2.53 shows two ways this might happen.

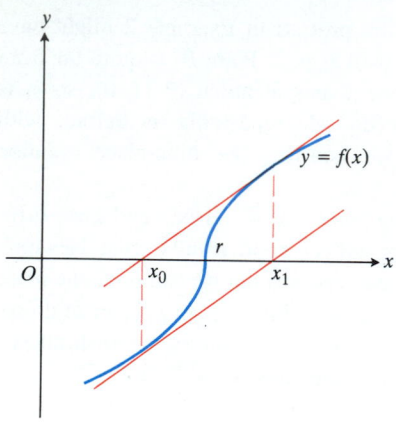

2.52   The graph of a function for which Newton's method fails to converge.

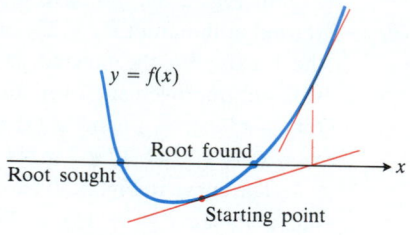

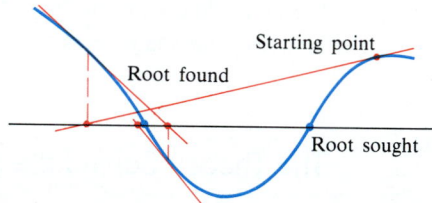

2.53   Newton's method may miss the root you want if you start too far away.

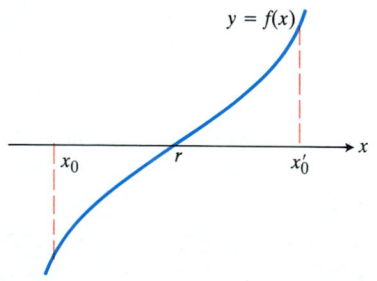

2.54   The curve $y = f(x)$ is convex toward the axis between x_0 and r and between x_0' and r. Newton's method will converge to r from either starting point.

When will Newton's method converge? A result from advanced calculus says that if the inequality

$$\left|\frac{f(x)f''(x)}{[f'(x)]^2}\right| < 1 \tag{7}$$

holds for all values of x in an interval about a root r of f, then the method will converge to r for any starting value x_0 in that interval. This is a *sufficient*, but not a necessary, condition. The method can (and does) converge in some cases when there is no interval about r in which the inequality (Eq. (7)) holds. Newton's method will always work if the curve $y = f(x)$ is convex ("bulges") toward the axis in the interval between x_0 and the root sought. See Fig. 2.54.

The speed with which Newton's method converges to a root r is expressed by the advanced-calculus formula

$$|r - x_{n+1}| \leq \frac{1}{2}\frac{\max|f''|}{\min|f'|}(r - x_n)^2, \tag{8}$$

where "max" and "min" refer to the maximum and minimum values of $|f''|$ and $|f'|$ in an interval surrounding r. The formula says that the error at step $n + 1$ is no greater than a constant times the square of the error at step n. That might not seem like much, but think of what it says. If the constant $(1/2)\max|f''|/\min|f'|$ is less than or equal to 1, and $|r - x_n| \leq 10^{-3}$, then $|r - x_{n+1}| \leq 10^{-6}$. *In a single step*, the method takes us to within less than one-millionth of the root. Little wonder that Newton's method, when it applies, is the method of choice for modern computers.

✳ Chaos in Newton's Method

The process of finding roots by Newton's method can be chaotic, meaning that for some equations the final outcome may be extremely sensitive to the starting value's location. The equation $4x^4 - 4x^2 = 0$ is a case in point (Fig. 2.55). Starting in either red zone on the x-axis leads to root A. Starting in the black zone leads to root B. Points in the blue zone lead to root C. The boundary points that separate the

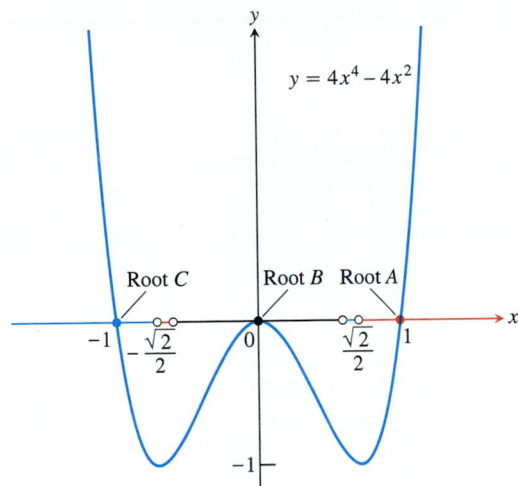

$$y = 4x^4 - 4x^2$$

Root C   Root B   Root A

2.55 The color coding on the x-axis shows which starting values for Newton's method lead to which roots. The process is chaotic near the color boundaries, where small changes in initial value can make dramatic changes in outcome.

zones (the four open circles on the x-axis) lead to horizontal tangents and no solution at all.

If you think of the roots as attractors of other points, the coloring shows the intervals of points they attract ("intervals of attraction"). You might think that points between roots A and B would be attracted either to A or to B, but that is not what happens. Between A and B there is an open interval whose points all go to root C. Similarly, the interval joining B and C contains an open interval of points attracted to A. On opposite sides of the boundaries of these intervals of attraction, we can find, *arbitrarily close together,* points whose ultimate destinations are far apart.

We encounter a truly dramatic example of chaotic behavior when we apply Newton's method to solve the complex-number equation

$$z^6 - 1 = 0. \tag{9}$$

It has six solutions: 1, -1, and the four numbers $\pm(1/2) \pm (\sqrt{3}/2)\,i$. As Fig. 2.56 suggests, each of the six roots has infinitely many "basins" of attraction in the complex plane (Appendix 6). Starting points in red basins are attracted to the root 1, those in the green basin to the root $(1/2) + (\sqrt{3}/2)i$, and so on. Each basin has a boundary whose complicated pattern repeats without end under successive magnifications.

2.56 This computer-generated initial value portrait uses color to show where different points in the complex plane end up when they are used as starting values in applying Newton's method to solve the equation $z^6 - 1 = 0$. Red points go to 1, green points to $(1/2) + (\sqrt{3}/2)\,i$, dark blue points to $(-1/2) + (\sqrt{3}/2)\,i$, and so on. Starting values that generate sequences that do not arrive within 0.1 units of a root after 32 steps are colored black.

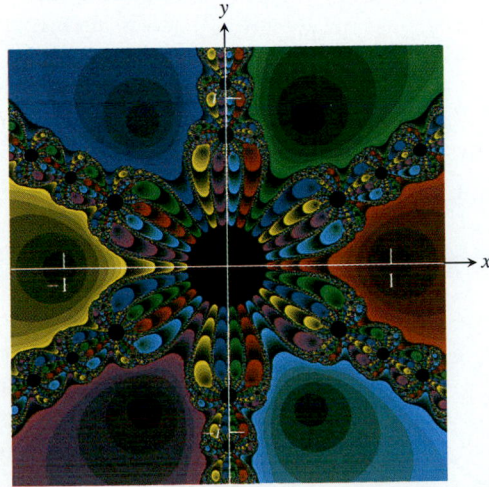

EXERCISES 2.7

1. Estimate the solutions of the equation $x^2 + x - 1 = 0$ (Fig. 2.57). Start with $x_0 = 1$ and $x_0 = -1$ and find x_2 in each case.

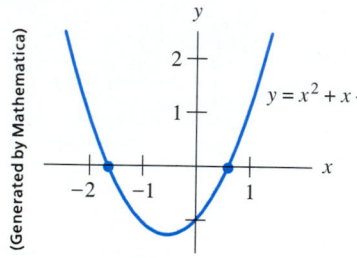

2.57   The graph of the equation in Exercise 1.

2. Estimate the one real solution of the equation $x^3 + 3x + 1 = 0$ (Fig. 2.58). Start with $x_0 = 1$ and find x_2. (In Section 3.2, we shall see how to use calculus to show that the equation has no other real solutions.)

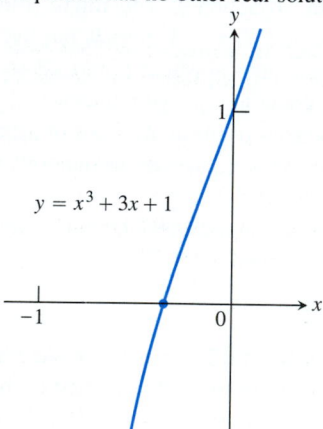

2.58   The graph of the equation in Exercise 2.

3. Estimate the two zeros of the function $f(x) = x^4 + x - 3$ (Fig. 2.59). Start with $x_0 = 1$ and $x_0 = -1$ and find x_2 in each case.

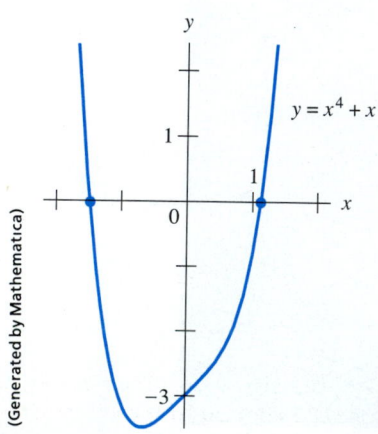

2.59   The graph of the function in Exercise 3.

4. Estimate the two zeros of the function $f(x) = 2x - x^2 + 1$ (Fig. 2.60). Start with $x_0 = 0$ and $x_0 = 2$ and find x_2 in each case.

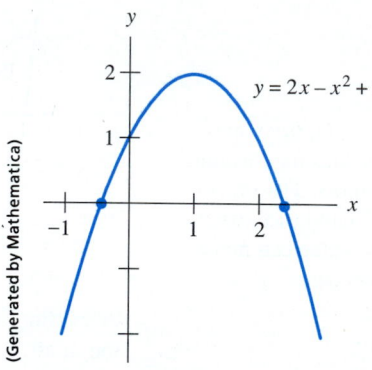

2.60   The graph of the function in Exercise 4.

5. Find the positive fourth root of 2 by solving the equation $x^4 - 2 = 0$. Start with $x_0 = 1$ and find x_2.

6. Find the negative fourth root of 2 by solving the equation $x^4 - 2 = 0$. Start with $x_0 = -1$ and find x_2.

7. Suppose your first guess is lucky, in the sense that x_0 is a root of $f(x) = 0$. What happens to x_1 and later approximations?

8. You plan to estimate $\pi/2$ to five decimal places by solving the equation $\cos x = 0$ by Newton's method. Does it matter what your starting value is? Explain.

9. *Oscillation.*   Show that if $h > 0$, applying Newton's method to

$$f(x) = \begin{cases} \sqrt{x}, & x \geq 0, \\ \sqrt{-x}, & x < 0, \end{cases}$$

leads to $x_1 = -h$ if $x_0 = h$ and to $x_1 = h$ if $x_0 = -h$. Draw a picture that shows what is going on.

10. *Approximations that get worse and worse.*   Apply Newton's method to $f(x) = x^{1/3}$ with $x_0 = 1$, and calculate x_1, x_2, x_3, and x_4. Find a formula for $|x_n|$. What happens to $|x_n|$ as $n \to \infty$? Draw a picture that shows what is going on.

11. CALCULATOR   Use the Intermediate Value Theorem from Section 1.10 to show that $f(x) = x^3 + 2x - 4$ has a root between $x = 1$ and $x = 2$. Then find the root to five decimal places.

12. a) Explain why the following three statements ask for the same information.

    1. Find the zeros of $f(x) = x^3 - 3x - 1$.

    2. Find the x-coordinates of the points where the curve $y = x^3$ crosses the line $y = 3x + 1$.

    3. Find the x-coordinates of the points where the graph of the function $g(x) = x^4/4 - 3x^2/2 - x + 5$ has horizontal tangents.

b) CALCULATOR Find the positive root of $f(x) = x^3 - 3x - 1$ to five decimal places.

c) CALCULATOR Find the two negative roots of $f(x) = x^3 - 3x - 1$ to five decimal places.

13. CALCULATOR Estimate π to five decimal places by applying Newton's method to the equation $\tan x = 0$ with $x_0 = 3$. Remember to use radians.

14. Solve the equation $\sec x = 4$ to five decimal places on the interval $[0, \pi/2]$.

15. CALCULATOR *Locating a planet.* To calculate a planet's space coordinates, we have to solve trigonometric equations like $x = 1 + 0.5 \sin x$. Graphing the function $f(x) = x - 1 - 0.5 \sin x$ suggests that the function has a root near $x = 1.5$. Use one application of Newton's method to improve this estimate. That is, start with $x_0 = 1.5$ and find x_1. (The value of the root is 1.49870 to five decimal places.) Remember to use radians.

16. CALCULATOR *Finding an ion concentration.* While trying to find the acidity of a saturated solution of magnesium hydroxide in hydrochloric acid, you derive the equation

$$\frac{3.64 \times 10^{-11}}{[OH_3^+]^2} = [OH_3^+] + 3.6 \times 10^{-4}$$

for the hydronium ion concentration $[OH_3^+]$. To find the value of $[OH_3^+]$, you set $x = 10^4 [OH_3^+]$ and convert the equation to

$$x^3 + 3.6x^2 - 36.4 = 0.$$

You then solve this equation by Newton's method. What do you get for x? (Make it good to two decimal places.) For $[OH_3^+]$?

Computer or Programmable Calculator

17. The curve $y = \tan x$ crosses the line $y = 2x$ somewhere between $x = 0$ and $x = \pi/2$. Where?

18. Find the two real solutions of the equation $x^4 - 2x^3 - x^2 - 2x + 2 = 0$ (Fig. 2.61).

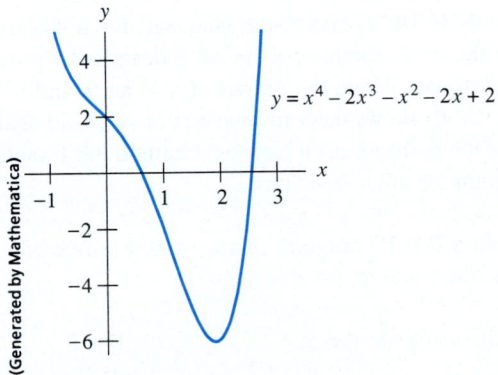

2.61 The graph of the equation in Exercise 18.

19. Find the four real solutions of the equation $2x^4 - 4x^2 + 1 = 0$ (Fig. 2.62).

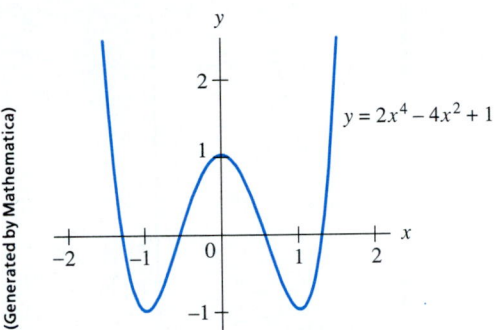

2.62 The graph of the equation in Exercise 19.

20. *The sonobuoy problem.* In submarine location problems it is often necessary to find the submarine's closest point of approach (CPA) to a sonobuoy (sound detector) in the water. Suppose that a submarine travels on a parabolic path $y = x^2$ and that the buoy is located at the point $(2, -1/2)$ (Fig. 2.63). As we shall see in Section 3.5, Exercise 20, the value of x that minimizes the distance between the point (x, x^2) and the point $(2, -1/2)$ is a solution of the equation

$$\frac{1}{x^2 + 1} = x.$$

Solve this equation by Newton's method to find the CPA to five decimal places. (*Source: The Contraction Mapping Principle* by C. O. Wilde, UMAP Unit 326 [Arlington, Mass.: COMAP, Inc.].)

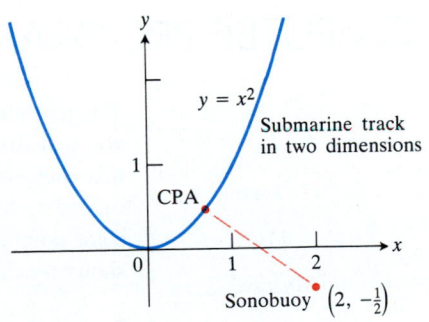

2.63 The submarine track in Exercise 20.

21. *Curves that are nearly flat at the root.* Some curves are so flat that Newton's method stops too far from the root to give a useful estimate. Try Newton's method on $f(x) = (x - 1)^{40}$ with a starting value of $x_0 = 2$ (Fig. 2.64). How close does the computer come to the root $x = 1$?

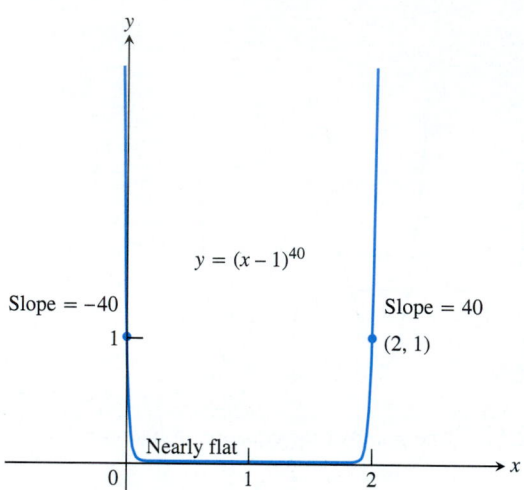

2.64   The graph of $y = (x - 1)^{40}$ (Exercise 21).

22. *Finding a root different from the one sought.*   All three zeros of $f(x) = 4x^4 - 4x^2$ can be found by starting Newton's method near $x = -\sqrt{2}/2$. (See Fig. 2.55.) Try it.

23. Find the approximate values of r_1 through r_4 in the factorization

$$8x^4 - 14x^3 - 9x^2 + 11x - 1$$
$$= 8(x - r_1)(x - r_2)(x - r_3)(x - r_4)$$

(Fig. 2.65).

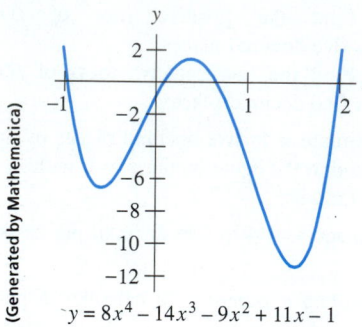

2.65   The graph of the polynomial $y = 8x^4 - 14x^3 - 9x^2 + 11x - 1$ in Exercise 23.

24. *Chaos in Newton's method.*   If you have a computer or a calculator that can be programmed to do complex-number arithmetic, experiment with Newton's method to solve the equation $z^6 - 1 = 0$. The recursion relation to use is

$$z_{n+1} = z_n - \frac{z_n^6 - 1}{6z_n^5} \qquad \text{or} \qquad z_{n+1} = \frac{5}{6}z_n + \frac{1}{6z_n^5}.$$

Try these starting values (among others): $2, i, \sqrt{3} + i$.

EXPLORER PROGRAM	
Bisections, Secants, Newton	Graphs any function you key in and finds its roots by the method you select.

✱ A COMPUTER PROGRAM FOR NEWTON'S METHOD

You may already be familiar with BASIC as a computer language, but if you are not you will still be able to follow the next examples. Example 3 describes a program that uses Newton's method to estimate where the graphs of $y = \sin x$ and $y = x^2$ intersect. A quick sketch (Fig. 2.66) shows them to intersect at zero and again at some point near $x = 1$. That is the point we need Newton's method for. Example 4 shows how to modify the program for other functions.

Example 3   Write and run a BASIC program that uses Newton's method to estimate the positive solution of the equation $\sin x = x^2$.

Solution   The calculations to be made are these:

1. Start with $x_0 = 1$.

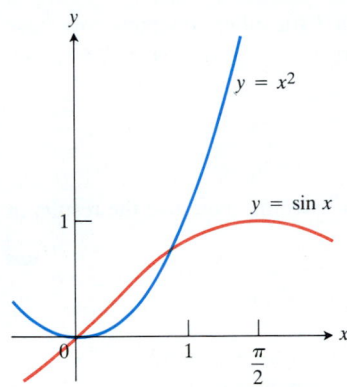

2.66 The curves $y = x^2$ and $y = \sin x$ cross at $x = 0$ and again near $x = 1$. Example 3 uses Newton's method to find this second solution of the equation $\sin x = x^2$.

2. We want to make $f(x) = \sin x - x^2$ small, zero if possible; its derivative is $f'(x) = \cos x - 2x$.

3. Given any x_n, the next value of x is

$$x_{n+1} = x_n - \frac{f(x_n)}{f'(x_n)}.$$

For convenience, we label the steps in the program 10, 20, 30, and so on, leaving numbers in between for any additional steps we may wish to include later. The computer does not recognize x^2 unless we write it as $x \,\hat{}\, 2$, as shown in line 30 of the program that follows. Likewise, $10\,\hat{}\,(-6)$ in line 80 is the notation for 10^{-6} in the program. The asterisk (*) in $2 * x$ indicates multiplication—the machine does not recognize $2x$ as 2 times x.

Now, let us look again at line 80. We do not know in advance how many iterations may be needed to find a "sufficiently accurate" estimate of the root of $f(x) = 0$, so we arbitrarily tell the computer to stop after it reaches a value of x for which the absolute value of f is less than 10^{-6}. The entire command in line 80 is known as an "IF...THEN" conditional. If the condition is satisfied, the machine goes to line 100; but if the condition is not satisfied, the machine proceeds to line 90, which sends it back to line 50 for another round of calculation. Here is the program.

	PROGRAM		COMMENT
10	INPUT "ENTER A STARTING VALUE ", X		Screen prompt for a starting value.
20	N = 0		We start counting the steps.
30	DEF FNF(X) = SIN(X) − X ˆ 2		The function is $f(x) = \sin x - x^2$.
40	DEF FNG(X) = COS(X) − 2 * X		We let G stand for f'.
50	PRINT N; X; FNF(X); FNF(X)/FNG(X)		So we can see the values at each step.
60	X = X − FNF(X)/FNG(X)		The next x will be $x_n - f(x_n)/f'(x_n)$.
70	N = N + 1		Numbers the new x.
80	IF ABS(FNF(X)) < 10 ˆ (−6) THEN 100		Conditional command discussed in text.
90	GOTO 50		If the condition ABS(FNF(X)) < 10 ˆ (−6) is not yet satisfied, do it again.
100	PRINT N; X; FNF(X); FNF(X)/FNG(X)		Print the final result.
110	END		The program stops only after ABS(FNF(X)) < 10 ˆ (−6).

Table 2.4 shows what happens when you run the program with the starting value $x = 1$.

TABLE 2.4
Estimates of the positive solution of $\sin x - x^2 = 0$

n	x_n	$f(x_n)$	$f(x_n)/f'(x_n)$
0	1	− 0.1585 29	0.1086 04
1	0.8913 961	− 1.6637 21E − 02	1.4411 18E − 02
2	0.8769 848	− 2.8818 85E − 04	2.5858 16E − 04
3	0.8767 262	0	0

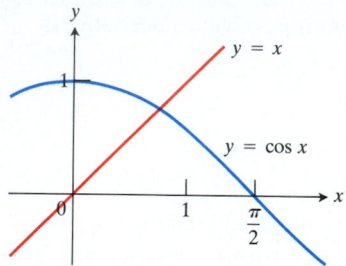

2.67   Example 4 shows how to modify the program in Example 3 to find where $\cos x = x$.

Example 4   To run the program of Example 3 for other functions, you have only to change lines 30 and 40. To estimate where $\cos x$ equals x (Fig. 2.67), you would use

30	DEF FNF(X) = X − COS(X)
40	DEF FNG(X) = 1 + SIN(X)

Running the program with a starting value of $x = 1$ would then give the results in Table 2.5.

TABLE 2.5
Estimates of the solution of $\cos x = x$

n	x_n	$f(x_n)$	$f(x_n)/f'(x_n)$
0	1	0.4596 977	0.2496 361
1	0.7503 639	0.0189 231	1.1250 99E − 02
2	0.7391 129	4.6372 42E − 05	2.7707 64E − 05
3	0.7390 851	0	0

REVIEW QUESTIONS

1. What rules do you know for calculating derivatives? Give examples.

2. Explain how the three formulas

   a) $\dfrac{d}{dx}(x^n) = nx^{n-1}$,

   b) $\dfrac{d}{dx}(cu) = c\dfrac{du}{dx}$,

   c) $\dfrac{d}{dx}(u_1 + u_2 + \cdots + u_n) = \dfrac{du_1}{dx} + \dfrac{du_2}{dx} + \cdots + \dfrac{du_n}{dx}$

   enable us to differentiate any polynomial.

3. What formula do we need, in addition to the three listed in Question 2, to differentiate rational functions?

4. What is a second derivative? A third derivative? How many derivatives do the functions you know have? Give examples.

5. When a body moves along a coordinate line and its position s is a differentiable function of t, how do you define the body's velocity, speed, and acceleration? Give an example.

6. Besides velocity, speed, and acceleration, what other rates of change are found with derivatives?

7. What are the derivatives of the six basic trigonometric functions? How does their calculation depend on radian measure?

8. When is the composite of two functions differentiable at a point? What do you need to know to evaluate its derivative there? Give examples.

9. What is implicit differentiation and what is it good for? Give examples.

10. What is the linearization $L(x)$ of a function $f(x)$ at a point $x = a$? What is required of f at a for the linearization to exist? How are linearizations used? Give examples.

11. If x moves from x_0 to a nearby value $x_0 + dx$, how do we estimate the corresponding change in the value of a differentiable function $f(x)$? How do we estimate the relative change? The percentage change? Give an example.

12. How are derivatives expressed in differential notation? Give examples.

13. Describe Newton's method for solving equations. Give an example. What is the theory behind the method? What are some of the things to watch out for when you use the method?

MISCELLANEOUS EXERCISES

In Exercises 1–8, find dy/dx.

1. $y = (x + 1)^4(x^2 + 2x)$

2. $y = (2x + 1)/(4 - 8x)$

3. $y = (x^2 + \sin x + 1)^3$

4. $y = \cos x/(1 + \sec x)$

5. $y = (\sec x + \tan x)^5$

6. $y = x^{-2}\sin(x^3)$

7. $x^2 + xy + y^2 - 5x = 2$

8. $5x^{4/5} + 10y^{6/5} = 15$

In Exercises 9–14, find ds/dt when s equals the given expression.

9. $(2t - 5)(4 - t)^{-1}$

10. $t/\sqrt{t + 1}$

11. $\left(-1 - \dfrac{t}{2} - \dfrac{t^2}{4}\right)^2$

12. $\sec^2\left(1 + \dfrac{t}{2}\right)$

13. $(t^2 - 8t)^{-1/2}$

14. $t^2 \csc 5t$

In Exercises 15 and 16, find dp/dq.

15. $p^3 + 4pq - 3q^2 = 2$

16. $q = (5p^2 + 2p)^{-3/2}$

In Exercises 17 and 18, find dr/ds.

17. $r \cos 2s + \sin^2 s = \pi$

18. $2rs - r - s + s^2 = -3$

19. The parabola $y = x^2 + C$ is to be tangent to the line $y = x$. Find C.

20. Show that the tangent to the curve $y = x^3$ at any point (a, a^3) meets the curve again at a point where the slope is four times the slope at (a, a^3).

21. Show that the function $y = \cos(2x - A)$ satisfies the equation

$$y'' = -4y$$

for every value of A. Equations like $y'' = -4y$ arise when we design loudspeakers and audio amplifiers, and their solutions can always be expressed as cosines. We shall say more about such equations in Chapter 15.

22. The position at time $t \geq 0$ of a particle moving along a coordinate line is

$$s = 10 \cos(t + \pi/4).$$

a) What is the particle's starting position $(t = 0)$?

b) What are the points farthest to the left and right of the origin reached by the particle?

c) Find the particle's velocity and acceleration at the points in (b).

d) When does the particle first reach the origin? What are its velocity, speed, and acceleration then?

23. On Earth, you can easily shoot a paper clip 64 ft into the air with a rubber band. In t sec after firing, the paper clip is $s = 64t - 16t^2$ ft above your hand.

a) How long does it take the paper clip to reach its maximum height? With what velocity does it leave your hand?

b) On the moon, the same force will send the paper clip to a height of $s = 64t - 2.6t^2$ ft in t sec. About how long will it take the paper clip to reach its maximum height and how high will it go?

24. At time t sec, the positions of two particles on a coordinate line are $s_1 = 3t^3 - 12t^2 + 18t + 5$ m and $s_2 = -t^3 + 9t^2 - 12t$ m. When do the particles have the same velocities?

25. The following data give the coordinates s of a moving body for various values of t. Plot s vs. t on coordinate paper and sketch a smooth curve through the given points. Assuming that this smooth curve represents the motion of the body, estimate the velocity at (a) $t = 1.0$; (b) $t = 2.5$; (c) $t = 2.0$.

s (ft)	10	38	58	70	74	70	58	38	10
t (sec)	0	0.5	1.0	1.5	2.0	2.5	3.0	3.5	4.0

26. The graphs in Fig. 2.68 show the distance traveled (mi), velocity (mph), and acceleration (mph/sec) for a 2-min automobile trip. Which graph shows (a) distance? (b) velocity? (c) acceleration?

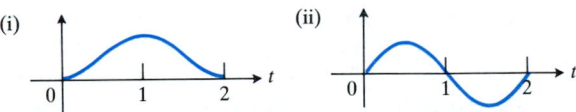

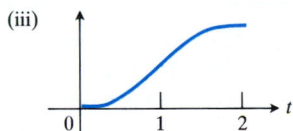

2.68   The graphs for Exercise 26.

27. *Marginal revenue.*   A bus will hold 60 people. The number x of people per trip who use the bus is related to the fare charged (p dollars) by the law $p = [3 - (x/40)]^2$. Write an expression for the total revenue $r(x)$ per trip received by the bus company. What number of people per trip will make the marginal revenue dr/dx equal to zero? What is the corresponding fare? (This is the fare that maximizes the revenue, so the bus company should probably rethink its fare policy.)

28. *Industrial production*

a) Economists often use the expression "rate of growth" in relative rather than absolute terms. For example, let

$u = f(t)$ be the number of people in the labor force at time t in a given industry. (We treat this function as though it were differentiable even though it is an integer-valued step function. We model its graph with a smooth curve.)

Let $v = g(t)$ be the average production per person in the labor force at time t. The total production is then $y = uv$. If the labor force is growing at the rate of 4% per year ($du/dt = 0.04u$) and the production per worker is growing at the rate of 5% per year ($dv/dt = 0.05v$), find the rate of growth of the total production, y.

b) Suppose that the labor force in (a) is decreasing at the rate of 2% per year while the production per person is increasing at the rate of 3% per year. Is the total production increasing, or is it decreasing, and at what rate?

29. A particle of constant mass m moves along the x-axis. Its velocity v and position x satisfy the equation

$$\frac{1}{2} m(v^2 - v_0^2) = \frac{1}{2} k(x_0^2 - x^2),$$

where k, v_0, and x_0 are constants. Show that whenever $v \neq 0$,

$$m \frac{dv}{dt} = -kx.$$

30. *The rate of a chemical reaction.* When two chemicals, A and B, combine to form an amount p of product, the rate dp/dt at which the product forms is called the **reaction rate**. In many reactions, one molecule of product is formed from one molecule of A and one molecule of B. Suppose that the initial molar masses of A and B are equal, both having value a. Under these conditions, the amount of product at any time t after mixing is given by the function $p = a^2 kt / (akt + 1)$. In this equation, k is a positive constant of proportionality from the chemical law of mass action, representing the affinity of the two chemicals.

a) Find dp/dt.

b) Find the time at which the reaction is proceeding at its fastest rate and the value of dp/dt at that time.

31. Find all values of the constants m and b for which the function

$$y = \begin{cases} \sin x & \text{for } x < \pi, \\ mx + b & \text{for } x \geq \pi, \end{cases}$$

is (a) continuous at $x = \pi$; (b) differentiable at $x = \pi$.

32. Does the function

$$f(x) = \begin{cases} \dfrac{1 - \cos x}{x} & \text{for } x \neq 0, \\ 0 & \text{for } x = 0, \end{cases}$$

have a derivative at $x = 0$? Explain.

33. The graph of $y = \sin(x - \sin x)$ in Fig. 2.69 suggests that the curve might have horizontal tangents at the x-axis. Does it?

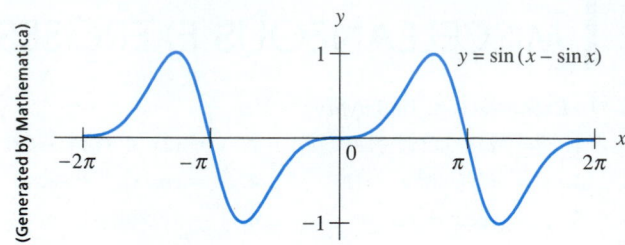

2.69 The curve in Exercise 33.

34. Figure 2.70 shows a boat 1 km offshore, sweeping the shore with a searchlight. The light turns at the constant rate $d\theta/dt = -3/5$ radians per second. (This rate is the light's *angular velocity.*)

a) Express x (see the figure) in terms of θ.

b) Differentiate both sides of the equation you obtained in (a) with respect to t. Then substitute $d\theta/dt = -3/5$. This will express dx/dt (the rate at which the light moves along the shore) as a function of θ.

c) How fast (m/sec) is the light moving along the shore when it reaches point A?

d) How many revolutions per minute is 0.6 radian per second?

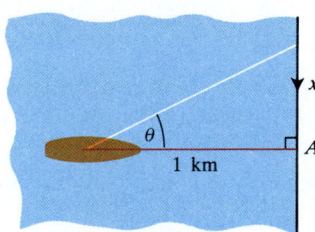

2.70 The boat in Exercise 34.

35. If the identity $\sin(x + a) = \sin x \cos a + \cos x \sin a$ is differentiated with respect to x, is the resulting equation also an identity? Does this principle apply to the equation $x^2 - 2x - 8 = 0$? Explain.

36. Find dy/dt at $t = 0$ if $y = 3 \sin 2x$ and $x = t^2 + \pi$.

37. Find ds/du at $u = 2$ if $s = t^2 + 5t$ and $t = (u^2 + 2u)^{1/3}$.

38. Find dw/ds at $s = 0$ if $w = \sin(\sqrt{r} - 2)$ and $r = 8 \sin(s + \pi/6)$.

39. Find the points where the tangent to the curve $y = \sqrt{x}$ at $x = 4$ crosses the coordinate axes.

40. What horizontal line crosses the curve $y = \sqrt{x}$ at a 45° angle?

41. Find the tangent to the curve $y = 2/\sqrt{x - 1}$ at the point where $x = 10$.

42. Write an equation for the line through $(2, 1)$ normal to the curve $x^2 = 4y$.

43. Find d^3y/dx^3 if

a) $y = \sqrt{2x - 1}$
b) $y = 1/(3x + 2)$
c) $y = ax^3 + bx^2 + cx + d$.

44. For what value of c is the curve $y = c/(x + 1)$ tangent to the line through the points $(0, 3)$ and $(5, -2)$?

45. Show that the normal line at any point of the circle $x^2 + y^2 = a^2$ passes through the origin.

46. Find the lines that are tangent and normal to the curve $(y - x)^2 = 2x + 4$ at the point $(6, 2)$.

47. Find the lines that are tangent and normal to the curve at the given point.

a) $x^2 + 2y^2 = 9$ at $(1, 2)$
b) $x^3 + y^2 = 2$ at $(1, 1)$
c) $xy + 2x - 5y = 2$ at $(3, 2)$

48. Which of the following statements could be true if $f''(x) = x^{1/3}$?

I. $f(x) = \dfrac{9}{28} x^{7/3} + 9$ II. $f'(x) = \dfrac{9}{28} x^{7/3} - 2$

III. $f'(x) = \dfrac{3}{4} x^{4/3} + 6$ IV. $f(x) = \dfrac{3}{4} x^{4/3} - 4$

a)   I only   b)   III only
c)   II and IV only   d)   I and III only

49. The designer of a 30-ft-diameter spherical hot-air balloon wishes to suspend the gondola 8 ft below the bottom of the balloon with suspension cables tangent to the surface of the balloon. Figure 2.71 shows two of the cables running from the top edges of the gondola to their points of tangency, $(-12, -9)$ and $(12, -9)$. How wide must the gondola be?

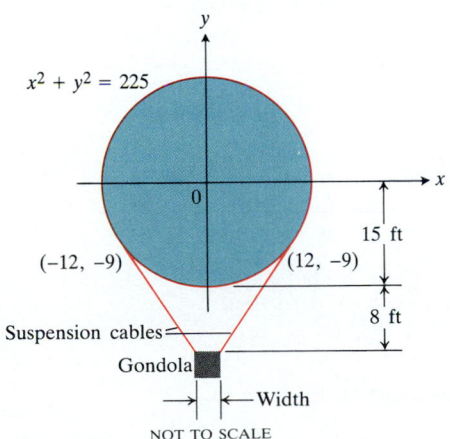

2.71   The balloon and gondola in Exercise 49.

50. *What determines the frequency of a vibrating piano string?*   We measure the frequencies at which wires vibrate in cycles (trips back and forth) per second. The unit of measure is a hertz (Hz): 1 cycle per second. Middle A on a piano has a frequency of 440 Hz. For any given wire, the fundamental frequency y is a function of four variables:

  r: the radius of the wire,
  l: the length,
  d: the density of the wire,
  T: the tension (force) holding the wire taut.

With r and l in centimeters, d in grams per cubic centimeter, and T in dynes (it takes about 100,000 dynes to lift an apple), the fundamental frequency of the wire is

$$y = \frac{1}{2rl} \sqrt{\frac{T}{\pi d}}.$$

If we keep all the variables fixed except one, then y can be alternately thought of as four different functions of one variable: $y(r)$, $y(l)$, $y(d)$, and $y(T)$. How would changing each variable then affect the string's fundamental frequency? To find out, calculate $y'(r)$, $y'(l)$, $y'(d)$, and $y'(T)$.

51. If $y^3 + y = 2 \cos x$, find the value of d^2y/dx^2 at the point $(0, 1)$.

52. If $x^{1/3} + y^{1/3} = 4$, find d^2y/dx^2 at the point $(8, 8)$.

53. a) By differentiating $x^2 - y^2 = 1$ implicitly, show that $dy/dx = x/y$.
b) Then show that $d^2y/dx^2 = -1/y^3$.

54. *An osculating circle.*   Find the values of h, k, and a that make the circle $(x - h)^2 + (y - k)^2 = a^2$ tangent to the parabola $y = x^2 + 1$ at the point $(1, 2)$ and that also make the second derivatives d^2y/dx^2 have the same value on both curves there. Circles like this one that are tangent to a curve and have the same second derivative as the curve at the point of tangency are called *osculating circles* (from the Latin *osculari* meaning "to kiss"). We shall encounter them again in Chapter 11.

55. We can obtain a useful linear approximation to $1/(1 + \tan x)$ at $x = 0$ by combining the approximations

$$\frac{1}{1 + x} \approx 1 - x \qquad \text{and} \qquad \tan x \approx x$$

to get

$$\frac{1}{1 + \tan x} \approx 1 - x.$$

Show that this is the standard linear approximation of $1/(1 + \tan x)$ at $x = 0$.

56. Find the linearization of $f(x) = \sqrt{1 + x} + \sin x - 0.5$ at $x = 0$.

57. Find the linearization of $f(x) = 2/(1 - x) + \sqrt{1 + x} - 3.1$ at $x = 0$.

58. Write a formula that estimates the change that occurs in
   a) the volume of a right circular cone when the radius changes from r_0 to $r_0 + dr$ and the height does not change;
   b) the lateral surface area $S = \pi r \sqrt{r^2 + h^2}$ of a right circular cone when the height changes from h_0 to $h_0 + dh$ and the radius does not change.

59. a) How accurately should you measure the edge of a cube to be reasonably sure of calculating the cube's surface area with an error of no more than 2%?
   b) Suppose the edge is measured with the accuracy required in (a). About how accurately can the cube's volume be calculated from the edge measurement? To find out, estimate the percentage error in the volume calculation that would result from using the edge measurement.

60. The circumference of the equator of a sphere is measured as 10 cm with a possible error of 0.4 cm. The measurement is then used to calculate the radius. The radius is then used to calculate the surface area and volume of the sphere. Estimate the percentage errors in the calculated values of (a) the radius, (b) the surface area, and (c) the volume.

61. To find the height of a tree above horizontal ground you measure the angle α from the ground to the treetop from a point 100 ft away from the tree's base and calculate the height with the formula $h = 100 \tan \alpha$. The best figure you can get for α with the equipment at hand is $30° \pm 1°$. About how much error could the tolerance of $\pm 1°$ create in your calculation of h? Remember to work in radians.

62. To find the height of a lamppost (Fig 2.72), you stand a 6-ft pole 20 ft from the lamp and measure the length of its shadow. The figure you get for a is 15 ft, give or take an inch. Calculate the height of the lamppost from the value $a = 15$ and estimate the possible error in the result.

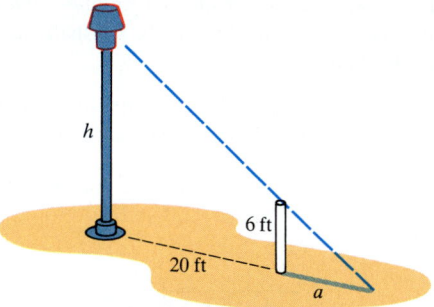

2.72 The lamppost in Exercise 62.

63. *The period of a clock pendulum.* The period T of a clock pendulum (time for one full swing and back) is given by the formula $T^2 = 4\pi^2 L/g$, where T is measured in seconds, $g = 32.2$ ft/sec², and L, the length of the pendulum, is measured in feet. Find approximately
   a) the length of a clock pendulum whose period is $T = 1$ sec;

   b) the change dT in T if the pendulum in (a) is lengthened 0.01 ft; and
   c) the amount the clock gains or loses in a day as a result of the period's changing by the amount dT found in (b).

64. *The linearization is the best linear approximation.* Suppose that $y = f(x)$ is differentiable at $x = a$ and that $g(x) = m(x - a) + c$ is a linear function in which m and c are constants (Fig. 2.73). If the error $E(x) = f(x) - g(x)$ were small enough near $x = a$, we might think of using g as a linear approximation of f instead of the linearization $L(x) = f(a) + f'(a)(x - a)$. Show that if we impose on g the conditions

1. $E(a) = 0$ $\qquad \left(\begin{array}{l} \text{The approximation error} \\ \text{is zero at } x = a. \end{array} \right)$

2. $\displaystyle \lim_{x \to a} \frac{E(x)}{x - a} = 0$ $\qquad \left(\begin{array}{l} \text{The error is negligible when} \\ \text{compared with } (x - a). \end{array} \right)$

then $g(x) = f(a) + f'(a)(x - a)$. Thus, the linearization gives the only linear approximation whose error is both zero at $x = a$ and negligible in comparison with $x - a$.

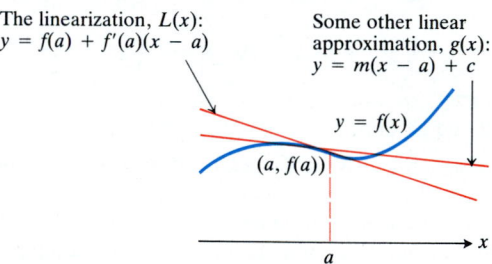

The linearization, $L(x)$:
$y = f(a) + f'(a)(x - a)$

Some other linear approximation, $g(x)$:
$y = m(x - a) + c$

$y = f(x)$

$(a, f(a))$

2.73 The linearization of f at a gives the best linear approximation of f at a (Exercise 64).

65. CALCULATOR Use Newton's method to find where the curve $y = -x^3 + 3x + 4$ crosses the x-axis (Fig 2.74). Give your answer to five decimal places.

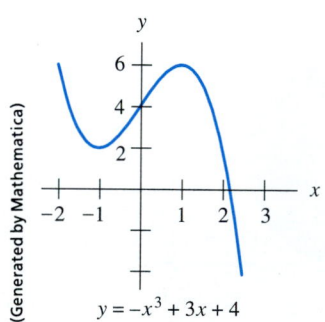

(Generated by Mathematica)

$y = -x^3 + 3x + 4$

2.74 The graph for Exercise 65.

66. COMPUTER OR PROGRAMMABLE CALCULATOR Solve the equation $2 \cos x - \sqrt{1 + x} = 0$ to five decimal places (Fig. 2.75).

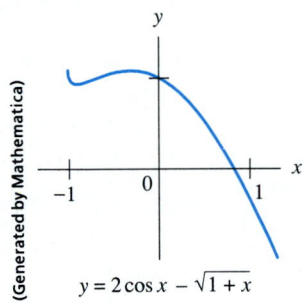

(Generated by Mathematica)

$y = 2 \cos x - \sqrt{1 + x}$

2.75   The graph for Exercise 66.

67. *A surprising result.*   Suppose that the functions f and g are defined throughout an open interval containing the point x_0, that f is differentiable at x_0, that $f(x_0) = 0$, and that g is continuous at x_0. Show that the product fg is differentiable at x_0. This shows, for example, that while $|x|$ is not differentiable at $x = 0$, the product $x|x|$ *is* differentiable at $x = 0$. (*Hint:* Write down the difference quotient for the product fg and see what its limit has to be.)

68. *Continuation of Exercise 67.*   Use the result of Exercise 67 to show that the following functions are differentiable at $x = 0$.

a) $|x| \sin x$ b) $x^{2/3} \sin x$ c) $\sqrt[3]{x}\,(1 - \cos x)$

d) $h(x) = \begin{cases} x^2 \sin(1/x), & x \neq 0 \\ 0, & x = 0 \end{cases}$

69. *The generalized product rule.*   Use mathematical induction (Appendix 5) to prove that if $y = u_1 u_2 \cdots u_n$ is a finite product of differentiable functions, then y is differen-

tiable on their common domain and

$$\frac{dy}{dx} = \frac{du_1}{dx} u_2 \cdots u_n + u_1 \frac{du_2}{dx} \cdots u_n + \cdots$$

$$+ u_1 u_2 \cdots u_{n-1} \frac{du_n}{dx}.$$

This is Eq. (14) in Exercises 2.1.

70. *Leibniz's rule for higher order derivatives of products.* Leibniz's rule for higher order derivatives of products of differentiable functions says that

a) $\dfrac{d^2(uv)}{dx^2} = \dfrac{d^2u}{dx^2} v + 2 \dfrac{du}{dx}\dfrac{dv}{dx} + u \dfrac{d^2v}{dx^2},$

b) $\dfrac{d^3(uv)}{dx^3} = \dfrac{d^3u}{dx^3} v + 3 \dfrac{d^2u}{dx^2}\dfrac{dv}{dx} + 3 \dfrac{du}{dx}\dfrac{d^2v}{dx^2} + u \dfrac{d^3v}{dx^3},$

c) $\dfrac{d^n(uv)}{dx^n} = \dfrac{d^nu}{dx^n} v + n \dfrac{d^{n-1}u}{dx^{n-1}}\dfrac{dv}{dx} + \cdots$

$$+ \frac{n(n-1)\cdots(n-k+1)}{k!}\frac{d^{n-k}u}{dx^{n-k}}\frac{d^kv}{dv^k}$$

$$+ \cdots + u\frac{d^nv}{dx^n}.$$

The equations in (a) and (b) are special cases of the equation in (c). Derive the equation in (c) by mathematical induction, using the fact that $\dbinom{m}{k} + \dbinom{m}{k+1} = \dfrac{m!}{k!\,(m-k)!} + \dfrac{m!}{(k+1)!(m-k-1)!}.$

71. Suppose that a function f satisfies the following conditions for all real values of x and y:

1. $f(x + y) = f(x) \cdot f(y);$

2. $f(x) = 1 + xg(x),$ where $\lim\limits_{x \to 0} g(x) = 1.$

Show that the derivative $f'(x)$ exists at every value of x and that $f'(x) = f(x).$

CHAPTER

3

APPLICATIONS OF DERIVATIVES

Overview   This chapter describes some of the most important applications of derivatives. It shows how to calculate rates of change that we cannot measure directly from rates of change that we can. It shows how to draw reliable graphs and how to find the largest and smallest values that a differentiable function assumes on a closed interval. It also shows how to use an ingenious rule to find limits of quotients when both numerator and denominator approach zero. It shows how to use a formula from the eighteenth century to tell how accurate linearizations are. The formula even provides a way to add quadratic terms to linearizations to make them more accurate.

The theorem that makes many of these accomplishments possible is the Mean Value Theorem of differential calculus. In Chapter 4 we shall see how the same theorem provides the gateway to integral calculus, the second main branch of the field.

3.1 Related Rates of Change

How rapidly will the fluid level inside a vertical cylindrical storage tank drop if we pump the fluid out at the rate of 3000 L/min?

A question like this asks us to calculate a rate that we cannot measure directly from a rate that we can. To do so, we write an equation that relates the variables involved and differentiate it to get an equation that relates the rate we seek to the rate we know.

Example 1   *Pumping Out a Tank.*   How rapidly will the fluid level inside a vertical cylindrical tank drop if we pump the fluid out at the rate of 3000 L/min?

Solution   We draw a picture of a partially filled vertical cylindrical tank, calling its radius r and the height of the fluid h (Fig. 3.1). Call the volume of the fluid V.

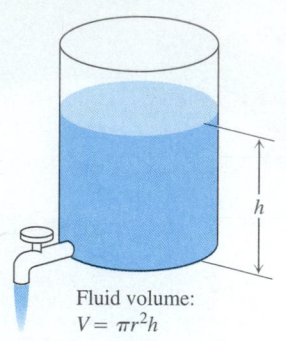

Fluid volume:
$V = \pi r^2 h$

3.1   The cylindrical tank in Example 1.

As time passes, the radius remains constant but V and h change. We think of V and h as differentiable functions of time and use t to represent time. We are told that

$$\frac{dV}{dt} = -3000. \qquad \text{(We pump at the rate of 3000 L/min.)}$$

We are asked to find

$$\frac{dh}{dt}. \qquad \text{(How fast will the fluid level drop?)}$$

To find dh/dt, we first write an equation that relates h to V. The equation depends on the units chosen for V, r, and h. With V in liters and r and h in meters, the appropriate equation is

$$V = 1000\pi r^2 h$$

because a cubic meter contains 1000 liters.

We then use the Chain Rule to differentiate both sides of the equation $V = 1000\pi r^2 h$ with respect to t to get an equation that relates dh/dt to dV/dt:

$$\frac{dV}{dt} = 1000\pi r^2 \frac{dh}{dt}. \qquad (1)$$

We substitute the known value $dV/dt = -3000$ and solve for dh/dt:

$$\frac{dh}{dt} = \frac{-3000}{1000\pi r^2} = -\frac{3}{\pi r^2}. \qquad (2)$$

The fluid level will drop at the rate of $3/\pi r^2$ m/min.

Equation (2) shows how the rate at which the fluid level drops depends on the tank's radius. If r is small, dh/dt will be large; if r is large, dh/dt will be small.

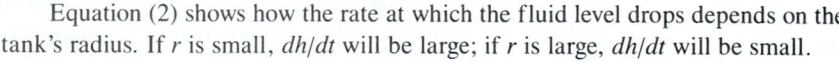

If $r = 1$ m: $\qquad \dfrac{dh}{dt} = -\dfrac{3}{\pi} \approx -0.95$ m/min $= -95$ cm/min.

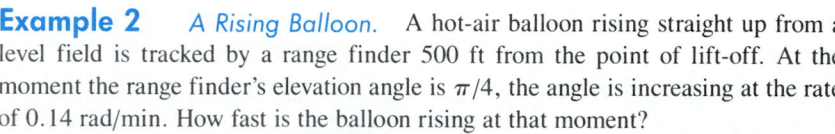

If $r = 10$ m: $\qquad \dfrac{dh}{dt} = -\dfrac{3}{100\pi} \approx -0.0095$ m/min $= -0.95$ cm/min.

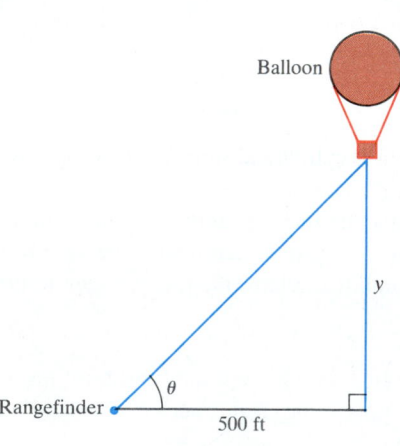

Balloon

y

Rangefinder

θ

500 ft

3.2   If $d\theta/dt = 0.14$ when $\theta = \pi/4$, what is the value of dy/dt when $\theta = \pi/4$? See Example 2.

Example 2   *A Rising Balloon.*   A hot-air balloon rising straight up from a level field is tracked by a range finder 500 ft from the point of lift-off. At the moment the range finder's elevation angle is $\pi/4$, the angle is increasing at the rate of 0.14 rad/min. How fast is the balloon rising at that moment?

Solution   We answer the question in six steps.

STEP 1: *Draw a picture and name the variables and constants* (Fig. 3.2). The variables in the picture are

$\qquad \theta = $ the angle the range finder makes with the ground (radians),

$\qquad y = $ the height of the balloon (feet).

We let t represent time and assume θ and y to be differentiable functions of t.

The one constant in the picture is the distance from the range finder to the point of lift-off (500 ft). There is no need to give it a special symbol.

STEP 2: *Write down the additional numerical information .*

$$\frac{d\theta}{dt} = 0.14 \text{ rad/min} \qquad \text{when} \quad \theta = \frac{\pi}{4}$$

STEP 3: *Write down what we are asked to find .* We want dy/dt when $\theta = \pi/4$.

STEP 4: *Write an equation that relates the variables y and θ .*

$$\frac{y}{500} = \tan \theta, \qquad \text{or} \qquad y = 500 \tan \theta$$

STEP 5: *Differentiate with respect to t to find how dy/dt* (which we want) *is related to $d\theta/dt$* (which we know).

$$\frac{dy}{dt} = 500 \sec^2\theta \, \frac{d\theta}{dt}$$

STEP 6: *Evaluate with $\theta = \pi/4$ and $d\theta/dt = 0.14$ to find dy/dt .*

$$\frac{dy}{dt} = 500(\sqrt{2})^2(0.14) = (1000)(0.14) = 140 \qquad \left(\sec\frac{\pi}{4} = \sqrt{2}\right)$$

At the moment in question, the balloon is rising at the rate of 140 ft/min.

Strategy for Solving Related Rate Problems

1. *Draw a picture and name the variables and constants.* Use t for time. Assume all variables are differentiable functions of t.

2. *Write down the numerical information* (in terms of the symbols you have chosen).

3. *Write down what you are asked to find* (usually a rate, expressed as a derivative).

4. *Write an equation that relates the variables.* You may have to combine two or more equations to get a single equation that relates the variable whose rate you want to the variable whose rate you know.

5. *Differentiate* with respect to t to express the rate you want in terms of the rate and variables whose values you know.

6. *Evaluate.*

Example 3   *Truck Convoys.*   Two truck convoys leave a depot, convoy A traveling east at 40 mph and convoy B traveling north at 30 mph. How fast is the distance between the convoys changing 6 min later, when convoy A is 4 mi from the depot and convoy B is 3 mi from the depot?

Solution   We carry out the steps of the basic strategy.

STEP 1: *Picture and variables.* We picture the convoys in the coordinate plane, using the positive x-axis as the eastbound highway and the positive y-axis as the

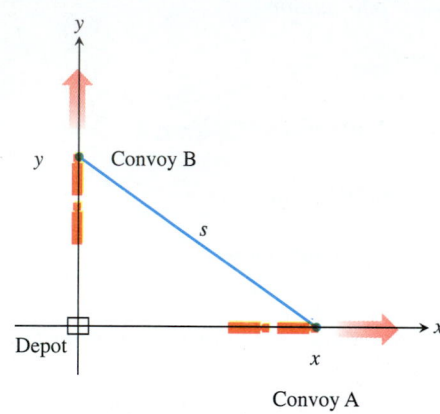

northbound highway (Fig. 3.3). We let t represent time and set

$$x = \text{position of convoy A at time } t,$$

$$y = \text{position of convoy B at time } t,$$

$$s = \text{distance between convoys at time } t.$$

We assume x, y, and s to be differentiable functions of t.

STEP 2: *Numerical information.* At the time in question,

$$x = 4 \text{ mi,} \qquad y = 3 \text{ mi,} \qquad \frac{dx}{dt} = 40 \text{ mph,} \qquad \frac{dy}{dt} = 30 \text{ mph.}$$

STEP 3: *To find:* $\dfrac{ds}{dt}$.

3.3 If you know where the convoys are and how fast they are moving, you can calculate how fast the distance between them is growing (Example 3).

STEP 4: *How the variables are related:* $s^2 = x^2 + y^2$.   (Pythagorean theorem)

(The equation $s = \sqrt{x^2 + y^2}$ would also work.)

STEP 5: *Differentiate with respect to t.*

$$2s\frac{ds}{dt} = 2x\frac{dx}{dt} + 2y\frac{dy}{dt}$$

$$\frac{ds}{dt} = \frac{1}{s}\left(x\frac{dx}{dt} + y\frac{dy}{dt}\right)$$

$$= \frac{1}{\sqrt{x^2 + y^2}}\left(x\frac{dx}{dt} + y\frac{dy}{dt}\right)$$

STEP 6: *Evaluate with $x = 4$, $y = 3$, $dx/dt = 40$, $dy/dt = 30$.*

$$\frac{ds}{dt} = \frac{1}{\sqrt{4^2 + 3^2}}\,(4(40) + 3(30)) = \frac{1}{5}\,(160 + 90) = \frac{250}{5} = 50$$

At the moment in question, the distance between the convoys is growing at the rate of 50 mph.

Example 4   Water runs into a conical tank at the rate of 9 ft³/min. The tank stands point down and has a height of 10 ft and a base radius of 5 ft. How fast is the water level rising when the water is 6 ft deep?

Solution We draw and label a picture of a partially filled conical tank (Fig. 3.4). The variables in the problem are

$$V = \text{the volume (ft}^3\text{) of water in the tank at time } t \text{ (min),}$$

$$x = \text{the radius (ft) of the surface of the water at time } t,$$

$$y = \text{the depth (ft) of water in the tank at time } t.$$

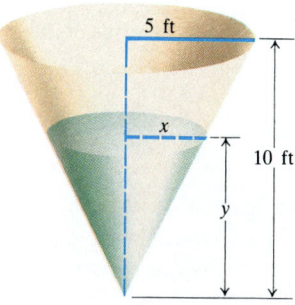

3.4 The conical tank in Example 4. To show that the water level is changing, we denote the depth of the water by a variable, in this case y. The rate at which the level changes is then dy/dt.

The constants are the dimensions of the tank, and the rate

$$\frac{dV}{dt} = 9 \text{ ft}^3/\text{min.}$$

We are asked for the value of dy/dt when $y = 6$.

The relation between V and y is expressed by the equation

$$V = \frac{1}{3}\pi x^2 y. \tag{3}$$

This equation involves x as well as V and y, but we can eliminate x in the following way. By similar triangles (Fig. 3.4),

$$\frac{x}{y} = \frac{5}{10} \quad \text{or} \quad x = \frac{1}{2}y.$$

Therefore,

$$V = \frac{1}{3}\pi\left(\frac{1}{2}y\right)^2 y = \frac{1}{12}\pi y^3. \tag{4}$$

We differentiate both sides of Eq. (4) to express dy/dt in terms of dV/dt:

$$\frac{dV}{dt} = \frac{1}{4}\pi y^2 \frac{dy}{dt} \quad \text{or} \quad \frac{dy}{dt} = \frac{4}{\pi y^2}\frac{dV}{dt}. \tag{5}$$

When $dV/dt = 9$ and $y = 6$,

$$\frac{dy}{dt} = \frac{4}{\pi \cdot 36} \cdot 9 = \frac{1}{\pi} = 0.32 \text{ ft/min.} \quad \text{(Rounded)}$$

At the moment in question, the water level is rising at the rate of 0.32 ft/min.

EXERCISES 3.1

1. Suppose that the radius r and area $A = \pi r^2$ of a circle are differentiable functions of t. Write an equation that relates dA/dt to dr/dt.

2. Suppose that the radius r and surface area $S = 4\pi r^2$ of a sphere are differentiable functions of t. Write an equation that relates dS/dt to dr/dt.

3. Suppose that the side length x and volume $V = x^3$ of a cube are differentiable functions of t. Write an equation that relates dV/dt to dx/dt.

4. Suppose that the radius r and volume $V = (1/3)\pi r^2 h$ of a right circular cone are differentiable functions of t. How is dV/dt related to dr/dt if h is constant?

5. Suppose that the height h and volume $V = (1/3)\pi r^2 h$ of a right circular cone are differentiable functions of t. How is dV/dt related to dh/dt if r is constant?

6. Let x and y be differentiable functions of t and let $s = \sqrt{x^2 + y^2}$ be the distance between the points $(x, 0)$ and $(0, y)$ in the xy-plane. How is ds/dt related to dx/dt and dy/dt?

7. *Heating a plate.* When a circular plate of metal is heated in an oven, its radius increases at the rate of 0.01 cm/min. At what rate is the plate's area increasing when the radius is 50 cm?

8. *Changing voltage.* Ohm's law for electrical circuits like the one in Fig. 3.5 states that $V = IR$, where V is the voltage, I is the current in amperes, and R is the resistance in ohms. Suppose that V is increasing at the rate of 1 volt/sec while I is decreasing at the rate of 1/3 amp/sec. Let t denote time in seconds.
   a) What is the value of dV/dt?
   b) What is the value of dI/dt?
   c) What equation relates dR/dt to dV/dt and dI/dt?
   d) Find the rate at which R is changing when $V = 12$ volts and $I = 2$ amp. Is R increasing, or decreasing?

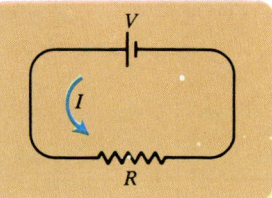

3.5 The circuit in Exercise 8.

9. *Changing dimensions in a rectangle.* The length l of a rectangle is decreasing at the rate of 2 cm/sec while the width w is increasing at the rate of 2 cm/sec. When $l = 12$ cm and $w = 5$ cm, find the rates of change of (a) the

area, (b) the perimeter, and (c) the lengths of the diagonals of the rectangle. Which of these quantities are decreasing and which are increasing?

10. *Commercial air traffic.*   Two commercial jets at 40,000 ft are flying at 520 mph along straight-line courses that cross at right angles. How fast is the distance between the planes closing when plane A is 5 mi from the intersection point and plane B is 12 mi from the intersection point?

11. *A sliding ladder.*   A 13-ft ladder resting on horizontal ground is leaning against a vertical wall when its base starts to slide away from the wall. By the time the base is 12 ft from the wall, the base is moving at the rate of 5 ft/sec. How fast is the top of the ladder sliding down the wall then? How fast is the area of the triangle formed by the ladder, wall, and ground changing?

12. *Boring a cylinder.*   The mechanics at Lincoln Automotive are reboring a 6-in.-deep cylinder to fit a new piston. The machine they are using increases the cylinder's radius one-thousandth of an inch every 3 min. How rapidly is the cylinder volume increasing when the bore (diameter) is 3.80 in.?

13. *A growing sand pile.*   Sand falls from a conveyor belt at the rate of 10 ft³/min onto a conical pile. The radius of the base of the pile is always equal to half the pile's height. How fast is the height growing when the pile is 5 ft high?

14. *A growing raindrop.*   Suppose that a drop of mist is a perfect sphere and that, through condensation, the drop picks up moisture at a rate proportional to its surface area. Show that under these circumstances the drop's radius increases at a constant rate.

15. *The radius of an inflating balloon.*   A spherical balloon is inflated with helium at the rate of 100π ft³/min. How fast is the balloon's radius increasing at the instant the radius is 5 ft? How fast is the surface area increasing?

16. *Hauling in a dinghy.*   A dinghy is pulled toward a dock by a rope from the bow through a ring on the dock 6 ft above the bow (Fig. 3.6). If the rope is hauled in at the rate of 2 ft/sec, how fast is the boat approaching the dock when 10 ft of rope are out?

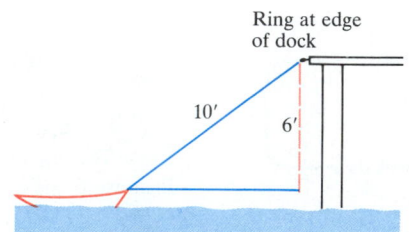

Ring at edge
of dock

10'

6'

3.6   The dinghy in Exercise 16.

17. *A balloon and a bicycle.*   A balloon is rising vertically above a level, straight road at a constant rate of 1 ft/sec. Just when the balloon is 65 ft above the ground, a bicycle passes

under it, going 17 ft/sec. How fast is the distance between the bicycle and balloon increasing 3 sec later?

18. *Making coffee.*   Coffee is draining from a conical filter into a cylindrical coffee pot at the rate of 10 in³/min (Fig. 3.7). How fast is the level in the pot rising when the coffee in the cone is 5 in. deep? How fast is the level in the cone falling then?

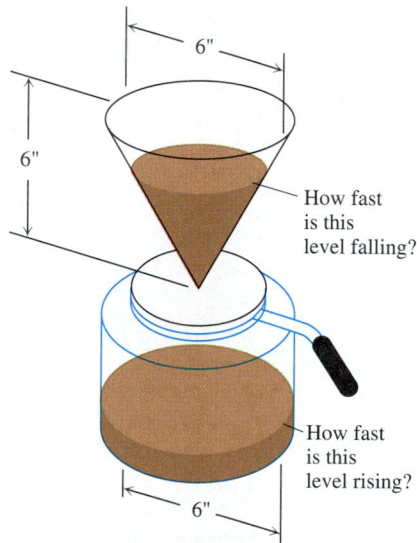

6"

6"

How fast
is this
level falling?

How fast
is this
level rising?

6"

3.7   The pot and filter in Exercise 18.

19. *Cardiac output.*   In the late 1860s, Adolf Fick, a professor of physiology in the faculty of medicine in Würzburg, Germany, developed the method we use today for measuring how much blood your heart pumps in a minute. Your cardiac output as you read this sentence is probably about 7 liters per minute. At rest it is likely to be a bit under 6 L/min. If you are a trained runner running a marathon, your cardiac output can be as high as 30 L/min.

Your cardiac output can be calculated with the formula

$$y = \frac{Q}{D},$$

where Q is the number of milliliters of CO_2 you exhale in a minute and D is the difference between the CO_2 concentration (mL/L) in the blood pumped to the lungs and the CO_2 concentration in the blood returning from the lungs. With $Q = 233$ mL/min and $D = 97 - 56 = 41$ mL/L,

$$y = \frac{233 \text{ mL/min}}{41 \text{ mL/L}} \approx 5.95 \text{ L/min},$$

close to the 6 L/min that most people have at rest. (Data courtesy of J. Kenneth Herd, M.D., Quillen College of Medicine, East Tennessee State University.)

Suppose that when $Q = 233$ and $D = 41$, we also know that D is decreasing at the rate of 2 units per minute but that Q remains unchanged. What is happening to the cardiac output?

20. Water is being pumped from an inverted right circular conical tank at the rate of 36 ft³/min. The tank, which stands vertex down and base up, has a height of 12 ft and a base diameter of 16 ft. How fast is the water level dropping when the water is 9 ft deep?

21. *Moving along a parabola.* A particle moves along the parabola $y = x^2$ in the first quadrant in such a way that its x-coordinate increases at a steady 10 m/sec. How fast is the angle of inclination θ of the line joining the particle to the origin changing when $x = 3$ m? What is the limiting value of $d\theta/dt$ as $x \to \infty$?

22. *Motion in the plane.* The coordinates of a particle in the metric xy-plane are differentiable functions of time t with $dx/dt = -1$ m/sec and $dy/dt = -5$ m/sec. How fast is the particle's distance from the origin changing as it passes through the point $(5, 12)$?

23. *Cost, revenue, and profit.* A company can manufacture x items at a cost of $c(x)$ dollars, a sales revenue of $r(x)$ dollars, and a profit of $p(x) = r(x) - c(x)$ dollars (everything in thousands). Find the rates of change of cost, revenue, and profit for the following values of x and dx/dt.
   a) $r(x) = 9x$, $c(x) = x^3 - 6x^2 + 15x$, and $dx/dt = 0.1$ when $x = 2$
   b) $r(x) = 70x$, $c(x) = x^3 - 6x^2 + 45/x$, and $dx/dt = 0.05$ when $x = 1.5$

24. *A moving shadow.* A man 6 ft tall walks at the rate of 5 ft/sec toward a streetlight that is 16 ft above the ground. At what rate is the tip of his shadow moving? At what rate is the length of his shadow changing?

25. *Another moving shadow.* A light shines from the top of a pole 50 ft high (Fig. 3.8). A ball is dropped from the same height from a point 30 ft away from the light. How fast is the shadow of the ball moving along the ground 1/2 sec later? (Assume the ball falls $s = 16t^2$ ft in t sec.)

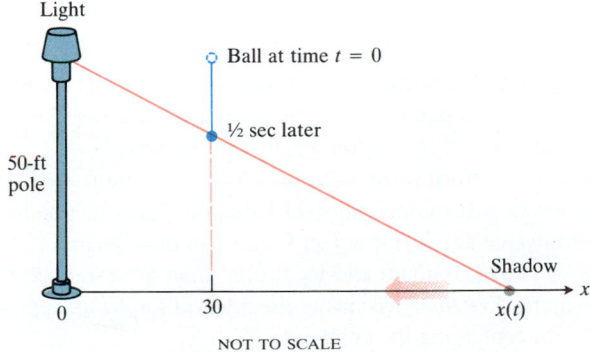

3.8   The streetlight and shadow in Exercise 25.

26. *A melting ice layer.* A spherical iron ball 8 in. in diameter is coated with a layer of ice of uniform thickness. If the ice melts at the rate of 10 in³/min, how fast is the thickness of the ice decreasing when it is 2 in. thick? How fast is the outer surface area of ice decreasing?

27. *Highway patrol.* A highway patrol plane flies 3 mi above a level, straight road at a steady ground speed of 120 mph. The pilot sees an oncoming car and determines with radar that the line-of-sight distance from the plane to the car is 5 mi and decreasing at the rate of 160 mph. Find the car's speed along the highway.

28. You are videotaping a race from a stand 132 ft from the track, following a car that is moving at 180 mph (264 ft/sec) (Fig. 3.9). How fast will your camera angle θ be changing when the car is right in front of you? A half second later?

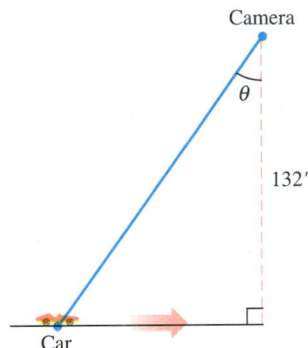

3.9   The camera angle in Exercise 28.

29. CALCULATOR *A building's shadow.* On the morning of a day when the sun will pass directly overhead, the shadow of an 80-ft building on level ground is 60 ft long (Fig. 3.10). At the moment in question, the angle θ the sun makes with the ground is increasing at the rate of 0.27°/min. At what rate is the length of the building's shadow decreasing? (Remember to use radians. Express your answer in inches per minute to the nearest tenth.)

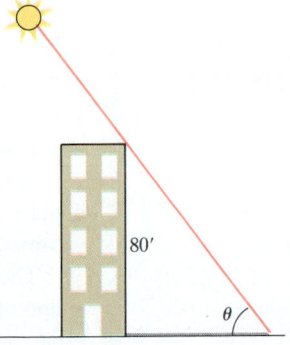

3.10   The building in Exercise 29.

30. CALCULATOR Two ships are steaming straight away from a point O along routes that make a 120° angle. Ship A moves at 14 knots (nautical miles per hour—a nautical mile is 2000 yd). Ship B moves at 21 knots. How fast is the distance between the ships increasing when $OA = 5$ and $OB = 3$ nautical miles?

3.2 Maxima, Minima, and the Mean Value Theorem

One of the useful things we do with derivatives is find where functions take on their maximum and minimum values. In this section we lay the theoretical ground for finding these extreme values, as they are called. This will enable us to graph polynomials and rational functions effectively (Sections 3.3 and 3.4) and to solve important optimization problems from other fields (Section 3.5).

We also introduce one of the most influential theorems in calculus, the Mean Value Theorem.

Maxima and Minima—Local vs. Absolute

Figure 3.11 shows a point c where a function $y = f(x)$ has a maximum value. If we move to either side of c, the function values get smaller and the curve falls away. When we take in more of the curve, however, we find that f assumes an even larger value at d. Thus, $f(c)$ is not the absolute maximum value of f on the interval $[a, b]$ but only a relative or local maximum value.

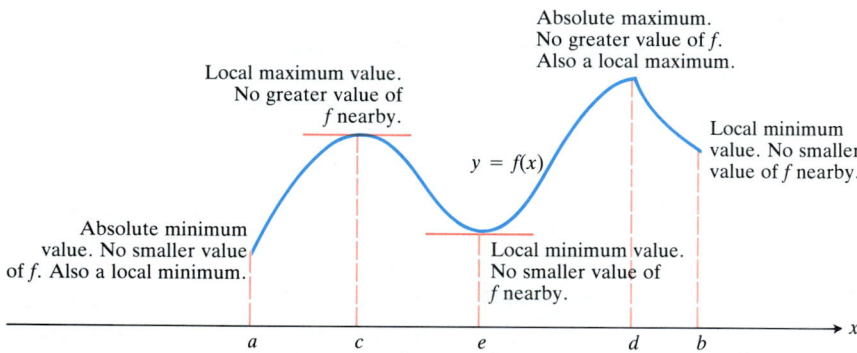

3.11   How to classify maxima and minima.

DEFINITIONS

A function f has a **local maximum** value at an interior point c of its domain if $f(x) \leq f(c)$ for all x in some open interval I about c. The function has an **absolute maximum** value at c if $f(x) \leq f(c)$ for all x in the domain.

Similarly, f has a **local minimum** value at an interior point c of its domain if $f(x) \geq f(c)$ for all x in an open interval I about c. The function has an **absolute minimum** value at c if $f(x) \geq f(c)$ for all x in the domain.

The definitions of local maximum and local minimum are extended to endpoints of the function's domain by requiring the interval I to be an appropriate half-open interval containing the endpoint.

Notice that an absolute maximum is also a local maximum: Being the largest value overall, it is also the largest value in its immediate neighborhood. Hence a list of all local maxima will include the absolute maximum if there is one. Similarly, an absolute minimum, when it exists, is also a local minimum. A list of all local minima will include the absolute minimum if there is one.

The First Derivative Theorem

In Fig. 3.11 two extreme values of f occur at endpoints of the function's domain, one occurs at a point where f' fails to exist, and two occur at interior points where $f' = 0$. This is typical for a function defined on a closed interval. As the following theorem says, a function's first derivative is always zero at an interior point where the function has a local extreme value. Hence the only places where a function f can ever have an extreme value are

1. interior points where f' is zero,

2. interior points where f' does not exist,

3. endpoints of the function's domain.

Points in the domain of f where f' is zero or fails to exist are called **critical points** of f. Thus, the only points worth considering in a search for a function's extreme values are critical points and endpoints. We shall see the importance of this observation as the chapter continues.

THEOREM 1

The First Derivative Theorem for Local Extreme Values

If a function f has a local maximum or a local minimum value at an interior point c of an interval where it is defined, and if f' is defined at c, then

$$f'(c) = 0.$$

Proof   You may not have seen an argument like the one we are about to use, so we shall explain its form first. We want to show that $f'(c) = 0$, and our plan is to do that indirectly by showing first that $f'(c)$ cannot be positive and second that $f'(c)$ cannot be negative either. Why does that show that $f'(c) = 0$? Because, in the entire real number system, only one number is neither positive nor negative, and that number is zero.

To be specific, suppose f has a local maximum value at $x = c$, so that $f(x) \leq f(c)$ for all values of x near c (Fig. 3.12). Since c is an interior point of f's domain, the limit

$$\lim_{x \to c} \frac{f(x) - f(c)}{x - c} \tag{1}$$

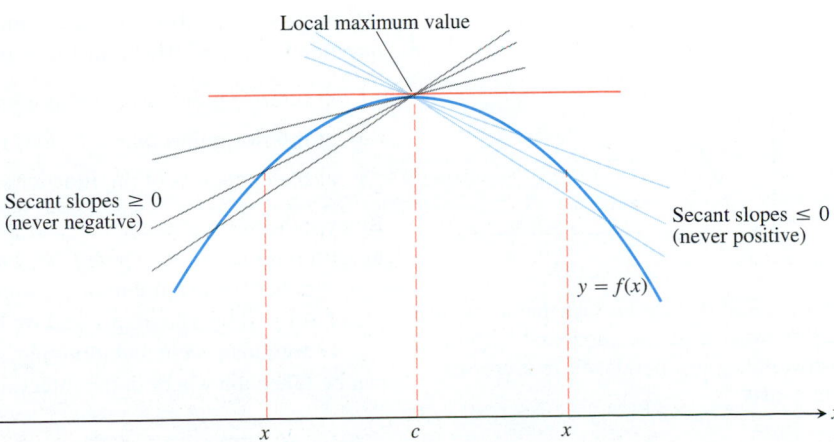

Local maximum value

Secant slopes ≥ 0
(never negative)

Secant slopes ≤ 0
(never positive)

$y = f(x)$

3.12   A curve with a local maximum value. The slope at c, simultaneously the limit of nonpositive numbers and nonnegative numbers, is zero.

When the French mathematician Michel Rolle published his theorem in 1691, his goal was to show that between every two zeros of a polynomial function there always lies a zero of the polynomial we now know to be the function's derivative. (The version of the theorem we prove here is not restricted to polynomials.)

Rolle distrusted the new methods of calculus, however, and spent a great deal of time and energy denouncing their use and attacking l'Hôpital's all too popular (he felt) calculus book. It is ironic that Rolle is known today only for his inadvertent contribution to a field he tried so hard to suppress.

defining $f'(c)$ is two-sided. (We used this formulation of the derivative in proving Theorem 8 in Section 1.10.) This means that the right-hand and left-hand limits both exist at $x = c$, and both equal $f'(c)$.

When we examine these limits separately, we find that

$$\lim_{x \to c^+} \frac{f(x) - f(c)}{x - c} \le 0 \qquad (2)$$

because, immediately to the right of c, $f(x) \le f(c)$ and $x - c > 0$. Similarly,

$$\lim_{x \to c^-} \frac{f(x) - f(c)}{x - c} \ge 0 \qquad (3)$$

because, immediately to the left of c, $f(x) \le f(c)$ and $x - c < 0$.

Inequality (2) says that $f'(c)$ cannot be greater than zero, whereas (3) says that $f'(c)$ cannot be less than zero. So $f'(c) = 0$.

This proves the theorem for local maximum values. To prove it for local minimum values, replace f by $-f$ and run through the argument again.

Rolle's Theorem

There is strong geometric evidence that between any two points where a smooth curve crosses the x-axis there is a point on the curve where the tangent is horizontal. A 300-year-old theorem of Michel Rolle (1652–1719) assures us that this is indeed the case.

THEOREM 2

Rolle's Theorem

Suppose that $y = f(x)$ is continuous at every point of the closed interval $[a, b]$ and differentiable at every point of its interior (a, b). If

$$f(a) = f(b) = 0,$$

then there is at least one number c between a and b at which

$$f'(c) = 0.$$

See Fig. 3.13.

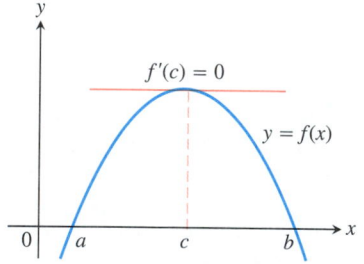

3.13 Rolle's theorem says that a smooth curve has at least one horizontal tangent between any two points where it crosses the x-axis.

Proof   We know from Section 1.10 that a continuous function defined on a closed interval assumes absolute maximum and minimum values on the interval. The question is, where? Theorem 1 tells us there are only three places to look:

1. at interior points where f' is zero,

2. at interior points where f' does not exist,

3. at the endpoints of the function's domain, in this case a and b.

By hypothesis, f has a derivative at every interior point. That rules out (2), leaving us with interior points where $f' = 0$ and with the two endpoints a and b.

If either the maximum or the minimum occurs at a point c inside the interval, then $f'(c) = 0$ by Theorem 1 and we have found a point for Rolle's theorem.

If both maximum and minimum are at a or b, then f is constant, $f' = 0$, and c can be taken anywhere in the interval. This completes the proof.

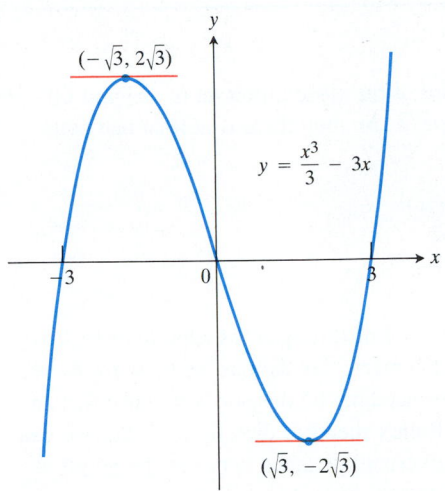

3.14   As predicted by Rolle's theorem, this smooth curve has horizontal tangents between the points where it crosses the x-axis.

Example 1   The polynomial function

$$f(x) = \frac{x^3}{3} - 3x$$

graphed in Fig. 3.14 is continuous at every point of $[-3, 3]$ and differentiable at every point of $(-3, 3)$. Since $f(-3) = f(3) = 0$, Rolle's theorem says that f' must be zero at least once in the open interval between $a = -3$ and $b = 3$. In fact, $f'(x) = x^2 - 3$ is zero twice in this interval, once at $x = -\sqrt{3}$ and again at $x = \sqrt{3}$.

Example 2   As the function $f(x) = 1 - |x|$ shows (Fig. 3.15), the differentiability of f is essential to Rolle's theorem. If we allow even one interior point where f is not differentiable, its graph may fail to have a horizontal tangent.

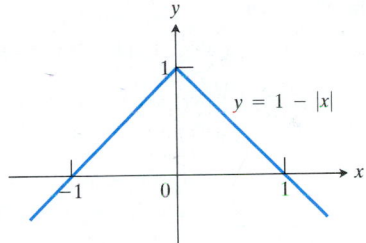

3.15   This curve has no horizontal tangent between the points where it crosses the x-axis.

The Mean Value Theorem

The Mean Value Theorem is a slanted version of Rolle's theorem. See Fig. 3.16. The figure shows the graph of a differentiable function f defined on an interval $a \leq x \leq b$. There is a point on the curve where the tangent is parallel to the chord AB. In Rolle's theorem, the line AB is the x-axis and $f'(c) = 0$. Here the line AB is a chord joining the endpoints of the curve above a and b, and $f'(c)$ is the slope of the chord.

3.16   Geometrically, the Mean Value Theorem says that somewhere between A and B the curve has at least one tangent parallel to chord AB.

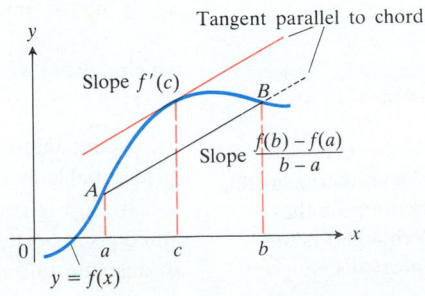

The Mean Value Theorem

If $y = f(x)$ is continuous at every point of the closed interval $[a, b]$ and differentiable at every point of its interior (a, b), then there is at least one number c between a and b at which

$$\frac{f(b) - f(a)}{b - a} = f'(c). \tag{4}$$

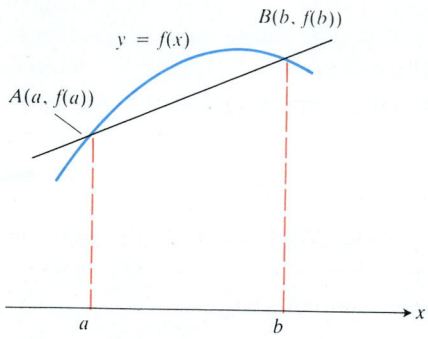

3.17   The graph of f and the chord AB over the interval $a \leq x \leq b$.

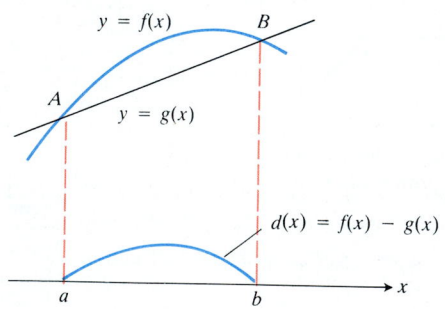

3.18   The chord AB in Fig. 3.17 is the graph of the function $g(x)$. The function $d(x) = f(x) - g(x)$ gives the vertical distance between the graphs of f and g at x.

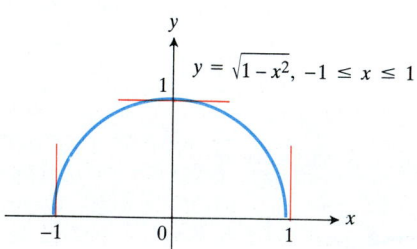

3.19   The function $f(x) = \sqrt{1 - x^2}$ satisfies the hypotheses (and conclusion) of the Mean Value Theorem on the interval $[-1, 1]$ even though f is not differentiable at the interval's endpoints.

Proof   If we graph f over $[a, b]$ and draw the line through the endpoints $A(a, f(a))$ and $B(b, f(b))$, the figure we get (Fig. 3.17) resembles the one we drew for Rolle's theorem. The difference is that the line AB need not be the x-axis because $f(a)$ and $f(b)$ may not be zero. We cannot apply Rolle's theorem directly to f, but we can apply it to the function that measures the vertical distance between the graph of f and the line AB. This, it turns out, will tell us what we want to know about the derivative of f.

The line AB is the graph of the function

$$g(x) = f(a) + \frac{f(b) - f(a)}{b - a}(x - a) \tag{5}$$

(point–slope equation), and the formula for the vertical distance between the graphs of f and g at x is

$$d(x) = f(x) - g(x) = f(x) - f(a) - \frac{f(b) - f(a)}{b - a}(x - a). \tag{6}$$

Figure 3.18 shows the graphs of f, g, and d together.

The function d satisfies the hypotheses of Rolle's theorem on the interval $[a, b]$. It is continuous on $[a, b]$ and differentiable in (a, b) because f and g are. Both $d(a)$ and $d(b)$ are zero because the graphs of f and g pass through A and B.

Therefore $d' = 0$ at some point c between a and b. To see what this says about f', we differentiate both sides of Eq. (6) with respect to x and set $x = c$. This gives

$$d'(x) = f'(x) - \frac{f(b) - f(a)}{b - a}, \qquad \text{(Derivative of Eq. (6)\ldots)}$$

$$d'(c) = f'(c) - \frac{f(b) - f(a)}{b - a}, \qquad (\ldots \text{with } x = c)$$

$$0 = f'(c) - \frac{f(b) - f(a)}{b - a}, \qquad (d'(c) = 0)$$

$$f'(c) = \frac{f(b) - f(a)}{b - a},$$

which is what we set out to prove.

Notice that the hypotheses of the Mean Value Theorem do not require f to be differentiable at either a or b. Continuity at a and b is enough (Fig. 3.19).

If $f'(x)$ is continuous on $[a, b]$, then the Max-Min Theorem for continuous functions in Section 1.10 tells us that f' has an absolute maximum value max f' and an absolute minimum value min f' on the interval. Since the number $f'(c)$ can

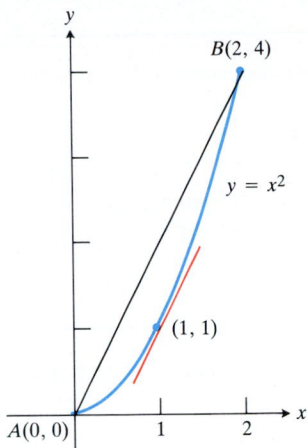

3.20   As we find in Example 3, $x = 1$ is where the tangent is parallel to the chord.

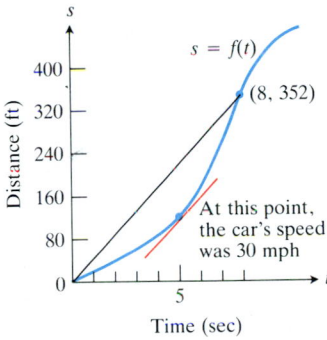

3.21   Distance vs. elapsed time for the car in Example 4.

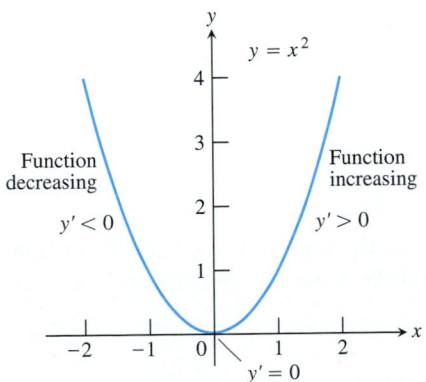

3.22   The function $y = x^2$ decreases on $(-\infty, 0)$, where $y' = 2x$ is negative, and increases on $(0, \infty)$, where y' is positive. In between these two intervals, $y' = 0$.

neither exceed max f' nor be less than min f', the equation

$$\frac{f(b) - f(a)}{b - a} = f'(c) \tag{7}$$

gives us the inequality

$$\min f' \leq \frac{f(b) - f(a)}{b - a} \leq \max f'. \tag{8}$$

The importance of the Mean Value Theorem lies in the estimates that sometimes come from Eq. (8) and in the mathematical conclusions that come from Eq. (7), one of which we shall see in a moment.

We usually do not know any more about the number c than the theorem tells us, which is that c exists. In a few cases we can satisfy our curiosity about the identity of c, as in the next example. Keep in mind, however, that our ability to identify c is the exception rather than the rule, and the importance of the theorem lies elsewhere.

Example 3   The function $f(x) = x^2$ (Fig. 3.20) is continuous for $0 \leq x \leq 2$ and differentiable for $0 < x < 2$. Since $f(0) = 0$ and $f(2) = 4$, the Mean Value Theorem says that at some point c in the interval the derivative $f'(x) = 2x$ must have the value $(4 - 0)/(2 - 0) = 2$. In this (exceptional) case we can identify c by solving the equation $2c = 2$ to get $c = 1$.

Physical Interpretations

If we think of the number $(f(b) - f(a))/(b - a)$ as the average change in f over $[a, b]$ and $f'(c)$ as an instantaneous change, then the Mean Value Theorem says that at some interior point the instantaneous change must equal the average change over the entire interval.

Example 4   If a car accelerating from zero takes 8 sec to go 352 ft, its average velocity for the 8-second interval is $352/8 = 44$ ft/sec. At some point during the acceleration, the Mean Value Theorem says, the speedometer must read exactly 30 mph (44 ft/sec) (Fig. 3.21).

The First Corollary

Among other things, the Mean Value Theorem is famous for three important corollaries. The first, which we shall see in a moment, says exactly when the graphs of differentiable functions rise and fall. The second, which we shall come to in Chapter 4, says that only constant functions can have zero derivatives. The third, also in Chapter 4, says that functions with identical derivatives throughout an interval can differ on that interval by at most a constant value.

You may have noticed that differentiable functions seem to increase when their derivatives are positive and decrease when their derivatives are negative (Fig. 3.22). The corollary we are about to prove says this is always the case. To prove the corollary, we need precise definitions of increasing and decreasing.

DEFINITIONS

A function f defined throughout an interval I is said to **increase** on I if, for any two points x_1 and x_2 in I,

$$x_2 > x_1 \implies f(x_2) > f(x_1).$$

Similarly, f is said to **decrease** on I if, for any two points x_1 and x_2 in I,

$$x_2 > x_1 \implies f(x_2) < f(x_1).$$

COROLLARY 1

The First Derivative Test for Increasing and Decreasing: f increases when $f' > 0$ and decreases when $f' < 0$.

Suppose that f is continuous at each point of the closed interval $[a, b]$ and differentiable at each point of its interior (a, b). If $f' > 0$ at each point of (a, b), then f increases throughout $[a, b]$. If $f' < 0$ at each point of (a, b), then f decreases throughout $[a, b]$. In either case, f is one-to-one on $[a, b]$.

Proof   Let x_1 and x_2 be any two numbers in $[a, b]$ with $x_1 < x_2$. Apply the Mean Value Theorem to f on $[x_1, x_2]$:

$$f(x_2) - f(x_1) = f'(c)(x_2 - x_1) \tag{9}$$

for some c between x_1 and x_2. The sign of the right-hand side of Eq. (9) is the same as the sign of $f'(c)$ because $x_2 - x_1$ is positive. Therefore

$$f(x_2) > f(x_1) \qquad \text{if } f'(x) \text{ is positive on } (a, b)$$

(f is increasing) and

$$f(x_2) < f(x_1) \qquad \text{if } f'(x) \text{ is negative on } (a, b)$$

(f is decreasing). In either case, $x_1 \neq x_2$ implies that $f(x_1) \neq f(x_2)$, so f is one-to-one.

Notice that we were also able to show that a function has to be one-to-one on any interval where its derivative is positive or its derivative is negative. Knowing this will pay off later, in Chapter 6.

Corollary 1 provides a useful tool for graphing, as we shall see in the next section.

Solving Equations

When we solve equations numerically, we usually want to know beforehand how many solutions to look for in a given interval. With the help of Corollary 1 we can sometimes find out.

Suppose, for example, that

1. f is continuous on $[a, b]$ and differentiable on (a, b);
2. $f(a)$ and $f(b)$ have opposite signs;
3. $f' > 0$ or $f' < 0$ throughout (a, b).

Then f has exactly one zero between a and b: It cannot have more than one because f is one-to-one on $[a, b]$. Yet it has at least one, by the Intermediate Value Theorem.

Example 5   The function $f(x) = x^3 + 3x + 1$ is continuous and differentiable on $[-1, 1]$, $f(-1) = -3$ and $f(1) = 5$ have opposite signs, and $f'(x) = 3x^2 + 3$ is always positive. The equation $x^3 + 3x + 1 = 0$ therefore has exactly one solution in the interval $[-1, 1]$ (Fig. 3.23).

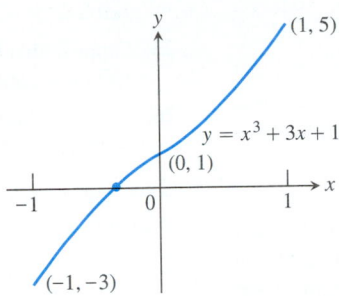

3.23   The only real zero of the polynomial $y = x^3 + 3x + 1$ is the one shown here between -1 and 0.

In this case we also know that the zero in $[-1, 1]$ is f's only zero because f is one-to-one throughout its entire domain.

EXERCISES 3.2

Show that the equations in Exercises 1–4 have exactly one solution in the given interval.

1. $x - \dfrac{2}{x} = 0, \quad 1 \le x \le 3$

2. $2x - \cos x = 0, \quad -\pi \le x \le \pi$

3. $x^4 + 3x + 1 = 0, \quad -2 \le x \le -1$

4. $-x^3 - 3x + 1 = 0, \quad 0 \le x \le 1$

For the functions and intervals in Exercises 5–8, find the value or values of c that satisfy the equation

$$\frac{f(b) - f(a)}{b - a} = f'(c)$$

in the conclusion of the Mean Value Theorem.

5. $f(x) = x^2 + 2x - 1, \quad 0 \le x \le 1$

6. $f(x) = x^{2/3}, \quad 0 \le x \le 1$

7. $f(x) = x + \dfrac{1}{x}, \quad \dfrac{1}{2} \le x \le 2$

8. $f(x) = \sqrt{x - 1}, \quad 1 \le x \le 3$

9. Show that at some instant during a 2-h automobile trip the car's speedometer reading will equal the average speed for the trip.

10. *Temperature change.*   It took 14 sec for a thermometer to rise from $-19°C$ to $100°C$ when it was taken from a freezer and placed in boiling water. Show that somewhere along the way the mercury was rising at exactly $8.5°C/sec$.

11. a) Plot the zeros of each polynomial on a line together with the zeros of its first derivative.

   i) $y = x^4 - 4$

   ii) $y = x^2 + 8x + 15$

   iii) $y = x^3 - 3x^2 + 4 = (x + 1)(x - 2)^2$

   iv) $y = x^3 - 33x^2 + 216x = x(x - 9)(x - 24)$

   What pattern do you see?

   b) Prove that between every two zeros of the polynomial $x^n + a_{n-1}x^{n-1} + \cdots + a_1x + a_0$ there lies a zero of the polynomial $nx^{n-1} + (n - 1)a_{n-1}x^{n-2} + \cdots + a_1$.

12. Which of the following functions satisfy the hypotheses of the Mean Value Theorem on the intervals given for them? Explain.

   a) $f(x) = x^{2/3}, \quad [-1, 8]$

   b) $f(x) = \sqrt{x(1 - x)}, \quad [0, 1]$

   c) $f(x) = x^{4/5}, \quad [0, 1]$

   d) $f(x) = \begin{cases} \dfrac{\sin x}{x}, & -\pi \le x < 0, \\ 0, & x = 0 \end{cases}$

13. For what values of a, m, and b does the function

$$f(x) = \begin{cases} 3, & x = 0, \\ -x^2 + 3x + a, & 0 < x < 1, \\ mx + b, & 1 \le x \le 2, \end{cases}$$

satisfy the hypotheses of the Mean Value Theorem on the interval $[0, 2]$?

14. Suppose that f'' is continuous on $[a, b]$ and that f has three zeros in the interval. Show that f'' has at least one zero in (a, b). Generalize this result.

15. The function

$$f(x) = \begin{cases} x & \text{if } 0 \le x < 1, \\ 0 & \text{if } x = 1, \end{cases}$$

is zero at $x = 0$ and again at $x = 1$. Yet its derivative $f' = 1$ is different from zero at every point between 0 and 1. Why doesn't this contradict the Mean Value Theorem?

16. Suppose the derivative of a differentiable function $f(x)$ is never zero on $[0, 1]$. Show that $f(0) \neq f(1)$.

17. Show that for any numbers a and b,
$$|\sin b - \sin a| \leq |b - a|.$$

18. Suppose that f is differentiable for $a \leq x \leq b$ and that $f(b) < f(a)$. Show that f' is negative at some point between a and b.

19. Show that
a) $y = 1/x$ decreases on any interval on which it is defined;
b) $y = 1/x^2$ increases on any interval to the left of the origin and decreases on any interval to the right of the origin.

20. Show that if $f'' > 0$ throughout an interval $[a, b]$, then f' has at most one zero in $[a, b]$. What if $f'' < 0$ throughout $[a, b]$ instead?

21. Suppose that f and g are differentiable on $[a, b]$ and that $f(a) = g(a)$ and $f(b) = g(b)$. Show that there is at least one point c in (a, b) where the tangents to the graphs of f and g are parallel.

22. Let f be differentiable at every value of x and suppose that $f(1) = 1$, that $f' < 0$ on $(-\infty, 1)$, and that $f' > 0$ on $(1, \infty)$.
a) Show that $f(x) \geq 1$ for all x.
b) Must $f'(1) = 0$? Explain. (*Hint:* See Section 1.10.)

23. Let $f(x) = px^2 + qx + r$ be a quadratic function defined on a closed interval $[a, b]$. Show that there is exactly one point c in (a, b) at which f satisfies the conclusion of the Mean Value Theorem.

24. Show that a cubic polynomial can have at most three real zeros.

25. Suppose that $f(0) = 3$ and that $f'(x) = 0$ for all x. Show that $f(x)$ must equal 3 for all x.

26. Suppose that $f'(x) = 2$ and that $f(0) = 5$. Show that $f(x) = 2x + 5$ at every value of x.

Make the estimates in Exercises 27 and 28 by applying the inequality
$$\min f' \leq \frac{f(b) - f(a)}{b - a} \leq \max f'.$$

27. CALCULATOR Suppose that $f'(x) = 1/(1 + x^4 \cos x)$ for $0 \leq x \leq 0.1$ and that $f(0) = 1$. Estimate $f(0.1)$.

28. CALCULATOR Suppose that $f'(x) = 1/(1 - x^4)$ for $0 \leq x \leq 0.1$ and that $f(0) = 2$. Estimate $f(0.1)$.

29. *The geometric mean of a and b.*   The **geometric mean** of two positive numbers a and b is the number $\sqrt{ab}$. Show that the value of c in the conclusion of the Mean Value Theorem for $f(x) = 1/x$ on an interval $[a, b]$ of positive numbers is $c = \sqrt{ab}$.

30. *The arithmetic mean of a and b.*   The **arithmetic mean** of two numbers a and b is the number $(a + b)/2$. Show that the value of c in the conclusion of the Mean Value Theorem for $f(x) = x^2$ on any interval $[a, b]$ is $c = (a + b)/2$.

3.3 Graphing with y' and y''

The problem with graphing a curve $y = f(x)$ without calculus is that we may have no way to know what the curve does between the points we plot. Even if we use a computer to plot hundreds of points and connect them with short line segments (which is all a computer can do), the verification that the graph looks like what is on the screen must still come from elsewhere. The computer can only suggest what *might* be true.

In this section we see how to use derivatives, when they exist, to tell how graphs behave between the points we plot.

The First Derivative

When we know that a function has a derivative at every point of an interval, we know that the function is continuous throughout the interval and that its graph is connected (Section 1.10). Thus the graphs of $\sin x$ and $\cos x$ remain unbroken, however far they are extended, as do the graphs of polynomials. The graphs of $\tan x$ and $1/x^2$ break only at the points where the functions are undefined. On every interval that avoids these points, the functions are differentiable, and hence continuous, and their graphs are connected.

3.24 The differentiable function $y = f(x)$ increases on (a, c) where $f' > 0$, decreases on (c, d) where $f' < 0$, and increases again on (d, b). The transitions are marked by horizontal tangents.

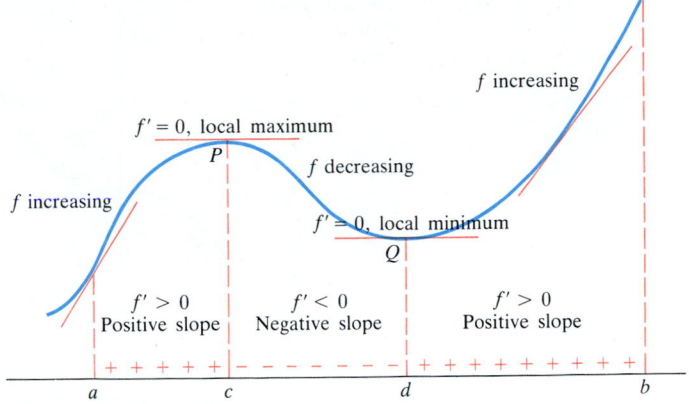

We gain additional information about a function's graph when we know where its first derivative is positive, negative, and zero. For, as we saw in Section 3.2, this tells us where the graph rises, falls, and has horizontal tangents (Fig. 3.24). We may also profit from identifying isolated points where the first derivative is undefined, for these may signal the presence of a corner or a cusp (Fig. 3.25). We shall say more about cusps at the end of the section.

There are two things to watch out for here, however. A curve may have a horizontal tangent without having a local maximum or minimum (Fig. 3.26). Also, a curve may have a local maximum or minimum without having a horizontal tangent (Fig. 3.25).

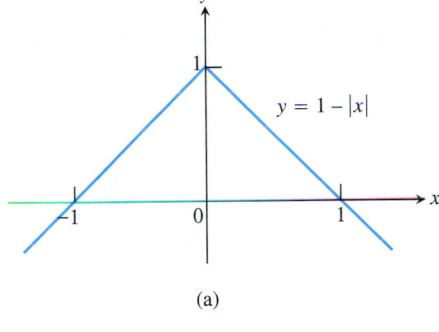

(a)

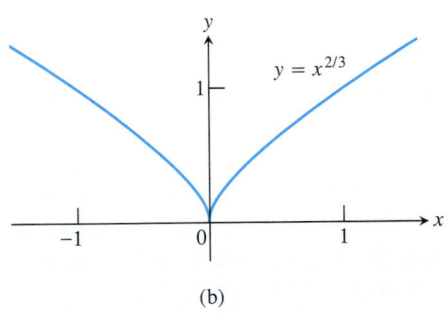

(b)

3.25 (a) The graph of $y = 1 - |x|$ has a corner at $x = 0$ where the function's derivative is undefined. The right-hand and left-hand derivatives exist there but they have different values. (b) The graph of $y = x^{2/3}$ has a cusp at $x = 0$. The derivative $y' = (2/3)x^{-1/3}$ approaches ∞ as $x \to 0^+$ and approaches $-\infty$ as $x \to 0^-$. The curve does have a tangent at the origin, but the tangent is vertical.

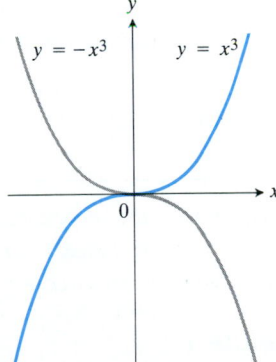

3.26 The curves $y = x^3$ and $y = -x^3$ have horizontal tangents at the origin without having maxima or minima there.

Concavity

As you can see in Fig. 3.27, the curve $y = x^3$ rises as x increases, but the portions defined on the intervals $(-\infty, 0)$ and $(0, \infty)$ turn in different ways. As we come in from the left toward the origin along the curve, the curve turns to our right and falls below its tangents. As we leave the origin, the curve turns to our left and rises above its tangents.

To put it another way, the slopes of the tangents decrease as the curve approaches the origin from the left and increase as the curve moves from the origin into the first quadrant.

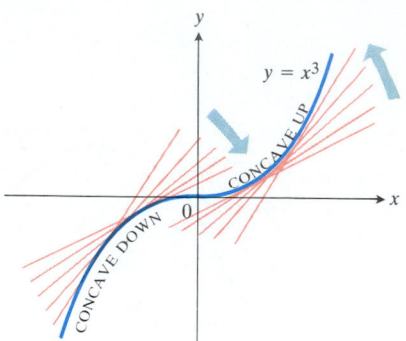

3.27   The graph of $y = x^3$ is concave down on the left, concave up on the right.

We say that the curve $y = x^3$ is concave down on $(-\infty, 0)$, where y' decreases, and concave up on $(0, \infty)$, where y' increases.

DEFINITIONS

The graph of a differentiable function is **concave down** on an interval where the slope y' decreases. It is **concave up** on an interval where the slope y' increases.

If a function $y = f(x)$ has a second derivative as well as a first (as do most of the functions in this text), we can apply Corollary 1 of the Mean Value Theorem (Section 3.2) to the function $f' = y'$ to conclude that y' decreases if $y'' < 0$ and increases if $y'' > 0$. This gives us what is called the second derivative test for concavity.

The Second Derivative Test for Concavity

The graph of $y = f(x)$ is concave down on any interval where $y'' < 0$ and concave up on any interval where $y'' > 0$.

The idea of the second derivative test is that if $y'' < 0$, then y' decreases as x increases and the tangent turns clockwise. If $y'' > 0$, then y' increases as x increases and the tangent turns counterclockwise.

Example 1

a) The curve $y = x^3$ (Fig. 3.27) is concave down on $(-\infty, 0)$ where $y'' = 6x$ is negative and concave up on $(0, \infty)$ where $y'' = 6x$ is positive.

b) The parabola $y = x^2$ is concave up on every interval because $y'' = 2$ is always positive.

Points of Inflection

To study the motion of a body moving on a line we often graph the body's position as a function of time. One reason for doing so is to reveal where the body's acceleration (given by the second derivative) changes sign. On the graph, these are the points where the concavity changes. In mathematics such points are called points of inflection.

DEFINITION

A point on the graph of a differentiable function where the concavity changes is called a **point of inflection**.

Thus a point of inflection on a twice-differentiable curve is a point where y'' is positive on one side and negative on the other. At such a point y'' is zero because derivatives have the intermediate value property.

At a point of inflection on the graph of a twice-differentiable function, $y'' = 0$.

Example 2   The graph of the simple harmonic motion $y = \sin t$ (Fig. 3.28) changes concavity at $t = 0$ and $t = \pi$, where the acceleration $y'' = -\sin t$ is zero.

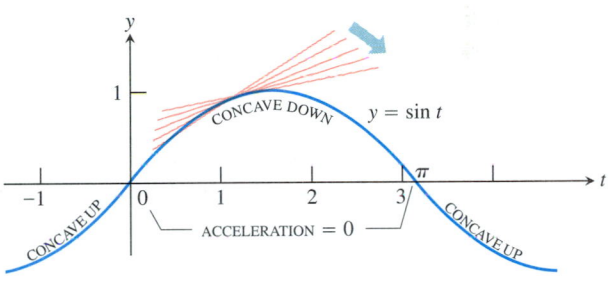

3.28   The simple harmonic motion in Example 2.

It is possible for y'' to be zero at a point that is *not* a point of inflection (Fig. 3.29a), and you have to watch out for this. Also, a point of inflection may occur where y'' fails to exist (Fig. 3.29b).

A General Procedure for Graphing

We now apply what we have learned to graph a cubic equation. The steps we take are those of a general procedure we recommend for graphing.

Example 3   Graph the function

$$y = x^3 - 3x^2 + 4 = (x + 1)(x - 2)^2.$$

Solution   STEP 1:   *Find y' and y''.*

$$y = x^3 - 3x^2 + 4,$$
$$y' = 3x^2 - 6x,$$
$$y'' = 6x - 6$$

STEP 2:   *Find where y' is positive, negative, and zero.* This will show where the curve may have local maxima and minima and where the curve is rising and falling.
   When factored, $y' = 3x(x - 2)$, so its zeros occur at $x = 0$ and $x = 2$. To find

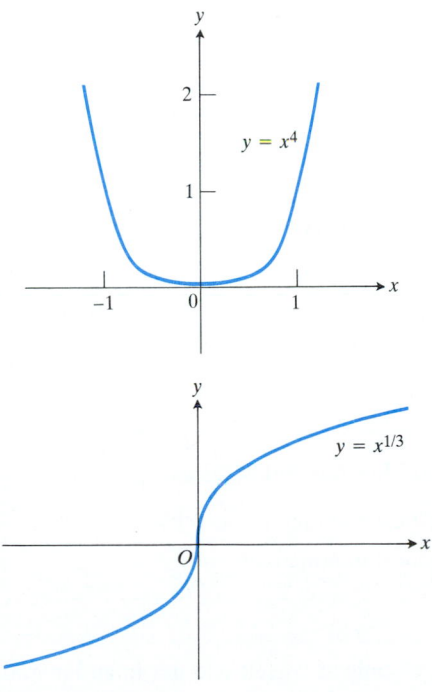

3.29   (a) The curve $y = x^4$ has no inflection point at $x = 0$. Even though $y'' = 12x^2$ is zero there, y'' does not change sign. (b) The curve $y = x^{1/3}$ has an inflection point at $x = 0$, but y'' does not exist there.

Sign of $3x$:

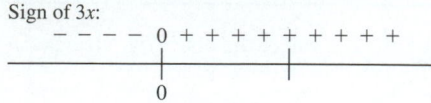

Sign of $(x - 2)$:

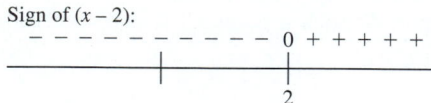

Sign of $3x(x - 2)$:

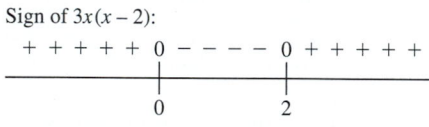

3.30 To find the sign pattern of $3x(x - 2)$, sketch the sign patterns of the factors $3x$ and $x - 2$ and "multiply."

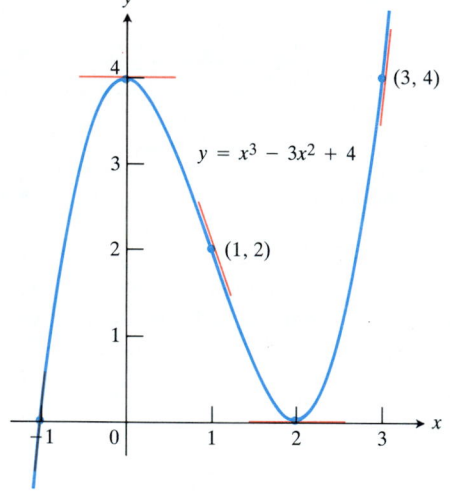

3.31 The graph of $y = x^3 - 3x^2 + 4$, based on the values of y' and y''.

the sign of y' elsewhere, we draw a picture like the one in Fig. 3.30. The sign pattern for y' tells us that the curve rises as it comes in from the left toward $x = 0$, falls from $x = 0$ to $x = 2$, and rises again to the right of $x = 2$. The curve has a local maximum at $x = 0$, where y' changes from $+$ to $-$, and a local minimum at $x = 2$, where y' changes from $-$ to $+$. There are no other extreme values: y is differentiable at all values of x, and its domain has no endpoints.

STEP 3: *Find where y'' is positive, negative, and zero.* This will tell us about concavity and possible inflection points.

When factored, $y'' = 6(x - 1)$, which is positive if $x > 1$, negative if $x < 1$, and zero if $x = 1$. The curve is concave down to the left of 1 and concave up to the right of 1. It has an inflection point at $x = 1$, where y'' changes sign.

STEP 4: *Make a summary table.* We include the values of y, y', and y'' at the intercepts and at the other important points. We summarize what we have learned about the curve's behavior.

x	y	y'	y''	Behavior
-1	0	9	-12	Rising, concave down
0	4	0	-6	Local maximum
1	2	-3	0	Falling, inflection point
2	0	0	6	Local minimum
3	4	9	12	Rising, concave up

STEP 5: *Draw the graph.* To do this, we plot the points from the table and sketch the tangents at these points. We then use the information about rise, fall, and concavity to draw the rest of the curve (Fig. 3.31).

A Useful Shortcut: The Second Derivative Test for Local Maxima and Minima

Instead of looking at how the sign of y' changes at a point where $y' = 0$, we can often use the following test to determine whether there is a local maximum or minimum at the point.

The Second Derivative Test for Local Maxima and Minima

If $f'(c) = 0$ and $f''(c) < 0$, then f has a local maximum at $x = c$.

If $f'(c) = 0$ and $f''(c) > 0$, then f has a local minimum at $x = c$.

Notice that the test requires finding y'' only at c itself and not in an interval about c. This makes the test relatively easy to apply. Unfortunately, the test fails if $y'' = 0$ or if y'' fails to exist or is hard to find. For instance, the test does not identify the local minimum of $y = x^4$ at $x = 0$ (Fig. 3.29a). The second derivative $y'' = 12x^2$ is zero when $x = 0$, so the test does not apply. (See Miscellaneous Exercises 22 and 23 at the end of the chapter.)

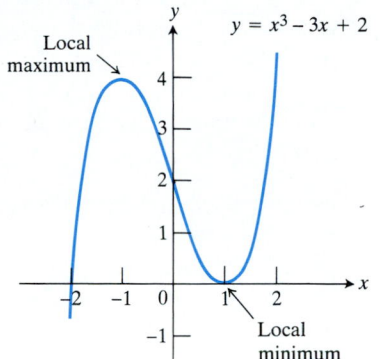

3.32   The graph of $y = x^3 - 3x + 2$.

Example 4   Find all maxima and minima of the function

$$y = x^3 - 3x + 2$$

on the interval $-\infty < x < \infty$.

Solution   The domain has no endpoints and the function is differentiable at every point. Therefore, extreme values can occur only where the first derivative,

$$y' = 3x^2 - 3 = 3(x - 1)(x + 1),$$

equals zero, which means at $x = 1$ and $x = -1$. The second derivative,

$$y'' = 6x,$$

is positive at $x = 1$ and negative at $x = -1$. Hence $y(1) = 0$ is a local minimum value and $y(-1) = 4$ is a local maximum value (Fig. 3.32).

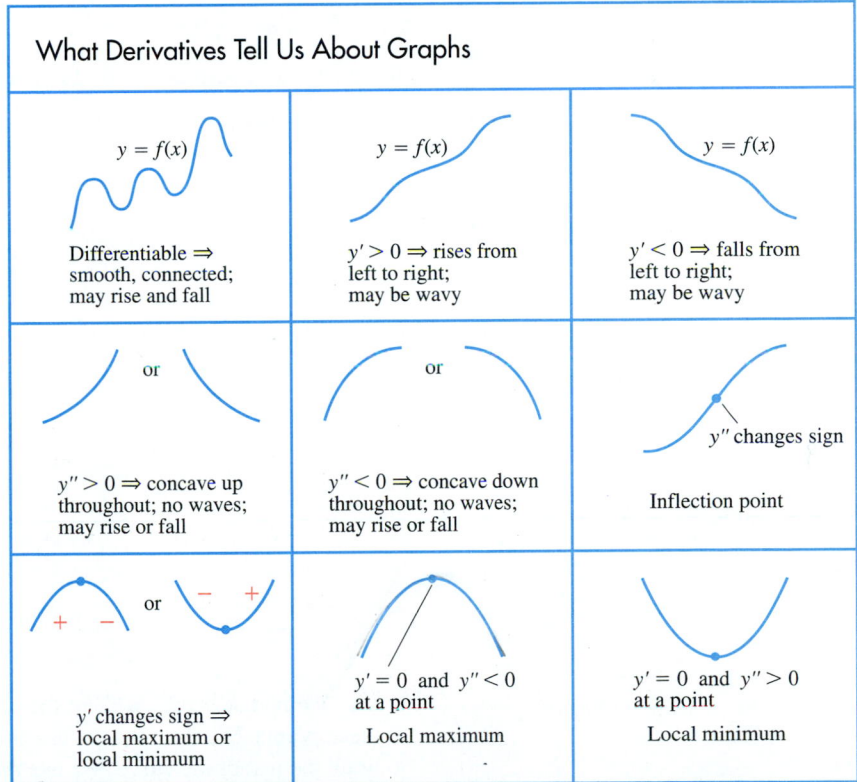

Cusps

The graph of a continuous function $y = f(x)$ is said to have a **cusp** at a point $x = x_0$ if $f'(x) \to \infty$ as x approaches x_0 from one side and $f'(x) \to -\infty$ as x approaches x_0 from the other side. The tangent to a graph at a cusp is vertical. When the first derivative of a function we want to graph is undefined at one or more isolated points, we modify our graphing procedure to include a search for cusps.

Steps in Graphing y = f(x)

1. Find y' and y''.
2. Find where y' is positive, negative, zero, or undefined.
3. Find where y'' is positive, negative, and zero.
4. Make a summary table.
5. Draw the graph.

Example 5   Graph the function $y = 1 - x^{2/3}$.

Solution

STEP 1: *Find y' and y''.*

$$y = 1 - x^{2/3},$$

$$y' = -\frac{2}{3}x^{-1/3},$$

$$y'' = \frac{2}{9}x^{-4/3}$$

STEP 2: *Find where y' is positive, negative, zero, or undefined.* The derivative $y' = -(2/3)x^{-1/3}$ is never zero. It is undefined at $x = 0$. It is positive for $x < 0$ and negative for $x > 0$. The curve rises as it approaches the origin from the left and falls as it moves away from the origin to the right. Since

$$\lim_{x \to 0^-} -(2/3)x^{-1/3} = \infty \quad \text{and} \quad \lim_{x \to 0^+} -(2/3)x^{-1/3} = -\infty,$$

the curve has a cusp at $x = 0$. The curve has a vertical tangent at $x = 0$.

STEP 3: *Find where y'' is positive, negative, and zero.* The second derivative

$$y'' = \frac{2}{9}x^{-4/3}$$

is positive for all $x \neq 0$. The curve is concave up on both sides of the origin.

STEP 4: *Make a summary table.* We summarize what we have learned about the curve so far.

x	y	y'	y''	**Behavior**
-1	0	$\frac{2}{3}$	$\frac{2}{9}$	Rising, concave up
0	1	∞		Cusp, vertical tangent
1	0	$-\frac{2}{3}$	$\frac{2}{9}$	Falling, concave up

STEP 5: *Draw the graph.* We plot the points from the table and sketch the tangents at these points. We then use the information about rise, fall, cusps, and concavity to draw the rest of the curve. We use the curve's symmetry as an additional guide: The function $y = 1 - x^{2/3}$ is even, so the graph is symmetric about the y-axis (Fig. 3.33).

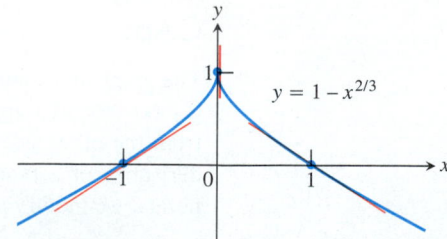

3.33   Like the graph of $x^{2/3}$ in Fig. 3.25(b), the graph of $y = 1 - x^{2/3}$ has a cusp at $x = 0$. The y-axis is the curve's tangent at this point.

EXERCISES 3.3

Find the local maximum and minimum values of the functions graphed in Exercises 1–4, and say at what values of x they are taken on. Find the coordinates of the inflection points. Identify the intervals on which the graphs are rising, falling, concave up, and concave down.

1. $y = \dfrac{x^3}{3} - \dfrac{x^2}{2} - 2x + \dfrac{1}{3}$

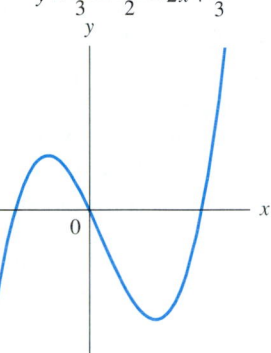

(Generated by Mathematica)

2. $y = \dfrac{x^4}{4} - 2x^2 + 4$

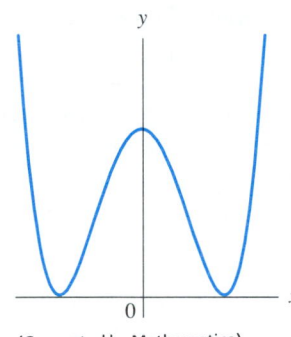

(Generated by Mathematica)

3. $y = \sin |x|$, $-2\pi \le x \le 2\pi$

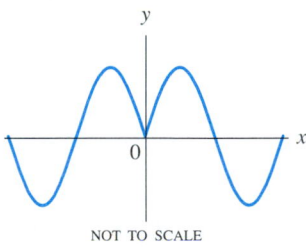

NOT TO SCALE

(Generated by Mathematica)

4. $y = x + \sin 2x$, $-\dfrac{2\pi}{3} \le x \le \dfrac{2\pi}{3}$

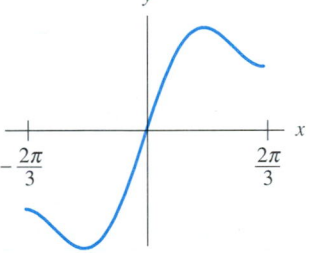

(Generated by Mathematica)

Use the steps listed in the text to graph the equations in Exercises 5–24. Indicate inflection points and local maxima and minima.

5. $y = x^2 - 4x + 3$

6. $y = 2x - x^2$

7. $y = 2x^3 - 3x^2$

8. $y = x^3 - x$

9. $y = 4 + 3x - x^3$

10. $y = -2x^3 + 6x^2 - 3$

11. $y = (x - 2)^3 + 1$

12. $y = x(6 - 2x)^2$

13. $y = 2x^4 - 4x^2 + 1$

14. $y = x^4 - 2x^2$

15. $y = x + \sin x$, $0 \le x \le 2\pi$

16. $y = x - \sin x$, $0 \le x \le 2\pi$

17. $y = (x + 1)^{3/2}$

18. $y = 3x^{5/3} - 5x$

19. $y = |x^2 - 1|$

20. $y = |x^3 - x|$

21. $x = y^2 - y + 1$

22. $x = 2y - y^2$

23. $x = y - y^3$

24. $x = y^3 - 6y^2 + 9y + 1$

Each of the curves in Exercises 25–28 has a vertical tangent at one or more points. Identify the points and sketch the curves.

25. $y = 5x^{1/5}$

26. $y = x^{3/4}$

27. $y = \sqrt{x + 1}$

28. $y = \sqrt{x(x - 1)}$

Each of the curves in Exercises 29–34 has one or more cusps. Locate the cusps and graph the curves.

29. $y = (x - 8)^{2/3}$

30. $y = (x + 1)^{2/5}$

31. $y = (x^2 - 1)^{2/3}$

32. $y = 16 - x^{4/5}$

33. $y = \sqrt{|x|}$

34. $y = \sqrt{|x - 4|}$

Velocity and acceleration. Each of the graphs in Exercises 35 and 36 is the graph of the position function $y = s(t)$ of a body moving back and forth on a coordinate line. At approximately what times is each body's (a) velocity equal to zero; (b) acceleration equal to zero?

35.

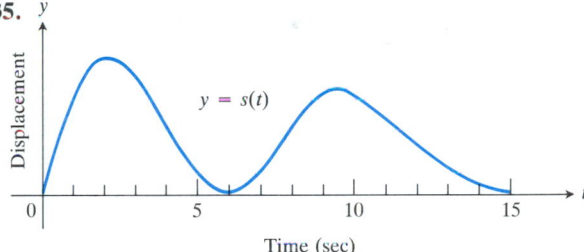

36.

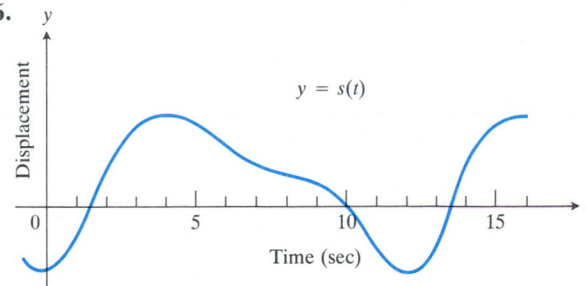

37. Figure 3.34 shows a portion of the graph of a twice-differentiable function $y = f(x)$. At each of the five points indicated, classify y' and y'' as positive, negative, or zero.

3.34 The graph for Exercise 37.

38. For what value of b will the curve $y = x^3 + bx^2 + cx + d$ (b, c, and d are constants) have a point of inflection at $x = 1$?

39. Sketch a smooth connected curve $y = f(x)$ through the origin with the properties that $f'(x) < 0$ for $x < 0$, and $f'(x) > 0$ for $x > 0$.

40. Sketch a smooth connected curve $y = f(x)$ through the origin with the properties that $f''(x) < 0$ for $x < 0$, and $f''(x) > 0$ for $x > 0$.

41. Sketch a smooth connected curve $y = f(x)$ with

$$f(-2) = 8, \qquad\qquad f'(2) = f'(-2) = 0,$$
$$f(0) = 4, \qquad\qquad f'(x) < 0 \quad \text{for} \quad |x| < 2,$$
$$f(2) = 0, \qquad\qquad f''(x) < 0 \quad \text{for} \quad x < 0,$$
$$f'(x) > 0 \quad \text{for} \quad |x| > 2, \qquad f''(x) > 0 \quad \text{for} \quad x > 0.$$

42. Sketch a smooth connected curve $y = f(x)$ with the following properties. Label coordinates where possible.

x	y	Curve
$x < 2$		Falling, concave up
2	1	Horizontal tangent
$2 < x < 4$		Rising, concave up
4	4	Inflection point
$4 < x < 6$		Rising, concave down
6	7	Horizontal tangent
$x > 6$		Falling, concave down

43. Sketch a connected curve $y = f(x)$ having $f'(x) > 0$ for $x < 2$ and $f'(x) < 0$ for $x > 2$ if
a) f' is continuous at $x = 2$,
b) $f'(x) = 1$ for $x < 2$ and $f'(x) = -1$ for $x > 2$,
c) $f'(x) \to 1^+$ as $x \to 2^-$ and $f'(x) \to -1^+$ as $x \to 2^+$.

44. For $x > 0$, sketch a smooth connected curve $y = f(x)$ that has $f(1) = 0$ and $f'(x) = 1/x$. Is such a curve necessarily concave down? Explain.

45. Suppose the derivative of the function $y = f(x)$ is

$$y' = (x - 1)^2(x - 2).$$

At what points, if any, does the graph of f have a local minimum, local maximum, or point of inflection? (*Hint:* Draw the sign pattern for y'.)

46. Suppose the derivative of the function $y = f(x)$ is

$$y' = (x - 1)^2(x - 2)(x - 4).$$

At what points, if any, does the graph of f have a local minimum, local maximum, or point of inflection?

47. If $f(x)$ is a differentiable function and $f'(c) = 0$ at an interior point c of f's domain, must f have a local maximum or minimum at $x = c$? Explain.

48. If $f(x)$ is a twice-differentiable function and $f''(c) = 0$ at an interior point c of f's domain, must the graph of f have an inflection point at $x = c$? Explain.

49. *Inflection points.*   True, or false? Explain.
a) A quadratic curve $y = ax^2 + bx + c$, $a \neq 0$, never has an inflection point.
b) A cubic curve $y = ax^3 + bx^2 + cx + d$, $a \neq 0$, always has exactly one inflection point.

50. *Horizontal tangents.*   True, or false? Explain.
a) The graph of every polynomial of even degree has at least one horizontal tangent.
b) The graph of every polynomial of odd degree has at least one horizontal tangent.

Computer Grapher or Graphing Calculator

51. Graph the curves $y = x^3 + cx + 1$ on the interval $-4 \le x \le 4$, for $c = -8, -4, 0,$ and 4. Use different colors if possible. Experiment with other values of c. What happens to the graph as $c \to -\infty$? As $c \to \infty$?

52. *Linearizations at inflection points.*   Linearizations fit particularly well at points of inflection. You will see what we mean if you graph the following examples. (You will see why the fit is so good if you do Exercise 23 in Section 3.7.)
a) $f(x) = \sin x$ and its linearization $L(x) = x$ at $x = 0$
b) *Newton's serpentine.* $f(x) = 4x/(x^2 + 1)$ and its linearization $L(x) = 4x$ at $x = 0$
c) *Newton's serpentine.* $f(x) = 4x/(x^2 + 1)$ and its linearization $L(x) = -(x/2) + 3\sqrt{3}/2$ at the point $(\sqrt{3}, \sqrt{3})$

In Exercises 53–56, find the inflection points (if any) on the graph of the function and the coordinates of the points on the graph where the function has a local maximum or local minimum value. Then graph the function in a region large enough to show all these points simultaneously. Add to your picture the graphs of the function's first and second derivatives. How are the values at which these graphs intersect the x-axis related to the graph of the function? In what other ways are the graphs of the derivatives related to the graph of the function?

53. $y = x^5 - 5x^4 - 240$ **54.** $y = x^3 - 12x^2$

55. $y = \dfrac{4}{5}x^5 + 16x^2 - 25$

56. $y = \dfrac{x^4}{4} - \dfrac{x^3}{3} - 4x^2 + 12x + 20$

57. a) Graph $y = x^{2/3}(x^2 - 2)$ for $-3 \le x \le 3$. Then use calculus to confirm what the screen shows about concavity, rise, and fall.
b) Does the curve have a cusp at $x = 0$, or does it just have a corner with different right-hand and left-hand derivatives?

58. a) Graph $y = 9x^{2/3}(x - 1)$ for $-0.5 \le x \le 1.5$. Then use calculus to confirm what the screen shows about concavity, rise, and fall. What concavity does the curve have to the left of the origin?
b) Does the curve have a cusp at $x = 0$, or does it just have a corner with different right-hand and left-hand derivatives?

To see how misleading a computer-generated graph can be, graph the functions in Exercises 59–64 for $-6 \le x \le 6$. If you can, vary the step size. The larger the step size, the worse things become, but decreasing the step size may not make the problems go away. What should the graphs really look like? (The notation $\lfloor x \rfloor$ denotes the greatest integer function.)

59. $y = \lfloor x \rfloor$ **60.** $y = \lfloor x \rfloor / 10$

61. $y = x - \lfloor x \rfloor$ **62.** $y = \sin(\lfloor x \rfloor)$

63. $y = \lfloor x \rfloor / x$ and $y = x / \lfloor x \rfloor$ together

64. $y = \tan(1/x)$

65. Graph $y = \lfloor 100x \rfloor / 100$ and $y = x$ together over the interval

$-6 \le x \le 6$, using as small a step size as you can. What difference do you see between the graphs?

EXPLORER PROGRAMS

Derivatives	Graphs the first and second derivatives of functions you key in
PowerGrapher	Graphs all the functions in this section
	Draws tangents to graphs

3.4 Graphing Rational Functions: Asymptotes and Dominant Terms

In Section 3.3, we graphed polynomials after looking at rise, fall, extreme values, concavity, and points of inflection. To graph rational functions reliably and effectively, we have to consider asymptotes, symmetry, and dominant terms as well. This section gives the details. As with other functions, computer-generated graphs of rational functions can sometimes mislead. We need confirmation from other sources to know whether to accept or reject the computer's description of how a function behaves.

Horizontal and Vertical Asymptotes

If the distance between the graph of a function and some fixed line approaches zero as the graph moves farther and farther from the origin, we say that the graph approaches the line asymptotically and that the line is an asymptote of the graph.

Example 1   The coordinate axes are asymptotes of the curve $y = 1/x$ (Fig. 3.35). The x-axis is an asymptote of the curve on the right because

$$\lim_{x \to \infty} \frac{1}{x} = 0$$

and on the left because

$$\lim_{x \to -\infty} \frac{1}{x} = 0.$$

The y-axis is an asymptote of the curve both above and below because

$$\lim_{x \to 0^+} \frac{1}{x} = \infty$$

and

$$\lim_{x \to 0^-} \frac{1}{x} = -\infty.$$

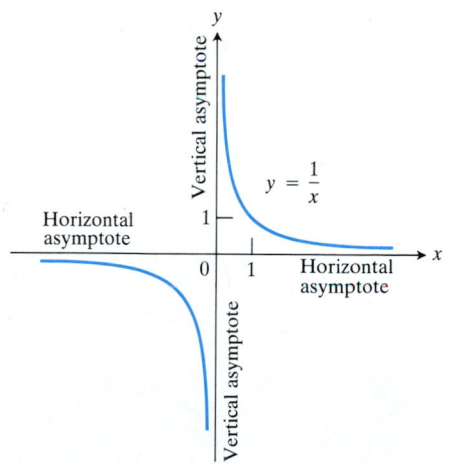

3.35   The coordinate axes are asymptotes of both branches of the hyperbola $y = 1/x$.

A line $y = b$ is a **horizontal asymptote** of the graph of a function $y = f(x)$ if either

$$\lim_{x \to \infty} f(x) = b \qquad \text{or} \qquad \lim_{x \to -\infty} f(x) = b.$$

A line $x = a$ is a **vertical asymptote** of the graph if either

$$\lim_{x \to a^+} f(x) = \pm\infty \qquad \text{or} \qquad \lim_{x \to a^-} f(x) = \pm\infty.$$

Example 2   Find the asymptotes of the curve

$$y = \frac{x}{x - 1}.$$

Solution   The asymptotes are quickly revealed if we divide $(x - 1)$ into x:

$$\begin{array}{r} 1 \\ x - 1 \overline{)\,x } \\ \underline{x - 1} \\ 1 \end{array}$$

This enables us to rewrite the formula for y as

$$y = 1 + \frac{1}{x - 1}.$$

From this we see that the curve in question is the graph of $y = 1/x$ shifted 1 unit up and 1 unit to the right. The asymptotes are now the lines $x = 1$ and $y = 1$ (Fig. 3.36). Indeed, if we apply the asymptote definitions, we find that

$$\lim_{x \to \infty} \left(1 + \frac{1}{x - 1}\right) = 1, \qquad \lim_{x \to -\infty} \left(1 + \frac{1}{x - 1}\right) = 1,$$

$$\lim_{x \to 1^+} \left(1 + \frac{1}{x - 1}\right) = \infty, \qquad \lim_{x \to 1^-} \left(1 + \frac{1}{x - 1}\right) = -\infty.$$

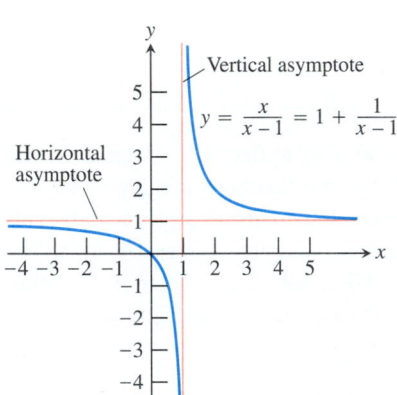

3.36   The lines $y = 1$ and $x = 1$ are asymptotes of the curve $y = x/(x - 1)$.

Example 3   Find the asymptotes of the curve

$$y = -\frac{8}{x^2 - 4}.$$

Solution   The line $y = 0$ is an asymptote on the right because $y \to 0$ as $x \to +\infty$ and again on the left because $y \to 0$ as $x \to -\infty$ (Fig. 3.37).
   The line $x = 2$ is an asymptote because

$$\lim_{x \to 2^-} \frac{-8}{x^2 - 4} = \infty$$

and again because

$$\lim_{x \to 2^+} \frac{-8}{x^2 - 4} = -\infty.$$

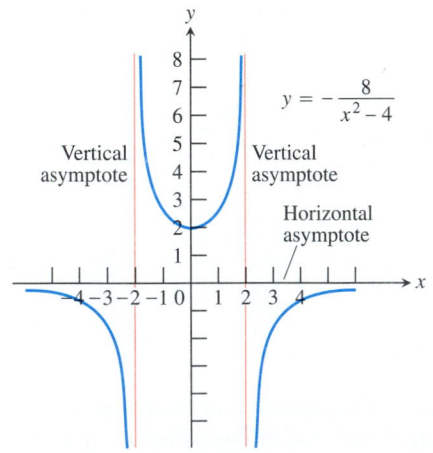

3.37   The graph of $y = -8/(x^2 - 4)$. Notice that the curve approaches the x-axis from only one side. Asymptotes do not have to be two-sided.

Similarly, the line $x = -2$ is an asymptote because $y \to \infty$ as $x \to -2^+$, and again because $y \to -\infty$ as $x \to -2^-$. There are no other asymptotes because y has a finite limit at every other point.

Oblique Asymptotes

If the degree of the numerator of a rational function is one greater than the degree of the denominator, the graph has an **oblique asymptote**, that is, an asymptote that is neither vertical nor horizontal.

Example 4   Find the asymptotes of the curve

$$y = \frac{x^2 - 3}{2x - 4}.$$

Solution   To find the asymptotes, oblique and otherwise, we divide $(2x - 4)$ into $(x^2 - 3)$:

$$
\begin{array}{r}
\frac{x}{2} + 1 \\
2x - 4 \overline{\smash{)}x^2 - 3} \\
\underline{x^2 - 2x} \\
2x - 3 \\
\underline{2x - 4} \\
1
\end{array}
$$

This tells us that

$$y = \frac{x^2 - 3}{2x - 4} = \underbrace{\frac{x}{2} + 1}_{\text{linear}} + \underbrace{\frac{1}{2x - 4}}_{\substack{\text{remainder} \\ \text{goes to } 0 \\ \text{as } x \to \pm\infty}}. \tag{1}$$

From this representation (see Fig. 3.38) we see that the line

$$y = \frac{x}{2} + 1$$

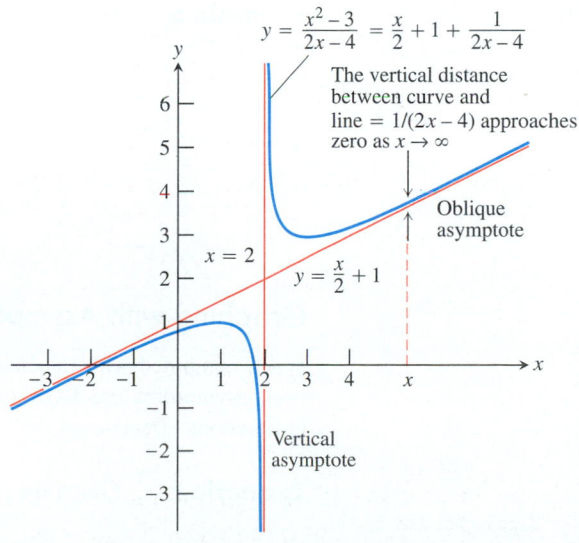

3.38   The graph of $y = (x^2 - 3)/(2x - 4)$.

is an asymptote of the curve:

$$\lim_{x \to \infty} \frac{1}{2x - 4} = 0, \qquad \text{so} \qquad y = \frac{x^2 - 3}{2x - 4} \to \frac{x}{2} + 1.$$

Equation (1) also reveals the presence of a vertical asymptote at $x = 2$. As $x \to 2^+$, $1/(2x - 4) \to \infty$, and as $x \to 2^-$, $1/(2x - 4) \to -\infty$.

Dominant Terms

Of all the observations we can make quickly about the function

$$y = \frac{x^2 - 3}{2x - 4},$$

probably the most useful is that

$$\frac{x^2 - 3}{2x - 4} = \frac{x}{2} + 1 + \frac{1}{2x - 4}. \qquad \text{(Obtained by dividing as in Eq. (1)).}$$

This tells us immediately that

$$y \approx \frac{x}{2} + 1 \qquad \text{for } x \text{ numerically large,}$$

$$y \approx \frac{1}{2x - 4} \qquad \text{for } x \text{ close to 2.}$$

If we want to know quickly how the function "goes," this is the way to find out. It goes like $y = (x/2) + 1$ when x is large and like $y = 1/(2x - 4)$ when x is close to 2.

We say that $(x/2) + 1$ **dominates** when x is large and that $1/(2x - 4)$ dominates when x is close to 2.

Dominant terms like these are the key to predicting the function's behavior over different portions of the x-axis.

Example 5   For

$$y = \frac{x + 3}{x + 2} = 1 + \frac{1}{x + 2}, \tag{2}$$

we have

$$y \approx 1, \quad |x| \text{ large,} \qquad y \approx \frac{1}{x + 2}, \quad x \text{ near } -2.$$

Graphing with Asymptotes and Dominant Terms

When combined with information about symmetry and about a function's derivatives, asymptotes and dominant terms tell us what we need to know to graph rational functions effectively.

Example 6   Graph the function

$$y = \frac{x^2 + 1}{x}.$$

Checklist for Graphing Rational Functions

1. Branches and signs
2. Symmetry
3. Dominant terms
4. Asymptotes
5. Rise, fall, extreme values
6. Concavity
7. Sketch dominant terms, asymptotes, and horizontal tangents
8. Graph the function

Solution   We learn what we can do about branches (separate connected pieces), symmetry, dominant terms and asymptotes, rise, fall, extreme values, and concavity. Then we draw the graph.

1. *Branches and signs.* The function is discontinuous at $x = 0$, so the graph has two branches. The function's values are positive if $x > 0$ and negative if $x < 0$, so the graph lies entirely in the first and third quadrants.

2. *Symmetry with respect to the coordinate axes and origin.* The function is odd, so its graph is symmetric with respect to the origin.

3. *Dominant terms.* We divide $x^2 + 1$ by x to express the function as

$$y = x + \frac{1}{x}. \tag{3}$$

This reveals that

$$y \approx x, \qquad x \text{ numerically large,}$$

$$y \approx \frac{1}{x}, \qquad x \text{ close to zero.}$$

4. *Asymptotes.* Equation (3) also reveals the curve's asymptotes:

$y = x,$ (Oblique asymptote, because $\frac{1}{x} \to 0$ as $x \to \infty$ or $x \to -\infty$)

$x = 0.$ (Vertical asymptote, because $\frac{1}{x} \to \infty$ as $x \to 0^+$ and $\frac{1}{x} \to -\infty$ as $x \to 0^-$)

5. *Rise and fall.* The first derivative (from Eq. (3)) is

$$y' = 1 - \frac{1}{x^2}.$$

The derivative is zero when

$$y' = 1 - \frac{1}{x^2} = 0, \qquad \text{or} \qquad x = \pm 1.$$

The derivative has the sign pattern in Fig. 3.39. The curve rises on $(-\infty, -1)$, falls on $(-1, 0)$ and $(0, 1)$, and rises again on $(1, \infty)$. There is a local maximum at $x = -1$ and a local minimum at $x = 1$. The curve has horizontal tangents at these points.

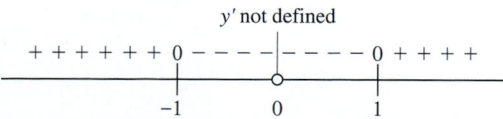

3.39   The sign pattern for $y' = 1 - (1/x^2)$.

6. *Concavity.* The second derivative $y'' = 2/x^3$ is negative if $x < 0$ and positive if $x > 0$. The curve is therefore concave down on $(-\infty, 0)$ and concave up on $(0, \infty)$.

7. *Graph the dominant terms and asymptotes, and sketch the horizontal tangents.* See Fig. 3.40(a).

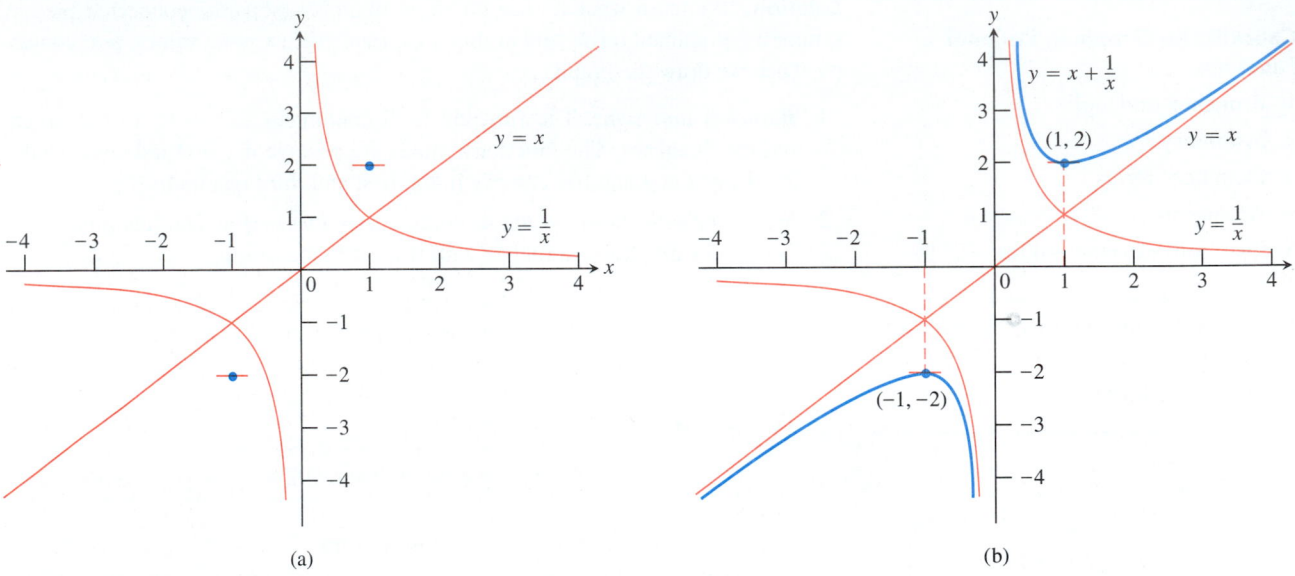

(a)

(b)

3.40   Stages in graphing $y = x + 1/x$:
(a) asymptotes, horizontal tangents, and
dominant terms; (b) the completed graph.

8.   *Graph the function.* We sketch a curve that fits the asymptotes, the
horizontal tangents, and the graphs of the dominant terms and that has the
other properties we discovered (Fig. 3.40b).

EXERCISES 3.4

Use the applicable steps in the checklist given with Example 6 to
graph the functions in Exercises 1–20.

1. $y = \dfrac{1}{x - 1}$

2. $y = \dfrac{1}{x + 1}$

3. $y = \dfrac{1}{2x + 4}$

4. $y = \dfrac{-3}{x - 3}$

5. $y = \dfrac{x + 3}{x + 2}$

6. $y = \dfrac{2x}{x + 1}$

7. $y = \dfrac{3x + 1}{2x - 1}$

8. $y = \dfrac{x - 4}{x - 5}$

9. $y = \dfrac{x^2 - 1}{x}$

10. $y = \dfrac{x^2 + 4}{2x}$

11. $y = \dfrac{x^4 + 1}{x^2}$

12. $y = \dfrac{x^3 + 1}{x^2}$

13. $y = \dfrac{1}{x^2 - 1}$

14. $y = \dfrac{x^2 - 4}{x^2 - 2}$

15. $y = \dfrac{x^2}{x - 1}$

16. $y = \dfrac{x^2 - 4}{x - 1}$

17. $y = \dfrac{x^2 - x + 1}{x - 1}$

18. $y = \dfrac{-x^2 + 2x - 2}{x - 1}$

19. $y = \dfrac{x}{x^2 - 1}$

20. $y = \dfrac{x - 1}{x^2(x - 2)}$

21. a) Find each critical point of

$$f(x) = \begin{cases} 0, & x = \pm 1, \\ \dfrac{x^{2/3}}{x^2 - 1}, & x \neq \pm 1, \end{cases}$$

and determine whether the function takes on an extreme
value there. If it does not, how does the function behave
as x approaches the critical point from the right and from
the left?

b) Graph the function.

22. Graph $y = (2 \sin x)/(1 + \cos x)$.

23. *Symmetry.*

a) Suppose an odd function is known to be increasing on
the interval $x > 0$. What can be said of its behavior on
the interval $x < 0$?

b) Suppose an even function is known to be increasing on
the interval $x < 0$. What can be said of its behavior on
the interval $x > 0$?

24. The word *asymptote* derives from an old Greek word for
"never touching." In practice, however, a curve may cross
one of its asymptotes a finite number of times or even
infinitely often, as does the curve $y = 2 + (\sin x)/x$, $x > 0$,
in Fig. 3.41. Show that the slope of the curve nevertheless
approaches the slope of the asymptote as $x \to \infty$.

$$y = 2 + \frac{\sin x}{x}$$

3.41 A curve that crosses its asymptote infinitely often (Exercise 24).

Computer Grapher or Graphing Calculator

Graph the functions in Exercises 25–30.

25. $y = x + 4/x^2$

26. $y = (x + 1)/(x^2 + 1)$

27. $y = (x^2 + 1)/(x^3 - 4x)$

28. $y = (x - 1)/(x^3 - 2x^2)$

29. $y = x + \sin 2x$

30. $y = x^2 + 3 \sin 2x$

Find the asymptotes and dominant terms of the rational functions in Exercises 31 and 32. Then graph the dominant terms and oblique asymptotes along with the function. Use different colors if you can. Try different window sizes.

31. $y = -\frac{x^2 - 4}{x + 1}$

32. $y = \frac{x^2 + x - 6}{2x - 2}$

33. *Pitfalls of computer graphing revisited.* Graph the function

$$y = -\frac{x^3 - 2}{x^2 + 1}$$

over the following intervals.

a) $-9 \le x \le 9$,

b) $-90 \le x \le 90$,

c) $-900 \le x \le 900$,

The graph in (a) should be good. The graph in (b) will probably indicate some activity near the origin but will not show what. The graph in (c) will look just like the graph of the line $y = -x$. Why?

34. *(Continuation of Exercise 33, for computer graphers in which you can control the step size).* Graph the function

$$y = -\frac{x^3 - 2}{x^2 + 1}$$

over the interval $-9 \le x \le 9$ with different step sizes. Start large, using only a few steps to cover the interval. Then decrease the step size to see how small it has to be to get good resolution. With large steps, the graph won't look anything like what it should be. With extremely small steps, the graph will be good but slow to develop. For best results, you want something in between.

35. Graph the function $y = x^{2/3}/(x^2 - 1)$ from Exercise 21(a) over the interval $-2 \le x \le 2$. The curve will appear to be concave down between $x = -1$ and $x = 1$, with no sign of a cusp at the origin. Zoom in on the origin and watch the true shape of the graph emerge. Why do you think the cusp fails to appear in the first view of the graph?

EXPLORER PROGRAM

PowerGrapher   Graphs functions together, so you can graph a function along with its dominant terms

3.5 Optimization

To optimize something means to make it as useful or effective as possible. In the mathematical models in which we use functions to describe the things that interest us, this usually means finding where some function has its greatest or smallest value. What is the most economical shape for an oil can? What is the stiffest beam we can cut from a 12-in. log?

In this section, we show how to answer such questions by finding the extreme values of differentiable functions. Our basic tool is the observation we made in Section 3.2 that the extreme values of a function f can occur only at the function's critical points (domain points where f' is zero or fails to exist) or at the endpoints of the function's domain. None of these points is necessarily the location of a local maximum or minimum value, but these are the only candidates.

Continuous Functions on Closed Intervals

Most optimization applications call for finding the absolute maximum value or absolute minimum value of a continuous function on a closed interval. The Max-

How to Find the Absolute Maximum and Minimum Values of a Continuous Function on a Closed Interval

1. Evaluate the function at its critical points and at the endpoints of its domain.

2. Take the largest and smallest of these values.

Min Theorem in Section 1.10 assures us they exist. In theory, the number of points where these values might occur could be infinite. In practice the number is usually so small that we can simply list the points and calculate the corresponding function values to identify the absolute maximum and minimum and see where they are taken on.

Example 1   Find the absolute maximum and minimum values of $y = x^{2/3}$ on the interval $-2 \leq x \leq 3$.

Solution   We evaluate the function at the critical points and endpoints and take the largest and smallest of these values.

The first derivative,

$$y' = \frac{2}{3} x^{-1/3}$$

$$= \frac{2}{3\sqrt[3]{x}},$$

has no zeros but is undefined at $x = 0$. The values of the function at this one critical point and at the endpoints are

Critical point value: $f(0) = 0$

Endpoint values: $f(-2) = (-2)^{2/3} = 4^{1/3},$

$$f(3) = (3)^{2/3} = 9^{1/3}.$$

We conclude that the function's maximum value is $9^{1/3}$, taken on at $x = 3$. The minimum value is 0, taken on at $x = 0$ (Fig. 3.42).

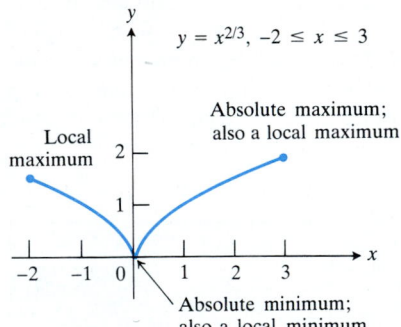

3.42   The extreme values of $y = x^{2/3}$ on the interval $-2 \leq x \leq 3$ (Example 1).

Example 2   *Geometry.*   A rectangle is to be fitted inside a semicircle of radius 2, with one side along the semicircle's diameter. What is the largest area the rectangle can have? What dimensions give this area?

Solution   To describe the dimensions of the rectangle, we place the circle and rectangle in the coordinate plane (Fig. 3.43). The length, height, and area of the rectangle can then be expressed in terms of the position x of the lower right-hand corner:

Length: $2x$

Height: $\sqrt{4 - x^2}$

Area: $2x \cdot \sqrt{4 - x^2}$

Our mathematical goal is to find the absolute maximum value of the continuous function

$$A(x) = 2x\sqrt{4 - x^2}$$

on the interval $0 \leq x \leq 2$. We do this by examining the values of A at the critical points and endpoints.

The derivative

$$\frac{dA}{dx} = \frac{-2x^2}{\sqrt{4 - x^2}} + 2\sqrt{4 - x^2}$$

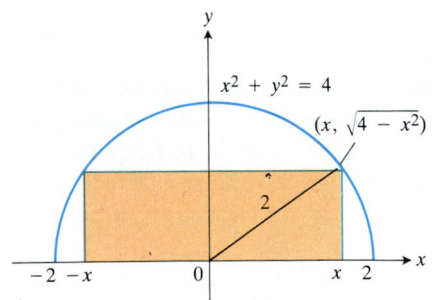

3.43   The rectangle and semicircle in Example 2.

is not defined when $x = 2$ and is equal to zero when

$$\frac{-2x^2}{\sqrt{4 - x^2}} + 2\sqrt{4 - x^2} = 0$$

$$-2x^2 + 2(4 - x^2) = 0$$

$$8 - 4x^2 = 0$$

$$x^2 = 2$$

$$x = \pm\sqrt{2}.$$

For $0 \le x \le 2$ we therefore have

Critical point value: $A(\sqrt{2}) = 2\sqrt{2}\sqrt{4 - 2} = 4,$

Endpoint values: $A(0) = 0,\quad A(2) = 0.$

The area has a maximum value of 4 when the rectangle is $2x = 2\sqrt{2}$ units long by $\sqrt{4 - x^2} = \sqrt{2}$ units high.

Applications in Industry, Business, and Optics

Example 3   *Metal Fabrication.*   An open-top box is to be made by cutting small congruent squares from the corners of a 12-by-12-in. sheet of tin and bending up the sides. How large should the squares cut from the corners be to make the box hold as much as possible?

Solution   We start with a picture, the way we do when we solve a problem in related rates (Fig. 3.44). In the figure the corner squares are x inches on a side. The volume of the box is a function of this variable:

$$V(x) = x(12 - 2x)^2 = 144x - 48x^2 + 4x^3.$$

The domain of V is the interval $0 \le x \le 6$.

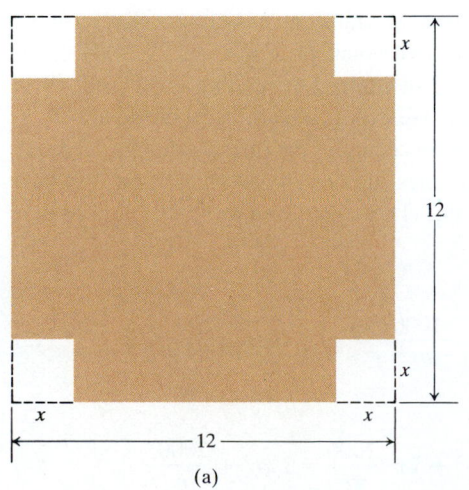

(a)

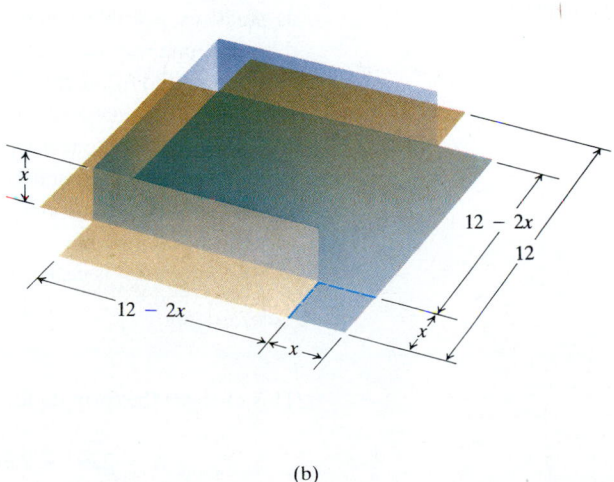

(b)

3.44   To make an open box, squares are cut from the corners of a square sheet of tin (a) and the sides are bent up (b). What value of x gives the largest volume?

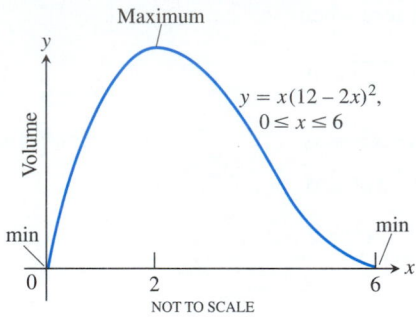

3.45   The volume of the box in Fig. 3.44, graphed as a function of x.

A graph of V (Fig. 3.45) suggests a minimum value of 0 at $x = 0$ and $x = 6$ and a maximum near $x = 2$. To learn more, we examine the first derivative of V with respect to x:

$$\frac{dV}{dx} = 144 - 96x + 12x^2 = 12(12 - 8x + x^2) = 12(2 - x)(6 - x).$$

Of the two zeros, $x = 2$ and $x = 6$, only $x = 2$ lies in the interior of the function's domain and makes the critical-point list. The values of V at this one critical point and two endpoints are

Critical point value: $V(2) = 128$

Endpoint values: $V(0) = 0, \quad V(6) = 0.$

The maximum volume is 128 in³. The cut-out squares should be 2 in. on a side.

Example 4   *Product Design.*   You have been asked to design a 1-L oil can shaped like a right circular cylinder. What dimensions will use the least material?

Solution   We picture the can as a right circular cylinder with height h and diameter $2r$ (Fig. 3.46). If r and h are measured in centimeters and the volume is expressed as 1000 cm³, then r and h are related by the equation

$$\pi r^2 h = 1000. \tag{1}$$

How shall we interpret the phrase "least material"? One possibility is to ignore the thickness of the material and the waste in manufacturing. Then we ask for dimensions r and h that make the total surface area

$$A = \underbrace{2\pi r^2}_{\substack{\text{cylinder} \\ \text{ends}}} + \underbrace{2\pi rh}_{\substack{\text{cylinder} \\ \text{wall}}} \tag{2}$$

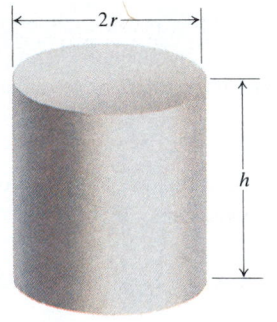

3.46   This 1-L can uses the least material when $h = 2r$ (Example 4).

as small as possible while satisfying the constraint $\pi r^2 h = 1000$. (Exercise 18 describes one way we might take waste into account.)

What kind of oil can do we expect? Not a tall, thin one like a 6-ft pipe, nor a short, wide one like a covered pizza pan. We expect something in between.

We are not quite ready to follow the procedure of the earlier examples because Eq. (2) gives A as a function of two variables and our procedure calls for A to be a function of a single variable. However, Eq. (1) can be solved to express either r or h in terms of the other.

Solving for h is easier, so we take

$$h = \frac{1000}{\pi r^2}. \tag{3}$$

This changes the formula for A to

$$A = 2\pi r^2 + 2\pi rh = 2\pi r^2 + 2\pi r \frac{1000}{\pi r^2} = 2\pi r^2 + \frac{2000}{r}. \tag{4}$$

Our mathematical goal is to find the minimum value of A on the open interval $r > 0$. Since A is differentiable throughout its domain, and its domain has no endpoints, it can have a minimum value only where its first derivative is zero.

$$A = 2\pi r^2 + \frac{2000}{r}$$

$$\frac{dA}{dr} = 4\pi r - \frac{2000}{r^2}$$

$$4\pi r - \frac{2000}{r^2} = 0$$

$$4\pi r^3 = 2000$$

$$r = \sqrt[3]{\frac{500}{\pi}}$$

So something happens at $r = \sqrt[3]{500/\pi}$, but what? To find out, we calculate d^2A/dr^2:

$$\frac{dA}{dr} = 4\pi r - \frac{2000}{r^2}$$

$$\frac{d^2A}{dr^2} = 4\pi + \frac{4000}{r^3}.$$

This derivative is positive throughout the domain of A. Hence, the graph of A is concave up throughout its entire domain. The value of A at $r = \sqrt[3]{500/\pi}$ is an absolute minimum (Fig. 3.47).

When $r = \sqrt[3]{500/\pi}$,

$$h = \frac{1000}{\pi r^2} = 2\sqrt[3]{500/\pi} = 2r. \qquad \text{(After some arithmetic)}$$

Thus the most efficient can is one in which the height equals the diameter. With a calculator we find

$$r = 5.42 \text{ cm}, \qquad h = 10.84 \text{ cm},$$

to the nearest hundredth.

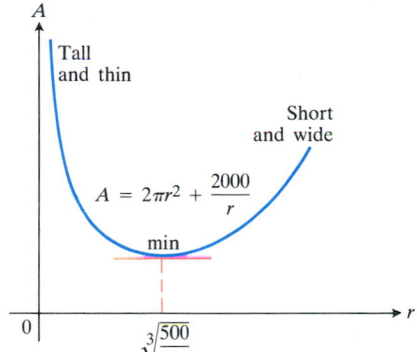

3.47 The graph of $A = 2\pi r^2 + 2000/r$ is concave up.

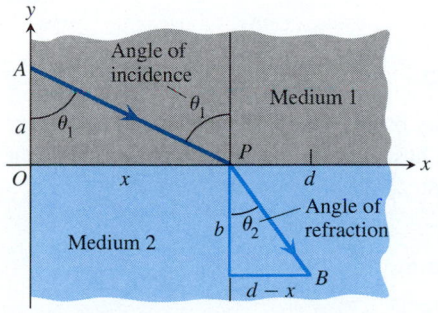

3.48 A light ray is refracted (deflected from its path) when it passes from one medium to another. θ_1 is the angle of incidence and θ_2 is the angle of refraction.

Example 5 *Fermat's Principle and Snell's Law of Refraction.* The speed of light depends on the medium through which it travels and tends to be slower in denser media. In a vacuum, light travels at the famous speed $c = 3 \times 10^8$ m/sec, but in the earth's atmosphere it travels slightly slower than that, and in glass slower still (about two-thirds as fast).

Fermat's principle in optics states that light always travels from one point to another along the quickest route. This observation enables us to predict the path light will take when it travels from a point in one medium across the boundary to a point in another medium. To see exactly what we mean, let us find the path that a ray of light will follow in going from point A in a medium where the speed of light is c_1 to point B in a medium where its speed is c_2.

Solution We assume that the two points lie in the xy-plane and that the x-axis separates the two media as in Fig. 3.48.

In either medium, where the speed of light remains constant, "shortest time" means "shortest path," and the ray of light will follow a straight line. Hence the path from A to B will consist of a line segment from A to a boundary point P, followed by another line segment from P to B. From the formula distance equals rate times time, we have

$$\text{time} = \frac{\text{distance}}{\text{rate}}.$$

The time required for light to travel from A to P is therefore

$$t_1 = \frac{AP}{c_1} = \frac{\sqrt{a^2 + x^2}}{c_1}.$$

From P to B the time is

$$t_2 = \frac{PB}{c_2} = \frac{\sqrt{b^2 + (d - x)^2}}{c_2}.$$

The time from A to B is the sum of these:

$$t = t_1 + t_2 = \frac{\sqrt{a^2 + x^2}}{c_1} + \frac{\sqrt{b^2 + (d - x)^2}}{c_2}. \tag{5}$$

Our mathematical goal is to find any value or values of x in the interval $0 \le x \le d$ at which t assumes its minimum value. We find

$$\frac{dt}{dx} = \frac{x}{c_1 \sqrt{a^2 + x^2}} - \frac{(d - x)}{c_2 \sqrt{b^2 + (d - x)^2}} \tag{6}$$

or, if we use the angles θ_1 and θ_2 in the figure,

$$\frac{dt}{dx} = \frac{\sin \theta_1}{c_1} - \frac{\sin \theta_2}{c_2}. \tag{7}$$

We can see from Eq. (6) that dt/dx is negative at $x = 0$ and positive at $x = d$, so it is zero at some point in between (Fig. 3.49). There is only one such point because dt/dx is an increasing function of x. At this point,

$$\frac{\sin \theta_1}{c_1} = \frac{\sin \theta_2}{c_2}. \tag{8}$$

This equation is **Snell's law** or the **law of refraction**.

We conclude that the path the ray of light follows is the one described by Snell's law. Figure 3.50 shows how this works out for air and water.

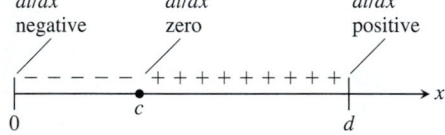

3.49 The sign pattern of dt/dx in Example 5.

3.50 For air and water, the light velocity ratio at room temperature is 1.33 and Snell's law becomes $\sin \theta_1 = 1.33 \sin \theta_2$. In this laboratory photograph, $\theta_1 = 35.5°$, $\theta_2 = 26°$, and $(\sin 35.5°/\sin 26°) = 0.581/0.438 \approx 1.33$, as predicted. (Photograph from *PSSC Physics*, Second Edition, 1965, D.C. Heath & Company with Education Development Center, Inc. NCFMF Book of Film Notes, 1974; The MIT Press with Education Development Center, Inc., Newton, Massachusetts. Reprinted by permission.)

Cost and Revenue in Economics

Here we want to point out one of the many places where calculus makes an important contribution to economic theory. It has to do with the relationship between profit, revenue (money received), and cost.

Suppose that

$r(x) = $ the revenue from selling x items,

$c(x) = $ the cost of producing the x items,

$p(x) = r(x) - c(x) = $ the profit from selling x items.

The marginal revenue and marginal cost at this production level (x items) are

$$\frac{dr}{dx} = \text{marginal revenue}, \qquad \frac{dc}{dx} = \text{marginal cost}.$$

The theorem we want to mention is the one that describes the relationship of p to these derivatives.

THEOREM 4

Maximum profit (if any) occurs at a production level at which marginal revenue equals marginal cost.

Developing a Physical Law

In developing a physical law, we typically observe an effect, measure values and list them in a table, and then try to find a rule by which one thing can be connected with another. The Alexandrian Greek Claudius Ptolemy (c.100–c.170 A.D.) tried to do this for the refraction of light by water. He made a table of angles of incidence and corresponding angles of refraction, with values very close to the ones we find for air and water today.

Angle in air (degrees)	Ptolemy's angle in water (degrees)	Modern angle in water (degrees)
10	8	7.5
20	15.5	15
30	22.5	22
40	28	29
50	35	35
60	40.5	40.5
70	45	45
80	50	47.6

The rule that connected these angles, however, eluded him, as it did everyone else for the next 1400 years. The Dutch mathematician Willebrord Snell (1580–1626) found it in 1621.

Finding a rule is nice, but the real glory of science is finding a way of thinking that makes the rule evident. Fermat discovered it around 1650. His idea was this: Of all the paths light might take to get from one point to another, it follows the path that takes the shortest time. In Example 5, you saw how this principle leads to Snell's law. The derivation we gave is Fermat's own.

Proof of Theorem 4   We assume that $r(x)$ and $c(x)$ are differentiable for all $x > 0$, so if $p(x) = r(x) - c(x)$ has a maximum value, it occurs at a production level at which $p'(x) = 0$. Since $p'(x) = r'(x) - c'(x)$, the equation $p'(x) = 0$ implies

$$r'(x) - c'(x) = 0, \quad \text{or} \quad r'(x) = c'(x).$$

This concludes the proof (Fig. 3.51).

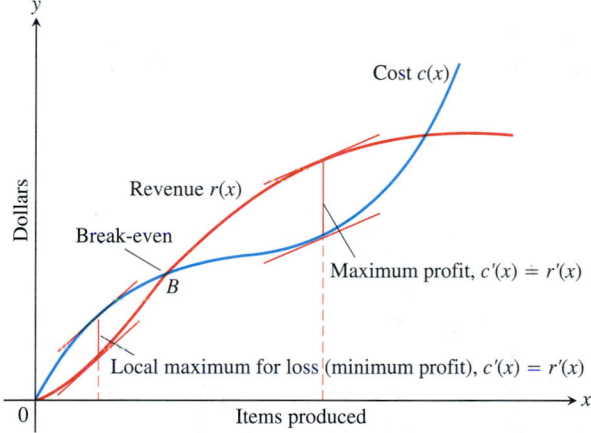

3.51   The graph of a typical cost function starts concave down and later turns concave up. It crosses the revenue curve at the break-even point B. To the left of B, the company operates at a loss. To the right, the company operates at a profit, with the maximum profit occurring where $c'(x) = r'(x)$. Farther to the right, cost exceeds revenue (perhaps because of a combination of market saturation and rising labor and material costs) and production levels become unprofitable again.

What guidance do we get from this theorem? We know that a production level at which $p'(x) = 0$ need not be a level of maximum profit. It might be a level of minimum profit, for example. But if we are making financial projections for a company, we should look for production levels at which marginal cost seems to equal marginal revenue. If there is a most profitable production level, it will be either an endpoint value or one of these.

Example 6   Suppose that $r(x) = 9x$ and $c(x) = x^3 - 6x^2 + 15x$, where x represents thousands of units. Is there a production level that maximizes profit? If so, what is it?

Solution

$$r(x) = 9x, \quad c(x) = x^3 - 6x^2 + 15x$$

$$r'(x) = 9, \quad c'(x) = 3x^2 - 12x + 15 \qquad \text{(Find } r'(x) \text{ and } c'(x).)$$

$$3x^2 - 12x + 15 = 9 \qquad \text{(Set them equal and solve for } x.)$$

$$3x^2 - 12x + 6 = 0$$

$$x^2 - 4x + 2 = 0$$

$$x = \frac{4 \pm \sqrt{16 - 4 \cdot 2}}{2} = \frac{4 \pm 2\sqrt{2}}{2} = 2 \pm \sqrt{2}$$

The possible production levels for maximum profit are $x = 2 + \sqrt{2}$ thousand units and $x = 2 - \sqrt{2}$ thousand units. A quick glance at the graphs in Fig. 3.52 or at the corresponding values of r and c shows $x = 2 + \sqrt{2}$ to be a point of maximum profit and $x = 2 - \sqrt{2}$ to be a local maximum for loss.

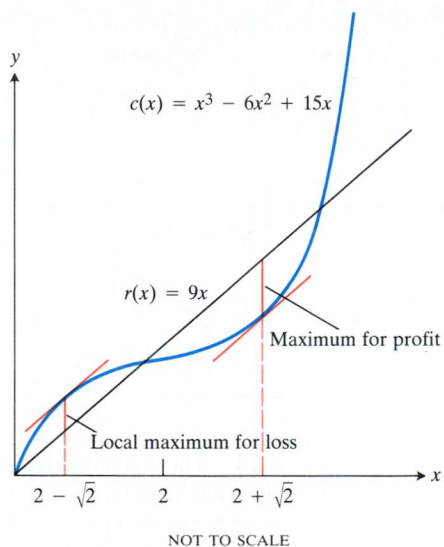

3.52   The cost and revenue curves for Example 6.

NOT TO SCALE

Modeling Discrete Phenomena with Differentiable Functions

In case you are wondering how we can use differentiable functions $c(x)$ and $r(x)$ to describe the cost and revenue that come from producing a number of items x that can only be an integer, here is the rationale.

When x is large, we can reasonably fit the cost and revenue data with smooth curves $c(x)$ and $r(x)$ that are defined not only at integer values of x but at the values in between. Once we have these differentiable functions, which are supposed to behave like the real cost and revenue when x is an integer, we can apply calculus to draw conclusions about their values. We then translate these mathematical conclusions into inferences about the real world that we hope will have predictive value. When they do, as in the case of the economic theory here, we say that the functions give a good model of reality.

What do we do when our calculus tells us that the best production level is a value of x that isn't an integer, as it did in Example 6 when it said that $x = 2 + \sqrt{2}$

thousand units would be the production level for maximum profit? The answer is to use the nearest convenient integer. For $x = 2 + \sqrt{2}$ thousand, we might use 3414, or perhaps 3410 or 3420 if we ship in boxes of 10.

Strategy for Solving Max-Min Problems

1. *Draw a picture.* Label the parts that are important in the problem.

2. *Write an equation.* Write an equation for the quantity whose maximum or minimum value you want. If you can, express the quantity as a function of a single variable, say $y = f(x)$. This may require some algebra and the use of information from the statement of the problem. Note the domain in which the values of x are to be found.

3. *Test the critical points and endpoints.* The extreme value of f will be found among the values f takes on at the endpoints of the domain and at the points where f' is zero or fails to exist. List the values of f at these points. If f has an absolute maximum or minimum on its domain, it will appear on the list. You may have to examine the sign pattern of f' or the sign of f'' to decide whether a given value represents a maximum, a minimum, or neither.

EXERCISES 3.5

1. If the perimeter of the circular sector in Fig. 3.53 is 100 m, what values of r and s will maximize the area?

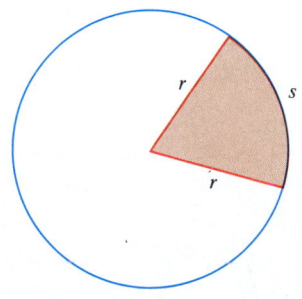

Area: $A = \dfrac{1}{2}rs$

Perimeter: $p = s + 2r$

3.53   The sector in Exercise 1.

2. What is the largest possible area for a right triangle whose hypotenuse is 5 cm long?

3. What is the smallest perimeter possible for a rectangle whose area is 16 in²?

4. Show that among all rectangles with a given perimeter, the one with the largest area is a square.

5. Figure 3.54 shows a rectangle inscribed in an isosceles right triangle whose hypotenuse is 2 units long.
   a) Express the y-coordinate of P in terms of x. (You might start by writing an equation for the line AB.)
   b) Express the area of the rectangle in terms of x.
   c) What is the largest area the rectangle can have?

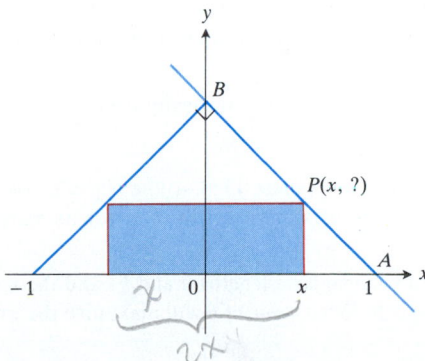

3.54   The rectangle in Exercise 5.

6. A rectangle has its base on the x-axis and its upper two vertices on the parabola $y = 12 - x^2$. What is the largest area the rectangle can have?

7. You are planning to make an open rectangular box from an 8-by-15-in. piece of cardboard by cutting squares from the corners and folding up the sides. What are the dimensions of the box of largest volume you can make this way?

8. You are planning to close off a corner of the first quadrant with a line segment 20 units long running from $(a, 0)$ to $(0, b)$. Show that the area of the triangle enclosed by the segment is largest when $a = b$.

9. A rectangular plot of farmland will be bounded on one side by a river and on the other three sides by a single-strand electric fence. With 800 m of wire at your disposal, what is the largest area you can enclose?

10. A 216-m² rectangular pea patch is to be enclosed by a fence and divided into two equal parts by another fence parallel to one of the sides. What dimensions for the outer rectangle will require the smallest total length of fence? How much fence will be needed?

11. *The lightest steel holding tank.* Your iron works has contracted to design and build a 500-ft³ square-based, open-top, rectangular steel holding tank for a paper company. The tank is to be made by welding half-inch-thick stainless steel plates together along their edges. As the production engineer, your job is to find dimensions for the base and height that will make the tank weigh as little as possible. What dimensions do you tell the shop to use?

12. *Catching rain water.* An 1125-ft³ open-top rectangular tank with a square base x ft on a side and y ft deep is to be built with its top flush with the ground to catch runoff water. The costs associated with the tank involve not only the material from which the tank is made but also an excavation charge proportional to the product xy. If the total cost is

$$c = 5(x^2 + 4xy) + 10xy,$$

what values of x and y will minimize it?

13. You are designing a rectangular poster to contain 50 in² of printing with margins of 4 in. each at top and bottom and 2 in. at each side. What overall dimensions will minimize the amount of paper used?

14. The height of an object moving vertically is given by

$$s = -16t^2 + 96t + 112,$$

with s in feet and t in seconds. Find (a) the object's velocity when $t = 0$, (b) its maximum height, and (c) its velocity when $s = 0$.

15. Two sides of a triangle have lengths a and b, and the angle between them is θ. What value of θ will maximize the triangle's area?

16. Find the largest possible value of $s = 2x + y$ if x and y are side lengths in a right triangle whose hypotenuse is $\sqrt{5}$ units long.

17. What are the dimensions of the lightest open-top right circular cylindrical can that will hold a volume of 1000 cm³? Compare the result here with the result in Example 4.

18. You are designing 1000-cm³ right circular cylindrical cans

whose manufacture will take waste into account. There is no waste in cutting the aluminum for the sides, but the tops and bottoms of radius r will be cut from squares that measure $2r$ units on a side. The total amount of aluminum used up by each can will therefore be

$$A = 8r^2 + 2\pi rh$$

rather than the $A = 2\pi r^2 + 2\pi rh$ in Example 4. In Example 4 the ratio of h to r for the most economical cans was 2 to 1. What is the ratio now?

19. a) The U.S. Postal Service will accept a box for domestic shipment only if the sum of its length and girth (distance around) does not exceed 108 in. What dimensions will give a box with a square end the largest possible volume (Fig. 3.55)?

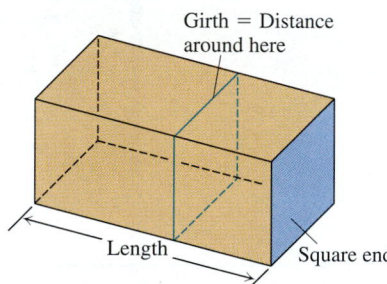

3.55 The box in Exercise 19.

b) GRAPHER Graph the volume of a 108-in. box (length plus girth equals 108 in.) as a function of its length and compare what you see with your answer in (a).

20. *Conclusion of the sonobuoy problem (Exercise 20, Section 2.7).*
a) Show that the value of x that minimizes the square of the distance, and hence the distance, between two points (x, x^2) and $(2, -1/2)$ in Fig. 3.56 is a solution of the equation $x = 1/(x^2 + 1)$.
b) GRAPHER Graph the function $f(x) = x - (1/(x^2 + 1))$.
c) CALCULATOR If you did not already do so in Section 2.7, solve the equation $x = 1/(x^2 + 1)$ to five decimal places.

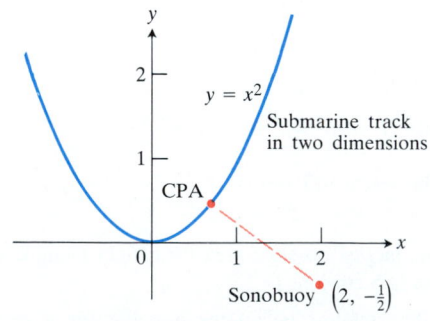

3.56 The submarine track and sonobuoy in Exercise 20. CPA = closest point of approach.

21. Compare the answers to the following two construction problems.

a) A rectangular sheet of perimeter 36 cm and dimensions x cm by y cm is to be rolled into the cylinder in Fig. 3.57(a). What values of x and y give the largest volume?

b) The rectangular sheet of perimeter 36 cm and dimensions x by y is to be revolved about one of the sides of length y to sweep out the cylinder in Fig. 3.57(b). What values of x and y give the largest volume?

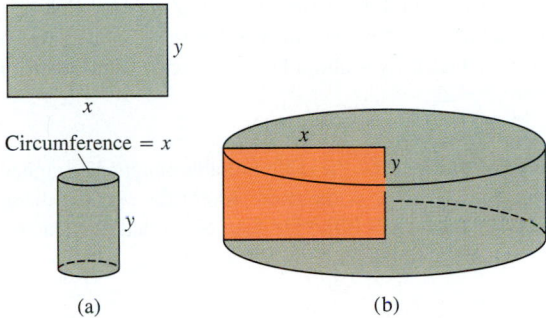

3.57 The rectangular sheet and cylinders in Exercise 21.

22. A right triangle whose hypotenuse is $\sqrt{3}$ m long is revolved about one of its legs to generate a right circular cone. Find the radius, height, and volume of the cone of greatest volume that can be made this way.

23. What value of a makes $f(x) = x^2 + (a/x)$ have (a) a local minimum at $x = 2$; (b) a point of inflection at $x = 1$?

24. Show that $f(x) = x^2 + (a/x)$ cannot have a local maximum for any value of a.

25. What values of a and b make

$$f(x) = x^3 + ax^2 + bx$$

have (a) a local maximum at $x = -1$ and a local minimum at $x = 3$; (b) a local minimum at $x = 4$ and a point of inflection at $x = 1$?

26. Find the volume of the largest right circular cone that can be inscribed in a sphere of radius 3 (Fig. 3.58).

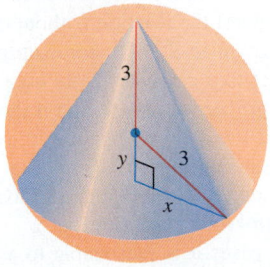

3.58 The sphere and cone in Exercise 26.

27. *The strength of a beam.* The strength S of a rectangular beam is proportional to its width times the square of its depth.

a) Find the dimensions of the strongest beam that can be cut from a 12-in.-diameter log (Fig. 3.59).

b) GRAPHER Graph S as a function of the beam's width w, assuming the proportionality constant to be $k = 1$. Reconcile what you see with your answer in (a).

c) GRAPHER On the same screen, or on a separate screen, graph S as a function of the beam's depth d, again taking $k = 1$. Compare the graphs with one another and with your answer in (a). What would be the effect of changing to some other value of k? Try it.

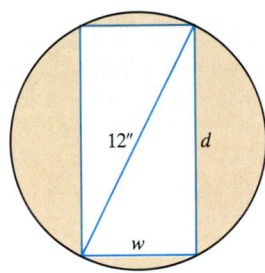

3.59 A typical cross section of the log and beam in Exercises 27 and 28.

28. *The stiffness of a beam.* The stiffness S of a rectangular beam is proportional to its width times the cube of its depth.

a) Find the dimensions of the stiffest beam that can be cut from a 12-in.-diameter log (Fig. 3.59).

b) GRAPHER Graph S as a function of the beam's width w, assuming the proportionality constant to be $k = 1$. Reconcile what you see with your answer in (a).

c) GRAPHER On the same screen, or on a separate screen, graph S as a function of the beam's depth d, again taking $k = 1$. Compare the graphs with one another and with your answer in (a). What would be the effect of changing to some other value of k? Try it.

29. a) The function $y = \cot x - \sqrt{2} \csc x$ has an absolute maximum value on the interval $0 < x < \pi$. Find it.

b) GRAPHER Graph the function and compare what you see with your answer in (a).

30. a) The function $y = \tan x + \sqrt{3} \cot x$ has an absolute minimum value on the interval $0 < x < \pi/2$. Find it.

b) GRAPHER Graph the function and compare what you see with your answer in (a).

31. How close does the curve $y = \sqrt{x}$ come to the point $(3/2, 0)$? (*Hint:* If you minimize the *square* of the distance, you can avoid square roots.)

32. How close does the semicircle $y = \sqrt{16 - x^2}$ come to the point $(1, \sqrt{3})$?

33. You are designing a trough with the dimensions shown in Fig. 3.60. Only the angle θ can be varied. What value of θ will maximize the trough's volume?

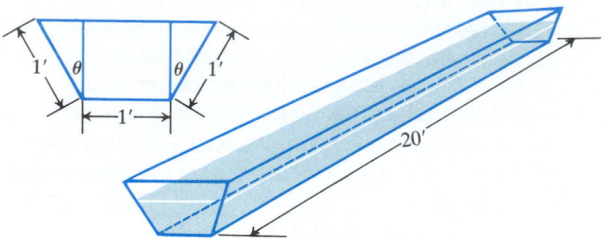

3.60   The trough in Exercise 33.

34. A rectangular sheet of $8\frac{1}{2}$-by-11-in. paper is placed on a flat surface, and one of the corners is lifted up and placed on the opposite longer edge. The other corners are held in their original positions. With all four corners now held fixed, the paper is smoothed flat (Fig. 3.61). The problem is to make the length of the crease as small as possible. Call the length L.

a) Try it with paper.

b) Show that $L^2 = 2x^3/(2x - 8.5)$.

c) Minimize L^2.

d) CALCULATOR   Find the minimum value of L to the nearest tenth of an inch.

e) GRAPHER   Graph L as a function of x and compare what you see with your answer in (d).

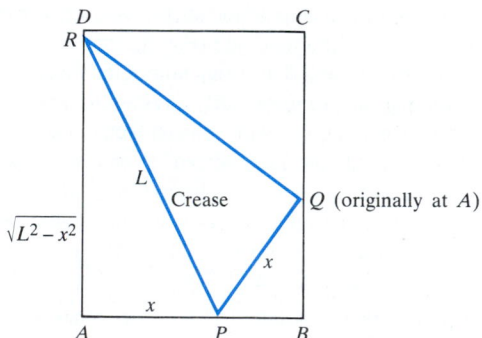

3.61   The paper in Exercise 34.

35. Fermat's principle in optics states that light always travels from one point to another along a path that minimizes the travel time. Figure 3.62 shows light from a source A reflected by a plane mirror to a receiver at point B. Show that for the light to obey Fermat's principle, the angle of incidence must equal the angle of reflection. (This result can also be derived without calculus. There is a purely geometric argument, which you may prefer.)

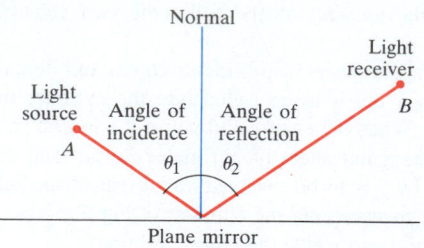

3.62   In studies of light reflection, the angles of incidence and reflection are measured from the line normal to the reflecting surface. Exercise 35 asks you to show that if light obeys Fermat's "least-time" principle, then $\theta_1 = \theta_2$.

36. Let $f(x)$ and $g(x)$ be the differentiable functions graphed in Fig. 3.63. Point c is the point where the vertical distance between the curves is the greatest. Show that the tangents to the curves at $x = c$ have to be parallel.

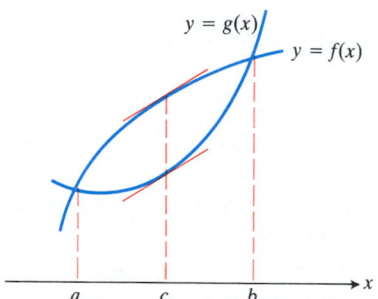

3.63   The graphs for Exercise 36.

37. *Tin pest.*   Metallic tin, when kept below 13°C for a while, becomes brittle and crumbles to a gray powder. Tin objects eventually crumble to this gray powder spontaneously if kept in a cold climate for years. The Europeans who saw the tin organ pipes in their churches crumble away years ago called the change *tin pest* because it seemed to be contagious. And indeed it was, for the gray powder is a catalyst for its own formation.

A *catalyst* for a chemical reaction is a substance that controls the rate of the reaction without undergoing any permanent change in itself. An *autocatalytic reaction* is one whose product is a catalyst for its own formation. Such a reaction may proceed slowly at first if the amount of catalyst present is small and slowly again at the end, when most of the original substance is used up. But in between, when both the substance and its catalyst product are abundant, the reaction proceeds at a faster pace.

In some cases it is reasonable to assume that the rate $v = dx/dt$ of the reaction is proportional both to the amount of the original substance present and to the amount of product. That is, v may be considered to be a function of x alone, and

$$v = kx(a - x) = kax - kx^2,$$

where

x = the amount of product,

a = the amount of substance at the beginning,

k = a positive constant.

At what value of x does the rate v have a maximum? What is the maximum value of v?

38. Suppose that at time $t \geq 0$ the position of a particle moving on the x-axis is $x = (t - 1)(t - 4)^4$.

a) When is the particle at rest?

b) During what time interval does the particle move to the left?

c) What is the fastest the particle goes while moving to the left?

d) GRAPHER Graph x as a function of t for $0 \leq t \leq 6$. Graph dx/dt over the same interval, in another color if possible. Compare the graphs with one another and with your answers in (a)–(c).

39. *Circle vs. square.*

a) A 4-m length of wire is available for making a circle and a square. How should the wire be distributed between the two shapes to maximize the sum of the enclosed areas?

b) GRAPHER Graph the total area enclosed by the wire as a function of the circle's radius. Reconcile what you see with your answer in (a).

c) GRAPHER Now graph the total area enclosed by the wire as a function of the square's side length. Again, reconcile what you see with your answer in (a).

40. At noon, ship A was 12 nautical miles due north of ship B. Ship A was sailing south at 12 knots (nautical miles per hour—a nautical mile is 2000 yd) and continued to do so all day. Ship B was sailing east at 8 knots and continued to do so all day.

a) Start counting time with $t = 0$ at noon and express the distance s between the ships as a function of t.

b) How rapidly was the distance between the ships changing at noon? One hour later?

c) CALCULATOR The visibility that day was 5 nautical miles. Did the ships ever sight each other?

d) GRAPHER Graph s and ds/dt together as functions of t for $-1 \leq t \leq 3$, using different colors if possible. Compare the graphs and reconcile what you see with your answers in (b) and (c).

e) The graph of ds/dt looks as if it might have a horizontal asymptote in the first quadrant. This in turn suggests that ds/dt approaches a limiting value as $t \to \infty$. What is this value? What is its relation to the ships' individual speeds?

41. Find the points on the curve $y = \sqrt{x}$ nearest the point $(c, 0)$ (a) if $c \geq 1/2$, (b) if $c < 1/2$.

42. CALCULATOR The 8-ft wall in Fig. 3.64 stands 27 ft from the building. Find the length of the shortest straight beam that will reach to the side of the building from the ground outside the wall.

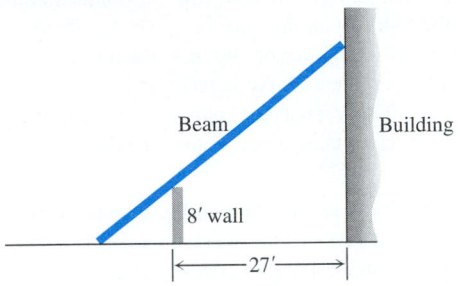

3.64 The wall and beam in Exercise 42.

43. Show that if a, b, c, and d are positive integers, then

$$\frac{(a^2 + 1)(b^2 + 1)(c^2 + 1)(d^2 + 1)}{abcd} \geq 16.$$

44. If the sum of the surface areas of a cube and a sphere is held constant, what ratio of an edge of the cube to the radius of the sphere will make the sum of the volumes (a) as small as possible, (b) as large as possible?

45. *How we cough.*

a) When we cough, the trachea (windpipe) contracts to increase the velocity of the air going out. This raises the questions of how much it should contract to maximize the velocity and whether it really contracts that much when we cough.

   Under reasonable assumptions about the elasticity of the tracheal wall and about how the air near the wall is slowed by friction, the average flow velocity v can be modeled by the equation

$$v = c(r_0 - r)r^2 \text{ cm/sec} \qquad \frac{r_0}{2} \leq r \leq r_0,$$

   where r_0 is the rest radius of the trachea in centimeters and c is a positive constant whose value depends in part on the length of the trachea.

   Show that v has its maximum value when $r = (2/3)r_0$, that is, when the trachea is about 33% contracted. The remarkable fact is that x-ray photographs confirm that the trachea contracts about this much during a cough.

b) GRAPHER Take r_0 to be 0.5 and c to be 1, and graph v over the interval $0 \leq r \leq 0.5$. Compare what you see to the claim that v is at a maximum when $r = (2/3)r_0$.

46. *Sensitivity to medicine (continuation of Exercise 52, Section 2.1).* Find the amount of medicine to which the body is most sensitive by finding the value of M that maximizes the derivative dR/dM.

47. It costs a manufacturer c dollars each to manufacture and distribute a certain item. If the items sell at x dollars each, the number sold is given by $n = a/(x - c) + b(100 - x)$, where a and b are positive constants. What selling price will bring a maximum profit?

48. You operate a tour service that offers the following rates:
   a) $200 per person if 50 people (the minimum number to book the tour) go on the tour.
   b) For each additional person, up to a maximum of 80 people total, everyone's charge is reduced by $2.

   It costs you $6000 (a fixed cost) plus $32 per person to conduct the tour. How many people does it take to maximize your profit?

49. *The best quantity to order.*   One of the formulas for inventory management says that the average weekly cost of ordering, paying for, and holding merchandise is

$$A(q) = \frac{km}{q} + cm + \frac{hq}{2},$$

where q is the quantity you order when things run low (shoes, radios, brooms, or whatever the item might be), k is the cost of placing an order (the same, no matter how often you order), c is the cost of one item (a constant), m is the number of items sold each week (a constant), and h is the weekly holding cost per item (a constant that takes into account things such as space, utilities, insurance, and security). Your job, as the inventory manager for your store, is to find the quantity that will minimize $A(q)$. What is it? (The formula you get for the answer is called the *Wilson lot size formula.*)

50. *Continuation of Exercise 49.*   Shipping costs sometimes depend on order size. When they do, it is more realistic to replace k by $k + bq$, the sum of k and a constant multiple of q. What is the most economical quantity to order now?

51. Use Theorem 4 to show that if $r(x) = 6x$ and $c(x) = x^3 - 6x^2 + 15x$ are your revenue and cost functions, then your operation will never be profitable and the best you can do is break even (have revenue equal cost).

EXPLORER PROGRAM

PowerGrapher   Evaluates functions at multiple points in an interval, enabling you to make informed guesses about zeros and extremes

Graphs functions singly or together

3.6 Indeterminate Forms and L'Hôpital's Rule

In the late 1600s, John Bernoulli discovered a rule for calculating limits of fractions whose numerators and denominators both approach zero. Today the rule is known as l'Hôpital's rule, after Guillaume François Antoine de l'Hôpital (1661–1704), Marquis de Sainte-Mesme, a French nobleman who wrote the introductory calculus text in which the rule first appeared. L'Hôpital's rule gives fast results and often applies when other methods fail.

The Indeterminate Form 0/0

If functions f and g are continuous at $x = a$ but $f(a) = g(a) = 0$, the limit

$$\lim_{x \to a} \frac{f(x)}{g(x)} \tag{1}$$

cannot be evaluated by substituting $x = a$. The substitution produces 0/0, a meaningless expression known as an **indeterminate form**.

As we have seen, the value of the limit in Eq. (1) is often hard to predict:

$$\lim_{x \to 2} \frac{x^2 - 4}{x - 2} = \lim_{x \to 2} (x + 2) = 4$$

$$\lim_{x \to 0} \frac{\sin x}{x} = 1$$

$$\lim_{x \to 0} \frac{1 - \cos x}{x} = 0.$$

The limit

$$f'(a) = \lim_{x \to a} \frac{f(x) - f(a)}{x - a}$$

from which we calculate derivatives always produces the indeterminate form 0/0. Our success in calculating derivatives suggests that we might turn things around and use derivatives to calculate limits that lead to indeterminate forms. For example, knowing the derivative of sin x would let us find

$$\lim_{x \to 0} \frac{\sin x}{x} = \lim_{x \to 0} \frac{\sin x - \sin 0}{x - 0} = \frac{d}{dx}(\sin x)\bigg|_{x=0} = \cos 0 = 1.$$

L'Hôpital's rule gives an explicit connection between derivatives and limits that lead to the indeterminate form 0/0.

THEOREM 5

L'Hôpital's Rule (First Form)

Suppose that $f(a) = g(a) = 0$, that $f'(a)$ and $g'(a)$ exist, and that $g'(a) \neq 0$. Then

$$\lim_{x \to a} \frac{f(x)}{g(x)} = \frac{f'(a)}{g'(a)}.$$

Proof   Working backward from $f'(a)$ and $g'(a)$, which are themselves limits, we have

$$\frac{f'(a)}{g'(a)} = \frac{\lim\limits_{x \to a} \dfrac{f(x) - f(a)}{x - a}}{\lim\limits_{x \to a} \dfrac{g(x) - g(a)}{x - a}} = \lim_{x \to a} \frac{\dfrac{f(x) - f(a)}{x - a}}{\dfrac{g(x) - g(a)}{x - a}}$$

$$= \lim_{x \to a} \frac{f(x) - f(a)}{g(x) - g(a)} = \lim_{x \to a} \frac{f(x) - 0}{g(x) - 0}$$

$$= \lim_{x \to a} \frac{f(x)}{g(x)}.$$

Example 1

a) $\lim\limits_{x \to 0} \dfrac{3x - \sin x}{x} = \dfrac{3 - \cos x}{1}\bigg|_{x=0} = 2$

b) $\lim\limits_{x \to 0} \dfrac{\sqrt{1 + x} - 1}{x} = \dfrac{\dfrac{1}{2\sqrt{1 + x}}}{1}\bigg|_{x=0} = \dfrac{1}{2}$

c) $\lim\limits_{x \to 0} \dfrac{x - \sin x}{x^3} = \dfrac{1 - \cos x}{3x^2}\bigg|_{x=0} = ?$

Notice that to apply l'Hôpital's rule to f/g we divide the derivative of f by the derivative of g. Do not fall into the trap of taking the derivative of f/g. The quotient to use is f'/g', not $(f/g)'$.

What can we do about the limit in Example 1(c)? The first form of l'Hôpital's rule does not tell us what the limit is because the derivative of $g(x) = x^3$ is zero at $x = 0$. However, a stronger form of l'Hôpital's rule says that whenever the rule gives 0/0 we can apply it again, repeating the process until we get a different result.

With this stronger rule we can finish the work begun in Example 1(c):

$$\lim_{x \to 0} \frac{x - \sin x}{x^3} = \lim_{x \to 0} \frac{1 - \cos x}{3x^2} \qquad \left(\text{Still } \frac{0}{0}; \text{ apply the rule again.}\right)$$

$$= \lim_{x \to 0} \frac{\sin x}{6x} \qquad \left(\text{Still } \frac{0}{0}; \text{ apply the rule again.}\right)$$

$$= \lim_{x \to 0} \frac{\cos x}{6} = \frac{1}{6}. \qquad \text{(A different result. Stop.)}$$

THEOREM 6

L'Hôpital's Rule (Stronger Form)

Suppose that $f(x_0) = g(x_0) = 0$ and that the functions f and g are both differentiable on an open interval (a,b) that contains the point x_0. Suppose also that $g' \neq 0$ at every point in (a, b) except possibly x_0. Then

$$\lim_{x \to x_0} \frac{f(x)}{g(x)} = \lim_{x \to x_0} \frac{f'(x)}{g'(x)}, \qquad (2)$$

provided the limit on the right exists.

You will find a proof of the stronger form in Appendix 7.

A Misnamed Rule and the First Differential Calculus Text

In 1694 John Bernoulli agreed to accept a retainer of 300 pounds per year from his former pupil l'Hôpital to solve problems for him and keep him up to date on calculus. One of the problems was the so-called 0/0 problem, which Bernoulli solved as agreed. When l'Hôpital published his notes on calculus in book form in 1696, the 0/0 rule appeared as a theorem. L'Hôpital acknowledged his debt to Bernoulli and, to avoid claiming authorship of the book's entire contents, had the book appear without bearing his own name. Bernoulli nevertheless accused l'Hôpital of plagiarism, an accusation inadvertently supported after l'Hôpital's death by the publisher's promotion of the book as l'Hôpital's. By 1721, Bernoulli, a man so jealous he threw his son Daniel out of the house for accepting a mathematics prize from the French Academy of Sciences, claimed to have been the author of the entire work. As puzzling and fickle as ever, history accepted Bernoulli's claim (until recently, at least) while naming the rule after l'Hôpital.

Example 2

$$\lim_{x \to 0} \frac{\sqrt{1 + x} - 1 - (x/2)}{x^2} \qquad \left(\frac{0}{0}\right)$$

$$= \lim_{x \to 0} \frac{(1/2)(1 + x)^{-1/2} - (1/2)}{2x} \qquad \left(\text{Still } \frac{0}{0}\right)$$

$$= \lim_{x \to 0} \frac{-(1/4)(1 + x)^{-3/2}}{2} = -\frac{1}{8}$$

When you apply l'Hôpital's rule, look for a change from 0/0 to something else. This is where the limit's value is revealed.

Example 3

$$\lim_{x \to 0} \frac{1 - \cos x}{x + x^2} \qquad \left(\frac{0}{0}\right)$$

$$= \lim_{x \to 0} \frac{\sin x}{1 + 2x} = \frac{0}{1} = 0$$

If we continue to differentiate in an attempt to apply l'Hôpital's rule once more, we get

$$\lim_{x \to 0} \frac{1 - \cos x}{x + x^2} = \lim_{x \to 0} \frac{\sin x}{1 + 2x} = \lim_{x \to 0} \frac{\cos x}{2} = \frac{1}{2},$$

which is wrong.

If you reach a point where one of the derivatives is zero and the other is not, then the limit in question is either zero, as in Example 3, or infinity, as in the next example.

Example 4

$$\lim_{x \to 0^+} \frac{\sin x}{x^2} \qquad \left(\frac{0}{0}\right)$$

$$= \lim_{x \to 0^+} \frac{\cos x}{2x} = \infty$$

The Forms ∞/∞, $\infty \cdot 0$, and $\infty - \infty$

In more advanced books it is proved that l'Hôpital's rule applies to the indeterminate form ∞/∞ as well as 0/0. If $f(x)$ and $g(x)$ both approach infinity as x approaches a, then

$$\lim_{x \to a} \frac{f(x)}{g(x)} = \lim_{x \to a} \frac{f'(x)}{g'(x)},$$

provided the limit on the right exists. In the notation $x \to a$, a may be either finite or infinite.

Example 5

a) $\displaystyle \lim_{x \to (\pi/2)^-} \frac{\tan x}{1 + \tan x} \qquad \left(\frac{\infty}{\infty}\right)$

$$= \lim_{x \to (\pi/2)^-} \frac{\sec^2 x}{\sec^2 x} = 1$$

b) $\displaystyle \lim_{x \to \infty} \frac{x - 2x^2}{3x^2 + 5} = \lim_{x \to \infty} \frac{1 - 4x}{6x} = \lim_{x \to \infty} \frac{-4}{6} = -\frac{2}{3}$

We can sometimes handle the forms $0 \cdot \infty$ and $\infty - \infty$ by using algebra to get 0/0 or ∞/∞ instead. Here again, we do not mean to suggest that there is a number $0 \cdot \infty$ or $\infty - \infty$ any more than we mean to suggest that there is a number 0/0 or ∞/∞. These forms are not numbers but descriptions of limits.

Example 6

$$\lim_{x \to 0^+} x \cot x \qquad (0 \cdot \infty)$$

$$= \lim_{x \to 0^+} \frac{x}{\tan x} \qquad \left(\frac{0}{0}\right)$$

$$= \lim_{x \to 0^+} \frac{1}{\sec^2 x} = \frac{1}{1} = 1$$

Example 7   Find $\displaystyle \lim_{x \to 0} \left(\frac{1}{\sin x} - \frac{1}{x}\right)$.

Solution   If $x \to 0^+$, then $\sin x \to 0^+$ and

$$\frac{1}{\sin x} - \frac{1}{x} \to \infty - \infty.$$

Similarly, if $x \to 0^-$, then $\sin x \to 0^-$ and

$$\frac{1}{\sin x} - \frac{1}{x} \to -\infty - (-\infty) = -\infty + \infty.$$

Neither form reveals what happens in the limit. To find out, we combine the original fractions,

$$\frac{1}{\sin x} - \frac{1}{x} = \frac{x - \sin x}{x \sin x},$$

and apply l'Hôpital's rule to the single fraction on the right:

$$\lim_{x \to 0} \left(\frac{1}{\sin x} - \frac{1}{x} \right) = \lim_{x \to 0} \frac{x - \sin x}{x \sin x} \qquad \left(\frac{0}{0} \right)$$

$$= \lim_{x \to 0} \frac{1 - \cos x}{\sin x + x \cos x} \qquad \left(\text{Still } \frac{0}{0} \right)$$

$$= \lim_{x \to 0} \frac{\sin x}{2 \cos x - x \sin x} = \frac{0}{2} = 0.$$

FLOWCHART 3.1   L'Hôpital's rule

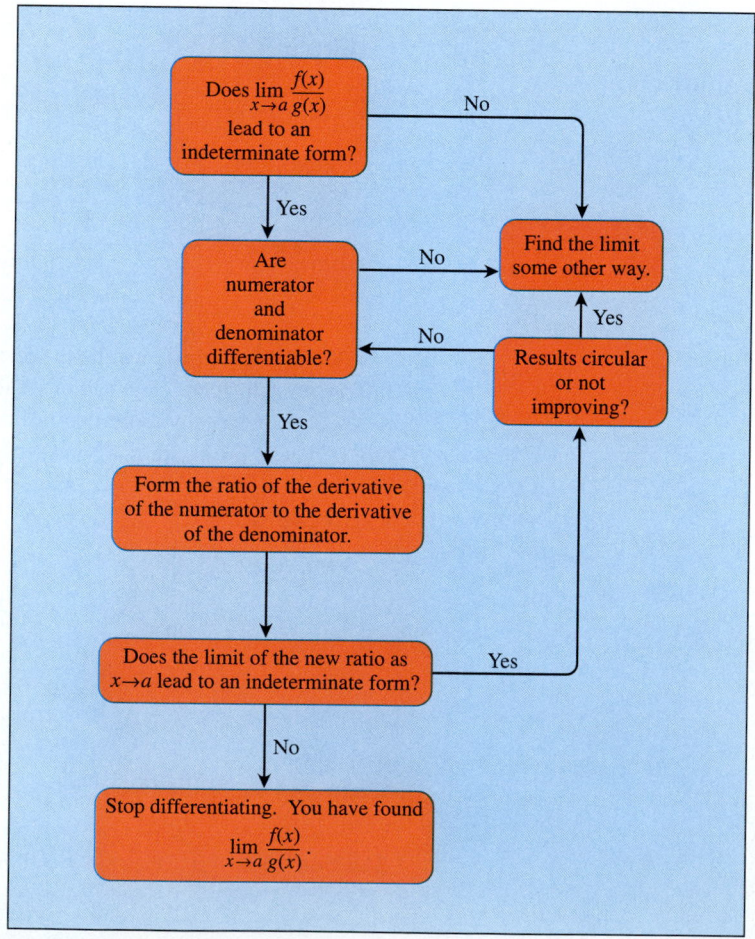

EXERCISES 3.6

Find the limits in Exercises 1–20.

1. $\lim_{x \to 2} \dfrac{x - 2}{x^2 - 4}$

2. $\lim_{t \to \infty} \dfrac{6t + 5}{3t - 8}$

3. $\lim_{x \to 1} \dfrac{x^3 - 1}{4x^3 - x - 3}$

4. $\lim_{x \to \pi/2} \dfrac{2x - \pi}{\cos x}$

5. $\lim_{t \to 0} \dfrac{\sin t^2}{t}$

6. $\lim_{x \to 0} \dfrac{\sin 5x}{x}$

7. $\lim_{x \to 0} \dfrac{\sqrt{ax + a^2} - a}{x}; \; a > 0$

8. $\lim_{x \to \infty} \dfrac{5x^2 - 3x}{7x^2 + 1}$

9. $\lim_{x \to 0} \dfrac{x \sin x}{1 - \cos x}$

10. $\lim_{t \to 0} \dfrac{\cos t - 1}{t^2}$

11. $\lim_{x \to \pi/2} \left(x - \dfrac{\pi}{2}\right)^2 \sec x$

12. $\lim_{x \to (\pi/2)^-} \left(\dfrac{\pi}{2} - x\right) \tan x$

13. $\lim_{x \to 0^+} \dfrac{2x}{x + 7\sqrt{x}}$

14. $\lim_{x \to \infty} \dfrac{x - 2x^2}{3x^2 + 5x}$

15. $\lim_{t \to 0} \dfrac{10(\sin t - t)}{t^3}$

16. $\lim_{x \to 0} \dfrac{x(1 - \cos x)}{x - \sin x}$

17. $\lim_{x \to 0^+} \left(\dfrac{1}{x} - \dfrac{1}{\sqrt{x}}\right)$

18. $\lim_{x \to \infty} 3\left(x - \sqrt{x^2 + x}\right)$

19. $\lim_{x \to -(\pi/2)^+} (\sec x + \tan x)$

20. $\lim_{x \to 0^+} (\csc x - \cot x + \cos x)$

21. L'Hôpital's rule does not seem to help with

$$\lim_{x \to \infty} \dfrac{\sqrt{9x + 1}}{\sqrt{x + 1}}.$$

Find the limit some other way.

22. L'Hôpital's rule does not seem to help with

$$\lim_{x \to (\pi/2)^-} \dfrac{\sec x}{\tan x}.$$

Try it—you just keep on going. Find the limit some other way.

23. The right triangle in Fig. 3.65 has one leg of length 1, another of length y, and a hypotenuse of length r. The angle opposite y has radian measure θ. Find the limits as $\theta \to (\pi/2)^-$ of (a) $r - y$, (b) $r^2 - y^2$, (c) $r^3 - y^3$.

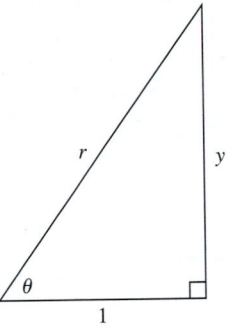

3.65 The triangle in Exercise 23.

24. Which is correct, (a) or (b)? Explain.

a) $\lim_{x \to 3} \dfrac{x - 3}{x^2 - 3} = \lim_{x \to 3} \dfrac{1}{2x} = \dfrac{1}{6}$

b) $\lim_{x \to 3} \dfrac{x - 3}{x^2 - 3} = \dfrac{0}{6} = 0$

25. Let

$$f(x) = \begin{cases} x + 2, & x \neq 0, \\ 0, & x = 0, \end{cases}$$

$$g(x) = \begin{cases} x + 1, & x \neq 0, \\ 0, & x = 0. \end{cases}$$

Show that

$$\lim_{x \to 0} \dfrac{f'(x)}{g'(x)} = 1, \quad \text{but} \quad \lim_{x \to 0} \dfrac{f(x)}{g(x)} = 2.$$

Doesn't this contradict l'Hôpital's rule? Explain.

26. Find a value of c that makes the function

$$f(x) = \begin{cases} \dfrac{9x - 3 \sin 3x}{5x^3}, & x \neq 0, \\ c, & x = 0, \end{cases}$$

continuous at $x = 0$.

27. Find a value of c that makes the function

$$f(x) = \begin{cases} \dfrac{(\tan x)^2}{\sin(4x^2/\pi)}, & 0 < x \leq \dfrac{\pi}{4}, \\ c, & x = 0, \end{cases}$$

satisfy the hypotheses of the Mean Value Theorem on the interval $[0, \pi/4]$.

28. CALCULATOR If you did not do so in Section 2.3, try to guess the value of

$$\lim_{h \to 0} \dfrac{1 - \cos h}{h^2}$$

by taking $h = 1, 0.1, 0.01$, and so on, as far as your calculator can go. Then confirm your guess by finding the limit with l'Hôpital's rule.

Graphing Calculator or Computer Grapher

In Exercises 29 and 30, graph the functions and use what you see to guess the limits. Then confirm your guess by finding the limit with l'Hôpital's rule.

29. $\lim\limits_{x \to 2} \dfrac{x^2 - 4}{\sqrt{x^2 + 5} - 3}$

30. $\lim\limits_{x \to 1} \dfrac{2x^2 - (3x + 1)\sqrt{x} + 2}{x - 1}$

EXPLORER PROGRAMS

Limit Problems — Offers practice in finding limits that lead to indeterminate forms

PowerGrapher — Graphs all the functions in this section

3.7 Approximation Errors, Quadratic Approximations, and the Mean Value Theorem

This section introduces two extensions of the Mean Value Theorem. The first tells how much error is incurred in replacing a differentiable function by one of its linearizations. The second tells how to improve a linear approximation by adding a quadratic term and tells how much error (usually less) to expect from the newly constructed quadratic approximation.

The Extended Mean Value Theorem (First Form)

In a slightly altered form, the Mean Value Theorem of Section 3.2 says that if a function f is continuous at every point of $[a, b]$ and differentiable at every point of (a, b), then there is at least one number c in (a, b) at which

$$f(b) = f(a) + f'(c)(b - a). \tag{1a}$$

When we use the theorem in this form we usually want to treat b as an independent variable. In such cases, we write x in place of b to get

$$f(x) = f(a) + f'(c)(x - a), \tag{1b}$$

now valid for the interval joining x and a, with c in between. The right-hand side of Eq. (1b) looks like the linearization

$$L(x) = f(a) + f'(a)(x - a).$$

In one, we have $f'(c)$, in the other $f'(a)$, but otherwise they are the same. For twice-differentiable functions, the resemblance is more than coincidental.

THEOREM 7

The Extended Mean Value Theorem (First Form)

If f and f' are continuous on $[a, b]$ and f' is differentiable on (a, b), then there is at least one number c_2 between a and b at which

$$f(b) = f(a) + f'(a)(b - a) + \frac{f''(c_2)}{2}(b - a)^2. \tag{2}$$

The significance of Eq. (2) is not merely that f satisfies an equation of the form

$$f(b) = f(a) + f'(a)(b - a) + k(b - a)^2 \tag{3}$$

for some number k. This is hardly news, for if we wanted to, we could always find k by solving Eq. (3) for it. The significance of Eq. (2) is that the value of k is half the value of f'' at some point between a and b. This, as we shall soon see, gives us the control we need over the errors in linear approximations.

Proof of Theorem 7   Equation (3) says that when $x = b$, the function $f(x)$ has the same value as the function

$$f(a) + f'(a)(x - a) + k(x - a)^2.$$

These two functions also have the same value when $x = a$ (namely, $f(a)$), but generally their difference,

$$F(x) = f(x) - f(a) - f'(a)(x - a) - k(x - a)^2,$$

is not zero.

The function $F(x)$ satisfies all the hypotheses of Rolle's theorem on the interval $[a, b]$: $F(a) = 0$; $F(b) = 0$; F is continuous on $[a, b]$ because f, $(x - a)$, and $(x - a)^2$ are; and F is differentiable on (a, b) for the same reason. Therefore, $F' = 0$ at some point c_1 between a and b:

$$F'(c_1) = 0, \qquad a < c_1 < b.$$

Because $F'(c_1) = 0$, the derivative

$$F'(x) = f'(x) - f'(a) - 2k(x - a) \tag{4}$$

satisfies the hypotheses of Rolle's theorem on the interval $[a, c_1]$: $F'(c_1) = 0$, and substituting $x = a$ in Eq. (4) shows that $F'(a) = 0$. Also, F' is continuous on $[a, c_1]$ and differentiable on (a, c_1) because both f' and $(x - a)$ are. Therefore, the derivative

$$F''(x) = f''(x) - 2k$$

is zero at some point c_2 between a and c_1, which means that

$$0 = f''(c_2) - 2k, \qquad \text{or} \qquad k = \frac{f''(c_2)}{2}.$$

Substituting this value of k in Eq. (3) gives Eq. (2), which is what we set out to prove. ∎

REMARK   Equation (2) also holds if b is less than a, provided f and f' continue to satisfy the hypotheses of continuity and differentiability on the interval joining b and a. To verify the equation for $b < a$, we can repeat the proof we just gave, referring to "the interval joining b and a" wherever we formerly referred to $[a, b]$ and (a, b). If we do this and take similar care with the other intervals in the proof, the arguments go through as before.

The Errors in Linear Approximations

We can now calculate the error in the linear approximation

$$f(x) \approx f(a) + f'(a)(x - a) \tag{5}$$

in Section 2.6. We begin by rewriting Eq. (2) with x in place of b. This gives

$$f(x) = \underbrace{f(a) + f'(a)(x - a)}_{\substack{\text{the linearization of} \\ f \text{ at } x = a}} + \underbrace{\frac{f''(c_2)}{2}(x - a)^2}_{\substack{\text{the error incurred} \\ \text{if we replace } f \text{ by} \\ \text{its linearization}}}, \tag{6}$$

with c_2 between x and a.

Equation (6) tells us that the error in the approximation $f(x) \approx f(a) + f'(a)(x-a)$ over the interval I joining x and a is

$$e_1(x) = \frac{f''(c_2)}{2}(x-a)^2. \tag{7}$$

If f'' is continuous, then it has a maximum value on I, and $|e_1(x)|$ satisfies the inequality

$$|e_1(x)| \le \frac{1}{2} \max |f''|(x-a)^2. \tag{8}$$

When we use this inequality to estimate the error, however, we usually cannot find the exact value of $\max |f''|$. We have to replace it with an upper bound or "worst-case" value instead. If M is *any* upper bound for $\max |f''|$, then

$$|e_1(x)| \le \frac{1}{2} M(x-a)^2. \tag{9}$$

This is the inequality we normally use in estimating $|e_1(x)|$. To make $|e_1(x)|$ small for a given M, we just make $(x-a)^2$ small by taking x close to a.

The Error in the Linear Approximation of $f(x)$ near $x = a$

If f, f', and f'' are continuous on the closed interval I joining x and a, then the error $e_1(x)$ incurred in replacing $f(x)$ by its linearization $L(x) = f(a) + f'(a)(x-a)$ on I satisfies the inequality

$$|e_1(x)| \le \frac{1}{2} M(x-a)^2, \tag{10}$$

where M is any upper bound for the values of $|f''|$ on I.

Example 1   The linearization of $f(x) = 1/(1-x)$ at $x = 0$ is $L(x) = 1 + x$ (Fig. 3.66). Estimate the error in the approximation

$$\frac{1}{1-x} \approx 1 + x$$

on the interval $|x| \le 0.1$.

Solution   We use inequality (10) to find an upper bound on $|e_1(x)|$ for the given values of x. We start by finding the second derivative of $f(x) = (1-x)^{-1}$:

$$f(x) = (1-x)^{-1}, \qquad f'(x) = (1-x)^{-2}, \qquad f''(x) = 2(1-x)^{-3}.$$

We then look for an upper bound M for the values of $|f''(x)|$ on the interval $-0.1 \le x \le 0.1$. On this interval,

$$|f''(x)| = \frac{2}{|1-x|^3}$$

$$\le \frac{2}{|1-0.1|^3} = \frac{2}{(0.9)^3} < 2.8.$$

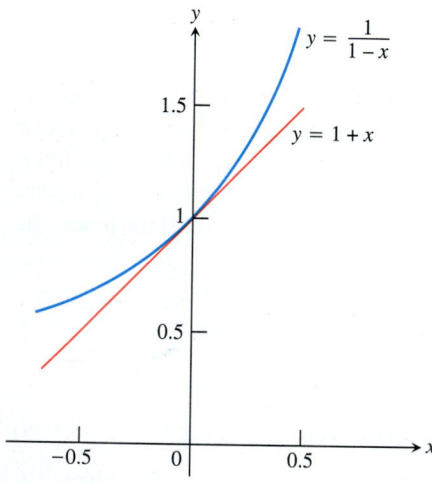

3.66   The function $f(x) = 1/(1-x)$ and its linearization $L(x) = 1 + x$ at $x = 0$. (The graph of f has another branch not shown here.) If $-0.1 \le x \le 0.1$, the values of L never differ from those of f by more than ± 0.014 (Example 1).

We may safely take $M = 2.8$. With this value of M and with $a = 0$, inequality (10) gives

$$|e_1(x)| \le \frac{1}{2}(2.8)(x - 0)^2$$

$$\le 1.4x^2$$

$$\le 1.4(0.1)^2 \qquad \text{(Because } |x| \le 0.1\text{)}$$

$$\le 0.014.$$

The error in the approximation $(1 - x)^{-1} \approx 1 + x$ is no more than 0.014 if $|x| \le 0.1$.

Quadratic Approximations

To get more accuracy in a linear approximation, we add a quadratic term. Typical quadratic approximations near $x = 0$ (Fig. 3.67) are

$$\frac{1}{1 - x} \approx 1 + x + x^2, \qquad \cos x \approx 1 - \frac{x^2}{2},$$

$$\sqrt{1 + x} \approx 1 + \frac{x}{2} - \frac{x^2}{8}, \qquad \sin x \approx x.$$

Others can be obtained from these with algebra. For $x \approx 0$ we have

$$\frac{1}{2 - x} = \frac{1}{2}\left(\frac{1}{1 - x/2}\right) \approx \frac{1}{2}\left[1 + \frac{x}{2} + \left(\frac{x}{2}\right)^2\right] = \frac{1}{2} + \frac{x}{4} + \frac{x^2}{8},$$

3.67 The functions (a) $y = \sqrt{1 + x}$ and (b) $y = \cos x$ with their linear and quadratic approximations at $x = 0$.

$$\frac{1 + x}{1 - x} \approx (1 + x)(1 + x + x^2) \approx 1 + 2x + 2x^2. \qquad \text{(Ignoring the } x^3 \text{ term)}$$

To combine approximations correctly, each approximation must be valid at the time

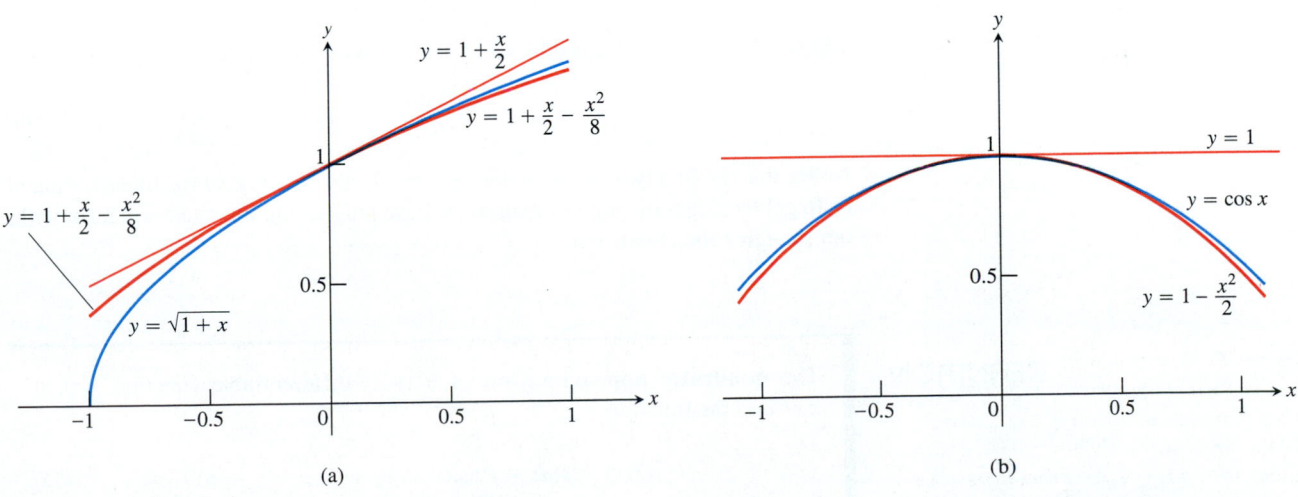

(a)   (b)

it is applied:

Right: $\dfrac{1}{1 - \sin x} \approx \dfrac{1}{1 - x} \approx 1 + x + x^2, \qquad x \approx 0$

Wrong: $\sin\left(\dfrac{1}{1 - x}\right) \underset{(1)}{\approx} \dfrac{1}{1 - x} \underset{(2)}{\approx} 1 + x + x^2.$

The second combination is wrong: approximation (1) requires $1/(1 - x) \approx 0$, and hence $|x|$ to be large, while approximation (2) requires $|x|$ to be small.

We can get a good estimate for the error in quadratic approximations by extending the Mean Value Theorem one step further.

THEOREM 8

The Extended Mean Value Theorem (Second Form)

If f, f', and f'' are continuous on $[a, b]$ and f'' is also differentiable on (a, b), then there exists a number c_3 between a and b for which

$$f(b) = f(a) + f'(a)(b - a) + \frac{f''(a)}{2}(b - a)^2 + \frac{f'''(c_3)}{6}(b - a)^3. \quad (11)$$

The theorem can be extended further still, as long as f has the necessary derivatives. You will find an outline of the general proof in Miscellaneous Exercise 43 at the end of the chapter.

In applications we usually write Eq. (11) with x in place of b:

$$f(x) = f(a) + f'(a)(x - a) + \frac{f''(a)}{2}(x - a)^2 + \frac{f'''(c_3)}{6}(x - a)^3, \quad (12)$$

with c_3 lying between a and x. From Eq. (12) we get the quadratic approximation

$$f(x) \approx f(a) + f'(a)(x - a) + \frac{f''(a)}{2}(x - a)^2, \quad (13)$$

which is valid on the interval between a and x with an error of

$$e_2(x) = \frac{f'''(c_3)}{6}(x - a)^3. \quad (14)$$

Notice that the first two terms on the right-hand side of (13) give the linearization of f. To get the quadratic approximation we have only to add the quadratic term without changing the linear part.

DEFINITION

The **quadratic approximation** of a twice-differentiable function $f(x)$ at $x = a$ is the function

$$Q(x) = f(a) + f'(a)(x - a) + \frac{f''(a)}{2}(x - a)^2. \quad (15)$$

If $f'''(x)$ is continuous on the closed interval from a to x, then it has a maximum value on the interval, and Eq. (14) tells us that

$$|e_2(x)| \leq \frac{1}{6}\max \left|f'''(x)\right| |x - a|^3. \tag{16}$$

We usually stand even less chance of finding $\max |f'''|$ than we do of finding $\max |f''|$. We therefore replace it by an upper bound M and estimate $|e_2(x)|$ with the inequality

$$|e_2(x)| \leq \frac{1}{6}M|x - a|^3. \tag{17}$$

The Error in the Quadratic Approximation of $f(x)$ near $x = a$

If f, f', f'', and f''' are continuous at every point of the closed interval I joining x and a, then the error $e_2(x)$ involved in replacing $f(x)$ by its quadratic approximation $Q(x)$ on I satisfies the inequality

$$|e_2(x)| \leq \frac{1}{6}M|x - a|^3, \tag{18}$$

where M is any upper bound for the values of $|f'''|$ on I.

Notice that we use "quadratic approximation of f" both for the function $Q(x)$ and for the approximation $Q(x) \approx f(x)$. It is conventional to do that.

Example 2   Find the quadratic approximation $Q(x)$ of the function $f(x) = 1/(1 - x)$ at $x = 0$. Estimate the error in the approximation $Q(x) \approx f(x)$ on the interval $|x| \leq 0.1$.

Solution   To find the quadratic approximation, we use Eq. (15) with $a = 0$:

$$f(x) = (1 - x)^{-1} \qquad f(0) = 1$$
$$f'(x) = (1 - x)^{-2} \qquad f'(0) = 1$$
$$f''(x) = 2(1 - x)^{-3} \qquad f''(0) = 2$$
$$f'''(x) = 6(1 - x)^{-4},$$

$$Q(x) = f(a) + f'(a)(x - a) + \frac{f''(a)}{2}(x - a)^2$$

$$= 1 + (1)(x - 0) + \frac{2}{2}(x - 0)^2$$

$$= 1 + x + x^2.$$

To estimate the approximation error, we first find an upper bound M for the values of $|f'''(x)|$ on the interval $-0.1 \leq x \leq 0.1$. On this interval,

$$|f'''(x)| = \frac{6}{|1 - x|^4} \leq \frac{6}{(0.9)^4} < 9.15,$$

so we may safely take $M = 9.15$. We then use inequality (18) with $a = 0$ and

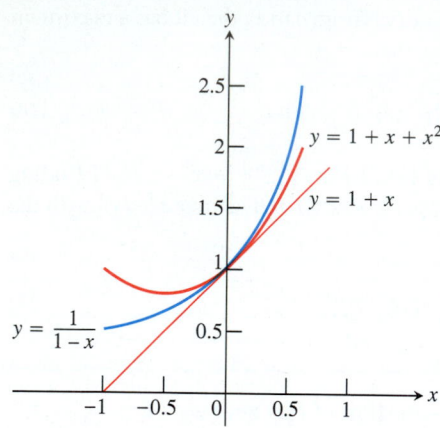

3.68   The function $f(x) = 1/(1 - x)$ and its linear and quadratic approximations at $x = 0$. The quadratic curve gives a better fit. (The graph of f has a branch not shown here.)

$M = 9.15$:

$$|e_2(x)| \leq \frac{1}{6}(9.15)|x - 0|^3$$

$$\leq \frac{1}{6}(9.15)(0.1)^3 \qquad \text{(Because } |x| \leq 0.1)$$

$$\leq 0.0016. \qquad \text{(Rounded up)}$$

The error in the approximation is no more than 0.0016 if $|x| \leq 0.1$. This is an improvement over the upper bound of 0.014 found for the linear approximation of $1/(1 - x)$ in Example 1. See Fig. 3.68.

Example 3   Find the quadratic approximation of $\sin x$ near $x = 0$. Give an upper bound for the error in the approximation on the interval $|x| \leq 0.3$.

Solution   To find the quadratic approximation, we use Eq. (15) with $f(x) = \sin x$ and $a = 0$:

$$f(x) = \sin x \qquad\qquad f(0) = 0$$

$$f'(x) = \cos x \qquad\qquad f'(0) = 1$$

$$f''(x) = -\sin x \qquad\qquad f''(0) = 0$$

$$f'''(x) = -\cos x,$$

$$Q(x) = f(a) + f'(a)(x - a) + f''(a)(x - a)^2$$

$$= 0 + (1)(x - 0) + 0$$

$$= x.$$

The quadratic approximation of $\sin x$ near $x = 0$ is $\sin x \approx x$ (Fig. 3.69).

How can the approximation be quadratic if there is no x^2 term? There *is* a quadratic term, but its coefficient is zero. The important fact is that no quadratic term is missing and we can estimate the approximation error with $e_2(x)$ instead of $e_1(x)$.

To find an upper bound for $|e_2(x)|$ on the interval $-0.3 \leq x \leq 0.3$, we first find an upper bound for $|f'''(x)|$. Since $|\cos x|$ never exceeds 1,

$$|f'''(x)| = |-\cos x| \leq 1$$

and we may safely take $M = 1$. With $M = 1$ and $a = 0$, inequality (18) gives

$$|e_2(x)| \leq \frac{1}{6}(1)|x - 0|^3$$

$$\leq \frac{1}{6}|x|^3$$

$$\leq \frac{1}{6}(0.3)^3 \qquad \text{(because } |x| \leq 0.3)$$

$$\leq 0.0045.$$

The error on the interval $-0.3 \leq x \leq 0.3$ will never exceed 0.0045.

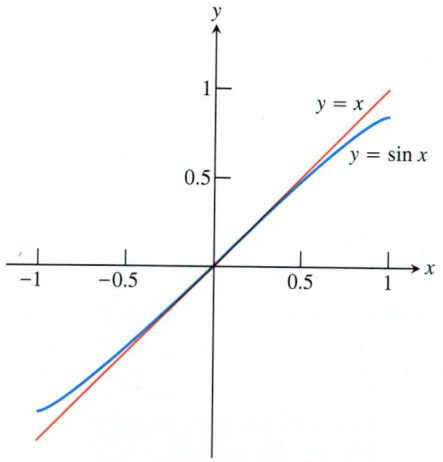

3.69   The function $y = x$ is the quadratic approximation of $y = \sin x$ at $x = 0$ as well as the linearization of $y = \sin x$ at $x = 0$. Example 3 explains why.

TABLE 3.1
Standard approximation formulas

Linear	Quadratic
Approximation at $x = a$	
$L(x) = f(a) + f'(a)(x - a)$	$Q(x) = f(a) + f'(a)(x - a) + \dfrac{f''(a)}{2}(x - a)^2$
Error bound	
$\|e_1(x)\| \le \dfrac{1}{2} M(x - a)^2$	$\|e_2(x)\| \le \dfrac{1}{6} M\|x - a\|^3$
M is any upper bound for $\max\|f''\|$ on the interval joining x and a.	M is any upper bound for $\max\|f'''\|$ on the interval joining x and a.
Common approximations near $x = 0$	
$\dfrac{1}{1 - x} \approx 1 + x$	$\dfrac{1}{1 - x} \approx 1 + x + x^2$
$\sqrt{1 + x} \approx 1 + \dfrac{x}{2}$	$\sqrt{1 + x} \approx 1 + \dfrac{x}{2} - \dfrac{x^2}{8}$
$\sin x \approx x$	$\sin x \approx x$
$\cos x \approx 1$	$\cos x \approx 1 - \dfrac{x^2}{2}$
$(1 + x)^k \approx 1 + kx$	$(1 + x)^k \approx 1 + kx + \dfrac{k(k - 1)}{2} x^2$
(any number k)	(any number k)

EXERCISES 3.7

Use formulas from Table 3.1 to find quadratic approximations for the functions in Exercises 1–6.

1. $x \sin x$ at $x = 0$ **2.** $\sqrt{1 + \sin x}$ at $x = 0$

3. $\cos \sqrt{1 + x}$ at $x = -1$ **4.** $\sqrt{x}$ at $x = 1$
(*Hint:* $x = 1 + (x - 1)$.)

5. $\dfrac{\sec x}{1 - x}$ at $x = 0$ **6.** $\dfrac{1}{\sqrt{1 - x^2}}$ at $x = 0$

Find upper bounds for the error $\|e_1(x)\|$ in the linear approximations in Exercises 7 and 8.

7. $\sqrt{1 + x} \approx 1$, $\|x\| \le 0.1$

8. $\cos x \approx 1$, $\|x\| \le 0.1$

In Exercises 9–12:
  a) Find the quadratic approximation of the function at $x = 0$.
  b) Find a numerical upper bound for $\|e_2(x)\|$ on the interval $\|x\| \le 0.1$.

9. $\sqrt{1 + x}$ **10.** $\cos x$ **11.** $\sec x$ **12.** $\tan x$

In Exercises 13–16:
  a) Find the quadratic approximation of the function.
  b) Find a numerical upper bound for $\|e_2(x)\|$ on the interval $\|x - a\| \le 0.1$.

13. $\sin x$ at $a = \pi/2$ **14.** $\sin x$ at $a = \pi$

15. $\cos x$ at $a = \pi/2$ **16.** $\cos x$ at $a = \pi$

17. For what values of x will the error $\|e_2(x)\|$ in the approximation $\sin x \approx x$ be (a) less than $1/100$; (b) less than 1% of $\|x\|$?

18. For what values of x will the error $\|e_2(x)\|$ in the approximation $\cos x \approx 1 - (x^2/2)$ be (a) less than $1/100$; (b) less than 1% of $\|x\|$?

19. Find the quadratic approximation at $x = 0$ of
  a) $f(x) = 2x + 3$,
  b) $f(x) = mx + b$ (m and b constant).

20. Show that every quadratic function $f(x) = c_0 + c_1 x + c_2 x^2$ (c_1, c_2, and c_3 constant) is its own quadratic approximation at $x = 0$.

21. Assume that the formula

$$\frac{d}{dx}(u^k) = k\,u^{k-1}\frac{du}{dx}$$

holds for any number k and show that

$$Q(x) = 1 + kx + \frac{k(k-1)}{2}x^2$$

is the quadratic approximation of $(1 + x)^k$.

22. CALCULATOR Convince yourself that the formulas in Table 3.1 are reasonable by testing them in special cases. For instance, find $(1 + 0.02)^{-2}$ first by multiplication and division and then by using the formula

$$(1 + x)^k \approx 1 + kx + \frac{k(k-1)}{2}x^2.$$

23. *Approximations at inflection points (continuation of Exercise 52, Section 3.3).*
   a) Show that if the graph of a twice-differentiable function

$f(x)$ has an inflection point at $x = a$, then the linearization of f at $x = a$ is also the quadratic approximation of f at $x = a$. This explains why tangent lines fit so well at inflection points.
   b) GRAPHER If you have not yet done so, do Exercise 52 in Section 3.3 now.

24. GRAPHER Graph each of the following functions together with its quadratic approximation at $x = 0$. If possible, use different colors for the function and approximation.
   a) $(1 + x)^{-1}$ b) $(1 + x)^2$ c) $(1 + x)^3$

EXPLORER PROGRAM	
PowerGrapher	Draws tangents to curves
	Graphs functions together

REVIEW QUESTIONS

1. Outline a general method for solving related rate problems. Illustrate with an example.

2. State the first derivative test for local extreme values.

3. What does it mean for a function $y = f(x)$ to have an absolute or local maximum or minimum value?

4. Explain how to find the local and absolute maximum and minimum values of a function $y = f(x)$.

5. What are the hypotheses and conclusion of Rolle's theorem? How does the theorem sometimes help you to tell how many solutions an equation has in a given interval?

6. What are the hypotheses and conclusion of the Mean Value Theorem? What physical interpretation does the theorem sometimes have? Give an example.

7. State the first derivative test for increasing and decreasing and prove it as a corollary of the Mean Value Theorem.

8. How do you test a twice-differentiable function to find out where its graph is concave down or concave up? What is an

inflection point? What physical significance do inflection points sometimes have?

9. List the steps you would take to graph a polynomial $y = f(x)$. How does calculus tell you the shape of the graph between the points you plot? Give an example.

10. List the steps you would take to graph a rational function $y = f(x)$. Illustrate with an example.

11. How do you know when the graph of a function has a cusp at a point $x = x_0$? Give an example.

12. Outline a general method for solving max-min problems. Illustrate with an example.

13. Describe l'Hôpital's rule. How do you know when to use the rule and when to stop? Give an example.

14. State the extended mean value theorems and explain how they enable us to estimate errors in linear and quadratic approximations of differentiable functions. Give examples.

MISCELLANEOUS EXERCISES

1. Points A and B move along the x- and y-axes, respectively, in such a way that the distance r (meters) along the perpendicular from the origin to line AB remains constant. How fast is OA changing, and is it increasing, or decreasing, when $OB = 2r$ and B is moving toward O at the rate of $0.3r$ m/sec?

2. The volume of a cube is increasing at the rate of 1200 cm^3/min at the instant its edges are 20 cm long. At what rate is the length of the edges changing at that instant?

3. *Relief from a heart attack.* A heart attack victim has been given a blood vessel dilator to lower the pressure against which the heart has to work. For a short while after the drug has been administered, the radii of the affected blood vessels will increase at about 1% per minute. According to Poiseuille's law, $V = kr^4$ (see Exercise 47, Section 2.6), what rate of increase (percent per minute) can we expect in the blood flow over the next few minutes, all other things being equal?

4. A sphere's volume is decreasing at the rate of $12\pi\,\text{m}^3/\text{min}$. Find the rates at which the radius and surface area are decreasing when the radius is 20 m. About how much should you expect the radius and surface area to change in the following 6 sec?

5. a) Show that the equation $x^4 + 2x^2 - 2 = 0$ has exactly one solution on $[0, 1]$.
   b) CALCULATOR Find the solution to as many decimal places as you can.

6. If $f'(x) \le 2$ for all x, what is the most the values of f can increase on $[0, 6]$?

7. a) Show that $f(x) = x/(x + 1)$ increases on every interval in its domain.
   b) Show that $f(x) = x^3 + 2x$ has no local maximum or minimum values.

8. CALCULATOR As a result of a heavy rain, the volume of water in a reservoir increased by 1400 acre-ft in 24 h. Show that at some instant during that period the reservoir's volume was increasing at a rate in excess of 225,000 gal/min. (An acre-foot is $43{,}560\,\text{ft}^3$, the volume that would cover one acre to the depth of one foot. A cubic foot holds 7.48 gal.)

9. Suppose that f is continuous on $[a, b]$ and that c is an interior point of the interval. Show that if $f'(x) \le 0$ on $[a, c]$ and $f'(x) \ge 0$ on $(c, b]$, then $f(x)$ is never less than $f(c)$ on $[a, b]$.

10. a) Show that $-1/2 \le x/(1 + x^2) \le 1/2$ for every value of x.
    b) Suppose that f is a function whose derivative is $f'(x) = x/(1 + x^2)$. Use the result in (a) to show that
$$|f(b) - f(a)| \le \frac{1}{2}|b - a|$$
    for any a and b.

Graph the functions in Exercises 11–14.

11. $y = 4x^3 - x^4$

12. $y = (1/8)(x - 3)(x + 3)^2$

13. $y = \dfrac{3x^2 - 6x - 5}{x^2 - 2x - 3}$

14. $y = \dfrac{3}{4}x^{4/3} + \dfrac{3}{2}x^{-2/3}$

15. Graph $y = x^2/(x^2 + 1)$ and $y = -1/(x^2 + 1)$ together.

16. a) Suppose that the first derivative of $y = f(x)$ is
$$y' = 6(x + 1)(x - 2)^2.$$

At what points, if any, does the graph of f have a local maximum, local minimum, or point of inflection?
   b) Suppose that the first derivative of $y = f(x)$ is
$$y' = 6x(x + 1)(x - 2).$$
   At what points, if any, does the graph of f have a local maximum, local minimum, or point of inflection?

17. Use the sign pattern for the derivative
$$\frac{df}{dx} = 6(x - 1)(x - 2)^2(x - 3)^3(x - 4)^4$$
to identify the points where f has local maximum and minimum values.

18. Use the following information to find the values of a, b, and c in the formula
$$f(x) = \frac{x + a}{bx^2 + cx + 2}.$$
   i) The values of a, b, and c are either 0 or 1.
   ii) The graph of f passes through the point $(-1, 0)$.
   iii) The line $y = 1$ is an asymptote of the graph of f.

19. Find the values of the right-hand and left-hand derivatives at the critical points of the function $f(x) = -|x^2 - 2x - 3|$. Then use these values along with information about rise, fall, and concavity to graph the function.

20. GRAPHER *Another pitfall of computer graphing.* A graph that is large enough to show a function's global behavior may fail to reveal important local features. The graph of $f(x) = (x^8/8) - (x^6/2) - x^5 + 5x^3$ is a case in point.
   a) Graph f over the interval $-2.5 \le x \le 2.5$. Where does the graph appear to have local extreme values or points of inflection?
   b) Now factor $f'(x)$ and show that f has a local maximum at $x = \sqrt[5]{5} \approx 1.70998$ and local minima at $x = \pm\sqrt{3} \approx \pm1.73205$.
   c) Zoom in on the graph to find a viewing window that shows clearly the presence of the extreme values at $x = \sqrt[5]{5}$ and $x = \sqrt{3}$.
      The moral here is that without calculus the existence of two of the three extreme values would probably have gone unnoticed. On any normal graph of the function, the values would lie close enough together to fall within the dimensions of a single pixel on the screen.
      (*Source: Uses of Technology in the Mathematics Curriculum* by Benny Evans and Jerry Johnson, Oklahoma State University, published in 1990 under National Science Foundation Grant USE-8950044)

21. GRAPHER *(Continuation of Exercise 20).*
   a) Graph $f(x) = (x^8/8) - (2/5)x^5 - 5x - (5/x^2) + 11$ over the interval $-2 \le x \le 2$. Where does the graph appear to have local extreme values or points of inflection?
   b) Show that f has a local maximum value at $x = \sqrt[7]{5} \approx 1.2585$ and a local minimum value at $x = \sqrt[4]{2} \approx 1.2599$.

c) Zoom in on the graph to find a viewing window that shows clearly the presence of the extreme values at $x = \sqrt[5]{5}$ and $x = \sqrt[3]{2}$.

22. *The (first) second derivative test.* The second derivative test for local maxima and minima (Section 3.3) says:
a) f has a local maximum value at $x = c$ if $f'(c) = 0$ and $f''(c) < 0$;
b) f has a local minimum value at $x = c$ if $f'(c) = 0$ and $f''(c) > 0$.

To prove statement (a), let $\epsilon = (1/2)|f''(c)|$. Then use the fact that

$$f''(c) = \lim_{h \to 0} \frac{f'(c + h) - f'(c)}{h} = \lim_{h \to 0} \frac{f'(c + h)}{h}$$

to conclude that for some $\delta > 0$,

$$0 < |h| < \delta \Rightarrow \frac{f'(c + h)}{h} < f''(c) + \epsilon < 0.$$

Thus $f'(c + h)$ is positive for $-\delta < h < 0$ and negative for $0 < h < \delta$. Prove statement (b) in a similar way.

23. *The (second) second derivative test.* Use the equation

$$f(x) = f(a) + f'(a)(x - a) + \frac{f''(c_2)}{2}(x - a)^2$$

from the Extended Mean Value Theorem in Section 3.7 to establish the following test.

Let f have continuous first and second derivatives and suppose that $f'(a) = 0$. Then
a) f has a local maximum at a if $f'' \leq 0$ throughout an interval whose interior contains a;
b) f has a local minimum at a if $f'' \geq 0$ throughout an interval whose interior contains a.

24. *Estimating reciprocals without division.* You can estimate the value of the reciprocal of a number a without ever dividing by a if you apply Newton's method to the function $f(x) = (1/x) - a$. For example, if $a = 3$, the function involved is $f(x) = (1/x) - 3$.
a) Graph $y = (1/x) - 3$. Where does the graph cross the x-axis?
b) Show that the recursion formula in this case is

$$x_{n+1} = x_n(2 - 3x_n),$$

so there is no need for division.

25. *(Continuation of Exercise 19, Section 3.5).* Suppose that instead of having a box with square ends you have a box with square sides (Fig. 3.70) so that its dimensions are h by

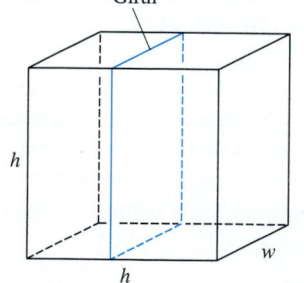

Girth

3.70 The box in Exercise 25.

h by w, its length is h, and its girth is $2h + 2w$. What dimensions will give the box its largest volume now? (In Exercise 19, Section 3.5, the dimensions were 36 by 18 by 18 in.)

26. *A max-min problem with a variable answer.* Sometimes the solution of a max-min problem depends on the proportions of the shapes involved. As a case in point, suppose that a right circular cylinder of radius r and height h is inscribed in a right circular cone of radius R and height H, as shown in Fig. 3.71. Find the value of r (in terms of R and H) that maximizes the total surface area of the cylinder (including top and bottom). As you will see, the answer depends on whether $H < R$, $H = R$, $R < H < 2R$, $H = 2R$, or $2R < H$.

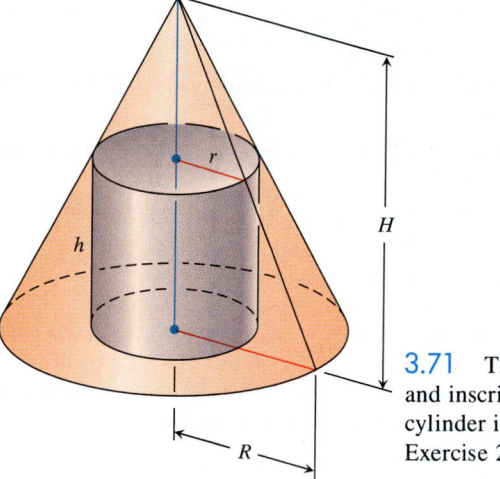

3.71 The cone and inscribed cylinder in Exercise 26.

27. *Nonnegative numbers with a fixed sum.* The sum of two nonnegative numbers is 20. Find the numbers
a) if their product is to be as large as possible;
b) if the sum of their squares is to be as large as possible;
c) if one number plus the square root of the other is to be as large as possible.

28. An isosceles triangle with its vertex pointing down has its base parallel to and above the x-axis. The base vertices lie on the parabola $y = 27 - x^2$. The triangle's remaining vertex lies at the origin. What is the largest area the triangle can have?

29. Find the height and radius of the largest right circular cylinder that will fit in a sphere of radius $\sqrt{3}$ (Fig. 3.72).

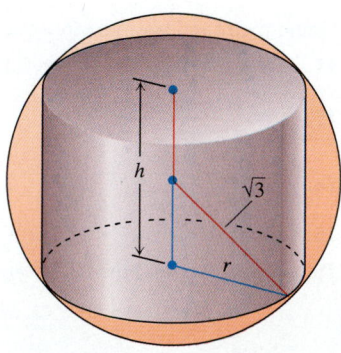

3.72 The cylinder and sphere in Exercise 29.

30. Figure 3.73 shows two right circular cones, one upside down in the other. The two bases are parallel. The vertex of the smaller cone lies at the center of the larger cone's base. What values of r and h will give the smaller cone the largest possible volume?

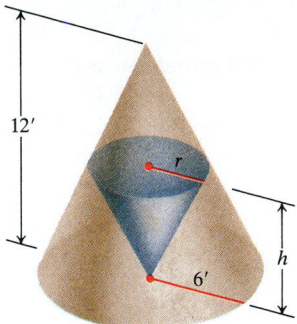

3.73   The cones in Exercise 30.

31. You are drawing up plans for the piping that will connect a drilling rig 12 mi offshore to a refinery on shore 20 mi down the coast (Fig. 3.74). What values of x and y will give the least expensive connection if underwater pipe costs $50,000 per mile and land-based pipe costs $30,000 per mile?

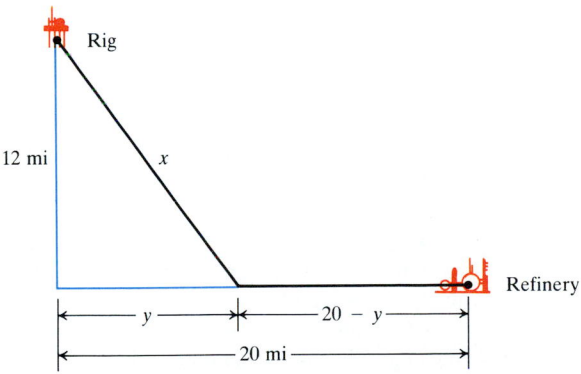

3.74   Diagram for the pipe in Exercise 31.

32. You are designing an athletic field in the shape of a rectangle x units long capped by semicircular regions of radius r at the two ends. The field is to be bounded by a 400-m track. What values of x and r will give the rectangle its largest area?

33. Your company can make x hundred grade A tires and y hundred grade B tires a day, where $0 \le x \le 4$ and $y = (40 - 10x)/(5 - x)$. The profit on grade A tires is twice the profit on grade B tires. Assuming all the tires sell, what are the most profitable numbers of tires to make?

34. The positions of two particles on the s-axis are $s_1 = \sin t$ and $s_2 = \sin(t + \pi/3)$. What is the farthest apart the particles ever get?

35. *The ladder problem.*   What is the approximate length (ft) of the longest ladder you can carry horizontally around the corner of the corridor in Fig. 3.75? Round your answer down to the nearest foot.

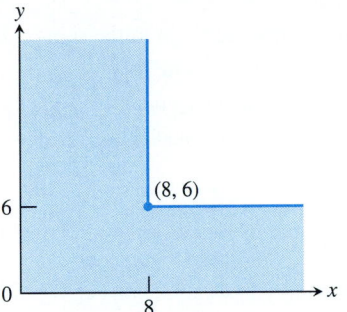

3.75   The corridor in Exercise 35.

36. You want to bore a hole in the side of the tank in Fig. 3.76 at a height that will make the stream of water coming out hit the ground as far from the tank as possible. If you drill the hole near the top, where the pressure is low, the water will exit slowly but spend a relatively long time in the air. If you drill the hole near the bottom, the water will exit at a higher velocity but have only a short time to fall. Where is the best place, if any, for the hole? (*Hint*: How long will it take an exiting particle of water to fall from height y to the ground?)

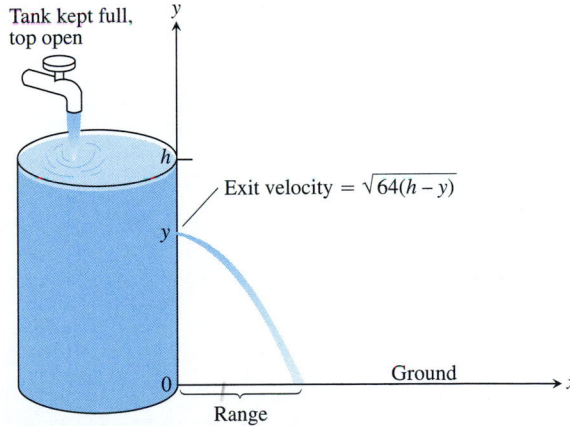

3.76   What height y gives the exit stream its greatest range? See Exercise 36.

37. Points A and B lie at the ends of a diameter of a unit circle and point C lies on the circumference. Is it true that the perimeter of triangle ABC is largest when the triangle is isosceles? How do you know?

38. Find the smallest value of the positive constant m that will make $mx - 1 + (1/x)$ greater than or equal to zero for all positive values of x.

39. *Schwarz's inequality.*

a) Show that if $a > 0$, then $f(x) = ax^2 + 2bx + c \geq 0$ for all (real) x if and only if $b^2 \leq ac$.

b) Derive **Schwarz's inequality**,

$$(a_1b_1 + a_2b_2 + \cdots + a_nb_n)^2 \leq (a_1^2 + a_2^2 + \cdots + a_n^2)(b_1^2 + b_2^2 + \cdots + b_n^2),$$

by applying what you learned in (a) to the sum

$$(a_1x + b_1)^2 + (a_2x + b_2)^2 + \cdots + (a_nx + b_n)^2.$$

c) Show that equality holds in Schwarz's inequality only if there exists a real number x that makes a_ix equal $-b_i$ for every value of i from 1 to n.

40. Let $h = fg$ be the product of two differentiable functions of x.

a) If f and g are positive, with local maxima at $x = a$, and if f' and g' change sign at a, does h have a local maximum at a?

b) If the graphs of f and g have inflection points at $x = a$, does the graph of h have an inflection point at a?

In either case, if the answer is yes, give a proof. If the answer is no, give a counterexample.

41. Evaluate the following limits.

a) $\lim\limits_{x \to 0} \dfrac{2 \sin 5x}{3x}$

b) $\lim\limits_{x \to 0} \sin 5x \cot 3x$

c) $\lim\limits_{x \to 0^+} x \csc^2 \sqrt{2x}$

d) $\lim\limits_{x \to \pi/2} (\sec x - \tan x)$

e) $\lim\limits_{x \to 0} \dfrac{x - \sin x}{x - \tan x}$

f) $\lim\limits_{x \to 0} \dfrac{\sin x^2}{x \sin x}$

g) $\lim\limits_{x \to 0} \dfrac{\sec x - 1}{x^2}$

h) $\lim\limits_{x \to 2} \dfrac{x^3 - 8}{x^2 - 4}$

i) $\lim\limits_{x \to \pi/2} \dfrac{1 - \sin x}{1 + \cos 2x}$

j) $\lim\limits_{x \to \pi/3} \dfrac{\cos x - 0.5}{x - (\pi/3)}$

42. L'Hôpital's rule does not help with the following limits. Find them some other way.

a) $\lim\limits_{x \to \infty} \dfrac{\sqrt{x + 5}}{\sqrt{x} + 5}$

b) $\lim\limits_{x \to \infty} \dfrac{2x}{x + 7\sqrt{x}}$

43. *Taylor's theorem.*   Suppose that f and its first n derivatives $f', f'', \ldots, f^{(n)}$ are continuous on $[a, b]$ and that $f^{(n)}$ is differentiable on (a, b). Let

$$F(x) = f(x) - f(a) - (x - a)f'(a) - (x - a)^2 f''(a)/2!$$

$$- \cdots - \frac{(x - a)^{(n)} f^{(n)}(a)}{n!} - k(x - a)^{n+1},$$

where k is chosen to make $F(b)$ zero. Show that

a) $F(a) = F(b) = 0$;

b) $F'(a) = F''(a) = \cdots = F^{(n)}(a) = 0$;

c) there exist numbers $c_1, c_2, \ldots, c_{n+1}$ with

$$a < c_{n+1} < c_n < \cdots < c_2 < c_1 < b$$

and with

$$F'(c_1) = 0 = F''(c_2) = F''(c_3) = \cdots$$
$$= F^{(n)}(c_n) = F^{(n+1)}(c_{n+1});$$

d) hence,

$$k = \frac{f^{(n+1)}(c_{n+1})}{(n+1)!}.$$

This, combined with the statement $F(b) = 0$, gives **Taylor's formula**:

$$f(b) = f(a) + f'(a)(b - a) + \frac{f''(a)}{2}(b - a)^2$$

$$+ \cdots + \frac{f^{(n)}(a)}{n!}(b - a)^n$$

$$+ \frac{f^{(n+1)}(c_{n+1})}{(n+1)!}(b - a)^{n+1},$$

for some number c_{n+1} between a and b. The derivation outlined here comes from James Wolfe's "A Proof of Taylor's Formula," *American Mathematical Monthly*, no. 60 (1953), p. 415.

44. *A cubic approximation.*   Write Taylor's formula (Exercise 43) with x in place of b. Then use it with $f(x) = 1/(1 - x)$, $a = 0$, and $n = 3$ to find the standard cubic approximation of $1/(1 - x)$. Give an upper bound for the magnitude of the error in the approximation when $|x| \leq 0.1$.

4

INTEGRATION

Overview   This chapter introduces the second main branch of calculus, called integral calculus. Integral calculus is the mathematics we use to find the lengths of curved paths and the cross-section areas of machine parts and airplane wings, to predict the future positions of moving bodies, and to calculate the effective voltages and currents in electrical circuits. In this chapter, we set the stage for these and other applications.

The development of integral calculus begins with the calculation of area as a limit of finite sums. The limits involved turn out to be special cases of a more general kind of limit called a definite integral. Presenting the properties of definite integrals is the central goal of this chapter.

The single most important concept in this chapter is the connection between definite integrals and derivatives. The discovery of this connection by Leibniz and Newton turned calculus into the most important mathematics for modeling the natural laws of science.

4.1 Indefinite Integrals

One of the early accomplishments of calculus was predicting the future position of a moving body from one of its known locations and a formula for its velocity function. Today we view this as one of a number of occasions on which we recover a function from one of its known values and a formula for its rate of change. It is a routine process today, thanks to calculus, to calculate how fast a space vehicle needs to be going at a certain point to escape the earth's gravitational field or to predict the useful life of a sample of radioactive polonium-210 from its present level of activity and its rate of decay.

The process of finding a function from one of its known values and its derivative $f(x)$ has two steps. The first is to find a formula that gives all the functions that could possibly have f as a derivative. These functions are the so-called antiderivatives of f, and the formula that gives them all is called the indefinite integral of f. The second step is to use the known function value to select the particular antiderivative we want from the indefinite integral. The first step is the subject of the present section; the second is the subject of the next.

Finding a formula that gives all of a function's antiderivatives might seem like an impossible task, or at least to require a little magic. But this is not the case at all. If we can find even one of a function's antiderivatives we can find them all, because of the second and third corollaries of the Mean Value Theorem of Section 3.2. We begin with these corollaries and then show how to "reverse" known differentiation formulas to find indefinite integrals.

The Mean Value Theorem's Second and Third Corollaries

The first corollary of the Mean Value Theorem gave the first derivative test for increasing and decreasing functions. The second corollary says that only constant functions have zero derivatives. The third says that functions with identical derivatives differ only by a constant value.

COROLLARY 2

> If $F'(x) = 0$ for all x in an interval I, then F has a constant value on I. That is, there is a constant C such that $F(x) = C$ for all x in I.

Corollary 2 is the converse of the rule that says the derivative of a constant function is zero. The derivative of a nonconstant function may be zero from time to time, but the only functions whose derivatives are zero throughout an entire interval are the functions that are constant on the interval.

Proof of Corollary 2   We want to show that F has a constant value throughout the interval I. We do so by showing that if x_1 and x_2 are any two points in I, then $F(x_1) = F(x_2)$.

Suppose that x_1 and x_2 are two points in I, numbered from left to right so that $x_1 < x_2$. Then F satisfies the hypotheses of the Mean Value Theorem on $[x_1, x_2]$: It is differentiable at every point of $[x_1, x_2]$ and hence continuous at every point of $[x_1, x_2]$. Therefore,

$$\frac{F(x_2) - F(x_1)}{x_2 - x_1} = F'(c)$$

at some point c between x_1 and x_2. Since $F' = 0$ throughout I, this equation translates successively into

$$\frac{F(x_2) - F(x_1)}{x_2 - x_1} = 0, \qquad F(x_2) - F(x_1) = 0, \qquad \text{and} \qquad F(x_2) = F(x_1).$$

COROLLARY 3

> If $F_1'(x) = F_2'(x)$ at each point of an interval I, then there is a constant C such that $F_1(x) = F_2(x) + C$ for all x in I.

Corollary 3 says that the only way two functions can have identical rates of change throughout an interval is for their values on the interval to differ by a constant. For example, we know that the derivative of the function x^2 is $2x$. Therefore, every other function with derivative $2x$ on an interval is given on that interval by the formula $F(x) = x^2 + C$ for some value of C (Fig. 4.1). No other functions there can have derivative $2x$.

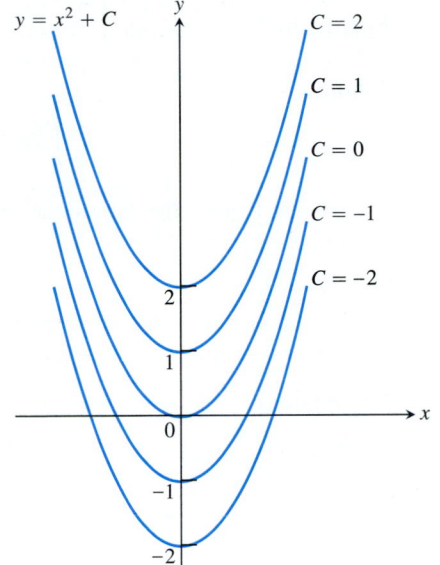

4.1 From a geometric point of view, Corollary 3 of the Mean Value Theorem says that the graphs of functions with identical derivatives can be superimposed by a vertical shift. The graphs of the functions with derivative $2x$ are the parabolas $y = x^2 + C$, shown here for selected values of C.

Proof of Corollary 3   Since $F_1'(x) = F_2'(x)$ at each point of I, the derivative of the difference function $F = F_1 - F_2$ at each point is

$$F'(x) = F_1'(x) - F_2'(x) = 0.$$

Hence, F has a constant value C throughout I (Corollary 2), so $F_1(x) - F_2(x) = C$. That is, $F_1(x) = F_2(x) + C$ at each point of I.

Finding Antiderivatives—Indefinite Integrals

As we mentioned in the introduction, a function $F(x)$ is an **antiderivative** of a function $f(x)$ if $F'(x) = f(x)$ at every point of f's domain. The set of all antiderivatives of f is called the indefinite integral of f with respect to x.

DEFINITION

If the function $f(x)$ is a derivative, then the set of all antiderivatives of f is the **indefinite integral of f with respect to x,** denoted by

$$\int f(x)\, dx.$$

The symbol $\int$ is an integral sign. The function f is the **integrand** of the integral and x is the **variable of integration.**

According to Corollary 3 of the Mean Value Theorem, once we have found an antiderivative $F(x)$ of a function $f(x)$, the other antiderivatives of f differ from F by a constant. We indicate this in integral notation by writing

$$\int f(x)\, dx = F(x) + C. \tag{1}$$

The constant C is the **constant of integration** or **arbitrary constant.** Equation (1) is read, "The indefinite integral of f with respect to x is $F(x) + C$." When we find $F(x) + C$, we say that we have **integrated** f and **evaluated** the integral.

Example 1   Evaluate $\int 2x\,dx$.

Solution

$$\int 2x\,dx = \underset{\displaystyle\text{an antiderivative of }2x}{x^2} + \underset{\displaystyle\text{the arbitrary constant}}{C}$$

As we saw before, the formula $x^2 + C$ generates all the antiderivatives of the function $2x$.

We find many of the indefinite integrals we need in scientific work by reversing derivative formulas. The next example shows what we mean.

Example 2   *Integral Formulas*

Indefinite integral	*Reversed derivative formula*
1. $\displaystyle\int x^n dx = \frac{x^{n+1}}{n+1} + C, \quad n \neq -1$	$\displaystyle\frac{d}{dx}\left(\frac{x^{n+1}}{n+1}\right) = x^n$
2. $\displaystyle\int \sin kx\,dx = -\frac{\cos kx}{k} + C$	$\displaystyle\frac{d}{dx}\left(-\frac{\cos kx}{k}\right) = \sin kx$
3. $\displaystyle\int \cos kx\,dx = \frac{\sin kx}{k} + C$	$\displaystyle\frac{d}{dx}\left(\frac{\sin kx}{k}\right) = \cos kx$
4. $\displaystyle\int \sec^2 x\,dx = \tan x + C$	$\displaystyle\frac{d}{dx}\tan x = \sec^2 x$
5. $\displaystyle\int \csc^2 x\,dx = -\cot x + C$	$\displaystyle\frac{d}{dx}(-\cot x) = \csc^2 x$
6. $\displaystyle\int \sec x \tan x\,dx = \sec x + C$	$\displaystyle\frac{d}{dx}\sec x = \sec x \tan x$
7. $\displaystyle\int \csc x \cot x\,dx = -\csc x + C$	$\displaystyle\frac{d}{dx}(-\csc x) = \csc x \cot x$

In case you are wondering why the integrals of the tangent, cotangent, secant, and cosecant do not appear here, the answer is that the usual formulas for them require logarithms. In Section 4.5, we shall show that these functions do have antiderivatives, but we shall have to wait until Chapters 6 and 7 to see what they are.

Example 3   *Selected Integrals from Example 2*

a) $\displaystyle\int x^5\,dx = \frac{x^6}{6} + C$ $\left(\begin{array}{l}\text{Formula 1}\\ \text{with } n = 5\end{array}\right)$

b) $\displaystyle\int \frac{1}{\sqrt{x}}\,dx = \int x^{-1/2}\,dx = 2x^{1/2} + C = 2\sqrt{x} + C$ $\left(\begin{array}{l}\text{Formula 1}\\ \text{with}\\ n = -1/2\end{array}\right)$

c) $\displaystyle\int \sin 2x\,dx = -\frac{\cos 2x}{2} + C$ $\left(\begin{array}{l}\text{Formula 2}\\ \text{with } k = 2\end{array}\right)$

d) $\displaystyle\int \cos \frac{x}{2}\,dx = \int \cos \frac{1}{2}x\,dx = \frac{\sin(1/2)x}{1/2} + C = 2\sin \frac{x}{2} + C$ $\left(\begin{array}{l}\text{Formula 3}\\ \text{with } k = 1/2\end{array}\right)$

The formulas in Example 2 hold because, in each case, the derivative of the function $F(x) + C$ on the right is the integrand $f(x)$ on the left. Finding an integral formula can sometimes be difficult, but checking it, once found, is relatively easy: Differentiate the right-hand side. If the derivative is the integrand, the formula is correct; otherwise it is wrong.

Example 4

Right: $$\int x \cos x \, dx = x \sin x + \cos x + C$$

Reason: The derivative of the right-hand side is the integrand:

$$\frac{d}{dx} (x \sin x + \cos x + C) = x \cos x + \sin x - \sin x + 0 = x \cos x.$$

Wrong: $$\int x \cos x \, dx = x \sin x + C$$

Reason: The derivative of the right-hand side is not the integrand:

$$\frac{d}{dx} (x \sin x + C) = x \cos x + \sin x + 0 \neq x \cos x.$$

Do not worry about how to derive the correct integral formula in Example 4 right now. We present a nice technique for doing so in Chapter 7.

Rules of Algebra for Antiderivatives

Among the things we know about antiderivatives are these:

1. A function is an antiderivative of a constant multiple $k\,f$ of a function f if and only if it is k times an antiderivative of f.

2. In particular, a function is an antiderivative of $-f$ if and only if it is the negative of an antiderivative of f.

3. A function is an antiderivative of a sum or difference $f \pm g$ if and only if it is the sum or difference of an antiderivative of f and an antiderivative of g.

When we express these observations in integral notation, we get the standard arithmetic rules for indefinite integration (Table 4.1).

TABLE 4.1
Rules for indefinite integration

1. *Constant Multiple Rule*: $$\int k\,f(x)\,dx = k \int f(x)\,dx$$
   (Does not work if k varies with x.)

2. *Rule for Negatives*: $$\int -f(x)\,dx = -\int f(x)\,dx$$
   (Rule 1 with $k = -1$)

3. *Sum and Difference Rule*: $$\int [f(x) \pm g(x)]\,dx = \int f(x)\,dx \pm \int g(x)\,dx$$

Example 5   *Rewriting the Constant of Integration*

$$\int 5 \sec x \tan x \, dx = 5 \int \sec x \tan x \, dx \qquad \text{(Table 4.1, rule 1)}$$

$$= 5(\sec x + C) \qquad \text{(Example 2, Formula 6)}$$

$$= 5 \sec x + 5C \qquad \text{(First form)}$$

$$= 5 \sec x + C' \qquad \text{(Shorter form)}$$

$$= 5 \sec x + C \qquad \text{(Usual form—no prime)}$$

What about all the different forms in Example 5? Each one of them gives all the antiderivatives of $f(x) = 5 \sec x \tan x$, so each answer is correct. But the least complicated of the three, and the usual choice, is

$$\int 5 \sec x \tan x \, dx = 5 \sec x + C.$$

The general rule, in practice, is this: If $F'(x) = f(x)$ and k is a constant, then

$$k \int f(x) \, dx = \int k f(x) \, dx = kF(x) + C. \qquad (2)$$

Just as the Sum and Difference Rule for differentiation enables us to differentiate expressions term by term, the Sum and Difference Rule for integration enables us to integrate expressions term by term. When we do so, we combine the individual constants of integration into a single arbitrary constant at the end.

Example 6   *Term-by-term Integration.*   Evaluate

$$\int (x^2 - 2x + 5) \, dx.$$

Solution If we recognize that $(x^3/3) - x^2 + 5x$ is an antiderivative of $x^2 - 2x + 5$, we can evaluate the integral as

$$\int (x^2 - 2x + 5) \, dx = \overbrace{\frac{x^3}{3} - x^2 + 5x}^{\text{antiderivative}} + \underset{\text{arbitrary constant}}{C}.$$

However, if we do not recognize the antiderivative right away, we can still generate it term by term with the Sum and Difference Rule:

$$\int (x^2 - 2x + 5) \, dx = \int x^2 \, dx - \int 2x \, dx + \int 5 \, dx$$

$$= \frac{x^3}{3} + C_1 - x^2 + C_2 + 5x + C_3.$$

The formula we obtain this way gives all the antiderivatives of $x^2 - 2x + 5$. But it is also more complicated than it needs to be to do that. If we were to combine C_1, C_2, and C_3 into a single constant $C = C_1 + C_2 + C_3$, the formula would simplify to

$$\frac{x^3}{3} - x^2 + 5x + C$$

and *still* give all the antiderivatives there are. For this reason we recommend that you go right to the final form even if you elect to integrate term by term. Write

$$\int (x^2 - 2x + 5) \, dx = \int x^2 \, dx - \int 2x \, dx + \int 5 \, dx = \frac{x^3}{3} - x^2 + 5x + C.$$

Find the simplest antiderivative you can for each part and then add the constant at the end.

The Integrals of sin²x and cos²x

We can sometimes use trigonometric identities to transform indefinite integrals we do not know how to evaluate into indefinite integrals we do know how to evaluate. Among the examples you should know about, important because of how frequently they arise in applications, are the integral formulas for $\sin^2 x$ and $\cos^2 x$. You need not remember the formulas themselves, but try to remember how they can be derived when you need them.

Example 7

$$\int \sin^2 x \, dx = \int \frac{1 - \cos 2x}{2} \, dx \qquad \left(\text{Because } \sin^2 x = \frac{1 - \cos 2x}{2}\right)$$

$$= \frac{1}{2} \int (1 - \cos 2x) \, dx = \frac{1}{2} \int dx - \frac{1}{2} \int \cos 2x \, dx$$

$$= \frac{1}{2} x - \frac{1}{2} \frac{\sin 2x}{2} + C = \frac{x}{2} - \frac{\sin 2x}{4} + C$$

Example 8

$$\int \cos^2 x \, dx = \int \frac{1 + \cos 2x}{2} \, dx \qquad \left(\cos^2 x = \frac{1 + \cos 2x}{2}\right)$$

$$= \frac{x}{2} + \frac{\sin 2x}{4} + C \qquad \left(\substack{\text{As in Example 7, but} \\ \text{with a sign change}}\right)$$

EXERCISES 4.1

Find antiderivatives for the functions in Exercises 1–16. Do as many as you can without writing anything down (except the answer).

1. a) $6x$ b) -2 c) $6x - 2$

2. a) $8x^7$ b) x^7 c) $x^7 - 6x + 8$

3. a) $-3x^{-4}$ b) x^{-4} c) $x^{-4} + 2x + 3$

4. a) $\dfrac{1}{x^2}$ b) $\dfrac{5}{x^2}$ c) $2 - \dfrac{5}{x^2}$

5. a) $\dfrac{3}{2}\sqrt{x}$ b) $4\sqrt{x}$ c) $x^2 - 4\sqrt{x}$

6. a) $\dfrac{1}{2}x^{-1/2}$ b) $-\dfrac{1}{2}x^{-3/2}$ c) $-\dfrac{3}{2}x^{-5/2}$

7. a) $\dfrac{2}{3}x^{-1/3}$ b) $\dfrac{1}{3}x^{-2/3}$ c) $-\dfrac{1}{3}x^{-4/3}$

8. a) $x^{-1/3}$ b) $x^{-2/3}$ c) $x^{-4/3}$

9. a) $-\sin 3x$ b) $3 \sin x$ c) $3 \sin x - \sin 3x$

10. a) $\pi \cos \pi x$ b) $\dfrac{\pi}{2} \cos \dfrac{\pi x}{2}$ c) $\cos \dfrac{\pi x}{2}$

11. a) $\sec^2 x$ b) $5 \sec^2 5x$ c) $\sec^2 5x$

12. a) $\csc^2 x$ b) $7 \csc^2 7x$ c) $\csc^2 7x$

13. a) $\sec x \tan x$ b) $2 \sec 2x \tan 2x$ c) $4 \sec 2x \tan 2x$

14. a) $\csc x \cot x$ b) $8 \csc 4x \cot 4x$ c) $\csc 4x \cot 4x$

15. $(\sin x - \cos x)^2$ (*Hint:* $2 \sin x \cos x = \sin 2x$)

16. $(1 + 2 \cos x)^2$

Evaluate the integrals in Exercises 17–50.

17. $\displaystyle\int 7 \, dx$

18. $\displaystyle\int (6 - 6x) \, dx$

19. $\displaystyle\int 3\sqrt{x} \, dx$

20. $\displaystyle\int \frac{4}{x^2} \, dx$

21. $\displaystyle\int t^{-1/3}\, dt$ **22.** $\displaystyle\int (1 - 4y^{-3})\, dy$

23. $\displaystyle\int (5v^2 + 2v)\, dv$ **24.** $\displaystyle\int \left(\frac{z^2}{2} + \frac{z^3}{3}\right) dz$

25. $\displaystyle\int (2s^3 - 5s + 7)\, ds$ **26.** $\displaystyle\int (1 - x^2 - 3x^5)\, dx$

27. $\displaystyle\int (x - 3)(4x^2 - 7)\, dx$ **28.** $\displaystyle\int \sqrt{25x^3}\, dx$

29. $\displaystyle\int \frac{t^3 - 1}{t - 1}\, dt$ **30.** $\displaystyle\int \left(\frac{1}{\sqrt{r}} + \sqrt{r}\right) dr$

31. $\displaystyle\int (x + 2 \cos x)\, dx$ **32.** $\displaystyle\int \left(\frac{1}{\sqrt{\theta}} + \sin \theta\right) d\theta$

33. $\displaystyle\int \sin \frac{t}{3}\, dt$ **34.** $\displaystyle\int 3 \cos 5y\, dy$

35. $\displaystyle\int 3 \csc^2 x\, dx$ **36.** $\displaystyle\int \frac{\sec^2 x}{3}\, dx$

37. $\displaystyle\int \frac{\csc x \cot x}{2}\, dx$ **38.** $\displaystyle\int \frac{2}{5} \sec \theta \tan \theta\, d\theta$

39. $\displaystyle\int \frac{\sec^3 \theta + \tan \theta}{\sec \theta}\, d\theta$ **40.** $\displaystyle\int \frac{1}{2} (\csc^2 x - \csc x \cot x)\, dx$

41. $\displaystyle\int (\sin 2y - \csc^2 y)\, dy$ **42.** $\displaystyle\int (2 \cos 2t - 3 \sin 3t)\, dt$

43. $\displaystyle\int 4 \sin^2 v\, dv$ **44.** $\displaystyle\int \frac{\cos^2 u}{7}\, du$

45. $\displaystyle\int \sin x \cos x\, dx$ **46.** $\displaystyle\int (1 - \cos^2 t)\, dt$

47. $\displaystyle\int (1 + \tan^2 \theta)\, d\theta$ **48.** $\displaystyle\int \frac{1 + \cot^2 x}{2}\, dx$

49. $\displaystyle\int \cos \theta (\tan \theta + \sec \theta)\, d\theta$

50. $\displaystyle\int \frac{\csc \theta}{\csc \theta - \sin \theta}\, d\theta$

Show that the integral formulas in Exercises 51–54 are correct by showing that the derivatives of the right-hand sides are the integrands in the integrals on the left-hand sides. (In Section 4.6 we shall see where formulas like these come from.)

51. $\displaystyle\int (7x - 2)^3\, dx = \frac{(7x - 2)^4}{28} + C$

52. $\displaystyle\int \sec^2 5x\, dx = \frac{\tan 5x}{5} + C$

53. $\displaystyle\int \frac{1}{(x + 1)^2}\, dx = -\frac{1}{x + 1} + C$

54. $\displaystyle\int \frac{1}{(x + 1)^2}\, dx = \frac{x}{x + 1} + C'$

55. Right, or wrong? Say which for each formula.

a) $\displaystyle\int x \sin x\, dx = \frac{x^2}{2} \sin x + C$

b) $\displaystyle\int x \sin x\, dx = -x \cos x + C$

c) $\displaystyle\int x \sin x\, dx = -x \cos x + \sin x + C$

56. Right, or wrong? Say which for each formula.

a) $\displaystyle\int (2x + 1)^2\, dx = \frac{(2x + 1)^3}{3} + C$

b) $\displaystyle\int 3(2x + 1)^2\, dx = (2x + 1)^3 + C$

c) $\displaystyle\int 6(2x + 1)^2\, dx = (2x + 1)^3 + C$

57. Suppose that

$$f(x) = \frac{d}{dx} (1 - \sqrt{x}) \quad\text{and}\quad g(x) = \frac{d}{dx} (x + 2).$$

Find:

a) $\displaystyle\int f(x)\, dx$ b) $\displaystyle\int g(x)\, dx$

c) $\displaystyle\int [-f(x)]\, dx$ d) $\displaystyle\int [-g(x)]\, dx$

e) $\displaystyle\int [f(x) + g(x)]\, dx$ f) $\displaystyle\int [f(x) - g(x)]\, dx$

g) $\displaystyle\int [x + f(x)]\, dx$ h) $\displaystyle\int [g(x) - 4]\, dx$

58. Repeat Exercise 57, assuming that

$$f(x) = \frac{d}{dx} e^x \quad\text{and}\quad g(x) = \frac{d}{dx} (x \sin x).$$

59. *A surprising graph*
a) GRAPHER   Graph $f(x) = \sin x \sin(x + 2) - \sin^2(x + 1)$.
b) Use calculus to verify what you see.

60. Show that if functions f and g have identical derivatives on $[a, b]$ and $f(c) = g(c)$ at some point c in $[a, b]$, then $f(x) = g(x)$ for all x in $[a, b]$.

EXPLORER PROGRAMS	
Derivatives	Graphs functions and their derivatives together
PowerGrapher	Graphs functions together

4.2 Initial Value Problems and Mathematical Modeling

As we mentioned earlier, one of the important things we can do with calculus is recover a function from its rate of change and a single known value. The recovery process has two steps. The first step is to evaluate the function's indefinite integral to obtain a formula that gives all the function's antiderivatives. The second is to use the known value to select from these antiderivatives the particular one we want. We showed how to take the first step in the preceding section. Now we show how to take the second. We also discuss mathematical modeling, the process by which we, as scientists, use mathematics to learn about reality.

Initial Value Problems

The problem of finding a function y of x when we know its derivative

$$\frac{dy}{dx} = f(x)$$

and its value y_0 at a particular point x_0 is called an **initial value problem.** We solve such a problem in two steps. First we integrate both sides of the equation with respect to x to find the indefinite integral of f:

$$y = F(x) + C. \tag{1}$$

Then we use the fact that y equals y_0 when x equals x_0 to find the right value for C. In this case $y_0 = F(x_0) + C$, so $C = y_0 - F(x_0)$. The solution of the initial value problem is the function

$$y = F(x) + [y_0 - F(x_0)]. \tag{2}$$

This function has the right derivative, because

$$\frac{dy}{dx} = \frac{d}{dx} F(x) + \frac{d}{dx} [y_0 - F(x_0)] = f(x) + 0 = f(x). \tag{3}$$

It also has the right value when $x = x_0$, because

$$y \Big|_{x = x_0} = F(x_0) + [y_0 - F(x_0)] = y_0. \tag{4}$$

An equation like

$$\frac{dy}{dx} = f(x) \tag{5}$$

that has a derivative in it is called a **differential equation.** Equation (5) gives dy/dx as a function of x. A more complicated differential equation might involve y on the right-hand side as well:

$$\frac{dy}{dx} = 2xy^2. \tag{6}$$

It might also involve higher-order derivatives:

$$\frac{d^2y}{dx^2} - \frac{dy}{dx} + 5y = 3. \tag{7}$$

We shall see how to solve equations like (6) and (7) when we get to Chapter 15. For the time being, we shall steer away from such complications.

The indefinite integral $y = F(x) + C$ of the function $f(x)$ is the **general solution** of the equation $dy/dx = f(x)$. It gives all the solutions of the equation (there are infinitely many, one for each value of C). We solve the equation by finding its general solution. We then solve the initial value problem by using the **initial condition** that y equals y_0 when x equals x_0 to find the **particular solution** $y = F(x) + [y_0 - F(x_0)]$.

Example 1   *Finding a Body's Velocity from its Acceleration and Initial Velocity.*   The acceleration of gravity near the surface of the earth is 9.8 m/sec^2. This means that the velocity v of a body falling freely in a vacuum changes at the rate of

$$\frac{dv}{dt} = 9.8 \text{ m/sec}^2.$$

If the body is dropped from rest, what will its velocity be t seconds after it is released?

Solution   In mathematical terms, we want to solve the initial value problem that consists of

The differential equation: $\dfrac{dv}{dt} = 9.8$

The initial condition: $v = 0$ when $t = 0$.

To solve it, we find the general solution of the differential equation for v by integrating both sides with respect to t:

$\dfrac{dv}{dt} = 9.8$ (Differential equation)

$\displaystyle\int \frac{dv}{dt}\, dt = \int 9.8\, dt$ (Integrate with respect to t.)

$v + C_1 = 9.8t + C_2$ (Integrals evaluated)

$v = 9.8t + C_2 - C_1$ (Solved for v)

$v = 9.8t + C$.   (Constants combined as one)

This last equation tells us that the body's velocity t seconds into the fall is $9.8t + C$ m/sec for some value of C. What value? We find out from the initial condition:

$v = 9.8t + C$

$0 = 9.8(0) + C$ ($v = 0$ when $t = 0$)

$C = 0$.

Conclusion: The body's velocity t seconds into the fall is

$$v = 9.8t + 0 = 9.8t \text{ m/sec}.$$

Notice what happened to the constants of integration when we solved the equation $dv/dt = 9.8$ in Example 1. They combined into a single constant at the end. For

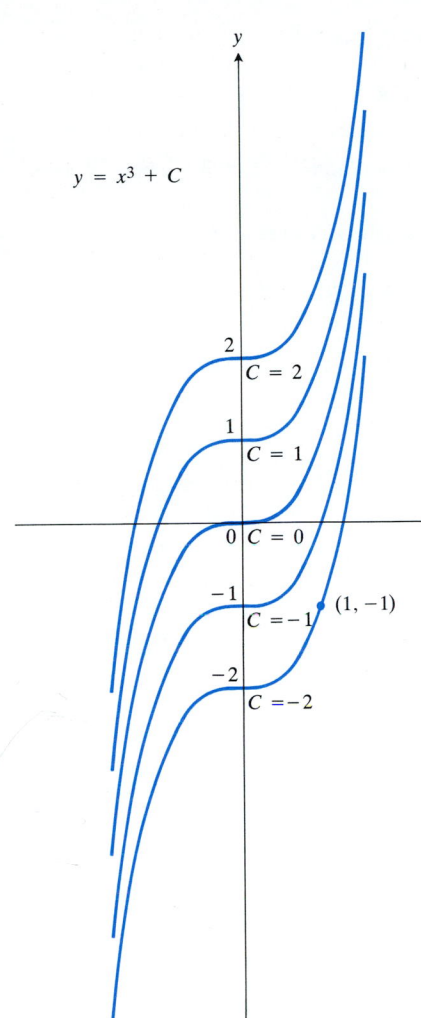

$y = x^3 + C$

$C = 2$

$C = 1$

$C = 0$

$C = -1$ (1, −1)

$C = -2$

4.2 The curves $y = x^3 + C$ fill the coordinate plane without overlapping. In Example 2 we identify the curve $y = x^3 - 2$ as the one that passes through the point $(1, -1)$.

this reason, we recommend that you go from $dv/dt = 9.8$ directly to the final form

$$v = \int 9.8 \, dt = 9.8t + C$$

without bothering to introduce separate constants for the two sides of the equation. We shall do this ourselves from now on.

Example 2   *Finding a Curve from its Slope Function and a Point*   Find the curve whose slope at the point (x, y) is $3x^2$ if the curve is required to pass through the point $(1, -1)$.

Solution   In mathematical language, we are asked to solve the initial value problem that consists of

The differential equation: $\dfrac{dy}{dx} = 3x^2$

The initial condition: $y = -1$ when $x = 1$.

To solve it, we integrate both sides of the differential equation with respect to x, obtaining

$$y = \int 3x^2 \, dx = x^3 + C.$$

Then we substitute $x = 1$ and $y = -1$ to find C:

$$y = x^3 + C,$$
$$-1 = (1)^3 + C,$$
$$C = -2.$$

The curve we want is $y = x^3 - 2$ (Fig. 4.2).

In the next example, we have to integrate a second derivative twice to find the function we are looking for. The first integration,

$$\int \frac{d^2s}{dt^2} \, dt = \frac{ds}{dt} + C, \tag{8}$$

gives the function's first derivative. The second integration gives the function.

Example 3   *Finding a Projectile's Height from its Acceleration, Initial Velocity, and Initial Position.*   A heavy projectile is fired straight up from a platform 3 m above the ground, with an initial velocity of 160 m/sec. Assume that the only force affecting the projectile during its flight is from gravity, which produces a downward acceleration of 9.8 m/sec². Find an equation for the projectile's height above the ground as a function of time t if $t = 0$ when the projectile is fired. How high above the ground is the projectile 3 sec later?

Solution   To model the problem, we draw a figure (Fig. 4.3) and let s denote the projectile's height above the ground at time t. We assume s to be a twice-differentiable function of t and represent the projectile's velocity and acceleration with the derivatives ds/dt and d^2s/dt^2.

Since gravity acts in the negative s-direction, the direction of decreasing s in

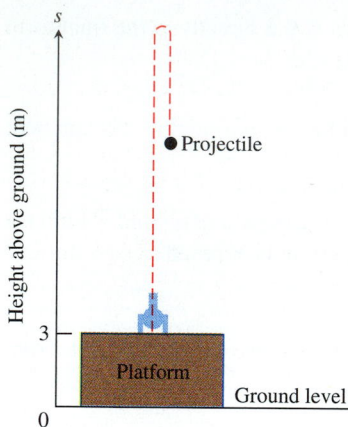

4.3 The sketch for modeling the projectile motion in Example 3.

our model, the initial value problem to solve is

ACCEL'

DOWNWARD (-)

The differential equation: $\dfrac{d^2 s}{dt^2} = -9.8$

The initial conditions: $\dfrac{ds}{dt} = 160$ and $s = 3$ when $t = 0.$

We integrate the differential equation with respect to t to find ds/dt:

$$\int \frac{d^2 s}{dt^2}\, dt = \int (-9.8)\, dt$$

$$\frac{ds}{dt} = -9.8t + C_1.$$

We apply the first initial condition to find C_1:

$$160 = -9.8(0) + C_1 \qquad \left(\frac{ds}{dt} = 160 \text{ when } t = 0\right)$$

$$C_1 = 160.$$

This completes the formula for ds/dt:

$$\frac{ds}{dt} = -9.8t + 160.$$

We integrate ds/dt with respect to t to find s:

$$\int \frac{ds}{dt}\, dt = \int (-9.8t + 160)\, dt$$

$$s = -4.9t^2 + 160t + C_2.$$

We apply the second initial condition to find C_2:

$$3 = -4.9(0)^2 + 160(0) + C_2 \qquad (s = 3 \text{ when } t = 0)$$

$$C_2 = 3.$$

This completes the formula for s as a function of t:

$$s = -4.9t^2 + 160t + 3.$$

To find the projectile's height 3 sec into its flight, we set $t = 3$ in the formula for s. The height is

$$s = -4.9(3)^2 + 160(3) + 3 = 438.9 \text{ m.}$$

When we find a function by integrating its first derivative, we have one constant of integration, as in the first two examples. When we find a function from its second derivative, we have to deal with two constants, one from each integration, as in Example 3. To find a function from its third derivative, we would have to find the values of three constants of integration, and so on. In each case, the values of the constants are determined by the problem's initial conditions. Each time we integrate, we need an initial condition to tell us the value of C.

Mathematical Modeling

The development of a mathematical model usually takes four steps: First we observe something in the real world (a ball bearing falling from rest or the trachea contracting during a cough, for example) and construct a system of mathematical variables and relationships that imitate some of its important features. We build a mathematical metaphor for what we see. Next we apply (usually) existing mathematics to the variables and relationships in the model to draw conclusions about them. After that we translate the mathematical conclusions into information about the system under study. Finally we check the information against observation to see if the model has predictive value. We also investigate the possibility that the model applies to other systems. The really good models are the ones that lead to conclusions that are consistent with observation, that have predictive value and broad application, and that are not too hard to use.

The natural cycle of mathematical imitation, deduction, interpretation, and comparison is shown in the diagrams in Fig. 4.4.

4.4 Diagrams of the modeling cycles for free fall, light refraction, melting ice, and escape velocity.

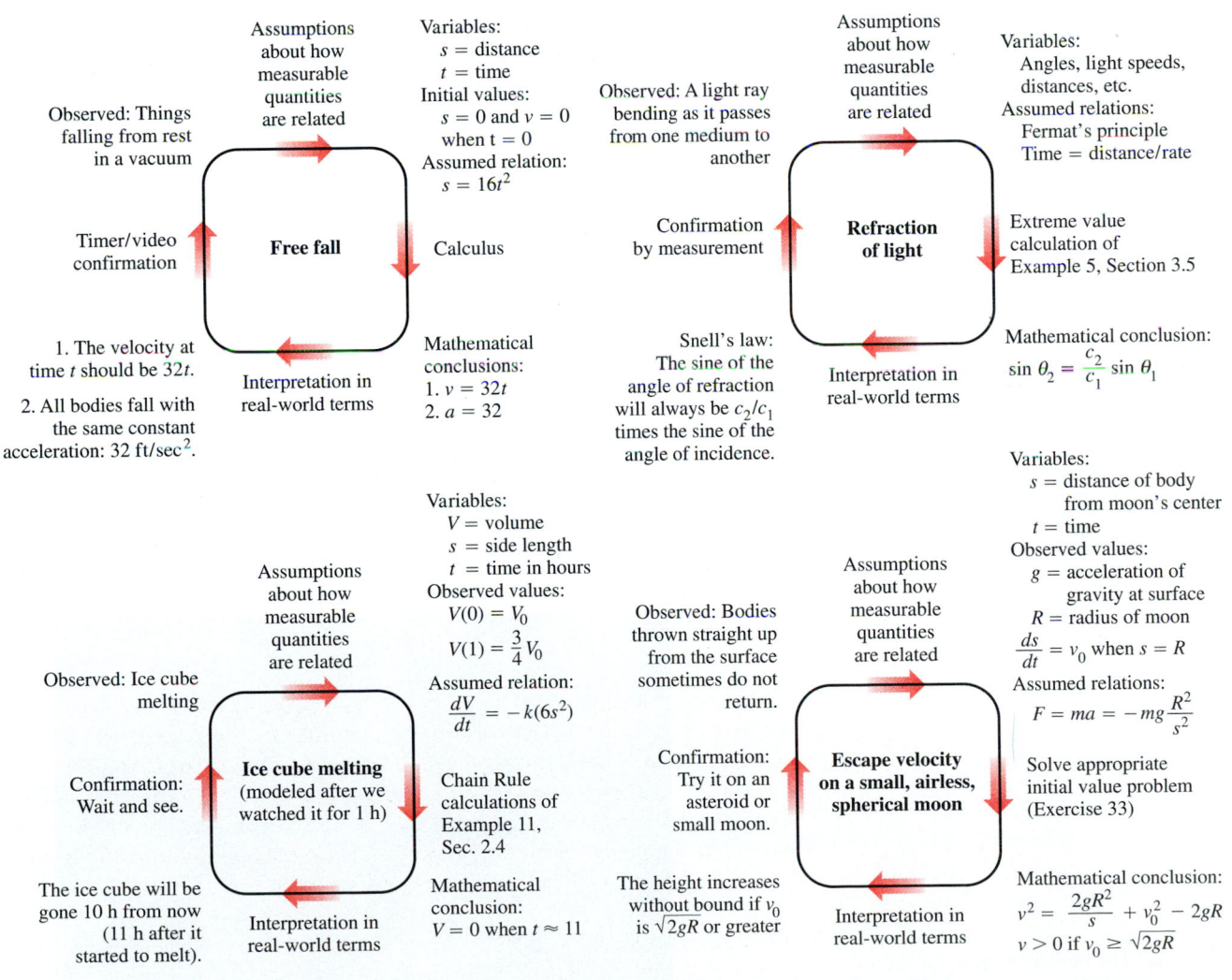

Free fall

Observed: Things falling from rest in a vacuum

Timer/video confirmation

1. The velocity at time t should be $32t$.

2. All bodies fall with the same constant acceleration: 32 ft/sec^2.

Assumptions about how measurable quantities are related

Variables:
s = distance
t = time
Initial values:
$s = 0$ and $v = 0$ when t = 0
Assumed relation:
$s = 16t^2$

Calculus

Mathematical conclusions:
1. $v = 32t$
2. $a = 32$

Interpretation in real-world terms

Refraction of light

Observed: A light ray bending as it passes from one medium to another

Confirmation by measurement

Snell's law: The sine of the angle of refraction will always be c_2/c_1 times the sine of the angle of incidence.

Assumptions about how measurable quantities are related

Variables:
Angles, light speeds, distances, etc.
Assumed relations:
Fermat's principle
Time = distance/rate

Extreme value calculation of Example 5, Section 3.5

Mathematical conclusion:
$\sin \theta_2 = \dfrac{c_2}{c_1} \sin \theta_1$

Interpretation in real-world terms

Ice cube melting (modeled after we watched it for 1 h)

Observed: Ice cube melting

Confirmation: Wait and see.

The ice cube will be gone 10 h from now (11 h after it started to melt).

Assumptions about how measurable quantities are related

Variables:
V = volume
s = side length
t = time in hours
Observed values:
$V(0) = V_0$
$V(1) = \dfrac{3}{4}V_0$
Assumed relation:
$\dfrac{dV}{dt} = -k(6s^2)$

Chain Rule calculations of Example 11, Sec. 2.4

Mathematical conclusion:
$V = 0$ when $t \approx 11$

Interpretation in real-world terms

Escape velocity on a small, airless, spherical moon

Observed: Bodies thrown straight up from the surface sometimes do not return.

Confirmation: Try it on an asteroid or small moon.

The height increases without bound if v_0 is $\sqrt{2gR}$ or greater

Assumptions about how measurable quantities are related

Variables:
s = distance of body from moon's center
t = time
Observed values:
g = acceleration of gravity at surface
R = radius of moon
$\dfrac{ds}{dt} = v_0$ when $s = R$
Assumed relations:
$F = ma = -mg\dfrac{R^2}{s^2}$

Solve appropriate initial value problem (Exercise 33)

Mathematical conclusion:
$v^2 = \dfrac{2gR^2}{s} + v_0^2 - 2gR$
$v > 0$ if $v_0 \geq \sqrt{2gR}$

Interpretation in real-world terms

Computer Simulation

When a system we want to study is complicated, we can sometimes experiment first to see how the system behaves under different circumstances. But if this is not possible (the experiments might be expensive, time-consuming, or dangerous), we might run a series of simulated experiments on a computer—experiments that behave like the real thing without the disadvantages. Thus we might model the effects of atomic war, the effect of waiting a year longer to harvest trees, the effect of crossing particular breeds of cattle, or the effect of increasing atmospheric ozone by 1%, all without having to pay the consequences or wait to see how things work out.

We also bring computers in when the model we want to use has too many calculations to be practical any other way. NASA's space flight models are run on computers—they have to be to generate course corrections on time. If you want to model the behavior of galaxies that contain billions of stars, a computer offers the only possible way. One of the most spectacular computer simulations in recent years, carried out by Alar Toomre at MIT, explained a peculiar galactic shape that was not consistent with our previous ideas about how galaxies are formed. The galaxies had acquired their odd shapes, Toomre concluded, by passing through one another (Fig. 4.5).

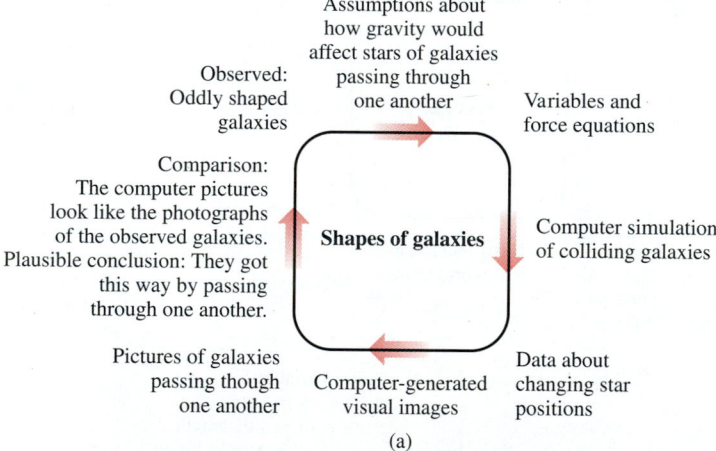

(a)

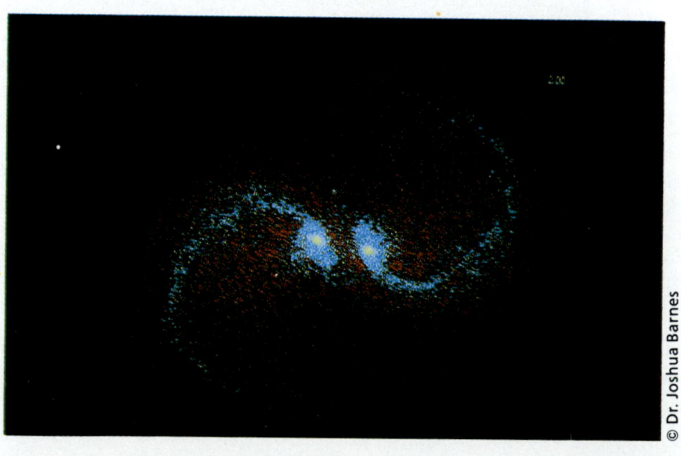

© Dr. Joshua Barnes

(b)

4.5   (a) The modeling cycle for the shapes of colliding galaxies. (b) The computer's image of how galaxies are reshaped by the collision.

EXERCISES 4.2

Solve the initial value problems in Exercises 1–16.

1. $\dfrac{dy}{dx} = 2x - 7$, $y = 0$ when $x = 2$

2. $\dfrac{dy}{dx} = 10 - x$, $y = -1$ when $x = 0$

3. $\dfrac{dy}{dx} = \dfrac{1}{x^2} + x$, $x > 0$; $y = 1$ when $x = 2$

4. $\dfrac{dy}{dx} = 9x^2 - 4x + 5$, $y = 0$ when $x = -1$

5. $\dfrac{dy}{dx} = 3\sqrt{x}$, $y = 4$ when $x = 9$

6. $\dfrac{dy}{dx} = \dfrac{1}{2\sqrt{x}}$, $y = 0$ when $x = 4$

7. $\dfrac{ds}{dt} = 1 + \cos t$, $s = 4$ when $t = 0$

8. $\dfrac{ds}{dt} = \cos t + \sin t$, $s = 1$ when $t = \pi$

9. $\dfrac{dv}{dt} = -\pi \sin \pi t$, $v = 0$ when $t = 0$

10. $\dfrac{dv}{dt} = \dfrac{1}{2} \sec t \tan t$, $v = 1$ when $t = 0$

11. $\dfrac{d^2y}{dx^2} = 2 - 6x$, $\dfrac{dy}{dx} = 4$ and $y = 1$ when $x = 0$

12. $\dfrac{d^2y}{dx^2} = 0$, $\dfrac{dy}{dx} = 2$ and $y = 0$ when $x = 0$

13. $\dfrac{d^2r}{dt^2} = \dfrac{2}{t^3}$, $\dfrac{dr}{dt} = 1$ and $r = 1$ when $t = 1$

14. $\dfrac{d^2s}{dt^2} = \dfrac{3t}{8}$, $\dfrac{ds}{dt} = 3$ and $s = 4$ when $t = 4$

15. $\dfrac{d^3y}{dx^3} = 6$; $\dfrac{d^2y}{dx^2} = -8$, $\dfrac{dy}{dx} = 0$,
and $y = 5$ when $x = 0$

16. $y^{(4)} = \cos x$; $y''' = 0$, $y'' = 1$, $y' = 2$,
and $y = 3$ when $x = 0$

Exercises 17 and 18 give the velocity $v = ds/dt$ and initial position of a body moving on a coordinate line. Find the body's position at time t.

17. $v = 9.8t$, $s = 10$ when $t = 0$

18. $v = \sin t$, $s = 0$ when $t = 0$

Exercises 19 and 20 give the acceleration $a = d^2s/dt^2$, initial velocity, and initial position of a body moving on a coordinate line. Find the body's position at time t.

19. $a = 32$; $v = 20$ and $s = 0$ when $t = 0$

20. $a = \sin t$; $v = -1$ and $s = 1$ when $t = 0$

21. Find the curve in the xy-plane that passes through the point $(9, 4)$ and whose slope at each point is $3\sqrt{x}$.

22. a) Find a function $y = f(x)$ with the following properties.

   i) $\dfrac{d^2y}{dx^2} = 6x$.

   ii) Its graph in the xy-plane passes through the point $(0, 1)$ and has a horizontal tangent there.

   b) How many functions like this are there? How do you know?

23. On the moon the acceleration of gravity is 1.6 m/sec². If a rock is dropped into a crevasse, how fast will it be going just before it hits bottom 30 sec later?

24. A rocket lifts off the surface of the earth with a constant acceleration of 20 m/sec². How fast will the rocket be going 1 min later?

25. With approximately what velocity do you enter the water if you dive from a 10-m platform? (Use $g = 9.8$ m/sec².)

26. CALCULATOR The acceleration of gravity near the surface of Mars is 3.72 m/sec². If a rock is blasted straight up from the surface with an initial velocity of 93 m/sec (about 208 mph), how high does it go? (*Hint:* When is the velocity zero?)

27. *How long will it take a tank to drain?*   If we open a valve to drain the water from a vertical cylindrical tank (Fig. 4.6), the water will flow fast when the tank is full but will slow down as the tank drains. The rate at which the water level drops is proportional to the square root of the water's depth. In the notation of Fig. 4.6, this means that

$$\frac{dy}{dt} = -k\sqrt{y}. \tag{9}$$

4.6   The draining tank in Exercises 27 and 28.

The value of the positive constant k depends on the acceleration of gravity and the cross-section area of the drain hole. Equation (9) has a minus sign because y decreases with time. Solve Eq. (9) by rewriting it as

$$\frac{1}{\sqrt{y}}\frac{dy}{dt} = -k \qquad (10)$$

and integrating both sides with respect to t. Combine the constants of integration into a single arbitrary constant. Then solve the resulting equation for y to express y as a function of t.

28. *Continuation of Exercise 27.* (a) Suppose t is measured in minutes and $k = 1/10$. Find y as a function of t if $y = 9$ ft when $t = 0$. (b) How long will it take the tank to drain if the water is 9 ft deep to start with?

29. *Stopping a car in time.* You are driving along a highway at a steady 60 mph (88 ft/sec) when you see an accident ahead and slam on the brakes. What constant deceleration is required to stop your car in 242 ft? To find out, carry out the following steps:

STEP 1: Solve the initial value problem

Differential equation: $\dfrac{d^2s}{dt^2} = -k$ (k constant)

Initial conditions: $\dfrac{ds}{dt} = 88$ and $s = 0$ when $t = 0$.

$\left(\begin{array}{l}\text{Measuring time and distance from}\\ \text{when the brakes are applied}\end{array}\right)$

STEP 2: Find the value of t that makes $ds/dt = 0$. (The answer will involve k.)

STEP 3: Find the value of k that makes $s = 242$ for the value of t you found in step 2.

30. *Motion along a coordinate line.* A particle moves along a coordinate line with acceleration $a = d^2s/dt^2 = 15\sqrt{t} - (3/\sqrt{t})$, subject to the conditions that $ds/dt = 4$ and $s = 0$ when $t = 1$. Find
a) the velocity $v = ds/dt$ in terms of t,
b) the position s in terms of t.

31. CALCULATOR *The hammer and the feather.* When *Apollo 15* astronaut David Scott dropped a hammer and a feather on the moon to demonstrate that in a vacuum all bodies fall with the same (constant) acceleration, he dropped them from about 4 ft above the ground. The television footage of the event shows the hammer and feather falling more slowly than on Earth, where, in a vacuum, they would have taken only half a second to fall the 4 ft. How long did it take the hammer and feather to fall 4 ft on the moon? To find out, solve the following initial value problem for s as a function of t. Then find the value of t that makes s equal 0.

Differential equation: $\dfrac{d^2s}{dt^2} = -5.2$ ft/sec^2

Initial conditions: $\dfrac{ds}{dt} = 0$ and $s = 4$ when $t = 0$

32. *The standard equation for free fall.* The standard equation for free fall near the surface of every planet is

$$s = \frac{1}{2}gt^2 + v_0 t + s_0,$$

where $s(t)$ is the body's position on the line of fall, g is the planet's (constant) acceleration of gravity, v_0 is the body's initial velocity, and s_0 is the body's initial position. Derive this equation by solving the following initial value problem.

Differential equation: $\dfrac{d^2s}{dt^2} = g$

Initial conditions: $\dfrac{ds}{dt} = v_0$ and $s = s_0$ when $t = 0$

33. *Escape velocity.* The gravitational attraction F exerted by an airless moon on a body of mass m at a distance s from the moon's center is given by the equation $F = -mg\,R^2 s^{-2}$, where g is the acceleration of gravity at the moon's surface and R is the moon's radius (Fig. 4.7). The force F is negative because it acts in the direction of decreasing s.

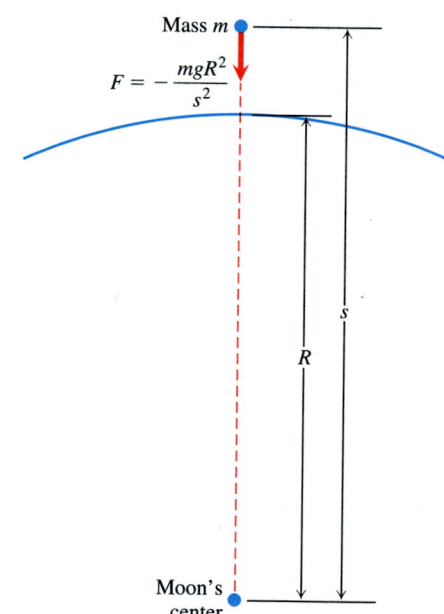

4.7   Diagram for Exercise 33.

a) If the body is projected vertically upward from the moon's surface with an initial velocity v_0 at time $t = 0$, show that the body's velocity at position s is given by the equation

$$v^2 = \frac{2gR^2}{s} + v_0^2 - 2gR.$$

Thus, the velocity remains positive as long as $v_0 \geq \sqrt{2gR}$. The velocity $v_0 = \sqrt{2gR}$ is called the moon's **escape** velocity. A body projected upward with this or a greater velocity will escape from the moon's gravitational pull.

b) Show that if $v_0 = \sqrt{2gR}$, then

$$s = R\left(1 + \frac{3v_0}{2R}\,t\right)^{2/3}.$$

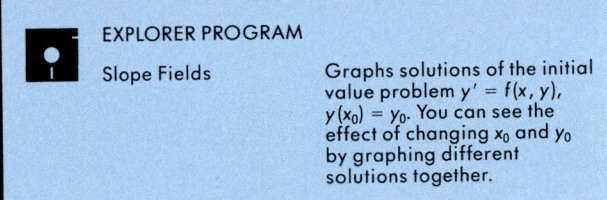

EXPLORER PROGRAM

Slope Fields

Graphs solutions of the initial value problem $y' = f(x, y)$, $y(x_0) = y_0$. You can see the effect of changing x_0 and y_0 by graphing different solutions together.

4.3 Definite Integrals

We now turn our attention to another kind of integration, the integration associated with summation. The new integrals are called definite integrals to distinguish them from the indefinite integrals we have worked with up to now. Definite integrals are numerical limits, not families of antiderivatives, and you may well wonder why two such different mathematical entities are both called integrals and what connection they could possibly have. The connection is beautiful, deep, and surprising, and its discovery by Newton and Leibniz in the mid-seventeenth century changed the mathematical world forever. We shall see for ourselves what that connection is when we get to Section 4.5.

The approach we take is first to define what we mean by the area of the region between the graph of a nonnegative continuous function $y = f(x)$ and an interval $a \leq x \leq b$ of the x-axis. We start by filling as much of the region as we can with vertical inscribed rectangles in the manner suggested by Fig. 4.8. The sum of the

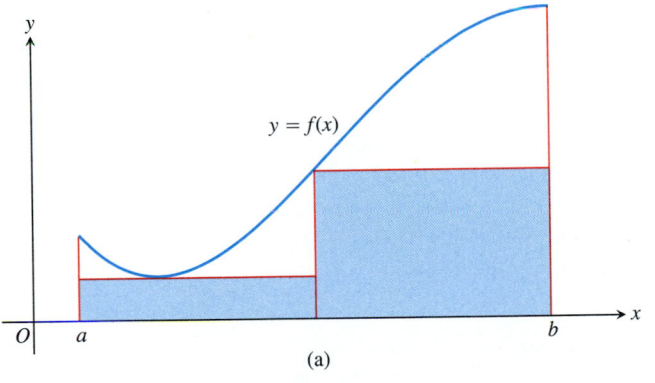

(a)

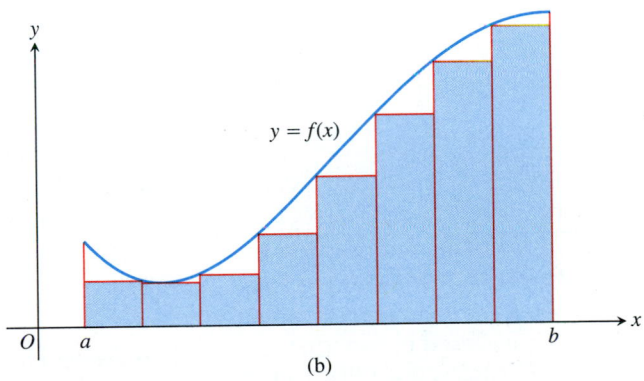

(b)

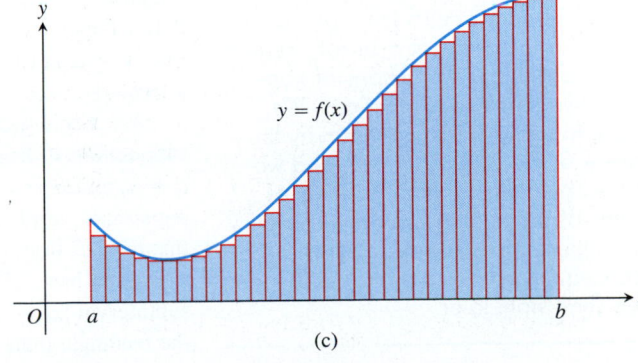

(c)

4.8 To define the area of the region beneath the graph of f from a to b, we approximate the region with inscribed rectangles and add the areas of the rectangles. As (a), (b), and (c) suggest, the approximation improves as the rectangles become narrower and more numerous.

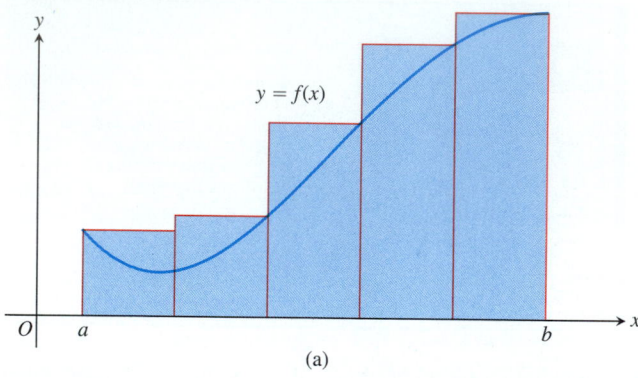

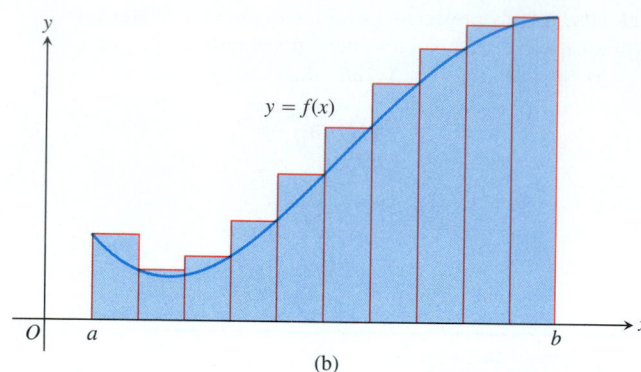

4.9   Circumscribed rectangles do as well as inscribed rectangles for estimating the area of a region under a curve.

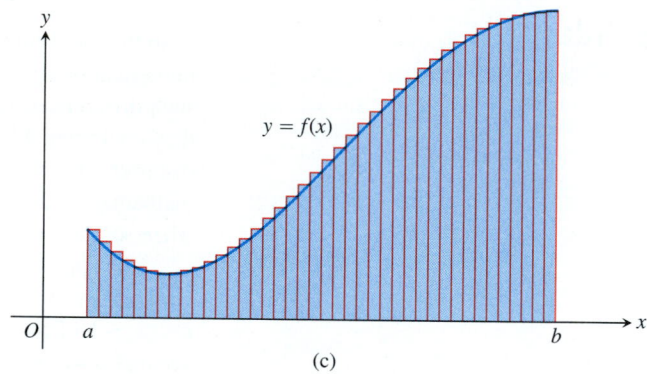

rectangle areas approximates the area of the region. The more rectangles we use, the better the approximation becomes. We define the area of the region to be the limit of the rectangle area sums as the rectangles become smaller and smaller and the number of rectangles we use approaches infinity. By the time we have said all this in precise mathematical terms, two additional features of our construction will become apparent. First, we get exactly the same limit if we use circumscribed rectangles (Fig. 4.9) or any other kinds of rectangles whose bases lie along the x-axis and whose tops touch the curve. Second, the limit of the sums of the rectangle areas exists for any continuous function whatever, not just for the nonnegative continuous functions with which we now begin.

The Area Under a Curve

Suppose $y = f(x)$ is nonnegative and continuous throughout $[a, b]$. We want to define the area of the region enclosed by the graph of f, the x-axis, and the vertical lines $x = a$ and $x = b$. We call this area **the area under the curve $y = f(x)$ from a to b.**

We begin with an approximation. We partition $[a, b]$ into n consecutive subintervals of length $\Delta x = (b - a)/n$ by choosing points $x_1, x_2, \ldots, x_{n-1}$ between $a = x_0$ and $b = x_n$ (Fig. 4.10). The vertical lines through the points x_i divide the region into vertical strips. We approximate each strip with an inscribed rectangle that reaches from the base of the strip on the x-axis to the lowest point of the curve above the base. If c_k is a point at which f assumes its minimum value on the kth subinterval $[x_{k-1}, x_k]$ (there *is* such a point because f is continuous), the height of the rectangle there is $f(c_k)$. The area of the kth rectangle, its height times its base

Defining the Area Under a Curve

We know general formulas for areas of regions enclosed by triangles, trapezoids, circles, and other shapes from classical Greek geometry. But there are no formulas from precalculus times for the areas of arbitrary regions like those under the graphs of non-negative continuous functions. We have to *define* what we mean by the areas of these new regions, and we do so by defining them as limits of sums of areas of inscribed rectangles. That these limits always exist is a consequence of a remarkable theorem we shall state later.

4.10 When we use inscribed rectangles to estimate the area under the graph of a continuous nonnegative function, the height of each rectangle is the minimum value f takes on the rectangle's base.

length, is

$$\text{area of } k\text{th rectangle} = f(c_k)\Delta x.$$

The area of the first rectangle in Fig. 4.10 is $f(c_1)\Delta x$, the area of the second is $f(c_2)\Delta x$, and so on down to the nth rectangle, whose area is $f(c_n)\Delta x$. The sum of these areas,

$$S_n = f(c_1)\Delta x + f(c_2)\Delta x + \cdots + f(c_n)\Delta x, \tag{1}$$

approximates the area under the curve from a to b.

Example 1   Estimate the area under the curve $y = x^2 + 1$ from $a = 0$ to $b = 1$ with $n = 4$ inscribed rectangles of equal width.

Solution   We graph $f(x) = x^2 + 1$ over $[0, 1]$ and partition $[0, 1]$ into subintervals of length $\Delta x = 1/4$ with the points

$$x_1 = \frac{1}{4}, \qquad x_2 = \frac{2}{4}, \qquad x_3 = \frac{3}{4}.$$

We sketch the inscribed rectangle for each subinterval (Fig. 4.11). The height of each rectangle (in this case) is the length of its left-hand edge, so the c_k's in Eq. (1) are the intervals' left-hand endpoints. The sum of the rectangle areas is

$$S_4 = f(0)\Delta x + f\left(\frac{1}{4}\right)\Delta x + f\left(\frac{1}{2}\right)\Delta x + f\left(\frac{3}{4}\right)\Delta x$$

$$= \left((0)^2 + 1\right)\left(\frac{1}{4}\right) + \left(\left(\frac{1}{4}\right)^2 + 1\right)\left(\frac{1}{4}\right) + \left(\left(\frac{1}{2}\right)^2 + 1\right)\left(\frac{1}{4}\right) +$$

$$\left(\left(\frac{3}{4}\right)^2 + 1\right)\left(\frac{1}{4}\right)$$

$$= \frac{1}{4} + \frac{17}{64} + \frac{5}{16} + \frac{25}{64} = 1.21875.$$

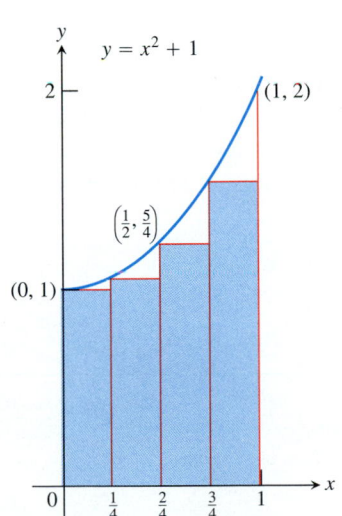

4.11 Rectangles for estimating the area under the graph of $f(x) = x^2 + 1$ from 0 to 1 (Example 1).

Since the rectangles do not cover the entire region under the curve, we expect the approximation to be too small. As you will see if you do Exercise 2 in Section 4.5, the area is 4/3, so the approximation falls short by about 8%.

As we just found, the area approximation given by S_n in Eq. (1) may not be particularly good if the number of rectangles is small. But we expect the approximation to improve as the rectangles become narrower and more numerous, so we define the area under the curve from a to b to be the limiting value of the sums S_n as n approaches infinity.

DEFINITION

> Let f be a nonnegative continuous function on the interval $[a, b]$. The **area under the curve** $y = f(x)$ from $x = a$ to $x = b$ is
>
> $$A = \lim_{n \to \infty} S_n = \lim_{n \to \infty} (f(c_1)\Delta x + f(c_2)\Delta x + \cdots + f(c_n)\Delta x). \qquad (2)$$

The limit in Eq. (2) always exists, as we shall explain shortly, and in Section 4.5 we shall find a straightforward technique for computing it. Before we can do that, however, we need to say what it means for sums to have limits, and to do that we need to develop an efficient notation for sums that involve large numbers of terms.

Sigma Notation for Finite Sums

We use the capital Greek letter Σ ("sigma") to write the sum

$$f(c_1)\Delta x + f(c_2)\Delta x + \cdots + f(c_n)\Delta x \qquad (3)$$

as

$$\sum_{k=1}^{n} f(c_k)\Delta x. \qquad (4)$$

The expression in (4) is read "the sum from k equals 1 to n of f of c_k delta x." When we write a sum this way, we say we have written it in sigma notation.

DEFINITION

> **Sigma Notation for Finite Sums**
>
> The symbol $\sum_{k=1}^{n} a_n$ denotes the sum $a_1 + a_2 + \cdots + a_n$. The a's are the **terms** of the sum: a_1 is the first term, a_2 is the second term, a_k is the k**th term**, and a_n is the nth and last term. The variable k is the **index of summation**. The values of k run through the integers from 1 to n. The number 1 is the **lower limit of summation**; the number n is the **upper limit of summation**.

Example 2

The sum in sigma notation	The sum written out—one term for each value of k	The value of the sum
$\displaystyle\sum_{k=1}^{5} k$	$1 + 2 + 3 + 4 + 5$	15
$\displaystyle\sum_{k=1}^{3} (-1)^k k$	$(-1)^1(1) + (-1)^2(2) + (-1)^3(3)$	$-1 + 2 - 3 = -2$
$\displaystyle\sum_{k=1}^{2} \frac{k}{k+1}$	$\dfrac{1}{1+1} + \dfrac{2}{2+1}$	$\dfrac{1}{2} + \dfrac{2}{3} = \dfrac{7}{6}$

Algebra Rules for Finite Sums

When we work with finite sums, we can always use the following rules.

Algebra Rules

1. **Sum Rule:**
$$\sum_{k=1}^{n} (a_k + b_k) = \sum_{k=1}^{n} a_k + \sum_{k=1}^{n} b_k$$

2. **Difference Rule:**
$$\sum_{k=1}^{n} (a_k - b_k) = \sum_{k=1}^{n} a_k - \sum_{k=1}^{n} b_k$$

3. **Constant Multiple Rule:**
$$\sum_{k=1}^{n} ca_k = c \cdot \sum_{k=1}^{n} a_k \quad \text{(Any number } c\text{)}$$

4. **Constant Value Rule:**
$$\sum_{k=1}^{n} a_k = n \cdot c \text{ if } a_k \text{ has the constant value } c$$

There are no surprises in this list of rules, but the formal proofs require mathematical induction (Appendix 5).

Example 3

a) $\displaystyle\sum_{k=1}^{n} (k - k^2) = \sum_{k=1}^{n} k - \sum_{k=1}^{n} k^2$ (Difference Rule)

b) $\displaystyle\sum_{k=1}^{n} (-a_k) = \sum_{k=1}^{n} (-1) \cdot a_k = -1 \cdot \sum_{k=1}^{n} a_k = -\sum_{k=1}^{n} a_k$ $\left(\begin{array}{c}\text{Constant} \\ \text{Multiple Rule}\end{array}\right)$

c) $\displaystyle\sum_{k=1}^{3} (k + 4) = \sum_{k=1}^{3} k + \sum_{k=1}^{3} 4$ (Sum Rule)

$\qquad\qquad = (1 + 2 + 3) + (3 \cdot 4) = 6 + 12 = 18$ $\left(\begin{array}{c}\text{Constant} \\ \text{Value Rule}\end{array}\right)$

Riemann Integrals

The existence of the limit in Eq. (2) is a consequence of a more general theorem that applies to any continuous function on a closed interval. In the more general theorem, the function may have negative values as well as positive ones. We shall introduce the theorem first and then discuss why it works, but we shall not go through the rigors of a complete proof.

We begin with an arbitrary continuous function f defined on an interval $[a, b]$ (Fig. 4.12). We partition the interval $[a, b]$ into n subintervals by choosing $n - 1$ points, $x_1, x_2, \ldots, x_{n-1}$, between a and b subject only to the condition that

$$a < x_1 < x_2 < \cdots < x_{n-1} < b.$$

To make the notation consistent, we denote a by x_0 and b by x_n. The set

$$P = \{x_0, x_1, \ldots, x_n\}$$

is called a **partition** of $[a, b]$.

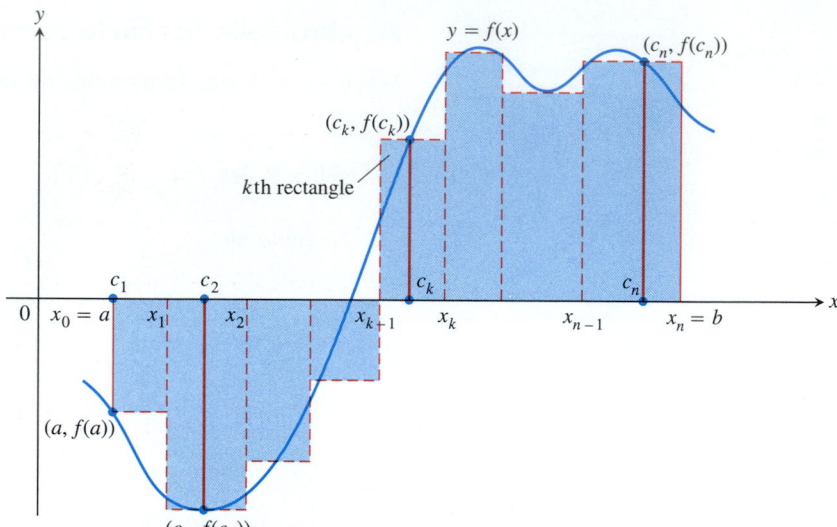

4.12 The graph of a typical function $y = f(x)$ over a closed interval $[a, b]$. The rectangles approximate the region between the graph of the function and the x-axis.

The partition P defines n closed subintervals, $[x_0, x_1], [x_1, x_2], \ldots, [x_{n-1}, x_n]$. The typical subinterval $[x_{k-1}, x_k]$ is called the **kth subinterval** of P. Its length is $\Delta x_k = x_k - x_{k-1}$. On each subinterval we stand a vertical rectangle that reaches from the x-axis to the curve $y = f(x)$. The exact height of the rectangle does not matter as long as its top or base touches the curve at some point $(c_k, f(c_k))$.

If $f(c_k)$ is positive, the product $f(c_k)\Delta x_k$ is the rectangle's area. If $f(c_k)$ is negative, $f(c_k)\Delta x_k$ is the negative of the area. In any case, we add the products to form the sum:

$$S_P = \sum_{k=1}^{n} f(c_k)\Delta \dot{x}_k. \tag{5}$$

This sum, which depends on P and the choice of the numbers c_k, is called a **Riemann sum for f on $[a, b]$,** in honor of the German mathematician Georg Friedrich Bernhard Riemann (*ree*-mahn) (1826–1866), who studied the limits of such sums for discontinuous as well as continuous functions.

Example 4   *Selected Riemann Sums for $f(x) = \sin \pi x$ on $[0, 3/2]$ (Fig. 4.13)*

Partition: $P = \{0, 1/2, 1, 3/2\}$

Subinterval lengths: $\Delta x_k = \dfrac{1}{2}, \quad k = 1, 2, 3$

Choice of c_k: The interval midpoints $c_1 = \dfrac{1}{4}, \quad c_2 = \dfrac{3}{4}, \quad c_3 = \dfrac{5}{4}$ (Fig. 4.13a)

The corresponding Riemann sum:

$$\sum_{k=1}^{3} f(c_k)\Delta x_k = \sum_{k=1}^{3} \sin(\pi c_k) \cdot \frac{1}{2} = \frac{1}{2} \sum_{k=1}^{3} \sin(\pi c_k)$$

$$= \frac{1}{2}\left(\frac{\sqrt{2}}{2} + \frac{\sqrt{2}}{2} - \frac{\sqrt{2}}{2}\right) = \frac{\sqrt{2}}{4} \approx 0.35.$$

If we choose each c_k to be the left-hand endpoint of its interval instead of the

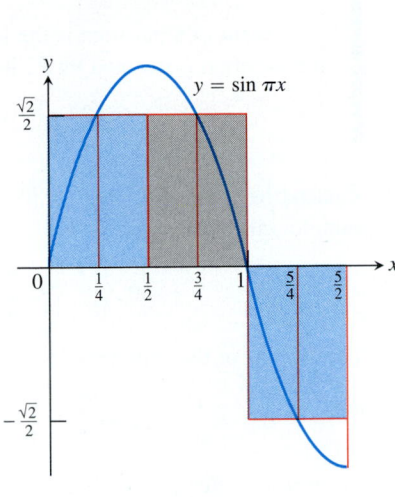

(a) **The first sum**

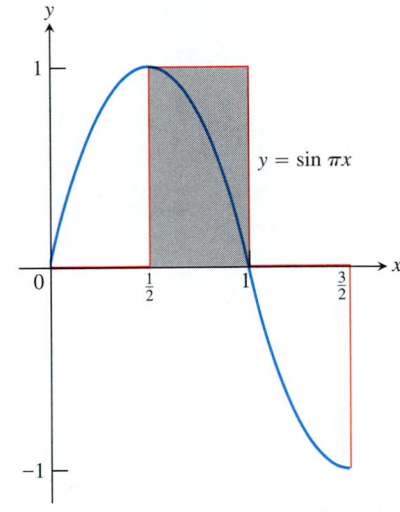

(b) **The second sum**

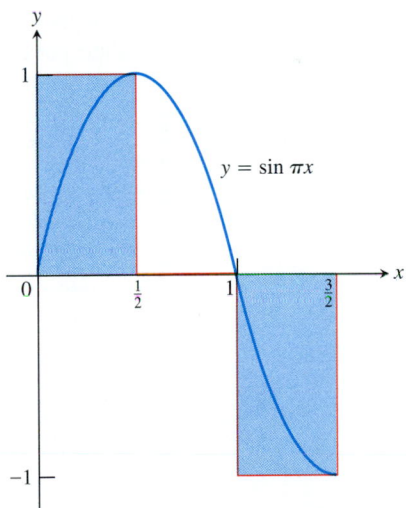

(c) **The third sum**

4.13 The approximating rectangles for the Riemann sums in Example 4.

midpoint, then $c_1 = 0$, $c_2 = 1/2$, $c_3 = 1$ (Fig. 4.13b), and the Riemann sum is

$$\sum_{k=1}^{3} \sin(\pi c_k)\Delta x_k = \frac{1}{2}\left(\sin 0 + \sin \frac{\pi}{2} + \sin \pi\right) = \frac{1}{2}(0 + 1 + 0) = \frac{1}{2}.$$

If we choose each c_k to be the right-hand endpoint of its subinterval, then $c_1 = 1/2$, $c_2 = 1$, $c_3 = 3/2$ (Fig. 4.13c), and the Riemann sum is

$$\sum_{k=1}^{3} \sin(\pi c_k)\Delta x_k = \frac{1}{2}\left(\sin \frac{\pi}{2} + \sin \pi + \sin \frac{3\pi}{2}\right) = \frac{1}{2}(1 + 0 - 1) = 0.$$

As the rectangles of a Riemann sum cover the region between the curve and the x-axis with increasing accuracy, we should find the sums approaching a limiting value of some kind. To make this idea precise, we need to say what it means for partitions to become finer and for Riemann sums to have a limit. We accomplish this with the following definitions.

DEFINITION

> The **norm** of a partition is the length of the partition's longest subinterval. If the partition is denoted by P, its norm is denoted by double bars:
>
> $$\|P\| \qquad \text{(read, "the norm of } P\text{")}$$

Example 5   The set $P = \{0, 1/4, 2/3, 1, 3/2, 2\}$ is a partition of $[0, 2]$. The subintervals of the partition are

$$\left[0, \frac{1}{4}\right], \quad \left[\frac{1}{4}, \frac{2}{3}\right], \quad \left[\frac{2}{3}, 1\right], \quad \left[1, \frac{3}{2}\right], \quad \left[\frac{3}{2}, 2\right].$$

The lengths of the subintervals are

$$\Delta x_1 = \frac{1}{4}, \quad \Delta x_2 = \frac{2}{3} - \frac{1}{4} = \frac{5}{12}, \quad \Delta x_3 = \frac{1}{3}, \quad \Delta x_4 = \frac{1}{2}, \quad \Delta x_5 = \frac{1}{2}.$$

The norm of the partition is 1/2, the longest of these lengths. As you can see, there can be more than one longest subinterval.

The way we say that successive partitions of an interval become finer is to say that their norms approach zero.

DEFINITION

> **The Definite Integral as a Limit of Riemann Sums**
>
> Let $f(x)$ be a function defined on a closed interval $[a, b]$. We say that the **limit** of the Riemann sums $\sum_{k=1}^{n} f(c_k)\Delta x_k$ on $[a, b]$ as $\|P\| \to 0$ is the number I if the following condition is satisfied:
>
> Given any positive number ϵ, there exists a positive number δ such that for every partition P of $[a, b]$,
>
> $$\|P\| < \delta \quad \Rightarrow \quad \left|\sum_{k=1}^{n} f(c_k)\Delta x_k - I\right| < \epsilon \qquad (6)$$
>
> for any choice of the numbers c_k in the subintervals $[x_{k-1}, x_k]$.
>
> If the limit exists, we write
>
> $$\lim_{\|P\| \to 0} \sum_{k=1}^{n} f(c_k)\Delta x_k = I. \qquad (7)$$
>
> We call I the **definite integral** of f over $[a, b]$, and we say that f is **integrable** over $[a, b]$ and that the Riemann sums of f on $[a, b]$ **converge** to I.

The amazing fact is that despite the potential for variety in the Riemann sums $\Sigma f(c_k)\Delta x_k$ as the partitions change and the c_k's are chosen at random in the intervals of each new partition, the sums always have the same limit as $\|P\| \to 0$ when f is continuous on $[a, b]$.

THEOREM 1

> **The Existence of Definite Integrals**
>
> All continuous functions are integrable. That is, if a function f is continuous on an interval $[a, b]$, then its definite integral over $[a, b]$ exists.

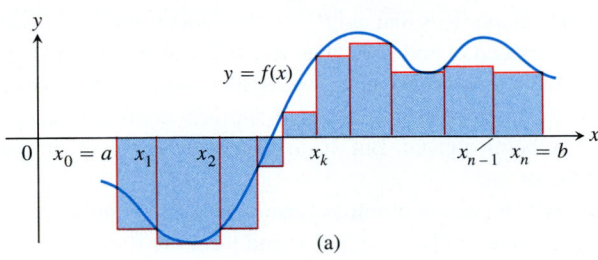

(a)

The lower sum $L = \sum\limits_{k=1}^{n} \min_k \Delta x_k$ is less than ...

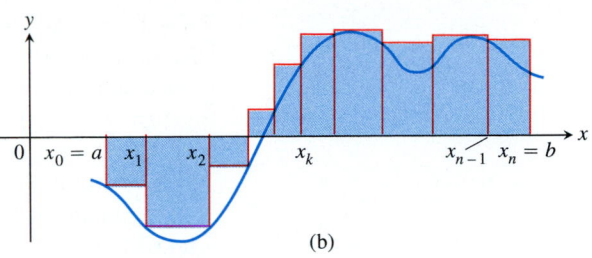

(b)

... the upper sum $U = \sum\limits_{k=1}^{n} \max_k \Delta x_k$.

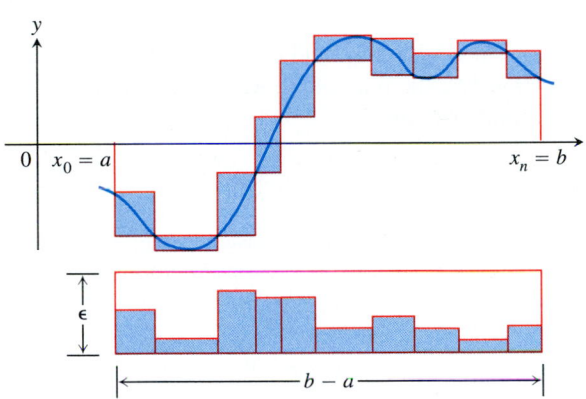

(c)

The difference $U - L$ can be made very small: less than $\epsilon \cdot (b - a)$.

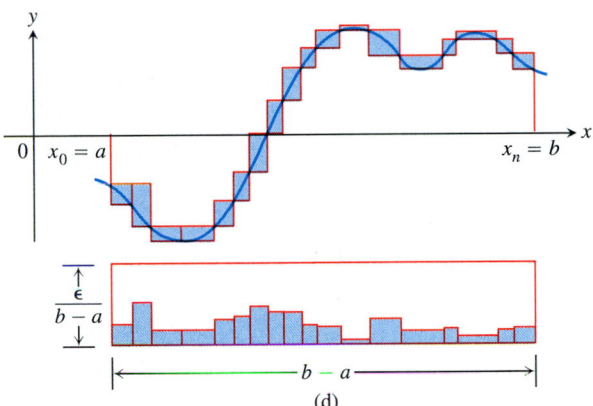

(d)

We can make $U - L$ smaller than any given positive ϵ by making $\|P\|$ small enough.

4.14 For each $\epsilon > 0$, there exists a $\delta > 0$ such that all the blocks in part (d) have heights less than $\epsilon/(b - a)$ if their maximum width is less than δ. This makes $U - L$, the sum of the blocks' areas, less than $(\epsilon/(b - a)) \cdot (b - a) = \epsilon$.

Why should we expect such a theorem to hold? Imagine a typical partition P of the interval $[a, b]$. The function f, being continuous, has a minimum value $\min_k$ ("min kay") and a maximum value $\max_k$ ("max kay") on each subinterval. The products $\min_k \Delta x_k$ associated with the minimum values (Fig. 4.14a) add up to what we call the **lower sum** for f on P:

$$L = \min_1 \Delta x_1 + \min_2 \Delta x_2 + \cdots + \min_n \Delta x_n.$$

The products $\max_k \Delta x_k$ obtained from the maximum values (Fig. 4.14b) add up to the **upper sum** for f on P:

$$U = \max_1 \Delta x_1 + \max_2 \Delta x_2 + \cdots + \max_n \Delta x_n.$$

The difference $U - L$ between the upper and lower sums is the sum of the areas of the shaded blocks in Fig. 4.14(c). As $\|P\| \to 0$, the blocks in Fig. 4.14(c) become more numerous, narrower, and shorter. As Fig. 4.14(d) suggests, we can make the nonnegative number $U - L$ less than any prescribed positive ϵ by taking $\|P\|$ close enough to zero. In other words,

$$\lim_{\|P\| \to 0} (U - L) = 0, \tag{8}$$

and, as shown in more advanced texts,

$$\lim_{\|P\| \to 0} L = \lim_{\|P\| \to 0} U. \tag{9}$$

The fact that Eqs. (8) and (9) hold for any continuous function is a consequence of a special property, called *uniform continuity*, that continuous functions have on closed intervals. This property guarantees that as $\|P\| \to 0$ the blocks in Fig. 4.14(c) that make up the difference between U and L become less tall as they become less wide and that we can make them all as short as we please by making them narrow enough. Passing over the ϵ-δ arguments associated with uniform continuity keeps our derivation of Eq. (9) from being a proof. But the argument is right in spirit and gives a faithful portrait of the proof.

Assuming that Eq. (9) holds for any continuous function on $[a, b]$, suppose we choose a point c_k from each subinterval $[x_{k-1}, x_k]$ of P and form the Riemann sum $\sum_{k=1}^{n} f(c_k) \Delta x_k$. Then $\min_k \leq f(c_k) \leq \max_k$ for each k, so

$$L \leq \sum_{k=1}^{n} f(c_k) \Delta x_k \leq U. \tag{10}$$

The Riemann sum for f is sandwiched between L and U. By a modified version of the Sandwich Theorem of Section 1.9, the limit of the Riemann sums as $\|P\| \to 0$ exists and equals the common limit of U and L:

$$\lim_{\|P\| \to 0} L = \lim_{\|P\| \to 0} \sum_{k=1}^{n} f(c_k) \Delta x_k = \lim_{\|P\| \to 0} U. \tag{11}$$

Pause for a moment to see how remarkable the conclusion in Eq. (11) really is. It says that no matter how we choose the points c_k to form the Riemann sums as $\|P\| \to 0$, the limit is always the same. We can take every $f(c_k)$ to be the minimum value of f on $[x_{k-1}, x_k]$. The limit is the same. We can take every $f(c_k)$ to be the maximum value of f on $[x_{k-1}, x_k]$. The limit is the same. We can choose every c_k at random. The limit is the same.

This fact was first discovered (without uniform continuity) by Cauchy in 1823 and later put on a solid logical foundation (with uniform continuity) by other mathematicians of the nineteenth century. The limit, which we have called the definite integral of f over $[a, b]$, is often called the **Riemann integral** of f over $[a, b]$ because it was Riemann's idea to trap the limit between upper and lower sums.

Although we stated the integral existence theorem specifically for continuous functions, many discontinuous functions are integrable, as Riemann discovered in his pioneering work in 1854. We treat the integration of bounded piecewise continuous functions in Miscellaneous Exercises 57–64 at the end of this chapter. We shall explore the integration of unbounded functions in Section 7.7.

Functions with No Riemann Integral

While many discontinuous functions are integrable, many others are not. The function

$$f(x) = \begin{cases} 1 & \text{when } x \text{ is rational,} \\ 0 & \text{when } x \text{ is irrational,} \end{cases}$$

for example, has no Riemann integral over $[0, 1]$. For any partition P of $[0, 1]$, the upper and lower sums are

$$U = \sum \max_k \Delta x_k = \sum 1 \cdot \Delta x_k = \sum \Delta x_k = 1, \qquad \left(\begin{array}{l} \text{Every subinterval} \\ \text{contains a rational} \\ \text{number.} \end{array} \right)$$

$$L = \sum \min_k \Delta x_k = \sum 0 \cdot \Delta x_k = 0. \qquad \left(\begin{array}{l} \text{Every subinterval} \\ \text{contains an irrational} \\ \text{number.} \end{array} \right)$$

For the integral of f to exist over $[0, 1]$, U and L would have to have the same limit as $\|P\| \to 0$. But they do not, for

$$\lim_{\|P\| \to 0} L = 0 \qquad \text{while} \qquad \lim_{\|P\| \to 0} U = 1.$$

Therefore, f has no integral on $[0, 1]$. No constant multiple kf has an integral either, unless k is zero.

Terminology

There is a fair amount of terminology to learn in connection with definite integrals.

The definite integral of a function $f(x)$ over an interval $[a, b]$ is usually denoted by the symbol

$$\int_a^b f(x) \, dx,$$

which is read, "the integral of f of x dx from a to b." As in indefinite integration, the symbol $\int$ is an **integral sign.** Leibniz chose it because it resembled the S in the German word for *summation*. We shall see the connection between definite and indefinite integration in Section 4.5.

When we find the value of $\int_a^b f(x) \, dx$, we say that we have **evaluated the integral** and that we have **integrated** f from a to b. We call $[a, b]$ the **interval of integration.** The numbers a and b are the **limits of integration,** a being the **lower limit of integration** and b the **upper limit of integration.** The function f is the **integrand** of the integral. The variable x is the **variable of integration.**

The value of the definite integral of a function over any particular interval depends on the function and not on the letter we choose to represent its independent variable. If we decide to use t or u instead of x, we simply write the integral as

$$\int_a^b f(t) \, dt \qquad \text{or} \qquad \int_a^b f(u) \, du \qquad \text{instead of} \qquad \int_a^b f(x) \, dx.$$

No matter how we write the integral, it is still the same number, defined as a limit of Riemann sums. Since it does not matter what letter we use, the variable of integration is called a **dummy variable.**

Example 6   Express the limit of Riemann sums

$$\lim_{\|P\| \to 0} \sum_{k=1}^n (3c_k^2 - 2c_k + 5)\Delta x_k$$

as an integral if P denotes a partition of the interval $[-1, 3]$.

Solution   The function being evaluated at c_k in each term of the sum is $f(x) = 3x^2 - 2x + 5$. The interval being partitioned is $[-1, 3]$. The limit is therefore the integral of f from -1 to 3:

$$\lim_{\|P\| \to 0} \sum_{k=1}^n (3c_k^2 - 2c_k + 5)\Delta x_k = \int_{-1}^3 (3x^2 - 2x + 5) \, dx.$$

Area

We can now see that if a function f is nonnegative and integrable over an interval $[a, b]$, then the area under the graph of f from a to b is the definite integral of f from a to b. This integral definition of area gives the same results as the classical formu-

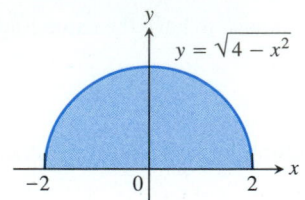

4.15   The semicircular region in Example 7.

las for the areas of triangles, trapezoids, semicircles, and other shapes to which the new and old calculation methods apply (a fact we shall not prove). We can sometimes use this observation to evaluate integrals.

Example 7   Evaluate

$$\int_{-2}^{2} \sqrt{4 - x^2} \, dx$$

by regarding the integral as the area under the graph of $f(x) = \sqrt{4 - x^2}$ from -2 to 2.

Solution   The graph of f from -2 to 2 is the curve $y = \sqrt{4 - x^2}$, a semicircle of radius 2 (Fig. 4.15). The area between the semicircle and the x-axis is

$$\text{Area} = \frac{1}{2} \pi r^2 = \frac{1}{2} \pi (2)^2 = 2\pi.$$

Since the area is the value of the integral,

$$\int_{-2}^{2} \sqrt{4 - x^2} \, dx = 2\pi.$$

If an integrable function is negative, the areas of the rectangles we use to approximate the region between the function's graph and the x-axis are the negatives of the terms $f(c_k)\Delta x_k$ in the associated Riemann sums. The area between the graph and the x-axis is therefore the negative of the function's integral from a to b (Fig. 4.16).

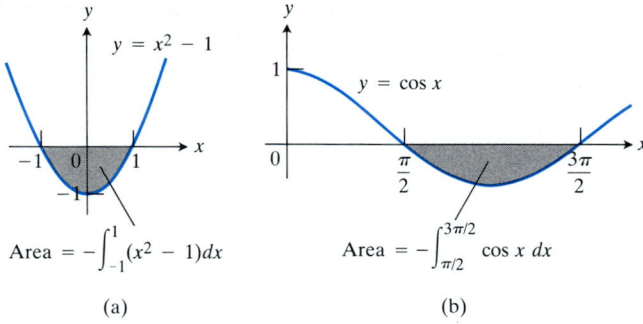

(a)   (b)

4.16   When a function's graph dips below the x-axis, the area between the axis and the portion of the graph below the axis is the negative of the function's integral.

If an integrable function $y = f(x)$ has both positive and negative values on an interval $[a, b]$, then the Riemann sums for f on $[a, b]$ add the areas of the rectangles that lie above the x-axis to the negatives of the areas of the rectangles that lie below the x-axis, as in Fig. 4.17. The resulting cancellation reduces the sums, so their limiting value is a number whose magnitude is less than the total area between the curve and the x-axis. The value of the integral is the area above the axis minus the area below the axis.

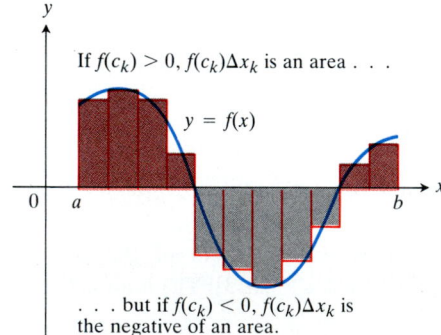

4.17   If f has negative as well as positive values, the Riemann sums for f add the areas of the rectangles that lie above the axis to the negatives of the areas of the rectangles that lie below the axis. The resulting cancellation reduces the value of the sums' limit to something less than the total area.

For any integrable function,

$$\int_{a}^{b} f(x) \, dx = (\text{area above } x\text{-axis}) - (\text{area below } x\text{-axis}). \qquad (12)$$

Example 8

$$\int_0^{2\pi} \sin x \, dx = 0$$

The integral is the area of region A above the x-axis minus the area of region B below the x-axis (Fig. 4.18). The areas are the same, so the integral is zero.

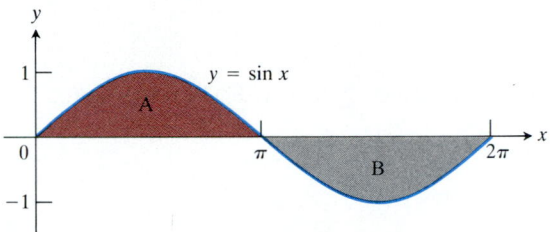

4.18   The areas of the shaded regions are equal, so the integral of the sine from 0 to 2π is zero.

Conclusion

In defining areas as definite integrals, we have vastly extended our ability to calculate the areas of regions in the plane. But area is only one of thousands of important quantities we define and calculate by definite integration. As our story unfolds, we shall learn how to use integrals to calculate lengths, surface areas, volumes, masses, hydrostatic forces against dams, electrical voltages and currents, and much more. The list of things we calculate with integrals is nearly endless.

Before we can do any of this, however, we need to learn how to evaluate integrals effectively. We shall learn this in the next four sections.

EXERCISES 4.3

Write the sums in Exercises 1–4 without sigma notation. Then evaluate them.

1. $\displaystyle\sum_{k=1}^{2} \frac{6k}{k+1}$

2. $\displaystyle\sum_{k=1}^{3} \frac{k-1}{k}$

3. $\displaystyle\sum_{k=1}^{4} (-1)^k \cos k\pi$

4. $\displaystyle\sum_{k=1}^{3} (-1)^{k+1} \sin \frac{\pi}{k}$

Express the sums in Exercises 5–10 in sigma notation.

5. $1 + 2 + 3 + 4 + 5 + 6$

6. $1 + 4 + 9 + 16$

7. $\dfrac{1}{2} + \dfrac{1}{4} + \dfrac{1}{8} + \dfrac{1}{16}$

8. $2 + 4 + 6 + 8 + 10$

9. $1 - \dfrac{1}{2} + \dfrac{1}{3} - \dfrac{1}{4} + \dfrac{1}{5}$

10. $-\dfrac{1}{5} + \dfrac{2}{5} - \dfrac{3}{5} + \dfrac{4}{5} - \dfrac{5}{5}$

The lower limit of summation does not have to be 1. It can be some other integer.

11. Which of the following express $1 + 2 + 4 + 8 + 16 + 32$ in sigma notation?

a) $\displaystyle\sum_{k=1}^{6} 2^{k-1}$

b) $\displaystyle\sum_{k=0}^{5} 2^k$

c) $\displaystyle\sum_{k=-1}^{4} 2^{k+1}$

12. Which formula is not equivalent to the others?

a) $\displaystyle\sum_{k=-1}^{1} \frac{(-1)^k}{k+2}$

b) $\displaystyle\sum_{k=0}^{2} \frac{(-1)^k}{k+1}$

c) $\displaystyle\sum_{k=1}^{3} \frac{(-1)^k}{k}$

d) $\displaystyle\sum_{k=2}^{4} \frac{(-1)^{k-1}}{k-1}$

13. Suppose that $\sum_{k=1}^{n} a_k = -5$ and $\sum_{k=1}^{n} b_k = 6$. Find the values of

a) $\sum_{k=1}^{n} 3a_k$

b) $\sum_{k=1}^{n} \frac{b_k}{6}$

c) $\sum_{k=1}^{n} (a_k + b_k)$

d) $\sum_{k=1}^{n} (a_k - b_k)$

e) $\sum_{k=1}^{n} (b_k - 2a_k)$

14. Suppose that $\sum_{k=1}^{n} a_k = 0$ and $\sum_{k=1}^{n} b_k = 1$. Find the values of

a) $\sum_{k=1}^{n} 8a_k$

b) $\sum_{k=1}^{n} 250b_k$

c) $\sum_{k=1}^{n} (a_k + 1)$

d) $\sum_{k=1}^{n} (b_k - 1)$

In Exercises 15 and 16, graph each function over the given interval. Partition the interval into four subintervals of equal length. Then (a) sketch the rectangles associated with $\sum_{k=1}^{4} \min_k \Delta x$ and add their areas; (b) sketch the rectangles associated with $\sum_{k=1}^{4} \max_k \Delta x$ and add their areas.

15. $f(x) = x^2$ on $[-1, 1]$

16. $f(x) = 1 + \sin x$ on $[-\pi/2, \pi/2]$

Express the limits in Exercises 17–26 as definite integrals.

17. $\lim_{\|P\| \to 0} \sum_{k=1}^{n} c_k^2 \Delta x_k$, where P is a partition of $[0, 2]$

18. $\lim_{\|P\| \to 0} \sum_{k=1}^{n} 2c_k^3 \Delta x_k$, where P is a partition of $[-1, 0]$

19. $\lim_{\|P\| \to 0} \sum_{k=1}^{n} (c_k^2 - 3c_k) \Delta x_k$, where P is a partition of $[-7, 5]$

20. $\lim_{\|P\| \to 0} \sum_{k=1}^{n} \frac{1}{c_k} \Delta x_k$, where P is a partition of $[1, 4]$

21. $\lim_{\|P\| \to 0} \sum_{k=1}^{n} \frac{1}{1 - c_k} \Delta x_k$, where P is a partition of $[2, 3]$

22. $\lim_{\|P\| \to 0} \sum_{k=1}^{n} \sqrt{4 - c_k^2} \Delta x_k$, where P is a partition of $[0, 1]$

23. $\lim_{\|P\| \to 0} \sum_{k=1}^{n} \cos c_k \Delta x_k$, where P is a partition of $[0, 4]$

24. $\lim_{\|P\| \to 0} \sum_{k=1}^{n} \tan c_k \Delta x_k$, where P is a partition of $[0, \pi/4]$

25. $\lim_{\|P\| \to 0} \sum_{k=1}^{n} \sec c_k \Delta x_k$, where P is a partition of $[-\pi/4, 0]$

26. $\lim_{\|P\| \to 0} \sum_{k=1}^{n} \sin^3 c_k \Delta x_k$, where P is a partition of $[-\pi, \pi]$

Express the areas of the shaded regions in Exercises 27–30 as integrals (but do not evaluate the integrals).

27.

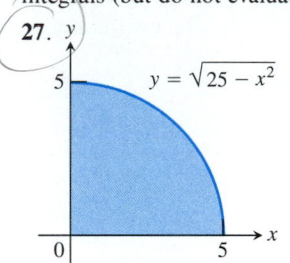

$y = \sqrt{25 - x^2}$

28.

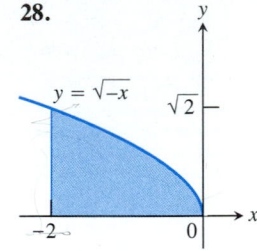

$y = \sqrt{-x}$

29.

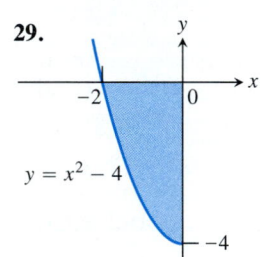

$y = x^2 - 4$

30.

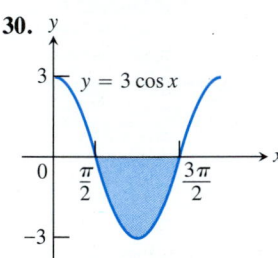

$y = 3 \cos x$

In Exercises 31 and 32, express the area of the shaded region between the curve and the x-axis as a difference of two integrals.

31.

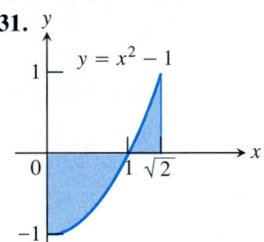

$y = x^2 - 1$

32.

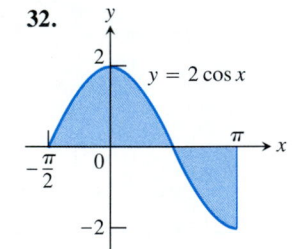

$y = 2 \cos x$

In Exercises 33–36, find the value of each integral by regarding it as the area under the graph of an appropriately chosen function and using an area formula from plane geometry.

33. $\int_{-1}^{1} \sqrt{1 - x^2} \, dx$

34. $\int_{0}^{2} \sqrt{4 - x^2} \, dx$

35. $\int_{-1}^{1} (1 - |x|) \, dx$

36. $\int_{-1}^{1} (1 + \sqrt{1 - x^2}) \, dx$

37. What values of a and b maximize the value of

$$\int_{a}^{b} (x - x^2) \, dx ?$$

(*Hint:* Where is the integrand positive?)

38. What values of a and b minimize the value of

$$\int_a^b (x^4 - 2x^2)\, dx\,?$$

39. *Upper and lower sums for increasing functions*
   a) Suppose the graph of a continuous function $f(x)$ rises steadily as x moves from left to right across an interval $[a, b]$. Let P be a partition of $[a, b]$ into n subintervals of length $\Delta x = (b - a)/n$. Show by referring to Fig. 4.19 that the difference between the upper and lower sums for f on this partition can be represented graphically as the area of a rectangle R whose dimensions are $[f(b) - f(a)]$ by Δx. (*Hint:* The difference $U - L$ is the sum of areas of rectangles whose diagonals $Q_0 Q_1, Q_1 Q_2, \ldots, Q_{n-1} Q_n$ lie along the curve. There is no overlapping when these rectangles are shifted horizontally onto R.)
   b) Suppose that instead of being equal, the lengths Δx_k of the subintervals of the partition of $[a, b]$ vary in size. Show that

$$U - L \le |f(b) - f(a)|\Delta x_{max},$$

   where Δx_{max} is the norm P, and hence that $\lim_{\|P\| \to 0} (U - L) = 0$.

40. *Upper and lower sums for decreasing functions (continuation of Exercise 39)*
   a) Draw a figure like Fig. 4.19 for a continuous function $f(x)$ whose values decrease steadily as x moves from left to right across the interval $[a, b]$. Let P be a partition of $[a, b]$ into subintervals of equal length. Find an expression for $U - L$ that is analogous to the one you found for $U - L$ in Exercise 39(a).
   b) Suppose that instead of being equal, the lengths Δx_k of the subintervals of P vary in size. Show that the inequality

$$U - L \le |f(b) - f(a)|\Delta x_{max}$$

   of Exercise 39(b) still holds and hence that $\lim_{\|P\| \to 0} (U - L) = 0$.

COMPUTER GRAPHER If you have access to a Riemann sum grapher, graph the upper and lower Riemann sums for the

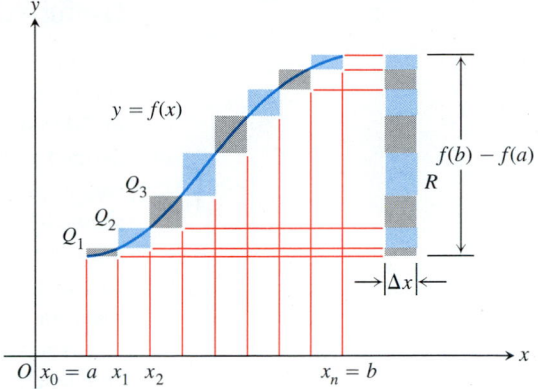

4.19 If f is an increasing continuous function, the blocks representing $U - L$ fill the rectangle on the right without overlapping (Exercises 39 and 40).

integrals in Exercises 41–46. In each case try it with $n = 4$, 10, 20, and 50 subintervals of equal length.

41. $\displaystyle\int_0^1 (1 - x)\, dx = \frac{1}{2}$ **42.** $\displaystyle\int_0^1 (x^2 + 1)\, dx = \frac{4}{3}$

43. $\displaystyle\int_{-\pi}^{\pi} \cos x\, dx = 0$ **44.** $\displaystyle\int_0^{\pi/4} \sec^2 x\, dx = 1$

45. $\displaystyle\int_{-1}^1 |x|\, dx = 1$

46. $\displaystyle\int_1^2 \frac{1}{x}\, dx = \ln 2 \approx 0.6931471806$

EXPLORER PROGRAM

Riemann Sums   Draws rectangles for upper-endpoint and lower-endpoint Riemann sums. Also gives the trapezoidal sum for comparison (see Section 4.7)

4.4 Evaluating Definite Integrals (First Steps)

In this section, we describe eight operating rules that facilitate our work with definite integrals. We also introduce the Mean Value Theorem for Definite Integrals and the associated notion of the average value of a function over a closed interval. We also show how to evaluate integrals of constant functions and examine two rare instances in which we can evaluate definite integrals of nonconstant functions directly from Riemann sums.

Useful Rules for Working with Integrals

We often want to add and subtract definite integrals, multiply their integrands by constants, and compare them with other definite integrals. We do this with the rules in Table 4.2. All the rules except the first two follow from the way integrals are defined as Riemann sums. You might think that this would make them relatively easy to prove. After all, we might argue, sums have these properties so their limits should have them, too. But when we get down to the details we find that most of the proofs require complicated ϵ-δ arguments with norms of subdivisions and are not easy at all. We shall omit all but two of the proofs. The remaining proofs can be found in more advanced texts.

Notice that Rule 1 is a definition. We want every integral over an interval of zero length to be zero. Rule 1 extends the definition of definite integral to allow for the case $a = b$. Rule 2, also a definition, extends the definition of definite integral to allow for the case $b < a$. Rules 3 and 4 are like the analogous rules for limits and indefinite integrals, and we use them the same way. Once we know the integrals of two functions, we automatically know the integrals of their linear combinations. We may also use Rules 3 and 4 repeatedly to evaluate integrals of arbitrary finite linear

TABLE 4.2
Rules of algebra for definite integrals

1. *Zero*: $\displaystyle\int_a^a f(x)\,dx = 0$ (A definition)

2. *Order of integration*: $\displaystyle\int_b^a f(x)\,dx = -\int_a^b f(x)\,dx$ (Also a definition)

3. *Constant multiples*: $\displaystyle\int_a^b kf(x)\,dx = k\int_a^b f(x)\,dx$ (Any number k)

   $\displaystyle\int_a^b -f(x)\,dx = -\int_a^b f(x)\,dx$ ($k = -1$)

4. *Sums and differences*: $\displaystyle\int_a^b (f(x) \pm g(x))\,dx = \int_a^b f(x)\,dx \pm \int_a^b g(x)\,dx$

5. *Domination*: $f(x) \geq g(x)$ on $[a, b]$ $\Rightarrow$ $\displaystyle\int_a^b f(x)\,dx \geq \int_a^b g(x)\,dx$

6. *Nonnegative functions*: $f(x) \geq 0$ on $[a, b]$ $\Rightarrow$ $\displaystyle\int_a^b f(x)\,dx \geq 0$

   (Special case of Rule 5)

7. *Max-min*:   If max f and min f are the maximum and minimum values of f on $[a, b]$, then

   $$\min f \cdot (b - a) \leq \int_a^b f(x)\,dx \leq \max f \cdot (b - a).$$

8. *Additivity*:   If f is integrable on the intervals joining a, b, and c, then

   $$\int_a^b f(x)\,dx + \int_b^c f(x)\,dx = \int_a^c f(x)\,dx.$$

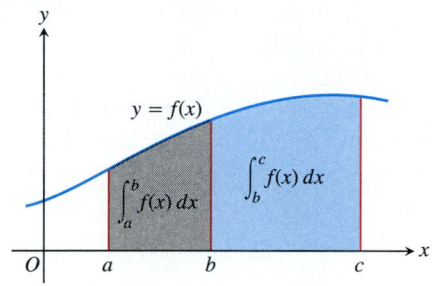

4.20   Additivity for definite integrals:

$$\int_a^b f(x)\,dx + \int_b^c f(x)\,dx = \int_a^c f(x)\,dx$$

$$\int_b^c f(x)\,dx = \int_a^c f(x)\,dx - \int_a^b f(x)\,dx.$$

combinations of integrable functions term by term. For any constants $c_1, \ldots, c_n$, regardless of sign, and functions $f_1(x), \ldots, f_n(x)$, integrable on $[a, b]$,

$$\int_a^b (c_1f_1(x) + \cdots + c_nf_n(x))\,dx = c_1\int_a^b f_1(x)\,dx + \cdots + c_n\int_a^b f_n(x)\,dx. \quad (1)$$

The proof, omitted, comes from mathematical induction.

Figure 4.20 illustrates Rule 8 with a positive function, but the rule applies to any integrable function.

Proof of Rule 3   Rule 3 says that the integral of k times a function is k times the integral of the function. This is true because

$$\int_a^b kf(x)\,dx = \lim_{\|P\|\to 0} \sum_{i=1}^n kf(c_i)\Delta x_i$$

$$= \lim_{\|P\|\to 0} k\sum_{i=1}^n f(c_i)\Delta x_i$$

$$= k\lim_{\|P\|\to 0} \sum_{i=1}^n f(c_i)\Delta x_i = k\int_a^b f(x)\,dx.$$

Proof of Rule 7   Rule 7 says that the integral of f over $[a, b]$ is never smaller than the minimum value of f times the length of the interval and never larger than the maximum value of f times the length of the interval. The reason is that for every partition of $[a, b]$ and for every choice of the points c_k,

$$\min f \cdot (b - a) = \min f \cdot \sum_{k=1}^n \Delta x_k \quad \left(\sum_{k=1}^n \Delta x_k = b - a\right)$$

$$= \sum_{k=1}^n \min f \cdot \Delta x_k$$

$$\leq \sum_{k=1}^n f(c_k)\Delta x_k \qquad (\min f \leq f(c_k))$$

$$\leq \sum_{k=1}^n \max f \cdot \Delta x_k \qquad (f(c_k) \leq \max f)$$

$$= \max f \cdot \sum_{k=1}^n \Delta x_k$$

$$= \max f \cdot (b - a).$$

In short, all Riemann sums for f on $[a, b]$ satisfy the inequality

$$\min f \cdot (b - a) \leq \sum_{k=1}^n f(c_k)\Delta x_k \leq \max f \cdot (b - a). \quad (2)$$

Hence their limit, the integral, does too.

Example 1   Suppose that f, g, and h are integrable, that

$$\int_{-1}^1 f(x)\,dx = 5, \qquad \int_1^4 f(x)\,dx = -2, \qquad \int_{-1}^1 h(x)\,dx = 7,$$

and that $g(x) \geq f(x)$ on $[-1, 1]$. Then

1. $\displaystyle\int_4^1 f(x)\,dx = -\int_1^4 f(x)\,dx = -(-2) = 2;$ \qquad (Rule 2)

2. $\displaystyle\int_{-1}^{1} [2f(x) + 3h(x)]\, dx = 2\int_{-1}^{1} f(x)\, dx + 3\int_{-1}^{1} h(x)\, dx$

$= 2(5) + 3(7) = 31;$ (Rules 3 and 4)

3. $\displaystyle\int_{-1}^{1} [f(x) - h(x)]\, dx = 5 - 7 = -2;$ (Rule 4)

4. $\displaystyle\int_{-1}^{1} g(x)\, dx \geq 5;$ $\begin{pmatrix} \text{Because} \\ g(x) \geq f(x) \\ \text{on } [-1, 1], \\ \text{Rule 5} \end{pmatrix}$

5. $\displaystyle\int_{-1}^{4} f(x)\, dx = \int_{-1}^{1} f(x)\, dx + \int_{1}^{4} f(x)\, dx = 5 + (-2) = 3.$ (Rule 8)

The Average Value of a Function

If we divide the inequality in the Max-Min Rule for definite integrals by $(b - a)$, we get

$$\min f \leq \frac{1}{b - a}\int_{a}^{b} f(x)\, dx \leq \max f. \tag{3}$$

If f is continuous, the Intermediate Value Theorem in Section 1.10 says that f must assume every value between $\min f$ and $\max f$. In particular, f must assume the value

$$\frac{1}{b - a}\int_{a}^{b} f(x)\, dx.$$

THEOREM 2

The Mean Value Theorem for Definite Integrals

If f is continuous on the closed interval $[a, b]$, then at some point c in the interval $[a, b]$,

$$f(c) = \frac{1}{b - a}\int_{a}^{b} f(x)\, dx. \tag{4}$$

The number on the right-hand side of Eq. (4) is called the average value or mean value of f on $[a, b]$.

DEFINITION

The **average value (mean value)** of an integrable function f on $[a, b]$ is

$$\frac{1}{b - a}\int_{a}^{b} f(x)\, dx. \tag{5}$$

Notice that the average value of f on $[a, b]$ is the integral of f divided by the length of the interval. (Also see Fig. 4.21.)

Example 2   Find the average value of $f(x) = \sqrt{4 - x^2}\, dx$ on the interval $[-2, 2]$.

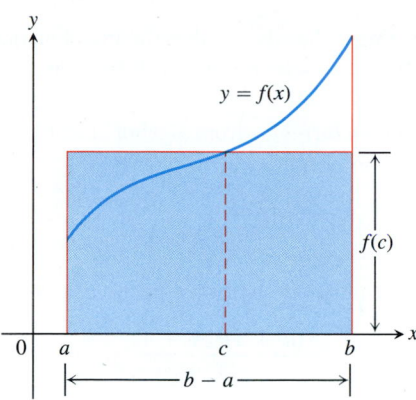

4.21 If f is continuous and positive on $[a, b]$, its average value $f(c)$ is the height of a rectangle whose area $f(c)(b - a)$ equals $\int_a^b f(x)\, dx$, the area under the graph of f.

Solution

$$\text{Avg. value of } f \text{ on } [-2, 2] = \frac{1}{2 - (-2)} \int_{-2}^{2} \sqrt{4 - x^2}\, dx \qquad \text{(Eq. 5)}$$

$$= \frac{1}{4}(2\pi) = \frac{\pi}{2} \qquad \begin{pmatrix} \text{Value from} \\ \text{Example 7,} \\ \text{Section 4.3} \end{pmatrix}$$

The average value of $f(x) = \sqrt{4 - x^2}$ on $[-2, 2]$ is $\pi/2$.

As you will see in the next section, we use average values to calculate the effective voltage and current in an alternating electric circuit.

Constant Functions

What integrals can we evaluate directly from the definition besides integrals that represent known areas? One answer is integrals of constant functions. The integral of a constant function over an interval $[a, b]$ is the function's value times the length of the interval.

THEOREM 3

If f has the constant value c on $[a, b]$, then

$$\int_a^b f(x)\, dx = \int_a^b c\, dx = c(b - a). \qquad (6)$$

Proof The Riemann sums for f on $[a, b]$ all have the constant value

$$\sum_{k=1}^{n} f(c_k)\Delta x_k = \sum_{k=1}^{n} c \cdot \Delta x_k = c \cdot \sum_{k=1}^{n} \Delta x_k = c(b - a).$$

The limit of these sums, the integral they converge to, therefore has this value too.

Example 3

$$\int_{-3}^{1} 5\, dx = 5(1 - (-3)) = 5(4) = 20$$

Formulas for Riemann Sums

We can occasionally evaluate the definite integral of a nonconstant function by finding an explicit formula for some of the Riemann sums that converge to the integral. This is how the mathematicians of the Renaissance found the areas under curves, and it seems to be a natural way to proceed. After all, integrals are defined as limits of Riemann sums. However, each calculation requires its own bit of ingenuity and the formulas developed to evaluate one integral do not readily apply to any other. This lack of generality is one of the features that distinguishes the calculus of the Renaissance from the methods of Leibniz and Newton that are the subject of the next section. Why bother with these calculations if we are about to learn in Section 4.5 a better way to evaluate integrals? One reason is that the forthcoming examples portray the state of the art in Renaissance times. Another is that the mathematics within the examples is still current and being acquainted with it will help later, in Chapter 8.

The two examples we shall study depend on formulas from algebra. The first is the formula for the sum of the first n positive integers:

$$\sum_{k=1}^{n} k = 1 + 2 + 3 + \cdots + n = \frac{n(n+1)}{2}. \tag{7}$$

The second is the formula for the sum of the squares of the first n positive integers:

$$\sum_{k=1}^{n} k^2 = 1^2 + 2^2 + 3^2 + \cdots + n^2 = \frac{n(n+1)(2n+1)}{2 \; 6}. \tag{8}$$

Both formulas can be derived by mathematical induction.

Example 4   Let a and b be positive numbers with $a < b$. Evaluate $\displaystyle\int_a^b x \, dx$.

Solution   We evaluate the integral as a limit of sums.

We graph the function $f(x) = x$ (Fig. 4.22) and partition the interval $[a, b]$ into n subintervals of length $\Delta x = (b - a)/n$ with the points

$$x_0 = a,$$
$$x_1 = a + \Delta x,$$
$$x_2 = a + 2\Delta x,$$
$$\vdots$$
$$x_{n-1} = a + (n-1)\Delta x,$$
$$x_n = b.$$

The inscribed rectangles defined by this partition have areas

$$f(a)\Delta x = a \cdot \Delta x,$$
$$f(x_1)\Delta x = (a + \Delta x) \cdot \Delta x,$$
$$f(x_2)\Delta x = (a + 2\Delta x) \cdot \Delta x,$$
$$\vdots \qquad \vdots$$
$$f(x_{n-1})\Delta x = (a + (n-1)\Delta x) \cdot \Delta x.$$

The sum of these areas is

$$S_n = (a + (a + \Delta x) + (a + 2\Delta x) + \cdots + (a + (n-1)\Delta x)) \cdot \Delta x$$
$$= (na + (1 + 2 + \cdots + (n-1))\Delta x)\Delta x$$
$$= \left(na + \frac{(n-1)n}{2}\Delta x\right)\Delta x \qquad \text{(From Eq. (7), with n replaced by $n-1$)}$$
$$= \left(a + \frac{(n-1)}{2}\Delta x\right)n\,\Delta x$$
$$= \left(a + \frac{b-a}{2} \cdot \frac{n-1}{n}\right)(b-a). \qquad \left(\Delta x = \frac{b-a}{n}\right)$$

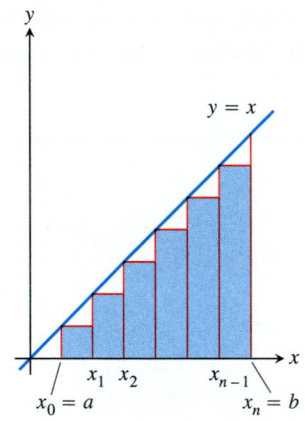

4.22   Rectangles for approximating the value of $\int_a^b x \, dx$ (Example 4).

The value of the integral is the limit of S_n as $n \to \infty$. The only place n appears in the final form is in the fraction

$$\frac{n-1}{n} = 1 - \frac{1}{n},$$

whose limit as $n \to \infty$ is $(1 - 0) = 1$. Therefore,

$$\int_a^b x \, dx = \lim_{n \to \infty} S_n = \left(a + \frac{b-a}{2} \cdot 1 \right)(b-a)$$

$$= \left(\frac{b+a}{2} \right)(b-a) = \frac{b^2}{2} - \frac{a^2}{2}.$$

(9)

For instance,

$$\int_2^3 4x \, dx = 4 \int_2^3 x \, dx = 4 \left(\frac{(3)^2}{2} - \frac{(2)^2}{2} \right) = 4 \left(\frac{9}{2} - 2 \right) = 4 \left(\frac{5}{2} \right) = 10.$$

Example 5   Evaluate $\displaystyle\int_0^b x^2 dx, \quad b > 0.$

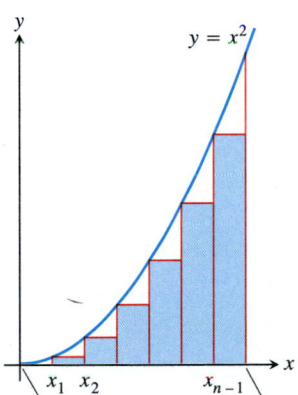

4.23   Rectangles for approximating the value of $\int_0^b x^2 \, dx$ (Example 5).

Solution   We evaluate the integral as a limit of sums.

We graph the function $f(x) = x^2$ (Fig. 4.23) and partition the interval $[0, b]$ into n subintervals of length $\Delta x = b/n$ with the points

$$x_0 = 0, \qquad x_1 = \Delta x, \qquad x_2 = 2\Delta x, \qquad \ldots, \qquad x_{n-1} = (n-1)\Delta x, \qquad x_n = b.$$

The inscribed rectangles defined by this partition have areas

$$f(x_0)\Delta x = (0)^2 \, \Delta x,$$
$$f(x_1)\Delta x = (\Delta x)^2 \, \Delta x,$$
$$f(x_2)\Delta x = (2\Delta x)^2 \, \Delta x,$$
$$\vdots$$
$$f(x_{n-1})\Delta x = ((n-1)\Delta x)^2 \, \Delta x.$$

The sum of these areas is

$$S_n = (1^2 + 2^2 + \cdots + (n-1)^2)(\Delta x)^3$$

$$= \frac{(n-1)(n)(2n-1)}{6} \left(\frac{b}{n} \right)^3 \qquad \text{(From Eq. (8) with } n \text{ replaced by } n - 1)$$

$$= \frac{b^3}{6} \cdot \frac{n-1}{n} \cdot \frac{n}{n} \cdot \frac{2n-1}{n}$$

$$= \frac{b^3}{6} \left(1 - \frac{1}{n} \right)\left(2 - \frac{1}{n} \right).$$

We now let n increase without bound and get

$$\int_0^b x^2 \, dx = \lim_{n \to \infty} S_n = \frac{b^3}{6}(1)(2) = \frac{b^3}{3}.$$

(10)

For instance,

$$\int_0^1 x^2 dx = \frac{1}{3}, \qquad \int_0^2 x^2 dx = \frac{8}{3}, \qquad \text{and} \qquad \int_0^3 x^2 dx = \frac{27}{3} = 9.$$

With the rules in Table 4.2 and the results in Examples 3–5, we can now evaluate many other integrals.

Example 6

a) $\displaystyle\int_0^2 (3x^2 - 2x + 5)\, dx = 3\int_0^2 x^2\, dx - 2\int_0^2 x\, dx + 5\int_0^2 dx$

$$= 3 \cdot \frac{8}{3} - 2\left(\frac{2^2}{2} - \frac{0^2}{2}\right) + 5(2 - 0)$$

$$= 8 - 4 + 10 = 14.$$

b) $\displaystyle\int_1^2 x^2\, dx = \int_0^2 x^2\, dx - \int_0^1 x^2\, dx = \frac{8}{3} - \frac{1}{3} = \frac{7}{3}$ (Values from Example 5)

c) $\displaystyle\int_3^1 (x + 2)\, dx = -\int_1^3 (x + 2)\, dx$

$$= -\int_1^3 x\, dx - \int_1^3 2\, dx$$

$$= -\left(\frac{(3)^2}{2} - \frac{(1)^2}{2}\right) - 2(3 - 1)$$

$$= -4 - 4 = -8$$

EXERCISES 4.4

1. Suppose that f and g are continuous and that

$$\int_1^2 f(x)\, dx = -4, \qquad \int_1^5 f(x)\, dx = 6, \qquad \int_1^5 g(x)\, dx = 8.$$

Use the rules in Table 4.2 to find

a) $\displaystyle\int_2^2 g(x)\, dx$

b) $\displaystyle\int_5^1 g(x)\, dx$

c) $\displaystyle\int_1^2 3 f(x)\, dx$

d) $\displaystyle\int_2^5 f(x)\, dx$

e) $\displaystyle\int_1^5 [f(x) - g(x)]\, dx$

f) $\displaystyle\int_1^5 [4f(x) - g(x)]\, dx$

2. Suppose that f and h are continuous and that

$$\int_1^9 f(x)\, dx = -1, \qquad \int_7^9 f(x)\, dx = 5, \qquad \int_7^9 h(x)\, dx = 4.$$

Use the rules in Table 4.2 to find

a) $\displaystyle\int_1^9 -2f(x)\, dx$

b) $\displaystyle\int_7^9 [f(x) + h(x)]\, dx$

c) $\displaystyle\int_7^9 [2f(x) - 3h(x)]\, dx$

d) $\displaystyle\int_9^1 f(x)\, dx$

e) $\displaystyle\int_1^7 f(x)\, dx$

f) $\displaystyle\int_9^7 [h(x) - f(x)]\, dx$

3. Suppose $\displaystyle\int_1^2 f(x)\, dx = 5$. Find

a) $\displaystyle\int_1^2 f(u)\, du$,

b) $\displaystyle\int_1^2 f(z)\, dz$,

c) $\displaystyle\int_2^1 f(t)\, dt$.

4. Suppose that f is continuous and that

$$\int_0^3 f(x)\, dx = 3, \qquad \int_0^4 f(z)\, dz = 7.$$

Find $\int_{3}^{4} f(y)\,dy$.

5. Use Rule 7 in Table 4.2 to find upper and lower bounds for the value of

$$\int_{0}^{1} \frac{1}{1+x^2}\,dx.$$

6. *Continuation of Exercise 5.*   Use Rule 7 in Table 4.2 to find upper and lower bounds for the values of

$$\int_{0}^{1/2} \frac{1}{1+x^2}\,dx \quad \text{and} \quad \int_{1/2}^{1} \frac{1}{1+x^2}\,dx.$$

Then add these to arrive at an improved estimate of

$$\int_{0}^{1} \frac{1}{1+x^2}\,dx.$$

Evaluate the integrals in Exercises 7–20.

7. $\int_{-2}^{2} 4.4\,dx$

8. $\int_{0}^{1/2} \pi\,dx$

9. $\int_{3}^{1} 7\,dx$

10. $\int_{0}^{-2} dx$

11. $\int_{0}^{2} 5x\,dx$

12. $\int_{3}^{5} \frac{x}{8}\,dx$

13. $\int_{0}^{2} (3 - 2x)\,dx$

14. $\int_{2}^{0} x\,dx$

15. $\int_{2}^{1} (1 - x)\,dx$

16. $\int_{\sqrt{2}/2}^{\sqrt{2}} (x - \sqrt{2})\,dx$

17. $\int_{1}^{2} x^2\,dx$

18. $\int_{0}^{3} (x^2 - 3)\,dx$

19. $\int_{0}^{1} (x - 1)^2\,dx$

20. $\int_{\sqrt{2}}^{0} (6x^2 + 3x)\,dx$

21. Find the area under the curve $y = x^2 + 1$ from $x = 0$ to $x = 3$.

22. Find the area under the curve $y = 6x^2 - 2$ from $x = 1$ to $x = 2$.

In Exercises 23–28, find the average value of f over the given interval.

23. $f(x) = 2|x|$, $[-1, 1]$

24. $f(x) = \cos x$, $[0, 2\pi]$

25. $f(x) = \sqrt{1 - x^2}$, $[-1, 1]$

26. $f(x) = x^2$, $[0, 2]$

27. $f(x) = 2x + 1$, $[0, 2\sqrt{2}]$

28. $f(x) = 3x^2 - 2x + 2$, $[0, 2]$

29. Suppose that f is continuous and that

$$\int_{1}^{2} f(x)\,dx = 4.$$

Show that $f(x) = 4$ at least once on the interval $[1, 2]$.

30. Show that the value of $\int_{0}^{1} \sin(x^2)\,dx$ cannot possibly be 2.

31. Suppose that f and g are continuous on $[a, b]$ and that $\int_{a}^{b} (f(x) - g(x))\,dx = 0$. Show that $f(x) = g(x)$ at least once in the interval.

32. Use the formula

$$\sum_{k=1}^{n} k^3 = 1^3 + 2^3 + \cdots + n^3 = \left(\frac{n(n+1)}{2}\right)^2$$

and the method of Example 5 to show that if $b > 0$, then

$$\int_{0}^{b} x^3\,dx = \frac{b^4}{4}.$$

Evaluate the integrals in Exercises 33–36 with the help of the result in Exercise 32.

33. $\int_{0}^{\sqrt{2}} 3x^3\,dx$

34. $\int_{\sqrt{5}}^{\sqrt{3}} x^3\,dx$

35. $\int_{0}^{1/2} \left(4x^3 - \frac{3}{2}x^2 + 5x - 7\right)\,dx$

36. $\int_{1}^{\sqrt{2}} (2x^3 + 6x^2 - 7x - 4)\,dx$

37. Evaluate the integral $\int_{a}^{b} x\,dx$ in Example 4 by carrying out the calculations of Example 4 with circumscribed rectangles (Fig. 4.24) instead of inscribed rectangles.

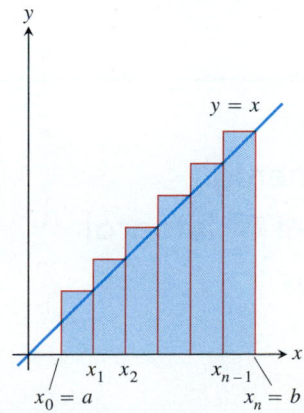

4.24   The circumscribed rectangles for Exercise 37.

38. Evaluate $\int_0^b x^2 \, dx$, $b > 0$, by carrying out the calculations of Example 5 with circumscribed rectangles (Fig. 4.25) instead of inscribed rectangles.

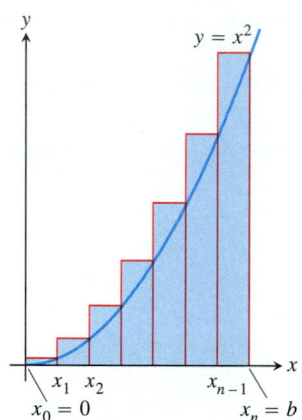

4.25   The circumscribed rectangles for Exercise 38.

39. Let

$$S_n = \frac{1}{n}\left[\frac{1}{n} + \frac{2}{n} + \frac{3}{n} + \cdots + \frac{n-1}{n}\right].$$

Calculate $\lim_{n\to\infty} S_n$ by showing that S_n is an approximating sum of the integral

$$\int_0^1 x \, dx,$$

whose value we know from Example 4. (*Hint:* Subdivide [0, 1] into n intervals of equal length and write out the approximating sum for inscribed rectangles.)

40. Let

$$S_n = \frac{1^2}{n^3} + \frac{2^2}{n^3} + \cdots + \frac{(n-1)^2}{n^3}.$$

To calculate $\lim_{n\to\infty} S_n$, show that

$$S_n = \frac{1}{n}\left[\left(\frac{1}{n}\right)^2 + \left(\frac{2}{n}\right)^2 + \cdots + \left(\frac{n-1}{n}\right)^2\right]$$

and interpret S_n as an approximating sum of the integral

$$\int_0^1 x^2 \, dx,$$

whose value we know from Example 5. (*Hint:* Subdivide [0, 1] into n intervals of equal length and write out the approximating sum for inscribed rectangles.)

41. Use the formula

$$\sin h + \sin 2h + \sin 3h + \cdots + \sin mh$$

$$= \frac{\cos(h/2) - \cos((m + (1/2))h)}{2\sin(h/2)}$$

to find the area under the curve $y = \sin x$ from $x = 0$ to $x = \pi/2$, in two steps:

a) Divide the interval $[0, \pi/2]$ into n equal subintervals and calculate the corresponding upper sum U; then

b) find the limit of U as $n \to \infty$ and

$$\Delta x = (b - a)/n \to 0.$$

EXPLORER PROGRAMS

Riemann Sums	Draws rectangles for upper-endpoint and lower-endpoint Riemann sums. Also gives the trapezoidal sum for comparison (Section 4.7)
Sequences and Series	Enables you to look for numerical and graphical indications of convergence of sums S_n given by formulas like those in Examples 4 and 5

4.5 The Fundamental Theorem of Integral Calculus

This section presents the fundamental theorem of integral calculus. The first part of the theorem says that the definite integral of a continuous function is a differentiable function of its upper limit of integration and tells us what the value of that derivative is. The second part of the theorem tells us that the definite integral of a continuous function from a to b can be found from any one of the function's antiderivatives F as the number $F(b) - F(a)$. The discovery by Newton and Leibniz of these astonishing connections between integration and differentiation started the mathematical development that fueled the scientific revolution for the next 200 years and constitutes what is still regarded as the most important computational discovery in the history of the Western world.

The Fundamental Theorem, Part 1

If $f(t)$ is an integrable function, its integral from any fixed number a to another number x defines a function F whose value at x is

$$F(x) = \int_a^x f(t)\, dt. \tag{1}$$

For example, if f is nonnegative and x lies to the right of a, the area under the graph of f from a to x is

$$A_a^x = \int_a^x f(t)\, dt. \tag{2}$$

The variable x in the function A_a^x is the upper limit of integration of an integral, but the function is just like any other real-valued function of a real variable. For each value of the input x there is a well defined numerical output A_a^x, in this case the value of the integral of f from a to x.

The formula

$$F(x) = \int_a^x f(t)\, dt$$

gives an important way to define new functions in science and engineering and provides an especially useful way to describe solutions of differential equations. (We shall say more about this later.) The reason for our mentioning the formula now is that it makes the connection between integrals and derivatives. For if f is any continuous function whatever, F is a differentiable function of x and, even more important, its derivative, dF/dx, is f itself. At every value of x,

$$\frac{d}{dx}\int_a^x f(t)\, dt = f(x). \tag{3}$$

This conclusion is beautiful, powerful, deep, and unexpected, and Eq. (3) may well be the single most important equation in calculus. It says that the differential equation $dF/dx = f$ has a solution for every continuous function f. It says that every continuous function f is the derivative of some other function, namely $\int_a^x f(t)\, dt$. It says that every continuous function has an antiderivative. (That was how we knew in Section 4.1 that $\tan x$, $\cot x$, $\sec x$, and $\csc x$ had antiderivatives.) And it says that the processes of integration and differentiation are inverses of one another. Equation (3) is so important that it is the first part of the Fundamental Theorem of Calculus.

THEOREM 4

The Fundamental Theorem of Calculus, Part 1

If f is continuous on $[a, b]$, then the function

$$F(x) = \int_a^x f(t)\, dt \tag{4}$$

has a derivative at every point of $[a, b]$ and

$$\frac{dF}{dx} = \frac{d}{dx}\int_a^x f(t)\, dt = f(x). \tag{5}$$

Proof   We prove Theorem 4 by applying the definition of derivative directly to the function $F(x)$. This means writing out Fermat's difference quotient,

$$\frac{F(x + h) - F(x)}{h},\tag{6}$$

and showing that its limit as $h \to 0$ is the number $f(x)$.

When we replace $F(x + h)$ and $F(x)$ by their defining integrals, the numerator in Eq. (6) becomes

$$F(x + h) - F(x) = \int_a^{x+h} f(t)\, dt - \int_a^x f(t)\, dt.\tag{7}$$

The Additivity Rule for integrals (Table 4.2 in Section 4.4) simplifies this to

$$\int_x^{x+h} f(t)\, dt,\tag{8}$$

so that Eq. (6) becomes

$$\frac{F(x + h) - F(x)}{h} = \frac{1}{h}[F(x + h) - F(x)]$$
$$= \frac{1}{h}\int_x^{x+h} f(t)\, dt.\tag{9}$$

According to the Mean Value Theorem for Definite Integrals (Theorem 2 in the preceding section), the value of the last expression in Eq. (9) is one of the values taken on by f in the interval joining x and $x + h$. That is, for some number c in this interval,

$$\frac{1}{h}\int_x^{x+h} f(t)\, dt = f(c).\tag{10}$$

We can therefore find out what happens to $(1/h)$ times the integral as $h \to 0$ by watching what happens to $f(c)$ as $h \to 0$.

What does happen to $f(c)$ as $h \to 0$? As $h \to 0$, the endpoint $x + h$ approaches x, pushing c ahead of it like a bead on a wire:

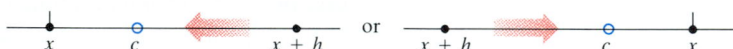

So c approaches x, and, since f is continuous at x, $f(c)$ approaches $f(x)$:

$$\lim_{h \to 0} f(c) = f(x).\tag{11}$$

Going back to the beginning, then, we have

$$\frac{dF}{dx} = \lim_{h \to 0} \frac{F(x + h) - F(x)}{h} \qquad \text{(definition of derivative)}$$

$$= \lim_{h \to 0} \frac{1}{h}\int_x^{x+h} f(t)\, dt \qquad \text{(Eq. 9)}$$

$$= \lim_{h \to 0} f(c) \qquad \text{(Eq. 10)}$$

$$= f(x). \qquad \text{(Eq. 11)}$$

This concludes the proof.

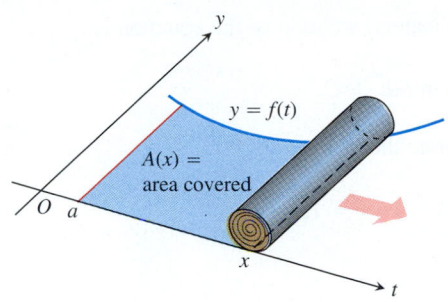

4.26   The rate at which the carpet covers the floor at the point x is the width of the carpet's leading edge as it rolls past x. In symbols, $dA/dx = f(x)$.

When the values of f are positive, the equation

$$\frac{d}{dx}\int_a^x f(t)\, dt = f(x)$$

has a nice geometric interpretation. For then the integral of f from a to x is the area $A(x)$ of the region between the graph of f and the x-axis from a to x. Imagine covering this region from left to right by unrolling a carpet of variable width $f(t)$ (Fig. 4.26). As the carpet rolls past x, the rate at which the floor is being covered is $f(x)$.

Example 1

$$\frac{d}{dx}\int_{-\pi}^x \cos t\, dt = \cos x \qquad \text{(Eq. (5) with } f(t) = \cos t)$$

$$\frac{d}{dx}\int_0^x \frac{1}{1 + t^2}\, dt = \frac{1}{1 + x^2} \qquad \left(\text{Eq. (5) with } f(t) = \frac{1}{1 + t^2}\right)$$

Example 2   Find dy/dx if

$$y = \int_1^{x^2} \cos t\, dt.$$

Solution   Notice that the upper limit of integration is not x but x^2. To find dy/dx we must therefore treat y as the composite of

$$y = \int_1^u \cos t\, dt \quad \text{and} \quad u = x^2$$

and apply the Chain Rule:

$$\frac{dy}{dx} = \frac{dy}{du}\frac{du}{dx} \qquad \text{(Chain Rule)}$$

$$= \frac{d}{du}\int_1^u \cos t\, dt \cdot \frac{du}{dx} \qquad \text{(Substitute the formula for } y.)$$

$$= \cos u \cdot \frac{du}{dx} \qquad \text{(Eq. (5) with } f(t) = \cos t)$$

$$= \cos x^2 \cdot 2x \qquad (u = x^2)$$

$$= 2x \cos x^2. \qquad \text{(Usual form)}$$

Example 3   Express the solution of the following initial value problem as an integral.

Differential equation: $\quad \dfrac{dy}{dx} = \tan x$

Initial condition: $\quad y = 5 \quad \text{when} \quad x = 1$

Solution   The function

$$F(x) = \int_1^x \tan t\, dt$$

is an antiderivative of $\tan x$. Hence the general solution of the equation is

$$y = \int_1^x \tan t \, dt + C.$$

As always, the initial conditions determine the right value for C:

$$5 = \int_1^1 \tan t \, dt + C \qquad (y = 5 \text{ when } x = 1)$$

$$5 = 0 + C \tag{12}$$

$$C = 5.$$

The solution of the initial value problem is

$$y = \int_1^x \tan t \, dt + 5.$$

How did we know where to start integrating when we constructed $F(x)$? We could have started anywhere, but the best value to start with is the initial value of x (in this case $x = 1$). Then the integral will be zero when we apply the initial condition (as it was in Eq. 12) and C will automatically be the initial value of y.

The Evaluation of Definite Integrals

We now come to the second part of the Fundamental Theorem of Calculus, the part that describes how to evaluate definite integrals.

THEOREM 5

The Fundamental Theorem of Calculus, Part 2

If f is continuous at every point of $[a, b]$ and F is any antiderivative of f on $[a, b]$, then

$$\int_a^b f(x) \, dx = F(b) - F(a). \tag{13}$$

To evaluate $\int_a^b f(x) \, dx$:

1. Find an antiderivative F of f. Any antiderivative will do, so pick the simplest one you can.
2. Calculate the number $F(b) - F(a)$.

This number will be $\int_a^b f(x) \, dx$.

Theorem 5 says that to evaluate the definite integral of a continuous function f from a to b, all we need do is find an antiderivative F of f and calculate the number $F(b) - F(a)$. The existence of the antiderivative is assured by the first part of the Fundamental Theorem.

Proof of Theorem 5 To prove Theorem 5, we use the fact that functions with identical derivatives differ only by a constant. We already know one function whose derivative equals f, namely,

$$G(x) = \int_a^x f(t) \, dt.$$

Therefore, if F is any other such function, then

$$F(x) = G(x) + C \tag{14}$$

throughout $[a, b]$ for some constant C. When we use Eq. (14) to calculate

Notation

The usual notation for the number $F(b) - F(a)$ is $F(x)]_a^b$, or $[F(x)]_a^b$, depending on whether F has one or more terms. As you will see, this notation provides a compact "recipe" for the evaluation.

Write $F(x)]_a^b$ for $F(b) - F(a)$ when $F(x)$ has a single term.

Write $[F(x)]_a^b$ for $F(b) - F(a)$ when $F(x)$ has more than one term.

$F(b) - F(a)$, we find that

$$F(b) - F(a) = [G(b) + C] - [G(a) + C] = G(b) - G(a)$$

$$= \int_a^b f(t)\, dt - \int_a^a f(t)\, dt$$

$$= \int_a^b f(t)\, dt - 0 = \int_a^b f(t)\, dt.$$

This establishes Eq. (13) and concludes the proof.

Example 4

a) $\displaystyle \int_0^\pi \cos x\, dx = \sin x \Big]_0^\pi = \sin \pi - \sin 0 = 0 - 0 = 0$

b) $\displaystyle \int_0^2 x^2\, dx = \frac{x^3}{3} \Big]_0^2 = \frac{2^3}{3} - \frac{0^3}{3} = \frac{8}{3}$

c) $\displaystyle \int_{-2}^2 (4 - x^2)\, dx = \left[4x - \frac{x^3}{3} \right]_{-2}^2$

$$= \left[4(2) - \frac{(2)^3}{3} \right] - \left[4(-2) - \frac{(-2)^3}{3} \right]$$

$$= \left[8 - \frac{8}{3} \right] - \left[-8 + \frac{8}{3} \right] = \frac{16}{3} - \left(-\frac{16}{3} \right) = \frac{32}{3}$$

Example 5

Find the area between the x-axis and the curves (a) $y = 4 - x^2$, (b) $y = x^2 - 4$ for $-2 \le x \le 2$.

Solution   We graph the curves over $[-2, 2]$ to see where the function values are positive and negative (Fig. 4.27).

a) Since $y = 4 - x^2 \ge 0$ on $[-2, 2]$, the area between the curve and the x-axis from -2 to 2 is

$$\text{Area} = \int_{-2}^2 (4 - x^2)\, dx = \left[4x - \frac{x^3}{3} \right]_{-2}^2 = \frac{32}{3}. \qquad \binom{\text{Value from}}{\text{Example 4}}$$

b) Since $y = x^2 - 4 \le 0$ on $[-2, 2]$, the area between the curve and the x-axis from -2 to 2 is the negative of the integral of $x^2 - 4$ from -2 to 2:

$$\text{Area} = -\int_{-2}^2 (x^2 - 4)\, dx$$

$$= \int_{-2}^2 -(x^2 - 4)\, dx$$

$$= \int_{-2}^2 (4 - x^2)\, dx = \frac{32}{3}. \qquad \text{(Value from Example 4)}$$

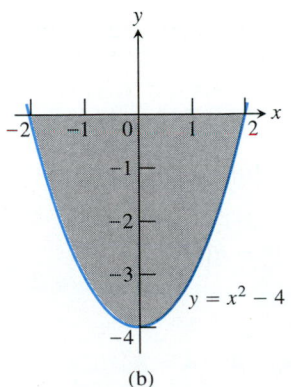

4.27   The graphs in (a) and (b) enclose the same amount of area with the x-axis, but the definite integrals of the functions from -2 to 2 differ in sign.

When the graph of $y = f(x)$ crosses the x-axis between $x = a$ and $x = b$, we find the area between the graph and the axis from a to b by taking the following steps.

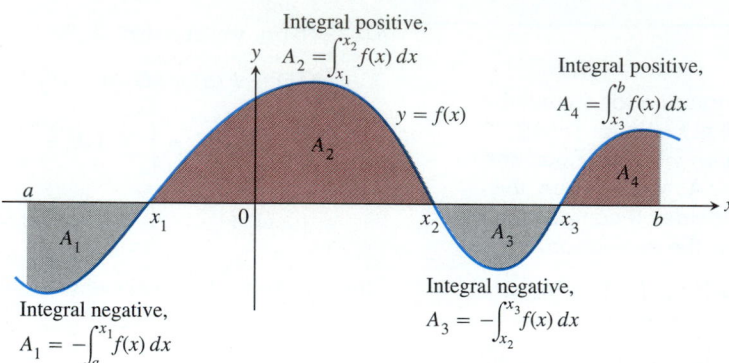

4.28 The area between the graph of f and the x-axis from a to b is found with four integrations. This is less work than it seems because we have to find the antiderivative of f only once.

Steps for Finding Area When f Has Both Positive and Negative Values on $[a, b]$ (Fig. 4.28)

STEP 1: Find the points where $f = 0$.

STEP 2: Use the zeros of f to partition $[a, b]$ into subintervals.

STEP 3: Integrate f over each subinterval.

STEP 4: Add the absolute values of the results.

Example 6   Find the area of the region between the x-axis and the curve

$$y = x^3 - 4x, \qquad -2 \le x \le 2.$$

Solution

STEP 1: *The zeros of y.* We factor the formula for y to find where y is zero:

$$y = x^3 - 4x = x(x^2 - 4) = x(x - 2)(x + 2).$$

The zeros occur at $x = -2, 0,$ and $2.$

STEP 2: *The intervals of integration.* The points $x = -2, 0,$ and 2 partition $[-2, 2]$ into two subintervals, $[-2, 0]$ and $[0, 2]$.

STEP 3: *The integrations*

$$\int_{-2}^{0} (x^3 - 4x)\, dx = \left[\frac{x^4}{4} - 2x^2\right]_{-2}^{0} = [0] - [4 - 8] = 4$$

$$\int_{0}^{2} (x^3 - 4x)\, dx = \left[\frac{x^4}{4} - 2x^2\right]_{0}^{2} = [4 - 8] - [0] = -4$$

STEP 4: *The absolute values added*

$$\text{Area of region} = |4| + |-4| = 4 + 4 = 8$$

Figure 4.29 shows the graph of $y = x^3 - 4x$ over $[-2, 2]$. The first integral in step 3 gives the area A_1. The second integral gives the negative of the area A_2. The sum of the integrals' absolute values gives $A_1 + A_2$.

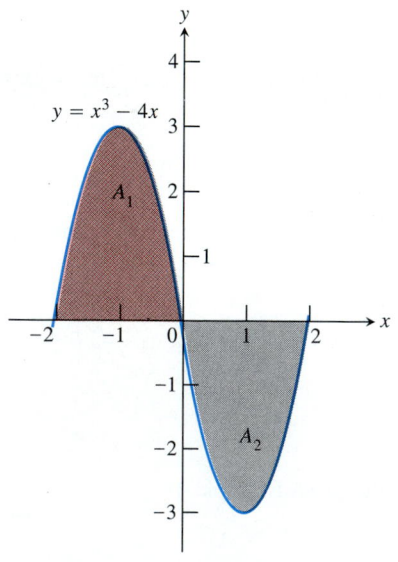

4.29 The graph of $y = x^3 - 4x$ from $x = -2$ to $x = 2$.

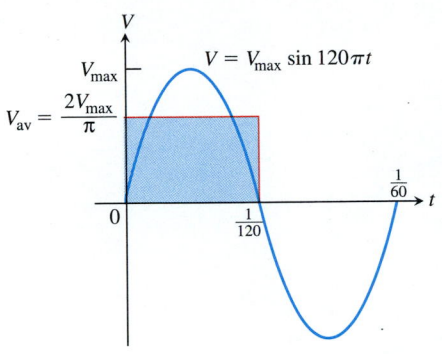

4.30 The graph of the household voltage $V = V_{max} \sin 120\pi t$ over a full cycle. Its average value over a half cycle is $2V_{max}/\pi$. Its average value over a full cycle is zero.

Example 7   *Household Electricity.*   We model the voltage in our home wiring with the sine function

$$V = V_{max} \sin 120\pi t,$$

which expresses the voltage V in volts as a function of time t in seconds. The function runs through 60 cycles each second. The positive number V_{max} ("vee max") is the **peak voltage.**

The average value of V over a half cycle (duration 1/120 sec; see Fig. 4.30) is

$$V_{av} = \frac{1}{(1/120) - 0} \int_0^{1/120} V_{max} \sin 120\pi t \, dt$$

$$= 120 V_{max} \left[-\frac{1}{120\pi} \cos 120\pi t \right]_0^{1/120}$$

$$= \frac{V_{max}}{\pi} \left[-\cos \pi + \cos 0 \right] = \frac{2 V_{max}}{\pi}.$$

The average value of the voltage over a full cycle, as we can see from Fig. 4.30, is zero. (Also see Exercise 53.) If we measured the voltage with a standard moving-coil galvanometer, the meter would read zero.

To measure the voltage effectively, we use an instrument that measures the square root of the average value of the square of the voltage, namely

$$V_{rms} = \sqrt{(V^2)_{av}}. \tag{15}$$

The subscript "rms" (read the letters separately) stands for "root mean square." Since the average value of $V^2 = (V_{max})^2 \sin^2 120\pi t$ over a cycle is

$$(V^2)_{av} = \frac{1}{(1/60) - 0} \int_0^{1/60} (V_{max})^2 \sin^2 120\pi t \, dt = \frac{(V_{max})^2}{2}, \tag{16}$$

(Exercise 53c), the rms voltage is

$$V_{rms} = \sqrt{\frac{(V_{max})^2}{2}} = \frac{V_{max}}{\sqrt{2}}. \tag{17}$$

The values given for household currents and voltages are always rms values. Thus, "115 volts ac" means that the rms voltage is 115. The peak voltage,

$$V_{max} = \sqrt{2}\, V_{rms} = \sqrt{2} \cdot 115 = 163 \text{ volts}, \qquad \text{(Calculator, rounded)}$$

obtained from Eq. (17), is considerably higher.

EXERCISES 4.5

Evaluate the integrals in Exercises 1–26.

1. $\displaystyle\int_1^2 (2x + 5)\, dx$

2. $\displaystyle\int_0^1 (1 + x^2)\, dx$

3. $\displaystyle\int_0^3 (4 - x^3)\, dx$

4. $\displaystyle\int_0^1 (x^2 - 2x + 3)\, dx$

5. $\displaystyle\int_0^1 (x^2 + \sqrt{x})\, dx$

6. $\displaystyle\int_0^5 x^{3/2}\, dx$

7. $\displaystyle\int_1^{32} x^{-6/5}\, dx$

8. $\displaystyle\int_{-2}^{-1} \frac{2}{x^2}\, dx$

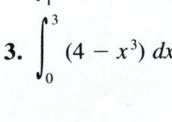

9. $\displaystyle\int_0^\pi \sin x \, dx$

10. $\displaystyle\int_0^\pi (1 + \cos x) \, dx$

11. $\displaystyle\int_0^{\pi/3} 2 \sec^2 x \, dx$

12. $\displaystyle\int_{\pi/6}^{5\pi/6} \csc^2 x \, dx$

13. $\displaystyle\int_{\pi/4}^{3\pi/4} \csc\theta \cot\theta \, d\theta$

14. $\displaystyle\int_0^{\pi/3} 4 \sec u \tan u \, du$

15. $\displaystyle\int_{\pi/2}^0 4 \sin^2 t \, dt$

16. $\displaystyle\int_{-\pi/3}^{\pi/3} 2 \cos^2 v \, dv$

17. $\displaystyle\int_{-\pi/2}^{\pi/2} (8y^2 + \sin y) \, dy$

18. $\displaystyle\int_{-\pi/4}^0 4(\sec^2 t + \tan^2 t) \, dt$

19. $\displaystyle\int_{-1}^1 (r + 1)^2 \, dr$

20. $\displaystyle\int_9^4 \frac{1 - \sqrt{u}}{\sqrt{u}} \, du$

21. $\displaystyle\int_{-1}^0 \left(\frac{u^7}{2} - u^{15}\right) du$

22. $\displaystyle\int_{-\sqrt{3}}^{\sqrt{3}} (t + 1)(t^2 + 4) \, dt$

23. $\displaystyle\int_1^{\sqrt{2}} \frac{s^2 + \sqrt{s}}{s^2} \, ds$

24. $\displaystyle\int_{1/2}^1 \left(\frac{1}{v^3} - \frac{1}{v^4}\right) dv$

25. $\displaystyle\int_{-4}^4 |x| \, dx$

26. $\displaystyle\int_0^\pi \frac{1}{2}(\cos x + |\cos x|) \, dx$

In Exercises 27–30, find the total area of the region between the curve and the x-axis.

27. $y = 2 - x, \quad 0 \le x \le 3$

28. $y = 3x^2 - 3, \quad -2 \le x \le 2$

29. $y = x^3 - 3x^2 + 2x, \quad 0 \le x \le 2$

30. $y = x^3 - 4x, \quad -2 \le x \le 2$

Find the areas of the shaded regions in Exercises 31–34.

31.
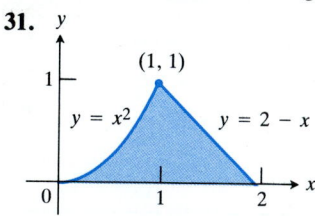

32.
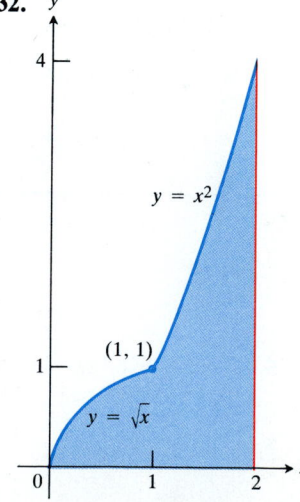

33.
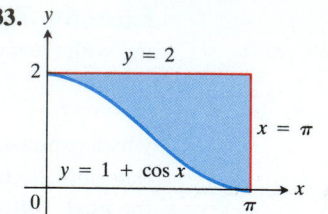

34.
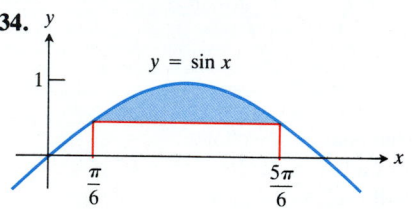

35. *Testing the consistency of the new definition of area.* Whenever we find a new way to calculate something, it is a good idea to be sure that the new and old ways agree on objects to which they both apply. If you use an integral to find the area of the triangle in Fig. 4.31, will you still get $A = (1/2)bh$?

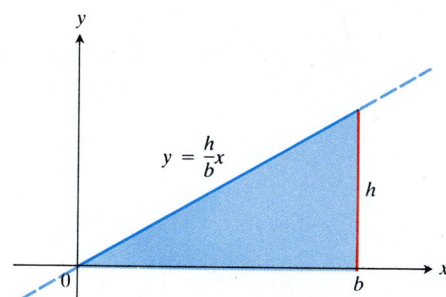

4.31 Is the area of this triangle still $(1/2) \, bh$? See Exercise 35.

36. Find $(d/dx) \int_0^x \cos t \, dt$ in two ways: (a) by evaluating the integral and differentiating the result with respect to x and (b) by applying the first part of the Fundamental Theorem directly to the integral.

Find dy/dx in Exercises 37–44.

37. $\displaystyle y = \int_0^x \sqrt{1 + t^2} \, dt$

38. $\displaystyle y = \int_1^x \frac{1}{t} \, dt, \quad x > 0$

39. $\displaystyle y = \int_0^{\sqrt{x}} \sin(t^2) \, dt$

40. $\displaystyle y = \int_0^{\tan x} \frac{dt}{1 + t^2}$

41. $\displaystyle y = \int_x^4 \sqrt{t^4 + 7} \, dt$

42. $\displaystyle y = \int_{x^2}^0 \cos \sqrt{t} \, dt$

43. $y = \int_{\sin x}^{0} \dfrac{dt}{\sqrt{1 - t^2}}$, $|x| < \dfrac{\pi}{2}$

44. $y = \int_{\sec x}^{0} \dfrac{dt}{t\sqrt{t^2 - 1}}$, $-\dfrac{\pi}{2} < x < 0$

Each of the following functions solves one of the initial value problems in Exercises 45–48. Which function solves which problem?

a) $y = \int_{1}^{x} \dfrac{1}{t}\, dt - 3$

b) $y = \int_{0}^{x} \sec t\, dt + 4$

c) $y = \int_{-1}^{x} \sec t\, dt + 4$

d) $y = \int_{\pi}^{x} \dfrac{1}{t}\, dt - 3$

45. $\dfrac{dy}{dx} = \dfrac{1}{x}$, $y(\pi) = -3$

46. $y' = \sec x$, $y(-1) = 4$

47. $y' = \sec x$, $y(0) = 4$

48. $y' = \dfrac{1}{x}$, $y(1) = -3$

Express the solutions of the initial value problems in Exercises 49 and 50 as integrals.

49. $\dfrac{dy}{dx} = \sec x$, $y = 3$ when $x = 2$

50. $\dfrac{ds}{dt} = f(t)$, $s = s_0$ when $t = t_0$

51. Show that if k is a positive constant, then the area between the x-axis and one arch of the curve $y = \sin kx$ is always $2/k$.

52. *Archimedes' area formula for parabolas.* Archimedes (287–212 B.C.), inventor, military engineer, physicist, and the greatest mathematician of classical times in the Western world, discovered that the area under a parabolic arch like the one in Fig. 4.32 is always two-thirds the base times the height.

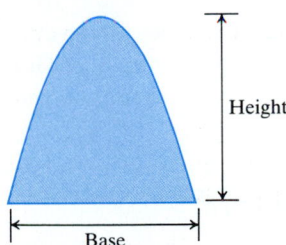

Height

Base

4.32   The parabolic arch in Exercise 52.

a) Find the area under the parabolic arch
$$y = 6 - x - x^2, \quad -3 \le x \le 2.$$

b) Find the height of the arch. (Where does y have its maximum value?)

c) Show that the area is two–thirds the base times the height.

d) Graph the parabolic arch $y = h - (4h/b^2)x^2$, $-b/2 \le x \le b/2$, assuming that h and b are positive. Then use

calculus to find the area of the region between the arch and the x-axis.

53. *Continuation of Example 7*

a) Show by evaluating the integral in the expression
$$\frac{1}{(1/60) - 0} \int_{0}^{1/60} V_{max} \sin 120\pi t\, dt$$
that the average value of $V = V_{max} \sin 120\pi t$ over a full cycle is zero.

b) The circuit that runs your electric stove is rated 240 volts rms. What is the peak value of the allowable voltage?

c) Verify Eq. (16).

54. Suppose that the position at time t (sec) of a particle moving along a coordinate axis is
$$s = \int_{0}^{t} f(x)\, dx$$
meters, where f is the differentiable function graphed in Fig. 4.33. Use the graph to answer the following questions.

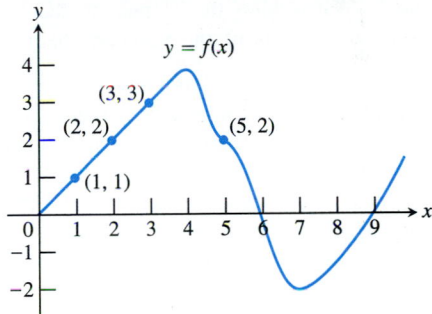

4.33   The graph of the function f in Exercise 54.

a) What is the particle's velocity at time $t = 5$?

b) Is the acceleration of the particle at time $t = 5$ positive, or negative?

c) What is the particle's position at time $t = 3$?

d) At what time during the first 9 sec does s have its largest value?

e) Approximately when is the acceleration zero?

f) When is the particle moving toward the origin? away from the origin?

55. Suppose $\int_{1}^{x} f(t)\, dt = x^2 - 2x + 1$. Find $f(x)$.

56. Find $f(4)$ if $\int_{0}^{x} f(t)\, dt = x \cos \pi x$.

57. Find
$$\lim_{x \to 0} \frac{1}{x^3} \int_{0}^{x} \frac{t^2}{t^4 + 1}\, dt.$$

58. Find (a) the linearization at $x = 0$ and (b) the quadratic approximation at $x = 0$ of the function
$$f(x) = 2 \int_{0}^{x} \frac{10}{1 + t^4}\, dt.$$

59. Suppose that f has a positive derivative for all values of x and that $f(1) = 0$. Which of the following statements must be true of the function

$$g(x) = \int_0^x f(t) \, dt?$$

a) g is a differentiable function of x.
b) g is a continuous function of x.
c) The graph of g has a horizontal tangent at $x = 1$.
d) g has a local maximum at $x = 1$.
e) g has a local minimum at $x = 1$.
f) The graph of g has an inflection point at $x = 1$.
g) The graph of dg/dx crosses the x-axis at $x = 1$.

Computer Grapher

If you have access to a program that will graph a function $f(x)$ and its integral $F(x) = \int_a^x f(t) \, dt$ together, try it on the functions in Exercises 60–63. Watch how the slopes of the curves $y = \int_0^x f(t) \, dt$ match the graphs of the functions being integrated. We know that

i) If F has a relative maximum at $x = c$, then $f(c) = 0$.
ii) The integral F is an increasing function of x on any interval on which f is positive.
iii) The integral F is a decreasing function of x on any interval on which f is negative.

Look for these relationships in the graphs.

60. $f(x) = \sin x$ **61.** $f(x) = x \cos \pi x$

62. $f(x) = x^2 - 4$ **63.** $f(x) = x^3 - 4x^2 + 3x$

64. *The Fundamental Theorem.*   If f is continuous, we expect

$$\lim_{h \to 0} \frac{1}{h} \int_x^{x+h} f(t) \, dt$$

to equal $f(x)$, as in the proof of Part 1 of the Fundamental Theorem. For instance, if $f(t) = \cos t$, then

$$\frac{1}{h} \int_x^{x+h} \cos t \, dt = \frac{\sin(x + h) - \sin x}{h}. \tag{18}$$

The right-hand side of Eq. (18) is the difference quotient for the derivative of the sine, and we expect its limit as $h \to 0$ to be $\cos x$.

Graph $\cos x$ for $-\pi \le x \le 2\pi$. Then, in a different color if possible, graph the right-hand side of Eq. (18) as a function of x for $h = 2, 1, 0.5$, and 0.1. Watch how the latter curves converge to the graph of the cosine as $h \to 0$.

65. Repeat Exercise 64 for $f(t) = 3t^2$. What is

$$\lim_{h \to 0} \frac{1}{h} \int_x^{x+h} 3t^2 \, dt = \lim_{h \to 0} \frac{(x + h)^3 - x^3}{h}?$$

Graph $f(x) = 3x^2$ for $-1 \le x \le 1$. Then graph the quotient $((x + h)^3 - x^3)/h$ as a function of x for $h = 1, 0.5, 0.2$, and 0.1. Use different colors if possible. Watch how the latter curves converge to the graph of $3x^2$ as $h \to 0$.

EXPLORER PROGRAMS

Derivatives	Graphs $f(x)$ and $\int_a^x f(t) \, dt$ together for your choice of f and a
PowerGrapher	Graphs functions together

4.6 Integration by Substitution: Running the Chain Rule Backward

A change of variable can often turn an unfamiliar integral into one we can evaluate. The method for doing this is called the substitution method of integration. It is the principal method by which integrals are evaluated. This section shows how and why the method works.

The Generalized Power Rule in Integral Form

When u is a differentiable function of x, and n is a rational number different from -1, the Chain Rule tells us that

$$\frac{d}{dx}\left(\frac{u^{n+1}}{n+1}\right) = u^n \frac{du}{dx}. \tag{1}$$

This same equation, from another point of view, says that $u^{n+1}/(n + 1)$ is one of the antiderivatives of the product $u^n(du/dx)$. The set of all antiderivatives of $u^n \, (du/dx)$ is therefore

$$\int \left(u^n \frac{du}{dx}\right) dx = \frac{u^{n+1}}{n+1} + C. \tag{2}$$

The integral on the left-hand side of this equation is usually written in the simpler "differential" form,

$$\int u^n \, du,$$
(3)

obtained by treating the dx's as differentials that cancel. Combining Eqs. (2) and (3) then gives the following rule.

If u is any differentiable function of x, then

$$\int u^n \, du = \frac{u^{n+1}}{n+1} + C, \qquad n \neq -1.$$
(4)

Whenever we can cast an integral in the form

$$\int u^n \, du,$$

with u a differentiable function of x, and du the differential of u, we can integrate with respect to u in the usual way to evaluate the integral as $[u^{n+1}/(n+1)] + C$.

Example 1   Evaluate $\int (x+2)^5 \, dx$.

Solution   We can put the integral in the form

$$\int u^5 \, du$$

by substituting

$$u = x + 2, \qquad du = d(x+2) = dx.$$

Then

$$\int (x+2)^5 \, dx = \int u^5 \, du$$

$$= \frac{u^6}{6} + C \qquad \text{(Integrate, using Eq. (4) with } n = 5.\text{)}$$

$$= \frac{(x+2)^6}{6} + C. \qquad \text{(Replace } u \text{ by } x+2.\text{)}$$

Example 2

$$\int \sqrt{1+x^2} \cdot 2x \, dx = \int u^{1/2} \, du \qquad \left(\begin{matrix}\text{Substitue } u = 1 + x^2, \\ du = 2x \, dx.\end{matrix}\right)$$

$$= \frac{u^{(1/2)+1}}{(1/2)+1} + C \qquad \left(\begin{matrix}\text{Integrate, using Eq. (4)} \\ \text{with } n = 1/2.\end{matrix}\right)$$

$$= \frac{2}{3} u^{3/2} + C \qquad \text{(Simpler form)}$$

$$= \frac{2}{3} (1+x^2)^{3/2} + C \qquad \text{(Replace } u \text{ by } 1 + x^2.\text{)}$$

Example 3 *Adjusting the Integrand by a Multiplicative Constant to Put it in Standard Form*

$$\int \sqrt{4x-1}\, dx = \int u^{1/2} \cdot \frac{1}{4}\, du \qquad \left(\begin{array}{l} \text{Substitute } u = 4x - 1, \\ du = 4\, dx, \frac{1}{4}\, du = dx. \end{array} \right)$$

$$= \frac{1}{4} \int u^{1/2}\, du \qquad \left(\begin{array}{l} \text{Constant Multiple Rule. The} \\ \text{integral is now in standard form.} \end{array} \right)$$

$$= \frac{1}{4} \frac{u^{3/2}}{3/2} + C \qquad \left(\begin{array}{l} \text{Integrate, using Eq. (4) with} \\ n = 1/2. \end{array} \right)$$

$$= \frac{1}{6} u^{3/2} + C \qquad \text{(Simpler form)}$$

$$= \frac{1}{6} (4x-1)^{3/2} + C \qquad \text{(Replace } u \text{ by } 4x - 1.)$$

Integrals of Trigonometric Functions

If u is a differentiable function of x, then $\sin u$ is a differentiable function of x. The Chain Rule gives the derivative of $\sin u$ with respect to x as

$$\frac{d}{dx} \sin u = \cos u \frac{du}{dx}. \tag{5}$$

From another point of view, this equation says that $\sin u$ is one of the antiderivatives of the product $\cos u \cdot (du/dx)$. The set of all antiderivatives of the product is therefore

$$\int \left(\cos u \frac{du}{dx} \right) dx = \sin u + C. \tag{6}$$

A formal cancellation of the dx's in the integral on the left leads to the following rule.

If u is a differentiable function of x, then

$$\int \cos u\, du = \sin u + C. \tag{7}$$

Equation (7) says that whenever we can cast an integral in the form

$$\int \cos u\, du,$$

we can integrate with respect to u in the usual way to evaluate the integral as $\sin u + C$.

The companion formula for the integral of $\sin u$ when u is a differentiable function of x is

---

$$\int \sin u\, du = -\cos u + C. \tag{8}$$

Example 4

$$\int \cos (7x + 5) \, dx = \int \cos u \cdot \frac{1}{7} \, du \qquad \left(\begin{array}{l} \text{Substitute } u = 7x + 5, \\ du = 7 \, dx, \frac{1}{7} \, du = dx. \end{array} \right)$$

$$= \frac{1}{7} \int \cos u \, du = \frac{1}{7} \sin u + C$$

$$= \frac{1}{7} \sin (7x + 5) + C$$

Example 5

$$\int x^2 \sin(x^3) \, dx = \int \sin u \cdot \frac{1}{3} \, du \qquad \left(\begin{array}{l} \text{Substitute } u = x^3, \\ du = 3x^2 \, dx, \\ \frac{1}{3} \, du = x^2 \, dx. \end{array} \right)$$

$$= \frac{1}{3} \int \sin u \, du = \frac{1}{3} (-\cos u) + C$$

$$= -\frac{1}{3} \cos(x^3) + C$$

The Chain Rule formulas for the derivatives of the tangent, cotangent, secant, and cosecant of a differentiable function u of x lead to the following integrals.

$$\int \sec^2 u \, du = \tan u + C \tag{9}$$

$$\int \csc^2 u \, du = -\cot u + C \tag{10}$$

$$\int \sec u \tan u \, du = \sec u + C \tag{11}$$

$$\int \csc u \cot u \, du = -\csc u + C \tag{12}$$

In each formula we assume u to be a differentiable function of x. Each formula can be checked by differentiating the right-hand side with respect to x. In each case the Chain Rule applies to produce the integrand on the left.

Example 6

$$\int \frac{1}{\cos^2 2x} \, dx = \int \sec^2 2x \, dx$$

$$= \int \sec^2 u \cdot \frac{1}{2} \, du \qquad \left(\text{Substitute } u = 2x, \, du = 2 \, dx, \, dx = \frac{1}{2} \, du. \right)$$

$$= \frac{1}{2} \int \sec^2 u \, du$$

$$= \frac{1}{2} \tan u + C \qquad \text{(Integrate, using Eq. (9).)}$$

$$= \frac{1}{2} \tan 2x + C \qquad \text{(Replace } u \text{ by } 2x.\text{)}$$

The Substitution Method of Integration

To evaluate the integral

$$\int f(g(x))g'(x)\,dx$$

when f and g' are continuous functions, carry out the following steps:

STEP 1: Substitute $u = g(x)$ and $du = g'(x)\,dx$ to obtain the integral

$$\int f(u)\,du.$$

STEP 2: Integrate with respect to u.

STEP 3: Replace u by $g(x)$ in the result.

The Substitution Method of Integration

The substitutions in the preceding examples are all instances of the following general rule.

$$\int f(g(x)) \cdot g'(x)\,dx = \int f(u)\,du \qquad \text{1. Substitute } u = g(x),\ du = g'(x)\,dx.$$

$$= F(u) + C \qquad \text{2. Evaluate by finding an antiderivative } F(u) \text{ of } f(u). \text{ (Any one will do.)}$$

$$= F(g(x)) + C \qquad \text{3. Substitute back.}$$

These three steps are the steps of the substitution method of integration.

The method works because $F(g(x))$ is an antiderivative of $f(g(x)) \cdot g'(x)$ whenever F is an antiderivative of f:

$$\frac{d}{dx} F(g(x)) = F'(g(x)) \cdot g'(x) \qquad \text{(Chain Rule)}$$

$$= f(g(x)) \cdot g'(x) \qquad \text{(Because } F' = f\text{)}$$

Example 7

$$\int (x^2 + 2x - 3)^2 (x + 1)\,dx = \int u^2 \cdot \frac{1}{2}\,du \qquad \left(\begin{array}{l} \text{Substitute} \\ u = x^2 + 2x - 3, \\ du = 2x\,dx + 2\,dx \\ \quad = 2(x + 1)\,dx, \\ \frac{1}{2}\,du = (x + 1)\,dx. \end{array} \right)$$

$$= \frac{1}{2} \int u^2\,du$$

$$= \frac{1}{2} \cdot \frac{u^3}{3} + C = \frac{1}{6} u^3 + C \qquad \left(\begin{array}{l} \text{Integrate with} \\ \text{respect to } u. \end{array} \right)$$

$$= \frac{1}{6} (x^2 + 2x - 3)^3 + C \qquad \text{(Replace } u.\text{)}$$

Example 8

$$\int \sin^4 x \cos x\,dx = \int u^4\,du \qquad \text{(Substitute } u = \sin x,\ du = \cos x\,dx.\text{)}$$

$$= \frac{u^5}{5} + C \qquad \text{(Integrate with respect to } u.\text{)}$$

$$= \frac{\sin^5 x}{5} + C \qquad \text{(Replace } u.\text{)}$$

There is often more than one way to make a successful substitution, as the next example shows.

Example 9   Evaluate $\displaystyle\int \frac{2z\,dz}{\sqrt[3]{z^2 + 1}}$.

Solution   We can use the substitution method of integration as an exploratory tool: Substitute for the most troublesome part of the integrand and see how things work out. For the integral here, we might try $u = z^2 + 1$ or we might even press our luck and take u to be the entire cube root. Here is what happens in each case.

SOLUTION 1:   Subsitute $u = z^2 + 1$.

$$\int \frac{2z\, dz}{\sqrt[3]{z^2 + 1}} = \int \frac{du}{u^{1/3}} \qquad \text{(Substitute } u = z^2 + 1,\ du = 2z\, dz.)$$

$$= \int u^{-1/3}\, du \qquad \left(\text{In the form } \int u^n\, du.\right)$$

$$= \frac{u^{2/3}}{2/3} + C \qquad \text{(Integrate with respect to } u.)$$

$$= \frac{3}{2} u^{2/3} + C$$

$$= \frac{3}{2} (z^2 + 1)^{2/3} + C \qquad \text{(Replace } u.)$$

SOLUTION 2:   Substitute $u = \sqrt[3]{z^2 + 1}$.

$$\int \frac{2z\, dz}{\sqrt[3]{z^2 + 1}} = \int \frac{3\, u^2\, du}{u} \qquad \left(\begin{aligned} &\text{Substitute } u = \sqrt[3]{z^2 + 1}, \\ &u^3 = z^2 + 1,\ 3u^2\, du = 2z\, dz. \end{aligned}\right)$$

$$= 3 \int u\, du$$

$$= 3 \cdot \frac{u^2}{2} + C \qquad \text{(Integrate with respect to } u.)$$

$$= \frac{3}{2} (z^2 + 1)^{2/3} + C \qquad \text{(Replace } u.)$$

Substitution in Definite Integrals

The formula for evaluating definite integrals by substitution first appeared in a book by Isaac Barrow (1630–1677), Newton's mathematics teacher at Cambridge University. It looks like this:

Substitution in Definite Integrals

THE FORMULA

$$\int_a^b f(g(x)) \cdot g'(x)\, dx = \int_{g(a)}^{g(b)} f(u)\, du \tag{13}$$

HOW TO USE IT

Substitute $u = g(x)$, $du = g'(x)\, dx$, and integrate from $g(a)$ to $g(b)$.

To use the formula, make the same u-substitution you would use to evaluate the corresponding indefinite integral. Then integrate with respect to u from the value of u at $x = a$ to the value of u at $x = b$.

Example 10

$$\int_0^{\pi/4} \tan x \sec^2 x \, dx = \int_0^1 u \, du \qquad \left(\begin{array}{l} \text{Substitute } u = \tan x,\, du = \sec^2 x \, dx, \\ \text{and integrate from } \tan 0 = 0 \\ \text{to } \tan (\pi/4) = 1. \end{array} \right)$$

$$= \frac{u^2}{2} \Big]_0^1 \qquad \text{(Evaluate the definite integal.)}$$

$$= \frac{(1)^2}{2} - \frac{(0)^2}{2} = \frac{1}{2}$$

We do not have to use Eq. (13) if we do not want to. We can always transform the integral as an indefinite integral, integrate, change back to x, and use the original x-limits. In the next example we evaluate a definite integral both ways—with Eq. (13) and without.

Example 11   Evaluate $\int_{-1}^1 3x^2\sqrt{x^3 + 1} \, dx$.

Solution

METHOD 1:   Transform the integral and evaluate it with transformed limits.

$$\int_{-1}^1 3x^2\sqrt{x^3 + 1} \, dx = \int_0^2 \sqrt{u} \, du \qquad \left(\begin{array}{l} \text{Substitute } u = x^3 + 1, \\ du = 3x^2 \, dx \text{ and integrate} \\ \text{from } u(-1) = 0 \text{ to } u(1) = 2. \end{array} \right)$$

$$= \frac{2}{3} u^{3/2} \Big]_0^2$$

$$= \frac{2}{3} \Big[2^{3/2} - 0^{3/2} \Big] = \frac{2}{3} [2\sqrt{2}] = \frac{4\sqrt{2}}{3}$$

METHOD 2:   Transform the integral as an indefinite integral, integrate, change back to x, and use the original x-limits.

$$\int 3x^2\sqrt{x^3 + 1} \, dx = \int \sqrt{u} \, du \qquad \left(\begin{array}{l} \text{Substitute} \\ u = x^3 + 1, \\ du = 3x^2 \, dx. \end{array} \right)$$

$$= \frac{2}{3} u^{3/2} + C$$

$$= \frac{2}{3} (x^3 + 1)^{3/2} + C$$

$$\int_{-1}^1 3x^2\sqrt{x^3 + 1} \, dx = \frac{2}{3} (x^3 + 1)^{3/2} \Big]_{-1}^1 \qquad \left(\begin{array}{l} \text{Use the} \\ \text{indefinite} \\ \text{integral just} \\ \text{found.} \end{array} \right)$$

$$= \frac{2}{3} \Big[((1)^3 + 1)^{3/2} - ((-1)^3 + 1)^{3/2} \Big]$$

$$= \frac{2}{3} \Big[2^{3/2} - 0^{3/2} \Big] = \frac{2}{3} [2\sqrt{2}] = \frac{4\sqrt{2}}{3}$$

Which method is better—evaluating the transformed integral with transformed limits or transforming back to use the original limits of integration? In Example 11 the first method seems easier, but that is not always the case. As a rule, it is best to know both methods and use whichever one seems better at the time.

EXERCISES 4.6

Evaluate the indefinite integrals in Exercises 1–8 by using the given substitutions to reduce the integrals to standard form.

1. $\int x \sin(2x^2)\, dx, \quad u = 2x^2$

2. $\int \sec 2x \tan 2x\, dx, \quad u = 2x$

3. $\int \left(1 - \cos \frac{t}{2}\right)^2 \sin \frac{t}{2}\, dt, \quad u = 1 - \cos \frac{t}{2}$

4. $\int 28(7x - 2)^3 dx, \quad u = 7x - 2$

5. $\int \frac{9r^2 dr}{\sqrt{1 - r^3}}, \quad u = 1 - r^3$

6. $\int 12(y^4 + 4y^2 + 1)^2 (y^3 + 2y)\, dy, \quad u = y^4 + 4y^2 + 1$

7. $\int \csc^2 2\theta \cot 2\theta\, d\theta, \quad$ (a) $u = \cot 2\theta$; (b) $u = \csc 2\theta$

8. $\int \frac{dt}{\sqrt{5t}}, \quad$ (a) $u = 5t$; (b) $u = \sqrt{5t}$

Evaluate the definite integrals in Exercises 9–12 by using the given substitutions.

9. $\int_0^{\pi/6} \frac{\sin 2x}{\cos^2 2x}\, dx, \quad u = \cos 2x$

10. $\int_{\pi/6}^{\pi/2} \sin^2 \theta \cos \theta\, d\theta, \quad u = \sin \theta$

11. $\int_{-\pi/2}^{\pi/2} \frac{\cos t}{(2 + \sin t)^2}\, dt, \quad u = 2 + \sin t$

12. $\int_{\pi^2/4}^{\pi^2} \frac{\sin\sqrt{y}}{\sqrt{y}}\, dy, \quad u = \sqrt{y}$

Evaluate the indefinite integrals in Exercises 13–22.

13. $\int \frac{dx}{(1 - x)^2}$

14. $\int \frac{4y}{\sqrt{2y^2 + 1}}\, dy$

15. $\int \sec^2(x + 2)\, dx$

16. $\int \sec^2\left(\frac{x}{4}\right)\, dx$

17. $\int 8r(r^2 - 1)^{1/3}\, dr$

18. $\int x^4(7 - x^5)^3\, dx$

19. $\int \sec\left(\theta + \frac{\pi}{2}\right) \tan\left(\theta + \frac{\pi}{2}\right)\, d\theta$

20. $\int \sqrt{\tan x} \sec^2 x\, dx$

21. $\int \frac{6x^3}{\sqrt[4]{1 + x^4}}\, dx$

22. $\int (s^3 + 2s^2 - 5s + 6)^2 (3s^2 + 4s - 5)\, ds$

Evaluate the definite integrals in Exercises 23–42.

23. a) $\int_0^3 \sqrt{y + 1}\, dy$ b) $\int_{-1}^0 \sqrt{y + 1}\, dy$

24. a) $\int_0^1 r\sqrt{1 - r^2}\, dr$ b) $\int_{-1}^1 r\sqrt{1 - r^2}\, dr$

25. a) $\int_0^{\pi/4} \tan x \sec^2 x\, dx$ b) $\int_{-\pi/4}^0 \tan x \sec^2 x\, dx$

26. a) $\int_0^1 x^3(1 + x^4)^3\, dx$ b) $\int_{-1}^1 x^3(1 + x^4)^3\, dx$

27. a) $\int_0^1 \frac{x^3}{\sqrt{x^4 + 9}}\, dx$ b) $\int_{-1}^0 \frac{x^3}{\sqrt{x^4 + 9}}\, dx$

28. a) $\int_{-1}^1 \frac{x}{(1 + x^2)^2}\, dx$ b) $\int_0^1 \frac{x}{(1 + x^2)^2}\, dx$

29. a) $\int_0^{\sqrt{7}} x(x^2 + 1)^{1/3}\, dx$ b) $\int_{-\sqrt{7}}^0 x(x^2 + 1)^{1/3}\, dx$

30. a) $\int_0^\pi 3\cos^2 x \sin x\, dx$ b) $\int_{2\pi}^{3\pi} 3\cos^2 x \sin x\, dx$

31. a) $\int_0^{\pi/6} (1 - \cos 3x) \sin 3x\, dx$

b) $\int_{\pi/6}^{\pi/3} (1 - \cos 3x) \sin 3x\, dx$

32. a) $\int_0^{\sqrt{3}} \frac{4x}{\sqrt{x^2 + 1}}\, dx$ b) $\int_{-\sqrt{3}}^{\sqrt{3}} \frac{4x}{\sqrt{x^2 + 1}}\, dx$

33. a) $\int_0^{2\pi} \frac{\cos x}{\sqrt{2 + \sin x}}\, dx$ b) $\int_{-\pi}^\pi \frac{\cos x}{\sqrt{2 + \sin x}}\, dx$

34. a) $\int_{-\pi/2}^0 \frac{\sin x}{(3 + \cos x)^2}\, dx$ b) $\int_0^{\pi/2} \frac{\sin x}{(3 + \cos x)^2}\, dx$

35. $\int_0^1 \sqrt{t^5 + 2t}\ (5t^4 + 2)\, dt$

36. $\int_1^4 \frac{dy}{2\sqrt{y}\ (1 + \sqrt{y})^2}$

37. $\int_0^{\pi/2} \cos^3 2x \sin 2x\, dx$

38. $\displaystyle\int_{-\pi/4}^{\pi/4} \tan^2 x \sec^2 x \, dx$

39. $\displaystyle\int_{0}^{\pi} \frac{8 \sin t}{\sqrt{5 - 4 \cos t}} \, dt$

40. $\displaystyle\int_{0}^{\pi/4} (1 - \sin 2t)^{3/2} \cos 2t \, dt$

41. $\displaystyle\int_{0}^{1} 15x^2 \sqrt{5x^3 + 4} \, dx$

42. $\displaystyle\int_{0}^{1} (y^3 + 6y^2 - 12y + 5)(y^2 + 4y - 4) \, dy$

Find the total areas of the shaded regions in Exercises 43 and 44.

43.

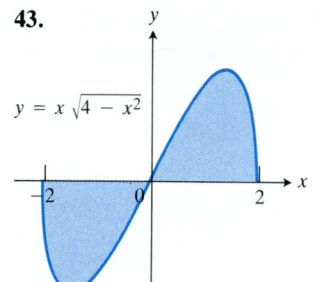

$y = x\sqrt{4 - x^2}$

44.

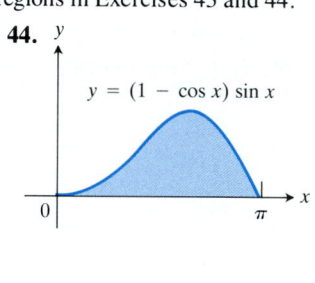

$y = (1 - \cos x) \sin x$

45. It looks as if we can integrate $2 \sin x \cos x$ with respect to x in three different ways:

a) $\displaystyle\int 2 \sin x \cos x \, dx = \int 2u \, du$ $\left(\begin{matrix}u = \sin x, \\ du = \cos x \, dx\end{matrix}\right)$

$\qquad = u^2 + C_1 = \sin^2 x + C_1;$

b) $\displaystyle\int 2 \sin x \cos x \, dx = \int -2u \, du$ $\left(\begin{matrix}u = \cos x, \\ du = -\sin x \, dx, \\ -du = \sin x \, dx\end{matrix}\right)$

$\qquad = -u^2 + C_2 = -\cos^2 x + C_2;$

c) $\displaystyle\int 2 \sin x \cos x \, dx = \int \sin 2x \, dx$ $\left(\begin{matrix}2 \sin x \cos x \\ = \sin 2x\end{matrix}\right)$

$\qquad = -\dfrac{\cos 2x}{2} + C_3.$

Can all three integrations be correct? Explain.

46. The substitution $u = \tan x$ gives

$$\int \sec^2 x \tan x \, dx = \int u \, du = \frac{u^2}{2} + C = \frac{\tan^2 x}{2} + C.$$

The substitution $u = \sec x$ gives

$$\int \sec^2 x \tan x \, dx = \int u \, du = \frac{u^2}{2} + C = \frac{\sec^2 x}{2} + C.$$

Can both integrations be correct? Explain.

Solve the initial value problems in Exercises 47–50.

47. $\dfrac{ds}{dt} = 24t(3t^2 - 1)^3, \quad s = 0$ when $t = 0$

48. $\dfrac{dy}{dx} = 4x(x^2 + 8)^{-1/3}, \quad y = 0$ when $x = 0$

49. $\dfrac{ds}{dt} = 6 \sin(t + \pi), \quad s = 0$ when $t = 0$

50. $\dfrac{d^2 s}{dt^2} = -4 \sin\left(2t - \dfrac{\pi}{2}\right), \quad \dfrac{ds}{dt} = 100$ and $s = 0$ when $t = 0$

51. Suppose that $F(x)$ is an antiderivative of $f(x) = (\sin x)/x$, $x > 0$. Express

$$\int_{1}^{3} \frac{\sin 2x}{x} \, dx$$

in terms of F.

52. Show that if f is continuous, then

$$\int_{0}^{1} f(x) \, dx = \int_{0}^{1} f(1 - x) \, dx.$$

53. Suppose that

$$\int_{0}^{1} f(x) \, dx = 3.$$

Find

$$\int_{-1}^{0} f(x) \, dx$$

if (a) f is odd, (b) f is even.

54. a) Show that

$$\int_{-a}^{a} h(x) \, dx = \begin{cases} 0 & \text{if } h \text{ is odd} \\ 2\displaystyle\int_{0}^{a} h(x) \, dx & \text{if } h \text{ is even.} \end{cases}$$

b) Test the result in part (a) with $h(x) = \sin x$ and with $h(x) = \cos x$, taking $a = \pi/2$ in each case.

Sequences of Substitution

If you do not know what substitution to make, try reducing the integral step by step, using a trial substitution to simplify the integral a bit and then another to simplify it some more. You will see what we mean if you try the sequences of substitutions in Exercises 55 and 56.

55. $\displaystyle\int_{0}^{\pi/4} \frac{18 \tan^2 x \sec^2 x}{(2 + \tan^3 x)^2} \, dx$

a) $u = \tan x$, followed by $v = u^3$, then by $w = 2 + v$

b) $u = \tan^3 x$ followed by $v = 2 + u$

c) $u = 2 + \tan^3 x$

56. $\displaystyle\int \sqrt{1 + \sin^2(x - 1)} \, \sin(x - 1) \cos(x - 1) \, dx$

a) $u = x - 1$, followed by $v = \sin u$, then by $w = 1 + v^2$

b) $u = \sin(x - 1)$ followed by $v = 1 + u^2$

c) $u = 1 + \sin^2(x - 1)$

Evaluate the integrals in Exercises 57–60.

57. $\displaystyle\int \frac{1}{x^2} \sin \frac{1}{x} \cos \frac{1}{x} \, dx$

58. $\displaystyle\int \sin^3\left(\frac{3t+1}{2}\right)\cos\left(\frac{3t+1}{2}\right)dt$

59. $\displaystyle\int \frac{\cos\sqrt{\theta}}{\sqrt{\theta}\,\sin^2\sqrt{\theta}}\,d\theta$

60. $\displaystyle\int \frac{(2r-1)\cos\sqrt{3(2r-1)^2+6}}{\sqrt{3(2r-1)^2+6}}\,dr$

The Shift Property for Definite Integrals

A basic property of definite integrals is their invariance under translation, as expressed by the equation

$$\int_a^b f(x)\,dx = \int_{a-c}^{b-c} f(x+c)\,dx. \tag{14}$$

The equation holds whenever f is integrable and defined for the necessary values of x. For example (Fig. 4.34),

$$\int_0^1 x^3\,dx = \int_{-2}^{-1}(x+2)^3\,dx = \int_2^3 (x-2)^3\,dx. \tag{15}$$

61. Use a substitution to verify Eq. (14).

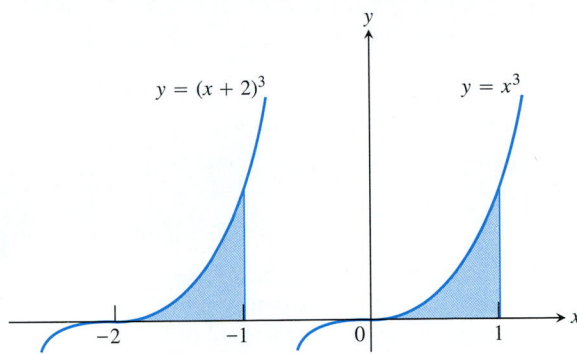

4.34   The first two integrations in Eq. (15). The shaded regions, being congruent, have equal areas.

62. For each of the following functions, graph $f(x)$ over $[a,b]$ and $f(x+c)$ over $[a-c, b-c]$ to convince yourself that Eq. (14) is reasonable.
a) $f(x) = x^2,\ a = 0,\ b = 1,\ c = 1$
b) $f(x) = \sin x,\ a = 0,\ b = \pi,\ c = \pi/2$
c) $f(x) = \sqrt{x-4},\ a = 4,\ b = 8,\ c = 5$

4.7 Numerical Integration

As we have seen, the ideal way to evaluate a definite integral

$$\int_a^b f(x)\,dx$$

is to find a formula $F(x)$ for one of the antiderivatives of $f(x)$ and calculate the number $F(b) - F(a)$. But some antiderivatives are hard to find, and still others, like the antiderivatives of $(\sin x)/x$ and $\sqrt{1+x^4}$, have no elementary formulas. We do not mean merely that no one has yet succeeded in finding simple expressions for evaluating the antiderivatives of $(\sin x)/x$ and $\sqrt{1+x^4}$. We mean it has been proved that no such expressions exist.

Whatever the reason, when we cannot evaluate a definite integral with an antiderivative, we turn to numerical methods such as the trapezoidal rule and Simpson's rule, described in this section. These rules enable us to estimate an integral's value to as many decimal places as we please whenever we want. They also enable us to calculate integrals with reasonable accuracy from numerical tables of function values. This comes in handy when the only information we have about a function is a set of values measured in the laboratory or in the field.

The Trapezoidal Rule

When we cannot find a workable antiderivative for a function f that we have to integrate, we partition the interval of integration, replace f by a closely fitting polynomial on each subinterval, integrate the polynomials, and add the results to approximate the integral of f. The higher the degrees of the polynomials for a given partition, the better the results. For a given degree, the finer the partition, the better the results, until we reach limits imposed by round-off and truncation errors.

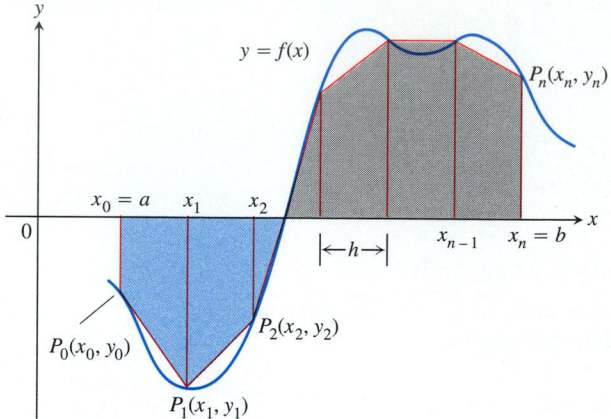

4.35 The trapezoidal rule approximates short stretches of the curve $y = f(x)$ with line segments. To estimate the integral of f from a to b, we add the "signed" areas of the trapezoids made by joining the ends of the segments to the x-axis.

Linear polynomials (degree 1) give the simplest useful approximation. We partition $[a, b]$ into n subintervals of length $h = (b - a)/n$ and join the corresponding points on the curve with line segments (Fig. 4.35). The vertical lines from the ends of the segments to the partition points create a collection of trapezoids that approximate the region between the curve and the x-axis. We add the areas of the trapezoids, counting area above the x-axis as positive and area below the axis as negative. The sum of the "signed" areas is

$$T = \frac{1}{2}(y_0 + y_1)h + \frac{1}{2}(y_1 + y_2)h + \cdots$$
$$+ \frac{1}{2}(y_{n-2} + y_{n-1})h + \frac{1}{2}(y_{n-1} + y_n)h$$
$$= h\left(\frac{1}{2}y_0 + y_1 + y_2 + \cdots + y_{n-1} + \frac{1}{2}y_n\right)$$
$$= \frac{h}{2}(y_0 + 2y_1 + 2y_2 + \cdots + 2y_{n-1} + y_n),$$

(1)

where

$$y_0 = f(a), \quad y_1 = f(x_1), \quad \cdots, \quad y_{n-1} = f(x_{n-1}), \quad y_n = f(b).$$

The trapezoidal rule says: Use T to estimate the integral of f from a to b.

The Trapezoidal Rule

To approximate

$$\int_a^b f(x)\, dx,$$

use

$$T = \frac{h}{2}(y_0 + 2y_1 + 2y_2 + \cdots + 2y_{n-1} + y_n)$$

(2)

(for n subintervals of length $h = (b - a)/n$).

TABLE 4.3

x	$y = x^2$
1	1
$\frac{5}{4}$	$\frac{25}{16}$
$\frac{6}{4}$	$\frac{36}{16}$
$\frac{7}{4}$	$\frac{49}{16}$
2	4

Example 1   Use the trapezoidal rule with $n = 4$ to estimate

$$\int_1^2 x^2 \, dx.$$

Compare the estimate with the exact value of the integral.

Solution   The exact value of the integral is

$$\int_1^2 x^2 dx = \frac{x^3}{3}\Big]_1^2 = \frac{8}{3} - \frac{1}{3} = \frac{7}{3}.$$

To find the trapezoidal approximation, we divide the interval of integration into four subintervals of equal length and list the values of $y = x^2$ at the endpoints and subdivision points (see Table 4.3). We then evaluate Eq. (2) with $n = 4$ and $h = 1/4$:

$$T = \frac{h}{2}(y_0 + 2y_1 + 2y_2 + 2y_3 + y_4)$$

$$= \frac{1}{8}\left(1 + 2\left(\frac{25}{16}\right) + 2\left(\frac{36}{16}\right) + 2\left(\frac{49}{16}\right) + 4\right) = \frac{75}{32} = 2.34375.$$

The approximation overestimates the area by about half a percent of its true value. Each trapezoid contains slightly more than the corresponding strip under the curve (Fig. 4.36).

Controlling the Error in the Trapezoidal Approximation

Pictures suggest that the error

$$E_T = \int_a^b f(x) \, dx - T \qquad (3)$$

in the trapezoidal approximation will go down as the **step size** h decreases, because the trapezoids fit the curve better as their number increases. A theorem from advanced calculus assures us that this will always be the case if f has a continuous second derivative.

The Error Estimate for the Trapezoidal Rule

If f'' is continuous and M is any upper bound for the values of $|f''|$ on $[a, b]$, then

$$|E_T| \le \frac{b - a}{12} h^2 M. \qquad (4)$$

Although theory tells us there will always be a smallest safe value of M, in practice we can hardly ever find it. Instead, we find the best value we can and go on to estimate $|E_T|$ from there. This may seem sloppy, but it works. To make $|E_T|$ small for a given M, we just make h small.

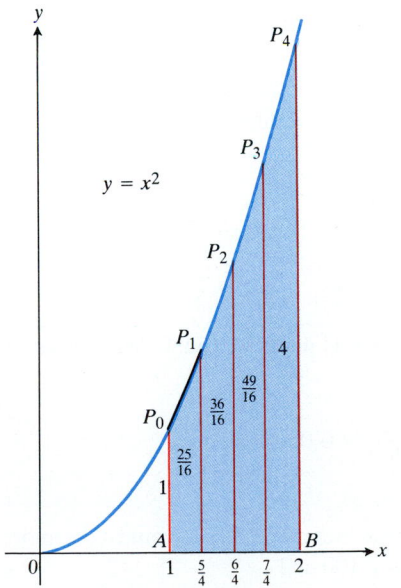

4.36   The trapezoidal approximation of the area under the graph of $y = x^2$ from $x = 1$ to $x = 2$ is a slight overestimate.

Example 2   Find an upper bound for the error in the approximation found in Example 1 for the value of

$$\int_1^2 x^2 \, dx.$$

Solution   We first find an upper bound M for the magnitude of the second derivative of $f(x) = x^2$ on the interval $1 \leq x \leq 2$. Since $f''(x) = 2$ for all x, we may safely take $M = 2$. With $b - a = 1$ and $h = 1/4$, Eq. (4) gives

$$|E_T| \leq \frac{b - a}{12} h^2 M = \frac{1}{12} \left(\frac{1}{4}\right)^2 (2) = \frac{1}{96}.$$

This is precisely what we find when we subtract $T = 75/32$ from $\int_1^2 x^2 \, dx = 7/3$, since $7/3 - 75/32 = -1/96$. Here we are able to give the error *exactly*, but this is exceptional.

Example 3   The trapezoidal rule is used to estimate the value of

$$\int_0^1 x \sin x \, dx$$

when $n = 10$ steps. Find an upper bound for the error in the estimate.

Solution   We use the formula

$$|E_T| \leq \frac{b - a}{12} h^2 M$$

with $b = 1$, $a = 0$, and $h = 1/n = 1/10$. This gives

$$|E_T| \leq \frac{1}{12} \left(\frac{1}{10}\right)^2 M = \frac{1}{1200} M.$$

The number M can be any upper bound for the values of $|f''|$ on $[0, 1]$. To choose a value for M, we calculate f'' to see how big it might be. A straightforward differentiation gives

$$f'' = 2 \cos x - x \sin x.$$

By the triangle inequality,

$$|f''| \leq 2|\cos x| + |x| \, |\sin x| \leq 2 + (1)(1) = 3$$

because $0 \leq x \leq 1$, and because $|\cos x|$ and $|\sin x|$ never exceed 1. We can safely take $M = 3$. Therefore,

$$|E_T| \leq \frac{1}{1200} (3) = \frac{1}{400} = 0.0025.$$

The error is no greater than 2.5×10^{-3}.

For greater accuracy we would not try to improve M but would take more steps. With $n = 100$ steps, for example, $h = 1/100$ and

$$|E_T| \leq \frac{1}{12} \left(\frac{1}{100}\right)^2 (2) < 1.67 \times 10^{-5}.$$

Example 4   As we shall see in Section 6.2, the value of ln 2 is given by the integral

$$\ln 2 = \int_1^2 \frac{1}{x}\, dx .$$

How many subintervals (steps) should be used in the trapezoidal rule to approximate the integral, and hence the value of ln 2, with an error of absolute value less than 10^{-4}?

Solution   To determine n, the number of subintervals, we use Eq. (4) with

$$b - a = 2 - 1 = 1, \qquad h = \frac{b - a}{n} = \frac{1}{n},$$

$$f''(x) = \frac{d^2}{dx^2}(x^{-1}) = 2x^{-3} = \frac{2}{x^3}.$$

Then

$$|E_T| \le \frac{b - a}{12} h^2 \max|f''(x)| = \frac{1}{12}\left(\frac{1}{n}\right)^2 \max\left|\frac{2}{x^3}\right|,$$

where max refers to the interval [1, 2].

This is one of the rare cases where we can find the exact value of $\max|f''|$. On [1, 2], $y = 2/x^3$ decreases steadily from a maximum of $y = 2$ to a minimum of $y = 1/4$ (Fig. 4.37). Therefore,

$$|E_T| \le \frac{1}{12}\left(\frac{1}{n}\right)^2 \cdot 2 = \frac{1}{6n^2}.$$

The error's absolute value will therefore be less than 10^{-4} if

$$\frac{1}{6n^2} < 10^{-4}, \qquad \frac{10^4}{6} < n^2, \qquad \frac{100}{\sqrt{6}} < n, \qquad \text{or} \qquad 40.83 < n .$$

The first integer beyond 40.83 is $n = 41$. With $n = 41$ subdivisions we can guarantee calculating ln 2 with an error of magnitude less than 10^{-4}. Any larger n will work, too. ∎

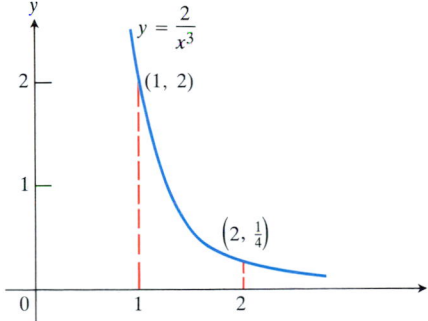

4.37   The continuous function $y = 2/x^3$ has its maximum value on [1, 2] at $x = 1$.

Simpson's Rule

Simpson's rule for approximating $\int_a^b f(x)\, dx$ is based on approximating f with quadratic polynomials instead of linear polynomials. We approximate the graph of f with parabolic arcs instead of line segments.

The area of the shaded region under the parabola in Fig. 4.38 is

$$A_p = \frac{h}{3}(y_0 + 4y_1 + y_2). \tag{5}$$

We can derive the formula for A_p in the following way. To simplify the algebra, we use the coordinate system shown in Fig. 4.39. The area under the parabola is the same no matter where the y-axis is, as long as we preserve the vertical scale. The parabola has an equation of the form

$$y = Ax^2 + Bx + C ,$$

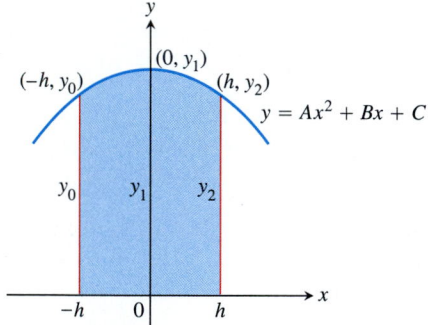

4.38   Simpson's rule approximates short stretches of curve with parabolic arcs.

so the area under it from $x = -h$ to $x = h$ is

$$A_p = \int_{-h}^{h} (Ax^2 + Bx + C)\, dx = \left[\frac{Ax^3}{3} + \frac{Bx^2}{2} + Cx\right]_{-h}^{h}$$

$$= \frac{2Ah^3}{3} + 2Ch = \frac{h}{3}(2Ah^2 + 6C). \tag{6}$$

Since the curve passes through the three points $(-h, y_0)$, $(0, y_1)$, and (h, y_2), we also have

$$y_0 = Ah^2 - Bh + C, \qquad y_1 = C, \qquad y_2 = Ah^2 + Bh + C,$$

from which we obtain

$$C = y_1,$$
$$Ah^2 - Bh = y_0 - y_1,$$
$$Ah^2 + Bh = y_2 - y_1, \tag{7}$$
$$2Ah^2 = y_0 + y_2 - 2y_1.$$

4.39   By integrating from $-h$ to h, we find the shaded area to be

$$A_p = \frac{h}{3}\Big(y_0 + 4y_1 + y_2\Big).$$

Hence, expressing the area A_p in terms of y_0, y_1, and y_2, we have

$$A_p = \frac{h}{3}(2Ah^2 + 6C) = \frac{h}{3}((y_0 + y_2 - 2y_1) + 6y_1) = \frac{h}{3}(y_0 + 4y_1 + y_2). \tag{8}$$

The formula for A_p counts area above the x-axis as positive and area below the axis as negative. Simpson's rule follows from partitioning $[a, b]$ into an even number of subintervals of equal length, applying the formula to successive interval pairs, and adding the results.

Simpson's Rule

To approximate

$$\int_a^b f(x)\, dx,$$

use

$$S = \frac{h}{3}(y_0 + 4y_1 + 2y_2 + 4y_3 + \cdots + 2y_{n-2} + 4y_{n-1} + y_n) \tag{9}$$

(n even, $h = (b - a)/n$).

The y's in Eq. (9) are the values of f at the partition points

$$x_0 = a, \quad x_1 = a + h, \quad x_2 = a + 2h, \quad \cdots, \quad x_{n-1} = a + (n-1)/h, \quad b = x_n$$

(Fig. 4.40).

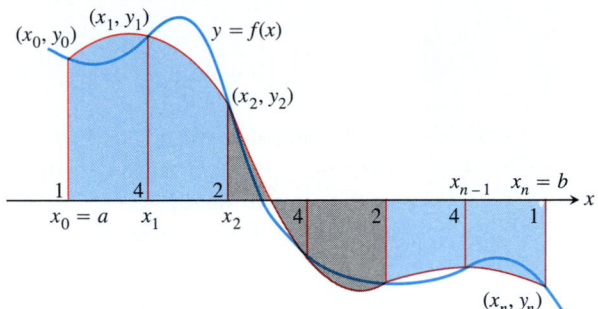

4.40 The y's in Eq. (9) are the values of f at the partition points. The subintervals have equal length, and the number of subintervals must be even.

Simpson's One-Third Rule

The rule

$$\text{Area} = \frac{h}{3}(y_0 + 4y_1 + y_2)$$

(Eq. (5) in the text), for which Thomas Simpson (1720–1761) became famous, was known long before he was born. It is another of history's beautiful quirks that one of the ablest mathematicians of eighteenth-century England is remembered not for his successful texts and his contributions to mathematical analysis but for a rule that was never his, that he never laid claim to, and that bears his name only because he happened to mention it in one of his books.

Error Control for Simpson's Rule

The Simpson's rule error,

$$E_S = \int_a^b f(x)\, dx - S, \tag{10}$$

decreases with the step size too. The inequality for controlling it, however, assumes f to have a continuous fourth derivative instead of merely a continuous second derivative. The formula, once again from advanced calculus, is this:

The Error Estimate for Simpson's Rule

If $f^{(4)}$ is continuous and M is any upper bound for the values of $\left|f^{(4)}\right|$ on $[a, b]$, then

$$|E_S| \leq \frac{b-a}{180} h^4 M. \tag{11}$$

As with the trapezoidal rule, we can almost never find the smallest possible value of M. We just find the best value we can and go on from there to estimate $|E_S|$.

Example 5   Use Simpson's rule with $n = 4$ to approximate

$$\int_0^1 5x^4\, dx.$$

What estimate does Eq. (11) give for the error in the approximation?

Solution   Again we have chosen an integral whose exact value we can calculate directly:

$$\int_0^1 5x^4\, dx = x^5 \Big]_0^1 = 1.$$

TABLE 4.4

x	$y = 5x^4$
0	0
$\dfrac{1}{4}$	$\dfrac{5}{256}$
$\dfrac{2}{4}$	$\dfrac{80}{256}$
$\dfrac{3}{4}$	$\dfrac{405}{256}$
1	5

To find the Simpson approximation, we partition the interval of integration into four subintervals and evaluate $f(x) = 5x^4$ at the partition points (Table 4.4). We then evaluate Eq. (9) with $n = 4$ and $h = 1/4$:

$$S = \frac{h}{3}(y_0 + 4y_1 + 2y_2 + 4y_3 + y_4)$$

$$= \frac{1}{12}\left(0 + 4\left(\frac{5}{256}\right) + 2\left(\frac{80}{256}\right) + 4\left(\frac{405}{256}\right) + 5\right) = 1.00260. \qquad \text{(Rounded)}$$

To estimate the error, we first find an upper bound M for the magnitude of the fourth derivative of $f(x) = 5x^4$ on the interval $0 \le x \le 1$. Since the fourth derivative has the constant value $f^{(4)}(x) = 120$, we may safely take $M = 120$. With $b - a = 1$ and $h = 1/4$, Eq. (11) then gives

$$|E_S| \le \frac{b-a}{180}h^4 M = \frac{1}{180}\left(\frac{1}{4}\right)^4(120) = \frac{1}{384} < 0.00261.$$

Calculus and Computers

Here is another example of calculus having something important to say about computation. It is a straightforward matter to implement the trapezoidal rule and Simpson's rule on a computer. But that in itself is not enough. We need to know how many steps to take to achieve the accuracy we want, and the guidance for *that* comes from calculus.

Which Rule Gives Better Results?

The answer lies in the error-control formulas for the two rules:

$$|E_T| \le \frac{b-a}{12}h^2 M, \qquad |E_S| \le \frac{b-a}{180}h^4 M. \qquad (12)$$

The M's of course mean different things, the first being an upper bound on $|f''|$ and the second an upper bound on $|f^{(4)}|$. But there is more. The factor $(b - a)/180$ in the Simpson formula is one-fifteenth of the factor $(b - a)/12$ in the trapezoidal formula. More important still, the Simpson formula has an h^4 while the trapezoidal formula has only an h^2. If h is one-tenth, then h^2 is one-hundredth but h^4 is only one ten-thousandth. If both M's are 1, for example, and $b - a = 1$, then, with $h = 1/10$,

$$|E_T| \le \frac{1}{12}\left(\frac{1}{10}\right)^2 \cdot 1 \le \frac{1}{1200}, \qquad (13)$$

while

$$|E_S| \le \frac{1}{180}\left(\frac{1}{10}\right)^4 \cdot 1 \le \frac{1}{1,800,000} = \frac{1}{1500} \cdot \frac{1}{1200}. \qquad (14)$$

For roughly the same amount of computational effort, we get better accuracy with Simpson's rule—at least in this case.

The h^2 versus h^4 is the key. If h is less than 1, then h^4 can be significantly smaller than h^2. On the other hand, if h equals 1, there is no difference between h^2 and h^4. If h is greater than 1, the value of h^4 may be significantly larger than the value of h^2. In the latter two cases, the error-control formulas offer little help. We have to go back to the geometry of the curve $y = f(x)$ to see whether trapezoids or parabolas, if either, are going to give the results we want.

Working with Numerical Data

The next example shows how to use Simpson's rule to estimate the integral of a function from values measured in the laboratory or in the field even when we have no formula for the function. The trapezoidal rule can be used the same way.

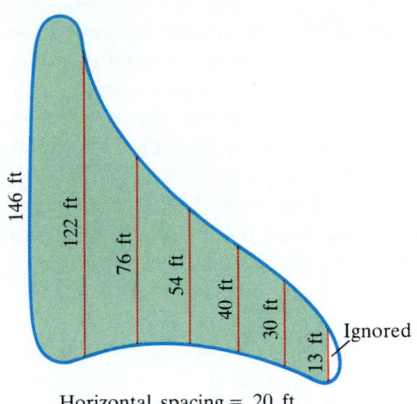

146 ft | 122 ft | 76 ft | 54 ft | 40 ft | 30 ft | 13 ft | Ignored

Horizontal spacing = 20 ft

4.41   The swamp in Example 6.

Example 6   A town wants to drain and fill a small polluted swamp (Fig. 4.41). The swamp averages 5 ft deep. About how many cubic yards of dirt will it take to fill the area after the swamp is drained?

Solution   To calculate the volume of the swamp, we estimate the surface area and multiply by 5. To estimate the area, we use Simpson's rule with $h = 20$ ft and the y's equal to the distances measured across the swamp, as shown in Fig. 4.41:

$$S = \frac{h}{3}(y_0 + 4y_1 + 2y_2 + 4y_3 + 2y_4 + 4y_5 + y_6)$$

$$= \frac{20}{3}(146 + 488 + 152 + 216 + 80 + 120 + 13) = 8100.$$

The volume is about $(8100)(5) = 40{,}500$ ft³ or 1500 yd³.

Round-off Errors

Although decreasing the step size h reduces the error in the Simpson and trapezoidal approximations in theory, it may fail to do so in practice. When h is very small, say $h = 10^{-5}$, the round-off errors in the arithmetic required to evaluate S and T may accumulate to such an extent that the error formulas no longer describe what is going on. Shrinking h below a certain size can actually make things worse. While this will not be an issue in the present book, you should consult a text on numerical analysis for alternative methods if you are having problems with round-off.

EXERCISES 4.7

Use (a) the trapezoidal rule and (b) Simpson's rule to approximate the integrals in Exercises 1–6 with $n = 4$. Then (c) find the integral's exact value for comparison.

1. $\int_0^2 x \, dx$

2. $\int_0^2 x^2 \, dx$

3. $\int_0^2 x^3 \, dx$

4. $\int_1^2 \frac{1}{x^2} \, dx$

5. $\int_0^4 \sqrt{x} \, dx$

6. $\int_0^\pi \sin x \, dx$

7. Use Eq. (4) to estimate the error in using the trapezoidal rule with $n = 10$ to estimate the value of

$$\ln 2 = \int_1^2 \frac{1}{x} \, dx.$$

For comparison, $\ln 2 = 0.69314718\ldots$.

8. Use Eq. (11) to estimate the error in using Simpson's rule with $n = 10$ to estimate the value of

$$\ln 2 = \int_1^2 \frac{1}{x} \, dx.$$

For comparison, $\ln 2 = 0.69314718\ldots$.

In Exercises 9–14, estimate the minimum number of subdivisions needed to approximate the integrals with an error of absolute value less than 10^{-4} by (a) the trapezoidal rule and (b) Simpson's rule.

9. $\int_0^2 x \, dx$

10. $\int_0^2 x^2 \, dx$

11. $\int_0^2 x^3 \, dx$

12. $\int_1^2 \frac{1}{x^2} \, dx$

13. $\int_1^4 \sqrt{x} \, dx$

14. $\int_0^\pi \sin x \, dx$

15. As the fish-and-game warden of your township, you are responsible for stocking the town pond (Fig. 4.42) with fish before fishing season. The average depth of the pond is 20 ft. You plan to start the season with one fish per 1000 ft³. You intend to have at least 25% of the opening day's fish population left at the end of the season. What is the maximum number of licenses the town can sell if the average seasonal catch is 20 fish per license?

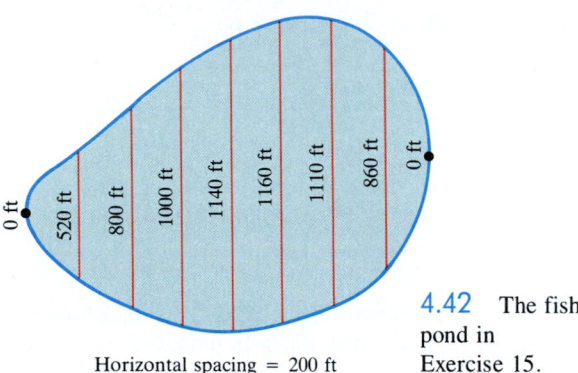

0 ft   520 ft   800 ft   1000 ft   1140 ft   1160 ft   1110 ft   860 ft   0 ft

Horizontal spacing = 200 ft

4.42   The fish pond in Exercise 15.

16. CALCULATOR The design of a new airplane requires a gasoline tank of constant cross-section area in each wing. A scale drawing of a cross section is shown in Fig. 4.43. The tank must hold 5000 lb of gasoline, which has a density of 42 lb/ft³. Estimate the length of the tank.

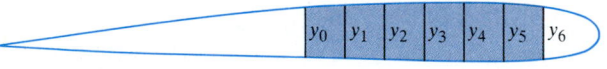

y_0 y_1 y_2 y_3 y_4 y_5 y_6

$y_0 = 1.5$ ft, $y_1 = 1.6$ ft, $y_2 = 1.8$ ft, $y_3 = 1.9$ ft,
$y_4 = 2.0$ ft, $y_5 = y_6 = 2.1$ ft   Horizontal spacing = 1 ft

4.43   The cross section of the gasoline tank in Exercise 16.

17. CALCULATOR A vehicle's aerodynamic drag is determined in part by its cross-section area and, all other things being equal, engineers try to make this area as small as possible. Use Simpson's rule to estimate the cross-section area of James Worden's solar-powered Solectria car at MIT, shown at right, from the diagram in Fig. 4.44.

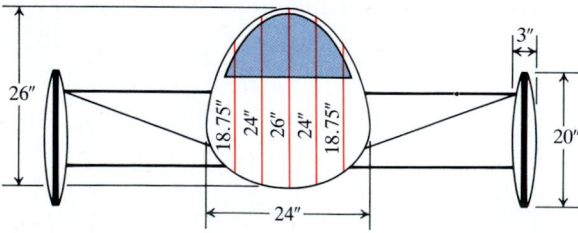

26″   18.75″   24″   26″   24″   18.75″   3″   20″

24″

4.44   The solar-powered car in Exercise 17.

18. CALCULATOR *The dye-dilution technique for measuring cardiac output.* Instead of measuring a patient's cardiac output with exhaled carbon dioxide, as in Exercise 19 in Section 3.1, a doctor may prefer to use the dye-dilution technique described here. You start by injecting 5–10 mg of dye in a main vein near the heart. The dye is drawn into the right side of the heart and then pumped through the lungs and out the left side of the heart into the aorta, where its concentration is measured each second as the blood flows past. The data in Table 4.5 and the plot in Fig. 4.45 show

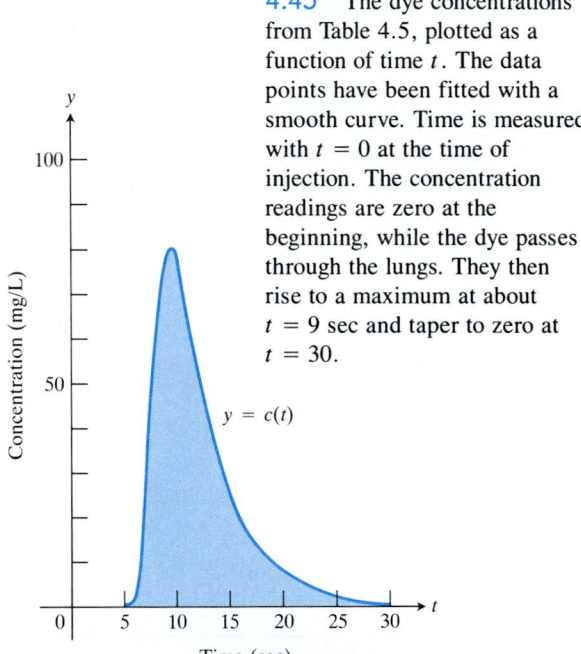

4.45   The dye concentrations from Table 4.5, plotted as a function of time t. The data points have been fitted with a smooth curve. Time is measured with $t = 0$ at the time of injection. The concentration readings are zero at the beginning, while the dye passes through the lungs. They then rise to a maximum at about $t = 9$ sec and taper to zero at $t = 30$.

Concentration (mg/L)

$y = c(t)$

Time (sec)

TABLE 4.5
Dye-dilution data for Exercise 18

Seconds after injection t	Dye concentration (adjusted for recirculation) $c(t)$	Seconds after injection t	Dye concentration (adjusted for recirculation) $c(t)$
1	0	16	18.5
2	0	17	14.5
3	0	18	11.5
4	0	19	9.1
5	0	20	7.3
6	1.5	21	5.7
7	38.0	22	4.5
8	67.0	23	3.6
9	80.0	24	2.8
10	73.0	25	2.3
11	61.0	26	1.8
12	48.0	27	1.4
13	36.0	28	1.1
14	29.0	29	0.9
15	23.0	30	0

the response of a healthy, resting patient to an injection of 5.6 mg of dye.

The patient's cardiac output is calculated by dividing the area under the concentration curve into the number of milligrams of dye and multiplying the result by 60:

$$\text{Cardiac output} = \frac{\text{milligrams of dye}}{\text{area under curve}} \times 60. \qquad (15)$$

You can see why if you check the units in which these quantities are measured. The dye is in milligrams, the area is in (milligrams/liter) × seconds, and

$$\frac{\text{mg}}{\frac{\text{mg}}{\text{L}} \cdot \text{sec}} \cdot 60 = \text{mg} \cdot \frac{\text{L}}{\text{mg} \cdot \text{sec}} \cdot 60 = \frac{\text{L}}{\text{sec}} \cdot 60 = \frac{\text{L}}{\text{min}}.$$

a) Use the trapezoidal rule and the data in Table 4.5 to calculate the area under the concentration curve in Fig. 4.45.

b) Then use Eq. (15) to calculate the patient's cardiac output.

19. CALCULATOR The rate at which flashbulbs give off light varies during the flash. For some bulbs, the rate at which light is produced, measured in lumens, reaches a peak and fades quickly (Fig. 4.46a). For others, the light, instead of reaching a peak, stays at a moderate level for a longer time (Fig. 4.46b).

To find out how much light reaches the film in a camera, we must know when the shutter opens and closes. A

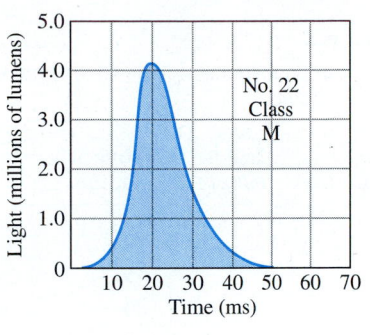

(a)

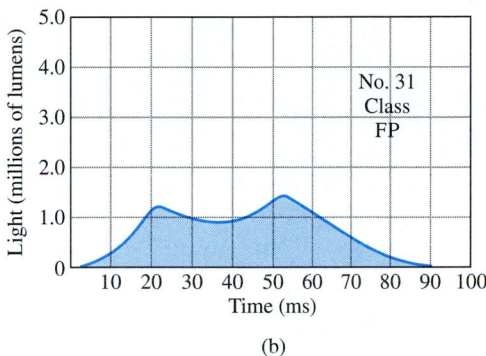

(b)

4.46 Flashbulb output data from Tables 4.6 and 4.7 plotted and connected by smooth curves.

typical focal-plane shutter opens 20 milliseconds (ms) after the button is pressed and stays open for 50 ms. The amount of light emitted by the flashbulb while the shutter is open is

$$A = \int_{20}^{70} L(t)\,dt \text{ lumen-ms,}$$

where $L(t)$ is the bulb's lumen rate as a function of time. Use the trapezoidal rule and the data from Tables 4.6 and 4.7 to estimate A and find out which bulb gets more light to

TABLE 4.6
Light output (in millions of lumens) vs. time (ms) for No. 22 flashbulb

Time after ignition	Light output	Time after ignition	Light output
0	0	30	1.7
5	0.2	35	0.7
10	0.5	40	0.35
15	2.6	45	0.2
20	4.2	50	0
25	3.0		

Data from *Photographic Lamp and Equipment Guide*, P4-15P, General Electric Company, Cleveland, Ohio.

TABLE 4.7
Light output (in millions of lumens) vs. time (ms) for
No. 31 flashbulb

Time after ignition	Light output	Time after ignition	Light output
0	0	50	1.3
5	0.1	55	1.4
10	0.3	60	1.3
15	0.7	65	1.0
20	1.0	70	0.8
25	1.2	75	0.6
30	1.0	80	0.3
35	0.9	85	0.2
40	1.0	90	0
45	1.1		

Data from *Photographic Lamp and Equipment Guide*, P4-15P, General Electric Company, Cleveland, Ohio.

the film. (From *Integration* by W. U. Walton et. al., Project CALC, Education Development Center, Inc., Newton, MA, 1975, p. 83.)

20. CALCULATOR *Usable values of the sine-integral function.* The sine-integral function,

$$\text{Si}(x) = \int_0^x \frac{\sin t}{t} \, dt, \qquad (\text{``Sine integral of } x\text{''})$$

is one of the many functions in engineering whose formulas cannot be simplified. There is no elementary formula for the antiderivative of $(\sin t)/t$. The values of $\text{Si}(x)$, however, are readily estimated by numerical integration.

Although the notation does not show it explicitly, the function being integrated is

$$f(t) = \begin{cases} \dfrac{\sin t}{t}, & t \neq 0, \\ 1, & t = 0, \end{cases}$$

the continuous extension of $(\sin t)/t$ to the interval $[0, x]$. The function has derivatives of all orders at every point of its domain. Its graph is smooth (Fig. 4.47) and you can expect good results from Simpson's rule.

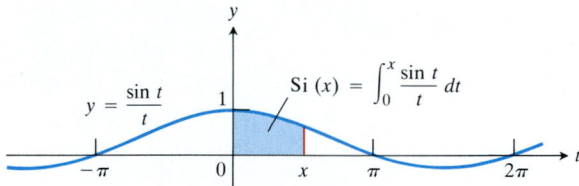

4.47   The continuous extension of $y = (\sin t)/t$. The sine-integral function $\text{Si}(x)$ is the subject of Exercises 20 and 25.

a) Use the fact that $|f^{(4)}| \leq 1$ on $[0, \pi/2]$ to give an upper bound for the error that will occur if

$$\text{Si}\left(\frac{\pi}{2}\right) = \int_0^{\pi/2} \frac{\sin t}{t} \, dt$$

is estimated by Simpson's rule with $n = 4$.
b) Estimate $\text{Si}(\pi/2)$ by Simpson's rule with $n = 4$.
c) Express the error bound you found in (a) as a percentage of the value you found in (b).

21. *Polynomials of low degree.* The magnitude of the error in the trapezoidal approximation of $\int_a^b f(x)\, dx$ is

$$|E_T| = \frac{b-a}{12} h^2 |f''(c)|,$$

where c is some point (usually unidentified) in (a, b). If f is a linear function of x, then $f''(c) = 0$, so $E_T = 0$ and T gives the exact value of the integral for any value of h. This is no surprise, really, for if f is linear, the line segments approximating the graph of f fit the graph exactly. The surprise comes with Simpson's rule. The magnitude of the error in Simpson's rule is

$$|E_S| = \frac{b-a}{180} h^4 |f^{(4)}(c)|,$$

where once again c lies in (a, b). If f is a polynomial of degree less than four, $f^{(4)} = 0$ no matter what c is, so $E_S = 0$ and S gives the integral's exact value—even if we use only two steps. As a case in point, use Simpson's rule with $n = 2$ to estimate

$$\int_0^2 x^3 \, dx.$$

Compare your answer with the integral's exact value.

22. CALCULATOR *The corrected trapezoidal rule.* It is possible (from advanced calculus) to write the error term for the trapezoidal rule in the form

$$E_T = \frac{h^2}{12} \int_a^b f''(x)\, dx - \frac{h^4}{720} \int_a^b f^{(4)}(x)\, dx + \cdots . \quad (16)$$

For sufficiently small h, we can improve T by subtracting the correction term

$$C(n) = \frac{h^2}{12} \int_a^b f''(x)\, dx = \frac{h^2}{12}\left(f'(b) - f'(a)\right)$$

$$= \frac{(b-a)^2}{12n^2}\left(f'(b) - f'(a)\right). \quad (17)$$

For the evaluation of $\ln 2 = \int_1^2 (1/x)\, dx$, for example, $f'(x) = -1/x^2$, and the correction is

$$C(n) = \frac{(b-a)^2}{12n^2}\left(f'(b) - f'(a)\right) = \frac{1}{12n^2}\left(\frac{3}{4}\right) = \frac{1}{16n^2}.$$

Subtracting $1/(16n^2)$ from $T(n)$ reduces the error in $T(n)$ to one-thousandth of its former size (Table 4.8).

TABLE 4.8
Corrected trapezoidal rule calculations of $\ln 2 = \int_1^2 \frac{1}{x} \, dx$

n	Estimate $T(n)$	Error $T(n) - \ln 2$	Correction $\frac{1}{16n^2}$	Corrected estimate $T(n) - \frac{1}{16n^2}$	Error $\left(T(n) - \frac{1}{16n^2}\right) - \ln 2$
10	0.6937 71403	0.0006 24222	0.0006 25	0.6931 46403	−0.0000 00776
20	0.6933 03382	0.0001 56201	0.0001 5625	0.6931 47132	−0.0000 00049
30	0.6932 16615	0.0000 69434	0.0000 69444	0.6931 47171	−0.0000 00010
40	0.6931 86240	0.0000 39059	0.0000 39062	0.6931 47181	0

Use the trapezoidal rule and then the corrected formula $T(n) - C(n)$ to estimate the value of

$$\int_0^1 \frac{4}{1 + x^2} \, dx$$

for $n = 10$. The integral's exact value is π.

Computer or Programmable Calculator

At the end of this exercise set you will find a BASIC program for evaluating integrals numerically with the trapezoidal rule and Simpson's rule. If you can implement the program, or if you have access to an integral evaluator of some other kind, estimate the values of the integrals in Exercises 23–26.

23. $\int_{-1}^1 2\sqrt{1 - x^2} \, dx$ (The exact value is π.)

24. $\int_0^1 \sqrt{1 + x^4} \, dx$ (A nonelementary integral that arose in Newton's research)

25. $\int_0^{\pi/2} \frac{\sin x}{x} \, dx$ (The integral from Exercise 20. To avoid division by zero, you may have to start the integration at a small positive number like 10^{-6} instead of 0.)

26. $\int_0^{\pi/2} \sin(x^2) \, dx$ (An integral associated with the diffraction of light)

EXPLORER PROGRAM

Integral Evaluation   Evaluates $\int_a^b f(x) \, dx$ by the trapezoidal rule, Simpson's rule, and Romberg's accelerated version of the trapezoidal rule.

✳ A COMPUTER PROGRAM FOR NUMERICAL INTEGRATION

Here is a BASIC computer program for approximating the value of

$$\int_a^b f(x) \, dx.$$

We begin with the trapezoidal rule and then show how to change the program for Simpson's rule.

PROGRAM LISTING

```
10      DEF FNF(X) = F(X)
20      INPUT "ENTER LOWER LIMIT     "; A
30      INPUT "ENTER UPPER LIMIT     "; B
```

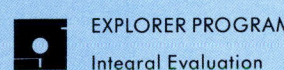

COMMENTS

Key in the formula for $f(x)$.
Enter the lower limit of integration.
Enter the upper limit.

PROGRAM LISTING

```
40        INPUT "ENTER NUMBER TO PRINT     "; C
50        FOR K = 1 TO C

60        N = 10 * K
70        S = 0
80        H = (B − A)/N
90        FOR X = A TO B − H + .001 STEP H


100       S = S + (FNF(X) + FNF(X + H))/2

110       NEXT X
120       PRINT N; H * S
130       NEXT K
140       END
```

COMMENTS

Enter the number of data to print.
Each value of K gives one trapezoidal
approximation.
Number of subintervals in the Kth approximation
Begin at 0 for each new K.
Width of each subinterval
The left-hand endpoints of the N subintervals
(CHANGE for Simpson's rule). The termination
point is adjusted to avoid a premature ending.
Cumulative trapezoidal sum (CHANGE for
Simpson's rule).
Returns to line 90 for next step.

Returns to line 50 with new K.

Table 4.9 shows the results for $f(x) = 1/x$, $a = 1$, $b = 2$, and $c = 5$. The numbers $T(n)$ are the trapezoidal approximations to

$$\ln 2 = \int_1^2 \frac{1}{x}\, dx$$

for $n = 10, 20, 30, 40,$ and 50. Notice that with $n = 30$ we already have the accuracy required in Example 4. The error-control formula in Eq. (4) always gives a safe worst-case value; the results in practice are often better.

To change the program for trapezoidal approximations to Simpson's rule, do the following:

REVISED COMMANDS:

```
90        FOR X = A TO B − 2*H STEP 2*H
100       LET S = S + (FNF(X) + 4*FNF(X + H) + FNF(X + 2*H))/3
```

COMMENT

We now group the subintervals in pairs.
Cumulative Simpson sum.

Table 4.10 shows the result of using the Simpson program with $f(x) = 1/x$, $a = 1$, $b = 2$, and $c = 5$ to approximate $\ln 2 = \int_1^2 (1/x)\, dx$ with $n = 10, 20, 30,$ $40,$ and 50 subintervals. Notice the improvement over results with the trapezoidal rule: The approximation for $n = 50$ rounds accurately to six places instead of three.

TABLE 4.9
Trapezoidal approximations of $\ln 2 = \int_1^2 (1/x)\, dx$ with error limits from Eq. (4)

n	$T(n)$	Error limit from Eq. (4)
10	0.6937 713	0.0006 242
20	0.6933 035	0.0001 564
30	0.6932 168	0.0000 697
40	0.6931 864	0.0000 393
50	0.6931 723	0.0000 252

For comparison: $\ln 2 = 0.6931471805...$

TABLE 4.10
Simpson's rule approximations of $\ln 2 = \int_1^2 (1/x)\, dx$ with error limits from Eq. (11)

n	$S(n)$	Error limit from Eq. (11)
10	0.6931 502	0.0000 031
20	0.6931 476	0.0000 005
30	0.6931 475	0.0000 004
40	0.6931 474	0.0000 003
50	0.6931 473	0.0000 002

For comparison: $\ln 2 = 0.6931471805...$

REVIEW QUESTIONS

1. Can a function have more than one antiderivative? If so, how are the antiderivatives related? What theorem in Chapter 3 is the key to the relationship? Explain the connection.

2. What is an indefinite integral? How do you evaluate one? What general formulas do you know for evaluating definite integrals?

3. What is an initial value problem? How do you solve one? Give an example.

4. If you know the acceleration of a body moving along a coordinate line as a function of time, what more do you need to know to find the body's position as a function of time? Give an example.

5. What is a partition of an interval? What is the norm of a partition? Give examples.

6. What is a Riemann sum? What does it mean for a number to be a limit of Riemann sums? Give an example.

7. How is $\int_a^b f(x)\, dx$ defined? What is it called? When does it exist?

8. What is the relation between definite integrals and area? Give an example. Describe some other interpretations of definite integrals.

9. State eight rules for working with definite integrals (Table 4.2). Give an example of each.

10. State the Mean Value Theorem for Definite Integrals.

11. What is the average value of an integrable function over a closed interval? How do average values arise in applications? Give an example.

12. State the fundamental theorem of integral calculus. What is it good for? Illustrate each part of the theorem with an example.

13. Must every continuous function be the derivative of some other function? Explain.

14. Write a formula for the area $A(x)$ of the region that is bounded above by the semicircle $y = \sqrt{4 - t^2}$, below by the t-axis, on the left by the line $t = -1$, and on the right by the line $t = x$, $-1 < x < 2$. Find dA/dx.

15. True, or false?
   a) If $\int_a^b f(x)\, dx$ exists, then f is differentiable on $[a, b]$.
   b) If f is differentiable on $[a, b]$, then $\int_a^b f(x)\, dx$ exists.
   c) If f is continuous on $[a, b]$, then $\int_a^b f(x)\, dx$ exists.
   d) If f is continuous on $[a, b]$, then $F(x) = \int_a^x f(t)\, dt$ is continuous on $[a, b]$.
   e) If f is continuous on $[a, b]$, then $F(x) = \int_a^x f(t)\, dt$ is differentiable on $[a, b]$.

16. How does integration by substitution work? What do you do about the limits in definite integrals? Give examples.

17. What numerical methods are available for estimating the values of definite integrals that cannot be evaluated directly with antiderivatives? How accurate are they? How can the accuracy sometimes be improved? Give examples.

MISCELLANEOUS EXERCISES

Solve the initial value problems in Exercises 1–6.

1. $\dfrac{dy}{dx} = \dfrac{x^2 + 1}{x^2}$, $y = -1$ when $x = 1$

2. $\dfrac{dy}{dx} = \left(x + \dfrac{1}{x}\right)^2$, $y = 1$ when $x = 1$

3. $\dfrac{ds}{dt} = \dfrac{4t}{(1 + t^2)^2}$, $s = 0$ when $t = 0$

4. $\dfrac{dr}{ds} = s\sqrt{s^2 - 4}$, $r = 3$ when $s = 2$

5. $\dfrac{d^2y}{dx^2} = 15\sqrt{x} + \dfrac{3}{\sqrt{x}}$, $\dfrac{dy}{dx} = 8$ and $y = 0$ when $x = 1$

6. $\dfrac{d^3u}{dr^3} = -\cos r$; $\dfrac{d^2u}{dr^2} = 0$, $\dfrac{du}{dr} = 0$,

   and $u = -1$ when $r = 0$

7. Suppose $F(x)$ is an antiderivative of $f(x) = \sqrt{1 + x^4}$. Express $\int_0^1 \sqrt{1 + x^4}\, dx$ in terms of F.

8. Show that $y = x^2 + \displaystyle\int_1^x \dfrac{1}{t}\, dt + 1$ solves the following initial value problem.

$$y'' = 2 - \dfrac{1}{x^2}, \quad y = 2 \text{ and } y' = 3 \quad \text{when } x = 1$$

9. Does any function $y = f(x)$ satisfy all of the following conditions? If so, what is it? If not, why not?
   i) $d^2y/dx^2 = 0$ for all x
   ii) $dy/dx = 1$ when $x = 0$
   iii) $y = 0$ when $x = 0$

10. Find an equation for the curve in the xy-plane that passes through the point $(1, -1)$ if its slope at x is always $3x^2 + 2$.

11. You sling a shovelful of dirt up from the bottom of a hole with an initial velocity of 32 ft/sec. The dirt must rise 17 ft above the release point to clear the edge of the hole. Is that enough speed to get the dirt out, or had you better duck?

12. The acceleration of a particle moving along a coordinate

line is $d^2s/dt^2 = 2 + 6t$ m/sec². At $t = 0$, the velocity is 4 m/sec. Find the velocity as a function of t. Then find how far the particle moves during the first second of its trip, from $t = 0$ to $t = 1$.

13. *Stopping a motorcycle.*   The State of Illinois Cycle Rider Safety Program requires riders to be able to brake from 30 mph (44 ft/sec) to 0 in 45 ft. What constant deceleration does it take to do that? To find out, carry out these steps:

STEP 1:   Solve the following initial value problem. The answer will involve k.

$$\frac{d^2s}{dt^2} = -k, \quad ds/dt = 44 \text{ and } s = 0 \text{ when } t = 0$$

STEP 2:   Find the time t^* ("tee star") when $ds/dt = 0$. The answer will still involve k.

STEP 3:   Solve the equation $s(t^*) = 45$ for k.

14. Which of the graphs in Fig. 4.48 shows the solution of the initial value problem

$$\frac{dy}{dx} = 2x, \qquad y = 4 \text{ when } x = 1?$$

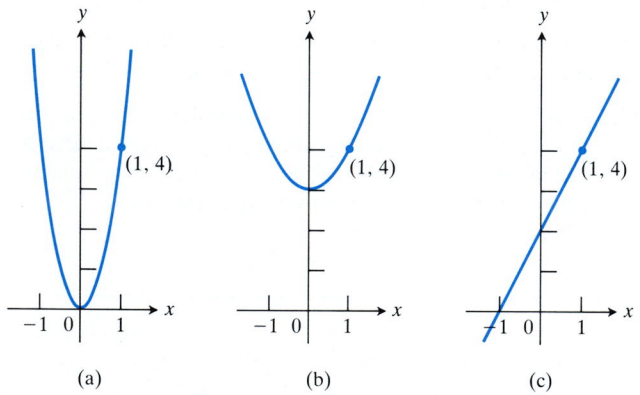

(a)   (b)   (c)

4.48   The graphs for Exercise 14.

15. Skydivers A and B are in a helicopter hovering at 6400 ft. Skydiver A jumps and descends for 4 sec before opening her parachute. The helicopter then climbs to 7000 ft and hovers there. Forty-five seconds after A leaves the aircraft, B jumps and descends for 13 sec before opening her parachute. Both skydivers descend at 16 ft/sec with parachute open. Assume that the skydivers fall freely (no effective air resistance) before their parachutes open.
a) At what altitude does A's parachute open?
b) At what altitude does B's parachute open?
c) Which skydiver lands first?

16. Show that

$$y = \frac{1}{a} \int_0^x f(t) \sin a(x - t) \, dt$$

solves the initial value problem

$$\frac{d^2y}{dx^2} + a^2 y = f(x), \quad \frac{dy}{dx} = 0 \text{ and } y = 0 \text{ when } x = 0.$$

(*Hint:* $\sin(ax - at) = \sin ax \cos at - \cos ax \sin at$.)

Express the limits in Exercises 17 and 18 as definite integrals.

17. $\displaystyle \lim_{\|P\| \to 0} \sum_{k=1}^{n} \frac{1}{c_k} \Delta x_k$, where P is a partition of $[1, 2]$

18. $\displaystyle \lim_{\|P\| \to 0} \sum_{k=1}^{n} e^{c_k} \Delta x_k$, where P is a partition of $[0, 1]$

19. Suppose $\displaystyle \int_{-2}^{2} f(x) \, dx = 4$, $\displaystyle \int_{2}^{5} f(x) \, dx = 3$, $\displaystyle \int_{-2}^{5} g(x) \, dx = 2$.

Which, if any, of the following statements are true?

a) $\displaystyle \int_{5}^{2} f(x) \, dx = -3$ b) $\displaystyle \int_{-2}^{5} (f(x) + g(x)) = 9$

c) $f(x) \le g(x)$ on the interval $-2 \le x \le 5$

20. Suppose $\displaystyle \int_{0}^{1} f(x) \, dx = \pi$. Find

a) $\displaystyle \int_{0}^{1} f(t) \, dt$,   b) $\displaystyle \int_{1}^{0} f(y) \, dy$,

c) $\displaystyle \int_{0}^{1} -3f(z) \, dz$.

Evaluate the integrals in Exercises 21–38.

21. $\displaystyle \int_{2}^{3} \left(t - \frac{2}{t}\right)\left(t + \frac{2}{t}\right) dt$ 22. $\displaystyle \int_{-1}^{0} (1 - 3w)^2 \, dw$

23. $\displaystyle \int_{-1}^{0} \frac{12 \, dx}{(2 - 3x)^2}$ 24. $\displaystyle \int_{1}^{4} \frac{(1 + \sqrt{u})^{1/2}}{\sqrt{u}} \, du$

25. $\displaystyle \int_{0}^{1} \frac{dr}{\sqrt[3]{(7 - 5r)^2}}$ 26. $\displaystyle \int_{0}^{1} t^{1/3} (1 + t^{4/3})^{-7} \, dt$

27. $\displaystyle \int_{1}^{25} \frac{dv}{\sqrt{v} \sqrt{2\sqrt{v} - 1}}$

28. $\displaystyle \int_{-1}^{0} \sqrt{u^4 + 4u^3 + 6u^2 + 4u + 1} \, du$

29. $\displaystyle \int_{\pi}^{3\pi} \cot^2 \frac{x}{6} \, dx$ 30. $\displaystyle \int_{0}^{\pi} \tan^2 \frac{\theta}{3} \, d\theta$

31. $\displaystyle \int_{0}^{\pi/2} 5 (\sin t)^{3/2} \cos t \, dt$ 32. $\displaystyle \int_{-1}^{1} 2y \sin(1 - y^2) \, dy$

33. $\displaystyle \int_{-\pi/2}^{\pi/2} 15 \sin^4 3x \cos 3x \, dx$

34. $\displaystyle\int_{0}^{\pi/2} \frac{3 \sin x \cos x}{\sqrt{1 + 3 \sin^2 x}}\, dx$

35. $\displaystyle\int_{0}^{1} \pi z^2 \sec^2 \left(\frac{\pi z^3}{3}\right) dz$ **36.** $\displaystyle\int_{0}^{\pi/4} \cot^2 \left(s + \frac{\pi}{4}\right) ds$

37. $\displaystyle\int_{0}^{\pi/3} \frac{\tan \theta}{\sqrt{2 \sec \theta}}\, d\theta$ **38.** $\displaystyle\int_{\pi^2/36}^{\pi^2/4} \frac{\cos \sqrt{t}}{\sqrt{t} \sin \sqrt{t}}\, dt$

39. CALCULATOR Compute the average value of the temperature function

$$f(x) = 37 \sin \left(\frac{2\pi}{365} (x - 101)\right) + 25$$

for a 365-day year. This is one way to estimate the annual mean air temperature in Fairbanks, Alaska. The National Weather Service's official figure, a numerical average of the daily normal mean air temperatures for the year, is 25.7°F, which is slightly higher than the average value of $f(x)$. Figure 2.28 shows why.

40. Let f be a function that is differentiable on $[a, b]$. In Chapter 2 we defined the average rate of change of f on $[a, b]$ to be

$$\frac{f(b) - f(a)}{b - a}$$

and the instantaneous rate of change of f at x to be $f'(x)$. In this chapter we defined the average value of a function. For the new definition of average to be consistent with the old one, we should have

$$\frac{f(b) - f(a)}{b - a} = \text{average value of } f' \text{ on } [a, b].$$

Show that this is the case.

41. The area of the region in the xy-plane enclosed by the x-axis, the curve $y = f(x)$, $f(x) \geq 0$, and the lines $x = 1$ and $x = b$ is equal to $\sqrt{b^2 + 1} - \sqrt{2}$ for all $b > 1$. Find $f(x)$.

42. Find

a) $\displaystyle\lim_{h \to 0} \frac{1}{h} \int_{x}^{x+h} \frac{du}{u + \sqrt{u^2 + 1}},$

b) $\displaystyle\lim_{x \to 2} \left(\frac{x}{x - 2} \int_{2}^{x} f(t)\, dt\right)$, assuming f is continuous.

43. Suppose x and y are related by the equation

$$x = \int_{0}^{y} \frac{1}{\sqrt{1 + 4t^2}}\, dt.$$

Show that d^2y/dx^2 is proportional to y and find the constant of proportionality.

44. Prove that

$$\int_{0}^{x} \left(\int_{0}^{u} f(t)\, dt\right) du = \int_{0}^{x} f(u)(x - u)\, du.$$

(*Hint:* Express the integral on the right-hand side as the difference of two integrals. Then show that both sides of the equation have the same derivative with respect to x.)

45. Find $f(4)$ if

a) $\displaystyle\int_{0}^{x^2} f(t)\, dt = x \cos \pi x$,   b) $\displaystyle\int_{0}^{f(x)} t^2\, dt = x \cos \pi x$.

46. Find $f(\pi/2)$ from the following information.

i) f is positive and continuous.

ii) The area under the curve $y = f(x)$ from $x = 0$ to $x = a$ is

$$\frac{a^2}{2} + \frac{a}{2} \sin a + \frac{\pi}{2} \cos a.$$

Calculator Exercises (Mostly)

47. What step size h would you use to be sure of estimating the value of

$$\ln 3 = \int_{1}^{3} \frac{1}{x}\, dx$$

by Simpson's rule with an error of no more than 10^{-4} in absolute value?

48. A brief calculation shows that if $0 \leq x \leq 1$, then the second derivative of $f(x) = \sqrt{1 + x^4}$ lies between 0 and 8. Based on this, about how many subintervals would you need to estimate the integral of f from 0 to 1 with an error no greater than 10^{-3} in absolute value?

49. A direct calculation shows that

$$\int_{0}^{\pi} 2 \sin^2 x\, dx = \pi.$$

How close do you come to this value by using the trapezoidal rule with $n = 6$? Simpson's rule with $n = 6$?

50. You are planning to use Simpson's rule to estimate the value of the integral

$$\int_{1}^{2} f(x)\, dx$$

with an error magnitude less than 10^{-5}. You have determined that $|f^{(4)}(x)| \leq 3$ throughout the interval of integration. How many subintervals should you use to ensure the required accuracy? (Remember that for Simpson's rule the number has to be even.)

51. *A new parking lot.* To meet the demand for parking, your town has allocated the area shown in Fig. 4.49. As the town

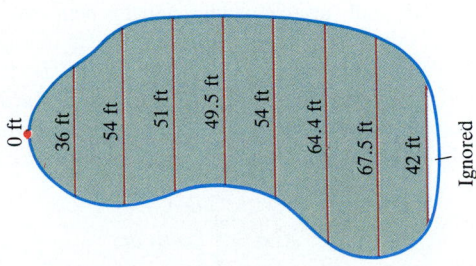

Horizontal spacing = 15 ft

4.49 The parking lot in Exercise 51.

engineer, you have been asked by the town council to find out if the lot can be built for $11,000. The cost to clear the land will be 10 cents a square foot, and the lot will cost 2 dollars a square foot to pave. Can the job be done for $11,000?

52. *The error function.*   The error function,

$$\text{erf}(x) = \frac{2}{\sqrt{\pi}} \int_0^x e^{-t^2} \, dt,$$

important in probability and in the theories of heat flow and signal transmission, must be evaluated numerically because there is no elementary expression for the antiderivative of e^{-t^2}.

a) Use Simpson's rule with $n = 10$ to estimate erf(1).

b) In [0, 1],

$$\left| \frac{d^4}{dt^4} \left(e^{-t^2} \right) \right| \le 12.$$

Give an upper bound for the magnitude of the error of the estimate in (a).

Leibniz's Rule

In applications, we sometimes encounter functions like

$$f(x) = \int_{\sin x}^{x^2} (1 + t) \, dt \quad \text{and} \quad g(x) = \int_{\sqrt{x}}^{2\sqrt{x}} \sin t^2 \, dt,$$

defined by integrals that have variable upper limits of integration and variable lower limits of integration at the same time. The first integral can be evaluated directly but the second cannot. We may find the derivative of either integral, however, by a formula called **Leibniz's rule:**

Leibniz's Rule

If f is continuous on $[a, b]$, and $u(x)$ and $v(x)$ are differentiable functions of x whose values lie in $[a, b]$, then

$$\frac{d}{dx} \int_{u(x)}^{v(x)} f(t) \, dt = f(v(x)) \frac{dv}{dx} - f(u(x)) \frac{du}{dx}.$$

Figure 4.50 gives a geometric interpretation of Leibniz's rule. It shows a carpet of variable width $f(t)$ that is being rolled up at the left at the same time x as it is being unrolled at the right. (In this interpretation time is x, not t.) At time x, the floor is covered from $u(x)$ to $v(x)$. The rate du/dx at which the carpet is being rolled up need not be the same as the rate dv/dx at which the carpet is being laid down. At any given time x, the area covered by carpet is

$$A(x) = \int_{u(x)}^{v(x)} f(t) \, dt.$$

At what rate is the covered area changing? At the instant x, $A(x)$ is increasing by the width $f(v(x))$ of the unrolling carpet times the rate dv/dx at which the carpet is being unrolled. That is, $A(x)$

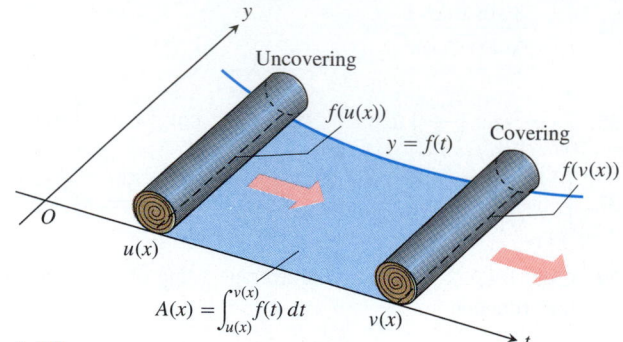

$$A(x) = \int_{u(x)}^{v(x)} f(t) \, dt$$

4.50   Rolling and unrolling a carpet: a geometric interpretation of Leibniz's rule:

$$\frac{dA}{dx} = f(v(x)) \frac{dv}{dx} - f(u(x)) \frac{du}{dx}.$$

is being increased at the rate

$$f(v(x)) \frac{dv}{dx}.$$

At the same time, A is being decreased at the rate

$$f(u(x)) \frac{du}{dx},$$

the width at the end that is being rolled up times the rate du/dx. The net rate of change in A is

$$\frac{dA}{dx} = f(v(x)) \frac{dv}{dx} - f(u(x)) \frac{du}{dx},$$

which is precisely Leibniz's rule.

To prove the rule, let F be an antiderivative of f on $[a, b]$. Then

$$\int_{u(x)}^{v(x)} f(t) \, dt = F(v(x)) - F(u(x)). \qquad (1)$$

Differentiating both sides of this equation with respect to x gives the equation we want:

$$\frac{d}{dx} \int_{u(x)}^{v(x)} f(t) \, dt = \frac{d}{dx} \Big[F(v(x)) - F(u(x)) \Big]$$

$$= F'(v(x)) \frac{dv}{dx} - F'(u(x)) \frac{du}{dx} \quad \text{(Chain Rule)}$$

$$= f(v(x)) \frac{dv}{dx} - f(u(x)) \frac{du}{dx}.$$

You will see another nice way to derive the rule if you do Miscellaneous Exercise 55 in Chapter 12.

Use Leibniz's rule to find the derivatives of the functions in Exercises 53–55.

53. $f(x) = \int_{1/x}^{x} \frac{1}{t} \, dt$

54. $f(x) = \int_{\cos x}^{\sin x} \frac{1}{1 - t^2} \, dt$

55. $g(y) = \int_{\sqrt{y}}^{2\sqrt{y}} \sin t^2 \, dt$

56. Use Leibniz's rule to find the value of x that maximizes the value of the integral

$$\int_{x}^{x+3} t(5 - t)\, dt.$$

Problems like this arise in the mathematical theory of political elections. See "The Entry Problem in a Political Race," by Steven J. Brams and Philip D. Straffin, Jr., in *Political Equilibrium,* Peter Ordeshook and Kenneth Shepfle, Editors, Kluwer-Nijhoff, Boston, 1982, pp. 181–95.

Bounded Piecewise Continuous Functions

Although we are mainly interested in continuous functions, many functions in applications are piecewise continuous. All bounded piecewise continuous functions are integrable (as are many unbounded functions, as we shall see in Chapter 7). **Bounded** on an interval I means that for some finite constant M, $|f(x)| \le M$ for all x in I. **Piecewise continuous** on I means that I can be partitioned into open subintervals on which f is continuous. To integrate a bounded piecewise continuous function that has a continuous extension to each closed subinterval of the partition, we integrate the individual extensions and add the results. The integral of the function

$$f(x) = \begin{cases} 1 - x, & -1 \le x < 0, \\ x^2, & 0 \le x < 2, \\ -1, & 2 \le x \le 3, \end{cases}$$

(Fig. 4.51) over $[-1, 3]$ is

$$\int_{-1}^{3} f(x)\, dx = \int_{-1}^{0} (1 - x)\, dx + \int_{0}^{2} x^2 dx + \int_{2}^{3} (-1)\, dx$$

$$= \left[x - \frac{x^2}{2} \right]_{-1}^{0} + \left[\frac{x^3}{3} \right]_{0}^{2} + \left[-x \right]_{2}^{3}$$

$$= \frac{3}{2} + \frac{8}{3} - 1 = \frac{19}{6}.$$

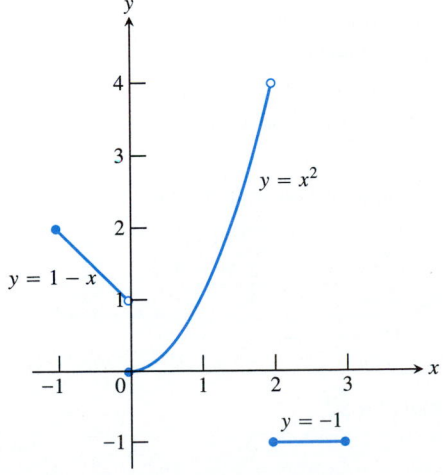

4.51   The graph of a piecewise continuous function. Functions like this are integrated piece by piece.

The Fundamental Theorem applies to bounded piecewise continuous functions with the restriction that the equation $(d/dx)\int_{a}^{x} f(t)\, dt = f(x)$ is expected to hold only at values of x at which f is continuous. There is a similar restriction on Leibniz's rule.

Graph the functions in Exercises 57–62 and integrate them over their domains.

57. $f(x) = \begin{cases} x^{2/3}, & -8 \le x < 0, \\ -4, & 0 \le x \le 3 \end{cases}$

58. $f(x) = \begin{cases} \sqrt{-x}, & -4 \le x < 0, \\ x^2 - 4, & 0 \le x \le 3 \end{cases}$

59. $g(t) = \begin{cases} t, & 0 \le t < 1, \\ \sin \pi t, & 1 \le t \le 2 \end{cases}$

60. $h(z) = \begin{cases} \sqrt{1 - z}, & 0 \le z < 1, \\ (7z - 6)^{-1/3}, & 1 \le z \le 2 \end{cases}$

61. $f(x) = \begin{cases} 1, & -2 \le x < -1, \\ 1 - x^2, & -1 \le x < 1, \\ 2, & 1 \le x \le 2 \end{cases}$

62. $h(r) = \begin{cases} r, & -1 \le r < 0, \\ 1 - r^2, & 0 \le r < 1, \\ 1, & 1 \le r \le 2 \end{cases}$

63. Find the average value of the function graphed in Fig. 4.52(a).

64. Find the average value of the function graphed in Fig. 4.52(b).

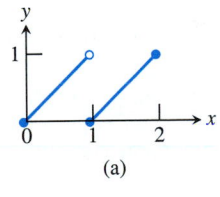

(a)

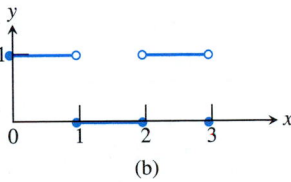

(b)

4.52   The graphs for Exercises 63 and 64.

Approximating Finite Sums with Integrals

In many applications of calculus, integrals are used to approximate finite sums—the reverse of the usual procedure of using finite sums to approximate integrals. Here is an example.

EXAMPLE   Estimate the sum of the square roots of the first n positive integers, $\sqrt{1} + \sqrt{2} + \cdots + \sqrt{n}$.

Solution See Fig. 4.53. The integral

$$\int_0^1 \sqrt{x}\, dx = \frac{2}{3} x^{3/2} \Big]_0^1 = \frac{2}{3}$$

is the limit of the sums

$$S_n = \sqrt{\frac{1}{n}} \cdot \frac{1}{n} + \sqrt{\frac{2}{n}} \cdot \frac{1}{n} + \cdots + \sqrt{\frac{n}{n}} \cdot \frac{1}{n}$$

$$= \frac{\sqrt{1} + \sqrt{2} + \cdots + \sqrt{n}}{n^{3/2}}.$$

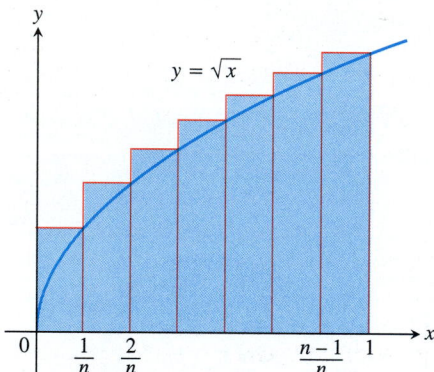

4.53 The relation of the circumscribed rectangles to the integral $\int_0^1 \sqrt{x}\, dx$ leads to an estimate of the sum $\sqrt{1} + \sqrt{2} + \sqrt{3} + \cdots + \sqrt{n}$.

Therefore, when n is large, S_n will be close to 2/3 and we will have

$$\text{Root sum} = \sqrt{1} + \sqrt{2} + \cdots + \sqrt{n} = S_n \cdot n^{3/2} \approx \frac{2}{3} n^{3/2}.$$

The following table shows how good the approximation can be.

n	Root sum	$(2/3)n^{3/2}$	Relative error
10	22.468	21.082	$1.386/\,22.468 \approx 6\%$
50	239.04	235.70	1.4%
100	671.46	666.67	0.7%
1000	21,097	21,081	0.07%

65. Evaluate

$$\lim_{n \to \infty} \frac{1^5 + 2^5 + 3^5 + \cdots + n^5}{n^6}$$

by showing that the limit is

$$\int_0^1 x^5 \, dx$$

and evaluating the integral.

66. See Exercise 65. Evaluate

$$\lim_{n \to \infty} \frac{1}{n^4} (1^3 + 2^3 + 3^3 + \cdots + n^3).$$

67. Let $f(x)$ be a continuous function. Express

$$\lim_{n \to \infty} \frac{1}{n}\left[f\!\left(\frac{1}{n}\right) + f\!\left(\frac{2}{n}\right) + \cdots + f\!\left(\frac{n}{n}\right) \right]$$

as a definite integral.

68. Use the result of Exercise 67 to evaluate

a) $\displaystyle\lim_{n \to \infty} \frac{1}{n^2} (2 + 4 + 6 + \cdots + 2n),$

b) $\displaystyle\lim_{n \to \infty} \frac{1}{n^{16}} (1^{15} + 2^{15} + 3^{15} + \cdots + n^{15}),$

c) $\displaystyle\lim_{n \to \infty} \frac{1}{n}\left(\sin \frac{\pi}{n} + \sin \frac{2\pi}{n} + \sin \frac{3\pi}{n} + \cdots + \sin \frac{n\pi}{n} \right).$

What can be said about the following limits?

d) $\displaystyle\lim_{n \to \infty} \frac{1}{n^{17}} (1^{15} + 2^{15} + 3^{15} + \cdots + n^{15})$

e) $\displaystyle\lim_{n \to \infty} \frac{1}{n^{15}} (1^{15} + 2^{15} + 3^{15} + \cdots + n^{15})$

69. a) Show that the area A_n of an n-sided regular polygon in a circle of radius r is

$$A_n = \frac{nr^2}{2} \sin \frac{2\pi}{n}.$$

b) Find the limit of A_n as $n \to \infty$. Is this answer consistent with what you know about the area of a circle?

70. Let

$$S_n = 1 + 2 + \cdots + n = \frac{n(n+1)}{2}$$

be the sum of the first n integers.

a) Use mathematical induction to show that

$$S_n^{(2)} = S_1 + S_2 + \cdots + S_n$$
$$= \frac{n(n+1)(n+2)}{2 \cdot 3}.$$

b) Use mathematical induction to show that

$$S_n^{(3)} = S_1^{(2)} + S_2^{(2)} + \cdots + S_n^{(2)}$$
$$= \frac{n(n+1)(n+2)(n+3)}{2 \cdot 3 \cdot 4}.$$

c) For any $k > 1$, let

$$S_n^{(k)} = S_1^{(k-1)} + \cdots + S_n^{(k-1)}.$$

Guess a formula for $S_n^{(k)}$.

CHAPTER

5

APPLICATIONS OF DEFINITE INTEGRALS

Overview Many things we want to know can be calculated with integrals: the areas between curves, the volumes and surface areas of solids, the lengths of curves, the amount of work it takes to pump liquids from below ground, the forces against floodgates, the coordinates of the points where solid objects will balance. We can define all of these in natural ways as limits of Riemann sums of continuous functions on closed intervals. Then we evaluate these limits by applying techniques of integral calculus.

There is a clear pattern to how we define and calculate the integrals in our applications, a pattern that, once learned, enables us to define new integrals whenever we need them. In this chapter we look at specific applications first; then, in the concluding section, we examine the pattern and show how it leads to integrals in new situations.

5.1 Areas of Regions Between Curves

This section shows how to find the areas of regions in the coordinate plane by integrating the functions that define the regions' boundaries.

The Basic Formula, Derived from Riemann Sums

Suppose that functions $f_1(x)$ and $f_2(x)$ are continuous and that $f_1(x) \geq f_2(x)$ throughout an interval $a \leq x \leq b$ (Fig. 5.1). The region between the curves $y = f_1(x)$ and $y = f_2(x)$ from a to b might accidentally have a shape that would let us find its area with a formula from geometry, but we usually have to find the area with an integral.

To see what that integral should be, we approximate the region with vertical rectangles. Figure 5.2 shows a typical approximation, based on a partition $P = \{x_0, x_1, \ldots, x_n\}$ of the interval $[a, b]$. The kth rectangle, shown in detail in Fig. 5.3, is Δx_k units wide and runs from the point $(c_k, f_2(c_k))$ on the lower curve to the point $(c_k, f_1(c_k))$ on the upper curve. Its area, height × width, is $(f_1(c_k) - f_2(c_k))\Delta x_k$.

We approximate the area of the region by adding the areas of the rectangles from a to b:

$$\text{Rectangle area sum} = \sum_{k=1}^{n} (f_1(c_k) - f_2(c_k))\, \Delta x_k. \tag{1}$$

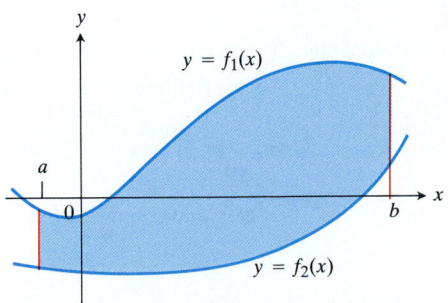

5.1 A typical region between two curves.

317

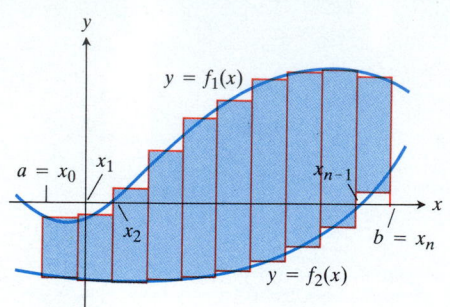

5.2   To derive a formula for the region's area, we think of approximating the region with rectangles perpendicular to the x-axis.

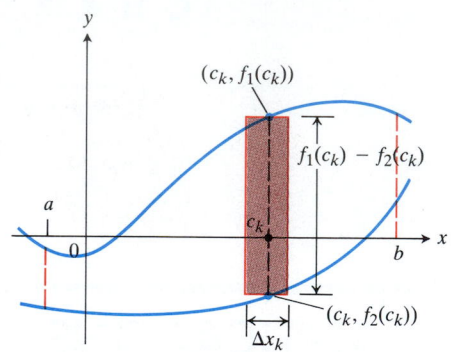

5.3   A typical rectangle is $f_1(c_k) - f_2(c_k)$ units high by Δx_k units wide.

This sum is a Riemann sum for the difference function $(f_1 - f_2)$ over the closed interval $[a, b]$.

Since f_1 and f_2 are continuous, two things happen as we subdivide $[a, b]$ more finely and let the norm of the partition go to zero. The rectangles approximate the region with increasing geometric accuracy, and the Riemann sums in Eq. (1) approach a limit. We therefore define this limit, the definite integral of $(f_1 - f_2)$ from a to b, to be the area of the region.

DEFINITION

> If f_1 and f_2 are continuous functions with $f_1(x) \geq f_2(x)$ throughout the interval $a \leq x \leq b$, then the **area of the region between the curves** $y = f_1(x)$ and $y = f_2(x)$ from a to b is the integral of $(f_1 - f_2)$ from a to b:
>
> $$\text{Area} = \int_a^b (f_1(x) - f_2(x))\, dx. \qquad (2)$$

Equation (2) is the basic formula for finding the areas of regions between curves. If the curves do not cross one another on the interval of integration, or if they cross only at one or both of the interval's endpoints, we can find the area of the region between the curves with a single application of Eq. (2). If the curves cross at one or more interior points of the interval of integration, we need to apply the formula more than once. Either way, we proceed with the following steps.

> **How to Find the Area of the Region Between Two Curves**
>
> STEP 1:   Sketch the bounding curves. Among other things, this will show how many integrations to use.
>
> STEP 2:   Set up and evaluate the appropriate integral or integrals.

Example 1   Find the area between the curves $y = \sec^2 x$ and $y = \sin x$ from 0 to $\pi/4$.

Solution   STEP 1:   *Sketch the bounding curves.* We sketch the curves together over the interval $[0, \pi/4]$ (Fig. 5.4a). The curves do not cross within the interval, so we

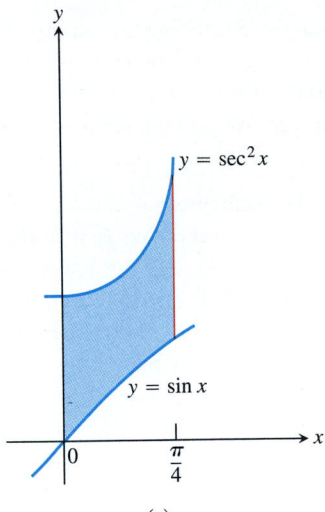

(a)

can find the area with a single application of Eq. (2), integrating from $a = 0$ to $b = \pi/4$.

STEP 2: *Set up and evaluate the integral.* The upper curve is $y = \sec^2 x$, so we take $f_1(x) = \sec^2 x$ in the area formula. The lower curve is $y = \sin x$, so $f_2(x) = \sin x$. The area between the curves is

$$\int_a^b (f_1(x) - f_2(x))\, dx = \int_0^{\pi/4} (\sec^2 x - \sin x)\, dx$$

$$= \Big[\tan x + \cos x\Big]_0^{\pi/4}$$

$$= \Big[1 + \frac{\sqrt{2}}{2}\Big] - \Big[0 + 1\Big] = \frac{\sqrt{2}}{2}.$$

Example 2 Find the area of the whale-shaped region enclosed by the parabola $y = 2 - x^2$ and the line $y = -x$.

Solution STEP 1: *Sketch the bounding curves.* We sketch the curves together (Fig. 5.4b). The curves cross only at the endpoints of the interval of integration, so we can find the area with a single application of Eq. (2).

STEP 2: *Set up and evaluate the integral.* The upper curve is $y = 2 - x^2$, so $f_1(x) = 2 - x^2$. The lower curve is $y = -x$, so $f_2(x) = -x$. The limits of integration are the x-coordinates of the points where the curves cross. We find them by solving the equations $y = 2 - x^2$ and $y = -x$ simultaneously for x:

$$2 - x^2 = -x$$

$$x^2 - x - 2 = 0$$

$$(x + 1)(x - 2) = 0$$

$$x = -1, \qquad x = 2.$$

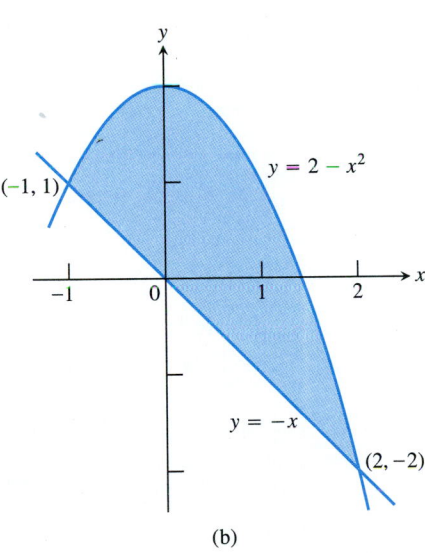

(b)

5.4 We find the areas of regions like the ones shown here with a single application of Eq. (2) (Examples 1 and 2).

The region runs from $x = -1$ on the left to $x = 2$ on the right. The area of the region enclosed by the curves is

$$\int_a^b (f_1(x) - f_2(x))\, dx = \int_{-1}^2 ((2 - x^2) - (-x))\, dx$$

$$= \int_{-1}^2 (2 + x - x^2)\, dx$$

$$= \Big[2x + \frac{x^2}{2} - \frac{x^3}{3}\Big]_{-1}^2$$

$$= \Big(4 + \frac{4}{2} - \frac{8}{3}\Big) - \Big(-2 + \frac{1}{2} + \frac{1}{3}\Big)$$

$$= 6 + \frac{3}{2} - \frac{9}{3} = \frac{9}{2}.$$

Example 3 Find the area of the region between the curves $y = 4 - x^2$ and $y = -x + 2$ from $x = -2$ to $x = 3$.

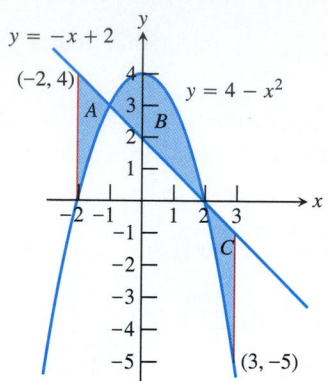

5.5 To find the area of the region between the parabola $y = 4 - x^2$ and the line $y = -x + 2$ from $x = -2$ to $x = 3$, we use Eq. (2) to find the area of each subregion and then add the results (Example 3).

Solution STEP 1: *Sketch the bounding curves.* Sketching the curves over the interval $[-2, 3]$ (Fig. 5.5) shows that the curves intersect at two points within the interval. The region whose area we wish to find is the union of three subregions (A, B, and C) whose areas we can find with Eq. (2). We add the areas of A, B, and C to find the area we want.

STEP 2: *Set up and evaluate the integrals.* For subregions A and C, $y = -x + 2$ is the upper curve and $y = 4 - x^2$ the lower. For subregion B it is the other way around: $y = 4 - x^2$ is the upper curve and $y = -x + 2$ the lower. The limits of integration are the endpoints of the interval $[-2, 3]$ and the x-coordinates of the two interior points where the curves cross. We find the latter two by solving the equations $y = 4 - x^2$ and $y = -x + 2$ simultaneously for x:

$$4 - x^2 = -x + 2$$
$$x^2 - x - 2 = 0$$
$$(x + 1)(x - 2) = 0$$
$$x = -1, \qquad x = 2.$$

The area we seek is the sum

$$\underbrace{\int_{-2}^{-1} ((-x + 2) - (4 - x^2))\, dx}_{\text{area of } A} + \underbrace{\int_{-1}^{2} ((4 - x^2) - (-x + 2))\, dx}_{\text{area of } B}$$

$$+ \underbrace{\int_{2}^{3} ((-x + 2) - (4 - x^2))\, dx}_{\text{area of } C}$$

$$= \int_{-2}^{-1} (x^2 - x - 2)\, dx + \int_{-1}^{2} (-x^2 + x + 2)\, dx + \int_{2}^{3} (x^2 - x - 2)\, dx$$

$$= \frac{11}{6} + \frac{9}{2} + \frac{11}{6} = \frac{49}{6}.$$

Boundaries with Changing Formulas

The formulas for the upper and lower boundaries of a region may change even if the curves themselves do not cross. When this happens, we break the region into subregions that correspond to the formula changes and apply Eq. (2) as before.

Example 4   Find the area of the region in the first quadrant that is bounded above by the curve $y = \sqrt{x}$ and below by the x-axis and the line $y = x - 2$.

Solution STEP 1: *Sketch the bounding curves.* A sketch (Fig. 5.6) shows that the region's upper boundary consists of the curve $y = \sqrt{x}$, so $f_1(x) = \sqrt{x}$. The lower boundary consists of two curves, first $y = 0$ for $0 \le x \le 2$ and then $y = x - 2$ for $2 \le x \le 4$. Hence the formula for f_2 changes from $f_2(x) = 0$ for $0 \le x \le 2$ to $f_2(x) = x - 2$ for $2 \le x \le 4$. We will need two integrals to find the area, one for $0 \le x \le 2$ and the other for $2 \le x \le 4$.

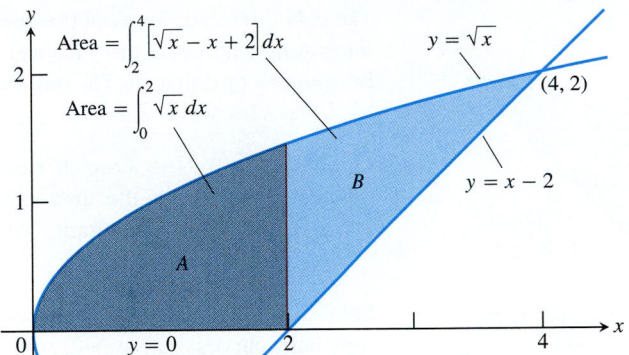

5.6 When the formula for a bounding curve changes, the area integral changes to match (Example 4).

STEP 2: *Set up and evaluate the integrals.* We add the areas of regions A and B (see Fig. 5.6) to find the total area:

$$\text{Total area} = \underbrace{\int_0^2 (\sqrt{x} - 0)\, dx}_{\text{area of } A} + \underbrace{\int_2^4 (\sqrt{x} - (x - 2))\, dx}_{\text{area of } B}$$

$$= \left[\frac{2}{3} x^{3/2}\right]_0^2 + \left[\frac{2}{3} x^{3/2} - \frac{x^2}{2} + 2x\right]_2^4$$

$$= \frac{2}{3}(2)^{3/2} - 0 + \left(\frac{2}{3}(4)^{3/2} - 8 + 8\right) - \left(\frac{2}{3}(2)^{3/2} - 2 + 4\right)$$

$$= \frac{2}{3}(8) - 2 = \frac{10}{3}.$$

Integrating with Respect to y

If a region's bounding curves are described by giving x as a function of y, the basic formula changes.

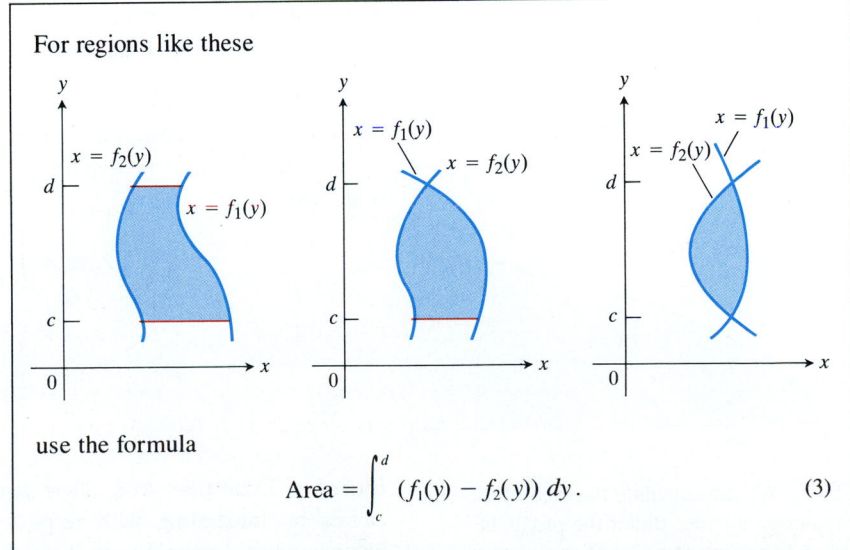

For regions like these

use the formula

$$\text{Area} = \int_c^d (f_1(y) - f_2(y))\, dy. \tag{3}$$

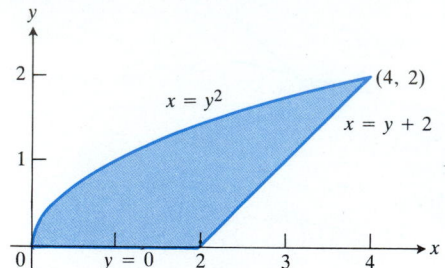

5.7   While it takes two integrations to find the area of this region if we integrate with respect to x, it takes only one if we integrate with respect to y (Example 5).

The only difference between this formula and the one in Eq. (2) is that we are now integrating with respect to y instead of x. We can sometimes reduce the number of integrations by doing so. The basic steps are the same as before.

Example 5   *The Area in Example 4 Found by a Single Integration with Respect to y.*   Find the area of the region between the curves $x = y^2$ and $x = y + 2$ in the first quadrant.

Solution   STEP 1:   *Sketch the bounding curves.* A sketch (Fig. 5.7) shows that the right-hand curve is $x = y + 2$, so $f_1(y) = y + 2$. The left-hand curve is $x = y^2$, so $f_2(y) = y^2$. The curves cross only at an endpoint of the interval $0 \le y \le 2$, so we can find the area with a single application of Eq. (3).

STEP 2:   *Set up and evaluate the integral.* The area of the region is

$$\int_c^d (f_1(y) - f_2(y))\, dy = \int_0^2 ((y + 2) - (y^2))\, dy$$

$$= \left[\frac{y^2}{2} + 2y - \frac{y^3}{3}\right]_0^2$$

$$= \left(2 + 4 - \frac{8}{3}\right) - (0) = \frac{10}{3}.$$

This is the result of Example 4, found with less work.

Combining Integrals with Formulas from Geometry

Sometimes the fastest way to find the area of a region is to combine calculus and geometry.

Example 6   *The Area of the Region in Examples 4 and 5 Found the Fastest Way.*   Find the area of the region in Examples 4 and 5.

Solution   The area we want to find is the area between the curve $y = \sqrt{x}$, $0 \le x \le 4$, and the x-axis, *minus* the area of a triangle with base 2 and height 2 (Fig. 5.8):

$$\text{Area} = \int_0^4 \sqrt{x}\, dx - \frac{1}{2}(2)(2)$$

$$= \frac{2}{3}x^{3/2}\Big]_0^4 - 2$$

$$= \frac{2}{3}(8) - 0 - 2 = \frac{10}{3}.$$

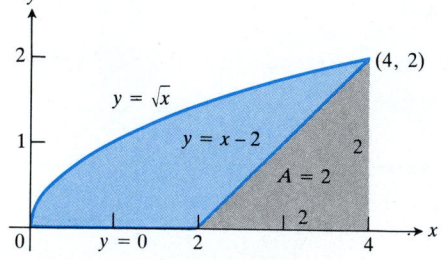

5.8   We can calculate the area of the blue region as the area under the parabola $y = \sqrt{x}$ minus the area of the gray triangular region.

Moral of Examples 4–6   It is sometimes easier to find the area between two curves by integrating with respect to y instead of x. Examine each region beforehand to determine which method, if either, is easier. Sketching the region may also reveal how to use geometry to simplify your work.

EXERCISES 5.1

Find the areas of the shaded regions in Exercises 1–6.

1.

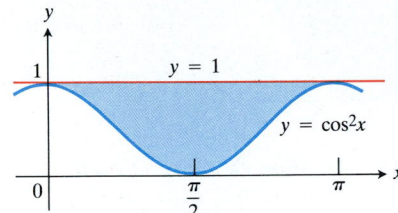

2.

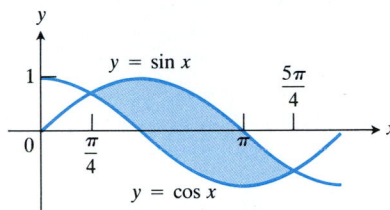

3.

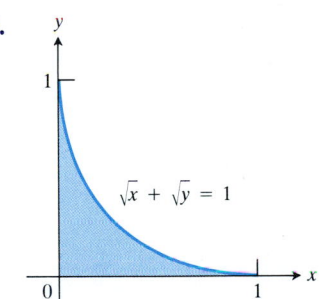

4.

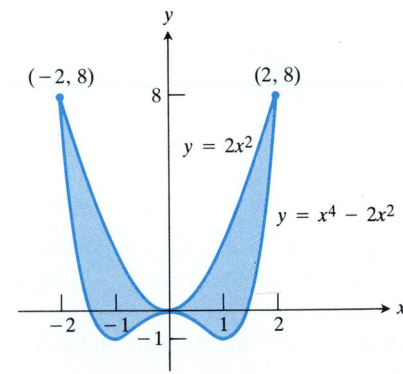

NOT TO SCALE

5.

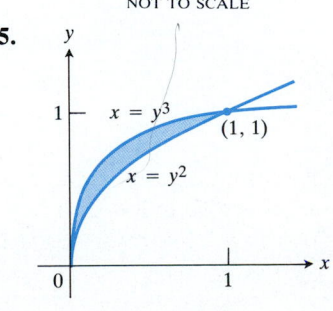

6.

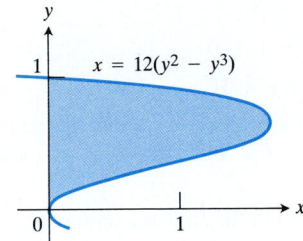

Find the areas of the regions enclosed by the lines and curves in Exercises 7–18.

7. $y = x^2 - 2$ and $y = 2$

8. $y = 2x - x^2$ and $y = -3$

9. $y = x^4$ and $y = 8x$

10. $y = x^2 - 2x$ and $y = x$

11. $y = x^2$ and $y = -x^2 + 4x$

12. $y = 7 - 2x^2$ and $y = x^2 + 4$

13. $y = 4 - 4x^2$ and $y = x^4 - 1$

14. $y = x^3$ and $y = 3x^2 - 4$

15. $y = \sqrt{|x|}$ and $5y = x + 6$

16. $y = x\sqrt{a^2 - x^2}$, $a > 0$, and $y = 0$

17. $y = x^4 - 4x^2 + 4$ and $y = x^2$

18. $y = |x^2 - 4|$ and $y = (x^2/2) + 4$

Find the areas of the regions enclosed by the lines and curves in Exercises 19–26.

19. $x = 2y^2$, $x = 0$, and $y = 3$

20. $x = y^2$ and $x = y + 2$

21. $4x = y^2 - 4$ and $4x = y + 16$

22. $x = y^2$ and $x = -2y^2 + 3$

23. $x = -y^2$ and $x = 2 - 3y^2$

24. $x = y^{2/3}$ and $x = 2 - y^4$

25. $x = y^2 - 1$ and $x = |y|\sqrt{1 - y^2}$

26. $x = y^3 - y^2$ and $x = 2y$

Find the areas of the regions enclosed by the lines and curves in Exercises 27–34.

27. $y = 2 \sin x$ and $y = \sin 2x$, $0 \le x \le \pi$

28. $y = 8 \cos x$ and $y = \sec^2 x$, $-\pi/3 \le x \le \pi/3$

29. $y = \sec^2 x$, $y = \tan^2 x$, $x = -\pi/4$, and $x = \pi/4$

30. $y = \sec^2(\pi x/3)$ and $y = x^{1/3}$, $-1 \le x \le 1$

31. $x = \cos(\pi y/2)$ and $x = 1 - y^2$

32. $x = \sin|y|$ and $x = -1$, $-2\pi \le y \le 2\pi$

33. $x = 3 \sin y\sqrt{\cos y}$ and $x = 0$, $0 \le y \le \pi/2$

34. $x = \tan^2 y$ and $x = -\tan^2 y$, $-\pi/4 \le y \le \pi/4$

Find the areas of the regions enclosed by the lines and curves in Exercises 35–38.

35. $y = 2 - (x - 2)^2$ and $y = x$

36. $y = x^2 - 2x - 3$ and $y = x - 3$

37. $x = 4 - y^2$ and $x = y + 2$

38. $x = 3y - y^2$ and $x + y = 3$

39. Find the area of the region in the first quadrant bounded by the line $y = x$, the line $x = 2$, the curve $y = 1/x^2$, and the x-axis.

40. Find the area of the region in the first quadrant bounded on the left by the y-axis, below by the line $y = x/4$, above left by the curve $y = 1 + \sqrt{x}$, and above right by the curve $y = 2/\sqrt{x}$.

41. Find the area of the "triangular" region bounded by the y-axis and the curves $y = \sin x$ and $y = \cos x$ in the first quadrant.

42. Find the area of the region between the curve $y = 3 - x^2$ and the line $y = -1$ by integrating with respect to (a) x; (b) y.

43. The area of the region between the curve $y = x^2$ and the line $y = 4$ is divided into two equal portions by the line $y = c$.
a) Find c by integrating with respect to y. (This puts c into the limits of integration.)
b) Find c by integrating with respect to x. (This puts c into the integrand as well.)

44. Figure 5.9 shows triangle AOC inscribed in the region cut from the parabola $y = x^2$ by the line $y = a^2$. Find the limit of the ratio of the area of the triangle to the area of the parabolic region as a approaches zero.

45. Suppose the area of the region between the graph of the continuous function f (Fig. 5.10) and the x-axis from $x = a$ to $x = b$ is 4 square units. Find the area between the curves $y = f(x)$ and $y = 2f(x)$ from $x = a$ to $x = b$.

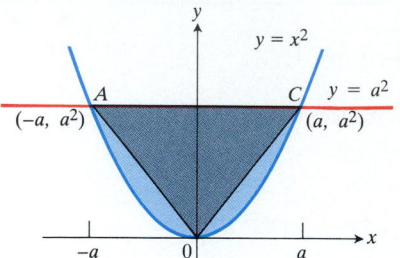

5.9   The triangle and parabola in Exercise 44.

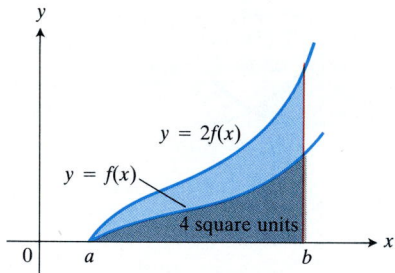

5.10   The curves and regions in Exercise 45.

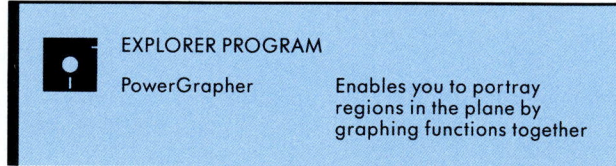

EXPLORER PROGRAM

PowerGrapher   Enables you to portray regions in the plane by graphing functions together

5.2 Volumes: Slicing, Disks, and Washers

Now that we have defined (and can calculate) the areas of a wide variety of regions bounded by smooth curves, we can define (and calculate) the volumes of an equally wide variety of cylinders based on these regions (Fig. 5.11). This enables us to define the volumes of more arbitrary solids. We approximate the solids with

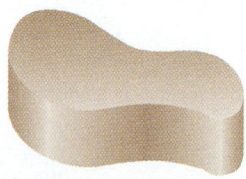

5.11   The volume of a cylinder is always defined to be its base area times its height.

Plane region whose area we know

Cylinder based on region
Volume =
base area × height

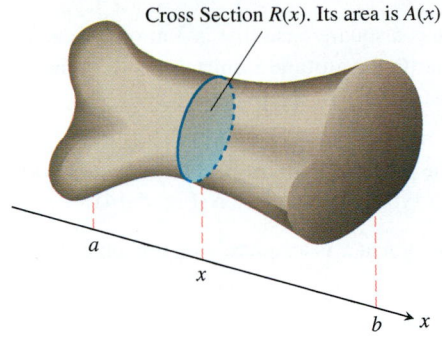

Cross Section $R(x)$. Its area is $A(x)$

5.12 If the area $A(x)$ of the cross section $R(x)$ is a continuous function of x, we can find the volume of the solid by integrating A from a to b.

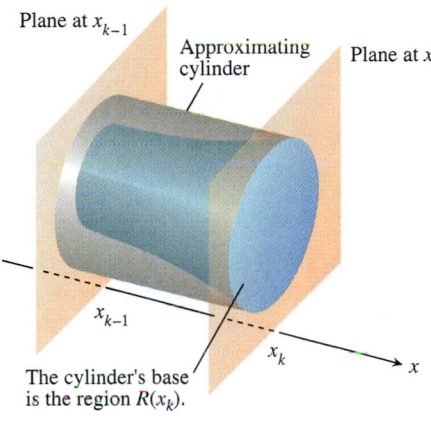

Plane at x_{k-1}

Approximating cylinder

Plane at x_k

x_{k-1}

x_k

The cylinder's base is the region $R(x_k)$.

NOT TO SCALE

5.13 Enlarged view of the slice of the solid between the planes at x_{k-1} and x_k and its approximating cylinder.

cylinders based on typical cross sections, add the volumes of the cylinders to form Riemann sums, and define and calculate the volumes of the solids as the limiting integrals of these sums.

Volumes of Arbitrary Solids: Slicing

Suppose we want to find the volume of a solid like the one shown in Fig. 5.12. The solid lies between planes perpendicular to the x-axis at $x = a$ and $x = b$. Each cross section of the solid by a plane perpendicular to the x-axis is a region whose area we know how to find. Specifically, at each point x in the closed interval $[a, b]$ the cross section of the solid is a region $R(x)$ whose area is $A(x)$. This makes A a real-valued function of x. If it is also a continuous function of x, we can use it to define and calculate the volume of the solid as an integral in the following way.

We partition the interval $[a, b]$ along the x-axis in the usual manner and slice the solid, as we would a loaf of bread, by planes perpendicular to the x-axis at the partition points. The kth slice, the one between the planes at x_{k-1} and x_k, has approximately the same volume as the cylinder between these two planes based on the region $R(x_k)$ (Fig. 5.13). The volume of this cylinder is

$$V_k = \text{base area} \times \text{height}$$

$$= A(x_k) \times (\text{distance between the planes at } x_{k-1} \text{ and } x_k) \tag{1}$$

$$= A(x_k)\Delta x_k.$$

The volume of the solid is therefore approximated by the cylinder volume sum

$$\sum_{k=1}^{n} A(x_k)\Delta x_k. \tag{2}$$

This is a Riemann sum for the function $A(x)$ on the closed interval $[a, b]$. We expect the approximations from sums like these to improve as the norm of the partition of $[a, b]$ goes to zero, so we define their limiting integral to be the volume of the solid.

DEFINITION

The **volume** of a solid of known integrable cross-section area $A(x)$ from $x = a$ to $x = b$ is the integral of A from a to b:

$$\text{Volume} = \int_a^b A(x)\,dx. \tag{3}$$

To apply Eq. (3), we take the following steps.

How to Find Volumes by the Method of Slicing

STEP 1: Sketch the solid and a typical cross section.

STEP 2: Find a formula for $A(x)$.

STEP 3: Find the limits of integration.

STEP 4: Integrate $A(x)$ to find the volume.

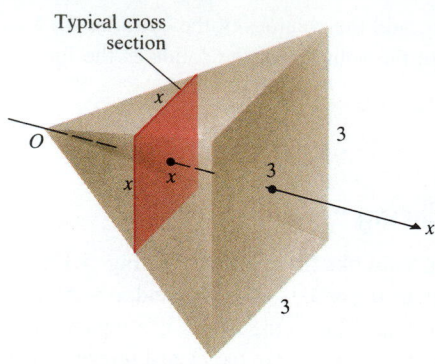

Typical cross section

5.14   The cross sections of the pyramid in Example 1 are squares.

Example 1   A pyramid 3 m high has a square base that is 3 m on a side. The cross section of the pyramid perpendicular to the altitude x units down from the vertex is a square x units on a side. Find the volume of the pyramid.

Solution   STEP 1:   *A sketch.* We draw the pyramid with its altitude along the x-axis and its vertex at the origin and include a typical cross section (Fig. 5.14).

STEP 2:   *A formula for $A(x)$.* The cross section at x is a square x meters on a side, so its area is

$$A(x) = x^2.$$

STEP 3:   *The limits of integration.* The squares go from $x = 0$ to $x = 3$.

STEP 4:   *The volume.*

$$\text{Volume} = \int_a^b A(x)\,dx \qquad (\text{Eq. (3)})$$

$$= \int_0^3 x^2\,dx = \frac{x^3}{3}\Big]_0^3 = 9.$$

The volume is 9 m^3.

Example 2   A curved wedge is cut from a cylinder of radius 3 by two planes. One plane is perpendicular to the axis of the cylinder. The second plane crosses the first plane at a 45° angle at the center of the cylinder. Find the volume of the wedge.

Solution   STEP 1:   *A sketch.* We draw the wedge and sketch a typical cross section perpendicular to the x-axis (Fig. 5.15).

STEP 2:   *The formula for $A(x)$.* The cross section at x is a rectangle of area

$$A(x) = (\text{height})(\text{width}) = (x)(2\sqrt{9 - x^2})$$

$$= 2x\sqrt{9 - x^2}.$$

STEP 3:   *The limits of integration.* The rectangles run from $x = 0$ to $x = 3$.

STEP 4:   *The volume.*

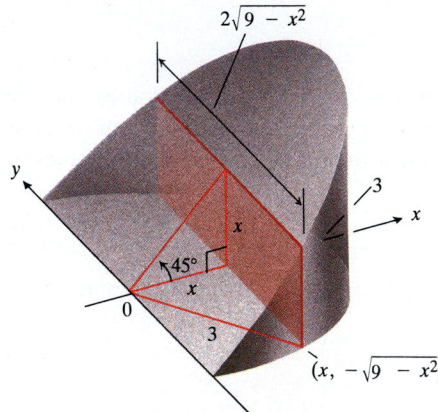

5.15   The wedge of Example 2, sliced perpendicular to the x-axis. The cross sections are rectangles.

$$\text{Volume} = \int_a^b A(x)\,dx = \int_0^3 2x\sqrt{9 - x^2}\,dx \qquad (\text{Eq. (3)})$$

$$= -\frac{2}{3}(9 - x^2)^{3/2}\Big]_0^3 = 0 + \frac{2}{3}(9)^{3/2} \qquad \left(\begin{array}{l}\text{Substitute } u = 9 - x^2,\\ du = -2x\,dx\text{, integrate,}\\ \text{and substitute back.}\end{array}\right)$$

$$= 18.$$

Example 3   *Cavalieri's Theorem.*   Cavalieri's theorem says that two solids with equal altitudes and identical parallel cross sections have the same volume (Fig. 5.16). We can see this immediately from Eq. (3) because the cross-section area function $A(x)$ is the same in each case.

Bonaventura Cavalieri (1598–1647)

Cavalieri, a student of Galileo's whom Galileo told to study calculus, discovered that if two plane regions can be arranged to lie over the same interval of the x-axis in such a way that they have identical vertical cross sections at every point, then the regions have the same area. The theorem was good enough to win Cavalieri a chair at the University of Bologna in 1629. The solid geometry version in Example 3, which Cavalieri never proved, was given his name by later geometers.

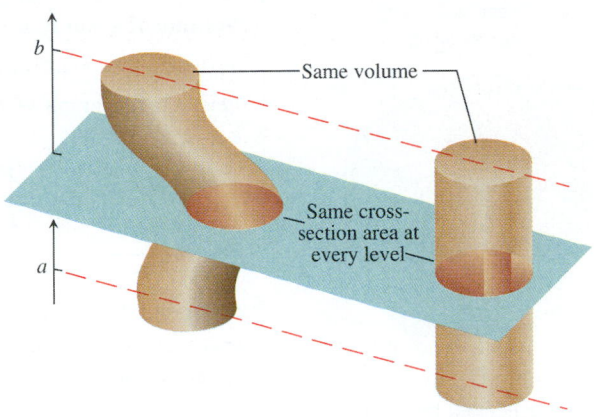

5.16 *Cavalieri's theorem*: These two solids have the same volume even though the one on the left looks bigger. You can illustrate this yourself with stacks of coins.

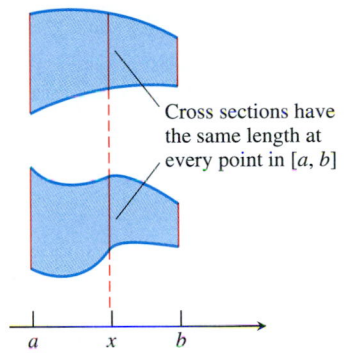

Cross sections have the same length at every point in [a, b]

Volumes of Solids of Revolution: Disks

The most common application of the method of slicing is to solids of revolution. **Solids of revolution** are solids whose shapes can be generated by revolving plane regions about axes. Thread spools are solids of revolution; so are hand weights and billiard balls (Fig. 5.17). Solids of revolution sometimes have volumes we can find with formulas from geometry, as in the case of a billiard ball. But when we want to find the volume of a blimp or to predict the weight of a part we are going to have turned on a lathe, formulas from geometry are of little help and we turn to calculus for the answers.

As we said, a solid of revolution is generated by revolving a plane region about an axis. If we can set things up so that the region is the region between the graph of a continuous function $y = f(x)$, $a \le x \le b$, and the x-axis, and the axis of revolution is the x-axis (Fig. 5.18), we can calculate the solid's volume in the following way.

The typical cross section of the solid perpendicular to the axis of revolution (in this case the x-axis) is a disk of radius $f(x)$ and area

$$A(x) = \pi(\text{radius})^2 = \pi(f(x))^2. \tag{4}$$

The solid's volume, being the integral of A from $x = a$ to $x = b$, is the integral of the quantity $\pi(f(x))^2$ from a to b.

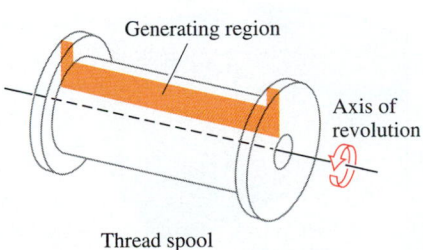

Thread spool

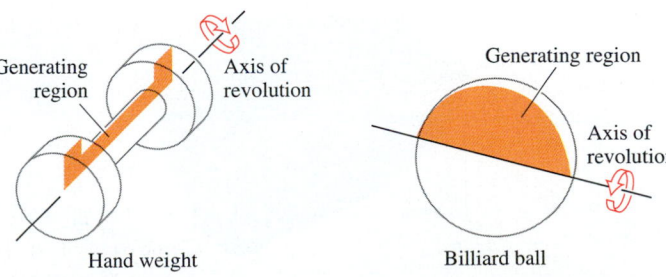

Hand weight   Billiard ball

5.17 Many familiar shapes can be modeled as solids of revolution.

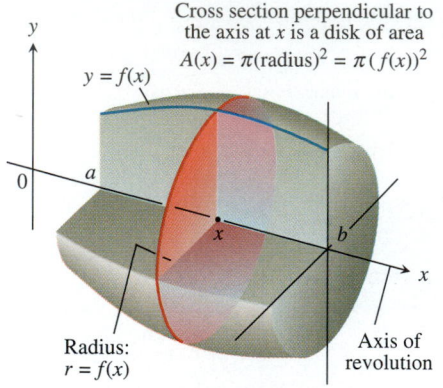

Cross section perpendicular to
the axis at x is a disk of area
$A(x) = \pi(\text{radius})^2 = \pi(f(x))^2$

$y = f(x)$

a

x

b

Radius:
$r = f(x)$

Axis of
revolution

5.18   The solid generated by revolving
the region between the curve $y = f(x)$ and
the x-axis from a to b about the x-axis.

5.19   The solid generated by revolving
the region bounded by the curve $y = \sqrt{x}$
and the lines $y = 0$ and $x = 4$ about the
x-axis (Example 4). Part (e) shows a typical
slice perpendicular to the x-axis. The
radius of the slice at x is $f(x)$. (Image
structure generated by Mathematica™.)

Volume of a Solid of Revolution (Rotation About the x-axis)

The volume of the solid generated by revolving about the x-axis the region
between the graph of the continuous function $y = f(x)$ and the x-axis from
$x = a$ to $x = b$ is

$$\text{Volume} = \int_a^b \pi(\text{radius function})^2 \, dx = \int_a^b \pi(f(x))^2 \, dx. \qquad (5)$$

Example 4   The region between the curve $y = \sqrt{x}$, $0 \le x \le 4$, and the
x-axis is revolved about the x-axis to generate the solid in Fig. 5.19. Find the volume of the solid.

Solution

$$\text{Volume} = \int_a^b \pi(\text{radius})^2 \, dx \qquad \text{(Eq. (5))}$$

$$= \int_0^4 \pi(\sqrt{x})^2 \, dx \qquad \left(\begin{array}{l}\text{The radius function is} \\ f(x) = \sqrt{x}, \ 0 \le x \le 4.\end{array}\right)$$

$$= \pi \int_0^4 x \, dx = \pi \left.\frac{x^2}{2}\right]_0^4 = \pi \frac{(4)^2}{2} = 8\pi$$

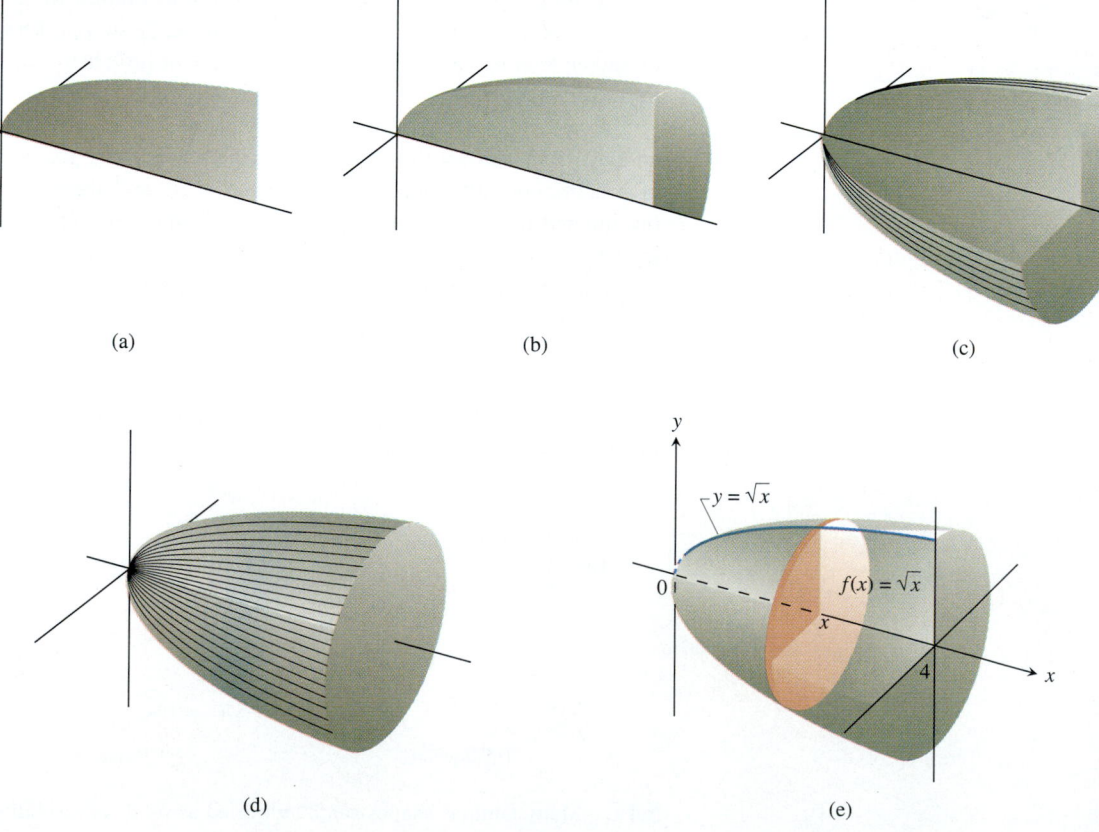

(a)

(b)

(c)

$y = \sqrt{x}$

0

$f(x) = \sqrt{x}$

x

4

y

x

(d)

(e)

How to Find Volumes Using Eq. (5)

1. Square the expression for the radius function $f(x)$.
2. Multiply by π.
3. Integrate from a to b.

The axis of revolution in the next example is not the x-axis, but the rule for calculating the volume is the same: Integrate $\pi(\text{radius})^2$ between appropriate limits.

Example 5   Find the volume of the solid generated by revolving the region bounded by $y = \sqrt{x}$ and the lines $y = 1$ and $x = 4$ about the line $y = 1$.

Solution   We draw a figure that shows the region and the radius at a typical point on the axis of revolution (Fig. 5.20(e)). The region runs from $x = 1$ to $x = 4$. At each x in the interval $1 \leq x \leq 4$, the cross-section radius is

$$\text{Radius at } x = \sqrt{x} - 1.$$

Therefore,

$$
\begin{aligned}
\text{Volume} &= \int_{1}^{4} \pi(\text{radius})^2 \, dx \\
&= \int_{1}^{4} \pi(\sqrt{x} - 1)^2 \, dx \\
&= \pi \int_{1}^{4} (x - 2\sqrt{x} + 1) \, dx \\
&= \pi \left[\frac{x^2}{2} - 2 \cdot \frac{2}{3} x^{3/2} + x \right]_{1}^{4} = \frac{7\pi}{6}.
\end{aligned}
$$

5.20   The solid generated by revolving the region bounded by $y = \sqrt{x}$, $x = 4$, and $y = 1$ about the line $y = 1$ (Example 5). Part (e) shows a typical slice perpendicular to the axis of revolution. The radius at x is $\sqrt{x} - 1$. The cross-section area at x is $A(x) = \pi(\sqrt{x} - 1)^2$. (Image structure generated by Mathematica™.)

To find the volume of the solid generated by revolving the region between a curve $x = f(y)$ and the y-axis from $y = c$ to $y = d$ about the y-axis, use Eq. (5) with x replaced by y.

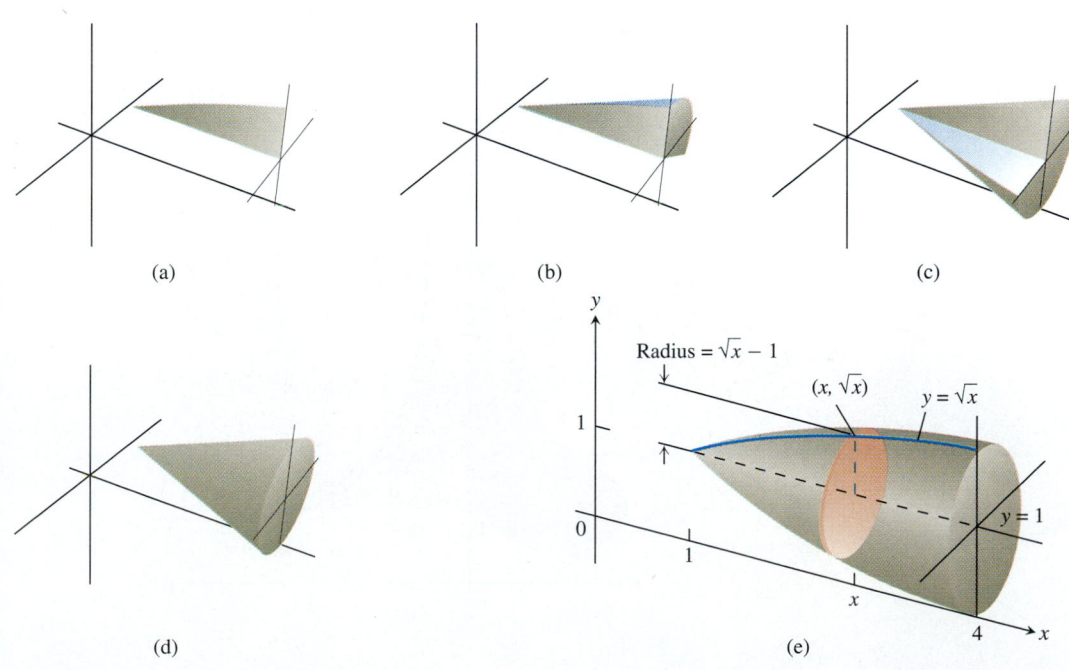

(a)   (b)   (c)

(d)   (e)

Radius $= \sqrt{x} - 1$

$(x, \sqrt{x})$

$y = \sqrt{x}$

$y = 1$

Volume of a Solid of Revolution (Rotation About the *y*-axis)

$$\text{Volume} = \int_c^d \pi(\text{radius})^2 \, dy = \int_c^d \pi(f(y))^2 \, dy \qquad (6)$$

Example 6   The region between the curve $x = 2/y$, $1 \leq y \leq 4$, and the *y*-axis is revolved about the *y*-axis to generate a solid (Fig. 5.21). Find the solid's volume.

Solution

$$\text{Volume} = \int_1^4 \pi(\text{radius})^2 \, dy = \int_1^4 \pi\left(\frac{2}{y}\right)^2 dy \qquad \text{(Eq. (6))}$$

$$= 4\pi \int_1^4 \frac{1}{y^2} \, dy = 4\pi\left[-\frac{1}{y}\right]_1^4 = 4\pi\left[\frac{3}{4}\right] = 3\pi$$

5.21   The solid in Example 6. (Image structure generated by Mathematica™.)

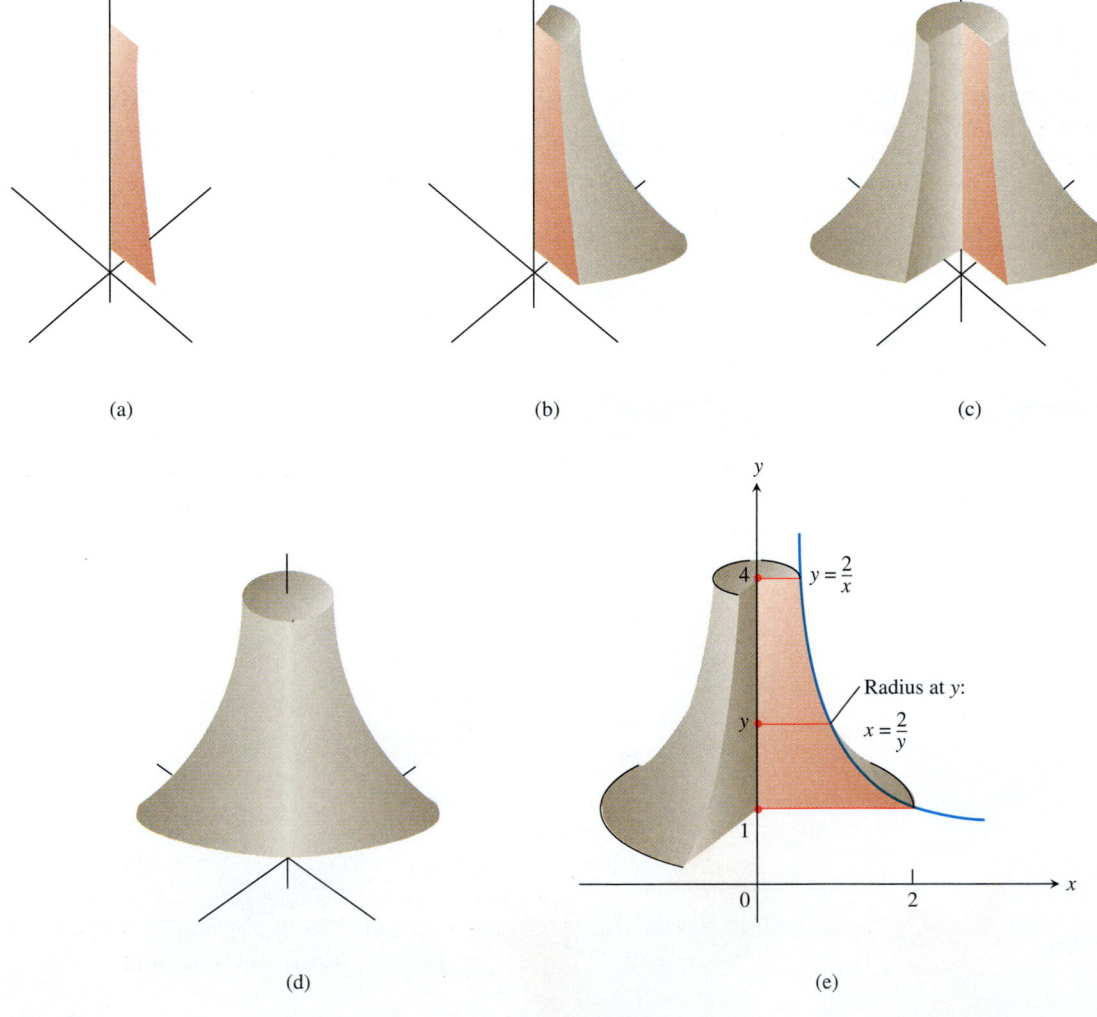

(a)   (b)   (c)

(d)   (e)

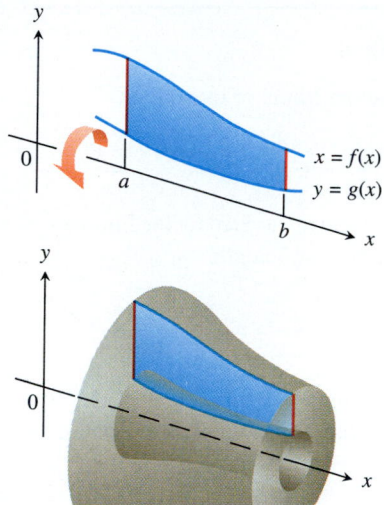

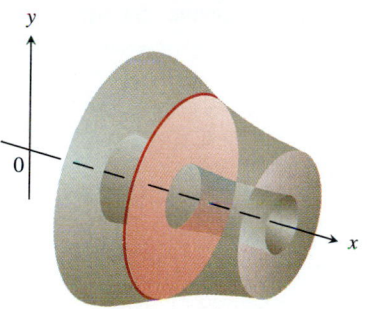

5.22 The cross sections of the solid of revolution generated here are washers, not disks, so the integral $\int_a^b A(x)\,dx$ leads to a slightly different formula.

The Washer Method

If the region we revolve to generate a solid of revolution does not border on the axis of revolution, the solid has a hole in it (Fig. 5.22). The cross sections perpendicular to the axis of revolution are washers instead of disks. The dimensions of a typical washer (Fig. 5.23) are

Outer washer radius: $R(x) = f(x),$

Inner washer radius: $r(x) = g(x).$

The washer's area is

$$A(x) = \pi(\text{outer radius})^2 - \pi(\text{inner radius})^2$$
$$= \pi(R(x))^2 - \pi(r(x))^2 \qquad (7)$$
$$= \pi(R^2(x) - r^2(x)).$$

The volume of the solid is the integral of A from $x = a$ to $x = b$.

When the region in Fig. 5.22 is revolved about the x-axis, this line segment generates the washer

$y = f(x)$

$y = g(x)$

$g(x)$

x

$f(x)$

5.23 The curves $y = f(x)$ and $y = g(x)$ determine the outer and inner radii of the cross-section washer at x.

Washer Method for Calculating Volumes

$$\text{Volume} = \int_a^b \pi(R^2(x) - r^2(x))\,dx, \qquad (8)$$

$R(x) = $ outer radius, $r(x) = $ inner radius.

Notice that the function being integrated in Eq. (8) is $\pi(R^2 - r^2)$, not $\pi(R - r)^2$. Notice also that Eq. (8) reduces to the disk method formula if the inner radius $r(x)$ is zero throughout $[a, b]$. The disk method is a special case of the washer method. To apply Eq. (8), we take the following steps.

How to Find Volumes by the Washer Method

STEP 1:   Sketch the generating region and find the limits of integration.

STEP 2:   Draw a line segment across the region perpendicular to the axis of revolution.

STEP 3:   Find the inner and outer radii of the washer generated by the line segment. These give the radius functions for the volume integral.

STEP 4:   Evaluate the volume integral.

Example 7   The region bounded by the curve $y = x^2 + 1$ and the line $y = -x + 3$ is revolved about the x-axis to generate a solid. Find the volume of the solid.

Solution   STEP 1:   *Sketch the generating region and find the limits of integration* (Fig. 5.24). The limits of integration are the x-coordinates of the points where the parabola and the line cross. We find the coordinates by solving the equations $y = x^2 + 1$ and $y = -x + 3$ simultaneously for x:

$$x^2 + 1 = -x + 3$$
$$x^2 + x - 2 = 0$$
$$(x + 2)(x - 1) = 0$$
$$x = -2, \qquad x = 1.$$

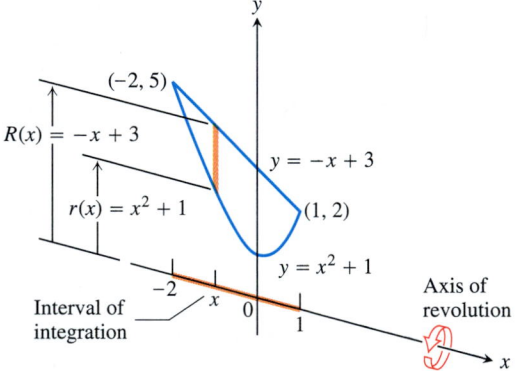

5.24   The region bounded by the curve $y = x^2 + 1$ and the line $y = -x + 3$, spanned by a line segment perpendicular to the axis of revolution (Example 7).

STEP 2:   *Draw a line segment across the region perpendicular to the axis of revolution* (the orange segment in Fig. 5.24).

STEP 3:   *Find the outer and inner radii of the washer generated by the line segment* (Fig. 5.25). We read these radii from the formulas for the bounding curves:

Outer radius: $R(x) = -x + 3,$

Inner radius: $r(x) = x^2 + 1.$

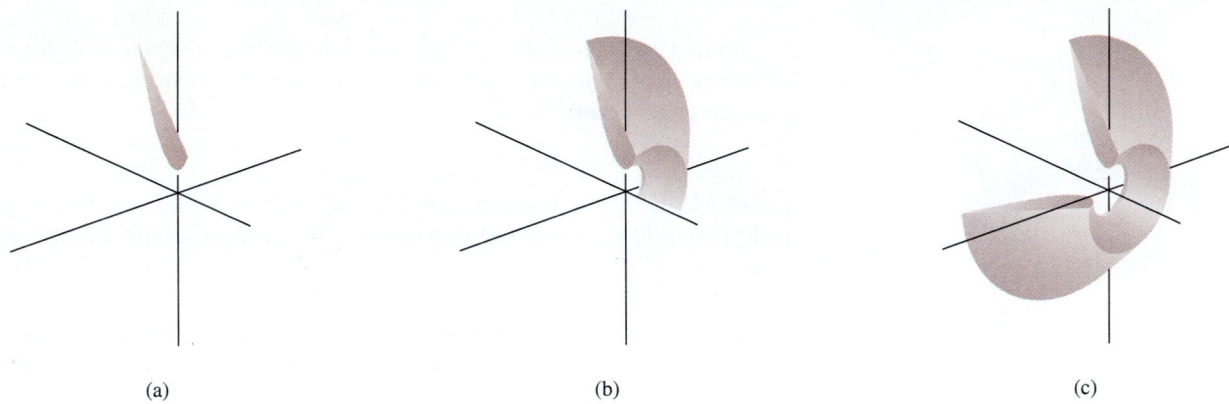

(a)   (b)   (c)

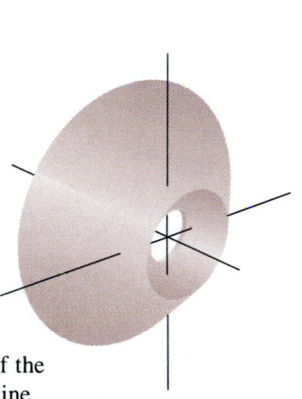

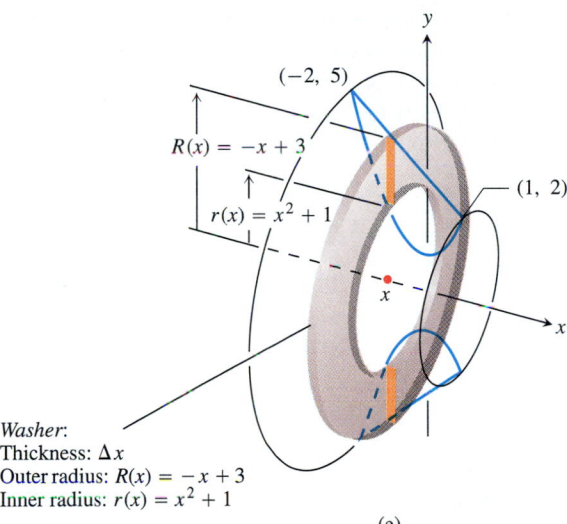

(−2, 5)

$R(x) = -x + 3$

$r(x) = x^2 + 1$

(1, 2)

x

y

x

Washer:
Thickness: Δx
Outer radius: $R(x) = -x + 3$
Inner radius: $r(x) = x^2 + 1$

(d)   (e)

5.25 The outer and inner radii of the washer swept out by revolving the line segment in Fig. 5.24 about the *x*-axis are the distances of the segments' endpoints from the axis.

STEP 4: *Evaluate the volume integral.*

$$\text{Volume} = \int_a^b \pi(R^2(x) - r^2(x))\,dx \qquad \text{(Eq. (8))}$$

$$= \int_{-2}^1 \pi((-x+3)^2 - (x^2+1)^2)\,dx \qquad \begin{pmatrix} \text{Values from} \\ \text{steps 1 and 3} \end{pmatrix}$$

$$= \int_{-2}^1 \pi(8 - 6x - x^2 - x^4)\,dx$$

$$= \pi\left[8x - 3x^2 - \frac{x^3}{3} - \frac{x^5}{5}\right]_{-2}^1 = \frac{117\pi}{5}$$

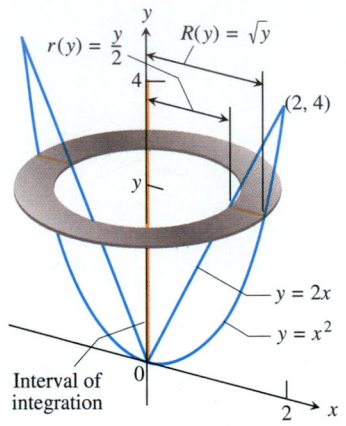

5.26 The region bounded by the parabola $y = x^2$ and the line $y = 2x$ revolved about the y-axis. The line segment perpendicular to the axis of revolution sweeps out a washer whose radii are given by the formulas for the curves.

To find the volume of a solid generated by revolving a region about the y-axis, we use the preceding steps and integrate with respect to y instead of x. The next example shows how to do this.

Example 8   The region bounded by the parabola $y = x^2$ and the line $y = 2x$ in the first quadrant is revolved about the y-axis to generate a solid. Find the volume of the solid.

Solution   STEP 1: *Sketch the region and find the limits of integration* (Fig. 5.26). The curves $y = x^2$ and $y = 2x$ cross at $(0, 0)$ and $(2, 4)$, so we integrate from $y = 0$ to $y = 4$.

STEP 2: *Draw a line segment across the region perpendicular to the axis of revolution* (the orange segment in Fig. 5.26).

STEP 3: *Find the inner and outer radii of the washer generated by the line segment.* From Fig. 5.26,

$$R(y) = \sqrt{y}, \qquad r(y) = \frac{y}{2}.$$

STEP 4: *Evaluate the volume integral.*

$$\text{Volume} = \int_c^d \pi(R^2(y) - r^2(y))\, dy \qquad \text{(Eq. (8) with } y \text{ in place of } x)$$

$$= \int_0^4 \pi\left((\sqrt{y})^2 - \left(\frac{y}{2}\right)^2\right) dy \qquad \text{(Values from steps 1 and 3)}$$

$$= \pi \int_0^4 \left(y - \frac{y^2}{4}\right) dy = \pi\left[\frac{y^2}{2} - \frac{y^3}{12}\right]_0^4 = \frac{8}{3}\pi$$

If the axis of revolution is not one of the coordinate axes, the rule for finding the volume is the same: Find formulas for the solid's outer and inner radii and integrate $\pi(R^2 - r^2)$ between appropriate limits. In the next example we revolve a region about a line parallel to the y-axis. We would handle a region revolved about a line parallel to the x-axis in a similar way.

Example 9   The region between the parabola $y = x^2$ and the line $y = 2x$ is revolved about the line $x = 2$ parallel to the y-axis. Find the volume swept out.

Solution   STEP 1: *Sketch the region and find the limits of integration* (Fig. 5.27). The line crosses the parabola at $(0, 0)$ and $(2, 4)$, so we integrate from $y = 0$ to $y = 4$.

STEP 2: *Draw a line segment across the region perpendicular to the axis of revolution* (the orange segment in Fig. 5.27).

STEP 3: *Find the inner and outer radii of the washer generated by the line segment.* From Fig. 5.27,

$$R(y) = 2 - \frac{y}{2}, \qquad r(y) = 2 - \sqrt{y}.$$

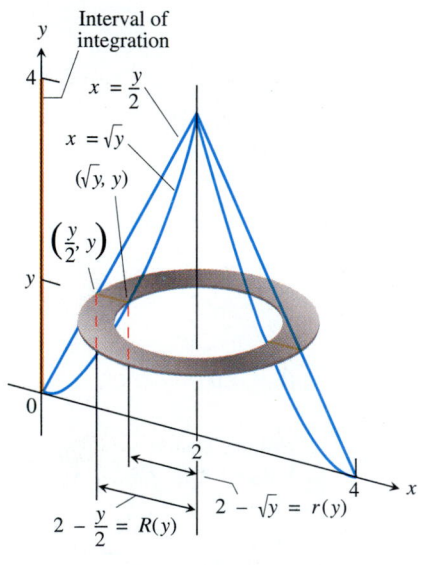

5.27 The inner and outer radii of the washer are measured, as always, as distances from the axis of revolution.

STEP 4:   *Integrate with respect to y to find the volume.*

$$\text{Volume} = \int_{c}^{d} \pi(R^2(y) - r^2(y))\, dy \qquad \text{(Eq. (8) with } y \text{ in place of } x)$$

$$= \int_{0}^{4} \pi\left(\left(2 - \frac{y}{2}\right)^2 - \left(2 - \sqrt{y}\right)^2\right) dy \qquad \text{(Values from steps 1 and 3)}$$

$$= \pi\int_{0}^{4} \left(\frac{y^2}{4} - 3y + 4\sqrt{y}\right) dy$$

$$= \pi\left[\frac{y^3}{12} - \frac{3y^2}{2} + \frac{8}{3}y^{3/2}\right]_{0}^{4} = \frac{8}{3}\pi$$

EXERCISES 5.2

Use the disk method to find the volumes of the solids generated by revolving about the x-axis the regions bounded by the lines and curves in Exercises 1–6.

1. $y = x^2$, $y = 0$, $x = 2$

2. $y = \sqrt{9 - x^2}$, $y = 0$

3. $y = x - x^2$, $y = 0$

4. $y = x^3$, $y = 0$, $x = 2$

5. $y = \sqrt{\cos x}$, $0 \le x \le \pi/2$, $y = 0$, $x = 0$

6. $y = \sec x$, $y = 0$, $x = -\pi/4$, $x = \pi/4$

Use the disk method to find the volumes of the solids generated by revolving about the y-axis the regions bounded by the lines and curves in Exercises 7–12.

7. $x = \sqrt{4 - y}$, $x = 0$, $y = 0$

8. $x = \sqrt{5}y^2$, $x = 0$, $y = -1$, $y = 1$

9. $x = 1 - y^2$, $x = 0$

10. $x = y^{3/2}$, $x = 0$, $y = 2$

11. $x = \sqrt{2 \sin 2y}$, $0 \le y \le \pi/2$, $x = 0$

12. $x = 2/(y + 1)$, $x = 0$, $y = 0$, $y = 1$

Use the washer method to find the volumes of the solids generated by revolving about the x-axis the regions bounded by the lines and curves in Exercises 13–20.

13. $y = x$, $y = 1$, $x = 0$ **14.** $y = 2x$, $y = x$, $x = 1$

15. $x = \sqrt{y}$, $y = 4$, $x = 0$ **16.** $y = x^2 + 3$, $y = 4$

17. $y = x^2 + 1$, $y = x + 3$

18. $y = 4 - x^2$, $y = 2 - x$

19. $y = \sec x$, $y = \sqrt{2}$, $-\pi/4 \le x \le \pi/4$

20. $y = \sec x$, $y = \tan x$, $x = 0$, $x = 1$

Use the washer method to find the volumes of the solids generated by revolving about the y-axis the regions bounded by the lines and curves in Exercises 21–26.

21. $y = x - 1$, $y = 1$, $x = 1$

22. $y = x - 1$, $y = 2$, $x = 1$

23. $y = x^2$, $y = 0$, $x = 2$

24. $y = x$, $y = \sqrt{x}$

25. The semicircle $x = \sqrt{25 - y^2}$ and the y-axis

26. The semicircle $x = \sqrt{25 - y^2}$ and the line $x = 4$

Find the volumes of the solids generated by revolving the shaded regions in Exercises 27 and 28 about the indicated axes.

27. The x-axis

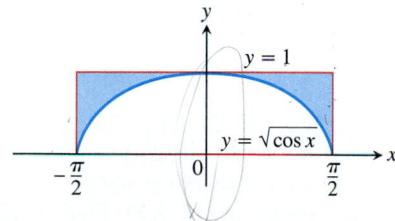

28. The y-axis

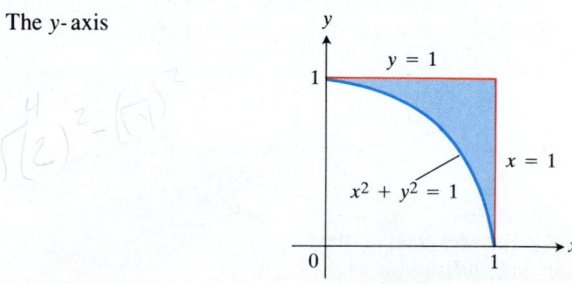

29. Find the volume of the solid generated by revolving the region bounded by $y = \sqrt{x}$ and the lines $y = 2$ and $x = 0$ about the

    a) x-axis,   b) y-axis,
    c) line $y = 2$,   d) line $x = 4$.

30. Find the volume of the solid generated by revolving the triangular region bounded by the lines $y = 2x$, $y = 0$, and $x = 1$

    a) about the line $x = 1$,
    b) about the line $x = 2$.

31. Find the volume of the solid generated by revolving the region bounded by the parabola $y = x^2$ and the line $y = 1$

    a) about the line $y = 1$,
    b) about the line $y = 2$,
    c) about the line $y = -1$.

32. By integration, find the volume of the solid generated by revolving the triangular region with vertices $(0, 0)$, $(b, 0)$, $(0, h)$

    a) about the x-axis,
    b) about the y-axis.

33. *Minimizing a volume.* The arch $y = \sin x$, $0 \le x \le \pi$, is revolved about the line $y = c$ to generate the solid shown in Fig. 5.28. Find the value of c that minimizes the volume of the solid.

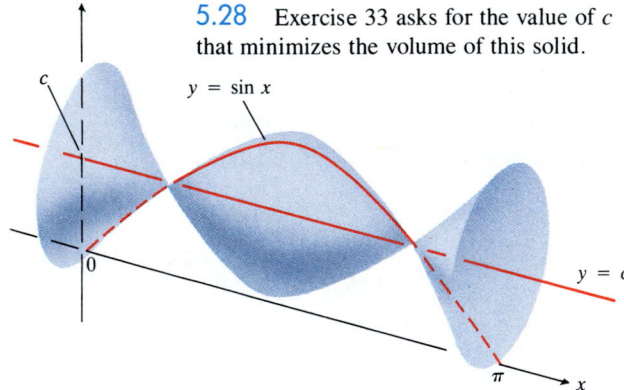

5.28   Exercise 33 asks for the value of c that minimizes the volume of this solid.

$y = \sin x$

$y = c$

34. CALCULATOR *Designing a plumb bob.* Having been asked to design a brass plumb bob that will weigh in the neighborhood of 190 g, you decide to shape it like the solid of revolution in Fig. 5.29. Find the plumb bob's volume. If you specify a brass that weighs 8.5 g/cm³, how much will the plumb bob weigh (to the nearest gram)?

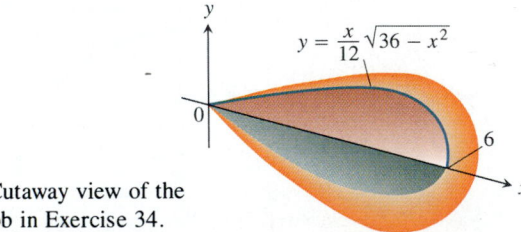

$y = \dfrac{x}{12}\sqrt{36 - x^2}$

5.29   Cutaway view of the plumb bob in Exercise 34.

35. CALCULATOR *Designing a wok.* You are designing a wok frying pan that will be shaped like a spherical bowl with handles. A bit of experimentation at home persuades you that you can get one that holds about 3 L if you make it 9 cm deep and give the sphere a radius of 16 cm. To be sure, you picture the wok as a solid of revolution (Fig. 5.30) and calculate its volume with an integral. What volume do you really get (to the nearest cubic centimeter)?

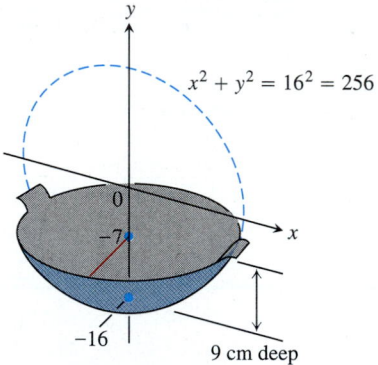

$x^2 + y^2 = 16^2 = 256$

-7

-16

9 cm deep

Dimensions in centimeters

5.30   The wok in Exercise 35.

36. CALCULATOR *An auxiliary fuel tank.* You are designing an auxiliary fuel tank that will fit under a helicopter's fuselage to extend its range. After some experimentation at your drawing board, you decide to shape the tank like the surface generated by revolving the curve $y = 1 - (x^2/16)$, $-4 \le x \le 4$, about the x-axis (dimensions in feet).

    a) How many cubic feet of fuel will the tank hold (to the nearest cubic foot)?
    b) A cubic foot holds 7.481 gal. If the helicopter gets 2 mi to the gallon, how many additional miles will the helicopter be able to fly once the tank is installed (to the nearest mile)?

37. *The volume of a torus.* The disk $x^2 + y^2 \le a^2$ is revolved about the line $x = b$ ($b > a$) to generate a solid shaped like a doughnut and called a *torus*. Find its volume. (*Hint:* $\int_{-a}^{a} \sqrt{a^2 - y^2}\, dy = \pi a^2/2$, since it is the area of a semicircle of radius a.)

38. a) A hemispherical bowl of radius a contains water to a depth h. Find the volume of water in the bowl.
    b) (Related rates) Water runs into a sunken concrete hemispherical bowl of radius 5 m at the rate of 0.2 m³/sec. How fast is the water level in the bowl rising when the water is 4 m deep?

39. *Testing the consistency of the calculus definition of volume.* The volume formulas in this section are all consistent with the standard formulas from geometry.
    a) As a case in point, show that if you revolve the region enclosed by the semicircle $y = \sqrt{a^2 - x^2}$ and the x-axis about the x-axis to generate a solid sphere, the disk for-

mula for volume (Eq. 6) will give $(4/3)\pi a^3$ just as it should.

b) Use calculus to find the volume of a right circular cone of height h and base radius r.

40. *Cavalieri's original theorem.* Prove Cavalieri's original theorem (marginal note, page 327) assuming each region to be the region between the graphs of two continuous functions on $[a, b]$.

41. *Cavalieri's theorem.* A solid lies between planes perpendicular to the x-axis at $x = 0$ and $x = 12$ (Fig. 5.31). The cross sections by planes perpendicular to the x-axis are circular disks whose diameters run from the line $y = x/2$ to the line $y = x$. Use Cavalieri's theorem (Example 3) to explain why the solid has the same volume as a right circular cone with base radius 3 and height 12.

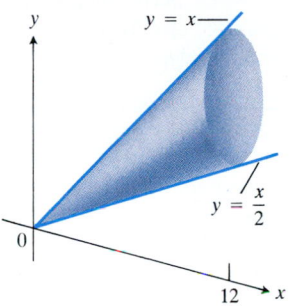

5.31   The cone in Exercise 41.

42. *The volume of a hemisphere (a classical application of Cavalieri's theorem).* Derive the formula for the volume of a solid hemisphere of radius R by comparing its cross sections with the cross sections of a solid right circular cylinder of radius R and height R from which a solid right circular cone of base radius R and altitude R has been removed.

Volumes by Slicing

Find the volumes of the solids in Exercises 43–50.

43. The solid lies between planes perpendicular to the x-axis at $x = 0$ and $x = 4$. The cross sections perpendicular to the axis on the interval $0 \le x \le 4$ are squares whose diagonals run from the parabola $y = -\sqrt{x}$ to the parabola $y = \sqrt{x}$.

44. The solid lies between planes perpendicular to the x-axis at $x = -1$ and $x = 1$. The cross sections perpendicular to the x-axis between these planes are squares whose diagonals run from the semicircle $y = -\sqrt{1 - x^2}$ to the semicircle $y = \sqrt{1 - x^2}$.

45. The solid lies between planes perpendicular to the x-axis at $x = -1$ and $x = 1$. The cross sections perpendicular to the axis between these planes are squares with edges running from the semicircle $y = -\sqrt{1 - x^2}$ to the semicircle $y = \sqrt{1 - x^2}$ (Fig. 5.32).

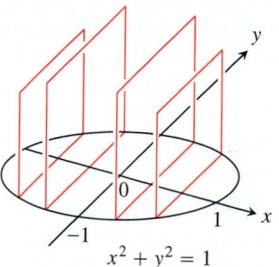

5.32   Wire frame outline of the solid in Exercise 45.

46. The solid lies between planes perpendicular to the x-axis at $x = -1$ and $x = 1$. The cross sections perpendicular to the x-axis are circular disks whose diameters run from the parabola $y = x^2$ to the parabola $y = 2 - x^2$.

47. The solid lies between planes perpendicular to the x-axis at $x = -\pi/3$ and $x = \pi/3$. The cross sections perpendicular to the x-axis are circular disks with diameters running from the curve $y = \tan x$ to the curve $y = \sec x$.

48. The solid lies between planes perpendicular to the x-axis at $x = 0$ and $x = 2$. The cross sections perpendicular to the x-axis are circular disks with diameters running from the x-axis up to the parabola $y = \sqrt{5}x^2$.

49. The base of the solid is the disk $x^2 + y^2 \le 1$. The cross sections by planes perpendicular to the y-axis between $y = -1$ and $y = 1$ are isosceles right triangles with one leg in the disk (Fig. 5.33).

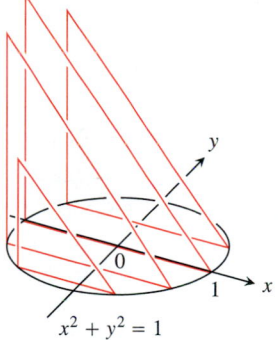

5.33   Wire frame outline of the solid in Exercise 49.

50. The base of the solid is the region between the curve $y = 2\sqrt{\sin x}$ and the interval $[0, \pi]$ on the x-axis. The cross sections perpendicular to the x-axis are equilateral triangles (Fig. 5.34).

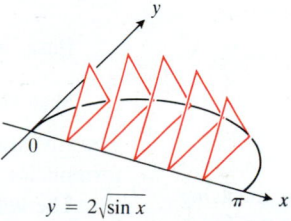

$y = 2\sqrt{\sin x}$

5.34   Wire frame outline of the solid in Exercise 50.

51. *A twisted solid.*   A square of side length s lies in a plane perpendicular to a line L. One vertex of the square lies on L. As this vertex moves a distance h along L, the square turns one revolution about L. Find the volume of the solid generated by this motion. What will the volume be if the square turns twice instead of once?

52. (Continuation of Example 2.) We could have found the volume of the wedge in Example 2 by drawing a picture like the one in Fig. 5.35, taking cross sections perpendicular to the y-axis, and integrating with respect to y. Find the volume this way.

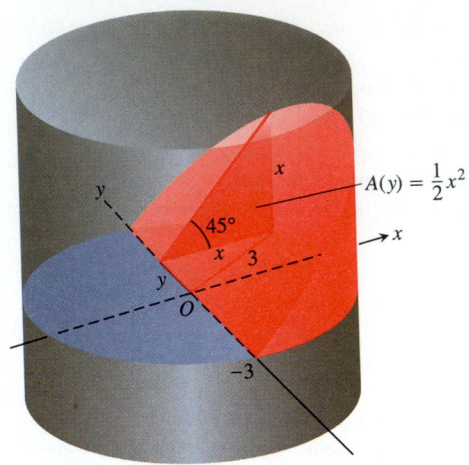

EXPLORER PROGRAM

PowerGrapher   Enables you to portray regions in the plane by graphing their bounding curves

5.35   The cross sections of the wedge perpendicular to the y-axis are isosceles right triangles. Exercise 52 asks you to use their areas to find the volume of the wedge.

5.3 Cylindrical Shells: An Alternative to Washers

When we need to find the volume of a shape that can be modeled as a solid of revolution, cylindrical shells sometimes work better than washers. In part, the reason is that the formula they lead to does not require squaring.

The Basic Shell Formula

We arrive at the cylindrical shell volume formula in the following way.

Suppose we revolve the tinted region $ABCD$ in Fig. 5.36 about the y-axis to generate a solid. To find the volume of the solid, we first approximate the region with vertical rectangles based on a partition of the closed interval $[a, b]$ on which the region stands. The rectangles run parallel to the y-axis, the axis of revolution.

Figure 5.36 also shows a typical approximating rectangle. Its dimensions are $f(c_k)$ by Δx_k. We choose the point c_k to be the midpoint of the interval from x_{k-1} to x_k. Since it does not matter where the c_k's are chosen in their intervals when we find limits of Riemann sums, we are free to choose each c_k as we please. The resulting formula is less cumbersome if we use midpoints. You will see why in just a moment.

Again with reference to Fig. 5.36, the cylindrical shell swept out by revolving the rectangle about the y-axis is a solid cylinder with these dimensions:

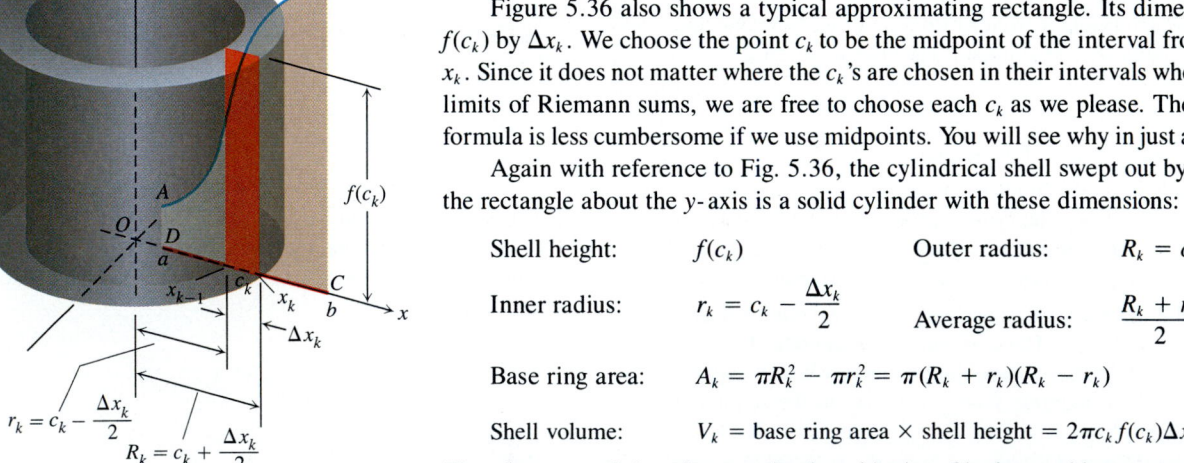

Shell height: $f(c_k)$ Outer radius: $R_k = c_k + \dfrac{\Delta x_k}{2}$

Inner radius: $r_k = c_k - \dfrac{\Delta x_k}{2}$ Average radius: $\dfrac{R_k + r_k}{2} = c_k$

Base ring area: $A_k = \pi R_k^2 - \pi r_k^2 = \pi (R_k + r_k)(R_k - r_k)$ (1)

Shell volume: $V_k =$ base ring area $\times$ shell height $= 2\pi c_k f(c_k)\Delta x_k$.

5.36   The solid swept out by revolving region $ABCD$ about the y-axis can be approximated with cylindrical shells like the one shown here.

The advantage of choosing c_k to be the midpoint of its interval becomes clear in the formula for A_k. With this choice, $R_k + r_k$ equals $2c_k$; without this choice, it doesn't.

The volumes of the shells generated by the partition of $[a, b]$ add up to

$$\text{Shell volume sum} = \sum_{k=1}^{n} 2\pi c_k f(c_k)\Delta x_k. \qquad (2)$$

We take the volume of the solid swept out by revolving region *ABCD* about the *y*-axis to be the limit of the shell volume sums as the norm of the partition of $[a, b]$ goes to zero. If *f* is continuous, the limit exists and can be found by integrating the product $2\pi x\, f(x)$ from $x = a$ to $x = b$.

The Shell Method (Axis the *y*-Axis)

Suppose $y = f(x)$ is continuous and nonnegative throughout an interval $a \le x \le b$ that does not cross the *y*-axis. Then the volume of the solid generated by revolving the region between the graph of *f* and the interval $a \le x \le b$ about the *y*-axis is found by integrating $2\pi x\, f(x)$ with respect to *x* from *a* to *b*.

$$\text{Volume} = \int_a^b 2\pi \binom{\text{shell}}{\text{radius}}\binom{\text{shell}}{\text{height}}\, dx = \int_a^b 2\pi x\, f(x)\, dx. \qquad (3)$$

One way to remember Eq. (3) is to imagine that a cylindrical shell of average circumference $2\pi x$, height $f(x)$, and thickness dx has been cut along a generating rectangle and rolled flat like a sheet of tin (Fig. 5.37). The sheet is almost a rectangular solid of dimensions $2\pi x$ by $f(x)$ by dx. Hence the shell's volume is about $2\pi x\, f(x)\, dx$. Equation (3) says that the volume of the complete solid is the integral of $2\pi x\, f(x)\, dx$ from *a* to *b*.

How to Find Volumes by the Shell Method

To apply Eq. (3), take these steps:

STEP 1: Sketch the region.

STEP 2: Draw a thin rectangle across the region parallel to the *y*-axis (the axis of revolution). The radius of the cylindrical shell swept out by the rectangle is *x*. The height of the shell is $f(x)$, the height of the rectangle. The width of the rectangle is dx. Add this information to the picture. (We also sketched the shell, but you need not do that in your own work.)

STEP 3: Find limits of integration *a* and *b* that include all possible rectangles like this from one end of the region to the other.

STEP 4: Integrate the product $2\pi x\, f(x)$ with respect to *x* from *a* to *b* to find the volume.

5.37 How to remember the integral formula for cylindrical shells.

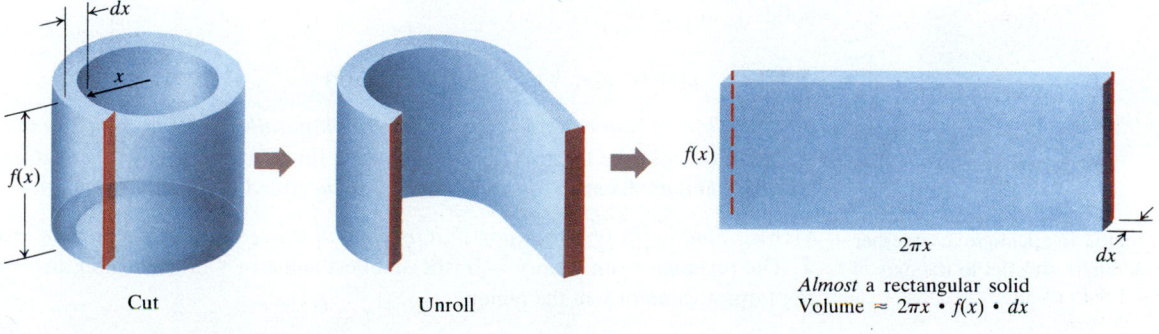

Cut   Unroll   *Almost* a rectangular solid
                             Volume $\approx 2\pi x \cdot f(x) \cdot dx$

Height: $f(x)$ Inner circumference: $2\pi x$ Thickness: dx

Example 1   The region bounded by the curve $y = \sqrt{x}$, the x-axis, and the line $x = 4$ is revolved about the y-axis to generate a solid. Find the volume of the solid.

Solution   STEP 1:   *Sketch the region* (Fig. 5.38).

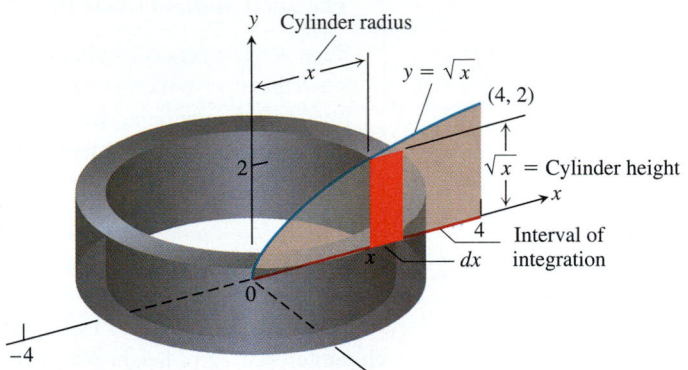

5.38   The region in Example 1, together with a thin rectangle parallel to the axis of revolution (the y-axis, in this case). The rectangle sweeps out a cylindrical shell.

STEP 2:   *Draw a thin rectangle across the region parallel to the y-axis.* Label the rectangle's height $\sqrt{x}$, its width dx, and its distance from the y-axis x. (We added all this to Fig. 5.38.)

STEP 3:   *Find limits of integration that include all rectangles like the one in step 2.* The limits in this case are $x = 0$ and $x = 4$.

STEP 4:   *Integrate to find the volume.*

$$\text{Volume} = \int_a^b 2\pi x\, f(x)\, dx = \int_0^4 2\pi x \sqrt{x}\, dx \qquad \left(\begin{array}{l}\text{Eq. (3) with values}\\\text{from steps 2 and 3}\end{array}\right)$$

$$= 2\pi \int_0^4 x^{3/2}\, dx = 2\pi \left[\frac{2}{5} x^{5/2}\right]_0^4 = \frac{128\pi}{5}$$

The volume is $128\pi/5$.

To use shells to find the volume of a solid generated by revolving a region about the x-axis instead of the y-axis, use Eq. (3) with y in place of x. Except for changes in notation, the steps we take to implement the new formula are the same as before.

Example 2   The region bounded by the curve $y = \sqrt{x}$, the x-axis, and the line $x = 4$ is revolved about the x-axis to generate a solid. Find the volume of the solid.

Solution   STEP 1:   *Sketch the region* (Fig. 5.39).

STEP 2:   *Draw a thin rectangle across the region parallel to the x-axis* (the axis of revolution). Describe the rectangle's height as a function of y. Label the rectangle's width dy and its distance from the x-axis y. (We added all this to Fig. 5.39.)

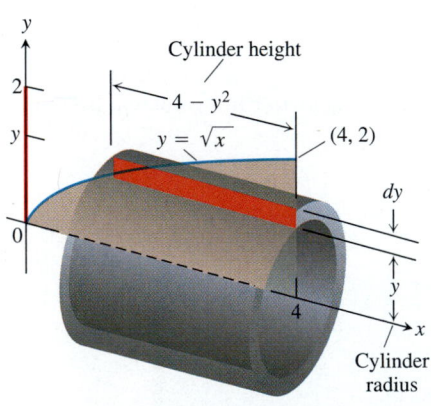

5.39   The region in Example 2, together with a thin rectangle parallel to the axis of revolution.

STEP 3:   *Find limits of integration that include all the rectangles like the one in step 2.* The rectangles run from $y = 0$, the smallest value of y in the region, to $y = 2$, the largest value of y in the region.

STEP 4:   *Integrate to find the volume.*

$$\text{Volume} = \int_c^d 2\pi \binom{\text{shell}}{\text{radius}}\binom{\text{shell}}{\text{height}}\, dy = \int_0^2 2\pi y(4 - y^2)\, dy$$

$$= 2\pi \int_0^2 (4y - y^3)\, dy = 2\pi \left[2y^2 - \frac{y^4}{4}\right]_0^2 = 8\pi,$$

in agreement with the disk method calculation in Example 4, Section 5.2.

If the axis of revolution is a line parallel to one of the coordinate axes, we use the same steps as before. The only added complication is that the expression for the radius of the typical cylinder is no longer simply x or y.

Example 3   The region in the first quadrant bounded by the parabola $y = x^2$, the y-axis, and the line $y = 1$ is revolved about the line $x = 2$ to generate a solid. Find the volume of the solid.

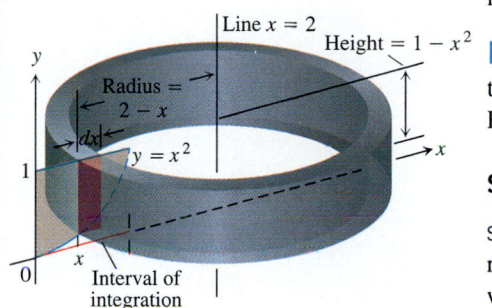

5.40   The region in Example 3. As always, the radius of the cylindrical shell is measured as a distance from the axis of revolution.

Solution   STEP 1:   *Sketch the region* (Fig. 5.40).

STEP 2:   *Draw a thin rectangle across the region parallel to the line $x = 2$ (the axis of revolution).* Describe the rectangle's height as a function of x. Label the rectangle's width dx. Describe the rectangle's distance from the line $x = 2$ as a function of x. This is the radius of the typical cylindrical shell the rectangle sweeps out. (We added all this to Fig. 5.40.)

STEP 3:   *Find limits of integration that take in all rectangles like the one in step 2.* The limits are $x = 0$ and $x = 1$, the extreme values of x in the region.

STEP 4:   *Integrate to find the volume.*

$$\text{Volume} = \int_a^b 2\pi \binom{\text{shell}}{\text{radius}}\binom{\text{shell}}{\text{height}}\, dx = \int_0^1 2\pi (2 - x)(1 - x^2)\, dx$$

$$= 2\pi \int_0^1 (2 - x - 2x^2 + x^3)\, dx = 2\pi \left[2x - \frac{x^2}{2} - \frac{2}{3}x^3 + \frac{x^4}{4}\right]_0^1 = \frac{13\pi}{6}$$

The disk, washer, and shell methods for calculating the volumes of solids of revolution always agree. In Section 6.1 (Exercise 22) we shall be able to prove the equivalence of the washer and shell methods for a broad class of solids. In the meantime, we illustrate the agreement with an example.

Example 4   The disk enclosed by the circle $x^2 + y^2 = 4$ is revolved about the y-axis to generate a solid sphere. A hole of diameter 2 is then bored through the sphere along the y-axis. Find the volume of the "cored" sphere.

Solution   The cored sphere could have been generated by revolving the shaded region in Fig. 5.41(a) about the y-axis. Thus there are three methods we might use to find the volume: disks, washers, and shells.

METHOD 1:   *Disks and subtraction.* Figure 5.41(b) shows the solid sphere with the core pulled out. The core is a circular cylinder with spherical end caps. Our plan is to subtract the volume of the core from the volume of the sphere.

5.41 (a) The tinted region generates a cored sphere. (b) An exploded view showing the sphere with the core removed. (c) A phantom view showing a cross-section slice of the sphere with the core removed. (d) Finding the volume with cylindrical shells.

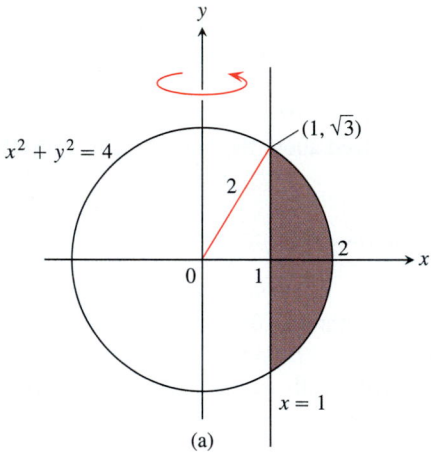

(a)

The volume of the cored sphere is the volume swept out by the shaded region as it revolves about the y-axis.

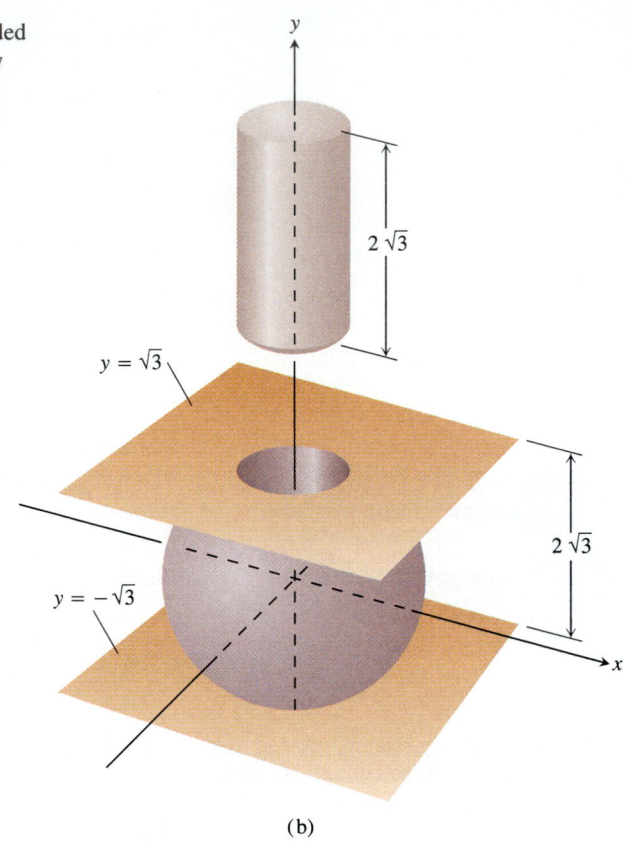

(b)

With the method of disks, we can calculate the volume of the hole left by the core and subtract it from the volume of the truncated sphere.

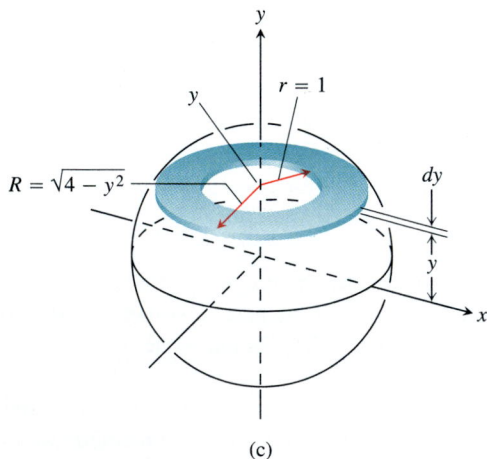

(c)

With the method of washers, we can calculate the volume of the cored sphere by modeling it as a stack of washers perpendicular to the y-axis.

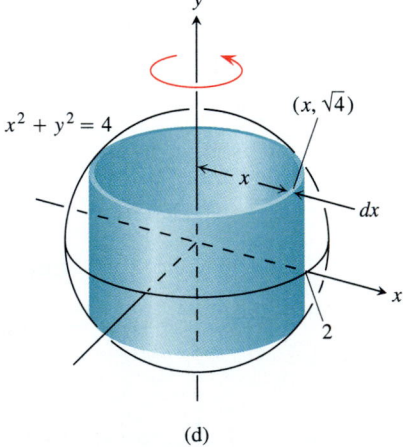

(d)

With the method of cylindrical shells, we can calculate the volume of the cored sphere by modeling it as a union of shells parallel to the y-axis.

We can simplify matters by imagining that the two caps have already been sliced off the sphere by planes perpendicular to the y-axis at $y = \sqrt{3}$ and $y = -\sqrt{3}$. With the caps removed, the truncated sphere has volume T, say. From this we subtract the volume of the hole, a right circular cylinder of radius 1 and height $2\sqrt{3}$.

The volume of the hole is

$$H = \pi(1)^2(2\sqrt{3}) = 2\pi\sqrt{3}.$$

The truncated sphere (before drilling) is a solid of revolution whose cross sections perpendicular to the y-axis are disks. The radius of a typical disk is $\sqrt{4 - y^2}$. Therefore,

$$T = \int_{-\sqrt{3}}^{\sqrt{3}} \pi(\text{radius})^2 \, dy = \int_{-\sqrt{3}}^{\sqrt{3}} \pi(4 - y^2) \, dy = \pi\left[4y - \frac{y^3}{3}\right]_{-\sqrt{3}}^{\sqrt{3}} = 6\pi\sqrt{3}.$$

The volume of the cored sphere is

$$T - H = 6\pi\sqrt{3} - 2\pi\sqrt{3} = 4\pi\sqrt{3}.$$

METHOD 2: *Washers.* The cored sphere is a solid of revolution whose cross sections perpendicular to the y-axis are washers (Fig. 5.41c). The radii of a typical washer are

Outer radius: $R = \sqrt{4 - y^2}$,

Inner radius: $r = 1$.

The volume of the cored sphere is therefore

$$V = \int_{-\sqrt{3}}^{\sqrt{3}} \pi(R^2 - r^2) \, dy = \int_{-\sqrt{3}}^{\sqrt{3}} \pi(4 - y^2 - 1) \, dy = \pi\left[3y - \frac{y^3}{3}\right]_{-\sqrt{3}}^{\sqrt{3}} = 4\pi\sqrt{3}.$$

METHOD 3: *Cylindrical shells.* We model the volume of the cored sphere with cylindrical shells like the one in Fig. 5.41(d). The typical shell has radius x, height $2\sqrt{4 - x^2}$, and thickness dx. The volume of the cored sphere is

$$V = \int_1^2 2\pi\binom{\text{shell}}{\text{radius}}\binom{\text{shell}}{\text{height}} dx = \int_1^2 4\pi x\sqrt{4 - x^2} \, dx$$

$$= 4\pi\left[-\frac{1}{3}(4 - x^2)^{3/2}\right]_1^2 \qquad \left(\begin{array}{l}\text{After substituting } u = 4 - x^2,\\ \text{integrating, and substituting back}\end{array}\right)$$

$$= 0 - 4\pi\left[-\frac{1}{3}(4 - 1)^{3/2}\right] = 4\pi\sqrt{3}.$$

Table 5.1 summarizes the methods of finding volumes with washers and shells.

EXERCISES 5.3

Use the shell method to find the volumes of the solids generated by revolving about the y-axis the regions bounded by the curves and lines in Exercises 1–6.

1. $y = x$, $y = -x/2$, $x = 2$

2. $y = \sqrt{x}$, $y = 0$, $x = 4$

3. $y = x^2 + 1$, $y = 0$, $x = 0$, $x = 1$

4. $y = 2x - 1$, $y = \sqrt{x}$, $x = 0$

5. $y = 1/x$, $\quad y = 0$, $\quad x = 1/2$, $\quad x = 2$

6. $y = 3/(2\sqrt{x})$, $\quad y = 0$, $\quad x = 1$, $\quad x = 4$

Use the shell method to find the volumes of the solids generated by revolving about the x-axis the regions bounded by the curves and lines in Exercises 7–12.

7. $y = |x|$, $\quad y = 1$

8. $y = x$, $\quad y = 1$, $\quad x = 2$

9. $y = \sqrt{x}$, $\quad y = 0$, $\quad y = x - 2$

10. $y = 2$, $\quad x = -y$, $\quad x = \sqrt{y}$

11. The parabola $x = 2y - y^2$ and the y-axis

12. The parabola $x = 2y - y^2$ and the line $y = x$

Find the volumes of the solids generated by revolving the shaded regions in Exercises 13–16 about the indicated axes.

13. The y-axis

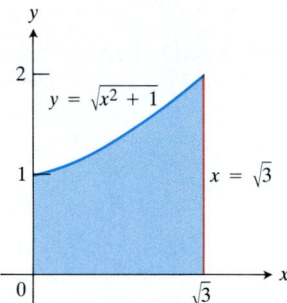

14. The y-axis

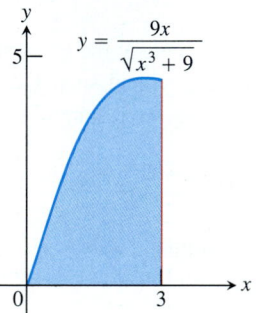

15. a) The x-axis   b) The line $y = 1$
c) The line $y = 8/5$ d) The line $y = -2/5$

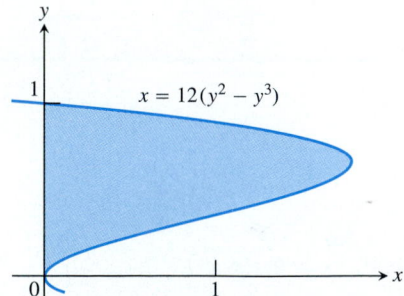

16. a) The x-axis
b) The line $y = 2$
c) The line $y = 5$
d) The line $y = -5/8$

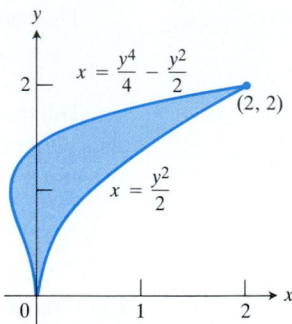

In Exercises 17–24, find the volumes of the solids generated by revolving the regions about the given axes.

17. The triangle with vertices $(1, 1)$, $(1, 2)$, and $(2, 2)$ about (a) the x-axis; (b) the y-axis; (c) the line $x = 10/3$; (d) the line $y = 1$

18. The region in the first quadrant bounded by the curve $x = y - y^3$ and the y-axis about (a) the x-axis; (b) the line $y = 1$

19. The region in the first quadrant bounded by $x = y - y^3$, $x = 1$, and $y = 1$ about (a) the x-axis; (b) the y-axis; (c) the line $x = 1$; (d) the line $y = 1$

20. The triangular region bounded by the lines $2y = x + 4$, $y = x$, and $x = 0$ about (a) the x-axis; (b) the y-axis; (c) the line $x = 4$; (d) the line $y = 8$

21. The region in the first quadrant bounded by $y = x^3$ and $y = 4x$ about (a) the x-axis; (b) the line $y = 8$

22. The region bounded by $y = \sqrt{x}$ and $y = x^2/8$ about (a) the x-axis; (b) the y-axis

23. The region bounded by $y = 2x - x^2$ and $y = x$ about (a) the y-axis; (b) the line $x = 1$

24. The region bounded by $y = \sqrt{x}$, $y = 2$, $x = 0$ about (a) the x-axis; (b) the y-axis; (c) the line $x = 4$; (d) the line $y = 2$

25. Suppose that the function $f(x)$ is nonnegative and continuous for $x \geq 0$. Suppose also that, for every positive number b, revolving the region enclosed by the graph of f, the coordinate axes, and the line $x = b$ about the y-axis generates a solid of volume $2\pi b^3$. Find $f(x)$.

EXPLORER PROGRAM

PowerGrapher   Enables you to portray regions in the plane by graphing their bounding curves

TABLE 5.1
Washers versus shells

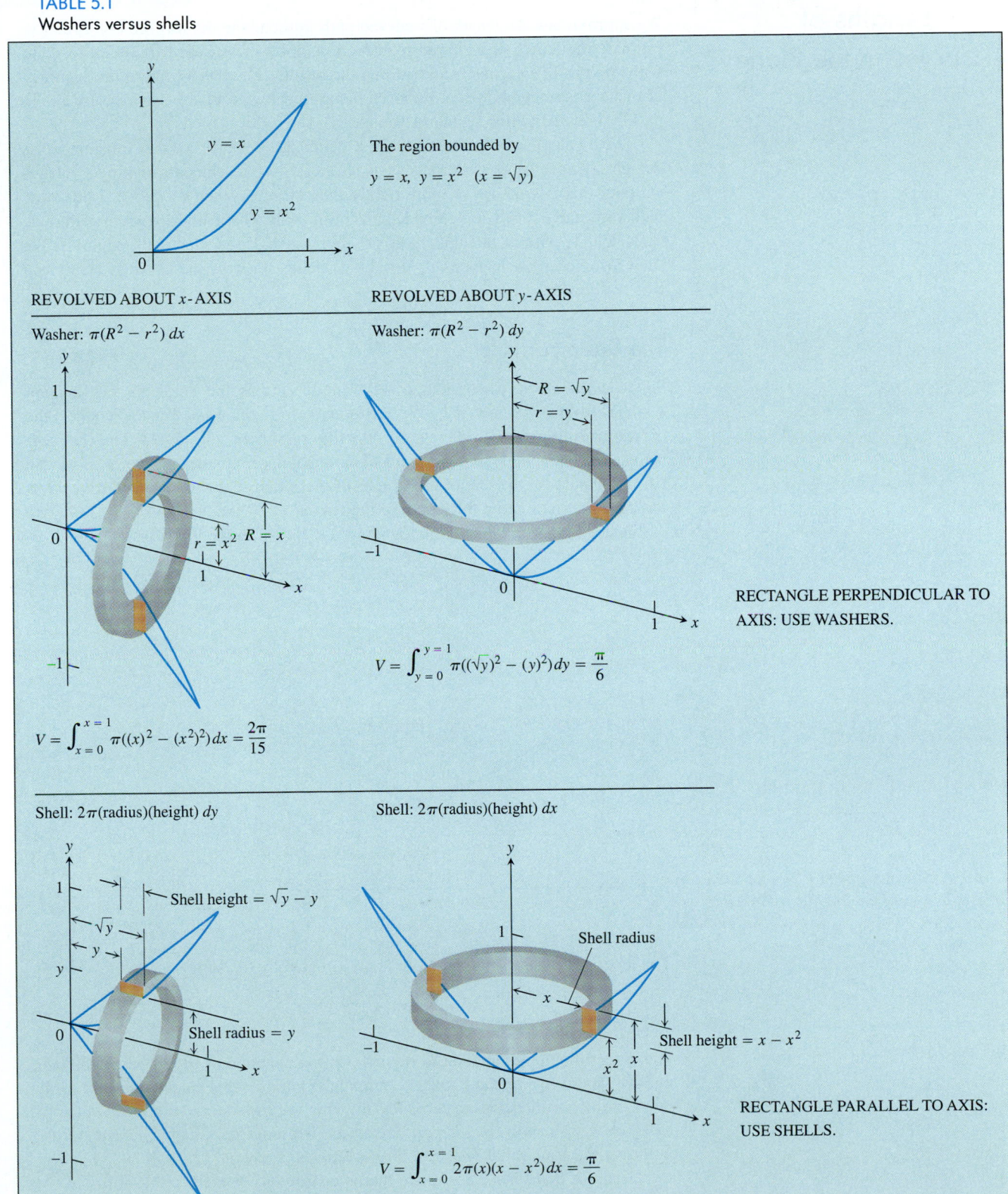

The region bounded by
$y = x$, $y = x^2$ $(x = \sqrt{y})$

REVOLVED ABOUT x-AXIS

Washer: $\pi(R^2 - r^2)\,dx$

$r = x^2$ $R = x$

$$V = \int_{x=0}^{x=1} \pi((x)^2 - (x^2)^2)\,dx = \frac{2\pi}{15}$$

REVOLVED ABOUT y-AXIS

Washer: $\pi(R^2 - r^2)\,dy$

$R = \sqrt{y}$
$r = y$

$$V = \int_{y=0}^{y=1} \pi((\sqrt{y})^2 - (y)^2)\,dy = \frac{\pi}{6}$$

RECTANGLE PERPENDICULAR TO AXIS: USE WASHERS.

Shell: $2\pi(\text{radius})(\text{height})\,dy$

Shell height $= \sqrt{y} - y$
$\sqrt{y}$
y
Shell radius $= y$

$$V = \int_{y=0}^{y=1} 2\pi(y)(\sqrt{y} - y)\,dy = \frac{2\pi}{15}$$

Shell: $2\pi(\text{radius})(\text{height})\,dx$

Shell radius
x
Shell height $= x - x^2$
x^2 x

$$V = \int_{x=0}^{x=1} 2\pi(x)(x - x^2)\,dx = \frac{\pi}{6}$$

RECTANGLE PARALLEL TO AXIS: USE SHELLS.

5.4 Lengths of Curves in the Plane

We approximate the length of a curved path in the plane the way we use a ruler to estimate the length of a curved road on a map, by measuring from point to point with straight-line segments and adding the results. There is a limit to the accuracy of such an estimate, however, imposed in part by how accurately we measure and in part by how many line segments we use.

With calculus we can usually do a better job because we can imagine using straight-line segments as short as we please, each set of segments making a polygonal path that fits the curve more tightly than before. When we proceed this way, with a smooth enough curve, the lengths of the polygonal paths approach a limit we can calculate with an integral. In this section we shall see what that integral is. We shall also see what happens if, instead of being smooth, the curve is Helga von Koch's snowflake curve.

The Basic Formula

Suppose the curve whose length we want to find is the graph of the function $y = f(x)$ from $x = a$ to $x = b$. We partition $[a, b]$ in the usual way and connect the corresponding points on the curve with line segments (Fig. 5.42). The line segments, taken together, form a polygonal path that approximates the curve. If we can find a formula for the sum of the lengths of these line segments, we will have a formula for approximating the length of the curve.

The length of a typical line segment PQ (shown in the figure) is

$$\sqrt{(\Delta x_k)^2 + (\Delta y_k)^2}. \tag{1}$$

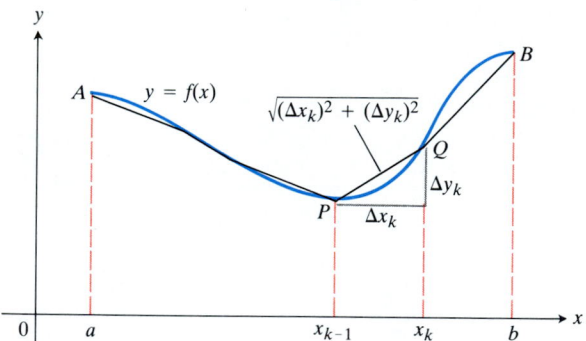

5.42   A typical segment PQ of a polygonal path approximating the curve AB.

The length of the curve is therefore approximated by the sum

$$\sum_{k=1}^{n} \sqrt{(\Delta x_k)^2 + (\Delta y_k)^2}. \tag{2}$$

We expect the approximation to improve as the partition of $[a, b]$ becomes finer, and we would like to show that the sums in (2) approach a calculable limit as the norm of the partition goes to zero. To show this, we rewrite the sum in (2) in a form to which we can apply the Integral Existence Theorem from Chapter 4. Our starting point is the Mean Value Theorem for derivatives.

Suppose that f has a derivative that is continuous at every point of $[a, b]$ (we

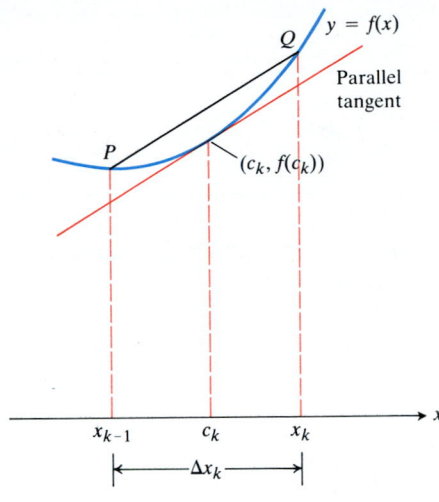

call such a function **smooth** on $[a, b]$ and call its graph a **smooth curve**). Then, by the Mean Value Theorem, there is a point $(c_k, f(c_k))$ on the curve between P and Q where the tangent is parallel to the segment PQ (Fig. 5.43). At this point

$$f'(c_k) = \frac{\Delta y_k}{\Delta x_k}, \tag{3}$$

$$\Delta y_k = f'(c_k) \Delta x_k. \tag{4}$$

With this substitution for Δy_k, the sums in (2) take the form

$$\sum_{k=1}^{n} \sqrt{(\Delta x_k)^2 + (f'(c_k) \Delta x_k)^2} = \sum_{k=1}^{n} \sqrt{1 + (f'(c_k))^2} \, \Delta x_k. \tag{5}$$

The sums on the right are Riemann sums for the continuous function $\sqrt{1 + (f'(x))^2}$ on the interval $[a, b]$. They therefore converge to the integral of this function as the norm of the partition of the interval goes to zero. We define this integral to be the length of the curve from a to b.

5.43   Enlargement of the arc PQ in Fig. 5.42.

DEFINITION

If the function f is smooth on $[a, b]$, the **length of the curve $y = f(x)$ from a to b** is the number

$$L = \int_a^b \sqrt{1 + \left(\frac{dy}{dx}\right)^2} \, dx. \tag{6}$$

Example 1   Find the length of the curve

$$y = \frac{4\sqrt{2}}{3} x^{3/2} - 1, \qquad 0 \le x \le 1.$$

Solution   We use Eq. (6) with $a = 0$, $b = 1$, and

$$y = \frac{4\sqrt{2}}{3} x^{3/2} - 1,$$

$$\frac{dy}{dx} = \frac{4\sqrt{2}}{3} \cdot \frac{3}{2} x^{1/2} = 2\sqrt{2} x^{1/2},$$

$$1 + \left(\frac{dy}{dx}\right)^2 = 1 + \left(2\sqrt{2} x^{1/2}\right)^2 = 1 + 8x.$$

The length of the curve from $x = 0$ to $x = 1$ is

$$L = \int_0^1 \sqrt{1 + \left(\frac{dy}{dx}\right)^2} \, dx = \int_0^1 \sqrt{1 + 8x} \, dx \qquad \left(\begin{array}{l}\text{Eq. (6) with } a = 0, \\ b = 1\end{array}\right)$$

$$= \frac{2}{3} \cdot \frac{1}{8} (1 + 8x)^{3/2} \Big]_0^1 = \frac{13}{6}. \qquad \left(\begin{array}{l}\text{Substitute } u = 1 + 8x, \text{ integrate,} \\ \text{and replace } u \text{ by } 1 + 8x.\end{array}\right)$$

Dealing With Discontinuities in dy/dx

At a point on a curve where dy/dx fails to exist, dx/dy may exist and we may be able to find the curve's length by interchanging x and y in Eq. (6). The revised formula looks like this:

$$L = \int_c^d \sqrt{1 + \left(\frac{dx}{dy}\right)^2} \, dy. \tag{7}$$

To use Eq. (7), we express x as a function of y, calculate dx/dy, and proceed as before to square, add 1, take the square root, and integrate.

Example 2   Find the length of the curve $y = (x/2)^{2/3}$ from $x = 0$ to $x = 2$.

Solution   The derivative

$$\frac{dy}{dx} = \frac{2}{3}\left(\frac{x}{2}\right)^{-1/3} = \frac{2}{3}\left(\frac{2}{x}\right)^{1/3}$$

is not defined at $x = 0$, so we cannot find the curve's length with Eq. (6).
We therefore rewrite the equation to express x in terms of y:

$$y = \left(\frac{x}{2}\right)^{2/3}$$

$$y^{3/2} = \frac{x}{2} \qquad \text{(Raise both sides to the power 3/2.)}$$

$$x = 2y^{3/2}. \qquad \text{(Solve for } x.\text{)}$$

From this we see that the curve whose length we want is also the graph of $x = 2y^{3/2}$ from $y = 0$ to $y = 1$ (Fig. 5.44).
The derivative

$$\frac{dx}{dy} = 2 \cdot \frac{3}{2} y^{1/2} = 3y^{1/2}$$

is continuous throughout the interval $0 \le y \le 1$. We may therefore find the curve's length by setting

$$\left(\frac{dx}{dy}\right)^2 = (3y^{1/2})^2 = 9y$$

in Eq. (7) and integrating from $y = 0$ to $y = 1$:

$$L = \int_c^d \sqrt{1 + \left(\frac{dx}{dy}\right)^2} \, dy = \int_0^1 \sqrt{1 + 9y} \, dy \qquad \text{(Eq. (7))}$$

$$= \frac{1}{9} \cdot \frac{2}{3} (1 + 9y)^{3/2} \Big]_0^1 \qquad \left(\begin{array}{l}\text{Substitute } u = 1 + 9y, \, du/9 = dy, \\ \text{integrate, and substitute back.}\end{array}\right)$$

$$= \frac{2}{27} (10\sqrt{10} - 1) = 2.27. \qquad \text{(To two places with a calculator)}$$

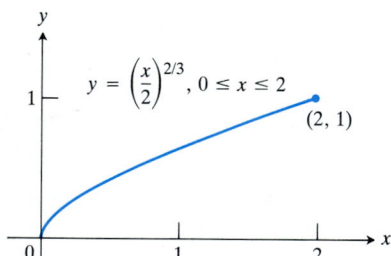

5.44   The graph of $y = (x/2)^{2/3}$ from $x = 0$ to $x = 2$ is also the graph of $x = 2y^{3/2}$ from $y = 0$ to $y = 1$, a function that has a continuous first derivative. We may therefore use Eq. (7) with $x = 2y^{3/2}$ to find the curve's length.

The Short Differential Formula

The equations

$$L = \int_a^b \sqrt{1 + \left(\frac{dy}{dx}\right)^2} \, dx \qquad \text{and} \qquad L = \int_c^d \sqrt{1 + \left(\frac{dx}{dy}\right)^2} \, dy \tag{8}$$

are often written with differentials instead of derivatives. This is done formally by thinking of the derivatives as quotients of differentials and bringing the dx and dy inside the radicals to cancel the denominators. In the first integral we have

$$\sqrt{1 + \left(\frac{dy}{dx}\right)^2}\,dx = \sqrt{1 + \frac{dy^2}{dx^2}}\,dx = \sqrt{dx^2 + \frac{dy^2}{dx^2}\,dx^2} = \sqrt{dx^2 + dy^2}. \quad (9)$$

In the second integral we have

$$\sqrt{1 + \left(\frac{dx}{dy}\right)^2}\,dy = \sqrt{1 + \frac{dx^2}{dy^2}}\,dy = \sqrt{dy^2 + \frac{dx^2}{dy^2}\,dy^2} = \sqrt{dx^2 + dy^2}. \quad (10)$$

Thus the integrals in (8) reduce to the same differential formula:

$$L = \int_a^b \sqrt{dx^2 + dy^2}. \quad (11)$$

Of course, dx and dy must be expressed in terms of a common variable, and appropriate limits of integration must be found before the integration in Eq. (11) is performed.

We can shorten Eq. (11) still further. Think of dx and dy as two sides of a small triangle whose hypotenuse is

$$ds = \sqrt{dx^2 + dy^2} \quad (12)$$

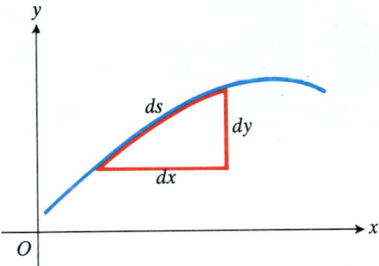

5.45 Diagram for remembering the equation $ds = \sqrt{dx^2 + dy^2}$.

(Fig. 5.45). The differential ds is then regarded as a differential of arc length that can be integrated between appropriate limits to give the length of the curve. With $\sqrt{dx^2 + dy^2}$ set equal to ds, the integral in Eq. (11) simply becomes the integral of ds.

DEFINITION

> **The Arc Length Differential and the Differential Formula for Arc Length**
>
> $$ds = \sqrt{dx^2 + dy^2} \qquad\qquad L = \int ds$$
>
> arc length differential   differential formula for arc length

Curves with Infinite Length

As you may recall from Section 2.5, Helga von Koch's snowflake curve K is the limit curve of an infinite sequence $C_1, C_2, \ldots, C_n, \ldots$ of "triangular" polygonal curves. Figure 5.46 shows the first four curves in the sequence. Each time we introduce a new vertex in the construction process, it remains as a vertex in all subsequent curves and becomes a point on the limit curve K. This means that each of the C's is itself a polygonal approximation of K—the endpoints of its sides all belonging to K. The length of K should therefore be the limit of the lengths of the curves C_n. At least, that is what it should be if we apply the definition of length we developed for smooth curves.

What, then, is the limit of the lengths of the curves C_n? If the original equilateral triangle C_1 has sides of length 1, the total length of C_1 is 3. To make C_2 from C_1, we replace each side of C_1 by four segments, each of which is one-third as long as the original side. The total length of C_2 is therefore 3(4/3). To get the length of C_3, we multiply by 4/3 again. We do so again to get the length of C_4. By the time we get

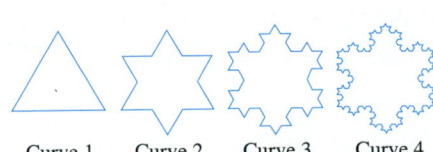

Curve 1   Curve 2   Curve 3   Curve 4

5.46 The first four polygonal approximations in the construction of Helga von Koch's snowflake.

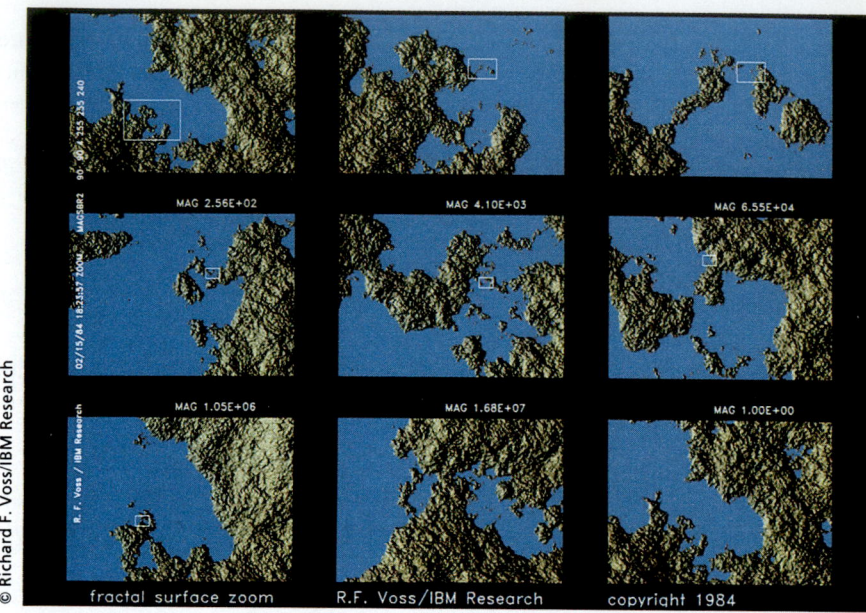

© Richard F. Voss/IBM Research

fractal surface zoom   R.F. Voss/IBM Research   copyright 1984

5.47 Repeated magnifications of a fractal coastline. Like Helga von Koch's snowflake curve, coasts like these are too rough to have a measurable length.

out to C_n, we have a curve of length $3(4/3)^{n-1}$.

Curve Number	1	2	3	$\cdots$	n	$\cdots$
Length	3	$3\left(\dfrac{4}{3}\right)$	$3\left(\dfrac{4}{3}\right)^2$	$\cdots$	$3\left(\dfrac{4}{3}\right)^{n-1}$	$\cdots$

The length of C_{10} is nearly 40 and the length of C_{100} is greater than 7,000,000,000,000. The lengths grow too rapidly to have a finite limit. Therefore the snowflake curve has no length, or, if you prefer, infinite length.

What went wrong? Nothing. The formulas we derived for length are for the graphs of smooth functions, curves that are smooth enough to have a continuously turning tangent at every point. Helga von Koch's snowflake curve is too rough for that, and our derivative-based formulas do not apply.

Benoit Mandelbrot's theory of fractals has proved to be a rich source of curves with infinite length, curves that when magnified prove to be as rough and varied as they looked before magnification. Like coastlines on an ocean, such curves cannot be smoothed out by magnification (Fig. 5.47).

EXERCISES 5.4

Set up, but do not evaluate, integrals for the lengths of the curves in Exercises 1–8.

1. $y = x^2$, $-1 \le x \le 2$

2. $y = \tan x$, $-\pi/3 \le x \le 0$

3. $y^2 + 2y = 2x + 1$ from $(-1, -1)$ to $(7, 3)$

4. $y = \sin x - x \cos x$, $0 \le x \le \pi$

5. $x = \sin y$, $0 \le y \le \pi$

6. $x = \sqrt{1 - y^2}$, $-1/2 \le y \le 1/2$

7. $y = \displaystyle\int_0^x \tan t \, dt$, $0 \le x \le \pi/6$

8. $x = \displaystyle\int_0^y \sqrt{\sec^2 t - 1} \, dt$, $-\pi/3 \le y \le \pi/4$

Find the lengths of the curves in Exercises 9–20.

9. $y = (1/3)(x^2 + 2)^{3/2}$ from $x = 0$ to $x = 3$

10. $y = x^{3/2}$ from $x = 0$ to $x = 4$

11. $9x^2 = 4y^3$ from (0, 0) to $(2\sqrt{3}, 3)$

12. $y = x^{2/3}$ from (0, 0) to $(4, \sqrt[3]{16})$

13. $y = (x^3/3) + 1/(4x)$ from $x = 1$ to $x = 3$ (*Hint:* $1 + (dy/dx)^2$ is a perfect square.)

14. $y = (x^{3/2}/3) - x^{1/2}$ from $x = 1$ to $x = 9$ (*Hint:* $1 + (dy/dx)^2$ is a perfect square.)

15. $x = (y^4/4) + 1/(8y^2)$ from $y = 1$ to $y = 2$ (*Hint:* $1 + (dx/dy)^2$ is a perfect square.)

16. $x = (y^3/6) + 1/(2y)$ from $y = 2$ to $y = 3$ (*Hint:* $1 + (dx/dy)^2$ is a perfect square.)

17. $y = (3/4)x^{4/3} - (3/8)x^{2/3} + 5$, $1 \le x \le 8$

18. $x = (y^3/3) + y^2 + y + 1/(4(y + 1))$, $0 \le y \le 2$

19. $x = \int_0^y \sqrt{\sec^4 t - 1}\, dt$, $-\pi/4 \le y \le \pi/4$

20. $y = \int_{-2}^x \sqrt{3t^4 - 1}\, dt$, $-2 \le x \le -1$

21. *The length of an astroid.* The graph of the equation $x^{2/3} + y^{2/3} = 1$ is one of a family of curves called *astroids* (not "asteroids") because of their starlike appearance (Fig. 5.48). Find the length of this particular astroid by finding the length of half the first-quadrant portion, $y = (1 - x^{2/3})^{3/2}$, $\sqrt{2}/4 \le x \le 1$, and multiplying by 8.

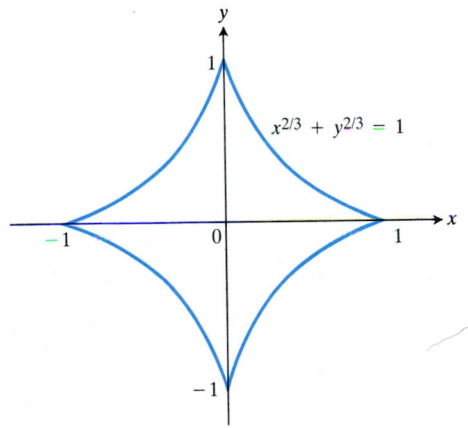

$x^{2/3} + y^{2/3} = 1$

5.48   The astroid in Exercise 21.

22. Find a curve through the point (1, 1) whose length integral (Eq. 6) is

$$L = \int_1^4 \sqrt{1 + \frac{1}{4x}}\, dx.$$

23. Find the length of the curve

$$y = \int_0^x \sqrt{\cos 2t}\, dt$$

from $x = 0$ to $x = \pi/4$. (*Hint:* For $0 \le x \le \pi/4$, $1 + \cos 2x = 2\cos^2 x$.)

24. Suppose that the length of the smooth curve $y = f(x)$ over the interval $0 \le x \le a$ is always $\sqrt{2}a$. Find f.

25. *Testing the definition of length.* The Pythagorean theorem tells us that the length of line segment OP in Fig. 5.49 is $\sqrt{a^2 + b^2}$. Show that the calculus definition of length gives the same result by using Eq. (6) to find the length of the curve $y = (b/a)\, x, 0 \le x \le a$.

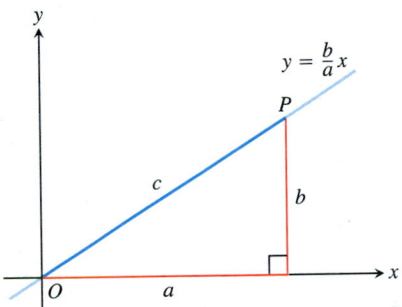

5.49   Is c still equal to $\sqrt{a^2 + b^2}$? To find out, do Exercise 25.

You may have wondered why the curves we have been working with have such unusual formulas. The reason is that the square root $\sqrt{1 + (dy/dx)^2}$ in the arc length integral almost never produces a function whose antiderivative we can find. In fact, this square root is a well-known source of nonelementary integrals. Most arc length integrals have to be evaluated numerically, as in the exercises that follow.

26. CALCULATOR   Your metal fabrication company is bidding for a contract to make sheets of corrugated iron roofing like the one shown in Fig. 5.50. The cross sections of the corrugated sheets are to conform to the curve

$$y = \sin \frac{3\pi}{20}\, x,   0 \le x \le 20 \text{ in.}$$

If the roofing is to be stamped from flat sheets by a process

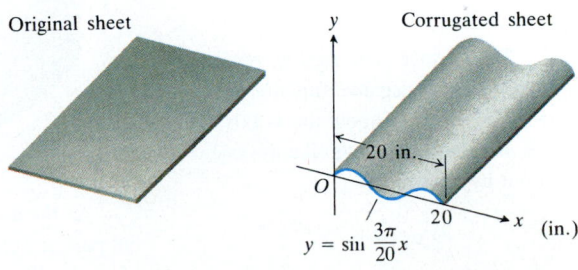

Original sheet   Corrugated sheet

20 in.

$y = \sin \dfrac{3\pi}{20} x$

5.50   How wide does the original sheet have to be? See Exercise 26.

that does not stretch the material, how wide should the original material be? To find out, use numerical integration to approximate the length of the sine curve.

27. CALCULATOR Your engineering firm is bidding for the contract to construct the tunnel shown in Fig. 5.51. The tunnel

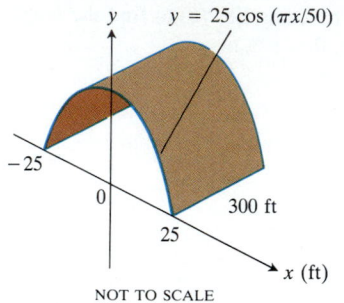

$y = 25 \cos(\pi x/50)$

300 ft

NOT TO SCALE

5.51   The tunnel in Exercise 27.

is 300 ft long and 50 ft wide at the base. The cross section is shaped like one arch of the curve $y = 25 \cos(\pi x/50)$. Upon completion, the tunnel's inside surface (excluding the roadway) will be treated with a waterproof sealer that costs $1.75 per square foot to apply. How much will it cost to apply the sealer? (*Hint:* Use numerical integration to find the length of the cosine curve.)

EXPLORER PROGRAMS

Integral Evaluation	Evaluates integrals by Simpson's rule and other numerical methods
PowerGrapher	Graphs any function you can key in

5.5 Areas of Surfaces of Revolution

When you jump rope, the rope sweeps out a surface in the space around you, a surface called a surface of revolution. As you can imagine, the area of this surface depends on the rope's length and on how far away each segment of the rope swings. This section explores the relation between the area of a surface of revolution and the length and reach of the curve that generates it.

The Basic Formula

Suppose we want to find the area of the surface swept out by revolving the graph of a nonnegative function $y = f(x)$, $a \le x \le b$, about the x-axis. We partition the closed interval $[a, b]$ in the usual way and use the points in the partition to divide the graph into short arcs. Figure 5.52 shows a typical arc PQ and the band it sweeps out as part of the graph of f.

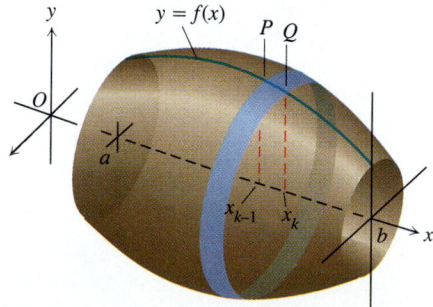

5.52   The surface generated by revolving the graph of a nonnegative function $y = f(x)$, $a \le x \le b$, about the x-axis. The surface is a union of bands like the one swept out by the arc PQ.

As the arc PQ revolves about the x-axis, the line segment joining P and Q sweeps out part of a cone whose axis lies along the x-axis (magnified view in Fig. 5.53). A piece of a cone like this is called a frustum of the cone, *frustum* being Latin for "piece." The surface area of the frustum approximates the surface area of the band swept out by the arc PQ.

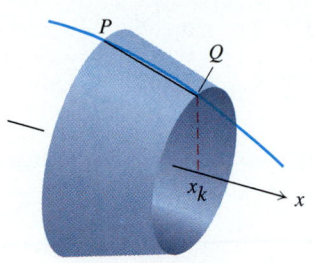

5.53 The line segment joining P and Q sweeps out a frustum of a cone.

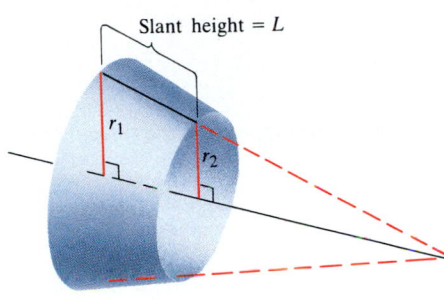

5.54 The important dimensions of the frustum in Fig. 5.53.

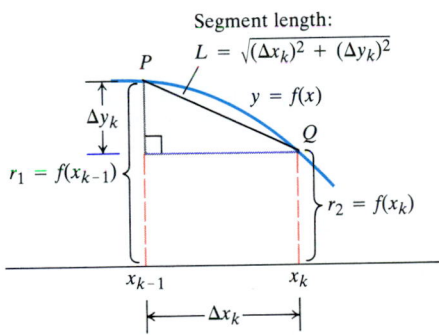

5.55 The important dimensions associated with the arc and segment PQ.

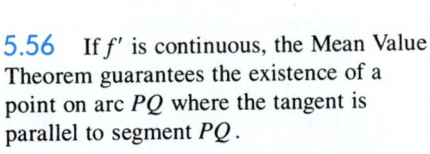

5.56 If f' is continuous, the Mean Value Theorem guarantees the existence of a point on arc PQ where the tangent is parallel to segment PQ.

The surface area of the frustum of a cone (see Fig. 5.54) is 2π times the average of the base radii times the slant height:

$$\text{Frustum surface area} = 2\pi \cdot \frac{r_1 + r_2}{2} \cdot L = \pi(r_1 + r_2)L. \tag{1}$$

For the frustum swept out by the segment PQ (Fig. 5.55), this works out to be

$$\text{Frustum surface area} = \pi(f(x_{k-1}) + f(x_k))\sqrt{(\Delta x_k)^2 + (\Delta y_k)^2}. \tag{2}$$

The area of the original surface, being the sum of the areas of the bands swept out by arcs like arc PQ, is approximated by the frustum area sum

$$\sum_{k=1}^{n} \pi(f(x_{k-1}) + f(x_k))\sqrt{(\Delta x_k)^2 + (\Delta y_k)^2}. \tag{3}$$

We expect the approximation to improve as the partition of $[a, b]$ becomes finer, and we would like to show that the sums in (3) approach a calculable limit as the norm of the partition goes to zero.

To show this, we try to rewrite the sum in (3) as the Riemann sum of some function over the interval from a to b. As in the calculation of arc length, we begin by appealing to the Mean Value Theorem for derivatives.

Suppose as before that f has a derivative that is continuous at every point of $[a, b]$. Then, by the Mean Value Theorem, there is a point $(c_k, f(c_k))$ on the curve between P and Q where the tangent is parallel to the segment PQ (Fig. 5.56). At this point,

$$f'(c_k) = \frac{\Delta y_k}{\Delta x_k}, \tag{4}$$

$$\Delta y_k = f'(c_k)\Delta x_k. \tag{5}$$

With this substitution for Δy_k, the sums in (3) take the form

$$\sum_{k=1}^{n} \pi(f(x_{k-1}) + f(x_k))\sqrt{(\Delta x_k)^2 + (f'(c_k)\Delta x_k)^2}$$
$$= \sum_{k=1}^{n} \pi(f(x_{k-1}) + f(x_k))\sqrt{1 + (f'(c_k))^2}\,\Delta x_k. \tag{6}$$

At this point there is both good news and bad.

The bad news is that the sums in (6) are not the Riemann sums of any function because the points x_{k-1}, x_k, and c_k are not the same and there is no way to make them the same. The good news is that this does not matter. A theorem called Bliss's theorem, from advanced calculus, assures us that as the norm of the subdivision of

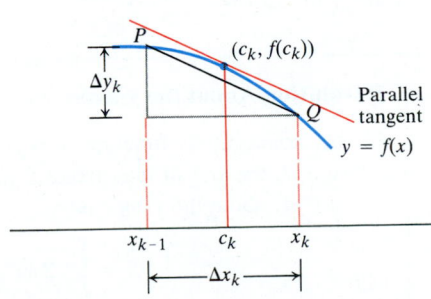

$[a, b]$ goes to zero, the sums in (6) converge to

$$\int_a^b 2\pi f(x)\sqrt{1 + (f'(x))^2}\, dx \qquad (7)$$

just the way we want them to. We therefore define this integral to be the area of the surface swept out by the graph of f from a to b.

DEFINITION

If the nonnegative function f is smooth throughout the interval $a \le x \le b$, the **area of the surface** generated by revolving the curve $y = f(x)$ about the x-axis is

$$S = \int_a^b 2\pi y\sqrt{1 + \left(\frac{dy}{dx}\right)^2}\, dx. \qquad (8)$$

Notice that the square root in Eq. (8) is the same one that appears in the formula for arc length. (More about that later.)

Example 1   Find the area of the surface generated by revolving the curve $y = 2\sqrt{x}$, $1 \le x \le 2$, about the x-axis (Fig. 5.57).

Solution   We evaluate the formula

$$\text{Surface area} = \int_a^b 2\pi y\sqrt{1 + \left(\frac{dy}{dx}\right)^2}\, dx \qquad \text{(Eq. (8))}$$

with

$$a = 1, \qquad b = 2, \qquad y = 2\sqrt{x}, \qquad \frac{dy}{dx} = \frac{1}{\sqrt{x}},$$

$$\sqrt{1 + \left(\frac{dy}{dx}\right)^2} = \sqrt{1 + \left(\frac{1}{\sqrt{x}}\right)^2} = \sqrt{1 + \frac{1}{x}} = \sqrt{\frac{x+1}{x}} = \frac{\sqrt{x+1}}{\sqrt{x}}.$$

With these substitutions,

$$\text{Surface area} = \int_1^2 2\pi \cdot 2\sqrt{x}\,\frac{\sqrt{x+1}}{\sqrt{x}}\, dx = 4\pi \int_1^2 \sqrt{x+1}\, dx$$

$$= 4\pi \cdot \frac{2}{3}(x+1)^{3/2}\Big]_1^2 = \frac{8\pi}{3}(3\sqrt{3} - 2\sqrt{2}).$$

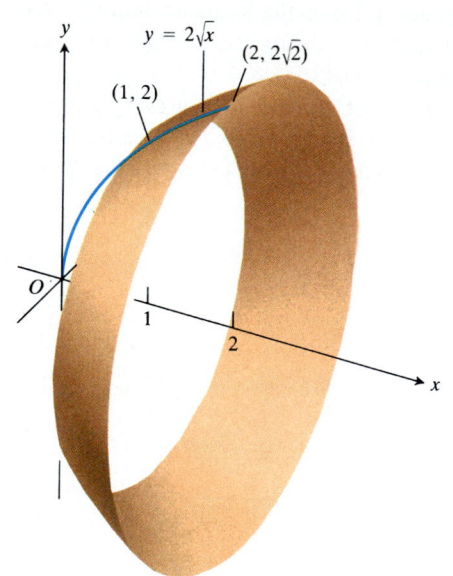

$y = 2\sqrt{x}$

$(2, 2\sqrt{2})$

$(1, 2)$

5.57   Example 1 calculates the area of this surface.

Revolution About the y-axis

If the axis of revolution is the y-axis, we use the formula we get from interchanging x and y in Eq. (8):

Revolution About the y-axis

If the nonnegative function $x = f(y)$ is smooth throughout the interval $c \le y \le d$, the area of the surface S generated by revolving the curve $x = f(y)$, $c \le y \le d$, about the y-axis is

$$S = \int_c^d 2\pi x\sqrt{1 + \left(\frac{dx}{dy}\right)^2}\, dy. \qquad (9)$$

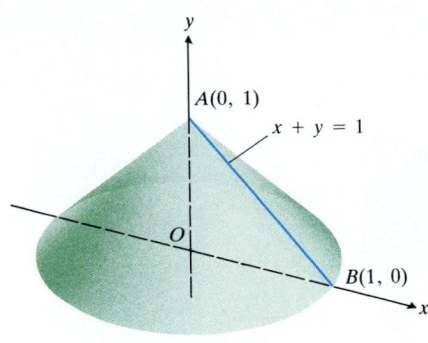

5.58 Revolving line segment AB about the y-axis generates a cone whose lateral surface area we can now calculate in two different ways. See Example 2.

Example 2   The line segment $x = 1 - y$, $0 \le y \le 1$, is revolved about the y-axis to generate the cone in Fig. 5.58. Find its lateral surface area.

Solution   Here we have a calculation we can check with a formula from geometry:

$$\text{Lateral surface area} = \frac{\text{base circumference}}{2} \times \text{slant height} = \pi\sqrt{2}.$$

To see how Eq. (9) gives the same result, we take

$$c = 0, \qquad d = 1, \qquad x = 1 - y, \qquad \frac{dx}{dy} = -1,$$

$$\sqrt{1 + \left(\frac{dx}{dy}\right)^2} = \sqrt{1 + (-1)^2} = \sqrt{2}$$

and calculate

$$\text{Surface area} = \int_c^d 2\pi x \sqrt{1 + \left(\frac{dx}{dy}\right)^2}\, dy = \int_0^1 2\pi(1 - y)\sqrt{2}\, dy$$

$$= 2\pi\sqrt{2}\left[y - \frac{y^2}{2}\right]_0^1 = 2\pi\sqrt{2}\left(1 - \frac{1}{2}\right) = \pi\sqrt{2}.$$

The results agree, as they should.

The Short Differential Form

The equations

$$S = \int_a^b 2\pi y \sqrt{1 + \left(\frac{dy}{dx}\right)^2}\, dx \qquad \text{and} \qquad S = \int_c^d 2\pi x \sqrt{1 + \left(\frac{dx}{dy}\right)^2}\, dy \quad (10)$$

are often written in terms of the arc length differential $ds = \sqrt{dx^2 + dy^2}$ as

$$S = \int_a^b 2\pi y\, ds \qquad \text{and} \qquad S = \int_c^d 2\pi x\, ds. \quad (11)$$

In the first of these, y is the distance from the x-axis to an element of arc length ds. In the second, x is the distance from the y-axis to an element of arc length ds. In both cases the integrals have the form

$$S = \int 2\pi(\text{radius})(\text{band width}) = \int 2\pi\rho\, ds, \quad (12)$$

where ρ is the radius from the axis of revolution to an element of arc length ds (Fig. 5.59).

If you wish to remember only one formula for surface area, you might make it the short differential form.

Short Differential Form

$$S = \int 2\pi\rho\, ds \quad (13)$$

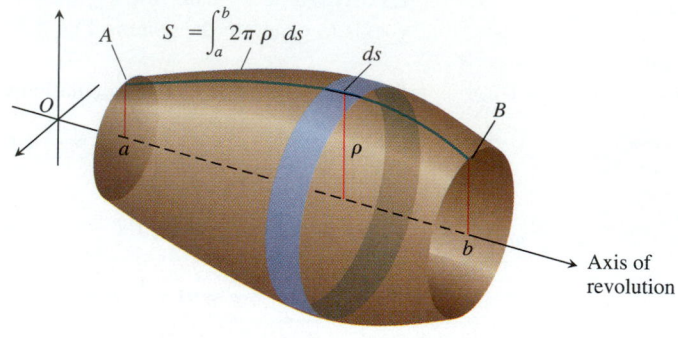

5.59 The area of the surface swept out by revolving arc AB about the axis shown here is $\int_a^b 2\pi\rho \, ds$. The exact expression depends on the formulas for ρ and ds.

In any particular problem, you would then express the radius function ρ and the arc length differential ds in terms of a common variable and supply limits of integration for that variable.

Example 3 Find the area of the surface generated by revolving the curve $y = x^3$, $0 \le x \le 1/2$, about the x-axis (Fig. 5.60).

Solution We start with the short differential form:

$$S = \int 2\pi\rho \, ds$$

$$= \int 2\pi y \, ds \qquad \left(\begin{array}{l}\text{For revolution about the } x\text{-axis,}\\ \text{the radius function is } \rho = y.\end{array}\right)$$

$$= \int 2\pi y \sqrt{dx^2 + dy^2}. \qquad (ds = \sqrt{dx^2 + dy^2})$$

We then decide whether to express dy in terms of dx or dx in terms of dy. The original form of the equation, $y = x^3$, makes it easier to express dy in terms of dx, so we continue the calculation with

$$y = x^3, \qquad dy = 3x^2 \, dx, \qquad \text{and} \qquad \sqrt{dx^2 + dy^2} = \sqrt{dx^2 + (3x^2 \, dx)^2}$$
$$= \sqrt{1 + 9x^4} \, dx.$$

With these substitutions, x becomes the variable of integration, and we continue with

$$S = \int_{x=0}^{x=1/2} 2\pi y \sqrt{dx^2 + dy^2} = \int_0^{1/2} 2\pi x^3 \sqrt{1 + 9x^4} \, dx$$

$$= 2\pi \cdot \frac{1}{36} \cdot \frac{2}{3} (1 + 9x^4)^{3/2} \Big]_0^{1/2} \qquad \left(\begin{array}{l}\text{Substitute } u = 1 + 9x^4,\ du/36 = x^3 dx,\\ \text{integrate, and substitute back.}\end{array}\right)$$

$$= \frac{\pi}{27}\left[\left(1 + \frac{9}{16}\right)^{3/2} - 1\right]$$

$$= \frac{\pi}{27}\left[\left(\frac{25}{16}\right)^{3/2} - 1\right] = \frac{\pi}{27}\left(\frac{125}{64} - 1\right) = \frac{61\pi}{1728}.$$

As with arc length calculations, even the simplest curves can provide a workout.

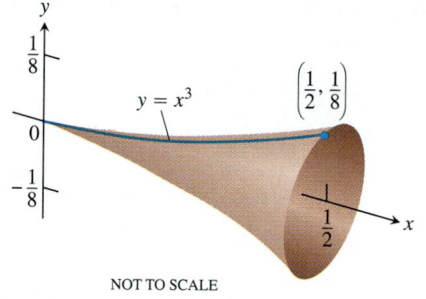

NOT TO SCALE

5.60 The surface generated by revolving the curve $y = x^3$, $0 \le x \le 1/2$, about the x-axis could be the design for a champagne glass. The surface area is calculated in Example 3.

EXERCISES 5.5

In Exercises 1–8, set up, but do not evaluate, an integral for the area of the surface generated by revolving the given curve about the given axis.

1. $y = \tan x$, $-\pi/4 \le x \le \pi/4$; x-axis
2. $y = x^2$, $0 \le x \le 2$; x-axis
3. $xy = 1$, $1 \le y \le 2$; y-axis
4. $x = \sin y$, $0 \le y \le \pi$; y-axis
5. $x^{1/2} + y^{1/2} = 3$ from (4, 1) to (1, 4); x-axis
6. $y + 2\sqrt{y} = x$, $1 \le y \le 2$; y-axis
7. $x = \displaystyle\int_0^y \tan t \, dt$, $0 \le y \le \pi/3$; y-axis
8. $y = \displaystyle\int_1^x \sqrt{t^2 - 1} \, dt$, $1 \le x \le \sqrt{5}$; x-axis

In Exercises 9–22, find the areas of the surfaces generated by revolving the curves about the indicated axes.

9. $y = x/2$, $0 \le x \le 4$, x-axis. Check your result with a formula from geometry, as in Example 2.
10. $y = x/2$, $0 \le x \le 4$, y-axis. Check your result with a formula from geometry, as in Example 2.
11. $y = (x/2) + (1/2)$, $1 \le x \le 3$, x-axis. Check your result with the geometry formula in Eq. (1).
12. $y = (x/2) + (1/2)$, $1 \le x \le 3$, y-axis. Check your result with the geometry formula in Eq. (1).
13. $y = x^3/9$, $0 \le x \le 2$, x-axis
14. $y = \sqrt{x}$, $3/4 \le x \le 15/4$, x-axis
15. $y = \sqrt{2x - x^2}$, $0 \le x \le 2$, x-axis
16. $y = \sqrt{x + 1}$, $1 \le x \le 5$, x-axis
17. $x = y^3/3$, $0 \le y \le 1$, y-axis
18. $x = (1/3)y^{3/2} - y^{1/2}$, $0 \le y \le 3$, y-axis
19. $x = 2\sqrt{4 - y}$, $0 \le y \le 15/4$, y-axis (Fig. 5.61)

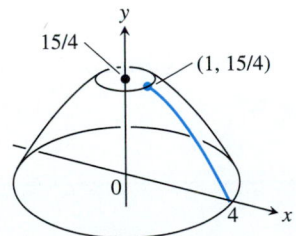

5.61   The surface in Exercise 19.

20. $x = \sqrt{2y - 1}$, $1/2 \le y \le 1$, y-axis (Fig. 5.62)

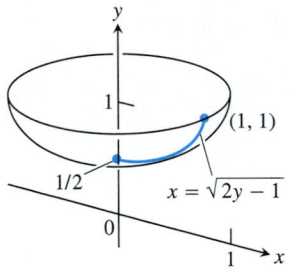

5.62   The surface in Exercise 20.

21. $x = (y^4/4) + 1/(8y^2)$, $1 \le y \le 2$, x-axis (*Hint:* Express $ds = \sqrt{dx^2 + dy^2}$ in terms of dy and evaluate the integral $S = \int 2\pi y \, ds$ with appropriate limits.)
22. $y = (1/3)(x^2 + 2)^{3/2}$, $0 \le x \le \sqrt{2}$, y-axis (*Hint:* Express $ds = \sqrt{dx^2 + dy^2}$ in terms of dx and evaluate the integral $S = \int 2\pi x \, ds$ with appropriate limits.)
23. *Testing the new definition.*   Show that the surface area of a sphere of radius a is still $4\pi a^2$ by using Eq. (8) to find the area of the surface generated by revolving the curve $y = \sqrt{a^2 - x^2}$, $-a \le x \le a$, about the x-axis.
24. *Testing the new definition.*   The lateral (side) surface area of a cone of height h and base radius r should be $\pi r \sqrt{r^2 + h^2}$, the semiperimeter of the base times the slant height. Show that this is still the case by finding the area of the surface generated by revolving the line segment $y = (r/h) x$, $0 \le x \le h$, about the x-axis (Fig. 5.63).

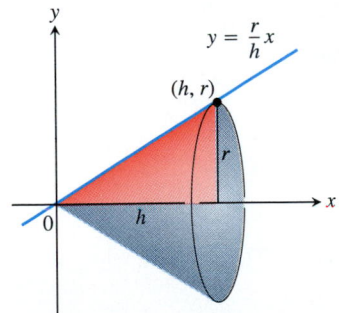

5.63   The cone in Exercise 24.

25. a) Write an integral for the area of the surface generated by revolving the curve $y = \cos x$, $-\pi/2 \le x \le \pi/2$, about the x-axis. In Sections 7.3 and 7.4 we shall see how to evaluate such integrals.

b) CALCULATOR   Find the surface area numerically.

26. *The surface of an astroid.*   Find the area of the surface generated by revolving about the x-axis the portion of the astroid $x^{2/3} + y^{2/3} = 1$ shown in Fig. 5.64. (*Hint:* Revolve the first-quadrant portion $y = (1 - x^{2/3})^{3/2}$, $0 \le x \le 1$, about the x-axis and double your result.)

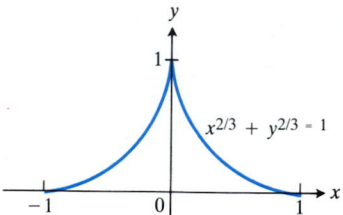

5.64   The generating curve in Exercise 26.

27. *Enameling woks.*   Your company decided to put out a deluxe version of the successful wok you designed in Exercise 35 of Section 5.2 (Fig. 5.65). The plan is to coat it inside with white enamel and outside with blue enamel. Each enamel will be sprayed on 0.5 mm thick before baking. Your manufacturing department wants to know how much enamel it will take for a production run of 5000 woks. What do you tell them? (Neglect waste and unused material, and give your answer in liters.)

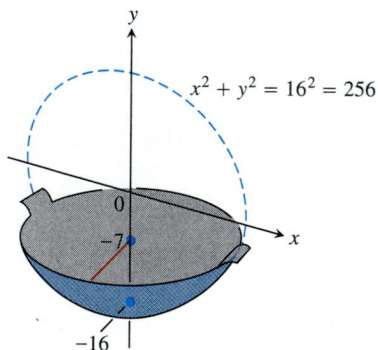

5.65   The wok in Exercise 27.

28. *Slicing bread.*   Did you know that if you cut a spherical loaf of bread into slices of equal width, each slice will have the same amount of crust? To see why, suppose the semicircle $y = \sqrt{r^2 - x^2}$ in Fig. 5.66 is revolved about the x-axis to generate a sphere. Let AB be an arc of the semicircle that lies above an interval of length h on the x-axis. Show that the area swept out by AB does not depend on the location of the interval. (It does depend on the length of the interval.)

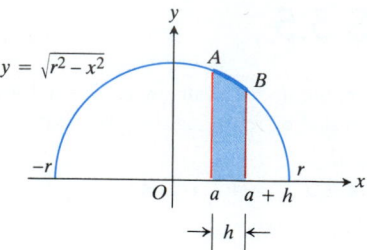

5.66   The semicircle in Exercise 28.

29. *Surfaces generated by curves that cross the axis of revolution.*   The surface area formula in Eq. (8) was developed under the assumption that the function f whose graph generated the surface was nonnegative over the interval $[a, b]$. For curves that cross the axis of revolution, we replace Eq. (8) with the absolute value formula

$$S = \int 2\pi\rho \, ds = \int 2\pi |f(x)| \, ds. \qquad (14)$$

Use Eq. (14) to find the surface area of the double cone generated by revolving the line segment $y = x$, $-1 \le x \le 2$, about the x-axis.

30. *(Exercise 29, continued.)*   Find the area of the surface generated by revolving the curve $y = x^3/9$, $-\sqrt{3} \le x \le \sqrt{3}$, about the x-axis. What do you think will happen if you drop the absolute values from Eq. (14) and attempt to find the surface area with the formula $S = \int 2\pi f(x) \, ds$ instead? Try it.

Numerical Integrator

If you have access to a computer or calculator with a numerical integration program, find, to two decimal places, the areas of the surfaces generated by revolving the curves in Exercises 31–34 about the x-axis.

31. $y = \sin x$, $0 \le x \le \pi$

32. $y = x^2/4$, $0 \le x \le 2$

33. $y = x + \sin 2x$, $-2\pi/3 \le x \le 2\pi/3$ (the curve in Section 3.3, Exercise 4)

34. $y = \dfrac{x}{12} \sqrt{36 - x^2}$, $0 \le x \le 6$ (the surface of the plumb bob in Section 5.2, Exercise 34)

EXPLORER PROGRAMS	
Integral Evaluation	Evaluates integrals numerically
PowerGrapher	Graphs any function you can key in

5.6 Moments and Centers of Mass

Many structures and mechanical systems behave as if their masses were concentrated at a single point called the center of mass (Fig. 5.67). It is important to know how to locate this point, and it turns out that doing so is basically a mathematical enterprise. For the moment we shall deal only with one- and two-dimensional objects. Three-dimensional objects are best done with the multiple integrals of Chapter 13.

Masses Along a Line

We develop our mathematical model in stages. The first stage is to imagine masses m_1, m_2, and m_3 placed on a rigid x-axis that is supported by a fulcrum at the origin.

The resulting system might balance, or it might not. It depends on how large the masses are and how they are arranged.

Each mass m_k exerts a downward force $m_k g$ equal to the magnitude of the mass times the acceleration of gravity. Each of these forces has a tendency to turn the axis about the origin, the way somebody's weight might turn a seesaw. This turning effect, called a **torque,** is measured by multiplying the force $m_k g$ by the signed distance x_k from the mass to the origin. Masses to the left of the origin exert a negative (counterclockwise) torque. Masses to the right of the origin exert a positive (clockwise) torque.

We use the sum of the torques to measure the tendency of a system to rotate about the origin. This sum is called the **system torque.**

$$\text{System torque} = m_1 g x_1 + m_2 g x_2 + m_3 g x_3 \tag{1}$$

The system will balance if and only if its net torque is zero.

If we factor out the g in Eq. (1), we see that the system torque is

$$g(m_1 x_1 + m_2 x_2 + m_3 x_3). \tag{2}$$

a feature of the environment

a feature of the system

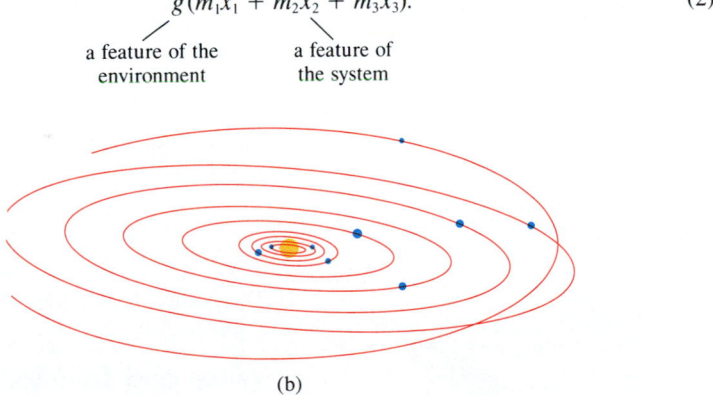

(b)

5.67 (a) The motion of this wrench gliding on ice seems haphazard until we notice that the wrench is simply turning about its center of mass as the center moves in a straight line. (b) The planets, asteroids, and comets of our solar system revolve about their collective center of mass. (It lies inside the sun.)

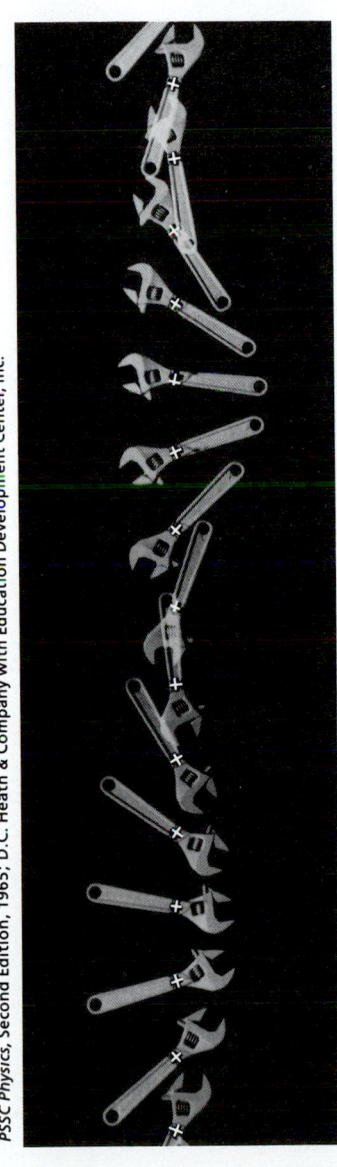

PSSC Physics, Second Edition, 1965; D.C. Heath & Company with Education Development Center, Inc.

(a)

Mass vs. Weight

Weight is the force that results from gravity pulling on a mass. If an object of mass m is placed in a location where the acceleration of gravity is g, the object's weight there is

$$F = mg$$

(a version of Newton's second law).

Thus the torque is the product of the gravitational acceleration g, which is a feature of the environment in which the system happens to reside, and the number $(m_1x_1 + m_2x_2 + m_3x_3)$, which is a feature of the system itself, a constant that stays the same no matter where the system is placed.

The number $(m_1x_1 + m_2x_2 + m_3x_3)$ is called the **moment of the system about the origin.**

$$M_O = \text{Moment of system about origin} = \sum m_k x_k \qquad (3)$$

(We shift to sigma notation here to allow for sums with more terms. For $\sum m_k x_k$, read "summation $m\ k\ x\ k$.")

We usually want to know where to place the fulcrum to make the system balance, that is, at what point $\bar{x}$ to place it to make the torque zero.

Special location
for balance

The torque of each mass about the fulcrum in this special location is

$$\text{Torque of } m_k \text{ about } \bar{x} = \begin{pmatrix} \text{signed distance} \\ \text{of } m_k \text{ from } \bar{x} \end{pmatrix} \cdot \begin{pmatrix} \text{downward} \\ \text{force} \end{pmatrix}$$

$$= (x_k - \bar{x}) \cdot m_k g. \qquad (4)$$

When we write the equation that says that the sum of these torques is zero, we get an equation we can solve for $\bar{x}$:

$$\sum (x_k - \bar{x})m_k g = 0 \qquad \text{(Sum of the torques equals zero)}$$

$$g \sum (x_k - \bar{x})m_k = 0 \qquad \text{(Constant Multiple Rule for Sums)}$$

$$\sum (m_k x_k - \bar{x}m_k) = 0 \qquad \text{(g divided out, m_k distributed)}$$

$$\sum m_k x_k - \sum \bar{x}m_k = 0 \qquad \text{(Difference Rule for Sums)}$$

$$\sum m_k x_k = \bar{x} \sum m_k \qquad \text{(Rearranged, Constant Multiple Rule again)}$$

$$\bar{x} = \frac{\sum m_k x_k}{\sum m_k}. \qquad \text{(Solved for } \bar{x})$$

This last equation tells us to find $\bar{x}$ by dividing the system's moment about the origin by the system's total mass:

$$\bar{x} = \frac{\sum x_k m_k}{\sum m_k} = \frac{\text{system moment about origin}}{\text{system mass}}. \qquad (5)$$

The point $\bar{x}$ is called the system's **center of mass.**

Wires and Thin Rods

In many applications, we want to know the center of mass of a rod or a thin strip of metal. In cases like these, where we can model the distribution of mass with a continuous function, the summation signs in our formulas become integrals in a manner we shall now describe.

Imagine a long, thin strip lying along the x-axis from $x = a$ to $x = b$ and cut into small pieces of mass Δm_k by a partition of the interval $[a, b]$.

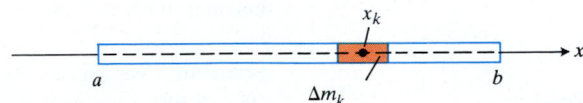

Each piece is Δx units long and lies approximately x_k units from the origin. Now observe three things.

First, the strip's center of mass $\bar{x}$ is nearly the same as the center of mass of the system of point masses we would get by attaching each mass Δm_k to the point x_k:

$$\bar{x} \approx \frac{\text{system moment}}{\text{system mass}}. \qquad (6)$$

Second, the moment of each piece of the strip about the origin is approximately $x_k \Delta m_k$, so the system moment is approximately the sum of the $x_k \Delta m_k$:

$$\text{system moment} \approx \sum x_k \Delta m_k. \qquad (7)$$

Third, if the density of the strip at x_k is $\delta(x_k)$, expressed in terms of mass per unit length, and δ is continuous, then Δm_k is approximately equal to $\delta(x_k)\Delta x$ (mass per unit length times length):

$$\Delta m_k \approx \delta(x_k)\Delta x. \qquad (8)$$

Combining these three observations gives

$$\bar{x} \approx \frac{\text{system moment}}{\text{system mass}} \approx \frac{\sum x_k \Delta m_k}{\sum \Delta m_k} \approx \frac{\sum x_k\, \delta(x_k)\Delta x}{\sum \delta(x_k)\Delta x} \qquad (9)$$

The sum in the numerator of the last quotient in (9) is a Riemann sum for the continuous function $x\,\delta(x)$ over the closed interval $[a, b]$. The sum in the denominator is a Riemann sum for the function $\delta(x)$ over this interval. We expect the approximations in (9) to improve as the strip is partitioned more finely and are led to the equation

$$\bar{x} = \frac{\displaystyle\int_a^b x\delta(x)\,dx}{\displaystyle\int_a^b \delta(x)\,dx}. \qquad (10)$$

This is the formula we use to calculate $\bar{x}$.

Density

A material's density is defined to be its mass per unit volume. In practice, however, we tend to use units we can conveniently measure. For wires, rods, and narrow strips we use mass per unit length. For flat sheets and plates we use mass per unit area.

Moment, Mass, and Center of Mass of a Thin Rod or Strip Along the x-Axis

Moment about the origin: $M_O = \displaystyle\int_a^b x\delta(x)\,dx$ (11a)

Mass: $M = \displaystyle\int_a^b \delta(x)\,dx$ (11b)

Center of mass: $\bar{x} = \dfrac{M_O}{M}$ (11c)

Equation (11c) says that to find the center of mass of a rod or a thin strip, we divide its moment about the origin by its mass.

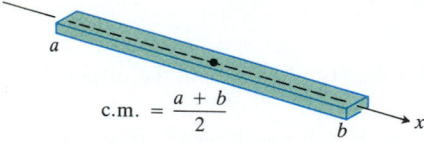

5.68 The center of mass of a straight, thin rod or strip of constant density lies halfway between its ends.

Example 1   Strips and Rods of Constant Density.   Show that the center of mass of a straight, thin strip or rod of constant density is always located halfway between its two ends.

Solution   We model the strip as a portion of the x-axis from $x = a$ to $x = b$ (Fig. 5.68). Our goal is to show that $\bar{x} = (a + b)/2$, the point halfway between a and b.

The key is the density's having a constant value. This enables us to regard the function $\delta(x)$ in the integrals in Eqs. (11) as a constant (call it δ), with the result that

$$M_O = \int_a^b \delta x \, dx = \delta \int_a^b x \, dx = \delta \left[\frac{1}{2} x^2 \right]_a^b = \frac{\delta}{2} (b^2 - a^2),$$

$$M = \int_a^b \delta \, dx = \delta \int_a^b dx = \delta \left[x \right]_a^b = \delta (b - a),$$

$$\bar{x} = \frac{M_O}{M} = \frac{\frac{\delta}{2}(b^2 - a^2)}{\delta(b - a)} = \frac{a + b}{2}. \qquad \text{(The } \delta\text{'s cancel.)}$$

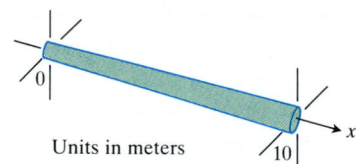

5.69 We can treat a rod of variable thickness as a rod of variable density. See Example 2.

Example 2   A Variable Density.   The 10-m-long rod in Fig. 5.69 thickens from left to right so that its density, instead of being constant, is $\delta(x) = 1 + (x/10)$ kg/m. Find the rod's center of mass.

Solution   The rod's moment about the origin (Eq. 11a) is

$$M_O = \int_0^{10} x\delta(x) \, dx = \int_0^{10} x\left(1 + \frac{x}{10}\right) dx = \int_0^{10} \left(x + \frac{x^2}{10}\right) dx$$

$$= \left[\frac{x^2}{2} + \frac{x^3}{30} \right]_0^{10} = 50 + \frac{100}{3} = \frac{250}{3} \text{ kg} \cdot \text{m}. \qquad \left(\begin{array}{l}\text{The units of a moment} \\ \text{are mass} \times \text{length.}\end{array}\right)$$

The rod's mass (Eq. 11b) is

$$M = \int_0^{10} \delta(x) \, dx = \int_0^{10} \left(1 + \frac{x}{10}\right) dx = \left[x + \frac{x^2}{20} \right]_0^{10} = 10 + 5 = 15 \text{ kg}.$$

The center of mass (Eq. 11c) is located at the point

$$\bar{x} = \frac{M_O}{M} = \frac{250}{3} \cdot \frac{1}{15} = \frac{50}{9} \approx 5.56 \text{ m}.$$

Masses Distributed over a Plane Region

Suppose we have a finite collection of masses located in the coordinate plane, the mass m_k being located at the point (x_k, y_k) (see Fig. 5.70). The total mass of the system is

System mass: $M = \sum m_k .$

Each mass m_k has a moment about each axis. Its moment about the x-axis is $m_k y_k$, and its moment about the y-axis is $m_k x_k$. The moments of the entire system about the two axes are

Moment about x-axis: $M_x = \sum m_k y_k ,$

Moment about y-axis: $M_y = \sum m_k x_k .$

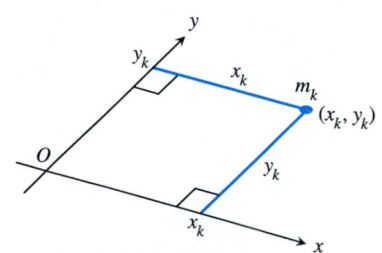

5.70 Each mass m_k has a moment about each axis.

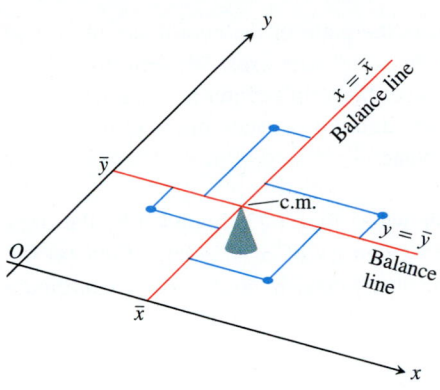

5.71 A two-dimensional array of masses balances on its center of mass.

The x-coordinate of the system's center of mass is defined to be

$$\bar{x} = \frac{M_y}{M} = \frac{\sum m_k x_k}{\sum m_k}.$$ (12)

With this choice of $\bar{x}$, as in the one-dimensional case, the system balances about the line $x = \bar{x}$ (Fig. 5.71).

The y-coordinate of the system's center of mass is defined to be

$$\bar{y} = \frac{M_x}{M} = \frac{\sum m_k y_k}{\sum m_k}.$$ (13)

With this choice of $\bar{y}$, the system balances about the line $y = \bar{y}$ as well. The torques exerted by the masses about the line $y = \bar{y}$ cancel out. Thus, as far as balance is concerned, the system behaves as if all its mass were at the single point $(\bar{x}, \bar{y})$. We call this point the system's center of mass.

Thin, Flat Plates

In many applications, we need to find the center of mass of a thin, flat plate: a disk of aluminum, say, or a triangular sheet of steel. In such cases we assume the distribution of mass to be continuous, and the formulas we use to calculate $\bar{x}$ and $\bar{y}$ contain integrals instead of finite sums. The integrals arise in the following way.

Imagine the plate occupying a region in the xy-plane, cut into thin strips parallel to one of the axes (in Fig. 5.72, the y-axis). The center of mass of a typical strip is $(\tilde{x}, \tilde{y})$. (The symbol ~ over the x and y is a *tilde*, pronounced to rhyme with "Hilda." Thus $\tilde{x}$ is read "x tilde.") We treat the strip's mass Δm as if it were concentrated at $(\tilde{x}, \tilde{y})$. The moment of the strip about the y-axis is then $\tilde{x} \Delta m$. The moment of the strip about the x-axis is $\tilde{y} \Delta m$. Equations (12) and (13) then become

$$\bar{x} = \frac{M_y}{M} = \frac{\sum \tilde{x} \Delta m}{\sum \Delta m}, \qquad \bar{y} = \frac{M_x}{M} = \frac{\sum \tilde{y} \Delta m}{\sum \Delta m}.$$ (14)

As in the one-dimensional case, the sums in the numerator and denominator are Riemann sums for integrals and approach these integrals as limiting values as the strips into which the plate is cut become narrower and narrower. We write these integrals symbolically as

$$\bar{x} = \frac{\int \tilde{x}\ dm}{\int dm} \qquad \text{and} \qquad \bar{y} = \frac{\int \tilde{y}\ dm}{\int dm}.$$ (15)

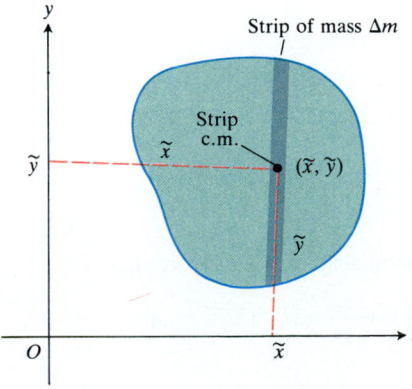

5.72 A plate cut into thin strips parallel to the y-axis. The moment exerted by a typical strip about each axis is the moment its mass Δm would exert if concentrated at the strip's center of mass $(\tilde{x}, \tilde{y})$.

Moments, Mass, and Center of Mass of a Thin Plate Covering a Region in the xy-Plane

Moment about the x-axis: $M_x = \displaystyle\int \tilde{y}\ dm$

Moment about the y-axis: $M_y = \displaystyle\int \tilde{x}\ dm$ (16)

Mass: $M = \displaystyle\int dm$

Center of mass: $\bar{x} = \dfrac{M_y}{M}, \qquad \bar{y} = \dfrac{M_x}{M}$

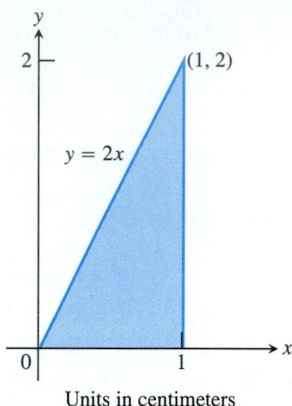

5.73 The plate in Example 3.

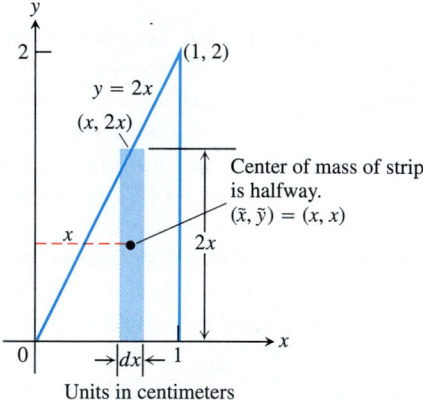

5.74 Modeling the plate in Example 3 with vertical strips.

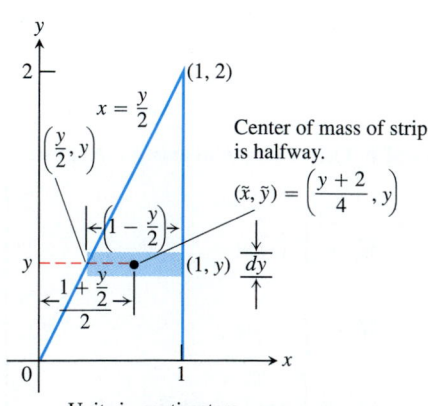

5.75 Modeling the plate in Example 3 with horizontal strips.

To evaluate these integrals, we picture the plate in the coordinate plane and stretch a strip of mass parallel to one of the coordinate axes. We then express the strip's mass dm and the coordinates $(\tilde{x}, \tilde{y})$ of the strip's center of mass in terms of x or y. Finally, we integrate $\tilde{y}\, dm$, $\tilde{x}\, dm$, and dm between limits of integration determined by the plate's location in the plane.

Example 3   The triangular plate shown in Fig. 5.73, bounded by the lines $y = 0$, $y = 2x$, and $x = 1$, has a constant density of $\delta = 3\text{g/cm}^2$. Find (a) the plate's moment M_y about the y-axis, (b) the plate's mass M, and (c) the x-coordinate of the plate's center of mass.

Solution   METHOD 1: *Vertical strips* (Fig. 5.74).

a) The moment M_y: The typical vertical strip has

center of mass (c.m.): $(\tilde{x}, \tilde{y}) = (x, x)$,

length: $2x$,

width: dx,

area: $dA = 2x\, dx$,

mass: $dm = \delta\, dA = 3 \cdot 2x\, dx = 6x\, dx$,

distance of c.m. from y-axis: $\tilde{x} = x$.

The moment of the strip about the y-axis is

$$\tilde{x}\, dm = x \cdot 6x\, dx = 6x^2\, dx.$$

The moment of the plate about the y-axis is therefore

$$M_y = \int \tilde{x}\, dm = \int_0^1 6x^2\, dx = 2x^3 \Big]_0^1 = 2\ \text{g} \cdot \text{cm}.$$

b) The plate's mass:

$$M = \int dm = \int_0^1 6x\, dx = 3x^2 \Big]_0^1 = 3\ \text{g}.$$

c) The x-coordinate of the plate's center of mass:

$$\tilde{x} = \frac{M_y}{M} = \frac{2\ \text{g} \cdot \text{cm}}{3\ \text{g}} = \frac{2}{3}\ \text{cm}.$$

By a similar computation we could find M_x and $\bar{y} = M_x/M$.

METHOD 2: *Horizontal strips* (Fig. 5.75).

a) The moment M_y: The typical horizontal strip (see the figure) has

center of mass (c.m.): $(\tilde{x}, \tilde{y}) = \left(\frac{1}{2}\left(1 + \frac{y}{2}\right), y\right) = \left(\frac{y+2}{4}, y\right),$

length: $1 - \dfrac{y}{2} = \dfrac{2-y}{2},$

width: dy,

area: $dA = \dfrac{2-y}{2}\, dy,$

How to Find a Plate's Center of Mass

1. Picture the plate in the xy-plane.
2. Sketch a strip of mass parallel to one of the coordinate axes and find its dimensions.
3. Find the strip's mass dm and center of mass $(\tilde{x}, \tilde{y})$.
4. Integrate $\tilde{y}\, dm$, $\tilde{x}\, dm$, and dm to find M_x, M_y, and M.
5. Divide the moments by the mass to calculate $\bar{x}$ and $\bar{y}$.

mass: $dm = \delta\, dA = 3 \cdot \dfrac{2 - y}{2}\, dy$,

distance of c.m. to y-axis: $\tilde{x} = \dfrac{y + 2}{4}$.

The moment of the strip about the y-axis is

$$\tilde{x}\, dm = \frac{y + 2}{4} \cdot 3 \cdot \frac{2 - y}{2}\, dy = \frac{3}{8}(4 - y^2)\, dy.$$

The moment of the plate about the y-axis is

$$M_y = \int \tilde{x}\, dm = \int_0^2 \frac{3}{8}(4 - y^2)\, dy = \frac{3}{8}\left[4y - \frac{y^3}{3}\right]_0^2 = \frac{3}{8}\left(\frac{16}{3}\right) = 2\ \text{g} \cdot \text{cm}.$$

b) The plate's mass:

$$M = \int dm = \int_0^2 \frac{3}{2}(2 - y)\, dy = \frac{3}{2}\left[2y - \frac{y^2}{2}\right]_0^2 = \frac{3}{2}(4 - 2) = 3\ \text{g}.$$

c) The x-coordinate of the plate's center of mass:

$$\bar{x} = \frac{M_y}{M} = \frac{2\ \text{g} \cdot \text{cm}}{3\ \text{g}} = \frac{2}{3}\ \text{cm}.$$

By a similar computation, we could find M_x and $\bar{y}$.

If the distribution of mass in a thin, flat plate has an axis of symmetry, the center of mass will lie on this axis. If there are two axes of symmetry, the center of mass will lie at their intersection. These facts often help to simplify our work.

Example 4   Find the center of mass of a thin plate of constant density δ covering the region bounded above by the parabola $y = 4 - x^2$ and below by the x-axis (Fig. 5.76).

Solution   Since the plate is symmetric about the y-axis and its density is constant, the distribution of mass is symmetric about the y-axis and the center of mass lies on the y-axis. This means that $\bar{x} = 0$. It remains to find $\bar{y} = M_x/M$.

A trial calculation with horizontal strips (Fig. 5.76a) leads to an inconvenient integration:

$$M_x = \int_0^4 2\delta y \sqrt{4 - y}\, dy.$$

We therefore model the distribution of mass with vertical strips instead (5.76b). The typical vertical strip has

center of mass (c.m.): $(\tilde{x},\ \tilde{y}) = \left(x,\ \dfrac{4 - x^2}{2}\right)$,

length: $4 - x^2$,

width: dx,

area: $dA = (4 - x^2)\, dx$,

mass: $dm = \delta\, dA = \delta(4 - x^2)\, dx$,

distance from c.m. to x-axis: $\tilde{y} = \dfrac{4 - x^2}{2}$.

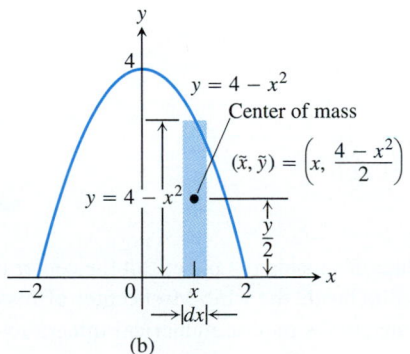

(a)

(b)

5.76   Modeling the plate in Example 4 with (a) horizontal strips leads to an inconvenient integration, so we model with (b) vertical strips instead.

The moment of the strip about the x-axis is

$$\tilde{y} \, dm = \frac{4 - x^2}{2} \cdot \delta(4 - x^2) \, dx = \frac{\delta}{2}(4 - x^2)^2 \, dx.$$

The moment of the plate about the x-axis is

$$M_x = \int \tilde{y} \, dm = \int_{-2}^{2} \frac{\delta}{2}(4 - x^2)^2 \, dx = \frac{256}{15} \delta. \qquad (17)$$

The mass of the plate is

$$M = \int dm = \int_{-2}^{2} \delta(4 - x^2) \, dx = \frac{32}{3} \delta. \qquad (18)$$

Therefore,

$$\bar{y} = \frac{M_x}{M} = \frac{\dfrac{256}{15}\delta}{\dfrac{32}{3}\delta} = \frac{8}{5}.$$

The plate's center of mass is the point

$$(\bar{x}, \bar{y}) = \left(0, \frac{8}{5}\right).$$

Example 5   *Variable Density.*   Find the center of mass of the plate in Example 4 if the density at any point (x, y) is $\delta = 2x^2$, twice the square of the distance from the point to the y-axis.

Solution   The mass distribution is still symmetric about the y-axis, so $\bar{x} = 0$. With $\delta = 2x^2$, Eqs. (17) and (18) become

$$M_x = \int \tilde{y} \, dm = \int_{-2}^{2} \frac{\delta}{2}(4 - x^2)^2 \, dx = \int_{-2}^{2} x^2(4 - x^2)^2 \, dx = \frac{2048}{105}, \qquad (17')$$

$$M = \int dm = \int_{-2}^{2} \delta(4 - x^2) \, dx = \int_{-2}^{2} 2x^2(4 - x^2) \, dx = \frac{256}{15}. \qquad (18')$$

Therefore,

$$\bar{y} = \frac{M_x}{M} = \frac{2048}{105} \cdot \frac{15}{256} = \frac{8}{7}.$$

The plate's new center of mass is

$$(\bar{x}, \bar{y}) = \left(0, \frac{8}{7}\right).$$

The center of mass of a flat convex plate lies inside the plate. But the center of mass of a bent wire or rod will probably not lie inside the wire. Most center-of-mass calculations for wires and rods that lie along curves require numerical integration. For wires or rods shaped like circular arcs, however, the antiderivatives we need are readily available.

Example 6   Find the center of mass of a wire of constant density δ shaped like a semicircle of radius a.

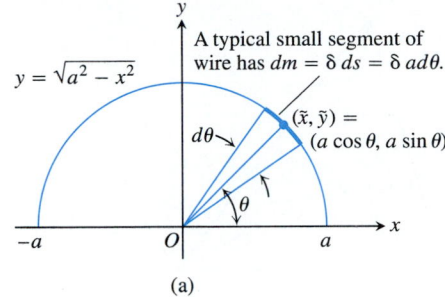

(a)

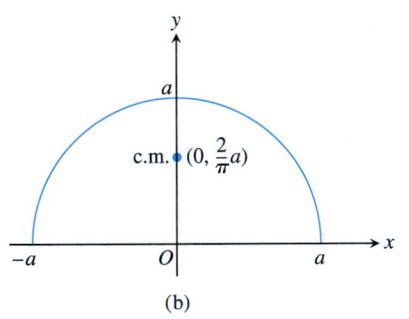

(b)

5.77 The semicircular wire in Example 6. (a) The dimensions and variables used in finding the center of mass. (b) The center of mass does not lie on the wire.

Solution We model the wire with the semicircle $y = \sqrt{a^2 - x^2}$ (Fig. 5.77). The distribution of mass is symmetric about the y-axis, so $\bar{x} = 0$. To find $\bar{y}$, we imagine the wire divided into short segments. The typical segment (Fig. 5.77a) has

length: $\quad ds = a\, d\theta$,

mass: $\quad dm = \delta\, ds = \delta a\, d\theta$, (mass per unit length times length)

distance of c.m. to x-axis: $\quad \tilde{y} = a \sin \theta$.

Hence,

$$\bar{y} = \frac{\int \tilde{y}\, dm}{\int dm} = \frac{\int_0^\pi a \sin \theta \cdot \delta a\, d\theta}{\int_0^\pi \delta a\, d\theta} = \frac{\delta a^2[-\cos \theta]_0^\pi}{\delta a \pi} = \frac{2}{\pi} a.$$

The center of mass lies on the axis of symmetry at the point $(0, 2a/\pi)$, about two-thirds of the way up from the origin (Fig. 5.77b).

Centroids

When the density function is constant, it cancels out of the numerator and denominator of the formulas for $\bar{x}$ and $\bar{y}$. This happened in nearly every example in this section. As far as $\bar{x}$ and $\bar{y}$ were concerned, δ might as well have been 1. Thus, when the density is constant, the location of the center of mass is a feature of the geometry of the object and not of the material from which it is made. In such cases engineers may call the center of mass the **centroid** of the shape, as in "Find the centroid of a triangle or a solid cone." To do so, just set δ equal to 1 and proceed to find $\bar{x}$ and $\bar{y}$ as before, by dividing moments by masses.

EXERCISES 5.6

1. Two children balance on a seesaw. The 40-lb child is 5 ft from the fulcrum. How far from the fulcrum is the 50-lb child?

2. You weld the ends of two steel rods of equal length into a right-angled frame (Fig. 5.78). Where is the frame's center of mass? (*Hint:* Where is the center of mass of each rod?)

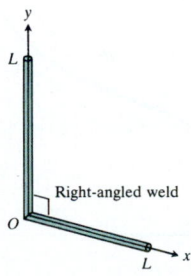

5.78 The welded rods in Exercise 2.

Exercises 3–6 give the density functions of thin rods lying along various intervals of the x-axis. Find each rod's moment about the origin and center of mass.

3. $\delta(x) = 4,\ 0 \le x \le 2$

4. $\delta(x) = 1 + (x/3),\ 0 \le x \le 3$

5. $\delta(x) = \left(1 + \dfrac{x}{4}\right)^2,\ 0 \le x \le 4$

6. $\delta(x) = \begin{cases} 2, & 0 \le x \le 3, \\ 1, & 3 \le x \le 6 \end{cases}$

In Exercises 7–18, find the center of mass of a thin plate of constant density δ covering the given region.

7. The region bounded by the parabola $y = x^2$ and the line $y = 4$

8. The region bounded by the parabola $y = 25 - x^2$ and the x-axis

9. The region bounded by the parabola $y = x - x^2$ and the line $y = -x$

10. The region enclosed by the parabolas $y = x^2 - 3$ and $y = -2x^2$

11. The region bounded by the y-axis and the curve $x = y - y^3$, $0 \le y \le 1$

12. The region bounded by the parabola $x = y^2 - y$ and the line $y = x$

13. The region bounded by the x-axis and the curve $y = \cos x$, $-\pi/2 \le x \le \pi/2$

14. The region between the x-axis and the curve $y = \sec x$, $-\pi/4 \le x \le \pi/4$

15. The region bounded by the parabolas $y = 2x^2 - 4x$ and $y = 2x - x^2$

16. a) The region cut from the first quadrant by the circle $x^2 + y^2 = 9$

   b) The region bounded by the x-axis and the semicircle $y = \sqrt{9 - x^2}$

   Compare your answer with the answer in (a).

17. The "triangular" region in the first quadrant between the circle $x^2 + y^2 = 9$ and the lines $x = 3$ and $y = 3$. (*Hint:* Use geometry to find the area.)

18. The region bounded above by the curve $y = 1/x^3$, below by the curve $y = -1/x^3$, and on the left and right by the lines $x = 1$ and $x = a > 1$. Also, find $\lim_{a \to \infty} \bar{x}$.

19. Find the moment about the x-axis of a wire of constant density that lies along the curve $y = \sqrt{x}$ from $x = 0$ to $x = 2$.

20. Find the moment about the x-axis of a wire of constant density that lies along the curve $y = x^3$ from $x = 0$ to $x = 1$.

21. Suppose the density of the wire in Example 6 is $\delta = k \sin \theta$ (k constant). Find the center of mass.

22. *The centroid of a triangle lies at the intersection of the triangle's medians (Fig. 5.79a).* You may recall that the point inside a triangle that lies one third of the way from each side toward the opposite vertex is the point where the triangle's three medians intersect. Show that the centroid lies at the intersection of the medians by showing that it too lies one third of the way from each side toward the opposite vertex. To do so, take the following steps.

  1. Stand one side of the triangle on the x-axis as in Fig. 5.79(b). Express dm in terms of L and dy.

  2. Use similar triangles to show that $L = (b/h)(h - y)$. Substitute this expression for L in your formula for dm.

  3. Show that $\bar{y} = h/3$.

  4. Extend the argument to the other sides.

Use the result in Exercise 22 to find the centroids of the triangles whose vertices appear in Exercises 23–26. (*Hint:* Draw each triangle first.)

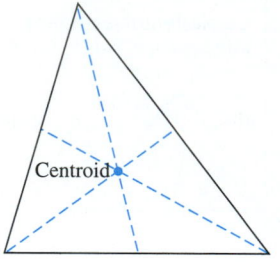

(a)

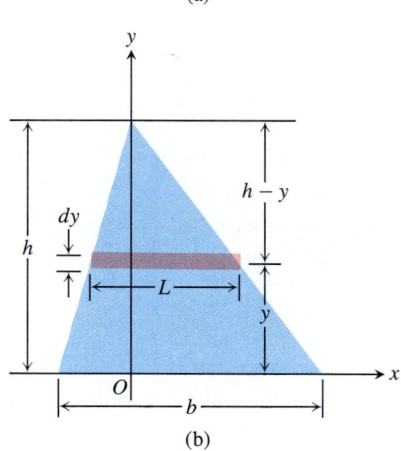

(b)

5.79 The triangle in Exercise 22. a) The centroid. (b) The dimensions and variables to use in locating the center of mass.

23. $(-1, 0), (1, 0), (0, 3)$ **24.** $(0, 0), (1, 0), (0, 1)$

25. $(0, 0), (a, 0), (0, a)$ **26.** $(0, 0), (a, 0), (0, b)$

Verify the statements and engineering formulas in Exercises 27–30.

27. *Centroid of any differentiable plane curve (Fig. 5.80).*

$$\bar{x} = \frac{\int x \, ds}{\text{length}}, \qquad \bar{y} = \frac{\int y \, ds}{\text{length}}$$

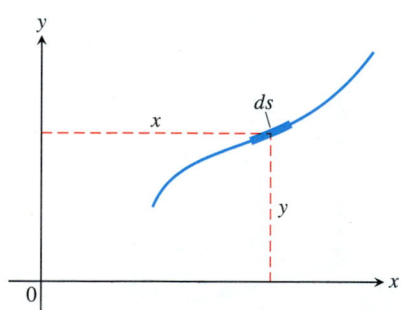

5.80 Diagram for Exercise 27.

28. *Centroid of a parabolic segment (shaded region in Fig. 5.81).* Whatever the value of $p > 0$ in the parabola's equation,

$$\bar{y} = \frac{3}{5}a.$$

29. *Centroid of a circular arc (Fig. 5.82).* For wires and thin rods of constant density shaped like circular arcs centered at the origin and symmetric about the y-axis,

$$\bar{y} = \frac{a \sin \alpha}{\alpha} = \frac{ac}{s}.$$

30. *(Continuation of Exercise 29.)*
   a) Show that when α is small, the distance d from the centroid to chord AB is about $2h/3$ (in the notation of Fig. 5.82) by taking the following steps.

   1. Show that

   $$\frac{d}{h} = \frac{\sin \alpha - \alpha \cos \alpha}{\alpha - \alpha \cos \alpha}. \tag{19}$$

   2. Apply l'Hôpital's rule to show that $\lim_{\alpha \to 0} (d/h) = 2/3$.

   b) CALCULATOR The error (difference between d and $2h/3$) is small even for angles greater than $45°$. See for yourself by evaluating the right-hand side of Eq. (19) for $\alpha = 0.2, 0.4, 0.6, 0.8$, and 1.0 rad.

EXPLORER PROGRAM

PowerGrapher   Graphs all the functions in this section. Makes tables of values of functions and differences of functions

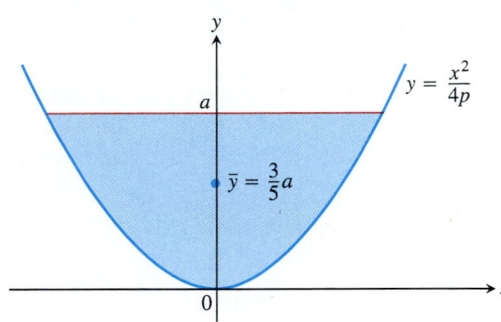

5.81 Diagram for Exercise 28.

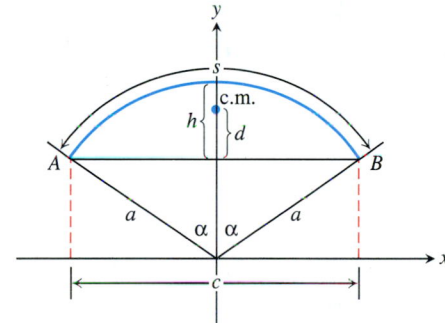

5.82 The circular arc in Exercises 29 and 30.

5.7 Work

In everyday life, *work* means any activity that requires muscular or mental effort. In science, however, the term is used in a narrower sense that involves the application of a force to a body and the body's subsequent displacement. This section shows how to calculate work. The applications run from stretching springs to pumping liquids out of subterranean tanks to lifting satellites into orbit.

The Constant-Force Formula for Work

We begin with a definition.

DEFINITION

When a body moves a distance d along a straight line as the result of being acted on by a force that has a constant magnitude F in the direction of the motion, the **work** W done by the force in moving the body is F times d.

$$W = Fd \tag{1}$$

We can see right away that there is a considerable difference between what we are used to calling work and what this formula says work is. If you push a car down a street, you are doing work, both by your own reckoning and according to Eq. (1). But if you push against the car and the car does not move, Eq. (1) says you are doing no work, no matter how hard or how long you push.

Work is measured in foot-pounds, newton-meters, or whatever force-distance unit is appropriate to the occasion.

Example 1   It takes a force of about 1 newton (1 N) to lift an apple from a table. If you lift it 1 m, you have done about 1 newton-meter (N · m) of work on the apple.

The Variable-Force Integral Formula for Work

If the force you apply varies along the way, as it will if you are lifting a leaking bucket or compressing a spring, the formula $W = Fd$ has to be replaced by an integral formula that takes the variation in F into account. It takes calculus to measure the work done by a variable force.

Suppose that the force performing the work varies continuously along a line that we can take to be the x-axis and that the force is represented by the function $F(x)$. We are interested in the work done along an interval from $x = a$ to $x = b$. We partition the closed interval $[a, b]$ in the usual way and choose an arbitrary point c_k in each subinterval $[x_{k-1}, x_k]$. If the subinterval is short enough, F, being continuous, will not vary much from x_{k-1} to x_k. The amount of work done by the force from x_{k-1} to x_k will be nearly equal to $F(c_k)$ times distance Δx_k, as it would be if we could apply Eq. (1). The total work done from a to b is thus approximated by the Riemann sum

$$\sum_{k=1}^{n} F(c_k)\, \Delta x_k. \tag{2}$$

We expect the approximations to improve as the norm of the partition goes to zero, so we define the work done by the force from a to b to be the integral of F from a to b.

DEFINITION

The **work** done by a continuous force $F(x)$ directed along the x-axis from $x = a$ to $x = b$ is

$$W = \int_a^b F(x)\, dx. \tag{3}$$

Example 2   A leaky 5-lb bucket is lifted from the ground into the air by pulling in 20 ft of rope at a constant speed (Fig. 5.83). The rope weighs 0.08 lb/ft. The bucket starts with 2 gal of water (16 lb) and leaks at a constant rate. It finishes draining just as it reaches the top. How much work was spent

a) lifting the water alone;

b) lifting the water and bucket together;

c) lifting the water, bucket, and rope?

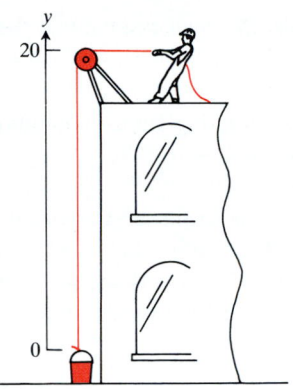

5.83 The leaky bucket in Example 2.

Solution

a) *The water alone.* The force required to lift the water is equal to the water's weight, which varies steadily from 16 to 0 lb over the 20-ft lift. When the bucket is x ft off the ground, the water weighs

$$F(x) = 16\left(\frac{20 - x}{20}\right) = 16\left(1 - \frac{x}{20}\right) = 16 - \frac{4x}{5} \text{ lb}.$$

original weight of water → / proportion left at elevation x ← \

The work done is

$$W = \int_a^b F(x)\, dx \qquad \text{(Use Eq. (3) for variable forces.)}$$

$$= \int_0^{20} \left(16 - \frac{4x}{5}\right) dx = \left[16x - \frac{2x^2}{5}\right]_0^{20} = 320 - 160 = 160 \text{ ft} \cdot \text{lb}.$$

b) *The water and bucket together.* According to Eq. (1), it takes $5 \times 20 = 100$ ft · lb to lift a 5-lb weight 20 ft. Therefore

$$160 + 100 = 260 \text{ ft} \cdot \text{lb}$$

of work were spent lifting the water and bucket together.

c) *The water, bucket, and rope.* Now the total weight at level x is

lb/ft   ft

$$F(x) = \underbrace{\left(16 - \frac{4x}{5}\right)}_{\substack{\text{variable} \\ \text{weight} \\ \text{of water}}} + \underbrace{5}_{\substack{\text{constant} \\ \text{weight of} \\ \text{bucket}}} + \underbrace{(0.08)(20 - x)}_{\substack{\text{weight of rope} \\ \text{paid out at} \\ \text{elevation } x}}.$$

The work lifting the rope is

$$\text{Work on rope} = \int_0^{20} (0.08)(20 - x)\, dx = \int_0^{20} (1.6 - 0.08x)\, dx$$

$$= \left[1.6x - 0.04x^2\right]_0^{20} = 32 - 16 = 16 \text{ ft} \cdot \text{lb}.$$

The total work for the water, bucket, and rope combined is

$$160 + 100 + 16 = 276 \text{ ft} \cdot \text{lb}.$$

Hooke's Law for Springs: $F = kx$

Hooke's law says that the amount of force F it takes to stretch or compress a spring x length units from its natural (unstressed) length is proportional to x. In symbols,

$$F = kx. \tag{4}$$

The number k, measured in force units per unit length, is a constant characteristic of the spring, called the **spring constant.** Hooke's law (Eq. 4) gives good results as

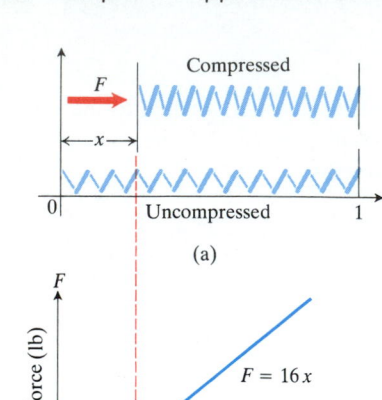

(a)

(b)

5.84   The force F required to hold a spring under compression increases linearly as the spring is compressed.

long as the force doesn't distort the metal in the spring. We shall assume that the forces in this section are too small to do that.

Example 3   Find the work required to compress a spring from its natural length of 1 ft to a length of 0.75 ft if the spring constant is $k = 16$ lb/ft.

Solution   We picture the uncompressed spring laid out along the x-axis with its movable end at the origin and its fixed end at $x = 1$ ft (Fig. 5.84). This enables us to describe the force required to compress the spring from 0 to x with the formula $F = 16x$. As the spring is compressed from 0 to 0.25 ft, the force varies from

$$F(0) = 16 \cdot 0 = 0 \text{ lb} \qquad \text{to} \qquad F(0.25) = 16 \cdot 0.25 = 4 \text{ lb.}$$

The work done by F over this interval is

$$W = \int_0^{0.25} 16x \, dx = 8x^2 \Big]_0^{0.25} = 0.5 \text{ ft} \cdot \text{lb.} \qquad \begin{pmatrix} \text{Eq. (3) with } a = 0, \\ b = 0.25, F(x) = 16x \end{pmatrix}$$

Example 4   A spring has a natural length of 1 m. A force of 24 N stretches the spring to a length of 1.8 m.

a) Find the spring constant k.

b) How much work will it take to stretch the spring 2 m beyond its natural length?

c) How far will a 45-N force stretch the spring?

Solution

a) *The spring constant.* We find the spring constant from Eq. (4). A force of 24 N stretches the spring 0.8 m, so

$$24 = k(0.8) \qquad\qquad \text{(Eq. (4) with } F = 24, x = 0.8)$$

$$k = 24/0.8 = 30 \text{ N/m.}$$

b) *The work to stretch the spring* 2 m. We imagine the unstressed spring hanging along the x-axis with its free end at $x = 0$ (Fig. 5.85). Then the force required to stretch the spring x m beyond its natural length is the force required to pull the free end of the spring x units from the origin. Hooke's law with $k = 30$ tells us that this force is

$$F(x) = 30x.$$

The work required to apply this force from $x = 0$ m to $x = 2$ m is

$$W = \int_0^2 30x \, dx = 15x^2 \Big]_0^2 = 60 \text{ N} \cdot \text{m.}$$

c) *How far will a 45-N force stretch the spring?* We substitute $F = 45$ in the equation $F = 30x$ to find

$$45 = 30x, \qquad \text{or} \qquad x = 1.5 \text{ m.}$$

A 45-N force will stretch the spring 1.5 m. No calculus is required to find this.

5.85   A 24-N weight stretches this spring 0.8 m beyond its unstressed length.

Pumping Liquids from Containers

To find how much work it takes to pump all or part of the liquid from a container, imagine lifting the liquid out one horizontal slab at a time and applying the equation $W = Fd$ to each slab. The integral we get each time depends on the weight of the liquid and the cross-section dimensions of the container, but the way we find the integral is the same for all containers. The next two examples show what to do.

Example 5   How much work does it take to pump the water from a full upright right circular cylindrical tank of radius 5 ft and height 10 ft to a level 4 ft above the top of the tank?

5.86   To find the work it takes to pump the water from a tank, think of lifting the water one thin slab at a time.

Solution   We draw coordinate axes (Fig. 5.86) and imagine the water divided into thin slabs by planes perpendicular to the y-axis at the points of a partition of the interval $[0, 10]$.

The typical slab between the planes at y and $y + \Delta y$ has a volume of approximately

$$\Delta V = \pi (\text{radius})^2 (\text{thickness}) = \pi (5)^2 \Delta y = 25\pi \, \Delta y \text{ ft}^3.$$

The force $F(y)$ required to lift this slab is equal to its weight,

$$F(y) = w\Delta V = 25 \, \pi w \Delta y \text{ lb},$$

where w is the weight of a cubic foot of water (we can substitute for w later).

The distance through which F must act is about $(14 - y)$ ft, so the work of lifting this slab 4 ft above the top of the tank is about

$$\Delta W = 25\pi w (14 - y) \, \Delta y \text{ ft} \cdot \text{lb}.$$

The work of lifting all the slabs of water is about

$$\sum_0^{10} \Delta W = \sum_0^{10} 25\pi w \, (14 - y)\Delta y \text{ ft} \cdot \text{lb}.$$

This is a Riemann sum for the function $25\pi w \, (14 - y)$ over the interval from $y = 0$ to $y = 10$. The work of pumping the tank dry is the limit of these sums as the norm of the partition goes to zero:

$$\text{Work} = \int_0^{10} 25\pi w \, (14 - y) \, dy = 25\pi w \int_0^{10} (14 - y) \, dy$$

$$= 25\pi w \left[14y - \frac{y^2}{2} \right]_0^{10} = 25\pi w [140 - 50]$$

$$= 2250\pi (62.5) \qquad \text{(Water weighs 62.5 lb/ft}^3\text{)}$$

$$= 441{,}786 \text{ ft} \cdot \text{lb}. \qquad \text{(Nearest ft} \cdot \text{lb, with a calculator)}$$

A 1-horsepower pump rated at 550 ft · lb per second could empty the tank in a little less than 14 min.

Weight-Density

Weight per unit volume is called **weight-density.** Here are some typical values (lb/ft³).

Gasoline	42
Mercury	849
Milk	64.5
Olive oil	57
Seawater	64
Water	62.5

Example 6   The inverted conical tank in Fig. 5.87 is filled to within 2 ft of the top with olive oil weighing 57 lb/ft³. How much work does it take to pump the oil to the rim of the tank?

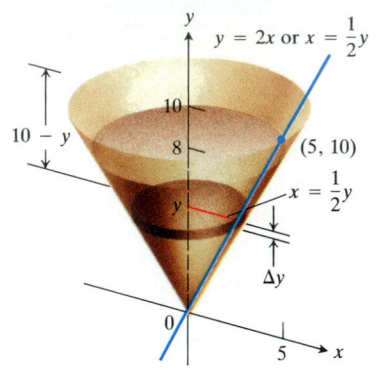

5.87   The olive oil in Example 6.

How to Find Work Done During Pumping

1. Draw a figure with a coordinate system.

2. Find the weight of a thin horizontal slab of liquid.

3. Find the work to lift the slab to its destination.

4. Integrate the work expression from the top to the bottom of the liquid.

Weight vs. Mass

Weight is the force that results from gravity pulling on a mass. The two are related by the equation in Newton's second law,

Weight = mass × acceleration.

Thus,

$$\text{Newtons} = \text{kilograms} \times \frac{\text{m}}{\text{sec}^2},$$

$$\text{Pounds} = \text{slugs} \times \frac{\text{ft}}{\text{sec}^2}.$$

To convert mass to weight, multiply by the acceleration of gravity. To convert weight to mass, divide by the acceleration of gravity.

Solution   We imagine the oil divided into thin slabs by planes perpendicular to the y-axis at the points of a partition of the interval $[0, 8]$.

The typical slab between the planes at y and $y + \Delta y$ has a volume of about

$$\Delta V = \pi (\text{radius})^2 (\text{thickness}) = \pi \left(\frac{1}{2}y\right)^2 \Delta y = \frac{\pi}{4}y^2 \, \Delta y \text{ ft}^3.$$

The force $F(y)$ required to lift this slab is equal to its weight,

$$F(y) = 57 \, \Delta V = \frac{57\pi}{4}y^2 \, \Delta y \text{ lb.} \qquad \begin{pmatrix}\text{Weight} = \text{weight per} \\ \text{unit volume} \times \text{volume}\end{pmatrix}$$

The distance through which $F(y)$ must act to lift this slab to the level of the rim of the cone is about $(10 - y)$ ft, so the work done lifting the slab is about

$$\Delta W = \frac{57\pi}{4}(10 - y)\, y^2 \Delta y \text{ ft} \cdot \text{lb.}$$

The work done lifting all the slabs from $y = 0$ to $y = 8$ to the rim is about

$$\sum_0^8 \frac{57\pi}{4}(10 - y)\, y^2 \, \Delta y \text{ ft} \cdot \text{lb.}$$

This is a Riemann sum for the function $(57\pi/4)(10 - y)y^2$ on the interval from $y = 0$ to $y = 8$. The work of pumping the oil to the rim is the limit of these sums as the norm of the partition goes to zero.

$$\text{Work} = \int_0^8 \frac{57\pi}{4}(10 - y)\, y^2 \, dy = \frac{57\pi}{4}\int_0^8 (10y^2 - y^3) \, dy$$

$$= \frac{57\pi}{4}\left[\frac{10y^3}{3} - \frac{y^4}{4}\right]_0^8 = 30{,}561 \text{ ft} \cdot \text{lb} \qquad \begin{pmatrix}\text{With a calculator,} \\ \text{rounded}\end{pmatrix}$$

✱ Work and Kinetic Energy

Equation (3) explains an important relationship between work and kinetic energy.

The Work–Kinetic Energy Relation

The net work done by a force acting on a body along its line of motion equals the change in the body's kinetic energy.

$$\text{Work} = \Delta \text{KE} \tag{5}$$

To derive this relation, we let m be the body's mass. We let t represent time and assume that the motion takes place along the x-axis. The body's velocity v can then be written as dx/dt. We also assume that the force is directed along the x-axis from x_1 to x_2, and we let $F(x)$ be the force's magnitude over this interval. We then bring together three ideas:

1. The equation for work: $W = \int_{x_1}^{x_2} F(x) \, dx$

2. The formula for kinetic energy: $K = \frac{1}{2}mv^2$

3. Newton's second law of motion: $F = m\dfrac{dv}{dt}$

Kinetic Energy and Space Debris Effects (comparisons for collisions at 10 km/sec)

- Less than 0.01 cm — Surface erosion

- Less than 0.03 cm — Possibly serious damage

- 0.3 cm at 10 km/sec (32,630 ft/sec) — Bowling ball at 60 mph (88 ft/sec)

- 1-cm aluminum sphere at 10 km/sec — 400-lb safe at 60 mph (88 ft/sec)

From *Station Break*, NASA's monthly newsletter on the Space Station Freedom Program, Vol. 2, No. 4, April 1990, p. 7.

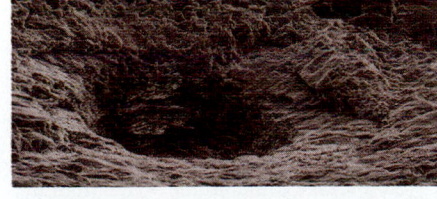

A scanning electron micrograph of a crater made in one of space shuttle *Challenger*'s windows by an orbiting paint chip. At the velocities encountered in space, even a collision with a small speck can cause significant damage. (NASA photo from the *Boston Globe*, Sci-Tech section, p. 31, August 22, 1988.)

We apply the Chain Rule to rewrite the work integral in the following way:

$$W = \int_{x_1}^{x_2} F(x)\, dx = \int_{x_1}^{x_2} m\frac{dv}{dt}\, dx \qquad \left(F = m\frac{dv}{dt}\right)$$

$$= \int_{x_1}^{x_2} mv\frac{dv}{dx}\, dx \qquad \left(\begin{array}{l}\text{Chain Rule:}\\ \dfrac{dv}{dt} = \dfrac{dv}{dx}\dfrac{dx}{dt} = v\dfrac{dv}{dx}\end{array}\right)$$

$$= \int_{v_1}^{v_2} mv\, dv \qquad \left(\begin{array}{l}v_1 \text{ and } v_2 \text{ are the}\\ \text{body's velocities}\\ \text{at } x_1 \text{ and } x_2.\end{array}\right)$$

$$= \frac{1}{2}mv^2\Big]_{v_1}^{v_2} = \frac{1}{2}mv_2^2 - \frac{1}{2}mv_1^2$$

$$= (\text{kinetic energy at } x_2) - (\text{kinetic energy at } x_1)$$

$$= \Delta KE. \qquad (\text{For short})$$

This establishes the relation $W = \Delta KE$ in Eq. (5) and gives it in coordinate form.

> **Work and Kinetic Energy**
> $$\text{Work} = \frac{1}{2}mv_2^2 - \frac{1}{2}mv_1^2 \qquad (6)$$

The virtue of this result is that if we know the body's velocities v_1 and v_2 we can find the work that caused the velocity to change even without knowing anything about the magnitude of the applied force or how long the force acted. Thus, we can find the work it takes to throw a baseball at 90 mph without knowing how long the ball is held or how the force applied to it by the pitcher's hand varies along the way. The same idea applies to tennis. If we know how fast the ball is going right after it is served, we can calculate how much work went into accelerating the ball without knowing exactly how much pressure the racket put on the ball or how long the racket and ball were in contact.

Example 7   A 2-oz tennis ball was served at 160 ft/sec (about 109 mph). How much work was done on the ball to make it go this fast?

Solution   We use Eq. (6) with $v_1 = 0$, $v_2 = 160$, and m equal to the mass of the ball. To find the ball's mass from its weight, we express the weight in pounds and divide by 32 ft/sec^2, the acceleration of gravity. The resulting engineering unit is called a *slug*.

$$\text{Weight} = 2\text{ oz} = \frac{1}{8}\text{ lb,}$$

$$\text{Mass} = \frac{1}{8}\cdot\frac{1}{32} = \frac{1}{256}\text{ slug,}$$

$$\text{Work} = \frac{1}{2}mv_2^2 - \frac{1}{2}mv_1^2 = \frac{1}{2}\left(\frac{1}{256}\right)(160)^2 - 0 = 50 \qquad (\text{Eq. 6})$$

It took 50 ft·lb of work to accelerate the ball from 0 to 160 ft/sec.

EXERCISES 5.7

1. The workers in Example 2 changed to a larger bucket that held 5 gal (40 lb) of water, but the new bucket had an even larger leak so that it, too, was empty by the time it reached the top. Assuming that the water leaked out at a steady rate, how much work was done lifting the water? (Do not include the rope and bucket.)

2. The bucket in Example 2 is hauled up twice as fast so that there is still 1 gal (8 lb) of water left when the bucket reaches the top. How much work is done lifting the water this time? (Do not include the rope and bucket.)

3. A mountain climber is about to haul up a 50-m length of hanging rope. How much work will it take if the rope weighs 0.74 N/m?

4. It took 1800 N · m of work to stretch a spring from its natural length of 2 m to a length of 5 m. What is the spring's spring constant?

5. An electric elevator with a motor at the top has a multistrand cable weighing 4.5 lb/ft. When the car is at the first floor, 180 ft of cable are paid out, and effectively 0 ft are out when the car is at the top floor. How much work does the motor do just lifting the cable when it takes the car from the first floor to the top?

6. A bag of sand originally weighing 144 lb was lifted at a constant rate. As it rose, sand also leaked out at a constant rate. The sand was half gone by the time the bag had been lifted 18 ft. How much work was done lifting the sand this far? (Neglect the weight of the bag and lifting equipment.)

7. A force of 2 N will stretch a rubber band 2 cm. Assuming Hooke's law applies, how far will a 4-N force stretch the rubber band? How much work does it take to stretch the rubber band this far?

8. If a force of 90 N stretches a spring 1 m beyond its natural length, how much work does it take to stretch the spring 5 m beyond its natural length?

9. *Subway car springs.* It takes a force of 21,714 lb to compress a coil spring assembly on a subway car from its free height of 8 in. to its fully compressed height of 5 in.
   a) What is the assembly's spring constant?
   b) How much work does it take to compress the assembly the first half inch? the second half inch?
   (Data courtesy of Bombardier, Inc., Mass Transit Division, for spring assemblies in subway cars delivered to the New York City Transit Authority from 1985 to 1987.)

10. A bathroom scale is compressed 1/16 in. when a 150-lb person stands on it. Assuming the scale behaves like a spring, how much does someone who compresses the scale 1/8 in. weigh? How much work does it take to compress the scale 1/8 in?

11. A rectangular tank with its top at ground level is used to catch run-off water (Fig. 5.88).

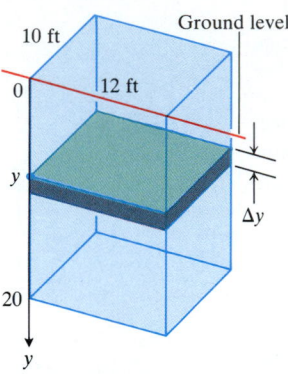

5.88 The tank in Exercise 11.

a) How much work does it take to empty the tank by pumping the water back to ground level once the tank is full?
b) If the water is pumped to ground level with a (5/11)-hp motor (work output 250 ft · lb/sec), how long will it take to empty the full tank?
c) Show that the pump in part (b) will lower the water level 10 ft (halfway) during the first 25 min of pumping.

12. A full rectangular cistern (rainwater storage tank) with its top 10 ft below ground level (Fig. 5.89) is to be emptied by pumping its contents to ground level.
a) How much work will it take to empty the cistern?
b) How long will it take a (1/2)-hp pump, rated at 275 ft · lb/sec, to pump the tank dry?
c) How long will it take the pump in part (b) to empty the tank halfway? (It will be less than half the time required to empty the tank completely.)

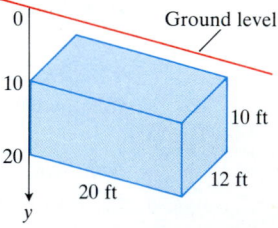

5.89 The cistern in Exercise 12.

13. How much work would it take to pump the water from the tank in Example 5 to the level of the top of the tank (instead of 4 ft higher)?

14. Suppose that instead of being completely full, the tank in Example 5 is only half full. How much work does it take to pump the water that's left to a level 4 ft above the top of the tank?

15. A vertical right circular cylindrical tank measures 30 ft high and 20 ft in diameter. It is full of kerosene weighing

51.2 lb/ft^3. How much work does it take to pump the kerosene to the level of the top of the tank?

16. The cylindrical tank in Fig. 5.90 can be filled by pumping water from a lake 15 ft below the bottom of the tank. There are two ways to go about it. One is to pump the water through a hose attached to a valve in the bottom of the tank. The other is to attach the hose to the rim of the tank and let the water pour in. Which way will be faster? Why?

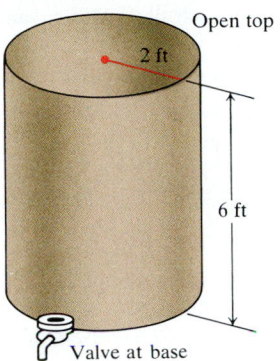

5.90   The tank in Exercise 16.

17. To design the interior surface of a huge stainless steel tank, you revolve the curve $y = x^2$, $0 \le x \le 4$, about the y-axis. The container, with dimensions in meters, is to be filled with seawater, which weighs 10,000 N/m^3. How much work will it take to empty the tank by pumping the water to the tank's top?

18. We model pumping from spherical containers the way we do from other containers, with the axis of integration along the vertical axis of the sphere. Use Fig. 5.91 to find out how much work it takes to pump the water from a full hemispherical bowl of radius 5 ft to a height 4 ft above the top of the bowl.

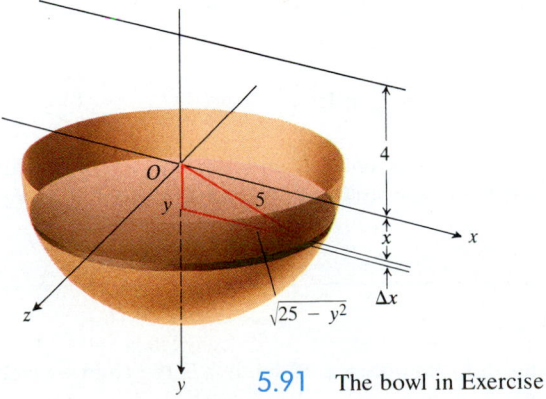

5.91   The bowl in Exercise 18.

19. You are in charge of the evacuation and repair of the storage tank shown in Fig. 5.92. The tank is a hemisphere of radius 10 ft and is full of benzene weighing 56 lb/ft^3. A firm you

contacted says it can empty the tank for 1/2¢ per foot-pound of work. Find the work required to empty the tank by pumping the benzene to an outlet 2 ft above the top of the tank. If you have $5000 budgeted for the job, can you afford to hire the firm?

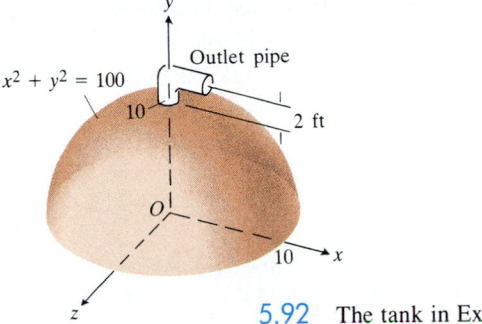

5.92   The tank in Exercise 19.

20. Your town has decided to drill a well to increase its water supply. As the town engineer, you have determined that a water tower will be necessary to provide the pressure needed for distribution, and you have designed the system in Fig. 5.93. The water is to be pumped from a 300-ft well through a vertical 4-in. pipe into the base of a cylindrical tank 20 ft in diameter and 25 ft high. The base of the tank will be 60 ft above ground. The pump is a 3-hp pump, rated at 1650 ft · lb/sec. How long will it take to fill the tank the first time? (Include the time it takes to fill the pipe.)

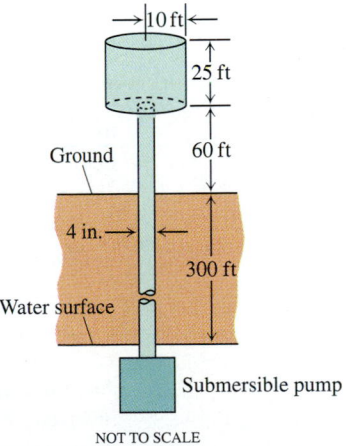

5.93   The water tower and well in Exercise 20.

21. *Work and kinetic energy in baseball.*   How many foot-pounds of work does it take to throw a baseball 90 mph? A baseball weighs 5 oz = 0.3125 lb.

22. *Work and kinetic energy in golf.*   A 1.6-oz golf ball is driven off the tee at a speed of 280 ft/sec (about 191 mph). How many foot-pounds of work are done getting the ball into the air?

23. CALCULATOR *The 1990 U.S. Open men's tennis championship.* During the match in which Peter Sampras won the 1990 U.S. Open, Sampras hit a serve that was clocked at a phenomenal 124 mph. How much work did Sampras have to do on the 2-oz ball to get it to that speed?

24. CALCULATOR A quarterback threw a 14.5-oz football 88 ft/sec (60 mph). How many foot-pounds of work were done on the ball to get it to this speed?

25. CALCULATOR How much work has to be performed on a 6.5-oz softball to pitch it 132 ft/sec (90 mph)?

26. A 2-oz steel ball bearing is placed on a vertical spring whose spring constant is $k = 18$ lb/ft. The spring is compressed 3 inches and released. About how high does the ball bearing go?

27. CALCULATOR *Putting a satellite in orbit.* The strength of the earth's gravitational field varies with the distance r from the earth's center, and the magnitude of the gravitational force experienced by a satellite of mass m during and after launch is

$$F(r) = \frac{m\,MG}{r^2}.$$

Here, $M = 5.975 \times 10^{24}$ kg is the earth's mass, $G = 6.6720 \times 10^{-11}$ Nm²kg^{-2} is the universal gravitational

constant, and r is measured in meters. The number of newton-meters of work it takes to lift a 1000-kg satellite from the earth's surface to a circular orbit 35,780 km above the earth's center is therefore given by the integral

$$\text{Work} = \int_{6,370,000}^{35,780,000} \frac{1000\,MG}{r^2}\,dr.$$

Evaluate the integral. The lower limit of integration is the earth's radius in meters at the launch site. (This calculation does not take into account energy spent by the launch vehicle or energy spent bringing the satellite to orbit velocity.)

28. CALCULATOR *Forcing electrons together.* Two electrons r meters apart repel each other with a force of

$$F = \frac{23 \times 10^{-29}}{r^2}$$

newtons.
a) Suppose one electron is held fixed at the point $(1, 0)$ on the x-axis (units in meters). How much work does it take to move a second electron along the x-axis from the point $(-1, 0)$ to the origin?
b) Suppose an electron is held fixed at each of the points $(-1, 0)$ and $(1, 0)$. How much work does it take to move a third electron along the x-axis from $(5, 0)$ to $(3, 0)$?

5.8 Fluid Pressures and Fluid Forces

We make dams thicker at the bottom than at the top (Fig. 5.94) because the pressure against them increases with depth. The deeper the water, the thicker the dam has to be.

It is a remarkable fact that the pressure at any point on the dam depends only on how far below the surface the point is and not on how much the surface happens to be tilted at that point. The pressure, in pounds per square foot at a point h feet below the surface, is always $62.5h$. The number 62.5 is the weight-density of water in pounds per cubic foot.

The formula, pressure $= 62.5h$, makes sense when you think of the units involved:

$$\frac{\text{lb}}{\text{ft}^2} = \frac{\text{lb}}{\text{ft}^3} \times \text{ft.} \tag{1}$$

As you can see, this equation depends only on units and not on what fluid is involved. The pressure h feet below the surface of any fluid is the fluid's weight-density times the depth.

5.94   To withstand the increasing pressure, dams are built thicker as they go down.   ·

> **The Pressure-Depth Equation**
>
> In a fluid that is standing still, the pressure at depth h is the fluid's weight-density times h:
>
> $$p = wh. \tag{2}$$

Weight-Density

A fluid's **weight-density** is its weight per unit volume. Here are some typical values (lb/ft³).

Gasoline	42
Mercury	849
Milk	64.5
Molasses	100
Olive oil	57
Seawater	64
Water	62.5

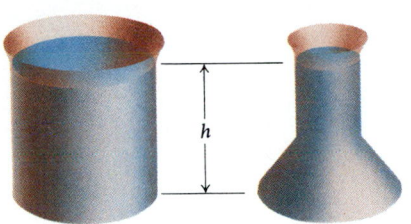

5.95 These two containers are filled with water to the same depth and have the same base area. The total force is therefore the same on the bottom of each container. The containers' shapes do not matter here. The only things that count are depth and base area.

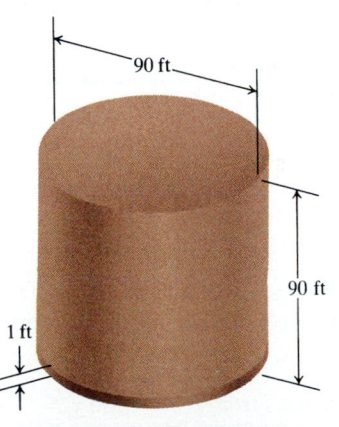

SHADED BAND NOT TO SCALE

5.96 Schematic drawing of the molasses tank in Example 1. How much force did the lowest foot of the vertical wall have to withstand when the tank was full? It takes an integral to find out. Notice that the proportions of the tank were ideal.

In this section we use the equation $p = wh$ to derive a formula for the total force exerted by a fluid against all or part of a vertical or horizontal containing wall.

The Constant-Depth Formula for Fluid Force

In a container of fluid with a flat horizontal base, the total force exerted by the fluid against the base can be calculated by multiplying the area of the base by the pressure at that level. We can do this because total force equals force per unit area (pressure) times area. (See Fig. 5.95.) If F, p, and A are the total force, pressure, and area, then

$$F = \text{total force} = \text{force per unit area} \times \text{area}$$

$$= \text{pressure} \times \text{area} = p\,A$$

$$= wh\,A. \qquad\qquad (p = wh \text{ from Eq. (2)})$$

Fluid Force on a Constant-Depth Surface

$$F = p\,A = wh\,A \qquad\qquad (3)$$

Example 1   *The Great Molasses Flood.*   At 1:00 P.M. on January 15, 1919, an unusually warm day, a 90-ft-high, 90-ft-diameter cylindrical metal tank in which the Puritan Distilling Company was storing molasses at the corner of Foster and Commercial streets in Boston's North End exploded. The molasses flooded into the streets, 30 ft deep, trapping pedestrians and horses, knocking down buildings, and oozing into homes. It was eventually tracked all over town and even carried into the suburbs on trolley cars and people's shoes. The cleanup went on for weeks.

Given that the molasses weighed 100 lb/ft³, what was the total force exerted by the molasses against the bottom of the tank at the time it blew? Assuming the tank was full, we can find out from Eq. (3):

$$\text{Total force} = wh\,A = (100)(90)(\pi(45)^2) = 57{,}255{,}526 \text{ lb.} \qquad \text{(Rounded)}$$

How about the force against the walls of the tank? For example, what was the total force against the bottom foot-wide band of tank wall (Fig. 5.96)?

The area of the band was

$$A = \pi\,d\,h = \pi(90)(1) = 90\pi \text{ ft}^2.$$

The tank was 90 ft deep, so the pressure near the bottom was about

$$p = wh = (100)(90) = 9000 \text{ lb/ft}^2.$$

Therefore the total force against the band was about

$$F = wh\,A = (9000)(90\pi) = 2{,}544{,}690 \text{ lb.} \qquad \text{(Rounded)}$$

But this is not exactly right. The top of the band was 89 ft below the surface, not 90, and the pressure there was less. To find out exactly what the force on the band was, we need to take into account the variation of the pressure across the band, and this means using calculus.

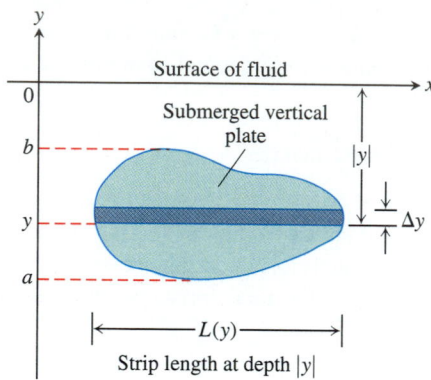

5.97 The force exerted by a fluid against one side of an approximating horizontal strip $|y|$ units beneath the surface is about

$$\Delta F = \text{pressure} \times \text{area} = w|y| \times L(y)\,\Delta y.$$

The plate shown here is flat, but it might have been curved instead, like the vertical wall of a tank. In either case, the strip length $L(y)$ is measured along the surface of the plate.

The Variable-Depth Integral for the Fluid Force Against a Submerged Vertical Wall

Suppose we want to know the force against one side of a vertical plate submerged in a fluid of weight-density w. To find it, we model the plate as a region extending from $y = a$ to $y = b$ below the x-axis in the xy-plane (Fig. 5.97). We partition $[a, b]$ in the usual way and imagine the region to be cut into thin horizontal strips by planes perpendicular to the y-axis at the partition points. The typical strip from y to $y + \Delta y$ is Δy units wide by $L(y)$ units long. We assume $L(y)$ to be a continuous function of y.

The pressure varies across the strip from top to bottom, just as it did in the molasses tank. But if the strip is narrow enough, the pressure will remain close to its bottom-edge value $w|y|$. The total force against one side of the strip will therefore be about

$$\Delta F = (\text{pressure along bottom edge}) \times (\text{area})$$
$$= w|y| \times L(y)\,\Delta y. \tag{4}$$

The force against the entire plate wall will be about

$$\sum_a^b \Delta F = \sum_a^b w|y|L(y)\,\Delta y. \tag{5}$$

The sum on the right-hand side of Eq. (5) is a Riemann sum for the continuous function $w|y|L(y)$ on $[a, b]$. We expect the approximations to improve as the norm of the partition goes to zero, so we define the total force against the plate wall to be the limit of these sums as the norm goes to zero.

DEFINITION

The Integral for Fluid Force

Suppose a submerged vertical plate runs from depth $|a|$ up to a depth $|b|$ in a fluid of weight density w, as in Fig. 5.97. Suppose also that at depth $|y|$ the plate is $L(y)$ units wide, as measured along the plate. Then the total force of the fluid against one side of the plate is

$$F = \int_a^b w|y|L(y)\,dy. \tag{6}$$

Example 2   A vertical triangular plate lies base up 2 ft below the surface of a swimming pool (Fig. 5.98). Find the fluid force against one side of the plate.

Solution   We see from Fig. 5.98 that the plate runs from $y = -5$ to $y = -2$ and that its width at level y is $L(y) = 2(y + 5)$. Therefore, Eq. (6) gives

$$\text{Force} = \int_a^b w|y|L(y)\,dy = \int_{-5}^{-2} (62.5)(-y)(2)(y + 5)\,dy \quad \left(\begin{matrix}\text{For water,}\\ w = 62.5.\end{matrix}\right)$$

$$= -125 \int_{-5}^{-2} (y^2 + 5y) = -125\left[\frac{y^3}{3} + \frac{5y^2}{2}\right]_{-5}^{-2}$$

$$= 1687.5 \text{ lb.}$$

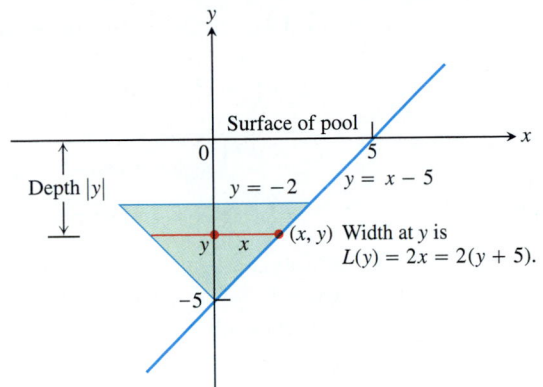

5.98 The important dimensions of the plate in Example 2.

> ### How to Find Fluid Force
>
> Whatever coordinate system you use, you can always find the fluid force against one side of a submerged vertical plate or wall by taking these steps:
>
> 1. Find expressions for the length and depth of a typical thin horizontal strip.
> 2. Multiply their product by the fluid's weight-density w and integrate over the interval of depths occupied by the plate or wall.

It is sometimes more natural to put the origin of the coordinate system at the bottom of the plate instead of at the fluid's surface. This simplifies the factor $|y|$ in the integrand to y, but otherwise the integral remains the same.

Example 3   We can now calculate exactly the force exerted by the molasses against the bottom 1-ft band of the Puritan Distilling Company's storage tank when the tank was full.

The tank was a right circular cylindrical tank 90 ft high and 90 ft in diameter. Using a coordinate system with the origin at the bottom of the tank and the y-axis pointing up (Fig. 5.99), we find that the typical horizontal strip at level y has

Strip depth: $(90 - y)$,

Strip length: $\pi \times$ tank diameter $= 90\pi$.

The force against the band is therefore

$$\text{Force} = \int_0^1 w\,(\text{depth})(\text{length})\,dy = \int_0^1 100(90 - y)(90\pi)\,dy \qquad \left(\begin{array}{l}\text{For molasses,} \\ w = 100\end{array}\right)$$

$$= 9000\pi \int_0^1 (90 - y)\,dy = 2{,}530{,}553 \text{ lb.} \qquad\qquad \text{(Rounded)}$$

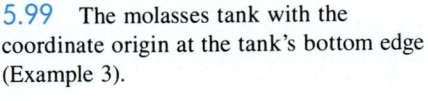

5.99 The molasses tank with the coordinate origin at the tank's bottom edge (Example 3).

As expected, the force is slightly less than the constant-depth estimate following Example 1.

Fluid Forces and Centroids of Flat Vertical Plates

When we know the location of the centroid of a submerged flat vertical plate, there is a shortcut we can take to find the force against one side of the plate. From Eq. (6),

$$\text{Force} = \int_a^b w|y|L(y)\,dy = w\int_a^b |y|L(y)\,dy \qquad \text{(See Fig. 5.97)}$$

$$= w \cdot \begin{pmatrix} \text{moment about } x\text{-axis of} \\ \text{region occupied by plate} \end{pmatrix}$$

$$= w \cdot (|\bar{y}| \times \text{area of plate}). \qquad (M_x/A = |\bar{y}|, \text{ so } M_x = |\bar{y}|\,A)$$

This derivation assumes that y is measured down from the liquid's surface so that the magnitude $|\bar{y}|$ of the y-coordinate of the plate's centroid gives the centroid's depth. To allow other coordinate systems, we need to cast our result in a coordinate free form:

Fluid Forces and Centroids

The force of a fluid of weight density w against one side of a submerged flat vertical plate is the product of w, the distance $\bar{h}$ from the plate's centroid to the fluid surface, and the plate's area:

$$F = w\bar{h}A. \tag{7}$$

Example 4   Use Eq. (7) to find the force in Example 2.

Solution   The centroid of the triangle (Fig. 5.98) lies on the y-axis, one third of the way from the base toward the opposite vertex, so

$$\bar{h} = |\bar{y}| = \left|-\left(2 + \frac{1}{3}\cdot 3\right)\right| = 3.$$

The triangle's area is

$$A = \frac{1}{2}(\text{base})(\text{height}) = \frac{1}{2}(6)(3) = 9.$$

Hence,

$$F = w\bar{h}A = (62.5)(3)(9) = 1687.5 \text{ lb.}$$

Equation (7) says that the fluid force on one side of a submerged flat vertical plate is the same as it would be if the plate's entire area lay $\bar{h}$ units beneath the surface. For many shapes, the location of the centroid can be found in a table, and Eq. (7) gives a practical way to find F. Of course, the centroid's location was found by someone who performed an integration equivalent to evaluating the integral in Eq. (6). We recommend for now that you practice your mathematical modeling by drawing pictures and thinking things through the way we did when we developed Eq. (6). Then check your results, when you conveniently can, with Eq. (7).

EXERCISES 5.8

1. Suppose the plate in Fig. 5.98 lies 4 ft below the surface instead of 2 ft. What is the fluid force on one side of the plate now?

2. What was the total fluid force against the cylindrical side wall of the molasses tank (Example 1) when the tank was full? half full?

3. The vertical ends of a watering trough are isosceles triangles like the one in Fig. 5.100 (dimensions in feet).
a) What is the fluid force against the ends when the tank is full? Does it matter how long the trough is?
b) CALCULATOR How many inches do you have to lower the water level to cut the force in half? (Answer to the nearest half inch.)

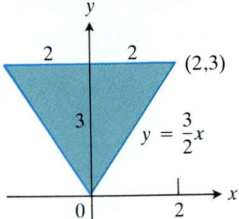

5.100   The end of the trough in Exercise 3.

4. The isosceles triangular plate in Fig. 5.101 is submerged vertically 1 ft below the surface of the water. Find the fluid force against one face of the plate.

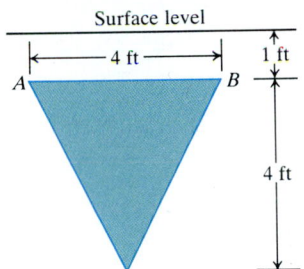

5.101   The plate in Exercise 4.

5. The plate in Exercise 4 is revolved 180° about line AB so that part of the plate sticks up above the water (Fig. 5.102). What force does the water exert on one face of the plate now?

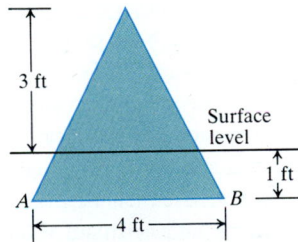

5.102   The plate from Exercise 4 repositioned as in Exercise 5.

6. The semicircular plate in Fig. 5.103 sticks straight down into the water with its diameter along the surface. How much force does the water exert against one side of the plate?

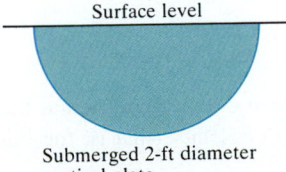

Submerged 2-ft diameter vertical plate

5.103   The plate in Exercise 6.

7. A horizontal rectangular fresh-water fish tank of interior dimensions $2 \times 2 \times 4$ ft is filled to within 2 in. of the top.
a) Find the fluid force against the sides and ends of the tank.
b) If the tank is sealed and stood on end (without spilling), so that one of the square ends is the base, what does that do to the fluid force on the rectangular sides?

8. The viewing portion of the rectangular glass window in a typical fish tank at the New England Aquarium in Boston is 63 in. wide and runs from 0.5 in. below the water's surface to 33.5 in. below the surface. Find the fluid force against this portion of the window. The weight-density of seawater is 64 lb/ft³. (In case you were wondering, the glass is 3/4 in. thick and the tank walls extend 4 in. above the water to keep the fish from jumping out.)

9. CALCULATOR A rectangular milk carton measures $3.75 \times 3.75 \times 7.75$ in. Find the force of the milk on one side when the carton is full.

10. A tank truck hauls milk in a 6-ft diameter horizontal right circular cylindrical tank. How much force does the milk exert on each end of the tank when the tank is half full?

11. The cubical metal tank in Fig. 5.104 is used to store liquids. It has a parabolic gate, held in place by bolts and designed to withstand a fluid force of 160 lb without rupturing. The liquid you plan to store has a weight-density of 50 lb/ft³.
a) What is the fluid force on the gate when the liquid is 2 ft deep?
b) What is the maximum height to which the container can be filled without exceeding its design limitation?

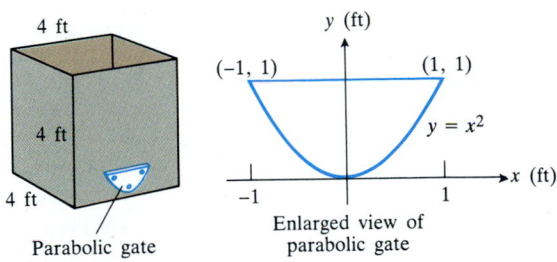

5.104   The tank in Exercise 11.

12. CALCULATOR The end plates in the trough in Fig. 5.105

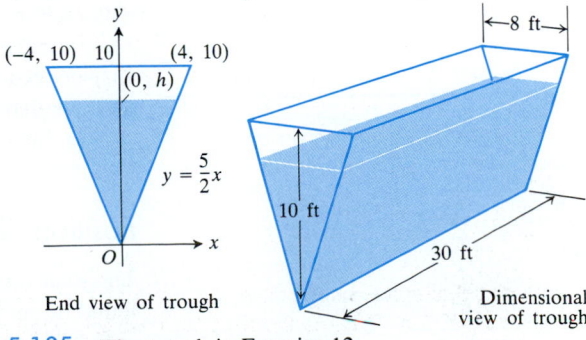

5.105   The trough in Exercise 12.

were designed to withstand a fluid force of 6667 lb. How many cubic feet of water can the tank hold without exceeding design limitations?

13. Water is running into the rectangular swimming pool in Fig. 5.106 at the rate of 1000 ft³/h.
   a) Find the fluid force against the triangular drain plate after 9 h of filling.
   b) The drain plate is designed to withstand a fluid force of 520 lb. How high can you fill the pool without exceeding this limitation?

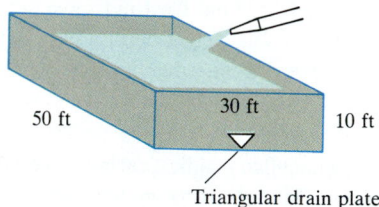

Triangular drain plate

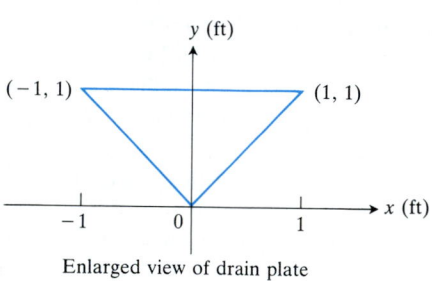

Enlarged view of drain plate

5.106   The drain plate in Exercise 13.

14. A vertical rectangular plate is submerged in a fluid with its top edge parallel to the fluid's surface. Show that the fluid force on one side of the plate is the average value of the pressure on the plate times the area of the plate.

15. Water pours into the tank in Fig. 5.107 at the rate of 4 ft³/min. The tank's cross sections are 4-ft-diameter semicircles. One end of the tank is movable, but moving it to increase the volume compresses a spring. The spring constant is $k = 100$ lb/ft. If the end of the tank moves 5 ft against the spring, the water will drain out of a safety hole in the bottom at the rate of 5 ft³/min. Will the movable end reach the hole before the tank overflows?

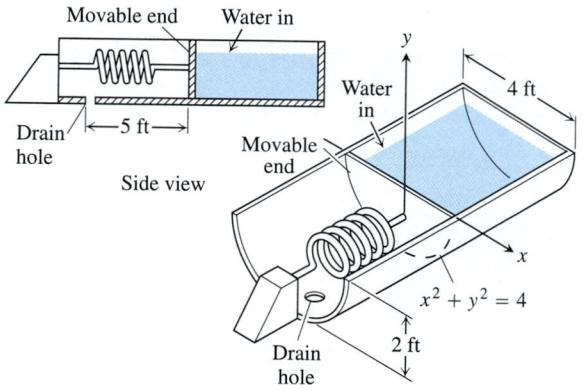

5.107   The tank in Exercise 15.

5.9 The Basic Pattern: Other Modeling Applications

There is a pattern to what we have been doing in the preceding sections. In each section we wanted to measure something that was modeled or described by one or more continuous functions. In Section 5.1 it was the area between the graphs of two continuous functions. In Section 5.2 it was the volume of a solid. In Section 5.7 it was the work done by a force directed along the x-axis whose magnitude was given by a continuous function, and so on. In each case we responded by partitioning the interval on which the function or functions were defined and approximating what we wanted to measure with Riemann sums over the interval. We then used the integral defined by the limit of the Riemann sums to define and calculate what we wanted to measure. Table 5.2 shows the pattern.

Literally thousands of things in biology, chemistry, economics, engineering, finance, geology, medicine, and other fields (the list would fill pages) are modeled and calculated by exactly this process.

In this section we review the process and look at a few more of the integrals it leads to.

Position Shift and Distance Traveled

The total distance traveled by a body moving up and down a coordinate line is found by integrating the absolute value of the body's velocity (that is, the body's speed) over the time interval of the motion.

If $s(t)$ is a body's position on a coordinate line at time t, then

$$\frac{ds}{dt} = v = \text{velocity}$$

and

$$\left|\frac{ds}{dt}\right| = |v| = \text{speed}.$$

TABLE 5.2
The phases of developing an integral to calculate something

Phase 1	Phase 2	Phase 3
We describe or model something we want to measure in terms of one or more continuous functions defined on a closed interval $[a, b]$.	We partition $[a, b]$ into subintervals of length Δx_k and choose a point c_k in each subinterval. We approximate what we want to measure with a finite sum. We identify the sum as a Riemann sum of a continuous function over $[a, b]$.	The approximations improve as the norm of the partition goes to zero. The Riemann sums approach a limiting integral. We use the integral to define and calculate what we originally wanted to measure.
The area between the curves $y = f_1(x)$, $y = f_2(x)$, on $[a, b]$ when $f_2(x) \leq f_1(x)$	$\sum (f_1(c_k) - f_2(c_k))\Delta x_k$	Area $= \displaystyle\int_a^b (f_1(x) - f_2(x))\, dx$
The volume of the solid with cross-section area $A(x)$, $a \leq x \leq b$	$\sum A(x_k)\Delta x_k$	Volume $= \displaystyle\int_a^b A(x)\, dx$
The length of a smooth curve $y = f(x)$, $a \leq x \leq b$	$\sum \sqrt{1 + (f'(c_k))^2}\, \Delta x_k$	Length $= \displaystyle\int_a^b \sqrt{1 + (f'(x))^2}\, dx$
The work done by a variable force of magnitude $F(x)$ directed along the x-axis from a to b	$\sum F(c_k)\Delta x_k$	Work $= \displaystyle\int_a^b F(x)\, dx$

To see why, we partition the time interval $a \leq t \leq b$ into subintervals in the usual way and let Δt_k denote the length of the kth interval. If Δt_k is small enough, the body's velocity $v(t)$ will not change much from t_{k-1} to t_k and the right-hand end-point value $v(t_k)$ will give a good approximation of the velocity throughout the interval. Accordingly, the change in the body's position coordinate during the kth time interval will be about

$$v(t_k)\Delta t_k. \tag{1}$$

The change will be positive if $v(t_k)$ is positive and negative if $v(t_k)$ is negative.

In either case, the distance traveled by the body during the kth interval will be about

$$|v(t_k)|\Delta t_k. \tag{2}$$

The total trip distance will be about

$$\sum_{k=1}^{n} |v(t_k)|\Delta t_k. \tag{3}$$

The sum in (3) is a Riemann sum for the speed $|v(t)|$ on the interval $[a, b]$. We expect the approximations we get from sums like these to improve as the norm of the partition of $[a, b]$ goes to zero. It therefore looks as if we should be able to calculate the total distance traveled by the body by integrating the body's speed from a to b. In practice, this turns out to be just the right thing to do. The mathematical model predicts the distance every time.

$$\text{Distance traveled} = \int_a^b |v(t)|\, dt \tag{4}$$

If we wish instead to predict how far up or down the line from its initial position a body will end up when a trip is over, we integrate v instead of its absolute value.

To understand why, we let $s(t)$ be the body's position at time t and let F be an antiderivative of v. Then

$$s(t) = F(t) + C$$

for some constant C. The shift in the body's position (**displacement**) caused by the trip from $t = a$ to $t = b$ is

$$s(b) - s(a) = (F(b) + C) - (F(a) + C) = F(b) - F(a) = \int_a^b v(t)\, dt. \qquad (5)$$

1. To find distance traveled, integrate speed.

2. To find displacement, integrate velocity.

$$\boxed{\text{Displacement} = \int_a^b v(t)\, dt \qquad (6)}$$

Example 1   The velocity of a body moving along a line from $t = 0$ to $t = 3\pi/2$ sec was

$$v(t) = 5 \cos t \text{ m/sec.}$$

Find the total distance traveled and the body's displacement.

Solution

$$\text{Distance traveled} = \int_0^{3\pi/2} |5 \cos t|\, dt \qquad \text{(Distance is the integral of speed.)}$$

$$= \int_0^{\pi/2} 5 \cos t\, dt + \int_{\pi/2}^{3\pi/2} (-5 \cos t)\, dt$$

$$= 5 \sin t \Big]_0^{\pi/2} - 5 \sin t \Big]_{\pi/2}^{3\pi/2}$$

$$= 5(1 - 0) - 5(-1 - 1) = 5 + 10 = 15 \text{ m}$$

$$\text{Displacement} = \int_0^{3\pi/2} 5 \cos t\, dt \qquad \left(\begin{array}{l}\text{Displacement is the integral of} \\ \text{velocity.}\end{array}\right)$$

$$= 5 \sin t \Big]_0^{3\pi/2} = 5(-1) - 5(0) = -5 \text{ m}$$

During the trip, the body traveled 5 m forward and 10 m backward for a total distance of 15 m. This displaced the body 5 m to the left (Fig. 5.108).

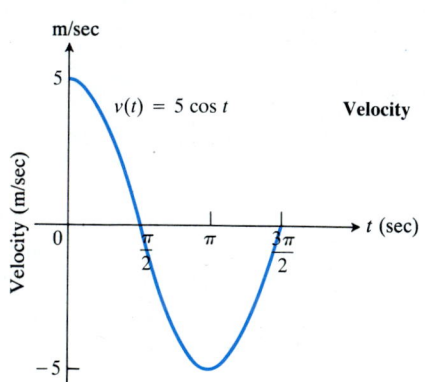

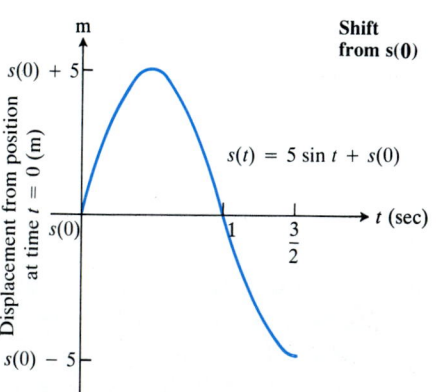

5.108   The position and velocity of the body in Example 1. The velocity is positive at first, and the corresponding displacement is positive. But the body stops at $t = \pi/2$ and reverses direction. By time $t = 3\pi/2$, the body has moved 5 m to the left of where it started.

Delesse's Rule

As you may know, the sugar in an apple starts turning into starch as soon as the apple is picked, and the longer the apple sits around, the starchier it becomes. You can tell fresh apples from stale by both flavor and consistency.

To find out how much starch is in a given apple, we can look at a thin slice

Delesse's Rule

Achille Ernest Delesse was a mid-nineteenth-century mining engineer interested in determining the composition of rocks. To find out how much of a particular mineral a rock contained, he cut it through, polished an exposed face, and covered the face with transparent waxed paper, trimmed to size. He then traced on the paper the exposed portions of the mineral that interested him. After weighing the paper, he cut out the mineral traces and weighed them. The ratio of the weights gave not only the proportion of the surface occupied by the mineral but, more important, the proportion of the entire rock occupied by the mineral. Delesse described his method in an article entitled "A Mechanical Procedure for Determining the Composition of Rocks," in *Annales des Mines*, 13, 1848, pp. 379–88. His method is still used by petroleum geologists today. A two-dimensional analogue of it is used to determine the porosities of the ceramic filters that extract organic molecules in chemistry laboratories and screen out microbes in water purifiers.

under a microscope. The cross sections of the starch granules will show up clearly, and it is easy to estimate the proportion of the viewing area they occupy. This two-dimensional proportion will be the same as the three-dimensional proportion of uncut starch granules in the apple itself. The apparently magical equality of these proportions was first discovered by a French geologist, Achille Ernest Delesse, in the 1840s. Its explanation lies in the notion of average value.

Suppose we want to find the proportion of some granular material in a solid and that the sample we have chosen to analyze is a cube whose edges have length L. We picture the cube with an x-axis along one edge and imagine slicing the cube with planes perpendicular to points of the interval $[0, L]$ (Fig. 5.109). Call the proportion of the area of the slice at x occupied by the granular material of interest (starch, in our apple example) $r(x)$ and assume r is a continuous function of x.

Now partition the interval $[0, L]$ into subintervals in the usual way. Imagine the cube sliced into thin slices by planes at the subdivision points. The length Δx_k of the kth subinterval is the distance between the planes at x_{k-1} and x_k. If the planes are close enough together, the sections cut from the grains by the planes will resemble cylinders with bases in the plane at x_k. The proportion of granular material between the planes will therefore be about the same as the proportion of cylinder base area in the plane at x_k, which in turn will be about $r(x_k)$. Thus the amount of granular material in the slab between the two planes will be about

$$\text{(Proportion)} \times \text{(slab volume)} = r(x_k)L^2 \, \Delta x_k. \tag{7}$$

The amount of granular material in the entire sample cube will be about

$$\sum_{k=1}^{n} r(x_k)L^2 \, \Delta x_k. \tag{8}$$

This sum is a Riemann sum for the function $r(x)L^2$ over the interval $[0, L]$. We expect the approximations by sums like these to improve as the norm of the subdivision of $[0, L]$ goes to zero and therefore expect the integral

$$\int_a^b r(x)L^2 \, dx \tag{9}$$

to give the amount of granular material in the sample cube.

We can obtain the proportion of granular material in the sample by dividing this amount by the cube's volume L^3. If we have chosen our sample well, this will also be the proportion of granular material in the solid from which the sample was

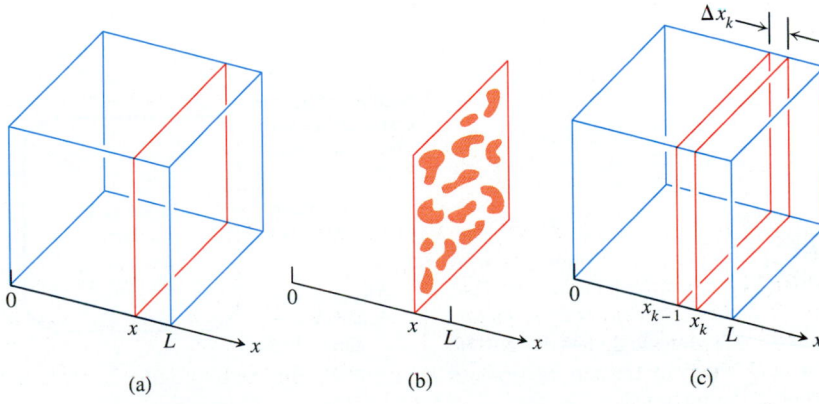

5.109 The steps leading to Delesse's rule: (a) a slice through a sample cube of material; (b) the granular material in the slice; (c) the slab between consecutive slices determined by a subdivision of $[0, L]$.

(a)   (b)   (c)

taken. Putting it all together, we get

$$\text{Proportion of granular material in solid} = \text{Proportion of granular material in the sample cube}$$

$$= \frac{\int_0^L r(x)L^2\,dx}{L^3} = \frac{L^2\int_0^L r(x)\,dx}{L^3} = \frac{1}{L}\int_0^L r(x)\,dx \quad (10)$$

$$= \text{average value of } r(x) \text{ over } [0, L]$$

$$= \text{proportion of area occupied by granular material in a typical cross section.}$$

This is Delesse's rule. Once we have found $\bar{r}$, the average of $r(x)$ over $[0, L]$, we have found the proportion of granular material in the solid.

In practice, $\bar{r}$ is found by averaging over a number of cross sections. There are several things to watch out for in the process. In addition to the possibility that the granules cluster in ways that make representative samples difficult to find, there is the possibility that we might not recognize a granule's trace for what it is. Some cross sections of normal red blood cells look like disks and ovals, for example, but others look surprisingly like outlines of dumbbells. We do not want to dismiss the dumbbells as experimental error the way one research group did a few years ago.

Useless Integrals

Some of the integrals we get from forming Riemann sums do what we want, but others do not. It all depends on how we choose to model the problems we want to solve. Some choices are good; others are not. Here is an example.

We use the surface area formula

$$\text{Surface area} = \int_a^b 2\pi f(x)\sqrt{1 + \left(\frac{df}{dx}\right)^2}\,dx \quad (11)$$

because it has predictive value and always gives results consistent with information from other sources. In other words, the model we used to derive the formula (Fig. 5.110) was a good one.

Why not find the surface area by approximating with cylindrical bands instead of conical bands, as suggested in Fig. 5.111? The Riemann sums we get this way converge just as nicely as the ones based on conical bands, and the resulting integral

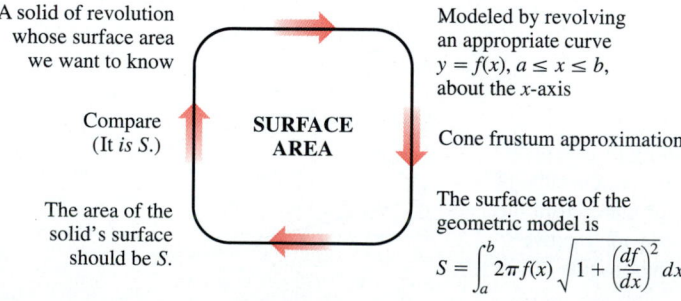

A solid of revolution whose surface area we want to know

Modeled by revolving an appropriate curve $y = f(x)$, $a \le x \le b$, about the x-axis

Compare (It *is* S.)

SURFACE AREA

Cone frustum approximation

The area of the solid's surface should be S.

The surface area of the geometric model is

$$S = \int_a^b 2\pi f(x)\sqrt{1 + \left(\frac{df}{dx}\right)^2}\,dx.$$

5.110 The modeling cycle for surface area.

is simpler. Instead of Eq. (11), we get

$$\text{Surface area candidate} = \int_a^b 2\pi f(x)\, dx. \qquad (12)$$

After all, we might argue, we used cylinders to derive good volume formulas, so why not use them again to derive surface area formulas?

The answer is that the formula in Eq. (12) has no predictive value and almost never gives results consistent with experience. The comparison step in the model fails for this formula.

There is a moral here: Just because we end up with a nice-looking integral does not mean it will do what we want. Constructing an integral is not enough—we have to test it too.

Pappus's Theorems

In the third century, an Alexandrian Greek named Pappus discovered two formulas that relate centroids to surfaces and solids of revolution. The formulas provide shortcuts to a number of otherwise lengthy calculations.

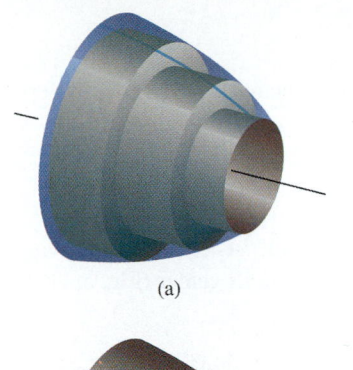

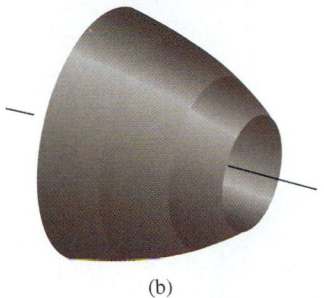

(a)

(b)

5.111   Why not use (a) cylindrical bands instead of (b) conical bands to approximate surface area?

THEOREM 1

Pappus's Theorem for Volumes

If a plane region is revolved once about a line in the plane that does not cut through the region's interior, then the volume of the solid it generates is equal to the region's area times the distance traveled by the region's centroid during the revolution. If ρ is the distance from the axis of revolution to the centroid, then

$$V = 2\pi \rho A. \qquad (13)$$

Naturally, we need to impose some restriction on the region R for the theorem to hold. The usual restriction is that the boundary of R be smooth enough to make the lengths of the cross sections of the region parallel to the axis of revolution vary continuously with their distance from the axis. (In Fig. 5.112, these are the lengths $L(y)$.)

Proof   We draw the axis of revolution as the x-axis with the region R in the first quadrant. We let $L(y)$ denote the length of the cross section of the region perpendicular to the y-axis at y (Fig. 5.112). By the method of cylindrical shells, the volume of the solid generated by revolving the region about the x-axis is

$$V = \int_c^d 2\pi\,(\text{shell radius})(\text{shell height})\, dy = 2\pi \int_c^d yL(y)\, dy. \qquad (14)$$

The y-coordinate of R's centroid is

$$\bar{y} = \frac{\int_c^d \tilde{y}\, dA}{A} = \frac{\int_c^d yL(y)\, dy}{A},$$

so that

$$\int_c^d yL(y)\, dy = A\bar{y}.$$

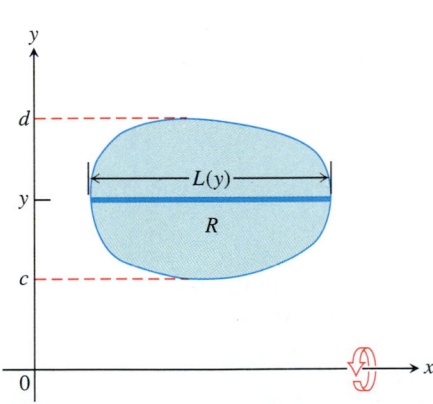

5.112   The region R is to be revolved (once) about the x-axis to generate a solid. A 1700-year-old theorem says that the solid's volume can be calculated by multiplying the region's area by the distance traveled by its centroid during the revolution.

Substituting $A\bar{y}$ for the last integral in (14) gives the result we set out to derive:

$$V = 2\pi\,\bar{y}\,A. \tag{15}$$

Example 2   The volume of the torus (doughnut) generated by revolving a circle of radius a about an axis at a distance $b \geq a$ from its center (Fig. 5.113) is

$$V = 2\pi(b)(\pi a^2) = 2\pi^2\,ba^2.$$

Distance from axis of revolution to centroid

b

a

Area: πa^2

5.113   With Pappus's first theorem, we can find the volume of a torus without having to integrate: $(2\pi b)(\pi a^2)$.

THEOREM 2

Pappus's Theorem for Surface Areas

If an arc of a smooth plane curve is revolved once about a line in the plane that does not cut through the arc's interior, then the area of the surface generated by the arc equals the length of the arc times the distance traveled by the arc's centroid during the revolution. If ρ is the distance from the axis of revolution to the centroid, then

$$S = 2\pi\rho L. \tag{16}$$

The proof we give assumes that we can model the axis of revolution as the x-axis and the arc as the graph of a smooth function of x.

Proof   We draw the axis of revolution as the x-axis with the arc extending from $x = a$ to $x = b$ in the first quadrant (Fig. 5.114). The area of the surface generated by the arc is

$$S = \int_{x=a}^{x=b} 2\pi y\,ds = 2\pi\int_{x=a}^{x=b} y\,ds. \tag{17}$$

The y-coordinate of the arc's centroid is

$$\bar{y} = \frac{\displaystyle\int_{x=a}^{x=b} \tilde{y}\,ds}{\displaystyle\int_{x=a}^{x=b} ds} = \frac{\displaystyle\int_{x=a}^{x=b} y\,ds}{L}. \qquad \left(\begin{array}{l} L = \int ds \text{ is the arc's} \\ \text{length and } \tilde{y} = y. \end{array}\right)$$

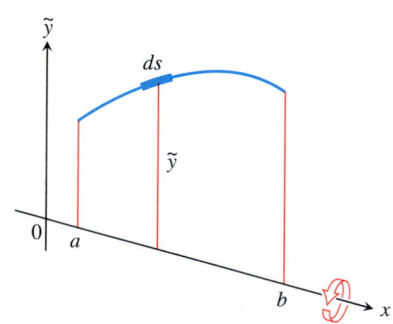

5.114   Pappus's second theorem tells how to calculate the area of the surface swept out by revolving an arc like this about the x-axis.

Hence,

$$\int_{x=a}^{x=b} y \, ds = \bar{y}L.$$

Substituting $\bar{y}L$ for the last integral in Eq. (17) gives the result we set out to derive:

$$S = 2\pi\bar{y}L. \qquad (18)$$

Example 3   The surface area of the torus in Example 1 is

$$S = 2\pi(b)(2\pi a) = 4\pi^2 \, ba.$$

The theorems of Pappus can be applied "backward" to locate an object's centroid.

Example 4   Locate the centroid of a semicircular region.

Solution   We model the region as the region enclosed by the x-axis and the semicircle $y = \sqrt{a^2 - x^2}$ and imagine revolving the region about the x-axis to generate a solid sphere (Fig. 1.115). We then have

$$\bar{y} = \frac{V}{2\pi A} \qquad \text{(Eq. (13))}$$

$$= \frac{\frac{4}{3}\pi a^3}{2\pi \cdot \frac{1}{2}\pi a^2} = \frac{4}{3\pi} a.$$

The centroid lies on the axis of symmetry at the point $\bar{y} = 4a/3\pi$.

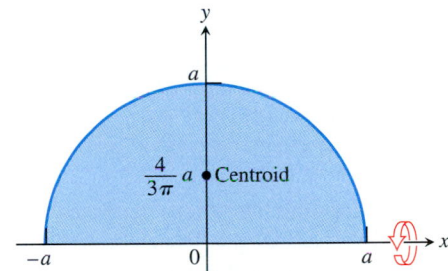

5.115   If we use one of Pappus's theorems, we can find the y-coordinate of the centroid of a semicircular region without having to integrate.

EXERCISES 5.9

In Exercises 1–8, the function $v(t)$ is the velocity in meters per second of a body moving along a coordinate line. (a) Graph v as a function of t to see where it is positive and negative. Then find (b) the total distance traveled by the body during the given time interval and (c) the body's displacement.

1. $v(t) = 5 \cos t, \quad 0 \le t \le 2\pi$

2. $v(t) = \sin \pi t, \quad 0 \le t \le 2$

3. $v(t) = 6 \sin 3t, \quad 0 \le t \le \pi/2$

4. $v(t) = 4 \cos 2t, \quad 0 \le t \le \pi$

5. $v(t) = 49 - 9.8t, \quad 0 \le t \le 10$

6. $v(t) = 8 - 1.6t, \quad 0 \le t \le 10$

7. $v(t) = 6t^2 - 18t + 12 = 6(t - 1)(t - 2), \quad 0 \le t \le 2$

8. $v(t) = 6t^2 - 18t + 12 = 6(t - 1)(t - 2), \quad 0 \le t \le 3$

9. The function $x = (1/3)t^3 - 3t^2 + 8t$ gives the position of a

body moving on the x-axis at time $t \ge 0$ (s in meters, t in seconds).

a) Show that the body is moving to the right at time $t = 0$.

b) When does the particle move to the left?

c) What is the particle's position at time $t = 3$?

d) When $t = 3$, what is the total distance the particle has traveled?

10. CALCULATOR   Table 5.3 shows the velocity of a model train engine moving back and forth on a track for 10 sec. Use Simpson's rule to find the resulting displacement and total distance traveled.

11. The photograph in Fig. 5.116 shows a grid superimposed on the polished face of a piece of granite. Use the grid and Delesse's rule to estimate the proportion of shrimp-colored granular material in the rock.

TABLE 5.3
Selected velocities of the model train engine in Exercise 10

Time (sec)	Velocity (in./sec)	Time (sec)	Velocity (in./sec)
0	0	6	−11
1	12	7	−6
2	22	8	2
3	10	9	6
4	−5	10	0
5	−13		

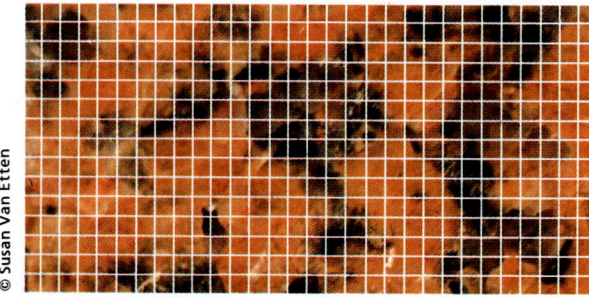

© Susan Van Etten

5.116   The granite in Exercise 11.

12. The photograph in Fig. 5.117 shows a grid superimposed on a microscopic view of a stained section of human lung tissue. The clear spaces between the cells are cross sections of the lung's air sacs (called *alveolae*—accent on the second syllable). Use the grid and Delesse's rule to estimate the proportion of air space in the lung.

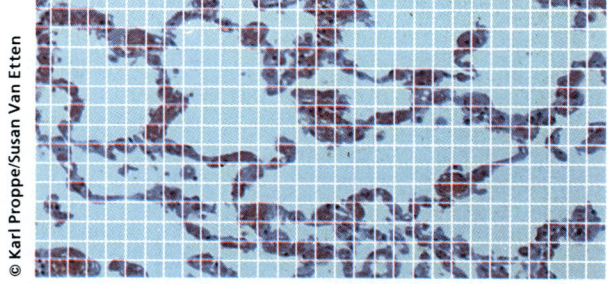

© Karl Proppe/Susan Van Etten

5.117   The lung tissue in Exercise 12.

13. *Modeling surface area.* The lateral surface area of the cone generated by revolving the line segment $y = x/\sqrt{3}$, $0 \le x \le \sqrt{3}$, about the x-axis (Fig. 5.118) should be (1/2)(base circumference)(slant height) = $(1/2)(2\pi)(2) =$

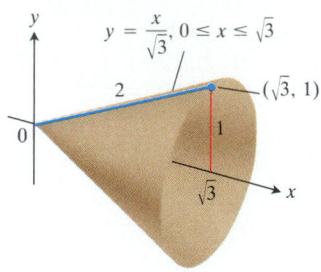

$y = \dfrac{x}{\sqrt{3}}, 0 \le x \le \sqrt{3}$

$(\sqrt{3}, 1)$

5.118   The cone in Exercise 13.

2π. What do you get if you use Eq. (12) with $f(x) = x/\sqrt{3}$?

14. *Modeling surface area.* The only surface for which Eq. (12) gives the surface area we want is a cylinder. Show that Eq. (12) gives $S = 2\pi rh$ for the cylinder generated by revolving the line segment $y = r$, $0 \le x \le h$, about the x-axis.

The Theorems of Pappus

15. The square region with vertices $(0, 2)$, $(2, 0)$, $(4, 2)$, and $(2, 4)$ is revolved about the x-axis to generate a solid. Find the volume and surface area of the solid.

16. Use a theorem of Pappus to find the volume generated by revolving about the line $x = 5$ the triangular region bounded by the coordinate axes and the line $2x + y = 6$. (As you saw in Exercise 22 of Section 5.6, the centroid of a triangle lies at the intersection of the medians, one third of the way from the midpoint of each side toward the opposite vertex.)

17. Find the volume of the torus generated by revolving the circle $(x - 2)^2 + y^2 = 1$ about the y-axis.

18. Use the theorems of Pappus to find the lateral surface area and the volume of a right circular cone.

19. Use the second theorem of Pappus and the fact that the surface area of a sphere of radius a is $4\pi a^2$ to find the centroid of the semicircle $y = \sqrt{a^2 - x^2}$.

20. As found in Exercise 19, the centroid of the semicircle $y = \sqrt{a^2 - x^2}$ lies at the point $(0, 2a/\pi)$. Find the area of the surface swept out by revolving the semicircle about the line $y = a$.

21. The area of the region R enclosed by the semiellipse $y = (b/a)\sqrt{a^2 - x^2}$ and the x-axis is $(1/2)\,\pi ab$ and the volume of the ellipsoid generated by revolving R about the x-axis is $(4/3)\,\pi ab^2$. Find the centroid of R. Notice the remarkable fact that the location is independent of a.

22. As found in Example 4, the centroid of the region enclosed by the x-axis and the semicircle $y = \sqrt{a^2 - x^2}$ lies at the point $(0, 4a/3\pi)$. Find the volume of the solid generated by revolving this region about the line $y = -a$.

23. The region of Exercise 22 is revolved about the line $y = x - a$ to generate a solid. Find the volume of the solid.

24. As found in Exercise 19, the centroid of the semicircle $y = \sqrt{a^2 - x^2}$ lies at the point $(0, 2a/\pi)$. Find the area of the surface generated by revolving the semicircle about the line $y = x - a$.

25. Find the moment about the x-axis of the semicircular region in Example 4. If you use results already known, you will not need to integrate.

REVIEW QUESTIONS

1. How do you define and calculate the area between the graphs of two continuous functions? Give an example.

2. How do you define and calculate the volumes of solids by the method of slicing? Give an example.

3. How are the disk and washer methods for calculating volumes derived from the method of slicing? Give examples of volume calculations by these methods.

4. Describe the method of cylindrical shells. Give an example.

5. How do you define and calculate the length of the graph of a smooth function over a closed interval? Give an example. What about functions that do not have continuous first derivatives?

6. How do you define and calculate the area of the surface swept out by revolving the graph of a smooth function $y = f(x)$, $a \le x \le b$, about the x-axis? Give an example.

7. How do you locate the center of mass of a straight, narrow rod or strip of material? Give an example. If the density of the material is constant, you can tell right away where the center of mass is. Where is it?

8. How do you locate the center of mass of a thin flat plate of material? Give an example.

9. How do you define and calculate the work done by a variable force directed along a portion of the x-axis? How do you calculate the work it takes to pump a liquid from a tank? Give examples.

10. How is work related to kinetic energy? How does this relationship enable us to calculate the work it takes to launch a ball? Give an example.

11. How do you calculate the force exerted by a liquid against a portion of a vertical wall? Give an example.

12. Suppose you know, as a function of time, the velocity of a body that will be moving back and forth along a coordinate line from time a to time b. How can you predict how much the motion will shift the body's position? How can you predict the total distance the body will travel?

13. What does Delesse's rule say? Give an example.

14. What do Pappus's two theorems say? Give examples of how they are used to calculate surface areas and volumes and to locate centroids.

15. There is a basic pattern to the way we constructed integrals in this chapter. What is it?

MISCELLANEOUS EXERCISES

Find the areas of the regions enclosed by the curves and lines in Exercises 1–4.

1. $y = x + 1$, $y = 3 - x^2$

2. $y^2 = 4x$, $y = 4x - 2$

3. $y = \sin x$, $y = x$, $0 \le x \le \pi/4$

4. $y = |\sin x|$, $y = 1$, $-\pi/2 \le x \le \pi/2$

5. Find the area of the "triangular" region bounded on the left by $x + y = 2$, on the right by $y = x^2$, and above by $y = 2$.

6. Find the area of the "triangular" region bounded on the left by $y = \sqrt{x}$, on the right by $y = 6 - x$, and below by $y = 1$.

7. Find the extreme values of $f(x) = x^3 - 3x^2$ and find the area of the region enclosed by the graph of f and the x-axis.

8. Find the area of the region cut from the first quadrant by the curve $x^{1/2} + y^{1/2} = a^{1/2}$.

9. Find the area of the region enclosed by the curves $x = y^{2/3}$, $x = y$, and $y = -1$.

10. Find the area of the region between the curves $y = \sin x$ and $y = \cos x$ for $0 \le x \le 3\pi/2$.

Find the volumes of the solids in Exercises 11–16.

11. The solid lies between planes perpendicular to the x-axis at $x = 0$ and $x = 1$. The cross sections perpendicular to the x-axis between these planes are circular disks whose diameters run from the parabola $y = x^2$ to the parabola $y = \sqrt{x}$.

12. The base of the solid is the region in the first quadrant between the line $y = x$ and the parabola $y = 2\sqrt{x}$. The cross sections of the solid perpendicular to the x-axis are equilateral triangles whose bases stretch from the line to the curve.

13. The solid lies between planes perpendicular to the x-axis at

$x = \pi/4$ and $x = 5\pi/4$. The cross sections between these planes are circular disks whose diameters run from the curve $y = 2 \cos x$ to the curve $y = 2 \sin x$.

14. The solid (Fig. 5.119) lies between planes perpendicular to the x-axis at $x = 0$ and $x = 6$. The cross sections between these planes are squares whose bases run from the x-axis up to the curve $x^{1/2} + y^{1/2} = \sqrt{6}$.

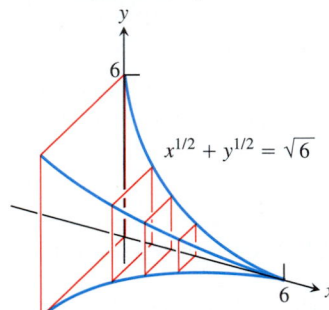

5.119 The solid in Exercise 14.

15. The solid lies between planes perpendicular to the x-axis at $x = 0$ and $x = 4$. The cross sections of the solid perpendicular to the x-axis between these planes are circular disks whose diameters run from the curve $x^2 = 4y$ to the curve $y^2 = 4x$.

16. The base of the solid is the region bounded by the parabola $y^2 = 4ax$, $a > 0$, and the line $x = a$ in the xy-plane. Each cross section perpendicular to the x-axis is an equilateral triangle with one edge in the plane. (The triangles all lie on the same side of the plane.)

17. Find the volume of the solid generated by revolving the region bounded by the x-axis, the curve $y = 3x^4$, and the lines $x = 1$ and $x = -1$ about (a) the x-axis; (b) the y-axis; (c) the line $x = 1$; (d) the line $y = 3$.

18. Find the volume of the solid generated by revolving the "triangular" region bounded by the curve $y = 4/x^3$ and the lines $x = 1$ and $y = 1/2$ about (a) the x-axis; (b) the y-axis; (c) the line $x = 2$; (d) the line $y = 4$.

19. Find the volume of the solid generated by revolving the region bounded on the left by the parabola $x = y^2 + 1$ and on the right by the line $x = 5$ about (a) the x-axis; (b) the y-axis; (c) the line $x = 5$.

20. Find the volume of the solid generated by revolving the region bounded by the parabola $y^2 = 4x$ and the line $y = x$ about (a) the x-axis; (b) the y-axis; (c) the line $x = 4$; (d) the line $y = 4$.

21. Find the volume of the solid generated by revolving the "triangular" region bounded by the x-axis, the line $x = \pi/3$, and the curve $y = \tan x$ in the first quadrant about the x-axis.

22. Find the volume of the solid generated by revolving the region bounded by the curve $y = \sin x$ and the lines $x = 0$, $x = \pi$, and $y = 2$ about the line $y = 2$.

23. Find the volume of the solid generated by revolving the region between the x-axis and the curve $y = x^2 - 2x$ about

(a) the x-axis, (b) the line $y = -1$, (c) the line $x = 2$, (d) the line $y = 2$.

24. Find the volume of the solid generated by revolving about the x-axis the region bounded by $y = 2 \tan x$, $y = 0$, $x = -\pi/4$, and $x = \pi/4$. (The region lies in the first and third quadrants and resembles a skewed bow tie.)

25. A round hole of radius $\sqrt{3}$ ft is bored through the center of a solid sphere of radius 2 ft. Find the volume of material removed from the sphere.

26. CALCULATOR The profile of a football resembles the ellipse in Fig. 5.120. Find the football's volume to the nearest cubic inch.

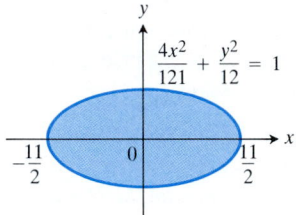

5.120 The profile of the elliptical football in Exercise 26.

27. A solid is generated by revolving about the x-axis the region bounded by the graph of the continuous function $y = f(x)$, the x-axis, and the lines $x = a$ and $x = b$. Its volume, for all $b > a$, is $b^2 - ab$. Find $f(x)$.

28. A solid is generated by revolving about the x-axis the region bounded by the graph of the continuous function $y = f(x)$, the x-axis, and the lines $x = 0$ and $x = a$. Its volume, for all $a > 0$, is $a^2 + a$. Find $f(x)$.

Find the lengths of the curves in Exercises 29–31.

29. $y = x^{1/2} - (1/3)x^{3/2}$, $\quad 0 \le x \le 3$

30. $x = y^{2/3}$, $\quad 1 \le y \le 8$

31. $y = (5/12)x^{6/5} - (5/8)x^{4/5}$, $\quad 1 \le x \le 32$

32. Suppose that the function $f(x)$ is smooth for $x \ge 0$ and that $f(0) = a$. Let $s(x)$ denote the length of the graph of f from $(0, a)$ to $(x, f(x))$, $x > 0$.

a) Find $f(x)$ if $s(x) = Cx$ for some constant C. What are the allowable values for C?

b) Could $s(x)$ equal x^n for any $n > 1$? Explain.

In Exercises 33–36, find the areas of the surfaces generated by revolving the curves about the given axes.

33. $y = \sqrt{2x + 1}$, $\quad 0 \le x \le 12$, $\quad x$-axis

34. $y = x^3/9$, $\quad -1 \le x \le 1$, $\quad x$-axis

35. $x = (1/3)y^{3/2} - y^{1/2}$, $\quad 4 \le y \le 9$, $\quad y$-axis

36. $y = x^2$, $\quad 2 \le y \le 6$, $\quad y$-axis

37. At points on the curve $y = 2\sqrt{x}$, line segments of length $h = y$ are drawn perpendicular to the xy-plane (Fig. 5.121). Find the area of the surface formed by these perpendiculars from $(0, 0)$ to $(3, 2\sqrt{3})$.

5.121   The surface in Exercise 37.

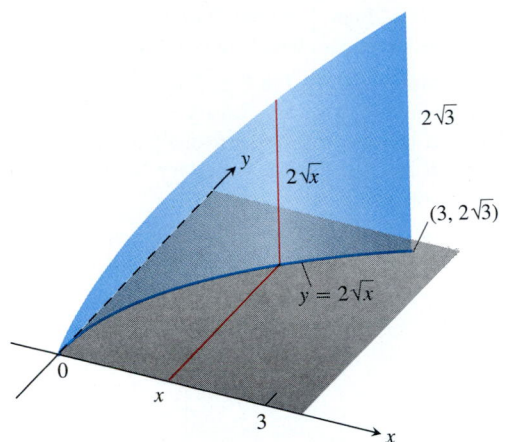

$2\sqrt{3}$

$2\sqrt{x}$

$(3, 2\sqrt{3})$

$y = 2\sqrt{x}$

0

3

38. At points on a circle of radius a, line segments are drawn perpendicular to the plane of the circle, the perpendicular at each point P being of length ks, where s is the length of the arc of the circle measured counterclockwise from $(a, 0)$ to P and k is a positive constant (Fig. 5.122). Find the area of the surface formed by the perpendiculars along the arc beginning at $(a, 0)$ and extending once around the circle.

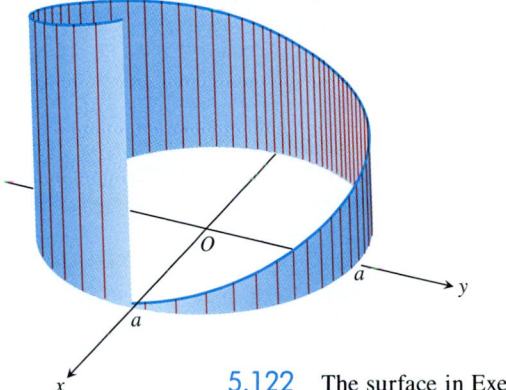

O

a

a

x

5.122   The surface in Exercise 38.

39. Find the center of mass of a thin, flat plate of constant density covering the region enclosed by the x-axis, the lines $x = 2$ and $x = -2$, and the parabola $y = x^2$.

40. Find the center of mass of a thin, flat plate of constant density covering the "triangular" region in the first quadrant bounded by the y-axis, the parabola $y = x^2/4$, and the line $y = 4$.

41. Find the center of mass of a thin, flat plate of density $\delta = 3$ covering the region enclosed by the parabola $y^2 = x$ and the line $x = 2y$.

42. a) Find the center of mass of a thin plate of constant density covering the region between the curve $y = 3/x^{3/2}$ and the x-axis from $x = 1$ to $x = 9$.
  b) Find the plate's center of mass if, instead of being constant, the density is $\delta(x) = x$.

43. Find the centroid of the region bounded below by the x-axis and above by the curve $y = 1 - x^n$, n an even positive integer. What is the limiting position of the centroid as $n \to \infty$?

44. CALCULATOR   When you haul a telephone pole on a two-wheeled carriage behind a truck, you want the wheels to be three feet or so behind the pole's center of mass to provide an adequate "tongue" weight. New England Telephone's class 1 40-ft wooden poles have a 27-in. circumference at the top and a 43.5-in. circumference at the base.
  a) About how far from the top is the center of mass?
  b) A typical class 1 pole weighs 1730 lb. What is its weight density?

45. A rock climber is about to haul up 10 kg of equipment that has been hanging beneath her on 40 m of rope that weighs 0.8 N/m. How much work will it take? (*Hint*: Solve for the rope and equipment separately; then add.)

46. You drove an 800-gal tank truck from the base to the summit of Mt. Washington and discovered on arrival that the tank was only half full. You started out with a full tank of water, climbed at a steady rate, and took 50 min to accomplish the 4750-ft elevation change. Assuming that the water leaked out at a steady rate, how much work was spent in carrying water to the top? Do not count the work done in getting yourself and the truck there. Water weighs 8 lb/U.S. gal.

47. If a force of 20 lb is required to hold a spring 1 ft beyond its unstressed length, how much work does it take to stretch the spring this far? How much work does it take to stretch the spring an additional foot?

48. A right circular conical tank, point down, with top radius 5 ft and height 10 ft is filled with a liquid whose weight-density is 60 lb/ft³. How much work does it take to pump the liquid to a point 2 ft above the tank? If the pump is driven by a motor rated at 275 ft · lb/sec (1/2-hp), how long will it take to empty the tank?

49. When a particle of mass m is at $(x, 0)$, it is attracted toward the origin with a force whose magnitude is k/x^2. If the particle starts from rest at $x = b$ and is acted on by no other forces, find the work done on it by the time it reaches $x = a$, $0 < a < b$.

50. Below the surface of the earth, the force of its gravitational attraction is directly proportional to the distance from its center. Find the work done in lifting an object, whose weight at the surface is w lb, from a distance r ft below the earth's surface to the surface.

51. A storage tank is a right circular cylinder 20 ft long and 8 ft in diameter with its axis horizontal. If the tank is half full of olive oil weighing 57 lb/ft³, find the work done in emptying it through a pipe that runs from the bottom of the tank to an outlet that is 6 ft above the top of the tank.

52. Suppose that the gas in a circular cylinder of cross-section area A is being compressed by a piston.
  a) If p is the pressure of the gas in pounds per square inch and V is the volume in cubic inches, show that the work

done in compressing the gas from state (p_1, V_1) to state (p_2, V_2) is given by the equation

$$\text{Work} = \int_{(p_1, V_1)}^{(p_2, V_2)} p \, dV.$$

(*Hint:* In the coordinates suggested in Fig. 5.123, $dV = A \, dx$. The force against the piston is pA.)

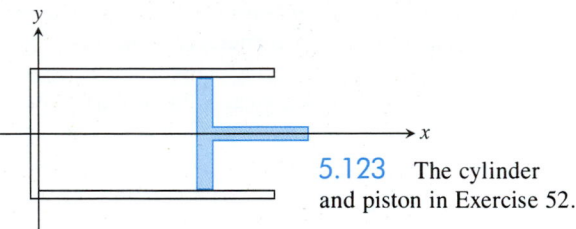

5.123   The cylinder and piston in Exercise 52.

b) Use the integral in (a) to find the work done in compressing the gas from $V_1 = 243$ in³ to $V_2 = 32$ in³ if $p_1 = 50$ lb/in³ and p and V obey the gas law $pV^{1.4} = $ constant for adiabatic processes.

53. A particle of mass m starts from rest at time $t = 0$ and is moved along the x-axis with constant acceleration a from $x = 0$ to $x = h$ against a variable force of magnitude $F(t) = t^2$. Find the work done.

54. Suppose a 1.6-oz golf ball is placed on a vertical spring with spring constant $k = 2$ lb/in. The spring is compressed 6 in. and released. About how high does the ball go (measured from the spring's rest position)?

55. The vertical triangular plate in Fig. 5.124 is the end plate of a trough full of water. What is the fluid force against the plate?

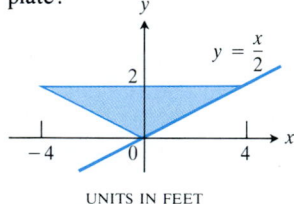

5.124   The plate in Exercise 55.

UNITS IN FEET

56. The vertical trapezoidal plate in Fig. 5.125 is the end plate of a trough of maple syrup weighing 75 lb/ft³. What is the force exerted by the syrup against the end plate of the trough when the syrup is 10 in. deep?

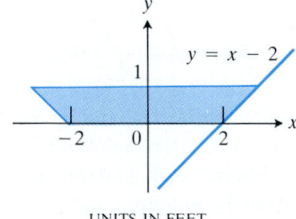

5.125   The plate in Exercise 56.

UNITS IN FEET

57. A flat vertical gate in the face of a dam is shaped like the parabolic region between the curve $y = 4x^2$ and the line $y = 4$, with measurements in feet. The top of the gate lies 5 ft below the surface of the water. Find the force exerted by the water against the gate.

58. CALCULATOR   A standard olive oil can measures 5.75 by 3.5 by 10 in. high. Find the fluid force against the base and each side of the can when the can is full.

59. The container profiled in Fig. 5.126 is filled with two non-mixing liquids of weight density w_1 and w_2. Find the fluid force on one side of the vertical square plate $ABCD$. The points B and D lie in the boundary layer and the square is $6\sqrt{2}$ ft on a side.

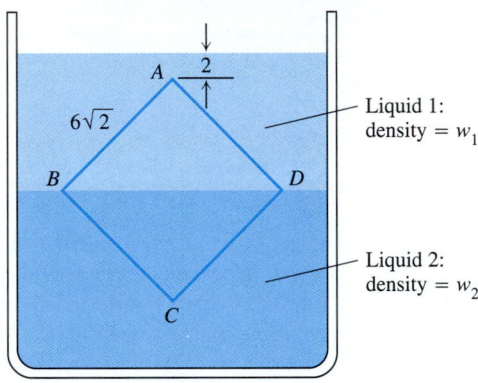

5.126   Profile of the container in Exercise 59.

60. The isosceles trapezoidal plate in Fig. 5.127 is submerged vertically in water with its upper edge 4 ft below the surface. Find the fluid force on one side of the plate in two different ways:
a) By evaluating an integral.
b) By dividing the plate into a parallelogram and an isosceles triangle, locating their centroids, and using the equation $F = w\bar{h}A$ from Section 5.8.

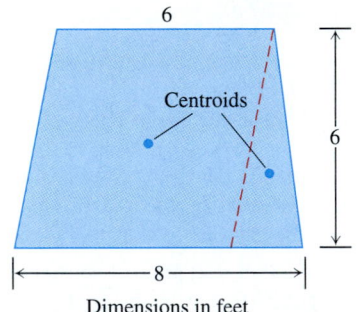

5.127   The trapezoidal plate in Exercise 60.

In Exercises 61–64, the function v is the velocity in meters per second of a body moving on a coordinate line. (a) Graph v as a function of t to see where it is positive and negative. Then find (b) the total distance the body travels during the time interval and (c) the body's displacement.

61. $v = 3t^2 - 15t + 18, \quad 0 \le t \le 3$

62. $v = t^3 - 3t^2 + 2t, \quad 0 \le t \le 2$

63. $v = 5 \cos t, \quad 0 \le t \le 3\pi/2$

64. $v = \pi \sin \pi t, \quad 0 \le t \le 3/2$

6

THE CALCULUS OF TRANSCENDENTAL FUNCTIONS

Overview Many of the functions we use in mathematics and science are inverses of one another. The functions $\ln x$ and e^x are probably the best-known function–inverse pair, but others are nearly as important. The trigonometric functions, when suitably restricted, have important inverses, and there are other useful pairs of logarithmic and exponential functions. Less widely known are the hyperbolic functions and their inverses, functions that arise when we study hanging cables, heat flow, and the friction encountered by objects falling through the air. We shall describe all of these functions in this chapter and look at the kinds of problems where they naturally occur.

6.1 Inverse Functions

We have reached the point in our development of calculus where we need the inverses of some of the functions we have been using. In this section, we review what it means for one function to be the inverse of another and look at the formulas, graphs, and derivatives of function–inverse pairs. We shall study the pair $\ln x$ and e^x in detail in the next two sections.

One-to-One Functions Have Inverses

As you know, a function is a rule that assigns a number from its range to each number in its domain. The numbers in the domain are called inputs and the numbers assigned to them, outputs. Functions like $y = \sin x$ and $y = x^2$ can assign the same output to different inputs. The sines of 0 and π are both 0 and the squares of -1 and 1 are both 1. Other functions, however, like $y = \sqrt{x}$ and $y = (x/4) + 3$, never assign an output number more than once. As you know, functions like these are called *one-to-one*.

Since each output of a one-to-one function comes from just one input, a one-to-one function can be reversed to send the outputs back to the inputs from which they came (Fig. 6.1).

The function defined by reversing a one-to-one function f is called the **inverse**

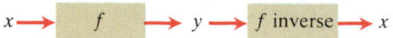

6.1 The inverse of a function f sends every output of f back to the input from which it came.

397

$$x \longrightarrow \boxed{f} \longrightarrow f(x) \longrightarrow \boxed{f^{-1}} \longrightarrow x = f^{-1}(f(x))$$

$$y \longrightarrow \boxed{f^{-1}} \longrightarrow f^{-1}(y) \longrightarrow \boxed{f} \longrightarrow y = f(f^{-1}(y))$$

6.2 If $y = f(x)$ is a one-to-one function, then $f^{-1}(f(x)) = x$ and $f(f^{-1}y)) = y$. Each of the composites $f^{-1} \circ f$ and $f \circ f^{-1}$ is the identity function on its own domain.

of f. The symbol for the inverse is f^{-1}, read "f inverse." The -1 in f^{-1} is *not* an exponent: $f^{-1}(x)$ does not mean $1/f(x)$.

The result of composing f and f^{-1} in either order is the **identity function,** the function that assigns each number to itself (Fig. 6.2). This gives us a way to test whether two functions f and g are inverses of one another. Compute $f \circ g$ and $g \circ f$. If both composites are identity functions, then f and g are inverses of one another; otherwise they are not. If f squares every number in its domain, g had better take square roots or it isn't the inverse of f.

Increasing functions have inverses and decreasing functions have inverses, as you will be asked to show in Exercise 20. In fact, these are the only continuous functions to have inverses, although we shall not prove this.

The Graphs of Inverses

How is the graph of the inverse of a function related to the graph of the function? If the function is increasing, say, its graph rises from left to right, like the graph in Fig. 6.3(a). To read the graph, we start at the point x on the x-axis, go up to the graph, and then move over to the y-axis to read the value of y. If we start with y and want to find the x from which it came, we reverse the process (Fig. 6.3b).

The graph of f is already the graph of f^{-1}, although the latter graph is not drawn in the usual way with the domain axis horizontal and the range axis vertical. The input–output pairs are reversed. To display the graph in the usual way, we have to reverse the pairs by reflecting the graph in the 45° line $y = x$ (Fig. 6.3c) and interchanging the letters x and y (Fig. 6.3d). This puts the independent variable,

6.3 (a) To find the value of f at x, we start at x and go up to the curve and over to the y-axis. (b) The graph of f is also the graph of f^{-1}. To find the x that gave y, we start at y and go over to the curve and down to the x-axis. Notice that the domain of f^{-1} is the range of f and the range of f^{-1} is the domain of f.

To draw the graph of f^{-1} in the usual way, (c) we reflect it in the line $y = x$ and then (d) interchange the letters x and y. We now have a normal-looking graph of f^{-1} as a function of x.

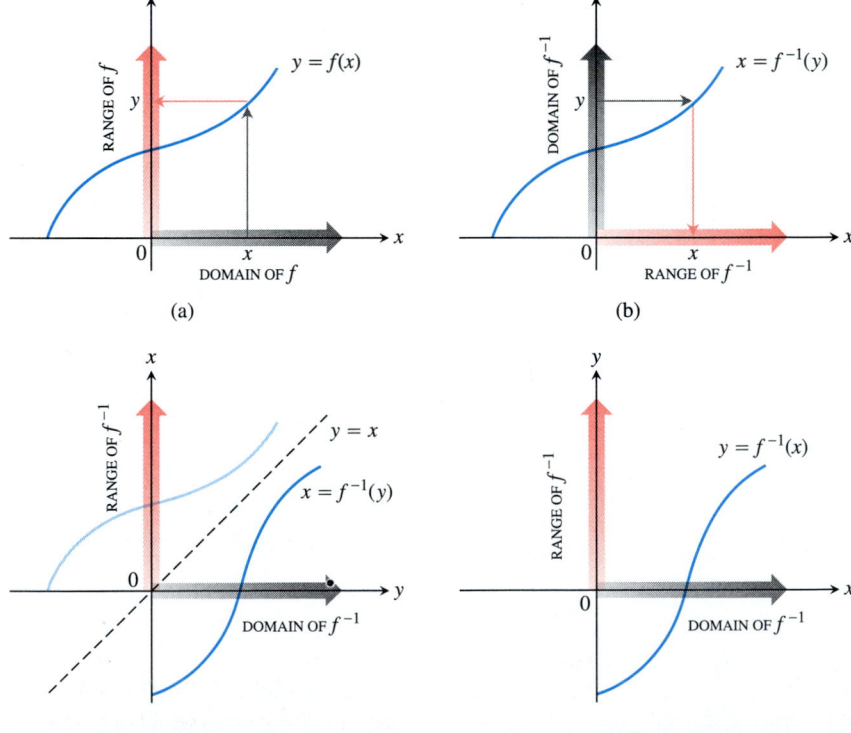

(a)   (b)

(c)   (d)

To Express f^{-1} as a Function of x:

1. Solve the equation $y = f(x)$ for x in terms of y.

2. Interchange x and y. The resulting formula will be $y = f^{-1}(x)$.

now called x, on the horizontal axis and the dependent variable, now called y, on the vertical axis.

The steps we take to draw a normal-looking graph of f^{-1} also tell us how to express f^{-1} as a function of x.

Example 1   Find the inverse of $y = \frac{1}{2}x + 1$, expressed as a function of x.

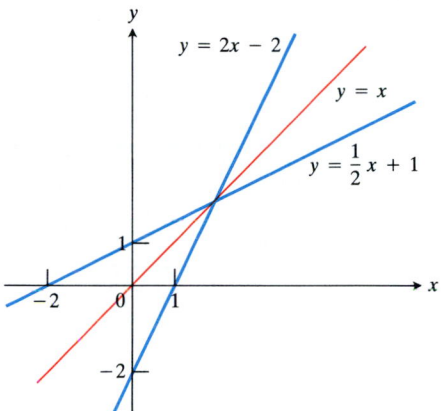

6.4   Graphing the functions $y = (1/2)x + 1$ and $y = 2x - 2$ together shows the graphs' symmetry with respect to the line $y = x$. The graphs of a function and its inverse always have this symmetry.

Solution

STEP 1: Solve for x in terms of y: $y = \frac{1}{2}x + 1$

$$2y = x + 2$$

$$x = 2y - 2$$

STEP 2: Interchange x and y: $y = 2x - 2$

The inverse of the function $f(x) = (1/2)x + 1$ is the function $f^{-1}(x) = 2x - 2$. To check, we verify that both composites give the identity function:

$$f^{-1}(f(x)) = 2\left(\frac{1}{2}x + 1\right) - 2 = x + 2 - 2 = x,$$

$$f(f^{-1}(x)) = \frac{1}{2}(2x - 2) + 1 = x - 1 + 1 = x.$$

See Fig. 6.4.

Example 2   Find the inverse of the function $y = x^2$, $x \geq 0$, expressed as a function of x.

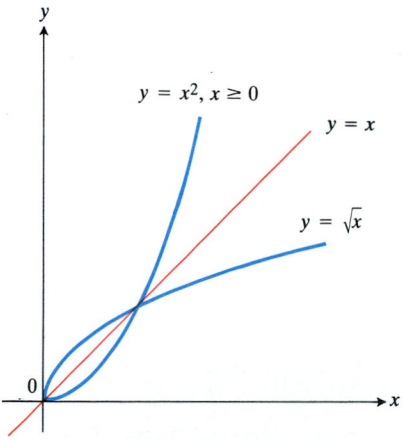

6.5   The functions $y = \sqrt{x}$ and $y = x^2$, $x \geq 0$, are inverses of one another. Again, notice the symmetry with respect to the line $y = x$. The graphs are mirror images of one another across the line.

Solution

STEP 1: Solve for x in terms of y: $y = x^2$

$$\sqrt{y} = \sqrt{x^2} = x \qquad (\sqrt{x^2} = x \text{ because } x \geq 0)$$

STEP 2: Interchange x and y: $y = \sqrt{x}$

The inverse of the function $y = x^2$, $x \geq 0$, is the function $y = \sqrt{x}$. See Fig. 6.5.

Notice that, unlike the restricted function $y = x^2$, $x \geq 0$, the unrestricted function $y = x^2$ is not one-to-one and therefore has no inverse.

Derivatives of Inverses of Differentiable Functions

The relation between the graphs of f and f^{-1} suggests an important relationship between their derivatives (Fig. 6.6).

If $f'(a) \neq 0$, then f^{-1} has a derivative at the point $f(a)$ and its value there is $1/f'(a)$. We have to impose some mathematical conditions on f to make sure this conclusion holds. The usual conditions, from advanced calculus, are stated in Theorem 1.

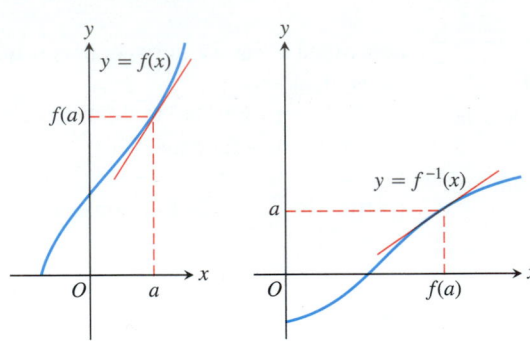

The slopes are reciprocal: $\left.\dfrac{df^{-1}}{dx}\right|_{f(a)} = \dfrac{1}{\left.\dfrac{df}{dx}\right|_a}$

6.6   The derivative of f^{-1} at the point $f(a)$ is the reciprocal of the derivative of f at a.

THEOREM 1

The Derivative Rule for Inverses

If f is differentiable at every point of an interval I, and df/dx is never zero on I, then f^{-1} is differentiable at every interior point of the interval $f(I)$. The value of df^{-1}/dx at any particular point $f(a)$ is the reciprocal of the value of df/dx at a.

$$\left.\frac{df^{-1}}{dx}\right|_{x=f(a)} = \frac{1}{\left(\left.\dfrac{df}{dx}\right|_{x=a}\right)} \tag{1}$$

Example 3   Verify Eq. (1) for $f(x) = \dfrac{1}{2}x + 1$ and its inverse $f^{-1}(x) = 2x - 2$.

Solution   We have

$$\frac{d}{dx}\left(f^{-1}(x)\right) = \frac{d}{dx}(2x - 2) = 2, \qquad \frac{d}{dx}(f(x)) = \frac{d}{dx}\left(\frac{1}{2}x + 1\right) = \frac{1}{2}.$$

Therefore, regardless of the values of a and $f(a)$, we have

$$\frac{df^{-1}}{dx} = \frac{1}{df/dx} = \frac{1}{1/2} = 2.$$

Example 4   Verify Eq. (1) for $f(x) = x^2$, $x \geq 0$, and its inverse $f^{-1}(x) = \sqrt{x}$ at the point $f(2) = 2^2 = 4$ in the domain of f^{-1} (Fig. 6.7).

Solution

$$\left.\frac{df^{-1}}{dx}\right|_{x=f(2)} = \left.\frac{d}{dx}\left(\sqrt{x}\right)\right|_{x=4} = \left.\left(\frac{1}{2\sqrt{x}}\right)\right|_{x=4} = \frac{1}{4},$$

$$\left.\frac{df}{dx}\right|_{x=2} = \left.\frac{d}{dx}(x^2)\right|_{x=2} = \left.(2x)\right|_{x=2} = 4,$$

$$\left.\frac{df^{-1}}{dx}\right|_{x=f(2)} = \frac{1}{\left.\dfrac{df}{dx}\right|_{x=2}}$$

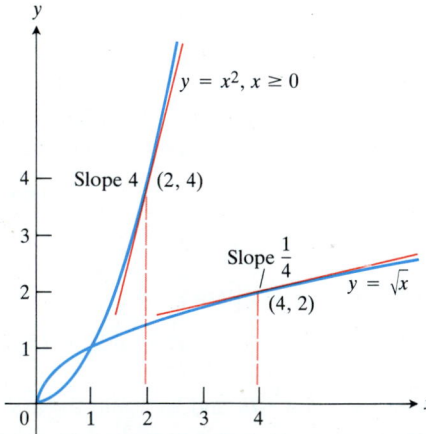

6.7   The graphs of inverse functions have reciprocal slopes at corresponding points.

Another Way to Look at Theorem 1

If $y = f(x)$ is differentiable at $x = a$ and we change x by a small amount dx, the corresponding change in y is approximately

$$dy = f'(a)\, dx.$$

This means that y changes about $f'(a)$ times as fast as x and that x changes about $1/f'(a)$ times as fast as y.

EXERCISES 6.1

In Exercises 1–4,
   a) Find the inverse f^{-1} of the function f, expressed as a function of x.
   b) Graph f and f^{-1} together.
   c) Verify Eq. (1) by evaluating df/dx at $x = a$ and df^{-1}/dx at $x = f(a)$.

1. $f(x) = 2x + 3$, $a = -1$
2. $f(x) = 5 - 4x$, $a = 1/2$
3. $f(x) = (1/5)x + 7$, $a = -1$
4. $f(x) = 2x^2$, $x \geq 0$, $a = 5$

In Exercises 5–8, find $f^{-1}(x)$.

5. $f(x) = x^2 + 1$, $x \geq 0$

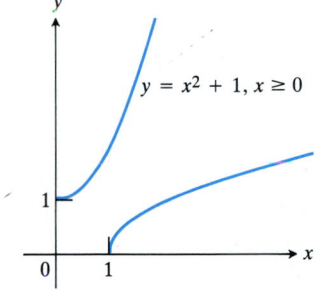

6. $f(x) = x^2$, $x \leq 0$

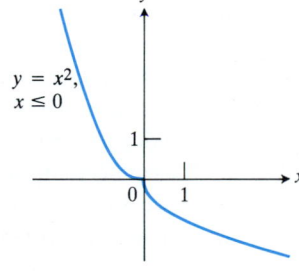

7. $f(x) = x^3 - 1$

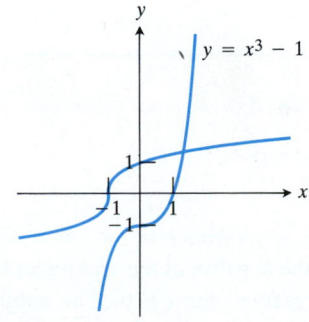

8. $f(x) = x^2 - 2x + 1$, $x \geq 1$

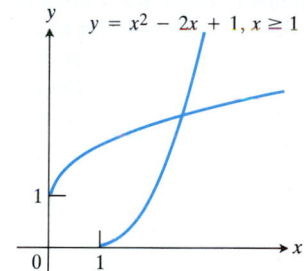

In Exercises 9–16, find $f^{-1}(x)$ and compose f and f^{-1} to show that $f(f^{-1}(x)) = f^{-1}(f(x)) = x$.

9. $f(x) = x^5$

10. $f(x) = x^4$, $x \geq 0$

11. $f(x) = x^3 + 1$

12. $f(x) = (1/2)x - 7/2$

13. $f(x) = 1/x^2$, $x > 0$

14. $f(x) = 1/x^3$, $x \neq 0$

15. $f(x) = (x + 1)^2$, $x \geq -1$

16. $f(x) = x^{2/3}$, $x \geq 0$

17. Graph $y = x^3$ and $y = x^{1/3}$ together over the interval $-2 \leq x \leq 2$ and sketch the tangents at $(1, 1)$ and $(-1, -1)$. What lines are tangent to the curves at $x = 0$?

18. Graph the curve $y = 1/x$, $x > 0$, and notice its symmetry about the line $y = x$. Find the inverse of the function $f(x) = 1/x$.

19. One of the virtues of Eq. (1) is that it enables us to find values of df^{-1}/dx even when we do not have an explicit formula for the derivative. As a case in point, let $f(x) = x^2 - 4x - 3$, $x > 2$, and find the value of df^{-1}/dx at the point $x = -3 = f(4)$.

20. *Increasing functions and decreasing functions.*   As we saw in Section 3.2, a function $f(x)$ increases on its domain if for any two points x_1 and x_2 in the domain,

$$x_2 > x_1 \quad \Rightarrow \quad f(x_2) > f(x_1).$$

Similarly, a function decreases on its domain if for any two

points x_1 and x_2 in the domain,

$$x_2 > x_1 \quad \Rightarrow \quad f(x_2) < f(x_1).$$

Show that increasing functions and decreasing functions are one-to-one. That is, show that $x_2 \ne x_1$ always implies $f(x_2) \ne f(x_1)$.

21. *Still another way to view Theorem 1.*   If we write $g(x)$ for $f^{-1}(x)$, Eq. (1) can be written as

$$g'(f(a)) = \frac{1}{f'(a)}, \qquad \text{or} \qquad g'(f(a)) \cdot f'(a) = 1.$$

If we then write x for a, we get

$$g'(f(x)) \cdot f'(x) = 1.$$

The latter equation may remind you of the Chain Rule, and indeed there is a connection.

   Assume that f and g are differentiable functions that are inverses of one another, so that $(g \circ f)(x) = x$. Differentiate both sides of this equation with respect to x, using the Chain Rule to express $(g \circ f)'(x)$ as a product of derivatives of g and f. What do you find? (This is not a proof of Theorem 1 because we assume here the theorem's conclusion that $g = f^{-1}$ is differentiable.)

22. *Equivalence of the washer and shell methods for finding volume.*   Let f be continuous on the interval $a \le x \le b$, with $a > 0$, and suppose that f has a continuous inverse, f^{-1}. Revolve about the y-axis the region bounded by the graph of

f and the lines $x = a$ and $y = f(b)$ to generate a solid. Then the values of the integrals given by the washer and shell methods for the volume have identical values:

$$\int_{f(a)}^{f(b)} \pi \left((f^{-1}(y))^2 - a^2 \right) dy = \int_a^b 2\pi x (f(b) - f(x)) \, dx.$$

To prove this equality, define

$$W(t) = \int_{f(a)}^{f(t)} \pi \left((f^{-1}(y))^2 - a^2 \right) dy,$$

$$S(t) = \int_a^t 2\pi x (f(t) - f(x)) \, dx.$$

Then show that the functions W and S agree at a point of $[a, b]$ and have identical derivatives on $[a, b]$. As you saw in Section 4.1, Exercise 60, this will guarantee $W(t) = S(t)$ for all t in $[a, b]$. In particular, $W(b) = S(b)$. (*Source:* "Disks and Shells Revisited," by Walter Carlip, *American Mathematical Monthly*, Vol. 98, No. 2, February 1991, pp. 154–156.)

EXPLORER PROGRAM

| Picard's Fixed Point | An equation solver that also enables you to toggle between the graphs of a function and of its inverse. |

6.2 Natural Logarithms

As we mentioned before, the most important function–inverse pair in mathematics and science is the pair consisting of the natural logarithm function $\ln x$ and the exponential function e^x. The key to understanding e^x is $\ln x$, so we introduce $\ln x$ first. The importance of logarithms came at first from the improvement they brought to arithmetic. The revolutionary properties of logarithms made possible the calculations of the great seventeenth-century advances in offshore navigation and celestial mechanics. Nowadays we do complicated arithmetic with calculators, but the properties of logarithms remain as important as ever.

The Natural Logarithm Function

The natural logarithm of a positive number x, written as $\ln x$, is the value of an integral.

DEFINITION

The Natural Logarithm Function

$$\ln x = \int_1^x \frac{1}{t} \, dt, \qquad x > 0 \tag{1}$$

   If $x > 1$, then $\ln x$ is the area under the curve $y = 1/t$ from $t = 1$ to $t = x$ (Fig. 6.8). If x is less than 1 (but still positive), $\ln x$ gives the negative of the area under the curve from $t = x$ to $t = 1$. The function is not defined for $x \le 0$. The natural

The Development of Logarithms

In the late 1500s, a Scottish baron, John Napier, invented a device called the **logarithm** that simplified arithmetic by replacing multiplication by addition. The equation that accomplished this was

$$\ln ax = \ln a + \ln x.$$

To multiply two positive numbers a and x, you looked up their logarithms in a table, added the logarithms, found the sum in the body of the table, and read the table backward to find the product ax.

Having the table was the key, of course, and Napier spent the last 20 years of his life working on a table he never finished (while the astronomer Tycho Brahe waited in vain for the information he needed to speed his calculations). The table was completed after Napier's death (and Brahe's) by Napier's friend Henry Briggs in London. Logarithms subsequently became widely known as Briggs's logarithms and some older books on navigation still refer to them by this name.

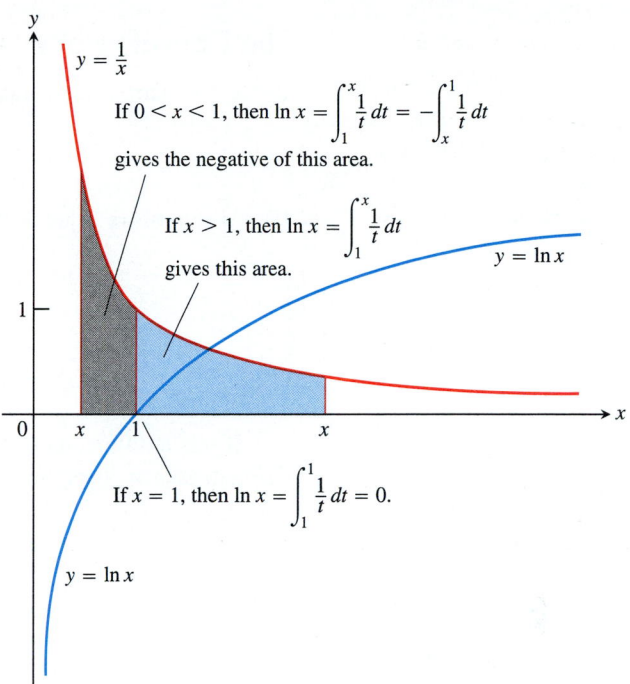

If $0 < x < 1$, then $\ln x = \int_1^x \frac{1}{t}\, dt = -\int_x^1 \frac{1}{t}\, dt$ gives the negative of this area.

If $x > 1$, then $\ln x = \int_1^x \frac{1}{t}\, dt$ gives this area.

If $x = 1$, then $\ln x = \int_1^1 \frac{1}{t}\, dt = 0.$

6.8   The graph of $y = \ln x$ and its relation to the function $y = 1/x$, $x > 0$. The graph of the logarithm rises above the x-axis as x moves from 1 to the right, and it falls below the axis as x moves from 1 to the left.

logarithm of 1 itself is zero because

$$\ln 1 = \int_1^1 \frac{1}{t}\, dt = 0. \qquad \text{(Upper and lower limits equal)} \qquad (2)$$

Notice that we show the graph of $y = 1/x$ in Fig. 6.8 but use the notation $y = 1/t$ in the integral. Using x for everything would have us writing

$$\ln x = \int_1^x \frac{1}{x}\, dx. \qquad (3)$$

This has too many x's to make sense, so we changed the variable of integration.

Example 1   You can find values of $\ln x$ to eight or ten digits on most scientific calculators by entering a value for x and pressing $\boxed{\ln x}$. Typical values, rounded except for $\ln 1$, are shown in Table 6.1. The natural logarithm of x grows so slowly as $x \to \infty$ that we use $\ln x$ as a standard for measuring the growth of other functions. (We discuss this in more detail in Section 6.6.)

TABLE 6.1
Selected values of ln x (rounded except for ln 1)

x	0.5	1	2	10	1000	10,000	100,000
$\ln x$	-0.693	0	0.693	2.303	6.908	9.210	11.513

The Derivative of $y = \ln x$

By the first part of the Fundamental Theorem of Calculus (in Section 4.5),

$$\frac{d}{dx} \ln x = \frac{d}{dx} \int_1^x \frac{1}{t} \, dt = \frac{1}{x}.$$

For every positive value of x, therefore,

$$\frac{d}{dx} \ln x = \frac{1}{x}. \tag{4}$$

If u is a differentiable function of x whose values are positive, so that $\ln u$ is defined, then applying the Chain Rule

$$\frac{dy}{dx} = \frac{dy}{du} \frac{du}{dx}$$

to the function $y = \ln u$ gives

$$\frac{d}{dx} \ln u = \frac{d}{du} \ln u \cdot \frac{du}{dx} = \frac{1}{u} \frac{du}{dx}. \tag{5}$$

$$\frac{d}{dx} \ln u = \frac{1}{u} \frac{du}{dx}, \qquad u > 0 \tag{6}$$

Example 2
$$\frac{d}{dx} \ln 2x = \frac{1}{2x} \frac{d}{dx} (2x) = \frac{1}{2x} (2) = \frac{1}{x}$$

Notice the remarkable occurrence in Example 2. The function $y = \ln 2x$ has the same derivative as the function $y = \ln x$. This is true of $y = \ln ax$ for any constant a:

$$\frac{d}{dx} \ln ax = \frac{1}{ax} \cdot \frac{d}{dx} (ax) = \frac{1}{ax} (a) = \frac{1}{x}. \tag{7}$$

Example 3   Equation (6) with $u = x^2 + 3$ gives

$$\frac{d}{dx} \ln(x^2 + 3) = \frac{1}{x^2 + 3} \cdot \frac{d}{dx} (x^2 + 3) = \frac{1}{x^2 + 3} \cdot 2x = \frac{2x}{x^2 + 3}.$$

The Rules of Arithmetic for Logarithms

The rules that made logarithms the single most important improvement in arithmetic before the advent of modern computers are listed in Table 6.2. As you can see, the rules made it possible to replace multiplication of positive numbers by addition, and division of positive numbers by subtraction. They also made it possible to replace exponentiation by multiplication. For the moment, we add the restriction that the exponent n in Rule 4 be a rational number. You will see why when we prove the rule.

TABLE 6.2
Rules of arithmetic for logarithms

For any positive numbers a and x and for any exponent n,

1. $\ln ax = \ln a + \ln x$

2. $\ln \dfrac{a}{x} = \ln a - \ln x$

3. $\ln \dfrac{1}{x} = -\ln x$ (Rule 2 with $a = 1$)

4. $\ln x^n = n \ln x$ (Exponent Rule)

Example 4

$\ln 6 = \ln(2 \cdot 3) = \ln 2 + \ln 3$ (Rule 1)

$\ln 4 - \ln 5 = \ln \dfrac{4}{5} = \ln 0.8$ (Rule 2)

$\ln \dfrac{1}{8} = -\ln 8$ (Rule 3)

      $= -\ln 2^3 = -3 \ln 2$ (Rule 4)

One useful consequence of Rule 4 is that $\ln \sqrt[m]{x} = \ln x^{1/m} = (1/m)\ln x$.

$$\ln \sqrt[m]{x} = \frac{1}{m} \ln x \tag{8}$$

Example 5

$$\ln \sqrt{\cos x} = \frac{1}{2} \ln \cos x, \qquad \ln \sqrt[3]{x+1} = \frac{1}{3} \ln(x+1)$$

The derivations of the rules in Table 6.2 go like this:

Proof that ln ax = ln a + ln x   The argument is unusual—and elegant. It starts by observing that $\ln ax$ and $\ln x$ have the same derivative (Eq. 7). According to Corollary 3 of the Mean Value Theorem, then, the functions must differ by a constant, which means that

$$\ln ax = \ln x + C \tag{9}$$

for some C. With this much accomplished, it remains only to show that C equals $\ln a$.

Equation (9) holds for all positive values of x, so it must hold for $x = 1$. Hence,

$\ln(a \cdot 1) = \ln 1 + C$

   $\ln a = 0 + C$ ($\ln 1 = 0$)

     $C = \ln a$.   (Rearranged)

Substituting $C = \ln a$ in Eq. (9) gives the equation we wanted to prove:

$$\ln ax = \ln a + \ln x. \tag{10}$$

Proof that ln (a/x) = ln a − ln x   We get this from Eq. (10) in two stages. Equation (10) with a replaced by $1/x$ gives

$$\ln \frac{1}{x} + \ln x = \ln\left(\frac{1}{x} \cdot x\right)$$
$$= \ln 1 = 0,$$

so that

$$\ln \frac{1}{x} = -\ln x. \tag{11}$$

Equation (10) with x replaced by $1/x$ then gives

$$\ln \frac{a}{x} = \ln\left(a \cdot \frac{1}{x}\right) = \ln a + \ln \frac{1}{x}$$
$$= \ln a - \ln x.$$

Proof that ln xⁿ = n ln x (assuming n rational)   We use the same-derivative argument again. For all positive values of x,

$$\frac{d}{dx} \ln x^n = \frac{1}{x^n} \frac{d}{dx} (x^n) \qquad \text{(Eq. (6) with } u = x^n)$$

$$= \frac{1}{x^n} n\, x^{n-1} \qquad \left(\begin{array}{l}\text{Here is where we need } n \text{ to be rational, at}\\ \text{least for now. We have proved the Power}\\ \text{Rule only for rational exponents.}\end{array}\right)$$

$$= n \cdot \frac{1}{x} = \frac{d}{dx} (n \ln x).$$

Since $\ln x^n$ and $n \ln x$ have the same derivative,

$$\ln x^n = n \ln x + C$$

for some constant C. Taking x to be 1 identifies C as zero, and we're done.

As for using the rule $\ln x^n = n \ln x$ for irrational values of n, you may go right ahead and do so. It does hold for all n, and there is no need to pretend otherwise. From the point of view of mathematical development, however, we want you to be aware that the rule is far from proved. The situation is much worse than not knowing how to take the derivative of x^n when n is irrational. We do not even know how to raise a number like 2 to an irrational power like $\sqrt{3}$. Even your calculator cannot do so. Calculators work only with finite decimals and only with a finite assortment of them at that.

There is hope, however, for besides 0 and 1 there is one other number we *can* raise to irrational powers in our current stage of mathematical development, and that is the number e. The function e^x is defined for all values of x, irrational as well as rational. This, as we shall see in the next section, saves the day.

The Graph of ln x; Domain and Range

The derivative $d(\ln x)/dx = 1/x$ is positive and continuous at every point in the function's domain, so $\ln x$ is an increasing function of x. The graph is connected and rises steadily from left to right with a continuously turning tangent. The second derivative

$$\frac{d^2(\ln x)}{dx^2} = \frac{d}{dx}\left(\frac{1}{x}\right) = -\frac{1}{x^2}$$

is always negative, so the graph is concave down throughout.

We know that $\ln 1$ is zero and we can estimate $\ln 2$ numerically to show that

$$\ln 2 = \int_1^2 \frac{1}{t}\, dt \approx 0.693.$$

The arithmetic rules in Table 6.2 then give

$$\ln 4 = \ln 2^2 = 2 \ln 2 \approx 1.38, \qquad\qquad \ln 8 = \ln 2^3 = 3 \ln 2 \approx 2.07,$$

$$\ln \frac{1}{2} = \ln 2^{-1} = -\ln 2 \approx -0.69, \qquad \ln \frac{1}{4} = \ln 4^{-1} = -\ln 4 \approx -1.38,$$

and so on. We plot the points and connect them with a smooth curve with the guidance of tangent lines. The results are shown in Fig. 6.9.

How high and how low does the graph go? Since $\ln 2 > 1/2$,

$$\ln 2^n = n \ln 2 > n\left(\frac{1}{2}\right) = \frac{n}{2}, \qquad \ln 2^{-n} = -n \ln 2 < -n\left(\frac{1}{2}\right) = -\frac{n}{2}.$$

The logarithm of x tends to ∞ as x tends to ∞ and tends to $-\infty$ as x approaches zero from above. In short,

$$\lim_{x \to \infty} \ln x = \infty \qquad \text{and} \qquad \lim_{x \to 0^+} \ln x = -\infty. \tag{12}$$

The domain of $\ln x$ is the set of positive real numbers. The range of $\ln x$ is the entire real line.

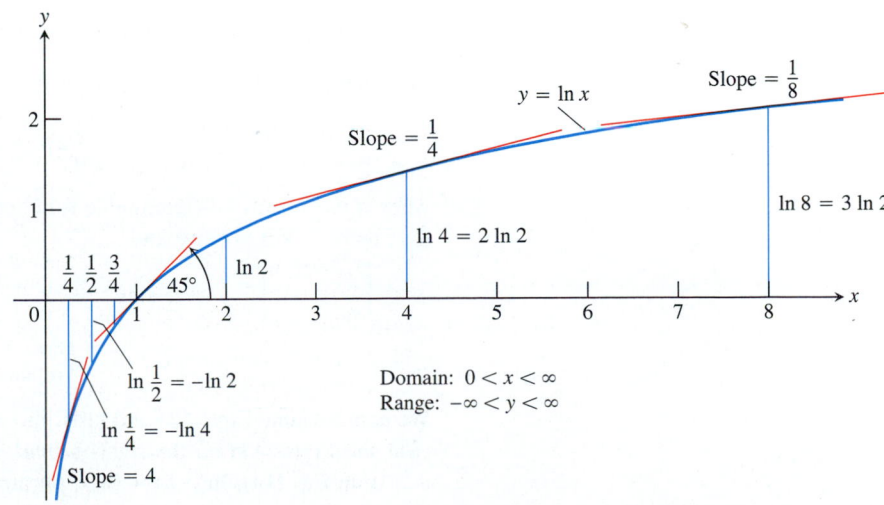

6.9   The graph of $y = \ln x$.

Logarithmic Differentiation

The derivatives of functions given by formulas that involve products, quotients, and powers can often be found more quickly if we take the natural logarithm of both sides before differentiating. This enables us to use the rules in Table 6.2 to simplify the formulas before differentiating. The process, called **logarithmic differentiation**, is illustrated in the next example.

Example 6   Find dy/dx if

$$y = \frac{(x^2 + 1)(x + 3)^{1/2}}{x - 1}, \qquad x > 1.$$

Logarithmic differentiation of $y = f(x)$ has four steps:

1. Take logs of both sides and simplify.
2. Differentiate implicitly.
3. Solve for dy/dx.
4. Substitute for y.

Solution   We take the natural logarithm of both sides and simplify the result with the rules in Table 6.2:

$$\ln y = \ln \frac{(x^2 + 1)(x + 3)^{1/2}}{x - 1}$$

$$= \ln(x^2 + 1)(x + 3)^{1/2} - \ln(x - 1) \qquad \text{(Rule 2)}$$

$$= \ln(x^2 + 1) + \ln(x + 3)^{1/2} - \ln(x - 1) \qquad \text{(Rule 1)}$$

$$= \ln(x^2 + 1) + \frac{1}{2}\ln(x + 3) - \ln(x - 1). \qquad \text{(Rule 4)}$$

We then take derivatives of both sides with respect to x, using implicit differentiation on the left:

$$\frac{1}{y}\frac{dy}{dx} = \frac{1}{x^2 + 1} \cdot 2x + \frac{1}{2} \cdot \frac{1}{x + 3} - \frac{1}{x - 1}.$$

Next we solve for dy/dx:

$$\frac{dy}{dx} = y\left(\frac{2x}{x^2 + 1} + \frac{1}{2x + 6} - \frac{1}{x - 1}\right).$$

Finally, we substitute for y:

$$\frac{dy}{dx} = \frac{(x^2 + 1)(x + 3)^{1/2}}{x - 1}\left(\frac{2x}{x^2 + 1} + \frac{1}{2x + 6} - \frac{1}{x - 1}\right).$$

The Integral $\int (1/u)\, du$

Equation (6) leads to the integral formula

$$\int \frac{1}{u}\, du = \ln u + C \qquad (13)$$

when u is a positive differentiable function, but what if u is negative? If u is negative, then $-u$ is positive and

$$\int \frac{1}{u}\, du = \int \frac{1}{(-u)}\, d(-u)$$

$$= \ln(-u) + C. \qquad \left(\begin{array}{l}\text{Eq. (13) with } u \\ \text{replaced by } -u\end{array}\right) \qquad (14)$$

We can combine Eqs. (13) and (14) into a single formula by noticing that in each case the expression on the right is $\ln|u| + C$. In Eq. (13), $\ln u = \ln|u|$ because $u > 0$; in Eq. (14), $\ln(-u) = \ln|u|$ because $u < 0$. Whether u is positive or negative, the integral of $(1/u)\, du$ is $\ln|u| + C$.

If u is a nonzero differentiable function,

$$\int \frac{1}{u}\, du = \ln|u| + C. \tag{15}$$

We know that

$$\int u^n\, du = \frac{u^{n+1}}{n+1} + C, \qquad n \neq -1.$$

Equation (15) explains what to do when n equals -1. With this gap filled, we can now integrate all rational powers of all nonzero differentiable functions.

In applications it is important to remember that the function u in Eq. (15) can be any nonzero differentiable function $f(x)$. The equation says that integrals of a certain *form* lead to logarithms. That is,

$$\int \frac{f'(x)}{f(x)}\, dx = \ln|f(x)| + C \tag{16}$$

whenever $f(x)$ is a differentiable function that maintains a constant sign on the domain given for it.

Example 7

$$\int_0^2 \frac{2x}{x^2 - 5}\, dx = \int_{-5}^{-1} \frac{du}{u} = \ln|u|\Big]_{-5}^{-1} \qquad \left(\begin{matrix} u = x^2 - 5, & du = 2x\, dx, \\ u(0) = -5, & u(2) = -1 \end{matrix} \right)$$

$$= \ln|-1| - \ln|-5| = \ln 1 - \ln 5 = -\ln 5$$

Example 8

$$\int_{-\pi/2}^{\pi/2} \frac{4 \cos \theta}{3 + 2 \sin \theta}\, d\theta = \int_1^5 \frac{2}{u}\, du \qquad \left(\begin{matrix} u = 3 + 2 \sin \theta, & du = 2 \cos \theta\, d\theta, \\ u(-\pi/2) = 1, & u(\pi/2) = 5 \end{matrix} \right)$$

$$= 2 \ln|u|\Big]_1^5$$

$$= 2 \ln|5| - 2 \ln|1| = 2 \ln 5$$

The Integrals of tan x and cot x

Equation (15) tells us at last how to integrate the tangent and cotangent functions. For the tangent,

$$\int \tan x\, dx = \int \frac{\sin x}{\cos x}\, dx = \int \frac{-du}{u} \qquad \left(\begin{matrix} u = \cos x, \\ du = -\sin x\, dx \end{matrix} \right)$$

$$= -\int \frac{du}{u} = -\ln|u| + C \qquad \text{(Eq. (15))}$$

$$= -\ln|\cos x| + C = \ln \frac{1}{|\cos x|} + C \qquad \text{(Table 6.2, Rule 3)}$$

$$= \ln|\sec x| + C.$$

For the cotangent,

$$\int \cot x \, dx = \int \frac{\cos x \, dx}{\sin x} = \int \frac{du}{u} \qquad \begin{pmatrix} u = \sin x, \\ du = \cos x \, dx \end{pmatrix}$$

$$= \ln|u| + C = \ln|\sin x| + C.$$

The general formulas are

$$\int \tan u \, du = -\ln|\cos u| + C = \ln|\sec u| + C \qquad (17)$$

$$\int \cot u \, du = \ln|\sin u| + C = -\ln|\csc x| + C \qquad (18)$$

Example 9

$$\int_0^{\pi/6} \tan 2x \, dx = \int_0^{\pi/3} \tan u \cdot \frac{du}{2} = \frac{1}{2} \int_0^{\pi/3} \tan u \, du \qquad \begin{pmatrix} \text{Substitute } u = 2x, \\ dx = du/2, \, u(0) = 0, \\ u(\pi/6) = \pi/3 \end{pmatrix}$$

$$= \frac{1}{2} \ln|\sec u| \Big]_0^{\pi/3} = \frac{1}{2}(\ln 2 - \ln 1) = \frac{1}{2}\ln 2 \qquad (\text{Eq. (17)})$$

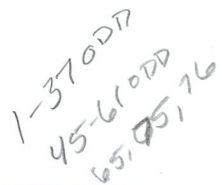

EXERCISES 6.2

1. Express the following logarithms in terms of $\ln 2$ and $\ln 3$.
 a) $\ln 0.75$ b) $\ln(4/9)$ c) $\ln(1/2)$
 d) $\ln\sqrt[3]{9}$ e) $\ln 3\sqrt{2}$ f) $\ln\sqrt{13.5}$

2. Express the following logarithms in terms of $\ln 5$ and $\ln 7$.
 a) $\ln(1/125)$ b) $\ln 9.8$ c) $\ln 7\sqrt{7}$
 d) $\ln 1225$ e) $\ln 0.056$ f) $(\ln 35 + \ln(1/7))/(\ln 25)$

Find the derivatives with respect to x of the functions whose formulas appear in Exercises 3–26.

3. $\ln 3x$

4. $\ln kx$, k constant

5. $\ln x^3$

6. $(\ln x)^3$

7. $\ln x^{3/2}$

8. $\ln(10/x)$

9. $\ln(x + 2)$

10. $\ln(x^2 + 2x)$

11. $\ln(2 - \cos x)$

12. $x \ln x - x$

13. $\ln(\ln x)$

14. $\ln(\ln(\ln x))$

15. $x(\sin(\ln x) + \cos(\ln x))$

16. $\ln(\sec x + \tan x)$

17. $\ln \dfrac{1}{x\sqrt{x + 1}}$

18. $\dfrac{1}{2} \ln \dfrac{1 + x}{1 - x}$

19. $\dfrac{1 + \ln x}{1 - \ln x}$

20. $\sqrt{\ln \sqrt{x}}$

21. $\ln(\sec(\ln x))$

22. $\ln\left(\dfrac{\sqrt{\sin x \cos x}}{1 + 2 \ln x}\right)$

23. $\ln\left(\dfrac{(x^2 + 1)^5}{\sqrt{1 - x}}\right)$

24. $\ln \sqrt{\dfrac{(x + 1)^5}{(x + 2)^{20}}}$

25. $\displaystyle\int_{x^2/2}^{x^2} \ln\sqrt{t} \, dt$

26. $\displaystyle\int_{\sqrt{x}}^{\sqrt[3]{x}} \ln t \, dt$

Use logarithmic differentiation to find dy/dx in Exercises 27–38.

27. $y = \sqrt{x(x + 1)}$

28. $y = \sqrt{\dfrac{x}{x + 1}}$

29. $y = \sqrt{x + 3} \sin x$

30. $y = \dfrac{\tan x}{\sqrt{2x + 1}}$

31. $y = x(x + 1)(x + 2)$

32. $y = \dfrac{1}{x(x + 1)(x + 2)}$

33. $y = \dfrac{x + 5}{x \cos x}$

34. $y = \dfrac{x \sin x}{\sqrt{\sec x}}$

35. $y = \dfrac{x\sqrt{x^2 + 1}}{(x + 1)^{2/3}}$

36. $y = \sqrt{\dfrac{(x + 1)^{10}}{(2x + 1)^5}}$

37. $y = \sqrt[3]{\dfrac{x(x - 2)}{x^2 + 1}}$

38. $y = \sqrt[3]{\dfrac{x(x + 1)(x - 2)}{(x^2 + 1)(2x + 3)}}$

Find the limits in Exercises 39–44.

39. $\displaystyle\lim_{x \to \infty} \ln \dfrac{1}{x}$

40. $\displaystyle\lim_{x \to \infty} \dfrac{\ln x}{x}$

41. $\displaystyle\lim_{x\to\infty}\int_x^{2x}\frac{1}{t}\,dt$

42. $\displaystyle\lim_{x\to\infty}\frac{1}{x\ln x}\int_1^x \ln t\,dt$

43. $\displaystyle\lim_{\theta\to 0^+}\frac{\ln(\sin\theta)}{\ln(\cot\theta)}$

44. $\displaystyle\lim_{t\to 0}\frac{\ln(1+2t)-2t}{t^2}$

Evaluate the integrals in Exercises 45–62.

45. $\displaystyle\int_{-3}^{-2}\frac{dx}{x}$

46. $\displaystyle\int_{-1}^{0}\frac{3\,dx}{3x-2}$

47. $\displaystyle\int\frac{2y\,dy}{y^2-25}$

48. $\displaystyle\int\frac{8r\,dr}{4r^2-5}$

49. $\displaystyle\int_0^{\pi}\frac{\sin t}{2-\cos t}\,dt$

50. $\displaystyle\int_0^{\pi/3}\frac{4\sin\theta}{1-4\cos\theta}\,d\theta$

51. $\displaystyle\int_1^{2}\frac{2\ln x}{x}\,dx$

52. $\displaystyle\int_2^{4}\frac{dx}{x\ln x}$

53. $\displaystyle\int_2^{4}\frac{dx}{x(\ln x)^2}$

54. $\displaystyle\int_2^{16}\frac{dx}{2x\sqrt{\ln x}}$

55. $\displaystyle\int\frac{3\sec^2 t}{6+3\tan t}\,dt$

56. $\displaystyle\int\frac{\sec y\tan y}{2+\sec y}\,dy$

57. $\displaystyle\int_0^{\pi/2}\tan\frac{x}{2}\,dx$

58. $\displaystyle\int_{\pi/4}^{\pi/2}\cot t\,dt$

59. $\displaystyle\int_{\pi/2}^{\pi}2\cot\frac{\theta}{3}\,d\theta$

60. $\displaystyle\int_0^{\pi/12}6\tan 3x\,dx$

61. $\displaystyle\int\frac{dx}{2\sqrt{x}+2x}$

62. $\displaystyle\int\frac{\sec x\,dx}{\sqrt{\ln(\sec x+\tan x)}}$

63. *The linearization and quadratic approximation of* $\ln(1+x)$ *at* $x=0$. Instead of approximating $\ln x$ near $x=1$, we approximate $\ln(1+x)$ near $x=0$. We get simpler formulas this way:

Linearization at $x=0$: $\ln(1+x)\approx x$

Quadratic approximation at $x=0$: $\ln(1+x)\approx x-\dfrac{x^2}{2}$.

a) Derive these formulas.

b) CALCULATOR Estimate to five decimal places the errors involved in replacing $\ln(1+x)$ by its linearization and quadratic approximation on the interval $[0, 0.1]$.

64. Locate and identify the extreme values of
a) $\ln(\cos x)$ on $[-\pi/4, \pi/3]$,
b) $\cos(\ln x)$ on $[1/2, 2]$.

65. Find the area between the curves $y=\ln x$ and $y=\ln 2x$ from $x=1$ to $x=5$.

66. Find the area between the curve $y=\tan x$ and the x-axis from $x=-\pi/4$ to $x=\pi/3$.

67. The region in the first quadrant bounded by the coordinate axes, the line $y=3$, and the curve $x=2/\sqrt{y+1}$ is

revolved about the y-axis to generate a solid. Find the volume of the solid.

68. The region between the curve $y=\sqrt{\cot x}$ and the x-axis from $x=\pi/6$ to $x=\pi/2$ is revolved about the x-axis to generate a solid. Find the volume of the solid.

69. The region between the curve $y=1/x^2$ and the x-axis from $x=1/2$ to $x=2$ is revolved about the y-axis to generate a solid. Find the volume of the solid.

70. In Section 5.3, Exercise 14, we revolved about the y-axis the region between the curve $y=9x/\sqrt{x^3+9}$ and the x-axis from $x=0$ to $x=2$ to generate a solid of volume 36π. What volume do you get if you revolve the region about the x-axis instead? (See Section 5.3, Exercise 14, for a graph.)

71. Find the lengths of the following curves.
a) $y=(x^2/8)-\ln x$, $4\le x\le 8$
b) $y=\sin x-(1/4)\ln|\sec x+\tan x|$, $0\le x\le\pi/3$

72. Find a curve through the point $(1, 0)$ whose length from $x=1$ to $x=2$ is

$$L=\int_1^2\sqrt{1+\frac{1}{x^2}}\,dx.$$

73. CALCULATOR
a) Find the centroid of the region between the curve $y=1/x$ and the x-axis from $x=1$ to $x=2$. Give the coordinates to two decimal places.
b) Sketch the region and show the centroid in your sketch.

74. a) Find the center of mass of a thin plate of constant density covering the region between the curve $y=1/\sqrt{x}$ and the x-axis from $x=1$ to $x=16$.
b) Find the center of mass if, instead of being constant, the density function is $\delta(x)=4/\sqrt{x}$.

Solve the initial value problems in Exercises 75 and 76.

75. $\dfrac{dy}{dx}=1+\dfrac{1}{x}$, $y=3$ when $x=1$

76. $\dfrac{d^2y}{dx^2}=\sec^2 x$, $y=0$ and $\dfrac{dy}{dx}=1$ when $x=0$

77. A body moves along a coordinate line with acceleration $d^2s/dt^2=4/(4-t)^2$. When $t=0$, the body's velocity is 2 m/sec. Find the total distance traveled by the body from time $t=1$ to $t=2$ sec.

78. *Estimating values of* $\ln x$ *with Simpson's rule.* Although linearizations are good for replacing the logarithmic function over short intervals, when it comes to estimating *particular* values of $\ln x$, Simpson's rule is better.

As a case in point, the values of $\ln(1.2)$ and $\ln(0.8)$ to five places are

$$\ln(1.2)=0.18232,\quad \ln(0.8)=-0.22314.$$

Estimate $\ln(1.2)$ and $\ln(0.8)$ first with the formula $\ln(1+x)\approx x$ and then use Simpson's rule with $n=2$. (Impressive, isn't it?)

Graphing Calculator or Computer Grapher

79. *The errors in the approximations* $\ln(1 + x) \approx x$ *and* $\ln(1 + x) \approx x - x^2/2$. Graph $\ln(1 + x)$, x, and $x - (x^2/2)$ together for $0 \le x \le 0.5$. Use different colors, if available. At what points do the approximations of $\ln(1 + x)$ seem best? Least good? Find as good an upper bound for the errors as your calculator will allow.

80. Graph $\ln x$, $\ln 2x$, $\ln 4x$, $\ln 8x$, and $\ln 16x$ (as many as you can) together for $0 \le x \le 10$. What is going on? Explain.

81. Graph $y = \ln|\sin x|$ for $0 \le x \le 22$. How would you change the formula to turn the arches upside down?

82. a) Graph $y = \sin x$ and the curves $y = \ln(a + \sin x)$ for

$a = 2, 4, 8, 20,$ and 50 together for $0 \le x \le 23$.

b) Why do the curves flatten as a increases? (*Hint:* Find an a-dependent upper bound for $|y'|$.)

83. Does the graph of $y = \sqrt{x} - \ln x$, $x > 0$, have an inflection point? Try to answer the question (a) by graphing, (b) by using calculus.

EXPLORER PROGRAM

PowerGrapher   Graphs all logarithmic functions

6.3 The Exponential Function

Whenever we have a quantity y whose rate of change over time is proportional to the amount of y present, we have a function that satisfies the differential equation

$$\frac{dy}{dt} = ky.$$

If, in addition, $y = y_0$ when $t = 0$, the function is none other than the exponential function $y = y_0 e^{kt}$. This section shows where the exponential function comes from (it is the inverse of $\ln x$) and explores the properties that account for the amazing frequency with which the function appears in mathematics and its applications. We shall look at some of these applications in Section 6.5.

The Exponential Function exp(x)

The natural logarithm function, being an increasing function of x with domain $(0, \infty)$ and range $(-\infty, \infty)$, has an inverse whose domain is $(-\infty, \infty)$ and whose range is $(0, \infty)$. We call the inverse the exponential function of x, abbreviated $\exp(x)$.

DEFINITION

> The function $y = \exp(x)$, **the exponential function of x,** is the inverse of $y = \ln x$.

Since $\exp(x)$ and $\ln x$ are inverses, their composites in either order give the identity function.

$$\exp(\ln x) = x, \qquad x > 0 \tag{1}$$

$$\ln(\exp(x)) = x, \qquad \text{all } x \tag{2}$$

We shall have more to say about these equations in a moment.

Because $\exp(x)$ is the inverse of $\ln x$, its graph is obtained by reflecting the graph of $\ln x$ across the line $y = x$ (Fig. 6.10). As you can see,

$$\lim_{x \to \infty} \exp(x) = \infty \quad \text{and} \quad \lim_{x \to -\infty} \exp(x) = 0. \tag{3}$$

The Number e

The number e is $\exp(1)$, the one number whose natural logarithm is 1. (See Fig. 6.10.)

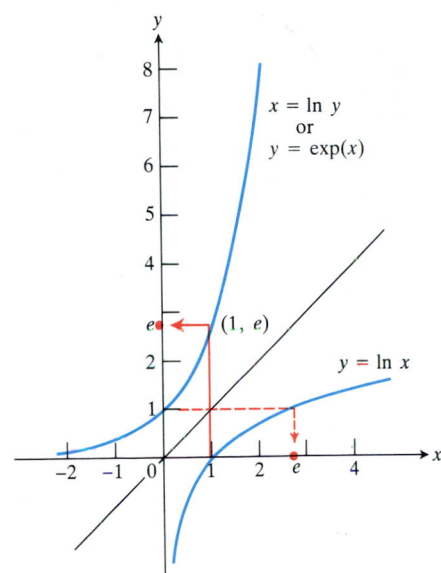

6.10 The graphs of $y = \ln x$ and its inverse $y = \exp(x)$. The number e is the one number whose natural logarithm is 1.

DEFINITION

$$e = \exp(1) = \ln^{-1}(1) \tag{4}$$

Although e is not a rational number, we shall see in Chapter 8 that it is possible to find its value with a computer to as many places as we want with the formula

$$e = \lim_{n \to \infty} \left(1 + 1 + \frac{1}{2} + \frac{1}{6} + \cdots + \frac{1}{n!}\right), \tag{5}$$

To fifteen places,

$$e = 2.7 \ 1828 \ 1828 \ 45 \ 90 \ 45.$$

(This arrangement makes the number easier to remember.)

The Function $y = e^x$

We can raise the number e to a rational power x in the usual way:

$$e^2 = e \cdot e, \qquad e^{-2} = \frac{1}{e^2}, \qquad e^{1/2} = \sqrt{e},$$

and so on. Since e is positive, e^x is positive too. This means that e^x has a logarithm. When we take the logarithm we find that

$$\ln e^x = x \ln e = x \cdot 1 = x. \tag{6}$$

But we also have

$$\ln(\exp(x)) = x. \qquad \text{(Eq. (2))}$$

This tells us that

$$e^x = \exp(x) \qquad \text{when } x \text{ is rational.} \qquad (7)$$

Equation (7) provides a way to complete the definition of e^x to include irrational values of x. The function $\exp(x)$ is defined for all x, so we can use it to give a value to e^x at points where e^x had no previous value.

DEFINITION

For every real number x, $\qquad e^x = \exp(x).$ $\hspace{6em}$ (8)

Example 1   You can find values of e^x on most scientific calculators to eight or ten digits by entering x and pressing $\boxed{e^x}$ or $\boxed{\text{INV}}\ \boxed{\ln x}$. Typical values (rounded) are

$$e^{-1} = 0.3679, \qquad e^2 = 7.3891,$$

$$e^{10} = 22026, \qquad e^{100} = 2.6881 \times 10^{43}.$$

The values of e^x grow very rapidly as $x \to \infty$, a point we shall return to in Section 6.6.

Equations Involving ln x and eˣ

When we replace $\exp(x)$ by e^x in Eqs. (1) and (2), we get the two most important rules for combining $\ln x$ and e^x.

For all positive x: $\qquad\qquad e^{\ln x} = x \hspace{4em} (9)$

For all x: $\qquad\qquad\qquad \ln(e^x) = x \hspace{4em} (10)$

If you did Exercises 5 and 6 in Section 1.3 on your calculator, these equations explain what you found. If you did not do them, you might want to do them now, along with the next example.

Example 2

a) $\ln e^2 = 2$

b) $\ln e^{-1} = -1$

c) $\ln\sqrt{e} = \dfrac{1}{2}$

d) $\ln e^{\sin x} = \sin x$

e) $e^{\ln 2} = 2$

f) $e^{\ln(x^2 + 1)} = x^2 + 1$

g) $e^{3\ln 2} = e^{\ln 2^3} = e^{\ln 8} = 8 \qquad$ (One way)

h) $e^{3\ln 2} = (e^{\ln 2})^3 = 2^3 = 8 \qquad$ (Another way)

Example 3   Find y if $\ln y = 3t + 5$.

Solution   Exponentiate:

$$e^{\ln y} = e^{3t + 5}$$

$$y = e^{3t + 5}. \qquad \text{(Eq. (9))}$$

Example 4   Find k if $e^{2k} = 10$.

Solution   Take the natural logarithm of both sides:

$$e^{2k} = 10$$

$$\ln e^{2k} = \ln 10$$

$$2k = \ln 10 \qquad \text{(Eq. (10))}$$

$$k = \frac{1}{2} \ln 10.$$

Useful Operating Rules

1. To remove logarithms from an equation, exponentiate both sides.
2. To remove exponentials, take the logarithm of both sides.

Laws of Exponents

Even though e^x is defined in a seemingly roundabout way as $\ln^{-1} x$, it obeys the familiar laws of exponents from algebra (Table 6.3).

Proof of Law 1   Let

$$y_1 = e^{x_1} \qquad \text{and} \qquad y_2 = e^{x_2}. \tag{11}$$

Then,

$$x_1 = \ln y_1 \quad \text{and} \quad x_2 = \ln y_2 \qquad \text{(Take logs of both sides of Eqs. (11))}$$

$$x_1 + x_2 = \ln y_1 + \ln y_2 = \ln y_1 y_2 \qquad \text{(Table 6.2, Rule 1)}$$

$$e^{x_1 + x_2} = e^{\ln y_1 y_2} \qquad \text{(Exponentiate)}$$

$$= y_1 y_2 \qquad (e^{\ln u} = u)$$

$$= e^{x_1} e^{x_2}.$$

TABLE 6.3
Laws of exponents for e^x

For all numbers x, x_1, and x_2,

1. $e^{x_1} \cdot e^{x_2} = e^{x_1 + x_2}$

2. $e^{-x} = \dfrac{1}{e^x}$

3. $\dfrac{e^{x_1}}{e^{x_2}} = e^{x_1 - x_2}$

Laws 2 and 3 follow from Law 1, as you will be asked to show in Exercise 74.

Example 5

a) $e^{x + \ln 2} = e^x \cdot e^{\ln 2} = 2e^x \qquad \text{(Law 1)}$

b) $e^{-\ln x} = \dfrac{1}{e^{\ln x}} = \dfrac{1}{x} \qquad \text{(Law 2)}$

c) $\dfrac{e^{2x}}{e} = e^{2x - 1} \qquad \text{(Law 3)}$

The Derivative and Integral of e^x

Transcendental Numbers and Transcendental Functions

Numbers that are solutions of polynomial equations with rational coefficients are called **algebraic:** -2 is algebraic because it satisfies the equation $x + 2 = 0$, and $\sqrt{3}$ is algebraic because it satisfies the equation $x^2 - 3 = 0$. Numbers that are not algebraic are called **transcendental,** a term coined by Euler to describe numbers, like e and π, that appeared to "transcend the power of algebraic methods." But it was not until a hundred years after Euler's death (1873) that Charles Hermite proved the transcendence of e in the sense that we describe. A few years later (1882), C. L. F. Lindemann proved the transcendence of π.

Today we call a function $y = f(x)$ algebraic if it satisfies an equation of the form

$$P_n y^n + \cdots + P_1 y + P_0 = 0$$

in which the P's are polynomials in x with rational coefficients. The function $y = 1/\sqrt{x + 1}$ is algebraic because it satisfies the equation $(x + 1)y^2 - 1 = 0$. Here the polynomials are $P_2 = x + 1$, $P_1 = 0$, and $P_0 = -1$. Polynomials and rational functions with rational coefficients are algebraic, as are all sums, products, quotients, rational powers, and rational roots of algebraic functions.

Functions that are not algebraic are called transcendental. The six basic trigonometric functions are transcendental, as are the inverses of the trigonometric functions and the exponential and logarithmic functions that are the main subject of the present chapter.

The exponential function is differentiable because it is the inverse of a differentiable function whose derivative is never zero. Starting with $y = e^x$, we have, in order,

$$y = e^x$$

$$\ln y = x \qquad \text{(Logarithms of both sides)}$$

$$\frac{1}{y}\frac{dy}{dx} = 1 \qquad \text{(Derivatives of both sides with respect to } x\text{)}$$

$$\frac{dy}{dx} = y$$

$$\frac{dy}{dx} = e^x. \qquad \text{(} y \text{ replaced by } e^x\text{)}$$

The startling conclusion we draw from this sequence of equations is that e^x is its own derivative. Never before have we encountered such a function. No matter how many times we differentiate it, we get the function back. As we shall see in Section 6.5, constant multiples of e^x are the only functions that behave this way. These functions are like the story of a disciple who asked a guru what was holding up the earth. "An elephant," replied the guru. The disciple naturally wanted to know what was holding up the elephant. "Another elephant," said the guru. "And what about *that* elephant?" asked the disciple. The guru paused a moment and then replied, "It's elephants all the way down."

$$\frac{d}{dx} e^x = e^x \tag{12}$$

Example 6

$$\frac{d}{dx}(5e^x) = 5\frac{d}{dx}e^x = 5e^x$$

The Chain Rule extends Eq. (12) in the usual way to a more general form.

If u is any differentiable function of x, then

$$\frac{d}{dx} e^u = e^u \frac{du}{dx}. \tag{13}$$

Example 7

a) $\dfrac{d}{dx} e^{-x} = e^{-x}\dfrac{d}{dx}(-x) = e^{-x}(-1) = -e^{-x}$ $\qquad \left(\begin{array}{l}\text{Eq. (13) with}\\ u = -x\end{array}\right)$

b) $\dfrac{d}{dx} e^{\sin x} = e^{\sin x}\dfrac{d}{dx}(\sin x) = e^{\sin x}\cdot \cos x$ $\qquad \left(\begin{array}{l}\text{Eq. (13) with}\\ u = \sin x\end{array}\right)$

The integral equivalent of Eq. (13) is

$$\int e^u \, du = e^u + C. \tag{14}$$

Example 8

$$\int_0^{\ln 2} e^{3x} \, dx = \int_0^{\ln 8} e^u \cdot \frac{1}{3} \, du \qquad \left(\begin{array}{l} u = 3x, \frac{1}{3} \, du = dx, u(0) = 0, \\ u(\ln 2) = 3 \ln 2 = \ln 2^3 = \ln 8 \end{array} \right)$$

$$= \frac{1}{3} \int_0^{\ln 8} e^u \, du$$

$$= \frac{1}{3} e^u \Big]_0^{\ln 8}$$

$$= \frac{1}{3} [8 - 1] = \frac{7}{3}$$

Example 9

$$\int_0^{\pi/2} e^{\sin x} \cos x \, dx = e^{\sin x} \Big]_0^{\pi/2} \qquad \left(\begin{array}{l} \text{Antiderivative} \\ \text{from Example 7} \end{array} \right)$$

$$= e^1 - e^0 = e - 1$$

Example 10   Solve the initial value problem

Differential equation: $\quad e^y \dfrac{dy}{dx} = 2x, \qquad x > \sqrt{3},$

Initial condition: $\quad y = 0 \quad \text{when} \quad x = 2.$

Solution   We integrate both sides of the differential equation with respect to x to obtain

$$e^y = x^2 + C.$$

We use the initial condition to determine C:

$$C = e^0 - (2)^2$$
$$= 1 - 4 = -3.$$

This completes the formula for e^y:

$$e^y = x^2 - 3. \tag{15}$$

To find y, we take logarithms of both sides:

$$\ln e^y = \ln(x^2 - 3)$$
$$y = \ln(x^2 - 3). \tag{16}$$

Notice that the solution is valid for $x > \sqrt{3}$.

It is always a good idea to check a solution in the original equation. From Eqs.

(15) and (16), we have

$$e^y \frac{dy}{dx} = e^y \frac{d}{dx} \ln(x^2 - 3) \qquad \text{(Eq. (16))}$$

$$= e^y \frac{2x}{x^2 - 3}$$

$$= (x^2 - 3)\frac{2x}{x^2 - 3} \qquad \text{(Eq. (15))}$$

$$= 2x.$$

The solution checks.

EXERCISES 6.3

Find simpler expressions for the numbers in Exercises 1 and 2.

1. a) $e^{\ln 7}$ b) $2 \ln\sqrt{e}$ c) $\ln(\ln e^e)$

2. a) $e^{-\ln 8}$ b) $e^{5 \ln(1/2)}$ c) $e^{\ln 2 + \ln 3}$

Simplify the expressions in Exercises 3 and 4.

3. a) $e^{-\ln(x^2)}$ b) $e^{\ln x + \ln y}$ c) $\ln(e^{1/x})$

4. a) $\ln(1/e^x)$ b) $\ln(e^{-x^2})$ c) $e^{\ln x - 2 \ln y}$

In Exercises 5 and 6, solve for k.

5. a) $e^{2k} = 4$ b) $e^{5k} = \frac{1}{4}$ c) $e^{(k/1000)} = a$

6. a) $100e^{10k} = 200$ b) $80e^k = 1$ c) $e^{(\ln 0.8)k} = 0.1$

In Exercises 7 and 8, solve for t.

7. a) $e^t = 1$ b) $4e^{-0.1t} = 20$ c) $e^{-0.3t} = 27$

8. a) $e^{-0.01t} = 1000$ b) $e^{kt} = \frac{1}{2}$ c) $e^{(\ln 0.2)t} = 0.4$

In Exercises 9–16, solve for y.

9. $\ln y = 2t + 4$ **10.** $\ln y = -t + 5$

11. $\ln (y - 40) = 5t$ **12.** $\ln(1 - 2y) = t$

13. $e^{\sqrt{y}} = x^2$ **14.** $e^{(x^2)}e^{(2x + 1)} = e^y$

15. $\ln(y - 1) = x + \ln x$

16. $\ln(y^2 - 1) - \ln(y + 1) = \sin x$

In Exercises 17–34, find dy/dx if y equals the given expression.

17. e^{-5x} **18.** $e^{2x/3}$

19. $e^{5 - 7x}$ **20.** $e^{(4\sqrt{x} + x^2)}$

21. $xe^x - e^x$ **22.** $(1 + 2x)e^{-2x}$

23. $(x^2 - 2x + 2)e^x$ **24.** $(9x^2 - 6x + 2)e^{3x}$

25. $e^x(\sin x + \cos x)$ **26.** $\ln(3x \, e^{-x})$

27. $\cos(e^{-x^2})$ **28.** $x^3 e^{-2x} \cos 5x$

29. $e^{(\cos x + \ln x)}$ **30.** $\ln(e^x/(1 + e^x))$

31. $x^3 e^x - 3x^2 e^x + 6xe^x - 6x + \sqrt{2}$

32. $\dfrac{ax - 1}{a^2} e^{ax}$ **33.** $\displaystyle\int_0^{\ln x} \sin e^t \, dt$

34. $\displaystyle\int_{e^{4\sqrt{x}}}^{e^{2x}} \ln t \, dt$

In Exercises 35–38, find dy/dx.

35. $\ln y = e^y \sin x$ **36.** $\ln xy = e^{x + y}$

37. $e^{2x} = \sin(x + 3y)$ **38.** $\tan y = e^x + \ln x$

Evaluate the integrals in Exercises 39–52.

39. $\displaystyle\int_1^{e^2} \frac{1}{x} \, dx$ **40.** $\displaystyle\int_{\ln 2}^{\ln 3} e^x \, dx$

41. $\displaystyle\int_{-\ln 2}^0 e^{-x} \, dx$ **42.** $\displaystyle\int_{-1}^1 e^{(t + 1)} \, dt$

43. $\displaystyle\int_{\ln 3}^{\ln 5} e^{2\theta} \, d\theta$ **44.** $\displaystyle\int_{-1}^1 2xe^{-x^2} \, dx$

45. $\displaystyle\int_1^4 \frac{e^{\sqrt{r}}}{2\sqrt{r}} \, dr$ **46.** $\displaystyle\int_{-1}^1 \frac{e^t}{1 + e^t} \, dt$

47. $\displaystyle\int_e^{e^e} \frac{dx}{x \ln x}$ **48.** $\displaystyle\int_0^{\ln 13} \frac{e^s \, ds}{1 + 2e^s}$

49. $\displaystyle\int_{\ln(\pi/6)}^{\ln(\pi/2)} 2e^u \cos e^u \, du$

50. $\int \dfrac{\tan(\ln v)}{v}\,dv$

51. $\int (1 + e^{\tan\theta})\sec^2\theta\,d\theta$

52. $\int \dfrac{dx}{1 + e^x}$

Find the limits in Exercises 53–56.

53. $\lim\limits_{\theta\to 0}\dfrac{\cos\theta - 1}{e^\theta - \theta - 1}$

54. $\lim\limits_{h\to 0}\dfrac{e^h - (1 + h)}{h^2}$

55. $\lim\limits_{t\to\infty}\dfrac{e^t + t^2}{e^t - t}$

56. $\lim\limits_{x\to\infty} x^2 e^{-x}$

57. *The linearization and quadratic approximation of e^x at $x = 0$.* The standard linear and quadratic approximations of e^x at $x = 0$ are

Linearization at $x = 0$: $\quad e^x \approx 1 + x$

Quadratic approximation at $x = 0$: $\quad e^x \approx 1 + x + \dfrac{x^2}{2}$.

a) Derive these formulas.
b) CALCULATOR Estimate to five decimal places the errors involved in replacing e^x by its linearization and quadratic approximation on $[0, 0.1]$.

58. a) Show that $y = Ce^{ax}$ solves the differential equation $dy/dx = ay$ for any choice of the constant C.
b) Solve the initial value problem

$$\frac{dy}{dx} = -2y,\;\; y = 3\quad\text{when}\quad x = 0.$$

59. Find the maximum value of $f(x) = x^2\ln(1/x)$.

60. Find the absolute maximum and minimum values of $f(x) = e^x - 2x$ on $[0, 1]$.

61. The curve $y = (x - 3)^2 e^x$ has a horizontal tangent at $x = 3$. Does it have a local extreme value there, or does it have a point of inflection?

62. Find the maximum and minimum values of the periodic function $f(x) = e^{\sin x}$ (Fig. 6.11).

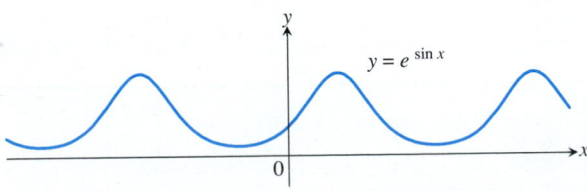

6.11 The graph of the function in Exercise 62.

Find the areas of the shaded regions in Exercises 63 and 64.

63.

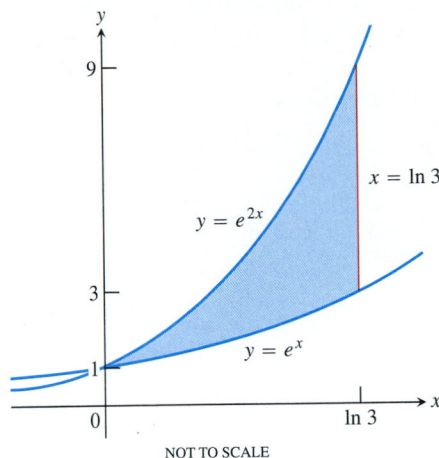

NOT TO SCALE

64.

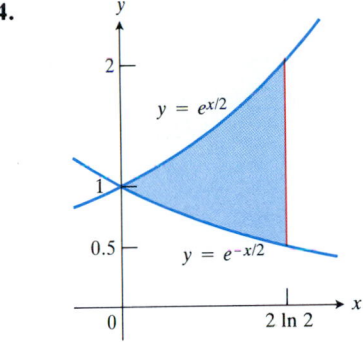

65. Let $A(t)$ be the area of the region in the first quadrant enclosed by the coordinate axes, the curve $y = e^{-x}$, and the vertical line $x = t$, $t > 0$ (Fig. 6.12). Let $V(t)$ be the volume of the solid generated by revolving the region about the x-axis. Find the following limits.
a) $\lim\limits_{t\to\infty} A(t)$
b) $\lim\limits_{t\to\infty} V(t)/A(t)$
c) $\lim\limits_{t\to 0^+} V(t)/A(t)$

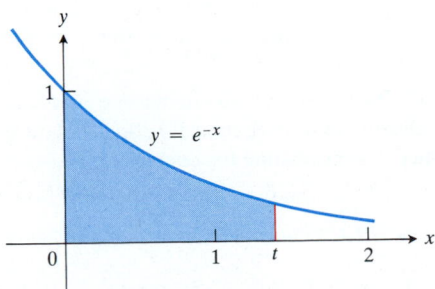

6.12 The region in Exercise 65.

66. Show that for any number $a > 1$,

$$\int_1^a \ln x \, dx + \int_0^{\ln a} e^y \, dy = a \ln a.$$

(*Hint:* Look at Fig. 6.13.)

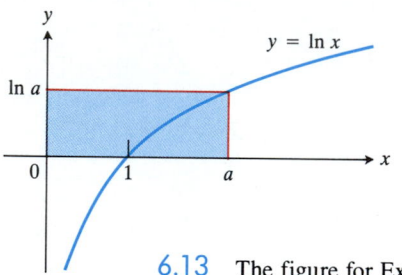

6.13   The figure for Exercise 66.

67. Find a curve through the origin in the xy-plane whose length from $x = 0$ to $x = 1$ is

$$L = \int_0^1 \sqrt{1 + \frac{1}{4} e^x} \, dx.$$

68. Find the area of the surface generated by revolving the curve $x = (e^y + e^{-y})/2$, $0 \le y \le \ln 2$, about the y-axis (Fig. 6.14).

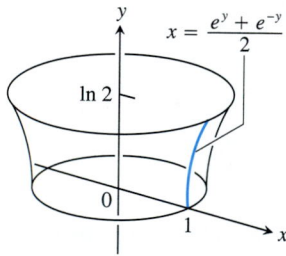

6.14   The surface in Exercise 68.

Solve the initial value problems in Exercises 69 and 70.

69. $\dfrac{dy}{dx} = e^x \sin(e^x - 2)$, $\quad y = 0$ when $x = \ln(\pi + 2)$

70. $\dfrac{1}{y + 1} \dfrac{dy}{dx} = \dfrac{1}{2x}$, $\quad x > 0$; $\quad y = 2$ when $x = 1$

71. CALCULATOR *A decimal representation of e.* Find e to as many decimal places as your calculator allows by solving the equation $\ln x = 1$.

72. CALCULATOR *The inverse relation between e^x and $\ln x$.* If you did not already do so in Section 1.3, find out how good your calculator is at evaluating the composites

$$e^{\ln x} \quad \text{and} \quad \ln(e^x).$$

73. GRAPHER *The linearization and quadratic approximation of e^x at $x = 0$.*
a) Graph e^x, $1 + x$, and $1 + x + x^2/2$ together on the interval $-2 \le x \le 2$. Use different colors, if available. On what intervals do the approximations appear to overestimate e^x? underestimate e^x?

b) CALCULATOR Now, without changing the display, restrict your attention to the interval $[-1, 1]$. Where do the errors of largest magnitude in the approximations occur? Estimate these magnitudes to five decimal places.

74. *Exponent laws.* Starting with the equation $e^{x_1} e^{x_2} = e^{x_1 + x_2}$, derived in the text, show that $e^{-x} = 1/e^x$ for any real number x. Then show that $e^{x_1}/e^{x_2} = e^{x_1 - x_2}$ for any numbers x_1 and x_2.

75. *The geometric, logarithmic, and arithmetic mean inequality.*
a) Show that the graph of e^x is concave up over every interval of x-values.
b) Show, by reference to Fig. 6.15, that if $0 < a < b$ then

$$e^{(\ln a + \ln b)/2} \cdot (\ln b - \ln a) < \int_{\ln a}^{\ln b} e^x \, dx <$$

$$\frac{e^{\ln a} + e^{\ln b}}{2} \cdot (\ln b - \ln a).$$

6.15   The figure for Exercise 75.

c) Use the inequality in (b) to conclude that

$$\sqrt{ab} < \frac{b - a}{\ln b - \ln a} < \frac{a + b}{2}.$$

This inequality says that the geometric mean of two positive numbers is less than their logarithmic mean, which in turn is less than their arithmetic mean.

(For more about this inequality, see "The Geometric, Logarithmic, and Arithmetic Mean Inequality" by Frank Burk, *American Mathematical Monthly,* Vol. 94, No. 6, June–July 1987, pp. 527–28.)

EXPLORER PROGRAMS

Bisections, Secants, Newton	Solves equations numerically
PowerGrapher	Graphs all functions in this section

6.4 Other Exponential and Logarithmic Functions

While we have not yet devised a way to raise positive numbers to any but rational powers, we have an exception in the number e. The definition $e^x = \ln^{-1}x$ defines e^x for every real value of x, irrational as well as rational. In this section, we show how this enables us to raise any other positive number to an arbitrary power and thus to define an exponential function $y = a^x$ for any positive number a. We also prove the power rule for differentiation in its final form (good for all exponents) and define functions like x^x and $(\sin x)^{\tan x}$ that involve raising the values of one function to powers given by another function.

Just as e^x is but one of many exponential functions, $\ln x$ is one of many logarithmic functions, the others being the inverses of the functions a^x. These logarithmic functions have important applications in science and engineering.

The Function a^x

Since $a = e^{\ln a}$ for any positive number a, we can think of a^x as $(e^{\ln a})^x = e^{x \ln a}$. We therefore make the following definition.

DEFINITION

> **The Function $y = a^x$**
>
> If a is a positive number and x is any number whatever, $a^x = e^{x \ln a}$.   (1)

Example 1 $2^{\sqrt{3}} = e^{\sqrt{3} \ln 2}, \qquad 2^\pi = e^{\pi \ln 2}$

Example 2   The number x^n can now be defined for any positive number x and any real number n:

$$x^n = e^{n \ln x}. \qquad \text{(Eq. (1) with } a = x \text{ and } x = n\text{)}$$

Example 3   The n in the equation $\ln x^n = n \ln x$ no longer has to be a rational number. It can be any real number as long as $x > 0$:

$$\ln x^n = \ln(e^{n \ln x}) \qquad \text{(Example 2)}$$

$$= n \ln x. \qquad \left(\begin{array}{l}\ln e^u = u \text{ for any } u \text{, in} \\ \text{particular for } u = n \ln x\end{array}\right)$$

The exponential function a^x obeys all the standard laws of exponents (Table 6.4). (We omit the proofs.)

Law 3, together with the definition of x^n as $e^{n \ln x}$, enables us to prove the Power Rule for differentiation in its final form. Differentiating x^n with respect to x gives

TABLE 6.4
Laws of exponents for a^x
($a > 0$, any x and y)

1. $a^x \cdot a^y = a^{x+y}$

2. $a^{-x} = \dfrac{1}{a^x}$

3. $\dfrac{a^x}{a^y} = a^{x-y}$

4. $(a^x)^y = a^{xy} = (a^y)^x$

$$\frac{d}{dx} x^n = \frac{d}{dx} e^{n \ln x} \qquad \text{(Definition of } x^n, x > 0\text{)}$$

$$= e^{n \ln x} \cdot \frac{d}{dx} (n \ln x) \qquad \text{(Chain Rule for } e^u\text{)}$$

$$= x^n \cdot \frac{n}{x} \qquad \text{(The definition again)}$$

$$= n x^{n-1}. \qquad \text{(Table 6.4, Law 3)}$$

In short, as long as $x > 0$,

$$\frac{d}{dx} x^n = n x^{n-1}. \tag{2}$$

The Chain Rule extends Eq. (2) to the Power Rule's final form.

Power Rule (Final Form)

If u is a positive differentiable function of x, and n is any real number, then u^n is a differentiable function of x and

$$\frac{d}{dx} u^n = nu^{n-1} \frac{du}{dx}. \tag{3}$$

Example 4

$$\frac{d}{dx} x^{\sqrt{2}} = \sqrt{2}x^{\sqrt{2}-1} \qquad (x > 0)$$

$$\frac{d}{dx} (\sin x)^\pi = \pi(\sin x)^{\pi-1}\cos x \qquad (\sin x > 0)$$

The Derivative of a^x

We find the derivative of a^x the way we found the derivative of x^n, starting with the definition $a^x = e^{x \ln a}$:

$$\frac{d}{dx} a^x = \frac{d}{dx} e^{x \ln a} = e^{x \ln a} \cdot \frac{d}{dx} (x \ln a) \qquad \text{(Chain Rule)}$$

$$= a^x \ln a.$$

If $a > 0$, then

$$\frac{d}{dx} a^x = a^x \ln a. \tag{4}$$

Equation (4) shows why the function e^x is the exponential function preferred in calculus. If $a = e$, then $\ln a = 1$ and Eq. (4) simplifies to

$$\frac{d}{dx} e^x = e^x \ln e = e^x.$$

Equation (4) also comes in a more general form, based on the Chain Rule.

If $a > 0$ and u is a differentiable function of x, then a^u is a differentiable function of x and

$$\frac{d}{dx} a^u = a^u \ln a \frac{du}{dx}. \tag{5}$$

Example 5

a) $\dfrac{d}{dx} 3^x = 3^x \ln 3$

b) $\dfrac{d}{dx} 3^{-x} = 3^{-x} \ln 3 \dfrac{d}{dx}(-x) = -3^{-x} \ln 3$

c) $\dfrac{d}{dx} 3^{\sin x} = 3^{\sin x} \ln 3 \dfrac{d}{dx}(\sin x) = 3^{\sin x} (\ln 3) \cos x$

Other Power Functions

The ability to raise positive numbers to real powers makes it possible to define functions like x^x and $x^{\ln x}$ for $x > 0$. We find the derivatives of such functions by rewriting them as powers of e.

Example 6   Find dy/dx if $y = x^x$, $x > 0$.

Solution   Write x^x as a power of e:

$$y = x^x = e^{x \ln x}. \qquad \text{(Eq. (1) with } a = x)$$

Then differentiate as usual:

$$\frac{dy}{dx} = \frac{d}{dx} e^{x \ln x}$$

$$= e^{x \ln x} \frac{d}{dx}(x \ln x)$$

$$= e^{x \ln x} \left(x \cdot \frac{1}{x} + \ln x \right)$$

$$= e^{x \ln x} (1 + \ln x).$$

L'Hôpital's Rule and the Forms 1^∞, 0^0, and ∞^0

The forms 1^∞, 0^0, and ∞^0 can sometimes be handled by taking logarithms first. The idea is to calculate the limit of the logarithm and exponentiate that limit.

If $\lim\limits_{x \to a} \ln f(x) = L$,

then $\lim\limits_{x \to a} f(x) = \lim\limits_{x \to a} e^{\ln f(x)} = e^L.$ (6)

Example 7   Show that

$$\lim_{x \to 0^+} (1 + x)^{1/x} = e. \qquad (7)$$

Solution   The limit leads to the indeterminate form 1^∞. We let $f(x) = (1 + x)^{1/x}$ and find $\lim_{x \to 0^+} \ln f(x)$. Since

$$\ln f(x) = \ln(1 + x)^{1/x} = \frac{1}{x} \ln(1 + x),$$

l'Hôpital's rule gives

$$\lim_{x \to 0^+} \ln f(x) = \lim_{x \to 0^+} \frac{\ln(1 + x)}{x} \qquad \left(\frac{0}{0}\right)$$

$$= \lim_{x \to 0^+} \frac{\dfrac{1}{1 + x}}{1} = 1.$$

Therefore,

$$\lim_{x \to 0^+} f(x) = e^1 = e. \qquad \text{(Eq. (6) with } L = 1\text{)}$$

An Alternate Definition of e

The limit

$$\lim_{n \to \infty} \left(1 + \frac{1}{n}\right)^n = e$$

is the special case of Eq. (7) obtained by restricting x to be the reciprocal of a positive integer. It gives a way to define e without logarithms. However, the proof that the limit exists then has to be different from the logarithm-based proof in Example 7.

Example 8   Find $\lim_{x \to \infty} x^{1/x}$.

Solution   The limit leads to the form ∞^0. We let $f(x) = x^{1/x}$ and find $\lim_{x \to \infty} \ln f(x)$. Since

$$\ln f(x) = \ln x^{1/x} = \frac{\ln x}{x},$$

l'Hôpital's rule gives

$$\lim_{x \to \infty} \ln f(x) = \lim_{x \to \infty} \frac{\ln x}{x} \qquad \left(\frac{\infty}{\infty}\right)$$

$$= \lim_{x \to \infty} \frac{1/x}{1} = 0.$$

Therefore,

$$\lim_{x \to \infty} f(x) = e^0 = 1. \qquad \text{(Eq. (6) with } L = 0\text{)}$$

The Graph of a^x

From the formula

$$\frac{d}{dx} a^x = a^x \ln a,$$

we see that the derivative of a^x is positive if $a > 1$ and negative if $0 < a < 1$. Thus, a^x is an increasing function of x if $a > 1$ and a decreasing function of x if $0 < a < 1$. In either case, a^x is one-to-one. The second derivative,

$$\frac{d^2}{dx^2} a^x = \frac{d}{dx} (a^x \ln a)$$

$$= \ln a \frac{d}{dx} (a^x)$$

$$= \ln a (a^x \ln a)$$

$$= (\ln a)^2 a^x,$$

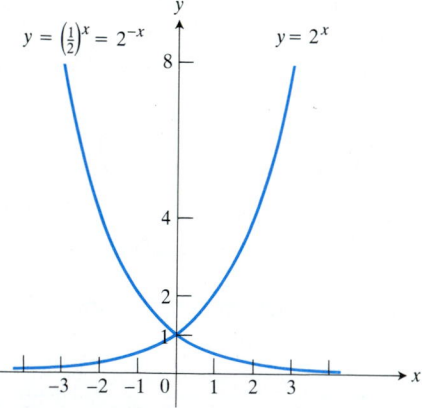

6.16   The graphs of 2^x and $(1/2)^x = 2^{-x}$. Both functions are one-to-one. As $x \to \infty$, $2^x \to \infty$ and $(1/2)^x \to 0$. Similarly, as $x \to -\infty$, $2^x \to 0$ and $(1/2)^x \to \infty$.

On calculators and computers that do not offer a^x as a standard function, you have to enter a^x as $\exp(x * \log(a))$. The generating functions for the graphs here are $\exp(x * \log(2))$ and $\exp(-x * \log(2))$.

is always positive, so the graph is concave up. Figure 6.16 shows the graphs of $y = 2^x$ (increasing and one-to-one) and $y = (1/2)^x$ (decreasing and one-to-one).

The Integral of a^u

If $a \neq 1$, so that $\ln a \neq 0$, we can divide both sides of Eq. (5) by $\ln a$ to obtain

$$a^u \frac{du}{dx} = \frac{1}{\ln a} \frac{d}{dx} (a^u). \tag{8}$$

Integrating with respect to x then gives

$$\int a^u \frac{du}{dx} \, dx = \int \frac{1}{\ln a} \frac{d}{dx} (a^u) \, dx = \frac{1}{\ln a} \int \frac{d}{dx} (a^u) \, dx = \frac{1}{\ln a} a^u + C. \tag{9}$$

Writing the first integral in differential form then gives

$$\int a^u \, du = \frac{a^u}{\ln a} + C. \tag{10}$$

Example 9

a) $\displaystyle\int 2^x \, dx = \frac{2^x}{\ln 2} + C$ (Eq. (10) with $a = 2$, $u = x$)

b) $\displaystyle\int 2^{\sin x} \cos x \, dx = \int 2^u \, du = \frac{2^u}{\ln 2} + C = \frac{2^{\sin x}}{\ln 2} + C$ ($u = \sin x$ in Eq. (10))

Base a Logarithms

As we saw earlier, if a is any positive number other than 1, the function a^x is one-to-one and has a nonzero derivative at every point. It therefore has a differentiable inverse. We call the inverse the **base a logarithm of x** and denote it by $\log_a x$.

DEFINITION

For any positive number $a \neq 1$,

$$\log_a x = \text{inverse of } a^x.$$

Because the logarithm and exponential are inverses, their composites in either order give the identity function.

$$\log_a (a^x) = x \qquad \text{(for all } x\text{)} \tag{11}$$

$$a^{(\log_a x)} = x \qquad \text{(for each positive } x\text{)} \tag{12}$$

Example 10

a) $\log_2(2^5) = 5,$ $\log_{10}(10^{-7}) = -7$

b) $2^{(\log_2 3)} = 3,$ $10^{(\log_{10} 4)} = 4$

The Evaluation of $\log_a x$

The values of $\log_a x$ can be calculated from the natural logarithms of a and x with the following formula.

$$\log_a x = \frac{\ln x}{\ln a} \qquad\qquad (13)$$

We derive this formula from Eq. (12):

$a^{(\log_a x)} = x$	(Eq. (12))
$\ln a^{(\log_a x)} = \ln x$	(Natural logarithms of both sides)
$\log_a x \cdot \ln a = \ln x$	(The Exponent Rule)
$\log_a x = \dfrac{\ln x}{\ln a}.$	(Solved for $\log_a x$)

TABLE 6.5
Rules of arithmetic for base a
logarithms

1. $\log_a uv = \log_a u + \log_a v$
2. $\log_a \dfrac{u}{v} = \log_a u - \log_a v$
3. $\log_a u^n = n \log_a u$

Example 11

$$\log_{10} 2 = \frac{\ln 2}{\ln 10} \approx \frac{0.69}{2.30} = 0.3$$

Example 12   The ratio of $\ln x$ to $\log_{10} x$ has a constant value:

$$\frac{\ln x}{\log_{10} x} = \ln 10 = 2.302585093. \qquad \text{(To ten digits)}$$

This explains what you discovered in Section 1.8, Exercise 48(d).

Example 13   *Calculators*   Most calculators have keys for $\log_{10} x$ and $\ln x$. To find logarithms to other bases, we use Eq. (13).

To find $\log_2 x$, find $\ln x$ and divide by $\ln 2$: $\log_2 5 = \dfrac{\ln 5}{\ln 2}.$

To find $\ln x$ given $\log_2 x$, multiply by $\ln 2$: $\ln 5 = \log_2 5 \cdot \ln 2.$

The rules of arithmetic are the same for $\log_a x$ as for $\ln x$ (Table 6.5). These rules come from dividing the corresponding rules for the natural logarithm by $\ln a$. For example,

$\ln uv = \ln u + \ln v$	(Rule 1 for natural logarithms . . .)
$\dfrac{\ln uv}{\ln a} = \dfrac{\ln u}{\ln a} + \dfrac{\ln v}{\ln a}$	(. . . divided by $\ln a$ . . .)
$\log_a uv = \log_a u + \log_a v.$	(. . . gives Rule 1 for base a logarithms.)

The Graph of $\log_a x$

The graph of $\log_a x$ is obtained by reflecting the graph of a^x across the line $y = x$ (Fig. 6.17). It looks very much like a scaled version of the graph of $\ln x$. Indeed it is, because of Eq. (13).

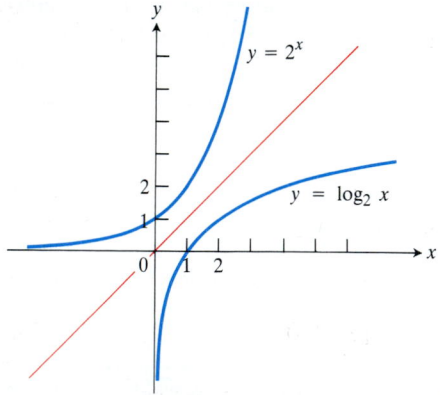

6.17   The graph of 2^x and its inverse, $\log_2 x$. As $x \to \infty$, $\log_2 x \to \infty$. As $x \to 0^+$, $\log_2 x \to -\infty$. Notice the symmetry in the line $y = x$.

On calculators and computers that do not offer $\log_a x$, use $(\log(x))/(\log(a))$ (on computers that use log for the natural logarithm). The generating function for the logarithmic function here is $(\log(x))/(\log(2))$.

The Derivative of $\log_a u$

To find the derivative of a base a logarithm, we first convert it to a natural logarithm. If u is a positive differentiable function of x, then

$$\frac{d}{dx}(\log_a u) = \frac{d}{dx}\left(\frac{\ln u}{\ln a}\right) = \frac{1}{\ln a}\frac{d}{dx}(\ln u) = \frac{1}{\ln a}\cdot\frac{1}{u}\frac{du}{dx}.$$

$$\frac{d}{dx}\log_a u = \frac{1}{\ln a}\cdot\frac{1}{u}\frac{du}{dx} \tag{14}$$

Example 14

$$\frac{d}{dx}\log_{10}(3x+1) = \frac{1}{\ln 10}\cdot\frac{1}{3x+1}\frac{d}{dx}(3x+1) = \frac{3}{(\ln 10)(3x+1)}$$

Integrals Involving $\log_a x$

To evaluate integrals involving base a logarithms, we convert them to natural logarithms.

Example 15

$$\int \frac{\log_2 x}{x}\,dx = \frac{1}{\ln 2}\int \frac{\ln x}{x}\,dx \qquad \left(\log_2 x = \frac{\ln x}{\ln 2}\right)$$

$$= \frac{1}{\ln 2}\int u\,du \qquad \left(u = \ln x,\ du = \frac{1}{x}\,dx\right)$$

$$= \frac{1}{\ln 2}\frac{u^2}{2} + C = \frac{1}{\ln 2}\frac{(\ln x)^2}{2} + C = \frac{(\ln x)^2}{2\ln 2} + C$$

Base 10 Logarithms

Base 10 logarithms, often called **common logarithms,** appear in many scientific formulas. For example, earthquake intensity is often reported on the **Richter scale.** Here the formula is

$$\text{Magnitude } R = \log_{10}\left(\frac{a}{T}\right) + B, \tag{15}$$

where a is the amplitude of the ground motion in microns at the receiving station, T is the period of the seismic wave in seconds, and B is an empirical factor that allows for the weakening of the seismic wave with increasing distance from the epicenter of the earthquake.

Example 16   For an earthquake 10,000 km from the receiving station, $B = 6.8$. If the recorded vertical ground motion is $a = 10$ microns and the period is $T = 1$ sec, the earthquake's magnitude is

$$R = \log_{10}\left(\frac{10}{1}\right) + 6.8 = 1 + 6.8 = 7.8.$$

An earthquake of this magnitude does great damage near its epicenter.

The pH scale for measuring the acidity of a solution is a logarithmic scale. The pH value (hydrogen potential) of the solution is the common logarithm of the reciprocal of the solution's hydronium ion concentration, $[H_3O^+]$:

$$pH = \log_{10}\frac{1}{[H_3O^+]} = -\log_{10}[H_3O^+]. \qquad (16)$$

The hydronium ion concentration is measured in moles per liter. Vinegar has a pH of 3, distilled water a pH of 7, seawater a pH of 8.15, and household ammonia a pH of 12. The total scale ranges from about 0.1 for normal hydrochloric acid to 14 for a normal (1 N) solution of sodium hydroxide.

Another example of the use of common logarithms is the db ("dee bee") scale for measuring loudness in decibels. If I is the intensity of sound in watts per square meter, the decibel level of the sound is

$$\text{Sound level} = 10\log_{10}(I \times 10^{12})\ \text{db}. \qquad (17)$$

If you ever wondered why doubling the power of your audio amplifier increases the sound level by only a few decibels, Eq. (17) provides the answer. As the following example shows, doubling I adds only about 3 db.

Example 17   Doubling I in Eq. (17) adds about 3 db. Writing log for $\log_{10}$ (a common practice), we have

$$\text{Sound level with } I \text{ doubled} = 10\log(2I \times 10^{12}) \qquad \left(\begin{matrix}\text{Eq. (17) with}\\ 2I \text{ for } I\end{matrix}\right)$$

$$= 10\log(2\cdot I \times 10^{12})$$

$$= 10\log 2 + 10\log(I \times 10^{12})$$

$$= \text{original sound level} + 10\log 2$$

$$\approx \text{original sound level} + 3. \qquad (\log_{10}2 \approx 0.30)$$

REMARK ON NOTATION   Many advanced texts and research publications in mathematics use $\log x$, with no base specified, to represent the natural logarithm $\ln x$. Most texts in the physical sciences use $\log x$ to represent $\log_{10}x$. Most calculators use $\ln x$ for the natural logarithm and $\log x$ for base 10 logarithms. Computers, however, may use LOG(X) for the natural logarithm. One then evaluates $\log_{10}x$ as (LOG(X))/(LOG(10)).

Most Foods Are Acidic (pH < 7).

Food	pH value
Bananas	4.5–4.7
Grapefruit	3.0–3.3
Oranges	3.0–4.0
Limes	1.8–2.0
Milk	6.3–6.6
Soft drinks	2.0–4.0
Spinach	5.1–5.7

Typical Sound Levels

Threshold of hearing	0 db
Rustle of leaves	10 db
Average whisper	20 db
Quiet automobile	50 db
Ordinary conversation	65 db
Pneumatic drill 10 feet away	90 db
Threshold of pain	120 db

EXERCISES 6.4

In Exercises 1–18, find the derivative of the expression with respect to x, r, s, t, v, or θ, whichever is appropriate.

1. x^{π}

2. $x^{1+\sqrt{2}}$

3. $r^{-\sqrt{2}}$

4. r^{1-e}

5. 2^x

6. 9^{-x}

7. $5^{\sqrt{s}}$

8. $2^{(s^2)}$

9. $2^{\sec\theta}$

10. $3^{\tan\theta}\ln 3$

11. $x^{\ln x}$

12. $x^{(x+1)}$

13. $(\sqrt{t})^t$

14. $(1-t)^t$

15. $(\sin x)^{\sin x}$

16. $(\cos v)^{\sqrt{v}}$

17. $(\ln x)^{\ln x}$

18. $x^{\log_2 x}$

Evaluate the integrals in Exercises 19–32.

19. $\displaystyle\int_0^1 3x^{\sqrt{3}}\, dx$

20. $\displaystyle\int_0^2 x^{\sqrt{2}-1}\, dx$

21. $\displaystyle\int_0^3 (\sqrt{2}+1)\, x^{\sqrt{2}}\, dx$

22. $\displaystyle\int_1^e x^{\ln 2 - 1}\, dx$

23. $\displaystyle\int_0^1 5^x\, dx$

24. $\displaystyle\int_{-1}^0 2^x\, dx$

25. $\displaystyle\int_0^1 \frac{1}{2^x}\, dx$

26. $\displaystyle\int_{-1}^1 2^{(x+1)}\, dx$

27. $\displaystyle\int_{-1}^0 4^{-x}\ln 2\, dx$

28. $\displaystyle\int_{-2}^0 5^{-x}\, dx$

29. $\displaystyle\int_1^{\sqrt{2}} x 2^{x^2}\, dx$

30. $\displaystyle\int_0^{\pi/2} 2^{\cos x}\sin x\, dx$

31. $\displaystyle\int_2^4 x^{2x}(1+\ln x)\, dx$

32. $\displaystyle\int_1^2 \frac{2^{\ln x}}{x}\, dx$

Express the numbers in Exercises 33 and 34 in terms of natural logarithms and simplify.

33. a) $\log_4 16$ b) $\log_{32} 8$ c) $\log_4 \sqrt{2}$

34. a) $\log_5 0.04$ b) $\log_{0.5} 4$ c) $\log_{0.2} \sqrt{5}$

Rewrite the expressions in Exercises 35 and 36 as expressions that involve neither exponentials nor logarithms.

35. a) $2^{\log_4 x}$ b) $9^{\log_3 x}$ c) $\log_2(e^{(\ln 2)(\sin x)})$

36. a) $25^{\log_5(3x^2)}$ b) $\log_e(e^x)$ c) $\log_4(2^{e^x \sin x})$

Express the ratios in Exercises 37 and 38 as ratios of natural logarithms and simplify.

37. a) $\dfrac{\log_2 x}{\log_3 x}$ b) $\dfrac{\log_2 x}{\log_8 x}$ c) $\dfrac{\log_x a}{\log_{x^2} a}$

38. a) $\dfrac{\log_9 x}{\log_3 x}$ b) $\dfrac{\log_{\sqrt{10}} x}{\log_{\sqrt{2}} x}$ c) $\dfrac{\log_a b}{\log_b a}$

Solve the equations in Exercises 39 and 40 for x.

39. $3^{\log_3 7} + 2^{\log_2 5} = 5^{\log_5 x}$

40. $8^{\log_8 3} - e^{\ln 5} = x^2 - 7^{\log_7 3x}$

In Exercises 41–56, find the derivative of the expression with respect to x, r, s, t, or θ, whichever is appropriate.

41. $\log_4 x + \log_4 x^2$

42. $\log_{25} e^x - \log_5 \sqrt{x}$

43. $\log_2 r \cdot \log_3 r$

44. $r \log_4 (1/r)$

45. $\log_{10} \sqrt{s+1}$

46. $\log_3 (3s^2 + 1)$

47. $1/(\log_6 t)$

48. $\log_2 (8t^{\ln 2})$

49. $\log_{\ln 2} x^{\ln 2}$

50. $x \sin (\log_7 x)$

51. $\log_5 e^\theta$

52. $3^{\log_2 \theta}$

53. $\log_2 \left(\dfrac{x^2 e^x}{2\sqrt{x+1}}\right)$

54. $\log_7 \left(\dfrac{\sin x \cos x}{e^x 2^x}\right)$

55. $3 \log_8 (\log_2 t)$

56. $t \log_3 (e^{(\sin t)(\ln 3)})$

Evaluate the integrals in Exercises 57–66.

57. $\displaystyle\int \frac{\log_{10} x}{x}\, dx$

58. $\displaystyle\int_1^4 \frac{\log_2 x}{x}\, dx$

59. $\displaystyle\int_1^4 \frac{\ln 2 \, \log_2 x}{x}\, dx$

60. $\displaystyle\int_1^e \frac{2 \ln 10 \, \log_{10} x}{x}\, dx$

61. $\displaystyle\int_0^2 \frac{\log_2 (x+2)}{x+2}\, dx$

62. $\displaystyle\int_{1/10}^{10} \frac{\log_{10} (10x)}{x}\, dx$

63. $\displaystyle\int_0^9 \frac{2 \log_{10} (x+1)}{x+1}\, dx$

64. $\displaystyle\int_2^3 \frac{2 \log_2 (x-1)}{x-1}\, dx$

65. $\displaystyle\int \frac{dx}{x \log_{10} x}$

66. $\displaystyle\int \frac{dx}{x (\log_8 x)^2}$

Find the limits in Exercises 67–74.

67. $\displaystyle\lim_{x\to 0} \frac{3^{\sin x} - 1}{x}$

68. $\displaystyle\lim_{x\to 0} \frac{(1/2)^x - 1}{x}$

69. $\displaystyle\lim_{x\to 1^+} x^{1/(1-x)}$

70. $\displaystyle\lim_{x\to 0^+} (e^x + x)^{1/x}$

71. $\displaystyle\lim_{x\to 0^+} x^x$

72. $\displaystyle\lim_{x\to 0^+} x^{1/\ln x}$

73. $\displaystyle\lim_{x\to\infty} (\ln x)^{1/x}$

74. $\displaystyle\lim_{x\to\infty} (1 + 2x)^{1/(2 \ln x)}$

75. CALCULATOR *Linear and quadratic approximations of $\log_3 x$.*
   a) Find the linearization and quadratic approximation of $f(x) = \log_3 x$ at $x = 3$. Give the coefficients to five decimal places.
   b) Estimate the errors in these approximations on the interval [2.9, 3.1], again to five decimal places.

76. CALCULATOR *Linear and quadratic approximations of 2^x.*
   a) Find the linearization and quadratic approximation of $f(x) = 2^x$ at $x = 0$. Give the coefficients to five decimal places.
   b) Estimate the errors in these approximations on the interval [-1, 1], again to five decimal places.

77. CALCULATOR The curves $y = x^2$ and $y = 2^x$ intersect at $x = 2$ and $x = 4$. There is also a third intersection between $x = -1$ and $x = 0$ (Fig. 6.18). Find the coordinates of that

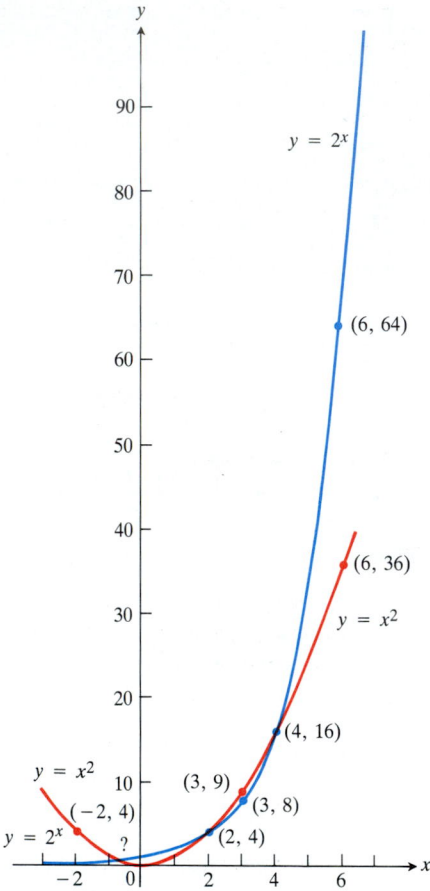

6.18   The curves $y = x^2$ and $y = 2^x$ intersect at $x = 2$, $x = 4$, and $x = ?$ See Exercise 77.

third intersection as accurately as your calculator will allow, by solving the equation $x^2 = 2^x$ numerically.

78. *Blood pH.*   The pH of human blood normally falls between 7.37 and 7.44. Find the corresponding bounds for the hydronium ion concentration $[H_3O^+]$.

79. In any solution, the product of the hydronium ion concentration $[H_3O^+]$ (moles/L) and the hydroxyl ion concentration $[OH^-]$ (moles/L) is about 10^{-14}.
   a) What value of $[H_3O^+]$ minimizes the sum of the concentrations, $S = [H_3O^+] + [OH^-]$? (*Hint:* Change notation. Let $x = [H_3O^+]$.)
   b) What is the pH of a solution in which S has this minimum value?
   c) What ratio of $[H_3O^+]$ to $[OH^-]$ minimizes S?

80. Given that $x > 0$, find the maximum value, if any, of (a) $x^{1/x}$; (b) x^{1/x^2}; (c) x^{1/x^n} (n a positive integer). (d) Show that $\lim_{x \to \infty} x^{1/x^n} = 1$ for every positive integer n.

81. *Audio amplifier.*   By what factor k do you have to multiply the intensity I of the sound from your audio amplifier to add 10 db to the sound level?

82. CALCULATOR   Find out how close you can come to
$$e = \lim_{n \to \infty} \left(1 + \frac{1}{n}\right)^n = 2.7 \ 1828 \ 1828 \ 45 \ 90 \ 45 \dots$$
with your calculator by taking $n = 10, 10^2, 10^3, \dots$. You can expect the approximations to approach e at first, but on some calculators they will move away again as round-off errors take their toll.

83. *The continuous extension of $(\sin x)^x$ to $[0, \pi]$.*
   a) GRAPHER   Graph $f(x) = (\sin x)^x$ on the interval $0 \leq x \leq \pi$. What value would you assign to f to make it continuous at $x = 0$?
   b) Verify your conclusion in (a) by finding $\lim_{x \to 0^+} f(x)$ with l'Hôpital's rule.
   c) GRAPHER   Returning to the graph, estimate the maximum value of f on $[0, \pi]$. About where is max f taken on?
   d) GRAPHER   Sharpen your estimate in (c) by graphing f' in the same window to see where its graph crosses the x-axis. To simplify your work, you might want to delete the exponential factor from the expression for f' and graph just the factor that has a zero.
   e) ROOT FINDER   Sharpen your estimate of the location of max f further still by solving the equation $f' = 0$ numerically.
   f) CALCULATOR   Estimate max f by evaluating f at the locations you found in (c), (d), and (e). What is your best value for max f?

84. *(Continuation of Exercise 83) The function $(\sin x)^{\tan x}$.*
   a) GRAPHER   Graph $f(x) = (\sin x)^{\tan x}$ on the interval $-7 \leq x \leq 7$. How do you account for the gaps in the graph? How wide are the gaps?
   b) GRAPHER   Now graph f on the interval $0 \leq x \leq \pi$. The function is not defined at $x = \pi/2$, but the graph has no break at this point. What is going on? What value does the graph appear to give for f at $x = \pi/2$? (*Hint:* Use l'Hôpital's rule to find $\lim f$ as $x \to (\pi/2)^-$ and $x \to (\pi/2)^+$.)
   c) GRAPHER AND ROOT FINDER   Continuing with the graph in (b), find max f and min f as accurately as you can and estimate the values of x at which they are taken on.

85. GRAPHER   *The place of $\ln x$ among the powers of x.*   The natural logarithm
$$\ln x = \int_1^x \frac{1}{t}\, dt$$
fills the gap in the set of formulas
$$\int t^{k-1}\, dt = \frac{t^k}{k} + C, \qquad k \neq 0, \qquad (18)$$
but the formulas themselves do not reveal how well the logarithm fits in. We can see the nice fit graphically if we select from Eq. (18) the specific antiderivatives
$$\int_1^x t^{k-1}\, dt = \frac{x^k - 1}{k}, \qquad x > 0,$$

and compare their graphs with the graph of $\ln x$.

a) Graph the functions $f(x) = (x^k - 1)/k$ together with $\ln x$ on the interval $0 \le x \le 50$ for $k = \pm 1, \pm 0.5, \pm 0.1,$ and ± 0.05.

b) Show that $\lim\limits_{k \to 0} \dfrac{x^k - 1}{k} = \ln x$.

(Based on "The Place of $\ln x$ Among the Powers of x" by Henry C. Finlayson, *American Mathematical Monthly,* Vol. 94, No. 5, May 1987, p. 450.)

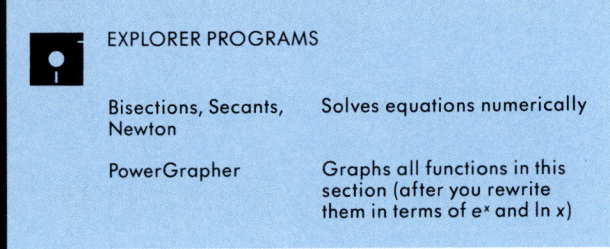

EXPLORER PROGRAMS	
Bisections, Secants, Newton	Solves equations numerically
PowerGrapher	Graphs all functions in this section (after you rewrite them in terms of e^x and $\ln x$)

6.5 Growth and Decay

In this section we derive the well-known law of exponential change and describe some of the applications of logarithmic and exponential functions that account for the importance of these functions in engineering and science.

The Law of Exponential Change

Suppose the quantity y we are interested in—velocity, temperature, electrical current, whatever—grows or decreases at a rate that at any given time t is proportional to the amount present. If we also know the amount present at time $t = 0$, call it y_0, we can find y as a function of t by solving the initial value problem

Differential equation: $\qquad \dfrac{dy}{dt} = ky,$ $\qquad\qquad$ (1)

Initial condition: $\qquad y = y_0$ when $t = 0$.

If y is increasing, then k is positive, and we use Eq. (1) to say that the rate of growth is proportional to what has already been accumulated. If y is decreasing, then k is negative, and we use Eq. (1) to say that the rate of decay is proportional to the amount still left.

We see right away that the constant function $y = 0$ is a solution of Eq. (1). To find what the nonzero solutions are, we divide by y to get, in order,

$$\frac{1}{y}\frac{dy}{dt} = k$$

$$\ln|y| = kt + C \qquad \left(\begin{array}{l}\text{Integrate with respect to } t, \text{ using the}\\ \text{equation } \int (1/u)\,du = \ln|u| + C.\end{array}\right)$$

$$|y| = e^{kt + C} \qquad \text{(Exponentiate.)}$$

$$|y| = e^C \cdot e^{kt} \qquad \text{(A law of exponents)}$$

$$y = \pm e^C e^{kt} \qquad \text{(Eliminate absolute values.)}$$

$$y = Ae^{kt}. \qquad \text{(} A \text{ is a more convenient name for } \pm e^C.\text{)}$$

By allowing A to take on the value 0 (as well as the arbitrary positive and negative values given by the expression $\pm e^C$), we can include the additional solution $y = 0$ in the formula.

We find the right value of A for the initial value problem by solving for A when $y = y_0$ and $t = 0$:

$$y_0 = A\, e^{k \cdot 0} = A.$$

The solution of the initial value problem is

$$y = y_0 \, e^{kt}.$$

We call the solution the Law of Exponential Change.

The Law of Exponential Change

$$y = y_0 \, e^{kt} \tag{2}$$

Growth: $k > 0$ Decay: $k < 0$

The number k is the **rate constant** of the equation.

The derivation of Eq. (2) explains why the only functions that are their own derivatives are constant multiples of the exponential function. The equation $dy/dt = y$ is Eq. (1) with $k = 1$. With $k = 1$, the solutions are all given by the formula $y = y_0 e^t$.

Example 1   *The Growth of a Cell.*   In an ideal environment, the mass m of a cell will grow exponentially, at least early on. Nutrients pass quickly through the cell wall, and growth is limited only by the metabolism within the cell, which in turn depends on the mass of participating molecules. If we make the reasonable assumption that, at each instant of time, the cell's growth rate dm/dt is proportional to the mass that has already been accumulated, then

$$\frac{dm}{dt} = km \qquad \text{and} \qquad m = m_0 \, e^{kt}.$$

There are limitations, of course, and in any particular case we would expect this equation to provide reliable information only for values of m below a certain size.

Example 2   *Birth Rates and Population Growth.*   Strictly speaking, the number of individuals in a population (of people, plants, foxes, or whatever) is a discontinuous function of time because it takes on discrete values. However, as soon as the number of individuals becomes large enough, it may safely be described with a continuous and even differentiable function. If we assume that the proportion of reproducing individuals remains constant and assume a constant fertility, then at any instant t the birth rate is proportional to the number $y(t)$ of individuals present. If, further, we neglect departures, arrivals, and deaths, the growth rate dy/dt will be the same as the birth rate ky. In other words,

$$\frac{dy}{dt} = ky.$$

Once again, we find that $y = y_0 \, e^{kt}.$

Example 3   One model for the way diseases spread assumes that the rate dy/dt at which the number of infected people changes is proportional to the number y itself. The more infected people there are, the faster the disease will spread. The fewer there are, the slower it will spread. Once again,

$$y = y_0 \, e^{kt}.$$

Suppose that in the course of any given year, the number y of cases of a disease is reduced by 20%. If there are 10,000 cases today, how many years will it take to reduce the number of cases to 1000?

Solution   The equation we use is $y = y_0 \, e^{kt}$, and there are three things to find:

1. the value of y_0,
2. the value of k,
3. the value of t that makes $y = 1000$.

STEP 1:   *The value of y_0.* We are free to count time beginning anywhere we want. If we start counting from today, then $y = 10,000$ when $t = 0$, so $y_0 = 10,000$. Our equation becomes

$$y = 10,000 \, e^{kt}.$$

STEP 2:   *The value of k.* When $t = 1$, the number of cases will be 80% of its present value, or 8000. Hence

$$10,000 \, e^{k(1)} = 8000$$

$$e^k = 0.8$$

$$\ln e^k = \ln 0.8$$

$$k = \ln 0.8.$$

At any given time t, therefore,

$$y = 10,000 \, e^{(\ln 0.8)t}.$$

STEP 3:   *The value of t that makes $y = 1000$.* Set y equal to 1000 and solve for t:

$$10,000 \, e^{(\ln 0.8)t} = 1000$$

$$e^{(\ln 0.8)t} = 0.1$$

$$(\ln 0.8)t = \ln 0.1 \qquad \text{(Logs of both sides)}$$

$$t = \frac{\ln 0.1}{\ln 0.8} = 10.32. \qquad \text{(Calculator, rounded)}$$

It will take a little more than 10 years to reduce the number of cases to 1000.

Continuously Compounded Interest

If you invest an amount A_0 of money at a fixed annual interest rate r and interest is added to your account k times a year, it turns out that the amount of money you will have at the end of t years is

$$A_t = A_0 \left(1 + \frac{r}{k}\right)^{kt}. \tag{3}$$

The money might be added ("compounded," bankers say) monthly ($k = 12$), weekly ($k = 52$), daily ($k = 365$), or even more frequently, say by the hour or by the minute. But there is still a limit to how much you will earn that way, and the limit is

$$\lim_{k \to \infty} A_t = \lim_{k \to \infty} A_0 \left(1 + \frac{r}{k}\right)^{kt} = A_0 \, e^{rt}, \tag{4}$$

as you will be asked to show in Exercise 16.

The resulting formula for the amount of money in your account after t years is $A(t) = A_0 \, e^{rt}$.

The Continuous Compound Interest Formula

$$A(t) = A_0 \, e^{rt} \tag{5}$$

Interest paid according to this formula is said to be **compounded continuously**. The number r is called the **continuous interest rate.**

Example 4   Suppose you deposit \$621 in a bank account that pays 6% compounded continuously. How much money will you have in the account 8 years later?

Solution   We use Eq. (5) with $A_0 = 621$, $r = 0.06$, and $t = 8$:

$$A(8) = 621 \, e^{(0.06)(8)} = 621 \, e^{0.48} = 1003.58. \qquad \text{(Calculator, nearest cent)}$$

Had the bank paid interest quarterly ($k = 4$ in Eq. 3), the amount in your account would have been an even \$1000. Thus the effect of continuous compounding, as compared with quarterly compounding, has been an addition of \$3.58. A bank might decide it would be worth this additional amount to be able to advertise, "We compound your money every second, night and day—better than that, we compound the interest continuously."

Radioactivity

For radon-222 gas, t is measured in days and $k = 0.18$. For radium-226, which used to be painted on watch dials to make them glow at night (a dangerous practice), t is measured in years and $k = 4.3 \times 10^{-2}$. The decay of radium in the earth's crust is the source of the radon we find in our basements.

When an atom emits some of its mass as radiation, the remainder of the atom re-forms to make an atom of some new element. This process of radiation and change is called **radioactive decay,** and an element whose atoms go spontaneously through this process is called **radioactive.** Thus, radioactive carbon-14 decays into nitrogen; radium, through a number of intervening radioactive steps, decays into lead.

Experiments have shown that at any given time the rate at which a radioactive element decays (as measured by the number of nuclei that change per unit time) is approximately proportional to the number of radioactive nuclei present. Thus, the decay of a radioactive element is described by the equation $dy/dt = -ky$. If y_0 is the number of radioactive nuclei present at time zero, the number still present at any later time t will be $y = y_0 \, e^{-kt}$.

Radioactive Decay Equation

$$y = y_0 \, e^{-kt} \tag{6}$$

Example 5 *The Half-life of a Radioactive Element.* The **half-life** of a radioactive element is the time required for half of the radioactive nuclei present in a sample to decay. It is a remarkable fact that the half-life is a constant that does not depend on the number of radioactive nuclei initially present in the sample.

To see why, let y_0 be the number of radioactive nuclei initially present in the sample. Then the number y present at any later time t will be

$$y = y_0 e^{-kt}.$$

We seek the value of t at which

$$y_0 e^{-kt} = \frac{1}{2} y_0,$$

for this will be the time when the number of radioactive nuclei present equals half the original number. The y_0's cancel in this equation to give

$$e^{-kt} = \frac{1}{2}, \qquad -kt = \ln\frac{1}{2} = -\ln 2, \qquad \text{and} \qquad t = \frac{\ln 2}{k}.$$

This value of t is the half-life of the element. It depends only on the value of k; the number y_0 does not enter in.

$$\text{Half-life} = \frac{\ln 2}{k} \qquad (7)$$

Example 6 *Polonium-210.* The effective radioactive lifetime of polonium-210 is so short we measure it in days rather than years. The number of radioactive atoms remaining after t days in a sample that starts with y_0 radioactive atoms is

$$y = y_0 e^{-4.95 \times 10^{-3} t}.$$

Find the element's half-life.

Solution

$$\text{Half-life} = \frac{\ln 2}{k} \qquad \text{(Eq. (7))}$$

$$= \frac{\ln 2}{4.95 \times 10^{-3}} \qquad \begin{pmatrix}\text{The } k \text{ from polonium's}\\ \text{decay equation}\end{pmatrix}$$

$$= 140 \text{ days} \qquad \text{(Calculator, rounded)}$$

Example 7 *Carbon-14.* People who do carbon-14 dating use a figure of 5700 years for its half-life (more about carbon-14 dating in the exercises). Find the age of a sample in which 10% of the radioactive nuclei originally present have decayed.

Solution We use the decay equation $y = y_0 e^{-kt}$. There are two things to find:

1. the value of k,
2. the value of t that makes $y_0 e^{-kt} = (9/10)y_0$ or $e^{-kt} = 9/10$.

Carbon-14 Dating

The half-lives of radioactive elements can sometimes be used to date events from the Earth's past. The ages of rocks more than 2 billion years old have been measured by the extent of the radioactive decay of uranium (half-life 4.5 billion years!). In a living organism, the ratio of radioactive carbon, carbon-14, to ordinary carbon stays fairly constant during the lifetime of the organism, being approximately equal to the ratio in the organism's surroundings at the time. After the organism's death, however, no new carbon is ingested, and the proportion of carbon-14 in the organism's remains decreases as the carbon-14 decays. Since the half-life of carbon-14 is known to be about 5700 years, it is possible to estimate the age of organic remains by comparing the proportion of carbon-14 they contain with the proportion assumed to have been in the organism's environment at the time it lived. Archeologists have dated shells (which contain $CaCO_3$), seeds, and wooden artifacts this way. The estimate of 15,500 years for the age of the cave paintings at Lascaux, France, is based on carbon-14 dating. After generations of controversy, the Shroud of Turin was shown by carbon-14 dating in 1988 to have been made later than 1200 A.D.

STEP 1: *The value of* k: We use the half-life equation:

$$k = \frac{\ln 2}{\text{half-life}}$$

$$= \frac{\ln 2}{5700} = 1.2 \times 10^{-4}$$

STEP 2: *The value of* t *that makes* $e^{-kt} = 9/10$:

$$e^{-1.2 \times 10^{-4}t} = 0.9$$

$$-1.2 \times 10^{-4}t = \ln(0.9) \qquad \text{(Logs of both sides)}$$

$$t = \frac{\ln(0.9)}{-1.2 \times 10^{-4}} = 878 \text{ years} \qquad \text{(Calculator, rounded)}$$

The sample is about 878 years old.

Heat Transfer: Newton's Law of Cooling

Soup left in a tin cup cools to the temperature of the surrounding air. A hot silver ingot immersed in water cools to the temperature of the surrounding water. In situations like these, the rate at which an object's temperature is changing at any given time is roughly proportional to the difference between its temperature and the temperature of the surrounding medium. This observation is called **Newton's law of cooling**, although it applies to warming as well. It can be written as an equation in the following way.

If T is the temperature of the object at time t, and T_s is the surrounding temperature, then

$$\frac{dT}{dt} = -k(T - T_s). \tag{8}$$

Since T_s is constant, Eq. (8) is the same as $dy/dt = -ky$ with $y = (T - T_s)$. Hence, the solution of Eq. (8) is

$$y = y_0\, e^{-kt},$$

or

$$T - T_s = (T_0 - T_s)e^{-kt}, \tag{9}$$

where T_0 is the value of T at time zero.

Newton's Law of Cooling

$$T - T_s = (T_0 - T_s)e^{-kt} \tag{10}$$

Example 8   A hard-boiled egg at 98°C is put in a sink of 18°C water to cool. After 5 minutes, the egg's temperature is found to be 38°C. Assuming that the water has not warmed appreciably, how much longer will it take the egg to reach 20°C?

Solution   We find how long it would take the egg to cool from 98°C to 20°C and subtract the 5 minutes that have already elapsed.

According to Eq. (10), the egg's temperature t minutes after it is put in the sink is

$$T - 18 = (98 - 18)e^{-kt}, \quad \text{or} \quad T = 18 + 80e^{-kt}.$$

To find k, we use the information that $T = 38$ when $t = 5$. This gives

$$38 = 18 + 80e^{-5k}$$

$$e^{-5k} = \frac{1}{4}$$

$$-5k = \ln\frac{1}{4} = -\ln 4$$

$$k = \frac{1}{5}\ln 4 = 0.28$$

(to two decimal places). The egg's temperature at time t is

$$T = 18 + 80e^{-0.28t}.$$

When will $T = 20$? When

$$20 = 18 + 80e^{-0.28t}$$

$$80e^{-0.28t} = 2$$

$$e^{-0.28t} = \frac{1}{40}$$

$$-0.28t = \ln\frac{1}{40} = -\ln 40$$

$$t = \frac{\ln 40}{0.28} = 13 \text{ min.} \qquad \text{(To two digits)}$$

The egg's temperature will reach 20°C 13 minutes after it is put in the water to cool. Since it took 5 minutes to reach 38°C, it will take 8 more to reach 20°C.

EXERCISES 6.5

CALCULATOR HELPFUL   The answers to most of the following exercises are in terms of logarithms and exponentials. A calculator will enable you to express the answers in decimal form.

1. *Human evolution continues.*   The analysis of tooth shrinkage by C. Loring Brace and colleagues at the University of Michigan's Museum of Anthropology indicates that human tooth size is continuing to decrease and that the evolutionary process did not come to a halt some 30,000 years ago as many scientists contend. In northern Europeans, for example, tooth size reduction now has a rate of 1% per 1000 years.

a) If t represents time in years and y represents tooth size, use the condition that $y = 0.99y_0$ when $t = 1000$ to find the value of k in the equation $y = y_0\, e^{kt}$. Then use this value of k to answer the following questions.

b) In about how many years will human teeth be 90% of their present size?

c) What will be our descendants' tooth size 20,000 years from now (as a percentage of our present tooth size)?

(Source: *LSA Magazine*, Spring 1989, Vol. 12, No. 2, p. 19, Ann Arbor, MI.)

2. *Atmospheric pressure.*   The earth's atmospheric pressure p

is often modeled by assuming that the rate dp/dh at which p changes with the altitude h above sea level is proportional to p. Suppose that the pressure at sea level is 1013 millibars (about 14.7 pounds per square inch) and that the pressure at an altitude of 20 km is 90 millibars.

a) Solve the equation $dp/dh = kp$ (k a constant) to express p in terms of h. Determine the values of k and the constant of integration from the given initial conditions.

b) What is the atmospheric pressure at $h = 50$ km?

c) At what altitude is the pressure equal to 900 millibars?

3. *First order chemical reactions.*   In some chemical reactions, the rate at which the amount of a substance changes with time is proportional to the amount present. For the change of δ-glucono lactone into gluconic acid, for example,

$$\frac{dy}{dt} = -0.6y$$

when t is measured in hours. If there are 100 grams of δ-glucono lactone present when $t = 0$, how many grams will be left after the first hour?

4. *The inversion of sugar.*   The processing of raw sugar has a step called "inversion" that changes the sugar's molecular structure. Once the process has begun, the rate of change of the amount of raw sugar is proportional to the amount of raw sugar remaining. If 1000 kg of raw sugar reduces to 800 kg of raw sugar during the first 10 hours, how much raw sugar will remain after another 14 hours?

5. *Working underwater.*   The intensity $L(x)$ of light x feet beneath the surface of the ocean satisfies the differential equation

$$\frac{dL}{dx} = -kL.$$

As a diver, you know from experience that diving to 18 ft in the Caribbean Sea cuts the intensity in half. You cannot work without artificial light when the intensity falls below one-tenth of the surface value. About how deep can you expect to work without artificial light?

6. *Voltage in a discharging capacitor.*   Suppose that electricity is draining from a capacitor at a rate that is proportional to the voltage V across its terminals and that, if t is measured in seconds,

$$\frac{dV}{dt} = -\frac{1}{40}V.$$

Solve this equation for V, using V_0 to denote the value of V when $t = 0$. How long will it take the voltage to drop to 10% of its original value?

7. *Cholera bacteria.*   Suppose that the bacteria in a colony can grow unchecked, by the law of exponential change. The colony starts with 1 bacterium and doubles every half hour. How many bacteria will the colony contain at the end of 24 hr? (Under favorable laboratory conditions, the number of cholera bacteria can double every 30 min. In an infected

person, many bacteria are destroyed, but this example helps explain why a person who feels well in the morning may be dangerously ill by evening.)

8. *Growth of bacteria.*   A colony of bacteria is grown under ideal conditions in a laboratory so that the population increases exponentially with time. At the end of 3 hr there are 10,000 bacteria. At the end of 5 hr there are 40,000. How many bacteria were present initially?

9. *The incidence of a disease (continuation of Example 3).* Suppose that in any given year the number of cases can be reduced by 25% instead of 20%.

a) How long will it take to reduce the number of cases to 1000?

b) How long will it take to eradicate the disease, that is, reduce the number of cases to less than 1?

10. *The U.S. population.*   The Museum of Science in Boston displays a running total of the U.S. population. On April 10, 1991, the total was increasing at the rate of 1 person every 21 sec. The displayed population figure for 3:30 P.M. that day was 252,360,611.

a) Assuming exponential growth at a constant rate, find the rate constant for the population's growth (people per year).

b) At this rate, what will the U.S. population be at 3:30 P.M. Boston time on April 10, 2001?

11. *Oil depletion.*   Suppose the amount of oil pumped from one of the canyon wells in Whittier, California, decreases at the continuous rate of 10% per year. When will the well's output fall to one-fifth of its present value?

12. *Continuous price discounting.*   To encourage buyers to place 100-unit orders, your firm's sales department applies a continuous discount that makes the unit price a function $p(x)$ of the number of units x ordered. The discount decreases the price at the rate of $0.01 per unit ordered. The price per unit for a 100-unit order is $p(100) = 20.09.

a) Find $p(x)$ by solving the following initial value problem:

   Differential equation: $\quad \dfrac{dp}{dx} = -\dfrac{1}{100}p,$

   Initial condition: $\qquad p(100) = 20.09.$

b) Find the unit price $p(10)$ for a 10-unit order and the unit price $p(90)$ for a 90-unit order.

c) The sales department has asked you to find out if it is discounting so much that the firm's revenue, $r(x) = x \cdot p(x)$, will actually be less for a 100-unit order than, say, for a 90-unit order. Reassure them by showing that r has its maximum value at $x = 100$.

d) GRAPHER   Graph the revenue function $r(x) = xp(x)$ for $0 \le x \le 200$.

13. *John Napier's question.*   John Napier was the first person to answer the question What happens if you invest an amount of money at 100% interest, compounded continuously?

a) What does happen?

b) How long does it take to triple your money?

c) How much can you earn in a year?

14. *Benjamin Franklin's will.* The Franklin Technical Institute of Boston owes its existence to a provision in a codicil to Benjamin Franklin's will. In part it reads:

I wish to be useful even after my Death, if possible, in forming and advancing other young men that may be serviceable to their Country in both Boston and Philadelphia. To this end I devote Two thousand Pounds Sterling, which I give, one thousand thereof to the Inhabitants of the Town of Boston in Massachusetts, and the other thousand to the inhabitants of the City of Philadelphia, in Trust and for the Uses, Interests and Purposes hereinafter mentioned and declared.

Franklin's plan was to lend money to young apprentices at 5% interest with the provision that each borrower should pay each year

. . . with the yearly Interest, one tenth part of the Principal, which sums of Principal and Interest shall be again let to fresh Borrowers. . . . If this plan is executed and succeeds as projected without interruption for one hundred Years, the Sum will then be one hundred and thirty-one thousand Pounds of which I would have the Managers of the Donation to the Inhabitants of the Town of Boston, then lay out at their discretion one hundred thousand Pounds in Public Works. . . . The remaining thirty-one thousand Pounds, I would have continued to be let out on Interest in the manner above directed for another hundred Years. . . . At the end of this second term if no unfortunate accident has prevented the operation the sum will be Four Millions and Sixty-one Thousand Pounds.

It was not always possible to find as many borrowers as Franklin had planned, but the managers of the trust did the best they could. At the end of 100 years from the reception of the Franklin gift, in January 1894, the fund had grown from 1000 pounds to almost exactly 90,000 pounds. In 100 years the original capital had multiplied about 90 times instead of the 131 times Franklin had imagined.

What rate of interest, compounded continuously for 100 years, would have multiplied Benjamin Franklin's original capital by 90?

15. *(Continuation of Exercise 14)* In Benjamin Franklin's estimate that the original 1000 pounds would grow to 131,000 in 100 years, he was using an annual rate of 5% and compounding once each year. What rate of interest per year when compounded continuously for 100 years would multiply the original amount by 131?

16. *The continuous compound interest formula.* Use the equation $\lim_{x \to 0^+} (1 + x)^{1/x} = e$ from Section 6.4 (Eq. 7) to confirm the statement that

$$\lim_{k \to \infty} A_0 \left(1 + \frac{r}{k} \right)^{kt} = A_0 e^{rt}.$$

17. *Radon-222.* The decay equation for radon-222 gas is $y = y_0 e^{-0.18t}$, with t in days. About how long will it take the radon in a sealed sample of air to fall to 90% of its original value?

18. *Polonium-210.* The half-life of polonium is 140 days, but your sample will not be useful to you after 95% of the radioactive nuclei present on the day the sample arrives has disintegrated. For about how many days after the sample arrives will you be able to use the polonium?

19. *The mean life of a radioactive nucleus.* Physicists using the radioactivity equation $y = y_0 e^{-kt}$ call the number $1/k$ the *mean life* of a radioactive nucleus. The mean life of a radon nucleus is about $1/0.18 = 5.6$ days. The mean life of a carbon-14 nucleus is more than 8000 years. Show that 95% of the radioactive nuclei originally present in a sample will disintegrate within three mean lifetimes, i.e., by time $t = 3/k$. Thus, the mean life of a nucleus gives a quick way to estimate how long the radioactivity of a sample will last.

20. *Californium-252.* What costs $27 million per gram and can be used to treat brain cancer, analyze coal for its sulfur content, and detect explosives in luggage? The answer is Californium-252, a radioactive isotope so rare that only 8 g of it have been made in the western world since its discovery by Glenn Seaborg in 1950. The half-life of the isotope is 2.645 years—long enough for a useful service life and short enough to have a high radioactivity per unit mass. One microgram of the isotope releases 170 million neutrons per second.

a) What is the value of k in the decay equation for this isotope?

b) What is the isotope's mean life? (See Exercise 19.)

c) How long will it take 95% of a sample's radioactive nuclei to disintegrate?

21. *Cooling soup.* Suppose that a cup of soup cooled from 90°C to 60°C after 10 minutes in a room whose temperature was 20°C. Use Newton's law of cooling to answer the following questions.

a) How much longer would it take the soup to cool to 35°C?

b) Instead of being left to stand in the room, the cup of 90°C soup is put in a freezer whose temperature is −15°C. How long will it take the soup to cool from 90°C to 35°C?

22. *A beam of unknown temperature.* An aluminum beam was brought into a machine shop where the temperature was held at 65°. After 10 minutes, the beam's temperature was 35°F and after another 10 minutes it was 50°F. Use Newton's law of cooling to estimate the beam's initial temperature.

23. *Surrounding medium of unknown temperature.* A pan of warm water (46°C) was put in a refrigerator. Ten minutes later, the water's temperature was 39°C; 10 minutes after that, it was 33°C. Use Newton's law of cooling to estimate how cold the refrigerator was.

24. *Silver cooling in air.*   The temperature of an ingot of silver is 60°C above room temperature right now. Twenty minutes ago, it was 70°C above room temperature. How far above room temperature will the silver be 15 minutes from now? Two hours from now? When will the silver be 10°C above room temperature?

25. The charcoal from a tree killed in the volcanic eruption that formed Crater Lake in Oregon contained 44.5% of the carbon-14 found in living matter. About how old is Crater Lake?

26. To see the effect of a relatively small error in the estimate of the amount of carbon-14 in a sample being dated, consider this hypothetical situation:

a) A fossilized bone found in central Illinois in the year 2000 A.D. contains 17% of its original carbon-14 content. Estimate the year the animal died.

b) Repeat (a) assuming 18% instead of 17%.

c) Repeat (a) assuming 16% instead of 17%.

27. *Art forgery.*   A painting attributed to Vermeer (1632–1675), which should contain no more than 96.2% of its original carbon-14, contains 99.5% instead. About how old is the forgery?

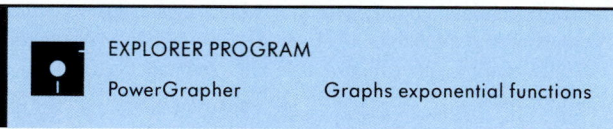

EXPLORER PROGRAM

PowerGrapher   Graphs exponential functions

6.6 The Rates at Which Functions Grow

This section shows how to compare the rates at which functions of x grow as x becomes large and introduces the so-called little-oh and big-oh notation sometimes used to describe the results of these comparisons. The exponential function e^x grows so rapidly as $x \to \infty$, and the logarithmic function $\ln x$ grows so slowly, that they set the standards by which all other functions are judged.

We shall restrict our attention to functions whose values eventually become and remain positive as $x \to \infty$. (Functions with negative values can be compared by taking absolute values first.)

Relative Rates of Growth

You may have noticed that exponential functions like 2^x and e^x seem to grow more rapidly as x gets large than the polynomials and rational functions we graphed in Chapter 3. These exponentials certainly grow more rapidly than x itself, as Figs. 6.10 and 6.16 show, and you can see 2^x outgrowing x^2 as x increases in Fig. 6.18. In fact, as $x \to \infty$, the functions 2^x and e^x grow faster than any power of x, even $x^{1,000,000}$, as you will be asked to confirm in Exercise 9.

To get a feeling for how rapidly the values of $y = e^x$ grow with increasing x, think of graphing the function on a large blackboard, with the axes scaled in centimeters. At $x = 1$ cm, the graph is $e^1 \approx 3$ cm above the x-axis. At $x = 6$ cm, the graph is $e^6 \approx 403$ cm ≈ 4 m high (it is about to go through the ceiling if it hasn't done so already). At $x = 10$ cm, the graph is $e^{10} \approx 22{,}026$ cm ≈ 220 m high, higher than most buildings. At $x = 24$ cm, the graph is more than halfway to the moon, and at $x = 43$ cm from the origin, the graph is high enough to reach past the nearest neighboring star, Proxima Centauri:

$$e^{43} \approx 4.7 \times 10^{18} \text{ cm}$$

$$= 4.7 \times 10^{13} \text{ km}$$

$$\approx 1.57 \times 10^8 \text{ light-seconds} \qquad \left(\begin{matrix}\text{In a vacuum, light travels at 300,000} \\ \text{km/sec.}\end{matrix}\right) \quad (1)$$

$$\approx 5.0 \text{ light-years}$$

The distance to Proxima Centauri is about 4.22 light-years. Yet with $x = 43$ cm from the origin, the graph is still less than 2 feet to the right of the y-axis.

In contrast, logarithmic functions like $y = \log_2 x$ and $y = \ln x$ grow more slowly as $x \to \infty$ than any positive power of x (see Exercise 11). With axes scaled in centimeters, you have to go nearly 5 light-years out on the x-axis to find a point where the graph of $y = \ln x$ is even $y = 43$ cm high.

These important comparisons of exponential, polynomial, and logarithmic functions can be made precise by defining what it means for a function $f(x)$ to grow faster than a function $g(x)$ as $x \to \infty$.

DEFINITION

Rates of Growth as $x \to \infty$

1. f **grows faster than** g as $x \to \infty$ if

$$\lim_{x \to \infty} \frac{f(x)}{g(x)} = \infty \tag{2}$$

or, equivalently, if

$$\lim_{x \to \infty} \frac{g(x)}{f(x)} = 0. \tag{3}$$

2. If f grows faster than g as $x \to \infty$, we also say that g **grows slower than** f as $x \to \infty$.

3. f and g **grow at the same rate** as $x \to \infty$ if

$$\lim_{x \to \infty} \frac{f(x)}{g(x)} = L \neq 0. \qquad (L \text{ finite and not zero}) \tag{4}$$

According to these definitions, $y = 2x$ does not grow faster than $y = x$. The two functions grow at the same rate because

$$\lim_{x \to \infty} \frac{2x}{x} = \lim_{x \to \infty} 2 = 2,$$

which is a finite, nonzero limit. The reason for this apparent disregard of common sense is that we want "f grows faster than g" to mean that for large x-values, g is negligible when compared with f.

Example 1 e^x grows faster than x^2 as $x \to \infty$ because

$$\lim_{x \to \infty} \frac{e^x}{x^2} = \lim_{x \to \infty} \frac{e^x}{2x} = \lim_{x \to \infty} \frac{e^x}{2} = \infty,$$

as we see by two applications of l'Hôpital's rule.

Example 2

a) 3^x grows faster than 2^x as $x \to \infty$ because

$$\lim_{x \to \infty} \frac{3^x}{2^x} = \lim_{x \to \infty} \left(\frac{3}{2} \right)^x = \infty.$$

b) As part (a) may have suggested, exponential functions with different bases never grow at the same rate as $x \to \infty$. If $a > b > 0$, then a^x grows faster than b^x. Since $(a/b) > 1$,

$$\lim_{x \to \infty} \frac{a^x}{b^x} = \lim_{x \to \infty} \left(\frac{a}{b}\right)^x = \infty.$$

Example 3 x^2 grows faster than $\ln x$ as $x \to \infty$ because

$$\lim_{x \to \infty} \frac{x^2}{\ln x} = \lim_{x \to \infty} \frac{2x}{1/x} \qquad \text{(l'Hôpital's rule)}$$

$$= \lim_{x \to \infty} 2x^2 = \infty.$$

Example 4 $\ln x$ grows slower than x as $x \to \infty$ because

$$\lim_{x \to \infty} \frac{\ln x}{x} = \lim_{x \to \infty} \frac{1/x}{1} \qquad \text{(l'Hôpital's rule)}$$

$$= \lim_{x \to \infty} \frac{1}{x} = 0.$$

Example 5   In contrast to exponential functions, logarithmic functions always grow at the same rate as $x \to \infty$. For any bases a and b,

$$\lim_{x \to \infty} \frac{\log_a x}{\log_b x} = \lim_{x \to \infty} \frac{\ln x / \ln a}{\ln x / \ln b} = \frac{\ln b}{\ln a}.$$

The limiting ratio is always finite and never zero.

If f grows at the same rate as g as $x \to \infty$, and g grows at the same rate as h as $x \to \infty$, then f grows at the same rate as h as $x \to \infty$. The reason is that

$$\lim_{x \to \infty} \frac{f}{g} = L_1 \qquad \text{and} \qquad \lim_{x \to \infty} \frac{g}{h} = L_2$$

together imply

$$\lim_{x \to \infty} \frac{f}{h} = \lim_{x \to \infty} \frac{f}{g} \cdot \frac{g}{h} = L_1 L_2.$$

If L_1 and L_2 are finite and nonzero, then so is $L_1 L_2$.

Example 6   Show that $\sqrt{x^2 + 5}$ and $(2\sqrt{x} - 1)^2$ grow at the same rate as $x \to \infty$.

Solution   We show that the functions grow at the same rate by showing that they both grow at the same rate as x:

$$\lim_{x \to \infty} \frac{\sqrt{x^2 + 5}}{x} = \lim_{x \to \infty} \sqrt{1 + \frac{5}{x^2}} = 1,$$

$$\lim_{x \to \infty} \frac{(2\sqrt{x} - 1)^2}{x} = \lim_{x \to \infty} \left(\frac{2\sqrt{x} - 1}{\sqrt{x}}\right)^2 = \lim_{x \to \infty} \left(2 - \frac{1}{\sqrt{x}}\right)^2 = 4.$$

Order of Magnitude as $x \to \infty$

Here we introduce the "little-oh" and "big-oh" notation invented by number theorists a hundred years ago and now commonplace in mathematical analysis and computer science.

DEFINITION

A function f is **of smaller order than** g as $x \to \infty$ if $\lim_{x \to \infty} \dfrac{f(x)}{g(x)} = 0$.

We indicate this by writing $f = o(g)$ ("f is little-oh of g").

Notice that saying $f = o(g)$ as $x \to \infty$ is another way of saying that f grows slower than g as $x \to \infty$.

Example 7

$$\ln x = o(x) \text{ as } x \to \infty \quad \text{because} \quad \lim_{x \to \infty} \frac{\ln x}{x} = 0$$

$$x^2 = o(x^3 + 1) \text{ as } x \to \infty \quad \text{because} \quad \lim_{x \to \infty} \frac{x^2}{x^3 + 1} = 0$$

If $f(x)/g(x)$ fails to approach zero as $x \to \infty$ but the ratio remains bounded, we say that f is at most the order of g as $x \to \infty$. What we mean, exactly, is that $f(x)/g(x)$ stays less than or equal to some integer M for x sufficiently large.

DEFINITION

A function f is **of at most the order of** g as $x \to \infty$ if there is a positive integer M for which

$$\frac{f(x)}{g(x)} \leq M,$$

for x sufficiently large. We indicate this by writing $f = O(g)$ ("f is big-oh of g").

Example 8

$$x + \sin x = O(x) \text{ as } x \to \infty \quad \text{because} \quad \frac{x + \sin x}{x} \leq 2 \text{ for } x \text{ sufficiently large.}$$

Example 9

$$e^x + x^2 = O(e^x) \text{ as } x \to \infty \quad \text{because} \quad \frac{e^x + x^2}{e^x} \to 1 \text{ as } x \to \infty,$$

$$x = O(e^x) \text{ as } x \to \infty \quad \text{because} \quad \frac{x}{e^x} \to 0 \text{ as } x \to \infty.$$

If you look at the definitions again, you will see that $f = o(g)$ implies $f = O(g)$. Also, if f and g grow at the same rate, then $f = O(g)$ and $g = O(f)$, as you will be asked to confirm in Exercise 15.

Sequential vs. Binary Search

Computer scientists sometimes measure the efficiency of an algorithm by counting the number of steps a computer must take to make the algorithm do something. There can be significant differences in how efficiently algorithms perform, even if they are designed to accomplish the same task. These differences are often described in big-oh notation. Here is an example.

Webster's *Third New International Dictionary* lists about 26,000 words that begin with the letter *a*. One way to look up a word, or to find out it is not there, is to read through the list one word at a time from the beginning until you either find the word or determine that it is not there. This method, called sequential search, makes no particular use of the words' alphabetical arrangement. You can be sure of getting an answer, but it might take 26,000 steps.

Another way to find the word or to determine it is not there is to go straight to the middle of the list (give or take a few words). If you do not find the word, then go to the middle of the half that contains it and forget about the half that does not. (You know which half contains it because you know the list is ordered alphabetically.) This method eliminates roughly 13,000 words in a single step. If you do not find your word on the second try, then jump to the middle of the half that contains it. Continue this way until you have either found your word or divided the list in half so many times that there are no words left. How many times do you have to divide the list to find the word or determine that it is not there? At most 15, because

$$2^{14} < 26{,}000 < 2^{15}.$$

That certainly beats a possible 26,000 steps.

For a list of length n, a sequential search algorithm takes on the order of n steps to find a word or determine that it is not in the list. A binary search, as the second algorithm is called, takes on the order of $\log_2 n$ steps. The reason is that if $2^{m-1} < n \leq 2^m$, then $m - 1 < \log_2 n \leq m$ and the number of bisections required to narrow the list down to one word will be no more than $m = \lceil \log_2 n \rceil$, the smallest integer greater than or equal to $\log_2 n$.

Big-oh notation provides a compact way to say all this. The number of steps in a sequential search of an ordered list is $O(n)$, while the number of steps in a binary search is $O(\log_2 n)$. In our example, there is a big difference between the two (26,000 vs. 15), and the difference can only increase with n because n grows faster than $\log_2 n$ as $n \to \infty$.

> To find an item in a list of length n:
>
> A sequential search takes $O(n)$ steps.
>
> A binary search takes $O(\log_2 n)$ steps.

EXERCISES 6.6

1. Which of the following functions grow slower than e^x as $x \to \infty$?

a) $x + 3$

b) $x^3 - 3x + 1$

c) $\sqrt{x}$

d) 4^x

e) $(5/2)^x$

f) $\ln x$

g) $\log_{10} x$

h) e^{-x}

i) e^{x+1}

j) $(1/2)e^x$

2. Which of the following functions grow faster than x^2 as $x \to \infty$? Which grow at the same rate as x^2? Which grow slower?

a) $x^2 + 4x$

b) $x^3 + 3$

c) x^5

d) $15x + 3$

e) $\sqrt{x^4 + 5x}$

f) $(x + 1)^2$

g) $\ln x$

h) $\ln(x^2)$

i) $\ln(10^x)$

j) 2^x

3. Which of the following functions grow at the same rate as $\ln x$ as $x \to \infty$?

a) $\log_3 x$

b) $\log_2 x^2$

c) $\log_{10}\sqrt{x}$

d) $1/x$

e) $1/\sqrt{x}$

f) e^{-x}

g) x

h) $5 \ln x$

i) 2

j) $\sin x$

4. Order the following functions from fastest growing to slowest growing as $x \to \infty$.

a) e^x

b) x^x

c) $(\ln x)^x$

d) $e^{x/2}$

5. Show that $\sqrt{10x + 1}$ and $\sqrt{x + 1}$ grow at the same rate as $x \to \infty$ by showing that they both grow at the same rate as $\sqrt{x}$ as $x \to \infty$.

6. Show that $\sqrt{x^4 + x}$ and $\sqrt{x^4 - x^3}$ grow at the same rate as $x \to \infty$ by showing that they both grow at the same rate as x^2 as $x \to \infty$.

7. True, or false? As $x \to \infty$,

a) $x = o(x)$

b) $x = o(x + 5)$

c) $x = O(x + 5)$

d) $x = O(2x)$

e) $e^x = o(e^{2x})$

f) $x + \ln x = O(x)$

g) $\ln x = o(\ln 2x)$

h) $\sqrt{x^2 + 5} = O(x)$

8. True, or false? As $x \to \infty$,

a) $\dfrac{1}{x + 3} = O\left(\dfrac{1}{x}\right)$

b) $\dfrac{1}{x} + \dfrac{1}{x^2} = O\left(\dfrac{1}{x}\right)$

c) $\dfrac{1}{x} - \dfrac{1}{x^2} = o\left(\dfrac{1}{x}\right)$

d) $2 + \cos x = O(2)$

e) $e^x + x = O(e^x)$

f) $x \ln x = o(x^2)$

g) $\ln(\ln x) = O(\ln x)$

h) $\ln(x) = o(\ln(x^2 + 1))$

9. Show that e^x grows faster as $x \to \infty$ than x^n for any positive integer n, even $x^{1,000,000}$. (*Hint:* What is the nth derivative of x^n?)

10. *The function e^x outgrows any polynomial.* Show that e^x grows faster as $x \to \infty$ than any polynomial

$$a_n x^n + a_{n-1} x^{n-1} + \cdots + a_1 x + a_0.$$

11. a) Show that $\ln x$ grows slower as $x \to \infty$ than $x^{1/n}$ for any positive integer n, even $x^{1/1,000,000}$.

b) CALCULATOR   Although the values of $x^{1/1,000,000}$ eventually overtake the values of $\ln x$, you have to go way out on the x-axis before this happens. Find a value of x greater than 1 for which $x^{1/1,000,000} > \ln x$. You might start by observing that when $x > 1$ the equality $\ln x = x^{1/1,000,000}$ is equivalent to $\ln \ln x = (\ln x)/1,000,000$.

c) CALCULATOR   Even $x^{1/10}$ takes a long time to overtake $\ln x$. Experiment with a calculator to find the value of x at which the graphs of $x^{1/10}$ and $\ln x$ cross, or, equivalently, where $\ln x = 10 \ln \ln x$. Bracket the crossing point between powers of 10 and then close in by successive halving.

d) GRAPHER or ROOT FINDER   *(Continuation of part c).* The value of x where $\ln x = 10 \ln \ln x$ is too far out for many graphers and root finders to identify. Try it on the equipment available to you and see what happens.

12. *The function $\ln x$ grows slower than any polynomial.*   Show that $\ln x$ grows slower as $x \to \infty$ than any nonconstant polynomial.

13. Suppose you have four different algorithms for solving the same problem and each algorithm takes a number of steps equal to one of the functions listed here:

$$n, \qquad n \log_2 n, \qquad n^2, \qquad n(\log_2 n)^2.$$

Which of the algorithms, if any, is the most efficient in the long run?

14. CALCULATOR   Suppose you are looking for an item in an ordered list one million items long. How many steps might it take to find that item with a sequential search? A binary search?

15. Show that if functions f and g grow at the same rate as $x \to \infty$ then $f = O(g)$ and $g = O(f)$.

16. a) What do the conclusions we drew in Section 1.8 about the limits of rational functions as $x \to \infty$ tell us about the relative growth rates of polynomials as $x \to \infty$?

b) When is a polynomial f of smaller order than a polynomial g as $x \to \infty$? of at most the order of g as $x \to \infty$?

Order of Magnitude as $x \to a$

We can broaden the application of the notion of order of magnitude if we lift the restriction $x \to \infty$ in our definitions and consider limits as $x \to a$ for any number a. The following exercises show what we mean.

17. *Simpson's rule and the trapezoidal rule.*   Show that the error E_S in the Simpson's rule approximation of a definite integral is $O(h^4)$ as $h \to 0$, while the error E_T in the trapezoidal rule approximation is $O(h^2)$. This gives another way to explain the relative accuracies of the two approximation methods.

18. *Linearization vs. quadratic approximation.* Suppose the function f has a continuous third derivative. Show that the magnitude $|e_2(x)|$ of the error in the quadratic approximation of f at $x = a$ is $O(|x - a|^3)$ as $x \to a$ while the magnitude $|e_1(x)|$ of the error in the linearization at $x = a$ is $O((x - a)^2)$. This gives another way to explain why we expect an improvement in accuracy when we shift from a linearization to a quadratic approximation.

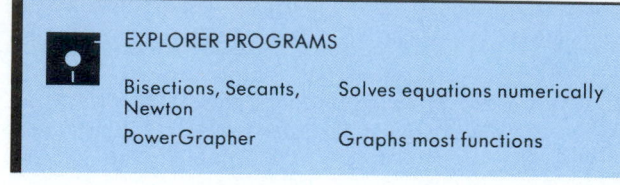

EXPLORER PROGRAMS

Bisections, Secants,   Solves equations numerically
Newton

PowerGrapher   Graphs most functions

6.7 The Inverse Trigonometric Functions

Inverse trigonometric functions arise in problems that require finding angles from side measurements in triangles. They also provide antiderivatives for a wide variety of functions and appear in solutions to a number of differential equations that arise in mathematics, engineering, and physics. In this section we show how the functions are defined, graphed, and evaluated. In the next section, we look at their derivatives and integrals.

The Arc Sine

The function $y = \sin x$ is not one-to-one; it runs through its full range of values from -1 to 1 twice on every interval of length 2π. However, if we restrict the domain of the sine to $[-\pi/2, \pi/2]$, we find that the restricted function

$$y = \sin x, \qquad -\pi/2 \le x \le \pi/2, \tag{1}$$

is one-to-one. It therefore has an inverse, which we denote by

$$y = \sin^{-1}x. \tag{2}$$

This equation is read "y equals the **arc sine** of x" or "y equals arc sine x" and is often written as

$$y = \arcsin x. \tag{3}$$

6.19 Arcs on the unit circle whose lengths represent $\sin^{-1}x$ and $\cos^{-1}x$.

In case you are wondering what the "arc" is doing there, look at Fig. 6.19. The figure gives the geometric interpretation of $y = \sin^{-1}x$ when y is positive. If $x = \sin y$, then y is the *arc* on the unit circle whose sine is x. For every value of x in the interval $[-1, 1]$, $y = \sin^{-1}x$ is the number in the interval $[-\pi/2, \pi/2]$ whose sine is x. For instance (Fig. 6.20),

$$\sin^{-1}0 = 0 \qquad \text{because} \quad \sin 0 = 0,$$

$$\sin^{-1}\sqrt{3}/2 = \pi/3 \qquad \text{because} \quad \sin \pi/3 = \sqrt{3}/2,$$

$$\sin^{-1}1 = \pi/2 \qquad \text{because} \quad \sin \pi/2 = 1,$$

$$\sin^{-1}(-\sqrt{2}/2) = -\pi/4 \qquad \text{because} \quad \sin(-\pi/4) = -\sqrt{2}/2.$$

Figure 6.21 shows the graph of $\sin^{-1}x$. The light blue curve, $x = \sin y$, is the reflection of the graph of $\sin x$ across the line $y = x$. The graph of $\sin^{-1}x$ is the portion running from $y = -\pi/2$ to $y = \pi/2$.

The -1 in $\sin^{-1}x$ means "inverse," not "reciprocal." The *reciprocal* of $\sin x$ is

$$(\sin x)^{-1} = \frac{1}{\sin x} = \csc x.$$

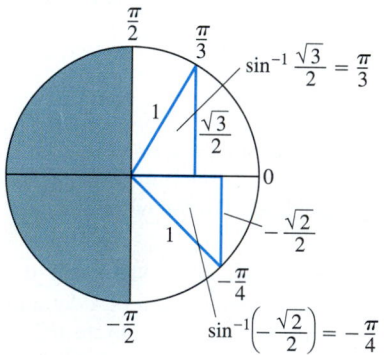

6.20 The angle whose measure is $y = \sin^{-1}x$ ranges from $-\pi/2$ to $\pi/2$.

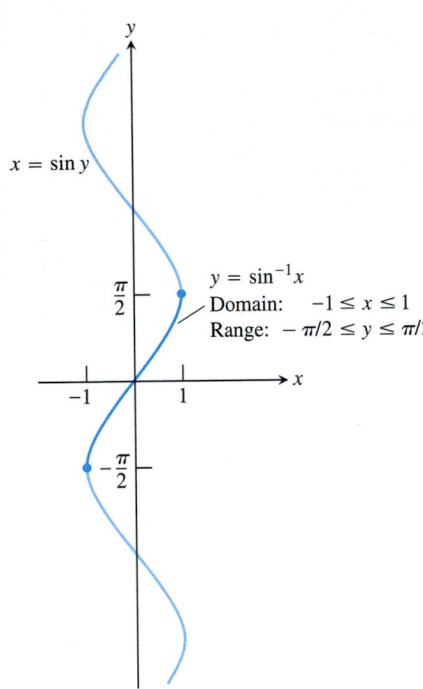

$x = \sin y$

$y = \sin^{-1}x$
Domain: $-1 \le x \le 1$
Range: $-\pi/2 \le y \le \pi/2$

6.21   The graph of $y = \sin^{-1}x$.

The graph of the arc sine in Fig. 6.21 is symmetric about the origin because the graph of $x = \sin y$ is symmetric about the origin. Algebraically, this means that

$$\sin^{-1}(-x) = -\sin^{-1}x \qquad (4)$$

for every x in the domain of the arc sine, which is another way to say that the function $y = \sin^{-1}x$ is odd.

The Arc Cosine

Like the sine function, the cosine function $y = \cos x$ is not one-to-one, but its restriction to the interval $[0, \pi]$,

$$y = \cos x, \qquad 0 \le x \le \pi, \qquad (5)$$

is one-to-one. The restricted function therefore has an inverse,

$$y = \cos^{-1}x, \qquad (6)$$

which we call the **arc cosine** of x. For each value of x in the interval $[-1, 1]$, $y = \cos^{-1}x$ is the number in the interval $[0, \pi]$ whose cosine is x. The graph of $y = \cos^{-1}x$ is shown in Fig. 6.22.

As we can see from Fig. 6.23, the arc cosine of x satisfies the identity

$$\cos^{-1}x + \cos^{-1}(-x) = \pi, \qquad (7)$$

or

$$\cos^{-1}(-x) = \pi - \cos^{-1}x. \qquad (8)$$

And we can see from the triangle in Fig. 6.24 that for $x > 0$,

$$\sin^{-1}x + \cos^{-1}x = \pi/2 \qquad (9)$$

because $\sin^{-1}x$ and $\cos^{-1}x$ are then complementary angles in a right triangle whose hypotenuse is 1 unit long and one of whose legs is x units long. Equation (9) holds for the other values of x in $[-1, 1]$ as well, but we cannot draw this conclusion from the geometry of the triangle in Fig. 6.24. It is, however, a consequence of Eqs. (4) and (8) (see Exercise 46).

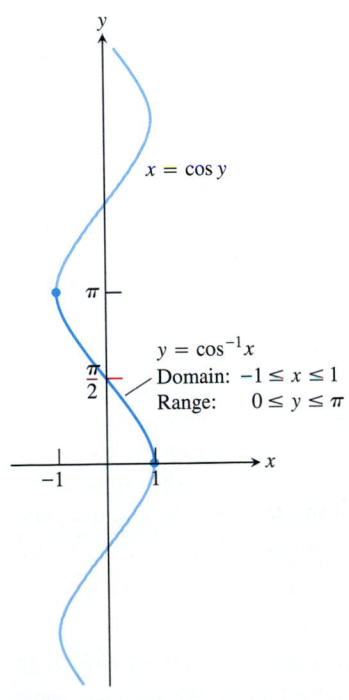

$x = \cos y$

$y = \cos^{-1}x$
Domain: $-1 \le x \le 1$
Range: $0 \le y \le \pi$

6.22   The graph of $\cos^{-1}x$.

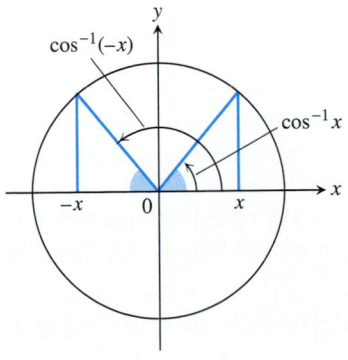

$\cos^{-1}(-x)$

$\cos^{-1}x$

6.23 $\cos^{-1}x + \cos^{-1}(-x) = \pi$

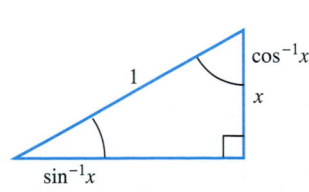

$\cos^{-1}x$

$\sin^{-1}x$

6.24   In this figure,
$$\sin^{-1}x + \cos^{-1}x = \pi/2.$$

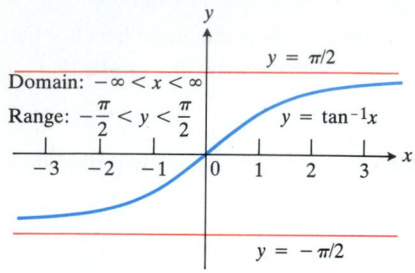

6.25 The branch chosen for $y = \tan^{-1}x$ is the one through the origin.

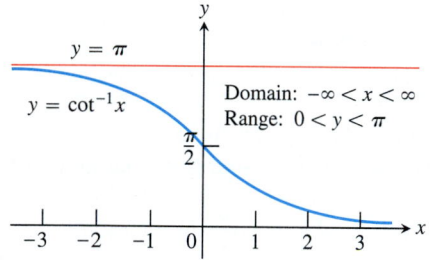

6.26 The graph of
$$y = \cot^{-1}x = \pi/2 - \tan^{-1}x.$$

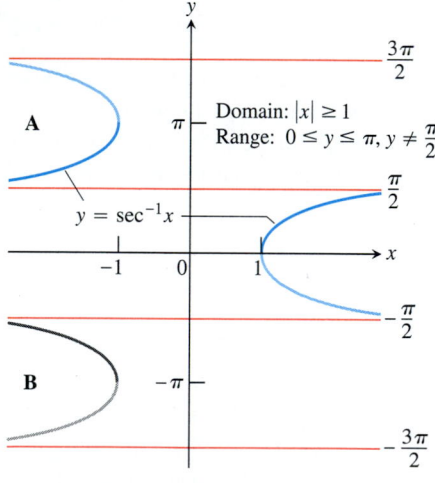

6.27 There are two logical choices for the left-hand branch of $y = \sec^{-1}x$. With choice **A**, Eq. (17) holds, but the formula for the derivative of the arc secant is complicated by absolute value bars. Choice **B** leads to a simpler derivative formula, but Eq. (17) no longer holds. The more frequent choice is **A**, the choice we have made.

Inverses of tan x, sec x, csc x, cot x

The other four basic trigonometric functions, $y = \tan x$, $y = \sec x$, $y = \csc x$, and $y = \cot x$, also have inverses when suitably restricted. The inverse of

$$y = \tan x, \qquad -\pi/2 < x < \pi/2, \tag{10}$$

is denoted by

$$y = \tan^{-1}x. \tag{11}$$

The domain of the arc tangent is the entire real line, and the range is the open interval $(-\pi/2, \pi/2)$. For every value of x, $y = \tan^{-1}x$ is the angle between $-\pi/2$ and $\pi/2$ whose tangent is x. The graph of $y = \tan^{-1}x$ is shown in Fig. 6.25.

The graph of $y = \tan^{-1}x$ is symmetric about the origin because it is a branch of the graph of $x = \tan y$ that is symmetric about the origin. Algebraically, this means that

$$\tan^{-1}(-x) = -\tan^{-1}x. \tag{12}$$

Like the arc sine, the arc tangent is an odd function of x.

The inverses of the (restricted) functions

$$y = \cot x, \qquad 0 < x < \pi, \tag{13}$$

$$y = \sec x, \qquad 0 \le x \le \pi, \quad x \ne \pi/2, \tag{14}$$

$$y = \csc x, \qquad -\pi/2 \le x \le \pi/2, \quad x \ne 0, \tag{15}$$

are chosen to be the functions graphed in Figs. 6.26, 6.27, and 6.28. They are chosen this way to satisfy the relationships

$$\cot^{-1}x = \pi/2 - \tan^{-1}x, \tag{16}$$

$$\sec^{-1}x = \cos^{-1}(1/x), \tag{17}$$

$$\csc^{-1}x = \sin^{-1}(1/x). \tag{18}$$

We shall not dwell on these relationships, but they are handy for graphing and for finding the values of $\cot^{-1}x$, $\sec^{-1}x$, and $\csc^{-1}x$ on a calculator that gives only $\tan^{-1}x$, $\cos^{-1}x$, and $\sin^{-1}x$.

WARNING ABOUT THE ARC SECANT Some people choose $\sec^{-1}x$ to lie between 0 and $\pi/2$ when x is positive and between $-\pi$ and $-\pi/2$ when x is negative (hence as a negative angle in the third quadrant, as shown by the gray curve in Fig. 6.27). This has the advantage of simplifying the formula for the derivative of $\sec^{-1}x$ but the disadvantage of failing to satisfy Eq. (17) when x is negative. Also, some mathematical tables give third-quadrant values for $\sec^{-1}x$ instead of the second-quadrant values used in this book. Watch out for this when you use tables.

Right-Triangle Interpretations

The right-triangle interpretations of the inverse trigonometric functions in Fig. 6.29 can be useful in integration problems that require substitutions. We shall use some of them in Chapter 7.

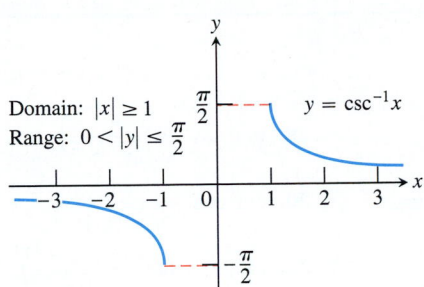

Domain: $|x| \geq 1$
Range: $0 < |y| \leq \dfrac{\pi}{2}$

$y = \csc^{-1}x$

6.28 The graph of
$$y = \csc^{-1}x = \sin^{-1}(1/x).$$

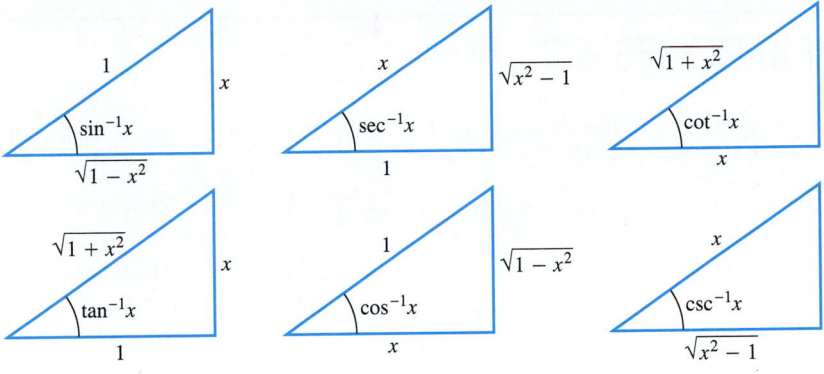

6.29 Right-triangle interpretations of the inverse trigonometric function angles (first-quadrant values).

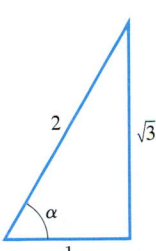

6.30 If $\alpha = \sin^{-1}(\sqrt{3}/2)$, then the values of the trigonometric functions of α can be read from this triangle (Example 1).

Example 1 Find $\csc\alpha$, $\cos\alpha$, $\sec\alpha$, $\tan\alpha$, and $\cot\alpha$ if

$$\alpha = \sin^{-1}\frac{\sqrt{3}}{2}. \tag{19}$$

Solution Equation (19) tells us that

$$\sin\alpha = \frac{\sqrt{3}}{2}.$$

We draw a reference triangle with side $\sqrt{3}$ and hypotenuse 2 (Fig. 6.30). The length of the remaining side is

$$\sqrt{(2)^2 - \left(\sqrt{3}\right)^2} = \sqrt{4 - 3} = 1.$$

The values we want can be read as ratios of side lengths from the completed triangle:

$$\csc\alpha = \frac{2}{\sqrt{3}} = \frac{2\sqrt{3}}{3}, \qquad \cos\alpha = \frac{1}{2}, \qquad \sec\alpha = 2,$$

$$\tan\alpha = \frac{\sqrt{3}}{1} = \sqrt{3}, \qquad \cot\alpha = \frac{1}{\sqrt{3}} = \frac{\sqrt{3}}{3}.$$

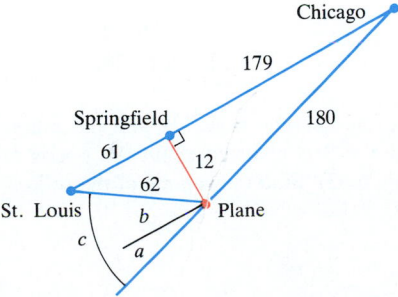

6.31 Diagram for drift correction (Example 2), with distances rounded to the nearest mile (drawing not to scale).

Example 2 *Drift Correction.* During an airplane flight from Chicago to St. Louis the navigator determines that the plane is 12 mi off course, as shown in Fig. 6.31. Find the angle a for a course parallel to the original, correct course, the angle b, and the correction angle $c = a + b$.

Solution

$$a = \sin^{-1}\frac{12}{180} \approx 0.067 \text{ radians} \approx 3.8°,$$

$$b = \sin^{-1}\frac{12}{62} \approx 0.195 \text{ radians} \approx 11.2°,$$

$$c = a + b \approx 15°.$$

EXERCISES 6.7

Use reference triangles like those in Figs. 6.29 and 6.30 to find the angles in Exercises 1–12.

1. a) $\tan^{-1}1$ b) $\tan^{-1}(-\sqrt{3})$ c) $\tan^{-1}\left(\dfrac{1}{\sqrt{3}}\right)$

2. a) $\tan^{-1}(-1)$ b) $\tan^{-1}\sqrt{3}$ c) $\tan^{-1}\left(\dfrac{-1}{\sqrt{3}}\right)$

3. a) $\sin^{-1}\left(\dfrac{-1}{2}\right)$ b) $\sin^{-1}\left(\dfrac{1}{\sqrt{2}}\right)$ c) $\sin^{-1}\left(\dfrac{-\sqrt{3}}{2}\right)$

4. a) $\sin^{-1}\left(\dfrac{1}{2}\right)$ b) $\sin^{-1}\left(\dfrac{-1}{\sqrt{2}}\right)$ c) $\sin^{-1}\left(\dfrac{\sqrt{3}}{2}\right)$

5. a) $\cos^{-1}\left(\dfrac{1}{2}\right)$ b) $\cos^{-1}\left(\dfrac{-1}{\sqrt{2}}\right)$ c) $\cos^{-1}\left(\dfrac{\sqrt{3}}{2}\right)$

6. a) $\cos^{-1}\left(\dfrac{-1}{2}\right)$ b) $\cos^{-1}\left(\dfrac{1}{\sqrt{2}}\right)$ c) $\cos^{-1}\left(\dfrac{-\sqrt{3}}{2}\right)$

7. a) $\sec^{-1}(-\sqrt{2})$ b) $\sec^{-1}\left(\dfrac{2}{\sqrt{3}}\right)$ c) $\sec^{-1}(-2)$

8. a) $\sec^{-1}\sqrt{2}$ b) $\sec^{-1}\left(\dfrac{-2}{\sqrt{3}}\right)$ c) $\sec^{-1}2$

9. a) $\csc^{-1}\sqrt{2}$ b) $\csc^{-1}\left(\dfrac{-2}{\sqrt{3}}\right)$ c) $\csc^{-1}2$

10. a) $\csc^{-1}(-\sqrt{2})$ b) $\csc^{-1}\left(\dfrac{2}{\sqrt{3}}\right)$ c) $\csc^{-1}(-2)$

11. a) $\cot^{-1}(-1)$ b) $\cot^{-1}\sqrt{3}$ c) $\cot^{-1}\left(\dfrac{-1}{\sqrt{3}}\right)$

12. a) $\cot^{-1}1$ b) $\cot^{-1}(-\sqrt{3})$ c) $\cot^{-1}\left(\dfrac{1}{\sqrt{3}}\right)$

13. Given that $\alpha = \sin^{-1}(1/2)$, find $\cos\alpha$, $\tan\alpha$, $\sec\alpha$, $\csc\alpha$.

14. Given that $\alpha = \cos^{-1}(-1/2)$, find $\sin\alpha$, $\tan\alpha$, $\sec\alpha$, $\csc\alpha$.

Evaluate the expressions in Exercises 15–34.

15. $\sin\left(\dfrac{\pi}{12} + \cos^{-1}\dfrac{\sqrt{2}}{2}\right)$ **16.** $\tan\left(\sin^{-1}\left(-\dfrac{1}{2}\right)\right)$

17. $\sin^{-1}\left(\dfrac{1}{2}\sec\left(\cos^{-1}\dfrac{1}{2}\right)\right)$

18. $\cot\left(\sin^{-1}\left(-\dfrac{1}{2}\right) - \sec^{-1}2\right)$

19. $\csc(\sec^{-1}2) + \cos(\tan^{-1}(-\sqrt{3}))$

20. $\tan(\sec^{-1}1) + \sin(\csc^{-1}1)$

21. $\sec^{-1}(\sqrt{2}\,(\cos(\cot^{-1}1)))$

22. $\sec^{-1}\left(\csc\left(\sin^{-1}\left(-\dfrac{\sqrt{2}}{2}\right)\right)\right)$

23. $\sec^{-1}\left(\sec\left(-\dfrac{\pi}{6}\right)\right)$ **24.** $\cot^{-1}\left(\cot\left(-\dfrac{\pi}{4}\right)\right)$

25. $\sin(\tan^{-1}2.4)$ **26.** $\cos(\sin^{-1}0.8)$

27. $\sin\left(\tan^{-1}\dfrac{x}{2}\right)$ **28.** $\tan\left(\cos^{-1}\dfrac{x+1}{\sqrt{2x^2+2}}\right)$

29. $\cot\left(\sin^{-1}\dfrac{3}{\sqrt{4x^2+9}}\right)$ **30.** $\cot\left(\cos^{-1}\dfrac{-1}{\sqrt{x^2+1}}\right)$

31. $\sin\left(\tan^{-1}\dfrac{x}{\sqrt{1-x^2}}\right)$ **32.** $\cos\left(\dfrac{\pi}{2} - \tan^{-1}\dfrac{x}{\sqrt{1-x^2}}\right)$

33. $\sec^{-1}\left(\sqrt{2}\cos\left(\dfrac{13\pi}{12} - \sin^{-1}\left(-\dfrac{1}{2}\sin\left(\sec^{-1}\sqrt{3}\right.\right.\right.\right.$
$\left.\left.\left.\left. + \csc^{-1}\sqrt{3}\right)\right)\right)\right)$

34. $\sec^{-1}\left(-4\cos\left(\dfrac{2}{3}\sin^{-1}\left(\tan\left(\dfrac{3\pi}{4} + \cot^{-1}0\right)\right)\right)\right)$

Find the limits in Exercises 35–38. (If in doubt, look at the inverse function's graph.)

35. a) $\lim\limits_{x\to 1^-}\sin^{-1}x$ b) $\lim\limits_{x\to -1^+}2\cos^{-1}x$

36. a) $\lim\limits_{x\to\infty}\tan^{-1}5x$ b) $\lim\limits_{x\to -\infty}\cot^{-1}(x/2)$

37. a) $\lim\limits_{x\to\infty}\sec^{-1}(\sqrt{x})$ b) $\lim\limits_{x\to -\infty}\sec^{-1}(x+5)$

38. a) $\lim\limits_{x\to\infty}\csc^{-1}\left(\dfrac{x+1}{x-1}\right)$ b) $\lim\limits_{x\to -\infty}\csc^{-1}(2x+3)$

39. You are sitting in a classroom next to the wall looking at the blackboard at the front of the room (Fig. 6.32). The blackboard is 12 ft long and starts 3 ft from the wall you are sitting next to. Show that your viewing angle is

$$\alpha = \cot^{-1}\dfrac{x}{15} - \cot^{-1}\dfrac{x}{3}$$

if you are x ft from the front wall.

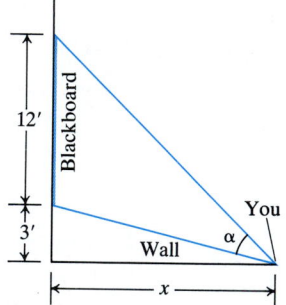

6.32 The classroom in Exercise 39.

40. The region between the curve $y = \sec^{-1}x$ and the x-axis between $x = 1$ and $x = 2$ is revolved about the y-axis to generate a solid (Fig. 6.33). Find the volume of the solid.

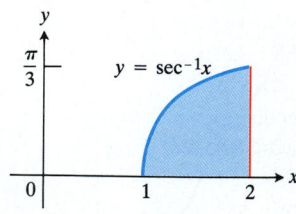

6.33 The region in Exercise 40.

41. A particle starts at the origin at $t = 0$ sec and moves along the s-axis in such a way that its velocity at position s is $ds/dt = \cos^2 \pi s$. How long will it take the particle to reach $s = 1/4$? Will it ever reach $s = 1/2$? Why?

42. CALCULATOR   Figure 6.34 shows a cone whose slant height is 3 m. How large should the indicated angle be to maximize the cone's volume?

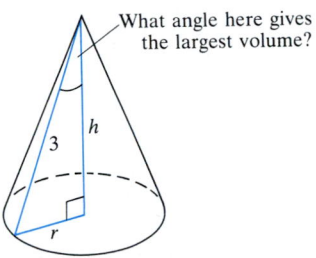
What angle here gives the largest volume?

6.34   The cone in Exercise 42.

43. CALCULATOR   Find the angle α in Fig. 6.35.

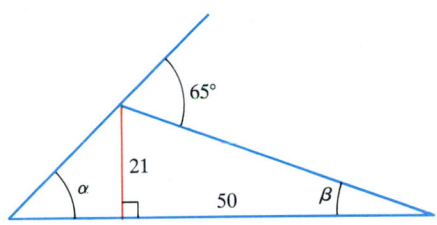

6.35   The figure for Exercise 43. (*Hint:* $\alpha + \beta = 65°$.)

44. Figure 6.36 offers an informal proof that
$$\tan^{-1}1 + \tan^{-1}2 + \tan^{-1}3 = \pi.$$
See if you can tell what is going on. (*Source:* "Behold! Sums of Arctan" by Edward M. Harris, *The College Mathematics Journal*, Vol. 18, No. 2, March 1987, p. 141.)

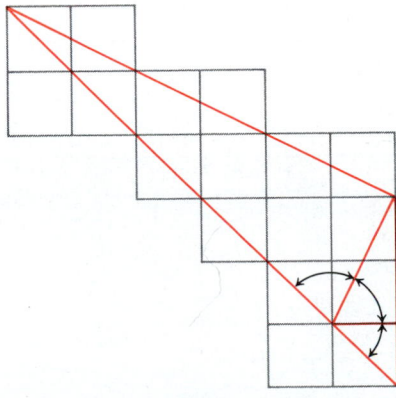

6.36   The informal proof for Exercise 44.

45. *Two derivations of the identity $\sec^{-1}(-x) = \pi - \sec^{-1}x$.*
a) *(Geometric)*   Figure 6.37 offers a pictorial proof that $\sec^{-1}(-x) = \pi - \sec^{-1}x$. See if you can tell what is going on.

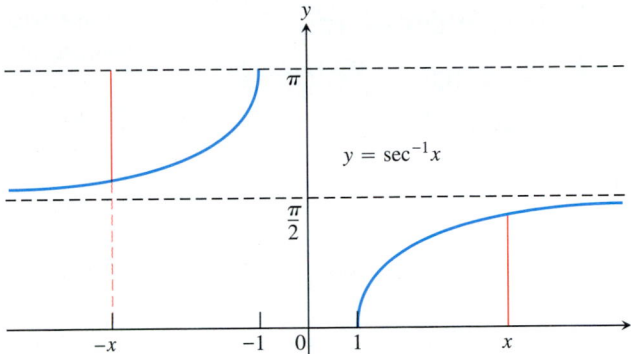

6.37   The figure for Exercise 45(a).

b) *(Algebraic)*   Derive the identity $\sec^{-1}(-x) = \pi - \sec^{-1}x$ by combining the following two equations from the text:
$$\cos^{-1}(-x) = \pi - \cos^{-1}x \qquad \text{(Eq. (8))}$$
$$\sec^{-1}x = \cos^{-1}(1/x) \qquad \text{(Eq. (17))}$$

46. *The identity $\sin^{-1}x + \cos^{-1}x = \pi/2$.*   Figure 6.24 establishes the identity for $0 < x < 1$. To establish it for the rest of $[-1, 1]$, verify by direct calculation that it holds for $x = 1, 0,$ and -1. Then, for values of x in $(-1, 0)$, let $x = -a$, $a > 0$, and apply Eqs. (4) and (8) to the sum $\sin^{-1}(-a) + \cos^{-1}(-a)$.

Graphing Calculator or Computer Grapher

47. *Newton's serpentine.*   Graph Newton's serpentine, $y = 4x/(x^2 + 1)$. Then graph $y = 2\sin(2\tan^{-1}x)$ in the same graphing window. What do you see? Explain.

48. Graph the rational function $y = (2 - x^2)/x^2$. Then graph $y = \cos(2\sec^{-1}x)$ in the same graphing window. What do you see? Explain.

Graph the composite functions in Exercises 49–51. What are the domains and ranges of these functions? Are the calculator's (computer's) graphs correct?

49. a) $y = \tan^{-1}(\tan x)$ b) $y = \tan(\tan^{-1}x)$

50. a) $y = \sin^{-1}(\sin x)$ b) $y = \sin(\sin^{-1}x)$

51. a) $y = \cos^{-1}(\cos x)$ b) $y = \cos(\cos^{-1}x)$

EXPLORER PROGRAM	
PowerGrapher	Graphs inverse trigonometric functions expressed in terms of ATN(X)

6.8 Derivatives of Inverse Trigonometric Functions; Related Integrals

In this section, we list the standard formulas for the derivatives of the inverse trigonometric functions (Table 6.6), show how they are derived, and discuss their companion integral formulas. As we shall see, the restrictions on the domains of the inverse trigonometric functions reappear in natural ways as restrictions on the domains of the derivatives.

Example 1

a) $\dfrac{d}{dx}\sin^{-1}x^2 = \dfrac{1}{\sqrt{1-(x^2)^2}} \cdot \dfrac{d}{dx}(x^2) = \dfrac{2x}{\sqrt{1-x^4}}$

b) $\dfrac{d}{dx}\tan^{-1}\sqrt{x+1} = \dfrac{1}{1+(\sqrt{x+1})^2} \cdot \dfrac{d}{dx}(\sqrt{x+1})$

$= \dfrac{1}{x+2} \cdot \dfrac{1}{2\sqrt{x+1}} = \dfrac{1}{2\sqrt{x+1}(x+2)}$

c) $\dfrac{d}{dx}\sec^{-1}(-3x) = \dfrac{1}{|-3x|\sqrt{(-3x)^2-1}} \cdot \dfrac{d}{dx}(-3x)$

$= \dfrac{-3}{|3x|\sqrt{9x^2-1}} = \dfrac{-1}{|x|\sqrt{9x^2-1}}.$

Example 2

$\displaystyle\int_0^1 \dfrac{e^{\tan^{-1}x}}{1+x^2}\,dx = \int_0^{\pi/4} e^u\,du \qquad \left(\begin{matrix}u = \tan^{-1}x, \quad du = \dfrac{dx}{1+x^2},\\ u(0)=0, \quad u(1)=\pi/4\end{matrix}\right)$

$= e^u\Big]_0^{\pi/4} = e^{\pi/4} - 1$

To show where the formulas in Table 6.6 come from, we derive Formulas 1 and 5.

The Derivative of $y = \sin^{-1}u$

We know that the function $x = \sin y$ is differentiable in the interval $-\pi/2 < y < \pi/2$ and that its derivative, the cosine, is positive there. The derivative rule for inverses in Section 6.1 therefore assures us that the inverse function $y = \sin^{-1}x$ is differentiable throughout the interval $-1 < x < 1$. We cannot expect it to be differentiable at $x = 1$ or $x = -1$, however, because the tangents to the graph are vertical at these points (see Fig. 6.38).

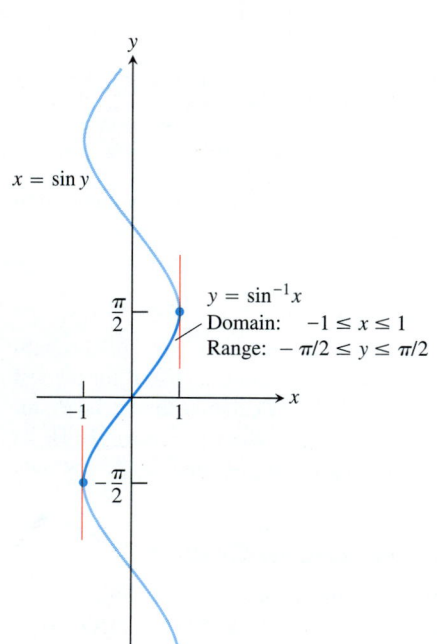

$x = \sin y$

$y = \sin^{-1}x$
Domain: $-1 \le x \le 1$
Range: $-\pi/2 \le y \le \pi/2$

6.38 The graph of $y = \sin^{-1}x$ has vertical tangents at $x = -1$ and $x = 1$.

TABLE 6.6
Derivatives of the inverse trigonometric functions

1. $\dfrac{d(\sin^{-1}u)}{dx} = \dfrac{du/dx}{\sqrt{1-u^2}}, \quad -1 < u < 1$

2. $\dfrac{d(\cos^{-1}u)}{dx} = -\dfrac{du/dx}{\sqrt{1-u^2}}, \quad -1 < u < 1$

3. $\dfrac{d(\tan^{-1}u)}{dx} = \dfrac{du/dx}{1+u^2}$

4. $\dfrac{d(\cot^{-1}u)}{dx} = -\dfrac{du/dx}{1+u^2}$

5. $\dfrac{d(\sec^{-1}u)}{dx} = \dfrac{du/dx}{|u|\sqrt{u^2-1}}, \quad |u| > 1$

6. $\dfrac{d(\csc^{-1}u)}{dx} = \dfrac{-du/dx}{|u|\sqrt{u^2-1}}, \quad |u| > 1$

To calculate the derivative of $y = \sin^{-1}x$, we differentiate both sides of the equation $\sin y = x$ with respect to x:

$$\sin y = x,$$

$$\frac{d}{dx} \sin y = 1, \tag{1}$$

$$\cos y \frac{dy}{dx} = 1.$$

We then divide through by $\cos y$ (> 0 for $-\pi/2 < y < \pi/2$) to get

$$\frac{dy}{dx} = \frac{1}{\cos y} = \frac{1}{\sqrt{1 - \sin^2 y}} = \frac{1}{\sqrt{1 - x^2}}. \tag{2}$$

The derivative of $y = \sin^{-1}x$ with respect to x is

$$\frac{d}{dx} \sin^{-1}x = \frac{1}{\sqrt{1 - x^2}}. \tag{3}$$

If u is a differentiable function of x with $|u| < 1$, we apply the Chain Rule in the form

$$\frac{dy}{dx} = \frac{dy}{du}\frac{du}{dx}$$

to $y = \sin^{-1}u$ to obtain

$$\frac{d}{dx} \sin^{-1}u = \frac{1}{\sqrt{1 - u^2}}\frac{du}{dx}. \tag{4}$$

The formulas for the derivatives of $\cos^{-1}u$ and $\tan^{-1}u$ can be derived in a similar way.

The Derivative of $y = \sec^{-1}u$

We begin by differentiating both sides of the equation $\sec y = x$ with respect to x:

$$\sec y = x,$$

$$\frac{d}{dx} \sec y = 1,$$

$$\sec y \tan y \frac{dy}{dx} = 1, \tag{5}$$

$$\frac{dy}{dx} = \frac{1}{\sec y \tan y}.$$

To express the result in terms of x, we use the relations

$$\sec y = x \quad \text{and} \quad \tan y = \pm\sqrt{\sec^2 y - 1} = \pm\sqrt{x^2 - 1}.$$

Hence, if $|x| > 1$,

$$\frac{dy}{dx} = \pm\frac{1}{x\sqrt{x^2 - 1}}.$$

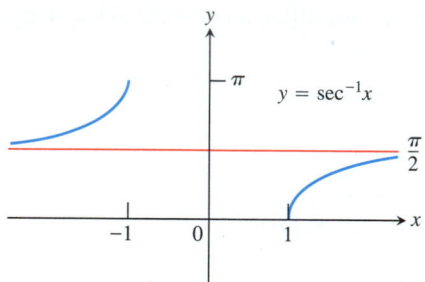

6.39 The slope of the curve $y = \sec^{-1}x$ is positive for $x < -1$ and $x > 1$.

What do we do about the sign? A glance at Fig. 6.39 shows that for $|x| > 1$ the slope of the graph of $y = \sec^{-1}x$ is always positive. Therefore,

$$\frac{d}{dx} \sec^{-1}x = \begin{cases} \dfrac{1}{x\sqrt{x^2-1}} & \text{if } x > 1, \\[3mm] -\dfrac{1}{x\sqrt{x^2-1}} & \text{if } x < -1. \end{cases} \tag{6}$$

With absolute values, we can write Eq. (6) as a single formula:

$$\frac{d}{dx} \sec^{-1}x = \frac{1}{|x|\sqrt{x^2-1}}, \qquad |x| > 1. \tag{7}$$

If u is a differentiable function of x with $|u| > 1$, we can then apply the Chain Rule to obtain

$$\frac{d}{dx} \sec^{-1}u = \frac{1}{|u|\sqrt{u^2-1}} \frac{du}{dx}. \tag{8}$$

Integration Formulas

We might expect the six derivative formulas in Table 6.6 to lead to six new integration formulas, but only three are in frequent use (Table 6.7).

TABLE 6.7
Integrals leading to inverse trigonometric functions

1. $\displaystyle\int \frac{du}{\sqrt{1-u^2}} = \sin^{-1}u + C$ (Valid for $u^2 < 1$)

3. $\displaystyle\int \frac{du}{1+u^2} = \tan^{-1}u + C$ (Valid for all u)

5. $\displaystyle\int \frac{du}{u\sqrt{u^2-1}} = \int \frac{d(-u)}{(-u)\sqrt{u^2-1}} = \sec^{-1}|u| + C = \cos^{-1}\left|\frac{1}{u}\right| + C$

 (Valid for $u^2 > 1$)

Example 3 Evaluate

a) $\displaystyle\int_0^1 \frac{dx}{1+x^2}$

b) $\displaystyle\int_{2/\sqrt{3}}^{\sqrt{2}} \frac{dx}{x\sqrt{x^2-1}}.$

Solution

a) $\displaystyle\int_0^1 \frac{dx}{1+x^2} = \tan^{-1}x \Big]_0^1 = \tan^{-1}1 - \tan^{-1}0 = \frac{\pi}{4} - 0 = \frac{\pi}{4}$

b) $\displaystyle\int_{2/\sqrt{3}}^{\sqrt{2}} \frac{dx}{x\sqrt{x^2-1}} = \sec^{-1}x \Big]_{2/\sqrt{3}}^{\sqrt{2}} = \frac{\pi}{4} - \frac{\pi}{6} = \frac{\pi}{12}$

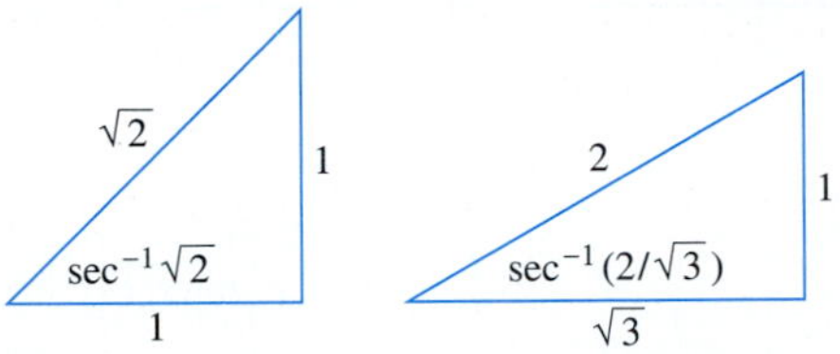

$\sqrt{2}$

1

$\sec^{-1}\sqrt{2}$

1

2

1

$\sec^{-1}(2/\sqrt{3})$

$\sqrt{3}$

An isosceles right
triangle:

A 30-60-90 triangle:

$\sec^{-1}\sqrt{2} = \dfrac{\pi}{4}$

$\sec^{-1}\left(\dfrac{2}{\sqrt{3}}\right) = \dfrac{\pi}{6}$

6.40 Triangles you can use for finding
$\sec^{-1}\sqrt{2}$ and $\sec^{-1}(2/\sqrt{3})$.

If you do not see right away how to evaluate $\sec^{-1}x$ at the limits of integration, draw triangles like the ones in Fig. 6.40.

Example 4   Evaluate

$$\int \frac{x^2\,dx}{\sqrt{1-x^6}}.$$

Solution   The resemblance between the given integral and the standard form

$$\int \frac{du}{\sqrt{1-u^2}} = \sin^{-1}u + C$$

suggests the substitution

$$u = x^3, \qquad du = 3x^2\,dx.$$

Indeed,

$$\int \frac{x^2\,dx}{\sqrt{1-x^6}} = \frac{1}{3}\int \frac{3x^2\,dx}{\sqrt{1-(x^3)^2}} = \frac{1}{3}\int \frac{du}{\sqrt{1-u^2}}$$

$$= \frac{1}{3}\sin^{-1}u + C = \frac{1}{3}\sin^{-1}(x^3) + C.$$

Example 5   Evaluate

$$\int \frac{dx}{\sqrt{9-x^2}}.$$

Solution   This is the integral for the arc sine of x with 9 in place of 1. To get the 1 back, we factor a 9 from $9 - x^2$:

$$\sqrt{9-x^2} = \sqrt{9\left(1 - \frac{x^2}{9}\right)} = 3\sqrt{1 - \left(\frac{x}{3}\right)^2}.$$

Then

$$\int \frac{dx}{\sqrt{9-x^2}} = \int \frac{dx}{3\sqrt{1-(x/3)^2}}.$$

We now substitute

$$u = \frac{x}{3} \quad \text{and} \quad du = \frac{dx}{3}, \quad \text{or} \quad dx = 3\,du.$$

This gives

$$\int \frac{dx}{\sqrt{9-x^2}} = \int \frac{dx}{3\sqrt{1-(x/3)^2}} = \int \frac{3\,du}{3\sqrt{1-u^2}}$$

$$= \int \frac{du}{\sqrt{1-u^2}} = \sin^{-1}u + C$$

$$= \sin^{-1}\left(\frac{x}{3}\right) + C.$$

EXERCISES 6.8

Find dy/dx in Exercises 1–18.

1. $y = \cos^{-1}x^2$

2. $y = \cos^{-1}(1/x)$

3. $y = 5\tan^{-1}3x$

4. $y = \cot^{-1}\sqrt{x}$

5. $y = \sin^{-1}(x/2)$

6. $y = \sin^{-1}(1 - x)$

7. $y = \sec^{-1}5x$

8. $y = (1/3)\tan^{-1}(x/3)$

9. $y = \csc^{-1}(x^2 + 1)$

10. $y = \cos^{-1}2x$

11. $y = \csc^{-1}\sqrt{x} + \sec^{-1}\sqrt{x}$

12. $y = \csc^{-1}\dfrac{1}{x}, \quad x > 0$

13. $y = \cot^{-1}\sqrt{x - 1}$

14. $y = x\sqrt{1 - x^2} - \cos^{-1}x$

15. $y = \sqrt{x^2 - 1} - \sec^{-1}x$

16. $y = \cot^{-1}\dfrac{1}{x} - \tan^{-1}x$

17. $y = x\sin^{-1}x + \sqrt{1 - x^2}$

18. $y = \ln(x^2 + 4) - x\tan^{-1}(x/2)$

Verify the integration formulas in Exercises 19–22.

19. $\displaystyle\int \frac{\tan^{-1}x}{x^2}\,dx = \ln x - \frac{1}{2}\ln(1 + x^2) - \frac{\tan^{-1}x}{x} + C$

20. $\displaystyle\int x^3\cos^{-1}5x\,dx = \frac{x^4}{4}\cos^{-1}5x + \frac{5}{4}\int \frac{x^4\,dx}{\sqrt{1 - 25x^2}}$

21. $\displaystyle\int (\sin^{-1}x)^2\,dx = x(\sin^{-1}x)^2 - 2x + 2\sqrt{1 - x^2}\,\sin^{-1}x + C$

22. $\displaystyle\int \ln(a^2 + x^2)\,dx = x\ln(a^2 + x^2) - 2x + 2a\tan^{-1}\frac{x}{a} + C$

Evaluate the integrals in Exercises 23–42.

23. $\displaystyle\int_0^{1/2} \frac{dx}{\sqrt{1 - x^2}}$

24. $\displaystyle\int_{-1}^{1} \frac{dx}{1 + x^2}$

25. $\displaystyle\int_{\sqrt{2}}^{2} \frac{dx}{x\sqrt{x^2 - 1}}$

26. $\displaystyle\int_{-2}^{-\sqrt{2}} \frac{dx}{x\sqrt{x^2 - 1}}$

27. $\displaystyle\int_0^{\sqrt{2}/2} \frac{y\,dy}{\sqrt{1 - y^4}}$

28. $\displaystyle\int_0^{1} \frac{4s\,ds}{\sqrt{4 - s^4}}$

29. $\displaystyle\int_0^{2} \frac{dx}{1 + (x - 1)^2}$

30. $\displaystyle\int_0^{\ln\sqrt{3}} \frac{e^x\,dx}{1 + e^{2x}}$

31. $\displaystyle\int_{-\pi/2}^{\pi/2} \frac{2\cos\theta\,d\theta}{1 + \sin^2\theta}$

32. $\displaystyle\int \frac{4\,dt}{t(1 + \ln^2 t)}$

33. $\displaystyle\int_1^{3} \frac{dy}{\sqrt{y}(1 + y)}$

34. $\displaystyle\int_1^{5/4} \frac{dr}{\sqrt{1 - 4(r - 1)^2}}$

35. $\displaystyle\int_{1/\sqrt{3}}^{1} \frac{dx}{x\sqrt{4x^2 - 1}}$

36. $\displaystyle\int \frac{6\sec^2\theta\,d\theta}{\sqrt{1 - \tan^2\theta}}$

37. $\displaystyle\int \frac{dx}{x\sqrt{x^4 - 1}}$

38. $\displaystyle\int_1^{\sqrt{3}} \frac{2\,dx}{(1 + x^2)\tan^{-1}x}$

39. $\displaystyle\int_2^{4} \frac{dy}{2y\sqrt{y - 1}}$

40. $\displaystyle\int_{1/2}^{3/4} \frac{ds}{\sqrt{s}\sqrt{1 - s}}$

41. $\displaystyle\int_{-2/3}^{-\sqrt{2}/3} \frac{dt}{t\sqrt{9t^2 - 1}}$

42. $\displaystyle\int \frac{12\,dx}{\sqrt{e^{2x} - 1}}$

Find the limits in Exercises 43 and 44.

43. $\displaystyle\lim_{x \to 0} \frac{\sin^{-1}x}{x}$

44. $\displaystyle\lim_{x \to 0} \frac{\tan^{-1}x}{x}$

45. Find the volume of the solid of revolution in Fig. 6.41.

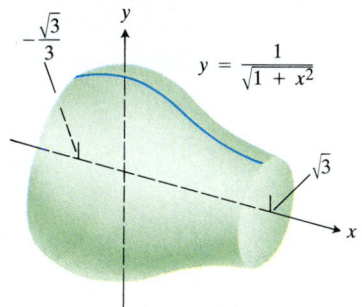

6.41   The solid in Exercise 45.

46. Find the length of the curve $y = \sqrt{1 - x^2}, \ -1/2 \le x \le 1/2$.

47. *Continuation of Exercise 39, Section 6.7.*   You want to move your chair to a position along the wall that will maximize your viewing angle α. About how far from the front of the room should you sit?

48. Find the center of mass of a thin plate of constant density covering the region bounded by the curves $y = 1/(1 + x^2)$ and $y = -1/(1 + x^2)$ and by the lines $x = 0$ and $x = 1$ (Fig. 6.42).

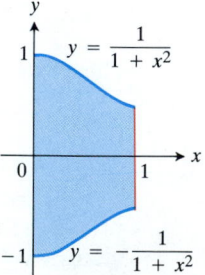

6.42   The region in Exercise 48.

49. What value of x maximizes θ in Fig. 6.43? How large is θ at that point? Begin by showing that $\theta = \pi - \cot^{-1}x - \cot^{-1}(2 - x)$.

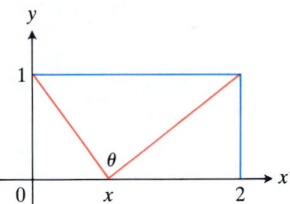

6.43   The angle in Exercise 49.

50. Find the linearizations of $\sin^{-1}x$, $\cos^{-1}x$, $\tan^{-1}x$, and $\cot^{-1}x$ at $x = 0$.

Solve the initial value problems in Exercises 51–54.

51. Differential equation: $\dfrac{dy}{dx} = \dfrac{1}{x\sqrt{x^2 - 1}}, \quad x > 1$

Initial condition: $y = \pi$ when $x = 2$

52. Differential equation: $\dfrac{dy}{dx} = \dfrac{1}{\sqrt{1 - x^2}}$

Initial condition: $y = 1$ when $x = 0$

53. Differential equation: $\dfrac{dy}{dx} = -\dfrac{1}{\sqrt{1 - x^2}}$

Initial condition: $y = \pi/2$ when $x = -\sqrt{2}/2$

54. Differential equation: $\dfrac{dy}{dx} = -\dfrac{1}{1 + x^2}$

Initial condition: $y = \pi/2$ when $x = 0$

55. CALCULATOR   Use numerical integration to estimate the value of

$$\sin^{-1}0.6 = \int_0^{0.6} \frac{dx}{\sqrt{1 - x^2}}.$$

For reference, $\sin^{-1}0.6 = 0.64350$ to five places.

56. CALCULATOR   Use numerical integration to estimate the value of

$$\pi = 4 \int_0^1 \frac{1}{1 + x^2}\, dx.$$

57. Can the integrations in (a) and (b) both be correct? Explain.

a) $\displaystyle \int \frac{dx}{\sqrt{1 - x^2}} = \sin^{-1}x + C$

b) $\displaystyle \int \frac{dx}{\sqrt{1 - x^2}} = -\int -\frac{dx}{\sqrt{1 - x^2}} = -\cos^{-1}x + C$

58. a) Show that the functions

$$f(x) = \sin^{-1}\frac{x - 1}{x + 1} \quad \text{and} \quad g(x) = 2\tan^{-1}\sqrt{x},$$

both defined for $x \geq 0$, have the same derivative and therefore that

$$f(x) = g(x) + C. \tag{9}$$

b) Find C. (*Hint:* Evaluate both sides of Eq. (9) for a particular value of x.)

Verify the derivative formulas in Exercises 59–62.

59. $\dfrac{d(\cos^{-1}u)}{dx} = -\dfrac{du/dx}{\sqrt{1 - u^2}}, \quad -1 < u < 1$

60. $\dfrac{d(\tan^{-1}u)}{dx} = \dfrac{du/dx}{1 + u^2}$

61. $\dfrac{d(\cot^{-1}u)}{dx} = -\dfrac{du/dx}{1 + u^2}$

62. $\dfrac{d(\csc^{-1}u)}{dx} = \dfrac{-du/dx}{|u|\sqrt{u^2 - 1}}, \quad |u| > 1$

Graphing Calculator or Computer Grapher

Graph the functions in Exercises 63–66 together with their derivatives.

63. $\sin^{-1}x$ **64.** $\cos^{-1}x$ **65.** $\tan^{-1}x$ **66.** $\sec^{-1}x$

EXPLORER PROGRAMS

Derivatives	Graphs functions and their derivatives
PowerGrapher	Graphs all functions in this section

6.9 Hyperbolic Functions

It can be shown that every function defined on an interval centered at the origin can be written in a unique way as the sum of one even function and one odd function. For an arbitrary function f, the decomposition is

$$f(x) = \underbrace{\frac{f(x) + f(-x)}{2}}_{\text{even part}} + \underbrace{\frac{f(x) - f(-x)}{2}}_{\text{odd part}}$$

If we write e^x this way, we get

$$e^x = \underbrace{\frac{e^x + e^{-x}}{2}}_{\text{even part}} + \underbrace{\frac{e^x - e^{-x}}{2}}_{\text{odd part}}$$

The even and odd parts of e^x, called the hyperbolic cosine and hyperbolic sine of x, respectively, turn out to be useful functions in their own right. They describe the motions of waves in elastic solids, the shapes of hanging electric power lines, and the temperature distributions in metal cooling fins. They even come up in the general theory of relativity. The designers of the Gateway Arch to the West in St. Louis used a hyperbolic cosine function to predict the arch's internal forces and then shaped the arch like an upside-down hyperbolic cosine curve.

This section introduces the calculus of hyperbolic functions.

Definitions and Identities

The hyperbolic cosine and hyperbolic sine functions are defined by the first two equations in Table 6.8. The table also lists the definitions of the hyperbolic tangent, cotangent, secant, and cosecant. As we shall see, the hyperbolic functions bear a number of similarities to the trigonometric functions after which they are named. (See Exercise 91 as well.)

TABLE 6.8
The six basic hyperbolic functions

Hyperbolic cosine of x:	$\cosh x = \dfrac{e^x + e^{-x}}{2}$
Hyperbolic sine of x:	$\sinh x = \dfrac{e^x - e^{-x}}{2}$
Hyperbolic tangent:	$\tanh x = \dfrac{\sinh x}{\cosh x} = \dfrac{e^x - e^{-x}}{e^x + e^{-x}}$
Hyperbolic cotangent:	$\coth x = \dfrac{\cosh x}{\sinh x} = \dfrac{e^x + e^{-x}}{e^x - e^{-x}}$
Hyperbolic secant:	$\operatorname{sech} x = \dfrac{1}{\cosh x} = \dfrac{2}{e^x + e^{-x}}$
Hyperbolic cosecant:	$\operatorname{csch} x = \dfrac{1}{\sinh x} = \dfrac{2}{e^x - e^{-x}}$

See Figs. 6.44 and 6.45 for graphs. The notation $\cosh x$ is often read "kosh x," rhyming with either "gosh x" or "gauche x," and $\sinh x$ is pronounced as if spelled "cinch x" or "shine x."

Identities

Hyperbolic functions satisfy the identities shown in Table 6.9. Except for differences in sign, these are the identities we already know for trigonometric functions.

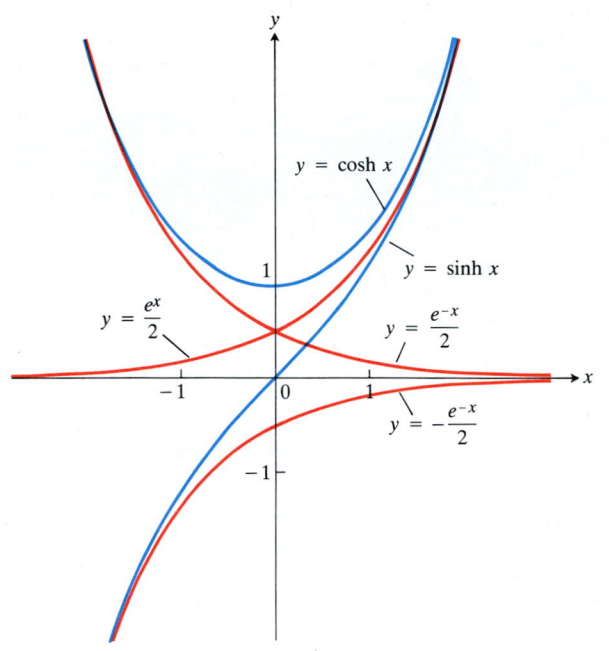

6.44   The graphs of the hyperbolic sine and cosine functions.

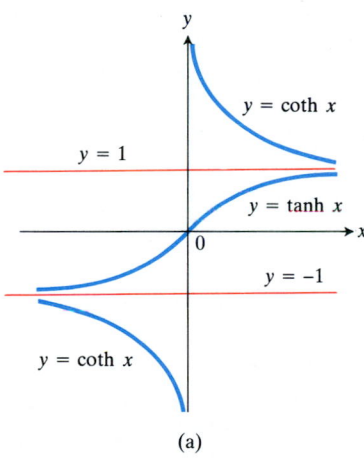

(a)

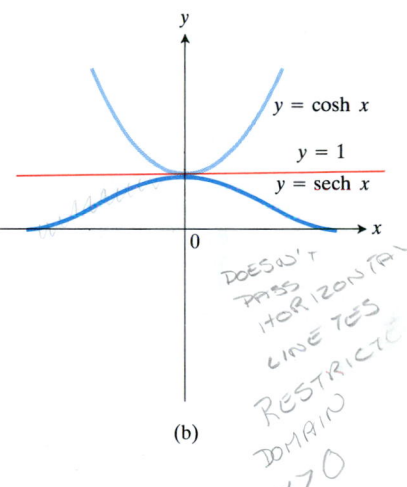

(b)

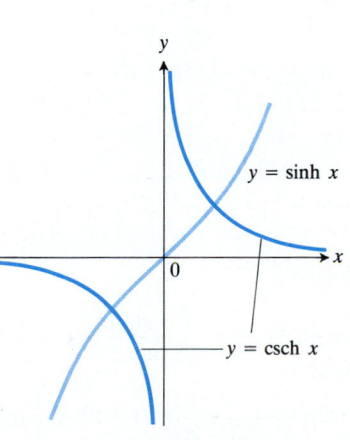

6.45   The graphs of the hyperbolic cotangent, secant, and cosecant of x, and their relations to the graphs of the hyperbolic targent, cosine, and sine of x.

(c)

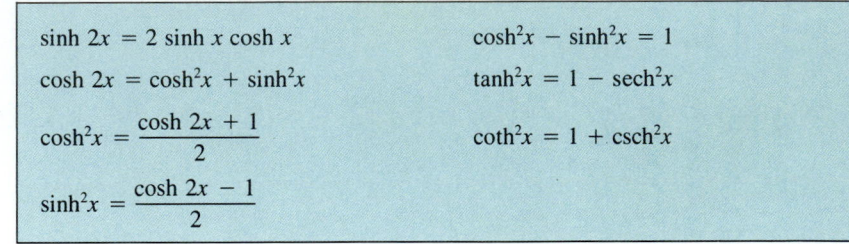

TABLE 6.9
Identities for hyperbolic functions

$\sinh 2x = 2 \sinh x \cosh x$	$\cosh^2 x - \sinh^2 x = 1$
$\cosh 2x = \cosh^2 x + \sinh^2 x$	$\tanh^2 x = 1 - \text{sech}^2 x$
$\cosh^2 x = \dfrac{\cosh 2x + 1}{2}$	$\coth^2 x = 1 + \text{csch}^2 x$
$\sinh^2 x = \dfrac{\cosh 2x - 1}{2}$	

Derivatives and Integrals

The six hyperbolic functions, being rational combinations of the differentiable functions e^x and e^{-x}, have derivatives at every point at which they are defined. Finding formulas for the derivatives is a straightforward exercise using the sum, difference, and quotient rules of differentiation. Again, there are similarities with trigonometric functions. These derivatives are listed in Table 6.10.

TABLE 6.10
Derivatives of hyperbolic functions

$\dfrac{d}{dx}(\sinh u) = \cosh u \dfrac{du}{dx}$	$\dfrac{d}{dx}(\coth u) = -\text{csch}^2 u \dfrac{du}{dx}$
$\dfrac{d}{dx}(\cosh u) = \sinh u \dfrac{du}{dx}$	$\dfrac{d}{dx}(\text{sech}\, u) = -\text{sech}\, u \tanh u \dfrac{du}{dx}$
$\dfrac{d}{dx}(\tanh u) = \text{sech}^2 u \dfrac{du}{dx}$	$\dfrac{d}{dx}(\text{csch}\, u) = -\text{csch}\, u \coth u \dfrac{du}{dx}$

These derivative formulas produce the integral formulas in Table 6.11.

TABLE 6.11
Integral formulas for hyperbolic functions

$\displaystyle\int \sinh u \, du = \cosh u + C$	$\displaystyle\int \text{csch}^2 u \, du = -\coth u + C$
$\displaystyle\int \cosh u \, du = \sinh u + C$	$\displaystyle\int \text{sech}\, u \tanh u \, du = -\text{sech}\, u + C$
$\displaystyle\int \text{sech}^2 u \, du = \tanh u + C$	$\displaystyle\int \text{csch}\, u \coth u \, du = -\text{csch}\, u + C$

Example 1

$$\int \coth 5x \, dx = \int \frac{\cosh 5x}{\sinh 5x} \, dx = \frac{1}{5}\int \frac{du}{u} \qquad (u = \sinh 5x)$$

$$= \frac{1}{5}\ln|u| + C = \frac{1}{5}\ln|\sinh 5x| + C$$

Example 2

$$\int_0^1 \sinh^2 x \, dx = \int_0^1 \frac{\cosh 2x - 1}{2} \, dx \qquad \text{(Table 6.9)}$$

$$= \frac{1}{2} \int_0^1 (\cosh 2x - 1) \, dx = \frac{1}{2} \left[\frac{\sinh 2x}{2} - x \right]_0^1$$

$$= \frac{\sinh 2}{4} - \frac{1}{2} = 0.40672 \qquad \left(\begin{matrix} \text{Calculator,} \\ \text{rounded} \end{matrix} \right)$$

As in Example 2, hyperbolic functions are usually evaluated with a calculator. Use the $\boxed{\text{hyp}}$, $\boxed{\text{cos}}$, $\boxed{\text{sin}}$, and $\boxed{\text{tan}}$ keys, or $\boxed{e^x}$.

Example 3

$$\int_0^{\ln 2} 4e^x \sinh x \, dx = \int_0^{\ln 2} 4 \, e^x \frac{e^x - e^{-x}}{2} \, dx = \int_0^{\ln 2} (2e^{2x} - 2) \, dx$$

$$= [e^{2x} - 2x]_0^{\ln 2} = (e^{2 \ln 2} - 2 \ln 2) - (1 - 0)$$

$$= 4 - 2 \ln 2 - 1 = 3 - 2 \ln 2$$

$$= 1.61371 \qquad \text{(Calculator, rounded)}$$

The Inverse Hyperbolic Functions

We use the inverses of the six basic hyperbolic functions in integration. Since

$$\frac{d}{dx} \sinh x = \cosh x \qquad (1)$$

is positive for every value of x, the hyperbolic sine is an increasing function of x and therefore has an inverse, which we denote by

$$y = \sinh^{-1} x. \qquad (2)$$

This equation is read "y equals the arc hyperbolic sine of x" or "y equals the inverse hyperbolic sine of x." For every value of x in the interval $-\infty < x < \infty$, the value of $y = \sinh^{-1} x$ is the number whose hyperbolic sine is x. The graphs of $y = \sinh x$ and $y = \sinh^{-1} x$ are shown in Fig. 6.46(a).

The function $y = \cosh x$ is not one-to-one, as we can see quickly enough from its graph in Fig. 6.44. But the restricted function

$$y = \cosh x, \qquad x \geq 0, \qquad (3)$$

is one-to-one and therefore has an inverse, denoted by

$$y = \cosh^{-1} x. \qquad (4)$$

For every value of $x \geq 1$, $y = \cosh^{-1} x$ is the number in the interval $0 \leq y < \infty$ whose hyperbolic cosine is x. The graphs of $y = \cosh x$, $x \geq 0$, and $y = \cosh^{-1} x$ are shown in Fig. 6.46(b).

Like $y = \cosh x$, the function $y = \text{sech } x = 1/\cosh x$ fails to be one-to-one,

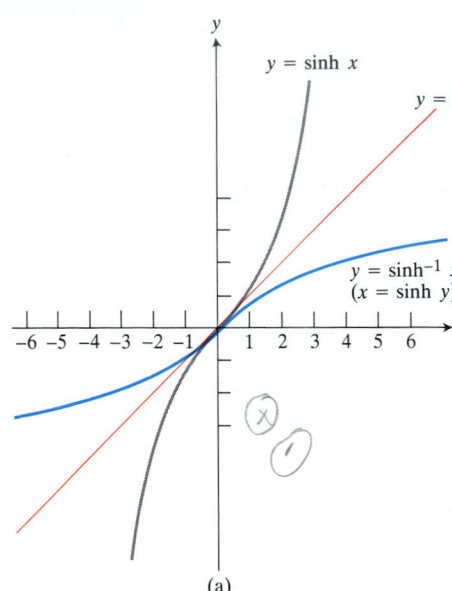

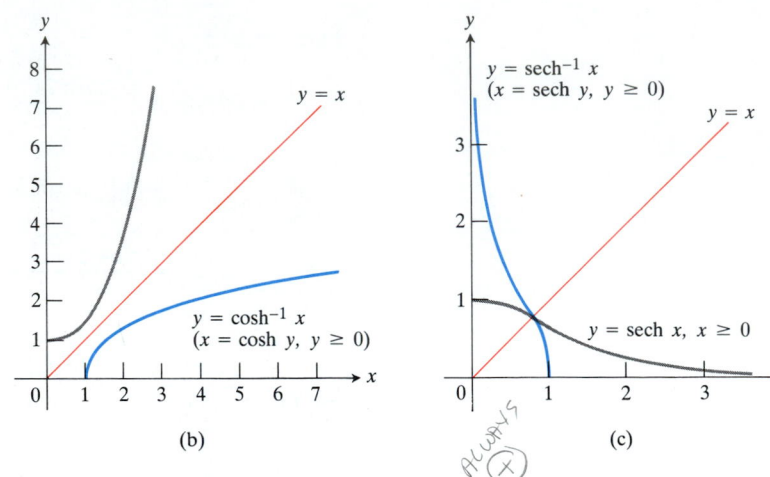

(b)

(c)

but its restriction to nonnegative values of x does have an inverse, denoted by

$$y = \text{sech}^{-1}x. \tag{5}$$

For every value of x in the interval $(0, 1]$, $y = \text{sech}^{-1}x$ is the nonnegative number whose hyperbolic secant is x. The graphs of $y = \text{sech } x$, $x \geq 0$, and $y = \text{sech}^{-1}x$ are shown in Fig. 6.46(c).

The hyperbolic tangent, cotangent, and cosecant are one-to-one on their domains and therefore have inverses, denoted by

$$y = \tanh^{-1}x, \qquad y = \coth^{-1}x, \qquad y = \text{csch}^{-1}x. \tag{6}$$

These functions are graphed in Fig. 6.47.

6.46 The graphs of the inverse hyperbolic sine, cosine, and secant of x. Notice the symmetries about the line $y = x$.

Useful Identities

The inverse hyperbolic secant, cosecant, and cotangent satisfy the identities shown in Table 6.12. These identities are handy when we have to calculate the values of $\text{sech}^{-1}x$, $\text{csch}^{-1}x$, and $\coth^{-1}x$ on a calculator that gives only $\cosh^{-1}x$, $\sinh^{-1}x$, and $\tanh^{-1}x$.

TABLE 6.12
Identities for inverse hyperbolic functions

$\text{sech}^{-1}x = \cosh^{-1}\dfrac{1}{x}$
$\text{csch}^{-1}x = \sinh^{-1}\dfrac{1}{x}$
$\coth^{-1}x = \tanh^{-1}\dfrac{1}{x}$

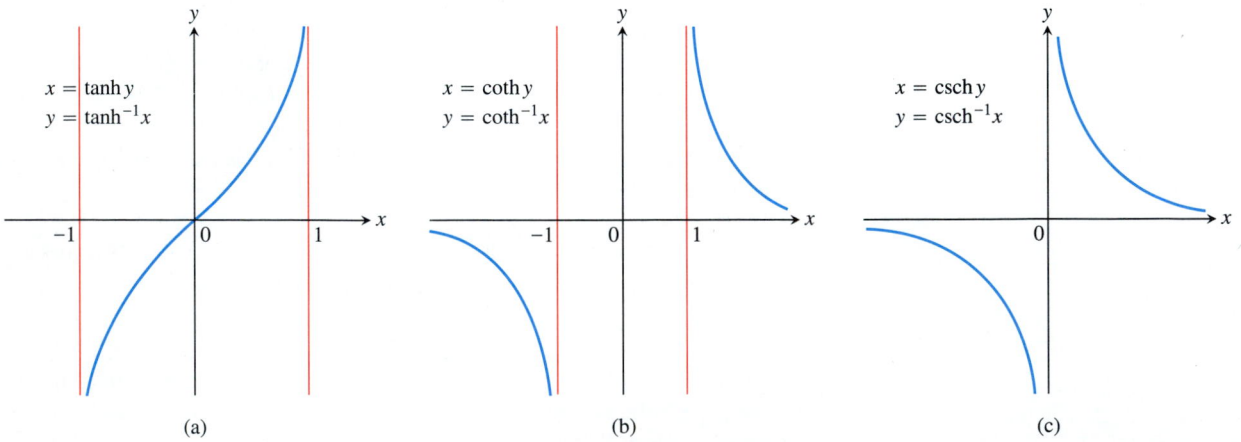

6.47 The graphs of the inverse hyperbolic tangent, cotangent, and cosecent of x.

TABLE 6.13
Derivatives of inverse hyperbolic functions

$$\frac{d(\sinh^{-1}u)}{dx} = \frac{1}{\sqrt{1 + u^2}}\frac{du}{dx}$$

$$\frac{d(\cosh^{-1}u)}{dx} = \frac{1}{\sqrt{u^2 - 1}}\frac{du}{dx}, \quad u > 1$$

$$\frac{d(\tanh^{-1}u)}{dx} = \frac{1}{1 - u^2}\frac{du}{dx}, \quad |u| < 1$$

$$\frac{d(\coth^{-1}u)}{dx} = \frac{1}{1 - u^2}\frac{du}{dx}, \quad |u| > 1$$

$$\frac{d(\text{sech}^{-1}u)}{dx} = \frac{-du/dx}{u\sqrt{1 - u^2}}, \quad 0 < u < 1$$

$$\frac{d(\text{csch}^{-1}u)}{dx} = \frac{-du/dx}{|u|\sqrt{1 + u^2}}, \quad u \neq 0$$

Derivatives and Integrals

The chief merit of the inverse hyperbolic functions lies in integration. You will see why after we derive the formulas for their derivatives, in Table 6.13.

The restrictions $|u| < 1$ and $|u| > 1$ on the derivative formulas for $\tanh^{-1}u$ and $\coth^{-1}u$ come from the natural restrictions on the values of these functions. (See Figs. 6.47a and b.) The distinction between $|u| < 1$ and $|u| > 1$ becomes important when we convert the derivative formulas into integral formulas. If $|u| < 1$, the integral of $1/(1 - u^2)$ is $\tanh u + C$. If $|u| > 1$, the integral is $\coth u + C$.

Example 4   Show that if u is a differentiable function of x whose values are greater than 1, then

$$\frac{d}{dx}\cosh^{-1}u = \frac{1}{\sqrt{u^2 - 1}}\frac{du}{dx}.$$

Solution   First we find the derivative of $y = \cosh^{-1}x$ for $x > 1$:

$$y = \cosh^{-1}x$$

$$x = \cosh y \qquad \text{(Inverse function property)}$$

$$1 = \sinh y \frac{dy}{dx} \qquad \text{(Differentiation with respect to } x)$$

$$\frac{dy}{dx} = \frac{1}{\sinh y} = \frac{1}{\sqrt{\cosh^2 y - 1}} \qquad \text{(Since } x > 1, y > 0 \text{ and } \sinh y > 0)$$

$$= \frac{1}{\sqrt{x^2 - 1}}. \qquad \text{(}\cosh y = x)$$

In short,

$$\frac{d}{dx}\cosh^{-1}x = \frac{1}{\sqrt{x^2 - 1}}.$$

The Chain Rule gives the final result:

$$\frac{d}{dx}\cosh^{-1}u = \frac{1}{\sqrt{u^2-1}}\frac{du}{dx}.$$

The derivative formulas in Table 6.13 lead to the integration formulas in Table 6.14.

TABLE 6.14
Integrals leading to inverse hyperbolic functions

$$\int \frac{du}{\sqrt{1+u^2}} = \sinh^{-1}u + C$$

$$\int \frac{du}{\sqrt{u^2-1}} = \cosh^{-1}u + C, \quad u>1$$

$$\int \frac{du}{1-u^2} = \begin{cases} \tanh^{-1}u + C & \text{if } |u|<1 \\ \coth^{-1}u + C & \text{if } |u|>1 \end{cases}$$

$$\int \frac{du}{u\sqrt{1-u^2}} = -\operatorname{sech}^{-1}|u| + C = -\cosh^{-1}\left(\frac{1}{|u|}\right) + C$$

$$\int \frac{du}{u\sqrt{1+u^2}} = -\operatorname{csch}^{-1}|u| + C = -\sinh^{-1}\left(\frac{1}{|u|}\right) + C$$

Example 5

$$\int_0^1 \frac{2\,dx}{\sqrt{1+4x^2}} = \int_0^2 \frac{du}{\sqrt{1+u^2}} = \sinh^{-1}u\Big]_0^2 \qquad (u=2x,\ du=2\,dx)$$

$$= \sinh^{-1}2 - 0 = 1.4436 \qquad \text{(Calculator, rounded)}$$

As in Example 5, inverse hyperbolic functions are usually evaluated with a calculator. On calculators without $\boxed{\text{inv}}$ and $\boxed{\text{hyp}}$ keys, we use formulas that express the inverse hyperbolic functions as combinations of natural logarithms. We have included these in Table 6.15 in the exercises.

EXERCISES 6.9

Each of Exercises 1–4 gives a value of $\sinh x$ or $\cosh x$. Use the definitions and the identity $\cosh^2x - \sinh^2x = 1$ to find the values of the remaining five hyperbolic functions.

1. $\sinh x = -\frac{3}{4}$

2. $\sinh x = \frac{4}{3}$

3. $\cosh x = \frac{17}{15}, \quad x>0$

4. $\cosh x = \frac{13}{5}, \quad x>0$

Rewrite the expressions in Exercises 5–10 in terms of exponentials and simplify the results as much as you can.

5. $2\cosh(\ln x)$

6. $\sinh(2\ln x)$

7. $\cosh 5x + \sinh 5x$

8. $(\sinh x + \cosh x)^4$

9. $\cosh 3x - \sinh 3x$

10. $\ln(\cosh x + \sinh x) + \ln(\cosh x - \sinh x)$

11. Use the identities

$$\sinh(x+y) = \sinh x \cosh y + \cosh x \sinh y$$
$$\cosh(x+y) = \cosh x \cosh y + \sinh x \sinh y,$$

to show that

a) $\sinh 2x = 2 \sinh x \cosh x$;

b) $\cosh 2x = \cosh^2 x + \sinh^2 x$.

12. Use the definitions of $\cosh x$ and $\sinh x$ to show that

$$\cosh^2 x - \sinh^2 x = 1.$$

In Exercises 13–24, find dy/dx if y equals the given expression.

13. $\sinh 3x$

14. $\dfrac{1}{2}\sinh(2x + 1)$

15. $2 \tanh \dfrac{x}{2}$

16. $x - \tanh x$

17. $\ln(\text{sech } x)$

18. $\ln(\text{csch } x)$

19. $\ln(\text{csch } x + \coth x)$

20. $x \cosh x - \sinh x$

21. $\dfrac{1}{2} \ln|\tanh x|$

22. $\tan^{-1}(\sinh x)$

23. a) $\cosh^2 x$ b) $\sinh^2 x$ c) $\dfrac{1}{2}\cosh 2x$

24. $(x^2 + 1) \text{ sech } (\ln x)$ (*Hint:* Express sech(ln x) in terms of exponentials and simplify before differentiating.)

In Exercises 25–36, find dy/dx if y equals the given expression.

25. $\sinh^{-1} 2x$

26. $2 \cosh^{-1} \sqrt{x}$

27. $(1 - x)\tanh^{-1} x$

28. $(1 - x^2)\coth^{-1} x$

29. $x \, \text{sech}^{-1} x$

30. $x^2 \text{csch}^{-1}(x^2)$

31. $\sinh^{-1}(\tan x)$

32. $\cosh^{-1}(\sec x), \quad 0 < x < \pi/2$

33. $\tanh^{-1}(\sin x), \quad -\pi/2 < x < \pi/2$

34. $\coth^{-1}(\sec x), \quad -\pi/2 < x < \pi/2$

35. $\text{sech}^{-1}(\sin x), \quad 0 < x < \pi/2$

36. $\text{csch}^{-1}(\tan x), \quad 0 < x < \pi/2$

Verify the integration formulas in Exercises 37–44.

37. a) $\displaystyle \int \text{sech } x \, dx = \tan^{-1}(\sinh x) + C$

   b) $\displaystyle \int \text{sech } x \, dx = \sin^{-1}(\tanh x) + C$

38. $\displaystyle \int x \sinh x \, dx = x \cosh x - \sinh x + C$

39. $\displaystyle \int \text{sech}^{-1} x \, dx = x \, \text{sech}^{-1} x + \sin^{-1} x + C$

40. $\displaystyle \int x \, \text{sech}^{-1} x \, dx = \frac{x^2}{2} \text{sech}^{-1} x - \frac{1}{2}\sqrt{1 - x^2} + C$

41. $\displaystyle \int x \coth^{-1} x \, dx = \frac{x^2 - 1}{2} \coth^{-1} x + \frac{x}{2} + C$

42. $\displaystyle \int \tanh^{-1} x \, dx = x \tanh^{-1} x + \frac{1}{2} \ln(1 - x^2) + C$

43. $\displaystyle \int x \cosh^{-1} x \, dx = \frac{2x^2 - 1}{4} \cosh^{-1} x - \frac{x}{4}\sqrt{x^2 - 1} + C$

44. $\displaystyle \int x \sinh^{-1} x \, dx = \frac{2x^2 + 1}{4} \sinh^{-1} x - \frac{x}{4}\sqrt{x^2 + 1} + C$

Evaluate the integrals in Exercises 45–60.

45. $\displaystyle \int_{-1}^{1} \cosh 5x \, dx$

46. $\displaystyle \int_{-1}^{0} \cosh(2x + 1)dx$

47. $\displaystyle \int_{-3}^{3} \sinh x \, dx$

48. $\displaystyle \int_{-\pi}^{\pi} \tanh 2x \, dx$

49. $\displaystyle \int_{0}^{1/2} 4e^x \cosh x \, dx$

50. $\displaystyle \int_{0}^{1/2} 4e^{-x} \sinh x \, dx$

51. $\displaystyle \int \frac{\cosh(\ln x)}{x} \, dx$

52. $\displaystyle \int \frac{\sinh x}{\cosh x} \, dx$

53. $\displaystyle \int_{0}^{\ln 3} \text{sech}^2 x \, dx$

54. $\displaystyle \int_{0}^{\ln 2} \tanh^2 x \, dx$

55. $\displaystyle \int_{1}^{4} \frac{\cosh \sqrt{x}}{\sqrt{x}} \, dx$

56. $\displaystyle \int_{\ln 2}^{\ln 3} \text{csch}^2 x \, dx$

57. $\displaystyle \int_{-\ln 2}^{0} \cosh^2 3x \, dx$

58. $\displaystyle \int_{-\ln 2}^{\ln 2} \sqrt{\cosh 2x - 1} \, dx$

59. $\displaystyle \int \text{sech}^3 5x \tanh 5x \, dx$

60. $\displaystyle \int \tanh^3 x \, dx$

Evaluating Inverse Hyperbolic Functions and Related Integrals

When hyperbolic function keys are not available on a calculator, it is still possible to evaluate the inverse hyperbolic functions by expressing them as logarithms. The conversion formulas are listed in Table 6.15.

TABLE 6.15
Logarithm formulas for evaluating inverse hyperbolic functions

$$\sinh^{-1} x = \ln(x + \sqrt{x^2 + 1}), \quad -\infty < x < \infty$$

$$\cosh^{-1} x = \ln(x + \sqrt{x^2 - 1}), \quad x \geq 1$$

$$\tanh^{-1} x = \frac{1}{2} \ln \frac{1 + x}{1 - x}, \quad |x| < 1$$

$$\text{sech}^{-1} x = \ln \left(\frac{1 + \sqrt{1 - x^2}}{x} \right), \quad 0 < x \leq 1$$

$$\text{csch}^{-1} x = \ln \left(\frac{1}{x} + \frac{\sqrt{1 + x^2}}{|x|} \right), \quad x \neq 0$$

$$\coth^{-1} x = \frac{1}{2} \ln \frac{x + 1}{x - 1}, \quad |x| > 1$$

Use the formulas in Table 6.15 to express the numbers in Exercises 61–72 in terms of natural logarithms.

61. $\sinh^{-1}(-5/12)$

62. $\sinh^{-1}(-4/3)$

63. $\cosh^{-1}(5/3)$

64. $\cosh^{-1}(2/\sqrt{3})$

65. $\tanh^{-1}(-1/2)$

66. $\tanh^{-1}(-3/5)$

67. $\coth^{-1}(5/4)$

68. $\coth^{-1}(-2)$

69. $\operatorname{sech}^{-1}(3/5)$

70. $\operatorname{sech}^{-1}(4/5)$

71. $\operatorname{csch}^{-1}(-1/\sqrt{3})$

72. $\operatorname{csch}^{-1}(5/12)$

Evaluate the integrals in Exercises 73–80 in terms of (a) inverse hyperbolic functions; (b) natural logarithms using the formulas in Table 6.15.

73. $\displaystyle\int_{0}^{1} \frac{dx}{\sqrt{1+x^2}}$

74. $\displaystyle\int_{4/5}^{12/13} \frac{dx}{x\sqrt{1-x^2}}$

75. $\displaystyle\int_{5/4}^{5/3} \frac{dx}{\sqrt{x^2-1}}$

76. $\displaystyle\int_{0}^{1/2} \frac{dx}{1-x^2}$

77. $\displaystyle\int_{5/4}^{2} \frac{dx}{1-x^2}$

78. $\displaystyle\int_{0}^{2\sqrt{3}} \frac{dx}{\sqrt{4+x^2}}$

79. $\displaystyle\int_{1}^{2} \frac{dx}{x\sqrt{4+x^2}}$

80. $\displaystyle\int_{0}^{\pi} \frac{\cos x\, dx}{\sqrt{1+\sin^2 x}}$

In Exercises 81 and 82, find the volumes of the solids generated by revolving the shaded regions about the x-axis.

81.

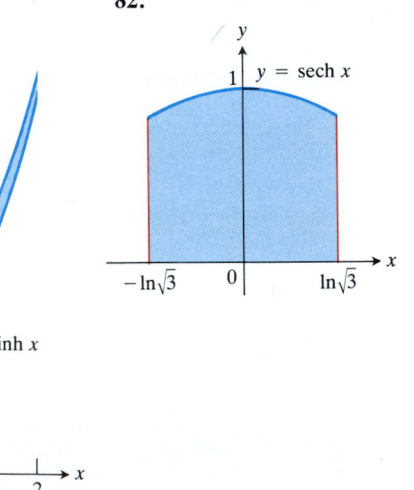

82.

83. Use the formula

$$\int \operatorname{sech} x\, dx = \sin^{-1}(\tanh x) + C$$

to find the centroid of the shaded region in Exercise 82.

84. Find the volume of the solid generated by revolving the shaded region in Fig. 6.48 about the line $y = 1$.

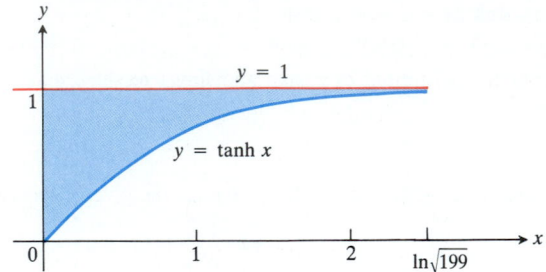

6.48 The region in Exercise 84.

85. a) Find the centroid of the curve $y = \cosh x$, $-\ln 2 \le x \le \ln 2$.

b) CALCULATOR Evaluate the coordinates to two decimal places. Then sketch the curve and plot the centroid to show its relation to the curve.

86. *Even and odd.* Show that if a function f is defined on an interval symmetric about the origin (so that f is defined at $-x$ whenever it is defined at x), then

$$f(x) = \frac{f(x) + f(-x)}{2} + \frac{f(x) - f(-x)}{2}. \tag{7}$$

Then show that $(f(x) + f(-x))/2$ is even and that $(f(x) - f(-x))/2$ is odd.

87. *Continuation of Exercise 86.* Equation (7) simplifies considerably if f itself is (a) even or (b) odd. What are the new equations?

88. *Resistance proportional to the square of the velocity.* If a body of mass m falling from rest under the action of gravity encounters an air resistance proportional to the square of the velocity, then the body's velocity t seconds into the fall satisfies the differential equation

$$m\frac{dv}{dt} = mg - kv^2,$$

where k is a constant that depends on the body's aerodynamic properties and the density of the air. (We assume that the fall is short enough so that the variation in the air's density will not affect the outcome significantly.)

a) Show that

$$v = \sqrt{\frac{mg}{k}} \tanh\left(\sqrt{\frac{gk}{m}}\, t\right)$$

satisfies the differential equation and the initial condition that $v = 0$ when $t = 0$.

b) Find the body's *limiting velocity*, $\lim_{t\to\infty} v$.

89. *Tractor trailers and the tractrix.* When a tractor trailer turns into a cross street or driveway, its rear wheels follow a curve like the one in Fig. 6.49. (This is why the rear wheels sometimes ride up over the curb.) We can find an equation for the curve if we picture the rear wheels as a mass M at the point $(1, 0)$ on the x-axis attached by a rod of unit length to a point P representing the cab at the origin. As the point P

moves up the y-axis, it drags M along behind it. The curve traced by M, called a *tractrix* from the Latin word *tractum* for "drag," can be shown to be the graph of the function $y = f(x)$ that solves the initial value problem

Differential equation: $\dfrac{dy}{dx} = -\dfrac{1}{x\sqrt{1 - x^2}} + \dfrac{x}{\sqrt{1 - x^2}}$,

Initial condition: $y = 0$ when $x = 1$.

Solve the initial value problem to find an equation for the curve. (You need an inverse hyperbolic function.)

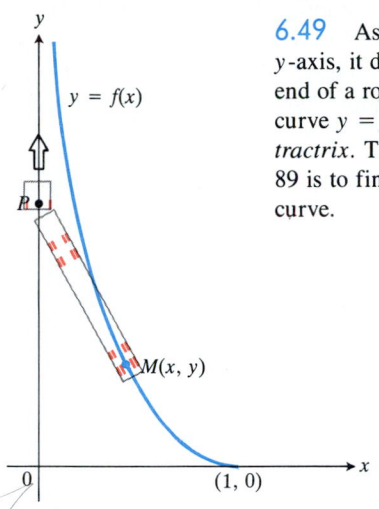

6.49 As P moves up the y-axis, it drags M after it at the end of a rod 1 unit long. The curve $y = f(x)$ traced by M is a *tractrix*. The goal of Exercise 89 is to find an equation for the curve.

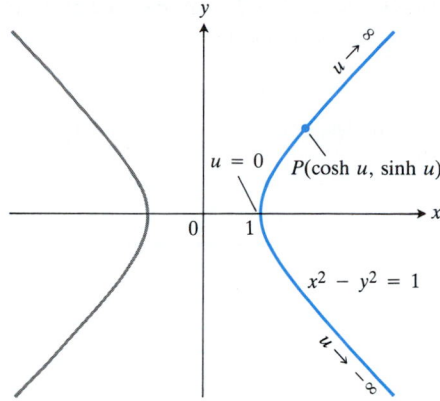

6.50 Since $\cosh^2 u - \sinh^2 u = 1$, the point ($\cosh u$, $\sinh u$) lies on the right-hand branch of the hyperbola $x^2 - y^2 = 1$ for every value of u (Exercise 91).

90. *Accelerations whose magnitudes are proportional to displacement.* Suppose that the position of a body moving along a coordinate line at time t is

a) $s = a \cos kt + b \sin kt$;

b) $s = a \cosh kt + b \sinh kt$.

Show that in both cases the acceleration d^2s/dt^2 is proportional to s but that in the first case it is always directed toward the origin while in the second case it is directed away from the origin.

91. *The hyperbolic in hyperbolic functions.* In case you are wondering where the name *hyperbolic* comes from, here is the answer: Just as $x = \cos u$ and $y = \sin u$ are identified with points (x, y) on the unit circle, the functions $x = \cosh u$ and $y = \sinh u$ are identified with points (x, y) on the right-hand branch of the unit hyperbola, $x^2 - y^2 = 1$ (Fig. 6.50).

Another analogy between hyperbolic and circular functions is that the variable u in the coordinates ($\cosh u$, $\sinh u$) for the points of the right-hand branch of the hyperbola $x^2 - y^2 = 1$ is twice the area of the sector AOP pictured in Fig. 6.51. To see why this is so, carry out the following steps.

a) Show that the area $A(u)$ of sector AOP is given by the formula

$$A(u) = \frac{1}{2} \cosh u \sinh u - \int_{1}^{\cosh u} \sqrt{x^2 - 1}\, dx.$$

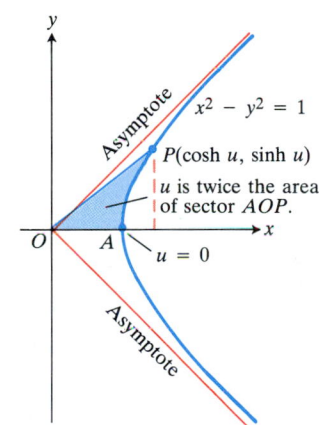

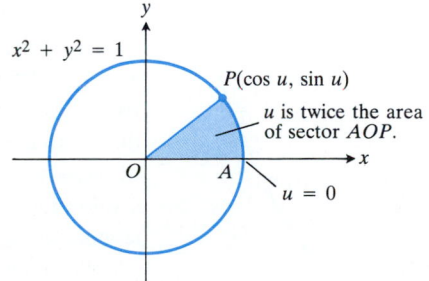

6.51 One of the analogies between hyperbolic and circular functions is revealed by these two diagrams (Exercise 91).

b) Differentiate both sides of the equation in (a) with respect to u to show that

$$A'(u) = \frac{1}{2}.$$

c) Solve this last equation for $A(u)$. What is the value of $A(0)$?

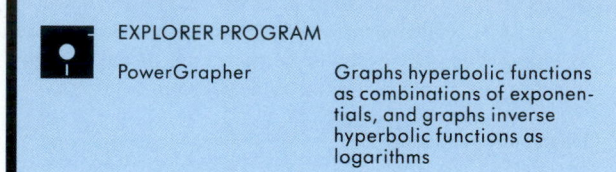

EXPLORER PROGRAM

PowerGrapher   Graphs hyperbolic functions as combinations of exponentials, and graphs inverse hyperbolic functions as logarithms

6.10 Hanging Cables

In this section, we show that chains, telephone lines, TV cables, and electric power lines that are strung from one support to another always hang in a hyperbolic cosine curve. In contrast, cables of suspension bridges, which do not hang freely but support a constant load per horizontal foot, hang in parabolas (Exercise 54, Section 9.1).

The Mathematical Model

The starting point for the mathematical description of a hanging cable's shape is the observation that the forces acting on any section of a cable hanging at rest must cancel each other out. We know this because the cable is not moving. If we think about the forces acting on a section that runs from the cable's lowest point A to a point P partway up one of the sides, we find that they can be described as

1. the horizontal tension at A,

2. the tangential tension at P, and

3. the downward force exerted on the cable section by gravity.

Figure 6.52 shows a typical section of cable with the forces acting on it. The horizontal tension at A is shown by the arrow of length H pointing to the left. The tangential tension at P is shown by the arrow of length T pointing up and to the right along the tangent to the cable at P. The force of gravity is shown by an arrow pointing straight down. Its length is ws, the weight of the cable section, expressed as a product of the cable's weight per unit length w and the section's length s. The tension at P has a horizontal component of magnitude $T \cos \phi$ and a vertical component of magnitude $T \sin \phi$.

To make full use of the information that the forces acting on section AP are in balance, we introduce a coordinate system in the plane in which the cable hangs (Fig. 6.53) and assume that the cable lies along a differentiable curve in that plane. This enables us to write equations for the relationships among H, T, and ws and, from these equations, to derive a formula for the curve. We take the x-axis perpendicular to the force of gravity and the y-axis vertical with the positive y-axis passing through A. For the moment we let the y-coordinate of A be y_0. Later we shall find a value for y_0 that makes the cable's equation in the coordinate system particularly simple.

Since the cable section lies in the first quadrant, the angle ϕ that the tangent to the cable at P makes with the horizontal is also the angle of inclination of the tangent line. The slope dy/dx of the curve along which the cable lies therefore has the value $\tan \phi$ at P. At A, the curve's lowest point, $dy/dx = 0$.

Since the cable section AP is moving neither left nor right, the magnitude of the horizontal component of the tension at P just equals the horizontal tension at A:

$$T \cos \phi = H. \tag{1}$$

Hanging cable

T

ϕ

P

$T \sin \phi$

$T \cos \phi$

s

A

H

ws

6.52   The forces on a typical section AP of a hanging cable. Each arrow in the diagram is labeled with its length.

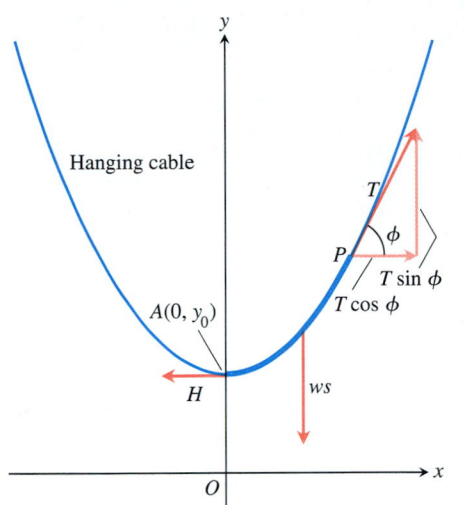

6.53 To find the shape of a hanging cable, we match it to a curve in a coordinate plane.

Similarly, the fact that section AP is moving neither up nor down tells us that the magnitude of the vertical component of the tension at P equals the weight of the cable section:

$$T \sin \phi = ws. \tag{2}$$

These are the only two facts we have about the hanging cable and, if we are to determine its shape by finding y as a function of x, this must be where we start. The only other information we have, an artifact of the coordinate system and our assumption that the curve along which the cable lies is differentiable, is that

$$\frac{dy}{dx} = \tan \phi. \tag{3}$$

The importance of Eq. (3) is that it gets x and y into the picture, for we can now write

$$\frac{dy}{dx} = \tan \phi = \frac{T \sin \phi}{T \cos \phi} = \frac{ws}{H}$$

or, simply,

$$\frac{dy}{dx} = \frac{ws}{H}. \tag{4}$$

While the equations we have so far do not give a way to express s directly in terms of x and y, we do know from our work with arc length that

$$\frac{ds}{dx} = \sqrt{1 + \left(\frac{dy}{dx}\right)^2}. \tag{5}$$

We can therefore eliminate s from Eq. (4) by differentiating both sides with respect to x and replacing the resulting ds/dx by $\sqrt{1 + (dy/dx)^2}$. This gives

$$\frac{d^2y}{dx^2} = \frac{w}{H}\sqrt{1 + \left(\frac{dy}{dx}\right)^2}. \tag{6}$$

We now have an equation that shows how the coordinates of the points on the hanging cable are related.

To find y as a function of x, we solve Eq. (6) subject to the conditions that

$$\frac{dy}{dx}(0) = 0, \quad \text{(The cable is horizontal at } x = 0.\text{)}$$

$$y(0) = y_0. \quad \text{(The cable crosses the } y\text{-axis at } y = y_0.\text{)}$$

To make the work easier, we can replace w/H temporarily with the single letter a. Equation (6) then becomes

$$\frac{d^2y}{dx^2} = a\sqrt{1 + \left(\frac{dy}{dx}\right)^2}. \tag{7}$$

While we have had relatively little experience with second-order differential equations, we can draw on our experience with first-order equations by observing that Eq. (7) is a first-order equation in the variable dy/dx. If we represent dy/dx by the single letter p, writing

$$\frac{dy}{dx} = p,$$

Why the Hyperbolic Substitution?

In case you are wondering why we substituted $p = \sinh u$ instead of $p = \tan u$ to evaluate the integral in Eq. (8), the answer is that the tangent substitution leads to a logarithmic expression for p that is harder to integrate. We shall study substitutions like these in the next chapter.

Eq. (7) becomes

$$\frac{dp}{dx} = a\sqrt{1 + p^2}.$$

From this we get

$$\frac{dp}{\sqrt{1 + p^2}} = a \, dx$$

and

$$\int \frac{dp}{\sqrt{1 + p^2}} = \int a \, dx = ax + C. \tag{8}$$

We evaluate the integral on the left by substituting

$$p = \sinh u, \qquad dp = \cosh u \, du,$$

$$1 + p^2 = 1 + \sinh^2 u = \cosh^2 u.$$

Then

$$\int \frac{dp}{\sqrt{1 + p^2}} = \int \frac{\cosh u \, du}{\sqrt{\cosh^2 u}}$$

$$= \int \frac{\cosh u \, du}{\cosh u} \qquad (\cosh u > 0) \tag{9}$$

$$= \int du = u + C'$$

$$= \sinh^{-1} p + C'.$$

This combines with Eq. (8) to give

$$\sinh^{-1} p = ax + C_1. \qquad (C_1 = C - C') \tag{10}$$

The initial condition

$$p(0) = \frac{dy}{dx}(0) = 0$$

determines the value of C_1:

$$C_1 = \sinh^{-1}(0) - a \cdot 0 = 0.$$

Therefore,

$$\sinh^{-1} p = ax$$

and

$$p = \sinh ax. \tag{11}$$

When restated in terms of x and y, Eq. (11) becomes

$$\frac{dy}{dx} = \sinh ax,$$

an equation readily integrated to give

$$y = \frac{1}{a} \cosh ax + C_2.$$

The remaining initial condition, $y(0) = y_0$, determines C_2:

$$y(0) = \frac{1}{a} \cosh 0 + C_2$$

$$y_0 = \frac{1}{a} + C_2$$

$$C_2 = \frac{1}{a} - y_0.$$

We now go back and adjust the coordinate system for the cable to set the value of y_0 at

$$y_0 = \frac{1}{a} = \frac{H}{w}.$$

This makes C_2 equal to zero, and the final equation for the curve of the hanging cable (Fig. 6.54) is

$$y = \frac{1}{a} \cosh ax = \frac{H}{w} \cosh \frac{w}{H} x. \tag{12}$$

To summarize, here is what we have learned about the shape of a hanging cable: If we choose a coordinate system for the plane of the cable in which the x-axis is horizontal, the force of gravity is straight down, the positive y-axis points straight up, and the lowest point of the cable lies at the point $y = H/w$ on the y-axis, then the cable lies along the graph of the hyperbolic cosine

$$y = \frac{H}{w} \cosh \frac{w}{H} x. \tag{13}$$

At least, this is true for the half of the cable that lies in the first quadrant, which was our assumption for the section we analyzed. What about the other half? It lies along the hyperbolic cosine too. The cable and hyperbolic cosine are both symmetric with respect to the y-axis, so if they match on one side they match on the other.

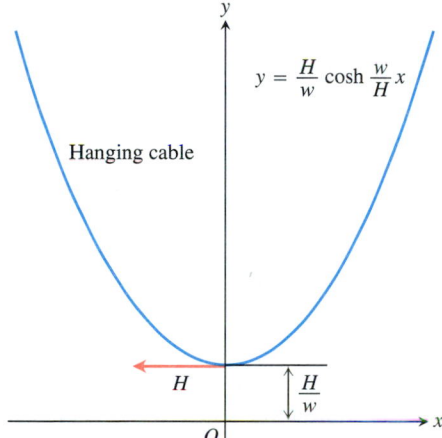

6.54 In a coordinate system chosen to match H and w in the manner shown, a hanging cable lies along the hyperbolic cosine $y = (H/w)\cosh(wx/H)$.

Catenaries and Chain Curves

The curve $y = (1/a)\cosh ax$ is sometimes called a **chain curve** or a **catenary,** the latter deriving from the Latin *catena*, meaning "chain."

The Gateway Arch to the West

Our calculations for hanging cables explain the shape of the Gateway Arch to the West in St. Louis. When a cable hangs freely, all its internal forces are in equilibrium and there are no transverse forces working to push the cable out of shape. If an arch were built hanging down in the shape of a hyperbolic cosine curve, there would be no transverse forces in the arch. If the arch were then turned right side up, all the forces, reversed, would still be in equilibrium, with no transverse forces present to encourage collapse. The inherent stability of the hyperbolic cosine shape was what led the architects to choose it for the Gateway Arch.

Example 1   Show that, in the coordinate system of Fig. 6.54, the magnitude T of the tension at any point $P(x, y)$ along a hanging cable is given by the equation

$$T = wy.$$

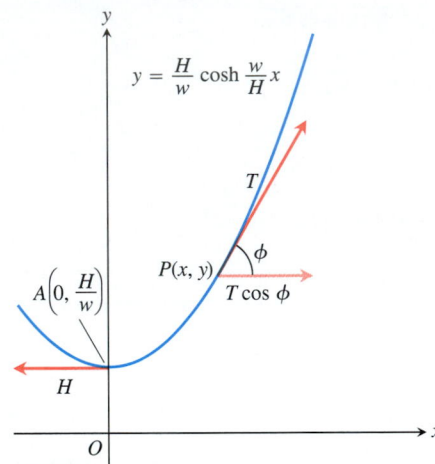

$$y = \frac{H}{w} \cosh \frac{w}{H} x$$

6.55   As explained in Example 1, $T = wy$ in this coordinate system.

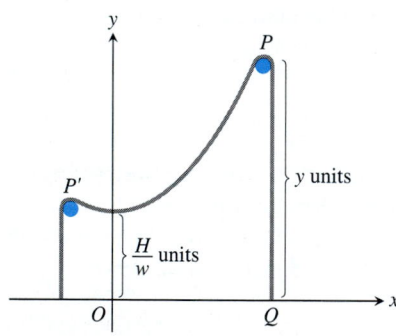

6.56   If the axes are positioned so that the lowest point of the cable is H/w units above the x-axis, then the tension at each point $P(x, y)$ on the cable is exactly equal to the weight of a piece of cable y units long.

Solution   We work in a coordinate system in which the cable lies along the curve

$$y = \frac{H}{w} \cosh \frac{w}{H} x.$$

We sketch the section of cable from the point $A(0, H/w)$ up to an arbitrary point $P(x, y)$ and show the tensions at A and P (Fig. 6.55). The cable's slope at P is

$$\tan \phi = \frac{dy}{dx} = \sinh \frac{w}{H} x.$$

The tension at P satisfies the equation

$$T \cos \phi = H,$$

from which we find

$$T = H \sec \phi = H\sqrt{1 + \tan^2\phi}$$

$$= H\sqrt{1 + \sinh^2 \frac{w}{H} x} = H\sqrt{\cosh^2 \frac{w}{H} x}$$

$$= H \cosh \frac{w}{H} x$$

$$= w \frac{H}{w} \cosh \frac{w}{H} x = wy.$$

What does it mean for T to equal wy at every point along the hanging cable in the chosen coordinate system? It means that the magnitude of the tension at $P(x, y)$ is exactly equal to the weight of y units of cable. Thus, if the cable to the right of P were allowed to hang straight down from P over a smooth peg and were cut off at the point Q where it crossed the x-axis, the weight of the cable length PQ would be just enough to keep the cable from slipping at P. The cable can be draped over two smooth pegs without slipping if the ends just reach the x-axis (Fig. 6.56).

Example 2   A typical electric power line of 4-gauge copper wire might span 350 ft between poles and have a tension of 690 lb at the poles (achieved by winching the wire over a pulley). The wire would then make an angle of about 2° with the horizontal. As we can see from the equation

$$H = T \cos \phi = T \cos 2°$$

$$= (690)(0.9994) = 689.6, \qquad \text{(Calculator, rounded)}$$

such a flat span would exhibit very little difference between T and H. The difference between T and H is more noticeable in wires that sag a lot. If the wire meets the pole at an angle of 60° from the horizontal, the tension at the wire's lowest point is only half the tension at the pole, as we can see from the equation

$$H = T \cos 60° = \frac{T}{2}.$$

For a fine article about how construction crews determine the tensions in the cables they string, read Thomas O'Neil's "Constructing Power Lines," *The UMAP Journal*, Vol. 4, No. 3, 1983, pp. 259–64.

EXERCISES 6.10

1. a) Find the length of the segment of the catenary $y = (1/2)\cosh 2x$ from $x = 0$ to $x = \ln \sqrt{5}$.
   b) Find the length of the segment of the catenary $y = (1/a)\cosh ax$ from $x = 0$ to $x = b > 0$.

2. Show that the slope, taken as positive, at a point P on a catenary is always w/H times the distance along the catenary from the bottom of the catenary up to P.

3. Show that the area of the region in the first quadrant enclosed by the catenary $y = (1/a)\cosh ax$, the coordinate axes, and the line $x = b$ is the same as the area of a rectangle of height $1/a$ and length s, where s is the length of the catenary from $x = 0$ to $x = b$ (Fig. 6.57).

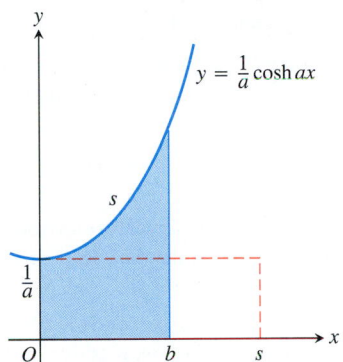

6.57   The region and rectangle in Exercise 3.

4. *A minimal surface.*   Find the area of the surface swept out by revolving about the x-axis the curve $y = 4 \cosh(x/4)$, $-\ln 16 \le x \le \ln 81$ (Fig. 6.58).

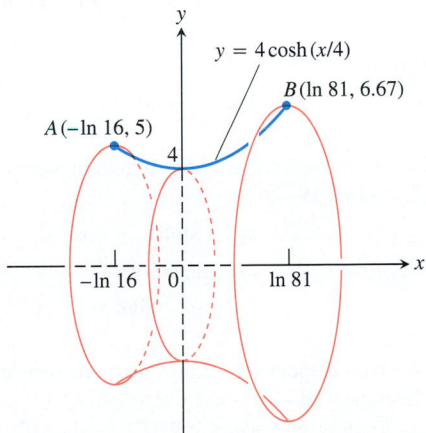

6.58   The minimal surface in Exercise 4.

It can be shown that, of all continuously differentiable curves joining the points A and B in Fig. 6.58, the catenary $y = 4 \cosh(x/4)$ generates the surface of least area. If you made a rigid wire frame of the end-circles through A and B and dipped them in a soap-film solution, the surface spanning the circles would be the one generated by the catenary.

5. The length of arc AP in Fig. 6.55 is $s = (1/a) \sinh ax$ (where $a = w/H$).
   a) Show that the coordinates of P may be expressed in terms of s as

   $$x = \frac{1}{a} \sinh^{-1} as, \qquad y = \sqrt{s^2 + \frac{1}{a^2}}.$$

   b) Use the equations in (a) to confirm that

   $$ds^2 = dx^2 + dy^2.$$

6. *The sag and horizontal tension in a cable.*   The ends of a cable 32 ft long and weighing 2 lb/ft are fastened at the same level to posts 30 ft apart.
   a) Model the cable with the equation

   $$y = \frac{1}{a} \cosh ax, \quad -15 \le x \le 15.$$

   Use information from Exercise 5(a) to show that a satisfies the equation

   $$16a = \sinh 15a. \tag{14}$$

   b) GRAPHER   Solve Eq.(14) graphically by estimating the coordinates of the points where the graphs of the equations $y = 16a$ and $y = \sinh 15a$ intersect in the ay-plane.
   c) EQUATION SOLVER or ROOT FINDER   Solve Eq. (14) for a numerically. Compare your solution with the value you found in (b).
   d) Estimate the horizontal tension in the cable at the cable's lowest point.
   e) GRAPHER   Graph the catenary

   $$y = \frac{1}{a} \cosh ax$$

   over the interval $-15 \le x \le 15$. Estimate the sag in the cable at its center.

REVIEW QUESTIONS

1. What is the inverse of a one-to-one function? How do you tell when functions are inverses of one another? Give examples of functions that are inverses of one another; that are not inverses of one another.

2. What continuous functions have inverses?

3. How are the graphs, domains, and ranges of functions and their inverses related? Give an example.

4. How are the derivatives of functions and their inverses related? Give an example.

5. What are the rules for doing arithmetic with logarithms?

6. What is logarithmic differentiation? Give an example.

7. What integrals lead naturally to logarithms? Give an example.

8. Graph the functions $\ln x$ and e^x together. What are the domains and ranges of these functions? Can e^x ever equal zero?

9. What laws of exponents do exponential functions obey?

10. How are the functions a^x and $\log_a x$ defined? What are their derivatives? What integrals are associated with these functions? Give examples. What are the restrictions on a?

11. How is $\log_a x$ evaluated in terms of $\ln x$? How is the graph of $\log_a x$ related to the graph of $\ln x$?

12. What does the graph of the function a^x look like if $a > 1$? If $0 < a < 1$?

13. Describe some of the applications of base 10 logarithms.

14. Where does the equation $y = y_0 e^{kt}$ come from? Describe some of the applications of this equation.

15. How do you compare the growth rates of functions of x as $x \to \infty$? Give examples.

16. Explain little-oh and big-oh notation. Give examples.

17. Graph the six basic inverse trigonometric functions. Indicate their domains and ranges.

18. What are the derivatives of the inverse trigonometric functions? What integrals are associated with these functions? Give examples.

19. How are the six basic hyperbolic functions defined? Graph them and indicate their domains and ranges.

20. What identities are associated with hyperbolic functions?

21. What are the derivatives of the six basic hyperbolic functions? What are the corresponding integrals? Give examples.

22. Graph the six inverse hyperbolic functions and give their domains and ranges.

23. What are the derivatives of the inverse hyperbolic functions? Give examples of integrals that lead naturally to inverse hyperbolic functions.

24. What shapes do hanging cables assume? How do the relations among the magnitudes of the forces acting on a section of cable lead to a differential equation for the cable's shape? What is the equation? How is it solved?

MISCELLANEOUS EXERCISES

Find the derivatives with respect to x of the functions whose formulas appear in Exercises 1–14.

1. $\ln (\cos^{-1}x)$

2. $\ln (\sin^{-1}x)$

3. $\log_2(x^2)$

4. $\log_5(x - 7)$

5. $(1 + x^2)e^{\tan^{-1}x}$

6. $x^2\sec^{-1}x - \sqrt{x^2 - 1}$

7. $(x + 2)^{x + 2}$

8. $\dfrac{1}{3 \ln 2} 8^{\sin^{-1}x}$

9. $\sin^{-1}(\sqrt{1 - x})$

10. $(1 + x^2) \cot^{-1}2x$

11. $x \tan^{-1}x - \dfrac{1}{2} \ln x$

12. $x \cos^{-1}x - \sqrt{1 - x^2}$

13. $2\sqrt{x - 1} \sec^{-1} \sqrt{x}$

14. $\csc^{-1} (\sec x), \quad 0 < x < \pi/2$

Verify the integration formulas in Exercises 15 and 16.

15. $\displaystyle \int \tan^{-1}t \, dt = t \tan^{-1} t - \dfrac{1}{2} \ln (1 + t^2) + C$

16. $\displaystyle \int \sec^{-1} \theta \, d\theta = \theta \sec^{-1} \theta - \ln \left(\theta + \sqrt{\theta^2 - 1} \right) + C, \theta > 1$

Find dy/dx in Exercises 17 and 18.

17. $x^{\ln y} = 2$

18. $\ln xy = e^{x + y}$

Find the derivatives with respect to s of the functions whose formulas appear in Exercises 19–24.

19. $s - \coth s$

20. $s \sinh s - \cosh s$

21. $\ln(\operatorname{csch} s) + s \coth s$

22. $\ln(\operatorname{sech} s) + s \tanh s$

23. $\sin^{-1}(\tanh s)$

24. $\tan^{-1}(\sinh s)$

Find the derivatives with respect to v of the functions whose formulas appear in Exercises 25–30.

25. $\sqrt{1 + v^2} \sinh^{-1} v$

26. $\sqrt{v^2 - 1} \cosh^{-1} v$

27. $1 - \tanh^{-1} (1/v), \quad |v| > 1$

28. $\coth^{-1} (\csc v), \quad 0 < v < \pi/2$

29. $\text{sech}^{-1}(\cos 2v), \quad 0 < v < \pi/4$

30. $\text{csch}^{-1}(\cot v), \quad 0 < v < \pi/2$

Evaluate the integrals in Exercises 31–48.

31. $\displaystyle\int \frac{dx}{3 - 2e^{-x}}$

32. $\displaystyle\int_1^e \frac{\sqrt{\ln x}}{x}\, dx$

33. $\displaystyle\int \frac{\tan(\ln t)}{t}\, dt$

34. $\displaystyle\int_0^{\pi/2} \frac{\cos\theta\, d\theta}{\sqrt{4 - \sin^2\theta}}$

35. $\displaystyle\int_1^e \frac{2\ln t}{t\,\ln^2 t + 31t}\, dt$

36. $\displaystyle\int_{-1}^0 \frac{dx}{x^2 + 2x + 2}$

37. $\displaystyle\int_1^e \frac{\pi\cos(\pi\ln z)}{z}\, dz$

38. $\displaystyle\int_1^8 \frac{\log_4 q}{q}\, dq$

39. $\displaystyle\int_1^e \frac{8\ln 3\,\log_3 x}{x}\, dx$

40. $\displaystyle\int_0^1 u\,3^{(u^2)}\, du$

41. $\displaystyle\int_0^{\pi/4} 2^{\tan\theta}\sec^2\theta\, d\theta$

42. $\displaystyle\int_0^{\pi/6} 4e^{-4\sin w}\cos w\, dw$

43. $\displaystyle\int \frac{3\,dx}{x\sqrt{\ln x - \ln^2 x}}$

44. $\displaystyle\int_1^{1 + (\sqrt{2}/2)} \frac{dt}{\sqrt{1 - (t - 1)^2}}$

45. $\displaystyle\int_1^{3\sqrt{3}} \frac{dy}{y^{2/3} + y^{4/3}}$

46. $\displaystyle\int \frac{dx}{\sqrt{1 - x^2}\left(\frac{\pi}{4} + \sin^{-1}x\right)}$

47. $\displaystyle\int \frac{dx}{x\sqrt{x^2 - 1}\,\sec^{-1}x}$

48. $\displaystyle\int_{\ln\log_4\ln 4}^{\ln\log_4\ln 16} 4^{(e^x)}e^x\, dx$

Evaluate the integrals in Exercises 49–54.

49. $\displaystyle\int_0^{\ln 2} 4\,e^x\cosh x\, dx$

50. $\displaystyle\int_0^{\ln 2} \frac{\sinh t}{1 + \cosh t}\, dt$

51. $\displaystyle\int_{-\ln 3}^{\ln 3} 3\sqrt{1 + \cosh 2\theta}\, d\theta$

52. $\displaystyle\int_1^2 \frac{5\,\text{sech}(\ln r)}{r}\, dr$

53. $\displaystyle\int 10\,\text{csch}^2 s\coth s\, ds$

54. $\displaystyle\int 4\,\text{sech}^4 x\tanh x\, dx$

Evaluate the integrals in Exercises 55–60 in terms of (a) inverse hyperbolic functions and (b) natural logarithms, using Table 6.15 in Exercises 6.9.

55. $\displaystyle\int_0^{\pi/2} \frac{\sin x\, dx}{\sqrt{1 + \cos^2 x}}$

56. $\displaystyle\int_{\sqrt{2}}^{\sqrt{17}} \frac{2x\, dx}{\sqrt{x^4 - 1}}$

57. $\displaystyle\int_{1/5}^{1/2} \frac{4\tanh^{-1}x}{1 - x^2}\, dx$

58. $\displaystyle\int_{\pi/6}^{\pi/4} \frac{\cos x\, dx}{\sin x\sqrt{1 + \sin^2 x}}$

59. $\displaystyle\int_{3/5}^{4/5} \frac{2\,\text{sech}^{-1}x}{x\sqrt{1 - x^2}}\, dx$

60. $\displaystyle\int_{\sqrt{8}}^{\sqrt{3}} \frac{e^{\coth^{-1}x}}{1 - x^2}\, dx$

Solve the initial value problems in Exercises 61 and 62.

61. $\displaystyle\frac{1}{y + 1}\frac{dy}{dx} = -\frac{1}{x^2}, \quad x > 0; \quad y = 0 \text{ when } x = 1$

62. $\displaystyle\frac{d^2y}{dx^2} = e^{-x}, \quad y = 0 \text{ and } \frac{dy}{dx} = -2 \text{ when } x = 0$

63. Solve for y.
  a) $e^{2y} = x^2$
  b) $3^y = 2^{y + 1}$

64. Solve for x.
  a) $\ln(x - 1) = y + \ln x$
  b) $4^{-x} = 3^{x + 2}$

Find the limits in Exercises 65–70.

65. $\displaystyle\lim_{x \to 0} \frac{2^{\sin x} - 1}{e^x - 1}$

66. $\displaystyle\lim_{x \to 0} \frac{xe^x}{4 - 4e^x}$

67. $\displaystyle\lim_{x \to 4} \frac{\sin^2(\pi x)}{e^{x - 4} + 3 - x}$

68. $\displaystyle\lim_{b \to 1^-} \int_0^b \frac{dx}{\sqrt{1 - x^2}}$

69. $\displaystyle\lim_{x \to \infty}\left(1 + \frac{3}{x}\right)^x$

70. $\displaystyle\lim_{x \to \infty} \frac{1}{x}\int_0^x \tan^{-1}t\, dt$

Find the limits in Exercises 71 and 72.

71. $\displaystyle\lim_{n \to \infty}\left(\frac{1}{n + 1} + \frac{1}{n + 2} + \cdots + \frac{1}{2n}\right)$

72. $\displaystyle\lim_{n \to \infty}\frac{1}{n}\left(e^{1/n} + e^{2/n} + \cdots + e^{(n - 1)/n} + e^{n/n}\right)$

73. The function $f(x) = e^x + x$, being differentiable and one-to-one, has a differentiable inverse $f^{-1}(x)$. Find the value of df^{-1}/dx at the point $f(\ln 2)$.

74. Find the inverse of the function $f(x) = 1 + (1/x), x \neq 0$. Then show that $f^{-1}(f(x)) = f(f^{-1}(x)) = x$ and that
$$\left.\frac{df^{-1}}{dx}\right|_{f(x)} = \frac{1}{f'(x)}.$$

75. Find the area between the curve $y = 2(\ln x)/x$ and the x-axis from $x = 1$ to $x = e$.

76. The functions $\ln 5x$ and $\ln 3x$ differ by a constant. What constant?

77. a) Show that the area between the curve $y = 1/x$ and the x-axis from $x = 10$ to $x = 20$ is the same as the area between the curve and the x-axis from $x = 1$ to $x = 2$.
  b) Show that the area between the curve $y = 1/x$ and the x-axis from ka to kb is the same as the area between the curve and the x-axis from $x = a$ to $x = b$ $(0 < a < b, k > 0)$.

78. A particle moves along the s-axis with acceleration $d^2s/dt^2 = 4/(4 - t)^2$. If $ds/dt = 2$ when $t = 0$, what distance does the particle travel from $t = 1$ to $t = 2$?

79. Find the length of the curve $x = (y/4)^2 - 2\ln(y/4)$, $4 \leq y \leq 12$.

80. Find the first four derivatives with respect to x of $f(x) = xe^x$. What pattern do you see? Verify it by mathematical induction.

81. Find the value of c that makes
$$f(x) = \begin{cases} \dfrac{x}{6^x - 5^x}, & x \neq 0 \\ c, & x = 0 \end{cases}$$
continuous at $x = 0$.

82. Find the quadratic approximation of
a) $f(x) = \ln(\sec x + \tan x)$ at $x = 0$;
b) $f(x) = \ln x$ at $x = 1$.

83. The region between the curve $y = 1/(2\sqrt{x})$ and the x-axis from $x = 1/4$ to $x = 4$ is revolved about the x-axis to generate a solid.
a) Find the volume of the solid.
b) Find the centroid of the region.

84. a) Use l'Hôpital's rule and mathematical induction to prove that for any positive integer n,
$$\lim_{x \to \infty} \frac{(\ln x)^n}{x} = 0.$$
b) Use mathematical induction and the rule $\ln ax = \ln a + \ln x$ to prove that at any value x where the functions $u_1(x), u_2(x), \ldots, u_n(x)$ are all positive,
$$\ln(u_1 \cdot u_2 \cdot \cdots \cdot u_n) = \ln u_1 + \ln u_2 + \cdots + \ln u_n.$$

85. Find the quadratic approximation at $x = 0$ of
$$\text{erf}(x) = \frac{2}{\sqrt{\pi}} \int_0^x e^{-t^2} \, dt.$$

86. GRAPHER   Graph the following functions and use what you see to locate and estimate the extreme values, identify the coordinates of the inflection points, and identify the intervals on which the graphs are concave up and concave down. Then confirm your estimates by working with the functions' derivatives.
a) $y = (\ln x)/\sqrt{x}$ b) $y = e^{-x^2}$
c) $y = (1 + x)e^{-x}$

87. Find $f'(2)$ if $f(x) = e^{g(x)}$ and
$$g(x) = \int_2^x \frac{t}{1 + t^4} \, dt.$$

88. Find the following limits.
a) $\lim_{x \to 0} \left(\frac{\sin x}{x} \right)^{1/x^2}$

b) $\lim_{x \to \infty} x \, e^{-x^2} \int_0^x e^{(t^2)} \, dt$

89. a) If $(\ln x)/x = (\ln 2)/2$, must x equal 2?
b) If $(\ln x)/x = -2 \ln 2$, must x equal 1/2?

90. *How to evaluate $\ln x$ with square roots.*
a) Show that $\lim_{h \to 0} \dfrac{e^h - 1}{h} = 1$.
b) Show that if x is any positive number, then
$$\lim_{n \to \infty} n\left(\sqrt[n]{x} - 1 \right) = \ln x.$$
(*Hint:* Take $x = e^{nh}$ and apply the result in (a).)
c) CALCULATOR   The result in (b) provides a way to find $\ln x$ to any desired finite number of decimal places (within the limitations of your calculator) using nothing fancier than a square root key. Take $n = 2^k$; then obtain $\sqrt[n]{x}$ from x by taking k successive square roots. See how close you can come this way to calculating $\ln 2 = 0.6931\ 47180\ 56$.

91. CALCULATOR   Find the following numbers as accurately as

your calculator will allow, using only $\boxed{\ln x}$ and $\boxed{\div}$.
a) $\log_{10} 5$ b) $\log_2 3$ c) $\log_7 2$

92. *Varying a logarithm's base.*
a) Find $\lim \log_a 2$ as $a \to 0^+$, 1^-, 1^+, and ∞.
b) GRAPHER   Graph $y = \log_a 2$ as a function of a over the interval $0 \le a \le 4$.

93. Find the areas between the curves $y = 2(\log_2 x)/x$ and $y = 2(\log_4 x)/x$ and the x-axis from $x = 1$ to $x = e$. What is the ratio of the larger area to the smaller?

94. GRAPHER   Graph $f(x) = x \ln x$. Does the function appear to have an absolute minimum value? Either way, confirm your observation with calculus.

95. $\pi^e < e^\pi$. Why does Fig. 6.59 "prove" that π^e is less than e^π? (*Source:* "Proof Without Words" by Fouad Nakhil, *Mathematics Magazine*, Vol. 60, No. 3, June 1987, p. 165.)

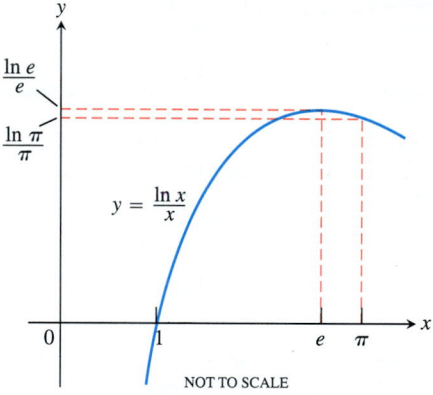

6.59   The figure for Exercise 95.

96. For what $x > 0$ does $x^{(x^x)} = (x^x)^x$?

97. GRAPHER   *Group blood testing.*   During World War II it was necessary to administer blood tests to large numbers of recruits. There are two standard ways to administer a blood test to N people. In method 1, each person is tested separately. In method 2, the blood samples of x people are pooled and tested as one large sample. If the test is negative, this one test is enough for all x people. If the test is positive, then each of the x people is tested separately, requiring a total of $x + 1$ tests. Using the second method and some probability theory it can be shown that, on the average, the total number of tests y will be
$$y = N\left(1 - q^x + \frac{1}{x} \right).$$
With $q = 0.99$ and $N = 1000$, find the integer value of x that minimizes y. Also find the integer value of x that maximizes y. (This second result is not important to the real-life situation.) The group testing method was used in World War II with a savings of 80% over the individual testing method, but not with the given value of q.

98. *Free fall in the fourteenth century.*   In the middle of the fourteenth century, Albert of Saxony (1316–1390) proposed a model of free fall that assumed that the velocity of a falling body was proportional to the distance fallen. It seemed reasonable to think that a body that had fallen 20 ft might be moving twice as fast as a body that had fallen 10 ft. And besides, none of the instruments in use at the time were accurate enough to prove otherwise. Today we can see just how far off Albert of Saxony's model was by solving the initial value problem implicit in his model. Solve the problem and compare your solution graphically with the equation $s = 16t^2$. You will see that it describes a motion that starts too slowly at first and then becomes too fast too soon to be realistic.

99. Find the age of a sample of charcoal in which 90% of the carbon-14 originally present has decayed.

100. CALCULATOR *Cooling a pie.* A deep-dish apple pie, whose internal temperature was 220°F when removed from the oven, was set out on a breezy 40° porch to cool. Fifteen minutes later, the pie's internal temperature was 180°F. How long did it take the pie to cool from there to 70°F?

101. *Transport through a cell membrane.* Under some conditions the result of the movement of a dissolved substance across a cell's membrane is described by the equation

$$\frac{dy}{dt} = k\frac{A}{V}(c - y).$$

In this equation, y is the concentration of the substance inside the cell and dy/dt is the rate with which y changes over time. The letters k, A, V, and c stand for constants, k being the *permeability coefficient* (a property of the membrane), A the surface area of the membrane, V the cell's volume, and c the concentration of the substance outside the cell. The equation says that the rate at which the concentration changes within the cell is proportional to the difference between it and the outside concentration.
a) Solve the equation for $y(t)$, using y_0 to denote $y(0)$.
b) Find the steady-state concentration, $\lim_{t\to\infty} y(t)$.
(Based on *Some Mathematical Models in Biology* by R. M. Thrall, J. A. Mortimer, K. R. Rebman, R. F. Baum, Eds., Revised Edition, December 1967, PB-202 364, pp. 101–103; distributed by N.T.I.S., U.S. Department of Commerce.)

102. Compare the rates of growth as $x\to\infty$ of
a) x and $5x$
b) $x + \frac{1}{x}$ and x
c) $x^2 + x$ and $x^2 - x$

103. Compare the rates of growth as $x\to\infty$ of
a) $\ln 2x$, $\ln x^2$, and $\ln(x + 2)$
b) $x^{\ln x}$ and $x^{\log_2 x}$ c) $\left(\frac{1}{2}\right)^x$ and $\left(\frac{1}{3}\right)^x$

104. True, or false?
a) $\frac{1}{x^2} + \frac{1}{x^4} = O\left(\frac{1}{x^2}\right)$ b) $\frac{1}{x^2} + \frac{1}{x^4} = O\left(\frac{1}{x^4}\right)$
c) $\sqrt{x^2 + 1} = O(x)$

105. True, or false?
a) $\ln x = o(x)$
b) $\ln \ln x = o(\ln x)$
c) $x = o(x + \ln x)$

106. Figure 6.60 shows an informal proof that

$$\tan^{-1}\frac{1}{2} + \tan^{-1}\frac{1}{3} = \frac{\pi}{4}.$$

See if you can tell what is going on. (*Source:* "Behold! Sums of Arctan" by Edward M. Harris, *The College Mathematics Journal*, Vol. 18, No. 2, March 1987, p. 141.)

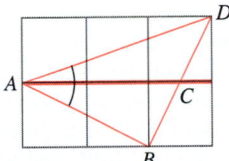

6.60 The figure for Exercise 106.

107. Show that

$$\tan^{-1}x + \tan^{-1}\frac{1}{x} = \text{const}.$$

Find the constant.

108. Use Fig. 6.61 to show that

$$\int_0^{\pi/2} \sin x \, dx = \frac{\pi}{2} - \int_0^1 \sin^{-1}x \, dx.$$

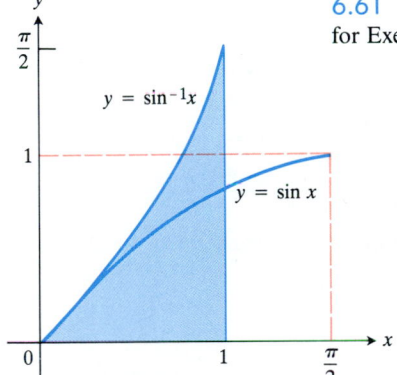

6.61 The figure for Exercise 108.

109. CALCULATOR *Locating a solar station.* You are under contract to build a solar station at ground level on the east–west line between the two buildings shown in Fig. 6.62. How far from the taller building should you place the station to maximize the number of hours it will be in the sun on a day when the sun passes directly overhead? Begin by

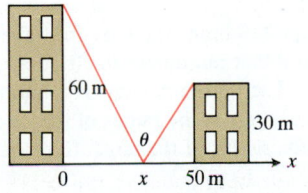

6.62 The figure for Exercise 109.

observing that

$$\theta = \pi - \cot^{-1}\frac{x}{60} - \cot^{-1}\frac{50 - x}{30}.$$

Then find the value of x that maximizes θ.

110. *The best branching angles for blood vessels and pipes.*   When a smaller pipe branches off from a larger one in a flow system, we may want it to run off at an angle that is best from some energy-saving point of view. We might require, for instance, that energy loss due to friction be minimized along the section AOB shown in Fig. 6.63. In this diagram, B is a given point to be reached by the smaller pipe, A is a point in the larger pipe upstream from B, and O is the point where the branching occurs. A law due to Poiseuille states that the loss of energy due to friction in nonturbulent flow is proportional to the length of the path and inversely proportional to the fourth power of the radius. Thus, the loss along AO is $(kd_1)/R^4$ and along OB is $(kd_2)/r^4$, where k is a constant, d_1 is the length of AO, d_2 is the length of OB, R is the radius of the larger pipe, and r is the radius of the smaller pipe. The angle θ is to be chosen to minimize the sum of these two losses:

$$L = k\frac{d_1}{R^4} + k\frac{d_2}{r^4}.$$

In our model, we assume that $AC = a$ and $BC = b$ are fixed. Thus we have the relations

$$d_1 + d_2\cos\theta = a, \qquad d_2\sin\theta = b,$$

so that

$$d_2 = b\csc\theta,$$
$$d_1 = a - d_2\cos\theta = a - b\cot\theta.$$

We can express the total loss L as a function of θ:

$$L = k\left(\frac{a - b\cot\theta}{R^4} + \frac{b\csc\theta}{r^4}\right).$$

a) Show that the critical value of θ for which $dL/d\theta$ equals zero is

$$\theta_c = \cos^{-1}\frac{r^4}{R^4}.$$

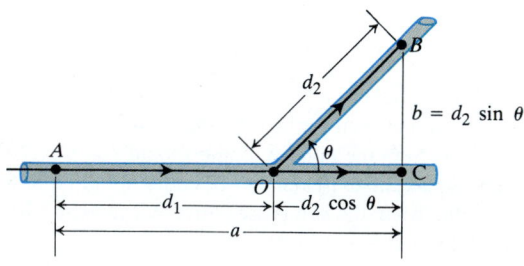

6.63   The smaller pipe OB branches away from the larger AOC at an angle θ that minimizes the friction loss along AO and OB. The optimum angle is found to be $\theta_c = \cos^{-1}(r^4/R^4)$, where r is the radius of the smaller pipe and R is the radius of the larger (Exercise 110).

b) CALCULATOR   If the ratio of the pipe radii is $r/R = 5/6$, estimate to the nearest degree the optimal branching angle given in part (a).

The mathematical analysis described here is also used to explain the angles at which arteries branch in an animal's body. (See *Introduction to Mathematics for Life Scientists*, 2nd Ed. by E. Batschelet [New York: Springer-Verlag, 1976].)

111. Suppose that g is an even function of x, and h is an odd function of x. Show that if $g(x) + h(x) = 0$ for all x, then $g(x) = 0$ for all x and $h(x) = 0$ for all x.

112. Use the result of Exercise 111 to show that if $f(x) = f_E(x) + f_O(x)$ is the sum of an even function $f_E(x)$ and an odd function $f_O(x)$, then

$$f_E(x) = (f(x) + f(-x))/2 \quad \text{and} \quad f_O(x) = (f(x) - f(-x))/2.$$

113. *A pursuit curve.*   A rabbit starts at the origin and runs at a constant speed up the y-axis. At the same time, a dog, running at the same speed as the rabbit, starts at point $(1, 0)$ and runs straight for the rabbit. It can be shown that the dog's path (Fig. 6.64) is the graph of the function $y = f(x)$ that solves the second order initial value problem

$$x\frac{d^2y}{dx^2} = \sqrt{1 + \left(\frac{dy}{dx}\right)^2},$$

$$y = 0 \quad \text{and} \quad dy/dx = 0 \quad \text{when} \quad x = 1.$$

Solve the initial value problem.

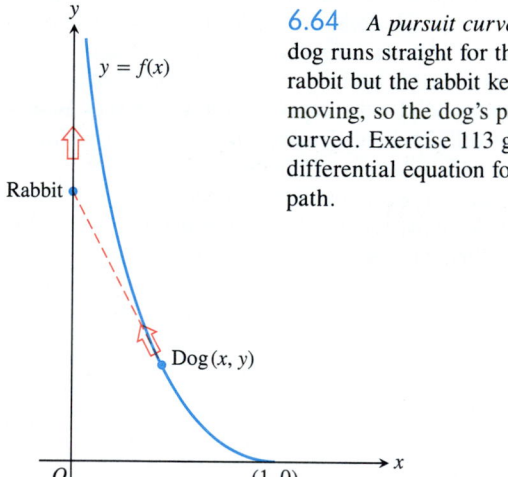

6.64   *A pursuit curve.* The dog runs straight for the rabbit but the rabbit keeps moving, so the dog's path is curved. Exercise 113 gives a differential equation for the path.

Use integral tables and the formulas in Table 6.15 in Exercises 6.9 to evaluate the integrals that arise in Exercises 114–116.

114. CALCULATOR   Find the length of the curve $y = x^2$, $0 \leq x \leq 1$.

115. CALCULATOR   Find the area of the surface generated by revolving the curve $y = x^2$, $0 \leq x \leq 1$, about the x-axis.

116. CALCULATOR   Find the length of the curve $y = \ln x$, $3/4 \leq x \leq 1$.

CHAPTER

7

TECHNIQUES OF INTEGRATION

Overview   We have seen how integrals arise in modeling real phenomena and in measuring objects in the world around us, and we know in theory how integrals are evaluated with antiderivatives. The more sophisticated our models become, however, the more involved our integrals become. We need to know how to change these integrals into forms we can work with. The goal of this chapter is to show how to change unfamiliar integrals into integrals we can recognize, find in a table, or evaluate with a computer.

7.1 Basic Integration Formulas

The ideal way to evaluate the indefinite integral of a continuous function f is to find an antiderivative of f and add a constant of integration to it to produce a formula that gives all the antiderivatives of f. Table 7.1 lists the integral formulas for the functions we have worked with so far, along with three new formulas. Together, these are the integral formulas that find the most frequent use. When an integral matches one of these formulas or can be changed into one of them in an appropriate way with a combination of algebra, trigonometry, and substitution, we have a ready-made solution for the problem at hand.

You will find a more extensive integral table at the back of this book. We recommend, however, that you do not consult this table before you are asked to do so in Section 7.6. The intervening sections are designed to train you in important algebraic and trigonometric procedures for changing complicated integrals into more workable forms, and using the extensive table prematurely will make the training less effective.

The three new formulas in Table 7.1 are

$$\int u \, dv = uv - \int v \, du; \qquad \begin{pmatrix} \text{Integration by parts,} \\ \text{Section 7.2} \end{pmatrix}$$

$$\int \sec u \, du = \ln|\sec u + \tan u| + C, \qquad \text{(Example 6, Eq. (1))}$$

$$\int \csc u \, du = -\ln|\csc u + \cot u| + C. \qquad \text{(Eq. (2))}$$

479

TABLE 7.1
Standard integral forms

Definition	$\displaystyle\int du = u + C$
Constant multiple	$\displaystyle\int a\,du = au + C$
Sum	$\displaystyle\int (du + dv) = \int du + \int dv$
Powers	$\displaystyle\int u^n du = \frac{u^{n+1}}{n+1} + C, \quad n \neq -1 \qquad \int \frac{du}{u} = \ln|u| + C$
Trigonometric	$\displaystyle\int \cos u\,du = \sin u + C \qquad \int \sin u\,du = -\cos u + C$
	$\displaystyle\int \tan u\,du = \ln|\sec u| + C \qquad \int \cot u\,du = \ln|\sin u| + C$
	$\displaystyle\int \sec u\,du = \ln|\sec u + \tan u| + C \qquad \int \csc u\,du = -\ln|\csc u + \cot u| + C$
Exponential	$\displaystyle\int e^u\,du = e^u + C \qquad \int a^u du = \frac{1}{\ln a}a^u + C; \quad a > 0,\ a \neq 1$
Special algebraic forms	$\displaystyle\int \frac{du}{1 + u^2} = \tan^{-1}u + C \qquad \int \frac{du}{\sqrt{1 - u^2}} = \sin^{-1}u + C$
	$\displaystyle\int \frac{du}{u\sqrt{u^2 - 1}} = \sec^{-1}|u| + C$
Integration by parts (from the Product Rule for derivatives)	$\displaystyle\int u\,dv = uv - \int v\,du$

The latter two are included to complete the list of integrals of the six trigonometric functions. The formula for integration by parts comes from integrating both sides of the formula $d(uv) = u\,dv + v\,du$ and rearranging the result. We shall study integration by parts in Section 7.2.

Substitution, Algebra, and Trigonometry

The most frequently used integration technique is *substitution*. We look at the function to be integrated and try to match it with a standard form.

Example 1   *Substitution.*   Evaluate $\displaystyle\int \frac{dx}{1 + 4x^2}$.

Solution   The nearest standard form is

$$\int \frac{du}{1 + u^2} = \tan^{-1}u + C.$$

Since $4x^2 = u^2$ if $u = 2x$, we substitute

$$u = 2x, \qquad du = 2\,dx, \qquad \text{and} \qquad dx = \frac{1}{2}\,du.$$

Then

$$\int \frac{dx}{1 + 4x^2} = \int \frac{(1/2)\,du}{1 + u^2} = \frac{1}{2}\int \frac{du}{1 + u^2}$$

$$= \frac{1}{2}\tan^{-1}u + C = \frac{1}{2}\tan^{-1}2x + C.$$

Example 2 *Completing the Square.* Evaluate $\displaystyle \int \frac{dx}{x^2 + 2x + 2}.$

Solution We complete the square (Appendix 1) to write the denominator in the form

$$x^2 + 2x + 2 = (x^2 + 2x + 1) + (2 - 1) = (x + 1)^2 + 1.$$

The substitution $u = x + 1$ then reduces the integral to

$$\int \frac{dx}{x^2 + 2x + 2} = \int \frac{dx}{(x + 1)^2 + 1} = \int \frac{du}{u^2 + 1}$$

$$= \tan^{-1}u + C = \tan^{-1}(x + 1) + C.$$

Example 3 *Expanding a Power and Using a Trigonometric Identity.*
Evaluate $\displaystyle \int (\sec x + \tan x)^2\,dx.$

Solution We expand the integrand and get

$$(\sec x + \tan x)^2 = \sec^2 x + 2\sec x \tan x + \tan^2 x.$$

The first two terms on the right-hand side of this equation are old friends; we can integrate them at once. What do we do about $\tan^2 x$? There is an identity that connects it with $\sec^2 x$:

$$\tan^2 x + 1 = \sec^2 x, \qquad \tan^2 x = \sec^2 x - 1.$$

We replace $\tan^2 x$ by $\sec^2 x - 1$ and get

$$\int (\sec x + \tan x)^2\,dx = \int (\sec^2 x + 2\sec x \tan x + \sec^2 x - 1)\,dx$$

$$= 2\int \sec^2 x\,dx + 2\int \sec x \tan x\,dx - \int 1\,dx$$

$$= 2\tan x + 2\sec x - x + C.$$

We shall do more with trigonometric techniques in Sections 7.3 and 7.4.

Example 4 *Reducing an Improper Fraction.* Evaluate $\displaystyle \int \frac{(x - 2)^3}{x^2 - 4}\,dx.$

Solution   We divide the numerator and denominator by the common factor, $x - 2$:

$$\frac{(x - 2)^3}{x^2 - 4} = \frac{(x - 2)^3}{(x - 2)(x + 2)}$$

$$= \frac{(x - 2)^2}{x + 2} = \frac{x^2 - 4x + 4}{x + 2}.$$

$$\begin{array}{r} x \phantom{{}^2} - 6 \phantom{{}+ 4} \\ x + 2\overline{\smash{)}x^2 - 4x + 4} \\ \underline{x^2 + 2x \phantom{{}+ 4}} \\ - 6x + 4 \\ \underline{- 6x - 12} \\ + 16 \end{array}$$

The result is an improper fraction, in which the degree of the numerator is equal to or greater than the degree of the denominator. To integrate such fractions, we first divide them, getting a quotient plus a remainder that is a proper fraction:

$$\frac{x^2 - 4x + 4}{x + 2} = x - 6 + \frac{16}{x + 2}.$$

Therefore,

$$\int \frac{(x - 2)^3}{x^2 - 4}\, dx = \int \left(x - 6 + \frac{16}{x + 2} \right) dx = \frac{x^2}{2} - 6x + 16 \ln|x + 2| + C.$$

We shall do more with integrating rational functions in Section 7.5.

Example 5   *Separating a Fraction.*   Evaluate $\displaystyle\int \frac{3x + 2}{\sqrt{1 - x^2}}\, dx.$

Solution   We first separate the integrand to get

$$\int \frac{3x + 2}{\sqrt{1 - x^2}}\, dx = 3 \int \frac{x\, dx}{\sqrt{1 - x^2}} + 2 \int \frac{dx}{\sqrt{1 - x^2}}.$$

In the first of these new integrals we substitute

$$u = 1 - x^2, \qquad du = -2x\, dx, \qquad \text{and} \qquad x\, dx = -\frac{1}{2}\, du:$$

$$3 \int \frac{x\, dx}{\sqrt{1 - x^2}} = 3 \int \frac{(-1/2)\, du}{\sqrt{u}} = -\frac{3}{2} \int u^{-1/2}\, du$$

$$= -\frac{3}{2} \left(\frac{u^{1/2}}{1/2} \right) + C_1 = -3\sqrt{1 - x^2} + C_1.$$

The second of the new integrals is a standard inverse trigonometric form,

$$\int \frac{dx}{\sqrt{1 - x^2}} = \sin^{-1}x + C_2.$$

Combining these results gives

$$\int \frac{3x + 2}{\sqrt{1 - x^2}}\, dx = -3\sqrt{1 - x^2} + 2 \sin^{-1}x + C.$$

Example 6   *The Secant Integral.*   Show that

$$\int \sec x\, dx = \ln|\sec x + \tan x| + C. \tag{1}$$

Solution

$$\int \sec x \, dx = \int (\sec x)(1) \, dx = \int \sec x \frac{\sec x + \tan x}{\sec x + \tan x} \, dx$$

$$= \int \frac{\sec^2 x + \sec x \tan x}{\sec x + \tan x} \, dx$$

$$= \int \frac{du}{u} \qquad \left(\begin{array}{l} u = \sec x + \tan x \\ du = \sec^2 x + \sec x \tan x \end{array} \right)$$

$$= \ln|u| + C = \ln|\sec x + \tan x| + C.$$

With cosecants and cotangents in place of secants and tangents, the method of Example 6 leads to the companion formula

$$\int \csc x \, dx = -\ln|\csc x + \cot x| + C. \tag{2}$$

See Exercise 60.

✳ Mercator's World Map

The integral of the secant plays an important role in making maps for compass navigation. The easiest course for a sailor or pilot to steer is a course whose compass heading is constant. This might be a course of 45° (northeast), for example, or a course of 225° (southwest), or whatever. Such a course will lie along a spiral that winds around the globe toward one of the poles (Fig. 7.1) unless the course runs due north or south or lies parallel to the equator.

In 1569, Gerhard Krämer (1512–1594), a Flemish surveyor and geographer known by his Latinized last name, Mercator, made a world map on which all spirals of constant compass heading appeared as straight lines (Fig. 7.2). A sailor could then read the compass heading for a voyage between any two points from the direction of a straight line connecting them on Mercator's map.

Figure 7.2 shows that the vertical lines of longitude that meet at the poles on the globe have been spread apart to lie parallel on the map. The horizontal lines of latitude are also parallel, as they are on the globe, but they are no longer evenly spaced. The spacing increases toward the poles.

The secant function plays a role in determining the correct spacing of all these lines. The scaling factor by which the horizontal distance must be increased at latitude $\alpha°$ to make the lines of longitude parallel is precisely $\sec \alpha$. There is no spreading at the equator, where $\sec 0° = 1$. At latitude 30° north and south, horizontal distances are multiplied by $\sec 30° \approx 1.15$. At 60°, they are multiplied by $\sec 60° = 2$, and so on. The farther from the equator you go, the larger the factor and the more the spread.

The lines of latitude must be spread apart toward the poles to match the spread in the lines of longitude. The scaling process, however, is complicated by the fact that the appropriate scaling factor changes with the latitude. We overcome this difficulty by integration. If R is the radius of the globe being modeled (Fig. 7.3), the map distance between the line representing the equator and the line representing the latitude $\alpha°$ is R times the integral of the secant from zero to α:

$$\text{Vertical map distance from } 0° \text{ to } \alpha° = R \int_0^\alpha \sec x \, dx. \tag{3}$$

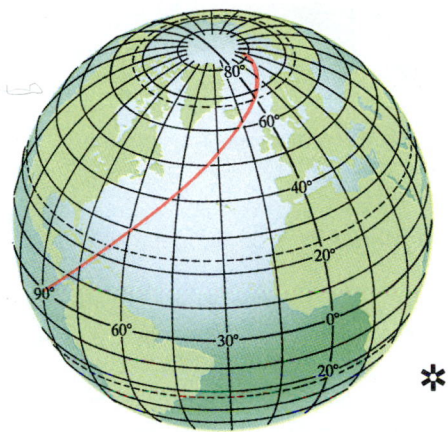

7.1 A flight with a constant bearing of 45° E of N from the Galápagos Islands in the Pacific to Franz Josef Land in the Arctic Ocean as it appears on a globe.

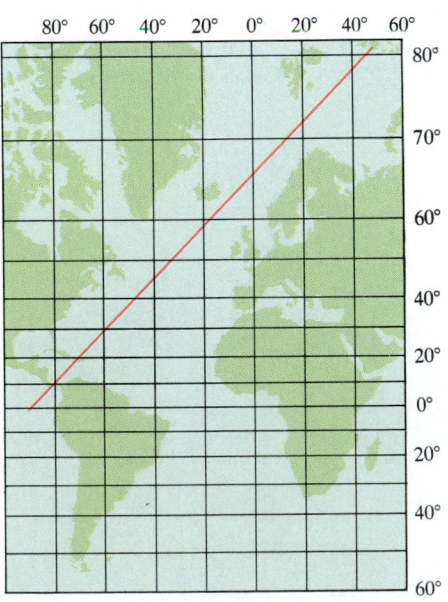

7.2 The flight of Fig. 7.1 traced on a Mercator map.

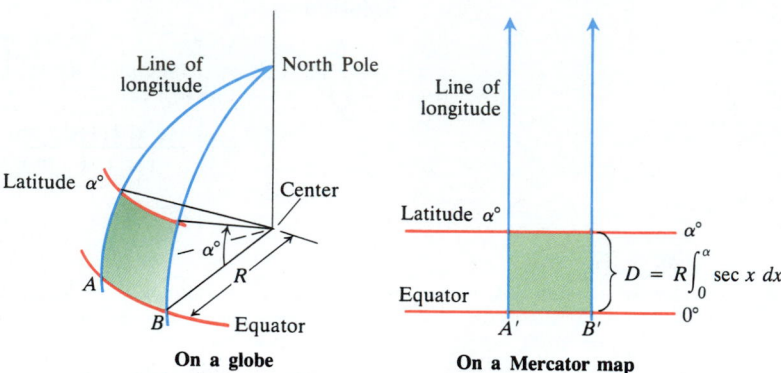

7.3 Lines of latitude and longitude on a globe and on a Mercator map.

On a globe

On a Mercator map

Therefore, the map distance between two latitude lines on the same side of the equator, say at $\alpha°$ and $\beta°$ $(\alpha < \beta)$ is

$$R\int_0^\beta \sec x \, dx - R\int_0^\alpha \sec x \, dx = R\int_\alpha^\beta \sec x \, dx = R \ln|\sec x + \tan x|\Big|_\alpha^\beta. \quad (4)$$

Example 7   Suppose that the equatorial length of a Mercator map just matches the equator of a globe of radius 25 cm. Then Eq. (4) gives the spacing on the map between the equator and latitude 20° north as

$$25\int_0^{20°} \sec x \, dx = 25 \ln|\sec x + \tan x|\Big|_0^{20°} \approx 9 \text{ cm.}$$

The spacing between 60° and 80° north is given by Eq. (4) as

$$25\int_{60°}^{80°} \sec x \, dx = 25 \ln|\sec x + \tan x|\Big|_{60°}^{80°} \approx 28 \text{ cm.}$$

As you can see, the map distance between latitude lines that are 20° apart is considerably greater near the pole than it is near the equator. The navigational properties of a Mercator map are achieved at the expense of a considerable distortion of distance.

EXERCISES 7.1

Evaluate the integrals in Exercises 1–36 by finding substitutions that reduce them to standard forms.

1. $\displaystyle\int \frac{16x \, dx}{\sqrt{8x^2 + 1}}$

2. $\displaystyle\int_0^{\pi/2} \frac{3 \cos x \, dx}{\sqrt{1 + 3 \sin x}}$

3. $\displaystyle\int_0^{\pi/2} 3\sqrt{\sin v} \cos v \, dv$

4. $\displaystyle\int_{\pi/6}^{\pi/2} \cot^3 y \, \csc^2 y \, dy$

5. $\displaystyle\int_0^1 \frac{16x \, dx}{8x^2 + 2}$

6. $\displaystyle\int_0^{\sqrt{\pi/3}} 4z \tan z^2 \, dz$

7. $\displaystyle\int \frac{dx}{x - \sqrt{x}}$

8. $\displaystyle\int_{\pi/6}^{\pi/2} \cot(\pi - \theta) \, d\theta$

9. $\displaystyle\int_{-\pi}^{\pi} \sec \frac{t}{3} \, dt$

10. $\displaystyle\int_{\sqrt{5}}^{\sqrt{6}} x \sec(x^2 - 5) \, dx$

11. $\displaystyle\int_{3\pi/2}^{7\pi/4} \csc(s - \pi) \, ds$

12. $\displaystyle\int_{3/\pi}^{2/\pi} \frac{1}{\theta^2} \csc \frac{1}{\theta} \, d\theta$

13. $\displaystyle\int_0^{\sqrt{\ln 2}} 2xe^{x^2} \, dx$

14. $\displaystyle\int_{\pi/2}^{\pi} \sin y e^{\cos y} \, dy$

15. $\displaystyle\int_0^{\pi/3} e^{\tan v} \sec^2 v \, dv$

16. $\displaystyle\int_{\ln 2}^{\ln 3} \frac{e^{\sqrt{t}} \, dt}{\sqrt{t}}$

17. $\displaystyle\int_{-1}^{0} 3^{x+1} \, dx$

18. $\displaystyle\int_1^2 \frac{2^{\ln x}}{x} \, dx$

19. $\displaystyle\int \frac{2^{\sqrt{w}}\,dw}{2\sqrt{w}}$

20. $\displaystyle\int_0^1 10^{2\theta}\,d\theta$

21. $\displaystyle\int_0^{\sqrt{3}/3} \frac{9\,du}{1 + 9u^2}$

22. $\displaystyle\int_{-1}^0 \frac{4\,dx}{1 + (2x + 1)^2}$

23. $\displaystyle\int \frac{6\,dy}{\sqrt{y}(1 + y)}$

24. $\displaystyle\int \frac{dx}{e^x + e^{-x}}$

25. $\displaystyle\int_0^{1/6} \frac{dx}{\sqrt{1 - 9x^2}}$

26. $\displaystyle\int_0^1 \frac{dt}{\sqrt{4 - t^2}}$

27. $\displaystyle\int_0^{1/\sqrt{2}} \frac{2s\,ds}{\sqrt{1 - s^4}}$

28. $\displaystyle\int \frac{2\,dx}{x\sqrt{1 - 4\ln^2 x}}$

29. $\displaystyle\int \frac{6\,dx}{x\sqrt{25x^2 - 1}}$

30. $\displaystyle\int_{-6}^{-3\sqrt{2}} \frac{dr}{r\sqrt{r^2 - 9}}$

31. $\displaystyle\int \frac{dy}{\sqrt{e^{2y} - 1}}$

32. $\displaystyle\int \frac{dx}{x\sqrt{4x^2 - 1}}$

33. $\displaystyle\int_1^{e^{\pi/3}} \frac{dx}{x\cos(\ln x)}$

34. $\displaystyle\int_{1/4}^{1/3} (\sec \pi x + \csc \pi x)\,dx$

35. $\displaystyle\int \frac{\ln x\,dx}{x + 4x\ln^2 x}$

36. $\displaystyle\int \frac{\tan \theta\,d\theta}{2\sec \theta + 1}$

Evaluate each integral in Exercises 37–42 by completing the square and using a substitution to reduce it to standard form.

37. $\displaystyle\int \frac{dx}{\sqrt{-x^2 + 4x - 3}}$

38. $\displaystyle\int \frac{dx}{\sqrt{2x - x^2}}$

39. $\displaystyle\int_1^2 \frac{8\,dx}{x^2 - 2x + 2}$

40. $\displaystyle\int_2^4 \frac{2\,dx}{x^2 - 6x + 10}$

41. $\displaystyle\int \frac{dx}{(x + 1)\sqrt{x^2 + 2x}}$

42. $\displaystyle\int \frac{dx}{(x - 2)\sqrt{x^2 - 4x + 3}}$

Evaluate the integrals in Exercises 43–46.

43. $\displaystyle\int_{\pi/4}^{3\pi/4} (\csc x - \cot x)^2\,dx$

44. $\displaystyle\int_0^{\pi/4} (\sec x + 4\cos x)^2\,dx$

45. $\displaystyle\int_{\pi/6}^{\pi/3} (\csc x - \sec x)(\sin x + \cos x)\,dx$

46. $\displaystyle\int (\sin 3x \cos 2x - \cos 3x \sin 2x)\,dx$

Evaluate the integrals in Exercises 47–50.

47. $\displaystyle\int \frac{x}{x + 1}\,dx$

48. $\displaystyle\int_0^1 \frac{x^2}{x^2 + 1}\,dx$

49. $\displaystyle\int_{\sqrt{2}}^3 \frac{2x^3}{x^2 - 1}\,dx$

50. $\displaystyle\int_{-1}^3 \frac{4x^2 - 7}{2x + 3}\,dx$

Evaluate the integrals in Exercises 51–54.

51. $\displaystyle\int_0^{\sqrt{3}/2} \frac{1 - x}{\sqrt{1 - x^2}}\,dx$

52. $\displaystyle\int \frac{x + 2\sqrt{x - 1}}{2x\sqrt{x - 1}}\,dx$

53. $\displaystyle\int_0^{\pi/4} \frac{1 + \sin x}{\cos^2 x}\,dx$

54. $\displaystyle\int_0^{1/2} \frac{2 - 8x}{1 + 4x^2}\,dx$

55. Find the length of the curve $y = \ln(\cos x)$, $0 \le x \le \pi/3$.

56. Find the length of the curve $y = \ln(\sec x)$, $0 \le x \le \pi/4$.

57. Find the centroid of the region bounded by the x-axis, the curve $y = \sec x$, and the lines $x = -\pi/4$, $x = \pi/4$.

58. Find the area of the region bounded above by $y = 2\cos x$ and below by $y = \sec x$, $-\pi/4 \le x \le \pi/4$.

59. Find the centroid of the region that is bounded above by the line $y = 2$ and below by the curve $y = \csc x$, $\pi/6 \le x \le 5\pi/6$.

60. *The integral of csc x.* Repeat the derivation in Example 6, using cofunctions, to show that

$$\int \csc x\,dx = -\ln|\csc x + \cot x| + C.$$

61. Show that the integral

$$\int \left((x^2 - 1)(x + 1)\right)^{-2/3} dx$$

can be evaluated with any of the following substitutions:
a) $u = 1/(x + 1)$
b) $u = ((x - 1)/(x + 1))^k$ for $k = $
    $1, 1/2, 1/3, -1/3, -2/3,$ and -1
c) $u = \tan^{-1}x$
d) $u = \tan^{-1}\sqrt{x}$ e) $u = \tan^{-1}((x - 1)/2)$
f) $u = \cos^{-1}x$ g) $u = \cosh^{-1}x$

What is the value of the integral? (From "Problems and Solutions," *College Mathematics Journal*, Vol. 21, No. 5, Nov. 1990, p. 425–426.)

62. Evaluate

$$\int_0^{\pi/2} \frac{\sin^n x}{\sin^n x + \cos^n x}\,dx,$$

where n is any integer. (*Hint:* Substitute $x = (\pi/2) - u$ and add the new and old integrals.)

CALCULATOR or TABLES How far apart should the lines of latitude in Exercises 63 and 64 be on the Mercator map in Example 7?

63. Latitudes 30° and 45° north (New Orleans, Louisiana, and Minneapolis, Minnesota)

64. Latitudes 45° and 60° north (Salem, Oregon, and Seward, Alaska)

7.2 Integration by Parts

Integration by parts is a technique used mainly to simplify integrals of the form

$$\int f(x)\, g(x)\, dx, \tag{1}$$

in which f can be differentiated repeatedly to become zero and g can be integrated repeatedly without difficulty. The integral

$$\int xe^x\, dx$$

is such an integral because $f(x) = x$ can be differentiated twice to become zero and $g(x) = e^x$ can be integrated repeatedly without difficulty. Integration by parts also applies to integrals like

$$\int e^x \sin x\, dx,$$

in which each part of the integrand appears again after repeated differentiation or integration.

In this section, we describe integration by parts and show how to apply it.

The Formula

The formula for integration by parts comes from the Product Rule,

$$\frac{d}{dx}(uv) = u\frac{dv}{dx} + v\frac{du}{dx}.$$

In its differential form, the rule becomes

$$d(uv) = u\, dv + v\, du,$$

which is then written as

$$u\, dv = d(uv) - v\, du$$

and integrated to give the following formula.

When and How to Use Integration by Parts

When: If substitution doesn't work, try integration by parts.

How: Start with an integral of the form

$$\int f(x)\, g(x)\, dx.$$

Match this with an integral of the form

$$\int u\, dv$$

by choosing dv to be part of the integrand including dx and possibly $f(x)$ or $g(x)$.

Guideline for choosing u and dv: The formula

$$\int u\, dv = uv - \int v\, du$$

gives a new integral on the right side of the equation. If the new integral is more complex than the original one, try a different choice.

The Integration by Parts Formula

$$\int u\, dv = uv - \int v\, du \tag{2}$$

The integration by parts formula expresses one integral, $\int u\, dv$, in terms of a second integral, $\int v\, du$. With a proper choice of u and v, the second integral may be easier to evaluate than the first. This is the reason for the importance of the formula. When faced with an integral we cannot handle, we can replace it by one with which we might have more success.

The equivalent formula for definite integrals is

$$\int_{v_1}^{v_2} u\, dv = (u_2 v_2 - u_1 v_1) - \int_{u_1}^{u_2} v\, du. \tag{3}$$

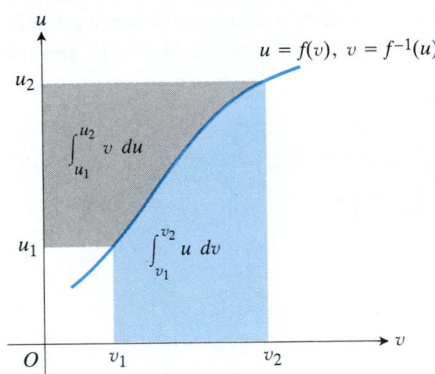

7.4 The area of the blue region, $\int_{v_1}^{v_2} u\, dv$, is equal to the area of the large rectangle, $u_2 v_2$, minus the areas of the small rectangle, $u_1 v_1$, and the gray region, $\int_{u_1}^{u_2} v\, du$. In symbols,

$$\int_{v_1}^{v_2} u\, dv = (u_2 v_2 - u_1 v_1) - \int_{u_1}^{u_2} v\, du.$$

Figure 7.4 shows how the different parts of Eq. (3) may be interpreted as areas.

Example 1   Evaluate $\displaystyle\int x \cos x\, dx$.

Solution   We use the formula $\int u\, dv = uv - \int v\, du$ with

$$u = x, \qquad dv = \cos x\, dx,$$

$$du = dx, \qquad v = \sin x. \qquad \text{(Simplest function with } dv = \cos x\, dx\text{)}$$

Then

$$\int x \cos x\, dx = x \sin x - \int \sin x\, dx = x \sin x + \cos x + C.$$

Before we go on to new examples, let us look at some of the other choices we might have made in Example 1.

Example 2   *Example 1 Revisited.*   To apply integration by parts to

$$\int x \cos x\, dx = \int f(x)\, g(x)\, dx = \int u\, dv$$

we have four immediate choices:

1. Let $u = 1$ and $dv = f(x)\, g(x)\, dx = x \cos x\, dx$.
2. Let $u = f(x) = x$ and $dv = g(x)\, dx = \cos x\, dx$.
3. Let $u = f(x)\, g(x) = x \cos x$ and $dv = dx$.
4. Let $u = g(x) = \cos x$ and $dv = x\, dx$.

Let's examine these one at a time.

Choice 1 will not do because we do not know how to integrate $dv = x \cos x\, dx$ to get v.

Choice 2 worked well. We used it in Example 1.

Choice 3 leads to $du = \dfrac{du}{dx}\, dx = (-x \sin x + \cos x)\, dx$, and $v = x$, with a new integral

$$\int v\, du = \int (-x^2 \sin x + x \cos x)\, dx.$$

This is worse than the integral we started with.

Choice 4 leads to $du = -\sin x\, dx$ and $v = \displaystyle\int x\, dx = x^2/2$, so the new integral is

$$\int v\, du = -\int \frac{x^2}{2} \sin x\, dx.$$

This is also harder than the original because we have gone from $x \cos x$ to $x^2 \sin x$ in the integrand.

Summary.   Keep in mind that the goal is to go from the given integral ($\int u\, dv$) to a new integral ($\int v\, du$) that is simpler. (Integration by parts does not always work, so we cannot expect to achieve the goal one hundred percent of the time.)

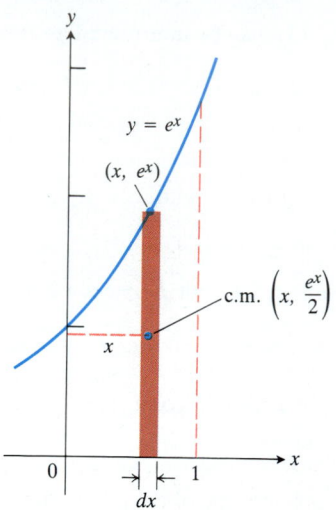

7.5 The moment of the strip about the y-axis is $x\, \delta\, dA = \delta\, xe^x dx$ (Example 3).

Example 3   Find the moment about the y-axis of a thin plate of constant density δ covering the region in the first quadrant bounded by the curve $y = e^x$ and the line $x = 1$ (Fig. 7.5).

Solution   A typical vertical strip has

center of mass (c.m.): $(\tilde{x},\, \tilde{y}) = \left(x,\, \dfrac{e^x}{2}\right),$

length: $e^x,$

width: $dx,$

area: $dA = e^x\, dx,$

mass: $dm = \delta\, dA = \delta\, e^x\, dx.$

The moment of the strip about the y-axis is therefore

$$\tilde{x}\, dm = x \cdot \delta\, e^x\, dx = \delta\, xe^x\, dx.$$

The moment of the plate about the y-axis is

$$M_y = \int \tilde{x}\, dm = \delta \int_0^1 xe^x\, dx.$$

To evaluate this integral we use the formula $\int u\, dv = uv - \int v\, du$ with

$$u = x, \qquad dv = e^x\, dx,$$

$$du = dx, \qquad v = e^x. \qquad \text{(Simplest function with } dv = e^x\, dx\text{)}$$

Then

$$\int xe^x\, dx = xe^x - \int e^x\, dx,$$

so

$$\int_0^1 xe^x\, dx = xe^x \Big]_0^1 - \int_0^1 e^x\, dx = e - \Big[e^x\Big]_0^1 = e - [e - 1] = 1.$$

The moment of the plate about the y-axis is

$$M_y = \delta \int_0^1 xe^x\, dx = \delta \cdot 1 = \delta.$$

Example 4   Evaluate $\displaystyle\int \ln x\, dx.$

Solution   We use the formula $\int u\, dv = uv - \int v\, du$ with

$$u = \ln x, \qquad \text{(Simplifies when differentiated)}$$

$$du = \frac{1}{x}\, dx,$$

$$dv = dx, \qquad \text{(Easy to integrate)}$$

$$v = x. \qquad \text{(Simplest function with } dv = dx\text{)}$$

Then

$$\int \ln x\, dx = x \ln x - \int x \cdot \frac{1}{x}\, dx = x \ln x - \int dx = x \ln x - x + C.$$

Repeated Use

Sometimes we have to use integration by parts more than once to obtain an answer.

Example 5   Evaluate $\displaystyle\int x^2 e^x\,dx$.

Solution   We use the formula $\int u\,dv = uv - \int v\,du$ with

$$u = x^2, \qquad dv = e^x\,dx, \qquad v = e^x, \qquad du = 2x\,dx.$$

This gives

$$\int x^2 e^x\,dx = x^2 e^x - 2\int x e^x\,dx.$$

It takes a second integration by parts to find the integral on the right. As in Example 3, its value is $x e^x - e^x + C'$. Hence

$$\int x^2 e^x\,dx = x^2 e^x - 2x e^x + 2e^x + C.$$

Solving for the Unknown Integral

Integrals like the next one occur in electrical engineering. Their evaluation requires two integrations by parts followed by solving for the unknown integral.

Example 6   Evaluate $\displaystyle\int e^x \cos x\,dx$.

Solution   We first use the formula $\int u\,dv = uv - \int v\,du$ with

$$u = e^x, \qquad dv = \cos x\,dx, \qquad v = \sin x, \qquad du = e^x\,dx.$$

Then

$$\int e^x \cos x\,dx = e^x \sin x - \int e^x \sin x\,dx. \tag{4}$$

The second integral is like the first, except that it has $\sin x$ in place of $\cos x$. To evaluate it, we use integration by parts with

$$u = e^x, \qquad dv = \sin x\,dx, \qquad v = -\cos x, \qquad du = e^x\,dx.$$

Then

$$\int e^x \cos x\,dx = e^x \sin x - \left(-e^x \cos x - \int (-\cos x)(e^x\,dx) \right)$$

$$= e^x \sin x + e^x \cos x - \int e^x \cos x\,dx.$$

The unknown integral now appears on both sides of the equation. Combining the two expressions gives

$$2\int e^x \cos x\,dx = e^x \sin x + e^x \cos x.$$

Dividing by 2 and adding a constant of integration gives

$$\int e^x \cos x \, dx = \frac{e^x \sin x + e^x \cos x}{2} + C.$$

Our choice of $u = e^x$ and $dv = \sin x \, dx$ in the second integration may have seemed arbitrary but it wasn't. In theory, we could have chosen $u = \sin x$ and $dv = e^x \, dx$. Doing so, however, would have turned Eq. (4) into

$$\int e^x \cos x \, dx = e^x \sin x - \left(e^x \sin x - \int e^x \cos x \, dx \right)$$

$$= \int e^x \cos x \, dx.$$

The resulting identity is correct but useless. *Moral:* Once you have decided on what to differentiate and integrate in circumstances like these, stick with them. General formulas for the integrals of $e^{ax} \cos bx$ and the closely related $e^{ax} \sin bx$ can be found in the integral table at the end of this book.

Tabular Integration

We have seen that integrals of the form $\int f(x) \, g(x) \, dx$, in which f can be differentiated repeatedly to become zero and g can be integrated repeatedly without difficulty, are natural candidates for integration by parts. However, if many repetitions are required, the calculations can be cumbersome. In situations like this, there is a way to organize the calculations that saves a great deal of work. It is called **tabular integration** and is illustrated in the following examples.

Example 7   Evaluate $\int x^2 e^x \, dx$ by tabular integration.

Solution   With $f(x) = x^2$ and $g(x) = e^x$, we list

$f(x)$ and its derivatives		$g(x)$ and its integrals
x^2	$(+)$	e^x
$2x$	$(-)$	e^x
2	$(+)$	e^x
0		e^x

We add the products of the functions connected by the arrows, with the middle sign changed, to obtain

$$\int x^2 e^x \, dx = x^2 e^x - 2xe^x + 2e^x + C.$$

Example 8   Evaluate $\int x^3 \sin x \, dx$ by tabular integration.

Solution   With $f(x) = x^3$ and $g(x) = \sin x$, we list

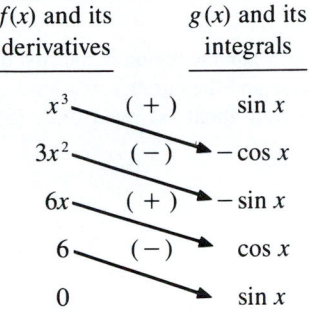

$f(x)$ and its derivatives		$g(x)$ and its integrals
x^3	$(+)$	$\sin x$
$3x^2$	$(-)$	$-\cos x$
$6x$	$(+)$	$-\sin x$
6	$(-)$	$\cos x$
0		$\sin x$

Again we add the products of the functions connected by the arrows, with every other sign changed, to obtain

$$\int x^3 \sin x \, dx = -x^3 \cos x + 3x^2 \sin x + 6x \cos x - 6 \sin x + C.$$

EXERCISES 7.2

Evaluate the integrals in Exercises 1–30.

1. $\int x \sin x \, dx$

2. $\int \theta \cos 2\theta \, d\theta$

3. $\int t^2 \cos t \, dt$

4. $\int x^2 \sin x \, dx$

5. $\int_1^2 x \ln x \, dx$

6. $\int x^3 \ln x \, dx$

7. $\int \tan^{-1} y \, dy$

8. $\int \sin^{-1} y \, dy$

9. $\int x \sec^2 x \, dx$

10. $\int 4x \sec^2 2x \, dx$

11. $\int x^3 e^x \, dx$

12. $\int p^4 e^{-p} \, dp$

13. $\int (x^2 - 5x)e^x \, dx$

14. $\int (r^2 + r + 1)e^r \, dr$

15. $\int x^5 e^x \, dx$

16. $\int t^2 e^{4t} \, dt$

17. $\int_0^{\pi/2} \theta^2 \sin 2\theta \, d\theta$

18. $\int_0^{\pi/2} x^3 \cos 2x \, dx$

19. $\int_{2/\sqrt{3}}^2 t \sec^{-1} t \, dt$

20. $\int_0^{1/\sqrt{2}} 2x \sin^{-1}(x^2) \, dx$

21. $\int e^\theta \sin \theta \, d\theta$

22. $\int e^{-y} \cos y \, dy$

23. $\int e^{2x} \cos 3x \, dx$

24. $\int e^{-2x} \sin 2x \, dx$

25. $\int e^{\sqrt{3s+9}} \, ds$

26. $\int_0^1 x\sqrt{1-x} \, dx$

27. $\int_0^{\pi/3} x \tan^2 x \, dx$

28. $\int \ln(x + x^2) \, dx$

29. $\int \sin(\ln x) \, dx$

30. $\int z(\ln z)^2 \, dz$

31. Find the area of the region enclosed by the x-axis and the curve $y = x \sin x$ for (a) $0 \le x \le \pi$, (b) $\pi \le x \le 2\pi$.

32. Use the method of cylindrical shells to find the volume swept out by revolving the region bounded by $x = 0$, $y = 0$, and $y = \cos x$, $0 \le x \le \pi/2$, about the y-axis.

33. Find the volume swept out by revolving about the y-axis the region in the first quadrant bounded by the coordinate axes, the curve $y = e^{-x}$, and the line $x = 1$.

34. a) Find the center of mass of a thin plate of constant density covering the region in the first quadrant enclosed by the curve $y = x^2 e^x$, the x-axis, and the line $x = 1$.

b) CALCULATOR Find the coordinates of the center of mass to two decimal places. Show the center of mass in a rough sketch of the plate.

35. a) Find the center of mass of a thin plate of constant density covering the region enclosed by the curve $y = \ln x$, the x-axis, and the line $x = e$.

b) CALCULATOR Find the coordinates of the center of mass to two decimal places. Show the center of mass in a rough sketch of the plate.

36. Find the volume swept out when the region in the first quadrant bounded by the x-axis and the curve $y = x \sin x$, $0 \le x \le \pi$ (Fig. 7.6), is revolved about (a) the x-axis, (b) the line $x = \pi$.

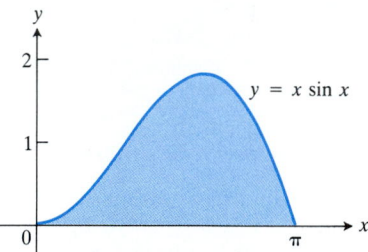

7.6   The region in Exercise 36.

37. Find the moment about the y-axis of a thin plate of density $\delta = 1 + x$ covering the region bounded by the x-axis and the curve $y = \sin x$, $0 \le x \le \pi$.

38. Although we usually drop the constant of integration in determining v as $\int dv$ in integration by parts, choosing the constant to be different from zero can occasionally be helpful. As a case in point, evaluate

$$\int x \tan^{-1} x \, dx,$$

with $u = \tan^{-1} x$ and $v = (x^2/2) + C$, and find a value of C that simplifies the resulting formula.

Integrating Inverses

Integration by parts leads to a rule for integrating inverses that usually gives good results:

$$\int f^{-1}(x) \, dx = \int y f'(y) \, dy \qquad \left(\begin{matrix} y = f^{-1}(x), \, x = f(y) \\ dx = f'(y) \, dy \end{matrix} \right)$$

$$= y f(y) - \int f(y) \, dy \quad \left(\begin{matrix} \text{Integration by parts with} \\ u = y, \, dv = f'(y) \, dy \end{matrix} \right)$$

$$= x f^{-1}(x) - \int f(y) \, dy$$

The idea is to take the most complicated part of the integral, in this case $f^{-1}(x)$, and simplify it first. For the integral of $\ln x$, we get

$$\int \ln x \, dx = \int y e^y \, dy \qquad \left(\begin{matrix} y = \ln x, \, x = e^y \\ dx = e^y \, dy \end{matrix} \right)$$

$$= y e^y - e^y + C \qquad \text{(Example 3)}$$

$$= x \ln x - x + C.$$

For the integral of $\cos^{-1} x$ we get

$$\int \cos^{-1} x \, dx = x \cos^{-1} x - \int \cos y \, dy \qquad (y = \cos^{-1} x)$$

$$= x \cos^{-1} x - \sin y + C$$

$$= x \cos^{-1} x - \sin(\cos^{-1} x) + C.$$

Use the formula

$$\int f^{-1}(x) \, dx = x f^{-1}(x) - \int f(y) \, dy \qquad (y = f^{-1}(x)) \qquad (5)$$

to evaluate the integrals in Exercises 39–42. Express your answers in terms of x.

39. $\int \sin^{-1} x \, dx$ 40. $\int \tan^{-1} x \, dx$

41. $\int \sec^{-1} x \, dx$ 42. $\int \log_2 x \, dx$

Another way to integrate $f^{-1}(x)$ (when f^{-1} is integrable, of course) is to use integration by parts with $u = f^{-1}(x)$ and $dv = dx$ to rewrite the integral of f^{-1} as

$$\int f^{-1}(x) \, dx = x f^{-1}(x) - \int x \left(\frac{d}{dx} f^{-1}(x) \right) dx. \qquad (6)$$

Exercises 43 and 44 compare the results of using Eqs. (5) and (6).

43. Equations (5) and (6) give different formulas for the integral of $\cos^{-1} x$:

a) $\int \cos^{-1} x \, dx = x \cos^{-1} x - \sin(\cos^{-1} x) + C$,   (Eq. (5))

b) $\int \cos^{-1} x \, dx = x \cos^{-1} x - \sqrt{1 - x^2} + C$.   (Eq. (6))

Can both integrations be correct? Explain.

44. Equations (5) and (6) lead to different formulas for the integral of $\tan^{-1} x$:

a) $\int \tan^{-1} x \, dx = x \tan^{-1} x - \ln \sec(\tan^{-1} x) + C$, (Eq. (5))

b) $\int \tan^{-1} x \, dx = x \tan^{-1} x - \ln \sqrt{1 + x^2} + C$.   (Eq. (6))

Can both integrations be correct? Explain.

Evaluate the integrals in Exercises 45 and 46 with both Eq. (5) and Eq. (6). In each case, check your work by differentiating your answer with respect to x.

45. $\int \sinh^{-1} x \, dx$ 46. $\int \tanh^{-1} x \, dx$

7.3 Trigonometric Integrals

Trigonometric integrals involve algebraic combinations of trigonometric functions. In principle, we can always express such integrals in terms of sines and cosines, but it is often simpler to work with other functions, as in the integral

$$\int \sec^2 x \, dx = \tan x + C.$$

The general idea is to use identities to transform the integrals we have into forms that are easier to integrate.

Products of Sines and Cosines

We begin with integrals of the form

$$\int \sin^m x \cos^n x \, dx,$$

where m and n are nonnegative integers (positive or zero). We can divide the work into three cases:

Case 1: m is odd. PEEL

Case 2: m is even and n is odd.

Case 3: m and n are both even. USE IDENTIES

In each case we can use a trigonometric identity to transform the integral into a more convenient form.

CASE 1: If m is odd, we write m as $2k + 1$ and use the identity $\sin^2 x = 1 - \cos^2 x$ to obtain

$$\sin^m x = \sin^{2k+1} x = (\sin^2 x)^k \sin x = (1 - \cos^2 x)^k \sin x. \qquad (1)$$

Then we combine the single $\sin x$ with dx in the integral and set $\sin x \, dx$ equal to $-d(\cos x)$.

Example 1 Evaluate $\displaystyle\int \sin^3 x \cos^2 x \, dx$.

Solution

$$\int \sin^3 x \cos^2 x \, dx = \int \sin^2 x \cos^2 x \sin x \, dx$$

$$= \int (1 - \cos^2 x) \cos^2 x (-d(\cos x))$$

$$= \int (1 - u^2)(u^2)(-du) \qquad (u = \cos x)$$

$$= \int (u^4 - u^2) \, du$$

$$= \frac{u^5}{5} - \frac{u^3}{3} + C$$

$$= \frac{\cos^5 x}{5} - \frac{\cos^3 x}{3} + C$$

CASE 2: If m is even and n is odd in $\int \sin^m x \cos^n x \, dx$, we write n as $2k + 1$ and use the identity $\cos^2 x = 1 - \sin^2 x$ to obtain

$$\cos^n x = \cos^{2k+1} x = (\cos^2 x)^k \cos x = (1 - \sin^2 x)^k \cos x.$$

We then combine the single $\cos x$ with dx and set $\cos x \, dx$ equal to $d(\sin x)$.

Example 2   Evaluate $\int \cos^5 x \, dx$.

Solution

$$\int \cos^5 x \, dx = \int \cos^4 x \cos x \, dx = \int (1 - \sin^2 x)^2 \, d(\sin x)$$

$$= \int (1 - u^2)^2 \, du \qquad (u = \sin x)$$

$$= \int (1 - 2u^2 + u^4) \, du$$

$$= u - \frac{2}{3}u^3 + \frac{1}{5}u^5 + C = \sin x - \frac{2}{3}\sin^3 x + \frac{1}{5}\sin^5 x + C$$

CASE 3:   If both m and n are even in $\int \sin^m x \cos^n x \, dx$, we substitute

$$\sin^2 x = \frac{1 - \cos 2x}{2}, \qquad \cos^2 x = \frac{1 + \cos 2x}{2} \tag{2}$$

to reduce the integrand to one in lower powers of $\cos 2x$.

Example 3   Evaluate $\int \sin^2 x \cos^4 x \, dx$.

Solution

$$\int \sin^2 x \cos^4 x \, dx = \int \left(\frac{1 - \cos 2x}{2}\right)\left(\frac{1 + \cos 2x}{2}\right)^2 dx$$

$$= \frac{1}{8}\int (1 - \cos 2x)(1 + 2\cos 2x + \cos^2 2x) \, dx$$

$$= \frac{1}{8}\int (1 + \cos 2x - \cos^2 2x - \cos^3 2x) \, dx$$

For the term involving $\cos^2 2x$ we use Eq. (2) again:

$$\int \cos^2 2x \, dx = \frac{1}{2}\int (1 + \cos 4x) \, dx$$

$$= \frac{1}{2}\left(x + \frac{1}{4}\sin 4x\right). \qquad \left(\begin{array}{l}\text{Omitting the constant of} \\ \text{integration until the final result}\end{array}\right)$$

For the $\cos^3 2x$ term we use Case 2:

$$\int \cos^3 2x \, dx = \int (1 - \sin^2 2x) \cos 2x \, dx \qquad \left(\begin{array}{l}u = \sin 2x, \\ du = 2 \cos 2x \, dx\end{array}\right)$$

$$= \frac{1}{2}\int (1 - u^2) \, du = \frac{1}{2}\left(\sin 2x - \frac{1}{3}\sin^3 2x\right). \qquad \left(\begin{array}{l}\text{Again} \\ \text{omitting } C\end{array}\right)$$

FLOWCHART 7.1 $\int \sin^m x \cos^n x\, dx$ (m, n nonnegative integers)

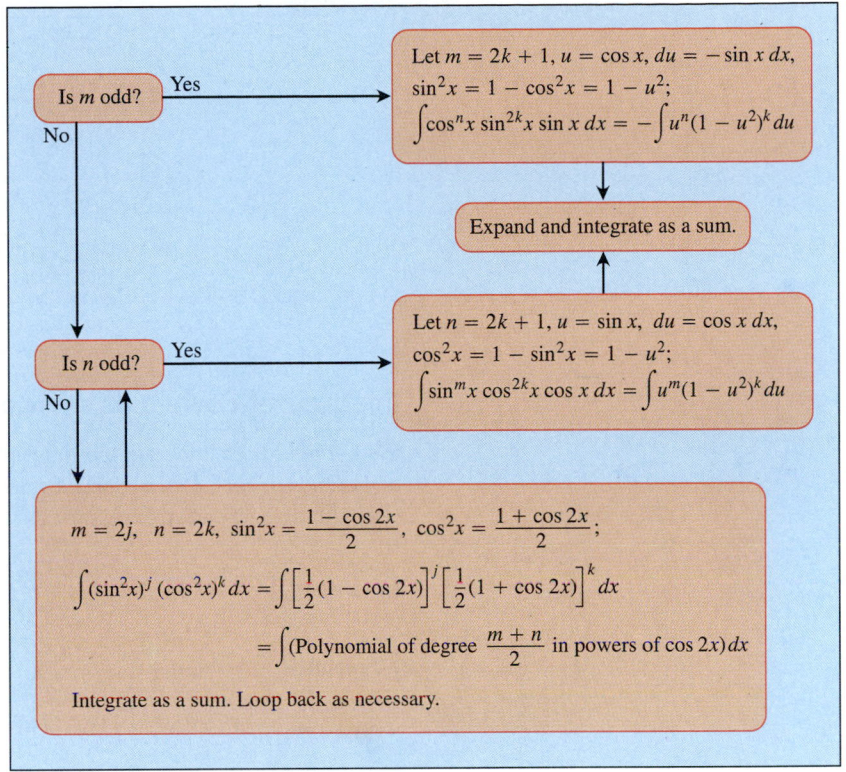

Combining everything and simplifying, we get

$$\int \sin^2 x \cos^4 x\, dx = \frac{1}{16}\left(x - \frac{1}{4}\sin 4x + \frac{1}{3}\sin^3 2x\right) + C.$$

Eliminating Square Roots

In the next example, we use the identity $\cos^2\theta = (1 + \cos 2\theta)/2$ to eliminate a square root.

Example 4   Evaluate $\displaystyle\int_0^{\pi/4} \sqrt{1 + \cos 4x}\, dx$.

Solution   To eliminate the square root, we use the identity

$$\cos^2\theta = \frac{1 + \cos 2\theta}{2}, \qquad \text{or} \qquad 1 + \cos 2\theta = 2\cos^2\theta.$$

With $\theta = 2x$, this becomes

$$1 + \cos 4x = 2\cos^2 2x.$$

Therefore,

$$\int_0^{\pi/4} \sqrt{1 + \cos 4x}\, dx = \int_0^{\pi/4} \sqrt{2 \cos^2 2x}\, dx = \int_0^{\pi/4} \sqrt{2}\sqrt{\cos^2 2x}\, dx$$

$$= \sqrt{2} \int_0^{\pi/4} |\cos 2x|\, dx$$

$$= \sqrt{2} \int_0^{\pi/4} \cos 2x\, dx \qquad \begin{pmatrix} \text{Because} \\ \cos 2x \geq 0 \\ \text{on } [0, \pi/4] \end{pmatrix}$$

$$= \sqrt{2} \left[\frac{\sin 2x}{2} \right]_0^{\pi/4} = \frac{\sqrt{2}}{2} [1 - 0] = \frac{\sqrt{2}}{2}.$$

Integrals of Powers of tan x and sec x

We know how to integrate the tangent and secant and their squares. To integrate higher powers we use the identities $\tan^2 x = \sec^2 x - 1$ and $\sec^2 x = \tan^2 x + 1$ and integrate by parts when necessary to reduce the higher powers to lower powers.

Example 5   Evaluate $\displaystyle\int \tan^4 x\, dx$.

Solution

$$\int \tan^4 x\, dx = \int \tan^2 x \cdot \tan^2 x\, dx = \int \tan^2 x \cdot (\sec^2 x - 1)\, dx$$

$$= \int \tan^2 x \sec^2 x\, dx - \int \tan^2 x\, dx$$

$$= \int \tan^2 x \sec^2 x\, dx - \int (\sec^2 x - 1)\, dx$$

$$= \int \tan^2 x \sec^2 x\, dx - \int \sec^2 x\, dx + \int dx$$

In the first integral, we let

$$u = \tan x, \qquad du = \sec^2 x\, dx$$

and have

$$\int u^2\, du = \frac{1}{3} u^3 + C'.$$

The remaining integrals are standard forms, so

$$\int \tan^4 x\, dx = \frac{1}{3}\tan^3 x - \tan x + x + C.$$

Example 6   Evaluate $\displaystyle\int \sec^3 x\, dx$.

Solution   We integrate by parts, using

$$u = \sec x, \qquad dv = \sec^2 x\, dx, \qquad v = \tan x, \qquad du = \sec x \tan x\, dx.$$

Then

$$\int \sec^3 x \, dx = \sec x \tan x - \int (\tan x)(\sec x \tan x \, dx)$$

$$= \sec x \tan x - \int (\sec^2 x - 1)\sec x \, dx \qquad (\tan^2 x = \sec^2 x - 1)$$

$$= \sec x \tan x + \int \sec x \, dx - \int \sec^3 x \, dx.$$

Combining the two secant-cubed integrals gives

$$2 \int \sec^3 x \, dx = \sec x \tan x + \int \sec x \, dx$$

and

$$\int \sec^3 x \, dx = \frac{1}{2} \sec x \tan x + \frac{1}{2} \ln |\sec x + \tan x| + C.$$

With cosecants and cotangents in place of secants and tangents, the method of Example 6 leads to the companion formula

$$\int \csc^3 x \, dx = -\frac{1}{2} \csc x \cot x - \frac{1}{2} \ln |\csc x + \cot x| + C \qquad (3)$$

(Exercise 61).

Products of Sines and Cosines

The integrals

$$\int \sin mx \sin nx \, dx, \qquad \int \sin mx \cos nx \, dx, \qquad \int \cos mx \cos nx \, dx,$$

$m \ne n$, arise in many mathematical and scientific applications that require the use of trigonometric functions. (The numbers m and n need not be integers, although they sometimes are.) We can evaluate these integrals with integration by parts, but two such integrations are required in each case. It is simpler to use the identities

$$\sin mx \sin nx = \frac{1}{2}[\cos(m - n)x - \cos(m + n)x], \qquad (4)$$

$$\sin mx \cos nx = \frac{1}{2}[\sin(m - n)x + \sin(m + n)x], \qquad (5)$$

$$\cos mx \cos nx = \frac{1}{2}[\cos(m - n)x + \cos(m + n)x]. \qquad (6)$$

These come from combining the identities

$$\cos(A + B) = \cos A \cos B - \sin A \sin B, \qquad (7)$$

$$\cos(A - B) = \cos A \cos B + \sin A \sin B, \qquad (8)$$

$$\sin(A + B) = \sin A \cos B + \cos A \sin B, \qquad (9)$$

$$\sin(A - B) = \sin A \cos B - \cos A \sin B. \qquad (10)$$

For example, if we take $A = mx$ and $B = nx$ in Eqs. (7) and (8), add, and divide by 2, we get Eq. (6). We get Eq. (4) by subtracting (7) from (8) and dividing by 2. To get Eq. (5), we add Eqs. (9) and (10) and divide by 2.

Example 7 Evaluate $\displaystyle\int \sin 3x \cos 5x \, dx$.

Solution From Eq. (5) with $m = 3$ and $n = 5$ we get

$$\int \sin 3x \cos 5x \, dx = \frac{1}{2}\int [\sin(-2x) + \sin 8x] \, dx$$

$$= \frac{1}{2}\int (\sin 8x - \sin 2x) \, dx$$

$$= -\frac{\cos 8x}{16} + \frac{\cos 2x}{4} + C.$$

Definite Integrals of Even Functions

The definite integral of an even function $f(x)$ over $[-a, a]$ is twice the value of the integral of f over $[0, a]$ (Fig. 7.7). This is because the integral of f from $-a$ to 0 has the same value as the integral of f from 0 to a:

$$\int_{-a}^{0} f(x) \, dx = \int_{a}^{0} f(-u)(-du) \qquad (x = -u, \, dx = -du)$$

$$= -\int_{a}^{0} f(u) \, du \qquad (f \text{ even} \Leftrightarrow f(-u) = f(u)) \qquad (11)$$

$$= \int_{0}^{a} f(u) \, du.$$

This observation saves time when the antiderivative of f is more easily evaluated at 0 than at $-a$.

What functions of x are even?

Constants, and even powers of x: $0, 1, -5, x^2, x^{-4}$

Cosines: $\cos ax$ (any number a)

Products and quotients of even functions: $x^2 \cos x$, $\dfrac{\sin^2 x}{\cos x}$

A product of two odd functions: $x \sin x$

Roots of even functions: $\sqrt{x^2 + 1}$, $\sqrt[3]{1 - x^2}$

7.7 The graph of an even function is symmetric about the y-axis, so

$$\int_{-a}^{0} f(x) \, dx = \int_{0}^{a} f(x) \, dx.$$

If f is even,

$$\int_{-a}^{a} f(x) \, dx = 2\int_{0}^{a} f(x) \, dx.$$

Example 8

$$\int_{-\pi/4}^{\pi/4} \cos x \, dx = 2\int_{0}^{\pi/4} \cos x \, dx = 2\Big[\sin x\Big]_{0}^{\pi/4} = 2\Big[\frac{\sqrt{2}}{2} - 0\Big] = \sqrt{2}.$$

Definite Integrals of Odd Functions

In many applications, we integrate functions over intervals that are symmetric about the origin. In an amazing number of cases, these integrals turn out to equal

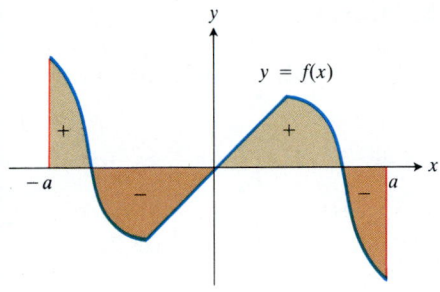

7.8 The graph of an odd function over an interval that is symmetric about the origin encloses as much area above the axis as below. If f is odd,

$$\int_{-a}^{a} f(x)\, dx = 0.$$

zero. For example,

$$\int_{-1}^{1} 2x\, dx = x^2\Big]_{-1}^{1} = 1 - 1 = 0,$$

$$\int_{-\pi}^{\pi} \sin x\, dx = -\cos x\Big]_{-\pi}^{\pi} = -\cos \pi + \cos(-\pi) = -(-1) + (-1) = 0,$$

$$\int_{-a}^{a} \frac{\sin x}{\cos^2 x}\, dx = \frac{1}{\cos x}\Big]_{-a}^{a} = \frac{1}{\cos a} - \frac{1}{\cos(-a)} = 0.$$

These three integrals are zero because they are integrals of odd functions over intervals that are symmetric about the origin. For each function, $f(-x) = -f(x)$. A calculation similar to the one in Eq. (11) shows that

$$\int_{-a}^{0} f(x)\, dx = -\int_{0}^{a} f(x)\, dx,$$

so the integral over $[-a, a]$ is zero (see Exercise 62). Geometrically (Fig. 7.8), the amount of area enclosed by the graph of f above the x-axis to the left of the origin equals the amount of area enclosed below the x-axis to the right of the origin and cancels it out in integration. Similarly, the area enclosed beneath the x-axis to the left of the origin cancels the area enclosed above the x-axis to the right of the origin.

What functions of x are odd?

Odd powers of x: $x,\quad x^{-1},\quad x^3,\quad x^{-3}$

Sines: $\sin ax$

Odd integer powers of odd functions: $\sin^3 x,\quad \dfrac{1}{\sin^5 x}$

The product of an odd function and an even function: $\cos x \sin 3x,\quad x^2\sin x$

The quotient of an odd function and an even function: $\tan x = \dfrac{\sin x}{\cos x}$

EXERCISES 7.3

Evaluate the integrals in Exercises 1–20.

1. $\displaystyle\int_{0}^{\pi/2} \sin^5 x\, dx$

2. $\displaystyle\int_{0}^{\pi} \sin^5 \frac{x}{2}\, dx$

3. $\displaystyle\int_{-\pi/2}^{\pi/2} \cos^3 x\, dx$

4. $\displaystyle\int_{0}^{\pi/6} 3\cos^5 3x\, dx$

5. $\displaystyle\int \sin^7 y\, dy$

6. $\displaystyle\int 7\cos^7 t\, dt$

7. $\displaystyle\int 8\sin^4 x\, dx$

8. $\displaystyle\int 8\cos^4 2\pi x\, dx$

9. $\displaystyle\int_{-\pi/4}^{\pi/4} \frac{4\sin^4 x}{\cos^2 x}\, dx$

10. $\displaystyle\int \frac{\cos^6 \theta}{\sin^2 \theta}\, d\theta$

11. $\displaystyle\int_{\pi/6}^{\pi/2} \frac{3\cos^3 t}{\sqrt{\sin^3 t}}\, dt$

12. $\displaystyle\int_{-\pi/3}^{\pi/3} \frac{\sin^3 x}{\sqrt{\cos x}}\, dx$

13. $\displaystyle\int_{0}^{2\pi} \sin^2 2\theta \cos^3 2\theta\, d\theta$

14. $\displaystyle\int_{0}^{\pi/2} \sin^{5/2} x \cos^3 x\, dx$

15. $\displaystyle\int_{-\pi/2}^{\pi/4} 16\sin^2 x \cos^2 x\, dx$

16. $\displaystyle\int 8\sin^4 y \cos^2 y\, dy$

17. $\displaystyle\int 35\sin^4 \theta \cos^3 \theta\, d\theta$

18. $\displaystyle\int 4\sin^4 t \cos^4 t\, dt$

19. $\displaystyle\int \frac{4\sin^2(\ln x) \cos^2(\ln x)}{x}\, dx$

20. $\displaystyle\int \frac{12\sin^3(\tan^{-1} y)}{1 + y^2}\, dy$

Evaluate the integrals in Exercises 21–32.

21. $\displaystyle\int_0^{2\pi} \sqrt{\frac{1 - \cos x}{2}}\, dx$

22. $\displaystyle\int_0^{\pi} \sqrt{1 - \cos 2x}\, dx$

23. $\displaystyle\int_0^{\pi} \sqrt{1 - \sin^2 t}\, dt$

24. $\displaystyle\int_0^{\pi} \sqrt{1 - \cos^2\theta}\, d\theta$

25. $\displaystyle\int_{-\pi/4}^{\pi/4} \sqrt{1 + \tan^2 x}\, dx$

26. $\displaystyle\int_{-\pi/4}^{\pi/4} \sqrt{\sec^2 x - 1}\, dx$

27. $\displaystyle\int \theta\sqrt{1 - \cos 2\theta}\, d\theta$

28. $\displaystyle\int_{-\pi}^{\pi} (1 - \cos^2 t)^{3/2}\, dt$

29. $\displaystyle\int_0^{\pi} \sqrt{1 + \sin x}\, dx$ $\quad\left(Hint: \sin x = \cos\left(x - \frac{\pi}{2}\right).\right)$

30. $\displaystyle\int_0^{\pi} \sqrt{1 - \sin x}\, dx$

31. $\displaystyle\int \cos x \ln\left(\frac{1 - \cos 2x}{2}\right) dx$

32. $\displaystyle\int \frac{\cos x}{\sqrt{1 - \cos 2x}}\, dx$

Evaluate the integrals in Exercises 33–50.

33. $\displaystyle\int_{-\pi/3}^{0} 2 \sec^3 x\, dx$

34. $\displaystyle\int e^x \sec^3 (e^x - 1)\, dx$

35. $\displaystyle\int_{\pi/2}^{\pi} \csc^3 \frac{x}{2}\, dx$

36. $\displaystyle\int \frac{\csc^3\sqrt{\theta}}{\sqrt{\theta}}\, d\theta$

37. $\displaystyle\int_0^{\pi/4} \tan^2 x \sec x\, dx$

38. $\displaystyle\int_{\pi/6}^{\pi/2} \cot^2 x \csc x\, dx$

39. $\displaystyle\int_0^{\pi/4} \sec^4\theta\, d\theta$

40. $\displaystyle\int_0^{\pi/12} 3 \sec^4 3x\, dx$

41. $\displaystyle\int \csc^4\theta\, d\theta$

42. $\displaystyle\int_{\pi/2}^{\pi} 3 \csc^4 \frac{\theta}{2}\, d\theta$

43. $\displaystyle\int_0^{\pi/4} 4 \tan^3 x\, dx$

44. $\displaystyle\int_{\pi/6}^{\pi/3} \cot^3 x\, dx$

45. $\displaystyle\int_{\pi/4}^{\pi/2} 8 \cot^4 t\, dt$

46. $\displaystyle\int 6 \tan^4 x\, dx$

47. $\displaystyle\int \sec^2 t \ln(\cos t)\, dt$

48. $\displaystyle\int_{-\pi/3}^{\pi/3} \frac{\sin^2\theta}{\cos^4\theta}\, d\theta$

49. $\displaystyle\int \tan^3 x (\sec x)^{3/2}\, dx$

50. $\displaystyle\int \cot^3 x (\csc x)^{-1/2}\, dx$

Evaluate the integrals in Exercises 51–56.

51. $\displaystyle\int_{-\pi}^{0} \sin 3x \cos 2x\, dx$

52. $\displaystyle\int_0^{\pi/2} \sin 2x \cos 3x\, dx$

53. $\displaystyle\int_{\pi/12}^{\pi/6} 8 \sin 4x \sin 2x\, dx$

54. $\displaystyle\int_{2\pi}^{3\pi} \sin \frac{x}{3} \cos \frac{x}{6}\, dx$

55. $\displaystyle\int_{2\pi}^{4\pi} \cos \frac{x}{3} \cos \frac{x}{4}\, dx$

56. $\displaystyle\int_0^{\pi/2} \cos \frac{x}{2} \cos 7x\, dx$

57. Which integrals are zero and which are not? (You can do most of these without writing anything down.)

a) $\displaystyle\int_{-\pi}^{\pi} \sin x \cos^2 x\, dx$
b) $\displaystyle\int_{-L}^{L} \sqrt[3]{\sin x}\, dx$

c) $\displaystyle\int_{-\pi/4}^{\pi/4} x \sec x\, dx$
d) $\displaystyle\int_{-\pi/2}^{\pi/2} x \sin x\, dx$

e) $\displaystyle\int_{-a}^{a} \sin mx \cos mx\, dx, \quad m \neq 0$

f) $\displaystyle\int_{-\pi/2}^{\pi/2} \cos^3 x\, dx$
g) $\displaystyle\int_{-\ln 2}^{\ln 2} x(e^x + e^{-x})\, dx$

h) $\displaystyle\int_{-\pi/2}^{\pi/2} \sin x \sin 2x\, dx$
i) $\displaystyle\int_{-a}^{a} (e^x \sin x + e^{-x}\sin x)\, dx$

58. Which integrals are zero and which are not? (You can do most of these without writing anything down.)

a) $\displaystyle\int_{-1}^{1} \sin 3x \cos 5x\, dx$
b) $\displaystyle\int_{-a}^{a} x\sqrt{a^2 - x^2}\, dx$

c) $\displaystyle\int_{-\pi/4}^{\pi/4} \tan^3 x\, dx$
d) $\displaystyle\int_{-\pi/2}^{\pi/2} x \cos x\, dx$

e) $\displaystyle\int_{-\pi}^{\pi} \sin^5 x\, dx$
f) $\displaystyle\int_{-\pi}^{\pi} \cos^5 x\, dx$

g) $\displaystyle\int_{-\pi/2}^{\pi/2} \sin^2 x \cos x\, dx$
h) $\displaystyle\int_{-\pi/4}^{\pi/4} \sec x \tan x\, dx$

i) $\displaystyle\int_{-1}^{1} \frac{\sin x\, dx}{e^x + e^{-x}}$

59. Which integrals in Exercise 57 have even integrands? Evaluate these integrals.

60. Which integrals in Exercise 58 have even integrands? Evaluate these integrals.

61. *The integral of $\csc^3 x$.* Use the method of Example 6, with cosecants and cotangents in place of secants and tangents, to show that

$$\int \csc^3 x\, dx = -\frac{1}{2} \csc x \cot x - \frac{1}{2} \ln|\csc x + \cot x| + C.$$

62. *The integral of an odd continuous function f(x) from x = −a to x = a is zero.* Use the substitution $u = -x$ together with the fact that f is odd to show that

$$\int_{-a}^{0} f(x)\,dx = -\int_{0}^{a} f(x)\,dx.$$

Then use this to show that

$$\int_{-a}^{a} f(x)\,dx = 0.$$

63. a) Show that the reciprocal of an odd function, when defined, is odd. Are reciprocals of even functions (when defined) even?

b) Show that the quotient of an odd function and an even function, when defined, is odd.

64. *Orthogonal functions.* Integrable functions f and g are said to be **orthogonal** on an interval $[a, b]$ if

$$\int_{a}^{b} f(x)\,g(x)\,dx = 0.$$

a) Prove that if m and n are integers and $m \neq \pm n$, then $\sin mx$ and $\sin nx$ are orthogonal on any interval of length 2π.

b) Prove the same for $\cos mx$ and $\cos nx$.

c) Prove the same for $\sin mx$ and $\cos nx$ even if $m = \pm n$.

These and other sets of orthogonal functions are used extensively in studies of oscillation, temperature distribution, and electrical potential.

7.4 Trigonometric Substitutions

We now embark on a three-step program that will enable us (in theory, at least) to integrate all rational functions of x. The first step is to study substitutions that change binomials like $a^2 + x^2$, $a^2 - x^2$, and $x^2 - a^2$ into single squared terms. The second step will be to simplify integrals involving $ax^2 + bx + c$ by completing the square and then replacing the resulting sums and differences of squares by single squared terms. The third and last step, taken in Section 7.5, will be to express rational functions of x as sums of polynomials (which we already know how to integrate), fractions with linear-factored denominators (which become logarithms or fractions when integrated), and fractions with quadratic denominators (which we shall be able to integrate by the techniques of the present section).

Trigonometric Substitutions for Combining Squares

Trigonometric substitutions enable us to replace the binomials

$$a^2 + x^2, \qquad a^2 - x^2, \qquad \text{and} \qquad x^2 - a^2$$

by single squared terms and thereby transform a number of important integrals into integrals we can recognize or can find in a table. The most commonly used substitutions, $x = a \tan \theta$, $x = a \sin \theta$, and $x = a \sec \theta$, come from the reference triangles in Fig. 7.9.

With $x = a \tan \theta$,

$$a^2 + x^2 = a^2 + a^2\tan^2\theta = a^2(1 + \tan^2\theta) = a^2\sec^2\theta. \tag{1}$$

7.9 Reference triangles for trigonometric substitutions that change binomials into single squared terms.

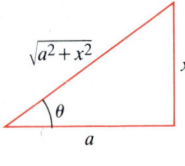

$x = a \tan \theta$
$\sqrt{a^2 + x^2} = a|\sec \theta|$

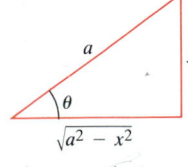

$x = a \sin \theta$
$\sqrt{a^2 - x^2} = a|\cos \theta|$

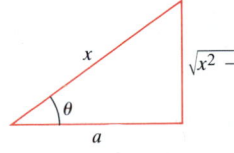

$x = a \sec \theta$
$\sqrt{x^2 - a^2} = a|\tan \theta|$

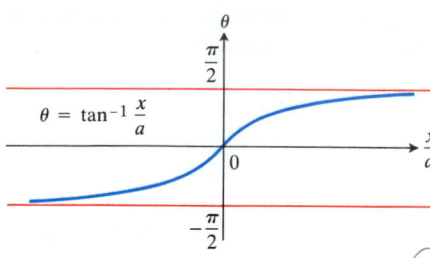

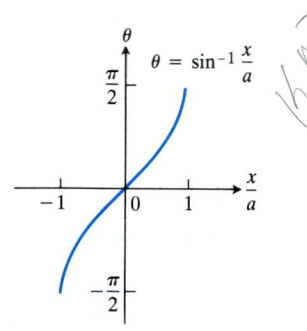

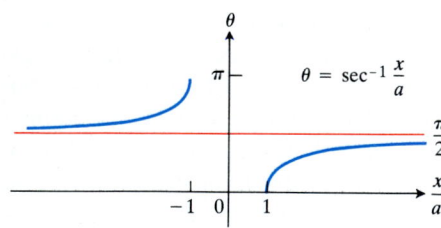

7.10 The arc tangent, arc sine, and arc secant of x/a, graphed as functions of x/a.

With $x = a \sin \theta$,

$$a^2 - x^2 = a^2 - a^2\sin^2\theta = a^2(1 - \sin^2\theta) = a^2\cos^2\theta. \tag{2}$$

With $x = a \sec \theta$,

$$x^2 - a^2 = a^2\sec^2\theta - a^2 = a^2(\sec^2\theta - 1) = a^2\tan^2\theta. \tag{3}$$

$x = a \tan \theta$	replaces	$a^2 + x^2$	by $a^2\sec^2\theta,$ $\qquad(4)$
$x = a \sin \theta$	replaces	$a^2 - x^2$	by $a^2\cos^2\theta,$ $\qquad(5)$
$x = a \sec \theta$	replaces	$x^2 - a^2$	by $a^2\tan^2\theta.$ $\qquad(6)$

We want any substitution we use in an integration to be reversible so that we can change back to the original variable afterward. For example, if $x = a \tan \theta$, we want to be able to set $\theta = \tan^{-1}(x/a)$ after the integration takes place. If $x = \sin \theta$, we want to be able to set $\theta = \sin^{-1}(x/a)$ when we're done, and similarly for $x = a \sec \theta$.

As we know from Section 6.7, the functions in these substitutions have inverses only for selected values of θ (Fig. 7.10). For reversibility,

$$x = a \tan \theta \quad \text{requires} \quad \theta = \tan^{-1}\frac{x}{a} \quad \text{with} \quad -\frac{\pi}{2} < \theta < \frac{\pi}{2},$$

$$x = a \sin \theta \quad \text{requires} \quad \theta = \sin^{-1}\frac{x}{a} \quad \text{with} \quad -\frac{\pi}{2} \le \theta \le \frac{\pi}{2},$$

$$x = a \sec \theta \quad \text{requires} \quad \theta = \sec^{-1}\frac{x}{a} \quad \text{with} \quad \begin{cases} 0 \le \theta < \dfrac{\pi}{2} & \text{if} \quad \dfrac{x}{a} \ge 1, \\[2mm] \dfrac{\pi}{2} < \theta \le \pi & \text{if} \quad \dfrac{x}{a} \le -1. \end{cases}$$

Example 1   Evaluate $\displaystyle\int \frac{dx}{\sqrt{4 + x^2}}.$

Solution   We set

$$x = 2 \tan \theta, \qquad dx = 2 \sec^2\theta \, d\theta, \qquad -\frac{\pi}{2} < \theta < \frac{\pi}{2},$$

$$4 + x^2 = 4 + 4 \tan^2\theta = 4(1 + \tan^2\theta) = 4 \sec^2\theta.$$

Then

$$\int \frac{dx}{\sqrt{4 + x^2}} = \int \frac{2 \sec^2\theta \, d\theta}{\sqrt{4 \sec^2\theta}} = \int \frac{\sec^2\theta \, d\theta}{|\sec \theta|} \qquad (\sqrt{\sec^2\theta} = |\sec \theta|)$$

$$= \int \sec \theta \, d\theta \qquad \left(\sec \theta > 0 \text{ for } -\frac{\pi}{2} < \theta < \frac{\pi}{2}\right)$$

$$= \ln|\sec \theta + \tan \theta| + C$$

$$= \ln\left|\frac{\sqrt{4 + x^2}}{2} + \frac{x}{2}\right| + C \qquad \text{(From Fig. 7.11)}$$

$$= \ln|\sqrt{4 + x^2} + x| + C'. \qquad \text{(Taking } C' = C - \ln 2)$$

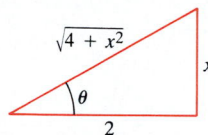

Notice how we expressed $\ln|\sec\theta + \tan\theta|$ in terms of x: We drew a reference triangle for the original substitution $x = 2\tan\theta$ (Fig. 7.11) and read the ratios from the triangle.

7.11   Reference triangle for $x = 2\tan\theta$ (Example 1):

$$\tan\theta = \frac{x}{2}$$

and

$$\sec\theta = \frac{\sqrt{4 + x^2}}{2}$$

Example 2   Evaluate $\displaystyle\int \frac{x^2\, dx}{\sqrt{9 - x^2}}$.

Solution   To replace $9 - x^2$ by a single squared term, we set

$$x = 3\sin\theta, \qquad dx = 3\cos\theta\, d\theta, \qquad -\frac{\pi}{2} < \theta < \frac{\pi}{2},$$

$$9 - x^2 = 9(1 - \sin^2\theta) = 9\cos^2\theta.$$

Then

$$\int \frac{x^2\, dx}{\sqrt{9 - x^2}} = \int \frac{9\sin^2\theta \cdot 3\cos\theta\, d\theta}{|3\cos\theta|}$$

$$= 9\int \sin^2\theta\, d\theta \qquad \left(\cos\theta > 0 \text{ for } -\frac{\pi}{2} < \theta < \frac{\pi}{2}\right)$$

$$= 9\int \frac{1 - \cos 2\theta}{2}\, d\theta$$

$$= \frac{9}{2}\left(\theta - \frac{\sin 2\theta}{2}\right) + C$$

$$= \frac{9}{2}(\theta - \sin\theta\cos\theta) + C$$

$$= \frac{9}{2}\left(\sin^{-1}\frac{x}{3} - \frac{x}{3} \cdot \frac{\sqrt{9 - x^2}}{3}\right) + C \quad \text{(Fig. 7.12)}$$

$$= \frac{9}{2}\sin^{-1}\frac{x}{3} - \frac{x}{2}\sqrt{9 - x^2} + C.$$

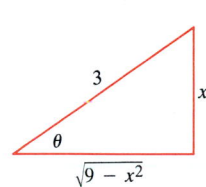

7.12   Reference triangle for $x = 3\sin\theta$ (Example 2):

$$\sin\theta = \frac{x}{3}$$

and

$$\cos\theta = \frac{\sqrt{9 - x^2}}{3}$$

Example 3   Evaluate $\displaystyle\int \frac{dx}{\sqrt{x^2 - 25}}$.

Solution   To replace $x^2 - 25$ by a single squared term, we set

$$x = 5\sec\theta, \qquad dx = 5\sec\theta\tan\theta\, d\theta, \qquad \theta = \sec^{-1}\frac{x}{5},$$

$$x^2 - 25 = 25\sec^2\theta - 25 = 25(\sec^2\theta - 1) = 25\tan^2\theta.$$

With this substitution, we have

$$0 < \theta < \frac{\pi}{2} \text{ for } \frac{x}{5} > 1 \qquad \text{and} \qquad \frac{\pi}{2} < \theta < \pi \text{ for } \frac{x}{5} < -1$$

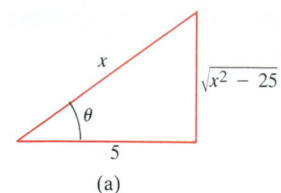

(a)

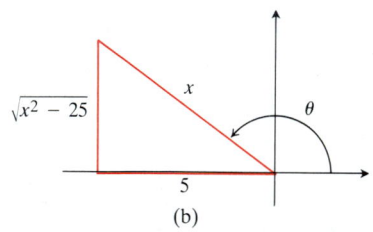

(b)

7.13 (a) Reference triangle for
$x = 5 \sec \theta$, $0 < \theta < \pi/2$, $x > 5$;
(b) reference triangle for
$x = 5 \sec \theta$, $\pi/2 < \theta < \pi$, $x < -5$
in Example 3.

(Fig. 7.10). Then

$$\int \frac{dx}{\sqrt{x^2 - 25}} = \int \frac{5 \sec \theta \tan \theta \, d\theta}{\sqrt{25 \tan^2\theta}}$$

$$= \int \frac{\sec \theta \tan \theta \, d\theta}{|\tan \theta|} \qquad \left(\sqrt{\tan^2\theta} = |\tan \theta|\right)$$

$$= \pm \int \sec \theta \, d\theta = \pm \ln|\sec \theta + \tan \theta| + C$$

$$= \pm \ln\left|\frac{x}{5} \pm \frac{\sqrt{x^2 - 25}}{5}\right| + C \qquad \text{(Fig. 7.13)}$$

$$= \pm \ln|x \pm \sqrt{x^2 - 25}| + C'. \qquad (C' = C \pm \ln 5)$$

What do we do about the signs? When $0 < \theta < \pi/2$, the tangent and secant are both positive and both signs are (+). When $\pi/2 < \theta < \pi$, the tangent and secant are both negative and both signs are (−). Therefore,

$$\int \frac{dx}{\sqrt{x^2 - 25}} = \begin{cases} \ln|x + \sqrt{x^2 - 25}| + C', \\ \text{or} \\ -\ln|x - \sqrt{x^2 - 25}| + C'. \end{cases} \qquad (7)$$

Fortunately, we do not have to work with this two-line formula because the two logarithmic expressions on the right-hand side differ only by a constant:

$$-\ln|x - \sqrt{x^2 - 25}| = \ln\left|\frac{1}{x - \sqrt{x^2 - 25}}\right|$$

$$= \ln\left|\frac{1}{x - \sqrt{x^2 - 25}} \cdot \frac{x + \sqrt{x^2 - 25}}{x + \sqrt{x^2 - 25}}\right|$$

$$= \ln\left|\frac{x + \sqrt{x^2 - 25}}{25}\right| \qquad \left(\begin{matrix} x^2 - (\sqrt{x^2 - 25})^2 \\ = x^2 - x^2 + 25 = 25 \end{matrix}\right)$$

$$= \ln|x + \sqrt{x^2 - 25}| - \ln 25.$$

Therefore,

$$\int \frac{dx}{\sqrt{x^2 - 25}} = \ln|x + \sqrt{x^2 - 25}| + C.$$

Example 4   Find the volume of the solid generated by revolving about the x-axis the region bounded by the curve $y = 4/(x^2 + 4)$, the x-axis, and the lines $x = 0$ and $x = 2$.

Solution   We sketch the region (Fig. 7.14) and use the disk method from Section 5.2:

$$\text{Volume} = \int_0^2 \pi (f(x))^2 \, dx = 16\pi \int_0^2 \frac{dx}{(x^2 + 4)^2}.$$

To evaluate the integral, we set

$$x = 2 \tan \theta, \qquad dx = 2 \sec^2\theta \, d\theta, \qquad \theta = \tan^{-1}\frac{x}{2},$$

$$x^2 + 4 = 4 \tan^2\theta + 4 = 4(\tan^2\theta + 1) = 4 \sec^2\theta$$

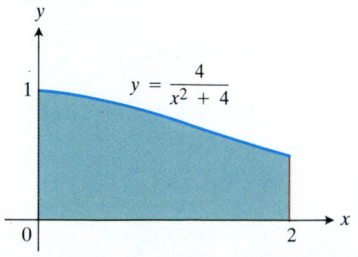

7.14   The region in Example 4.

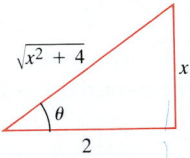

7.15 Reference triangle for $x = 2 \tan \theta$ (Example 4). We have $\theta = 0$ when $x = 0$ and $\theta = \pi/4$ when $x = 2$.

(Fig. 7.15). With these substitutions, we have $\theta = 0$ when $x = 0$ and $\theta = \pi/4$ when $x = 2$.

$$\text{Volume} = 16\pi \int_{x=0}^{x=2} \frac{dx}{(x^2 + 4)^2} = 16\pi \int_{\theta=0}^{\theta=\pi/4} \frac{2 \sec^2\theta \, d\theta}{(4 \sec^2\theta)^2}$$

$$= 16\pi \int_0^{\pi/4} \frac{2 \sec^2\theta \, d\theta}{16 \sec^4\theta} = \pi \int_0^{\pi/4} 2 \cos^2\theta \, d\theta$$

$$= \pi \int_0^{\pi/4} (1 + \cos 2\theta) \, d\theta = \pi\left[\theta + \frac{\sin 2\theta}{2}\right]_0^{\pi/4} \quad \left(\begin{array}{l} 2\cos^2\theta = \\ 1 + \cos 2\theta \end{array}\right)$$

$$= \pi\left[\frac{\pi}{4} + \frac{1}{2}\right] = 4.04. \qquad \text{(Calculator, rounded)}$$

Integrals Involving $ax^2 + bx + c,\ a \neq 0$

As in Section 7.1, we handle these by first completing the square (Appendix 1):

$$ax^2 + bx + c = a\left(x^2 + \frac{b}{a}x + \left(\frac{b}{2a}\right)^2\right) + \left(c - \frac{b^2}{4a}\right) = a\left(x + \frac{b}{2a}\right)^2 + \left(c - \frac{b^2}{4a}\right).$$

We then substitute

$$u = x + \frac{b}{2a}, \qquad x = u - \frac{b}{2a}, \qquad dx = du.$$

Example 5   Evaluate $\displaystyle\int \frac{dx}{\sqrt{2x - x^2}}$.

Solution   First we do the necessary algebra:

$$2x - x^2 = -(x^2 - 2x + 1) + 1 = 1 - (x - 1)^2.$$

Then we substitute $u = x - 1$ and $du = dx$ to get

$$\int \frac{dx}{\sqrt{2x - x^2}} = \int \frac{dx}{\sqrt{1 - (x - 1)^2}} = \int \frac{du}{\sqrt{1 - u^2}}$$

$$= \sin^{-1}u + C = \sin^{-1}(x - 1) + C.$$

Example 6   Evaluate $\displaystyle\int \frac{dx}{4x^2 + 4x + 2}$.

Solution

$$4x^2 + 4x + 2 = 4\left(x^2 + x + \frac{1}{4}\right) + 2 - \frac{4}{4} = 4\left(x + \frac{1}{2}\right)^2 + 1 = (2x + 1)^2 + 1$$

Hence if we let $u = 2x + 1$, we get

$$\int \frac{dx}{4x^2 + 4x + 2} = \int \frac{dx}{(2x + 1)^2 + 1} = \frac{1}{2}\int \frac{du}{u^2 + 1}$$

$$= \frac{1}{2}\tan^{-1}u + C = \frac{1}{2}\tan^{-1}(2x + 1) + C.$$

Two Useful Formulas

Integrals of the form $\int du/(u^2 + a^2)$ and $\int du/\sqrt{a^2 - u^2}$ arise so often in applications of integration that many people find it saves time to memorize formulas for evaluating them.

$$\int \frac{du}{u^2 + a^2} = \frac{1}{a} \tan^{-1} \frac{u}{a} + C \qquad (8)$$

$$\int \frac{du}{\sqrt{a^2 - u^2}} = \sin^{-1} \frac{u}{a} + C \qquad (9)$$

We can derive Eq. (8) by substituting $u = a \tan \theta$ and Eq. (9) by substituting $u = a \sin \theta$ (Exercise 50).

Example 7

a) $\displaystyle\int \frac{dx}{(x + 1)^2 + 4} = \frac{1}{2} \tan^{-1} \frac{x + 1}{2} + C \qquad \left(\begin{matrix}\text{Eq. (8) with } u = x + 1, \\ du = dx, \, a = 2\end{matrix}\right)$

b) $\displaystyle\int \frac{dx}{\sqrt{3 - 4x^2}} = \frac{1}{2} \int \frac{du}{\sqrt{a^2 - u^2}} \qquad (u = 2x, \frac{du}{2} = dx, a = \sqrt{3})$

$\displaystyle\qquad = \frac{1}{2} \sin^{-1} \frac{u}{a} + C \qquad \text{(Eq. (9))}$

$\displaystyle\qquad = \frac{1}{2} \sin^{-1} \frac{2x}{\sqrt{3}} + C$

EXERCISES 7.4

Evaluate the integrals in Exercises 1–16.

1. $\displaystyle\int_{-2}^{2} \frac{dx}{4 + x^2}$

2. $\displaystyle\int_{0}^{2} \frac{dx}{8 + 2x^2}$

3. $\displaystyle\int_{0}^{3/2} \frac{dx}{\sqrt{9 - x^2}}$

4. $\displaystyle\int_{0}^{1/2\sqrt{2}} \frac{2 \, dx}{\sqrt{1 - 4x^2}}$

5. $\displaystyle\int \frac{dx}{\sqrt{x^2 - 4}}$

6. $\displaystyle\int \frac{3 \, dx}{\sqrt{9x^2 - 1}}$

7. $\displaystyle\int \frac{2 \, dt}{\sqrt{t} + 4t\sqrt{t}}$

8. $\displaystyle\int_{0}^{1/3} \frac{3 \, dy}{\sqrt{1 + 9y^2}}$

9. $\displaystyle\int_{0}^{3\sqrt{2}/4} \frac{dx}{\sqrt{9 - 4x^2}}$

10. $\displaystyle\int_{0}^{5} \sqrt{25 - \theta^2} \, d\theta$

11. $\displaystyle\int_{1/\sqrt{3}}^{1} \frac{2 \, dz}{z\sqrt{4z^2 - 1}}$

12. $\displaystyle\int_{8/\sqrt{3}}^{8} \frac{24 \, dx}{x\sqrt{x^2 - 16}}$

13. $\displaystyle\int \frac{dy}{y\sqrt{4 + \ln^2 y}}$

14. $\displaystyle\int_{0}^{1} \frac{x^3 \, dx}{\sqrt{x^2 + 1}}$

15. $\displaystyle\int_{1}^{2} \frac{6 \, dx}{\sqrt{4 - (x - 1)^2}}$

16. $\displaystyle\int \frac{\sqrt{1 - w^2}}{w^2} \, dw$

Evaluate the integrals in Exercises 17–40.

17. $\displaystyle\int_{1}^{3} \frac{dy}{y^2 - 2y + 5}$

18. $\displaystyle\int_{1}^{4} \frac{dy}{y^2 - 2y + 10}$

19. $\displaystyle\int_{1}^{3/2} \frac{(x - 1) \, dx}{\sqrt{2x - x^2}}$

20. $\displaystyle\int \frac{(x - 2) \, dx}{\sqrt{5 + 4x - x^2}}$

21. $\displaystyle\int \frac{dx}{\sqrt{x^2 - 2x}}$

22. $\displaystyle\int \frac{\cos t\, dt}{\sqrt{\sin^2 t + 2 \sin t}}$

23. $\displaystyle\int_{-2}^{2} \frac{(x+2)\, dx}{\sqrt{x^2 + 4x + 13}}$

24. $\displaystyle\int_{0}^{1} \frac{(1-x)\, dx}{\sqrt{8 + 2x - x^2}}$

25. $\displaystyle\int \frac{ds}{\sqrt{s^2 - 2s + 5}}$

26. $\displaystyle\int_{1}^{3} \frac{2z\, dz}{z^2 - 2z + 5}$

27. $\displaystyle\int \frac{3\, dx}{9x^2 - 6x + 5}$

28. $\displaystyle\int_{-1}^{0} \frac{6\, dt}{\sqrt{3 - 2t - t^2}}$

29. $\displaystyle\int \frac{dr}{\sqrt{r^2 - 2r - 3}}$

30. $\displaystyle\int_{\sqrt{2}-1}^{1} \frac{dx}{(x+1)\sqrt{x^2 + 2x}}$

31. $\displaystyle\int_{-2}^{3} \frac{2\, d\theta}{\theta^2 + 4\theta + 5}$

32. $\displaystyle\int_{1}^{5/2} \frac{v\, dv}{\sqrt{v^2 - 2v + 5}}$

33. $\displaystyle\int \frac{dx}{\sqrt{9x^2 - 6x + 5}}$

34. $\displaystyle\int \frac{3y\, dy}{9y^2 - 6y + 5}$

35. $\displaystyle\int_{-2}^{1} \frac{dx}{\sqrt{x^2 + 4x + 13}}$

36. $\displaystyle\int \frac{(z-1)\, dz}{\sqrt{z^2 - 4z + 3}}$

37. $\displaystyle\int_{5}^{6} \frac{dt}{\sqrt{t^2 - 2t - 8}}$

38. $\displaystyle\int \frac{x\, dx}{\sqrt{9x^2 - 6x + 5}}$

39. $\displaystyle\int \frac{r\, dr}{\sqrt{r^2 + 4r + 5}}$

40. $\displaystyle\int \frac{(2x+3)\, dx}{4x^2 + 4x + 5}$

Evaluate the integrals in Exercises 41 and 42.

41. $\displaystyle\int \frac{4x^2\, dx}{(1 - x^2)^{3/2}}$

42. $\displaystyle\int_{0}^{1} \frac{4\, dx}{(4 - x^2)^{3/2}}$

43. Find the area of the region cut from the first quadrant by the elliptical curve $3y = \sqrt{9 - x^2}$.

44. Find the average value of the function $f(x) = 4/(x^2 - 4x + 8)$ over the interval $[2, 4]$.

45. CALCULATOR   To three decimal places, find the length of the curve $y = x^2$, $0 \le x \le \sqrt{3}/2$.

46. CALCULATOR   To three decimal places, find the area of the region in Fig. 7.16.

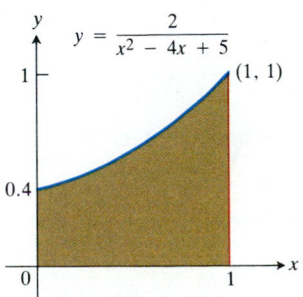

7.16   The region in Exercise 46.

47. CALCULATOR   To three decimal places, find the volume of the solid of revolution in Fig. 7.17.

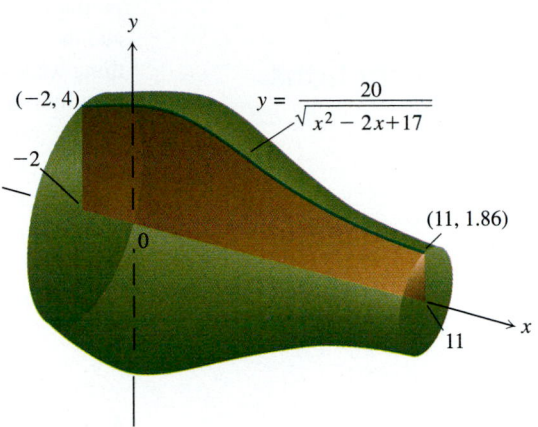

7.17   The solid in Exercise 47.

48. Find the area of the surface generated by revolving the arc in Fig. 7.18 about the x-axis.

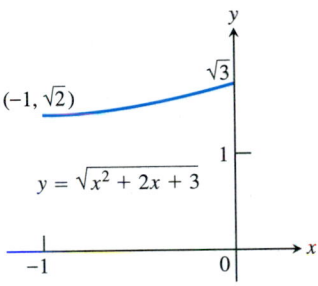

7.18   The arc in Exercise 48.

49. CALCULATOR   To two decimal places, find the coordinates of the centroid of the region in Fig. 7.19.

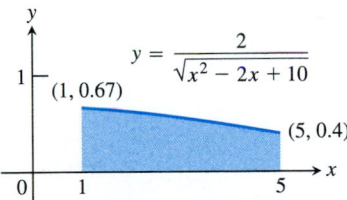

7.19   The region in Exercise 49.

50. Derive the formula in (a) Eq. (8); (b) Eq. (9).

Evaluate the limits in Exercises 51 and 52.

51. $\displaystyle\lim_{a \to 1^+} \int_{a}^{3} \frac{dx}{(x+1)\sqrt{x^2 + 2x - 3}}$

52. $\displaystyle\lim_{a \to -5^+} \int_{a}^{-4} \frac{dx}{\sqrt{-x^2 - 8x - 15}}$

7.5 Rational Functions and Partial Fractions

In this section, we describe the method of partial fractions, a powerful tool for integrating rational functions of the form $f(x)/g(x)$ where $f(x)$ and $g(x)$ are polynomials and the degree of f is less than the degree of g. Such a proper fraction may be something we know how to integrate, such as $1/(x + 2)$ or $2x/(x^2 + 4)$, but it may also be a more complicated expression. A theorem from advanced algebra (mentioned later in more detail) says that the more complicated expressions can all be replaced by sums of simpler fractions that we can integrate with techniques we already know. The first step toward integrating rational functions is therefore to find these replacement formulas. The examples in this section show how this is done.

Example 1   *The Kind of Addition We Want to Reverse.*

$$\frac{2}{x + 1} + \frac{3}{x - 3} = \frac{2(x - 3) + 3(x + 1)}{(x + 1)(x - 3)} = \frac{5x - 3}{x^2 - 2x - 3} \tag{1}$$

Adding the fractions on the left produces the fraction on the right. The reverse process consists of finding constants A and B such that

$$\frac{5x - 3}{x^2 - 2x - 3} = \frac{A}{x + 1} + \frac{B}{x - 3}. \tag{2}$$

(Pretend for a moment that we do not know that $A = 2$ and $B = 3$ will work.) We call the fractions $A/(x + 1)$ and $B/(x - 3)$ **partial fractions** because their denominators are only part of the original denominator $x^2 - 2x - 3$. We call A and B **undetermined coefficients** until proper values for them have been found.

To find A and B, we first clear Eq. (2) of fractions, obtaining

$$5x - 3 = A(x - 3) + B(x + 1) = (A + B)x - 3A + B.$$

This will be an identity in x if and only if the coefficients of like powers of x on the two sides are equal:

$$A + B = 5, \qquad -3A + B = -3.$$

These equations determine that the values of A and B are $A = 2$, $B = 3$.

To evaluate undetermined coefficients,

1. clear the equation of fractions;
2. equate the coefficients of like powers of x;
3. solve for the coefficients.

Example 2   *Two Linear Factors in the Denominator.*   Evaluate

$$\int \frac{5x - 3}{(x + 1)(x - 3)} \, dx.$$

Solution   From Example 1,

$$\int \frac{5x - 3}{(x + 1)(x - 3)} \, dx = \int \frac{2}{x + 1} \, dx + \int \frac{3}{x - 3} \, dx$$

$$= 2 \ln|x + 1| + 3 \ln|x - 3| + C.$$

Example 3   *A Repeated Linear Factor in the Denominator.*   Express

$$\frac{6x + 7}{(x + 2)^2}$$

as a sum of partial fractions.

Solution   Since the denominator has a repeated linear factor, $(x + 2)^2$, we must express the fraction in the form

$$\frac{6x + 7}{(x + 2)^2} = \frac{A}{x + 2} + \frac{B}{(x + 2)^2}.$$ (3)

Clearing Eq. (3) of fractions gives

$$6x + 7 = A(x + 2) + B = Ax + (2A + B).$$

Matching coefficients of like terms gives $A = 6$ and

$$7 = 2A + B = 12 + B, \quad \text{or} \quad B = -5.$$

Hence,

$$\frac{6x + 7}{(x + 2)^2} = \frac{6}{x + 2} - \frac{5}{(x + 2)^2}.$$

Example 4   *An Improper Fraction.*   **Express**

$$\frac{2x^3 - 4x^2 - x - 3}{x^2 - 2x - 3}$$

as a sum of partial fractions.

Solution   First we divide the denominator into the numerator to get a polynomial plus a proper fraction. Then we write the proper fraction as a sum of partial fractions. Long division gives

$$x^2 - 2x - 3 \overline{\smash{\big)}\, \begin{array}{l} 2x \\ 2x^3 - 4x^2 - x - 3 \\ \underline{2x^3 - 4x^2 - 6x} \\ 5x - 3 \end{array}}$$

Hence,

$$\frac{2x^3 - 4x^2 - x - 3}{x^2 - 2x - 3} = 2x + \frac{5x - 3}{x^2 - 2x - 3} \qquad \text{(Result of the division)}$$

$$= 2x + \frac{2}{x + 1} + \frac{3}{x - 3}. \qquad \left(\begin{array}{l} \text{Proper fraction expanded} \\ \text{as in Example 1} \end{array} \right)$$

Example 5   *An Irreducible Quadratic Factor in the Denominator.*
Express

$$\frac{-2x + 4}{(x^2 + 1)(x - 1)^2}$$

as a sum of partial fractions.

Solution   The denominator has an irreducible quadratic factor as well as a repeated linear factor, so we write

$$\frac{-2x + 4}{(x^2 + 1)(x - 1)^2} = \frac{Ax + B}{x^2 + 1} + \frac{C}{x - 1} + \frac{D}{(x - 1)^2}.$$ (4)

Notice the numerator over $x^2 + 1$: For quadratic factors, we use first-degree numerators, not constant numerators. Clearing the equation of fractions gives

$$- 2x + 4 = (Ax + B)(x - 1)^2 + C(x - 1)(x^2 + 1) + D(x^2 + 1)$$
$$= (A + C)x^3 + (-2A + B - C + D)x^2$$
$$+ (A - 2B + C)x + (B - C + D).$$

Equating coefficients of like terms gives

Coefficients of x^3: $\quad 0 = A + C$,

Coefficients of x^2: $\quad 0 = -2A + B - C + D$,

Coefficients of x^1: $\quad -2 = A - 2B + C$

Coefficients of x^0: $\quad 4 = B - C + D$.

We solve these equations simultaneously to find the values of A, B, C, and D:

$$-4 = -2A, \quad A = 2 \qquad \text{(Subtract fourth equation from second)}$$
$$C = -A = -2 \qquad \text{(From the first equation)}$$
$$B = 1 \qquad \text{($A = 2$ and $C = -2$ in third equation)}$$
$$D = 4 - B + C = 1. \qquad \text{(From the fourth equation)}$$

We substitute these values into Eq. (4), obtaining

$$\frac{-2x + 4}{(x^2 + 1)(x - 1)^2} = \frac{2x + 1}{x^2 + 1} - \frac{2}{x - 1} + \frac{1}{(x - 1)^2}.$$

Example 6 Evaluate $\displaystyle\int \frac{-2x + 4}{(x^2 + 1)(x - 1)^2} \, dx$.

Solution We expand the integrand by partial fractions, as in Example 5, and integrate the terms of the expansion:

$$\int \frac{-2x + 4}{(x^2 + 1)(x - 1)^2} \, dx = \int \left(\frac{2x + 1}{x^2 + 1} - \frac{2}{x - 1} + \frac{1}{(x - 1)^2} \right) dx \qquad \text{(Example 5)}$$

$$= \int \left(\frac{2x}{x^2 + 1} + \frac{1}{x^2 + 1} - \frac{2}{x - 1} + \frac{1}{(x - 1)^2} \right) dx$$

$$= \ln(x^2 + 1) + \tan^{-1}x - 2 \ln|x - 1| - \frac{1}{x - 1} + C.$$

General Description of the Method

Success in writing a rational function $f(x)/g(x)$ as a sum of partial fractions depends on two things:

1. *The degree of $f(x)$ must be less than the degree of $g(x)$.* (If it is not, divide and work with the remainder term.)

2. *We must know the factors of $g(x)$.* (In theory, any polynomial with real coefficients can be written as a product of real linear factors and real quadratic factors. In practice, the factors may be hard to find.)

A theorem from advanced algebra says that if these two conditions are met, we may write $f(x)/g(x)$ as a sum of partial fractions by taking the following steps.

STEP 1: Let $x - r$ be a linear factor of $g(x)$. Suppose $(x - r)^m$ is the highest power of $x - r$ that divides $g(x)$. Then assign the sum of m partial fractions to this factor, as follows:

$$\frac{A_1}{x - r} + \frac{A_2}{(x - r)^2} + \cdots + \frac{A_m}{(x - r)^m}.$$

Do this for each distinct linear factor of $g(x)$.

STEP 2: Let $x^2 + px + q$ be a quadratic factor of $g(x)$. Suppose

$$(x^2 + px + q)^n$$

is the highest power of this factor that divides $g(x)$. Then to this factor assign the sum of the n partial fractions:

$$\frac{B_1x + C_1}{x^2 + px + q} + \frac{B_2x + C_2}{(x^2 + px + q)^2} + \cdots + \frac{B_nx + C_n}{(x^2 + px + q)^n}.$$

Do this for each distinct quadratic factor of $g(x)$ that cannot be factored into linear factors with real coefficients.

STEP 3: Set the original fraction $f(x)/g(x)$ equal to the sum of all these partial fractions. Clear the resulting equation of fractions and arrange the terms in decreasing powers of x.

STEP 4: Equate the coefficients of corresponding powers of x and solve the resulting equations for the undetermined coefficients.

EXERCISES 7.5

Expand the quotients in Exercises 1–8 by partial fractions.

1. $\dfrac{5x - 13}{(x - 3)(x - 2)}$

2. $\dfrac{5x - 7}{x^2 - 3x + 2}$

3. $\dfrac{x + 4}{(x + 1)^2}$

4. $\dfrac{2x + 2}{x^2 - 2x + 1}$

5. $\dfrac{x + 1}{x^2(x - 1)}$

6. $\dfrac{z}{z^3 - z^2 - 6z}$

7. $\dfrac{x^2 + 8}{x^2 - 5x + 6}$

8. $\dfrac{x^4 + 9}{x^2(x^2 + 9)}$

Evaluate the integrals in Exercises 9–50.

9. $\displaystyle\int_0^{1/2} \frac{dx}{1 - x^2}$

10. $\displaystyle\int_1^2 \frac{dx}{x^2 + 2x}$

11. $\displaystyle\int \frac{x + 4}{x^2 + 5x - 6}\, dx$

12. $\displaystyle\int \frac{2x + 1}{x^2 - 7x + 12}\, dx$

13. $\displaystyle\int_4^8 \frac{y\, dy}{y^2 - 2y - 3}$

14. $\displaystyle\int \frac{y^3 + 4y^2}{y^3 + y}\, dy$

15. $\displaystyle\int \frac{dt}{t^3 + t^2 - 2t}$

16. $\displaystyle\int \frac{x + 3}{2x^3 - 8x}\, dx$

17. $\displaystyle\int \frac{x^3\, dx}{x^2 + 2x + 1}$

18. $\displaystyle\int_2^6 \frac{x^3\, dx}{x^2 - 2x + 1}$

19. $\displaystyle\int \frac{dx}{(x^2 - 1)^2}$

20. $\displaystyle\int \frac{x^2\, dx}{(x - 1)(x^2 + 2x + 1)}$

21. $\displaystyle\int_0^{\ln 2} \frac{e^t\, dt}{e^{2t} + 3e^t + 2}$

22. $\displaystyle\int \frac{(x + 1)^2 \tan^{-1}x + x^3 + x}{(x + 1)^2(x^2 + 1)}\, dx$

23. $\displaystyle\int \frac{\cos y\, dy}{\sin^2 y + \sin y - 6}$

24. $\displaystyle\int_{\pi/3}^{\pi/2} \frac{\sin \theta\, d\theta}{\cos^2\theta + \cos \theta - 2}$

25. $\displaystyle\int \frac{1 - \sqrt{x}}{1 + \sqrt{x}}\, dx$

26. $\displaystyle\int \frac{d\theta}{\sqrt{\theta} + \sqrt[3]{\theta}}$

27. $\displaystyle\int t \ln(t + 5)\, dt$

28. $\displaystyle\int_0^1 \ln(r^2 + 1)\, dr$

29. $\displaystyle\int_0^{2\sqrt{2}} \frac{x^3\, dx}{x^2 + 1}$

30. $\displaystyle\int_0^1 \frac{x^4 + 2x}{x^2 + 1}\, dx$

31. $\displaystyle\int_1^2 \frac{dy}{y^3 + y}$

32. $\displaystyle\int_0^1 \frac{dx}{(x + 1)(x^2 + 1)}$

33. $\displaystyle\int_0^{\sqrt{3}} \frac{5x^2\, dx}{x^2 + 1}$

34. $\displaystyle\int \frac{3t^2 + t + 4}{t^3 + t}\, dt$

35. $\displaystyle\int \frac{4x + 4}{x^2(x^2 + 1)}\, dx$

36. $\displaystyle\int_{-1}^0 \frac{x^3 - x}{(x^2 + 1)(x - 1)^2}\, dx$

37. $\displaystyle\int_0^1 \frac{x^3 + 1}{x^2 + 1}\, dx$

38. $\displaystyle\int_0^1 \frac{x^4 + 2x}{x^2 + 1}\, dx$

39. $\displaystyle\int_0^1 \frac{y^2 + 2y + 1}{(y^2 + 1)^2}\, dy$

40. $\displaystyle\int_{-1/2}^{1/2} \frac{8x^2 + 8x + 2}{(4x^2 + 1)^2}\, dx$

41. $\displaystyle\int_0^1 \frac{2r^3 + 3r^2 + 5r + 2}{r^2 + r + 1}\, dr$

42. $\displaystyle\int_0^1 \frac{dx}{(x^2 + 1)^2}$

43. $\displaystyle\int_{-1}^0 \frac{2x + 2}{(x^2 + 1)(x - 1)^3}\, dx$

44. $\displaystyle\int_1^2 \frac{3x^3 - 5x^2 - 4x + 3}{x^3(x^2 + x - 1)}\, dx$

45. $\displaystyle\int \frac{w - 1}{4w^2 - 4w - 1}\, dw$

46. $\displaystyle\int \frac{2x - 1}{4x^2 + 8x + 1}\, dx$

47. $\displaystyle\int \frac{3\theta^2 - \theta + 1}{\theta^3 - 1}\, d\theta$

48. $\displaystyle\int \frac{y^2 - 4}{y^3 - 5y + 2}\, dy$

49. $\displaystyle\int \frac{2x^3 + 5x^2 + 8x + 4}{(x^2 + 2x + 2)^2}\, dx$

50. $\displaystyle\int \frac{x^4 - 4x^3 + 2x^2 - 3x + 1}{x^6 + 3x^4 + 3x^2 + 1}\, dx$

51. Find the volume of the solid generated by revolving about the x-axis the region bounded by the x-axis, the lines $x = 0.5$ and $x = 2.5$, and the curve $y = 3/\sqrt{3x - x^2}$ (Fig. 7.20).

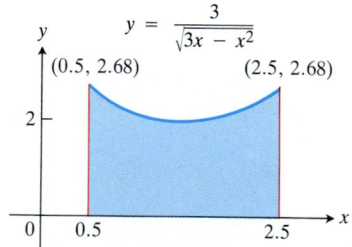

7.20   The region in Exercise 51.

52. Find the length of the curve $y = \ln(1 - x^2)$, $0 \le x \le 1/2$.

53. CALCULATOR Find, to two decimal places, the x-coordinate of the centroid of the region in the first quadrant bounded by the x-axis, the curve $y = \tan^{-1} x$, and the line $x = \sqrt{3}$.

54. CALCULATOR Find, to two decimal places, the x-coordinate of the centroid of the region in Fig. 7.21.

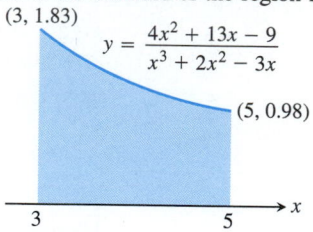

7.21   The region in Exercise 54.

55. CALCULATOR *Social diffusion.* Sociologists sometimes use the phrase "social diffusion" to describe the way information spreads through a population. The information might be a rumor, a cultural fad, or news about a technical innovation. In a sufficiently large population, the number of people x who have the information is treated as a differentiable function of time t, and the rate of diffusion, dx/dt, is assumed to be proportional to the number of people who have the information times the number of people who do not. This leads to the equation

$$\frac{dx}{dt} = kx(N - x),$$

where N is the number of people in the population.

Suppose t is measured in days, $k = 1/250$, and two people start a rumor at time $t = 0$ in a population of $N = 1000$ people.

a) Find x as a function of t.

b) When will half the population have heard the rumor? (This is when the rumor will be spreading the fastest.)

56. *Second-order chemical reactions.* Many chemical reactions are the result of the interaction of two molecules that undergo a change to produce a new product. The rate of the reaction typically depends on the concentrations of the two kinds of molecules. If a is the amount of substance A and b is the amount of substance B at time $t = 0$, and if x is the amount of product at time t, then the rate of formation of x may be given by the differential equation

$$\frac{dx}{dt} = k(a - x)(b - x),$$

or

$$\frac{1}{(a - x)(b - x)}\frac{dx}{dt} = k,$$

where k is a constant for the reaction. Integrate both sides of this equation to obtain a relation between x and t

a) if $a = b$, and

b) if $a \ne b$. Assume in each case that $x = 0$ when $t = 0$.

57. *An integral that connects π to its approximation 22/7.* Evaluate

$$\int_0^1 \frac{x^4(x - 1)^4}{x^2 + 1}\, dx.$$

58. *Autocatalytic reactions.* The equation that describes the autocatalytic reaction in Exercise 37 in Section 3.5 can be rewritten as

$$\frac{dx}{dt} = kx(x - a).$$

Read Exercise 37 for background (there is no need to do the exercise). Then solve the equation above to find x as a function of t. Assume that $x = x_0$ when $t = 0$.

7.6 Using Integral Tables; Reduction Formulas

The numbered integration formulas at the back of this book are stated in terms of constants a, b, c, m, n, and so on. These constants can usually assume any real value and need not be integers. Occasional limitations on their values are stated with the formulas. Formula 5 requires $n \neq -1$, for example, and Formula 11 requires $n \neq -2$.

The formulas also assume that the constants do not take on values that require dividing by zero or taking even roots of negative numbers. For example, Formula 8 assumes $a \neq 0$, and Formula 13(a) cannot be used unless b is negative.

The examples in this section show how the formulas are commonly used.

Example 1   Evaluate $\displaystyle\int x(2x + 5)^{-1}\, dx$.

Solution   We use Formula 8 (not 7, which requires $n \neq -1$):

$$\int x(ax + b)^{-1}\, dx = \frac{x}{a} - \frac{b}{a^2} \ln |ax + b| + C.$$

With $a = 2$ and $b = 5$, we have

$$\int x(2x + 5)^{-1}\, dx = \frac{x}{2} - \frac{5}{4} \ln |2x + 5| + C.$$

Example 2   Evaluate $\displaystyle\int \frac{dx}{x\sqrt{2x + 4}}$.

Solution   We use Formula 13(b):

$$\int \frac{dx}{x\sqrt{ax + b}} = \frac{1}{\sqrt{b}} \ln \left| \frac{\sqrt{ax + b} - \sqrt{b}}{\sqrt{ax + b} + \sqrt{b}} \right| + C, \qquad \text{if } b > 0.$$

With $a = 2$ and $b = 4$, we have

$$\int \frac{dx}{x\sqrt{2x + 4}} = \frac{1}{\sqrt{4}} \ln \left| \frac{\sqrt{2x + 4} - \sqrt{4}}{\sqrt{2x + 4} + \sqrt{4}} \right| + C$$

$$= \frac{1}{2} \ln \left| \frac{\sqrt{2x + 4} - 2}{\sqrt{2x + 4} + 2} \right| + C.$$

Formula 13(a), which requires $b < 0$, would not have been appropriate here. It *is* appropriate, however, in the next example.

Example 3 Evaluate $\displaystyle\int \frac{dx}{x\sqrt{2x-4}}$.

Solution We use Formula 13(a):

$$\int \frac{dx}{x\sqrt{ax+b}} = \frac{2}{\sqrt{-b}}\tan^{-1}\sqrt{\frac{ax+b}{-b}} + C, \quad \text{if } b < 0.$$

With $a = 2$ and $b = -4$, we have

$$\int \frac{dx}{x\sqrt{2x-4}} = \frac{2}{\sqrt{-(-4)}}\tan^{-1}\sqrt{\frac{2x-4}{-(-4)}} + C = \tan^{-1}\sqrt{\frac{x-2}{2}} + C.$$

Example 4 Evaluate $\displaystyle\int \frac{dx}{x^2\sqrt{2x-4}}$.

Solution We begin with Formula 15:

$$\int \frac{dx}{x^2\sqrt{ax+b}} = -\frac{\sqrt{ax+b}}{bx} - \frac{a}{2b}\int \frac{dx}{x\sqrt{ax+b}} + C.$$

With $a = 2$ and $b = -4$, we have

$$\int \frac{dx}{x^2\sqrt{2x-4}} = -\frac{\sqrt{2x-4}}{-4x} + \frac{2}{2\cdot4}\int \frac{dx}{x\sqrt{2x-4}} + C.$$

We then use Formula 13(a) to evaluate the integral on the right (Example 3) to obtain

$$\int \frac{dx}{x^2\sqrt{2x-4}} = \frac{\sqrt{2x-4}}{4x} + \frac{1}{4}\tan^{-1}\sqrt{\frac{x-2}{2}} + C.$$

Example 5 Evaluate $\displaystyle\int x\sin^{-1}x\, dx$.

Solution We use Formula 99:

$$\int x^n\sin^{-1}ax\, dx = \frac{x^{n+1}}{n+1}\sin^{-1}ax - \frac{a}{n+1}\int \frac{x^{n+1}\,dx}{\sqrt{1-a^2x^2}}, \quad n \neq -1.$$

With $n = 1$ and $a = 1$, we have

$$\int x\sin^{-1}x\, dx = \frac{x^2}{2}\sin^{-1}x - \frac{1}{2}\int \frac{x^2\,dx}{\sqrt{1-x^2}}.$$

The integral on the right is found in the table as Formula 33:

$$\int \frac{x^2}{\sqrt{a^2-x^2}}\, dx = \frac{a^2}{2}\sin^{-1}\frac{x}{a} - \frac{1}{2}x\sqrt{a^2-x^2} + C.$$

With $a = 1$,

$$\int \frac{x^2\,dx}{\sqrt{1-x^2}} = \frac{1}{2}\sin^{-1}x - \frac{1}{2}x\sqrt{1-x^2} + C.$$

The combined result is

$$\int x \sin^{-1}x \, dx = \frac{x^2}{2} \sin^{-1}x - \frac{1}{2}\left(\frac{1}{2} \sin^{-1}x - \frac{1}{2}x\sqrt{1 - x^2}\right) + C$$

$$= \left(\frac{x^2}{2} - \frac{1}{4}\right)\sin^{-1}x + \frac{1}{4}x\sqrt{1 - x^2} + C.$$

Example 6   The region in the first quadrant enclosed by the x-axis, the line $x = 1$, and the curve $y = \sin^{-1}x$ is revolved about the y-axis to generate a solid. Find the volume of the solid.

Solution   We sketch the region (Fig. 7.22) and decide on the method of cylindrical shells. The volume is

$$V = \int_0^1 2\pi(\text{radius})(\text{height}) \, dx$$

$$= \int_0^1 2\pi x \sin^{-1}x \, dx$$

$$= 2\pi\left[\left(\frac{x^2}{2} - \frac{1}{4}\right)\sin^{-1}x + \frac{1}{4}x\sqrt{1 - x^2}\right]_0^1 \quad \text{(Example 5)}$$

$$= 2\pi\left[\left(\frac{1}{4}\right)\left(\frac{\pi}{2}\right) + 0\right] - 2\pi[0 + 0] = \frac{\pi^2}{4}.$$

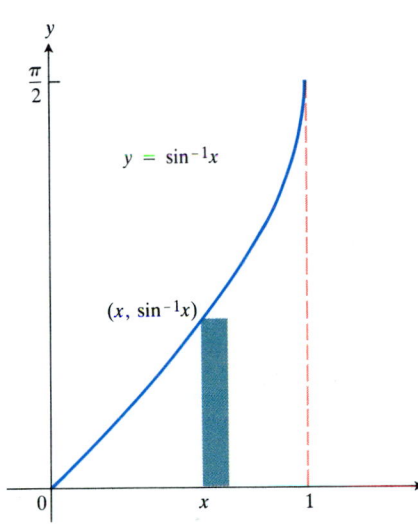

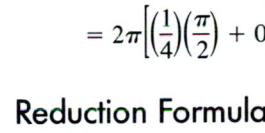

7.22   The region in Example 6.

Reduction Formulas

The time required to evaluate integrals by repeated applications of integration by parts can sometimes be shortened by using reduction formulas such as

$$\int (\ln x)^n \, dx = x(\ln x)^n - n\int (\ln x)^{n-1} \, dx, \tag{1}$$

$$\int \sec^n x \, dx = \frac{1}{n-1} \sec^{n-2}x \tan x - \frac{n-2}{n-1}\int \sec^{n-2}x \, dx, \tag{2}$$

$$\int \tan^n x \, dx = \frac{1}{n-1} \tan^{n-1}x - \int \tan^{n-2}x \, dx. \tag{3}$$

A **reduction formula** replaces an integral containing some power of a function with an integral of the same form in which the power has been decreased or reduced. By applying such a formula repeatedly, we can eventually express the original integral in terms of a power low enough to be evaluated directly.

Example 7   Evaluate $\int \tan^5 x \, dx$.

Solution   We apply Eq. (3) with $n = 5$ to get

$$\int \tan^5 x \, dx = \frac{1}{4} \tan^4 x - \int \tan^3 x \, dx.$$

We then apply Eq. (3) again, with $n = 3$, to evaluate the remaining integral:

$$\int \tan^3 x \, dx = \frac{1}{2} \tan^2 x - \int \tan x \, dx = \frac{1}{2} \tan^2 x + \ln|\cos x| + C.$$

The combined result is

$$\int \tan^5 x \, dx = \frac{1}{4} \tan^4 x - \frac{1}{2} \tan^2 x - \ln|\cos x| + C.$$

As their form suggests, reduction formulas are derived by integration by parts.

Example 8   Show that for any positive integer n,

$$\int (\ln x)^n \, dx = x(\ln x)^n - n \int (\ln x)^{n-1} \, dx.$$

Solution   We use the integration by parts formula

$$\int u \, dv = uv - \int v \, du$$

with

$$u = (\ln x)^n, \qquad du = n(\ln x)^{n-1} \frac{dx}{x}, \qquad dv = dx, \qquad v = x,$$

to obtain

$$\int (\ln x)^n \, dx = x(\ln x)^n - n \int (\ln x)^{n-1} \, dx.$$

Nonelementary Integrals

The development of computers and calculators that find antiderivatives by symbolic manipulation has led to a renewed interest in determining which antiderivatives can be expressed as finite combinations of elementary functions (the functions we have been studying) and which cannot. Integrals of functions that do not have elementary antiderivatives are called **nonelementary** integrals. They require infinite series (Chapter 8) or numerical methods for their evaluation. Examples of the latter include the error function

$$\operatorname{erf}(x) = \frac{2}{\sqrt{\pi}} \int_0^x e^{-t^2} \, dt$$

and integrals such as

$$\int \sin x^2 \, dx \qquad \text{and} \qquad \int \sqrt{1 + x^4} \, dx$$

that arise in engineering and physics. These and a number of others, such as

$$\int \frac{e^x}{x} \, dx, \qquad \int e^{(e^x)} \, dx, \qquad \int \frac{1}{\ln x} \, dx, \qquad \int \ln(\ln x) \, dx, \qquad \int \frac{\sin x}{x} \, dx,$$

$$\int \sqrt{1 - k^2 \sin^2 x} \, dx, \qquad 0 < k < 1,$$

look so easy they tempt us to try them just to see how they turn out. It can be proved, however, that there is no way to express these integrals as finite combinations of elementary functions. The same applies to integrals that can be changed into these by substitution. The integrands all have antiderivatives—they are, after all, continuous—but none of the antiderivatives is elementary.

None of the integrals you are asked to evaluate in the present chapter falls into this category, but you may encounter nonelementary integrals from time to time in your other work.

A General Procedure for Indefinite Integration

While there is no surefire way to evaluate all indefinite integrals, the procedure in Flowchart 7.2 may help.

FLOWCHART 7.2   Procedure for Indefinite Integration

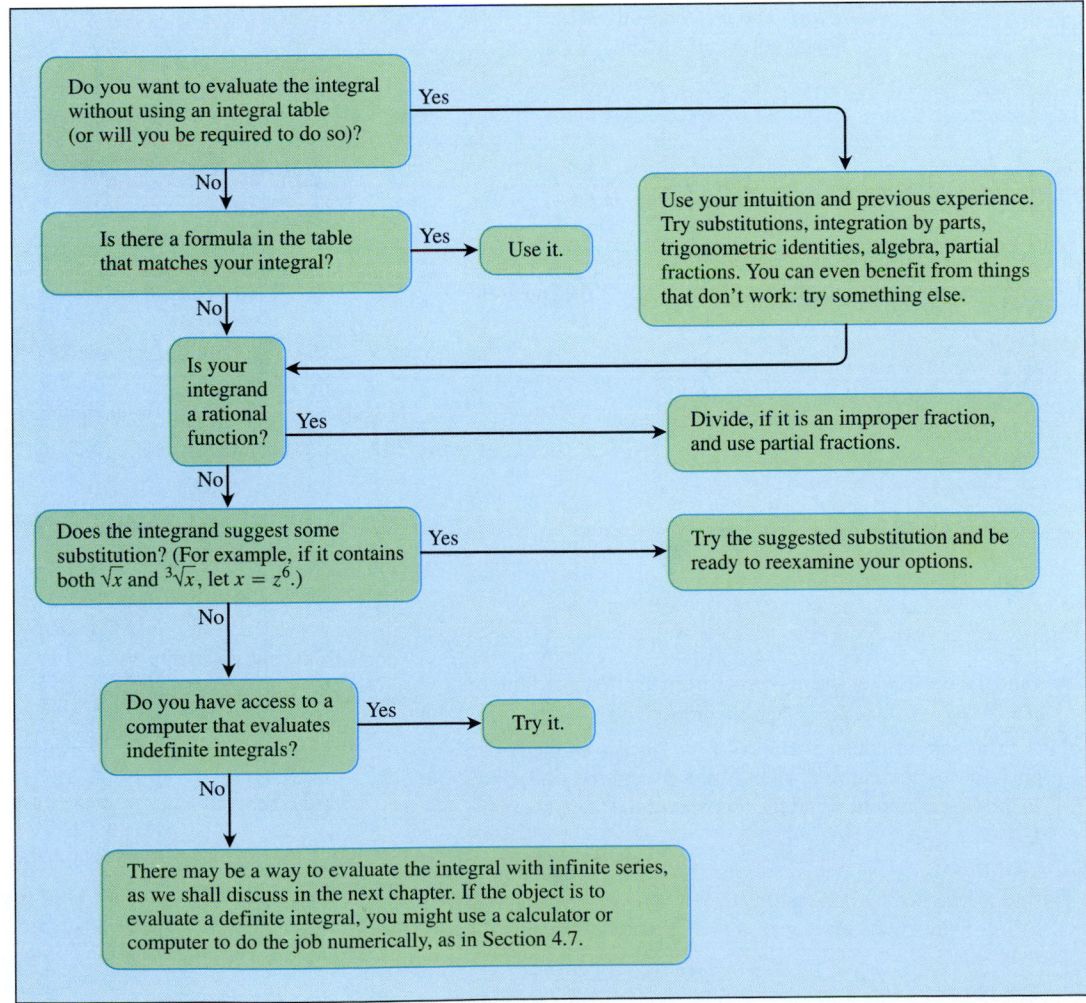

EXERCISES 7.6

Use the integral formulas at the back of this book to evaluate the integrals in Exercises 1–14.

1. $\displaystyle\int x \cos^{-1}x \, dx$

2. $\displaystyle\int_{6}^{9} \frac{dx}{x\sqrt{x-3}}$

3. $\displaystyle\int_{0}^{1/2} x \tan^{-1} 2x \, dx$

4. $\displaystyle\int \frac{dx}{(9-x^2)^2}$

5. $\displaystyle\int_{4}^{10} \frac{\sqrt{4x+9}}{x^2} \, dx$

6. $\displaystyle\int_{3}^{11} \frac{dx}{x^2\sqrt{7+x^2}}$

7. $\displaystyle\int \frac{dx}{x^2\sqrt{7-x^2}}$

8. $\displaystyle\int_{-2}^{-\sqrt{2}} \frac{\sqrt{x^2-2}}{x} \, dx$

9. $\displaystyle\int_{-\pi/12}^{\pi/4} \frac{dx}{5+4\sin 2x}$

10. $\displaystyle\int \frac{dx}{4+5\sin 2x}$

11. $\displaystyle\int_{3}^{6} \frac{x}{\sqrt{x-2}} \, dx$ (*Hint:* The n in Formula 7 need not be an integer.)

12. $\displaystyle\int x\sqrt{2x-3} \, dx$

13. $\displaystyle\int \frac{\sqrt{3x-4}}{x} \, dx$

14. $\displaystyle\int_{0}^{1} x^2\tan^{-1}x \, dx$

In Exercises 15–18, use the substitution to change the integral into one you can find in the tables. Then evaluate the integral.

15. $\displaystyle\int_{0}^{1} \sin^{-1}\sqrt{x} \, dx, \quad u = \sqrt{x}$

16. $\displaystyle\int_{3/4}^{1} \frac{\cos^{-1}\sqrt{x}}{\sqrt{x}} \, dx, \quad u = \sqrt{x}$

17. $\displaystyle\int_{0}^{1/2} \frac{\sqrt{x}}{\sqrt{1-x}} \, dx, \quad u = \sqrt{x}$

18. $\displaystyle\int_{\pi/4}^{\pi/2} \cot x\sqrt{1-\sin^2x} \, dx, \quad u = \sin x$

19. Find the centroid of the region cut from the first quadrant by the curve $y = 1/\sqrt{x+1}$ and the line $x = 3$.

20. A thin plate of constant density $\delta = 1$ occupies the region enclosed by the curve $y = 36/(2x+3)$ and the line $x = 3$ in the first quadrant. Find the moment of the plate about the y-axis.

21. CALCULATOR Use the integral table and a calculator to find to two decimal places the area of the surface generated by revolving the curve $y = x^2$, $-1 \le x \le 1$, about the x-axis.

22. The head of your firm's accounting department has asked you to find a formula she can use in a computer program to calculate the year-end inventory of gasoline in the company's tanks. A typical tank is shaped like a right circular cylinder of radius r and length L, mounted horizontally, as shown in Fig. 7.23. The data come to the accounting office as depth measurements taken with a vertical measuring stick marked in centimeters.

a) Show, in the notation of Fig. 7.23, that the volume of gasoline that fills the tank to a depth d is
$$V = 2L\int_{-r}^{-r+d} \sqrt{r^2-y^2} \, dy.$$

b) Evaluate the integral to produce the requested formula.

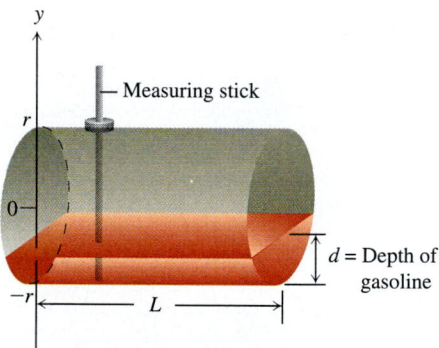

7.23   The tank in Exercise 22.

Exercises 23–26 refer to the formulas at the back of this book.

23. Verify Formula 55 for $x > 0$ by differentiating the right side.

24. Verify Formula 76 by differentiating the right side.

25. Verify Formula 9 by integrating
$$\int \frac{x}{(ax+b)^2} \, dx$$
with the substitution $u = ax + b$.

26. Verify Formula 46 by integrating
$$\int \frac{dx}{x^2\sqrt{x^2-a^2}}$$
with the substitution $x = a \sec u$.

Use reduction formulas to evaluate the integrals in Exercises 27–40.

27. $\displaystyle\int_{-\pi}^{\pi} \cos^4x \, dx$

28. $\displaystyle\int_{\pi/6}^{\pi/3} \cos^63x \, dx$

29. $\displaystyle\int_{0}^{\pi} \sin^4x \, dx$

30. $\displaystyle\int_{0}^{\pi/4} \sin^52x \, dx$

31. $\displaystyle\int_{0}^{\pi/8} \tan^3 2x \, dx$

32. $\displaystyle\int_{0}^{\pi/4} \tan^4\!\left(\frac{x}{2}\right) dx$

33. $\displaystyle\int_{\pi/4}^{3\pi/4} \cot^4 x \, dx$

34. $\displaystyle\int_{\pi/8}^{3\pi/8} \cot^4 2x \, dx$

35. $\displaystyle\int_{-\pi/3}^{\pi/3} \sec^4 x \, dx$

36. $\displaystyle\int_{0}^{\pi/4} \sec^5 x \, dx$

37. $\displaystyle\int_{\pi/4}^{\pi/2} \csc^4 x \, dx$

38. $\displaystyle\int_{\pi/4}^{\pi/2} \csc^5 x \, dx$

39. $\displaystyle\int_{1}^{e} 16x^3 (\ln x)^2 \, dx$

40. $\displaystyle\int_{1}^{e} (\ln x)^3 \, dx$

Evaluate the integrals in Exercises 41–44 by making a trigonometric substitution and then applying a reduction formula.

41. $\displaystyle\int_{0}^{1} (x^2 + 1)^{-3/2} \, dx$

42. $\displaystyle\int_{0}^{1} (x^2 + 1)^{3/2} \, dx$

43. $\displaystyle\int_{0}^{3/5} \frac{dx}{(1 - x^2)^3}$

44. $\displaystyle\int_{1}^{2} \frac{(x^2 - 1)^{3/2}}{x} \, dx$

45. Verify the formula

$$\int \sin^n x \, dx = -\frac{\sin^{n-1}x \cos x}{n} + \frac{n-1}{n} \int \sin^{n-2}x \, dx$$

by differentiating the right-hand side and combining the results to get $\sin^n x$.

46. Verify the formula

$$\int \csc^n x \, dx = -\frac{\csc^{n-2}x \cot x}{n-1} + \frac{n-2}{n-1} \int \csc^{n-2}x \, dx$$

by differentiating the right-hand side and combining the results to get $\csc^n x$.

47. a) Derive the formula

$$\int x^n \sin x \, dx = -x^n \cos x + n \int x^{n-1} \cos x \, dx.$$

   b) What is the corresponding formula for the integral of $x^n \sin ax$?

48. a) Derive the formula

$$\int x^n \cos x \, dx = x^n \sin x - n \int x^{n-1} \sin x \, dx.$$

   b) What is the corresponding formula for the integral of $x^n \cos ax$?

7.7 Improper Integrals

In many applications we want to integrate a function over an interval that is not closed. To do so, we first integrate the function over a closed interval inside the nonclosed interval; then we take the limit of this integral as the closed interval expands to fill the nonclosed interval. Figure 7.24 shows how to do this for a half-open interval $[a, c)$. It is conventional to call the resulting limit, whether it exists or not, the **improper integral** of the function over the nonclosed interval.

How can we tell if the improper integral exists? That is the basic question and the one we address in this section. Once the integral is known to exist, its value, if not immediately apparent, can be estimated by numerical methods.

Convergence and Divergence

In lifting-line theory in aerodynamics, the function

$$f(x) = \sqrt{\frac{1 + x}{1 - x}} \tag{1}$$

needs to be integrated over the interval from $x = 0$ to $x = 1$. The function is not defined at $x = 1$, although it is defined and continuous everywhere else in $[0, 1]$. To integrate f from 0 to 1, we integrate f from 0 to a positive number b less than 1 and take the limit of the resulting definite integral as b approaches 1. If the limit exists, we define the integral of f from 0 to 1 to be this value and write

$$\int_{0}^{1} \sqrt{\frac{1 + x}{1 - x}} \, dx = \lim_{b \to 1^-} \int_{0}^{b} \sqrt{\frac{1 + x}{1 - x}} \, dx. \tag{2}$$

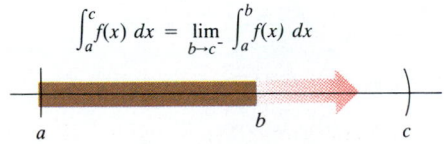

$$\int_{a}^{c} f(x) \, dx = \lim_{b \to c^-} \int_{a}^{b} f(x) \, dx$$

7.24 To integrate over a half-open interval $[a, c)$, we integrate over a closed interval $[a, b]$ inside $[a, c)$ and then take the limit as $[a, b]$ expands to fill $[a, c)$.

In this case, we say that the integral

$$\int_0^1 \sqrt{\frac{1+x}{1-x}}\, dx$$

converges and that the area under the curve $y = \sqrt{(1+x)/(1-x)}$ from 0 to 1 is the value of the integral. If the limit in (2) does not exist, we say that the integral **diverges.**

Example 1   Determine whether

$$\int_0^1 \sqrt{\frac{1+x}{1-x}}\, dx$$

converges.

Solution   See Fig. 7.25(a). Multiplying numerator and denominator by $\sqrt{1+x}$ gives

$$\int \sqrt{\frac{1+x}{1-x}}\, dx = \int \frac{1+x}{\sqrt{1-x^2}}\, dx = \int \frac{1}{\sqrt{1-x^2}}\, dx + \int \frac{x}{\sqrt{1-x^2}}\, dx$$

$$= \sin^{-1}x - \sqrt{1-x^2} + C.$$

Therefore,

$$\lim_{b \to 1^-} \int_0^b \sqrt{\frac{1+x}{1-x}}\, dx = \lim_{b \to 1^-} \left[\sin^{-1}x - \sqrt{1-x^2} \right]_0^b$$

$$= \lim_{b \to 1^-} [\sin^{-1}b - \sqrt{1-b^2} + 1] \qquad (3)$$

$$= \sin^{-1}1 - 0 + 1 = (\pi/2) + 1.$$

The integral converges to $(\pi/2) + 1$.

The preceding computation integrates vertical strips with respect to x. We can get the same result by integrating horizontal strips with respect to y (Fig. 7.25b). In

7.25   (a) To evaluate

$$\int_0^1 \sqrt{(1+x)/(1-x)}\, dx,$$

we evaluate

$$\int_0^b \sqrt{(1+x)/(1-x)}\, dx$$

and let b approach 1 from below. (b) If we treat the curve as the graph of the function $x = (y^2 - 1)/(y^2 + 1)$, we can make an equivalent calculation by integrating from 1 to c with respect to y and letting c approach infinity. We then add 1 to the result to include the area of the square (Example 1).

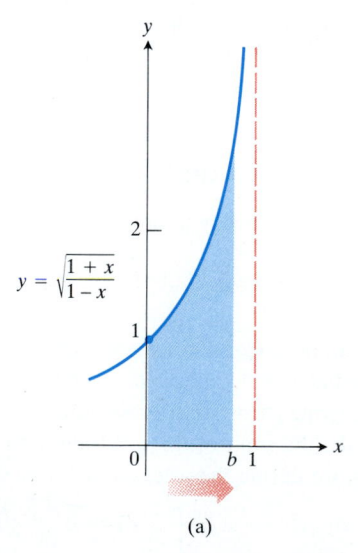

(a)

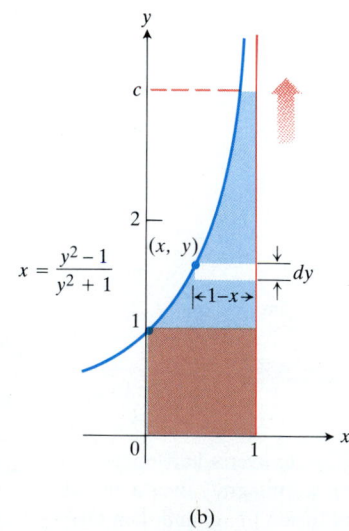

(b)

this case,

$$dA = (1 - x)\, dy = \frac{2}{y^2 + 1}\, dy, \qquad \left(\text{Using } x = \frac{y^2 - 1}{y^2 + 1}\right)$$

and the portion of the area above the shaded square is

$$\lim_{c \to \infty} \int_1^c \frac{2}{y^2 + 1}\, dy = \lim_{c \to \infty} \left[2 \tan^{-1} y\right]_1^c$$

$$= \lim_{c \to \infty} 2 \tan^{-1} c - 2 \cdot \frac{\pi}{4} = 2 \cdot \frac{\pi}{2} - \frac{\pi}{2} = \frac{\pi}{2}. \tag{4}$$

Including the shaded square, the area is $(\pi/2) + 1$, in agreement with our first calculation.

We call the limit in (4) the integral of $2/(y^2 + 1)$ from 1 to ∞:

$$\int_1^\infty \frac{2}{y^2 + 1}\, dy = \lim_{c \to \infty} \int_1^c \frac{2}{y^2 + 1}\, dy = \frac{\pi}{2}.$$

More generally, if f is integrable over every interval $[a, b]$ and $\lim_{b \to \infty} \int_a^b f(x)\, dx$ exists, we call the limit the integral of f from a to ∞:

$$\int_a^\infty f(x)\, dx = \lim_{b \to \infty} \int_a^b f(x)\, dx. \tag{5}$$

The notation

$$\int_a^b f(x)\, dx$$

for improper integrals is the same as the notation for definite integrals. In any given case it is usually a simple matter to tell whether a particular integral is to be calculated as an ordinary definite integral or as a limit. If a and b are finite and f is continuous at every point of $[a, b]$, the integral is an ordinary definite integral. If f becomes infinite at one or more points in the interval of integration, or one or both of the limits of integration are infinite, then the designated integral is improper and is to be calculated as a limit.

DEFINITION

$\int_a^b f(x)\, dx$ denotes an **improper integral** if

1. f becomes infinite at one or more points of the interval of integration and/or

2. one or both of the limits of integration is infinite.

Example 2   In the integral $$\int_0^1 \frac{dx}{x},$$ the function

$$f(x) = \frac{1}{x}$$

becomes infinite at $x = 0$. We cut off the point $x = 0$ and start our integration at

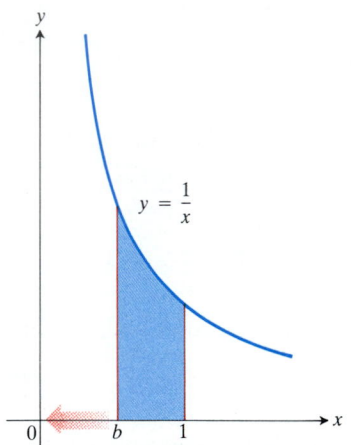

7.26   Example 2 shows that

$$\int_0^1 \frac{dx}{x} = \lim_{b \to 0^+} \int_b^1 \frac{dx}{x}$$

diverges.

some positive number $b < 1$ (Fig. 7.26):

$$\int_b^1 \frac{dx}{x} = \ln x \Big]_b^1 = \ln 1 - \ln b = \ln \frac{1}{b}.$$

We then investigate the behavior of the integral as b approaches 0 from the right. Since

$$\lim_{b \to 0^+} \int_b^1 \frac{dx}{x} = \lim_{b \to 0^+} \left(\ln \frac{1}{b} \right) = +\infty,$$

the integral from $x = 0$ to $x = 1$ diverges.

The method to be used when the function f becomes infinite at an interior point of the interval of integration is illustrated in the following example.

Example 3   In the integral $\displaystyle\int_0^3 \frac{dx}{(x-1)^{2/3}},$ the function

$$f(x) = \frac{1}{(x-1)^{2/3}}$$

becomes infinite at $x = 1$, which lies between the limits of integration 0 and 3. In such a case, we again cut out the point where $f(x)$ becomes infinite. This time we integrate from 0 to b, where b is slightly less than 1, and start again on the other side of 1 at c and integrate from c to 3 (Fig. 7.27). This gives two integrals to investigate:

$$\int_0^b \frac{dx}{(x-1)^{2/3}} \quad \text{and} \quad \int_c^3 \frac{dx}{(x-1)^{2/3}}.$$

If the first of these has a definite limit as $b \to 1^-$ and if the second also has a definite limit as $c \to 1^+$, then we say that the integral of f from 0 to 3 *converges* and that its value is

$$\int_0^3 \frac{dx}{(x-1)^{2/3}} = \lim_{b \to 1^-} \int_0^b \frac{dx}{(x-1)^{2/3}} + \lim_{c \to 1^+} \int_c^3 \frac{dx}{(x-1)^{2/3}}.$$

If either limit fails to exist, we say that the integral of f from zero to 3 *diverges*. For this example,

$$\lim_{b \to 1^-} \int_0^b (x-1)^{-2/3} dx = \lim_{b \to 1^-} [3(b-1)^{1/3} - 3(0-1)^{1/3}] = 3$$

and

$$\lim_{c \to 1^+} \int_c^3 (x-1)^{-2/3} dx = \lim_{c \to 1^+} [3(3-1)^{1/3} - 3(c-1)^{1/3}] = 3\sqrt[3]{2}.$$

Since both limits exist and are finite, the integral of f converges and its value is $3 + 3\sqrt[3]{2}$.

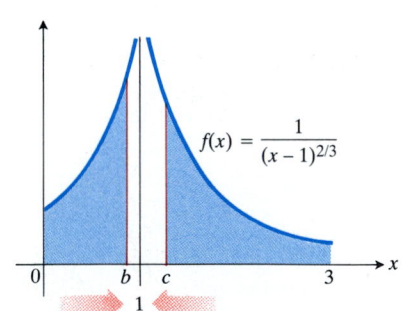

7.27   Example 3 shows that

$$\int_0^3 f(x)\, dx = \lim_{b \to 1^-} \int_0^b f(x)\, dx + \lim_{c \to 1^+} \int_c^3 f(x)\, dx$$

converges to $3 + 3\sqrt[3]{2}$.

Example 4   Each cross section of the solid infinite horn in Fig. 7.28, cut by a plane perpendicular to the x-axis for $-\infty < x \le \ln 2$, is a circular disk with one diameter reaching from the x-axis to the curve $y = e^x$. Find the volume of the horn.

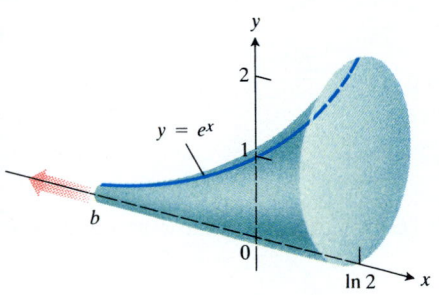

$y = e^x$

7.28   The calculation in Example 4 shows that this infinite solid horn has a finite volume.

Solution   The area of a typical cross section is

$$A(x) = \pi(\text{radius})^2 = \pi\left(\frac{1}{2}y\right)^2 = \frac{\pi}{4}e^{2x}.$$

We define the volume of the horn to be the limit as $b \to -\infty$ of the volume of the portion from b to $\ln 2$. As in Section 5.2 (the method of slicing), the volume of this portion is

$$V = \int_b^{\ln 2} A(x)\, dx = \int_b^{\ln 2} \frac{\pi}{4}e^{2x}\, dx = \frac{\pi}{8}e^{2x}\Big]_b^{\ln 2}$$

$$= \frac{\pi}{8}\left(e^{\ln 4} - e^{2b}\right) = \frac{\pi}{8}\left(4 - e^{2b}\right).$$

As $b \to -\infty$, $e^{2b} \to 0$ and $V \to (\pi/8)(4 - 0) = \pi/2$. The volume of the horn is $\pi/2$.

Example 5   Evaluate $\displaystyle\int_2^\infty \frac{x+3}{(x-1)(x^2+1)}\, dx.$

Solution

$$\int_2^\infty \frac{x+3}{(x-1)(x^2+1)}\, dx = \lim_{b\to\infty} \int_2^b \frac{x+3}{(x-1)(x^2+1)}\, dx$$

$$= \lim_{b\to\infty} \int_2^b \left(\frac{2}{x-1} - \frac{2x+1}{x^2+1}\right) dx \qquad \text{(Partial fractions)}$$

$$= \lim_{b\to\infty} \left[2\ln(x-1) - \ln(x^2+1) - \tan^{-1}x\right]_2^b$$

$$= \lim_{b\to\infty} \left[\ln\frac{(x-1)^2}{x^2+1} - \tan^{-1}x\right]_2^b \qquad \left(\begin{array}{l}\text{Combine the}\\\text{logarithms.}\end{array}\right)$$

$$= \lim_{b\to\infty} \left[\ln\left(\frac{(b-1)^2}{b^2+1}\right) - \tan^{-1}b\right] - \ln\left(\frac{1}{5}\right) + \tan^{-1}2$$

$$= 0 - \frac{\pi}{2} + \ln 5 + \tan^{-1}2 \approx 1.1458$$

Notice that we combined the logarithms in the antiderivative *before* we calculated the limit as $b \to \infty$. Had we not done so, we would have encountered the indeterminate form

$$\lim_{b\to\infty} (2\ln(b-1) - \ln(b^2+1)) = \infty - \infty.$$

The way to evaluate the indeterminate form, of course, is to combine the logarithms, so we would have arrived at the same answer in the end. But our original route was shorter.

The Integral $\displaystyle\int_1^\infty dx/x^p$

The convergence of the integral $\int_1^\infty dx/x^p$ depends on the value of the exponent p. The next example illustrates this with $p = 1$ and $p = 2$.

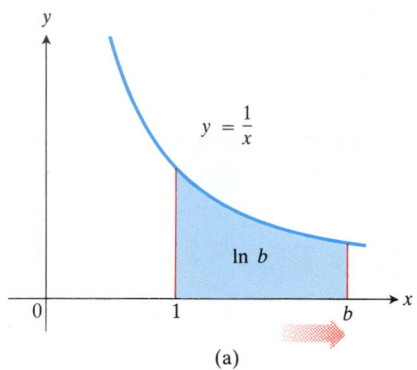

(a)

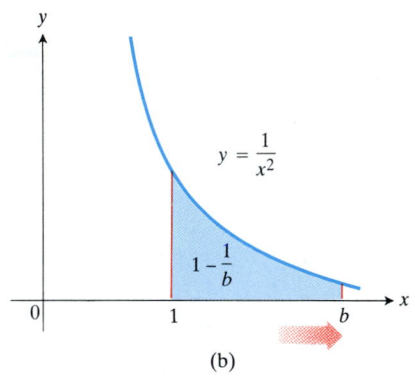

(b)

7.29   (a) The shaded area, $\ln b$, does not approach a finite limit as $b \rightarrow \infty$. (b) The shaded area, $1 - (1/b)$, approaches 1 as $b \rightarrow \infty$ (Example 6).

Example 6   Do the integrals

$$\int_1^{\infty} \frac{dx}{x} \quad \text{and} \quad \int_1^{\infty} \frac{dx}{x^2}$$

converge?

Solution   The curves $y = 1/x$ and $y = 1/x^2$ both approach the x-axis as $x \rightarrow \infty$ (Fig. 7.29). In the first case,

$$\int_1^b \frac{dx}{x} = \ln x \Big]_1^b = \ln b, \quad \text{so that} \quad \lim_{b \to \infty} \int_1^b \frac{dx}{x} = \lim_{b \to \infty} \ln b = \infty.$$

Therefore,

$$\int_1^{\infty} \frac{dx}{x} = \infty$$

and the integral *diverges*.

In the second case,

$$\int_1^b \frac{dx}{x^2} = -\frac{1}{x} \Big]_1^b = 1 - \frac{1}{b}, \quad \text{so that} \quad \int_1^b \frac{dx}{x^2} = \lim_{b \to \infty} \int_1^b \frac{dx}{x^2} = \lim_{b \to \infty} \left(1 - \frac{1}{b}\right) = 1.$$

The integral *converges* and its value is 1.

Generally, the integral $\int_1^{\infty} dx/x^p$ converges when $p > 1$ but diverges when $p \leq 1$ (Exercise 63).

Tests for Convergence and Divergence

The improper integrals that arise in practice usually cannot be evaluated directly, and we have to settle for the two-step procedure of first establishing the fact of convergence and then approximating the sum numerically. The principal methods of determining convergence are the domination and limit comparison tests.

Example 7   Determine whether the improper integral $\int_1^{\infty} e^{-x^2} dx$ converges.

Solution   Even though we cannot find any simpler expression for

$$I(b) = \int_1^b e^{-x^2} dx$$

(the integral is nonelementary), we can show that $I(b)$ has a finite limit as $b \rightarrow \infty$.

The function $I(b)$ represents the area between the x-axis and the curve $y = e^{-x^2}$ from $x = 1$ to $x = b$. It is an increasing function of b. Therefore, it either becomes infinite as $b \rightarrow \infty$ or has a finite limit as $b \rightarrow \infty$. It does not become infinite, because for every value of $b \geq 1$,

$$I(b) = \int_1^b e^{-x^2} dx \leq \int_1^b e^{-x} dx \quad \text{(Fig. 7.30)}$$

$$= e^{-1} - e^{-b} < e^{-1} < 0.37. \quad \text{(Calculator, rounded)}$$

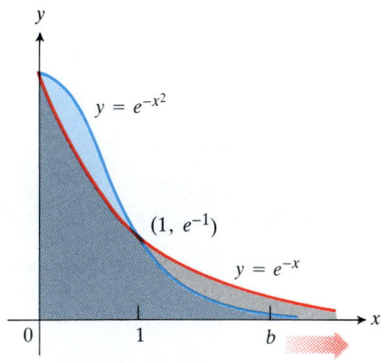

7.30   The graphs of $y = e^{-x^2}$ and $y = e^{-x}$ (Example 7).

Hence,

$$\int_1^\infty e^{-x^2}\, dx = \lim_{b\to\infty} \int_1^b e^{-x^2}\, dx$$

converges to a definite finite value. We have not calculated what the value is, but we know that it is less than 0.37.

A positive function f **dominates** a positive function g as $x \to \infty$ if $g(x) \le f(x)$ for all values of x beyond some point a. For instance, $f(x) = e^{-x}$ dominates $g(x) = e^{-x^2}$ as $x \to \infty$ because $e^{-x^2} \le e^{-x}$ for all $x > a = 1$.

If g is dominated by f as $x \to \infty$, then

$$\int_a^b g(x)\, dx \le \int_a^b f(x)\, dx, \qquad b > a,$$

and from this it can be argued as in Example 7 that

$$\int_a^\infty g(x)\, dx \quad \text{converges if} \quad \int_a^\infty f(x)\, dx \quad \text{converges.}$$

Turning this around says that

$$\int_a^\infty f(x)\, dx \quad \text{diverges if} \quad \int_a^\infty g(x)\, dx \quad \text{diverges.}$$

We state these results as a theorem and then give examples.

THEOREM 1

Domination Test for Convergence and Divergence of Improper Integrals

If f and g are integrable over every finite interval $[a, b]$ and if $0 \le g(x) \le f(x)$ for all $x > a$, then

1. $\displaystyle\int_a^\infty g(x)\, dx$ converges if $\displaystyle\int_a^\infty f(x)\, dx$ converges,

2. $\displaystyle\int_a^\infty f(x)\, dx$ diverges if $\displaystyle\int_a^\infty g(x)\, dx$ diverges.

Example 8

a) $\displaystyle\int_1^\infty \frac{1}{e^{2x}}\, dx$ converges because $\dfrac{1}{e^{2x}} < \dfrac{1}{e^x}$ and $\displaystyle\int_1^\infty \frac{1}{e^x}\, dx$ converges.

b) $\displaystyle\int_1^\infty \frac{1}{\sqrt{x}}\, dx$ diverges because $\dfrac{1}{\sqrt{x}} \ge \dfrac{1}{x}$ for $x \ge 1$ and $\displaystyle\int_1^\infty \frac{1}{x}\, dx$ diverges.

c) $\displaystyle\int_1^\infty \left(\frac{1}{x} + \frac{1}{x^2}\right) dx$ diverges because $\dfrac{1}{x} + \dfrac{1}{x^2} > \dfrac{1}{x}$ and $\displaystyle\int_1^\infty \frac{1}{x}\, dx$ diverges.

Another useful result, which we shall not prove, is the limit comparison test for the convergence and divergence of improper integrals. (You will find a similar result for infinite series in Chapter 8.) It goes like this:

THEOREM 2

Limit Comparison Test for Convergence and Divergence of Improper Integrals

Suppose that f and g are positive functions that are integrable over every finite interval $[a, b]$ and that

$$\lim_{x \to \infty} \frac{f(x)}{g(x)} = L, \qquad 0 < L < \infty.$$

Then $\displaystyle\int_a^\infty f(x)\, dx$ and $\displaystyle\int_a^\infty g(x)\, dx$ both converge or both diverge.

In the language of Section 6.6, Theorem 2 says that if two functions grow at the same rate as $x \to \infty$, then their integrals from a to ∞ behave alike: they both converge or both diverge. This does not mean that their integrals must have the same value, however, as the next example shows.

Example 9   Compare

$$\int_1^\infty \frac{dx}{x^2} \qquad \text{and} \qquad \int_1^\infty \frac{dx}{1 + x^2}$$

with the limit comparison test.

Solution   With

$$f(x) = \frac{1}{x^2} \qquad g(x) = \frac{1}{1 + x^2},$$

we have

$$\lim_{x \to \infty} \frac{1}{x^2} \frac{1 + x^2}{1} = \lim_{x \to \infty} \frac{1 + x^2}{x^2} = \lim_{x \to \infty} \left(\frac{1}{x^2} + 1 \right) = 0 + 1 = 1$$

as a positive finite limit. Therefore,

$$\int_1^\infty \frac{dx}{1 + x^2} \quad \text{converges because} \quad \int_1^\infty \frac{dx}{x^2} \quad \text{converges.}$$

The integrals converge to different values, however:

$$\int_1^\infty \frac{dx}{x^2} = 1, \qquad \text{(Example 6)}$$

and

$$\int_1^\infty \frac{dx}{1 + x^2} = \lim_{b \to \infty} \int_1^b \frac{dx}{1 + x^2} = \lim_{b \to \infty} [\tan^{-1} b - \tan^{-1} 1] = \frac{\pi}{2} - \frac{\pi}{4} = \frac{\pi}{4}.$$

Example 10

$$\int_1^\infty \frac{3}{e^x + 5} dx \quad \text{converges because} \quad \int_1^\infty \frac{1}{e^x} dx \quad \text{converges}$$

and

$$\lim_{x \to \infty} \frac{1}{e^x} \cdot \frac{e^x + 5}{3} = \lim_{x \to \infty} \frac{e^x + 5}{3e^x} = \lim_{x \to \infty} \left(\frac{1}{3} + \frac{5}{3e^x} \right) = \frac{1}{3} + 0 = \frac{1}{3},$$

a positive finite limit. As far as the convergence of the improper integral is concerned, $3/(e^x + 5)$ behaves like $1/e^x$.

In the mathematics underlying studies of light, electricity, and sound, we sometimes encounter integrals that have two infinite limits of integration instead of just one. The next definition addresses the convergence of such integrals.

DEFINITION

If $\displaystyle\int_{-\infty}^{a} f(x)\, dx$ and $\displaystyle\int_{a}^{\infty} f(x)\, dx$ both converge, we say that $\displaystyle\int_{-\infty}^{\infty} f(x)\, dx$ **converges** and we define its value to be

$$\int_{-\infty}^{\infty} f(x)\, dx = \int_{-\infty}^{a} f(x)\, dx + \int_{a}^{\infty} f(x)\, dx. \tag{6}$$

If either or both of the integrals on the right diverge, we say that $\displaystyle\int_{-\infty}^{\infty} f(x)\, dx$ **diverges**.

The choice of a in Eq. (6) in unimportant (Exercise 71) and we can evaluate or determine the convergence of $\int_{-\infty}^{\infty} f(x)\, dx$ with any convenient choice.

The integral of f from $-\infty$ to ∞ is not necessarily the same as $\lim_{b \to \infty} \int_{-b}^{b} f(x)\, dx$, which may exist even if $\int_{-\infty}^{\infty} f(x)\, dx$ does not converge (Exercise 81).

Example 11

$$\int_{-\infty}^{\infty} \frac{dx}{1 + x^2} = \int_{-\infty}^{0} \frac{dx}{1 + x^2} + \int_{0}^{\infty} \frac{dx}{1 + x^2} \qquad \text{(Eq. (6) with } a = 0\text{)}$$

$$= \lim_{b \to -\infty} \left[\tan^{-1} x \right]_{b}^{0} + \lim_{c \to \infty} \left[\tan^{-1} x \right]_{0}^{c}$$

$$= \lim_{b \to -\infty} \left[\tan^{-1} 0 - \tan^{-1} b \right] + \lim_{c \to \infty} \left[\tan^{-1} c - \tan^{-1} 0 \right]$$

$$= 0 - \left(-\frac{\pi}{2} \right) + \frac{\pi}{2} - 0 = \pi.$$

We interpret the value of this integral as the area of the infinite region between the curve $y = 1/(1 + x^2)$ and the x-axis.

Concluding Remarks

We know that

$$\int_{1}^{\infty} \frac{1}{x^2}\, dx$$

converges, but what about integrals like

$$\int_{2}^{\infty} \frac{1}{x^2}\, dx \qquad \text{and} \qquad \int_{100}^{\infty} \frac{1}{x^2}\, dx ?$$

The answer is that they converge, too. The existence of the limits

$$\lim_{b \to \infty} \int_2^b \frac{1}{x^2} \, dx \qquad \text{and} \qquad \lim_{b \to \infty} \int_{100}^b \frac{1}{x^2} \, dx$$

does not depend on the starting points $a = 2$ and $a = 100$, but only on the values of $1/x^2$ as x approaches ∞. Indeed, we find for any positive a that

$$\lim_{b \to \infty} \int_a^b \frac{1}{x^2} \, dx = \lim_{b \to \infty} \left(-\frac{1}{b} + \frac{1}{a} \right) = \frac{1}{a}.$$

The *value* of the limit depends on the value of a, but the *existence* of the limit does not. The convergence or divergence of integrals that are improper only because one limit of integration is infinite never depends on the other limit of integration.

Our final example shows how an improper integral may diverge without becoming infinite.

Example 12   The integral

$$\int_0^b \cos x \, dx = \sin b$$

takes all values between -1 and $+1$ as b varies between $2n\pi - \pi/2$ and $2n\pi + \pi/2$, where n is any integer. Hence,

$$\lim_{b \to \infty} \int_0^b \cos x \, dx$$

does not exist and $\int_0^\infty \cos x \, dx$ diverges. We might say that the integral "diverges by oscillation."

EXERCISES 7.7

Evaluate the integrals in Exercises 1–18.

1. $\displaystyle\int_0^\infty \frac{dx}{x^2 + 1}$

2. $\displaystyle\int_0^1 \frac{dx}{\sqrt{x}}$

3. $\displaystyle\int_{-1}^1 \frac{dx}{x^{2/3}}$

4. $\displaystyle\int_1^\infty \frac{dx}{x^{1.001}}$

5. $\displaystyle\int_0^4 \frac{dx}{\sqrt{4 - x}}$

6. $\displaystyle\int_0^1 \frac{dx}{\sqrt{1 - x^2}}$

7. $\displaystyle\int_0^1 \frac{dr}{r^{0.999}}$

8. $\displaystyle\int_{-\infty}^{-2} \frac{2dx}{x^2 - 1}$

9. $\displaystyle\int_2^\infty \frac{2}{v^2 - v} \, dv$

10. $\displaystyle\int_0^\infty \frac{dx}{(1 + x)\sqrt{x}}$

11. $\displaystyle\int_0^{\ln 2} \frac{\sinh x}{\cosh x - 1} \, dx$

12. $\displaystyle\int_{-\infty}^\infty \operatorname{sech} x \, dx$

13. $\displaystyle\int_0^1 \frac{\theta + 1}{\sqrt{\theta^2 + 2\theta}} \, d\theta$

14. $\displaystyle\int_0^2 \frac{s + 1}{\sqrt{4 - s^2}} \, ds$

15. $\displaystyle\int_0^1 \ln x \, dx$

16. $\displaystyle\int_0^1 x \ln x \, dx$

17. $\displaystyle\int_0^\infty x^4 e^{-x} \, dx$

18. $\displaystyle\int_0^\infty 2e^{-\theta} \sin \theta \, d\theta$

In Exercises 19–50, use direct integration, the domination test, or the limit comparison test, as appropriate, to determine whether the integrals diverge or converge. If more than one method applies, use whatever method you prefer.

19. $\displaystyle\int_1^\infty \frac{dx}{x^3 + 1}$

20. $\displaystyle\int_0^\infty \frac{dx}{x^3}$

21. $\displaystyle\int_0^\infty \frac{dx}{x^{3/2} + 1}$

22. $\displaystyle\int_0^\infty \frac{dx}{1 + e^x}$

23. $\displaystyle\int_0^{\pi/2} \tan \theta \, d\theta$

24. $\displaystyle\int_{-1}^1 \frac{dx}{x^2}$

25. $\displaystyle\int_{-1}^{1} \frac{dx}{x^{2/5}}$

26. $\displaystyle\int_{0}^{\infty} \frac{dt}{\sqrt{t}}$

27. $\displaystyle\int_{2}^{\infty} \frac{dv}{\sqrt{v-1}}$

28. $\displaystyle\int_{1}^{\infty} \frac{5}{s} \, ds$

29. $\displaystyle\int_{0}^{2} \frac{dx}{1-x^2}$

30. $\displaystyle\int_{2}^{\infty} \frac{dx}{(x+1)^2}$

31. $\displaystyle\int_{1}^{\infty} \frac{dr}{(r-2)^2}$

32. $\displaystyle\int_{1}^{3} \frac{2 \, dx}{\sqrt{|x-2|}}$

33. $\displaystyle\int_{0}^{\infty} \frac{dx}{\sqrt{x^6+1}}$

34. $\displaystyle\int_{-1}^{1} \frac{dx}{\sqrt[3]{x}}$

35. $\displaystyle\int_{0}^{\pi} \frac{\sin\theta}{\sqrt{\pi-\theta}} \, d\theta$

36. $\displaystyle\int_{1}^{\infty} \frac{\sqrt{x+1}}{x^2} \, dx$

37. $\displaystyle\int_{\pi}^{\infty} \frac{2+\cos x}{x} \, dx$

38. $\displaystyle\int_{1}^{\infty} \frac{\ln u}{u} \, du$

39. $\displaystyle\int_{6}^{\infty} \frac{1}{\sqrt{\theta+5}} \, d\theta$

40. $\displaystyle\int_{1}^{\infty} \frac{dx}{\sqrt{2x+10}}$

41. $\displaystyle\int_{2}^{\infty} \frac{2}{t^2-1} \, dt$

42. $\displaystyle\int_{1}^{\infty} \frac{1}{e^{\ln x}} \, dx$

43. $\displaystyle\int_{2}^{\infty} \frac{1}{\ln x} \, dx$

44. $\displaystyle\int_{1}^{\infty} \frac{1}{\sqrt{e^x - x}} \, dx$

45. $\displaystyle\int_{1}^{\infty} \frac{1}{e^x - 2^x} \, dx$

46. $\displaystyle\int_{2}^{\infty} \frac{1}{y^3 - 5} \, dy$

47. $\displaystyle\int_{0}^{\infty} \frac{dx}{\sqrt{x + x^4}}$

(*Hint:* Compare the integral with $\int dx/\sqrt{x}$ for x near zero and with $\int dx/x^2$ for large x.)

48. $\displaystyle\int_{0}^{\infty} \frac{2x-3}{(x^2 - 3x + 2)^{4/3}} \, dx$

49. $\displaystyle\int_{1}^{\infty} \frac{d\theta}{\theta^2(|\sin\theta| + |\cos\theta|)}$

(*Hint:* $|\sin\theta| + |\cos\theta| \geq \sin^2\theta + \cos^2\theta$.)

50. $\displaystyle\int_{0}^{\infty} \frac{dr}{2r^5 + 7r^4 + r^3 + 5r^2 + 2r + 1}$

Evaluate the integrals in Exercises 51–60.

51. $\displaystyle\int_{-1}^{\infty} \frac{dx}{x^2 + 5x + 6}$

52. $\displaystyle\int_{0}^{\infty} \frac{dx}{(x+1)(x^2+1)}$

53. $\displaystyle\int_{0}^{\infty} \frac{dy}{(y^2+1)^2}$

54. $\displaystyle\int_{-1}^{\infty} \frac{v^2 - 2v + 1}{(1+v^2)^2} \, dv$

55. $\displaystyle\int_{3}^{\infty} \frac{x+3}{2x^3 - 8x} \, dx$

56. $\displaystyle\int_{2}^{\infty} \frac{2-x^2}{2x(x^2-1)} \, dx$

57. $\displaystyle\int_{0}^{1} \frac{dx}{\sqrt{x^2 + 2x}}$

58. $\displaystyle\int_{1}^{\infty} \frac{1}{x\sqrt{x^2-1}} \, dx$

59. $\displaystyle\int_{0}^{4} \frac{1-x}{\sqrt{8 + 2x - x^2}} \, dx$

60. $\displaystyle\int_{-2}^{-1} \frac{dx}{\sqrt{-x^2 - 4x - 3}}$

61. *Estimating the value of a convergent improper integral whose domain is infinite.* Show that

$$\int_{3}^{\infty} e^{-3x} \, dx = \frac{1}{3} e^{-9} < 0.000042,$$

and hence that $\int_{3}^{\infty} e^{-x^2} \, dx < 0.000042$. Therefore, $\int_{0}^{\infty} e^{-x^2} \, dx$ can be replaced by $\int_{0}^{3} e^{-x^2} \, dx$ without introducing an error of magnitude greater than 0.000042. Evaluate this last integral numerically.

62. *The infinite paint can or Gabriel's horn*
As Example 6 shows, the integral $\int_{1}^{\infty} (dx/x)$ diverges. This means that the integral

$$\int_{1}^{\infty} 2\pi \frac{1}{x}\sqrt{1 + \frac{1}{x^4}} \, dx,$$

which measures the *surface area* of the solid of revolution traced out by revolving the curve $y = 1/x$, $1 \leq x$, about the x-axis, diverges also. By comparing the two integrals, we see that, for every finite value $b > 1$,

$$\int_{1}^{b} 2\pi \frac{1}{x}\sqrt{1 + \frac{1}{x^4}} \, dx > \int_{1}^{b} \frac{1}{x} \, dx.$$

However, the integral

$$\int_{1}^{\infty} \pi \left(\frac{1}{x}\right)^2 dx$$

for the *volume* of the solid converges. Calculate it. This solid of revolution is sometimes described as a can that does not hold enough paint to cover its own exterior (Fig. 7.31).

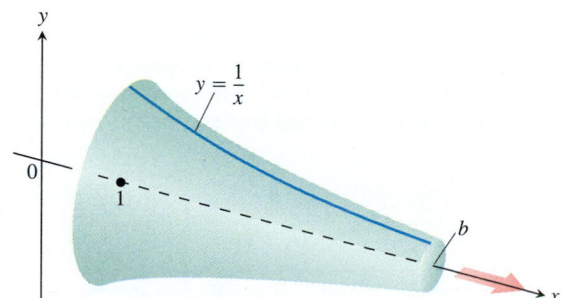

7.31 Extending this solid indefinitely to the right produces an "infinite" solid with a finite volume but an infinite surface area. To learn why, read Exercise 62.

63. Show that

$$\int_1^\infty \frac{dx}{x^p} = \frac{1}{p-1} \quad \text{when} \quad p > 1$$

but that the integral is infinite when $p < 1$. Example 6 shows what happens when $p = 1$.

64. Find the values of p for which each integral converges:

a) $\int_1^2 \frac{dx}{x(\ln x)^p}$,   b) $\int_2^\infty \frac{dx}{x(\ln x)^p}$.

Exercises 65–68 are about the infinite region in the first quadrant between the curve $y = e^{-x}$ and the x-axis.

65. Find the area of the region.

66. Find the centroid of the region.

67. Find the volume swept out by revolving the region about the y-axis.

68. Find the volume swept out by revolving the region about the x-axis.

69. Find the area of the region that lies between the curves $y = \sec x$ and $y = \tan x$ for $0 \le x \le \pi/2$.

70. Show that the area of the region between the curve $y = 1/(1 + x^2)$ and the entire x-axis is the same as the area of the unit disk, $x^2 + y^2 \le 1$.

71. Show that if $f(x)$ is integrable on every interval of real numbers and a and b are real numbers with $a < b$, then
a) $\int_{-\infty}^a f(x)\,dx$ and $\int_a^\infty f(x)\,dx$ both converge if and only if $\int_{-\infty}^b f(x)\,dx$ and $\int_b^\infty f(x)\,dx$ both converge.
b) $\int_{-\infty}^a f(x)\,dx + \int_a^\infty f(x)\,dx = \int_{-\infty}^b f(x)\,dx + \int_b^\infty f(x)\,dx$ when the integrals involved converge.

72. a) Show that if f is even and the necessary integrals exist, then

$$\int_{-\infty}^\infty f(x)\,dx = 2\int_0^\infty f(x)\,dx.$$

b) Show that if f is odd and the necessary integrals exist, then

$$\int_{-\infty}^\infty f(x)\,dx = 0.$$

Use direct evaluation, the domination and limit comparison tests, and the results in Exercise 72, as appropriate, to determine the convergence or divergence of the integrals in Exercises 73–80. If more than one method applies, use whatever method you prefer.

73. $\int_{-\infty}^\infty \frac{dx}{\sqrt{x^2 + 1}}$ **74.** $\int_{-\infty}^\infty \frac{dx}{\sqrt{x^6 + 1}}$

75. $\int_{-\infty}^\infty \frac{dx}{e^x + e^{-x}}$ **76.** $\int_{-\infty}^\infty \frac{e^{-x}\,dx}{x^2 + 1}$

77. $\int_{-\infty}^\infty e^{-|x|}\,dx$ **78.** $\int_{-\infty}^\infty \frac{dx}{(x + 1)^2}$

79. $\int_{-\infty}^\infty \frac{|\sin x| + |\cos x|}{|x| + 1}\,dx$ **80.** $\int_{-\infty}^\infty \frac{x\,dx}{(x^2 + 1)(x^2 + 2)}$

(*Hint:* $|\sin \theta| + |\cos \theta| \ge \sin^2 \theta + \cos^2 \theta$.)

81. $\int_{-\infty}^\infty f(x)\,dx$ **may not equal** $\lim_{b\to\infty} \int_{-b}^b f(x)\,dx$. Show that

$$\int_0^\infty \frac{2x\,dx}{x^2 + 1}$$

diverges and hence that

$$\int_{-\infty}^\infty \frac{2x\,dx}{x^2 + 1}$$

diverges. Then show that

$$\lim_{b\to\infty} \int_{-b}^b \frac{2x\,dx}{x^2 + 1} = 0.$$

82. Here is an argument that ln 3 equals $\infty - \infty$. Where does the argument go wrong? Explain.

$$\ln 3 = \ln 1 + \ln 3 = \ln 1 - \ln \frac{1}{3}$$

$$= \lim_{b\to\infty} \ln\left(\frac{b-2}{b}\right) - \ln\frac{1}{3}$$

$$= \lim_{b\to\infty} \left[\ln \frac{x-2}{x}\right]_3^b$$

$$= \lim_{b\to\infty} \left[\ln (x - 2) - \ln x\right]_3^b$$

$$= \lim_{b\to\infty} \int_3^b \left(\frac{1}{x-2} - \frac{1}{x}\right)dx$$

$$= \int_3^\infty \left(\frac{1}{x-2} - \frac{1}{x}\right)dx$$

$$= \int_3^\infty \frac{1}{x-2}\,dx - \int_3^\infty \frac{1}{x}\,dx$$

$$= \lim_{b\to\infty} \left[\ln(x - 2)\right]_3^b - \lim_{b\to\infty} \left[\ln x\right]_3^b$$

$$= \infty - \infty.$$

83. Find
a) $\displaystyle\lim_{b\to\infty} \int_{-b}^b \sin x\,dx$,

b) $\displaystyle\lim_{x\to 0^+} x \int_x^1 \frac{\cos t}{t^2}\,dt$

84. For what value or values of a does

$$\int_1^\infty \left(\frac{ax}{x^2 + 1} - \frac{1}{2x}\right)dx$$

converge? Evaluate the corresponding integral(s).

REVIEW QUESTIONS

1. What are the general methods for finding indefinite integrals?

2. What substitution(s) would you consider if the integrand contained the following terms?

a) $\sqrt{x^2 + 9}$ b) $\sqrt{x^2 - 9}$

c) $\sqrt{9 - x^2}$ d) $\sin^3 x \cos^2 x$

e) $\sin^2 x \cos^2 x$

3. What method(s) would you try if the integrand contained the following terms?

a) $\sin^{-1} x$ b) $\ln x$

c) $\sqrt{1 + 2x - x^2}$ d) $x \sin x$

e) $\dfrac{2x + 3}{x^2 - 5x + 6}$ f) $\sin 5x \cos 3x$

g) $\dfrac{1 - \sqrt{x}}{1 + \sqrt[4]{x}}$ h) $x\sqrt{2x + 3}$

4. What is a reduction formula? How are reduction formulas used? Illustrate by using the formula

$$\int \tan^n ax = \frac{\tan^{n-1} ax}{a(n - 1)} - \int \tan^{n-2} ax\, dx$$

to evaluate

$$\int_0^{\pi/2} \tan^3\left(\frac{x}{2}\right) dx.$$

5. Discuss two types of improper integrals. Define convergence and divergence of each type. How do you test for convergence and divergence of improper integrals? Give examples.

6. How is $\int_{-\infty}^{\infty} f(x)\, dx$ defined? Is it the same as $\lim_{b \to \infty} \int_{-b}^{b} f(x)\, dx$? Give examples.

MISCELLANEOUS EXERCISES

Evaluate the following integrals.

1. $\displaystyle\int \frac{\cos x\, dx}{\sqrt{1 + \sin x}}$

2. $\displaystyle\int_0^{\sqrt{2}/2} \frac{\sin^{-1} x\, dx}{\sqrt{1 - x^2}}$

3. $\displaystyle\int \frac{\tan x\, dx}{\cos^2 x}$

4. $\displaystyle\int \frac{y}{y^4 + 1}\, dy$

5. $\displaystyle\int_0^9 e^{\ln \sqrt{x}}\, dx$

6. $\displaystyle\int \frac{\cos \sqrt{x}}{\sqrt{x}}\, dx$

7. $\displaystyle\int \frac{dx}{\sqrt{x^2 + 2x + 2}}$

8. $\displaystyle\int_4^5 \frac{(3x - 7)\, dx}{(x - 1)(x - 2)(x - 3)}$

9. $\displaystyle\int x^2 e^x\, dx$

10. $\displaystyle\int \sqrt{z^2 + 1}\, dz$

11. $\displaystyle\int \frac{dr}{1 + \sqrt{r}}$

12. $\displaystyle\int_5^{25} \sqrt{\frac{25}{x} - 1}\, dx$

13. $\displaystyle\int \frac{4x^3 - 20x}{x^4 - 10x^2 + 9}\, dx$

14. $\displaystyle\int_0^{64} \frac{dx}{\sqrt{1 + \sqrt{x}}}$

15. $\displaystyle\int t^{2/3}(t^{5/3} + 1)^{2/3}\, dt$

16. $\displaystyle\int_{\pi/6}^{\pi/2} \frac{\cot x\, dx}{\ln(e \sin x)}$

17. $\displaystyle\int \frac{dt}{\sqrt{e^t + 1}}$

18. $\displaystyle\int \frac{(\sin x)\, e^{\sec x}}{\cos^2 x}\, dx$

19. $\displaystyle\int_0^{\pi/2} \frac{\cos x\, dx}{1 + \sin^2 x}$

20. $\displaystyle\int_{1/2}^{(1 + \sqrt{2})/2} \frac{dx}{\sqrt{2x - x^2}}$

21. $\displaystyle\int_0^{\pi/2} \frac{\sin x\, dx}{1 + \cos^2 x}$

22. $\displaystyle\int_0^{\pi/4} \frac{\cos 2t}{1 + \sin 2t}\, dt$

23. $\displaystyle\int_{\pi/4}^{\pi/3} \frac{dx}{\sin x \cos x}$

24. $\displaystyle\int \sqrt{1 + \sin x}\, dx$

25. $\displaystyle\int_{-\pi/2}^0 \sqrt{1 - \sin u}\, du$

26. $\displaystyle\int \frac{dx}{\sqrt{(a^2 + x^2)^3}}$

27. $\displaystyle\int \frac{e^{2x}\, dx}{\sqrt[3]{1 + e^x}}$

28. $\displaystyle\int_0^1 \ln \sqrt{1 + x^2}\, dx$

29. $\displaystyle\int_{5/4}^{5/3} \frac{dx}{(x^2 - 1)^{3/2}}$

30. $\displaystyle\int_0^{3/5} \frac{x^3\, dx}{\sqrt{1 - x^2}}$

31. $\displaystyle\int \frac{dx}{x(2 + \ln x)}$

32. $\displaystyle\int \frac{\cos 2x - 1}{\cos 2x + 1}\, dx$

33. $\displaystyle\int_0^{\ln 2} \frac{e^{2x}\, dx}{\sqrt[3]{e^x + 1}}$

34. $\displaystyle\int \frac{2 \sin \sqrt{x}\, dx}{\sqrt{x} \sec \sqrt{x}}$

35. $\displaystyle\int_0^3 (16 + x^2)^{-3/2}\, dx$

36. $\displaystyle\int_0^{(e - 1)^2} \frac{dw}{\sqrt{w} + 1}$

37. $\displaystyle\int \sin \sqrt{v + 1}\, dv$

38. $\displaystyle\int \cos \sqrt{1 - x}\, dx$

39. $\displaystyle\int_0^1 \frac{dx}{4 - x^2}$

40. $\displaystyle\int_0^1 \frac{dx}{x^3 + 1}$

41. $\displaystyle\int \frac{dy}{y(2y^3 + 1)^2}$

42. $\displaystyle\int \frac{x\,dx}{1 + \sqrt{x}}$

43. $\displaystyle\int \frac{dx}{x(x^2 + 1)^2}$

44. $\displaystyle\int \ln\sqrt{x - 1}\,dx$

45. $\displaystyle\int \frac{dx}{e^x - 1}$

46. $\displaystyle\int \frac{(x + 1)\,dx}{x^2(x - 1)}$

47. $\displaystyle\int \frac{x\,dx}{x^2 + 4x + 3}$

48. $\displaystyle\int_{\ln 2}^{\ln 3} \frac{15\,du}{(e^u - e^{-u})^2}$

49. $\displaystyle\int \frac{4\,dx}{x^3 + 4x}$

50. $\displaystyle\int \frac{da}{5a^2 + 8a + 5}$

51. $\displaystyle\int \frac{\sqrt{x^2 - 1}}{x}\,dx$

52. $\displaystyle\int e^x \cos 2x\,dx$

53. $\displaystyle\int \frac{dx}{x(3\sqrt{x} + 1)}$

54. $\displaystyle\int \frac{dx}{x(1 + \sqrt[3]{x})}$

55. $\displaystyle\int_{\pi/6}^{\pi/2} \frac{\cot\theta\,d\theta}{1 + \sin^2\theta}$

56. $\displaystyle\int \frac{z^5\,dz}{\sqrt{1 + z^2}}$

57. $\displaystyle\int \frac{e^{4t}\,dt}{(1 + e^{2t})^{2/3}}$

58. $\displaystyle\int \frac{dx}{x^{1/5}\sqrt{1 + x^{4/5}}}$

59. $\displaystyle\int \frac{(x^3 + x^2)\,dx}{x^2 + x - 2}$

60. $\displaystyle\int_2^3 \frac{x^3 + 1}{x^3 - x}\,dx$

61. $\displaystyle\int \frac{x\,dx}{(x - 1)^2}$

62. $\displaystyle\int \frac{(x + 1)\,dx}{(x^2 + 2x - 3)^{2/3}}$

63. $\displaystyle\int \frac{dy}{(2y + 1)\sqrt{y^2 + y}}$

64. $\displaystyle\int \frac{ds}{s^2\sqrt{a^2 - s^2}}$

65. $\displaystyle\int (1 - x^2)^{3/2}\,dx$

66. $\displaystyle\int \ln(x + \sqrt{1 + x^2})\,dx$

67. $\displaystyle\int p\tan^2 p\,dp$

68. $\displaystyle\int x\cos^2 x\,dx$

69. $\displaystyle\int_0^\pi x^2\sin x\,dx$

70. $\displaystyle\int x\sin^2 x\,dx$

71. $\displaystyle\int_0^1 \frac{dt}{t^4 + 4t^2 + 3}$

72. $\displaystyle\int \frac{du}{e^{4u} + 4e^{2u} + 3}$

73. $\displaystyle\int_0^2 x\ln\sqrt{x + 2}\,dx$

74. $\displaystyle\int (x + 1)^2 e^x\,dx$

75. $\displaystyle\int \sec^{-1}q\,dq$

76. $\displaystyle\int \frac{8\,dx}{x^4 + 2x^3}$

77. $\displaystyle\int \frac{x\,dx}{x^4 - 16}$

78. $\displaystyle\int_0^{\pi/2} \frac{\cos x\,dx}{\sqrt{1 + \cos x}}$

79. $\displaystyle\int \frac{\cos x\,dx}{\sin^3 x - \sin x}$

80. $\displaystyle\int_0^{\ln 2} \frac{du}{(e^u + e^{-u})^2}$

81. $\displaystyle\int_0^1 \frac{u\,du}{1 + \sqrt{u} + u}$

82. $\displaystyle\int \frac{\sec^2 t\,dt}{\sec^2 t - 3\tan t + 1}$

83. $\displaystyle\int \frac{dt}{\sec^2 t + \tan^2 t}$

84. $\displaystyle\int_0^{\tan^{-1}\sqrt{2}} \frac{dx}{1 + \cos^2 x}$

85. $\displaystyle\int e^{2t}\cos(e^t)\,dt$

86. $\displaystyle\int_0^1 \ln\sqrt{x^2 + 1}\,dx$

87. $\displaystyle\int x\ln(x^3 + x)\,dx$

88. $\displaystyle\int_0^1 x^3 e^{x^2}\,dx$

89. $\displaystyle\int \frac{\cos x\,dx}{\sqrt{4 - \cos^2 x}}$

90. $\displaystyle\int_0^{\pi/4} \frac{\sec^2 x\,dx}{\sqrt{4 - \sec^2 x}}$

91. $\displaystyle\int x^2\sin(1 - x)\,dx$

92. $\displaystyle\int_0^1 \frac{dx}{(x^2 + 1)(2 + \tan^{-1}x)}$

93. $\displaystyle\int \frac{dx}{\cot^3 x}$

94. $\displaystyle\int_0^{1/3} \theta\ln\sqrt[3]{3\theta + 1}\,d\theta$

95. $\displaystyle\int \frac{x^3\,dx}{(x^2 + 1)^2}$

96. $\displaystyle\int_0^{3/4} \frac{x\,dx}{\sqrt{1 - x}}$

97. $\displaystyle\int_0^4 x\sqrt{2x + 1}\,dx$

98. $\displaystyle\int \ln(x + \sqrt{x^2 - 1})\,dx$

99. $\displaystyle\int \ln(x - \sqrt{x^2 - 1})\,dx$

100. $\displaystyle\int_0^{\ln\sqrt{3}} e^{-x}\tan^{-1}(e^x)\,dx$

101. $\displaystyle\int \ln(x + \sqrt{x})\,dx$

102. $\displaystyle\int \tan^{-1}\sqrt{x}\,dx$

103. $\displaystyle\int_2^4 4\sec^{-1}\sqrt{x}\,dx$

104. $\displaystyle\int_0^{\pi^2} \cos\sqrt{x}\,dx$

105. $\displaystyle\int_0^{\pi^2/4} \sin\sqrt{x}\,dx$

106. $\displaystyle\int_0^2 \tan^{-1}\sqrt{x + 1}\,dx$

107. $\displaystyle\int \sqrt{1 - x^2}\,\sin^{-1}x\,dx$

108. $\displaystyle\int_0^\pi x\sin^2(2x)\,dx$

109. $\displaystyle\int \frac{\tan x\,dx}{\tan x + \sec x}$

110. $\displaystyle\int \frac{dt}{\sqrt{e^{2t} + 1}}$

111. $\displaystyle\int \frac{dx}{(\cos^2 x + 4\sin x - 5)\cos x}$

112. $\displaystyle\int \frac{dt}{a + be^{ct}}, \quad a,\,b,\,c \neq 0$

113. $\int \sqrt{\dfrac{1 - \cos x}{\cos \alpha - \cos x}} \, dx, \quad \alpha \text{ constant}, \ 0 < \alpha < x < \pi$

114. $\int \dfrac{dx}{9x^4 + x^2}$

115. $\int_0^1 \ln(2x^2 + 4) \, dx$

116. $\int \dfrac{\sin x \, dx}{\cos^2 x - 5 \cos x + 4}$

117. $\int \dfrac{dt}{\sqrt{1 - e^{-t}}}$

118. $\int \dfrac{\tan^{-1}x}{x^2} \, dx$

119. $\int_0^1 3(x - 1)^2 \left(\int_0^x \sqrt{1 + (t - 1)^4} \, dt \right) dx$

120. $\int \left(\dfrac{1}{t} + \dfrac{1}{t + t \ln t} \right) dt$

121. $\int_3^\infty \dfrac{8x + 24}{x(x^2 - 4)} \, dx$

122. $\int_2^\infty \dfrac{4v^3 + v - 1}{v^2(v - 1)(v^2 + 1)} \, dv$

123. $\int_0^\infty y^3 e^{-y^2} \, dy$

124. $\int_1^\infty \dfrac{\ln \theta}{\theta^2} \, d\theta$

We dare you to evaluate the integrals in Exercises 125–134.

125. $\int (\sin^{-1}x)^2 \, dx$

126. $\int \dfrac{dx}{x(x + 1)(x + 2) \cdots (x + m)}$

127. $\int x \sin^{-1}x \, dx$

128. $\int \sin^{-1}\sqrt{y} \, dy$

129. $\int \dfrac{d\theta}{1 - \tan^2\theta}$

130. $\int \ln(\sqrt{x} + \sqrt{1 + x}) \, dx$

131. $\int \dfrac{dt}{t - \sqrt{1 - t^2}}$

132. $\int \dfrac{(2e^{2x} - e^x) \, dx}{\sqrt{3e^{2x} - 6e^x - 1}}$

133. $\int \dfrac{dx}{x^4 + 4}$

134. $\int \dfrac{dx}{x^6 - 1}$

In Exercises 135 and 136, evaluate the limit by identifying it with a definite integral and evaluating the integral.

135. $\displaystyle\lim_{n \to \infty} \sum_{k=1}^n \ln \sqrt[n]{1 + \dfrac{k}{n}}$

136. $\displaystyle\lim_{n \to \infty} \sum_{k=0}^{n-1} \dfrac{1}{\sqrt{n^2 - k^2}}$

137. Find the volume of the solid generated by revolving the triangular region enclosed by the curve $y = \ln x$ and the lines $x = 1$ and $y = 1$ about (a) the x-axis; (b) the y-axis; (c) the line $x = 1$; (d) the line $y = 1$.

138. Find the volume of the solid generated by revolving the region in Fig. 7.32 about the y-axis.

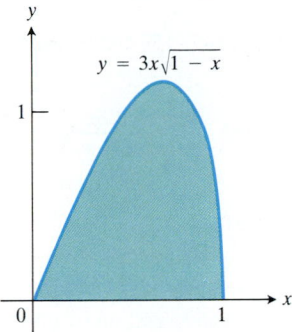

7.32   The region in Exercise 138.

139. Find the centroid of the region bounded by the curves $y = \pm(1 - x^2)^{-1/2}$ and the lines $x = 0$ and $x = 1$.

140. The infinite region bounded by the coordinate axes and the curve $y = -\ln x$ in the first quadrant is revolved about the x-axis to generate a solid. Find the volume of the solid.

141. The region between the curve $y = x \ln x$ and the x-axis from $x = 0$ to $x = 2$ is revolved about the x-axis to generate a solid (Fig. 7.33). Find the volume of the solid.

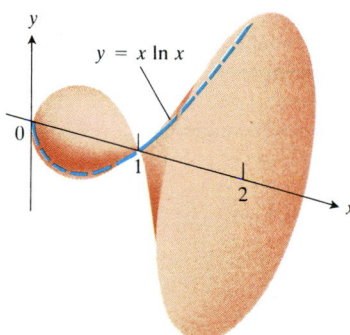

7.33   The solid of revolution in Exercise 141.

142. Find the length of the curve $y = \ln x$ from $x = 1$ to $x = e$.

143. Find the area of the surface generated by revolving the curve in Exercise 142 about the y-axis.

144. Find the area of the surface generated by revolving the curve $y = \cos x$, $-\pi/2 \le x \le \pi/2$ about the x-axis.

145. Show that if $f(x)$ is a polynomial of degree n, then

$$\int e^x f(x) \, dx = e^x(f(x) - f'(x) + f''(x) - \cdots + (-1)^n f^{(n)}(x)).$$

146. Without evaluating either integral, show why

$$2\int_{-1}^1 \sqrt{1 - x^2} \, dx = \int_{-1}^1 \dfrac{1}{\sqrt{1 - x^2}} \, dx.$$

(*Hint:* Interpret one integral as an area and the other as a length.) (*Source:* Peter A. Lindstrom, *Mathematics Magazine,* Vol. 45, No. 1, January 1972, p. 47.)

Which of the integrals in Exercises 147–150 converge and which diverge?

147. $\displaystyle\int_{-\infty}^{\infty} \frac{4x\,dx}{x^2 + 1}$ **148.** $\displaystyle\int_{-\infty}^{\infty} \frac{8\,dx}{2x^{4/3} + 3}$

149. $\displaystyle\int_{-\infty}^{\infty} \frac{e^{-x}\,dx}{e^{-x} + e^x}$ **150.** $\displaystyle\int_{-\infty}^{\infty} \frac{x^3\,dx}{1 + e^x}$

151. a) GRAPHER Graph the function $e^{(x - e^x)}$ for $-5 \le x \le 3$.
 b) Show that $\int_{-\infty}^{\infty} e^{(x - e^x)}\,dx$ converges, and find the integral's value.

152. Find $\displaystyle\lim_{n\to\infty} \int_0^1 \frac{ny^{n-1}}{1 + y}\,dy$.

The Substitution $z = \tan(x/2)$

The substitution

$$z = \tan\frac{x}{2} \tag{1}$$

reduces the problem of integrating any rational function of $\sin x$ and $\cos x$ to a problem involving a rational function of z. This in turn can be integrated by partial fractions. Thus the substitution (1) is a powerful tool. It is cumbersome, however, and is used only when simpler methods fail.

Figure 7.34 shows how $\tan\,(x/2)$ expresses a rational function of $\sin x$ and $\cos x$. To see the effect of the substitution, we calculate

$$\cos x = 2\cos^2\frac{x}{2} - 1 = \frac{2}{\sec^2(x/2)} - 1$$

$$= \frac{2}{1 + \tan^2(x/2)} - 1 = \frac{2}{1 + z^2} - 1,$$

$$\cos x = \frac{1 - z^2}{1 + z^2}, \tag{2}$$

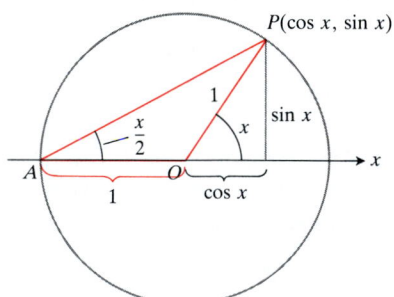

7.34   From this figure, we can read the relation

$$\tan\frac{x}{2} = \frac{\sin x}{1 + \cos x}$$

and

$$\sin x = 2\sin\frac{x}{2}\cos\frac{x}{2} = 2\,\frac{\sin(x/2)}{\cos(x/2)}\cdot\cos^2\frac{x}{2}$$

$$= 2\tan\frac{x}{2}\cdot\frac{1}{\sec^2(x/2)} = \frac{2\tan(x/2)}{1 + \tan^2(x/2)},$$

$$\sin x = \frac{2z}{1 + z^2}. \tag{3}$$

Finally, $x = 2\tan^{-1}z$, so

$$dx = \frac{2\,dz}{1 + z^2}. \tag{4}$$

EXAMPLE 1

$$\int\frac{1}{1 + \cos x}\,dx = \int\frac{1 + z^2}{2}\,\frac{2\,dz}{1 + z^2}$$

$$= \int dz = z + C = \tan\frac{x}{2} + C$$

EXAMPLE 2

$$\int\frac{1}{2 + \sin x}\,dx = \int\frac{1 + z^2}{2 + 2z + 2z^2}\,\frac{2\,dz}{1 + z^2}$$

$$= \int\frac{dz}{z^2 + z + 1} = \int\frac{dz}{(z + (1/2))^2 + 3/4}$$

$$= \int\frac{du}{u^2 + a^2}$$

$$= \frac{1}{a}\tan^{-1}\frac{u}{a} + C$$

$$= \frac{2}{\sqrt{3}}\tan^{-1}\frac{2z + 1}{\sqrt{3}} + C$$

$$= \frac{2}{\sqrt{3}}\tan^{-1}\frac{1 + 2\tan(x/2)}{\sqrt{3}} + C$$

Use the substitutions in Eqs. (1)–(4) to evaluate the integrals in Exercises 153–160. Integrals like these arise in calculating the average angular velocity of the output shaft of a universal joint when the input and output shafts are not aligned.

153. $\displaystyle\int_0^{\pi/2} \frac{dx}{1 + \sin x}$ **154.** $\displaystyle\int_{\pi/3}^{\pi/2} \frac{dx}{1 - \cos x}$

155. $\displaystyle\int \frac{dx}{1 - \sin x}$ **156.** $\displaystyle\int_0^{\pi/2} \frac{dx}{2 + \cos x}$

157. $\displaystyle\int \frac{\cos x\,dx}{1 - \cos x}$ **158.** $\displaystyle\int \frac{dx}{1 + \sin x + \cos x}$

159. $\displaystyle\int \frac{dx}{\sin x - \cos x}$ **160.** $\displaystyle\int_{\pi/2}^{2\pi/3} \frac{dx}{\sin x + \tan x}$

The Gamma Function and Stirling's Formula

Euler's gamma function $\Gamma(x)$ ("gamma of x"; Γ is a Greek capital g) uses an integral to extend the factorial function from the nonnegative integers to other real values. The formula is

$$\Gamma(x) = \int_0^\infty t^{x-1}e^{-t}\,dt, \quad x > 0. \tag{5}$$

For each positive x, the number $\Gamma(x)$ is the integral of $t^{x-1}e^{-t}$ with respect to t from 0 to ∞. Figure 7.35 shows the graph of Γ near the origin. You will see how to calculate $\Gamma(1/2)$ if you do Miscellaneous Exercise 30 in Chapter 13.

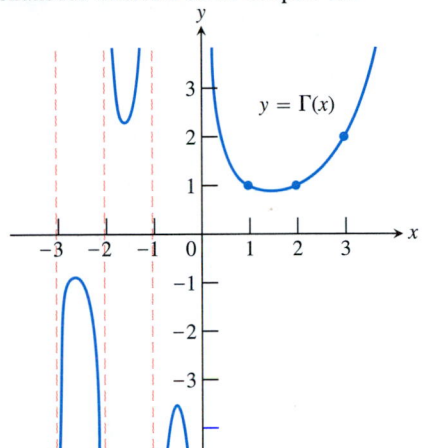

7.35 $\Gamma(x)$ is a continuous function of x whose value at each positive integer $n + 1$ is $n!$. The defining integral formula for Γ (Eq. 5) is valid only for $x > 0$, but we can extend Γ to negative noninteger values of x with the formula $\Gamma(x) = (\Gamma(x + 1))/x$, which is the subject of Exercise 161.

161. *If n is a nonnegative integer, $\Gamma(n + 1) = n!$.*
   a) Show that $\Gamma(1) = 1$.
   b) Then apply integration by parts to the integral for $\Gamma(x + 1)$ to show that $\Gamma(x + 1) = x\Gamma(x)$. This gives

$$\Gamma(2) = 1\Gamma(1) = 1$$
$$\Gamma(3) = 2\Gamma(2) = 2$$
$$\Gamma(4) = 3\Gamma(3) = 6$$
$$\vdots$$
$$\Gamma(n + 1) = n\Gamma(n) = n! \tag{6}$$

   c) Use mathematical induction to verify Eq. (6) for every nonnegative integer n.

162. *Stirling's formula.* Scottish mathematician James Stirling (1692–1770) showed that

$$\lim_{x \to \infty} \left(\frac{e}{x}\right)^x \sqrt{\frac{x}{2\pi}}\; \Gamma(x) = 1,$$

so for large x,

$$\Gamma(x) = \left(\frac{x}{e}\right)^x \sqrt{\frac{2\pi}{x}}\,(1 + \epsilon(x)), \quad \epsilon(x) \to 0 \text{ as } x \to \infty.$$

Dropping $\epsilon(x)$ leads to the approximation

$$\Gamma(x) \approx \left(\frac{x}{e}\right)^x \sqrt{\frac{2\pi}{x}}\;. \quad \textbf{(Stirling's formula)} \tag{7}$$

   a) *Stirling's approximation for $n!$.* Use Eq. (7) and the fact that $n! = n\Gamma(n)$ to show that

$$n! \approx \left(\frac{n}{e}\right)^n \sqrt{2n\pi}\;. \quad \textbf{(Stirling's approximation)} \tag{8}$$

As you will see if you do Exercise 82 in Section 8.1, Eq. (8) leads to the approximation

$$\sqrt[n]{n!} \approx \frac{n}{e}. \tag{9}$$

   b) CALCULATOR Compare your calculator's value for $n!$ with the value given by Stirling's approximation for $n = 10, 20, 30, \ldots$, as far as your calculator can go.
   c) CALCULATOR A refinement of Eq. (6) gives

$$\Gamma(x) = \left(\frac{x}{e}\right)^x \sqrt{\frac{2\pi}{x}}\; e^{1/(12x)}\,(1 + \epsilon(x)),$$

or

$$\Gamma(x) \approx \left(\frac{x}{e}\right)^x \sqrt{\frac{2\pi}{x}}\; e^{1/(12x)},$$

which tells us that

$$n! \approx \left(\frac{n}{e}\right)^n \sqrt{2n\pi}\; e^{1/(12n)}. \tag{10}$$

Compare the values given for 10! by your calculator, Stirling's approximation, and Eq. (10).

(The authors wish to thank Richard Askey for sharing his classroom notes on this material.)

CHAPTER

8

INFINITE SERIES

Overview   In this chapter, we study infinite polynomials called power series and, as a product of our study, develop one of the most remarkable formulas in all of mathematics. The formula, known as Taylor's formula, does two things for us. It shows how to calculate the value of an infinitely differentiable function like e^x at any point, just from its value and the values of its derivatives at the origin. And, as if that were not enough, the formula gives us polynomial approximations of differentiable functions of any order we want, along with their error formulas, all in a single equation (subject to the number of derivatives available).

Power series have many additional uses. They provide an efficient way to evaluate nonelementary integrals and they solve differential equations that give insight into heat flow, vibration, chemical diffusion, and signal transmission. What you will learn here sets the stage for the roles played by series of functions of all kinds in science and mathematics.

8.1 Limits of Sequences of Numbers

This section describes what it means for an infinite sequence of numbers to have a limit and shows how to find the limits of many of the sequences that arise in mathematics and applied fields.

Informally, a sequence is an ordered collection of things, but in this chapter the things will usually be numbers. We have seen sequences before, such as sequences $x_0, x_1, \ldots, x_n, \ldots$ of numerical approximations generated by Newton's method and the sequence $A_3, A_4, \ldots, A_n, \ldots$ of areas of n-sided regular polygons used to define the area of a circle. These sequences have limits, but many equally important sequences do not. The sequence $1, 2, 3, \ldots, n, \ldots$ of positive integers has no limit, nor does the sequence $2, 3, 5, 7, 11, 13, \ldots$ of prime numbers. We need to know when sequences do and do not have limits and how to find the limits when they exist. As with functions, we usually find out with theorems based on a formal definition. You will see a close parallel between what we do here and what we did with limits of functions in Chapter 1.

Definitions and Notation

Our starting point is a formal definition of sequence.

DEFINITION

> An **infinite sequence** (or **sequence**) of numbers is a function whose domain is the set of integers greater than or equal to some integer n_0.

Usually n_0 is 1 and the domain is the set of all positive integers. But sometimes we want to start our sequences elsewhere. We might take $n_0 = 0$, for instance, when we begin Newton's method, or take $n_0 = 3$ when we work with n-sided polygons.

Sequences are defined the way other functions are, some typical rules being

$$a(n) = n - 1, \qquad a(n) = (-1)^{n+1} \frac{1}{n}, \qquad a(n) = \frac{n-1}{n} \qquad (1)$$

(Example 1 and Fig. 8.1).

To indicate that the domains are sets of integers, we use a letter like n from the middle of the alphabet for the independent variable, instead of the x, y, z, and t used so widely in other contexts. The formulas in the defining rules, however, like those in (1), are often valid for domains much larger than the set of positive integers. This can be an advantage, as we shall see.

The number $a(n)$ is the **nth term** of the sequence, or the **term with index n**. For example, if $a(n) = (n-1)/n$, then the terms are

First term	Second term	Third term	nth term
$a(1) = 0$	$a(2) = \frac{1}{2},$	$a(3) = \frac{2}{3}, \ldots,$	$a(n) = \frac{n-1}{n}.$ (2)

When we use the subscript notation a_n for $a(n)$, the sequence in (2) becomes

$$a_1 = 0, \qquad a_2 = \frac{1}{2}, \qquad a_3 = \frac{2}{3}, \qquad \ldots, \qquad a_n = \frac{n-1}{n}. \qquad (3)$$

To describe sequences, we often write the first few terms as well as a formula for the nth term.

Example 1

We write	For the sequence whose defining rule is
$0, 1, 2, \ldots, n-1, \ldots$	$a_n = n - 1$
$1, \dfrac{1}{2}, \dfrac{1}{3}, \ldots, \dfrac{1}{n}, \ldots$	$a_n = \dfrac{1}{n}$
$1, -\dfrac{1}{2}, \dfrac{1}{3}, -\dfrac{1}{4}, \ldots, (-1)^{n+1}\dfrac{1}{n}, \ldots$	$a_n = (-1)^{n+1}\dfrac{1}{n}$
$0, \dfrac{1}{2}, \dfrac{2}{3}, \dfrac{3}{4}, \ldots, \dfrac{n-1}{n}, \ldots$	$a_n = \dfrac{n-1}{n}$
$0, -\dfrac{1}{2}, \dfrac{2}{3}, -\dfrac{3}{4}, \ldots, (-1)^{n+1}\left(\dfrac{n-1}{n}\right), \ldots$	$a_n = (-1)^{n+1}\left(\dfrac{n-1}{n}\right)$
$3, 3, 3, \ldots, 3, \ldots$	$a_n = 3$

8.1 The sequences of Example 1 are graphed here in two different ways: by plotting the numbers a_n on a horizontal axis and by plotting the points (n, a_n) in the coordinate plane.

The terms $a_n = n - 1$ eventually surpass every integer, so the sequence $\{a_n\}$ diverges, . . .

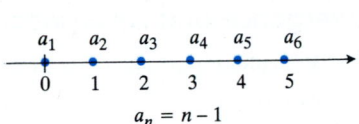

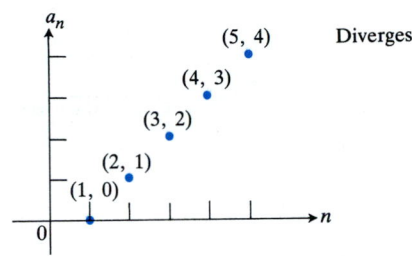

. . . but the terms $a_n = 1/n$ decrease steadily and get arbitrarily close to 0 as n increases, so the sequence $\{a_n\}$ converges to 0.

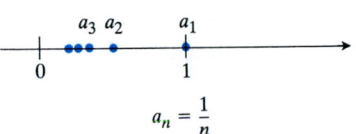

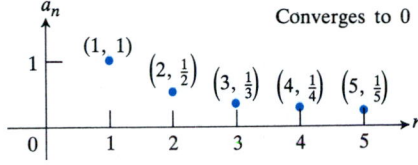

The terms $a_n = (-1)^{n+1}(1/n)$ alternate in sign but still converge to 0.

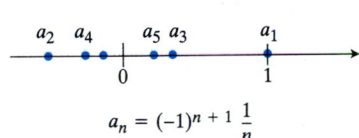

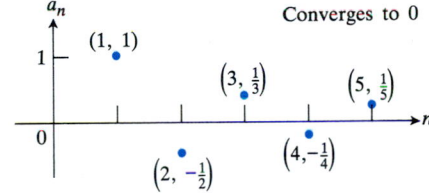

The terms $a_n = (n-1)/n$ approach 1 steadily and get arbitrarily close as n increases, so the sequence $\{a_n\}$ converges to 1.

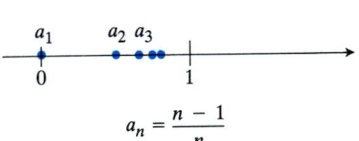

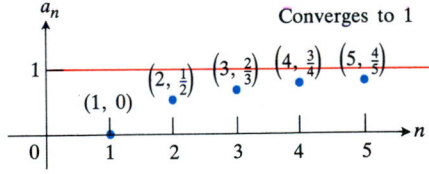

The terms $a_n = (-1)^{n+1}[(n-1)/n]$ alternate in sign. The positive terms approach 1. But the negative terms approach -1 as n increases, so the sequence $\{a_n\}$ diverges.

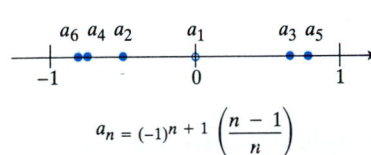

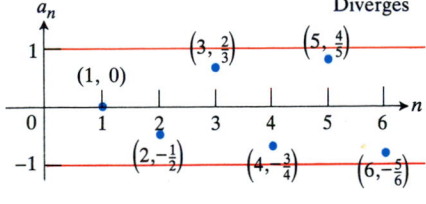

The terms in the sequence of constants $a_n = 3$ have the same value regardless of n; so we say the sequence $\{a_n\}$ converges to 3.

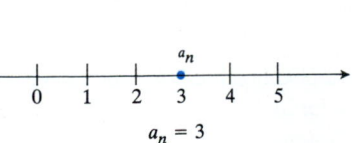

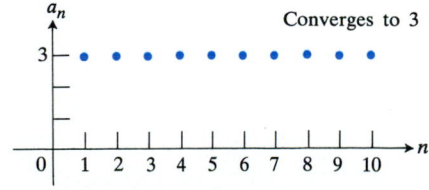

Notation   We refer to the sequence whose nth term is a_n with the notation $\{a_n\}$ ("the sequence a sub n"). The second sequence in Example 1 is $\{1/n\}$ ("the sequence 1 over n"); the last sequence is $\{3\}$ ("the sequence 3").

Convergence and Divergence

As Fig. 8.1 shows, the sequences of Example 1 do not behave the same way. The sequences $\{1/n\}$, $\{(-1)^{n+1}(1/n)\}$, and $\{(n-1)/n\}$ each seem to approach a single limiting value as n increases, and $\{3\}$ is at a limiting value from the very first. On the other hand, terms of $\{(-1)^{n+1}(n-1)/n\}$ seem to accumulate near two different values, -1 and 1, while the terms of $\{n-1\}$ become increasingly large and do not accumulate anywhere.

To distinguish sequences that approach a unique limiting value L, as n increases, from those that do not, we say that the former sequences *converge,* according to the following definition.

DEFINITIONS

> The sequence $\{a_n\}$ **converges** to the number L if to every positive number ϵ there corresponds an integer N such that for all n,
>
> $$n > N \quad \Rightarrow \quad |a_n - L| < \epsilon. \tag{4}$$
>
> If no such number and integer exist, we say that $\{a_n\}$ **diverges.**
> If $\{a_n\}$ converges to L, we write $\lim\limits_{n \to \infty} a_n = L$, or simply $a_n \to L$, and call L the **limit** of the sequence (Fig. 8.2).

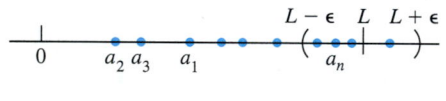

8.2 $a_n \to L$ if L is a horizontal asymptote of the sequence of points $\{(n, a_n)\}$. In this figure, all the a_n's after a_N lie within ϵ of L.

Example 2   Show that $\{1/n\}$ converges to 0.

Solution   We set $a_n = 1/n$ and $L = 0$ in the definition of convergence. To show that $1/n \to 0$, we must show that for any $\epsilon > 0$, there exists an integer N such that for all n,

$$n > N \quad \Rightarrow \quad \left| \frac{1}{n} - 0 \right| < \epsilon. \tag{5}$$

This implication will hold for all n for which

$$\frac{1}{n} < \epsilon \quad \text{or, equivalently,} \quad n > \frac{1}{\epsilon}.$$

Pick an integer N greater than $1/\epsilon$. Then any n greater than N will automatically be greater than $1/\epsilon$ and the implication in (5) will hold.

Example 3   Show that if k is any number, then $\{k\}$ converges to k.

Solution   We set $a_n = k$ and $L = k$ in the definition of convergence. To show that $a_n \to k$, we must show that for any $\epsilon > 0$ there exists an integer N such that for all n,

$$n > N \quad \Rightarrow \quad |k - k| < \epsilon.$$

This implication holds for any integer N because $|k - k| = 0$ is less than every positive ϵ for all n.

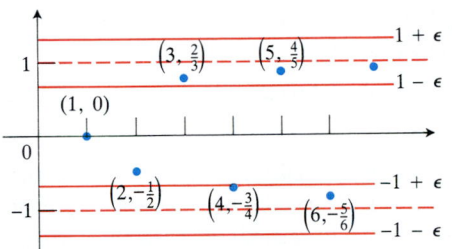

$$a_n = (-1)^{n+1}\left(\frac{n-1}{n}\right)$$

Neither the ϵ-interval about 1 nor the ϵ-interval about -1 contains a complete tail of the sequence.

8.3 The sequence $\{(-1)^{n+1}[(n-1)/n]\}$ diverges.

Example 4 Show that $\{(-1)^{n+1}(n-1)/n\}$ diverges.

Solution Take a positive ϵ smaller than 1 so that the bands shown in Fig. 8.3 about the lines $y = 1$ and $y = -1$ do not overlap. Any $\epsilon < 1$ will do. Convergence to 1 would require every point of the graph beyond a certain index N to lie inside the upper band, but this will never happen. As soon as a point (n, a_n) lies within the upper band, every alternate point starting with $(n + 1, a_{n+1})$ will lie within the lower band. Hence the sequence cannot converge to 1. Likewise, it cannot converge to -1. On the other hand, because the terms of the sequence get alternately closer to 1 and -1, they never accumulate near any other value. Therefore, the sequence diverges.

A **tail** of a sequence $\{a_n\}$ is the collection of all terms whose indices are greater than some integer N. In other words, a tail is one of the sets $\{a_n \mid n > N\}$. Another way to say that $a_n \to L$ is to say that every ϵ-interval about L contains a tail of the sequence. The convergence or divergence of a sequence has nothing to do with how a sequence starts out. It depends only on how the tails behave.

The behavior of $\{(-1)^{n+1}[(n-1)/n]\}$ is qualitatively different from that of $\{n - 1\}$, which diverges because it outgrows every real number L. To describe the behavior of $\{n - 1\}$ we write

$$\lim_{n \to \infty} (n - 1) = \infty.$$

In speaking of infinity as a limit of a sequence $\{a_n\}$, we do not mean that the difference between a_n and infinity becomes small as n increases. We mean that a_n becomes numerically large as n increases.

Limits Are Unique

A sequence cannot converge to two different limits.

If a sequence $\{a_n\}$ converges, then its limit is unique.

The argument goes like this: If $\{a_n\}$ converged to two different limits L_1 and L_2, we could take ϵ to be the positive number $|L_1 - L_2|/2$. Because $a_n \to L_1$, there would exist an integer N_1 such that for all n,

$$n > N_1 \quad \Rightarrow \quad |a_n - L_1| < \epsilon.$$

There would also exist an integer N_2 such that for all n,

$$n > N_2 \quad \Rightarrow \quad |a_n - L_2| < \epsilon.$$

For all n greater than both N_1 and N_2, we would then have

$$|L_1 - L_2| = |L_1 - a_n + a_n - L_2| \le |L_1 - a_n| + |a_n - L_2| \qquad \begin{pmatrix}\text{Triangle} \\ \text{inequality}\end{pmatrix}$$
$$< \epsilon + \epsilon = 2(|L_1 - L_2|/2)$$
$$< |L_1 - L_2|.$$

Factorial Notation

The notation $n!$ ("n factorial") means the product $1 \cdot 2 \cdot 3 \cdot \cdots \cdot n$ of the integers from 1 to n. Notice that $(n + 1)! = (n + 1) \cdot n!$. Thus, $4! = 1 \cdot 2 \cdot 3 \cdot 4 = 24$ and $5! = 1 \cdot 2 \cdot 3 \cdot 4 \cdot 5 = 5 \cdot 4! = 120$. We also define $0!$ to be 1. Factorials grow even faster than exponentials, as the following table suggests.

n	e^n(rounded)	$n!$
1	3	1
5	148	120
10	22,026	3,628,800
20	4.8×10^8	2.4×10^{18}

But this is absurd; no number is less than itself. Hence, if a sequence converges, its limit is unique.

Recursive Definitions

So far, we have calculated each a_n directly from the value of n. But sequences are often defined **recursively** by giving

1. the value of the first term (or first few terms);

2. a rule, called a **recursion formula,** for calculating any later term from terms that precede it.

The initial value $a_1 = 1$ and the recursion formula $a_{n+1} = a_n + 1$ defines the sequence of positive integers recursively. The terms of the factorial sequence $1, 2, 6, 24, \ldots, n!, \ldots$ are usually calculated recursively with the formula $(n + 1)! = (n + 1)n!$. The initial value $x_0 = 1$ and the recursion formula $x_{n+1} = x_n - [(\sin x_n - x_n^2)/(\cos x_n - 2x_n)]$ define a sequence that converges to the solution of the equation $\sin x = x^2$ (Section 2.7, Example 3). Recursion formulas arise regularly in computer programming, and we shall encounter them again when we preview numerical methods for solving differential equations (Section 15.9).

Subsequences

If the terms of one sequence occur in their given order among the terms of a second sequence, we call the first sequence a **subsequence** of the second.

Example 5   *Subsequences of the Sequence of Positive Integers*

a) The sequence $2, 4, 6, \ldots, 2n, \ldots$ of even integers

b) The sequence $1, 3, 5, \ldots, 2n + 1, \ldots$ of odd integers

c) The sequence $2, 3, 5, 7, 11, \ldots$ of primes

Subsequences are important for two reasons. First, if a sequence $\{a_n\}$ converges to a limit L, then all subsequences also converge to L. If we know a sequence converges, it may be quicker for us to find or estimate its limit by choosing a rapidly convergent subsequence. The second reason is related to the first: If any subsequence of the original sequence diverges, or if two subsequences have different limits, then the original sequence diverges. The sequence $\{(-1)^{n-1}\}$ diverges because the subsequence $1, 1, 1, \ldots$ of odd-numbered terms converges to 1 while the subsequence $-1, -1, -1, \ldots$ of even-numbered terms converges to -1, which is a different limit.

Useful Theorems

The study of limits would be cumbersome if we had to answer every question about convergence by applying the definition directly. Fortunately, three theorems will make this largely unnecessary from now on. The first is the sequence version of Theorem 1 in Chapter 1.

THEOREM 1

The following rules hold if $\lim_{n \to \infty} a_n = A$ and $\lim_{n \to \infty} b_n = B$

1. *Sum Rule:* $\lim \{a_n + b_n\} = A + B$

2. *Difference Rule:* $\lim \{a_n - b_n\} = A - B$

3. *Product Rule:* $\lim \{a_n \cdot b_n\} = A \cdot B$

4. *Constant Multiple Rule:* $\lim \{k \cdot b_n\} = k \cdot B$ (Any number k)

5. *Quotient Rule:* $\lim \dfrac{a_n}{b_n} = \dfrac{A}{B}$ if $B \neq 0$

The limits are all taken as $n \to \infty$, and A and B are real numbers.

By combining Theorem 1 with Examples 2 and 3, we can proceed immediately to

$$\lim_{n \to \infty} \left(-\frac{1}{n} \right) = -1 \cdot \lim_{n \to \infty} \frac{1}{n} = -1 \cdot 0 = 0,$$

$$\lim_{n \to \infty} \left(\frac{n-1}{n} \right) = \lim_{n \to \infty} \left(1 - \frac{1}{n} \right) = \lim_{n \to \infty} 1 - \lim_{n \to \infty} \frac{1}{n} = 1 - 0 = 1,$$

$$\lim_{n \to \infty} \frac{5}{n^2} = 5 \cdot \lim_{n \to \infty} \frac{1}{n} \cdot \lim_{n \to \infty} \frac{1}{n} = 5 \cdot 0 \cdot 0 = 0,$$

$$\lim_{n \to \infty} \frac{4 - 7n^6}{n^6 + 3} = \lim_{n \to \infty} \frac{(4/n^6) - 7}{1 + (3/n^6)} = \frac{0 - 7}{1 + 0} = -7.$$

One general consequence of Theorem 1 is that every nonzero multiple of a divergent sequence $\{a_n\}$ diverges. For, suppose $\{ca_n\}$ were to converge for some number $c \neq 0$. Then $(1/c)\{ca_n\} = \{a_n\}$ would converge by the Constant Multiple Rule—but it does not.

The next theorem is the sequence version of the Sandwich Theorem.

THEOREM 2

The Sandwich Theorem for Sequences

If $a_n \leq b_n \leq c_n$ for all n beyond some index N, and if $\lim a_n = \lim c_n = L$, then $\lim b_n = L$ also.

An immediate consequence of Theorem 2 is that, if $|b_n| \leq c_n$ and $c_n \to 0$, then $b_n \to 0$ because $-c_n \leq b_n \leq c_n$. We use this fact in the next example.

Example 6   Since $1/n \to 0$, we know that

a) $\dfrac{\cos n}{n} \to 0$ because $0 \leq \left| \dfrac{\cos n}{n} \right| = \dfrac{|\cos n|}{n} \leq \dfrac{1}{n}$;

b) $\dfrac{1}{2^n} \to 0$ because $0 \leq \dfrac{1}{2^n} \leq \dfrac{1}{n}$;

c) $(-1)^n \dfrac{1}{n} \to 0$ because $0 \leq \left| (-1)^n \dfrac{1}{n} \right| \leq \dfrac{1}{n}$.

The application of Theorems 1 and 2 is broadened by a theorem stating that applying a continuous function to a convergent sequence produces a convergent sequence. We state the theorem without proof.

THEOREM 3

If $a_n \to L$ and if f is a function that is continuous at L and defined at all a_n, then $f(a_n) \to f(L)$.

Example 7

Show that $\sqrt{(n+1)/n} \to 1$.

Solution We know that $(n+1)/n \to 1$. Taking $f(x) = \sqrt{x}$ and $L = 1$ in Theorem 3 gives $\sqrt{(n+1)/n} \to \sqrt{1} = 1$.

Example 8

Show that $2^{1/n} \to 1$ (Fig. 8.4).

Solution We know that $1/n \to 0$. Taking $f(x) = 2^x$ and $L = 0$ in Theorem 3 therefore gives $2^{1/n} \to 2^0 = 1$. Some sample values:

n	$2^{1/n}$
1	2
2	1.4142 13562
4	1.1892 07115
10	1.0717 73463
100	1.0069 55555
1000	1.0006 93387
10000	1.0000 69317

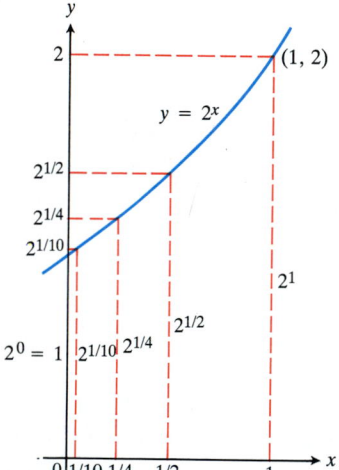

8.4 As $n \to \infty$, $x = 1/n \to 0$ and $y = 2^{1/n} \to 2^0 = 1$ (Example 8).

In Section 1.3, we saw that entering 2 in a calculator and pressing the square root key repeatedly produces a succession of numbers that approach 1. The result of Example 8 now tells us why. The successive square roots form a subsequence $2^{1/2}$, $2^{1/4}$, $2^{1/8}$, $2^{1/16}$, ... of the sequence $\{2^{1/n}\}$. The sequence converges to 1, so the subsequence does too.

The next theorem enables us to use l'Hôpital's rule to find the limits of some sequences. We state and prove the theorem first and then show how to apply it.

THEOREM 4

Suppose that $f(x)$ is a function defined for all $x \geq n_0$ and $\{a_n\}$ is a sequence such that $a_n = f(n)$ when $n \geq n_0$. If

$$\lim_{x \to \infty} f(x) = L, \qquad \text{then} \qquad \lim_{n \to \infty} a_n = L.$$

Proof Suppose that $\lim_{x \to \infty} f(x) = L$. Then for each positive number ϵ there is a number M such that for all x,

$$x > M \quad \Rightarrow \quad |f(x) - L| < \epsilon.$$

Let N be an integer greater than M and greater than or equal to n_0. Then

$$n > N \quad \Rightarrow \quad a_n = f(n) \quad \text{and} \quad |a_n - L| = |f(n) - L| < \epsilon.$$

Example 9   Show that $\lim_{n\to\infty}(\ln n)/n = 0$.

Solution   The function $(\ln x)/x$ is defined for all $x \geq 1$ and agrees with the given sequence at positive integers. Therefore $\lim_{n\to\infty}(\ln n)/n$ will equal $\lim_{x\to\infty}(\ln x)/x$ if the latter exists. A single application of l'Hôpital's rule shows that

$$\lim_{x\to\infty}\frac{\ln x}{x} = \lim_{x\to\infty}\frac{1/x}{1} = \frac{0}{1} = 0.$$

We conclude that $\lim_{n\to\infty}(\ln n)/n = 0$.

When we use l'Hôpital's rule to find the limit of a sequence, we often treat n as a continuous real variable and differentiate directly with respect to n. This saves us from having to rewrite the formula for a_n as we did in Example 9.

Example 10   Find $\lim_{n\to\infty}(2^n/5n)$.

Solution   By l'Hôpital's rule,

$$\lim_{n\to\infty}\frac{2^n}{5n} = \lim_{n\to\infty}\frac{2^n \cdot \ln 2}{5} = \infty.$$

Limits That Arise Frequently

The limits in Table 8.1 are useful and arise frequently. The first limit is from Example 9. The others are derived in Appendix 8.

TABLE 8.1

1. $\lim_{n\to\infty}\dfrac{\ln n}{n} = 0$ 　　　　　　2. $\lim_{n\to\infty}\sqrt[n]{n} = 1$

3. $\lim_{n\to\infty}x^{1/n} = 1$ 　　$(x > 0)$ 　　4. $\lim_{n\to\infty}x^n = 0$ 　　$(|x| < 1)$

5. $\lim_{n\to\infty}\left(1 + \dfrac{x}{n}\right)^n = e^x$ 　$(\text{Any } x)$ 　　6. $\lim_{n\to\infty}\dfrac{x^n}{n!} = 0$ 　$(\text{Any } x)$

In formulas (3)–(6), x remains fixed while $n \to \infty$.

Example 11   *Limits from Table 8.1*

1. $\dfrac{\ln(n^2)}{n} = \dfrac{2 \ln n}{n} \to 2 \cdot 0 = 0$ 　　　　(Formula 1)

2. $\sqrt[n]{n^2} = n^{2/n} = (n^{1/n})^2 \to (1)^2 = 1$ 　　(Formula 2)

3. $\sqrt[n]{3n} = 3^{1/n}(n^{1/n}) \to 1 \cdot 1 = 1$ 　　　(Formula 3 with $x = 3$, and Formula 2)

4. $\left(-\dfrac{1}{2}\right)^n \to 0$ 　　　　　　　　$\left(\text{Formula 4 with } x = -\dfrac{1}{2}\right)$

5. $\left(\dfrac{n-2}{n}\right)^n = \left(1 - \dfrac{2}{n}\right)^n \to e^{-2}$ 　　(Formula 5 with $x = -2$)

6. $\dfrac{100^n}{n!} \to 0$ 　　　　　　　　　(Formula 6 with $x = 100$)

Example 12   Does the sequence whose nth term is

$$a_n = \left(\frac{n+1}{n-1}\right)^n$$

converge? If so, find $\lim_{n \to \infty} a_n$.

Solution 1   *(L'Hôpital's Rule)*   The limit leads to the indeterminate form 1^∞. To apply l'Hôpital's rule, we first change the form to $\infty \cdot 0$ by taking the natural logarithm of a_n (as in Section 6.4, Example 7):

$$\ln a_n = \ln \left(\frac{n+1}{n-1}\right)^n = n \ln \left(\frac{n+1}{n-1}\right).$$

Then,

$$\lim_{n \to \infty} \ln a_n = \lim_{n \to \infty} n \ln \left(\frac{n+1}{n-1}\right) \qquad (\infty \cdot 0)$$

$$= \lim_{n \to \infty} \frac{\ln \left(\frac{n+1}{n-1}\right)}{1/n} \qquad \left(\frac{0}{0}\right)$$

$$= \lim_{n \to \infty} \frac{-2/(n^2-1)}{-1/n^2} \qquad \text{(l'Hôpital's rule)}$$

$$= \lim_{n \to \infty} \frac{2n^2}{n^2-1} = 2.$$

Therefore,

$$a_n = e^{\ln a_n} \to e^2.$$

The limit exists and equals e^2.

Solution 2   (Less general, but works well in this case.) We have

$$\left(\frac{n+1}{n-1}\right)^n = \frac{\left(\frac{n+1}{n}\right)^n}{\left(\frac{n-1}{n}\right)^n} = \frac{\left(1+\frac{1}{n}\right)^n}{\left(1+\frac{-1}{n}\right)^n}.$$

Hence,

$$\lim_{n \to \infty} \left(\frac{n+1}{n-1}\right)^n = \frac{\lim_{n \to \infty} \left(1+\frac{1}{n}\right)^n}{\lim_{n \to \infty} \left(1+\frac{-1}{n}\right)^n} = \frac{e^1}{e^{-1}} \qquad \left(\begin{array}{l}\text{Table 8.1,} \\ \text{Formula 5,} \\ \text{with } x = 1 \\ \text{and with } x = -1\end{array}\right)$$

$$= e^2$$

✳ Picard's Method for Finding Roots

The problem of finding the roots of the equation

$$f(x) = 0 \qquad (6)$$

is equivalent to that of finding the solutions of the equation

$$g(x) = f(x) + x = x, \qquad (7)$$

obtained by adding x to both sides of Eq. (6). Any value of x that satisfies Eq. (7) satisfies (6), and conversely. By this simple change we cast Eq. (6) into a form that may render it solvable on a computer by a powerful method called **Picard's method** (after the French mathematician Charles Émile Picard, 1856–1941).

If the domain of g contains the range of g, we can start with a point x_0 in the domain and apply g repeatedly to get

$$x_1 = g(x_0), \qquad x_2 = g(x_1), \qquad x_3 = g(x_2), \qquad \ldots . \qquad (8)$$

Under simple restrictions that we shall describe shortly, the sequence generated by the recursion formula $x_{n+1} = g(x_n)$ will converge to a point x for which $g(x) = x$. This point solves the equation $f(x) = 0$ because

$$f(x) = g(x) - x = x - x = 0. \qquad (9)$$

A point x for which $g(x) = x$ is a **fixed point** of g. We see in Eq. (9) that the fixed points of g are precisely the roots of f.

We begin with an example whose outcome we know so that we can see how the successive approximations work. You will find a BASIC program for Picard's method after the exercises at the end of this section.

Example 13   Solve the equation $(1/4)x + 3 = x$.

Solution By algebra we know that the solution is $x = 4$. To apply Picard's method, we take

$$g(x) = \frac{1}{4}x + 3,$$

choose a starting point, say $x_0 = 1$, and calculate the initial terms of the sequence $x_{n+1} = g(x_n)$. Table 8.2 lists the results. In ten steps, the solution of the original equation is found with an error of magnitude less than 3×10^{-6}. Figure 8.5 shows the geometry of the solution.

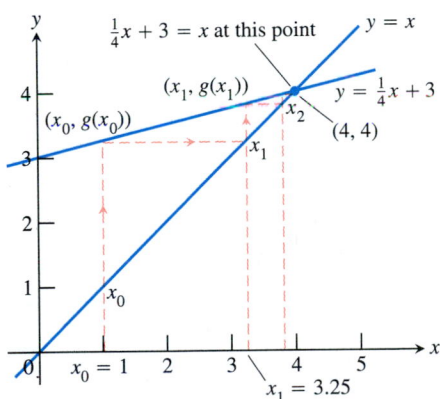

8.5 The geometric interpretation of the Picard solution of the equation $g(x) = (1/4)x + 3 = x$ in Example 13. We start with $x_0 = 1$ and calculate the first y-value, $g(x_0)$. This becomes the second x-value, x_1. The second y-value, $g(x_1)$, becomes the third x-value, x_2, and so on. The process is shown here as a path (called the *iteration path*) that starts at $x_0 = 1$, moves up to $(x_0, g(x_0)) = (x_0, x_1)$, over to (x_1, x_1), up to $(x_1, g(x_1))$, and so on. The path converges to the point where the graph of g meets the line $y = x$, the point where $g(x) = x$.

TABLE 8.2
Successive iterates of $g(x) = (1/4)x + 3$, starting with $x_0 = 1$

x_n	$x_{n+1} = g(x_n) = (1/4)x_n + 3$
$x_0 = 1$	$x_1 = g(x_0) = (1/4)(1) + 3 = 3.25$
$x_1 = 3.25$	$x_2 = g(x_1) = (1/4)(3.25) + 3 = 3.8125$
$x_2 = 3.8125$	$x_3 = g(x_2) = 3.9531\ 25$
$x_3 = 3.9531\ 25$	$x_4 = 3.9882\ 8125$
$\vdots$	$x_5 = 3.9970\ 70313$
	$x_6 = 3.9992\ 67578$
	$x_7 = 3.9998\ 16895$
	$x_8 = 3.9999\ 54224$
	$x_9 = 3.9999\ 88556$
	$x_{10} = 3.9999\ 97139$
	$\vdots$

Example 14   Solve the equation $\cos x = x$.

Solution   We take $g(x) = \cos x$, choose $x_0 = 1$ as a starting value, and use the recursion formula $x_{n+1} = g(x_n)$ to find

$$x_0 = 1, \qquad x_1 = \cos 1, \qquad x_2 = \cos (x_1), \ldots.$$

We can approximate the first 50 terms or so on a calculator in radian mode by entering 1 and pressing $\boxed{\cos}$ repeatedly. The display stops changing when $\cos x = x$ to the number of decimal places in the display.

Try it for yourself. As you press the key, notice that the successive approximations lie alternately above and below the fixed point

$$x = 0.7390 \ 85133 \ldots.$$

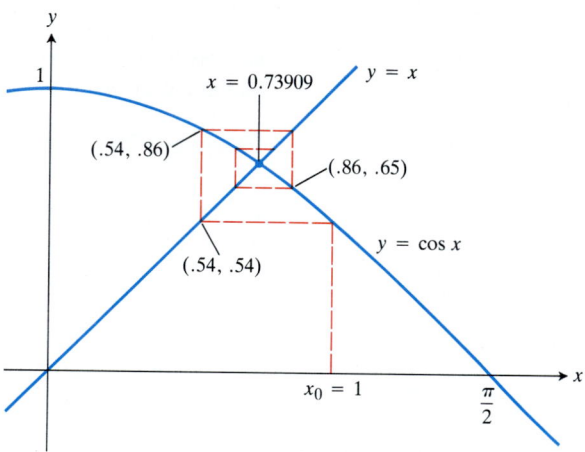

8.6   The solution of $\cos x = x$ by Picard's method starting at $x_0 = 1$ (Example 14).

Figure 8.6 shows that the values oscillate this way because the path of the procedure spirals around the fixed point.

Example 15   Picard's method will not solve the equation

$$g(x) = 4x - 12 = x.$$

As Fig. 8.7 shows, any choice of x_0 except $x_0 = 4$, the solution itself, generates a divergent sequence that moves away from the solution.

The difficulty in Example 15 can be traced to the fact that the slope of the line $y = 4x - 12$ exceeds 1, the slope of the line $y = x$. Conversely, the process worked in Example 13 because the slope of the line $y = (1/4)x + 3$ was numerically less than 1. A theorem from advanced calculus tells us that if $g'(x)$ is continuous on a closed interval I whose interior contains a solution of the equation $g(x) = x$, and if $|g'(x)| < 1$ on I, then any choice of x_0 in the interior of I will lead to the solution. (See Exercise 99 about what to do if $|g'(x)| > 1$.)

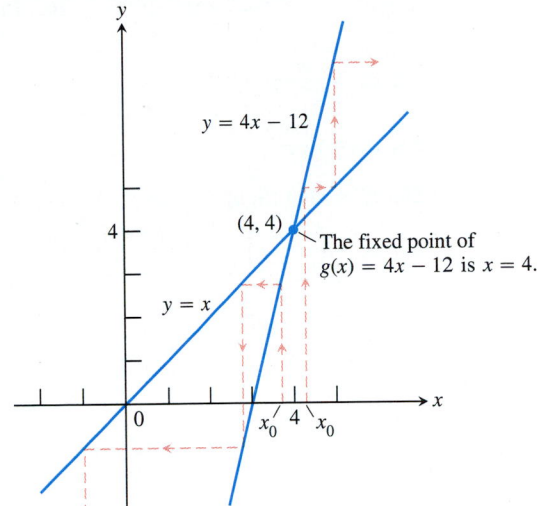

8.7 Applying the Picard method to $g(x) = 4x - 12$ will not find the fixed point unless x_0 is 4 itself (Example 15).

EXERCISES 8.1

Each of Exercises 1–4 gives a formula for the nth term a_n of a sequence $\{a_n\}$. Find the values of a_1, a_2, a_3, and a_4.

1. $a_n = \dfrac{1-n}{n^2}$ **2.** $a_n = \dfrac{1}{n!}$

3. $a_n = \dfrac{(-1)^{n+1}}{2n-1}$ **4.** $a_n = 2 + (-1)^n$

Each of Exercises 5–10 gives the first term or two of a sequence and a recursion formula for the remaining terms. Write out the first ten terms of each sequence.

5. $x_1 = 1$, $x_{n+1} = x_n + (1/2^n)$

6. $x_1 = 1$, $x_{n+1} = x_n/(n+1)$

7. $x_1 = 2$, $x_{n+1} = (-1)^{n+1}x_n/2$

8. $x_1 = -2$, $x_{n+1} = nx_n/(n+1)$

9. $x_1 = x_2 = 1$, $x_{n+2} = x_{n+1} - x_n$

10. $x_1 = 2$, $x_2 = -1$, $x_{n+2} = x_{n+1}/x_n$

Which of the sequences $\{a_n\}$ in Exercises 11–70 converge and which diverge? Find the limit of each convergent sequence.

11. $a_n = 2 + (0.1)^n$ **12.** $a_n = \dfrac{n + (-1)^n}{n}$

13. $a_n = \dfrac{1 - 2n}{1 + 2n}$ **14.** $a_n = \dfrac{2n+1}{1 - 3\sqrt{n}}$

15. $a_n = \dfrac{1 - 5n^4}{n^4 + 8n^3}$ **16.** $a_n = \dfrac{n+3}{n^2 + 5n + 6}$

17. $a_n = \dfrac{n^2 - 2n + 1}{n - 1}$ **18.** $a_n = \dfrac{1 - n^3}{70 - 4n^2}$

19. $a_n = 1 + (-1)^n$ **20.** $a_n = (-1)^n\left(1 - \dfrac{1}{n}\right)$

21. $a_n = \left(\dfrac{n+1}{2n}\right)\left(1 - \dfrac{1}{n}\right)$

22. $a_n = \left(2 - \dfrac{1}{2^n}\right)\left(3 + \dfrac{1}{2^n}\right)$

23. $a_n = \dfrac{(-1)^{n+1}}{2n - 1}$ **24.** $a_n = \left(-\dfrac{1}{2}\right)^n$

25. $a_n = \dfrac{\sin n}{n}$ **26.** $a_n = \dfrac{\sin^2 n}{2^n}$

27. $a_n = \sqrt{\dfrac{2n}{n+1}}$ **28.** $a_n = \sin\left(\dfrac{\pi}{2} + \dfrac{1}{n}\right)$

29. $a_n = \tan^{-1} n$ **30.** $a_n = \ln n - \ln(n+1)$

31. $a_n = \dfrac{n}{2^n}$ **32.** $a_n = \dfrac{3^n}{n^3}$

33. $a_n = \dfrac{\ln(n+1)}{\sqrt{n}}$ **34.** $a_n = \dfrac{\ln n}{\ln 2n}$

35. $a_n = 8^{1/n}$ **36.** $a_n = (0.03)^{1/n}$

37. $a_n = \left(1 + \dfrac{7}{n}\right)^n$ **38.** $a_n = \left(1 - \dfrac{1}{n}\right)^n$

39. $a_n = \dfrac{1}{(0.9)^n}$ **40.** $a_n = n\pi \cos n\pi$

41. $a_n = \sqrt[n]{10n}$ **42.** $a_n = \sqrt[n]{n^2}$

43. $a_n = \left(\dfrac{3}{n}\right)^{1/n}$ **44.** $a_n = (n+4)^{1/(n+4)}$

45. $a_n = \dfrac{\ln n}{n^{1/n}}$ **46.** $a_n = \sqrt[n]{4^n n}$

47. $a_n = \left(\dfrac{1}{3}\right)^n + \dfrac{1}{\sqrt{2^n}}$ **48.** $a_n = \sqrt[n]{3^{2n+1}}$

49. $a_n = \dfrac{n!}{n^n}$ (*Hint:* Compare the quotient with $1/n$.)

50. $a_n = \dfrac{(-4)^n}{n!}$

51. $a_n = \left(\dfrac{1}{n}\right)^{1/\ln n}$

52. $a_n = \dfrac{n!}{2^n \cdot 3^n}$

53. $a_n = \dfrac{n!}{10^{6n}}$

54. $a_n = \dfrac{3^n \cdot 6^n}{2^{-n} \cdot n!}$

55. $a_n = \tanh n$

56. $a_n = \sinh(\ln n)$

57. $a_n = \ln\left(1 + \dfrac{1}{n}\right)^n$

58. $a_n = \left(\dfrac{n}{n+1}\right)^n$

59. $a_n = \left(\dfrac{3n+1}{3n-1}\right)^n$

60. $a_n = \left(1 - \dfrac{1}{n^2}\right)^n$

61. $a_n = \sqrt[n]{\dfrac{x^n}{2n+1}}, \quad x > 0$

62. $a_n = \dfrac{\left(\dfrac{10}{11}\right)^n}{\left(\dfrac{9}{10}\right)^n + \left(\dfrac{11}{12}\right)^n}$

63. $a_n = \dfrac{1}{n}\displaystyle\int_1^n \dfrac{1}{x}\,dx$

64. $a_n = \displaystyle\int_1^n \dfrac{1}{x^p}\,dx, \quad p > 1$

65. $a_n = \sqrt[n]{n^2 + n}$

66. $a_n = \dfrac{(\ln n)^{200}}{n}$

67. $a_n = \dfrac{n^2}{2n-1}\sin\dfrac{1}{n}$

68. $a_n = n\left(1 - \cos\dfrac{1}{n}\right)$

69. $a_n = n - \sqrt{n^2 - n}$

70. $a_n = \dfrac{1}{\sqrt{n^2 - 1} - \sqrt{n^2 + n}}$

CALCULATOR In Exercises 71–74, experiment with a calculator to identify a value of N that will make the inequality hold for $n \geq N$.

71. $\left|\sqrt[n]{0.5} - 1\right| < 10^{-3}$

72. $\left|\sqrt[n]{n} - 1\right| < 10^{-3}$

73. $(0.9)^n < 10^{-3}$

74. $2^n/n! < 10^{-7}$

75. CALCULATOR *A recursive definition of $\pi/2$.* If you start with $x_1 = 1$ and define the subsequent terms of $\{x_n\}$ by the rule $x_n = x_{n-1} + \cos x_{n-1}$, you generate a sequence that converges rapidly to $\pi/2$. Try it. Figure 8.8 explains what is going on.

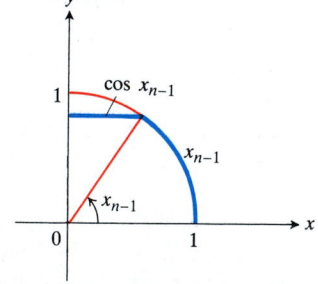

8.8 The length $\pi/2$ of the circular arc is approximated by $x_{n-1} + \cos x_{n-1}$ (Exercise 75).

76. The first term of a sequence is $x_1 = 1$. Each succeeding term is the sum of all those that come before it:

$$x_{n+1} = x_1 + x_2 + \cdots + x_n.$$

Write out enough early terms of the sequence to deduce a general formula for x_n that holds for $n \geq 2$.

77. A sequence of rational numbers is described as follows:

$$\frac{1}{1}, \frac{3}{2}, \frac{7}{5}, \frac{17}{12}, \cdots, \frac{a}{b}, \frac{a+2b}{a+b}, \cdots$$

Here the numerators form one sequence, the denominators form a second sequence, and their ratios form a third sequence. Let x_n and y_n be, respectively, the numerator and the denominator of the nth fraction $r_n = x_n/y_n$.

a) Verify that $x_1^2 - 2y_1^2 = -1$, $x_2^2 - 2y_2^2 = +1$ and, more generally, that if $a^2 - 2b^2 = -1$ or $+1$, then

$$(a + 2b)^2 - 2(a + b)^2 = +1 \quad \text{or} \quad -1,$$

respectively.

b) The fractions $r_n = x_n/y_n$ approach a limit as n increases. What is that limit? (*Hint:* Use part (a) to show that $r_n^2 - 2 = \pm(1/y_n)^2$ and that y_n is not less than n.)

78. The **Fibonacci sequence** is defined recursively as follows: $x_1 = 1$, $x_2 = 1$, and $x_{n+2} = x_n + x_{n+1}$. It can also be described in another way that amounts to "zippering" two sequences together. (Although it may seem unnecessary, we found it easier to write a BASIC program doing it this way.) Let $a_n = x_{2n-1}$ and $b_n = x_{2n}$. Then the Fibonacci sequence can be written as

$$a_1, b_1, a_2, b_2, a_3, b_3, \ldots, a_k, b_k, a_{k+1}, b_{k+1}, \ldots.$$

The recursion formulas now become $a_1 = b_1 = 1$ and $a_{n+1} = a_n + b_n$, $b_{n+1} = a_{n+1} + b_n$. Verify that the following BASIC program will thus give the first 12 terms of the Fibonacci sequence, and fill in the rest of the table. (You do not need a computer to do the calculations.) The sequence is named for Leonardo Fibonacci (c. 1170–1240), who used the sequence to model successive generations of a rabbit population.

PROGRAM		n	a_n	b_n
10	a = 1	1	1	1
20	b = 1	2	2	
30	FOR n = 1 to 6	3		
40	PRINT n,a,b	4		
50	a = a + b	5		
60	b = a + b	6		
70	NEXT n			
80	END			

Note: The a on the right in line 60 comes from the a on the left in line 50.

79. *Sequences generated by Newton's method.* Newton's method, applied to a differentiable function $f(x)$, begins with a starting value x_0 and constructs from it a sequence of numbers $\{x_n\}$ that under favorable circumstances converges to a zero of f. The recursion formula for the sequence is

$$x_{n+1} = x_n - \frac{f(x_n)}{f'(x_n)}.$$

Do the following sequences converge? If so, to what value? In each case, begin by identifying the function f that generates the sequence.

a) $x_0 = 1, \quad x_{n+1} = x_n - \dfrac{x_n^2 - 2}{2x_n} = \dfrac{x_n}{2} + \dfrac{1}{x_n}$

b) $x_0 = 1, \quad x_{n+1} = x_n - \dfrac{\tan x_n - 1}{\sec^2 x_n}$

c) $x_0 = 1, \quad x_{n+1} = x_n - 1$

80. CALCULATOR Newton's method uses the formula $x_{n+1} = (x_n + a/x_n)/2$ to generate a sequence of approximations to the positive solution of the equation $x^2 - a = 0$, $a > 0$. Starting with $x_0 = 1$ and $a = 3$, calculate the successive terms of the sequence until you have approximated $\sqrt{3}$ as accurately as your calculator permits.

81. *Pythagorean triples.*   A triple of positive integers a, b, and c is called a **Pythagorean triple** if $a^2 + b^2 = c^2$. Let a be an odd positive integer and let

$$b = \left\lfloor \frac{a^2}{2} \right\rfloor \quad \text{and} \quad c = \left\lceil \frac{a^2}{2} \right\rceil$$

be, respectively, the integer floor and ceiling for $a^2/2$.

a) Show that $a^2 + b^2 = c^2$ (Fig. 8.9). (*Hint:* Let $a = 2n + 1$ and express b and c in terms of n.)

b) By direct calculation, or by appealing to Fig. 8.9, find

$$\lim_{a \to \infty} \frac{\left\lfloor \dfrac{a^2}{2} \right\rfloor}{\left\lceil \dfrac{a^2}{2} \right\rceil}.$$

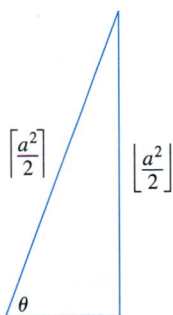

8.9   The right triangle for the Pythagorean triple in Exercise 81.

82. *The nth root of n!*

a) Show that $\lim_{n \to \infty} (2n\pi)^{1/(2n)} = 1$

and hence, using Stirling's approximation (Chapter 7, Miscellaneous Exercise 162a), that

$$\sqrt[n]{n!} \approx \frac{n}{e} \quad \text{for large values of } n.$$

b) **CALCULATOR** Test the approximation in (a) for $n = 40$, $50, 60, \ldots$, as far as your calculator will allow.

83. Suppose that $f(x)$ is defined for all $0 \le x \le 1$, that f is differentiable, and that $f(0) = 0$. Define a sequence $\{a_n\}$ by the rule $a_n = nf(1/n)$. Show that $\lim_{n \to \infty} a_n = f'(0)$.

Use the result of Exercise 83 to find the limits of the sequences whose nth terms appear in Exercises 84–86.

84. $a_n = n \tan^{-1}\dfrac{1}{n}$

85. $a_n = n(e^{1/n} - 1)$

86. $a_n = n \ln\left(1 + \dfrac{2}{n}\right)$

87. a) Assuming that $\lim_{n \to \infty} (1/n^c) = 0$ if c is any positive constant, show that

$$\lim_{n \to \infty} \frac{\ln n}{n^c} = 0$$

if c is any positive constant.

b) Prove that $\lim_{n \to \infty} (1/n^c) = 0$ if c is any positive constant. (*Hint:* If $\epsilon = 0.001$ and $c = 0.04$, how large should N be to ensure that $|1/n^c - 0| < \epsilon$ if $n > N$?)

88. Prove Theorem 2.

89. Prove Theorem 3. (*Outline of proof:* Assume the hypotheses of the theorem and let ϵ be any positive number. For this ϵ there exists a $\delta > 0$ such that, for all x,

$$|x - L| < \delta \Rightarrow |f(x) - f(L)| < \epsilon.$$

For such a $\delta > 0$, there exists a positive integer N such that, for all n,

$$n > N \Rightarrow |a_n - L| < \delta.$$

What is the conclusion?)

90. *The zipper theorem.* Prove the "zipper theorem" for sequences: If $\{a_n\}$ and $\{b_n\}$ both converge to L, then the sequence

$$a_1, \quad b_1, \quad a_2, \quad b_2, \quad \ldots, \quad a_n, \quad b_n, \quad \ldots$$

converges to L.

✱ Picard's Method

CALCULATOR   Use Picard's method to solve the equations in Exercises 91–96 to five decimal places.

91. $\sqrt{x} = x$

92. $x^2 = x$

93. $\cos x + x = 0$

94. $\cos x = x + 1$

95. $x - \sin x = 0.1$

96. $\sqrt{x} = 4 - \sqrt{1 + x}$ (*Hint:* Square both sides first.)

97. Solving the equation $\sqrt{x} = x$ by Picard's method finds the solution $x = 1$ but not the solution $x = 0$. Why? (*Hint:* Graph $y = x$ and $y = \sqrt{x}$ together.)

98. Solving the equation $x^2 = x$ by Picard's method finds the solution $x = 0$ but not the solution $x = 1$. Why? (*Hint:* Graph $y = x^2$ and $y = x$ together.)

99. *If $|g'(x)| > 1$, try g^{-1} instead.*   There is more to be learned from Examples 13 and 15. Example 15 showed that we cannot apply Picard's method to find a fixed point of

$g(x) = 4x - 12$. But we can apply the method to find a fixed point of $g^{-1}(x) = (1/4)x + 3$ because the derivative of g^{-1} is $1/4$, whose value is less than 1 in magnitude on any interval. In Example 13, we found the fixed point of g^{-1} to be $x = 4$. Now notice that 4 is also a fixed point of g, since

$$g(4) = 4(4) - 12 = 4.$$

In finding the fixed point of g^{-1}, we found the fixed point of g.

A function and its inverse always have the same fixed points. The graphs of the functions are symmetric about the line $y = x$ and therefore intersect the line at the same points.

We now see that the application of Picard's method is quite broad. For suppose g is one-to-one, with a continuous first derivative whose magnitude is greater than 1 on a closed interval I whose interior contains a fixed point of g. Then the derivative of g^{-1}, being the reciprocal of g', has magnitude less than 1 on I. Picard's method applied to g^{-1} on I will find the fixed point of g. As cases in point, find the fixed points of

a) $g(x) = 2x + 3$ b) $g(x) = 1 - 4x$

100. COMPUTER *The zeros of $x^2 - 2x - 3$.* Using a computer programmed for Picard's method (there is a program at the end of the exercises), try solving the equation $x^2 - 2x - 3 = 0$ in three different ways:

a) Write the equation as $x = \pm\sqrt{2x + 3}$ and take $g(x) = \pm\sqrt{2x + 3}$.

b) Write the equation as $x = (x^2 - 3)/2$ and take $g(x) = (x^2 - 3)/2$.

c) Write the equation as $x = 3/(x - 2)$ and take $g(x) = 3/(x - 2)$.

Explain what happens in each case.

EXPLORER PROGRAMS

Picard's Fixed Point Method	Generates graphs and iteration paths like those in Figs. 8.5, 8.6, and 8.7. In addition to being a powerful equation solver, the program enables you to toggle back and forth between the graphs of a function and its inverse.
Sequences and Series	Generates terms of one or two sequences and graphs them. You may define the sequences recursively or by giving formulas for their nth terms. Enables you to look for graphical and numerical evidence of convergence or divergence

---

✳ A COMPUTER PROGRAM FOR PICARD'S METHOD

The BASIC program listed below uses Picard's iteration method to solve the equation $g(x) = x$. You enter a formula for $g(x)$, a starting value for x, and a numerical tolerance (called TL) for how close $g(x)$ and x need to be before the iterations stop. The program also contains a limit (currently set at 100, but changeable) on how many iterations the computer should take before reporting nonconvergence.

Program Listing

Here is a listing of the program with $g(x) = \sqrt{2x + 3}$, to find the positive root of $x^2 - 2x - 3 = 0$.

LINE		COMMENT
10	DEF FNG(X) = SQR(2 * X + 3)	Enters $g(x) = \sqrt{2x + 3}$.
20	LIMIT = 100	The program will count 100 steps before reporting nonconvergence.
30	INPUT "ENTER STARTING VALUE (X) "; X	Prompt for entering the starting value.
40	INPUT "ENTER VALUE FOR TERMINATION "; TL	Tells the computer when it has come close enough to stop; 10^{-6} might be good.
50	FOR ITER = 1 TO LIMIT	Iterations start at 1, and end at LIMIT if TL is not reached

(continued)

LINE		COMMENT
60	PRINT "AT ITER NO ";ITER;	
	" X = "; X; " G(X) = "; FNG(X)	
70	IF ABS(X − FNG(X)) < TL THEN 120	
80	X = FNG(X)	Redefines x for the next step.
90	NEXT ITER	Closes the FOR NEXT loop.
100	PRINT "DID NOT CONVERGE IN ";	
	LIMIT; " ITERATIONS"	
110	STOP	
120	PRINT "CONVERGED IN "; ITER; " ITERATIONS"	
130	END	

8.2 Infinite Series

In mathematics and science we often use infinite polynomials like

$$1 + x + x^2 + x^3 + \cdots + x^n + \cdots$$

to represent functions (the series above represents $1/(1 − x)$ for $|x| < 1$), evaluate nonelementary integrals, and solve differential equations. For any particular value of x, such a polynomial is calculated as an infinite sum of constants, a sum we call an *infinite series*. The goal of this and the next three sections is to learn to work with infinite series. Then, in Sections 8.6–8.8, we shall build on what we have learned to study infinite polynomials as infinite series of powers of x.

We begin by asking how to assign meaning to an expression like

$$1 + \frac{1}{2} + \frac{1}{4} + \frac{1}{8} + \frac{1}{16} + \cdots.$$

The way to do so is not to try to add all the terms at once (we cannot) but rather to add the terms one at a time from the beginning and look for a pattern in how these partial sums grow. When we do this, we find the following.

Partial sum		Value
first:	$s_1 = 1$	1
second:	$s_2 = 1 + \frac{1}{2}$	$2 − \frac{1}{2}$
third:	$s_3 = 1 + \frac{1}{2} + \frac{1}{4}$	$2 − \frac{1}{4}$
$\vdots$		
nth:	$s_n = 1 + \frac{1}{2} + \frac{1}{4} + \cdots + \frac{1}{2^{n-1}}$	$2 − \frac{1}{2^{n-1}}$ $\left(\begin{array}{c}\text{After some}\\\text{algebra}\end{array}\right)$

Indeed there is a pattern. The partial sums form a sequence whose nth term is

$$s_n = 2 − \frac{1}{2^{n-1}},$$

and this sequence converges to 2. We say

"the sum of the series $1 + \frac{1}{2} + \frac{1}{4} + \cdots + \frac{1}{2^{n-1}} + \cdots$ is 2."

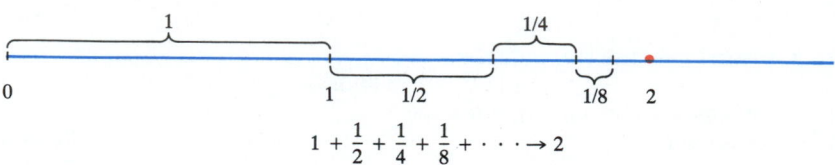

8.10 As the lengths 1, 1/2, 1/4, $1/8, \ldots,$ are added one by one, the sum approaches 2.

$$1 + \frac{1}{2} + \frac{1}{4} + \frac{1}{8} + \cdots \to 2$$

Is the sum of any finite number of terms in the series 2? No. Can we actually add an infinite number of terms one by one? No. But we can still define their sum by defining it to be the limit of the sequence of partial sums as $n \to \infty$, in this case 2 (Fig. 8.10). Our knowledge of sequences and limits enables us to break away from the confines of finite sums.

DEFINITIONS

Given a sequence of numbers $\{a_n\}$, an expression of the form

$$a_1 + a_2 + a_3 + \cdots + a_n + \cdots \tag{1}$$

is an **infinite series**. The number a_n is the **nth term** of the series. The sequence $\{s_n\}$ defined by

$$s_1 = a_1$$
$$s_2 = a_1 + a_2$$
$$\vdots \tag{2}$$
$$s_n = a_1 + a_2 + \cdots + a_n = \sum_{k=1}^{n} a_k$$

is the **sequence of partial sums** of the series, the number s_n being the **nth partial sum.** If the sequence of partial sums converges to a limit L, we say that the series **converges** and that its **sum** is L. In this case, we also write

$$a_1 + a_2 + \cdots + a_n + \cdots = \sum_{n=1}^{\infty} a_n = L. \tag{3}$$

If the sequence of partial sums of the series does not converge, we say that the series **diverges.**

When we begin to study a given series $a_1 + a_2 + \cdots + a_n + \cdots$, we might not know whether it converges or diverges. In either case, it is convenient to use sigma notation to write the series as

$$\sum_{n=1}^{\infty} a_n, \quad \sum_{k=1}^{\infty} a_k, \quad \text{or} \quad \sum a_n.$$

The first of these is read "summation from n equals 1 to infinity of a_n"; the second as "summation from k equals 1 to infinity of a_k"; and the third as "summation a_n."

Geometric Series

Geometric series are series of the form

$$a + ar + ar^2 + \cdots + ar^{n-1} + \cdots = \sum_{n=1}^{\infty} ar^{n-1} \tag{4}$$

in which a and r are fixed real numbers and $a \neq 0$. The **ratio** r can be positive, as in

$$1 + \frac{1}{2} + \frac{1}{4} + \cdots + \frac{1}{2^{n-1}} + \cdots, \tag{5}$$

or negative, as in

$$1 - \frac{1}{3} + \frac{1}{9} - \cdots + (-1)^{n-1}\frac{1}{3^{n-1}} + \cdots. \tag{6}$$

If $r = 1$, the nth partial sum of the geometric series in (4) is

$$s_n = a + a(1) + a(1)^2 + \cdots + a(1)^{n-1} = na,$$

and the series diverges because $\lim_{n\to\infty} s_n = \pm\infty$, depending on the sign of a. If $r \neq 1$, we can determine the convergence or divergence of the series in the following way. We multiply the nth partial sum

$$s_n = a + ar + ar^2 + \cdots + ar^{n-1}$$

by r, obtaining

$$rs_n = ar + ar^2 + \cdots + ar^{n-1} + ar^n.$$

We then subtract rs_n from s_n. Most of the terms on the right cancel when we do this, leaving only

$$s_n - rs_n = a - ar^n \qquad \text{or} \qquad s_n(1-r) = a(1-r^n). \tag{7}$$

We solve for s_n, obtaining

$$s_n = \frac{a(1-r^n)}{(1-r)} \qquad (r \neq 1). \tag{8}$$

If $|r| < 1$, then $r^n \to 0$ as $n \to \infty$ (as we saw in Section 8.1), and $s_n \to a/(1-r)$. In other words, the series converges to $a/(1-r)$. If $|r| > 1$, then $|r^n| \to \infty$ and the series diverges.

If $|r| < 1$, the geometric series converges and

$$\sum_{n=1}^{\infty} ar^{n-1} = \frac{a}{1-r}. \tag{9}$$

If $|r| \geq 1$, the series diverges.

Example 1 The geometric series with $a = 1/9$ and $r = 1/3$ is

$$\frac{1}{9} + \frac{1}{27} + \frac{1}{81} + \cdots = \frac{1}{9}\left(1 + \frac{1}{3} + \frac{1}{3^2} + \cdots\right) = \frac{1/9}{1-(1/3)} = \frac{1}{6}.$$

Example 2 The geometric series with $a = 4$ and $r = -1/2$ is

$$4 - 2 + 1 - \frac{1}{2} + \frac{1}{4} - \cdots = 4\left(1 - \frac{1}{2} + \frac{1}{4} - \frac{1}{8} + \frac{1}{16} - \cdots\right)$$

$$= \frac{4}{1+(1/2)} = \frac{8}{3}.$$

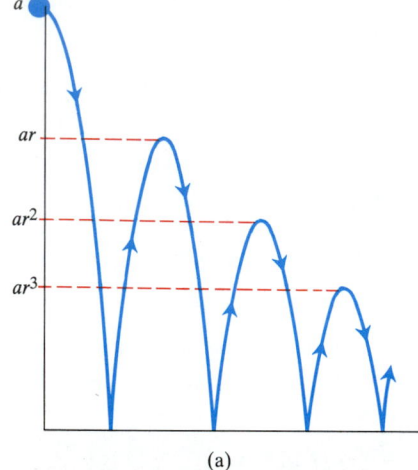

a

ar

ar^2

ar^3

(a)

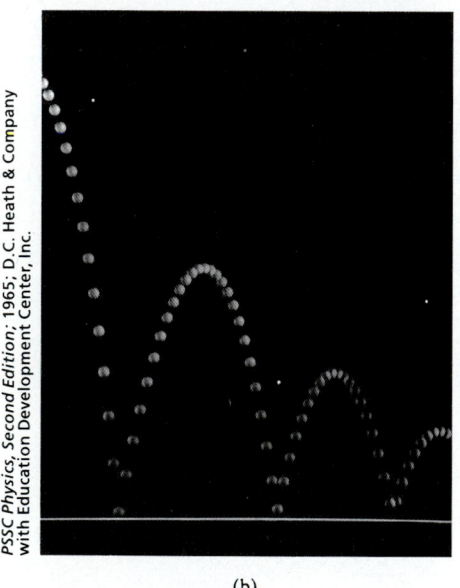

PSSC Physics, Second Edition; 1965; D.C. Heath & Company with Education Development Center, Inc.

(b)

8.11   (a) Example 3 shows how to use a geometric series to calculate the total vertical distance traveled by a bouncing ball if the height of each rebound is reduced by the factor r. (b) A stroboscopic photo of a bouncing ball.

Example 3   You drop a ball from a meters above a flat surface. Each time the ball hits the surface after falling a distance h, it rebounds a distance rh, where r is positive but less than 1. Find the total distance the ball travels up and down (Fig. 8.11).

Solution   The total distance is

$$s = a + \underbrace{2ar + 2ar^2 + 2ar^3 + \cdots}_{\text{This sum is } 2ar/(1 - r).} = a + \frac{2ar}{1 - r} = a\frac{1 + r}{1 - r}.$$

If $a = 6$ m and $r = 2/3$, for instance, the distance is

$$s = 6\frac{1 + (2/3)}{1 - (2/3)} = 6\left(\frac{5/3}{1/3}\right) = 30 \text{ m}.$$

Repeating Decimals

We use geometric series to explain why repeating decimals represent rational numbers.

Example 4   Express the repeating decimal 5.23 23 23 . . . as the ratio of two integers.

Solution

$$5.23\ 23\ 23\ldots = 5 + \frac{23}{100} + \frac{23}{(100)^2} + \frac{23}{(100)^3} + \cdots$$

$$= 5 + \frac{23}{100}\underbrace{\left(1 + \frac{1}{100} + \left(\frac{1}{100}\right)^2 + \cdots\right)}_{1/(1 - 0.01)}$$

$$= 5 + \frac{23}{100}\left(\frac{1}{0.99}\right) = 5 + \frac{23}{99} = \frac{518}{99}.$$

Sums of Other Convergent Series

For geometric series we get a closed form for the nth partial sum and, from that, the formula $s = a/(1 - r)$. Unfortunately, such formulas are rare. What we usually have to do is test the series for convergence and then find some way to estimate the sums of the series that pass the test. Sometimes a convergent series represents a known function like $\sin x$ or $\ln x$. In other cases we might need a computer or calculator to estimate the sum.

The next example, however, is another of those rare cases in which we can find the series' sum from a formula for s_n.

Example 5   Determine whether $\sum_{n=1}^{\infty} [1/(n(n + 1))]$ converges. If it does, find the sum.

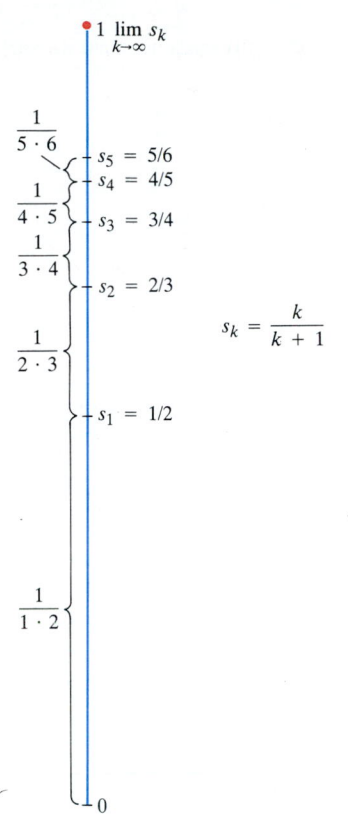

8.12   The sum of the first k terms of the series

$$\sum_{n=1}^{\infty} \frac{1}{n(n+1)}$$

in Example 5 is $k/(k+1)$, and the sum of the series is

$$\lim_{k \to \infty} \frac{k}{k+1} = 1.$$

Solution   We look for a pattern in the sequence of partial sums that might lead us to a closed expression for s_k. The key to success here, as in the integration

$$\int \frac{dx}{x(x+1)} = \int \frac{dx}{x} - \int \frac{dx}{x+1},$$

is partial fractions. The observation that

$$\frac{1}{k(k+1)} = \frac{1}{k} - \frac{1}{k+1} \tag{10}$$

permits us to write the partial sum

$$\sum_{n=1}^{k} \frac{1}{n(n+1)} = \frac{1}{1 \cdot 2} + \frac{1}{2 \cdot 3} + \cdots + \frac{1}{k \cdot (k+1)}$$

as

$$s_k = \left(\frac{1}{1} - \frac{1}{2}\right) + \left(\frac{1}{2} - \frac{1}{3}\right) + \cdots + \left(\frac{1}{k} - \frac{1}{k+1}\right). \tag{11}$$

Removing parentheses and canceling the terms of opposite sign collapses the sum to

$$s_k = 1 - \frac{1}{k+1}. \tag{12}$$

We then see that $s_k \to 1$ as $k \to \infty$. The series converges, and its sum is 1 (Fig. 8.12).

$$\sum_{n=1}^{\infty} \frac{1}{n(n+1)} = 1.$$

Divergent Series

Geometric series with $|r| \geq 1$ are not the only series to diverge.

Example 6   The series

$$\sum_{n=1}^{\infty} n^2 = 1 + 4 + 9 + \cdots + n^2 + \cdots$$

diverges because the partial sums grow beyond every number L. After $n = 1$, the number $s_n = 1 + 4 + 9 + \cdots + n^2$ is greater than n^2.

Example 7   The series

$$\sum_{n=1}^{\infty} \frac{n+1}{n} = \frac{2}{1} + \frac{3}{2} + \frac{4}{3} + \cdots + \frac{n+1}{n} + \cdots$$

diverges because the partial sums eventually outgrow every preassigned number. Each term is greater than 1, so the sum of n terms is greater than n.

A series can diverge without its partial sums becoming large. The partial sums can oscillate between two extremes, for instance, as they do in the next example.

Example 8 The geometric series $\sum_{n=1}^{\infty} (-1)^{n+1}$ diverges because its partial sums alternate between 1 and 0:

$$s_1 = (-1)^2 = 1,$$

$$s_2 = (-1)^2 + (-1)^3 = 1 - 1 = 0,$$

$$s_3 = (-1)^2 + (-1)^3 + (-1)^4 = 1 - 1 + 1 = 1,$$

and so on.

The *n*th-Term Test for Divergence

We present a test for detecting the kind of divergence that occurs in Examples 6, 7, and 8.

The *n*th-Term Test for Divergence

If $\lim_{n \to \infty} a_n \neq 0$, or if $\lim_{n \to \infty} a_n$ fails to exist, then $\sum_{n=1}^{\infty} a_n$ diverges.

When we apply the *n*th-Term Test to the series in Examples 6, 7, and 8, we find that

$$\sum_{n=1}^{\infty} n^2 \quad \text{diverges because } n^2 \to \infty,$$

$$\sum_{n=1}^{\infty} \frac{n+1}{n} \quad \text{diverges because } \frac{n+1}{n} \to 1,$$

$$\sum_{n=1}^{\infty} (-1)^{n+1} \quad \text{diverges because } \lim_{n \to \infty} (-1)^{n+1} \text{ does not exist.}$$

The reason the *n*th-Term Test works is that $\lim_{n \to \infty} a_n$ must equal zero if $\sum a_n$ converges. To see why, let

$$s_n = a_1 + a_2 + \cdots + a_n$$

and suppose that $\sum a_n$ converges to S; that is, $s_n \to S$. When n is large, so is $n - 1$, and both s_n and s_{n-1} are close to S. Their difference, a_n, must then be close to zero. More formally, $a_n = s_n - s_{n-1} \to S - S = 0$.

Example 9 Determine whether each series converges or diverges. If it converges, find its sum.

a) $\displaystyle\sum_{n=1}^{\infty} \frac{n}{2n+5}$ b) $\displaystyle\sum_{n=1}^{\infty} \frac{5(-1)^n}{4^n}$

Solution

a) $\lim_{n \to \infty} \dfrac{n}{2n+5} = \dfrac{1}{2} \neq 0$. The series diverges by the *n*th-Term Test.

b) This is a geometric series with $a = -5/4$ and $r = -1/4$. It converges to

$$\frac{a}{1-r} = \frac{-5/4}{1 + (1/4)} = -1.$$

A Necessary (but Not Sufficient) Condition for Convergence

We often state the nth-Term Test for divergence a shorter way.

If $\sum_{n=1}^{\infty} a_n$ converges, then $a_n \to 0$.

CAUTION   This does *not* mean that Σa_n converges if $a_n \to 0$. A series Σa_n may diverge even though $a_n \to 0$. Thus, $\lim a_n = 0$ is a *necessary* but *not a sufficient* condition for the series Σa_n to converge.

Example 10   The series

$$1 + \underbrace{\frac{1}{2} + \frac{1}{2}}_{2 \text{ terms}} + \underbrace{\frac{1}{4} + \frac{1}{4} + \frac{1}{4} + \frac{1}{4}}_{4 \text{ terms}} + \cdots + \underbrace{\frac{1}{2^n} + \frac{1}{2^n} + \cdots + \frac{1}{2^n}}_{2^n \text{ terms}} + \cdots$$

diverges even though its terms form a sequence that converges to 0.

Whenever we have two convergent series, we can add them, subtract them, and multiply them by constants to make other convergent series. The next theorem gives the details.

THEOREM 5

If $\Sigma a_n = A$ and $\Sigma b_n = B$, then

1. *Sum Rule:* $\sum (a_n + b_n) = A + B$

2. *Difference Rule:* $\sum (a_n - b_n) = A - B$

3. *Constant Multiple Rule:* $\sum k a_n = k \sum a_n = kA$ (Any number k)

As corollaries of Theorem 5 we have the following.

1. Every nonzero constant multiple of a divergent series diverges.

2. If Σa_n converges and Σb_n diverges, then $\Sigma(a_n + b_n)$ and $\Sigma(a_n - b_n)$ both diverge.

The proofs resemble the proofs of similar theorems discussed earlier, and we omit them.

Example 11

a) $\displaystyle\sum_{n=1}^{\infty} \frac{4}{2^{n-1}} = 4 \sum_{n=1}^{\infty} \frac{1}{2^{n-1}} = 4 \frac{1}{1 - (1/2)} = 8$

b) $\displaystyle\sum_{n=1}^{\infty} \frac{3^{n-1} - 1}{6^{n-1}} = \sum_{n=1}^{\infty} \frac{1}{2^{n-1}} - \sum_{n=1}^{\infty} \frac{1}{6^{n-1}} = 2 - \frac{1}{1 - (1/6)} = 2 - \frac{6}{5} = \frac{4}{5}$

We can always add a finite number of terms to a series or delete a finite number of terms from a series without altering its convergence or divergence. If $\sum_{n=1}^{\infty} a_n$

converges, then $\sum_{n=k}^{\infty} a_n$ converges for any $k > 1$ and

$$\sum_{n=1}^{\infty} a_n = a_1 + a_2 + \cdots + a_{k-1} + \sum_{n=k}^{\infty} a_n. \tag{13}$$

Conversely, if $\sum_{n=k}^{\infty} a_n$ **converges for any** $k > 1$, then $\sum_{n=1}^{\infty} a_n$ converges. Thus,

$$\sum_{n=1}^{\infty} \frac{1}{5^n} = \frac{1}{5} + \frac{1}{25} + \frac{1}{125} + \sum_{n=4}^{\infty} \frac{1}{5^n} \tag{14}$$

and

$$\sum_{n=4}^{\infty} \frac{1}{5^n} = \left(\sum_{n=1}^{\infty} \frac{1}{5^n} \right) - \frac{1}{5} - \frac{1}{25} - \frac{1}{125}. \tag{15}$$

Reindexing

As long as we preserve the order of its terms, we can reindex any series without altering its convergence. To raise the starting value of the index h units, replace the n in the formula for a_n by $n - h$:

$$\sum_{n=1}^{\infty} a_n = \sum_{n=1+h}^{\infty} a_{n-h} = a_1 + a_2 + a_3 + \cdots.$$

To lower the starting value of **the index** h **units, replace the** n **in the formula for** a_n **by** $n + h$:

$$\sum_{n=1}^{\infty} a_n = \sum_{n=1-h}^{\infty} a_{n+h} = a_1 + a_2 + a_3 + \cdots.$$

It works like translation.

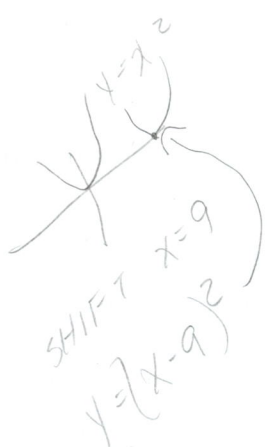

Example 12   We can write the geometric series that starts with

$$1 + \frac{1}{2} + \frac{1}{4} + \cdots$$

as

$$\sum_{n=0}^{\infty} \frac{1}{2^n}, \qquad \sum_{n=5}^{\infty} \frac{1}{2^{n-5}}, \qquad \text{or even} \qquad \sum_{n=-4}^{\infty} \frac{1}{2^{n+4}}.$$

The partial sums **remain the same** no matter what indexing we choose.

We usually give preference to **indexings that** lead to simple expressions. We chose the indexing in Example 11(b) **to begin** with $n = 1$ to match the indexing in the geometric series formulas in Eqs. **(4) and (9)**. Beginning with $n = 0$ instead would have produced the simpler formula

$$\sum_{n=0}^{\infty} \frac{3^n - 1}{6^n}.$$

EXERCISES 8.2

In Exercises 1–6, find a formula for the nth partial sum of each series and use it to find the series' sum if the series converges.

1. $2 + \dfrac{2}{3} + \dfrac{2}{9} + \dfrac{2}{27} + \cdots + \dfrac{2}{3^{n-1}} + \cdots$

2. $\dfrac{9}{100} + \dfrac{9}{100^2} + \dfrac{9}{100^3} + \cdots + \dfrac{9}{100^n} + \cdots$

3. $1 - \dfrac{1}{2} + \dfrac{1}{4} - \dfrac{1}{8} + \cdots + (-1)^{n-1}\dfrac{1}{2^{n-1}} + \cdots$

4. $1 - 2 + 4 - 8 + \cdots + (-1)^{n-1}2^{n-1} + \cdots$

5. $\dfrac{1}{2\cdot 3} + \dfrac{1}{3\cdot 4} + \dfrac{1}{4\cdot 5} + \cdots + \dfrac{1}{(n+1)(n+2)} + \cdots$

6. $\dfrac{5}{1\cdot 2} + \dfrac{5}{2\cdot 3} + \dfrac{5}{3\cdot 4} + \cdots + \dfrac{5}{n(n+1)} + \cdots$

In Exercises 7–14, write out the first few terms of each series to show how the series starts. Then find the sum of the series.

7. $\displaystyle\sum_{n=0}^{\infty} \dfrac{(-1)^n}{4^n}$

8. $\displaystyle\sum_{n=2}^{\infty} \dfrac{1}{4^n}$

9. $\displaystyle\sum_{n=1}^{\infty} \dfrac{7}{4^n}$

10. $\displaystyle\sum_{n=0}^{\infty} (-1)^n \dfrac{5}{4^n}$

11. $\displaystyle\sum_{n=0}^{\infty} \left(\dfrac{5}{2^n} + \dfrac{1}{3^n}\right)$

12. $\displaystyle\sum_{n=0}^{\infty} \left(\dfrac{5}{2^n} - \dfrac{1}{3^n}\right)$

13. $\displaystyle\sum_{n=0}^{\infty} \left(\dfrac{1}{2^n} + \dfrac{(-1)^n}{5^n}\right)$

14. $\displaystyle\sum_{n=0}^{\infty} \left(\dfrac{2^{n+1}}{5^n}\right)$

Use partial fractions to find the sum of each series in Exercises 15–18.

15. $\displaystyle\sum_{n=1}^{\infty} \dfrac{4}{(4n-3)(4n+1)}$

16. $\displaystyle\sum_{n=1}^{\infty} \dfrac{1}{(4n-3)(4n+1)}$

17. $\displaystyle\sum_{n=3}^{\infty} \dfrac{4}{(4n-3)(4n+1)}$

18. $\displaystyle\sum_{n=1}^{\infty} \dfrac{2n+1}{n^2(n+1)^2}$

Which series in Exercises 19–38 converge and which diverge? If a series converges, find its sum.

19. $\displaystyle\sum_{n=0}^{\infty} \left(\dfrac{1}{\sqrt{2}}\right)^n$

20. $\displaystyle\sum_{n=1}^{\infty} \ln \dfrac{1}{n}$

21. $\displaystyle\sum_{n=1}^{\infty} (-1)^{n+1} \dfrac{3}{2^n}$

22. $\displaystyle\sum_{n=1}^{\infty} (\sqrt{2})^n$

23. $\displaystyle\sum_{n=0}^{\infty} \cos n\pi$

24. $\displaystyle\sum_{n=0}^{\infty} \dfrac{\cos n\pi}{5^n}$

25. $\displaystyle\sum_{n=0}^{\infty} e^{-2n}$

26. $\displaystyle\sum_{n=1}^{\infty} \dfrac{n^2+1}{n}$

27. $\displaystyle\sum_{n=1}^{\infty} (-1)^{n+1} n$

28. $\displaystyle\sum_{n=1}^{\infty} \dfrac{2}{10^n}$

29. $\displaystyle\sum_{n=0}^{\infty} \dfrac{2^n-1}{3^n}$

30. $\displaystyle\sum_{n=1}^{\infty} \left(1 - \dfrac{1}{n}\right)^n$

31. $\displaystyle\sum_{n=0}^{\infty} \dfrac{n!}{1000^n}$

32. $\displaystyle\sum_{n=0}^{\infty} \dfrac{1}{x^n}, \quad |x| > 1$

33. $\displaystyle\sum_{n=0}^{\infty} \left(\dfrac{e}{\pi}\right)^n$

34. $\displaystyle\sum_{n=0}^{\infty} \left(\dfrac{\ln(3/2)}{\ln 2}\right)^n$

35. $\displaystyle\sum_{n=1}^{\infty} \ln\left(\dfrac{n}{n+1}\right)$

36. $\displaystyle\sum_{n=1}^{\infty} \ln\left(\dfrac{n}{2n+1}\right)$

37. $\displaystyle\sum_{n=1}^{\infty} \left(\dfrac{1}{\sqrt{n}} - \dfrac{1}{\sqrt{n+1}}\right)$

38. $\displaystyle\sum_{n=0}^{\infty} \dfrac{e^{n\pi}}{\pi^{ne}}$

The series in Exercises 39–42 are geometric series. Find a and r in each case. Express the inequality $|r| < 1$ in terms of x and find the values of x for which the inequality holds and the series converges.

39. $\dfrac{1}{1+x} = \displaystyle\sum_{n=0}^{\infty} (-1)^n x^n$

40. $\dfrac{1}{1+x^2} = \displaystyle\sum_{n=0}^{\infty} (-1)^n x^{2n}$

41. $\dfrac{6}{3-x} = \displaystyle\sum_{n=0}^{\infty} 3\left(\dfrac{x-1}{2}\right)^n$

42. $\dfrac{2+\sin x}{8+2\sin x} = \displaystyle\sum_{n=0}^{\infty} \dfrac{(-1)^n}{2}\left(\dfrac{1}{3+\sin x}\right)^n$

In Exercises 43–48, find the values of x for which the given geometric series converges. Also, find the sum of the series (as a function of x) for those values of x.

43. $\displaystyle\sum_{n=0}^{\infty} 2^n x^n$

44. $\displaystyle\sum_{n=0}^{\infty} (-1)^n x^{2n}$

45. $\displaystyle\sum_{n=0}^{\infty} (-1)^n (x+1)^n$

46. $\displaystyle\sum_{n=0}^{\infty} \left(-\dfrac{1}{2}\right)^n (x-3)^n$

47. $\displaystyle\sum_{n=0}^{\infty} \sin^n x$

48. $\displaystyle\sum_{n=0}^{\infty} (\ln x)^n$

Express the numbers in Exercises 49–56 as ratios of integers.

49. $0.\overline{23} = 0.23\ 23\ 23\ldots$

50. $0.\overline{234} = 0.234\ 234\ 234\ldots$

51. $0.\overline{7} = 0.7777\ldots$

52. $0.\overline{d} = 0.dddd\ldots, \quad$ where d is a digit

53. $0.0\overline{6} = 0.06666\ldots$

54. $1.\overline{414} = 1.414\ 414\ 414\ldots$

55. $1.24\overline{123} = 1.24\ 123\ 123\ 123\ldots$

56. $3.\overline{142857} = 3.142857\ 142857\ldots$

57. A ball is dropped from a height of 4 m. Each time it strikes the pavement after falling from a height of h m it rebounds to a height of $0.75h$ m. Find the total distance the ball travels up and down.

58. *Continuation of Exercise 57.* Find the total number of seconds the ball in Exercise 57 is traveling. (*Hint:* The formula $s = 4.9t^2$ gives $t = \sqrt{s/4.9}$.)

59. The series in Exercise 5 can also be written as

$$\sum_{n=1}^{\infty} \frac{1}{(n+1)(n+2)} \quad \text{and} \quad \sum_{n=-1}^{\infty} \frac{1}{(n+3)(n+4)}.$$

Write it as a sum beginning with (a) $n = -2$, (b) $n = 0$, (c) $n = 5$.

60. The series in Exercise 6 can also be written as

$$\sum_{n=1}^{\infty} \frac{5}{n(n+1)} \quad \text{and} \quad \sum_{n=0}^{\infty} \frac{5}{(n+1)(n+2)}.$$

Write it as a sum beginning with (a) $n = -1$, (b) $n = 3$, (c) $n = 20$.

61. Find a closed-form expression for the nth partial sum of the series

$$\sum_{n=1}^{\infty} (-1)^{n+1}.$$

62. Find a closed-form expression for the nth partial sum of the series

$$\sum_{n=1}^{\infty} \ln\left(\frac{n}{n+1}\right).$$

(*Hint:* Write out the first few partial sums to see what is going on.)

63. Figure 8.13 shows the first five of an infinite sequence of squares. The outermost square has an area of 4 m². Each of the other squares is obtained by joining the midpoints of the sides of the square before it. Find the sum of the areas of all the squares.

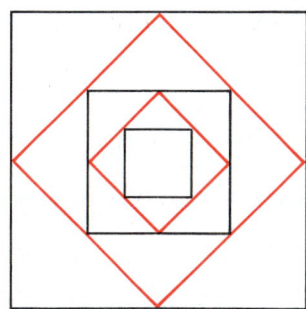

8.13 The first five squares in Exercise 63.

64. Figure 8.14 shows the first three rows of a sequence of rows of semicircles. There are 2^n semicircles in the nth row,

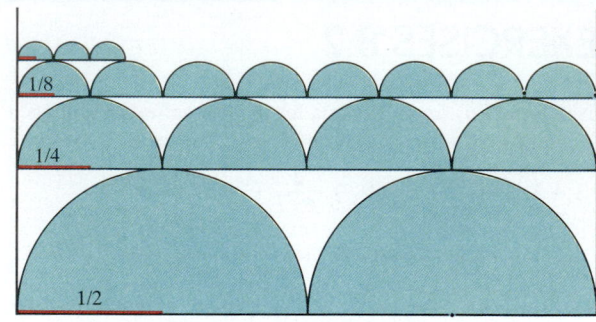

8.14 The semicircles in Exercise 64.

each of radius $1/2^n$. Find the sum of the areas of all the semicircles.

65. *Helga von Koch's snowflake curve.* Helga von Koch's snowflake (p. 144) is a curve of infinite length that encloses a region of finite area. To see why this is so, suppose the curve is generated by starting with an equilateral triangle whose sides have length 1.
   a) Find the length L_n of the nth curve C_n and show that $\lim_{n\to\infty} L_n = \infty$.
   b) Find the area A_n of the region enclosed by C_n and calculate $\lim_{n\to\infty} A_n$.

66. Make up an example of two divergent series whose term-by-term sum converges.

67. Show by example that $\Sigma (a_n/b_n)$ may diverge even though Σa_n and Σb_n converge and no b_n equals 0.

68. Show by example that $\Sigma (a_n/b_n)$ may converge to something other than A/B even when $A = \Sigma a_n$, $B = \Sigma b_n \neq 0$, and no b_n equals 0.

69. Find convergent geometric series $A = \Sigma a_n$ and $B = \Sigma b_n$ that illustrate the fact that $\Sigma a_n b_n$ may converge without being equal to AB.

70. Show that if Σa_n converges, and $a_n \neq 0$ for all n, then $\Sigma (1/a_n)$ diverges.

EXPLORER PROGRAM

Sequences and Series — Enables you to look for numerical and graphical indications of convergence and divergence of a sequence or series. Also plots the initial terms of a series and the series' partial sums in a common graph. You may define your sequence or the terms of your series either recursively or by giving a formula for the nth term.

8.3 Series Without Negative Terms: The Comparison and Integral Tests

Given a series $\Sigma\, a_n$, we have two questions:

1. Does the series converge?
2. If it converges, what is its sum?

Much of the rest of this chapter is devoted to answering the first question. But as a practical matter, the second question is just as important; we come back to it later.

In this section and the next we study series that do not have negative terms. The reason for this restriction is that the partial sums of these series form nondecreasing sequences, and nondecreasing sequences *that are bounded from above* always converge. To show that a series of nonnegative terms converges, we need only show that there is some number beyond which the partial sums never go.

It may at first seem to be a drawback that this approach establishes the fact of convergence without producing the sum of the series in question. Surely it would be better to compute sums of series directly from nice formulas for their partial sums. But in most cases such formulas are not available, and in their absence we have to turn instead to the two-step procedure, as we said earlier, of first establishing convergence and then approximating the sum.

Nondecreasing Sequences

Suppose that $\Sigma\, a_n$ is an infinite series and that $a_n \geq 0$ for every n. Then, when we calculate the partial sums s_1, s_2, s_3, and so on, we see that each one is greater than or equal to its predecessor because $s_{n+1} = s_n + a_n$:

$$s_1 \leq s_2 \leq s_3 \leq \cdots \leq s_n \leq s_{n+1} \leq \cdots . \tag{1}$$

A sequence $\{s_n\}$ with the property that $s_n \leq s_{n+1}$ for every n is called a **nondecreasing sequence.**

There are two kinds of nondecreasing sequences—those that increase beyond any finite bound and those that don't. The former diverge to infinity, so we turn our attention to the other kind: those that do not grow beyond all bounds. Such a sequence is said to be **bounded from above,** and any number M such that $s_n \leq M$ for all n is called an **upper bound** of the sequence.

Example 1   If $s_n = n/(n + 1)$, then 1 is an upper bound and so is any number greater than 1. No number smaller than 1 is an upper bound, so for this sequence 1 is the **least upper bound.**

A nondecreasing sequence that is bounded from above always has a least upper bound, but we shall not prove this fact. Instead, we shall prove that if L is the least upper bound, then the sequence converges to L. The following argument shows why L is the limit.

Suppose we plot the points $(1, s_1), (2, s_2), \ldots, (n, s_n)$ in the xy-plane. If M is an upper bound of the sequence, all these points will lie on or below the line $y = M$ (Fig. 8.15). The line $y = L$ is the lowest such line. None of the points (n, s_n) lies above $y = L$, but some do lie above any lower line $y = L - \epsilon$, if ϵ is a positive number. The sequence converges to L because

a) $s_n \leq L$ for *all* values of n and

b) given any $\epsilon > 0$, there exists at least one integer N for which $s_N > L - \epsilon$.

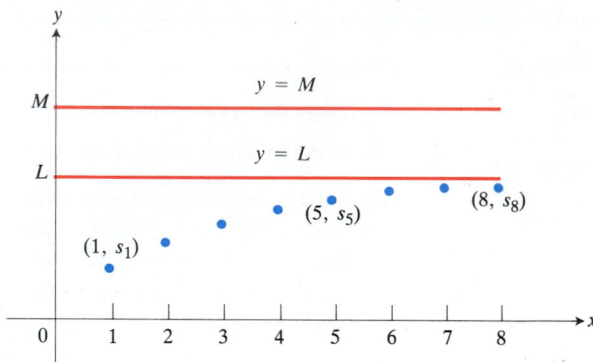

8.15 If the terms of a nondecreasing sequence have an upper bound M, they have a limit $L \le M$.

The fact that $\{s_n\}$ is a nondecreasing sequence tells us further that

$$s_n \ge s_N > L - \epsilon \qquad \text{for all } n \ge N.$$

This means that *all* the numbers s_n beyond the Nth number lie within ϵ of L. This is precisely the condition for L to be the limit of the sequence s_n.

The facts for nondecreasing sequences are summarized in the following theorem. A similar result holds for nonincreasing sequences (Exercise 39).

THEOREM 6

The Nondecreasing Sequence Theorem

A nondecreasing sequence converges if and only if its terms are bounded from above. If all the terms are less than or equal to M, then the limit of the sequence is less than or equal to M as well.

Theorem 6 tells us that we can show that a series $\Sigma \, a_n$ of nonnegative terms converges if we can show that its partial sums are bounded from above. The question is how to find out in any particular instance whether the s_n's have an upper bound.

Sometimes we can show that the s_n's are bounded above by showing that each one is less than or equal to the corresponding partial sum of a series that is already known to converge. The next example shows how this can happen.

Example 2   The series

$$\sum_{n=0}^{\infty} \frac{1}{n!} = 1 + \frac{1}{1!} + \frac{1}{2!} + \frac{1}{3!} + \cdots \tag{2}$$

converges because its terms are all positive and less than or equal to the corresponding terms of

$$1 + \sum_{n=0}^{\infty} \frac{1}{2^n} = 1 + 1 + \frac{1}{2} + \frac{1}{2^2} + \cdots. \tag{3}$$

To see how this relationship leads to an upper bound for the partial sums of $\Sigma_{n=0}^{\infty} (1/(n!))$, let

$$s_n = 1 + \frac{1}{1!} + \frac{1}{2!} + \cdots + \frac{1}{n!}$$

and observe that, for each n,

$$s_n \leq 1 + 1 + \frac{1}{2} + \frac{1}{2^2} + \cdots + \frac{1}{2^n} < 1 + \sum_{n=0}^{\infty} \frac{1}{2^n} = 1 + \frac{1}{1 - (1/2)} = 3.$$

Thus the partial sums of $\sum_{n=0}^{\infty} (1/(n!))$ are all less than 3, so $\sum_{n=0}^{\infty} (1/(n!))$ converges.

The fact that 3 is an upper bound for the partial sums of $\sum_{n=0}^{\infty} (1/(n!))$ does not necessarily mean that the series converges to 3. As we shall see in Section 8.7, the series converges to e.

Nicole Oresme (1320–1382)

The argument we use to show the divergence of the harmonic series was devised by the French theologian, mathematician, physicist, and bishop Nicole Oresme (pronounced "or-rem"). Oresme was a vigorous opponent of astrology, a dynamic preacher, an adviser of princes, a friend of King Charles V, a popularizer of science, and a skillful translator of Latin into French. Oresme did not believe in Albert of Saxony's generally accepted model of free fall (Chapter 6, Miscellaneous Exercise 98) but preferred Aristotle's constant-acceleration model, the model that became popular among Oxford scholars in the 1330s and that Galileo eventually used three hundred years later.

Example 3   *The Harmonic Series.*   The series

$$\sum_{1}^{\infty} \frac{1}{n} = 1 + \frac{1}{2} + \frac{1}{3} + \cdots + \frac{1}{n} + \cdots$$

is called the **harmonic series**. It diverges because there is no upper bound for its sequence of partial sums. To see why, imagine grouping the terms of the series in the following way:

$$1 + \frac{1}{2} + \underbrace{\left(\frac{1}{3} + \frac{1}{4}\right)}_{>\frac{2}{4} = \frac{1}{2}} + \underbrace{\left(\frac{1}{5} + \frac{1}{6} + \frac{1}{7} + \frac{1}{8}\right)}_{>\frac{4}{8} = \frac{1}{2}} + \underbrace{\left(\frac{1}{9} + \frac{1}{10} + \cdots + \frac{1}{16}\right)}_{>\frac{8}{16} = \frac{1}{2}} + \cdots.$$

The sum of the first two terms is 1.5. The sum of the next two terms is $1/3 + 1/4$, which is greater than $1/4 + 1/4 = 1/2$. The sum of the next four terms is $1/5 + 1/6 + 1/7 + 1/8$, which is greater than $1/8 + 1/8 + 1/8 + 1/8 = 1/2$. The sum of the next eight terms is $1/9 + 1/10 + 1/11 + 1/12 + 1/13 + 1/14 + 1/15 + 1/16$, which is greater than $8/16 = 1/2$. The sum of the next 16 terms is greater than $16/32 = 1/2$, and so on. In general, the sum of 2^n terms ending with $1/2^{n+1}$ is greater than $2^n/2^{n+1} = 1/2$. The sequence of partial sums is not bounded: If $n = 2^k$, the partial sum s_n is greater than $k/2$. The harmonic series diverges. (We shall see later that the nth partial sum is slightly greater than $\ln(n + 1)$.)

Notice that the nth-Term Test for divergence does not detect the divergence of the harmonic series. The nth term, $1/n$, goes to zero but the series still diverges.

Comparison Test for Convergence

We established the convergence of the series in Example 2 by comparing it with a series that was already known to converge. This kind of comparison is typical of a procedure called the Comparison Test for convergence of series of nonnegative terms.

> **Comparison Test for Series of Nonnegative Terms**
>
> Let $\Sigma\, a_n$ be a series with no negative terms.
>
> a) **Test for convergence.** The series $\Sigma\, a_n$ converges if there is a convergent series $\Sigma\, c_n$ with $a_n \leq c_n$ for all $n > n_0$, for some positive integer n_0.
>
> b) **Test for divergence.** The series $\Sigma\, a_n$ diverges if there is a divergent series of nonnegative terms $\Sigma\, d_n$ with $a_n \geq d_n$ for all $n > n_0$.

In part (a), the partial sums of the series $\Sigma\, a_n$ are bounded above by

$$M = a_1 + a_2 + \cdots + a_{n_0} + \sum_{n = n_0 + 1}^{\infty} c_n.$$

They therefore form a nondecreasing sequence with a limit L that is less than or equal to M.

In part (b), the partial sums for $\Sigma\, a_n$ are not bounded from above. If they were, the partial sums for $\Sigma\, d_n$ would be bounded by

$$M' = d_1 + d_2 + \cdots + d_{n_0} + \sum_{n = n_0 + 1}^{\infty} a_n$$

and $\Sigma\, d_n$ would have to converge instead of diverge.

To apply the Comparison Test to a series, we do not have to include the early terms of the series. We can start the test with any index N, provided we include all the terms of the series being tested from there on.

Example 4   We can establish the convergence of the series

$$5 + \frac{2}{3} + 1 + \frac{1}{7} + \frac{1}{2} + \frac{1}{3!} + \frac{1}{4!} + \cdots + \frac{1}{k!} + \cdots$$

by ignoring the first four terms and comparing the remainder with the convergent geometric series

$$\sum_{n = 1}^{\infty} \frac{1}{2^n} = \frac{1}{2} + \frac{1}{4} + \frac{1}{8} + \cdots.$$

To apply the Comparison Test, we need to have on hand a list of series we already know about. Here is what we know so far:

Convergent series	Divergent series
Geometric series with $|r| < 1$	Geometric series with $|r| \geq 1$
Telescoping series like $\displaystyle\sum_{n = 1}^{\infty} \frac{1}{n(n + 1)}$	The harmonic series $\displaystyle\sum_{n = 1}^{\infty} \frac{1}{n}$
The series $\displaystyle\sum_{n = 0}^{\infty} \frac{1}{n!}$	Any series $\Sigma\, a_n$ with $\lim a_n \neq 0$

The next test, the Integral Test, will add some series to these lists.

The Integral Test

We introduce the Integral Test with a specific example, a series that is related to the harmonic series, but in which the nth term is $1/n^2$ instead of $1/n$.

Example 5   Does the series

$$\sum_{n = 1}^{\infty} \frac{1}{n^2} = 1 + \frac{1}{4} + \frac{1}{9} + \frac{1}{16} + \cdots + \frac{1}{n^2} + \cdots \tag{4}$$

converge, or does it diverge?

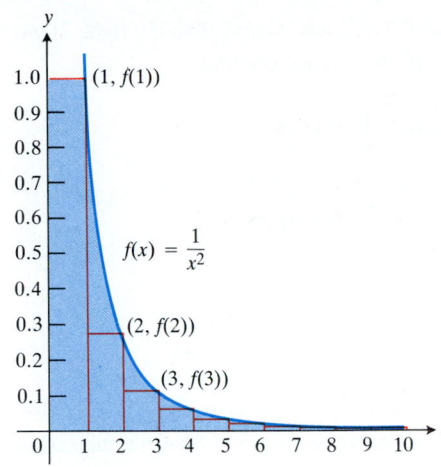

8.16   When $f(x) = 1/x^2$,

$$s_n = f(1) + f(2) + \cdots + f(n)$$

$$< f(1) + \int_1^n f(x)\, dx$$

$$= 1 + \left(\frac{1}{1} - \frac{1}{n}\right) < 2.$$

Solution   When we studied improper integrals in Chapter 7, we learned that

$$\int_1^\infty \frac{1}{x}\, dx \text{ diverges} \qquad \text{and} \qquad \int_1^\infty \frac{1}{x^2}\, dx \text{ converges.}$$

If we can show that the sequence of partial sums of series (4) is bounded above, we can conclude that the series converges. Figure 8.16 suggests how we can find an upper bound for

$$s_n = \frac{1}{1^2} + \frac{1}{2^2} + \frac{1}{3^2} + \cdots + \frac{1}{n^2} = f(1) + f(2) + f(3) + \cdots + f(n),$$

where $f(x) = 1/x^2$.

The first term of the series is $f(1) = 1$, which we can interpret as the area of a rectangle of height $1/1^2$ and base the length of the interval $[0, 1]$ on the x-axis. The next rectangle, over the interval $[1, 2]$, has area $f(2) = 1/2^2$. That rectangle lies below the curve $y = 1/x^2$, so

$$f(2) = \frac{1}{2^2} < \int_1^2 \frac{1}{x^2}\, dx.$$

In the same way,

$$f(3) = \frac{1}{3^2} < \int_2^3 \frac{1}{x^2}\, dx, \ldots, f(n) = \frac{1}{n^2} < \int_{n-1}^n \frac{1}{x^2}\, dx. \tag{5}$$

Adding these inequalities, we get

$$f(2) + f(3) + \cdots + f(n) < \int_1^n (1/x^2)\, dx = -\frac{1}{x}\Big]_1^n = 1 - \frac{1}{n}.$$

To get s_n we must add $f(1) = 1$. When we do so, we have

$$s_n = f(1) + f(2) + f(3) + \cdots + f(n) < 2 - \frac{1}{n} < 2.$$

The sequence of partial sums $\{s_n\}$ is an increasing sequence that is bounded above, so it has a limit, and the given series converges. Its sum is known to be $\pi^2/6$, which is about 1.64493.

We now state and prove the Integral Test in more general terms.

Integral Test

Let $a_n = f(n)$ where $f(x)$ is a continuous, positive, decreasing function of x for all $x \geq 1$. Then the series $\Sigma\, a_n$ and the integral $\int_1^\infty f(x)\, dx$ both converge or both diverge.

Proof   We start with the assumption that f is a decreasing function with $f(n) = a_n$ for every n. This leads us to observe that the rectangles in Fig. 8.17(a), which have areas $a_1, a_2, \ldots, a_n$, collectively enclose more area than that under the curve $y = f(x)$ from $x = 1$ to $x = n + 1$. That is,

$$\int_1^{n+1} f(x)\, dx \leq a_1 + a_2 + \cdots + a_n.$$

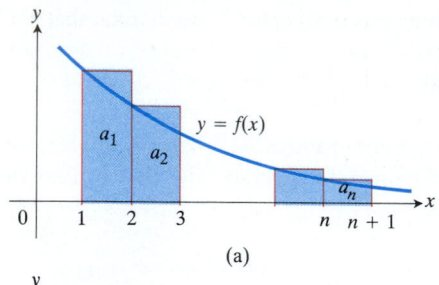

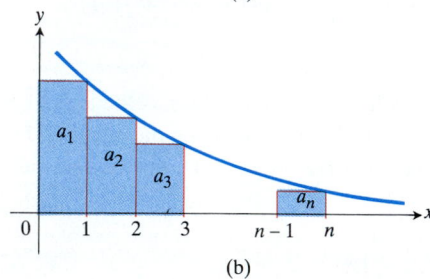

8.17   Subject to the conditions of the Integral Test, the series $\sum_{n=1}^{\infty} a_n$ and the integral $\int_1^{\infty} f(x)\, dx$ both converge or both diverge.

In Fig. 8.17(b) the rectangles have been faced to the left instead of to the right. If we momentarily disregard the first rectangle, of area a_1, we see that

$$a_2 + a_3 + \cdots + a_n \le \int_1^n f(x)\, dx.$$

If we include a_1, we have

$$a_1 + a_2 + \cdots + a_n \le a_1 + \int_1^n f(x)\, dx.$$

Combining these results gives

$$\int_1^{n+1} f(x)\, dx \le a_1 + a_2 + \cdots + a_n \le a_1 + \int_1^n f(x)\, dx. \tag{6}$$

If the integral $\int_1^{\infty} f(x)\, dx$ is finite, the right-hand inequality shows that $\Sigma\, a_n$ is also finite. But if $\int_1^{\infty} f(x)\, dx$ is infinite, then the left-hand inequality shows that the series is also infinite.

Hence the series and the integral are both finite or both infinite.

Example 6   *The p-series.*   If p is a real constant, the series

$$\sum_{n=1}^{\infty} \frac{1}{n^p} = \frac{1}{1^p} + \frac{1}{2^p} + \frac{1}{3^p} + \cdots + \frac{1}{n^p} + \cdots \tag{7}$$

converges if $p > 1$ and diverges if $p \le 1$. To prove this, let $f(x) = 1/x^p$. Then, if $p > 1$, we have $-p + 1 < 0$ and

$$\int_1^{\infty} x^{-p}\, dx = \lim_{b \to \infty} \frac{x^{-p+1}}{-p+1} \Bigg]_1^b = \frac{1}{1-p} \lim_{b \to \infty} (b^{-p+1} - 1) = \frac{1}{p-1},$$

which is finite. Hence the p-series converges if $p > 1$.

If $p = 1$, we have

$$1 + \frac{1}{2} + \frac{1}{3} + \cdots + \frac{1}{n} + \cdots,$$

which we already know diverges. Or, by the Integral Test,

$$\int_1^{\infty} x^{-1} dx = \lim_{b \to \infty} \ln x \Bigg]_1^b = \infty,$$

and, since the integral diverges, the series diverges.

Finally, if $p < 1$, then the terms of the p-series are greater than the corresponding terms of the divergent harmonic series. Hence the p-series diverges, by the Comparison Test.

We have convergence for $p > 1$ but divergence for every other value of p.

The Limit Comparison Test

We now present a more powerful form of the Comparison Test, known as the Limit Comparison Test. It is particularly handy when we deal with series in which a_n is a rational function of n. The next example will show you what we mean.

Example 7   Do the following series converge, or diverge?

a) $\displaystyle\sum_{n=2}^{\infty} \frac{2n}{n^2 - n + 1}$

b) $\displaystyle\sum_{n=2}^{\infty} \frac{2n^3 + 100n^2 + 1000}{(1/8)n^6 - n + 2}$

Solution   In determining convergence or divergence, only the tails count. And when n is very large, the highest powers of n in numerator and denominator are what count the most. So in (a), we reason this way:

$$a_n = \frac{2n}{n^2 - n + 1}$$

behaves approximately like $2n/n^2 = 2/n$, and, by comparing it with $\Sigma\ 1/n$, we guess that $\Sigma\ a_n$ diverges. In (b), we reason that a_n will behave approximately like $2n^3/(1/8)n^6 = 16/n^3$ and, by comparing it with $\Sigma\ 1/n^3$, a p-series with $p = 3$, we guess that the series converges.

To be more precise, in part (a) we take

$$a_n = \frac{2n}{n^2 - n + 1} \qquad \text{and} \qquad d_n = \frac{1}{n}$$

and look at the ratio

$$\frac{a_n}{d_n} = \frac{2n^2}{n^2 - n + 1} = \frac{2}{1 - \left(\frac{1}{n}\right) + \left(\frac{1}{n^2}\right)}.$$

Clearly, as $n \to \infty$, the limit is 2: $\lim\ (a_n/d_n) = 2$.

This means that, in particular, if we take $\epsilon = 1$ in the definition of limit, we know there is an integer N such that a_n/d_n is within 1 unit of this limit for all $n \geq N$:

$$2 - 1 \leq a_n/d_n \leq 2 + 1 \qquad \text{for} \quad n \geq N.$$

Thus $a_n \geq d_n$ for $n \geq N$. Therefore, by the Comparison Test, $\Sigma\ a_n$ diverges because $\Sigma\ d_n$ diverges.

In part (b), if we let $c_n = 1/n^3$, we can show that $\lim(a_n/c_n) = 16$.

Taking $\epsilon = 1$ in the definition of limit, we can conclude that there is an index N' such that a_n/c_n is between 15 and 17 when $n \geq N'$. Since $\Sigma\ c_n$ converges, so also does $\Sigma\ 17c_n$ and thus $\Sigma\ a_n$.

Our rather rough guesswork paved the way for successful choices of comparison series. We make all of this more precise in the following Limit Comparison Test.

Limit Comparison Test

a) **Test for convergence.** If $a_n \geq 0$ for $n \geq n_0$ and there is a convergent series $\Sigma\ c_n$ such that $c_n > 0$ and

$$\lim \frac{a_n}{c_n} < \infty, \tag{8}$$

then $\Sigma\ a_n$ converges.

b) **Test for divergence.** If $a_n \geq 0$ for $n \geq n_0$ and there is a divergent series $\Sigma\ d_n$ such that $d_n > 0$ and

$$\lim \frac{a_n}{d_n} > 0, \tag{9}$$

then $\Sigma\ a_n$ diverges.

A simpler version of the Limit Comparison Test combines parts (a) and (b) in the following way.

Simplified Limit Comparison Test

If the terms of the two series $\Sigma\, a_n$ and $\Sigma\, b_n$ are positive for $n \geq n_0$, and the limit of a_n/b_n is finite and positive, then both series converge or both diverge.

The Simplified Limit Comparison Test is the one we use most often.

Example 8   Which of the following series converge and which diverge?

a) $\dfrac{3}{4} + \dfrac{5}{9} + \dfrac{7}{16} + \dfrac{9}{25} + \cdots = \displaystyle\sum_{n=1}^{\infty} \dfrac{2n + 1}{(n + 1)^2}$

b) $\dfrac{101}{3} + \dfrac{102}{10} + \dfrac{103}{29} + \cdots = \displaystyle\sum_{n=1}^{\infty} \dfrac{100 + n}{n^3 + 2}$

c) $\dfrac{1}{1} + \dfrac{1}{3} + \dfrac{1}{7} + \cdots = \displaystyle\sum_{n=1}^{\infty} \dfrac{1}{2^n - 1}$

Solution

a) Let $a_n = (2n + 1)/(n^2 + 2n + 1)$ and $d_n = 1/n$. Then

$$\sum d_n \text{ diverges} \qquad \text{and} \qquad \lim \frac{a_n}{d_n} = \lim \frac{2n^2 + n}{n^2 + 2n + 1} = 2,$$

so $\Sigma\, a_n$ diverges.

b) Let $a_n = (100 + n)/(n^3 + 2)$. When n is large, this ought to compare with $n/n^3 = 1/n^2$, so we let $c_n = 1/n^2$ and apply the Limit Comparison Test:

$$\sum c_n \text{ converges} \qquad \text{and} \qquad \lim \frac{a_n}{c_n} = \lim \frac{n^3 + 100n^2}{n^3 + 2} = 1,$$

so $\Sigma\, a_n$ converges.

c) Let $a_n = 1/(2^n - 1)$ and $c_n = 1/2^n$. (We reason that $2^n - 1$ behaves somewhat like 2^n when n is large.) Then

$$\frac{a_n}{c_n} = \frac{2^n}{2^n - 1} = \frac{1}{1 - (1/2)^n} \to 1 \qquad \text{as} \quad n \to \infty.$$

Because $\Sigma\, c_n$ converges, we conclude that $\Sigma\, a_n$ does also.

EXERCISES 8.3

Which series in Exercises 1–32 converge and which diverge?
Give reasons for your answers.

1. $\displaystyle\sum_{n=1}^{\infty} \dfrac{1}{10^n}$ **2.** $\displaystyle\sum_{n=1}^{\infty} \dfrac{n}{n + 2}$ **3.** $\displaystyle\sum_{n=1}^{\infty} \dfrac{\sin^2 n}{2^n}$ **4.** $\displaystyle\sum_{n=1}^{\infty} \dfrac{5}{n}$ **5.** $\displaystyle\sum_{n=1}^{\infty} \dfrac{1 + \cos n}{n^2}$ **6.** $\displaystyle\sum_{n=1}^{\infty} -\dfrac{1}{8^n}$

7. $\displaystyle\sum_{n=2}^{\infty} \frac{\ln n}{n}$

8. $\displaystyle\sum_{n=1}^{\infty} \frac{1}{n\sqrt{n}}$

9. $\displaystyle\sum_{n=1}^{\infty} \frac{2^n}{3^n}$

10. $\displaystyle\sum_{n=0}^{\infty} \frac{-2}{n+1}$

11. $\displaystyle\sum_{n=1}^{\infty} \frac{1}{1+\ln n}$

12. $\displaystyle\sum_{n=1}^{\infty} \frac{1}{2n-1}$

13. $\displaystyle\sum_{n=1}^{\infty} \frac{2^n}{n+1}$

14. $\displaystyle\sum_{n=1}^{\infty} \left(\frac{n}{3n+1}\right)^n$

15. $\displaystyle\sum_{n=1}^{\infty} \frac{1}{\sqrt{n^3+2}}$

16. $\displaystyle\sum_{n=2}^{\infty} \frac{\sqrt{n}}{\ln n}$

17. $\displaystyle\sum_{n=1}^{\infty} \frac{n}{n^2+1}$

18. $\displaystyle\sum_{n=1}^{\infty} \frac{1}{n\sqrt[n]{n}}$

19. $\displaystyle\sum_{n=1}^{\infty} \left(1+\frac{1}{n}\right)^n$

20. $\displaystyle\sum_{n=1}^{\infty} \frac{\sqrt{n}}{n^2+1}$

21. $\displaystyle\sum_{n=1}^{\infty} \frac{1-n}{n\cdot 2^n}$

22. $\displaystyle\sum_{n=1}^{\infty} \frac{1}{(\ln 2)^n}$

23. $\displaystyle\sum_{n=1}^{\infty} \frac{1}{3^{n-1}+1}$

24. $\displaystyle\sum_{n=1}^{\infty} \frac{10n+1}{n(n+1)(n+2)}$

25. $\displaystyle\sum_{n=1}^{\infty} \frac{\tan^{-1}n}{n^{1.1}}$

26. $\displaystyle\sum_{n=2}^{\infty} n^2 e^{-n}$

27. $\displaystyle\sum_{n=1}^{\infty} \frac{1}{n(1+\ln^2 n)}$

28. $\displaystyle\sum_{n=3}^{\infty} \frac{1}{n(\ln n)\sqrt{(\ln n)^2-1}}$

29. $\displaystyle\sum_{n=1}^{\infty} \frac{1}{1+2+3+\cdots+n}$

30. $\displaystyle\sum_{n=1}^{\infty} \frac{n}{1+2^2+3^2+\cdots+n^2}$

31. $\displaystyle\sum_{n=1}^{\infty} \operatorname{sech} n$

32. $\displaystyle\sum_{n=1}^{\infty} \operatorname{sech}^2 n$

33. Determine whether the following series converges.

$$\sum_{n=1}^{\infty} \frac{1\cdot 3\cdot 5\cdot\ \cdots\ \cdot(2n-1)}{2\cdot 4\cdot 6\cdot\ \cdots\ \cdot(2n)}$$

You will see where this series comes from if you do Exercise 24 in Section 8.8.

34. For what value or values of a, if any, does the following series converge?

$$\sum_{n=1}^{\infty} \left(\frac{a}{n+2}-\frac{1}{n+4}\right)$$

35. CALCULATOR   There is absolutely no empirical evidence for the divergence of the harmonic series even though we know it diverges. The partial sums, which satisfy the inequality

$$\ln(n+1) = \int_1^{n+1} \frac{1}{x}\,dx \le 1+\frac{1}{2}+\cdots+\frac{1}{n}$$

$$\le 1+\int_1^n \frac{1}{x}\,dx = 1+\ln n$$

(Eq. 6), just grow too slowly. To see what we mean, suppose you had started with $s_1 = 1$ the day the universe was formed, thirteen billion years ago, and added a new term every *second*. About how large would s_n be today?

36. There are no values of x for which $\sum_{n=1}^{\infty}(1/nx)$ converges. Why?

37. Show that if $\sum_{n=1}^{\infty} a_n$ is a convergent series of nonnegative numbers, then $\sum_{n=1}^{\infty}(a_n/n)$ converges.

38. Show that if $\sum a_n$ and $\sum b_n$ are convergent series with $a_n \ge 0$ and $b_n \ge 0$, then $\sum a_n b_n$ converges. (*Hint:* From some integer on, $0 \le a_n$ and $b_n < 1$, so $a_n b_n \le a_n$.)

39. *Nonincreasing sequences.*   A sequence of numbers $\{s_n\}$ in which $s_n \ge s_{n+1}$ for every n is called a **nonincreasing sequence**. A sequence $\{s_n\}$ is bounded from below if there is a finite constant M with $M \le s_n$ for every n. Such a number M is called a lower bound for the sequence. Deduce from Theorem 6 that a nonincreasing sequence that is bounded from below converges and that a nonincreasing sequence that is not bounded from below diverges.

40. *The Cauchy condensation test.*   The Cauchy condensation test says: Let $\{a_n\}$ be a nonincreasing sequence ($a_n \ge a_{n+1}$ for all n) of positive terms that converges to 0. Then $\sum a_n$ converges if and only if $\sum 2^n a_{2^n}$ converges. For example, $\sum(1/n)$ diverges because $\sum 2^n \cdot (1/2^n) = \sum 1$ diverges. Show why the test works.

41. Use the Cauchy condensation test from Exercise 40 to show that

a) $\displaystyle\sum_{n=2}^{\infty} \frac{1}{n\ln n}$ diverges;

b) $\displaystyle\sum_{n=1}^{\infty} \frac{1}{n^p}$ converges if $p > 1$ and diverges if $p \le 1$.

42. *Logarithmic p–series.*

a) Show that

$$\int_2^{\infty} \frac{dx}{x\,(\ln x)^p} \qquad (p \text{ a positive constant})$$

converges if and only if $p > 1$.

Knowing this about the integral, what can you deduce about the convergence or divergence of the following series?

b) $\displaystyle\sum_{n=2}^{\infty} \frac{1}{n\ln n}$

c) $\displaystyle\sum_{n=2}^{\infty} \frac{1}{n\,(\ln n)^{1.01}}$

d) $\displaystyle\sum_{n=5}^{\infty} \frac{n^{1/2}}{(\ln n)^3}$

e) $\displaystyle\sum_{n=3}^{\infty} \frac{1}{n\ln(n^3)}$

f) $\displaystyle\sum_{n=2}^{\infty} \frac{1}{n(\ln n)^{(n+1)/2}}$

43. *Euler's constant.*   Graphs like those in Fig. 8.17 suggest that as n increases there is little change in the difference between the sum

$$1+\frac{1}{2}+\cdots+\frac{1}{n}$$

and the integral

$$\ln n = \int_1^n \frac{1}{x}\,dx.$$

To explore this idea, carry out the following steps.

a) By taking $f(x) = 1/x$ in inequality (6), show that

$$\ln(n + 1) \le 1 + \frac{1}{2} + \cdots + \frac{1}{n} \le 1 + \ln n$$

or

$$0 < \ln(n + 1) - \ln n$$

$$\le 1 + \frac{1}{2} + \cdots + \frac{1}{n} - \ln n \le 1.$$

Thus, the sequence

$$a_n = 1 + \frac{1}{2} + \cdots + \frac{1}{n} - \ln n$$

is bounded from below and from above.

b) Show that

$$\frac{1}{n + 1} < \int_n^{n+1} \frac{1}{x}\, dx = \ln(n + 1) - \ln n,$$

and use this result to show that the sequence $\{a_n\}$ in part (a) is decreasing.

Since a decreasing sequence that is bounded from below converges (Exercise 39), the numbers a_n defined in (a) converge:

$$1 + \frac{1}{2} + \cdots + \frac{1}{n} - \ln n \to \gamma.$$

The number γ, whose value is $0.5772\ldots$, is called *Euler's constant*. In contrast to other special numbers like π and e, no other expression with a simple law of formulation has ever been found for γ.

44. *Prime numbers.* The prime numbers form a sequence $\{p_n\} = \{2, 3, 5, 7, 11, 13, 17, 19, \ldots\}$. It is known that $\lim_{n \to \infty} [(n \ln n)/p_n] = 1$. Using this fact, show that

$$\sum_{n=1}^{\infty} \frac{1}{p_n} = \frac{1}{2} + \frac{1}{3} + \frac{1}{5} + \frac{1}{7} + \frac{1}{11} + \cdots + \frac{1}{p_n} + \cdots$$

diverges. (See Exercise 42.)

EXPLORER PROGRAM

Sequences and Series

Generates terms of one or two sequences and graphs them while you watch. You may define the sequences recursively or by giving formulas for their nth terms. Enables you to look for graphical and numerical evidence of convergence or divergence.

8.4 Series Without Negative Terms: The Ratio and Root Tests

Convergence tests that depend on comparing one series with another series or with an integral are called *extrinsic* tests. They are useful, but there are reasons to look for tests that do not require comparison. As a practical matter, we may not be able to find the series or function we need to make a comparison work. And, in principle, all the information about a given series should be contained in its own terms. We therefore turn our attention to *intrinsic* tests—those that depend only on the series at hand.

The Ratio Test

Our first intrinsic test, the Ratio Test, measures the rate of growth (or decline) of a series by examining the ratio a_{n+1}/a_n. For a geometric series, this rate of growth is a constant, and the series converges if and only if its ratio is less than 1 in absolute value. But even if the ratio is not constant, we may be able to find a geometric series for comparison, as in Example 1.

Example 1   Let $a_1 = 1$ and define a_{n+1} to be $a_{n+1} = \dfrac{n}{2n + 1} a_n$.

Does the series $\Sigma\, a_n$ converge, or diverge?

Solution   We begin by writing out a few terms of the series:

$$a_1 = 1, \qquad a_2 = \frac{1}{3} a_1 = \frac{1}{3}, \qquad a_3 = \frac{2}{5} a_2 = \frac{1 \cdot 2}{3 \cdot 5}, \qquad a_4 = \frac{3}{7} a_3 = \frac{1 \cdot 2 \cdot 3}{3 \cdot 5 \cdot 7}.$$

The series in Example 1 converges rapidly, as the following computer data suggest.

n	s_n
5	1.5607 5
10	1.5705 5
15	1.5707 89894
20	1.5707 96149
25	1.5707 96322
30	1.5707 96327
35	1.5707 96327

Each term is somewhat less than 1/2 the term before it, because $n/(2n + 1)$ is less than 1/2. Therefore the terms of the given series are less than or equal to the terms of the geometric series

$$1 + \left(\frac{1}{2}\right) + \left(\frac{1}{4}\right) + \cdots + \left(\frac{1}{2}\right)^{n-1} + \cdots$$

that converges to 2. So our series also converges, and its sum is less than 2. Computer data show that the sum is approximately 1.5707 96327.

In proving the Ratio Test, we shall make a comparison with appropriate geometric series as in Example 1, but when we *apply* it we do not actually make a direct comparison.

The Ratio Test

Let $\Sigma\, a_n$ be a series with positive terms, and suppose that

$$\lim_{n \to \infty} \frac{a_{n+1}}{a_n} = \rho.$$

Then

a) the series *converges* if $\rho < 1$,

b) the series *diverges* if $\rho > 1$,

c) the series *may converge or it may diverge* if $\rho = 1$. (The test provides no information.)

Proof

a) $\rho < 1$. Let r be a number between ρ and 1. Then the number $\epsilon = r - \rho$ is positive. Since

$$\frac{a_{n+1}}{a_n} \to \rho,$$

a_{n+1}/a_n must lie within ϵ of ρ when n is large enough, say for all $n \geq N$. In particular,

$$\frac{a_{n+1}}{a_n} < \rho + \epsilon = r, \qquad \text{when } n > N.$$

That is,

$$a_{N+1} < ra_N,$$

$$a_{N+2} < ra_{N+1} < r^2 a_N,$$

$$a_{N+3} < ra_{N+2} < r^3 a_N,$$

$$\vdots$$

$$a_{N+m} < ra_{N+m-1} < r^m a_N.$$

These inequalities show that the terms of our series, after the Nth term, approach zero more rapidly than the terms in a geometric series with ratio

$r < 1$. More precisely, consider the series $\Sigma\, c_n$, where $c_n = a_n$ for $n = 1$, $2, \dots, N$ and $c_{N+1} = ra_N$, $c_{N+2} = r^2a_N, \dots, c_{N+m} = r^m a_N, \dots$. Now $a_n \le c_n$ for all n, and

$$\sum_{n=1}^{\infty} c_n = a_1 + a_2 + \cdots + a_{N-1} + a_N + ra_N + r^2 a_N + \cdots$$

$$= a_1 + a_2 + \cdots + a_{N-1} + a_N(1 + r + r^2 + \cdots).$$

The geometric series $1 + r + r^2 + \cdots$ converges because $|r| < 1$, so $\Sigma\, c_n$ converges. Since $a_n \le c_n$, $\Sigma\, a_n$ also converges.

b) $\rho > 1$. From some index M on,

$$\frac{a_{n+1}}{a_n} > 1 \qquad \text{and} \qquad a_M < a_{M+1} < a_{M+2} < \cdots.$$

The terms of the series do not approach zero as n becomes infinite, and the series diverges by the nth-Term Test.

c) $\rho = 1$. The two series

$$\sum_{n=1}^{\infty} \frac{1}{n} \qquad \text{and} \qquad \sum_{n=1}^{\infty} \frac{1}{n^2}$$

show that some other test for convergence must be used when $\rho = 1$.

$$\text{For } \sum_{n=1}^{\infty} \frac{1}{n}: \qquad \frac{a_{n+1}}{a_n} = \frac{1/(n+1)}{1/n} = \frac{n}{n+1} \to 1.$$

$$\text{For } \sum_{n=1}^{\infty} \frac{1}{n^2}: \qquad \frac{a_{n+1}}{a_n} = \frac{1/(n+1)^2}{1/n^2} = \left(\frac{n}{n+1}\right)^2 \to 1^2 = 1.$$

In both cases $\rho = 1$, yet the first series diverges while the second converges (Section 8.3, Examples 3 and 6).

The Ratio Test is often effective when the terms of the series contain factorials of expressions involving n or expressions raised to the nth power or combinations of the two, as in the next example.

Example 2   Use the Ratio Test to investigate the convergence of the following series.

a) $\displaystyle\sum_{n=1}^{\infty} \frac{n!\,n!}{(2n)!}$
b) $\displaystyle\sum_{n=1}^{\infty} \frac{4^n n!\,n!}{(2n)!}$
c) $\displaystyle\sum_{n=0}^{\infty} \frac{2^n + 5}{3^n}$

Solution

a) If $a_n = n!\,n!/(2n)!$, then $a_{n+1} = (n+1)!(n+1)!/(2n+2)!$, and

$$\frac{a_{n+1}}{a_n} = \frac{(n+1)!(n+1)!(2n)!}{n!\,n!\,(2n+2)(2n+1)(2n)!}$$

$$= \frac{(n+1)(n+1)}{(2n+2)(2n+1)} = \frac{n+1}{4n+2} \to \frac{1}{4}.$$

The series converges because $\rho = 1/4$ is less than 1.

b) If $a_n = 4^n n!n!/(2n)!$, then

$$\frac{a_{n+1}}{a_n} = \frac{4^{n+1}(n+1)!(n+1)!}{(2n+2)(2n+1)(2n)!} \cdot \frac{(2n)!}{4^n n!n!}$$

$$= \frac{4(n+1)(n+1)}{(2n+2)(2n+1)} = \frac{2(n+1)}{2n+1} \to 1.$$

Because the limit is $\rho = 1$, we cannot decide on the basis of the Ratio Test alone whether the series converges or diverges. However, when we note that $a_{n+1}/a_n = (2n+2)/(2n+1)$, we conclude that a_{n+1} is always greater than a_n because $(2n+2)/(2n+1)$ is always greater than 1. Therefore, all terms are greater than or equal to $a_1 = 2$, and the nth term does not approach zero as $n \to \infty$. Hence, by the nth-Term Test, the series diverges.

c) For the series $\sum_{n=0}^{\infty} (2^n + 5)/3^n$,

$$\frac{a_{n+1}}{a_n} = \frac{(2^{n+1} + 5)/3^{n+1}}{(2^n + 5)/3^n} = \frac{1}{3} \cdot \frac{2^{n+1} + 5}{2^n + 5} = \frac{1}{3} \cdot \left(\frac{2 + 5 \cdot 2^{-n}}{1 + 5 \cdot 2^{-n}}\right) \to \frac{1}{3} \cdot \frac{2}{1} = \frac{2}{3}.$$

The series converges because $\rho = 2/3$ is less than 1.

This does *not* mean that 2/3 is the sum of the series. In fact,

$$\sum_{n=0}^{\infty} \frac{2^n + 5}{3^n} = \sum_{n=0}^{\infty} \left(\frac{2}{3}\right)^n + \sum_{n=0}^{\infty} \frac{5}{3^n} = \frac{1}{1 - (2/3)} + \frac{5}{1 - (1/3)} = \frac{21}{2}.$$

The nth-Root Test

We return to the question "Does $\Sigma\, a_n$ converge?" When there is a simple formula for a_n, we can try one of the tests we already have. But consider the following example.

Example 3   Let $a_n = f(n)/2^n$, where

$$f(n) = \begin{cases} n & \text{if } n \text{ is a prime number,} \\ 1 & \text{otherwise.} \end{cases}$$

Does $\Sigma\, a_n$ converge?

Solution   We write out several terms of the series:

$$\sum a_n = \frac{1}{2} + \frac{2}{4} + \frac{3}{8} + \frac{1}{16} + \frac{5}{32} + \frac{1}{64} + \frac{7}{128} + \cdots + \frac{f(n)}{2^n} + \cdots.$$

Clearly, this is not a geometric series. The nth term approaches zero as $n \to \infty$, so we do not know if the series diverges. The Integral Test does not look promising. The Ratio Test produces

$$\frac{a_{n+1}}{a_n} = \frac{1}{2}\frac{f(n+1)}{f(n)} = \begin{cases} \dfrac{1}{2} & \text{if neither } n \text{ nor } n+1 \text{ is a prime,} \\[2mm] \dfrac{1}{2n} & \text{if } n \text{ is a prime} \ge 3, \\[2mm] \dfrac{n+1}{2} & \text{if } n+1 \text{ is a prime} \ge 5. \end{cases}$$

The ratio sometimes is close to zero, sometimes is very large, and sometimes is 1/2. It has no limit because there are infinitely many primes. A test that will answer the question (affirmatively—yes, the series does converge) is the nth-Root Test. To apply it, we consider the following:

$$\sqrt[n]{a_n} = \frac{\sqrt[n]{f(n)}}{2} = \begin{cases} \dfrac{\sqrt[n]{n}}{2} & \text{if } n \text{ is a prime,} \\[2ex] \dfrac{1}{2} & \text{otherwise.} \end{cases}$$

Therefore,

$$\frac{1}{2} \le \sqrt[n]{a_n} \le \frac{\sqrt[n]{n}}{2}$$

and $\lim \sqrt[n]{a_n} = 1/2$ by the Sandwich Theorem. Because this limit is less than 1, the nth-Root Test tells us that the given series converges, as we shall now see.

The nth-Root Test

Let $\Sigma\, a_n$ be a series with $a_n \ge 0$ for $n \ge n_0$, and suppose that $\sqrt[n]{a_n} \to \rho$. Then

a) the series *converges* if $\rho < 1$,

b) the series *diverges* if $\rho > 1$,

c) the test is *not conclusive* if $\rho = 1$.

Proof

a) $\rho < 1$.   Choose an $\epsilon > 0$ so small that $\rho + \epsilon < 1$ also. Since $\sqrt[n]{a_n} \to \rho$, the terms $\sqrt[n]{a_n}$ eventually get closer than ϵ to ρ. In other words, there exists an index $N \ge n_0$ such that

$$\sqrt[n]{a_n} < \rho + \epsilon \qquad \text{when } n \ge N.$$

Then it is also true that

$$a_n < (\rho + \epsilon)^n \qquad \text{for } n \ge N.$$

Now, $\Sigma_{n=N}^{\infty} (\rho + \epsilon)^n$, a geometric series with ratio $(\rho + \epsilon) < 1$, converges. By comparison, $\Sigma_{n=N}^{\infty} a_n$ converges, from which it follows that

$$\sum_{n=1}^{\infty} a_n = a_1 + \cdots + a_{N-1} + \sum_{n=N}^{\infty} a_n$$

converges.

b) $\rho > 1$.   For all indices beyond some integer M, we have $\sqrt[n]{a_n} > 1$, so that $a_n > 1$ for $n > M$. The terms of the series do not converge to zero. The series diverges by the nth-Term Test.

c) $\rho = 1$.   The series $\Sigma_{n=1}^{\infty} (1/n)$ and $\Sigma_{n=1}^{\infty} (1/n^2)$ show that the test is not conclusive when $\rho = 1$. The first series diverges and the second converges, but in both cases $\sqrt[n]{a_n} \to 1$.

Example 4   Which of the following series converges and which diverges?

a) $\displaystyle\sum_{n=1}^{\infty} \frac{n^2}{2^n}$

b) $\displaystyle\sum_{n=1}^{\infty} \frac{2^n}{n^2}$

Solution Series (a) converges because $\sqrt[n]{a_n} = \dfrac{\sqrt[n]{n^2}}{2} \to \dfrac{1}{2} < 1$. But series (b) diverges because $\sqrt[n]{b_n} = \dfrac{2}{\sqrt[n]{n^2}} \to 2 > 1$.

1-30 EVERY 6TH PROB

EXERCISES 8.4

Which series in Exercises 1–30 converge and which diverge? Give reasons for your answers.

1. $\displaystyle\sum_{n=1}^{\infty} \dfrac{n^2}{2^n}$

2. $\displaystyle\sum_{n=1}^{\infty} \dfrac{n!}{10^n}$

3. $\displaystyle\sum_{n=1}^{\infty} \dfrac{n^{10}}{10^n}$

4. $\displaystyle\sum_{n=1}^{\infty} n^2 e^{-n}$

5. $\displaystyle\sum_{n=1}^{\infty} \left(\dfrac{n-2}{n}\right)^n$

6. $\displaystyle\sum_{n=1}^{\infty} \dfrac{2+(-1)^n}{1.25^n}$

7. $\displaystyle\sum_{n=1}^{\infty} n! e^{-n}$

8. $\displaystyle\sum_{n=1}^{\infty} \dfrac{(-2)^n}{3^n}$

9. $\displaystyle\sum_{n=1}^{\infty} \left(1 - \dfrac{3}{n}\right)^n$

10. $\displaystyle\sum_{n=1}^{\infty} \left(1 - \dfrac{1}{n^2}\right)^n$

11. $\displaystyle\sum_{n=1}^{\infty} \dfrac{\ln n}{n^3}$

12. $\displaystyle\sum_{n=1}^{\infty} \left(\dfrac{1}{n} - \dfrac{1}{n^2}\right)$

13. $\displaystyle\sum_{n=1}^{\infty} \dfrac{\ln n}{n}$

14. $\displaystyle\sum_{n=1}^{\infty} \dfrac{n \ln n}{2^n}$

15. $\displaystyle\sum_{n=1}^{\infty} \dfrac{(n+1)(n+2)}{n!}$

16. $\displaystyle\sum_{n=1}^{\infty} e^{-n}(n^3)$

17. $\displaystyle\sum_{n=1}^{\infty} \dfrac{(n+3)!}{3! n! 3^n}$

18. $\displaystyle\sum_{n=1}^{\infty} -\dfrac{n^2}{2^n}$

19. $\displaystyle\sum_{n=1}^{\infty} \dfrac{n!}{(2n+1)!}$

20. $\displaystyle\sum_{n=1}^{\infty} \dfrac{n!}{n^n}$

21. $\displaystyle\sum_{n=2}^{\infty} \dfrac{n}{(\ln n)^n}$

22. $\displaystyle\sum_{n=2}^{\infty} \dfrac{1}{(\ln n)^2}$

23. $\displaystyle\sum_{n=1}^{\infty} \dfrac{n! \ln n}{n(n+2)!}$

24. $\displaystyle\sum_{n=1}^{\infty} \dfrac{3^n}{n^3 2^n}$

25. $\displaystyle\sum_{n=1}^{\infty} \dfrac{(n!)^n}{(n^n)^2}$

26. $\displaystyle\sum_{n=1}^{\infty} \dfrac{(n!)^n}{n^{(n^2)}}$

27. $\displaystyle\sum_{n=1}^{\infty} \dfrac{n^n}{2^{(n^2)}}$

28. $\displaystyle\sum_{n=1}^{\infty} \dfrac{n^n}{(2^n)^2}$

29. $\displaystyle\sum_{n=1}^{\infty} \dfrac{1 \cdot 3 \cdot \cdots \cdot (2n-1)}{4^n 2^n n!}$

30. $\displaystyle\sum_{n=1}^{\infty} \dfrac{1 \cdot 3 \cdot \cdots \cdot (2n-1)}{[2 \cdot 4 \cdot \cdots \cdot (2n)](3n+1)}$

Which of the series $\Sigma_{n=1}^{\infty} a_n$ defined by the formulas in Exercises 31–42 converge and which diverge? Give reasons for your answers.

31. $a_1 = 2, \quad a_{n+1} = \dfrac{1 + \sin n}{n} a_n$

32. $a_1 = \dfrac{1}{3}, \quad a_{n+1} = \dfrac{3n-1}{2n+5} a_n$

33. $a_1 = 3, \quad a_{n+1} = \dfrac{n}{n+1} a_n$

34. $a_1 = 2, \quad a_{n+1} = \dfrac{2}{n} a_n$

35. $a_1 = -1, \quad a_{n+1} = \dfrac{1 + \ln n}{n} a_n$

36. $a_1 = \dfrac{1}{2}, \quad a_{n+1} = \dfrac{n + \ln n}{n + 10} a_n$

37. $a_n = \dfrac{2^n n! n!}{(2n)!}$

38. $a_n = \dfrac{(3n)!}{n!(n+1)!(n+2)!}$

39. $a_1 = 1, \ a_{n+1} = \dfrac{n(n+1)}{(n+2)(n+3)} a_n$

(*Hint:* Write out several terms, see which factors cancel, and then generalize.)

40. $a_1 = 1, \quad a_{n+1} = \dfrac{n}{(n-1)(n+1)} a_n$

41. $a_1 = a_2 = 1, \quad a_{n+1} = \dfrac{1}{1 + a_n}$

42. $a_n = 1/3^n$ if n is odd, $\quad a_n = n/3^n$ if n is even

43. Neither the Ratio nor the nth-Root Test helps with p-series, $p > 0$. Try each test on $\Sigma_{n=1}^{\infty} (1/n^p)$ to see what happens.

8.5 Alternating Series and Absolute Convergence

A series in which the terms are alternately positive and negative is called an **alternating series.** Here are three examples:

$$1 - \frac{1}{2} + \frac{1}{3} - \frac{1}{4} + \frac{1}{5} - \cdots + \frac{(-1)^{n+1}}{n} + \cdots \tag{1}$$

$$-2 + 1 - \frac{1}{2} + \frac{1}{4} - \frac{1}{8} + \cdots + \frac{(-1)^{n}4}{2^{n}} + \cdots \tag{2}$$

$$1 - 2 + 3 - 4 + 5 - 6 + \cdots + (-1)^{n-1}n + \cdots \tag{3}$$

Series (1), called the **alternating harmonic series,** converges, as we shall see in a moment. Series (2), a geometric series with ratio $r = -1/2$, converges to $-2/[1 + (1/2)] = -4/3$. Series (3) diverges because the nth term does not approach zero.

We prove the convergence of the alternating harmonic series by applying a general result known as the Alternating Series Theorem.

THEOREM 7

> ### The Alternating Series Theorem (Leibniz's Theorem)
>
> The series
>
> $$\sum_{n=1}^{\infty} (-1)^{n+1}a_n = a_1 - a_2 + a_3 - a_4 + \cdots \tag{4}$$
>
> converges if all three of the following conditions are satisfied:
>
> 1. The a_n's are all positive.
> 2. $a_n \geq a_{n+1}$ for all n.
> 3. $a_n \to 0$.

Proof If n is an even integer, say $n = 2m$, then the sum of the first n terms is

$$
\begin{aligned}
s_{2m} &= (a_1 - a_2) + (a_3 - a_4) + \cdots + (a_{2m-1} - a_{2m}) \\
&= a_1 - (a_2 - a_3) - (a_4 - a_5) - \cdots - (a_{2m-2} - a_{2m-1}) - a_{2m}.
\end{aligned}
\tag{5}
$$

The first equality shows that s_{2m} is the sum of m nonnegative terms, since each term in parentheses is positive or zero. Hence $s_{2m+2} \geq s_{2m}$, and the sequence $\{s_{2m}\}$ is nondecreasing. The second equality shows that $s_{2m} \leq a_1$. Since $\{s_{2m}\}$ is nondecreasing and bounded from above, it has a limit, say

$$\lim_{m \to \infty} s_{2m} = L. \tag{6}$$

If n is an odd integer, say $n = 2m + 1$, then the sum of the first n terms is

$$s_{2m+1} = s_{2m} + a_{2m+1}.$$

Since $a_n \to 0$,

$$\lim_{m \to \infty} a_{2m+1} = 0$$

and, as $m \to \infty$,

$$s_{2m+1} = s_{2m} + a_{2m+1} \to L + 0 = L. \tag{7}$$

When we combine the results of (6) and (7), we get

$$\lim_{n \to \infty} s_n = L.$$

Example 1   The alternating harmonic series

$$\sum_{n=1}^{\infty} (-1)^{n+1} \frac{1}{n} = 1 - \frac{1}{2} + \frac{1}{3} - \frac{1}{4} + \cdots$$

satisfies the three requirements of Theorem 7; therefore it converges.

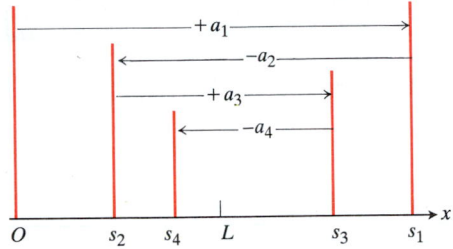

8.18   The partial sums of an alternating series that satisfies the hypotheses of Theorem 7 straddle their limit.

We can use a graphical interpretation of the partial sums to gain added insight into the way an alternating series converges to its limit L when the three conditions of Theorem 7 are satisfied. Starting from the origin of the x-axis (Fig. 8.18), we lay off the positive distance $s_1 = a_1$. To find the point corresponding to $s_2 = a_1 - a_2$, we back up a distance equal to a_2. Since $a_2 \le a_1$, we do not back up any farther than O at most. Next we go forward a distance a_3 and mark the point corresponding to $s_3 = a_1 - a_2 + a_3$. Since $a_3 \le a_2$, we go forward by an amount no greater than the previous backward step; that is, $s_3 \le s_1$. We continue in this seesaw fashion, backing up or going forward as the signs in the series demand. But each forward or backward step is shorter than (or at most the same size as) the preceding step, because $a_{n+1} \le a_n$. And since the nth term approaches zero as n increases, the size of step we take forward or backward gets smaller and smaller. We oscillate across the limit L, and the amplitude of oscillation approaches zero. The even-numbered partial sums $s_2, s_4, s_6, \ldots, s_{2m}$ increase toward L, while the odd-numbered sums $s_1, s_3, s_5, \ldots, s_{2m+1}$ decrease toward L. The limit L lies between any two successive sums s_n and s_{n+1} and hence differs from s_n by an amount less than a_{n+1}.

Because

$$|L - s_n| < a_{n+1} \qquad \text{for every } n, \tag{8}$$

we can make useful estimates of the sums of convergent alternating series.

THEOREM 8

The Alternating Series Estimation Theorem

If the alternating series $\displaystyle\sum_{n=1}^{\infty} (-1)^{n+1} a_n$

satisfies the three conditions of Theorem 7, then

$$s_n = a_1 - a_2 + \cdots + (-1)^{n+1} a_n$$

approximates the sum L of the series with an error whose absolute value is less than a_{n+1}, the numerical value of the first unused term. Furthermore, the remainder, $L - s_n$, has the same sign as the first unused term.

We leave the determination of the sign of the remainder for Exercise 53.

Example 2   We first try Theorem 8 on an alternating series whose sum we already know, namely the geometric series

$$\sum_{n=0}^{\infty} (-1)^n \frac{1}{2^n} = 1 - \frac{1}{2} + \frac{1}{4} - \frac{1}{8} + \frac{1}{16} - \frac{1}{32} + \frac{1}{64} - \frac{1}{128} \Bigm| + \frac{1}{256} - \cdots.$$

The theorem says that if we truncate the series after the eighth term, we throw away a total that is positive and less than $1/256$. The sum of the first eight terms is 0.6640 625. The sum of the series is

$$\frac{1}{1 - (-1/2)} = \frac{1}{3/2} = \frac{2}{3}.$$

The difference,

$$\frac{2}{3} - 0.6640\ 625 = 0.0026\ 04166\ 6\ldots,$$

is positive and less than

$$\frac{1}{256} = 0.0039\ 0625.$$

Absolute Convergence

DEFINITION

A series $\Sigma\ a_k$ **converges absolutely** (is **absolutely convergent**) if the corresponding series of absolute values, $\Sigma|a_k|$, converges.

The geometric series

$$1 - \frac{1}{2} + \frac{1}{4} - \frac{1}{8} + \cdots$$

converges absolutely because the corresponding series of absolute values

$$1 + \frac{1}{2} + \frac{1}{4} + \frac{1}{8} + \cdots$$

converges. But the alternating harmonic series, although it converges, does not converge absolutely. The corresponding series of absolute values is the (divergent) harmonic series.

DEFINITION

A series that converges but does not converge absolutely **converges conditionally.**

The alternating harmonic series converges conditionally.

Absolute convergence is important because, first, we have many good tests for convergence of series of positive terms. Second, if a series converges absolutely, then it converges. That is the thrust of the next theorem.

THEOREM 9

The Absolute Convergence Theorem

If $\sum_{n=1}^{\infty} |a_n|$ converges, then $\sum_{n=1}^{\infty} a_n$ converges.

Proof   For each n,

$$-|a_n| \le a_n \le |a_n|, \quad \text{so} \quad 0 \le a_n + |a_n| \le 2|a_n|.$$

If $\sum_{n=1}^{\infty} |a_n|$ converges, then $\sum_{n=1}^{\infty} 2|a_n|$ converges and, by the Comparison Test, the nonnegative series $\sum_{n=1}^{\infty} (a_n + |a_n|)$ converges. The equality $a_n = (a_n + |a_n|)$

$-|a_n|$ now lets us express $\sum_{n=1}^{\infty} a_n$ as the difference of two convergent series:

$$\sum_{n=1}^{\infty} a_n = \sum_{n=1}^{\infty} (a_n + |a_n| - |a_n|) = \sum_{n=1}^{\infty} (a_n + |a_n|) - \sum_{n=1}^{\infty} |a_n|.$$

Therefore, $\sum_{n=1}^{\infty} a_n$ converges. $\quad\blacksquare$

We can rephrase Theorem 9 to say that *every absolutely convergent series converges.* However, the converse statement is false. Many convergent series do not converge absolutely. The convergence of many series depends on the series' having infinitely many positive and negative terms arranged in a particular order.

Example 3   For $\sum_{n=1}^{\infty} (-1)^{n+1} \dfrac{1}{n^2} = 1 - \dfrac{1}{4} + \dfrac{1}{9} - \dfrac{1}{16} + \cdots$, the corresponding series of absolute values is

$$\sum_{n=1}^{\infty} \frac{1}{n^2} = 1 + \frac{1}{4} + \frac{1}{9} + \frac{1}{16} + \cdots.$$

The series converges because it is a p-series with $p = 2 > 1$. Therefore

$$\sum_{n=1}^{\infty} (-1)^{n+1} \frac{1}{n^2}$$

converges absolutely and

$$\sum_{n=1}^{\infty} (-1)^{n+1} \frac{1}{n^2}$$

converges. $\quad\blacksquare$

Example 4   For $\sum_{n=1}^{\infty} \dfrac{\sin n}{n^2} = \dfrac{\sin 1}{1} + \dfrac{\sin 2}{4} + \dfrac{\sin 3}{9} + \cdots$, the corresponding series of absolute values is

$$\sum_{n=1}^{\infty} \left| \frac{\sin n}{n^2} \right| = \frac{|\sin 1|}{1} + \frac{|\sin 2|}{4} + \cdots,$$

which converges by comparison with $\sum_{n=1}^{\infty} (1/n^2)$ because $|\sin n| \leq 1$ for every n. The original series converges absolutely; therefore it converges. $\quad\blacksquare$

Alternating p-series

When p is a positive constant, the sequence $\{1/n^p\}$ is a decreasing sequence with limit zero. Therefore the alternating p-series

$$\sum_{n=1}^{\infty} \frac{(-1)^{n-1}}{n^p} = 1 - \frac{1}{2^p} + \frac{1}{3^p} - \frac{1}{4^p} + \cdots, \qquad p > 0$$

converges.

For $p > 1$, the series converges absolutely. For $0 < p \leq 1$, the series converges conditionally.

Conditional convergence: $1 - \dfrac{1}{\sqrt{2}} + \dfrac{1}{\sqrt{3}} - \dfrac{1}{\sqrt{4}} + \cdots$

Absolute convergence: $1 - \dfrac{1}{2^{3/2}} + \dfrac{1}{3^{3/2}} - \dfrac{1}{4^{3/2}} + \cdots$

Rearranging Series

One other important fact about absolutely convergent series is the following theorem.

THEOREM 10

The Rearrangement Theorem for Absolutely Convergent Series

If $\sum_{n=1}^{\infty} a_n$ converges absolutely, and $b_1, b_2, \ldots, b_n, \ldots$ is any arrangement of the sequence $\{a_n\}$, then $\sum b_n$ converges and

$$\sum_{n=1}^{\infty} b_n = \sum_{n=1}^{\infty} a_n.$$

(For an outline of the proof, see Exercise 60.)

Example 5   As we saw in Example 3, the series

$$1 - \frac{1}{4} + \frac{1}{9} - \frac{1}{16} + \cdots + (-1)^{n-1}\frac{1}{n^2} + \cdots$$

converges absolutely. A possible rearrangement of the terms of the series might start with a positive term, then two negative terms, then three positive terms, then four negative terms, and so on: After k terms of one sign, take $k + 1$ terms of the opposite sign. The first ten terms of such a series look like this:

$$1 - \frac{1}{4} - \frac{1}{16} + \frac{1}{9} + \frac{1}{25} + \frac{1}{49} - \frac{1}{36} - \frac{1}{64} - \frac{1}{100} - \frac{1}{144} + \cdots.$$

The Rearrangement Theorem says that both series converge to the same value. In this example, if we had the second series to begin with, we would probably be glad to exchange it for the first, if we knew that we could. We can do even better: The sum of either series is also equal to

$$\sum_{n=1}^{\infty} \frac{1}{(2n-1)^2} - \sum_{n=1}^{\infty} \frac{1}{(2n)^2}.$$

(See Exercise 57.)

CAUTION ABOUT REARRANGEMENTS   If we rearrange infinitely many terms of a conditionally convergent series, we can get results that are far different from the sum of the original series. The next example illustrates some of the things that can happen.

Example 6   Consider the alternating harmonic series

$$\frac{1}{1} - \frac{1}{2} + \frac{1}{3} - \frac{1}{4} + \frac{1}{5} - \frac{1}{6} + \frac{1}{7} - \frac{1}{8} + \frac{1}{9} - \frac{1}{10} + \frac{1}{11} - \cdots.$$

Here, the series of terms $\sum [1/(2n-1)]$ diverges to $+\infty$ and the series of terms $\sum(-1/2n)$ diverges to $-\infty$. No matter how far out in the sequence of odd-numbered terms we begin, we can always add enough positive terms to get an arbitrarily large sum. Similarly, with the negative terms, no matter how far out we start, we can add enough consecutive even-numbered terms to get a negative sum of arbitrarily large absolute value. If we wished to do so, we could start adding odd-numbered terms until we had a sum greater than $+3$, say, and then follow that with enough consecutive negative terms to make the new total less than -4. We could then add enough positive terms to make the total greater than $+5$ and follow with consecutive unused negative terms to make a new total less than -6, and so on. In this way, we could make the swings arbitrarily large in either direction.

Another possibility, with the same series, is to focus on a particular limit. Suppose we try to get sums that converge to 1. We start with the first term, $1/1$, and then subtract $1/2$. Next we add $1/3$ and $1/5$, which brings the total back to 1 or above. Then we add consecutive negative terms until the total is less than 1. We continue in this manner: When the sum is less than 1, add positive terms until the total is 1 or more; then subtract (add negative) terms until the total is again less than 1. This process can be continued indefinitely. Because both the odd-numbered terms and the even-numbered terms of the original series approach zero as $n \rightarrow \infty$, the amount by which our partial sums exceed 1 or fall below it approaches zero. So the new series converges to 1. The rearranged series starts like this:

$$\frac{1}{1} - \frac{1}{2} + \frac{1}{3} + \frac{1}{5} - \frac{1}{4} + \frac{1}{7} + \frac{1}{9} - \frac{1}{6} + \frac{1}{11} + \frac{1}{13} - \frac{1}{8} + \frac{1}{15} + \frac{1}{17} - \frac{1}{10}$$

$$+ \frac{1}{19} + \frac{1}{21} - \frac{1}{12} + \frac{1}{23} + \frac{1}{25} - \frac{1}{14} + \frac{1}{27} - \frac{1}{16} + \cdots.$$

FLOWCHART 8.1   Procedure for Determining Convergence

The kind of behavior illustrated by this example is typical of what can happen with any conditionally convergent series. Moral: Add the terms of a conditionally convergent series in the order given.

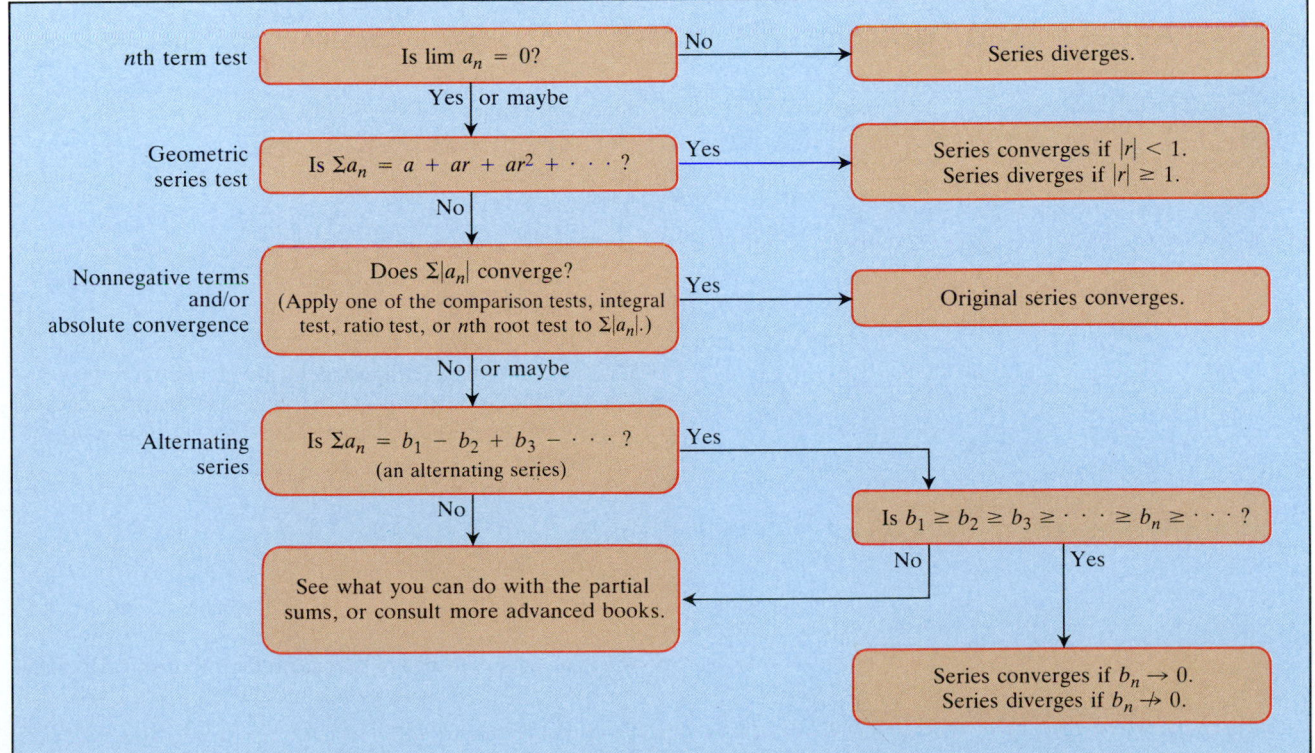

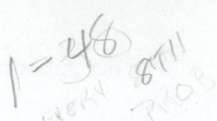

EXERCISES 8.5

Which of the alternating series in Exercises 1–12 converge and which diverge?

1. $\displaystyle\sum_{n=1}^{\infty} (-1)^{n+1} \frac{1}{n^{3/2}}$

2. $\displaystyle\sum_{n=2}^{\infty} (-1)^{n+1} \frac{1}{\ln n}$

3. $\displaystyle\sum_{n=1}^{\infty} (-1)^{n+1}$

4. $\displaystyle\sum_{n=1}^{\infty} (-1)^{n+1} \frac{10^n}{n^{10}}$

5. $\displaystyle\sum_{n=1}^{\infty} (-1)^{n+1} \frac{\sqrt{n}+1}{n+1}$

6. $\displaystyle\sum_{n=1}^{\infty} (-1)^{n+1} \frac{\ln n}{n}$

7. $\displaystyle\sum_{n=2}^{\infty} (-1)^n \log_n 2$

8. $\displaystyle\sum_{n=2}^{\infty} (-1)^{n+1} \frac{\ln n}{\ln n^2}$

9. $\displaystyle\sum_{n=1}^{\infty} (-1)^{n+1} \frac{3\sqrt{n}+1}{\sqrt{n}+1}$

10. $\displaystyle\sum_{n=1}^{\infty} (-1)^n \ln\left(1+\frac{1}{n}\right)$

11. $\displaystyle\sum_{n=1}^{\infty} (-1)^n \left(\sqrt{n+\sqrt{n}} - \sqrt{n}\right)$

12. $\displaystyle\sum_{n=1}^{\infty} (-1)^n \sqrt{n} \sin\frac{1}{n}$

Which of the series in Exercises 13–44 converge absolutely, which converge conditionally, and which diverge?

13. $\displaystyle\sum_{n=1}^{\infty} (-1)^{n+1} (0.1)^n$

14. $\displaystyle\sum_{n=1}^{\infty} (-1)^{n+1} \frac{1}{\sqrt{n}}$

15. $\displaystyle\sum_{n=1}^{\infty} (-1)^{n+1} \frac{n}{n^3+1}$

16. $\displaystyle\sum_{n=1}^{\infty} \frac{n!}{2^n}$

17. $\displaystyle\sum_{n=1}^{\infty} (-1)^n \frac{1}{n+3}$

18. $\displaystyle\sum_{n=1}^{\infty} (-1)^n \frac{\sin n}{n^2}$

19. $\displaystyle\sum_{n=1}^{\infty} (-1)^{n+1} \frac{3+n}{5+n}$

20. $\displaystyle\sum_{n=2}^{\infty} (-1)^n \frac{1}{\ln n^3}$

21. $\displaystyle\sum_{n=1}^{\infty} (-1)^{n+1} \frac{1+n}{n^2}$

22. $\displaystyle\sum_{n=1}^{\infty} \frac{(-2)^{n+1}}{n+5^n}$

23. $\displaystyle\sum_{n=1}^{\infty} n^2 (2/3)^n$

24. $\displaystyle\sum_{n=1}^{\infty} (-1)^{n+1} (\sqrt[n]{10})$

25. $\displaystyle\sum_{n=1}^{\infty} (-1)^n \frac{\tan^{-1} n}{n^2+1}$

26. $\displaystyle\sum_{n=2}^{\infty} (-1)^{n+1} \frac{1}{n \ln n}$

27. $\displaystyle\sum_{n=1}^{\infty} \left(\frac{1}{n} - \frac{1}{2n}\right)$

28. $\displaystyle\sum_{n=1}^{\infty} (-1)^{n+1} \frac{(0.1)^n}{n}$

29. $\displaystyle\sum_{n=1}^{\infty} (-1)^n \frac{n}{n+1}$

30. $\displaystyle\sum_{n=1}^{\infty} \frac{(-1)^n}{1+\sqrt{n}}$

31. $\displaystyle\sum_{n=1}^{\infty} \frac{-1}{n^2+2n+1}$

32. $\displaystyle\sum_{n=1}^{\infty} (5)^{-n}$

33. $\displaystyle\sum_{n=1}^{\infty} \frac{(-100)^n}{n!}$

34. $\displaystyle\sum_{n=2}^{\infty} (-1)^n \left(\frac{\ln n}{\ln n^2}\right)^n$

35. $\displaystyle\sum_{n=1}^{\infty} \frac{\cos n\pi}{n\sqrt{n}}$

36. $\displaystyle\sum_{n=1}^{\infty} \frac{\cos n\pi}{n}$

37. $\displaystyle\sum_{n=1}^{\infty} \frac{(-1)^n}{\sqrt{n}+\sqrt{n+1}}$

38. $\displaystyle\sum_{n=1}^{\infty} \frac{(-1)^{n+1}(n!)^2}{(2n)!}$

39. $\displaystyle\sum_{n=1}^{\infty} (-1)^n \frac{(2n)!}{2^n n! \, n}$

40. $\displaystyle\sum_{n=1}^{\infty} (-1)^n \frac{(n!)^2 \, 3^n}{(2n+1)!}$

41. $\displaystyle\sum_{n=1}^{\infty} (-1)^n (\sqrt{n+1} - \sqrt{n})$

42. $\displaystyle\sum_{n=1}^{\infty} (-1)^n (\sqrt{n^2+n} - n)$

43. $\displaystyle\sum_{n=1}^{\infty} (-1)^n \operatorname{sech} n$

44. $\displaystyle\sum_{n=1}^{\infty} (-1)^n \operatorname{csch} n$

In Exercises 45–48, estimate the error involved in using the sum of the first four terms to approximate the sum of the entire series.

45. $\displaystyle\sum_{n=1}^{\infty} (-1)^{n+1} \frac{1}{n}$

46. $\displaystyle\sum_{n=1}^{\infty} (-1)^{n+1} \frac{1}{10^n}$

47. $\ln(1.01) = \displaystyle\sum_{n=1}^{\infty} (-1)^{n+1} \frac{(0.01)^n}{n}$

(In Section 8.8, you will see why the series adds to $\ln(1.01)$.)

48. $\dfrac{1}{1+t} = \displaystyle\sum_{n=0}^{\infty} (-1)^n t^n, \quad 0 < t < 1$

CALCULATOR Approximate the sums in Exercises 49 and 50 with an error of magnitude less than 5×10^{-6}.

49. $\displaystyle\sum_{n=0}^{\infty} (-1)^n \frac{1}{(2n)!}$ (As you will see in Section 8.7, the sum is cos 1, the cosine of 1 radian.)

50. $\displaystyle\sum_{n=0}^{\infty} (-1)^n \frac{1}{n!}$ (The sum is $\frac{1}{e}$.)

51. a) The series

$$\frac{1}{3} - \frac{1}{2} + \frac{1}{9} - \frac{1}{4} + \frac{1}{27} - \frac{1}{8} + \cdots + \frac{1}{3^n} - \frac{1}{2^n} + \cdots$$

does not meet one of the conditions of Theorem 7. Which one?

b) Find the sum of the series in (a).

52. CALCULATOR The limit L of an alternating series that satisfies the conditions of Theorem 7 lies between the values of any two consecutive partial sums. This suggests using the average

$$\frac{s_n + s_{n+1}}{2} = s_n + \frac{1}{2}(-1)^{n+2} a_{n+1}$$

to estimate L. Compute

$$s_{20} + \frac{1}{2} \cdot \frac{1}{21}$$

as an approximation to the sum of the alternating harmonic series. The exact sum is $\ln 2 = 0.6931\ldots$.

53. *The sign of the remainder of an alternating series that satis-*

fies the conditions of Theorem 7. Prove the assertion in Theorem 8 that whenever an alternating series satisfying the conditions of Theorem 7 is approximated with one of its partial sums, then the remainder (sum of the unused terms) has the same sign as the first unused term. (*Hint:* Group the remainder's terms in consecutive pairs.)

54. Show that the sum of the first $2n$ terms of the series

$$1 - \frac{1}{2} + \frac{1}{2} - \frac{1}{3} + \frac{1}{3} - \frac{1}{4} + \frac{1}{4} - \frac{1}{5} + \frac{1}{5} - \frac{1}{6} + \cdots$$

is the same as the sum of the first n terms of the series

$$\frac{1}{1 \cdot 2} + \frac{1}{2 \cdot 3} + \frac{1}{3 \cdot 4} + \frac{1}{4 \cdot 5} + \frac{1}{5 \cdot 6} + \cdots.$$

Do these series converge? What is the sum of the first $2n + 1$ terms of the first series? If the series converge, what is their sum?

55. Show by example that $\sum_{n=1}^{\infty} a_n b_n$ may diverge even if $\sum_{n=1}^{\infty} a_n$ and $\sum_{n=1}^{\infty} b_n$ both converge.

56. Show that the alternating series

$$\sum_{n=2}^{\infty} \frac{(-1)^n}{\sqrt{n} + (-1)^n}$$

diverges. Which of the three conditions of Leibniz's theorem is not satisfied? (*Hint:* Show that

$$\frac{(-1)^n}{\sqrt{n} + (-1)^n} = \frac{(-1)^n}{\sqrt{n}} - \frac{1}{n + (-1)^n \sqrt{n}}$$

and write the original series as the difference of the two series.)

57. *Unzipping absolutely convergent series.*

a) Show that if $\sum_{n=1}^{\infty} |a_n|$ converges and

$$b_n = \begin{cases} a_n & \text{if } a_n \geq 0, \\ 0 & \text{if } a_n < 0, \end{cases}$$

then $\sum_{n=1}^{\infty} b_n$ converges.

b) Use the results in (a) to show likewise that if $\sum_{n=1}^{\infty} |a_n|$ converges and

$$c_n = \begin{cases} 0 & \text{if } a_n \geq 0, \\ a_n & \text{if } a_n < 0, \end{cases}$$

then $\sum_{n=1}^{\infty} c_n$ converges.

In other words, if a series converges absolutely, its positive terms form a convergent series, and so do its negative terms. Furthermore,

$$\sum_{n=1}^{\infty} a_n = \sum_{n=1}^{\infty} b_n + \sum_{n=1}^{\infty} c_n$$

because $b_n = (a_n + |a_n|)/2$ and $c_n = (a_n - |a_n|)/2$.

58. What is wrong here:

Multiply both sides of the alternating harmonic series

$$S = 1 - \frac{1}{2} + \frac{1}{3} - \frac{1}{4} + \frac{1}{5} - \frac{1}{6} +$$

$$\frac{1}{7} - \frac{1}{8} + \frac{1}{9} - \frac{1}{10} + \frac{1}{11} - \frac{1}{12} + \cdots$$

by 2 to get

$$2S = 2 - 1 +$$

$$\frac{2}{3} - \frac{1}{2} + \frac{2}{5} - \frac{1}{3} + \frac{2}{7} - \frac{1}{4} + \frac{2}{9} - \frac{1}{5} + \frac{2}{11} - \frac{1}{6} + \cdots.$$

Collect terms with the same denominator, as the arrows indicate, to arrive at

$$2S = 1 - \frac{1}{2} + \frac{1}{3} - \frac{1}{4} + \frac{1}{5} - \frac{1}{6} + \cdots.$$

The series on the right-hand side of this equation is the series we started with. Therefore, $2S = S$, and dividing by S gives $2 = 1$. (*Source: Riemann's Rearrangement Theorem* by Stewart Galanor, *Mathematics Teacher*, Vol. 80, No. 8, 1987, pp. 675–81.)

59. CALCULATOR In Example 6, suppose the goal is to arrange the terms to get a new series that converges to $-1/2$. Start the new arrangement with the first negative term, which is $-1/2$. Whenever you have a sum that is less than or equal to $-1/2$, start introducing positive terms, taken in order, until the new total is greater than $-1/2$. Then add negative terms until the total is less than or equal to $-1/2$ again. Continue this process until your partial sums have been above the target at least three times and finish at or below it. If s_n is the sum of the first n terms of your new series, plot the points (n, s_n) to illustrate how the sums are behaving.

60. *Outline of the proof of the Rearrangement Theorem (Theorem 10).* Let ϵ be a positive real number, let $L = \sum_{n=1}^{\infty} a_n$, and let $s_k = \sum_{n=1}^{k} a_n$. Show that for some index N_1 and for some index $N_2 \geq N_1$,

$$\sum_{n=N_1}^{\infty} |a_n| < \frac{\epsilon}{2} \qquad \text{and} \qquad |s_{N_2} - L| < \frac{\epsilon}{2}.$$

Since all the terms $a_1, a_2, \ldots, a_{N_2}$ appear somewhere in the sequence $\{b_n\}$, there is an index $N_3 \geq N_2$ such that if $n \geq N_3$, then $(\sum_{k=1}^{n} b_k) - s_{N_2}$ is at most a sum of terms a_m with $m \geq N_1$. Therefore, if $n \geq N_3$,

$$\left| \sum_{k=1}^{n} b_k - L \right| \leq \left| \sum_{k=1}^{n} b_k - s_{N_2} \right| + |s_{N_2} - L|$$

$$\leq \sum_{k=N_1}^{\infty} |a_k| + |s_{N_2} - L| < \epsilon.$$

EXPLORER PROGRAM

Sequences and Series

Generates terms of one or two sequences and graphs them. You may define the sequences recursively or by giving formulas for their nth terms. Enables you to look for graphical and numerical evidence of convergence or divergence.

8.6 Power Series

Now that we know how to test infinite series for convergence, we can study the infinite polynomials we mentioned at the beginning of Section 8.2. We call these infinite polynomials *power series* because they are defined as infinite series of powers of some variable, in our case x. Like polynomials, power series can be added, subtracted, multiplied, differentiated, and integrated to give new power series. They can be divided, too, but we shall not go into that. Almost any function with infinitely many derivatives can be represented by a power series, as long as the derivatives do not become too large. The series for e^x, $\sin x$, $\cos x$, $\ln(1 + x)$, $\tan^{-1}x$, and so on enable us to approximate the values of these functions as accurately as we please. We shall show you what these series are as the chapter continues.

Power Series and Convergence

We begin with a formal definition.

DEFINITIONS

A **power series** is a series of the form

$$\sum_{n=0}^{\infty} c_n x^n = c_0 + c_1 x + c_2 x^2 + \cdots + c_n x^n + \cdots \tag{1}$$

A POWER SERIES CENTERED AT $x=0$

or

$$\sum_{n=0}^{\infty} c_n(x - a)^n = c_0 + c_1(x - a) + c_2(x - a)^2 + \cdots + c_n(x - a)^n + \cdots \tag{2}$$

CENTERED @ $x=a$

in which the **center** a and the **coefficients** $c_0, c_1, c_2, \ldots, c_n, \ldots$ are constants.

Equation (1) is the special case obtained by taking $a = 0$ in Eq. (2).

Example 1   Taking all the coefficients to be 1 in Eq. (1) gives the geometric power series

$$\sum_{n=0}^{\infty} x^n = 1 + x + x^2 + \cdots + x^n + \cdots.$$

The series converges to $1/(1 - x)$ when $|x| < 1$. We express this fact by writing

$$\frac{1}{1 - x} = 1 + x + x^2 + \cdots + x^n + \cdots \qquad \text{for} \quad -1 < x < 1. \tag{3}$$

Up to now we have used Eq. (3) as a formula to give us the sum of the series on the right. We now change the focus: We think of the partial sums of the series on the right as polynomials $P_n(x)$ that approximate the function on the left. For values of x near zero, we need take only a few terms of the series to get a good approximation. As we move toward $x = 1$ or -1, we must take more terms. Figure 8.19 shows the graphs of $y = 1/(1 - x)$, and the approximating polynomials $y = P_n(x)$ for $n = 0$, 1, and 2.

Example 2   The power series

$$1 - \frac{1}{2}(x - 2) + \frac{1}{4}(x - 2)^2 + \cdots + \left(-\frac{1}{2}\right)^n(x - 2)^n + \cdots \tag{4}$$

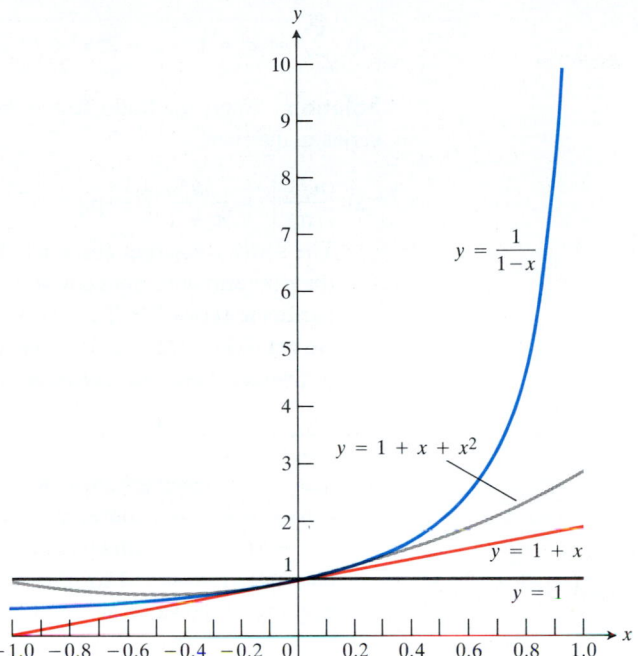

8.19 The graphs of $y = 1/(1 - x)$ and the approximating polynomials $P_0(x) = 1$, $P_1(x) = 1 + x$, and $P_2(x) = 1 + x + x^2$.

matches Eq. (2) with $a = 2$, $c_0 = 1$, $c_1 = -1/2$, $c_2 = 1/4, \ldots, c_n = (-1/2)^n$. This is a geometric series with first term 1 and ratio $r = -(x - 2)/2$. The series converges for $|(x - 2)/2| < 1$ or $0 < x < 4$. The sum is

$$\frac{1}{1 - r} = \frac{1}{1 + \dfrac{x - 2}{2}} = \frac{2}{x},$$

so we write

$$\frac{2}{x} = 1 - \frac{(x - 2)}{2} + \frac{(x - 2)^2}{4} - \cdots + \left(-\frac{1}{2}\right)^n (x - 2)^n + \cdots, \qquad 0 < x < 4.$$

We use series (4) to generate polynomial approximations to $2/x$ (or $1/x$ if we divide by 2) for values of x near 2. We keep the powers of $(x - 2)$, instead of expanding them, and write the approximating polynomials as

$$P_0(x) = 1, \quad P_1(x) = 1 - \frac{1}{2}(x - 2), \quad P_2(x) = 1 - \frac{1}{2}(x - 2) + \frac{1}{4}(x - 2)^2,$$

and so on. The higher powers of $(x - 2)$ decrease rapidly as n increases as long as $|x - 2|$ is small.

Example 3   For what values of x do the following series converge?

a) $\displaystyle\sum_{n=1}^{\infty} (-1)^{n-1} \frac{x^n}{n} = x - \frac{x^2}{2} + \frac{x^3}{3} - \cdots$

b) $\displaystyle\sum_{n=1}^{\infty} (-1)^{n-1} \frac{x^{2n-1}}{2n - 1} = x - \frac{x^3}{3} + \frac{x^5}{5} - \cdots$

c) $\displaystyle\sum_{n=0}^{\infty} \frac{x^n}{n!} = 1 + x + \frac{x^2}{2!} + \frac{x^3}{3!} + \cdots$

d) $\displaystyle\sum_{n=0}^{\infty} n!\, x^n = 1 + x + 2!\, x^2 + 3!\, x^3 + \cdots$

Solution   Apply the Ratio Test to the series $\Sigma|u_n|$, where u_n is the nth term of the series in question.

a) $\left|\dfrac{u_{n+1}}{u_n}\right| = \dfrac{n}{n+1}\,|x| \to |x|$

The series converges absolutely for $|x| < 1$. It diverges if $|x| > 1$ because the nth term does not converge to zero. At $x = 1$, we get the alternating harmonic series $1 - 1/2 + 1/3 - 1/4 + \cdots$, which converges. At $x = -1$ we get $-1 - 1/2 - 1/3 - 1/4 - \cdots$, the negative of the harmonic series; it diverges. Series (a) converges for $-1 < x \le 1$ and diverges elsewhere.

b) $\left|\dfrac{u_{n+1}}{u_n}\right| = \dfrac{2n-1}{2n+1}\,x^2 \to x^2$

The series converges absolutely for $x^2 < 1$. It diverges for $x^2 > 1$ because the nth term does not converge to zero. At $x = 1$ the series becomes $1 - 1/3 + 1/5 - 1/7 + \cdots$, which converges by the Alternating Series Theorem. It converges at $x = -1$ for the same reason. The value at $x = -1$ is the negative of the value at $x = 1$.

c) $\left|\dfrac{u_{n+1}}{u_n}\right| = \left|\dfrac{x^{n+1}}{(n+1)!} \cdot \dfrac{n!}{x^n}\right| = \dfrac{|x|}{n+1} \to 0$ for every x

The series converges absolutely for all x.

d) $\left|\dfrac{u_{n+1}}{u_n}\right| = \left|\dfrac{(n+1)!\, x^{n+1}}{n!\, x^n}\right| = (n+1)\,|x| \to \infty$ unless $x = 0$

The series diverges for all values of x except $x = 0$. The nth term does not approach zero unless $x = 0$.

Example 3 illustrates how we usually test a power series for convergence and the kinds of results we get.

How to Test a Power Series for Convergence

STEP 1:   Use the Ratio Test (or nth-Root Test) to find the interval where the series converges absolutely. Ordinarily, this is an open interval

$$|x - a| < h \qquad \text{or} \qquad a - h < x < a + h.$$

In some instances, as in Example 3(c), the series converges for all values of x. These instances are not uncommon. In rare cases, the series may converge only at a single point, as in Example 3(d).

STEP 2:   If the interval of absolute convergence is finite, test for convergence or divergence at each of the two endpoints, as in Example 2 and Examples 3(a) and (b). Neither the Ratio Test nor the nth-Root Test is useful at these points. Use a Comparison Test, the Integral Test, or the Alternating Series Theorem.

STEP 3:   If the interval of absolute convergence is $a - h < x < a + h$, the series diverges (it does not even converge conditionally) for $|x - a| > h$, because for those values of x the nth term does not approach zero.

In the next section we shall see how series in powers of $(x - a)$ are generated by the values of a function f and its derivatives at $x = a$. Since $1/x$, $\sqrt{x}$, and $\ln x$ exist and have derivatives of all orders at $x = 1$, the power series we use to represent them can be in powers of $x - 1$. But they do not have power series representations in powers of x because they have no derivatives at $x = 0$.

To simplify the notation, the next theorem deals with the convergence of series of the form $\Sigma a_n x^n$. For series of the form $\Sigma a_n (x - a)^n$ we can replace $(x - a)$ by x' and apply the results to the series $\Sigma a_n (x')^n$.

THEOREM 11

The Convergence Theorem for Power Series

If

$$\sum_{n=0}^{\infty} a_n x^n = a_0 + a_1 x + a_2 x^2 + \cdots \tag{5}$$

converges for $x = c$ $(c \neq 0)$, then it converges absolutely for all $|x| < |c|$. If the series diverges for $x = d$, then it diverges for all $|x| > |d|$.

Proof Suppose the series

$$\sum_{n=0}^{\infty} a_n c^n \tag{6}$$

converges. Then

$$\lim_{n \to \infty} a_n c^n = 0.$$

Hence, there is an integer N such that $|a_n c^n| < 1$ for all $n \geq N$. That is,

$$|a_n| < \frac{1}{|c|^n} \qquad \text{for } n \geq N. \tag{7}$$

Now take any x such that $|x| < |c|$ and consider

$$|a_0| + |a_1 x| + \cdots + |a_{N-1} x^{N-1}| + |a_N x^N| + |a_{N+1} x^{N+1}| + \cdots.$$

There is only a finite number of terms prior to $|a_N x^N|$, and their sum is finite. Starting with $|a_N x^N|$ and beyond, the terms are less than

$$\left| \frac{x}{c} \right|^N + \left| \frac{x}{c} \right|^{N+1} + \left| \frac{x}{c} \right|^{N+2} + \cdots \tag{8}$$

because of (7). But the series in (8) is a geometric series with ratio $r = |x/c|$, which is less than 1 since $|x| < |c|$. Hence the series (8) converges, so the original series (6) converges absolutely. This proves the first half of the theorem.

The second half of the theorem follows from the first. If the series diverges at $x = d$ and converges at a value x_0 with $|x_0| > |d|$, we may take $c = x_0$ in the first half of the theorem and conclude that the series converges absolutely at d. But the series cannot converge absolutely and diverge at one and the same time. Hence, if it diverges at d, it diverges for all $|x| > |d|$. $\blacksquare$

The Radius and Interval of Convergence

The examples we have looked at and the theorem we just proved lead to the conclusion that a power series always behaves in exactly one of the following three ways.

Possible Behavior of $\Sigma\, c_n\,(x - a)^n$

1. The series converges at $x = a$ and **diverges elsewhere.**

2. There is a positive number h such that **the series diverges for $|x - a| > h$** but converges absolutely for $|x - a| < h$. The series may or may not converge at either of the endpoints $x = a - h$ and $x = a + h$.

3. The series converges absolutely for every x.

In case 2, the set of points at which the series converges is a finite interval, called the **interval of convergence.** We know from past examples that the interval may be open, half-open, or closed, depending on the particular series. But no matter which kind of interval it is, h is called the **radius of convergence** of the series, and $a + h$ is the least upper bound of the set of points at which the series converges. The convergence is absolute at every point in the interior of the interval. If $a = 0$, the interval is centered at the origin. If a power series converges absolutely for all values of x, we say that its radius of convergence is infinite. If it converges only at $x = a$, we say that the radius of convergence is zero.

Term-by-Term Differentiation of Power Series

A theorem from advanced calculus tells us that a power series can be differentiated term by term at each point in the interior of its interval of convergence.

THEOREM 12

The Term-by-Term Differentiation Theorem

If $\sum_{n=0}^{\infty} c_n(x - a)^n$ converges for $a - h < x < a + h$ for some $h > 0$, it defines a function f:

$$f(x) = \sum_{n=0}^{\infty} c_n(x - a)^n, \qquad a - h < x < a + h.$$

Such a function f has derivatives of all orders inside the interval of convergence. We can obtain the derivatives by differentiating the original series term by term:

$$f'(x) = \sum_{n=1}^{\infty} n\, c_n(x - a)^{n-1},$$

$$f''(x) = \sum_{n=2}^{\infty} n(n-1)c_n(x - a)^{n-2},$$

and so on. Each of these derived series converges at every interior point of the interval of convergence of the original series.

Here is an example of how to apply term-by-term differentiation.

Example 4   Is there a more familiar name for the function f defined by the power series

$$f(x) = \sum_{n=0}^{\infty} \frac{(-1)^n x^{2n+1}}{2n+1} = x - \frac{x^3}{3} + \frac{x^5}{5} - \cdots, \qquad -1 \le x \le 1?$$

A Word of Caution

Term-by-term differentiation might not work for other kinds of series. For example, the trigonometric series

$$\sum_{n=1}^{\infty} \frac{\sin(n!\,x)}{n^2}$$

converges for every x. But if we differentiate term by term we get the series

$$\sum_{n=1}^{\infty} \frac{n!\cos(n!\,x)}{n^2},$$

which diverges for all x.

Solution   We differentiate the original series term by term and get

$$f'(x) = \sum_{n=0}^{\infty} \frac{(2n+1)(-1)^n x^{2n}}{2n+1} = \sum_{n=0}^{\infty} (-1)^n x^{2n} = 1 - x^2 + x^4 - x^6 + \cdots.$$

This is a geometric series with first term 1 and ratio $-x^2$, so

$$f'(x) = \frac{1}{1-(-x^2)} = \frac{1}{1+x^2}.$$

We can now integrate $f'(x) = 1/(1+x^2)$ to get

$$\int f'(x)\, dx = \int \frac{dx}{1+x^2} = \tan^{-1}x + C.$$

The series for $f(x)$ is zero when $x = 0$, so $C = 0$. Hence

$$f(x) = x - \frac{x^3}{3} + \frac{x^5}{5} - \frac{x^7}{7} + \cdots = \tan^{-1}x, \qquad -1 \le x \le 1. \tag{9}$$

The function is the restriction of $\tan^{-1}x$ to the interval $[-1, 1]$.

Notice that the original series in **Example 4 converges at both** endpoints of the original interval of convergence, **but the differentiated series converges** only inside the interval. This is all that the theorem **guarantees.**

Term-by-Term Integration of Power Series

Another theorem from **advanced calculus** states that a power series can be integrated term by term **throughout its** interval of convergence.

THEOREM 13

The Term-by-Term Integration Theorem

If $\sum_{n=0}^{\infty} c_n(x-a)^n$ converges when $a - h < x < a + h$ for some $h > 0$ and $f(x) = \sum_{n=0}^{\infty} c_n(x-a)^n$ for $a - h < x < a + h$, then the series

$$\sum_{n=0}^{\infty} c_n \frac{(x-a)^{n+1}}{n+1}$$ converges for $a - h < x < a + h$ and

$$\int f(x)\, dx = \sum_{n=0}^{\infty} c_n \frac{(x-a)^{n+1}}{n+1} + C \qquad \text{for} \quad a - h < x < a + h.$$

Example 5   The series

$$\frac{1}{1+t} = 1 - t + t^2 - t^3 \cdots$$

converges on the open interval $-1 < t < 1$. Therefore,

$$\ln(1+x) = \int_0^x \frac{1}{1+t}\, dt = t - \frac{t^2}{2} + \frac{t^3}{3} - \frac{t^4}{4} \cdots \Big]_0^x$$

$$= x - \frac{x^2}{2} + \frac{x^3}{3} - \frac{x^4}{4} + \cdots, \qquad -1 < x < 1.$$

The latter series also converges at $x = 1$, but that was not guaranteed by the theorem.

Multiplication of Power Series

Still another theorem from advanced calculus states that power series can be multiplied term by term on any common interval of absolute convergence.

THEOREM 14

The Series Multiplication Theorem for Power Series

If $A(x) = \sum_{n=0}^{\infty} a_n x^n$ and $B(x) = \sum_{n=0}^{\infty} b_n x^n$ converge absolutely for $|x| < h$, and

$$c_n = a_0 b_n + a_1 b_{n-1} + a_2 b_{n-2} + \cdots + a_{n-1} b_1 + a_n b_0 = \sum_{k=0}^{n} a_k b_{n-k}, \quad (10)$$

then $\sum_{n=0}^{\infty} c_n x^n$ converges absolutely to $A(x) B(x)$ for $|x| < h$:

$$\left(\sum_{n=0}^{\infty} a_n x^n \right) \cdot \left(\sum_{n=0}^{\infty} b_n x^n \right) = \sum_{n=0}^{\infty} c_n x^n. \quad (11)$$

Example 6

$$\sum_{n=0}^{\infty} x^n = 1 + x + x^2 + \cdots + x^n + \cdots = \frac{1}{1-x} \qquad \text{for } |x| < 1$$

by itself to get the power series for $1/(1-x)^2$ for $|x| < 1$.

Solution Let

$$A(x) = \sum_{n=0}^{\infty} a_n x^n = 1 + x + x^2 + \cdots + x^n + \cdots = 1/(1-x),$$

$$B(x) = \sum_{n=0}^{\infty} b_n x^n = 1 + x + x^2 + \cdots + x^n + \cdots = 1/(1-x),$$

and

$$c_n = \underbrace{a_0 b_n + a_1 b_{n-1} + \cdots + a_k b_{n-k} + \cdots + a_n b_0}_{n+1 \text{ terms, each with value } 1} = n + 1.$$

Then, by the Series Multiplication Theorem,

$$A(x) \cdot B(x) = \sum_{n=0}^{\infty} c_n x^n = \sum_{n=0}^{\infty} (n+1) x^n$$

$$= 1 + 2x + 3x^2 + 4x^3 + \cdots + (n+1) x^n + \cdots$$

is the series for $1/(1-x)^2$. The series all converge absolutely for $|x| < 1$.

EXERCISES 8.6

Each of Exercises 1–34 gives a formula for the nth term of a series. (a) Find the series' radius and interval of convergence. For what values of x does the series converge (b) absolutely; (c) conditionally?

1. x^n

2. $(x + 5)^n$

3. $(-1)^n (x + 1)^n$

4. $\dfrac{(x - 2)^n}{n}$

5. $\dfrac{(x - 2)^n}{10^n}$

6. $(2x)^n$

7. $\dfrac{nx^n}{n+2}$

8. $\dfrac{(-1)^n(x+2)^n}{n}$

9. $\dfrac{x^n}{n\sqrt{n}\,3^n}$

10. $\dfrac{(x-1)^n}{\sqrt{n}}$

11. $\dfrac{(-1)^n x^n}{n!}$

12. $\dfrac{3^n x^n}{n!}$

13. $\dfrac{x^{2n+1}}{n!}$

14. $\dfrac{(2x-3)^{2n+1}}{n!}$

15. $\dfrac{x^n}{\sqrt{n^2+3}}$

16. $\dfrac{(-1)^n x^n}{\sqrt{n^2+3}}$

17. $\dfrac{n\,x^n}{4^n(n^2+1)}$

18. $\dfrac{n(x-3)^n}{5^n}$

19. $\dfrac{\sqrt{n}\,x^n}{3^n}$

20. $\sqrt[n]{n}\,(x-1)^n$

21. $\left(1+\dfrac{1}{n}\right)^n x^n$

22. $(\ln n)x^n$

23. $n^n x^n$

24. $n!(x-4)^n$

25. $\dfrac{(-1)^{n+1}(x-2)^n}{n\,2^n}$

26. $(-2)^n(n+1)(x-1)^n$

27. $\dfrac{x^n}{n\ln n}$ (*Hint:* See Section 8.3, Exercise 41.)

28. $\dfrac{(x-1)^{2n+1}}{n^{3/2}}$

29. $\left(\dfrac{1}{|x|}\right)^n$

30. $\left(\dfrac{x^2+1}{3}\right)^n$

31. $(\cosh n)x^n$

32. $(\sinh n)x^n$

33. $(\tanh n)x^n$

34. $(\operatorname{sech} n)x^n$

In Exercises 35–38, find the series' interval of convergence and, within this interval, the sum of the series as a function of x.

35. $\displaystyle\sum_{n=0}^{\infty}\dfrac{(x-1)^{2n}}{2^n}$

36. $\displaystyle\sum_{n=1}^{\infty}(\ln x)^n$

37. $\displaystyle\sum_{n=0}^{\infty}\left(\dfrac{x^2-1}{2}\right)^n$

38. $\displaystyle\sum_{n=0}^{\infty}\left(\dfrac{\sqrt{x}}{2}-1\right)^n$

The geometric series in Exercises 39 and 40 have two intervals of convergence. Identify the intervals and find the sum of each series as a function of x.

39. $\displaystyle\sum_{n=0}^{\infty}\left(\dfrac{3}{2-x^2}\right)^n$

40. $\displaystyle\sum_{n=0}^{\infty}\left(\dfrac{1}{x-1}\right)^n$

41. For what values of x does the series

$$1-\frac{1}{2}(x-3)+\frac{1}{4}(x-3)^2+\cdots+\left(-\frac{1}{2}\right)^n(x-3)^n+\cdots$$

converge? What is its sum? What series do you get if you differentiate the given series term by term? For what values of x does the new series converge? What is its sum?

42. If you integrate the series in Exercise 41 term by term, what new series do you get? For what values of x does the new series converge and what is another name for its sum?

43. The series for $\tan x$,

$$\tan x = x+\frac{x^3}{3}+\frac{2x^5}{15}+\frac{17x^7}{315}+\frac{62x^9}{2835}+\cdots,$$

converges for $-\pi/2<x<\pi/2$.
a) Find the first five terms of the series for $\ln|\sec x|$. For what values of x should the series converge?
b) Find the first five terms of the series for $\sec^2 x$. For what values of x should this series converge?
c) Check your result in (b) by squaring the series given for $\sec x$ in Exercise 44.

44. The series for $\sec x$,

$$\sec x = 1+\frac{x^2}{2}+\frac{5}{24}x^4+\frac{61}{720}x^6+\frac{277}{8064}x^8+\cdots,$$

converges for $-\pi/2<x<\pi/2$.
a) Find the first five terms of a series for the function $\ln|\sec x+\tan x|$. For what values of x should the series converge?
b) Find the first four terms of a series for $\sec x\tan x$. For what values of x should the series converge?
c) Check your result in (b) by multiplying the series for $\sec x$ by the series given for $\tan x$ in Exercise 43.

45. *Equality of convergent power series.*
a) Show that if two power series $\sum_{n=0}^{\infty}a_n x^n$ and $\sum_{n=0}^{\infty}b_n x^n$ are convergent and equal for all values of x in an open interval $(-c,c)$, then $a_n=b_n$ for every n. (*Hint:* Let $f(x)=\sum_{n=0}^{\infty}a_n x^n=\sum_{n=0}^{\infty}b_n x^n$. Differentiate term by term to show that a_n and b_n both equal $f^{(n)}(0)/(n!)$.)
b) Show that if $\sum_{n=0}^{\infty}a_n x^n=0$ for all x in an open interval $(-c,c)$, then $a_n=0$ for every n.

46. *The sum of the series $\sum_{n=0}^{\infty}(n^2/2^n)$.*   To find the sum of this series, express $1/(1-x)$ as a geometric series, differentiate both sides of the resulting equation with respect to x, multiply both sides of the result by x, differentiate again, multiply by x again, and set x equal to $1/2$. What do you get? (*Source:* David E. Dobbs' letter to the editor, *Illinois Mathematics Teacher,* Vol. 33, Issue 4, 1982, p. 27.)

47. *Convergence at endpoints.*   Show by examples that the convergence of a power series at an endpoint of its interval of convergence may be either conditional or absolute.

8.7 Taylor Series and Maclaurin Series

This section shows how to find power series called Taylor series for functions that have infinitely many derivatives and shows how to control the errors involved in using the partial sums of these series as polynomial approximations of the functions they represent. Our work here continues our earlier work with linearizations and quadratic approximations. One of the important features of Taylor series is that they enable us to extend the domains of functions to include complex numbers. We shall go into that briefly at the end of the section.

DEFINITIONS

Let f be a function with derivatives of all orders throughout some interval containing a as an interior point. Then the **Taylor series generated by f at a** is

$$\sum_{k=0}^{\infty} \frac{f^{(k)}(a)}{k!}(x-a)^k = f(a) + f'(a)(x-a) + \frac{f''(a)}{2!}(x-a)^2$$

$$+ \cdots + \frac{f^{(n)}(a)}{n!}(x-a)^n + \cdots . \tag{1}$$

The **Maclaurin series generated by f** is

$$\sum_{k=0}^{\infty} \frac{f^{(k)}(0)}{k!}x^k = f(0) + f'(0)x + \frac{f''(0)}{2!}x^2 + \cdots + \frac{f^{(n)}(0)}{n!}x^n + \cdots , \tag{2}$$

the Taylor series generated by f at $x = 0$.

Once we have found the Taylor series generated by a function f at a particular a, we can apply our usual tests to find where the series converges: usually in some interval $(a - h, a + h)$ or for all x. When the series does converge, we ask, "Does it converge to $f(x)$?"

Example 1   Find the Taylor series generated by $f(x) = 1/x$ at $a = 2$. Where, if anywhere, does the series converge to $1/x$?

Solution   We need to compute $f(2)$, $f'(2)$, $f''(2)$, and so on. Taking derivatives we get

$$f(x) = x^{-1}, \qquad\qquad f(2) = 2^{-1} = \frac{1}{2},$$

$$f'(x) = -x^{-2}, \qquad\qquad f'(2) = -\frac{1}{2^2},$$

$$f''(x) = 2!\, x^{-3}, \qquad\qquad \frac{f''(2)}{2!} = 2^{-3} = \frac{1}{2^3},$$

$$f'''(x) = -3!\, x^{-4}, \qquad\qquad \frac{f'''(2)}{3!} = -\frac{1}{2^4},$$

$$\vdots$$

$$f^{(n)}(x) = (-1)^n n!\, x^{-(n+1)}, \qquad \frac{f^{(n)}(2)}{n!} = \frac{(-1)^n}{2^{n+1}}.$$

The Taylor series is

$$f(2) + f'(2)(x - 2) + \frac{f''(2)}{2!}(x - 2)^2 + \frac{f'''(2)}{3!}(x - 2)^3 +$$

$$\cdots + \frac{f^{(n)}(2)}{n!}(x - 2)^n + \cdots$$

$$= \frac{1}{2} - \frac{(x - 2)}{2^2} + \frac{(x - 2)^2}{2^3} - \cdots + (-1)^n \frac{(x - 2)^n}{2^{n+1}} + \cdots.$$

This is a geometric series with first term 1/2 and ratio $r = -(x - 2)/2$. It converges absolutely for $|x - 2| < 2$ and its sum is

$$\frac{1/2}{1 + (x - 2)/2} = \frac{1}{2 + (x - 2)} = \frac{1}{x}.$$

In this example the Taylor series generated by $f(x) = 1/x$ at $a = 2$ converges to $1/x$ for $|x - 2| < 2$ or $0 < x < 4$.

Taylor Polynomials

The linearization and quadratic approximation of a twice-differentiable function f at a point a are the polynomials

$$P_1(x) = f(a) + f'(a)(x - a),$$

$$P_2(x) = f(a) + f'(a)(x - a) + \frac{f''(a)}{2}(x - a)^2.$$

If f has derivatives of higher order at a, then it has higher order polynomial approximations as well, one for each available derivative. These polynomials are called the Taylor polynomials of f.

DEFINITION

Let f be a function with derivatives of order k for $k = 1, 2, \ldots, N$ in some interval containing a as an interior point. Then for any integer n from 0 through N, the **Taylor polynomial** of order n generated by f at a is the polynomial

$$P_n(x) = f(a) + f'(a)(x - a) + \frac{f''(a)}{2!}(x - a)^2 + \cdots + \frac{f^{(k)}(a)}{k!}(x - a)^k$$

$$+ \cdots + \frac{f^{(n)}(a)}{n!}(x - a)^n. \tag{3}$$

We speak of a Taylor polynomial of *order n* rather than *degree n* because $f^{(n)}(a)$ may be zero. The first two Taylor polynomials of cos x at $x = 0$, for example, are $P_0(x) = 1$ and $P_1(x) = 1$. The first-order polynomial has degree zero, not one.

Special Property of Taylor Polynomials

What is so special about Taylor polynomials? We answer by looking for polynomials whose values at $x = a$ are equal to $f(a)$ and whose derivatives at $x = a$ are $f'(a)$, $f''(a)$, and so on. With that object in mind, we start with a polynomial of order n in powers of $(x - a)$ with undetermined coefficients $c_0, c_1, c_2, \ldots, c_n$:

$$P(x) = c_0 + c_1(x - a) + c_2(x - a)^2 + c_3(x - a)^3 + \cdots + c_n(x - a)^n. \tag{4}$$

Its derivatives of order $1, 2, \ldots, n$ are

$$P'(x) = c_1 + 2c_2(x - a) + 3c_3(x - a)^2 + \cdots \qquad + nc_n(x - a)^{n-1}$$

$$P''(x) = \qquad 2c_2 + \quad (3)(2)c_3(x - a) + \cdots + n(n - 1)c_n(x - a)^{n-2}$$

$$P'''(x) = \qquad\qquad\qquad (3!)c_3 + \cdots + n(n - 1)(n - 2)c_n(x - a)^{n-3}$$

$$\vdots$$

$$P^{(n)}(x) = \qquad\qquad\qquad\qquad\qquad\qquad\qquad (n!)c_n$$

When we substitute $x = a$, the terms with $(x - a)$ become 0. We also want $P(a) = f(a)$, $P'(a) = f'(a)$, $P''(a) = f''(a), \ldots, P^{(n)}(a) = f^{(n)}(a)$. This leads to the equations

$$P(a) = f(a) = c_0, \qquad\qquad P'(a) = f'(a) = c_1,$$

$$P''(a) = f''(a) = (2!)c_2, \qquad P'''(a) = f'''(a) = (3!)c_3,$$

$$\vdots$$

$$P^{(n)}(a) = f^{(n)}(a) = (n!)c_n.$$

When we determine the coefficients c_k in this way, we get

$$c_0 = f(a), \quad c_1 = f'(a), \quad c_2 = \frac{f''(a)}{2!}, \quad c_3 = \frac{f'''(a)}{3!}, \quad \cdots, \quad c_n = \frac{f^{(n)}(a)}{n!}.$$

Substituting these values for $c_0, c_1, \ldots, c_n$ in Eq. (4) gives the Taylor polynomial $P_n(x)$ in Eq. (3). The special property of the Taylor polynomials is just this:

> The Taylor polynomial of order n and its first n derivatives have the same values that f and its first n derivatives have at $x = a$.

A function that has derivatives of all orders at $x = a$ generates a Taylor polynomial for every $n \geq 0$.

Example 2   Find the Taylor polynomials generated by $f(x) = e^x$ at $a = 0$.

Solution   Expressed in terms of x, the given function and its derivatives are

$$f(x) = e^x, \qquad f'(x) = e^x, \qquad \cdots, \qquad f^{(n)}(x) = e^x,$$

so

$$f(0) = e^0 = 1, \qquad f'(0) = 1, \qquad \cdots, \qquad f^{(n)}(0) = 1,$$

and

$$P_n(x) = 1 + x + \frac{x^2}{2!} + \frac{x^3}{3!} + \cdots + \frac{x^n}{n!}.$$

See Fig. 8.20.

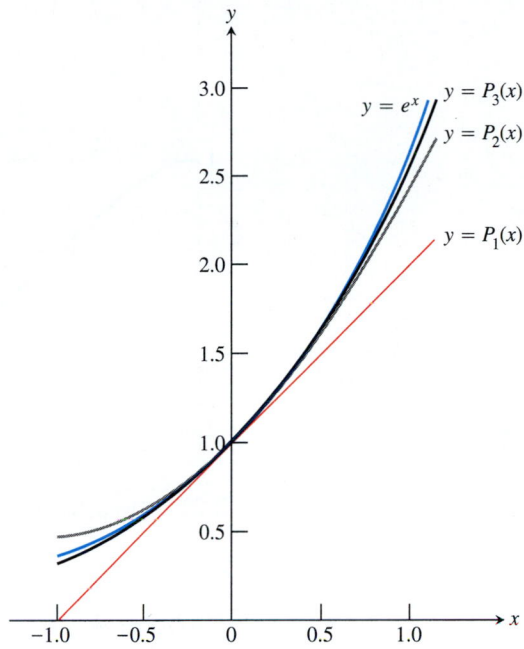

8.20 The graph of $f(x) = e^x$ and its Taylor polynomials $P_1(x) = 1 + x$, $P_2(x) = 1 + x + (x^2/2!)$, and $P_3(x) = 1 + x + (x^2/2!) + (x^3/3!)$. Notice the close agreement near the center $x = 0$.

Example 3   Find the Taylor polynomials generated by $f(x) = \cos x$ at $a = 0$.

Solution   The cosine and its derivatives are

$$f(x) = \cos x, \qquad f'(x) = -\sin x,$$

$$f''(x) = -\cos x, \qquad f^{(3)}(x) = \sin x,$$

$$\vdots \qquad\qquad \vdots$$

$$f^{(2n)}(x) = (-1)^n \cos x, \qquad f^{(2n+1)}(x) = (-1)^{n+1} \sin x.$$

When $x = 0$, the cosines are 1 and the sines are 0, so

$$f^{(2n)}(0) = (-1)^n, \qquad f^{(2n+1)}(0) = 0.$$

Notice that the Taylor polynomials of order $2n$ and of order $2n + 1$ are identical:

$$P_{2n}(x) = P_{2n+1}(x) = 1 - \frac{x^2}{2!} + \frac{x^4}{4!} - \cdots + (-1)^n \frac{x^{2n}}{(2n)!}.$$

Figure 8.21 shows how well these polynomials approximate $y = \cos x$ near $x = 0$. Only the right-hand portions of the graphs are shown because the graphs are symmetric about the y-axis.

8.21 The polynomials

$$P_{2n}(x) = \sum_{k=0}^{n}[(-1)^k x^{2k}/(2k)!]$$

converge to $\cos x$ as $n \to \infty$. Notice how we can deduce the behavior of $\cos x$ arbitrarily far away solely from the values of the cosine and its derivatives at $a = 0$.

Taylor's Theorem with Remainder

The preceding examples involved Taylor polynomials of order n for $n = 0, 1, 2, \ldots$. As $n \to \infty$, the Taylor polynomials become the partial sums of the Taylor series generated by f at a. The next theorem helps us find out whether these sums converge to $f(x)$ in some interval $(a - h, a + h)$ or, perhaps, for all x.

THEOREM 15

> **Taylor's Theorem**
>
> If f and its first n derivatives $f', f'', \ldots, f^{(n)}$ are continuous on $[a, b]$ or on $[b, a]$, and $f^{(n)}$ is differentiable on (a, b) or on (b, a), then there exists a number c between a and b such that
>
> $$f(b) = f(a) + f'(a)(b - a) + \frac{f''(a)}{2!}(b - a)^2 + \cdots$$
>
> $$+ \frac{f^{(n)}(a)}{n!}(b - a)^n + \frac{f^{(n+1)}(c)}{(n+1)!}(b - a)^{n+1}.$$

We prove the theorem assuming $a < b$. The proof for the case $a > b$ is nearly the same.

Proof (for $a < b$)   The Taylor polynomial

$$P_n(x) = f(a) + f'(a)(x - a) + \frac{f''(a)}{2!}(x - a)^2 + \cdots + \frac{f^{(n)}(a)}{n!}(x - a)^n$$

and its first n derivatives match the function f and its first n derivatives at $x = a$. We do not disturb that matching if we add another term of the form $K(x - a)^{n+1}$, where K is any constant, because such a term and its first n derivatives are all equal to zero at $x = a$. The new function

$$\phi_n(x) = P_n(x) + K(x - a)^{n+1}$$

and its first n derivatives still agree with f and its first n derivatives at $x = a$.

We now choose the particular value of K that makes the curve $y = \phi_n(x)$ agree with the original curve $y = f(x)$ at $x = b$. In symbols,

$$f(b) = P_n(b) + K(b - a)^{n+1}, \quad \text{or} \quad K = \frac{f(b) - P_n(b)}{(b - a)^{n+1}}. \tag{5}$$

With K defined by Eq. (5), the function

$$F(x) = f(x) - \phi_n(x) \tag{6}$$

measures the difference between the original function f and the approximating function ϕ_n for each x in $[a, b]$.

We now use Rolle's theorem. First, because $F(a) = F(b) = 0$ and both F and F' are continuous on $[a, b]$, we know that

$$F'(c_1) = 0 \quad \text{for some } c_1 \text{ in } (a, b).$$

Next, because $F'(a) = F'(c_1) = 0$ and both F' and F'' are continuous on $[a, c_1]$, we know that

$$F''(c_2) = 0 \quad \text{for some } c_2 \text{ in } (a, c_1).$$

Rolle's theorem, applied successively to F'', F''', ..., $F^{(n-1)}$ implies the existence of

$$c_3 \text{ in } (a, c_2) \qquad \text{such that } F'''(c_3) = 0,$$
$$c_4 \text{ in } (a, c_3) \qquad \text{such that } F^{(4)}(c_4) = 0,$$
$$\vdots$$
$$c_n \text{ in } (a, c_{n-1}) \qquad \text{such that } F^{(n)}(c_n) = 0.$$

Finally, because $F^{(n)}$ is continuous on $[a, c_n]$ and differentiable on (a, c_n), and $F^{(n)}(a) = F^{(n)}(c_n) = 0$, Rolle's theorem implies that there is a number c_{n+1} in (a, c_n) such that

$$F^{(n+1)}(c_{n+1}) = 0. \tag{7}$$

If we differentiate

$$F(x) = f(x) - P_n(x) - K(x - a)^{n+1}$$

$n + 1$ times, we get

$$F^{(n+1)}(x) = f^{(n+1)}(x) - 0 - (n + 1)! \, K. \tag{8}$$

Equations (7) and (8) together give

$$K = \frac{f^{(n+1)}(c)}{(n + 1)!} \qquad \text{for some number } c = c_{n+1} \text{ in } (a, b). \tag{9}$$

Equations (5) and (9) give

$$f(b) = P_n(b) + \frac{f^{(n+1)}(c)}{(n+1)!}(b-a)^{n+1}.$$

This concludes the proof.

When we apply Taylor's theorem, we usually want to hold a fixed and treat b as an independent variable. Taylor's formula is easier to use in circumstances like these if we change b to x. Here is how the theorem reads with this change.

Taylor's Formula

If f has derivatives of all orders in an open interval I containing a, then for each positive integer n and for each x in I,

$$f(x) = f(a) + f'(a)(x-a) + \frac{f''(a)}{2!}(x-a)^2 + \cdots$$

$$+ \frac{f^{(n)}(a)}{n!}(x-a)^n + R_n(x), \tag{10}$$

where

$$R_n(x) = \frac{f^{(n+1)}(c)}{(n+1)!}(x-a)^{n+1} \qquad \text{for some } c \text{ between } a \text{ and } x. \tag{11}$$

When we state Taylor's theorem this way, it says that for each x in I,

$$f(x) = P_n(x) + R_n(x). \tag{12}$$

Pause for a moment to think about how remarkable this equation is. For any value of n we want, the equation gives both a polynomial approximation of f of that order and a formula for the error involved in using that approximation over the interval I.

Equation (10) is called **Taylor's formula.** The function $R_n(x)$ is called the **remainder of order** n or the **error term** for the approximation of f by $P_n(x)$ over I. When $R_n(x) \to 0$ as $n \to \infty$ for all x in I, we say that the Taylor series generated by f at $x = a$ **converges** to f on I and we write

$$f(x) = \sum_{k=0}^{\infty} \frac{f^{(k)}(a)}{k!}(x-a)^k. \tag{13}$$

Example 4   *The Maclaurin Series for e^x.*   Show that the Taylor series generated by $f(x) = e^x$ at $x = 0$ converges to $f(x)$ for every real value of x.

Solution   The function has derivatives of all orders throughout the interval $-\infty < x < \infty$. Equations (10) and (11) with $f(x) = e^x$ and $a = 0$ give

$$e^x = 1 + x + \frac{x^2}{2!} + \cdots + \frac{x^n}{n!} + R_n(x), \qquad \text{(Polynomial from Example 2)}$$

and

$$R_n(x) = \frac{e^c}{(n+1)!} x^{n+1} \qquad \text{for some } c \text{ between } 0 \text{ and } x.$$

Since e^x is an increasing function of x, e^c lies between $e^0 = 1$ and e^x. When x is negative, so is c, and $e^c < 1$. When x is zero, $e^x = 1$ and $R_n(x) = 0$. When x is positive, so is c, and $e^c < e^x$. Thus,

$$|R_n(x)| \leq \frac{|x|^{n+1}}{(n+1)!} \qquad \text{when } x \leq 0,$$

and

$$|R_n(x)| < e^x \frac{x^{n+1}}{(n+1)!} \qquad \text{when } x > 0.$$

Finally, because

$$\lim_{n \to \infty} \frac{x^{n+1}}{(n+1)!} = 0 \qquad \text{for every } x, \qquad \text{(Section 8.1)}$$

$\lim_{n \to \infty} R_n(x) = 0$, and the series converges to e^x for every x.

$$e^x = \sum_{k=0}^{\infty} \frac{x^k}{k!} = 1 + x + \frac{x^2}{2!} + \cdots + \frac{x^k}{k!} + \cdots. \tag{14}$$

Estimating the Remainder

It is often possible to estimate $R_n(x)$ as we did in Example 4. This method of estimation is so convenient that we state it as a theorem for future reference.

THEOREM 16

The Remainder Estimation Theorem

If there are positive constants M and r such that $|f^{(n+1)}(t)| \leq Mr^{n+1}$ for all t between a and x, inclusive, then the remainder term $R_n(x)$ in Taylor's theorem satisfies the inequality

$$|R_n(x)| \leq M \frac{r^{n+1}|x - a|^{n+1}}{(n+1)!}.$$

If these conditions hold for every n and all the other conditions of Taylor's theorem are satisfied by f, then the series converges to $f(x)$.

In the simplest examples, we can take $r = 1$ provided f and all its derivatives are bounded in magnitude by some constant M. But if $f(x) = 2 \cos(3x)$, each time we differentiate we get a factor of 3 and r needs to be greater than 1. In this particular case, we can take $r = 3$ along with $M = 2$.

We are now ready to look at some examples of how the Remainder Estimation Theorem and Taylor's theorem can be used together to settle questions of convergence. As you will see, they can also be used to determine the accuracy with which a function is approximated by one of its Taylor polynomials.

Example 5   *The Maclaurin Series for* sin x.   Show that the Maclaurin series for sin x converges to sin x for all x.

Solution   The function and its derivatives are

$$f(x) = \sin x, \qquad f'(x) = \cos x,$$
$$f''(x) = -\sin x, \qquad f'''(x) = -\cos x,$$
$$\vdots$$
$$f^{(2k)}(x) = (-1)^k \sin x, \qquad f^{(2k+1)}(x) = (-1)^k \cos x,$$

so

$$f^{(2k)}(0) = 0 \qquad \text{and} \qquad f^{(2k+1)}(0) = (-1)^k.$$

The series has only odd-power terms and, for $n = 2k + 1$, Taylor's theorem gives

$$\sin x = x - \frac{x^3}{3!} + \frac{x^5}{5!} - \cdots + \frac{(-1)^k x^{2k+1}}{(2k+1)!} + R_{2k+1}(x).$$

All the derivatives of sin x have absolute values less than or equal to 1, so we can apply the Remainder Estimation Theorem with $M = 1$ and $r = 1$ to obtain

$$|R_{2k+1}(x)| \le 1 \cdot \frac{|x|^{2k+2}}{(2k+2)!}.$$

Since $(|x|^{2k+2}/(2k+2)!) \to 0$ as $k \to \infty$, whatever the value of x, $R_{2k+1}(x) \to 0$, and the Maclaurin series for sin x converges to sin x for every x.

$$\sin x = \sum_{k=0}^{\infty} \frac{(-1)^k x^{2k+1}}{(2k+1)!} = x - \frac{x^3}{3!} + \frac{x^5}{5!} - \frac{x^7}{7!} + \cdots. \qquad (15)$$

Example 6   *The Maclaurin Series for* cos x.   Show that the Maclaurin series for cos x converges to cos x for every value of x.

Solution   We add the remainder term to the Taylor polynomial for cos x in Example 3 to obtain Taylor's formula for cos x with $n = 2k$:

$$\cos x = 1 - \frac{x^2}{2!} + \frac{x^4}{4!} - \cdots + (-1)^k \frac{x^{2k}}{(2k)!} + R_{2k}(x).$$

Because the derivatives of the cosine have absolute value less than or equal to 1, the Remainder Estimation Theorem with $M = 1$ and $r = 1$ gives

$$|R_{2k}(x)| \le 1 \cdot \frac{|x|^{2k+1}}{(2k+1)!}.$$

For every value of x, $R_{2k} \to 0$ as $k \to \infty$. Therefore, the series converges to cos x for every value of x.

Who Invented Taylor Series?

Brook Taylor (1685–1731) did not invent Taylor series, and Maclaurin series were not developed by Colin Maclaurin (1698–1746). James Gregory was working with Taylor series when Taylor was only a few years old, and he published the Maclaurin series for tan x, sec x, tan⁻¹x, and sec⁻¹x ten years before Maclaurin was born. At about the same time, Nicolaus Mercator discovered the Maclaurin series for ln(1 + x).

Taylor was not aware of Gregory's work when he published his book *Methodus incrementorum directa et inversa* in 1715, which contained what we now call Taylor series. Maclaurin quoted Taylor's work in a calculus book he wrote in 1742. The book popularized series representations of functions and although Maclaurin never claimed to have discovered them, Taylor series centered at $a = 0$ became known as Maclaurin series. History evened things up in the end. Maclaurin, a brilliant mathematician, was the original discoverer of the rule for solving systems of equations that we now call Cramer's rule.

$$\cos x = \sum_{k=0}^{\infty} \frac{(-1)^k x^{2k}}{(2k)!} = 1 - \frac{x^2}{2!} + \frac{x^4}{4!} - \frac{x^6}{6!} + \cdots. \tag{16}$$

Example 7   Find the Maclaurin series for $\cos 2x$ and show that it converges to $\cos 2x$ for every value of x.

Solution   We replace the x in Eq. (16) by $2x$, to obtain

$$\cos 2x = \sum_{k=0}^{\infty} \frac{(-1)^k (2x)^{2k}}{(2k)!} = 1 - \frac{(2x)^2}{2!} + \frac{(2x)^4}{4!} - \frac{(2x)^6}{6!} + \cdots.$$

The Maclaurin series for $\cos x$ converges to $\cos x$ for every value of x and therefore converges for every value of $2x$.

Truncation Error

The Maclaurin series for e^x converges to e^x for all x. But we still need to decide how many terms to use to approximate e^x to a given degree of accuracy.

Here are some examples of how to use the Remainder Estimation Theorem to estimate truncation error.

Example 8   Calculate e with an error of less than 10^{-6}.

Solution   We can use the result of Example 4 with $x = 1$ to write

$$e = 1 + 1 + \frac{1}{2!} + \cdots + \frac{1}{n!} + R_n(1),$$

with

$$R_n(1) = e^c \frac{1}{(n+1)!} \qquad \text{for some } c \text{ between 0 and 1.}$$

For the purposes of this example, we assume that we know that $e < 3$. Hence, we are certain that

$$\frac{1}{(n+1)!} < R_n(1) < \frac{3}{(n+1)!}$$

because $1 < e^c < 3$.

By experiment we find that $1/9! > 10^{-6}$, while $3/10! < 10^{-6}$. Thus we should take $(n + 1)$ to be at least 10, or n to be at least 9. With an error of less than 10^{-6},

$$e = 1 + 1 + \frac{1}{2} + \frac{1}{3!} + \cdots + \frac{1}{9!} = 2.7182\ 82. \qquad \text{(rounded)}$$

Example 9   For what values of x can we replace $\sin x$ by $x - (x^3/3!)$ with an error of magnitude no greater than 3×10^{-4}?

Solution   Here we can take advantage of the fact that the Maclaurin series for $\sin x$ is an alternating series for every nonzero value of x. According to the Alternating Series Estimation Theorem (Section 8.5), the error in truncating

$$\sin x = x - \frac{x^3}{3!} \; \Big| + \frac{x^5}{5!} - \cdots$$

after $(x^3/3!)$ is no greater than

$$\left| \frac{x^5}{5!} \right| = \frac{|x|^5}{120}.$$

Therefore the error will be less than or equal to 3×10^{-4} if

$$\frac{|x|^5}{120} < 3 \times 10^{-4} \qquad \text{or} \qquad |x| < \sqrt[5]{360 \times 10^{-4}} \approx 0.514. \qquad \begin{pmatrix} \text{Rounded} \\ \text{down, to be} \\ \text{safe} \end{pmatrix}$$

The Alternating Series Estimation Theorem tells us something that the Remainder Estimation Theorem does not: namely, that the estimate $x - (x^3/3!)$ for $\sin x$ is an underestimate when x is positive because then $x^5/120$ is positive.

Figure 8.22 shows the graph of $\sin x$, along with the graphs of a number of its approximating Taylor polynomials. Notice that the graph of $P_3(x) = x - (x^3/3!)$ is almost indistinguishable from the sine curve when $-1 \le x \le 1$. However, it crosses the x-axis at $\pm \sqrt{6} \approx \pm 2.45$, whereas the sine curve crosses the axis at $\pm \pi \approx \pm 3.14$.

You might wonder how the estimate given by the Remainder Estimation Theorem compares with the one just obtained from the Alternating Series Estimation Theorem. If we write

$$\sin x = x - \frac{x^3}{3!} + R_3,$$

then the Remainder Estimation Theorem gives

$$|R_3| \le 1 \cdot \frac{|x|^4}{4!} = \frac{|x|^4}{24},$$

which is not very good. But when we recognize that $x - (x^3/3!) = 0 + x + 0x^2 - (x^3/3!) + 0x^4$ is the Taylor polynomial of order 4 as well as of order

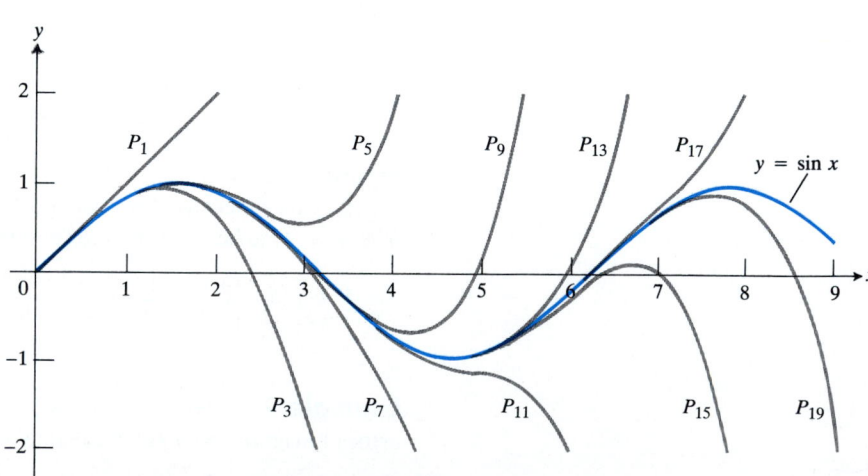

8.22   The polynomials

$$P_{2n+1}(x) = \sum_{k=0}^{n} [(-1)^k x^{2k+1}/(2k+1)!]$$

converge to $\sin x$ as $n \to \infty$.

3, then we have

$$\sin x = x - \frac{x^3}{3!} + 0 + R_4,$$

and the Remainder Estimation Theorem with $M = r = 1$ gives

$$|R_4| \le 1 \cdot \frac{|x|^5}{5!} = \frac{|x|^5}{120}.$$

This is what we had from the Alternating Series Estimation Theorem.

Combining Taylor Series

On common intervals of convergence, Taylor series can be added, subtracted, and multiplied by constants just as other series can, and the results are once again Taylor series. The Taylor series for $f(x) + g(x)$ is the sum of the Taylor series for $f(x)$ and $g(x)$ because the nth derivative of $f + g$ is $f^{(n)} + g^{(n)}$, and so on. Thus we obtain the Maclaurin series for $(1 + \cos 2x)/2$ by adding 1 to the Maclaurin series for $\cos 2x$ and dividing the combined results by 2, and the Maclaurin series for $\sin x + \cos x$ is the term-by-term sum of the Maclaurin series for $\sin x$ and $\cos x$.

$e^{i\theta} = \cos \theta + i \sin \theta$

As you may recall, a complex number is a number of the form $a + bi$, where a and b are real numbers and $i = \sqrt{-1}$. If we substitute $x = i\theta$ (θ real) in the Maclaurin series for e^x and use the relations

$$i^2 = -1, \qquad i^3 = i^2 i = -i, \qquad i^4 = i^2 i^2 = 1, \qquad i^5 = i^4 i = i,$$

and so on, to simplify the result, we obtain

$$e^{i\theta} = 1 + \frac{i\theta}{1!} + \frac{i^2\theta^2}{2!} + \frac{i^3\theta^3}{3!} + \frac{i^4\theta^4}{4!} + \frac{i^5\theta^5}{5!} + \frac{i^6\theta^6}{6!} + \cdots$$

$$= \left(1 - \frac{\theta^2}{2!} + \frac{\theta^4}{4!} - \frac{\theta^6}{6!} + \cdots\right) + i\left(\theta - \frac{\theta^3}{3!} + \frac{\theta^5}{5!} - \cdots\right) = \cos \theta + i \sin \theta.$$

(17)

This does not *prove* that $e^{i\theta} = \cos \theta + i \sin \theta$ because we have not yet defined what it means to raise e to an imaginary power. But it does say how we ought to define $e^{i\theta}$ to be consistent with other things we know.

DEFINITION

For any real number θ, $\qquad e^{i\theta} = \cos \theta + i \sin \theta.$ (18)

Equation (18), called **Euler's formula**, enables us to define e^{a+bi} to be $e^a \cdot e^{bi}$ for any complex number $a + bi$.

One of the amazing consequences of Euler's formula is the equation

$$e^{i\pi} = -1.$$

When written in the form $e^{i\pi} + 1 = 0$, this equation combines the five most important constants in mathematics.

EXERCISES 8.7

In Exercises 1–8, find the Taylor polynomials of orders 0, 1, 2, and 3 generated by f at a.

1. $f(x) = \ln x, \quad a = 1$
2. $f(x) = \ln(1 + x), \quad a = 0$
3. $f(x) = 1/x, \quad a = 2$
4. $f(x) = 1/(x + 2), \quad a = 0$
5. $f(x) = \sin x, \quad a = \pi/4$
6. $f(x) = \cos x, \quad a = \pi/4$
7. $f(x) = \sqrt{x}, \quad a = 4$
8. $f(x) = \sqrt{x + 4}, \quad a = 0$

Find the Maclaurin series for the functions in Exercises 9–18.

9. e^{-x}
10. $e^{x/2}$
11. $\sin 3x$
12. $5 \cos \pi x$
13. $\cos(-x)$
14. $x \sin x$
15. $\cosh x = (e^x + e^{-x})/2$
16. $\sinh x = (e^x - e^{-x})/2$
17. $(x^2/2) - 1 + \cos x$
18. $\cos^2 x$ (*Hint:* $\cos^2 x = (1 + \cos 2x)/2$.)

Quadratic Approximations

Write out Taylor's formula (Eq. 10) with $n = 2$ and $a = 0$ for the functions in Exercises 19–24. This will give you the quadratic approximations of these functions at $x = 0$ and the associated error terms.

19. $e^{\tan x}$
20. $e^{\sin x}$
21. $\ln(\cos x)$
22. $\dfrac{1}{\sqrt{1 - x^2}}$
23. $\sinh x$
24. $\cosh x$

25. Use the Taylor series generated by e^x at $x = a$ to show that
$$e^x = e^a \left[1 + (x - a) + \frac{(x - a)^2}{2!} + \cdots \right].$$

26. Find the Taylor series generated by e^x at $a = 1$. Compare your answer with the formula in Exercise 25.

27. For approximately what values of x can you replace $\sin x$ by $x - (x^3/6)$ with an error of magnitude no greater than 5×10^{-4}?

28. If $\cos x$ is replaced by $1 - (x^2/2)$ and $|x| < 0.5$, what estimate can be made of the error? Does $1 - (x^2/2)$ tend to be too large, or too small?

29. How close is the approximation $\sin x = x$ when $|x| < 10^{-3}$? For which of these values of x is $x < \sin x$?

30. The estimate $\sqrt{1 + x} = 1 + (x/2)$ is used when x is small. Estimate the error when $|x| < 0.01$.

31. The approximation $e^x = 1 + x + (x^2/2)$ is used when x is small. Use the Remainder Estimation Theorem to estimate the error when $|x| < 0.1$.

32. When $x < 0$, the series for e^x is an alternating series. Use the Alternating Series Estimation Theorem to estimate the error that results from replacing e^x by $1 + x + (x^2/2)$ when $-0.1 < x < 0$. Compare with Exercise 31.

33. Estimate the error in the approximation $\sinh x = x + (x^3/3!)$ when $|x| < 0.5$. (*Hint:* Use R_4, not R_3.)

34. When $0 \le h \le 0.01$, show that e^h may be replaced by $1 + h$ with an error of magnitude no greater than 0.6% of h. Use $e^{0.01} = 1.01$.

Each of the series in Exercises 35 and 36 is the value of the Maclaurin series of a function $f(x)$ at some point. What function and what point? What is the sum of the series?

35. $(0.1) - \dfrac{(0.1)^3}{3!} + \dfrac{(0.1)^5}{5!} - \cdots + \dfrac{(-1)^k(0.1)^{2k+1}}{(2k+1)!} + \cdots$

36. $1 - \dfrac{\pi^2}{4^2 \cdot 2!} + \dfrac{\pi^4}{4^4 \cdot 4!} - \cdots + \dfrac{(-1)^k(\pi)^{2k}}{4^{2k} \cdot (2k!)} - \cdots$

37. Differentiate the Maclaurin series for $\sin x$, $\cos x$, and e^x term by term and compare your results with the Maclaurin series for $\cos x$, $\sin x$, and e^x.

38. Integrate the Maclaurin series for $\sin x$, $\cos x$, and e^x term by term and compare your results with the Maclaurin series for $\cos x$, $\sin x$, and e^x.

39. Multiply the Maclaurin series for $2 \cos x$ and $\sin x$ together to find the first five nonzero terms of the product series. Confirm that this is the beginning of the Maclaurin series for $\sin 2x$.

40. Multiply the Maclaurin series for e^x and $\cos x$ together to find the first five nonzero terms of the Maclaurin series for $e^x \cos x$.

41. Use the Maclaurin series for $\cos x$ and the Alternating Series Estimation Theorem to show that
$$\frac{1}{2} - \frac{x^2}{24} < \frac{1 - \cos x}{x^2} < \frac{1}{2} \quad \text{for} \quad x \ne 0.$$

(This is the inequality in Section 1.9, Exercise 21.)

42. Use series to verify that $y = y_0 e^x$ is a solution of the equation $y' = y$ for any constant y_0.

43. Use the identity $\sin^2 x = (1 - \cos 2x)/2$ to obtain a series for $\sin^2 x$. Then differentiate this series to obtain a series for $2 \sin x \cos x$. Check that this is the series for $\sin 2x$.

44. *A convergent Taylor series that converges to its generating function only at its center.* It can be shown (though not simply) that the function f defined by the rule

$$f(x) = \begin{cases} 0 & \text{if } x = 0 \\ e^{-1/x^2} & \text{if } x \neq 0 \end{cases}$$

has derivatives of all orders at $x = 0$ (Fig. 8.23) and that $f^{(n)}(0) = 0$ for all n.
a) Find the Maclaurin series generated by f. At what values of x does the series converge? At what values of x does the series converge to $f(x)$?
b) Write out Taylor's formula (Eq. 10) for f, taking $a = 0$ and assuming $x \neq 0$. What does the formula tell you about the value of $R_n(x)$?

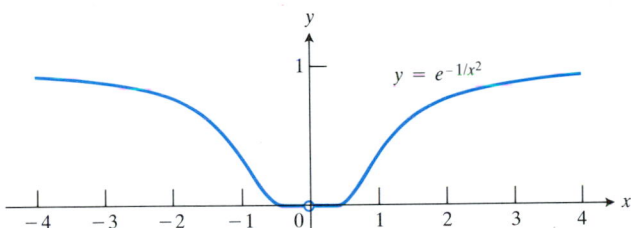

8.23 The graph of the continuous extension of $y = e^{-1/x^2}$ is so flat at the origin that all of its derivatives there are zero (Exercise 44).

45. Use Eq. (18) to write the following powers of e in the form $a + bi$.
a) $e^{-i\pi}$ b) $e^{i\pi/4}$ c) $e^{-i\pi/2}$

46. *Euler's identities.* Use Eq. (18) to show that

$$\cos \theta = \frac{e^{i\theta} + e^{-i\theta}}{2} \quad \text{and} \quad \sin \theta = \frac{e^{i\theta} - e^{-i\theta}}{2i}.$$

47. Establish the equations in Exercise 46 by combining the formal Maclaurin series for $e^{i\theta}$ and $e^{-i\theta}$.

48. Show that
a) $\cosh i\theta = \cos \theta$,
b) $\sinh i\theta = i \sin \theta$.

49. By multiplying the Maclaurin series for e^x and $\sin x$, find the terms through x^5 of the Maclaurin series for $e^x \sin x$. This series is the imaginary part of the series for

$$e^x \cdot e^{ix} = e^{(1 + i)x}.$$

Use this fact to check your answer. For what values of x should the series for $e^x \sin x$ converge?

50. When a and b are real, we define $e^{(a + ib)x}$ to be e^{ax} $(\cos bx + i \sin bx)$. From this definition, show that

$$\frac{d}{dx} e^{(a + ib)x} = (a + ib) e^{(a + ib)x}.$$

51. Use the definition of $e^{i\theta}$ to show that for any real numbers θ, θ_1, and θ_2,
a) $e^{i\theta_1} e^{i\theta_2} = e^{i(\theta_1 + \theta_2)}$,
b) $e^{-i\theta} = 1/e^{i\theta}$.

52. Two complex numbers $a + ib$ and $c + id$ are equal if and only if $a = c$ and $b = d$. Use this fact to evaluate

$$\int e^{ax} \cos bx \, dx \quad \text{and} \quad \int e^{ax} \sin bx \, dx$$

from

$$\int e^{(a + ib)x} \, dx = \frac{a - ib}{a^2 + b^2} e^{(a + ib)x} + C,$$

where $C = C_1 + iC_2$ is a complex constant of integration.

53. *The Maclaurin series generated by $f(x) = \sum_{n=0}^{\infty} a_n x^n$ is $\sum_{n=0}^{\infty} a_n x^n$.* A function defined by a power series $\sum_{n=0}^{\infty} a_n x^n$ with a radius of convergence $c > 0$ has a Maclaurin series that converges to the function at every point of $(-c, c)$. Show this by showing that the Maclaurin series generated by $f(x) = \sum_{n=0}^{\infty} a_n x^n$ is the series $\sum_{n=0}^{\infty} a_n x^n$ itself.
An immediate consequence of this is that series like

$$x \sin x = x^2 - \frac{x^4}{3!} + \frac{x^6}{5!} - \frac{x^8}{7!} + \cdots$$

and

$$x^2 e^x = x^2 + x^3 + \frac{x^4}{2!} + \frac{x^5}{3!} + \cdots,$$

obtained by multiplying Maclaurin series by powers of x, as well as series obtained by integration and differentiation of convergent power series, are themselves the Maclaurin series generated by the functions they represent.

54. *Maclaurin series for even functions and odd functions (continuation of Section 8.6, Exercise 45).* Suppose that $f(x) = \sum_{n=0}^{\infty} a_n x^n$ converges for all x in an open interval $(-c, c)$. Show that
a) If f is even, then $a_1 = a_3 = a_5 = \cdots = 0$, i.e., the series for f contains only even powers of x.
b) If f is odd, then $a_0 = a_2 = a_4 = \cdots = 0$, i.e., the series for f contains only odd powers of x.

55. *Taylor polynomials of periodic functions*
a) Show that every continuous periodic function $f(x)$, $-\infty < x < \infty$, is bounded in magnitude by showing that there exists a positive constant M such that $|f(x)| \leq M$ for all x.
b) Show that the graph of every Taylor polynomial generated by $f(x) = \cos x$ must eventually move away from the

graph of cos x as $|x|$ increases. You can see this happening in Fig. 8.21. The Taylor polynomials of sin x behave in a similar way (Fig. 8.22).

56. GRAPHER

a) Graph the curves $y = 1 - (x^2/6)$ and $y = (\sin x)/x$ together with the line $y = 1$.

b) Use the Maclaurin series for sin x and the Alternating Series Estimation Theorem to show that

$$1 - \frac{x^2}{6} < \frac{\sin x}{x} < 1 \quad \text{for} \quad x \neq 0.$$

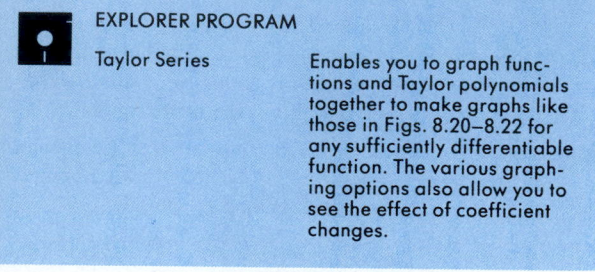

EXPLORER PROGRAM

Taylor Series

Enables you to graph functions and Taylor polynomials together to make graphs like those in Figs. 8.20–8.22 for any sufficiently differentiable function. The various graphing options also allow you to see the effect of coefficient changes.

8.8 Calculations with Taylor Series

This section introduces the binomial series, discusses how to choose useful centers for Taylor series, and shows how series are sometimes used to evaluate nonelementary integrals. The section concludes with a table of frequently used Maclaurin series.

The Binomial Series

The Maclaurin series generated by $f(x) = (1 + x)^m$, when m is constant, is

$$1 + mx + \frac{m(m - 1)}{2!} x^2 + \frac{m(m - 1)(m - 2)}{3!} x^3 + \cdots$$

$$+ \frac{m(m - 1)(m - 2) \cdots (m - k + 1)}{k!} x^k + \cdots. \qquad (1)$$

This series, called the **binomial series,** converges absolutely for $|x| < 1$. To derive the series, we first list the function and its derivatives:

$$f(x) = (1 + x)^m,$$

$$f'(x) = m(1 + x)^{m - 1},$$

$$f''(x) = m(m - 1)(1 + x)^{m - 2},$$

$$f'''(x) = m(m - 1)(m - 2)(1 + x)^{m - 3},$$

$$\vdots$$

$$f^{(k)}(x) = m(m - 1)(m - 2) \cdots (m - k + 1)(1 + x)^{m - k}.$$

We then evaluate these at $x = 0$ and substitute in the Maclaurin series formula to obtain

$$1 + mx + \frac{m(m - 1)}{2!} x^2 + \cdots$$

$$+ \frac{m(m - 1)(m - 2) \cdots (m - k + 1)}{k!} x^k + \cdots. \qquad (2)$$

If m is an integer greater than or equal to zero, the series stops after $(m + 1)$ terms because the coefficients from $k = m + 1$ on are zero.

If m is not a positive integer or zero, the series is infinite and converges for $|x| < 1$. To see why, let u_k be the term involving x^k. Then apply the ratio test for absolute convergence to see that

$$\left| \frac{u_{k+1}}{u_k} \right| = \left| \frac{m-k}{k+1} x \right| \to |x| \qquad \text{as } k \to \infty.$$

Our derivation of the binomial series shows only that it is generated by $(1 + x)^m$ and converges for $|x| < 1$. The derivation does not show that the series actually converges to $(1 + x)^m$. The series does, but we shall assume that part without proof.

For $-1 < x < 1$,

$$(1 + x)^m = 1 + \sum_{k=1}^{\infty} \binom{m}{k} x^k, \tag{3}$$

where

$$\binom{m}{1} = m, \qquad \binom{m}{2} = \frac{m(m-1)}{2!},$$

and

$$\binom{m}{k} = \frac{m(m-1)(m-2)\cdots(m-k+1)}{k!} \qquad \text{for } k \geq 3.$$

Example 1   Show that when $m = -1$, Eq. (3) gives the geometric series

$$\frac{1}{1+x} = 1 - x + x^2 - x^3 + \cdots + (-1)^k x^k + \cdots. \tag{4}$$

Solution   When $m = -1$,

$$\binom{-1}{1} = -1, \qquad \binom{-1}{2} = \frac{-1(-2)}{2!} = 1,$$

and

$$\binom{-1}{k} = \frac{-1(-2)(-3)\cdots(-1-k+1)}{k!} = (-1)^k \left(\frac{k!}{k!} \right) = (-1)^k.$$

With these coefficient values, Eq. (3) becomes

$$(1 + x)^{-1} = 1 + \sum_{k=1}^{\infty} (-1)^k x^k = 1 - x + x^2 - x^3 + \cdots + (-1)^k x^k + \cdots,$$

which is Eq. (4).

Choosing Centers for Taylor Series

Taylor's formula,

$$f(x) = f(a) + f'(a)(x-a) + \frac{f''(a)}{2!}(x-a)^2 + \cdots$$

$$+ \frac{f^{(n)}(a)}{n!}(x-a)^n + \frac{f^{(n+1)}(c)}{(n+1)!}(x-a)^{n+1}, \tag{5}$$

expresses the value of f at x in terms of f and its derivatives at a. In numerical computations, we therefore need a to be a point where we know the values of f and its derivatives. We also need a to be close enough to the values of x we are interested in to make $(x - a)^{n+1}$ so small we can neglect the remainder.

Example 2   What value of a might we choose in Taylor's formula (Eq. 5) to compute sin 35° efficiently?

Solution   The radian measure for 35° is $35\pi/180$. We could choose $a = 0$ and use the series

$$\sin x = x - \frac{x^3}{3!} + \frac{x^5}{5!} - \cdots + (-1)^n \frac{x^{2n+1}}{(2n+1)!} + 0 \cdot x^{2n+2} + R_{2n+2}(x). \quad (6)$$

Alternatively, we could choose $a = \pi/6$ (which corresponds to 30°) and use the series

$$\sin x = \sin \frac{\pi}{6} + \cos \frac{\pi}{6} \left(x - \frac{\pi}{6}\right) - \sin \frac{\pi}{6} \frac{(x - \pi/6)^2}{2!} - \cos \frac{\pi}{6} \frac{(x - \pi/6)^3}{3!}$$

$$+ \cdots + \sin \left(\frac{\pi}{6} + n \frac{\pi}{2}\right) \frac{(x - \pi/6)^n}{n!} + R_n(x).$$

The remainder in Eq. (6) satisfies the inequality

$$|R_{2n+2}(x)| \leq \frac{|x|^{2n+3}}{(2n+3)!},$$

which tends to zero as $n \to \infty$, no matter how large $|x|$ may be. We could therefore calculate sin 35° by placing

$$x = \frac{35\pi}{180} \approx 0.6108 \ 652$$

in the approximation

$$\sin x \approx x - \frac{x^3}{6} + \frac{x^5}{120} - \frac{x^7}{5040}.$$

This gives a truncation error of magnitude no greater than 3.3×10^{-8}, since

$$\left|R_8 \left(\frac{35\pi}{180}\right)\right| < \frac{(0.611)^9}{9!} < 3.3 \times 10^{-8}.$$

By using the series with $a = \pi/6$, we could obtain equal accuracy with a smaller exponent n, but at the expense of introducing cos $\pi/6 = \sqrt{3}/2$ as one of the coefficients. In this series, with $a = \pi/6$, we would take $x = 35\pi/180$, and the quantity $(x - a)$ would be

$$x - \frac{\pi}{6} = \frac{35\pi}{180} - \frac{30\pi}{180}$$

$$= \frac{5\pi}{180} \approx 0.0872 \ 665,$$

which decreases rapidly as it is raised to higher powers.

Evaluating Nonelementary Integrals

Maclaurin series are often used to express nonelementary integrals in terms of series.

Example 3   Express $\int \sin x^2 \, dx$ as a power series.

Solution   From the series for $\sin x$ we obtain

$$\sin x^2 = x^2 - \frac{x^6}{3!} + \frac{x^{10}}{5!} - \frac{x^{14}}{7!} + \frac{x^{18}}{9!} - \cdots .$$

Therefore,

$$\int \sin x^2 \, dx = C + \frac{x^3}{3} - \frac{x^7}{7 \cdot 3!} + \frac{x^{11}}{11 \cdot 5!} - \frac{x^{15}}{15 \cdot 7!} + \frac{x^{19}}{19 \cdot 9!} - \cdots .$$

Example 4   Estimate $\int_0^1 \sin x^2 \, dx$ with an error of less than 0.001.

Solution   From the indefinite integral in Example 3,

$$\int_0^1 \sin x^2 \, dx = \frac{1}{3} - \frac{1}{7 \cdot 3!} + \frac{1}{11 \cdot 5!} - \frac{1}{15 \cdot 7!} + \frac{1}{19 \cdot 9!} - \cdots .$$

The series alternates, and we find by experiment that

$$\frac{1}{11 \cdot 5!} \approx 0.0007\ 6$$

is the first term to be numerically less than 0.001. The sum of the preceding two terms gives

$$\int_0^1 \sin x^2 \, dx \approx \frac{1}{3} - \frac{1}{42} \approx 0.310.$$

With two more terms we could estimate

$$\int_0^1 \sin x^2 \, dx \approx 0.3102\ 68$$

with an error of less than 10^{-6}, and with only one term beyond that we have

$$\int_0^1 \sin x^2 \, dx \approx \frac{1}{3} - \frac{1}{42} + \frac{1}{1320} - \frac{1}{75600} + \frac{1}{6894720} \approx 0.3102\ 68303,$$

with an error of less than 10^{-9}. To guarantee this accuracy with the error formula for the trapezoidal rule would require using about 13,000 subintervals.

Arctangents

In Section 8.6, Example 4, we found a series for $\tan^{-1} x$ by differentiating to get

$$\frac{d}{dx} \tan^{-1} x = \frac{1}{1 + x^2} = 1 - x^2 + x^4 - x^6 + \cdots$$

and integrating to get

$$\tan^{-1}x = x - \frac{x^3}{3} + \frac{x^5}{5} - \frac{x^7}{7} + \cdots.$$

However, we did not prove the term-by-term integration theorem on which this conclusion depended. We now derive the series again by integrating both sides of the finite formula

$$\frac{1}{1 + t^2} = 1 - t^2 + t^4 - t^6 + \cdots + (-1)^n t^{2n} + \frac{(-1)^{n+1} t^{2n+2}}{1 + t^2}, \tag{7}$$

in which the last term comes from adding the remaining terms as a geometric series with first term $a = t^{2n+2}$ and ratio $r = -t^2$. Integrating both sides of Eq. (7) from $t = 0$ to $t = x$ gives

$$\tan^{-1}x = x - \frac{x^3}{3} + \frac{x^5}{5} - \frac{x^7}{7} + \cdots + (-1)^n \frac{x^{2n+1}}{2n+1} + R(n, x),$$

where

$$R(n, x) = \int_0^x \frac{(-1)^{n+1} t^{2n+2}}{1 + t^2} \, dt.$$

The denominator of the integrand is greater than or equal to 1; hence

$$|R(n, x)| \leq \int_0^{|x|} t^{2n+2} \, dt = \frac{|x|^{2n+3}}{2n+3}.$$

If $|x| \leq 1$, the right side of this inequality approaches zero as $n \to \infty$. Therefore $\lim_{n \to \infty} R(n, x) = 0$ if $|x| \leq 1$ and

$$\tan^{-1}x = \sum_{n=0}^{\infty} \frac{(-1)^n x^{2n+1}}{2n+1}, \qquad |x| \leq 1.$$

We take this route instead of finding the Maclaurin series directly because the formulas for the higher order derivatives of $\tan^{-1}x$ are unmanageable.

$$\boxed{\tan^{-1}x = x - \frac{x^3}{3} + \frac{x^5}{5} - \frac{x^7}{7} + \cdots, \qquad |x| \leq 1} \tag{8}$$

When we put $x = 1$ and $\tan^{-1}1 = \pi/4$ in Eq. (8), we get **Leibniz's formula:**

$$\frac{\pi}{4} = 1 - \frac{1}{3} + \frac{1}{5} - \frac{1}{7} + \frac{1}{9} - \cdots + \frac{(-1)^n}{2n+1} + \cdots.$$

This series converges too slowly to be a useful source of decimal approximations of π. It is better to use a formula like

$$\pi = 48 \tan^{-1}\frac{1}{18} + 32 \tan^{-1}\frac{1}{57} - 20 \tan^{-1}\frac{1}{239}, \tag{9}$$

which uses values of x closer to zero.

Frequently Used Maclaurin Series

$$\frac{1}{1-x} = 1 + x + x^2 + \cdots + x^n + \cdots = \sum_{n=0}^{\infty} x^n, \qquad |x| < 1$$

$$\frac{1}{1+x} = 1 - x + x^2 - \cdots + (-x)^n + \cdots = \sum_{n=0}^{\infty} (-1)^n x^n, \qquad |x| < 1$$

$$e^x = 1 + x + \frac{x^2}{2!} + \cdots + \frac{x^n}{n!} + \cdots = \sum_{n=0}^{\infty} \frac{x^n}{n!}, \qquad |x| < \infty$$

$$\sin x = x - \frac{x^3}{3!} + \frac{x^5}{5!} - \cdots + (-1)^n \frac{x^{2n+1}}{(2n+1)!} + \cdots = \sum_{n=0}^{\infty} \frac{(-1)^n x^{2n+1}}{(2n+1)!}, \qquad |x| < \infty$$

$$\cos x = 1 - \frac{x^2}{2!} + \frac{x^4}{4!} - \cdots + (-1)^n \frac{x^{2n}}{(2n)!} + \cdots = \sum_{n=0}^{\infty} \frac{(-1)^n x^{2n}}{(2n)!}, \qquad |x| < \infty$$

$$\ln(1+x) = x - \frac{x^2}{2} + \frac{x^3}{3} - \cdots + (-1)^{n-1} \frac{x^n}{n} + \cdots = \sum_{n=1}^{\infty} \frac{(-1)^{n-1} x^n}{n}, \qquad -1 < x \le 1$$

$$\ln \frac{1+x}{1-x} = 2 \tanh^{-1} x = 2 \left(x + \frac{x^3}{3} + \frac{x^5}{5} + \cdots + \frac{x^{2n+1}}{2n+1} + \cdots \right) = 2 \sum_{n=0}^{\infty} \frac{x^{2n+1}}{2n+1}, \qquad |x| < 1$$

$$\tan^{-1} x = x - \frac{x^3}{3} + \frac{x^5}{5} - \cdots + (-1)^n \frac{x^{2n+1}}{2n+1} + \cdots = \sum_{n=0}^{\infty} \frac{(-1)^n x^{2n+1}}{2n+1}, \qquad |x| \le 1$$

Binomial Series

$$(1+x)^m = 1 + mx + \frac{m(m-1)x^2}{2!} + \frac{m(m-1)(m-2)x^3}{3!} + \cdots + \frac{m(m-1)(m-2)\cdots(m-k+1)x^k}{k!} + \cdots$$

$$= 1 + \sum_{k=1}^{\infty} \binom{m}{k} x^k, \qquad |x| < 1,$$

where

$$\binom{m}{1} = m, \qquad \binom{m}{2} = \frac{m(m-1)}{2!}, \qquad \binom{m}{k} = \frac{m(m-1)\ldots(m-k+1)}{k!} \qquad \text{for } k \ge 3.$$

Note: It is customary to define $\binom{m}{0}$ to be 1 and to take $x^0 = 1$ (even in the usually excluded case where $x = 0$) to write the binomial series compactly as

$$(1+x)^m = \sum_{k=0}^{\infty} \binom{m}{k} x^k, \qquad |x| < 1.$$

If m is a *positive integer,* the series terminates at x^m, and the result converges for all x.

EXERCISES 8.8

What Taylor series would you use to represent the functions in Exercises 1–6 near the given values of x? (There may be more than one good answer.) Write out the first four nonzero terms of the series you choose.

1. $\cos x$ near $x = 1$ **2.** $\sin x$ near $x = 6.3$

3. e^x near $x = 0.4$ **4.** $\ln x$ near $x = 1.3$

5. $\cos x$ near $x = 69$ **6.** $\tan^{-1} x$ near $x = 2$

CALCULATOR In Exercises 7–14, use series and a calculator to estimate the value of each integral with an error of magnitude less than 10^{-3}.

7. $\displaystyle\int_0^{0.2} \sin x^2 \, dx$ **8.** $\displaystyle\int_0^{0.1} \tan^{-1} x \, dx$

9. $\displaystyle\int_0^{0.1} x^2 e^{-x^2} \, dx$ **10.** $\displaystyle\int_0^{0.1} \frac{\tan^{-1} x}{x} \, dx$

11. $\displaystyle\int_0^{0.4} \frac{1 - e^{-x}}{x} \, dx$ **12.** $\displaystyle\int_0^{0.1} \frac{\ln(1 + x)}{x} \, dx$

13. $\displaystyle\int_0^{0.1} \frac{1}{\sqrt{1 + x^4}} \, dx$ **14.** $\displaystyle\int_0^{0.25} \sqrt[3]{1 + x^2} \, dx$

CALCULATOR Use series to evaluate the integrals in Exercises 15–18 as accurately as your calculator will allow.

15. $\displaystyle\int_0^{0.1} \frac{\sin x}{x} \, dx$ **16.** $\displaystyle\int_0^{0.1} e^{-x^2} \, dx$

17. $\displaystyle\int_0^{0.1} \sqrt{1 + x^4} \, dx$ **18.** $\displaystyle\int_0^1 \frac{1 - \cos x}{x^2} \, dx$

19. Replace x by $-x$ in the Maclaurin series for $\ln(1 + x)$ to obtain a series for $\ln(1 - x)$. Then subtract this from the Maclaurin series for $\ln(1 + x)$ to show that for $|x| < 1$,

$$\ln \frac{1 + x}{1 - x} = 2\left(x + \frac{x^3}{3} + \frac{x^5}{5} + \cdots\right).$$

20. How many terms of the Maclaurin series for $\ln(1 + x)$ should you add to be sure of calculating $\ln(1.1)$ with an error of magnitude less than 10^{-8}?

21. According to the Alternating Series Estimation Theorem, how many terms of the Maclaurin series for $\tan^{-1} 1$ would you have to add to be sure of finding $\pi/4$ with an error of magnitude less than 10^{-3}?

22. Show that the Maclaurin series for $\tan^{-1} x$ diverges for $|x| > 1$.

23. CALCULATOR About how many terms of the Maclaurin series for $\tan^{-1} x$ would you have to use to evaluate each term on the right-hand side of the equation

$$\pi = 48 \tan^{-1} \frac{1}{18} + 32 \tan^{-1} \frac{1}{57} - 20 \tan^{-1} \frac{1}{239}$$

with an error of magnitude less than 10^{-6}? In contrast, the convergence of $\sum_{n=1}^{\infty} (1/n^2)$ to $\pi^2/6$ is so slow that even 50 terms will not yield two-place accuracy.

24. Integrate the binomial series for $(1 - x^2)^{-1/2}$ to show that for $|x| < 1$

$$\sin^{-1} x = x + \sum_{n=1}^{\infty} \frac{1 \cdot 3 \cdot 5 \cdot \cdots \cdot (2n - 1)}{2 \cdot 4 \cdot 6 \cdot \cdots \cdot (2n)} \frac{x^{2n+1}}{2n + 1}.$$

25. a) Use the binomial series and the fact that

$$\frac{d}{dx} \sin^{-1} x = (1 - x^2)^{-1/2}$$

to generate the first four nonzero terms of the Maclaurin series for $\sin^{-1} x$. What is the radius of convergence?

b) Use your result in (a) to find the first five nonzero terms of the Maclaurin series for $\cos^{-1} x$.

26. a) Find the first four nonzero terms of the Maclaurin series for

$$\sinh^{-1} x = \int_0^x \frac{dt}{\sqrt{1 + t^2}}.$$

b) CALCULATOR Use the first three terms of the series in (a) to estimate $\sinh^{-1} 0.25$. Give an upper bound for the magnitude of the estimation error.

27. Obtain the Maclaurin series for $1/(1 + x)^2$ from the series for $-1/(1 + x)$.

28. Use the Maclaurin series for $1/(1 - x^2)$ to obtain a series for $2x/(1 - x^2)^2$.

29. Integrate the first three nonzero terms of the Maclaurin series for $\tan t$ from 0 to x to obtain the first three nonzero terms of the Maclaurin series for $\ln \sec x$.

30. *The series for* $\tanh^{-1}(x) = (1/2)\ln((1 + x)/(1 - x))$.
a) Show that

$$\int_0^x \frac{dt}{1 - t^2} = \int_0^x \left(1 + t^2 + t^4 + \cdots + t^{2n} + \frac{t^{2n+2}}{1 - t^2}\right) dt$$

or, put another way, that

$$\tanh^{-1} x = x + \frac{x^3}{3} + \frac{x^5}{5} + \cdots + \frac{x^{2n+1}}{2n + 1} + R(n, x),$$

where

$$R(n, x) = \int_0^x \frac{t^{2n+2}}{1 - t^2} \, dt.$$

b) Show that

$$|R(n, x)| \le \frac{1}{1 - x^2} \frac{|x|^{2n+3}}{2n + 3} \quad \text{if} \quad |x| < 1.$$

This shows that $\lim_{n \to \infty} R(n, x) = 0$ if $|x| < 1$ and hence that

$$\tanh^{-1} x = \sum_{n=0}^{\infty} \frac{x^{2n+1}}{2n + 1}, \quad |x| < 1.$$

c) Show that the series for $\tanh^{-1} x$ diverges if $|x| \ge 1$.

✳ 8.9 A Computer Mystery

You may never have to actually use a series to compute the value of π, $\ln 2$, e, $\sin 35°$, or the like, because if you need any of these numbers you can either look up appropriate approximations in a table or (more likely) find them with a calculator. But it was an exciting new idea in mathematics (and still can be an exciting idea) to realize that we can get *all* the information about e^x, for example, from just knowing that $e^0 = 1$ and that the derivative of e^x is e^x. That is enough to generate the Maclaurin series for e^x, which converges to e^x for all x. By multiplying the series for e^x by the series for e^y, we can also show that $e^x \cdot e^y = e^{x+y}$; by differentiation, we can show that the derivative of the series for e^x is that same series, and so on.

Those of you who are interested in computer science may find the following "detective story" challenging. We wanted to compare, for purposes of illustration, three different ways of calculating e.

1. e is the root of the equation $\ln x = 1$. (Newton's method produced 2.7 1828 1828 on the third iteration, starting with $x_0 = 3$.)

2. e is the sum of the series

$$1 + \frac{1}{1} + \frac{1}{2!} + \frac{1}{3!} + \frac{1}{4!} + \frac{1}{5!} + \cdots .$$

(This produced the same degree of precision when we included 14 or more terms.)

3. $e = \lim_{n \to \infty} \left(1 + \frac{1}{n}\right)^n$

This is often taken as the definition of e. With a computer, one can just pick any large value of n and tell the computer to print the value of $a_n = [1 + (1/n)]^n$. Of course, it is also instructive to see what happens for various values of n along the way. When we did that, we got a surprise. Here is a portion of the table of values we got, starting at $n = 10^5$.

n	$\left(1 + \frac{1}{n}\right)^n$	
10^5	2.7 1826 8237	
2×10^5	2.7 1827 5305	
3×10^5	2.7 1827 6664	
4×10^5	2.7 1827 7479	
5×10^5	2.7 1827 7751	
6×10^5	2.7 1827 9110	
7×10^5	2.7 1827 9110	(What's this? A repetition?)
8×10^5	2.7 1828 1828	(Hooray!)
9×10^5	2.7 1827 9110	(The machine likes this number?)
10^6	2.7 1828 1828	(Well, that's better.)
10^{13}	1	(What's going on?)
$\vdots$		
10^{19}	1	

We offer this as a puzzle. What is the machine doing? (Don't read on until you give it your best try!) Want a clue? Recall the definition of a^b.

A Computer Mystery Solved

We have just presented some data for $[1 + (1/n)]^n$ for selected values of n between 10^5 and 10^{19}. How can we account for the obvious errors in some of these answers?

Let's start with the definition of a^b as $\exp(b \ln a)$:

$$\left(1 + \frac{1}{n}\right)^n = \exp\left[n \ln\left(1 + \frac{1}{n}\right)\right].$$ (1)

For large values of n, $1/n$ is small, and we look at the series for $\ln(1 + x)$ with $x = 1/n$. The result is

$$\ln\left(1 + \frac{1}{n}\right) = \frac{1}{n} - \frac{1}{2n^2} + \frac{1}{3n^3} - \cdots.$$ (2)

Clearly, this series approaches zero as $n \to \infty$. But Eq. (1) requires that we multiply this series by n. Doing so, we get

$$n \ln\left(1 + \frac{1}{n}\right) = 1 - \frac{1}{2n} + \frac{1}{3n^2} - \cdots.$$ (3)

For large values of n, the terms from $1/3n^2$ and beyond are very small compared to the first two terms. So, for a first approximation, we have

$$\left(1 + \frac{1}{n}\right)^n \approx \exp\left(1 - \frac{1}{2n}\right).$$ (4)

How about the irregular data that we got from the computer? We guessed that it might come from inaccurate values for $\ln[1 + (1/n)]^n$, so we programmed the computer to give those data. At the same time, we had the computer give the values of $1 - (1/2n)$ and $1 - (1/2n) + (1/3n^2)$. These latter values were the same for all values of n shown in the table. And they are certainly more accurate than the machine's values of $n \ln[1 + (1/n)]$.

n	$n \ln[1 + (1/n)]$	$1 - (1/2n) + (1/3n^2)$
10^5	0.99999 5	0.99999 5
2×10^5	0.99999 76	0.99999 75
3×10^5	0.99999 81	0.99999 8333
4×10^5	0.99999 84	0.99999 875
5×10^5	0.99999 85	0.99999 90
6×10^5	0.99999 90	0.99999 91667
7×10^5	0.99999 90	0.99999 92857
8×10^5	1	0.99999 94444
9×10^5	0.99999 9	0.99999 95
10^6	1	0.99999 95455
1.1×10^6	1.00000 01	0.99999 95833
1.2×10^6	0.99999 6	0.99999 96154
1.3×10^6	0.99999 0	0.99999 96429
1.4×10^6	1.00000 04	0.99999 96667
1.5×10^6	1.00000 05	0.99999 96875
2×10^6	1	0.99999 97500

Naturally, if the computer has wrong values for $n \ln[1 + (1/n)]$, it will give wrong answers for $\exp(n \ln[1 + (1/n)])$. So, we have tracked down the source of the trouble; the values of $\ln[1 + (1/n)]$ are not accurate enough to give right answers when multiplied by n, if $n \geq 3 \times 10^5$. For $n = 10^{13}$, the computer says $\ln(1.0000\ 00000\ 0001) = 0$. When it multiplies this by 10^{13}, it still says "zero" and $\exp(0) = 1$. We have gone beyond the machine's realm of reliability. It isn't the computer's fault—or ours. But we have gained some insight by solving the mystery.

One more remark: If we use the approximation (4) and the fact that

$$\exp\left(1 - \frac{1}{2n}\right) = \exp(1) \cdot \exp\left(-\frac{1}{2n}\right), \tag{5}$$

we can go a step further. Remember that we are talking about $n \geq 3 \times 10^5$, so $1/2n \leq (1/6) \times 10^{-5}$. If we put $h = -1/2n$ in the Maclaurin series for $\exp(h) = e^h$, we get

$$\exp\left(-\frac{1}{2n}\right) = 1 - \frac{1}{2n} \tag{6}$$

with an error less than $(1/2)(1/2n)^2$ by the Alternating Series Estimation Theorem. This error is less than 0.3×10^{-11} for $n \geq 3 \times 10^5$. Therefore, for large values of n, the combined results of Eqs. (4), (5), and (6) yield

$$\left(1 + \frac{1}{n}\right)^n \approx e \cdot \left(1 - \frac{1}{2n}\right) \approx e - \frac{e}{2n}.$$

For $n = 10^6$, the machine should *not* give $2.7\ 1828\ 1828$ (which seems so very accurate), but $2.7\ 1828\ 0469$ to nine decimals.

REVIEW QUESTIONS

1. Define infinite sequence (sequence), infinite series (series), and sequence of partial sums of a series.

2. Define convergence for (a) sequences, (b) series.

3. Describe Picard's method for finding roots. When can it be expected to work?

4. Which of the following statements are true, and which are false?
a) If a sequence does not converge, then it diverges.
b) If a sequence $\{a_n\}$ does not converge, then a_n tends to infinity as n does.
c) If a series does not converge, then its nth term does not approach zero as n tends to infinity.
d) If the nth term of a series does not approach zero as n tends to infinity, then the series diverges.
e) If a sequence $\{a_n\}$ converges, then there is a number L such that a_n lies within 1 unit of L (i) for all values of n, (ii) for all but a finite number of values of n.
f) If all partial sums of a series are less than some constant L, then the series converges.
g) If a series converges, then its partial sums s_n are bounded (that is, $m \leq s_n \leq M$ for some constants m and M).

5. What tests do you know for the convergence and divergence of infinite series?

6. Under what circumstances do you know for sure that a bounded sequence converges?

7. Define absolute convergence and conditional convergence for infinite series. Give examples of series that (a) converge absolutely, (b) converge conditionally, (c) diverge.

8. Under what circumstances can you guarantee that rearranging infinitely many terms of an infinite series will not alter the series' sum? When might it alter the series' sum? Give examples.

9. What are geometric series? Under what circumstances do they converge? diverge?

10. What test would you apply to decide whether an alternating series converges? Give examples of convergent and of divergent alternating series. How can you estimate the error involved in using a partial sum to estimate the sum of a convergent alternating series?

11. What is a power series? How do you test a power series for convergence? What kinds of results can you get? Give examples.

12. What are the basic facts about term-by-term differentiation and integration of power series? Give examples.

13. What is the Taylor series generated by a function $f(x)$ at a point $x = a$? What do you need to know about f to construct the series? Give an example.

14. What is the Maclaurin series generated by a function $f(x)$? Give an example.

15. What are Taylor polynomials and what good are they? Give examples.

16. What is Taylor's formula and what does Taylor's theorem say about it? When does the Taylor series generated by f at $x = a$ converge to f?

17. What does Taylor's formula tell us about the error in a linearization?

18. What are the Maclaurin series for e^x, $\sin x$, $\cos x$, $1/(1 + x)$, $1/(1 - x)$, $1/(1 + x^2)$, $\tan^{-1}x$, and $\ln(1 + x)$? How do you estimate the errors involved in replacing these series by their partial sums? Give examples.

19. What do you take into account when choosing a center for a Taylor series? Illustrate with examples.

20. How can you sometimes use series to evaluate nonelementary integrals? Give an example.

MISCELLANEOUS EXERCISES

Determine which of the sequences $\{a_n\}$ in Exercises 1–14 converge and which diverge. Find the limit of each convergent sequence.

1. $a_n = 1 + \dfrac{(-1)^n}{n}$

2. $a_n = \dfrac{1 - 2^n}{2^n}$

3. $a_n = \cos \dfrac{n\pi}{2}$

4. $a_n = \left(\dfrac{3}{n}\right)^{1/n}$

5. $a_n = \dfrac{n}{\ln(n^2)}$

6. $a_n = \left(\dfrac{n + 5}{n}\right)^n$

7. $a_n = \sqrt[2n]{\dfrac{3^n}{n}}$

8. $a_n = \dfrac{1}{3^{2n - 1}}$

9. $a_n = \dfrac{(-4)^n}{n!}$

10. $a_n = \dfrac{\ln(2n^3 + 1)}{n}$

11. $a_n = n(2^{1/n} - 1)$

12. $a_n = \sqrt[n]{2n + 1}$

13. $a_n = \left(1 + \dfrac{1}{n}\right)^{-n}$

14. $a_n = \dfrac{\left(\dfrac{10}{11}\right)^n + \dfrac{1}{n}}{\left(\dfrac{9}{10}\right)^n + \left(\dfrac{11}{12}\right)^n}$

In Exercises 15–18, find the sums of the series that converge.

15. $\displaystyle\sum_{n=1}^{\infty} \ln\left(\dfrac{n}{n + 1}\right)$

16. $\displaystyle\sum_{n=2}^{\infty} \dfrac{-2}{n(n + 1)}$

17. $\displaystyle\sum_{n=0}^{\infty} e^{-n}$

18. $\displaystyle\sum_{n=1}^{\infty} (-1)^n \dfrac{3}{4^n}$

Which of the series in Exercises 19–40 converge absolutely, which converge conditionally, and which diverge? Give reasons for your answers.

19. $\displaystyle\sum_{n=1}^{\infty} \dfrac{1}{\sqrt{n}}$

20. $\displaystyle\sum_{n=1}^{\infty} \dfrac{-5}{n}$

21. $\displaystyle\sum_{n=1}^{\infty} \dfrac{(-1)^n}{\sqrt{n}}$

22. $\displaystyle\sum_{n=1}^{\infty} \dfrac{1}{2n^3}$

23. $\displaystyle\sum_{n=1}^{\infty} \dfrac{(-1)^n}{\ln(n + 1)}$

24. $\displaystyle\sum_{n=2}^{\infty} \dfrac{1}{n(\ln n)^2}$

25. $\displaystyle\sum_{n=1}^{\infty} \dfrac{(-1)^n}{n\sqrt{n^2 + 1}}$

26. $\displaystyle\sum_{n=1}^{\infty} \dfrac{(-1)^n 3n^2}{n^3 + 1}$

27. $\displaystyle\sum_{n=1}^{\infty} \dfrac{n + 1}{n!}$

28. $\displaystyle\sum_{n=1}^{\infty} \dfrac{(-1)^n (n^2 + 1)}{2n^2 + n - 1}$

29. $\displaystyle\sum_{n=1}^{\infty} \dfrac{(-3)^n}{n!}$

30. $\displaystyle\sum_{n=1}^{\infty} \dfrac{2^n 3^n}{n^n}$

31. $\displaystyle\sum_{n=1}^{\infty} \dfrac{1}{\sqrt{n(n + 1)(n + 2)}}$

32. $\displaystyle\sum_{n=2}^{\infty} \dfrac{1}{n\sqrt{n^2 - 1}}$

33. $\displaystyle\sum_{n=1}^{\infty} \dfrac{1}{(3n - 2)^{n + (1/2)}}$

34. $\displaystyle\sum_{n=1}^{\infty} \dfrac{(\tan^{-1}n)^2}{n^2 + 1}$

35. $\displaystyle\sum_{n=1}^{\infty} (-1)^n \tanh n$

36. $\displaystyle\sum_{n=1}^{\infty} \tan^{-1}\left(\dfrac{\cos n\,\pi}{n}\right)$

37. $\displaystyle\sum_{n=1}^{\infty} (-1)^n \ln(1 + e^{-n})$

38. $\displaystyle\sum_{n=2}^{\infty} \dfrac{\log_n(n!)}{n^3}$ (*Hint:* First show that $\log_n(n!) < n$.)

39. $\displaystyle\sum_{n=1}^{\infty} \dfrac{\ln(n!)}{n^4}$ (*Hint:* First show that $n! \le n^n$.)

40. $\displaystyle\sum_{n=1}^{\infty} \dfrac{\Gamma(n)}{n^n}$ (See Chapter 7, Miscellaneous Exercise 161.)

Each of Exercises 41–50 gives a formula for the nth term of a series. (a) Find the series' radius and interval of convergence. For what values of x does the series converge (b) absolutely; (c) conditionally?

41. $\dfrac{(x + 2)^n}{n\,3^n}$

42. $\dfrac{(x - 1)^{2n - 2}}{(2n - 1)!}$

43. $\dfrac{x^n}{n^n}$

44. $\dfrac{n + 1}{2n + 1} \dfrac{(x - 1)^n}{2^n}$

45. $\dfrac{(-1)^{n - 1}(x - 1)^n}{n^2}$

46. $\dfrac{x^n}{\sqrt{n}}$

47. $(\operatorname{csch} n) x^n$

48. $(\coth n) x^n$

49. $\dfrac{(n + 1) x^{2n - 1}}{3^n}$

50. $\dfrac{(\cos x)^n}{n}$

51. Find the radius of convergence of the series

$$\sum_{n=1}^{\infty} \frac{2 \cdot 5 \cdot 8 \cdot \cdots \cdot (3n-1)}{2 \cdot 4 \cdot 6 \cdot \cdots \cdot (2n)} x^n.$$

52. a) Show that the series

$$\sum_{n=1}^{\infty} \left(\sin \frac{1}{2n} - \sin \frac{1}{2n+1} \right)$$

converges.

b) CALCULATOR Estimate the error involved in using the sum of the first twenty terms of the series ($n = 20$) to estimate the sum of the series. Is the estimate too large, or too small?

Each of the series in Exercises 53–58 is the value of the Maclaurin series of a function $f(x)$ at a particular point. What function and what point? What is the sum of the series?

53. $1 - \dfrac{1}{4} + \dfrac{1}{16} - \cdots + (-1)^n \dfrac{1}{4^n} + \cdots$

54. $\dfrac{2}{3} - \dfrac{4}{18} + \dfrac{8}{81} - \cdots + (-1)^{n-1} \dfrac{2^n}{n3^n} + \cdots$

55. $\pi - \dfrac{\pi^3}{3!} + \dfrac{\pi^5}{5!} - \cdots + (-1)^n \dfrac{\pi^{2n+1}}{(2n+1)!} + \cdots$

56. $1 - \dfrac{\pi^2}{9 \cdot 2!} + \dfrac{\pi^4}{81 \cdot 4!} - \cdots + (-1)^n \dfrac{\pi^{2n}}{3^{2n}(2n)!} + \cdots$

57. $1 + \ln 2 + \dfrac{(\ln 2)^2}{2!} + \cdots + \dfrac{(\ln 2)^n}{n!} + \cdots$

58. $\dfrac{1}{\sqrt{3}} - \dfrac{1}{9\sqrt{3}} + \dfrac{1}{45\sqrt{3}} - \cdots$

$$+ (-1)^{n-1} \frac{1}{(2n-1)(\sqrt{3})^{2n-1}} + \cdots$$

In Exercises 59 and 60, find the first four nonzero terms of the Taylor series for the function $f(x)$ at the center $x = a$.

59. $f(x) = \sqrt{3 + x^2}$ at $x = -1$

60. $f(x) = 1/(1 - x)$ at $x = 2$

61. CALCULATOR Start with $x_0 = 1$ and press $\boxed{\sin}$ and $\boxed{\sqrt{x}}$ alternately to start generating a sequence of terms with the recursion formula $x_{n+1} = \sqrt{\sin(x_n)}$. Does the sequence appear to be convergent? If so, what equation does the limit satisfy?

62. Figure 8.24 provides an informal proof that $\sum_{n=1}^{\infty} (1/n^2)$ is less than 2. Explain what is going on. (*Source:* "Convergence with Pictures" by P. J. Rippon, *American Mathematical Monthly*, Vol. 93, No. 6, 1986, pp. 476–78.)

63. Find a closed-form formula for the nth partial sum of the series $\sum_{n=2}^{\infty} \ln[1 - (1/n^2)]$ and use it to determine whether the series converges.

64. Evaluate $\sum_{k=2}^{\infty} 1/(k^2 - 1)$ by finding the limit as $n \to \infty$ of the nth partial sum.

65. Prove that the sequence $\{x_n\}$ and the series $\sum_{k=1}^{\infty} (x_{k+1} - x_k)$ both converge or both diverge.

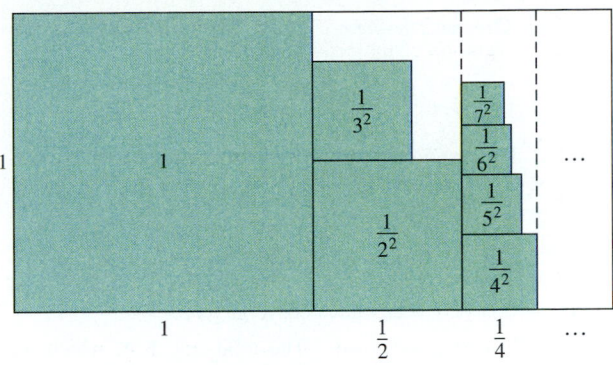

8.24 The figure for Exercise 62.

66. Assuming $|x| > 1$, show that

$$\frac{1}{1-x} = -\frac{1}{x} - \frac{1}{x^2} - \frac{1}{x^3} - \cdots.$$

67. *Generalizing Euler's constant.* Figure 8.25 shows the graph of a positive twice-differentiable decreasing function f whose second derivative is positive on $(0, \infty)$. For each n, the number A_n is the area of the lunar region between the curve and the line segment joining the points $(n, f(n))$ and $(n+1, f(n+1))$.

a) Use the figure to show that $\sum_{n=1}^{\infty} A_n < (1/2)$ $(f(1) - f(2))$.

b) Then show the existence of

$$\lim_{n \to \infty} \left[\sum_{k=1}^{n} f(k) - \frac{1}{2}(f(1) + f(n)) - \int_1^n f(x)\, dx \right].$$

c) Then show the existence of

$$\lim_{n \to \infty} \left[\sum_{k=1}^{n} f(k) - \int_1^n f(x)\, dx \right].$$

If $f(x) = 1/x$, the limit in (c) is Euler's constant. (*Source:* "Convergence with Pictures" by P. J. Rippon, *American Mathematical Monthly*, Vol. 93, No. 6, 1986, pp. 476–78.)

8.25 The figure for Exercise 67.

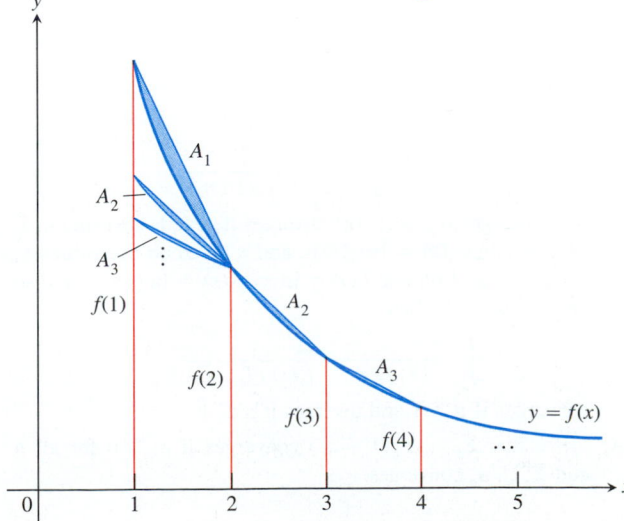

68. *Nicole Oresme's theorem.* Prove Nicole Oresme's theorem that

$$1 + \frac{1}{2} \cdot 2 + \frac{1}{4} \cdot 3 + \cdots + \frac{n}{2^{n-1}} + \cdots = 4.$$

$\left(\textit{Hint:}\ \text{Differentiate both sides of the equation}\ 1/(1-x) = 1 + \sum_{n=1}^{\infty} x^n.\right)$

69. Prove that the sequence of approximations $x_0, x_1, x_2, \ldots$ ($x_0 \neq 1$) generated by Newton's method to find the zero of $f(x) = (x-1)^{40}$ really does converge to 1.

70. *Raabe's (or Gauss's) test.* The following test, which we state without proof, is an extension of the Ratio Test.

Raabe's test: If $\sum_{n=1}^{\infty} u_n$ is a series of positive constants and there exist constants C, K, and N such that

$$\frac{u_{n+1}}{u_n} = 1 - \frac{C}{n} + \frac{f(n)}{n^2}, \tag{1}$$

where $|f(n)| < K$ for $n \geq N$, then $\sum_{n=1}^{\infty} u_n$ converges if $C > 1$ and diverges if $C \leq 1$.

Show that the results of Raabe's test agree with what you know about the series $\sum_{n=1}^{\infty} (1/n^2)$ and $\sum_{n=1}^{\infty} (1/n)$.

71. *(Continuation of Exercise 70.)* Suppose that the terms of $\sum_{n=1}^{\infty} u_n$ are defined recursively by the formulas

$$u_1 = 1, \quad u_{n+1} = \frac{(2n-1)^2}{(2n)(2n+1)} u_n.$$

Apply Raabe's test to determine whether the series converges.

72. a) Suppose $a_1, a_2, a_3, \ldots, a_n$ are positive numbers satisfying the following conditions:
   i) $a_1 \geq a_2 \geq a_3 \geq \cdots$;
   ii) the series $a_2 + a_4 + a_8 + a_{16} + \cdots$ diverges.

   Show that the series

$$\frac{a_1}{1} + \frac{a_2}{2} + \frac{a_3}{3} + \cdots$$

   diverges.

b) Use the result in (a) to show that

$$1 + \sum_{n=2}^{\infty} \frac{1}{n \ln n}$$

   diverges.

73. If p is a constant, show that the series

$$1 + \sum_{n=3}^{\infty} \frac{1}{n \cdot \ln n \cdot [\ln(\ln n)]^p}$$

(a) converges if $p > 1$, (b) diverges if $p \leq 1$. In general, if $f_1(x) = x$, $f_{n+1}(x) = \ln(f_n(x))$, and n takes on the values 1, 2, 3, ..., we find that $f_2(x) = \ln x$, $f_3(x) = \ln(\ln x)$, and so on. If $f_n(a) > 1$, then

$$\int_a^{\infty} \frac{dx}{f_1(x)\, f_2(x) \ldots f_n(x)\, (f_{n+1}(x))^p}$$

converges if $p > 1$ and diverges if $p \leq 1$.

74. Prove that $\sum_{n=1}^{\infty} a_n/(1 + a_n)$ converges if $a_n > 0$ for all n and $\sum_{n=1}^{\infty} a_n$ converges.

75. If $\sum_{n=1}^{\infty} a_n$ converges, and if $a_n \neq 1$ and $a_n > 0$ for all n,
a) Show that $\sum_{n=1}^{\infty} a_n^2$ converges.
b) Does $\sum_{n=1}^{\infty} a_n/(1 - a_n)$ converge? Explain.

76. *(Continuation of Exercise 75.)* If $\sum_{n=1}^{\infty} a_n$ converges, and if $1 > a_n > 0$ for all n, show that $\sum_{n=1}^{\infty} \ln(1 - a_n)$ converges. (*Hint:* First show that $|\ln(1 - a_n)| \leq a_n/(1 - a_n)$.)

77. a) Find the interval of convergence of the series

$$y = 1 + \frac{1}{6}x^3 + \frac{1}{180}x^6 + \cdots + \frac{1 \cdot 4 \cdot 7 \cdot \cdots \cdot (3n-2)}{(3n)!} x^{3n} + \cdots.$$

b) Show that the function defined by the series satisfies the differential equation of the form

$$\frac{d^2 y}{dx^2} = x^a y + b$$

and find the values of the constants a and b.

78. a) Find the Maclaurin series for the function $x^2/(1 + x)$.
b) Does the series converge at $x = 1$? Explain.

79. Find the first three nonzero terms of the Maclaurin series for $\ln(\cos x)$
a) by substituting the series for $y = 1 - \cos x$ into the series for $\ln(1 - y)$.
b) by applying the definition to the function $f(x) = \ln(\cos x)$.
Which method is faster?

80. Find the first three nonzero terms of the expansion of $f(x) = \sqrt{1 + x^2}$ in powers of $(x - 1)$.

81. Expand $f(x) = 1/(1 - x)$ in powers of $(x - 2)$. Find the series' interval and radius of convergence.

82. Expand $f(x) = 1/(x + 1)$ in powers of $(x - 3)$. Find the series' interval and radius of convergence.

83. Expand $f(x) = \cos x$ in powers of $(x - \pi)$. Find the series' radius and interval of convergence.

84. Expand $f(x) = 1/x$ in powers of $(x - a)$. Find the interval of convergence if (a) $a > 0$, (b) $a < 0$.

85. *A fast estimate of $\pi/2$.* As you saw if you did Exercise 75 in Section 8.1, the sequence generated by starting with $x_0 = 1$ and applying the recursion formula $x_{n+1} = x_n + \cos x_n$ converges rapidly to $\pi/2$. To explain the speed of the convergence, let $\epsilon_n = (\pi/2) - x_n$ (Fig. 8.26). Then

$$\epsilon_{n+1} = \frac{\pi}{2} - x_n - \cos x_n$$

$$= \epsilon_n - \cos\left(\frac{\pi}{2} - \epsilon_n\right)$$

$$= \epsilon_n - \sin \epsilon_n$$

$$= \frac{1}{3!}\left(\epsilon_n\right)^3 - \frac{1}{5!}\left(\epsilon_n\right)^5 + \cdots.$$

Use this equality to show that

$$0 < \epsilon_{n+1} < \frac{1}{6}\left(\epsilon_n\right)^3.$$

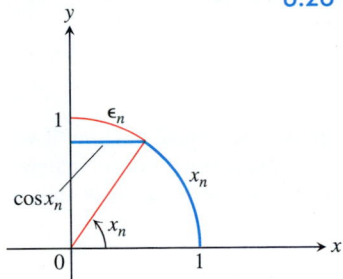

8.26 The figure for Exercise 85.

86. Suppose $\sum_{n=1}^{\infty} a_n$ is a convergent series of positive numbers. Does $\sum_{n=1}^{\infty} \ln(1 + a_n)$ converge? Explain.

CALCULATOR Estimate the values of the integrals in Exercises 87 and 88 with an error of magnitude less than 10^{-8}.

87. $\int_{0}^{1} x \sin(x^3)\, dx$ **88.** $\int_{0}^{1/64} \dfrac{\tan^{-1} x}{\sqrt{x}}\, dx$

89. *Series for $\tan^{-1} x$ for $|x| > 1$.* Derive the series

$$\tan^{-1}x = \frac{\pi}{2} - \frac{1}{x} + \frac{1}{3x^3} - \frac{1}{5x^5} + \cdots, \quad x > 1,$$

$$\tan^{-1}x = -\frac{\pi}{2} - \frac{1}{x} + \frac{1}{3x^3} - \frac{1}{5x^5} + \cdots, \quad x < -1,$$

by integrating the series

$$\frac{1}{1 + t^2} = \frac{1}{t^2} \cdot \frac{1}{1 + (1/t^2)} = \frac{1}{t^2} - \frac{1}{t^4} + \frac{1}{t^6} - \frac{1}{t^8} + \cdots$$

in the first case from x to ∞ and in the second case from $-\infty$ to x.

90. Use series multiplication to find the terms through x^5 of the Maclaurin series for (a) $(\tan^{-1}x)[\ln(1 + x)]$; (b) $(\tan^{-1}x)/(1 - x)$.

91. *An infinite product.* The infinite product

$$\prod_{n=1}^{\infty} (1 + a_n) = (1 + a_1)(1 + a_2)(1 + a_3)\ldots$$

is said to converge if the series

$$\sum_{n=1}^{\infty} \ln(1 + a_n),$$

obtained by taking the natural logarithm of the product, converges. Prove that the product converges if $a_n > -1$ for every n and if $\sum_{n=1}^{\infty}|a_n|$ converges. (*Hint:* Show that

$$|\ln(1 + a_n)| \le \frac{|a_n|}{1 - |a_n|} < 2|a_n|$$

when $|a_n| < 1/2$.)

92. *The value of $\sum_{n=1}^{\infty} \tan^{-1}(2/n^2)$.*

a) Use the formula for the tangent of the difference of two angles to show that

$$\tan(\tan^{-1}(n + 1) - \tan^{-1}(n - 1)) = \frac{2}{n^2}$$

and hence that

$$\tan^{-1}\frac{2}{n^2} = \tan^{-1}(n + 1) - \tan^{-1}(n - 1).$$

b) Show that

$$\sum_{n=1}^{N} \tan^{-1}\frac{2}{n^2} = \tan^{-1}(N + 1) + \tan^{-1}N - \frac{\pi}{4}.$$

c) Find the value of $\sum_{n=1}^{\infty} \tan^{-1}\dfrac{2}{n^2}$.

93. *Quality control*

a) Differentiate the series

$$\frac{1}{1 - x} = 1 + x + x^2 + \cdots + x^n + \cdots$$

to obtain a series for $1/(1 - x)^2$.

b) In one throw of two dice, the probability of getting a roll of 7 is $p = 1/6$. If you throw the dice repeatedly, the probability that a 7 will appear for the first time at the nth throw is $q^{n-1}p$, where $q = 1 - p = 5/6$. The expected number of throws until a 7 first appears is $\sum_{n=1}^{\infty} n\,q^{n-1}p$. Find the sum of this series.

c) As an engineer applying statistical control to an industrial operation, you inspect items taken at random from the assembly line. You classify each sampled item as either "good" or "bad." If the probability of an item's being good is p and of an item being bad is $q = 1 - p$, the probability that the first bad item is found is the nth one inspected is $p^{n-1}q$. The average number inspected up to and including the first bad item found is $\sum_{n=1}^{\infty} n\,p^{n-1}q$. Evaluate this sum, assuming $0 < p < 1$.

94. *Expected value.* Suppose that a random variable X may assume the values 1, 2, 3, ... , with probabilities $p_1, p_2, p_3, \ldots$, where p_k is the probability that X equals k ($k = 1, 2, 3, \ldots$). Suppose also that $p_k \ge 0$ and that $\sum_{k=1}^{\infty} p_k = 1$. The **expected value** of X, denoted by $E(X)$, is the number $\sum_{k=1}^{\infty} k p_k$, provided the series converges. In each of the following cases, show that $\sum_{k=1}^{\infty} p_k = 1$ and find $E(X)$ if it exists. (*Hint:* See Exercise 93.)

a) $p_k = 2^{-k}$ b) $p_k = \dfrac{5^{k-1}}{6^k}$

c) $p_k = \dfrac{1}{k(k + 1)} = \dfrac{1}{k} - \dfrac{1}{k + 1}$

95. a) Prove the following theorem: If $\{c_n\}$ is a sequence of numbers such that every sum $t_n = \sum_{k=1}^{n} c_k$ is bounded, then the series $\sum_{n=1}^{\infty} c_n/n$ converges and is equal to $\sum_{n=1}^{\infty} t_n/(n(n + 1))$.

Outline of proof: Replace c_1 by t_1 and c_n by $t_n - t_{n-1}$ for $n \ge 2$. If $s_{2n+1} = \sum_{k=1}^{2n+1} c_k/k$, show that

$$s_{2n+1} = t_1\left(1 - \frac{1}{2}\right) + t_2\left(\frac{1}{2} - \frac{1}{3}\right)$$

$$+ \cdots + t_{2n}\left(\frac{1}{2n} - \frac{1}{2n + 1}\right) + \frac{t_{2n+1}}{2n + 1}$$

$$= \sum_{k=1}^{2n} \frac{t_k}{k(k + 1)} + \frac{t_{2n+1}}{2n + 1}.$$

Because $|t_k| < M$ for some constant M, the series $\sum_{k=1}^{\infty} t_k/(k(k + 1))$ converges absolutely and s_{2n+1} has a limit as $n \to \infty$. Finally, if $s_{2n} = \sum_{k=1}^{2n} c_k/k$, then $s_{2n+1} - s_{2n} = c_{2n+1}/(2n + 1)$ approaches zero as

$n \to \infty$ because $|c_{2n+1}| = |t_{2n+1} - t_{2n}| < 2M$. Hence the sequence of partial sums of the series $\Sigma c_k/k$ converges and the limit is $\Sigma_{k=1}^{\infty} t_k/(k(k+1))$.

b) Show how the foregoing theorem applies to the alternating harmonic series

$$1 - \frac{1}{2} + \frac{1}{3} - \frac{1}{4} + \frac{1}{5} - \frac{1}{6} + \cdots .$$

c) Show that the series

$$1 - \frac{1}{2} - \frac{1}{3} + \frac{1}{4} + \frac{1}{5} - \frac{1}{6} - \frac{1}{7} + \cdots$$

converges. (After the first term, the signs are two negative, two positive, two negative, two positive, and so on in that pattern.)

96. *The convergence of $\Sigma_{n=1}^{\infty} [(-1)^{n-1} x^n]/n$ to $\ln(1 + x)$ for $-1 < x \leq 1$.*

a) Show by long division or otherwise that

$$\frac{1}{1+t} = 1 - t + t^2 - t^3 + \cdots + (-1)^n t^n + \frac{(-1)^{n+1} t^{n+1}}{1+t}.$$

b) By integrating the equation of part (a) with respect to t from 0 to x, show that

$$\ln(1 + x) = x - \frac{x^2}{2} + \frac{x^3}{3} - \frac{x^4}{4} + \cdots$$
$$+ (-1)^n \frac{x^{n+1}}{n+1} + R_{n+1}$$

where

$$R_{n+1} = (-1)^{n+1} \int_0^x \frac{t^{n+1}}{1+t} \, dt.$$

c) If $x \geq 0$, show that

$$|R_{n+1}| \leq \int_0^x t^{n+1} \, dt = \frac{x^{n+2}}{n+2}.$$

$\left(\text{*Hint:* As } t \text{ varies from 0 to } x, \right.$

$$1 + t \geq 1 \quad \text{and} \quad t^{n+1}/(1+t) \leq t^{n+1},$$

and

$$\left. \left| \int_0^x f(t) \, dt \right| \leq \int_0^x |f(t)| \, dt . \right)$$

d) If $-1 < x < 0$, show that

$$|R_{n+1}| \leq \left| \int_0^x \frac{t^{n+1}}{1-|x|} \, dt \right| = \frac{|x|^{n+2}}{(n+2)(1-|x|)}.$$

$\left(\text{*Hint:* If } x < t \leq 0, \text{ then } |1 + t| \geq 1 - |x| \text{ and} \right.$

$$\left. \left| \frac{t^{n+1}}{1+t} \right| \leq \frac{|t|^{n+1}}{1-|x|} . \right)$$

e) Use the foregoing results to prove that the series

$$x - \frac{x^2}{2} + \frac{x^3}{3} - \frac{x^4}{4} + \cdots + \frac{(-1)^n x^{n+1}}{n+1} + \cdots$$

converges to $\ln(1 + x)$ for $-1 < x \leq 1$.

Power Series and Indeterminate Forms

In considering

$$\lim_{x \to a} \frac{f(x)}{g(x)}$$

when $f(a) = g(a) = 0$, we can quickly determine the limit if we have Taylor series that converge to these functions on an interval centered at $x = a$. The following example shows how this is done.

EXAMPLE 1 Evaluate $\displaystyle\lim_{x \to 0} \frac{\sin x - \tan x}{x^3}$.

SOLUTION To terms in x^5, the Maclaurin series for $\sin x$ and $\tan x$ are

$$\sin x = x - \frac{x^3}{3!} + \frac{x^5}{5!} - \cdots, \quad \tan x = x + \frac{x^3}{3} + \frac{2x^5}{15} + \cdots.$$

Hence,

$$\sin x - \tan x = -\frac{x^3}{2} - \frac{x^5}{8} - \cdots = x^3 \left(-\frac{1}{2} - \frac{x^2}{8} - \cdots \right)$$

and

$$\lim_{x \to 0} \frac{\sin x - \tan x}{x^3} = \lim_{x \to 0} \left(-\frac{1}{2} - \frac{x^2}{8} - \cdots \right) = -\frac{1}{2}.$$

Use series to find the limits in Exercises 97–113.

97. $\displaystyle\lim_{x \to 0} \frac{e^x - (1 + x)}{x^2}$

98. $\displaystyle\lim_{t \to 0} \frac{1 - \cos t - (t^2/2)}{t^4}$

99. $\displaystyle\lim_{\theta \to \infty} \theta \sin \frac{1}{\theta}$

100. $\displaystyle\lim_{x \to 0} \frac{1 - \cos x}{\sin x}$

101. $\displaystyle\lim_{x \to 0} \frac{\sin x}{e^x - 1}$

102. $\displaystyle\lim_{y \to 0} \frac{\sin y - y + (y^3/6)}{y^5}$

103. $\displaystyle\lim_{u \to 0} \frac{e^u - e^{-u} - 2u}{u - \sin u}$

104. $\displaystyle\lim_{x \to 0} \frac{x - \tan^{-1} x}{x^3}$

105. $\displaystyle\lim_{\theta \to 0} \frac{\tan \theta - \sin \theta}{\theta^3 \cos \theta}$

106. $\displaystyle\lim_{x \to \infty} x^2 \left(e^{-1/x^2} - 1 \right)$

107. $\displaystyle\lim_{h \to 0} \frac{\ln(1 + h^2)}{1 - \cos h}$

108. $\displaystyle\lim_{t \to 0} \frac{\tan 3t}{t}$

109. $\displaystyle\lim_{x \to 1} \frac{\ln x}{x - 1}$

110. $\displaystyle\lim_{x \to 1} \frac{\ln x^2}{x - 1}$

111. $\displaystyle\lim_{x \to 0} \left(\frac{1}{\sin x} - \frac{1}{x} \right)$

112. $\displaystyle\lim_{x \to 0} \left(\frac{1}{2 - 2 \cos x} - \frac{1}{x^2} \right)$

113. $\displaystyle\lim_{x \to 0} \frac{\ln(1 - x) - \sin x}{1 - \cos^2 x}$

114. Find values of r and s that make

$$\lim_{x \to 0} \left(\frac{\sin 3x}{x^3} + \frac{r}{x^2} + s \right) = 0.$$

9

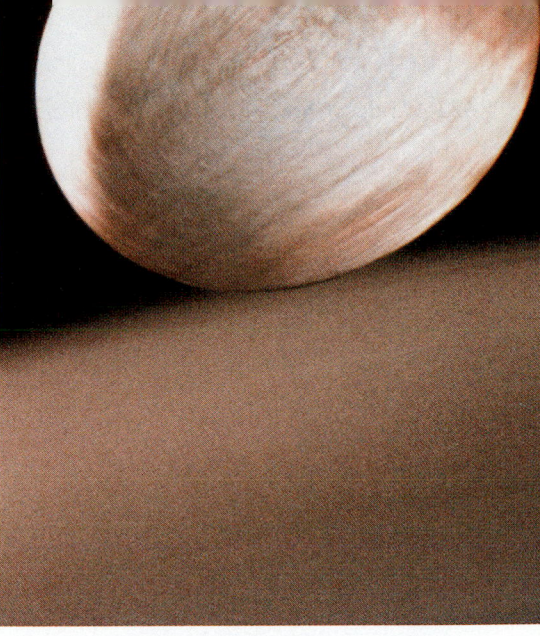

CONIC SECTIONS, PARAMETRIZED CURVES, AND POLAR COORDINATES

Overview   The study of motion has been important since ancient times, and, as you know, calculus gives us the mathematics we need to describe motion. In this chapter, we extend our ability to analyze motion by showing how to keep track of the position of a moving body as a function of time. We do this with parametric equations. We study them in the coordinate plane here and then extend our work to three dimensions in Chapters 10 and 11. We begin our study by developing equations for conic sections, since these are the paths traveled by planets, satellites, and other bodies (even electrons) whose motions are driven by inverse square forces. Planetary motion is best described in polar coordinates (another of Newton's inventions, although James Bernoulli usually gets the credit because he published first), so we spend our remaining time finding out what curves, derivatives, and integrals look like in this new coordinate system.

9.1 Equations for Conic Sections

The conic sections that originated in Greek geometry are described today as the graphs of quadratic equations in the coordinate plane. The Greeks of Plato's time described these curves as the curves formed by cutting a double cone with a plane (Fig. 9.1); hence the name "conic section." We begin by reviewing briefly the equations for parabolas and circles and then continue on to ellipses and hyperbolas.

   We use the mathematics of conic sections to describe the paths of planets, comets, moons, asteroids, and satellites that are moved through space by gravitational forces. Once we know that the path of a freely moving body is a conic section, we immediately have useful information about its velocity and the force that drives it, as we shall see in Chapter 11.

Equations from the Distance Formula: Circles and Parabolas

As you know, the distance between two points (x_1, y_1) and (x_2, y_2) in the coordinate plane is

$$d = \sqrt{(x_2 - x_1)^2 + (y_2 - y_1)^2}.$$

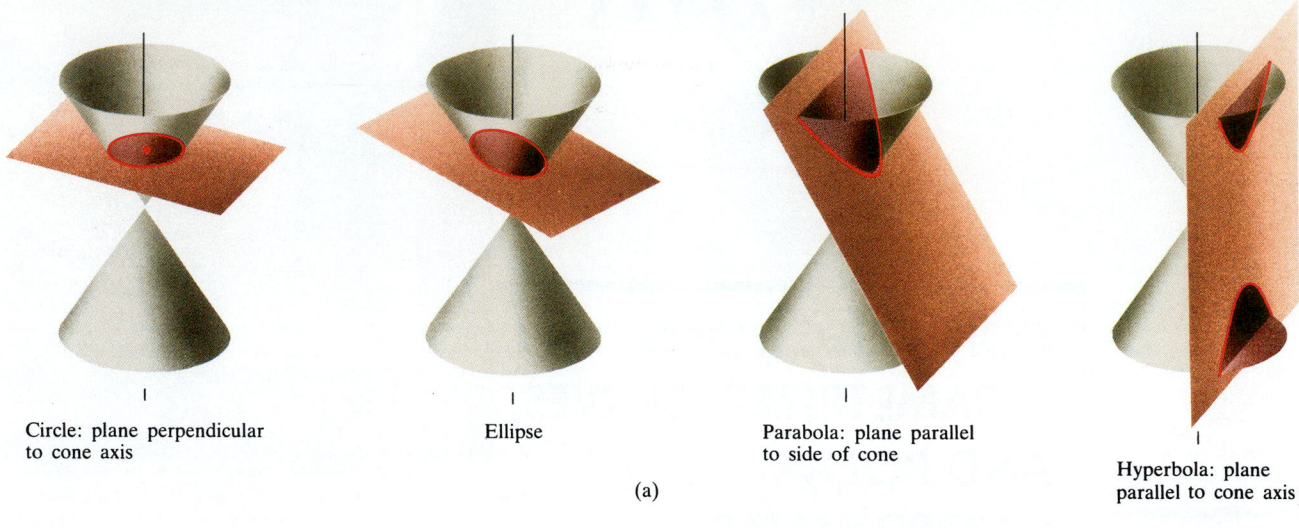

Circle: plane perpendicular
to cone axis

Ellipse

Parabola: plane parallel
to side of cone

Hyperbola: plane
parallel to cone axis

(a)

9.1   The standard conic sections (a) are
the curves in which a plane cuts a double
cone. Hyperbolas come in two parts, called
branches. The point and lines obtained by
passing the plane through the cone's vertex
(b) are *degenerate* conic sections.

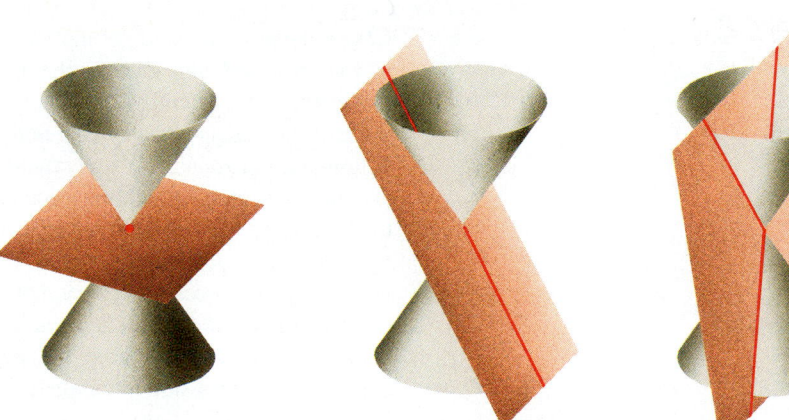

Point: plane through
cone vertex only

Single line: plane
tangent to cone

Pair of intersecting lines

(b)

In Section 1.5, we used this formula to derive the standard equation for circles centered at the origin and the standard equations for parabolas with vertices at the
origin (see Table 9.1). The horizontal and vertical shift formulas then gave us equations for circles and a variety of parabolas in other locations.

**The Standard-Form Equation for a Circle of Radius *a* Centered
at the Origin**

$$x^2 + y^2 = a^2$$

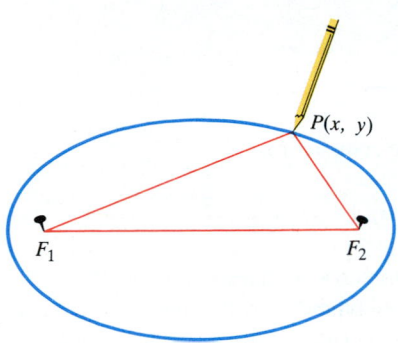

9.2 How to draw an ellipse.

TABLE 9.1
The standard-form equations for parabolas with vertices at the origin ($p > 0$)

Equation	Focus	Directrix	Axis	Direction
$x^2 = 4py$	$(0, p)$	$y = -p$	y-axis	Opens up
$x^2 = -4py$	$(0, -p)$	$y = p$	y-axis	Opens down
$y^2 = 4px$	$(p, 0)$	$x = -p$	x-axis	Opens to right
$y^2 = -4px$	$(-p, 0)$	$x = p$	x-axis	Opens to left

Ellipses

The equations we use for ellipses come from the distance formula, too.

DEFINITIONS

An **ellipse** is the set of points in a plane whose distances from two fixed points in the plane have a constant sum. The two fixed points are the **foci** of the ellipse.

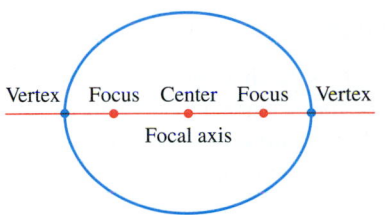

9.3 The center, vertices, and foci of an ellipse lie on the focal axis.

The quickest way to construct an ellipse uses the definition. Put a loop of string around two tacks F_1 and F_2, pull the string taut with a pencil point P, and move the pencil around to trace a closed curve (Fig. 9.2). The curve is an ellipse because the sum $PF_1 + PF_2$, being equal to the length of the loop minus the distance between the tacks, remains constant. The ellipse's foci lie at F_1 and F_2.

DEFINITIONS

The line through the foci of an ellipse is the ellipse's **focal axis.** The point on the axis halfway between the foci is the ellipse's **center.** The points where the focal axis crosses the ellipse are the ellipse's **vertices** (Fig. 9.3).

If the foci are $F_1(-c, 0)$ and $F_2(c, 0)$ (Fig. 9.4), and the sum of the distances $PF_1 + PF_2$ is denoted by $2a$, then the coordinates of a point P on the ellipse satisfy the equation

$$\sqrt{(x + c)^2 + y^2} + \sqrt{(x - c)^2 + y^2} = 2a. \tag{1}$$

To simplify this equation, we move the second radical to the right-hand side, square, isolate the remaining radical, and square again, obtaining

$$\frac{x^2}{a^2} + \frac{y^2}{a^2 - c^2} = 1. \tag{2}$$

Since the sum $PF_1 + PF_2$ is greater than the length F_1F_2 (triangle inequality for triangle PF_1F_2), the number $2a$ is greater than $2c$. Accordingly, a is greater than c and the number $a^2 - c^2$ in Eq. (2) is positive.

The algebraic steps taken to arrive at Eq. (2) can be reversed to show that every point P whose coordinates satisfy an equation of this form with $0 < c < a$ also satisfies the equation $PF_1 + PF_2 = 2a$. Thus, a point lies on the ellipse if and only if its coordinates satisfy Eq. (2).

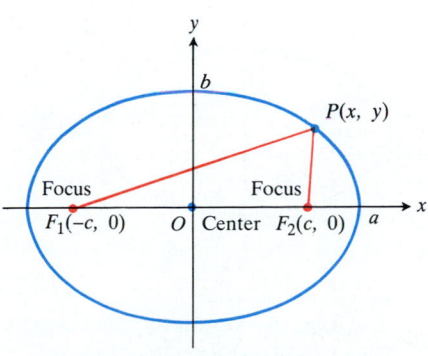

9.4 The ellipse defined by the equation $PF_1 + PF_2 = 2a$ is the graph of the equation $(x^2/a^2) + (y^2/b^2) = 1$.

If

$$b = \sqrt{a^2 - c^2}, \tag{3}$$

then $a^2 - c^2 = b^2$ and Eq. (2) takes the more compact form

$$\frac{x^2}{a^2} + \frac{y^2}{b^2} = 1. \tag{4}$$

Equation (4) reveals that this ellipse is symmetric with respect to the origin and both coordinate axes. It lies inside the rectangle bounded by the lines $x = \pm a$ and $y = \pm b$. It crosses the axes at the points $(\pm a, 0)$ and $(0, \pm b)$. The tangents at these points are perpendicular to the axes because the slope

$$\frac{dy}{dx} = -\frac{b^2 x}{a^2 y}$$

is zero when $x = 0$ and infinite when $y = 0$. These observations are the basis of the drawing lesson below.

The Major and Minor Axes of an Ellipse

The **major axis** of the ellipse described by Eq. (4) is the line segment of length $2a$ joining the points $(\pm a, 0)$. The **minor axis** is the line segment of length $2b$ joining the points $(0, \pm b)$. The number a itself is called the **semimajor axis** and the number b the **semiminor axis.** The number c, which can be found from Eq. (3) as

$$c = \sqrt{a^2 - b^2}, \tag{5}$$

is the **center-to-focus distance** of the ellipse.

DRAWING   LESSON

How to Graph the Ellipse $\dfrac{x^2}{a^2} + \dfrac{y^2}{b^2} = 1$

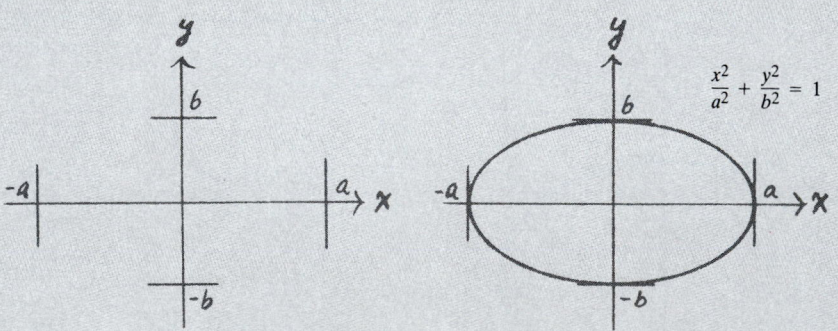

1. Mark the points $(\pm a, 0)$ and $(0, \pm b)$ with line segments perpendicular to the coordinate axes.

2. Use the segments as tangent lines to guide your drawing.

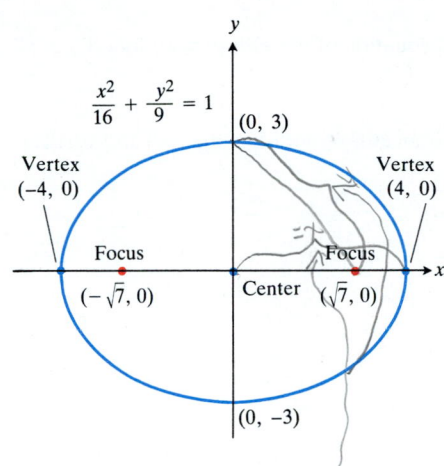

9.5 The major axis of $(x^2/16) + (y^2/9)$ $= 1$ is horizontal (Example 1).

Example 1   *Major Axis Horizontal.*   The ellipse

$$\frac{x^2}{16} + \frac{y^2}{9} = 1 \tag{6}$$

(Fig. 9.5) has

Semimajor axis: $a = \sqrt{16} = 4$,   Semiminor axis: $b = \sqrt{9} = 3$,

Center-to-focus distance: $c = \sqrt{16 - 9} = \sqrt{7}$,

Foci: $(\pm c, 0) = (\pm\sqrt{7}, 0)$,

Vertices: $(\pm a, 0) = (\pm 4, 0)$,

Center: $(0, 0)$.

Example 2   *Major Axis Vertical.*   The ellipse

$$\frac{x^2}{9} + \frac{y^2}{16} = 1, \tag{7}$$

obtained by interchanging x and y in Eq. (6), represents an ellipse with its major axis vertical instead of horizontal (Fig. 9.6). With a^2 still equal to 16 and b^2 equal to 9, we have

Semimajor axis: $a = \sqrt{16} = 4$,   Semiminor axis: $b = \sqrt{9} = 3$,

Center-to-focus distance: $c = \sqrt{16 - 9} = \sqrt{7}$,

Foci: $(0, \pm c) = (0, \pm\sqrt{7})$,

Vertices: $(0, \pm a) = (0, \pm 4)$,

Center: $(0, 0)$.

There is never any cause for confusion in analyzing equations like (6) and (7). We simply find the intercepts on the coordinate axes; then we know which way the major axis runs because it is the longer of the two axes. The center always lies at the origin and the foci always lie on the major axis.

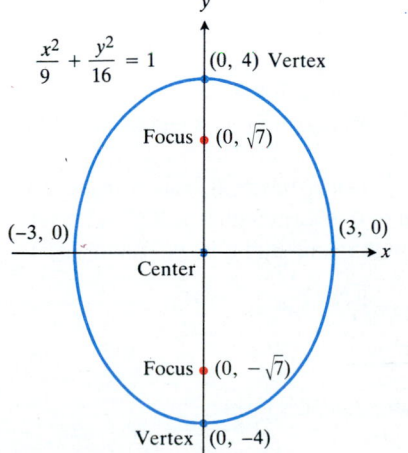

9.6 The major axis of $(x^2/9) + (y^2/16)$ $= 1$ is vertical (Example 2).

Standard-Form Equations for Ellipses Centered at the Origin

Foci on the x-axis: $\dfrac{x^2}{a^2} + \dfrac{y^2}{b^2} = 1$ $(a > b)$

Center-to-focus distance: $c = \sqrt{a^2 - b^2}$

Foci: $(\pm c, 0)$

Vertices: $(\pm a, 0)$

Foci on the y-axis: $\dfrac{x^2}{b^2} + \dfrac{y^2}{a^2} = 1$ $(a > b)$

Center-to-focus distance: $c = \sqrt{a^2 - b^2}$

Foci: $(0, \pm c)$

Vertices: $(0, \pm a)$

In each case, a is the semimajor axis and b is the semiminor axis.

Example 3   Find the standard-form equation of the ellipse with foci $(0, \pm 3)$ and vertices $(0, \pm 4)$.

Solution   The standard-form equation for an ellipse with foci $(0, \pm c)$ and vertices $(0, \pm a)$ is

$$\frac{x^2}{b^2} + \frac{y^2}{a^2} = 1,$$

where $c = \sqrt{a^2 - b^2}$. In the ellipse at hand, $c = 3$ and $a = 4$, so

$$3 = \sqrt{(4)^2 - b^2},$$

$$9 = 16 - b^2,$$

$$b^2 = 7.$$

The equation we seek is

$$\frac{x^2}{7} + \frac{y^2}{16} = 1.$$

Hyperbolas

DEFINITIONS

A **hyperbola** is the set of points in a plane whose distances from two fixed points in the plane have a constant difference. The two fixed points are the **foci** of the hyperbola.

If the foci are $F_1(-c, 0)$ and $F_2(c, 0)$ (Fig. 9.7) and the constant difference is $2a$, then a point (x, y) lies on the hyperbola if and only if

$$\sqrt{(x + c)^2 + y^2} - \sqrt{(x - c)^2 + y^2} = \pm 2a. \tag{8}$$

To simplify this equation, we move the second radical to the right-hand side, square, isolate the remaining radical, and square again, obtaining

$$\frac{x^2}{a^2} + \frac{y^2}{a^2 - c^2} = 1. \tag{9}$$

So far, this looks just like the equation for an ellipse. But now $a^2 - c^2$ is negative because $2a$, being the difference of two sides of triangle PF_1F_2, is less than $2c$, the third side.

The algebraic steps taken to arrive at Eq. (9) can be reversed to show that every point P whose coordinates satisfy an equation of this form with $0 < a < c$ also satisfies Eq. (8). Thus, a point lies on the hyperbola if and only if its coordinates satisfy Eq. (9).

If we let b denote the positive square root of $c^2 - a^2$,

$$b = \sqrt{c^2 - a^2}, \tag{10}$$

then $a^2 - c^2 = -b^2$ and Eq. (9) takes the more compact form

$$\frac{x^2}{a^2} - \frac{y^2}{b^2} = 1. \tag{11}$$

The only difference between Eq. (11) and the equation for an ellipse (Eq. 4) is the

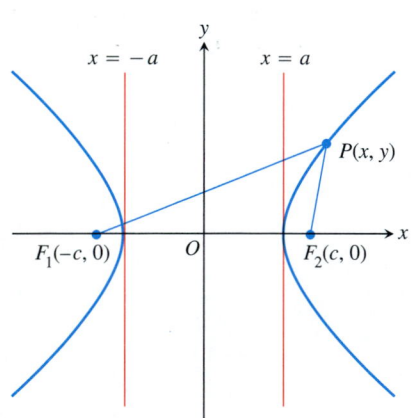

9.7   Hyperbolas have two branches. For points on the right-hand branch of the hyperbola shown here, $PF_1 - PF_2 = 2a$. For points on the left-hand branch, $PF_2 - PF_1 = 2a$.

minus sign in the equation and the new relation

$$c^2 = a^2 + b^2 \tag{12}$$

given by Eq. (10).

Like the ellipse, the hyperbola is symmetric with respect to the origin and both coordinate axes. It crosses the x-axis at the points $(\pm a, 0)$. The tangents at these points are vertical because the derivative

$$\frac{dy}{dx} = \frac{b^2 x}{a^2 y}$$

is infinite when $y = 0$. The hyperbola has no y-intercepts; in fact, no part of the curve lies between the lines $x = -a$ and $x = a$.

DEFINITIONS

The line through the foci of a hyperbola is the hyperbola's **focal axis.** The point on the axis halfway between the foci is the hyperbola's **center.** The points where the focal axis crosses the hyperbola arc the hyperbola's **vertices** (Fig. 9.8).

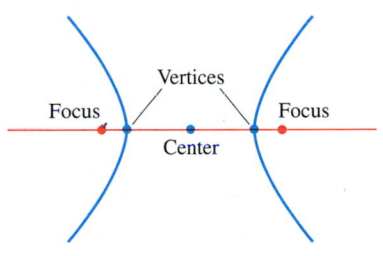

9.8 The center, vertices, and foci of a hyperbola lie on the focal axis.

Asymptotes of Hyperbolas—Graphing

As we saw in Section 3.4, the distance between a curve and some fixed line may approach zero as the curve moves farther and farther from the origin. If this happens, we call the line an **asymptote** of the curve. The hyperbola

$$\frac{x^2}{a^2} - \frac{y^2}{b^2} = 1 \tag{13}$$

has two asymptotes, the lines

$$y = \pm \frac{b}{a} x \tag{14}$$

(Exercise 67). The asymptotes give us the guidance we need to graph hyperbolas quickly. (See the drawing lesson on page 630.) The fastest way to find the equations of the asymptotes is to replace the 1 in Eq. (13) by 0 and solve the resulting equation for y:

$$\underbrace{\frac{x^2}{a^2} - \frac{y^2}{b^2} = 1}_{\text{hyperbola}} \Rightarrow \underbrace{\frac{x^2}{a^2} - \frac{y^2}{b^2} = 0}_{\text{0 for 1}} \Rightarrow \underbrace{y = \pm \frac{b}{a} x.}_{\text{asymptotes}} \tag{15}$$

Example 4 *Foci on the x-axis.* The equation

$$\frac{x^2}{4} - \frac{y^2}{5} = 1 \tag{16}$$

is Eq. (11) with $a^2 = 4$ and $b^2 = 5$ (Fig. 9.9a). We have

Center-to-focus distance: $c = \sqrt{a^2 + b^2} = \sqrt{4 + 5} = 3,$

Foci: $(\pm c, 0) = (\pm 3, 0),$ Vertices: $(\pm a, 0) = (\pm 2, 0),$

Center: $(0, 0),$

Asymptotes: $\dfrac{x^2}{4} - \dfrac{y^2}{5} = 0$ or $y = \pm \dfrac{\sqrt{5}}{2} x.$

How to Graph the Hyperbola $\dfrac{x^2}{a^2} - \dfrac{y^2}{b^2} = 1$

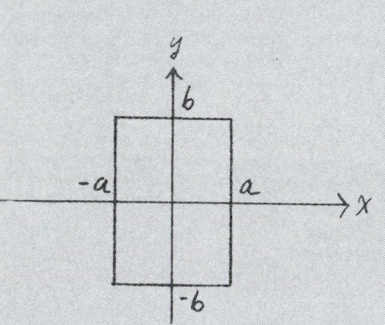

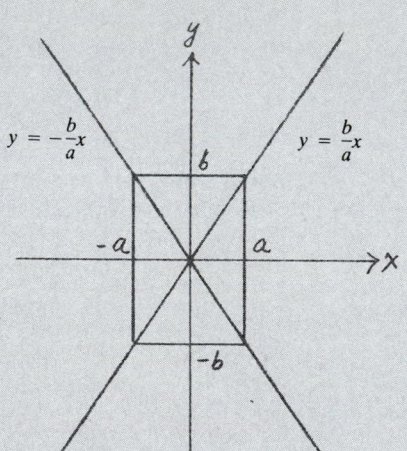

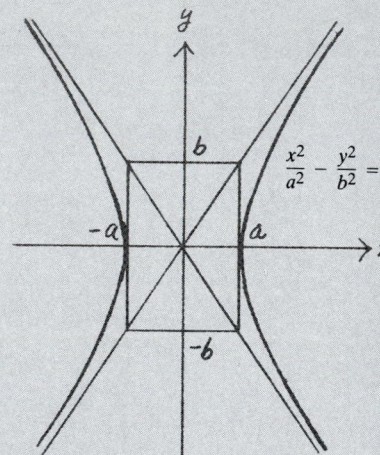

1. Mark the points $(\pm a, 0)$ and $(0, \pm b)$ with line segments and complete the rectangle they determine.

2. Sketch the asymptotes by extending the rectangle's diagonals.

3. Use the rectangle and asymptotes to guide your drawing.

Standard-Form Equations for Hyperbolas Centered at the Origin

Foci on the x-axis: $\dfrac{x^2}{a^2} - \dfrac{y^2}{b^2} = 1$

Center-to-focus distance: $c = \sqrt{a^2 + b^2}$

Foci: $(\pm c, 0)$

Vertices: $(\pm a, 0)$

Asymptotes: $\dfrac{x^2}{a^2} - \dfrac{y^2}{b^2} = 0$ or $y = \pm\dfrac{b}{a}x$

Foci on the y-axis: $\dfrac{y^2}{a^2} - \dfrac{x^2}{b^2} = 1$

Center-to-focus distance: $c = \sqrt{a^2 + b^2}$

Foci: $(0, \pm c)$

Vertices: $(0, \pm a)$

Asymptotes: $\dfrac{y^2}{a^2} - \dfrac{x^2}{b^2} = 0$ or $y = \pm\dfrac{a}{b}x$

Notice the difference in the asymptote equations (b/a in the first, a/b in the second).

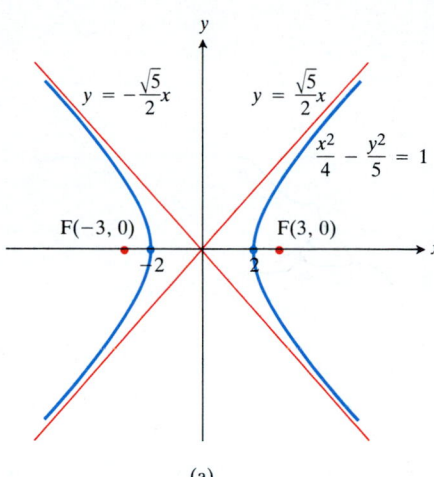

(a)

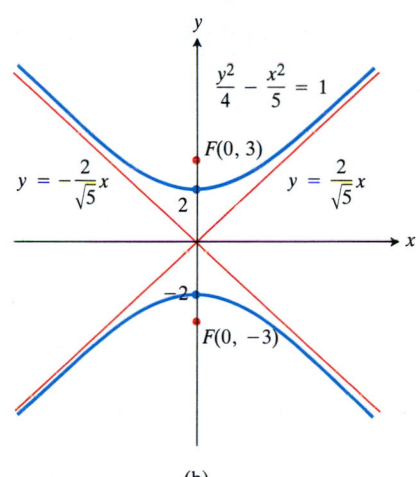

(b)

9.9   (a) The hyperbola in Example 4.
(b) The hyperbola in Example 5.

Example 5   *Foci on the y-axis.*   The hyperbola

$$\frac{y^2}{4} - \frac{x^2}{5} = 1, \tag{17}$$

obtained by interchanging x and y in Eq. (16), represents a hyperbola with vertices on the y-axis (Fig. 9.9b). With a^2 still equal to 4 and b^2 equal to 5, we have

Center-to-focus distance: $c = \sqrt{a^2 + b^2} = \sqrt{4 + 5} = 3,$

Foci: $(0, \pm c) = (0, \pm 3),$ Vertices: $(0, \pm a) = (0, \pm 2),$

Center: $(0, 0),$

Asymptotes: $\frac{y^2}{4} - \frac{x^2}{5} = 0$ or· $y = \pm \frac{2}{\sqrt{5}} x.$

Example 6   Find an equation for the hyperbola with asymptotes $y = \pm (4/3)x$ and foci $(\pm 10, 0)$.

Solution   The standard-form equation for a hyperbola with foci $(\pm c, 0)$ on the x-axis is

$$\frac{x^2}{a^2} - \frac{y^2}{b^2} = 1,$$

where $c = \sqrt{a^2 + b^2}$. From the asymptote equation $y = \pm (b/a)x$, we learn that

$$\frac{b}{a} = \frac{4}{3}, \quad \text{or} \quad b = \frac{4}{3} a.$$

Hence,

$$c^2 = a^2 + b^2 = a^2 + \frac{16}{9} a^2 = \frac{25}{9} a^2,$$

and we have

$$a^2 = \frac{9}{25} c^2 = \frac{9}{25} (10)^2 = 36,$$

$$b^2 = c^2 - a^2 = 100 - 36 = 64. \quad \left(\begin{array}{l} \text{The foci are } (\pm 10, 0), \\ \text{so } c = 10 \text{ and } c^2 = 100. \end{array} \right)$$

The equation we seek is

$$\frac{x^2}{36} - \frac{y^2}{64} = 1.$$

Classifying Conic Sections by Eccentricity: The Focus–Directrix Equation

Although the center-to-focus distance c does not appear in the standard equation for an ellipse,

$$\frac{x^2}{a^2} + \frac{y^2}{b^2} = 1 \quad (a > b),$$

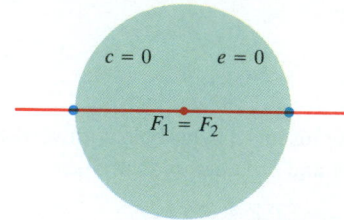 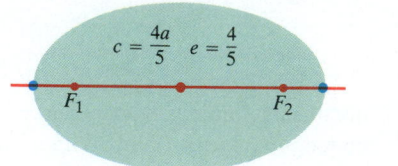 

9.10 The ellipse changes from a circle to a line segment as c increases from 0 to a.

we may still determine the value of c from the equation

$$c = \sqrt{a^2 - b^2}.$$

If we keep a fixed and vary c over the interval $0 \le c \le a$, the resulting ellipses will vary in shape (Fig. 9.10). They are circles if $c = 0$ (so that $a = b$) and flatten as c increases. In the extreme case $c = a$, the foci and vertices overlap and the ellipse degenerates into a line segment.

We use the ratio of c to a to describe the various shapes the ellipse can take. We call this ratio the ellipse's eccentricity.

DEFINITION

> The **eccentricity** of the ellipse $(x^2/a^2) + (y^2/b^2) = 1$ $(a > b)$ is the number
>
> $$e = \frac{c}{a} = \frac{\sqrt{a^2 - b^2}}{a}. \qquad (18)$$

TABLE 9.2
Eccentricities of planetary orbits

Mercury	0.21	Saturn	0.06
Venus	0.01	Uranus	0.05
Earth	0.02	Neptune	0.01
Mars	0.09	Pluto	0.25
Jupiter	0.05		

The planets in the solar system revolve around the sun in elliptical orbits with the sun at one focus. Most of the planets, including Earth, have orbits that are nearly circular, as can be seen from the eccentricities in Table 9.2. Pluto, however, has a fairly eccentric orbit, with $e = 0.25$, as does Mercury, with $e = 0.21$. Other members of the solar system have orbits that are even more eccentric. Icarus, an asteroid about 1 mile wide that revolves around the sun every 409 Earth days, has an orbital eccentricity of 0.83 (Fig. 9.11).

Example 7   The orbit of Halley's comet is an ellipse 36.18 astronomical units long by 9.12 astronomical units wide. (One *astronomical unit* [AU] is the semi-major axis of the earth's orbit, about 92,600,000 miles.) Its eccentricity is

$$e = \frac{\sqrt{a^2 - b^2}}{a} = \frac{\sqrt{(36.18/2)^2 - (9.12/2)^2}}{(36.18/2)} = \frac{\sqrt{(18.09)^2 - (4.56)^2}}{18.09}$$

$$= 0.97. \qquad \text{(Rounded, with a calculator)}$$

Whereas a parabola has one focus and one directrix, each ellipse has two foci and two directrices. These are the lines perpendicular to the major axis at distances $\pm a/e$ from the center. The parabola has the property that

$$PF = 1 \cdot PD \qquad (19)$$

for any point P on it, where F is the focus and D is the point nearest P on the directrix. For an ellipse, it can be shown that the equations that replace (19) are

$$PF_1 = e \cdot PD_1, \qquad PF_2 = e \cdot PD_2. \qquad (20)$$

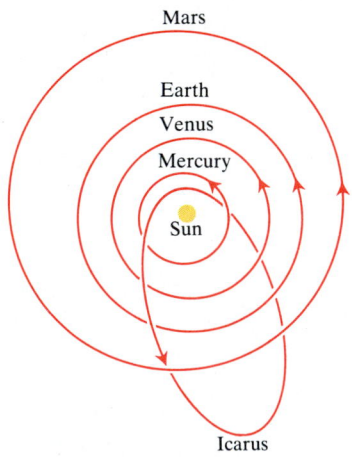

9.11 The orbit of the asteroid Icarus is highly eccentric. The earth's orbit is so nearly circular that both its foci lie inside the sun.

Halley's Comet

Edmund Halley (1656–1742; pronounced *haw*-ley), British biologist, geologist, sea captain, pirate, spy, Antarctic voyager, astronomer, adviser on fortifications, company founder and director, and author of the first actuarial mortality tables, was also the mathematician who pushed and harried Newton into writing his *Principia*. Despite his accomplishments, Halley is known today chiefly as the man who calculated the orbit of the great comet of 1682: "wherefore if according to what we have already said [the comet] should return again about the year 1758, candid posterity will not refuse to acknowledge that this was first discovered by an Englishman." Indeed, candid posterity did not refuse—ever since the comet's return in 1758, it has been known as Halley's comet.

Last seen rounding the sun during the winter and spring of 1985–86, the comet is due to return again in the year 2062. A recent study indicates that the comet has made about 2000 cycles so far with about the same number to go before the sun erodes it away completely.

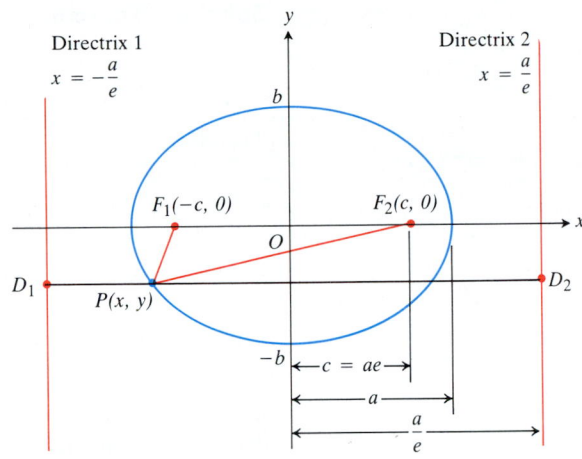

9.12 The foci and directrices of the ellipse $(x^2/a^2) + (y^2/b^2) = 1$. Directrix 1 corresponds to focus F_1, and directrix 2 to focus F_2.

Here, e is the eccentricity, P is any point on the ellipse, F_1 and F_2 are the foci, and D_1 and D_2 are the points on the directrices nearest P (Fig. 9.12).

In each equation in (20) the directrix and focus must correspond; that is, if we use the distance from P to F_1, we must also use the distance from P to the directrix at the same end of the ellipse. The directrix $x = -a/e$ corresponds to $F_1(-c, 0)$, and the directrix $x = a/e$ corresponds to $F_2(c, 0)$.

We define the eccentricity of a hyperbola with the same formula we use for the ellipse, $e = c/a$, only in this case c equals $\sqrt{a^2 + b^2}$ instead of $\sqrt{a^2 - b^2}$. In contrast to the eccentricity of an ellipse, the eccentricity of a hyperbola is always greater than 1.

DEFINITION

The **eccentricity** of the hyperbola $(x^2/a^2) - (y^2/b^2) = 1$ is the number

$$e = \frac{c}{a} = \frac{\sqrt{a^2 + b^2}}{a}. \tag{21}$$

In both ellipse and hyperbola, the eccentricity is the ratio of the distance between the foci to the distance between the vertices (because $c/a = 2c/2a$).

$$\text{Eccentricity} = \frac{\text{distance between foci}}{\text{distance between vertices}} \tag{22}$$

In an ellipse, the foci are closer together than the vertices and the ratio is less than 1. In a hyperbola, the foci are farther apart than the vertices and the ratio is greater than 1.

Example 8   Locate the vertices of an ellipse of eccentricity 0.8 whose foci lie at the points $(0, \pm 7)$.

Solution   The vertices are the points $(0, \pm a)$, where

$$\frac{c}{a} = \frac{7}{a} = e = 0.8.$$

Therefore,

$$a = \frac{7}{0.8} = 8.75$$

and the vertices are the points $(0, \pm 8.75)$.

Example 9   Find the eccentricity of the hyperbola $9x^2 - 16y^2 = 144$.

Solution   We divide both sides of the hyperbola's equation by 144 to put it in standard form, obtaining

$$\frac{9x^2}{144} - \frac{16y^2}{144} = 1 \qquad \text{and thus} \qquad \frac{x^2}{16} - \frac{y^2}{9} = 1.$$

With $a^2 = 16$ and $b^2 = 9$, we find that

$$c = \sqrt{a^2 + b^2} = \sqrt{16 + 9} = 5,$$

so

$$e = \frac{c}{a} = \frac{5}{4}.$$

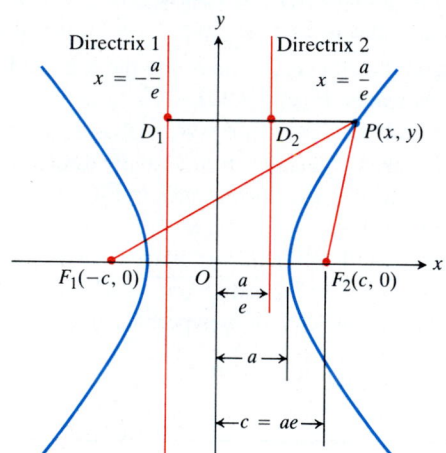

9.13   The foci and directrices of the hyperbola $(x^2/a^2) - (y^2/b^2) = 1$. No matter where P lies on the hyperbola, $PF_1 = e \cdot PD_1$ and $PF_2 = e \cdot PD_2$.

As with the ellipse, it can be shown that the lines $x = \pm a/e$ act as directrices for the hyperbola and that

$$PF_1 = e \cdot PD_1 \qquad \text{and} \qquad PF_2 = e \cdot PD_2. \tag{23}$$

Here P is any point on the hyperbola, F_1 and F_2 are the foci, and D_1 and D_2 are the points nearest P on the directrices (Fig. 9.13).

To complete the picture, we now define the eccentricity of a parabola to be $e = 1$. Equations (19), (20), and (23) then have the common form $PF = e \cdot PD$.

DEFINITION

The **eccentricity** of a parabola is $e = 1$.

The "focus–directrix" equation $PF = e \cdot PD$ unites the parabola, ellipse, and hyperbola in the following way. Suppose that a point P moves in a plane in such a way that its distance PF from a fixed point F in the plane (the focus) is a constant multiple of its distance PD from a fixed line in the plane (the directrix). That is, suppose that

$$PF = e \cdot PD, \tag{24}$$

where e is the constant of proportionality. Then the path traced by P is

a) a *parabola* if $e = 1$,

b) an *ellipse* of eccentricity e if $e < 1$, and

c) a *hyperbola* of eccentricity e if $e > 1$.

Equation (24) may not look like much to get excited about. There are no coordinates in it and when we try to translate it into coordinate form it translates in different ways, depending on the numerical size of e. At least, that is what happens in the Cartesian plane. However, in the polar coordinate plane, as we shall see in Section 9.7, the equation $PF = e \cdot PD$ translates into a single equation regardless of the size of e, an equation so simple that it has been the equation of choice for astronomers and space scientists for nearly three hundred years.

Given the focus and corresponding directrix of a hyperbola centered at the origin and with foci on the x-axis, we can use the dimensions shown in Fig. 9.13 to find e. Knowing e, we can derive a Cartesian equation for the hyperbola from the equation $PF = e \cdot PD$, as in the next example. We can find equations for ellipses centered at the origin and with foci on the x-axis in a similar way, using the dimensions shown in Fig. 9.12.

Example 10   Find the standard-form equation for the hyperbola centered at the origin that has a focus at $(3, 0)$ and has the line $x = 1$ as the corresponding directrix.

Solution   We first use the dimensions shown in Fig. 9.13 to find the hyperbola's eccentricity. The focus is

$$(c, 0) = (3, 0), \qquad \text{so} \qquad c = 3.$$

The directrix is the line

$$x = \frac{a}{e} = 1, \qquad \text{so} \qquad a = e.$$

When combined with the equation $e = c/a$ that defines eccentricity, these results give

$$e = \frac{c}{a} = \frac{3}{e}, \qquad \text{so} \qquad e^2 = 3 \quad \text{and} \quad e = \sqrt{3}.$$

Knowing e, we can now derive the equation we want from the equation $PF = e \cdot PD$. In the notation of Fig. 9.14, we have

$$PF = e \cdot PD \qquad \text{(Eq. (24))}$$
$$\sqrt{(x - 3)^2 + (y - 0)^2} = \sqrt{3}\,|x - 1| \qquad (e = \sqrt{3})$$
$$x^2 - 6x + 9 + y^2 = 3(x^2 - 2x + 1)$$
$$2x^2 - y^2 = 6$$
$$\frac{x^2}{3} - \frac{y^2}{6} = 1.$$

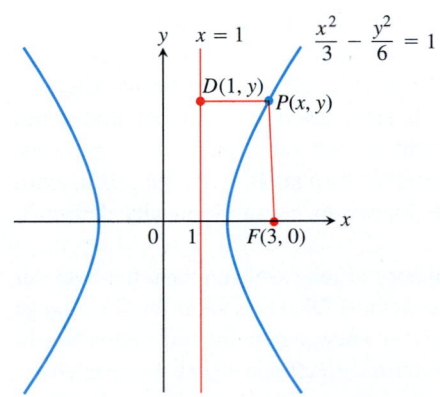

9.14   The hyperbola in Example 10.

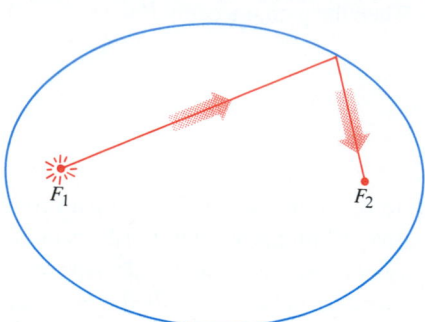

9.15   An elliptical mirror (shown here in profile) reflects light from one focus to the other.

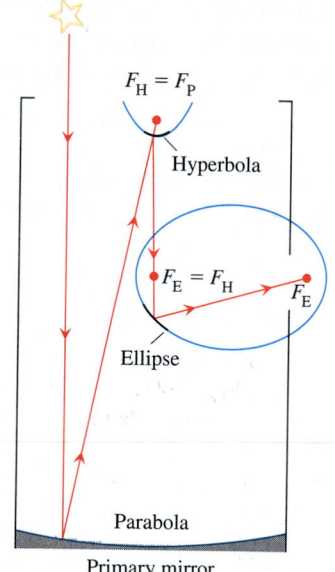

9.17   In this schematic drawing of a reflecting telescope, starlight reflects off a primary parabolic mirror toward the mirror's focus F_P. It is then reflected by a small hyperbolic mirror, whose focus is $F_H = F_P$, toward the second focus of the hyperbola, $F_E = F_H$. Since this focus is shared by an ellipse, the light is reflected by the elliptical mirror to the ellipse's second focus to be seen by an observer.

As recent experience with NASA's Hubble space telescope shows, the mirrors have to be nearly perfect to focus properly. The aberration causing the malfunction in Hubble's primary mirror amounts to about half a wavelength of visible light, no more than 1/50 the width of a human hair.

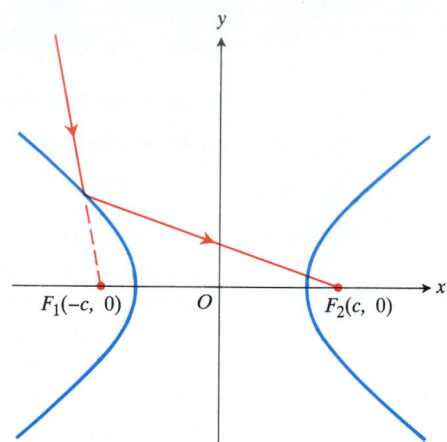

9.16   In this profile of a hyperbolic mirror, light coming toward focus F_1 is reflected toward focus F_2.

Reflective Properties

Like parabolas, ellipses and hyperbolas have reflective properties that are important in science and engineering. If an ellipse is revolved about its major axis to generate a surface (the surface is called an *ellipsoid*), and the interior is silvered to produce a mirror, light from one focus will be reflected to the other focus (Fig. 9.15). Ellipsoids reflect sound the same way, and this property is used to construct *whispering galleries,* rooms in which a person standing at one focus can hear a whisper from the other focus. Statuary Hall in the U.S. Capitol building is a whispering gallery. Ellipsoids also appear in instruments used to study aircraft noise in wind tunnels (sound at one focus can be received at the other focus with relatively little interference from other sources).

Light directed toward one focus of a hyperbolic mirror is reflected toward the other focus (Fig. 9.16). This property of hyperbolas is combined with the reflective properties of parabolas and ellipses in designing modern telescopes (Fig. 9.17).

Other Applications

Ellipses appear in airplane wings (British Spitfire) and sometimes in gears designed for racing bicycles. Stereo systems often have elliptical styli, and water pipes are sometimes designed with elliptical cross sections to allow for expansion when the water freezes. The triggering mechanisms in some lasers are elliptical, and stones on a beach become more and more elliptical as they are ground down by waves. There are also applications of ellipses to fossil formation. The ellipsolith, once thought to be a separate species, is now known to be an elliptically deformed nautilus.

Hyperbolic paths arise in Einstein's theory of relativity and form the basis for the (unrelated) LORAN radio navigation system. (LORAN is short for "long range navigation.") Hyperbolas also form the basis for a new system the Burlington Northern Railroad is developing for using synchronized electronic signals from satellites to track freight trains. A few years ago, computers aboard Burlington Northern locomotives in Minnesota were able to track trains to within one mile per hour of their speed and to within 150 feet of their actual location.

EXERCISES 9.1

Match the parabolas in Exercises 1–4 with the following equations:

$$x^2 = 2y, \quad x^2 = -6y, \quad y^2 = 8x, \quad y^2 = -4x.$$

Then find the parabola's focus and directrix.

1.
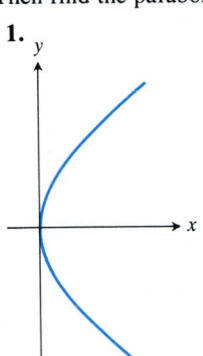

2.
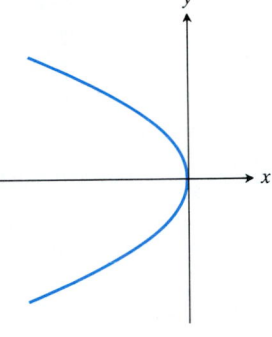

3.
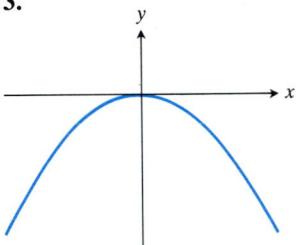

4.
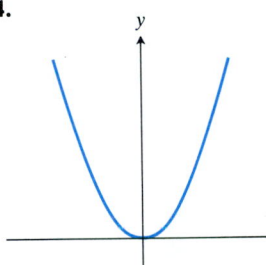

Match the conic sections in Exercises 5–8 with the following equations:

$$\frac{x^2}{4} + \frac{y^2}{9} = 1, \quad \frac{x^2}{2} + y^2 = 1, \quad \frac{y^2}{4} - x^2 = 1, \quad \frac{x^2}{4} - \frac{y^2}{9} = 1.$$

Then find the conic section's foci, eccentricity, and directrices. If the conic section is a hyperbola, find its asymptotes as well.

5.
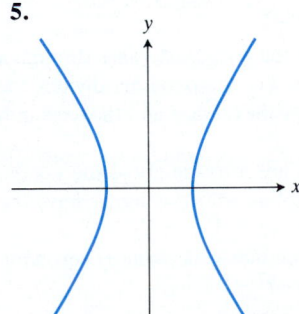

6.
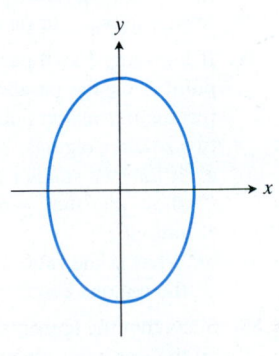

7.
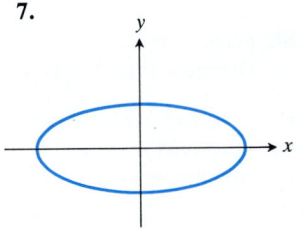

8.
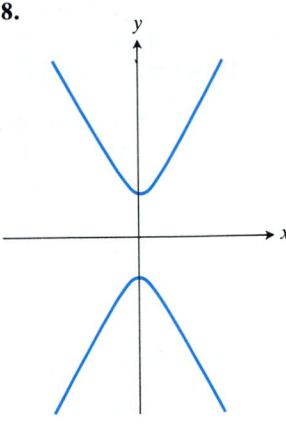

Exercises 9–16 give equations for ellipses. Put each equation in standard form and find the ellipse's eccentricity. Then sketch the ellipse. Include the foci in your sketch.

9. $16x^2 + 25y^2 = 400$

10. $7x^2 + 16y^2 = 112$

11. $2x^2 + y^2 = 2$

12. $2x^2 + y^2 = 4$

13. $3x^2 + 2y^2 = 6$

14. $9x^2 + 10y^2 = 90$

15. $6x^2 + 9y^2 = 54$

16. $169x^2 + 25y^2 = 4225$

Exercises 17–24 give equations for hyperbolas. Put each equation in standard form and find the hyperbola's eccentricity and asymptotes. Then sketch the hyperbola. Include the asymptotes and foci in your sketch.

17. $x^2 - y^2 = 1$

18. $9x^2 - 16y^2 = 144$

19. $y^2 - x^2 = 8$

20. $y^2 - x^2 = 4$

21. $8x^2 - 2y^2 = 16$

22. $y^2 - 3x^2 = 3$

23. $8y^2 - 2x^2 = 16$

24. $64x^2 - 36y^2 = 2304$

Exercises 25–30 give information about the foci, vertices, and eccentricity of ellipses centered at the origin of the xy-plane. In each case, find the ellipse's standard-form equation from the information given.

25. Foci: $(\pm\sqrt{2}, 0)$
Vertices: $(\pm 2, 0)$

26. Foci: $(0, \pm 4)$
Vertices: $(0, \pm 5)$

27. Foci: $(0, \pm 3)$
Eccentricity: 0.5

28. Foci: $(\pm 8, 0)$
Eccentricity: 0.2

29. Vertices: $(0, \pm 70)$
Eccentricity: 0.1

30. Vertices: $(\pm 10, 0)$
Eccentricity: 0.24

Exercises 31–34 give information about the foci and corresponding directrices of ellipses centered at the origin of the xy-plane. In each case, use the dimensions in Fig. 9.12 to find the eccentricity of the ellipse. Then use the equation $PF = e \cdot PD$ to find the ellipse's standard-form equation.

31. Focus: $(\sqrt{5}, 0)$
Directrix: $x = \dfrac{9}{\sqrt{5}}$

32. Focus: $(4, 0)$
Directrix: $x = \dfrac{16}{3}$

33. Focus: $(-4, 0)$
Directrix: $x = -16$

34. Focus: $(-\sqrt{2}, 0)$
Directrix: $x = -2\sqrt{2}$

Exercises 35–42 give information about the foci, vertices, eccentricities, and asymptotes of hyperbolas centered at the origin of the xy-plane. In each case, find the hyperbola's standard-form equation from the information given.

35. Foci: $(0, \pm\sqrt{2})$
Asymptotes: $y = \pm x$

36. Foci: $(\pm 2, 0)$
Asymptotes: $y = \pm\dfrac{1}{\sqrt{3}}x$

37. Vertices: $(\pm 3, 0)$
Asymptotes: $y = \pm\dfrac{4}{3}x$

38. Vertices: $(0, \pm 2)$
Asymptotes: $y = \pm\dfrac{1}{2}x$

39. Vertices: $(0, \pm 1)$
Eccentricity: 3

40. Vertices: $(\pm 2, 0)$
Eccentricity: 2

41. Foci: $(\pm 3, 0)$
Eccentricity: 3

42. Foci: $(0, \pm 5)$
Eccentricity: 1.25

Exercises 43–46 give information about the foci and corresponding directrices of hyperbolas centered at the origin of the xy-plane. In each case, use the dimensions in Fig. 9.13 to find the eccentricity of the hyperbola. Then use the equation $PF = e \cdot PD$ to find the hyperbola's standard-form equation.

43. Focus: $(4, 0)$
Directrix: $x = 2$

44. Focus: $(\sqrt{10}, 0)$
Directrix: $x = \sqrt{2}$

45. Focus: $(-2, 0)$
Directrix: $x = -\dfrac{1}{2}$

46. Focus: $(-6, 0)$
Directrix: $x = -2$

Sketch the regions whose points satisfy the inequalities or sets of inequalities in Exercises 47–52.

47. $9x^2 + 16y^2 \le 144$

48. $x^2 + y^2 \ge 1$ and $4x^2 + y^2 \le 4$

49. $x^2 + 4y^2 \ge 4$ and $4x^2 + 9y^2 \le 36$

50. $(x^2 + y^2 - 4)(x^2 + 9y^2 - 9) \le 0$

51. $4y^2 - x^2 \ge 4$

52. $|x^2 - y^2| \le 1$

53. *Archimedes' formula for the volume of a parabolic solid.* The region enclosed by the parabola $y = (4h/b^2)x^2$ and the line $y = h$ is revolved about the y-axis to generate a solid (Fig. 9.18). Show that the volume of the solid is 3/2 the volume of the corresponding cone.

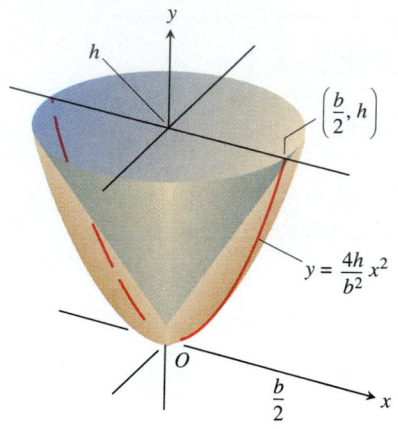

9.18   The cone and parabolic solid in Exercise 53.

54. *Suspension bridge cables hang in parabolas.*   Figure 9.19 shows a cable of a suspension bridge supporting a uniform load of w pounds per horizontal foot. It can be shown that if H is the horizontal tension in the cable at the origin, then the curve of the cable satisfies the differential equation

$$\frac{dy}{dx} = \frac{w}{H}x.$$

Show that the cable hangs in a parabola by solving this equation with the condition that $y = 0$ when $x = 0$.

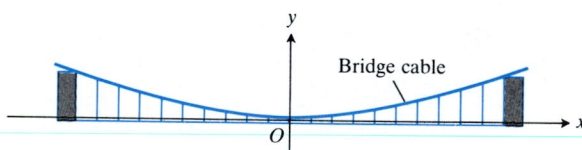

9.19   The suspension bridge cable in Exercise 54.

55. Find equations for the lines that are tangent to the circle $(x - 2)^2 + (y - 1)^2 = 5$ at the points where the circle crosses the coordinate axes. (*Hint:* Use implicit differentiation.)

56. Find an equation for the circle centered at $(-2, 1)$ that passes through the point $(1, 3)$. Is the point $(1.1, 2.8)$ inside, outside, or on the circle?

57. If lines are drawn parallel to the coordinate axes through a point P on the parabola $y^2 = kx$, the parabola divides the rectangular region bounded by these lines and the axes into two smaller regions.
   a) If the two smaller regions are revolved about the y-axis, show that they generate solids whose volumes have the ratio 4:1.
   b) What is the ratio of the volumes generated by revolving the regions about the x-axis?

58. Show that the tangents to the curve $y^2 = 4px$ from any point on the line $x = -p$ are perpendicular.

59. Find the dimensions of the rectangle of largest area that can be inscribed in the ellipse $x^2 + 4y^2 = 4$ with its sides parallel to the coordinate axes. What is the area of the rectangle?

60. Find the center of mass of a thin homogeneous plate that is bounded below by the x-axis and above by the ellipse $(x^2/9) + (y^2/16) = 1$.

61. Find the volume of the solid generated by revolving the region enclosed by the ellipse $9x^2 + 4y^2 = 36$ about the (a) x-axis, (b) y-axis.

62. The "triangular" region in the first quadrant bounded by the x-axis, the line $x = 4$, and the hyperbola $9x^2 - 4y^2 = 36$ is revolved about the x-axis to generate a solid. Find the volume of the solid.

63. The region bounded on the left by the y-axis, on the right by the hyperbola $x^2 - y^2 = 1$, and above and below by the lines $y = \pm 3$ is revolved about the y-axis to generate a solid. Find the volume of the solid.

64. The curve $y = \sqrt{x^2 + 1}$, $0 \le x \le \sqrt{2}$, which is part of the upper branch of the hyperbola $y^2 - x^2 = 1$, is revolved about the x-axis to generate a surface. Find the area of the surface.

65. The circular waves in Fig. 9.20 were made by touching the surface of the water in a ripple tank, first at A and then at B. As the circular waves expanded, their point of intersection seemed to trace a hyperbola. Did it really do that?

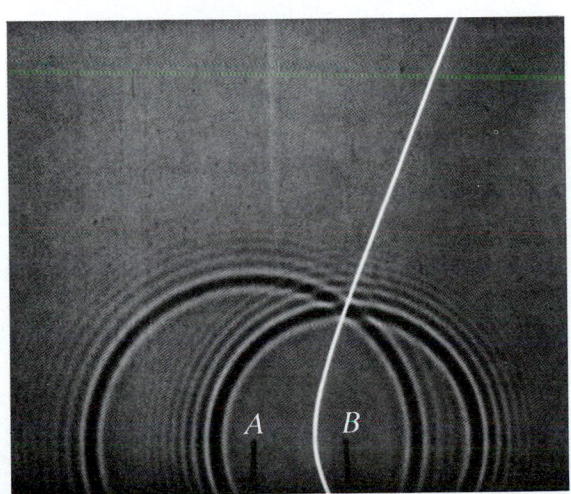

9.20 The expanding circles of ripples in Exercise 65. Photograph from *PSSC Physics,* Second Edition, 1965, D. C. Heath & Company with Education Development Center, Inc. *NCFMF Book of Film Notes,* 1974, The MIT Press with Education Development Center, Inc., Newton, Massachusetts.

To find out, we can model the waves with circles centered at A and B (Fig. 9.21). At time t, the point P is $r_A(t)$ units

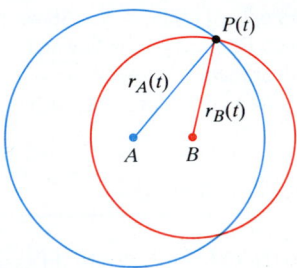

9.21 The model for the waves in Exercise 65.

from A and $r_B(t)$ units from B. Since the radii of the circles increase at a constant rate, the rate at which the waves are traveling is

$$\frac{dr_A}{dt} = \frac{dr_B}{dt}.$$

Conclude from this equation that $r_A - r_B$ has a constant value, so that P must lie on a hyperbola with foci at A and B.

66. *The reflective property of parabolas.* Figure 9.22 shows a typical point $P(x_0, y_0)$ on the parabola $y^2 = 4px$. The line L is tangent to the parabola at P. The parabola's focus lies at $F(p, 0)$. The ray L' extending from P to the right is parallel to the x-axis. We show that light from F to P will be reflected out along L' by showing that β equals α. Establish this equality by taking the following steps.

1. Show that $\tan \beta = 2p/y_0$.
2. Show that $\tan \phi = y_0/(x_0 - p)$.
3. Use the identity

$$\tan \alpha = \frac{\tan \phi - \tan \beta}{1 + \tan \phi \tan \beta}$$

to show that $\tan \alpha = 2p/y_0$.

Since the angles involved are both acute, $\tan \beta = \tan \alpha$ implies $\beta = \alpha$.

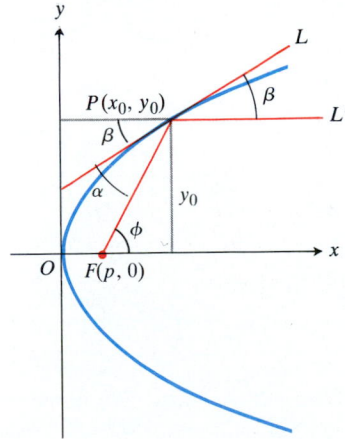

9.22 A parabolic reflector sends all light from the focus out parallel to the parabola's axis (Exercise 66).

67. *The asymptotes of $(x^2/a^2) - (y^2/b^2) = 1$.* Show that the vertical distance between the line $y = (b/a)x$ and the upper half of the right-hand branch $y = (b/a)\sqrt{x^2 - a^2}$ of the hyperbola $(x^2/a^2) - (y^2/b^2) = 1$ approaches 0 by showing that

$$\lim_{x \to \infty}\left(\frac{b}{a}x - \frac{b}{a}\sqrt{x^2 - a^2}\right) = \frac{b}{a}\lim_{x \to \infty}\left(x - \sqrt{x^2 - a^2}\right) = 0.$$

Similar results hold for the remaining portions of the hyperbola and the lines $y = \pm(b/a)x$.

9.2 The Graphs of Quadratic Equations in x and y; Rotations About the Origin

In this section, we establish one of the most amazing results in analytic geometry, which is that the Cartesian graph of any equation of the form

$$Ax^2 + Bxy + Cy^2 + Dx + Ey + F = 0, \tag{1}$$

in which A, B, and C are not all zero, is nearly always a conic section. The only exceptions are the case in which the graph consists of two parallel lines and the case in which there is no graph at all (Table 9.3). It is conventional to call all graphs of Eq. (1), curved or not, **quadratic curves.**

The Cross-Product Term

You may have noticed that the term Bxy did not appear in the equations for the conic sections in Section 9.1. This happened because the axes of the conic sections ran parallel to (in fact, coincided with) the coordinate axes.

To see what happens when the parallelism is absent, let us write an equation for a hyperbola with $a = 3$ and foci at $F_1(-3, -3)$ and $F_2(3, 3)$ (Fig. 9.23). The equa-

TABLE 9.3
Examples of quadratic curves

$Ax^2 + Bxy + Cy^2 + Dx + Ey + F = 0$							
	A	B	C	D	E	F	Equation	Remarks
Circle	1		1			-4	$x^2 + y^2 = 4$	$A = C$
Parabola			1	-9			$y^2 = 9x$	Quadratic in y, linear in x
Ellipse	4		9			-36	$4x^2 + 9y^2 = 36$	A, C have same sign, $A \neq C$
Hyperbola	1		-1			-1	$x^2 - y^2 = 1$	A, C have opposite signs
One line (still a conic section)	1						$x^2 = 0$	y-axis
Intersecting lines (still a conic section)		1		1	-1	-1	$xy + x - y - 1 = 0$	Factors to $(x - 1)(y + 1) = 0$, so $x = 1$, $y = -1$
Parallel lines (not a conic section)	1			-3		2	$x^2 - 3x + 2 = 0$	Factors to $(x - 1)(x - 2) = 0$, so $x = 1$, $x = 2$
Point	1		1				$x^2 + y^2 = 0$	The origin
No graph	1					1	$x^2 = -1$	No graph

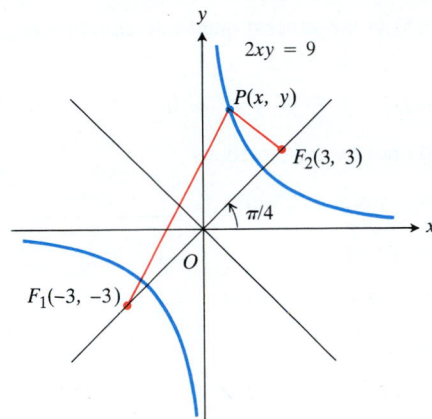

9.23 The focal axis of the hyperbola $2xy = 9$ makes an angle of $\pi/4$ radians with the positive x-axis.

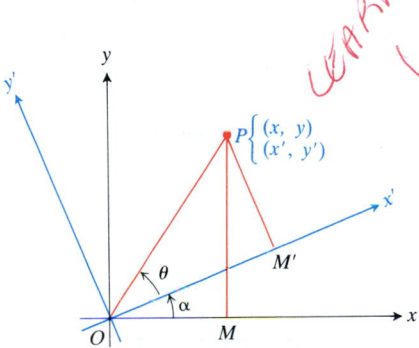

9.24 A counterclockwise rotation through angle α about the origin.

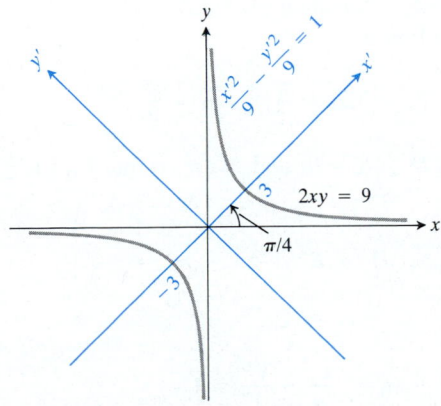

9.25 The hyperbola in Example 1.

tion $|PF_1 - PF_2| = 2a$ then becomes $|PF_1 - PF_2| = 2(3) = 6$ and

$$\sqrt{(x + 3)^2 + (y + 3)^2} - \sqrt{(x - 3)^2 + (y - 3)^2} = \pm 6.$$

When we transpose one radical, square, solve for the radical that still appears, and square again, this reduces to

$$2xy = 9, \tag{2}$$

which is a special case of Eq. (1) in which the cross-product term is present. The asymptotes of the hyperbola in Eq. (2) are the x- and y-axes, and the focal axis makes an angle of $\pi/4$ radians with the positive x-axis. As in this example, the cross-product term is present in Eq. (1) only when the axes of the conic are tilted.

Rotating the Coordinate Axes to Eliminate the Cross-Product Term

To eliminate the xy-term from the equation of a conic, we rotate the coordinate axes to eliminate the "tilt" in the axes of the conic. The equations for the rotations we use are derived in the following way. In the notation of Fig. 9.24, which shows a counterclockwise rotation about the origin through an angle α,

$$x = OM = OP \cos(\theta + \alpha) = OP \cos\theta \cos\alpha - OP \sin\theta \sin\alpha,$$
$$y = MP = OP \sin(\theta + \alpha) = OP \cos\theta \sin\alpha + OP \sin\theta \cos\alpha. \tag{3}$$

Since

$$OP \cos\theta = OM' = x' \qquad \text{and} \qquad OP \sin\theta = M'P = y', \tag{4}$$

Eqs. (3) reduce to the following.

> **Equations for Rotating the Coordinate Axes**
>
> $$x = x' \cos\alpha - y' \sin\alpha,$$
> $$y = x' \sin\alpha + y' \cos\alpha \tag{5}$$

Example 1   The x- and y-axes are rotated through an angle of $\pi/4$ radians about the origin. Find an equation for the hyperbola $2xy = 9$ in the new coordinates.

Solution   Since $\cos \pi/4 = \sin \pi/4 = 1/\sqrt{2}$, we substitute

$$x = \frac{x' - y'}{\sqrt{2}}, \qquad y = \frac{x' + y'}{\sqrt{2}}$$

from Eqs. (5) into the equation $2xy = 9$ and obtain

$$2\left(\frac{x' - y'}{\sqrt{2}}\right)\left(\frac{x' + y'}{\sqrt{2}}\right) = 9$$

$$x'^2 - y'^2 = 9$$

$$\frac{x'^2}{9} - \frac{y'^2}{9} = 1.$$

See Fig. 9.25.

If we apply the rotation equations in (5) to the general quadratic equation (1), we obtain a new quadratic equation

$$A'x'^2 + B'x'y' + C'y'^2 + D'x' + E'y' + F' = 0. \tag{6}$$

The new coefficients are related to the old ones by the equations

$$A' = A\cos^2\alpha + B\cos\alpha\sin\alpha + C\sin^2\alpha,$$
$$B' = B\cos 2\alpha + (C - A)\sin 2\alpha,$$
$$C' = A\sin^2\alpha - B\sin\alpha\cos\alpha + C\cos^2\alpha,$$
$$D' = D\cos\alpha + E\sin\alpha, \tag{7}$$
$$E' = -D\sin\alpha + E\cos\alpha,$$
$$F' = F.$$

These equations show, among other things, that if we start with a quadratic equation for a curve in which the cross-product term is present ($B \neq 0$), we can find a rotation angle α that produces a quadratic equation in which no cross-product term appears ($B' = 0$). To find α, we put $B' = 0$ in the second equation in (7) and solve the resulting equation,

$$B\cos 2\alpha + (C - A)\sin 2\alpha = 0,$$

for α. In practice, this means finding α from one of the two equations

$$\cot 2\alpha = \frac{A - C}{B} \qquad \text{or} \qquad \tan 2\alpha = \frac{B}{A - C}. \tag{8}$$

Example 2   The coordinate axes are to be rotated through an angle α to produce an equation for the curve

$$x^2 + xy + y^2 - 6 = 0 \tag{9}$$

that has no cross-product term. Find a suitable value for α and the corresponding new equation.

Solution 1   *Using Eqs. (8), (7), and (6).*   Equation (9) has $A = B = C = 1$. We substitute these values into Eq. (8) to find α:

$$\cot 2\alpha = \frac{A - C}{B} = \frac{1 - 1}{1} = 0, \qquad 2\alpha = \frac{\pi}{2}, \qquad \alpha = \frac{\pi}{4}.$$

Substituting $\alpha = \pi/4$, $A = B = C = 1$, $D = E = 0$, and $F = -6$ into Eqs. (7) gives

$$A' = \frac{3}{2}, \qquad B' = 0, \qquad C' = \frac{1}{2}, \qquad D' = E' = 0, \qquad F' = -6.$$

Equation (6) then gives

$$\frac{3}{2}x'^2 + \frac{1}{2}y'^2 - 6 = 0, \qquad \text{or} \qquad \frac{x'^2}{4} + \frac{y'^2}{12} = 1.$$

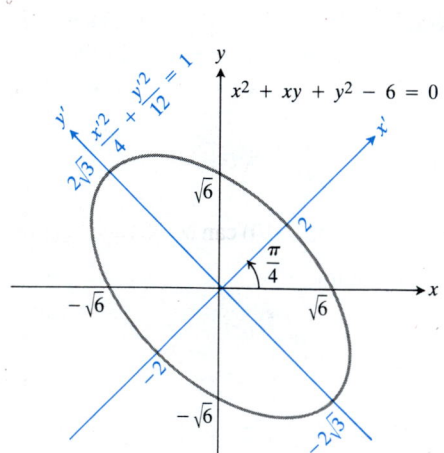

9.26   The ellipse in Example 2.

This is the equation of an ellipse with foci on the new y'-axis (Fig. 9.26).

Solution 2   *Using the Rotation Equations Directly.*   This method brings in fewer formulas but requires more arithmetic. We begin by finding α as in Solution 1 and substitute the value found, $\alpha = \pi/4$, into the rotation equations to get

$$x = x' \cos \alpha - y' \sin \alpha = \frac{\sqrt{2}}{2} x' - \frac{\sqrt{2}}{2} y',$$

$$y = x' \sin \alpha + y' \cos \alpha = \frac{\sqrt{2}}{2} x' + \frac{\sqrt{2}}{2} y'.$$

We then substitute the primed expressions for x and y in the original equation, $x^2 + xy + y^2 - 6 = 0$. This gives

$$\left(\frac{\sqrt{2}}{2} x' - \frac{\sqrt{2}}{2} y'\right)^2 + \left(\frac{\sqrt{2}}{2} x' - \frac{\sqrt{2}}{2} y'\right)\left(\frac{\sqrt{2}}{2} x' + \frac{\sqrt{2}}{2} y'\right)$$

$$+ \left(\frac{\sqrt{2}}{2} x' + \frac{\sqrt{2}}{2} y'\right)^2 - 6 = 0,$$

or, skipping over the arithmetic,

$$\frac{x'^2}{4} + \frac{y'^2}{12} = 1.$$

The Graphs of Quadratic Equations

We now return to our analysis of the graph of the general quadratic equation.

Since axes may always be rotated to eliminate the cross-product term, there is no loss of generality in assuming that this has been done, and our equation has the form

$$Ax^2 + Cy^2 + Dx + Ey + F = 0. \tag{10}$$

Equation (10) represents

a) a *circle* if $A = C \neq 0$ (special cases: the graph is a point or there is no graph at all);

b) a *parabola* if Eq. (10) is quadratic in one variable and linear in the other;

c) an *ellipse* if A and C are both positive or both negative (special cases: a single point or no graph at all);

d) a *hyperbola* if A and C have opposite signs (special case: a pair of intersecting lines);

e) a *straight line* if A and C are zero and at least one of D and E is different from zero;

f) *one or two straight lines* if the left-hand side of Eq. (10) can be factored into the product of two linear factors.

If $A = C \neq 0$, for example, we can complete the squares on the x-terms and y-terms to rewrite Eq. (10) in the form

$$A(x - h)^2 + A(y - k)^2 = F'' \tag{11}$$

and divide by A to obtain the equivalent equation

$$(x - h)^2 + (y - k)^2 = F''/A. \tag{12}$$

The form of Eq. (12) reveals the graph to be a circle of radius $\sqrt{F''/A}$ centered at (h, k). The exceptions (a point, no graph at all) arise when $F'' = 0$ or $F''/A < 0$.

Similarly, if A and C have opposite signs, completing the squares gives an equivalent equation of the form

$$\pm |A|(x - h)^2 \mp |C|(y - k)^2 = F''. \tag{13}$$

If $F'' \neq 0$, we may divide through by F'' and put the equation in the form

$$\pm \frac{(x - h)^2}{a^2} \mp \frac{(y - k)^2}{b^2} = 1. \tag{14}$$

The graph is a hyperbola centered at (h, k), i.e., a hyperbola that has been shifted to place its center at the point (h, k). The exception (two intersecting lines) occurs when $F'' = 0$, for then Eq. (13) reduces to

$$y = \pm \sqrt{\frac{|A|}{|C|}} (x - h) + k. \tag{15}$$

The Discriminant

There is a quick way to tell whether the graph of the equation

$$Ax^2 + Bxy + Cy^2 + Dx + Ey + F = 0 \tag{16}$$

is a parabola, an ellipse, or a hyperbola. The test does not require us to eliminate the xy-term first.

As we have seen, if B is not zero, then rotating the axes through the angle α determined by the equation

$$\cot 2\alpha = \frac{A - C}{B} \tag{17}$$

will change Eq. (16) into the equivalent form

$$A'x'^2 + C'y'^2 + D'x' + E'y' + F' = 0 \tag{18}$$

without a cross-product term.

Now, the graph of Eq. (18) is a (real or degenerate)

a) *parabola* if A' or $C' = 0$, that is, if $A'C' = 0$;

b) *ellipse* if A' and C' have the same sign, that is, if $A'C' > 0$;

c) *hyperbola* if A' and C' have opposite signs, that is, if $A'C' < 0$.

It can also be verified, by using Eqs. (7), that for any rotation of axes,

$$B^2 - 4AC = B'^2 - 4A'C'. \tag{19}$$

This means that the quantity $B^2 - 4AC$ is not changed by a rotation. But when we rotate through the angle α given by Eq. (17), B' becomes zero, so that

$$B^2 - 4AC = -4A'C'$$

(Exercise 46). Since the curve is a parabola if $A'C' = 0$, an ellipse if $A'C' > 0$, and a hyperbola if $A'C' < 0$, the curve must be

a) a *parabola* if $B^2 - 4AC = 0$,

b) an *ellipse* if $B^2 - 4AC < 0$, $\qquad\qquad$ (20)

c) a *hyperbola* if $B^2 - 4AC > 0$.

The number $B^2 - 4AC$ is called the **discriminant** of Eq. (16). What we have just seen is that the graph of Eq. (16) is a parabola if the discriminant is zero, an ellipse if the discriminant is negative, and a hyperbola if the discriminant is positive (with the understanding that occasional degenerate cases may arise).

Example 3

a) $3x^2 - 6xy + 3y^2 + 2x - 7 = 0$ represents a parabola because

$$B^2 - 4AC = (-6)^2 - 4 \cdot 3 \cdot 3 = 36 - 36 = 0.$$

b) $x^2 + xy + y^2 - 1 = 0$ represents an ellipse because

$$B^2 - 4AC = (1)^2 - 4 \cdot 1 \cdot 1 = -3 < 0.$$

c) $xy - y^2 - 5y + 1 = 0$ represents a hyperbola because

$$B^2 - 4AC = (1)^2 - 4(0)(-1) = 1 > 0.$$

✳ How Calculators Use Rotations to Evaluate Sines and Cosines

Some calculators use rotations to calculate sines and cosines of arbitrary angles. The procedure goes something like this: The calculator has, stored,

1. ten angles or so, say

$$\alpha_1 = \sin^{-1}(10^{-1}), \quad \alpha_2 = \sin^{-1}(10^{-2}), \quad \ldots, \quad \alpha_{10} = \sin^{-1}(10^{-10}),$$

and

2. twenty numbers, the sines and cosines of the angles $\alpha_1, \alpha_2, \ldots, \alpha_{10}$.

To calculate the sine and cosine of an arbitrary angle θ, we enter θ (in radians) into the calculator. The calculator subtracts or adds multiples of 2π to θ to replace θ by the angle between 0 and 2π that has the same sine and cosine as θ (we shall continue to call the angle θ). The calculator then "writes" θ as a sum of multiples of α_1 (as many as possible, without overshooting) plus multiples of α_2 (again, as many as possible), and so on, working its way to α_{10}. This gives

$$\theta \approx m_1\alpha_1 + m_2\alpha_2 + \cdots + m_{10}\alpha_{10}.$$

The calculator then rotates the point $(1, 0)$ through m_1 copies of α_1 (through α_1, m_1 times in succession), plus m_2 copies of α_2, and so on, finishing off with m_{10} copies of α_{10} (Fig. 9.27). The coordinates of the final position of $(1, 0)$ on the unit circle are the values the calculator gives for $(\cos \theta, \sin \theta)$.

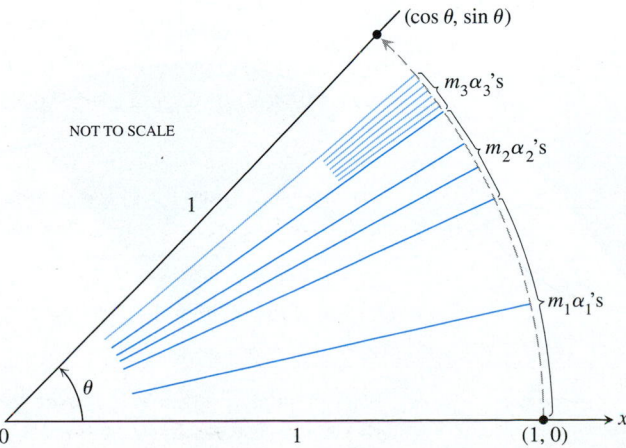

9.27 To calculate the sine and cosine of an angle θ between 0 and 2π, the calculator rotates the point $(1, 0)$ to an appropriate location on the unit circle and displays the resulting coordinates.

✷ Sections of a Cone

We have defined ellipses, parabolas, and hyperbolas in the coordinate plane algebraically, with coordinate equations derived from the distance formula. We have also seen that plane curves whose points P satisfy equations of the form $PF = e \cdot PD$ for suitably chosen points F and D in the coordinate plane are "algebraic" parabolas, ellipses, and hyperbolas in this sense. In this section, we show that every nondegenerate curve obtained by slicing a double right circular cone with a plane that is not perpendicular to the axis of the cone satisfies an equation of the form $PF = e \cdot PD$ for suitably chosen points F and P in the cutting plane. Thus, every geometric ellipse, parabola, or hyperbola is also an algebraic ellipse, parabola, or hyperbola in its own plane.

As you know, circles do not satisfy a focus–directrix equation, but geometric circles are algebraic circles nonetheless. If we slice a right circular cone with a plane perpendicular to the cone's axis at a point O different from the cone's vertex, the points on the circle of intersection have a constant distance r from O. If we then coordinatize the plane with the origin at O, the circle is the graph of the equation $x^2 + y^2 = r^2$. Thus, every nondegenerate geometric conic section is an algebraic conic section in the coordinates of the plane that cuts the section from the cone.

The Basic Argument   Suppose that the cutting plane makes an acute angle α with the axis of the cone and that the acute angle between the side and axis of the cone is β (Fig. 9.28). Then the section is

a) an ellipse if $\beta < \alpha < 90°$;

b) a parabola if $\alpha = \beta$;

c) a hyperbola if $0 \leq \alpha < \beta$.

9.28   The intersection of a plane and a right circular double cone in (a) an ellipse, (b) a parabola, (c) a hyperbola. The angle β is the angle between the side and axis of the cone. The angle α is the acute angle between the plane and the cone's axis.

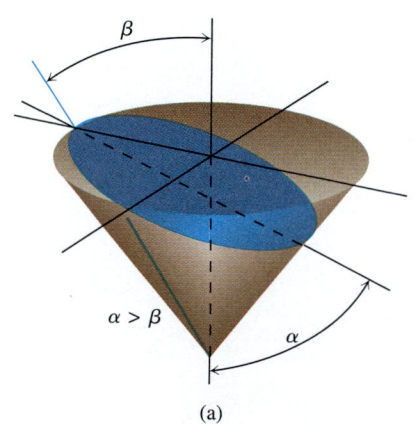

(a)

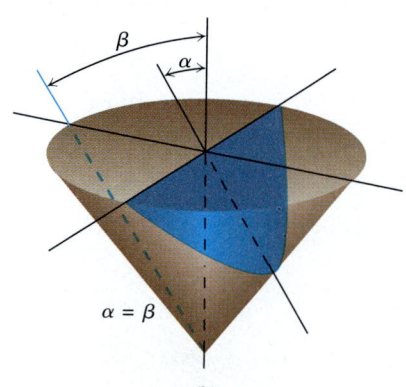

(b)

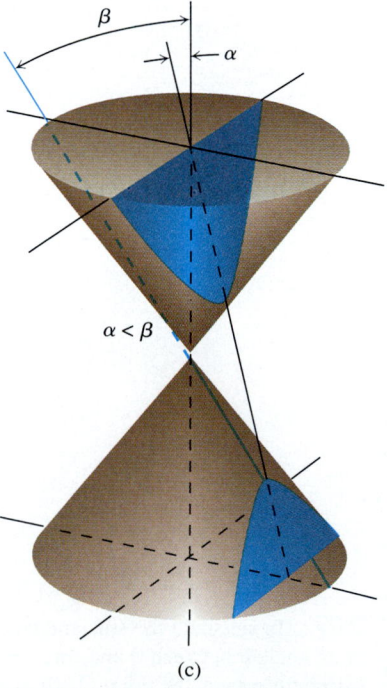

(c)

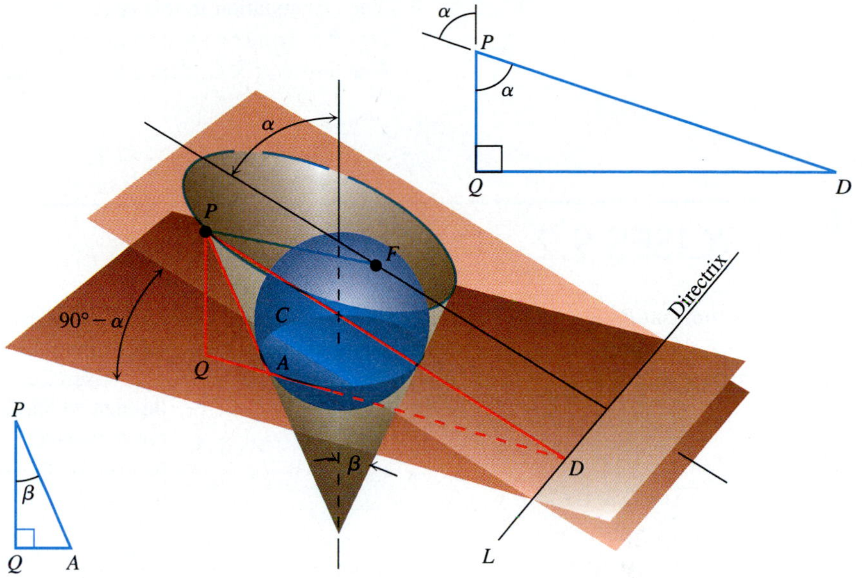

9.29   Every nondegenerate geometric ellipse satisfies a focus–directrix equation in the cutting plane. Line L is the directrix that corresponds to the focus F.

The construction we are about to make assumes we are working with an ellipse, but the argument works for the other two cases as well.

Inscribe a sphere that is tangent to the cone along a circle C and tangent to the cutting plane at a point F (Fig. 9.29). Let P be any point on the conic section. We shall see that F is a focus of the conic section and that the corresponding directrix is the line L in which the cutting plane intersects the plane of circle C.

To this end, let Q be the point where the line through P parallel to the axis of the cone intersects the plane of C. Let A be the point where the line joining P to the vertex of the cone intersects C. Let line PD be the perpendicular from P to L. Then $\overline{PA}$ and $\overline{PF}$, being tangents to the sphere from a common point P, have the same length:

$$PA = PF.$$

From right triangle PQA, we have

$$PQ = PA \cos \beta.$$

From right triangle PQD, we also have

$$PQ = PD \cos \alpha.$$

Equating the formulas for PQ gives, in sequence,

$$PA \cos \beta = PD \cos \alpha,$$

$$\frac{PA}{PD} = \frac{\cos \alpha}{\cos \beta},$$

$$\frac{PF}{PD} = \frac{\cos \alpha}{\cos \beta}, \qquad (PA = PF) \tag{21}$$

$$PF = \frac{\cos \alpha}{\cos \beta} \cdot PD,$$

$$PF = e \cdot PD. \qquad \left(e = \frac{\cos \alpha}{\cos \beta}\right)$$

The last equation in this sequence is the one we have been looking for. It characterizes P as lying on an algebraically defined ellipse, parabola, or hyperbola with focus F and directrix L, depending on whether $e < 1$, $e = 1$, or $e > 1$.

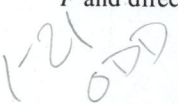

EXERCISES 9.2

Use the discriminant $B^2 - 4AC$ to decide whether the equations in Exercises 1–16 represent parabolas, ellipses, or hyperbolas.

1. $x^2 - 3xy + y^2 - x = 0$

2. $3x^2 - 18xy + 27y^2 - 5x + 7y = -4$

3. $3x^2 - 7xy + \sqrt{17}\, y^2 = 1$

4. $2x^2 - \sqrt{15}\, xy + 2y^2 + x + y = 0$

5. $x^2 + 2xy + y^2 + 2x - y + 2 = 0$

6. $2x^2 - y^2 + 4xy - 2x + 3y = 6$

7. $x^2 + 4xy + 4y^2 - 3x = 6$

8. $x^2 + y^2 + 3x - 2y = 10$

9. $xy + y^2 - 3x = 5$

10. $3x^2 + 6xy + 3y^2 - 4x + 5y = 12$

11. $3x^2 - 5xy + 2y^2 - 7x - 14y = -1$

12. $2x^2 - 4.9xy + 3y^2 - 4x = 7$

13. $x^2 - 3xy + 3y^2 + 6y = 7$

14. $25x^2 + 21xy + 4y^2 - 350x = 0$

15. $6x^2 + 3xy + 2y^2 + 17y + 2 = 0$

16. $3x^2 + 12xy + 12y^2 + 435x - 9y + 72 = 0$

In Exercises 17–26, rotate the coordinate axes to change the given equation into an equation that has no cross-product (xy) term. Then identify the graph of the equation. (The new equations will vary with the size and direction of the rotation you choose.)

17. $xy = 2$

18. $x^2 + xy + y^2 = 1$

19. $3x^2 + 2\sqrt{3}xy + y^2 - 8x + 8\sqrt{3}y = 0$

20. $x^2 - \sqrt{3}xy + 2y^2 = 1$

21. $x^2 - 2xy + y^2 = 2$

22. $x^2 - 3xy + y^2 = 5$

23. $\sqrt{2}x^2 + 2\sqrt{2}xy + \sqrt{2}y^2 - 8x + 8y = 0$

24. $xy - y - x + 1 = 0$

25. $3x^2 + 2xy + 3y^2 = 19$

26. $3x^2 + 4\sqrt{3}xy - y^2 = 7$

CALCULATOR The conic sections in Exercises 17–26 were chosen to have rotation angles that were "nice" in the sense that once we knew $\cot 2\alpha$ or $\tan 2\alpha$ we could identify 2α and find $\sin\alpha$ and $\cos\alpha$ from familiar triangles. The conic sections we encounter in practice may not have such nice rotation angles,

and we may have to use a calculator to determine α from the value of $\cot 2\alpha$ or $\tan 2\alpha$.

In Exercises 27–32, use a calculator to find an angle α through which the coordinate axes can be rotated to change the given equation into a quadratic equation that has no cross-product term. Then find $\sin\alpha$ and $\cos\alpha$ to two decimal places and use Eqs. (7) to find the coefficients of the new equation to the nearest decimal place. In each case, say whether the conic section is an ellipse, a hyperbola, or a parabola.

27. $x^2 - xy + 3y^2 + x - y - 3 = 0$

28. $2x^2 + xy - 3y^2 + 3x - 7 = 0$

29. $x^2 - 4xy + 4y^2 - 5 = 0$

30. $2x^2 - 12xy + 18y^2 - 49 = 0$

31. $3x^2 + 5xy + 2y^2 - 8y - 1 = 0$

32. $2x^2 + 7xy + 9y^2 + 20x - 86 = 0$

33. What effect does a 90° rotation about the origin have on the equations of the following conic sections? Give the new equation in each case.
 a) The ellipse $(x^2/a^2) + (y^2/b^2) = 1$ $(a > b)$
 b) The hyperbola $(x^2/a^2) - (y^2/b^2) = 1$
 c) The circle $x^2 + y^2 = a^2$
 d) The line $y = mx$
 e) The line $y = mx + b$

34. What effect does a 180° rotation about the origin have on the equations of the following conic sections? Give the new equation in each case.
 a) The ellipse $(x^2/a^2) + (y^2/b^2) = 1$ $(a > b)$
 b) The hyperbola $(x^2/a^2) - (y^2/b^2) = 1$
 c) The circle $x^2 + y^2 = a^2$
 d) The line $y = mx$
 e) The line $y = mx + b$

35. The hyperbola $xy = a$. The hyperbola $xy = 1$ is one of many hyperbolas of the form $xy = a$ that appear in science and mathematics.
 a) Rotate the coordinate axes through an angle of 45° to change the equation $xy = 1$ into an equation with no xy-term. What is the new equation?
 b) Do the same for the equation $xy = a$.

36. Find the eccentricity of the hyperbola $xy = 2$.

37. Show that the equation $x^2 + y^2 = a^2$ becomes $x'^2 +$

$y'^2 = a^2$ for every choice of the angle α in the rotation equations.

38. Show that rotating the axes through an angle of $\pi/4$ radians will eliminate the xy-term from Eq. (1) whenever $A = C$.

39. a) What kind of conic section is the curve $xy + 2x - y = 0$?

b) Solve the equation $xy + 2x - y = 0$ for y and sketch the curve as the graph of a rational function of x.

c) Find equations for the lines parallel to the line $y = -2x$ that are normal to the curve. Add the lines to your sketch.

40. What values of a, b, and c make the ellipse

$$4x^2 + y^2 + ax + by + c = 0$$

tangent to the x-axis at the origin and also pass through the point $(-1, 2)$?

41. Starting with the equation

$$Ax^2 + Bxy + Cy^2 + Dx + Ey + F = 0,$$

find an equation for the conic section that has all of the following properties:

1. It is symmetric about the origin.

2. It passes through the point $(1, 0)$.

3. The line $y = 1$ is tangent to it at the point $(-2, 1)$.

What kind of conic section is it?

42. Find an equation for the curve $x^2 + 2xy + y^2 - 1 = 0$ after a rotation that eliminates the xy-term. Identify the graph.

43. Prove or find counterexamples to the following statements about the graph of the equation $Ax^2 + Bxy + Cy^2 + Dx + Ey + F = 0$.

a) If $AC > 0$, the graph is an ellipse.

b) If $AC > 0$, the graph is a hyperbola.

c) If $AC < 0$, the graph is a hyperbola.

44. *A nice area formula for ellipses.* When $B^2 - 4AC$ is negative, the equation

$$Ax^2 + Bxy + Cy^2 = 1$$

represents an ellipse. If the ellipse's semi-axes are a and b,

its area is πab (a standard formula). Show that the area is also given by the formula $2\pi/\sqrt{4AC - B^2}$. (*Hint:* Rotate the coordinate axes to eliminate the xy-term and apply Eq. (19) to the new equation.)

45. *Other invariants.* We describe the fact that $B'^2 - 4A'C'$ equals $B^2 - 4AC$ after a rotation about the origin by saying that the discriminant of a quadratic equation is an **invariant** of the equation. Use Eqs. (7) to show that the numbers (a) $A + C$ and (b) $D^2 + E^2$ are also invariants, in the sense that

$$A' + C' = A + C \quad \text{and} \quad D'^2 + E'^2 = D^2 + E^2.$$

We can use these equalities to check against numerical errors when we rotate axes. They can also be helpful in shortening the work required to find values for the new coefficients.

46. *A proof that $B'^2 - 4A'C' = B^2 - 4AC$.* Use Eqs. (7) to show that $B'^2 - 4A'C' = B^2 - 4AC$ for any rotation of axes about the origin. The calculation works out nicely but requires patience.

***47.** Sketch a figure similar to Fig. 9.29 for the case in which the conic section is a parabola. Then derive Eq. (21) on the basis of your figure.

***48.** Sketch a figure similar to Fig. 9.29 for the case in which the conic section is a hyperbola. Then derive Eq. (21) on the basis of your figure.

***49.** Which parts of the geometric construction described in the last subsection ("Sections of a Cone") become impossible if the conic section is a circle?

EXPLORER PROGRAM

Conic Sections

Shows how the coefficients in the equation $Ax^2 + Bxy + Cy^2 + Dx + Ey + F = 0$ change as the conic section it describes is rotated in the xy-plane

9.3 Parametrizations of Curves

When the path of a particle moving in the plane looks like the curve in Fig. 9.30, we cannot hope to describe it with a Cartesian formula that expresses y directly in terms of x or x directly in terms of y. Instead, we express each of the particle's coordinates as a function of time t and describe the path with a pair of equations, $x = f(t)$ and $y = g(t)$. Indeed, for studying motion, equations like these are preferable to a Cartesian formula for the path because they immediately tell us the particle's position at any time t. They become equations for the motion as well as equations for the path along which the motion takes place. They also enable us to calculate the particle's velocity and acceleration at any time t, as we shall see in

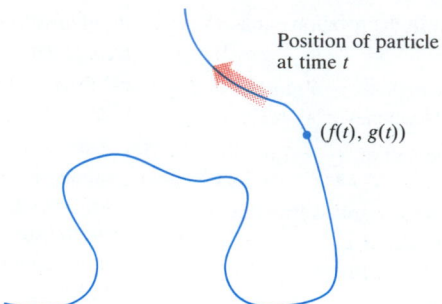

9.30 The path traced by a particle moving in the xy-plane is not always the graph of a function of x or a function of y.

Chapter 11. In the present section, we focus on the geometry of curves defined by parametric equations.

DEFINITIONS

If x and y are given as continuous functions

$$x = f(t), \qquad y = g(t) \qquad (1)$$

over an interval of t-values, then the set of points $(x, y) = (f(t), g(t))$ defined by these equations is called a **curve** in the coordinate plane. The equations are **parametric equations** for the curve. The variable t is a **parameter** for the curve and its domain I is called the **parameter interval**. If I is a closed interval, $a \le t \le b$, the point $(f(a), g(a))$ is the **initial point** of the curve and $(f(b), g(b))$ is the **terminal point** of the curve. When we give parametric equations and a parameter interval for a curve in the plane, we say that we have **parametrized** the curve. The equations and interval constitute a **parametrization** of the curve.

In many applications, t denotes time, but it might instead denote an angle (as in some of the following examples) or the distance a particle has traveled along its path from its starting point (as it sometimes will when we later study motion).

Example 1   *The Circle* $x^2 + y^2 = 1$.   The equations and parameter interval

$$x = \cos t, \qquad y = \sin t, \qquad 0 \le t \le 2\pi,$$

describe the position $P(x, y)$ of a particle that moves counterclockwise around the circle $x^2 + y^2 = 1$ as t increases (Fig. 9.31).

We know that the point lies on this circle for every value of t because

$$x^2 + y^2 = \cos^2 t + \sin^2 t = 1.$$

But how much of the circle does the point $P(x, y)$ actually traverse? To find out, we track the motion as t runs from 0 to 2π. The parameter t is the radian measure of the angle that radius OP makes with the positive x-axis. The particle starts at $(1, 0)$, moves up and to the left as t approaches $\pi/2$, and continues around the circle to stop again at $(1, 0)$ when $t = 2\pi$. The particle traces the circle exactly once. ◻

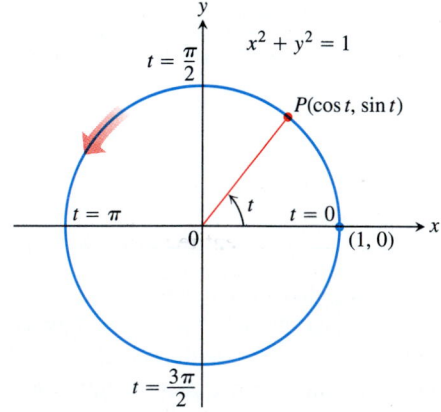

9.31   The equations $x = \cos t$ and $y = \sin t$ describe motion on the unit circle $x^2 + y^2 = 1$. The arrow shows the direction of increasing t (Example 1).

Example 2   *A Semicircle.*   The equations and parameter interval

$$x = \cos t, \qquad y = -\sin t, \qquad 0 \le t \le \pi,$$

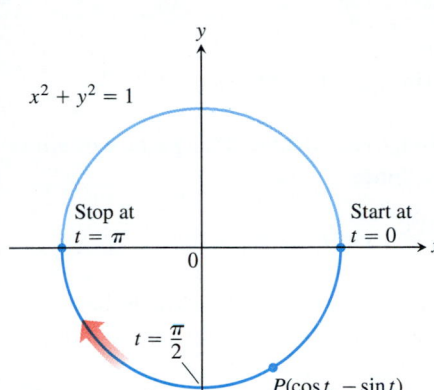

9.32 The point $P(\cos t, -\sin t)$ moves clockwise as t increases from 0 to π (Example 2).

describe the position $P(x, y)$ of a particle that moves clockwise around the circle $x^2 + y^2 = 1$ as t increases from 0 to π.

We know that the point P lies on this circle for all t because its coordinates satisfy the circle's equation. How much of the circle does the particle traverse? To find out, we track the motion as t runs from 0 to π. As in Example 1, the particle starts at $(1, 0)$. But now as t increases, y becomes negative, decreasing to -1 when $t = \pi/2$ and then increasing back to 0 as t approaches π. The motion stops at $t = \pi$ with only the lower half of the circle covered (Fig. 9.32).

Example 3   *Half a Parabola.*   The position $P(x, y)$ of a particle moving in the xy-plane is given by the equations and parameter interval

$$x = \sqrt{t}, \qquad y = t, \qquad t \geq 0.$$

Identify the path traced by the particle and describe the motion.

Solution   We try to identify the path by eliminating t between the equations $x = \sqrt{t}$ and $y = t$. With any luck, this will produce a recognizable algebraic relation between x and y. We find that

$$y = t = (\sqrt{t})^2 = x^2.$$

This means the particle's position coordinates satisfy the equation $y = x^2$, so the particle moves along the parabola $y = x^2$.

It would be a mistake, however, to conclude that the particle's path is the entire parabola $y = x^2$—it is only half the parabola. The particle's x-coordinate is never negative. The particle starts at $(0, 0)$ when $t = 0$ and rises into the first quadrant as t increases (Fig. 9.33).

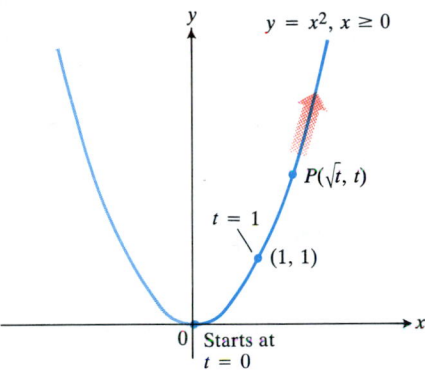

9.33 The equations $x = \sqrt{t}$, $y = t$ and interval $t \geq 0$ describe the motion of a particle that traces the right-hand half of the parabola $y = x^2$ (Example 3).

Example 4   *A Complete Parabola.*   The position $P(x, y)$ of a particle moving in the xy-plane is given by the equations and parameter interval

$$x = t, \qquad y = t^2, \qquad -\infty < t < \infty.$$

Identify the particle's path and describe the motion.

Solution   We identify the path by eliminating t between the equations $x = t$ and $y = t^2$, obtaining

$$y = (t)^2 = x^2.$$

The particle's position coordinates satisfy the equation $y = x^2$, so the particle moves along this curve.

In contrast to Example 3, the particle now traverses the entire parabola. As t increases from $-\infty$ to ∞, the particle comes down the left-hand side, passes through the origin, and moves up the right-hand side (Fig. 9.34).

As Example 4 illustrates, the graph of a function $y = f(x)$, $x \, \varepsilon \, I$, has the automatic parametrization $x = t$, $y = f(t)$, $t \, \varepsilon \, I$. This is so simple we usually do not use it, but the point of view is occasionally helpful.

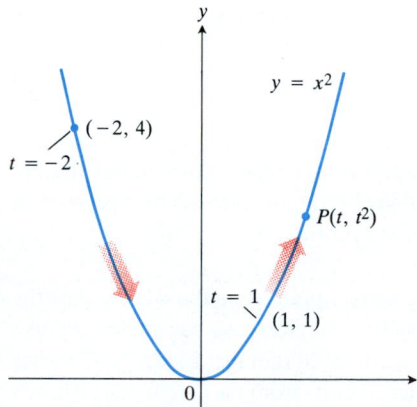

9.34 The path defined by the parametric equations $x = t$, $y = t^2$ and interval $-\infty < t < \infty$ (Example 4).

Example 5   *A Parametrization of the Ellipse $x^2/a^2 + y^2/b^2 = 1$.*   Describe the motion of a particle whose position $P(x, y)$ at time t is given by the

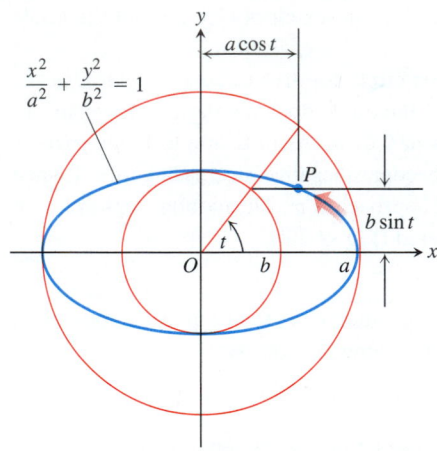

9.35   The coordinates of P are $x = a \cos t$, $y = b \sin t$ (Example 5).

equations and parameter interval

$$x = a \cos t, \qquad y = b \sin t, \qquad 0 \le t \le 2\pi.$$

Solution   We find a Cartesian equation for the coordinates of the particle by eliminating t between the equations for x and y. Since

$$\frac{x^2}{a^2} + \frac{y^2}{b^2} = \frac{a^2 \cos^2 t}{a^2} + \frac{b^2 \sin^2 t}{b^2} = \cos^2 t + \sin^2 t = 1,$$

the motion takes place on the ellipse $x^2/a^2 + y^2/b^2 = 1$. The particle begins at $(a, 0)$ when t equals zero and moves counterclockwise around the ellipse, traversing it exactly once as t moves from 0 to 2π (Fig. 9.35).

Example 6   *A Parametrization of the Circle $x^2 + y^2 = a^2$.*   The equations and parameter interval

$$x = a \cos t, \qquad y = a \sin t, \qquad 0 \le t \le 2\pi,$$

obtained by taking $b = a$ in Example 5, describe the circle $x^2 + y^2 = a^2$.

Example 7   *A Parametrization of the Right-hand Branch of the Hyperbola $x^2 - y^2 = 1$.*   Describe the motion of the particle whose position $P(x, y)$ at time t is given by the equations and parameter interval

$$x = \sec t, \qquad y = \tan t, \qquad -\frac{\pi}{2} < t < \frac{\pi}{2}.$$

Solution   We find a Cartesian equation for the coordinates of P by eliminating t between the equations for x and y. Since

$$x^2 - y^2 = \sec^2 t - \tan^2 t = 1,$$

we see that the motion takes place somewhere on the hyperbola $x^2 - y^2 = 1$. Since $x = \sec t$ is always positive for the parameter values $-\pi/2 < t < \pi/2$, the motion takes place on the hyperbola's right-hand branch. As t moves from $-\pi/2$ to $\pi/2$, the particle comes in along the lower half of the right-hand branch, reaching the vertex $(1, 0)$ at $t = 0$. It then moves into the first quadrant to complete the coverage of the right-hand branch as t approaches $\pi/2$ (Fig. 9.36).

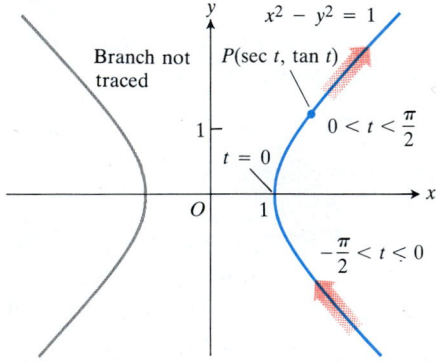

9.36   The equations $x = \sec t$, $y = \tan t$ and interval $-\pi/2 < t < \pi/2$ describe the right-hand branch of the hyperbola $x^2 - y^2 = 1$ (Example 7).

Example 8   *Cycloids.*   A wheel of radius a rolls along a horizontal straight line without slipping. Find parametric equations for the path traced by a point P on the wheel's circumference. The path is called a **cycloid.**

Solution   We take the line to be the x-axis, mark a point P on the wheel, start the wheel with P at the origin, and roll the wheel to the right. As parameter, we use the angle t through which the wheel turns, measured in radians. Figure 9.37 shows the wheel a short while later, when its base lies at units from the origin. The wheel's center C lies at (at, a) and the coordinates of P are

$$x = at + a \cos \theta, \qquad y = a + a \sin \theta. \qquad (2)$$

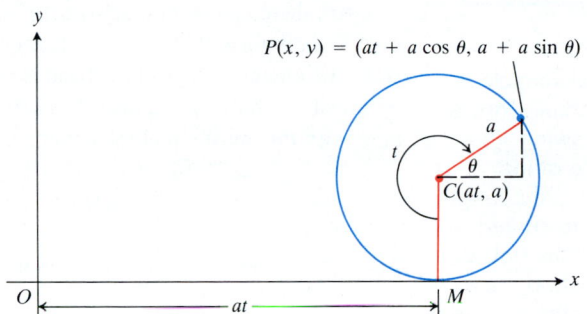

9.37 The position of $P(x, y)$ on the rolling wheel at time t (Example 8).

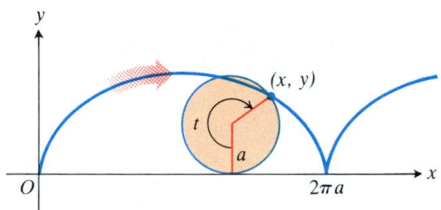

9.38 The cycloid $x = a(t - \sin t)$, $y = a(1 - \cos t)$, shown for $t \geq 0$.

To express θ in terms of t, we observe that $t + \theta = 3\pi/2$, so that

$$\theta = \frac{3\pi}{2} - t. \tag{3}$$

This makes

$$\cos \theta = \cos\left(\frac{3\pi}{2} - t\right) = -\sin t, \qquad \sin \theta = \sin\left(\frac{3\pi}{2} - t\right) = -\cos t. \tag{4}$$

The equations we seek are

$$x = at - a \sin t, \qquad y = a - a \cos t. \tag{5}$$

These are usually written with the a factored out:

$$x = a(t - \sin t), \qquad y = a(1 - \cos t). \tag{6}$$

Figure 9.38 shows the first arch of the cylcoid and part of the next.

✳ Brachistochrones and Tautochrones

If we turn Fig. 9.38 upside down, Eqs. (6) still apply and the resulting curve (Fig. 9.39) has two interesting physical properties. The first relates to the origin O and the point B at the bottom of the first arch. Among all smooth curves joining these points, the cycloid is the curve along which a frictionless bead, subject only to the force of gravity, will slide from O to B the fastest. This makes the cycloid a **brachistochrone** (brah-*kiss*-toe-krone), or shortest time curve for these points. The second property is that even if you start the bead partway down the curve toward B, it will still take the bead the same amount of time to reach B. This makes the cycloid a **tautochrone** (*taw*-toe-krone), or same-time curve for O and B.

9.39 To study motion along an upside-down cycloid under the influence of gravity, we turn Fig. 9.38 upside down. This points the y-axis in the direction of the gravitational force and makes the downward y-coordinates positive. The equations and parameter interval for the cycloid are still

$$x = a(t - \sin t),$$
$$y = a(1 - \cos t), \quad t \geq 0.$$

The arrow shows the direction of increasing t.

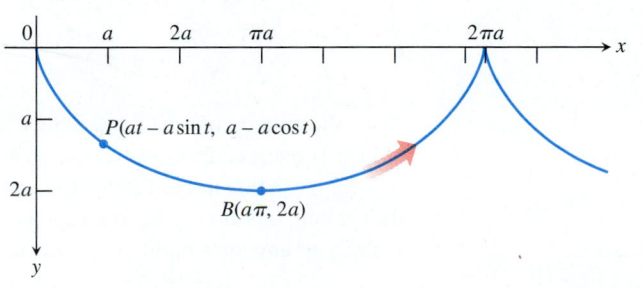

Huygens's Clock

One problem with a pendulum clock whose bob swings in a circular arc is that the frequency of the swing depends on the amplitude of the swing. The wider the swing, the longer it takes the bob to return to center.

This does not happen if the bob can be made to swing in a cycloid, as you will see if you read about the cycloid's tautochrone property at the end of the section. In 1673, Christiaan Huygens (1629–1695), the Dutch mathematician, physicist, and astronomer who discovered the rings of Saturn, driven by a need to make accurate determinations of longitude at sea, designed a pendulum clock whose bob would swing in a cycloid. He hung the bob from a fine wire constrained by guards that caused it to draw up as it swung. How were the guards shaped? They were cycloids, too.

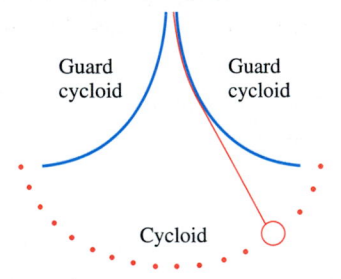

Are there any other brachistochrones joining O and B, or is the cycloid the only one? We can formulate this as a mathematical question in the following way. At the start, the kinetic energy of the bead is zero, since its velocity is zero. The work done by gravity in moving the bead from $(0, 0)$ to any other point (x, y) in the plane is mgy, and this must equal the change in kinetic energy. That is,

$$mgy = \frac{1}{2} mv^2 - \frac{1}{2} m(0)^2. \tag{7}$$

Thus, the velocity of the bead when it reaches (x, y) has to be

$$v = \sqrt{2gy}. \tag{8}$$

That is,

$$\frac{ds}{dt} = \sqrt{2gy} \qquad (ds \text{ is the arc length differential along the bead's path.})$$

or

$$dt = \frac{ds}{\sqrt{2gy}} = \frac{\sqrt{1 + (dy/dx)^2}\, dx}{\sqrt{2gy}}. \tag{9}$$

The time T_f it takes the bead to slide along a particular path $y = f(x)$ from O to $B(a\pi, 2a)$ is

$$T_f = \int_{x=0}^{x=a\pi} \sqrt{\frac{1 + (dy/dx)^2}{2gy}}\, dx. \tag{10}$$

What curves $y = f(x)$, if any, minimize the value of this integral?

At first sight, we might guess that the straight line joining O and B would give the shortest time, but perhaps not. There might be some advantage in having the bead fall vertically at first to build up its velocity faster. With a higher velocity, the bead could travel a longer path and still reach B first. Indeed, this is the right idea. The solution, from a branch of mathematics known as the calculus of variations, is that the original cycloid from O to B is the one and only brachistochrone for O and B.

While the solution of the brachistochrone problem is beyond our present reach, we can still show why the cycloid is a tautochrone. For the cycloid, Eq. (10) takes the form

$$\begin{aligned} T_{\text{cycloid}} &= \int_{x=0}^{x=a\pi} \sqrt{\frac{dx^2 + dy^2}{2gy}} \\ &= \int_{t=0}^{t=\pi} \sqrt{\frac{a^2(2 - 2\cos t)}{2ga(1 - \cos t)}}\, dt \qquad \left(\begin{matrix} \text{From Eqs. (6), } dx = a(1 - \cos t)\, dt, \\ dy = a\sin t\, dt, \text{ and } y = a(1 - \cos t) \end{matrix}\right) \\ &= \int_0^\pi \sqrt{\frac{a}{g}}\, dt = \pi \sqrt{\frac{a}{g}}. \end{aligned} \tag{11}$$

Thus, the amount of time it takes the frictionless bead to slide down the cycloid to B after it is released from rest at O is $\pi\sqrt{a/g}$.

Suppose that instead of starting the bead at O we start it at some lower point on the cycloid, a point (x_0, y_0) corresponding to the parameter value $t_0 > 0$. The bead's velocity at any later point (x, y) on the cycloid is

$$v = \sqrt{2g(y - y_0)} = \sqrt{2ga(\cos t_0 - \cos t)}. \qquad (y = a(1 - \cos t))$$

The Witch of Agnesi

Although l'Hôpital wrote the first text on differential calculus, the first text to include differential and integral calculus along with analytic geometry, infinite series, and differential equations was written in the 1740s by the Italian mathematician Maria Gaetana Agnesi (1718–1799). Agnesi, a gifted scholar and linguist whose Latin essay defending higher education for women was published when she was only nine years old, was a well-published scientist by age 20 and an honorary faculty member of the University of Bologna by age 30.

Today, Agnesi is remembered chiefly for a bell-shaped curve called *the witch of Agnesi*. This name, found only in English texts, is the result of a mistranslation. Agnesi's own name for the curve was *versiera* or "turning curve." John Colson, a noted Cambridge mathematician who felt Agnesi's text so important that he learned Italian to translate it "for the benefit of British youth" (he particularly had in mind young women, for whom he hoped Agnesi would be a role model), probably confused *versiera* with *avversiera*, which means "wife of the devil" and translates into "witch." You can find out more about the witch by doing Exercise 28.

Accordingly, the time required for the bead to slide down to B is

$$T = \int_{t_0}^{\pi} \sqrt{\frac{a^2(2 - \cos t)}{2ag(\cos t_0 - \cos t)}}\, dt = \sqrt{\frac{a}{g}} \int_{t_0}^{\pi} \sqrt{\frac{1 - \cos t}{\cos t_0 - \cos t}}\, dt$$

$$= \sqrt{\frac{a}{g}} \int_{t_0}^{\pi} \sqrt{\frac{2 \sin^2(t/2)}{[2\cos^2(t_0/2) - 1] - [2\cos^2(t/2) - 1]}}\, dt$$

$$= \sqrt{\frac{a}{g}} \int_{t_0}^{\pi} \frac{\sin(t/2)\, dt}{\sqrt{\cos^2(t_0/2) - \cos^2(t/2)}}$$

$$= \sqrt{\frac{a}{g}} \int_{t=t_0}^{t=\pi} \frac{-2\, du}{\sqrt{a^2 - u^2}} \qquad \left(\begin{array}{l} u = \cos(t/2) \\ -2du = \sin(t/2)\, dt \\ a = \cos(t_0/2) \end{array} \right)$$

$$= 2\sqrt{\frac{a}{g}} \left[-\sin^{-1}\frac{u}{a} \right]_{t=t_0}^{t=\pi}$$

$$= 2\sqrt{\frac{a}{g}} \left[-\sin^{-1}\frac{\cos\,(t/2)}{\cos\,(t_0/2)} \right]_{t_0}^{\pi} = 2\sqrt{\frac{a}{g}}(-\sin^{-1}0 + \sin^{-1}1) = \pi\sqrt{\frac{a}{g}}.$$

This is precisely the time it takes the bead to slide to B from O. It takes the bead the same amount of time to reach B no matter where it starts. Beads starting simultaneously from O, A, and C in Fig. 9.40, for instance, will all reach B at the same time.

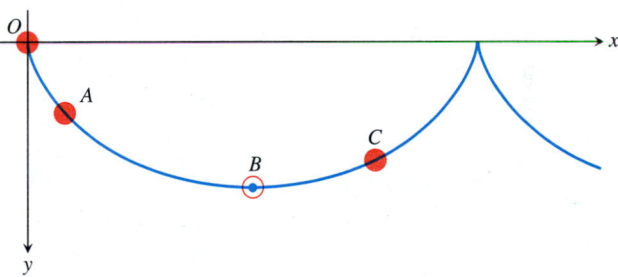

9.40 Beads released simultaneously on the cycloid at O, A, and C will reach B at the same time.

Standard Parametrizations

Circle $x^2 + y^2 = a^2$:

$$x = a \cos t$$
$$y = a \sin t$$
$$0 \le t \le 2\pi$$

Ellipse $\dfrac{x^2}{a^2} + \dfrac{y^2}{b^2} = 1$:

$$x = a \cos t$$
$$y = b \sin t$$
$$0 \le t \le 2\pi$$

Cycloid generated by a circle of radius a:

$$x = a(t - \sin t), \qquad y = a(1 - \cos t)$$

EXERCISES 9.3

Exercises 1–24 give parametric equations and parameter intervals for the motion of a particle in the xy-plane. Identify the particle's path by finding a Cartesian equation for it. Graph the Cartesian equation. Indicate the portion of the graph traced by the particle and the direction of motion.

1. $x = \cos t$, $y = \sin t$, $0 \le t \le \pi$

2. $x = \cos 2t$, $y = \sin 2t$, $0 \le t \le \pi$

3. $x = \sin 2\pi t$, $y = \cos 2\pi t$, $0 \le t \le 1$

4. $x = \cos(\pi - t)$, $y = \sin(\pi - t)$, $0 \le t \le \pi$

5. $x = 4 \cos t$, $y = 2 \sin t$, $0 \le t \le 2\pi$

6. $x = 4 \sin t$, $y = 2 \cos t$, $0 \le t \le \pi$

7. $x = 4 \cos t$, $y = 5 \sin t$, $0 \le t \le \pi$

8. $x = 4 \sin t$, $y = 5 \cos t$, $0 \le t \le 2\pi$

9. $x = 3t$, $y = 9t^2$, $-\infty < t < \infty$

10. $x = -\sqrt{t}$, $y = t$, $t \ge 0$

11. $x = t$, $y = \sqrt{t}$, $t \ge 0$

12. $x = \sec^2 t - 1$, $y = \tan t$, $-\pi/2 < t < \pi/2$

13. $x = -\sec t$, $y = \tan t$, $-\pi/2 < t < \pi/2$

14. $x = \csc t$, $y = \cot t$, $0 < t < \pi$

15. $x = 2t - 5$, $y = 4t - 7$, $-\infty < t < \infty$

16. $x = 1 - t$, $y = 1 + t$, $-\infty < t < \infty$

17. $x = t$, $y = 1 - t$, $0 \le t \le 1$

18. $x = 3t$, $y = 2 - 2t$, $0 \le t \le 1$

19. $x = t$, $y = \sqrt{1 - t^2}$, $-1 \le t \le 1$

20. $x = t$, $y = \sqrt{4 - t^2}$, $0 \le t \le 2$

21. $x = t^2$, $y = \sqrt{t^4 + 1}$, $t \ge 0$

22. $x = \sqrt{t + 1}$, $y = \sqrt{t}$, $t \ge 0$

23. $x = \cosh t$, $y = \sinh t$, $-\infty < t < \infty$

24. $x = 2 \sinh t$, $y = 2 \cosh t$, $-\infty < t < \infty$

25. Find parametric equations and a parameter interval for the motion of a particle that starts at $(a, 0)$ and traces the circle $x^2 + y^2 = a^2$
    a) once clockwise,
    b) once counterclockwise,
    c) twice clockwise,
    d) twice counterclockwise.
    (There are many correct ways to do these, so your answers may not be the same as the ones in the back of the book.)

26. Find parametric equations and a parameter interval for the motion of a particle that starts at $(a, 0)$ and traces the ellipse $(x^2/a^2) + (y^2/b^2) = 1$
    a) once clockwise,
    b) once counterclockwise,
    c) twice clockwise,

d) twice counterclockwise.
(As in Exercise 25, there are many correct answers.)

27. *The involute of a circle.*   If a string wound around a fixed circle is unwound while held taut in the plane of the circle, its end P traces an *involute* of the circle. In Fig. 9.41, the circle in question is the unit circle in the xy-plane and the initial position of the tracing point is the point $(1, 0)$ on the x-axis. The unwound portion of the string is tangent to the circle at Q, and t is the radian measure of the angle from the positive x-axis to segment OQ. Derive parametric equations for the involute by expressing the coordinates x and y of P in terms of t for $t \ge 0$.

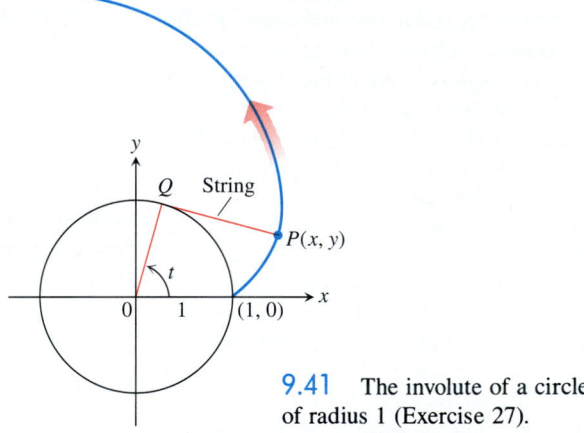

9.41   The involute of a circle of radius 1 (Exercise 27).

28. *The witch of Maria Agnesi.*   The bell-shaped witch of Maria Agnesi can be constructed in the following way. Start with a circle of radius 1, centered at the point $(0, 1)$ on the y-axis (Fig. 9.42). Choose a point A on the line $y = 2$ and connect it to the origin with a line segment. Call the point where the segment crosses the circle B. Let P be the point where the vertical line through A crosses the horizontal line through B. The witch is the curve traced by P as A moves along the line $y = 2$. Find parametric equations and a parameter interval for the witch by expressing the coordi-

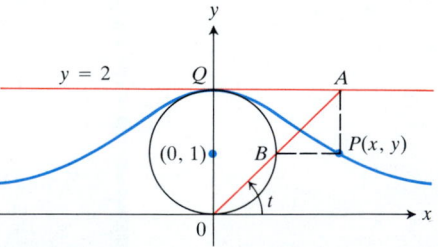

9.42   The witch of Maria Agnesi (Exercise 28).

nates of P in terms of t, the radian measure of the angle that segment OA makes with the positive x-axis. The following equalities (which you may assume) will help:

1. $x = AQ$

2. $y = 2 - AB \sin t$

3. $AB \cdot OA = (AQ)^2$

29. Find the point on the parabola $x = t$, $y = t^2$, $-\infty < t < \infty$, closest to the point $(2, 1/2)$. (*Hint:* Minimize the square of the distance as a function of t.)

30. Find the point on the ellipse $x = 2 \cos t$, $y = \sin t$, $0 \le t \le 2\pi$ closest to the point $(3/4, 0)$. (*Hint:* Minimize the square of the distance as a function of t.)

31. *Parametrizations of lines in the plane (Fig. 9.43)*
   a) Show that the equations and parameter interval $x = x_0 + (x_1 - x_0)t$, $y = y_0 + (y_1 - y_0)t$, $-\infty < t < \infty$, describe the line through the points (x_0, y_0) and (x_1, y_1).
   b) Using the same parameter interval, write parametric equations for the line through a point (x_1, y_1) and the origin.
   c) Using the same parameter interval, write parametric equations for the line through $(-1, 0)$ and $(0, 1)$.

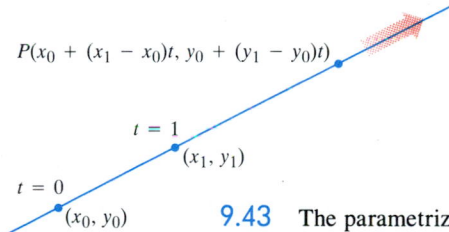

$P(x_0 + (x_1 - x_0)t, y_0 + (y_1 - y_0)t)$

$t = 1$

(x_1, y_1)

$t = 0$

(x_0, y_0)

9.43   The parametrized line in Exercise 31. The arrow shows the direction of increasing t.

Computer Graphing Exercises

If you have access to a parametric equation grapher, graph the following equations over the given parameter intervals.

32. *Ellipse.* $x = 4 \cos t$, $y = 2 \sin t$, over
   a) $0 \le t \le 2\pi$,
   b) $0 \le t \le \pi$,
   c) $-\pi/2 \le t \le \pi/2$.

33. *Hyperbola branch.* $x = \sec t$ (enter as $1/\cos(t)$), $y = \tan t$ (enter as $(\sin (t)/\cos (t))$), over
   a) $-1.5 \le t \le 1.5$,
   b) $-0.5 \le t \le 0.5$,
   c) $-0.1 \le t \le 0.1$.

34. *Parabola.* $x = 2t + 3$, $y = t^2 - 1$, $-2 \le t \le 2$

35. *Cycloid.* $x = t - \sin t$, $y = 1 - \cos t$, over
   a) $0 \le t \le 2\pi$,
   b) $0 \le t \le 4\pi$,
   c) $\pi \le t \le 3\pi$.

36. *Astroid.* $x = \cos^3 t$, $y = \sin^3 t$, over
   a) $0 \le t \le 2\pi$,   b) $-\pi/2 \le t \le \pi/2$.

37. *A nice curve (a deltoid)*

$$x = 2 \cos t + \cos 2t,$$
$$y = 2 \sin t - \sin 2t,$$
$$0 \le t \le 2\pi$$

What happens if you replace 2 with -2 in the equations for x and y? Graph the new equations and find out.

38. *An even nicer curve*

$$x = 3 \cos t + \cos 3t,$$
$$y = 3 \sin t - \sin 3t,$$
$$0 \le t \le 2\pi$$

What happens if you replace 3 with -3 in the equations for x and y? Graph the new equations and find out.

39. *Projectile motion.*   Graph

$$x = (64 \cos \alpha)t,$$
$$y = -16t^2 + (64 \sin \alpha)t,$$
$$0 \le t \le 4 \sin \alpha,$$

for the following firing angles:
   a) $\alpha = \pi/4$,   b) $\alpha = \pi/6$,   c) $\alpha = \pi/3$,
   d) $\alpha = \pi/2$ (watch out—here it comes!)

40. *Three beautiful curves*
   a) *Epicycloid:*

$$x = 9 \cos t - \cos 9t$$
$$y = 9 \sin t - \sin 9t$$
$$0 \le t \le 2\pi$$

   b) *Hypocycloid:*

$$x = 8 \cos t + 2 \cos 4t$$
$$y = 8 \sin t - 2 \sin 4t$$
$$0 \le t \le 2\pi$$

   c) *Hypotrochoid:*

$$x = \cos t + 5 \cos 3t$$
$$y = 6 \cos t - 5 \sin 3t$$
$$0 \le t \le 2\pi$$

EXPLORER PROGRAM

PowerGrapher

Traces the curves for $x(t)$, $y(t)$, and $P(x, y)$ in side-by-side displays as t increases through the parameter interval. Also graphs $P(x, y)$ in a separate display.

Graphs different sets of parametric equations in a common display.

9.4 The Calculus of Parametrized Curves

This section shows how to find slopes, lengths, centroids, and surface areas associated with parametrized curves.

Slopes of Parametrized Curves

DEFINITIONS

A parametrized curve $x = f(t)$, $y = g(t)$ is said to be **differentiable at $t = t_0$** if f and g are differentiable at $t = t_0$. The curve is **differentiable** if it is differentiable at every parameter value. The curve is **smooth** if f and g are smooth, i.e., if f and g have continuous first derivatives.

At a point on a differentiable parametrized curve where y is also a differentiable function of x, the derivatives dx/dt, dy/dt, and dy/dx are related by the Chain Rule equation

$$\frac{dy}{dt} = \frac{dy}{dx}\frac{dx}{dt} \tag{1}$$

(Fig. 9.44). If $dx/dt \neq 0$, we may divide both sides of this equation by dx/dt to solve for dy/dx.

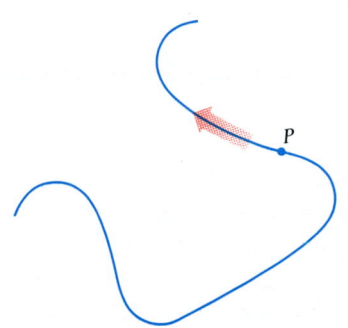

9.44 When the three first derivatives dx/dt, dy/dt, and dy/dx exist at a point P on a parametrized curve, they are related by the Chain Rule equation

$$\frac{dy}{dt} = \frac{dy}{dx}\frac{dx}{dt}.$$

Formula for Finding dy/dx from dy/dt and dx/dt $(dx/dt \neq 0)$

$$\frac{dy}{dx} = \frac{dy/dt}{dx/dt} \tag{2}$$

Example 1   Find the tangent to the right-hand hyperbola branch

$$x = \sec t, \qquad y = \tan t, \qquad -\frac{\pi}{2} < t < \frac{\pi}{2},$$

at the point $(\sqrt{2}, 1)$, where $t = \pi/4$ (Fig. 9.45).

Solution   The slope of the curve at t is

$$\frac{dy}{dx} = \frac{dy/dt}{dx/dt} = \frac{\sec^2 t}{\sec t \tan t} = \frac{\sec t}{\tan t}. \qquad \text{(Eq.(2))}$$

Setting t equal to $\pi/4$ gives

$$\left.\frac{dy}{dx}\right|_{t=\pi/4} = \frac{\sec(\pi/4)}{\tan(\pi/4)} = \frac{\sqrt{2}}{1} = \sqrt{2}.$$

The point–slope equation of the tangent is

$$y - y_0 = m(x - x_0)$$
$$y - 1 = \sqrt{2}(x - \sqrt{2})$$
$$y = \sqrt{2}x - 2 + 1$$
$$y = \sqrt{2}x - 1.$$

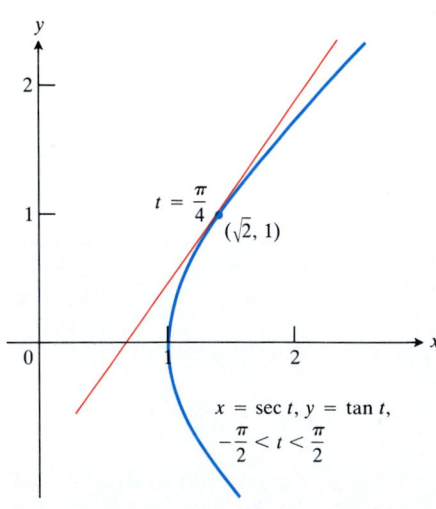

$$x = \sec t,\ y = \tan t,$$
$$-\frac{\pi}{2} < t < \frac{\pi}{2}$$

9.45   The hyperbola branch in Example 1.

The Parametric Formula for d^2y/dx^2

If the parametric equations for a curve define y as a twice-differentiable function of x, we may calculate d^2y/dx^2 as a function of t in the following way:

$$\frac{d^2y}{dx^2} = \frac{d}{dx}(y')$$

$$= \frac{dy'/dt}{dx/dt}. \qquad \text{(Eq. (2) with } y \text{ replaced by } y')$$

Formula for Finding d^2y/dx^2 from dx/dt and $y' = dy/dx$ $(dx/dt \neq 0)$

$$\frac{d^2y}{dx^2} = \frac{dy'/dt}{dx/dt} \qquad (3)$$

Notice the lack of symmetry in Eq. (3). To find d^2y/dx^2, we divide the derivative of y' by the derivative of x, not by the derivative of x'.

Example 2   Find d^2y/dx^2 if $x = t - t^2$ and $y = t - t^3$.

Solution   STEP 1:   Express y' in terms of t:

To find d^2y/dx^2 in terms of t,
1. Express $y' = dy/dx$ in terms of t
2. Find dy'/dt
3. Divide dy'/dt by dx/dt.

$$y' = \frac{dy}{dx} = \frac{dy/dt}{dx/dt} = \frac{1 - 3t^2}{1 - 2t} \qquad \text{(Eq. (2) with } x = t - t^2, y = t - t^3)$$

STEP 2:   Differentiate y' with respect to t:

$$\frac{dy'}{dt} = \frac{d}{dt}\left(\frac{1 - 3t^2}{1 - 2t}\right) = \frac{2 - 6t + 6t^2}{(1 - 2t)^2}$$

STEP 3:   Divide dy'/dt by dx/dt:

$$\frac{dx}{dt} = \frac{d}{dt}(t - t^2) = 1 - 2t$$

$$\frac{dy'/dt}{dx/dt} = \frac{2 - 6t + 6t^2}{(1 - 2t)^2} \cdot \frac{1}{1 - 2t}$$

$$= \frac{2 - 6t + 6t^2}{(1 - 2t)^3}.$$

Lengths of Parametrized Curves; Centroids

We can find an integral for the length of a smooth curve

$$x = f(t), \qquad y = g(t), \qquad a \leq t \leq b,$$

by rewriting the integral $L = \int ds$ from Section 5.4 in the following way:

$$\text{Length} = \int_{t=a}^{t=b} ds = \int_a^b \sqrt{dx^2 + dy^2} \qquad (ds = \sqrt{dx^2 + dy^2})$$

$$= \int_a^b \sqrt{\left(\frac{dx^2}{dt^2} + \frac{dy^2}{dt^2}\right) dt^2} = \int_a^b \sqrt{\left(\frac{dx}{dt}\right)^2 + \left(\frac{dy}{dt}\right)^2} \, dt. \qquad (4)$$

The only requirement besides the continuity of the integrand is that the point $P(x, y)$ not trace out any portion of the curve more than once as t moves from a to b.

Parametric Formula for Arc Length

If the functions $x = f(t)$ and $y = g(t)$ have continuous first derivatives with respect to t for $a \leq t \leq b$, and if the point $P(x, y)$ traces the curve defined by these equations exactly once as t moves from $t = a$ to $t = b$, then the length of the curve is given by the formula

$$\text{Length} = \int_a^b \sqrt{\left(\frac{dx}{dt}\right)^2 + \left(\frac{dy}{dt}\right)^2} \, dt. \tag{5}$$

What if we have two different parametrizations for a curve whose length we want to find—does it matter which one we use? The answer, from advanced calculus, is no. As long as the parametrization we choose meets the conditions preceding Eq. (5), the formula gives the correct length.

Example 3   Find the length of the astroid (Fig. 9.46)

$$x = \cos^3 t, \qquad y = \sin^3 t, \qquad 0 \leq t \leq 2\pi.$$

Solution   Because of the curve's symmetry with respect to the coordinate axes, its length is four times the length of the first-quadrant portion. We have

$$x = \cos^3 t, \qquad y = \sin^3 t,$$

$$\left(\frac{dx}{dt}\right)^2 = [3\cos^2 t(-\sin t)]^2 = 9\cos^4 t \sin^2 t,$$

$$\left(\frac{dy}{dt}\right)^2 = [3\sin^2 t(\cos t)]^2 = 9\sin^4 t \cos^2 t,$$

$$\sqrt{\left(\frac{dx}{dt}\right)^2 + \left(\frac{dy}{dt}\right)^2} = \sqrt{9\cos^2 t \sin^2 t \underbrace{(\cos^2 t + \sin^2 t)}_{1}},$$

$$= \sqrt{9\cos^2 t \sin^2 t}$$

$$= 3|\cos t \sin t|$$

$$= 3\cos t \sin t. \qquad (\cos t \sin t \geq 0 \text{ for } 0 \leq t \leq \pi/2)$$

Therefore,

$$\text{Length of first-quadrant portion} = \int_0^{\pi/2} 3\cos t \sin t \, dt$$

$$= \frac{3}{2}\int_0^{\pi/2} \sin 2t \, dt$$

$$= -\frac{3}{4}\cos 2t \Big]_0^{\pi/2} = \frac{3}{2}.$$

The length of the astroid is four times this: $4(3/2) = 6$.

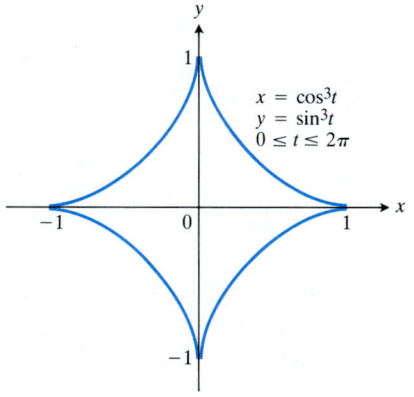

$x = \cos^3 t$
$y = \sin^3 t$
$0 \leq t \leq 2\pi$

9.46   The astroid in Example 3.

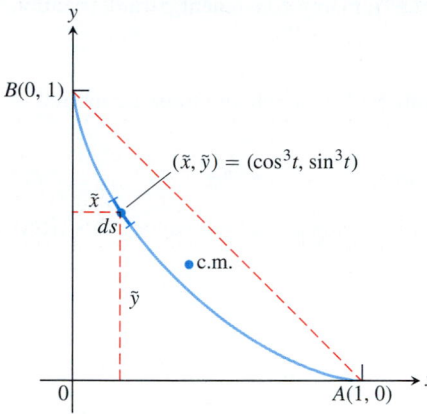

9.47   A typical segment of the arc in Example 4. The centroid (c.m.) of the curve lies about a third of the way toward chord AB.

Example 4   Find the centroid of the first-quadrant arc of the astroid in Example 3.

Solution   We take the curve's density to be $\delta = 1$ and calculate the curve's mass and moments about the coordinate axes as we did at the end of Section 5.6.

The distribution of mass is symmetric about the line $y = x$, so $\bar{x} = \bar{y}$. A typical segment of the curve (Fig. 9.47) has mass

$$dm = 1 \cdot ds = \sqrt{\left(\frac{dx}{dt}\right)^2 + \left(\frac{dy}{dx}\right)^2}\, dt = 3 \cos t \sin t\, dt \qquad \text{(From Example 3)}$$

The curve's mass is

$$M = \int_0^{\pi/2} dm = \int_0^{\pi/2} 3 \cos t \sin t\, dt = \frac{3}{2}. \qquad \text{(Again from Example 3)}$$

The curve's moment about the x-axis is

$$M_x = \int \tilde{y}\, dm = \int_0^{\pi/2} \sin^3 t \cdot 3 \cos t \sin t\, dt$$

$$= 3 \int_0^{\pi/2} \sin^4 t \cos t\, dt = 3 \cdot \frac{\sin^5 t}{5}\Big]_0^{\pi/2} = \frac{3}{5}.$$

Hence,

$$\bar{y} = \frac{M_x}{M} = \frac{3/5}{3/2} = \frac{2}{5}.$$

The centroid is the point $(2/5, 2/5)$ (Fig. 9.47).

The Area of a Surface of Revolution

The formula $S = \int 2\pi\, \rho\, ds$ developed in Section 5.5 for the area of the surface swept out by revolving a smooth curve about an axis translates into $S = \int 2\pi\, y\, ds$ if the axis is the x-axis and into $S = \int 2\pi\, x\, ds$ if the axis is the y-axis. With $ds = \sqrt{(dx/dt)^2 + (dy/dt)^2}\, dt$, these lead to the following formulas.

Parametric Formulas for the Area of a Surface of Revolution

If the functions $x = f(t)$ and $y = g(t)$ are nonnegative and have continuous first derivatives with respect to t for $a \le t \le b$, and if the point $P(x, y)$ traces the curve defined by these equations exactly once as t moves from $t = a$ to $t = b$, then the areas of the surfaces generated by revolving the curve about the coordinate axes are as follows.

1. Revolution about the x-axis ($y \ge 0$)

$$\text{Surface area} = \int_a^b 2\pi\, y\, \sqrt{\left(\frac{dx}{dt}\right)^2 + \left(\frac{dy}{dt}\right)^2}\, dt \qquad (6)$$

2. Revolution about the y-axis ($x \ge 0$)

$$\text{Surface area} = \int_a^b 2\pi\, x\, \sqrt{\left(\frac{dx}{dt}\right)^2 + \left(\frac{dy}{dt}\right)^2}\, dt \qquad (7)$$

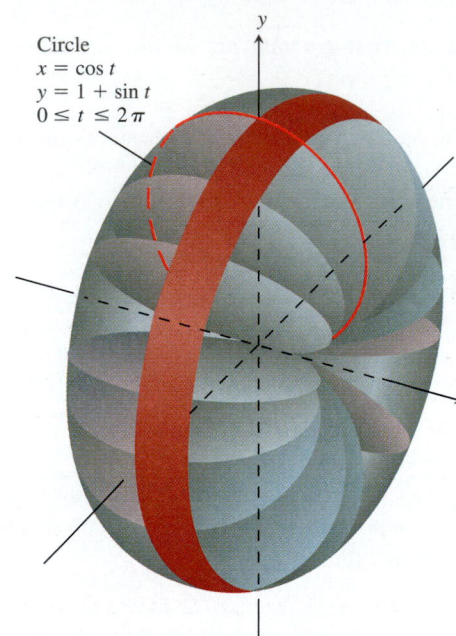

Circle
$x = \cos t$
$y = 1 + \sin t$
$0 \le t \le 2\pi$

9.48   The surface in Example 5.

As with length, we can calculate surface area from any convenient parametrization that meets the criteria stated above.

Example 5   The standard parametrization of the circle of radius 1 centered at the point $(0, 1)$ in the xy-plane is

$$x = \cos t, \qquad y = 1 + \sin t, \qquad 0 \le t \le 2\pi.$$

Use these equations to find the area of the surface swept out by revolving this circle about the x-axis (Fig. 9.48).

Solution   We use Eq. (6) with

$$y = 1 + \sin t, \qquad \frac{dx}{dt} = -\sin t, \qquad \frac{dy}{dt} = \cos t$$

to obtain

$$\text{Area} = \int_a^b 2\pi y \sqrt{\left(\frac{dx}{dt}\right)^2 + \left(\frac{dy}{dt}\right)^2} \, dt \qquad \text{(Eq.(6))}$$

$$= \int_0^{2\pi} 2\pi(1 + \sin t)\underbrace{\sqrt{(-\sin t)^2 + (\cos t)^2}}_{1} \, dt$$

$$= 2\pi \int_0^{2\pi} (1 + \sin t) \, dt = 2\pi \Big[t - \cos t\Big]_0^{2\pi}$$

$$= 2\pi[(2\pi - 1) - (0 - 1)] = 4\pi^2.$$

EXERCISES 9.4

In Exercises 1–12, find an equation for the line tangent to the curve at the point defined by the given value of t. Also, find the value of d^2y/dx^2 at this point.

1. $x = 2 \cos t, \quad y = 2 \sin t; \quad t = \pi/4$

2. $x = \sin 2\pi t, \quad y = \cos 2\pi t; \quad t = -1/6$

3. $x = 4 \sin t, \quad y = 2 \cos t; \quad t = \pi/4$

4. $x = \cos t, \quad y = \sqrt{3} \cos t; \quad t = 2\pi/3$

5. $x = t, \quad y = \sqrt{t}; \quad t = 1/4$

6. $x = \sec^2 t - 1, \quad y = \tan t; \quad t = -\pi/4$

7. $x = \sec t, \quad y = \tan t; \quad t = \pi/6$

8. $x = -\sqrt{t + 1}, \quad y = \sqrt{3t}; \quad t = 3$

9. $x = 2t^2 + 3, \quad y = t^4; \quad t = -1$

10. $x = 1/t, \quad y = -2 + \ln t; \quad t = 1$

11. $x = t - \sin t, \quad y = 1 - \cos t; \quad t = \pi/3$

12. $x = \cos t, \quad y = 1 + \sin t; \quad t = \pi/2$

Find the lengths of the curves in Exercises 13–18.

13. $x = \cos t, \quad y = t + \sin t, \quad 0 \le t \le \pi$

14. $x = t^3, \quad y = 3t^2/2, \quad 0 \le t \le \sqrt{3}$

15. $x = t^2/2, \quad y = (2t + 1)^{3/2}/3, \quad 0 \le t \le 4$

16. $x = (2t + 3)^{3/2}/3, \quad y = t + t^2/2, \quad 0 \le t \le 3$

17. $x = 8 \cos t + 8t \sin t,$
$\quad y = 8 \sin t - 8t \cos t, \quad 0 \le t \le \pi/2$

18. $x = \ln(\sec t + \tan t) - \sin t,$
$\quad y = \cos t, \quad 0 \le t \le \pi/3$ (See Fig. 9.49.)

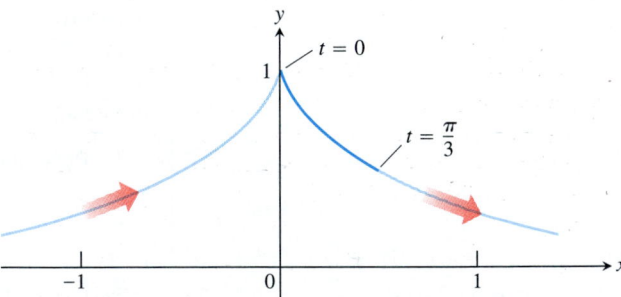

9.49   The curve in Exercise 18 is a portion of a curve that covers the entire x-axis as t runs between $-\pi/2$ and $\pi/2$. The arrows show the direction of increasing t.

Find the areas of the surfaces generated by revolving the curves in Exercises 19–22 about the indicated axes.

19. $x = \cos t$, $y = 2 + \sin t$, $0 \le t \le 2\pi$, about the x-axis

20. $x = (2/3)t^{3/2}$, $y = 2\sqrt{t}$, $0 \le t \le \sqrt{3}$, about the y-axis

21. $x = t + \sqrt{2}$, $y = (t^2/2) + \sqrt{2}\,t$, $-\sqrt{2} \le t \le \sqrt{2}$, about the y-axis

22. $x = \ln(\sec t + \tan t) - \sin t$, $y = \cos t$, $0 \le t \le \pi/3$, about the x-axis

23. *A cone frustum.* The line segment joining the points $(0, 1)$ and $(2, 2)$ is revolved about the x-axis to generate a frustum of a cone. Find the surface area of the frustum with the parametrization

$$x = 2t, \quad y = t + 1, \quad 0 \le t \le 1.$$

Check your result with the geometry formula

$$\text{Area} = \pi(r_1 + r_2)(\text{slant height}).$$

24. *A cone.* The line segment joining the origin to the point (h, r) is revolved about the x-axis to generate a cone of height h and base radius r. Find the cone's surface area with the parametrization

$$x = ht, \quad y = rt, \quad 0 \le t \le 1.$$

Check your result with the geometry formula

$$\text{Area} = \pi(r)(\text{slant height}).$$

25. a) Find the coordinates of the centroid of the curve

$$x = \cos t + t \sin t, \quad y = \sin t - t \cos t, \quad 0 \le t \le \pi/2.$$

b) CALCULATOR The curve is a portion of the involute in Fig. 9.41. Sketch the curve. Find the centroid's coordinates to the nearest tenth and add the centroid to your sketch.

26. a) Find the coordinates of the centroid of the curve

$$x = e^t \cos t, \quad y = e^t \sin t, \quad 0 \le t \le \pi.$$

b) CALCULATOR Sketch the curve. Find the centroid's coordinates to the nearest tenth and add the centroid to your sketch.

27. a) Find the coordinates of the centroid of the curve

$$x = \cos t, \quad y = t + \sin t, \quad 0 \le t \le \pi.$$

b) Sketch the curve and add the centroid to your sketch.

28. INTEGRAL EVALUATOR Most centroid calculations for curves are done with a calculator or computer that has an integral evaluation program. As a case in point, find, to the nearest hundredth, the coordinates of the centroid of the curve

$$x = t^3, \quad y = 3t^2/2, \quad 0 \le t \le \sqrt{3}.$$

29. *Length is independent of parametrization.* To illustrate the fact that the numbers we get for length do not depend on the way we parametrize our curves (except for the mild restrictions mentioned earlier), calculate the length of the semicircle $y = \sqrt{1 - x^2}$ with these two different parametrizations:

a) $x = \cos 2t$, $y = \sin 2t$, $0 \le t \le \pi/2$
b) $x = \sin \pi t$, $y = \cos \pi t$, $-1/2 \le t \le 1/2$

30. *Elliptic integrals.* The length of the ellipse

$$x = a \cos t, \quad y = b \sin t, \quad 0 \le t \le 2\pi$$

turns out to be

$$\text{Length} = 4a \int_0^{\pi/2} \sqrt{1 - e^2 \cos^2 t}\ dt,$$

where e is the ellipse's eccentricity. The integral in this formula, called an *elliptic integral,* is nonelementary except when $e = 0$ or 1.

a) CALCULATOR Use the Trapezoidal Rule with $n = 10$ to estimate the length of the ellipse when $a = 1$ and $e = 1/2$.

b) Use the fact that the absolute value of the second derivative of $f(t) = \sqrt{1 - e^2 \cos^2 t}$ is less than 1 to find an upper bound for the error in the estimate you obtained in (a).

The curves drawn by computer in Exercises 31 and 32 are called *Bowditch curves* or *Lissajous figures.* In each case, find the point in the first quadrant where the tangent to the curve is horizontal and find the equations of the two tangents at the origin.

31.

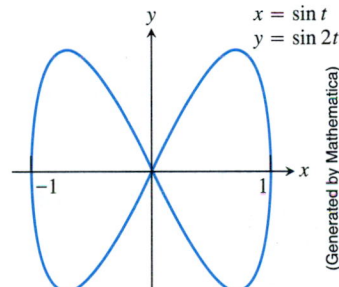

$x = \sin t$
$y = \sin 2t$

(Generated by Mathematica)

32.

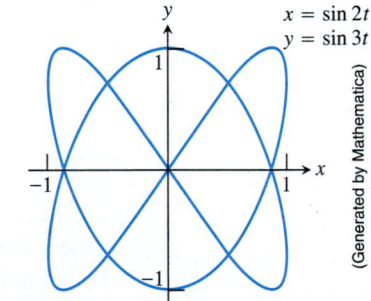

$x = \sin 2t$
$y = \sin 3t$

(Generated by Mathematica)

Computer Grapher

Graph the parametric curves in Exercises 33–39 over parameter intervals of your choice. The curves are all Bowditch curves (Lissajous figures), their general formula being

$$x = a \sin(mt + d), \qquad y = b \sin nt,$$

with m and n integers.

33. $x = \sin 2t, \quad y = \sin t$

34. $x = \sin 3t, \quad y = \sin 4t$

35. $x = \sin t, \quad y = \sin 4t$

36. $x = \sin t, \quad y = \sin 5t$

37. $x = \sin 3t, \quad y = \sin 5t$

38. $x = \sin(3t + \pi/2), \quad y = \sin 5t$

39. $x = \sin(3t + \pi/4), \quad y = \sin 5t$

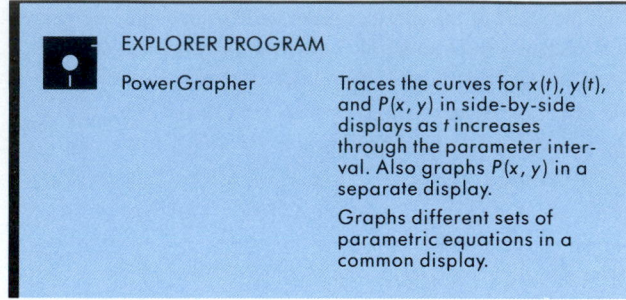

EXPLORER PROGRAM

PowerGrapher — Traces the curves for $x(t)$, $y(t)$, and $P(x, y)$ in side-by-side displays as t increases through the parameter interval. Also graphs $P(x, y)$ in a separate display.

Graphs different sets of parametric equations in a common display.

9.5 Polar Coordinates

In this section, we define polar coordinates and study their relation to Cartesian coordinates. One of the distinctions between polar and Cartesian coordinates is that while a point in the plane has just one pair of Cartesian coordinates, it has infinitely many pairs of polar coordinates. This has interesting consequences for graphing, as we shall see in the next section. Polar coordinates enable us to describe all conic sections with a single equation, as we shall see in Section 9.7, and the calculus we have done in rectangular coordinates carries over to this new system as well.

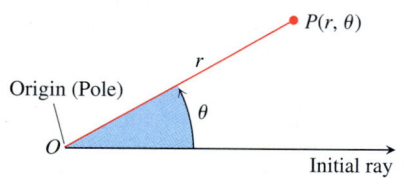

9.50 To define polar coordinates for the plane, we start with an origin called the pole, and an initial ray.

Definition of Polar Coordinates

To define polar coordinates, we first fix an **origin** O (called the **pole**) and an **initial ray** from O (Fig. 9.50). Then each point P can be located by assigning to it a **polar coordinate pair** (r, θ), in which the first number, r, gives the directed distance from O to P and the second number, θ, gives the directed angle from the initial ray to the segment OP:

Polar Coordinates

$$P(r, \theta) \tag{1}$$

Directed distance from O to P Directed angle from initial ray to OP

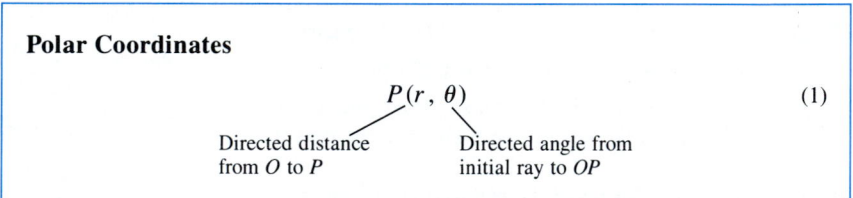

As in trigonometry, the angle θ is positive when measured counterclockwise and negative when measured clockwise. But the angle associated with a given point is not unique. For instance, the point 2 units from the origin along the ray $\theta = 30°$ has polar coordinates $r = 2$, $\theta = 30°$. It also has coordinates $r = 2$, $\theta = -330°$ (Fig. 9.51). As we shall see in Example 1, it has infinitely many coordinate pairs.

9.51 Polar coordinates are not unique. Since the ray $\theta = -330°$ is the same as the ray $\theta = 30°$, the point P defined by the coordinates $(2, 30°)$ also has coordinates $(2, -330°)$.

Negative Values of r; Changing to Radian Measure

There are occasions when we wish to allow r to be negative. That is why we say "directed distance" in (1). The ray $\theta = 30°$ and the ray $\theta = 210°$ together make a complete line through O (Fig. 9.52). The point P $(2, 210°)$ 2 units from O on the ray

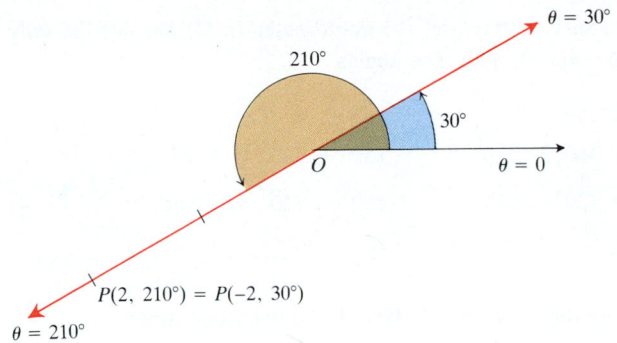

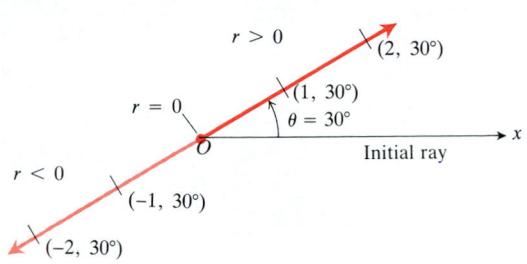

9.53 The ray $\theta = 30°$ and its opposite, the ray $\theta = 210°$, make a straight line.

9.52 Points can have polar coordinates with negative r-values.

$\theta = 210°$ has polar coordinates $r = 2$, $\theta = 210°$. It can be reached by turning 210° counterclockwise from the initial ray and going forward 2 units. It can also be reached by turning 30° counterclockwise from the initial ray and going *backward* two units. So we say that the point also has polar coordinates $r = -2$, $\theta = 30°$.

Whenever the angle between two rays is 180°, the rays make a straight line. We then say that each ray is the **opposite** of the other. Points on the ray $\theta = \alpha$ have polar coordinates (r, α) with $r \geq 0$. Points on the opposite ray, the ray $\theta = \alpha + 180°$, have coordinates (r, α) with $r \leq 0$ (Fig. 9.53).

Example 1 Find all the polar coordinates of the point $(2, 30°)$. Express the angles in radians as well as in degrees.

Solution We sketch the initial ray of the coordinate system, draw the ray through the origin that makes a 30° angle with the initial ray, and mark the point $(2, 30°)$ (Fig. 9.54). We then find formulas for the coordinate pairs in which $r = 2$ and $r = -2$ and convert the formulas to radian measure.

For $r = 2$: The angles

$$30° + 1 \cdot 360° = 390° \qquad 30° - 1 \cdot 360° = -330°$$

$$30° + 2 \cdot 360° = 750° \qquad 30° - 2 \cdot 360° = -690°$$

$$30° + 3 \cdot 360° = 1110° \qquad 30° - 3 \cdot 360° = -1050° \qquad (2)$$

$$\vdots \qquad\qquad\qquad \vdots$$

all end in the same ray as the angle 30°. Thus, the polar coordinates

$$(2, 30° + n \cdot 360°), \qquad n = 0, \pm 1, \pm 2, \ldots \qquad (3)$$

all identify the point $(2, 30°)$.

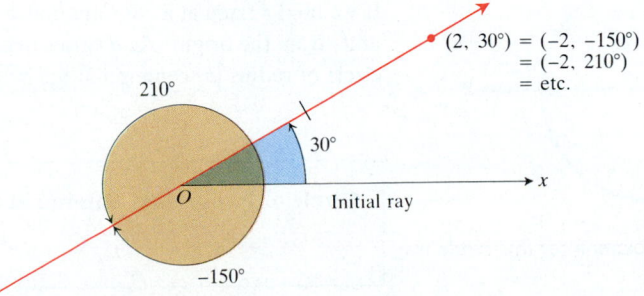

9.54 The point $P(2, 30°)$ has many different polar coordinates.

For r = −2: Numerous as they are, the coordinates in (3) are not the only polar coordinates of the point (2, 30°). The angles

$$
\begin{array}{ll}
-150° & -150° \\
-150° + 360° = 210° & -150° - 360° = -510° \\
-150° + 720° = 570° & -150° - 720° = -870°
\end{array} \tag{4}
$$

$$\vdots \qquad\qquad \vdots$$

all define the ray opposite the ray $\theta = 30°$. Hence, the polar coordinates

$$(-2, -150° + n \cdot 360°), \qquad n = 0, \pm 1, \pm 2, \ldots \tag{5}$$

represent the point (2, 30°) as well.

Radian measure: The radian formulas that correspond to (3) and (5) are

$$\left(2, \frac{\pi}{6} + 2n\pi\right), \qquad n = 0, \pm 1, \pm 2, \ldots \tag{6}$$

and

$$\left(-2, -\frac{5\pi}{6} + 2n\pi\right), \qquad n = 0, \pm 1, \pm 2, \ldots. \tag{7}$$

When $n = 0$, these formulas give

$$(2, \pi/6) \qquad \text{and} \qquad (-2, -5\pi/6).$$

When $n = 1$, they give

$$(2, 13\pi/6) \qquad \text{and} \qquad (-2, 7\pi/6),$$

and so on.

The Use of Radian Measure

Although nothing in the definition of polar coordinates requires the use of radian measure, we shall need to have all angles in radians when we differentiate and integrate trigonometric functions of θ. We shall therefore use radian measure almost exclusively from now on.

Elementary Coordinate Equations and Inequalities

If we hold r fixed at a constant nonzero value $r = a$, then the point $P(r, \theta)$ lies $|a|$ units from the origin. As θ varies over any interval of length 2π radians, P traces a circle of radius $|a|$ centered at the origin (Fig. 9.55).

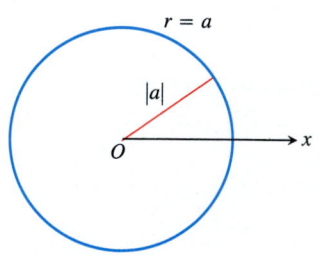

9.55 The polar equation for this circle is $r = a$.

Circle of Radius $|a|$ Centered at the Origin

$$r = a \tag{8}$$

(a)

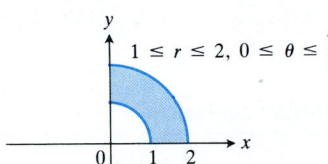

(b)

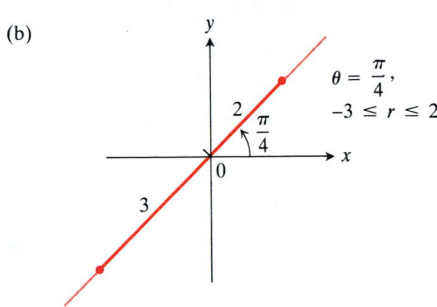

(c)

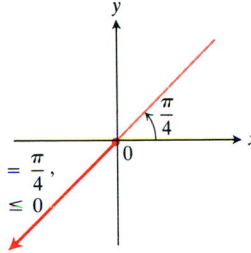

(d)

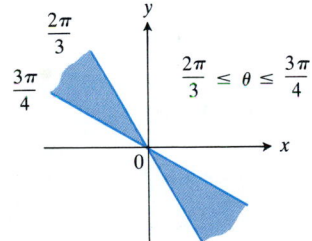

9.56   The graphs of typical inequalities in r and θ (Example 4).

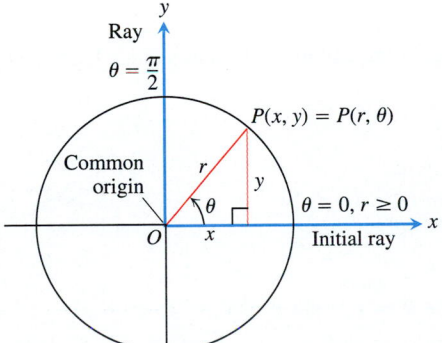

9.57   The usual way to relate polar and Cartesian coordinates.

Example 2   The equation $r = 1$ is an equation for the circle of radius 1 centered at the origin. So is the equation $r = -1$.

If we hold θ fixed at a constant value $\theta = \theta_0$ and let r run between $-\infty$ and ∞, the point $P(r, \theta)$ traces a line through the origin that makes an angle of measure θ_0 with the initial ray. The line therefore consists of all the points in the plane that have coordinates of the form (r, θ_0).

> **Equation for Lines Through the Origin**
>
> $$\theta = \theta_0 \tag{9}$$

Example 3   One equation for the line in Fig. 9.53 is $\theta = \pi/6$. The equations $\theta = 7\pi/6$ and $\theta = -5\pi/6$ are also equations for this line.

Equations of the form $r = a$ and $\theta = \theta_0$ can be combined to define regions, segments, and rays.

Example 4   Graph the sets of points whose polar coordinates satisfy the following conditions.

a) $1 \le r \le 2$ and $0 \le \theta \le \dfrac{\pi}{2}$

b) $-3 \le r \le 2$ and $\theta = \dfrac{\pi}{4}$

c) $r \le 0$ and $\theta = \dfrac{\pi}{4}$

d) $\dfrac{2\pi}{3} \le \theta \le \dfrac{3\pi}{4}$ (no restriction on r)

Solution   The graphs are shown in Fig. 9.56.

Cartesian Versus Polar Coordinates

When we use both polar and Cartesian coordinates in a plane, we place the two origins together and take the initial polar ray to be the positive x-axis. The ray $\theta = \pi/2$, $r > 0$, becomes the positive y-axis (Fig. 9.57). The two sets of coordinates are then related by the following equations.

> **Equations Relating Polar and Cartesian Coordinates**
>
> $$x = r \cos \theta, \qquad y = r \sin \theta, \qquad x^2 + y^2 = r^2, \qquad \frac{y}{x} = \tan \theta \tag{10}$$

Equations (10) are the equations we use to rewrite polar equations in Cartesian form and vice versa.

Example 5

Polar equation	Cartesian equivalent
$r \cos \theta = 2$	$x = 2$
$r^2 \cos \theta \sin \theta = 4$	$xy = 4$
$r^2 \cos^2 \theta - r^2 \sin^2 \theta = 1$	$x^2 - y^2 = 1$
$r = 1 + 2r \cos \theta$	$y^2 - 3x^2 - 4x - 1 = 0$
$r = 1 - \cos \theta$	$x^4 + y^4 + 2x^2 y^2 + 2x^3 + 2xy^2 - y^2 = 0$

With some curves, we are better off with polar coordinates; with others, we aren't.

Example 6

Find a Cartesian equation for the curve

$$r \cos\left(\theta - \frac{\pi}{3}\right) = 3.$$

Solution   We use the identity

$$\cos(A - B) = \cos A \cos B + \sin A \sin B$$

with $A = \theta$ and $B = \pi/3$:

$$r \cos\left(\theta - \frac{\pi}{3}\right) = 3,$$

$$r\left(\cos \theta \cos \frac{\pi}{3} + \sin \theta \sin \frac{\pi}{3}\right) = 3,$$

$$r \cos \theta \cdot \frac{1}{2} + r \sin \theta \cdot \frac{\sqrt{3}}{2} = 3,$$

$$\frac{1}{2} x + \frac{\sqrt{3}}{2} y = 3,$$

$$x + \sqrt{3}y = 6.$$

Example 7

Replace the following polar equations by equivalent Cartesian equations, and identify their graphs.

a) $r \cos \theta = -4$

b) $r^2 = 4r \cos \theta$

c) $r = \dfrac{4}{2 \cos \theta - \sin \theta}$

Solution   We use the substitutions $r \cos \theta = x$, $r \sin \theta = y$, $r^2 = x^2 + y^2$.

a) $r \cos \theta = -4$

The Cartesian equation: $r \cos \theta = -4$
$$x = -4$$

The graph: Vertical line through $x = -4$ on the x-axis

b) $r^2 = 4r \cos \theta$

The Cartesian equation:
$$r^2 = 4r \cos \theta$$
$$x^2 + y^2 = 4x$$
$$x^2 - 4x + y^2 = 0$$
$$x^2 - 4x + 4 + y^2 = 4 \quad \left(\begin{array}{c}\text{Completing} \\ \text{the square}\end{array}\right)$$
$$(x - 2)^2 + y^2 = 4$$

The graph: Circle, radius 2, center $(h, k) = (2, 0)$

c) $r = \dfrac{4}{2 \cos \theta - \sin \theta}$

The Cartesian equation:
$$r(2 \cos \theta - \sin \theta) = 4$$
$$2r \cos \theta - r \sin \theta = 4$$
$$2x - y = 4$$
$$y = 2x - 4$$

The graph: Line, slope $m = 2$, y-intercept $b = -4$

EXERCISES 9.5

NOTE: *All angles are in radians.*

1. Pick out the polar coordinate pairs that label the same point.

a) $(3, 0)$ b) $(-3, 0)$

c) $(-3, \pi)$ d) $(-3, 2\pi)$

e) $(2, 2\pi/3)$ f) $(2, -2\pi/3)$

g) $(2, 7\pi/3)$ h) $(-2, \pi/3)$

i) $(2, -\pi/3)$ j) $(2, \pi/3)$

k) $(-2, -\pi/3)$ l) $(-2, 2\pi/3)$

m) (r, θ) n) $(r, \theta + \pi)$

o) $(-r, \theta + \pi)$ p) $(-r, \theta)$

2. Find the Cartesian coordinates of the points whose polar coordinates are given in parts (a)–(l) of Exercise 1.

3. Plot the following points (given in polar coordinates). Then find all the polar coordinates of each point.

a) $(2, \pi/2)$ b) $(2, 0)$

c) $(-2, \pi/2)$ d) $(-2, 0)$

4. Plot the following points (given in polar coordinates). Then find all the polar coordinates of each point.

a) $(3, \pi/4)$ b) $(-3, \pi/4)$

c) $(3, -\pi/4)$ d) $(-3, -\pi/4)$

5. Find the Cartesian coordinates of the following points (given in polar coordinates).

a) $(\sqrt{2}, \pi/4)$ b) $(1, 0)$

c) $(0, \pi/2)$ d) $(-\sqrt{2}, \pi/4)$

e) $(-3, 5\pi/6)$ f) $(5, \tan^{-1}(4/3))$

g) $(-1, 7\pi)$ h) $(2\sqrt{3}, 2\pi/3)$

6. Find all polar coordinates of the origin.

Graph the sets of points whose polar coordinates satisfy the equations and inequalities in Exercises 7–22.

7. $r = 2$ 8. $0 \le r \le 2$

9. $r \ge 1$ 10. $1 \le r \le 2$

11. $0 \le \theta \le \pi/6, \quad r \ge 0$ 12. $\theta = 2\pi/3, \quad r \le -2$

13. $\theta = \pi/3, \quad -1 \le r \le 3$

14. $\theta = 11\pi/4, \quad r \ge -1$

15. $\theta = \pi/2, \quad r \ge 0$ 16. $\theta = \pi/2, \quad r \le 0$

17. $0 \le \theta \le \pi, \quad r = 1$

18. $0 \le \theta \le \pi, \quad r = -1$

19. $\pi/4 \le \theta \le 3\pi/4, \quad 0 \le r \le 1$

20. $-\pi/4 \le \theta \le \pi/4, \quad -1 \le r \le 1$

21. $-\pi/2 \le \theta \le \pi/2, \quad 1 \le r \le 2$

22. $0 \le \theta \le \pi/2, \quad 1 \le |r| \le 2$

Replace the polar equations in Exercises 23–42 by equivalent Cartesian equations. Then identify the graph.

23. $r \cos \theta = 2$ 24. $r \sin \theta = -1$

25. $r \sin \theta = 4$ 26. $r \cos \theta = 0$

27. $r \sin \theta = 0$ 28. $r \cos \theta = -3$

29. $r \cos \theta + r \sin \theta = 1$ 30. $r \sin \theta = r \cos \theta$

31. $r^2 = 1$ 32. $r^2 = 4r \sin \theta$

33. $r = \dfrac{5}{\sin\theta - 2\cos\theta}$

34. $r^2\sin 2\theta = 2$

35. $r = \cot\theta\csc\theta$

36. $r = 4\tan\theta\sec\theta$

37. $r = \csc\theta\, e^{r\cos\theta}$

38. $r\sin\theta = \ln r + \ln\cos\theta$

39. $r^2 + 2r^2\cos\theta\sin\theta = 1$

40. $\cos^2\theta = \sin^2\theta$

41. $r = 2\cos\theta + 2\sin\theta$

42. $r = 2\cos\theta - \sin\theta$

Replace the Cartesian equations in Exercises 43–52 by equivalent polar equations.

43. $x = 7$

44. $y = 1$

45. $x = y$

46. $x - y = 3$

47. $x^2 + y^2 = 4$

48. $x^2 - y^2 = 1$

49. $\dfrac{x^2}{9} + \dfrac{y^2}{4} = 1$

50. $xy = 2$

51. $y^2 = 4x$

52. $x^2 - y^2 = 25\sqrt{x^2 + y^2}$

9.6 Graphing in Polar Coordinates

The graph of a polar coordinate equation $F(r, \theta) = 0$ consists of the points whose polar coordinates in some form satisfy the equation. We say "in some form" because some coordinate pairs of a point on the graph may not satisfy the equation even when others do.

To speed our work, we look for symmetries, for values of θ at which the curve passes through the origin, and for points at which r takes on extreme values. When the curve passes through the origin, we also try to calculate the curve's slope there. This section gives the details.

Symmetry

The three parts of Fig. 9.58 illustrate the standard polar coordinate tests for symmetry.

Symmetry Tests for Polar Graphs

1. *Symmetry about the x-axis:* If the point (r, θ) lies on the graph, the point $(r, -\theta)$ or $(-r, \pi - \theta)$ lies on the graph (Fig. 9.58a).

2. *Symmetry about the y-axis:* If the point (r, θ) lies on the graph, the point $(r, \pi - \theta)$ or $(-r, -\theta)$ lies on the graph (Fig. 9.58b).

3. *Symmetry about the origin:* If the point (r, θ) lies on the graph, the point $(-r, \theta)$ or $(r, \theta + \pi)$ lies on the graph (Fig. 9.58c).

If a curve has any two of the symmetries listed here, it also has the third (as you will be invited to show in Exercise 34). If two of the tests are positive there is no need to apply the third.

Slope

The slope of a polar curve $r = f(\theta)$ is not given by the derivative $r' = df/d\theta$, but by a different formula. To see why, and what the formula is, think of the graph of f as the graph of the parametric equations

$$x = r\cos\theta = f(\theta)\cos\theta, \qquad y = r\sin\theta = f(\theta)\sin\theta. \tag{1}$$

If f is a differentiable function of θ, then so are x and y and, when $dx/d\theta \neq 0$, we

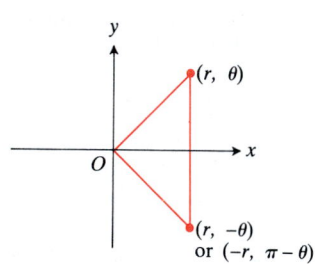

About the x-axis

(a)

About the y-axis

(b)

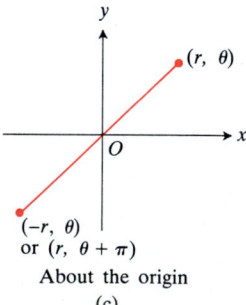

About the origin

(c)

9.58   Three tests for symmetry.

may calculate dy/dx from the parametric formula

$$\frac{dy}{dx} = \frac{dy/d\theta}{dx/d\theta} \qquad \text{(Section 9.4, Eq. (2) with } t = \theta)$$

$$= \frac{\dfrac{d}{d\theta}(f(\theta) \cdot \sin \theta)}{\dfrac{d}{d\theta}(f(\theta) \cdot \cos \theta)}$$

$$= \frac{\dfrac{df}{d\theta}\sin \theta + f(\theta) \cos \theta}{\dfrac{df}{d\theta}\cos \theta - f(\theta) \sin \theta} \qquad \text{(Product Rule for Derivatives)}$$

$$= \frac{r' \sin \theta + r \cos \theta}{r' \cos \theta - r \sin \theta}. \qquad (r = f, \ r' = df/d\theta)$$

Slope of a Polar Curve

If $r = f(\theta)$ is differentiable and $dx/d\theta \neq 0$, then the slope dy/dx at the point (r, θ) on the graph of f is given by the formula

$$\text{Slope at } (r, \theta) = \frac{r' \sin \theta + r \cos \theta}{r' \cos \theta - r \sin \theta}. \qquad (2)$$

If $r = 0$ when $\theta = \theta_0$, then Eq. (2) reduces to

$$\text{Slope at } (0, \theta_0) = \frac{r' \sin \theta_0}{r' \cos \theta_0} = \frac{\sin \theta_0}{\cos \theta_0} = \tan \theta_0. \qquad (3)$$

Slopes at the Origin

If the graph of $r = f(\theta)$ passes through the origin at the value $\theta = \theta_0$, the slope of the curve there is

$$\text{Slope at } (0, \theta_0) = \tan \theta_0. \qquad (4)$$

The reason we say "slope at $(0, \theta_0)$" and not just "slope at the origin" is that a polar curve may pass through the origin more than once, with different slopes at different θ-values. This is not the case in our first example, however.

Example 1   *A Cardioid.*   Graph the curve

$$r = 1 - \cos \theta.$$

Solution   The curve is symmetric about the x-axis because

$$(r, \theta) \text{ on the graph} \quad \Rightarrow \quad r = 1 - \cos \theta$$

$$\Rightarrow \quad r = 1 - \cos(-\theta) \qquad \begin{pmatrix} \cos \theta = \\ \cos(-\theta) \end{pmatrix}$$

$$\Rightarrow \quad (r, -\theta) \text{ on the graph}.$$

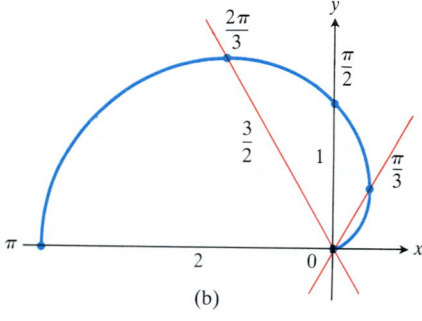

θ	$r = 1 - \cos\theta$
0	0
$\frac{\pi}{3}$	$\frac{1}{2}$
$\frac{\pi}{2}$	1
$\frac{2\pi}{3}$	$\frac{3}{2}$
π	2

(a)

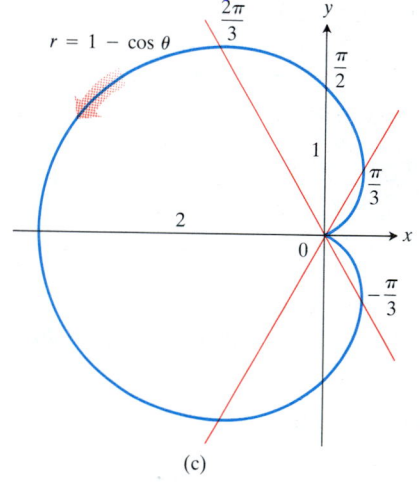

(b)

(c)

9.59 The steps in graphing the cardioid $r = 1 - \cos\theta$ (Example 1). The arrow shows the direction of increasing θ.

As θ increases from 0 to π, $\cos\theta$ decreases from 1 to -1, and $r = 1 - \cos\theta$ increases from a minimum value of 0 to a maximum value of 2. As θ continues on from π to 2π, $\cos\theta$ increases from -1 back to 1 and r decreases from 2 back to 0. The curve starts to repeat when $\theta = 2\pi$ because the cosine has period 2π.

The curve leaves the origin with slope $\tan(0) = 0$ and returns to the origin with slope $\tan(2\pi) = 0$.

We make a table of values from $\theta = 0$ to $\theta = \pi$, plot the points, draw a smooth curve through them with a horizontal tangent at the origin, and reflect the curve across the x-axis to complete the graph (Fig. 9.59). The curve is called a *cardioid* because of its heart shape.

Example 2   Graph the curve $r^2 = 4\cos\theta$.

Solution   Although $\cos\theta$ has period 2π, the equation $r^2 = 4\cos\theta$ requires $\cos\theta \geq 0$, so we get the entire graph by running θ from $-\pi/2$ to $\pi/2$. The curve is symmetric about the x-axis because

$$(r, \theta) \text{ on the graph} \quad \Rightarrow \quad r^2 = 4\cos\theta$$
$$\Rightarrow \quad r^2 = 4\cos(-\theta) \qquad \begin{pmatrix} \cos\theta = \\ \cos(-\theta) \end{pmatrix}$$
$$\Rightarrow \quad (r, -\theta) \text{ on the graph.}$$

The curve is also symmetric about the origin because

$$(r, \theta) \text{ on the graph} \quad \Rightarrow \quad r^2 = 4\cos\theta$$
$$\Rightarrow \quad (-r)^2 = 4\cos\theta$$
$$\Rightarrow \quad (-r, \theta) \text{ on the graph.}$$

Together, these two symmetries imply symmetry about the y-axis.

The curve passes through the origin when $\theta = -\pi/2$ and $\theta = \pi/2$. It has a vertical tangent both times because $\tan\theta$ is infinite.

For each value of θ in the interval between $-\pi/2$ and $\pi/2$, the formula $r^2 = 4\cos\theta$ gives two values of r:

$$r = \pm 2\sqrt{\cos\theta}.$$

We make a short table of values, plot the corresponding points, and use information about symmetry and tangents to guide us in connecting the points with a smooth curve (Fig. 9.60).

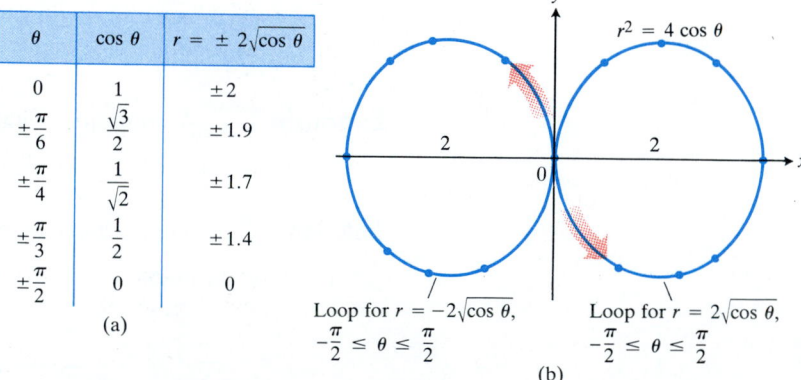

θ	$\cos\theta$	$r = \pm 2\sqrt{\cos\theta}$
0	1	± 2
$\pm\frac{\pi}{6}$	$\frac{\sqrt{3}}{2}$	± 1.9
$\pm\frac{\pi}{4}$	$\frac{1}{\sqrt{2}}$	± 1.7
$\pm\frac{\pi}{3}$	$\frac{1}{2}$	± 1.4
$\pm\frac{\pi}{2}$	0	0

(a)

Loop for $r = -2\sqrt{\cos\theta}$, $-\frac{\pi}{2} \leq \theta \leq \frac{\pi}{2}$

Loop for $r = 2\sqrt{\cos\theta}$, $-\frac{\pi}{2} \leq \theta \leq \frac{\pi}{2}$

(b)

9.60 The graph of $r^2 = 4\cos\theta$ (Example 2). The arrows show the direction of increasing θ. The values of r in the table were found with a calculator and rounded.

STEP 1: First graph $r = f(\theta)$ in the *Cartesian* $r\theta$-plane (that is, plot the values of θ on a horizontal axis and the corresponding values of r along a vertical axis).

STEP 2: Then use the Cartesian graph as a "table" and guide to sketch the *polar* coordinate graph.

Faster Graphing

One way to graph a polar equation $r = f(\theta)$ is to make a table of (r, θ) values, plot the corresponding points, and connect them in order of increasing θ. This can work well if enough points have been plotted to reveal all the loops and dimples in the graph. In this section we describe another method of graphing that is usually quicker and more reliable. The steps in the new method are described in the sidelight.

This method is better than simple point plotting because the Cartesian graph, even when hastily drawn, shows at a glance where r is positive, negative, and nonexistent, as well as where r is increasing and decreasing. As examples, we graph $r = 1 + \cos(\theta/2)$ and $r^2 = \sin 2\theta$.

Example 3   Graph the curve

$$r = 1 + \cos \frac{\theta}{2}.$$

Solution   We first graph r as a function of θ in the Cartesian $r\theta$-plane. Since the cosine has period 2π, we must let θ run from 0 to 4π to produce the entire graph (Fig. 9.61a). The arrows from the θ-axis to the curve give the radius vectors for graphing $r = 1 + \cos(\theta/2)$ in the polar plane (Fig. 9.61b).

Example 4   *A Lemniscate.*   Graph the curve $r^2 = \sin 2\theta$.

Solution   Here we begin by plotting r^2 (not r) as a function of θ in the Cartesian $r^2\theta$-plane, treating r^2 as a variable that may have negative as well as positive values (Fig. 9.62a). We pass from there to the graph of $r = \pm\sqrt{\sin 2\theta}$ in the $r\theta$-plane (Fig. 9.62b) and then draw the polar graph (Fig. 9.62c). The graph in Fig. 9.62(b) "covers" the final polar graph in Fig. 9.62(c) twice. We could have managed with either loop alone, with the two upper halves, or with the two lower halves. The double covering does no harm, however, and we actually learn a little more about the behavior of the function this way.

Finding the Points Where Curves Intersect

The fact that we can represent a point in different ways in polar coordinates makes extra care necessary in deciding when a point lies on the graph of a polar equation and in determining the points at which the graphs of polar equations intersect. The problem is that a point of intersection may satisfy the equation of one curve with polar coordinates that are different from the ones with which it satisfies the equation of another curve. Thus, solving the equations of two curves simultaneously may not identify all their points of intersection. The only sure way to identify all the points of intersection is to graph the equations.

Example 5   Show that the point $(2, \pi/2)$ lies on the curve $r = 2 \cos 2\theta$.

Solution   It may seem at first that the point $(2, \pi/2)$ does not lie on the curve because substituting the given coordinates into the equation gives

$$2 = 2 \cos 2\left(\frac{\pi}{2}\right) = 2 \cos \pi = -2,$$

How to Use Cartesian Graphs to Draw Polar Graphs

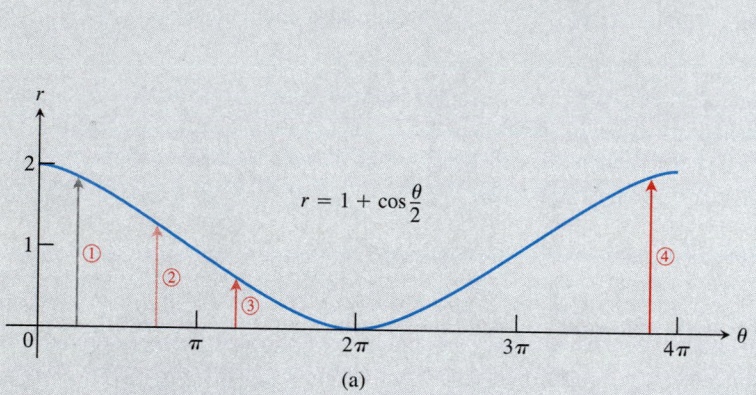

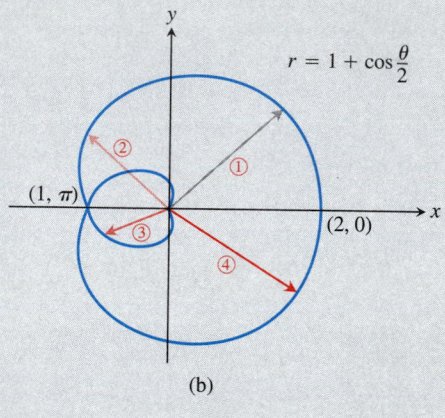

9.61 (a) The graph of $r = 1 + \cos(\theta/2)$ in the Cartesian $r\theta$-plane gives us radius vectors from the θ-axis to the curve. (b) The radius vectors help us draw the graph in the polar $r\theta$-plane.

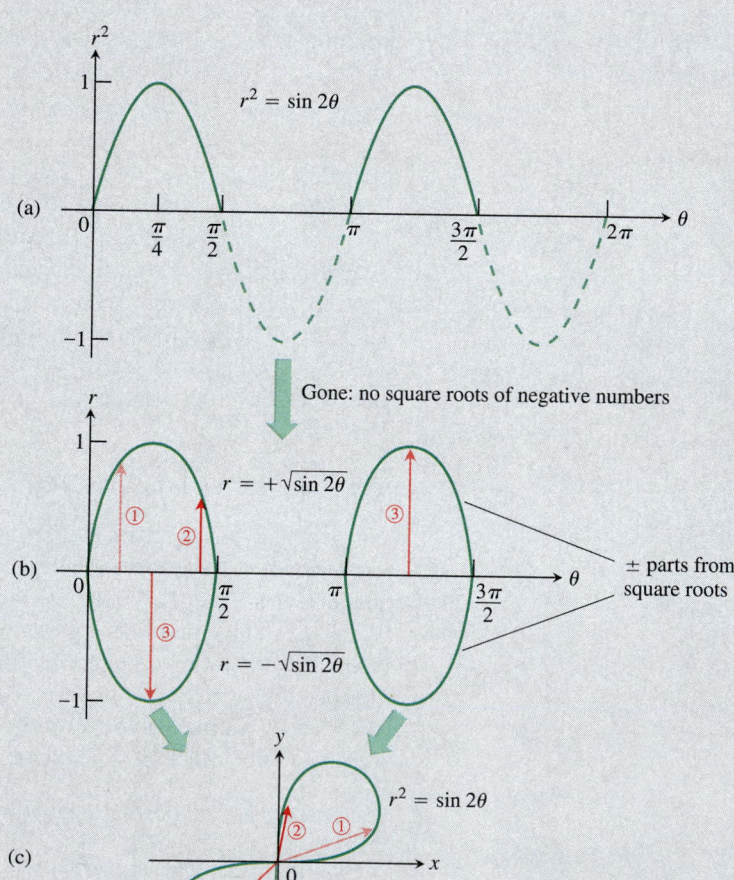

9.62 (a) The graph of $r^2 = \sin 2\theta$ in the Cartesian $r^2\theta$-plane includes negative values of the dependent variable r^2 as well as positive values. (b) When we graph r vs. θ in the Cartesian $r\theta$-plane, we ignore the points where r is imaginary but plot $+$ and $-$ parts from the points where r^2 is positive. (c) In the polar $r\theta$-plane, the radius vectors from the previous sketch cover the final graph twice.

which is not a true equality. The magnitude is right, but the sign is wrong. This suggests looking for a pair of coordinates for the given point in which r is negative, for example,

$$\left(-2, -\frac{\pi}{2}\right).$$

When we try these in the equation $r = 2 \cos 2\theta$, we find

$$-2 = 2 \cos 2\left(-\frac{\pi}{2}\right) = 2(-1) = -2,$$

and the equation is satisfied. The point $(2, \pi/2)$ does lie on the curve.

Example 6   Find the points of intersection of the curves

$$r^2 = 4 \cos \theta \qquad \text{and} \qquad r = 1 - \cos \theta.$$

Solution   In Cartesian coordinates, we can always find the points where two curves cross by solving their equations simultaneously. In polar coordinates, the story is different. Simultaneous solution may reveal some intersection points without revealing others. In this example, simultaneous solution reveals only two of the four intersection points. The others may be found by graphing. (Also, see Exercise 33.)

If we substitute $\cos \theta = r^2/4$ in the equation $r = 1 - \cos \theta$, we get

$$r = 1 - \cos \theta = 1 - \frac{r^2}{4}$$

$$4r = 4 - r^2$$

$$r^2 + 4r - 4 = 0$$

$$r = -2 \pm 2\sqrt{2}. \qquad \text{(Quadratic Formula)}$$

The value $r = -2 - 2\sqrt{2}$ has too large an absolute value to belong to either curve. The values of θ corresponding to $r = -2 + 2\sqrt{2}$ are

$$\theta = \cos^{-1}(1 - r) \qquad \text{(From } r = 1 - \cos \theta\text{)}$$

$$= \cos^{-1}(1 - (2\sqrt{2} - 2)) \qquad \text{(Set } r = 2\sqrt{2} - 2\text{)}$$

$$= \cos^{-1}(3 - 2\sqrt{2})$$

$$= \pm 80°. \qquad \text{(With a calculator, rounded)}$$

We have thus identified two intersection points:

$$(r, \theta) = (2\sqrt{2} - 2, \pm 80°).$$

If we graph the equations $r^2 = 4 \cos \theta$ and $r = 1 - \cos \theta$ together (Fig. 9.63), as we can now do by combining the graphs in Figs. 9.59 and 9.60, we see that the curves also intersect at the origin and at the point $(2, \pi)$. Why weren't the r-values of these points revealed by the simultaneous solution? The answer is that the points $(0, 0)$ and $(2, \pi)$ are not on the curves "simultaneously." They are not reached at the same value of θ. On the curve $r = 1 - \cos \theta$, the point $(2, \pi)$ is reached when $\theta = \pi$. On the curve $r^2 = 4 \cos \theta$, it is reached when $\theta = 0$, where it is identified not by the coordinates $(2, \pi)$, which do not satisfy the equation, but by the coordi-

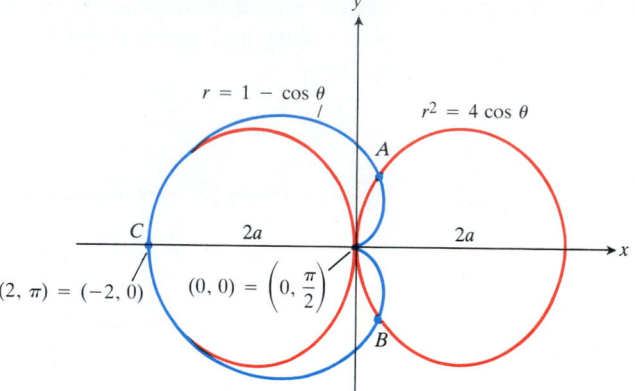

9.63 The four points of intersection of the curves $r = 1 - \cos\theta$ and $r^2 = 4\cos\theta$ (Example 6). Only A and B were found by simultaneous solution. The other two were disclosed by graphing.

nates $(-2, 0)$, which do. Similarly, the cardioid reaches the origin when $\theta = 0$, but the curve $r^2 = 4\cos\theta$ reaches the origin when $\theta = \pi/2$.

EXERCISES 9.6

Sketch the curves in Exercises 1–10.

1. $r = 1 + \cos\theta$ **2.** $r = 2 - 2\cos\theta$

3. $r = 1 - \sin\theta$ **4.** $r = 1 + \sin\theta$

5. $r = 2 + \sin\theta$ **6.** $r = 1 + 2\sin\theta$

7. $r^2 = 4\cos 2\theta$ **8.** $r^2 = 4\sin\theta$

9. $r = \theta$ **10.** $r = \sin(\theta/2)$

Use Eq. (2) to find the slopes of the curves in Exercises 11–14 at the given points. Sketch the curves along with their tangents at these points.

11. *Cardioid.* $r = -1 + \cos\theta;$ $\theta = \pm\pi/2$

12. *Cardioid.* $r = -1 + \sin\theta;$ $\theta = 0, \pi/2, \pi$

13. *Four-leaved rose.* $r = \sin 2\theta;$ $\theta = \pm\pi/4, \pm 3\pi/4$, and the values of θ at which the curve passes through the origin

14. *Four-leaved rose.* $r = \cos 2\theta;$ $\theta = 0, \pm\pi/2, \pi$, and the values of θ at which the curve passes through the origin

Graph the lemniscates in Exercises 15 and 16.

15. $r^2 = 4\cos 2\theta$ **16.** $r^2 = 4\sin 2\theta$

Graph the limaçons in Exercises 17–20. Limaçon (*"lee*-ma-sahn") is Old French for "snail." You will see why the name is appropriate when you graph the limaçons in Exercise 17. Equations for limaçons have the form $r = a \pm b\cos\theta$ or $r = a \pm b\sin\theta$. There are four basic shapes.

17. *Limaçons with an inner loop*

  a) $r = \dfrac{1}{2} + \cos\theta$ b) $r = \dfrac{1}{2} + \sin\theta$

18. *Cardioids*

  a) $r = 1 - \cos\theta$ b) $r = -1 + \sin\theta$

19. *Dimpled limaçons*

  a) $r = \dfrac{3}{2} + \cos\theta$ b) $r = \dfrac{3}{2} - \sin\theta$

20. *Cardioid-like limaçons*

  a) $r = 2 + \cos\theta$ b) $r = -2 + \sin\theta$

21. Sketch the region defined by the inequality $0 \le r \le 2 - 2\cos\theta$.

22. Sketch the region defined by the inequality $0 \le r^2 \le \cos\theta$.

23. Show that the point $(2, 3\pi/4)$ lies on the curve $r = 2\sin 2\theta$.

24. Show that the point $(1/2, 3\pi/2)$ lies on the curve $r = -\sin(\theta/3)$.

Find the points of intersection of the pairs of curves in Exercises 25–32.

25. $r = 1 + \cos\theta, \quad r = 1 - \cos\theta$

26. $r = 1 + \sin\theta, \quad r = 1 - \sin\theta$

27. $r = 1 - \sin\theta, \quad r^2 = 4\sin\theta$

28. $r^2 = \sqrt{2}\sin\theta, \quad r^2 = \sqrt{2}\cos\theta$

29. $r = 1, \quad r = 2\sin 2\theta$

30. $r = 1, \quad r^2 = 2\sin 2\theta$

31. $r = \sqrt{2}\cos 2\theta, \quad r = \sqrt{2}\sin 2\theta$

32. $r^2 = \sqrt{2}\cos 2\theta, \quad r^2 = \sqrt{2}\sin 2\theta$

33. *Continuation of Example 6.* The simultaneous solution of the equations

$$r^2 = 4\cos\theta, \tag{5}$$

$$r = 1 - \cos\theta, \tag{6}$$

in the text did not reveal the points $(0, 0)$ and $(2, \pi)$ in which their graphs intersected.

a) We could have found the point $(2, \pi)$, however, by replacing the (r, θ) in Eq. (5) by the equivalent $(-r, \theta + \pi)$ to obtain

$$r^2 = 4 \cos \theta$$
$$(-r)^2 = 4 \cos(\theta + \pi) \qquad (7)$$
$$r^2 = -4 \cos \theta.$$

Solve Eqs. (6) and (7) simultaneously to show that $(2, \pi)$ is a common solution. (This will still not reveal that the graphs intersect at $(0, 0)$.)

b) The origin is still a special case. (It often is.) Here is one way to handle it: Set $r = 0$ in Eqs. (5) and (6) and solve each equation for a corresponding value of θ. Since $(0, \theta)$ is the origin for *any* θ, this will show that both curves pass through the origin even if they do so for different θ-values.

34. Show that a curve with any two of the symmetries listed at the beginning of the section automatically has the third.

Computer Grapher

Find the points of intersection of the pairs of curves in Exercises 35–38.

35. $r^2 = \sin 2\theta, \quad r^2 = \cos 2\theta$

36. $r = 1 + \cos \dfrac{\theta}{2}, \quad r = 1 - \sin \dfrac{\theta}{2}$

37. $r = 1, \quad r = 2 \sin 2\theta$ **38.** $r = 1, \quad r^2 = 2 \sin 2\theta$

39. *A rose within a rose.* Graph the equation $r = 1 - 2 \sin 3\theta$.

40. Graph the *nephroid of Freeth:*

$$r = 1 + 2 \sin \frac{\theta}{2}$$

41. Graph the *roses* $r = \cos m\theta$ for $m = 1/3, 2, 3,$ and 7.

42. *Spirals.* Polar coordinates are just the thing for defining spirals. Graph the following spirals:

a) *A logarithmic spiral:* $r = e^{\theta/10}$
b) *A hyperbolic spiral:* $r = 8/\theta$
c) *Equilateral hyperbola:* $r = \pm 10/\sqrt{\theta}$ (Try using different colors for the two branches.)

EXPLORER PROGRAM

PowerGrapher   Graphs equations of the form $r = f(\theta)$ singly or together in a common display

---

9.7 Polar Equations for Conic Sections

Polar coordinates are important in astronomy and astronautical engineering because the ellipses, parabolas, and hyperbolas along which satellites, moons, planets, and comets move can all be described with one general polar equation. In Cartesian coordinates, the equations of conics have different forms, but not so here. This section develops the general equation, along with special equations for lines and circles.

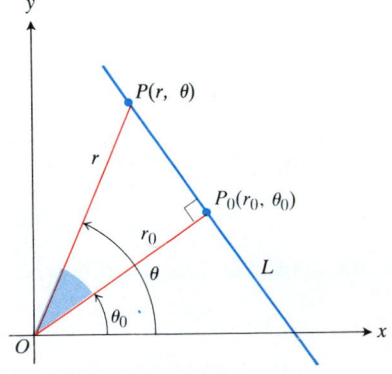

9.64 We can obtain a polar equation for line L by reading the relation $r_0/r = \cos(\theta - \theta_0)$ from triangle OP_0P.

Lines

Suppose the perpendicular from the origin to line L meets L at the point $P_0(r_0, \theta_0)$, with $r_0 \geq 0$ (Fig. 9.64). Then, if $P(r, \theta)$ is any other point on L, the points P, P_0, and O are the vertices of a right triangle, from which we can establish the trigonometric relation

$$\frac{r_0}{r} = \cos(\theta - \theta_0) \qquad (1)$$

or

$$r \cos(\theta - \theta_0) = r_0. \qquad (2)$$

The coordinates of P_0 satisfy this equation as well.

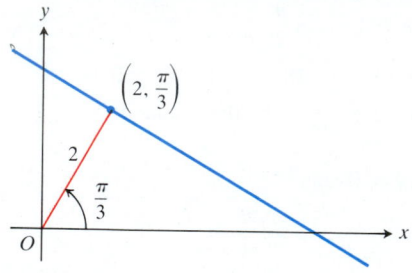

9.65 The standard polar equation for the line shown here is

$$r \cos\left(\theta - \frac{\pi}{3}\right) = 2$$

(Example 1).

The Standard Polar Equation for Lines

If the point $P_0(r_0, \theta_0)$ is the foot of the perpendicular from the origin to the line L, and $r_0 \geq 0$, then an equation for L is

$$r \cos(\theta - \theta_0) = r_0. \tag{3}$$

Example 1 Use the identity $\cos(A - B) = \cos A \cos B + \sin A \sin B$ to find a Cartesian equation for the line in Fig. 9.65.

Solution

$$r \cos\left(\theta - \frac{\pi}{3}\right) = 2$$

$$r\left(\cos\theta \cos\frac{\pi}{3} + \sin\theta \sin\frac{\pi}{3}\right) = 2$$

$$\frac{1}{2} r \cos\theta + \frac{\sqrt{3}}{2} r \sin\theta = 2$$

$$\frac{1}{2} x + \frac{\sqrt{3}}{2} y = 2$$

$$x + \sqrt{3}y = 4$$

Circles

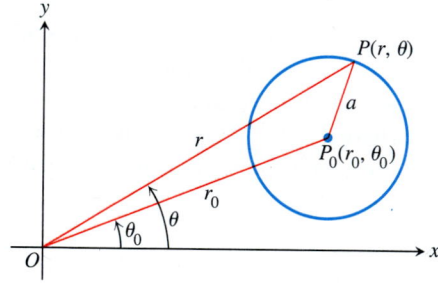

9.66 We can get a polar equation for this circle by applying the Law of Cosines to triangle OP_0P.

To find a polar equation for the circle of radius a centered at $P_0(r_0, \theta_0)$, we let $P(r,\theta)$ be a point on the circle and apply the Law of Cosines to triangle OP_0P (Fig. 9.66). This gives

$$a^2 = r_0^2 + r^2 - 2r_0 r \cos(\theta - \theta_0). \tag{4}$$

If the circle passes through the origin, then $r_0 = a$ and Eq. (4) simplifies somewhat:

$$a^2 = a^2 + r^2 - 2ar \cos(\theta - \theta_0) \qquad \text{(Eq.(4) with } r_0 = a)$$

$$r^2 = 2ar \cos(\theta - \theta_0)$$

$$r = 2a \cos(\theta - \theta_0). \tag{5}$$

If the circle's center lies on the positive x-axis, so that $\theta_0 = 0$, Eq. (5) becomes

$$r = 2a \cos\theta. \tag{6}$$

If, instead, the circle's center lies on the positive y-axis, $\theta = \pi/2$, $\cos(\theta - \pi/2) = \sin\theta$, and Eq. (5) becomes

$$r = 2a \sin\theta. \tag{7}$$

We obtain equations for circles through the origin centered on the negative x- and y-axes from Eqs. (6) and (7) by replacing r with $-r$.

Polar Equations for Circles Through the Origin Centered on the x- and y-Axes, Radius a

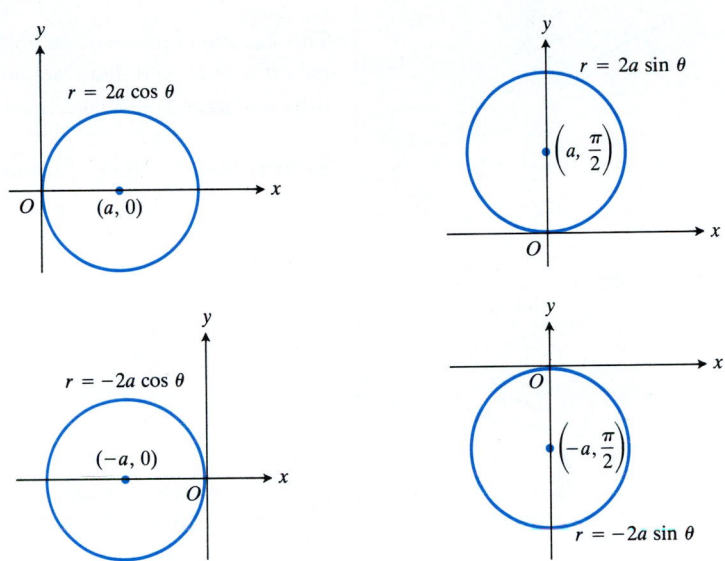

Example 2   *Circles Through the Origin*

Radius	Center (polar coordinates)	Equation
3	$(3, 0)$	$r = 6 \cos \theta$
2	$(2, \pi/2)$	$r = 4 \sin \theta$
1/2	$(-1/2, 0)$	$r = -\cos \theta$
1	$(-1, \pi/2)$	$r = -2 \sin \theta$

Ellipses, Parabolas, and Hyperbolas

To find polar equations for ellipses, parabolas, and hyperbolas, we first assume that the conic has one focus at the origin (for the parabola, its only focus) and that the corresponding directrix is the vertical line $x = k$ lying to the right of the origin (Fig. 9.67). This makes

$$PF = r \tag{8}$$

and

$$PD = k - FB = k - r \cos \theta. \tag{9}$$

The conic's focus–directrix equation $PF = e \cdot PD$ then becomes

$$r = e(k - r \cos \theta), \tag{10}$$

9.67 If a conic section is put in this position, then $PF = r$ and $PD = k - r \cos \theta$.

TABLE 9.4
Equations for conic sections ($e > 0$)

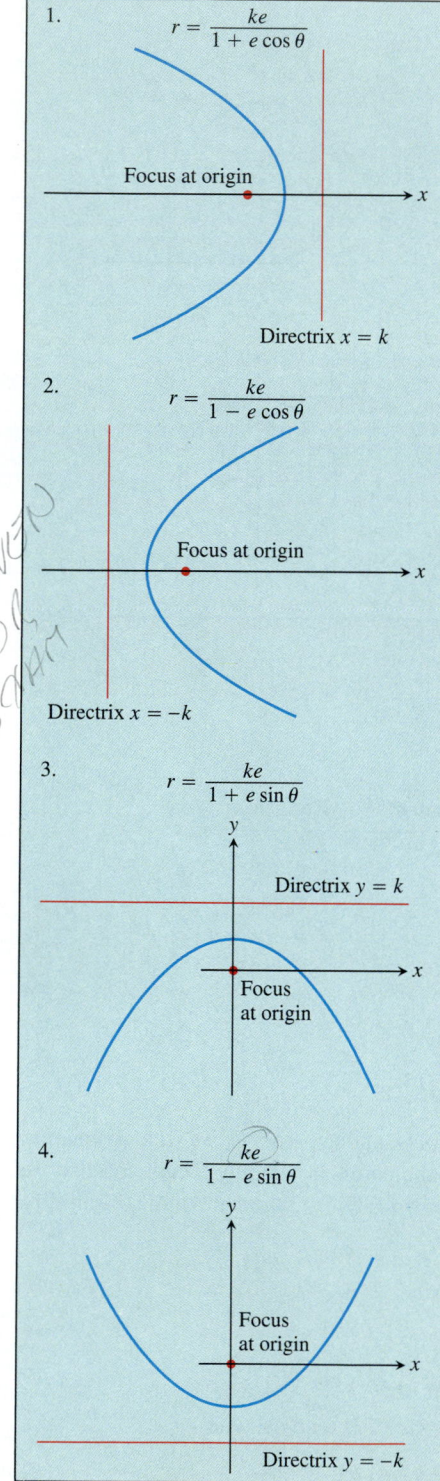

1.
$$r = \frac{ke}{1 + e \cos \theta}$$

Focus at origin

Directrix $x = k$

2.
$$r = \frac{ke}{1 - e \cos \theta}$$

Directrix $x = -k$

Focus at origin

3.
$$r = \frac{ke}{1 + e \sin \theta}$$

Directrix $y = k$

Focus at origin

4.
$$r = \frac{ke}{1 - e \sin \theta}$$

Focus at origin

Directrix $y = -k$

which can be solved for r to obtain

$$r = \frac{ke}{1 + e \cos \theta}. \tag{11}$$

This equation represents an ellipse if $0 < e < 1$, a parabola if $e = 1$, and a hyperbola if $e > 1$. And there we have it—ellipses, parabolas, and hyperbolas all with the same basic coordinate equation.

Example 3   *Typical Conics from Eq. (11)*

$$e = \frac{1}{2}: \qquad \text{ellipse} \qquad r = \frac{k}{2 + \cos \theta}$$

$$e = 1: \qquad \text{parabola} \qquad r = \frac{k}{1 + \cos \theta}$$

$$e = 2: \qquad \text{hyperbola} \qquad r = \frac{2k}{1 + 2 \cos \theta}$$

From time to time, you may see variations of Eq. (11) that come from relocating the directrix. If the directrix is the line $x = -k$ to the left of the origin (the origin is still the focus), the equation you get in place of Eq. (11) is

$$r = \frac{ke}{1 - e \cos \theta}. \tag{12}$$

The denominator now has a ($-$) instead of a ($+$). If the directrix is either of the lines $y = k$ or $y = -k$, the equations have sines in them instead of cosines, as shown in Table 9.4.

Example 4   Find an equation for the hyperbola with eccentricity 3/2, directrix $x = 2$, and focus at the origin.

Solution   We use Eq. (1) in Table 9.4 with $k = 2$ and $e = 3/2$ to get

$$r = \frac{2(3/2)}{1 + (3/2) \cos \theta}, \qquad \text{or} \qquad r = \frac{6}{2 + 3 \cos \theta}.$$

Example 5   Find the directrix of the parabola

$$r = \frac{25}{10 + 10 \cos \theta}.$$

Solution   We divide the numerator and denominator by 10 to put the equation in standard form:

$$r = \frac{5/2}{1 + \cos \theta}.$$

This is the equation

$$r = \frac{ke}{1 + e \cos \theta}$$

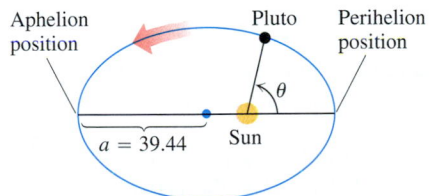

Directrix
$x = k$

Center   Focus at origin

$\longrightarrow x$

ea

a

$\dfrac{a}{e}$

9.68 In an ellipse with semimajor axis a, the focus–directrix distance is $k = (a/e) - ea$, so $ke = a(1 - e^2)$.

GIVEN ➝

with $k = 5/2$ and $e = 1$. The equation of the directrix is $x = 5/2$.

From the ellipse diagram in Fig. 9.68, we see that k is related to the eccentricity e and the semimajor axis a by the equation

$$k = \frac{a}{e} - ea.$$ (13)

$ke = a$?

From this, we find that $ke = a(1 - e^2)$. Replacing ke by $a(1 - e^2)$ in Eq. (11) gives the standard polar equation for an ellipse with eccentricity e and semimajor axis a.

> **Ellipse with Eccentricity e and Semimajor Axis a**
>
> $$r = \frac{a(1 - e^2)}{1 + e \cos \theta}$$ (14)

Notice that when $e = 0$, Eq. (14) becomes $r = a$, which represents a circle.

Equation (14) is the starting point for calculating planetary orbits in astronomy.

Example 6   Find a polar equation for an ellipse with semimajor axis 39.44 AU (astronomical units) and eccentricity 0.25. This is the approximate size of Pluto's orbit around the sun.

Solution   We use Eq. (14) with $a = 39.44$ and $e = 0.25$ to find

$$r = \frac{39.44(1 - (0.25)^2)}{1 + 0.25 \cos \theta}$$

$$= \frac{147.9}{4 + \cos \theta}.$$

At its point of closest approach (perihelion), Pluto is

$$r_{\min} = \frac{147.9}{4 + 1} = 29.58 \text{ AU}$$

from the sun. At its most distant point (aphelion), Pluto is

$$r_{\max} = \frac{147.9}{4 - 1} = 49.3 \text{ AU}$$

from the sun (Fig. 9.69).

Aphelion position

Pluto

Perihelion position

θ

$a = 39.44$ Sun

9.69   The orbit of Pluto (Example 6).

Example 7   Find the distance from one focus of the ellipse in Example 6 to the associated directrix.

Solution   We use Eq. (13) with $a = 39.44$ and $e = 0.25$ to find

$$k = 39.44\left(\frac{1}{0.25} - 0.25\right) = 147.9 \text{ AU}.$$

EXERCISES 9.7

Find polar and Cartesian equations for the lines in Exercises 1–4.

1.

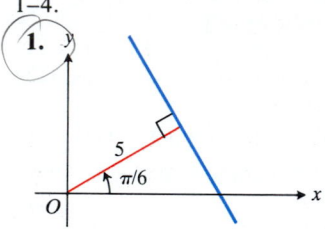

2.

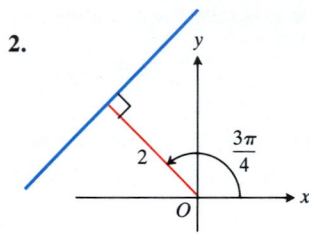

3. **4.**

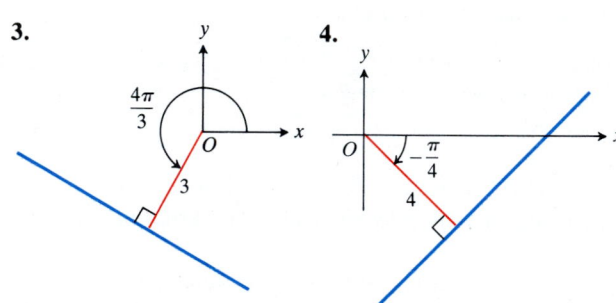

Sketch the lines in Exercises 5–8 and find Cartesian equations for them.

5. $r \cos\left(\theta - \dfrac{\pi}{4}\right) = \sqrt{2}$ **6.** $r \cos\left(\theta - \dfrac{2\pi}{3}\right) = 3$

7. $r \cos\left(\theta - \dfrac{3\pi}{2}\right) = 1$ **8.** $r \cos\left(\theta + \dfrac{\pi}{3}\right) = 2$

Find polar equations for the circles in Exercises 9–12.

9. **10.**

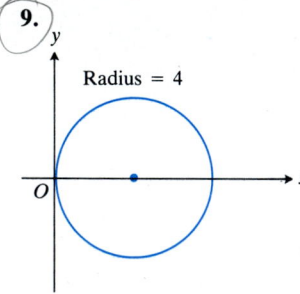

Radius = 4

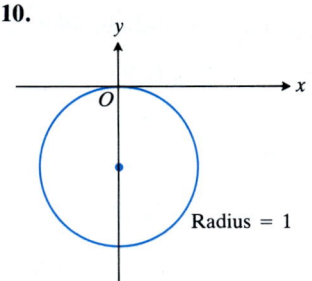

Radius = 1

11. **12.**

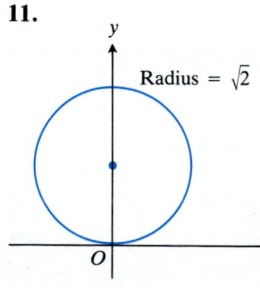

Radius = $\sqrt{2}$

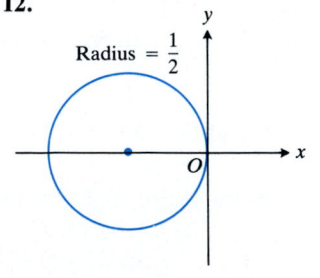

Radius = $\dfrac{1}{2}$

Sketch the circles in Exercises 13–16. Give polar coordinates for their centers and identify their radii.

13. $r = 4 \cos \theta$ **14.** $r = 6 \sin \theta$

15. $r = -2 \cos \theta$ **16.** $r = -8 \sin \theta$

Exercises 17–24 give the eccentricities of conic sections with one focus at the origin, along with the directrix corresponding to that focus. Find a polar equation for each conic section.

17. $e = 1, \quad x = 2$ **18.** $e = 1, \quad y = 2$

19. $e = 2, \quad x = 4$ **20.** $e = 5, \quad y = -6$

21. $e = 1/2, \quad x = 1$ **22.** $e = 1/4, \quad x = -2$

23. $e = 1/5, \quad y = -10$ **24.** $e = 1/3, \quad y = 6$

Sketch the parabolas and ellipses in Exercises 25–32. Include the directrix that corresponds to the focus at the origin. Label the vertices with appropriate polar coordinates. Label the centers of the ellipses as well.

25. $r = \dfrac{1}{1 + \cos \theta}$ **26.** $r = \dfrac{6}{2 + \cos \theta}$

27. $r = \dfrac{25}{10 - 5 \cos \theta}$ **28.** $r = \dfrac{4}{2 - 2 \cos \theta}$

29. $r = \dfrac{400}{16 + 8 \sin \theta}$ **30.** $r = \dfrac{12}{3 + 3 \sin \theta}$

31. $r = \dfrac{8}{2 - 2 \sin \theta}$ **32.** $r = \dfrac{4}{2 - \sin \theta}$

Sketch the regions defined by the inequalities in Exercises 33 and 34.

33. $0 \le r \le 2 \cos \theta$ **34.** $-3 \cos \theta \le r \le 0$

35. *Perihelion and aphelion.* Suppose that a planet travels about its sun in an ellipse whose semimajor axis has length a.

a) Show that $r = a(1 - e)$ when the planet is closest to the sun and that $r = a(1 + e)$ when the planet is farthest from the sun.

b) Use the data in Table 9.5 to find how close each planet in our solar system comes to the sun and how far away each planet gets from the sun.

36. *Planetary orbits.* In Example 6, we found a polar equation for the orbit of Pluto. Use the data in Table 9.5 to find polar equations for the orbits of the other planets.

37. a) Find Cartesian equations for the curves $r = 2 \sin \theta$ and $r = \csc \theta$.

b) Sketch the curves together and label their points of intersection in both Cartesian and polar coordinates.

38. Repeat Exercise 37 for $r = 2 \cos \theta$ and $r = \sec \theta$.

39. Find a polar equation for the parabola whose focus lies at the origin and whose directrix is the line $r \cos \theta = 4$.

TABLE 9.5
Semimajor axes and eccentricities of the planets in our solar system

Planet	Semimajor axis (astronomical units)	Eccentricity
Mercury	0.3871	0.2056
Venus	0.7233	0.0068
Earth	1.000	0.0167
Mars	1.524	0.0934
Jupiter	5.203	0.0484
Saturn	9.539	0.0543
Uranus	19.18	0.0460
Neptune	30.06	0.0082
Pluto	39.44	0.2481

40. Find a polar equation for the parabola whose focus lies at the origin and whose directrix is the line $r \cos(\theta - \pi/2) = 2$.

41. *The space engineer's formula for eccentricity.* The space engineer's formula for the eccentricity of an elliptical orbit is

$$e = \frac{r_{max} - r_{min}}{r_{max} + r_{min}},$$

where r is the distance from the space vehicle to the focus of the ellipse along which it travels. Why does the formula work? (*Hint:* You do not need calculus. Just think about the geometric description of eccentricity.)

42. *Halley's comet (Continuation of Example 7, Section 9.1)*
   a) Write an equation for the orbit of Halley's comet in a coordinate system in which the sun lies at the origin and the other focus lies on the negative x-axis, scaled in astronomical units.
   b) How close does the comet come to the sun in astronomical units? In miles?

c) What is the farthest the comet gets from the sun in astronomical units? In miles?

43. *Drawing ellipses with string (Continuation of Exercise 41).* You have a string with a knot in each end that can be pinned to a drawing board. The string is 10 in. long from the center of one knot to the center of the other. How far apart should the pins be to use the method illustrated in Fig. 9.2 (Section 9.1) to draw an ellipse of eccentricity 0.2? The resulting ellipse would resemble the orbit of Mercury.

44. a) Show that the equations $x = r \cos \theta$, $y = r \sin \theta$ transform the polar equation

$$r = \frac{k}{1 + e \cos \theta}$$

into the Cartesian equation

$$(1 - e^2)x^2 + y^2 + 2k\,ex - k^2 = 0.$$

   b) Then apply the criteria of Section 9.2 to show that

$$
\begin{aligned}
e &= 0 & &\Rightarrow & &\text{circle} \\
0 &< e < 1 & &\Rightarrow & &\text{ellipse} \\
e &= 1 & &\Rightarrow & &\text{parabola} \\
e &> 1 & &\Rightarrow & &\text{hyperbola.}
\end{aligned}
$$

Computer Grapher

Graph the lines and conic sections in Exercises 45–54.

45. $r = 3 \sec(\theta - \pi/3)$ **46.** $r = 4 \sec(\theta + \pi/6)$

47. $r = 4 \sin \theta$ **48.** $r = -2 \cos \theta$

49. $r = 8/(4 + \cos \theta)$ **50.** $r = 8/(4 + \sin \theta)$

51. $r = 1/(1 - \sin \theta)$ **52.** $r = 1/(1 + \cos \theta)$

53. $r = 1/(1 + 2 \sin \theta)$ **54.** $r = 1/(1 + 2 \cos \theta)$

EXPLORER PROGRAM

PowerGrapher   Graphs all conic sections in polar coordinates

---

9.8 Integration in Polar Coordinates

This section shows how to calculate areas of plane regions, lengths of curves, and areas of surfaces of revolution in polar coordinates. The general methods for setting up the integrals are the same as for Cartesian coordinates but the resulting formulas are somewhat different. Formulas for finding the centroids of fan-shaped regions are derived at the end of the exercises.

Area in the Plane

The region TOS in Fig. 9.70 is bounded by the rays $\theta = \alpha$ and $\theta = \beta$ and the curve $r = f(\theta)$. We partition angle TOS into n subangles of measure $\Delta\theta$ and approximate a typical sector POQ by a circular sector of radius $r_k = f(\theta_k)$ and central angle $\Delta\theta$

9.70   To derive a formula for the area swept out by the radius OP as P moves from T to S along the curve $r = f(\theta)$, we partition region TOS into sectors.

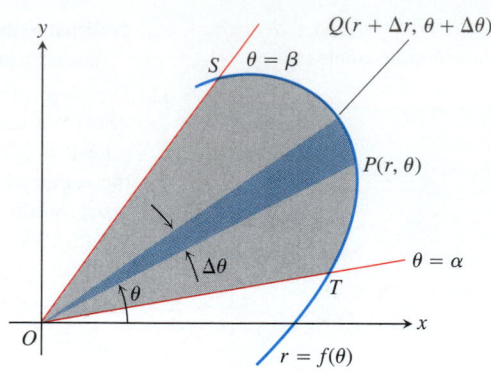

9.71   For some θ_k between θ and $\theta + \Delta\theta$, the area of the shaded circular sector just equals the area of the sector POQ bounded by the curve shown in Fig. 9.70.

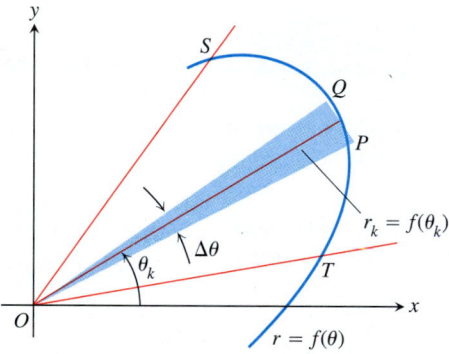

(Fig. 9.71). If f is continuous for $\alpha \le \theta \le \beta$, we can choose θ_k to make the area $(1/2)r_k^2\,\Delta\theta$ of the circular sector equal the area of sector POQ. If we choose such a radius for each of the n sectors determined by the partition of angle TOS, the area of region TOS will equal the sum

$$A = \sum_{k=1}^{n} \frac{1}{2} r_k^2\,\Delta\theta = \sum_{k=1}^{n} \frac{1}{2}\Big(f(\theta_k)\Big)^2\Delta\theta.$$

Letting $\Delta\theta \to 0$ then shows that

$$A = \lim_{\Delta\theta\to0} \sum \frac{1}{2}(f(\theta_k))^2\,\Delta\theta = \int_{\alpha}^{\beta} \frac{1}{2}f^2(\theta)\,d\theta = \int_{\alpha}^{\beta} \frac{1}{2}r^2\,d\theta.$$

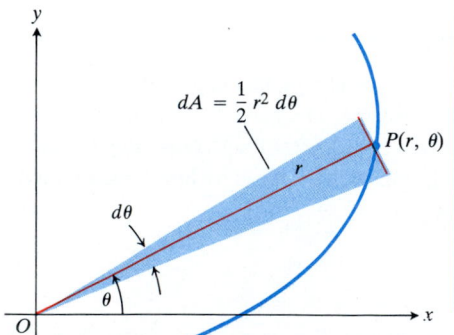

9.72   The area differential dA.

Area of the Fan-shaped Region Between the Origin and the Curve
$r = f(\theta), \quad \alpha \le \theta \le \beta$

$$A = \int_{\alpha}^{\beta} \frac{1}{2} r^2\,d\theta. \tag{1}$$

This is the integral of the **area differential** (Fig. 9.72)

$$dA = \frac{1}{2} r^2\,d\theta. \tag{2}$$

Example 1   Find the area of the region enclosed by the cardioid $r = 2(1 + \cos\theta)$.

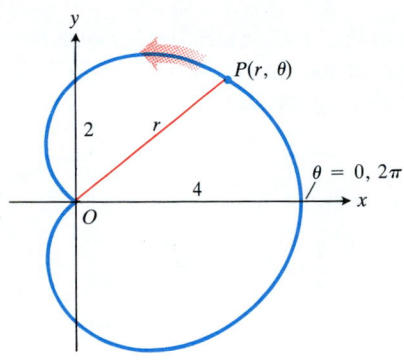

9.73 The cardioid $r = 2(1 + \cos \theta)$ (Example 1).

Solution We graph the cardioid (Fig. 9.73) and determine that the radius OP sweeps out the region exactly once as θ runs from 0 to 2π. The area is therefore

$$\int_{\theta = 0}^{\theta = 2\pi} \frac{1}{2} r^2 \, d\theta = \int_0^{2\pi} \frac{1}{2} \cdot 4(1 + \cos \theta)^2 \, d\theta$$

$$= \int_0^{2\pi} 2(1 + 2 \cos \theta + \cos^2 \theta) \, d\theta$$

$$= \int_0^{2\pi} \left(2 + 4 \cos \theta + 2 \frac{1 + \cos 2\theta}{2} \right) d\theta$$

$$= \int_0^{2\pi} (3 + 4 \cos \theta + \cos 2\theta) \, d\theta$$

$$= 3\theta + 4 \sin \theta + \frac{\sin 2\theta}{2} \Big]_0^{2\pi}$$

$$= 6\pi - 0 = 6\pi.$$

Example 2   Find the area inside the smaller loop of the limaçon

$$r = 2 \cos \theta + 1.$$

Solution   After sketching the curve (Fig. 9.74), we see that the smaller loop is traced out by the point (r, θ) as θ increases from $\theta = 2\pi/3$ to $\theta = 4\pi/3$. Since the curve is symmetric about the x-axis (the equation is unaltered when we replace θ by $-\theta$), we may calculate the area of the shaded half of the inner loop by integrating from $\theta = 2\pi/3$ to $\theta = \pi$. The area A we seek will be twice the value of the resulting integral:

$$A = 2 \int_{2\pi/3}^{\pi} \frac{1}{2} r^2 \, d\theta = \int_{2\pi/3}^{\pi} r^2 \, d\theta.$$

Since

$$r^2 = (2 \cos \theta + 1)^2$$

$$= 4 \cos^2 \theta + 4 \cos \theta + 1$$

$$= 4 \cdot \frac{1 + \cos 2\theta}{2} + 4 \cos \theta + 1$$

$$= 2 + 2 \cos 2\theta + 4 \cos \theta + 1$$

$$= 3 + 2 \cos 2\theta + 4 \cos \theta,$$

we have

$$A = \int_{2\pi/3}^{\pi} (3 + 2 \cos 2\theta + 4 \cos \theta) \, d\theta$$

$$= \left[3\theta + \sin 2\theta + 4 \sin \theta \right]_{2\pi/3}^{\pi}$$

$$= (3\pi) - \left(2\pi - \frac{\sqrt{3}}{2} + 4 \cdot \frac{\sqrt{3}}{2} \right)$$

$$= \pi - \frac{3\sqrt{3}}{2}.$$

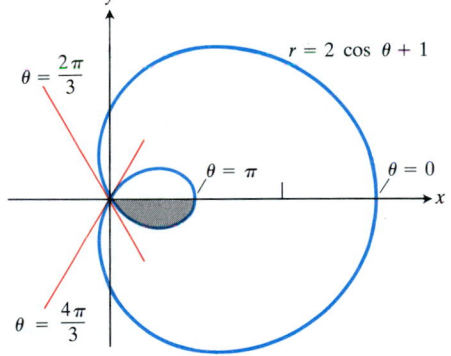

9.74   The limaçon in Example 2.

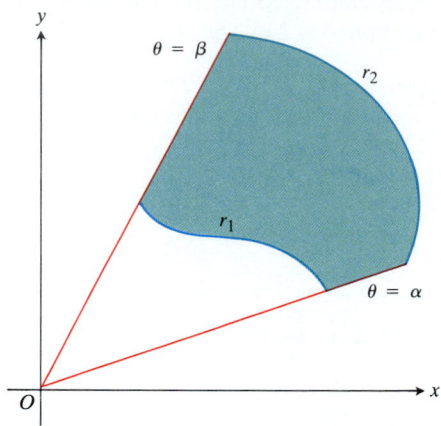

9.75 The area of the shaded region is calculated by subtracting the area of the region between r_1 and the origin from the area of the region between r_2 and the origin.

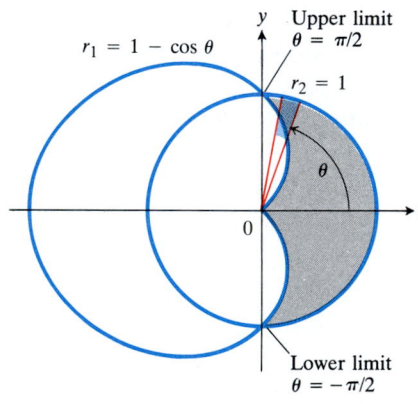

9.76 The region and limits of integration in Example 3.

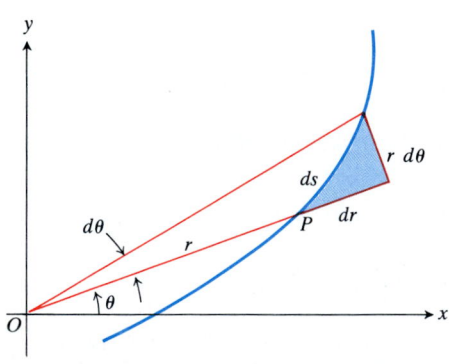

9.77 For arc length, $ds^2 = r^2\, d\theta^2 + dr^2$.

To find the area of a region like the one in Fig. 9.75, which lies between two polar curves from $\theta = \alpha$ to $\theta = \beta$, we subtract the integral of $(1/2)r_1^2\, d\theta$ from the integral of $(1/2)r_2^2\, d\theta$. This leads to the following formula.

Area of the Region $0 \le r_1(\theta) \le r \le r_2(\theta), \quad \alpha \le \theta \le \beta$

$$A = \int_\alpha^\beta \frac{1}{2} r_2^2\, d\theta - \int_\alpha^\beta \frac{1}{2} r_1^2\, d\theta = \int_\alpha^\beta \frac{1}{2}\,(r_2^2 - r_1^2)\, d\theta \qquad (3)$$

Example 3   Find the area of the region that lies inside the circle $r = 1$ and outside the cardioid $r = 1 - \cos\theta$.

Solution   We sketch the region to determine its boundaries and find the limits of integration (Fig. 9.76). The outer curve is $r_2 = 1$, the inner curve is $r_1 = 1 - \cos\theta$, and θ runs from $-\pi/2$ to $\pi/2$. The area, from Eq. (3), is

$$A = \int_{-\pi/2}^{\pi/2} \frac{1}{2}\,(r_2^2 - r_1^2)\, d\theta$$

$$= 2\int_0^{\pi/2} \frac{1}{2}\,(r_2^2 - r_1^2)\, d\theta \qquad \text{(Symmetry)}$$

$$= \int_0^{\pi/2} (1 - (1 - 2\cos\theta + \cos^2\theta))\, d\theta$$

$$= \int_0^{\pi/2} (2\cos\theta - \cos^2\theta)\, d\theta = \int_0^{\pi/2} \left(2\cos\theta - \frac{\cos 2\theta + 1}{2}\right) d\theta$$

$$= \left[2\sin\theta - \frac{\sin 2\theta}{4} - \frac{\theta}{2}\right]_0^{\pi/2} = 2 - \frac{\pi}{4}.$$

The Length of a Curve

We calculate the length of a curve $r = f(\theta)$, $\alpha \le \theta \le \beta$, by expressing the differential $ds = \sqrt{dx^2 + dy^2}$ in terms of θ and integrating from α to β. To express ds in terms of θ, we first write dx and dy as

$$dx = d(r\cos\theta) = -r\sin\theta\, d\theta + \cos\theta\, dr,$$
$$dy = d(r\sin\theta) = r\cos\theta\, d\theta + \sin\theta\, dr. \qquad (4)$$

We then square and add (arithmetic omitted) to obtain

$$ds = \sqrt{dx^2 + dy^2}$$
$$= \sqrt{r^2\, d\theta^2 + dr^2}. \qquad (5)$$

Think of ds as the hypotenuse of a right triangle whose sides are $r\, d\theta$ and dr (Fig. 9.77). For the purpose of evaluation, Eq. (5) is usually written with $d\theta$ factored out:

$$ds = \sqrt{r^2 + \left(\frac{dr}{d\theta}\right)^2}\, d\theta. \qquad (6)$$

Length of a Curve

If $r = f(\theta)$ has a continuous first derivative for $\alpha \le \theta \le \beta$ and if the point $P(r, \theta)$ traces the curve $r = f(\theta)$ exactly once as θ runs from α to β, then the length of the curve is given by the formula

$$S = \text{Length} = \int_\alpha^\beta \sqrt{r^2 + \left(\frac{dr}{d\theta}\right)^2}\, d\theta. \qquad (7)$$

$$\left(\frac{DR}{D\Theta}\right) = R'$$

Example 4   Find the length of the cardioid $r = 1 - \cos\theta$.

Solution   We sketch the cardioid to determine the limits of integration (Fig. 9.78). The point $P(r, \theta)$ starts at the origin and traces the curve once, counterclockwise as θ runs from 0 to 2π, so these are the values we take for α and β.
With

$$r = 1 - \cos\theta, \qquad \frac{dr}{d\theta} = \sin\theta,$$

we have

$$r^2 + \left(\frac{dr}{d\theta}\right)^2 = (1 - \cos\theta)^2 + (\sin\theta)^2$$

$$= 1 - 2\cos\theta + \underbrace{\cos^2\theta + \sin^2\theta}_{1}$$

$$= 2 - 2\cos\theta$$

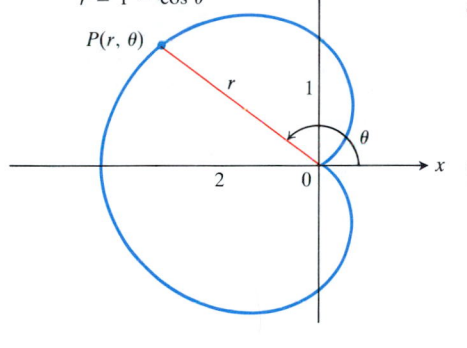

$r = 1 - \cos\theta$

$P(r, \theta)$

9.78   Example 4 calculates the length of this cardioid. The point P traces the curve once as θ runs from 0 to 2π.

and

$$\text{Length} = \int_\alpha^\beta \sqrt{r^2 + \left(\frac{dr}{d\theta}\right)^2}\, d\theta = \int_0^{2\pi} \sqrt{2 - 2\cos\theta}\, d\theta$$

$$= \int_0^{2\pi} \sqrt{4\sin^2\frac{\theta}{2}}\, d\theta \qquad \left(1 - \cos\theta = 2\sin^2\frac{\theta}{2}\right)$$

$$= \int_0^{2\pi} 2\left|\sin\frac{\theta}{2}\right| d\theta$$

$$= \int_0^{2\pi} 2\sin\frac{\theta}{2}\, d\theta \qquad \left(\sin\frac{\theta}{2} \ge 0 \text{ for } 0 \le \theta \le 2\pi\right)$$

$$= \left[-4\cos\frac{\theta}{2}\right]_0^{2\pi} = 4 + 4 = 8.$$

The Area of a Surface of Revolution

The formula for the area of a surface of revolution is $S = \int 2\pi\rho\, ds$, just as in rectangular coordinates, but now we express the radius function ρ and the arc length differential ds in terms of r and θ.

Area of a Surface of Revolution

If $r = f(\theta)$ has a continuous first derivative for $\alpha \leq \theta \leq \beta$ and if the point $P(r, \theta)$ traces the curve $r = f(\theta)$ exactly once as θ runs from α to β, then the areas of the surfaces generated by revolving the curve about the x- and y-axes are given by the following formulas:

1. Revolution about the x-axis ($y \geq 0$)

$$\text{Surface area} = \int_{\alpha}^{\beta} 2\pi y \, ds = \int_{\alpha}^{\beta} 2\pi r \sin\theta \sqrt{r^2 + \left(\frac{dr}{d\theta}\right)^2} \, d\theta \qquad (8)$$

2. Revolution about the y-axis ($x \geq 0$)

$$\text{Surface area} = \int_{\alpha}^{\beta} 2\pi x \, ds = \int_{\alpha}^{\beta} 2\pi r \cos\theta \sqrt{r^2 + \left(\frac{dr}{d\theta}\right)^2} \, d\theta \qquad (9)$$

Example 5   Find the area of the surface generated by revolving the right-hand loop of the lemniscate $r^2 = \cos 2\theta$ about the y-axis.

Solution   We sketch the loop to determine the limits of integration (Fig. 9.79). The point $P(r, \theta)$ traces the curve once, counterclockwise as θ runs from $-\pi/4$ to $\pi/4$, so these are the values we take for α and β.

We evaluate the surface area integrand

$$2\pi r \cos\theta \sqrt{r^2 + \left(\frac{dr}{d\theta}\right)^2} = 2\pi \cos\theta \sqrt{r^4 + \left(r\frac{dr}{d\theta}\right)^2} \qquad (10)$$

in stages. First of all, $r^2 = \cos 2\theta$, so

$$2r\frac{dr}{d\theta} = -2\sin 2\theta$$

$$r\frac{dr}{d\theta} = -\sin 2\theta$$

$$\left(r\frac{dr}{d\theta}\right)^2 = \sin^2 2\theta.$$

9.79   The right-hand half of a lemniscate (a) is revolved about the y-axis to generate a surface (b), whose area is calculated in Example 5.

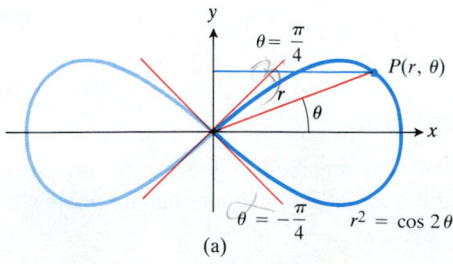

(a)

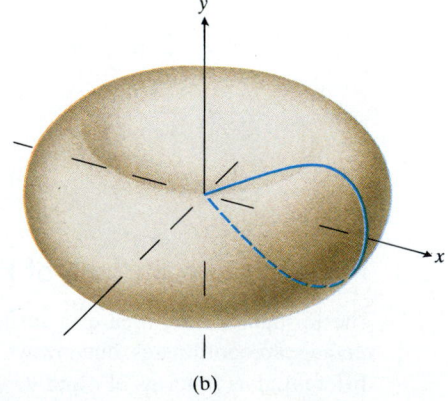

(b)

Since

$$r^4 = (r^2)^2 = \cos^2 2\theta,$$

the square root on the right-hand side of Eq. (10) simplifies to

$$\sqrt{r^4 + \left(r\frac{dr}{d\theta}\right)^2} = \sqrt{\cos^2 2\theta + \sin^2 2\theta} = 1.$$

Hence,

$$\text{Surface area} = \int_\alpha^\beta 2\pi\, r \cos\theta \sqrt{r^2 + \left(\frac{dr}{d\theta}\right)^2}\, d\theta$$

$$= \int_{-\pi/4}^{\pi/4} 2\pi \cos\theta \cdot (1)\, d\theta$$

$$= 2\pi\Big[\sin\theta\Big]_{-\pi/4}^{\pi/4} = 2\pi\left[\frac{\sqrt{2}}{2} + \frac{\sqrt{2}}{2}\right] = 2\pi\sqrt{2}.$$

EXERCISES 9.8

UP TO 31

Find the areas of the regions in Exercises 1–16.

1. Inside the convex limaçon $r = 4 + 2\cos\theta$

2. Inside the cardioid $r = a(1 + \cos\theta)$, $a > 0$

3. Inside one leaf of the four-leaved rose $r = \cos 2\theta$

4. Inside the lemniscate $r^2 = 2a^2\cos 2\theta$, $a > 0$

5. Inside one loop of the lemniscate $r^2 = 4\sin 2\theta$

6. Inside the six-leaved rose $r^2 = 2\sin 3\theta$ (Fig. 9.80)

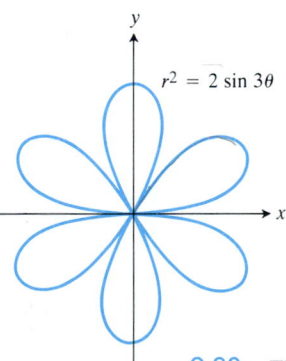

9.80   The rose in Exercise 6.

7. Shared by the circles $r = 2\cos\theta$ and $r = 2\sin\theta$

8. Shared by the circles $r = 1$ and $r = 2\sin\theta$

9. Shared by the circle $r = 2$ and the cardioid $r = 2(1 - \cos\theta)$

10. Shared by the cardioids $r = 2(1 + \cos\theta)$ and $r = 2(1 - \cos\theta)$

11. Inside the lemniscate $r^2 = 6\cos 2\theta$ and outside the circle $r = \sqrt{3}$

12. Inside the circle $r = 3a\cos\theta$ and outside the cardioid $r = a(1 + \cos\theta)$, $a > 0$

13. Inside the circle $r = -2\cos\theta$ and outside the circle $r = 1$

14. a) Inside the outer loop of the limaçon $r = 2\cos\theta + 1$ in Example 2
    b) Inside the outer loop and outside the inner loop

15. Inside the circle $r = 6$ above the line $r = 3\csc\theta$

16. Inside the lemniscate $r^2 = 6\cos 2\theta$ to the right of the line $r = (3/2)\sec\theta$.

17. a) Find the area of the shaded region in Fig. 9.81.
    b) It looks as if the graph of $r = \tan\theta$ could be asymptotic to the lines $x = 1$ and $x = -1$. Is it?

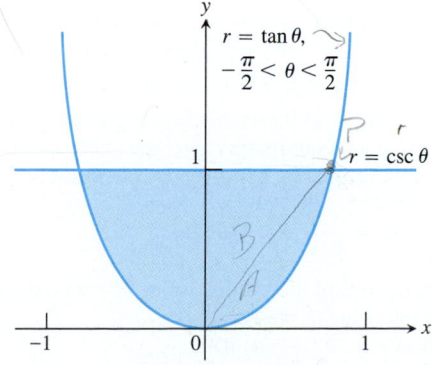

9.81   The region in Exercise 17.

18. The area of the region that lies inside the cardioid $r = \cos \theta + 1$ and outside the circle $r = \cos \theta$ is not

$$\frac{1}{2} \int_0^{2\pi} [(\cos \theta + 1)^2 - \cos^2 \theta] \, d\theta = \pi.$$

Why not? What *is* the area?

Find the lengths of the curves in Exercises 19–25.

19. The spiral $r = \theta^2, \quad 0 \le \theta \le \sqrt{5}$

20. The spiral $r = e^\theta/\sqrt{2}, \quad 0 \le \theta \le \pi$

21. The cardioid $r = 1 + \cos \theta$

22. The curve $r = a \sin^2 (\theta/2), \quad 0 \le \theta \le \pi, \quad a > 0$

23. The curve $r = \cos^3 (\theta/3), \quad 0 \le \theta \le \pi/4$

24. The curve $r = \sqrt{1 + \sin 2\theta}, \quad 0 \le \theta \le \pi\sqrt{2}$

25. The curve $r = \sqrt{\cos 2\theta}, \quad 0 \le \theta \le \pi/6$

26. *Circumferences of circles.*   As usual, when faced with a new formula, it is a good idea to try it out on familiar objects to be sure it gives results consistent with past experience. Use the length formula in Eq. (7) to calculate the circumferences of the following circles:

a) $r = a$ b) $r = a \cos \theta$ c) $r = a \sin \theta$

Find the areas of the surfaces generated by revolving the curves in Exercises 27–30 about the indicated axes.

27. $r = \sqrt{\cos 2\theta}, \quad 0 \le \theta \le \pi/4, \quad y$-axis

28. $r = \sqrt{2}e^{\theta/2}, \quad 0 \le \theta \le \pi/2, \quad x$-axis

29. $r^2 = \cos 2\theta, \quad x$-axis

30. $r = 2a \cos \theta, \quad y$-axis

31. $r = f(\theta)$ vs. $r = 2f(\theta)$.   Suppose that the function $r = f(\theta)$ satisfies the criteria set forth for Eqs. (7) and (8) on the interval $\alpha \le \theta \le \beta$.

a) Show that the length of the curve $r = 2f(\theta), \alpha \le \theta \le \beta$, is twice the length of the curve $r = f(\theta), \alpha \le \theta \le \beta$.

b) Show that the area of the surface generated by revolving the curve $r = 2f(\theta), \alpha \le \theta \le \beta$, about the x-axis is four times the area of the surface generated by revolving the curve $r = f(\theta), \alpha \le \theta \le \beta$, about the x-axis.

32. *Average value.*   If f is an integrable function of θ, the average value of the polar coordinate r over the curve $r = f(\theta)$, $\alpha \le \theta \le \beta$, with respect to θ is given by the formula

$$r_{av} = \frac{1}{\beta - \alpha} \int_\alpha^\beta f(\theta) \, d\theta.$$

Use this formula to find the average value of r with respect to θ over the following curves.

a) The cardioid $r = a(1 - \cos \theta)$

b) The circle $r = a$

c) The circle $r = a \cos \theta, \quad -\pi/2 \le \theta \le \pi/2$

Centroids of Fan-Shaped Regions

Since the centroid of a triangle is located on each median, two-thirds of the way from the vertex to the opposite base, the lever arm for the moment about the x-axis of the thin triangular region in Fig. 9.82 is about $(2/3)r \sin \theta$. Similarly, the lever arm for the moment of the triangular region about the y-axis is about $(2/3)r \cos \theta$. These approximations improve as $\Delta\theta \to 0$ and lead to the following formulas for the coordinates of the centroid of region AOB:

$$\bar{x} = \frac{\int \frac{2}{3} r \cos \theta \cdot \frac{1}{2} r^2 \, d\theta}{\int \frac{1}{2} r^2 \, d\theta} = \frac{\frac{2}{3} \int r^3 \cos \theta \, d\theta}{\int r^2 \, d\theta},$$

$$\bar{y} = \frac{\int \frac{2}{3} r \sin \theta \cdot \frac{1}{2} r^2 \, d\theta}{\int \frac{1}{2} r^2 \, d\theta} = \frac{\frac{2}{3} \int r^3 \sin \theta \, d\theta}{\int r^2 \, d\theta},$$

with limits $\theta = \alpha$ to $\theta = \beta$ on all integrals.

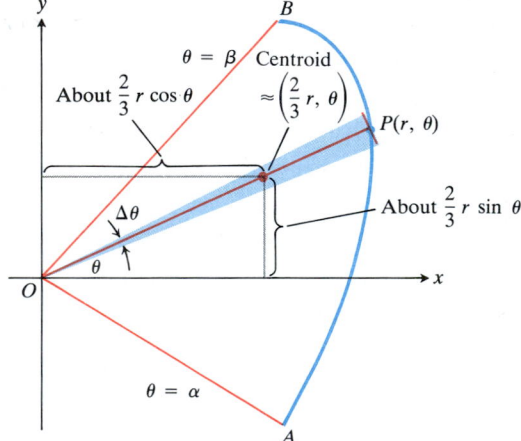

9.82   The moment of the thin triangular sector about the x-axis is approximately

$$\frac{2}{3} r \sin \theta \, dA = \frac{2}{3} r \sin \theta \cdot \frac{1}{2} r^2 d\theta = \frac{1}{3} r^3 \sin \theta \, d\theta.$$

33. Find the centroid of the region enclosed by the cardioid $r = a(1 + \cos \theta)$.

34. Find the centroid of the semicircular region $0 \le r \le a$, $0 \le \theta \le \pi$.

EXPLORER PROGRAM

PowerGrapher   Graphs curves in polar coordinates in color

REVIEW QUESTIONS

1. What reflective properties do parabolas, ellipses, and hyperbolas have? What applications use these properties?

2. How are ellipses defined in terms of distance? Give typical equations for ellipses. Graph one of the equations and include the ellipse's vertices and foci. How is the eccentricity of an ellipse defined? Sketch an ellipse whose eccentricity is close to 1 and another whose eccentricity is close to 0.

3. How are hyperbolas defined in terms of distance? Give typical equations for hyperbolas. Graph one of the equations and include the hyperbola's vertices, foci, axes, and asymptotes. What values can the eccentricities of hyperbolas have?

4. Explain the equation $PF = e \cdot PD$.

5. What can be said about the graph of the equation
$$Ax^2 + Bxy + Cy^2 + Dx + Ey + F = 0$$
if A, B, and C are not all zero? What can you tell from the number $B^2 - 4AC$?

6. How do you find a coordinate system in which the new equation for a conic section has no xy-term? Give an example.

7. Give parametric equations for a circle, parabola, and ellipse. In each case, give a parameter domain that covers the conic exactly once.

8. Give parametric equations for one branch of a hyperbola. What is the appropriate domain for the parameter if the curve is to be traced out exactly once by your equations?

9. What is a cycloid? What are typical parametric equations for a cycloid? What physical problems have cycloids for solutions?

10. How do you find dy/dx and d^2y/dx^2 when a curve in the xy-plane is given by parametric equations? Give examples.

11. How do you find the length of a parametrized curve? Give an example.

12. How do you find the area of a surface of revolution generated by a parametrized curve? Give an example.

13. Make a diagram to show the standard relations between Cartesian coordinates (x, y) and polar coordinates (r, θ). Express each set of coordinates in terms of the other kind.

14. If a point has polar coordinates (r_0, θ_0), what other polar coordinates does the point have?

15. How do you test the graph of the equation $F(r, \theta) = 0$ for symmetry about the origin? About the x-axis? About the y-axis? Give examples.

16. Describe a technique for graphing the equation $r = f(\theta)$ that involves the Cartesian $r\theta$-plane. Give an example.

17. What are the standard equations for lines, circles, ellipses, parabolas, and hyperbolas in polar coordinates? Give examples.

18. How do you find the areas of plane regions in polar coordinates? Give examples.

19. How do you find the length of a curve in polar coordinates? Give an example.

20. How do you find the area of a surface of revolution in polar coordinates? Give an example.

MISCELLANEOUS EXERCISES

1. Find an equation for the parabola with focus (4, 0) and directrix $x = 3$. Sketch the parabola together with its vertex, focus, and directrix.

2. Find the vertex, focus, and directrix of the parabola
$$x^2 - 6x - 12y + 9 = 0.$$

3. Find an equation for the curve traced by the point $P(x, y)$ if the distance from P to the vertex of the parabola $x^2 = 4y$ is twice the distance from P to the focus. Identify the curve.

4. A line segment of length $a + b$ runs from the x-axis to the y-axis. The point P on the segment lies a units from one end and b units from the other end. Show that P traces an ellipse as the ends of the segment slide along the axes.

5. The vertices of an ellipse of eccentricity 0.5 lie at the points $(0, \pm 2)$. Where do the foci lie?

6. Find an equation for the ellipse of eccentricity 2/3 that has the line $x = 2$ as a directrix and the point (4, 0) as the corresponding focus.

7. One focus of a hyperbola lies at the point $(0, -7)$ and the corresponding directrix is the line $y = -1$. Find an equation for the hyperbola if its eccentricity is (a) 2; (b) 5.

8. Find an equation for the hyperbola with foci $(0, -2)$ and (0, 2) that passes through the point (12, 7).

Two curves are said to be *orthogonal* if their tangents cross at right angles at every point where the curves intersect. Exercises 9–12 are about orthogonal conic sections.

9. Sketch the curves $xy = 2$ and $x^2 - y^2 = 3$ together and show that they are orthogonal.

10. Sketch the curves $y^2 = 4x + 4$ and $y^2 = 64 - 16x$ together and show that they are orthogonal.

11. Show that the curves $2x^2 + 3y^2 = a^2$ and $ky^2 = x^3$ are orthogonal for all values of the constants a and $k(a \neq 0, k \neq 0)$. Sketch the four curves corresponding to $a = 2$, $a = 4$, $k = 1/2$, $k = -2$ in one diagram.

12. Show that the parabolas

$$y^2 = 4a(a - x)$$

and

$$y^2 = 4b(b + x),$$

$a > 0$ and $b > 0$, have a common focus, the same for any a and b. Show that the parabolas intersect at the points $(a - b, \pm 2\sqrt{ab})$ and that each a-parabola is orthogonal to every b-parabola. (By varying a and b, we obtain two families of confocal parabolas. Each family is said to be a set of *orthogonal trajectories* of the other family. See Fig. 9.83.)

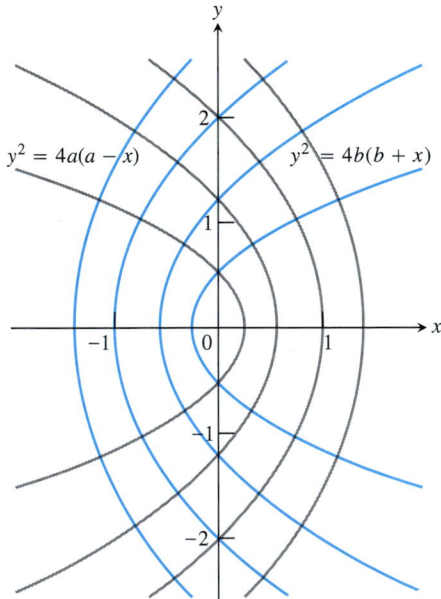

9.83 Some of the parabolas in Exercise 12.

13. A comet moves in a parabolic orbit with the sun at the focus. When the comet is 4×10^7 miles from the sun, the line from the comet to the sun makes a 60° angle with the orbit's axis (Fig. 9.84). How close will the comet come to the sun?

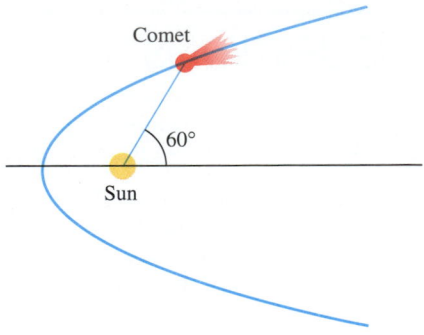

9.84 The comet in Exercise 13.

14. A ripple tank is made by bending a strip of tin around the perimeter of an ellipse for the wall of the tank and soldering a flat bottom onto this. An inch or two of water is put in the tank and the experimenter pokes a finger into it, right at one focus of the ellipse. Ripples radiate outward through the water, reflect from the strip around the edge of the tank, and in a short time a drop of water spurts up at the second focus. Why?

15. Set up the integrals that give (a) the area of a quadrant of the circle $x^2 + y^2 = a^2$, (b) the area of a quadrant of the ellipse $b^2x^2 + a^2y^2 = a^2b^2$. Show that the integral in (b) is b/a times the integral in (a), and deduce the area of the ellipse from the known area of the circle.

16. A rope with a ring in one end is looped over two pegs in a horizontal line. The free end, after being passed through the ring, has a weight suspended from it to make the rope hang taut. If the rope slips freely over the pegs and through the ring, the weight will descend as far as possible. Assume that the length of the rope is at least four times as great as the distance between the pegs and that the configuration of the rope is symmetric with respect to the line of the vertical part of the rope.

a) Find the angle formed at the bottom of the loop (Fig. 9.85).

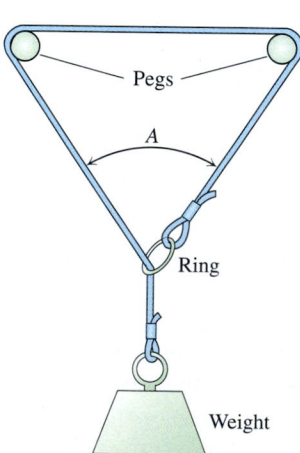

9.85 Exercise 16 asks how large the angle A will be when the frictionless rope shown here is pulled tight by the weight.

b) Show that for each fixed position of the ring on the rope, the possible locations of the ring in space lie on an ellipse with foci at the pegs.

c) Justify the original symmetry assumption by combining the result in (b) with the assumption that the rope and weight will take a rest position of minimal potential energy.

17. Two radar stations lie 20 km apart along an east–west line. A low-flying plane traveling from west to east is known to have a speed of v_0 km/sec. At $t = 0$ a signal is sent from the station at $(-10, 0)$, bounces off the plane, and is received at $(10, 0)$ $30/c$ seconds later (c is the velocity of the signal). When $t = 10/v_0$, another signal is sent out from the station at $(-10, 0)$, reflects off the plane, and is once again received $30/c$ seconds later by the other station. Find the position of the plane when it reflects the second signal under the assumption that v_0 is much less than c.

18. *LORAN.* A radio signal was sent simultaneously from towers A and B located several hundred miles apart on the California coast. A ship offshore received the signal from A 1400 microseconds before it received the signal from B.

a) Assume that the radio signals traveled at 980 ft per microsecond. What can be said about the approximate location of the ship relative to the two towers?

b) Find out what you can about LORAN and other hyperbolic radio navigation systems. (See, for example, *American Practical Navigator,* by Nathaniel Bowditch, Vol. I, U.S. Defense Mapping Agency Hydrographic Center, Publication No. 9, 1977, Chapter 43.)

19. On a level plane, at the same instant, you hear the sound of a rifle and that of the bullet hitting the target. What is your location in relation to the rifle and target?

20. Show that no tangent can be drawn from the origin to the hyperbola $x^2 - y^2 = 1$. (*Hint:* If the tangent to a curve at a point $P(x, y)$ on the curve passes through the origin, then the slope of the curve at P is y/x.)

21. *Constructing tangents to parabolas.* Show that the tangent to the parabola $y^2 = 4px$ at the point $P(x_1, y_1) \neq (0, 0)$ on the parabola meets the axis of symmetry x_1 units to the left of the vertex. This provides an accurate way to construct a tangent to the parabola at any point other than the origin (where we already have the y-axis): Mark the point $P(x_1, y_1)$ in question, drop a perpendicular from P to the x-axis, measure $2x_1$ units to the left, mark that point, and draw a line from there through P.

22. *The reflective property of ellipses.* An ellipsoid is generated by rotating an ellipse about its major axis. The inside surface of the ellipsoid is silvered to produce a mirror. Show that a light ray emanating from one focus will be reflected to the other focus. (*Hint:* Put the ellipse in standard position in the plane, as in Fig. 9.86, and show that the lines from the foci to a point P on the ellipse make equal angles with the tangent at P.)

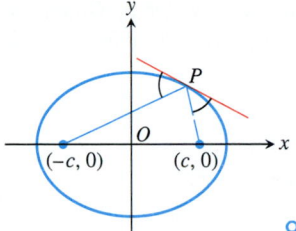

9.86 The ellipse in Exercise 22.

What points in the xy-plane satisfy the equations and inequalities in Exercises 23–30? Draw a figure for each exercise.

23. $(x^2 - y^2 - 1)(x^2 + y^2 - 25)(x^2 + 4y^2 - 4) = 0$

24. $(x + y)(x^2 + y^2 - 1) = 0$

25. $(x^2/9) + (y^2/16) \leq 1$

26. $(x^2/9) - (y^2/16) \leq 1$

27. $(9x^2 + 4y^2 - 36)(4x^2 + 9y^2 - 16) \leq 0$

28. $(9x^2 + 4y^2 - 36)(4x^2 + 9y^2 - 16) > 0$

29. $x^4 - (y^2 - 9)^2 = 0$

30. $x^2 + xy + y^2 < 3$

Use the discriminant to decide whether the equations in Exercises 31–34 represent parabolas, ellipses, or hyperbolas.

31. $x^2 + xy + y^2 + x + y + 1 = 0$

32. $x^2 + 3xy + 2y^2 + x + y + 1 = 0$

33. $x^2 + 4xy + 4y^2 + x + y + 1 = 0$

34. $x^2 + 2xy - 2y^2 + x + y + 1 = 0$

Identify the conic sections in Exercises 35 and 36. Then rotate the coordinate axes to find a new equation for the conic section that has no xy-term. (The new equations will vary with the rotations you choose.)

35. $2x^2 + xy + 2y^2 - 15 = 0$

36. $x^2 + 2\sqrt{3}xy - y^2 + 4 = 0$

37. Find the points on the parabola $x = 2t$, $y = t^2$, $-\infty < t < \infty$, closest to the point $(0, 3)$.

38. Is the curve $\sqrt{x} + \sqrt{y} = 1$ part of a conic section? If so, what kind of conic section? If not, why not?

39. CALCULATOR Find the eccentricity of the ellipse $x^2 + xy + y^2 = 1$ to the nearest hundredth.

40. Show that any tangent to the hyperbola $xy = a^2$ makes a triangle of area $2a^2$ with the hyperbola's asymptotes.

41. Find the eccentricity of the hyperbola $xy = 1$.

42. Show that the curve $2xy - \sqrt{2}y + 2 = 0$ is a hyperbola. Find the hyperbola's center, vertices, foci, axes, and asymptotes.

Exercises 43–48 give parametric equations and parameter intervals for the motion of a particle in the xy-plane. Identify the particle's path by finding a Cartesian equation for it. Graph the Cartesian equation and indicate the direction of motion and the portion traced by the particle.

43. $x = t/2, \quad y = t + 1; \quad -\infty < t < \infty$

44. $x = \sqrt{t}, \quad y = 1 - \sqrt{t}; \quad t \geq 0$

45. $x = (1/2)\tan t, \quad y = (1/2)\sec t; \quad -\pi/2 < t < \pi/2$

46. $x = -2 \cos t, \quad y = 2 \sin t; \quad 0 \leq t \leq \pi$

47. $x = -\cos t, \quad y = \cos^2 t; \quad 0 \leq t \leq \pi$

48. $x = 4 \cos t, \quad y = 9 \sin t; \quad 0 \leq t \leq 2\pi$

49. Find parametric equations and a parameter interval for the motion of a particle in the xy-plane that traces the ellipse $16x^2 + 9y^2 = 144$ once counterclockwise. (There are many ways to do this, so your answer may not be the same as the one in the back of the book.)

50. Find parametric equations and a parameter interval for the motion of a particle that starts at the point $(-2, 0)$ in the xy-plane and traces the circle $x^2 + y^2 = 4$ three times clockwise. (There are many ways to do this.)

51. *Epicycloids.* When a circle rolls externally along the circumference of a second, fixed circle, any point P on the circumference of the rolling circle describes an *epicycloid* (Fig. 9.87). Let the fixed circle have its center at the origin O and have radius a. Let the radius of the rolling circle be b and let the initial position of the tracing point P be $A(a, 0)$. Find parametric equations for the epicycloid, using as the parameter the angle θ from the positive x-axis to the line through the circles' centers.

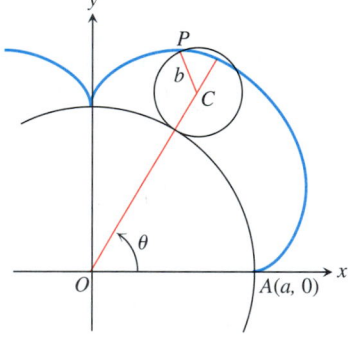

9.87   An epicycloid with $b = a/4$ (Exercise 51).

52. *Hypocycloids.* When a circle rolls on the inside of a fixed circle, any point P on the circumference of the rolling circle describes a *hypocycloid*. Let the fixed circle be $x^2 + y^2 = a^2$, let the radius of the rolling circle be b, and let the initial position of the tracing point P be $A(a, 0)$. Find parametric equations for the hypocycloid, using as the parameter the angle θ from the positive x-axis to the line joining the circles' centers. In particular, if $b = a/4$, as in Fig. 9.88, show that the hypocycloid is the astroid

$$x = a \cos^3 \theta, \quad y = a \sin^3 \theta.$$

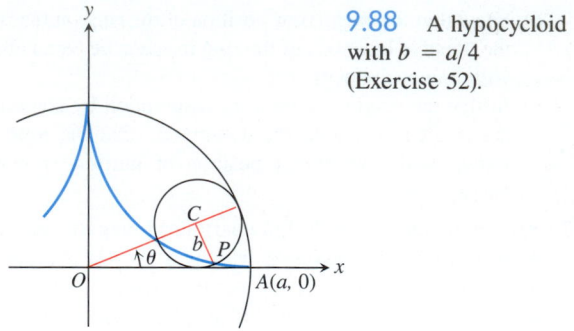

9.88   A hypocycloid with $b = a/4$ (Exercise 52).

53. *More about hypocycloids.* Figure 9.89 shows a circle of radius a tangent to the inside of a circle of radius $2a$. The point P, shown as the point of tangency in the figure, is attached to the smaller circle. What path does P trace as the smaller circle rolls around the inside of the larger circle?

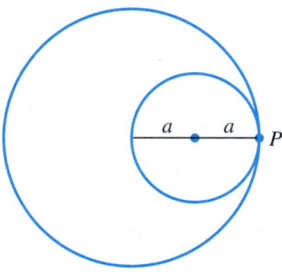

9.89   The circles in Exercise 53.

54. *Generating a cardioid with circles.* Cardioids are special epicycloids (Exercise 51). Show that if you roll a circle of radius a about another circle of radius a in the polar coordinate plane, as in Fig. 9.90, the original point of contact P will trace a cardioid. (*Hint:* Start by showing that angles OBC and PAD both have measure θ.)

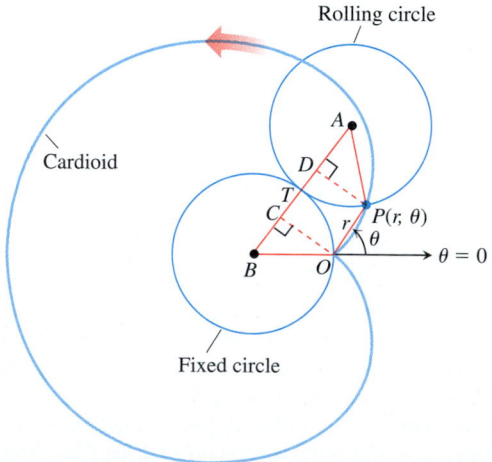

9.90   As the circle centered at A rolls around the circle centered at B, the point P traces a cardioid (Exercise 54).

In Exercises 55 and 56, find an equation for the line in the *xy*-plane that is tangent to the curve at the point corresponding to the given value of *t*. Also, find the value of d^2y/dx^2 at this point.

55. $x = (1/2)\tan t, \quad y = (1/2)\sec t; \quad t = \pi/3$

56. $x = 1 + 1/t^2, \quad y = 1 - 3/t; \quad t = 2$

Find the lengths of the curves in Exercises 57 and 58.

57. $x = e^{2t} - \dfrac{t}{8}, \quad y = e^t; \quad 0 \le t \le \ln 2$.

58. The closed loop in Fig. 9.91.

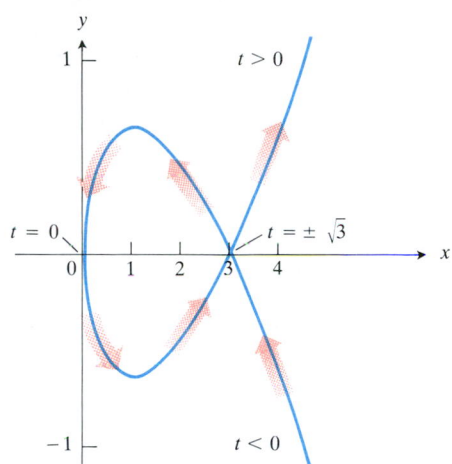

9.91   Exercise 58 refers to the curve $x = t^2$, $y = (t^3/3) - t$ shown here. The loop starts at $t = -\sqrt{3}$ and ends at $t = \sqrt{3}$.

Find the areas of the surfaces generated by revolving the curves in Exercises 59 and 60 about the axes indicated.

59. $x = t^2/2, \quad y = 2t, \quad 0 \le t \le \sqrt{5}; \quad x\text{-axis}$

60. $x = t^2 + 1/(2t), \quad y = 4\sqrt{t}, \quad 1/\sqrt{2} \le t \le 1; \quad y\text{-axis}$

61. Find the centroid of the region enclosed by the *x*-axis and the cycloid arch

$$x = a(t - \sin t), \quad y = a(1 - \cos t), \quad 0 \le t \le 2\pi.$$

(*Hint:* Express $\tilde{x}$, $\tilde{y}$, and dx in terms of *t* and *dt*.)

62. (Continuation of Exercise 61.) Use a theorem of Pappus to find the volume of the solid generated by revolving the region in Exercise 61 about the *x*-axis.

63. A point $P(x, y)$ moves in the plane in such a way that for $t \ge 0$,

$$\frac{dx}{dt} = \frac{1}{t + 2} \quad \text{and} \quad \frac{dy}{dt} = 2t.$$

a) Express *x* and *y* as functions of *t* if $x = \ln 2$ and $y = 1$ when $t = 0$.

b) Express *y* in terms of *x*.

c) Express *x* in terms of *y*.

d) Find the average rate of change in *y* with respect to *x* as *t* varies from 0 to 2.

e) Find dy/dx when $t = 1$.

64. Find parametric equations and a Cartesian equation for the curve traced by the point $P(x, y)$ if its coordinates satisfy the differential equations

$$\frac{dx}{dt} = -2y, \quad \frac{dy}{dt} = \cos t,$$

subject to the conditions that $x = 3$ and $y = 0$ when $t = 0$. Identify the curve.

65. Find the first moments about the coordinate axes of the curve

$$x = (2/3)t^{3/2}, \quad y = 2\sqrt{t}, \quad 0 \le t \le \sqrt{3}.$$

66. *Pythagorean triples.* Suppose that the coordinates of a particle $P(x, y)$ moving in the plane are

$$x = \frac{1 - t^2}{1 + t^2} \quad \text{and} \quad y = \frac{2t}{1 + t^2}$$

for $-\infty < t < \infty$. Show that $x^2 + y^2 = 1$ and hence that the motion takes place on the unit circle. What one point of the circle is not covered by the motion? Sketch the circle and indicate the direction of motion for increasing *t*. For what values of *t* does $(x, y) = (0, -1)$? $(1, 0)$? $(0, 1)$?

From $x^2 + y^2 = 1$, we obtain

$$(t^2 - 1)^2 + (2t)^2 = (t^2 + 1)^2,$$

an equation of interest in number theory because it generates *Pythagorean triples* of integers. When *t* is an integer greater than 1, $a = t^2 - 1$, $b = 2t$, and $c = t^2 + 1$ are positive integers that satisfy the equation $a^2 + b^2 = c^2$.

Each of the graphs in Exercises 67–74 is the graph of one of the equations (a)–(l) listed below. Find the equation for each graph.

a) $r = \cos 2\theta$ b) $r \cos \theta = 1$

c) $r = \dfrac{6}{1 - 2\cos \theta}$ d) $r = \sin 2\theta$

e) $r = \theta$ f) $r^2 = \cos 2\theta$

g) $r = 1 + \cos \theta$ h) $r = 1 - \sin \theta$

i) $r = \dfrac{2}{1 - \cos \theta}$ j) $r^2 = \sin 2\theta$

k) $r = -\sin \theta$ l) $r = 2\cos \theta + 1$

67. Four-leaved rose   **68.** Spiral

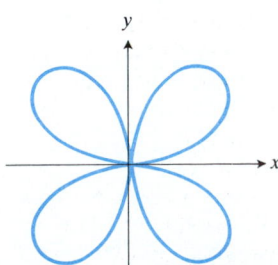

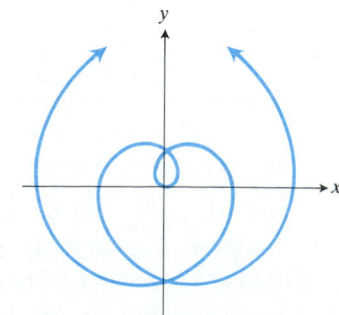

69. Limaçon

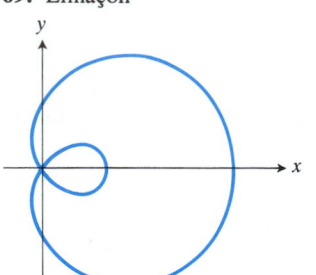

70. Lemniscate

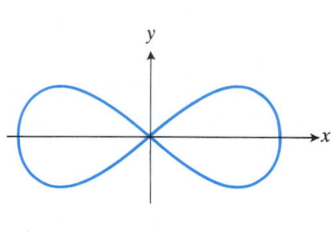

71. Circle

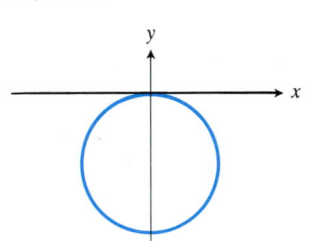

72. Cardioid

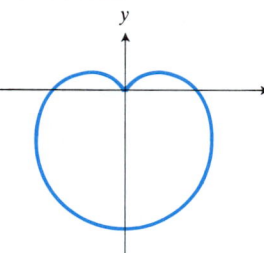

73. Parabola   **74.** Lemniscate

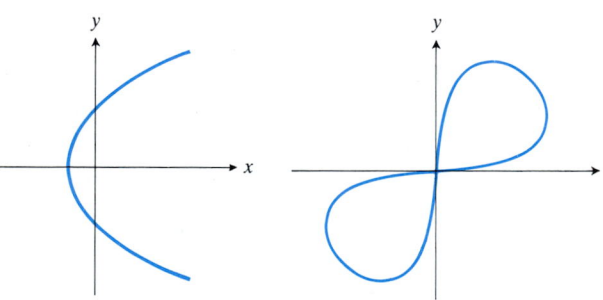

Sketch the regions defined by the polar coordinate inequalities in Exercises 75 and 76.

75. $0 \le r \le 6 \cos \theta$

76. $-4 \sin \theta \le r \le 0$

Find the points of intersection of the curves given by the polar coordinate equations in Exercises 77–84.

77. $r = \sin \theta, \quad r = 1 + \sin \theta$

78. $r = \cos \theta, \quad r = 1 - \cos \theta$

79. $r = 1 + \sin \theta, \quad r = -1 + \sin \theta$

80. $r = 1 + \cos \theta, \quad r = -1 - \cos \theta$

81. $r = a, \quad r = a(1 - \sin \theta), \quad a > 0$

82. $r = a \sec \theta, \quad r = 2a \sin \theta, \quad a > 0$

83. $r = a \cos \theta, \quad r = a(1 + \cos \theta), \quad a > 0$

84. $r = a(1 + \cos 2\theta), \quad r = a \cos 2\theta, \quad a > 0$

Find equations for the lines that are tangent to the polar coordinate curves in Exercises 85 and 86 at the origin.

85. The lemniscate $r^2 = \cos 2\theta$

86. The limaçon $r = 2 \cos \theta + 1$

87. Find polar coordinate equations for the lines that are tangent to the tips of the petals of the four-leaved rose $r = \sin 2\theta$.

88. Find polar coordinate equations for the lines that are tangent to the cardioid $r = 1 + \sin \theta$ at the points where it crosses the x-axis.

Sketch the conic sections whose polar coordinate equations are given in Exercises 89–92. Give polar coordinates for the vertices and, in the case of ellipses, for the centers as well.

89. $r = \dfrac{2}{1 + \cos \theta}$ **90.** $r = \dfrac{8}{2 + \cos \theta}$

91. $r = \dfrac{6}{1 - 2 \cos \theta}$ **92.** $r = \dfrac{12}{3 + \sin \theta}$

Exercises 93–96 give the eccentricities of conic sections with one focus at the origin of the polar coordinate plane, along with the directrix for that focus. Find a polar equation for each conic section.

93. $e = 2, \quad r \cos \theta = 2$ **94.** $e = 1, \quad r \cos \theta = -4$

95. $e = 1/2, \quad r \sin \theta = 2$ **96.** $e = 1/3, \quad r \sin \theta = -6$

97. Find a polar coordinate equation for

a) the parabola with focus at the origin and vertex at $(a, \pi/4)$;

b) the ellipse with foci at the origin and $(2, 0)$ and one vertex at $(4, 0)$;

c) the hyperbola with one focus at the origin, center at $(2, \pi/2)$, and a vertex at $(1, \pi/2)$.

98. *A satellite orbit.*   A satellite is in an orbit that passes over the North and South Poles of the earth. When it is over the South Pole it is at the highest point of its orbit, 1000 miles above the earth's surface. Above the North Pole it is at the lowest point of its orbit, 300 miles above the earth's surface.

a) Assuming that the orbit (with reference to the earth) is an ellipse with one focus at the center of the earth, find its eccentricity. (Take the diameter of the earth to be 8000 miles.)

b) Using the north–south axis of the earth as the x-axis and the center of the earth as origin, find a polar coordinate equation for the orbit.

Find the areas of the regions in the polar coordinate plane described in Exercises 99–102.

99. Enclosed by the limaçon $r = 2 - \cos \theta$

100. Enclosed by one leaf of the three-leaved rose $r = \sin 3\theta$

101. Inside the two-leaved rose $r = 1 + \cos 2\theta$ and outside the circle $r = 1$

102. Inside the cardioid $r = 2(1 + \sin \theta)$ and outside the circle $r = 2 \sin \theta$

Find the areas of the surfaces generated by revolving the polar coordinate curves in Exercises 103 and 104 about the indicated axes.

103. $r = \sqrt{\cos 2\theta}$, $0 \le \theta \le \pi/4$, about the x-axis

104. $r^2 = \sin 2\theta$, about the y-axis

105. a) Find an equation in polar coordinates for the curve

$$x = e^{2t} \cos t, \quad y = e^{2t} \sin t, \quad -\infty < t < \infty.$$

b) Sketch the curve.

c) Find the length of the curve from $t = 0$ to $t = 2\pi$.

106. Find the length of the curve $r = 2 \sin^3(\theta/3)$, $0 \le \theta \le 3\pi$, in the polar coordinate plane.

107. Find the area of the surface generated by revolving the first-quadrant portion of the cardioid $r = 1 + \cos \theta$ about the x-axis. (*Hint:* Use the identities $1 + \cos \theta = 2 \cos^2(\theta/2)$ and $\sin \theta = 2 \sin(\theta/2) \cos(\theta/2)$ to simplify the integral.)

108. Sketch the regions enclosed by the curves $r = 2a \cos^2(\theta/2)$ and $r = 2a \sin^2(\theta/2)$, $a > 0$, in the polar coordinate plane and find the area of the portion of the plane they have in common.

The Angle Between the Radius Vector and the Tangent Line to a Polar Coordinate Curve

In Cartesian coordinates, when we want to discuss the direction of a curve at a point, we use the angle ϕ measured counterclockwise from the positive x-axis to the tangent line. In polar coordinates, it is more convenient to calculate the angle ψ from the *radius vector* to the tangent line (Fig. 9.92). The angle ϕ can then be calculated from the relation

$$\phi = \theta + \psi, \tag{1}$$

which comes from applying the exterior angle theorem to the triangle in Fig. 9.92.

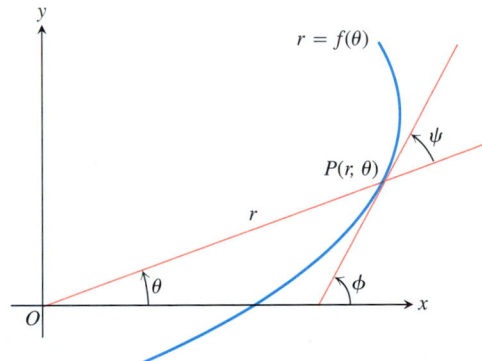

9.92   The angle ψ between the tangent line and the radius vector.

Suppose the equation of the curve is given in the form $r = f(\theta)$, where $f(\theta)$ is a differentiable function of θ. Then

$$x = r \cos \theta \quad \text{and} \quad y = r \sin \theta \tag{2}$$

are differentiable functions of θ with

$$\frac{dx}{d\theta} = -r \sin \theta + \cos \theta \frac{dr}{d\theta},$$

$$\frac{dy}{d\theta} = r \cos \theta + \sin \theta \frac{dr}{d\theta}. \tag{3}$$

Since $\psi = \phi - \theta$ from (1),

$$\tan \psi = \tan(\phi - \theta) = \frac{\tan \phi - \tan \theta}{1 + \tan \phi \tan \theta}.$$

Furthermore,

$$\tan \phi = \frac{dy}{dx} = \frac{dy/d\theta}{dx/d\theta}$$

because $\tan \phi$ is the slope of the curve at P. Also,

$$\tan \theta = \frac{y}{x}.$$

Hence

$$\tan \psi = \frac{\dfrac{dy/d\theta}{dx/d\theta} - \dfrac{y}{x}}{1 + \dfrac{y}{x} \dfrac{dy/d\theta}{dx/d\theta}}$$

$$= \frac{x \dfrac{dy}{d\theta} - y \dfrac{dx}{d\theta}}{x \dfrac{dx}{d\theta} + y \dfrac{dy}{d\theta}} \tag{4}$$

The numerator in the last expression in Eq. (4) is found from Eqs. (2) and (3) to be

$$x \frac{dy}{d\theta} - y \frac{dx}{d\theta} = r^2.$$

Similarly, the denominator is

$$x \frac{dx}{d\theta} + y \frac{dy}{d\theta} = r \frac{dr}{d\theta}.$$

When we substitute these into Eq. (4), we obtain

$$\tan \psi = \frac{r}{dr/d\theta}. \tag{5}$$

This is the equation we use for finding ψ.

109. Show, by reference to a figure, that the angle β between the tangents to two curves at a point of intersection may be found from the formula

$$\tan \beta = \frac{\tan \psi_2 - \tan \psi_1}{1 + \tan \psi_2 \tan \psi_1}. \tag{6}$$

When will the two curves intersect at right angles?

110. Find the value of $\tan \psi$ for the curve $r = \sin^4(\theta/4)$.

111. Find the angle between the curve $r = 2a \sin 3\theta$ and its tangent when $\theta = \pi/3$.

112. For the hyperbolic spiral $r\theta = a$ show that $\psi = 3\pi/4$ when $\theta = 1$ radian, and that $\psi \to \pi/2$ as the spiral winds around the origin. Sketch the curve and indicate ψ for $\theta = 1$ radian.

113. The circles $r = \sqrt{3} \cos \theta$ and $r = \sin \theta$ intersect at the point $(\sqrt{3}/2, \pi/3)$. Show that their tangents are perpendicular there.

114. Sketch the cardioid $r = a(1 + \cos \theta)$ and circle $r = 3a \cos \theta$ in one diagram and find the angle between their tangents at the point of intersection that lies in the first quadrant.

115. Find the points of intersection of the parabolas

$$r = \frac{1}{1 - \cos \theta} \quad \text{and} \quad r = \frac{3}{1 + \cos \theta}$$

and the angles between their tangents at these points.

116. Find points on the cardioid $r = a(1 + \cos \theta)$ where the tangent line is (a) horizontal, (b) vertical.

117. Show that the parabolas $r = a/(1 + \cos \theta)$ and $r = b/(1 - \cos \theta)$ are orthogonal at each point of intersection $(ab \neq 0)$.

118. Find the angle at which the cardioid $r = a(1 - \cos \theta)$ crosses the ray $\theta = \pi/2$.

119. Find the angle between the line $r = 3 \sec \theta$ and the cardioid $r = 4(1 + \cos \theta)$ at one of their intersections.

120. Find the slope of the tangent line to the curve $r = a \tan(\theta/2)$ at $\theta = \pi/2$.

121. Find the angle at which the parabolas $r = 1/(1 - \cos \theta)$ and $r = 1/(1 - \sin \theta)$ intersect in the first quadrant.

122. The equation $r^2 = 2 \csc 2\theta$ represents a curve in polar coordinates.
a) Sketch the curve.
b) Find an equivalent Cartesian equation for the curve.
c) Find the angle at which the curve intersects the ray $\theta = \pi/4$.

123. Suppose that the angle ψ from the radius vector to the tangent line of the curve $r = f(\theta)$ has the constant value α.
a) Show that the area bounded by the curve and two rays $\theta = \theta_1$, $\theta = \theta_2$, is proportional to $r_2^2 - r_1^2$, where (r_1, θ_1) and (r_2, θ_2) are polar coordinates of the ends of the arc of the curve between these rays. Find the factor of proportionality.
b) Show that the length of the arc of the curve in part (a) is proportional to $r_2 - r_1$ and find the proportionality constant.

124. Let P be a point on the hyperbola $r^2 \sin 2\theta = 2a^2$. Show that the triangle formed by OP, the tangent at P, and the initial line is isosceles.

VECTORS AND ANALYTIC GEOMETRY IN SPACE

Overview   This chapter introduces vectors and three-dimensional coordinate systems. Just as the coordinate plane is the natural place to study functions of a single variable, coordinate space is the place to study functions of two variables (or more). We establish coordinates in space by adding a third axis that measures distance above and below the xy-plane. This builds on what we already know without forcing us to start over again.

Equations in three variables define surfaces in space the way equations in two variables define curves in the plane. We use these surfaces to graph functions of two variables (not in this chapter, but later), define regions, bound solids, describe walls of containers, and so on. In short, we use them to do all the things we do in the plane, but stepped up one dimension.

Once in space, we can model motion in three dimensions and track the positions of moving bodies with vectors. We can also calculate the directions and magnitudes of their velocities and accelerations and predict the effects of the forces that are driving them. As we shall see in Chapter 11, coordinates and vectors make a powerful combination. Coordinates tell us where moving bodies are, and vectors tell us what is happening to them as they go along.

10.1 Vectors in the Plane

Some of the things we measure are completely determined by their magnitudes. To record mass, length, or time, for example, we need only write down a number and name an appropriate unit of measure. But we need more than that to describe a force, displacement, or velocity, for these quantities have direction as well as magnitude. To describe a force, we need to record the direction in which it acts as well as how large it is. To describe a body's displacement, we have to say in what direction it moves as well as how far. To describe a body's velocity at any given time, we have to know where the body is headed as well as how fast it is going.

Quantities that have direction as well as magnitude are usually represented by arrows that point in the direction of the action and whose lengths give the magnitude of the action in terms of a suitably chosen unit. Arrows with the same length and direction are regarded as equivalent. We can draw the arrow that represents a force or velocity wherever we want as long as it has the right length and direction.

When we work with arrows in mathematics, we think of them as directed line segments and we call the sets of equivalent segments *vectors*.

DEFINITIONS

> Directed line segments (arrows) in the plane are **equivalent** if they have the same length and direction. Each directed line segment is also equivalent to itself. The collection of all directed line segments equivalent to a given directed line segment is the **vector** determined by the segment. Each directed line segment in the vector **represents** the vector. When we draw a directed line segment as an arrow, we say that we **draw** the vector it represents. The **length (magnitude)** of a vector is the common length of the directed line segments that represent it. The **direction** of a vector is the common direction of the directed line segments that represent it.

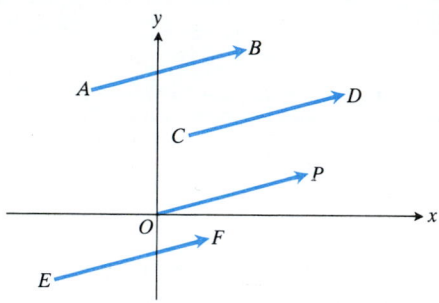

10.1   Arrows with the same length and direction represent the same vector (Example 1).

Thus, the arrows we use when we draw vectors are understood to represent the same vector if they have the same length, are parallel, and point the same way.

In print, vectors are usually described with single boldface roman letters, as in **v** ("vector vee"). The vector defined by the directed line segment from point A to point B is written as $\overrightarrow{AB}$ ("vector ab"). In handwritten work it is customary to draw small arrows above letters that represent vectors. Thus the equation appearing as $\mathbf{v} = \overrightarrow{AB}$ in print would appear as $\vec{v} = \overrightarrow{AB}$ when written by hand.

Example 1   The four arrows in Fig. 10.1 have the same length and direction. They therefore represent the same vector, and we write

$$\overrightarrow{AB} = \overrightarrow{CD} = \overrightarrow{OP} = \overrightarrow{EF}.$$

Scalars and Scalar Multiples

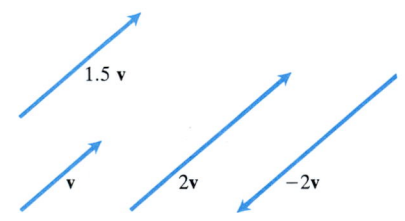

10.2   Scalar multiples of **v**.

We multiply a vector by a positive real number by multiplying its length by the number (Fig. 10.2). To multiply a vector by 2, we double its length. To multiply a vector by 1.5, we increase its length by 50%, and so on. We multiply a vector by a negative number by reversing the vector's direction and multiplying the length by the number's absolute value. To multiply a vector by -2, we reverse the vector's direction and double its length.

If c is a nonzero real number and **v** a vector, the direction of $c\mathbf{v}$ agrees with that of **v** if c is positive and is opposite to that of **v** if c is negative. Since real numbers work like scaling factors in this context, we call them **scalars** and call multiples like $c\mathbf{v}$ scalar multiples of **v**.

To include multiplication by zero, we adopt the convention that multiplying a vector by zero produces the **zero vector 0**, consisting of points that are degenerate line segments of zero length. Unlike other vectors, the vector **0** has no direction.

Geometric Addition: The Parallelogram Law

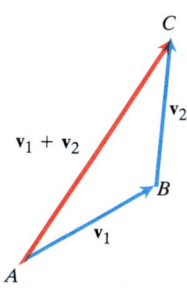

10.3   The sum of $\mathbf{v}_1$ and $\mathbf{v}_2$.

Two vectors $\mathbf{v}_1$ and $\mathbf{v}_2$ may be added geometrically by drawing a representative of $\mathbf{v}_1$, say from A to B as in Fig. 10.3, and then a representative of $\mathbf{v}_2$ starting from the

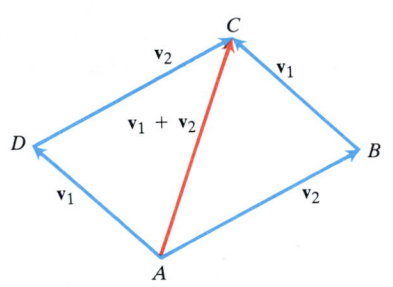

10.4 The Parallelogram Law of Addition. Quadrilateral $ABCD$ is a parallelogram because both pairs of opposite sides have equal lengths. The law was used by Aristotle to describe the combined action of two forces.

terminal point B of $\mathbf{v}_1$. In Fig. 10.3, $\mathbf{v}_2 = \overrightarrow{BC}$. The sum $\mathbf{v}_1 + \mathbf{v}_2$ is then the vector represented by the arrow from the initial point A of $\mathbf{v}_1$ to the terminal point C of $\mathbf{v}_2$. That is, if

$$\mathbf{v}_1 = \overrightarrow{AB} \qquad \text{and} \qquad \mathbf{v}_2 = \overrightarrow{BC},$$

then

$$\begin{aligned}.\mathbf{v}_1 + \mathbf{v}_2 &= \overrightarrow{AB} + \overrightarrow{BC} \\ &= \overrightarrow{AC}.\end{aligned}$$

This description of addition is sometimes called the **Parallelogram Law** of addition because $\mathbf{v}_1 + \mathbf{v}_2$ is given by the diagonal of the parallelogram determined by $\mathbf{v}_1$ and $\mathbf{v}_2$ (Fig. 10.4).

Components

Whenever a vector $\mathbf{v}$ can be written as a sum

$$\mathbf{v} = \mathbf{v}_1 + \mathbf{v}_2,$$

the vectors $\mathbf{v}_1$ and $\mathbf{v}_2$ are said to be **components** of $\mathbf{v}$. We also say that we have **represented** or **resolved** $\mathbf{v}$ in terms of $\mathbf{v}_1$ and $\mathbf{v}_2$.

The most common algebra of vectors is based on representing each vector in terms of components parallel to the Cartesian coordinate axes and writing each component as an appropriate multiple of a **basic** vector of unit length. The basic vector in the positive x-direction is the vector $\mathbf{i}$ determined by the directed line segment that runs from $(0, 0)$ to $(1, 0)$. The basic vector in the positive y-direction is the vector $\mathbf{j}$ determined by the directed line segment from $(0, 0)$ to $(0, 1)$. Then $a\mathbf{i}$, a being a scalar, represents a vector of length $|a|$ parallel to the x-axis, pointing to the right if a is positive and to the left if a is negative. Similarly, $b\mathbf{j}$ is a vector of length $|b|$ parallel to the y-axis, pointing up if b is positive and down if b is negative. Figure 10.5 shows a vector $\mathbf{v} = \overrightarrow{AC}$ resolved into its $\mathbf{i}$- and $\mathbf{j}$-components as the sum

$$\mathbf{v} = a\mathbf{i} + b\mathbf{j}.$$

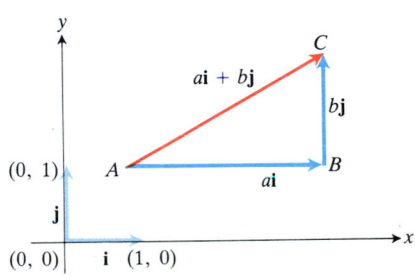

10.5 The basic vectors $\mathbf{i}$ and $\mathbf{j}$. Any vector $\overrightarrow{AC}$ in the plane can be expressed as a scalar multiple of $\mathbf{i}$ plus a scalar multiple of $\mathbf{j}$.

DEFINITION

If $\mathbf{v} = a\mathbf{i} + b\mathbf{j}$, the vectors $a\mathbf{i}$ and $b\mathbf{j}$ are the **vector components of v in the directions of i and j.** The numbers a and b are the **scalar components of v in the directions of i and j.**

Components give us a way to define the equality of vectors algebraically.

DEFINITION

Equality of Vectors (Algebraic Definition)

$$a\mathbf{i} + b\mathbf{j} = a'\mathbf{i} + b'\mathbf{j} \qquad \Leftrightarrow \qquad a = a' \text{ and } b = b' \tag{1}$$

That is, two vectors are equal if and only if their scalar components in the directions of $\mathbf{i}$ and $\mathbf{j}$ are identical.

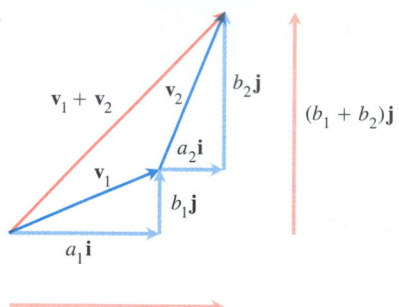

10.6 If $\mathbf{v}_1 = a_1 \mathbf{i} + b_1 \mathbf{j}$ and $\mathbf{v}_2 = a_2 \mathbf{i} + b_2 \mathbf{j}$, then $\mathbf{v}_1 + \mathbf{v}_2 = (a_1 + a_2) \mathbf{i} + (b_1 + b_2) \mathbf{j}$.

Algebraic Addition

Two vectors may be added algebraically by adding their corresponding scalar components, as shown in Fig. 10.6.

> If $\mathbf{v}_1 = a_1 \mathbf{i} + b_1 \mathbf{j}$, and $\mathbf{v}_2 = a_2 \mathbf{i} + b_2 \mathbf{j}$, then
> $$\mathbf{v}_1 + \mathbf{v}_2 = (a_1 + a_2) \mathbf{i} + (b_1 + b_2) \mathbf{j}. \qquad (2)$$

Example 2

$$(2\mathbf{i} - 4\mathbf{j}) + (5\mathbf{i} + 3\mathbf{j}) = (2 + 5)\mathbf{i} + (-4 + 3)\mathbf{j} = 7\mathbf{i} - \mathbf{j}.$$

Subtraction

The negative of a vector $\mathbf{v}$ is the vector $-\mathbf{v}$ that has the same length as $\mathbf{v}$ but points in the opposite direction. To subtract a vector $\mathbf{v}_2$ from a vector $\mathbf{v}_1$, we add $-\mathbf{v}_2$ to $\mathbf{v}_1$. This may be done geometrically by drawing $-\mathbf{v}_2$ from the tip of $\mathbf{v}_1$ and then drawing the vector from the initial point of $\mathbf{v}_1$ to the tip of $-\mathbf{v}_2$, as shown in Fig. 10.7(a), where

$$\overrightarrow{AD} = \overrightarrow{AB} + \overrightarrow{BD} = \mathbf{v}_1 + (-\mathbf{v}_2) = \mathbf{v}_1 - \mathbf{v}_2.$$

Another way to draw $\mathbf{v}_1 - \mathbf{v}_2$ is to draw $\mathbf{v}_1$ and $\mathbf{v}_2$ with a common initial point and then draw $\mathbf{v}_1 - \mathbf{v}_2$ as the vector from the tip of $\mathbf{v}_2$ to the tip of $\mathbf{v}_1$. This is illustrated in Fig. 10.7(b), where

$$\overrightarrow{CB} = \overrightarrow{CA} + \overrightarrow{AB} = -\mathbf{v}_2 + \mathbf{v}_1 = \mathbf{v}_1 - \mathbf{v}_2.$$

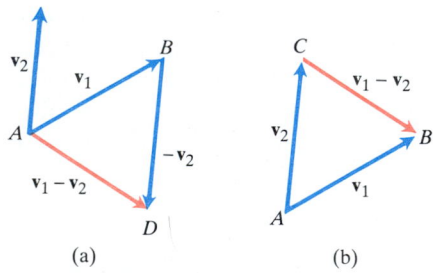

10.7 Two ways to draw $\mathbf{v}_1 - \mathbf{v}_2$: (a) as $\mathbf{v}_1 + (-\mathbf{v}_2)$ and (b) as the vector from the tip of $\mathbf{v}_2$ to the tip of $\mathbf{v}_1$.

Thus, $\overrightarrow{CB}$ is the vector that when added to $\mathbf{v}_2$ gives $\mathbf{v}_1$:

$$\overrightarrow{CB} + \mathbf{v}_2 = (\mathbf{v}_1 - \mathbf{v}_2) + \mathbf{v}_2 = \mathbf{v}_1.$$

In terms of components, vector subtraction follows the algebraic law

> $$\mathbf{v}_1 - \mathbf{v}_2 = (a_1 - a_2) \mathbf{i} + (b_1 - b_2) \mathbf{j}, \qquad (3)$$

which says that corresponding scalar components are subtracted.

Example 3

$$(6\mathbf{i} + 2\mathbf{j}) - (3\mathbf{i} - 5\mathbf{j}) = (6 - 3)\mathbf{i} + (2 - (-5))\mathbf{j} = 3\mathbf{i} + 7\mathbf{j}.$$

We find the components of the vector from a point $P_1(x_1, y_1)$ to a point $P_2(x_2, y_2)$ by subtracting the components of $\overrightarrow{OP_1} = x_1 \mathbf{i} + y_1 \mathbf{j}$ from the components of $\overrightarrow{OP_2} = x_2 \mathbf{i} + y_2 \mathbf{j}$.

> The vector from $P_1(x_1, y_1)$ to $P_2(x_2, y_2)$ is
> $$\overrightarrow{P_1P_2} = (x_2 - x_1) \mathbf{i} + (y_2 - y_1) \mathbf{j}. \qquad (4)$$

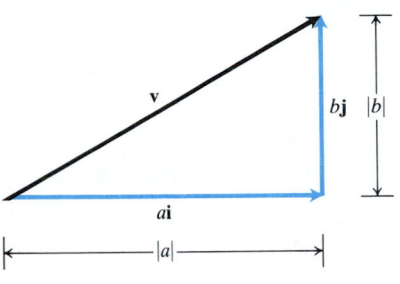

10.8 The length of **v** is $\sqrt{|a|^2 + |b|^2} = \sqrt{a^2 + b^2}$.

Example 4   The vector from P_1 (3, 4) to P_2 (5, 1) is

$$\overrightarrow{P_1P_2} = (5 - 3)\,\mathbf{i} + (1 - 4)\,\mathbf{j} = 2\,\mathbf{i} - 3\,\mathbf{j}.$$

Length

We calculate the length of $\mathbf{v} = a\mathbf{i} + b\mathbf{j}$ by representing **v** as the hypotenuse of a right triangle with sides $|a|$ and $|b|$ (Fig. 10.8) and applying the Pythagorean theorem to get

$$\text{Length of } \mathbf{v} = \sqrt{|a|^2 + |b|^2} = \sqrt{a^2 + b^2}.$$

The usual symbol for the length is $|\mathbf{v}|$, which is read "the length of **v**" or "the magnitude of **v**," the latter being more common in applied fields. The bars are the same as the ones we use for absolute values.

> The length or magnitude of $\mathbf{v} = a\mathbf{i} + b\mathbf{j}$ is
> $$|\mathbf{v}| = \sqrt{a^2 + b^2}. \tag{5}$$

Scalar Multiplication

Scalar multiplication can be accomplished component by component.

> If c is a scalar and $\mathbf{v} = a\mathbf{i} + b\mathbf{j}$ is a vector, then
> $$c\mathbf{v} = c(a\mathbf{i} + b\mathbf{j}) = (ca)\,\mathbf{i} + (cb)\,\mathbf{j}. \tag{6}$$

To check that the length of $c\mathbf{v}$ is still $|c|$ times the length of **v** when we do scalar multiplication this way, we can calculate the length with Eq. (5):

$$
\begin{aligned}
|c\mathbf{v}| &= |(ca)\,\mathbf{i} + (cb)\,\mathbf{j}| && \text{(Eq. 6)}\\
&= \sqrt{(ca)^2 + (cb)^2} && \text{(Eq. (5) with } ca \text{ and } cb \text{ in place of } a \text{ and } b)\\
&= \sqrt{c^2(a^2 + b^2)}\\
&= \sqrt{c^2}\,\sqrt{a^2 + b^2}\\
&= |c|\,|\mathbf{v}|.
\end{aligned}
$$

> If c is a scalar and **v** is a vector, then
> $$|c\mathbf{v}| = |c|\,|\mathbf{v}|. \tag{7}$$

Example 5   If $c = -2$ and $\mathbf{v} = -3\,\mathbf{i} + 4\,\mathbf{j}$, then

$$|\mathbf{v}| = |-3\,\mathbf{i} + 4\,\mathbf{j}| = \sqrt{(-3)^2 + (4)^2} = \sqrt{9 + 16} = \sqrt{25} = 5$$

$$|-2\mathbf{v}| = |(-2)(-3\,\mathbf{i} + 4\,\mathbf{j})| = |6\,\mathbf{i} - 8\,\mathbf{j}| = \sqrt{(6)^2 + (-8)^2} = \sqrt{36 + 64}$$

$$= \sqrt{100} = 10 = |-2|5 = |c|\,|\mathbf{v}|.$$

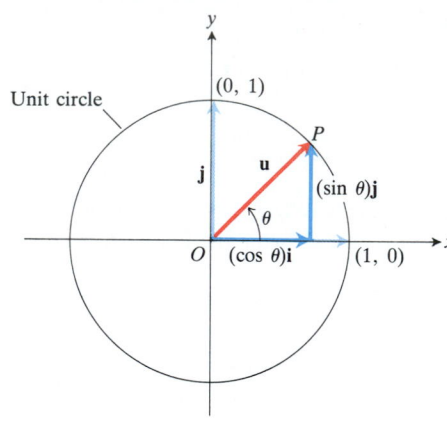

10.9 The unit vector that makes an angle of measure θ with the positive x-axis. Every unit vector has the form

$$\mathbf{u} = (\cos\theta)\,\mathbf{i} + (\sin\theta)\,\mathbf{j}$$

for some θ.

The Zero Vector

In terms of components, the zero vector is the vector

$$\mathbf{0} = 0\,\mathbf{i} + 0\,\mathbf{j}. \tag{8}$$

It is the only vector whose length is zero, as we can see from the fact that

$$|a\,\mathbf{i} + b\,\mathbf{j}| = \sqrt{a^2 + b^2} = 0 \qquad \Leftrightarrow \qquad a = b = 0.$$

Unit Vectors

Any vector $\mathbf{u}$ whose length is equal to the unit of length used along the coordinate axes is called a **unit vector.** The vectors $\mathbf{i}$ and $\mathbf{j}$ are unit vectors:

$$|\,\mathbf{i}\,| = |1\,\mathbf{i} + 0\,\mathbf{j}| = \sqrt{1^2 + 0^2} = 1, \qquad |\,\mathbf{j}\,| = |0\,\mathbf{i} + 1\,\mathbf{j}| = \sqrt{0^2 + 1^2} = 1.$$

If $\mathbf{u}$ is the unit vector obtained by rotating $\mathbf{i}$ through an angle θ in the positive direction, then $\mathbf{u}$ has a horizontal component $\cos\theta$ and a vertical component $\sin\theta$ (Fig. 10.9), so that

$$\mathbf{u} = (\cos\theta)\,\mathbf{i} + (\sin\theta)\,\mathbf{j}. \tag{9}$$

If we allow the angle θ in Eq. (9) to vary from 0 to 2π, the point P in Fig. 10.9 traces the unit circle $x^2 + y^2 = 1$ once counterclockwise. Since this includes all possible directions, every unit vector in the plane is given by Eq. (9) for some value of θ.

In handwritten work it is common to denote unit vectors with small "hats," as in $\hat{\mathbf{u}}$ (pronounced "u hat"). In hat notation, $\mathbf{i}$ and $\mathbf{j}$ become $\hat{\mathbf{i}}$ and $\hat{\mathbf{j}}$.

Direction as a Vector

It is common in subjects like classical electricity and magnetism, which use vectors a great deal, to define the direction of a nonzero vector $\mathbf{v}$ to be the unit vector obtained by dividing $\mathbf{v}$ by its own length.

DEFINITION   If $\mathbf{v} \neq \mathbf{0}$, the **direction** of $\mathbf{v}$ is the vector $\dfrac{\mathbf{v}}{|\mathbf{v}|}$. $\tag{10}$

Notice that instead of just saying that $\mathbf{v}/|\mathbf{v}|$ *represents* the direction of $\mathbf{v}$, we say that it *is* the direction of $\mathbf{v}$.

To see that $\mathbf{v}/|\mathbf{v}|$ really is a unit vector, we can calculate its length directly:

$$\text{Length of } \frac{\mathbf{v}}{|\mathbf{v}|} = \left|\frac{\mathbf{v}}{|\mathbf{v}|}\right|$$

$$= \frac{1}{|\mathbf{v}|}\,|\mathbf{v}| \qquad \left(\text{Eq. (7) with } c = \frac{1}{|\mathbf{v}|}\right)$$

$$= 1.$$

Any nonzero vector can be expressed in terms of its length and direction by using the equation

$$v = |v| \cdot \frac{v}{|v|} = (\text{length of } v) \cdot (\text{direction of } v). \qquad (11)$$

Example 6   Express $v = 3\,i - 4\,j$ in terms of its length and direction.

Solution   Length of v: $|v| = \sqrt{(3)^2 + (-4)^2} = \sqrt{9 + 16} = 5$

Direction of v: $\dfrac{v}{|v|} = \dfrac{3\,i - 4\,j}{5} = \dfrac{3}{5}\,i - \dfrac{4}{5}\,j$

$$v = 3\,i - 4\,j = 5\left(\frac{3}{5}\,i - \frac{4}{5}\,j\right)$$

 length of v direction of v

It follows from the definition of direction as a vector that vectors A and B have the same direction if

$$\frac{A}{|A|} = \frac{B}{|B|} \qquad \text{or} \qquad A = \frac{|A|}{|B|}\,B. \qquad (12)$$

Thus, if A and B have the same direction, A is a positive scalar multiple of B. Conversely, if $A = kB,\ k > 0$, then

$$\frac{A}{|A|} = \frac{k\,B}{|k\,B|} = \frac{k}{|k|}\,\frac{B}{|B|} = \frac{k}{k}\,\frac{B}{|B|} = \frac{B}{|B|}. \qquad (13)$$

Therefore, two nonzero vectors A and B have the same direction if and only if A is a positive scalar multiple of B.

We say that two nonzero vectors A and B have *opposite* directions if their directions are opposite in sign:

$$\frac{A}{|A|} = -\frac{B}{|B|}. \qquad (14)$$

From this it follows that A and B have opposite directions if and only if A is a negative scalar multiple of B.

Example 7

a) Same direction: $A = 3\,i - 4\,j$ and $B = \dfrac{3}{2}\,i - 2\,j = \dfrac{1}{2}A$

 (B is a positive scalar multiple of A.)

b) Opposite directions: $A = 3\,i - 4\,j$ and $B = -9\,i + 12\,j = -3A$

 (B is a negative scalar multiple of A.)

Slopes, Tangents, and Normals

Two vectors are said to be **parallel** if they are either positive or negative scalar multiples of one another or, equivalently, if the line segments representing them are parallel. Similarly, a vector is parallel to a line if the segments that represent the vector are parallel to the line. The **slope** of a vector that is not parallel to the y-axis is the slope shared by the lines parallel to the vector. Thus, when $a \neq 0$, the vector

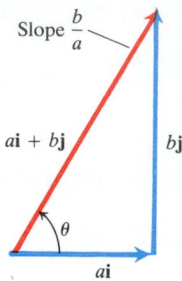

10.10   If $a \neq 0$, the vector $a\mathbf{i} + b\mathbf{j}$ has slope $b/a = \tan \theta$.

$\mathbf{v} = a\mathbf{i} + b\mathbf{j}$ has a well-defined slope, which can be calculated from the components of $\mathbf{v}$ as the number b/a (Fig. 10.10).

When we say that a vector is **tangent** or **normal** to a curve at a point, we mean that the vector is parallel or normal to the line that is tangent to the curve at the point. The next example shows how to find such vectors.

Example 8   Find unit vectors tangent and normal to the curve

$$y = \frac{x^3}{2} + \frac{1}{2}$$

at the point $(1, 1)$.

Solution   We find the unit vectors that are parallel and normal to the curve's tangent line at the point $(1, 1)$, shown in Fig. 10.11.

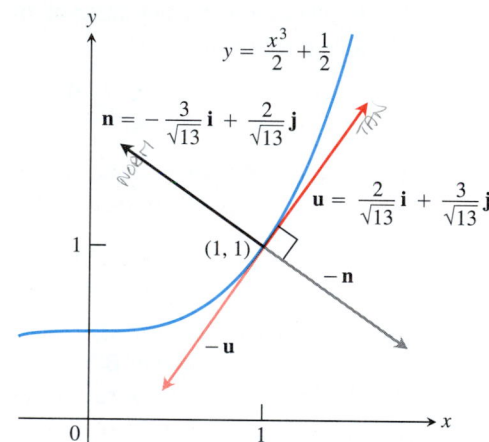

10.11   The unit tangent and normal vectors at the point $(1, 1)$ on the curve $y = (x^3/2) + 1/2$.

The slope of the line tangent to the curve at $(1, 1)$ is

$$y' = \left. \frac{3x^2}{2} \right|_{x=1} = \frac{3}{2}.$$

We find a unit vector with this slope. The vector $\mathbf{v} = 2\mathbf{i} + 3\mathbf{j}$ has slope $3/2$, as does every nonzero multiple of $\mathbf{v}$. To find a multiple of $\mathbf{v}$ that is a unit vector, we divide $\mathbf{v}$ by its length,

$$|\mathbf{v}| = \sqrt{2^2 + 3^2} = \sqrt{13}.$$

This produces the unit vector

$$\mathbf{u} = \frac{\mathbf{v}}{|\mathbf{v}|} = \frac{2}{\sqrt{13}}\mathbf{i} + \frac{3}{\sqrt{13}}\mathbf{j}.$$

The vector $\mathbf{u}$ is tangent to the curve at $(1, 1)$ because it has the same direction as $\mathbf{v}$. Of course, the vector

$$-\mathbf{u} = -\frac{2}{\sqrt{13}}\mathbf{i} - \frac{3}{\sqrt{13}}\mathbf{j},$$

which points in the opposite direction, is also tangent to the curve at $(1, 1)$. Without some additional requirement, there is no reason to prefer one of these vectors to the other.

If $\mathbf{v} = a\mathbf{i} + b\mathbf{j}$, then $\mathbf{p} = -b\mathbf{i} + a\mathbf{j}$ and $\mathbf{q} = b\mathbf{i} - a\mathbf{j}$ are perpendicular to $\mathbf{v}$ because their slopes are both $-a/b$, the negative reciprocal of $\mathbf{v}$'s slope.

To find unit vectors normal to the curve at $(1, 1)$, we look for unit vectors whose slopes are the negative reciprocal of the slope of $\mathbf{u}$. This is quickly done by interchanging the scalar components of $\mathbf{u}$ and changing the sign of one of them. We obtain

$$\mathbf{n} = -\frac{3}{\sqrt{13}}\,\mathbf{i} + \frac{2}{\sqrt{13}}\,\mathbf{j} \quad \text{and} \quad -\mathbf{n} = \frac{3}{\sqrt{13}}\,\mathbf{i} - \frac{2}{\sqrt{13}}\,\mathbf{j}.$$

Again, either one will do. The vectors have opposite directions but both are normal to the curve at the point $(1, 1)$.

EXERCISES 10.1

1. The vectors **A**, **B**, and **C** in Fig. 10.12 lie in a plane. Copy them on a sheet of paper. Then, by arranging the vectors head to tail, as in Figs. 10.3 and 10.6, sketch
   a) $\mathbf{A} + \mathbf{B}$
   b) $\mathbf{A} + \mathbf{B} + \mathbf{C}$
   c) $\mathbf{A} - \mathbf{B}$
   d) $\mathbf{A} - \mathbf{C}$

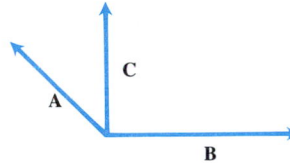

10.12 The vectors for Exercise 1.

2. The vectors **A**, **B**, and **C** in Fig. 10.13 lie in a plane. Copy them on a sheet of paper. Then, by arranging the vectors head to tail, as in Figs. 10.3 and 10.6, sketch
   a) $\mathbf{A} - \mathbf{B}$
   b) $\mathbf{A} - \mathbf{B} + \mathbf{C}$
   c) $2\mathbf{A} - \mathbf{B}$
   d) $\mathbf{A} + \mathbf{B} + \mathbf{C}$

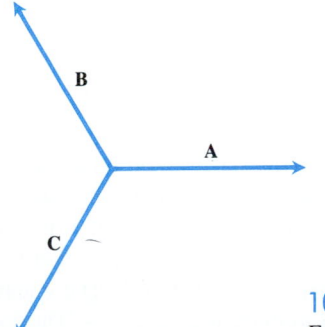

10.13 The vectors for Exercise 2.

Write the sums and differences in Exercises 3–8 in the form $a\,\mathbf{i} + b\,\mathbf{j}$.

3. $(2\,\mathbf{i} - 7\,\mathbf{j}) + (\mathbf{i} + 6\,\mathbf{j})$

4. $(\sqrt{3}\,\mathbf{i} - 3\,\mathbf{j}) + 6\,\mathbf{i}$

5. $(-2\,\mathbf{i} + 6\,\mathbf{j}) - 2(\mathbf{i} + \mathbf{j}) + 3\,\mathbf{i} - 4\,\mathbf{j}$

6. $3\!\left(\frac{\mathbf{i}}{2} - \frac{\mathbf{j}}{3}\right) - \frac{1}{2}(5\,\mathbf{i} - 2\,\mathbf{j})$

7. $2((\ln 2)\,\mathbf{i} + \mathbf{j}) - ((\ln 8)\,\mathbf{i} + \pi\mathbf{j})$

8. $(\mathbf{i} + \sqrt{2}\,\mathbf{j}) - 7(\mathbf{i} - \mathbf{j}) - \sqrt{2}\,(2\,\mathbf{i} + \mathbf{j})$

Express the vectors in Exercises 9–16 in the form $a\mathbf{i} + b\mathbf{j}$ and sketch them as arrows in the coordinate plane.

9. $\overrightarrow{P_1P_2}$ if P_1 is the point $(1, 3)$ and P_2 is the point $(2, -1)$

10. $\overrightarrow{OP_3}$ if O is the origin and P_3 is the midpoint of the vector $\overrightarrow{P_1P_2}$ joining $P_1(2, -1)$ and $P_2(-4, 3)$

11. The vector from the point $A(2, 3)$ to the origin

12. The sum of the vectors $\overrightarrow{AB}$ and $\overrightarrow{CD}$, given the four points $A(1, -1)$, $B(2, 0)$, $C(-1, 3)$, and $D(-2, 2)$

13. The unit vectors $\mathbf{u} = (\cos \theta)\,\mathbf{i} + (\sin \theta)\,\mathbf{j}$ for $\theta = \pi/6$ and $\theta = 2\pi/3$. Include the circle $x^2 + y^2 = 1$ in your sketch.

14. The unit vectors $\mathbf{u} = (\cos \theta)\,\mathbf{i} + (\sin \theta)\,\mathbf{j}$ for $\theta = -\pi/4$ and $\theta = -3\pi/4$. Include the circle $x^2 + y^2 = 1$ in your sketch.

15. The unit vector obtained by rotating $\mathbf{j}$ $120°$ clockwise about the origin

16. The unit vector obtained by rotating $\mathbf{i}$ $135°$ counterclockwise about the origin

In Exercises 17–20, find the unit vectors that are tangent and normal to the curve at the given point (four vectors in all). Then sketch the vectors and curve together.

17. $y = x^2$, $(2, 4)$

18. $x^2 + 2y^2 = 6$, $(2, 1)$

19. $y = \tan^{-1}x$, $(1, \pi/4)$

20. $y = \displaystyle\sum_{n=0}^{\infty} \frac{x^n}{n!}$, $(0, 1)$

In Exercises 21–24, find the unit vectors that are tangent and normal to the curve at the given point (four vectors in all).

21. $3x^2 + 8xy + 2y^2 - 3 = 0$, $(1, 0)$

22. $x^2 - 6xy + 8y^2 - 2x - 1 = 0$, $(1, 1)$

23. $y = \displaystyle\int_0^x \sqrt{3 + t^4}\; dt$, $(0, 0)$

24. $y = \displaystyle\int_e^x \ln(\ln t)\; dt$, $(e, 0)$

Use Eq. (11) to express the vectors in Exercises 25–30 as products of their lengths and directions.

25. $\mathbf{i} + \mathbf{j}$ 26. $2\mathbf{i} - 3\mathbf{j}$

27. $\sqrt{3}\,\mathbf{i} + \mathbf{j}$ 28. $-2\mathbf{i} + 3\mathbf{j}$

29. $5\mathbf{i} + 12\mathbf{j}$ 30. $-5\mathbf{i} - 12\mathbf{j}$

31. Show that $\mathbf{A} = 3\mathbf{i} + 6\mathbf{j}$ and $\mathbf{B} = -\mathbf{i} - 2\mathbf{j}$ have opposite directions. Sketch $\mathbf{A}$ and $\mathbf{B}$ together.

32. Show that $\mathbf{A} = 3\mathbf{i} + 6\mathbf{j}$ and $\mathbf{B} = (1/2)\mathbf{i} + \mathbf{j}$ have the same direction.

33. Find a vector 2 units long in the direction of $\mathbf{A} = -\mathbf{i} - \mathbf{j}$. How many such vectors are there?

34. Find a vector 5 units long in the direction opposite to the direction of $\mathbf{A} = (3/5)\mathbf{i} + (4/5)\mathbf{j}$. How many such vectors are there?

35. Let $\mathbf{v}$ be a vector in the plane not parallel to the y-axis. Is the slope of $-\mathbf{v}$ the same as the slope of $\mathbf{v}$, or is it the negative of the slope of $\mathbf{v}$? Explain.

10.2 Cartesian (Rectangular) Coordinates and Vectors in Space

Our goal now is to describe the three-dimensional Cartesian coordinate system and learn our way around in space. This means defining distance, practicing with the arithmetic of vectors (the rules are the same as in the plane but with an extra term), and making connections between sets of points in space and equations and inequalities. Everyone we know finds it harder to draw in three dimensions than in two, so we have included some drawing tips as well (pages 714–715). The Cartesian coordinates for space are often called *rectangular coordinates* because the axes that define them meet at right angles.

Cartesian Coordinates

To locate points in space, we use three mutually perpendicular coordinate axes, arranged as in Fig. 10.14. The axes Ox, Oy, and Oz shown there make what is known as a *right-handed* coordinate frame. When you hold your right hand so that the fingers curl from the positive x-axis toward the positive y-axis, your thumb points along the positive z-axis.

The Cartesian coordinates (x, y, z) of a point P in space are the numbers at which the planes through P perpendicular to the axes cut the axes.

Points that lie on the x-axis have y- and z-coordinates equal to zero. That is, they have coordinates of the form $(x, 0, 0)$. Similarly, points on the y-axis have coordinates of the form $(0, y, 0)$. Points on the z-axis have coordinates of the form $(0, 0, z)$.

The points in a plane perpendicular to the x-axis all have the same x-coordinate, this being the number at which that plane cuts the x-axis. Similarly, the points in a plane perpendicular to the y-axis have a common y-coordinate and the points in a plane perpendicular to the z-axis have a common z-coordinate. To write equations for these planes, we name the common coordinate's value. The equation $x = 2$ is an equation for the plane perpendicular to the x-axis at $x = 2$. The equation $y = 3$ is an equation for the plane perpendicular to the y-axis at $y = 3$. The equation $z = 5$ is an equation for the plane perpendicular to the z-axis at $z = 5$. Figure 10.15 shows the planes $x = 2$, $y = 3$, and $z = 5$, together with their intersection point $(2, 3, 5)$.

The planes $x = 2$ and $y = 3$ in Fig. 10.15 intersect in a line that runs parallel to the z-axis. This line is described by the *pair* of equations $x = 2$, $y = 3$. A point (x, y, z) lies on the line if and only if $x = 2$ and $y = 3$. Similarly, the line of intersection of the planes $y = 3$ and $z = 5$ is described by the equation pair $y = 3$, $z = 5$. This line runs parallel to the x-axis. The line of intersection of the

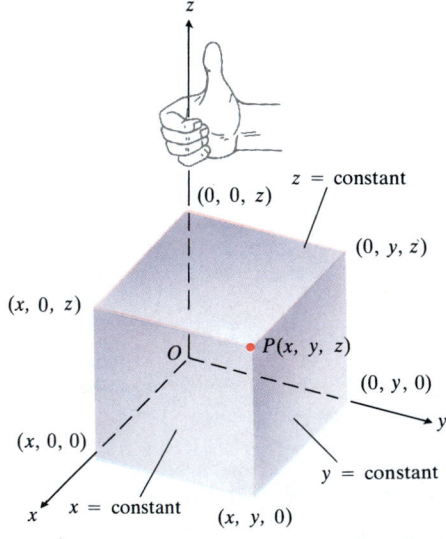

10.14   The Cartesian coordinate system is right-handed.

10.15 The planes $x = 2$, $y = 3$, and $z = 5$ determine three lines through the point $(2, 3, 5)$.

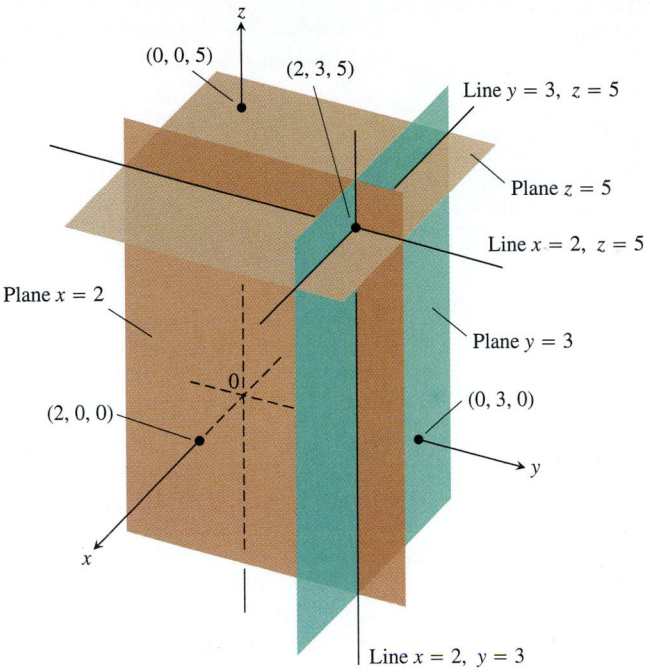

planes $x = 2$ and $z = 5$, parallel to the y-axis, is described by the equation pair $x = 2, z = 5$.

The planes determined by the three coordinate axes are the **xy-plane,** whose standard equation is $z = 0$; the **yz-plane,** whose standard equation is $x = 0$; and the **xz-plane,** whose standard equation is $y = 0$. They meet in the point $(0, 0, 0)$, which is called the **origin** of the coordinate system (Fig. 10.16).

The three **coordinate planes** $x = 0$, $y = 0$, and $z = 0$ divide space into eight cells called **octants.** The octant in which the point coordinates are all positive is called the **first octant,** but there is no conventional numbering for the remaining seven octants.

In the following examples, we match coordinate equations and inequalities with the sets of points they define in space.

Example 1

Defining equations and inequalities	Verbal description
$z \geq 0$	The half-space consisting of the points on and above the xy-plane.
$x = -3$	The plane perpendicular to the x-axis at $x = -3$. This plane lies parallel to the yz-plane and 3 units behind it.
$z = 0, x \leq 0, y \geq 0$	The second quadrant of the xy-plane.
$x \geq 0, y \geq 0, z \geq 0$	The first octant.
$-1 \leq y \leq 1$	The slab between the planes $y = -1$ and $y = 1$ (planes included).
$y = -2, z = 2$	The line in which the planes $y = -2$ and $z = 2$ intersect. Alternatively, the line through the point $(0, -2, 2)$ parallel to the x-axis.

10.16 The planes $x = 0$, $y = 0$, and $z = 0$ are the planes determined by the coordinate axes. They divide space into eight cells called octants.

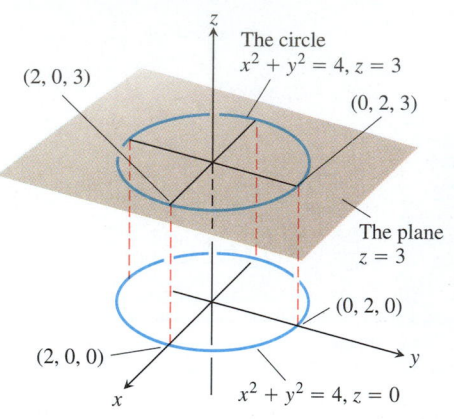

10.17 The circle $x^2 + y^2 = 4$, $z = 3$.

Example 2 Identify the set of points $P(x, y, z)$ whose coordinates satisfy the two equations

$$x^2 + y^2 = 4 \quad \text{and} \quad z = 3.$$

Solution The points lie in the horizontal plane $z = 3$ and, in this plane, make up the circle $x^2 + y^2 = 4$. We call this set of points "the circle $x^2 + y^2 = 4$ in the plane $z = 3$" or, more simply, "the circle $x^2 + y^2 = 4$, $z = 3$" (Fig. 10.17).

Vectors in Space

The sets of equivalent directed line segments we use to represent forces, displacements, and velocities in space are called vectors, just as they are in the plane. The same rules of addition, subtraction, and scalar multiplication apply.

The vectors represented by the directed line segments from the origin to the points $(1, 0, 0)$, $(0, 1, 0)$, and $(0, 0, 1)$ are the **basic vectors.** We denote them by **i**, **j**, and **k**. The **position vector r** from the origin O to the typical point $P(x, y, z)$ is

$$\mathbf{r} = \overrightarrow{OP} = x\mathbf{i} + y\mathbf{j} + z\mathbf{k}. \tag{1}$$

Addition, Subtraction, and Scalar Multiplication for Vectors in Space

For any vectors $\mathbf{A} = a_1\mathbf{i} + a_2\mathbf{j} + a_3\mathbf{k}$ and $\mathbf{B} = b_1\mathbf{i} + b_2\mathbf{j} + b_3\mathbf{k}$, and for any scalar c,

$$\mathbf{A} + \mathbf{B} = (a_1 + b_1)\mathbf{i} + (a_2 + b_2)\mathbf{j} + (a_3 + b_3)\mathbf{k}$$

$$\mathbf{A} - \mathbf{B} = (a_1 - b_1)\mathbf{i} + (a_2 - b_2)\mathbf{j} + (a_3 - b_3)\mathbf{k}$$

$$c\mathbf{A} = (ca_1)\mathbf{i} + (ca_2)\mathbf{j} + (ca_3)\mathbf{k}.$$

The Vector Between Two Points

Often we want to express the vector $\overrightarrow{P_1P_2}$ from the point $P_1(x_1, y_1, z_1)$ to the point $P_2(x_2, y_2, z_2)$ in terms of the coordinates of P_1 and P_2. To do so, we observe (Fig. 10.18) that

$$\overrightarrow{P_1P_2} = \overrightarrow{OP_2} - \overrightarrow{OP_1}$$

$$= (x_2\mathbf{i} + y_2\mathbf{j} + z_2\mathbf{k}) - (x_1\mathbf{i} + y_1\mathbf{j} + z_1\mathbf{k}) \tag{2}$$

$$= (x_2 - x_1)\mathbf{i} + (y_2 - y_1)\mathbf{j} + (z_2 - z_1)\mathbf{k}.$$

The vector from $P_1(x_1, y_1, z_1)$ to $P_2(x_2, y_2, z_2)$ is

$$\overrightarrow{P_1P_2} = (x_2 - x_1)\mathbf{i} + (y_2 - y_1)\mathbf{j} + (z_2 - z_1)\mathbf{k}. \tag{3}$$

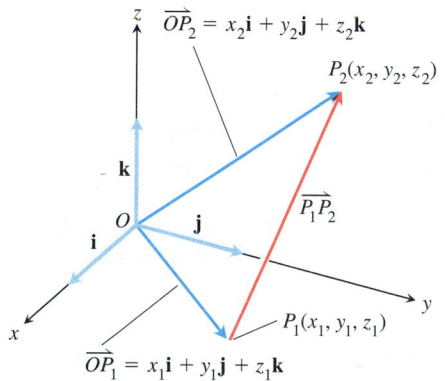

10.18 The vector from P_1 to P_2 is $\overrightarrow{P_1P_2} = (x_2 - x_1)\mathbf{i} + (y_2 - y_1)\mathbf{j} + (z_2 - z_1)\mathbf{k}$.

Length and Direction

As always, the important features of a vector are its length and direction. The length of a vector $a\mathbf{i} + b\mathbf{j} + c\mathbf{k}$ is calculated by applying the Pythagorean theorem twice.

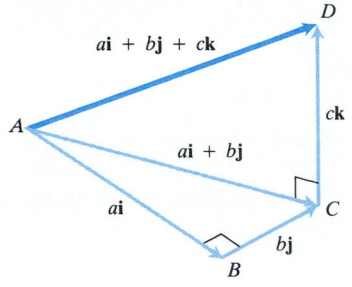

10.19   The length of the vector $\overrightarrow{AD}$ can be determined by applying the Pythagorean theorem to the right triangles ABC and ACD.

In the notation of Fig. 10.19,

$$|\overrightarrow{AC}| = |a\mathbf{i} + b\mathbf{j}| = \sqrt{a^2 + b^2},$$

from triangle ABC, and

$$|a\mathbf{i} + b\mathbf{j} + c\mathbf{k}| = |\overrightarrow{AD}| = \sqrt{|\overrightarrow{AC}|^2 + |\overrightarrow{CD}|^2}$$
$$= \sqrt{a^2 + b^2 + c^2},$$

from triangle ACD.

The **length** or **magnitude** of $\mathbf{A} = a\mathbf{i} + b\mathbf{j} + c\mathbf{k}$ is

$$|\mathbf{A}| = |a\mathbf{i} + b\mathbf{j} + c\mathbf{k}| = \sqrt{a^2 + b^2 + c^2}. \tag{4}$$

Example 3   The length of $\mathbf{A} = \mathbf{i} - 2\mathbf{j} + 3\mathbf{k}$ is

$$|\mathbf{A}| = \sqrt{(1)^2 + (-2)^2 + (3)^2} = \sqrt{1 + 4 + 9} = \sqrt{14}.$$

If we multiply a vector $\mathbf{A} = a_1\mathbf{i} + a_2\mathbf{j} + a_3\mathbf{k}$ by a scalar c, the length of $c\mathbf{A}$ is $|c|$ times the length of $\mathbf{A}$, just as in the plane. The reason is the same, as well:

$$c\mathbf{A} = ca_1\mathbf{i} + ca_2\mathbf{j} + ca_3\mathbf{k},$$
$$|c\mathbf{A}| = \sqrt{(ca_1)^2 + (ca_2)^2 + (ca_3)^2} = \sqrt{c^2a_1^2 + c^2a_2^2 + c^2a_3^2} \tag{5}$$
$$= |c|\sqrt{a_1^2 + a_2^2 + a_3^2} = |c||\mathbf{A}|.$$

Example 4   If $\mathbf{A}$ is the vector of Example 3, then the length of

$$2\mathbf{A} = 2(\mathbf{i} - 2\mathbf{j} + 3\mathbf{k}) = 2\mathbf{i} - 4\mathbf{j} + 6\mathbf{k}$$

is

$$\sqrt{(2)^2 + (-4)^2 + (6)^2} = \sqrt{4 + 16 + 36} = \sqrt{56}$$
$$= \sqrt{4 \cdot 14} = 2\sqrt{14} = 2|\mathbf{A}|.$$

As with vectors in the plane, vectors of unit length are called **unit vectors**. The vectors $\mathbf{i}$, $\mathbf{j}$, and $\mathbf{k}$ are unit vectors because

$$|\mathbf{i}| = |1\mathbf{i} + 0\mathbf{j} + 0\mathbf{k}| = \sqrt{1^2 + 0^2 + 0^2} = 1,$$
$$|\mathbf{j}| = |0\mathbf{i} + 1\mathbf{j} + 0\mathbf{k}| = \sqrt{0^2 + 1^2 + 0^2} = 1,$$
$$|\mathbf{k}| = |0\mathbf{i} + 0\mathbf{j} + 1\mathbf{k}| = \sqrt{0^2 + 0^2 + 1^2} = 1.$$

The direction of a nonzero vector $\mathbf{A}$ is the unit vector obtained by dividing $\mathbf{A}$ by its length $|\mathbf{A}|$.

The **direction** of $\mathbf{A}$ is $\dfrac{\mathbf{A}}{|\mathbf{A}|}$. $\tag{6}$

As in the plane, we can use the equation

$$\mathbf{A} = |\mathbf{A}| \cdot \frac{\mathbf{A}}{|\mathbf{A}|} \tag{7}$$

to express any nonzero vector as a product of its length and direction.

Example 5   Express $\mathbf{A} = \mathbf{i} - 2\,\mathbf{j} + 3\,\mathbf{k}$ as a product of its length and direction.

Solution

$$\mathbf{A} = |\mathbf{A}| \cdot \frac{\mathbf{A}}{|\mathbf{A}|} \qquad \text{(Eq. 7)}$$

$$= \sqrt{14} \cdot \frac{\mathbf{i} - 2\,\mathbf{j} + 3\,\mathbf{k}}{\sqrt{14}} \qquad \text{(Values from Example 3)}$$

$$= \sqrt{14}\left(\frac{1}{\sqrt{14}}\,\mathbf{i} - \frac{2}{\sqrt{14}}\,\mathbf{j} + \frac{3}{\sqrt{14}}\,\mathbf{k}\right) = (\text{length of } \mathbf{A}) \cdot (\text{direction of } \mathbf{A})$$

Example 6   Find a unit vector $\mathbf{u}$ in the direction of the vector from $P_1(1, 0, 1)$ to $P_2(3, 2, 0)$.

Solution   The vector we want is the direction of $\overrightarrow{P_1P_2}$. To find it, we divide $\overrightarrow{P_1P_2}$ by its own length:

$$\overrightarrow{P_1P_2} = (3 - 1)\,\mathbf{i} + (2 - 0)\,\mathbf{j} + (0 - 1)\,\mathbf{k} = 2\,\mathbf{i} + 2\,\mathbf{j} - \mathbf{k},$$

$$|\overrightarrow{P_1P_2}| = \sqrt{(2)^2 + (2)^2 + (-1)^2} = \sqrt{4 + 4 + 1} = \sqrt{9} = 3,$$

$$\mathbf{u} = \frac{\overrightarrow{P_1P_2}}{|\overrightarrow{P_1P_2}|} = \frac{2\,\mathbf{i} + 2\,\mathbf{j} - \mathbf{k}}{3} = \frac{2}{3}\,\mathbf{i} + \frac{2}{3}\,\mathbf{j} - \frac{1}{3}\,\mathbf{k}.$$

Example 7   Find a vector 6 units long in the direction of $\mathbf{A} = 2\,\mathbf{i} + 2\,\mathbf{j} - \mathbf{k}$.

Solution   The vector we want is

$$6\,\frac{\mathbf{A}}{|\mathbf{A}|} = 6\,\frac{2\,\mathbf{i} + 2\,\mathbf{j} - \mathbf{k}}{\sqrt{2^2 + 2^2 + (-1)^2}} = 6\,\frac{2\,\mathbf{i} + 2\,\mathbf{j} - \mathbf{k}}{3} = 4\,\mathbf{i} + 4\,\mathbf{j} - 2\,\mathbf{k}.$$

Distance in Space

To find the distance between two points P_1 and P_2 in space, we find the length of $\overrightarrow{P_1P_2}$. Equation (8) gives the resulting formula.

The Distance Between $P_1(x_1, y_1, z_1)$ and $P_2(x_2, y_2, z_2)$

$$|\overrightarrow{P_1P_2}| = \sqrt{(x_2 - x_1)^2 + (y_2 - y_1)^2 + (z_2 - z_1)^2} \tag{8}$$

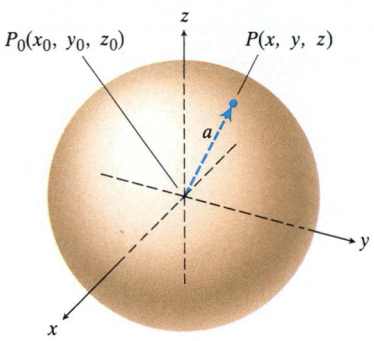

10.20 The standard equation of the sphere of radius a centered at the point (x_0, y_0, z_0) is

$$(x - x_0)^2 + (y - y_0)^2 + (z - z_0)^2 = a^2.$$

Example 8 The distance between $P_1(2, 1, 5)$ and $P_2(-2, 3, 0)$ is

$$|\overrightarrow{P_1 P_2}| = \sqrt{(-2 - 2)^2 + (3 - 1)^2 + (0 - 5)^2}$$

$$= \sqrt{16 + 4 + 25}$$

$$= \sqrt{45} = 3\sqrt{5}.$$

Spheres

We use Eq. (8) to write equations for spheres (Fig. 10.20). Since a point $P(x, y, z)$ lies on the sphere of radius a centered at $P_0(x_0, y_0, z_0)$ if and only if it lies a units from P_0, it lies on the sphere if and only if

$$|\overrightarrow{P_0 P}| = a$$

or

$$(x - x_0)^2 + (y - y_0)^2 + (z - z_0)^2 = a^2.$$

The Standard Equation for the Sphere of Radius a and Center (x_0, y_0, z_0)

$$(x - x_0)^2 + (y - y_0)^2 + (z - z_0)^2 = a^2 \qquad (9)$$

Example 9 Find the center and radius of the sphere

$$x^2 + y^2 + z^2 + 2x - 4y = 0.$$

Solution Complete the squares on the x-terms and y-terms to obtain

$$x^2 + 2x + 1 + y^2 - 4y + 4 + z^2 = 0 + 1 + 4$$

$$(x + 1)^2 + (y - 2)^2 + z^2 = 5.$$

This is Eq. (9) with $x_0 = -1$, $y_0 = 2$, $z_0 = 0$, and $a = \sqrt{5}$. The center is $(-1, 2, 0)$ and the radius is $\sqrt{5}$.

Example 10 *Sets Bounded by Spheres or Portions of Spheres*

Defining equations and inequalities	Description
a) $x^2 + y^2 + z^2 < 4$	The interior of the sphere $x^2 + y^2 + z^2 = 4$.
b) $x^2 + y^2 + z^2 \leq 4$	The solid ball bounded by the sphere $x^2 + y^2 + z^2 = 4$. Alternatively, the sphere $x^2 + y^2 + z^2 = 4$ together with its interior.
c) $x^2 + y^2 + z^2 > 4$	The exterior of the sphere $x^2 + y^2 + z^2 = 4$.
d) $x^2 + y^2 + z^2 = 4, \quad z \leq 0$	The lower hemisphere cut from the sphere $x^2 + y^2 + z^2 = 4$ by the xy-plane (the plane $z = 0$).

How to Draw Three-dimensional Objects to Look Three-dimensional

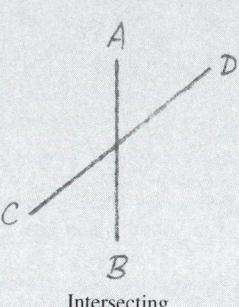
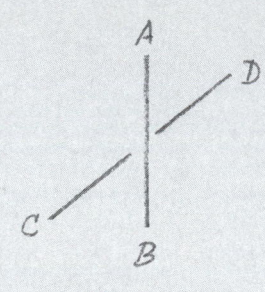
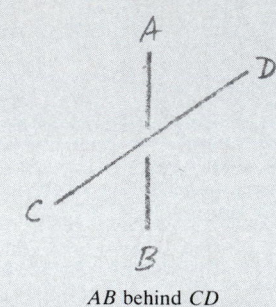

1. Break lines. When one line passes behind another, break it to show that it doesn't touch and that part of it is hidden.

Intersecting   *CD* behind *AB*   *AB* behind *CD*

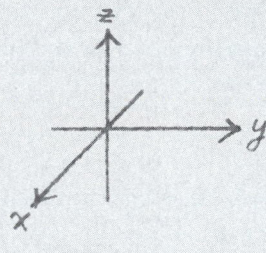
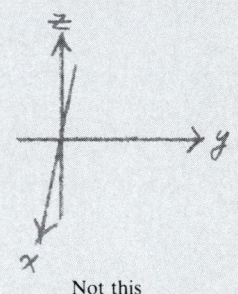

2. Make the angle between the positive *x*-axis and the positive *y*-axis large enough.

This   Not this

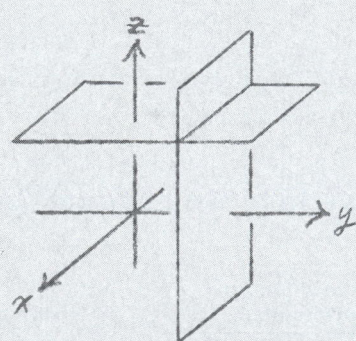

3. Draw planes parallel to the coordinate planes as if they were rectangles with sides parallel to the coordinate axes.

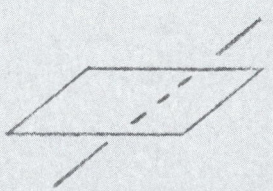
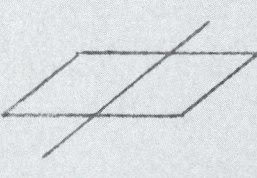

4. Dash or omit hidden portions of lines. Don't let the line touch the boundary of the parallelogram that represents the plane, unless the line lies in the plane.

Line below plane   Line above plane   Line *in* plane

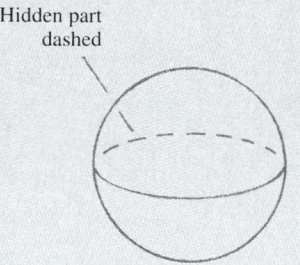

Hidden part dashed

Sphere first

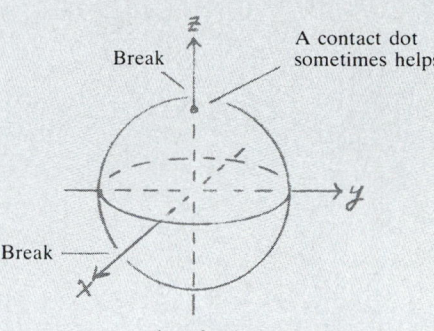

Break

A contact dot sometimes helps

Break

Axes later

5. Spheres: Draw the sphere first (outline and equator); draw axes, if any, later. Use line breaks and dashed lines.

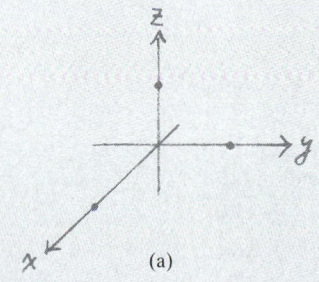

Advice ignored

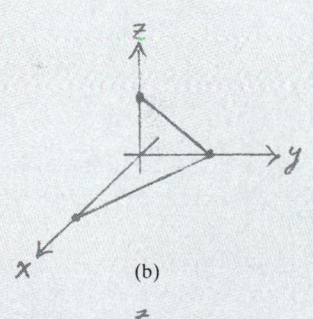

Advice followed

6. A general rule for perspective: Draw the object as if it lies some distance away, below, and to the left.

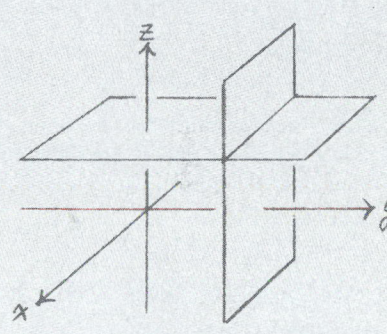

(a)

(b)

7. To draw a plane that crosses all three coordinate axes, follow the steps shown here: (a) Sketch the axes and mark the intercepts. (b) Connect the intercepts to form two sides of a parallelogram. (c) Complete the parallelogram and enlarge it by drawing lines parallel to its sides. (d) Darken the exposed parts, break hidden lines, and, if desired, dash hidden portions of the axes. You may wish to erase the smaller parallelogram at this point.

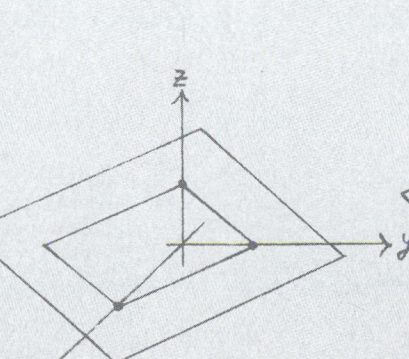

(c)

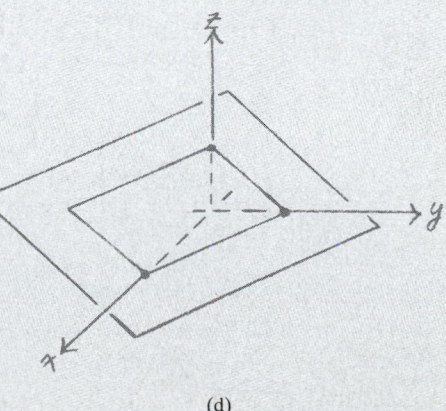

(d)

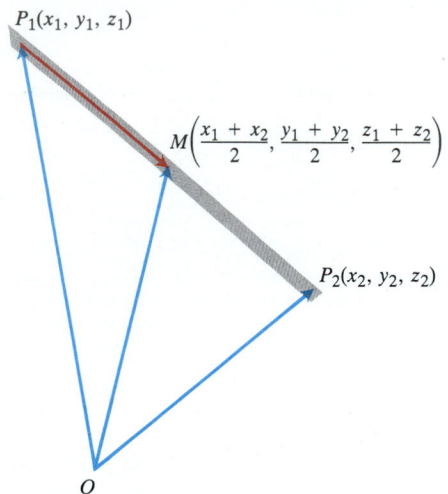

Midpoints of Line Segments

The coordinates of the midpoint M of the line segment joining two points $P_1(x_1, y_1, z_1)$ and $P_2(x_2, y_2, z_2)$ are found by averaging the coordinates of P_1 and P_2:

$$M = \left(\frac{x_1 + x_2}{2}, \ \frac{y_1 + y_2}{2}, \ \frac{z_1 + z_2}{2}\right). \tag{10}$$

To see why, observe that these coordinates are the scalar components of the position vector $\overrightarrow{OM}$ (Fig. 10.21) and that

$$\overrightarrow{OM} = \overrightarrow{OP_1} + \frac{1}{2}\,(\overrightarrow{P_1P_2}) = \overrightarrow{OP_1} + \frac{1}{2}\,(\overrightarrow{OP_2} - \overrightarrow{OP_1})$$

$$= \frac{1}{2}\,(\overrightarrow{OP_1} + \overrightarrow{OP_2})$$

$$= \frac{x_1 + x_2}{2}\,\mathbf{i} + \frac{y_1 + y_2}{2}\,\mathbf{j} + \frac{z_1 + z_2}{2}\,\mathbf{k}.$$

10.21   The coordinates of the point halfway between P_1 and P_2 are found by averaging the coordinates of P_1 and P_2.

Example 11   The midpoint of the segment joining $P_1(3, \ -2, \ 0)$ and $P_2(7, 4, 4)$ is

$$\left(\frac{3 + 7}{2}, \ \frac{-2 + 4}{2}, \ \frac{0 + 4}{2}\right) = (5, 1, 2).$$

EXERCISES 10.2

In Exercises 1–12, give a geometric description of the set of points in space whose coordinates satisfy the given pairs of equations.

1. $x = 2, \quad y = 3$
2. $x = -1, \quad z = 0$
3. $y = 0, \quad z = 0$
4. $x = 1, \quad y = 0$
5. $x^2 + y^2 = 4, \quad z = 0$
6. $x^2 + y^2 = 4, \quad z = -2$
7. $x^2 + z^2 = 4, \quad y = 0$
8. $y^2 + z^2 = 1, \quad x = 0$
9. $x^2 + y^2 + z^2 = 1, \quad x = 0$
10. $x^2 + y^2 + z^2 = 25, \quad y = -4$
11. $x^2 + y^2 + (z + 3)^2 = 25, \quad z = 0$
12. $x^2 + (y - 1)^2 + z^2 = 4, \quad y = 0$

In Exercises 13–18, describe the sets of points in space whose coordinates satisfy the given inequalities or combinations of equations and inequalities.

13. a) $x \geq 0, \quad y \geq 0, \quad z = 0$
b) $x \geq 0, \quad y \leq 0, \quad z = 0$

14. a) $0 \leq x \leq 1$
b) $0 \leq x \leq 1, \quad 0 \leq y \leq 1$
c) $0 \leq x \leq 1, \quad 0 \leq y \leq 1, \quad 0 \leq z \leq 1$

15. a) $x^2 + y^2 + z^2 \leq 1$
b) $x^2 + y^2 + z^2 > 1$

16. a) $x^2 + y^2 \leq 1, \quad z = 0$
b) $x^2 + y^2 \leq 1, \quad z = 3$
c) $x^2 + y^2 \leq 1, \quad$ no restriction on z

17. a) $x^2 + y^2 + z^2 = 1, \quad z \geq 0$
b) $x^2 + y^2 + z^2 \leq 1, \quad z \geq 0$

18. a) $x = y, \quad z = 0$
b) $x = y, \quad$ no restriction on z

In Exercises 19–28, describe the given set with a single equation or with a pair of equations.

19. The plane perpendicular to the
a) x-axis at $(3, 0, 0)$
b) y-axis at $(0, \ -1, \ 0)$
c) z-axis at $(0, 0, \ -2)$

20. The plane through the point $(3, \ -1, \ 2)$ perpendicular to the
a) x-axis   b) y-axis   c) z-axis

21. The plane through the point $(3, \ -1, \ 1)$ parallel to the
a) xy-plane   b) yz-plane   c) xz-plane

22. The circle of radius 2 centered at $(0, 0, 0)$ and lying in the
  a) xy-plane   b) yz-plane   c) xz-plane

23. The circle of radius 2 centered at $(0, 2, 0)$ and lying in the
  a) xy-plane   b) yz-plane   c) plane $y = 2$

24. The circle of radius 1 centered at $(-3, 4, 1)$ and lying in a plane parallel to the
  a) xy-plane   b) yz-plane   c) xz-plane

25. The line through the point $(1, 3, -1)$ parallel to the
  a) x-axis   b) y-axis   c) z-axis

26. The set of points in space equidistant from the origin and the point $(0, 2, 0)$

27. The circle in which the plane through the point $(1, 1, 3)$ perpendicular to the z-axis meets the sphere of radius 5 centered at the origin

28. The set of points in space that lie 2 units from the point $(0, 0, 1)$ and, at the same time, 2 units from the point $(0, 0, -1)$

Write inequalities to describe the sets in Exercises 29–34.

29. The slab bounded by the planes $z = 0$ and $z = 1$ (planes included)

30. The solid cube in the first octant bounded by the planes $x = 2$, $y = 2$, and $z = 2$

31. The half-space consisting of the points on and below the xy-plane

32. The upper hemisphere of the sphere of radius 1 centered at the origin

33. The (a) interior and (b) exterior of the sphere of radius 1 centered at the point $(1, 1, 1)$

34. The closed region bounded by the spheres of radius 1 and radius 2 centered at the origin. (*Closed* means the spheres are to be included. Had we wanted the spheres left out, we would have asked for the *open* region bounded by the spheres. This is analogous to the way we use "closed" and "open" to describe intervals: "closed" means endpoints included, "open" means endpoints left out. Closed sets include boundaries; open sets leave them out.)

In Exercises 35–46, express each vector as a product of its length and direction.

35. $2\mathbf{i} + \mathbf{j} - 2\mathbf{k}$

36. $3\mathbf{i} - 6\mathbf{j} + 2\mathbf{k}$

37. $\mathbf{i} + 4\mathbf{j} - 8\mathbf{k}$

38. $9\mathbf{i} - 2\mathbf{j} + 6\mathbf{k}$

39. $5\mathbf{k}$

40. $6\mathbf{i}$

41. $-4\mathbf{j}$

42. $\dfrac{3}{5}\mathbf{i} + \dfrac{4}{5}\mathbf{k}$

43. $-\dfrac{1}{3}\mathbf{j} + \dfrac{1}{4}\mathbf{k}$

44. $\dfrac{1}{\sqrt{2}}\mathbf{i} - \dfrac{1}{\sqrt{2}}\mathbf{k}$

45. $\dfrac{1}{\sqrt{6}}\mathbf{i} - \dfrac{1}{\sqrt{6}}\mathbf{j} - \dfrac{1}{\sqrt{6}}\mathbf{k}$

46. $\dfrac{\mathbf{i}}{\sqrt{3}} + \dfrac{\mathbf{j}}{\sqrt{3}} + \dfrac{\mathbf{k}}{\sqrt{3}}$

In Exercises 47–52, find the length and direction of the vector from point P_1 to point P_2. Also find the midpoint of the line segment joining P_1 and P_2.

47. $P_1(1, 1, 1)$, $P_2(3, 3, 0)$

48. $P_1(-1, 1, 5)$, $P_2(2, 5, 0)$

49. $P_1(1, 4, 5)$, $P_2(4, -2, 7)$

50. $P_1(3, 4, 5)$, $P_2(2, 3, 4)$

51. $P_1(0, 0, 0)$, $P_2(2, -2, -2)$

52. $P_1(5, 3, -2)$, $P_2(0, 0, 0)$

53. Find the vectors whose lengths and directions are given. Try to answer without writing anything down.

Length	Direction
a) 2	$\mathbf{i}$
b) -4	$\mathbf{j}$
c) $\sqrt{3}$	$\mathbf{k}$
d) $\dfrac{1}{2}$	$\dfrac{3}{5}\mathbf{j} + \dfrac{4}{5}\mathbf{k}$
e) 7	$\dfrac{6}{7}\mathbf{i} - \dfrac{2}{7}\mathbf{j} + \dfrac{3}{7}\mathbf{k}$
f) a	$u_1\mathbf{i} + u_2\mathbf{j} + u_3\mathbf{k}$

54. Find a vector of magnitude 7 in the direction of $\mathbf{A} = 12\mathbf{i} - 5\mathbf{k}$.

55. Find a vector $\sqrt{5}$ units long in the direction of $\mathbf{A} = \mathbf{i} + \mathbf{j} + \mathbf{k}$.

56. Find a vector 1/2 unit long in the direction of $\mathbf{A} = \mathbf{i} - \mathbf{j}$.

Spheres and Distance

57. Find the centers and radii of the following spheres.
  a) $(x + 2)^2 + y^2 + (z - 2)^2 = 8$
  b) $\left(x + \dfrac{1}{2}\right)^2 + \left(y + \dfrac{1}{2}\right)^2 + \left(z + \dfrac{1}{2}\right)^2 = \dfrac{21}{4}$
  c) $(x - \sqrt{2})^2 + (y - \sqrt{2})^2 + (z + \sqrt{2})^2 = 2$
  d) $x^2 + \left(y + \dfrac{1}{3}\right)^2 + \left(z - \dfrac{1}{3}\right)^2 = \dfrac{29}{9}$

58. Find equations for the spheres whose centers and radii are given here.

Center	Radius
a) $(1, 2, 3)$	$\sqrt{14}$
b) $(0, -1, 5)$	2
c) $(-2, 0, 0)$	$\sqrt{3}$
d) $(0, -7, 0)$	7

Find the centers and radii of the spheres in Exercises 59–62.

59. $x^2 + y^2 + z^2 + 4x - 4z = 0$

60. $2x^2 + 2y^2 + 2z^2 + x + y + z = 9$

61. $x^2 + y^2 + z^2 - 2az = 0$

62. $3x^2 + 3y^2 + 3z^2 + 2y - 2z = 9$

63. Find a formula for the distance from the point $P(x, y, z)$ to the

   a) x-axis,   b) y-axis,   c) z-axis.

64. Find a formula for the distance from the point $P(x, y, z)$ to the

   a) xy-plane,   b) yz-plane,   c) xz-plane.

Geometry with Vectors

65. Suppose that A, B, and C are the corner points of the thin triangular plate of constant density shown in Fig. 10.22.

   a) Find the vector from C to the midpoint M of side AB.

   b) Find the vector from C to the point that lies two-thirds of the way from C to M on the median CM.

   c) Find the coordinates of the point in which the medians of $\triangle ABC$ intersect. According to Exercise 22, Section 5.6, this point is the plate's center of mass.

66. Find the vector from the origin to the point of intersection of the medians of the triangle whose vertices are

   $A(1, -1, 2)$, $B(2, 1, 3)$, and $C(-1, 2, -1)$.

67. Let $ABCD$ be a general, not necessarily planar, quadrilateral in space. Show that the two segments joining the midpoints of opposite sides of $ABCD$ bisect each other. (*Hint:* Show that the segments have the same midpoint.)

68. Vectors are drawn from the center of a regular n-sided polygon in the plane to the vertices of the polygon. Show that the sum of the vectors is zero. (*Hint:* What happens to the sum if you rotate the polygon about its center?)

69. Suppose that A, B, and C are vertices of a triangle and that a, b, and c are, respectively, the midpoints of the opposite sides. Show that $\overrightarrow{Aa} + \overrightarrow{Bb} + \overrightarrow{Cc} = 0$.

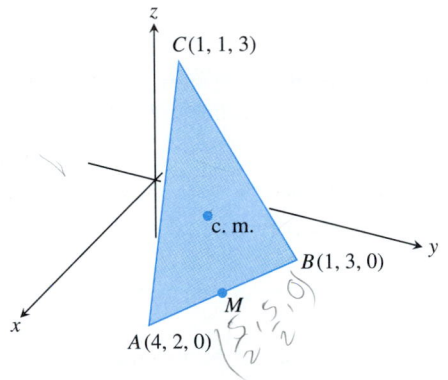

10.22   The triangular plate in Exercise 65.

10.3 Dot Products

We now introduce the dot product of two vectors, the first of two methods we shall learn for multiplying vectors together. Our motivation is the need to calculate the work done by a constant force in displacing a mass. If we can represent the force and displacement as vectors, we can calculate the work by finding their dot product. Dot products are also called *scalar products* because the result of the multiplication is a scalar and not a vector. Cross products, which we shall study in the next section, are always vectors.

Scalar products have applications in mathematics as well as in engineering and physics. This section describes the geometric and algebraic properties on which many of these applications depend.

DEFINITION

> The **scalar product A · B** ("A dot B") or **dot product** of two vectors **A** and **B** is the number
>
> $$\mathbf{A} \cdot \mathbf{B} = |\mathbf{A}||\mathbf{B}| \cos \theta, \qquad (1)$$
>
> where θ measures the smaller angle made by **A** and **B** when their initial points coincide (as in Fig. 10.23).

In words, the scalar product of **A** and **B** is the length of **A** times the length of **B** times the cosine of the angle between **A** and **B**. The product is a scalar, not a vector. It is called the dot product because of the dot in the notation **A · B**.

10.23   **A · B** is the number $|\mathbf{A}||\mathbf{B}| \cos \theta$.

From Eq. (1) we see that the scalar product of two vectors is positive when the angle between them is acute, negative when the angle is obtuse.

Since the angle a vector **A** makes with itself is zero, and the cosine of zero is 1,

$$\mathbf{A} \cdot \mathbf{A} = |\mathbf{A}||\mathbf{A}| \cos 0 = |\mathbf{A}||\mathbf{A}|(1) = |\mathbf{A}|^2, \qquad \text{or} \qquad |\mathbf{A}| = \sqrt{\mathbf{A} \cdot \mathbf{A}}. \qquad (2)$$

As we shall see, this provides a handy way to calculate a vector's length.

Calculation

To calculate $\mathbf{A} \cdot \mathbf{B}$ from the components of **A** and **B**, we let

$$\mathbf{A} = a_1 \mathbf{i} + a_2 \mathbf{j} + a_3 \mathbf{k},$$

$$\mathbf{B} = b_1 \mathbf{i} + b_2 \mathbf{j} + b_3 \mathbf{k},$$

and

$$\mathbf{C} = \mathbf{B} - \mathbf{A} = (b_1 - a_1) \mathbf{i} + (b_2 - a_2) \mathbf{j} + (b_3 - a_3) \mathbf{k}.$$

Then we apply the law of cosines to a triangle whose sides represent the vectors **A**, **B**, and **C** (Fig. 10.24) and obtain

$$|\mathbf{C}|^2 = |\mathbf{A}|^2 + |\mathbf{B}|^2 - 2|\mathbf{A}||\mathbf{B}| \cos \theta,$$

$$|\mathbf{A}||\mathbf{B}| \cos \theta = \frac{|\mathbf{A}|^2 + |\mathbf{B}|^2 - |\mathbf{C}|^2}{2}.$$

The left side of this equation is $\mathbf{A} \cdot \mathbf{B}$, and we may evaluate the right side by applying Eq. (4) of Section 10.2 to find the lengths of **A**, **B**, and **C**. The result of this algebra is the formula

$$\mathbf{A} \cdot \mathbf{B} = a_1 b_1 + a_2 b_2 + a_3 b_3. \qquad (3)$$

Thus, to find the scalar product of two given vectors we multiply their corresponding **i**, **j**, and **k** components and add the results. In particular, from Eq. (2) we have

$$|\mathbf{A}| = \sqrt{\mathbf{A} \cdot \mathbf{A}} = \sqrt{a_1^2 + a_2^2 + a_3^2}. \qquad (4)$$

When we solve Eq. (1) for θ, we get a formula for finding angles between nonzero vectors.

The Angle Between Two Nonzero Vectors

The angle between two nonzero vectors **A** and **B** is

$$\theta = \cos^{-1}\left(\frac{\mathbf{A} \cdot \mathbf{B}}{|\mathbf{A}||\mathbf{B}|}\right). \qquad (5)$$

Since the values of the arc cosine lie in $[0, \pi]$, Eq. (5) automatically gives the smaller of the two angles made by **A** and **B**.

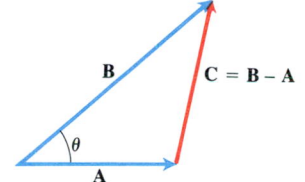

10.24 Equation (3) is obtained by applying the law of cosines to a triangle whose sides represent **A**, **B**, and **C = B − A**.

Example 1   Find the angle between $\mathbf{A} = \mathbf{i} - 2\mathbf{j} - 2\mathbf{k}$ and $\mathbf{B} = 6\mathbf{i} + 3\mathbf{j} + 2\mathbf{k}$.

Solution   We use Eq. (5):

$$\mathbf{A} \cdot \mathbf{B} = (1)(6) + (-2)(3) + (-2)(2) = 6 - 6 - 4 = -4,$$

$$|\mathbf{A}| = \sqrt{\mathbf{A} \cdot \mathbf{A}} = \sqrt{(1)^2 + (-2)^2 + (-2)^2} = \sqrt{9} = 3,$$

$$|\mathbf{B}| = \sqrt{\mathbf{B} \cdot \mathbf{B}} = \sqrt{(6)^2 + (3)^2 + (2)^2} = \sqrt{49} = 7,$$

$$\theta = \cos^{-1}\left(\frac{\mathbf{A} \cdot \mathbf{B}}{|\mathbf{A}||\mathbf{B}|}\right) \qquad \text{(Eq. (5))}$$

$$= \cos^{-1}\left(\frac{-4}{(3)(7)}\right) = \cos^{-1}\left(-\frac{4}{21}\right)$$

$$= 101°. \qquad \text{(Calculator, rounded)}$$

Laws of Multiplication

From the equation $\mathbf{A} \cdot \mathbf{B} = a_1 b_1 + a_2 b_2 + a_3 b_3$, we can see right away that

$$\mathbf{A} \cdot \mathbf{B} = \mathbf{B} \cdot \mathbf{A}. \tag{6}$$

In other words, scalar multiplication is commutative. We can also see from Eq. (3) that if c is any number, then

$$(c\mathbf{A}) \cdot \mathbf{B} = \mathbf{A} \cdot (c\mathbf{B}) = c(\mathbf{A} \cdot \mathbf{B}). \tag{7}$$

If $\mathbf{C} = c_1 \mathbf{i} + c_2 \mathbf{j} + c_3 \mathbf{k}$ is any third vector, then

$$\mathbf{A} \cdot (\mathbf{B} + \mathbf{C}) = a_1(b_1 + c_1) + a_2(b_2 + c_2) + a_3(b_3 + c_3)$$

$$= (a_1 b_1 + a_2 b_2 + a_3 b_3) + (a_1 c_1 + a_2 c_2 + a_3 c_3)$$

$$= \mathbf{A} \cdot \mathbf{B} + \mathbf{A} \cdot \mathbf{C}.$$

Hence scalar products obey the distributive law:

$$\mathbf{A} \cdot (\mathbf{B} + \mathbf{C}) = \mathbf{A} \cdot \mathbf{B} + \mathbf{A} \cdot \mathbf{C}. \tag{8}$$

If we combine this with the commutative law, Eq. (6), it is also evident that

$$(\mathbf{A} + \mathbf{B}) \cdot \mathbf{C} = \mathbf{A} \cdot \mathbf{C} + \mathbf{B} \cdot \mathbf{C}. \tag{9}$$

Equations (8) and (9) together permit us to multiply sums of vectors by the familiar laws of algebra. For example,

$$(\mathbf{A} + \mathbf{B}) \cdot (\mathbf{C} + \mathbf{D}) = \mathbf{A} \cdot \mathbf{C} + \mathbf{A} \cdot \mathbf{D} + \mathbf{B} \cdot \mathbf{C} + \mathbf{B} \cdot \mathbf{D}. \tag{10}$$

Orthogonal Vectors

Two vectors whose scalar product is zero are said to be *orthogonal*. From the equation $\mathbf{A} \cdot \mathbf{B} = |\mathbf{A}||\mathbf{B}| \cos \theta$, we can see that when neither $|\mathbf{A}|$ nor $|\mathbf{B}|$ is zero, $\mathbf{A} \cdot \mathbf{B}$ is zero if and only if cos θ is zero, or θ equals 90°. Hence, for the vectors we are dealing with, "orthogonal" means the same as "perpendicular." In other scientific contexts in which the word is used, there may be no such geometric interpretation.

The zero vector $\mathbf{0} = 0\,\mathbf{i} + 0\,\mathbf{j} + 0\,\mathbf{k}$ is orthogonal to every vector because its dot product with every vector is zero.

DEFINITION

> Vectors $\mathbf{A}$ and $\mathbf{B}$ are **orthogonal** if $\mathbf{A} \cdot \mathbf{B} = 0$.   (11)

Example 2   The vectors $\mathbf{A} = 3\,\mathbf{i} - 2\,\mathbf{j} + \mathbf{k}$ and $\mathbf{B} = 2\,\mathbf{j} + 4\,\mathbf{k}$ are orthogonal because

$$\mathbf{A} \cdot \mathbf{B} = (3)(0) + (-2)(2) + (1)(4) = 0.$$

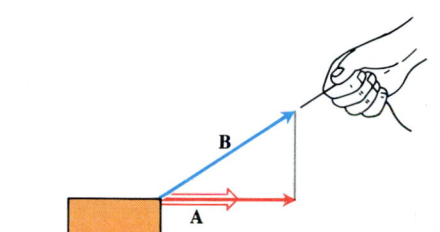

10.25 We may pull on the box with force $\mathbf{B}$ to move the box in the direction of $\mathbf{A}$. The effective force in this direction is represented by the vector projection of $\mathbf{B}$ onto $\mathbf{A}$.

Vector Projections and Scalar Components

The vector we get by projecting a vector $\mathbf{B}$ onto the line through a vector $\mathbf{A}$ is called the **vector projection of B onto A,** sometimes denoted

$$\text{proj}_{\mathbf{A}}\mathbf{B} \qquad (\text{"the vector projection of } \mathbf{B} \text{ onto } \mathbf{A}\text{"}).$$

If $\mathbf{B}$ represents a force, then the vector projection of $\mathbf{B}$ onto $\mathbf{A}$ represents the effective force in the direction of $\mathbf{A}$ (Fig. 10.25).

If the angle between $\mathbf{B}$ and $\mathbf{A}$ is acute, the length of the vector projection of $\mathbf{B}$ onto $\mathbf{A}$ is $|\mathbf{B}| \cos \theta$. If the angle is obtuse, its cosine is negative and the length of the vector projection of $\mathbf{B}$ onto $\mathbf{A}$ is $-|\mathbf{B}| \cos \theta$. In either case, the number $|\mathbf{B}| \cos \theta$ itself is called the **scalar component of B in the direction of A** (Fig. 10.26).

The scalar component of $\mathbf{B}$ in the direction of $\mathbf{A}$ can be found by dividing both sides of the equation $\mathbf{A} \cdot \mathbf{B} = |\mathbf{A}||\mathbf{B}| \cos \theta$ by $|\mathbf{A}|$. This gives

$$|\mathbf{B}| \cos \theta = \frac{\mathbf{A} \cdot \mathbf{B}}{|\mathbf{A}|} = \mathbf{B} \cdot \frac{\mathbf{A}}{|\mathbf{A}|}.   (12)$$

Equation (12) says that the scalar component of $\mathbf{B}$ in the direction of $\mathbf{A}$ can be obtained by "dotting" $\mathbf{B}$ with the direction of $\mathbf{A}$.

The vector projection of $\mathbf{B}$ onto $\mathbf{A}$ is the scalar component of $\mathbf{B}$ in the direction of $\mathbf{A}$ times the direction of $\mathbf{A}$. If the angle between $\mathbf{A}$ and $\mathbf{B}$ is acute, the vector projection has length $|\mathbf{B}| \cos \theta$ and direction $\mathbf{A}/|\mathbf{A}|$. If the angle is obtuse, the vector projection has length $-|\mathbf{B}| \cos \theta$ and direction $-\mathbf{A}/|\mathbf{A}|$. In either case,

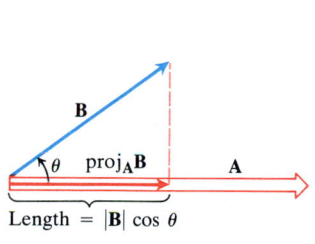

$$\text{proj}_{\mathbf{A}}\mathbf{B} = (|\mathbf{B}| \cos \theta) \frac{\mathbf{A}}{|\mathbf{A}|}.   (13)$$

Equations (12) and (13) together give a useful way to calculate $\text{proj}_{\mathbf{A}}\mathbf{B}$:

$$\text{proj}_{\mathbf{A}}\mathbf{B} = (|\mathbf{B}| \cos \theta) \frac{\mathbf{A}}{|\mathbf{A}|} \qquad (\text{Eq. (13)})$$

$$= \left(\frac{\mathbf{A} \cdot \mathbf{B}}{|\mathbf{A}|}\right) \frac{\mathbf{A}}{|\mathbf{A}|} \qquad (\text{Eq. (12)})$$

$$= \frac{\mathbf{A} \cdot \mathbf{B}}{\mathbf{A} \cdot \mathbf{A}} \mathbf{A}. \qquad (|\mathbf{A}|^2 = \mathbf{A} \cdot \mathbf{A})   (14)$$

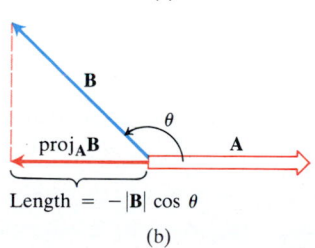

10.26 (a) When θ is acute, the length of the vector projection of $\mathbf{B}$ onto $\mathbf{A}$ is $|\mathbf{B}| \cos \theta$, the scalar component of $\mathbf{B}$ in the direction of $\mathbf{A}$. (b) When θ is obtuse, the scalar component of $\mathbf{B}$ in the direction of $\mathbf{A}$ is negative and the length of the vector projection of $\mathbf{B}$ onto $\mathbf{A}$ is $-|\mathbf{B}| \cos \theta$.

Example 3   Find the vector projection of $\mathbf{B} = 6\,\mathbf{i} + 3\,\mathbf{j} + 2\,\mathbf{k}$ onto $\mathbf{A} = \mathbf{i} - 2\,\mathbf{j} - 2\,\mathbf{k}$ and the scalar component of $\mathbf{B}$ in the direction of $\mathbf{A}$.

Solution   We find $\text{proj}_{\mathbf{A}}\mathbf{B}$ from Eq. (14):

$$\text{proj}_{\mathbf{A}}\mathbf{B} = \frac{\mathbf{A} \cdot \mathbf{B}}{\mathbf{A} \cdot \mathbf{A}}\,\mathbf{A} = \frac{6 - 6 - 4}{1 + 4 + 4}\,(\mathbf{i} - 2\,\mathbf{j} - 2\,\mathbf{k})$$

$$= -\frac{4}{9}(\mathbf{i} - 2\,\mathbf{j} - 2\,\mathbf{k}) = -\frac{4}{9}\,\mathbf{i} + \frac{8}{9}\,\mathbf{j} + \frac{8}{9}\,\mathbf{k}.$$

We find the scalar component of $\mathbf{B}$ in the direction of $\mathbf{A}$ from Eq. (12):

$$|\mathbf{B}| \cos \theta = \mathbf{B} \cdot \frac{\mathbf{A}}{|\mathbf{A}|} = (6\,\mathbf{i} + 3\,\mathbf{j} + 2\,\mathbf{k}) \cdot \left(\frac{1}{3}\,\mathbf{i} - \frac{2}{3}\,\mathbf{j} - \frac{2}{3}\,\mathbf{k}\right)$$

$$= 2 - 2 - \frac{4}{3} = -\frac{4}{3}.$$

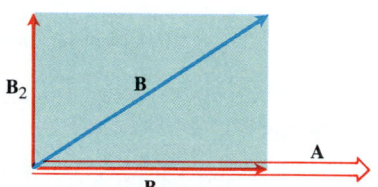

10.27   The vector $\mathbf{B}$ as the sum of vectors parallel and orthogonal to $\mathbf{A}$.

Writing a Vector as a Sum of Orthogonal Vectors

In mechanics, we often want to express a vector $\mathbf{B}$ as a sum of a vector $\mathbf{B}_1$ parallel to a vector $\mathbf{A}$ and a vector $\mathbf{B}_2$ orthogonal to $\mathbf{A}$ (Fig. 10.27). We can do this by writing $\mathbf{B}$ as a sum of its vector projection on $\mathbf{A}$ plus whatever is left over, because the left-over part will automatically be orthogonal to $\mathbf{A}$.

Where Vectors Came From

Although Aristotle used vectors to describe the effects of forces, the idea of resolving vectors into geometric components parallel to the coordinate axes came from Descartes. The algebra of vectors we use today was developed simultaneously and independently in the 1870s by Josiah Willard Gibbs (1839–1903), a mathematical physicist at Yale University, and by the English mathematical physicist Oliver Heaviside (1850–1925), the Heaviside of Heaviside layer fame. The works of Gibbs and Heaviside grew out of more complicated mathematical theories developed some years earlier by the Irish mathematician William Hamilton (1805–1865) and the German linguist, physicist, and geometer Hermann Grassman (1809–1877). Hamilton's quaternions and Grassman's algebraic forms are still in use but tend to appear in more theoretical work.

> **Formula for Writing B as a Vector Parallel to A Plus a Vector Orthogonal to A**
>
> GIVEN →
>
> $$\mathbf{B} = \underbrace{\frac{\mathbf{A} \cdot \mathbf{B}}{\mathbf{A} \cdot \mathbf{A}}\,\mathbf{A}}_{\substack{\mathbf{B}_1 \\ \text{parallel to } \mathbf{A}}} + \underbrace{\left(\mathbf{B} - \frac{\mathbf{A} \cdot \mathbf{B}}{\mathbf{A} \cdot \mathbf{A}}\,\mathbf{A}\right)}_{\substack{\mathbf{B}_2 \\ \text{orthogonal to } \mathbf{A}}}. \tag{15}$$
>
>

The vector $\mathbf{B}_1$, being the vector projection of $\mathbf{B}$ onto $\mathbf{A}$, is parallel to $\mathbf{A}$, while $\mathbf{B}_2$ can be seen to be orthogonal to $\mathbf{A}$ because $\mathbf{A} \cdot \mathbf{B}_2$ is zero:

$$\mathbf{A} \cdot \mathbf{B}_2 = \mathbf{A} \cdot \left(\mathbf{B} - \frac{\mathbf{A} \cdot \mathbf{B}}{\mathbf{A} \cdot \mathbf{A}}\,\mathbf{A}\right) = \mathbf{A} \cdot \mathbf{B} - \frac{\mathbf{A} \cdot \mathbf{B}}{\mathbf{A} \cdot \mathbf{A}}\,\mathbf{A} \cdot \mathbf{A} = \mathbf{A} \cdot \mathbf{B} - \mathbf{A} \cdot \mathbf{B} = 0. \tag{16}$$

Example 4   Express $\mathbf{B} = 2\,\mathbf{i} + \mathbf{j} - 3\,\mathbf{k}$ as the sum of a vector parallel to $\mathbf{A} = 3\,\mathbf{i} - \mathbf{j}$ and a vector orthogonal to $\mathbf{A}$.

Solution   We use Eq. (15). With

$$\mathbf{A} \cdot \mathbf{B} = 6 - 1 = 5 \qquad \text{and} \qquad \mathbf{A} \cdot \mathbf{A} = 9 + 1 = 10,$$

Eq. (15) gives

$$\mathbf{B} = \frac{\mathbf{A} \cdot \mathbf{B}}{\mathbf{A} \cdot \mathbf{A}}\,\mathbf{A} + \left(\mathbf{B} - \frac{\mathbf{A} \cdot \mathbf{B}}{\mathbf{A} \cdot \mathbf{A}}\,\mathbf{A}\right) = \frac{5}{10}(3\,\mathbf{i} - \mathbf{j}) + \left(2\,\mathbf{i} + \mathbf{j} - 3\,\mathbf{k} - \frac{5}{10}(3\,\mathbf{i} - \mathbf{j})\right)$$

$$= \left(\frac{3}{2}\,\mathbf{i} - \frac{1}{2}\,\mathbf{j}\right) + \left(\frac{1}{2}\,\mathbf{i} + \frac{3}{2}\,\mathbf{j} - 3\,\mathbf{k}\right).$$

Check: The first vector in the sum is parallel to **A** because it is $(1/2)$**A**. The second vector in the sum is orthogonal to **A** because

$$\left(\frac{1}{2}\mathbf{i} + \frac{3}{2}\mathbf{j} - 3\mathbf{k}\right) \cdot (3\mathbf{i} - \mathbf{j}) = \frac{3}{2} - \frac{3}{2} = 0.$$

Lines in the Plane and Distances from Points to Lines

Dot products give a new understanding of the equations we write for lines in the plane and a quick way to calculate distances from points to lines.

Suppose L is a line through the point $P_0(x_0, y_0)$ perpendicular to a vector $\mathbf{N} = A\mathbf{i} + B\mathbf{j}$ (Fig. 10.28). Then a point $P(x, y)$ lies on L if and only if

$$\mathbf{N} \cdot \overrightarrow{P_0P} = 0 \qquad \text{or} \qquad A(x - x_0) + B(y - y_0) = 0. \tag{17}$$

When rearranged, the second equation becomes

$$Ax + By = Ax_0 + By_0. \tag{18}$$

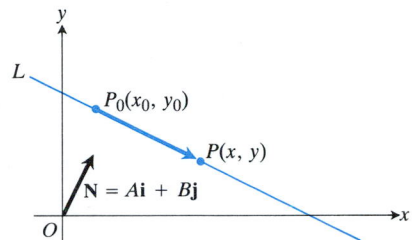

10.28 A point $P(x, y)$ lies on the line through P_0 perpendicular to **N** if and only if $\mathbf{N} \cdot \overrightarrow{P_0P} = 0$.

Line Through $P(x_0, y_0)$ Perpendicular to $\mathbf{N} = A\mathbf{i} + B\mathbf{j}$

$$Ax + By = C, \qquad C = Ax_0 + By_0 \tag{19}$$

Notice how the components of **N** become the coefficients in the equation $Ax + By = C$.

Example 5   Find an equation for the line through $P_0(3, 5)$ perpendicular to $\mathbf{N} = \mathbf{i} + 2\mathbf{j}$.

Solution   We use Eqs. (19) with $A = 1$, $B = 2$, and $C = (1)(3) + (2)(5) = 13$ to get

$$x + 2y = 13.$$

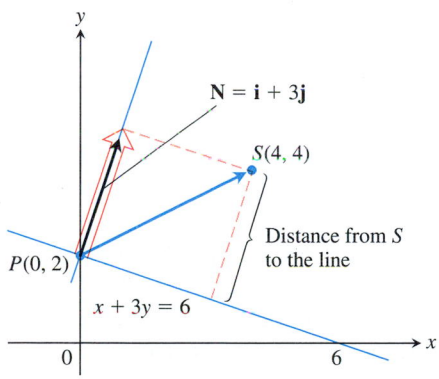

10.29   The distance from point S to the line is the length of the vector projection of $\overrightarrow{PS}$ onto **N** (Example 6).

Example 6   Find the distance from the point $S(4, 4)$ to the line L: $x + 3y = 6$.

Solution   We find a point P on the line and calculate the distance as the length of the vector projection of $\overrightarrow{PS}$ onto a vector **N** perpendicular to the line (Fig. 10.29). Any point on the line will do for P and we can find **N** from the coefficients of $x + 3y = 6$ as $\mathbf{N} = \mathbf{i} + 3\mathbf{j}$. With $P = (0, 2)$, say, we then have

$$\overrightarrow{PS} = (4 - 0)\mathbf{i} + (4 - 2)\mathbf{j} = 4\mathbf{i} + 2\mathbf{j},$$

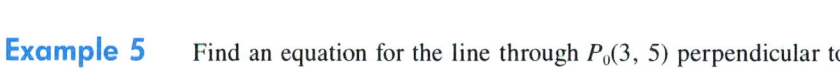

$$\text{distance from } S \text{ to } L = |\text{proj}_{\mathbf{N}} \overrightarrow{PS}| = \left|\overrightarrow{PS} \cdot \frac{\mathbf{N}}{|\mathbf{N}|}\right|$$

$$= \left|(4\mathbf{i} + 2\mathbf{j}) \cdot \frac{\mathbf{i} + 3\mathbf{j}}{\sqrt{(1)^2 + (3)^2}}\right| = \frac{4 + 6}{\sqrt{10}} = \sqrt{10}.$$

To find the distance from a point S to the line L: $Ax + By = C$, find

1. a point P on L,
2. $\overrightarrow{PS}$,
3. the direction of $\mathbf{N} = A\mathbf{i} + B\mathbf{j}$.

Then, calculate the distance as

$$\left|\overrightarrow{PS} \cdot \frac{\mathbf{N}}{|\mathbf{N}|}\right|.$$

Work

As we mentioned at the beginning of the section, the work done by a constant force $\mathbf{F}$ when the point of application undergoes a displacement $\overrightarrow{PQ}$ (Fig. 10.30) is defined to be the dot product of $\mathbf{F}$ with $\overrightarrow{PQ}$.

DEFINITION

> The **work** done by a constant force $\mathbf{F}$ acting through a displacement $\overrightarrow{PQ}$ is
>
> $$\text{Work} = \mathbf{F} \cdot \overrightarrow{PQ} = |\mathbf{F}||\overrightarrow{PQ}| \cos \theta. \qquad (20)$$

10.30   The work done by a constant force $\mathbf{F}$ during a displacement $\overrightarrow{PQ}$ is $(|\mathbf{F}| \cos \theta)|\overrightarrow{PQ}|$.

Example 7   If $|\mathbf{F}| = 40$ newtons (about 9 pounds), $|\overrightarrow{PQ}| = 3$ m, and $\theta = 60°$, the work done by $\mathbf{F}$ in acting from P to Q is

$$
\begin{aligned}
\text{Work} &= |\mathbf{F}||\overrightarrow{PQ}| \cos \theta && \text{(Eq. 20)} \\
&= (40)(3) \cos 60° && \text{(Given values)} \\
&= (120)(1/2) \\
&= 60 \text{ newton-meters.}
\end{aligned}
$$

We shall encounter work problems of greater interest in Chapter 14, where we show how to use integration to calculate the work done in moving a mass or a charged particle along a curve in space against a variable force.

EXERCISES 10.3

In Exercises 1–12, find $\mathbf{A} \cdot \mathbf{B}$, $|\mathbf{A}|$, $|\mathbf{B}|$, the cosine of the angle between $\mathbf{A}$ and $\mathbf{B}$, the scalar component of $\mathbf{B}$ in the direction of $\mathbf{A}$, and the vector projection of $\mathbf{B}$ onto $\mathbf{A}$.

1. $\mathbf{A} = 3\mathbf{i} + 2\mathbf{j}$, $\mathbf{B} = 5\mathbf{j} + \mathbf{k}$

2. $\mathbf{A} = \mathbf{i}$, $\mathbf{B} = 5\mathbf{j} - 3\mathbf{k}$

3. $\mathbf{A} = 3\mathbf{i} - 2\mathbf{j} - \mathbf{k}$, $\mathbf{B} = -2\mathbf{j}$

4. $\mathbf{A} = -2\mathbf{i} + 7\mathbf{j}$, $\mathbf{B} = \mathbf{k}$

5. $\mathbf{A} = 5\mathbf{j} - 3\mathbf{k}$, $\mathbf{B} = \mathbf{i} + \mathbf{j} + \mathbf{k}$

6. $\mathbf{A} = \dfrac{1}{\sqrt{2}}\mathbf{i} + \dfrac{1}{\sqrt{3}}\mathbf{j} + \dfrac{1}{\sqrt{6}}\mathbf{k}$, $\mathbf{B} = \dfrac{1}{\sqrt{2}}\mathbf{j} - \mathbf{k}$

7. $\mathbf{A} = -\mathbf{i} + \mathbf{j}$, $\mathbf{B} = \sqrt{2}\mathbf{i} + \sqrt{3}\mathbf{j} + 2\mathbf{k}$

8. $\mathbf{A} = \mathbf{i} + \mathbf{k}$, $\mathbf{B} = \mathbf{i} + \mathbf{j} + \mathbf{k}$

9. $\mathbf{A} = 2\mathbf{i} - 4\mathbf{j} + \sqrt{5}\mathbf{k}$, $\mathbf{B} = -2\mathbf{i} + 4\mathbf{j} - \sqrt{5}\mathbf{k}$

10. $\mathbf{A} = -5\mathbf{i} + \mathbf{j}$, $\mathbf{B} = 2\mathbf{i} + \sqrt{17}\mathbf{j} + 10\mathbf{k}$

11. $\mathbf{A} = 10\mathbf{i} + 11\mathbf{j} - 2\mathbf{k}$, $\mathbf{B} = 3\mathbf{j} + 4\mathbf{k}$

12. $\mathbf{A} = 2\mathbf{i} + 10\mathbf{j} - 11\mathbf{k}$, $\mathbf{B} = 2\mathbf{i} + 2\mathbf{j} + \mathbf{k}$

13. Write $\mathbf{B} = 3\mathbf{j} + 4\mathbf{k}$ as the sum of a vector parallel to $\mathbf{A} = \mathbf{i} + \mathbf{j}$ and a vector orthogonal to $\mathbf{A}$.

14. Write $\mathbf{B} = \mathbf{j} + \mathbf{k}$ as the sum of a vector parallel to $\mathbf{A} = \mathbf{i} + \mathbf{j}$ and a vector orthogonal to $\mathbf{A}$.

15. Write $\mathbf{B} = 8\mathbf{i} + 4\mathbf{j} - 12\mathbf{k}$ as the sum of a vector parallel to $\mathbf{A} = \mathbf{i} + 2\mathbf{j} - \mathbf{k}$ and a vector orthogonal to $\mathbf{A}$.

16. $\mathbf{B} = \mathbf{i} + (\mathbf{j} + \mathbf{k})$ is already the sum of a vector parallel to $\mathbf{i}$ and a vector orthogonal to $\mathbf{i}$. If you use Eq. (15) with $\mathbf{A} = \mathbf{i}$, do you get $\mathbf{B}_1 = \mathbf{i}$ and $\mathbf{B}_2 = \mathbf{j} + \mathbf{k}$? Try it and find out.

In Exercises 17–20, find an equation for the line in the xy-plane that passes through the given point perpendicular to the given vector. Then sketch the line. Include the vector in your sketch as a vector starting at the origin (as in Fig. 10.28).

17. $P(2, 1)$, $\mathbf{i} + 2\mathbf{j}$

18. $P(-2, 1)$, $\mathbf{i} - \mathbf{j}$

19. $P(-1, 2)$, $-2\mathbf{i} - \mathbf{j}$

20. $P(-1, 2)$, $2\mathbf{i} - 3\mathbf{j}$

In Exercises 21–24, find the distance in the xy-plane from the point to the line.

21. $S(2, 8)$, $x + 3y = 6$

22. $S(0, 0)$, $x + 3y = 6$

23. $S(2, 1)$, $x + y = 1$

24. $S(1, 3)$, $y = -2x$

25. Show that the vectors

$$\mathbf{A} = \frac{1}{\sqrt{3}}\,(\mathbf{i} - \mathbf{j} + \mathbf{k}), \quad \mathbf{B} = \frac{1}{\sqrt{2}}\,(\mathbf{j} + \mathbf{k}),$$

$$\mathbf{C} = \frac{1}{\sqrt{6}}\,(-2\,\mathbf{i} - \mathbf{j} + \mathbf{k})$$

are orthogonal to one another.

26. Find the vector projections of $\mathbf{D} = \mathbf{i} + \mathbf{j} + \mathbf{k}$ on the vectors $\mathbf{A}, \mathbf{B},$ and $\mathbf{C}$ of Exercise 25. Then show that $\mathbf{D}$ is the sum of these vector projections.

27. *Cancellation in dot products is not valid.* In real-number multiplication, if $ab_1 = ab_2$ and a is not zero, we can safely cancel the a and conclude that $b_1 = b_2$. Not so for vector multiplication: If $\mathbf{A} \cdot \mathbf{B}_1 = \mathbf{A} \cdot \mathbf{B}_2$ and $\mathbf{A} \neq \mathbf{0}$, it is not safe to conclude that $\mathbf{B}_1 = \mathbf{B}_2$. See if you can come up with an example. Keep it simple: Experiment with $\mathbf{i}, \mathbf{j},$ and $\mathbf{k}$.

28. *Sums and differences.* In Fig. 10.31, it looks as if $\mathbf{v}_1 + \mathbf{v}_2$ and $\mathbf{v}_1 - \mathbf{v}_2$ are orthogonal. Is this mere coincidence, or are there circumstances under which we may expect the sum of two vectors to be orthogonal to their difference? Find out by expanding the left-hand side of the equation

$$(\mathbf{v}_1 + \mathbf{v}_2) \cdot (\mathbf{v}_1 - \mathbf{v}_2) = 0.$$

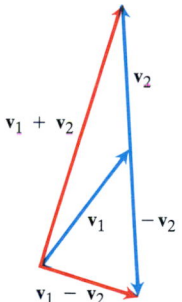

10.31 The vectors in Exercise 28.

29. Show that the diagonals of a rhombus are perpendicular.

30. Show that squares are the only rectangles with perpendicular diagonals.

CALCULATOR Use a calculator to find the angles in Exercises 31–34 to the nearest tenth of a degree.

31. Find the interior angles of the triangle ABC whose vertices are $A(-1, 0, 2)$, $B(2, 1, -1)$, and $C(1, -2, 2)$.

32. Find the angle between $\mathbf{A} = 2\,\mathbf{i} + 2\,\mathbf{j} + \mathbf{k}$ and $\mathbf{B} = 2\,\mathbf{i} + 10\,\mathbf{j} - 11\,\mathbf{k}$.

33. Find the angle between the diagonal of a cube and the diagonal of one of its faces. (*Hint:* Use a cube whose edges represent $\mathbf{i}, \mathbf{j},$ and $\mathbf{k}$.)

34. Find the angle between the diagonal of a cube and one of the edges it meets at a vertex.

Work

35. Find the work done by a force $\mathbf{F} = -5\,\mathbf{k}$ (magnitude 5 newtons) in moving an object along the line from the origin to the point $(1, 1, 1)$ (distance in meters).

36. A locomotive exerted a constant force of 60,000 newtons on a freight train while pulling it 1 km along a straight track. How much work did the locomotive do?

37. How much work does it take to slide a crate 20 m along a loading dock by pulling on it with a 200-newton force at an angle of 30° from the horizontal?

38. The wind passing over a boat's sail exerted a 1000-lb-magnitude force $\mathbf{F}$ as shown in Fig. 10.32. If the force vector made a 60° angle with the line of the boat's forward motion, how much work did the wind perform in moving the boat forward 1 mi? Answer in foot-pounds.

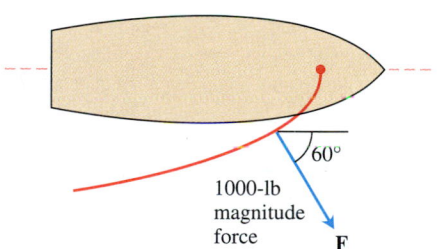

10.32 The boat in Exercise 38.

Angles Between Curves

The angles between two differentiable curves at a point of intersection are the angles between the curves' tangents at these points. Find the angles between the curves in Exercises 39–43 at their points of intersection. (You will not need a calculator.)

39. $3x + y = 5, \quad 2x - y = 4$

40. $y = \sqrt{3}x - 1, \quad y = -\sqrt{3}x + 2$

41. $y = \sqrt{(3/4) + x}, \quad y = \sqrt{(3/4) - x}$

42. $y = x^3, \quad y = \sqrt{x}$ (two points of intersection)

43. $y = x^2, \quad y = \sqrt[3]{x}$ (two points of intersection)

10.4 Cross Products

When we turn a bolt by applying a force to a wrench (Fig. 10.33), the torque we produce acts along the axis of the bolt to drive the bolt forward. The magnitude of the torque depends on how far out on the wrench the force is applied and on how

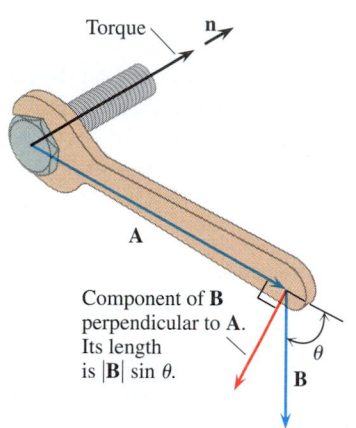

10.33   The torque vector describes the tendency of the force **B** to drive the bolt forward.

much of the force is actually perpendicular to the wrench at that point. The number we use to measure the torque's magnitude is a product made up of the length of the lever arm **A** and the scalar component of **B** perpendicular to **A**. In the notation of Fig. 10.33,

$$\text{magnitude of the torque vector} = |\mathbf{A}||\mathbf{B}|\sin\theta.$$

If we let **n** be a unit vector along the axis of the bolt in the direction of the torque, then the complete description of the torque vector is

$$\text{Torque vector} = \mathbf{n}|\mathbf{A}||\mathbf{B}|\sin\theta.$$

In mathematics, we call the vector $\mathbf{n}|\mathbf{A}||\mathbf{B}|\sin\theta$ the vector product of **A** and **B**.

Vector products are widely used to describe the effects of forces in studies of electricity, magnetism, fluid mechanics, and planetary motion. The goal of this section is to present the mathematical properties of vector products that account for their use in these fields. We shall use these properties ourselves in Chapter 11, where we study motion in space, and in Chapter 14, where we study the vector integrals of fluid flow. In the next section we shall also see how to combine vector products with scalar products to produce equations for lines and planes in space.

The Vector Product of Two Vectors in Space

When we define vector products in mathematics, we start with two nonzero vectors **A** and **B** in space without requiring them to have any particular physical interpretation. If **A** and **B** are not parallel, they determine a plane. We select a unit vector **n** perpendicular to the plane by the **right-hand rule.** This means we choose **n** to be the unit (normal) vector that points the way your right thumb points when your fingers curl through the angle θ from **A** to **B** (Fig. 10.34). We then define the **vector product A × B** ("A cross B") of **A** and **B** to be the vector

$$\mathbf{A} \times \mathbf{B} = \mathbf{n}|\mathbf{A}||\mathbf{B}|\sin\theta. \tag{1}$$

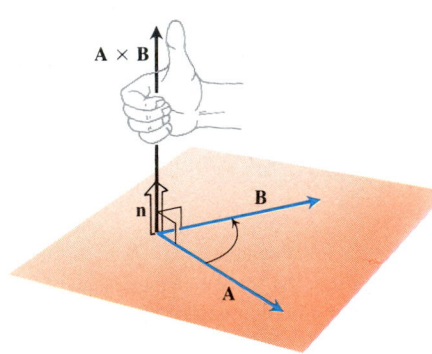

10.34   The construction of **A × B**.

Since **A × B** is a scalar multiple of **n**, it is perpendicular to both **A** and **B**.

If θ approaches 0° or 180° in Eq. (1), the length of **A × B** approaches zero. We therefore define **A × B** to be **0** if **A** and **B** are parallel (and fail to determine a plane). This is consistent with our torque interpretation as well. If the force **B** in Fig. 10.33 is parallel to the wrench, meaning that we are trying to turn the bolt by pushing or pulling straight along the handle of the wrench, the torque produced is **0**.

If one or both of **A** and **B** is zero, we define **A × B** to be zero. This way, the cross product of two vectors **A** and **B** is zero if and only if **A** and **B** are parallel or one or both of them is the zero vector.

The vector product of **A** and **B** is often called the **cross product** of **A** and **B** because of the cross in the notation **A × B**. In contrast to the dot product **A · B**, which is a scalar, the cross product **A × B** is a vector.

A × B Versus B × A

Reversing the order of the factors in a nonzero cross product reverses the direction of the resulting vector. When the fingers of our right hand curl through the angle θ from **B** to **A**, our thumb points the opposite way and the unit vector we choose in

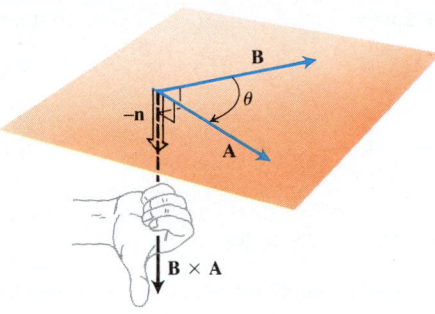

10.35   The construction of **B** × **A**.

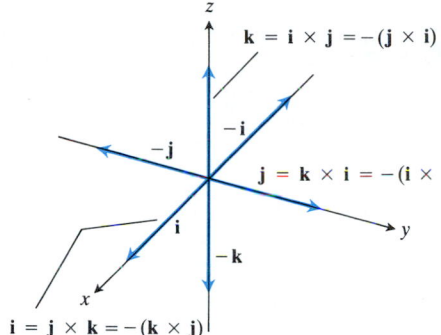

10.36   The pairwise cross products of **i**, **j**, and **k** can be read from this diagram.

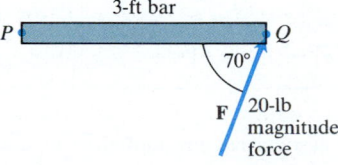

10.37   The parallelogram determined by **A** and **B**. Its area is |**A** × **B**|.

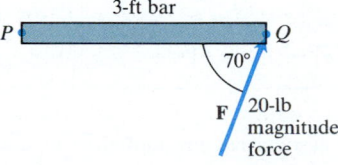

10.38   The magnitude of the torque exerted by **F** about *P* is about 56.4 ft · lb (Example 1).

forming **B** × **A** is the negative of the one we choose in forming **A** × **B** (Fig. 10.35). Thus,

$$\mathbf{B} \times \mathbf{A} = -(\mathbf{A} \times \mathbf{B}) \qquad (2)$$

for all vectors **A** and **B**. Unlike the dot product, the cross product is not commutative.

When we apply the definition to calculate the pairwise cross products of the unit vectors **i**, **j**, and **k**, we find (Fig. 10.36)

$$\mathbf{i} \times \mathbf{j} = -(\mathbf{j} \times \mathbf{i}) = \mathbf{k},$$
$$\mathbf{j} \times \mathbf{k} = -(\mathbf{k} \times \mathbf{j}) = \mathbf{i}, \qquad (3)$$
$$\mathbf{k} \times \mathbf{i} = -(\mathbf{i} \times \mathbf{k}) = \mathbf{j},$$

and

$$\mathbf{i} \times \mathbf{i} = \mathbf{j} \times \mathbf{j} = \mathbf{k} \times \mathbf{k} = \mathbf{0}.$$

|**A** × **B**| Is the Area of a Parallelogram

Because **n** is a unit vector, the magnitude of **A** × **B** is

$$AREA = |\mathbf{A} \times \mathbf{B}| = |\mathbf{n}||\mathbf{A}||\mathbf{B}||\sin \theta| = |\mathbf{A}||\mathbf{B}||\sin \theta|. \qquad (4)$$

This is the area of the parallelogram determined by **A** and **B** (Fig. 10.37), |**A**| being the base of the parallelogram and |**B**||sin θ| the height.

The Magnitude of a Torque

Equation (4) is the equation we use to calculate magnitudes of torques.

Example 1   The magnitude of the torque exerted by force **F** about the pivot point *P* in Fig. 10.38 is

$$|\overrightarrow{PQ} \times \mathbf{F}| = |\overrightarrow{PQ}||\mathbf{F}| \sin 70° \qquad \text{(Eq. (4))}$$
$$= (3)(20)(0.94) \qquad \text{(Calculator, rounded)}$$
$$= 56.4 \text{ ft-lb.}$$

The Associative and Distributive Laws

As a rule, cross-product multiplication is not associative because (**A** × **B**) × **C** lies in the plane of **A** and **B** whereas **A** × (**B** × **C**) lies in the plane of **B** and **C**. However, the **Scalar Distributive Law**

$$(r\mathbf{A}) \times (s\mathbf{B}) = (rs)\mathbf{A} \times \mathbf{B} \qquad (5)$$

does hold, as do the **Vector Distributive Laws**

$$\mathbf{A} \times (\mathbf{B} + \mathbf{C}) = \mathbf{A} \times \mathbf{B} + \mathbf{A} \times \mathbf{C} \tag{6}$$

and

$$(\mathbf{B} + \mathbf{C}) \times \mathbf{A} = \mathbf{B} \times \mathbf{A} + \mathbf{C} \times \mathbf{A}. \tag{7}$$

As a special case of (5) we have

$$(-\mathbf{A}) \times \mathbf{B} = \mathbf{A} \times (-\mathbf{B}) = -(\mathbf{A} \times \mathbf{B}).$$

The Scalar Distributive Law can be verified by applying Eq. (1) to the products on both sides of Eq. (5) and comparing the results. The Vector Distributive Law in Eq. (6) is not easy to prove, however. We shall assume it here and leave the proof to Appendix 9. Equation (7) follows from Eq. (6): multiply both sides of Eq. (6) by -1 and reverse the orders of the products.

The Determinant Formula for $\mathbf{A} \times \mathbf{B}$

Our next objective is to show how to calculate the components of $\mathbf{A} \times \mathbf{B}$ from the components of $\mathbf{A}$ and $\mathbf{B}$.

Suppose

$$\mathbf{A} = a_1\,\mathbf{i} + a_2\,\mathbf{j} + a_3\,\mathbf{k}, \qquad \mathbf{B} = b_1\,\mathbf{i} + b_2\,\mathbf{j} + b_3\,\mathbf{k}.$$

Then the distributive laws and the rules for multiplying $\mathbf{i}$, $\mathbf{j}$, and $\mathbf{k}$ tell us that

$$
\begin{aligned}
\mathbf{A} \times \mathbf{B} &= (a_1\,\mathbf{i} + a_2\,\mathbf{j} + a_3\,\mathbf{k}) \times (b_1\,\mathbf{i} + b_2\,\mathbf{j} + b_3\,\mathbf{k}) \\
&= a_1 b_1\,\mathbf{i} \times \mathbf{i} + a_1 b_2\,\mathbf{i} \times \mathbf{j} + a_1 b_3\,\mathbf{i} \times \mathbf{k} \\
&\quad + a_2 b_1\,\mathbf{j} \times \mathbf{i} + a_2 b_2\,\mathbf{j} \times \mathbf{j} + a_2 b_3\,\mathbf{j} \times \mathbf{k} \\
&\quad + a_3 b_1\,\mathbf{k} \times \mathbf{i} + a_3 b_2\,\mathbf{k} \times \mathbf{j} + a_3 b_3\,\mathbf{k} \times \mathbf{k} \\
&= (a_2 b_3 - a_3 b_2)\,\mathbf{i} + (a_3 b_1 - a_1 b_3)\,\mathbf{j} + (a_1 b_2 - a_2 b_1)\,\mathbf{k}.
\end{aligned}
\tag{8}
$$

The terms at the end of Eq. (8) are the same as the terms in the expansion of the symbolic determinant

$$
\begin{vmatrix}
\mathbf{i} & \mathbf{j} & \mathbf{k} \\
a_1 & a_2 & a_3 \\
b_1 & b_2 & b_3
\end{vmatrix}
$$

We therefore have the following rule.

> If $\mathbf{A} = a_1\,\mathbf{i} + a_2\,\mathbf{j} + a_3\,\mathbf{k}$ and $\mathbf{B} = b_1\,\mathbf{i} + b_2\,\mathbf{j} + b_3\,\mathbf{k}$, then
>
> $$\mathbf{A} \times \mathbf{B} = \begin{vmatrix} \mathbf{i} & \mathbf{j} & \mathbf{k} \\ a_1 & a_2 & a_3 \\ b_1 & b_2 & b_3 \end{vmatrix}. \tag{9}$$

Equation (9) is remarkable, given that neither the associative nor commutative law holds for cross-product multiplication.

Example 2   Find $\mathbf{A} \times \mathbf{B}$ and $\mathbf{B} \times \mathbf{A}$ if

$$\mathbf{A} = 2\,\mathbf{i} + \mathbf{j} + \mathbf{k}, \qquad \mathbf{B} = -4\,\mathbf{i} + 3\,\mathbf{j} + \mathbf{k}.$$

Determinant Formulas

(See Appendix 10.)

$$\begin{vmatrix} a & b \\ c & d \end{vmatrix} = ad - bc$$

Example

$$
\begin{aligned}
\begin{vmatrix} 2 & 1 \\ -4 & 3 \end{vmatrix} &= (2)(3) - (-4)(1) \\
&= 6 + 4 = 10
\end{aligned}
$$

$$
\begin{vmatrix}
a_{11} & a_{12} & a_{13} \\
a_{21} & a_{22} & a_{23} \\
a_{31} & a_{32} & a_{33}
\end{vmatrix}
$$

$$
= a_{11} \begin{vmatrix} a_{22} & a_{23} \\ a_{32} & a_{33} \end{vmatrix} - a_{12} \begin{vmatrix} a_{21} & a_{23} \\ a_{31} & a_{33} \end{vmatrix} + a_{13} \begin{vmatrix} a_{21} & a_{22} \\ a_{31} & a_{32} \end{vmatrix}
$$

Example

$$
\begin{vmatrix}
-5 & 3 & 1 \\
2 & 1 & 1 \\
-4 & 3 & 1
\end{vmatrix}
$$

$$
\begin{aligned}
&= (-5)\begin{vmatrix} 1 & 1 \\ 3 & 1 \end{vmatrix} - (3)\begin{vmatrix} 2 & 1 \\ -4 & 1 \end{vmatrix} + (1)\begin{vmatrix} 2 & 1 \\ -4 & 3 \end{vmatrix} \\
&= -5(1 - 3) - 3(2 + 4) + 1(6 + 4) \\
&= 10 - 18 + 10 = 2
\end{aligned}
$$

Solution   We use Eq. (9) to find $\mathbf{A} \times \mathbf{B}$:

$$\mathbf{A} \times \mathbf{B} = \begin{vmatrix} \mathbf{i} & \mathbf{j} & \mathbf{k} \\ 2 & 1 & 1 \\ -4 & 3 & 1 \end{vmatrix} = \begin{vmatrix} 1 & 1 \\ 3 & 1 \end{vmatrix} \mathbf{i} - \begin{vmatrix} 2 & 1 \\ -4 & 1 \end{vmatrix} \mathbf{j} + \begin{vmatrix} 2 & 1 \\ -4 & 3 \end{vmatrix} \mathbf{k}$$

$$= -2\,\mathbf{i} - 6\,\mathbf{j} + 10\,\mathbf{k}.$$

Equation (2) then gives

$$\mathbf{B} \times \mathbf{A} = -\mathbf{A} \times \mathbf{B} = 2\,\mathbf{i} + 6\,\mathbf{j} - 10\,\mathbf{k}.$$

Example 3   Find a vector perpendicular to the plane of $P(1,\ -1,\ 0)$, $Q(2,\ 1,\ -1)$, and $R(-1,\ 1,\ 2)$.

Solution   The vector $\overrightarrow{PQ} \times \overrightarrow{PR}$ is perpendicular to the plane because it is perpendicular to both vectors. In terms of components,

$$\overrightarrow{PQ} = (2 - 1)\,\mathbf{i} + (1 + 1)\,\mathbf{j} + (-1 - 0)\,\mathbf{k} = \mathbf{i} + 2\,\mathbf{j} - \mathbf{k},$$

$$\overrightarrow{PR} = (-1 - 1)\,\mathbf{i} + (1 + 1)\,\mathbf{j} + (2 - 0)\,\mathbf{k} = -2\,\mathbf{i} + 2\,\mathbf{j} + 2\,\mathbf{k},$$

$$\overrightarrow{PQ} \times \overrightarrow{PR} = \begin{vmatrix} \mathbf{i} & \mathbf{j} & \mathbf{k} \\ 1 & 2 & -1 \\ -2 & 2 & 2 \end{vmatrix}$$

$$= \begin{vmatrix} 2 & -1 \\ 2 & 2 \end{vmatrix} \mathbf{i} - \begin{vmatrix} 1 & -1 \\ -2 & 2 \end{vmatrix} \mathbf{j} + \begin{vmatrix} 1 & 2 \\ -2 & 2 \end{vmatrix} \mathbf{k}$$

$$= 6\,\mathbf{i} + 6\,\mathbf{k}.$$

Example 4   Find the area of the triangle with vertices $P(1,\ -1,\ 0)$, $Q(2,\ 1,\ -1)$, and $R(-1,\ 1,\ 2)$ (Fig. 10.39).

Solution   The area of the parallelogram determined by P, Q, and R is

$$|\overrightarrow{PQ} \times \overrightarrow{PR}| = |6\,\mathbf{i} + 6\,\mathbf{k}| \qquad \text{(Values from Example 3)}$$

$$= \sqrt{(6)^2 + (6)^2} = \sqrt{2 \cdot 36} = 6\sqrt{2}.$$

The triangle's area is half of this, or $3\sqrt{2}$.

Example 5   Find a unit vector perpendicular to the plane of $P(1,\ -1,\ 0)$, $Q(2,\ 1,\ -1)$, and $R(-1,\ 1,\ 2)$.

Solution   Since $\overrightarrow{PQ} \times \overrightarrow{PR}$ is perpendicular to the plane, its direction $\mathbf{n}$ is a unit vector perpendicular to the plane. In component form,

$$\mathbf{n} = \frac{\overrightarrow{PQ} \times \overrightarrow{PR}}{|\overrightarrow{PQ} \times \overrightarrow{PR}|}$$

$$= \frac{6\,\mathbf{i} + 6\,\mathbf{k}}{6\sqrt{2}} \qquad \text{(Values from Examples 3 and 4)}$$

$$= \frac{1}{\sqrt{2}}\,\mathbf{i} + \frac{1}{\sqrt{2}}\,\mathbf{k}.$$

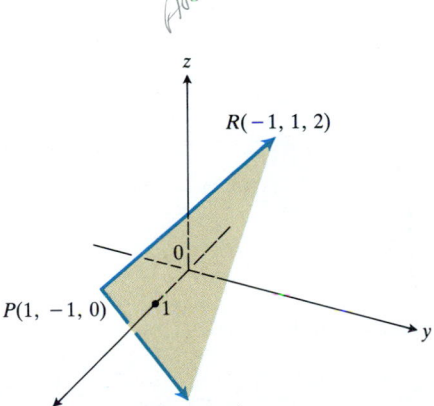

10.39   The area of triangle PQR is half of $|\overrightarrow{PQ} \times \overrightarrow{PR}|$ (Example 4).

EXERCISES 10.4

In Exercises 1–8, find the length and direction (when defined) of $A \times B$ and $B \times A$.

1. $A = 2\,i - 2\,j - k$, $B = i - k$

2. $A = 2\,i + 3\,j$, $B = -i + j$

3. $A = 2\,i - 2\,j + 4\,k$, $B = -i + j - 2\,k$

4. $A = i + j - k$, $B = 0$

5. $A = 2\,i$, $B = -3\,j$

6. $A = i \times j$, $B = j \times k$

7. $A = -8\,i - 2\,j - 4\,k$, $B = 2\,i + 2\,j + k$

8. $A = \frac{3}{2}\,i - \frac{1}{2}\,j + k$, $B = i + j + 2\,k$

In Exercises 9–14, sketch the coordinate axes and then include the vectors A, B, and $A \times B$ as vectors coming out from the origin.

9. $A = i$, $B = j$

10. $A = i + k$, $B = j$

11. $A = i - k$, $B = j + k$

12. $A = 2\,i - j$, $B = i + 2\,j$

13. $A = i + 3\,j + 2\,k$, $B = k$

14. $A = i + 2\,j$, $B = 2\,j + k$

In Exercises 15–18:
a) Find a vector N perpendicular to the plane of the points P, Q, and R.
b) Find the area of triangle PQR.
c) Find a unit vector perpendicular to plane PQR.

15. $P(1, -1, 2)$, $Q(2, 0, -1)$, $R(0, 2, 1)$

16. $P(1, 1, 1)$, $Q(2, 1, 3)$, $R(3, -1, 1)$

17. $P(2, -2, 1)$, $Q(3, -1, 2)$, $R(3, -1, 1)$

18. $P(-2, 2, 0)$, $Q(0, 1, -1)$, $R(-1, 2, -2)$

19. Let $A = 5\,i - j + k$, $B = j - 5\,k$, $C = -15\,i + 3\,j - 3\,k$. Which vectors, if any, are (a) perpendicular, (b) parallel?

20. Let $A = i + 2\,j - k$, $B = -i + j + k$, $C = i + k$, $D = -(\pi/2)\,i - \pi\,j + (\pi/2)\,k$. Which vectors, if any, are (a) perpendicular, (b) parallel?

In Exercises 21 and 22, find the magnitude of the torque exerted by F on the bolt at P if $\overrightarrow{PQ} = 8$ in. and $|F| = 30$ lb. Answer in foot-pounds.

21.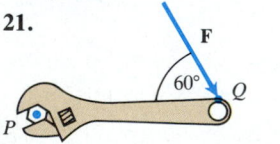

22.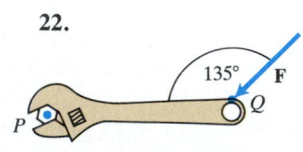

23. If $A = 2\,i - j$ and $B = i + 3\,j - 2\,k$, find $A \times B$. Then calculate $(A \times B) \cdot A$ and $(A \times B) \cdot B$.

24. Is $(A \times B) \cdot A$ always zero? Explain. What about $(A \times B) \cdot B$?

25. Given vectors A, B, and C, use dot-product and cross-product notation, as appropriate, to describe the following:
a) The vector projection of A onto B
b) A vector orthogonal to A and B
c) A vector with the length of A and the direction of B
d) A vector orthogonal to $A \times B$ and C
e) A vector in the plane of B and C perpendicular to A

26. *Cancellation is not valid in cross products either.* Find an example to show that $A \times B = A \times C$ need not imply that B equals C even if $A \neq 0$.

27. If $\overrightarrow{AB} \times \overrightarrow{AC} = 2\,i - 4\,j + 4\,k$, find the area of triangle ABC.

28. a) Suppose that on each triangular face of a tetrahedron $ABCD$ (Fig. 10.40) we construct an outward normal vector whose magnitude equals the area of the face. Show that the sum of these four vectors is 0.

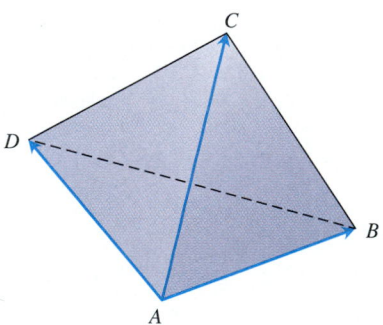

10.40   The tetrahedron in Exercise 28.

b) Formulate and prove an analogous statement for triangles in the coordinate plane.
c) The solid in (a) need not be a tetrahedron. Show that the conclusion holds for any starlike polyhedron with triangular faces. (A polyhedron is *starlike* if it has an interior point from which it is possible to view all the faces simultaneously without other faces getting in the way.) Do not give a technical argument. Just describe the basic idea behind the proof.

   The statement in (a) is actually true for any polyhedron in space. It need not be starlike and its faces need not be triangular.

10.5 Lines and Planes in Space

This section shows how to use scalar and vector products to write equations for lines, line segments, and planes in space.

Equations for Lines and Line Segments

Suppose L is a line in space that passes through a point $P_0(x_0, y_0, z_0)$ and lies parallel to a vector $\mathbf{v} = A\mathbf{i} + B\mathbf{j} + C\mathbf{k}$. Then L is the set of all points $P(x, y, z)$ for which the vector P_0P is parallel to $\mathbf{v}$ (Fig. 10.41). That is, P lies on L if and only if

$$\overrightarrow{P_0P} = t\mathbf{v} \tag{1}$$

for some number t, $-\infty < t < \infty$.

When we write Eq. (1) in terms of $\mathbf{i}$-, $\mathbf{j}$-, and $\mathbf{k}$-components and equate the corresponding components of the two sides, we get three equations involving the parameter t:

$$(x - x_0)\mathbf{i} + (y - y_0)\mathbf{j} + (z - z_0)\mathbf{k} = t(A\mathbf{i} + B\mathbf{j} + C\mathbf{k}), \quad \begin{pmatrix} \text{Eq. (1) spelled} \\ \text{out} \end{pmatrix}$$

$$x - x_0 = tA, \qquad y - y_0 = tB, \qquad z - z_0 = tC. \quad \text{(Components equated)}$$

When rearranged, these last three equations give us the standard parametrization of the line for the parameter interval $-\infty < t < \infty$.

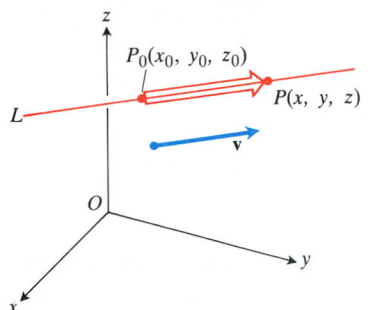

10.41 A point P lies on the line through P_0 parallel to $\mathbf{v}$ if and only if $\overrightarrow{P_0P}$ is a scalar multiple of $\mathbf{v}$.

[handwritten: know]

Standard Parametrization of the Line Through $P_0(x_0, y_0, z_0)$ Parallel to $\mathbf{v} = A\mathbf{i} + B\mathbf{j} + C\mathbf{k}$

$$x = x_0 + tA, \qquad y = y_0 + tB, \qquad z = z_0 + tC, \qquad -\infty < t < \infty \tag{2}$$

We rarely use any t-interval other than $(-\infty, \infty)$ when we parametrize a line, so we call the equations in (2) the **standard parametric equations** for the line through P_0 parallel to $\mathbf{v}$. When we want parametric equations for this line, these are the ones we produce, the interval $(-\infty, \infty)$ being understood.

Example 1 Find parametric equations for the line through $(-2, 0, 4)$ parallel to $\mathbf{v} = 2\mathbf{i} + 4\mathbf{j} - 2\mathbf{k}$ (Fig. 10.42).

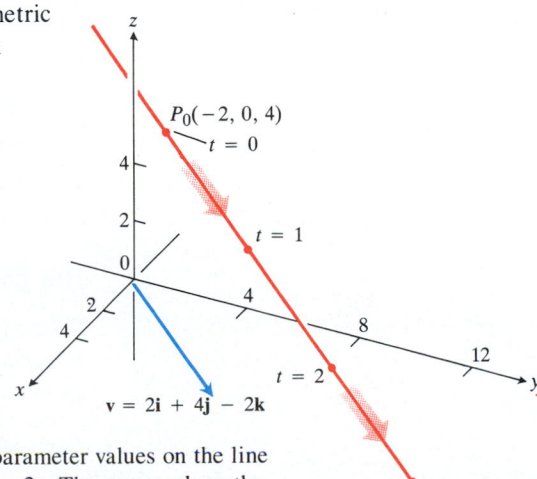

10.42 Selected points and parameter values on the line $x = -2 + 2t$, $y = 4t$, $z = 4 - 2t$. The arrows show the direction of increasing t (Example 1).

Solution With $P_0(x_0, y_0, z_0)$ equal to $(-2, 0, 4)$ and $A\mathbf{i} + B\mathbf{j} + C\mathbf{k}$ equal to $2\mathbf{i} + 4\mathbf{j} - 2\mathbf{k}$, Eqs. (2) become

$$x = -2 + 2t, \qquad y = 4t, \qquad z = 4 - 2t.$$

Example 2   Find parametric equations for the line through the points $P(-3, 2, -3)$ and $Q(1, -1, 4)$.

Solution The vector

$$\overrightarrow{PQ} = 4\mathbf{i} - 3\mathbf{j} + 7\mathbf{k}$$

is parallel to the line, and Eqs. (2) with $(x_0, y_0, z_0) = (-3, 2, -3)$ give

$$x = -3 + 4t, \qquad y = 2 - 3t, \qquad z = -3 + 7t. \qquad (3)$$

We could equally well have chosen $Q(1, -1, 4)$ as the "base point" and written

$$x = 1 + 4t, \qquad y = -1 - 3t, \qquad z = 4 + 7t \qquad (4)$$

as equations for the line. The parametrizations in (3) and (4) place you at different points for a given value of t, but each set of equations covers the line completely as t runs from $-\infty$ to ∞.

To parametrize a line segment joining two points, we first parametrize the line through the points. We then find the t-values for the points and restrict t to lie in the closed interval bounded by these values. The line equations together with this added restriction give a parametrization for the segment.

Example 3   Parametrize the line segment joining the points $P(-3, 2, -3)$ and $Q(1, -1, 4)$ (Fig. 10.43).

Solution We begin with equations for the line through P and Q, taking them, in this case, from Example 2:

$$x = -3 + 4t, \qquad y = 2 - 3t, \qquad z = -3 + 7t. \qquad (5)$$

We observe that the point

$$(x, y, z) = (-3 + 4t, 2 - 3t, -3 + 7t)$$

passes through $P(-3, 2, -3)$ at $t = 0$ and $Q(1, -1, 4)$ at $t = 1$. We add the restriction $0 \le t \le 1$ to Eqs. (5) to parametrize the segment:

$$x = -3 + 4t, \qquad y = 2 - 3t, \qquad z = -3 + 7t, \qquad 0 \le t \le 1.$$

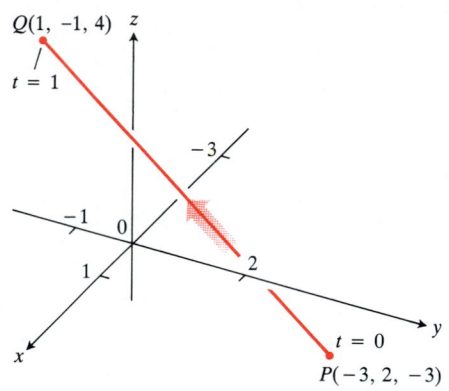

10.43 Example 3 derives a parametrization of the line segment joining P and Q. The arrow shows the direction of increasing t for this parametrization.

The Distance from a Point to a Line in Space

To find the distance from a point S to a line L that passes through a point P parallel to a vector $\mathbf{v}$, we find the length of the component of $\overrightarrow{PS}$ normal to the line (Fig. 10.44). In the notation of Fig. 10.44, the length is

$$|\overrightarrow{PS}| \sin \theta = |\overrightarrow{PS}| \frac{|\overrightarrow{PS} \times \mathbf{v}|}{|\overrightarrow{PS}| |\mathbf{v}|} \qquad \left(|\overrightarrow{PS} \times \mathbf{v}| = |\overrightarrow{PS}| |\mathbf{v}| \sin \theta\right)$$

$$= \frac{|\overrightarrow{PS} \times \mathbf{v}|}{|\mathbf{v}|}. \qquad (6)$$

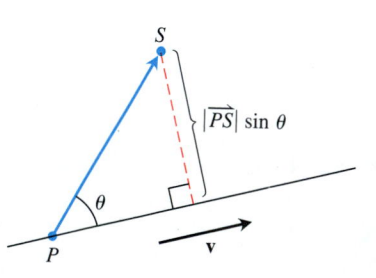

10.44 The distance from S to the line through P parallel to $\mathbf{v}$ is $|\overrightarrow{PS}| \sin \theta$, where θ is the angle between $\overrightarrow{PS}$ and $\mathbf{v}$.

GIVEN →

Distance from a Point S to a Line Through P Parallel to v

$$\text{Distance} = \frac{|\overrightarrow{PS} \times \mathbf{v}|}{|\mathbf{v}|} \tag{7}$$

Example 4   Find the distance from the point $S(1, 1, 5)$ to the line

$$L: \quad x = 1 + t, \quad y = 3 - t, \quad z = 2t.$$

Solution   We see from the equations for L that L passes through $P(1, 3, 0)$ parallel to $\mathbf{v} = \mathbf{i} - \mathbf{j} + 2\mathbf{k}$. With

$$\overrightarrow{PS} = (1 - 1)\mathbf{i} + (1 - 3)\mathbf{j} + (5 - 0)\mathbf{k} = -2\mathbf{j} + 5\mathbf{k}$$

and

$$\overrightarrow{PS} \times \mathbf{v} = \begin{vmatrix} \mathbf{i} & \mathbf{j} & \mathbf{k} \\ 0 & -2 & 5 \\ 1 & -1 & 2 \end{vmatrix} = \mathbf{i} + 5\mathbf{j} + 2\mathbf{k},$$

Eq. (7) gives

$$\text{Distance from } S \text{ to } L = \frac{|\overrightarrow{PS} \times \mathbf{v}|}{|\mathbf{v}|} = \frac{\sqrt{1 + 25 + 4}}{\sqrt{1 + 1 + 4}} = \frac{\sqrt{30}}{\sqrt{6}} = \sqrt{5}.$$

Equations for Planes

Suppose M is a plane in space that passes through a point $P_0(x_0, y_0, z_0)$ and is normal (perpendicular) to the nonzero vector $\mathbf{N} = A\mathbf{i} + B\mathbf{j} + C\mathbf{k}$. Then M consists of all points $P(x, y, z)$ for which the vector P_0P is orthogonal to $\mathbf{N}$ (Fig. 10.45). That is, P lies on M if and only if

$$\mathbf{N} \cdot \overrightarrow{P_0P} = 0 \tag{8}$$

or

$$A(x - x_0) + B(y - y_0) + C(z - z_0) = 0.$$

When rearranged, this becomes

$$Ax + By + Cz = Ax_0 + By_0 + Cz_0.$$

10.45   The standard equation for a plane in space is defined in terms of a vector normal to the plane: A point P lies in the plane through P_0 normal to $\mathbf{N}$ if and only if $\mathbf{N} \cdot \overrightarrow{P_0P} = 0$.

Standard Equation for the Plane Through $P_0(x_0, y_0, z_0)$ Normal to $\mathbf{N} = A\mathbf{i} + B\mathbf{j} + C\mathbf{k}$.

$$Ax + By + Cz = D, \qquad D = Ax_0 + By_0 + Cz_0 \tag{9}$$

Notice how the components of $\mathbf{N}$ become coefficients in the equation

$$Ax + By + Cz = D,$$

just the way they did for lines in two dimensions.

Example 5   Find an equation for the plane through $P_0(-3, 0, 7)$ perpendicular to $\mathbf{N} = 5\mathbf{i} + 2\mathbf{j} - \mathbf{k}$.

Solution   We use Eqs. (9) to get

$$D = 5(-3) + 2(0) - 1(7) = -15 - 7 = -22$$

and

$$5x + 2y - z = -22.$$

Example 6   Find an equation for the plane through the points $A(0, 0, 1)$, $B(2, 0, 0)$, and $C(0, 3, 0)$.

Solution   We find a vector normal to the plane and use it with one of the points (it does not matter which one) to write an equation for the plane.

The cross product

$$\overrightarrow{AB} \times \overrightarrow{AC} = \begin{vmatrix} \mathbf{i} & \mathbf{j} & \mathbf{k} \\ 2 & 0 & -1 \\ 0 & 3 & -1 \end{vmatrix} = 3\mathbf{i} + 2\mathbf{j} + 6\mathbf{k}$$

is normal to the plane. We substitute the components of this vector and the coordinates of the point $(0, 0, 1)$ into Eqs. (9) to get $D = 3(0) + 2(0) + 6(1) = 6$ and

$$3x + 2y + 6z = 6$$

as an equation for the plane.

Example 7   Find the point in which the line

$$x = \frac{8}{3} + 2t, \qquad y = -2t, \qquad z = 1 + t$$

meets the plane $3x + 2y + 6z = 6$.

Solution   The point

$$\left(\frac{8}{3} + 2t, -2t, 1 + t \right)$$

lies in the plane if its coordinates satisfy the equation of the plane; that is, if

$$3\left(\frac{8}{3} + 2t \right) + 2(-2t) + 6(1 + t) = 6$$

$$8 + 6t - 4t + 6 + 6t = 6$$

$$8t = -8$$

$$t = -1.$$

The point of intersection is

$$(x, y, z)\Big|_{t = -1} = \left(\frac{8}{3} - 2, 2, 1 - 1 \right) = \left(\frac{2}{3}, 2, 0 \right).$$

To find the distance from a point S to a plane $Ax + By + Cz = D$, find

1. a point P on the plane,
2. $\overrightarrow{PS}$,
3. the direction of $\mathbf{N} = A\mathbf{i} + B\mathbf{j} + C\mathbf{k}$.

Then calculate the distance as

$$\left| \overrightarrow{PS} \cdot \frac{\mathbf{N}}{|\mathbf{N}|} \right|.$$

GIVEN

Example 8   Find the distance from the point $S(1, 1, 3)$ to the plane $3x + 2y + 6z = 6$.

Solution   We use the same approach we used in Section 10.3 to find the distance from a point to a line: We find a point P in the plane and calculate the length of the vector projection of $\overrightarrow{PS}$ onto a vector $\mathbf{N}$ normal to the plane (Fig. 10.46).

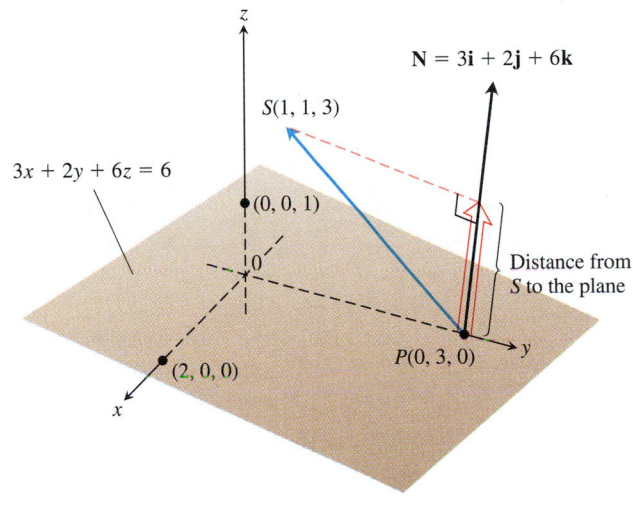

The coefficients in the equation $3x + 2y + 6z = 6$ give

$$\mathbf{N} = 3\mathbf{i} + 2\mathbf{j} + 6\mathbf{k}.$$

The points on the plane easiest to find from the plane's equation are the intercepts. If we take P to be the y-intercept $(0, 3, 0)$, then

$$\overrightarrow{PS} = (1 - 0)\mathbf{i} + (1 - 3)\mathbf{j} + (3 - 0)\mathbf{k}$$

$$= \mathbf{i} - 2\mathbf{j} + 3\mathbf{k},$$

$$|\mathbf{N}| = \sqrt{(3)^2 + (2)^2 + (6)^2}$$

$$= \sqrt{49} = 7,$$

Distance from S to the plane $= \left| \overrightarrow{PS} \cdot \frac{\mathbf{N}}{|\mathbf{N}|} \right|$

$$= \left| (\mathbf{i} - 2\mathbf{j} + 3\mathbf{k}) \cdot \left(\frac{3}{7}\mathbf{i} + \frac{2}{7}\mathbf{j} + \frac{6}{7}\mathbf{k} \right) \right|$$

$$= \left| \frac{3}{7} - \frac{4}{7} + \frac{18}{7} \right| = \frac{17}{7}.$$

10.46   The distance from S to the plane is the length of the vector projection of $\overrightarrow{PS}$ onto $\mathbf{N}$ (Example 8).

Angles Between Planes; Lines of Intersection

The angle between two intersecting planes is defined to be the (acute) angle made by their normal vectors (Fig. 10.47).

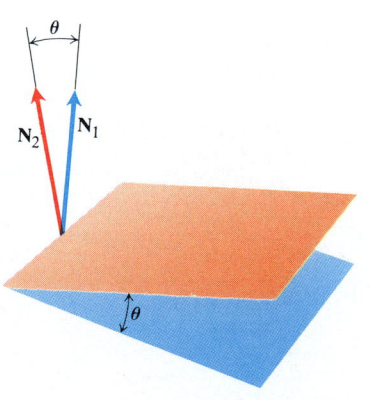

10.47   The angle between two planes is obtained from the angle between their normals.

Example 9   Find the angle between the planes $3x - 6y - 2z = 15$ and $2x + y - 2z = 5$.

Solution   The vectors

$$\mathbf{N}_1 = 3\,\mathbf{i} - 6\,\mathbf{j} - 2\,\mathbf{k}, \qquad \mathbf{N}_2 = 2\,\mathbf{i} + \mathbf{j} - 2\,\mathbf{k}$$

are normals to the planes. The angle between them is

$$\theta = \cos^{-1}\!\left(\frac{\mathbf{N}_1 \cdot \mathbf{N}_2}{|\mathbf{N}_1||\mathbf{N}_2|}\right) \qquad \text{(Eq. (5), Section 10.3)}$$

KNOW

$$= \cos^{-1}\frac{4}{21}$$

$$= 79°. \qquad \text{(Calculator, rounded)}$$

Example 10   Find a vector parallel to the line of intersection of the planes $3x - 6y - 2z = 15$ and $2x + y - 2z = 5$.

Solution   Any vector parallel to the line of intersection will be parallel to both planes and therefore perpendicular to their normals. Conversely, any vector perpendicular to the planes' normals will be parallel to both planes and hence parallel to their line of intersection. These requirements are met by the vector

$$\mathbf{v} = \mathbf{N}_1 \times \mathbf{N}_2 = \begin{vmatrix} \mathbf{i} & \mathbf{j} & \mathbf{k} \\ 3 & -6 & -2 \\ 2 & 1 & -2 \end{vmatrix} = 14\,\mathbf{i} + 2\,\mathbf{j} + 15\,\mathbf{k}.$$

Any nonzero scalar multiple of $\mathbf{v}$ will do as well.

Example 11   Find parametric equations for the line in which the planes $3x - 6y - 2z = 15$ and $2x + y - 2z = 5$ intersect.

Solution   We find a vector parallel to the line and a point on the line and use Eqs. (2).

Example 10 gives a vector parallel to the line, $\mathbf{v} = 14\,\mathbf{i} + 2\,\mathbf{j} + 15\,\mathbf{k}$. To find a point on the line, we can take any point common to the two planes. Substituting $z = 0$ in the plane equations and solving for x and y simultaneously gives the point $(3, -1, 0)$. The line is

$$x = 3 + 14t, \qquad y = -1 + 2t, \qquad z = 15t.$$

FIND PT ON LINE
OF INTERSECTION
$z_0 = 0$, SOLVE FOR x_0, y_0

$3x_0 - 6y_0 - 0 = 15$
$2x_0 - y_0 - 0 = 5$

EXAM 3 UP TO 10.5

EXERCISES 10.5

Find parametric equations for the lines in Exercises 1–12.

1. The line through $P(3, -4, -1)$ parallel to the vector $\mathbf{i} + \mathbf{j} + \mathbf{k}$

2. The line through $P(1, 2, -1)$ and $Q(-1, 0, 1)$

3. The line through $P(-2, 0, 3)$ and $Q(3, 5, -2)$

4. The line through $P(1, 2, 0)$ and $Q(1, 1, -1)$

5. The line through the origin parallel to the vector $2\,\mathbf{j} + \mathbf{k}$

6. The line through the point $(3, -2, 1)$ parallel to the line $x = 1 + 2t$, $y = 2 - t$, $z = 3t$

7. The line through $(1, 1, 1)$ parallel to the z-axis

8. The line through $(2, 4, 5)$ perpendicular to the plane $3x + 7y - 5z = 21$

9. The line through $(0, -7, 0)$ perpendicular to the plane $x + 2y + 2z = 13$

10. The line through $(2, 3, 0)$ perpendicular to the vectors $\mathbf{A} = \mathbf{i} + 2\mathbf{j} + 3\mathbf{k}$ and $\mathbf{B} = 3\mathbf{i} + 4\mathbf{j} + 5\mathbf{k}$

11. The x-axis

12. The z-axis

Find parametrizations for the line segments joining the points in Exercises 13–20. Draw coordinate axes and sketch each segment, indicating the direction of increasing t for your parametrization.

13. $(0, 0, 0)$, $(1, 1, 1)$
14. $(0, 0, 0)$, $(1, 0, 0)$
15. $(1, 0, 0)$, $(1, 1, 0)$
16. $(1, 1, 0)$, $(1, 1, 1)$
17. $(0, -1, 1)$, $(0, 1, 1)$
18. $(3, 0, 0)$, $(0, 2, 0)$
19. $(2, 2, 0)$, $(1, 2, -2)$
20. $(1, -1, -2)$, $(0, 2, 1)$

Find equations for the planes in Exercises 21–26.

21. The plane through $P_0(0, 2, -1)$ normal to $\mathbf{N} = 3\mathbf{i} - 2\mathbf{j} - \mathbf{k}$

22. The plane through $(1, -1, 3)$ parallel to the plane $3x + y + z = 7$.

23. The plane through $(1, 1, -1)$, $(2, 0, 2)$, and $(0, -2, 1)$

24. The plane through $(2, 4, 5)$, $(1, 5, 7)$, and $(-1, 6, 8)$

25. The plane through $P_0(2, 4, 5)$ perpendicular to the line
$$x = 5 + t, \quad y = 1 + 3t, \quad z = 4t$$

26. The plane through $A(1, -2, 1)$ perpendicular to the vector from the origin to A

27. Use Eqs. (2) to generate a parametrization of the line through $P(2, -4, 7)$ parallel to $\mathbf{v}_1 = 2\mathbf{i} - \mathbf{j} + 3\mathbf{k}$. Then generate another parametrization of the line using the point $P_2(-2, -2, 1)$ and the vector $\mathbf{v}_2 = -\mathbf{i} + (1/2)\mathbf{j} - (3/2)\mathbf{k}$.

28. Use Eqs. (9) to generate an equation for the plane through $P_1(4, 1, 5)$ normal to $\mathbf{N}_1 = \mathbf{i} - 2\mathbf{j} + \mathbf{k}$. Then generate another equation for the plane using the point $P_2(3, -2, 0)$ and the normal vector $\mathbf{N} = -\sqrt{2}\,\mathbf{i} + 2\sqrt{2}\,\mathbf{j} - \sqrt{2}\,\mathbf{k}$.

In Exercises 29–34, find the distance from the point to the line.

29. $x = 4t$, $y = -2t$, $z = 2t$; $S(0, 0, 12)$

30. $x = 5 + 3t$, $y = 5 + 4t$, $z = -3 - 5t$; $S(0, 0, 0)$

31. $x = 2 + 2t$, $y = 1 + 6t$, $z = 3$; $S(2, 1, 3)$

32. $x = 2t$, $y = 1 + 2t$, $z = 2t$; $S(2, 1, -1)$

33. $x = 4 - t$, $y = 3 + 2t$, $z = -5 + 3t$; $S(3, -1, 4)$

34. $x = 10 + 4t$, $y = -3$, $z = 4t$; $S(-1, 4, 3)$

In Exercises 35–40, find the distance from the point to the plane.

35. $(2, -3, 4)$, $x + 2y + 2z = 13$

36. $(0, 0, 0)$, $3x + 2y + 6z = 6$

37. $(0, 1, 1)$, $4y + 3z = -12$

38. $(2, 2, 3)$, $2x + y + 2z = 4$

39. $(0, -1, 0)$, $2x + y + 2z = 4$

40. $(1, 0, -1)$, $-4x + y + z = 4$

In Exercises 41–44, find the point in which the line meets the plane.

41. $x = 1 - t$, $y = 3t$, $z = 1 + t$; $2x - y + 3z = 6$

42. $x = 2$, $y = 3 + 2t$, $z = -2 - 2t$;
$$6x + 3y - 4z = -12$$

43. $x = 1 + 2t$, $y = 1 + 5t$, $z = 3t$; $x + y + z = 2$

44. $x = -1 + 3t$, $y = -2$, $z = 5t$; $2x - 3z = 7$

Find the angles between the planes in Exercises 45 and 46 (calculator not needed).

45. $x + y = 1$, $2x + y - 2z = 2$

46. $5x + y - z = 10$, $x - 2y + 3z = -1$

CALCULATOR   Use a calculator to find the angles between the planes in Exercises 47–50 to the nearest tenth of a degree.

47. $2x + 2y + 2z = 3$, $2x - 2y - z = 5$

48. $x + y + z = 1$, $z = 0$ (the xy-plane)

49. $2x + 2y - z = 3$, $x + 2y + z = 2$

50. $4y + 3z = -12$, $3x + 2y + 6z = 6$

Find parametrizations for the lines in which the planes in Exercises 51–54 intersect.

51. $x + y + z = 1$, $x + y = 2$

52. $3x - 6y - 2z = 3$, $2x + y - 2z = 2$

53. $x - 2y + 4z = 2$, $x + y - 2z = 5$

54. $5x - 2y = 11$, $4y - 5z = -17$

55. Find the points in which the line $x = 1 + 2t$, $y = -1 - t$, $z = 3t$ meets the coordinate planes.

56. Find equations for the line in the plane $z = 3$ that makes a $30°$-angle with $\mathbf{i}$ and a $60°$-angle with $\mathbf{j}$.

57. *Perspective in computer graphics.*   In computer graphics and perspective drawing we need to represent objects seen by the eye in space as images on a two-dimensional plane. Suppose the eye is at $E(x_0, 0, 0)$ (Fig. 10.48) and that we want to represent a point $P_1(x_1, y_1, z_1)$ as a point on the yz-plane. We do this by projecting P_1 onto the plane with a ray from E. The point P_1 will be portrayed as the point P $(0, y, z)$. The problem for us as graphics designers is to find y and z given E and P_1.

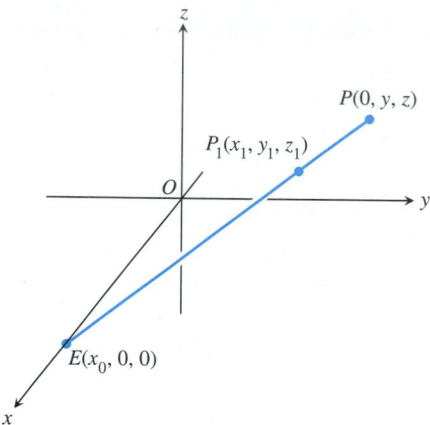

10.48   The figure for Exercise 57.

a) Write a vector equation that holds between $\overrightarrow{EP}$ and $\overrightarrow{EP_1}$. Use the equation to express y and z in terms of x_0, x_1, y_1, and z_1.

b) Test the formulas obtained for y and z in part (a) by investigating their behavior at $x_1 = 0$ and $x_1 = x_0$ and by seeing what happens as $x_0 \rightarrow \infty$.

58. *Hidden lines.*   Here is another typical problem in computer graphics. Your eye is at $(4, 0, 0)$. You are looking at a triangular plate whose vertices are at $(1, 0, 1)$, $(1, 1, 0)$, and $(-2, 2, 2)$. The line segment from $(1, 0, 0)$ to $(0, 2, 2)$ passes through the plate. What portion of the line segment is hidden from your view by the plate? (This is an exercise in finding intersections of lines and planes.)

COMPUTER   Exercises 59–62 give the coordinates of points A, B, C, and S. What outputs do you get when you use them as inputs for the vector utility program listed in this section?

	A	B	C	S
59.	$(1, 0, -1)$	$(0, 2, 0)$	$(-1, 0, 1)$	$(0, 0, 3)$
60.	$(1, 0, 1)$	$(0, 1, 1)$	$(1, 1, 0)$	$(2, 2, 2)$
61.	$(-2, 0, 4)$	$(2, 8, 0)$	$(2, 4, -2)$	$(0, 4, 2)$
62.	$(1, 3, 2)$	$(2, -1, 4)$	$(3, 0, -1)$	$(0, 0, 0)$

✳ COMPUTER PROGRAM: VECTOR OPERATIONS

The following BASIC program enables you to calculate the results of most of the vector operations you have seen so far. If you enter the coordinates of points A, B, and C, and later, the coordinates of a point S, the program will display the following information on the screen.

- The lengths of $\overrightarrow{OA}$, $\overrightarrow{OB}$, and $\overrightarrow{AB}$
- The coordinates of the midpoint of segment AB
- $\overrightarrow{OA} \cdot \overrightarrow{OB}$
- The angle between $\overrightarrow{OA}$ and $\overrightarrow{OB}$ in both radians and degrees (when the vectors are not parallel)
- The components of $\overrightarrow{OA} \times \overrightarrow{OB}$
- Parametric equations for line AB
- Parametric equations for the line through C parallel to $\overrightarrow{AB}$
- The distance from C to line AB
- The components of a vector **N** normal to plane ABC
- An equation for plane ABC
- The distance from S to plane ABC

Most of the computations follow the formulas in the text. The program uses subscripted variables called "arrays" to hold the components of vectors. In this way, the variable $A(I)$ can refer to any of the components of vector $\overrightarrow{OA}$ by setting $I = 1$, 2, or 3. This allows FOR loops to compute component values without using several statements.

Another use of FOR loops with subscripted variables is to get a sum, as in getting the dot product. The SUM must be set to zero before the loop begins; the successive values are added into SUM within the loop.

BASIC does not have an arc cosine function. Its only inverse trigonometric function is the arc tangent, ATN(X). To get the arc cosine, we used the identity

$$\cos^{-1}(x) = \frac{\pi}{2} - \tan^{-1}\left(\frac{x}{\sqrt{1-x^2}}\right).$$

This relation holds as long as x is different from 1, which it will be if vectors $\overrightarrow{OA}$ and $\overrightarrow{OB}$ are not parallel.

To find the distance from a point S to a plane, the program finds a line through S normal to the plane, calculates the parameter value of the point of intersection, finds the coordinates of the point, and calculates its distance from S. If $N_1 x + N_2 y + N_3 z = D$ is the plane and $\mathbf{N}$ is the normal vector, the normal line through (S_1, S_2, S_3) is

$$x = S_1 + N_1 t, \qquad y = S_2 + N_2 t, \qquad z = S_3 + N_3 t.$$

At the intersection of this line with the plane, the values of x, y, and z satisfy the plane's equation, so

$$N_1(S_1 + N_1 t) + N_2(S_2 + N_2 t) + N_3(S_3 + N_3 t) = D.$$

Solving for t (the program calls the solution T) gives

$$T = \frac{D - (N_1 S_1 + N_2 S_2 + N_3 S_3)}{N_1^2 + N_2^2 + N_3^2} = \frac{D - \mathbf{N} \cdot \overrightarrow{OS}}{|\mathbf{N}|}.$$

The distance from the intersection to S is

$$\text{Distance} = \sqrt{(S_1 - (S_1 + N_1 T))^2 + (S_2 - (S_2 + N_2 T))^2 + (S_3 - (S_3 + N_3 T))^2}$$
$$= \sqrt{N_1^2 T^2 + N_2^2 T^2 + N_3^2 T^2} = |T||\mathbf{N}|.$$

To find the distance from C to line AB, the program finds the vector projection $\overrightarrow{PB}$ of $\overrightarrow{CB}$ onto $\overrightarrow{AB}$ and calculates $|\overrightarrow{CB} - \overrightarrow{PB}|$ (Fig. 10.49).

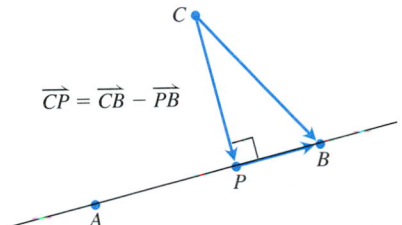

$$\overrightarrow{CP} = \overrightarrow{CB} - \overrightarrow{PB}$$

10.49   Another way to find the distance from a point C to a line AB is to calculate it as $|\overrightarrow{CB} - \overrightarrow{PB}|$.

PROGRAM		COMMENTS
10	DIM A(3), B(3), C(3), AB(3), AC(3), XPAB(3), Q(3), N(3)	Define arrays (subscripted variables) for vectors.
20	PI = 3.141593	Define π.
30	PRINT "ENTER THE COORDINATES OF THREE POINTS"	Give instructions to the user.
40	PRINT "RESPOND TO EACH PROMPT WITH 3 NUMBERS,"	
50	PRINT "USE A COMMA BETWEEN THE VALUES"	
60	PRINT	
70	INPUT "ENTER COMPONENTS OF POINT A "; A(1), A(2), A(3)	Get components of $\overrightarrow{OA}$,
80	INPUT "ENTER COMPONENTS OF POINT B "; B(1), B(2), B(3)	and of $\overrightarrow{OB}$.
90	INPUT "ENTER COORDINATES OF POINT C "; C(1), C(2), C(3)	Get coordinates of point C.
100	PRINT	
110	SUM = 0: FOR I = 1 TO 3: SUM = SUM + A(I) ^ 2: NEXT I	Get sum for length of $\overrightarrow{OA}$.
120	LA = SQR(SUM)	Get square root.
130	PRINT "LENGTH OF VECTOR OA IS "; LA	Print length of $\overrightarrow{OA}$.
140	SUM = 0: FOR I = 1 TO 3: SUM = SUM + B(I) ^ 2: NEXT I	Get sum for length of $\overrightarrow{OB}$.
150	LB = SQR(SUM)	Get square root.
160	PRINT "LENGTH OF VECTOR OB IS "; LB	Print length of $\overrightarrow{OB}$.
170	FOR I = 1 TO 3: AB(I) = B(I) − A(I): NEXT I	Get components of $\overrightarrow{AB}$.
180	SUM = 0: FOR I = 1 TO 3: SUM = SUM + AB(I) ^ 2: NEXT I	Get sum for length of $\overrightarrow{AB}$.

PROGRAM	COMMENTS
190 LAB = SQR(SUM)	Get square root.
200 PRINT "LENGTH OF VECTOR AB IS "; LAB	Print length of $\overrightarrow{AB}$.
210 PRINT "MIDPOINT OF AB IS "; (A(1) + B(1))/2, (A(2) + B(2))/2, (A(3) + B(3))/2	Compute and print coordinates of midpoint of segment AB.
220 DPAB = 0: FOR I = 1 TO 3: DPAB = DPAB + A(I) * B(I): NEXT I	Get sum for dot product of $\overrightarrow{OA}$ and $\overrightarrow{OB}$.
230 PRINT "THE DOT PRODUCT OF OA AND OB IS "; DPAB	Print dot product.
240 X = DPAB/ABS(LA * LB)	Put a value into temporary variable x.
250 ANGAB = 0: IF ABS(X) < 1 THEN ANGAB = PI/2 − ATN(X/SQR(1 − X ^ 2))	Get angle between $\overrightarrow{OA}$ and $\overrightarrow{OB}$—use trigonometric identity.
260 PRINT "ANGLE BETWEEN OA AND OB IS "; ANGAB; " RADIANS"	Print angle in radian measure.
270 PRINT " THIS IS EQUAL TO "; ANGAB/PI * 180; " DEGREES"	Print angle in degrees.
280 XPAB(1) = A(2) * B(3) − A(3) * B(2)	Get components of $\overrightarrow{OA} \times \overrightarrow{OB}$ using the determinant rule.
290 XPAB(2) = A(3) * B(1) − A(1) * B(3)	
300 XPAB(3) = A(1) * B(2) − A(2) * B(1)	
310 PRINT "COMPONENTS OF CROSS PRODUCT, OA X OB: "; XPAB(1); XPAB(2); XPAB(3)	Print components of $\overrightarrow{OA} \times \overrightarrow{OB}$.
320 PRINT "LINE THROUGH A AND B (PARAMETRIC FORM):"	Print first part of output for line through A and B,
330 PRINT " X = "; A(1);" + ";AB(1);"(t)"	then the equation for x,
340 PRINT " Y = "; A(2); " + ";AB(2);"(t)"	next the equation for y,
350 PRINT " Z = "; A(3); " + ";AB(3);"(t)"	and lastly the equation for z.
360 PRINT "LINE THROUGH C PARALLEL TO AB:"	Print the first part of output for parallel line through C,
370 PRINT " X = "; C(1); " + ";AB(1);"(t)"	then the equation for x,
380 PRINT " Y = "; C(2); " + ";AB(2);"(t)"	next the equation for y,
390 PRINT " Z = "; C(3); " + ";AB(3);"(t)"	and lastly the equation for z.
400 FOR I = 1 TO 3: CB(I) = B(I) − C(I): NEXT I	Compute components of $\overrightarrow{CB}$
410 DPABCB = 0: FOR I = 1 TO 3: DPABCB = DPABCB + AB(I) * CB(I): NEXT I	$\overrightarrow{AB} \cdot \overrightarrow{CB}$.
420 FOR I = 1 TO 3: P(I) = DPABCB/LAB ^ 2* AB(I): NEXT I	Components of $\overrightarrow{PB} = \text{proj}_{\overrightarrow{AB}} \overrightarrow{CB}$
430 SUM = 0: FOR I = 1 TO 3: SUM = SUM + (CB(I) − P(I)) ^ 2: NEXT I	$\overrightarrow{CP} \cdot \overrightarrow{CP}$
440 PRINT "DISTANCE FROM C TO LINE AB IS ": SQR(SUM)	The distance from C to line AB is $\sqrt{\overrightarrow{CP} \cdot \overrightarrow{CP}}$.
450 FOR I = 1 TO 3: AC(I) = A(I) − C(I): NEXT I	The components of $\overrightarrow{AC}$
460 N(1) = AB(2) * AC(3) − AB(3) * AC(2)	Get components of a vector **N** normal to plane through A, B, and C using the determinant rule.
470 N(2) = AB(3) * AC(1) − AB(1) * AC(3)	
480 N(3) = AB(1) * AC(2) − AB(2) * AC(1)	
490 PRINT "COMPONENTS OF A NORMAL TO PLANE ABC ARE";	Print first part of output for normal vector,
500 PRINT " "; N(1); N(2); N(3)	and then the components.
510 SUM = 0: FOR I = 1 TO 3: SUM = SUM + N(I) * C(I): NEXT I	Get product of components of **N** and coordinates of C.
520 PRINT "EQUATION OF PLANE ABC IS "	Print first part of output for equation of plane,
530 PRINT " "; N(1);"X + ";N(2);"Y + ";N(3);"Z = ";SUM	and then the equation.
540 PRINT "TO FIND THE DISTANCE FROM POINT S TO THE PLANE"	Prompt user and get coordinates of
550 INPUT " ENTER COORDINATES OF S: ", S(1),S(2),S(3)	point S.
560 S2 = 0: FOR I = 1 TO 3: S2 = S2 + (S(I) − C(I)) * N(I): NEXT I	Find $\mathbf{N} \cdot \overrightarrow{OS}$.
570 S3 = 0: FOR I = 1 TO 3: S3 = S3 + N(I) ^ 2: NEXT I	Add squares of components of **N**.
580 DIST = S2/SQR(S3)	Compute distance to the plane.
590 PRINT "DISTANCE IS ";ABS(DIST)	Print the distance.
600 END	

Here is the output for

$$A = (2, 0, 0), \qquad B = (0, 3, 0), \qquad C = (0, 0, 1), \qquad S = (1, 1, 3).$$

LENGTH OF VECTOR OA IS   2
LENGTH OF VECTOR OB IS   3
LENGTH OF VECTOR AB IS   3.605551
MIDPOINT OF AB IS   1   1.5   0
THE DOT PRODUCT OF OA AND OB IS 0
ANGLE BETWEEN OA AND OB IS 1.570797 RADIANS
  THIS IS EQUAL TO 90 DEGREES
COMPONENTS OF CROSS PRODUCT, OA X OB: 0 0 6
LINE THROUGH A AND B (PARAMETRIC FORM):
  X = 2 − 2(t)
  Y = 0 + 3(t)
  Z = 0 + 0(t)

LINE THROUGH C PARALLEL TO AB:
  X = 0 + − 2(t)
  Y = 0 + 3(t)
  Z = 1 + 0(t)
DISTANCE FROM C TO LINE AB IS 1.941451
COMPONENTS OF A NORMAL TO PLANE ABC ARE − 3 − 2 − 6
EQUATION OF PLANE ABC IS − 3X + − 2Y + − 6Z = −6
TO FIND THE DISTANCE FROM POINT S TO THE PLANE
  ENTER COORDINATES OF S: 1, 1, 3
DISTANCE IS   2.428572

10.6 Products of Three Vectors or More

The triple products $(\mathbf{A} \times \mathbf{B}) \cdot \mathbf{C}$ and $(\mathbf{A} \times \mathbf{B}) \times \mathbf{C}$ appear frequently in engineering and physics in the calculations associated with electricity, magnetism, and fluid flow. In this section, we study the three-dimensional geometry of the triple scalar product $(\mathbf{A} \times \mathbf{B}) \cdot \mathbf{C}$ (we shall use it to derive the surface area formulas in Section 14.4) and derive the identities that are used to evaluate triple products from the components of the vectors being multiplied.

The Triple Scalar or Box Product

The product $(\mathbf{A} \times \mathbf{B}) \cdot \mathbf{C}$ is called the **triple scalar product** of $\mathbf{A}$, $\mathbf{B}$, and $\mathbf{C}$ (in that order). As you can see from the formula

$$|(\mathbf{A} \times \mathbf{B}) \cdot \mathbf{C}| = |\mathbf{A} \times \mathbf{B}||\mathbf{C}||\cos \theta|, \tag{1}$$

the absolute value of the product is the volume of the parallelepiped (parallelogram-sided box) determined by $\mathbf{A}$, $\mathbf{B}$, and $\mathbf{C}$ (Fig. 10.50). The number $|\mathbf{A} \times \mathbf{B}|$ is the area of the base parallelogram. The number $|\mathbf{C}||\cos \theta|$ is the parallelepiped's height. Because of the geometry here, $(\mathbf{A} \times \mathbf{B}) \cdot \mathbf{C}$ is often called the **box product** of $\mathbf{A}$, $\mathbf{B}$, and $\mathbf{C}$.

By treating the planes of $\mathbf{B}$ and $\mathbf{C}$ and of $\mathbf{C}$ and $\mathbf{A}$ as the base planes of the parallelepiped determined by $\mathbf{A}$, $\mathbf{B}$, and $\mathbf{C}$, we see that

$$(\mathbf{A} \times \mathbf{B}) \cdot \mathbf{C} = (\mathbf{B} \times \mathbf{C}) \cdot \mathbf{A} = (\mathbf{C} \times \mathbf{A}) \cdot \mathbf{B}. \tag{2}$$

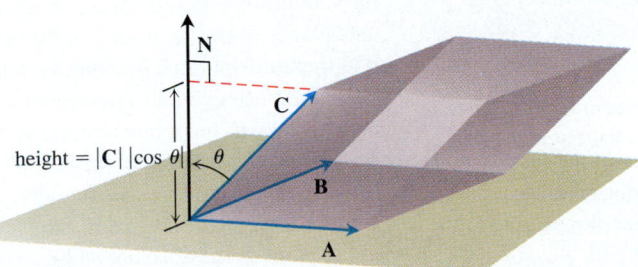

10.50   The volume of the parallelepiped shown here is $|(\mathbf{A} \times \mathbf{B}) \cdot \mathbf{C}|$. The absolute value bars correct the sign when θ is greater than 90°.

Since the dot product is commutative, Eq. (2) also gives

$$(\mathbf{A} \times \mathbf{B}) \cdot \mathbf{C} = \mathbf{A} \cdot (\mathbf{B} \times \mathbf{C}). \tag{3}$$

The dot and cross may be interchanged in a triple scalar product without altering its value.

The triple scalar product can be evaluated as a determinant in the following way:

$$\mathbf{A} \cdot (\mathbf{B} \times \mathbf{C}) = \mathbf{A} \cdot \left[\begin{vmatrix} b_2 & b_3 \\ c_2 & c_3 \end{vmatrix} \mathbf{i} - \begin{vmatrix} b_1 & b_3 \\ c_1 & c_3 \end{vmatrix} \mathbf{j} + \begin{vmatrix} b_1 & b_2 \\ c_1 & c_2 \end{vmatrix} \mathbf{k} \right]$$

$$= a_1 \begin{vmatrix} b_2 & b_3 \\ c_2 & c_3 \end{vmatrix} - a_2 \begin{vmatrix} b_1 & b_3 \\ c_1 & c_3 \end{vmatrix} + a_3 \begin{vmatrix} b_1 & b_2 \\ c_1 & c_2 \end{vmatrix}$$

$$= \begin{vmatrix} a_1 & a_2 & a_3 \\ b_1 & b_2 & b_3 \\ c_1 & c_2 & c_3 \end{vmatrix}.$$

KNOW →
ALTERNATE
FORMULA

The number $\mathbf{A} \cdot (\mathbf{B} \times \mathbf{C})$ can be calculated with the formula

$$\mathbf{A} \cdot (\mathbf{B} \times \mathbf{C}) = (\mathbf{A} \times \mathbf{B}) \cdot \mathbf{C} = \begin{vmatrix} a_1 & a_2 & a_3 \\ b_1 & b_2 & b_3 \\ c_1 & c_2 & c_3 \end{vmatrix}. \tag{4}$$

Example 1   Find the volume of the box (parallelepiped) determined by $\mathbf{A} = \mathbf{i} + 2\mathbf{j} - \mathbf{k}$, $\mathbf{B} = -2\mathbf{i} + 3\mathbf{k}$, and $\mathbf{C} = 7\mathbf{j} - 4\mathbf{k}$.

Solution   The volume is the absolute value of

$$\mathbf{A} \cdot (\mathbf{B} \times \mathbf{C}) = \begin{vmatrix} 1 & 2 & -1 \\ -2 & 0 & 3 \\ 0 & 7 & -4 \end{vmatrix}$$

$$= \begin{vmatrix} 0 & 3 \\ 7 & -4 \end{vmatrix} - 2 \begin{vmatrix} -2 & 3 \\ 0 & -4 \end{vmatrix} - \begin{vmatrix} -2 & 0 \\ 0 & 7 \end{vmatrix}$$

$$= -21 - 16 + 14 = -23,$$

or 23.

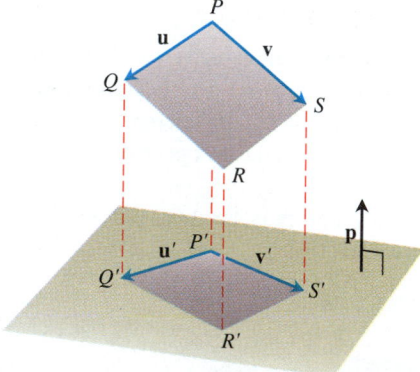

10.51   The parallelogram determined by two vectors **u** and **v** in space and the orthogonal projection of the parallelogram onto a plane. The projection lines, orthogonal to the plane, lie parallel to the unit normal vector **p**.

The Area of a Parallelogram's Projection on a Plane

Our next goal is to establish a result we shall need in Section 14.4 when we derive the standard integral formula for the area of a curved surface lying above a coordinate plane. At the beginning of the derivation, we approximate the surface by tiling it with parallelograms (in context, a natural thing to do). We then project each parallelogram orthogonally (straight down) onto the plane and work backward from the area of the parallelogram's image to find the area of the parallelogram itself. How do we find the area of the image? The answer, as we shall see in a moment, is that if the parallelogram's sides are determined by the vectors **u** and **v** (Fig. 10.51) and if **p** is a unit vector normal to the plane, then the area of the parallelogram's image in the plane is the absolute value of $(\mathbf{u} \times \mathbf{v}) \cdot \mathbf{p}$.

THEOREM 1

> The area of the orthogonal projection of the parallelogram determined by two vectors **u** and **v** in space onto a plane with unit normal vector **p** is given by the formula
>
> $$\text{Area} = |(\mathbf{u} \times \mathbf{v}) \cdot \mathbf{p}|. \tag{5}$$

Proof In the notation of Fig. 10.51, which shows a typical parallelogram determined by vectors **u** and **v** and its orthogonal projection onto a plane with unit normal vector **p**,

$$\mathbf{u} = \overrightarrow{PP'} + \mathbf{u}' + \overrightarrow{Q'Q}$$

$$= \mathbf{u}' + \overrightarrow{PP'} - \overrightarrow{QQ'} \quad \left(\overrightarrow{Q'Q} = -\overrightarrow{QQ'}\right) \tag{6}$$

$$= \mathbf{u}' + s\mathbf{p}. \qquad \text{(For some scalar } s \text{ because } \overrightarrow{PP'} - \overrightarrow{QQ'} \text{ is parallel to } \mathbf{p}\text{)}$$

Similarly,

$$\mathbf{v} = \mathbf{v}' + t\mathbf{p} \tag{7}$$

for some scalar t. Hence,

$$\mathbf{u} \times \mathbf{v} = (\mathbf{u}' + s\mathbf{p}) \times (\mathbf{v}' + t\mathbf{p})$$

$$= (\mathbf{u}' \times \mathbf{v}') + s(\mathbf{p} \times \mathbf{v}') + t(\mathbf{u}' \times \mathbf{p}) + \underbrace{st(\mathbf{p} \times \mathbf{p})}_{\mathbf{0}}. \tag{8}$$

The vectors $\mathbf{p} \times \mathbf{v}'$ and $\mathbf{u}' \times \mathbf{p}$ are both orthogonal to **p**. Hence, when we dot both sides of Eq. (8) with **p**, the only surviving term on the right is $(\mathbf{u}' \times \mathbf{v}') \cdot \mathbf{p}$. We are left with

$$(\mathbf{u} \times \mathbf{v}) \cdot \mathbf{p} = (\mathbf{u}' \times \mathbf{v}') \cdot \mathbf{p}. \tag{9}$$

In particular,

$$|(\mathbf{u} \times \mathbf{v}) \cdot \mathbf{p}| = |(\mathbf{u}' \times \mathbf{v}') \cdot \mathbf{p}|. \tag{10}$$

The absolute value on the right is the volume of the box determined by $\mathbf{u}'$, $\mathbf{v}'$, and **p**. The height of this particular box is $|\mathbf{p}| = 1$, so the box's volume is the same as its base area, the area of parallelogram $P'Q'R'S'$. Combining this observation with Eq. (10) gives

$$\text{Area of } P'Q'R'S' = |(\mathbf{u}' \times \mathbf{v}') \cdot \mathbf{p}| = |(\mathbf{u} \times \mathbf{v}) \cdot \mathbf{p}|. \tag{11}$$

This says that the area of the orthogonal projection of the parallelogram determined by **u** and **v** onto a plane with unit normal vector **p** is $|(\mathbf{u} \times \mathbf{v}) \cdot \mathbf{p}|$, which is what we set out to prove. ∎

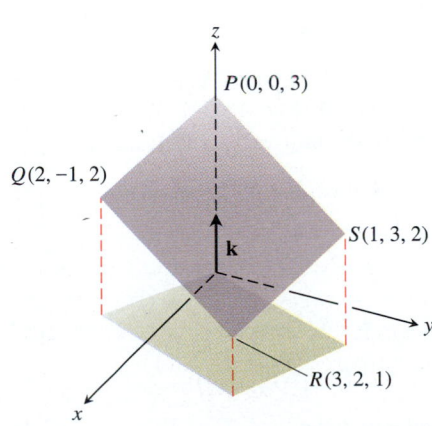

10.52 Example 2 calculates the area of the orthogonal projection of parallelogram *PQRS* on the *xy*-plane.

Example 2 Find the area of the orthogonal projection onto the *xy*-plane of the parallelogram determined by the points $P(0, 0, 3)$, $Q(2, -1, 2)$, $R(3, 2, 1)$, and $S(1, 3, 2)$ (Fig. 10.52).

Solution With

$$\mathbf{u} = \overrightarrow{PQ} = 2\mathbf{i} - \mathbf{j} - \mathbf{k}, \qquad \mathbf{v} = \overrightarrow{PS} = \mathbf{i} + 3\mathbf{j} - \mathbf{k}, \qquad \text{and} \qquad \mathbf{p} = \mathbf{k},$$

Eq. (5) gives

$$\text{Area} = (\mathbf{u} \times \mathbf{v}) \cdot \mathbf{p} = \begin{vmatrix} 2 & -1 & -1 \\ 1 & 3 & -1 \\ 0 & 0 & 1 \end{vmatrix} = \begin{vmatrix} 2 & -1 \\ 1 & 3 \end{vmatrix} = 7.$$

Triple Vector Products

The **triple vector products** $(\mathbf{A} \times \mathbf{B}) \times \mathbf{C}$ and $\mathbf{A} \times (\mathbf{B} \times \mathbf{C})$ are usually not equal, although the formulas for evaluating them from components are similar.

$$(\mathbf{A} \times \mathbf{B}) \times \mathbf{C} = (\mathbf{A} \cdot \mathbf{C})\mathbf{B} - (\mathbf{B} \cdot \mathbf{C})\mathbf{A} \qquad (12)$$

$$\mathbf{A} \times (\mathbf{B} \times \mathbf{C}) = (\mathbf{A} \cdot \mathbf{C})\mathbf{B} - (\mathbf{A} \cdot \mathbf{B})\mathbf{C} \qquad (13)$$

The first product is a multiple of $\mathbf{B}$ minus a multiple of $\mathbf{A}$. The second is the same multiple of $\mathbf{B}$ minus a multiple of $\mathbf{C}$. Take a moment to compare the two formulas.

Equation (13) follows from (12) by permuting the letters $\mathbf{A}$, $\mathbf{B}$, and $\mathbf{C}$:

$$(\mathbf{A} \times \mathbf{B}) \times \mathbf{C} = (\mathbf{A} \cdot \mathbf{C})\mathbf{B} - (\mathbf{B} \cdot \mathbf{C})\mathbf{A} \qquad \text{(Eq. (12))}$$

$$(\mathbf{B} \times \mathbf{C}) \times \mathbf{A} = (\mathbf{B} \cdot \mathbf{A})\mathbf{C} - (\mathbf{C} \cdot \mathbf{A})\mathbf{B} \qquad (\mathbf{A} \rightarrow \mathbf{B} \rightarrow \mathbf{C} \rightarrow \mathbf{A})$$

$$= (\mathbf{A} \cdot \mathbf{B})\mathbf{C} - (\mathbf{A} \cdot \mathbf{C})\mathbf{B} \qquad \left(\text{Dot multiplication is commutative.}\right)$$

$$\mathbf{A} \times (\mathbf{B} \times \mathbf{C}) = (\mathbf{A} \cdot \mathbf{C})\mathbf{B} - (\mathbf{A} \cdot \mathbf{B})\mathbf{C} \qquad \left(\begin{array}{l}\text{Reverse the order of the} \\ \text{product on the left; change} \\ \text{the signs on the right.}\end{array}\right)$$

To derive Eq. (12), we first eliminate two degenerate cases. If any of the vectors is zero, Eq. (12) holds because both sides are zero. If $\mathbf{B}$ is a scalar multiple of $\mathbf{A}$, both sides of Eq. (12) are zero again (see Exercise 12).

Suppose now that none of the vectors involved is zero and that $\mathbf{A}$ and $\mathbf{B}$ are not parallel. The vector $(\mathbf{A} \times \mathbf{B}) \times \mathbf{C}$ is orthogonal to $\mathbf{A} \times \mathbf{B}$ and consequently parallel to the plane of $\mathbf{A}$ and $\mathbf{B}$. This means that

$$(\mathbf{A} \times \mathbf{B}) \times \mathbf{C} = m\mathbf{A} + n\mathbf{B} \qquad (14)$$

for some scalars m and n. To evaluate m and n, we let $\mathbf{I}$ equal $\mathbf{A}/|\mathbf{A}|$ and let $\mathbf{J}$ be a unit vector orthogonal to $\mathbf{I}$ in the plane of $\mathbf{A}$ and $\mathbf{B}$ (Fig. 10.53). We then let $\mathbf{K}$ denote $\mathbf{I} \times \mathbf{J}$ and write $\mathbf{A}$, $\mathbf{B}$, and $\mathbf{C}$ in terms of the unit vectors $\mathbf{I}$, $\mathbf{J}$, and $\mathbf{K}$:

$$\mathbf{A} = a_1 \mathbf{I}, \qquad \mathbf{B} = b_1 \mathbf{I} + b_2 \mathbf{J}, \qquad \mathbf{C} = c_1 \mathbf{I} + c_2 \mathbf{J} + c_3 \mathbf{K}. \qquad (15)$$

Then

$$\mathbf{A} \times \mathbf{B} = a_1 b_2 \mathbf{K},$$

$$(\mathbf{A} \times \mathbf{B}) \times \mathbf{C} = a_1 b_2 c_1 \mathbf{J} - a_1 b_2 c_2 \mathbf{I}, \qquad (16)$$

$$m(a_1 \mathbf{I}) + n(b_1 \mathbf{I} + b_2 \mathbf{J}) = a_1 b_2 c_1 \mathbf{J} - a_1 b_2 c_2 \mathbf{I}. \qquad \text{(Eq. (14))}$$

This final equation is equivalent to the scalar equations

$$m a_1 + n b_1 = -a_1 b_2 c_2,$$

$$n b_2 = a_1 b_2 c_1.$$

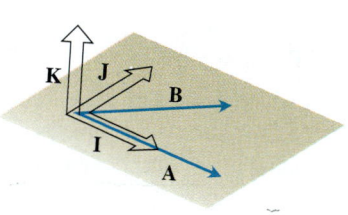

10.53 The orthogonal vectors $\mathbf{I}$ and $\mathbf{J}$ in the plane of $\mathbf{A}$ and $\mathbf{B}$, together with their cross product $\mathbf{K}$.

If b_2 were zero, $\mathbf{A}$ and $\mathbf{B}$ would be parallel, contrary to our assumption. Hence b_2 is not zero and we may divide both sides of the equation $nb_2 = a_1 b_2 c_1$ by b_2 to obtain

$$n = a_1 c_1 = \mathbf{A} \cdot \mathbf{C}.$$

Then, by substitution,

$$ma_1 = -nb_1 - a_1 b_2 c_2$$

$$= -a_1 c_1 b_1 - a_1 b_2 c_2.$$

Since $|\mathbf{A}| = a_1 \neq 0$, we may divide by a_1 to obtain

$$m = -(b_1 c_1 + b_2 c_2) = -(\mathbf{B} \cdot \mathbf{C}).$$

Substituting these values for m and n in Eq. (14) now gives Eq. (12). This concludes the derivation.

Example 3   Verify Eq. (12) for

$$\mathbf{A} = \mathbf{i} - \mathbf{j} + 2\,\mathbf{k}, \qquad \mathbf{B} = 2\,\mathbf{i} + \mathbf{j} + \mathbf{k}, \qquad \mathbf{C} = \mathbf{i} + 2\,\mathbf{j} - \mathbf{k}.$$

Solution   Since

$$\mathbf{A} \cdot \mathbf{C} = -3 \qquad \text{and} \qquad \mathbf{B} \cdot \mathbf{C} = 3,$$

the right-hand side of Eq. (12) is

$$(\mathbf{A} \cdot \mathbf{C})\mathbf{B} - (\mathbf{B} \cdot \mathbf{C})\mathbf{A} = -3\mathbf{B} - 3\mathbf{A} = -3(3\,\mathbf{i} + 3\,\mathbf{k}) = -9\,\mathbf{i} - 9\,\mathbf{k}.$$

As for the left-hand side,

$$\mathbf{A} \times \mathbf{B} = \begin{vmatrix} \mathbf{i} & \mathbf{j} & \mathbf{k} \\ 1 & -1 & 2 \\ 2 & 1 & 1 \end{vmatrix} = -3\,\mathbf{i} + 3\,\mathbf{j} + 3\,\mathbf{k},$$

so

$$(\mathbf{A} \times \mathbf{B}) \times \mathbf{C} = \begin{vmatrix} \mathbf{i} & \mathbf{j} & \mathbf{k} \\ -3 & 3 & 3 \\ 1 & 2 & -1 \end{vmatrix} = -9\,\mathbf{i} - 9\,\mathbf{k}.$$

We can sometimes use Eq. (12) to replace a string of cross multiplications with an expression that is less cumbersome to evaluate.

Example 4   Express $(\mathbf{A} \times \mathbf{B}) \times (\mathbf{C} \times \mathbf{D})$ in terms of cross products with fewer factors.

Solution   For convenience, set $\mathbf{C} \times \mathbf{D}$ equal to $\mathbf{V}$. Then,

$$(\mathbf{A} \times \mathbf{B}) \times \mathbf{V} = (\mathbf{A} \cdot \mathbf{V})\mathbf{B} - (\mathbf{B} \cdot \mathbf{V})\mathbf{A} \qquad \text{(Eq. 12)}$$

and

$$(\mathbf{A} \times \mathbf{B}) \times (\mathbf{C} \times \mathbf{D}) = (\mathbf{A} \cdot \mathbf{C} \times \mathbf{D})\mathbf{B} - (\mathbf{B} \cdot \mathbf{C} \times \mathbf{D})\mathbf{A}. \qquad (\mathbf{V} = \mathbf{C} \times \mathbf{D})$$

This expresses the original product as a scalar multiple of $\mathbf{B}$ minus a scalar multiple of $\mathbf{A}$. Had we wanted to do so, we could have used Eq. (13) instead to express it as a scalar multiple of $\mathbf{C}$ minus a scalar multiple of $\mathbf{D}$ (Exercise 10).

Notice that we tried to keep the answer as uncluttered as possible by omitting the usual parentheses from the expressions $\mathbf{A} \cdot \mathbf{C} \times \mathbf{D}$ and $\mathbf{B} \cdot \mathbf{C} \times \mathbf{D}$. There is only one way these products make sense, so we do not need parentheses to tell us what to do.

FINAL UP TO 10.6

EXERCISES 10.6

In Exercises 1–4, verify that $(\mathbf{A} \times \mathbf{B}) \cdot \mathbf{C} = (\mathbf{B} \times \mathbf{C}) \cdot \mathbf{A} = (\mathbf{C} \times \mathbf{A}) \cdot \mathbf{B}$. Find the volume of the parallelepiped determined by $\mathbf{A}$, $\mathbf{B}$, and $\mathbf{C}$. Also, find $(\mathbf{A} \times \mathbf{B}) \times \mathbf{C}$ and $\mathbf{A} \times (\mathbf{B} \times \mathbf{C})$.

A	B	C
1. $2\,\mathbf{i}$	$2\,\mathbf{j}$	$2\,\mathbf{k}$
2. $\mathbf{i} - \mathbf{j} + \mathbf{k}$	$2\,\mathbf{i} + \mathbf{j} - 2\,\mathbf{k}$	$-\mathbf{i} + 2\,\mathbf{j} - \mathbf{k}$
3. $2\,\mathbf{i} + \mathbf{j}$	$2\,\mathbf{i} - \mathbf{j} + \mathbf{k}$	$\mathbf{i} + 2\,\mathbf{k}$
4. $\mathbf{i} + \mathbf{j} - 2\,\mathbf{k}$	$-\mathbf{i} - \mathbf{k}$	$2\,\mathbf{i} + 4\,\mathbf{j} - 2\,\mathbf{k}$

5. Which of the following are always true and which are not always true?
   a) $|\mathbf{A}| = \sqrt{\mathbf{A} \cdot \mathbf{A}}$
   b) $\mathbf{A} \cdot \mathbf{A} = |\mathbf{A}|$
   c) $\mathbf{A} \times \mathbf{0} = \mathbf{0} \times \mathbf{A} = \mathbf{0}$
   d) $\mathbf{A} \times (-\mathbf{A}) = \mathbf{0}$
   e) $\mathbf{A} \times \mathbf{B} = \mathbf{B} \times \mathbf{A}$
   f) $\mathbf{A} \times (\mathbf{B} + \mathbf{C}) = \mathbf{A} \times \mathbf{B} + \mathbf{A} \times \mathbf{C}$
   g) $(\mathbf{A} \times \mathbf{B}) \cdot \mathbf{B} = 0$
   h) $(\mathbf{A} \times \mathbf{B}) \cdot \mathbf{C} = \mathbf{A} \cdot (\mathbf{B} \times \mathbf{C})$
   i) $(\mathbf{A} \times \mathbf{B}) \times \mathbf{C} = \mathbf{A} \times (\mathbf{B} \times \mathbf{C})$

6. Which of the following are always true and which are not always true?
   a) $\mathbf{A} \cdot \mathbf{B} = \mathbf{B} \cdot \mathbf{A}$
   b) $(\mathbf{A} \times \mathbf{B}) = -(\mathbf{B} \times \mathbf{A})$
   c) $(-\mathbf{A}) \times \mathbf{B} = -(\mathbf{A} \times \mathbf{B})$
   d) $(c\mathbf{A}) \cdot \mathbf{B} = \mathbf{A} \cdot (c\mathbf{B}) = c(\mathbf{A} \cdot \mathbf{B})$ (Any number c)
   e) $(c\mathbf{A}) \times \mathbf{B} = \mathbf{A} \times (c\mathbf{B}) = c(\mathbf{A} \times \mathbf{B})$ (Any number c)
   f) $\mathbf{A} \cdot \mathbf{A} = |\mathbf{A}|^2$
   g) $(\mathbf{A} \times \mathbf{A}) \cdot \mathbf{A} = 0$
   h) $(\mathbf{A} \times \mathbf{B}) \cdot \mathbf{A} = \mathbf{B} \cdot (\mathbf{A} \times \mathbf{B})$
   i) $(\mathbf{A} \times \mathbf{B}) \times (\mathbf{C} \times \mathbf{A}) = k\mathbf{A}$ for some scalar k

7. Suppose
$$\mathbf{A} \cdot \mathbf{A} = 4, \quad \mathbf{B} \cdot \mathbf{B} = 4, \quad \mathbf{A} \cdot \mathbf{B} = 0,$$
$$(\mathbf{A} \times \mathbf{B}) \times \mathbf{C} = \mathbf{0}, \quad (\mathbf{A} \times \mathbf{B}) \cdot \mathbf{C} = 8.$$

Find
   a) $\mathbf{A} \cdot \mathbf{C}$
   b) $|\mathbf{C}|$
   c) $|\mathbf{B} \times \mathbf{C}|$
   (*Hint*: Picture the vectors and think geometrically. Use basic, coordinate-free definitions. Avoid long calculations.)

8. Show that any vector $\mathbf{A}$ satisfies the identity
$$\mathbf{i} \times (\mathbf{A} \times \mathbf{i}) + \mathbf{j} \times (\mathbf{A} \times \mathbf{j}) + \mathbf{k} \times (\mathbf{A} \times \mathbf{k}) = 2\mathbf{A}.$$

9. Use Eq. (4) and facts about determinants (Appendix 10) to show that
   a) $\mathbf{A} \cdot (\mathbf{C} \times \mathbf{B}) = -\mathbf{A} \cdot (\mathbf{B} \times \mathbf{C})$
   b) $\mathbf{A} \cdot (\mathbf{A} \times \mathbf{B}) = 0$
   c) $(\mathbf{A} + \mathbf{D}) \cdot (\mathbf{B} \times \mathbf{C}) = \mathbf{A} \cdot (\mathbf{B} \times \mathbf{C}) + \mathbf{D} \cdot (\mathbf{B} \times \mathbf{C})$

10. Show that
$$(\mathbf{A} \times \mathbf{B}) \times (\mathbf{C} \times \mathbf{D}) = (\mathbf{A} \times \mathbf{B} \cdot \mathbf{D})\mathbf{C} - (\mathbf{A} \times \mathbf{B} \cdot \mathbf{C})\mathbf{D}.$$

11. Let $P(1, 2, -1)$, $Q(3, -1, 4)$, and $R(2, 6, 2)$ be three vertices of a parallelogram $PQRS$.
   a) Find the coordinates of S.
   b) Find the area of $PQRS$.
   c) Find the area of the orthogonal projection of $PQRS$ on each coordinate plane.

12. Show that if $\mathbf{B}$ is a scalar multiple of $\mathbf{A}$, then both sides of the equation
$$(\mathbf{A} \times \mathbf{B}) \times \mathbf{C} = (\mathbf{A} \cdot \mathbf{C})\mathbf{B} - (\mathbf{B} \cdot \mathbf{C})\mathbf{A}$$
are zero.

13. Show that the area of a parallelogram in space equals the square root of the sum of the squares of the areas of the parallelogram's orthogonal projections on the coordinate planes.

14. Show that $(\mathbf{A} \times \mathbf{B}) \times (\mathbf{C} \times \mathbf{D}) = \mathbf{0}$ if $\mathbf{A}$, $\mathbf{B}$, $\mathbf{C}$, and $\mathbf{D}$ are coplanar.

10.7 Surfaces in Space

Just as we call the graph of an equation $F(x, y) = 0$ in the plane a curve, we call the graph of an equation $F(x, y, z) = 0$ in space a **surface.** We use surfaces to describe boundaries of solids, to model membranes across which fluids flow, to describe plates over which electrical charges are distributed, and to define the walls of containers that are subjected to pressures of various kinds. We shall see all this and more in the chapters to come.

Our goal in the present section is to become acquainted with the surfaces most commonly used in the theory and application of the calculus of functions of more than one variable. This means finding out what the surfaces look like, what their equations are, and how to draw them.

Cylinders

The simplest surfaces to draw and write equations for, besides planes, are cylinders.

DEFINITIONS

> A **cylinder** is a surface composed of all the lines parallel to a given line that pass through a given plane curve. The curve is a **generating curve** for the cylinder.

In solid geometry, *cylinder* usually means *circular cylinder,* but that is not the case here. We allow our cylinders to have cross sections of any kind. The cross sections of the cylinder in our first example are parabolas, not circles. Cylinders, as we now define them, need not even be closed.

Example 1   *The Parabolic Cylinder $y = x^2$.*   Find an equation for the cylinder made by the lines parallel to the z-axis that pass through the parabola $y = x^2$, $z = 0$ (Fig. 10.54).

Solution   Suppose that the point $P_0(x_0, x_0^2, 0)$ lies on the parabola $y = x^2$ in the xy-plane. Then, for any value of z, the point $Q(x_0, x_0^2, z)$ will lie on the cylinder because it lies on the line $x = x_0$, $y = x_0^2$ through P_0 parallel to the z-axis. Conversely, any point $Q(x_0, x_0^2, z)$ whose y-coordinate is the square of its x-coordinate lies on the cylinder because it lies on the line $x = x_0$, $y = x_0^2$ through P_0 parallel to the z-axis (Fig. 10.55).

Regardless of the value of z, therefore, the points on the surface are the points whose coordinates satisfy the equation $y = x^2$. This makes $y = x^2$ an equation for the cylinder. Because of this, we call the cylinder "the cylinder $y = x^2$."

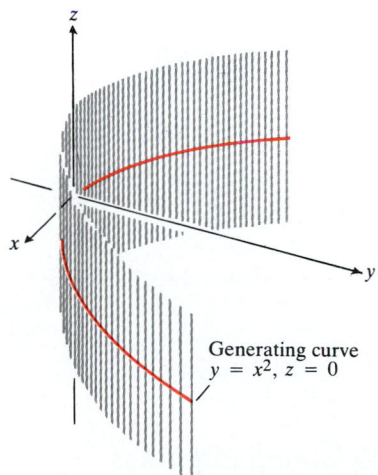

Generating curve
$y = x^2$, $z = 0$

10.54   The cylinder generated by lines parallel to the z-axis and passing through the parabola $y = x^2$ in the xy-plane (Example 1).

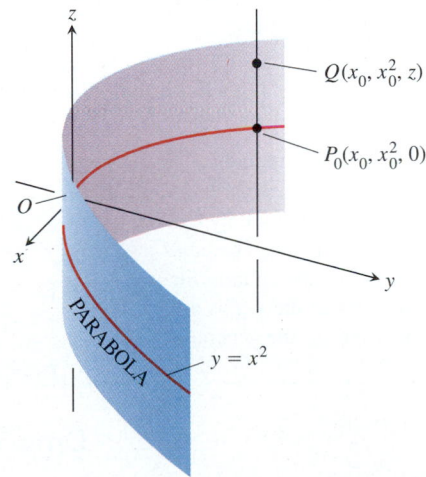

$Q(x_0, x_0^2, z)$

$P_0(x_0, x_0^2, 0)$

PARABOLA

$y = x^2$

10.55   Every point of the cylinder in Fig. 10.54 has coordinates of the form (x_0, x_0^2, z). We call the cylinder "the cylinder $y = x^2$."

As Example 1 suggests, any curve $f(x, y) = c$ in the xy-plane defines a cylinder parallel to the z-axis whose equation is also $f(x, y) = c$. The equation $x^2 + y^2 = 1$ defines the circular cylinder made by the lines parallel to the z-axis that pass through the circle $x^2 + y^2 = 1$ in the xy-plane. The equation $x^2 + 4y^2 = 9$ defines the elliptical cylinder made by the lines parallel to the z-axis that pass through the ellipse $x^2 + 4y^2 = 9$ in the xy-plane.

In a similar way, any curve $g(x, z) = c$ in the xz-plane defines a cylinder parallel to the y-axis whose space equation is also $g(x, z) = c$. Any curve $h(y, z) = c$ defines a cylinder parallel to the x-axis whose space equation is also $h(y, z) = c$. An equation in any two of the three Cartesian coordinates defines a cylinder parallel to the axis of the third coordinate. See Figs. 10.56 and 10.57.

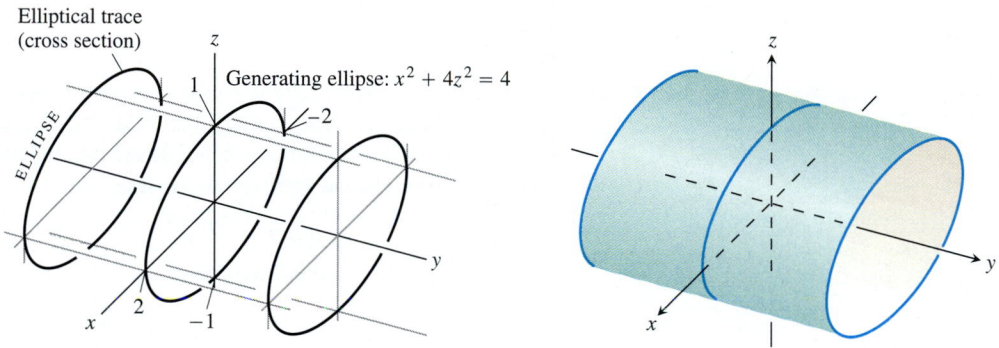

10.56 The elliptic cylinder $x^2 + 4z^2 = 4$ is made of lines parallel to the y-axis and passing through the ellipse $x^2 + 4z^2 = 4$ in the xz-plane. The cross sections or "traces" of the cylinder in planes perpendicular to the y-axis are ellipses congruent to the generating ellipse. The cylinder extends along the entire y-axis but we can draw only a finite portion of it.

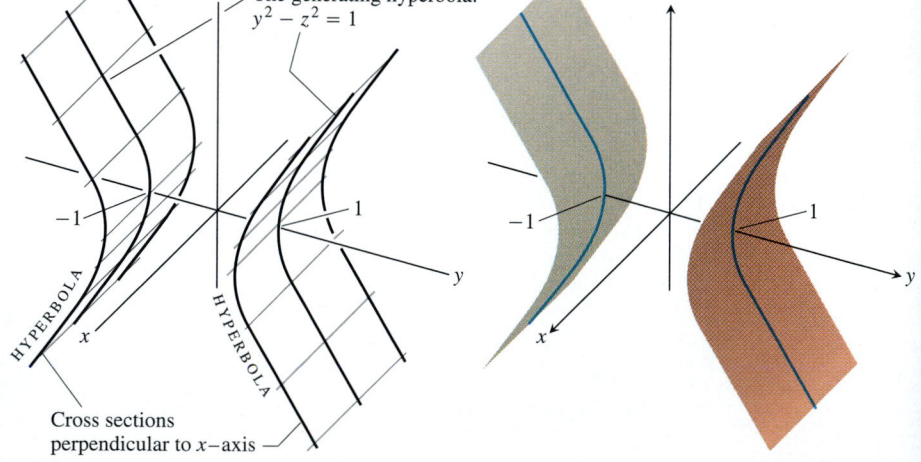

10.57 The hyperbolic cylinder $y^2 - z^2 = 1$ is made of lines parallel to the x-axis and passing through the hyperbola $y^2 - z^2 = 1$ in the yz-plane. The cross sections or "traces" of the cylinder in planes perpendicular to the x-axis are hyperbolas congruent to the generating hyperbola.

Drawing Cylinders Parallel to Coordinate Axes

See page 749 for some advice about drawing cylinders. As always, pencil is safer than pen because you can erase, but the advice applies no matter what medium you choose. Just determine which axes the cylinder is parallel to and carry out the steps shown.

How to Draw Cylinders Parallel to the Coordinate Axes

$$x^2 + y^2 = 1 \qquad z = y^2$$

1. Sketch all three coordinate axes *very lightly*.

2. Sketch the trace of the cylinder in the coordinate plane of the two variables that appear in the cylinder's equation. Sketch *very lightly*.

3. Sketch traces in parallel planes on either side (again, lightly).

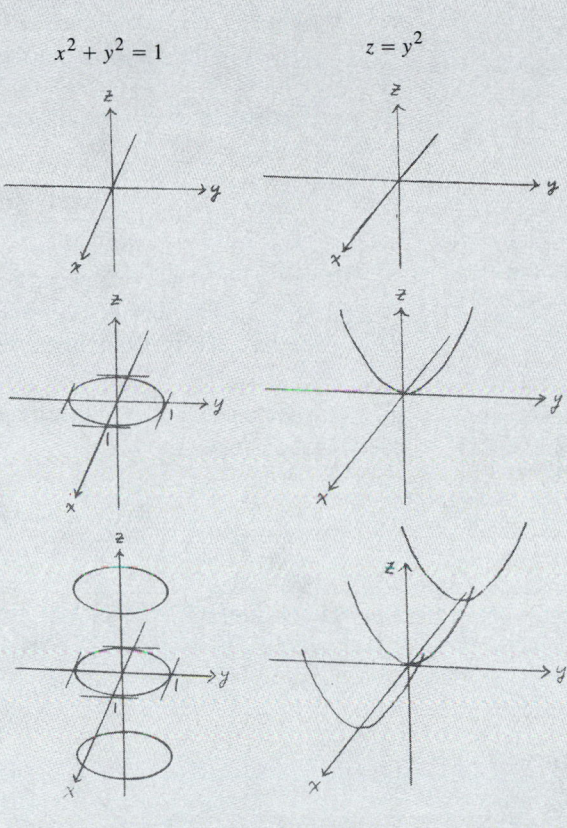

4. Add parallel outer edges to give the shape definition.

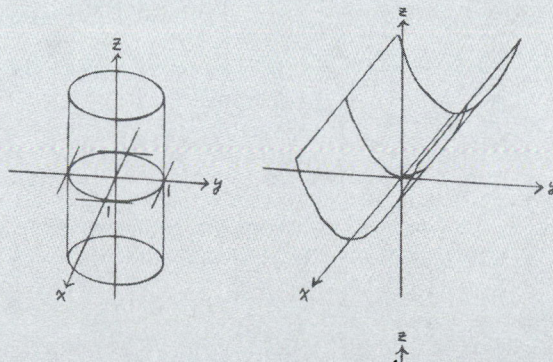

5. If more definition is required, darken the parts of the lines that are exposed to view. Leave the hidden parts light. Use line breaks when you can.

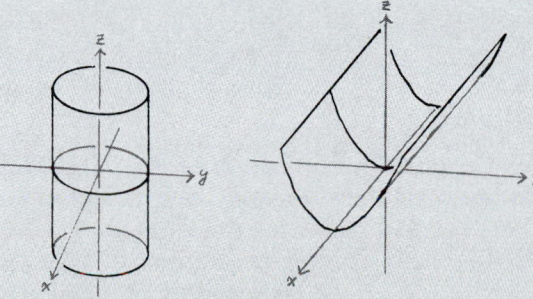

Quadric Surfaces

The cylinders in Figs. 10.55–10.57 are all **quadric surfaces,** surfaces whose equations combine quadratic terms with linear terms and constants. The examples that follow describe the other important quadric surfaces: ellipsoids (these include spheres), paraboloids, cones, and hyperboloids. The numbers a, b, and c that appear in the equations for these surfaces are assumed to be positive.

Example 2   The **ellipsoid**

$$\frac{x^2}{a^2} + \frac{y^2}{b^2} + \frac{z^2}{c^2} = 1 \tag{1}$$

(Fig. 10.58) cuts the coordinate axes at $(\pm a, 0, 0)$, $(0, \pm b, 0)$, and $(0, 0, \pm c)$. It lies within the rectangular box defined by the inequalities $|x| \le a$, $|y| \le b$, and $|z| \le c$. The surface is symmetric with respect to each of the coordinate planes because the variables in the defining equation are squared.

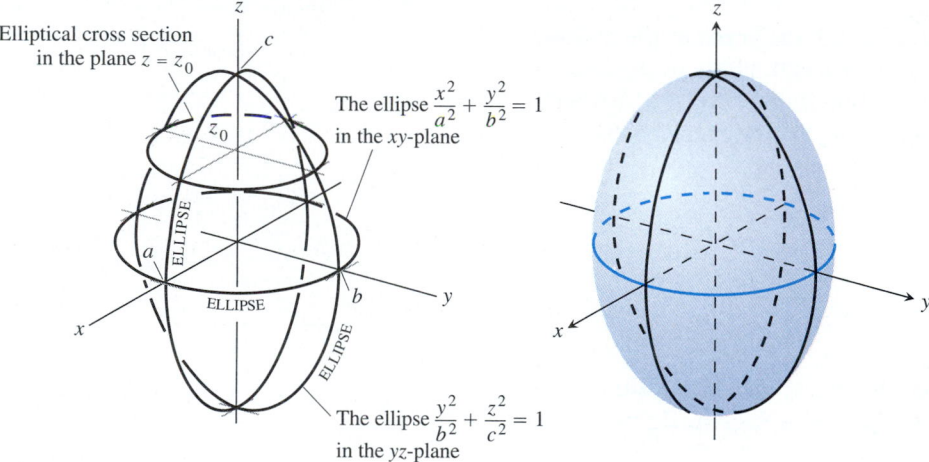

10.58   The ellipsoid
$$\frac{x^2}{a^2} + \frac{y^2}{b^2} + \frac{z^2}{c^2} = 1$$
in Example 2.

The curves in which the three coordinate planes cut the surface are ellipses. For example,

$$\frac{x^2}{a^2} + \frac{y^2}{b^2} = 1 \qquad \text{when} \qquad z = 0.$$

The section cut from the surface by the plane $z = z_0$, $|z_0| < c$, is the ellipse

$$\frac{x^2}{a^2(1 - (z_0/c)^2)} + \frac{y^2}{b^2(1 - (z_0/c)^2)} = 1 \tag{2}$$

(Exercise 70).

If any two of the semiaxes a, b, and c are equal, the surface is an ellipsoid of revolution. If all three are equal, the surface is a sphere.

Example 3   The **elliptic paraboloid**

$$\frac{x^2}{a^2} + \frac{y^2}{b^2} = \frac{z}{c} \tag{3}$$

is symmetric with respect to the planes $x = 0$ and $y = 0$ (Fig. 10.59). The only

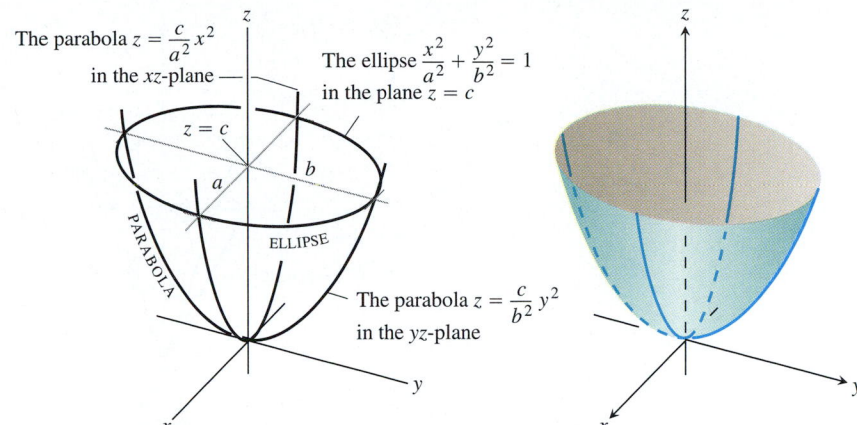

10.59 The elliptic paraboloid $(x^2/a^2) + (y^2/b^2) = z/c$ in Example 3. The cross sections perpendicular to the z-axis above the xy-plane are ellipses. The cross sections in the planes that contain the z-axis are parabolas.

intercept on the axes is the origin. Except for this point, the surface lies above the xy-plane because z is positive whenever either x or y is different from zero. The sections cut by the coordinate planes are

$$x = 0: \qquad \text{the parabola } z = \frac{c}{b^2}y^2,$$

$$y = 0: \qquad \text{the parabola } z = \frac{c}{a^2}x^2, \tag{4}$$

$$z = 0: \qquad \text{the point } (0, 0, 0).$$

Each plane $z = z_0$ above the xy-plane cuts the surface in the ellipse

$$\frac{x^2}{a^2} + \frac{y^2}{b^2} = \frac{z_0}{c}.$$

Example 4   The **circular paraboloid** or **paraboloid of revolution**

$$\frac{x^2}{a^2} + \frac{y^2}{a^2} = \frac{z}{c} \tag{5}$$

is obtained by taking $b = a$ in Eq. (3) for the elliptic paraboloid. The cross sections of the surface by planes perpendicular to the z-axis are circles centered on the z-axis. The cross sections by planes containing the z-axis are congruent parabolas with a common focus at the point $(0, 0, a^2/4c)$.

Shapes cut from circular paraboloids are used for antennas in radio telescopes, satellite trackers, and microwave radio links (Fig. 10.60).

10.60 Many antennas are shaped like pieces of paraboloids of revolution. (a) Radio telescopes use the same principles as optical telescopes. (b) A "rectangular-cut" radar reflector. (c) The profile of a horn antenna in a microwave radio link.

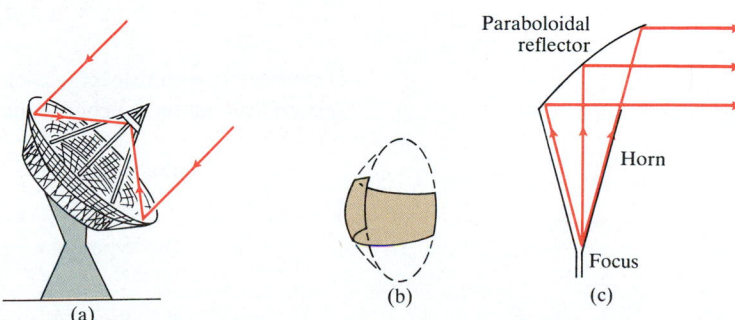

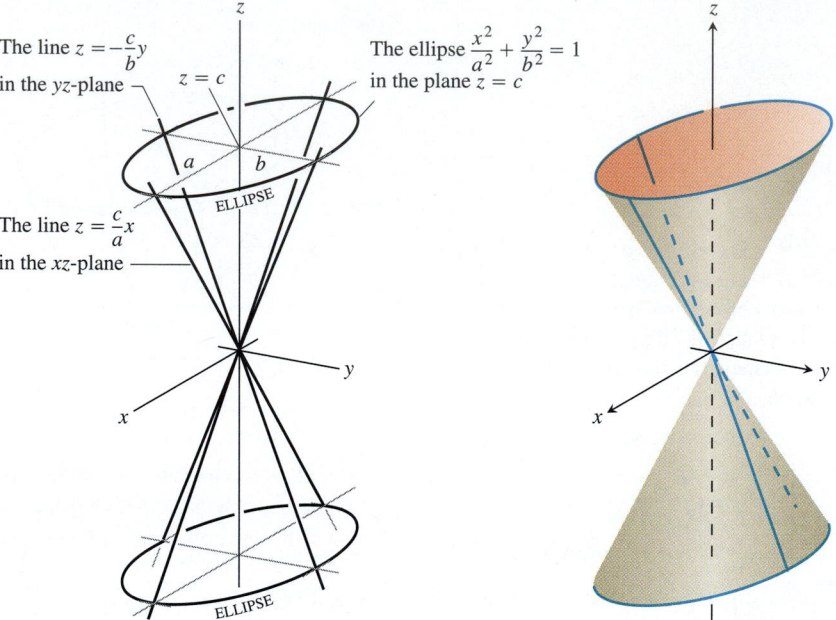

10.61 The elliptic cone $(x^2/a^2) + (y^2/b^2) = z^2/c^2$ in Example 5. Planes perpendicular to the z-axis cut the cone in ellipses above and below the xy-plane. Vertical planes that contain the z-axis cut it in pairs of intersecting lines.

Example 5   The **elliptic cone**

$$\frac{x^2}{a^2} + \frac{y^2}{b^2} = \frac{z^2}{c^2} \qquad (6)$$

is symmetric with respect to the three coordinate planes (Fig. 10.61). The sections cut by the coordinate planes are

$$x = 0: \qquad \text{the lines } z = \pm\frac{c}{b}\, y, \qquad (7)$$

$$y = 0: \qquad \text{the lines } z = \pm\frac{c}{a}\, x, \qquad (8)$$

$$z = 0: \qquad \text{the point } (0, 0, 0).$$

The sections cut by planes $z = z_0$ above and below the xy-plane are ellipses whose centers lie on the z-axis and whose vertices lie on the lines in Eqs. (7) and (8).

If $a = b$, the cone is a right circular cone.

Example 6   The **hyperboloid of one sheet**

$$\frac{x^2}{a^2} + \frac{y^2}{b^2} - \frac{z^2}{c^2} = 1 \qquad (9)$$

is symmetric with respect to each of the three coordinate planes (Fig. 10.62). The sections cut out by the coordinate planes are

$$x = 0: \qquad \text{the hyperbola } \frac{y^2}{b^2} - \frac{z^2}{c^2} = 1,$$

$$y = 0: \qquad \text{the hyperbola } \frac{x^2}{a^2} - \frac{z^2}{c^2} = 1, \qquad (10)$$

$$z = 0: \qquad \text{the ellipse } \frac{x^2}{a^2} + \frac{y^2}{b^2} = 1.$$

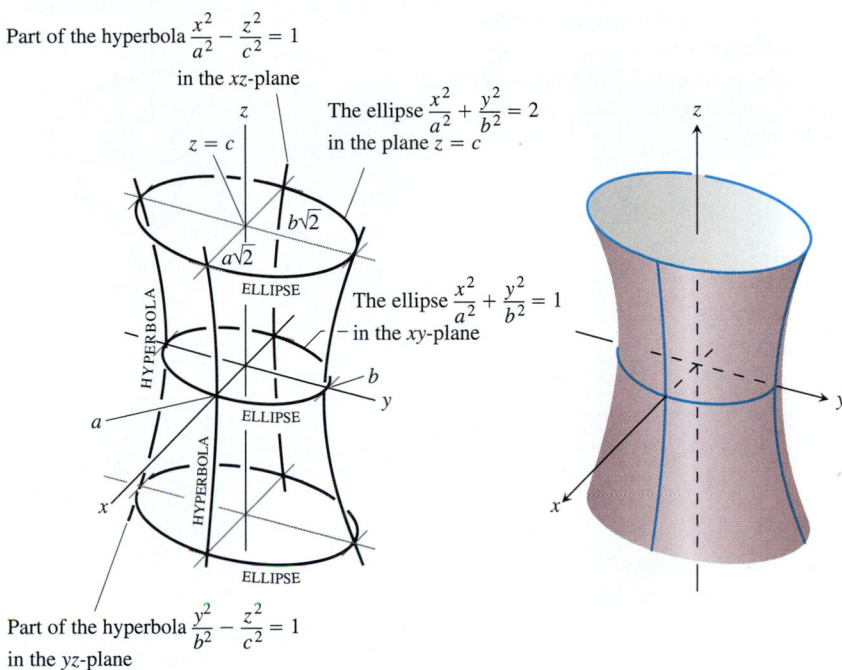

Part of the hyperbola $\dfrac{x^2}{a^2} - \dfrac{z^2}{c^2} = 1$ in the xz-plane

The ellipse $\dfrac{x^2}{a^2} + \dfrac{y^2}{b^2} = 2$ in the plane $z = c$

$z = c$

$b\sqrt{2}$

$a\sqrt{2}$

ELLIPSE

The ellipse $\dfrac{x^2}{a^2} + \dfrac{y^2}{b^2} = 1$ in the xy-plane

HYPERBOLA

b

ELLIPSE

a

HYPERBOLA

ELLIPSE

Part of the hyperbola $\dfrac{y^2}{b^2} - \dfrac{z^2}{c^2} = 1$ in the yz-plane

10.62 The hyperboloid $(x^2/a^2) + (y^2/b^2) - (z^2/c^2) = 1$ in Example 6. Planes perpendicular to the z-axis cut it in ellipses. Vertical planes containing the z-axis cut it in hyperbolas.

The plane $z = z_0$ cuts the surface in an ellipse with center on the z-axis and vertices on one of the hyperbolas in (10).

The surface is connected, meaning that it is possible to travel from one point on it to any other without leaving the surface. For this reason, it is said to have *one* sheet, in contrast to the hyperboloid in the next example, which has two sheets.

If $a = b$, the hyperboloid is a surface of revolution.

Example 7   The **hyperboloid of two sheets**

$$\frac{z^2}{c^2} - \frac{x^2}{a^2} - \frac{y^2}{b^2} = 1 \tag{11}$$

is symmetric with respect to the three coordinate planes (Fig. 10.63). The plane $z = 0$ does not intersect the surface; in fact, for a horizontal plane to intersect the surface, we must have $|z| \geq c$. The hyperbolic sections

$$x = 0: \quad \frac{z^2}{c^2} - \frac{y^2}{b^2} = 1, \qquad y = 0: \quad \frac{z^2}{c^2} - \frac{x^2}{a^2} = 1,$$

have their vertices and foci on the z-axis. The surface is separated into two portions, one above the plane $z = c$ and the other below the plane $z = -c$. This accounts for its name.

Equations (9) and (11) have different numbers of negative terms. The number in each case is the same as the number of sheets of the hyperboloid. If we compare with Eq. (6), we see that replacing the 1 on the right side of either Eq. (9) or (11) by zero gives the equation of a cone. If we replace the 1 on the right side of either Eq. (9) or Eq. (11) by 0, we obtain the equation

$$\frac{x^2}{a^2} + \frac{y^2}{b^2} - \frac{z^2}{c^2} = 0$$

10.63   The hyperboloid $(z^2/c^2) - (x^2/a^2) - (y^2/b^2) = 1$ in Example 7. Planes perpendicular to the z-axis above and below the vertices cut it in ellipses. Vertical planes containing the z-axis cut it in hyperbolas.

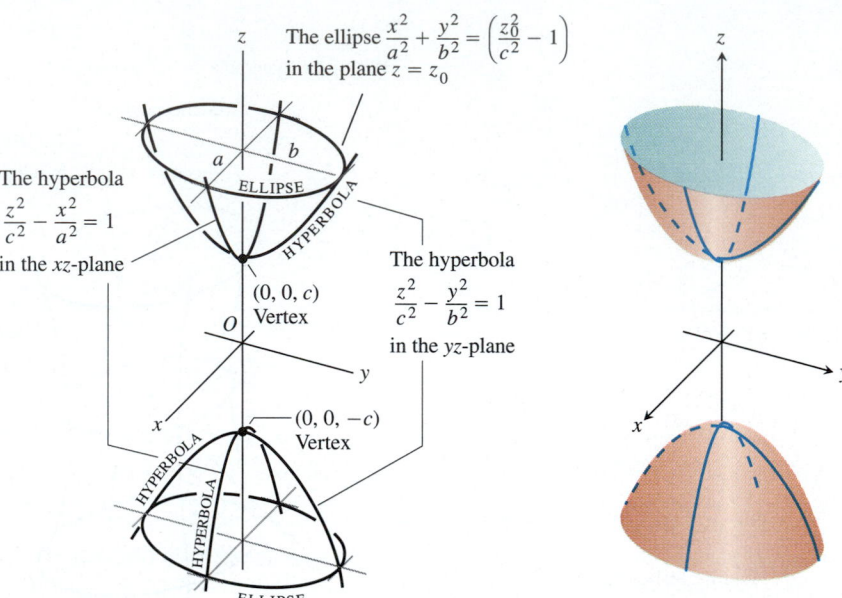

The ellipse $\dfrac{x^2}{a^2} + \dfrac{y^2}{b^2} = \left(\dfrac{z_0^2}{c^2} - 1\right)$ in the plane $z = z_0$

The hyperbola $\dfrac{z^2}{c^2} - \dfrac{x^2}{a^2} = 1$ in the xz-plane

ELLIPSE

HYPERBOLA

The hyperbola $\dfrac{z^2}{c^2} - \dfrac{y^2}{b^2} = 1$ in the yz-plane

$(0, 0, c)$ Vertex

$(0, 0, -c)$ Vertex

HYPERBOLA

HYPERBOLA

ELLIPSE

for an elliptic cone (Eq. 6). The hyperboloids are asymptotic to this cone (Fig. 10.64) in the same way that the hyperbolas

$$\frac{x^2}{a^2} - \frac{y^2}{b^2} = \pm 1$$

are asymptotic to the lines

$$\frac{x^2}{a^2} - \frac{y^2}{b^2} = 0$$

in the xy-plane.

Example 8   The **hyperbolic paraboloid**

$$\frac{y^2}{b^2} - \frac{x^2}{a^2} = \frac{z}{c} \tag{12}$$

has symmetry with respect to the planes $x = 0$ and $y = 0$ (Fig. 10.65). The sections in these planes are

$$x = 0: \qquad \text{the parabola } z = \frac{c}{b^2} y^2, \tag{13}$$

$$y = 0: \qquad \text{the parabola } z = -\frac{c}{a^2} x^2. \tag{14}$$

In the plane $x = 0$, the parabola opens upward from the origin. The parabola in the plane $y = 0$ opens downward.

If we cut the surface by a plane $z = z_0 > 0$, the section is a hyperbola,

$$\frac{y^2}{b^2} - \frac{x^2}{a^2} = \frac{z_0}{c}, \tag{15}$$

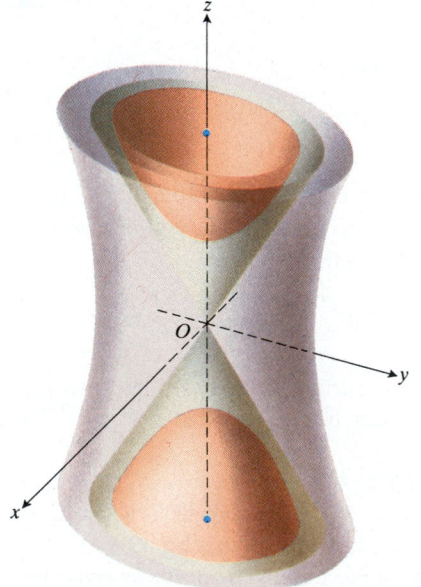

10.64   Both hyperboloids are asymptotic to the cone (Example 7).

with its focal axis parallel to the y-axis and its vertices on the parabola in (13). If z_0

DRAWING LESSON

How to Draw Quadric Surfaces

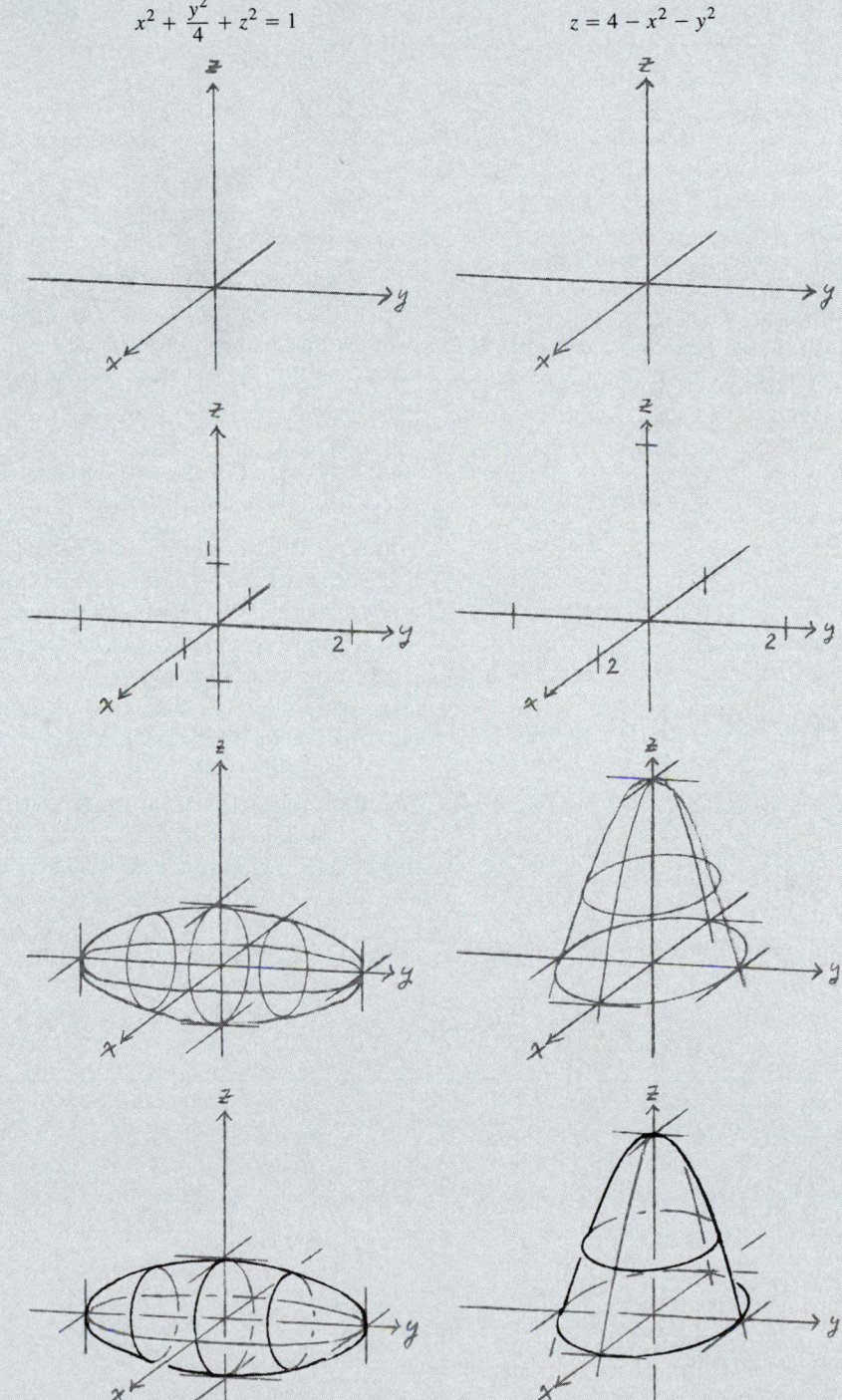

$$x^2 + \frac{y^2}{4} + z^2 = 1 \qquad\qquad z = 4 - x^2 - y^2$$

1. Lightly sketch the three coordinate axes.

2. Decide on a scale and mark the intercepts on the axes.

3. Sketch cross sections in the coordinate planes and in a few parallel planes, but don't clutter the picture. Use tangent lines as guides.

4. If more is required, darken the parts exposed to view. Leave the rest light. Use line breaks when you can.

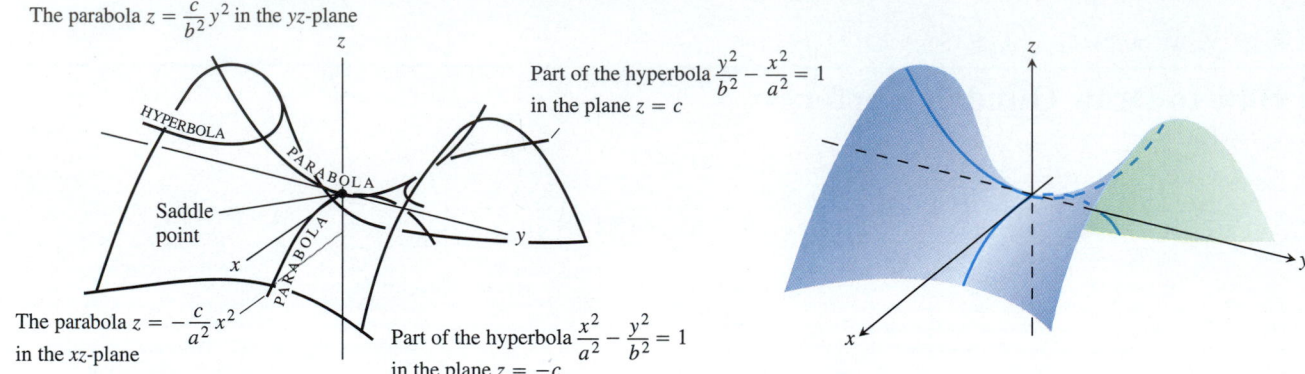

The parabola $z = \dfrac{c}{b^2} y^2$ in the yz-plane

HYPERBOLA

PARABOLA

Saddle point

PARABOLA

The parabola $z = -\dfrac{c}{a^2} x^2$ in the xz-plane

Part of the hyperbola $\dfrac{y^2}{b^2} - \dfrac{x^2}{a^2} = 1$ in the plane $z = c$

Part of the hyperbola $\dfrac{x^2}{a^2} - \dfrac{y^2}{b^2} = 1$ in the plane $z = -c$

10.65 The hyperbolic paraboloid $(y^2/b^2) - (x^2/a^2) = z/c$ in Example 8. The cross sections in planes perpendicular to the z-axis above and below the xy-plane are hyperbolas. The cross sections in planes perpendicular to the other axes are parabolas.

is negative, the focal axis is parallel to the x-axis and the vertices lie on the parabola in (14).

Near the origin, the surface is shaped like a saddle. To a person traveling along the surface in the yz-plane, the origin looks like a minimum. To a person traveling in the xz-plane, the origin looks like a maximum. Such a point is called a **minimax** or **saddle point** of a surface. We shall discuss maximum and minimum points on surfaces in Chapter 12.

Liquid Mirror Telescopes

When a circular pan of liquid is rotated about its vertical axis, the surface of the liquid does not stay flat. Instead, it assumes the shape of a paraboloid of revolution, exactly what is needed for the primary mirror of a reflecting telescope. At the turn of the century, attempts to make reliable mirrors with revolving mercury failed because of surface ripples and focus losses caused by variations in the speed of rotation. Today, these difficulties can be overcome with synchronous motors driven by oscillator-stabilized power supplies and checked against constant clocks.

Using the same idea, astronomers at the Steward Observatory's Mirror Laboratory in Tucson, Arizona, have used a large heated spinning turntable to cast borosilicate glass blanks for lightweight mirrors. Spincast mirrors cost less than traditionally cast mirrors and can be made larger. Their shorter focal lengths also allow them to be installed in compact telescope frames that are less expensive to house and less likely to flex in strong winds.

EXERCISES 10.7

Sketch the surfaces in Exercises 1–64.

Cylinders

1. $x^2 + y^2 = 4$

2. $x^2 + z^2 = 4$

3. $z = y^2 - 1$

4. $x = y^2$

5. $x^2 + 4z^2 = 16$

6. $4x^2 + y^2 = 36$

7. $z^2 - y^2 = 1$

8. $yz = 1$

Ellipsoids

9. $9x^2 + y^2 + z^2 = 9$

10. $4x^2 + 4y^2 + z^2 = 16$

11. $4x^2 + 9y^2 + 4z^2 = 36$

12. $9x^2 + 4y^2 + 36z^2 = 36$

Paraboloids

13. $z = x^2 + 4y^2$

14. $z = x^2 + 9y^2$

15. $z = 8 - x^2 - y^2$

16. $z = 18 - x^2 - 9y^2$

17. $x = 4 - 4y^2 - z^2$ **18.** $y = 1 - x^2 - z^2$

Cones

19. $x^2 + y^2 = z^2$ **20.** $y^2 + z^2 = x^2$

21. $4x^2 + 9z^2 = 9y^2$ **22.** $9x^2 + 4y^2 = 36z^2$

Hyperboloids

23. $x^2 + y^2 - z^2 = 1$ **24.** $y^2 + z^2 - x^2 = 1$

25. $(y^2/4) + (z^2/9) - (x^2/4) = 1$

26. $(x^2/4) + (y^2/4) - (z^2/9) = 1$

27. $z^2 - x^2 - y^2 = 1$ **28.** $(y^2/4) - (x^2/4) - z^2 = 1$

29. $x^2 - y^2 - (z^2/4) = 1$ **30.** $(x^2/4) - y^2 - (z^2/4) = 1$

Hyperbolic Paraboloids

31. $y^2 - x^2 = z$ **32.** $x^2 - y^2 = z$

Assorted

33. $x^2 + y^2 + z^2 = 4$ **34.** $4x^2 + 4y^2 = z^2$

35. $z = 1 + y^2 - x^2$ **36.** $y^2 - z^2 = 4$

37. $y = -(x^2 + z^2)$ **38.** $z^2 - 4x^2 - 4y^2 = 4$

39. $16x^2 + 4y^2 = 1$ **40.** $z = x^2 + y^2 + 1$

41. $x^2 + y^2 - z^2 = 4$ **42.** $x = 4 - y^2$

43. $x^2 + z^2 = y$ **44.** $z^2 - (x^2/4) - y^2 = 1$

45. $x^2 + z^2 = 1$ **46.** $4x^2 + 4y^2 + z^2 = 4$

47. $16y^2 + 9z^2 = 4x^2$ **48.** $z = x^2 - y^2 - 1$

49. $9x^2 + 4y^2 + z^2 = 36$ **50.** $4x^2 + 9z^2 = y^2$

51. $x^2 + y^2 - 16z^2 = 16$ **52.** $z^2 + 4y^2 = 9$

53. $z = -(x^2 + y^2)$ **54.** $y^2 - x^2 - z^2 = 1$

55. $x^2 - 4y^2 = 1$ **56.** $z = 4x^2 + y^2 - 4$

57. $4y^2 + z^2 - 4x^2 = 4$ **58.** $z = 1 - x^2$

59. $x^2 + y^2 = z$ **60.** $(x^2/4) + y^2 - z^2 = 1$

61. $yz = -1$ **62.** $36x^2 + 9y^2 + 4z^2 = 36$

63. $9x^2 + 16y^2 = 4z^2$ **64.** $4z^2 - x^2 - y^2 = 4$

65. a) Express the area A of the cross section cut from the ellipsoid

$$x^2 + \frac{y^2}{4} + \frac{z^2}{9} = 1$$

by the plane $z = c$ as a function of c. (The area of an ellipse with semiaxes a and b is πab.)

b) Use slices perpendicular to the z-axis to find the volume of the ellipsoid in (a).

c) Now find the volume of the ellipsoid

$$\frac{x^2}{a^2} + \frac{y^2}{b^2} + \frac{z^2}{c^2} = 1.$$

Does your formula give the volume of a sphere of radius a if $a = b = c$?

66. A barrel has the shape of an ellipsoid with equal pieces cut from the ends by planes perpendicular to the z-axis (Fig. 10.66). The barrel is $2h$ units high, its midsection radius is R, and its end radii are both r. Find a formula for the barrel's volume. Then check two things. First, suppose the sides of the barrel are straightened to turn the barrel into a cylinder of radius R and height $2h$. Does your formula give the cylinder's volume? Second, suppose $r = 0$ and $h = R$ so the barrel is a sphere. Does your formula give the sphere's volume?

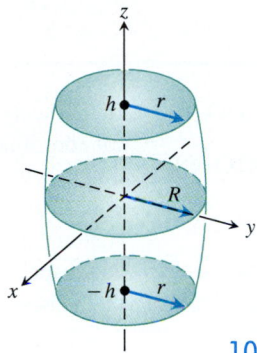

10.66   The barrel in Exercise 66.

67. Show that the volume of the segment cut from the paraboloid

$$\frac{x^2}{a^2} + \frac{y^2}{b^2} = \frac{z}{c}$$

by the plane $z = h$ equals half the segment's base times its altitude. (Figure 10.59 shows the segment for the special case $h = c$.)

68. a) Find the volume of the solid bounded by the hyperboloid

$$\frac{x^2}{a^2} + \frac{y^2}{b^2} - \frac{z^2}{c^2} = 1$$

and the planes $z = 0$ and $z = h$, $h > 0$.

b) Express your answer in (a) in terms of h and the areas A_0 and A_h of the regions cut by the hyperboloid from the planes $z = 0$ and $z = h$.

c) Show that the volume in (a) is also given by the formula

$$V = \frac{h}{6}\left(A_0 + 4A_m + A_h\right),$$

where A_m is the area of the region cut by the hyperboloid from the plane $z = h/2$.

69. *Cross sections of quadric surfaces cut by planes perpendicular to the coordinate axes are always conic sections.* The general equation for a quadric surface is

$$Gx^2 + Hy^2 + Iz^2 + Jxy + Kyz + Lxz +$$
$$Mx + Ny + Pz + Q = 0. \quad (16)$$

Use this equation together with information from Section 9.2 to show that the cross section of any quadric surface by a plane perpendicular to one of the coordinate axes is a conic section.

70. Verify Eq. (2).

Computer Grapher

If you have access to a 3-D computer grapher, try graphing the surfaces in Exercises 71–77.

71. $z = y^2$

72. $z = 1 - y^2$

73. $z = x^2 + y^2$

74. $z = x^2 + 2y^2$

75. $z = \sqrt{1 - x^2}$ (upper half of a circular cylinder)

76. $z = \sqrt{1 - (y^2/4)}$ (upper half of an elliptical cylinder)

77. $z = \sqrt{x^2 + 2y^2 + 4}$ (one sheet of an elliptic hyperboloid)

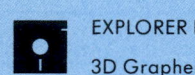

EXPLORER PROGRAM

3D Grapher   Graphs equations of the form
                    $z = f(x, y)$

10.8 Cylindrical and Spherical Coordinates

In this section we introduce two new systems of coordinates for space: the cylindrical coordinate system and the spherical coordinate system. In the cylindrical coordinate system, cylinders whose axes lie along the z-axis and planes that contain the z-axis have especially simple equations. In the spherical coordinate system, spheres centered at the origin and single cones at the origin whose axes lie along the z-axis have especially simple equations. When your work involves these shapes, these may be the best coordinate systems to use, as we shall see in later chapters.

Cylindrical Coordinates

We obtain cylindrical coordinates for space by combining polar coordinates in the xy-plane with the usual z-axis. This assigns to every point in space one or more coordinate triples of the form (r, θ, z), as shown in Fig. 10.67.

The values of x, y, r, and θ in cylindrical coordinates are related by the usual equations:

$$x = r \cos \theta, \qquad r^2 = x^2 + y^2, \qquad y = r \sin \theta, \qquad \tan \theta = y/x. \qquad (1)$$

We shall use cylindrical coordinates to study planetary motion in Section 11.5.

In cylindrical coordinates, the equation $r = a$ describes not just a circle in the xy-plane but an entire cylinder about the z-axis (Fig. 10.68). The z-axis itself is given by the equation $r = 0$. The equation $\theta = \theta_0$ describes the plane that contains the z-axis and makes an angle of θ_0 radians with the positive x-axis.

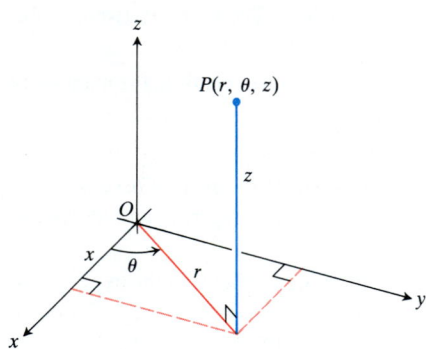

10.67 The cylindrical coordinates of a point in space are r, θ, and z.

Example 1 Describe the points in space whose cylindrical coordinates satisfy the equations

$$r = 2, \qquad \theta = \frac{\pi}{4}.$$

Solution These points make up the line in which the cylinder $r = 2$ cuts the portion of the plane $\theta = \pi/4$ in which r is positive (Fig. 10.69). This is the line through the point $(2, \pi/4, 0)$ parallel to the z-axis.

Example 2 Sketch the surface $r = 1 + \cos \theta$.

Solution The equation involves only r and θ; the coordinate variable z is missing. Therefore, the surface is a cylinder of lines that pass through the cardioid $r = 1 + \cos \theta$ in the $r\theta$-plane and lie parallel to the z-axis. The rules for sketching

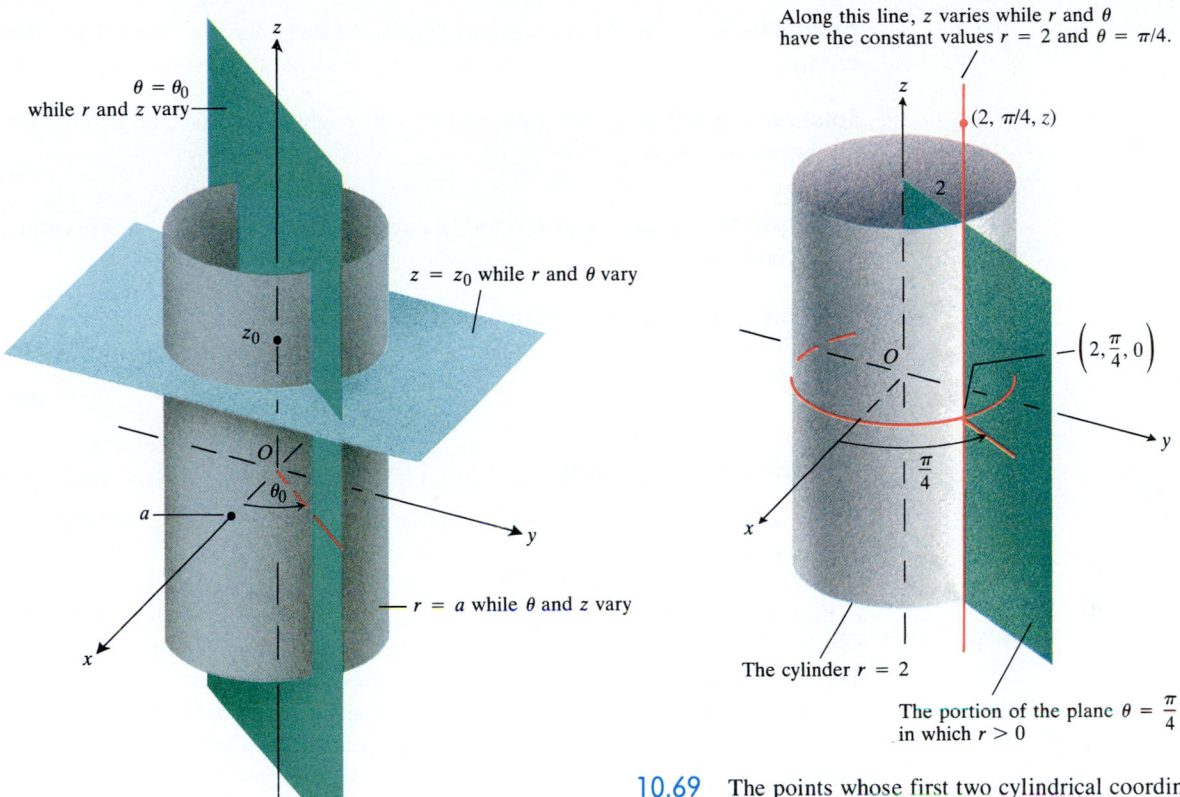

$\theta = \theta_0$ while r and z vary

$z = z_0$ while r and θ vary

z_0

θ_0

a

$r = a$ while θ and z vary

Along this line, z varies while r and θ have the constant values $r = 2$ and $\theta = \pi/4$.

$(2, \pi/4, z)$

$\left(2, \dfrac{\pi}{4}, 0\right)$

$\dfrac{\pi}{4}$

The cylinder $r = 2$

The portion of the plane $\theta = \dfrac{\pi}{4}$ in which $r > 0$

10.69   The points whose first two cylindrical coordinates are $r = 2$ and $\theta = \pi/4$ form a line parallel to the z-axis (Example 1).

10.68   Planes and cylinders that have constant-coordinate equations in cylindrical coordinates.

the cylinder are the same as always: sketch the x-, y-, and z-axes, draw a few perpendicular cross sections, connect the cross sections with parallel lines, and darken the exposed parts (Fig. 10.70).

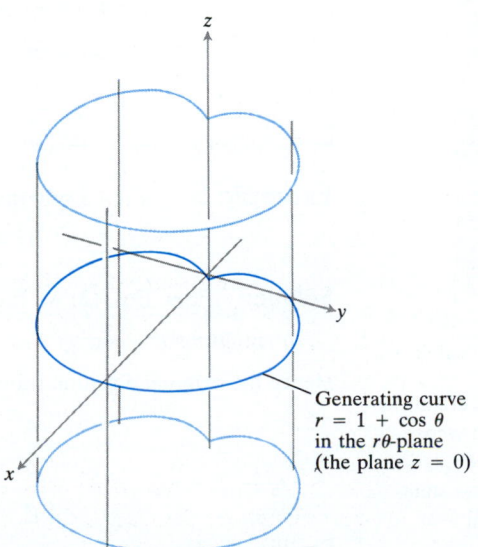

Generating curve $r = 1 + \cos\theta$ in the $r\theta$-plane (the plane $z = 0$)

10.70   The cylindrical-coordinate equation $r = 1 + \cos\theta$ defines a cylinder in space whose cross sections perpendicular to the z-axis are cardioids (Example 2).

Example 3
Find a Cartesian equation for the surface $z = r^2$ and identify the surface.

Solution   From Eqs. (1) we have $z = r^2 = x^2 + y^2$. The surface is the circular paraboloid $x^2 + y^2 = z$.

Example 4
Find an equation for the circular cylinder $4x^2 + 4y^2 = 9$ in cylindrical coordinates.

Solution   The cylinder consists of the points whose distance from the z-axis is $\sqrt{x^2 + y^2} = 3/2$. The corresponding equation in cylindrical coordinates is $r = 3/2$.

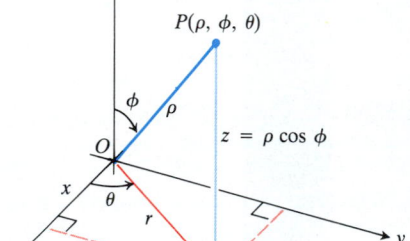

10.71   The spherical coordinates ρ, ϕ, and θ and their relation to x, y, z, and r.

Spherical Coordinates

Spherical coordinates locate points in space with two angles and a distance, as shown in Fig. 10.71.

The first coordinate, $\rho = |\overrightarrow{OP}|$, is the point's distance from the origin. Unlike r, the variable ρ is never negative. The second coordinate, ϕ, is the angle the vector $\overrightarrow{OP}$ makes with the positive z-axis. It is required to lie in the interval from 0 to π. The third coordinate is the angle θ from cylindrical coordinates.

The equation $\rho = a$ describes the sphere of radius a centered at the origin (Fig. 10.72). The equation $\phi = \phi_0$ describes a single cone whose vertex lies at the origin and whose axis lies along the z-axis. (We have to broaden our interpretation here to include the xy-plane as the cone $\phi = \pi/2$.) If ϕ_0 is greater than $\pi/2$, the cone $\phi = \phi_0$ opens downward.

A few books give spherical coordinates in the order (ρ, θ, ϕ), with the θ and ϕ reversed. Watch out for this when you read elsewhere.

Selected Equations Relating Cartesian (Rectangular), Cylindrical, and Spherical Coordinates

$$r = \rho \sin \phi, \qquad x = r \cos \theta = \rho \sin \phi \cos \theta,$$

$$z = \rho \cos \phi, \qquad y = r \sin \theta = \rho \sin \phi \sin \theta, \qquad (2)$$

$$\rho = \sqrt{x^2 + y^2 + z^2} = \sqrt{r^2 + z^2}$$

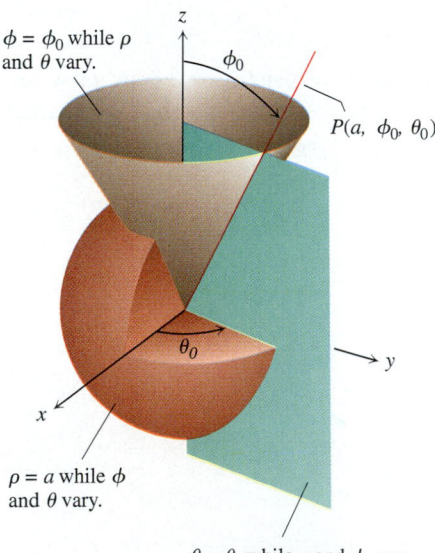

$\phi = \phi_0$ while ρ and θ vary.

$P(a, \phi_0, \theta_0)$

$\rho = a$ while ϕ and θ vary.

$\theta = \theta_0$ while ρ and ϕ vary.

10.72   Spheres whose centers are at the origin, single cones at the origin whose axes lie along the z-axis, and half-planes "hinged" along the z-axis have constant-coordinate equations in spherical coordinates.

Example 5
Find a spherical coordinate equation for the sphere
$$x^2 + y^2 + (z - 1)^2 = 1.$$

Solution   From Eqs. (2) we find that the left side of the equation is
$$\rho^2 \sin^2 \phi (\cos^2 \theta + \sin^2 \theta) + \rho^2 \cos^2 \phi - 2\rho \cos \phi + 1 = \rho^2 - 2\rho \cos \phi + 1.$$
Hence the original equation transforms into
$$\rho^2 - 2\rho \cos \phi + 1 = 1,$$
$$\rho^2 = 2\rho \cos \phi,$$
$$\rho = 2 \cos \phi.$$
See Fig. 10.73.

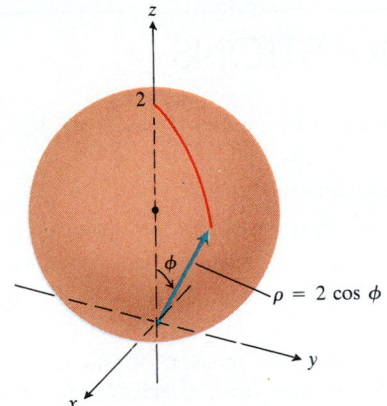

10.73 The sphere $\rho = 2 \cos \phi$ in Example 5. Notice that the equation restricts ϕ to lie in the interval $0 \leq \phi \leq \pi/2$ because the spherical coordinate ρ is not allowed to be negative.

EXERCISES 10.8

The following table gives the coordinates of specific points in space in one of three coordinate systems. In Exercises 1–10, find coordinates for each point in the other two systems. There may be more than one right answer because points in cylindrical and spherical coordinates may have more than one coordinate triple.

Rectangular (x, y, z)	Cylindrical (r, θ, z)	Spherical (ρ, ϕ, θ)
1. $(0, 0, 0)$		
2. $(1, 0, 0)$		
3. $(0, 1, 0)$		
4. $(0, 0, 1)$		
5.	$(1, 0, 0)$	
6.	$(\sqrt{2}, 0, 1)$	
7.	$(1, \pi/2, 1)$	
8.		$(\sqrt{3}, \pi/3, -\pi/2)$
9.		$(2\sqrt{2}, \pi/2, 3\pi/2)$
10.		$(\sqrt{2}, \pi, 3\pi/2)$

In Exercises 11–26, translate the equations from the given coordinate system (rectangular, cylindrical, spherical) into equations in the other two systems. Also, identify the set of points defined by the equation.

11. $r = 0$

12. $x^2 + y^2 = 5$

13. $z = 0$

14. $z = -2$

15. $\rho \cos \phi = 3$

16. $\sqrt{x^2 + y^2} = z$

17. $\rho \sin \phi \cos \theta = 0$

18. $\tan^2 \phi = 1$

19. $x^2 + y^2 + z^2 = 4$

20. $\rho = 6 \cos \phi$

21. $z = r^2 \cos 2\theta$

22. $z^2 - r^2 = 1$

23. $r = \csc \theta$

24. $z = x^2 + y^2$

25. $3 \tan^2 \phi = 1$

26. $\rho^2 \cos 2\phi = -1$

In Exercises 27–34, describe the sets of points in space whose cylindrical coordinates satisfy the given equations or pairs of equations. Sketch.

27. $r = 4$

28. $r^2 + z^2 = 1$

29. $r = 1 - \cos \theta$

30. $r = 2 \cos \theta$

31. $r = 2, \quad z = 3$

32. $\theta = \pi/6, \quad z = r$

33. $r = 3, \quad z = \theta/2$

34. $r^2 = \cos 2\theta$

35. Find the rectangular coordinates of the center of the sphere
$$r^2 + z^2 = 4r \cos \theta + 6r \sin \theta + 2z.$$

36. What symmetry will you find in a surface whose spherical coordinate equation has the form $\rho = f(\phi)$ (independent of θ)?

In Exercises 37–44, describe the sets of points in space whose spherical coordinates satisfy the given equations or pairs of equations. Sketch.

37. $\phi = \pi/6$

38. $\rho = 6, \quad \phi = \pi/6$

39. $\rho = 5, \quad \theta = \pi/4$

40. $\theta = \pi/4, \quad \phi = \pi/4$

41. $\rho = \cos \phi$

42. $\rho = 1 - \cos \phi$

43. $\rho = \sin \phi$

44. $\theta = \pi/2, \quad \rho = 4 \sin \phi$

REVIEW QUESTIONS

1. When do two directed line segments represent the same vector?

2. How are vectors added and subtracted?

3. How are the length and direction of a vector calculated?

4. If a vector is multiplied by a positive scalar, how is the result related to the original vector? What if the scalar is zero? Negative?

5. Define the *scalar product (dot product)* of two vectors. Which algebraic laws (commutative, associative, distributive) are satisfied by dot products and which, if any, are not? Give examples. When is the scalar product of two vectors equal to zero?

6. What is the vector projection of a vector **B** onto a vector **A**? How do you write **B** as the sum of a vector parallel to **A** and a vector orthogonal to **A**?

7. Define the *vector product (cross product)* of two vectors. Which algebraic laws (commutative, associative, distributive) are satisfied by cross products and which are not? Give examples. When is the vector product of two vectors equal to zero?

8. What is the determinant formula for evaluating the cross product of two vectors? Use it in an example.

9. How are vector and scalar products used to find equations for lines, line segments, planes? Give examples.

10. How can vectors be used to calculate the distance between a point and a line? A point and a plane? Give examples.

11. What is the geometric interpretation of $|(\mathbf{A} \times \mathbf{B}) \cdot \mathbf{C}|$ as a volume? How may the product be calculated from the components of **A**, **B**, and **C**? Give an example.

12. What theorem do you know about the areas of orthogonal projections of parallelograms on planes? Give an example.

13. What formulas are available for expressing $(\mathbf{A} \times \mathbf{B}) \times \mathbf{C}$ and $\mathbf{A} \times (\mathbf{B} \times \mathbf{C})$ as scalar multiples of **A**, **B**, and **C**? Give examples.

14. What is a cylinder? Give examples of equations that define cylinders in Cartesian coordinates; in cylindrical coordinates. What advice can you give about drawing cylinders?

15. Give examples of ellipsoids, paraboloids, cones, and hyperboloids (equations and sketches). What advice can you give about sketching these surfaces?

16. How are cylindrical and spherical coordinates defined? Draw diagrams that show how cylindrical and spherical coordinates are related to rectangular coordinates. What sets have constant-coordinate equations (like $x = 1$, $r = 1$, or $\phi = \pi/3$) in the three coordinate systems?

MISCELLANEOUS EXERCISES

1. Draw the unit vectors $\mathbf{u} = (\cos \theta)\,\mathbf{i} + (\sin \theta)\,\mathbf{j}$ for $\theta = 0$, $\pi/2$, $2\pi/3$, $5\pi/4$, and $5\pi/3$, together with the coordinate axes and unit circle.

2. Find the unit vector obtained by rotating
   a) **i** clockwise 45°   b) **j** counterclockwise 120°

In Exercises 3 and 4, find the unit vectors that are tangent and normal to the curve at point P.

3. $y = \tan x$, $P(\pi/4, 1)$ 4. $x^2 + y^2 = 25$, $P(3, 4)$

Express the vectors in Exercises 5–8 in terms of their lengths and directions.

5. $\sqrt{2}\,\mathbf{i} + \sqrt{2}\,\mathbf{j}$ 6. $-\mathbf{i} - \mathbf{j}$

7. $2\,\mathbf{i} - 3\,\mathbf{j} + 6\,\mathbf{k}$ 8. $\mathbf{i} + 2\,\mathbf{j} - \mathbf{k}$

9. Use vectors to prove that
$$(a^2 + b^2)(c^2 + d^2) \geq (ac + bd)^2$$
   for any four numbers a, b, c, and d. (*Hint:* Let $\mathbf{A} = a\,\mathbf{i} + b\,\mathbf{j}$ and $\mathbf{B} = c\,\mathbf{i} + d\,\mathbf{j}$.)

10. In Fig. 10.74, D is the midpoint of side AB of triangle ABC, and E is one-third of the way between C and B. Use vectors to prove that F is the midpoint of line segment CD.

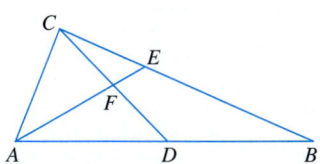

10.74   The triangle in Exercise 10.

11. The initial points of the vectors $2\,\mathbf{i} + 3\,\mathbf{j}$, $4\,\mathbf{i} + \mathbf{j}$, and $5\,\mathbf{i} + y\,\mathbf{j}$ lie at the origin. Find the value of y that makes them terminate on a common straight line.

12. Suppose that x and y are positive numbers whose sum is 1 and that $\mathbf{A} = \overrightarrow{OA}$ and $\mathbf{B} = \overrightarrow{OB}$ are vectors from the origin to

points A and B. Show that the point P determined by the vector $\overrightarrow{OP} = x\mathbf{A} + y\mathbf{B}$ lies on the line segment AB.

In Exercises 13 and 14, find $|\mathbf{A}|$, $|\mathbf{B}|$, $\mathbf{A} \cdot \mathbf{B}$, $\mathbf{B} \cdot \mathbf{A}$, $\mathbf{A} \times \mathbf{B}$, $\mathbf{B} \times \mathbf{A}$, $|\mathbf{A} \times \mathbf{B}|$, the acute angle between $\mathbf{A}$ and $\mathbf{B}$, the scalar component of $\mathbf{B}$ in the direction of $\mathbf{A}$, and the vector projection of $\mathbf{B}$ onto $\mathbf{A}$.

13. $\mathbf{A} = \mathbf{i} + \mathbf{j}$,
$\mathbf{B} = 2\mathbf{i} + \mathbf{j} - 2\mathbf{k}$

14. $\mathbf{A} = 5\mathbf{i} + \mathbf{j} + \mathbf{k}$,
$\mathbf{B} = \mathbf{i} - 2\mathbf{j} + 3\mathbf{k}$

In Exercises 15 and 16, write $\mathbf{B}$ as the sum of a vector parallel to $\mathbf{A}$ and a vector orthogonal to $\mathbf{A}$.

15. $\mathbf{A} = 2\mathbf{i} + \mathbf{j} - \mathbf{k}$,
$\mathbf{B} = \mathbf{i} + \mathbf{j} - 5\mathbf{k}$

16. $\mathbf{A} = \mathbf{i} - 2\mathbf{j}$,
$\mathbf{B} = \mathbf{i} + \mathbf{j} + \mathbf{k}$

In Exercises 17 and 18, draw coordinate axes and then sketch $\mathbf{A}$, $\mathbf{B}$, and $\mathbf{A} \times \mathbf{B}$ as vectors at the origin.

17. $\mathbf{A} = \mathbf{i}$, $\mathbf{B} = \mathbf{i} + \mathbf{j}$

18. $\mathbf{A} = \mathbf{i} - \mathbf{j}$, $\mathbf{B} = \mathbf{i} + \mathbf{j}$

19. *Work.*   Find the work done in pushing a car 800 ft with a force of magnitude 70 lb directed 30° downward from the horizontal against the back of the car.

20. *Torque.*   The operator's manual for the Toro 21-in. lawn mower says "tighten the spark plug to 15 ft · lb (20.4 N · m)." If you are installing the plug with an 11-in. socket wrench that places the center of your hand 9 in. from the axis of the spark plug, about how hard should you pull? Answer in pounds.

21. Show that $|\mathbf{A} + \mathbf{B}| \leq |\mathbf{A}| + |\mathbf{B}|$ for any vectors $\mathbf{A}$ and $\mathbf{B}$.

22. Suppose that vectors $\mathbf{A}$ and $\mathbf{B}$ are not parallel and that $\mathbf{A} = \mathbf{C} + \mathbf{D}$, where $\mathbf{C}$ is parallel to $\mathbf{B}$ and $\mathbf{D}$ is orthogonal to $\mathbf{B}$. Express $\mathbf{C}$ and $\mathbf{D}$ in terms of $\mathbf{A}$ and $\mathbf{B}$.

23. Show that $\mathbf{C} = |\mathbf{B}|\mathbf{A} + |\mathbf{A}|\mathbf{B}$ bisects the angle between $\mathbf{A}$ and $\mathbf{B}$.

24. Show that $|\mathbf{B}|\mathbf{A} + |\mathbf{A}|\mathbf{B}$ and $|\mathbf{B}|\mathbf{A} - |\mathbf{A}|\mathbf{B}$ are orthogonal.

25. *Dot multiplication is positive definite.*   Show that dot multiplication of vectors is *positive definite;* that is, show that $\mathbf{A} \cdot \mathbf{A} \geq 0$ for every vector $\mathbf{A}$ and that $\mathbf{A} \cdot \mathbf{A} = 0$ if and only if $\mathbf{A} = \mathbf{0}$.

26. By forming the cross product of two appropriate vectors, derive the trigonometric identity
$$\sin(A - B) = \sin A \cos B - \cos A \sin B.$$

In Exercises 27 and 28, find the distance from the point to the line in the xy-plane.

27. $(3, 2)$, $3x + 4y = 2$

28. $(-1, 1)$, $5x - 12y = 9$

In Exercises 29 and 30, find the distance from the point to the line.

29. $(2, 2, 0)$; $x = -t$, $y = t$, $z = -1 + t$

30. $(0, 4, 1)$; $x = 2 + t$, $y = 2 + t$, $z = t$

31. Find parametric equations for the line that passes through the point $(1, 2, 3)$ parallel to the vector $\mathbf{v} = -3\mathbf{i} + 7\mathbf{k}$.

32. Find parametric equations for the line segment joining the points $P(1, 2, 0)$ and $Q(1, 3, -1)$.

In Exercises 33 and 34, find the distance from the point to the plane.

33. $(6, 0, -6)$, $x - y = 4$

34. $(3, 0, 10)$, $2x + 3y + z = 2$

35. Find an equation for the plane that passes through the point $(3, -2, 1)$ normal to $\mathbf{N} = 2\mathbf{i} + \mathbf{j} - \mathbf{k}$.

36. Find an equation for the plane that passes through the point $(-1, 6, 0)$ perpendicular to the line $x = -1 + t$, $y = 6 - 2t$, $z = 3t$.

37. The equation $\mathbf{N} \cdot \overrightarrow{P_0P} = 0$ represents the plane through P_0 normal to $\mathbf{N}$. What set does the inequality $\mathbf{N} \cdot \overrightarrow{P_0P} > 0$ represent?

38. Find the angle between the planes $x = 7$ and $x + y + \sqrt{2}z = -3$.

39. Find the angle between the planes $x + y = 1$ and $y + z = 1$.

40. Find the distance from the point $P(1, 4, 0)$ to the plane through the points $A(0, 0, 0)$, $B(2, 0, -1)$, and $C(2, -1, 0)$.

41. Find the distance from the point $(2, 2, 3)$ to the plane $2x + 3y + 5z = 0$.

42. Find a vector parallel to the plane $2x - y - z = 4$ and orthogonal to the vector $\mathbf{i} + \mathbf{j} + \mathbf{k}$.

43. Find a unit vector orthogonal to $\mathbf{A}$ in the plane of $\mathbf{B}$ and $\mathbf{C}$ if $\mathbf{A} = 2\mathbf{i} - \mathbf{j} + \mathbf{k}$, $\mathbf{B} = \mathbf{i} + 2\mathbf{j} + \mathbf{k}$, and $\mathbf{C} = \mathbf{i} + \mathbf{j} - 2\mathbf{k}$.

44. Find a vector of magnitude 2 parallel to the line of intersection of the planes $x + 2y + z - 1 = 0$ and $x - y + 2z + 7 = 0$.

45. Find the point in which the line through the origin perpendicular to the plane $2x - y - z = 4$ meets the plane $3x - 5y + 2z = 6$.

46. Find parametric equations for the line in which the planes $x + 2y + z = 1$ and $x - y + 2z = -8$ intersect.

47. Show that the line in which the planes
$$x + 2y - 2z = 5 \quad \text{and} \quad 5x - 2y - z = 0$$
intersect is parallel to the line
$$x = -3 + 2t, \quad y = 3t, \quad z = 1 + 4t.$$

48. Find the point in which the line through $P(3, 2, 1)$ normal to the plane $2x - y + 2z = -2$ meets the plane.

49. What angle does the line of intersection of the planes $2x + y - z = 0$ and $x + y + 2z = 0$ make with the positive x-axis?

50. The line

$$L: \quad x = 3 + 2t, \quad y = 2t, \quad z = t$$

intersects the plane $x + 3y - z = -4$ in a point P. Find the coordinates of P and find equations for the line through P perpendicular to L.

51. Show that for every real number k the plane

$$x - 2y + z + 3 + k(2x - y - z + 1) = 0$$

contains the line of intersection of the planes

$$x - 2y + z + 3 = 0 \quad \text{and} \quad 2x - y - z + 1 = 0.$$

52. Find an equation for the plane through $A(-2, 0, -3)$ and $B(1, -2, 1)$ that lies parallel to the line through $C(-2, -13/5, 26/5)$ and $D(16/5, -13/5, 0)$.

53. *Submarine hunting.* Two surface ships on maneuvers are trying to determine a submarine's course and speed to prepare for an aircraft intercept. As shown in Fig. 10.75, ship A is located at $(4, 0, 0)$ while ship B is located at $(0, 5, 0)$. All coordinates are given in thousands of feet. Ship A locates the submarine in the direction of the vector $2\mathbf{i} + 3\mathbf{j} - (1/3)\mathbf{k}$, and ship B locates it in the direction of the vector $18\mathbf{i} - 6\mathbf{j} - \mathbf{k}$. Four minutes ago, the submarine was located at $(2, -1, -1/3)$. The aircraft is due in 20 min. Assuming the submarine moves in a straight line at a constant speed, to what position should the surface ships direct the aircraft?

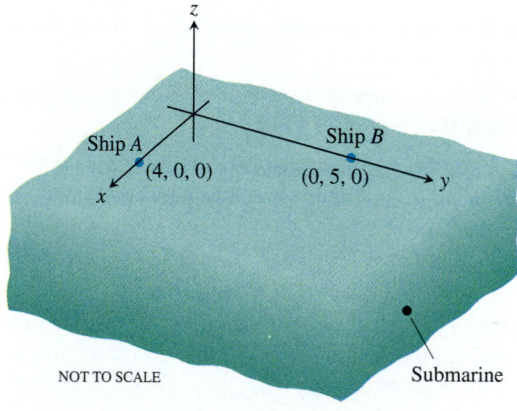

10.75   The ships and submarine in Exercise 53.

54. Two helicopters, H_1 and H_2, are traveling together. At time $t = 0$ hours, they separate and follow different straight-line paths given by

H_1: $\quad x = 6 + 40t, \quad y = -3 + 10t, \quad z = -3 + 2t$

H_2: $\quad x = 6 + 110t, \quad y = -3 + 4t, \quad z = -3 + t,$

all coordinates measured in miles. Due to system malfunctions, H_2 stops its flight at $(446, 13, 1)$ and, in a negligible amount of time, lands at $(446, 13, 0)$. Two hours later, H_1 is advised of this fact and heads toward H_2 at 150 mph. How long will it take H_1 to reach H_2?

55. *The distance between two lines.* Find the distance between the line L_1 through the points $A(1, 0, -1)$ and $B(-1, 1, 0)$ and the line L_2 through the points $C(3, 1, -1)$ and $D(4, 5, -2)$. The distance is to be measured along the line perpendicular to the two lines. First find a vector $\mathbf{N}$ perpendicular to both lines. Then project $\overrightarrow{AC}$ onto $\mathbf{N}$.

56. *(Continuation of Exercise 55.)* Find the distance between the line through $A(4, 0, 2)$ and $B(2, 4, 1)$ and the line through $C(1, 3, 2)$ and $D(2, 2, 4)$.

57. The parallelogram in Fig. 10.76 has vertices at $A(2, -1, 4)$, $B(1, 0, -1)$, $C(1, 2, 3)$, and D. Find
a) the coordinates of D,
b) the cosine of the interior angle at B,
c) the vector projection of $\overrightarrow{BA}$ onto $\overrightarrow{BC}$,
d) the area of the parallelogram,
e) an equation for the plane of the parallelogram,
f) the areas of the orthogonal projections of the parallelogram on the three coordinate planes.

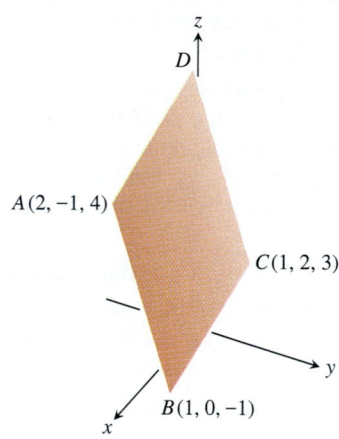

10.76   The parallelogram in Exercise 57.

58. Which of the following are equations for the plane through the points $P(1, 1, -1)$, $Q(3, 0, 2)$, and $R(-2, 1, 0)$?
a) $(2\mathbf{i} - 3\mathbf{j} + 3\mathbf{k}) \cdot ((x + 2)\mathbf{i} + (y - 1)\mathbf{j} + z\mathbf{k}) = 0$
b) $x = 3 - t, \quad y = -11t, \quad z = 2 - 3t$
c) $(x + 2) + 11(y - 1) = 3z$
d) $(2\mathbf{i} - 3\mathbf{j} + 3\mathbf{k}) \times ((x + 2)\mathbf{i} + (y - 1)\mathbf{j} + z\mathbf{k}) = \mathbf{0}$
e) $(2\mathbf{i} - \mathbf{j} + 3\mathbf{k}) \times (-3\mathbf{i} + \mathbf{k}) \cdot ((x + 2)\mathbf{i} + (y - 1)\mathbf{j} + z\mathbf{k}) = 0$

59. a) Show that

$$\begin{vmatrix} x_1 - x & y_1 - y & z_1 - z \\ x_2 - x & y_2 - y & z_2 - z \\ x_3 - x & y_3 - y & z_3 - z \end{vmatrix} = 0$$

is an equation for the plane through the three non-collinear points $P_1(x_1, y_1, z_1)$, $P_2(x_2, y_2, z_2)$, and $P_3(x_3, y_3, z_3)$.

b) What set of points in space is described by the equation

$$\begin{vmatrix} x & y & z & 1 \\ x_1 & y_1 & z_1 & 1 \\ x_2 & y_2 & z_2 & 1 \\ x_3 & y_3 & z_3 & 1 \end{vmatrix} = 0?$$

60. Show that the lines

$$x = a_1 s + b_1, \quad y = a_2 s + b_2, \quad z = a_3 s + b_3,$$
$$-\infty < s < \infty,$$

and

$$x = c_1 t + d_1, \quad y = c_2 t + d_2, \quad z = c_3 t + d_3,$$
$$-\infty < t < \infty,$$

intersect or are parallel if and only if

$$\begin{vmatrix} a_1 & c_1 & b_1 - d_1 \\ a_2 & c_2 & b_2 - d_2 \\ a_3 & c_3 & b_3 - d_3 \end{vmatrix} = 0.$$

61. Use vectors to show that the distance from $P_1(x_1, y_1)$ to the line $ax + by = c$ is

$$d = \frac{|ax_1 + by_1 - c|}{\sqrt{a^2 + b^2}}.$$

62. a) Use vectors to show that the distance from $P_1(x_1, y_1, z_1)$ to the plane $Ax + By + Cz = D$ is

$$d = \frac{|Ax_1 + By_1 + Cz_1 - D|}{\sqrt{A^2 + B^2 + C^2}}.$$

b) Find an equation for the sphere that is tangent to the planes $x + y + z = 3$ and $x + y + z = 9$ if the planes $2x - y = 0$ and $3x - z = 0$ pass through the center of the sphere.

63. a) Show that the distance between the parallel planes $Ax + By + Cz = D_1$ and $Ax + By + Cz = D_2$ is

$$d = \frac{|D_1 - D_2|}{|A\mathbf{i} + B\mathbf{j} + C\mathbf{k}|}. \qquad (1)$$

b) Use Eq. (1) to find the distance between the planes $2x + 3y - z = 6$ and $2x + 3y - z = 12$.

c) Find an equation for the plane parallel to the plane $2x - y + 2z = -4$ if the point $(3, 2, -1)$ is equidistant from the two planes.

d) Write equations for the planes that lie parallel to and 5 units away from the plane $x - 2y + z = 3$.

64. Show that if $\mathbf{A}$, $\mathbf{B}$, $\mathbf{C}$, and $\mathbf{D}$ are any vectors, then

a) $\mathbf{A} \times (\mathbf{B} \times \mathbf{C}) + \mathbf{B} \times (\mathbf{C} \times \mathbf{A}) + \mathbf{C} \times (\mathbf{A} \times \mathbf{B}) = \mathbf{0}$,

b) $\mathbf{A} \times \mathbf{B} = (\mathbf{A} \cdot \mathbf{B} \times \mathbf{i})\mathbf{i} + (\mathbf{A} \cdot \mathbf{B} \times \mathbf{j})\mathbf{j} + (\mathbf{A} \cdot \mathbf{B} \times \mathbf{k})\mathbf{k}$,

c) $(\mathbf{A} \times \mathbf{B}) \cdot (\mathbf{C} \times \mathbf{D}) = \begin{vmatrix} \mathbf{A} \cdot \mathbf{C} & \mathbf{B} \cdot \mathbf{C} \\ \mathbf{A} \cdot \mathbf{D} & \mathbf{B} \cdot \mathbf{D} \end{vmatrix}$.

65. Prove that four points A, B, C, and D are coplanar (lie in a common plane) if and only if $\overrightarrow{AD} \cdot (\overrightarrow{AB} \times \overrightarrow{BC}) = 0$.

66. Prove or disprove the formula

$$\mathbf{A} \times [\mathbf{A} \times (\mathbf{A} \times \mathbf{B})] \cdot \mathbf{C} = -|\mathbf{A}|^2 \mathbf{A} \cdot \mathbf{B} \times \mathbf{C}.$$

The equations in Exercises 67–76 define sets both in the plane and in three-dimensional space. Identify the sets.

Rectangular Coordinates

67. $x = 0$

68. $x + y = 1$

69. $x^2 + y^2 = 4$

70. $x^2 + 4y^2 = 16$

71. $x = y^2$

72. $y^2 - x^2 = 1$

Cylindrical Coordinates

73. $r = 1 - \cos\theta$

74. $r = \sin\theta$

75. $r^2 = 2\cos 2\theta$

76. $r = \cos 2\theta$

Spherical Coordinates

Describe the sets defined by the spherical-coordinate equations and inequalities in Exercises 77–82.

77. $\rho = 2$

78. $\theta = \pi/4$

79. $\phi = \pi/6$

80. $\rho = 1, \quad \phi = \pi/2$

81. $\rho = 1, \quad 0 \le \phi \le \pi/2$

82. $1 \le \rho \le 2$

In Exercises 83–88, translate the equations from the given coordinate system (rectangular, cylindrical, spherical) into the other two systems. Identify the set of points defined by the equation.

Rectangular

83. $z = 2$

84. $y^2 = x^2 + z^2$

Cylindrical

85. $z = r^2\cos^2 2\theta$

86. $r = \cos\theta$

Spherical

87. $\rho = 4\sec\phi$

88. $\rho\tan\phi\sin\phi = -1$

Areas in the Plane

To find the area of a parallelogram in a plane, picture the plane as the xy-plane in space. Then write vectors for a pair of adjacent sides (their $\mathbf{k}$-components will be zero) and calculate the magnitude of their cross product.

Find the areas of the parallelograms whose vertices are given in Exercises 89–92.

89. $A(1, 0)$, $B(0, 1)$, $C(-1, 0)$, $D(0, -1)$

90. $A(0, 0)$, $B(7, 3)$, $C(9, 8)$, $D(2, 5)$

91. $A(-1, 2)$, $B(2, 0)$, $C(7, 1)$, $D(4, 3)$

92. $A(-6, 0)$, $B(1, -4)$, $C(3, 1)$, $D(-4, 5)$

Find the areas of the triangles whose vertices are given in Exercises 93–96.

93. $A(0, 0)$, $B(-2, 3)$, $C(3, 1)$

94. $A(-1, -1)$, $B(3, 3)$, $C(2, 1)$

95. $A(-5, 3)$, $B(1, -2)$, $C(6, -2)$

96. $A(-6, 0)$, $B(10, -5)$, $C(-2, 4)$

97. Show that the area of a triangle in the xy-plane with one vertex at the origin and the other two vertices at (a_1, a_2) and (b_1, b_2) is

$$\pm \frac{1}{2} \begin{vmatrix} a_1 & a_2 \\ b_1 & b_2 \end{vmatrix}.$$

What controls the sign?

11

VECTOR-VALUED FUNCTIONS AND MOTION IN SPACE

Overview   When a body travels through space, the three equations $x = f(t)$, $y = g(t)$, and $z = h(t)$ that give the body's coordinates as functions of time serve as parametric equations for the body's motion and path. With vector notation, we can condense these three equations into a single equation $\mathbf{r} = f(t)\,\mathbf{i} + g(t)\,\mathbf{j} + h(t)\,\mathbf{k}$ that gives the body's position as a vector function of time.

In this chapter, we show how to use calculus to differentiate and integrate vector functions to study the paths, velocities, and accelerations of moving bodies. As we go along, we shall see how our work answers the standard questions about the paths and motions of projectiles, planets, and satellites. In the final section, we combine our new vector calculus with what we know about vector algebra and geometry, derivatives, the solutions of differential equations and initial value problems, and the equations of conic sections in polar coordinates to derive Kepler's laws of planetary motion from Newton's laws of motion and gravitation.

11.1 Vector-Valued Functions and Space Curves

To track a particle moving in space, we run a vector $\mathbf{r}$ from the origin to the particle (Fig. 11.1) and study the changes in $\mathbf{r}$. If the particle's position coordinates are twice-differentiable functions of time, then so is $\mathbf{r}$ and we can find the particle's velocity and acceleration vectors at any time by differentiating $\mathbf{r}$. Conversely, if we know either the particle's velocity vector or acceleration vector as a continuous function of time, and if we have enough information about the particle's initial velocity and initial position, we can find $\mathbf{r}$ as a function of time by integration. This section shows how the calculations go.

Definitions

When a particle moves through space during a time interval I, we think of the particle's coordinates as functions defined on I:

$$x = f(t), \qquad y = g(t), \qquad z = h(t), \qquad t \, \varepsilon \, I. \tag{1}$$

767

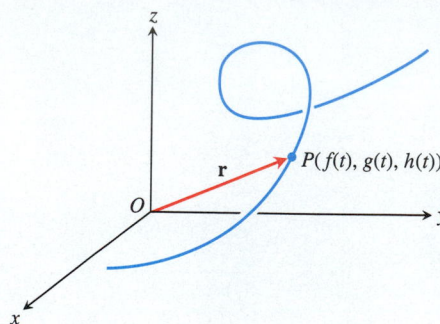

11.1   The position vector $\mathbf{r} = \overrightarrow{OP}$ of a particle moving through space is a function of time.

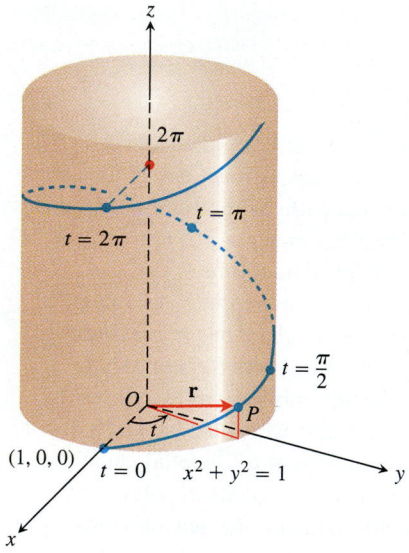

11.2   The upper half of the helix $\mathbf{r} = (\cos t)\,\mathbf{i} + (\sin t)\,\mathbf{j} + t\,\mathbf{k}$.

The points $(x, y, z) = (f(t), g(t), h(t))$, $t \, \varepsilon \, I$, make up the **curve** in space that we call the particle's path. The equations and interval in (1) **parametrize** the curve. The vector

$$\mathbf{r} = \overrightarrow{OP} = f(t)\,\mathbf{i} + g(t)\,\mathbf{j} + h(t)\,\mathbf{k} \tag{2}$$

from the origin to the particle's **position** $P(f(t), g(t), h(t))$ at time t is the particle's **position vector.** The functions f, g, and h are the **component functions (components)** of the position vector. We think of the particle's path as the **curve traced by** $\mathbf{r}$ during the time interval I.

Equation (2) defines $\mathbf{r}$ as a vector function of the real variable t on the interval I. More generally, a **vector function** or **vector-valued function** on a domain set D is a rule that assigns a vector in space to each element in D. For now, the domains will be intervals of real numbers. Later, in Chapter 14, the domains will be regions in the plane or in space. Vector functions will then be called "vector fields."

When we need to distinguish them from vector functions, we refer to real-valued functions as **scalar functions.** The components of $\mathbf{r}$ are scalar functions of t. When we define a vector-valued function by giving its component functions, we assume the vector function's domain to be the common domain of its component functions.

Example 1   *A Helix.*   The vector function

$$\mathbf{r} = (\cos t)\,\mathbf{i} + (\sin t)\,\mathbf{j} + t\,\mathbf{k}$$

is defined for all real values of t. The curve traced by $\mathbf{r}$ is a helix (from an old Greek word for spiral) that winds around the circular cylinder $x^2 + y^2 = 1$ (Fig. 11.2). The curve lies on the cylinder because the **i**- and **j**-components of $\mathbf{r}$, being the x- and y-coordinates of the tip of $\mathbf{r}$, satisfy the cylinder's equation:

$$x^2 + y^2 = (\cos t)^2 + (\sin t)^2 = 1.$$

The curve rises as the **k**-component $z = t$ increases. Each time t increases by 2π, the curve completes one turn around the cylinder. The equations

$$x = \cos t, \qquad y = \sin t, \qquad z = t$$

parametrize the helix, the interval $-\infty < t < \infty$ being understood. You will find more helices in Fig. 11.3.

11.3   Helices drawn by computer.

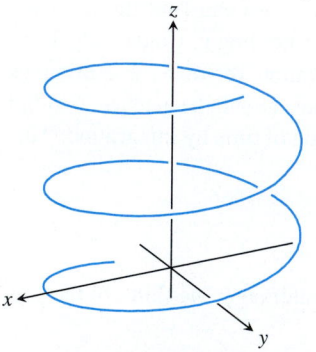

$\mathbf{r} = (\cos t)\mathbf{i} + (\sin t)\mathbf{j} + t\mathbf{k}$

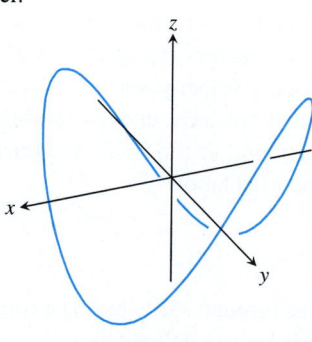

$\mathbf{r} = (\cos t)\mathbf{i} + (\sin t)\mathbf{j} + (\sin 2t)\mathbf{k}$

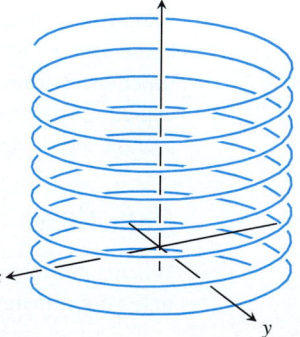

$\mathbf{r} = (\cos t)\mathbf{i} + (\sin t)\mathbf{j} + 0.3t\mathbf{k}$

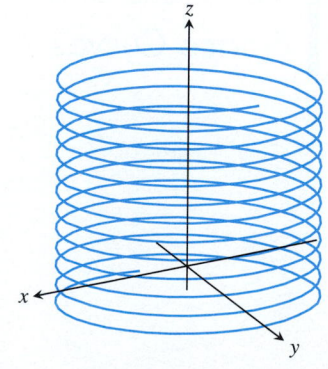

$\mathbf{r} = (\cos 5t)\mathbf{i} + (\sin 5t)\mathbf{j} + t\mathbf{k}$

(Generated by Mathematica)

Limits and Continuity

We define limits of vector functions in terms of components.

DEFINITION

> A vector function $\mathbf{f}$ defined by the rule $\mathbf{f}(t) = f(t)\,\mathbf{i} + g(t)\,\mathbf{j} + h(t)\,\mathbf{k}$ has **limit** $\mathbf{L} = L_1\,\mathbf{i} + L_2\,\mathbf{j} + L_3\,\mathbf{k}$ as t **approaches** t_0 if
>
> $$\lim_{t \to t_0} f(t) = L_1, \qquad \lim_{t \to t_0} g(t) = L_2, \qquad \text{and} \qquad \lim_{t \to t_0} h(t) = L_3. \qquad (3)$$
>
> We denote the existence of the limit in the usual way by writing
>
> $$\lim_{t \to t_0} \mathbf{f}(t) = \mathbf{L}.$$

Example 2   If $\mathbf{f}(t) = (\cos t)\,\mathbf{i} + (\sin t)\,\mathbf{j} + t\mathbf{k}$, then

$$\lim_{t \to \pi/4} \mathbf{f}(t) = \left(\lim_{t \to \pi/4} \cos t\right)\mathbf{i} + \left(\lim_{t \to \pi/4} \sin t\right)\mathbf{j} + \left(\lim_{t \to \pi/4} t\right)\mathbf{k}$$

$$= \frac{\sqrt{2}}{2}\,\mathbf{i} + \frac{\sqrt{2}}{2}\,\mathbf{j} + \frac{\pi}{4}\,\mathbf{k}.$$

We define continuity for vector functions the same way we define continuity for scalar functions.

DEFINITION

> A vector function $\mathbf{f}$ is **continuous at a point** $t = t_0$ in its domain if $\lim_{t \to t_0} \mathbf{f}(t) = \mathbf{f}(t_0)$. The function is **continuous** if it is continuous at every point in its domain.

Since limits are defined in terms of components, we can test vector functions for continuity by examining their components (Exercise 36).

> **Component Test for Continuity at a Point**
>
> A vector function $\mathbf{f}$ defined by the rule $\mathbf{f}(t) = f(t)\,\mathbf{i} + g(t)\,\mathbf{j} + h(t)\,\mathbf{k}$ is continuous at $t = t_0$ if and only if f, g, and h are continuous at t_0.

Example 3   a) The function

$$\mathbf{r} = (\cos t)\,\mathbf{i} + (\sin t)\,\mathbf{j} + t\mathbf{k}$$

is continuous because $\cos t$, $\sin t$, and t are continuous.

b) The function

$$\mathbf{r} = (\cos t)\,\mathbf{i} + (\sin t)\,\mathbf{j} + \left\lfloor \frac{t}{2\pi} \right\rfloor \mathbf{k}$$

is discontinuous at every integer multiple of 2π.

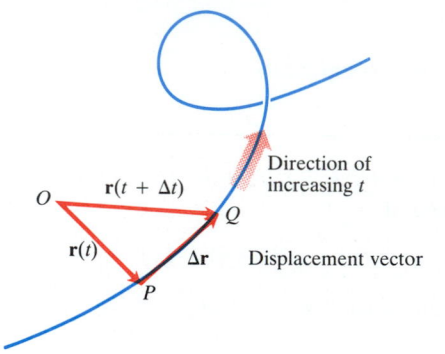

11.4   Between time t and time $t + \Delta t$, the particle moving along the path shown here undergoes the displacement $\overrightarrow{PQ} = \Delta \mathbf{r}$. The vector sum $\mathbf{r}(t) + \Delta \mathbf{r}$ gives the new position, $\mathbf{r}(t + \Delta t)$.

Derivatives and Motion

Suppose that $\mathbf{r} = f(t)\,\mathbf{i} + g(t)\,\mathbf{j} + h(t)\,\mathbf{k}$ is the position vector of a particle moving along a curve in space and that f, g, and h are differentiable functions of t. Then the difference between the particle's positions at time t and a nearby time $t + \Delta t$ can be expressed as the vector difference

$$\Delta \mathbf{r} = \mathbf{r}(t + \Delta t) - \mathbf{r}(t) \qquad (4)$$

(Fig. 11.4). In terms of components,

$$\begin{aligned}
\Delta \mathbf{r} &= \mathbf{r}(t + \Delta t) - \mathbf{r}(t) \\
&= [f(t + \Delta t)\,\mathbf{i} + g(t + \Delta t)\,\mathbf{j} + h(t + \Delta t)\,\mathbf{k}] \\
&\quad - [f(t)\,\mathbf{i} + g(t)\,\mathbf{j} + h(t)\,\mathbf{k}] \qquad (5) \\
&= [f(t + \Delta t) - f(t)]\,\mathbf{i} + [g(t + \Delta t) - g(t)]\,\mathbf{j} + [h(t + \Delta t) - h(t)]\,\mathbf{k}.
\end{aligned}$$

As Δt approaches zero, three things seem to happen simultaneously. First, Q approaches P along the curve. Second, the secant line PQ seems to approach a limiting position tangent to the curve at P. Third, the quotient $\Delta \mathbf{r}/\Delta t$ approaches the limit

$$\begin{aligned}
\lim_{\Delta t \to 0} \frac{\Delta \mathbf{r}}{\Delta t} &= \left[\lim_{\Delta t \to 0} \frac{f(t + \Delta t) - f(t)}{\Delta t}\right]\mathbf{i} + \left[\lim_{\Delta t \to 0} \frac{g(t + \Delta t) - g(t)}{\Delta t}\right]\mathbf{j} \\
&\quad + \left[\lim_{\Delta t \to 0} \frac{h(t + \Delta t) - h(t)}{\Delta t}\right]\mathbf{k} \qquad (6) \\
&= \left[\frac{df}{dt}\right]\mathbf{i} + \left[\frac{dg}{dt}\right]\mathbf{j} + \left[\frac{dh}{dt}\right]\mathbf{k}.
\end{aligned}$$

We are therefore led by past experience to the following definition.

DEFINITION

> The vector function $\mathbf{r} = f(t)\,\mathbf{i} + g(t)\,\mathbf{j} + h(t)\,\mathbf{k}$ is **differentiable at $t = t_0$** if f, g, and h are differentiable at t_0. Also, $\mathbf{r}$ is said to be **differentiable** if it is differentiable at every point of its domain. At any point t at which $\mathbf{r}$ is differentiable, its **derivative** is the vector
>
> $$\frac{d\mathbf{r}}{dt} = \lim_{\Delta t \to 0} \frac{\mathbf{r}(t + \Delta t) - \mathbf{r}(t)}{\Delta t} = \frac{df}{dt}\,\mathbf{i} + \frac{dg}{dt}\,\mathbf{j} + \frac{dh}{dt}\,\mathbf{k}.$$
>
> The curve traced by $\mathbf{r}$ is **smooth** if $d\mathbf{r}/dt$ is continuous, i.e., if f, g, and h have continuous first derivatives.

Look once again at Fig. 11.4. We drew the figure for Δt positive, so $\Delta \mathbf{r}$ points forward, in the direction of the motion. The vector $\Delta \mathbf{r}/\Delta t$, having the same direction as $\Delta \mathbf{r}$, points forward too. Had Δt been negative, $\Delta \mathbf{r}$ would have pointed backward, against the direction of motion. The quotient $\Delta \mathbf{r}/\Delta t$, however, being a negative scalar multiple of $\Delta \mathbf{r}$, would once again have pointed forward. No matter how $\Delta \mathbf{r}$ points, $\Delta \mathbf{r}/\Delta t$ points forward and we expect the vector $d\mathbf{r}/dt = \lim_{\Delta t \to 0} \Delta \mathbf{r}/\Delta t$, when different from $\mathbf{0}$, to do the same. This means that the derivative $d\mathbf{r}/dt$ is just what we want for modeling a particle's velocity. It points in the direction of motion and gives the rate of change of position with respect to time.

DEFINITIONS

If the position vector **r** of a particle moving in space is a differentiable function of time t, then the vector

$$\mathbf{v} = \frac{d\mathbf{r}}{dt}$$

is the particle's **velocity vector.** At any time t, the direction of **v** is the **direction of motion,** the magnitude of **v** is the particle's **speed,** and the derivative $\mathbf{a} = d\mathbf{v}/dt$, when it exists, is the particle's **acceleration vector.** In short,

1. Velocity is the derivative of position: $\mathbf{v} = \dfrac{d\mathbf{r}}{dt}$

2. Speed is the magnitude of velocity: Speed $= |\mathbf{v}|$

3. Acceleration is the derivative of velocity: $\mathbf{a} = \dfrac{d\mathbf{v}}{dt} = \dfrac{d^2\mathbf{r}}{dt^2}$

4. The vector $\mathbf{v}/|\mathbf{v}|$ is the direction of motion at time t.

We use the formula $\mathbf{A} = |\mathbf{A}| \cdot (\mathbf{A}/|\mathbf{A}|)$ to express the velocity of a moving particle as the product of its speed and direction.

$$\text{Velocity} = |\mathbf{v}| \cdot \frac{\mathbf{v}}{|\mathbf{v}|} = (\text{speed}) \cdot (\text{direction}) \tag{7}$$

Example 4   The vector

$$\mathbf{r} = (3 \cos t)\,\mathbf{i} + (3 \sin t)\,\mathbf{j} + t^2\,\mathbf{k}$$

gives the position of a moving body at time t. Find the body's speed and direction when $t = 2$. At what times, if any, are the body's velocity and acceleration orthogonal?

Solution

$$\mathbf{r} = (3 \cos t)\,\mathbf{i} + (3 \sin t)\,\mathbf{j} + t^2\,\mathbf{k}$$

$$\mathbf{v} = \frac{d\mathbf{r}}{dt} = -(3 \sin t)\,\mathbf{i} + (3 \cos t)\,\mathbf{j} + 2t\,\mathbf{k}$$

$$\mathbf{a} = \frac{d^2\mathbf{r}}{dt^2} = -(3 \cos t)\,\mathbf{i} - (3 \sin t)\,\mathbf{j} + 2\,\mathbf{k}$$

At $t = 2$, the body's speed and direction are

Speed: $|\mathbf{v}(2)| = \sqrt{(-3 \sin 2)^2 + (3 \cos 2)^2 + (2 \cdot 2)^2} = 5$

Direction: $\dfrac{\mathbf{v}(2)}{|\mathbf{v}(2)|} = -\left(\dfrac{3}{5} \sin 2\right)\mathbf{i} + \left(\dfrac{3}{5} \cos 2\right)\mathbf{j} + \dfrac{4}{5}\,\mathbf{k}.$

To find when **v** and **a** are orthogonal, we look for values of t for which

$$\mathbf{v} \cdot \mathbf{a} = 9 \sin t \cos t - 9 \cos t \sin t + 4t = 4t = 0.$$

The only value is $t = 0$.

Differentiation Rules

Because the derivatives of vector functions are defined component by component, the rules for differentiating vector functions have the same form as the rules for differentiating scalar functions.

Differentiation Rules for Vector Functions

Constant Function Rule: $\dfrac{d}{dt}\mathbf{C} = \mathbf{0}$ (any constant vector $\mathbf{C}$)

If $\mathbf{u}$ and $\mathbf{v}$ are differentiable functions of t, then

Scalar Multiple Rule: $\dfrac{d}{dt}(c\mathbf{u}) = c\dfrac{d\mathbf{u}}{dt}$ (any number c)

Sum Rule: $\dfrac{d}{dt}(\mathbf{u} + \mathbf{v}) = \dfrac{d\mathbf{u}}{dt} + \dfrac{d\mathbf{v}}{dt}$

Difference Rule: $\dfrac{d}{dt}(\mathbf{u} - \mathbf{v}) = \dfrac{d\mathbf{u}}{dt} - \dfrac{d\mathbf{v}}{dt}$

Dot-Product Rule: $\dfrac{d}{dt}(\mathbf{u} \cdot \mathbf{v}) = \dfrac{d\mathbf{u}}{dt} \cdot \mathbf{v} + \mathbf{u} \cdot \dfrac{d\mathbf{v}}{dt}$

Cross-Product Rule: $\dfrac{d}{dt}(\mathbf{u} \times \mathbf{v}) = \dfrac{d\mathbf{u}}{dt} \times \mathbf{v} + \mathbf{u} \times \dfrac{d\mathbf{v}}{dt}$

Chain Rule (Short Form):   If $\mathbf{r}$ is a differentiable function of t and t is a differentiable function of s, then

$$\frac{d\mathbf{r}}{ds} = \frac{d\mathbf{r}}{dt}\frac{dt}{ds}.$$

When using the Cross-Product Rule, remember to preserve the order of the factors. If $\mathbf{u}$ comes first on the left, it must come first on the right or the signs will be wrong.

We shall prove the product rules and the Chain Rule but leave the rules for constants, scalar multiples, sums, and differences as exercises.

Proof of the Dot-Product Rule

Suppose that $\mathbf{u} = u_1(t)\,\mathbf{i} + u_2(t)\,\mathbf{j} + u_3(t)\,\mathbf{k}$

and $\mathbf{v} = v_1(t)\,\mathbf{i} + v_2(t)\,\mathbf{j} + v_3(t)\,\mathbf{k}.$

Then

$$\frac{d}{dt}(\mathbf{u} \cdot \mathbf{v}) = \frac{d}{dt}(u_1v_1 + u_2v_2 + u_3v_3)$$

$$= \underbrace{u_1'v_1 + u_2'v_2 + u_3'v_3}_{\mathbf{u}' \cdot \mathbf{v}} + \underbrace{u_1v_1' + u_2v_2' + u_3v_3'}_{\mathbf{u} \cdot \mathbf{v}'}.$$

Proof of the Cross-Product Rule We model the proof after the proof of the product rule for scalar functions. According to the definition of derivative,

$$\frac{d}{dt}(\mathbf{u} \times \mathbf{v}) = \lim_{h \to 0} \frac{\mathbf{u}(t + h) \times \mathbf{v}(t + h) - \mathbf{u}(t) \times \mathbf{v}(t)}{h}.$$

To change this fraction into an equivalent one that contains the difference quotients for the derivatives of $\mathbf{u}$ and $\mathbf{v}$, we subtract and add $\mathbf{u}(t) \times \mathbf{v}(t + h)$ in the numerator. Then

$$\frac{d}{dt}(\mathbf{u} \times \mathbf{v})$$

$$= \lim_{h \to 0} \frac{\mathbf{u}(t + h) \times \mathbf{v}(t + h) - \mathbf{u}(t) \times \mathbf{v}(t + h) + \mathbf{u}(t) \times \mathbf{v}(t + h) - \mathbf{u}(t) \times \mathbf{v}(t)}{h}$$

$$= \lim_{h \to 0} \left[\frac{\mathbf{u}(t + h) - \mathbf{u}(t)}{h} \times \mathbf{v}(t + h) + \mathbf{u}(t) \times \frac{\mathbf{v}(t + h) - \mathbf{v}(t)}{h} \right] \tag{8}$$

$$= \lim_{h \to 0} \frac{\mathbf{u}(t + h) - \mathbf{u}(t)}{h} \times \lim_{h \to 0} \mathbf{v}(t + h) + \lim_{h \to 0} \mathbf{u}(t) \times \lim_{h \to 0} \frac{\mathbf{v}(t + h) - \mathbf{v}(t)}{h}.$$

The last of these equalities holds because the limit of the cross product of two vector functions is the cross product of their limits if the latter exist (Exercise 37). As h approaches zero, $\mathbf{v}(t + h)$ approaches $\mathbf{v}(t)$ because $\mathbf{v}$, being differentiable at t, is continuous at t (Exercise 38). The two fractions approach the values of $d\mathbf{u}/dt$ and $d\mathbf{v}/dt$ at t. In short,

$$\frac{d}{dt}(\mathbf{u} \times \mathbf{v}) = \frac{d\mathbf{u}}{dt} \times \mathbf{v} + \mathbf{u} \times \frac{d\mathbf{v}}{dt}.$$

Proof of the Chain Rule Suppose that $\mathbf{r} = f(t)\,\mathbf{i} + g(t)\,\mathbf{j} + h(t)\,\mathbf{k}$ is a differentiable vector function of t and that t is a differentiable scalar function of some other variable s. Then f, g, and h are differentiable functions of s, and the Chain Rule for differentiable real-valued functions gives

$$\frac{d\mathbf{r}}{ds} = \frac{df}{ds}\,\mathbf{i} + \frac{dg}{ds}\,\mathbf{j} + \frac{dh}{ds}\,\mathbf{k}$$

$$= \frac{df}{dt}\frac{dt}{ds}\,\mathbf{i} + \frac{dg}{dt}\frac{dt}{ds}\,\mathbf{j} + \frac{dh}{dt}\frac{dt}{ds}\,\mathbf{k}$$

$$= \left(\frac{df}{dt}\,\mathbf{i} + \frac{dg}{dt}\,\mathbf{j} + \frac{dh}{dt}\,\mathbf{k}\right)\frac{dt}{ds}$$

$$= \frac{d\mathbf{r}}{dt}\frac{dt}{ds}.$$

Derivatives of Vectors of Constant Length

We might at first think that a vector whose length remains constant as time passes has to have a zero derivative, but this need not be the case. Think of a moving clock hand. Its length remains constant as time passes, but its direction still changes. What we *can* say about the derivative of a vector of constant length is that it is always orthogonal to the vector. Direction changes take place at right angles.

If **u** is a differentiable vector function of t of constant length, then

$$\mathbf{u} \cdot \frac{d\mathbf{u}}{dt} = 0. \tag{9}$$

To see why Eq. (9) holds, suppose **u** is a differentiable function of t and that $|\mathbf{u}|$ is constant. Then $\mathbf{u} \cdot \mathbf{u} = |\mathbf{u}|^2$ is constant and we may differentiate both sides of this equation to get

$$\frac{d}{dt}(\mathbf{u} \cdot \mathbf{u}) = 0$$

$$\frac{d\mathbf{u}}{dt} \cdot \mathbf{u} + \mathbf{u} \cdot \frac{d\mathbf{u}}{dt} = 0 \qquad \text{(Dot-Product Rule with } \mathbf{v} = \mathbf{u}\text{)}$$

$$2\mathbf{u} \cdot \frac{d\mathbf{u}}{dt} = 0 \qquad \text{(Dot multiplication is commutative.)}$$

$$\mathbf{u} \cdot \frac{d\mathbf{u}}{dt} = 0.$$

Example 5   Show that

$$\mathbf{u} = (\sin t)\,\mathbf{i} + (\cos t)\,\mathbf{j} + \sqrt{3}\,\mathbf{k}$$

has constant length and is orthogonal to its derivative.

Solution $\mathbf{u} = (\sin t)\,\mathbf{i} + (\cos t)\,\mathbf{j} + \sqrt{3}\,\mathbf{k}$

$$|\mathbf{u}| = \sqrt{(\sin t)^2 + (\cos t)^2 + (\sqrt{3})^2} = \sqrt{1 + 3} = 2$$

$$\frac{d\mathbf{u}}{dt} = (\cos t)\,\mathbf{i} - (\sin t)\,\mathbf{j}$$

$$\mathbf{u} \cdot \frac{d\mathbf{u}}{dt} = \sin t \cos t - \cos t \sin t = 0$$

Integrals of Vector Functions

A differentiable vector function $\mathbf{F}(t)$ is an **antiderivative** of a vector function $\mathbf{f}(t)$ on an interval I if $d\mathbf{F}/dt = \mathbf{f}$ at each point of I. If $\mathbf{F}$ is an antiderivative of $\mathbf{f}$ on I, it can be shown, working one component at a time, that every antiderivative of $\mathbf{f}$ on I has the form $\mathbf{F} + \mathbf{C}$ for some constant vector $\mathbf{C}$ (Exercise 41). The set of all antiderivatives of $\mathbf{f}$ on I is the **indefinite integral** of $\mathbf{f}$ on I; we denote it in the usual way and write

$$\int \mathbf{f}(t)\, dt = \mathbf{F}(t) + \mathbf{C}. \tag{10}$$

The usual arithmetic rules for indefinite integrals apply.

Example 6

$$\int ((\cos t)\,\mathbf{i} + \mathbf{j} - 2t\,\mathbf{k})\,dt$$

$$= \left(\int \cos t\,dt\right)\mathbf{i} + \left(\int dt\right)\mathbf{j} - \left(\int 2t\,dt\right)\mathbf{k} \qquad (11)$$

$$= (\sin t + C_1)\,\mathbf{i} + (t + C_2)\,\mathbf{j} - (t^2 + C_3)\,\mathbf{k} \qquad (12)$$

$$= (\sin t)\,\mathbf{i} + t\,\mathbf{j} - t^2\,\mathbf{k} + \mathbf{C} \qquad (\mathbf{C} = C_1\,\mathbf{i} + C_2\,\mathbf{j} + C_3\,\mathbf{k})$$

As in the integration of scalar functions, we recommend that you skip the steps in (11) and (12) and go directly to the final form. Find an antiderivative for each component and add a constant vector at the end.

Definite integrals of vector functions are defined in terms of components.

DEFINITION

If the components of $\mathbf{f}(t) = f(t)\,\mathbf{i} + g(t)\,\mathbf{j} + h(t)\,\mathbf{k}$ are integrable over the interval $a \le t \le b$, then $\mathbf{f}$ is **integrable** over $[a, b]$ and the **definite integral** of $\mathbf{f}$ from a to b is

$$\int_a^b \mathbf{f}(t)\,dt = \left(\int_a^b f(t)\,dt\right)\mathbf{i} + \left(\int_a^b g(t)\,dt\right)\mathbf{j} + \left(\int_a^b h(t)\,dt\right)\mathbf{k}. \qquad (13)$$

All of the usual arithmetic rules for definite integrals apply (Exercise 39).

Example 7

$$\int_0^\pi ((\cos t)\,\mathbf{i} + \mathbf{j} - 2t\,\mathbf{k})\,dt = \left(\int_0^\pi \cos t\,dt\right)\mathbf{i} + \left(\int_0^\pi dt\right)\mathbf{j} - \left(\int_0^\pi 2t\,dt\right)\mathbf{k}$$

$$= \Big[\sin t\Big]_0^\pi \mathbf{i} + \Big[t\Big]_0^\pi \mathbf{j} - \Big[t^2\Big]_0^\pi \mathbf{k}$$

$$= [0 - 0]\,\mathbf{i} + [\pi - 0]\,\mathbf{j} - [\pi^2 - 0^2]\,\mathbf{k}$$

$$= \pi\,\mathbf{j} - \pi^2\,\mathbf{k}$$

Example 8   *Finding a Particle's Position Function from its Velocity Function and Initial Position.*   The velocity of a particle moving in space is

$$\frac{d\mathbf{r}}{dt} = (\cos t)\,\mathbf{i} - (\sin t)\,\mathbf{j} + \mathbf{k}.$$

Find the particle's position as a function of t if $\mathbf{r} = 2\,\mathbf{i} + \mathbf{k}$ when $t = 0$.

Solution   Our goal is to solve the initial value problem that consists of

The differential equation: $\dfrac{d\mathbf{r}}{dt} = (\cos t)\,\mathbf{i} - (\sin t)\,\mathbf{j} + \mathbf{k},$

The initial condition: $\mathbf{r} = 2\,\mathbf{i} + \mathbf{k}$ when $t = 0.$

To solve it, we first use what we know about antiderivatives to find the general solution of the differential equation. Integrating both sides with respect to t gives

$$\mathbf{r} = (\sin t)\,\mathbf{i} + (\cos t)\,\mathbf{j} + t\mathbf{k} + \mathbf{C}.$$

We then use the initial condition to find the right value for $\mathbf{C}$:

$$(\sin 0)\,\mathbf{i} + (\cos 0)\,\mathbf{j} + (0)\,\mathbf{k} + \mathbf{C} = 2\,\mathbf{i} + \mathbf{k} \qquad \left(\begin{matrix}\mathbf{r} = 2\,\mathbf{i} + \mathbf{k}\\ \text{when } t = 0\end{matrix}\right)$$

$$\mathbf{j} + \mathbf{C} = 2\,\mathbf{i} + \mathbf{k}$$

$$\mathbf{C} = 2\,\mathbf{i} - \mathbf{j} + \mathbf{k}$$

The particle's position as a function of t is

$$\mathbf{r} = (\sin t + 2)\,\mathbf{i} + (\cos t - 1)\,\mathbf{j} + (t + 1)\,\mathbf{k}.$$

To check (always a good idea), we can see from this formula that

$$\frac{d\mathbf{r}}{dt} = (\cos t + 0)\,\mathbf{i} + (-\sin t - 0)\,\mathbf{j} + (1 + 0)\,\mathbf{k} = (\cos t)\,\mathbf{i} - (\sin t)\,\mathbf{j} + \mathbf{k}$$

and that when $t = 0$,

$$\mathbf{r} = (\sin 0 + 2)\,\mathbf{i} + (\cos 0 - 1)\,\mathbf{j} + (0 + 1)\,\mathbf{k} = 2\,\mathbf{i} + \mathbf{k}.$$

Motion in the Plane

When a particle moves in the xy-plane, its z-coordinate is zero and its position vector reduces to

$$\mathbf{r} = f(t)\,\mathbf{i} + g(t)\,\mathbf{j}. \tag{14}$$

Example 9   The position vector of a particle moving along the parabola $y = x^2$ in the xy-plane is

$$\mathbf{r} = t\mathbf{i} + t^2\,\mathbf{j}.$$

Find the particle's velocity and acceleration vectors. Sketch them on the parabola together with the position vector at time $t = 1$.

Solution

$$\mathbf{r} = t\mathbf{i} + t^2\,\mathbf{j}$$

$$\mathbf{v} = \frac{d\mathbf{r}}{dt} = \mathbf{i} + 2t\mathbf{j}$$

$$\mathbf{a} = \frac{d\mathbf{v}}{dt} = 0\,\mathbf{i} + 2\,\mathbf{j} = 2\,\mathbf{j}.$$

At time $t = 1$,

$$\mathbf{r} = \mathbf{i} + \mathbf{j}, \qquad \mathbf{v} = \mathbf{i} + 2\,\mathbf{j}, \qquad \mathbf{a} = 2\,\mathbf{j}.$$

We sketch the parabola in the xy-plane and draw in the vectors (Fig. 11.5).

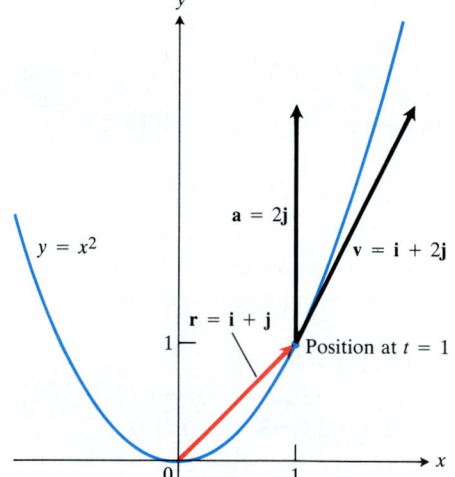

11.5   The position, velocity, and acceleration of the particle in Example 9 at $t = 1$. Notice the convention of drawing the velocity and acceleration as vectors at the point on the curve instead of at the origin.

EXERCISES 11.1

In Exercises 1–6, $\mathbf{r}$ is the position of a particle in space at time t. Find the particle's velocity and acceleration vectors. Then find the particle's speed and direction of motion at the given value of t. Write the velocity at that time as the product of its speed and direction.

1. $\mathbf{r} = (2 \cos t)\,\mathbf{i} + (3 \sin t)\,\mathbf{j} + 4t\,\mathbf{k}, \quad t = \pi/2$

2. $\mathbf{r} = (t + 1)\,\mathbf{i} + (t^2 - 1)\,\mathbf{j} + 2t\,\mathbf{k}, \quad t = 1$

3. $\mathbf{r} = (\cos 2t)\,\mathbf{j} + (2 \sin t)\,\mathbf{k}, \quad t = 0$

4. $\mathbf{r} = e^t\,\mathbf{i} + \dfrac{2}{9}e^{2t}\,\mathbf{j}, \quad t = \ln 3$

5. $\mathbf{r} = (\sec t)\,\mathbf{i} + (\tan t)\,\mathbf{j} + \dfrac{4}{3}t\,\mathbf{k}, \quad t = \pi/6$

6. $\mathbf{r} = (2 \ln(t + 1))\,\mathbf{i} + t^2\,\mathbf{j} + \dfrac{t^2}{2}\,\mathbf{k}, \quad t = 1$

In Exercises 7–10, $\mathbf{r}$ is the position of a particle in space at time t. Find the angle between the velocity and acceleration vectors at time $t = 0$.

7. $\mathbf{r} = (3t + 1)\,\mathbf{i} + \sqrt{3}t\,\mathbf{j} + t^2\,\mathbf{k}$

8. $\mathbf{r} = \left(\dfrac{\sqrt{2}}{2}t\right)\mathbf{i} + \left(\dfrac{\sqrt{2}}{2}t - 16t^2\right)\mathbf{j}$

9. $\mathbf{r} = (\ln(t^2 + 1))\,\mathbf{i} + (\tan^{-1}t)\,\mathbf{j} + \sqrt{t^2 + 1}\,\mathbf{k}$

10. $\mathbf{r} = \dfrac{4}{9}(1 + t)^{3/2}\,\mathbf{i} + \dfrac{4}{9}(1 - t)^{3/2}\,\mathbf{j} + \dfrac{1}{3}t\,\mathbf{k}$

In Exercises 11 and 12, $\mathbf{r}$ is the position vector of a particle moving in space. Find the time or times in the given time interval when the velocity and acceleration vectors are orthogonal.

11. $\mathbf{r} = (t - \sin t)\,\mathbf{i} + (1 - \cos t)\,\mathbf{j}, \quad 0 \le t \le 2\pi$

12. $\mathbf{r} = (\sin t)\,\mathbf{i} + t\,\mathbf{j} + (\cos t)\,\mathbf{k}, \quad t \ge 0$

Evaluate the integrals in Exercises 13–16.

13. $\displaystyle\int_0^1 (t^3\,\mathbf{i} + 7\,\mathbf{j} + (t + 1)\,\mathbf{k})\,dt$

14. $\displaystyle\int_1^4 \left(\dfrac{1}{t}\,\mathbf{i} + \dfrac{1}{5 - t}\,\mathbf{j} + \dfrac{1}{2t}\,\mathbf{k}\right)dt$

15. $\displaystyle\int_{-\pi/4}^{\pi/4} ((\sin t)\,\mathbf{i} + (1 + \cos t)\,\mathbf{j} + (\sec^2 t)\,\mathbf{k})\,dt$

16. $\displaystyle\int_0^{\pi/3} ((\sec t \tan t)\,\mathbf{i} + (\tan t)\,\mathbf{j} + (2 \sin t \cos t)\,\mathbf{k})\,dt$

Exercises 17–20 give the position vectors of particles moving in the xy-plane. In each case, find the particle's velocity and acceleration vectors at the given times. Sketch them as vectors on the curve, together with the position vector.

17. Motion on the circle $x^2 + y^2 = 1$
$$\mathbf{r} = (\sin t)\,\mathbf{i} + (\cos t)\,\mathbf{j}; \quad t = \pi/4 \text{ and } \pi/2$$

18. Motion on the circle $x^2 + y^2 = 16$
$$\mathbf{r} = \left(4 \cos \dfrac{t}{2}\right)\mathbf{i} + \left(4 \sin \dfrac{t}{2}\right)\mathbf{j}; \quad t = \pi \text{ and } 3\pi/2$$

19. Motion on the cycloid $x = t - \sin t,\ y = 1 - \cos t$
$$\mathbf{r} = (t - \sin t)\,\mathbf{i} + (1 - \cos t)\,\mathbf{j}; \quad t = \pi \text{ and } 3\pi/2$$

20. Motion on the parabola $y = x^2 + 1$
$$\mathbf{r} = t\,\mathbf{i} + (t^2 + 1)\,\mathbf{j}; \quad t = -1, 0, \text{ and } 1$$

Exercises 21–24 give the velocity function and initial position of a particle moving in space. Find the particle's position function.

21. Velocity function: $\dfrac{d\mathbf{r}}{dt} = -t\,\mathbf{i} - t\,\mathbf{j} - t\,\mathbf{k}$
    Initial position: $\mathbf{r} = \mathbf{i} + 2\,\mathbf{j} + 3\,\mathbf{k} \quad$ when $t = 0$

22. Velocity function: $\dfrac{d\mathbf{r}}{dt} = (180t)\,\mathbf{i} + (180t - 16t^2)\,\mathbf{j}$
    Initial position: $\mathbf{r} = 100\,\mathbf{j} \quad$ when $t = 0$

23. Velocity function: $\dfrac{d\mathbf{r}}{dt} = \dfrac{3}{2}(t + 1)^{1/2}\,\mathbf{i} + e^{-t}\,\mathbf{j} + \dfrac{1}{t + 1}\,\mathbf{k}$
    Initial position: $\mathbf{r} = \mathbf{k} \quad$ when $t = 0$

24. Velocity function: $\dfrac{d\mathbf{r}}{dt} = (t^3 + 4t)\,\mathbf{i} + t\,\mathbf{j} + 2t^2\,\mathbf{k}$
    Initial position: $\mathbf{r} = \mathbf{i} + \mathbf{j} \quad$ when $t = 0$

Exercises 25 and 26 give the acceleration function and the initial position and velocity of a particle moving in space. Find the particle's position function.

25. Acceleration function: $\dfrac{d^2\mathbf{r}}{dt^2} = -32\,\mathbf{k}$
    Initial position and velocity: $\mathbf{r} = 100\,\mathbf{k} \quad$ and
    $$\dfrac{d\mathbf{r}}{dt} = 8\,\mathbf{i} + 8\,\mathbf{j} \quad \text{when } t = 0$$

26. Acceleration function: $\dfrac{d^2\mathbf{r}}{dt^2} = -(\mathbf{i} + \mathbf{j} + \mathbf{k})$
    Initial position and velocity: $\mathbf{r} = 10\,\mathbf{i} + 10\,\mathbf{j} + 10\,\mathbf{k} \quad$ and
    $$\dfrac{d\mathbf{r}}{dt} = 0 \quad \text{when } t = 0$$

27. A particle moves on a cycloid in the xy-plane in such a way that its position at time t is
$$\mathbf{r} = (t - \sin t)\,\mathbf{i} + (1 - \cos t)\,\mathbf{j}.$$
Find the maximum and minimum values of $|\mathbf{v}|$ and $|\mathbf{a}|$. (*Hint:* Find the extreme values of $|\mathbf{v}|^2$ and $|\mathbf{a}|^2$ first and take square roots later.)

28. A particle moves around the ellipse $(y/3)^2 + (z/2)^2 = 1$ in the yz-plane in such a way that its position at time t is

$$\mathbf{r} = (3 \cos t)\,\mathbf{j} + (2 \sin t)\,\mathbf{k}.$$

Find the maximum and minimum values of $|\mathbf{v}|$ and $|\mathbf{a}|$. (See the hint in Exercise 27.)

29. Suppose a particle moves on a sphere centered at the origin in such a way that its position vector $\mathbf{r}$ is a differentiable function of time (Fig. 11.6). Show that the particle's velocity is always orthogonal to $\mathbf{r}$.

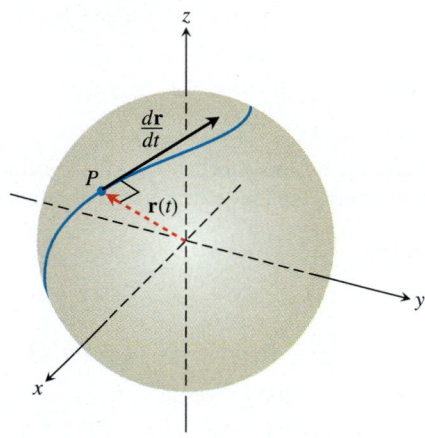

11.6 The velocity vector of a particle P that moves on the surface of a sphere is tangent to the sphere (Exercise 29).

30. Let $\mathbf{v}$ be a differentiable vector function of t. Show that if $\mathbf{v} \cdot (d\mathbf{v}/dt) = 0$ for all t, then $|\mathbf{v}|$ is constant.

31. *Derivatives of triple scalar products*
  a) Show that if $\mathbf{u}$, $\mathbf{v}$, and $\mathbf{w}$ are differentiable vector functions of t, then

$$\frac{d}{dt}(\mathbf{u} \cdot \mathbf{v} \times \mathbf{w}) \tag{15}$$

$$= \frac{d\mathbf{u}}{dt} \cdot \mathbf{v} \times \mathbf{w} + \mathbf{u} \cdot \frac{d\mathbf{v}}{dt} \times \mathbf{w} + \mathbf{u} \cdot \mathbf{v} \times \frac{d\mathbf{w}}{dt}.$$

  b) Show that Eq. (15) is equivalent to

$$\frac{d}{dt}\begin{vmatrix} u_1 & u_2 & u_3 \\ v_1 & v_2 & v_3 \\ w_1 & w_2 & w_3 \end{vmatrix} = \begin{vmatrix} \dfrac{du_1}{dt} & \dfrac{du_2}{dt} & \dfrac{du_3}{dt} \\ v_1 & v_2 & v_3 \\ w_1 & w_2 & w_3 \end{vmatrix} + \begin{vmatrix} u_1 & u_2 & u_3 \\ \dfrac{dv_1}{dt} & \dfrac{dv_2}{dt} & \dfrac{dv_3}{dt} \\ w_1 & w_2 & w_3 \end{vmatrix}$$

$$+ \begin{vmatrix} u_1 & u_2 & u_3 \\ v_1 & v_2 & v_3 \\ \dfrac{dw_1}{dt} & \dfrac{dw_2}{dt} & \dfrac{dw_3}{dt} \end{vmatrix} \tag{16}$$

Equation (16) says that the derivative of a 3 by 3 determinant of differentiable functions is the sum of the three determinants obtained from the original by differentiating one row at a time. The result extends to determinants of any order.

32. *(Continuation of Exercise 31)* Suppose that $\mathbf{r} = f(t)\,\mathbf{i} + g(t)\,\mathbf{j} + h(t)\,\mathbf{k}$ and that f, g, and h have derivatives through order three. Use Eq. (15) or (16) to show that

$$\frac{d}{dt}\left(\mathbf{r} \cdot \frac{d\mathbf{r}}{dt} \times \frac{d^2\mathbf{r}}{dt^2}\right) = \mathbf{r} \cdot \left(\frac{d\mathbf{r}}{dt} \times \frac{d^3\mathbf{r}}{dt^3}\right). \tag{17}$$

(Hint: Differentiate on the left and look for vectors whose products are zero.)

33. *The Constant Function Rule.* Prove that if $\mathbf{f}$ is the vector function with the constant value $\mathbf{C}$, then $d\mathbf{f}/dt = \mathbf{0}$.

34. *The Scalar Multiple Rule.* Prove that if $\mathbf{f}$ is a differentiable function of t and c is any real number, then

$$\frac{d(c\mathbf{f})}{dt} = c\,\frac{d\mathbf{f}}{dt}.$$

35. *The Sum and Difference Rules.* Prove that if $\mathbf{u}$ and $\mathbf{v}$ are differentiable functions of t, then

$$\frac{d}{dt}(\mathbf{u} + \mathbf{v}) = \frac{d\mathbf{u}}{dt} + \frac{d\mathbf{v}}{dt}$$

and

$$\frac{d}{dt}(\mathbf{u} - \mathbf{v}) = \frac{d\mathbf{u}}{dt} - \frac{d\mathbf{v}}{dt}.$$

36. *The component test for continuity at a point.* Show that the vector function $\mathbf{f}$ defined by the rule $\mathbf{f}(t) = f(t)\mathbf{i} + g(t)\,\mathbf{j} + h(t)\,\mathbf{k}$ is continuous at $t = t_0$ if and only if f, g, and h are continuous at t_0.

37. *Limits of cross products of vector functions.* Suppose that $\mathbf{f}(t) = f_1(t)\,\mathbf{i} + f_2(t)\,\mathbf{j} + f_3(t)\,\mathbf{k}$, $\mathbf{g}(t) = g_1(t)\,\mathbf{i} + g_2(t)\,\mathbf{j} + g_3(t)\,\mathbf{k}$, $\lim_{t \to t_0} \mathbf{f}(t) = \mathbf{A}$, and $\lim_{t \to t_0} \mathbf{g}(t) = \mathbf{B}$. Use the determinant formula for cross products and the Limit Product Rule for scalar functions to show that

$$\lim_{t \to t_0} (\mathbf{f}(t) \times \mathbf{g}(t)) = \left(\lim_{t \to t_0} \mathbf{f}(t)\right) \times \left(\lim_{t \to t_0} \mathbf{g}(t)\right).$$

38. *Differentiable vector functions are continuous.* Show that if $\mathbf{f}(t) = f(t)\,\mathbf{i} + g(t)\,\mathbf{j} + h(t)\,\mathbf{k}$ is differentiable at $t = t_0$, then it is continuous at t_0 as well.

39. Establish the following properties of integrable vector functions.
  a) The *Constant Scalar Multiple Rule:*

$$\int_a^b k\mathbf{f}(t)\,dt = k \int_a^b \mathbf{f}(t)\,dt \quad \text{(any scalar } k)$$

  The *Rule for Negatives,*

$$\int_a^b (-\mathbf{f}(t))\,dt = -\int_a^b \mathbf{f}(t)\,dt,$$

  is obtained by taking $k = -1$.
  b) The *Sum and Difference Rules:*

$$\int_a^b (\mathbf{f}(t) \pm \mathbf{g}(t))\,dt = \int_a^b \mathbf{f}(t)\,dt \pm \int_a^b \mathbf{g}(t)\,dt$$

c) The *Constant Vector Multiple Rules:*

$$\int_a^b \mathbf{C} \cdot \mathbf{f}(t)\, dt = \mathbf{C} \cdot \int_a^b \mathbf{f}(t)\, dt \qquad \text{(any constant vector } \mathbf{C})$$

and

$$\int_a^b \mathbf{C} \times \mathbf{f}(t)\, dt = \mathbf{C} \times \int_a^b \mathbf{f}(t)\, dt \qquad \text{(any constant vector } \mathbf{C})$$

40. *Products of scalar and vector functions.* Suppose that the scalar function $u(t)$ and the vector function $\mathbf{f}(t)$ are both defined for $a \leq t \leq b$.

a) Show that $u\mathbf{f}$ is continuous on $[a, b]$ if u and $\mathbf{f}$ are continuous on $[a, b]$.

b) If u and $\mathbf{f}$ are both differentiable on $[a, b]$, show that $u\mathbf{f}$ is differentiable on $[a, b]$ and that

$$\frac{d}{dt}(u\mathbf{f}) = u\frac{d\mathbf{f}}{dt} + \mathbf{f}\frac{du}{dt}.$$

41. *Antiderivatives of vector functions*

a) Use Corollary 3 of the Mean Value Theorem for scalar functions to show that if two vector functions $\mathbf{F}_1(t)$ and $\mathbf{F}_2(t)$ have identical derivatives on an interval I, then the functions differ by a constant vector value throughout I.

b) Use the result in (a) to show that if $\mathbf{F}(t)$ is any antiderivative of $\mathbf{f}(t)$ on I, then every other antiderivative of $\mathbf{f}$ on I equals $\mathbf{F}(t) + \mathbf{C}$ for some constant vector $\mathbf{C}$.

42. *The Fundamental Theorem of Calculus.* The Fundamental Theorem of Calculus for scalar functions of a real variable holds for vector functions of a real variable as well. Prove this by using the theorem for scalar functions to show first that if a vector function $\mathbf{f}(t)$ is continuous for $a \leq t \leq b$, then

$$\frac{d}{dt}\int_a^t \mathbf{f}(\tau)\, d\tau = \mathbf{f}(t)$$

at every point t of $[a, b]$. Then use the conclusion in part (b) of Exercise 41 to show that if $\mathbf{F}$ is any antiderivative of $\mathbf{f}$ on $[a, b]$ then

$$\int_a^b \mathbf{f}(t)\, dt = \mathbf{F}(b) - \mathbf{F}(a).$$

11.2 Modeling Projectile Motion

When we shoot a projectile into the air we usually want to know beforehand how far it will go (will it reach the target?), how high it will rise (will it clear the hill?), and when it will land (when do we get results?). We get this information from equations that calculate the answers we want from the direction and magnitude of the projectile's initial velocity vector, described in terms of the angle and speed at which the projectile is fired. The equations come from combining calculus and Newton's second law of motion in vector form. In this section, we derive these equations and show how to use them to get the information we want about projectile motion.

The Vector and Parametric Equations for Ideal Projectile Motion

To derive equations for projectile motion, we assume that the projectile behaves like a particle moving in a vertical coordinate plane and that the only force acting on the projectile during its flight is the constant force of gravity, which always points straight down. In practice, none of these assumptions really holds. The ground moves beneath the projectile as the earth turns, the air creates a frictional force that varies with the projectile's speed and altitude, and the force of gravity changes as the projectile moves along. All this must be taken into account by applying corrections to the predictions of the ideal equations we are about to derive. The corrections, however, are not the subject of this section.

We assume that our projectile is launched from the origin at time $t = 0$ into the first quadrant with an initial velocity $\mathbf{v}_0$ (Fig. 11.7). If $\mathbf{v}_0$ makes an angle α with the horizontal, then

$$\mathbf{v}_0 = (|\mathbf{v}_0| \cos \alpha)\, \mathbf{i} + (|\mathbf{v}_0| \sin \alpha)\, \mathbf{j}. \tag{1}$$

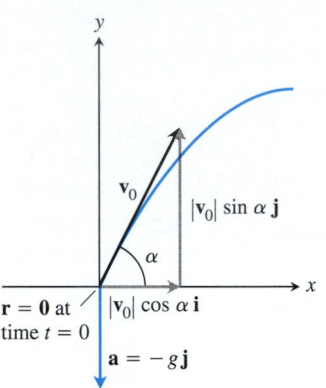

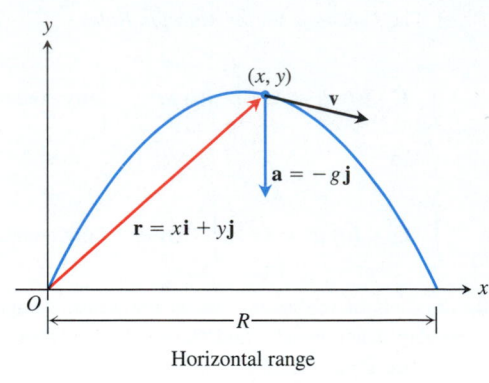

(a) Position, velocity, acceleration, and launch angle at $t = 0$.

(b) Position, velocity, and acceleration at a later time t.

11.7   The flight of an ideal projectile.

If we use the simpler notation v_0 for the initial speed $|\mathbf{v}_0|$, then

$$\mathbf{v}_0 = (v_0 \cos \alpha)\, \mathbf{i} + (v_0 \sin \alpha)\, \mathbf{j}. \qquad (2)$$

The projectile's initial position is

$$\mathbf{r}_0 = 0\, \mathbf{i} + 0\, \mathbf{j} = \mathbf{0}. \qquad (3)$$

If the only force acting on the projectile during its flight is the constant downward acceleration of gravity of magnitude g, then

$$\frac{d^2\mathbf{r}}{dt^2} = -g\mathbf{j}. \qquad (4)$$

We find the projectile's position as a function of time t by solving the following initial value problem:

Differential equation: $\dfrac{d^2\mathbf{r}}{dt^2} = -g\mathbf{j}$

Initial conditions: $\mathbf{r} = \mathbf{0}$ and $\dfrac{d\mathbf{r}}{dt} = \mathbf{v}_0$ when $t = 0$

The first integration gives

$$\frac{d\mathbf{r}}{dt} = -(gt)\, \mathbf{j} + \mathbf{v}_0. \qquad (5)$$

A second integration gives

$$\mathbf{r} = -\frac{1}{2}gt^2\, \mathbf{j} + \mathbf{v}_0 t + \mathbf{r}_0. \qquad (6)$$

Substituting the values of $\mathbf{v}_0$ and $\mathbf{r}_0$ from Eqs. (2) and (3) gives

$$\mathbf{r} = -\frac{1}{2}gt^2\, \mathbf{j} + \underbrace{(v_0 \cos \alpha)t\mathbf{i} + (v_0 \sin \alpha)t\mathbf{j}}_{\mathbf{v}_0 t} + \mathbf{0} \qquad (7)$$

$$= (v_0 \cos \alpha)t\; \mathbf{i} + \left((v_0 \sin \alpha)t - \frac{1}{2}gt^2\right) \mathbf{j}.$$

Equations of Motion for an Ideal Projectile Launched from the Origin at $t = 0$

Vector Form: $\mathbf{r} = (v_0 \cos \alpha)t \, \mathbf{i} + \left((v_0 \sin \alpha)t - \dfrac{1}{2}gt^2\right)\mathbf{j}$ (8)

Parametric Form: $x = (v_0 \cos \alpha)t, \quad y = (v_0 \sin \alpha)t - \dfrac{1}{2}gt^2$ (9)

where α is the **launch angle (firing angle, angle of elevation)** and v_0 is the **initial speed.**

If we measure time in seconds and distance in meters, g is 9.8 m/sec^2 and the equations in (9) give x and y in meters. If we measure time in seconds and distance in feet, then g is 32 ft/sec^2 and the equations in (9) give x and y in feet.

Example 1   A projectile is fired from the origin over horizontal ground at an initial speed of 500 m/sec at a launch angle of 60°. Where will the projectile be 10 sec later?

Solution   We use Eqs. (9) with $v_0 = 500$, $\alpha = 60°$, $g = 9.8$, and $t = 10$ to find the projectile's coordinates to the nearest meter 10 sec after firing:

$$x = (v_0 \cos \alpha)t = 500 \cdot \frac{1}{2} \cdot 10 = 2500 \text{ m}$$

$$y = (v_0 \sin \alpha)t - \frac{1}{2}gt^2$$

$$= 500 \cdot \frac{\sqrt{3}}{2} \cdot 10 - \frac{1}{2} \cdot 9.8 \cdot (10)^2$$

$$= 2500\sqrt{3} - 490$$

$$= 3840 \text{ m.} \qquad\qquad \text{(Calculator, rounded)}$$

Ten seconds after firing, the projectile is 3840 m in the air and 2500 m downrange.

Height, Flight Time, and Range

Equations (9) enable us to answer most questions about an ideal projectile fired from the origin.

The projectile reaches its highest point when its vertical velocity component is zero, that is, when

$$\frac{dy}{dt} = v_0 \sin \alpha - gt = 0, \qquad \text{or} \qquad t = \frac{v_0 \sin \alpha}{g}.$$

For this value of t, the value of y is

$$y_{\max} = (v_0 \sin \alpha)\left(\frac{v_0 \sin \alpha}{g}\right) - \frac{1}{2}g\left(\frac{v_0 \sin \alpha}{g}\right)^2 = \frac{(v_0 \sin \alpha)^2}{2g}.$$ (10)

To find when the projectile lands, when fired over horizontal ground, we set y equal to zero in Eqs. (9) and solve for t:

$$(v_0 \sin \alpha)t - \frac{1}{2} gt^2 = 0$$

$$t\left(v_0 \sin \alpha - \frac{1}{2} gt\right) = 0 \tag{11}$$

$$t = 0, \quad t = \frac{2v_0 \sin \alpha}{g}.$$

Since 0 is the time the projectile is fired, $(2v_0 \sin \alpha)/g$ must be the time when the projectile strikes the ground.

To find the projectile's range R, the distance from the origin to the point of impact on horizontal ground, we find the value of x when $t = (2v_0 \sin \alpha)/g$:

$$x = (v_0 \cos \alpha)t$$

$$R = (v_0 \cos \alpha)\left(\frac{2v_0 \sin \alpha}{g}\right)$$

$$= \frac{v_0^2}{g}(2 \sin \alpha \cos \alpha) = \frac{v_0^2}{g} \sin 2\alpha. \tag{12}$$

The range is largest when $\sin 2\alpha = 1$ or $\alpha = 45°$.

Height, Flight Time, and Range

For an ideal projectile fired from the origin over horizontal ground:

Maximum height:	$y_{max} = \dfrac{(v_0 \sin \alpha)^2}{2g}$	(13)
Flight time (time to impact):	$t = \dfrac{2v_0 \sin \alpha}{g}$	(14)
Range (distance to point of impact):	$R = \dfrac{v_0^2}{g} \sin 2\alpha$	(15)

Example 2   Find the maximum height, flight time, and range of a projectile fired from the origin over horizontal ground at an initial speed of 500 m/sec and a launch angle of 60° (same projectile as in Example 1).

Solution   (With a calculator, to the nearest meter and second)

Maximum height (Eq. 13): $y_{max} = \dfrac{(v_0 \sin \alpha)^2}{2g}$

$$= \frac{(500 \sin 60°)^2}{2(9.8)} = 9566 \text{ m}$$

Flight time (Eq. 14): $t = \dfrac{2v_0 \sin \alpha}{g}$

$$= \frac{2(500) \sin 60°}{9.8} = 88 \text{ sec}$$

Range (Eq. 15): $R = \dfrac{v_0^2}{g} \sin 2\alpha$

$$= \frac{(500)^2 \sin 120°}{9.8} = 22{,}092 \text{ m}$$

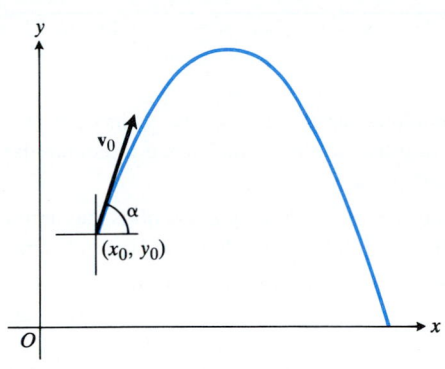

11.8 The path of a projectile fired from (x_0, y_0) with an initial velocity $\mathbf{v}_0$ at an angle of α degrees with the horizontal.

Firing from (x_0, y_0)

If we fire our ideal projectile from the point (x_0, y_0) instead of the origin (Fig. 11.8), the equations that replace Eqs. (9) are

$$x = x_0 + (v_0 \cos \alpha)t, \qquad y = y_0 + (v_0 \sin \alpha)t - \frac{1}{2} gt^2, \tag{16}$$

as you will be invited to show in Exercise 20.

Example 3 An athlete throws a 16-lb shot at an angle of 45° to the horizontal from 8 ft above the ground at an initial speed of 44 ft/sec (Fig. 11.9). How far from the inner edge of the stopboard does the shot land?

Solution We use a coordinate system in which the x-axis lies along the ground and the shot's coordinates at the time of launch ($t = 0$) are

$$x_0 = 0, \qquad y_0 = 8.$$

With these values for x_0 and y_0, and with $\alpha = 45°$ and $g = 32$ ft/sec^2, Eqs. (16) tell us that

$$x = 0 + (44 \cos 45°)t = 22\sqrt{2}\,t,$$

$$y = 8 + (44 \sin 45°)t - 16t^2 = 8 + 22\sqrt{2}\,t - 16t^2.$$

We find *when* the shot lands by setting y equal to zero and using the quadratic formula to solve for t, obtaining the positive solution

$$t = \frac{22\sqrt{2} + \sqrt{968 + 512}}{32} = 2.174 \text{ sec.} \qquad \text{(Calculator, rounded)}$$

We then find the value of x at this time from the equation $x = 22\sqrt{2}\,t$:

$$x = 22\sqrt{2}(2.174) = 67.64 \text{ ft.} \qquad \text{(Calculator, rounded)}$$

The shot lands about 67 ft 8 in. from the stopboard. Air resistance, variations in gravity, and the rotation of the earth have little effect on a dense object, like a shot, that moves relatively slowly and for only a few seconds. In such cases, we expect good results from the ideal model. ⌐

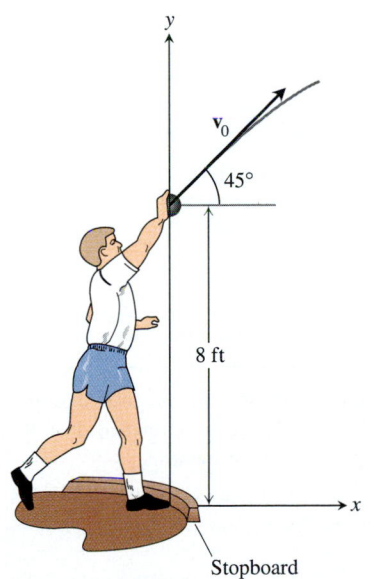

11.9 The shot put in Example 3. How far will the shot go?

Ideal Trajectories Are Parabolic

It is often claimed that water from a hose traces a parabola in the air, but anyone who looks closely enough will see this is not so. The air slows the water down and its forward progress is too slow at the end to match the rate at which it falls.

What is really being claimed is that ideal projectiles move along parabolas, and this we can see from Eqs. (9). If we substitute $t = x/(v_0 \cos \alpha)$ from the first equation into the second, we obtain the Cartesian coordinate equation

$$y = -\left(\frac{g}{2v_0^2 \cos^2\alpha}\right) x^2 + (\tan \alpha)\, x. \tag{17}$$

This equation is quadratic in x and linear in y, so its graph is a parabola.

EXERCISES 11.2

CALCULATOR   The projectiles in the following exercises are to be treated as ideal projectiles whose behavior is faithfully portrayed by the equations derived in the text. Most of the arithmetic, however, is realistic and is best done with a calculator. All launch angles are assumed to be measured from the horizontal. All projectiles are assumed to be fired from the origin over horizontal ground, unless stated otherwise.

1. A projectile is fired at a speed of 840 m/sec at an angle of 60°. How long will it take to get 21 km downrange?

2. Find the muzzle speed of a gun whose maximum range is 24.5 km.

3. A projectile is fired over level ground with an initial speed of 500 m/sec at an angle of elevation of 45°.
    a) When and how far away will the projectile strike?
    b) How high overhead will the projectile be when it is 5 km downrange?
    c) What is the highest the projectile will go?

4. A baseball is thrown from the stands 32 ft above the field at an angle of 30°. When and how far away will the ball strike the ground if its initial speed is 32 ft/sec?

5. Show that a projectile fired at an angle of α degrees, $0 < \alpha < 90$, has the same range as a projectile fired at the same speed at an angle of $(90 - \alpha)$ degrees. (In models that take air resistance into account, this symmetry is lost.)

6. What two angles of elevation will enable a projectile to reach a target 16 km downrange on the same level as the gun if the projectile's initial speed is 400 m/sec?

7. A spring gun at ground level fires a golf ball at an angle of 45°. The ball lands 10 m away. What was the ball's initial speed? For the same initial speed, find the two firing angles that make the range 6 m.

8. An electron in a TV tube is beamed horizontally at a speed of 5×10^6 m/sec toward the face of the tube 40 cm away. About how far will the electron drop before it hits?

9. Laboratory tests designed to find how far golf balls of different hardness go when hit with a driver showed that a 100-compression, Surlyn-covered, two-piece ball hit with a club-head speed of 100 mph at a launch angle of 9° carried 248.8 yd.
    a) What was the launch speed of the ball? (It was more than 100 mph. At the same time the club head was moving forward, the compressed ball was kicking away from the club face, adding to the ball's forward speed. The data are from "Does Compression Really Matter?" by Lew Fishman, *Golf Digest*, August 1986, pp. 35–37.)
    b) How much work was done on the ball getting it into the air? (Golf balls weigh 1.6 oz.)

10. Show that doubling a projectile's initial speed at a given launch angle multiplies its range by 4. By about what percentage should you increase the initial speed to double the height and range?

11. Show that a projectile attains three-quarters of its maximum height in half the time it takes to reach the maximum height.

12. A human cannonball is to be fired with an initial speed of $v_0 = 80\sqrt{10/3}$ ft/sec. The circus performer (of the right caliber, naturally) hopes to land on a special cushion located 200 ft downrange. The circus is being held in a large room with a flat ceiling 75 ft high. Can the performer be fired to the cushion without striking the ceiling? If so, what should the cannon's angle of elevation be?

13. A golf ball leaves the ground at a 30° angle at a speed of 90 ft/sec. Will it clear the top of a 30-ft tree 135 ft away?

14. A golf ball is hit from the tee to an elevated green with an initial speed of 116 ft/sec at an angle of elevation of 45°, as shown in Fig. 11.10. Will the ball reach the pin?

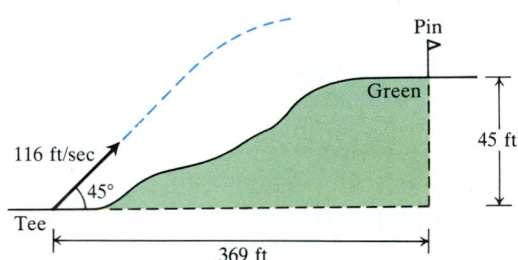

11.10   The tee and green in Exercise 14.

15. All other things being equal, the shot in Example 3 would have gone slightly farther if it had been launched at a 40° angle. How much farther? Answer in inches.

16. In Moscow, in 1987, Natalya Lisouskaya of the USSR set a women's world record by putting an 8-lb 13-oz shot 73 ft 10 in.
    a) Assuming that she launched the shot at a 40° angle to the horizontal 7 ft above the ground, what was the shot's initial speed?
    b) How much work was done on the shot to launch it at that speed? (Do not include the work it took to lift the shot 7 ft off the ground.)

17. A baseball hit by a Boston Red Sox player at a 20° angle from 3 ft above the ground just cleared the left end of the "Green Monster," the left-field wall in Fenway Park (Fig. 11.11). The wall there is 37 ft high and 315 ft from home plate. About how fast was the ball going? How long did it take the ball to reach the wall?

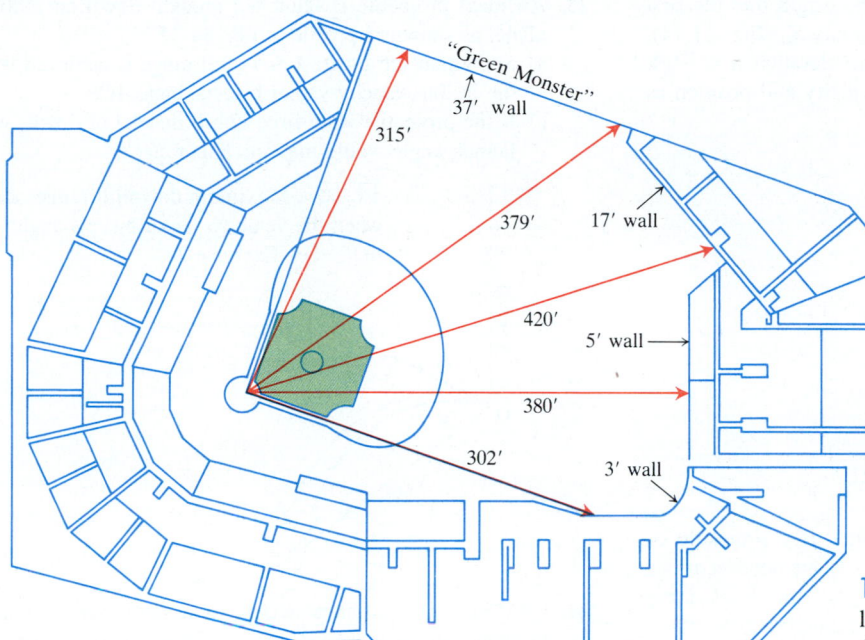

11.11 The "Green Monster," the left-field wall at Fenway Park in Boston (Exercise 17).

18. The multiflash photograph in Fig. 11.12 shows a model train engine moving at a constant speed on a straight track. As the engine moved along, a marble was fired into the air by a spring in the engine's smokestack. The marble, which continued to move with the same forward speed as the engine, rejoined the engine 1 sec after it was fired. Measure the angle the marble's path made with the horizontal and use the information to find how high the marble went and how fast the engine was moving.

PSSC Physics, Second Edition, 1965; D.C. Heath & Company with Education Development Center, Inc.

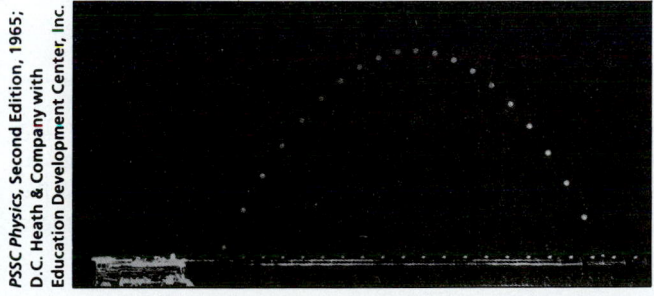

11.12 The train in Exercise 18.

19. Figure 11.13 shows an experiment with two marbles. Marble A was launched toward marble B with launch angle α and initial speed v_0. At the same instant, marble B was released to fall from rest $R \tan \alpha$ units directly above a spot

R units downrange from A. The marbles were found to collide regardless of the value of v_0. Was this mere coincidence, or must this happen? How do you know?

11.13 The marbles in Exercise 19.

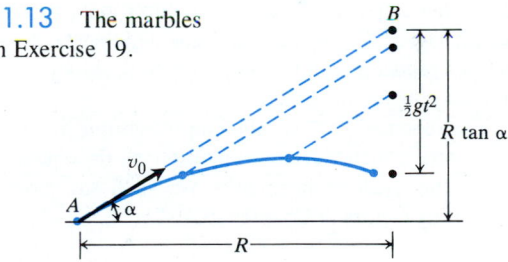

20. Derive the equations

$$x = x_0 + (v_0 \cos \alpha)t,$$

$$y = y_0 + (v_0 \sin \alpha)t - \frac{1}{2} gt^2$$

(Eqs. 16 in the text) by solving the following initial value problem for a vector $\mathbf{r}$ in the plane.

Differential equation: $\dfrac{d^2 \mathbf{r}}{dt^2} = -g \mathbf{j}$

Initial conditions: $\mathbf{r} = x_0 \mathbf{i} + y_0 \mathbf{j}$ and

$$\frac{d\mathbf{r}}{dt} = (v_0 \cos \alpha) \mathbf{i} + (v_0 \sin \alpha) \mathbf{j}$$

when $t = 0$

21. An ideal projectile, launched from the origin into the first octant at time $t = 0$ with initial velocity $\mathbf{v}_0$ (Fig. 11.14), experiences a constant downward acceleration $\mathbf{a} = -g\mathbf{k}$ from gravity. Find the projectile's velocity and position as functions of t.

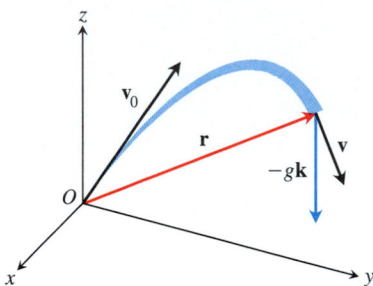

11.14   The projectile in Exercise 21.

22. *Air resistance proportional to velocity.*   If a projectile of mass m launched with initial velocity $\mathbf{v}_0$ encounters an air resistance proportional to its velocity, the total force $m(d^2\mathbf{r}/dt^2)$ on the projectile satisfies the equation

$$m\frac{d^2\mathbf{r}}{dt^2} = -mg\,\mathbf{j} - k\frac{d\mathbf{r}}{dt},$$

where k is the proportionality constant. Show that one integration of this equation gives

$$\frac{d\mathbf{r}}{dt} + \frac{k}{m}\mathbf{r} = \mathbf{v}_0 - gt\,\mathbf{j}.$$

Solve this equation. To do so, multiply both sides of the equation by $e^{(k/m)t}$. The left-hand side will then be the derivative of a product, with the result that both sides of the equation can now be integrated.

(The function $e^{(k/m)t}$ is called an *integrating factor* for the differential equation because multiplying the equation by it makes the equation integrable. We shall say more about integrating factors in Chapter 15.)

23. An ideal projectile is launched straight down an inclined plane, as shown in profile in Fig. 11.15.
a) Show that the greatest downhill range is achieved when the initial velocity vector bisects angle AOR.
b) If the projectile were fired uphill instead of down, what launch angle would maximize its range?

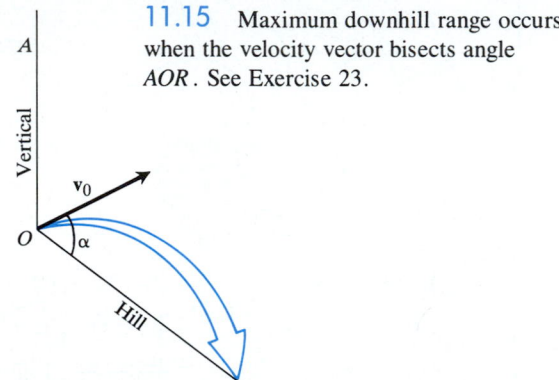

11.15   Maximum downhill range occurs when the velocity vector bisects angle AOR. See Exercise 23.

24. GRAPHER   If you have access to a parametric equation grapher and have not yet done Exercise 39 in Section 9.3, do it now. It is about ideal projectile motion.

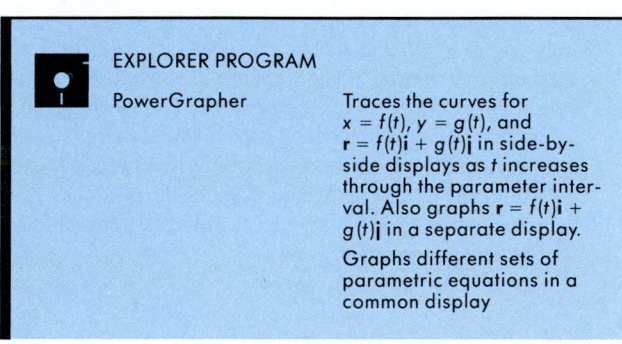

EXPLORER PROGRAM

PowerGrapher — Traces the curves for $x = f(t)$, $y = g(t)$, and $\mathbf{r} = f(t)\mathbf{i} + g(t)\mathbf{j}$ in side-by-side displays as t increases through the parameter interval. Also graphs $\mathbf{r} = f(t)\mathbf{i} + g(t)\mathbf{j}$ in a separate display.

Graphs different sets of parametric equations in a common display

11.3 Directed Distance and the Unit Tangent Vector **T**

As you can imagine, differentiable curves, especially those with continuous first and second derivatives, have been subjects of intense study, for their mathematical interest as well as for their applications to motion in space. In this section and the next, we describe some of the features that account for the importance of these curves.

Distance Along a Curve

One of the special features of space curves whose coordinate functions have continuous first derivatives is that, like plane curves, they are smooth enough to have a measurable length. This enables us to locate points along these curves by giving their directed distance s along the curve from some **base point,** the way we locate points on coordinate axes by giving their directed distance from the origin

11.16 Smooth curves can be scaled like coordinate axes or like tape measures that include negative numbers as well as positive numbers. The coordinate given to each point is its directed "highway" distance from the base point.

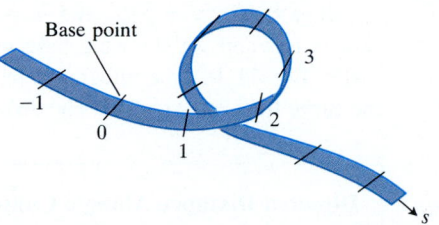

(Fig. 11.16). Although time is the natural parameter for describing a moving body's velocity and acceleration, the **directed distance coordinate** s is the natural parameter for studying a curve's geometry. The relationships between these parameters play an important role in calculations of space flight.

To calculate the distances along parametrized curves in space, we add a z-term to the length formula we use for curves in the plane.

DEFINITION

> If $x = f(t)$, $y = g(t)$, and $z = h(t)$ have continuous first derivatives with respect to t for $a \le t \le b$, and the position vector $\mathbf{r} = f(t)\,\mathbf{i} + g(t)\,\mathbf{j} + h(t)\,\mathbf{k}$ defined by these equations traces its curve exactly once as t moves from $t = a$ to $t = b$, then the **length** of the curve is given by the formula
>
> $$\text{Length} = \int_a^b \sqrt{\left(\frac{dx}{dt}\right)^2 + \left(\frac{dy}{dt}\right)^2 + \left(\frac{dz}{dt}\right)^2}\, dt. \tag{1}$$

Just as for plane curves, we can calculate the length of a curve in space from any convenient parametrization that meets the stated conditions. Again, we shall omit the proof.

Notice that the square root in Eq. (1) is $|\mathbf{v}|$, the length of the velocity vector $d\mathbf{r}/dt$. This lets us write the formula for length a shorter way.

Length Formula (Short Form)

$$\text{Length} = \int_a^b |\mathbf{v}|\, dt \tag{2}$$

As in linear motion, distance traveled is the integral of speed.

Example 1   Find the length of one turn of the helix

$$\mathbf{r} = (\cos t)\,\mathbf{i} + (\sin t)\,\mathbf{j} + t\mathbf{k}.$$

Solution   The helix makes one full turn as t runs from 0 to 2π (Fig. 11.17). We find the length of this portion of the curve from Eq. (2):

$$\text{Length} = \int_a^b |\mathbf{v}|\, dt = \int_0^{2\pi} \sqrt{(-\sin t)^2 + (\cos t)^2 + (1)^2}\, dt$$

$$= \int_0^{2\pi} \sqrt{2}\, dt = 2\pi\sqrt{2}.$$

This is $\sqrt{2}$ times the length of the circle in the xy-plane over which the helix stands.

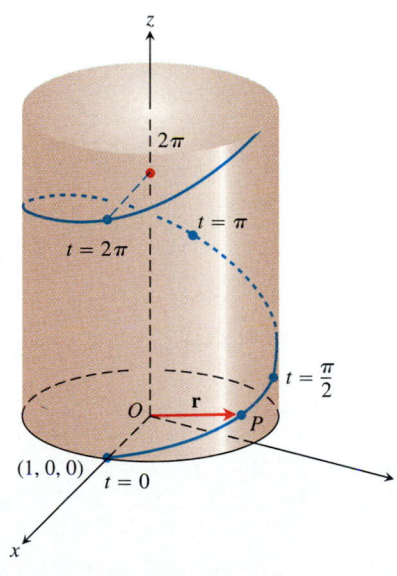

11.17   The helix $\mathbf{r} = (\cos t)\,\mathbf{i} + (\sin t)\,\mathbf{j} + t\,\mathbf{k}$ makes a complete turn during the time interval from $t = 0$ to $t = 2\pi$. We calculate the distance traveled by P by integrating $|\mathbf{v}|$ from $t = 0$ to $t = 2\pi$ (Example 1).

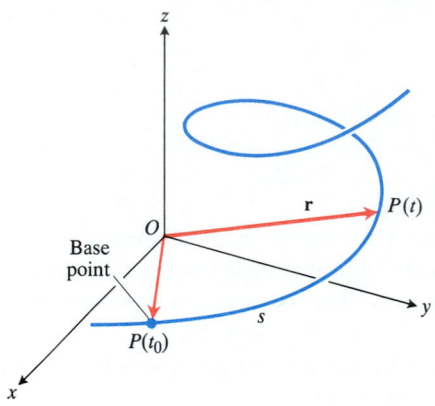

11.18 The directed distance along the curve from $P(t_0)$ to any point $P(t)$ is

$$s = \int_{t_0}^{t} |\mathbf{v}(\tau)| \, d\tau.$$

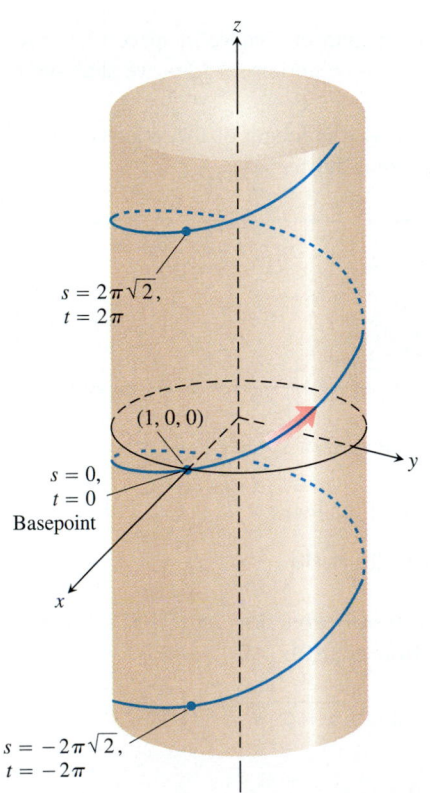

11.19 Directed distances along the helix $\mathbf{r} = (\cos t)\,\mathbf{i} + (\sin t)\,\mathbf{j} + t\,\mathbf{k}$ from the base point $(1, 0, 0)$ (Example 2).

If $x = f(t)$, $y = g(t)$, and $z = h(t)$ have continuous first derivatives with respect to t and we choose a base point $P(t_0)$ on the curve $\mathbf{r} = f(t)\,\mathbf{i} + g(t)\,\mathbf{j} + h(t)\,\mathbf{k}$ (Fig. 11.18), the integral of $|\mathbf{v}|$ from t_0 to t gives the directed distance along the curve from $P(t_0)$ to $P(t)$. The distance is a function of t, and we denote it by s.

Directed Distance Along a Curve from t_0 to t

$$s = \int_{t_0}^{t} \sqrt{x'(\tau)^2 + y'(\tau)^2 + z'(\tau)^2}\, d\tau = \int_{t_0}^{t} |\mathbf{v}(\tau)|\, d\tau \qquad (3)$$

The value of s is positive if t is greater than t_0 and negative if t is less than t_0.

Example 2   If $t_0 = 0$, the directed distance along the helix

$$\mathbf{r} = (\cos t)\,\mathbf{i} + (\sin t)\,\mathbf{j} + t\mathbf{k}$$

from t_0 to t is

$$
\begin{aligned}
s &= \int_{t_0}^{t} |\mathbf{v}(\tau)|\, d\tau && \text{(Eq. (3))} \\[2mm]
&= \int_{0}^{t} \sqrt{2}\, d\tau && \text{(Value from Example 1)} \\[2mm]
&= \sqrt{2}\,t.
\end{aligned}
$$

Thus, $s = 2\pi\sqrt{2}$ when $t = 2\pi$, $s = -2\pi\sqrt{2}$ when $t = -2\pi$, and so on (Fig. 11.19).

Example 3   *Distance Along a Line.*   Show that if $\mathbf{u} = u_1\,\mathbf{i} + u_2\,\mathbf{j} + u_3\,\mathbf{k}$ is a unit vector, then the directed distance along the line

$$\mathbf{r} = (x_0 + tu_1)\,\mathbf{i} + (y_0 + tu_2)\,\mathbf{j} + (z_0 + tu_3)\,\mathbf{k}$$

from the point $P_0(x_0, y_0, z_0)$ where $t = 0$ is t itself.

Solution

$$\mathbf{v} = \frac{d}{dt}(x_0 + tu_1)\,\mathbf{i} + \frac{d}{dt}(y_0 + tu_2)\,\mathbf{j} + \frac{d}{dt}(z_0 + tu_3)\,\mathbf{k} = u_1\,\mathbf{i} + u_2\,\mathbf{j} + u_3\,\mathbf{k} = \mathbf{u},$$

so

$$s(t) = \int_{0}^{t} |\mathbf{v}|\, d\tau = \int_{0}^{t} |\mathbf{u}|\, d\tau = \int_{0}^{t} 1\, d\tau = t.$$

If the derivatives beneath the radical in Eq. (3) are continuous, the Fundamental Theorem of Calculus tells us that s is a differentiable function of t whose derivative is

$$\frac{ds}{dt} = |\mathbf{v}(t)|. \qquad (4)$$

As we expect, the speed with which a particle moves along a smooth path is the magnitude of $\mathbf{v}$.

Notice that while the base point $P(t_0)$ plays a role in defining s in Eq. (3) it plays no role in Eq. (4). The rate at which a moving particle covers the distance along its path has nothing to do with how far away the base point is.

Notice also that as long as $|\mathbf{v}|$ is different from zero, *as we shall assume it to be in all examples from now on*, ds/dt is positive and s is an increasing function of t.

The Unit Tangent Vector **T**

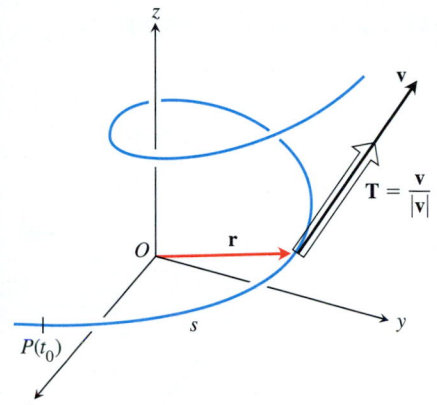

11.20 We find the unit tangent vector **T** by dividing **v** by $|\mathbf{v}|$.

Since ds/dt is positive for the curves we are considering from now on, s is one-to-one and has an inverse that gives t as a differentiable function of s (Section 6.1). The derivative of the inverse is

$$\frac{dt}{ds} = \frac{1}{ds/dt} = \frac{1}{|\mathbf{v}|}. \tag{5}$$

This makes **r** a differentiable function of s whose derivative can be calculated with the Chain Rule for vector-valued functions to be

$$\frac{d\mathbf{r}}{ds} = \frac{d\mathbf{r}}{dt}\frac{dt}{ds} = \mathbf{v} \cdot \frac{1}{|\mathbf{v}|} = \frac{\mathbf{v}}{|\mathbf{v}|}. \tag{6}$$

This tells us that $d\mathbf{r}/ds$ has the constant length

$$\left|\frac{d\mathbf{r}}{ds}\right| = \frac{1}{|\mathbf{v}|}|\mathbf{v}| = 1. \tag{7}$$

Together, Eqs. (6) and (7) say that $d\mathbf{r}/ds$ is a unit vector that points in the direction of **v**. We call $d\mathbf{r}/ds$ the unit tangent vector of the curve traced by **r** and denote it by **T** (Fig. 11.20).

DEFINITION

> The **unit tangent vector** of a differentiable curve $\mathbf{r} = \mathbf{f}(t)$ is
>
> $$\mathbf{T} = \frac{d\mathbf{r}}{ds} = \frac{d\mathbf{r}/dt}{ds/dt} = \frac{\mathbf{v}}{|\mathbf{v}|}. \tag{8}$$

The unit tangent vector **T** is a differentiable function of t whenever **v** is a differentiable function of t. As we shall see in the next section, **T** is one of three unit vectors in a traveling reference frame that is used to describe the motion of space vehicles and other bodies moving in three dimensions.

Example 4   Find the unit tangent vector of the helix

$$\mathbf{r} = (\cos t)\,\mathbf{i} + (\sin t)\,\mathbf{j} + t\,\mathbf{k}.$$

Solution

$$\mathbf{v} = (-\sin t)\,\mathbf{i} + (\cos t)\,\mathbf{j} + \mathbf{k}$$

$$|\mathbf{v}| = \sqrt{(-\sin t)^2 + (\cos t)^2 + (1)^2} = \sqrt{2}$$

$$\mathbf{T} = \frac{\mathbf{v}}{|\mathbf{v}|} = -\frac{\sin t}{\sqrt{2}}\,\mathbf{i} + \frac{\cos t}{\sqrt{2}}\,\mathbf{j} + \frac{1}{\sqrt{2}}\,\mathbf{k}$$

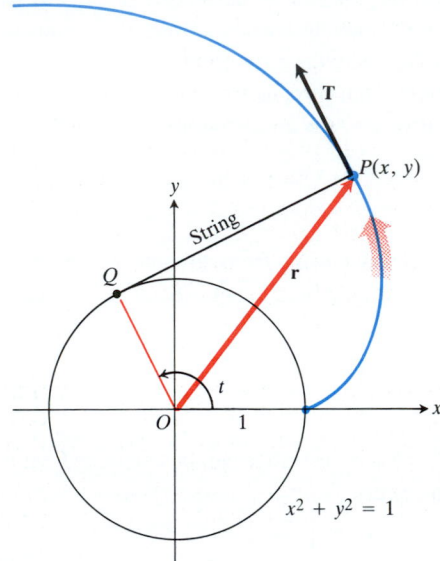

11.21   The *involute* of a circle is the path traced by the endpoint P of a string unwinding from a circle, here the unit circle in the xy-plane. The position vector of P can be shown to be
$\mathbf{r} = (\cos t + t \sin t)\,\mathbf{i} + (\sin t - t \cos t)\,\mathbf{j}$,
where t is the angle from the positive x-axis to Q. Example 5 derives a formula for the curve's unit tangent vector $\mathbf{T}$.

Example 5   *The Involute of a Circle (Fig. 11.21).*   Find the unit tangent vector of the curve

$$\mathbf{r} = (\cos t + t \sin t)\,\mathbf{i} + (\sin t - t \cos t)\,\mathbf{j}, \qquad t > 0.$$

Solution

$$\mathbf{v} = \frac{d\mathbf{r}}{dt} = (-\sin t + \sin t + t \cos t)\,\mathbf{i} + (\cos t - \cos t + t \sin t)\,\mathbf{j}$$

$$= (t \cos t)\,\mathbf{i} + (t \sin t)\,\mathbf{j}$$

$$|\mathbf{v}| = \sqrt{t^2\cos^2 t + t^2\sin^2 t} = \sqrt{t^2} = |t| = t \qquad (|t| = t \text{ because } t > 0)$$

$$\mathbf{T} = \frac{\mathbf{v}}{|\mathbf{v}|} = \frac{\mathbf{v}}{t} = (\cos t)\,\mathbf{i} + (\sin t)\,\mathbf{j}$$

Example 6   For the counter-clockwise motion

$$\mathbf{r} = (\cos t)\,\mathbf{i} + (\sin t)\,\mathbf{j}$$

around the unit circle,

$$\mathbf{v} = (-\sin t)\,\mathbf{i} + (\cos t)\,\mathbf{j}$$

is already a unit vector, so $\mathbf{T} = \mathbf{v}$ (Fig. 11.22).

11.22   For the counterclockwise motion $\mathbf{r} = (\cos t)\,\mathbf{i} + (\sin t)\,\mathbf{j}$ about the unit circle, the unit tangent vector is $\mathbf{v}$ itself (Example 6).

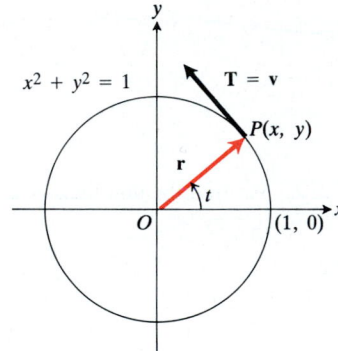

EXERCISES 11.3

In Exercises 1–8, find the curve's unit tangent vector. Also, find the length of the indicated portion of the curve.

1. $\mathbf{r} = (2 \cos t)\,\mathbf{i} + (2 \sin t)\,\mathbf{j} + \sqrt{5}t\,\mathbf{k}$,   from $t = 0$ to $t = \pi$

2. $\mathbf{r} = (6 \sin 2t)\,\mathbf{i} + (6 \cos 2t)\,\mathbf{j} + 5t\,\mathbf{k}$,   from $t = 0$ to $t = \pi$

3. $\mathbf{r} = t\mathbf{i} + (2/3)t^{3/2}\,\mathbf{k}$,   from $t = 0$ to $t = 8$

4. $\mathbf{r} = (\cos^3 t)\,\mathbf{j} + (\sin^3 t)\,\mathbf{k}$,   from $t = 0$ to $t = \pi/2$

5. $\mathbf{r} = (2 + t)\,\mathbf{i} - (t + 1)\,\mathbf{j} + t\,\mathbf{k}$,   from $t = 0$ to $t = 3$

6. $\mathbf{r} = 6t^3\,\mathbf{i} - 2t^3\,\mathbf{j} - 3t^3\,\mathbf{k}$,   from $t = -1$ to $t = 1$

7. $\mathbf{r} = (t \cos t)\,\mathbf{i} + (t \sin t)\,\mathbf{j} + (2\sqrt{2}/3)t^{3/2}\,\mathbf{k}$,   from $t = 0$ to $t = \pi$

8. $\mathbf{r} = (t \sin t + \cos t)\,\mathbf{i} + (t \cos t - \sin t)\,\mathbf{j}$,   from $t = 0$ to $t = \sqrt{2}$

In Exercises 9–12, find the directed distance along the curve from the point where $t = 0$ by evaluating the integral

$$s = \int_0^t |\mathbf{v}(\tau)|\,d\tau$$

from Eq. (3). Then find the length of the indicated portion of the curve.

9. $\mathbf{r} = (4 \cos t)\,\mathbf{i} + (4 \sin t)\,\mathbf{j} + 3t\,\mathbf{k}$,   from $t = 0$ to $t = \pi/2$

10. $\mathbf{r} = (\cos t + t \sin t)\,\mathbf{i} + (\sin t - t \cos t)\,\mathbf{j}$,   from $t = \pi/2$ to $t = \pi$

11. $\mathbf{r} = (e^t\cos t)\,\mathbf{i} + (e^t\sin t)\,\mathbf{j} + e^t\,\mathbf{k}$,   from $t = 0$ to $t = -\ln 4$

12. $\mathbf{r} = (1 + 2t)\,\mathbf{i} + (1 + 3t)\,\mathbf{j} + (6 - 6t)\,\mathbf{k}$,   from $t = 0$ to $t = -1$

13. Find the length of the curve

$$\mathbf{r} = (\sqrt{2}t)\,\mathbf{i} + (\sqrt{2}t)\,\mathbf{j} + (1 - t^2)\,\mathbf{k}$$

from $(0, 0, 1)$ to $(\sqrt{2}, \sqrt{2}, 0)$.

14. The length $2\pi\sqrt{2}$ of the turn of the helix in Example 1 is

also the length of the diagonal of a square 2π units on a side. Show how to obtain this square by cutting away and flattening a portion of the cylinder around which the helix winds.

15. a) Show that the curve

   $$\mathbf{r} = (\cos t)\,\mathbf{i} + (\sin t)\,\mathbf{j} + (1 - \cos t)\,\mathbf{k},\ 0 \le t \le 2\pi$$

   is an ellipse by showing that it is the curve of intersection of a cylinder and a plane. Find equations for the cylinder and plane.

   b) Sketch the ellipse on the cylinder. Add to your sketch the unit tangent vectors at $t = 0$, $\pi/2$, π, and $3\pi/2$.

   c) Show that the acceleration vector always lies parallel to the plane (orthogonal to a vector normal to the plane). This means that if you draw the acceleration as a vector attached to the ellipse, it will lie in the plane of the ellipse. Add the acceleration vectors for $t = 0$, $\pi/2$, π, and $3\pi/2$ to your sketch.

   d) Write an integral for the length of the ellipse. Do not try to evaluate the integral—it is nonelementary.

e) NUMERICAL INTEGRATOR Estimate the length of the ellipse to two decimal places.

16. *Length is independent of parametrization.*   To illustrate the fact that the length of a smooth space curve does not depend on the parametrization we use to compute it as long as the parametrization meets the conditions given with Eq. (1), calculate the length of one turn of the helix in Example 1 with the following parametrizations.

   a) $\mathbf{r} = (\cos 4t)\,\mathbf{i} + (\sin 4t)\,\mathbf{j} + 4t\,\mathbf{k},\quad 0 \le t \le \pi/2$
   b) $\mathbf{r} = (\cos(t/2))\,\mathbf{i} + (\sin(t/2))\,\mathbf{j} + (t/2)\,\mathbf{k},\quad 0 \le t \le 4\pi$
   c) $\mathbf{r} = (\cos t)\,\mathbf{i} - (\sin t)\,\mathbf{j} - t\,\mathbf{k},\quad -2\pi \le t \le 0$

EXPLORER PROGRAM

| Integral Evaluation | Evaluates definite integrals numerically |

11.4 Curvature, Torsion, and the Frenet Frame

In this section we define a frame of mutually orthogonal unit vectors that always travels with a body moving along a curve in space (Fig. 11.23). The frame has three vectors. The first is $\mathbf{T}$, the unit tangent vector. The second is $\mathbf{N}$, the unit vector that gives the direction of $d\mathbf{T}/ds$. The third is $\mathbf{B} = \mathbf{T} \times \mathbf{N}$. These vectors and their derivatives, when available, give useful information about a vehicle's orientation in space and about how the vehicle's path turns and twists as the vehicle moves along.

For example, the magnitude of the derivative $d\mathbf{T}/ds$ tells how much a vehicle's path turns to the left or right as it moves along; it is called the *curvature* of the vehicle's path. The number $-(d\mathbf{B}/ds) \cdot \mathbf{N}$ tells how much a vehicle's path rotates or twists as the vehicle moves along; it is called the *torsion* of the vehicle's path. Look at Fig. 11.23 again. If P were a train climbing a curved banked track, the rate at which the headlight turned from side to side per unit distance would be the curvature of the track. The rate at which the engine rotated about its longitudinal axis (the line of $\mathbf{T}$) would be the torsion.

We begin with curves in the plane and then move into space.

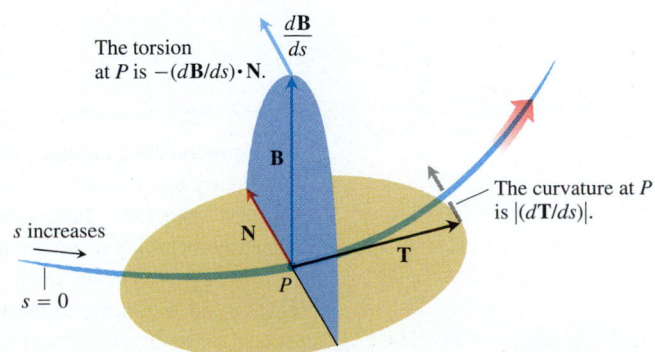

11.23 Every moving body travels with a frame of mutually orthogonal unit vectors (**T**, **N**, and **B**) that describes how the body moves.

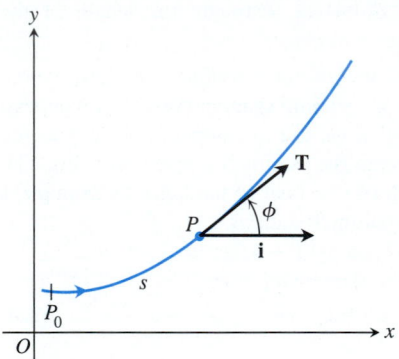

11.24   The value of $|d\phi/ds|$ at P is the curvature of the curve at P.

The Curvature of a Plane Curve

As we move along a differentiable curve in the plane, the unit tangent vector **T** turns as the curve bends. We measure the rate at which **T** turns by measuring the change in the angle ϕ that **T** makes with **i** (Fig. 11.24). At each point P, the absolute value of $d\phi/ds$, stated in radians per unit of length along the curve, is called the **curvature** at P. If $|d\phi/ds|$ is large, **T** turns sharply as we pass through P and the curvature at P is large. If $|d\phi/ds|$ is close to zero, **T** turns more slowly and the curvature at P is small. On circles and lines, the curvature is constant, as we shall see in Examples 1 and 2. On other curves, the curvature can vary from place to place. The traditional symbol for the curvature function is the Greek letter κ (kappa).

If $x = f(t)$ and $y = g(t)$ are twice-differentiable functions of t, we can derive a formula for the curvature of the curve $\mathbf{r} = f(t)\mathbf{i} + g(t)\mathbf{j}$ in the following way. In Newton's dot notation, in which $\dot{y}$ ("y dot") means dy/dt, $\ddot{y}$ ("y double dot") means d^2y/dt^2, and so on, we have

$$\tan \phi = \frac{dy}{dx} = \frac{dy/dt}{dx/dt} = \frac{\dot{y}}{\dot{x}} \qquad \text{and} \qquad \phi = \tan^{-1}\left(\frac{\dot{y}}{\dot{x}}\right). \tag{1}$$

Hence,

$$\frac{d\phi}{ds} = \frac{d\phi}{dt} \cdot \frac{dt}{ds} = \frac{1}{1 + (\dot{y}/\dot{x})^2} \frac{d}{dt}\left(\frac{\dot{y}}{\dot{x}}\right) \cdot \frac{1}{(\dot{x}^2 + \dot{y}^2)^{1/2}} \qquad \left(\frac{dt}{ds} = \frac{1}{|\mathbf{v}|}\right)$$

$$= \frac{\dot{x}^2}{(\dot{x}^2 + \dot{y}^2)^{3/2}} \frac{\dot{x}\ddot{y} - \dot{y}\ddot{x}}{\dot{x}^2} \tag{2}$$

$$= \frac{\dot{x}\ddot{y} - \dot{y}\ddot{x}}{|\mathbf{v}|^3}.$$

The curvature, therefore, is

$$\left|\frac{d\phi}{ds}\right| = \frac{|\dot{x}\ddot{y} - \dot{y}\ddot{x}|}{|\mathbf{v}|^3}. \tag{3}$$

The observation that $|\dot{x}\ddot{y} - \dot{y}\ddot{x}|$ is the magnitude of

$$\mathbf{v} \times \mathbf{a} = \begin{vmatrix} \mathbf{i} & \mathbf{j} & \mathbf{k} \\ \dot{x} & \dot{y} & 0 \\ \ddot{x} & \ddot{y} & 0 \end{vmatrix} \tag{4}$$

enables us to write Eq. (3) in a compact vector form.

Curvature

$$\kappa = \frac{|\mathbf{v} \times \mathbf{a}|}{|\mathbf{v}|^3} \tag{5}$$

Equation (5) calculates the curvature, a geometric property of the curve, from the velocity and acceleration of any vector representation of the curve in which $|\mathbf{v}|$ is different from zero. Take a moment to think about how remarkable this really is: From any formula for motion along a curve, no matter how variable the motion may be (as long as **v** is never zero), we can calculate a physical property of the curve that seems to have nothing to do with the way the curve is traversed.

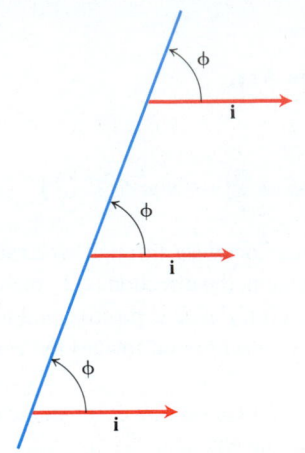

11.25 Along a line, the angle ϕ stays the same from point to point and the curvature, $d\phi/ds$, is zero (Example 1).

Example 1   *The Curvature of a Straight Line Is Zero.*   On a straight line, ϕ is constant (Fig. 11.25), so $d\phi/ds = 0$.

Example 2   *The Curvature of a Circle of Radius a Is $1/a$.*   To see why, parametrize the circle with the equation

$$\mathbf{r} = (a \cos t)\, \mathbf{i} + (a \sin t)\, \mathbf{j}.$$

Then

$$\mathbf{v} = -(a \sin t)\, \mathbf{i} + (a \cos t)\, \mathbf{j},$$

$$\mathbf{a} = -(a \cos t)\, \mathbf{i} - (a \sin t)\, \mathbf{j}.$$

Hence

$$\mathbf{v} \times \mathbf{a} = \begin{vmatrix} \mathbf{i} & \mathbf{j} & \mathbf{k} \\ -a \sin t & a \cos t & 0 \\ -a \cos t & -a \sin t & 0 \end{vmatrix} = (a^2\sin^2 t + a^2\cos^2 t)\, \mathbf{k} = a^2\, \mathbf{k},$$

$$|\mathbf{v}|^3 = \left[\sqrt{(-a \sin t)^2 + (a \cos t)^2}\right]^3 = a^3,$$

and

$$\kappa = \frac{|\mathbf{v} \times \mathbf{a}|}{|\mathbf{v}|^3} = \frac{|a^2\, \mathbf{k}|}{a^3} = \frac{a^2|\mathbf{k}|}{a^3} = \frac{1}{a}.$$

The larger the circle, the more gradually it curves.

Circle of Curvature and Radius of Curvature

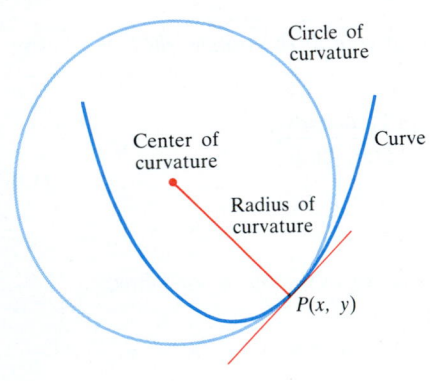

11.26 The osculating circle or circle of curvature at $P(x, y)$ lies toward the inner side of the curve.

The **circle of curvature** or **osculating circle** at a point P on a plane curve where $\kappa \neq 0$ is the circle in the plane of the curve that

1. is tangent to the curve at P (has the same tangent the curve has);
2. has the same curvature the curve has at P; and
3. lies toward the concave or inner side of the curve (as in Fig. 11.26).

The **radius of curvature** of the curve at P is the radius of the circle of curvature, which, according to Example 2, is

$$\text{Radius of curvature} = \rho = \frac{1}{\kappa}. \tag{6}$$

To calculate ρ, we calculate κ and take its reciprocal.

The **center of curvature** of the curve at P is the center of the circle of curvature.

The Principal Unit Normal Vector for Curves in the Plane

The vectors $d\mathbf{T}/ds$ and $d\mathbf{T}/d\phi$ are related by the Chain Rule equation

$$\frac{d\mathbf{T}}{ds} = \frac{d\mathbf{T}}{d\phi} \frac{d\phi}{ds}. \tag{7}$$

Furthermore, since $\mathbf{T}$ is a vector of constant length, $d\mathbf{T}/ds$ and $d\mathbf{T}/d\phi$ are both orthogonal to $\mathbf{T}$ (Section 11.1). They have the same direction if $d\phi/ds$ is positive

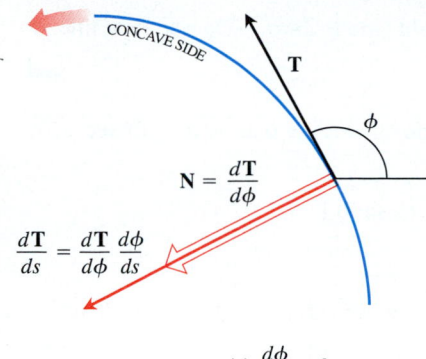

$$\frac{d\mathbf{T}}{ds} = \frac{d\mathbf{T}}{d\phi}\frac{d\phi}{ds}$$

(a) $\dfrac{d\phi}{ds} > 0$

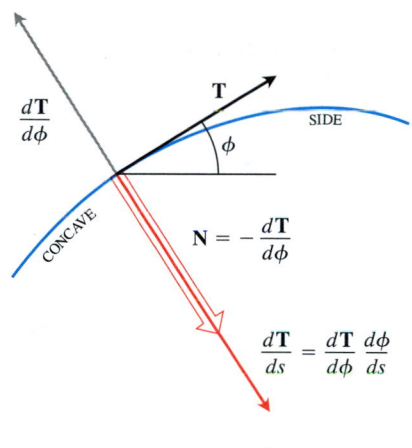

(b) $\dfrac{d\phi}{ds} < 0$

11.27 The vector $d\mathbf{T}/ds$, normal to the curve, always points inward. The principal unit normal vector $\mathbf{N}$ is the direction of $d\mathbf{T}/ds$.

and opposite directions if $d\phi/ds$ is negative. Now

$$\mathbf{T} = (\cos\phi)\,\mathbf{i} + (\sin\phi)\,\mathbf{j}, \tag{8}$$

and

$$\frac{d\mathbf{T}}{d\phi} = -(\sin\phi)\,\mathbf{i} + (\cos\phi)\,\mathbf{j} = \cos\!\left(\phi + \frac{\pi}{2}\right)\mathbf{i} + \sin\!\left(\phi + \frac{\pi}{2}\right)\mathbf{j} \tag{9}$$

is the unit vector obtained by rotating $\mathbf{T}$ counterclockwise through $\pi/2$ radians. Therefore, if we stand at a point on the curve facing in the direction of $\mathbf{T}$, the vector $d\mathbf{T}/ds = (d\mathbf{T}/d\phi)(d\phi/ds)$ will point toward the left if $d\phi/ds$ is positive and toward the right if $d\phi/ds$ is negative. In other words, $d\mathbf{T}/ds$ will point toward the concave side of the curve (Fig. 11.27).

Since $d\mathbf{T}/d\phi$ is a unit vector, the magnitude of $d\mathbf{T}/ds$ at any point on the curve is the curvature at that point, as we can see from the equation

$$\left|\frac{d\mathbf{T}}{ds}\right| = \left|\frac{d\mathbf{T}}{d\phi}\right|\left|\frac{d\phi}{ds}\right| = (1)(\kappa) = \kappa. \tag{10}$$

When $d\mathbf{T}/ds \neq \mathbf{0}$, its direction is given by the unit vector

$$\mathbf{N} = \frac{d\mathbf{T}/ds}{|d\mathbf{T}/ds|} = \frac{1}{\kappa}\frac{d\mathbf{T}}{ds}. \tag{11}$$

Since $\mathbf{N}$ points the same way $d\mathbf{T}/ds$ does, $\mathbf{N}$ is always orthogonal to $\mathbf{T}$ and directed toward the concave side of the curve. The vector $\mathbf{N}$ is the **principal unit normal vector** of the curve.

Because the directed distance is defined with ds/dt positive, dt/ds is positive and the Chain Rule gives

$$\begin{aligned}\mathbf{N} &= \frac{d\mathbf{T}/ds}{|d\mathbf{T}/ds|} = \frac{(d\mathbf{T}/dt)(dt/ds)}{|d\mathbf{T}/dt|\,|dt/ds|}\\[2mm] &= \frac{d\mathbf{T}/dt}{|d\mathbf{T}/dt|}.\end{aligned} \tag{12}$$

This formula enables us to find $\mathbf{N}$ without having to find ϕ, κ, or s first.

Example 3   Find $\mathbf{T}$ and $\mathbf{N}$ for the circular motion

$$\mathbf{r} = (\cos 2t)\,\mathbf{i} + (\sin 2t)\,\mathbf{j}.$$

Solution   We first find $\mathbf{T}$:

$$\mathbf{v} = -(2\sin 2t)\,\mathbf{i} + (2\cos 2t)\,\mathbf{j},$$

$$|\mathbf{v}| = \sqrt{4\sin^2 2t + 4\cos^2 2t} = 2,$$

$$\mathbf{T} = \frac{\mathbf{v}}{|\mathbf{v}|} = -(\sin 2t)\,\mathbf{i} + (\cos 2t)\,\mathbf{j}.$$

From this we find

$$\frac{d\mathbf{T}}{dt} = -(2\cos 2t)\,\mathbf{i} - (2\sin 2t)\,\mathbf{j},$$

$$\left|\frac{d\mathbf{T}}{dt}\right| = \sqrt{4\cos^2 2t + 4\sin^2 2t} = 2,$$

and

$$\mathbf{N} = \frac{d\mathbf{T}/dt}{|d\mathbf{T}/dt|} = -(\cos 2t)\,\mathbf{i} - (\sin 2t)\,\mathbf{j}. \qquad \text{(Eq. (12))}$$

Curvature for Curves in Space

In space there is no natural way to define an angle like ϕ with which to measure the change in $\mathbf{T}$ along a differentiable curve. But we still have s, the directed distance along the curve, and can define the curvature to be

$$\kappa = \left|\frac{d\mathbf{T}}{ds}\right|,$$

as it worked out to be for curves in the plane. The formula

$$\kappa = \frac{|\mathbf{v} \times \mathbf{a}|}{|\mathbf{v}|^3} \qquad (13)$$

still holds, as you will see if you do the calculation in Exercise 26.

Example 4   How do the values of a and b control the curvature of the helix (Fig. 11.28)

$$\mathbf{r} = (a\cos t)\,\mathbf{i} + (a\sin t)\,\mathbf{j} + bt\mathbf{k}? \qquad (a, b \ge 0)$$

Solution   We calculate the curvature with Eq. (13),

$$\mathbf{v} = -(a\sin t)\,\mathbf{i} + (a\cos t)\,\mathbf{j} + b\mathbf{k},$$

$$\mathbf{a} = -(a\cos t)\,\mathbf{i} - (a\sin t)\,\mathbf{j},$$

$$\mathbf{v} \times \mathbf{a} = \begin{vmatrix} \mathbf{i} & \mathbf{j} & \mathbf{k} \\ -a\sin t & a\cos t & b \\ -a\cos t & -a\sin t & 0 \end{vmatrix} = (ab\sin t)\,\mathbf{i} - (ab\cos t)\,\mathbf{j} + a^2\,\mathbf{k},$$

$$\kappa = \frac{|\mathbf{v} \times \mathbf{a}|}{|\mathbf{v}|^3} = \frac{\sqrt{a^2b^2 + a^4}}{(a^2 + b^2)^{3/2}} = \frac{a\sqrt{a^2 + b^2}}{(a^2 + b^2)^{3/2}} = \frac{a}{a^2 + b^2}. \qquad (14)$$

From Eq. (14) we see that increasing b for a fixed a decreases the curvature. Decreasing a for a fixed b eventually decreases the curvature as well. In other words, stretching a spring tends to straighten it.

If $b = 0$, the helix reduces to a circle of radius a and its curvature reduces to $1/a$, as it should. If $a = 0$, the helix becomes the z-axis, and its curvature reduces to 0, again as it should.

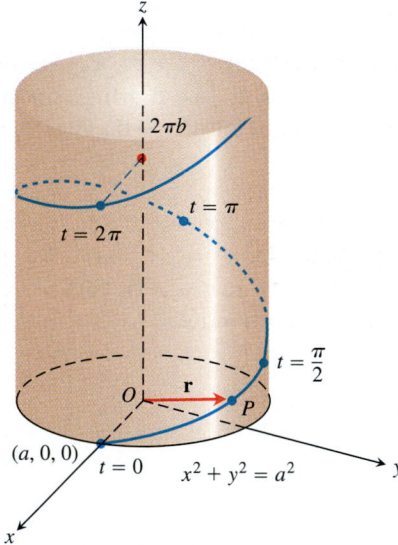

11.28   The helix $\mathbf{r} = (a\cos t)\,\mathbf{i} + (a\sin t)\,\mathbf{j} + bt\,\mathbf{k}$, drawn with a and b positive (Example 4).

N for Curves in Space

To define the principal unit normal vector of a curve in space, we use the same definition we use for a curve in the plane:

$$\mathbf{N} = \frac{d\mathbf{T}/ds}{|d\mathbf{T}/ds|} = \frac{1}{\kappa}\frac{d\mathbf{T}}{ds} = \frac{d\mathbf{T}/dt}{|d\mathbf{T}/dt|}. \qquad (15)$$

As before, $d\mathbf{T}/ds$ and its unit vector direction $\mathbf{N}$ are orthogonal to $\mathbf{T}$ because $\mathbf{T}$ has constant length, in this case 1.

Example 5   Find $\mathbf{N}$ for the helix in Example 4.

Solution   Using values from Example 4, we have

$$\mathbf{T} = \frac{\mathbf{v}}{|\mathbf{v}|} = \frac{-(a\sin t)\,\mathbf{i} + (a\cos t)\,\mathbf{j} + b\,\mathbf{k}}{\sqrt{a^2 + b^2}}$$

$$\frac{d\mathbf{T}}{dt} = -\frac{a}{\sqrt{a^2 + b^2}}\,((\cos t)\,\mathbf{i} + (\sin t)\,\mathbf{j})$$

$$\left|\frac{d\mathbf{T}}{dt}\right| = \frac{a}{\sqrt{a^2 + b^2}}\sqrt{\cos^2 t + \sin^2 t} = \frac{a}{\sqrt{a^2 + b^2}}$$

$$\mathbf{N} = \frac{d\mathbf{T}/dt}{|d\mathbf{T}/dt|} = -(\cos t)\,\mathbf{i} - (\sin t)\,\mathbf{j}.$$

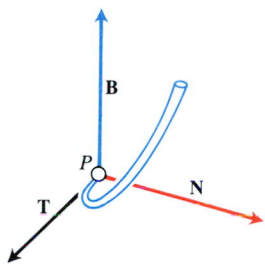

11.29   The vectors $\mathbf{T}$, $\mathbf{N}$, and $\mathbf{B}$ (in that order) make a right-handed frame of mutually orthogonal unit vectors in space. We call it the **Frenet** (fre-*nay*) **frame** (after Jean-Frédéric Frenet, 1816–1900).

Torsion and the Binormal Vector

The **binormal vector** of a curve in space is the vector $\mathbf{B} = \mathbf{T} \times \mathbf{N}$, a unit vector orthogonal to both $\mathbf{T}$ and $\mathbf{N}$ (Fig. 11.29). Together, $\mathbf{T}$, $\mathbf{N}$, and $\mathbf{B}$ define a moving right-handed vector frame that plays a significant role in calculating the flight paths of space vehicles.

The three planes determined by $\mathbf{T}$, $\mathbf{N}$, and $\mathbf{B}$ are shown in Fig. 11.30. The curvature $\kappa = |d\mathbf{T}/ds|$ can be thought of as the rate at which the normal plane turns as the point P moves along the curve. Similarly, the **torsion** $\tau = -(d\mathbf{B}/ds)\cdot\mathbf{N}$ is the rate at which the osculating plane twists about $\mathbf{T}$ as P moves along the curve. It gives a measure of how much the curve twists to the left or right.

The most widely used formula for torsion, derived in more advanced texts, is

$$\tau = \frac{\begin{vmatrix} \dot{x} & \dot{y} & \dot{z} \\ \ddot{x} & \ddot{y} & \ddot{z} \\ \dddot{x} & \dddot{y} & \dddot{z} \end{vmatrix}}{|\mathbf{v} \times \mathbf{a}|^2} \qquad (\mathbf{v} \times \mathbf{a} \neq \mathbf{0}). \tag{16}$$

This formula calculates the torsion directly from the derivatives of the component functions $x = f(t)$, $y = g(t)$, $z = h(t)$, that make up $\mathbf{r}$. The determinant's first row comes from $\mathbf{v}$, the second row comes from $\mathbf{a}$, and the third row comes from $\dot{\mathbf{a}}$.

Example 6   Find the torsion of the helix

$$\mathbf{r} = (\cos t)\,\mathbf{i} + (\sin t)\,\mathbf{j} + t\,\mathbf{k}.$$

Solution   We evaluate Eq. (16). We find the entries in the determinant by differentiating $\mathbf{r}$:

$$\mathbf{v} = -(\sin t)\,\mathbf{i} + (\cos t)\,\mathbf{j} + \mathbf{k}$$

$$\mathbf{a} = -(\cos t)\,\mathbf{i} - (\sin t)\,\mathbf{j}$$

$$\dot{\mathbf{a}} = (\sin t)\,\mathbf{i} - (\cos t)\,\mathbf{j}$$

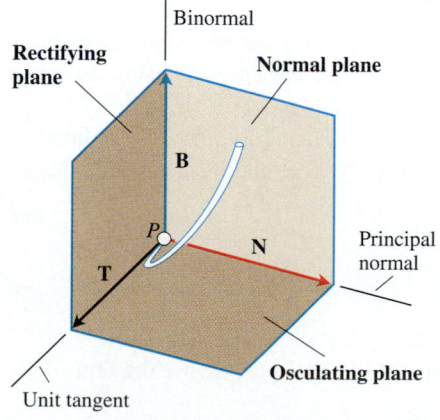

11.30   The three planes determined by $\mathbf{T}$, $\mathbf{N}$, and $\mathbf{B}$ have special names.

Then,

$$\tau = \frac{\begin{vmatrix} \dot{x} & \dot{y} & \dot{z} \\ \ddot{x} & \ddot{y} & \ddot{z} \\ \dddot{x} & \dddot{y} & \dddot{z} \end{vmatrix}}{|\mathbf{v} \times \mathbf{a}|^2} = \frac{\begin{vmatrix} -\sin t & \cos t & 1 \\ -\cos t & -\sin t & 0 \\ \sin t & -\cos t & 0 \end{vmatrix}}{\left| \begin{vmatrix} \mathbf{i} & \mathbf{j} & \mathbf{k} \\ -\sin t & \cos t & 1 \\ -\cos t & -\sin t & 0 \end{vmatrix} \right|^2}$$

$$= \frac{\cos^2 t + \sin^2 t}{|(\sin t)\,\mathbf{i} - (\cos t)\,\mathbf{j} + \mathbf{k}|^2} = \frac{1}{2}.$$

The Tangential and Normal Components of Acceleration

When a moving body is accelerated by gravity, brakes, a combination of rocket motors, or whatever, we usually want to know how much of the acceleration acts to move the body straight ahead in the direction of motion, that is, in the tangential direction $\mathbf{T}$. We can find out if we use the Chain Rule to rewrite $\mathbf{v}$ as

$$\mathbf{v} = \frac{d\mathbf{r}}{dt} = \frac{d\mathbf{r}}{ds}\frac{ds}{dt} = \mathbf{T}\frac{ds}{dt} \tag{17}$$

and differentiate both ends of this string of equalities to get

$$\mathbf{a} = \frac{d\mathbf{v}}{dt} = \frac{d}{dt}\left(\mathbf{T}\frac{ds}{dt}\right) = \frac{d^2 s}{dt^2}\,\mathbf{T} + \frac{ds}{dt}\frac{d\mathbf{T}}{dt}$$

$$= \frac{d^2 s}{dt^2}\,\mathbf{T} + \frac{ds}{dt}\left(\frac{d\mathbf{T}}{ds}\frac{ds}{dt}\right) = \frac{d^2 s}{dt^2}\,\mathbf{T} + \frac{ds}{dt}\left(\kappa \mathbf{N}\frac{ds}{dt}\right) \tag{18}$$

$$= \frac{d^2 s}{dt^2}\,\mathbf{T} + \kappa\left(\frac{ds}{dt}\right)^2 \mathbf{N}.$$

This is a remarkable equation. There is no $\mathbf{B}$ in it. No matter how the path of the moving body we are watching may appear to twist and turn, the acceleration $\mathbf{a}$ always lies in the plane of $\mathbf{T}$ and $\mathbf{N}$ orthogonal to $\mathbf{B}$. The equation also tells us exactly how much of the acceleration takes place tangent to the motion ($d^2 s/dt^2$) and how much takes place normal to the motion ($\kappa\,(ds/dt)^2$).

The **tangential** and **normal** scalar components of acceleration are

$$a_{\mathrm{T}} = \frac{d^2 s}{dt^2} = \frac{d}{dt}|\mathbf{v}| \qquad \text{and} \qquad a_{\mathrm{N}} = \kappa\left(\frac{ds}{dt}\right)^2 = \kappa|\mathbf{v}|^2. \tag{19}$$

That is,

$$\mathbf{a} = \frac{d^2 s}{dt^2}\,\mathbf{T} + \kappa\left(\frac{ds}{dt}\right)^2 \mathbf{N}. \tag{20}$$

Notice that the normal scalar component of the acceleration is the curvature times the square of the speed. This explains why you have to hold on when your car makes a sharp (large κ), high-speed (large $|\mathbf{v}|$) turn.

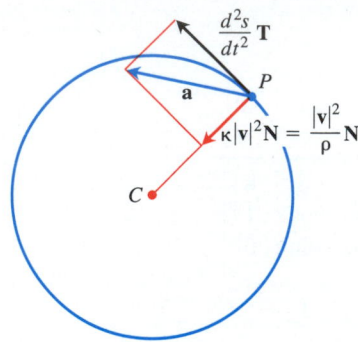

11.31 The tangential and normal components of the acceleration of a body speeding up as it moves counterclockwise around a circle of radius ρ.

If a body moves in a circle at a constant speed, d^2s/dt^2 is zero and all the acceleration points along $\mathbf{N}$ toward the circle's center. If the body is speeding up or slowing down, $\mathbf{a}$ has a nonzero tangential component (Fig. 11.31).

To calculate a_N we usually use the formula

$$a_\mathrm{N} = \sqrt{|\mathbf{a}|^2 - a_\mathrm{T}^2}, \tag{21}$$

which comes from solving the equation $|\mathbf{a}|^2 = \mathbf{a} \cdot \mathbf{a} = a_\mathrm{T}^2 + a_\mathrm{N}^2$ for a_N. With this formula we can find a_N without having to calculate κ first.

Example 7   Without finding $\mathbf{T}$ and $\mathbf{N}$, write the acceleration of the motion

$$\mathbf{r} = (\cos t + t \sin t)\,\mathbf{i} + (\sin t - t \cos t)\,\mathbf{j}, \qquad t > 0$$

in the form $\mathbf{a} = a_\mathrm{T}\mathbf{T} + a_\mathrm{N}\mathbf{N}$. (The path of the motion is the involute of the circle in Fig. 11.32.)

Solution   We use the first of Eqs. (19) to find a_T:

$$\mathbf{v} = (t \cos t)\,\mathbf{i} + (t \sin t)\,\mathbf{j} \qquad \left(\begin{array}{l}\text{Value from Section 11.3,}\\ \text{Example 5}\end{array}\right)$$

$$|\mathbf{v}| = \sqrt{t^2\cos^2 t + t^2\sin^2 t} = \sqrt{t^2} = |t| = t \qquad (t > 0)$$

$$a_\mathrm{T} = \frac{d}{dt}|\mathbf{v}| = \frac{d}{dt}(t) = 1 \qquad \text{(Eq. 19)}$$

Knowing a_T, we use Eq. (21) to find a_N:

$$\mathbf{a} = (\cos t - t \sin t)\,\mathbf{i} + (\sin t + t \cos t)\,\mathbf{j}$$

$$|\mathbf{a}|^2 = t^2 + 1 \qquad \left(\begin{array}{l}\text{after some}\\ \text{algebra}\end{array}\right)$$

$$a_\mathrm{N} = \sqrt{|\mathbf{a}|^2 - a_\mathrm{T}^2} = \sqrt{(t^2 + 1) - (1)} = \sqrt{t^2} = t.$$

We then use Eq. (20) to find $\mathbf{a}$:

$$\mathbf{a} = a_\mathrm{T}\mathbf{T} + a_\mathrm{N}\mathbf{N} = (1)\mathbf{T} + (t)\mathbf{N} = \mathbf{T} + t\mathbf{N}.$$

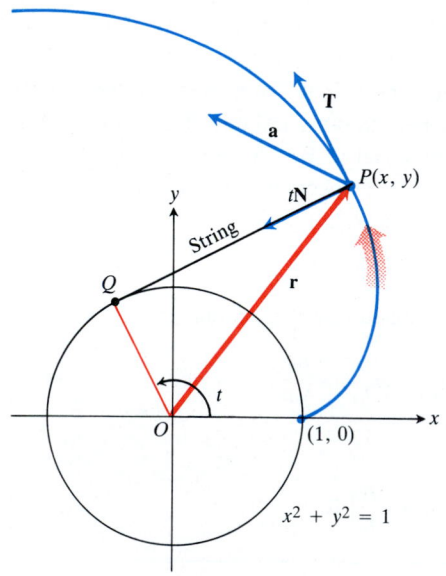

11.32 The tangential and normal components of the motion

$$\mathbf{r} = (\cos t + t \sin t)\,\mathbf{i} + (\sin t - t \cos t)\,\mathbf{j}$$

shown here are $\mathbf{T}$ and $t\mathbf{N}$ (Example 7).

Motion Formulas ($|\mathbf{v}| \neq 0$)

Unit tangent vector:	$\mathbf{T} = \dfrac{\mathbf{v}}{|\mathbf{v}|}$
Principal unit normal vector:	$\mathbf{N} = \dfrac{d\mathbf{T}/dt}{|d\mathbf{T}/dt|}$
Binormal vector:	$\mathbf{B} = \mathbf{T} \times \mathbf{N}$
Curvature:	$\kappa = \left|\dfrac{d\mathbf{T}}{ds}\right| = \dfrac{|\mathbf{v} \times \mathbf{a}|}{|\mathbf{v}|^3}$
Torsion:	$\tau = -\dfrac{d\mathbf{B}}{ds} \cdot \mathbf{N} = \dfrac{\begin{vmatrix} \dot{x} & \dot{y} & \dot{z} \\ \ddot{x} & \ddot{y} & \ddot{z} \\ \dddot{x} & \dddot{y} & \dddot{z} \end{vmatrix}}{|\mathbf{v} \times \mathbf{a}|^2}$
Tangential and normal scalar components of acceleration:	$\mathbf{a} = a_\mathrm{T}\mathbf{T} + a_\mathrm{N}\mathbf{N}$ $a_\mathrm{T} = \dfrac{d}{dt}|\mathbf{v}|$ $a_\mathrm{N} = \sqrt{|\mathbf{a}|^2 - a_\mathrm{T}^2}$

EXERCISES 11.4

Find **T**, **N**, and κ for the plane curves in Exercises 1–4.

1. $\mathbf{r} = t\mathbf{i} + (\ln \cos t)\mathbf{j}, \quad -\pi/2 < t < \pi/2$
2. $\mathbf{r} = (\ln \sec t)\mathbf{i} + t\mathbf{j}, \quad -\pi/2 < t < \pi/2$
3. $\mathbf{r} = (2t + 3)\mathbf{i} + (5 - t^2)\mathbf{j}$
4. $\mathbf{r} = (\cos t + t \sin t)\mathbf{i} + (\sin t - t \cos t)\mathbf{j}, \quad t > 0$

Find **T**, **N**, **B**, κ, and τ for the space curves in Exercises 5–12.

5. $\mathbf{r} = (3 \sin t)\mathbf{i} + (3 \cos t)\mathbf{j} + 4t\mathbf{k}$
6. $\mathbf{r} = (\cos t + t \sin t)\mathbf{i} + (\sin t - t \cos t)\mathbf{j} + 3\mathbf{k}$
7. $\mathbf{r} = (e^t \cos t)\mathbf{i} + (e^t \sin t)\mathbf{j} + 2\mathbf{k}$
8. $\mathbf{r} = (6 \sin 2t)\mathbf{i} + (6 \cos 2t)\mathbf{j} + 5t\mathbf{k}$
9. $\mathbf{r} = (t^3/3)\mathbf{i} + (t^2/2)\mathbf{j}, \quad t > 0$
10. $\mathbf{r} = (\cos^3 t)\mathbf{i} + (\sin^3 t)\mathbf{j}, \quad 0 < t < \pi/2$
11. $\mathbf{r} = t\mathbf{i} + a(\cosh(t/a))\mathbf{j}, \quad a > 0$
12. $\mathbf{r} = (\cosh t)\mathbf{i} - (\sinh t)\mathbf{j} + t\mathbf{k}$

In Exercises 13–16, write **a** in the form $\mathbf{a} = a_\mathrm{T}\mathbf{T} + a_\mathrm{N}\mathbf{N}$ without finding **T** and **N**.

13. $\mathbf{r} = (2t + 3)\mathbf{i} + (t^2 - 1)\mathbf{j}$
14. $\mathbf{r} = \ln(t^2 + 1)\mathbf{i} + (t - 2\tan^{-1}t)\mathbf{j}$
15. $\mathbf{r} = (a \cos t)\mathbf{i} + (a \sin t)\mathbf{j} + bt\mathbf{k}$
16. $\mathbf{r} = (1 + 3t)\mathbf{i} + (t - 2)\mathbf{j} - 3t\mathbf{k}$

In Exercises 17–20, write **a** in the form $\mathbf{a} = a_\mathrm{T}\mathbf{T} + a_\mathrm{N}\mathbf{N}$ at the given value of t without finding **T** and **N**. You can save yourself some work by evaluating the vectors **v** and **a** at the given value of t *before* finding their lengths.

17. $\mathbf{r} = (t + 1)\mathbf{i} + 2t\mathbf{j} + t^2\mathbf{k}, \quad t = 1$
18. $\mathbf{r} = (t \cos t)\mathbf{i} + (t \sin t)\mathbf{j} + t^2\mathbf{k}, \quad t = 0$
19. $\mathbf{r} = t^2\mathbf{i} + (t + (1/3)t^3)\mathbf{j} + (t - (1/3)t^3)\mathbf{k}, \quad t = 0$
20. $\mathbf{r} = (e^t \cos t)\mathbf{i} + (e^t \sin t)\mathbf{j} + \sqrt{2}e^t\mathbf{k}, \quad t = 0$

In Exercises 21 and 22, find **r**, **T**, **N**, and **B** at the given value of t. Then find equations for the osculating, normal, and rectifying planes at that value of t.

21. $\mathbf{r} = (\cos t)\mathbf{i} + (\sin t)\mathbf{j} - \mathbf{k}, \quad t = \pi/4$
22. $\mathbf{r} = (\cos t)\mathbf{i} + (\sin t)\mathbf{j} + t\mathbf{k}, \quad t = 0$

23. The speedometer on your car reads a steady 35 mph. Could you be accelerating? Explain.

24. Show that if a particle's speed is constant its acceleration is either zero or normal to its path.

25. Show that a particle's speed will be constant if the acceleration is always perpendicular to the velocity.

26. *How to derive the formula* $\kappa = |\mathbf{v} \times \mathbf{a}|/|\mathbf{v}|^3$ *for space curves.* To derive the formula, carry out these steps.

STEP 1: Use equations $\mathbf{v} = \mathbf{T}(ds/dt)$ and $\mathbf{a} = (d^2s/dt^2)\mathbf{T} + \kappa(ds/dt)^2\mathbf{N}$ to find a formula for $|\mathbf{v} \times \mathbf{a}|$.

STEP 2: Solve the resulting equation for κ, assuming $\mathbf{v} \times \mathbf{a} \neq 0$.

27. If a_N and $|\mathbf{v}|$ are known, the equation $a_\mathrm{N} = \kappa|\mathbf{v}|^2$ gives a convenient way to find curvature. Use it to find the curvature and radius of curvature of the curve

$$\mathbf{r} = (\cos t + t \sin t)\mathbf{i} + (\sin t - t \cos t)\mathbf{j}, \quad t > 0$$

in Example 7.

28. Show that the torsion of any sufficiently differentiable plane curve is zero when defined.

29. Show that a moving particle will move in a straight line if the normal component of its acceleration is zero.

30. Show that κ and τ are both zero for the line

$$\mathbf{r} = (x_0 + At)\mathbf{i} + (y_0 + Bt)\mathbf{j} + (z_0 + Ct)\mathbf{k}.$$

31. *Maximizing the curvature of a helix.* In Example 4, we found the curvature of the helix $\mathbf{r} = (a \cos t)\mathbf{i} + (a \sin t)\mathbf{j} + bt\mathbf{k}$ $(a, b \geq 0)$ to be $\kappa = a/(a^2 + b^2)$. What is the largest value κ can have for a given value of b?

32. *The torsion of a helix.* Find the torsion of the helix

$$\mathbf{r} = (a \cos t)\mathbf{i} + (a \sin t)\mathbf{j} + bt\mathbf{k} \quad (a, b > 0).$$

What is the largest value τ can have for a given value of a?

33. *Differentiable curves with zero torsion lie in planes.* That a sufficiently differentiable curve with zero torsion lies in a plane is a special case of the fact that a particle whose velocity remains normal to a fixed vector **C** moves in a plane perpendicular to **C**. This, in turn, can be viewed as the solution of the following problem in calculus.

Suppose $\mathbf{r} = f(t)\mathbf{i} + g(t)\mathbf{j} + h(t)\mathbf{k}$ is twice differentiable for all t in an interval $[a, b]$, that $\mathbf{r} = 0$ when $t = a$, and that $\mathbf{v} \cdot \mathbf{k} = 0$ for all t in $[a, b]$. Then $h(t) = 0$ for all t in $[a, b]$.

Solve this problem. (*Hint:* Start with $\mathbf{a} = d^2\mathbf{r}/dt^2$ and apply the initial conditions in reverse order.)

34. *A formula that calculates torsion from* **B** *and* **v**. If we start with the definition $\tau = -(d\mathbf{B}/ds) \cdot \mathbf{N}$ and apply the Chain Rule to rewrite $d\mathbf{B}/ds$ as

$$\frac{d\mathbf{B}}{ds} = \frac{d\mathbf{B}}{dt}\frac{dt}{ds} = \frac{d\mathbf{B}}{dt}\frac{1}{|\mathbf{v}|},$$

we arrive at the formula

$$\tau = -\frac{1}{|\mathbf{v}|}\left(\frac{d\mathbf{B}}{dt} \cdot \mathbf{N}\right).$$

The advantage of this formula over the one in Eq. (16) is that it is easier to derive and state. The disadvantage is that it can take a lot of work to evaluate without a computer. Use the new formula to find the torsion of the helix in Example 6.

35. Find an equation for the circle of curvature of the curve $\mathbf{r} = t\mathbf{i} + (\sin t)\mathbf{j}$ at the point $(\pi/2, 1)$. (The curve parametrizes the graph of $y = \sin x$ in the xy-plane.)

36. a) GRAPHER Graph the curve $\mathbf{r} = (2 \ln t)\mathbf{i} - (t + (1/t))\mathbf{j}$ for $0 < t < 10$.

b) Find an equation for the circle of curvature at the point $(0, -2)$.

37. Equation (2) says that the curvature of a plane curve $\mathbf{r} = f(t)\mathbf{i} + g(t)\mathbf{j}$ defined by twice differentiable functions $x = f(t)$ and $y = g(t)$ is given by the formula

$$\kappa = \frac{|\dot{x}\ddot{y} - \dot{y}\ddot{x}|}{(\dot{x}^2 + \dot{y}^2)^{3/2}}.$$

Apply the formula to find the curvatures of the following curves.

a) $\mathbf{r} = t\mathbf{i} + (\ln \sin t)\mathbf{j}, \quad 0 < t < \pi$

b) $\mathbf{r} = (\tan^{-1}(\sinh t))\mathbf{i} + (\ln \cosh t)\mathbf{j}$.

38. *A formula for the curvature of the graph of a function in the xy-plane.*

a) The graph $y = f(x)$ in the xy-plane automatically has the parametrization $x = x$, $y = f(x)$, and the vector formula $\mathbf{r} = x\mathbf{i} + f(x)\mathbf{j}$. Use this formula to show that if f is a twice-differentiable function of x, then

$$\kappa = \frac{|f''(x)|}{(1 + (f'(x))^2)^{3/2}}.$$

b) Use the formula in (a) to find the curvature of $y = \ln (\cos x)$, $-\pi/2 < x < \pi/2$. Compare your answer with the answer in Exercise 1.

39. a) Use the formula in Exercise 38 to find the curvature of the curve $y = e^x$ at the point $(0, 1)$. Then find an equation for the osculating circle at this point. Sketch the curve and circle together.

b) Use the circle's equation to find the values of dy/dx and d^2y/dx^2 for the circle at the point $(1, 0)$. Show that these derivatives have the same values as the corresponding derivatives of the function $y = e^x$ at this point.

40. *Smooth curves in the plane with zero curvature are linear.* Suppose that $x = f(t)$ and $y = g(t)$ have continuous first derivatives with respect to t and that

$$\frac{d\phi}{ds} = \frac{d}{ds}\tan^{-1}\left(\frac{\dot{y}}{\dot{x}}\right) = 0$$

for all t in some interval I. Show that $y = C_1 x + C_2$ for some constants C_1 and C_2. Thus, the curve $\mathbf{r} = f(t)\mathbf{i} + g(t)\mathbf{j}$, $t \, \varepsilon \, I$, lies along a straight line if its curvature is zero.

✱ COMPUTER PROGRAM: v, a, |v|, T, N, B, AND κ

The following BASIC program computes the velocity, acceleration, speed, the unit tangent, normal, and binormal vectors, and the curvature of the curve

$$\mathbf{r} = f(t)\mathbf{i} + g(t)\mathbf{j} + h(t)\mathbf{k} \text{ at any value of } t \text{ you select.}$$

There is a major difference between a numerical method like the one implemented here and the analytic technique by which we find the functions that define the quantities evaluated by the method. When we solve a problem analytically, we find a function that we can evaluate for any value of the parameter. The numerical method implemented by the program listed below must be run again with each new parameter value. The analytical result has another advantage in that we can often recognize critical values of the parameter where important changes will occur.

Another important difference between the two approaches is that a numerical solution almost always involves approximations. For example, a derivative is found from the ratio of two finite differences. It is not found as the limiting value of this ratio that is the definition of the derivative. The accuracy of such derivative approximations depends on the proper choice of the size of Δt. You may want to vary the value for DT in line 60 to investigate this point.

PROGRAM		COMMENT
10	DIM F(3), G(3), H(3), DR(3), DDR(3)	Define arrays (subscripted variables)
20	DIM TV(3), NV(3), VXA(3), BV(3)	for vectors
30	DEF FNF(T) = 3*COS(T)	Define parameter function for x
40	DEF FNG(T) = 3*SIN(T)	and for y
50	DEF FNH(T) = T^2	and for z

PROGRAM	COMMENT
60 DT = .01	Δt for derivative computations
70 INPUT "ENTER A VALUE FOR T ", TT	Get a t-value from the user in variable TT
80 T = TT	Save this in variable T
90 FOR I = −1 TO 1: F(I + 2) = FNF(T + I*DT): NEXT I	Compute three values for $x = f(t)$
100 FOR I = −1 TO 1: G(I + 2) = FNG(T + I*DT): NEXT I	and three for $y = g(t)$
110 FOR I = −1 TO 1: H(I + 2) = FNH(T + I*DT): NEXT I	and three for $z = h(t)$
120 DR(1) = (F(3) − F(1))/2/DT	Compute first component of velocity
130 DR(2) = (G(3) − G(1))/2/DT	and the second
140 DR(3) = (H(3) − H(1))/2/DT	and the third
150 DDR(1) = (F(3) − 2*F(2) + F(1))/DT^2	Do the same for
160 DDR(2) = (G(3) − 2*G(2) + G(1))/DT^2	the second
170 DDR(3) = (H(3) − 2*H(2) + H(1))/DT^2	derivative
180 PRINT "AT T = ";T;" THE COMPONENTS OF V AND A ARE"	Begin the output line for **V** and **A**,
190 PRINT DR(1); DR(2); DR(3)	now the velocity
200 PRINT DDR(1); DDR(2); DDR(3)	and the acceleration
210 SUM = 0: FOR I = 1 TO 3: SUM = SUM + DR(I)^2: NEXT I	Get sum of squares components of **V**
220 SPD = SQR(SUM)	and compute the speed
230 PRINT "SPEED = "; SPD	Print the speed
240 FOR I = 1 TO 3: TV(I) = DR(I)/SPD: NEXT I	Compute components of **T**
250 PRINT "COMPONENTS OF T VECTOR: "; TV(1); TV(2); TV(3)	and print them
260 SUM = 0: FOR I = 1 TO 3: SUM = SUM + DDR(I)^2: NEXT I	Sum squares of components of **A**
270 MAGNA = SQR(SUM)	and find |**A**|
280 FOR I = 1 TO 3: NV(I) = DDR(I)/MAGNA: NEXT I	Get components of **N**
290 PRINT "COMPONENTS OF N VECTOR: "; NV(1); NV(2); NV(3)	and print them
300 VXA(1) = DR(2)*DDR(3) − DR(3)*DDR(2)	Get first component of **V** × **A**
310 VXA(2) = DR(3)*DDR(1) − DR(1)*DDR(3)	and then the second
320 VXA(3) = DR(1)*DDR(2) − DR(2)*DDR(1)	and the third
330 FOR I = 1 TO 3: BV (I) = VXA(I)/SPD/MAGNA: NEXT I	Get components of **B**
340 PRINT "COMPONENTS OF B VECTOR: "; BV(1); BV(2); BV(3)	and print them
350 SUM = 0: FOR I = 1 TO 3: SUM = SUM + VXA(I)^2: NEXT I	Sum squares of components of **V** × **A**
360 PRINT "CURVATURE = "; SQR(SUM)/SPD^3	and compute and print the curvature
370 END	

Here is the output for $t = 2$.

```
AT T = 2 COMPONENTS OF V AND A ARE:
 − 2.727842   − 1.248431   4.000008
    1.249313   − 2.737045   2.000332
SPEED = 4.999977
COMPONENTS OF T VECTOR:  − 0.545571   − 0.2496874   0.8000053
COMPONENTS OF N VECTOR:    0.345786   − 0.7575618   0.5536536
COMPONENTS OF B VECTOR:    0.4678131    0.578688    0.4996421
CURVATURE = 0.1295347
```

Computer

Use the computer program just listed (or any similar program) to find **v**, **a**, |**v**|, **T**, **N**, **B**, and κ for the curves in Exercises 41–44 at the given values of t. (Change lines 30, 40, and 50 for each exercise.)

41. $\mathbf{r} = (t \cos t)\,\mathbf{i} + (t \sin t)\,\mathbf{j} + t\mathbf{k}, \quad t = \sqrt{3}$

42. $\mathbf{r} = (e^t \cos t)\,\mathbf{i} + (e^t \sin t)\,\mathbf{j} + e^t\,\mathbf{k}, \quad t = \ln 2$

43. $\mathbf{r} = (t - \sin t)\,\mathbf{i} + (1 - \cos t)\,\mathbf{j} + \sqrt{-t}\,\mathbf{k}, \quad t = -3\pi$

44. $\mathbf{r} = (3t - t^2)\,\mathbf{i} + (3t^2)\,\mathbf{j} + (3t + t^3)\,\mathbf{k}, \quad t = 1$

11.5 Planetary Motion and Satellites

In this section, we derive Kepler's laws of planetary motion from Newton's laws of motion and gravitation and discuss the orbits of Earth satellites. The derivation of Kepler's laws from Newton's is one of the triumphs of mathematical modeling with calculus. It draws on almost everything we have studied so far, including the algebra and geometry of vectors in space, the calculus of vector functions, the solutions of differential equations and initial value problems, and the polar coordinate description of conic sections.

Vector Equations for Motion in Polar and Cylindrical Coordinates

When a particle moves along a curve in the polar coordinate plane, we express its position, velocity, and acceleration in terms of the moving unit vectors

$$\mathbf{u}_r = (\cos\theta)\mathbf{i} + (\sin\theta)\mathbf{j}, \qquad \mathbf{u}_\theta = -(\sin\theta)\mathbf{i} + (\cos\theta)\mathbf{j} \qquad (1)$$

(Fig. 11.33). The vector $\mathbf{u}_r$ is the direction of the particle's position vector $\mathbf{r} = \overrightarrow{OP}$ and is related to $\mathbf{r}$ by the equations

$$\mathbf{u}_r = \frac{\mathbf{r}}{|\mathbf{r}|} = \frac{\mathbf{r}}{r} \qquad \text{and} \qquad \mathbf{r} = r\mathbf{u}_r, \qquad (2)$$

where r, the length of the position vector, is the positive polar distance coordinate of P. The vector $\mathbf{u}_r$ points in the direction of increasing r. The vector $\mathbf{u}_\theta$, orthogonal to $\mathbf{u}_r$, points in the direction of increasing θ. We find from (1) that

$$\frac{d\mathbf{u}_r}{d\theta} = -(\sin\theta)\mathbf{i} + (\cos\theta)\mathbf{j} = \mathbf{u}_\theta,$$

$$\frac{d\mathbf{u}_\theta}{d\theta} = -(\cos\theta)\mathbf{i} - (\sin\theta)\mathbf{j} = -\mathbf{u}_r. \qquad (3)$$

If we differentiate $\mathbf{u}_r$ and $\mathbf{u}_\theta$ with respect to t to find out how they change with time, the Chain Rule gives

$$\dot{\mathbf{u}}_r = \frac{d\mathbf{u}_r}{d\theta}\dot{\theta} = \dot{\theta}\mathbf{u}_\theta, \qquad \dot{\mathbf{u}}_\theta = \frac{d\mathbf{u}_\theta}{d\theta}\dot{\theta} = -\dot{\theta}\mathbf{u}_r. \qquad (4)$$

As in the previous section, we use Newton's dot notation for time derivatives to keep the formulas as simple as we can: $\dot{\mathbf{u}}_r$ means $d\mathbf{u}_r/dt$, $\dot{\theta}$ means $d\theta/dt$, and so on.

When we express the particle's velocity vector in terms of coordinate unit vectors (Fig. 11.34), we find that

$$\mathbf{v} = \frac{d\mathbf{r}}{dt} = \frac{d}{dt}\left(r\mathbf{u}_r\right) = \dot{r}\mathbf{u}_r + r\dot{\mathbf{u}}_r = \dot{r}\mathbf{u}_r + r\dot{\theta}\mathbf{u}_\theta. \qquad (5)$$

The acceleration is

$$\mathbf{a} = \dot{\mathbf{v}} = (\ddot{r}\mathbf{u}_r + \dot{r}\dot{\mathbf{u}}_r) + (r\ddot{\theta}\mathbf{u}_\theta + \dot{r}\dot{\theta}\mathbf{u}_\theta + r\dot{\theta}\dot{\mathbf{u}}_\theta). \qquad (6)$$

When we use Eqs. (4) to evaluate $\dot{\mathbf{u}}_r$ and $\dot{\mathbf{u}}_\theta$ and group the components, Eq. (6) becomes

$$\mathbf{a} = (\ddot{r} - r\dot{\theta}^2)\mathbf{u}_r + (r\ddot{\theta} + 2\dot{r}\dot{\theta})\mathbf{u}_\theta. \qquad (7)$$

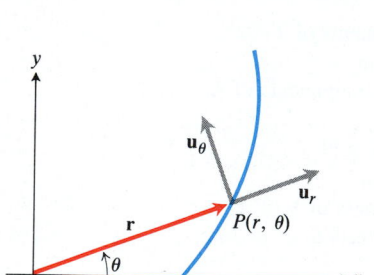

11.33 The length of the position vector $\mathbf{r}$ is the positive polar coordinate r of the point P. Thus, $\mathbf{u}_r$, which is $\mathbf{r}/|\mathbf{r}|$, is also $\mathbf{r}/r$. Equations (1) express $\mathbf{u}_r$ and $\mathbf{u}_\theta$ in terms of $\mathbf{i}$ and $\mathbf{j}$.

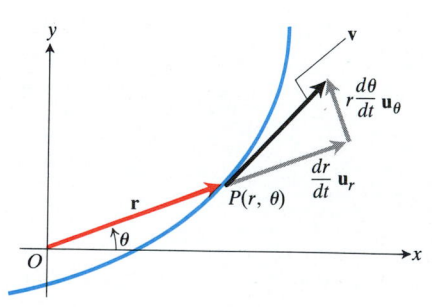

11.34 In polar coordinates, the velocity vector is

$$\mathbf{v} = (dr/dt)\mathbf{u}_r + r(d\theta/dt)\mathbf{u}_\theta.$$

Planets Move in Planes

Let us set our new coordinate system aside for a moment and turn our attention to the physics of a planet moving about a sun.

Newton's law of gravitation says that if **r** is the radius vector from the center of a sun of mass M to the center of a planet of mass m, then the force **F** of the gravitational attraction between the planet and sun is given by the equation

$$\mathbf{F} = -\frac{GmM}{|\mathbf{r}|^2}\frac{\mathbf{r}}{|\mathbf{r}|} \tag{8}$$

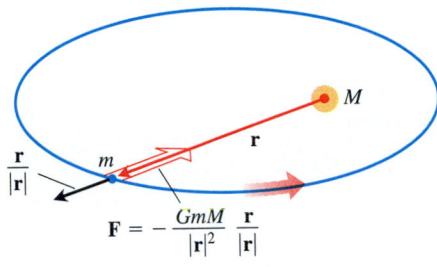

(Fig. 11.35). The number G is the (universal) **gravitational constant.** If we measure mass in kilograms, force in newtons, and distance in meters, G is about $6.6720 \times 10^{-11}\ \mathrm{Nm^2kg^{-2}}$.

Combining Eq. (8) with the equation $\mathbf{F} = m\ddot{\mathbf{r}}$, Newton's second law of motion, gives

$$m\ddot{\mathbf{r}} = -\frac{GmM}{|\mathbf{r}|^2}\frac{\mathbf{r}}{|\mathbf{r}|},$$

$$\ddot{\mathbf{r}} = -\frac{GM}{|\mathbf{r}|^2}\frac{\mathbf{r}}{|\mathbf{r}|}. \tag{9}$$

11.35 The force of gravity is directed along the line joining the centers of mass.

The planet is accelerated toward the sun's center at all times.

Equation (9) says that $\ddot{\mathbf{r}}$ is a scalar multiple of **r** and hence that

$$\mathbf{r} \times \ddot{\mathbf{r}} = \mathbf{0}. \tag{10}$$

A routine vector calculation shows $\mathbf{r} \times \ddot{\mathbf{r}}$ to be the derivative of $\mathbf{r} \times \dot{\mathbf{r}}$:

$$\frac{d}{dt}(\mathbf{r} \times \dot{\mathbf{r}}) = \underbrace{\dot{\mathbf{r}} \times \dot{\mathbf{r}}}_{\mathbf{0}} + \mathbf{r} \times \ddot{\mathbf{r}} = \mathbf{r} \times \ddot{\mathbf{r}}. \tag{11}$$

Hence Eq. (10) is equivalent to

$$\frac{d}{dt}(\mathbf{r} \times \dot{\mathbf{r}}) = \mathbf{0}, \tag{12}$$

which integrates to

$$\mathbf{r} \times \dot{\mathbf{r}} = \mathbf{C} \tag{13}$$

for some constant vector **C**.

The mathematical significance of Eq. (13) is that it tells us that **r** and $\dot{\mathbf{r}}$ always lie in a plane normal to a fixed vector **C**. The real-world interpretation is that the center of mass of the planet tracked by **r** moves in a plane through the center of the sun (Fig. 11.36).

Coordinates and Initial Conditions

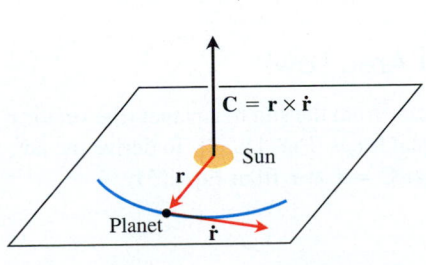

11.36 A planet that obeys Newton's laws of gravitation and motion travels in the plane through the sun's center of mass normal to $\mathbf{C} = \mathbf{r} \times \dot{\mathbf{r}}$.

We now introduce cylindrical coordinates in a way that places the origin at the sun's center of mass and makes the plane of the planet's motion the polar coordinate plane. This makes **r** the planet's polar coordinate position vector and makes $|\mathbf{r}|$ equal to r and $\mathbf{r}/|\mathbf{r}|$ equal to $\mathbf{u}_r$. We also position the z-axis in a way that makes **k** the direction of **C**. Thus, **k** has the same right-hand relation to $\mathbf{r} \times \dot{\mathbf{r}}$ that **C** does, and the planet's motion is counterclockwise when viewed from the positive z-axis. This

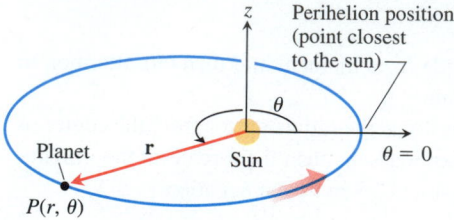

11.37 The coordinate system for planetary motion. The motion is counter-clockwise when viewed from above, as it is here, and $d\theta/dt$ is positive.

makes θ increase with t, so that $d\theta/dt$ is positive for all t. Finally, we rotate the polar coordinate plane about the z-axis, if necessary, to make the initial ray coincide with the direction $\mathbf{r}$ has when the planet is closest to the sun. This runs the ray through the planet's **perihelion** position (Fig. 11.37).

If we now measure time so that $t = 0$ at perihelion, we have the following initial conditions for the planet's motion.

1. $r = r_0$, the minimum radius, when $t = 0$
2. $\dot{r} = 0$ when $t = 0$ (because r has a minimum value then)
3. $\theta = 0$ when $t = 0$
4. $|\mathbf{v}| = v_0$ when $t = 0$

Since

$$
\begin{aligned}
v_0 &= |\mathbf{v}|_{t=0} & \text{(Standard notation)} \\
&= \left| \dot{r}\mathbf{u}_r + r\dot{\theta}\mathbf{u}_\theta \right|_{t=0} & \text{(Eq. (5))} \\
&= \left| r\dot{\theta}\mathbf{u}_\theta \right|_{t=0} & (\dot{r} = 0 \text{ when } t = 0) \\
&= \left(|r\dot{\theta}| \|\mathbf{u}_\theta\| \right)_{t=0} & \\
&= |r\dot{\theta}|_{t=0} & (\|\mathbf{u}_\theta\| = 1) \\
&= (r\dot{\theta})_{t=0}, & (r \text{ and } \dot{\theta} \text{ are both positive})
\end{aligned}
$$

we also know that

5. $r\dot{\theta} = v_0$ when $t = 0$.

Statement of Kepler's First Law (The Conic Section Law)

Kepler's first law says that a planet's path is a conic section with the sun at one focus. The eccentricity of the conic is

$$
e = \frac{r_0 v_0^2}{GM} - 1 \tag{14}
$$

and the polar equation is

$$
r = \frac{(1+e)r_0}{1 + e\cos\theta}. \tag{15}
$$

The derivation uses Kepler's second law, so we shall state and prove the second law before proving the first law.

Kepler's Second Law (The Equal Area Law)

Kepler's second law says that the radius vector from the sun to a planet (the vector $\mathbf{r}$ in our model) sweeps out equal areas in equal times (Fig. 11.38). To derive the law, we use Eq. (5) to evaluate the cross product $\mathbf{C} = \mathbf{r} \times \dot{\mathbf{r}}$ from Eq. (13):

$$
\begin{aligned}
\mathbf{C} = \mathbf{r} \times \dot{\mathbf{r}} &= \mathbf{r} \times \mathbf{v} \\
&= r\mathbf{u}_r \times (\dot{r}\mathbf{u}_r + r\dot{\theta}\mathbf{u}_\theta) & \text{(Eq. (5))} \\
&= r\dot{r}\underbrace{(\mathbf{u}_r \times \mathbf{u}_r)}_{\mathbf{0}} + r(r\dot{\theta})\underbrace{(\mathbf{u}_r \times \mathbf{u}_\theta)}_{\mathbf{k}} & (16) \\
&= r(r\dot{\theta})\,\mathbf{k}.
\end{aligned}
$$

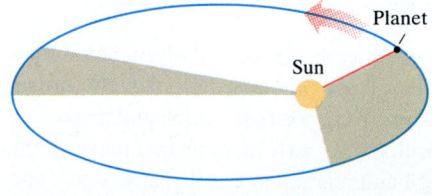

11.38 The line joining a planet to its sun sweeps over equal areas in equal times.

The German astronomer, mathematician, and physicist Johannes Kepler (1571–1630) was the first, and until Descartes the only, scientist to demand physical (as opposed to theological) explanations of celestial phenomena. His three laws of motion, the results of a lifetime of work, changed the course of astronomy forever and played a crucial role in the development of Newton's physics.

Setting t equal to zero shows that

$$\mathbf{C} = [r(r\dot{\theta})]_{t=0}\,\mathbf{k} = r_0 v_0\,\mathbf{k}. \tag{17}$$

Substituting this value for $\mathbf{C}$ in Eq. (16) gives

$$r_0 v_0\,\mathbf{k} = r^2\dot{\theta}\,\mathbf{k}, \quad \text{or} \quad r^2\dot{\theta} = r_0 v_0. \tag{18}$$

This is where the area comes in. The area differential in polar coordinates is

$$dA = \frac{1}{2}\,r^2\,d\theta$$

(Section 9.8). Accordingly, dA/dt has the constant value

$$\frac{dA}{dt} = \frac{1}{2}\,r^2\dot{\theta} = \frac{1}{2}\,r_0 v_0, \tag{19}$$

which is Kepler's second law.

For Earth, r_0 is about 150,000,000 km, v_0 is about 30 km/sec, and dA/dt is about 2,250,000,000 km²/sec. Every time your heart beats, Earth advances 30 km along its orbit and the radius joining Earth to the sun sweeps out 2,250,000,000 km² of area.

Proof of Kepler's First Law

To prove that a planet moves along a conic section with one focus at its sun, we need to express the planet's radius r as a function of θ. This requires a long sequence of calculations and some substitutions that are not altogether obvious.

We begin with the equation that comes from equating the coefficients of $\mathbf{u}_r = \mathbf{r}/|\mathbf{r}|$ in Eqs. (7) and (9):

$$\ddot{r} - r\dot{\theta}^2 = -\frac{GM}{r^2}. \tag{20}$$

We eliminate $\dot{\theta}$ temporarily by replacing it with $r_0 v_0/r^2$ from Eq. (18) and rearrange the resulting equation to get

$$\ddot{r} = \frac{r_0^2 v_0^2}{r^3} - \frac{GM}{r^2}. \tag{21}$$

We change this into a first-order equation by a change of variable. With

$$p = \frac{dr}{dt}, \quad \frac{d^2r}{dt^2} = \frac{dp}{dt} = \frac{dp}{dr}\frac{dr}{dt} = p\frac{dp}{dr}, \quad \text{(Chain Rule)}$$

Eq. (21) becomes

$$p\frac{dp}{dr} = \frac{r_0^2 v_0^2}{r^3} - \frac{GM}{r^2}. \tag{22}$$

Multiplying through by 2 and integrating with respect to r gives

$$p^2 = (\dot{r})^2 = -\frac{r_0^2 v_0^2}{r^2} + \frac{2GM}{r} + C_1. \tag{23}$$

The initial conditions that $r = r_0$ and $\dot{r} = 0$ when $t = 0$ determine the value of C_1 to be

$$C_1 = v_0^2 - \frac{2GM}{r_0}.$$

Accordingly, Eq. (23), after a suitable rearrangement, becomes

$$\dot{r}^2 = v_0^2 \left(1 - \frac{r_0^2}{r^2}\right) + 2GM\left(\frac{1}{r} - \frac{1}{r_0}\right). \tag{24}$$

The effect of going from Eq. (20) to Eq. (24) has been to replace a second order differential equation in r by a first order differential equation in r. Our goal is still to express r in terms of θ, so we now bring θ back into the picture. To accomplish this, we divide both sides of Eq. (24) by the squares of the corresponding sides of the equation $r^2\dot{\theta} = r_0 v_0$ (Eq. 18) and use the fact that $\dot{r}/\dot{\theta} = (dr/dt)/(d\theta/dt) = dr/d\theta$ to get

$$\frac{1}{r^4}\left(\frac{dr}{d\theta}\right)^2 = \frac{1}{r_0^2} - \frac{1}{r^2} + \frac{2GM}{r_0^2 v_0^2}\left(\frac{1}{r} - \frac{1}{r_0}\right)$$

$$= \frac{1}{r_0^2} - \frac{1}{r^2} + 2h\left(\frac{1}{r} - \frac{1}{r_0}\right). \qquad \left(h = \frac{GM}{r_0^2 v_0^2}\right) \tag{25}$$

To simplify further, we substitute

$$u = \frac{1}{r}, \qquad u_0 = \frac{1}{r_0}, \qquad \frac{du}{d\theta} = -\frac{1}{r^2}\frac{dr}{d\theta}, \qquad \left(\frac{du}{d\theta}\right)^2 = \frac{1}{r^4}\left(\frac{dr}{d\theta}\right)^2,$$

obtaining

$$\left(\frac{du}{d\theta}\right)^2 = u_0^2 - u^2 + 2\,hu - 2\,hu_0 = (u_0 - h)^2 - (u - h)^2, \tag{26}$$

$$\frac{du}{d\theta} = \pm\sqrt{(u_0 - h)^2 - (u - h)^2}. \tag{27}$$

Which sign do we take? We know that $\dot{\theta} = r_0 v_0/r^2$ is positive. Also, r starts from a minimum value at $t = 0$, so it cannot immediately decrease, and $\dot{r} \geq 0$, at least for early positive values of t. Therefore,

$$\frac{dr}{d\theta} = \frac{\dot{r}}{\dot{\theta}} \geq 0 \qquad \text{and} \qquad \frac{du}{d\theta} = -\frac{1}{r^2}\frac{dr}{d\theta} \leq 0.$$

The correct sign for Eq. (27) is the negative sign. With this determined, we rearrange Eq. (27) and integrate both sides with respect to θ:

$$\frac{-1}{\sqrt{(u_0 - h)^2 - (u - h)^2}}\frac{du}{d\theta} = 1$$

$$\cos^{-1}\left(\frac{u - h}{u_0 - h}\right) = \theta + C_2. \tag{28}$$

The constant of integration is zero because $u = u_0$ when $\theta = 0$ and because $\cos^{-1}(1) = 0$. Therefore,

$$\frac{u - h}{u_0 - h} = \cos\theta$$

and

$$\frac{1}{r} = u = h + (u_0 - h)\cos\theta. \tag{29}$$

A few more algebraic maneuvers produce the final equation

$$r = \frac{(1 + e)r_0}{1 + e \cos \theta},\tag{30}$$

where

$$e = \frac{1}{r_0 h} - 1 = \frac{r_0 v_0^2}{GM} - 1.\tag{31}$$

Together, Eqs. (30) and (31) say that the path of the planet is a conic section with one focus at the sun and with eccentricity $(r_0 v_0^2/GM) - 1$. This is the modern formulation of Kepler's first law.

Statement of Kepler's Third Law (The Time–Distance Law)

The time T it takes a planet to go around its sun once is the planet's **orbital period.** The semimajor axis a of the planet's orbit is the planet's **mean distance** from the sun. *Kepler's third law* says that T and a are related by the equation

$$\frac{T^2}{a^3} = \frac{4\pi^2}{GM}.\tag{32}$$

Since the right-hand side of this equation is constant within a given solar system, the ratio of T^2 to a^3 is the same for every planet in the system.

Kepler's third law is the starting point for working out the size of our solar system. It allows the semimajor axis of each planetary orbit to be expressed in astronomical units, Earth's semimajor axis being one unit. The distance between any two planets at any time can then be predicted in astronomical units and all that remains is to find one of these distances in kilometers. This can be done by bouncing radar waves off Venus, for example. The astronomical unit is now known, after a series of such measurements, to be 149,597,870 km.

We derive Kepler's third law by combining two different formulas for the area enclosed by the planet's elliptical orbit:

Formula 1:   Area $= \pi ab$ $\left(\begin{array}{l}\text{The geometry formula in which } a \text{ is the}\\ \text{semimajor axis and } b \text{ is the semiminor axis}\end{array}\right)$

Formula 2:   Area $= \displaystyle\int_0^T dA$

$$= \int_0^T \frac{1}{2} r_0 v_0 \, dt \qquad \text{(Eq. (19))}$$

$$= \frac{1}{2} T r_0 v_0.$$

Equating these gives

$$T = \frac{2\pi ab}{r_0 v_0} = \frac{2\pi a^2}{r_0 v_0} \sqrt{1 - e^2}. \qquad \text{(For any ellipse, } b = a\sqrt{1 - e^2}.)\tag{33}$$

It remains only to express a and e in terms of r_0, v_0, G, and M. Equation (31) does this for e. For a, we observe that setting θ equal to π in Eq. (30) gives

$$r_{\max} = r_0 \frac{1 + e}{1 - e}.$$

PSSC Physics, Second Edition, 1965; D.C. Heath & Company with Education Development Center, Inc.

11.39 This multiflash photograph shows a body being deflected by an inverse square law force. It moves along a hyperbola.

Hence,

$$2a = r_0 + r_{max} = \frac{2r_0}{1 - e} = \frac{2r_0\, GM}{2GM - r_0 v_0^2}. \tag{34}$$

Squaring both sides of Eq. (33) and substituting the results of Eqs. (31) and (34) now produces Kepler's third law, as you will see if you do Exercise 14.

Orbit Data

Although Kepler discovered his laws empirically and stated them only for the six planets known at the time, the modern derivations of Kepler's laws show that they apply to any body driven by a force that obeys an inverse square law. They apply to Halley's comet and the asteroid Icarus. They apply to the moon's orbit about Earth, and they applied to the orbit of the spacecraft Apollo 8 about the moon. They also applied to the air puck shown in Fig. 11.39 being deflected by an inverse square law force—its path is a hyperbola. Charged particles fired at the nuclei of atoms scatter along hyperbolic paths.

Tables 11.1–11.3 give additional data for planetary orbits and for the orbits of seven of Earth's artificial satellites (Fig. 11.40). Vanguard 1 sent back data that revealed differences between the levels of Earth's oceans and provided the first determination of the precise locations of some of the more isolated Pacific islands. The data also verified that the gravitation of the sun and moon would affect the orbits of Earth's satellites and that solar light could exert enough pressure to deform an orbit.

Syncom 3 is one of a series of U.S. Department of Defense telecommunications satellites. Tiros 11 (for "television infrared observation satellite") is one of a series of weather satellites. GOES 4 (for "geostationary operational environmental satellite") is one of a series of satellites designed to gather information about Earth's atmosphere. Its orbital period, 1436.2 minutes, is nearly the same as Earth's rotational period of 1436.1 minutes, and its orbit is nearly circular ($e = 0.0003$). Intelsat 5 is a heavy-capacity commercial telecommunications satellite.

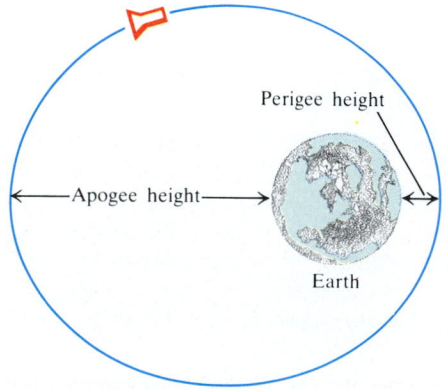

11.40 The orbit of an Earth satellite: $2a$ = diameter of Earth + perigee height + apogee height.

Circular Orbits

For circular orbits, e is zero, $r = r_0$ is a constant, and Eq. (14) gives

$$r = r_0 = \frac{GM}{v_0^2}, \tag{35}$$

which reduces to

$$r = \frac{GM}{v^2} \tag{36}$$

because v is constant as well (Exercise 12). Kepler's third law becomes

$$\frac{T^2}{r^3} = \frac{4\pi^2}{GM} \tag{37}$$

because $a = r$.

TABLE 11.1
Values of a, e, and T for the major planets

Planet	Semimajor axis $a^\dagger$	Eccentricity e	Period T
Mercury	57.95	0.2056	87.967 days
Venus	108.11	0.0068	224.701 days
Earth	149.57	0.0167	365.256 days
Mars	227.84	0.0934	1.8808 years
Jupiter	778.14	0.0484	11.8613 years
Saturn	1427.0	0.0543	29.4568 years
Uranus	2870.3	0.0460	84.0081 years
Neptune	4499.9	0.0082	164.784 years
Pluto	5909	0.2481	248.35 years

$\dagger$Millions of kilometers

TABLE 11.3
Numerical data

Gravitational constant: $G = 6.6720 \times 10^{-11}$ Nm^{2}kg^{-2}
(When you use this value of G in a calculation, remember to express force in newtons, distance in meters, mass in kilograms, and time in seconds.)
Sun's mass: 1.99×10^{30} kg
Earth's mass: 5.975×10^{24} kg
Equatorial radius of Earth: 6378.533 km
Polar radius of Earth: 6356.912 km
Earth's rotational period: 1436.1 min
Earth's orbital period: 1 year = 365.256 days

TABLE 11.2
Data on Earth's satellites

Name	Launch date	Time or expected time aloft	Weight at launch (kg)	Period (min)	Perigee height (km)	Apogee height (km)	Semimajor axis a (km)	Eccentricity
Sputnik 1	Oct. 1957	57.6 days	83.6	96.2	215	939	6,955	0.052
Vanguard 1	March 1958	300 years	1.47	138.5	649	4,340	8,872	0.208
Syncom 3	Aug. 1964	$> 10^6$ years	39	1436.2	35,718	35,903	42,189	0.002
Skylab 4	Nov. 1973	84.06 days	13,980	93.11	422	437	6,808	0.001
Tiros 11	Oct. 1978	500 years	734	102.12	850	866	7,236	0.001
GOES 4	Sept. 1980	$> 10^6$ years	627	1436.2	35,776	35,800	42,166	0.0003
Intelsat 5	Dec. 1980	$> 10^6$ years	1,928	1417.67	35,143	35,707	41,803	0.007

EXERCISES 11.5

Reminder: When a calculation involves the gravitational constant G, express distance in meters, mass in kilograms, and time in seconds.

Calculator Exercises

1. Since the orbit of Skylab 4 had a semimajor axis of $a = 6808$ km, Kepler's third law with M equal to the earth's mass should give the period. Calculate it. Compare your result with the value in Table 11.2.

2. Earth's distance from the sun at perihelion is approximately 149,577,000 km, and the eccentricity of the earth's orbit about the sun is 0.0167. Compute the velocity v_0 of the earth in its orbit at perihelion. (Use Eq. 14.)

3. In July 1965, the USSR launched Proton 1, weighing 12,200 kg (at launch), with a perigee height of 183 km, an apogee height of 589 km, and a period of 92.25 minutes. Using the relevant data for the mass of the earth and the gravitational constant G, compute the semimajor axis a of the orbit from Eq. (32). Compare your answer with the number you get by adding the perigee and apogee heights to the diameter of the earth.

4. a) The Viking 1 orbiter, which surveyed Mars from August 1975 to June 1976, had a period of 1639 min. Use this and the fact that the mass of Mars is 6.418×10^{23} kg to find the semimajor axis of the Viking 1 orbit.

   b) The Viking 1 orbiter was 1499 km from the surface of

Mars at its closest point and 35,800 km from the surface at its farthest point. Use this information together with the value you obtained in part (a) to estimate the average diameter of Mars.

5. The Viking 2 orbiter, which surveyed Mars from September 1975 to August 1976, moved in an ellipse whose semimajor axis was 22,030 km. What was the orbital period? (Express your answer in minutes.)

6. If a satellite is to hold a geostationary orbit, what must the semimajor axis of its orbit be? Compare your result with the semimajor axes of the satellites in Table 11.2.

7. The mass of Mars is 6.418×10^{23} kg. If a satellite revolving about Mars is to hold a stationary orbit (have the same period as the period of Mars's rotation, which is 1477.4 min), what must the semimajor axis of its orbit be?

8. The period of the moon's rotation about the earth is 2.36055×10^6 sec. About how far away is the moon?

9. A satellite moves around the earth in a circular orbit. Express the satellite's speed as a function of the orbit's radius.

10. If T is measured in seconds and a in meters, what is the value of T^2/a^3 for planets in our solar system? For satellites orbiting the earth? For satellites orbiting the moon? (The moon's mass is 7.354×10^{22} kg.)

Noncalculator Exercises

11. For what values of v_0 in Eq. (14) is the orbit in Eq. (15) a circle? An ellipse? A hyperbola?

12. Show that a planet in a circular orbit moves with a constant speed. (*Hint:* This is a consequence of one of Kepler's laws.)

13. Suppose that **r** is the position vector of a particle moving along a plane curve and dA/dt is the rate at which the vector sweeps out area. Without introducing coordinates, and assuming the necessary derivatives exist, give a geometric argument based on increments and limits for the validity of the equation

$$\frac{dA}{dt} = \frac{1}{2} |\mathbf{r} \times \dot{\mathbf{r}}|.$$

14. Complete the derivation of Kepler's third law (the part following Eq. 33).

Motion in the Polar Coordinate Plane

15. A particle moves in the plane in such a way that its polar coordinates at time t are $r = t$, $\theta = t$. Find **v**, **a**, and the curvature of the particle's path as functions of t.

16. At time t, the polar coordinates of a particle moving in the plane are

$$r = e^{\omega t} + e^{-\omega t}, \qquad \theta = t,$$

where ω ("omega") is a constant. Find the acceleration vector at time $t = 0$.

17. A ball is placed in a long frictionless tube that is pivoted at one end and rotates with constant angular velocity $d\theta/dt = 2$. The position of the ball at time t in polar coordinates is $r = \cosh 2t$, $\theta = 2t$. Show that the $\mathbf{u}_r$-component of the acceleration is always zero.

REVIEW QUESTIONS

1. State the rules for differentiating and integrating vector functions. Give examples.

2. How do you define and calculate the velocity, speed, direction of motion, and acceleration of a body moving along a sufficiently differentiable space curve? Give an example.

3. What is special about the derivatives of vectors of constant length? Give an example.

4. What are the vector and parametric equations for ideal projectile motion? How do you find a projectile's maximum height, flight time, and range? Give examples.

5. How do you define and calculate the length of a segment of a space curve? Give an example. What mathematical assumptions are involved in the definition?

6. How do you measure distance along a smooth curve in space from a preselected base point? Give an example of a directed-distance function.

7. What is a differentiable curve's unit tangent vector? Give an example.

8. Define curvature, circle of curvature (osculating circle), center of curvature, and radius of curvature for twice-differentiable curves in the plane. Give examples. What curves have zero curvature? constant curvature?

9. What is a plane curve's principal normal vector? Which way does it point? Give an example.

10. How do you define **N** and κ for curves in space? How are these quantities related? Give examples.

11. What is a curve's binormal vector? Give an example. What is the relation of this vector to the curve's torsion? Give an example.

12. What formulas are available for writing a moving body's acceleration as a sum of its tangential and normal components? Give an example. Why might one want to write the acceleration this way? What if the body moves at a constant speed? At a constant speed around a circle?

13. State Kepler's laws. To what do they apply?

MISCELLANEOUS EXERCISES

Graph the curves in Exercises 1 and 2, and sketch their velocity and acceleration vectors at the given values of t.

1. $\mathbf{r} = (4 \cos t)\,\mathbf{i} + (\sqrt{2} \sin t)\,\mathbf{j}, \quad t = 0$ and $\pi/4$

2. $\mathbf{r} = (\sqrt{3} \sec t)\,\mathbf{i} + (\sqrt{3} \tan t)\,\mathbf{j}, \quad t = 0$ and $\pi/6$

Evaluate the integrals in Exercises 3 and 4.

3. $\displaystyle\int_0^1 [(3 + 6t)\,\mathbf{i} + (4 + 8t)\,\mathbf{j} + (6\pi \cos \pi t)\,\mathbf{k}]\, dt$

4. $\displaystyle\int_e^{e^2} \left[\frac{2 \ln t}{t}\,\mathbf{i} + \frac{1}{t \ln t}\,\mathbf{j} + \frac{1}{t}\,\mathbf{k} \right]\, dt$

Exercises 5 and 6 give the velocity and initial position of a particle moving in space. Find the particle's position function.

5. $\dfrac{d\mathbf{r}}{dt} = -(\sin t)\,\mathbf{i} + (\cos t)\,\mathbf{j} + \mathbf{k}, \quad \mathbf{r} = \mathbf{j}$ when $t = 0$

6. $\dfrac{d\mathbf{r}}{dt} = \dfrac{1}{t^2 + 1}\,\mathbf{i} - \dfrac{1}{\sqrt{1 - t^2}}\,\mathbf{j} + \dfrac{1}{\sqrt{t^2 + 1}}\,\mathbf{k}, \quad \mathbf{r} = \mathbf{j} + \mathbf{k}$
   when $t = 0$.

Exercises 7 and 8 give the acceleration and the initial position and velocity of a particle moving in space. Find the particle's position function.

7. $\dfrac{d^2\mathbf{r}}{dt^2} = 2\,\mathbf{j}, \quad \mathbf{r} = \mathbf{i}$ and $\dfrac{d\mathbf{r}}{dt} = \mathbf{k}$ when $t = 0$

8. $\dfrac{d^2\mathbf{r}}{dt^2} = -2\,\mathbf{i} - 4\,\mathbf{j}, \quad \mathbf{r} = 3\,\mathbf{i} + 3\,\mathbf{j}$ and $\dfrac{d\mathbf{r}}{dt} = 4\,\mathbf{i}$ when $t = 1$

Find the lengths of the curves in Exercises 9 and 10.

9. $\mathbf{r} = (2 \cos t)\,\mathbf{i} + (2 \sin t)\,\mathbf{j} + t^2\,\mathbf{k}, \quad 0 \le t \le \pi/4$

10. $\mathbf{r} = (3 \cos t)\,\mathbf{i} + (3 \sin t)\,\mathbf{j} + 2t^{3/2}\,\mathbf{k}, \quad 0 \le t \le 3$

In Exercises 11–14, find $\mathbf{T}$, $\mathbf{N}$, $\mathbf{B}$, κ, and τ at the given value of t. Then find equations for the osculating, normal, and rectifying planes at that value of t.

11. $\mathbf{r} = \dfrac{4}{9}(1 + t)^{3/2}\,\mathbf{i} + \dfrac{4}{9}(1 - t)^{3/2}\,\mathbf{j} + \dfrac{1}{3} t\,\mathbf{k}, \quad t = 0$

12. $\mathbf{r} = (e^t \sin 2t)\,\mathbf{i} + (e^t \cos 2t)\,\mathbf{j} + 2e^t\,\mathbf{k}, \quad t = 0$

13. $\mathbf{r} = t\,\mathbf{i} + \dfrac{1}{2} e^{2t}\,\mathbf{j}, \quad t = \ln 2$

14. $\mathbf{r} = (3 \cosh 2t)\,\mathbf{i} + (3 \sinh 2t)\,\mathbf{j} + 6t\,\mathbf{k}, \quad t = \ln 2$

In Exercises 15 and 16, write $\mathbf{a}$ in the form $\mathbf{a} = a_T\mathbf{T} + a_N\mathbf{N}$ at $t = 0$ without finding $\mathbf{T}$ and $\mathbf{N}$.

15. $\mathbf{r} = (2 + 3t + 3t^2)\,\mathbf{i} + (4t + 4t^2)\,\mathbf{j} - (6 \cos t)\,\mathbf{k}$

16. $\mathbf{r} = (2 + t)\,\mathbf{i} + (t + 2t^2)\,\mathbf{j} + (1 + t^2)\,\mathbf{k}$

17. Find $\mathbf{T}$, $\mathbf{N}$, $\mathbf{B}$, κ, and τ as functions of t if $\mathbf{r} = (\sin t)\,\mathbf{i} + (\sqrt{2} \cos t)\,\mathbf{j} + (\sin t)\,\mathbf{k}$.

18. The position of a particle in the plane at time t is
$$\mathbf{r} = \frac{1}{\sqrt{1 + t^2}}\,\mathbf{i} + \frac{t}{\sqrt{1 + t^2}}\,\mathbf{j}.$$
Find the particle's highest speed.

19. Suppose $\mathbf{r} = (e^t \cos t)\,\mathbf{i} + (e^t \sin t)\,\mathbf{j}$. Show that the angle between $\mathbf{r}$ and $\mathbf{a}$ never changes. What *is* the angle?

20. At what times in the interval $0 \le t \le \pi$ are the velocity and acceleration vectors of the motion $\mathbf{r} = \mathbf{i} + (5 \cos t)\,\mathbf{j} + (3 \sin t)\,\mathbf{k}$ orthogonal?

21. The position of a particle moving in space at time $t \ge 0$ is
$$\mathbf{r} = 2\,\mathbf{i} + \left(4 \sin \frac{t}{2}\right)\mathbf{j} + \left(3 - \frac{t}{\pi}\right)\mathbf{k}.$$
Find the first time $\mathbf{r}$ is perpendicular to the vector $\mathbf{i} - \mathbf{j}$.

22. A particle moves around the unit circle in the xy-plane. Its position at time t is $\mathbf{r} = x\,\mathbf{i} + y\,\mathbf{j}$, where x and y are differentiable functions of t. Find dy/dt if $\mathbf{v} \cdot \mathbf{i} = y$. Is the motion clockwise, or counterclockwise?

23. You send a message through a pneumatic tube that follows the curve $9y = x^3$ (distance in meters). At the point $(3, 3)$, $\mathbf{v} \cdot \mathbf{i} = 4$ and $\mathbf{a} \cdot \mathbf{i} = -2$. Find the values of $\mathbf{v} \cdot \mathbf{j}$ and $\mathbf{a} \cdot \mathbf{j}$ at $(3, 3)$.

24. A particle moves in the plane so that its velocity and position vectors are always orthogonal. Show that the particle moves in a circle centered at the origin.

25. A circular wheel with radius 1 ft and center C rolls to the right along the x-axis at a half turn per second (Fig. 11.41). At time t seconds, the position vector of the point P on the wheel's circumference is
$$\mathbf{r} = (\pi t - \sin \pi t)\,\mathbf{i} + (1 - \cos \pi t)\,\mathbf{j}.$$
a) Sketch the curve traced by P during the interval $0 \le t \le 3$.

b) Find $\mathbf{v}$ and $\mathbf{a}$ at $t = 0, 1, 2$, and 3 and add these vectors to your sketch.

c) At any given time, what is the forward speed of the topmost point of the wheel? of C?

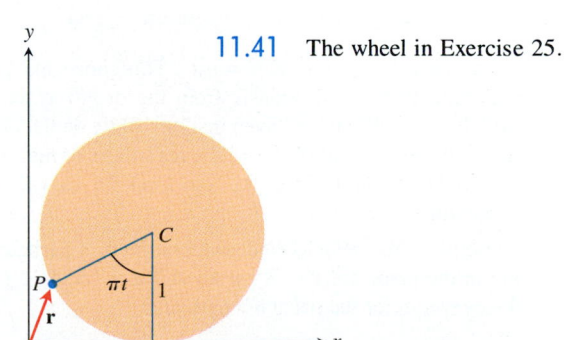

11.41   The wheel in Exercise 25.

26. The curve

$$\mathbf{r} = (2\sqrt{t}\cos t)\,\mathbf{i} + (3\sqrt{t}\sin t)\,\mathbf{j} + \sqrt{1-t}\,\mathbf{k}, \quad 0 \le t \le 1,$$

lies on a quadric surface. Describe the surface and find an equation for it.

27. Using the spherical coordinate θ as parameter, find a formula $\mathbf{r} = f(\theta)\,\mathbf{i} + g(\theta)\,\mathbf{j} + h(\theta)\,\mathbf{k}$ for the curve in which the plane $y + z = 0$ cuts the sphere $\rho = a$. Then find the length of the curve.

28. The line through the origin and the point $A(1, 1, 1)$ is the axis of rotation of a rigid body rotating with a constant angular speed of 6 rad/sec. The rotation appears to be clockwise when we look toward the origin from A. Find the velocity of the point of the body that is at the position $(1, 3, 2)$.

29. CALCULATOR *Javelin.*   In Potsdam in 1988, Petra Felke of (then) East Germany set a women's world record by throwing a javelin 262 ft 5 in.
a) Assuming that Felke launched the javelin at a 40° angle to the horizontal 7 ft above the ground, what was the javelin's initial speed?
b) How high did the javelin go?
c) A women's regulation javelin is 7 ft 2 1/2 in. (220 cm) long and weighs 1.32 lb (600 gm). How much work did Felke do on the javelin to launch it at that speed? (Do not include the work it took to lift the javelin the 7 ft.)

30. A golf ball is hit with an initial speed v_0 at an angle α to the horizontal from a point that lies at the foot of a straight-sided hill that is inclined at an angle ϕ to the horizontal, where

$$0 < \phi < \alpha < \frac{\pi}{2}.$$

Show that the ball lands at a distance

$$\frac{2v_0^2\cos\alpha}{g\cos^2\phi}\sin(\alpha - \phi),$$

measured up the face of the hill. Hence, show that the greatest range that can be achieved for a given v_0 occurs when $\alpha = (\phi/2) + (\pi/4)$, i.e., when the initial velocity vector bisects the angle between the vertical and the hill.

31. *Synchronous curves.*   By eliminating α from the ideal projectile equations

$$x = v_0(\cos\alpha)t, \quad y = (v_0\sin\alpha)t - \frac{1}{2}gt^2,$$

show that $x^2 + (y + gt^2/2)^2 = v_0^2 t^2$. This shows that projectiles launched simultaneously from the origin at the same initial speed will, at any given instant, all lie on the circle of radius $v_0 t$ centered at $(0, -gt^2/2)$, regardless of their launch angle. These circles are the *synchronous curves* of the launching.

32. At point P, the velocity and acceleration of a particle moving in the plane are $\mathbf{v} = 3\,\mathbf{i} + 4\,\mathbf{j}$ and $\mathbf{a} = 5\,\mathbf{i} + 15\,\mathbf{j}$. Find the curvature of the particle's path at P.

33. Show that the radius of curvature of a twice-differentiable plane curve $\mathbf{r} = f(t)\,\mathbf{i} + g(t)\,\mathbf{j}$ is given by the formula

$$\rho = \frac{\dot{x}^2 + \dot{y}^2}{\sqrt{\dot{x}^2 + \dot{y}^2 - \ddot{s}^2}}, \quad \text{where } \ddot{s} = \frac{d}{dt}\sqrt{\dot{x}^2 + \dot{y}^2}.$$

34. Find the point on the curve $y = e^x$ where the curvature is greatest.

35. Express the curvature of the curve

$$\mathbf{r} = \left(\int_0^t \cos\left(\frac{1}{2}\pi\theta^2\right)d\theta\right)\mathbf{i} + \left(\int_0^t \sin\left(\frac{1}{2}\pi\theta^2\right)d\theta\right)\mathbf{j}$$

as a function of the directed distance s measured along the curve from the origin (Fig. 11.42).

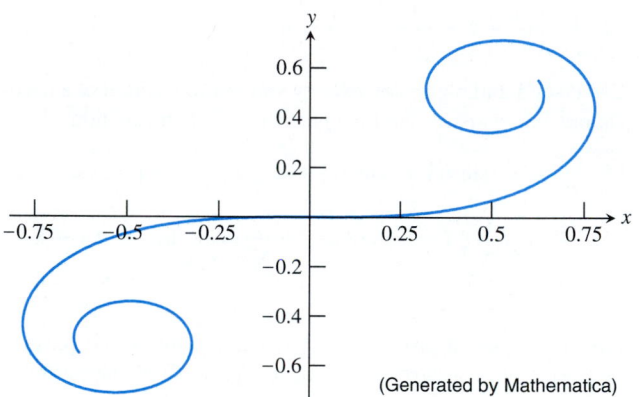

(Generated by Mathematica)

11.42   The curve in Exercise 35.

36. Find equations for the osculating, normal, and rectifying planes of the curve $\mathbf{r} = t\,\mathbf{i} + t^2\,\mathbf{j} + t^3\,\mathbf{k}$ at the point $(1, 1, 1)$.

37. Figure 11.43 shows the distance s measured counterclockwise around the circle $x^2 + y^2 = a^2$ from the point $(a, 0)$ to a point P. It also shows the angle ϕ that the tangent at P makes with the x-axis. Use the equations $s = a\theta$ and $\phi = \theta + \pi/2$ to calculate the circle's curvature directly from the definition $\kappa = |d\phi/ds|$.

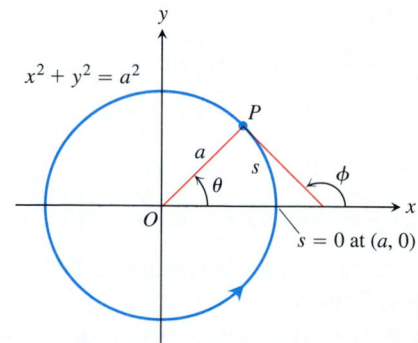

11.43   The circle in Exercise 37.

38. A frictionless particle P, starting from rest at time $t = 0$ at the point $(a, 0, 0)$, slides down the helix

$$\mathbf{r} = (a \cos \theta) \mathbf{i} + (a \sin \theta) \mathbf{j} + b\theta \mathbf{k} \qquad (a, b > 0)$$

under the influence of gravity, as in Fig. 11.44. The θ in this equation is the cylindrical coordinate θ and the helix is the curve $r = a$, $z = b\theta$, $\theta \geq 0$, in cylindrical coordinates. We assume θ to be a differentiable function of t for the motion. The law of conservation of energy tells us that the particle's speed after it has fallen a distance z is $\sqrt{2gz}$, where g is the constant acceleration of gravity.

a) Find the angular velocity $d\theta/dt$ when $\theta = 2\pi$.

b) Express the particle's θ- and z-coordinates as functions of t.

c) Express the tangential and normal components of the velocity $d\mathbf{r}/dt$ and acceleration $d^2\mathbf{r}/dt^2$ as functions of t. Does the acceleration have any nonzero component in the direction of the binormal vector $\mathbf{B}$?

11.44 The circular helix in Exercise 38.

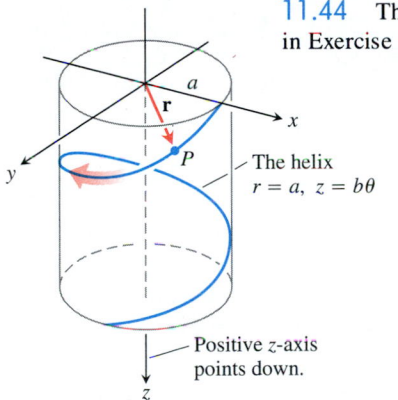

a

$\mathbf{r}$

x

y

P

The helix
$r = a$, $z = b\theta$

Positive z-axis
points down.

z

39. Suppose the curve in Exercise 38 is replaced by the conical helix $r = a\theta$, $z = b\theta$ shown in Fig. 11.45.

a) Express the angular velocity $d\theta/dt$ as a function of θ.

b) Express the distance the particle travels along the helix as a function of θ.

11.45 The conical helix in Exercise 39.

x

y

P

Conical helix
$r = a\theta$, $z = b\theta$

Cone $z = \dfrac{b}{a} r$

Positive z-axis points down.

z

40. *The view from Skylab 4.* What percentage of Earth's surface area could the astronauts see when Skylab 4 was at its apogee height, 437 km above the surface? To find out, model the visible surface as the surface generated by revolving the circular arc GT in Fig. 11.46 about the y-axis. Then carry out these steps:

1. Use similar triangles to show that $y_0/6380 = 6380/(6380 + 437)$. Solve for y_0.

2. Calculate the visible area as

$$VA = \int_{y_0}^{6380} 2\pi x \, ds.$$

3. Express the result as a percentage of Earth's surface area.

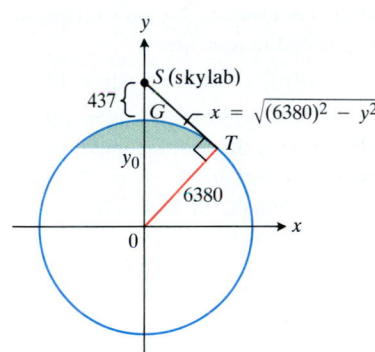

y

S (skylab)

$437 \{$

G

$x = \sqrt{(6380)^2 - y^2}$

T

y_0

6380

0

x

11.46 How much of the earth can you see from space at any one time? Exercise 40 shows how to find out.

41. Deduce from the orbit equation

$$r = \frac{(1 + e)r_0}{1 + e \cos \theta}$$

that a planet is closest to its sun when $\theta = 0$ and show that $r = r_0$ at that time.

42. GRAPHER or ROOT FINDER *A Kepler equation.* The problem of locating a planet in its orbit at a given time and date eventually leads to solving "Kepler" equations of the form

$$f(x) = x - 1 - \frac{1}{2} \sin x = 0.$$

a) Show that this particular equation has a solution between $x = 0$ and $x = 2$.

b) Find the solution to as many places as you can.

43. Express the curvature of a twice-differentiable curve $r = f(\theta)$ in the polar coordinate plane in terms of f and its derivatives.

44. A slender rod through the origin of the polar coordinate plane rotates (in the plane) about the origin at the rate of 3 rad/min. A beetle starting from the point $(2, 0)$ crawls along the rod toward the origin at the rate of 1 in./min.

a) Find the beetle's acceleration and velocity in polar form when it is halfway to (1 in. from) the origin.

b) CALCULATOR   To the nearest tenth of an inch, what will be the length of the path the beetle has traveled by the time it reaches the origin?

45. In Section 11.5, we found the velocity of a particle moving in the plane to be

$$\mathbf{v} = \dot{x}\,\mathbf{i} + \dot{y}\,\mathbf{j} = \dot{r}\,\mathbf{u}_r + r\dot{\theta}\,\mathbf{u}_\theta.$$

a) Express $\dot{x}$ and $\dot{y}$ in terms of $\dot{r}$ and $r\dot{\theta}$ by evaluating the dot products $\mathbf{v} \cdot \mathbf{i}$ and $\mathbf{v} \cdot \mathbf{j}$.

b) Express $\dot{r}$ and $r\dot{\theta}$ in terms of $\dot{x}$ and $\dot{y}$ by evaluating the dot products $\mathbf{v} \cdot \mathbf{u}_r$ and $\mathbf{v} \cdot \mathbf{u}_\theta$.

46. *Unit vectors for position and motion in cylindrical coordinates.*   When the position of a particle moving in space is given in cylindrical coordinates, the unit vectors we use to describe its position and motion are

$$\mathbf{u}_r = (\cos\theta)\,\mathbf{i} + (\sin\theta)\,\mathbf{j}, \qquad \mathbf{u}_\theta = -(\sin\theta)\,\mathbf{i} + (\cos\theta)\,\mathbf{j},$$

and $\mathbf{k}$ (Fig. 11.47). The particle's position vector is then $\mathbf{r} = r\mathbf{u}_r + z\mathbf{k}$, where r is the positive polar distance coordinate of the particle's position.

a) Show that $\mathbf{u}_r$, $\mathbf{u}_\theta$, and $\mathbf{k}$, in this order, form a right-handed frame of unit vectors.

b) Show that

$$\frac{d\mathbf{u}_r}{d\theta} = \mathbf{u}_\theta \qquad \text{and} \qquad \frac{d\mathbf{u}_\theta}{d\theta} = -\mathbf{u}_r.$$

c) Assuming that the necessary derivatives with respect to t exist, express $\mathbf{v} = \dot{\mathbf{r}}$ and $\mathbf{a} = \ddot{\mathbf{r}}$ in terms of $\mathbf{u}_r$, $\mathbf{u}_\theta$, $\mathbf{k}$, $\dot{r}$, and $\dot{\theta}$. (The dots indicate derivatives with respect to t: $\dot{\mathbf{r}}$ means $d\mathbf{r}/dt$, $\ddot{\mathbf{r}}$ means $d^2\mathbf{r}/dt^2$, and so on.) Section 11.5 shows how the vectors mentioned here are used in describing planetary motion.

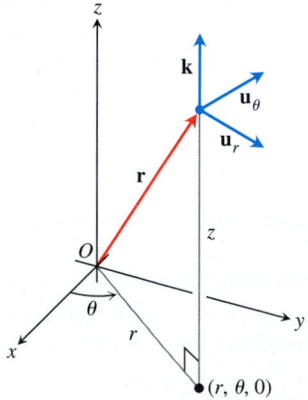

11.47   The unit vectors for describing motion in cylindrical coordinates (Exercise 46).

47. *Unit vectors for position and motion in spherical coordinates.*   Hold two of the three spherical coordinates ρ, ϕ, θ of a point P in space constant while letting the remaining coordinate increase. Let $\mathbf{u}$, with a subscript corresponding to the increasing coordinate, be the unit vector that points in the direction in which P then starts to move. The three resulting unit vectors, $\mathbf{u}_\rho$, $\mathbf{u}_\phi$, and $\mathbf{u}_\theta$ at P are shown in Fig. 11.48.

a) Express $\mathbf{u}_\rho$, $\mathbf{u}_\phi$, and $\mathbf{u}_\theta$ in terms of $\mathbf{i}$, $\mathbf{j}$, and $\mathbf{k}$.

b) Show that $\mathbf{u}_\rho \cdot \mathbf{u}_\phi = 0$.

c) Show that $\mathbf{u}_\theta = \mathbf{u}_\rho \times \mathbf{u}_\phi$.

d) Show that $\mathbf{u}_\rho$, $\mathbf{u}_\phi$, and $\mathbf{u}_\theta$, in that order, make a right-handed frame of mutually orthogonal vectors.

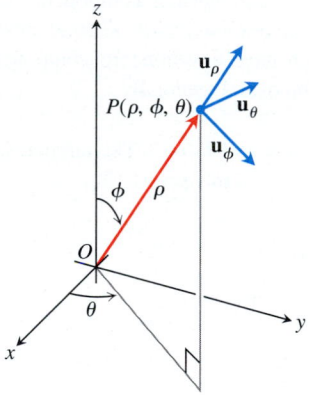

11.48   The unit vectors for describing motion in spherical coordinates (Exercise 47).

48. *(Continuation of Exercise 47)*   A particle P moves in space in such a way that its spherical coordinates are differentiable functions of time t. Express the particle's position and velocity vectors in terms of ρ, ϕ, and θ, the derivatives of ρ, ϕ, and θ, and the vectors $\mathbf{u}_\rho$, $\mathbf{u}_\phi$, and $\mathbf{u}_\theta$.

49. *Arc length in cylindrical coordinates*

a) Show that when you express $ds^2 = dx^2 + dy^2 + dz^2$ in terms of cylindrical coordinates, you get $ds^2 = dr^2 + r^2 d\theta^2 + dz^2$.

b) Interpret this result geometrically in terms of the edges and a diagonal of a box. Sketch the box.

c) Use the result in (a) to find the length of the curve $r = e^\theta$, $z = e^\theta$, $0 \le \theta \le \ln 8$.

50. *Arc length in spherical coordinates*

a) Show that when you express $ds^2 = dx^2 + dy^2 + dz^2$ in terms of spherical coordinates, you get $ds^2 = d\rho^2 + \rho^2 d\phi^2 + \rho^2 \sin^2\phi\, d\theta^2$.

b) Interpret this result geometrically in terms of the edges and a diagonal of a box. Sketch the box.

c) Use the result in (a) to find the length of the curve $\rho = 2e^\theta$, $\phi = \pi/6$, $0 \le \theta \le \ln 8$.

12

FUNCTIONS OF TWO OR MORE VARIABLES AND THEIR DERIVATIVES

Overview   We now begin our study of functions that have more than one independent variable. Functions of two or more variables appear more often in science than functions of a single variable, and their calculus is richer. Their derivatives are more varied because of the different ways in which the variables can interact. Their integrals lead to a greater variety of applications. The studies of probability, statistics, fluid dynamics, and electricity, to mention only a few, all lead in natural ways to functions of more than one variable.

In the present chapter, we introduce real-valued functions of two or more variables, define and calculate their derivatives, learn how to write the Chain Rule, and find out how to calculate the rates at which functions change as we move in different directions through their domains. We also learn how to estimate values of functions of two or more variables and how to calculate their extreme values in a given region.

The remaining chapters of the book build on what we do here. Chapter 13 deals with integrals of functions of two and three variables. These are the integrals with which we calculate the volumes and centers of mass of three-dimensional solids. In Chapter 14, we once more combine calculus with vectors, this time to derive the integral theorems that provide the mathematical foundation for the studies of fluid flow and electricity in physics and engineering. Finally, in Chapter 15, we examine various ways to solve equations of the form $dy/dx = f(x, y)$, equations in which the derivative of an unknown function y of x is given as a function of the two variables x and y.

12.1 Functions of Two or More Independent Variables

The values of many functions are determined by more than one independent variable. For example, the function $V = \pi r^2 h$ calculates the volume of a circular cylinder from its radius and height. The function $f(x, y) = x^2 + y^2$ calculates the height of the paraboloid $z = x^2 + y^2$ above the point (x, y) in the xy-plane. The function

$$w = \cos(1.7 \times 10^{-2} t - 0.2x)\, e^{-0.2x},$$

in Example 9, calculates the temperature x feet belowground on the tth day of the year as a fraction of the surface temperature on that day.

In this section, we define functions of more than one independent variable and discuss two ways to display their values graphically.

Functions and Variables

Real-valued functions of two or more real variables are defined much the way you would imagine from the single-variable case. The domains are sets of ordered pairs (triples, quadruples, whatever) of real numbers, and the ranges are sets of real numbers.

DEFINITIONS

> Suppose D is a set of n-tuples of real numbers $(x_1, x_2, \ldots, x_n)$. A **real-valued function** f on D is a rule that assigns a real number
>
> $$w = f(x_1, x_2, \ldots, x_n)$$
>
> to each element in D. The set D is the function's **domain.** The set of w-values taken on by f is the function's **range.** The symbol w is the **dependent variable** of f and f is said to be a function of the n **independent variables** x_1 to x_n. We call the x's the function's **input variables** and call w the function's **output variable.**

If f is a function of two independent variables, we usually call the variables x and y and picture the domain of f as a region in the xy-plane. If f is a function of three independent variables, we call the variables x, y, and z and picture the domain of f as a region in space.

When functions of more than one variable arise in applications, we tend to use letters that remind us of what the variables stand for. To say that the volume of a right circular cylinder is a function of its radius and height, we might write $V = f(r, h)$. To be even more specific, we might replace the notation $f(r, h)$ by the formula that calculates the value of V from the values of r and h and write $V = \pi r^2 h$. In either case, r and h would be the independent variables and V the dependent variable of the function.

As usual, we evaluate functions defined by formulas by substituting the values of the independent variables in the formula and calculating the corresponding value of the dependent variable.

Example 1   The value of the function $f(x, y, z) = \sqrt{x^2 + y^2 + z^2}$ at the point $(3, 0, 4)$ is

$$f(3, 0, 4) = \sqrt{(3)^2 + (0)^2 + (4)^2} = \sqrt{25} = 5.$$

Domains

In defining functions of more than one variable, we continue the practice of excluding inputs that lead to complex numbers or division by zero. If $f(x, y) = \sqrt{y - x^2}$, we do not allow y to be less than x^2. If $f(x, y) = 1/xy$, we do not allow the product xy to be zero.

These are the only restrictions, however, and the domains of functions are otherwise assumed to be the largest sets for which the defining rules generate real numbers.

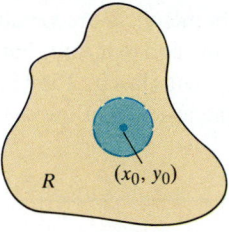

(a) Interior point

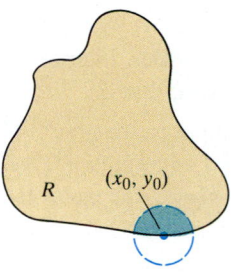

(b) Boundary point

12.1 Interior points and boundary points of a plane region R. An interior point is necessarily a point of R. A boundary point of R need not belong to R.

Example 2 *Functions of Two Variables*

Function	Domain	Range
$w = \sqrt{y - x^2}$	$y \geq x^2$	$w \geq 0$
$w = \dfrac{1}{xy}$	$xy \neq 0$	$w \neq 0$
$w = \sin xy$	Entire plane	$-1 \leq w \leq 1$
$w = -\dfrac{1}{x^2 + y^2}$	$(x, y) \neq (0, 0)$	$-\infty < w < 0$

Example 3 *Functions of Three Variables*

Function	Domain	Range
$w = \sqrt{x^2 + y^2 + z^2}$	Entire space	$w \geq 0$
$w = \dfrac{1}{x^2 + y^2 + z^2}$	$(x, y, z) \neq (0, 0, 0)$	$0 < w < \infty$
$w = xy \ln z$	Half-space $z > 0$	$-\infty < w < \infty$

The domains of functions defined on portions of the plane can have interior points and boundary points just as the domains of functions defined on intervals of the real line can.

DEFINITIONS

A point (x_0, y_0) in a region (set) R in the xy-plane is an **interior point** of R if it is the center of a disk that lies entirely in R (Fig. 12.1). A point (x_0, y_0) is a **boundary point** of R if every disk centered at (x_0, y_0) contains points that lie outside of R as well as points that lie in R. (The boundary point itself need not belong to R.)

The interior points of a region, as a set, make up the **interior** of the region. The region's boundary points make up its **boundary.** A region is **open** if it consists entirely of interior points. A region is **closed** if it contains all of its boundary points (Fig. 12.2).

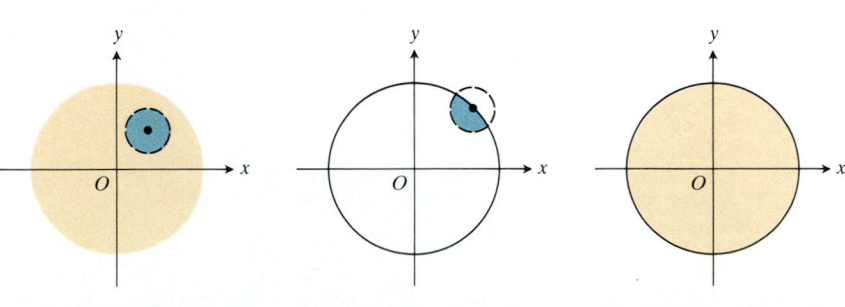

$\{(x, y) \mid x^2 + y^2 < 1\}$
Open unit disk. Every point an interior point.

$\{(x, y) \mid x^2 + y^2 = 1\}$
Boundary of unit disk. (The unit circle.)

$\{(x, y) \mid x^2 + y^2 \leq 1\}$
Closed unit disk. Contains all boundary points.

12.2 Interior points and boundary points of the unit disk in the plane.

As with intervals of real numbers, some regions in the plane are neither open nor closed. If you start with the open disk in Fig. 12.2 and add to it some but not all of its boundary points, the resulting set is neither open nor closed. The boundary points that *are* there keep the set from being open. The absence of the remaining boundary points keeps the set from being closed.

DEFINITIONS

A region in the plane is **bounded** if it lies inside a disk of fixed radius. A region is **unbounded** if it is not bounded.

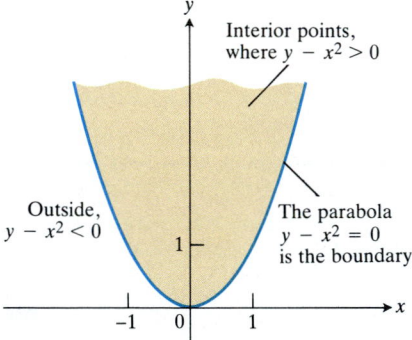

12.3 The domain of $f(x, y) = \sqrt{y - x^2}$ consists of the shaded region and its bounding parabola $y = x^2$.

Example 4

Bounded sets in the plane: Line segments, triangles, interiors of triangles, rectangles, disks

Unbounded sets in the plane: Lines, coordinate axes, the graphs of functions defined on infinite intervals, quadrants, half-planes, the plane itself

Example 5 The domain of the function $f(x, y) = \sqrt{y - x^2}$ is closed and unbounded (Fig. 12.3). The parabola $y = x^2$ is the boundary of the domain. The points inside the parabola make up the domain's interior.

The definitions of interior, boundary, open, closed, bounded, and unbounded for regions in space are similar to those for regions in the plane. To accommodate the extra dimension, we use solid spheres instead of disks.

DEFINITIONS

A point (x_0, y_0, z_0) in a region D in space is an **interior point** of D if it is the center of a solid sphere that lies entirely in D (Fig. 12.4). A point (x_0, y_0, z_0) is a **boundary point** of D if every sphere centered at (x_0, y_0, z_0) encloses points that lie outside of D as well as points that lie inside D. The **interior** of D is the set of interior points of D. The **boundary** of D is the set of boundary points of D.

A region D is **open** if it consists entirely of interior points. A region is **closed** if it contains its boundary.

12.4 Interior points and boundary points of a region in space.

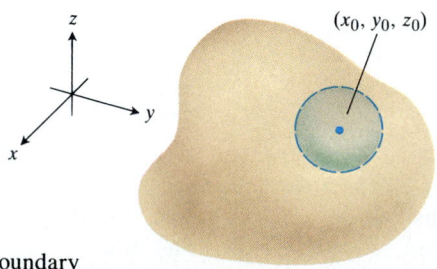

(a) Interior point

(b) Boundary point

Example 6

Open sets in space:	Interior of a solid sphere; the open half-space $z > 0$; the first octant (bounding planes absent); space itself
Closed sets in space:	Lines; planes; the closed half-space $z \geq 0$; the first octant together with its bounding planes; space itself
Neither open nor closed:	A solid sphere with part of its boundary removed; a solid cube with a missing face, edge, or corner point

Graphs and Level Curves of Functions of Two Variables

There are two standard ways to picture the values of a function $f(x, y)$. One is to draw and label curves in the domain on which f has a constant value. The other is to sketch the surface $z = f(x, y)$ in space.

DEFINITIONS

The set of points in the plane where a function $f(x, y)$ has a constant value $f(x, y) = c$ is called a **level curve** of f. The set of all points $(x, y, f(x, y))$ in space, for (x, y) in the domain of f, is called the **graph** of f. The graph of f is also called the **surface** $z = f(x, y)$.

Example 7

Graph the function

$$f(x, y) = 100 - x^2 - y^2$$

in space and plot the level curves $f(x, y) = 0$, $f(x, y) = 51$, and $f(x, y) = 75$, in the domain of f in the plane.

Solution The domain of f is the entire xy-plane and the range of f is the set of real numbers less than or equal to 100. The graph is the paraboloid $z = 100 - x^2 - y^2$, a portion of which is shown in Fig. 12.5.

The level curve $f(x, y) = 0$ is the set of points in the xy-plane at which

$$f(x, y) = 100 - x^2 - y^2 = 0, \qquad \text{or} \qquad x^2 + y^2 = 100,$$

which is the circle of radius 10 centered at the origin. Similarly, the level curves $f(x, y) = 51$ and $f(x, y) = 75$ are the circles

$$f(x, y) = 100 - x^2 - y^2 = 75, \qquad \text{or} \qquad x^2 + y^2 = 25,$$

$$f(x, y) = 100 - x^2 - y^2 = 51, \qquad \text{or} \qquad x^2 + y^2 = 49.$$

The level curve $f(x, y) = 100$ consists of the origin alone. (It is still called a level curve.)

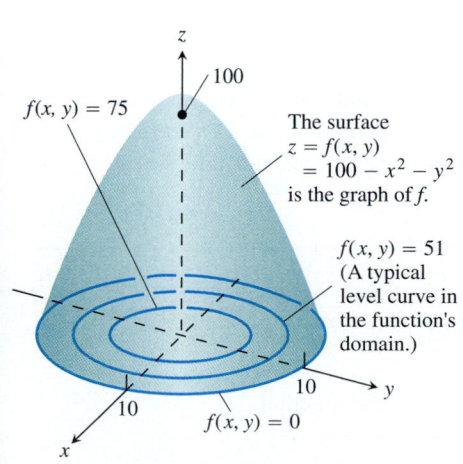

The surface
$z = f(x, y)$
$= 100 - x^2 - y^2$
is the graph of f.

$f(x, y) = 51$
(A typical level curve in the function's domain.)

12.5 The graph and selected level curves of the function $f(x, y) = 100 - x^2 - y^2$.

Contour Lines

The curve in space in which the plane $z = c$ cuts a surface $z = f(x, y)$ consists of the points that represent the function value $f(x, y) = c$. It is called the **contour**

The contour line $f(x, y) = 100 - y^2 - x^2 = 75$ is the circle $x^2 + y^2 = 25$ in the plane $z = 75$.

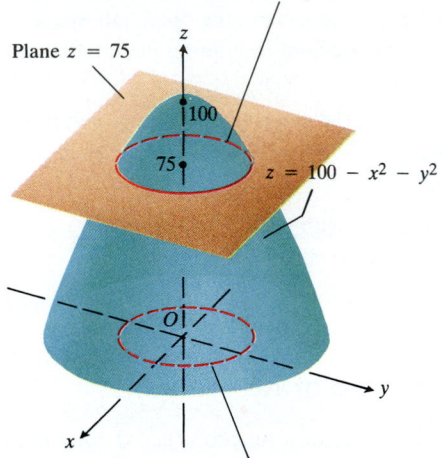

The level curve $f(x, y) = 100 - y^2 - x^2 = 75$ is the circle $x^2 + y^2 = 25$ in the xy-plane.

12.6   The graph of $f(x, y) = 100 - x^2 - y^2$ and its intersection with the plane $z = 75$.

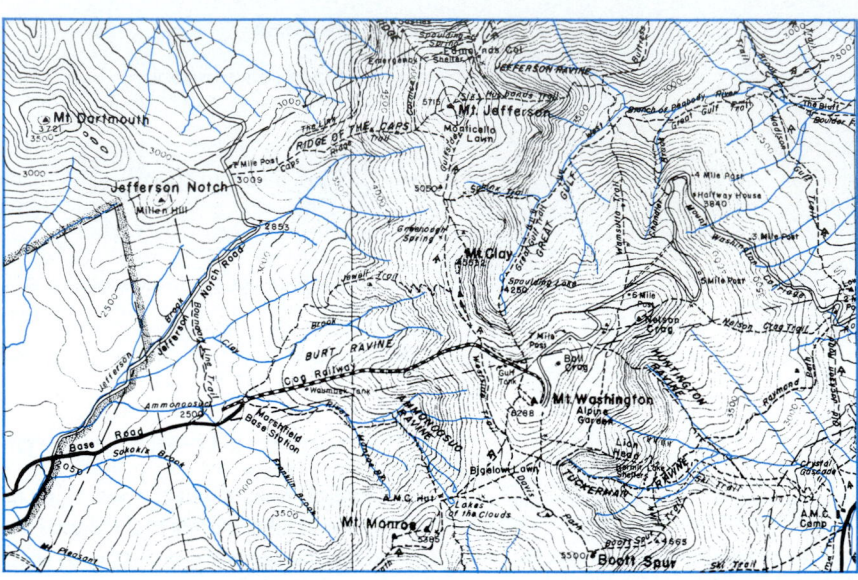

12.7   Contours on Mt. Washington in north-central New Hampshire. The streams, which follow paths of steepest descent, run perpendicular to the contours. So does the Cog Railway. (Based on a map by Louis F. Cutter, © 1987 Appalachian Mountain Club, Boston, Mass., by permission.)

line $f(x, y) = c$ to distinguish it from the level curve $f(x, y) = c$ in the domain of f. Figure 12.6 shows the contour line $f(x, y) = 75$ on the surface $z = 100 - x^2 - y^2$ defined by the function $f(x, y) = 100 - x^2 - y^2$. The contour line lies directly above the circle $x^2 + y^2 = 25$, which is the level curve $f(x, y) = 75$ in the function's domain.

Not everyone makes this distinction, however, and you may wish to call both kinds of curves by a single name and rely on context to convey which one you have in mind. On most maps, for example, the curves that represent constant elevation (height above sea level) are called contours, not level curves (Fig. 12.7).

Level Surfaces of Functions of Three Variables

In the plane, the points at which a function of two independent variables has a constant value $f(x, y) = c$ make a curve in the function's domain. In space, the points where a function of three independent variables has a constant value $f(x, y, z) = c$ make a surface in the function's domain.

DEFINITION

The set of points in space where a function of three independent variables has a constant value $f(x, y, z) = c$ is called a **level surface** of f.

Example 8   Describe the level surfaces of the function

$$f(x, y, z) = \sqrt{x^2 + y^2 + z^2}.$$

Solution   The value of f is the distance from the origin to the point (x, y, z). Each level surface

$$\sqrt{x^2 + y^2 + z^2} = c, \qquad c > 0,$$

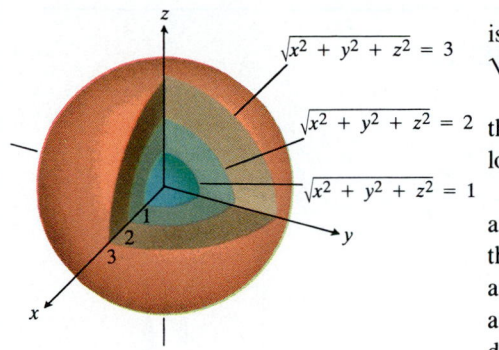

12.8 The level surfaces of the function $f(x, y, z) = \sqrt{x^2 + y^2 + z^2}$ are concentric spheres.

is a sphere of radius c centered at the origin (Fig. 12.8). The level surface $\sqrt{x^2 + y^2 + z^2} = 0$ consists of the origin alone.

We are not graphing the function here. The graph of the function, made up of the points $\left(x, y, z, \sqrt{x^2 + y^2 + z^2}\right)$, lies in a four-variable space. Instead, we are looking at level surfaces in the function's domain.

The function's level surfaces show how the function's values change as we move around in its domain. If we remain on a sphere of radius c centered at the origin, the function maintains a constant value, namely c. If we move from one sphere to another, the function's value changes. It increases if we move away from the origin and decreases if we move toward the origin. The way the function's values change depends on the direction we take. The dependence of change on direction is important. We shall return to it in Section 12.7.

Computer Graphing

The three-dimensional graphing programs for computers make it possible to graph functions of two variables with only a few keystrokes. We can often get information more quickly from a graph than we can from a formula.

Example 9   Figure 12.9 shows a computer-generated graph of the function

$$w = \cos(1.7 \times 10^{-2}t - 0.2x)\, e^{-0.2x},$$

where t is in days and x is in feet. The graph shows how the temperature beneath the earth's surface varies with time. The variation is given as a fraction of the variation at the surface. At a depth of 15 ft, the variation (change in vertical amplitude in the figure) is about 5 percent of the surface variation. At 30 ft, there is almost no variation in temperature during the year.

Another thing the graph shows that is not immediately apparent from the equation for f is that the temperature 15 ft below the surface is about half a year out of phase with the surface temperature. When the temperature is lowest on the surface (late January, say) it is at its highest 15 ft below. Fifteen feet below the ground, the seasons are reversed.

12.9   This computer-generated graph of
$$w = \cos(1.7 \times 10^{-2}t - 0.2x)e^{-0.2x}$$

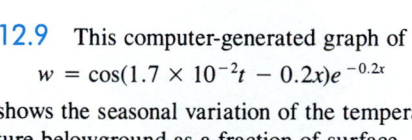

shows the seasonal variation of the temperature belowground as a fraction of surface temperature. $\Delta x = 0.375$ ft, $\Delta t = 15.625$ days. At $x = 15$ ft the variation is only 5% of the variation at the surface. At $x = 30$ ft the variation is less than 0.25% of the surface variation. (Adapted from art provided by Norton Starr for G. C. Berresford's "Differential Equations and Root Cellars," *The UMAP Journal*, Vol. 2, No. 3, (1981), pp. 53–75.)

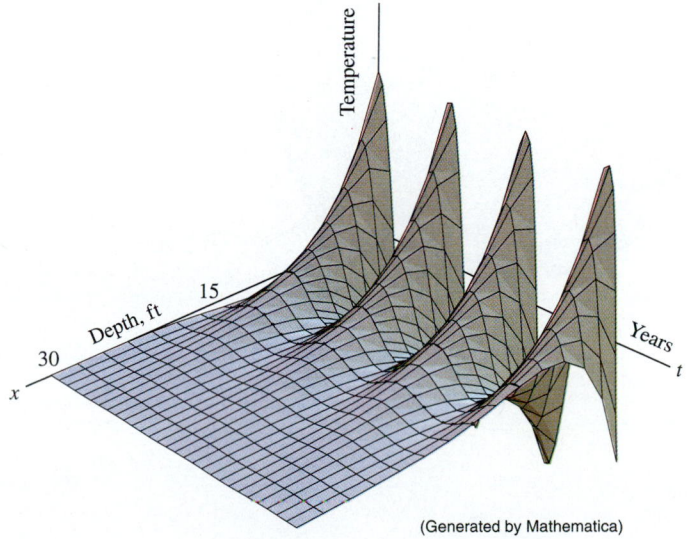

(Generated by Mathematica)

Surfaces Defined by Functions of Two Variables

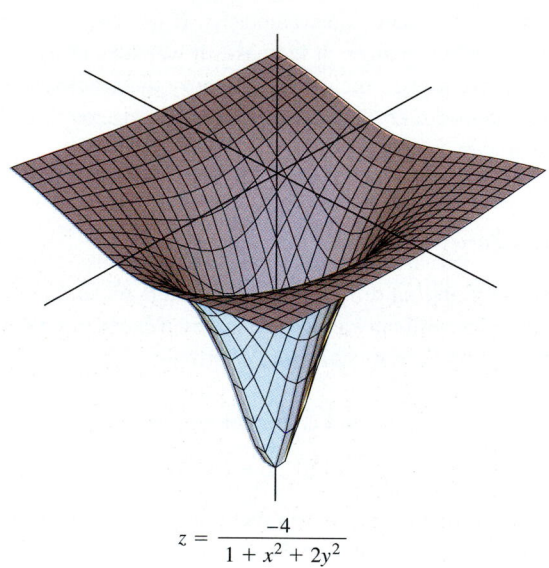

$$z = \frac{-4}{1 + x^2 + 2y^2}$$

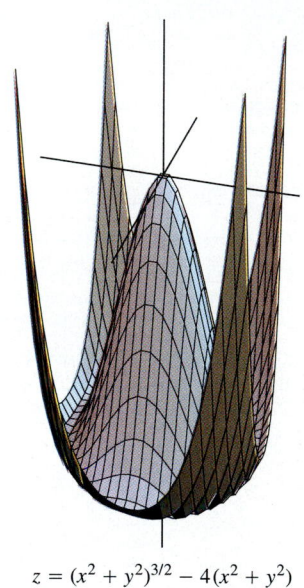

$$z = (x^2 + y^2)^{3/2} - 4(x^2 + y^2)$$

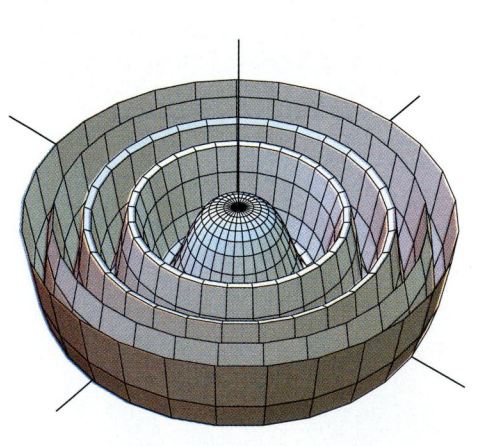

$$z = 1 + \cos(x^2 + y^2)$$

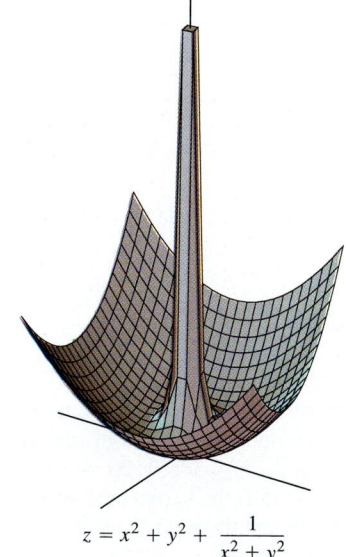

$$z = x^2 + y^2 + \frac{1}{x^2 + y^2}$$

(Generated by Mathematica)

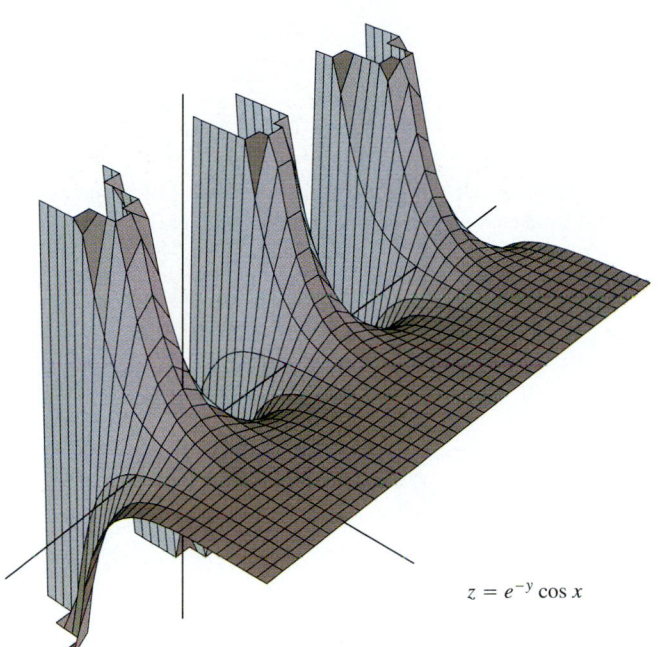

$$z = e^{-y} \cos x$$

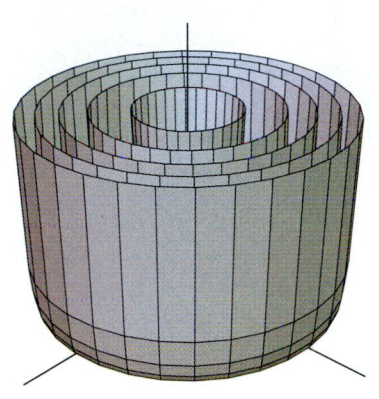

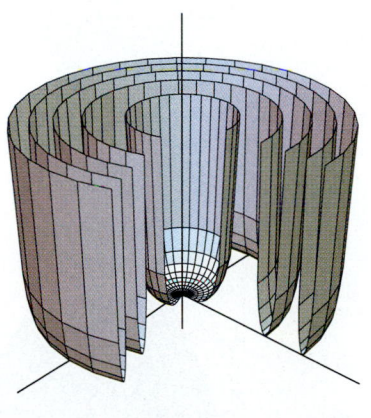

$$z = \ln(\sec(x^2 + y^2))$$

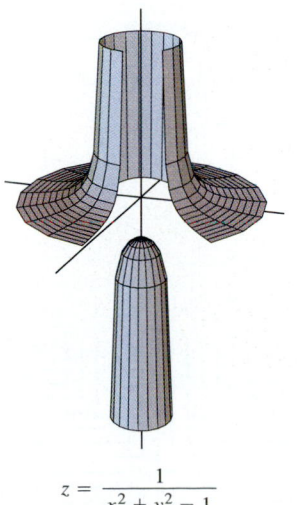

$$z = \frac{1}{x^2 + y^2 - 1}$$

(Generated by Mathematica)

EXERCISES 12.1

In Exercises 1–10, find the function's domain and range and describe the function's level curves.

1. $f(x, y) = \sqrt{y - x}$ **2.** $f(x, y) = 4x^2 + 9y^2$

3. $f(x, y) = y/x^2$ **4.** $f(x, y) = x^2 - y^2$

5. $f(x, y) = e^{\cos xy}$ **6.** $f(x, y) = \ln(x^2 + y^2)$

7. $f(x, y) = \sin^{-1}(y - x)$ **8.** $f(x, y) = \tan^{-1}(y/x)$

9. $f(x, y) = \ln x - \ln y$ **10.** $f(x, y) = \sqrt{\dfrac{x - y}{x + y}}$

Display the values of the functions in Exercises 11–20 in two ways: (a) by sketching the surface $z = f(x, y)$, and (b) by drawing an assortment of level curves in the function's domain. Label each level curve with the appropriate function value.

11. $f(x, y) = y^2$ **12.** $f(x, y) = 4 - y^2$

13. $f(x, y) = 1 + x^2 + y^2$ **14.** $f(x, y) = 4 - x^2 - y^2$

15. $f(x, y) = \sqrt{x^2 + y^2}$ **16.** $f(x, y) = 4x^2 + y^2$

17. $f(x, y) = 1 - |y|$ **18.** $f(x, y) = 1 - |x| - |y|$

19. $f(x, y) = e^{-(x^2 + y^2)}$ **20.** $f(x, y) = 1/(x^2 + y^2)$

Exercises 21–26 show level curves for the functions graphed in (a)–(f). Match each set of level curves with the appropriate function.

21.

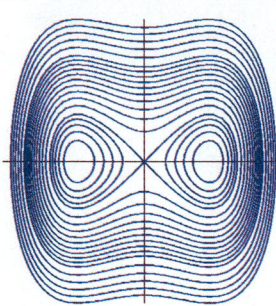

22.

23.

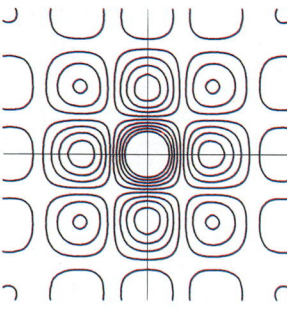

24.

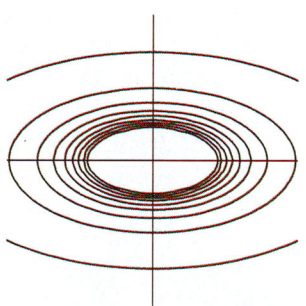

25.

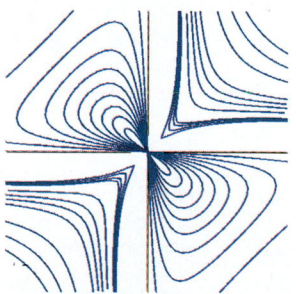

26.

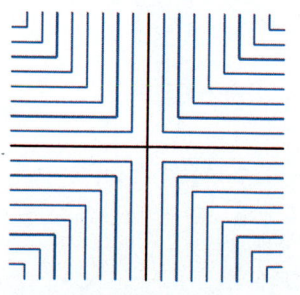

(a)

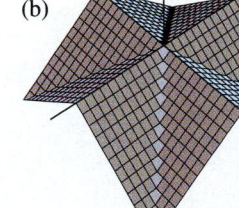

$z = (\cos x)(\cos y) \, e^{-\sqrt{x^2 + y^2}/4}$

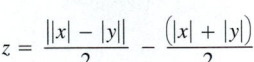

(b)

$z = \dfrac{||x| - |y||}{2} - \dfrac{(|x| + |y|)}{2}$

(c)

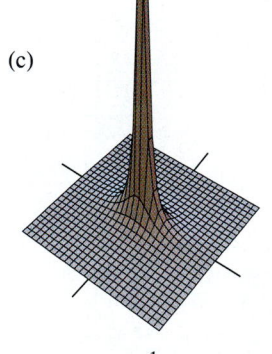

$z = \dfrac{1}{(x^2 + 4y^2)}$

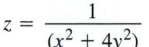

(d)

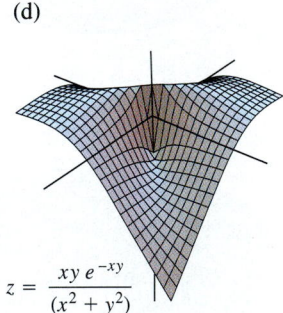

$z = \dfrac{xy \, e^{-xy}}{(x^2 + y^2)}$

(e)

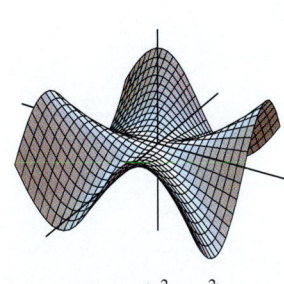

$z = \dfrac{xy(x^2 - y^2)}{(x^2 + y^2)}$

(f)

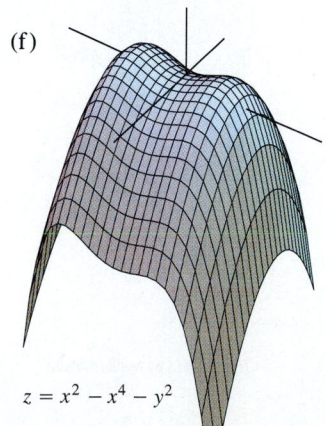

$z = x^2 - x^4 - y^2$

Describe the domains, ranges, and level surfaces of the functions in Exercises 27–34. Sketch a typical level surface for each function and label the surface with its function value.

27. $f(x, y, z) = \ln (x^2 + y^2 + z^2)$

28. $f(x, y, z) = \cosh(x^2 + z^2)$

29. $f(x, y, z) = \tan^{-1}(z - x^2 - y^2)$

30. $f(x, y, z) = \sin^{-1}z$

31. $f(x, y, z) = e^{-|z|}$

32. $f(x, y, z) = (x^2/25) + (y^2/16) + (z^2/9)$

33. $f(x, y, z) = \log_2 \left(z - \sqrt{x^2 + y^2}\right)$

34. $f(x, y, z) = |x| + |y| + |z|$

In Exercises 35–38, find an equation for the level curve of the function $f(x, y)$ that passes through the given point.

35. $f(x, y) = 16 - x^2 - y^2, \quad \left(2\sqrt{2}, \sqrt{2}\right)$

36. $f(x, y) = \sqrt{x^2 - 1}, \quad (1, 0)$

37. $f(x, y) = \int_x^y \dfrac{dt}{1 + t^2}, \quad \left(-\sqrt{2}, \sqrt{2}\right)$

38. $f(x, y) = \sum_{n=0}^{\infty} \left(\dfrac{x}{y}\right)^n, \quad (1, 2)$

In Exercises 39–42, find an equation for the level surface of the function $f(x, y, z)$ that passes through the given point.

39. $f(x, y, z) = \sqrt{x - y} - \ln z, \quad (3, -1, 1)$

40. $f(x, y, z) = \cos (x^2 - y + z^2), \quad (-1, 2, 1)$

41. $f(x, y, z) = \sum_{n=0}^{\infty} \dfrac{(x + y)^n}{n! \, z^n}, \quad (\ln 2, \ln 4, 3)$

42. $f(x, y, z) = \int_x^y \dfrac{d\theta}{\sqrt{1 - \theta^2}} + \int_{\sqrt{2}}^z \dfrac{dt}{t\sqrt{t^2 - 1}}, \quad (0, 1/2, 2)$

43. *The Concorde's sonic booms.*   The width w of the region in which people on the ground hear the Concorde's sonic boom directly, not reflected from a layer in the atmosphere, is a function of

$T = $ air temperature at ground level (in degrees Kelvin),

$h = $ the Concorde's altitude (in km),

$d = $ the vertical temperature gradient (temperature drop in degrees Kelvin per km).

The formula for w is

$$w = 4(Th/d)^{1/2}.$$

See Fig. 12.10.

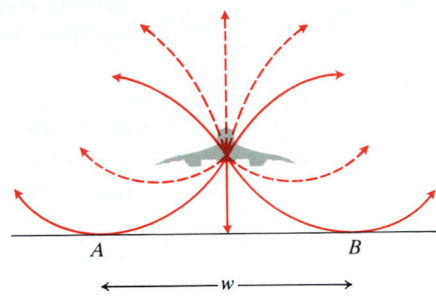

Sonic boom carpet

12.10   Sound waves from the Concorde bend as the temperature changes above and below the altitude at which the plane flies. The sonic boom carpet is the region on the ground that receives shock waves directly from the plane, not reflected from the atmosphere or diffracted along the ground. The carpet is determined by the grazing rays striking the ground from the point directly under the plane (Exercise 43).

The Washington-bound Concorde approaches the United States from Europe on a course that takes it south of Nantucket Island at an altitude of 16.8 km. If the surface temperature is 290 K and the vertical temperature gradient is 5 K/km, how far south of Nantucket must the plane be flown to keep its sonic boom carpet away from the island? (From "Concorde Sonic Booms as an Atmospheric Probe" by N. K. Balachandra, W. L. Donn, and D. H. Rind, *Science*, July 1, 1977, Vol. 197, p. 47.)

44. *The maximum value of a function on a line in space.*   Does the function $f(x, y, z) = xyz$ have a maximum value on the line $x = 20 - t$, $y = t$, $z = 20$? If so, what is it? (*Hint:* Along the line, $w = f(x, y, z)$ is a differentiable function of t.)

EXPLORER PROGRAM

3D Grapher — Graphs surfaces defined by equations of the form $z = f(x, y)$ over rectangular regions in the xy-plane

12.2 Limits and Continuity

This section deals with limits and continuity of functions of more than one variable. The definitions are analogous to those for functions of a single variable, except that there are now more variables to watch.

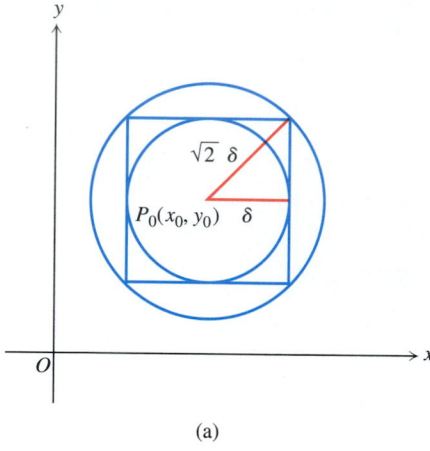

(a)

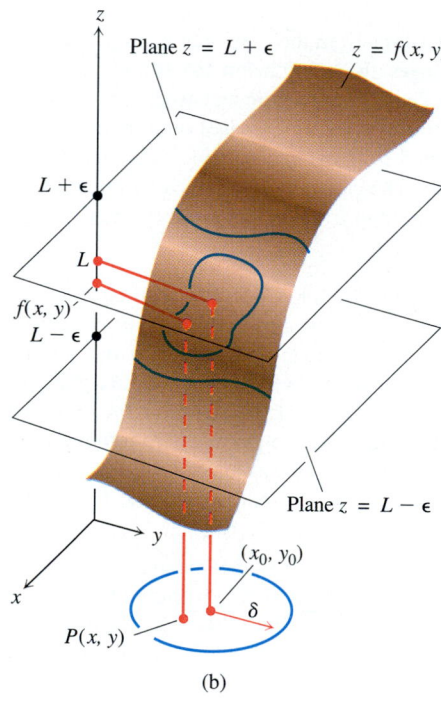

(b)

12.11 (a) The open square $|x - x_0| < \delta$, $|y - y_0| < \delta$ lies inside the open disk $\sqrt{(x - x_0)^2 + (y - y_0)^2} < \sqrt{2}\delta$ and contains the open disk $\sqrt{(x - x_0)^2 + (y - y_0)^2} < \delta$. Thus, we can make both $|x - x_0|$ and $|y - y_0|$ small by making $\sqrt{(x - x_0)^2 + (y - y_0)^2}$ small, and conversely. (b) For every point (x, y) within δ of (x_0, y_0), the value of $f(x, y)$ lies within ϵ of L.

Limits

If the values of a real-valued function $f(x, y)$ lie arbitrarily close to a fixed real number L for all points (x, y) sufficiently close to a point (x_0, y_0) but not equal to (x_0, y_0), we say that L is the limit of f as (x, y) approaches (x_0, y_0). In symbols, we write

$$\lim_{(x, y)\to(x_0, y_0)} f(x, y) = L, \tag{1}$$

and we say "the limit of f as (x, y) approaches (x_0, y_0) equals L." This is like the limit of a function of one variable, except that two independent variables are involved instead of one. In addition, if (x_0, y_0) is an interior point of f's domain, (x, y) can approach (x_0, y_0) from any direction. The direction of approach may turn out to be an issue, as some of the following examples show.

For (x, y) to be "close" to (x_0, y_0) means that the Cartesian distance $\sqrt{(x - x_0)^2 + (y - y_0)^2}$ is small in some sense. Since

$$|x - x_0| = \sqrt{(x - x_0)^2} \leq \sqrt{(x - x_0)^2 + (y - y_0)^2} \tag{2}$$

and

$$|y - y_0| = \sqrt{(y - y_0)^2} \leq \sqrt{(x - x_0)^2 + (y - y_0)^2}, \tag{3}$$

the inequality

$$\sqrt{(x - x_0)^2 + (y - y_0)^2} < \delta$$

for any value of δ implies that

$$|x - x_0| < \delta \qquad \text{and} \qquad |y - y_0| < \delta.$$

Conversely, if for some $\delta > 0$ both $|x - x_0| < \delta$ and $|y - y_0| < \delta$, then

$$\sqrt{(x - x_0)^2 + (y - y_0)^2} < \sqrt{\delta^2 + \delta^2} = \sqrt{2}\delta,$$

which is small if δ is small (Fig. 12.11). Thus, in calculating limits we may think either in terms of the distance in the plane or in terms of differences in individual coordinates.

> The **limit** of $f(x, y)$ as $(x, y)\to(x_0, y_0)$ is the number L if for any $\epsilon > 0$ there exists a $\delta > 0$ such that for all points $(x, y) \neq (x_0, y_0)$ in the domain of f, either
>
> 1. $\sqrt{(x - x_0)^2 + (y - y_0)^2} < \delta \quad \Rightarrow \quad |f(x, y) - L| < \epsilon$ or
> 2. $|x - x_0| < \delta$ and $|y - y_0| < \delta \quad \Rightarrow \quad |f(x, y) - L| < \epsilon.$

These equivalent definitions apply to boundary points as well as interior points of the domain of f. The only requirement is that the point (x, y) remain in the domain at all times.

It can be shown, as for functions of a single variable, that

$$\lim_{(x, y)\to(x_0, y_0)} x = x_0,$$

$$\lim_{(x, y)\to(x_0, y_0)} y = y_0, \tag{4}$$

$$\lim_{(x, y)\to(x_0, y_0)} k = k. \qquad \text{(Any number } k\text{)}$$

It can also be shown that the limit of the sum of two functions is the sum of their limits (when the latter exist), with similar results for the limits of the differences, products, constant multiples, and quotients of functions.

THEOREM 1

Properties of Limits of Functions of Two Variables

The following rules hold if $\lim\limits_{(x,y)\to(x_0,y_0)} f(x,y) = L_1$ and $\lim\limits_{(x,y)\to(x_0,y_0)} g(x,y) = L_2$.

1. *Sum Rule:* $\qquad\qquad\qquad$ $\lim [f(x,y) + g(x,y)] = L_1 + L_2$

2. *Difference Rule:* $\qquad\qquad$ $\lim [f(x,y) - g(x,y)] = L_1 - L_2$

3. *Product Rule:* $\qquad\qquad$ $\lim f(x,y) \cdot g(x,y) = L_1 \cdot L_2$

4. *Constant Multiple Rule:* $\quad$ $\lim kg(x,y) = kL_2$ $\quad$ (Any number k)

5. *Quotient Rule:* $\qquad\qquad$ $\lim \dfrac{f(x,y)}{g(x,y)} = \dfrac{L_1}{L_2}$ $\quad$ if $L_2 \neq 0$.

All limits are to be taken as $(x,y) \to (x_0, y_0)$, and L_1 and L_2 are to be real numbers.

When we combine Theorem 1 with the functions and limits in (4) we obtain the useful result that the limits of polynomials and rational functions as $(x,y) \to (x_0, y_0)$ may be calculated by evaluating the functions at the point (x_0, y_0). The only requirement is that the functions be defined at (x_0, y_0).

Example 1

a) $\lim\limits_{(x,y)\to(3,-4)} (x^2 + y^2) = (3)^2 + (-4)^2 = 9 + 16 = 25$

b) $\lim\limits_{(x,y)\to(0,1)} \dfrac{x - xy + 3}{x^2 y + 5xy - y^3} = \dfrac{0 - 0(1) + 3}{(0)^2(1) + 5(0)(1) - (1)^3} = -3$

Continuity

The definition of continuity for functions of two variables is analogous to the definition for functions of a single variable.

DEFINITIONS

A function $f(x,y)$ is said to be **continuous at the point (x_0, y_0)** if

1. f is defined at (x_0, y_0),

2. $\lim\limits_{(x,y)\to(x_0,y_0)} f(x,y)$ exists,

3. $\lim\limits_{(x,y)\to(x_0,y_0)} f(x,y) = f(x_0, y_0)$.

A function is said to be **continuous** if it is continuous at every point of its domain.

As with the definition of limit, the definition of continuity applies at boundary points as well as interior points of the domain of f. The only requirement is that the point (x,y) remain in the domain at all times.

As you may have guessed, one of the consequences of Theorem 1 is that algebraic combinations of continuous functions are continuous at every point at which all the functions involved are defined. This means that sums, differences, products, and constant multiples of continuous functions are continuous at any shared domain point. Also, the quotient of two continuous functions is continuous at any point at which the quotient is defined. In particular, polynomials and rational functions of two variables are continuous at every point at which they are defined.

If $z = f(x, y)$ is a continuous function of x and y, and $w = g(z)$ is a continuous function of z, then the composite $w = g(f(x, y))$ is continuous. Thus, composites like

$$e^{x-y}, \qquad \cos \frac{xy}{x^2 + 1}, \qquad \ln(1 + x^2 y^2)$$

are continuous at every point (x, y).

As with functions of a single variable, the general rule is that composites of continuous functions are continuous. The only requirement is that each function be continuous where it is applied.

Example 2   Show that the function

$$f(x, y) = \begin{cases} \dfrac{2xy}{x^2 + y^2}, & (x, y) \neq (0, 0), \\ 0, & (x, y) = (0, 0), \end{cases}$$

is continuous at every point except the origin (Fig. 12.12).

Solution   The function f is continuous at any point $(x, y) \neq (0, 0)$ because its values are then given by a rational function of x and y.

At $(0, 0)$ the value of f is defined, but f, we claim, has no limit as $(x, y) \rightarrow (0, 0)$. The reason is that different paths of approach to the origin can lead to different results, as we shall now see.

For every value of m, the line $y = mx$ is a level curve of f because the value

12.12   a) Stages in a computer's construction of the graph of

$$f(x, y) = \begin{cases} \dfrac{2xy}{x^2 + y^2}, & (x, y) \neq (0, 0), \\ 0, & (x, y) = (0, 0). \end{cases}$$

The function is continuous at every point except the origin. b) The level curves of f are the open rays from the origin together with the origin itself (as a degenerate curve). The rays in quadrants I and IV have been labeled with function values.

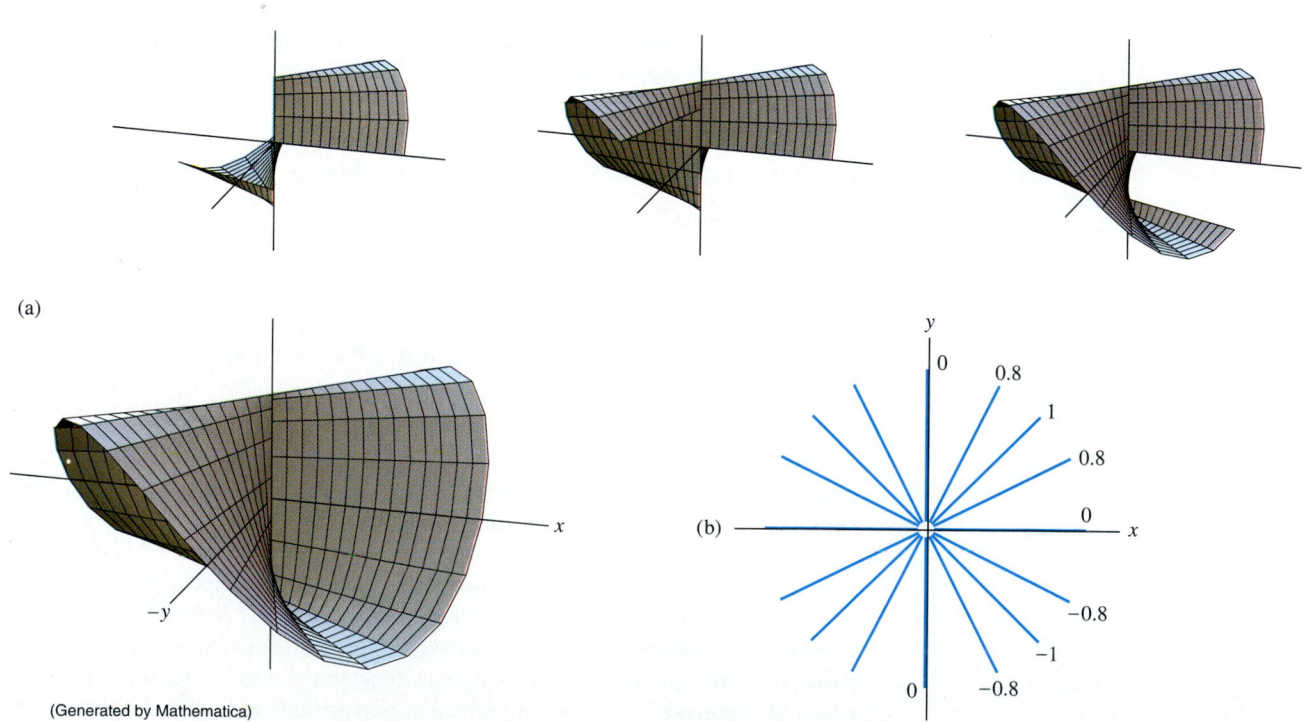

(a)

(b)

(Generated by Mathematica)

of f on the line has the constant value

$$f(x, y)\Big|_{y = mx} = \frac{2xy}{x^2 + y^2}\Big|_{y = mx} = \frac{2x(mx)}{x^2 + (mx)^2} = \frac{2mx^2}{x^2 + m^2x^2} = \frac{2m}{1 + m^2}.$$

Therefore, f has this number as its limit as (x, y) approaches the origin along the line:

$$\lim_{\substack{(x, y)\to(0, 0) \\ \text{along } y = mx}} f(x, y) = \lim_{(x, y)\to(0, 0)}\left[f(x, y)\Big|_{y = mx}\right] = \frac{2m}{1 + m^2}.$$

This limit changes with m. There is therefore no single number we may call the limit of f as (x, y) approaches the origin. The limit fails to exist, and the function is not continuous.

Example 2 illustrates an important point about limits of functions of two variables (or even more variables, for that matter). For a limit to exist at a point, the limit must be the same along every approach path. Therefore, if we ever find paths with different limits, we know the function has no limit at the point they approach.

The Two-Path Test for Discontinuity

If a function $f(x, y)$ has different limits along two different paths as (x, y) approaches (x_0, y_0), then $\lim_{(x, y)\to(x_0, y_0)} f(x, y)$ does not exist.

Example 3   Show that the function

$$f(x, y) = \frac{2x^2y}{x^4 + y^2}$$

(Fig. 12.13) has no limit as (x, y) approaches $(0, 0)$.

Solution   Along the curve $y = kx^2$, $x \neq 0$, the function has the constant value

$$f(x, y)\Big|_{y = kx^2} = \frac{2x^2y}{x^4 + y^2}\Big|_{y = kx^2} = \frac{2x^2(kx^2)}{x^4 + (kx^2)^2} = \frac{2kx^4}{x^4 + k^2x^4} = \frac{2k}{1 + k^2}.$$

Therefore,

$$\lim_{\substack{(x, y)\to(0, 0) \\ \text{along } y = kx^2}} f(x, y) = \lim_{(x, y)\to(0, 0)}\left[f(x, y)\Big|_{y = kx^2}\right] = \frac{2k}{1 + k^2}.$$

This limit varies with the path of approach. If (x, y) approaches $(0, 0)$ along the parabola $y = x^2$, for instance, $k = 1$ and the limit is 1. If (x, y) approaches $(0, 0)$ along the x-axis, $k = 0$ and the limit is 0. By the two-path test, f has no limit as (x, y) approaches $(0, 0)$.

The language we use here may seem contradictory. You might well ask, "What do you mean f has no limit as (x, y) approaches the origin—it has lots of limits." But that is the point. There is no path-independent limit, and therefore, by our definition, $\lim_{(x, y)\to(0, 0)} f(x, y)$ does not exist. It is our translating this formal statement

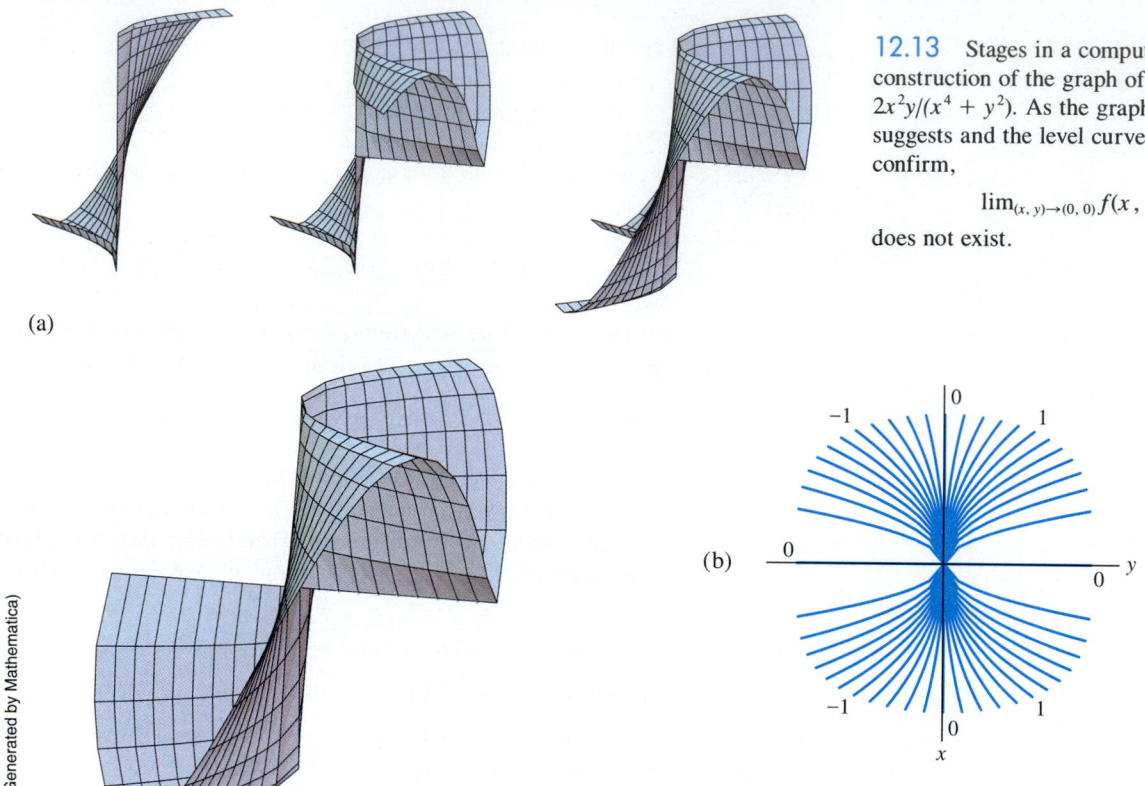

(a)

(b)

(Generated by Mathematica)

12.13 Stages in a computer's construction of the graph of $f(x, y) = 2x^2y/(x^4 + y^2)$. As the graph in (a) suggests and the level curve values in (b) confirm,

$$\lim_{(x, y) \to (0, 0)} f(x, y)$$

does not exist.

into the more colloquial "has no limit" that creates the apparent contradiction. The mathematics is fine. The problem arises in how we tend to talk about it. In the crunch, we need formality to keep things straight.

Functions of More Than Two Variables

The definitions of limit and continuity for functions of two variables and the conclusions about limits and continuity for sums, products, quotients, and composites of functions of two variables all extend to functions of three or more variables. Thus, functions like

$$\ln (x + y + z) \qquad \text{and} \qquad \frac{y \sin z}{x - 1}$$

are continuous throughout their domains and limits like

$$\lim_{P \to (1, 0, -1)} \frac{e^{x + z}}{z^2 + \cos \sqrt{xy}} = \frac{e^{1 - 1}}{(-1)^2 + \cos 0} = \frac{1}{2},$$

where P denotes the point (x, y, z), may be found by direct substitution.

Continuous Functions Defined on Closed, Bounded Regions

As we know, a function of a single variable that is continuous throughout a closed, bounded interval $[a, b]$ takes on an absolute maximum value and an absolute minimum value at least once in $[a, b]$. The same is true of a function $z = f(x, y)$ that is continuous on a closed, bounded set R in the plane (like a line segment, disk, or filled-in triangle). The function takes on an absolute maximum value at some point in R and an absolute minimum value at some point in R.

Similar theorems hold for functions of three or more variables. A continuous function $w = f(x, y, z)$, for example, must take on absolute maximum and minimum values on any closed, bounded set (solid ball or cube, spherical shell, rectangular plate) on which it is defined.

It is important to know how to find these extreme values and we shall show how to do so in Section 12.8. But first, we need to know about derivatives, and that will be the topic of the next section.

EXERCISES 12.2

Find the limits in Exercises 1–12.

1. $\displaystyle\lim_{(x, y)\to(0, 0)} \frac{3x^2 - y^2 + 5}{x^2 + y^2 + 2}$

2. $\displaystyle\lim_{(x, y)\to(1, 1)} \ln|1 + x^2y^2|$

3. $\displaystyle\lim_{(x, y)\to(0, \ln 2)} e^{x-y}$

4. $\displaystyle\lim_{(x, y)\to(0, 4)} \frac{x}{\sqrt{y}}$

5. $\displaystyle\lim_{P\to(1, 3, 4)} \sqrt{x^2 + y^2 + z^2} - 1$

6. $\displaystyle\lim_{P\to(1, 2, 6)} \left(\frac{1}{x} + \frac{1}{y} + \frac{1}{z}\right)$

7. $\displaystyle\lim_{(x, y)\to(0, \pi/4)} \sec x \tan y$

8. $\displaystyle\lim_{(x, y)\to(0, 0)} \cos \frac{x^2 + y^2}{x + y + 1}$

9. $\displaystyle\lim_{(x, y)\to(1, 1)} \cos \sqrt[3]{|xy| - 1}$

10. $\displaystyle\lim_{(x, y)\to(1, 0)} \frac{x \sin y}{x^2 + 1}$

11. $\displaystyle\lim_{(x, y)\to(0, 0)} \frac{e^y \sin x}{x}$

12. $\displaystyle\lim_{(x, y)\to(0, 0)} \tan^{-1}\left(1/\sqrt{x^2 + y^2}\right)$

Find the limits in Exercises 13–16 by rewriting the fractions first.

13. $\displaystyle\lim_{\substack{(x, y)\to(1, 1)\\ x \ne y}} \frac{x^2 - 2xy + y^2}{x - y}$

14. $\displaystyle\lim_{\substack{(x, y)\to(1, 1)\\ x \ne y}} \frac{x^2 - y^2}{x - y}$

15. $\displaystyle\lim_{\substack{(x, y)\to(1, 1)\\ x \ne 1}} \frac{xy - y - 2x + 2}{x - 1}$

16. $\displaystyle\lim_{\substack{(x, y)\to(2, -4)\\ y \ne -4, \, x \ne x^2}} \frac{y + 4}{x^2 y - xy + 4x^2 - 4x}$

Find the limits in Exercises 17–22. In each exercise, P stands for the point (x, y, z).

17. $\displaystyle\lim_{P\to(2, 3, -6)} \sqrt{x^2 + y^2 + z^2}$

18. $\displaystyle\lim_{P\to(0, -2, 0)} \ln\sqrt{x^2 + y^2 + z^2}$

19. $\displaystyle\lim_{P\to(3, 3, 0)} (\sin^2 x + \cos^2 y + \sec^2 z)$

20. $\displaystyle\lim_{P\to(\pi, 0, 3)} ze^{-2y} \cos 2x$

21. $\displaystyle\lim_{P\to(-1/4, \pi/2, 2)} \tan^{-1} xyz$

22. $\displaystyle\lim_{P\to(1, -1, -1)} \frac{2xy + yz}{x^2 + z^2}$

At what points (x, y) in the plane are the functions given by the formulas in Exercises 23–26 continuous?

23. a) $\sin(x + y)$ b) $\ln(x^2 + y^2)$

24. a) $\dfrac{x + y}{x - y}$ b) $\dfrac{y}{x^2 + 1}$

25. a) $\sin \dfrac{1}{xy}$ b) $\dfrac{x + y}{2 + \cos x}$

26. a) $\dfrac{x^2 + y^2}{x^2 - 3x + 2}$ b) $\dfrac{1}{x^2 - y}$

At what points (x, y, z) in space are the functions given by the formulas in Exercises 27–30 continuous?

27. a) $x^2 + y^2 - 2z^2$ b) $\sqrt{x^2 + y^2 - 1}$

28. a) $\ln xyz$ b) $e^{x+y} \cos z$

29. a) $xy \sin \dfrac{1}{z}$ b) $\dfrac{1}{x^2 + y^2 + z^2 - 1}$

30. a) $\dfrac{1}{|x| + |y| + |z|}$ b) $\dfrac{1}{|xy| + |z|}$

By considering different paths of approach, show that the functions in Exercises 31–38 have no limit as $(x, y)\to(0, 0)$.

31. $f(x, y) = \dfrac{x}{\sqrt{x^2 + y^2}}$

32. $f(x, y) = \dfrac{x^4}{x^4 + y^2}$

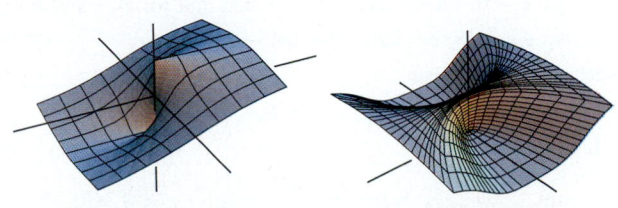

(Generated by Mathematica)   (Generated by Mathematica)

33. $f(x, y) = \dfrac{x^4 - y^2}{x^4 + y^2}$

34. $f(x, y) = \dfrac{xy}{|xy|}$

35. $f(x, y) = \dfrac{x - y}{x + y}$

36. $f(x, y) = \dfrac{x + y}{x - y}$

37. $f(x, y) = \dfrac{x^2 + y}{y}$

38. $f(x, y) = \dfrac{x^2}{x^2 - y}$

39. *Continuation of Example 2*

a) Reread Example 2. Then substitute $m = \tan \theta$ into the formula

$$f(x, y)\Big|_{y = mx} = \frac{2m}{1 + m^2}$$

and simplify the result to show how the value of f varies with the line's angle of inclination.

b) Use the formula you obtained in (a) to show that the limit of f as $(x, y) \to (0, 0)$ along the line $y = mx$ varies from -1 to 1 depending on the angle of approach.

40. Define $f(0, 0)$ in a way that extends

$$f(x, y) = xy\frac{x^2 - y^2}{x^2 + y^2}$$

to be continuous at the origin.

The Sandwich Theorem for functions of two variables states that if $g(x, y) \leq f(x, y) \leq h(x, y)$ for all $(x, y) \neq (x_0, y_0)$ in a disk centered at (x_0, y_0) and if g and h have the same finite limit L as $(x, y) \to (x_0, y_0)$, then $\lim_{(x, y) \to (x_0, y_0)} f(x, y)$ exists and equals L as well. Use this result to find the limits in Exercises 41 and 42.

41. Find $\lim\limits_{(x, y) \to (0, 0)} \dfrac{\tan^{-1} xy}{xy}$ if

$$1 - \frac{x^2 y^2}{3} < \frac{\tan^{-1} xy}{xy} < 1.$$

42. Find $\lim\limits_{(x, y) \to (0, 0)} \dfrac{4 - 4\cos\sqrt{|xy|}}{|xy|}$ if

$$2|xy| - \frac{x^2 y^2}{6} < 4 - 4\cos\sqrt{|xy|} < 2|xy|.$$

Changing to Polar Coordinates

If you cannot make any headway with $\lim_{(x, y) \to (0, 0)} f(x, y)$ in rectangular coordinates, try changing to polar coordinates. Substitute $x = r\cos\theta$, $y = r\sin\theta$, and investigate the limit of the resulting expression as $r \to 0$. In other words, try to decide whether there exists a number L satisfying the following criterion:

Given $\epsilon > 0$, there exists a $\delta > 0$ such that for all r and θ,

$$|r| < \delta \quad \Rightarrow \quad |f(r, \theta) - L| < \epsilon. \tag{5}$$

If such an L exists, then

$$\lim_{(x, y) \to (0, 0)} f(x, y) = \lim_{r \to 0} f(r, \theta) = L.$$

For instance,

$$\lim_{(x, y) \to (0, 0)} \frac{x^3}{x^2 + y^2} = \lim_{r \to 0} \frac{r^3 \cos^3\theta}{r^2} = \lim_{r \to 0} r\cos^3\theta = 0.$$

To verify the last of these equalities, we need to show that implication (5) in the criterion for limit is satisfied with $f(r, \theta) = r\cos^3\theta$ and $L = 0$. That is, we need to show that given any $\epsilon > 0$ there exists a $\delta > 0$ such that for all r and θ,

$$|r| < \delta \quad \Rightarrow \quad |r\cos^3\theta - 0| < \epsilon.$$

Since

$$|r\cos^3\theta| = |r||\cos^3\theta| \leq |r| \cdot 1 = |r|,$$

the implication holds for all r and θ if we take $\delta = \epsilon$.

In contrast,

$$\frac{x^2}{x^2 + y^2} = \frac{r^2 \cos^2\theta}{r^2} = \cos^2\theta$$

takes on all values from 0 to 1 regardless of how small $|r|$ is, so that $\lim_{(x, y) \to (0, 0)} x^2/(x^2 + y^2)$ does not exist.

In each of these instances, the existence or nonexistence of the limit as $r \to 0$ is fairly clear. Shifting to polar coordinates does not always help, however, and may even tempt us to false conclusions. For example, the limit may exist along every straight line (or ray) $\theta = $ constant and yet fail to exist in the broader sense. Example 3 illustrates this point. In polar coordinates, $f(x, y) = (2x^2 y)/(x^4 + y^2)$ becomes

$$f(r\cos\theta, r\sin\theta) = \frac{r\cos\theta\sin 2\theta}{r^2\cos^4\theta + \sin^2\theta}$$

for $r \neq 0$. If we hold θ constant and let $r \to 0$, the limit is 0. On the path $y = x^2$, however, we have $r\sin\theta = r^2\cos^2\theta$ and

$$f(r\cos\theta, r\sin\theta) = \frac{r\cos\theta\sin 2\theta}{r^2\cos^4\theta + (r\cos^2\theta)^2}$$

$$= \frac{2r\cos^2\theta\sin\theta}{2r^2\cos^4\theta} = \frac{r\sin\theta}{r^2\cos^2\theta} = 1.$$

In Exercises 43–48, find the limit of f as $(x, y) \to (0, 0)$ or show that the limit does not exist.

43. $f(x, y) = \dfrac{x^3 - xy^2}{x^2 + y^2}$

44. $f(x, y) = \cos\left(\dfrac{x^3 - y^3}{x^2 + y^2}\right)$

45. $f(x, y) = \dfrac{y^2}{x^2 + y^2}$

46. $f(x, y) = \dfrac{2x}{x^2 + x + y^2}$

47. $f(x, y) = \tan^{-1}\left(\dfrac{|x| + |y|}{x^2 + y^2}\right)$

48. $f(x, y) = \dfrac{x^2 - y^2}{x^2 + y^2}$

In Exercises 49 and 50, define $f(0, 0)$ in a way that extends f to be continuous at the origin.

49. $f(x, y) = \ln\left(\dfrac{y^2 + 3x}{3x^2 + x}\right)$

50. $f(x, y) = \dfrac{xy - x - y}{x + y}$

The Definitions of Limit and Continuity

Each of Exercises 51–54 gives a function $f(x, y)$ and a positive number ϵ. In each exercise, either show that there exists a $\delta > 0$ such that for all (x, y),

$$\sqrt{x^2 + y^2} < \delta \quad \Rightarrow \quad |f(x, y) - f(0, 0)| < \epsilon$$

or show that there exists a $\delta > 0$ such that for all (x, y),

$$|x| < \delta \quad \text{and} \quad |y| < \delta \quad \Rightarrow \quad |f(x, y) - f(0, 0)| < \epsilon.$$

Do either one or the other, whichever seems more convenient. There is no need to do both.

51. $f(x, y) = x^2 + y^2, \quad \epsilon = 0.01$

52. $f(x, y) = y/(x^2 + 1), \quad \epsilon = 0.05$

53. $f(x, y) = (x + y)/(x^2 + 1), \quad \epsilon = 0.01$

54. $f(x, y) = (x + y)/(2 + \cos x), \quad \epsilon = 0.02$

Each of Exercises 55–58 gives a function $f(x, y, z)$ and a positive number ϵ. In each exercise, either show that there exists a $\delta > 0$ such that for all (x, y, z),

$$\sqrt{x^2 + y^2 + z^2} < \delta \quad \Rightarrow \quad |f(x, y, z) - f(0, 0, 0)| < \epsilon$$

or show that there exists a $\delta > 0$ such that for all (x, y, z),

$$|x| < \delta, \quad |y| < \delta, \quad \text{and}$$
$$|z| < \delta \quad \Rightarrow \quad |f(x, y, z) - f(0, 0, 0)| < \epsilon.$$

Do either one or the other, whichever seems more convenient. There is no need to do both.

55. $f(x, y, z) = x^2 + y^2 + z^2, \quad \epsilon = 0.015$

56. $f(x, y, z) = xyz, \quad \epsilon = 0.008$

57. $f(x, y, z) = \dfrac{x + y + z}{x^2 + y^2 + z^2 + 1}, \quad \epsilon = 0.015$

58. $f(x, y, z) = \tan^2 x + \tan^2 y + \tan^2 z, \quad \epsilon = 0.03$

59. Show that $f(x, y, z) = x + y - z$ is continuous at every point (x_0, y_0, z_0).

60. Show that $f(x, y, z) = x^2 + y^2 + z^2$ is continuous at the origin.

12.3 Partial Derivatives

Partial derivatives are what we get when we hold all but one of the independent variables of a function constant and differentiate with respect to that one variable. In this section, we show how partial derivatives arise geometrically and how we can calculate them by applying the rules for differentiating functions of a single variable. We begin with functions of two independent variables.

Definitions and Notation

If (x_0, y_0) is a point in the domain of a function $f(x, y)$, the vertical plane $y = y_0$ will cut the surface $z = f(x, y)$ in the curve $z = f(x, y_0)$ (Fig. 12.14). This curve is the graph of the function $z = f(x, y_0)$ in the plane $y = y_0$. The horizontal coordinate in this plane is x; the vertical coordinate is z.

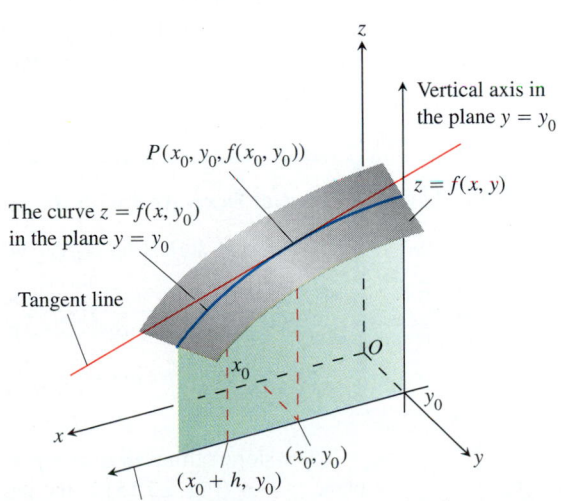

12.14 The intersection of the plane $y = y_0$ with the surface $z = f(x, y)$, viewed from a point above the first quadrant of the xy-plane.

We define the partial derivative of f with respect to x at the point (x_0, y_0) as the ordinary derivative of $f(x, y_0)$ with respect to x at the point $x = x_0$.

DEFINITION

Partial Derivative with Respect to x

The partial derivative of $f(x, y)$ with respect to x at the point (x_0, y_0) is

$$\frac{\partial f}{\partial x}\bigg|_{(x_0, y_0)} = \frac{d}{dx} f(x, y_0)\bigg|_{x = x_0} = \lim_{h \to 0} \frac{f(x_0 + h, y_0) - f(x_0, y_0)}{h}, \quad (1)$$

provided the limit exists. (The letter ∂ is a special round d.)

The slope of the curve $z = f(x, y_0)$ at the point $P(x_0, y_0, f(x_0, y_0))$ in the plane $y = y_0$ is the value of the partial derivative of f with respect to x at (x_0, y_0). The tangent to the curve at P is the line in the plane $y = y_0$ that passes through P with this slope. The partial derivative $\partial f/\partial x$ at (x_0, y_0) gives the rate of change of f with respect to x when y is held fixed at the value y_0.

The notation we use for the partial derivative of f with respect to x depends on what aspect of the derivative we want to emphasize. The usual list looks like this:

$\dfrac{\partial f}{\partial x}(x_0, y_0)$ or $f_x(x_0, y_0)$ "Partial derivative of f with respect to x at (x_0, y_0)" or "f sub x at (x_0, y_0)." Convenient for stressing the point (x_0, y_0).

$\dfrac{\partial z}{\partial x}\bigg|_{(x_0, y_0)}$ or $\left(\dfrac{\partial z}{\partial x}\right)_{(x_0, y_0)}$ "Partial derivative of z with respect to x at (x_0, y_0)." Common in science and engineering when you are dealing with variables and do not mention the function explicitly.

$f_x, \dfrac{\partial f}{\partial x}, z_x,$ or $\dfrac{\partial z}{\partial x}$ "Partial derivative of f (or z) with respect to x." Convenient when you regard the partial derivative as a function in its own right.

The definition of the partial derivative of $f(x, y)$ with respect to y at a point (x_0, y_0) is similar to the definition of the partial derivative of f with respect to x. We hold x fixed at the value x_0 and take the ordinary derivative of $f(x_0, y)$ with respect to y at y_0.

DEFINITION

Partial Derivative with Respect to y

The partial derivative of $f(x, y)$ with respect to y at the point (x_0, y_0) is

$$\frac{\partial f}{\partial y}\bigg|_{(x_0, y_0)} = \frac{d}{dy} f(x_0, y)\bigg|_{y = y_0} = \lim_{k \to 0} \frac{f(x_0, y_0 + k) - f(x_0, y_0)}{k}, \quad (2)$$

provided the limit exists.

The slope of the curve $z = f(x_0, y)$ at the point $P(x_0, y_0, f(x_0, y_0))$ in the vertical plane $x = x_0$ (Fig. 12.15) is the partial derivative of f with respect to y at (x_0, y_0). The tangent to the curve at P is the line in the plane $x = x_0$ that passes through P

Vertical axis
in the plane $x = x_0$

Tangent line

$P(x_0, y_0, f(x_0, y_0))$

$z = f(x, y)$

x_0

y_0

x

(x_0, y_0)

$(x_0, y_0 + k)$

12.15   The intersection of the plane $x = x_0$ with the surface $z = f(x, y)$, viewed from above the first quadrant of the xy-plane.

The curve $z = f(x_0, y)$
in the plane $x = x_0$

y

Horizontal axis
in the plane $x = x_0$

with this slope. The partial derivative gives the rate of change of f with respect to y at (x_0, y_0) when x is held fixed at the value x_0.

The partial derivative with respect to y is denoted the same way as the partial derivative with respect to x. We have

$$\frac{\partial f}{\partial y}(x_0, y_0), \qquad f_y(x_0, y_0), \qquad \frac{\partial f}{\partial y}, \qquad f_y, \tag{3}$$

and so on.

Notice that we now have two tangent lines associated with the surface $z = f(x, y)$ at the point $P(x_0, y_0, f(x_0, y_0))$ (Fig. 12.16). Is the plane they determine tangent to the surface at P? It would be nice if it were, but we shall have to learn more about partial derivatives to find out.

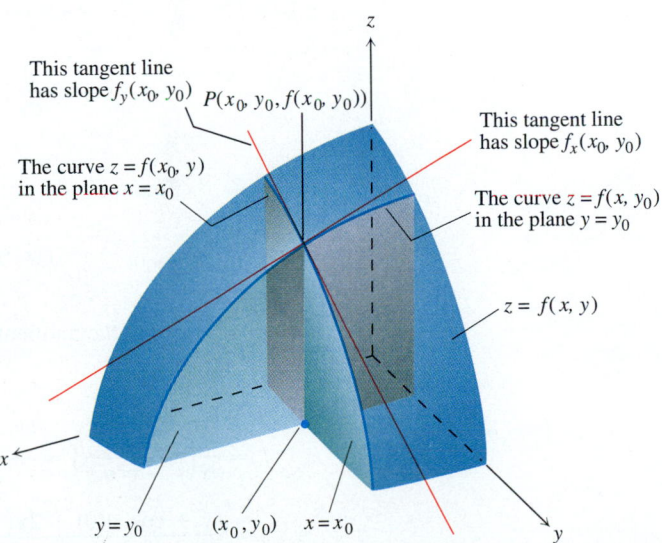

This tangent line
has slope $f_y(x_0, y_0)$ $P(x_0, y_0, f(x_0, y_0))$

This tangent line
has slope $f_x(x_0, y_0)$

The curve $z = f(x_0, y)$
in the plane $x = x_0$

The curve $z = f(x, y_0)$
in the plane $y = y_0$

$z = f(x, y)$

x

$y = y_0$ (x_0, y_0) $x = x_0$

y

12.16   Figures 12.14 and 12.15 combined. The tangent lines at $(x_0, y_0, f(x_0, y_0))$ determine a plane that, in this picture at least, appears to be tangent to the surface.

Calculations

As Eq. (1) shows, we calculate $\partial f/\partial x$ by differentiating f with respect to x in the usual way while treating y as a constant. Similarly, from Eq. (2) we see that we can calculate $\partial f/\partial y$ by differentiating f with respect to y in the usual way while holding x constant.

Example 1   Find the value of $\partial f/\partial x$ at the point $(4, -5)$ if

$$f(x, y) = x^2 + 3xy + y - 1.$$

Solution   We regard y as a constant and differentiate with respect to x:

$$\frac{\partial f}{\partial x} = \frac{\partial}{\partial x}(x^2 + 3xy + y - 1) = 2x + 3y + 0 - 0 = 2x + 3y.$$

The value of $\partial f/\partial x$ at $(4, -5)$ is $2(4) + 3(-5) = -7$.

Example 2   Find the value of $\partial f/\partial y$ at the point $(4, -5)$ if

$$f(x, y) = x^2 + 3xy + y - 1.$$

Solution   We regard x as a constant and differentiate with respect to y:

$$\frac{\partial f}{\partial y} = \frac{\partial}{\partial y}(x^2 + 3xy + y - 1) = 0 + 3x + 1 - 0 = 3x + 1.$$

The value of $\partial f/\partial y$ at $(4, -5)$ is $3(4) + 1 = 13$.

Example 3   Find $\partial f/\partial y$ if

$$f(x, y) = y \sin xy.$$

Solution   We treat x as a constant and f as a product of y and $\sin xy$:

$$\frac{\partial f}{\partial y} = \frac{\partial}{\partial y}(y \sin xy) = y\frac{\partial}{\partial y}\sin xy + \sin xy \frac{\partial}{\partial y}(y)$$

$$= y \cos xy \frac{\partial}{\partial y}(xy) + \sin xy = xy \cos xy + \sin xy.$$

Example 4   Find f_x if

$$f(x, y) = \frac{2y}{y + \cos x}.$$

Solution   We treat f as the quotient of $2y$ divided by $(y + \cos x)$. With y held constant, we get

$$f_x = \frac{\partial}{\partial x}\left(\frac{2y}{y + \cos x}\right) = \frac{(y + \cos x)\frac{\partial}{\partial x}(2y) - 2y\frac{\partial}{\partial x}(y + \cos x)}{(y + \cos x)^2}$$

$$= \frac{(y + \cos x)(0) - 2y(-\sin x)}{(y + \cos x)^2} = \frac{2y \sin x}{(y + \cos x)^2}.$$

Implicit differentiation works for partial derivatives the way it works for ordinary derivatives.

Example 5   Find $\partial z/\partial x$ if the equation

$$yz - \ln z = x + y$$

defines z as a function of the two independent variables x and y and the partial derivative exists.

Solution   We differentiate both sides of the equation with respect to x, holding y fixed and treating z as a differentiable function of x:

$$\frac{\partial}{\partial x}(yz) - \frac{\partial}{\partial x}\ln z = \frac{\partial x}{\partial x} + \frac{\partial y}{\partial x}$$

$$y\frac{\partial z}{\partial x} - \frac{1}{z}\frac{\partial z}{\partial x} = 1 + 0$$

$$\left(y - \frac{1}{z}\right)\frac{\partial z}{\partial x} = 1$$

$$\frac{\partial z}{\partial x} = \frac{z}{yz - 1}.$$

Functions of More Than Two Variables

The definitions of the partial derivatives of functions of more than two independent variables are like the definitions for functions of two variables. They are ordinary derivatives with respect to one variable, taken while the other independent variables are held constant.

Example 6   If x, y, and z are independent variables and

$$f(x, y, z) = x\sin(y + 3z),$$

then

$$\frac{\partial f}{\partial z} = \frac{\partial}{\partial z}(x\sin(y + 3z)) = x\frac{\partial}{\partial z}\sin(y + 3z)$$

$$= x\cos(y + 3z)\frac{\partial}{\partial z}(y + 3z) = 3x\cos(y + 3z).$$

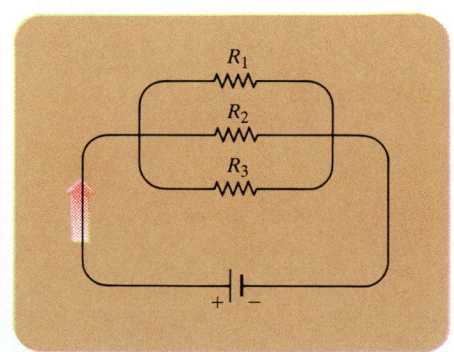

12.17   Resistors arranged this way are said to be connected in parallel (Example 7). Each resistor lets a portion of the current through. Their combined resistance R is calculated with the formula

$$\frac{1}{R} = \frac{1}{R_1} + \frac{1}{R_2} + \frac{1}{R_3}.$$

Example 7   *Electrical Resistors in Parallel.*   If resistors of R_1, R_2, and R_3 ohms are connected in parallel to make an R-ohm resistor, the value of R can be found from the equation

$$\frac{1}{R} = \frac{1}{R_1} + \frac{1}{R_2} + \frac{1}{R_3} \tag{4}$$

(Fig. 12.17). Find the value of $\partial R/\partial R_2$ when $R_1 = 30$, $R_2 = 45$, and $R_3 = 90$ ohms.

Solution   To find $\partial R/\partial R_2$, we treat R_1 and R_3 as constants and differentiate both sides of Eq. (4) with respect to R_2:

$$\frac{\partial}{\partial R_2}\left(\frac{1}{R}\right) = \frac{\partial}{\partial R_2}\left(\frac{1}{R_1} + \frac{1}{R_2} + \frac{1}{R_3}\right)$$

$$-\frac{1}{R^2}\frac{\partial R}{\partial R_2} = 0 - \frac{1}{R_2^2} + 0$$

$$\frac{\partial R}{\partial R_2} = \frac{R^2}{R_2^2} = \left(\frac{R}{R_2}\right)^2.$$

When $R_1 = 30$, $R_2 = 45$, and $R_3 = 90$,

$$\frac{1}{R} = \frac{1}{30} + \frac{1}{45} + \frac{1}{90} = \frac{3 + 2 + 1}{90} = \frac{6}{90} = \frac{1}{15},$$

so $R = 15$ and

$$\frac{\partial R}{\partial R_2} = \left(\frac{15}{45}\right)^2 = \left(\frac{1}{3}\right)^2 = \frac{1}{9}.$$

Second Order Partial Derivatives

When we differentiate a function $f(x, y)$ twice, we produce its second order derivatives. These derivatives are usually denoted by

$$\frac{\partial^2 f}{\partial x^2} \qquad \text{"}d \text{ squared } f\, d\, x \text{ squared"} \qquad \text{or} \qquad f_{xx} \qquad \text{"}f \text{ sub } x\, x\text{"}$$

$$\frac{\partial^2 f}{\partial y^2} \qquad \text{"}d \text{ squared } f\, d\, y \text{ squared"} \qquad f_{yy} \qquad \text{"}f \text{ sub } y\, y\text{"}$$

$$\frac{\partial^2 f}{\partial x\, \partial y} \qquad \text{"}d \text{ squared } f\, d\, x\, d\, y\text{"} \qquad f_{yx} \qquad \text{"}f \text{ sub } y\, x\text{"}$$

$$\frac{\partial^2 f}{\partial y\, \partial x} \qquad \text{"}d \text{ squared } f\, d\, y\, d\, x\text{"} \qquad f_{xy} \qquad \text{"}f \text{ sub } x\, y\text{"}$$

The defining equations are

$$\frac{\partial^2 f}{\partial x^2} = \frac{\partial}{\partial x}\left(\frac{\partial f}{\partial x}\right), \qquad \frac{\partial^2 f}{\partial x\, \partial y} = \frac{\partial}{\partial x}\left(\frac{\partial f}{\partial y}\right),$$

and so on. Notice the order in which the derivatives are taken:

$$\frac{\partial^2 f}{\partial x\, \partial y} \qquad \text{(Differentiate first with respect to } y \text{, then with respect to } x.)$$

$$f_{yx} = (f_y)_x \qquad \text{(Means the same thing.)}$$

Example 8   If $f(x, y) = x \cos y + y e^x$, then

$$\frac{\partial f}{\partial x} = \cos y + y e^x,$$

$$\frac{\partial^2 f}{\partial y\, \partial x} = \frac{\partial}{\partial y}\left(\frac{\partial f}{\partial x}\right) = -\sin y + e^x,$$

$$\frac{\partial^2 f}{\partial x^2} = \frac{\partial}{\partial x}\left(\frac{\partial f}{\partial x}\right) = y e^x.$$

Also,

$$\frac{\partial f}{\partial y} = -x \sin y + e^x$$

$$\frac{\partial^2 f}{\partial x \, \partial y} = \frac{\partial}{\partial x}\left(\frac{\partial f}{\partial y}\right) = -\sin y + e^x$$

$$\frac{\partial^2 f}{\partial y^2} = \frac{\partial}{\partial y}\left(\frac{\partial f}{\partial y}\right) = -x \cos y.$$

The Mixed Derivative Theorem

You may have noticed that the "mixed" second order derivatives

$$\frac{\partial^2 f}{\partial y \, \partial x} \qquad \text{and} \qquad \frac{\partial^2 f}{\partial x \, \partial y}$$

in Example 8 were equal. This was no mere coincidence. They have to be equal whenever f, f_x, f_y, f_{xy}, and f_{yx} are continuous.

THEOREM 2

The Mixed Derivative Theorem

If $f(x, y)$ and its partial derivatives f_x, f_y, f_{xy}, and f_{yx} are defined throughout an open region containing a point (a, b) and are all continuous at (a, b), then

$$f_{xy}(a, b) = f_{yx}(a, b). \tag{5}$$

You can find a proof of Theorem 2 in Appendix 11.

Theorem 2 says that when we want to calculate a mixed second order derivative we may differentiate in either order. This can work to our advantage, as the next example shows.

Example 9   Find $\partial^2 w/\partial x \partial y$ if

$$w = xy + \frac{e^y}{y^2 + 1}.$$

Solution   The symbol $\partial^2 w/\partial x \partial y$ tells us to differentiate first with respect to y and then with respect to x. However, if we postpone the differentiation with respect to y and differentiate first with respect to x, we can get the answer more quickly. In two steps,

$$\frac{\partial w}{\partial x} = y \qquad \text{and} \qquad \frac{\partial^2 w}{\partial y \, \partial x} = 1.$$

We are in for more work if we differentiate first with respect to y. (Just try it.)

Partial Derivatives of Still Higher Order

Although we shall deal mostly with first and second order partial derivatives because these appear most frequently in applications, there is no theoretical limit to how many times we can differentiate a function as long as the derivatives involved exist. Thus we get third and fourth order derivatives denoted by symbols like

$$\frac{\partial^3 f}{\partial x \, \partial y^2} = f_{yyx},$$

$$\frac{\partial^4 f}{\partial x^2 \, \partial y^2} = f_{yyxx},$$

and so on. As with second order derivatives, the order of differentiation is immaterial as long as the function and its derivatives through the order in question are all defined throughout an open region containing the point at which the derivatives are taken and are continuous at that point.

EXERCISES 12.3

In Exercises 1–22, find $\partial f / \partial x$ and $\partial f / \partial y$.

1. $f(x, y) = 2x^2 - 3y - 4$ **2.** $f(x, y) = x^2 - xy + y^2$

3. $f(x, y) = (x^2 - 1)(y + 2)$

4. $f(x, y) = 5xy - 7x^2 - y^2 + 3x - 6y + 2$

5. $f(x, y) = (xy - 1)^2$ **6.** $f(x, y) = (2x - 3y)^3$

7. $f(x, y) = \sqrt{x^2 + y^2}$ **8.** $f(x, y) = (x^3 + (y/2))^{2/3}$

9. $f(x, y) = 1/(x + y)$ **10.** $f(x, y) = x/(x^2 + y^2)$

11. $f(x, y) = (x + y)/(xy - 1)$ **12.** $f(x, y) = \tan^{-1}(y/x)$

13. $f(x, y) = \ln(x + y)$ **14.** $f(x, y) = e^{-x} \sin(x + y)$

15. $f(x, y) = e^{xy} \ln y$ **16.** $f(x, y) = e^{(x + y + 1)}$

17. $f(x, y) = \cos^2(3x - y^2)$ **18.** $f(x, y) = \sin^{-1} \sqrt{xy}$

19. $f(x, y) = x^y$ **20.** $f(x, y) = \log_y x$

21. $f(x, y) = \displaystyle\int_x^y f(t) \, dt$ (f continuous for all t)

22. $f(x, y) = \displaystyle\sum_{n=0}^{\infty} (xy)^n$ $(|xy| < 1)$

In Exercises 23–34, find f_x, f_y, and f_z.

23. $f(x, y, z) = 1 + xy^2 - 2z^2$

24. $f(x, y, z) = xy + yz + xz$

25. $f(x, y, z) = x - \sqrt{y^2 + z^2}$

26. $f(x, y, z) = (x^2 + y^2 + z^2)^{-1/2}$

27. $f(x, y, z) = \sin^{-1}(xyz)$

28. $f(x, y, z) = \sec^{-1}(x + yz)$

29. $f(x, y, z) = \ln(x + 2y + 3z)$

30. $f(x, y, z) = yz \ln(xy)$

31. $f(x, y, z) = e^{-(x^2 + y^2 + z^2)}$

32. $f(x, y, z) = e^{-xyz}$

33. $f(x, y, z) = \tanh(x + 2y + 3z)$

34. $f(x, y, z) = \sinh(xy - z^2)$

In Exercises 35–40, find the partial derivative of the function with respect to each variable.

35. $f(t, \alpha) = \cos(2\pi t - \alpha)$

36. $g(u, v) = v^2 e^{(2u/v)}$

37. $h(\rho, \phi, \theta) = \rho \sin \phi \cos \theta$

38. $g(r, \theta, z) = r(1 - \cos \theta) - z$

39. **Work done by the heart**   (Section 2.6, Exercise 48)

$$W(P, V, p, v, g) = PV + \frac{Vpv^2}{2g}$$

40. *Wilson lot size formula*   (Section 3.5, Exercise 49)

$$A(c, h, k, m, q) = \frac{km}{q} + cm + \frac{hq}{2}$$

Find all the second order partial derivatives of the functions in Exercises 41–46.

41. $f(x, y) = x + y + xy$ **42.** $f(x, y) = \sin xy$

43. $g(x, y) = x^2 y + \cos y + y \sin x$

44. $h(x, y) = xe^y + y + 1$

45. $r(x, y) = \ln(x + y)$ **46.** $s(x, y) = \tan^{-1}(y/x)$

In Exercises 47–50, verify that $w_{xy} = w_{yx}$.

47. $w = \ln(2x + 3y)$ **48.** $w = e^x + x \ln y + y \ln x$

49. $w = xy^2 + x^2 y^3 + x^3 y^4$

50. $w = x \sin y + y \sin x + xy$

51. Which order of differentiation will calculate f_{xy} faster: x first, or y first? Try to answer without writing anything down.

a) $f(x, y) = x \sin y + e^y$

b) $f(x, y) = 1/x$

c) $f(x, y) = y + (x/y)$

d) $f(x, y) = y + x^2y + 4y^3 - \ln(y^2 + 1)$

e) $f(x, y) = x^2 + 5xy + \sin x + 7e^x$

f) $f(x, y) = x \ln xy$

52. The fifth order partial derivative $\partial^5 f/\partial x^2 \partial y^3$ is zero for each of the following functions. To show this as quickly as possible, which variable would you differentiate with respect to first: x, or y? Try to answer without writing anything down.

a) $f(x, y) = y^2 x^4 e^x + 2$

b) $f(x, y) = y^2 + y(\sin x - x^4)$

c) $f(x, y) = x^2 + 5xy + \sin x + 7e^x$

d) $f(x, y) = xe^{y^2/2}$

53. Find the value of $\partial z/\partial x$ at the point $(1, 1, 1)$ if the equation

$$xy + z^3 x - 2yz = 0$$

defines z as a function of the two independent variables x and y and the partial derivative exists.

54. Find the value of $\partial x/\partial z$ at the point $(1, -1, -3)$ if the equation

$$xz + y \ln x - x^2 + 4 = 0$$

defines x as a function of the two independent variables y and z and the partial derivative exists.

Exercises 55 and 56 are about the angles and sides of the triangle in Fig. 12.18.

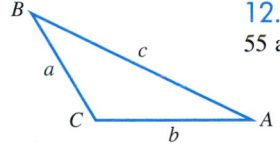

12.18 The triangle in Exercises 55 and 56.

55. Express A implicitly as a function of a, b, and c and calculate $\partial A/\partial a$ and $\partial A/\partial b$.

56. Express a implicitly as a function of A, b, and B and calculate $\partial a/\partial A$ and $\partial a/\partial B$.

57. Express v_x in terms of u and v if the equations $x = v \ln u$ and $y = u \ln v$ define u and v as functions of the independent variables x and y, and if v_x exists. (*Hint:* Differentiate both equations with respect to x and solve for v_x with Cramer's rule.)

58. Find $\partial x/\partial u$ and $\partial y/\partial u$ if the equations $u = x^2 - y^2$ and $v = x^2 - y$ define x and y as functions of the independent variables u and v, and the partial derivatives exist. (See the hint in Exercise 57.) Then let $s = x^2 + y^2$ and find $\partial s/\partial u$.

Laplace Equations

The *three-dimensional Laplace equation*

$$\frac{\partial^2 f}{\partial x^2} + \frac{\partial^2 f}{\partial y^2} + \frac{\partial^2 f}{\partial z^2} = 0 \tag{6}$$

is satisfied by steady-state temperature distributions $T = f(x, y, z)$ in space, by gravitational potentials, and by electro-

static potentials. The *two-dimensional Laplace equation*

$$\frac{\partial^2 f}{\partial x^2} + \frac{\partial^2 f}{\partial y^2} = 0, \tag{7}$$

obtained by dropping the $\partial^2 f/\partial z^2$ term from (6), describes potentials and steady-state temperature distributions in a plane (Fig. 12.19).

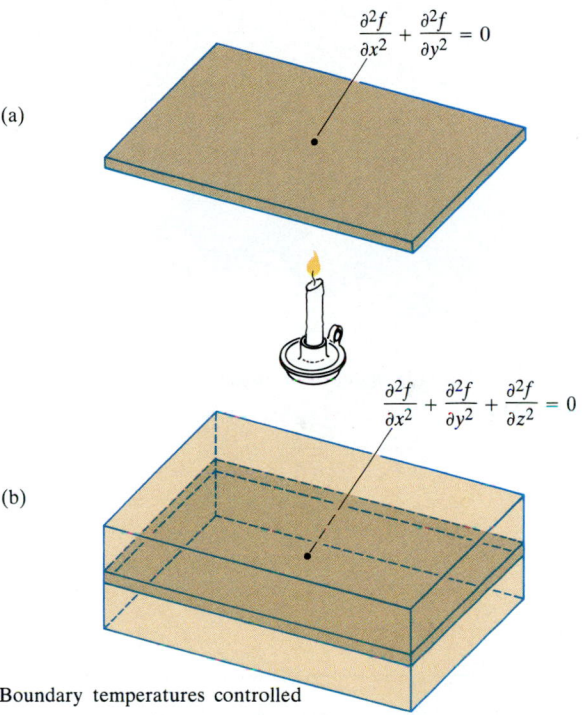

Boundary temperatures controlled

12.19 Steady-state temperature distributions in planes and solids satisfy Laplace equations. The plane (a) may be treated as a thin slice of the solid (b) perpendicular to the z-axis.

Show that each function in Exercises 59–64 satisfies a Laplace equation.

59. $f(x, y, z) = x^2 + y^2 - 2z^2$

60. $f(x, y, z) = e^{-2y} \cos 2x$

61. $f(x, y, z) = \ln \sqrt{x^2 + y^2}$

62. $f(x, y, z) = (x^2 + y^2 + z^2)^{-1/2}$

63. $f(x, y, z) = e^{3x + 4y} \cos 5z$

64. $f(x, y, z) = \cos 3x \cos 4y \sinh 5z$

65. For what values of n does

$$f(x, y, z) = (x^2 + y^2 + z^2)^n$$

satisfy the three-dimensional Laplace equation?

66. Find all solutions of the two-dimensional Laplace equation of the form

a) $f(x, y) = ax^2 + bxy + cy^2$

b) $f(x, y) = ax^3 + bx^2y + cxy^2 + dy^3$

The Wave Equation

If we stand on an ocean shore and take a snapshot of the waves, the picture shows a regular pattern of peaks and valleys in an instant of time (Fig. 12.20). We see periodic vertical motion in space, with respect to distance. If we stand in the water, we can feel the rise and fall of the water as the waves go by. We see periodic vertical motion in time. In physics, this beautiful symmetry

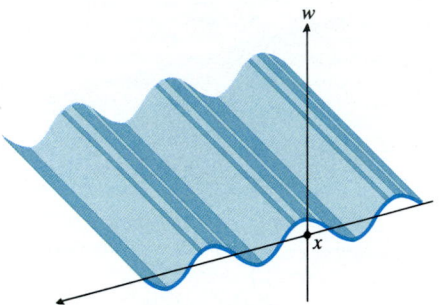

12.20   Waves in water at an instant in time. As time passes,

$$\frac{\partial^2 w}{\partial t^2} = c^2 \frac{\partial^2 w}{\partial x^2}.$$

is expressed by the *one-dimensional wave equation*

$$\frac{\partial^2 w}{\partial t^2} = c^2 \frac{\partial^2 w}{\partial x^2}, \tag{8}$$

where w is the wave height, x is the distance variable, t is the time variable, and c is the velocity with which the waves are propagated.

In our example, x is the distance across the ocean's surface, but in other applications x might be the distance along a vibrating string, distance through air (sound waves), or distance through space (light waves). The number c varies with the medium and type of wave.

Equations with partial derivatives are normally harder to solve than equations with ordinary derivatives because the solutions of a partial differential equation may have so many different forms. As a case in point, show that the functions in Exercises 67–73 are all solutions of the wave equation.

67. $w = \sin(x + ct)$ **68.** $w = \cos(2x + 2ct)$

69. $w = \sin(x + ct) + \cos(2x + 2ct)$

70. $w = \ln(2x + 2ct)$ **71.** $w = \tan(2x - 2ct)$

72. $w = 5\cos(3x + 3ct) + e^{x + ct}$

73. $w = f(u)$, where f is a differentiable function of u, $u = a(x + ct)$, and a is a constant.

12.4 Differentiability, Linearization, and Differentials

We know what it means for a function of two variables to have partial derivatives at a point, but we have yet to say what it means for such a function to be differentiable at a point. In this section, we define differentiability and proceed from there to linearizations and differentials. The mathematical results of the section stem from the Increment Theorem. As we shall see in the next section, this theorem also underlies the Chain Rule for functions of two variables.

Differentiability

Surprising as it may seem, the starting point for differentiability is not Fermat's difference quotient but rather the idea of increment. You may recall from our work with functions of a single variable that if $y = f(x)$ is differentiable at $x = x_0$, then the change in the value of f that results from changing x from x_0 to $x_0 + \Delta x$ is given by an equation of the form

$$\Delta y = f'(x_0)\Delta x + \epsilon \, \Delta x \tag{1}$$

in which $\epsilon \to 0$ as $\Delta x \to 0$. We used this property in Section 2.6 to derive the Chain Rule for functions of a single variable. For functions of two variables, the analogous property becomes the definition of differentiability. The Increment Theorem tells us when we may expect the property to hold.

THEOREM 3

The Increment Theorem for Functions of Two Variables

Suppose that the first partial derivatives of $f(x, y)$ are defined throughout an open region R containing the point (x_0, y_0) and that f_x and f_y are continuous at (x_0, y_0). Then the change $\Delta z = f(x_0 + \Delta x, y_0 + \Delta y) - f(x_0, y_0)$ in the value of f that results from moving from (x_0, y_0) to another point $(x_0 + \Delta x, y_0 + \Delta y)$ in R satisfies an equation of the form

$$\Delta z = f_x(x_0, y_0)\Delta x + f_y(x_0, y_0)\Delta y + \epsilon_1 \Delta x + \epsilon_2 \Delta y, \qquad (2)$$

in which $\epsilon_1, \epsilon_2 \to 0$ as $\Delta x, \Delta y \to 0$.

You will see where the epsilons come from if you read the proof in Appendix 11. You will also see that analogous theorems hold for functions of more than two variables.

The definition of differentiability for functions of two variables is based on Eq. (2).

DEFINITION

A function $f(x, y)$ is **differentiable at (x_0, y_0)** if $f_x(x_0, y_0)$ and $f_y(x_0, y_0)$ exist and Eq. (2) holds for f at (x_0, y_0). We call f **differentiable** if it is differentiable at every point in its domain.

In light of this definition, we have the immediate corollary of Theorem 3 that a function is differentiable if its first partial derivatives are continuous.

COROLLARY OF THEOREM 3

If the partial derivatives f_x and f_y of a function $f(x, y)$ are continuous throughout an open region R, then f is differentiable at every point of R.

If we replace the Δz in Eq. (2) by the expression $f(x, y) - f(x_0, y_0)$ and rewrite the equation as

$$f(x, y) = f(x_0, y_0) + f_x(x_0, y_0)\Delta x + f_y(x_0, y_0)\Delta y + \epsilon_1 \Delta x + \epsilon_2 \Delta y, \qquad (3)$$

we see that the right-hand side of the new equation approaches $f(x_0, y_0)$ as Δx and Δy approach 0. This tells us that a function $f(x, y)$ is continuous at every point where it is differentiable.

THEOREM 4

If a function $f(x, y)$ is differentiable at (x_0, y_0), then f is continuous at (x_0, y_0).

We can see from Theorems 3 and 4 that a function $f(x, y)$ is continuous at a point (x_0, y_0) if its first partial derivatives are continuous throughout some open region R containing the point. Our experience with functions of a single variable might tempt us to think that the mere existence of f_x and f_y at (x_0, y_0) would be enough to ensure the continuity of f at (x_0, y_0). After all, the existence of the derivative of a single-variable function at a point does ensure continuity. But here we have a departure from the single-variable theory. A function of two variables can be

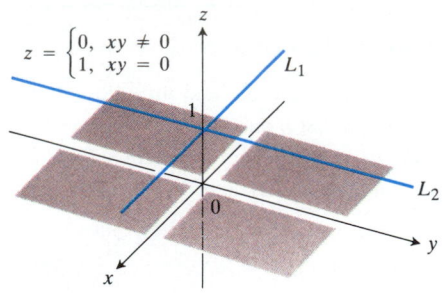

12.21   The function

$$f(x, y) = \begin{cases} 0, & xy \neq 0, \\ 1, & xy = 0, \end{cases}$$

is not continuous at $(0, 0)$. The limit of f as (x, y) approaches $(0, 0)$ along the line $y = x$ is 0, but $f(0, 0)$ itself is 1. The first partial derivatives of f at $(0, 0)$ both exist, $f_x(0, 0) = 0$ being the slope of line L_1 and $f_y(0, 0) = 0$ being the slope of line L_2.

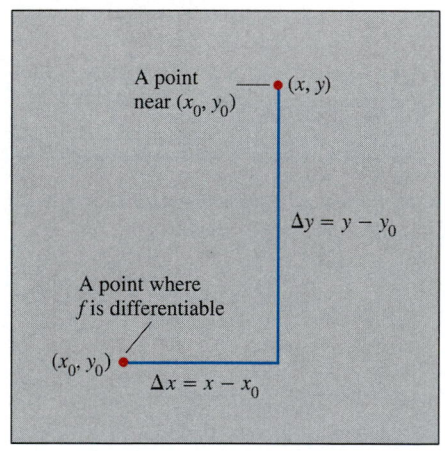

12.22   If f is differentiable at (x_0, y_0), then the value of f at any point (x, y) nearby is given by Eq. (5).

discontinuous at a point where its first partial derivatives exist (Fig. 12.21). Existence alone is not enough.

The Differentiability of Functions of More Than Two Variables

A function $w = f(x, y, z)$ of three independent variables is differentiable at a point $P_0(x_0, y_0, z_0)$ if $f_x, f_y,$ and f_z exist there and f satisfies an equation of the form

$$\Delta w = f_x \Delta x + f_y \Delta y + f_z \Delta z + \epsilon_1 \Delta x + \epsilon_2 \Delta y + \epsilon_3 \Delta z \tag{4}$$

in which $f_x, f_y,$ and f_z are evaluated at P_0 and $\epsilon_1, \epsilon_2, \epsilon_3 \to 0$ as $\Delta x, \Delta y, \Delta z \to 0$. A generalization of the Increment Theorem tells us that f is differentiable at a point P_0 if its first partial derivatives are continuous throughout an open region containing P_0. Similar results hold for functions of any finite number of independent variables.

Linearization

The functions of two variables that arise in science and mathematics can be complicated and we sometimes need to replace them with simpler functions that give the accuracy required for specific applications without being so hard to work with. We do this in a way that is similar to the way we find linear replacements for functions of a single variable (Section 2.6). The source of the linear replacements, or linearizations as we again call them, is Eq. (3).

Suppose the function we wish to replace is $z = f(x, y)$ and that we want the replacement to be effective near a point (x_0, y_0) at which we know the values of $f, f_x,$ and f_y and at which f is differentiable. Since f is differentiable, Eq. (3) holds for f at (x_0, y_0). Therefore, if we move from (x_0, y_0) to any point (x, y) by increments $\Delta x = x - x_0$ and $\Delta y = y - y_0$ (Fig. 12.22), the new value of f will be

$$f(x, y) = f(x_0, y_0) + f_x(x_0, y_0)(x - x_0) \qquad \begin{pmatrix} \text{Eq. (3), with} \\ \Delta x = x - x_0 \\ \text{and } \Delta y = y - y_0 \end{pmatrix} \tag{5}$$
$$+ f_y(x_0, y_0)(y - y_0) + \epsilon_1 \Delta x + \epsilon_2 \Delta y,$$

where $\epsilon_1, \epsilon_2 \to 0$ as $\Delta x, \Delta y \to 0$. If the increments Δx and Δy are small, the products $\epsilon_1 \Delta x$ and $\epsilon_2 \Delta y$ will be smaller still and we will have

$$f(x, y) \approx \underbrace{f(x_0, y_0) + f_x(x_0, y_0)(x - x_0) + f_y(x_0, y_0)(y - y_0)}_{L(x, y)}. \tag{6}$$

In other words, as long as Δx and Δy are small, f will have approximately the same value as the linear function L. When f is too hard to use, and our work can tolerate the error involved, we can safely replace f by L.

DEFINITIONS

The **linearization** of a function $f(x, y)$ at a point (x_0, y_0) where f is differentiable is the function

$$L(x, y) = f(x_0, y_0) + f_x(x_0, y_0)(x - x_0) + f_y(x_0, y_0)(y - y_0). \tag{7}$$

The approximation

$$L(x, y) \approx f(x, y) \tag{8}$$

is the **standard linear approximation** of f at (x_0, y_0).

In Section 12.7 we shall see that the plane $z = L(x, y)$ is tangent to the surface $z = f(x, y)$ at the point (x_0, y_0). Thus, the linearization of a function of two variables is a tangent-plane approximation to the function in the same way that the linearization of a function of a single variable is a tangent-line approximation. In the meantime, here is an example of a linearization.

Example 1   Find the linearization of

$$f(x, y) = x^2 - xy + \frac{1}{2}y^2 + 3$$

at the point $(3, 2)$.

Solution   We evaluate Eq. (7) with

$$f(x_0, y_0) = \left(x^2 - xy + \frac{1}{2}y^2 + 3\right)_{(3, 2)} = 8,$$

$$f_x(x_0, y_0) = \frac{\partial}{\partial x}\left(x^2 - xy + \frac{1}{2}y^2 + 3\right)_{(3, 2)} = (2x - y)_{(3, 2)} = 4,$$

$$f_y(x_0, y_0) = \frac{\partial}{\partial y}\left(x^2 - xy + \frac{1}{2}y^2 + 3\right)_{(3, 2)} = (-x + y)_{(3, 2)} = -1,$$

getting

$$L(x, y) = f(x_0, y_0) + f_x(x_0, y_0)(x - x_0) + f_y(x_0, y_0)(y - y_0) \qquad \text{(Eq. (7))}$$
$$= 8 + (4)(x - 3) + (-1)(y - 2) = 4x - y - 2.$$

The linearization of f at $(3, 2)$ is $L(x, y) = 4x - y - 2$.

How Accurate Is the Standard Linear Approximation?

To find the error in the approximation $f(x, y) \approx L(x, y)$, we need the second order partial derivatives of f. Suppose that the first and second order partial derivatives of f are continuous throughout an open set containing a closed rectangular region R centered at (x_0, y_0) and given by the inequalities

$$|x - x_0| \le h, \qquad |y - y_0| \le k \tag{9}$$

(Fig. 12.23). Since R is closed and bounded, the second partial derivatives all have maximum values on R. If B is the largest of these values, then, as explained in Section 12.10, the error $E(x, y) = f(x, y) - L(x, y)$ in the standard linear approximation satisfies the inequality

$$|E(x, y)| \le \frac{1}{2}B\left(|x - x_0| + |y - y_0|\right)^2 \tag{10}$$

throughout R.

When we use this inequality to estimate E, we usually cannot find the values of f_{xx}, f_{yy}, and f_{xy} that determine B and we have to replace B itself with an upper-bound or "worst-case" value instead. If M is any common upper bound for $|f_{xx}|$, $|f_{yy}|$, and $|f_{xy}|$ on R, then B will be less than or equal to M and we will know that

$$|E(x, y)| \le \frac{1}{2}M\left(|x - x_0| + |y - y_0|\right)^2. \tag{11}$$

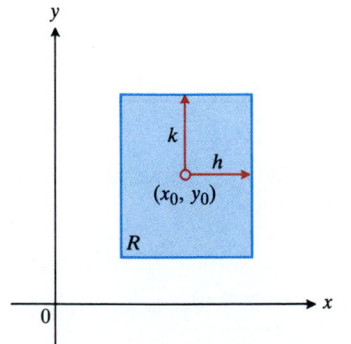

12.23   The rectangular region R: $|x - x_0| \le h$, $|y - y_0| \le k$ in the xy-plane. On this kind of region, we can find error bounds for our approximations.

This is the inequality we normally use in estimating E. When we need to make $|E(x, y)|$ small for a given M, we just make $|x - x_0|$ and $|y - y_0|$ small.

The Error in the Standard Linear Approximation of $f(x, y)$ near (x_0, y_0)

If f has continuous first and second partial derivatives throughout an open set containing a rectangle R centered at (x_0, y_0) and if M is any upper bound for the values of $|f_{xx}|$, $|f_{yy}|$, and $|f_{xy}|$ on R, then the error $E(x, y)$ incurred in replacing $f(x, y)$ on R by its linearization

$$L(x, y) = f(x_0, y_0) + f_x(x_0, y_0)(x - x_0) + f_y(x_0, y_0)(y - y_0) \tag{12}$$

satisfies the inequality

$$|E(x, y)| \leq \frac{1}{2} M \big(|x - x_0| + |y - y_0| \big)^2. \tag{13}$$

Example 2   In Example 1, we found the linearization of

$$f(x, y) = x^2 - xy + \frac{1}{2} y^2 + 3$$

at $(3, 2)$ to be

$$L(x, y) = 4x - y - 2.$$

Find an upper bound for the error in the approximation $f(x, y) \approx L(x, y)$ over the rectangle

$$R: \quad |x - 3| \leq 0.1, \quad |y - 2| \leq 0.1.$$

Express the upper bound as a percentage of $f(3, 2)$, the value of f at the center of the rectangle.

Solution   We use the inequality

$$|E(x, y)| \leq \frac{1}{2} M \big(|x - x_0| + |y - y_0| \big)^2. \quad \text{(Eq. (13))}$$

To find a suitable value for M, we calculate $f_{xx}, f_{xy},$ and f_{yy}, finding, after a routine differentiation, that all three derivatives are constant, with values

$$|f_{xx}| = |2| = 2, \qquad |f_{xy}| = |-1| = 1, \qquad |f_{yy}| = |1| = 1.$$

The largest of these is 2, so we may safely take M to be 2. With $(x_0, y_0) = (3, 2)$, we then know that, throughout R,

$$|E(x, y)| \leq \frac{1}{2} (2) \big(|x - 3| + |y - 2| \big)^2 = \big(|x - 3| + |y - 2| \big)^2.$$

Finally, since $|x - 3| \leq 0.1$ and $|y - 2| \leq 0.1$ on R, we have

$$|E(x, y)| \leq (0.1 + 0.1)^2 = 0.04.$$

As a percentage of $f(3, 2) = 8$, the error is no greater than

$$\frac{0.04}{8} \times 100 = 0.5\%.$$

As long as (x, y) stays in R, the approximation $f(x, y) \approx L(x, y)$ will be in error by no more than 0.04, which is 1/2% of the value of f at the center of R.

Predicting Change with Differentials

Suppose we know the values of a differentiable function $f(x, y)$ and its first partial derivatives at a point (x_0, y_0) and we want to predict how much this value will change if we move to a point $(x_0 + \Delta x, y_0 + \Delta y)$ nearby. If Δx and Δy are small, f and its linearization will change by nearly the same amount, so calculating the change in L gives a practical way to estimate the change in f.

The change in f is

$$\Delta f = f(x_0 + \Delta x, y_0 + \Delta y) - f(x_0, y_0).$$

A straightforward calculation with Eq. (7), using the notation $x - x_0 = \Delta x$ and $y - y_0 = \Delta y$, shows that the corresponding change in L is

$$\Delta L = L(x_0 + \Delta x, y_0 + \Delta y) - L(x_0, y_0)$$
$$= f_x(x_0, y_0)\Delta x + f_y(x_0, y_0)\Delta y. \qquad (14)$$

The formula for Δf is usually as hard to work with as the formula for f. The change in L, however, is just a known constant times Δx plus a known constant times Δy.

The change ΔL is usually described in the more suggestive notation

$$df = f_x(x_0, y_0)\, dx + f_y(x_0, y_0)\, dy, \qquad (15)$$

in which df denotes the change in the linearization that results from the changes dx and dy in x and y. As usual, we call dx and dy differentials of x and y, and we call df the corresponding differential of f.

DEFINITION

If we move (x_0, y_0) to a nearby point $(x_0 + dx, y_0 + dy)$, the resulting differential in f is

$$df = f_x(x_0, y_0)\, dx + f_y(x_0, y_0)\, dy. \qquad (16)$$

This change in the linearization of f is called the **total differential of f.**

Example 3   *Sensitivity to Change.*   Your company manufactures right circular cylindrical molasses storage tanks that are 25 ft high with a radius of 5 ft (Fig. 12.24). How sensitive are the tanks' volumes to small variations in height and radius?

Solution   As a function of radius r and height h, the typical tank's volume is

$$V = \pi r^2 h.$$

The change in volume caused by small changes dr and dh in radius and height is approximately

$$dV = V_r(5, 25)\, dr + V_h(5, 25)\, dh \qquad \left(\begin{matrix}\text{Eq. (16) with } f = V \text{ and}\\ (x_0, y_0) = (5, 25)\end{matrix}\right)$$
$$= (2\pi rh)_{(5, 25)}\, dr + (\pi r^2)_{(5, 25)}\, dh$$
$$= 250\pi\, dr + 25\pi\, dh.$$

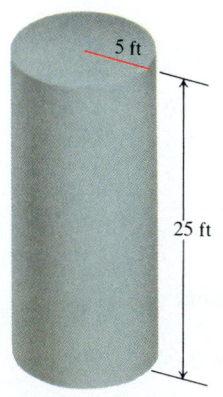

5 ft

25 ft

12.24   A small change in this tank's radius will change the volume 10 times as much as a change of the same size in the tank's height (Example 3).

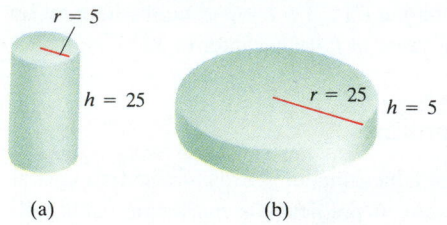

12.25   The volume of cylinder (a) is more sensitive to a small change in r than it is to an equally small change in h. The volume of cylinder (b) is more sensitive to small changes in h than it is to small changes in r.

Thus, a 1-unit change in r will change V by about 250π units. A 1-unit change in h will change V by about 25π units. The tank's volume is 10 times more sensitive to a small change in r than it is to a small change of equal size in h. As a quality control engineer concerned with being sure the tanks have the correct volume, you would want to pay special attention to their radii.

In contrast, if the values of r and h are reversed to make $r = 25$ and $h = 5$, then the total differential in V becomes

$$dV = (2\pi rh)_{(25, 5)} \, dr + (\pi r^2)_{(25, 5)} \, dh = 250\pi \, dr + 625\pi \, dh.$$

Now the volume is more sensitive to changes in h than to changes in r (Fig. 12.25).

The general rule to be learned from this example is that functions are most sensitive to small changes in the variables that generate the largest partial derivatives.

Absolute, Relative, and Percentage Change

When we move from (x_0, y_0) to a nearby point, we can describe the corresponding change in the value of a function $f(x, y)$ in three different ways.

	True	**Estimate**
Absolute change:	Δf	df
Relative change:	$\dfrac{\Delta f}{f(x_0, y_0)}$	$\dfrac{df}{f(x_0, y_0)}$
Percentage change:	$\dfrac{\Delta f}{f(x_0, y_0)} \times 100$	$\dfrac{df}{f(x_0, y_0)} \times 100$

(17)

Example 4   Suppose that the variables r and h change from the initial values of $(r_0, h_0) = (1, 5)$ by the amounts $dr = 0.03$ and $dh = -0.1$. Estimate the resulting absolute, relative, and percentage changes in the values of the function $V = \pi r^2 h$.

Solution   To estimate the absolute change in V, we evaluate

$$dV = V_r(r_0, h_0) \, dr + V_h(r_0, h_0) \, dh \tag{18}$$

to get

$$dV = 2\pi r_0 h_0 \, dr + \pi r_0^2 \, dh = 2\pi(1)(5)(0.03) + \pi(1)^2(-0.1)$$

$$= 0.3\pi - 0.1\pi = 0.2\pi.$$

We divide this by $V(r_0, h_0)$ to estimate the relative change:

$$\frac{dV}{V(r_0, h_0)} = \frac{0.2\pi}{\pi r_0^2 h_0} = \frac{0.2\pi}{\pi(1)^2(5)} = 0.04.$$

We multiply this by 100 to estimate the percentage change:

$$\frac{dV}{V(r_0, h_0)} \times 100 = 0.04 \times 100 = 4\%.$$

Example 5   The volume $V = \pi r^2 h$ of a right circular cylinder is to be calculated from measured values of r and h. Suppose that r is measured with an error of no more than 2% and h with an error of no more than 0.5%. Estimate the resulting possible percentage error in the calculation of V.

Solution   We are told that

$$\left| \frac{dr}{r} \times 100 \right| \le 2 \quad \text{and} \quad \left| \frac{dh}{h} \times 100 \right| \le 0.5.$$

Since

$$\frac{dV}{V} = \frac{2\pi rh \, dr + \pi r^2 \, dh}{\pi r^2 h} = \frac{2 \, dr}{r} + \frac{dh}{h},$$

we have

$$\left| \frac{dV}{V} \times 100 \right| = \left| 2 \frac{dr}{r} \times 100 + \frac{dh}{h} \times 100 \right|$$

$$\le 2 \left| \frac{dr}{r} \times 100 \right| + \left| \frac{dh}{h} \times 100 \right|$$

$$\le 2(2) + 0.5 = 4.5.$$

We estimate the error in the volume calculation to be at most 4.5%.

How accurately do we have to measure r and h to have a reasonable chance of calculating $V = \pi r^2 h$ with an error, say, of less than 2%? Questions like this are hard to answer for functions of more than one variable because there is usually no single right answer. Since

$$\frac{dV}{V} = 2 \frac{dr}{r} + \frac{dh}{h},$$

we see that dV/V is controlled by a combination of dr/r and dh/h. If we can measure h with great accuracy, we might come out all right even if we are sloppy about measuring r. On the other hand, our measurement of h might have so large a dh that the resulting dV/V would be too crude an estimate of $\Delta V/V$ to be useful even if dr were zero.

What we do in such cases is look for a reasonable square about the measured values (r_0, h_0) in which V will not vary by more than the allowed amount from $V_0 = \pi r_0^2 h_0$. The next example shows how this is done.

Example 6   Find a reasonable square about the point $(r_0, h_0) = (5, 12)$ in which the value of $V = \pi r^2 h$ will not vary by more than ± 0.1.

Solution   We approximate the variation ΔV by the differential

$$dV = 2\pi r_0 h_0 \, dr + \pi r_0^2 \, dh = 2\pi(5)(12) \, dr + \pi(5)^2 \, dh = 120\pi \, dr + 25\pi \, dh.$$

Since the region to which we are restricting our attention is a square (Fig. 12.26), we may set $dh = dr$ to get

$$dV = 120\pi \, dr + 25\pi \, dr = 145\pi \, dr.$$

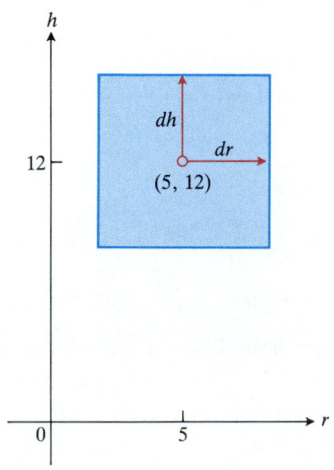

12.26   A small square about the point $(5, 12)$ in the rh-plane (Example 6).

We then ask, How small must we take dr to be sure that $|dV|$ is no larger than 0.1? To answer, we start with the inequality

$$|dV| \leq 0.1,$$

express dV in terms of dr,

$$|145\pi \, dr| \leq 0.1,$$

and find a corresponding upper bound for dr:

$$|dr| \leq \frac{0.1}{145\pi} \approx 2.1 \times 10^{-4}. \quad \left(\begin{array}{l}\text{Rounding down to make sure} \\ dr \text{ won't accidentally be too big}\end{array}\right)$$

With $dh = dr$, then, the square we want is described by the inequalities

$$|r - 5| \leq 2.1 \times 10^{-4}, \qquad |h - 12| \leq 2.1 \times 10^{-4}.$$

As long as (r, h) stays in this square, we may expect $|dV|$ to be less than or equal to 0.1 and we may expect $|\Delta V|$ to be approximately the same size.   ∎

Functions of More Than Two Variables

Analogous results hold for differentiable functions of more than two variables.

1. The **linearization** of a differentiable function $f(x, y, z)$ at a point $P_0(x_0, y_0, z_0)$ is

$$L(x, y, z) = f(P_0) + f_x(P_0)(x - x_0) + f_y(P_0)(y - y_0) + f_z(P_0)(z - z_0). \quad (19)$$

2. Suppose that R is a closed rectangular solid centered at P_0 and lying in an open region on which the second partial derivatives of f are continuous. Suppose also that $|f_{xx}|$, $|f_{yy}|$, $|f_{zz}|$, $|f_{xy}|$, $|f_{xz}|$, and $|f_{yz}|$ are all less than or equal to M throughout R. Then the **error** $E(x, y, z) = f(x, y, z) - L(x, y, z)$ in the approximation of f by L is bounded throughout R by the inequality

$$|E| \leq \frac{1}{2}M\big(|x - x_0| + |y - y_0| + |z - z_0|\big)^2. \quad (20)$$

3. If the second partial derivatives of f are continuous and if x, y, and z change from x_0, y_0, and z_0 by small amounts dx, dy, and dz, the **total differential**

$$df = f_x(P_0) \, dx + f_y(P_0) \, dy + f_z(P_0) \, dz \quad (21)$$

gives a good approximation of the resulting change in f.

Example 7   Find the linearization $L(x, y, z)$ of

$$f(x, y, z) = x^2 - xy + 3 \sin z$$

at the point $(x_0, y_0, z_0) = (2, 1, 0)$. Find an upper bound for the error incurred in replacing f by L on the rectangle

$$R: \quad |x - 2| \leq 0.01, \quad |y - 1| \leq 0.02, \quad |z| \leq 0.01.$$

Solution   A routine evaluation gives

$$f(2, 1, 0) = 2, \qquad f_x(2, 1, 0) = 3, \qquad f_y(2, 1, 0) = -2, \qquad f_z(2, 1, 0) = 3.$$

With these values, Eq. (19) becomes

$$L(x, y, z) = 2 + 3(x - 2) + (-2)(y - 1) + 3(z - 0) = 3x - 2y + 3z - 2.$$

Equation (20) gives an upper bound for the error incurred by replacing f by L on R. Since

$$f_{xx} = 2, \qquad f_{yy} = 0, \qquad f_{zz} = -3 \sin z,$$

$$f_{xy} = -1, \qquad f_{xz} = 0, \qquad f_{yz} = 0,$$

we may safely take M to be $\max|-3 \sin z| = 3$. Hence

$$|E| \leq \frac{1}{2}(3)(0.01 + 0.02 + 0.01)^2 = 0.0024.$$

The error will be no greater than 0.0024.

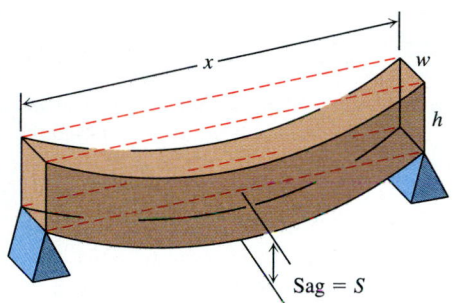

Example 8   *Controlling Sag in Uniformly Loaded Beams.*   A rectangular beam, supported at both ends, will sag when subjected to a uniform load (constant weight per linear foot). The amount S of sag (Fig. 12.27) is calculated with the formula

$$S = C \frac{px^4}{wh^3}.$$

In this equation,

$p =$ the load (newtons per meter of beam length),

$x =$ the length between supports (m),

$w =$ the width of the beam (m),

$h =$ the height of the beam (m),

$C =$ a constant that depends on the units of measurement and on the material from which the beam is made.

Find dS for a beam 4 m long, 10 cm wide, and 20 cm high that is subjected to a load of 100 N/m (Fig. 12.28). What conclusions can be drawn about the beam from the expression for dS?

12.27   A beam supported at its two ends before and after loading. Example 8 shows how the sag S is related to the weight of the load and the dimensions of the beam.

Solution   Since S is a function of the four independent variables p, x, w, and h, its differential dS is given by the equation

$$dS = S_p \, dp + S_x \, dx + S_w \, dw + S_h \, dh.$$

When we write this out for a particular set of values p_0, x_0, w_0, and h_0 and simplify the result, we find that

$$dS = S_0 \left(\frac{dp}{p_0} + \frac{4dx}{x_0} - \frac{dw}{w_0} - \frac{3dh}{h_0} \right), \tag{22}$$

where $S_0 = S(p_0, x_0, w_0, h_0) = Cp_0 x_0^4 / w_0 h_0^3$.

If $p_0 = 100$ N/m, $x_0 = 4$ m, $w_0 = 0.1$ m, and $h_0 = 0.2$ m, then

$$dS = S_0 \left(\frac{dp}{100} + dx - 10dw - 15dh \right). \tag{23}$$

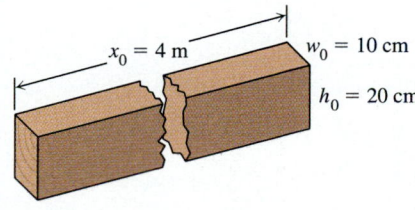

12.28   The dimensions of the beam in Example 8.

Here is what we can learn from Eq. (23). Since dp and dx appear with positive coefficients, increases in p and x will increase the sag. But dw and dh appear with negative coefficients, so increases in w and h will decrease the sag (make the beam stiffer). The sag is not very sensitive to changes in load because the coefficient of dp is $1/100$. The magnitude of the coefficient of dh is greater than the magnitude of the coefficient of dw. Making the beam 1 cm higher will therefore decrease the sag more than making the beam 1 cm wider.

EXERCISES 12.4

In Exercises 1–6, find the linearization $L(x, y)$ of the given function at each point.

1. $f(x, y) = x^2 + y^2 + 1$ at   (a) $(0, 0)$,   (b) $(1, 1)$

2. $f(x, y) = x^3 y^4$ at   (a) $(1, 1)$,   (b) $(0, 0)$

3. $f(x, y) = e^x \cos y$ at   (a) $(0, 0)$,   (b) $(0, \pi/2)$

4. $f(x, y) = (x + y + 2)^2$ at   (a) $(0, 0)$,   (b) $(1, 2)$

5. $f(x, y) = 3x - 4y + 5$ at   (a) $(0, 0)$,   (b) $(1, 1)$

6. $f(x, y) = e^{2y - x}$ at   (a) $(0, 0)$,   (b) $(1, 2)$

In Exercises 7–12, find the linearization $L(x, y)$ of the function $f(x, y)$ at P_0. Then use Eq. (13) to find an upper bound for the magnitude $|E|$ of the error in the approximation $f(x, y) \approx L(x, y)$ over the rectangle R.

7. $f(x, y) = x^2 - 3xy + 5$ at $P_0(2, 1)$,
R: $|x - 2| \le 0.1$, $|y - 1| \le 0.1$

8. $f(x, y) = (1/2)x^2 + xy + (1/4)y^2 + 3x - 3y + 4$ at P_0 $(2, 2)$,
R: $|x - 2| \le 0.1$, $|y - 2| \le 0.1$

9. $f(x, y) = 1 + y + x \cos y$ at $P_0(0, 0)$,
R: $|x| \le 0.2$, $|y| \le 0.2$
(Use $|\cos y| \le 1$ and $|\sin y| \le 1$ in estimating E.)

10. $f(x, y) = \ln x + \ln y$ at $P_0(1, 1)$,
R: $|x - 1| \le 0.2$, $|y - 1| \le 0.2$

11. $f(x, y) = e^x \cos y$ at $P_0(0, 0)$,
R: $|x| \le 0.1$, $|y| \le 0.1$
(Use $e^x \le 1.11$ and $|\cos y| \le 1$ in estimating E.)

12. $f(x, y) = xy^2 + y \cos(x - 1)$ at $P_0(1, 2)$,
R: $|x - 1| \le 0.1$, $|y - 2| \le 0.1$

13. You plan to calculate the area of a long, thin rectangle from measurements of its length and width. Which measurement should you be more careful with? Why?

14. a) Around the point $(1, 0)$, is $f(x, y) = x^2(y + 1)$ more sensitive to changes in x, or to changes in y?
b) What ratio of dx to dy will make df equal zero at $(1, 0)$?

15. Suppose T is to be calculated from the formula $T = x(e^y + e^{-y})$ and that x and y are known to be 2 and $\ln 2$ with maximum possible errors of $|dx| = 0.1$ and $|dy| = 0.02$. Estimate the maximum possible error in the computed value of T.

16. About how accurately may $V = \pi r^2 h$ be calculated from measurements of r and h that are in error by 1%?

17. If $r = 5.0$ cm and $h = 12.0$ cm to the nearest millimeter, what should we expect the maximum percentage error in calculating $V = \pi r^2 h$ to be?

18. To estimate the volume of a cylinder of radius about 2 m and height about 3 m, approximately how accurately should the radius and height be measured so that the error in the volume estimate will not exceed 0.1 m³? Assume that the possible error dr in measuring r is equal to the possible error dh in measuring h.

19. Give a reasonable square centered at $(1, 1)$ over which the value of $f(x, y) = x^3 y^4$ will not vary by more than ± 0.1.

20. *Variation in electrical resistance.*   The resistance R produced by wiring resistors of R_1 and R_2 ohms in parallel (Fig. 12.29) can be calculated from the formula

$$\frac{1}{R} = \frac{1}{R_1} + \frac{1}{R_2}.$$

a) Show that

$$dR = \left(\frac{R}{R_1}\right)^2 dR_1 + \left(\frac{R}{R_2}\right)^2 dR_2.$$

b) You have designed a two-resistor circuit like the one in Fig. 12.29 to have resistances of $R_1 = 100$ ohms and $R_2 = 400$ ohms, but there is always some variation in manufacturing and the resistors received by your firm will probably not have these exact values. Will the value of R be more sensitive to variation in R_1, or to variation in R_2?

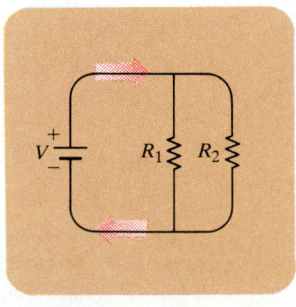

12.29   The circuit in Exercises 20 and 21.

21. *Continuation of Exercise 20.*   In another circuit like the one in Fig. 12.29, you plan to change R_1 from 20 to 20.1 ohms and R_2 from 25 to 24.9 ohms. By about what percentage will this change R?

22. *Error carry-over in coordinate changes*
a) If $x = 3 \pm 0.01$ and $y = 4 \pm 0.01$ (Fig. 12.30), with approximately what accuracy can you calculate the polar coordinates r and θ of the point $P(x, y)$ from the formulas $r^2 = x^2 + y^2$ and $\theta = \tan^{-1}(y/x)$? Express your estimates as percentage changes of the values that r and θ have at the point $(x_0, y_0) = (3, 4)$.
b) At the point $(x_0, y_0) = (3, 4)$, are the values of r and θ more sensitive to changes in x, or to changes in y?

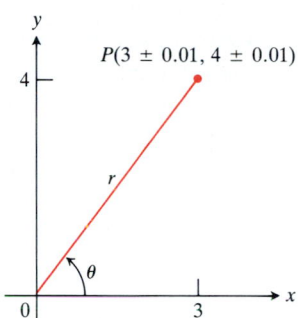

$P(3 \pm 0.01, 4 \pm 0.01)$

12.30   How much do the errors in the Cartesian coordinates of P affect the polar coordinates of P? See Exercise 22.

23. Does a function $f(x, y)$ with continuous first partial derivatives throughout an open region R have to be continuous on R? Explain.

24. If a function $f(x, y)$ has continuous second partial derivatives throughout an open region R, must the first order partial derivatives of f be continuous on R? Explain.

Functions of More Than Two Variables

Find the linearizations $L(x, y, z)$ of the functions in Exercises 25–30 at the given points.

25. $f(x, y, z) = xy + yz + xz$ at
   a) $(1, 1, 1)$ b) $(1, 0, 0)$ c) $(0, 0, 0)$

26. $f(x, y, z) = x^2 + y^2 + z^2$ at
   a) $(1, 1, 1)$ b) $(0, 1, 0)$ c) $(1, 0, 0)$

27. $f(x, y, z) = \sqrt{x^2 + y^2 + z^2}$ at
   a) $(1, 0, 0)$ b) $(1, 1, 0)$ c) $(1, 2, 2)$

28. $f(x, y, z) = (\sin xy)/z$ at
   a) $(\pi/2, 1, 1)$ b) $(2, 0, 1)$

29. $f(x, y, z) = e^x + \cos(y + z)$ at
   a) $(0, 0, 0)$ b) $\left(0, \frac{\pi}{2}, 0\right)$ c) $\left(0, \frac{\pi}{4}, \frac{\pi}{4}\right)$

30. $f(x, y, z) = \tan^{-1}(xyz)$ at
   a) $(1, 0, 0)$ b) $(1, 1, 0)$ c) $(1, 1, 1)$

In Exercises 31–34, find the linearization $L(x, y, z)$ of the function $f(x, y, z)$ at P_0. Then use Eq. (20) to find an upper bound for the magnitude of the error E in the approximation $f(x, y, z) \approx L(x, y, z)$ over the region R.

31. $f(x, y, z) = xz - 3yz + 2$ at $P_0(1, 1, 2)$
   R: $|x - 1| \le 0.01$, $|y - 1| \le 0.01$, $|z - 2| \le 0.02$

32. $f(x, y, z) = x^2 + xy + yz + (1/4)z^2$ at $P_0(1, 1, 2)$
   R: $|x - 1| \le 0.01$, $|y - 1| \le 0.01$, $|z - 2| \le 0.08$

33. $f(x, y, z) = xy + 2yz - 3xz$ at $P_0(1, 1, 0)$
   R: $|x - 1| \le 0.01$, $|y - 1| \le 0.01$, $|z| \le 0.01$

34. $f(x, y, z) = \sqrt{2} \cos x \sin(y + z)$ at $P_0(0, 0, \pi/4)$
   R: $|x| \le 0.01$, $|y| \le 0.01$, $|z - \pi/4| \le 0.01$

35. The beam of Example 8 is tipped on its side so that $h = 0.1$ m and $w = 0.2$ m.
   a) What is the value of dS now?
   b) Compare the sensitivity of the newly positioned beam to a small change in height with its sensitivity to an equally small change in width.

36. If $|a|$ is much greater than $|b|$, $|c|$, and $|d|$, to which of a, b, c, and d is the value of the determinant
$$f(a, b, c, d) = \begin{vmatrix} a & b \\ c & d \end{vmatrix}$$
most sensitive?

37. Estimate how strongly simultaneous errors of 2% in a, b, and c might affect the calculation of the product
$$p(a, b, c) = abc.$$

38. Estimate how much wood it takes to make a hollow rectangular box whose inside measurements are 5 ft long by 3 ft wide by 2 ft deep if the box is made of lumber 1/2-in. thick and the box has no top.

39. The area of a triangle is $(1/2)ab \sin C$, where a and b are the lengths of two sides of the triangle and C is the measure of the included angle. In surveying a triangular plot, you have measured a, b, and C to be 150 ft, 200 ft, and 60°, respectively. By about how much might your area calculation be in error if your values of a and b are off by half a foot each and your measurement of C is off by 2°? See Fig. 12.31. Remember to use radians.

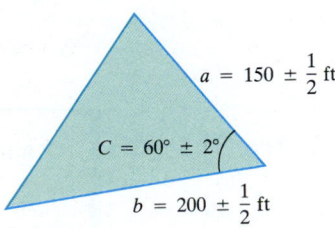

$a = 150 \pm \frac{1}{2}$ ft

$C = 60° \pm 2°$

$b = 200 \pm \frac{1}{2}$ ft

12.31   How accurately can you calculate the triangle's area from the measurement shown? See Exercise 39.

40. Suppose that $u = xe^y + y \sin z$ and that x, y, and z can be measured with maximum possible errors of ± 0.2, ± 0.6, and $\pm \pi/180$, respectively. Estimate the maximum possible error in calculating u from the measured values $x = 2$, $y = \ln 3$, $z = \pi/2$.

41. *The Wilson lot size formula.*   The Wilson lot size formula in economics says that the most economical quantity Q of goods (radios, shoes, brooms, whatever) for a store to order is given by the formula $Q = \sqrt{2KM/h}$, where K is the cost of placing the order, M is the number of items sold per week, and h is the weekly holding cost for each item (cost of space, utilities, security, and so on). To which of the variables K, M, and h is Q most sensitive near the point $(K_0, M_0, h_0) = (2, 20, 0.05)$?

12.5 The Chain Rule

When we are interested in the temperature $w = f(x, y, z)$ at points along a curve $x = g(t)$, $y = h(t)$, $z = k(t)$ in space, or in the pressure or density along a path through a gas or fluid, we may think of f as a function of the single variable t. For each value of t, the temperature at the point $(g(t), h(t), k(t))$ is the value of the composite $f(g(t), h(t), k(t))$. If we then wish to know the rate at which f changes with respect to t along the path, we have only to differentiate this composite with respect to t, provided, of course, the derivative exists.

Sometimes we can find the derivative by substituting the formulas for g, h, and k into the formula for f and differentiating directly with respect to t. But we often have to work with functions whose formulas are too complicated for convenient substitution or for which formulas are not readily available. To find a function's derivatives under circumstances like these, we use the Chain Rule. The form the Chain Rule takes depends on how many variables are involved but, except for the presence of additional variables, it works just like the Chain Rule from Chapter 2. The basic idea remains the same: Composites of differentiable functions are differentiable.

The Chain Rule for Functions of Two Variables

In Chapter 2, we used the Chain Rule when $w = f(x)$ was a differentiable function of x and $x = g(t)$ was a differentiable function of t. This made w a differentiable function of t and the Chain Rule said that dw/dt could be calculated with the formula

$$\frac{dw}{dt} = \frac{dw}{dx}\frac{dx}{dt}.$$

The analogous formula for a function $w = f(x, y)$ of two independent variables is given in Theorem 5.

THEOREM 5

Chain Rule for Functions of Two Independent Variables

If $w = f(x, y)$ is differentiable and x and y are differentiable functions of t, then w is a differentiable function of t and

$$\frac{dw}{dt} = \frac{\partial f}{\partial x}\frac{dx}{dt} + \frac{\partial f}{\partial y}\frac{dy}{dt}. \tag{1}$$

The way to remember the Chain Rule is to picture the diagram below. To find dw/dt, start at w and read down each route to t, multiplying derivatives along the way. Then add the products.

Chain Rule

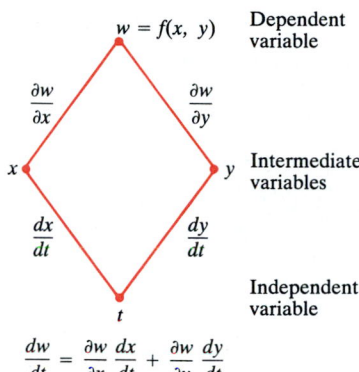

$$\frac{dw}{dt} = \frac{\partial w}{\partial x}\frac{dx}{dt} + \frac{\partial w}{\partial y}\frac{dy}{dt}$$

Proof   The proof consists of showing that if x and y are differentiable at $t = t_0$, then w is differentiable at t_0 and

$$\left(\frac{dw}{dt}\right)_{t_0} = \left(\frac{\partial w}{\partial x}\right)_{t_0}\left(\frac{dx}{dt}\right)_{t_0} + \left(\frac{\partial w}{\partial y}\right)_{t_0}\left(\frac{dy}{dt}\right)_{t_0}. \tag{2}$$

Let Δx, Δy, and Δw be the increments that result from changing t from t_0 to $t_0 + \Delta t$. Since f is differentiable,

$$\Delta w = \left(\frac{\partial w}{\partial x}\right)_{t_0}\Delta x + \left(\frac{\partial w}{\partial y}\right)_{t_0}\Delta y + \epsilon_1\Delta x + \epsilon_2\Delta y, \tag{3}$$

where ϵ_1, $\epsilon_2 \to 0$ as Δx, $\Delta y \to 0$. We divide Eq. (3) through by Δt and let Δt approach zero. The division gives

$$\frac{\Delta w}{\Delta t} = \left(\frac{\partial w}{\partial x}\right)_{t_0}\frac{\Delta x}{\Delta t} + \left(\frac{\partial w}{\partial y}\right)_{t_0}\frac{\Delta y}{\Delta t} + \epsilon_1\frac{\Delta x}{\Delta t} + \epsilon_2\frac{\Delta y}{\Delta t}, \tag{4}$$

and letting Δt approach zero gives

$$\left(\frac{dw}{dt}\right)_{t_0} = \lim_{\Delta t \to 0}\frac{\Delta w}{\Delta t}$$

$$= \left(\frac{\partial w}{\partial x}\right)_{t_0}\left(\frac{dx}{dt}\right)_{t_0} + \left(\frac{\partial w}{\partial y}\right)_{t_0}\left(\frac{dy}{dt}\right)_{t_0} + 0\cdot\left(\frac{dx}{dt}\right)_{t_0} + 0\cdot\left(\frac{dy}{dt}\right)_{t_0}. \tag{5}$$

This completes the proof.

Example 1   Use the Chain Rule to find the derivative of

$$w = xy$$

with respect to t along the path $x = \cos t$, $y = \sin t$. What is the derivative's value at $t = \pi/2$?

Solution   We evaluate the right-hand side of Eq. (1) with $w = xy$, $x = \cos t$, and $y = \sin t$:

$$\frac{\partial w}{\partial x} = y = \sin t, \quad \frac{\partial w}{\partial y} = x = \cos t, \quad \frac{dx}{dt} = -\sin t, \quad \frac{dy}{dt} = \cos t,$$

$$\frac{dw}{dt} = \frac{\partial w}{\partial x}\frac{dx}{dt} + \frac{\partial w}{\partial y}\frac{dy}{dt} = (\sin t)(-\sin t) + (\cos t)(\cos t) \quad \begin{pmatrix}\text{Eq. (1) with}\\ \text{values from}\\ \text{above}\end{pmatrix}$$

$$= -\sin^2 t + \cos^2 t = \cos 2t.$$

In this case we can check the result with a more direct calculation. As a function of t,

$$w = xy = \cos t \sin t = \frac{1}{2}\sin 2t,$$

so

$$\frac{dw}{dt} = \frac{d}{dt}\left(\frac{1}{2}\sin 2t\right) = \frac{1}{2}\cdot 2\cos 2t = \cos 2t.$$

Here we have three routes from w to t instead of two. But the rule for evaluating dw/dt is still the same. Read down each route, multiplying derivatives along the way; then add.

Chain Rule

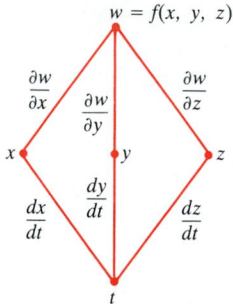

$w = f(x, y, z)$

$$\frac{dw}{dt} = \frac{\partial w}{\partial x}\frac{dx}{dt} + \frac{\partial w}{\partial y}\frac{dy}{dt} + \frac{\partial w}{\partial z}\frac{dz}{dt}$$

In either case,

$$\left(\frac{dw}{dt}\right)_{t = \pi/2} = \cos\left(2 \cdot \frac{\pi}{2}\right) = \cos \pi = -1.$$

The Chain Rule for Functions of Three Variables

To get the Chain Rule for functions of three variables, we add a term to Eq. (1).

Chain Rule for Functions of Three Independent Variables

If $w = f(x, y, z)$ is differentiable and x, y, and z are differentiable functions of t, then w is a differentiable function of t and

$$\frac{dw}{dt} = \frac{\partial f}{\partial x}\frac{dx}{dt} + \frac{\partial f}{\partial y}\frac{dy}{dt} + \frac{\partial f}{\partial z}\frac{dz}{dt}. \tag{6}$$

The derivation is identical with the derivation of Eq. (1) except that there are now three intermediate variables instead of two. The diagram we use for remembering the new equation is similar as well.

Example 2   *Changes in a Function's Values Along a Helix.*   Find dw/dt if

$$w = xy + z, \qquad x = \cos t, \qquad y = \sin t, \qquad z = t$$

(Fig. 12.32). What is the derivative's value at $t = 0$?

Solution

$$\frac{dw}{dt} = \frac{\partial w}{\partial x}\frac{dx}{dt} + \frac{\partial w}{\partial y}\frac{dy}{dt} + \frac{\partial w}{\partial z}\frac{dz}{dt} \qquad \text{(Eq. (6))}$$

$$= (y)(-\sin t) + (x)(\cos t) + (1)(1)$$

$$= (\sin t)(-\sin t) + (\cos t)(\cos t) + 1$$

$$= -\sin^2 t + \cos^2 t + 1 = 1 + \cos 2t,$$

$$\left(\frac{dw}{dt}\right)_{t = 0} = 1 + \cos(0) = 2.$$

The Chain Rule for Functions Defined on Surfaces

If we are interested in the temperature $w = f(x, y, z)$ at points (x, y, z) on a globe in space, we might prefer to think of x, y, and z as functions of the variables r and s that give the points' longitudes and latitudes. If $x = g(r, s)$, $y = h(r, s)$, and $z = k(r, s)$, we could then express the temperature as a function of r and s with the composite function

$$w = f(g(r, s), h(r, s), k(r, s)).$$

Under the right conditions, w would have partial derivatives with respect to both r and s that could be calculated in the following way.

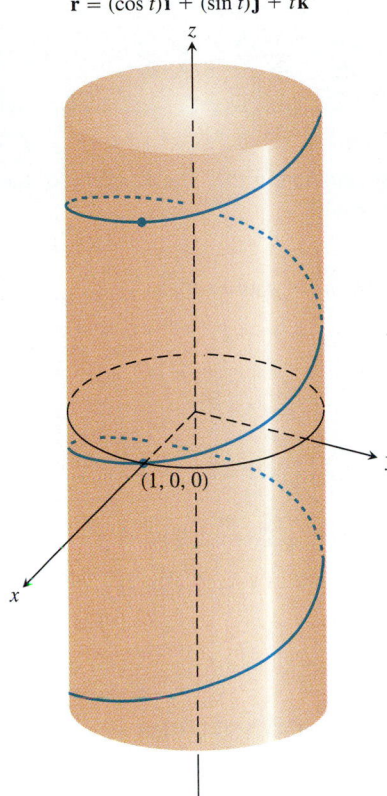

The helix
$\mathbf{r} = (\cos t)\mathbf{i} + (\sin t)\mathbf{j} + t\mathbf{k}$

$(1, 0, 0)$

12.32 Example 2 shows how the values of the function $w = xy + z$ vary with t along this helix.

Chain Rule for Two Independent Variables and Three Intermediate Variables

If $w = f(x, y, z)$, $x = g(r, s)$, $y = h(r, s)$, and $z = k(r, s)$ are differentiable, then w is a differentiable function of r and s and its partial derivatives are given by the equations

$$\frac{\partial w}{\partial r} = \frac{\partial w}{\partial x}\frac{\partial x}{\partial r} + \frac{\partial w}{\partial y}\frac{\partial y}{\partial r} + \frac{\partial w}{\partial z}\frac{\partial z}{\partial r}, \tag{7}$$

$$\frac{\partial w}{\partial s} = \frac{\partial w}{\partial x}\frac{\partial x}{\partial s} + \frac{\partial w}{\partial y}\frac{\partial y}{\partial s} + \frac{\partial w}{\partial z}\frac{\partial z}{\partial s}. \tag{8}$$

Equation (7) can be derived from Eq. (6) by holding s fixed and setting r equal to t. Similarly, Eq. (8) can be derived from Eq. (6) by holding r fixed and setting s equal to t. The tree diagrams are shown in Fig. 12.33.

Example 3   Express $\partial w/\partial r$ and $\partial w/\partial s$ in terms of r and s if

$$w = x + 2y + z^2, \qquad x = \frac{r}{s}, \qquad y = r^2 + \ln s, \qquad z = 2r.$$

Solution

$$\frac{\partial w}{\partial r} = \frac{\partial w}{\partial x}\frac{\partial x}{\partial r} + \frac{\partial w}{\partial y}\frac{\partial y}{\partial r} + \frac{\partial w}{\partial z}\frac{\partial z}{\partial r} \qquad \text{(Eq. (7))}$$

$$= (1)\left(\frac{1}{s}\right) + (2)(2r) + (2z)(2) = \frac{1}{s} + 4r + (4r)(2) = \frac{1}{s} + 12r,$$

$$\frac{\partial w}{\partial s} = \frac{\partial w}{\partial x}\frac{\partial x}{\partial s} + \frac{\partial w}{\partial y}\frac{\partial y}{\partial s} + \frac{\partial w}{\partial z}\frac{\partial z}{\partial s} \qquad \text{(Eq. (8))}$$

$$= (1)\left(-\frac{r}{s^2}\right) + 2\left(\frac{1}{s}\right) + (2z)(0) = \frac{2}{s} - \frac{r}{s^2}.$$

If f is a function of two variables instead of three, Eqs. (7) and (8) become one term shorter.

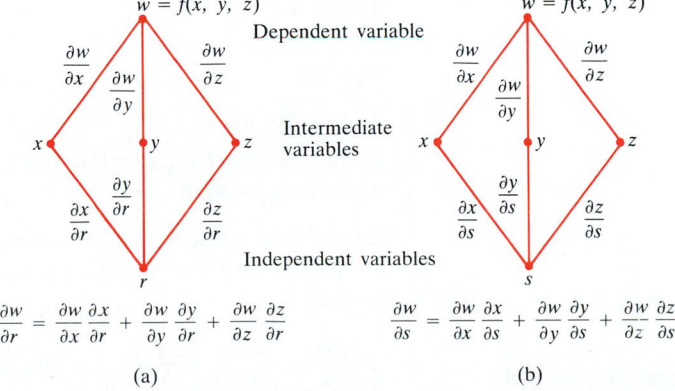

12.33 Tree diagrams for Eqs. (7) and (8).

Chain Rule

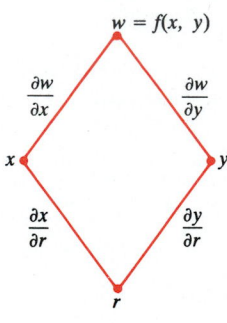

$$\frac{\partial w}{\partial r} = \frac{\partial w}{\partial x}\frac{\partial x}{\partial r} + \frac{\partial w}{\partial y}\frac{\partial y}{\partial r}$$

12.34 Tree diagram for Eqs. (9).

If $w = f(x, y)$, $x = g(r, s)$, and $y = h(r, s)$, then

$$\frac{\partial w}{\partial r} = \frac{\partial w}{\partial x}\frac{\partial x}{\partial r} + \frac{\partial w}{\partial y}\frac{\partial y}{\partial r} \qquad \text{and} \qquad \frac{\partial w}{\partial s} = \frac{\partial w}{\partial x}\frac{\partial x}{\partial s} + \frac{\partial w}{\partial y}\frac{\partial y}{\partial s}. \tag{9}$$

Figure 12.34 shows a tree diagram for the first equation in (9). The diagram for the second equation in (9) is similar—just replace r with s.

Example 4   Express $\partial w/\partial r$ and $\partial w/\partial s$ in terms of r and s if

$$w = x^2 + y^2, \qquad x = r - s, \qquad y = r + s.$$

Solution   We use Eqs. (9):

$$\frac{\partial w}{\partial r} = \frac{\partial w}{\partial x}\frac{\partial x}{\partial r} + \frac{\partial w}{\partial y}\frac{\partial y}{\partial r} \qquad\qquad \frac{\partial w}{\partial s} = \frac{\partial w}{\partial x}\frac{\partial x}{\partial s} + \frac{\partial w}{\partial y}\frac{\partial y}{\partial s}$$

$$= (2x)(1) + (2y)(1) \qquad\qquad = (2x)(-1) + (2y)(1)$$

$$= 2(r - s) + 2(r + s) \qquad\qquad = -2(r - s) + 2(r + s)$$

$$= 4r, \qquad\qquad\qquad\qquad = 4s.$$

If f is a function of x alone, Eqs. (7) and (8) simplify still further.

If $w = f(x)$ and $x = g(r, s)$, then

$$\frac{\partial w}{\partial r} = \frac{dw}{dx}\frac{\partial x}{\partial r} \qquad \text{and} \qquad \frac{\partial w}{\partial s} = \frac{dw}{dx}\frac{\partial x}{\partial s}. \tag{10}$$

Here dw/dx is the ordinary (single-variable) derivative (Fig. 12.35).

Chain Rule

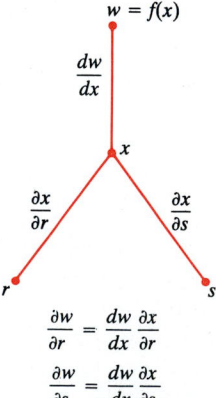

$$\frac{\partial w}{\partial r} = \frac{dw}{dx}\frac{\partial x}{\partial r}$$

$$\frac{\partial w}{\partial s} = \frac{dw}{dx}\frac{\partial x}{\partial s}$$

12.35 Tree diagram for Eqs. (10).

Implicit Differentiation (Continued from Chapter 2)

The two-variable Chain Rule in Eq. (1) leads to a formula that takes most of the work out of implicit differentiation. The formula arises in the following way. Suppose that

1. the function $F(x, y)$ is differentiable and
2. the equation $F(x, y) = 0$ defines y implicitly as a differentiable function of x, say $y = h(x)$.

Now, x is also a differentiable function of x, so what we have, in effect, is a function $w = F(u, v)$ in which u and v are both differentiable functions of x. By Eq. (1), therefore,

$$\frac{dw}{dx} = \frac{\partial w}{\partial u}\frac{du}{dx} + \frac{\partial w}{\partial v}\frac{dv}{dx}. \qquad \text{(Eq. (1) with } x = u, \, y = v, \, t = x)$$

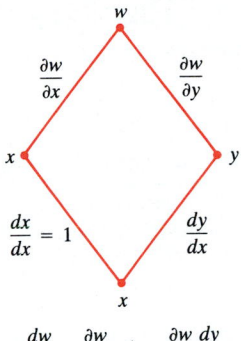

$$\frac{dw}{dx} = \frac{\partial w}{\partial x} \cdot 1 + \frac{\partial w}{\partial y}\frac{dy}{dx}$$

12.36   Tree diagram for Eq. (11).

With $u = x$, and $v = y = h(x)$,

$$\frac{dw}{dx} = \frac{\partial w}{\partial x}\frac{dx}{dx} + \frac{\partial w}{\partial y}\frac{dy}{dx} = \frac{\partial w}{\partial x} \cdot 1 + \frac{\partial w}{\partial y}\frac{dy}{dx} \tag{11}$$

(Fig. 12.36). Since $w = F(x, h(x)) = 0$ for all values of x (because h is defined by setting $F(x, y) = 0$), $dw/dx = 0$ and Eq. (11) reduces still further to

$$0 = \frac{\partial w}{\partial x} + \frac{\partial w}{\partial y}\frac{dy}{dx}. \tag{12}$$

If $\partial w/\partial y \neq 0$, we can solve Eq. (12) for dy/dx to get

$$\frac{dy}{dx} = -\frac{\partial w/\partial x}{\partial w/\partial y} = -\frac{F_x}{F_y}. \tag{13}$$

Suppose that $F(x, y)$ is differentiable and that the equation $F(x, y) = 0$ defines y as a differentiable function of x. Then, at any point where $F_y \neq 0$,

$$\frac{dy}{dx} = -\frac{F_x}{F_y}. \tag{14}$$

Example 5   Find dy/dx if $x^2 + \sin y - 2y = 0$.

Solution   Take $F(x, y) = x^2 + \sin y - 2y$. Then

$$\frac{dy}{dx} = -\frac{F_x}{F_y} = -\frac{2x}{\cos y - 2}. \qquad \text{(Eq. (14))}$$

This calculation is significantly shorter than the single-variable calculation with which we found dy/dx in Section 2.5, Example 3.   ∎

Remembering the Different Forms of the Chain Rule

How are we to remember all the different forms of the Chain Rule? The answer is that there is no need to remember them all. The best thing to do is to remember a few key equations, say (1), (6), and (7). Construct the others, when you need them, from tree diagrams. You can always draw the appropriate diagram by placing the dependent variable on top, the intermediate variables in the middle, and the selected independent variable at the bottom.

The Chain Rule for Functions of Many Variables

Suppose $w = f(x, y, \ldots, v)$ is a differentiable function of the variables $x, y, \ldots, v$ (a finite set) and $x, y, \ldots, v$ are themselves differentiable functions of the variables $p, q, \ldots, t$ (another finite set). Then w is a differentiable function of the variables p through t and the partial derivatives of w with respect to these variables are given by equations of the form

$$\frac{\partial w}{\partial p} = \frac{\partial w}{\partial x}\frac{\partial x}{\partial p} + \frac{\partial w}{\partial y}\frac{\partial y}{\partial p} + \cdots + \frac{\partial w}{\partial v}\frac{\partial v}{\partial p}. \tag{15}$$

The other equations are obtained by replacing p by $q, \ldots, t$, one at a time.

One way to remember Eq. (15) is to think of the right-hand side as the dot product of two vectors with components

$$\underbrace{\left(\frac{\partial w}{\partial x}, \frac{\partial w}{\partial y}, \ldots, \frac{\partial w}{\partial v}\right)}_{\substack{\text{derivatives of } w \text{ with} \\ \text{respect to the} \\ \text{intermediate variables}}} \quad \text{and} \quad \underbrace{\left(\frac{\partial x}{\partial p}, \frac{\partial y}{\partial p}, \ldots, \frac{\partial v}{\partial p}\right).}_{\substack{\text{derivatives of the intermediate} \\ \text{variables with respect to the} \\ \text{selected independent variable}}} \qquad (16)$$

EXERCISES 12.5

In Exercises 1–6, (a) express dw/dt as a function of t, both by using the Chain Rule and by expressing w in terms of t and differentiating directly with respect to t. Then (b) evaluate dw/dt at the given value of t.

1. $w = x^2 + y^2$, $x = \cos t$, $y = \sin t$; $t = \pi$

2. $w = x^2 + y^2$, $x = \cos t + \sin t$, $y = \cos t - \sin t$; $t = 0$

3. $w = \frac{x}{z} + \frac{y}{z}$, $x = \cos^2 t$, $y = \sin^2 t$, $z = 1/t$; $t = 3$

4. $w = \ln(x^2 + y^2 + z^2)$, $x = \cos t$, $y = \sin t$, $z = 4\sqrt{t}$; $t = 3$

5. $w = 2ye^x - \ln z$, $x = \ln(t^2 + 1)$, $y = \tan^{-1} t$, $z = e^t$; $t = 1$

6. $w = z - \sin xy$, $x = t$, $y = \ln t$, $z = e^{t-1}$; $t = 1$

In Exercises 7 and 8, (a) express $\partial z/\partial r$ and $\partial z/\partial \theta$ as functions of r and θ both by using the Chain Rule and by expressing z directly in terms of r and θ before differentiating. Then (b) evaluate $\partial z/\partial r$ and $\partial z/\partial \theta$ at the given point (r, θ).

7. $z = 4 e^x \ln y$, $x = \ln(r \cos \theta)$, $y = r \sin \theta$; $(r, \theta) = (2, \pi/4)$

8. $z = \tan^{-1}(x/y)$, $x = r \cos \theta$, $y = r \sin \theta$; $(r, \theta) = (1.3, \pi/6)$

In Exercises 9 and 10, (a) express $\partial w/\partial u$ and $\partial w/\partial v$ as functions of u and v both by using the Chain Rule and by expressing w directly in terms of u and v before differentiating. Then (b) evaluate $\partial w/\partial u$ and $\partial w/\partial v$ at the given point (u, v).

9. $w = xy + yz + xz$, $x = u + v$, $y = u - v$, $z = uv$; $(u, v) = (1/2, 1)$

10. $w = \ln(x^2 + y^2 + z^2)$, $x = ue^v \sin u$, $y = ue^v \cos u$, $z = ue^v$; $(u, v) = (-2, 0)$

In Exercises 11 and 12, (a) express $\partial u/\partial x$, $\partial u/\partial y$, and $\partial u/\partial z$ as functions of x, y, and z both by using the Chain Rule and by expressing u directly in terms of p, q, and r before differentiating. Then (b) evaluate $\partial u/\partial x$, $\partial u/\partial y$, and $\partial u/\partial z$ at the given point (x, y, z).

11. $u = \frac{p - q}{q - r}$, $p = x + y + z$, $q = x - y + z$, $r = x + y - z$; $(x, y, z) = (\sqrt{3}, 2, 1)$

12. $u = e^{qr} \sin^{-1} p$, $p = \sin x$, $q = z^2 \ln y$, $r = 1/z$; $(x, y, z) = (\pi/4, 1/2, -1/2)$

In Exercises 13–24, draw a tree diagram and write a Chain Rule formula for each derivative.

13. $\dfrac{dz}{dt}$ for $z = f(x, y)$, $x = g(t)$, $y = h(t)$

14. $\dfrac{dz}{dt}$ for $z = f(u, v, w)$, $u = g(t)$, $v = h(t)$, $w = k(t)$

15. $\dfrac{\partial w}{\partial u}$ and $\dfrac{\partial w}{\partial v}$ for $w = h(x, y, z)$, $x = f(u, v)$, $y = g(u, v)$, $z = k(u, v)$

16. $\dfrac{\partial w}{\partial x}$ and $\dfrac{\partial w}{\partial y}$ for $w = f(r, s, t)$, $r = g(x, y)$, $s = h(x, y)$, $t = k(x, y)$

17. $\dfrac{\partial w}{\partial u}$ and $\dfrac{\partial w}{\partial v}$ for $w = g(x, y)$, $x = h(u, v)$, $y = k(u, v)$

18. $\dfrac{\partial w}{\partial x}$ and $\dfrac{\partial w}{\partial y}$ for $w = g(u, v)$, $u = h(x, y)$, $v = k(x, y)$

19. $\dfrac{\partial z}{\partial t}$ and $\dfrac{\partial z}{\partial s}$ for $z = f(x, y)$, $x = g(t, s)$, $y = h(t, s)$

20. $\dfrac{\partial y}{\partial r}$ for $y = f(u)$, $u = g(r, s)$

21. $\dfrac{\partial w}{\partial s}$ and $\dfrac{\partial w}{\partial t}$ for $w = g(u)$, $u = h(s, t)$

22. $\dfrac{\partial w}{\partial p}$ for $w = f(x, y, z, v)$, $x = g(p, q)$, $y = h(p, q)$, $z = j(p, q)$, $v = k(p, q)$

23. $\dfrac{\partial w}{\partial r}$ and $\dfrac{\partial w}{\partial s}$ for $w = f(x, y)$, $x = g(r)$, $y = h(s)$

24. $\dfrac{\partial w}{\partial s}$ for $w = g(x, y)$, $x = h(r, s, t)$, $y = k(r, s, t)$

Assuming that the equations in Exercises 25–28 define y as a differentiable function of x, use Eq. (14) to find the value of dy/dx at the given point.

25. $x^3 - 2y^2 + xy = 0$, (1, 1)

26. $xy + y^2 - 3x - 3 = 0$, (−1, 1)

27. $x^2 + xy + y^2 - 7 = 0$, (1, 2)

28. $xe^y + \sin xy + y - \ln 2 = 0$, (0, ln 2)

Equation (14) can be generalized to functions of three variables and even more. The three-variable version goes like this:

If the equation $F(x, y, z) = 0$ determines z as a differentiable function of x and y, then, at points where $F_z \neq 0$,

$$\frac{\partial z}{\partial x} = -\frac{F_x}{F_z} \quad \text{and} \quad \frac{\partial z}{\partial y} = -\frac{F_y}{F_z}. \qquad (17)$$

Use these equations to find the values of $\partial z/\partial x$ and $\partial z/\partial y$ at the points in Exercises 29–32,

29. $z^3 - xy + yz + y^3 - 2 = 0$, (1, 1, 1)

30. $\frac{1}{x} + \frac{1}{y} + \frac{1}{z} - 1 = 0$, (2, 3, 6)

31. $\sin(x + y) + \sin(y + z) + \sin(x + z) = 0$, ($\pi, \pi, \pi$)

32. $xe^y + ye^z + 2 \ln x - 2 - 3 \ln 2 = 0$, (1, ln 2, ln 3)

33. Find $\partial w/\partial r$ when $r = 1$, $s = -1$ if $w = (x + y + z)^2$, $x = r - s$, $y = \cos(r + s)$, $z = \sin(r + s)$.

34. Find $\partial w/\partial v$ when $u = -1$, $v = 2$ if $w = xy + \ln z$, $x = v^2/u$, $y = u + v$, $z = \sin u$.

35. Find $\partial w/\partial v$ when $u = 0$, $v = 0$ if $w = x^2 + (y/x)$, $x = u - 2v + 1$, $y = 2u + v - 2$.

36. Find $\partial z/\partial u$ when $u = 0$, $v = 1$ if $z = \sin xy + x \sin y$, $x = u^2 + v^2$, $y = uv$.

37. Find $\partial z/\partial u$ and $\partial z/\partial v$ when $u = \ln 2$, $v = 1$ if $z = 5 \tan^{-1}x$ and $x = e^u + \ln v$.

38. Find $\partial w/\partial q$ when $p = 1$, $q = 4$ if $w = xy + yz + zu + uv$, $x = pq$, $y = q \tan^{-1}p$, $z = \ln(q - 3p)$, $u = \sqrt{q + 5p}, v = p + q + 1$.

39. If $f(u, v, w)$ is differentiable and $u = x - y, v = y - z$, and $w = z - x$, show that

$$\frac{\partial f}{\partial x} + \frac{\partial f}{\partial y} + \frac{\partial f}{\partial z} = 0.$$

40. a) Show that if we substitute polar coordinates $x = r \cos \theta$ and $y = r \sin \theta$ in a differentiable function $w = f(x, y)$, then

$$\frac{\partial w}{\partial r} = f_x \cos \theta + f_y \sin \theta$$

and

$$\frac{1}{r} \frac{\partial w}{\partial \theta} = -f_x \sin \theta + f_y \cos \theta.$$

b) Solve the equations in (a) to express f_x and f_y in terms of $\partial w/\partial r$ and $\partial w/\partial \theta$.

c) Show that

$$(f_x)^2 + (f_y)^2 = \left(\frac{\partial w}{\partial r}\right)^2 + \frac{1}{r^2}\left(\frac{\partial w}{\partial \theta}\right)^2.$$

41. *Changing voltage in a circuit.* The voltage V in a circuit that satisfies the law $V = IR$ (Fig. 12.37) is slowly dropping as the battery wears out. At the same time, the resistance R is increasing as the resistor heats up. Use the equation

$$\frac{dV}{dt} = \frac{\partial V}{\partial I}\frac{dI}{dt} + \frac{\partial V}{\partial R}\frac{dR}{dt}$$

to find how the current is changing at the instant when $R = 600$ ohms, $I = 0.04$ amp, $dR/dt = 0.5$ ohm/sec, and $dV/dt = -0.01$ volt/sec.

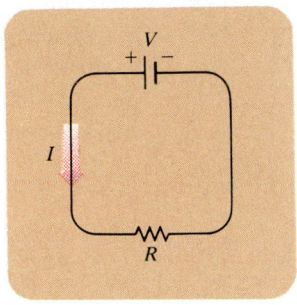

12.37   The battery-driven circuit in Exercise 41.

42. *Changing dimensions in a box.* The dimensions a, b, and c of the box in Fig. 12.38 are changing with time. At the instant in question, $a = 13$ cm, $b = 9$ cm, $c = 5$ cm, $da/dt = dc/dt = 2$ cm/sec, and $db/dt = -5$ cm/sec.
a) How fast are the volume $V = abc$ and the surface area $S = 2ab + 2ac + 2bc$ changing at that instant?
b) Is the box's diagonal increasing, or decreasing?

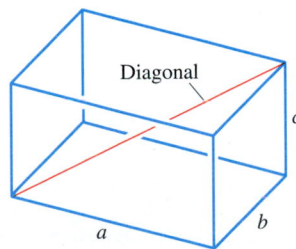

12.38   The box in Exercise 42.

43. Show that if $w = f(u, v)$ satisfies the Laplace equation $f_{uu} + f_{vv} = 0$, and if $u = (x^2 - y^2)/2$ and $v = xy$, then w satisfies the Laplace equation $w_{xx} + w_{yy} = 0$.

44. Let $w = f(u) + g(v)$, where $u = x + iy$ and $v = x - iy$ and $i = \sqrt{-1}$. Show that w satisfies the Laplace equation $w_{xx} + w_{yy} = 0$ if all the necessary functions are differentiable.

Changes in Functions Along Curves

45. Suppose that the partial derivatives of a function $f(x, y, z)$ at points on the helix $x = \cos t$, $y = \sin t$, $z = t$ are

$$f_x = \cos t, \quad f_y = \sin t, \quad f_z = t^2 + t - 2.$$

At what points on the curve, if any, can f take on extreme values?

46. Let $w = x^2 e^{2y} \cos 3z$. Find the value of dw/dt at the point $(1, \ln 2, 0)$ on the curve $x = \cos t$, $y = \ln(t + 2)$, $z = t$.

47. Let $T = f(x, y)$ be the temperature at the point (x, y) on the circle $x = \cos t$, $y = \sin t$ and suppose that

$$\frac{\partial T}{\partial x} = 8x - 4y, \quad \frac{\partial T}{\partial y} = 8y - 4x.$$

a) Find where the maximum and minimum temperatures on the circle occur by examining the derivatives dT/dt and d^2T/dt^2.

b) Suppose $T = 4x^2 - 4xy + 4y^2$. Find the maximum and minimum values of t on the circle.

12.6 Partial Derivatives with Constrained Variables*

In finding partial derivatives of functions like $w = f(x, y)$, we have assumed x and y to be independent. But in many applications this is not the case. For example, the internal energy U of a gas may be expressed as a function $U = f(p, v, T)$ of pressure, volume, and temperature. If the individual molecules of the gas do not interact, however, p, v, and T obey the ideal gas law

$$pv = nRT \qquad (n \text{ and } R \text{ constant})$$

and therefore fail to be independent. Finding partial derivatives in situations like these can be complicated. But it is better to face the complication now than to meet it for the first time while you are also trying to learn economics, engineering, or physics.

Decide Which Variables Are Dependent and Which Are Independent

If the variables in a function $w = f(x, y, z)$ are constrained by a relation like the one imposed on x, y, and z by the equation $z = x^2 + y^2$, the geometric meanings and the numerical values of the partial derivatives of f will depend on which variables are chosen to be dependent and which are chosen to be independent. To see how this choice can affect the outcome, we consider the calculation of $\partial w/\partial x$ when $w = x^2 + y^2 + z^2$ and $z = x^2 + y^2$.

Example 1   Find $\partial w/\partial x$ if

$$w = x^2 + y^2 + z^2 \qquad \text{and} \qquad z = x^2 + y^2.$$

Solution   We are given two equations in the four unknowns x, y, z, and w. Like many such systems, this one can be solved for two of the unknowns (the dependent variables) in terms of the others (the independent variables). In being asked for $\partial w/\partial x$, we are told that w is to be a dependent variable and x an independent variable. The possible choices for the other variables come down to

Dependent	Independent
w, z	x, y
w, y	x, z

*This section is based on notes written for MIT by Arthur P. Mattuck.

In either case, we can express w explicitly in terms of the selected independent variables. We do this by using the second equation to eliminate the remaining dependent variable in the first equation.

In the first case, the remaining dependent variable is z. We eliminate it from the first equation by replacing it by $x^2 + y^2$. The resulting expression for w is

$$w = x^2 + y^2 + z^2 = x^2 + y^2 + (x^2 + y^2)^2$$
$$= x^2 + y^2 + x^4 + 2x^2y^2 + y^4$$

and

$$\frac{\partial w}{\partial x} = 2x + 4x^3 + 4xy^2. \qquad (1)$$

This is the formula for $\partial w / \partial x$ when x and y are the independent variables.

In the second case, where the independent variables are x and z and the remaining dependent variable is y, we eliminate the dependent variable y in the expression for w by replacing y^2 by $z - x^2$. This gives

$$w = x^2 + y^2 + z^2 = x^2 + (z - x^2) + z^2 = z + z^2$$

and

$$\frac{\partial w}{\partial x} = 0. \qquad (2)$$

This is the formula for $\partial w / \partial x$ when x and z are the independent variables.

The formulas for $\partial w / \partial x$ in Eqs. (1) and (2) are genuinely different. We cannot change either formula into the other by using the relation $z = x^2 + y^2$. There is not just one $\partial w / \partial x$, there are two, and we see that the original instruction to find $\partial w / \partial x$ was incomplete. *Which* $\partial w / \partial x$? we ask.

The geometric interpretations of Eqs. (1) and (2) help to explain why the equations differ. The function $w = x^2 + y^2 + z^2$ measures the square of the distance from the point (x, y, z) to the origin. The condition $z = x^2 + y^2$ says that the point (x, y, z) lies on the surface of the paraboloid of revolution shown in Fig. 12.39. What does it mean to calculate $\partial w / \partial x$ at a point $P(x, y, z)$ that can move only on this surface? What is the value of $\partial w / \partial x$ when the coordinates of P are, say, $(1, 0, 1)$?

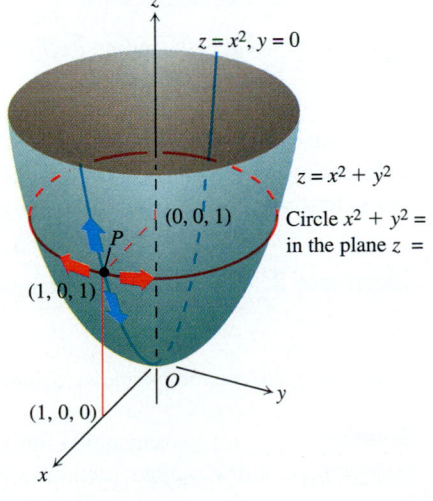

12.39 If P is constrained to lie on the paraboloid $z = x^2 + y^2$, the value of the partial derivative of $w = x^2 + y^2 + z^2$ with respect to x at P depends on the direction of motion (Example 1). (a) As x changes, with $y = 0$, P moves up or down the surface on the parabola $z = x^2$ in the xz-plane and $\partial w / \partial x = 2x + 4x^3$. (b) As x changes, with $z = 1$, P moves on the circle $x^2 + y^2 = 1$, $z = 1$, and $\partial w / \partial x = 0$.

Figure labels: $z = x^2, y = 0$; $z = x^2 + y^2$; $(0, 0, 1)$; Circle $x^2 + y^2 = 1$ in the plane $z = 1$; $(1, 0, 1)$; P; O; $(1, 0, 0)$

If we take x and y to be independent, then we find $\partial w/\partial x$ by holding y fixed (at $y = 0$ in this case) and letting x vary. This means that P moves along the parabola $z = x^2$ in the xz-plane. As P moves on this parabola, w, which is the square of the distance from P to the origin, changes. We calculate $\partial w/\partial x$ in this case (our first solution above) to be

$$\frac{\partial w}{\partial x} = 2x + 4x^3 + 4xy^2.$$

At the point $P(1, 0, 1)$, the value of this derivative is

$$\frac{\partial w}{\partial x} = 2 + 4 + 0 = 6.$$

If we take x and z to be independent, then we find $\partial w/\partial x$ by holding z fixed while x varies. Since the z-coordinate of P is 1, varying x moves P along a circle in the plane $z = 1$. As P moves along this circle, its distance from the origin remains constant, and w, being the square of this distance, does not change. That is,

$$\frac{\partial w}{\partial x} = 0,$$

as we found in our second solution.

How to Find $\partial w/\partial x$ When the Variables in $w = f(x, y, z)$ Are Constrained by Another Equation

As we saw in Example 1, a typical routine for finding $\partial w/\partial x$ when the variables in the function $w = f(x, y, z)$ are related by another equation has three steps.

STEP 1: Decide which variables are to be dependent and which are to be independent. (In practice, the decision is based on the physical or theoretical context of our work. In the exercises at the end of this section, we say which variables are which.)

STEP 2: Eliminate the other dependent variable(s) in the expression for w.

STEP 3: Differentiate as usual.

These steps apply to finding $\partial w/\partial y$ and $\partial w/\partial z$ as well.

If we cannot carry out step 2 after deciding which variables are dependent, we differentiate the equations as they are and try to solve for $\partial w/\partial x$ afterward. The next example shows how this is done.

Example 2   Find $\partial w/\partial x$ at the point $(x, y, z) = (2, -1, 1)$ if

$$w = x^2 + y^2 + z^2, \qquad z^3 - xy + yz + y^3 = 0,$$

and x and y are the independent variables.

Solution   It is not convenient to eliminate z in the expression for w. We therefore differentiate both equations implicitly with respect to x, treating x and y as independent variables and w and z as dependent variables. This gives

$$\frac{\partial w}{\partial x} = 2x + 2z\frac{\partial z}{\partial x} \qquad (3)$$

and

$$3z^2\frac{\partial z}{\partial x} - y + y\frac{\partial z}{\partial x} + 0 = 0. \qquad (4)$$

These equations may now be combined to express $\partial w/\partial x$ in terms of x, y, and z. We solve Eq. (4) for $\partial z/\partial x$ to get

$$\frac{\partial z}{\partial x} = \frac{y}{y + 3z^2}$$

and substitute into Eq. (3) to get

$$\frac{\partial w}{\partial x} = 2x + \frac{2yz}{y + 3z^2}.$$

The value of this derivative at $(x, y, z) = (2, -1, 1)$ is

$$\left(\frac{\partial w}{\partial x}\right)_{(2, -1, 1)} = 2(2) + \frac{2(-1)(1)}{-1 + 3(1)^2}$$

$$= 4 + \frac{-2}{2} = 3.$$

To show what variables are assumed to be independent in calculating a derivative, we can use the following notation:

$$\left(\frac{\partial w}{\partial x}\right)_y \qquad \partial w/\partial x \text{ with } x \text{ and } y \text{ independent,} \qquad (5)$$

$$\left(\frac{\partial w}{\partial x}\right)_z \qquad \partial w/\partial x \text{ with } x \text{ and } z \text{ independent,} \qquad (6)$$

$$\left(\frac{\partial f}{\partial y}\right)_{x, t} \qquad \partial f/\partial y \text{ with } y, x, \text{ and } t \text{ independent.} \qquad (7)$$

Example 3   Find

a) $\left(\dfrac{\partial w}{\partial x}\right)_{y, z}$ and   b) $\left(\dfrac{\partial w}{\partial x}\right)_{t, z}$

if $w = x^2 + y - z + \sin t$ and $x + y = t$.

Solution

a) With x, y, z independent, we have

$$t = x + y, \qquad w = x^2 + y - z + \sin(x + y),$$

$$\left(\frac{\partial w}{\partial x}\right)_{y, z} = 2x + 0 - 0 + \cos(x + y)\frac{\partial}{\partial x}(x + y)$$

$$= 2x + \cos(x + y).$$

b) With x, t, z independent, we have

$$y = t - x, \qquad w = x^2 + (t - x) - z + \sin t,$$

$$\left(\frac{\partial w}{\partial x}\right)_{t,\,z} = 2x - 1 - 0 + 0 = 2x - 1.$$

Arrow Diagrams

In solving problems like the one in Example 3, it often helps to start with an arrow diagram that shows how the variables and functions are related. If

$$w = x^2 + y - z + \sin t \qquad \text{and} \qquad x + y = t$$

and we are asked to find $\partial w / \partial x$ when x, y, and z are independent, the appropriate diagram is one like this:

$$
\begin{pmatrix} x \\ y \\ z \end{pmatrix}
\quad \longrightarrow \quad
\begin{pmatrix} x \\ y \\ z \\ t \end{pmatrix}
\quad \longrightarrow \quad w
\tag{8}
$$

$$
\begin{array}{ccc}
\text{independent} & \text{intermediate} & \text{dependent} \\
\text{variables} & \text{variables and} & \text{variable} \\
& \text{relations} & \\
& x = x & \\
& y = y & \\
& z = z & \\
& t = x + y &
\end{array}
$$

The diagram shows the independent variables on the left, the intermediate variables and their relation to the independent variables in the middle, and the dependent variable on the right.

To find $\partial w / \partial x$, we first apply the four-variable form of the Chain Rule to w, getting

$$\frac{\partial w}{\partial x} = \frac{\partial w}{\partial x}\frac{\partial x}{\partial x} + \frac{\partial w}{\partial y}\frac{\partial y}{\partial x} + \frac{\partial w}{\partial z}\frac{\partial z}{\partial x} + \frac{\partial w}{\partial t}\frac{\partial t}{\partial x}. \tag{9}$$

We then use the formula for w,

$$w = x^2 + y - z + \sin t,$$

to evaluate the partial derivatives of w that appear on the right-hand side of Eq. (9). This gives

$$\begin{aligned}
\frac{\partial w}{\partial x} &= 2x\frac{\partial x}{\partial x} + (1)\frac{\partial y}{\partial x} + (-1)\frac{\partial z}{\partial x} + \cos t\frac{\partial t}{\partial x} \\
&= 2x\frac{\partial x}{\partial x} + \frac{\partial y}{\partial x} - \frac{\partial z}{\partial x} + \cos t\frac{\partial t}{\partial x}.
\end{aligned} \tag{10}$$

To calculate the remaining partial derivatives, we apply what we know about the dependence and independence of the variables involved. As shown in the diagram (8), the variables x, y, and z are independent and $t = x + y$. Hence,

$$\frac{\partial x}{\partial x} = 1, \quad \frac{\partial y}{\partial x} = 0, \quad \frac{\partial z}{\partial x} = 0, \quad \frac{\partial t}{\partial x} = \frac{\partial}{\partial x}(x + y) = (1 + 0) = 1. \tag{11}$$

We substitute these values into Eq. (10) to find $\partial w/\partial x$:

$$\left(\frac{\partial w}{\partial x}\right)_{y,z} = 2x(1) + 0 - 0 + (\cos t)(1)$$

$$= 2x + \cos t$$

$$= 2x + \cos(x + y). \qquad \left(\begin{array}{l}\text{In terms of the independent}\\ \text{variables}\end{array}\right)$$

Working Without Specific Formulas

In applications, we often have to find an expression for the derivative of a function when neither the function nor the relation among its variables is given explicitly. Two observations can help us.

First, if a differentiable function $F(r, s)$ has a constant value throughout some region R of the rs-plane, and r and s are independent variables, then

$$\frac{\partial F}{\partial r} = \frac{\partial F}{\partial s} = 0 \qquad (12)$$

throughout R. In particular, Eq. (12) holds if $F(r, s) = 0$ throughout R.

Second, suppose that

$$f(x, y, z) = f(x(r, s), y(r, s), z(r, s)) = F(r, s) \qquad (13)$$

throughout some region R of the rs-plane, where all the functions are differentiable and r and s are independent variables. To display this information, we draw an arrow diagram:

$$
\begin{pmatrix} r \\ s \end{pmatrix} \longrightarrow \begin{pmatrix} x \\ y \\ z \end{pmatrix} \xrightarrow{f} w \qquad (14)
$$

$$F(r, s)$$

independent variables	intermediate variables and relations	dependent variable
	$x = x(r, s)$	$w = f(x, y, z)$
	$y = y(r, s)$	$= F(r, s)$
	$z = z(r, s)$	

Suppose further that $F(r, s) = 0$ throughout R. Then

$$\frac{\partial f}{\partial x}\frac{\partial x}{\partial r} + \frac{\partial f}{\partial y}\frac{\partial y}{\partial r} + \frac{\partial f}{\partial z}\frac{\partial z}{\partial r} = 0 \qquad (15)$$

and

$$\frac{\partial f}{\partial x}\frac{\partial x}{\partial s} + \frac{\partial f}{\partial y}\frac{\partial y}{\partial s} + \frac{\partial f}{\partial z}\frac{\partial z}{\partial s} = 0. \qquad (16)$$

The left-hand sides of these equations are $\partial F/\partial r$ and $\partial F/\partial s$, and these derivatives are zero because $F(r, s)$ has a constant value throughout R.

Example 4   Suppose that the equation $f(x, y, z) = 0$ determines z as a differentiable function of the independent variables x and y and that $\partial f/\partial z \neq 0$. Show that

$$\left(\frac{\partial z}{\partial x}\right)_y = -\frac{\partial f/\partial x}{\partial f/\partial z}.$$

Solution We regard all three of x, y, and z to be functions of the two independent variables x and y as in the diagram

$$\begin{pmatrix} x \\ y \end{pmatrix} \quad \longrightarrow \quad \begin{pmatrix} x \\ y \\ z \end{pmatrix} \quad \xrightarrow{f} \quad w \tag{17}$$

$$x = x$$
$$y = y$$
$$z = z(x, y)$$

This is the diagram in (14) with $r = x$, $s = y$, and a few details left out. From Eq. (15) with $r = x$, we have

$$\frac{\partial f}{\partial x}\frac{\partial x}{\partial x} + \frac{\partial f}{\partial y}\frac{\partial y}{\partial x} + \frac{\partial f}{\partial z}\frac{\partial z}{\partial x} = 0$$

$$\frac{\partial f}{\partial x}\cdot 1 + \frac{\partial f}{\partial y}\cdot 0 + \frac{\partial f}{\partial z}\frac{\partial z}{\partial x} = 0$$

$$\frac{\partial z}{\partial x} = -\frac{\partial f/\partial x}{\partial f/\partial z}.$$

EXERCISES 12.6

In Exercises 1–3, begin by drawing a diagram that shows the relations among the variables.

1. If $w = x^2 + y^2 + z^2$ and $z = x^2 + y^2$, find

a) $\left(\dfrac{\partial w}{\partial y}\right)_z$ b) $\left(\dfrac{\partial w}{\partial z}\right)_x$ c) $\left(\dfrac{\partial w}{\partial z}\right)_y$

2. If $w = x^2 + y - z + \sin t$ and $x + y = t$, find

a) $\left(\dfrac{\partial w}{\partial y}\right)_{x,z}$ b) $\left(\dfrac{\partial w}{\partial y}\right)_{z,t}$ c) $\left(\dfrac{\partial w}{\partial z}\right)_{x,y}$

d) $\left(\dfrac{\partial w}{\partial z}\right)_{y,t}$ e) $\left(\dfrac{\partial w}{\partial t}\right)_{x,z}$ f) $\left(\dfrac{\partial w}{\partial t}\right)_{y,z}$

3. Let $U = f(p, v, T)$ be the internal energy of a gas that obeys the ideal gas law $pv = nRT$ (n and R constant). Find

a) $\left(\dfrac{\partial U}{\partial p}\right)_v$ b) $\left(\dfrac{\partial U}{\partial T}\right)_v$

4. Find

a) $\left(\dfrac{\partial w}{\partial x}\right)_y$ b) $\left(\dfrac{\partial w}{\partial z}\right)_y$

at the point $(x, y, z) = (0, 1, \pi)$ if

$$w = x^2 + y^2 + z^2 \quad \text{and} \quad y \sin z + z \sin x = 1.$$

5. Find

a) $\left(\dfrac{\partial w}{\partial y}\right)_x$ b) $\left(\dfrac{\partial w}{\partial y}\right)_z$

at the point $(w, x, y, z) = (4, 2, 1, -1)$ if

$$w = x^2y^2 + yz - z^3 \quad \text{and} \quad x^2 + y^2 + z^2 = 6.$$

6. Find $\left(\dfrac{\partial u}{\partial y}\right)_x$ at the point $(u, v) = \left(\sqrt{2}, 1\right)$ if $x = u^2 + v^2$ and $y = uv$.

7. Suppose that $x^2 + y^2 = r^2$ and $x = r \cos \theta$, as in polar coordinates. Find

$$\left(\frac{\partial x}{\partial r}\right)_\theta \quad \text{and} \quad \left(\frac{\partial r}{\partial x}\right)_y.$$

8. Suppose that the equation $g(x, y, z) = 0$ determines z as a differentiable function of the independent variables x and y and that $g_z \neq 0$. Show that

$$\left(\frac{\partial z}{\partial y}\right)_x = -\frac{\partial g/\partial y}{\partial g/\partial z}.$$

9. Establish the fact, widely used in hydrodynamics, that if $f(x, y, z) = 0$, then

$$\left(\frac{\partial x}{\partial y}\right)_z\left(\frac{\partial y}{\partial z}\right)_x\left(\frac{\partial z}{\partial x}\right)_y = -1.$$

(*Hint*: Express all the derivatives in terms of the formal partial derivatives $\partial f/\partial x$, $\partial f/\partial y$, and $\partial f/\partial z$.)

10. If $z = x + f(u)$, where $u = xy$, show that

$$x\frac{\partial z}{\partial x} - y\frac{\partial z}{\partial y} = x.$$

11. Suppose

$$w = x^2 - y^2 + 4z + t \quad \text{and} \quad x + 2z + t = 25.$$

Show that the equations

$$\frac{\partial w}{\partial x} = 2x - 1 \quad \text{and} \quad \frac{\partial w}{\partial x} = 2x - 2$$

each give $\partial w/\partial x$, depending on which variables are chosen to be dependent and which are chosen to be independent. Identify the independent variables in each case.

12.7 Directional Derivatives, Gradient Vectors, and Tangent Planes

We saw in Section 12.4 that if a function $f(x, y, z)$ is differentiable, the rate at which the values of f change with respect to t along a differentiable curve $x = g(t)$, $y = h(t)$, $z = k(t)$ can be found from the formula

$$\frac{df}{dt} = \frac{\partial f}{\partial x}\frac{dx}{dt} + \frac{\partial f}{\partial y}\frac{dy}{dt} + \frac{\partial f}{\partial z}\frac{dz}{dt}. \tag{1}$$

At any particular point $P_0\,(g\,(t_0), h\,(t_0), k\,(t_0))$, this formula gives the rate of change of f with respect to increasing t and therefore depends, among other things, on the direction of motion along the curve. This observation is particularly important when the curve is a straight line and the parameter t is the distance along the line measured from P_0 in the direction of a given unit vector $\mathbf{u}$. For then df/dt is the rate of change of f with respect to distance in its domain in the direction of $\mathbf{u}$. By varying $\mathbf{u}$, we can find the rates at which f changes with respect to distance as we move through P_0 in different directions. These "directional derivatives" are extremely useful in science and engineering as well as in mathematics. The first goal of this section is to develop a formula for calculating them. The second goal is to show how to use vectors called gradients to define tangent planes and normal lines at points on level surfaces of functions of three variables.

Directional Derivatives of Functions of Three Variables

Suppose that the function $f(x, y, z)$ is differentiable throughout some region D in space, that $P_0(x_0, y_0, z_0)$ is a point in D, and that $\mathbf{u} = u_1\,\mathbf{i} + u_2\,\mathbf{j} + u_3\,\mathbf{k}$ is a unit vector (Fig. 12.40). Then

$$x = x_0 + su_1, \qquad y = y_0 + su_2, \qquad z = z_0 + su_3 \tag{2}$$

are parametric equations for the line through P_0 parallel to $\mathbf{u}$, and the parameter s measures the directed distance along this line from P_0 to the point $P_S(x_0 + su_1,\ y_0 + su_2, z_0 + su_3)$, as we saw in Example 3, Section 11.3. To calculate df/ds, the rate at which f changes with respect to distance in the direction of $\mathbf{u}$ at P_0, we substitute the derivatives

$$\frac{dx}{ds} = u_1, \qquad \frac{dy}{ds} = u_2, \qquad \frac{dz}{ds} = u_3$$

into the equation

$$\frac{df}{ds} = \frac{\partial f}{\partial x}\frac{dx}{ds} + \frac{\partial f}{\partial y}\frac{dy}{ds} + \frac{\partial f}{\partial z}\frac{dz}{ds}$$

and evaluate the partial derivatives of f at P_0. This gives

$$\frac{df}{ds} = \left(\frac{\partial f}{\partial x}\right)_{P_0} u_1 + \left(\frac{\partial f}{\partial y}\right)_{P_0} u_2 + \left(\frac{\partial f}{\partial z}\right)_{P_0} u_3. \tag{3}$$

The expression on the right-hand side of Eq. (3) is the dot product of $\mathbf{u}$ and the

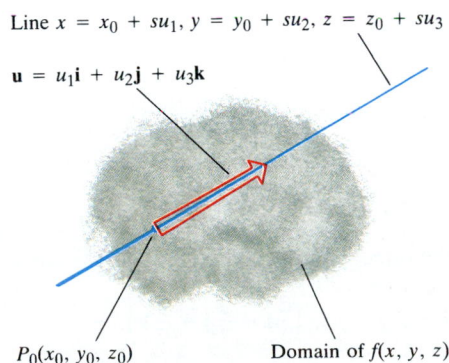

Line $x = x_0 + su_1, y = y_0 + su_2, z = z_0 + su_3$

$\mathbf{u} = u_1\mathbf{i} + u_2\mathbf{j} + u_3\mathbf{k}$

$P_0(x_0, y_0, z_0)$　　　Domain of $f(x, y, z)$

12.40　The derivative of f in the direction of the unit vector $\mathbf{u}$ at a point P_0 in its domain is the value of df/ds at P_0.

The notation ∇f is read "grad f" as well as "gradient of f" and "del f." The symbol ∇ by itself is read "del." Another notation for the gradient is grad f, read the way it is written.

vector

$$\nabla f = \left(\frac{\partial f}{\partial x}\right)_{P_0} \mathbf{i} + \left(\frac{\partial f}{\partial y}\right)_{P_0} \mathbf{j} + \left(\frac{\partial f}{\partial z}\right)_{P_0} \mathbf{k}.$$

This vector is called the *gradient* of f at the point P_0.

It is customary to picture the gradient as a vector in the domain of f. Its components are calculated by evaluating the three partial derivatives of f at (x_0, y_0, z_0). The derivative on the left-hand side of Eq. (3) is called the *derivative of f at the point P_0 in the direction of $\mathbf{u}$*. It is often denoted by

$$(D_{\mathbf{u}}f)_{P_0} \qquad \text{or} \qquad \left(\frac{df}{ds}\right)_{\mathbf{u}, P_0}$$

DEFINITIONS

If the partial derivatives of $f(x, y, z)$ are defined at $P_0(x_0, y_0, z_0)$, then the **gradient** of f at P_0 is the vector

$$\nabla f = \frac{\partial f}{\partial x}\mathbf{i} + \frac{\partial f}{\partial y}\mathbf{j} + \frac{\partial f}{\partial z}\mathbf{k} \qquad (4)$$

obtained by evaluating the partial derivatives of f at P_0.

If $\mathbf{u}$ is a unit vector, then the **derivative of f at P_0 in the direction of $\mathbf{u}$** is the number

$$(D_{\mathbf{u}}f)_{P_0} = (\nabla f)_{P_0} \cdot \mathbf{u}, \qquad (5)$$

which is the scalar product of the gradient f at P_0 and $\mathbf{u}$.

Example 1   Find the derivative of

$$f(x, y, z) = x^3 - xy^2 - z$$

at $P_0(1, 1, 0)$ in the direction of the vector $\mathbf{A} = 2\mathbf{i} - 3\mathbf{j} + 6\mathbf{k}$.

Solution   The direction of $\mathbf{A}$ is obtained by dividing $\mathbf{A}$ by its length:

$$|\mathbf{A}| = \sqrt{(2)^2 + (-3)^2 + (6)^2} = \sqrt{49} = 7,$$

$$\mathbf{u} = \frac{\mathbf{A}}{|\mathbf{A}|} = \frac{2}{7}\mathbf{i} - \frac{3}{7}\mathbf{j} + \frac{6}{7}\mathbf{k}.$$

The partial derivatives of f at P_0 are

$$f_x = 3x^2 - y^2|_{(1, 1, 0)} = 2, \qquad f_y = -2xy|_{(1, 1, 0)} = -2, \qquad f_z = -1|_{(1, 1, 0)} = -1.$$

The gradient of f at P_0 is

$$\nabla f|_{(1, 1, 0)} = 2\mathbf{i} - 2\mathbf{j} - \mathbf{k}$$

(Fig. 12.41). The derivative of f at P_0 in the direction of $\mathbf{A}$ is therefore

$$(D_{\mathbf{u}}f)|_{(1, 1, 0)} = \nabla f|_{(1, 1, 0)} \cdot \mathbf{u} = (2\mathbf{i} - 2\mathbf{j} - \mathbf{k}) \cdot \left(\frac{2}{7}\mathbf{i} - \frac{3}{7}\mathbf{j} + \frac{6}{7}\mathbf{k}\right)$$

$$= \frac{4}{7} + \frac{6}{7} - \frac{6}{7} = \frac{4}{7}.$$

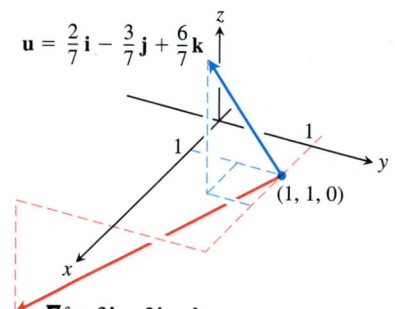

$$\mathbf{u} = \frac{2}{7}\mathbf{i} - \frac{3}{7}\mathbf{j} + \frac{6}{7}\mathbf{k}$$

$$\nabla f = 2\mathbf{i} - 2\mathbf{j} - \mathbf{k}$$

12.41   The change in f in the direction of $\mathbf{u}$ is $\nabla f \cdot \mathbf{u} = 4/7$ (Example 1).

Increments and Distance

The directional derivative plays the role of an ordinary derivative when we want to estimate how much a function f changes if we move a small distance ds from a point P_0 to a nearby point. If f were a function of a single variable, we would have

$$df = f'(P_0)\, ds. \qquad \text{(Ordinary derivative $\times$ increment)}$$

For a function of two or more variables, we use the formula

$$df = (\nabla f|_{P_0} \cdot \mathbf{u})\, ds, \qquad \text{(Directional derivative $\times$ increment)}$$

where $\mathbf{u}$ is the direction of motion away from P_0. The product on the right-hand side of this formula is the total differential of f in disguise:

$$df = f_x\, dx + f_y\, dy + f_z\, dz \qquad \text{(Total differential of f)}$$

$$= \nabla f \cdot (dx\,\mathbf{i} + dy\,\mathbf{j} + dz\,\mathbf{k})$$

$$= \nabla f \cdot \underbrace{\frac{dx\,\mathbf{i} + dy\,\mathbf{j} + dz\,\mathbf{k}}{\sqrt{dx^2 + dy^2 + dz^2}}}_{\substack{\text{unit vector } \mathbf{u} \text{ in the} \\ \text{direction of change}}} \underbrace{\sqrt{dx^2 + dy^2 + dz^2}}_{ds} \qquad (6)$$

$$= (\nabla f \cdot \mathbf{u})\, ds.$$

Estimating the Change in f in a Direction $\mathbf{u}$

To estimate the change in the value of a function f when we move a small distance ds from a point P_0 in a particular direction $\mathbf{u}$, use the formula

$$df = \underbrace{(\nabla f|_{P_0} \cdot \mathbf{u})}_{\substack{\text{directional} \\ \text{derivative}}} \cdot \underbrace{ds}_{\substack{\text{distance} \\ \text{increment}}}. \qquad (7)$$

Example 2 Estimate how much the value of

$$f(x, y, z) = xe^y + yz$$

will change if the point $P(x, y, z)$ is moved from $P_0(2, 0, 0)$ straight toward $P_1(4, 1, -2)$ a distance of $ds = 0.1$ units (Fig. 12.42).

Solution We first find the derivative of f at P_0 in the direction of the vector

$$\overrightarrow{P_0P_1} = 2\,\mathbf{i} + \mathbf{j} - 2\,\mathbf{k}.$$

The direction of this vector is

$$\mathbf{u} = \frac{\overrightarrow{P_0P_1}}{|\overrightarrow{P_0P_1}|} = \frac{\overrightarrow{P_0P_1}}{3} = \frac{2}{3}\,\mathbf{i} + \frac{1}{3}\,\mathbf{j} - \frac{2}{3}\,\mathbf{k}.$$

The gradient of f at P_0 is

$$\nabla f|_{(2,0,0)} = (e^y\,\mathbf{i} + (xe^y + z)\,\mathbf{j} + y\,\mathbf{k})|_{(2,0,0)} = \mathbf{i} + 2\,\mathbf{j}.$$

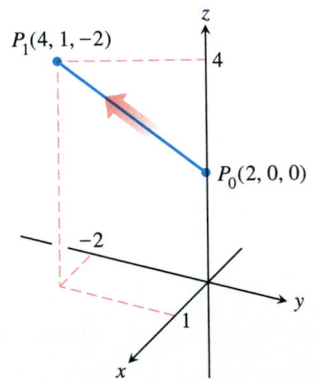

12.42 Example 2 shows how to estimate the change in the value of $f(x, y, z) = xe^y + yz$ as (x, y, z) moves a short distance $ds = 0.1$ units from P_0 along the line toward P_1.

Therefore,

$$\nabla f|_{P_0} \cdot \mathbf{u} = (\mathbf{i} + 2\,\mathbf{j}) \cdot \left(\frac{2}{3}\mathbf{i} + \frac{1}{3}\mathbf{j} - \frac{2}{3}\mathbf{k}\right) = \frac{2}{3} + \frac{2}{3} = \frac{4}{3}.$$

The change df in f that results from moving $ds = 0.1$ units away from P_0 in the direction of $\mathbf{u}$ is approximately

$$df = \nabla f|_{P_0} \cdot \mathbf{u} \; ds = \left(\frac{4}{3}\right)(0.1) \approx 0.13.$$

If we evaluate the dot product in the formula for the directional derivative, to obtain

$$D_{\mathbf{u}} f = \nabla f \cdot \mathbf{u} = |\nabla f|\,|\mathbf{u}| \cos \theta = |\nabla f| \cos \theta, \tag{8}$$

the following facts come to light.

Properties of the Directional Derivative $D_{\mathbf{u}} f = \nabla f \cdot \mathbf{u} = |\nabla f| \cos \theta$

1. The directional derivative has its largest positive value when $\cos \theta = 1$, or when $\mathbf{u}$ is the direction of the gradient. That is, f increases most rapidly at any point in its domain in the direction of ∇f. The derivative in this direction is $D_{\mathbf{u}} f = |\nabla f| \cos(0) = |\nabla f|$.

2. Similarly, f decreases most rapidly in the direction of $-\nabla f$. The derivative in this direction is $D_{\mathbf{u}} f = |\nabla f| \cos(\pi) = -|\nabla f|$.

3. $D_{-\mathbf{u}} f = \nabla f \cdot (-\mathbf{u}) = -\nabla f \cdot \mathbf{u} = -D_{\mathbf{u}} f$. That is, the derivative of f in the direction of $-\mathbf{u}$ is the negative of the derivative of f in the direction of $\mathbf{u}$.

4. The relationships of the partial derivatives of f to the directional derivative are

$$D_{\mathbf{i}} f = \nabla f \cdot \mathbf{i} = f_x, \qquad D_{\mathbf{j}} f = \nabla f \cdot \mathbf{j} = f_y, \qquad D_{\mathbf{k}} f = \nabla f \cdot \mathbf{k} = f_z.$$

Thus,

$$f_x = \text{derivative of } f \text{ in the } \mathbf{i}\text{-direction,}$$
$$f_y = \text{derivative of } f \text{ in the } \mathbf{j}\text{-direction,}$$
$$f_z = \text{derivative of } f \text{ in the } \mathbf{k}\text{-direction.}$$

Combining these results with property 3 gives

$$D_{-\mathbf{i}} f = -f_x, \qquad D_{-\mathbf{j}} f = -f_y, \qquad D_{-\mathbf{k}} f = -f_z.$$

5. Any direction $\mathbf{u}$ normal (perpendicular) to the gradient is a direction of zero change in f because

$$D_{\mathbf{u}} f = |\nabla f| \cos(\pi/2) = |\nabla f| \cdot 0 = 0.$$

Example 3   Find the directions in which $f(x, y, z) = x^2 + y^2 + z^2$ increases and decreases most rapidly at the point $(1, 1, 1)$ in its domain. At what rates does f change in these directions?

Solution   The most rapid increase at (x, y, z) is in the direction of

$$\nabla f = 2x\mathbf{i} + 2y\mathbf{j} + 2z\mathbf{k}.$$

At $(1, 1, 1)$ we have

$$\nabla f = 2\,\mathbf{i} + 2\,\mathbf{j} + 2\,\mathbf{k}, \qquad |\nabla f| = \sqrt{4 + 4 + 4} = 2\sqrt{3}.$$

The direction of ∇f is

$$\mathbf{u} = \frac{\nabla f}{|\nabla f|} = \frac{1}{\sqrt{3}}\,(\mathbf{i} + \mathbf{j} + \mathbf{k}).$$

The derivative of f in this direction is $|\nabla f| = 2\sqrt{3}$.

The direction of most rapid decrease at $(1, 1, 1)$ is

$$-\mathbf{u} = -\frac{1}{\sqrt{3}}\,(\mathbf{i} + \mathbf{j} + \mathbf{k})$$

and the derivative of f in this direction is $-|\nabla f| = -2\sqrt{3}$.

Results for Functions of Two Variables

For functions of two variables, we get results much like the ones for functions of three variables. We obtain the two-variable formulas by dropping the z-terms from the three-variable formulas. Thus, for a function $f(x, y)$ and a unit vector $\mathbf{u} = u_1\,\mathbf{i} + u_2\,\mathbf{j}$,

$$\nabla f = \frac{\partial f}{\partial x}\,\mathbf{i} + \frac{\partial f}{\partial y}\,\mathbf{j}, \tag{9}$$

and

$$D_{\mathbf{u}} f = \nabla f \cdot \mathbf{u} = \frac{\partial f}{\partial x}\,u_1 + \frac{\partial f}{\partial y}\,u_2. \tag{10}$$

The five properties listed earlier for directional derivatives of functions of three variables hold equally well in the two-variable case. In addition, property 5, which says that any direction normal to the gradient is a direction of zero change for the function, can now be turned around to say that the gradient of a function $f(x, y)$ is always normal to the function's level curves (Fig. 12.43). (See Exercise 51 for additional details.)

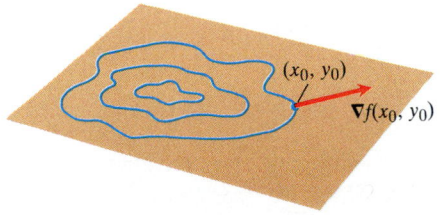

12.43   The gradient of a differentiable function of two variables is always normal to the function's level curves.

> At every point (x_0, y_0) in the domain of $f(x, y)$, the gradient of f is normal to the level curve through (x_0, y_0).

Example 4

a) Find the derivative of

$$f(x, y) = x^2 + y^2$$

at the point $P_0(1, 1)$ in the direction of the unit vector $\mathbf{u} = u_1\,\mathbf{i} + u_2\,\mathbf{j}$.

b) In what direction in its domain (the xy-plane) does f increase most rapidly at $(1, 1)$? What is the derivative of f in this direction?

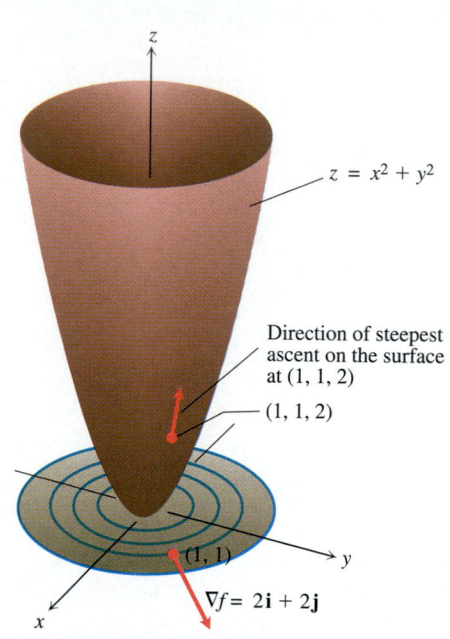

12.44   The gradient of $f(x, y) = x^2 + y^2$ at the point $(1, 1)$ is the vector $\nabla f = 2\mathbf{i} + 2\mathbf{j}$ in the xy-plane. This vector points in the direction in which f increases most rapidly at the point $(1, 1)$ and is normal to the level curve there. The corresponding direction on the surface $z = f(x, y)$ is the direction of steepest ascent at the point $(1, 1, 2)$ (Example 4).

Solution

a) We have

$$f(x, y) = x^2 + y^2, \qquad f_x(1, 1) = 2x|_{(1, 1)} = 2, \qquad f_y(1, 1) = 2y|_{(1, 1)} = 2,$$

$$(\nabla f)_{(1, 1)} = f_x(1, 1)\,\mathbf{i} + f_y(1, 1)\,\mathbf{j} = 2\,\mathbf{i} + 2\,\mathbf{j},$$

and

$$(D_{\mathbf{u}} f)_{(1, 1)} = (\nabla f)_{(1, 1)} \cdot \mathbf{u} = (2\,\mathbf{i} + 2\,\mathbf{j}) \cdot \mathbf{u} = 2u_1 + 2u_2.$$

The derivative of f at $(1, 1)$ in the direction of $\mathbf{u}$ is $2u_1 + 2u_2$.

b) The function increases most rapidly at $(1, 1)$ in the direction of ∇f at $(1, 1)$, which is

$$\left(\frac{\nabla f}{|\nabla f|}\right)_{(1, 1)} = \frac{2\,\mathbf{i} + 2\,\mathbf{j}}{\sqrt{2^2 + 2^2}} = \frac{2}{\sqrt{8}}\,\mathbf{i} + \frac{2}{\sqrt{8}}\,\mathbf{j} = \frac{1}{\sqrt{2}}\,(\mathbf{i} + \mathbf{j})$$

(Fig. 12.44). The derivative of f in this direction at $(1, 1)$ is

$$|\nabla f|_{(1, 1)} = 2\sqrt{2}.$$

Algebra Rules for Gradients

If we know the gradients of two functions f and g, we automatically know the gradients of their constant multiples, sum, difference, and product.

Algebra Rules for Gradients

1. *Constant Multiple Rule:* $\nabla(kf) = k\nabla f$ (Any number k)

2. *Sum Rule:* $\nabla(f + g) = \nabla f + \nabla g$

3. *Difference Rule:* $\nabla(f - g) = \nabla f - \nabla g$

4. *Product Rule:* $\nabla(fg) = f\nabla g + g\nabla f$

These rules have the same form as the corresponding rules for derivatives, as they should. You will see where the rules come from if you do Exercise 52.

Example 5   With

$$f(x, y, z) = e^x, \qquad g(x, y, z) = y - z,$$

$$\nabla f = e^x\,\mathbf{i}, \qquad \nabla g = \mathbf{j} - \mathbf{k},$$

we find

1. $\nabla(2f) = \nabla(2e^x) = 2e^x\,\mathbf{i} = 2\nabla f,$

2. $\nabla(f + g) = \nabla(e^x + y - z) = e^x\,\mathbf{i} + \mathbf{j} - \mathbf{k} = (e^x\,\mathbf{i}) + (\mathbf{j} - \mathbf{k}) = \nabla f + \nabla g,$

3. $\nabla(f - g) = \nabla(e^x - y + z) = e^x\,\mathbf{i} - \mathbf{j} + \mathbf{k} = (e^x\,\mathbf{i}) - (\mathbf{j} - \mathbf{k}) = \nabla f - \nabla g,$

4. $\nabla(fg) = \nabla(ye^x - ze^x) = (ye^x - ze^x)\,\mathbf{i} + e^x\,\mathbf{j} - e^x\,\mathbf{k}$

   $= (y - z)(e^x\,\mathbf{i}) + e^x(\mathbf{j} - \mathbf{k}) = g\nabla f + f\nabla g.$

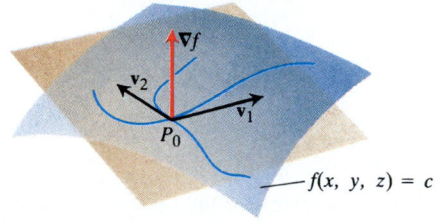

12.45 ∇f is orthogonal to the velocity vector of every differentiable curve in the surface through P_0. The velocity vectors at P_0 therefore lie in a common plane, which we call the tangent plane at P_0.

Equations for Tangent Planes and Normal Lines to Level Surfaces

To find equations for tangent planes and normal lines, we suppose that $P_0(x_0, y_0, z_0)$ is a point on the level surface $f(x, y, z) = c$ of a differentiable function f. If $x = g(t)$, $y = h(t)$, $z = k(t)$ is any differentiable curve on the surface, then

$$f(g(t), h(t), k(t)) = c \tag{11}$$

for every value of t. Differentiating both sides of this equation with respect to t gives

$$\frac{\partial f}{\partial x}\frac{dx}{dt} + \frac{\partial f}{\partial y}\frac{dy}{dt} + \frac{\partial f}{\partial z}\frac{dz}{dt} = 0. \tag{12}$$

The left-hand side of this equation is the dot product of ∇f with the curve's velocity vector $\mathbf{v}$, so the equation can be written in the form

$$\nabla f \cdot \mathbf{v} = 0. \tag{13}$$

What Eq. (13) shows more clearly than Eq. (12) is that the gradient of f is orthogonal to the velocity vector at every point of the curve.

Now let us restrict our attention to the differentiable curves that lie on the surface $f(x, y, z) = c$ and pass through P_0. All the velocity vectors at P_0 are orthogonal to ∇f at P_0, and hence all the tangent lines to these curves lie in the plane through P_0 normal to ∇f. We call this plane the tangent plane of the surface at P_0 (Fig. 12.45). We call the line through P_0 perpendicular to the plane the surface's normal line at P_0.

DEFINITIONS

The **tangent plane** at the point $P_0(x_0, y_0, z_0)$ on the level surface $f(x, y, z) = c$ is the plane

$$f_x(P_0)(x - x_0) + f_y(P_0)(y - y_0) + f_z(P_0)(z - z_0) = 0. \tag{14}$$

The **normal line** of the surface at P_0 is the line

$$x = x_0 + f_x(P_0)t, \qquad y = y_0 + f_y(P_0)t, \qquad z = z_0 + f_z(P_0)t. \tag{15}$$

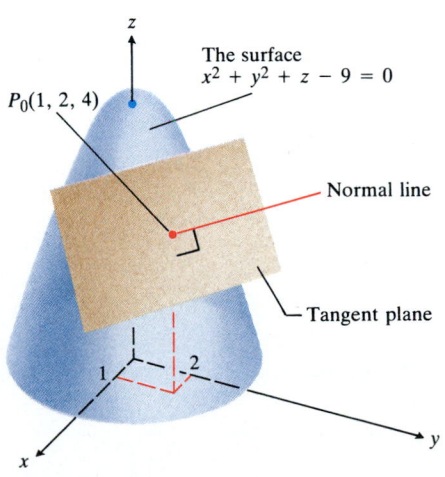

12.46 The tangent plane and normal line to the surface $x^2 + y^2 + z - 9 = 0$ at $P_0(1, 2, 4)$.

Example 6   Find equations for the tangent plane and normal line of the surface

$$f(x, y, z) = x^2 + y^2 + z - 9 = 0 \qquad \text{(A circular paraboloid)}$$

at the point $P_0(1, 2, 4)$.

Solution   The surface is shown in Fig. 12.46.

The tangent plane is the plane through P_0 perpendicular to the gradient of f at P_0. The gradient is

$$\nabla f|_{P_0} = (2x\mathbf{i} + 2y\mathbf{j} + \mathbf{k})_{(1, 2, 4)} = 2\mathbf{i} + 4\mathbf{j} + \mathbf{k}.$$

The plane is therefore the plane

$$2(x - 1) + 4(y - 2) + (z - 4) = 0 \qquad \text{or} \qquad 2x + 4y + z = 14.$$

The line normal to the surface at P_0 is

$$x = 1 + 2t, \qquad y = 2 + 4t, \qquad z = 4 + t.$$

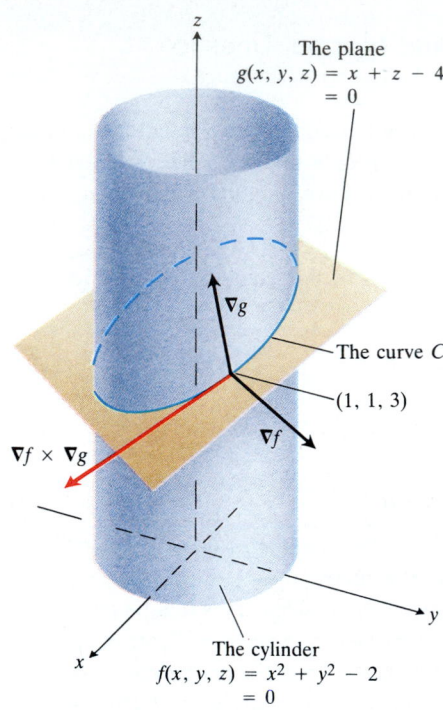

12.47 The cylinder $f(x, y, z) =$ $x^2 + y^2 - 2 = 0$ and the plane $g(x, y, z) =$ $x + z - 4 = 0$ intersect in a curve C. Any vector tangent to C will automatically be orthogonal to ∇f and ∇g and hence parallel to $\nabla f \times \nabla g$. This information enables us to write an equation for the tangent line in Example 7.

Example 7   The surfaces

$$f(x, y, z) = x^2 + y^2 - 2 = 0 \qquad \text{(A cylinder)}$$

and

$$g(x, y, z) = x + z - 4 = 0 \qquad \text{(A plane)}$$

meet in a curve C (Fig. 12.47). Find parametric equations for the line tangent to C at the point $P_0(1, 1, 3)$.

Solution   The tangent line is perpendicular to both ∇f and ∇g at P_0, so $\mathbf{v} = \nabla f \times \nabla g$ is parallel to the line. The components of $\mathbf{v}$ and the coordinates of P_0 give us equations for the line. We have

$$\nabla f_{(1, 1, 3)} = (2x\mathbf{i} + 2y\mathbf{j})_{(1, 1, 3)} = 2\mathbf{i} + 2\mathbf{j}$$

$$\nabla g_{(1, 1, 3)} = (\mathbf{i} + \mathbf{k})_{(1, 1, 3)} = \mathbf{i} + \mathbf{k}$$

$$\mathbf{v} = (2\mathbf{i} + 2\mathbf{j}) \times (\mathbf{i} + \mathbf{k}) = \begin{vmatrix} \mathbf{i} & \mathbf{j} & \mathbf{k} \\ 2 & 2 & 0 \\ 1 & 0 & 1 \end{vmatrix} = 2\mathbf{i} - 2\mathbf{j} - 2\mathbf{k}.$$

The line is

$$x = 1 + 2t, \qquad y = 1 - 2t, \qquad z = 3 - 2t.$$

Equations for Lines Tangent to Level Curves

The fact that the gradient of a function $f(x, y)$ is normal to the function's level curves in the plane enables us to use ∇f to write equations for tangent lines to level curves.

> The tangent to the level curve $f(x, y) = c$ at the point (x_0, y_0) is the line
>
> $$f_x(x_0, y_0)(x - x_0) + f_y(x_0, y_0)(y - y_0) = 0. \qquad (16)$$

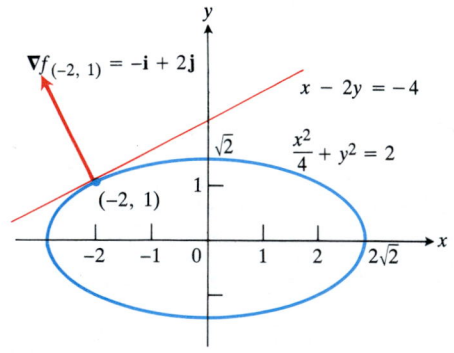

12.48   We can find the tangent to the ellipse $(x^2/4) + y^2 = 2$ by treating the ellipse as a level curve of the function $f(x, y) = (x^2/4) + y^2$ (Example 8).

Example 8   Find an equation for the tangent to the ellipse

$$\frac{x^2}{4} + y^2 = 2$$

(Fig. 12.48) at the point $(-2, 1)$.

Solution   The ellipse is a level curve of the function

$$f(x, y) = \frac{x^2}{4} + y^2.$$

The gradient of f at $(-2, 1)$ is

$$\nabla f|_{(-2, 1)} = \left(\frac{x}{2}\mathbf{i} + 2y\mathbf{j}\right)_{(-2, 1)} = -\mathbf{i} + 2\mathbf{j}.$$

The tangent is the line

$$(-1)(x + 2) + (2)(y - 1) = 0 \qquad \text{(Eq. (16))}$$

$$x - 2y = -4.$$

EXERCISES 12.7

In Exercises 1–4, find ∇f at the given point.

1. $f(x, y, z) = x^2 + y^2 - 2z^2 + z \ln x, \quad (1, 1, 1)$

2. $f(x, y, z) = 2z^3 - 3(x^2 + y^2)z + \tan^{-1} xz, \quad (1, 1, 1)$

3. $f(x, y, z) = (x^2 + y^2 + z^2)^{-1/2} + \ln(xyz), \quad (-1, 2, -2)$

4. $f(x, y, z) = e^{x+y}\cos z + (y + 1)\sin^{-1} x, \quad (0, 0, \pi/6)$

In Exercises 5–8, sketch the surface $f(x, y, z) = c$ together with ∇f at the given point.

5. $x^2 + y^2 - z = 0, \quad (1, 1, 2)$

6. $(y^2/2) + z^2 = 2, \quad (0, \sqrt{2}, 1)$

7. $(x^2/2) + (y^2/2) - (z^2/2) = 0, \quad (1, 1, -\sqrt{2})$

8. $z^2 - y^2 = 3, \quad (0, 1, 2)$

In Exercises 9–12, find ∇f at the given point. Then sketch ∇f together with the level curve through the point.

9. $f(x, y) = y - x, \quad (2, 1)$

10. $f(x, y) = \ln(x^2 + y^2), \quad (1, 1)$

11. $f(x, y) = y - x^2, \quad (-1, 0)$

12. $f(x, y) = x^2 - y^2, \quad (2, \sqrt{3})$

In Exercises 13–18, find the derivative of f at P_0 in the direction of $\mathbf{A}$.

13. $f(x, y, z) = xy + yz + zx, \quad P_0(1, -1, 2)$
    $\mathbf{A} = 3\mathbf{i} + 6\mathbf{j} - 2\mathbf{k}$

14. $f(x, y, z) = x^2 + 2y^2 - 3z^2, \quad P_0(1, 1, 1)$
    $\mathbf{A} = \mathbf{i} + \mathbf{j} + \mathbf{k}$

15. $f(x, y, z) = 3e^x \cos yz, \quad P_0(0, 0, 0)$
    $\mathbf{A} = -2\mathbf{i} + \mathbf{j} + 2\mathbf{k}$

16. $f(x, y, z) = \cos xy + e^{yz} + \ln zx, \quad P_0(1, 0, 1/2)$
    $\mathbf{A} = \mathbf{i} + \mathbf{j} - \sqrt{2}\mathbf{k}$

17. $f(x, y) = x - (y^2/x) + \sqrt{3} \sec^{-1}(2xy), \quad P_0(1, 1),$
    $\mathbf{A} = 12\mathbf{i} + 5\mathbf{j}$

18. $f(x, y) = \tan^{-1}(y/x) + \sqrt{3} \sin^{-1}(xy/2), \quad P_0(1, 1),$
    $\mathbf{A} = 3\mathbf{i} - 2\mathbf{j}$

In Exercises 19–24, find the directions in which f increases and decreases most rapidly at P_0. Then find the derivatives of f in these directions.

19. $f(x, y) = x^2 y + e^{xy} \sin y, \quad P_0(1, 0)$

20. $f(x, y) = x^{2/3} + xy + y^{2/3}, \quad P_0(-1, 1)$

21. $f(x, y, z) = xe^y + 4z^2 \cos^{-1}(\ln x), \quad P_0(1, \ln 2, 1/2)$

22. $f(x, y, z) = (x/y) - yz - 4\sqrt{xyz}, \quad P_0(4, 1, 1)$

23. $f(x, y, z) = \ln xy + \ln yz + \ln xz, \quad P_0(1, 1, 1)$

24. $f(x, y, z) = \ln(x^2 + y^2 - 1) + \sinh(xyz), \quad P_0(1, 1, 0)$

In Exercises 25–30, find equations for (a) the tangent plane and (b) the normal line at the point P_0 on the surface.

25. $2z - x^2 = 0, \quad P_0(2, 0, 2)$

26. $z - \ln(x^2 + y^2) = 0, \quad P_0(1, 0, 0)$

27. $x^2 + 2xy - y^2 + z^2 = 7, \quad P_0(1, -1, 3)$

28. $x - y^3 + zy - 2z^2 = 0, \quad P_0(-4, -2, 1)$

29. $\cos \pi x - x^2 y + e^{xz} + yz = 4, \quad P_0(0, 1, 2)$

30. $x^2 + y^2 - 2xy - x + 3y - z = -4, \quad P_0(2, -3, 18)$

In Exercises 31–34, find parametric equations for the line that is tangent to the curve of intersection of the surfaces at the given point.

31. Surfaces: $x + y^2 + 2z = 4, \quad x = 1$
    Point: $(1, 1, 1)$

32. Surfaces: $xyz = 1, \quad x^2 + 2y^2 + 3z^2 = 6$
    Point: $(1, 1, 1)$

33. Surfaces: $x^3 + 3x^2 y^2 + y^3 + 4xy - z^2 = 0,$
    $x^2 + y^2 + z^2 = 11$
    Point: $(1, 1, 3)$

34. Surfaces: $x^2 + y^2 = 4, \quad x^2 + y^2 - z = 0$
    Point: $(\sqrt{2}, \sqrt{2}, 4)$

In Exercises 35–38, sketch the curve $f(x, y) = c$ together with ∇f and the tangent line at the given point. Then write an equation for the tangent line.

35. $x^2 + y^2 = 4, \quad (\sqrt{2}, \sqrt{2})$

36. $x^2 - y = 1, \quad (\sqrt{2}, 1)$

37. $xy = -4, \quad (2, -2)$

38. $x^2 - xy + y^2 = 7, \quad (-1, 2)$ (This is the curve in Section 2.5, Example 4.)

39. By about how much will the value of $\cos \pi xy + xy^2$ change if (x, y) is moved a distance of $ds = 0.1$ unit from $(-1, -1)$ in the direction of $\mathbf{i} + \mathbf{j}$?

40. By about how much will the value of $\ln\sqrt{x^2 + y^2 + z^2}$ change if the point (x, y, z) is moved a distance of $ds = 0.1$ unit from $(3, 4, 12)$ in the direction of $3\mathbf{i} + 6\mathbf{j} - 2\mathbf{k}$?

41. By about how much will

$$f(x, y, z) = e^x\cos yz$$

change if the point $P(x, y, z)$ moves a distance of $ds = 0.1$ unit from the origin toward the point $(2, 1, -2)$?

42. By about how much will

$$f(x, y, z) = x + x\cos z - y\sin z + y$$

change if the point $P(x, y, z)$ moves from $P_0(2, -1, 0)$ a distance of $ds = 0.2$ unit toward the point $P_1(0, 1, 2)$?

43. In what two directions is the derivative of $f(x, y) = xy + y^2$ equal to zero at the point $(3, 2)$? (Fig. 12.49.)

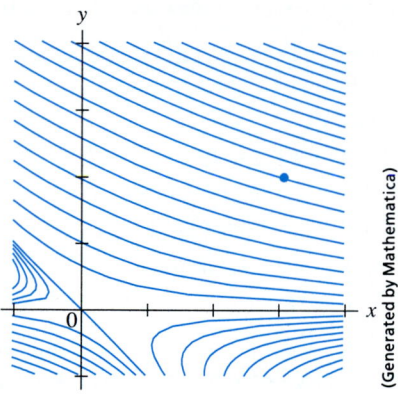

12.49   Level curves of $f(x, y) = xy + y^2$ (Exercise 43).

(Generated by Mathematica)

44. In what two directions is the derivative of $f(x, y) = (x^2 - y^2)/(x^2 + y^2)$ equal to zero at the point $(1, 1)$? (Fig. 12.50.)

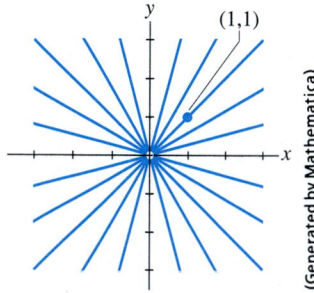

12.50   Level curves of $f(x, y) = (x^2 - y^2)/(x^2 + y^2)$ (Exercise 44).

(Generated by Mathematica)

45. *Change along a helix.*   Find the derivative of $f(x, y, z) = x^2 + y^2 + z^2$ in the direction of the unit tangent vector of the helix $\mathbf{r} = (\cos t)\mathbf{i} + (\sin t)\mathbf{j} + t\mathbf{k}$ at the points where

$t = -\pi/4, 0$, and $\pi/4$. The function f gives the square of the distance from a point $P(x, y, z)$ on the helix to the origin. The derivatives calculated here give the rates at which the square of the distance is changing with respect to t as P moves through the points where $t = -\pi/4$, 0, and $\pi/4$.

46. *Change along the involute of a circle.*   Find the derivative of $f(x, y) = x^2 + y^2$ in the direction of the unit tangent vector of the curve

$$\mathbf{r} = (\cos t + t\sin t)\mathbf{i} + (\sin t - t\cos t)\mathbf{j}, \quad t > 0$$

(Fig. 12.51).

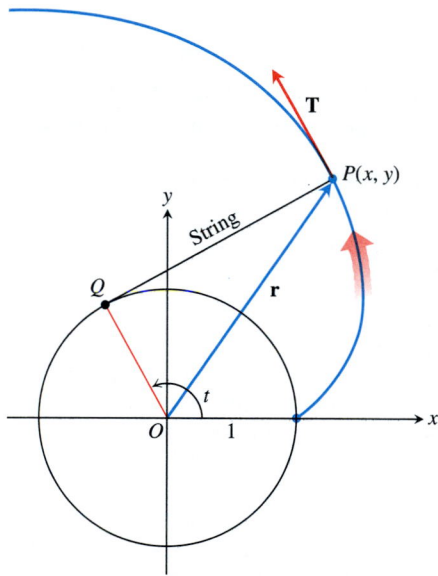

12.51   The involute of the unit circle from Section 11.3, Example 5. If you move out along the involute, covering distance along the curve at a steady rate, your distance from the origin will increase at a constant rate as well. (This is how to interpret the result of your calculation in Exercise 46.)

47. The derivative of $f(x, y)$ at $P_0(1, 2)$ in the direction of $\mathbf{i} + \mathbf{j}$ is $2\sqrt{2}$ and in the direction of $-2\mathbf{j}$ is -3. What is the derivative of f in the direction of $-\mathbf{i} - 2\mathbf{j}$?

48. The derivative of $f(x, y, z)$ at a point P is greatest in the direction of $\mathbf{A} = \mathbf{i} + \mathbf{j} - \mathbf{k}$. In this direction the value of the derivative is $2\sqrt{3}$.
a) Find ∇f at P.
b) Find the derivative of f at P in the direction of $\mathbf{i} + \mathbf{j}$.

49. *Normal curves and tangent curves.*   A curve is **normal** to a surface $f(x, y, z) = c$ at a point of intersection if the curve's velocity vector is a scalar multiple of ∇f at the point. The curve is **tangent** to the surface at a point of intersection if its velocity vector is orthogonal to ∇f there.

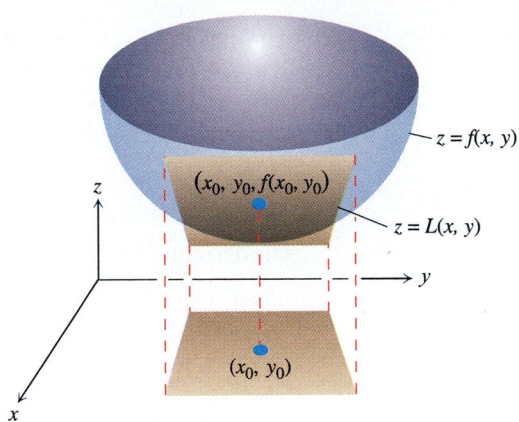

12.52 The graph of a function $z = f(x, y)$ and its linearization at a point (x_0, y_0). The plane defined by L is tangent to the surface at the point above the point (x_0, y_0). This furnishes a geometric explanation of why the values of L lie close to those of f in the immediate neighborhood of (x_0, y_0) (Exercise 50).

a) Show that the curve

$$\mathbf{r} = \sqrt{t}\,\mathbf{i} + \sqrt{t}\,\mathbf{j} - \frac{1}{4}(t + 3)\,\mathbf{k}$$

is normal to the surface $x^2 + y^2 - z = 3$ when $t = 1$.

b) Show that the curve

$$\mathbf{r} = \sqrt{t}\,\mathbf{i} + \sqrt{t}\,\mathbf{j} + (2t - 1)\,\mathbf{k}$$

is tangent to the surface $x^2 + y^2 - z = 1$ when $t = 1$.

50. *The linearization of $f(x, y)$ is a tangent–plane approximation.* Show that the tangent plane at the point $P_0(x_0, y_0, f(x_0, y_0))$ on the surface $z = f(x, y)$ defined by a differentiable function f is the plane

$$f_x(x_0, y_0)(x - x_0) + f_y(x_0, y_0)(y - y_0) - (z - f(x_0, y_0)) = 0$$

or

$$z = f(x_0, y_0) + f_x(x_0, y_0)(x - x_0) + f_y(x_0, y_0)(y - y_0).$$

Thus the tangent plane at P_0 is the graph of the linearization of f at P_0 (Fig. 12.52).

51. *Another way to see why gradients are normal to level curves.* Suppose that a differentiable function $f(x, y)$ has a constant value c along the differentiable curve $x = g(t)$, $y = h(t)$ for all values of t. Differentiate both sides of the equation $f(g(t), h(t)) = c$ with respect to t to show that ∇f is normal to the curve's tangent vector at every point.

52. *The algebra rules for gradients.* Given a constant k and the gradients

$$\nabla f = \frac{\partial f}{\partial x}\,\mathbf{i} + \frac{\partial f}{\partial y}\,\mathbf{j} + \frac{\partial f}{\partial z}\,\mathbf{k} \quad \text{and} \quad \nabla g = \frac{\partial g}{\partial x}\,\mathbf{i} + \frac{\partial g}{\partial y}\,\mathbf{j} + \frac{\partial g}{\partial z}\,\mathbf{k},$$

use the facts that

$$\frac{\partial}{\partial x}(kf) = k\frac{\partial f}{\partial x}, \qquad \frac{\partial}{\partial x}(f \pm g) = \frac{\partial f}{\partial x} \pm \frac{\partial g}{\partial x},$$

and so on, to establish the following rules:

a) $\nabla(kf) = k\nabla f$
b) $\nabla(f + g) = \nabla f + \nabla g$
c) $\nabla(f - g) = \nabla f - \nabla g$
d) $\nabla(fg) = f\nabla g + g\nabla f$

COMPUTER   Exercises 53–56 each give a function $f(x, y, z)$, a point P_0, and a vector $\mathbf{v}$. Use the gradient utility program at the end of the section to find ∇f at P_0, the derivative of f in the direction of $\mathbf{v}$, and equations for the tangent plane and normal line at P_0 on the level surface of f through P_0.

$f(x, y, z)$	P_0	$\mathbf{v}$
53. $4x^2y + y^2z$	$(1, 0, 1)$	$\mathbf{i} + \mathbf{j} + \mathbf{k}$
54. $xyz + y^3 - z$	$(1, 2, -1)$	$2\mathbf{i} - \mathbf{j} + 2\mathbf{k}$
55. $z\cos(x^2 + y^2)$	$(1, 1, 1)$	$2\mathbf{i} + 2\mathbf{j} - 2\mathbf{k}$
56. $x^3z + x^2y^2 + \sin yz$	$(1, 0, -1)$	$3\mathbf{i} - \mathbf{j} + 5\mathbf{k}$

✳ COMPUTER PROGRAM

A Gradient Utility Program

For versions of BASIC that accept multivariable functions, the following BASIC program carries out the standard calculations associated with gradients. Key in the program, including a formula for a differentiable function $f(x, y, z)$ in line 20. (To apply the program to a new function, you have only to change this one line.) Then run the program and enter, as prompted, the coordinates of a point P_0 and the components of a vector $\mathbf{v}$. In response to your entries, the program will display the following:

The gradient of f at P_0

The derivative of f at P_0 in the direction of $\mathbf{v}$

An equation for the plane tangent to the level surface of f through P_0

Equations for the line normal to the surface at P_0

The vector you enter for the direction need not be a unit vector. Any vector with the direction you want will do.

Here is a listing of the program with $f(x, y, z) = x^2 + y^2 + z$ in line 20.

PROGRAM	COMMENTS
10 DIM P(3), V(3), DF(3)	Define arrays to hold components of vectors
20 DEF FNF(X,Y, Z) = X^2 + Y^2 + Z	Define $f(x, y, z)$
30 PRINT "ENTER X − , Y − , Z − COORDINATES OF THE POINT P0."	Prompt the user for coordinates of a point
40 INPUT " USE COMMAS BETWEEN ", P(1),P(2),P(3)	and accept the input
50 PRINT "ENTER COMPONENTS OF A VECTOR THAT"	Prompt the user for components of a vector
60 INPUT" DEFINES THE DIRECTION ", V(1),V(2),V(3)	and accept the input
70 DEL = .01	Define delta value for computing derivatives
80 DF(1) = (FNF(P(1) + DEL,P(2),P(3)) − FNF(P(1) − DEL,P(2),P(3)))/DEL/2	Compute $\partial f/\partial x$
90 DF(2) = (FNF(P(1),P(2) + DEL,P(3)) − FNF(P(1),P(2) − DEL,P(3)))/DEL/2	then $\partial y/\partial x$
100 DF(3) = (FNF(P(1),P(2),P(3) + DEL) − FNF(P(1),P(2),P(3) − DEL))/DEL/2	and $\partial z/\partial x$
110 PRINT "COMPONENTS OF GRADIENT AT THE POINT"	Output line for components of
120 PRINT " ARE: ";DF(1);DF(2);DF(3)	gradient vector
130 PRINT	Print a blank line, cosmetic only
140 SUM = 0: FOR I = 1 TO 3: SUM = SUM + V(I)^2: NEXT I	Sum squares for
150 LENV = SQR(SUM)	the length of the vector
160 SUM = 0: FOR I = 1 TO 3: SUM = SUM + DF(I)*V(I)/LENV: NEXT I	Dot gradient with unit vector
170 PRINT "DERIVATIVE OF F AT P0 IN"	Output line for
180 PRINT " THE DIRECTION OF YOUR VECTOR IS ";SUM	directional derivative
190 PRINT	Print a blank line
200 SUM = 0: FOR I = 1 TO 3: SUM = SUM + P(I)*DF(I): NEXT I	Sum products of gradient components and coordinates
210 PRINT "THE PLANE TANGENT TO THE LEVEL"	
220 PRINT "SURFACE THROUGH P0 AT POINT PO IS"	
230 PRINT " ";DF(1);"X + ";DF(2);"Y + ";DF(3);"Z = ";SUM	
240 PRINT "THE LINE NORMAL TO THE LEVEL"	
250 PRINT "SURFACE THROUGH P0 AT POINT PO IS"	
260 PRINT " X = ";P(1);" + ";DF(1);"T"	
270 PRINT " Y = ";P(2);" + ";DF(2);"T"	
280 PRINT " Z = ";P(3);" + ";DF(3);"T"	
290 END	

With $f(x, y, z) = x^2 + y^2 + z$ in line 20, here is the screen output for $P_0 = (1, 2, 4)$, $\mathbf{v} = 2\,\mathbf{i} - 3\,\mathbf{j} + 5\,\mathbf{k}$ and $c = f(1, 2, 4) = (1)^2 + (2)^2 + 4 = 9$. The inputs have been highlighted.

```
ENTER X−, Y−, Z-COORDINATES OF THE POINT, P0
   USE COMMAS BETWEEN      1, 2, 4
ENTER COMPONENTS OF A VECTOR THAT
   DEFINES THE DIRECTION     2, −3, 5
COMPONENTS OF GRADIENT AT THE POINT
   ARE:     1.99995   3.999996   1.000023
DERIVATIVE OF F AT P0 IN
   THE DIRECTION OF YOUR VECTOR IS  − .4866601
TO FIND THE NORMAL LINE AND TANGENT PLANE
   AT P0 ON THE LEVEL SURFACE THROUGH P0,
   ENTER THE VALUE C OF F AT P0     9
THE EQUATION FOR THE PLANE IS
   1.99995 X + 3.999996 Y + 1.000023 Z = 14.00003
THE NORMAL LINE IS
   X = 1 + 1.99995 T
   Y = 2 + 3.999996 T
   Z = 4 + 1.000023 T
```

12.8 Maxima, Minima, and Saddle Points

As we mentioned at the end of Section 12.2, continuous functions defined on closed bounded regions in the xy-plane take on absolute maximum and minimum values on their domains (Figs. 12.53 and 12.54). It is important to be able to find these values and to know where they occur. For example, what is the highest temperature on a heated metal plate and where is it taken on? Where does a given surface attain its highest point above a given patch of the xy-plane? As we shall see in a moment, we can often answer questions like these by examining the partial derivatives of some function.

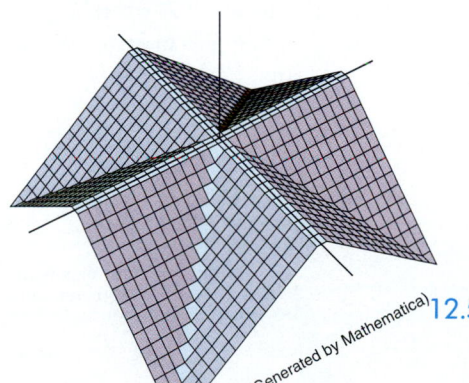

(Generated by Mathematica)

12.53 The function
$$z = (\cos x)(\cos y)e^{-\sqrt{x^2 + y^2}}$$
has a maximum value of 1 and a minimum value of about -0.067 on the square region $|x| \le 3\pi/2$, $|y| \le 3\pi/2$.

(Generated by Mathematica)

12.54 The "ridge" surface
$$z = \frac{1}{2}\left(\|x| - |y\| - |x| - |y|\right)$$
viewed from the point (10, 15, 20). The defining function has a maximum value of 0 and a minimum value of $-a$ on the square region $|x| \le a$, $|y| \le a$.

Our goal in this section is to show how to use partial derivatives to find the local maximum and minimum values of a continuous function $f(x, y)$ on a region R in the xy-plane. The first step is to use the function's first partial derivatives to make a (usually) short and comprehensive list of points where f can assume its local extreme values. What we do then depends on whether R is closed and bounded. If it is, we appeal to a theorem from advanced calculus that a continuous function on a closed, bounded region R assumes an absolute maximum value and an absolute minimum value on R, and we look through the list to find what these values are. If R is not closed and bounded, the function may not have absolute maximum and minimum values on R. However, we can still try to use the function's second partial derivatives to tell which points on the list, if any, give local maximum and minimum values. The routine is much the same as the routine for functions of a single variable. The main difference is that the first and second derivative tests now involve more derivatives.

The Derivative Tests

To find the extreme values of a continuous function of a single variable, we first look for points where the graph has a horizontal tangent line. At such a point we then look for a local maximum, a local minimum, or a point of inflection. We also examine the values of the function at the boundary points of its domain and at any points where the first derivative does not exist.

For a continuous function $f(x, y)$ of two independent variables, we look for points where the surface $z = f(x, y)$ has a horizontal tangent plane. At such points we then look for a local maximum, a local minimum, or a **saddle point** (Fig. 12.55). We also examine the values of the function at the boundary points in its domain and at any points where either f_x or f_y fails to exist.

We organize the search into three steps:

STEP 1:   *Make a list that includes the points where f has its local maxima and minima and evaluate f at all the points on the list.*

As we shall see later, the local maxima and minima of f can occur only at

i) boundary points of R,

ii) interior points of R where $f_x = f_y = 0$ (**the first derivative test**) and points where f_x or f_y fails to exist. (We call these the **critical points** of f.)

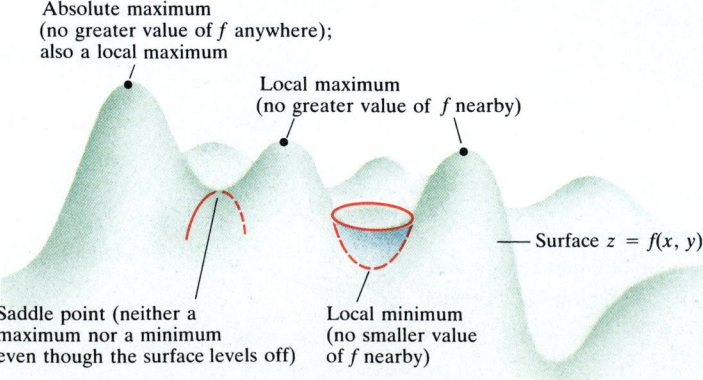

12.55 How values of $f(x, y)$ are classified. Saddle points are the three-dimensional analogues of points of inflection. Local values are often called *relative* values.

Absolute maximum
(no greater value of f anywhere);
also a local maximum

Local maximum
(no greater value of f nearby)

Surface $z = f(x, y)$

Saddle point (neither a maximum nor a minimum even though the surface levels off)

Local minimum
(no smaller value of f nearby)

Therefore, to make the list, we find the maximum and minimum values of f on the boundary of R and find the values of f at the critical points.

In theory, such a list could be long and might even include infinitely many points, but that is usually not the case in practice and definitely not the case in this book. The lists in the examples and exercises will be short.

STEP 2:   *If R is closed and bounded, look through the list for the maximum and minimum values of f. These will be the absolute maximum and minimum values of f on R.*

As we mentioned in the introduction, a function that is continuous on a closed bounded region of the plane has an absolute maximum value on the region and an absolute minimum value on the region. Since absolute maxima and minima are also local maxima and minima, the absolute maximum and minimum values of f already appear somewhere in the list we made in step 1. We have only to glance at the list to see what they are. If we then wish to learn which of the remaining values, if any, are local maxima and minima, we can go on to step 3.

STEP 3:   *If R is not closed and bounded, try the following second derivative test.*

The fact that $f_x = f_y = 0$ at an interior point (a, b) of R does not in itself guarantee that f will have an extreme value there. However, if f and its first and second partial derivatives are continuous on R, there is a second derivative test that may identify the behavior of f at (a, b). The **second derivative test** goes like this:

If $f_x(a, b) = f_y(a, b) = 0$, then

i) f has a **local maximum** at (a, b) if $f_{xx} < 0$ and $f_{xx}f_{yy} - f_{xy}^2 > 0$ at (a, b);

ii) f has a **local minimum** at (a, b) if $f_{xx} > 0$ and $f_{xx}f_{yy} - f_{xy}^2 > 0$ at (a, b);

iii) f has a **saddle point** at (a, b) if $f_{xx}f_{yy} - f_{xy}^2 < 0$ at (a, b).

iv) The test is *inconclusive* at (a, b) if $f_{xx}f_{yy} - f_{xy}^2 = 0$ at (a, b). We must find some other way to determine the behavior of f at (a, b).

The expression $f_{xx}f_{yy} - f_{xy}^2$ is called the **discriminant** of f. It is sometimes easier to remember in the determinate form

$$f_{xx}f_{yy} - f_{xy}^2 = \begin{vmatrix} f_{xx} & f_{xy} \\ f_{xy} & f_{yy} \end{vmatrix}.$$

The second derivative test is derived in Section 12.10.

We now look at examples that show these tests at work. After that, we show why the condition $f_x = f_y = 0$ is a necessary condition for having an extreme value at an interior point of the domain of a differentiable function.

The Tests at Work

In the first example we look at the function $f(x, y) = x^2 + y^2$, whose behavior we already know from looking at the formula: Its value is zero at the origin and increases steadily as (x, y) moves away from the origin. The point of Example 1 is to show how the derivative tests reveal this behavior.

Example 1   Find the extreme values of $f(x, y) = x^2 + y^2$.

Solution   The domain of f has no boundary points, for it is the entire plane (Fig. 12.56). The derivatives $f_x = 2x$ and $f_y = 2y$ exist everywhere. Therefore, local maxima and minima can occur only where

$$f_x = 2x = 0 \quad \text{and} \quad f_y = 2y = 0.$$

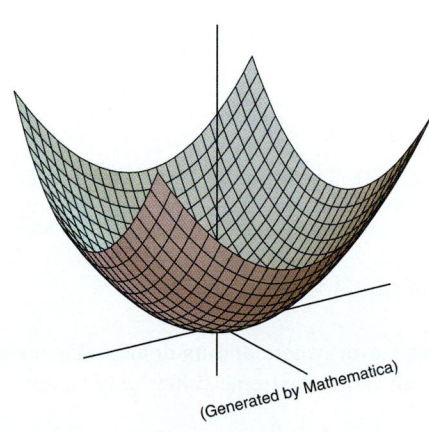

(Generated by Mathematica)

12.56   The graph of the function $f(x, y) = x^2 + y^2$ is the paraboloid $z = x^2 + y^2$. The function has only one extreme value, an absolute minimum value of 0 at the origin (Example 1).

The only possibility is the origin, where the value of f is zero. Since f is never negative, we see that zero is an absolute minimum.

We have not needed the second derivative test at all. Had we used it, we would have found

$$f_{xx} = 2, \qquad f_{yy} = 2, \qquad f_{xy} = 0$$

and

$$f_{xx} f_{yy} - f_{xy}^2 = (2)(2) - (0)^2 = 4 > 0,$$

identifying $(0, 0)$ as a local minimum. This in itself does not identify $(0, 0)$ as an absolute minimum. It takes more information to do that. ⌐

One virtue of the procedure we have described for finding extreme values is that it applies even to functions whose graphs are too complicated to draw. The next example illustrates this point.

Example 2   Find the extreme values of the function

$$f(x, y) = xy - x^2 - y^2 - 2x - 2y + 4.$$

Solution   The function is defined and differentiable for all x and y and its domain has no boundary points. The function therefore has extreme values only at the points where f_x and f_y are simultaneously zero. This leads to

$$f_x = y - 2x - 2 = 0, \qquad f_y = x - 2y - 2 = 0,$$

or

$$x = y = -2.$$

Therefore, the point $(-2, -2)$ is the only point where f may take on an extreme value. To see if it does so, we calculate

$$f_{xx} = -2, \qquad f_{yy} = -2, \qquad f_{xy} = 1.$$

The discriminant of f at $(a, b) = (-2, -2)$ is

$$f_{xx} f_{yy} - f_{xy}^2 = (-2)(-2) - (1)^2 = 4 - 1 = 3.$$

The combination

$$f_{xx} < 0 \qquad \text{and} \qquad f_{xx} f_{yy} - f_{xy}^2 > 0$$

tells us that f has a local maximum at $(-2, -2)$. The value of f at this point is $f(-2, -2) = 8$. ⌐

Example 3   Find the extreme values of $f(x, y) = xy$.

Solution   Since the function is differentiable everywhere and its domain has no boundary points (Fig. 12.57), the function can assume extreme values only where

$$f_x = y = 0 \qquad \text{and} \qquad f_y = x = 0.$$

Thus, the origin is the only point where f might have an extreme value. To see what happens there, we calculate

$$f_{xx} = 0, \qquad f_{yy} = 0, \qquad f_{xy} = 1.$$

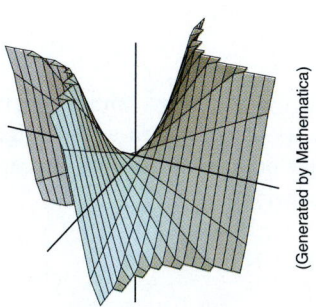

(Generated by Mathematica)

12.57   The surface $z = xy$ has a saddle point at the origin (Example 3).

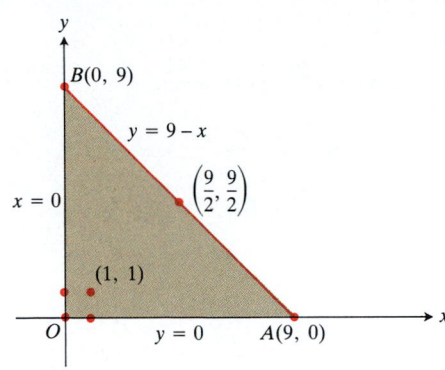

12.58 This triangular plate is the domain of the function in Example 4.

The discriminant,

$$f_{xx}f_{yy} - f_{xy}^2 = -1,$$

is negative. Therefore the function has a saddle point at $(0, 0)$. We conclude that $f(x, y) = xy$ assumes no extreme values at all.

Example 4   Find the absolute maximum and minimum values of

$$f(x, y) = 2 + 2x + 2y - x^2 - y^2$$

on the triangular plate in the first quadrant bounded by the lines $x = 0$, $y = 0$, $y = 9 - x$.

Solution   Since f is differentiable, the only places where f can assume extreme values are points on the boundary of the triangle (Fig. 12.58) and points inside the triangle where $f_x = f_y = 0$.

Boundary points. We take the triangle one side at a time:

1. On the segment OA, $y = 0$. The function

    $$f(x, y) = f(x, 0) = 2 + 2x - x^2$$

    may now be regarded as a function of x defined on the closed interval $0 \le x \le 9$. Its extreme values (we know from Chapter 3) may occur at the endpoints.

    $$x = 0 \qquad \text{where } f(0, 0) = 2,$$
    $$x = 9 \qquad \text{where } f(9, 0) = 2 + 18 - 81 = -61,$$

    and at the interior points where $f'(x, 0) = 2 - 2x = 0$. The only interior point where $f'(x, 0) = 0$ is $x = 1$, where

    $$f(x, 0) = f(1, 0) = 3.$$

2. On the segment OB, $x = 0$ and

    $$f(x, y) = f(0, y) = 2 + 2y - y^2.$$

    We know from the symmetry of f in x and y and from the analysis we just carried out that the candidates on this segment are

    $$f(0, 0) = 2, \qquad f(0, 9) = -61, \qquad f(0, 1) = 3.$$

3. We have already accounted for the values of f at the endpoints of AB, so we have only to look at the interior points of AB. With

    $$y = 9 - x,$$

    we have

    $$f(x, y) = 2 + 2x + 2(9 - x) - x^2 - (9 - x)^2 = -61 + 18x - 2x^2.$$

    Setting $f'(x, 9 - x) = 18 - 4x = 0$ gives

    $$x = \frac{18}{4} = \frac{9}{2}.$$

At this value of x,

$$y = 9 - \frac{9}{2} = \frac{9}{2}, \quad \text{and} \quad f(x, y) = f\left(\frac{9}{2}, \frac{9}{2}\right) = -\frac{41}{2}.$$

Interior points. For these we have

$$f_x = 2 - 2x = 0, \quad f_y = 2 - 2y = 0,$$

or

$$(x, y) = (1, 1),$$

where

$$f(1, 1) = 4.$$

Summary. We list all the candidates:

$$4, \quad 2, \quad -61, \quad 3, \quad -\frac{41}{2}.$$

The maximum is 4, which f assumes at $(1, 1)$. The minimum is -61, which f assumes at $(0, 9)$ and $(9, 0)$.

The Condition $f_x(a, b) = f_y(a, b) = 0$

The assertion that a function $f(x, y)$ with defined first partial derivatives can have an extreme value at an interior point of its domain only if f_x and f_y are both zero at that point is called the first derivative test for local extreme values.

THEOREM 6

The First Derivative Test for Local Extreme Values

If $f(x, y)$ has a local maximum or local minimum value at an interior point (a, b) of its domain where f_x and f_y are both defined, then f_x and f_y are both zero at (a, b).

Proof Suppose the value of f at (a, b) is a local maximum. Then

1. $x = a$ is an interior point of the domain of the curve $z = f(x, b)$ in which the plane $y = b$ cuts the surface $z = f(x, y)$ (Fig. 12.59);

2. the function $z = f(x, b)$ has a local maximum value at $x = a$;

3. the value of the derivative of $z = f(x, b)$ at $x = a$ is therefore zero (Theorem 1, Section 3.2).

Since this derivative is precisely $f_x(a, b)$, we conclude that $f_x(a, b) = 0$.

A similar argument with the function $z = f(a, y)$ shows that $f_y(a, b) = 0$.

This proves the theorem for local maximum values. To prove it for local minimum values, replace f by $-f$ and run through the argument again.

Despite the power of Theorem 6, we urge you to remember its limitations. It does not apply to points where either f_x or f_y fails to exist. It does not apply to boundary points of a function's domain. At a boundary point, a function can have an extreme value even if one or both of f_x and f_y are different from zero.

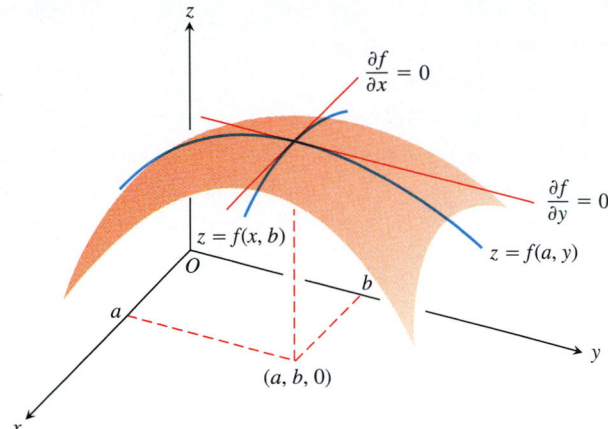

$$\frac{\partial f}{\partial x} = 0$$

$$\frac{\partial f}{\partial y} = 0$$

$z = f(x, b)$

$z = f(a, y)$

$(a, b, 0)$

12.59   The maximum of f occurs at $x = a$, $y = b$.

Summary of Max–Min Tests

The extreme values of $f(x, y)$ can occur only at

i) **boundary points** of the domain of f,

ii) **critical points** (interior points where $f_x = f_y = 0$ or points where f_x or f_y fails to exist).

If the first and second order partial derivatives of f are continuous throughout an open region containing a point (a, b) and $f_x(a, b) = f_y(a, b) = 0$, you may be able to classify (a, b) with the **second derivative test:**

i) $f_{xx} < 0$ and $f_{xx}f_{yy} - f_{xy}^2 > 0$ at (a, b) $\Rightarrow$ *local maximum,*

ii) $f_{xx} > 0$ and $f_{xx}f_{yy} - f_{xy}^2 > 0$ at (a, b) $\Rightarrow$ *local minimum,*

iii) $f_{xx}f_{yy} - f_{xy}^2 < 0$ at (a, b) $\Rightarrow$ *saddle point,*

iv) $f_{xx}f_{yy} - f_{xy}^2 = 0$ at (a, b) $\Rightarrow$ *test is inconclusive.*

EXERCISES 12.8

Find all maxima, minima, and saddle points of the functions in Exercises 1–30. Which, if any, of the maxima and minima are absolute?

1. $f(x, y) = x^2 + xy + y^2 + 3x - 3y + 4$

2. $f(x, y) = x^2 + 3xy + 3y^2 - 6x + 3y - 6$

3. $f(x, y) = 5xy - 7x^2 + 3x - 6y + 2$

4. $f(x, y) = 2xy - 5x^2 - 2y^2 + 4x + 4y - 4$

5. $f(x, y) = x^2 + xy + 3x + 2y + 5$

6. $f(x, y) = y^2 + xy - 2x - 2y + 2$

7. $f(x, y) = 2xy - 5x^2 - 2y^2 + 4x - 4$

8. $f(x, y) = 2xy - x^2 - 2y^2 + 3x + 4$

9. $f(x, y) = x^2 - 4xy + y^2 + 6y + 2$

10. $f(x, y) = 3x^2 + 6xy + 7y^2 - 2x + 4y$

11. $f(x, y) = 2x^2 + 3xy + 4y^2 - 5x + 2y$

12. $f(x, y) = 4x^2 - 6xy + 5y^2 - 20x + 26y$

13. $f(x, y) = x^2 - 4xy + y^2 + 5x - 2y$

14. $f(x, y) = x^2 - y^2 - 2x + 4y + 6$

15. $f(x, y) = x^2 - 2xy + 2y^2 - 2x + 2y + 1$

16. $f(x, y) = x^2 + 2xy$

17. $f(x, y) = 3 + 2x + 2y - 2x^2 - 2xy - y^2$

18. $f(x, y) = x^2 + xy + y^2 + x - 4y + 5$

19. $f(x, y) = x^3 - y^3 - 2xy + 6$

20. $f(x, y) = x^3 + y^3 + 3x^2 - 3y^2 - 8$

21. $f(x, y) = 6x^2 - 2x^3 + 3y^2 + 6xy$

22. $f(x, y) = 9x^3 + y^3/3 - 4xy$

23. $f(x, y) = x^3 + 3xy + y^3$

24. $f(x, y) = 4xy - x^4 - y^4$

25. $f(x, y) = \dfrac{1}{x^2 + y^2 - 1}$ **26.** $f(x, y) = \dfrac{1}{x} + xy + \dfrac{1}{y}$

27. $f(x, y) = \dfrac{-4}{1 + x^2 + y^2}$ **28.** $f(x, y) = \displaystyle\int_{x^2+2}^{y^2+5} \sqrt{t - 1}\, dt$

29. $f(x, y) = y \sin x$ **30.** $f(x, y) = e^{2x}\cos y$

In Exercises 31–38, find the absolute maxima and minima of the functions on the given domains.

31. $f(x, y) = 2x^2 - 4x + y^2 - 4y + 1$ on the closed triangular plate bounded by the lines $x = 0$, $y = 2$, $y = 2x$ in the first quadrant

32. $D(x, y) = x^2 - xy + y^2 + 1$ on the closed triangular plate in the first quadrant bounded by the lines $x = 0$, $y = 4$, $y = x$

33. $f(x, y) = x^2 + y^2$ on the closed triangular plate bounded by the lines $x = 0$, $y = 0$, $y + 2x = 2$ in the first quadrant

34. $T(x, y) = x^2 + xy + y^2 - 6x$ on the rectangular plate $0 \le x \le 5$, $-3 \le y \le 3$

35. $T(x, y) = x^2 + xy + y^2 - 6x + 2$ on the rectangular plate $0 \le x \le 5$, $-3 \le y \le 0$

36. $f(x, y) = 48xy - 32x^3 - 24y^2$ on the rectangular plate $0 \le x \le 1$, $0 \le y \le 1$

37. $f(x, y) = (4x - x^2)\cos y$ on the rectangular plate $1 \le x \le 3$, $-\pi/4 \le y \le \pi/4$ (Fig. 12.60)

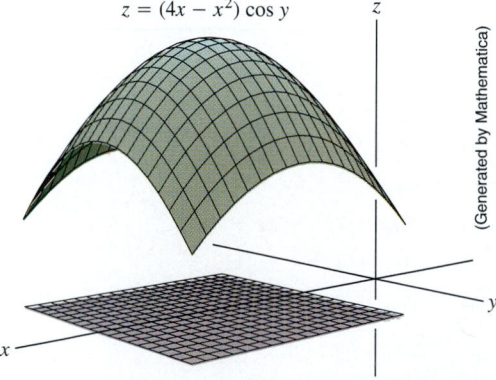

$z = (4x - x^2) \cos y$

12.60   The function and domain in Exercise 37.

38. $f(x, y) = 4x - 8xy + 2y + 1$ on the triangular plate bounded by the lines $x = 0$, $y = 0$, $x + y = 1$ in the first quadrant

39. *Temperatures.*   A flat circular plate has the shape of the region $x^2 + y^2 \le 1$. The plate, including the boundary where $x^2 + y^2 = 1$, is heated so that the temperature at the point (x, y) is

$$T(x, y) = x^2 + 2y^2 - x.$$

Find the temperatures at the hottest and coldest points on the plate (Fig. 12.61).

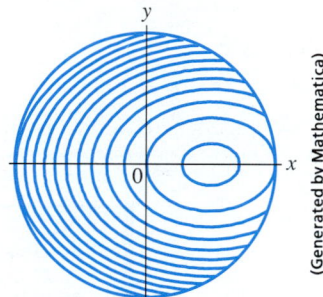

12.61   Curves of constant temperature are called isotherms. The figure shows isotherms of the temperature function $T(x, y) = x^2 + 2y^2 - x$ on the disk $x^2 + y^2 \le 1$ in the xy-plane. Exercise 39 asks you to locate the extreme temperatures.

40. Find the critical point of

$$f(x, y) = xy + 2x - \ln x^2 y$$

in the open first quadrant ($x > 0$, $y > 0$) and show that f takes on a minimum there (Fig. 12.62).

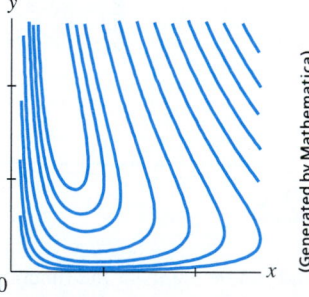

12.62   The function $f(x, y) = xy + 2x - \ln x^2 y$ takes on a minimum value somewhere in the open first quadrant $x > 0$, $y > 0$. Exercise 40 asks you to find where.

41. Find the maxima, minima, and saddle points of $f(x, y)$, if any, given that
   a) $f_x = 2x - 4y$ and $f_y = 2y - 4x$
   b) $f_x = 2x - 2$ and $f_y = 2y - 4$
   c) $f_x = 9x^2 - 9$ and $f_y = 2y + 4$

42. The discriminant $f_{xx}f_{yy} - f_{xy}^2$ is zero at the origin for each of the following functions, so the second derivative test fails there. Determine whether the function has a maximum, a minimum, or neither at the origin by imagining what the surface $z = f(x, y)$ looks like.
a) $f(x, y) = x^2y^2$ b) $f(x, y) = 1 - x^2y^2$
c) $f(x, y) = xy^2$ d) $f(x, y) = x^3y^2$
e) $f(x, y) = x^3y^3$ f) $f(x, y) = x^4y^4$

Extreme Values on Parametrized Curves

To find the extreme values of a function $f(x, y)$ on a curve $x = g(t)$, $y = h(t)$, we treat f as a function of the single variable t and use the Chain Rule to find where df/dt is zero. As in any other single-variable case, the extreme values of f are then found among the values at the
a) critical points (points where df/dt is zero or fails to exist), and
b) endpoints of the parameter domain.

43. Find the absolute maximum and minimum values of the following functions on
i) the quarter-circle $x^2 + y^2 = 4$ in the first quadrant,
ii) the half-circle $x^2 + y^2 = 4$, $y \geq 0$, and
iii) the full circle $x^2 + y^2 = 4$.
a) $f(x, y) = xy$ b) $f(x, y) = x + y$
c) $f(x, y) = 2x^2 + y^2$
Use the parametrization $x = 2 \cos t$, $y = 2 \sin t$.

44. Find the absolute maximum and minimum values, if any, of the function $f(x, y) = xy$ on the following.
a) The line $x = 2t$, $y = t + 1$
b) The line segment $x = 2t$, $y = t + 1$, $\quad -1 \leq t \leq 0$
c) The line segment $x = 2t$, $y = t + 1$, $\quad 0 \leq t \leq 1$

45. Find the absolute maximum and minimum values of the following functions on
i) the quarter-ellipse $(x^2/9) + (y^2/4) = 1$ in the first quadrant,
ii) the half-ellipse $(x^2/9) + (y^2/4) = 1$, $\quad y \geq 0$, and
iii) the full ellipse $(x^2/9) + (y^2/4) = 1$.
a) $f(x, y) = x^2 + 3y^2$ b) $f(x, y) = 2x + 3y$

46. Find the absolute maximum and minimum values of $f(x, y) = xy$ on the ellipse $x^2 + 4y^2 = 8$.

Least Squares and Regression Lines

When we try to fit a line $y = mx + b$ to a set of numerical data points (x_1, y_1), (x_2, y_2), . . . ,(x_n, y_n) (Fig. 12.63), we usually choose the line that minimizes the sum of the squares of the vertical distances from the points to the line. In theory, this means finding the values of m and b that minimize the value of the function

$$w = (mx_1 + b - y_1)^2 + \cdots + (mx_n + b - y_n)^2. \quad (1)$$

The values of m and b that do this are found with the first and

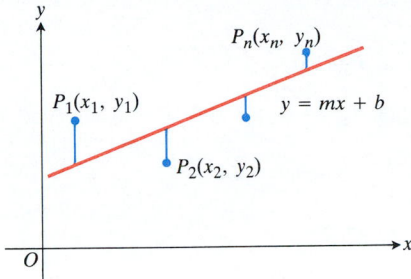

12.63   To fit a line to noncollinear points, we choose the line that minimizes the sum of the squares of the deviations.

second derivative tests to be

$$m = \frac{\left(\sum x_k\right)\left(\sum y_k\right) - n\sum x_k y_k}{\left(\sum x_k\right)^2 - n\sum x_k^2}, \quad (2)$$

$$b = \frac{1}{n}\left(\sum y_k - m\sum x_k\right), \quad (3)$$

with all sums running from $k = 1$ to $k = n$. Many scientific calculators have these formulas built in, enabling you to find m and b with only a few key presses after you have entered the data.

The line $y = mx + b$ determined by these values of m and b is called the **least squares line** or **regression line** for the data under study. Finding a least squares line lets you

1. summarize data with a simple expression,

2. predict values of y for other, experimentally untried values of x,

3. handle data analytically.

EXAMPLE   Find the least squares line for the points $(0, 1)$, $(1, 3)$, $(2, 2)$, $(3, 4)$, $(4, 5)$.

Solution   We organize the calculations in a table:

k	x_k	y_k	x_k^2	$x_k y_k$
1	0	1	0	0
2	1	3	1	3
3	2	2	4	4
4	3	4	9	12
5	4	5	16	20
Σ	10	15	30	39

Then we find

$$m = \frac{(10)(15) - 5(39)}{(10)^2 - 5(30)} = 0.9 \quad \binom{\text{Eq.(2) with } n = 5 \text{ and}}{\text{data from the table}}$$

and use the value of m to find

$$b = \frac{1}{5}(15 - (0.9)(10)) = 1.2. \quad \binom{\text{Eq. (3) with } n = 5,}{m = 0.9}$$

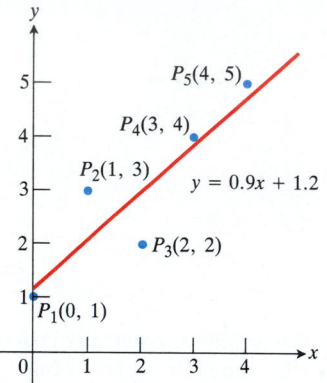

12.64   The least squares line for the data in the example.

The least squares line is $y = 0.9x + 1.2$ (Fig. 12.64).

In Exercises 47–50, use Eqs. (2) and (3) to find the least squares line for each set of data points. Then use the linear equation you obtain to predict the value of y that would correspond to $x = 4$.

47. $(-1, 2)$, $(0, 1)$, $(3, -4)$

48. $(-2, 0)$, $(0, 2)$, $(2, 3)$

49. $(0, 0)$, $(1, 2)$, $(2, 3)$

50. $(0, 1)$, $(2, 2)$, $(3, 2)$

51. CALCULATOR   Write a linear equation for the effect of irrigation on the yield of alfalfa by fitting a least squares line to the data in Table 12.1 (from the University of California Experimental Station, *Bulletin* No. 450, p. 8). Plot the data and draw the line.

52. CALCULATOR   *Craters on Mars.*   One theory of crater formation suggests that the frequency of large craters should fall off as the square of the diameter (Marcus, *Science,* June 21, 1968, p. 1334). Pictures from Mariner IV show the frequencies listed in Table 12.2. Fit a line of the form $F = m(1/D^2) + b$ to the data. Plot the data and draw the line.

TABLE 12.2
Crater sizes on Mars

Diameter in km, D	$1/D^2$ (for left value of class interval)	Frequency, F
32–45	0.001	53
45–64	0.0005	22
64–90	0.00025	14
90–128	0.000125	3

53. CALCULATOR   *Köchel numbers.*   In 1862, the German musicologist Ludwig von Köchel made a chronological list of the musical works of Wolfgang Amadeus Mozart. This list is the source of the Köchel numbers, or "K numbers," that now accompany the titles of Mozart's pieces (Sinfonia Concertante in E-flat major, K.364, for example). Table 12.3 gives the Köchel numbers and composition dates (y) of ten of Mozart's works.

a) Plot y vs. K to show that y is close to being a linear function of K.

b) Find a least squares line $y = mK + b$ for the data and add the line to your plot in (a).

c) Use the least squares line to estimate the year in which the Sinfonia Concertante was composed.

TABLE 12.1
Growth of alfalfa

x (total seasonal depth of water applied, in.)	y (average alfalfa yield, tons/acre)
12	5.27
18	5.68
24	6.25
30	7.21
36	8.20
42	8.71

TABLE 12.3
Compositions by Mozart

Köchel number K	Year Composed y
1	1761
75	1771
155	1772
219	1775
271	1777
351	1780
425	1783
503	1786
575	1789
626	1791

54. CALCULATOR *Submarine sinkings.*   The data in Table 12.4 show the results of an historical study of German submarines sunk by the U.S. Navy during 16 consecutive months of World War II. The data given for each month are the number of reported sinkings and the number of actual sinkings. The number of submarines sunk was slightly greater than the Navy's reports implied. Find a least squares line for estimating the number of actual sinkings from the number of reported sinkings.

TABLE 12.4
Sinkings of German submarines by U.S. during 16 consecutive months of WWII

Month	Guesses by U.S. (reported sinkings) x	Actual number y
1	3	3
2	2	2
3	4	6
4	2	3
5	5	4
6	5	3
7	9	11
8	12	9
9	8	10
10	13	16
11	14	13
12	3	5
13	4	6
14	13	19
15	10	15
16	16	15
	123	140

12.9 Lagrange Multipliers

As we saw in Section 12.8, we sometimes need to find the maximum and minimum values of functions whose domains are constrained to lie within some particular subset of the plane—a disk, for example, or a closed triangular region. But, as Fig. 12.65 suggests, functions may be subject to other kinds of constraints as well.

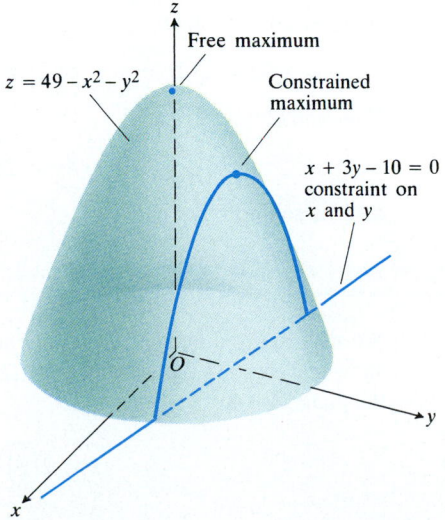

12.65   The function $f(x, y) = 49 - x^2 - y^2$, subject to the constraint $g(x, y) = x + 3y - 10 = 0$.

In this section, we explore a powerful method for finding the maxima and minima of constrained functions: the method of **Lagrange multipliers.** Lagrange developed the method in 1755 to solve sophisticated max–min problems in geometry. Today the method is important in economics, in engineering (where it is used in designing multistage rockets, for example), and in mathematics itself.

We begin with two examples and then describe the method in general terms and look at more examples.

Constrained Maxima and Minima

Example 1   Find the point $P(x, y, z)$ on the plane

$$2x + y - z - 5 = 0$$

that lies closest to the origin.

Solution   The problem asks us to find the minimum value of the function

$$|\overrightarrow{OP}| = \sqrt{(x - 0)^2 + (y - 0)^2 + (z - 0)^2} = \sqrt{x^2 + y^2 + z^2}$$

subject to the constraint that

$$2x + y - z - 5 = 0.$$

Since $|\overrightarrow{OP}|$ has a minimum value wherever the function

$$f(x, y, z) = x^2 + y^2 + z^2$$

has a minimum value, we may solve the problem by finding the minimum value of $f(x, y, z)$ subject to the constraint $2x + y - z - 5 = 0$. If we regard x and y as the independent variables in this equation and write z as

$$z = 2x + y - 5,$$

our problem reduces to one of finding the points (x, y) at which the function

$$h(x, y) = f(x, y, 2x + y - 5) = x^2 + y^2 + (2x + y - 5)^2$$

has its minimum value or values. Since the domain of h is the entire xy-plane, the first derivative test of Section 12.8 tells us that any minima that h might have must occur at points where

$$h_x = 2x + 2(2x + y - 5)(2) = 0, \qquad h_y = 2y + 2(2x + y - 5) = 0.$$

This leads to

$$10x + 4y = 20, \qquad 4x + 4y = 10,$$

and the solution

$$x = \frac{5}{3}, \qquad y = \frac{5}{6}.$$

We may apply a geometric argument together with the second derivative test to show that these values minimize h. The z-coordinate of the corresponding point on the plane $z = 2x + y - 5$ is

$$z = 2\left(\frac{5}{3}\right) + \frac{5}{6} - 5 = -\frac{5}{6}.$$

Therefore, the point we seek is

$$\text{Closest point:} \qquad P\left(\frac{5}{3}, \frac{5}{6}, -\frac{5}{6}\right).$$

Attempts to solve a constrained maximum or minimum problem by substitution, as we might call the method of Example 1, do not always go smoothly, as the next example shows. This is one of the reasons for learning the new method of this section, which does not require us to decide in advance which of the constrained variables to regard as independent.

Example 2

Find the points on the hyperbolic cylinder

$$x^2 - z^2 - 1 = 0$$

closest to the origin.

Solution 1 The cylinder is shown in Fig. 12.66. We seek the points on the cylinder closest to the origin. These are the points whose coordinates minimize the value of the function

$$f(x, y, z) = x^2 + y^2 + z^2 \qquad \text{(Square of the distance)}$$

subject to the constraint that $x^2 - z^2 - 1 = 0$. If we regard x and y as independent variables in the constraint equation, then

$$z^2 = x^2 - 1$$

and the values of $f(x, y, z) = x^2 + y^2 + z^2$ on the cylinder are given by the function

$$h(x, y) = x^2 + y^2 + (x^2 - 1) = 2x^2 + y^2 - 1.$$

To find the points on the cylinder whose coordinates minimize f, we look for the points in the xy-plane whose coordinates minimize h. The only extreme value of h occurs where

$$h_x = 4x = 0 \qquad \text{and} \qquad h_y = 2y = 0,$$

that is, at the point $(0, 0)$. But now we're in trouble—there are no points on the cylinder where both x and y are zero. What went wrong?

What happened was that the first derivative test found (as it should have) the point *in the domain of* h where h has a minimum value. We, on the other hand, want the points *on the cylinder* where h has a minimum value. While the domain of h is the entire xy-plane, the domain from which we can select the first two coordinates of the points (x, y, z) on the cylinder is restricted to the "shadow" of the cylinder on the xy-plane; it does not include the band between the lines $x = -1$ and $x = 1$ (Fig. 12.67).

We can avoid this problem if we treat y and z as independent variables (instead of x and y) and express x in terms of y and z as

$$x^2 = z^2 + 1.$$

With this substitution, $f(x, y, z) = x^2 + y^2 + z^2$ becomes

$$k(y, z) = (z^2 + 1) + y^2 + z^2 = 1 + y^2 + 2z^2$$

and we look for the points where k takes on its smallest value. The domain of k in

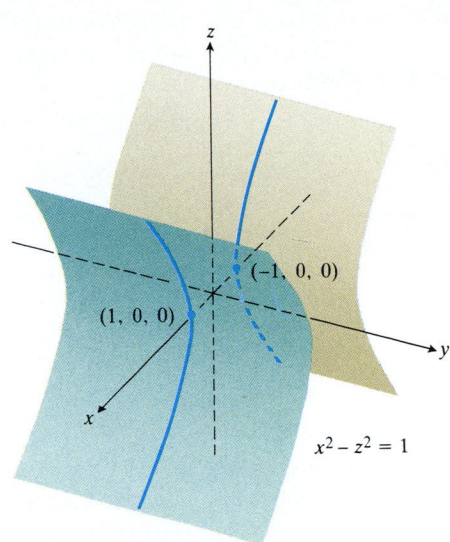

12.66 The hyperbolic cylinder $x^2 - z^2 - 1 = 0$ in Example 2.

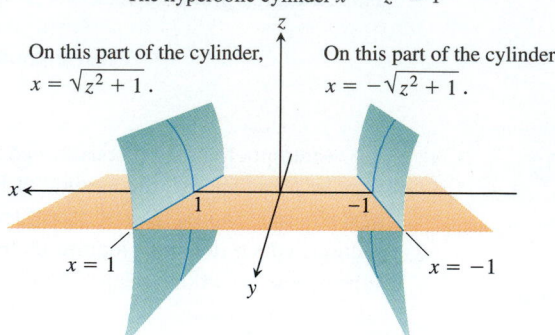

The hyperbolic cylinder $x^2 - z^2 = 1$

On this part of the cylinder, $x = \sqrt{z^2 + 1}$.

On this part of the cylinder, $x = -\sqrt{z^2 + 1}$.

$x = 1$ $x = -1$

12.67   The region in the xy-plane from which the first two coordinates of the points (x, y, z) on the hyperbolic cylinder $x^2 - z^2 = 1$ are selected excludes the band $-1 < x < 1$ in the xy-plane.

the yz-plane now matches the domain from which we select the y- and z-coordinates of the points (x, y, z) on the cylinder. Hence, the points that minimize k in the plane will have corresponding points on the cylinder. The smallest values of k occur where

$$k_y = 2y = 0 \qquad \text{and} \qquad k_z = 4z = 0,$$

or where $y = z = 0$. This leads to

$$x^2 = z^2 + 1 = 1, \qquad x = \pm 1.$$

The corresponding points on the cylinder are $(\pm 1, 0, 0)$. We can see from the inequality

$$k(y, z) = 1 + y^2 + 2z^2 \geq 1$$

that the points $(\pm 1, 0, 0)$ give a minimum value for k. We can also see that the minimum distance from the origin to a point on the cylinder is 1 unit.

Solution 2   Another way to find the points on the cylinder closest to the origin is to imagine a small sphere centered at the origin expanding like a soap bubble until it just touches the cylinder (Fig. 12.68). At each point of contact, the cylinder and sphere have the same tangent plane and normal line. Therefore, if the sphere and cylinder are represented as the level surfaces obtained by setting

$$f(x, y, z) = x^2 + y^2 + z^2 - a^2 \qquad \text{and} \qquad g(x, y, z) = x^2 - z^2 - 1$$

equal to 0, then the gradients ∇f and ∇g will be parallel where the surfaces touch. At any point of contact we will therefore be able to find a scalar λ ("lambda") such that

$$\nabla f = \lambda \nabla g,$$

or

$$2x\mathbf{i} + 2y\mathbf{j} + 2z\mathbf{k} = \lambda(2x\mathbf{i} - 2z\mathbf{k}).$$

Thus, the coordinates $x, y,$ and z of any point of tangency will have to satisfy the three scalar equations

$$2x = 2\lambda x, \qquad 2y = 0, \qquad 2z = -2\lambda z. \tag{1}$$

For what values of λ will a point (x, y, z) whose coordinates satisfy the equations in (1) also lie on the surface $x^2 - z^2 - 1 = 0$? To answer this question, we use

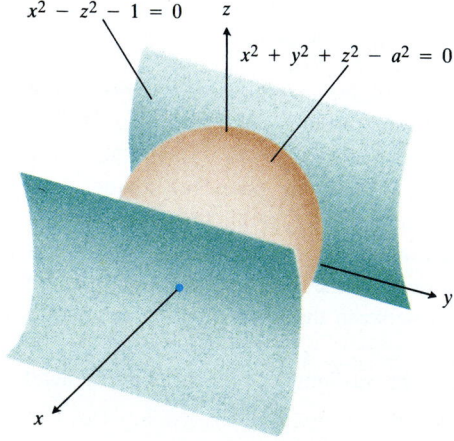

$x^2 - z^2 - 1 = 0$

$x^2 + y^2 + z^2 - a^2 = 0$

12.68   The sphere obtained by expanding a soap bubble centered at the origin until it just touches the hyperbolic cylinder $x^2 - z^2 - 1 = 0$. See Solution 2 of Example 2.

the fact that no point on the surface has a zero x-coordinate to conclude that $x \neq 0$ in the first equation in (1). This means that $2x = 2\lambda x$ only if

$$2 = 2\lambda \qquad \text{or} \qquad \lambda = 1.$$

For $\lambda = 1$, the equation $2z = -2\lambda z$ becomes $2z = -2z$. If this equation is to be satisfied as well, z must be zero. Since $y = 0$ also (from the equation $2y = 0$), we conclude that the points we seek all have coordinates of the form

$$(x, 0, 0).$$

What points on the surface $x^2 - z^2 = 1$ have coordinates of this form? The points $(x, 0, 0)$ for which

$$x^2 - (0)^2 = 1, \qquad x^2 = 1, \qquad \text{or} \qquad x = \pm 1.$$

The points on the cylinder closest to the origin are the points $(\pm 1, 0, 0)$.

The Method of Lagrange Multipliers

In Solution 2 of Example 2, we solved the problem by the **method of Lagrange multipliers.** In general terms, the method says that the extreme values of a function $f(x, y, z)$ whose variables are subject to a constraint $g(x, y, z) = 0$ are to be found on the surface $g = 0$ at the points where

$$\nabla f = \lambda \nabla g$$

for some scalar λ (called a **Lagrange multiplier**).

To explore the method further and see why it works, we first make the following observation, which we state as a theorem.

THEOREM 7

The Orthogonal Gradient Theorem

Suppose that $f(x, y, z)$ is differentiable in a region whose interior contains a differentiable curve

$$C: \quad \mathbf{r} = g(t)\,\mathbf{i} + h(t)\,\mathbf{j} + k(t)\,\mathbf{k}.$$

If P_0 is a point on C where f has a local maximum or minimum relative to its values on C, then ∇f is orthogonal to C at P_0.

Proof   We show that ∇f is orthogonal to the curve's velocity vector at P_0. The values of f on C are given by the composite $f(g(t), h(t), k(t))$, whose derivative with respect to t is

$$\frac{df}{dt} = \frac{\partial f}{\partial x}\frac{dg}{dt} + \frac{\partial f}{\partial y}\frac{dh}{dt} + \frac{\partial f}{\partial z}\frac{dk}{dt} = \nabla f \cdot \mathbf{v}.$$

At any point P_0 where f has a local maximum or minimum relative to its values on the curve, $df/dt = 0$, so

$$\nabla f \cdot \mathbf{v} = 0.$$

By dropping the z-terms in Theorem 7, we obtain a similar result for functions of two variables.

COROLLARY OF THEOREM 7

At the points on a differentiable curve $\mathbf{r} = g(t)\,\mathbf{i} + h(t)\,\mathbf{k}$ where a differentiable function $f(x, y)$ takes on its local maxima and minima relative to its values on the curve, $\nabla f \cdot \mathbf{v} = 0$.

Theorem 7 is the key to why the method of Lagrange multipliers works, as we shall now see. Suppose that $f(x, y, z)$ and $g(x, y, z)$ are differentiable and that P_0 is a point on the surface $g(x, y, z) = 0$ where f has a local maximum or minimum value relative to its other values on the surface. Then f takes on a local maximum or minimum at P_0 relative to its values on every differentiable curve through P_0 on the surface $g(x, y, z) = 0$. Therefore, ∇f is orthogonal to the velocity vector of every such differentiable curve through P_0. But so is ∇g (because ∇g is orthogonal to the level surface $g = 0$, as we saw in Section 12.7). Therefore, at P_0, ∇f is some scalar multiple λ of ∇g.

The Method of Lagrange Multipliers

Suppose that $f(x, y, z)$ and $g(x, y, z)$ are differentiable. To find the local maximum and minimum values of f subject to the constraint $g(x, y, z) = 0$, find the values of x, y, z, and λ that simultaneously satisfy the equations

$$\nabla f = \lambda \nabla g \qquad \text{and} \qquad g(x, y, z) = 0. \tag{2}$$

For functions of two independent variables, the appropriate equations are

$$\nabla f = \lambda \nabla g \qquad \text{and} \qquad g(x, y) = 0. \tag{3}$$

Example 3   Find the greatest and smallest values that the function

$$f(x, y) = xy$$

takes on the ellipse (Fig. 12.69)

$$\frac{x^2}{8} + \frac{y^2}{2} = 1.$$

Solution   We are asked to find the extreme values of $f(x, y) = xy$ subject to the constraint

$$g(x, y) = \frac{x^2}{8} + \frac{y^2}{2} - 1 = 0.$$

To do so, we first find the values of x, y, and λ for which

$$\nabla f = \lambda \nabla g \qquad \text{and} \qquad g(x, y) = 0.$$

The gradient equation gives

$$y\mathbf{i} + x\mathbf{j} = \frac{\lambda}{4} x\mathbf{i} + \lambda y\mathbf{j},$$

from which we find

$$y = \frac{\lambda}{4} x, \qquad x = \lambda y, \qquad \text{and} \qquad y = \frac{\lambda}{4}(\lambda y) = \frac{\lambda^2}{4} y,$$

so that $y = 0$ or $\lambda = \pm 2$. We now consider these two cases.

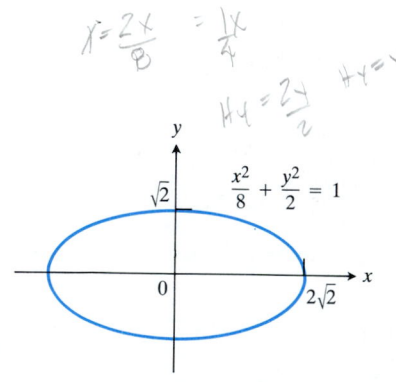

12.69   Example 3 shows how to find the largest and smallest values of the product xy on this ellipse.

CASE 1:   If $y = 0$, then $x = y = 0$, but the point $(0, 0)$ is not on the ellipse. Hence, $y \neq 0$.

CASE 2:   If $y \neq 0$, then $\lambda = \pm 2$ and $x = \pm 2y$. Substituting this in the equation $g(x, y) = 0$ gives

$$\frac{(\pm 2y)^2}{8} + \frac{y^2}{2} = 1, \qquad 4y^2 + 4y^2 = 8, \qquad \text{and} \qquad y = \pm 1.$$

The function $f(x, y) = xy$ therefore takes on its extreme values on the ellipse at the four points $(\pm 2, 1)$, $(\pm 2, -1)$. The extreme values are $xy = 2$ and $xy = -2$.

The Geometry of the Solution   The level curves of the function $f(x, y) = xy$ are the hyperbolas $xy = c$ (Fig. 12.70). The farther the hyperbolas lie from the origin, the larger the absolute value of f. We want to find the extreme values of $f(x, y)$, given that the point (x, y) also lies on the ellipse $x^2 + 4y^2 = 8$. Which hyperbolas intersecting the ellipse lie farthest from the origin? The hyperbolas that just graze the ellipse, the ones that are tangent to it. At these points, any vector normal to the hyperbola is normal to the ellipse, so the gradient $\nabla f = y\mathbf{i} + x\mathbf{j}$ is a multiple $(\lambda = \pm 2)$ of the gradient $\nabla g = (x/4)\mathbf{i} + y\mathbf{j}$. At the point $(2, 1)$, for example,

$$\nabla f = \mathbf{i} + 2\mathbf{j}, \qquad \nabla g = \frac{1}{2}\mathbf{i} + \mathbf{j}, \qquad \text{and} \qquad \nabla f = 2\nabla g.$$

At the point $(-2, 1)$,

$$\nabla f = \mathbf{i} - 2\mathbf{j}, \qquad \nabla g = -\frac{1}{2}\mathbf{i} + \mathbf{j}, \qquad \text{and} \qquad \nabla f = -2\nabla g.$$

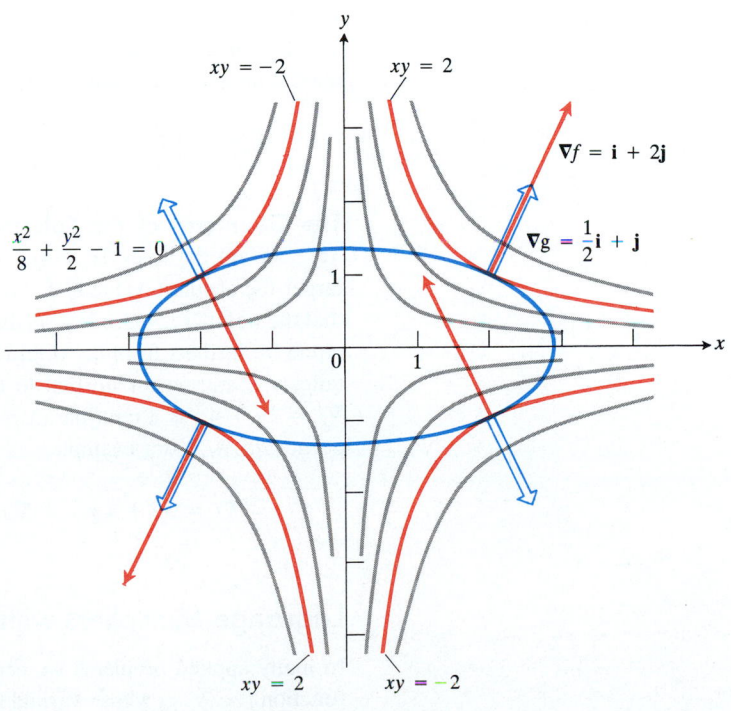

12.70   When subjected to the constraint $g(x, y) = x^2/8 + y^2/2 - 1 = 0$, the function $f(x, y) = xy$ takes on extreme values at the four points $(\pm 2, \pm 1)$. These are the points on the ellipse where ∇f (red) is a scalar multiple of ∇g (blue).

Example 4 Find the maximum and minimum values of the function $f(x, y) = 3x + 4y$ on the circle $x^2 + y^2 = 1$.

Solution We model this as a Lagrange multiplier problem with

$$f(x, y) = 3x + 4y, \qquad g(x, y) = x^2 + y^2 - 1$$

and look for the values of x, y, and λ that satisfy the equations

$$\nabla f = \lambda \nabla g: \quad 3\mathbf{i} + 4\mathbf{j} = 2x\lambda\mathbf{i} + 2y\lambda\mathbf{j},$$

$$g(x, y) = 0: \quad x^2 + y^2 - 1 = 0.$$

(4)

The gradient equation implies that $\lambda \neq 0$ and gives

$$x = \frac{3}{2\lambda}, \qquad y = \frac{2}{\lambda}.$$

These equations tell us, among other things, that x and y have the same sign. With these values for x and y, the equation $g(x, y) = 0$ gives

$$\left(\frac{3}{2\lambda}\right)^2 + \left(\frac{2}{\lambda}\right)^2 - 1 = 0,$$

so

$$\frac{9}{4\lambda^2} + \frac{4}{\lambda^2} = 1, \qquad 9 + 16 = 4\lambda^2, \qquad 4\lambda^2 = 25, \qquad \text{and} \qquad \lambda = \pm\frac{5}{2}.$$

Thus,

$$x = \frac{3}{2\lambda} = \pm\frac{3}{5}, \qquad y = \frac{2}{\lambda} = \pm\frac{4}{5},$$

and $f(x, y) = 3x + 4y$ has extreme values at $(x, y) = \pm(3/5, 4/5)$. (There are two points instead of four because x and y have the same sign.)

By calculating the value of $3x + 4y$ at the points $\pm(3/5, 4/5)$, we see that its maximum and minimum values on the circle $x^2 + y^2 = 1$ are

$$3\left(\frac{3}{5}\right) + 4\left(\frac{4}{5}\right) = \frac{25}{5} = 5 \qquad \text{and} \qquad 3\left(-\frac{3}{5}\right) + 4\left(-\frac{4}{5}\right) = -\frac{25}{5} = -5.$$

The Geometry of the Solution (Fig. 12.71) The level curves of $f(x, y) = 3x + 4y$ are the lines $3x + 4y = c$. The farther the lines lie from the origin, the larger the absolute value of f. We want to find the extreme values of $f(x, y)$ given that the point (x, y) also lies on the circle $x^2 + y^2 = 1$. Which lines intersecting the circle lie farthest from the origin? The lines tangent to the circle. At the points of tangency, any vector normal to the line is normal to the circle, so the gradient $\nabla f = 3\mathbf{i} + 4\mathbf{j}$ is a multiple ($\lambda = \pm 5/2$) of the gradient $\nabla g = 2x\mathbf{i} + 2y\mathbf{j}$. At the point (3/5, 4/5), for example,

$$\nabla f = 3\mathbf{i} + 4\mathbf{j}, \qquad \nabla g = \frac{6}{5}\mathbf{i} + \frac{8}{5}\mathbf{j}, \qquad \text{and} \qquad \nabla f = \frac{5}{2}\nabla g.$$

Lagrange Multipliers with Two Constraints

In many applied problems we need to find the extreme values of a differentiable function $f(x, y, z)$ whose variables are subject to two constraints.

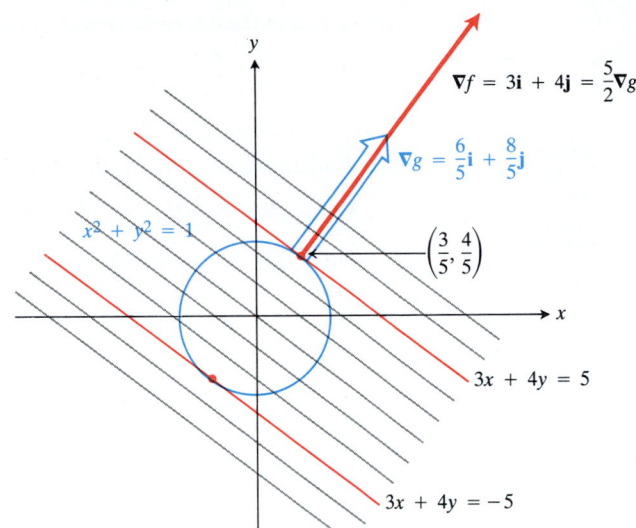

$$\nabla f = 3\mathbf{i} + 4\mathbf{j} = \tfrac{5}{2}\nabla g$$

$$\nabla g = \tfrac{6}{5}\mathbf{i} + \tfrac{8}{5}\mathbf{j}$$

$$x^2 + y^2 = 1$$

$$\left(\tfrac{3}{5}, \tfrac{4}{5}\right)$$

$$3x + 4y = 5$$

$$3x + 4y = -5$$

12.71 The function $f(x, y) = 3x + 4y$ takes on its largest value on the unit circle $g(x, y) = x^2 + y^2 - 1 = 0$ at the point $(3/5, 4/5)$ and its smallest value at the point $(-3/5, -4/5)$. At each of these points, ∇f is a scalar multiple of ∇g. The figure shows the gradients at the first point but not the second.

If the constraints on x, y, and z are

$$g_1(x, y, z) = 0 \qquad \text{and} \qquad g_2(x, y, z) = 0$$

and g_1 and g_2 are differentiable with ∇g_1 not parallel to ∇g_2, we find the constrained local maxima and minima of f by introducing two Lagrange multipliers λ and μ (mu, pronounced "mew"). That is, we locate the points $P(x, y, z)$ where f takes on its constrained extreme values by finding the values of x, y, z, λ, and μ that simultaneously satisfy the equations

$$\nabla f = \lambda\nabla g_1 + \mu\nabla g_2, \qquad g_1(x, y, z) = 0, \qquad g_2(x, y, z) = 0. \tag{5}$$

The equations in (5) have a nice geometric interpretation. The surfaces $g_1 = 0$ and $g_2 = 0$ (usually) intersect in a differentiable curve, say C (Fig. 12.72), and along this curve we seek the points where f has local maximum and minimum values relative to its other values on the curve. These are the points where ∇f is normal to C, as we saw in Theorem 7. But ∇g_1 and ∇g_2 are also normal to C at these points because C lies in the surfaces $g_1 = 0$ and $g_2 = 0$. Therefore ∇f lies in the plane determined by ∇g_1 and ∇g_2, which means that

$$\nabla f = \lambda\nabla g_1 + \mu\nabla g_2$$

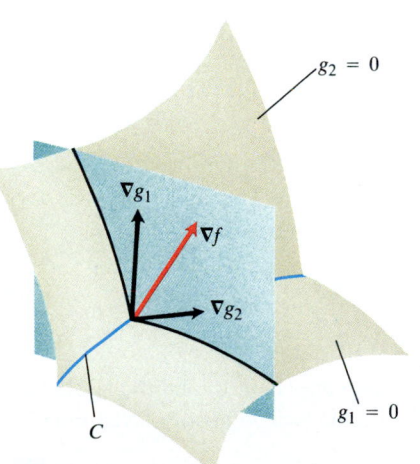

12.72 The vectors ∇g_1 and ∇g_2 lie in a plane perpendicular to the curve C because ∇g_1 is normal to the surface $g_1 = 0$ and ∇g_2 is normal to the surface $g_2 = 0$.

for some λ and μ. Since the points we seek also lie in both surfaces, their coordinates must satisfy the equations $g_1(x, y, z) = 0$ and $g_2(x, y, z) = 0$, which are the remaining requirements in Eq. (5).

Example 5 The plane $x + y + z = 1$ cuts the cylinder $x^2 + y^2 = 1$ in an ellipse (Fig. 12.73). Find the points on the ellipse that lie closest to and farthest from the origin.

Solution We model this as a Lagrange multiplier problem in which we find the extreme values of

$$f(x, y, z) = x^2 + y^2 + z^2$$

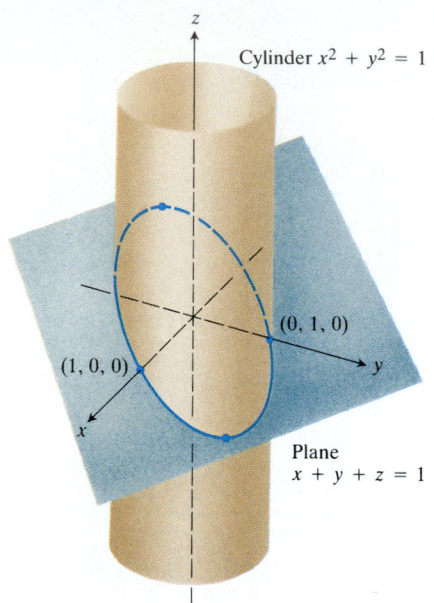

Cylinder $x^2 + y^2 = 1$

$(0, 1, 0)$

$(1, 0, 0)$

Plane
$x + y + z = 1$

12.73   On the ellipse where the plane and cylinder meet, what are the points closest to and farthest from the origin (Example 5)?

(the square of the distance from (x, y, z) to the origin) subject to the two constraints

$$g_1(x, y, z) = x^2 + y^2 - 1 = 0, \tag{6}$$

$$g_2(x, y, z) = x + y + z - 1 = 0. \tag{7}$$

The gradient equation in (5) then gives

$$\nabla f = \lambda \nabla g_1 + \mu \nabla g_2 \qquad \text{(Eq. (5))}$$

$$2x\mathbf{i} + 2y\mathbf{j} + 2z\mathbf{k} = \lambda(2x\mathbf{i} + 2y\mathbf{j}) + \mu(\mathbf{i} + \mathbf{j} + \mathbf{k}) \tag{8}$$

$$2x\mathbf{i} + 2y\mathbf{j} + 2z\mathbf{k} = (2\lambda x + \mu)\mathbf{i} + (2\lambda y + \mu)\mathbf{j} + \mu\mathbf{k} \tag{9}$$

or

$$2x = 2\lambda x + \mu, \qquad 2y = 2\lambda y + \mu, \qquad 2z = \mu. \tag{10}$$

The scalar equations in (10) yield

$$\begin{aligned} 2x = 2\lambda x + 2z \quad &\Rightarrow \quad (1 - \lambda)x = z, \\ 2y = 2\lambda y + 2z \quad &\Rightarrow \quad (1 - \lambda)y = z. \end{aligned} \tag{11}$$

Equations (11) are satisfied simultaneously if either $\lambda = 1$ and $z = 0$ or if $\lambda \neq 1$ and $x = y = z/(1 - \lambda)$.

If $z = 0$, then solving Eqs. (6) and (7) simultaneously to find the corresponding points on the ellipse gives the two points $(1, 0, 0)$ and $(0, 1, 0)$. This makes sense when you look at Fig. 12.73.

If $x = y$, then Eqs. (6) and (7) give

$$\begin{aligned} x^2 + x^2 - 1 &= 0 \qquad & x + x + z - 1 &= 0 \\ 2x^2 &= 1 \qquad & z &= 1 - 2x \tag{12} \\ x &= \pm\frac{\sqrt{2}}{2} \qquad & z &= 1 \mp \sqrt{2}. \end{aligned}$$

The corresponding points on the ellipse are

$$\left(\frac{\sqrt{2}}{2}, \frac{\sqrt{2}}{2}, 1 - \sqrt{2}\right) \qquad \text{and} \qquad \left(-\frac{\sqrt{2}}{2}, -\frac{\sqrt{2}}{2}, 1 + \sqrt{2}\right). \tag{13}$$

Again this makes sense when you look at Fig. 12.73.

The points on the ellipse closest to the origin are $(1, 0, 0)$ and $(0, 1, 0)$. The points on the ellipse farthest from the origin are the two points displayed in (13).

EXERCISES 12.9

1. Find the points on the ellipse $x^2 + 2y^2 = 1$ where $f(x, y) = xy$ has its extreme values.

2. Find the extreme values of $f(x, y) = xy$ subject to the constraint $g(x, y) = x^2 + y^2 - 10 = 0$.

3. Find the maximum value of $f(x, y) = 49 - x^2 - y^2$ on the line $x + 3y = 10$ (Fig. 12.65).

4. How close does the line $y = x + 1$ come to the parabola $y^2 = x$?

5. Find the extreme values of $f(x, y) = x^2y$ on the line $x + y = 3$.

6. Find the points on the curve $x^2y = 2$ nearest the origin.

7. Use the method of Lagrange multipliers to find

a) the minimum value of $x + y$, subject to the constraints $xy = 16$, $x > 0$, $y > 0$;

b) the maximum value of xy, subject to the constraint $x + y = 16$.

Comment on the geometry of each solution.

8. Find the points on the curve $x^2 + xy + y^2 = 1$ in the xy-plane that are nearest to and farthest from the origin.

9. Find the dimensions of the closed right circular cylindrical can of smallest surface area whose volume is 16π cm^3.

10. Use the method of Lagrange multipliers to find the dimensions of the rectangle of greatest area that can be inscribed in the ellipse $x^2/16 + y^2/9 = 1$ with sides parallel to the coordinate axes.

11. The temperature at a point (x, y) on a metal plate is $T(x, y) = 4x^2 - 4xy + y^2$. An ant on the plate walks around the circle of radius 5 centered at the origin. What are the highest and lowest temperatures encountered by the ant?

12. Your firm has been asked to design a storage tank for liquid petroleum gas. The customer's specifications call for a cylindrical tank with hemispherical ends, and the tank is to hold 8000 m^3 of gas. The customer also wants to use the smallest amount of material possible in building the tank. What radius and height do you recommend for the cylindrical portion of the tank?

13. Find the maximum and minimum values of $x^2 + y^2$ subject to the constraint $x^2 - 2x + y^2 - 4y = 0$.

14. Find the point on the plane $x + 2y + 3z = 13$ closest to the point $(1, 1, 1)$.

15. Find the maximum and minimum values of

$$f(x, y, z) = x - 2y + 5z$$

on the sphere

$$x^2 + y^2 + z^2 = 30.$$

16. Find the minimum distance from the surface $x^2 + y^2 - z^2 = 1$ to the origin.

17. Find the point on the surface $z = xy + 1$ nearest the origin.

18. Find the points on the surface $z^2 = xy + 4$ closest to the origin.

19. Find the points on the sphere $x^2 + y^2 + z^2 = 25$ where $f(x, y, z) = x + 2y + 3z$ has its maximum and minimum values.

20. Find three real numbers whose sum is 9 and the sum of whose squares is as small as possible.

21. Find the largest product the positive numbers x, y, and z can have if $x + y + z^2 = 16$.

22. A space probe in the shape of the ellipsoid

$$4x^2 + y^2 + 4z^2 = 16$$

enters the earth's atmosphere and its surface begins to heat. After one hour, the temperature at the point (x, y, z) on the probe's surface is

$$T(x, y, z) = 8x^2 + 4yz - 16z + 600.$$

Find the hottest point on the probe's surface.

23. *An example from economics.* In economics, the usefulness or *utility* of amounts x and y of two capital goods G_1 and G_2 is sometimes measured by a function $U(x, y)$. For example, G_1 and G_2 might be two chemicals a pharmaceutical company needs to have on hand and $U(x, y)$ the gain from manufacturing a product whose synthesis requires different amounts of the chemicals depending on the process used. If G_1 costs a dollars per kilogram, G_2 costs b dollars per kilogram, and the total amount allocated for the purchase of G_1 and G_2 together is c dollars, then the company's managers want to maximize $U(x, y)$ given that $ax + by = c$. Thus, they need to solve a typical Lagrange multiplier problem.

Suppose that

$$U(x, y) = xy + 2x$$

and that the equation $ax + by = c$ becomes

$$2x + y = 30$$

when reduced to lowest terms. Find the maximum value of U and the corresponding values of x and y subject to this latter constraint.

24. You are in charge of erecting a radio telescope on a newly discovered planet. To minimize interference, you want to place it where the magnetic field of the planet is weakest. The planet is spherical, with a radius of 6 units. Based on a coordinate system whose origin is at the center of the planet, the strength of the magnetic field is given by $M(x, y, z) = 6x - y^2 + xz + 60$. Where should you locate the radio telescope?

25. *The condition $\nabla f = \lambda \nabla g$ is not sufficient.* While $\nabla f = \lambda \nabla g$ is a necessary condition for the occurrence of an extreme value of $f(x, y)$ subject to the condition $g(x, y) = 0$, it does not in itself guarantee that one exists. As a case in point, try using the method of Lagrange multipliers to find a maximum value of $f(x, y) = x + y$ subject to the constraint that $xy = 16$. The method will identify the two points $(4, 4)$ and $(-4, -4)$ as candidates for the location of extreme values. Yet the sum $(x + y)$ has no maximum value on the hyperbola $xy = 16$. The farther you go from the origin on this hyperbola in the first quadrant, the larger the sum $f(x, y) = x + y$ becomes.

26. *A least squares plane.* The plane

$$z = Ax + By + C$$

is to be "fitted" to the following points (x_k, y_k, z_k):

$$(0, 0, 0), \quad (0, 1, 1), \quad (1, 1, 1), \quad (1, 0, -1).$$

Find the values of A, B, and C that minimize the sum of squares of the deviations

$$\sum_{k=1}^{4} (Ax_k + By_k + C - z_k)^2.$$

Lagrange Multipliers with Two Constraints

27. Find the point closest to the origin on the line of intersection of the planes $y + 2z = 12$ and $x + y = 6$.

28. Find the maximum value that $f(x, y, z) = x^2 + 2y - z^2$ can have on the line of intersection of the planes $2x - y = 0$ and $y + z = 0$.

29. Find the extreme values of $f(x, y, z) = x^2yz + 1$ on the intersection of the plane $z = 1$ with the sphere $x^2 + y^2 + z^2 = 10$.

30. a) Find the maximum value of $w = xyz$ on the line of intersection of the two planes $x + y + z = 40$ and $x + y - z = 0$.

b) Give a geometric argument to support your claim that you have found a maximum, and not a minimum, value of w.

31. Find the extreme values of the function $f(x, y, z) = xy + z^2$ on the circle in which the plane $y - x = 0$ intersects the sphere $x^2 + y^2 + z^2 = 4$.

12.10 Taylor's Formula, Second Derivatives, and Error Estimates

In this section, we derive the second derivative test for extreme values and the formula for estimating errors in linearizations of differentiable functions of two variables. The derivations are based on Taylor's formula from Section 8.7. The use of Taylor's formula in this context leads to a remarkable extension of the formula to functions of two variables. The extended formula provides polynomial approximations of all orders for functions of two variables, along with explicit formulas for the approximation errors.

The Derivation of the Second Derivative Test

Our first application of Taylor's formula is the derivation of the second derivative test for extreme values of a twice-differentiable function $f(x, y)$ at a point $P(a, b)$ where $f_x = f_y = 0$. We assume that the first and second order derivatives of f are continuous in an open region R about P (Fig. 12.74). We let h and k be increments small enough to put the point $S(a + h, b + k)$ and the line segment joining it to P inside R. We parametrize the segment PS with the equations and parameter interval

$$x = a + th, \qquad y = b + tk, \qquad 0 \le t \le 1.$$

This makes the value of f along the segment a function of t. We denote the function by F. Its defining formula is

$$F(t) = f(a + th, b + tk).$$

The function F is a differentiable function of t because f is differentiable and x and y are differentiable functions of t. The Chain Rule gives the derivative of F with respect to t as

$$F'(t) = f_x \frac{dx}{dt} + f_y \frac{dy}{dt} = hf_x + kf_y. \tag{1}$$

Since f_x and f_y are continuous throughout R, the function F' is continuous on the closed interval $0 \le t \le 1$.

Since f_x and f_y are differentiable (they have continuous partial derivatives), F' is a differentiable function of t. The Chain Rule gives the derivative of F' with respect to t as

$$F'' = \frac{\partial F'}{\partial x}\frac{dx}{dt} + \frac{\partial F'}{\partial y}\frac{dy}{dt}$$

$$= \frac{\partial}{\partial x}\left(hf_x + kf_y\right) \cdot h + \frac{\partial}{\partial y}\left(hf_x + kf_y\right) \cdot k \qquad \text{(From Eq. (1))} \tag{2}$$

$$= h^2 f_{xx} + 2hk f_{xy} + k^2 f_{yy}. \qquad (f_{xy} = f_{yx})$$

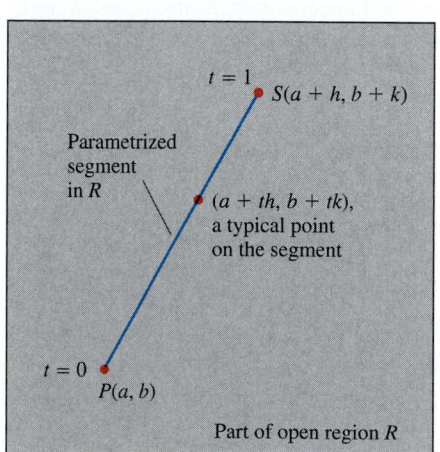

12.74 We begin the derivation of the second derivative test at $P(a, b)$ by parametrizing a typical line segment from P to a point S nearby.

In the figure: $t = 1$, $S(a + h, b + k)$; Parametrized segment in R; $(a + th, b + tk)$, a typical point on the segment; $t = 0$, $P(a, b)$; Part of open region R

Since F and F' are continuous on $[0, 1]$ and F' is differentiable on $(0, 1)$, we can apply Taylor's formula with $n = 2$ and $a = 0$ to obtain

$$F(1) = F(0) + F'(0)(1 - 0) + F''(c)\frac{(1 - 0)^2}{2}$$

$$F(1) = F(0) + F'(0) + \frac{1}{2}F''(c) \tag{3}$$

for some parameter value c between 0 and 1.

In terms of f,

$$F(1) = f(a + h, b + k),$$
$$F(0) = f(a, b),$$
$$F'(0) = hf_x(a, b) + kf_y(a, b), \tag{4}$$
$$F''(c) = (h^2 f_{xx} + 2hk f_{xy} + k^2 f_{yy})\big|_{(a + ch, b + ck)}.$$

With these values, Eq. (3) becomes

$$f(a + h, b + k) = f(a, b) + h f_x(a, b) + k f_y(a, b)$$
$$+ \frac{1}{2}\left(h^2 f_{xx} + 2hk f_{xy} + k^2 f_{yy}\right)\bigg|_{(a + ch, b + ck)}, \tag{5}$$

which reduces to

$$f(a + h, b + k) - f(a, b) = \frac{1}{2}\left(h^2 f_{xx} + 2hk f_{xy} + k^2 f_{yy}\right)\bigg|_{(a + ch, b + ck)} \tag{6}$$

since $f_x(a, b) = f_y(a, b) = 0$. We derive the second derivative test from this equation.

The presence of a maximum or minimum value of f at (a, b) is determined by the sign of $f(a + h, b + k) - f(a, b)$. By Eq. (6), this is the same as the sign of the quadratic form

$$Q(c) = (h^2 f_{xx} + 2hk f_{xy} + k^2 f_{yy})\big|_{(a + ch, b + ck)}. \tag{7}$$

The derivatives in Eq. (7) are evaluated at a point on segment PQ. Since these derivatives are continuous throughout R, their values at $(a + ch, b + ck)$ will be nearly the same as their values at (a, b) when h and k are small. In particular, if $Q(0) \neq 0$, the sign of $Q(c)$ will be the same as the sign of $Q(0)$ for sufficiently small values of h and k.

We can predict the sign of

$$Q(0) = h^2 f_{xx}(a, b) + 2hk f_{xy}(a, b) + k^2 f_{yy}(a, b) \tag{8}$$

and hence the sign of $Q(c)$ from the signs of f_{xx} and $f_{xx}f_{yy} - f_{xy}^2$ at (a, b). To do so, we multiply both sides of Eq. (8) by f_{xx} and rearrange the right-hand side to get

$$f_{xx}Q(0) = (h f_{xx} + k f_{xy})^2 + (f_{xx}f_{yy} - f_{xy}^2)k^2. \tag{9}$$

If $f_{xx} \neq 0$, the sign of $Q(0)$ can be determined from this equation, and we are led to the following criteria for the behavior of $f(x, y)$ at (a, b).

1. If $f_{xx} < 0$ and $f_{xx}f_{yy} - f_{xy}^2 > 0$ at (a, b), then $Q(0) < 0$ for all sufficiently small nonzero values of h and k, and f has a *local maximum* value at (a, b).

2. If $f_{xx} > 0$ and $f_{xx}f_{yy} - f_{xy}^2 > 0$ at (a, b), then $Q(0) > 0$ for all sufficiently small nonzero values of h and k, and f has a *local minimum* value at (a, b).

3. If $f_{xx}f_{yy} - f_{xy}^2 < 0$ at (a, b), then it can be shown that there are combinations of arbitrarily small nonzero values of h and k for which $Q(0) > 0$, and other values for which $Q(0) < 0$. Thus, arbitrarily close to the point $P_0(a, b, f(a, b))$ on the surface $z = f(x, y)$ there are points above P_0 and points below P_0. The function f has a *saddle point* at (a, b).

4. Finally, if $f_{xx}f_{yy} - f_{xy}^2 = 0$, we can draw *no conclusion* about the sign of $Q(c)$. Another test is needed. It does follow from Eq. (9) that $f_{xx}Q(0) \geq 0$ for all choices of h and k but the possibility that $Q(0)$ equals zero prevents us from drawing any reliable conclusion about the sign of $Q(c)$.

This completes the derivation of the second derivative test.

The Error Formula for Linear Approximations

Our second application of Taylor's formula is the verification that the difference $E(x, y)$ between the values of a function $f(x, y)$ and those of its linearization $L(x, y)$ at (x_0, y_0) satisfies the inequality

$$|E(x, y)| \leq \frac{1}{2} B (|x - x_0| + |y - y_0|)^2. \tag{10}$$

The function f is assumed to have continuous first and second partial derivatives throughout an open set containing a closed rectangular region R centered at (x_0, y_0). The number B is the largest value that any of $|f_{xx}|$, $|f_{yy}|$, and $|f_{xy}|$ take on R.

The inequality we want comes from Eq. (5). We substitute x_0 and y_0 for a and b, and $x - x_0$ and $y - y_0$ for h and k, respectively, and rearrange the equation in the form

$$f(x, y) = \underbrace{f(x_0, y_0) + f_x(x_0, y_0)(x - x_0) + f_y(x_0, y_0)(y - y_0)}_{\text{linearization } L(x, y)} \tag{11}$$

$$\underbrace{+ \frac{1}{2} \left((x - x_0)^2 f_{xx} + 2(x - x_0)(y - y_0)f_{xy} + (y - y_0)^2 f_{yy} \right) \Big|_{(x_0 + c(x - x_0), y_0 + c(y - y_0))}}_{\text{error } E(x, y)}.$$

This remarkable equation shows how to calculate E directly from the derivatives of f. In particular, it tells us that

$$|E| \leq \frac{1}{2} \left((x - x_0)^2 |f_{xx}| + 2|x - x_0| \, |y - y_0| \, |f_{xy}| + (y - y_0)^2 |f_{yy}| \right). \tag{12}$$

Hence, if B is an upper bound for the values of $|f_{xx}|$, $|f_{xy}|$, and $|f_{yy}|$ on R,

$$|E| \leq \frac{1}{2} \left((x - x_0)^2 B + 2|x - x_0| \, |y - y_0| B + (y - y_0)^2 B \right)$$

$$\leq \frac{1}{2} B \left((x - x_0)^2 + 2|x - x_0| \, |y - y_0| + (y - y_0)^2 \right) \tag{13}$$

$$\leq \frac{1}{2} B (|x - x_0| + |y - y_0|)^2.$$

Combining these last inequalities gives the inequality

$$|E| \leq \frac{1}{2} B \left(|x - x_0| + |y - y_0| \right)^2 \qquad (14)$$

that we set out to derive.

Taylor's Formula for Functions of Two Variables

The formula

$$F'(t) = \left(h \frac{\partial f}{\partial x} + k \frac{\partial f}{\partial y} \right)$$

from Eq. (1) says that applying the operator d/dt to $F(t) = f(a + ht, b + kt)$ gives the same result as applying the operator

$$\left(h \frac{\partial}{\partial x} + k \frac{\partial}{\partial y} \right)$$

to $f(x, y)$. Similarly, the formula

$$F''(t) = h^2 \frac{\partial^2 f}{\partial x^2} + 2hk \frac{\partial^2 f}{\partial x \, \partial y} + k^2 \frac{\partial^2 f}{\partial y^2}$$

from Eq. (2) says that applying d^2/dt^2 to $F(t)$ gives the same result as applying

$$\left(h \frac{\partial}{\partial x} + k \frac{\partial}{\partial y} \right)^2 = h^2 \frac{\partial^2}{\partial x^2} + 2hk \frac{\partial^2}{\partial x \, \partial y} + k^2 \frac{\partial^2}{\partial y^2}$$

to $f(x, y)$. These are the first two instances of a more general formula,

$$F^{(n)}(t) = \frac{d^n}{dt^n} F(t) = \left(h \frac{\partial}{\partial x} + k \frac{\partial}{\partial y} \right)^n f(x, y), \qquad (15)$$

which says that applying d^n/dt^n to $F(t)$ gives the same result as applying the operator

$$\left(h \frac{\partial}{\partial x} + k \frac{\partial}{\partial y} \right)^n,$$

after expanding it by the binomial theorem, to $f(x, y)$.

Suppose now that $f(x, y)$ and its partial derivatives through order $n + 1$ are continuous throughout a rectangular region centered at the point (a, b). Then we may extend the Taylor formula for $F(t)$ to more terms,

$$F(t) = F(0) + F'(0) \cdot t + \frac{F''(0)}{2!} t^2 + \cdots + \frac{F^{(n)}(0)}{n!} t^n + \text{remainder},$$

and take $t = 1$ to obtain

$$F(1) = F(0) + F'(0) + \frac{F''(0)}{2!} + \cdots + \frac{F^{(n)}(0)}{n!} + \text{remainder}.$$

When we replace the first n derivatives on the right of this last series by their equivalent expressions from Eq. (15) evaluated at $t = 0$ and add the appropriate remainder term, we arrive at the following formula.

Taylor's Formula for $f(x, y)$ at the Point (a, b)

Suppose $f(x, y)$ and its partial derivatives through order $n + 1$ are continuous throughout an open rectangular region R centered at a point (a, b). Then, throughout R,

$$f(a + h, b + k) = f(a, b) + (hf_x + kf_y)\big|_{(a, b)}$$

$$+ \frac{1}{2!} (h^2 f_{xx} + 2hk f_{xy} + k^2 f_{yy})\big|_{(a, b)}$$

$$+ \frac{1}{3!} (h^3 f_{xxx} + 3h^2 k f_{xxy} + 3hk^2 f_{xyy} + k^3 f_{yyy})\big|_{(a, b)} \qquad (16)$$

$$+ \cdots + \frac{1}{n!} \left(h \frac{\partial}{\partial x} + k \frac{\partial}{\partial y} \right)^n f \bigg|_{(a, b)}$$

$$+ \frac{1}{(n + 1)!} \left(h \frac{\partial}{\partial x} + k \frac{\partial}{\partial y} \right)^{n+1} f \bigg|_{(a + ch, b + ck)}$$

The derivatives in the first n derivative terms are evaluated at (a, b). The derivatives in the last term are evaluated at some point $(a + ch, b + ck)$ on the line segment joining (a, b) and $(a + h, b + k)$.

If $(a, b) = (0, 0)$ and we treat h and k as independent variables (denoting them now by x and y), then Eq. (16) assumes the following simpler form.

Taylor's Formula for $f(x, y)$ at the Origin

$$f(x, y) = f(0, 0) + xf_x + yf_y$$

$$+ \frac{1}{2!} (x^2 f_{xx} + 2xy f_{xy} + y^2 f_{yy})$$

$$+ \frac{1}{3!} (x^3 f_{xxx} + 3x^2 y f_{xxy} + 3xy^2 f_{xyy} + y^3 f_{yyy}) \qquad (17)$$

$$+ \cdots + \frac{1}{n!} \left(x \frac{\partial}{\partial x} + y \frac{\partial}{\partial y} \right)^n f$$

$$+ \frac{1}{(n + 1)!} \left(x \frac{\partial}{\partial x} + y \frac{\partial}{\partial y} \right)^{n+1} f \bigg|_{(cx, cy)}$$

The partial derivatives in the first n derivative terms are evaluated at $(0, 0)$. The partial derivatives in the final term are evaluated at (cy, cx), a point on the line segment joining the origin and (x, y).

Taylor's formula is the standard source of approximations of two-variable functions by polynomials in x and y. The first n derivative terms give the approximating polynomial; the last term gives the approximation error. The first three terms of Taylor's formula give the function's linearization. To improve on the linearization, we add higher power terms, the way we do for functions of a single variable.

Example 1   Use Taylor's formula to find a quadratic polynomial that approximates

$$f(x, y) = \sin x \sin y$$

near the origin. How accurate is the approximation if $|x| \le 0.1$ and $|y| \le 0.1$?

Solution   We take $n = 2$ in Eq. (17):

$$f(x, y) = f(0, 0) + (x f_x + y f_y)$$

$$+ \frac{1}{2}(x^2 f_{xx} + 2xy f_{xy} + y^2 f_{yy})$$

$$+ \frac{1}{6}(x^3 f_{xxx} + 3x^2 y f_{xxy} + 3xy^2 f_{xyy} + y^3 f_{yyy})_{(cx, cy)}$$

The first and second partial derivatives are evaluated at the origin and the third partial derivatives are evaluated at (cx, cy). With

$$f(x, y) = \sin x \sin y,$$

$$f(0, 0) = \sin x \sin y|_{(0, 0)} = 0,$$

$$f_x(0, 0) = \cos x \sin y|_{(0, 0)} = 0,$$

$$f_y(0, 0) = \sin x \cos y|_{(0, 0)} = 0,$$

$$f_{xx}(0, 0) = -\sin x \sin y|_{(0, 0)} = 0,$$

$$f_{xy}(0, 0) = \cos x \cos y|_{(0, 0)} = 1,$$

$$f_{yy}(0, 0) = -\sin x \sin y|_{(0, 0)} = 0,$$

we have

$$\sin x \sin y \approx 0 + 0 + 0 + \frac{1}{2}(x^2(0) + 2xy(1) + y^2(0))$$

$$\sin x \sin y \approx xy.$$

This is the standard quadratic approximation of $\sin x \sin y$ at the origin.

The error in the approximation is

$$E(x, y) = \frac{1}{6}(x^3 f_{xxx} + 3x^2 y f_{xxy} + 3xy^2 f_{xyy} + y^3 f_{yyy})|_{(cx, cy)}.$$

The third derivatives never exceed 1 in absolute value because they are products of sines and cosines. Also, $|x| \le 0.1$ and $|y| \le 0.1$. Hence

$$|E(x, y)| \le \frac{1}{6}((0.1)^3 + 3(0.1)^3 + 3(0.1)^3 + (0.1)^3)$$

$$\le \frac{8}{6}(0.1)^3$$

$$\le 0.00134. \qquad \text{(Rounded up, for safety)}$$

The error in the approximation will not exceed 0.00134 as long as $|x| \le 0.1$ and $|y| \le 0.1$.

EXERCISES 12.10

In Exercises 1–6, use Taylor's formula for $f(x, y)$ at the origin to find quadratic and cubic polynomial approximations of f near the origin.

1. $f(x, y) = e^x \cos y$

2. $f(x, y) = e^x \ln(1 + y)$

3. $f(x, y) = \sin(x^2 + y^2)$

4. $f(x, y) = \cos(x^2 + y^2)$

5. $f(x, y) = \dfrac{1}{1 - x - y}$

6. $f(x, y) = \dfrac{1}{1 - x - y + xy}$

7. Use Taylor's formula to find a quadratic approximation of $f(x, y) = \cos x \cos y$ at the origin. Estimate the error in the approximation if $|x| \le 0.1$ and $|y| \le 0.1$.

8. Use Taylor's formula to find a quadratic approximation of $e^x \sin y$ at the origin. Estimate the error in the approximation if $|x| \le 0.1$ and $|y| \le 0.1$.

9. *Error bounds for linear approximations of functions of three variables*

a) Suppose that R is a rectangular solid centered at $P(x_0, y_0, z_0)$ and lying in an open region on which $f(x, y, z)$ and its first and second order derivatives are continuous. Suppose also that h, k, and m are increments small enough for the line segment joining P to the point $S(x_0 + h, y_0 + k, z_0 + m)$ to lie in R. Parametrize PS with the equations and interval

$$x = x_0 + th, \quad y = y_0 + tk, \quad z = z_0 + tm, \quad 0 \le t \le 1.$$

Apply Taylor's formula for functions of a single variable to the function $F(t)$ whose defining formula is

$$F(t) = f(x_0 + th, y_0 + tk, z_0 + tm)$$

to show that

$$
\begin{aligned}
f(x_0 &+ h, y_0 + k, z_0 + m) \\
&= f(x_0, y_0, z_0) + hf_x(x_0, y_0, z_0) + kf_y(x_0, y_0, z_0) \\
&\quad + mf_z(x_0, y_0, z_0) \\
&\quad + \frac{1}{2}\Big(h^2 f_{xx} + k^2 f_{yy} + m^2 f_{zz} + 2hk f_{xy} \\
&\qquad + 2hm f_{xz} + 2km f_{yz}\Big)\Big|_{(x_0 + ch,\, y_0 + ck,\, z_0 + cm)}
\end{aligned}
\tag{5'}
$$

for some parameter value c between 0 and 1. This is the three-variable version of Eq. (5) in the text.

b) Use Eq. (5') to derive the inequality

$$|E| \le \frac{1}{2} M\big(|x - x_0| + |y - y_0| + |z - z_0|\big)^2$$

in Section 12.4, Eq. (20).

REVIEW QUESTIONS

1. Give examples of functions of two and three variables, and describe their domains and ranges.

2. Describe two ways to display the values of a function $f(x, y)$ graphically. Give examples.

3. What is a level surface of a function $f(x, y, z)$ of three independent variables? Give an example.

4. Give two equivalent definitions of

$$\lim_{(x, y) \to (x_0, y_0)} f(x, y) = L.$$

What is the basic theorem for calculating limits of sums, differences, products, constant multiples, and quotients of functions of two or more variables?

5. When is a function of two (three) variables continuous at a point in its domain? Give examples of functions that are continuous at some points but not continuous at others.

6. How are the partial derivatives $\partial f / \partial x$ and $\partial f / \partial y$ of a function $f(x, y)$ defined? What is their geometric meaning? What rates do they describe? How are they calculated? Give examples.

7. Give examples of second order partial derivatives of func-

tions of two variables. What is the basic theorem about mixed second order derivatives? What about mixed partial derivatives of higher order? Give examples.

8. What does it mean for a function $f(x, y)$ to be differentiable? The Increment Theorem states conditions under which f will be differentiable. What are they?

9. Why are differentiable functions of two independent variables continuous?

10. What is the linearization $L(x, y)$ of a function $f(x, y)$ at a point (x_0, y_0)? How are linearizations used? Give an upper bound for the error incurred by approximating f by L in the neighborhood of (x_0, y_0). Give examples.

11. If (x, y) moves from (x_0, y_0) to a point $(x_0 + dx, y_0 + dy)$ nearby, how do we estimate the corresponding change in a differentiable function $f(x, y)$? How do we estimate the relative change? The percentage change? Give an example.

12. Define the gradient of a function $f(x, y, z)$. Describe the role the gradient plays in defining directional derivatives, tangent planes, and normal lines. What is the relation between ∇f and the directions in which f changes most rapidly? least rapidly? What are the analogous results for functions of two variables? Give examples.

13. Describe how to find the extreme values (if any) of a function $f(x, y)$ that has continuous first and second order partial derivatives. Give examples.

14. How do you find the extreme values of a differentiable function along a differentiable curve in its domain? Give an example.

15. Describe the method of Lagrange multipliers and its geometric interpretation as it applies to a problem of maximizing or minimizing (a) a differentiable function $f(x, y)$ subject to a differentiable constraint $g(x, y) = 0$, (b) a differentiable function $f(x, y, z)$ subject to simultaneous differentiable constraints $g_1(x, y, z) = 0$ and $g_2(x, y, z) = 0$ when ∇g_1 is not parallel to ∇g_2. Give examples.

16. State Taylor's formula for a (sufficiently) differentiable function $f(x, y)$. Show how to use it to generate a quadratic polynomial approximation together with an appropriate error estimate.

MISCELLANEOUS EXERCISES

In Exercises 1–4, find the domain and range of f and identify the level curves. Sketch a typical level curve.

1. $f(x, y) = 9x^2 + y^2$

2. $f(x, y) = \sin(y - x)$

3. $f(x, y) = 1/xy$

4. $f(x, y) = \sqrt{x^2 - y}$

In Exercises 5–8, find the domain and range of f and identify the level surfaces. Sketch a typical level surface.

5. $f(x, y, z) = x^2 + y^2 - z$

6. $f(x, y, z) = x^2 + 4y^2 + 9z^2$

7. $f(x, y, z) = \dfrac{1}{x^2 + y^2 + z^2}$

8. $f(x, y, z) = \dfrac{1}{x^2 + y^2 + z^2 + 1}$

Find the limits in Exercises 9–14.

9. $\lim\limits_{(x, y) \to (\pi, \ln 2)} e^y \cos x$

10. $\lim\limits_{(x, y) \to (0, 0)} \dfrac{2 + y}{x + \cos y}$

11. $\lim\limits_{\substack{(x, y) \to (1, 1) \\ x \neq y}} \dfrac{x^2 - y^2}{x - y}$

12. $\lim\limits_{P \to (1, -1, e)} \ln |x + y + z|$

13. $\lim\limits_{(x, y) \to (0, 0)} y \sin \dfrac{1}{x}$

14. $\lim\limits_{(x, y) \to (1, 1)} \dfrac{x^3 y^3 - 1}{xy - 1}$

Show that the limits in Exercises 15 and 16 do not exist.

15. $\lim\limits_{(x, y) \to (0, 0)} \dfrac{y}{x^2 - y}$

16. $\lim\limits_{(x, y) \to (0, 0)} \dfrac{x^2 + y^2}{xy}$

17. a) Let $f(x, y) = (x^2 - y^2)/(x^2 + y^2)$ for $(x, y) \neq (0, 0)$. Is it possible to define $f(0, 0)$ in a way that makes f continuous at the origin? Why?

b) Let
$$f(x, y) = \begin{cases} \dfrac{\sin(x - y)}{|x| + |y|}, & |x| + |y| \neq 0, \\ 0, & (x, y) = (0, 0). \end{cases}$$

Is f continuous at the origin? Why?

18. Let
$$f(r, \theta) = \begin{cases} \dfrac{\sin 6r}{6r}, & r \neq 0, \\ 1, & r = 0 \end{cases}$$

(Fig. 12.75). Find

a) $\lim\limits_{r \to 0} f(r, \theta)$ b) $f_r(0, 0)$ c) $f_\theta(r, \theta)$, $r \neq 0$

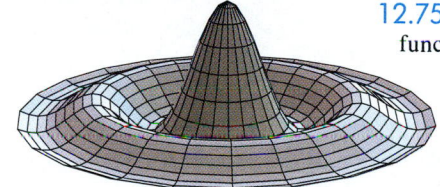

12.75 The graph of the function in Exercise 18.

(Generated by Mathematica)

In Exercises 19–24, find the partial derivative of the function with respect to each variable.

19. $g(r, \theta) = r \cos \theta + r \sin \theta$

20. $f(x, y) = \dfrac{1}{2} \ln(x^2 + y^2) + \tan^{-1} \dfrac{y}{x}$

21. $f(R_1, R_2, R_3) = \dfrac{1}{R_1} + \dfrac{1}{R_2} + \dfrac{1}{R_3}$

22. $h(x, y, z) = \sin(2\pi x + y - 3z)$

23. $P(n, R, T, V) = \dfrac{nRT}{V}$ (the Ideal Gas Law)

24. $f(r, l, T, d) = \dfrac{1}{2rl} \sqrt{\dfrac{T}{\pi d}}$

(the frequency of a struck piano string, Chapter 2, Miscellaneous Exercise 50)

Find the second order partial derivatives of the functions in Exercises 25–28.

25. $f(x, y) = y + \dfrac{x}{y}$

26. $f(x, y) = e^x + y \sin x$

27. $f(x, y) = x + xy - 5x^3 + \ln(x^2 + 1)$

28. $f(x, y) = y^2 - 3xy + \cos y + 7e^y$

29. If you did Exercise 40 in Section 12.2, you know that the function

$$f(x, y) = \begin{cases} xy \dfrac{x^2 - y^2}{x^2 + y^2}, & (x, y) \neq (0, 0), \\ 0, & (x, y) = (0, 0) \end{cases}$$

is continuous at the origin (Fig. 12.76). Find $f_{xy}(0, 0)$ and $f_{yx}(0, 0)$.

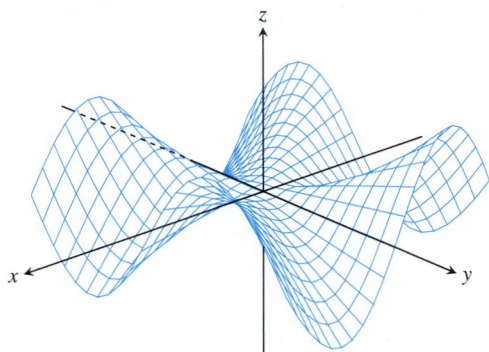

(Generated by Mathematica)

12.76   The graph of the function in Exercise 29.

30. Find a function $w = f(x, y)$ whose first partial derivatives are $\partial w/\partial x = 1 + e^x \cos y$ and $\partial w/\partial y = 2y - e^x \sin y$ and whose value at $(\ln 2, 0)$ is $\ln 2$.

In Exercises 31 and 32, find the linearization $L(x, y)$ of the function $f(x, y)$ at the point P_0. Then find an upper bound for the magnitude of the error E in the approximation $f(x, y) \approx L(x, y)$ over the rectangle R.

31. $f(x, y) = \sin x \cos y$, $P_0(\pi/4, \pi/4)$

$R: \quad \left| x - \dfrac{\pi}{4} \right| \leq 0.1, \quad \left| y - \dfrac{\pi}{4} \right| \leq 0.1$

32. $f(x, y) = xy - 3y^2 + 2$, $P_0(1, 1)$

$R: \quad |x - 1| \leq 0.1, \quad |y - 1| \leq 0.2$

Find the linearizations of the functions in Exercises 33 and 34 at the given points.

33. $f(x, y, z) = xy + 2yz - 3xz$ at $(1, 0, 0)$ and $(1, 1, 0)$

34. $f(x, y, z) = \sqrt{2} \cos x \, \sin(y + z)$ at $(0, 0, \pi/4)$ and $(\pi/4, \pi/4, 0)$.

35. You plan to calculate the volume inside a stretch of pipeline that is about 36 in. in diameter and 1 mi long. With which measurement should you be more careful—the length, or the diameter? Why?

36. Near the point $(1, 2)$, is $f(x, y) = x^2 - xy + y^2 - 3$ more sensitive to changes in x, or to changes in y? How do you know?

37. Suppose that the current I (amperes) in an electrical circuit is related to the voltage V (volts) and the resistance R (ohms)

by the equation $I = V/R$. If the voltage drops from 24 to 23 volts and the resistance drops from 100 to 80 ohms, will I increase, or decrease? By about how much? Express the changes in V and R and the estimated change in I as percentages of their original values.

38. If $a = 10$ cm and $b = 16$ cm to the nearest millimeter, what should you expect the maximum percentage error to be in the calculated area $A = \pi ab$ of the ellipse $x^2/a^2 + y^2/b^2 = 1$?

39. Let $y = uv$ and $z = u + v$, where u and v are positive independent variables.
   a) If u is measured with an error of 2% and v with an error of 3%, about what is the percentage error in the calculated value of y?
   b) Show that the percentage error in the calculated value of z is less than the percentage error in the value of y.

40. CALCULATOR   *Cardiac index.*   To make different people comparable in studies of cardiac output (Section 3.1, Exercise 19), researchers divide the measured cardiac output by the body surface area to find the *cardiac index* C:

$$C = \frac{\text{cardiac output}}{\text{body surface area}}.$$

The body surface area B is calculated with the formula

$$B = 71.84 w^{0.425} h^{0.725},$$

which gives B in square centimeters when w is measured in kilograms and h in centimeters. You are about to calculate the cardiac index of a person with the following measurements:

Cardiac output:	7 L/min
Weight:	70 kg
Height:	180 cm

Which will have a greater effect on the calculation, a 1-kg error in measuring the weight, or a 1-cm error in measuring the height?

41. Find dw/dt at $t = 0$ if $w = \sin(xy + \pi)$, $x = e^t$, $y = \ln(t + 1)$.

42. Find dw/dt at $t = 1$ if $w = xe^y + y \sin z - \cos z$, $x = 2\sqrt{t}$, $y = t - 1 + \ln t$, $z = \pi t$.

43. Find $\partial w/\partial r$ and $\partial w/\partial s$ when $r = \pi$ and $s = 0$ if $w = \sin(2x - y)$, $x = r + \sin s$, $y = rs$.

44. Find $\partial w/\partial u$ and $\partial w/\partial v$ when $u = v = 0$ if $w = \ln \sqrt{1 + x^2} - \tan^{-1} x$ and $x = 2e^u \cos v$.

Assuming that the equations in Exercises 45 and 46 define y as a differentiable function of x, find the value of dy/dx at the given point.

45. $1 - x - y^2 - \sin xy = 0$, $(0, 1)$

46. $2xy + e^{x+y} - 2 = 0$, $(0, \ln 2)$

47. Find the value of the derivative of $f(x, y, z) = xy + yz + xz$ with respect to t on the curve $x = \cos t, y = \sin t, z = \cos 2t$ at the point where $t = 1$ (Fig. 12.77).

12.77   The curve in Exercise 47.

48. Show that if $w = f(s)$ is any differentiable function of s whatever, and $s = y + 5x$, then

$$\frac{\partial w}{\partial x} - 5\frac{\partial w}{\partial y} = 0.$$

49. Let $w = f(r, \theta)$, $r = \sqrt{x^2 + y^2}$, and $\theta = \tan^{-1}(y/x)$. Find $\partial w/\partial x$ and $\partial w/\partial y$ and express your answers in terms of r and θ.

50. Let $z = f(u, v)$, $u = ax + by$, and $v = ax - by$. Express z_x and z_y in terms of f_u, f_v, and the constants a and b.

51. If a and b are constants, $w = u^3 + \tanh u + \cos u$, and $u = ax + by$, show that

$$a\frac{\partial w}{\partial y} = b\frac{\partial w}{\partial x}.$$

52. If $w = \ln(x^2 + y^2 + 2z)$, $x = r + s$, $y = r - s$, and $z = 2rs$, find w_r and w_s by the Chain Rule. Then check your answer another way.

53. The equations $e^u \cos v - x = 0$ and $e^u \sin v - y = 0$ define u and v as differentiable functions of x and y. Show that the angle between the vectors

$$\frac{\partial u}{\partial x}\mathbf{i} + \frac{\partial u}{\partial y}\mathbf{j} \quad \text{and} \quad \frac{\partial v}{\partial x}\mathbf{i} + \frac{\partial v}{\partial y}\mathbf{j}$$

is constant.

54. Introducing polar coordinates $x = r \cos \theta$ and $y = r \sin \theta$ changes $f(x, y)$ to $g(r, \theta)$. Find the value of $\partial^2 g/\partial\theta^2$ at the point $(r, \theta) = (2, \pi/2)$, given that

$$\frac{\partial f}{\partial x} = \frac{\partial f}{\partial y} = \frac{\partial^2 f}{\partial x^2} = \frac{\partial^2 f}{\partial y^2} = 1$$

at that point.

55. *A proof of Leibniz's rule.*   Leibniz's rule says that if f is continuous on $[a, b]$ and if $u(x)$ and $v(x)$ are differentiable functions of x whose values lie in $[a, b]$, then

$$\frac{d}{dx}\int_{u(x)}^{v(x)} f(t)\,dt = f(v(x))\frac{dv}{dx} - f(u(x))\frac{du}{dx}.$$

Prove the rule by setting

$$g(u, v) = \int_u^v f(t)\,dt, \quad u = u(x), \quad v = v(x)$$

and calculating dg/dx with the Chain Rule.

56. Suppose that f is a twice-differentiable function of r, that $r = \sqrt{x^2 + y^2 + z^2}$, and that

$$f_{xx} + f_{yy} + f_{zz} = 0.$$

Show that for some constants a and b,

$$f(r) = \frac{a}{r} + b.$$

57. A function $f(x, y)$ is *homogeneous of degree n* (n a nonnegative integer) if $f(tx, ty) = t^n f(x, y)$ for all t, x, and y. For such a function (sufficiently differentiable), prove that

a) $x\dfrac{\partial f}{\partial x} + y\dfrac{\partial f}{\partial y} = nf(x, y)$,

b) $x^2\left(\dfrac{\partial^2 f}{\partial x^2}\right) + 2xy\left(\dfrac{\partial^2 f}{\partial x\,\partial y}\right) + y^2\left(\dfrac{\partial^2 f}{\partial y^2}\right) = n(n-1)f.$

58. Let $\mathbf{r} = x\mathbf{i} + y\mathbf{j} + z\mathbf{k}$. Express x, y, and z as functions of the spherical coordinates ρ, ϕ, and θ and calculate $\partial\mathbf{r}/\partial\rho$, $\partial\mathbf{r}/\partial\phi$, and $\partial\mathbf{r}/\partial\theta$. Then express these derivatives in terms of the unit vectors

$$\mathbf{u}_\rho = (\sin\phi\cos\theta)\,\mathbf{i} + (\sin\phi\sin\theta)\,\mathbf{j} + (\cos\phi)\,\mathbf{k}$$
$$\mathbf{u}_\phi = (\cos\phi\cos\theta)\,\mathbf{i} + (\cos\phi\sin\theta)\,\mathbf{j} - (\sin\phi)\,\mathbf{k}$$
$$\mathbf{u}_\theta = -(\sin\theta)\,\mathbf{i} + (\cos\theta)\,\mathbf{j}.$$

In Exercises 59–62, find the directions in which f increases and decreases most rapidly at P_0 and find the derivative of f in each direction. Also, find the derivative of f at P_0 in the direction of the vector $\mathbf{A}$.

59. $f(x, y) = \cos x \cos y$, $P_0(\pi/4, \pi/4)$, $\mathbf{A} = 3\mathbf{i} + 4\mathbf{j}$

60. $f(x, y) = x^2 e^{-2y}$, $P_0(1, 0)$, $\mathbf{A} = \mathbf{i} + \mathbf{j}$

61. $f(x, y, z) = \ln(2x + 3y + 6z)$, $P_0(-1, -1, 1)$, $\mathbf{A} = 2\mathbf{i} + 3\mathbf{j} + 6\mathbf{k}$

62. $f(x, y, z) = x^2 + 3xy - z^2 + 2y + z + 4$, $P_0(0, 0, 0)$, $\mathbf{A} = \mathbf{i} + \mathbf{j} + \mathbf{k}$

In Exercises 63 and 64, sketch the surface $f(x, y, z) = c$ together with ∇f at the given points.

63. $x^2 + y + z^2 = 0$; $(0, -1, \pm 1)$, $(0, 0, 0)$

64. $y^2 + z^2 = 4$; $(2, \pm 2, 0)$, $(2, 0, \pm 2)$

In Exercises 65 and 66, find an equation for the plane tangent to the level surface $f(x, y, z) = c$ at the point P_0. Also find parametric equations for the line that is normal to the surface at P_0.

65. $x^2 - y - 5z = 0$, $P_0(2, -1, 1)$

66. $x^2 + y^2 + z = 4$, $P_0(1, 1, 2)$

In Exercises 67 and 68, find equations for the lines that are tangent and normal to the level curve $f(x, y) = c$ at the point P_0. Then sketch the lines and level curve together with ∇f at P_0.

67. $y - \sin x = 1$, $P_0(\pi, 1)$

68. $\dfrac{y^2}{2} - \dfrac{x^2}{2} = \dfrac{3}{2}$, $P_0(1, 2)$

In Exercises 69 and 70, find parametric equations for the line that is tangent to the curve of intersection of the surfaces at the given point.

69. Surfaces: $x^2 + 2y + 2z = 4,$ $y = 1$
    Point: $(1, 1, 1/2)$

70. Surfaces: $x + y^2 + z = 2,$ $y = 1$
    Point: $(1/2, 1, 1/2)$

71. Find the points on the surface

$$(y + z)^2 + (z - x)^2 = 16$$

where the normal line is parallel to the yz-plane.

72. Find the points on the surface

$$xy + yz + zx - x - z^2 = 0$$

where the tangent plane is parallel to the xy-plane.

73. Find the derivative of $f(x, y, z) = xyz$ in the direction of the velocity vector of the helix

$$\mathbf{r} = (\cos 3t)\,\mathbf{i} + (\sin 3t)\,\mathbf{j} + 3t\,\mathbf{k}$$

at time $t = \pi/3$.

74. What is the largest value that the directional derivative of $f(x, y, z) = xyz$ can have at the point $(1, 1, 1)$?

75. At the point $(1, 2)$ the function $f(x, y)$ has a derivative of 2 in the direction toward $(2, 2)$ and a derivative of -2 in the direction toward $(1, 1)$. Find the derivative in the direction toward $(4, 6)$.

76. Suppose that $\nabla f(x, y, z)$ is always parallel to the position vector $x\mathbf{i} + y\mathbf{j} + z\mathbf{k}$. Show that $f(0, 0, a) = f(0, 0, -a)$ for any a.

77. Show that the directional derivative of $f(x, y, z) = \sqrt{x^2 + y^2 + z^2}$ at the origin equals 1 in any direction but that f has no gradient vector at the origin.

78. Let $\mathbf{r} = x\mathbf{i} + y\mathbf{j} + z\mathbf{k}$ and let $r = |\mathbf{r}|$.
a) Show that $\nabla r = \mathbf{r}/r$.
b) Show that $\nabla(r^n) = nr^{n-2}\,\mathbf{r}$.
c) Find a function whose gradient equals $\mathbf{r}$.
d) Show that $\mathbf{r} \cdot d\mathbf{r} = r\,dr$.
e) Show that $\nabla(\mathbf{A} \cdot \mathbf{r}) = \mathbf{A}$ for any constant vector $\mathbf{A}$.

79. Show that the line normal to the surface $xy + z = 2$ at the point $(1, 1, 1)$ passes through the origin.

80. a) Sketch the surface $x^2 - y^2 + z^2 = 4$.
b) Find a vector normal to the surface at $(2, -3, 3)$. Add the vector to your sketch.
c) Find equations for the tangent plane and normal line at $(2, -3, 3)$.

81. Show that the curve

$$\mathbf{r} = (\ln t)\,\mathbf{i} + (t\ln t)\,\mathbf{j} + t\,\mathbf{k}$$

is tangent to the surface

$$xz^2 - yz + \cos xy = 1$$

at $(0, 0, 1)$.

82. Show that the curve

$$\mathbf{r} = \left(\frac{t^3}{4} - 2\right)\mathbf{i} + \left(\frac{4}{t} - 3\right)\mathbf{j} + \cos(t - 2)\,\mathbf{k}$$

is tangent to the surface

$$x^3 + y^3 + z^3 - xyz = 0$$

at $(0, -1, 1)$.

83. Which of the following statements are true if $f(x, y)$ is differentiable at (x_0, y_0)?
a) If $\mathbf{u}$ is a unit vector, the derivative of f at (x_0, y_0) in the direction of $\mathbf{u}$ is $(f_x(x_0, y_0)\,\mathbf{i} + f_y(x_0, y_0)\,\mathbf{j}) \cdot \mathbf{u}$.
b) The derivative of f at (x_0, y_0) in the direction of $\mathbf{u}$ is a vector.
c) The directional derivative of f at (x_0, y_0) has its greatest value in the direction of ∇f.
d) At (x_0, y_0), the gradient of f is normal to the curve $f(x, y) = f(x_0, y_0)$.

84. Suppose that a differentiable function $f(x, y)$ has the constant value c along the differentiable curve $x = g(t)$, $y = h(t)$; that is,

$$f(g(t), h(t)) = c$$

for all values of t. Differentiate both sides of this equation with respect to t to show that ∇f is orthogonal to the curve's tangent vector at every point on the curve.

85. Suppose cylindrical coordinates r, θ, z are introduced into a function $w = f(x, y, z)$ to yield $w = F(r, \theta, z)$. Show that

$$\nabla w = \frac{\partial w}{\partial r}\,\mathbf{u}_r + \frac{1}{r}\frac{\partial w}{\partial \theta}\,\mathbf{u}_\theta + \frac{\partial w}{\partial z}\,\mathbf{k}, \qquad (1)$$

where

$$\mathbf{u}_r = (\cos \theta)\,\mathbf{i} + (\sin \theta)\,\mathbf{j}$$
$$\mathbf{u}_\theta = -(\sin \theta)\,\mathbf{i} + (\cos \theta)\,\mathbf{j}.$$

(*Hint:* Express the right-hand side of Eq. (1) in terms of $\mathbf{i}, \mathbf{j}$, and $\mathbf{k}$ and use the Chain Rule to express the components of $\mathbf{i}, \mathbf{j}$, and $\mathbf{k}$ in rectangular coordinates.)

86. Suppose spherical coordinates ρ, ϕ, θ are introduced into a function $w = f(x, y, z)$ to yield a function $w = F(\rho, \phi, \theta)$. Show that

$$\nabla w = \frac{\partial w}{\partial \rho}\,\mathbf{u}_\rho + \frac{1}{\rho}\frac{\partial w}{\partial \phi}\,\mathbf{u}_\phi + \frac{1}{\rho \sin \phi}\frac{\partial w}{\partial \theta}\,\mathbf{u}_\theta, \qquad (2)$$

where

$$\mathbf{u}_\rho = (\sin \phi \cos \theta)\,\mathbf{i} + (\sin \phi \sin \theta)\,\mathbf{j} + (\cos \phi)\,\mathbf{k}$$
$$\mathbf{u}_\phi = (\cos \phi \cos \theta)\,\mathbf{i} + (\cos \phi \sin \theta)\,\mathbf{j} - (\sin \phi)\,\mathbf{k}$$
$$\mathbf{u}_\theta = -(\sin \theta)\,\mathbf{i} + (\cos \theta)\,\mathbf{j}.$$

(*Hint:* Express the right-hand side of Eq. (2) in terms of $\mathbf{i}, \mathbf{j}$, and $\mathbf{k}$ and use the Chain Rule to express the components of $\mathbf{i}, \mathbf{j}$, and $\mathbf{k}$ in rectangular coordinates.)

Test the functions in Exercises 87–92 for maxima, minima, and saddle points. Find the functions' values at these points.

87. $f(x, y) = x^2 - xy + y^2 + 2x + 2y - 4$

88. $f(x, y) = 5x^2 + 4xy - 2y^2 + 4x - 4y$

89. $f(x, y) = 6xy - x^3 - y^2$

90. $f(x, y) = 2x^3 + 3xy + y^3$

91. $f(x, y) = x^3 + y^3 - 3xy + 15$

92. $f(x, y) = x^3 + y^3 + 3x^2 - 3y^2$

In Exercises 93–100, find the absolute maximum and minimum values of f on the region R.

93. $f(x, y) = x^2 + xy + y^2 - 3x + 3y$
    R: The triangular region cut from the first quadrant by the line $x + y = 4$

94. $f(x, y) = x^2 - y^2 - 2x + 4y + 1$
    R: The rectangular region in the first quadrant bounded by the coordinate axes and the lines $x = 4$ and $y = 2$

95. $f(x, y) = y^2 - xy - 3y + 2x$
    R: The square region enclosed by the lines $x = \pm 2$ and $y = \pm 2$

96. $f(x, y) = 2x + 2y - x^2 - y^2$
    R: The square bounded by the coordinate axes and the lines $x = 2$, $y = 2$ in the first quadrant

97. $f(x, y) = x^2 - y^2 - 2x + 4y$
    R: The triangular region bounded below by the x-axis, above by the line $y = x + 2$, and on the right by the line $x = 2$

98. $f(x, y) = 4xy - x^4 - y^4 + 16$
    R: The triangular region bounded below by the line $y = -2$, above by the line $y = x$, and on the right by the line $x = 2$

99. $f(x, y) = x^3 + y^3 + 3x^2 - 3y^2$
    R: The square region enclosed by the lines $x = \pm 1$ and $y = \pm 1$

100. $f(x, y) = x^3 + 3xy + y^3 + 1$
    R: The square region enclosed by the lines $x = \pm 1$ and $y = \pm 1$

101. Show that the only possible maxima and minima of z on the surface $z = x^3 + y^3 - 9xy + 27$ occur at $(0, 0)$ and $(3, 3)$. Show that neither a maximum nor a minimum occurs at $(0, 0)$. Determine whether z has a maximum or a minimum at $(3, 3)$.

102. Find the maximum value of $f(x, y) = 6xye^{-(2x + 3y)}$ in the closed first quadrant (includes the nonnegative axes).

103. Find the points nearest the origin on the curve $xy^2 = 54$.

104. Find the extreme values of $f(x, y) = x^3 + y^2$ on the cylinder $x^2 + y^2 = 1$.

105. Suppose that the temperature T (degrees) at the point (x, y, z) on the sphere $x^2 + y^2 + z^2 = 1$ is $T = 400 xyz^2$. Locate the highest and lowest temperatures on the sphere.

106. Find the extreme values of $f(x, y) = x^2 + 3y^2 + 2y$ on the unit disk $x^2 + y^2 \leq 1$.

107. Find the point(s) on the surface $xyz = 1$ closest to the origin.

108. A closed rectangular box is to have a volume of V in^3. The cost of the material used in the box is a cents/in^2 for top and bottom, b cents/in^2 for front and back, and c cents/in^2 for the remaining two sides. What dimensions minimize the total cost of materials?

109. Find the minimum volume for a region bounded by the planes $x = 0$, $y = 0$, $z = 0$ and a plane tangent to the ellipsoid

$$\frac{x^2}{a^2} + \frac{y^2}{b^2} + \frac{z^2}{c^2} = 1$$

at a point in the first octant.

110. Prove the following theorem: If $f(x, y)$ is defined in a region R, and f_x and f_y exist and are bounded in R, then $f(x, y)$ is continuous in R. (The assumption of boundedness is essential.)

The One-Dimensional Heat Equation

If $w(x, t)$ represents the temperature at position x at time t in a uniform conducting rod with perfectly insulated sides (Fig. 12.78), then the partial derivatives w_{xx} and w_t satisfy a differential equation of the form

$$w_{xx} = \frac{1}{c^2} w_t. \tag{3}$$

This equation is called the **one-dimensional heat equation.** The value of the positive constant c^2 is determined by the material from which the rod is made. It has been determined experimentally for a broad range of materials. For a given application one finds the appropriate value in a table. For dry soil, for example, $c^2 = 0.19$ ft^2/day.

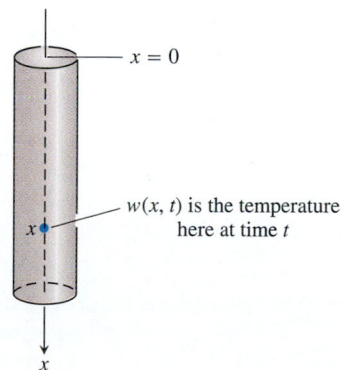

12.78   The temperature distribution in an insulated heat-conducting rod satisfies the equation

$$w_{xx} = \frac{1}{c^2} w_t.$$

In chemistry and biochemistry, the heat equation is known as the **diffusion equation.** In this context, $w(x, t)$ represents the concentration of a dissolved substance, a salt for instance, diffusing along a tube filled with liquid. The value of $w(x, t)$ is the concentration at point x at time t. In other applications, $w(x, t)$ represents the diffusion of a gas down a long, thin pipe.

In electrical engineering, the heat equation appears in the forms

$$v_{xx} = RCv_t \tag{4}$$

and

$$i_{xx} = RCi_t, \tag{5}$$

which are known as the **telegraph equations.** These equations describe the voltage v and the flow of current i in a coaxial cable or in any other cable in which leakage and inductance are negligible. The functions and constants in these equations are

$v(x, t)$ = voltage at point x at time t,

R = resistance per unit length,

C = capacitance to ground per unit of cable length,

$i(x, t)$ = current at point x at time t.

111. Find all solutions of the one-dimensional heat equation of the form $w = e^{rt}\sin \pi x$, r a constant.

112. Find all solutions of the one-dimensional heat equation that have the form $w = e^{rt} \sin kx$ and satisfy the conditions that $w(0, t) = 0$ and $w(L, t) = 0$. What happens to these solutions as $t \to \infty$?

The Cover Photograph

113. You plan to shape a test piece from a solid cylindrical blank of high-speed M-2 steel. The blank is 3.25 in. long, with a radius of 0.375 in. You will cut the piece by machining a disk from one end of the blank, shaping the other end into a right circular cone, and removing two cylindrical bands (of different depths) from the sides. Figure 12.79 shows the dimensions involved. You can control the radius and the horizontal length of your cuts with an accuracy of 1/2000 in. About how much error should you expect there to be in the mass of the finished piece? Express your answer as a percentage of the piece's ideal mass.

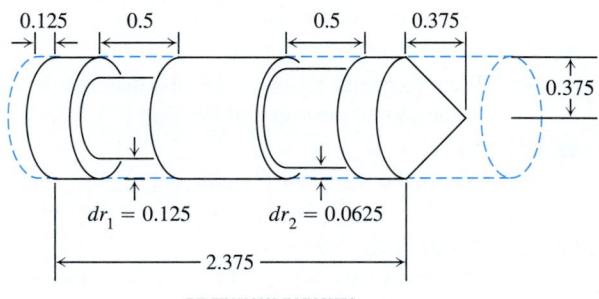

DIMENSIONS IN INCHES

12.79   The cylindrical blank (outline) and machined piece in Exercise 113.

CHAPTER

13

MULTIPLE INTEGRALS

Overview   This chapter introduces multiple integration, the complement of partial differentiation. The problems we can solve by integrating functions of two and three variables are similar to, but more general than, the problems we solve by integrating functions of a single variable. In Chapter 5, for example, we calculated the volumes of solids of revolution. Now we shall be able to calculate volumes of asymmetric solids in a variety of coordinate systems. Earlier, we calculated moments and centers of mass of rods and thin plates. Now we shall be able to handle plates more easily, work with more general density functions, and treat solids as well. As in the previous chapter, we perform the necessary calculations by drawing on our experience with functions of a single variable.

In the next chapter, we shall combine multiple integrals with vector functions to derive the magnificent vector integral theorems with which we calculate surface area and fluid flow. Multiple integrals also play an important part in statistics, electrical engineering, and the theory of electromagnetism.

13.1 Double Integrals

We now show how to integrate a continuous function $f(x, y)$ over a bounded region in the xy-plane. We begin with rectangular regions and then proceed to bounded regions of a more general nature. There are many similarities between the "double" integrals we define here and the "single" integrals we defined in Chapter 4 for functions of a single variable. Indeed, the connection is very strong. The basic theorem for evaluating double integrals says that every double integral can be evaluated in stages, using the single-integration methods already at our command.

Double Integrals Over Rectangles

Suppose that $f(x, y)$ is defined on a rectangular region R given by

$$R: \quad a \le x \le b, \quad c \le y \le d.$$

We imagine R to be covered by a network of lines parallel to the x- and y-axes

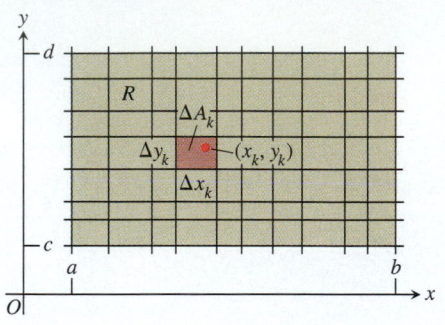

13.1   Rectangular grid partitioning the region R into small rectangles of area $\Delta A_k = \Delta x_k \Delta y_k$.

(Fig. 13.1). These lines partition R into small pieces of area

$$\Delta A = \Delta x \, \Delta y.$$

We number these in some order $\Delta A_1, \Delta A_2, \ldots, \Delta A_n$, choose a point (x_k, y_k) in each piece ΔA_k, and form the sum

$$S_n = \sum_{k=1}^{n} f(x_k, y_k) \, \Delta A_k. \tag{1}$$

If f is continuous throughout R, then, as we refine the mesh width to make both Δx and Δy go to zero, the sums in (1) approach a limit called the **double integral** of f over R. The notation for it is

$$\iint\limits_{R} f(x, y) \, dA.$$

Thus,

$$\iint\limits_{R} f(x, y) \, dA = \lim_{\Delta A \to 0} \sum_{k=1}^{n} f(x_k, y_k) \, \Delta A_k. \tag{2}$$

As with continuous functions of a single variable, the sums approach this limit no matter how the intervals $[a, b]$ and $[c, d]$ that determine R are partitioned, as long as the norms of the partitions both go to zero. The limit in (2) is also independent of the order in which the areas ΔA_k are numbered and independent of the choice of the point (x_k, y_k) within each ΔA_k. The values of the individual approximating sums S_n depend on these choices, but the sums approach the same limit in the end. The proof of the existence and uniqueness of this limit for a continuous function f is given in more advanced texts. The continuity of f is a sufficient, but not a necessary, condition for the existence of the double integral, and the limit in question exists for many discontinuous functions as well.

Properties of Double Integrals

Like single integrals, double integrals of continuous functions have algebraic properties that are useful in computations and applications. Among these properties are the following.

1. $\displaystyle\iint\limits_{R} k f(x, y) \, dA = k \iint\limits_{R} f(x, y) \, dA$ (any number k)

2. $\displaystyle\iint\limits_{R} (f(x, y) + g(x, y)) \, dA = \iint\limits_{R} f(x, y) \, dA + \iint\limits_{R} g(x, y) \, dA$

3. $\displaystyle\iint\limits_{R} (f(x, y) - g(x, y)) \, dA = \iint\limits_{R} f(x, y) \, dA - \iint\limits_{R} g(x, y) \, dA$

4. $\displaystyle\iint\limits_{R} f(x, y) \, dA \geq 0$ if $f(x, y) \geq 0$ on R

5. $\displaystyle\iint\limits_{R} f(x, y) \, dA \geq \iint\limits_{R} g(x, y) \, dA$ if $f(x, y) \geq g(x, y)$ on R

These are like the single-integral properties in Section 4.4. In addition, double inte-

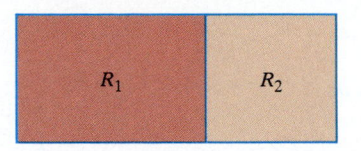

$$\iint_{R_1 \cup R_2} f(x, y)\ dA = \iint_{R_1} f(x, y)\ dA + \iint_{R_2} f(x, y)\ dA$$

13.2 Double integrals have the same kind of domain additivity property that single integrals have.

grals have a "domain additivity" property:

$$6. \iint_{R} f(x, y)\ dA = \iint_{R_1} f(x, y)\ dA + \iint_{R_2} f(x, y)\ dA.$$

The property holds when R is the union of two nonoverlapping rectangles R_1 and R_2 (Fig. 13.2). We shall omit the proofs.

Interpreting Double Integrals as Volumes

When $f(x, y)$ is positive, we may interpret the double integral of f over a rectangular region R as the volume of the solid prism bounded below by R and above by the surface $z = f(x, y)$ (Fig. 13.3). Each term $f(x_k, y_k)\ \Delta A_k$ in the sum $S_n = \Sigma f(x_k, y_k)\ \Delta A_k$ is the volume of a vertical rectangular prism that approximates the volume of the portion of the solid that stands directly above the base ΔA_k. The sum S_n thus approximates what we want to call the volume of the solid. We *define* this volume to be

$$\text{Volume} = \lim S_n = \iint_{R} f(x, y)\ dA. \tag{3}$$

As you might expect, this more general method of calculating volume agrees with the methods in Chapter 5 but we shall not prove this here.

Fubini's Theorem for Evaluating Double Integrals

We are now ready to evaluate our first double integral.

Suppose we wish to calculate the volume under the plane $z = 4 - x - y$ over the rectangular region $R: 0 \le x \le 2$, $0 \le y \le 1$ in the xy-plane. If we apply the method of slicing from Section 5.2, with slices perpendicular to the x-axis

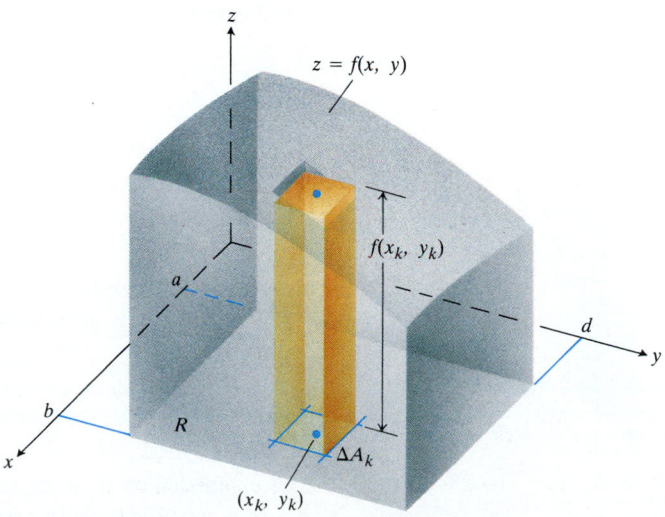

13.3 Approximating solids with rectangular prisms leads us to define the volumes of more general prisms as double integrals. The volume of the prism shown here is the double integral of $f(x, y)$ over the base region R.

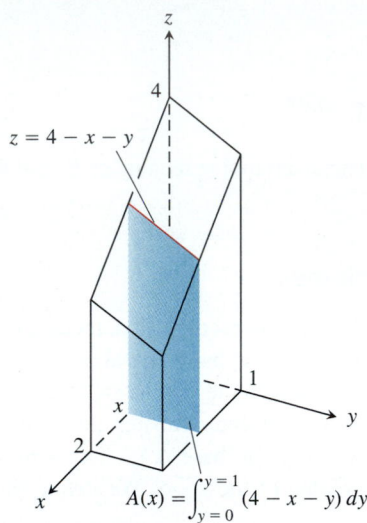

13.4 To obtain the cross-section area $A(x)$, we hold x fixed and integrate with respect to y.

(Fig. 13.4), then the volume is

$$\int_{x=0}^{x=2} A(x)\, dx, \tag{4}$$

where $A(x)$ is the cross-section area at x. For each value of x we may calculate $A(x)$ as the integral

$$A(x) = \int_{y=0}^{y=1} (4 - x - y)\, dy, \tag{5}$$

which is the area under the curve $z = 4 - x - y$ in the plane of the cross section at x. In calculating $A(x)$, x is held fixed and the integration takes place with respect to y. Combining (4) and (5), we see that the volume of the entire solid is

$$
\begin{aligned}
\text{Volume} &= \int_{x=0}^{x=2} A(x)\, dx = \int_{x=0}^{x=2} \left(\int_{y=0}^{y=1} (4 - x - y)\,dy \right) dx \\
&= \int_{x=0}^{x=2} \left[4y - xy - \frac{y^2}{2} \right]_{y=0}^{y=1} dx = \int_{x=0}^{x=2} \left(\frac{7}{2} - x \right) dx = \left[\frac{7}{2}x - \frac{x^2}{2} \right]_0^2 = 5.
\end{aligned} \tag{6}
$$

If we had just wanted to write instructions for calculating the volume, without carrying out any of the integrations, we could have written

$$\text{Volume} = \int_0^2 \int_0^1 (4 - x - y)\, dy\, dx.$$

The expression on the right, called an **iterated** or **repeated integral,** says that the volume is obtained by integrating $4 - x - y$ with respect to y from $y = 0$ to $y = 1$, holding x fixed, and then by integrating the resulting expression in x with respect to x from $x = 0$ to $x = 2$.

What would have happened if we had calculated the volume by slicing with planes perpendicular to the y-axis (Fig. 13.5)? As a function of y, the typical cross-section area is

$$A(y) = \int_{x=0}^{x=2} (4 - x - y)\, dx = \left[4x - \frac{x^2}{2} - xy \right]_{x=0}^{x=2} = 6 - 2y. \tag{7}$$

The volume of the entire solid is therefore

$$\text{Volume} = \int_{y=0}^{y=1} A(y)\, dy = \int_{y=0}^{y=1} (6 - 2y)\, dy = \left[6y - y^2 \right]_0^1 = 5,$$

in agreement with our earlier calculation.

Again, we may give instructions for calculating the volume as an iterated integral by writing

$$\text{Volume} = \int_0^1 \int_0^2 (4 - x - y)\, dx\, dy.$$

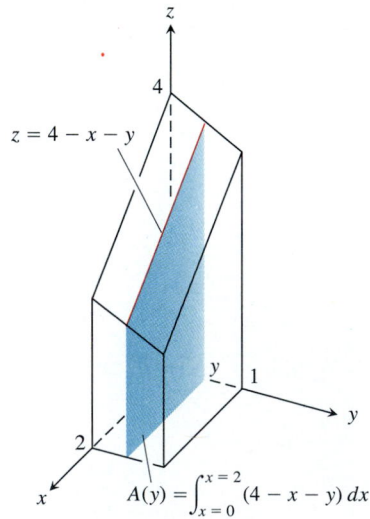

13.5 To obtain the cross-section area $A(y)$, we hold y fixed and integrate with respect to x.

The expression on the right says that the volume may be obtained by integrating $4 - x - y$ with respect to x from $x = 0$ to $x = 2$ (as we did in Eq. 7) and by integrating the result with respect to y from $y = 0$ to $y = 1$. In this iterated integral the order of integration is first x and then y, the reverse of the order we used in Eq. (6).

What do these two volume calculations with iterated integrals have to do with the double integral

$$\iint\limits_R (4 - x - y)\, dA$$

over the rectangle $R: 0 \le x \le 2$, $0 \le y \le 1$? The answer is that they both give the value of the double integral. A theorem proved by Guido Fubini (1879–1943) and published in 1907 says that the double integral of any continuous function over a rectangle can always be calculated as an iterated integral in either order of integration. (Fubini proved his theorem in greater generality, but this is how it translates into what we are doing now.)

THEOREM 1

Fubini's Theorem (First Form)

If $f(x, y)$ is continuous on the rectangular region $R: a \le x \le b$, $c \le y \le d$, then

$$\iint\limits_R f(x, y)\, dA = \int_c^d \int_a^b f(x, y)\, dx\, dy = \int_a^b \int_c^d f(x, y)\, dy\, dx.$$

Fubini's theorem says that double integrals over rectangles can always be calculated as iterated integrals. This means that we can evaluate a double integral by integrating one variable at a time, using the integration techniques we already know for functions of a single variable.

Fubini's theorem also says that we may calculate the double integral by integrating in *either* order, a genuine convenience, as we shall see in Example 3. In particular, when we calculate a volume by slicing, we may use either planes perpendicular to the x-axis or planes perpendicular to the y-axis. We get the same answer both ways.

Even more important is the fact that Fubini's theorem holds for *any* continuous function $f(x, y)$. In particular, f may have negative values as well as positive values on R, and, as we shall see later, the integrals we evaluate with Fubini's theorem may represent other things besides volumes.

Example 1   Calculate $\iint_R f(x, y)\, dA$ for

$$f(x, y) = 1 - 6x^2 y \qquad \text{and} \qquad R: \quad 0 \le x \le 2, \quad -1 \le y \le 1.$$

Solution By Fubini's theorem,

$$\iint\limits_R f(x, y)\, dA = \int_{-1}^1 \int_0^2 (1 - 6x^2 y)\, dx\, dy = \int_{-1}^1 \Big[x - 2x^3 y \Big]_{x=0}^{x=2} dy$$

$$= \int_{-1}^1 (2 - 16y)\, dy = \Big[2y - 8y^2 \Big]_{-1}^1 = 4.$$

Reversing the order of integration gives the same answer:

$$\int_0^2 \int_{-1}^1 (1 - 6x^2 y)\, dy\, dx = \int_0^2 \Big[y - 3x^2 y^2 \Big]_{y=-1}^{y=1} dx$$

$$= \int_0^2 \Big[(1 - 3x^2) - (-1 - 3x^2) \Big] dx = \int_0^2 2\, dx = 4.$$

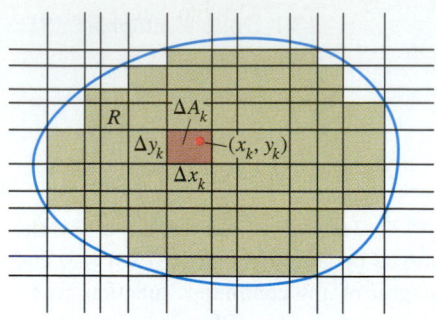

13.6 A rectangular grid partitioning a bounded nonrectangular region into cells.

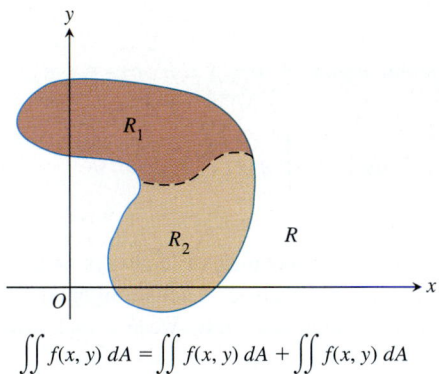

$$\iint\limits_{R} f(x, y)\, dA = \iint\limits_{R_1} f(x, y)\, dA + \iint\limits_{R_2} f(x, y)\, dA$$

13.7 The domain additivity property stated earlier for rectangular regions also holds for regions bounded by continuous curves.

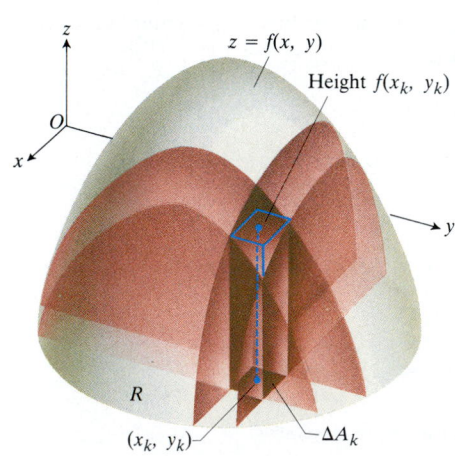

$$\text{Volume} = \lim \Sigma f(x_k, y_k)\Delta A_k = \iint f(x, y)\, dA$$

13.8 We define the volumes of solids with curved bases the same way we define the volumes of solids with rectangular bases.

Double Integrals over Bounded Nonrectangular Regions

To define the double integral of a function $f(x, y)$ over a bounded nonrectangular region, like the one shown in Fig. 13.6, we again imagine R to be covered by a rectangular grid, but we include in the partial sum only the small pieces of area $\Delta A = \Delta x\, \Delta y$ that lie entirely within the region (shaded in the figure). We number the pieces in some order, choose an arbitrary point (x_k, y_k) in each ΔA_k, and form the sum

$$S_n = \sum_{k=1}^{n} f(x_k, y_k)\, \Delta A_k.$$

The only difference between this sum and the one in Eq. (1) for rectangular regions is that now the areas ΔA_k may not cover all of R. But as the mesh becomes increasingly fine and the number of terms in S_n increases, more and more of R is included. If f is continuous and the boundary of R is made from the graphs of a finite number of continuous functions of x and/or continuous functions of y joined end to end, then the sums S_n will have a limit as the norms of the partitions that define the rectangular grid independently approach zero. We call the limit the **double integral** of f over R:

$$\iint\limits_{R} f(x, y)\, dA = \lim_{\Delta A \to 0} \sum f(x_k, y_k)\, \Delta A_k.$$

This limit may also exist under less restrictive circumstances, but we shall not pursue the point here.

Double integrals of continuous functions over nonrectangular regions have all the algebraic properties listed earlier for integrals over rectangular regions. The domain additivity property corresponding to property 6 says that if R is decomposed into nonoverlapping regions R_1 and R_2 with boundaries that are again made of a finite number of line segments or smooth curves (see Fig. 13.7), then

$$6'. \iint\limits_{R} f(x, y)\, dA = \iint\limits_{R_1} f(x, y)\, dA + \iint\limits_{R_2} f(x, y)\, dA.$$

If $f(x, y)$ is positive and continuous over R (Fig. 13.8), we define the volume of the solid region between R and the surface $z = f(x, y)$ to be $\iint_R f(x, y)\, dA$, as before.

If R is a region like the one shown in the xy-plane in Fig. 13.9, bounded "above" and "below" by the curves $y = g_2(x)$ and $y = g_1(x)$ and on the sides by the lines $x = a$, $x = b$, we may again calculate the volume by the method of slicing. We first calculate the cross-section area

$$A(x) = \int_{y=g_1(x)}^{y=g_2(x)} f(x, y)\, dy$$

and then integrate $A(x)$ from $x = a$ to $x = b$ to get the volume as an iterated integral:

$$V = \int_{a}^{b} A(x)\, dx = \int_{a}^{b} \int_{g_1(x)}^{g_2(x)} f(x, y)\, dy\, dx. \tag{8}$$

Similarly, if R is a region like the one shown in Fig. 13.10, bounded by the curves $x = h_2(y)$ and $x = h_1(y)$ and the lines $y = c$ and $y = d$, then the volume calculated by slicing is given by the iterated integral

$$\text{Volume} = \int_{c}^{d} \int_{h_1(y)}^{h_2(y)} f(x, y)\, dx\, dy. \tag{9}$$

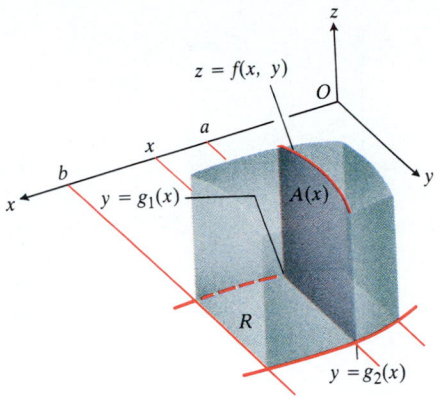

13.9 The area of the vertical slice shown here is

$$A(x) = \int_{g_1(x)}^{g_2(x)} f(x, y)\, dy.$$

To calculate the volume of the solid we integrate this area from $x = a$ to $x = b$.

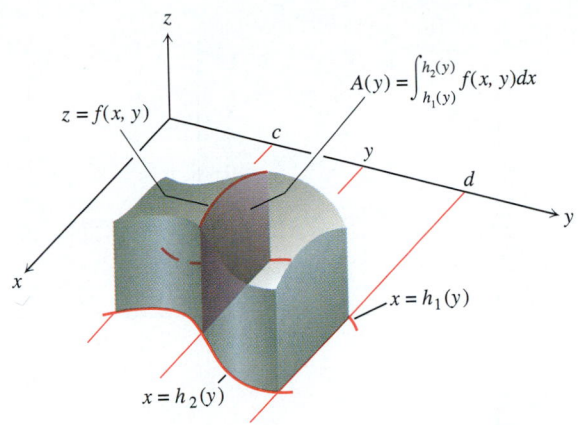

13.10 The volume of the solid shown here is

$$\int_c^d A(y)\, dy = \int_c^d \int_{h_1(y)}^{h_2(y)} f(x, y)\, dx\, dy.$$

The fact that the iterated integrals in Eqs. (8) and (9) both give the volume that we defined to be the double integral of f over R is a consequence of the following stronger form of Fubini's theorem.

THEOREM 2

Fubini's Theorem (Stronger Form)

Let $f(x, y)$ be continuous on a region R.

1. If R is defined by $a \le x \le b$, $g_1(x) \le y \le g_2(x)$, with g_1 and g_2 continuous on $[a, b]$, then

$$\iint_R f(x, y)\, dA = \int_a^b \int_{g_1(x)}^{g_2(x)} f(x, y)\, dy\, dx.$$

2. If R is defined by $c \le y \le d$, $h_1(y) \le x \le h_2(y)$, with h_1 and h_2 continuous on $[c, d]$, then

$$\iint_R f(x, y)\, dA = \int_c^d \int_{h_1(y)}^{h_2(y)} f(x, y)\, dx\, dy.$$

Example 2   Find the volume of the prism whose base is the triangle in the xy-plane bounded by the x-axis and the lines $y = x$ and $x = 1$ and whose top lies in the plane

$$z = f(x, y) = 3 - x - y.$$

Solution   See Fig. 13.11(a). For any x between 0 and 1, y may vary from $y = 0$ to $y = x$ (Fig. 13.11b). Hence,

$$V = \int_0^1 \int_0^x (3 - x - y)\, dy\, dx = \int_0^1 \left[3y - xy - \frac{y^2}{2}\right]_{y=0}^{y=x} dx$$

$$= \int_0^1 \left(3x - \frac{3x^2}{2}\right) dx = \left[\frac{3x^2}{2} - \frac{x^3}{2}\right]_{x=0}^{x=1} = 1.$$

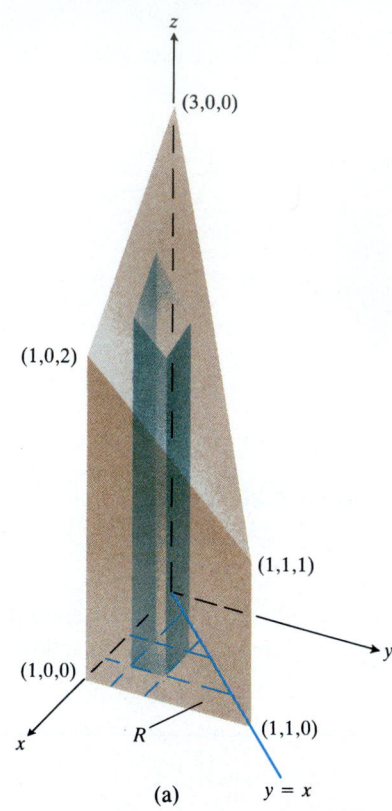

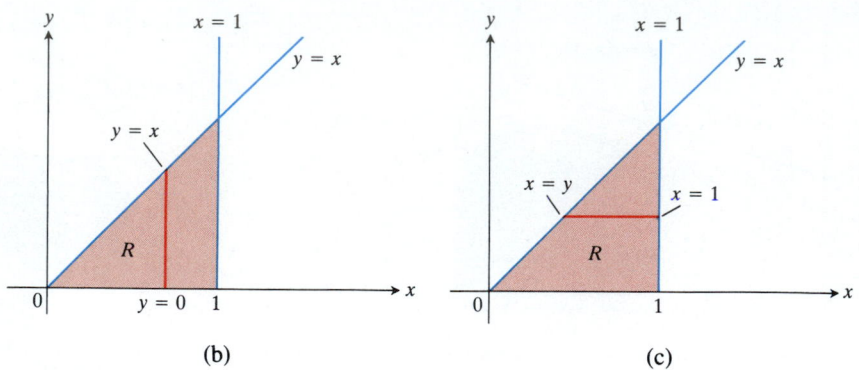

(b)

(c)

13.11   (a) Prism with a triangular base in the xy-plane. The volume of this prism is defined as a double integral over R. To evaluate it as an iterated integral, we may integrate first with respect to y and then with respect to x, or the other way around (Example 2).

(b) Integration limits of

$$\int_{x=0}^{x=1} \int_{y=0}^{y=x} f(x, y)\, dy\, dx.$$

If we integrate first with respect to y, we integrate along a vertical line through R and then integrate from left to right to include all the vertical lines in R.

(c) Integration limits of

$$\int_{y=0}^{y=1} \int_{x=y}^{x=1} f(x, y)\, dx\, dy.$$

If we integrate first with respect to x, we integrate along a horizontal line through R and then integrate from bottom to top to include all the horizontal lines in R.

When the order of integration is reversed (Fig. 13.11c), the integral for the volume is

$$V = \int_0^1 \int_y^1 (3 - x - y)\, dx\, dy = \int_0^1 \left[3x - \frac{x^2}{2} - xy \right]_{x=y}^{x=1} dy$$

$$= \int_0^1 \left(3 - \frac{1}{2} - y - 3y + \frac{y^2}{2} + y^2 \right) dy$$

$$= \int_0^1 \left(\frac{5}{2} - 4y + \frac{3}{2} y^2 \right) dy$$

$$= \left[\frac{5}{2} y - 2y^2 + \frac{y^3}{2} \right]_{y=0}^{y=1} = 1.$$

The two integrals are equal, as they should be.

While Fubini's theorem assures us that a double integral may be calculated as an iterated integral in either order of integration, the value of one integral may be easier to find than the value of the other. The next example shows how this can happen.

Example 3   Calculate

$$\iint_R \frac{\sin x}{x}\, dA,$$

where R is the triangle in the xy-plane bounded by the x-axis, the line $y = x$, and the line $x = 1$.

Solution   The region of integration is shown in Fig. 13.12. If we integrate first with respect to y and then with respect to x, we find

$$\int_0^1 \left(\int_0^x \frac{\sin x}{x}\, dy \right) dx = \int_0^1 \left(y \frac{\sin x}{x} \Big]_{y=0}^{y=x} \right) dx$$

$$= \int_0^1 \sin x\, dx = -\cos(1) + 1 \approx 0.46.$$

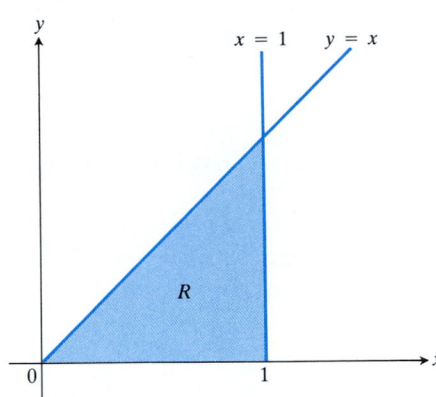

If we reverse the order of integration and attempt to calculate

$$\int_0^1 \int_y^1 \frac{\sin x}{x}\, dx\, dy,$$

we are stopped by the fact that $\int ((\sin x)/x)\, dx$ cannot be expressed in terms of elementary functions.

There is no general rule for predicting which order of integration (if either) will be a good one in circumstances like these, so do not worry about how to start your integrations. Just forge ahead and if the order you choose first does not work, try the other.

Finding the Limits of Integration

13.12   The region of integration in Example 3.

The hardest part of evaluating a double integral can be finding the limits of integration. Fortunately, there is a good procedure to follow.

Procedure for Finding Limits of Integration

A. To evaluate $\iint_R f(x, y)\, dA$ over a region R, integrating first with respect to y and then with respect to x, take the following steps:

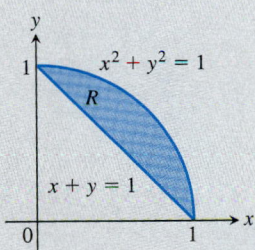   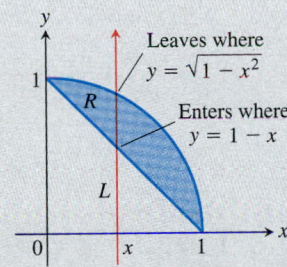   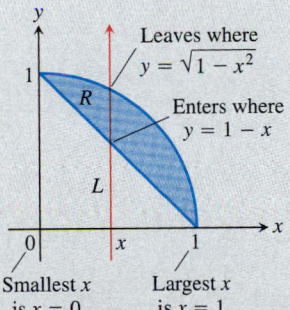

1. Sketch the region of integration and label the bounding curves.

2. Imagine a vertical line L cutting through R in the direction of increasing y. Mark the y-values where L enters and leaves. These are the y-limits of integration.

3. Choose x-limits that include all the vertical lines through R. The integral is

$$\iint_R f(x, y)\, dA =$$

$$\int_{x=0}^{x=1} \int_{y=1-x}^{y=\sqrt{1-x^2}} f(x, y)\, dy\, dx.$$

B. To evaluate the same double integral as an iterated integral with the order of integration reversed, use horizontal lines instead of vertical lines. The integral is

$$\iint_R f(x, y)\, dA = \int_0^1 \int_{1-y}^{\sqrt{1-y^2}} f(x, y)\, dx\, dy.$$

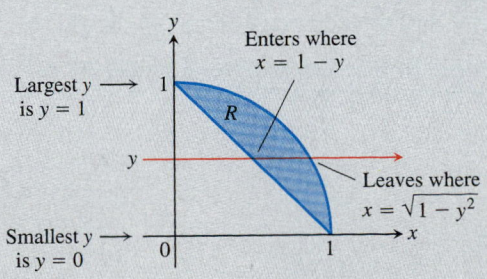

Example 4   Sketch the region over which the integration

$$\int_0^2 \int_{x^2}^{2x} (4x + 2) \, dy \, dx$$

takes place and write an equivalent integral with the order of integration reversed.

Solution   The region of integration is given by the inequalities $x^2 \leq y \leq 2x$ and $0 \leq x \leq 2$. It is therefore the region bounded by the curves $y = x^2$ and $y = 2x$ between $x = 0$ and $x = 2$ (Fig. 13.13a).

To find the limits for integrating in the reverse order, we imagine a horizontal line passing from left to right through the region. It enters at $x = y/2$ and leaves at $x = \sqrt{y}$. To include all such lines, we let y run from $y = 0$ to $y = 4$ (Fig. 13.13b). The integral is

$$\int_0^4 \int_{y/2}^{\sqrt{y}} (4x + 2) \, dx \, dy.$$

The common value of these integrals is 8.

13.13   To write

$$\int_0^2 \int_{x^2}^{2x} (4x + 2) \, dy \, dx$$

as a double integral with the order of integration reversed, we (a) sketch the region of integration and (b) work out what the new limits of integration have to be. Example 4 gives the details.

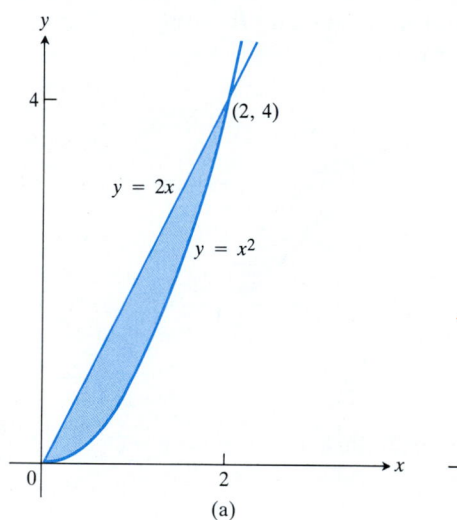

(a)

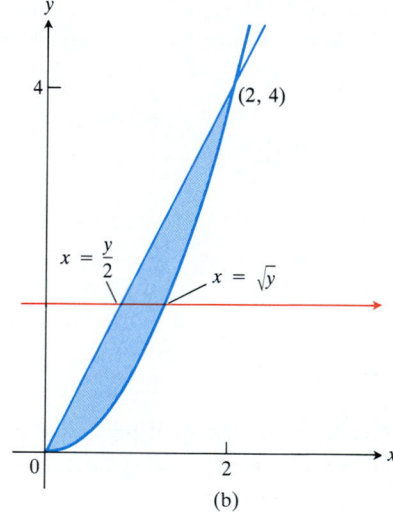

(b)

EXERCISES 13.1

In Exercises 1–10 sketch the region of integration and then evaluate the integral.

1. $\int_0^3 \int_0^2 (4 - y^2) \, dy \, dx$

2. $\int_0^3 \int_{-2}^0 (x^2 y - 2xy) \, dy \, dx$

3. $\int_{-1}^0 \int_{-1}^1 (x + y + 1) \, dx \, dy$

4. $\int_\pi^{2\pi} \int_0^\pi (\sin x + \cos y) \, dx \, dy$

5. $\int_0^\pi \int_0^x x \sin y \, dy \, dx$

6. $\int_0^\pi \int_0^{\sin x} y \, dy \, dx$

7. $\int_1^{\ln 8} \int_0^{\ln y} e^{x + y} \, dx \, dy$

8. $\int_1^2 \int_y^{y^2} dx \, dy$

9. $\displaystyle\int_{10}^{1}\int_{0}^{1/y} ye^{xy}\,dx\,dy$

10. $\displaystyle\int_{0}^{1}\int_{0}^{x^3} e^{y/x}\,dy\,dx$

28. $\displaystyle\int_{-1}^{1}\int_{-1/\sqrt{1-x^2}}^{1/\sqrt{1-x^2}} (2y+1)\,dy\,dx$

29. $\displaystyle\int_{-\infty}^{\infty}\int_{-\infty}^{\infty} \frac{1}{(x^2+1)(y^2+1)}\,dx\,dy$

30. $\displaystyle\int_{0}^{\infty}\int_{0}^{\infty} xe^{-(x+2y)}\,dx\,dy$

In Exercises 11–16, integrate f over the given region.

11. $f(x, y) = x/y$ over the region in the first quadrant bounded by the lines $y = x$, $y = 2x$, $x = 1$, $x = 2$

12. $f(x, y) = x^2 + y^2$ over the triangular region with vertices $(0, 0)$, $(1, 0)$, and $(0, 1)$

13. $f(x, y) = 1/xy$ over the square $1 \le x \le 2$, $1 \le y \le 2$

14. $f(x, y) = y \cos xy$ over the rectangle $0 \le x \le \pi$, $0 \le y \le 1$

15. $f(u, v) = v - \sqrt{u}$ over the triangular region cut from the first quadrant of the uv-plane by the line $u + v = 1$

16. $f(s, t) = e^s \ln t$ over the region in the first quadrant of the st-plane that lies under the curve $s = \ln t$ from $t = 1$ to $t = 2$

Each of Exercises 17–20 gives an integral over a region in a Cartesian coordinate plane. Sketch the region and evaluate the integral.

17. $\displaystyle\int_{-2}^{0}\int_{v}^{-v} 2\,dp\,dv$ (the pv-plane)

18. $\displaystyle\int_{0}^{1}\int_{0}^{\sqrt{1-s^2}} 8t\,dt\,ds$ (the st-plane)

19. $\displaystyle\int_{-\pi/3}^{\pi/3}\int_{0}^{\sec t} 3\cos t\,du\,dt$ (the tu-plane)

20. $\displaystyle\int_{0}^{3}\int_{-2}^{4-2u} \frac{4-2u}{v^2}\,dv\,du$ (the uv-plane)

In Exercises 21–26, sketch the region of integration and write an equivalent integral with the order of integration reversed. Then evaluate both integrals to confirm their equality.

21. $\displaystyle\int_{0}^{1}\int_{2}^{4-2x} dy\,dx$

22. $\displaystyle\int_{0}^{1}\int_{y}^{\sqrt{y}} dx\,dy$

23. $\displaystyle\int_{0}^{1}\int_{\sqrt{y}}^{1} dx\,dy$

24. $\displaystyle\int_{0}^{1}\int_{1}^{e^x} dy\,dx$

25. $\displaystyle\int_{0}^{3/2}\int_{0}^{9-4x^2} 16x\,dy\,dx$

26. $\displaystyle\int_{0}^{1}\int_{-\sqrt{1-y^2}}^{\sqrt{1-y^2}} 3y\,dx\,dy$

Evaluate the improper integrals in Exercises 27–30 as iterated integrals.

27. $\displaystyle\int_{1}^{\infty}\int_{e^{-x}}^{1} \frac{1}{x^3 y}\,dy\,dx$

In Exercises 31–38, sketch the region of integration and evaluate the integral.

31. $\displaystyle\int_{0}^{\pi}\int_{x}^{\pi} \frac{\sin y}{y}\,dy\,dx$

32. $\displaystyle\int_{0}^{1}\int_{y}^{1} x^2 e^{xy}\,dx\,dy$

33. $\displaystyle\int_{0}^{2}\int_{x}^{2} 2y^2\sin xy\,dy\,dx$

34. $\displaystyle\int_{0}^{2}\int_{0}^{4-x^2} \frac{xe^{2y}}{4-y}\,dy\,dx$

35. $\displaystyle\int_{0}^{2\sqrt{\ln 3}}\int_{y/2}^{\sqrt{\ln 3}} e^{x^2}\,dx\,dy$

36. $\displaystyle\int_{0}^{1/16}\int_{y^{1/4}}^{1/2} \cos(16\pi x^5)\,dx\,dy$

37. $\displaystyle\int_{0}^{3}\int_{\sqrt{x/3}}^{1} e^{y^3}\,dy\,dx$

38. $\displaystyle\int_{0}^{8}\int_{\sqrt[3]{x}}^{2} \frac{dy\,dx}{y^4+1}$

39. Find the volume of the region that lies under the paraboloid $z = x^2 + y^2$ and above the triangle enclosed by the lines $y = x$, $x = 0$, and $x + y = 2$ in the xy-plane.

40. Find the volume of the solid that is bounded above by the cylinder $z = x^2$ and below by the region enclosed by the parabola $y = 4 - x^2$ and the line $y = x$ in the xy-plane.

41. Find the volume of the solid whose base is the region in the xy-plane that is bounded by the parabola $y = 4 - x^2$ and the line $y = 3x$, while the top of the solid is bounded by the plane $z = x + 4$.

42. Find the volume of the solid in the first octant bounded by the coordinate planes, the cylinder $x^2 + y^2 = 4$, and the plane $z + y = 3$.

43. Find the volume of the solid in the first octant bounded by the coordinate planes, the plane $x = 3$, and the parabolic cylinder $z = 4 - y^2$.

44. Find the volume of the solid cut from the first octant by the surface $z = 4 - x^2 - y$.

45. Find the volume of the wedge cut from the first octant by the cylinder $z = 12 - 3y^2$ and the plane $x + y = 2$.

46. Find the volume of the solid cut from the square column $|x| + |y| \le 1$ by the planes $z = 0$ and $3x + z = 3$.

47. Find the volume of the solid that is bounded on the front and back by the planes $x = 2$ and $x = 1$, on the sides by the cylinders $y = \pm 1/x$, and above and below by the planes $z = x + 1$ and $z = 0$.

48. Find the volume of the solid that is bounded on the front and back by the planes $x = \pm\pi/3$, on the sides by the cylinders $y = \pm\sec x$, above by the cylinder $z = 1 + y^2$, and below by the xy-plane.

49. Integrate $f(x, y) = \sqrt{4 - x^2}$ over the smaller sector cut from the disk $x^2 + y^2 \le 4$ by the rays $\theta = \pi/6$ and $\theta = \pi/2$.

50. Integrate $f(x, y) = 1/(x^2 - x)(y - 1)^{2/3}$ over the infinite rectangle $2 \le x < \infty$, $0 \le y \le 2$.

51. A solid right (noncircular) cylinder has its base R in the xy-plane and is bounded from above by the paraboloid $z = x^2 + y^2$. The cylinder's volume is

$$V = \int_0^1 \int_0^y (x^2 + y^2)\, dx\, dy + \int_1^2 \int_0^{2-y} (x^2 + y^2)\, dx\, dy.$$

Sketch the base region R and express the cylinder's volume as a single iterated integral with the order of integration reversed. Then evaluate the integral to find the volume.

52. Evaluate the integral

$$\int_0^2 (\tan^{-1}\pi x - \tan^{-1}x)\, dx.$$

(*Hint:* Write the integrand as an integral.)

53. Over what region R in the xy-plane does

$$\iint_R (4 - x^2 - 2y^2)\, dA$$

have its maximum value?

54. Over what region R in the xy-plane does

$$\iint_R \left[(x^2 + y^2)^2 - 5(x^2 + y^2) + 4 \right] dA$$

have its minimum value?

Numerical Evaluation

Just as there are methods for evaluating single integrals numerically, there are corresponding methods for the numerical evaluation of double integrals. You will find a computer program for the two-variable trapezoidal rule immediately following this exercise set. If you can implement this program or if you have access to a double-integral evaluator of some other kind, estimate the values of the integrals in Exercises 55–58.

55. $\displaystyle \int_1^3 \int_1^x \frac{1}{xy}\, dy\, dx$

56. $\displaystyle \int_0^1 \int_0^1 e^{-(x^2 + y^2)}\, dy\, dx$

57. $\displaystyle \int_0^1 \int_0^1 \tan^{-1}xy\, dy\, dx$

58. $\displaystyle \int_{-1}^1 \int_0^{\sqrt{1-x^2}} 3\sqrt{1 - x^2 - y^2}\, dy\, dx$

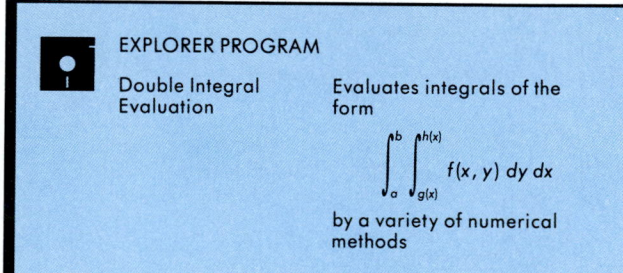

EXPLORER PROGRAM

Double Integral Evaluation

Evaluates integrals of the form

$$\int_a^b \int_{g(x)}^{h(x)} f(x, y)\, dy\, dx$$

by a variety of numerical methods

✳ A COMPUTER PROGRAM FOR EVALUATING DOUBLE INTEGRALS NUMERICALLY

The BASIC program listed here uses the trapezoidal rule to generate numerical approximations of integrals of the form

$$\int_a^b \int_{g_1(x)}^{g_2(x)} f(x, y)\, dy\, dx.$$

In the notation of the program, a is X0, b is XF (X final), and the interval $[a, b]$ is partitioned into NX subintervals of equal length. The vertical line segments that run from the curve $y = g_1(x)$ to the curve $y = g_2(x)$ above the partition points are each partitioned into NY subintervals of equal length (Fig. 13.14). These latter lengths vary with the length of the segment. The program calculates a trapezoidal sum of f-values along each vertical segment (as indicated in red in the figure) and then calculates a trapezoidal sum of these sums (indicated in black) to give the approximation for the region.

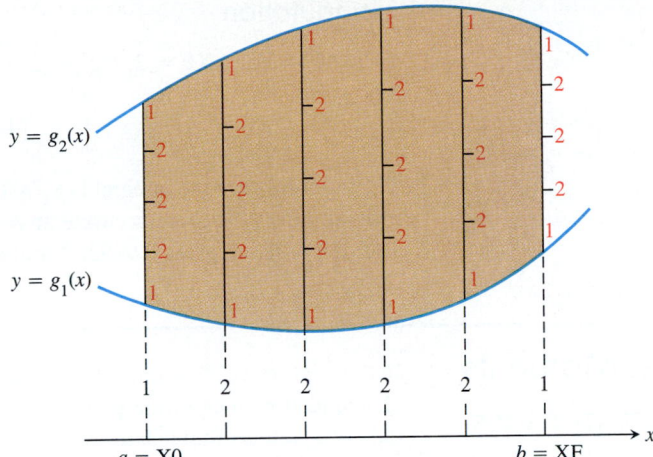

13.14   The partitions used in a typical trapezoidal rule estimate of the value of

$$\int_{a}^{b} \int_{y=g_1(x)}^{y=g_2(x)} f(x, y)\, dy\, dx.$$

For more accuracy, you can use Simpson's rule in place of the trapezoidal rule. You can find detailed descriptions of these and other methods in any standard text on numerical methods.

Program Listing

Here is a listing of the program with $f(x, y) = 4x + 2$, $g_1(x) = x^2$, and $g_2(x) = 2x$.

PROGRAM	COMMENT
10 NX = 20	NX is the number of x-intervals
20 NY = 20	NY is the number of y-intervals
30 DIM WX(NX + 1), WY(NY + 1)	Dimension arrays to hold weighting factors
40 DEF FN F(X, Y) = 4*X + 2	Define the function to be integrated
50 DEF FN G1(X) = X^2	Function G1(x) is the lower boundary
60 DEF FN G2(X) = 2*X	and G2(x) is the upper boundary
70 INPUT "ENTER X0, XF ", X0, XF	The left and right limits for x
80 DX = (XF − X0)/NX	Compute the value for delta x
90 FOR I = 2 TO NX: WX(I) = 2: NEXT I	Put the weights for x into
100 WX(1) = 1: WX(NX + 1) = 1	the WX array
110 FOR I = 2 TO NY: WY(I) = 2: NEXT I	Do the same for
120 WY(1) = 1: WY(NY + 1) = 1	the WY array
130 SUM = 0	Initialize the sum at zero
140 FOR I = 1 TO NX + 1	Begin a loop through all points in the region
150 X = X0 + (I − 1)*DX	Compute the x-value
160 Y0 = FN G1(X)	Compute the lower y-value
170 DY = (FN G2(X) − FN G1(X))/NY	and the delta-y at this x-value
180 FOR J = 1 TO NY + 1	Begin an inner loop at a constant value for x
190 Y = Y0 + (J − 1)*DY	moving up for the y's and
200 F = FN F(X, Y)	compute a function value at each point
210 SUM = SUM + WX(I)*WY(J)*F*DX/2*DY/2	and add to the sum after weighting it
220 NEXT J	End the inner loop
230 NEXT I	End the outer loop
240 PRINT "THE VALUE OF THE INTEGRAL IS ABOUT ";SUM	Print the answer
250 END	

Computation

With X0 = 0 and XF = 2, the program returns the value 7.980003 for

$$\int_0^2 \int_{x^2}^{2x} (4x + 2)\, dy\, dx.$$

The exact value of the integral is 8. With NX and NY equal to 50 instead of 20, the program gives the more accurate answer of 7.996799. With NX and NY equal to 100, the program gives 7.999367, but the calculation takes considerably longer.

13.2 Areas, Moments, and Centers of Mass

In this section we show how to use double integrals to define and calculate the areas of bounded regions in the plane and the masses, moments, centers of mass, and radii of gyration of thin plates covering these regions. The calculations are similar to the ones in Chapter 5, but we can now handle a greater variety of shapes.

Areas of Bounded Regions in the Plane

If we take $f(x, y) = 1$ in the definition of the double integral over a region R in the preceding section, the partial sums reduce to

$$S_n = \sum_{k=1}^{n} f(x_k, y_k)\Delta A_k = \sum_{k=1}^{n} \Delta A_k. \tag{1}$$

This approximates what we would like to call the area of R. As the increments in x and y independently approach zero, the coverage of R by the ΔA_k's (Fig. 13.15) becomes more nearly complete, and we define the area of R to be the limit

$$\text{Area} = \lim \sum \Delta A_k = \int\!\!\int_R dA. \tag{2}$$

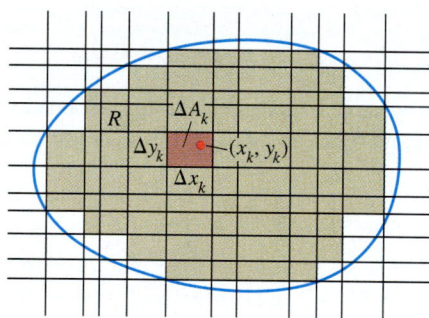

13.15   The first step in defining the area of a region is to partition the interior of the region into cells.

> **DEFINITION**
>
> The **area** of a closed bounded plane region R is the value of the integral
>
> $$\text{Area} = \int\!\!\int_R dA. \tag{3}$$

As with the other definitions in this chapter, the definition of area applies to a greater variety of regions than does the earlier single-variable definition of area, but it agrees with the earlier definition on regions to which they both apply.

To evaluate the area integral in (3), we integrate the constant function $f(x, y) = 1$ over R.

Example 1   Find the area of the region R bounded by $y = x$ and $y = x^2$ in the first quadrant.

Solution   We sketch the region (Fig. 13.16) and calculate the area as

$$A = \int_0^1 \int_{x^2}^{x} dy\, dx = \int_0^1 (x - x^2)\, dx = \left[\frac{x^2}{2} - \frac{x^3}{3}\right]_0^1 = \frac{1}{6}.$$

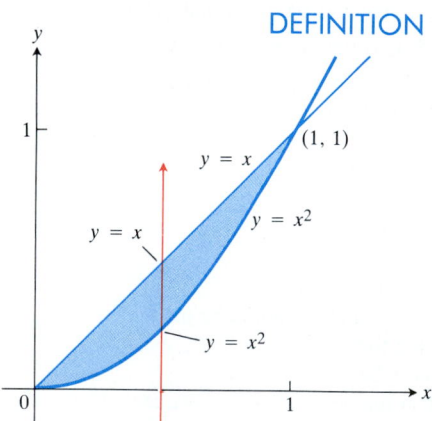

13.16   The area of the region between the parabola and the line in Example 1 is given by the double integral

$$\int_0^1 \int_{x^2}^{x} dy\, dx.$$

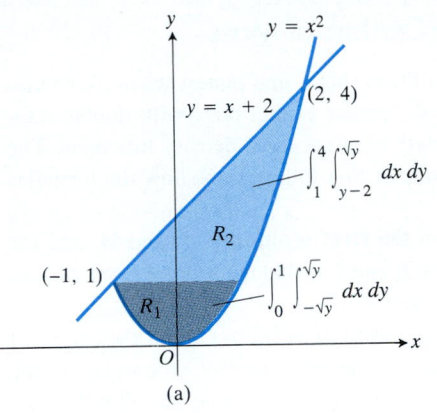

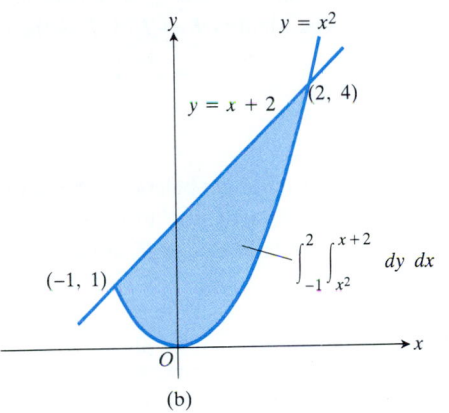

13.17 Calculating the area shown here takes (a) two integrals if the first integration is with respect to x, but (b) only one if the first integration is with respect to y (Example 2).

Example 2   Find the area of the region R enclosed by the parabola $y = x^2$ and the line $y = x + 2$.

Solution   If we divide R into the regions R_1 and R_2 shown in Fig. 13.17(a), we may calculate the area as

$$A = \int\!\!\int_{R_1} dA + \int\!\!\int_{R_2} dA = \int_0^1\!\int_{-\sqrt{y}}^{\sqrt{y}} dx\,dy + \int_1^4\!\int_{y-2}^{\sqrt{y}} dx\,dy.$$

On the other hand, reversing the order of integration (Fig. 13.17b) gives

$$A = \int_{-1}^2\!\int_{x^2}^{x+2} dy\,dx.$$

Clearly, this result is simpler and is the only one we would bother to write down in practice. Evaluation of this integral leads to the result

$$A = \int_{-1}^2 \Big[y\Big]_{x^2}^{x+2} dx = \int_{-1}^2 (x + 2 - x^2)\,dx = \left[\frac{x^2}{2} + 2x - \frac{x^3}{3}\right]_{-1}^2 = \frac{9}{2}.$$

Average Value

The average value of an integrable function of a single variable on a closed interval is the integral of the function over the interval divided by the length of the interval. For an integrable function of two variables defined on a closed and bounded region that has a measurable area, the average value is the integral over the region divided by the area of the region. If f is the function and R the region, then

$$\text{Average value of } f \text{ over } R = \frac{1}{\text{area of } R}\int\!\!\int_R f\,dA. \tag{4}$$

If f is the density of a thin plate covering R, then the double integral of f over R divided by the area of R is the plate's average density in units of mass per unit area. If $f(x, y)$ is the distance from the point (x, y) to a fixed point P, then the average value of f over R is the average distance of points in R from P.

Example 3   Find the average value of

$$f(x, y) = x \cos xy$$

over the rectangle $R: 0 \le x \le \pi, 0 \le y \le 1$.

Solution   The value of the integral of f over R is

$$\int_0^\pi\!\int_0^1 x \cos xy\,dy\,dx = \int_0^\pi \Big[\sin xy\Big]_{y=0}^{y=1} dx$$

$$= \int_0^\pi (\sin x - 0)\,dx$$

$$= -\cos x \Big]_0^\pi$$

$$= 1 + 1 = 2.$$

The area of R is π. The average value of f over R is $2/\pi$.

First and Second Moments and Centers of Mass

To find the moments and centers of mass of thin sheets and plates, we use formulas similar to those in Chapter 5. The main difference is that now, with double integrals, we can accommodate a greater variety of shapes and density functions. The formulas are given in Table 13.1 The examples that follow show how the formulas are used.

The mathematical difference between the **first moments** M_x and M_y and the **moments of inertia**, or **second moments,** I_x and I_y is that the second moments use the *squares* of the "lever-arm" distances x and y.

The moment I_0 is also called the **polar moment** of inertia about the origin. It is calculated by integrating the density $\delta(x, y)$ times $r^2 = x^2 + y^2$, the square of the distance from a representative point (x, y) to the origin. Notice that $I_0 = I_x + I_y$; once we find two, we get the third automatically. (The moment I_0 is sometimes called I_z, for moment of inertia about the z-axis. The identity $I_z = I_x + I_y$ is then called the **Perpendicular Axis Theorem.**)

The **radius of gyration** R_x is defined by the equation

$$I_x = MR_x^2.$$

It tells how far from the x-axis the entire mass of the plate might be concentrated to give the same I_x. The radius of gyration gives a convenient way to express the moment of inertia in terms of a mass and a length. The radii R_y and R_0 are defined

TABLE 13.1

Mass and moment formulas for thin plates covering regions in the *xy*-plane

Density: $\delta(x, y)$

Mass: $M = \iint \delta(x, y)\, dA$

First moments: $M_x = \iint y\delta(x, y)\, dA, \quad M_y = \iint x\delta(x, y)\, dA$

Center of mass: $\bar{x} = \dfrac{M_y}{M}, \quad \bar{y} = \dfrac{M_x}{M}$

Moments of inertia (second moments):

About the x-axis: $I_x = \iint y^2\delta(x, y)\, dA$

About the y-axis: $I_y = \iint x^2\delta(x, y)\, dA$

About the origin: $I_0 = \iint (x^2 + y^2)\delta(x, y)\, dA = I_x + I_y$

About a line L: $I_L = \iint r^2(x, y)\delta(x, y)\, dA$

   $r(x, y) = $ distance from (x, y) to L

Radii of gyration: About the x-axis: $R_x = \sqrt{I_x/M}$
About the y-axis: $R_y = \sqrt{I_y/M}$
About the origin: $R_0 = \sqrt{I_0/M}$

in a similar way, with

$$I_y = MR_y^2 \qquad \text{and} \qquad I_0 = MR_0^2.$$

We take square roots to get the formulas in Table 13.1.

Why the interest in moments of inertia? A body's first moments tell us about balance and about the torque the body exerts about different axes in a gravitational field. But if the body is a rotating shaft, we are more likely to be interested in how much energy is stored in the shaft or about how much energy it will take to accelerate the shaft to a particular angular velocity. This is where the second moment or moment of inertia comes in.

Think of partitioning the shaft into small blocks of mass Δm_k and let r_k denote the distance from the kth block's center of mass to the axis of rotation (Fig. 13.18). If the shaft rotates at an angular velocity of $\omega = d\theta/dt$ radians per second, the block's center of mass will trace its orbit at a linear speed of

$$v_k = \frac{d}{dt}\left(r_k\theta\right) = r_k \frac{d\theta}{dt} = r_k\,\omega. \tag{5}$$

The block's kinetic energy will be approximately

$$\frac{1}{2}\,\Delta m_k\,v_k^2 = \frac{1}{2}\,\Delta m_k\left(r_k\,\omega\right)^2 = \frac{1}{2}\,\omega^2\,r_k^2\,\Delta m_k. \tag{6}$$

The kinetic energy of the shaft will be approximately

$$\sum \frac{1}{2}\,\omega^2\,r_k^2\Delta m_k. \tag{7}$$

The integral approached by these sums as the shaft is partitioned into smaller and smaller blocks gives the shaft's kinetic energy:

$$\text{KE}_{\text{shaft}} = \int \frac{1}{2}\,\omega^2 r^2\,dm = \frac{1}{2}\,\omega^2 \int r^2\,dm. \tag{8}$$

The factor

$$I = \int r^2\,dm \tag{9}$$

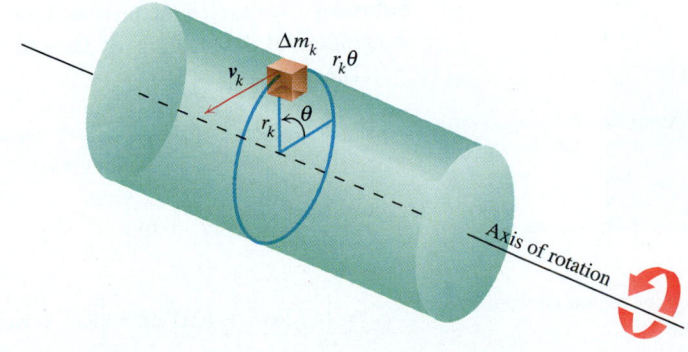

13.18 To find an integral for the amount of energy stored in a rotating shaft, we first imagine the shaft to be partitioned into small blocks. Each block has its own kinetic energy. We add the contributions of the individual blocks to find the kinetic energy of the shaft.

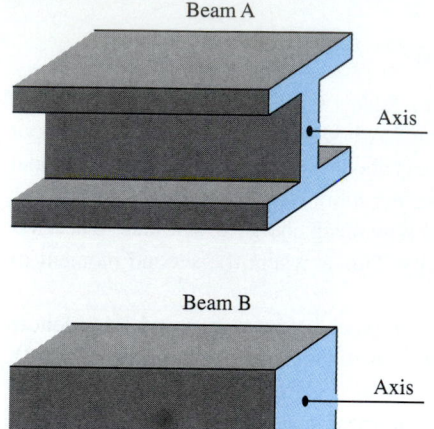

13.19 The greater the polar moment of inertia of the cross section of a beam about the beam's longitudinal axis, the stiffer the beam. Beams A and B have the same cross-section area, but beam A is stiffer.

First moments are "balancing" moments. Second moments are "turning" moments.

is the moment of inertia of the shaft about its axis of rotation, and we see from Eq. (8) that the shaft's kinetic energy is

$$KE_{shaft} = \frac{1}{2}I\omega^2. \tag{10}$$

To start a shaft of inertial moment I rotating at an angular velocity ω, we need to provide a kinetic energy of $KE = (1/2)I\omega^2$. To stop the shaft, we have to take this amount of energy back out. To start a locomotive with mass m moving at a linear velocity v, we need to provide a kinetic energy of $KE = (1/2)mv^2$. To stop the locomotive, we have to remove this amount of energy. The shaft's moment of inertia is analogous to the locomotive's mass. What makes the locomotive hard to start or stop is its mass. What makes the shaft hard to start or stop is its moment of inertia. The moment of inertia takes into account not only the mass but also its distribution.

The moment of inertia also plays a role in determining how much a horizontal metal beam will bend under a load. The stiffness of the beam is a constant times I, the polar moment of inertia of a typical cross section of the beam perpendicular to the beam's longitudinal axis. The greater the value of I, the stiffer the beam and the less it will bend under a given load. That is why we use I-beams instead of beams whose cross sections are square. The flanges at the top and bottom of the beam hold most of the beam's mass away from the longitudinal axis to maximize the value of I (Fig. 13.19).

If you want to see the moment of inertia at work, try the following experiment. Tape two coins to the ends of a pencil and twiddle the pencil about the center of mass. The moment of inertia accounts for the resistance you feel each time you change the direction of motion. Now move the coins an equal distance toward the center of mass and twiddle the pencil again. The system has the same mass and the same center of mass but now offers less resistance to the changes in motion. The moment of inertia has been reduced. The moment of inertia is what gives a baseball bat, golf club, or tennis racket its "feel." Tennis rackets that weigh the same, look the same, and have identical centers of mass will feel different and behave differently if their masses are not distributed the same way.

Example 4   A thin plate covers the triangular region bounded by the x-axis and the lines $x = 1$ and $y = 2x$ in the first quadrant. The plate's density at the point (x, y) is $\delta(x, y) = 6x + 6y + 6$. Find the plate's mass, first moments, center of mass, moments of inertia, and radii of gyration about the coordinate axes.

Solution   We sketch the plate and put in enough detail to determine the limits of integration for the integrals we have to evaluate (Fig. 13.20).

The plate's mass is

$$M = \int_0^1 \int_0^{2x} \delta(x, y)\, dy\, dx = \int_0^1 \int_0^{2x} (6x + 6y + 6)\, dy\, dx$$

$$= \int_0^1 \left[6xy + 3y^2 + 6y \right]_{y=0}^{y=2x} dx$$

$$= \int_0^1 (24x^2 + 12x)\, dx = \left[8x^3 + 6x^2 \right]_0^1 = 14.$$

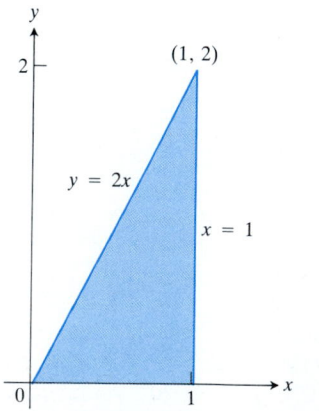

13.20 The triangular region covered by the plate in Example 4.

The first moment about the x-axis is

$$M_x = \int_0^1 \int_0^{2x} y\delta(x, y)\, dy\, dx = \int_0^1 \int_0^{2x} (6xy + 6y^2 + 6y)\, dy\, dx$$

$$= \int_0^1 \left[3xy^2 + 2y^3 + 3y^2 \right]_{y=0}^{y=2x} dx = \int_0^1 (28x^3 + 12x^2)\, dx$$

$$= \left[7x^4 + 4x^3 \right]_0^1 = 11.$$

A similar calculation gives

$$M_y = \int_0^1 \int_0^{2x} x\delta(x, y)\, dy\, dx = 10.$$

The coordinates of the center of mass are therefore

$$\bar{x} = \frac{M_y}{M} = \frac{10}{14} = \frac{5}{7}, \qquad \bar{y} = \frac{M_x}{M} = \frac{11}{14}.$$

The moment of inertia about the x-axis is

$$I_x = \int_0^1 \int_0^{2x} y^2\delta(x, y)\, dy\, dx = \int_0^1 \int_0^{2x} (6xy^2 + 6y^3 + 6y^2)\, dy\, dx$$

$$= \int_0^1 \left[2xy^3 + \frac{3}{2}y^4 + 2y^3 \right]_{y=0}^{y=2x} dx = \int_0^1 (40x^4 + 16x^3)\, dx = \left[8x^5 + 4x^4 \right]_0^1 = 12.$$

Similarly, the moment of inertia about the y-axis is

$$I_y = \int_0^1 \int_0^{2x} x^2\delta(x, y)\, dy\, dx = \frac{39}{5}.$$

Since we know I_x and I_y, we do not need to evaluate an integral to find I_0; we can use the equation $I_0 = I_x + I_y$ instead:

$$I_0 = 12 + \frac{39}{5} = \frac{60 + 39}{5} = \frac{99}{5}.$$

The three radii of gyration are

$$R_x = \sqrt{I_x/M} = \sqrt{12/14} = \sqrt{6/7},$$

$$R_y = \sqrt{I_y/M} = \sqrt{\left(\frac{39}{5}\right)/14} = \sqrt{39/70},$$

$$R_0 = \sqrt{I_0/M} = \sqrt{\left(\frac{99}{5}\right)/14} = \sqrt{99/70}.$$

Centroids of Geometric Figures

When the density of an object is constant, it cancels out of the numerator and denominator of the formulas for $\bar{x}$ and $\bar{y}$. As far as $\bar{x}$ and $\bar{y}$ are concerned, δ might as well be 1. Thus, when δ is constant, the location of the center of mass becomes a feature of the object's shape and not of the material of which it is made. In such cases, engineers may call the center of mass the **centroid** of the shape. To find a centroid, we set δ equal to 1 and proceed to find $\bar{x}$ and $\bar{y}$ as before, by dividing first moments by masses.

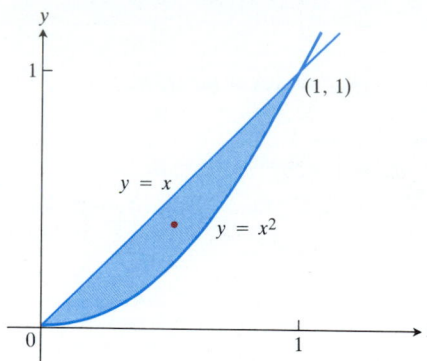

13.21 Example 5 calculates the coordinates of the centroid of the region shown here.

Example 5   Find the centroid of the region in the first quadrant that is bounded above by the line $y = x$ and below by the parabola $y = x^2$.

Solution   We sketch the region and include enough detail to determine the limits of integration (Fig. 13.21). We then set δ equal to 1 and evaluate the appropriate formulas from Table 13.1:

$$M = \int_0^1 \int_{x^2}^x 1\, dy\, dx = \int_0^1 \Big[y\Big]_{y=x^2}^{y=x} dx = \int_0^1 (x - x^2)\, dx = \left[\frac{x^2}{2} - \frac{x^3}{3}\right]_0^1 = \frac{1}{6},$$

$$M_x = \int_0^1 \int_{x^2}^x y\, dy\, dx = \int_0^1 \left[\frac{y^2}{2}\right]_{y=x^2}^{y=x} dx = \int_0^1 \left(\frac{x^2}{2} - \frac{x^4}{2}\right) dx = \left[\frac{x^3}{6} - \frac{x^5}{10}\right]_0^1 = \frac{1}{15},$$

$$M_y = \int_0^1 \int_{x^2}^x x\, dy\, dx = \int_0^1 \Big[xy\Big]_{y=x^2}^{y=x} dx = \int_0^1 (x^2 - x^3)\, dx = \left[\frac{x^3}{3} - \frac{x^4}{4}\right]_0^1 = \frac{1}{12}.$$

From these values of M, M_x, and M_y, we find

$$\bar{x} = \frac{M_y}{M} = \frac{1/12}{1/6} = \frac{1}{2} \qquad \text{and} \qquad \bar{y} = \frac{M_x}{M} = \frac{1/15}{1/6} = \frac{2}{5}.$$

The centroid is the point $\left(\frac{1}{2}, \frac{2}{5}\right)$.

EXERCISES 13.2

In Exercises 1–8, sketch the region bounded by the given lines and curves. Then find the region's area by double integration.

1. The coordinate axes and the line $x + y = 2$

2. The x-axis, the curve $y = e^x$, and the lines $x = 0$, $x = \ln 2$

3. The y-axis and the lines $y = 2x$, $y = 4$

4. The parabola $x = -y^2$ and the line $y = x + 2$

5. The parabolas $x = y^2$ and $x = 2y - y^2$

6. The parabola $x = y - y^2$ and the line $x + y = 0$

7. The semiellipse $y = 2\sqrt{1 - x^2}$ and the lines $x = \pm 1$, $y = -1$

8. Above by $y = x^2$, below by $y = -1$, on the left by $x = -2$, and on the right by $y = 2x - 1$

The integrals and sums of integrals in Exercises 9–14 give the areas of regions in the xy-plane. Sketch each region, label each bounding curve with its equation, and give the coordinates of the points where the curves intersect. Then find the area of the region.

9. $\int_0^6 \int_{y^2/3}^{2y} dx\, dy$

10. $\int_0^3 \int_{-x}^{x(2-x)} dy\, dx$

11. $\int_0^{\pi/4} \int_{\sin x}^{\cos x} dy\, dx$

12. $\int_{-1}^2 \int_{y^2}^{y+2} dx\, dy$

13. $\int_{-1}^0 \int_{-2x}^{1-x} dy\, dx + \int_0^2 \int_{-x/2}^{1-x} dy\, dx$

14. $\int_0^2 \int_{x^2-4}^0 dy\, dx + \int_0^4 \int_0^{\sqrt{x}} dy\, dx$

Constant Density

15. Find the center of mass of a thin plate of density $\delta = 3$ bounded by the lines $x = 0$, $y = x$, and the parabola $y = 2 - x^2$ in the first quadrant.

16. Find the moments of inertia and radii of gyration about the coordinate axes of a thin rectangular plate of constant density δ bounded by the coordinate axes and the lines $x = 3$ and $y = 3$ in the first quadrant.

17. Find the centroid of the region in the first quadrant bounded by the x-axis, the parabola $y^2 = 2x$, and the line $x + y = 4$.

18. Find the centroid of the triangular region cut from the first quadrant by the line $x + y = 3$.

19. Find the centroid of the semicircular region bounded by the x-axis and the curve $y = \sqrt{1 - x^2}$.

20. The area of the region in the first quadrant bounded by the parabola $y = 6x - x^2$ and the line $y = x$ is 125/6 square units. Find the centroid.

21. Find the centroid of the region cut from the first quadrant by the circle $x^2 + y^2 = a^2$.

22. Find the moment of inertia about the x-axis of a thin plate of density $\delta = 1$ bounded by the circle $x^2 + y^2 = 4$. Then use your result to find I_y and I_0 for the plate.

23. Find the centroid of the region between the x-axis and the arch $y = \sin x$, $0 \le x \le \pi$.

24. Find the moment of inertia with respect to the y-axis of a thin sheet of constant density $\delta = 1$ bounded by the curve $y = (\sin^2 x)/x^2$ and the interval $\pi \le x \le 2\pi$ of the x-axis.

25. Find the polar moment of inertia about the center of a thin rectangular plate of constant density covering the region $|x| \le a$, $|y| \le b$ in the xy-plane. (*Hint:* Find I_x. Then use the formula for I_x to find I_y.)

26. Find the moment of inertia and radius of gyration about the x-axis of a thin plate of constant density enclosed by the ellipse $(x^2/a^2) + (y^2/b^2) = 1$. (The ellipse's area is πab.)

27. *The centroid of an infinite region.*   Find the centroid of the infinite region in the second quadrant enclosed by the coordinate axes and the curve $y = e^x$. (Use improper integrals in the mass–moment formulas.)

28. *The first moment of an infinite plate.*   Find the first moment about the y-axis of a thin plate of density $\delta(x, y) = 1$ covering the infinite region under the curve $y = e^{-x^2/2}$ in the first quadrant.

Variable Density

29. Find the moment of inertia and radius of gyration about the x-axis of a thin plate bounded by the parabola $x = y - y^2$ and the line $x + y = 0$ if $\delta(x, y) = x + y$.

30. Find the mass of a thin plate occupying the smaller region cut from the ellipse $x^2 + 4y^2 = 12$ by the parabola $x = 4y^2$ if $\delta(x, y) = 5x$.

31. Find the center of mass of a thin triangular plate bounded by the y-axis and the lines $y = x$ and $y = 2 - x$ if $\delta(x, y) = 6x + 3y + 3$.

32. Find the center of mass and moment of inertia about the x-axis of a thin plate bounded by the curves $x = y^2$

and $x = 2y - y^2$ if the density at the point (x, y) is $\delta(x, y) = y + 1$.

33. Find the center of mass and the moment of inertia and radius of gyration about the y-axis of a thin rectangular plate cut from the first quadrant by the lines $x = 6$ and $y = 1$ if $\delta(x, y) = x + y + 1$.

34. Find the center of mass and the moment of inertia and radius of gyration about the y-axis of a thin plate bounded by the line $y = 1$ and the parabola $y = x^2$ if the density is $\delta(x, y) = y + 1$.

35. Find the center of mass and the moment of inertia and radius of gyration about the y-axis of a thin plate bounded by the x-axis, the lines $x = \pm 1$, and the parabola $y = x^2$ if $\delta(x, y) = 7y + 1$.

36. Find the center of mass and moment of inertia and radius of gyration about the x-axis of a thin rectangular plate bounded by the lines $x = 0$, $x = 20$, $y = -1$, and $y = 1$ if $\delta(x, y) = 1 + (x/20)$.

37. Find the center of mass, the moments of inertia, and radii of gyration about the coordinate axes and the polar moment of inertia and radius of gyration of a thin triangular plate bounded by the lines $y = x$, $y = -x$, and $y = 1$ if $\delta(x, y) = y + 1$.

38. Repeat Exercise 37 for $\delta(x, y) = 3x^2 + 1$.

Additional Exercises

39. Find the average value of $f(x, y) = \sin(x + y)$ over
   a) the rectangle $0 \le x \le \pi$, $0 \le y \le \pi$;
   b) the rectangle $0 \le x \le \pi$, $0 \le y \le \pi/2$.

40. Which do you think will be larger, the average value of $f(x, y) = xy$ over the square $0 \le x \le 1$, $0 \le y \le 1$, or the average value of f over the quarter circle $x^2 + y^2 \le 1$ in the first quadrant? Calculate them to find out.

41. Find the average height of the paraboloid $z = x^2 + y^2$ over the square $0 \le x \le 2$, $0 \le y \le 2$.

42. Find the average value of $f(x) = 1/(x^3 y)$ over the infinite region in the first quadrant that is bounded on the left by the line $x = 1$, above by the curve $y = e^{-x}$, and below by the curve $y = e^{-2x}$.

43. *Appliance design.*   When we design an appliance, one of the concerns is how hard the appliance will be to tip over. When tipped, it will right itself as long as its center of mass lies on the correct side of the *fulcrum*, the point on which the appliance is riding as it tips. Suppose the profile of an appliance of approximately constant density is parabolic, like an old fashioned radio. It fills the region $0 \le y \le a(1 - x^2)$, $-1 \le x \le 1$, in the xy-plane (Fig. 13.22). What values of a will guarantee that the appliance will have to be tipped more than 45° to fall over?

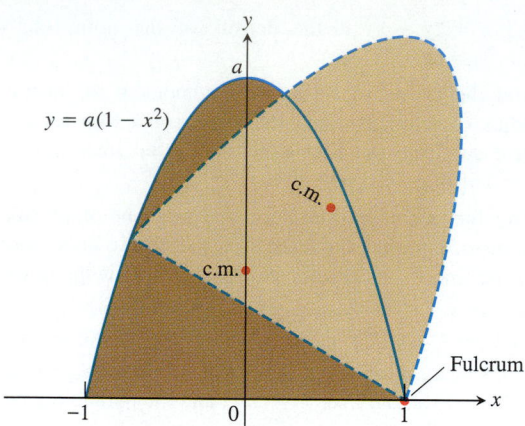

$y = a(1 - x^2)$

Fulcrum

13.22 The profile of the appliance in Exercise 43.

44. A rectangular plate of constant density $\delta(x, y) = 1$ occupies the region bounded by the lines $x = 4$ and $y = 2$ in the first quadrant. The moment of inertia I_a of the rectangle about the line $y = a$ is given by the integral

$$I_a = \int_0^4 \int_0^2 (y - a)^2 \, dy \, dx.$$

Find the value of a that minimizes I_a.

45. Find the centroid of the region in the xy-plane bounded by the curves $y = 1/\sqrt{1 - x^2}$, $y = -1/\sqrt{1 - x^2}$, and the lines $x = 0$, $x = 1$.

46. Find the radius of gyration of a slender rod of constant linear density δ gm/cm and length L cm with respect to an axis

a) through the rod's center of mass perpendicular to the rod's axis;

b) perpendicular to the rod's axis at one end of the rod.

The Parallel Axis Theorem

Let $L_{c.m.}$ be a line in the xy-plane that runs through the center of mass of a thin plate of mass m covering a region in the plane. Let L be a line in the plane parallel to and h units away from $L_{c.m.}$. The **Parallel Axis Theorem** says that under these conditions the moments of inertia I_L and $I_{c.m.}$ of the plate about L and $L_{c.m.}$ satisfy the equation

$$I_L = I_{c.m.} + mh^2. \qquad (11)$$

This equation gives a quick way to calculate one moment when the other moment and the mass are known.

47. *Proof of the Parallel Axis Theorem*

a) Show that the first moment of a thin flat plate about any line in the plane of the plate through the plate's center of mass is zero. (*Hint:* Place the center of mass at the origin with the line along the y-axis. What does the formula $\bar{x} = M_y/M$ then tell you?)

b) Use the result in (a) to derive the Parallel Axis Theorem. Assume that the plane is coordinatized in a way that makes $L_{c.m.}$ the y-axis and L the line $x = h$. Then expand the integrand of the integral for I_L to rewrite the integral as the sum of integrals whose values you recognize.

48. a) Use the Parallel Axis Theorem and the results of Example 4 to find the moments of inertia of the plate in Example 4 about the vertical and horizontal lines through the plate's center of mass.

b) Use the results in (a) to find the plate's moments of inertia about the lines $x = 1$ and $y = 2$.

Pappus's Formula

In addition to stating the centroid theorems in Section 5.9, Pappus knew that the centroid of the union of two nonoverlapping plane regions lies on the line segment joining their individual centroids. More specifically, suppose that m_1 and m_2 are the masses of thin plates P_1 and P_2 that cover nonoverlapping regions in the xy-plane. Let $\mathbf{c}_1$ and $\mathbf{c}_2$ be the vectors from the origin to the respective centers of mass of P_1 and P_2. Then the center of mass of the union $P_1 \cup P_2$ of the two plates is determined by the vector

$$\mathbf{c} = \frac{m_1 \mathbf{c}_1 + m_2 \mathbf{c}_2}{m_1 + m_2}. \qquad (12)$$

Equation (12) is known as **Pappus's formula.** For more than two nonoverlapping plates, as long as their number is finite, the formula generalizes to

$$\mathbf{c} = \frac{m_1 \mathbf{c}_1 + m_2 \mathbf{c}_2 + \cdots + m_n \mathbf{c}_n}{m_1 + m_2 + \cdots + m_n}. \qquad (13)$$

This formula is especially useful for finding the centroid of a plate of irregular shape that is made up of pieces of constant density whose centroids we know from geometry. We find the centroid of each piece and apply Eq. (13) to find the centroid of the plate.

49. Derive Pappus's formula (Eq. 12). (*Hint:* Sketch the plates as regions in the first quadrant and label their centers of mass as $(\bar{x}_1, \bar{y}_1)$ and $(\bar{x}_2, \bar{y}_2)$. What are the moments of $P_1 \cup P_2$ about the coordinate axes?)

50. Use Eq. (12) and mathematical induction to show that Eq. (13) holds for any positive integer $n > 2$.

51. Let A, B, and C be the shapes indicated in Fig. 13.23(a). Use Pappus's formula to find the centroid of

a) $A \cup B$ b) $A \cup C$ c) $B \cup C$

d) $A \cup B \cup C$

52. Locate the center of mass of the carpenter's square in Fig. 13.23(b).

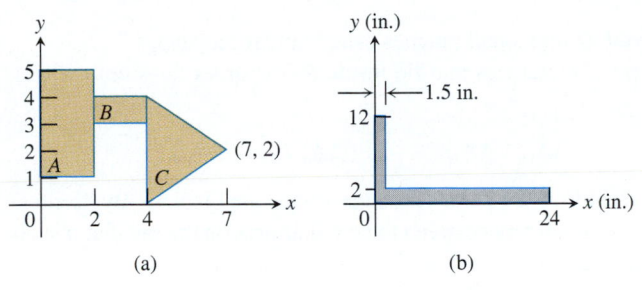

(a)

(b)

13.23   The figures for Exercises 51 and 52.

53. An isosceles triangle T has base $2a$ and altitude h. The base lies along the diameter of a semicircular disk D of radius a so that the two together make a shape resembling an ice cream cone. What relation must hold between a and h to place the centroid of $T \cup D$ on the common boundary of T and D? inside T?

54. An isosceles triangle T of altitude h has as its base one side of a square Q whose edges have length s. (The square and triangle do not overlap.) What relation must hold between h and s to place the centroid of $T \cup Q$ on the base of the triangle? Compare your answer with the answer to Exercise 53.

13.3 Double Integrals in Polar Form

Integrals are sometimes easier to evaluate if we change to polar coordinates. This section shows how to accomplish the change and how to evaluate integrals over regions whose boundaries are given by polar equations.

Integrals in Polar Coordinates

When we defined the double integral of a function over a region R in the xy-plane, we began by cutting R into rectangles whose sides were parallel to the coordinate axes. These were the natural shapes to use because their sides have either constant x-values or constant y-values. In polar coordinates, the natural shape is a "polar rectangle" whose sides have constant r- and θ values.

Suppose that a function $f(r, \theta)$ is defined over a region R that is bounded by the rays $\theta = \alpha$ and $\theta = \beta$ and by the continuous curves $r = g_1(\theta)$ and $r = g_2(\theta)$. Suppose also that $0 \leq g_1(\theta) \leq g_2(\theta) \leq a$ for every value of θ between α and β. Then R lies in a fan-shaped region Q defined by the inequalities $0 \leq r \leq a$ and $\alpha \leq \theta \leq \beta$. See Fig. 13.24.

We cover Q by a grid of circular arcs and rays. The arcs are cut from circles centered at the origin, with radii $\Delta r, 2\Delta r, \ldots, m\Delta r$, where $\Delta r = a/m$. The rays

13.24   The region R: $g_1(\theta) \leq r \leq g_2(\theta)$, $\alpha \leq \theta \leq \beta$ is contained in the fan-shaped region Q: $0 \leq r \leq a$, $\alpha \leq \theta \leq \beta$. The partition of Q by circular arcs and rays induces a partition of R.

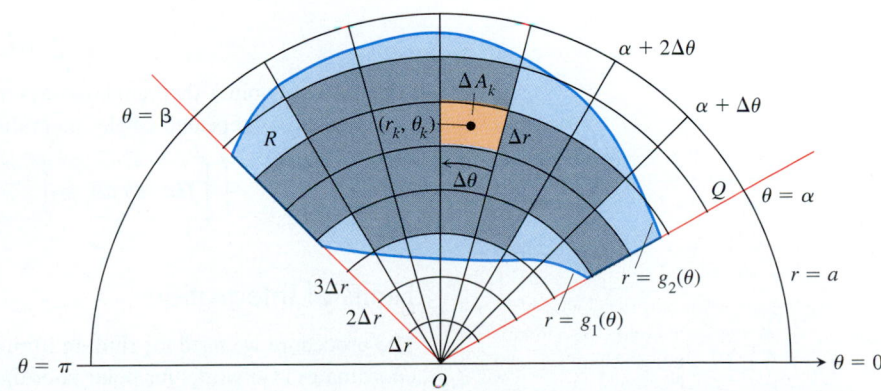

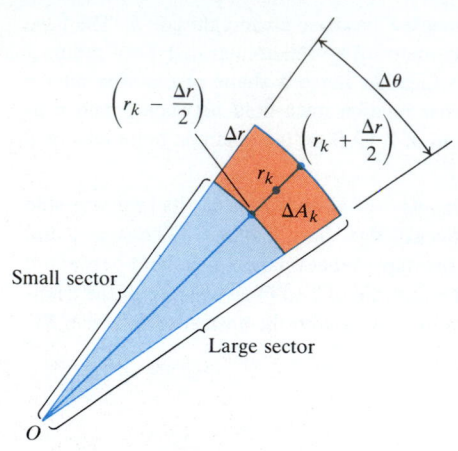

13.25 The observation that

$$\Delta A_k = \begin{pmatrix} \text{area of} \\ \text{large sector} \end{pmatrix} - \begin{pmatrix} \text{area of} \\ \text{small sector} \end{pmatrix}$$

leads to the formula $\Delta A_k = r_k \Delta r \Delta \theta$. The text explains why.

are given by

$$\theta = \alpha, \quad \theta = \alpha + \Delta\theta, \quad \theta = \alpha + 2\Delta\theta, \quad \ldots, \quad \theta = \alpha + m'\Delta\theta = \beta.$$

The arcs and rays divide Q into small patches called "polar rectangles."

We number the polar rectangles that lie inside R (the order does not matter), calling their areas

$$\Delta A_1, \quad \Delta A_2, \quad \ldots, \quad \Delta A_n.$$

We let (r_k, θ_k) be the center of the polar rectangle whose area is ΔA_k. By "center" we mean the point that lies halfway between the circular arcs on the ray that bisects the arcs. We then form the sum

$$S_n = \sum_{k=1}^{n} f(r_k, \theta_k)\Delta A_k. \tag{1}$$

If f is continuous throughout R, this sum will approach a limit as we refine the grid to make Δr and $\Delta\theta$ go to zero. The limit is called the double integral of f over R. In symbols,

$$\lim S_n = \iint\limits_{R} f(r, \theta)\, dA.$$

To evaluate this limit, we first have to write the sum S_n in a way that expresses ΔA_k in terms of Δr and $\Delta\theta$. The radius of the inner arc bounding ΔA_k is $r_k - (\Delta r/2)$ (Fig. 13.25). The area of the circular sector subtended by this arc at the origin is therefore

$$\frac{1}{2}\left(r_k - \frac{\Delta r}{2}\right)^2 \Delta\theta. \tag{2}$$

Similarly, the radius of the outer boundary of ΔA_k is $r_k + (\Delta r/2)$. The area of the sector it subtends is

$$\frac{1}{2}\left(r_k + \frac{\Delta r}{2}\right)^2 \Delta\theta. \tag{3}$$

Therefore,

$$\Delta A_k = \text{area of large sector} - \text{area of small sector}$$

$$= \frac{\Delta\theta}{2}\left[\left(r_k + \frac{\Delta r}{2}\right)^2 - \left(r_k - \frac{\Delta r}{2}\right)^2\right] = \frac{\Delta\theta}{2}[2r_k \Delta r] = r_k \Delta r \Delta\theta.$$

Combining this result with Eq. (1) gives

$$S_n = \sum_{k=1}^{n} f(r_k, \theta_k) r_k \Delta r \Delta\theta. \tag{4}$$

A version of Fubini's theorem now says that the limit approached by these sums can be evaluated by repeated single integrations with respect to r and θ as

$$\iint\limits_{R} f(r, \theta)\, dA = \int_{\theta=\alpha}^{\theta=\beta} \int_{r=g_1(\theta)}^{r=g_2(\theta)} f(r, \theta)\, r\, dr\, d\theta. \tag{5}$$

Limits of Integration

The procedure we used for finding limits of integration for integrals in rectangular coordinates also works for polar coordinates.

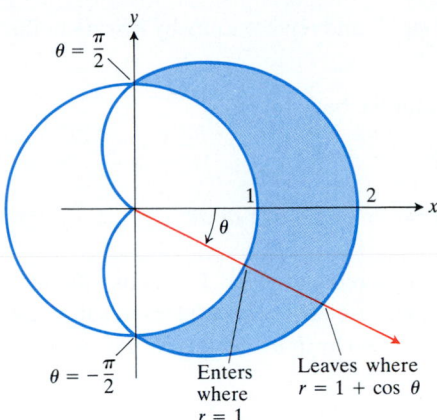

13.26 Example 1 describes the process by which we find limits of integration for calculating the area of the shaded region between the circle $r = 1$ and the cardioid $r = 1 + \cos \theta$.

Example 1 *How to Find Limits of Integration.* Find the limits of integration for integrating a function $f(r, \theta)$ over the region R that lies inside the cardioid $r = 1 + \cos \theta$ and outside the circle $r = 1$.

Solution We graph the cardioid and circle (Fig. 13.26) and carry out the following steps:

STEP 1: Holding θ fixed, let r increase to trace a ray out from the origin.

STEP 2: Integrate from the r-value where the ray enters R to the r-value where the ray leaves R.

STEP 3: Choose θ-limits to include all the rays from the origin that intersect R.

The result is the integral

$$\int_{-\pi/2}^{\pi/2} \int_{r=1}^{r=1+\cos\theta} f(r, \theta) \, r \, dr \, d\theta.$$

If $f(r, \theta)$ is the constant function whose value is 1, then the value of the integral of f over a region R is the area of R:

Area in Polar Coordinates

The area of a closed and bounded region R in the polar coordinate plane is given by the formula

$$\text{Area of } R = \iint_R r \, dr \, d\theta. \tag{6}$$

As you might expect, this formula for area is consistent with all earlier formulas, although we shall not prove the fact.

Example 2 Find the area enclosed by the lemniscate $r^2 = 4 \cos 2\theta$.

Solution We graph the lemniscate to determine the limits of integration (Fig. 13.27) and see that the total area is four times the first-quadrant portion.

$$\text{Area} = 4 \int_0^{\pi/4} \int_0^{\sqrt{4\cos 2\theta}} r \, dr \, d\theta = 4 \int_0^{\pi/4} \left[\frac{r^2}{2}\right]_{r=0}^{r=\sqrt{4\cos 2\theta}} d\theta$$

$$= 4 \int_0^{\pi/4} 2 \cos 2\theta \, d\theta = 4 \sin 2\theta \Big]_0^{\pi/4} = 4.$$

Changing Cartesian Integrals into Polar Integrals

The procedure for changing a Cartesian integral

$$\iint_R f(x, y) \, dx \, dy$$

into a polar integral has two steps:

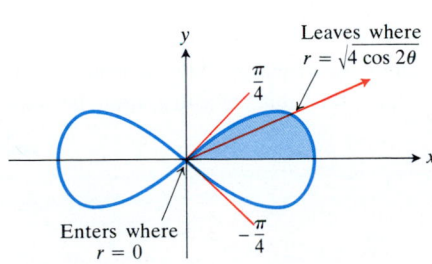

13.27 To integrate over the shaded region bounded by the lemniscate $r^2 = 4 \cos 2\theta$, we run r from 0 to $\sqrt{4 \cos 2\theta}$ and θ from 0 to $\pi/4$ (Example 2).

STEP 1:   Substitute $x = r \cos \theta$ and $y = r \sin \theta$, and replace $dx\, dy$ by $r\, dr\, d\theta$ in the Cartesian integral.

STEP 2:   Supply polar limits of integration for the boundary of R.

The Cartesian integral then becomes

$$\iint\limits_{R} f(x, y)\, dx\, dy = \iint\limits_{G} f(r \cos \theta, r \sin \theta)\, r\, dr\, d\theta, \tag{7}$$

where G denotes the region of integration in polar coordinates. This is like the substitution method in Chapter 4 except that there are now two variables to substitute for instead of one. Notice that $dx\, dy$ is not replaced by $dr\, d\theta$ but by $r\, dr\, d\theta$. We shall go into the reasons for that briefly in Section 13.7.

Example 3   Find the polar moment of inertia about the origin of a thin plate of density $\delta(x, y) = 1$ bounded by the quarter circle $x^2 + y^2 = 1$ in the first quadrant.

Solution   We sketch the region of integration to determine the limits of integration (Fig. 13.28).

In Cartesian coordinates, the polar moment is the value of the integral

$$\int_{0}^{1} \int_{0}^{\sqrt{1-x^2}} (x^2 + y^2)\, dy\, dx.$$

Integration with respect to y gives

$$\int_{0}^{1} \left(x^2 \sqrt{1 - x^2} + \frac{(1 - x^2)^{3/2}}{3} \right) dx,$$

an integral difficult to evaluate without tables.

Things go better if we change the original integral to polar coordinates. Substituting $x = r \cos \theta$, $y = r \sin \theta$, and replacing $dx\, dy$ by $r\, dr\, d\theta$, we get

$$\int_{0}^{1} \int_{0}^{\sqrt{1-x^2}} (x^2 + y^2)\, dy\, dx = \int_{0}^{\pi/2} \int_{0}^{1} (r^2)\, r\, dr\, d\theta$$

$$= \int_{0}^{\pi/2} \left[\frac{r^4}{4} \right]_{r=0}^{r=1} d\theta$$

$$= \int_{0}^{\pi/2} \frac{1}{4}\, d\theta = \frac{\pi}{8}.$$

Why was the polar coordinate transformation so effective? One reason is that $x^2 + y^2$ was simplified to r^2. Another is that the limits of integration became constants.

Example 4   Evaluate

$$\iint\limits_{R} e^{x^2 + y^2}\, dy\, dx,$$

where R is the semicircular region bounded by the x-axis and the curve $y = \sqrt{1 - x^2}$ (Fig. 13.29).

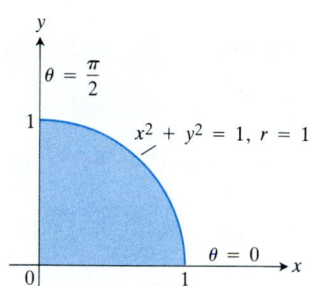

13.28   Example 3 evaluates the integral of a function over the region enclosed by the quarter circle $x^2 + y^2 = 1$ in the first quadrant. In polar coordinates, this region is described by simple inequalities:

$$0 \le r \le 1 \quad \text{and} \quad 0 \le \theta \le \pi/2.$$

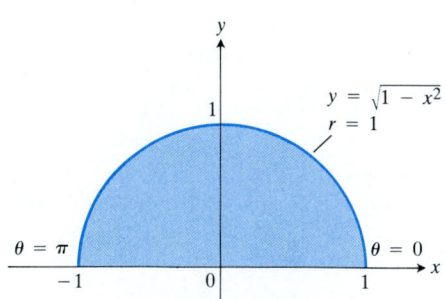

13.29   The semicircular region in Example 4 is described by the polar coordinate inequalities

$$0 \le r \le 1 \quad \text{and} \quad 0 \le \theta \le \pi.$$

Solution In Cartesian coordinates, the integral in question is a nonelementary integral and there is no direct way to integrate $e^{x^2+y^2}$ with respect to either x or y. Yet this integral and others like it are important in mathematics—in statistics, for example—and we must find a way to evaluate it. Polar coordinates save the day. Substituting $x = r \cos \theta$, $y = r \sin \theta$, and replacing $dx\, dy$ by $r\, dr\, d\theta$ enables us to evaluate the integral as

$$\iint_R e^{x^2+y^2}\, dy\, dx = \int_0^\pi \int_0^1 e^{r^2} r\, dr\, d\theta$$

$$= \int_0^\pi \left[\frac{1}{2} e^{r^2}\right]_0^1 d\theta$$

$$= \int_0^\pi \frac{1}{2}(e-1)\, d\theta = \frac{\pi}{2}(e-1).$$

The r in the $r\, dr\, d\theta$ was just what we needed to integrate e^{r^2}. Without it we would have been stuck, as we were at the beginning.

When we work in polar coordinates, we often have to integrate powers of sines and cosines. We accomplish this with reduction formulas, as in the next example. Reduction formulas were discussed in Section 7.6. You will find a number of useful reduction formulas in the integral table at the end of the book.

Example 5 Find the moment of inertia about the y-axis of a thin plate of constant density $\delta(x, y) = 1$ covering the region R enclosed by the cardioid $r = 1 - \cos \theta$.

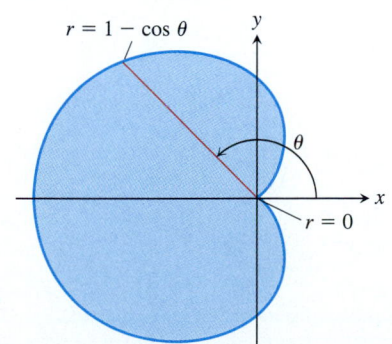

13.30 Example 5 shows how to find the moment of inertia about the y-axis of a thin plate covering the shaded region. To integrate over the region, we first run r from 0 to $1 - \cos \theta$ and then run θ from 0 to 2π.

Solution The required moment is

$$I_y = \iint_R x^2\, \delta(x, y)\, dA = \iint_R x^2\, dA.$$

We substitute $x = r \cos \theta$, replace dA by $r\, dr\, d\theta$, and sketch the region (Fig. 13.30) to determine the limits of integration:

$$I_y = \int_0^{2\pi} \int_0^{1-\cos\theta} r^3 \cos^2\theta\, dr\, d\theta$$

$$= \int_0^{2\pi} \frac{1}{4}(1 - \cos \theta)^4 \cos^2\theta\, d\theta.$$

We use the following reduction formula (Formula 61 in the integral table at the end of the book)

$$\int_0^{2\pi} \cos^n\theta\, d\theta = \frac{\cos^{n-1}\theta \sin \theta}{n}\bigg]_0^{2\pi} + \frac{n-1}{n}\int_0^{2\pi} \cos^{n-2}\theta\, d\theta$$

$$= \frac{n-1}{n}\int_0^{2\pi} \cos^{n-2}\theta\, d\theta$$

to evaluate the powers of cos θ that arise when we expand the integrand:

$$\int_0^{2\pi} \cos^2\theta \, d\theta = \frac{1}{2}\int_0^{2\pi} d\theta = \pi,$$

$$\int_0^{2\pi} \cos^3\theta \, d\theta = \frac{2}{3}\int_0^{2\pi} \cos\theta \, d\theta = \frac{2}{3}\sin\theta\Big]_0^{2\pi} = 0,$$

$$\int_0^{2\pi} \cos^4\theta \, d\theta = \frac{3}{4}\int_0^{2\pi} \cos^2\theta \, d\theta = \frac{3\pi}{4},$$

$$\int_0^{2\pi} \cos^5\theta \, d\theta = \frac{4}{5}\int_0^{2\pi} \cos^3\theta \, d\theta = 0,$$

$$\int_0^{2\pi} \cos^6\theta \, d\theta = \frac{5}{6}\int_0^{2\pi} \cos^4\theta \, d\theta = \frac{5}{6}\frac{3\pi}{4} = \frac{5\pi}{8}.$$

Therefore

$$I_y = \frac{1}{4}\int_0^{2\pi} (\cos^2\theta - 4\cos^3\theta + 6\cos^4\theta - 4\cos^5\theta + \cos^6\theta) \, d\theta$$

$$= \frac{1}{4}\Big[1 + \frac{18}{4} + \frac{5}{8}\Big]\pi = \frac{49\pi}{32}.$$

EXERCISES 13.3

In Exercises 1–15, change the Cartesian integral into an equivalent polar integral and evaluate the polar integral.

1. $\int_{-1}^{1}\int_{0}^{\sqrt{1-x^2}} dy \, dx$

2. $\int_{-1}^{1}\int_{-\sqrt{1-x^2}}^{\sqrt{1-x^2}} dy \, dx$

3. $\int_{0}^{1}\int_{0}^{\sqrt{1-y^2}} (x^2 + y^2) \, dx \, dy$

4. $\int_{-1}^{1}\int_{-\sqrt{1-y^2}}^{\sqrt{1-y^2}} (x^2 + y^2) \, dx \, dy$

5. $\int_{-a}^{a}\int_{-\sqrt{a^2-x^2}}^{\sqrt{a^2-x^2}} dy \, dx$

6. $\int_{0}^{2}\int_{0}^{\sqrt{4-y^2}} (x^2 + y^2) \, dx \, dy$

7. $\int_{0}^{1}\int_{y}^{\sqrt{2-y^2}} x \, dx \, dy$

8. $\int_{0}^{2}\int_{0}^{x} y \, dy \, dx$

9. $\int_{0}^{3}\int_{0}^{\sqrt{3}x} \frac{dy \, dx}{\sqrt{x^2 + y^2}}$

10. $\int_{0}^{2}\int_{0}^{\sqrt{4-x^2}} \frac{xy}{\sqrt{x^2 + y^2}} \, dy \, dx$

11. $\int_{0}^{1}\int_{0}^{\sqrt{1-x^2}} 5\sqrt{x^2 + y^2} \, dy \, dx$

12. $\int_{0}^{1}\int_{0}^{\sqrt{1-x^2}} e^{-(x^2 + y^2)} \, dy \, dx$

13. $\int_{0}^{2}\int_{0}^{\sqrt{1-(x-1)^2}} \frac{x + y}{x^2 + y^2} \, dy \, dx$

14. $\int_{-1}^{1}\int_{-\sqrt{1-y^2}}^{\sqrt{1-y^2}} \ln(x^2 + y^2 + 1) \, dx \, dy$

15. $\int_{-1}^{1}\int_{-\sqrt{1-x^2}}^{\sqrt{1-x^2}} \frac{2 \, dy \, dx}{(1 + x^2 + y^2)^2}$

16. a) The usual way to evaluate the improper integral $I = \int_0^\infty e^{-x^2} \, dx$ is first to calculate its square:

$$I^2 = \Big(\int_0^\infty e^{-x^2} dx\Big)\Big(\int_0^\infty e^{-y^2} dy\Big) = \int_0^\infty\int_0^\infty e^{-(x^2 + y^2)} \, dx \, dy.$$

Evaluate the last integral using polar coordinates and solve the resulting equation to show that $I = \sqrt{\pi}/2$.

b) Evaluate the integral

$$\int_0^\infty \frac{e^{-x}}{\sqrt{x}}\, dx.$$

17. Find the area of the region cut from the first quadrant by the curve $r = 2(2 - \sin 2\theta)^{1/2}$.

18. Find the area of the region that lies inside the cardioid $r = 1 + \cos\theta$ and outside the circle $r = 1$.

19. Find the area enclosed by one leaf of the rose $r = 12\cos 3\theta$.

20. Find the area of the region enclosed by the positive x-axis and spiral $r = 4\theta/3$, $0 \le \theta \le 2\pi$. The region looks like a snail shell.

21. Find the area of the region cut from the first quadrant by the cardioid $r = 1 + \sin\theta$.

22. Find the area of the region common to the interiors of the cardioids $r = 1 + \cos\theta$ and $r = 1 - \cos\theta$.

23. Find the area of the region that is bounded on the left by the y-axis and on the right by the parabola $r = \sec^2(\theta/2)$.

24. Find the area of the region inside the cardioid $r = 2 + 2\cos\theta$ that lies to the right of the parabola $r = \sec^2(\theta/2)$.

25. Integrate the function $f(x, y) = 1/(1 - x^2 - y^2)$ over the disk $x^2 + y^2 \le 3/4$.

26. Integrate the function $f(x, y) = (\ln(x^2 + y^2))/(x^2 + y^2)$ over the region between the circles $x^2 + y^2 = 1$ and $x^2 + y^2 = e^2$.

27. Find the first moment about the x-axis of a thin plate of constant density $\delta(x, y) = 3$, bounded below by the x-axis and above by the cardioid $r = 1 - \cos\theta$.

28. Find the moment of inertia about the x-axis and the polar moment of inertia about the origin of a thin disk bounded by the circle $x^2 + y^2 = a^2$ if the density is $\delta(x, y) = k(x^2 + y^2)$, k a constant.

29. Find the centroid of the region enclosed by the cardioid $r = 1 + \cos\theta$.

30. Find the polar moment of inertia about the origin of a thin plate bounded by the cardioid $r = 1 + \cos\theta$ if the density is $\delta(x, y) = 1$.

31. Find the mass of a thin plate that covers that region that lies outside the curve $r = a$ and inside the curve $r = 2a\sin\theta$ if the density is inversely proportional to the distance from the origin.

32. Find the polar moment of inertia about the origin of a thin plate of density $\delta(x, y) = 1$ covering the region that lies inside the cardioid $r = 1 - \cos\theta$ and outside the circle $r = 1$.

33. The region that lies inside the cardioid $r = 1 + \cos\theta$ and outside the circle $r = 1$ is the base of a solid right cylinder. The top of the cylinder lies in the plane $z = x$. Find the cylinder's volume.

34. The region enclosed by the lemniscate $r^2 = 2\cos 2\theta$ is the base of a solid right cylinder whose top is bounded by the sphere $z = \sqrt{2 - r^2}$. Find the cylinder's volume.

35. Find the average height of the hemisphere $z = \sqrt{1 - x^2 - y^2}$ above the disk $x^2 + y^2 \le 1$.

36. Find the average value of the square of the distance from the points inside a unit disk to a point on the boundary.

37. Let P_0 be a point inside a circle of radius a and let h denote the distance from P_0 to the center of the circle. Let d denote the distance from an arbitrary point P to P_0. Find the average value of d^2 over the region enclosed by the circle. (*Hint:* Simplify your work by placing the center of the circle at the origin and P_0 on the x-axis.)

38. Suppose that the area of a region in the polar coordinate plane is

$$A = \int_{\pi/4}^{3\pi/4} \int_{\csc\theta}^{2\sin\theta} r\, dr\, d\theta.$$

a) Sketch the region and find its area.

b) Use one of Pappus's theorems together with the centroid information in Exercise 22 of Section 5.9 to find the volume of the solid generated by revolving the region about the x-axis.

13.4 Triple Integrals in Rectangular Coordinates: Volumes and Average Values

Triple integrals are used to calculate the volumes of irregular three-dimensional shapes and the average values of functions over three-dimensional regions. They are also the integrals we use to calculate masses and moments of three-dimensional solids. When combined with vectors, as they will be in Chapter 14, they are the integrals we use to describe many of the phenomena associated with fluid flow and electromagnetism.

In the present section, we define triple integrals and use them to calculate volumes and average values.

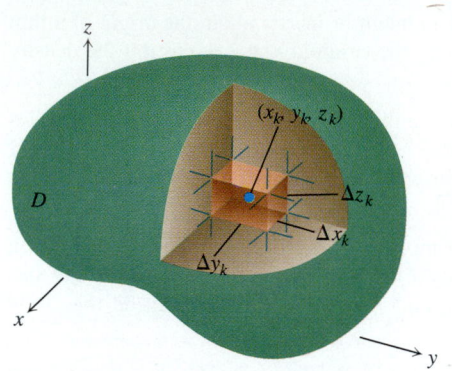

13.31   Partitioning a solid with rectangular cells of volume ΔV_k.

Triple Integrals

If $F(x, y, z)$ is a function defined on a closed bounded region D in space—the region occupied by a solid ball, for example, or a lump of clay—then the integral of F over D may be defined in the following way. We partition a rectangular region about D into rectangular cells by planes parallel to the coordinate planes (Fig. 13.31). We number the cells that lie inside D from 1 to n in some order, a typical cell having dimensions Δx_k by Δy_k by Δz_k and volume ΔV_k. We choose a point (x_k, y_k, z_k) in each cell and form the sum

$$S_n = \sum_{k=1}^{n} F(x_k, y_k, z_k)\, \Delta V_k. \tag{1}$$

If F is continuous and the bounding surface of D is made of smooth surfaces joined along continuous curves, then as Δx_k, Δy_k, and Δz_k approach zero independently the sums S_n approach a limit

$$\lim_{n \to \infty} S_n = \iiint_D F(x, y, z)\, dV. \tag{2}$$

We call this limit the **triple integral of F over D.** The limit also exists for some discontinuous functions.

Properties of Triple Integrals

Triple integrals have the same algebraic properties as double integrals and single integrals. If $F = F(x, y, z)$ and $G = G(x, y, z)$ are both integrable, then

1. $\displaystyle \iiint_D kF \, dV = k \iiint_D F \, dV$ (any number k),

2. $\displaystyle \iiint_D (F + G) \, dV = \iiint_D F \, dV + \iiint_D G \, dV,$

3. $\displaystyle \iiint_D (F - G) \, dV = \iiint_D F \, dV - \iiint_D G \, dV,$

4. $\displaystyle \iiint_D F \, dV \geq 0$ if $F \geq 0$ on D,

5. $\displaystyle \iiint_D F \, dV \geq \iiint_D G \, dV$ if $F \geq G$ on D.

Triple integrals also have a domain additivity property that proves useful in physics and engineering as well as in mathematics. If the domain D of a continuous function F is partitioned by smooth surfaces into a finite number of nonoverlapping cells $D_1, D_2, \ldots, D_n$, then

6. $\displaystyle \iiint_D F \, dV = \iiint_{D_1} F \, dV + \iiint_{D_2} F \, dV + \cdots + \iiint_{D_n} F \, dV.$

Volume of a Region in Space

If F is the constant function whose value is 1, then the sums in Eq. (1) reduce to

$$S_n = \sum F(x_k, y_k, z_k)\Delta V_k = \sum 1 \cdot \Delta V_k = \sum \Delta V_k. \tag{3}$$

As Δx, Δy, and Δz all approach zero, the cells ΔV_k become smaller and more numerous and fill up more and more of D. We therefore define the volume of D to be the triple integral

$$\lim_{} \sum_{k=1}^{n} \Delta V_k = \iiint_D dV.$$

DEFINITION

> The **volume** of a closed bounded region D in space is the value of the integral
>
> $$\text{Volume of } D = \iiint_D dV. \tag{4}$$

As we shall see in a moment, this integral enables us to calculate the volumes of solids enclosed by curved surfaces.

Evaluation of Triple Integrals

The triple integral is seldom evaluated directly from its definition as a limit. Instead, we apply a three-dimensional version of Fubini's theorem to evaluate the integral by repeated single integrations.

For example, suppose we want to integrate a continuous function $F(x, y, z)$ over a region D that is bounded below by a surface $z = f_1(x, y)$, above by the surface $z = f_2(x, y)$, and on the side by a cylinder C parallel to the z-axis (Fig. 13.32). Let R denote the vertical projection of D onto the xy-plane, which is the region in the xy-plane enclosed by C. The integral of F over D is then evaluated as

$$\iiint_D F(x, y, z)\, dV = \iint_R \left(\int_{z=f_1(x, y)}^{z=f_2(x, y)} F(x, y, z)\, dz \right) dy\, dx,$$

or

$$\iiint_D F(x, y, z)\, dV = \iint_R \int_{z=f_1(x, y)}^{z=f_2(x, y)} F(x, y, z)\, dz\, dy\, dx. \tag{5}$$

The z-limits of integration indicate that for every (x, y) in the region R, z may extend from the lower surface $z = f_1(x, y)$ to the upper surface $z = f_2(x, y)$. The y- and x-limits of integration have not been given explicitly in Eq. (5) but are to be determined in the usual way from the boundaries of R.

In case the lateral surface of the cylinder reduces to zero, as in Fig. 13.33 and Example 1, we may find the equation of the boundary of R by eliminating z between the two equations $z = f_1(x, y)$ and $z = f_2(x, y)$. This gives

$$f_1(x, y) = f_2(x, y),$$

an equation that contains no z and that defines the boundary of R in the xy-plane.

To supply the z-limits of integration in any particular instance we may use a procedure like the one for double integrals. We imagine a line L through a point

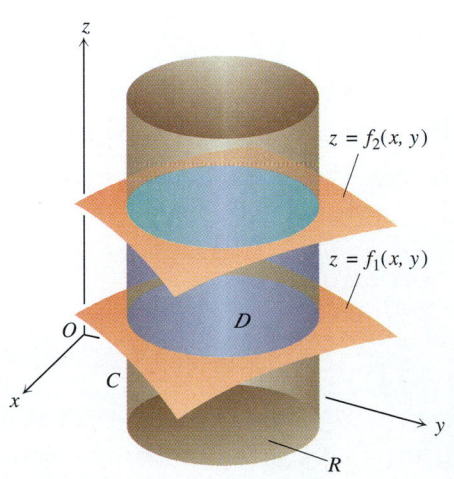

13.32 The enclosed volume can be found by evaluating

$$V = \iiint_R \int_{z=f_1(x, y)}^{z=f_2(x, y)} dz\, dy\, dx.$$

The curve in which the upper and lower bounding surfaces intersect. Along this curve, $f_1(x, y) = f_2(x, y)$.

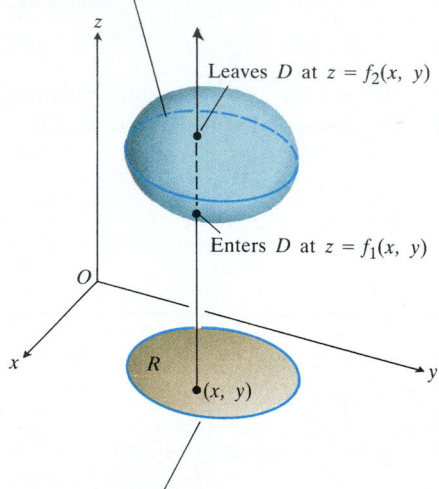

Leaves D at $z = f_2(x, y)$

Enters D at $z = f_1(x, y)$

The vertical projection of the curve of intersection onto the xy-plane. Along this curve, $f_1(x, y) = f_2(x, y)$ as well, and we can use this equation as an equation for the boundary of R.

13.33 A schematic diagram for finding the limits of integration for a triple integral of a function F over a three-dimensional region D enclosed by two surfaces. For the region here,

$$\iiint_D F \, dV = \iint_R \int_{z = f_1(x, y)}^{z = f_2(x, y)} F \, dz \, dy \, dx.$$

(x, y) in R and parallel to the z-axis. As z increases, the line enters D at $z = f_1(x, y)$ and leaves D at $z = f_2(x, y)$. These give the lower and upper limits of the integration with respect to z. The result of this integration is now a function of x and y alone, which we integrate over R, supplying limits in the usual way.

Example 1   Find the volume of the three-dimensional region enclosed by the surfaces $z = x^2 + 3y^2$ and $z = 8 - x^2 - y^2$.

Solution   The two surfaces (Fig. 13.34) intersect on the elliptical cylinder

$$x^2 + 3y^2 = 8 - x^2 - y^2, \quad \text{or} \quad x^2 + 2y^2 = 4.$$

The three-dimensional region projects into the two-dimensional region R in the xy-plane that is enclosed by the ellipse having this same equation. In the double integral with respect to y and x over R, if we integrate first with respect to y, holding x fixed, y varies from $-\sqrt{(4 - x^2)/2}$ to $+\sqrt{(4 - x^2)/2}$. Then x varies from -2 to $+2$. Thus we have

$$V = \int_{-2}^{2} \int_{-\sqrt{(4 - x^2)/2}}^{\sqrt{(4 - x^2)/2}} \int_{x^2 + 3y^2}^{8 - x^2 - y^2} dz \, dy \, dx = \int_{-2}^{2} \int_{-\sqrt{(4 - x^2)/2}}^{\sqrt{(4 - x^2)/2}} (8 - 2x^2 - 4y^2) \, dy \, dx$$

$$= \int_{-2}^{2} \left[(8 - 2x^2)y - \frac{4}{3} y^3 \right]_{y = -\sqrt{(4 - x^2)/2}}^{y = \sqrt{(4 - x^2)/2}} dx$$

$$= \int_{-2}^{2} \left(2(8 - 2x^2) \sqrt{\frac{4 - x^2}{2}} - \frac{8}{3} \left(\frac{4 - x^2}{2} \right)^{3/2} \right) dx$$

$$= \int_{-2}^{2} \left(8 \left(\frac{4 - x^2}{2} \right)^{3/2} - \frac{8}{3} \left(\frac{4 - x^2}{2} \right)^{3/2} \right) dx = \frac{4\sqrt{2}}{3} \int_{-2}^{2} (4 - x^2)^{3/2} \, dx$$

$$= 8\pi\sqrt{2}. \qquad \begin{pmatrix} \text{After integration with the} \\ \text{substitution } x = 2 \sin u \end{pmatrix}$$

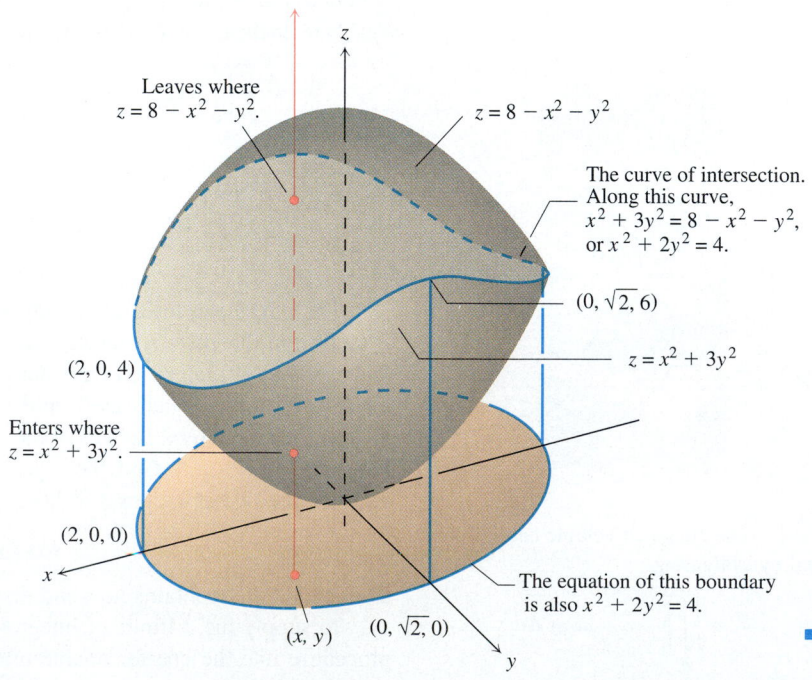

Leaves where $z = 8 - x^2 - y^2$.

$z = 8 - x^2 - y^2$

The curve of intersection. Along this curve, $x^2 + 3y^2 = 8 - x^2 - y^2$, or $x^2 + 2y^2 = 4$.

$(0, \sqrt{2}, 6)$

$z = x^2 + 3y^2$

$(2, 0, 4)$

Enters where $z = x^2 + 3y^2$.

$(2, 0, 0)$

The equation of this boundary is also $x^2 + 2y^2 = 4$.

(x, y) $(0, \sqrt{2}, 0)$

13.34   The volume of the region enclosed by these two paraboloids is calculated in Example 1.

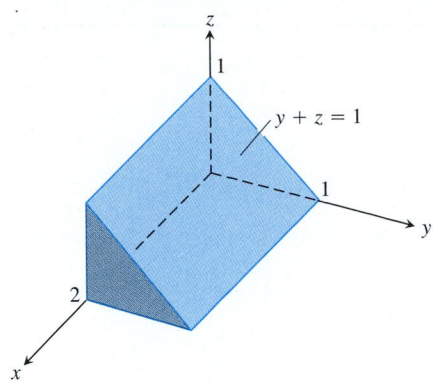

13.35 Example 2 shows how to calculate the volume of this prism with six different iterated triple integrals.

As we know, there are sometimes (but not always) two different orders in which the single integrations that evaluate a double integral may be worked. For triple integrals, there are sometimes (but not always) as many as *six* workable orders of integration. In the next example, all six are workable.

Example 2   Each of the following integrals gives the volume of the solid shown in Fig. 13.35.

a) $\int_0^1 \int_0^{1-z} \int_0^2 dx\, dy\, dz$
b) $\int_0^1 \int_0^{1-y} \int_0^2 dx\, dz\, dy$

c) $\int_0^1 \int_0^2 \int_0^{1-z} dy\, dx\, dz$
d) $\int_0^2 \int_0^1 \int_0^{1-z} dy\, dz\, dx$

e) $\int_0^1 \int_0^2 \int_0^{1-y} dz\, dx\, dy$
f) $\int_0^2 \int_0^1 \int_0^{1-y} dz\, dy\, dx$

Average Value of a Function in Space

The average value of a function F over a region D in space is defined by the formula

$$\text{Average value of } F \text{ over } D = \frac{1}{\text{volume of } D} \iiint_D F\, dV. \tag{6}$$

If $F(x, y, z) = \sqrt{x^2 + y^2 + z^2}$, then the average value of F over D is the average distance of points in D from the origin. If $F(x, y, z)$ is the density of a solid that occupies a region D in space, then the average value of F over D is the average density of the solid in units of mass per unit volume.

Example 3   Find the average value of $F(x, y, z) = xyz$ over the cube bounded by the coordinate planes and the planes $x = 2$, $y = 2$, and $z = 2$ in the first octant.

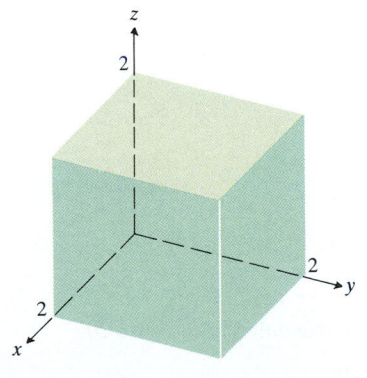

13.36 The solid cube bounded by the planes $x = 2$, $y = 2$, and $z = 2$ is the region of integration in Example 3.

Solution   We sketch the cube with enough detail to show the limits of integration (Fig. 13.36). We then use Eq. (6) to calculate the average value of F over the cube.

The volume of the cube is $(2)(2)(2) = 8$.

The value of the integral of F over the cube is

$$\int_0^2 \int_0^2 \int_0^2 xyz\, dx\, dy\, dz = \int_0^2 \int_0^2 \left[\frac{x^2}{2} yz\right]_{x=0}^{x=2} dy\, dz = \int_0^2 \int_0^2 2yz\, dy\, dz$$

$$= \int_0^2 \left[y^2 z\right]_{y=0}^{y=2} dz = \int_0^2 4z\, dz = \left[2z^2\right]_0^2 = 8.$$

With these values, Eq. (6) gives

$$\text{Average value of } xyz \text{ over the cube} = \frac{1}{\text{volume}} \iiint_{\text{cube}} xyz\, dV = \left(\frac{1}{8}\right)(8) = 1.$$

In evaluating the integral, we chose the order dx, dy, dz, but any of the other five possible orders would have done as well.

EXERCISES 13.4

1. Find the common value of the integrals in Example 2.

2. Write six different iterated triple integrals for the volume of the rectangular solid in the first octant bounded by the coordinate planes and the planes $x = 1$, $y = 2$, and $z = 3$. Evaluate one of the integrals.

3. Write six different iterated triple integrals for the volume of the tetrahedron cut from the first octant by the plane $6x + 3y + 2z = 6$. Evaluate one of the integrals.

4. Write six different iterated triple integrals for the volume of the region in the first octant enclosed by the cylinder $x^2 + z^2 = 4$ and the plane $y = 3$. Evaluate one of them.

Evaluate the integrals in Exercises 5–18.

5. $\displaystyle\int_0^1 \int_0^1 \int_0^1 (x^2 + y^2 + z^2)\, dz\, dy\, dx$

6. $\displaystyle\int_0^{\sqrt{2}} \int_0^{3y} \int_{x^2 + 3y^2}^{8 - x^2 - y^2} dz\, dx\, dy$

7. $\displaystyle\int_1^e \int_1^e \int_1^e \frac{1}{xyz}\, dx\, dy\, dz$

8. $\displaystyle\int_0^1 \int_0^{3 - 3x} \int_0^{3 - 3x - y} dz\, dy\, dx$

9. $\displaystyle\int_0^1 \int_0^{\pi} \int_0^{\pi} y \sin z\, dx\, dy\, dz$

10. $\displaystyle\int_{-1}^1 \int_{-1}^1 \int_{-1}^1 (x + y + z)\, dy\, dx\, dz$

11. $\displaystyle\int_0^3 \int_0^{\sqrt{9 - x^2}} \int_0^{\sqrt{9 - x^2}} dz\, dy\, dx$

12. $\displaystyle\int_0^2 \int_{-\sqrt{4 - y^2}}^{\sqrt{4 - y^2}} \int_0^{2x + y} dz\, dx\, dy$

13. $\displaystyle\int_0^1 \int_0^{2 - x} \int_0^{2 - x - y} dz\, dy\, dx$

14. $\displaystyle\int_0^1 \int_0^{1 - x^2} \int_3^{4 - x^2 - y} x\, dz\, dy\, dx$

15. $\displaystyle\int_0^{\pi} \int_0^{\pi} \int_0^{\pi} \cos(u + v + w)\, du\, dv\, dw$ (uvw-space)

16. $\displaystyle\int_1^e \int_1^e \int_1^e \ln r \ln s \ln t\, dt\, dr\, ds$ (rst-space)

17. $\displaystyle\int_0^{\pi/4} \int_0^{\ln \sec v} \int_{-\infty}^{2t} e^x\, dx\, dt\, dv$ (tvx-space)

18. $\displaystyle\int_0^7 \int_0^2 \int_0^{\sqrt{4 - q^2}} \frac{q}{r + 1}\, dp\, dq\, dr$ (pqr-space)

19. Figure 13.37 shows the region of integration of the integral
$$\int_{-1}^1 \int_{x^2}^1 \int_0^{1 - y} dz\, dy\, dx.$$

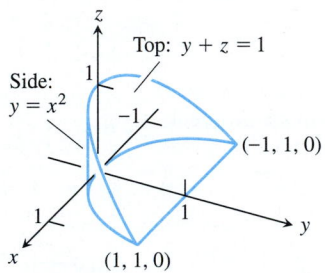

13.37   The region in Exercise 19.

Rewrite the integral as an equivalent iterated integral in the order
a) $dy\, dz\, dx$,   b) $dy\, dx\, dz$,
c) $dx\, dy\, dz$,   d) $dx\, dz\, dy$,
e) $dz\, dx\, dy$.

20. Figure 13.38 shows the region of integration of the integral
$$\int_0^1 \int_{-1}^0 \int_0^{y^2} dz\, dy\, dx.$$

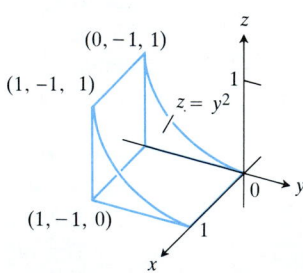

13.38   The region of integration in Exercise 20.

Rewrite the integral as an equivalent iterated integral in the order
a) $dy\, dz\, dx$,   b) $dy\, dx\, dz$,
c) $dx\, dy\, dz$,   d) $dx\, dz\, dy$,
e) $dz\, dx\, dy$.

Evaluate the integrals in Exercises 21–24 by changing the order of integration in an appropriate way.

21. $\displaystyle\int_0^4 \int_0^1 \int_{2y}^2 \frac{4 \cos(x^2)}{2\sqrt{z}}\, dx\, dy\, dz$

22. $\int_0^1 \int_0^1 \int_{x^2}^1 12\, xz\, e^{zy^2}\, dy\, dx\, dz$

23. $\int_0^1 \int_{\sqrt[3]{z}}^1 \int_0^{\ln 3} \dfrac{\pi e^{2x} \sin \pi y^2}{y^2}\, dx\, dy\, dz$

24. $\int_0^2 \int_0^{4-x^2} \int_0^x \dfrac{\sin 2z}{4-z}\, dy\, dz\, dx$

In Exercises 25–28, find the average value of $F(x, y, z)$ over the given region.

25. $F(x, y, z) = x^2 + 9$ over the cube in the first octant bounded by the coordinate planes and the planes $x = 2$, $y = 2$, and $z = 2$

26. $F(x, y, z) = x + y - z$ over the rectangular solid in the first octant bounded by the coordinate planes and the planes $x = 1$, $y = 1$, and $z = 2$

27. $F(x, y, z) = x^2 + y^2 + z^2$ over the cube in the first octant bounded by the coordinate planes and the planes $x = 1$, $y = 1$, and $z = 1$

28. $F(x, y, z) = xyz$ over the cube in the first octant bounded by the coordinate planes and the planes $x = a$, $y = a$, and $z = a$ ($a > 0$)

Find the volumes of the regions in Exercises 29–42.

29. The region between the cylinder $z = y^2$ and the xy-plane that is bounded by the planes $x = 0$, $x = 1$, $y = -1$, $y = 1$ (Fig. 13.39)

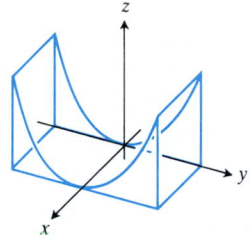

13.39   The region in Exercise 29.

30. The region cut from the first octant by the planes $x + z = 1$ and $y + 2z = 2$ (Fig. 13.40)

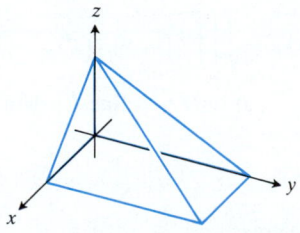

13.40   The region in Exercise 30.

31. The region in the first octant bounded by the coordinate planes, the plane $y + z = 2$, and the cylinder $x = 4 - y^2$ (Fig. 13.41)

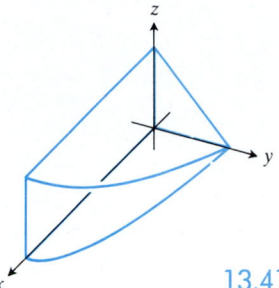

13.41   The region in Exercise 31.

32. The region in the first octant bounded by the coordinate planes, the plane $y = 1 - x$, and the surface $z = \cos(\pi x/2)$, $0 \le x \le 1$ (Fig. 13.42)

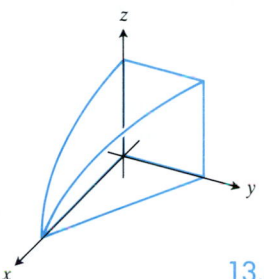

13.42   The region in Exercise 32.

33. The region in the first octant bounded by the coordinate planes and the surface $z = 4 - x^2 - y$ (Fig. 13.43)

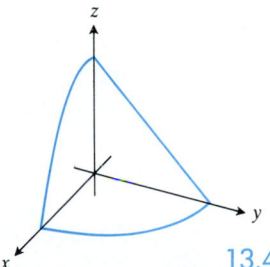

13.43   The region in Exercise 33.

34. The region in the first octant bounded by the coordinate planes, the plane $x + y = 4$, and the cylinder $y^2 + 4z^2 = 16$ (Fig. 13.44)

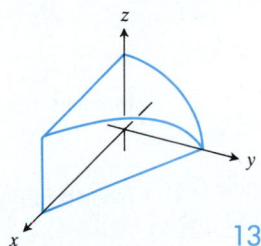

13.44   The region in Exercise 34.

35. The region common to the interiors of the cylinders $x^2 + y^2 = 1$ and $x^2 + z^2 = 1$ (Fig. 13.45)

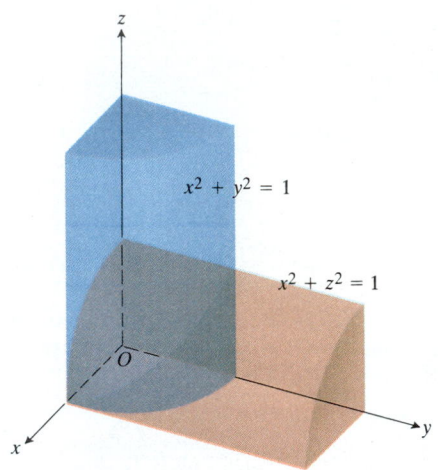

13.45   One-eighth of the region common to the cylinders $x^2 + y^2 = 1$ and $x^2 + z^2 = 1$ in Exercise 35.

36. The tetrahedron cut from the first octant by the plane $x + y/2 + z/3 = 1$

37. The region between the planes $x + y + 2z = 2$ and $2x + 2y + z = 4$ in the first octant

38. The wedge cut from the solid cylinder $x^2 + y^2 \le 1$ by the half-planes $z = -y$, $y \le 0$, and $z = 0$, $y \le 0$

39. The region cut from the solid cylinder $x^2 + y^2 \le 4$ by the plane $z = 0$ and the plane $x + z = 3$

40. The region cut from the solid elliptical cylinder $x^2 + 4y^2 \le 4$ by the xy-plane and the plane $z = x + 2$

41. The region bounded in back by the plane $x = 0$, on the front and sides by the parabolic cylinder $x = 1 - y^2$, on the top by the paraboloid $z = x^2 + y^2$, and on the bottom by the xy-plane

42. The wedge-shaped region enclosed on the side by the cylinder $x = -\cos y$, $-\pi/2 \le y \le \pi/2$, on the top by the plane $z = -2x$, and on the bottom by the xy-plane

43. Solve for a:

$$\int_0^1 \int_0^{4-a-x^2} \int_a^{4-x^2-y} dz \, dy \, dx = \frac{4}{15}.$$

44. For what value of c is the volume of the ellipsoid $x^2 + (y/2)^2 + (z/c)^2 = 1$ equal to 8π?

45. What domain D in space maximizes the value of the integral

$$\iiint_D (18 - x^2 - y^2 - 2z^2) \, dV?$$

46. What domain D in space minimizes the value of the integral

$$\iiint_D (36 - 9x^2 - 9y^2 - 4z^2) \, dV?$$

13.5 Masses and Moments in Three Dimensions

This section shows how we can calculate the masses and moments of three-dimensional objects in Cartesian coordinates. The formulas we use are similar to the ones we use for two-dimensional objects, the only difference being that we now have three coordinates instead of two. In Section 13.6, we shall discuss moment and mass calculations in spherical and cylindrical coordinates.

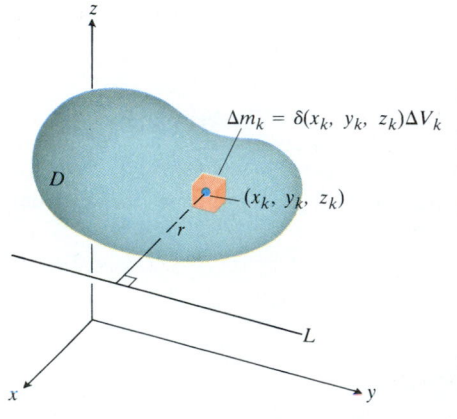

13.46   To define I_L we first imagine D to be subdivided into a finite number of mass elements Δm_k.

Masses and Moments

If $\delta(x, y, z)$ is the density function of an object occupying a region D in space, then the integral of δ over D gives the mass of the object. To see why, imagine subdividing D as in Fig. 13.46. The object's mass is the limit

$$M = \lim \sum_k \Delta m_k = \lim \sum_k \delta(x_k, y_k, z_k)\Delta V_k = \iiint_D \delta(x, y, z) \, dV. \qquad (1)$$

If $r(x, y, z)$ is the distance from the point (x, y, z) in D to a line L, then the moment of inertia of the mass

$$\Delta m_k = \delta(x_k, y_k, z_k)\Delta V_k$$

(shown in Fig. 13.46) about the line L is approximately

$$\Delta I_k = r^2(x_k, y_k, z_k)\Delta m_k.$$

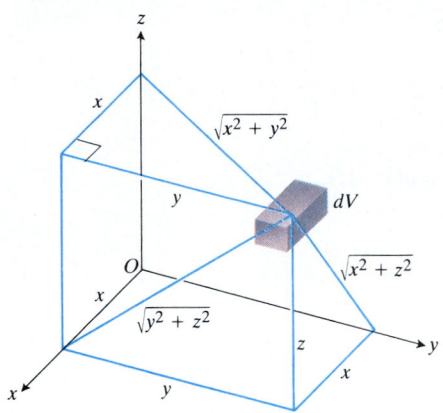

13.47 Distances from dV to the coordinate planes and axes.

The moment of inertia of the entire object about L is

$$I_L = \lim \sum_k \Delta I_k = \lim \sum_k r^2(x_k, y_k, z_k)\delta(x_k, y_k, z_k)\Delta V_k = \iiint_D r^2\delta\, dV.$$

If L is the x-axis, then $r^2 = y^2 + z^2$ (Fig. 13.47) and

$$I_x = \iiint_D (y^2 + z^2)\delta\, dV.$$

Similarly,

$$I_y = \iiint_D (x^2 + z^2)\delta\, dV \qquad \text{and} \qquad I_z = \iiint_D (x^2 + y^2)\delta\, dV.$$

These and other useful formulas are summarized in Table 13.2.

TABLE 13.2
Mass and moment formulas for objects in space

Mass: $\displaystyle M = \iiint_D \delta\, dV$ $(\delta(x, y, z) = \text{density})$

First moments about the coordinate planes:

$$M_{yz} = \iiint_D x\,\delta\, dV, \qquad M_{xz} = \iiint_D y\,\delta\, dV, \qquad M_{xy} = \iiint_D z\,\delta\, dV$$

Center of mass:

$$\bar{x} = \frac{\iiint x\,\delta\, dV}{M}, \qquad \bar{y} = \frac{\iiint y\,\delta\, dV}{M}, \qquad \bar{z} = \frac{\iiint z\,\delta\, dV}{M}$$

Moments of inertia (second moments):

$$I_x = \iiint (y^2 + z^2)\delta\, dV, \qquad I_y = \iiint (x^2 + z^2)\delta\, dV,$$

$$I_z = \iiint (x^2 + y^2)\delta\, dV,$$

Moment of inertia about a line L:

$$I_L = \iiint r^2\,\delta\, dV \qquad (r(x, y, z) = \text{distance from point } (x, y, z) \text{ to line } L)$$

Radius of gyration about a line L:

$$R_L = \sqrt{I_L/M}$$

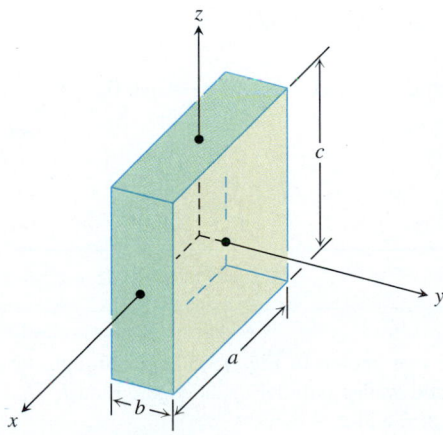

13.48 Example 1 calculates I_x, I_y, and I_z for the block shown here. The origin lies at the center of the block.

Example 1   Find I_x, I_y, I_z for the rectangular solid of constant density δ shown in Fig. 13.48.

Solution   The preceding formula for I_x gives

$$I_x = \int_{-c/2}^{c/2} \int_{-b/2}^{b/2} \int_{-a/2}^{a/2} (y^2 + z^2)\delta\, dx\, dy\, dz. \tag{2}$$

We can avoid some of the work of integration by observing that $(y^2 + z^2)\delta$ is an even function of x, y, and z and therefore

$$I_x = 8 \int_0^{c/2} \int_0^{b/2} \int_0^{a/2} (y^2 + z^2)\delta \, dx \, dy \, dz = 4a\delta \int_0^{c/2} \int_0^{b/2} (y^2 + z^2) \, dy \, dz$$

$$= 4a\delta \int_0^{c/2} \left[\frac{y^3}{3} + z^2 y\right]_{y=0}^{y=b/2} dz = 4a\delta \int_0^{c/2} \left(\frac{b^3}{24} + \frac{z^2 b}{2}\right) dz$$

$$= 4a\delta\left(\frac{b^3 c}{48} + \frac{c^3 b}{48}\right) = \frac{abc\delta}{12}(b^2 + c^2) = \frac{M}{12}(b^2 + c^2).$$

Similarly,

$$I_y = \frac{M}{12}(a^2 + c^2) \qquad \text{and} \qquad I_z = \frac{M}{12}(a^2 + b^2).$$

Example 2   Find the center of mass of a solid of constant density δ bounded below by the disk $R: x^2 + y^2 \leq 4$ in the plane $z = 0$ and above by the paraboloid $z = 4 - x^2 - y^2$ (Fig. 13.49).

Solution   By symmetry, $\bar{x} = \bar{y} = 0$. To find $\bar{z}$ we first calculate

$$M_{xy} = \iiint_R \int_{z=0}^{z=4-x^2-y^2} z \, \delta \, dz \, dy \, dx = \iint_R \left[\frac{z^2}{2}\right]_{z=0}^{z=4-x^2-y^2} \delta \, dy \, dx$$

$$= \frac{\delta}{2} \iint_R (4 - x^2 - y^2)^2 \, dy \, dx$$

$$= \frac{\delta}{2} \int_0^{2\pi} \int_0^2 (4 - r^2)^2 \, r \, dr \, d\theta \qquad \binom{\text{Polar}}{\text{coordinates}}$$

$$= \frac{\delta}{2} \int_0^{2\pi} \left[-\frac{1}{6}(4 - r^2)^3\right]_{r=0}^{r=2} d\theta = \frac{16\delta}{3} \int_0^{2\pi} d\theta = \frac{32\pi\delta}{3}.$$

A similar calculation gives

$$M = \iiint_R \int_0^{4-x^2-y^2} \delta \, dz \, dy \, dx = 8\pi\delta.$$

Therefore, $\bar{z} = M_{xy}/M = 4/3$, and the center of mass is $(\bar{x}, \bar{y}, \bar{z}) = (0, 0, 4/3)$.

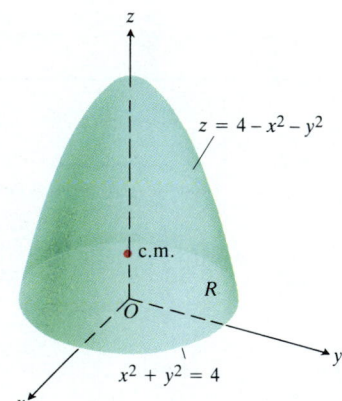

13.49   Example 2 calculates the coordinates of the center of mass of this solid.

EXERCISES 13.5

Constant Density

The solids in Exercises 1–12 all have constant density $\delta = 1$.

1. Evaluate the integral for I_x in Eq. (2) directly to show that the shortcut in Example 1 gives the same answer. Use the results in Example 1 to find the radius of gyration of the rectangular solid about each coordinate axis.

2. The coordinate axes shown in Fig. 13.50 run through the centroid of a solid wedge parallel to its edges. Find I_x, I_y, and I_z if $a = b = 6$ and $c = 4$.

3. Find the moments of inertia of the rectangular solid shown in Fig. 13.51 with respect to its edges by calculating I_x, I_y, and I_z.

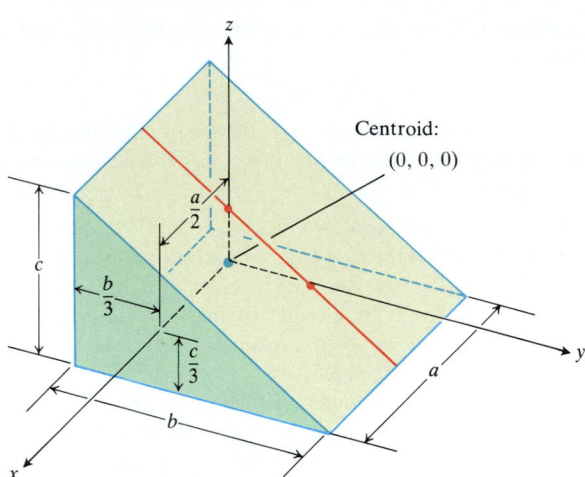

13.50   The figure for Exercise 2.

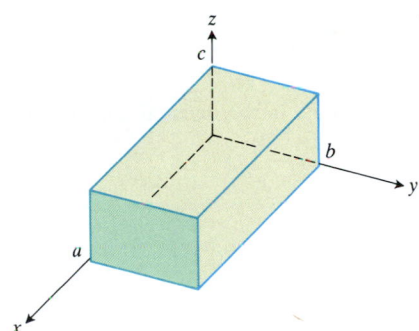

13.51   The rectangular solid in Exercise 3.

4. a) Find the centroid and the moments of inertia I_x, I_y, and I_z of the tetrahedron whose vertices are the points $(0, 0, 0)$, $(1, 0, 0)$, $(0, 1, 0)$ and $(0, 0, 1)$.
   b) Find the radius of gyration of the tetrahedron about the x-axis. Compare it with the distance from the centroid to the x-axis.

5. A solid "trough" of constant density is bounded below by the surface $z = 4y^2$, above by the plane $z = 4$, and on the ends by the planes $x = 1$ and $x = -1$. Find the center of mass and the moments of inertia with respect to the three axes.

6. A solid of constant density is bounded below by the plane $z = 0$, on the sides by the elliptical cylinder $x^2 + 4y^2 = 4$, and above by the plane $z = 2 - x$ (Fig. 13.52).
   a) Find $\bar{x}$ and $\bar{y}$.
   b) Evaluate the integral
   $$M_{xy} = \int_{-2}^{2} \int_{-(1/2)\sqrt{4-x^2}}^{(1/2)\sqrt{4-x^2}} \int_{0}^{2-x} z \, dz \, dy \, dx,$$

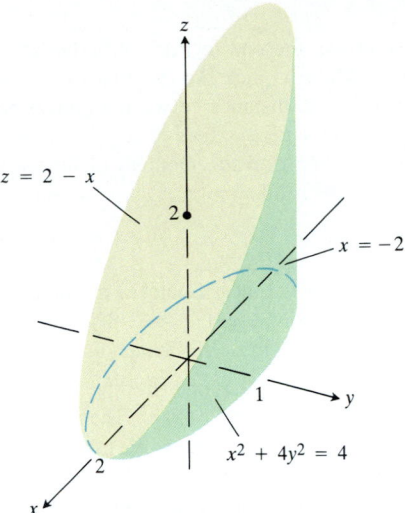

13.52   The solid in Exercise 6.

using integral tables to carry out the final integration with respect to x. Then divide M_{xy} by M to verify that $\bar{z} = 5/4$.

7. a) Find the center of mass of a solid of constant density bounded below by the paraboloid $z = x^2 + y^2$ and above by the plane $z = 4$.
   b) Find the plane $z = c$ that divides the solid into two parts of equal volume. This plane does not pass through the center of mass.

8. A solid cube 2 units on a side is bounded by the planes $x = \pm 1$, $z = \pm 1$, $y = 3$, and $y = 5$. Find the center of mass and the moments of inertia and radii of gyration about the coordinate axes. Suppose that the cube's mass is concentrated at the center of mass. How, if at all, does that change the moments of inertia and radii of gyration about the coordinate axes?

9. A wedge shaped like the one in Fig. 13.50 has $a = 4$, $b = 6$, and $c = 3$. Make a quick sketch to check for yourself that the square of the distance from a typical point (x, y, z) of the wedge to the line $L: z = 0$, $y = 6$ is $r^2 = (y - 6)^2 + z^2$. Then calculate the moment of inertia and radius of gyration of the wedge about L.

10. A wedge shaped like the one in Fig. 13.50 has $a = 4$, $b = 6$, and $c = 3$. Make a quick sketch to check for yourself that the square of the distance from a typical point (x, y, z) of the wedge to the line $L: x = 4$, $y = 0$ is $r^2 = (x - 4)^2 + y^2$. Then calculate the moment of inertia and radius of gyration of the wedge about L.

11. A rectangular solid like the one in Fig. 13.51 has $a = 4$, $b = 2$, and $c = 1$. Make a quick sketch to check for yourself that the square of the distance between a typical point (x, y, z) of the solid and the line $L: y = 2$, $z = 0$ is $r^2 = (y - 2)^2 + z^2$. Then find the moment of inertia and radius of gyration of the solid about L.

12. A rectangular solid like the one in Fig. 13.51 has $a = 4$, $b = 2$, and $c = 1$. Make a quick sketch to check for yourself that the square of the distance between a typical point (x, y, z) of the solid and the line $L: x = 4$, $y = 0$ is $r^2 = (x - 4)^2 + y^2$. Then find the moment of inertia and radius of gyration of the solid about L.

Variable Density

13. A solid region in the first octant is bounded by the coordinate planes and the plane $x + y + z = 2$. The density of the solid is $\delta(x, y, z) = 2x$. Find the center of mass.

14. A solid wedge shaped like the one in Fig. 13.50 has dimensions $a = 4$, $b = 6$, and $c = 3$. The density is $\delta(x, y, z) = x + 1$. Find the center of mass and the moments of inertia and radii of gyration about the coordinate axes. Notice that if the density is constant, the center of mass will be $(0, 0, 0)$.

15. A solid cube in the first octant is bounded by the coordinate planes and by the planes $x = 1$, $y = 1$, and $z = 1$. Find its center of mass and the moments of inertia and radii of gyration about the coordinate axes if

$$\delta(x, y, z) = x + y + z + 1.$$

16. A solid in the first octant is bounded by the planes $y = 0$ and $z = 0$ and by the surfaces $z = 4 - x^2$ and $x = y^2$ (Fig. 13.53). Its density function is $\delta(x, y, z) = kxy$.
a) Find the solid's mass.   b) Find $\bar{x}$.
c) Find $\bar{y}$.   d) Find $\bar{z}$.

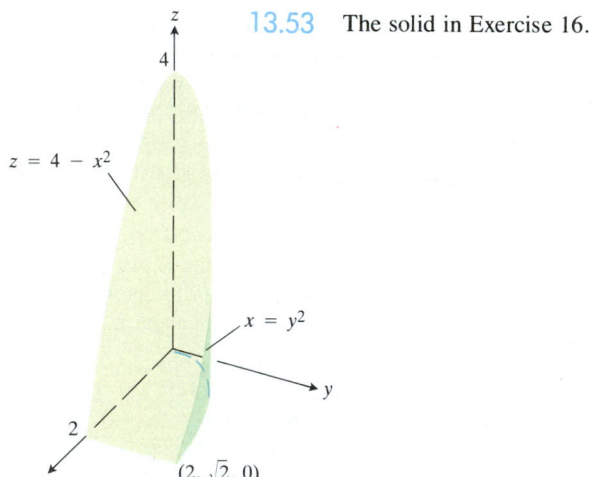

13.53   The solid in Exercise 16.

$z = 4 - x^2$

$x = y^2$

$(2, \sqrt{2}, 0)$

The Parallel Axis Theorem

The Parallel Axis Theorem (Exercises 13.2) holds in three dimensions as well as in two. Let $L_{c.m.}$ be a line through the center of mass of a body of mass m and let L be a parallel line h units away from $L_{c.m.}$. The **Parallel Axis Theorem** says that the moments of inertia $I_{c.m.}$ and I_L of the body about $L_{c.m.}$ and L satisfy the equation

$$I_L = I_{c.m.} + mh^2. \tag{3}$$

As in the two-dimensional case, the theorem gives a quick way to calculate one moment when the other moment and the mass are known.

17. *Proof of the Parallel Axis Theorem*
a) Show that the first moment of a body in space about any plane through the body's center of mass is zero. (*Hint:* Place the body's center of mass at the origin and let the plane be the yz-plane. What does the formula $\bar{x} = M_{yz}/M$ then tell you?)
b) To prove the Parallel Axis Theorem, place the body with its center of mass at the origin, with the line $L_{c.m.}$ along the z-axis and with the line L perpendicular to the xy-plane at the point $(h, 0, 0)$. Let D be the region of space occupied by the body (Fig. 13.54). Then, in the notation of the figure,

$$I_L = \iiint\limits_{D} |\mathbf{v} - h\mathbf{i}|^2 \, dm. \tag{4}$$

Expand the integrand in this integral and complete the proof.

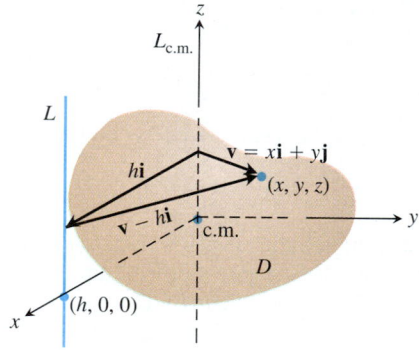

$L_{c.m.}$

L

$h\mathbf{i}$

$\mathbf{v} = x\mathbf{i} + y\mathbf{j}$

(x, y, z)

$\mathbf{v} - h\mathbf{i}$

c.m.

D

$(h, 0, 0)$

13.54   The diagram for the proof of the Parallel Axis Theorem in Exercise 17.

18. The moment of inertia about a diameter of a solid sphere of constant density and radius a is $(2/5)ma^2$, where m is the mass of the sphere. Find the moment of inertia about a line tangent to the sphere.

19. The moment of inertia of the rectangular solid in Fig. 13.51 about the z-axis is $I_z = abc(a^2 + b^2)/3$.
a) Use Eq. (3) to find the moment of inertia and radius of gyration of the solid about the line parallel to the z-axis through the solid's center of mass.
b) Use Eq. (3) and the result in (a) to find the moment of

inertia and radius of gyration of the solid about the line $x = 0$, $y = 2b$.

20. If $a = b = 6$ and $c = 4$, the moment of inertia of the solid wedge in Fig. 13.50 about the x-axis is $I_x = 208$. Find the moment of inertia of the wedge about the line $y = 4$, $z = -4/3$ (the edge line of the wedge's narrow end).

Pappus's Formula

Pappus's formula (Exercises 13.2) holds in three dimensions as well as in two. Suppose that bodies B_1 and B_2 of mass m_1 and m_2, respectively, occupy nonoverlapping regions in space and that $\mathbf{c}_1$ and $\mathbf{c}_2$ are the vectors from the origin to the bodies' respective centers of mass. Then the center of mass of the union $B_1 \cup B_2$ of the two bodies is determined by the vector

$$\mathbf{c} = \frac{m_1 \mathbf{c}_1 + m_2 \mathbf{c}_2}{m_1 + m_2}. \tag{5}$$

As before, this formula is called **Pappus's formula.** As in the two-dimensional case, the formula generalizes to

$$\mathbf{c} = \frac{m_1 \mathbf{c}_1 + m_2 \mathbf{c}_2 + \cdots + m_n \mathbf{c}_n}{m_1 + m_2 + \cdots + m_n} \tag{6}$$

for n bodies.

21. Derive Pappus's formula (Eq. 5). (*Hint:* Sketch B_1 and B_2 as nonoverlapping regions in the first octant and label their centers of mass $(\bar{x}_1, \bar{y}_1, \bar{z}_1)$ and $(\bar{x}_2, \bar{y}_2, \bar{z}_2)$. Express the moments of $B_1 \cup B_2$ about the coordinate planes in terms of the masses m_1 and m_2 and the coordinates of these centers.)

22. Figure 13.55 shows a solid made from three rectangular solids of constant density $\delta = 1$. Use Pappus's formula to find the center of mass of

a) $A \cup B$,   b) $A \cup C$,

c) $B \cup C$,   d) $A \cup B \cup C$.

23. a) Suppose that a solid right circular cone C of base radius a and altitude h is constructed on the circular base of a solid hemisphere S of radius a so that the union of the

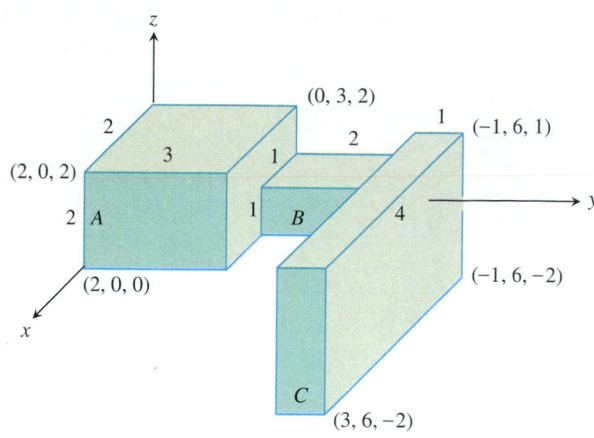

13.55   The solid in Exercise 22.

two solids resembles an ice cream cone. The centroid of a solid cone lies one-fourth of the way from the base toward the vertex. The centroid of a solid hemisphere lies three-eighths of the way from the base to the top. What relation must hold between h and a to place the centroid of $C \cup S$ in the common base of the two solids?

b) If you have not already done so, answer the analogous question about a triangle and a semicircle (Section 13.2, Exercise 53). The answers are not the same.

24. A solid pyramid P with height h and four congruent sides is built with its base as one face of a solid cube C whose edges have length s. The centroid of a solid pyramid lies one-fourth of the way from the base toward the vertex. What relation must hold between h and s to place the centroid of $P \cup C$ in the base of the pyramid? Compare your answer with the answer to Exercise 23. Also compare it to the answer to Exercise 54 in Section 13.2.

13.6 Triple Integrals in Cylindrical and Spherical Coordinates

If we are working with a solid like a cone or a cylindrical shell that has an axis of symmetry, we can often simplify our calculations by taking the axis to be the z-axis in a cylindrical coordinate system. In like manner, if we are working with a shape like a ball or spherical shell that has a center of symmetry, we can usually save time by choosing the center to be the origin in a spherical coordinate system.

This section shows how to calculate masses, moments, and volumes in cylindrical and spherical coordinates.

Cylindrical Coordinates

Cylindrical coordinates (Fig. 13.56) are good for describing cylinders whose axes run along the z-axis and planes that either contain the z-axis or lie perpendicular to

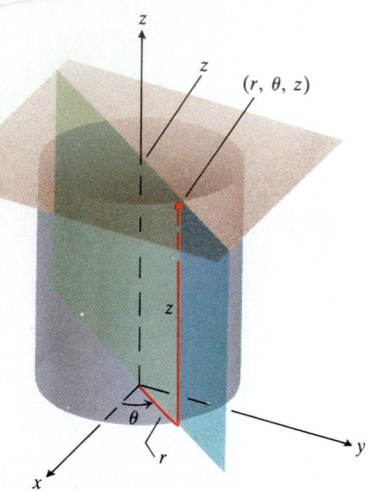

13.56 Cylindrical coordinates and typical surfaces of constant coordinate value.

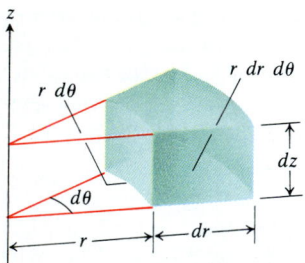

13.57 The volume element in cylindrical coordinates is $dV = dz\, r\, dr\, d\theta$.

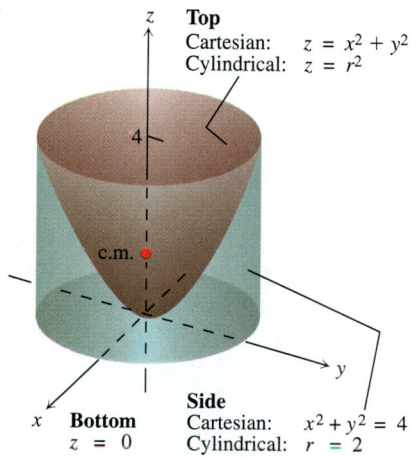

13.58 Example 1 shows how to locate the centroid of this region.

the z-axis. As we saw in Section 10.8, surfaces like these have equations of constant coordinate value:

$$r = 4 \qquad \text{(cylinder, radius 4, axis the } z\text{-axis),}$$

$$\theta = \frac{\pi}{3} \qquad \text{(plane containing the } z\text{-axis),}$$

$$z = 2 \qquad \text{(plane perpendicular to the } z\text{-axis).}$$

The volume element for subdividing a region in space with cylindrical coordinates is

$$dV = dz\, r\, dr\, d\theta \qquad (1)$$

(Fig. 13.57). Triple integrals in cylindrical coordinates are then evaluated as iterated integrals, as in the following example.

How to Integrate in Cylindrical Coordinates

To integrate a continuous function $f(r, \theta, z)$ over a region given by inequalities

$$g_1(r, \theta) \le z \le g_2(r, \theta),$$

$$h_1(\theta) \le r \le h_2(\theta),$$

$$\theta_1 \le \theta \le \theta_2,$$

evaluate the iterated integral

$$\int_{\theta = \theta_1}^{\theta = \theta_2} \int_{r = h_1(\theta)}^{r = h_2(\theta)} \int_{z = g_1(r, \theta)}^{z = g_2(r, \theta)} f(r, \theta, z)\, dz\, r\, dr\, d\theta. \qquad (2)$$

Integrate first with respect to z. Multiply by r and integrate with respect to r. Then integrate with respect to θ.

Example 1   Find the centroid of the region enclosed by the cylinder $x^2 + y^2 = 4$, bounded above by the paraboloid $z = x^2 + y^2$, and bounded below by the xy-plane.

Solution   We sketch the figure (Fig. 13.58) and find the cylindrical coordinate equations for the bounding surfaces. The coordinate inequalities for the region are

$$g_1(r, \theta) \le z \le g_2(r, \theta): \qquad 0 \le z \le r^2$$

$$h_1(\theta) \le r \le h_2(\theta): \qquad 0 \le r \le 2$$

$$\theta_1 \le \theta \le \theta_2: \qquad 0 \le \theta \le 2\pi.$$

The centroid lies on its axis of symmetry, in this case the z-axis, so $\bar{x} = \bar{y} = 0$.

To find $\bar{z}$, we divide the moment M_{xy} by the mass M. The value of M_{xy} is

$$M_{xy} = \int_0^{2\pi} \int_0^2 \int_0^{r^2} z \, dz \, r \, dr \, d\theta = \int_0^{2\pi} \int_0^2 \left[\frac{z^2}{2}\right]_0^{r^2} r \, dr \, d\theta$$

$$= \int_0^{2\pi} \int_0^2 \frac{r^5}{2} \, dr \, d\theta = \int_0^{2\pi} \left[\frac{r^6}{12}\right]_0^2 d\theta$$

$$= \int_0^{2\pi} \frac{16}{3} \, d\theta = \frac{32\pi}{3}.$$

The value of M is

$$M = \int_0^{2\pi} \int_0^2 \int_0^{r^2} dz \, r \, dr \, d\theta = \int_0^{2\pi} \int_0^2 \left[z\right]_0^{r^2} r \, dr \, d\theta$$

$$= \int_0^{2\pi} \int_0^2 r^3 \, dr \, d\theta = \int_0^{2\pi} \left[\frac{r^4}{4}\right]_0^2 d\theta$$

$$= \int_0^{2\pi} 4 \, d\theta = 8\pi.$$

Therefore,

$$\bar{z} = \frac{M_{xy}}{M} = \frac{32\pi}{3} \frac{1}{8\pi} = \frac{4}{3},$$

and the centroid is the point $(0, 0, 4/3)$.

Spherical Coordinates

Spherical coordinates (Fig. 13.59) are good for describing spheres centered at the origin, half-planes hinged along the z-axis, and single-napped cones whose vertices lie at the origin and whose axes lie along the z-axis. Surfaces like these have equations of constant coordinate value:

$\rho = 4$ (sphere, radius 4, center at origin),

$\phi = \dfrac{\pi}{3}$ $\begin{pmatrix}\text{cone opening up from the origin, making} \\ \text{an angle of } \pi/3 \text{ radians with the positive } z\text{-axis}\end{pmatrix}$,

$\theta = \dfrac{\pi}{3}$ $\begin{pmatrix}\text{half-plane, hinged along the } z\text{-axis, making} \\ \text{an angle of } \pi/3 \text{ radians with the positive } x\text{-axis}\end{pmatrix}$.

The volume element in spherical coordinates is

$$dV = \rho^2 \sin \phi \, d\rho \, d\phi \, d\theta \tag{3}$$

(Fig. 13.60), and triple integrals take the form

$$\iiint f(\rho, \phi, \theta) \, dV = \iiint f(\rho, \phi, \theta) \rho^2 \sin \phi \, d\rho \, d\phi \, d\theta. \tag{4}$$

To evaluate these integrals we usually integrate first with respect to ρ. The procedure for finding the limits of integration for a region D in space is shown in the following box.

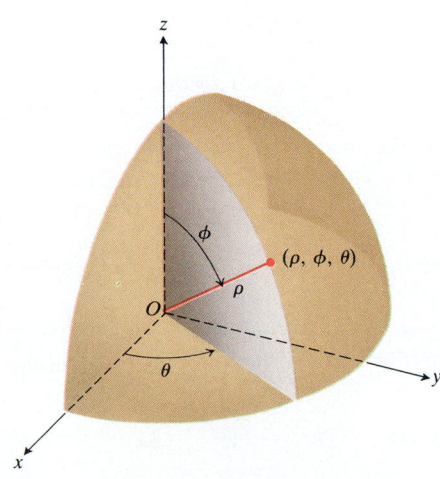

13.59 Spherical coordinates are measured with a distance and two angles.

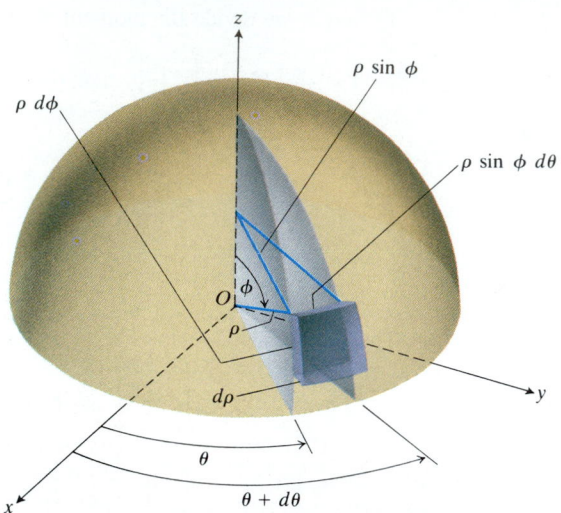

13.60 The volume element in spherical coordinates is

$$dV = d\rho \cdot \rho \, d\phi \cdot \rho \sin \phi \, d\theta$$
$$= \rho^2 \sin \phi \, d\rho \, d\phi \, d\theta.$$

How to Integrate in Spherical Coordinates

To integrate a continuous function $f(\rho, \phi, \theta)$ over a region given by the inequalities

$$g_1(\phi, \theta) \le \rho \le g_2(\phi, \theta),$$

$$h_1(\theta) \le \phi \le h_2(\theta),$$

$$\theta_1 \le \theta \le \theta_2,$$

evaluate the iterated integral

$$\int_{\theta = \theta_1}^{\theta = \theta_2} \int_{\phi = h_1(\theta)}^{\phi = h_2(\theta)} \int_{\rho = g_1(\phi, \theta)}^{\rho = g_2(\phi, \theta)} f(\rho, \phi, \theta) \rho^2 \sin \phi \, d\rho \, d\phi \, d\theta. \tag{5}$$

To do so, multiply f by ρ^2 and integrate with respect to ρ. Multiply the result by $\sin \phi$ and integrate with respect to ϕ. Then integrate with respect to θ.

Example 2   Find the volume of the upper region cut from the solid sphere $\rho \le 1$ by the cone $\phi = \pi/3$.

Solution   We sketch the cone and sphere (Fig. 13.61). The coordinate inequalities for the region occupied by the solid are

$$g_1(\phi, \theta) \le \rho \le g_2(\phi, \theta): \qquad 0 \le \rho \le 1,$$

$$h_1(\theta) \le \phi \le h_2(\theta): \qquad 0 \le \phi \le \frac{\pi}{3},$$

$$\theta_1 \le \theta \le \theta_2: \qquad 0 \le \theta \le 2\pi.$$

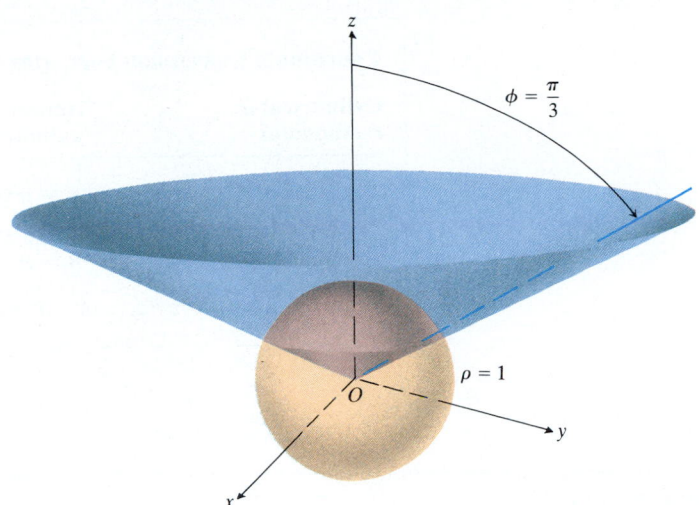

13.61   The region cut from the solid sphere $\rho \leq 1$ by the cone $\phi = \pi/3$ (Example 2).

The volume is the integral of $f(\rho, \phi, \theta) = 1$ over this region:

$$V = \int_0^{2\pi} \int_0^{\pi/3} \int_0^1 \rho^2 \sin \phi \, d\rho \, d\phi \, d\theta = \int_0^{2\pi} \int_0^{\pi/3} \left[\frac{\rho^3}{3} \right]_0^1 \sin \phi \, d\phi \, d\theta$$

$$= \int_0^{2\pi} \int_0^{\pi/3} \frac{1}{3} \sin \phi \, d\phi \, d\theta = \int_0^{2\pi} \left[-\frac{1}{3} \cos \phi \right]_0^{\pi/3} d\theta$$

$$= \int_0^{2\pi} \left(-\frac{1}{6} + \frac{1}{3} \right) d\theta = \frac{1}{6} (2\pi) = \frac{\pi}{3}.$$

Example 3   Find the moment of inertia about the z-axis of the region in Example 2.

Solution   In rectangular coordinates, the moment is

$$I_z = \int \int \int (x^2 + y^2) \, dV.$$

In spherical coordinates, $x^2 + y^2 = (\rho \sin \phi \cos \theta)^2 + (\rho \sin \phi \sin \theta)^2 = \rho^2 \sin^2 \phi$. Hence,

$$I_z = \int \int \int (\rho^2 \sin^2 \phi) \, \rho^2 \sin \phi \, d\rho \, d\phi \, d\theta = \int \int \int \rho^4 \sin^3 \phi \, d\rho \, d\phi \, d\theta.$$

For the region in Example 2, this becomes

$$I_z = \int_0^{2\pi} \int_0^{\pi/3} \int_0^1 \rho^4 \sin^3 \phi \, d\rho \, d\phi \, d\theta = \int_0^{2\pi} \int_0^{\pi/3} \left[\frac{\rho^5}{5} \right]_0^1 \sin^3 \phi \, d\phi \, d\theta$$

$$= \frac{1}{5} \int_0^{2\pi} \int_0^{\pi/3} (1 - \cos^2 \phi) \sin \phi \, d\phi \, d\theta = \frac{1}{5} \int_0^{2\pi} \left[-\cos \phi + \frac{\cos^3 \phi}{3} \right]_0^{\pi/3} d\theta$$

$$= \frac{1}{5} \int_0^{2\pi} \left(-\frac{1}{2} + 1 + \frac{1}{24} - \frac{1}{3} \right) d\theta = \frac{1}{5} \int_0^{2\pi} \frac{5}{24} \, d\theta = \frac{1}{24} (2\pi) = \frac{\pi}{12}.$$

> **Coordinate Conversion Formulas (from Section 10.8)**
>
> | **Cylindrical to rectangular** | **Spherical to cylindrical** | **Spherical to rectangular** |
> |---|---|---|
> | $x = r \cos\theta$ | $r = \rho \sin\phi$ | $x = \rho \sin\phi \cos\theta$ |
> | $y = r \sin\theta$ | $z = \rho \cos\phi$ | $y = \rho \sin\phi \sin\theta$ |
> | $z = z$ | $\theta = \theta$ | $z = \rho \cos\phi$ |
>
> Volume: $dV = dx\,dy\,dz$ or $dz\,r\,dr\,d\theta$ or $\rho^2 \sin\phi\,d\rho\,d\phi\,d\theta$

EXERCISES 13.6

Evaluate the cylindrical coordinate integrals in Exercises 1–6.

1. $\displaystyle\int_0^{2\pi}\int_0^1\int_r^{\sqrt{2-r^2}} dz\,r\,dr\,d\theta$

2. $\displaystyle\int_0^{2\pi}\int_0^3\int_{r^2/3}^{\sqrt{18-r^2}} dz\,r\,dr\,d\theta$

3. $\displaystyle\int_0^{2\pi}\int_0^{\theta/2\pi}\int_0^{3+24r^2} dz\,r\,dr\,d\theta$

4. $\displaystyle\int_0^{\pi}\int_0^{\theta/\pi}\int_{-\sqrt{4-r^2}}^{3\sqrt{4-r^2}} z\,dz\,r\,dr\,d\theta$

5. $\displaystyle\int_0^{2\pi}\int_0^1\int_r^{1/\sqrt{2-r^2}} 3\,dz\,r\,dr\,d\theta$

6. $\displaystyle\int_0^{2\pi}\int_0^1\int_{-1/2}^{1/2} (r^2\sin^2\theta + z^2)\,dz\,r\,dr\,d\theta$

Evaluate the spherical coordinate integrals in Exercises 7–12.

7. $\displaystyle\int_0^{\pi}\int_0^{\pi}\int_0^{2\sin\phi} \rho^2\sin\phi\,d\rho\,d\phi\,d\theta$

8. $\displaystyle\int_0^{2\pi}\int_0^{\pi/4}\int_0^2 (\rho\cos\phi)\,\rho^2\sin\phi\,d\rho\,d\phi\,d\theta$

9. $\displaystyle\int_0^{2\pi}\int_0^{\pi}\int_0^{(1-\cos\phi)/2} \rho^2\sin\phi\,d\rho\,d\phi\,d\theta$

10. $\displaystyle\int_0^{3\pi/2}\int_0^{\pi}\int_0^1 5\rho^3\sin^3\phi\,d\rho\,d\phi\,d\theta$

11. $\displaystyle\int_0^{2\pi}\int_0^{\pi/3}\int_{\sec\phi}^2 3\rho^2\sin\phi\,d\rho\,d\phi\,d\theta$

12. $\displaystyle\int_0^{2\pi}\int_0^{\pi/4}\int_0^{\sec\phi} \rho^3\cos^2\phi\,\sin\phi\,d\rho\,d\phi\,d\theta$

The integrals we have seen so far suggest that there are preferred orders of integration for cylindrical and spherical coordinates, but this is not really the case. The other orders usually work well and are occasionally better. Evaluate the integrals in Exercises 13–20.

13. $\displaystyle\int_0^{2\pi}\int_0^3\int_0^{z/3} r^3\,dr\,dz\,d\theta$

14. $\displaystyle\int_{-1}^1\int_0^{2\pi}\int_0^{1+\cos\theta} 4r\,dr\,d\theta\,dz$

15. $\displaystyle\int_0^1\int_0^{\sqrt{z}}\int_0^{2\pi} (r^2\cos^2\theta + z^2)r\,d\theta\,dr\,dz$

16. $\displaystyle\int_0^2\int_{r-2}^{\sqrt{4-r^2}}\int_0^{2\pi} (r\sin\theta + 1)r\,d\theta\,dz\,dr$

17. $\displaystyle\int_0^2\int_{-\pi}^0\int_{\pi/4}^{\pi/2} \rho^3\sin 2\phi\,d\phi\,d\theta\,d\rho$

18. $\displaystyle\int_{\pi/6}^{5\pi/6}\int_{\csc\phi}^{2\csc\phi}\int_0^{2\pi} \rho\sin\phi\,d\theta\,d\rho\,d\phi$

19. $\displaystyle\int_0^1\int_0^{\pi}\int_0^{\pi/4} 12\rho\sin^3\phi\,d\phi\,d\theta\,d\rho$

20. $\displaystyle\int_{\pi/6}^{\pi/2}\int_{-\pi/2}^{\pi/2}\int_{\csc\phi}^2 5\rho^4\sin^3\phi\,d\rho\,d\theta\,d\phi$

21. Set up an iterated triple integral for the volume of the sphere

$x^2 + y^2 + z^2 = 4$ in (a) spherical, (b) cylindrical, and (c) rectangular coordinates.

22. Let D denote the region in the first octant that is bounded below by the cone $z = \sqrt{x^2 + y^2}$ and above by the sphere $x^2 + y^2 + z^2 = 9$. Express the volume of D as an iterated triple integral in (a) cylindrical and (b) spherical coordinates. Then (c) find V.

23. Give the limits of integration for evaluating the integral

$$\int\int\int f(r, \theta, z)\, dz\, r\, dr\, d\theta$$

as an iterated integral over the region that is bounded below by the plane $z = 0$, on the side by the cylinder $r = \cos\theta$, and on top by the paraboloid $z = 3r^2$.

24. Convert the integral

$$\int_{-1}^{1}\int_{0}^{\sqrt{1-y^2}}\int_{0}^{x} (x^2 + y^2)\, dz\, dx\, dy$$

to an equivalent integral in cylindrical coordinates and evaluate the result.

25. Let D be the smaller cap cut from a solid ball of radius 2 units by a plane 1 unit from the center of the sphere. Express the volume of D as an iterated triple integral in (a) spherical, (b) cylindrical, and (c) rectangular coordinates. Then (d) find the volume by evaluating one of the three triple integrals.

26. Express the moment of inertia I_z of the solid hemisphere $x^2 + y^2 + z^2 \leq 1, z \geq 0$, as an iterated integral in (a) cylindrical and (b) spherical coordinates. Then (c) find I_z by evaluating one of these integrals.

Cylindrical Coordinates

In exercises that ask for masses and moments, assume that the density is $\delta = 1$ unless stated otherwise.

27. Find the volume of the region bounded below by the plane $z = 0$, laterally by the cylinder $x^2 + y^2 = 1$, and above by the paraboloid $z = x^2 + y^2$.

28. Find the volume of the region bounded below by the paraboloid $z = x^2 + y^2$, laterally by the cylinder $x^2 + y^2 = 1$, and above by the paraboloid $z = x^2 + y^2 + 1$.

29. Find the volume of the region bounded below by the plane $z = 0$ and above by the paraboloid $z = 4 - x^2 - y^2$.

30. Find the volume of the region enclosed by the cylinder $x^2 + y^2 = 4$ and the planes $z = 0$ and $y + z = 4$. (*Hint:* In cylindrical coordinates, $z = 4 - y$ becomes $z = 4 - r\sin\theta$.)

31. Find the volume of the region bounded above by the paraboloid $z = 5 - x^2 - y^2$ and below by the paraboloid $z = 4x^2 + 4y^2$.

32. Find the volume of the region bounded above by the paraboloid $z = 9 - x^2 - y^2$ and below by the xy-plane and lying *outside* the cylinder $x^2 + y^2 = 1$.

33. Find the volume of the region cut from the solid sphere $x^2 + y^2 + z^2 \leq 4$ by the cylinder $x^2 + y^2 = 1$.

34. Find the volume of the region bounded above by the sphere $x^2 + y^2 + z^2 = 2$ and below by the circular paraboloid $z = x^2 + y^2$.

35. Find the average value of the function $f(r, \theta, z) = r$ over the region bounded by the cylinder $r = 1$ between the planes $z = -1$ and $z = 1$.

36. Find the average value of the function $f(r, \theta, z) = r$ over the solid ball bounded by the sphere $r^2 + z^2 = 1$. (This is the sphere $x^2 + y^2 + z^2 = 1$.)

37. A solid is bounded below by the plane $z = 0$, above by the cone $z = r, r \geq 0$, and on the sides by the cylinder $r = 1$. Find the center of mass.

38. Find the centroid of the region that is bounded below by the surface $z = \sqrt{r}$ and above by the plane $z = 2$.

39. Find the moment of inertia and radius of gyration about the z-axis of a thick-walled right circular cylinder bounded on the inside by the cylinder $r = 1$, on the outside by the cylinder $r = 2$, and on the top and bottom by the planes $z = 4$ and $z = 0$.

40. A solid of constant density in the first octant is bounded above by the cone $z = \sqrt{x^2 + y^2}$, below by the plane $z = 0$, and on the sides by the cylinder $x^2 + y^2 = 4$ and the planes $x = 0$ and $y = 0$. Find the center of mass. (*Hint:* $\bar{x} = \bar{y}$.)

41. Find the moment of inertia of a solid circular cylinder of radius 1 and height 2 (a) about the axis of the cylinder, (b) about a line through the centroid perpendicular to the axis of the cylinder.

42. Find the moment of inertia of a right circular cone of base radius 1 and height 1 about an axis through the vertex parallel to the base.

43. Find the moment of inertia of a solid sphere of radius a about a diameter.

44. Find the centroid of the smaller region cut from the solid ball $r^2 + z^2 \leq 1$ by the half-planes $\theta = -\pi/3, r \geq 0$, and $\theta = \pi/3, r \geq 0$.

45. Show that the centroid of the solid semi-ellipsoid of revolution $(r^2/a^2) + (z^2/h^2) \leq 1, z \geq 0$, lies on the z-axis three-eighths of the way from the base to the top. The special case $h = a$ gives a solid hemisphere. Thus the centroid of a solid hemisphere lies on the axis of symmetry three-eighths of the way from the base to the top.

46. Find the moment of inertia of a right circular cone of base radius a and height h about its axis. (*Hint:* Place the cone with its vertex at the origin and its axis along the z-axis.)

47. Show that the centroid of a solid right circular cone is one-fourth of the way from the base to the vertex. (In general, the centroid of a solid cone or pyramid is one-fourth of the way from the centroid of the base to the vertex.)

48. A solid is bounded on the top by the paraboloid $z = r^2$, on the bottom by the plane $z = 0$, and on the sides by the cylinder $r = 1$. Find the center of mass and the moment of inertia and radius of gyration about the z-axis if the density is (a) $\delta(r, \theta, z) = z$, (b) $\delta(r, \theta, z) = r$.

49. A solid is bounded below by the cone $z = \sqrt{x^2 + y^2}$ and above by the plane $z = 1$. Find the center of mass and the moment of inertia and radius of gyration about the z-axis if the density is (a) $\delta(r, \theta, z) = z$, (b) $\delta(r, \theta, z) = z^2$.

50. A solid right circular cylinder is bounded by the cylinder $r = a$ and the planes $z = 0$ and $z = h$, $h > 0$. Find the center of mass and the moment of inertia and radius of gyration about the z-axis if the density is $\delta(r, \theta, z) = z + 1$.

Spherical Coordinates

In exercises that ask for masses and moments, assume that the density is $\delta = 1$ unless stated otherwise.

51. Find the volume of the portion of the solid sphere $\rho \le a$ that lies between the cones $\phi = \pi/3$ and $\phi = 2\pi/3$.

52. Find the volume of the region cut from the solid sphere $\rho \le a$ by the half-planes $\theta = 0$ and $\theta = \pi/6$ in the first octant.

53. Find the volume of the smaller region cut from the solid sphere $\rho \le 2$ by the plane $z = 1$.

54. Find the volume of the region enclosed by the surface $\rho = 1 - \cos\phi$ (Fig. 13.62).

55. Find the volume of the solid cut from the thick-walled cylinder $1 \le x^2 + y^2 \le 2$ by the cones $z = \pm\sqrt{x^2 + y^2}$.

56. Find the volume of the region that lies inside the sphere $x^2 + y^2 + z^2 = 2$ and outside the cylinder $x^2 + y^2 = 1$.

57. Find the average value of the function $f(\rho, \phi, \theta) = \rho$ over the solid ball $\rho \le 1$.

58. Figure 13.63 shows a solid that was made by drilling a conical hole in a solid hemisphere. The equation of the cone is

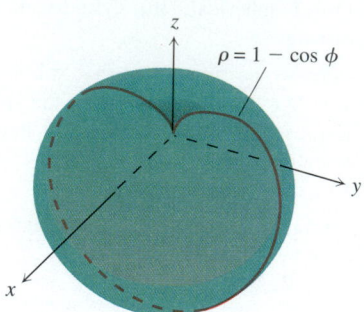

13.62   In spherical coordinates, the surface $\rho = 1 - \cos\phi$ is a cardioid of revolution (Exercise 54).

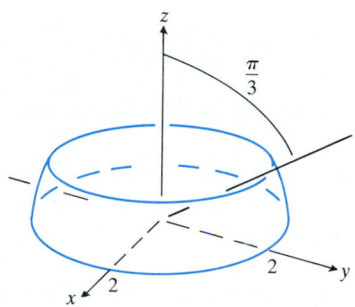

13.63   The solid in Exercise 58.

$\phi = \pi/3$. The radius of the hemisphere is $\rho = 2$. Find the center of mass.

59. Find the centroid of the solid bounded above by the sphere $\rho = a$ and below by the cone $\phi = \pi/4$.

60. A solid ball is bounded by the sphere $\rho = a$. Find the moment of inertia and radius of gyration about the z-axis if the density is

a) $\delta(\rho, \phi, \theta) = \rho^2$,   b) $\delta(\rho, \phi, \theta) = r = \rho\sin\phi$.

13.7 Substitutions in Multiple Integrals

This section shows how to evaluate multiple integrals by substitution. As in single integration, the goal of substitution is to replace complicated integrals by ones that are easier to evaluate. Substitutions can accomplish this by simplifying the integrand, by simplifying the limits of integration, or both.

Substitutions in Double Integrals

The polar coordinate substitution is a special case of a more general substitution method for double integrals, a method that pictures changes in variables as transformations of two-dimensional regions.

Suppose that a region G in the uv-plane is transformed one-to-one into the

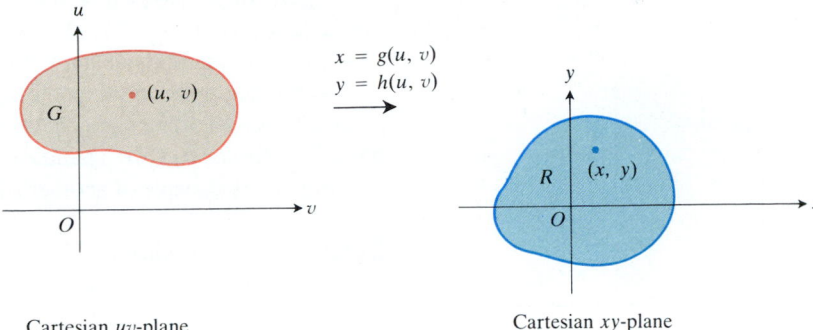

13.64   The equations $x = g(u, v)$ and $y = h(u, v)$ allow us to write an integral over a region R in the xy-plane as an integral over a region G in the uv-plane.

Cartesian uv-plane

Cartesian xy-plane

region R in the xy-plane by differentiable equations of the form

$$x = g(u, v), \qquad y = h(u, v),$$

as suggested in Fig. 13.64. Then any function $f(x, y)$ defined on R can be thought of as a function $f(g(u, v), h(u, v))$ defined on G as well. How is the integral of $f(x, y)$ over R related to the integral of $f(g(u, v), h(u, v))$ over G?

The answer is: If g, h, and f have continuous partial derivatives and $J(u, v)$ (to be discussed in a moment) is zero, if at all, only at isolated points, then

$$\iint_R f(x, y)\, dx\, dy = \iint_G f(g(u, v), h(u, v))\, |J(u, v)|\, du\, dv. \tag{1}$$

Notice the "Reversed" Order

The transforming equations $x = g(u, v)$ and $y = h(u, v)$ go from G to R, but we use them to change an integral over R into an integral over G.

The factor $J(u, v)$, whose absolute value appears in this formula, is the determinant

$$J(u, v) = \begin{vmatrix} \dfrac{\partial x}{\partial u} & \dfrac{\partial x}{\partial v} \\[2mm] \dfrac{\partial y}{\partial u} & \dfrac{\partial y}{\partial v} \end{vmatrix} = \frac{\partial(x, y)}{\partial(u, v)}. \tag{2}$$

It is called the **Jacobian determinant** or **Jacobian** of the coordinate transformation $x = g(u, v)$, $y = h(u, v)$, named after the mathematician Carl Jacobi. The alternative notation $\partial(x, y)/\partial(u, v)$ may help you to remember how the determinant is constructed from the partial derivatives of x and y. The derivation of Eq. (1) is intricate and we shall not give it here.

For polar coordinates, we have r and θ in place of u and v. With $x = r \cos \theta$ and $y = r \sin \theta$, the Jacobian is

Carl Gustav Jacob Jacobi

Jacobi (1804–1851), one of nineteenth-century Germany's most accomplished scientists, developed the theory of determinants and transformations into a powerful tool for evaluating multiple integrals and solving differential equations. He also applied transformation methods to study nonelementary integrals like the ones that arise in the calculation of arc length. Like Euler, Jacobi was a prolific writer and an even more prolific calculator and worked in a variety of mathematical and applied fields.

$$J(r, \theta) = \begin{vmatrix} \dfrac{\partial x}{\partial r} & \dfrac{\partial x}{\partial \theta} \\[2mm] \dfrac{\partial y}{\partial r} & \dfrac{\partial y}{\partial \theta} \end{vmatrix} = \begin{vmatrix} \cos \theta & -r \sin \theta \\[2mm] \sin \theta & r \cos \theta \end{vmatrix} = r(\cos^2\theta + \sin^2\theta) = r.$$

Hence, Eq. (1) becomes

$$\iint_R f(x, y)\, dx\, dy = \iint_G f(r \cos \theta, r \sin \theta)\, |r|\, dr\, d\theta$$

$$= \iint_G f(r \cos \theta, r \sin \theta)\, r\, dr\, d\theta, \quad \text{(If } r \geq 0) \tag{3}$$

which is Eq. (7) in Section 13.3.

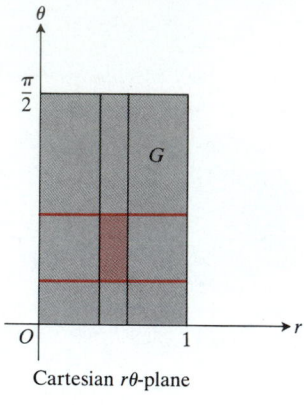

Cartesian $r\theta$-plane

$$x = r \cos \theta$$
$$y = r \sin \theta$$

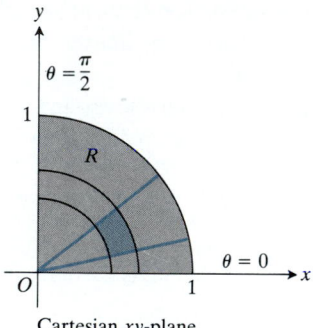

Cartesian xy-plane

13.65 The equations $x = r \cos \theta$, $y = r \sin \theta$ transform G into R.

Figure 13.65 shows how the equations $x = r \cos \theta$, $y = r \sin \theta$ transform the rectangle G: $0 \le r \le 1$, $0 \le \theta \le \pi/2$ into the quarter circle R bounded by $x^2 + y^2 = 1$ in the first quadrant of the xy-plane.

Notice that the integral on the right-hand side of Eq. (3) is not the integral of $f(r \cos \theta, r \sin \theta)$ over a region in the polar coordinate plane. It is the integral of the product of $f(r \cos \theta, r \sin \theta)$ and r over a region G in the *Cartesian $r\theta$-plane*.

Here is an example of another substitution.

Example 1   Evaluate

$$\int_0^4 \int_{x=y/2}^{x=(y/2)+1} \frac{2x-y}{2} \, dx \, dy$$

by applying the transformation

$$u = \frac{2x-y}{2}, \qquad v = \frac{y}{2} \tag{4}$$

and integrating over an appropriate region in the uv-plane.

Solution   We sketch the region R of integration in the xy-plane and identify its boundaries (Fig. 13.66).

To apply Eq. (1), we need to find the corresponding uv-region G and the Jacobian of the transformation. To find them, we first solve Eqs. (4) for x and y in terms of u and v. Routine algebra gives

$$x = u + v, \qquad y = 2v. \tag{5}$$

We then find the boundaries of G by substituting these expressions into the equations for the boundaries of R (Fig. 13.66):

xy-equations for the boundary of R	Corresponding uv-equations for the boundary of G	Simplified uv-equations
$x = y/2$	$u + v = 2v/2 = v$	$u = 0$
$x = (y/2) + 1$	$u + v = (2v/2) + 1 = v + 1$	$u = 1$
$y = 0$	$2v = 0$	$v = 0$
$y = 4$	$2v = 4$	$v = 2$

The Jacobian of the transformation (again from Eqs. 5) is

$$J(u, v) = \begin{vmatrix} \dfrac{\partial x}{\partial u} & \dfrac{\partial x}{\partial v} \\[2mm] \dfrac{\partial y}{\partial u} & \dfrac{\partial y}{\partial v} \end{vmatrix} = \begin{vmatrix} \dfrac{\partial}{\partial u}(u+v) & \dfrac{\partial}{\partial v}(u+v) \\[2mm] \dfrac{\partial}{\partial u}(2v) & \dfrac{\partial}{\partial v}(2v) \end{vmatrix} = \begin{vmatrix} 1 & 1 \\ 0 & 2 \end{vmatrix} = 2.$$

We now have everything we need to apply Eq. (1):

$$\int_0^4 \int_{x=y/2}^{x=(y/2)+1} \frac{2x-y}{2} \, dx \, dy = \int_{v=0}^{v=2} \int_{u=0}^{u=1} u \, |J(u, v)| \, du \, dv$$

$$= \int_0^2 \int_0^1 (u)(2) \, du \, dv = \int_0^2 \Big[u^2 \Big]_0^1 dv = \int_0^2 dv = 2.$$

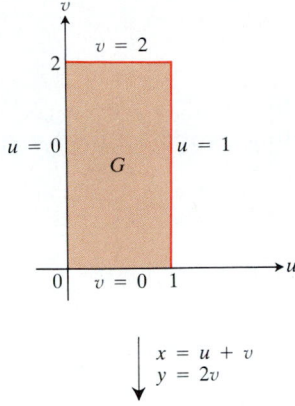

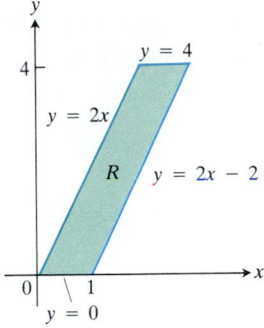

13.66 The equations $x = u + v$ and $y = 2v$ transform G into R. Reversing the transformation by the equations $u = (2x - y)/2$ and $v = y/2$ transforms R into G. See Example 1.

Substitutions in Triple Integrals

The cylindrical and spherical coordinate substitutions are special cases of a substitution method that pictures changes of variables in triple integrals as transformations of three-dimensional regions. The method is just like the method for double integrals, except that now we work in three dimensions instead of two.

Suppose that a region G in uvw-space is transformed one-to-one into the region D in xyz-space by differentiable equations of the form

$$x = g(u, v, w), \qquad y = h(u, v, w), \qquad z = k(u, v, w),$$

as suggested in Fig. 13.67. Then any function $F(x, y, z)$ defined on D can be thought of as a function

$$F(g(u, v, w), h(u, v, w), k(u, v, w)) = H(u, v, w)$$

defined on G. If g, h, and k have continuous first partial derivatives and the Jacobian of the transformation (see below) is never zero or is zero only at isolated points, then the integral of $F(x, y, z)$ over D is related to the integral of $H(u, v, w)$ over G by the equation

$$\iiint_D F(x, y, z)\, dx\, dy\, dz = \iiint_G H(u, v, w)\, |J(u, v, w)|\, du\, dv\, dw. \tag{6}$$

The factor $J(u, v, w)$, whose absolute value appears in this equation, is the **Jacobian determinant**

$$J(u, v, w) = \begin{vmatrix} \dfrac{\partial x}{\partial u} & \dfrac{\partial x}{\partial v} & \dfrac{\partial x}{\partial w} \\[2mm] \dfrac{\partial y}{\partial u} & \dfrac{\partial y}{\partial v} & \dfrac{\partial y}{\partial w} \\[2mm] \dfrac{\partial z}{\partial u} & \dfrac{\partial z}{\partial v} & \dfrac{\partial z}{\partial w} \end{vmatrix} = \frac{\partial(x, y, z)}{\partial(u, v, w)}. \tag{7}$$

As in the two-dimensional case, the derivation of Eq. (6) is complicated and we shall not go into it here.

For cylindrical coordinates, we have r, θ, and z in place of u, v, and w. The transformation from Cartesian $r\theta z$-space to Cartesian xyz-space is given by the equations

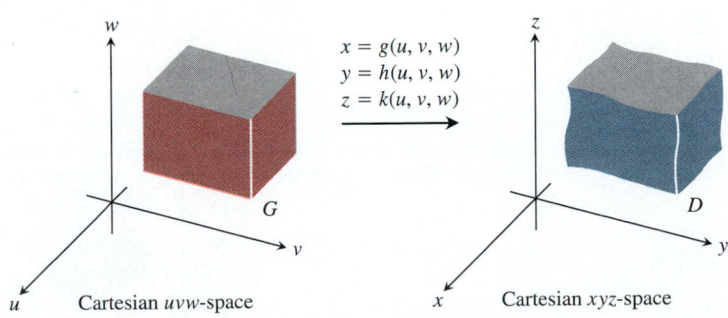

13.67 The equations $x = g(u, v, w)$, $y = h(u, v, w)$, and $z = k(u, v, w)$ allow us to write an integral over a region D in Cartesian xyz-space as an integral over a region G in Cartesian uvw-space.

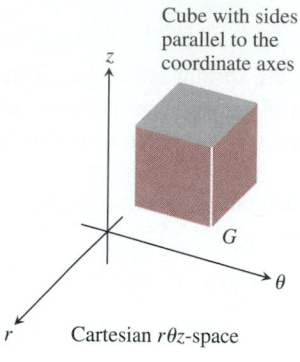

Cube with sides parallel to the coordinate axes

Cartesian $r\theta z$-space

$x = r\cos\theta$
$y = r\sin\theta$
$z = z$

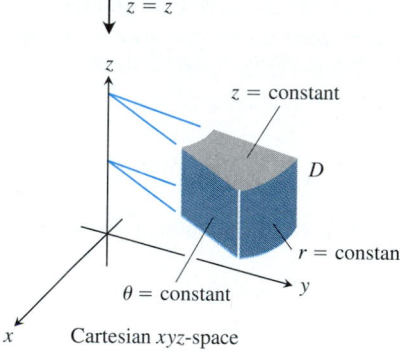

$z = $ constant

D

$r = $ constant

$\theta = $ constant

Cartesian xyz-space

13.68   The equations $x = r\cos\theta$, $y = r\sin\theta$, and $z = z$ transform G into D.

(Fig. 13.68). The Jacobian of the transformation is

$$J(r, \theta, z) = \begin{vmatrix} \dfrac{\partial x}{\partial r} & \dfrac{\partial x}{\partial \theta} & \dfrac{\partial x}{\partial z} \\[2mm] \dfrac{\partial y}{\partial r} & \dfrac{\partial y}{\partial \theta} & \dfrac{\partial y}{\partial z} \\[2mm] \dfrac{\partial z}{\partial r} & \dfrac{\partial z}{\partial \theta} & \dfrac{\partial z}{\partial z} \end{vmatrix} = \begin{vmatrix} \cos\theta & -r\sin\theta & 0 \\ \sin\theta & r\cos\theta & 0 \\ 0 & 0 & 1 \end{vmatrix} = r\cos^2\theta + r\sin^2\theta = r.$$

The corresponding version of Eq. (6) is

$$\iiint_D F(x, y, z)\, dx\, dy\, dz = \iiint_G H(r, \theta, z)\, |r|\, dr\, d\theta\, dz, \qquad (8)$$

and we can drop the absolute value signs whenever $r \geq 0$.

For spherical coordinates, we have ρ, ϕ, and θ in place of u, v, and w. The transformation from Cartesian $\rho\phi\theta$-space to Cartesian xyz-space is given by the equations

$$x = \rho\sin\phi\cos\theta, \qquad y = \rho\sin\phi\sin\theta, \qquad z = \rho\cos\phi$$

(Fig. 13.69). The Jacobian of the transformation is

$$J(\rho, \phi, \theta) = \begin{vmatrix} \dfrac{\partial x}{\partial \rho} & \dfrac{\partial x}{\partial \phi} & \dfrac{\partial x}{\partial \theta} \\[2mm] \dfrac{\partial y}{\partial \rho} & \dfrac{\partial y}{\partial \phi} & \dfrac{\partial y}{\partial \theta} \\[2mm] \dfrac{\partial z}{\partial \rho} & \dfrac{\partial z}{\partial \phi} & \dfrac{\partial z}{\partial \theta} \end{vmatrix} = \rho^2\sin\phi, \qquad (9)$$

as you will see if you do Exercise 9. The corresponding version of Eq. (6) is

$$\iiint_D F(x, y, z)\, dx\, dy\, dz = \iiint_G H(\rho, \phi, \theta)\, |\rho^2\sin\phi|\, d\rho\, d\phi\, d\theta, \qquad (10)$$

and we can drop the absolute value signs because $\sin\phi$ is never negative. Here is an example of another substitution.

Example 2   Evaluate

$$\int_0^3 \int_0^4 \int_{x = y/2}^{x = (y/2) + 1} \left(\frac{2x - y}{2} + \frac{z}{3} \right) dx\, dy\, dz$$

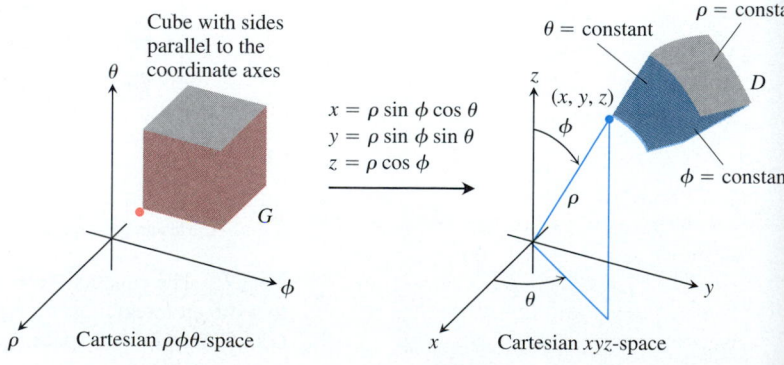

Cube with sides parallel to the coordinate axes

Cartesian $\rho\phi\theta$-space

$x = \rho\sin\phi\cos\theta$
$y = \rho\sin\phi\sin\theta$
$z = \rho\cos\phi$

$\theta = $ constant $\rho = $ consta

D

(x, y, z)

$\phi = $ constan

Cartesian xyz-space

13.69   The equations $x = \rho\sin\phi\cos\theta$, $y = \rho\sin\phi\sin\theta$, and $z = \rho\cos\phi$ transform G into D.

by applying the transformation

$$u = \frac{2x - y}{2}, \qquad v = \frac{y}{2}, \qquad w = \frac{z}{3} \tag{11}$$

and integrating over an appropriate region in uvw-space.

Solution We sketch the region D of integration in xyz-space and identify its boundaries (Fig. 13.70). In this case, the surfaces that bound D are all planes.

To apply Eq. (6), we need to find the corresponding uvw-region G and the Jacobian of the transformation. To find them, we first solve Eqs. (11) for x, y, and z in terms of u, v, and w. Routine algebra gives

$$x = u + v, \qquad y = 2v, \qquad z = 3w. \tag{12}$$

We then find the boundaries of G by substituting these expressions into the equations for the boundaries of D:

xyz-equations for the boundary of D	Corresponding uvw-equations for the boundary of G	Simplified uvw-equations
$x = y/2$	$u + v = 2v/2 = v$	$u = 0$
$x = (y/2) + 1$	$u + v = (2v/2) + 1 = v + 1$	$u = 1$
$y = 0$	$2v = 0$	$v = 0$
$y = 4$	$2v = 4$	$v = 2$
$z = 0$	$3w = 0$	$w = 0$
$z = 3$	$3w = 3$	$w = 1$

See Fig. 13.70.

The Jacobian of the transformation, again from Eqs. (12), is

$$J(u, v, w) = \begin{vmatrix} \dfrac{\partial x}{\partial u} & \dfrac{\partial x}{\partial v} & \dfrac{\partial x}{\partial w} \\[2mm] \dfrac{\partial y}{\partial u} & \dfrac{\partial y}{\partial v} & \dfrac{\partial y}{\partial w} \\[2mm] \dfrac{\partial z}{\partial u} & \dfrac{\partial z}{\partial v} & \dfrac{\partial z}{\partial w} \end{vmatrix} = \begin{vmatrix} 1 & 1 & 0 \\ 0 & 2 & 0 \\ 0 & 0 & 3 \end{vmatrix} = 6.$$

We now have everything we need to apply Eq. (6):

$$\int_0^3 \int_0^4 \int_{x = y/2}^{x = (y/2) + 1} \left(\frac{2x - y}{2} + \frac{z}{3} \right) dx\, dy\, dz$$

$$= \int_0^1 \int_0^2 \int_0^1 (u + w) |J(u, v, w)|\, du\, dv\, dw$$

$$= \int_0^1 \int_0^2 \int_0^1 (u + w)(6)\, du\, dv\, dw = 6 \int_0^1 \int_0^2 \left[\frac{u^2}{2} + uw \right]_0^1 dv\, dw$$

$$= 6 \int_0^1 \int_0^2 \left(\frac{1}{2} + w \right) dv\, dw = 6 \int_0^1 (1 + 2w)\, dw$$

$$= 6 \left[w + w^2 \right]_0^1 = 6(2) = 12.$$

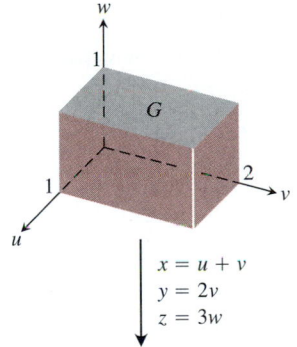

$x = u + v$
$y = 2v$
$z = 3w$

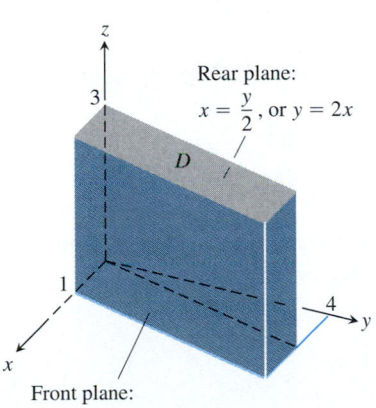

Rear plane:
$x = \dfrac{y}{2}$, or $y = 2x$

Front plane:
$x = \dfrac{y}{2} + 1$, or $y = 2x - 2$

13.70 The equations $x = u + v$, $y = 2v$, and $z = 3w$ transform G into D. Reversing the transformation by the equations $u = (2x - y)/2$, $v = y/2$, and $w = z/3$ transforms D into G. See Example 2.

Exercise 10 asks you to evaluate the original integral directly to show that its value is also 12.

EXERCISES 13.7

Double Integrals

1. Evaluate the integral

$$\int_0^4 \int_{x=y/2}^{x=(y/2)+1} \frac{2x-y}{2}\, dx\, dy$$

from Example 1 directly by integration with respect to x and y to confirm that its value is 2.

2. Find the Jacobian $\partial(x, y)/\partial(u, v)$ for the transformation
a) $x = u \cos v$, $\quad y = u \sin v$,
b) $x = u \sin v$, $\quad y = u \cos v$.

3. a) Solve the system

$$u = x - y, \quad v = 2x + y$$

for x and y in terms of u and v. Then find the value of the Jacobian $\partial(x, y)/\partial(u, v)$.

b) Let R be the region in the first quadrant bounded by the lines $y = -2x + 4$, $y = -2x + 7$, $y = x - 2$, and $y = x + 1$. Evaluate

$$\iint_R (2x^2 - xy - y^2)\, dx\, dy$$

by changing variables with the equations in (a) and integrating over a region G in the uv-plane.

4. a) Find the Jacobian of the transformation $x = u$, $y = uv$, and sketch the region $G\colon 1 \le u \le 2$, $1 \le uv \le 2$ in the uv-plane.

b) Then use Eq. (1) to transform the integral

$$\int_1^2 \int_1^2 \frac{y}{x}\, dy\, dx$$

into an integral over G, and evaluate both integrals.

5. Let R be the region in the first quadrant of the xy-plane bounded by the hyperbolas $xy = 1$, $xy = 9$ and the lines $y = x$, $y = 4x$ (Fig. 13.71). Use the transformation

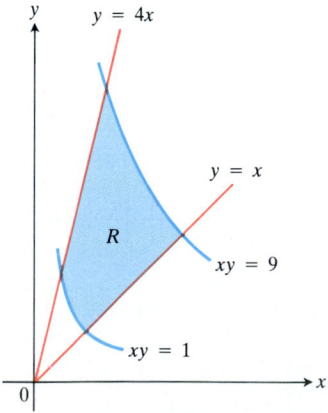

13.71 The xy-region in Exercise 5.

$x = u/v$, $y = uv$ with $u > 0$ and $v > 0$ to rewrite

$$\iint_R \left(\sqrt{\frac{y}{x}} + \sqrt{xy} \right) dx\, dy$$

as an integral over an appropriate region G in the uv-plane. Then evaluate the uv-integral over G.

6. The area πab of the ellipse $x^2/a^2 + y^2/b^2 = 1$ can be found by integrating the function $f(x, y) = 1$ over the region bounded by the ellipse in the xy-plane. Evaluating the integral directly requires a trigonometric substitution. An easier way to evaluate the integral is to use the transformation $x = au$, $y = bv$ and evaluate the transformed integral over the disk $G\colon u^2 + v^2 \le 1$ in the uv-plane. Find the area this way.

7. A thin plate of constant density covers the region bounded by the ellipse $x^2/a^2 + y^2/b^2 = 1$ in the xy-plane. Find the moment of the plate about the origin. (*Hint:* Use the transformation $x = ar \cos \theta$, $y = br \sin \theta$.)

8. Use the transformation $x = u + v/2$, $y = v$, to evaluate the integral

$$\int_0^2 \int_{y/2}^{(y+4)/2} y^3 (2x - y) e^{(2x-y)^2}\, dx\, dy$$

by first writing it as an integral over a region G in the uv-plane.

Triple Integrals

9. Evaluate the determinant in Eq. (9) to show that the Jacobian of the transformation from Cartesian $\rho\phi\theta$-space to Cartesian xyz-space is $\rho^2 \sin \phi$.

10. Evaluate the integral in Example 2 by integrating with respect to x, y, and z.

11. Find the volume of the ellipsoid

$$\frac{x^2}{a^2} + \frac{y^2}{b^2} + \frac{z^2}{c^2} = 1.$$

(*Hint:* Let $x = au$, $y = bv$, and $z = cw$. Then find the volume of an appropriate region in uvw-space.)

12. Evaluate

$$\iiint |xyz|\, dx\, dy\, dz$$

over the solid ellipsoid

$$\frac{x^2}{a^2} + \frac{y^2}{b^2} + \frac{z^2}{c^2} \le 1.$$

(*Hint:* Let $x = au$, $y = bv$, and $z = cw$. Then integrate over an appropriate region in uvw-space.)

13. Let D be the region in xyz-space defined by the inequalities

$$1 \le x \le 2, \quad 0 \le xy \le 2, \quad \text{and} \quad 0 \le z \le 1.$$

Evaluate

$$\iiint_D (x^2 y + 3xyz) \, dx \, dy \, dz$$

by applying the transformation

$$u = x, \quad v = xy, \quad w = 3z$$

and integrating over an appropriate region G in uvw-space.

14. Assuming the result that the center of mass of a solid hemisphere lies on the axis of symmetry three-eighths of the way from the base toward the top, show, by transforming the appropriate integrals, that the center of mass of a solid semi-ellipsoid $(x^2/a^2) + (y^2/b^2) + (z^2/c^2) \le 1$, $z \ge 0$ lies on the z-axis three-eighths of the way from the base toward the top. (You can do this without evaluating any of the integrals.)

REVIEW QUESTIONS

1. Define the double integral of a function of two variables over a bounded region in the coordinate plane.

2. How are double integrals evaluated as iterated integrals? Does the order of integration matter? How are the limits of integration determined? Give examples.

3. How are double integrals used to calculate areas, average values, masses, moments, centers of mass, and radii of gyration? Give examples.

4. How can you change a double integral in rectangular coordinates into a double integral in polar coordinates? Why might it be worthwhile to do so? Give an example.

5. Define the triple integral of a function $f(x, y, z)$ over a bounded region in space.

6. How are triple integrals in rectangular coordinates evaluated? How are the limits of integration determined? Give an example.

7. How are triple integrals in rectangular coordinates used to calculate volumes, average values, masses, moments, centers of mass, and radii of gyration? Give examples.

8. How are triple integrals defined in cylindrical and spherical coordinates? Why might one prefer working in one of these coordinate systems to working in rectangular coordinates?

9. How are triple integrals in cylindrical and spherical coordinates evaluated? How are the limits of integration found? Give examples.

10. How are substitutions in double integrals pictured as transformations of two-dimensional regions? Give a sample calculation.

11. How are substitutions in triple integrals pictured as transformations of three-dimensional regions? Give a sample calculation.

MISCELLANEOUS EXERCISES

1. Each of the following integrals represents the area of a region in a Cartesian coordinate plane. Sketch the region. Express the area of the region as a double integral with the order of integration reversed. Then evaluate both integrals.

a) $\displaystyle\int_0^4 \int_{-\sqrt{4-y}}^{(y-4)/2} dx \, dy$

b) $\displaystyle\int_{-1}^1 \int_{u^2}^1 dv \, du$

c) $\displaystyle\int_0^{3/2} \int_{-\sqrt{9-4t^2}}^{\sqrt{9-4t^2}} t \, ds \, dt$

d) $\displaystyle\int_0^2 \int_0^{4-w^2} 2 \, w \, dz \, dw$

2. Sketch the region over which the integral

$$\int_0^1 \int_{\sqrt{y}}^{2-\sqrt{y}} x \, y \, dx \, dy$$

is to be evaluated and find its value.

Evaluate the integrals in Exercises 3 and 4.

3. $\displaystyle\int_1^\infty \int_0^{1/(x\sqrt{x^2-1})} 2 \, dy \, dx$

4. $\displaystyle\int_0^1 \int_0^y e^{-x^2} \, dx \, dy$

5. Evaluate the integral

$$\int_0^\infty \frac{e^{-ax} - e^{-bx}}{x} \, dx .$$

(*Hint:* Use the relation

$$\frac{e^{-ax} - e^{-bx}}{x} = \int_a^b e^{-xy} \, dy$$

to form a double integral and evaluate the integral by changing the order of integration.)

6. Show that

$$\iint \frac{\partial^2 F(x, y)}{\partial x \, \partial y} \, dx \, dy$$

over the rectangle $x_0 \le x \le x_1$, $y_0 \le y \le y_1$, is

$$F(x_1, y_1) - F(x_0, y_1) - F(x_1, y_0) + F(x_0, y_0).$$

7. Evaluate

$$\int_0^a \int_0^b e^{\max(b^2 x^2, \, a^2 y^2)} \, dy \, dx ,$$

where a and b are positive numbers and

$$\max(b^2x^2, a^2y^2) = \begin{cases} b^2x^2 & \text{if } b^2x^2 \geq a^2y^2 \\ a^2y^2 & \text{if } b^2x^2 < a^2y^2 \end{cases}$$

8. a) Show, by changing to polar coordinates, that

$$\int_0^{a\sin\beta} \int_{y\cot\beta}^{\sqrt{a^2-y^2}} \ln(x^2 + y^2)\, dx\, dy = a^2\beta\left(\ln a - \frac{1}{2}\right),$$

where $a > 0$ and $0 < \beta < \pi/2$.

b) Rewrite the Cartesian integral with the order of integration reversed.

9. By changing the order of integration, show that the following double integral can be reduced to a single integral:

$$\int_0^x \int_0^u e^{m(x-t)}f(t)\, dt\, du = \int_0^x (x-t)e^{m(x-t)}f(t)\, dt.$$

Similarly, it can be shown that

$$\int_0^x \int_0^v \int_0^u e^{m(x-t)}f(t)\, dt\, du\, dv = \int_0^x \frac{(x-t)^2}{2}e^{m(x-t)}f(t)\, dt.$$

10. Sometimes a multiple integral with variable limits can be changed into one with constant limits. By changing the order of integration, show that

$$\int_0^1 f(x)\left(\int_0^x g(x-y)f(y)\, dy\right) dx$$

$$= \int_0^1 f(y)\left(\int_y^1 g(x-y)f(x)\, dx\right) dy$$

$$= \frac{1}{2}\int_0^1 \int_0^1 g(|x-y|)f(x)f(y)\, dx\, dy.$$

11. The base of a sand pile covers the region in the xy-plane that is bounded by the parabola $x^2 + y = 6$ and the line $y = x$. The height of the sand above the point (x, y) is x^2. Express the volume of sand as (a) a double integral, (b) a triple integral. Then (c) find the volume.

12. Suppose that $f(x, y)$ can be written as a product $f(x, y) = F(x)G(y)$ of a function of x and a function of y. Then the integral of f over the rectangle $R: a \leq x \leq b, c \leq y \leq d$ can be evaluated as a product as well, by the formula

$$\iint_R f(x, y)\, dA = \left(\int_a^b F(x)\, dx\right)\left(\int_c^d G(y)\, dy\right). \quad (1)$$

The argument is that

$$\iint_R f(x, y)\, dA = \int_c^d \left(\int_a^b F(x)G(y)\, dx\right) dy \quad \text{(Fubini's theorem)}$$

$$= \int_c^d \left(G(y)\int_a^b F(x)\, dx\right) dy \quad \left(\begin{array}{c}\text{Treating } G(y)\\ \text{as a constant}\end{array}\right)$$

$$= \int_c^d \left(\int_a^b F(x)\, dx\right)G(y)\, dy \quad \left(\begin{array}{c}\text{Algebraic}\\ \text{rearrangement}\end{array}\right)$$

$$= \left(\int_a^b F(x)\, dx\right)\int_c^d G(y)\, dy. \quad \left(\begin{array}{c}\text{The definite}\\ \text{integral}\\ \text{is a constant.}\end{array}\right)$$

When it applies, Eq. (1) can be a time saver. Use it to evaluate the following integrals.

a) $\displaystyle\int_0^{\ln 2} \int_0^{\pi/2} e^x \cos y\, dy\, dx$

b) $\displaystyle\int_1^2 \int_{-1}^1 \frac{x}{y^2}\, dx\, dy$

13. Find the centroid of the triangular region enclosed by the lines $x = 2$, $y = 2$ and the hyperbola $xy = 2$ in the xy-plane.

14. Find the polar moment of inertia about the origin of a thin triangular plate of constant density $\delta = 3$ bounded by the y-axis and the lines $y = 2x$ and $y = 4$ in the xy-plane.

15. Find the centroid of the region between the parabola $x + y^2 - 2y = 0$ and the line $x + 2y = 0$.

16. Find the centroid of the boomerang-shaped region between the parabolas $y^2 = -4(x - 1)$ and $y^2 = -2(x - 2)$.

17. Find the mass and first moments about the coordinate axes of a thin square plate bounded by the lines $x = \pm 1$, $y = \pm 1$ in the xy-plane if the density is $\delta(x, y) = x^2 + y^2 + 1/3$.

18. Find the center of mass and the moments of inertia and radii of gyration about the coordinate axes of a thin plate bounded by the line $y = x$ and the parabola $y = x^2$ in the xy-plane if the density is $\delta(x, y) = x + 1$.

19. Find the moment of inertia and radius of gyration about the x-axis of a thin triangular plate of constant density δ whose base lies along the interval $[0, b]$ on the x-axis and whose vertex lies on the line $y = h$ above the x-axis. As you will see, it does not matter where on the line this vertex lies. All such triangles have the same moment of inertia and radius of gyration.

20. Let $D_\mathbf{u} f$ denote the derivative of $f(x, y) = (x^2 + y^2)/2$ in the direction of the unit vector $\mathbf{u} = u_1\mathbf{i} + u_2\mathbf{j}$.
a) Find the average value of $D_\mathbf{u} f$ over the triangular region cut from the first quadrant by the line $x + y = 1$.
b) Show in general that the average value of $D_\mathbf{u} f$ over a region in the xy-plane is the value of $D_\mathbf{u} f$ at the centroid of the region.

21. a) Find the centroid of the region in the polar coordinate plane that lies inside the cardioid $r = 1 + \cos\theta$ and outside the circle $r = 1$.
b) CALCULATOR Sketch the region and show the centroid in your sketch.

22. a) Find the centroid of the plane region defined by the polar coordinate inequalities $0 \leq r \leq a$, $-\alpha \leq \theta \leq \alpha$ $(0 < \alpha \leq \pi)$. How does the centroid move as $\alpha \to \pi^-$?
b) CALCULATOR Sketch the region for $\alpha = 5\pi/6$ and show the centroid in your sketch.

23. Find the centroid of the region in the first quadrant bounded by the rays $\theta = 0$ and $\theta = \pi/2$ and the circles $r = 1$ and $r = 3$.

24. In Section 9.8 we derived the formula

$$A = \int_{\alpha}^{\beta} \frac{1}{2} f^2(\theta)\, d\theta$$

for the area of the fan-shaped region R between the origin and a curve $r = f(\theta)$, $\alpha \le \theta \le \beta$. Show that this formula is also a consequence of the equation

$$\text{Area of } R = \iint_{R} r\, dr\, d\theta$$

in Section 13.3.

25. The counterweight of a flywheel has the form of the smaller segment cut from a circle of radius a by a chord at a distance b from the center ($b < a$). Find the area of the counterweight and its polar moment of inertia about the center of the wheel.

26. Find the radii of gyration about the x- and y-axes of a thin plate of density $\delta = 1$ enclosed by one loop of the lemniscate $r^2 = 2a^2 \cos 2\theta$.

27. Evaluate $\displaystyle\int_{0}^{a \sin \beta} \int_{y \cot \beta}^{\sqrt{a^2 - y^2}} \ln (x^2 + y^2)\, dx\, dy$.

28. Integrate the function $f(x, y) = 1/(1 + x^2 + y^2)^2$ over the region enclosed by one loop of the lemniscate $(x^2 + y^2)^2 - (x^2 - y^2) = 0$.

29. The electrical charge distribution on a circular plate of radius R meters is $\sigma(r, \theta) = kr(1 - \sin \theta)$ coulomb/m^2 (k a constant). Integrate σ over the plate to find the total charge Q.

30. *The value of $\Gamma(1/2)$.* As we saw in the Miscellaneous Exercises in Chapter 7, the gamma function,

$$\Gamma(x) = \int_{0}^{\infty} t^{x-1} e^{-t}\, dt,$$

extends the factorial function from the nonnegative integers to other real values. Of particular interest in the theory of differential equations is the number

$$\Gamma\left(\frac{1}{2}\right) = \int_{0}^{\infty} t^{(1/2)-1} e^{-t}\, dt = \int_{0}^{\infty} \frac{e^{-t}}{\sqrt{t}}\, dt. \tag{2}$$

a) If you have not yet done Exercise 16(a) in Section 13.3, do it now to show that

$$I = \int_{0}^{\infty} e^{-y^2} = \frac{\sqrt{\pi}}{2}.$$

b) Substitute $y = \sqrt{t}$ in Eq. (2) to show that $\Gamma(1/2) = 2I = \sqrt{\pi}$.

31. Find the volume of the solid that is bounded above by the cylinder $z = 4 - x^2$, on the sides by the cylinder $x^2 + y^2 = 4$, and below by the xy-plane (Fig. 13.72).

32. Find the average value of $f(x, y, z) = 30\, xz\sqrt{x^2 + y}$ over the rectangular solid in the first octant bounded by the coordinate planes and the planes $x = 1$, $y = 3$, $z = 1$.

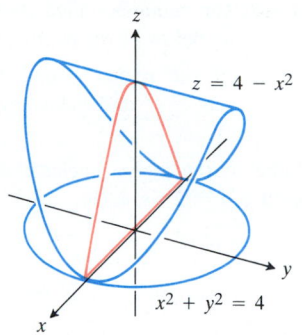

13.72 The solid in Exercise 31.

33. Convert

$$\int_{0}^{2\pi} \int_{0}^{1} \int_{0}^{\sqrt{4 - r^2}} r^2\, dz\, dr\, d\theta$$

to (a) rectangular coordinates, (b) spherical coordinates.

34. Write an iterated triple integral for the integral of $f(x, y, z) = 6 + 4y$ over the region in the first octant bounded by the cone $z = \sqrt{x^2 + y^2}$, the cylinder $x^2 + y^2 = 1$, and the coordinate planes in (a) rectangular coordinates, (b) cylindrical coordinates, (c) spherical coordinates. Then (d) find the integral of f by evaluating one of the triple integrals.

35. Set up an integral in rectangular coordinates equivalent to the integral

$$\int_{0}^{\pi/2} \int_{1}^{\sqrt{3}} \int_{1}^{\sqrt{4 - r^2}} r^3 \sin \theta \cos \theta\, z^2\, dz\, dr\, d\theta.$$

Arrange the order of integration to be z first, then y, then x.

36. The volume of a solid is

$$\int_{0}^{2} \int_{0}^{\sqrt{2x - x^2}} \int_{-\sqrt{4 - x^2 - y^2}}^{\sqrt{4 - x^2 - y^2}} dz\, dy\, dx.$$

a) Describe the solid by giving equations for the surfaces that form its boundary.

b) Convert the integral to cylindrical coordinates but do not evaluate the integral.

37. Find the volume of the portion of the solid cylinder $x^2 + y^2 \le 1$ that lies between the planes $z = 0$ and $x + y + z = 2$.

38. A hemispherical bowl of radius 5 cm is filled with water to within 3 cm of the top. Find the volume of water in the bowl.

39. Find the volume of the region in the first octant that lies between the cylinders $r = 1$ and $r = 2$ and that is bounded below by the xy-plane and above by the surface $z = xy$.

40. Find the volume of the region bounded above by the sphere $x^2 + y^2 + z^2 = 2$ and below by the paraboloid $z = x^2 + y^2$.

41. Find the volume of the region bounded above by the paraboloid $z = 3 - x^2 - y^2$ and below by the paraboloid $z = 2x^2 + 2y^2$.

42. Find the volume of the region enclosed by the spherical coordinate surface $\rho = 2 \sin \phi$ (Fig. 13.73).

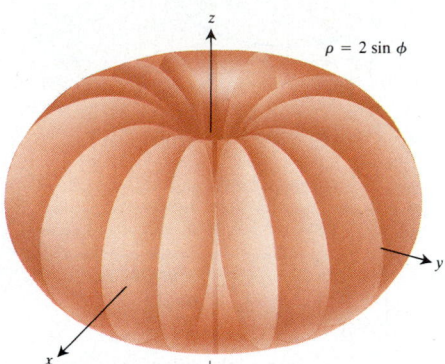
$\rho = 2 \sin \phi$

13.73 The spherical coordinate surface $\rho = 2 \sin \phi$ in Exercise 42 is the surface generated by revolving a circle of radius 1 about one of its tangents (in this case, the z-axis).

43. A circular cylindrical hole is bored through a solid sphere, the axis of the hole being a diameter of the sphere. The volume of the remaining solid is

$$V = 2 \int_0^{2\pi} \int_0^{\sqrt{3}} \int_1^{\sqrt{4-z^2}} r \, dr \, dz \, d\theta.$$

a) Find the radius of the hole and the radius of the sphere.
b) Evaluate the integral. (See Example 4, Section 5.3.)

44. Find the volume of material cut from the solid sphere $r^2 + z^2 \le 9$ by the cylinder $r = 3 \sin \theta$.

45. Find the volume of the region enclosed by the surfaces $z = x^2 + y^2$ and $z = (x^2 + y^2 + 1)/2$.

46. *Spherical vs. cylindrical coordinates.* Triple integrals involving spherical shapes do not always require spherical coordinates for convenient evaluation. Some calculations may be accomplished more easily with cylindrical coordinates. As a case in point, find the volume of the region bounded above by the sphere $x^2 + y^2 + z^2 = 8$ and below by the plane $z = 2$ by using (a) cylindrical coordinates, (b) spherical coordinates.

47. Find the moment of inertia about the z-axis of a solid of constant density $\delta = 1$ that is bounded above by the sphere $\rho = 2$ and below by the cone $\phi = \pi/3$ (spherical coordinates).

48. Find the moment of inertia of a solid of constant density δ bounded by two concentric spheres of radii a and b ($a < b$) about a diameter.

49. Find the moment of inertia about the z-axis of a solid of density $\delta = 1$ enclosed by the spherical coordinate surface $\rho = 1 - \cos \phi$ (Fig. 13.62).

50. *A challenge.* Setting up the following integral is straightforward but evaluating it can take hours. Some computer symbolic manipulation programs can evaluate the integral but others cannot. Some can evaluate the integral in one form but not another. One program we know evaluated the integral in less than a minute. Another took 20 minutes.

A square hole of side $2b$ is cut symmetrically through a solid sphere of radius a ($a > b\sqrt{2}$). Find the volume of material removed.

51. Show that if $u = x - y$ and $v = y$, then

$$\int_0^\infty \int_0^x e^{-sx} f(x - y, y) \, dy \, dx = \int_0^\infty \int_0^\infty e^{-s(u+v)} f(u, v) \, du \, dv.$$

52. What relationship must hold between the constants a, b, and c to make

$$\int_{-\infty}^\infty \int_{-\infty}^\infty e^{-(ax^2 + 2bxy + cy^2)} \, dx \, dy = 1?$$

(*Hint:* Let $s = \alpha x + \beta y$ and $t = \gamma x + \delta y$, where $(\alpha\delta - \beta\gamma)^2 = ac - b^2$. Then $ax^2 + 2bxy + cy^2 = s^2 + t^2$.)

CHAPTER

14

INTEGRATION IN VECTOR FIELDS

Overview   This chapter brings together all our previous work with derivatives, integrals, and vector functions to study integration in vector fields. The field concept has proved to be one of the most useful ideas in all of physical science. The mathematics in this chapter is the mathematics we use today to describe the properties of electric charge; explain the behavior of electromagnetic waves, including radio waves, x-rays, and visible light; predict the flow of air around airplanes and rockets; explain the flow of heat in a room or in a star; calculate the amount of work it takes to put a satellite into orbit; and model tropical storms and ocean currents.

14.1 Line Integrals

When a curve $\mathbf{r} = g(t)\,\mathbf{i} + h(t)\,\mathbf{j} + k(t)\,\mathbf{k}$, $a \leq t \leq b$, passes through the domain of a function $f(x, y, z)$ in space, the values of f along the curve are given by the composite function $f(g(t), h(t), k(t))$. If we integrate this composite with respect to arc length from $t = a$ to $t = b$, we calculate the so-called line integral of f along the curve. Despite the three-dimensional geometry, the line integral is an ordinary integral of a real-valued function over an interval of real numbers. We have evaluated such integrals ever since Chapter 4. The usual techniques apply and nothing new is needed.

The importance of line integrals lies in their application. These are the integrals with which we calculate the moments and masses of springs and curved wires. We also combine them with vectors to calculate the work done by variable forces along paths in space and the rates at which fluids flow along curves and across boundaries. The present section shows how line integrals are defined, evaluated, and used to calculate moments and masses. The next section brings line integrals together with vectors to calculate work and measure fluid flow.

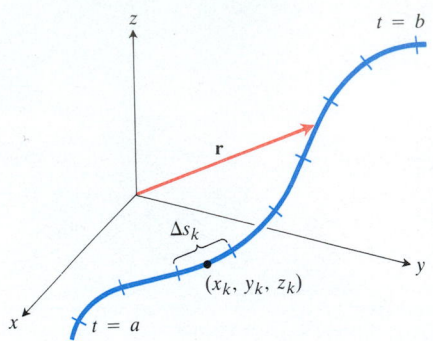

14.1 The curve $\mathbf{r} = g(t)\,\mathbf{i} + h(t)\,\mathbf{j} + k(t)\,\mathbf{k}$, partitioned into small arcs from $t = a$ to $t = b$. The length of a typical subarc is Δs_k.

Definitions and Notation

Line integrals are integrals of functions over curves. We define them in the following way. Suppose that $f(x, y, z)$ is a function whose domain contains the curve

$$\mathbf{r} = g(t)\,\mathbf{i} + h(t)\,\mathbf{j} + k(t)\,\mathbf{k}, \quad a \le t \le b.$$

We partition the curve into a finite number of subarcs (Fig. 14.1). The typical subarc has length Δs_k. In each subarc we choose a point (x_k, y_k, z_k) and form the sum

$$S_n = \sum_{k=1}^{n} f(x_k, y_k, z_k)\,\Delta s_k. \tag{1}$$

If f is continuous and $\mathbf{r}$ is smooth (the functions g, h, and k have continuous first derivatives), then the sums in (1) approach a limit as n increases and the lengths Δs_k approach zero. We call this limit the **line integral of f over the curve from a to b.** If the curve is denoted by a single letter, C for example, the notation for the integral is

$$\int_C f(x, y, z)\, ds \qquad \text{(``The integral of } f \text{ over } C\text{'')} \tag{2}$$

Evaluation

To evaluate the line integral in Eq. (2), we express ds and the integrand in terms of the curve's parameter t. A remarkable theorem from advanced calculus tells us that we can find the value from any smooth parametrization $x = g(t)$, $y = h(t)$, $z = k(t)$, $a \le t \le b$, for which ds/dt is positive by evaluating the integral

$$\int_{t=a}^{t=b} f(g(t), h(t), k(t))\,\frac{ds}{dt}\,dt = \int_a^b f(g(t), h(t), k(t))\sqrt{\left(\frac{dx}{dt}\right)^2 + \left(\frac{dy}{dt}\right)^2 + \left(\frac{dz}{dt}\right)^2}\,dt. \tag{3}$$

Since the square root in Eq. (3) is the length $|\mathbf{v}(t)|$ of the curve's velocity vector, the integral on the right-hand side can be simplified to give the following result.

THEOREM 1

> **The Evaluation Theorem for Line Integrals**
>
> The integral of a continuous function $f(x, y, z)$ over a space curve C can be calculated from any smooth parametrization $\mathbf{r} = g(t)\,\mathbf{i} + h(t)\,\mathbf{j} + k(t)\,\mathbf{k}$, $a \le t \le b$, of C for which $|\mathbf{v}| \ne 0$ as
>
> $$\int_C f(x, y, z)\, ds = \int_{t=a}^{t=b} f(g(t), h(t), k(t))\,|\mathbf{v}(t)|\,dt. \tag{4}$$

Notice that if f has the constant value 1, then the integral of f over C from a to b gives the length of C from a to b. We calculated lengths in Section 11.3 and will not repeat the calculations here.

In our first example, we integrate a function over two different paths from $(0, 0, 0)$ to $(1, 1, 1)$.

Example 1   Figure 14.2 shows two different paths from the origin to the point $(1, 1, 1)$. One path is the union of the line segments C_1 and C_2. The other path is the

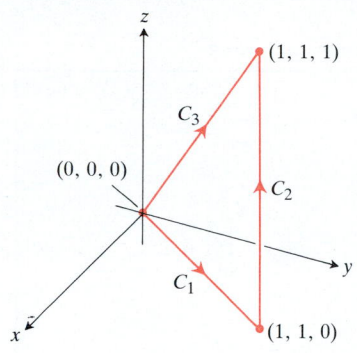

14.2 The paths of integration in Example 1. The arrows show the direction of increasing t in the parametrizations.

line segment C_3. Integrate the function

$$f(x, y, z) = x - 3y^2 + z$$

along each path.

Solution   We first parametrize the segments that make up the paths. According to the Evaluation Theorem, we can use any parametrization we want as long as $|\mathbf{v}(t)|$ is never zero and we integrate in the direction of increasing t so that $|\mathbf{v}(t)|\,dt$ is positive. We choose the simplest parametrizations we can think of, checking the lengths of the velocity vectors as we go along:

$$C_1: \quad \mathbf{r} = t\mathbf{i} + t\mathbf{j}, \quad 0 \le t \le 1; \qquad |\mathbf{v}| = \sqrt{1^2 + 1^2} = \sqrt{2}$$

$$C_2: \quad \mathbf{r} = \mathbf{i} + \mathbf{j} + t\mathbf{k}, \quad 0 \le t \le 1; \qquad |\mathbf{v}| = \sqrt{0^2 + 0^2 + 1^2} = 1$$

$$C_3: \quad \mathbf{r} = t\mathbf{i} + t\mathbf{j} + t\mathbf{k}, \quad 0 \le t \le 1; \qquad |\mathbf{v}| = \sqrt{1^2 + 1^2 + 1^2} = \sqrt{3}$$

Having chosen the parametrizations, we use Eq. (4) to integrate $f(x, y, z) = x - 3y^2 + z$ over each path in the direction of increasing t.

Path 1:
$$\int_{C_1 \cup C_2} f(x, y, z)\, ds = \int_{C_1} f(x, y, z)\, ds + \int_{C_2} f(x, y, z)\, ds$$

$$= \int_0^1 f(t, t, 0)\, \sqrt{2}\, dt + \int_0^1 f(1, 1, t)(1)\, dt$$

$$= \int_0^1 (t - 3t^2 + 0)\sqrt{2}\, dt + \int_0^1 (1 - 3 + t)(1)\, dt$$

$$= \sqrt{2} \int_0^1 (t - 3t^2)\, dt + \int_0^1 (t - 2)\, dt$$

$$= \sqrt{2} \left[\frac{t^2}{2} - t^3 \right]_0^1 + \left[\frac{t^2}{2} - 2t \right]_0^1 = -\frac{\sqrt{2}}{2} - \frac{3}{2}.$$

Path 2:
$$\int_{C_3} f(x, y, z)\, ds = \int_0^1 f(t, t, t)(\sqrt{3})\, dt = \int_0^1 (t - 3t^2 + t)(\sqrt{3})\, dt$$

$$= \sqrt{3} \int_0^1 (2t - 3t^2)\, dt = \sqrt{3} \left[t^2 - t^3 \right]_0^1 = 0$$

There are three things to notice about the integrations in Example 1. First, as soon as the components of the appropriate curve are substituted in the formula for f and ds is replaced by the appropriate $|\mathbf{v}(t)|\,dt$ for each segment, the integration becomes a straightforward integration with respect to t. Second, the integral of f over path 1 is obtained by integrating f over each section of the path and adding the results. Third, the integrals of f over path 1 and path 2 have different values. For most functions, the value of the line integral along a path joining two points changes when you change the path. For some functions, however, the value of the integral is the same for all paths joining the two points, as we shall see in Section 14.7.

Additivity

Line integrals have all the usual algebraic properties. The line integral of a sum of two functions is the sum of their line integrals, the line integral of a constant times a function is the constant times the line integral of the function, the line integral of

TABLE 14.1
Mass and moment formulas for coil springs, thin rods, and wires

Mass: $M = \displaystyle\int_C \delta(x, y, z)\, ds$

First moments about the coordinate planes:

$$M_{yz} = \int_C x\,\delta\,ds, \qquad M_{xz} = \int_C y\,\delta\,ds, \qquad M_{xy} = \int_C z\,\delta\,ds$$

Coordinates of the center of mass:

$$\bar{x} = M_{yz}/M, \qquad \bar{y} = M_{xz}/M, \qquad \bar{z} = M_{xy}/M$$

Moments of inertia:

$$I_x = \int_C (y^2 + z^2)\,\delta\,ds, \qquad I_y = \int_C (x^2 + z^2)\,\delta\,ds,$$

$$I_z = \int_C (x^2 + y^2)\,\delta\,ds, \qquad I_L = \int_C r^2\,\delta\,ds,$$

$$r(x, y, z) = \text{distance from point } (x, y, z) \text{ to line } L$$

Radius of gyration about a line L: $R_L = \sqrt{I_L/M}$.

a function from b to a is the negative of the line integral from a to b, and so on. Included in this list is the property that if a curve C is made by joining together a finite number of curves $C_1, C_2, \ldots, C_n$ end to end, then the line integral of a function over C is the sum of the line integrals over the curves that make it up:

$$\int_C f\,ds = \int_{C_1} f\,ds + \int_{C_2} f\,ds + \cdots + \int_{C_n} f\,ds. \tag{5}$$

We used Eq. (5) in Example 1 to calculate the integral of f over the first path by adding the integrals over C_1 and C_2.

Mass and Moment Calculations

We can model coil springs and wires as if they were masses distributed along continuously differentiable curves in space. The distribution of mass is modeled by a continuous density function $\delta(x, y, z)$. The spring's or wire's mass, center-of-mass coordinates, and moments are then calculated with the formulas in Table 14.1. The formulas apply to thin rods as well as springs and wires.

Example 2   A coil spring lies along the helix

$$\mathbf{r} = (\cos 4t)\,\mathbf{i} + (\sin 4t)\,\mathbf{j} + t\,\mathbf{k}, \qquad 0 \le t \le 2\pi.$$

The spring's density is a constant, $\delta = 1$. Find the spring's mass, the coordinates of the spring's center of mass, and the spring's moment of inertia and radius of gyration about the z-axis.

Solution   We sketch the spring (Fig. 14.3). Because of the symmetries involved, the center of mass lies at the point $(0, 0, \pi)$ on the z-axis.

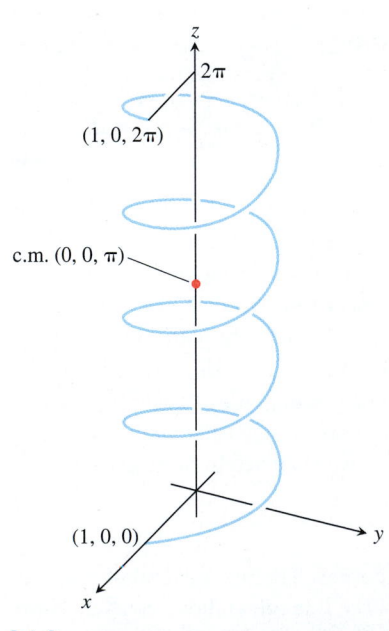

14.3   The helical spring in Example 2.

For the remaining calculations, we first express ds in terms of t:

$$ds = \sqrt{\left(\frac{dx}{dt}\right)^2 + \left(\frac{dy}{dt}\right)^2 + \left(\frac{dz}{dt}\right)^2} \, dt$$

$$= \sqrt{(-4 \sin 4t)^2 + (4 \cos 4t)^2 + 1} \, dt = \sqrt{17} \, dt.$$

The formulas in Table 14.1 then give

$$M = \int_{\text{helix}} \delta \, ds = \int_0^{2\pi} (1)\sqrt{17} \, dt = 2\pi\sqrt{17},$$

$$I_z = \int_{\text{helix}} (x^2 + y^2)\delta \, ds = \int_0^{2\pi} (\cos^2 4t + \sin^2 4t)(1)\sqrt{17} \, dt$$

$$= \int_0^{2\pi} \sqrt{17} \, dt = 2\pi\sqrt{17},$$

$$R_z = \sqrt{I_z/M} = \sqrt{2\pi\sqrt{17}/2\pi\sqrt{17}} = 1.$$

Notice that the radius of gyration about the z-axis is the radius of the cylinder around which the helix winds.

Example 3 A slender metal arch, thicker at the bottom than at the top, lies along the semicircle $y^2 + z^2 = 1$, $z \geq 0$, in the yz-plane (Fig. 14.4). Find the center of the arch's mass if the density at the point (x, y, z) on the arch is $\delta(x, y, z) = 2 - z$.

Solution We know that $\bar{x} = 0$ and $\bar{y} = 0$ because the arch lies in the yz-plane with its mass distributed symmetrically about the z-axis. To find $\bar{z}$, we parametrize the circle as

$$\mathbf{r} = (\cos t)\,\mathbf{j} + (\sin t)\,\mathbf{k}, \qquad 0 \leq t \leq \pi,$$

and express ds in terms of dt:

$$ds = \sqrt{\left(\frac{dx}{dt}\right)^2 + \left(\frac{dy}{dt}\right)^2 + \left(\frac{dz}{dt}\right)^2} \, dt = \sqrt{(0)^2 + (-\sin t)^2 + (\cos t)^2} \, dt = dt.$$

The formulas in Table 14.1 then give

$$M = \int_C \delta \, ds = \int_C (2 - z) \, ds = \int_0^{\pi} (2 - \sin t) \, dt = 2\pi - 2,$$

$$M_{xy} = \int_C z\,\delta \, ds = \int_C z(2 - z) \, ds = \int_0^{\pi} \sin t\,(2 - \sin t) \, dt$$

$$= \int_0^{\pi} (2 \sin t - \sin^2 t) \, dt = \frac{8 - \pi}{2},$$

$$\bar{z} = \frac{M_{xy}}{M} = \frac{8 - \pi}{2} \cdot \frac{1}{2\pi - 2} = \frac{8 - \pi}{4\pi - 4} = 0.57. \qquad \begin{pmatrix}\text{Calculator,}\\ \text{rounded}\end{pmatrix}$$

To the nearest hundredth, the center of mass is $(0, 0, 0.57)$.

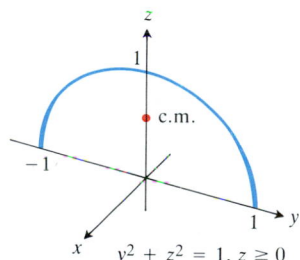

14.4 Example 3 shows how to find the center of mass of a circular arch of variable density.

EXERCISES 14.1

Match the vector equations in Exercises 1–8 with the graphs in Fig. 14.5.

(a)

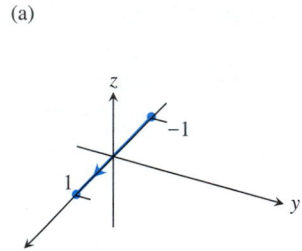

(b)

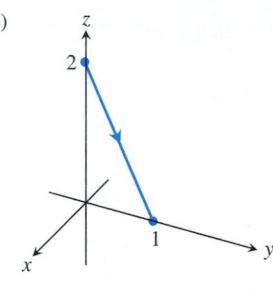

(c)

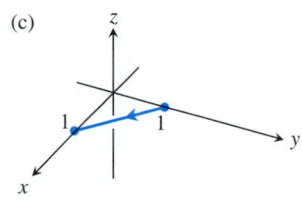

(d)

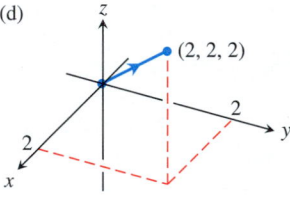

(e)

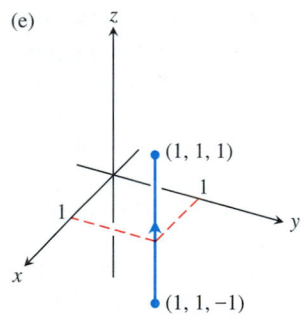

(f)

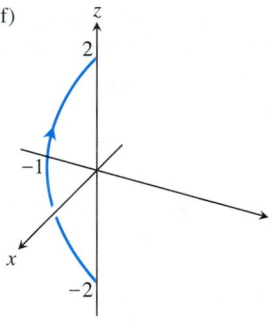

(g)

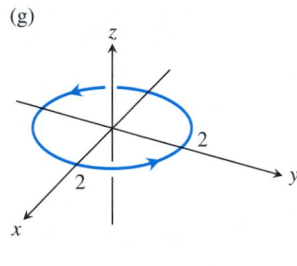

(h)

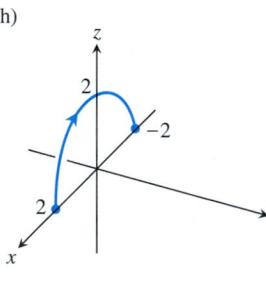

14.5   The graphs for Exercises 1–8.

1. $\mathbf{r} = t\mathbf{i} + (1 - t)\mathbf{j}, \quad 0 \le t \le 1$
2. $\mathbf{r} = \mathbf{i} + \mathbf{j} + t\mathbf{k}, \quad -1 \le t \le 1$
3. $\mathbf{r} = (2\cos t)\mathbf{i} + (2\sin t)\mathbf{j}, \quad 0 \le t \le 2\pi$
4. $\mathbf{r} = t\mathbf{i}, \quad -1 \le t \le 1$
5. $\mathbf{r} = t\mathbf{i} + t\mathbf{j} + t\mathbf{k}, \quad 0 \le t \le 2$
6. $\mathbf{r} = t\mathbf{j} + (2 - 2t)\mathbf{k}, \quad 0 \le t \le 1$

7. $\mathbf{r} = (t^2 - 1)\mathbf{j} + 2t\mathbf{k}, \quad -1 \le t \le 1$
8. $\mathbf{r} = (2\cos t)\mathbf{i} + (2\sin t)\mathbf{k}, \quad 0 \le t \le \pi$

In Exercises 9–14, evaluate the line integral of f along the given curve for the given parameter interval.

9. $f(x, y, z) = x + y$
   $\mathbf{r} = t\mathbf{i} + (1 - t)\mathbf{j}, \quad 0 \le t \le 1$
10. $f(x, y, z) = x - y + z - 2$
    $\mathbf{r} = t\mathbf{i} + (1 - t)\mathbf{j} + \mathbf{k}, \quad 0 \le t \le 1$
11. $f(x, y, z) = xy + y + z$
    $\mathbf{r} = 2t\mathbf{i} + t\mathbf{j} + (2 - 2t)\mathbf{k}, \quad 0 \le t \le 1$
12. $f(x, y, z) = \sqrt{x^2 + y^2}$
    $\mathbf{r} = (4\cos t)\mathbf{i} + (4\sin t)\mathbf{j} + 3t\mathbf{k}, \quad -2\pi \le t \le 2\pi$
13. $f(x, y, z) = 3z\sqrt{3x^2 + y^2 + z^2}$
    $\mathbf{r} = \mathbf{i} + \mathbf{j} + t\mathbf{k}, \quad -1 \le t \le 1$
14. $f(x, y, z) = \sqrt{3}/(x^2 + y^2 + z^2)$
    $\mathbf{r} = t\mathbf{i} + t\mathbf{j} + t\mathbf{k}, \quad 1 \le t < \infty$
15. Integrate $f(x, y, z) = x + \sqrt{y} - z^2$ over the path from $(0, 0, 0)$ to $(1, 1, 1)$ (Fig. 14.6a) given by
    C_1: $\mathbf{r} = t\mathbf{i} + t^2\mathbf{j}, \quad 0 \le t \le 1$
    C_2: $\mathbf{r} = \mathbf{i} + \mathbf{j} + t\mathbf{k}, \quad 0 \le t \le 1$
16. Integrate $f(x, y, z) = x + \sqrt{y} - z^2$ over the path from $(0, 0, 0)$ to $(1, 1, 1)$ (Fig. 14.6b) given by
    C_1: $\mathbf{r} = t\mathbf{k}, \quad 0 \le t \le 1$
    C_2: $\mathbf{r} = t\mathbf{j} + \mathbf{k}, \quad 0 \le t \le 1$
    C_3: $\mathbf{r} = t\mathbf{i} + \mathbf{j} + \mathbf{k}, \quad 0 \le t \le 1$

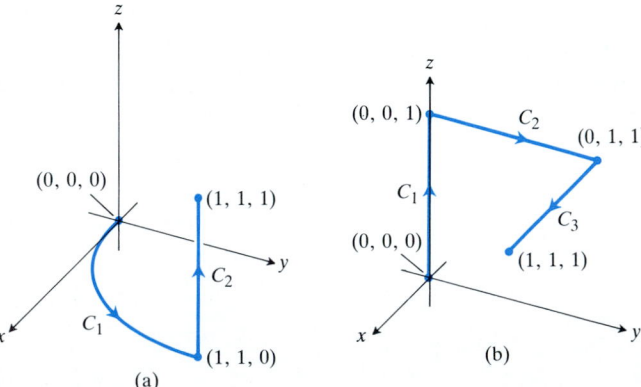

14.6   The paths of integration for Exercises 15 and 16.

17. Integrate $f(x, y, z) = (x + y + z)/(x^2 + y^2 + z^2)$ over the path $\mathbf{r} = t\mathbf{i} + t\mathbf{j} + t\mathbf{k}, a \le t \le b$.
18. Integrate $f(x, y, z) = -\sqrt{x^2 + z^2}$ over the circle $\mathbf{r} = (a\cos t)\mathbf{j} + (a\sin t)\mathbf{k}, 0 \le t \le 2\pi$, in the direction of increasing t.

19. Find I_x and R_x for the arch in Example 3.

20. Find the mass of a wire that lies along the curve $\mathbf{r} = (t^2 - 1)\mathbf{j} + 2t\mathbf{k}$, $0 \le t \le 1$, if the density is $\delta = (3/2)t$.

21. A circular wire hoop of constant density δ lies along the circle $x^2 + y^2 = a^2$ in the xy-plane. Find the hoop's moment of inertia and radius of gyration about the z-axis.

22. A slender rod of constant density lies along the line segment $\mathbf{r} = t\mathbf{j} + (2 - 2t)\mathbf{k}$, $0 \le t \le 1$ in the yz-plane. Find the moments of inertia and radii of gyration of the rod about the three coordinate axes.

23. A spring of constant density δ lies along the helix $\mathbf{r} = (\cos t)\mathbf{i} + (\sin t)\mathbf{j} + t\mathbf{k}$, $0 \le t \le 2\pi$.
a) Find I_z and R_z.
b) Suppose you have another spring of constant density δ that is twice as long as the spring in (a) and lies along the helix for $0 \le t \le 4\pi$. Do you expect I_z and R_z for the longer spring to be the same as those for the shorter one,

or should they be different? Check your predictions by calculating I_z and R_z for the longer spring.

24. A wire of density $\delta(x, y, z) = 15\sqrt{y + 2}$ lies along the curve $\mathbf{r} = (t^2 - 1)\mathbf{j} + 2t\mathbf{k}$, $-1 \le t \le 1$. Find its center of mass. Then sketch the curve and center of mass together.

25. Find the mass of a thin wire lying along the curve $\mathbf{r} = \sqrt{2}\,t\mathbf{i} + \sqrt{2}\,t\mathbf{j} + (4 - t^2)\mathbf{k}$, $0 \le t \le 1$, if the density is (a) $\delta = 3t$, (b) $\delta = 1$.

26. Find the center of mass of a thin wire lying along the curve $\mathbf{r} = t\mathbf{i} + 2t\mathbf{j} + (2/3)t^{3/2}\mathbf{k}$, $0 \le t \le 2$, if the density is $\delta = 3\sqrt{5 + t}$.

27. Find the center of mass, and the moments of inertia and radii of gyration about the coordinate axes of a thin wire lying along the curve

$$\mathbf{r} = t\mathbf{i} + \frac{2\sqrt{2}}{3}t^{3/2}\mathbf{j} + \frac{t^2}{2}\mathbf{k}, \quad 0 \le t \le 2,$$

if the density is $\delta = 1/(t + 1)$.

14.2 Vector Fields, Work, Circulation, and Flux

In studies of physical phenomena represented by vectors, we introduce the idea of a vector field on a domain in space, a function that assigns a vector quantity to each point in the domain. For example, we might be studying the velocity field of a moving fluid or the electric field created by a distribution of electric charges. In situations like these, integrals of real-valued functions over closed intervals are replaced by line integrals of functions over paths through the vector field, and the function being integrated is a scalar product of vectors. We can use these line integrals in vector fields to calculate the work done in moving an object along a path against a variable force. (The object might be a vehicle sent into space against Earth's gravitational field.) Or we might want to calculate the work done by a vector field in moving an object along a path through the field (for example, the work done by a particle accelerator in raising the energy of a particle). We can also use line integrals to calculate the rates at which fluids flow along and across curves.

Vector Fields

A **vector field** on a domain in the plane or in space is a function that assigns a vector to each point in the domain. A field of three-dimensional vectors might have a formula like

$$\mathbf{F} = M(x, y, z)\mathbf{i} + N(x, y, z)\mathbf{j} + P(x, y, z)\mathbf{k}. \tag{1}$$

The field is **continuous** if the **component functions** M, N, and P are continuous, **differentiable** if M, N, and P are differentiable, and so on. A field of two-dimensional vectors might have a formula like

$$\mathbf{F} = M(x, y)\mathbf{i} + N(x, y)\mathbf{j}. \tag{2}$$

If we attach a projectile's velocity vector to each point of the projectile's trajectory in the plane of motion, we have a two-dimensional field defined along the trajectory. If

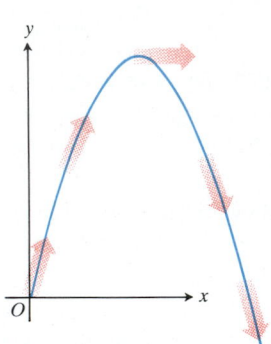

14.7 The velocity vectors $\mathbf{v}(t)$ of a projectile's motion make a vector field along the trajectory.

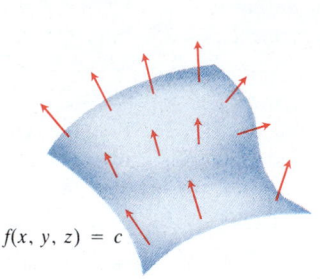

$f(x, y, z) = c$

14.8 The field of gradient vectors ∇f on a surface $f(x, y, z) = c$.

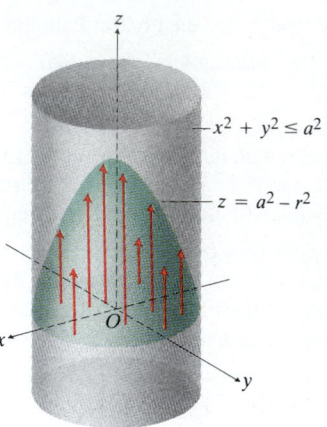

$x^2 + y^2 \leq a^2$

$z = a^2 - r^2$

14.9 The flow of fluid in a long cylindrical pipe. The vectors $\mathbf{v} = (a^2 - r^2)\,\mathbf{k}$ inside the cylinder that have their bases in the xy-plane have their tips on the paraboloid $z = a^2 - r^2$.

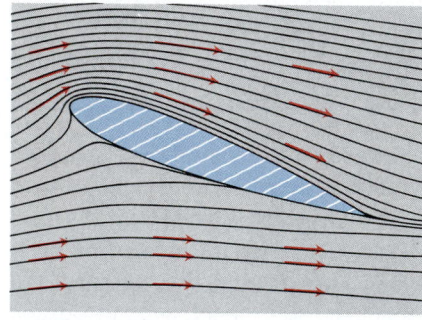

14.10 Velocity vectors of a flow around an airfoil in a wind tunnel. The streamlines were made visible by kerosene smoke. (Adapted from *NCFMF Book of Film Notes,* 1974, the MIT Press with Education Development Center, Inc., Newton, Massachusetts.)

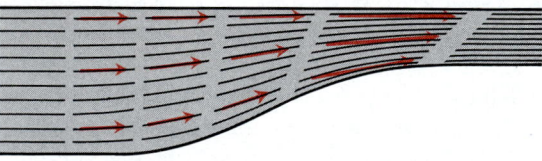

14.11 Streamlines in a contracting channel. The water speeds up as the channel narrows and the velocity vectors increase in length. (Adapted from *NCFMF Book of Film Notes,* 1974, the MIT Press with Education Development Center, Inc., Newton, Massachusetts.)

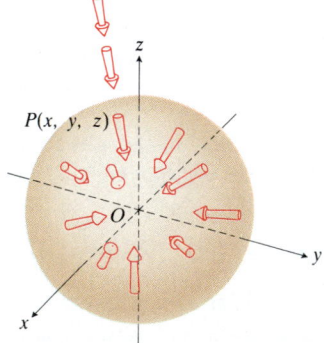

$P(x, y, z)$

14.12 Vectors in the gravitational field
$$\mathbf{F} = -\frac{GM(x\mathbf{i} + y\mathbf{j} + z\mathbf{k})}{(x^2 + y^2 + z^2)^{3/2}}$$

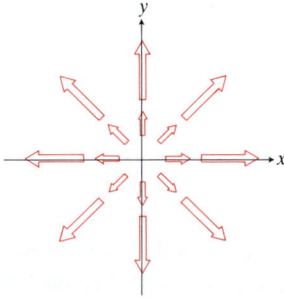

14.13 The radial field $\mathbf{F} = x\mathbf{i} + y\mathbf{j}$ of position vectors of points in the plane. Notice the convention that an arrow is drawn with its tail, not its head, at the point where $\mathbf{F}$ is evaluated.

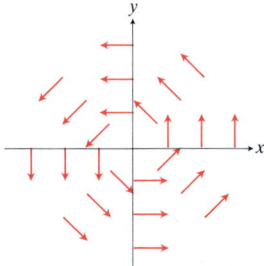

14.14 The circumferential or "spin" field of unit vectors
$$\mathbf{F} = (-y\mathbf{i} + x\mathbf{j})/(x^2 + y^2)^{1/2}$$
in the plane. The field is not defined at the origin.

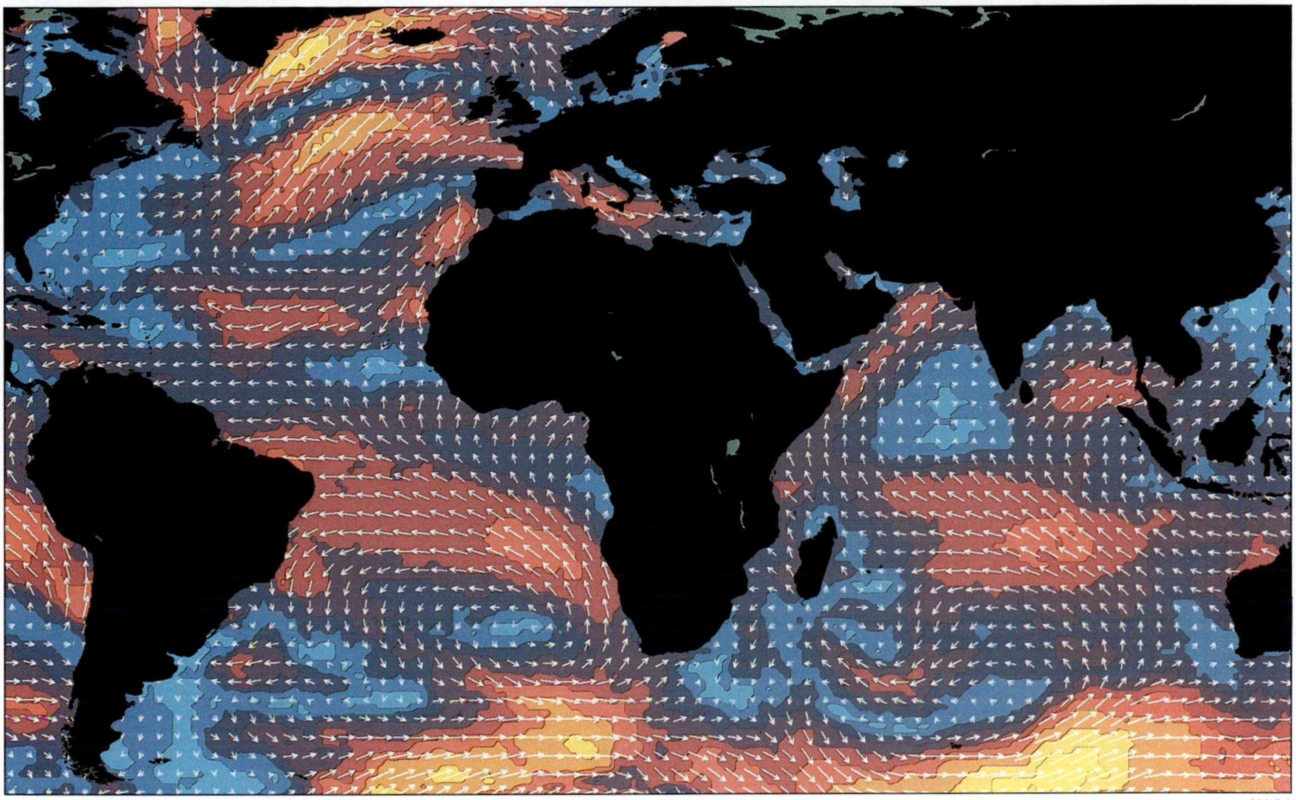

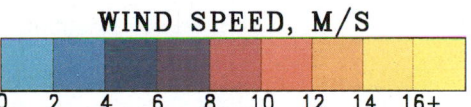

NASA

14.15 NASA's Seasat used radar during a 3-day period in September 1978 to take 350,000 wind measurements over the world's oceans. The arrows show wind direction; their length and the color contouring indicate speed. Notice the heavy storm south of Greenland.

we attach the gradient vector of a scalar function to each point of a level surface of the function, we have a three-dimensional field on the surface. If we attach the velocity vector to each point of a flowing fluid, we have a three-dimensional field defined on a region in space. These and other fields are illustrated in Figures 14.7–14.15. Some of the illustrations give formulas for the fields as well.

To sketch the fields that had formulas, we picked a representative selection of domain points and sketched the vectors attached to them. Notice the convention that the arrows representing the vectors are drawn with their tails, not their heads, at the points where the vector functions are evaluated. This is different from the way we drew the position vectors of the planets and projectiles we studied in Chapter 11, with their tails at the origin and their heads at the planet's and projectile's locations.

The Work Done by a Force over a Curve in Space

Suppose that the vector field $\mathbf{F} = M(x, y, z)\,\mathbf{i} + N(x, y, z)\,\mathbf{j} + P(x, y, z)\,\mathbf{k}$ represents a force throughout a region in space (it might be the force of gravity or an electromagnetic force of some kind) and that

$$\mathbf{r} = g(t)\,\mathbf{i} + h(t)\,\mathbf{j} + k(t)\,\mathbf{k}, \quad a \le t \le b,$$

is a curve in the region. Then the line integral of $\mathbf{F} \cdot \mathbf{T}$, the scalar component of $\mathbf{F}$ in the direction of the curve's unit tangent vector, over the curve is called the work done by **F** over the curve from a to b (Fig. 14.16).

DEFINITION

The **work** done by a continuous force $\mathbf{F} = M(x, y, z)\,\mathbf{i} + N(x, y, z)\,\mathbf{j} + P(x, y, z)\,\mathbf{k}$ over a smooth curve $\mathbf{r} = g(t)\,\mathbf{i} + h(t)\,\mathbf{j} + k(t)\,\mathbf{k}$ from $t = a$ to $t = b$ is the value of the line integral

$$\text{Work} = \int_{t=a}^{t=b} \mathbf{F} \cdot \mathbf{T}\, ds. \tag{3}$$

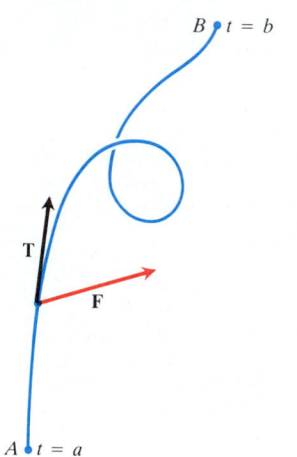

14.16 The work done by a continuous field **F** over a smooth path $\mathbf{r} = g(t)\,\mathbf{i} + h(t)\,\mathbf{j} + k(t)\,\mathbf{k}$ from A to B is the line integral of $\mathbf{F} \cdot \mathbf{T}$ from $t = a$ to $t = b$.

We derive Eq. (3) with the same kind of reasoning we used in Section 5.7 to derive the formula $W = \int_a^b F(x)\, dx$ for the work done by a continuous force of magnitude $F(x)$ directed along an interval of the x-axis. We divide the curve into short segments, apply the constant-force-times-distance formula for work to approximate the work over each curved segment, add the results to approximate the work over the entire curve, and calculate the work as the limit of the approximating sums as the segments become shorter and more numerous. To find exactly what the limiting integral should be, we partition the parameter interval $I = [a, b]$ in the usual way and choose a point c_k in each subinterval $[t_k, t_{k+1}]$. The partition of I determines ("induces," we say) a partition of the curve, with the point P_k being the tip of the position vector $\mathbf{r}$ at $t = t_k$ and Δs_k being the length of the curve segment $P_k P_{k+1}$ (Fig. 14.17). If $\mathbf{F}_k$ denotes the value of $\mathbf{F}$ at the point on the curve corresponding to $t = c_k$, and $\mathbf{T}_k$ denotes the curve's tangent vector at this point, then $\mathbf{F}_k \cdot \mathbf{T}_k$ is the scalar component of $\mathbf{F}$ in the direction of $\mathbf{T}$ at $t = c_k$ (Fig. 14.18). The work done by $\mathbf{F}$ along the curve segment $P_k P_{k+1}$ will be approximately

$$\begin{pmatrix} \text{force component in} \\ \text{direction of motion} \end{pmatrix} \times \begin{pmatrix} \text{distance} \\ \text{applied} \end{pmatrix} = \mathbf{F}_k \cdot \mathbf{T}_k \Delta s_k. \tag{4}$$

The work done by $\mathbf{F}$ along the curve from $t = a$ to $t = b$ will be approximately

$$\sum_{k=1}^{n} \mathbf{F}_k \cdot \mathbf{T}_k \Delta s_k. \tag{5}$$

As the norm of the partition of $[a, b]$ approaches zero, the norm of the induced partition of the curve approaches zero and the sums in (5) approach the line integral

$$\int_{t=a}^{t=b} \mathbf{F} \cdot \mathbf{T}\, ds. \tag{6}$$

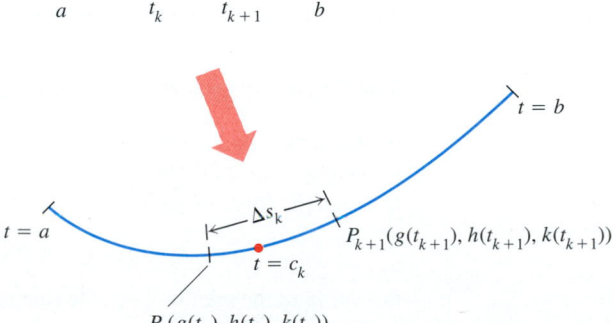

14.17 Each partition of the parameter interval $a \leq t \leq b$ induces a partition of the curve $\mathbf{r} = g(t)\,\mathbf{i} + h(t)\,\mathbf{j} + k(t)\,\mathbf{k}$.

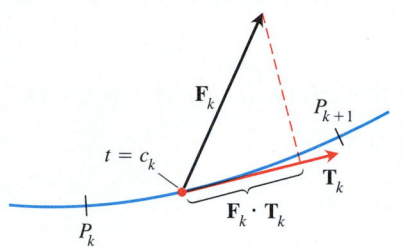

14.18 An enlarged view of the curve segment $P_k P_{k+1}$ in Fig. 14.17, showing the force vector and unit tangent vector at the point on the curve where $t = c_k$.

Notice how the numerical sign of the number we calculate with this integral depends on the direction in which the curve is traversed as t increases. If we reverse the direction of motion, we reverse the direction of $\mathbf{T}$ and change the signs of $\mathbf{F} \cdot \mathbf{T}$ and its integral.

Notation and Evaluation

There are six standard ways to write the work integral in Eq. (3).

Equivalent Ways to Write the Work Integral

$$\text{Work} = \int_{t=a}^{t=b} \mathbf{F} \cdot \mathbf{T} \, ds \qquad \text{(The definition)}$$

$$= \int_{t=a}^{t=b} \mathbf{F} \cdot d\mathbf{r} \qquad \text{(Compact differential form)}$$

$$= \int_{a}^{b} \mathbf{F} \cdot \frac{d\mathbf{r}}{dt} \, dt \qquad \left(\begin{array}{l}\text{Expanded to include } dt; \\ \text{emphasizes the} \\ \text{velocity vector } d\mathbf{r}/dt\end{array}\right)$$

$$= \int_{a}^{b} \left(M \frac{dg}{dt} + N \frac{dh}{dt} + P \frac{dk}{dt} \right) dt \qquad \left(\begin{array}{l}\text{Emphasizes the} \\ \text{component functions}\end{array}\right)$$

$$= \int_{a}^{b} \left(M \frac{dx}{dt} + N \frac{dy}{dt} + P \frac{dz}{dt} \right) dt \qquad \left(\begin{array}{l}\text{Abbreviates the} \\ \text{components of } \mathbf{r}\end{array}\right)$$

$$= \int_{a}^{b} M \, dx + N \, dy + P \, dz \qquad \left(\begin{array}{l}dt\text{'s canceled; the most} \\ \text{common differential form}\end{array}\right)$$

Despite their variety, these formulas are all evaluated the same way.

Evaluation

To evaluate the work integral, take these steps:

1. Evaluate $\mathbf{F}$ on the curve as a function of t.
2. Find $d\mathbf{r}/dt$.
3. Dot $\mathbf{F}$ with $d\mathbf{r}/dt$.
4. Integrate from $t = a$ to $t = b$.

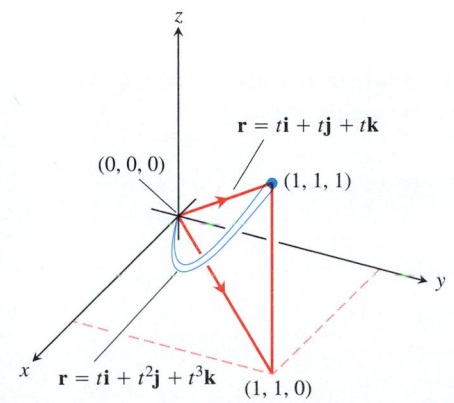

14.19 The curve in Example 1.

Example 1   Find the work done by the vector field

$$\mathbf{F} = (y - x^2) \, \mathbf{i} + (z - y^2) \, \mathbf{j} + (x - z^2) \, \mathbf{k}$$

over the curve

$$\mathbf{r} = t\mathbf{i} + t^2\mathbf{j} + t^3\mathbf{k}, \quad 0 \le t \le 1,$$

from $(0, 0, 0)$ to $(1, 1, 1)$ (Fig. 14.19).

Solution

STEP 1: *Evaluate* **F** *on the curve.*

$$\mathbf{F} = (y - x^2)\,\mathbf{i} + (z - y^2)\,\mathbf{j} + (x - z^2)\,\mathbf{k} = \underbrace{(t^2 - t^2)}_{0}\,\mathbf{i} + (t^3 - t^4)\,\mathbf{j} + (t - t^6)\,\mathbf{k}$$

STEP 2: *Find* $d\mathbf{r}/dt$.

$$\frac{d\mathbf{r}}{dt} = \frac{d}{dt}(t\mathbf{i} + t^2\,\mathbf{j} + t^3\,\mathbf{k}) = \mathbf{i} + 2t\mathbf{j} + 3t^2\,\mathbf{k}$$

STEP 3: *Dot* **F** *with* $d\mathbf{r}/dt$.

$$\mathbf{F} \cdot \frac{d\mathbf{r}}{dt} = ((t^3 - t^4)\,\mathbf{j} + (t - t^6)\,\mathbf{k}) \cdot (\mathbf{i} + 2t\mathbf{j} + 3t^2\,\mathbf{k})$$

$$= (t^3 - t^4)(2t) + (t - t^6)(3t^2) = 2t^4 - 2t^5 + 3t^3 - 3t^8$$

STEP 4: *Integrate from* $t = 0$ *to* $t = 1$.

$$\text{Work} = \int_0^1 (2t^4 - 2t^5 + 3t^3 - 3t^8)\,dt$$

$$= \left[\frac{2}{5}t^5 - \frac{2}{6}t^6 + \frac{3}{4}t^4 - \frac{3}{9}t^9\right]_0^1 = \frac{29}{60}.$$

Flow Integrals and Circulation

If instead of being a force field, the vector field $\mathbf{F} = M\mathbf{i} + N\mathbf{j} + P\mathbf{k}$ represents the velocity field of a fluid flowing through a region in space (a tidal basin, a riverbed, or the turbine chamber of a hydroelectric generator, for example), then the line integral of $\mathbf{F} \cdot \mathbf{T}$ along a curve in the region is called the fluid's flow along the curve.

DEFINITIONS

If $\mathbf{r} = g(t)\,\mathbf{i} + h(t)\,\mathbf{j} + k(t)\,\mathbf{k}$, $a \le t \le b$, is a smooth curve in the domain of a continuous velocity field $\mathbf{F} = M(x, y, z)\,\mathbf{i} + N(x, y, z)\,\mathbf{j} + P(x, y, z)\,\mathbf{k}$, the **flow** along the curve from $t = a$ to $t = b$ is the line integral of $\mathbf{F} \cdot \mathbf{T}$ over the curve from a to b,

$$\text{Flow} = \int_a^b \mathbf{F} \cdot \mathbf{T}\,ds. \tag{7}$$

The integral in this case is called a **flow integral.** If the curve is a closed loop, the flow is called the **circulation** around the curve.

We evaluate flow integrals the same way we evaluate work integrals.

Example 2   A fluid's velocity field is

$$\mathbf{F} = x\mathbf{i} + z\mathbf{j} + y\mathbf{k}.$$

Find the flow along the helix

$$\mathbf{r} = (\cos t)\,\mathbf{i} + (\sin t)\,\mathbf{j} + t\mathbf{k}, \quad 0 \le t \le \pi/2.$$

Solution

STEP 1:   *Evaluate* **F** *on the curve.*

$$\mathbf{F} = x\mathbf{i} + z\mathbf{j} + y\mathbf{k} = (\cos t)\,\mathbf{i} + t\mathbf{j} + (\sin t)\,\mathbf{k}$$

STEP 2:   *Find* $d\mathbf{r}/dt$.

$$\frac{d\mathbf{r}}{dt} = (-\sin t)\,\mathbf{i} + (\cos t)\,\mathbf{j} + \mathbf{k}$$

STEP 3:   *Find* $\mathbf{F} \cdot (d\mathbf{r}/dt)$.

$$\mathbf{F} \cdot \frac{d\mathbf{r}}{dt} = (\cos t)(-\sin t) + (t)(\cos t) + (\sin t)(1)$$

$$= -\sin t \cos t + t \cos t + \sin t$$

STEP 4:   *Integrate from* $t = a$ *to* $t = b$.

$$\text{Flow} = \int_{t=a}^{t=b} \mathbf{F} \cdot \frac{d\mathbf{r}}{dt}\, dt = \int_{0}^{\pi/2} (-\sin t \cos t + t \cos t + \sin t)\, dt$$

$$= \left[\frac{\cos^2 t}{2} + t \sin t\right]_{0}^{\pi/2} = \left(0 + \frac{\pi}{2}\right) - \left(\frac{1}{2} + 0\right) = \frac{\pi}{2} - \frac{1}{2}$$

Example 3   Find the circulation of a fluid around the circle

$$\mathbf{r} = (\cos t)\,\mathbf{i} + (\sin t)\,\mathbf{j}, \quad 0 \le t \le 2\pi,$$

if the velocity field is $\mathbf{F} = (x - y)\,\mathbf{i} + x\mathbf{j}$.

Solution

1. On the circle, $\mathbf{F} = (x - y)\,\mathbf{i} + x\mathbf{j} = (\cos t - \sin t)\,\mathbf{i} + (\cos t)\,\mathbf{j}$.

2. $\dfrac{d\mathbf{r}}{dt} = (-\sin t)\,\mathbf{i} + (\cos t)\,\mathbf{j}$

3. $\mathbf{F} \cdot \dfrac{d\mathbf{r}}{dt} = -\sin t \cos t + \underbrace{\sin^2 t + \cos^2 t}_{1}$

4. $\text{Circulation} = \displaystyle\int_{0}^{2\pi} \mathbf{F} \cdot \frac{d\mathbf{r}}{dt}\, dt = \int_{0}^{2\pi} (1 - \sin t \cos t)\, dt$

$$= \left[t - \frac{\sin^2 t}{2}\right]_{0}^{2\pi} = 2\pi$$

Flux Across a Plane Curve

To find the rate at which a fluid is entering or leaving a region enclosed by a curve C in the xy-plane, we calculate the line integral over C of $\mathbf{F} \cdot \mathbf{n}$, the scalar component of the fluid's velocity field in the direction of the curve's outward-pointing normal vector $\mathbf{n}$. The value of this integral is called the **flux** of $\mathbf{F}$ across C. Flux is Latin for *flow*, but many flux calculations involve no motion at all. If $\mathbf{F}$ were an electric field or a magnetic field, for instance, the integral of $\mathbf{F} \cdot \mathbf{n}$ would still be called the flux of the field across C.

DEFINITION

If C is a smooth closed curve in the domain of a continuous vector field $\mathbf{F} = M(x, y)\,\mathbf{i} + N(x, y)\,\mathbf{j}$ in the plane, and if $\mathbf{n}$ is the outward-pointing unit normal vector on C, the **flux** of $\mathbf{F}$ across C is given by the following line integral:

$$\text{Flux of } \mathbf{F} \text{ across } C = \int_C \mathbf{F} \cdot \mathbf{n} \, ds. \qquad (8)$$

Notice the difference between flux and circulation. The flux of $\mathbf{F}$ across C is the line integral with respect to arc length of $\mathbf{F} \cdot \mathbf{n}$, the scalar component of $\mathbf{F}$ in the direction of the outward normal. The circulation of $\mathbf{F}$ around C is the line integral with respect to arc length of $\mathbf{F} \cdot \mathbf{T}$, the scalar component of $\mathbf{F}$ in the direction of the unit tangent vector. Flux is the integral of the normal component of $\mathbf{F}$; circulation is the integral of the tangential component of $\mathbf{F}$.

To evaluate the integral in (8), we begin with a parametrization

$$x = g(t), \qquad y = h(t), \qquad a \le t \le b,$$

that traces the curve C exactly once as t increases from a to b. We can find the outward unit normal vector $\mathbf{n}$ by crossing the curve's unit tangent vector $\mathbf{T}$ with the vector $\mathbf{k}$. But which order do we choose, $\mathbf{T} \times \mathbf{k}$ or $\mathbf{k} \times \mathbf{T}$? Which one points outward? It depends on which way C is traversed as the parameter t increases. If the motion is clockwise, then $\mathbf{k} \times \mathbf{T}$ points outward; if the motion is counterclockwise, then $\mathbf{T} \times \mathbf{k}$ points outward (Fig. 14.20). The usual choice is $\mathbf{n} = \mathbf{T} \times \mathbf{k}$, the choice that assumes counterclockwise motion. Thus, while the value of the arc length integral in the definition of flux in Eq. (8) does not depend on which way C is traversed, the formulas we are about to derive for evaluating the integral in Eq. (8) will assume counterclockwise motion.

In terms of components,

$$\mathbf{n} = \mathbf{T} \times \mathbf{k} = \left(\frac{dx}{ds}\,\mathbf{i} + \frac{dy}{ds}\,\mathbf{j}\right) \times \mathbf{k} = \frac{dy}{ds}\,\mathbf{i} - \frac{dx}{ds}\,\mathbf{j}. \qquad (9)$$

If $\mathbf{F} = M(x, y)\,\mathbf{i} + N(x, y)\,\mathbf{j}$, then

$$\mathbf{F} \cdot \mathbf{n} = M(x, y)\frac{dy}{ds} - N(x, y)\frac{dx}{ds}.$$

Hence,

$$\int_C \mathbf{F} \cdot \mathbf{n} \, ds = \int_C \left(M\frac{dy}{ds} - N\frac{dx}{ds}\right) ds$$

$$= \oint_C M \, dy - N \, dx. \qquad (10)$$

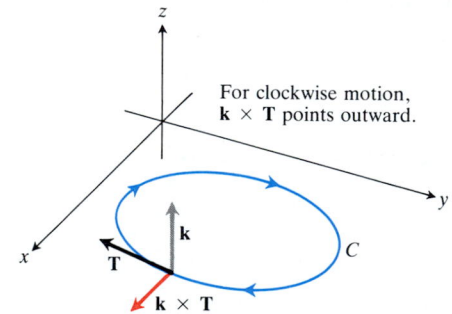

For clockwise motion, $\mathbf{k} \times \mathbf{T}$ points outward.

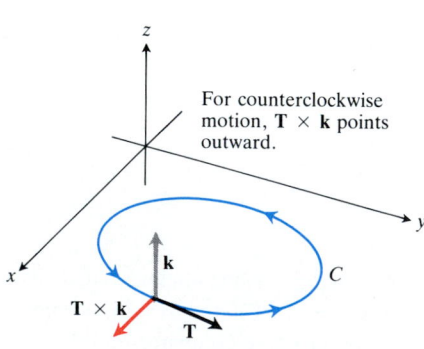

For counterclockwise motion, $\mathbf{T} \times \mathbf{k}$ points outward.

14.20   To find a unit outward normal vector for a curve C in the xy-plane that is traversed counterclockwise as t increases, we take $\mathbf{n} = \mathbf{T} \times \mathbf{k}$.

We put a directed circle $\circlearrowleft$ on the last integral as a reminder that the integration around the closed curve C is to be in the counterclockwise direction. To evaluate this integral, we express M, dy, N, and dx in terms of t and integrate from $t = a$ to $t = b$. Strange as it may seem, we do not need to know either $\mathbf{n}$ or ds to find the flux.

The Formula for Calculating Flux Across a Closed Plane Curve

$$\text{Flux of } \mathbf{F} = M\mathbf{i} + N\mathbf{j} \text{ across } C = \oint_C M \, dy - N \, dx \qquad (11)$$

The integral can be evaluated from any parametrization $x = g(t)$, $y = h(t)$, $a \le t \le b$, that traces C counterclockwise exactly once. To evaluate the integral, express M, dy, N, and dx in terms of t and integrate from $t = a$ to $t = b$.

Example 4   Find the flux of $\mathbf{F} = (x - y)\mathbf{i} + x\mathbf{j}$ across the circle $x^2 + y^2 = 1$ in the xy-plane.

Solution   The parametrization

$$\mathbf{r} = (\cos t)\mathbf{i} + (\sin t)\mathbf{j}, \quad 0 \le t \le 2\pi,$$

traces the circle counterclockwise exactly once. We can therefore use this parametrization in Eq. (11). With

$$M = x - y = \cos t - \sin t \qquad dy = d(\sin t) = \cos t \, dt$$

$$N = x = \cos t \qquad dx = d(\cos t) = -\sin t \, dt,$$

we find

$$\text{Flux} = \int_C M \, dy - N \, dx = \int_0^{2\pi} (\cos^2 t - \sin t \cos t + \cos t \sin t) \, dt \quad \text{(Eq. (11))}$$

$$= \int_0^{2\pi} \cos^2 t \, dt = \int_0^{2\pi} \frac{1 + \cos 2t}{2} \, dt = \left[\frac{t}{2} + \frac{\sin 2t}{4} \right]_0^{2\pi} = \pi.$$

The flux of $\mathbf{F}$ across the circle is π. Since the answer is positive, the net flow across the curve is outward. A net inward flow would give a negative flux.

EXERCISES 14.2

1. Give a formula $\mathbf{F} = M(x, y)\mathbf{i} + N(x, y)\mathbf{j}$ for the vector field in the plane that has the property that $\mathbf{F}$ points toward the origin with magnitude inversely proportional to the square of the distance from (x, y) to the origin. (The field is not defined at $(0, 0)$.)

2. Give a formula $\mathbf{F} = M(x, y)\mathbf{i} + N(x, y)\mathbf{j}$ for the vector field in the plane that has the properties that $\mathbf{F} = \mathbf{0}$ at $(0, 0)$ and that at any other point (a, b), $\mathbf{F}$ is tangent to the circle $x^2 + y^2 = a^2 + b^2$ and points in the clockwise direction with magnitude $|\mathbf{F}| = \sqrt{a^2 + b^2}$.

3. Sketch the sets of points in the plane where the field

$$\mathbf{F} = \left(\frac{x^2 + 2y^2 - 4}{4} \right)\mathbf{i} + \left(\frac{y - x^2}{4} \right)\mathbf{j}$$

is (a) vertical, (b) horizontal. Then sketch the field on these sets.

4. Draw the (a) vertical and (b) horizontal components of the spin field

$$\mathbf{F} = \frac{(-y\mathbf{i} + x\mathbf{j})}{(x^2 + y^2)^{1/2}}$$

along the circle $x^2 + y^2 = 4$.

In Exercises 5–10, find the work done by the force $\mathbf{F}$ from $(0, 0, 0)$ to $(1, 1, 1)$ over each of the following paths:

a) the line segment
   $$\mathbf{r} = t\mathbf{i} + t\mathbf{j} + t\mathbf{k},$$
   $$0 \le t \le 1,$$

b) (Fig. 14.21) the curve
   $$\mathbf{r} = t\mathbf{i} + t^2\mathbf{j} + t^4\mathbf{k},$$
   $$0 \le t \le 1,$$

c) the path consisting of the line segment from $(0, 0, 0)$ to $(1, 1, 0)$ followed by the line segment from $(1, 1, 0)$ to $(1, 1, 1)$.

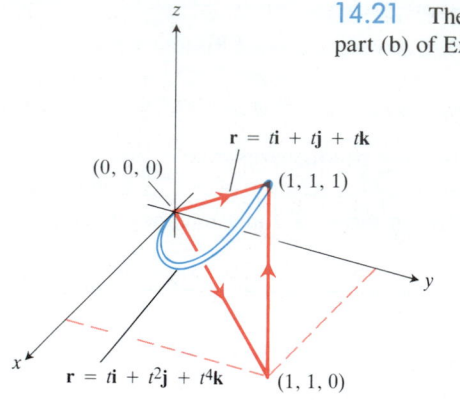

14.21 The curve in part (b) of Exercises 5–10.

5. $\mathbf{F} = 3y\mathbf{i} + 2x\mathbf{j} + 4z\mathbf{k}$

6. $\mathbf{F} = (1/(x^2 + 1))\,\mathbf{j}$

7. $\mathbf{F} = \sqrt{z}\,\mathbf{i} - 2x\mathbf{j} + \sqrt{y}\,\mathbf{k}$

8. $\mathbf{F} = xy\mathbf{i} + yz\mathbf{j} + xz\mathbf{k}$

9. $\mathbf{F} = (3x^2 - 3x)\,\mathbf{i} + 3z\mathbf{j} + \mathbf{k}$

10. $\mathbf{F} = (y + z)\,\mathbf{i} + (z + x)\,\mathbf{j} + (x + y)\,\mathbf{k}$

In Exercises 11–14, find the work done by $\mathbf{F}$ over the curve in the direction of increasing t.

11. $\mathbf{F} = xy\mathbf{i} + y\mathbf{j} - yz\mathbf{k}$
$\mathbf{r} = t\mathbf{i} + t^2\mathbf{j} + t\mathbf{k}, \quad 0 \le t \le 1$

12. $\mathbf{F} = 2y\mathbf{i} + 3x\mathbf{j} + (x + y)\,\mathbf{k}$
$\mathbf{r} = (\cos t)\,\mathbf{i} + (\sin t)\,\mathbf{j} + (t/6)\,\mathbf{k}, \quad 0 \le t \le 2\pi$

13. $\mathbf{F} = z\mathbf{i} + x\mathbf{j} + y\mathbf{k}$
$\mathbf{r} = (\sin t)\,\mathbf{i} + (\cos t)\,\mathbf{j} + t\mathbf{k}, \quad 0 \le t \le 2\pi$

14. $\mathbf{F} = 6z\mathbf{i} + y^2\mathbf{j} + 12x\mathbf{k}$
$\mathbf{r} = (\sin t)\,\mathbf{i} + (\cos t)\,\mathbf{j} + (t/6)\,\mathbf{k}, \quad 0 \le t \le 2\pi$

In Exercises 15–18, $\mathbf{F}$ is the velocity field of a fluid flowing through a region in space. Find the flow along the given curve in the direction of increasing t.

15. $\mathbf{F} = -4xy\mathbf{i} + 8y\mathbf{j} + 2\,\mathbf{k}$
$\mathbf{r} = t\mathbf{i} + t^2\mathbf{j} + \mathbf{k}, \quad 0 \le t \le 2$

16. $\mathbf{F} = x^2\mathbf{i} + yz\mathbf{j} + y^2\mathbf{k}$
$\mathbf{r} = 3t\mathbf{j} + 4t\mathbf{k}, \quad 0 \le t \le 1$

17. $\mathbf{F} = (x - z)\,\mathbf{i} + x\mathbf{k}$
$\mathbf{r} = (\cos t)\,\mathbf{i} + (\sin t)\,\mathbf{k}, \quad 0 \le t \le \pi$

18. $\mathbf{F} = -y\mathbf{i} + x\mathbf{j} + 2\mathbf{k}$
$\mathbf{r} = (-2 \cos t)\,\mathbf{i} + (2 \sin t)\,\mathbf{j} + 2t\mathbf{k}, \quad 0 \le t \le 2\pi$

19. Find the circulation and flux of the fields
$$\mathbf{F}_1 = x\mathbf{i} + y\mathbf{j} \quad \text{and} \quad \mathbf{F}_2 = -y\mathbf{i} + x\mathbf{j}$$

around and across each of the following curves.
a) The circle $\mathbf{r} = (\cos t)\,\mathbf{i} + (\sin t)\,\mathbf{j}, \quad 0 \le t \le 2\pi$
b) The ellipse $\mathbf{r} = (\cos t)\,\mathbf{i} + (4 \sin t)\,\mathbf{j}, \quad 0 \le t \le 2\pi$

20. Find the flux of the fields
$$\mathbf{F}_1 = 2x\mathbf{i} - 3y\mathbf{j} \quad \text{and} \quad \mathbf{F}_2 = 2x\mathbf{i} + (x - y)\,\mathbf{j}$$
across the circle
$$\mathbf{r} = (a \cos t)\,\mathbf{i} + (a \sin t)\,\mathbf{j}, \quad 0 \le t \le 2\pi.$$

In Exercises 21–24, find the circulation and flux of the field $\mathbf{F}$ around and across the closed semicircular path that consists of the semicircular arch $\mathbf{r}_1 = (a \cos t)\,\mathbf{i} + (a \sin t)\,\mathbf{j}, \; 0 \le t \le \pi$, followed by the line segment $\mathbf{r}_2 = t\mathbf{i}, \; -a \le t \le a$.

21. $\mathbf{F} = x\mathbf{i} + y\mathbf{j}$

22. $\mathbf{F} = x^2\mathbf{i} + y^2\mathbf{j}$

23. $\mathbf{F} = -y\mathbf{i} + x\mathbf{j}$

24. $\mathbf{F} = -y^2\mathbf{i} + x^2\mathbf{j}$

25. Evaluate the flow integral of the velocity field $\mathbf{F} = (x + y)\,\mathbf{i} - (x^2 + y^2)\,\mathbf{j}$ along each of the following paths from $(1, 0)$ to $(-1, 0)$ in the xy-plane.
a) The upper half of the circle $x^2 + y^2 = 1$
b) The line segment from $(1, 0)$ to $(-1, 0)$
c) The line segment from $(1, 0)$ to $(0, -1)$ followed by the line segment from $(0, -1)$ to $(-1, 0)$

26. The field $\mathbf{F} = xy\mathbf{i} + y\mathbf{j} - yz\mathbf{k}$ is the velocity field of a flow in space. Find the flow from $(0, 0, 0)$ to $(1, 1, 1)$ along the curve of intersection of the cylinder $y = x^2$ and the plane $z = x$.

27. Find the circulation of $\mathbf{F} = 2x\mathbf{i} + 2z\mathbf{j} + 2y\mathbf{k}$ around the closed path that consists of the helix $\mathbf{r}_1 = (\cos t)\,\mathbf{i} + (\sin t)\,\mathbf{j} + t\mathbf{k}, \; 0 \le t \le \pi/2$, and the line segments $\mathbf{r}_2 = \mathbf{j} + (\pi/2)(1 - t)\,\mathbf{k}, \; 0 \le t \le 1$, and $\mathbf{r}_3 = t\mathbf{i} + (1 - t)\,\mathbf{j}, \; 0 \le t \le 1$, traversed in the direction of increasing t.

28. Suppose that a function $f(t)$ is differentiable and positive for $a \le t \le b$, that $\mathbf{r} = t\mathbf{i} + f(t)\,\mathbf{j}, \; a \le t \le b$, and that $\mathbf{F} = y\mathbf{i}$. Show that
$$\int_{t=a}^{t=b} \mathbf{F} \cdot d\mathbf{r}$$
is the area of the region between the graph of f and the t-axis from $t = a$ to $t = b$.

EXPLORER PROGRAM

Vector Fields

Pictures fields of the form $\mathbf{F} = M(x, y)\,\mathbf{i} + N(x, y)\,\mathbf{j}$ and integrates $\mathbf{F} \cdot \mathbf{n}$ over city-block paths in the plane

14.3 Green's Theorem in the Plane

We now come to a theorem that describes the relationship between the way a fluid flows along or across the boundary of a plane region and the way the fluid moves around inside the region. We assume the fluid is incompressible—like water, for example, and not like a gas. The connection between the fluid's boundary behavior and its internal behavior is made possible by the notions of divergence and curl. The divergence of a fluid's velocity field is a measure of the rate at which fluid is being piped into or out of the region at any given point. The curl of the velocity field is a measure of the fluid's rate of rotation at each point.

Green's theorem states that, under conditions usually met in practice, the outward flux of a vector field across the boundary of a plane region equals the double integral of the divergence of the field over the interior of the region. In another form, Green's theorem states that the counterclockwise circulation of a vector field around the boundary of a region equals the double integral of the curl of the vector field over the region.

Green's theorem is one of the great theorems of calculus. It is deep and surprising and has far-reaching consequences. In pure mathematics, it ranks in importance with the Fundamental Theorem of Calculus. In applied mathematics the generalizations of Green's theorem to three dimensions provide the foundation for important theorems about electricity, magnetism, and fluid flow. We shall explore the three-dimensional forms of Green's theorem in Sections 14.5 and 14.6.

Throughout our discussion of Green's theorem, we talk in terms of velocity fields of fluid flows. We do so because fluid flows are easy to picture and the notions of flux and circulation are easy to interpret. We would like you to be aware, however, that Green's theorem applies to any vector field satisfying certain mathematical conditions. It does not depend for its validity on the field's having a particular physical interpretation.

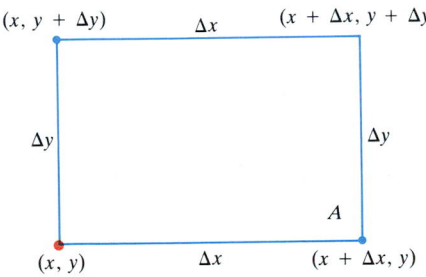

14.22 The rectangle for defining the flux density (divergence) of a vector field at a point (x, y).

Flux Density at a Point: Divergence

We need two new ideas for Green's theorem. The first is the idea of the flux density of a vector field at a point, which in mathematics is called the *divergence* of the vector field. We obtain it in the following way.

Suppose that

$$\mathbf{F}(x, y) = M(x, y)\,\mathbf{i} + N(x, y)\,\mathbf{j}$$

is the velocity field of a fluid flow in the plane and that the first partial derivatives of M and N are continuous at each point of a region R. Let (x, y) be a point in R and let A be a small rectangle with one corner at (x, y) that, along with its interior, lies entirely in R (Fig. 14.22). The sides of the rectangle, parallel to the coordinate axes, have lengths Δx and Δy. The rate at which fluid leaves the rectangle across the bottom edge is approximately

$$\mathbf{F}(x, y) \cdot (-\mathbf{j})\, \Delta x = -N(x, y)\, \Delta x. \tag{1}$$

This is the scalar component of the velocity at (x, y) in the direction of the outward normal times the length of the segment. If the velocity is in meters per second, for example, the exit rate will be in meters per second times meters or square meters per second. The rates at which the fluid crosses the other three sides in the directions of their outward normals can be estimated in a similar way. All told, we have

Top:	$\mathbf{F}(x, y + \Delta y) \cdot \mathbf{j}\, \Delta x = N(x, y + \Delta y)\, \Delta x,$	
Bottom:	$\mathbf{F}(x, y) \cdot (-\mathbf{j})\, \Delta x = -N(x, y)\, \Delta x,$	
Right:	$\mathbf{F}(x + \Delta x, y) \cdot \mathbf{i}\, \Delta y = M(x + \Delta x, y)\, \Delta y,$	(2)
Left:	$\mathbf{F}(x, y) \cdot (-\mathbf{i})\, \Delta y = -M(x, y)\, \Delta y.$	

Combining opposite pairs gives

Top and bottom: $(N(x, y + \Delta y) - N(x, y))\, \Delta x \approx \left(\dfrac{\partial N}{\partial y}\Delta y\right) \Delta x,$ (3)

Right and left: $(M(x + \Delta x, y) - M(x, y))\, \Delta y \approx \left(\dfrac{\partial M}{\partial x}\Delta x\right) \Delta y.$ (4)

Adding (3) and (4) gives

$$\text{Flux across rectangle boundary} \approx \left(\frac{\partial M}{\partial x} + \frac{\partial N}{\partial y}\right)\Delta x\, \Delta y. \tag{5}$$

We now divide by $\Delta x\, \Delta y$ to estimate the total flux per unit area or flux density for the rectangle:

$$\frac{\text{Flux across rectangle boundary}}{\text{Rectangle area}} \approx \left(\frac{\partial M}{\partial x} + \frac{\partial N}{\partial y}\right). \tag{6}$$

Finally, we let Δx and Δy approach zero to define what we call the *flux density* of **F** at the point (x, y).

In mathematics, we call the flux density the *divergence* of **F**. The symbol for it is div **F**, pronounced "divergence of **F**" or "div **F**."

DEFINITION

> The **flux density** or **divergence** of a vector field $\mathbf{F} = M\mathbf{i} + N\mathbf{j}$ at the point (x, y) is
>
> $$\text{div } \mathbf{F} = \frac{\partial M}{\partial x} + \frac{\partial N}{\partial y}. \tag{7}$$

If water were flowing into a region through a small hole at the point (x_0, y_0), the lines of flow would diverge there (hence the name) and div $\mathbf{F}(x_0, y_0)$ would be positive. If the water were draining out instead of flowing in, the divergence would be negative. See Fig. 14.23.

Example 1   Find the divergence of the vector field

$$\mathbf{F}(x, y) = (x^2 - y)\,\mathbf{i} + (xy - y^2)\,\mathbf{j}.$$

Solution   We use the formula in Eq. (7):

$$\text{div } \mathbf{F} = \frac{\partial M}{\partial x} + \frac{\partial N}{\partial y} = \frac{\partial}{\partial x}(x^2 - y) + \frac{\partial}{\partial y}(xy - y^2)$$

$$= 2x + x - 2y = 3x - 2y.$$

div $\mathbf{F}(x_0, y_0) > 0$

14.23   In the flow of an incompressible fluid across a plane region, the divergence is positive at a "source," a point where fluid enters the system, and negative at a "sink," a point where the fluid leaves the system.

Fluid arrives through a small hole (x_0, y_0).   div $\mathbf{F}(x_0, y_0) > 0$ Fluid leaves through a small hole (x_0, y_0).

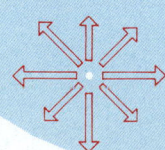

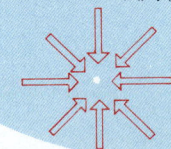

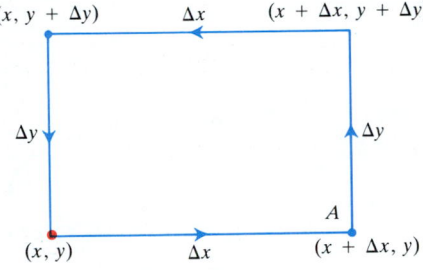

14.24 The rectangle for defining the circulation density (curl) of a vector field at a point (x, y).

Circulation Density at a Point: The Curl

The second of the two new ideas we need for Green's theorem is the idea of circulation density of a vector field $\mathbf{F}$ at a point, which in mathematics is called the *curl* of $\mathbf{F}$. To obtain it, we return to the velocity field

$$\mathbf{F}(x, y) = M(x, y)\,\mathbf{i} + N(x, y)\,\mathbf{j} \tag{8}$$

and the rectangle A. The rectangle is redrawn here as Fig. 14.24.

The counterclockwise circulation of $\mathbf{F}$ around the boundary of A is the sum of flow rates along the sides. For the bottom edge, the flow rate is approximately

$$\mathbf{F}(x, y) \cdot \mathbf{i}\,\Delta x = M(x, y)\,\Delta x. \tag{9}$$

This is the scalar component of the velocity $\mathbf{F}(x, y)$ in the direction of the tangent vector $\mathbf{i}$ times the length of the segment. The rates of flow along the other sides in the counterclockwise direction are expressed in a similar way. In all, we have

Top: $\qquad \mathbf{F}(x, y + \Delta y) \cdot (-\mathbf{i})\,\Delta x = -M(x, y + \Delta y)\,\Delta x,$

Bottom: $\qquad \mathbf{F}(x, y) \cdot \mathbf{i}\,\Delta x = M(x, y)\,\Delta x,$

Right: $\qquad \mathbf{F}(x + \Delta x, y) \cdot \mathbf{j}\,\Delta y = N(x + \Delta x, y)\,\Delta y,$

Left: $\qquad \mathbf{F}(x, y) \cdot (-\mathbf{j})\,\Delta y = -N(x, y)\,\Delta y.$
$$\tag{10}$$

We add opposite pairs to get

Top and bottom:

$$-(M(x, y + \Delta y) - M(x, y))\,\Delta x \approx -\left(\frac{\partial M}{\partial y}\,\Delta y\right)\Delta x, \tag{11}$$

Right and left:

$$(N(x + \Delta x, y) - N(x, y))\,\Delta y \approx \left(\frac{\partial N}{\partial x}\,\Delta x\right)\Delta y. \tag{12}$$

Adding (11) and (12) and dividing by $\Delta x\,\Delta y$ gives an estimate of the circulation density for the rectangle:

$$\frac{\text{Circulation around rectangle}}{\text{Rectangle area}} \approx \frac{\partial N}{\partial x} - \frac{\partial M}{\partial y}. \tag{13}$$

Finally, we let Δx and Δy approach zero to define what we call the *circulation density* of $\mathbf{F}$ at the point (x, y).

DEFINITION

> The **circulation density** or **curl** of a vector field $\mathbf{F} = M\mathbf{i} + N\mathbf{j}$ at the point (x, y) is
>
> $$\text{curl } \mathbf{F} = \frac{\partial N}{\partial x} - \frac{\partial M}{\partial y}. \tag{14}$$

If water is moving about a region in the xy-plane in a thin layer, then the circulation, or curl, at a point (x_0, y_0) gives a way to measure how fast and in what direction a small paddle wheel will spin if it is put into the water at (x_0, y_0) with its axis perpendicular to the plane (Fig. 14.25).

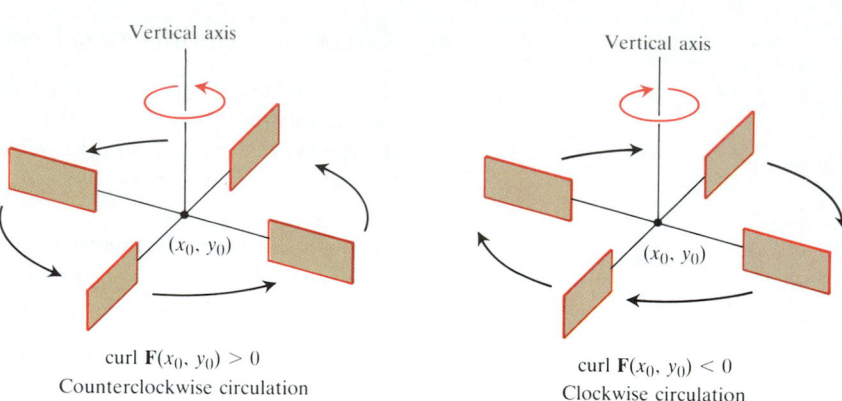

14.25 In the flow of an incompressible fluid over a plane region, the curl measures the rate of the fluid's rotation at a point. The curl is positive at points where the rotation is counterclockwise and negative where the rotation is clockwise.

curl $\mathbf{F}(x_0, y_0) > 0$
Counterclockwise circulation

curl $\mathbf{F}(x_0, y_0) < 0$
Clockwise circulation

Example 2   Find the curl of the vector field

$$\mathbf{F}(x, y) = (x^2 - y)\,\mathbf{i} + (xy - y^2)\,\mathbf{j}.$$

Solution   We use the formula in Eq. (14):

$$\text{curl } \mathbf{F} = \frac{\partial N}{\partial x} - \frac{\partial M}{\partial y} = \frac{\partial}{\partial x}(xy - y^2) - \frac{\partial}{\partial y}(x^2 - y) = y + 1.$$

Green's Theorem in the Plane

In one form, Green's theorem says that under suitable conditions the outward flux of a vector field across a simple closed curve in the plane (Fig. 14.26) equals the double integral of the divergence of the field over the region enclosed by the curve.

14.26 In proving Green's theorem, we distinguish between two kinds of closed curves, simple and not simple. Simple curves do not cross themselves. A circle is simple but a figure eight is not.

Simple   Simple

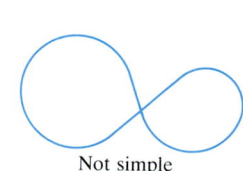

Not simple

THEOREM 2

Green's Theorem (Flux–Divergence Form)

The outward flux of a vector field $\mathbf{F} = M\mathbf{i} + N\mathbf{j}$ across a simple closed curve C equals the double integral of div $\mathbf{F}$ over the region R enclosed by C:

$$\underbrace{\oint_C M\,dy - N\,dx}_{\text{outward flux}} = \underbrace{\int\int_R \left(\frac{\partial M}{\partial x} + \frac{\partial N}{\partial y}\right) dx\,dy}_{\text{divergence integral}}.$$ (15)

In another form, Green's theorem says that the counterclockwise circulation of a vector field around a simple closed curve is the double integral of the curl of the field over the region enclosed by the curve.

THEOREM 3

Green's Theorem (Circulation–Curl Form)

The counterclockwise circulation of a vector field $\mathbf{F} = M\mathbf{i} + N\mathbf{j}$ around a simple closed curve C in the plane equals the double integral of curl $\mathbf{F}$ over the region enclosed by C:

$$\oint_C M\,dx + N\,dy = \iint_R \left(\frac{\partial N}{\partial x} - \frac{\partial M}{\partial y}\right) dx\,dy \qquad (16)$$

counterclockwise circulation   curl integral

The two forms of Green's theorem are equivalent. Applying Eq. (15) to the field $\mathbf{G}_1 = N\mathbf{i} - M\mathbf{j}$ gives Eq. (16), and applying Eq. (16) to $\mathbf{G}_2 = -N\mathbf{i} + M\mathbf{j}$ gives Eq. (15). We do not need to prove them both. We shall prove the curl form shortly.

We need two kinds of assumptions for Green's theorem to hold. First, we need conditions on M and N to ensure the existence of the integrals. The usual assumptions are that M, N, and their first partial derivatives are continuous at every point of some open region containing C and R. Second, we need geometric conditions on the curve C. It must be simple, closed, and made up of pieces along which we can integrate M and N. The usual assumptions are that C is **piecewise smooth** (consists of a finite number of smooth curves connected end to end). The proof we give for Green's theorem, however, assumes things about the shape of R as well. You can find less restrictive proofs in more advanced texts.

Example 3 Verify both forms of Green's theorem for the field

$$\mathbf{F}(x, y) = (x - y)\mathbf{i} + x\mathbf{j}$$

and the region R bounded by the circle

$$C: \quad \mathbf{r} = (\cos t)\mathbf{i} + (\sin t)\mathbf{j}, \quad 0 \le t \le 2\pi.$$

Solution We first express all functions, derivatives, and differentials in terms of t:

$$M = \cos t - \sin t, \qquad dx = d(\cos t) = -\sin t\,dt,$$

$$N = \cos t, \qquad dy = d(\sin t) = \cos t\,dt, \qquad (17)$$

$$\frac{\partial M}{\partial x} = 1, \qquad \frac{\partial M}{\partial y} = -1, \qquad \frac{\partial N}{\partial x} = 1, \qquad \frac{\partial N}{\partial y} = 0.$$

The two sides of Eq. (15):

$$\oint_C M\,dy - N\,dx = \int_{t=0}^{t=2\pi} (\cos t - \sin t)(\cos t\,dt) - (\cos t)(-\sin t\,dt)$$

$$= \int_0^{2\pi} \cos^2 t\,dt = \pi,$$

$$\iint_R \left(\frac{\partial M}{\partial x} + \frac{\partial N}{\partial y}\right) dx\,dy = \iint_R (1 + 0)\,dx\,dy = \iint_R dx\,dy = \pi.$$

The two sides of Eq. (16):

$$\oint_C M\,dx + N\,dy = \int_{t=0}^{t=2\pi} (\cos t - \sin t)(-\sin t\,dt) + (\cos t)(\cos t\,dt)$$

$$= \int_0^{2\pi} (-\sin t \cos t + 1)\,dt = 2\pi,$$

$$\iint_R \left(\frac{\partial N}{\partial x} - \frac{\partial M}{\partial y}\right) dx\,dy = \iint_R (1 - (-1))\,dx\,dy = 2\iint_R dx\,dy = 2\pi.$$

Using Green's Theorem to Evaluate Line Integrals

If we construct a closed curve C by piecing a number of different curves end to end, the process of evaluating a line integral over C can be lengthy because there are so many different integrals to evaluate. However, if C bounds a region R to which Green's theorem applies, we can use Green's theorem to change the line integral around C into one double integral over R.

Example 4   Evaluate the integral

$$\oint xy\,dy - y^2\,dx$$

around the square C cut from the first quadrant by the lines $x = 1$ and $y = 1$.

Solution   We can use either form of Green's theorem to change the line integral into a double integral over the square.

1. *With Eq.* (15):   Taking $M = xy$, $N = y^2$, and C and R as the square's boundary and interior gives

$$\oint_C xy\,dy - y^2\,dx = \iint_R (y + 2y)\,dx\,dy = \int_0^1 \int_0^1 3y\,dx\,dy$$

$$= \int_0^1 \Big[3xy\Big]_{x=0}^{x=1} dy$$

$$= \int_0^1 3y\,dy = \frac{3}{2}y^2\Big|_0^1 = \frac{3}{2}.$$

2. *With Eq.* (16):   Taking $M = -y^2$ and $N = xy$ gives the same result:

$$\oint_C -y^2\,dx + xy\,dy = \iint_R (y - (-2y))\,dx\,dy = \frac{3}{2}.$$

Example 5   Calculate the outward flux of the field $\mathbf{F}(x, y) = x\mathbf{i} + y^2\mathbf{j}$ across the square bounded by the lines $x = \pm 1$ and $y = \pm 1$.

Solution   Calculating the flux with a line integral would take four integrations, one for each side of the square. With Green's theorem, we can change the line integral to one double integral. With $M = x$, $N = y^2$, C the square, and R the square's interior, we have

$$\text{Flux} = \oint_C M \, dy - N \, dx = \int\int_R \left(\frac{\partial M}{\partial x} + \frac{\partial N}{\partial y}\right) dx \, dy \qquad \text{(Green's theorem)}$$

$$= \int_{-1}^{1}\int_{-1}^{1} (1 + 2y) \, dx \, dy = \int_{-1}^{1} \Big[x + 2xy\Big]_{x=-1}^{x=1} dy$$

$$= \int_{-1}^{1} (2 + 4y) \, dy = \Big[2y + 2y^2\Big]_{-1}^{1} = 4.$$

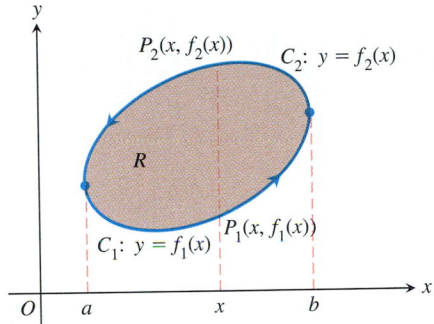

14.27 The boundary curve C is made up of C_1, the graph of $y = f_1(x)$, and C_2, the graph of $y = f_2(x)$.

A Proof of Green's Theorem (Special Regions)

Let C be a smooth simple closed curve in the xy-plane with the property that lines parallel to the axes cut it in no more than two points. Let R be the region enclosed by C and suppose that M, N, and their first partial derivatives are continuous at every point of some open region containing C and R. We want to show that

$$\oint_C M \, dx + N \, dy = \int\int_R \left(\frac{\partial N}{\partial x} - \frac{\partial M}{\partial y}\right) dx \, dy. \qquad (18)$$

Figure 14.27 shows C made up of two parts:

$$C_1: \quad y = f_1(x), \quad a \le x \le b, \qquad C_2: \quad y = f_2(x), \quad a \le x \le b.$$

For any x between a and b, we can integrate $\partial M / \partial y$ with respect to y from $y = f_1(x)$ to $y = f_2(x)$ and obtain

$$\int_{f_1(x)}^{f_2(x)} \frac{\partial M}{\partial y} \, dy = M(x, y)\Big]_{y=f_1(x)}^{y=f_2(x)} = M(x, f_2(x)) - M(x, f_1(x)). \qquad (19)$$

We can then integrate this with respect to x from a to b:

$$\int_a^b \int_{f_1(x)}^{f_2(x)} \frac{\partial M}{\partial y} \, dy \, dx = \int_a^b [M(x, f_2(x)) - M(x, f_1(x))] \, dx$$

$$= -\int_b^a M(x, f_2(x)) \, dx - \int_a^b M(x, f_1(x)) \, dx$$

$$= -\int_{C_2} M \, dx - \int_{C_1} M \, dx$$

$$= -\oint_C M \, dx.$$

Therefore

$$\oint_C M \, dx = \int\int_R \left(-\frac{\partial M}{\partial y}\right) dx \, dy. \qquad (20)$$

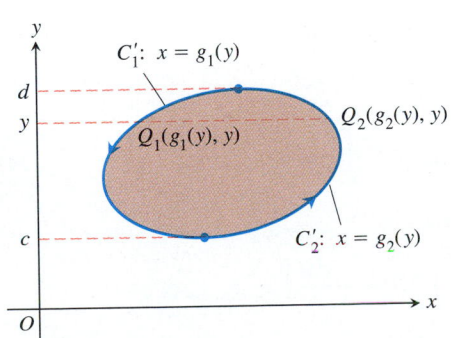

14.28 The boundary curve C is made up of C_1', the graph of $x = g_1(y)$, and C_2', the graph of $x = g_2(y)$.

Equation (20) is half the result we need for Eq. (18). Exercise 29 asks you to derive the other half by integrating $\partial N / \partial x$ first with respect to x and then with respect to y, as suggested by Fig. 14.28. Figure 14.28 shows the curve C of Fig. 14.27 decomposed into the two directed parts

$$C_1': \quad x = g_1(y), \quad d \ge y \ge c, \qquad C_2': \quad x = g_2(y), \quad c \le y \le d.$$

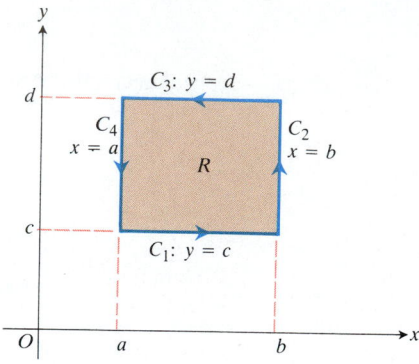

14.29 To prove Green's theorem for the rectangular region shown here, we divide the boundary into four directed line segments.

The result of this double integration is

$$\oint_C N \, dy = \iint_R \frac{\partial N}{\partial x} \, dx \, dy. \tag{21}$$

Combining Eqs. (20) and (21) gives Eq. (18). This concludes the proof.

Extending the Proof to Other Regions

The argument with which we just derived Green's equation does not apply directly to the rectangular region in Fig. 14.29 because the lines $x = a$, $x = b$, $y = c$, and $y = d$ meet the region's boundary in more than two points. However, if we divide the boundary C into four directed line segments,

$$C_1: \quad y = c, \quad a \le x \le b, \qquad C_2: \quad x = b, \quad c \le y \le d,$$

$$C_3: \quad y = d, \quad b \ge x \ge a, \qquad C_4: \quad x = a, \quad d \ge y \ge c,$$

we can modify the argument in the following way.

Proceeding as in the proof of Eq. (21), we have

$$\int_c^d \int_a^b \frac{\partial N}{\partial x} \, dx \, dy = \int_c^d [N(b, y) - N(a, y)] \, dy$$

$$= \int_c^d N(b, y) \, dy + \int_d^c N(a, y) \, dy \tag{22}$$

$$= \int_{C_2} N \, dy + \int_{C_4} N \, dy.$$

Because y is constant along C_1 and C_3,

$$\int_{C_1} N \, dy = \int_{C_3} N \, dy = 0,$$

so we can add

$$\int_{C_1} N \, dy + \int_{C_3} N \, dy$$

to the right-hand side of Eq. (22) without changing the equality. Doing so, we have

$$\int_c^d \int_a^b \frac{\partial N}{\partial x} \, dx \, dy = \oint_C N \, dy. \tag{23}$$

Similarly, we can show (Exercise 30) that

$$\int_a^b \int_c^d \frac{\partial M}{\partial y} \, dy \, dx = -\oint_C M \, dx. \tag{24}$$

Subtracting (24) from (23), we again arrive at

$$\oint_C M \, dx + N \, dy = \iint_R \left(\frac{\partial N}{\partial x} - \frac{\partial M}{\partial y} \right) dx \, dy.$$

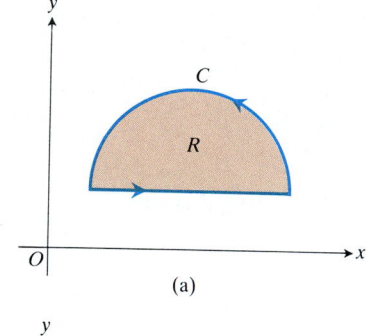

(a)

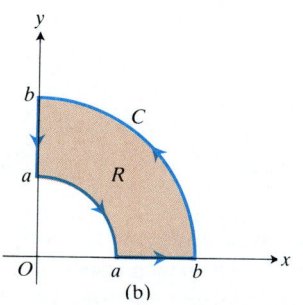

(b)

14.30 Other regions to which Green's theorem applies.

Regions like those in Fig. 14.30 can be handled with no greater difficulty. Equation (18) still applies. It also applies to the horseshoe-shaped region R shown in Fig. 14.31, as we see by putting together the regions R_1 and R_2 and their boundaries.

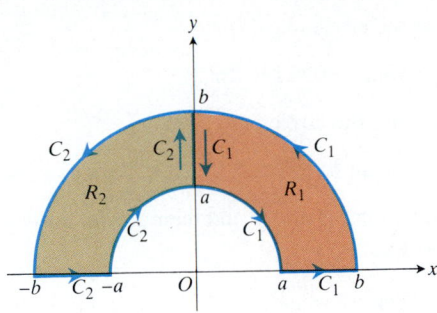

14.31   A region R that combines regions R_1 and R_2.

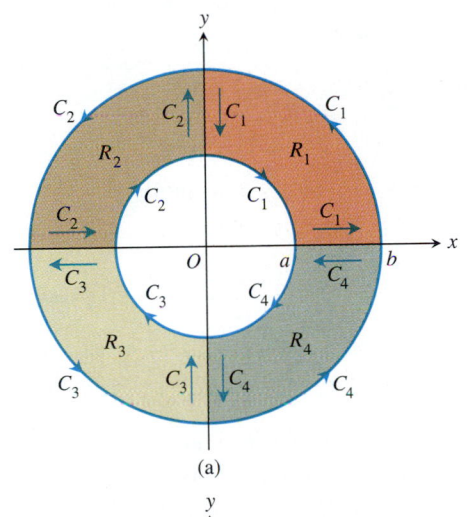

(a)

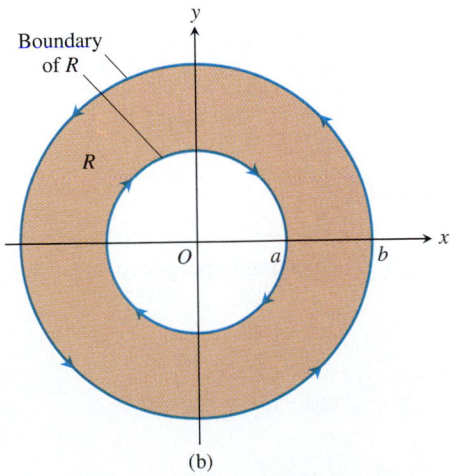

(b)

14.32   The annular region R shown here combines four smaller regions. In polar coordinates, we have $r = a$ for the inner circle, $r = b$ for the outer circle, and $a \leq r \leq b$ for the region itself.

Green's theorem applies to C_1, R_1, and to C_2, R_2, yielding

$$\int_{C_1} M\, dx + N\, dy = \int\int_{R_1} \left(\frac{\partial N}{\partial x} - \frac{\partial M}{\partial y} \right) dx\, dy,$$

$$\int_{C_2} M\, dx + N\, dy = \int\int_{R_2} \left(\frac{\partial N}{\partial x} - \frac{\partial M}{\partial y} \right) dx\, dy.$$

When we add these two equations, the line integral along the y-axis from b to a for C_1 cancels the integral over the same segment but in the opposite direction for C_2. Hence

$$\oint_C M\, dx + N\, dy = \int\int_R \left(\frac{\partial N}{\partial x} - \frac{\partial M}{\partial y} \right) dx\, dy,$$

where C consists of the two segments of the x-axis from $-b$ to $-a$ and from a to b and of the two semicircles, and where R is the region inside C.

The device of adding line integrals over separate boundaries to build up an integral over a single boundary can be extended to any finite number of subregions. In Fig. 14.32(a), let C_1 be the boundary of the region R_1 in the first quadrant. Similarly for the other three quadrants: C_i is the boundary of the region R_i, $i = 1, 2, 3, 4$. By Green's theorem,

$$\oint_{C_i} M\, dx + N\, dy = \int\int_{R_i} \left(\frac{\partial N}{\partial x} - \frac{\partial M}{\partial y} \right) dx\, dy. \tag{25}$$

We add Eqs. (25) for $i = 1, 2, 3, 4$, and get

$$\oint_{r=b} (M\, dx + N\, dy) + \oint_{r=a} (M\, dx + N\, dy) = \int\int_{a \leq r \leq b} \left(\frac{\partial N}{\partial x} - \frac{\partial M}{\partial y} \right) dx\, dy. \tag{26}$$

Equation (26) says that the double integral of $(\partial N/\partial x) - (\partial M/\partial y)$ over the annular ring R equals the line integral of $M\, dx + N\, dy$ over the complete boundary of R in the direction that keeps R on our left as we progress (Fig. 14.32b).

Example 6   Verify the circulation form of Green's theorem (Eq. 16) on the annular ring $R: h^2 \leq x^2 + y^2 \leq 1, 0 < h < 1$ (Fig. 14.33), if

$$M = \frac{-y}{x^2 + y^2}, \qquad N = \frac{x}{x^2 + y^2}.$$

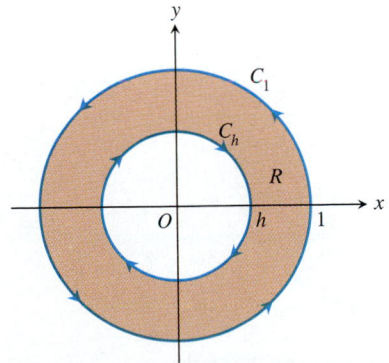

14.33   Green's theorem may be applied to the annular region R by integrating along the boundaries as shown (Example 6).

Solution   The boundary of R consists of the circle

$$C_1: \quad x = \cos t, \quad y = \sin t, \quad 0 \le t \le 2\pi,$$

traversed counterclockwise as t increases, and the circle

$$C_h: \quad x = h \cos \theta, \quad y = -h \sin \theta, \quad 0 \le \theta \le 2\pi,$$

traversed clockwise as θ increases. The functions M and N and their partial derivatives are continuous throughout R. Moreover,

$$\frac{\partial M}{\partial y} = \frac{(x^2 + y^2)(-1) + y(2y)}{(x^2 + y^2)^2} = \frac{y^2 - x^2}{(x^2 + y^2)^2} = \frac{\partial N}{\partial x},$$

so

$$\iint_R \left(\frac{\partial N}{\partial x} - \frac{\partial M}{\partial y} \right) dx \, dy = \iint_R 0 \, dx \, dy = 0.$$

The integral of $M \, dx + N \, dy$ over the boundary of R is

$$\int_C M \, dx + N \, dy = \oint_{C_1} \frac{x \, dy - y \, dx}{x^2 + y^2} + \oint_{C_h} \frac{x \, dy - y \, dx}{x^2 + y^2}$$

$$= \int_0^{2\pi} (\cos^2 t + \sin^2 t) \, dt - \int_0^{2\pi} \frac{h^2(\cos^2 \theta + \sin^2 \theta)}{h^2} \, d\theta$$

$$= 2\pi - 2\pi = 0.$$

The functions M and N in Example 6 are discontinuous at $(0, 0)$, so we cannot apply Green's theorem to the circle C_1 and the region inside it. We must exclude the origin. We do so by excluding the points inside C_h.

We could replace the circle C_1 in Example 6 by an ellipse or any other simple closed curve K surrounding C_h (Fig. 14.34). The result would still be

$$\oint_K (M \, dx + N \, dy) + \oint_{C_h} (M \, dx + N \, dy) = \iint_R \left(\frac{\partial N}{\partial x} - \frac{\partial M}{\partial y} \right) dy \, dx = 0,$$

which leads to the surprising conclusion that

$$\oint_K (M \, dx + N \, dy) = 2\pi$$

for any such curve K. We can explain this result by changing to polar coordinates. With

$$x = r \cos \theta, \qquad\qquad y = r \sin \theta,$$

$$dx = -r \sin \theta \, d\theta + \cos \theta \, dr, \qquad dy = r \cos \theta \, d\theta + \sin \theta \, dr,$$

we have

$$\frac{x \, dy - y \, dx}{x^2 + y^2} = \frac{r^2(\cos^2 \theta + \sin^2 \theta) \, d\theta}{r^2} = d\theta,$$

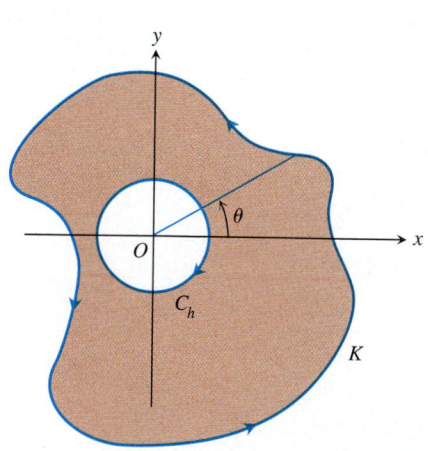

14.34   The region bounded by the circle C_h and the curve K.

and θ increases by 2π as we traverse K once counterclockwise.

The Normal and Tangential Forms of Green's Theorem

As we learned in Section 14.2, the flux of a two-dimensional vector field **F** across a closed curve C in the direction of its outer unit normal vector **n** is

$$\int_C \mathbf{F} \cdot \mathbf{n} \, ds. \quad \text{(Flux)}$$

Similarly, the circulation around C in the direction of its unit tangent vector **T** is

$$\int_C \mathbf{F} \cdot \mathbf{T} \, ds. \quad \text{(Circulation)}$$

Hence, the flux form of Green's theorem (Eq. (15)) can be written as

$$\oint_C \mathbf{F} \cdot \mathbf{n} \, ds = \int\int_R \left(\frac{\partial M}{\partial x} + \frac{\partial N}{\partial y} \right) dx \, dy \quad \text{(Normal form)} \tag{27}$$

and the circulation form of Green's theorem (Eq. (16)) can be written as

$$\oint_C \mathbf{F} \cdot \mathbf{T} \, ds = \int\int_R \left(\frac{\partial N}{\partial x} - \frac{\partial M}{\partial y} \right) dx \, dy. \quad \text{(Tangential form)} \tag{28}$$

The left-hand sides of these equations are now in vector form. How about the right-hand sides? Can they be expressed in vector form too? The answer is yes, as we shall see in Sections 14.5 and 14.6.

EXERCISES 14.3

In Exercises 1–4, verify Green's theorem by evaluating both sides of Eqs. (15) and (16) for the field $\mathbf{F} = M\mathbf{i} + N\mathbf{j}$. Take the domains of integration in each case to be the disk $R: x^2 + y^2 \le a^2$ and its bounding circle $C: \mathbf{r} = (a \cos t)\mathbf{i} + (a \sin t)\mathbf{j}, \, 0 \le t \le 2\pi$.

1. $\mathbf{F} = -y\mathbf{i} + x\mathbf{j}$

2. $\mathbf{F} = y\mathbf{i}$

3. $\mathbf{F} = 2x\mathbf{i} - 3y\mathbf{j}$

4. $\mathbf{F} = -x^2 y\mathbf{i} + xy^2\mathbf{j}$

In Exercises 5–10, use Green's theorem to find the counterclockwise circulation and outward flux for the field **F** and curve C.

5. $\mathbf{F} = (x - y)\mathbf{i} + (y - x)\mathbf{j}$
   C: The square bounded by $x = 0, x = 1, y = 0, y = 1$

6. $\mathbf{F} = (x^2 + 4y)\mathbf{i} + (x + y^2)\mathbf{j}$
   C: The square bounded by $x = 0, x = 1, y = 0, y = 1$

7. $\mathbf{F} = (y^2 - x^2)\mathbf{i} + (x^2 + y^2)\mathbf{j}$
   C: The triangle bounded by $y = 0, x = 3$, and $y = x$

8. $\mathbf{F} = (x + y)\mathbf{i} - (x^2 + y^2)\mathbf{j}$
   C: The triangle bounded by $y = 0, x = 1$, and $y = x$

9. $\mathbf{F} = (x + e^x \sin y)\mathbf{i} + (x + e^x \cos y)\mathbf{j}$
   C: The right-hand loop of the lemniscate $r^2 = \cos 2\theta$

10. $\mathbf{F} = \left(\tan^{-1} \dfrac{y}{x} \right)\mathbf{i} + \ln(x^2 + y^2)\mathbf{j}$

    C: The boundary of the region defined by the polar-coordinate inequalities $1 \le r \le 2, 0 \le \theta \le \pi$.

11. Find the counterclockwise circulation and outward flux of the field $\mathbf{F} = xy\mathbf{i} + y^2\mathbf{j}$ around and over the boundary of the region enclosed by the parabola $y = x^2$ and the line $y = x$ in the first quadrant.

12. Find the counterclockwise circulation and the outward flux of the field $\mathbf{F} = (-\sin y)\mathbf{i} + (x \cos y)\mathbf{j}$ around and over the square cut from the first quadrant by the lines $x = \pi/2$ and $y = \pi/2$.

13. Find the outward flux of the field

    $$\mathbf{F} = \left(3xy - \frac{x}{1 + y^2} \right)\mathbf{i} + (e^x + \tan^{-1} y)\mathbf{j}$$

    across the cardioid $r = a(1 + \cos \theta), a > 0$.

14. Find the counterclockwise circulation of the field $\mathbf{F} = (y + e^x \ln y)\mathbf{i} + (e^x/y)\mathbf{j}$ around the boundary of the region that is bounded above by the curve $y = 2 - x^2$ and below by the curve $y = x^4 + 1$.

Apply Green's theorem in one of its two forms to evaluate the line integrals in Exercises 15–20.

15. $\oint_C (y^2\, dx + x^2\, dy)$

C: The triangle bounded by $x = 0$, $x + y = 1$, $y = 0$

16. $\oint_C (3y\, dx + 2x\, dy)$

C: The boundary of $0 \le x \le \pi$, $0 \le y \le \sin x$

17. $\oint_C (6y + x)\, dx + (y + 2x)\, dy$

C: The circle $(x - 2)^2 + (y - 3)^2 = 4$

18. $\oint_C (2x + y^2)\, dx + (2xy + 3y)\, dy$

C: Any simple closed curve in the plane for which Green's theorem holds

19. $\oint_C 2xy^3\, dx + 4x^2 y^2\, dy$

C: The boundary of the "triangular" region in the first quadrant enclosed by the x-axis, the line $x = 1$, and the curve $y = x^3$

20. $\oint_C (4x - 2y)\, dx + (2x - 4y)\, dy$

C: The circle $(x - 2)^2 + (y - 2)^2 = 4$

21. Let C be the boundary of a region on which Green's theorem holds. Use Green's theorem to calculate

a) $\oint_C f(x)\, dx + g(y)\, dy$,

b) $\oint_C ky\, dx + hx\, dy$ (k and h constants).

22. Show that

$$\oint_C 4x^3 y\, dx + x^4\, dy = 0$$

for any closed curve to which Green's theorem applies.

23. Show that

$$\oint_C -y^3\, dx + x^3\, dy$$

is positive for any closed curve C to which Green's theorem applies.

24. Show that the value of

$$\oint_C xy^2\, dx + (x^2 y + 2x)\, dy$$

around any square depends only on the area of the square and not on its location in the plane.

25. *Green's theorem and Laplace's equation.* Assuming that all the necessary derivatives exist and are continuous, show that if $f(x, y)$ satisfies the Laplace equation

$$\frac{\partial^2 f}{\partial x^2} + \frac{\partial^2 f}{\partial y^2} = 0,$$

then

$$\oint_C \frac{\partial f}{\partial y}\, dx - \frac{\partial f}{\partial x}\, dy = 0$$

for all closed curves C to which Green's theorem applies. (The converse is also true: If the line integral is always zero, then f satisfies the Laplace equation.)

26. Among all smooth simple closed curves in the plane, oriented counterclockwise, find the one along which the work done by

$$\mathbf{F} = \left(\frac{1}{4} x^2 y + \frac{1}{3} y^3\right)\mathbf{i} + x\mathbf{j}$$

is greatest. (*Hint:* Where is curl $\mathbf{F}$ positive?)

27. Green's theorem holds for a region R with any finite number of holes as long as the bounding curves are smooth, simple, and closed and we integrate over each component of the boundary in the direction that keeps R on our immediate left as we go along (Fig. 14.35).

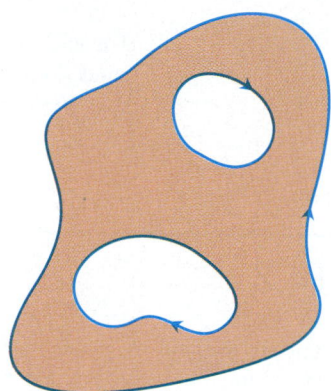

14.35   Green's theorem holds for regions with more than one hole (Exercise 27).

a) Let $f(x, y) = \ln(x^2 + y^2)$ and let C be the circle $x^2 + y^2 = a^2$. Evaluate the flux integral

$$\oint_C \nabla f \cdot \mathbf{n}\, ds.$$

b) Let K be an arbitrary smooth simple closed curve in the plane that does not pass through $(0, 0)$. Use Green's theorem to show that

$$\oint_K \nabla f \cdot \mathbf{n}\, ds$$

has two possible values, depending on whether $(0, 0)$ lies inside K or outside K.

28. *Bendixson's criterion.* The **streamlines** of a planar fluid flow are the smooth curves traced by the fluid's individual particles. The vectors $\mathbf{F} = M(x, y)\mathbf{i} + N(x, y)\mathbf{j}$ of the flow's velocity field are the tangent vectors of the streamlines. Show that if the flow takes place over a *simply connected* region R (no holes or missing points) and that if

$M_x + N_y \neq 0$ throughout R, then none of the streamlines in R is closed. In other words, no particle of fluid ever has a closed trajectory in R. The criterion $M_x + N_y \neq 0$ is called **Bendixson's criterion** for the nonexistence of closed trajectories.

29. Establish Eq. (21) to finish the proof of the special case of Green's theorem.

30. Establish Eq. (24) to complete the argument for the extension of Green's theorem.

Calculating Area with Green's Theorem

If a simple closed curve C in the plane and the region R it encloses satisfy the hypotheses of Green's theorem, then the area of R is given by the following formula.

> **Green's Theorem Area Formula**
>
> $$\text{Area of } R = \frac{1}{2} \oint_C x \, dy - y \, dx. \qquad (29)$$

The reason is that by Eq. (15), run backward,

$$\text{Area of } R = \iint_R dy \, dx = \iint_R \left(\frac{1}{2} + \frac{1}{2} \right) dy \, dx$$

$$= \oint_C \frac{1}{2} x \, dy - \frac{1}{2} y \, dx.$$

31. Show that if R is a region in the plane bounded by a piecewise smooth simple closed curve C, then

$$\text{Area of } R = \oint_C x \, dy = - \oint_C y \, dx.$$

32. Suppose that a nonnegative function $y = f(x)$ has a continu-

ous first derivative on $[a, b]$. Let C be the boundary of the region in the xy-plane that is bounded below by the x-axis, above by the graph of f, and on the sides by the lines $x = a$ and $x = b$. Show that

$$\int_a^b f(x) \, dx = - \oint_C y \, dx.$$

33. Let A be the area and $\bar{x}$ the x-coordinate of the centroid of a region R that is bounded by a piecewise smooth simple closed curve C in the xy-plane. Show that

$$\frac{1}{2} \oint_C x^2 \, dy = - \oint_C xy \, dx = \frac{1}{3} \oint_C x^2 \, dy - xy \, dx = A\bar{x}.$$

34. Let I_y be the moment of inertia about the y-axis of the region in Exercise 33. Show that

$$\frac{1}{3} \oint_C x^3 \, dy = - \oint_C x^2 y \, dx = \frac{1}{4} \oint_C x^3 \, dy - x^2 y \, dx = I_y.$$

Use the Green's theorem area formula (Eq. 29) to find the areas of the regions enclosed by the curves in Exercises 35–38.

35. The circle $\mathbf{r} = (a \cos t) \mathbf{i} + (a \sin t) \mathbf{j}, \quad 0 \leq t \leq 2\pi$

36. The ellipse $\mathbf{r} = (a \cos t) \mathbf{i} + (b \sin t) \mathbf{j}, \quad 0 \leq t \leq 2\pi$

37. The astroid (Fig. 9.46) $\mathbf{r} = (\cos^3 t) \mathbf{i} + (\sin^3 t) \mathbf{j}, \quad 0 \leq t \leq 2\pi$

38. The curve (Fig. 9.91) $\mathbf{r} = t^2 \mathbf{i} + ((t^3/3) - t) \mathbf{j}, \quad -\sqrt{3} \leq t \leq \sqrt{3}$

EXPLORER PROGRAM

| Vector Fields | Displays two-dimensional vector fields and integrates the curl over rectangular regions. Also calculates flux across rectangular boundaries |

14.4 Surface Area and Surface Integrals

We know how to integrate a function over a flat region in a plane, but what if the function is defined over a curved surface instead? How do we calculate its integral then? The trick to evaluating one of these so-called surface integrals is to rewrite it as a double integral over a region in a coordinate plane beneath the surface (Fig. 14.36). This changes the surface integral into the kind of integral we already know how to evaluate.

Our first step is to find a double integral formula for calculating the area of a curved surface. We will then see how the ideas that arise can be used again to define and evaluate surface integrals. With surface integrals under control, we shall be able to calculate the flux of a three-dimensional vector field through a surface and the masses and moments of thin shells of material. In Sections 14.5 and 14.6 we shall see how surface integrals provide just what we need to generalize the two forms of Green's theorem to three dimensions. One of these generalizations will

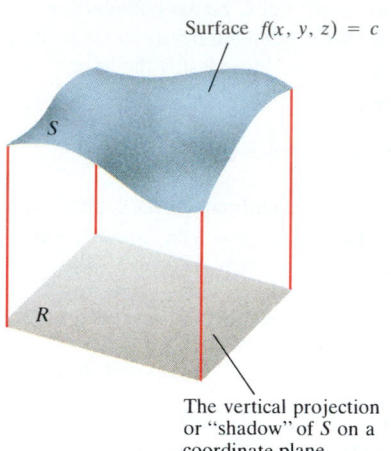

Surface $f(x, y, z) = c$

The vertical projection or "shadow" of S on a coordinate plane

14.36 As we shall soon see, the integral of a function $g(x, y, z)$ over a surface S in space can be calculated by evaluating a related double integral over the vertical projection or "shadow" of S on a coordinate plane.

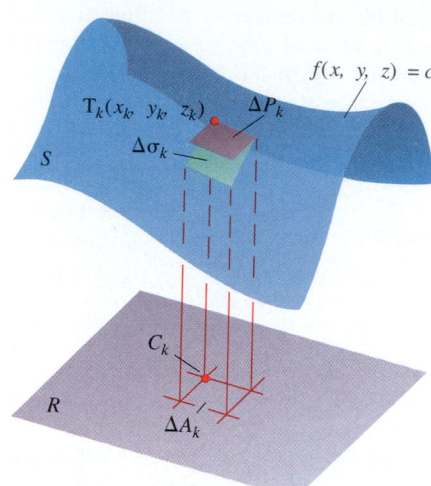

14.37 A surface S and its vertical projection onto a plane beneath it. You can think of R as the shadow of S on the plane. The tangent plate ΔP_k approximates the surface patch $\Delta\sigma_k$ above ΔA_k.

enable us to express the flux of a vector field through a closed surface as a triple integral over the three-dimensional region enclosed by the surface. The other generalization will enable us to express the circulation of a vector field around a closed curve in space as an integral over a surface bounded by the curve. These results have far-reaching consequences in mathematics as well as in the theories of electromagnetism and fluid flow.

The Definition of Surface Area

Figure 14.37 shows a surface S lying above its "shadow" region R in a plane beneath it. The surface is defined by the equation $f(x, y, z) = c$. If the surface is **smooth** (∇f is continuous and never vanishes on S), we can define and calculate its area as a double integral over R. It takes a while to derive the integral, but the integral itself is easy enough to work with.

The first step in defining the area of S is to partition the region R into small rectangles ΔA_k of the kind we would use if we were defining an integral over R. Directly above each ΔA_k lies a patch of surface $\Delta\sigma_k$ that we may approximate with a portion ΔP_k of the tangent plane. To be specific, we suppose that ΔP_k is a portion of the plane that is tangent to the surface at the point $T_k(x_k, y_k, z_k)$ directly above the back corner C_k of ΔA_k. If the tangent plane is parallel to R, then ΔP_k will be congruent to ΔA_k. Otherwise, it will be a parallelogram whose area is somewhat larger than the area of ΔA_k.

Figure 14.38 gives a magnified view of $\Delta\sigma_k$ and ΔP_k, showing the gradient vector $\nabla f(x_k, y_k, z_k)$ at T_k and a unit vector $\mathbf{p}$ that is normal to R. The figure also shows the angle γ_k between ∇f and $\mathbf{p}$. The other vectors in the picture, $\mathbf{u}$ and $\mathbf{v}$, lie along the edges of the patch ΔP_k in the tangent plane. Thus, both $\mathbf{u} \times \mathbf{v}$ and ∇f are normal to the tangent plane.

We now need the fact from Section 10.6 that $|(\mathbf{u} \times \mathbf{v}) \cdot \mathbf{p}|$ is the area of the projection of the parallelogram determined by $\mathbf{u}$ and $\mathbf{v}$ onto any plane whose normal is $\mathbf{p}$. In our case, this translates into the statement

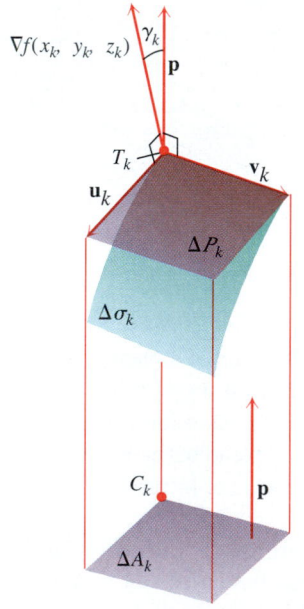

14.38 Magnified view from the preceding figure. The vector $\mathbf{u} \times \mathbf{v}$ (not shown) is parallel to the vector ∇f because both vectors are normal to the plane of ΔP_k.

$$|(\mathbf{u} \times \mathbf{v}) \cdot \mathbf{p}| = \Delta A_k. \tag{1}$$

Now, $|\mathbf{u} \times \mathbf{v}|$ itself is the area ΔP_k (standard fact about cross products) so Eq. (1)

becomes

$$\underbrace{|\mathbf{u} \times \mathbf{v}|}_{\Delta P_k} \underbrace{|\mathbf{p}|}_{1} \underbrace{|\cos(\text{angle between } \mathbf{u} \times \mathbf{v} \text{ and } \mathbf{p})|}_{\substack{\text{same as } |\cos\gamma_k| \text{ because} \\ \boldsymbol{\nabla}f \text{ and } \mathbf{u} \times \mathbf{v} \text{ are both} \\ \text{normal to the tangent plane}}} = \Delta A_k \qquad (2)$$

or

$$\Delta P_k |\cos \gamma_k| = \Delta A_k$$

or

$$\Delta P_k = \frac{\Delta A_k}{|\cos\gamma_k|},$$

provided $\cos \gamma_k \neq 0$. We will have $\cos \gamma_k \neq 0$ as long as $\boldsymbol{\nabla}f$ is not parallel to the ground plane and $\boldsymbol{\nabla}f \cdot \mathbf{p} \neq 0$.

Since the patches ΔP_k approximate the surface patches $\Delta\sigma_k$ that fit together to make S, the sum

$$\sum \Delta P_k = \sum \frac{\Delta A_k}{|\cos \gamma_k|} \qquad (3)$$

looks like an approximation of what we might like to call the surface area of S. It also looks as if the approximation would improve if we refined the partition of R. In fact, the sums on the right-hand side of Eq. (3) are approximating sums for the double integral

$$\iint_R \frac{1}{|\cos \gamma|} \, dA . \qquad (4)$$

We therefore define the **area** of S to be the value of this integral whenever it exists.

A Practical Formula

For any particular surface $f(x, y, z) = c$,

$$|\boldsymbol{\nabla}f \cdot \mathbf{p}| = |\boldsymbol{\nabla}f| \, |\mathbf{p}| \, |\cos \gamma|,$$

so

$$\frac{1}{|\cos \gamma|} = \frac{|\boldsymbol{\nabla}f|}{|\boldsymbol{\nabla}f \cdot \mathbf{p}|}.$$

This combines with Eq. (4) to give a practical formula for area.

The Formula for Surface Area

The area of the surface $f(x, y, z) = c$ over a closed and bounded plane region R is

$$\text{Surface area} = \iint_R \frac{|\boldsymbol{\nabla}f|}{|\boldsymbol{\nabla}f \cdot \mathbf{p}|} \, dA , \qquad (5)$$

where $\mathbf{p}$ is a unit vector normal to R and $\boldsymbol{\nabla}f \cdot \mathbf{p} \neq 0$.

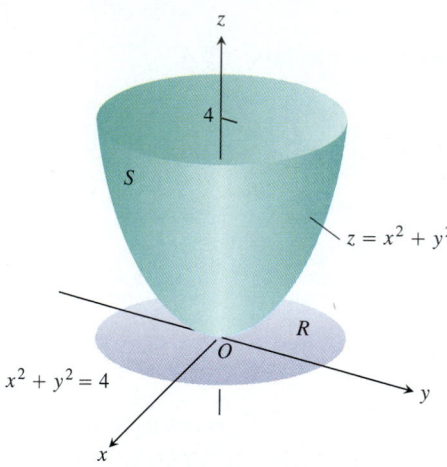

14.39   The area of this parabolic surface is calculated in Example 1.

Thus, the area is the double integral over R of the magnitude of ∇f divided by the magnitude of the scalar component of ∇f normal to R.

We reached Eq. (5) under the assumption that $\nabla f \cdot \mathbf{p} \neq 0$ throughout R and that ∇f is continuous. Whenever the integral exists, however, we may define its value to be the area of the portion of the surface $f(x, y, z) = c$ that lies over R.

Equation (5) agrees with the formulas for surface area in Section 5.5, although we shall not prove this fact.

Example 1   Find the area of the surface cut from the bottom of the paraboloid $x^2 + y^2 - z = 0$ by the plane $z = 4$.

Solution   We sketch the surface S and the region R below it in the xy-plane (Fig. 14.39). The surface S is part of the level surface $f(x, y, z) = x^2 + y^2 - z = 0$, and R is the disk $x^2 + y^2 \leq 4$ in the xy-plane. To get a unit vector normal to the plane of R, we can take $\mathbf{p} = \mathbf{k}$.

At any point (x, y, z) on the surface, we have

$$f(x, y, z) = x^2 + y^2 - z,$$

$$\nabla f = 2x\mathbf{i} + 2y\mathbf{j} - \mathbf{k},$$

$$|\nabla f| = \sqrt{(2x)^2 + (2y)^2 + (-1)^2}$$

$$= \sqrt{4x^2 + 4y^2 + 1},$$

$$|\nabla f \cdot \mathbf{p}| = |\nabla f \cdot \mathbf{k}| = |-1| = 1.$$

In the region R, $dA = dx\, dy$. Therefore,

$$\text{Surface area} = \iint_R \frac{|\nabla f|}{|\nabla f \cdot \mathbf{p}|} \, dA \qquad \text{(Eq. (5))}$$

$$= \iint_{x^2 + y^2 \leq 4} \sqrt{4x^2 + 4y^2 + 1} \, dx\, dy$$

$$= \int_0^{2\pi} \int_0^2 \sqrt{4r^2 + 1} \; r \, dr \, d\theta \qquad \text{(Polar coordinates)}$$

$$= \int_0^{2\pi} \left[\frac{1}{12} (4r^2 + 1)^{3/2} \right]_0^2 d\theta$$

$$= \int_0^{2\pi} \frac{1}{12} (17^{3/2} - 1) \, d\theta$$

$$= \frac{\pi}{6} (17\sqrt{17} - 1).$$

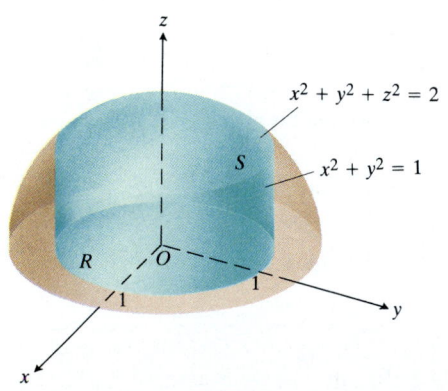

14.40   The cap cut from the hemisphere by the cylinder projects vertically onto the disk $R: x^2 + y^2 \leq 1$ (Example 2).

Example 2   Find the area of the cap cut from the hemisphere $x^2 + y^2 + z^2 = 2$, $z \geq 0$, by the cylinder $x^2 + y^2 = 1$ (Fig. 14.40).

Solution   The cap S is part of the level surface $f(x, y, z) = x^2 + y^2 + z^2 = 2$. It projects one-to-one onto the disk $R: x^2 + y^2 \leq 1$ in the xy-plane. The vector $\mathbf{p} = \mathbf{k}$ is normal to the plane of R.

At any point on the surface,

$$f(x, y, z) = x^2 + y^2 + z^2,$$

$$\nabla f = 2x\mathbf{i} + 2y\mathbf{j} + 2z\mathbf{k}$$

$$|\nabla f| = 2\sqrt{x^2 + y^2 + z^2} = 2\sqrt{2}, \quad \left(\begin{array}{l}\text{Because } x^2 + y^2 + z^2 = 2 \\ \text{at points of } S\end{array}\right)$$

$$|\nabla f \cdot \mathbf{p}| = |\nabla f \cdot \mathbf{k}| = |2z| = 2z.$$

Therefore,

$$\text{Surface area} = \iint_R \frac{|\nabla f|}{|\nabla f \cdot \mathbf{p}|} \, dA = \iint_R \frac{2\sqrt{2}}{2z} \, dA = \sqrt{2} \iint_R \frac{dA}{z}. \tag{6}$$

What do we do about the z?

Since z is the z-coordinate of a point on the sphere, we can express it in terms of x and y as

$$z = \sqrt{2 - x^2 - y^2}.$$

We continue the work of Eq. (6) with this substitution:

$$\begin{aligned}
\text{Surface area} &= \sqrt{2} \iint_R \frac{dA}{z} = \sqrt{2} \iint_{x^2 + y^2 \le 1} \frac{dA}{\sqrt{2 - x^2 - y^2}} \\
&= \sqrt{2} \int_0^{2\pi} \int_0^1 \frac{r \, dr \, d\theta}{\sqrt{2 - r^2}} \qquad \text{(Polar coordinates)} \\
&= \sqrt{2} \int_0^{2\pi} \left[-(2 - r^2)^{1/2}\right]_{r=0}^{r=1} d\theta = \sqrt{2} \int_0^{2\pi} \left(\sqrt{2} - 1\right) d\theta \\
&= 2\pi\left(2 - \sqrt{2}\right).
\end{aligned}$$

Surface Integrals

We now show how to integrate a function over a surface, using the ideas we just developed for calculating surface area.

Suppose, for example, that we have an electrical charge distributed over a surface $f(x, y, z) = c$ like the one shown in Fig. 14.41 and that the function $g(x, y, z)$ gives the charge per unit area (charge density) at each point on S. Then we may calculate the total charge on S as an integral in the following way.

We partition the shadow region R on the ground plane beneath the surface into small rectangles of the kind we would use if we were defining the surface area of S. Then directly above each ΔA_k lies a patch of surface $\Delta\sigma_k$ that we approximate with a parallelogram-shaped portion of tangent plane, ΔP_k.

Up to this point the construction proceeds as in the definition of surface area, but now we take one additional step: We evaluate g at (x_k, y_k, z_k) and approximate the total charge on the surface patch $\Delta\sigma_k$ by the product

$$g(x_k, y_k, z_k)\Delta P_k.$$

The rationale is that when the partition of R is sufficiently fine, the value of g throughout $\Delta\sigma_k$ is nearly constant and ΔP_k is nearly the same as $\Delta\sigma_k$. The total charge over S is then approximated by the sum

$$\text{Total charge} \approx \sum g(x_k, y_k, z_k) \, \Delta P_k = \sum g(x_k, y_k, z_k) \frac{\Delta A_k}{|\cos \gamma_k|}. \tag{7}$$

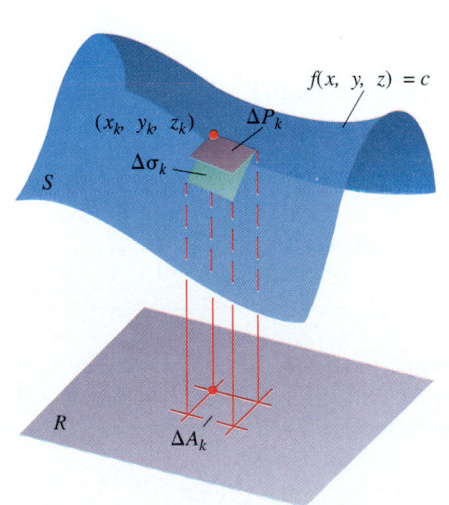

14.41 If we know how an electrical charge is distributed over a surface, we can find the total charge with a suitably modified surface integral.

If f, the function defining the surface S, and its first partial derivatives are continuous, and if g is continuous over S, then the sums on the right-hand side of Eq. (7) approach the limit

$$\iint_R g(x, y, z) \frac{dA}{|\cos \gamma|} = \iint_R g(x, y, z) \frac{|\nabla f|}{|\nabla f \cdot \mathbf{p}|} dA \tag{8}$$

as the rectangular subdivision of R is refined in the usual way. This limit is called the integral of g over the surface S and is calculated as a double integral over R. The value of the integral is the total charge on the surface S.

As you might expect, the formula in Eq. (8) defines the integral of *any* function g over the surface S as long as the integral exists.

DEFINITIONS

If R is the shadow region of a surface S defined by the equation $f(x, y, z) = c$, and g is a continuous function defined at the points of S, then the **integral of g over S** is the integral

$$\iint_R g(x, y, z) \frac{|\nabla f|}{|\nabla f \cdot \mathbf{p}|} dA, \tag{9}$$

where $\mathbf{p}$ is a unit vector normal to R and $\nabla f \cdot \mathbf{p} \neq 0$. The integral itself is called a **surface integral.**

The surface integral in (9) takes on different meanings in different applications. If g has the constant value 1, the integral gives the area of S. If g gives the mass density of a thin shell of material modeled by S, the integral gives the mass of the shell.

Algebraic Properties: The Surface Area Differential

We often abbreviate the integral in (9) by writing $d\sigma$ for $(|\nabla f|/|\nabla f \cdot \mathbf{p}|) \, dA$.

The Surface Area Differential and the Differential Form for Surface Integrals

$$d\sigma = \frac{|\nabla f|}{|\nabla f \cdot \mathbf{p}|} dA \qquad\qquad \iint_S g \, d\sigma \tag{10}$$

surface area   differential formula
differential   for surface integrals

Surface integrals have all the usual properties of double integrals, the surface integral of the sum of two functions being the sum of their surface integrals and so on. The domain additivity property takes the form

$$\iint_S g \, d\sigma = \iint_{S_1} g \, d\sigma + \iint_{S_2} g \, d\sigma + \cdots + \iint_{S_n} g \, d\sigma.$$

The idea is that if S is partitioned by smooth curves into a finite number of nonoverlapping smooth patches (i.e., if S is **piecewise smooth**), then the integral of a function g over S is the sum of the integrals over the patches. Thus, the integral of a

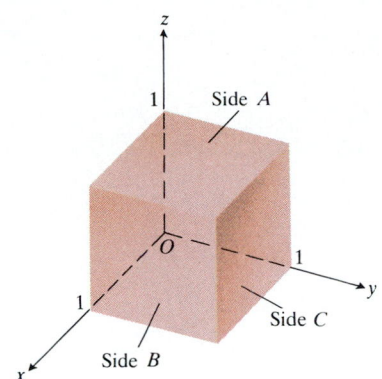

14.42 To integrate a function over the surface of a cube, we integrate over each face and add the results (Example 3).

function over the surface of a cube is the sum of the integrals over the faces of the cube. We integrate over a turtle shell of welded plates by integrating one plate at a time and adding the results.

Example 3   Integrate $g(x, y, z) = xyz$ over the surface of the cube cut from the first octant by the planes $x = 1$, $y = 1$, and $z = 1$ (Fig. 14.42).

Solution   We integrate xyz over each of the six sides and add the results. Since $xyz = 0$ on the sides that lie in the coordinate planes, the integral over the surface of the cube reduces to

$$\iint\limits_{\substack{\text{cube}\\\text{surface}}} xyz \, d\sigma = \iint\limits_{\text{side } A} xyz \, d\sigma + \iint\limits_{\text{side } B} xyz \, d\sigma + \iint\limits_{\text{side } C} xyz \, d\sigma.$$

Side A is the surface $f(x, y, z) = z = 1$ over the square region $R_{xy}: 0 \le x \le 1$, $0 \le y \le 1$, in the xy-plane. For this surface and region,

$$\mathbf{p} = \mathbf{k}, \qquad \nabla f = \mathbf{k}, \qquad |\nabla f| = 1, \qquad |\nabla f \cdot \mathbf{p}| = |\mathbf{k} \cdot \mathbf{k}| = 1,$$

$$d\sigma = \frac{|\nabla f|}{|\nabla f \cdot \mathbf{p}|} \, dA = \frac{1}{1} \, dx \, dy = dx \, dy,$$

$$xyz = xy(1) = xy,$$

and

$$\iint\limits_{\text{side } A} xyz \, d\sigma = \iint\limits_{R_{xy}} xy \, dx \, dy = \int_0^1 \int_0^1 xy \, dx \, dy = \int_0^1 \frac{y}{2} \, dy = \frac{1}{4}.$$

Symmetry tells us that the integrals of xyz over sides B and C are also 1/4. Hence,

$$\iint\limits_{\substack{\text{cube}\\\text{surface}}} xyz \, d\sigma = \frac{1}{4} + \frac{1}{4} + \frac{1}{4} = \frac{3}{4}.$$

Orientation

We call a smooth surface S **orientable** or **two-sided** if it is possible to define a field $\mathbf{n}$ of unit normal vectors on S that varies continuously with position. Any patch or subportion of an orientable surface is orientable. Spheres and other smooth closed surfaces in space (smooth surfaces that enclose solids) are orientable. By convention, we choose $\mathbf{n}$ on a closed surface to point outward.

Once $\mathbf{n}$ has been chosen, we say that we have **oriented** the surface, and we call the surface together with its normal field an **oriented surface.** The vector $\mathbf{n}$ at any point is called the **positive direction** at that point (Fig. 14.43).

The Möbius band in Fig. 14.44 is not orientable. No matter where we start to construct a continuous unit normal field (shown as the shaft of a thumbtack in the figure), moving the vector continuously around the surface in the manner shown will return it to the starting point with a direction opposite to the one it had when it started out. The vector at that point cannot point both ways and yet it must if the field is to be continuous. We conclude that no such field exists.

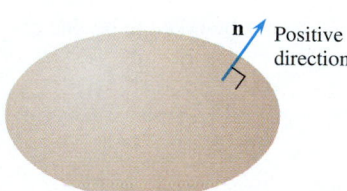

14.43 Smooth closed surfaces in space are orientable. The outward unit normal vector defines the positive direction at each point.

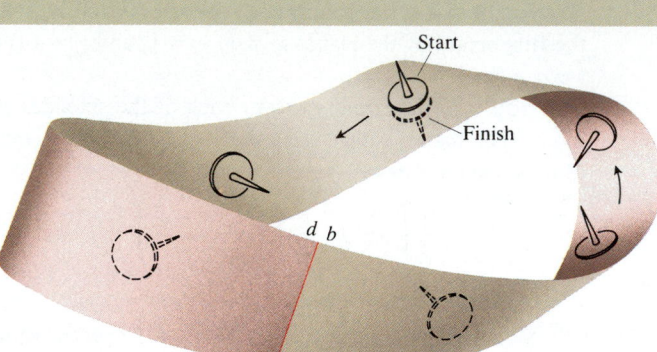

14.44 The Möbius band can be constructed by taking a rectangular strip of paper $abcd$, giving the end bc a single twist to interchange the positions of the vertices b and c, and then pasting the ends of the strip together to match a with c and b with d. The Möbius band is a nonorientable or one-sided surface.

The Surface Integral for Flux

Suppose that $\mathbf{F}$ is a continuous vector field defined over an oriented surface S and that $\mathbf{n}$ is the chosen unit normal field on the surface. We call the integral of $\mathbf{F} \cdot \mathbf{n}$ over S the flux across S in the positive direction. Thus, the flux is the integral over S of the scalar component of $\mathbf{F}$ in the direction of $\mathbf{n}$.

DEFINITION

> The **flux** of a three-dimensional vector field $\mathbf{F}$ across an oriented surface S in the direction of $\mathbf{n}$ is given by the formula
>
> $$\text{Flux} = \iint_S \mathbf{F} \cdot \mathbf{n} \, d\sigma. \tag{11}$$

This definition is analogous to the flux of a two-dimensional field $\mathbf{F}$ across a plane curve C. In the plane (Section 14.2), the flux is

$$\int_C \mathbf{F} \cdot \mathbf{n} \, ds,$$

the integral of the scalar component of $\mathbf{F}$ normal to the curve.

If $\mathbf{F}$ is the velocity field of a three-dimensional fluid flow, the flux of $\mathbf{F}$ across S is the net rate at which fluid is crossing S in the chosen positive direction. We shall discuss such flows in more detail in Section 14.5.

If S is part of a level surface $g(x, y, z) = c$, then $\mathbf{n}$ may be taken to be one of the two fields

$$\mathbf{n} = \pm \frac{\boldsymbol{\nabla} g}{|\boldsymbol{\nabla} g|}, \tag{12}$$

depending on which field gives the preferred direction.

Example 4   Find the flux of $\mathbf{F} = yz\mathbf{j} + z^2\mathbf{k}$ outward through the surface S cut from the semicircular cylinder $y^2 + z^2 = 1$, $z \geq 0$, by the planes $x = 0$ and $x = 1$.

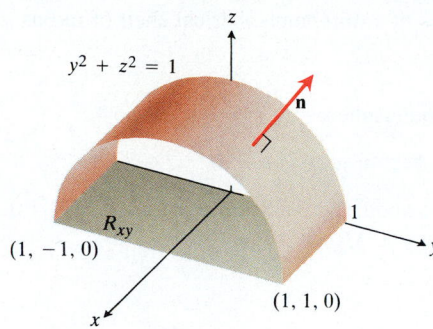

$y^2 + z^2 = 1$

n

R_{xy}

$(1, -1, 0)$

$(1, 1, 0)$

14.45 Example 4 calculates the flux of a vector field outward through this surface. The area of the shadow region R_{xy} is 2.

Solution The outward normal field on S (Fig. 14.45) may be calculated from the gradient of $g(x, y, z) = y^2 + z^2$ to be

$$\mathbf{n} = +\frac{\nabla g}{|\nabla g|} = \frac{2y\mathbf{j} + 2z\mathbf{k}}{\sqrt{4y^2 + 4z^2}} = \frac{2y\mathbf{j} + 2z\mathbf{k}}{2\sqrt{1}} = y\mathbf{j} + z\mathbf{k}.$$

With $\mathbf{p} = \mathbf{k}$, we also have

$$d\sigma = \frac{|\nabla g|}{|\nabla g \cdot \mathbf{k}|} dA = \frac{2}{|2z|} dA = \frac{1}{z} dA.$$

We can drop the absolute value bars because $z \geq 0$ on S.

The value of $\mathbf{F} \cdot \mathbf{n}$ on the surface is given by the formula

$$\mathbf{F} \cdot \mathbf{n} = (yz\mathbf{j} + z^2\mathbf{k}) \cdot (y\mathbf{j} + z\mathbf{k})$$

$$= y^2z + z^3 = z(y^2 + z^2)$$

$$= z. \qquad\qquad (y^2 + z^2 = 1 \text{ on } S)$$

Therefore, the flux of $\mathbf{F}$ outward through S is

$$\iint_S \mathbf{F} \cdot \mathbf{n}\, d\sigma = \iint_S (z)\left(\frac{1}{z} dA\right) = \iint_{R_{xy}} dA = \text{area}\, (R_{xy}) = 2.$$

Moments and Masses of Thin Shells

In engineering and physics, thin shells of material like bowls, metal drums, and domes are modeled with surfaces. Their moments and masses are calculated with the surface integral formulas in Table 14.2.

TABLE 14.2
Mass and moment formulas for very thin shells

Mass: $\quad M = \iint_S \delta(x, y, z)\, d\sigma \qquad (\delta(x, y, z) = \text{density at } (x, y, z))$

First moments about the coordinate planes:

$$M_{yz} = \iint_S x\, \delta\, d\sigma, \qquad M_{xz} = \iint_S y\, \delta\, d\sigma, \qquad M_{xy} = \iint_S z\, \delta\, d\sigma$$

Coordinates of center of mass:

$$\bar{x} = M_{yz}/M, \qquad \bar{y} = M_{xz}/M, \qquad \bar{z} = M_{xy}/M$$

Moments of inertia:

$$I_x = \iint_S (y^2 + z^2)\, \delta\, d\sigma, \qquad I_y = \iint_S (x^2 + z^2)\, \delta\, d\sigma,$$

$$I_z = \iint_S (x^2 + y^2)\, \delta\, d\sigma, \qquad I_L = \iint_S r^2\, \delta\, d\sigma$$

$$r(x, y, z) = \text{distance from point } (x, y, z) \text{ to line } L$$

Radius of gyration about a line L: $\quad R_L = \sqrt{I_L/M}$

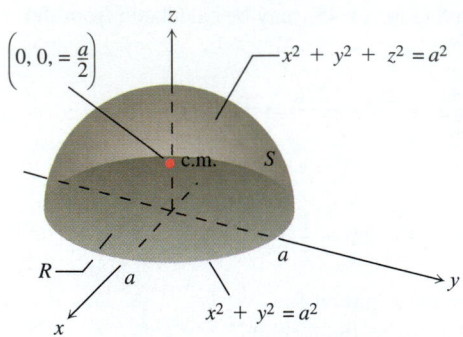

14.46   The center of mass of a thin hemispherical shell of constant density lies on the axis of symmetry halfway from the base to the top (Example 5).

Example 5   Find the center of mass of a thin hemispherical shell of radius a and constant density δ.

Solution   We model the shell with the hemisphere

$$f(x, y, z) = x^2 + y^2 + z^2 = a^2, \qquad z \geq 0$$

(Fig. 14.46). The symmetry of the surface about the z-axis tells us that $\bar{x} = \bar{y} = 0$. It remains only to find $\bar{z}$ from the formula $\bar{z} = M_{xy}/M$.

The mass of the shell is

$$M = \iint_S \delta \, d\sigma = \delta \iint_S d\sigma = (\delta)(\text{area of } S) = 2\pi a^2 \delta.$$

To evaluate the integral for M_{xy}, we take $\mathbf{p} = \mathbf{k}$ and calculate

$$|\nabla f| = |2x\mathbf{i} + 2y\mathbf{j} + 2z\mathbf{k}| = 2\sqrt{x^2 + y^2 + z^2} = 2a,$$

$$|\nabla f \cdot \mathbf{p}| = |\nabla f \cdot \mathbf{k}| = |2z| = 2z,$$

$$d\sigma = \frac{|\nabla f|}{|\nabla f \cdot \mathbf{p}|} \, dA = \frac{a}{z} \, dA.$$

Then

$$M_{xy} = \iint_S z \, \delta \, d\sigma = \delta \iint_R z \frac{a}{z} \, dA = \delta a \iint_R dA = \delta a (\pi a^2) = \delta \pi a^3,$$

$$\bar{z} = \frac{M_{xy}}{M} = \frac{\pi a^3 \delta}{2\pi a^2 \delta} = \frac{a}{2}.$$

The shell's center of mass is the point $(0, 0, a/2)$.

EXERCISES 14.4

Surface Area

1. Find the area of the surface cut from the paraboloid $x^2 + y^2 - z = 0$ by the plane $z = 2$.

2. Find the area of the band cut from the paraboloid $x^2 + y^2 - z = 0$ by the planes $z = 2$ and $z = 6$.

3. Find the area of the region cut from the plane $x + 2y + 2z = 5$ by the cylinder whose walls are $x = y^2$ and $x = 2 - y^2$.

4. Find the area of the portion of the surface $x^2 - 2z = 0$ that lies above the triangle bounded by the lines $x = \sqrt{3}$, $y = 0$, and $y = x$ in the xy-plane.

5. Find the area of the surface $x^2 - 2y - 2z = 0$ that lies above the triangle bounded by the lines $x = 2$, $y = 0$, and $y = 3x$ in the xy-plane.

6. Find the area of the surface $x^2 - 2 \ln x + \sqrt{15}y - z = 0$ above the square $R: 0 \leq y \leq 1$, $1 \leq x \leq 2$, in the xy-plane.

7. Find the area of the cap cut from the sphere $x^2 + y^2 + z^2 = 2$ by the cone $z = \sqrt{x^2 + y^2}$.

8. Find the area of the ellipse cut from the plane $z = cx$ by the cylinder $x^2 + y^2 = 1$.

9. Find the area of the portion of the cylinder $x^2 + z^2 = 1$ that lies between the planes $x = \pm 1/2$ and $y = \pm 1/2$.

10. Find the area of the portion of the surface $x = 9 - y^2 - z^2$ that lies above the ring $1 \leq y^2 + z^2 \leq 9$ in the yz-plane.

11. Find the area of the surface cut from the paraboloid $x^2 + y + z^2 = 1$ by the plane $y = 0$. (*Hint:* Project the surface on the xz-plane.)

12. Whenever we replace an old definition with a new one, it is a good idea to try out the new one on familiar objects to see that it gives the answers we expect. For instance, is the surface area of a circular cylinder of base radius a and height h still $2\pi ah$? Find out by using Eq. (5) with $\mathbf{p} = \mathbf{k}$ to calcu-

late the area of the surface cut from the cylinder $y^2 + z^2 = a^2$ by the planes $x = 0$ and $x = h > 0$. (Double the area above the xy-plane.)

Surface Integrals

13. Integrate $g(x, y, z) = x + y + z$ over the surface of the cube cut from the first octant by the planes $x = a$, $y = a$, $z = a$.

14. Integrate $g(x, y, z) = y + z$ over the surface of the wedge in Fig. 14.47.

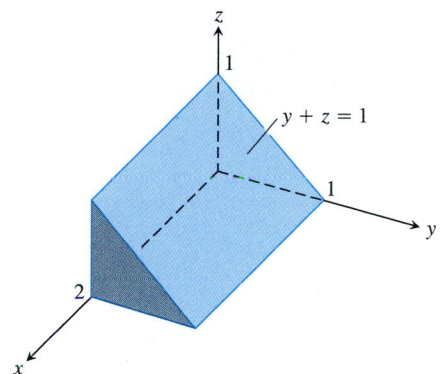

14.47   The wedge in Exercise 14.

15. Integrate $g(x, y, z) = xyz$ over the surface of the rectangular solid cut from the first octant by the planes $x = a$, $y = b$, and $z = c$.

16. Integrate $g(x, y, z) = xyz$ over the surface of the rectangular solid bounded by the planes $x = \pm a$, $y = \pm b$, and $z = \pm c$.

17. Integrate $g(x, y, z) = x + y + z$ over the portion of the plane $2x + 2y + z = 2$ that lies in the first octant.

18. Integrate $g(x, y, z) = x\sqrt{y^2 + 1}$ over the surface cut from the paraboloid $y^2 + 4z = 16$ by the planes $x = 0$, $x = 1$, and $z = 0$.

Flux Across a Surface

In Exercises 19–24, find the flux of the field $\mathbf{F}$ across the portion of the sphere $x^2 + y^2 + z^2 = a^2$ in the first octant in the direction away from the origin.

19. $\mathbf{F}(x, y, z) = z\mathbf{k}$

20. $\mathbf{F}(x, y, z) = -y\mathbf{i} + x\mathbf{j}$

21. $\mathbf{F}(x, y, z) = y\mathbf{i} - x\mathbf{j} + \mathbf{k}$

22. $\mathbf{F}(x, y, z) = zx\mathbf{i} + zy\mathbf{j} + z^2\mathbf{k}$

23. $\mathbf{F}(x, y, z) = x\mathbf{i} + y\mathbf{j} + z\mathbf{k}$

24. $\mathbf{F}(x, y, z) = \dfrac{x\mathbf{i} + y\mathbf{j} + z\mathbf{k}}{\sqrt{x^2 + y^2 + z^2}}$

25. Find the flux of the field $\mathbf{F}(x, y, z) = z^2\mathbf{i} + x\mathbf{j} - 3z\mathbf{k}$ upward through the surface cut from the parabolic cylinder $z = 4 - y^2$ by the planes $x = 0$, $x = 1$, and $z = 0$.

26. Find the flux of the field $\mathbf{F}(x, y, z) = 4x\mathbf{i} + 4y\mathbf{j} + 2\mathbf{k}$ outward through the surface cut from the bottom of the paraboloid $z = x^2 + y^2$ by the plane $z = 1$.

27. Let S be the portion of the cylinder $y = e^x$ in the first octant that projects parallel to the x-axis onto the rectangle R_{yz}: $1 \le y \le 2$, $0 \le z \le 1$ in the yz-plane (Fig. 14.48). Let $\mathbf{n}$ be the unit vector normal to S that points away from the yz-plane. Find the flux of the field $\mathbf{F}(x, y, z) = -2\mathbf{i} + 2y\mathbf{j} + z\mathbf{k}$ across S in the direction of $\mathbf{n}$.

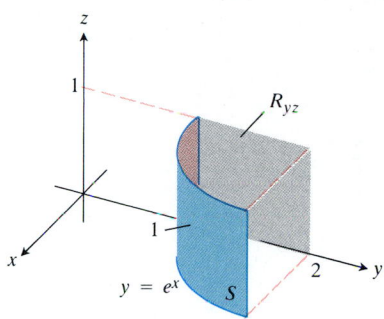

14.48   The surface and region in Exercise 27.

28. Let S be the portion of the cylinder $y = \ln x$ in the first octant whose projection parallel to the y-axis onto the xz-plane is the rectangle R_{xz}: $1 \le x \le e$, $0 \le z \le 1$. Let $\mathbf{n}$ be the unit vector normal to S that points away from the xz-plane. Find the flux of $\mathbf{F} = 2y\mathbf{j} + z\mathbf{k}$ through S in the direction of $\mathbf{n}$.

29. Find the outward flux of the field $\mathbf{F} = 2xy\mathbf{i} + 2yz\mathbf{j} + 2xz\mathbf{k}$ across the surface of the cube cut from the first octant by the planes $x = a$, $y = a$, $z = a$.

30. Find the outward flux of the field $\mathbf{F} = xz\mathbf{i} + yz\mathbf{j} + \mathbf{k}$ across the surface of the upper cap cut from the solid sphere $x^2 + y^2 + z^2 \le 25$ by the plane $z = 3$.

Moments and Masses

31. Find the centroid of the portion of the sphere $x^2 + y^2 + z^2 = a^2$ that lies in the first octant.

32. Find the centroid of the surface cut from the cylinder $y^2 + z^2 = 9$, $z \ge 0$, by the planes $x = 0$ and $x = 3$ (resembles the surface in Example 4).

33. Find the center of mass and the moment of inertia and radius of gyration about the z-axis of a thin shell of constant density δ cut from the cone $x^2 + y^2 - z^2 = 0$ by the planes $z = 1$ and $z = 2$.

34. Find the moment of inertia about the z-axis of a thin shell of constant density δ cut from the cone $4x^2 + 4y^2 - z^2 = 0$, $z \geq 0$, by the circular cylinder $x^2 + y^2 = 2x$ (Fig. 14.49).

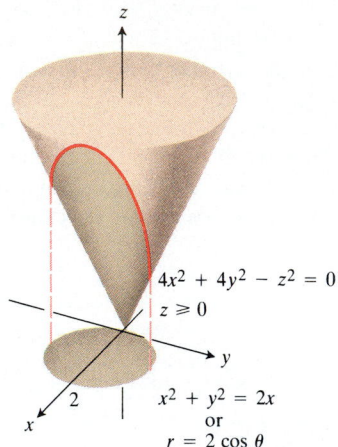

$4x^2 + 4y^2 - z^2 = 0$
$z \geq 0$

$x^2 + y^2 = 2x$
or
$r = 2 \cos \theta$

14.49   The surface in Exercise 34.

35. a) Find the moment of inertia about a diameter of a thin spherical shell of radius a and constant density δ. (Work with a hemispherical shell and double the result.)
b) Use the Parallel Axis Theorem (Exercises 13.5) and the result in (a) to find the moment of inertia about a line tangent to the shell.

36. a) Find the centroid of the lateral surface of a solid cone of base radius a and height h (cone surface minus the base).
b) Use Pappus's formula (Exercises 13.5) and the result in (a) to find the centroid of the complete surface of a solid cone (side plus base).
c) A cone of radius a and height h is joined to a hemisphere of radius a to make a surface S that resembles an ice cream cone. Use Pappus's formula and the results in (a) and Example 5 to find the centroid of S. How high does the cone have to be to place the centroid in the plane shared by the bases of the hemisphere and cone?

Special Formulas for Surface Area

If S is the surface defined by a function $z = f(x, y)$ that has continuous first partial derivatives throughout a region R_{xy} in the xy-plane (Fig. 14.50), then S is also the level surface $F(x, y, z) = 0$ of the function $F(x, y, z) = f(x, y) - z$. Taking the unit normal to R_{xy} to be $\mathbf{p} = \mathbf{k}$ then gives

$$|\nabla F| = |f_x \mathbf{i} + f_y \mathbf{j} - \mathbf{k}| = \sqrt{f_x^2 + f_y^2 + 1},$$
$$|\nabla F \cdot \mathbf{p}| = |(f_x \mathbf{i} + f_y \mathbf{j} - \mathbf{k}) \cdot \mathbf{k}| = |-1| = 1,$$

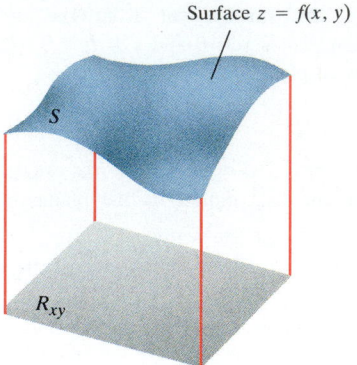

Surface $z = f(x, y)$

S

R_{xy}

14.50   For a surface $z = f(x, y)$, the surface area formula in Eq. (5) takes the form
$$S = \iint_{R_{xy}} \sqrt{f_x^2 + f_y^2 + 1} \, dx \, dy.$$

and

$$\frac{\text{Area}}{\text{of } S} = \iint_{R_{xy}} \frac{|\nabla F|}{|\nabla F \cdot \mathbf{p}|} \, dA = \iint_{R_{xy}} \sqrt{f_x^2 + f_y^2 + 1} \, dx \, dy. \quad (13)$$

Similarly, the area of a smooth surface $x = f(y, z)$ over a region R_{yz} in the yz-plane is

$$\text{Area of } S = \iint_{R_{yz}} \sqrt{f_y^2 + f_z^2 + 1} \, dy \, dz, \quad (14)$$

and the area of a smooth surface $y = f(x, z)$ over a region R_{xz} in the xz-plane is

$$\text{Area of } S = \iint_{R_{xz}} \sqrt{f_x^2 + f_z^2 + 1} \, dx \, dz. \quad (15)$$

Use Eqs. (13)–(15) to find the areas of the surfaces in Exercises 37–42.

37. The surface cut from the bottom of the paraboloid $z = x^2 + y^2$ by the plane $z = \sqrt{3}$.

38. The portion of the cone $z = \sqrt{x^2 + y^2}$ that lies over the region between the circle $x^2 + y^2 = 1$ and the ellipse $9x^2 + 4y^2 = 36$ in the xy-plane. (*Hint:* Use formulas from geometry to find the area of the region.)

39. The surface cut from the bottom of the paraboloid $x = 1 - y^2 - z^2$ by the yz-plane.

40. The triangle cut from the plane $2x + 6y + 3z = 6$ by the bounding planes of the first octant. Calculate the area three ways, once with each area formula.

41. The surface in the first octant cut from the cylinder $2y = z^2$ by the planes $x = 1$ and $y = 2$.

42. The portion of the plane $y + z = 4$ that lies above the region cut from the first quadrant of the xz-plane by the parabola $x = 4 - z^2$.

14.5 The Divergence Theorem

The divergence form of Green's theorem in the plane states that the net outward flux of a vector field across a simple closed curve in the plane can be calculated by integrating the divergence of the field over the region enclosed by the curve. The corresponding theorem in three dimensions, called the Divergence Theorem, states that the net outward flux of a vector field across a closed surface in space can be calculated by integrating the divergence of the field over the region enclosed by the surface. In this section, we prove the Divergence Theorem and show how it simplifies the calculation of flux. We also derive Gauss's law for flux in an electric field. The three-dimensional version of the circulation form of Green's theorem will be treated in the next section.

Divergence in Three Dimensions

The **divergence** of a vector field $\mathbf{F} = M(x, y, z)\,\mathbf{i} + N(x, y, z)\,\mathbf{j} + P(x, y, z)\,\mathbf{k}$ is the scalar function

$$\operatorname{div} \mathbf{F} = \frac{\partial M}{\partial x} + \frac{\partial N}{\partial y} + \frac{\partial P}{\partial z}. \tag{1}$$

Div $\mathbf{F}$ has the same physical interpretation as the divergence of a two-dimensional vector field. When $\mathbf{F}$ is the velocity field of a fluid flow, the value of div $\mathbf{F}$ at a point (x, y, z) is the rate at which fluid is being piped in or drained away at (x, y, z). The divergence is the flux per unit volume or flux density at the point. We shall say more about this at the end of the section.

Del Notation

The divergence of a three-dimensional vector field $\mathbf{F}$ is usually expressed in terms of the symbolic operator $\boldsymbol{\nabla}$ ("del"), defined by the equation

$$\boldsymbol{\nabla} = \mathbf{i}\,\frac{\partial}{\partial x} + \mathbf{j}\,\frac{\partial}{\partial y} + \mathbf{k}\,\frac{\partial}{\partial z}. \tag{2}$$

This operator can be applied to any differentiable vector field $\mathbf{F} = M\mathbf{i} + N\mathbf{j} + P\mathbf{k}$ to give the divergence of $\mathbf{F}$ as a symbolic dot product:

$$\boldsymbol{\nabla} \cdot \mathbf{F} = \frac{\partial M}{\partial x} + \frac{\partial N}{\partial y} + \frac{\partial P}{\partial z} = \operatorname{div} \mathbf{F}. \tag{3}$$

The notation $\boldsymbol{\nabla} \cdot \mathbf{F}$ is read "del dot $\mathbf{F}$."

When applied to a differentiable scalar function $f(x, y, z)$, the del operator gives the gradient of f:

$$\boldsymbol{\nabla} f = \frac{\partial f}{\partial x}\,\mathbf{i} + \frac{\partial f}{\partial y}\,\mathbf{j} + \frac{\partial f}{\partial z}\,\mathbf{k}. \tag{4}$$

This may now be read as "del f" as well as "grad f."

Since

$$\frac{\partial^2 f}{\partial x^2} + \frac{\partial^2 f}{\partial y^2} + \frac{\partial^2 f}{\partial z^2} = \boldsymbol{\nabla} \cdot \boldsymbol{\nabla} f = \boldsymbol{\nabla}^2 f, \tag{5}$$

The Divergence Theorem

Mikhail Vassilievich Ostrogradsky (1801–1862) was the first mathematician to publish a proof of the Divergence Theorem. Having been denied his degree at Kharkhov University by the minister for religious affairs and national education (for being an atheist), Ostrogradsky left Russia for Paris in 1822. There he met Laplace, Legendre, Fourier, Poisson, and Cauchy. While working on the theory of heat in the mid-1820s, he formulated the Divergence Theorem as a tool for turning volume integrals into surface integrals.

Carl Friedrich Gauss (1777–1855) had already proved the theorem while working on the theory of gravitation, but his notebooks were not published until many years later. The theorem is sometimes called Gauss's theorem. The list of Gauss's accomplishments in science and mathematics is truly astonishing, ranging from the invention of the electric telegraph (with Wilhelm Weber in 1833) to the development of a wonderfully accurate theory of planetary orbits and to work in non-Euclidean geometry that later became fundamental to Einstein's general theory of relativity.

Laplace's equation,

$$\frac{\partial^2 f}{\partial x^2} + \frac{\partial^2 f}{\partial y^2} + \frac{\partial^2 f}{\partial z^2} = 0,$$

(Exercises 12.3) can now be written as

$$\text{div grad } f = 0 \qquad \text{or} \qquad \nabla^2 f = 0. \tag{6}$$

If we think of a two-dimensional vector field

$$\mathbf{F}(x, y) = M(x, y)\,\mathbf{i} + N(x, y)\,\mathbf{j}$$

as a three-dimensional field whose **k**-component is zero, then

$$\nabla \cdot \mathbf{F} = \frac{\partial M}{\partial x} + \frac{\partial N}{\partial y} = \text{div } \mathbf{F}, \tag{7}$$

and the normal form of Green's theorem for two-dimensional fields (about flux and divergence; see Eq. 27 in Section 14.3) becomes

$$\int_C \mathbf{F} \cdot \mathbf{n}\, ds = \iint_R \nabla \cdot \mathbf{F}\, dA. \tag{8}$$

The tangential form of Green's theorem for two-dimensional fields can also be expressed in vector form in del notation, as we shall see in Section 14.6.

As these equations suggest, del notation provides an easy way to express and remember some of the most important relationships in mathematics, engineering, and physics. We shall see more of this as the chapter continues.

The Divergence Theorem

The Divergence Theorem says that under suitable conditions the outward flux of a vector field across a closed surface (oriented outward) equals the triple integral of the divergence of the field over the region enclosed by the surface.

THEOREM 4

The Divergence Theorem

The flux of a vector field $\mathbf{F} = M\mathbf{i} + N\mathbf{j} + P\mathbf{k}$ across a closed oriented surface S in the direction of the surface's outward unit normal field $\mathbf{n}$ equals the integral of $\nabla \cdot \mathbf{F}$ over the region D enclosed by the surface:

$$\underbrace{\iint_S \mathbf{F} \cdot \mathbf{n}\, d\sigma}_{\substack{\text{outward} \\ \text{flux}}} = \underbrace{\iiint_D \nabla \cdot \mathbf{F}\, dV}_{\substack{\text{divergence} \\ \text{integral}}}. \tag{9}$$

As you can see, Eq. (9) has the same form as Eq. (8), which gives the flux–divergence form of Green's theorem for two-dimensional fields. The only difference is that we are now working in three dimensions instead of two.

Proof for Special Regions

To prove the Divergence Theorem, we assume that the components of $\mathbf{F}$ have continuous partial derivatives. We also assume that D is a convex region with no holes

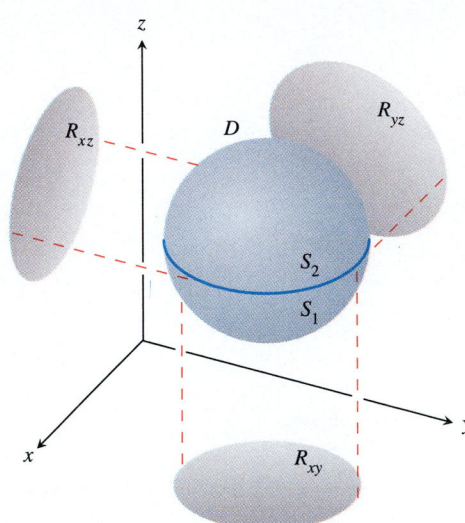

14.51 We first prove the Divergence Theorem for the kind of three-dimensional region shown here. We then extend the theorem to other regions.

or bubbles, such as a solid sphere, cube, or ellipsoid, and that S is a piecewise smooth surface. In addition, we assume that any line perpendicular to the xy-plane at an interior point of the region R_{xy} that is the projection of D on the xy-plane intersects the surface S in exactly two points, producing surfaces S_1 and S_2:

$$S_1: \qquad z = f_1(x,y), \qquad (x,y) \text{ in } R_{xy},$$

$$S_2: \qquad z = f_2(x,y), \qquad (x,y) \text{ in } R_{xy},$$

with $f_1 \leq f_2$. We make similar assumptions about the projection of D onto the other coordinate planes. See Fig. 14.51.

The components of the unit normal vector $\mathbf{n} = n_1\,\mathbf{i} + n_2\,\mathbf{j} + n_3\,\mathbf{k}$ are the cosines of the angles α, β, and γ that $\mathbf{n}$ makes with $\mathbf{i}$, $\mathbf{j}$, and $\mathbf{k}$ (Fig. 14.52). This is true because all the vectors involved are unit vectors. We have

$$n_1 = \mathbf{n} \cdot \mathbf{i} = |\mathbf{n}|\,|\mathbf{i}|\cos\alpha = \cos\alpha,$$

$$n_2 = \mathbf{n} \cdot \mathbf{j} = |\mathbf{n}|\,|\mathbf{j}|\cos\beta = \cos\beta, \qquad (10)$$

$$n_3 = \mathbf{n} \cdot \mathbf{k} = |\mathbf{n}|\,|\mathbf{k}|\cos\gamma = \cos\gamma.$$

Thus,

$$\mathbf{n} = (\cos\alpha)\,\mathbf{i} + (\cos\beta)\,\mathbf{j} + (\cos\gamma)\,\mathbf{k}$$

and

$$\mathbf{F} \cdot \mathbf{n} = M\cos\alpha + N\cos\beta + P\cos\gamma.$$

In component form, the Divergence Theorem states that

$$\iint_S (M\cos\alpha + N\cos\beta + P\cos\gamma)\,d\sigma = \iiint_D \left(\frac{\partial M}{\partial x} + \frac{\partial N}{\partial y} + \frac{\partial P}{\partial z}\right) dx\,dy\,dz. \qquad (11)$$

We prove the theorem by proving three separate equalities that add up to Eq. (11):

$$\iint M\cos\alpha\,d\sigma = \iiint \frac{\partial M}{\partial x}\,dx\,dy\,dz, \qquad (12)$$

$$\iint N\cos\beta\,d\sigma = \iiint \frac{\partial N}{\partial y}\,dx\,dy\,dz, \qquad (13)$$

$$\iint P\cos\gamma\,d\sigma = \iiint \frac{\partial P}{\partial z}\,dx\,dy\,dz. \qquad (14)$$

We prove Eq. (14) by converting the surface integral on the left to a double integral over the projection R_{xy} of D on the xy-plane (Fig. 14.53). The surface S consists of an upper part S_2 whose equation is $z = f_2(x,y)$ and a lower part S_1 whose equation is $z = f_1(x,y)$. On S_2, the outer normal $\mathbf{n}$ has a positive $\mathbf{k}$-component and

$$\cos\gamma\,d\sigma = dx\,dy \quad \text{because} \quad d\sigma = \frac{dA}{|\cos\gamma|} = \frac{dx\,dy}{\cos\gamma}. \qquad (15)$$

See Fig. 14.54. On S_1, the outer normal $\mathbf{n}$ has a negative $\mathbf{k}$-component and

$$\cos\gamma\,d\sigma = -dx\,dy. \qquad (16)$$

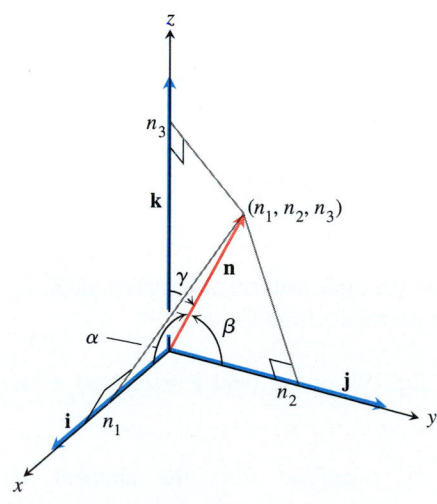

14.52 The scalar components of a unit normal vector $\mathbf{n}$ are the cosines of the angles α, β, and γ that it makes with $\mathbf{i}$, $\mathbf{j}$, and $\mathbf{k}$.

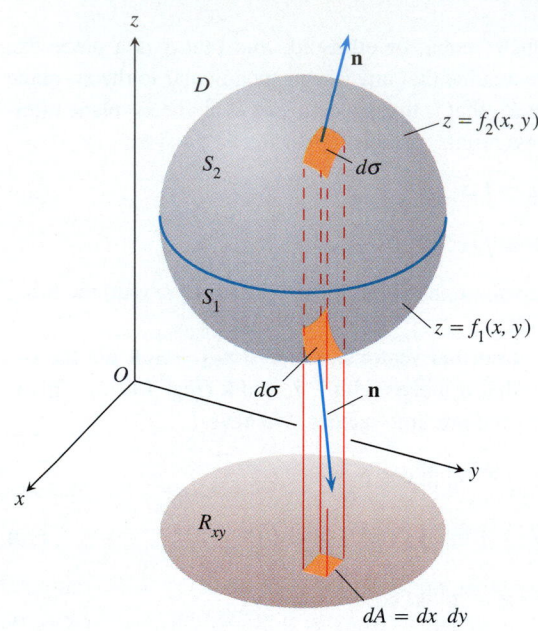

14.53 The three-dimensional region D enclosed by the surfaces S_1 and S_2 shown here projects vertically onto a two-dimensional region R_{xy} in the xy-plane.

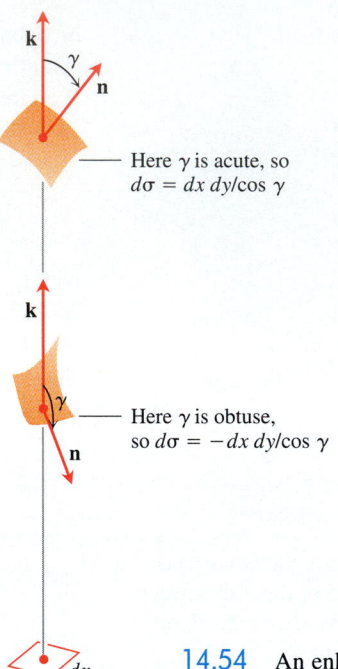

Here γ is acute, so $d\sigma = dx\, dy/\cos\gamma$

Here γ is obtuse, so $d\sigma = -dx\, dy/\cos\gamma$

14.54 An enlarged view of the area patches in Fig. 14.53. The relations $d\sigma = \pm dx\, dy/\cos\gamma$ are derived in Section 14.4.

Therefore,

$$\iint_S P \cos\gamma\, d\sigma = \iint_{S_2} P \cos\gamma\, d\sigma + \iint_{S_1} P \cos\gamma\, d\sigma$$

$$= \iint_{R_{xy}} P(x,y,f_2(x,y))\, dx\, dy - \iint_{R_{xy}} P(x,y,f_1(x,y))\, dx\, dy \tag{17}$$

$$= \iint_{R_{xy}} [P(x,y,f_2(x,y)) - P(x,y,f_1(x,y))]\, dx\, dy$$

$$= \iint_{R_{xy}} \left[\int_{f_1(x,y)}^{f_2(x,y)} \frac{\partial P}{\partial z}\, dz \right] dx\, dy = \iiint_D \frac{\partial P}{\partial z}\, dz\, dx\, dy.$$

This proves Eq. (14).

The proofs for Eqs. (12) and (13) follow the same pattern; or just permute x, y, z; M, N, P; α, β, γ, in order, and get those results from Eq. (14).

Example 1   Evaluate both sides of Eq. (9) for the field $\mathbf{F} = x\mathbf{i} + y\mathbf{j} + z\mathbf{k}$ over the sphere $x^2 + y^2 + z^2 = a^2$.

Solution   The outer unit normal to S, calculated from the gradient of $f(x,y,z) = x^2 + y^2 + z^2 - a^2$, is

$$\mathbf{n} = \frac{2(x\mathbf{i} + y\mathbf{j} + z\mathbf{k})}{\sqrt{4(x^2 + y^2 + z^2)}} = \frac{x\mathbf{i} + y\mathbf{j} + z\mathbf{k}}{a}.$$

Hence,

$$\mathbf{F} \cdot \mathbf{n} \, d\sigma = \frac{x^2 + y^2 + z^2}{a} \, d\sigma = \frac{a^2}{a} \, d\sigma = a \, d\sigma$$

because $x^2 + y^2 + z^2 = a^2$ on the surface. Therefore

$$\iint_S \mathbf{F} \cdot \mathbf{n} \, d\sigma = \iint_S a \, d\sigma = a \iint_S d\sigma = a(4\pi a^2) = 4\pi a^3.$$

The divergence of $\mathbf{F}$ is

$$\nabla \cdot \mathbf{F} = \frac{\partial}{\partial x}(x) + \frac{\partial}{\partial y}(y) + \frac{\partial}{\partial z}(z) = 3,$$

so

$$\iiint_D \nabla \cdot \mathbf{F} \, dV = \iiint_D 3 \, dV = 3\left(\frac{4}{3}\pi a^3\right) = 4\pi a^3.$$

Example 2   Calculate the flux of the field $\mathbf{F} = xy\mathbf{i} + yz\mathbf{j} + xz\mathbf{k}$ outward through the surface of the cube cut from the first octant by the planes $x = 1$, $y = 1$, and $z = 1$.

Solution   The Divergence Theorem says that instead of calculating the flux as a sum of six separate integrals, one for each face of the cube, we can calculate the flux by integrating the divergence

$$\nabla \cdot \mathbf{F} = \frac{\partial}{\partial x}(xy) + \frac{\partial}{\partial y}(yz) + \frac{\partial}{\partial z}(xz) = y + z + x$$

over the cube's interior:

$$\text{Flux} = \iint_{\substack{\text{cube} \\ \text{surface}}} \mathbf{F} \cdot \mathbf{n} \, d\sigma$$

$$= \iiint_{\substack{\text{cube} \\ \text{interior}}} \nabla \cdot \mathbf{F} \, dV \qquad \text{(The Divergence Theorem)}$$

$$= \int_0^1 \int_0^1 \int_0^1 (x + y + z) \, dx \, dy \, dz$$

$$= \frac{3}{2}. \qquad \text{(Routine integration)}$$

The Divergence Theorem for Other Regions

The Divergence Theorem can be extended to regions that can be split up into a finite number of simple regions of the type just discussed and to regions that can be defined as limits of simpler regions in certain ways. For example, suppose that D is the region between two concentric spheres and that $\mathbf{F}$ has continuously differentiable components throughout D and on the bounding surfaces. Split D by an equatorial plane and apply the Divergence Theorem to each half separately. The bottom half, D_1, is shown in Fig. 14.55. The surface that bounds D_1 consists of an outer

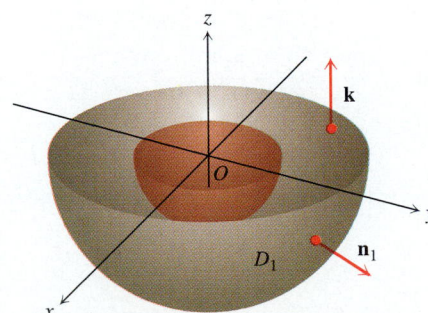

14.55   The lower half of the solid region between two concentric spheres.

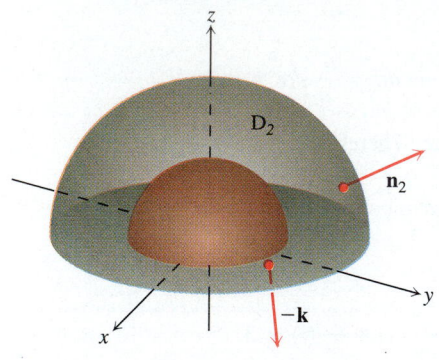

14.56 The upper half of the solid region between two concentric spheres.

hemisphere, a plane washer-shaped base, and an inner hemisphere. The Divergence Theorem says that

$$\iint_{S_1} \mathbf{F} \cdot \mathbf{n}_1 \, d\sigma_1 = \iiint_{D_1} \boldsymbol{\nabla} \cdot \mathbf{F} \, dV_1. \tag{18}$$

The unit normal $\mathbf{n}_1$ that points outward from D_1 points away from the origin along the outer surface, equals $\mathbf{k}$ along the flat base, and points toward the origin along the inner surface. Next apply the Divergence Theorem to D_2, as shown in Fig. 14.56:

$$\iint_{S_2} \mathbf{F} \cdot \mathbf{n}_2 \, d\sigma_2 = \iiint_{D_2} \boldsymbol{\nabla} \cdot \mathbf{F} \, dV_2. \tag{19}$$

As we follow $\mathbf{n}_2$ over S_2, pointing outward from D_2, we see that $\mathbf{n}_2$ equals $-\mathbf{k}$ along the washer-shaped base in the xy-plane, points away from the origin on the outer sphere, and points toward the origin on the inner sphere. When we add Eqs. (18) and (19), the surface integrals over the flat base cancel because of the opposite signs of $\mathbf{n}_1$ and $\mathbf{n}_2$. We thus arrive at the result

$$\iint_{S} \mathbf{F} \cdot \mathbf{n} \, d\sigma = \iiint_{D} \boldsymbol{\nabla} \cdot \mathbf{F} \, dV, \tag{20}$$

with D the region between the spheres, S the boundary of D consisting of two spheres, and $\mathbf{n}$ the unit normal to S directed outward from D.

Example 3   Find the net outward flux of the field

$$\mathbf{F} = \frac{x\mathbf{i} + y\mathbf{j} + z\mathbf{k}}{\rho^3}, \qquad \rho = \sqrt{x^2 + y^2 + z^2}$$

across the boundary of the region $D: 0 < a^2 \le x^2 + y^2 + z^2 \le b^2$.

Solution   The flux can be calculated by integrating $\boldsymbol{\nabla} \cdot \mathbf{F}$ over D. We have

$$\frac{\partial \rho}{\partial x} = \frac{1}{2}(x^2 + y^2 + z^2)^{-1/2}(2x) = \frac{x}{\rho}$$

and

$$\frac{\partial M}{\partial x} = \frac{\partial}{\partial x}(x\rho^{-3}) = \rho^{-3} - 3x\rho^{-4}\frac{\partial \rho}{\partial x} = \frac{1}{\rho^3} - \frac{3x^2}{\rho^5}.$$

Similarly,

$$\frac{\partial N}{\partial y} = \frac{1}{\rho^3} - \frac{3y^2}{\rho^5} \qquad \text{and} \qquad \frac{\partial P}{\partial z} = \frac{1}{\rho^3} - \frac{3z^2}{\rho^5}.$$

Hence,

$$\text{div } \mathbf{F} = \frac{3}{\rho^3} - \frac{3}{\rho^5}(x^2 + y^2 + z^2) = \frac{3}{\rho^3} - \frac{3\rho^2}{\rho^5} = 0$$

and

$$\iiint_{D} \boldsymbol{\nabla} \cdot \mathbf{F} \, dV = 0.$$

So the integral of $\boldsymbol{\nabla} \cdot \mathbf{F}$ over D is zero and the net outward flux across the boundary of D is zero. But there is more to learn from this example. The flux leaving D across the inner sphere S_a is the negative of the flux leaving D across the outer sphere S_b (because the sum of these fluxes is zero). This means that the flux of $\mathbf{F}$ across S_a in the direction away from the origin equals the flux of $\mathbf{F}$ across S_b in the direction away from the origin. Thus, the flux of $\mathbf{F}$ across a sphere centered at the origin is independent of the radius of the sphere. What is this flux?

To find it, we evaluate the flux integral directly. The outward unit normal on the sphere of radius a is

$$\mathbf{n} = \frac{x\mathbf{i} + y\mathbf{j} + z\mathbf{k}}{\sqrt{x^2 + y^2 + z^2}} = \frac{x\mathbf{i} + y\mathbf{j} + z\mathbf{k}}{a}.$$

Hence, on the sphere,

$$\mathbf{F} \cdot \mathbf{n} = \frac{x\mathbf{i} + y\mathbf{j} + z\mathbf{k}}{a^3} \cdot \frac{x\mathbf{i} + y\mathbf{j} + z\mathbf{k}}{a} = \frac{x^2 + y^2 + z^2}{a^4} = \frac{a^2}{a^4} = \frac{1}{a^2}$$

and

$$\iint\limits_{S_a} \mathbf{F} \cdot \mathbf{n}\, d\sigma = \frac{1}{a^2} \iint\limits_{S_a} d\sigma = \frac{1}{a^2}(4\pi a^2) = 4\pi.$$

The outward flux of $\mathbf{F}$ across any sphere centered at the origin is 4π.

Gauss's Law—One of the Four Great Laws of Electromagnetic Theory

There is still more to be learned from Example 3. In electromagnetic theory, the electric field created by a point charge q located at the origin is the inverse square field

$$\mathbf{E}(x, y, z) = q\frac{\mathbf{r}}{|\mathbf{r}|^3} = q\frac{x\mathbf{i} + y\mathbf{j} + z\mathbf{k}}{\rho^3},$$

where $\mathbf{r}$ is the position vector of the point (x, y, z) and $\rho = |\mathbf{r}| = \sqrt{x^2 + y^2 + z^2}$. In the notation of Example 3,

$$\mathbf{E} = q\mathbf{F}.$$

The calculations in Example 3 show that the outward flux of $\mathbf{E}$ across any sphere centered at the origin is $4\pi q$. But this result is not confined to spheres. The outward flux of $\mathbf{E}$ across any surface S that encloses the origin (and to which the Divergence Theorem applies) is also $4\pi q$. To see why, we have only to imagine a large sphere S_a centered at the origin and enclosing the surface S. Since

$$\nabla \cdot \mathbf{E} = \nabla \cdot q\mathbf{F} = q\nabla \cdot \mathbf{F} = 0$$

when $\rho > 0$, the integral of $\nabla \cdot \mathbf{E}$ over the region D between S and S_a is zero. Hence, by the Divergence Theorem,

$$\iint\limits_{\substack{\text{boundary} \\ \text{of } D}} \mathbf{E} \cdot \mathbf{n}\, d\sigma = 0,$$

and the flux of $\mathbf{E}$ across S in the direction away from the origin must be the same as the flux of $\mathbf{E}$ across S_a in the direction away from the origin, which is $4\pi q$. This statement, called *Gauss's law*, applies to much more general charge distributions than the one we have assumed here, as you will see in nearly any college physics text.

Gauss's Law

$$\iint_S \mathbf{E} \cdot \mathbf{n} \, d\sigma = 4\pi q \tag{21}$$

The Continuity Equation of Hydrodynamics

If $\mathbf{v}(x, y, z)$ is the velocity field of a fluid flowing smoothly through a region in space, $\delta = \delta(x, y, z, t)$ is the fluid's density at (x, y, z) at time t, and $\mathbf{F} = \delta\mathbf{v}$, then the **continuity equation** of hydrodynamics is the statement that

$$\nabla \cdot \mathbf{F} + \frac{\partial \delta}{\partial t} = 0. \tag{22}$$

The equation evolves naturally from the Divergence Theorem,

$$\iint_S \mathbf{F} \cdot \mathbf{n} \, d\sigma = \iiint_D \nabla \cdot \mathbf{F} \, dV, \tag{23}$$

if the functions involved have continuous first partial derivatives, as we shall now see.

First, the integral

$$\iint_S \mathbf{F} \cdot \mathbf{n} \, d\sigma$$

is the rate at which mass leaves D across S (leaves because $\mathbf{n}$ is the outer normal). To see why, consider a patch of area $\Delta\sigma$ on the surface (Fig. 14.57). In a short time interval Δt, the volume ΔV of fluid that flows across the patch is approximately equal to the volume of a cylinder with base area $\Delta\sigma$ and height $(\mathbf{v}\,\Delta t) \cdot \mathbf{n}$, where $\mathbf{v}$ is a velocity vector rooted at a point of the patch:

$$\Delta V \approx \mathbf{v} \cdot \mathbf{n} \, \Delta\sigma \, \Delta t.$$

The mass of this volume of fluid is about

$$\Delta m \approx \delta\mathbf{v} \cdot \mathbf{n} \, \Delta\sigma \Delta t,$$

so the rate at which mass is flowing out of D across the patch is about

$$\frac{\Delta m}{\Delta t} \approx \delta\mathbf{v} \cdot \mathbf{n} \, \Delta\sigma.$$

This leads to the approximation

$$\frac{\Sigma \, \Delta m}{\Delta t} \approx \sum \delta\mathbf{v} \cdot \mathbf{n} \, \Delta\sigma \tag{24}$$

as an estimate of the average rate at which mass flows across S. Finally, letting $\Delta\sigma \to 0$ and $\Delta t \to 0$ gives the instantaneous rate at which mass leaves D across S as

$$\frac{dm}{dt} = \iint_S \delta\mathbf{v} \cdot \mathbf{n} \, d\sigma, \tag{25}$$

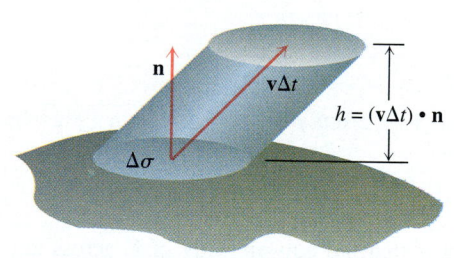

14.57 The fluid that flows upward through the patch $\Delta\sigma$ in a short time Δt fills a "cylinder" whose volume is approximately base × height = $\mathbf{v} \cdot \mathbf{n} \, \Delta\sigma \, \Delta t$.

which for our particular flow is

$$\frac{dm}{dt} = \int\int_S \mathbf{F} \cdot \mathbf{n} \, d\sigma. \tag{26}$$

Now let B be a solid sphere centered at a point Q in the flow. The average value of $\nabla \cdot \mathbf{F}$ over B is

$$\frac{1}{\text{volume of } B} \int\int\int_B \nabla \cdot \mathbf{F} \, dV. \tag{27}$$

It is a consequence of the continuity of the divergence that $\nabla \cdot \mathbf{F}$ actually takes on this value at some point P in B. Thus,

$$(\nabla \cdot \mathbf{F})_P = \frac{1}{\text{volume of } B} \int\int\int_B \nabla \cdot \mathbf{F} \, dV = \frac{\int\int_S \mathbf{F} \cdot \mathbf{n} \, d\sigma}{\text{volume of } B} \tag{28}$$

$$= \frac{\text{rate at which mass leaves } B \text{ across its surface } S}{\text{volume of } B}.$$

The fraction on the right describes decrease in mass per unit volume.

Now let the radius of B approach zero while the center Q stays fixed. The left-hand side of Eq. (28) converges to $(\nabla \cdot \mathbf{F})_Q$, the right side to $(-\partial\delta/\partial t)_Q$. The equality of these two limits is the continuity equation

$$\nabla \cdot \mathbf{F} = -\frac{\partial\delta}{\partial t}. \tag{29}$$

The continuity equation "explains" $\nabla \cdot \mathbf{F}$: The divergence of $\mathbf{F}$ at a point is the rate at which the density of the fluid is decreasing there.

The Divergence Theorem

$$\int\int_S \mathbf{F} \cdot \mathbf{n} \, d\sigma = \int\int\int_D \nabla \cdot \mathbf{F} \, dV$$

now says that the net decrease in density of the fluid in region D is accounted for by the mass transported across the surface S. In a way, the theorem is a statement about conservation of mass.

EXERCISES 14.5

In Exercises 1–12, use the Divergence Theorem to find the outward flux of $\mathbf{F}$ across the boundary of the region D.

1. $\mathbf{F} = (y - x)\mathbf{i} + (z - y)\mathbf{j} + (y - x)\mathbf{k}$
   D: The cube bounded by the planes $x = \pm 1$, $y = \pm 1$, and $z = \pm 1$

2. $\mathbf{F} = x^2\mathbf{i} + y^2\mathbf{j} + z^2\mathbf{k}$
   a) D: The cube cut from the first octant by the planes $x = 1$, $y = 1$, and $z = 1$
   b) D: The cube bounded by the planes $x = \pm 1$, $y = \pm 1$, and $z = \pm 1$
   c) D: The region cut from the solid cylinder $x^2 + y^2 \le 4$ by the planes $z = 0$ and $z = 1$

3. $\mathbf{F} = y\mathbf{i} + xy\mathbf{j} - z\mathbf{k}$
   D: The region inside the solid cylinder $x^2 + y^2 \le 4$ between the plane $z = 0$ and the paraboloid $z = x^2 + y^2$

4. $\mathbf{F} = x^2\mathbf{i} + xz\mathbf{j} - 3z\mathbf{k}$
   D: The solid sphere $x^2 + y^2 + z^2 \le 4$

5. $\mathbf{F} = x^2\mathbf{i} - 2xy\mathbf{j} + 3xz\mathbf{k}$
   D: The region cut from the first octant by the sphere $x^2 + y^2 + z^2 = 4$

6. $\mathbf{F} = (6x^2 + 2xy)\mathbf{i} + (2y + x^2z)\mathbf{j} + 4x^2y^3\mathbf{k}$
   D: The region cut from the first octant by the cylinder $x^2 + y^2 = 4$ and the plane $z = 3$

7. $\mathbf{F} = 2xz\mathbf{i} - xy\mathbf{j} - z^2\mathbf{k}$
   D: The wedge cut from the first octant by the plane $y + z = 4$ and the elliptical cylinder $4x^2 + y^2 = 16$

8. $\mathbf{F} = x^3\mathbf{i} + y^3\mathbf{j} + z^3\mathbf{k}$
   D: The solid sphere $x^2 + y^2 + z^2 \le a^2$

9. $\mathbf{F} = \sqrt{x^2 + y^2 + z^2}\,(x\mathbf{i} + y\mathbf{j} + z\mathbf{k})$
   D: The region $1 \le x^2 + y^2 + z^2 \le 2$

10. $\mathbf{F} = (x\mathbf{i} + y\mathbf{j} + z\mathbf{k})/\sqrt{x^2 + y^2 + z^2}$
    D: The region $1 \le x^2 + y^2 + z^2 \le 4$

11. $\mathbf{F} = (5x^3 + 12xy^2)\mathbf{i} + (y^3 + e^y \sin z)\mathbf{j} + (5z^3 + e^y\cos z)\mathbf{k}$
    D: The solid region between the spheres $x^2 + y^2 + z^2 = 1$ and $x^2 + y^2 + z^2 = 2$

12. $\mathbf{F} = \ln(x^2 + y^2)\,\mathbf{i} - \left(\dfrac{2z}{x}\tan^{-1}\dfrac{y}{x}\right)\mathbf{j} + z\sqrt{x^2 + y^2}\,\mathbf{k}$
    D: The thick-walled cylinder $1 \le x^2 + y^2 \le 2, -1 \le z \le 2$

13. Let S be the closed cubelike surface in Fig. 14.58, with its base the unit square in the xy-plane, its four sides lying in the planes $x = 0$, $x = 1$, $y = 0$, $y = 1$, and its top an arbitrary smooth surface whose identity is unknown. Let $\mathbf{F} = x\mathbf{i} - 2y\mathbf{j} + (z + 3)\mathbf{k}$. Suppose the outward flux of $\mathbf{F}$ through side A is 1 and through side B is -3. Find the outward flux of $\mathbf{F}$ through the top.

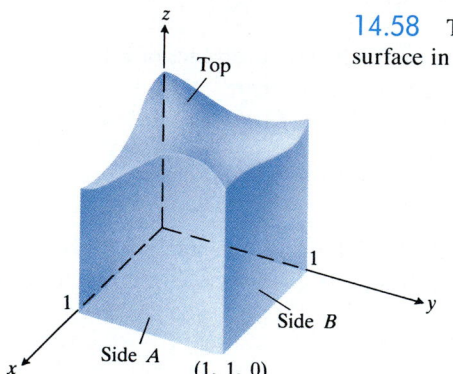

14.58   The cubelike surface in Exercise 13.

14. Let $\mathbf{F}$ be a field whose components have continuous first partial derivatives throughout a portion of space containing a closed and bounded region D and its smooth bounding surface S. Suppose that the length of the vector $\mathbf{F}$ never exceeds 1 on S. What bound can be placed on the numerical size of the integral

$$\iiint_D \nabla \cdot \mathbf{F}\, dV\,?$$

How do you know?

15. a) Show that the flux of the position vector field $\mathbf{F} = x\mathbf{i} + y\mathbf{j} + z\mathbf{k}$ outward through a smooth closed surface S is three times the volume of the region enclosed by the surface.

    b) Let $\mathbf{n}$ be the outward unit normal vector field on S. Show that it is not possible for $\mathbf{F}$ to be orthogonal to $\mathbf{n}$ at every point of S.

16. Among all rectangular solids defined by the inequalities $0 \le x \le a$, $0 \le y \le b$, $0 \le z \le 1$, find the one for which the total flux of $\mathbf{F} = (-x^2 - 4xy)\mathbf{i} - 6yz\mathbf{j} + 12z\mathbf{k}$ outward through the six sides is greatest.

17. Let $\mathbf{F} = x\mathbf{i} + y\mathbf{j} + z\mathbf{k}$ and suppose that the surface S and region D satisfy the hypotheses of the Divergence Theorem. Show that the volume of D is given by the formula

$$\text{Volume of } D = \frac{1}{3}\iint_S \mathbf{F}\cdot\mathbf{n}\,d\sigma.$$

18. Show that the outward flux of a constant vector field $\mathbf{F} = \mathbf{C}$ across any closed surface to which the Divergence Theorem applies is zero.

Harmonic Functions and Normal Derivatives

19. A function $f(x, y, z)$ is **harmonic** in a region D in space if throughout D it satisfies the Laplace equation

$$\nabla^2 f = \nabla \cdot \nabla f = \frac{\partial^2 f}{\partial x^2} + \frac{\partial^2 f}{\partial y^2} + \frac{\partial^2 f}{\partial z^2} = 0.$$

Suppose that f is harmonic throughout a bounded region D enclosed by a smooth surface S and that $\mathbf{n}$ is the outward unit normal vector field on S. Let $\partial f/\partial n = \nabla f \cdot \mathbf{n}$ be the derivative of f in the direction of $\mathbf{n}$. This derivative, a real-valued function defined on S, is called the **normal derivative** of f on S.

    a) Show that the integral of $\partial f/\partial n$ over S is zero. (*Hint:* Let $\mathbf{F} = \nabla f$.)

    b) Show that

$$\iint_S f\frac{\partial f}{\partial n}\,d\sigma = \iiint_D |\nabla f|^2\,dV.$$

(*Hint:* Let $\mathbf{F} = f\,\nabla f$.)

20. Let S be the surface of the portion of the solid sphere $x^2 + y^2 + z^2 \le a^2$ that lies in the first octant, let $f(x, y, z) = \ln\sqrt{x^2 + y^2 + z^2}$, and let $\partial f/\partial n$ be the normal derivative of f on S defined in Exercise 19. Calculate

$$\iint_S \frac{\partial f}{\partial n}\,d\sigma.$$

21. *Green's first formula.*   Suppose that f and g are scalar functions with continuous first- and second-order partial derivatives throughout a closed region D that is bounded by a piecewise smooth surface S. Show that

$$\iint_S f\frac{\partial g}{\partial n}\,d\sigma = \iiint_D (f\,\nabla^2 g + \nabla f \cdot \nabla g)\,dV. \qquad (30)$$

Equation (30) is **Green's first formula.** (*Hint:* Apply the Divergence Theorem to the field $\mathbf{F} = f\,\nabla g$.)

22. *Green's second formula*   (continuation of Exercise 21). Interchange f and g in Eq. (30) to obtain a similar formula. Then subtract this formula from Eq. (30) to show that

$$\iint_S \left(f\frac{\partial g}{\partial n} - g\frac{\partial f}{\partial n}\right)d\sigma = \iiint_D (f\,\nabla^2 g - g\,\nabla^2 f)\,dV. \qquad (31)$$

This equation is **Green's second formula.**

14.6 Stokes's Theorem

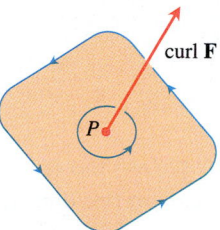

14.59 The circulation vector at a point P in a plane in a three-dimensional fluid flow. Notice its right-hand relation to the circulation line.

As we saw in Section 14.3, the circulation density or curl of a two-dimensional field $\mathbf{F} = M\mathbf{i} + N\mathbf{j}$ at a point (x, y) is described by the scalar quantity $(\partial N/\partial x - \partial M/\partial y)$. In three dimensions, the circulation around a point P in a plane is described not with a scalar but with a vector. This vector is normal to the plane of the circulation (Fig. 14.59) and points in the direction that gives it a right-hand relation to the circulation line. The length of the vector gives the rate of the fluid's rotation, which usually varies as the circulation plane is tilted about P from one position to another. It turns out that the vector of greatest circulation in a flow with velocity field $\mathbf{F} = M\mathbf{i} + N\mathbf{j} + P\mathbf{k}$ is the vector

$$\text{curl } \mathbf{F} = \left(\frac{\partial P}{\partial y} - \frac{\partial N}{\partial z}\right)\mathbf{i} + \left(\frac{\partial M}{\partial z} - \frac{\partial P}{\partial x}\right)\mathbf{j} + \left(\frac{\partial N}{\partial x} - \frac{\partial M}{\partial y}\right)\mathbf{k}. \tag{1}$$

We get this information from Stokes's theorem, the generalization of the circulation–curl form of Green's theorem to space.

In this section we present Stokes's theorem and investigate what it says about circulation. We shall also see how Stokes's theorem sometimes enables us to evaluate a line integral around a loop in space by calculating an easier surface integral. This is similar to the way we used Green's theorem to evaluate a line integral over a complicated curve in the plane by calculating an easier double integral over the region enclosed by the curve. In addition, we shall state the circulation–curl form of Green's theorem in its final vector form. Stokes's theorem has important applications in electromagnetic theory, but we shall not go into them here.

Stokes's Theorem

In del notation, the curl of a three-dimensional vector field $\mathbf{F} = M\mathbf{i} + N\mathbf{j} + P\mathbf{k}$ is written as the symbolic cross product

$$\text{curl } \mathbf{F} = \nabla \times \mathbf{F} = \begin{vmatrix} \mathbf{i} & \mathbf{j} & \mathbf{k} \\ \dfrac{\partial}{\partial x} & \dfrac{\partial}{\partial y} & \dfrac{\partial}{\partial z} \\ M & N & P \end{vmatrix}$$

$$= \left(\frac{\partial P}{\partial y} - \frac{\partial N}{\partial z}\right)\mathbf{i} + \left(\frac{\partial M}{\partial z} - \frac{\partial P}{\partial x}\right)\mathbf{j} + \left(\frac{\partial N}{\partial x} - \frac{\partial M}{\partial y}\right)\mathbf{k}. \tag{2}$$

Stokes's theorem says that, under conditions normally met in practice, the circulation of a vector field around the boundary of an oriented surface in space in the direction counterclockwise with respect to the surface's unit normal vector field $\mathbf{n}$ (Fig. 14.60) is equal to the double integral of the normal component of the curl of the field over the surface.

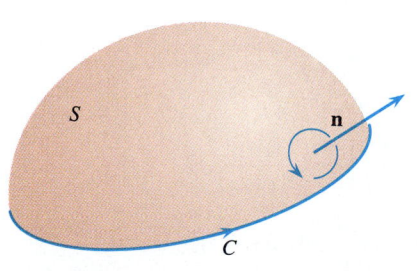

14.60 The orientation of the bounding curve C gives it a right-handed relation to the normal field $\mathbf{n}$.

THEOREM 5

Stokes's Theorem

The circulation of $\mathbf{F} = M\mathbf{i} + N\mathbf{j} + P\mathbf{k}$ around the boundary C of an oriented surface S in the direction counterclockwise with respect to the surface's unit normal vector $\mathbf{n}$ equals the integral of $\nabla \times \mathbf{F} \cdot \mathbf{n}$ over S.

$$\underbrace{\oint_C \mathbf{F} \cdot d\mathbf{r}}_{\substack{\text{counterclockwise} \\ \text{circulation}}} = \underbrace{\iint_S \nabla \times \mathbf{F} \cdot \mathbf{n} \, d\sigma}_{\text{curl integral}} \tag{3}$$

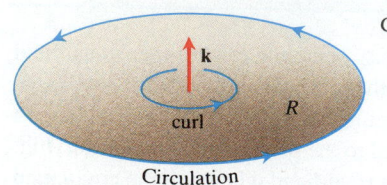

Green

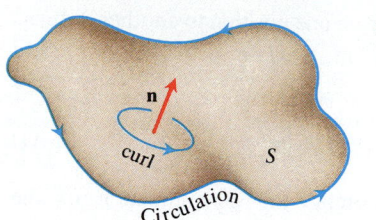

Stokes

14.61   Green's theorem vs. Stokes's theorem.

Naturally, we need some mathematical restrictions on **F**, C, and S to ensure the existence of the integrals in Stokes's equation. The usual restrictions are that all the functions and derivatives involved be continuous.

If C is a curve in the xy-plane, oriented counterclockwise, and R is the region bounded by C in the xy-plane, then $d\sigma = dx\,dy$ and

$$(\nabla \times \mathbf{F}) \cdot \mathbf{n} = (\nabla \times \mathbf{F}) \cdot \mathbf{k} = \left(\frac{\partial N}{\partial x} - \frac{\partial M}{\partial y}\right). \tag{4}$$

Under these conditions, Stokes's equation becomes

$$\oint_C \mathbf{F} \cdot d\mathbf{r} = \int\int_R \left(\frac{\partial N}{\partial x} - \frac{\partial M}{\partial y}\right) dx\,dy,$$

which is the circulation–curl form of the equation in Green's theorem. Conversely, by reversing these steps we can rewrite Green's theorem for two-dimensional fields in del notation as

$$\oint_C \mathbf{F} \cdot d\mathbf{r} = \int\int_R \nabla \times \mathbf{F} \cdot \mathbf{k}\,dA. \tag{5}$$

See Fig. 14.61

Example 1

Evaluate the two integrals in Stokes's theorem for the hemisphere

$$x^2 + y^2 + z^2 = 9, \qquad z \geq 0,$$

its bounding circle

$$x^2 + y^2 = 9, \qquad z = 0,$$

and the field

$$\mathbf{F} = y\mathbf{i} - x\mathbf{j}.$$

Solution   The integrand of the circulation integral is

$$\mathbf{F} \cdot d\mathbf{r} = (y\mathbf{i} - x\mathbf{j}) \cdot (dx\mathbf{i} + dy\mathbf{j} + dz\mathbf{k}) = y\,dx - x\,dy.$$

By the circulation form of Green's theorem for the plane,

$$\oint_C \mathbf{F} \cdot d\mathbf{r} = \oint_C (y\,dx - x\,dy) = \int\int_R \left(\frac{\partial N}{\partial x} - \frac{\partial M}{\partial y}\right) dx\,dy$$

$$= \int\int_{x^2+y^2 \leq 9} -2\,dx\,dy = (-2)9\pi = -18\pi.$$

For the curl integral of $\mathbf{F} = y\mathbf{i} - x\mathbf{j}$, we have

$$\nabla \times \mathbf{F} = \left(\frac{\partial P}{\partial y} - \frac{\partial N}{\partial z}\right)\mathbf{i} + \left(\frac{\partial M}{\partial z} - \frac{\partial P}{\partial x}\right)\mathbf{j} + \left(\frac{\partial N}{\partial x} - \frac{\partial M}{\partial y}\right)\mathbf{k}$$

$$= (0 - 0)\mathbf{i} + (0 - 0)\mathbf{j} + (-1 - 1)\mathbf{k} = -2\mathbf{k},$$

$$\mathbf{n} = \frac{x\mathbf{i} + y\mathbf{j} + z\mathbf{k}}{\sqrt{x^2 + y^2 + z^2}} = \frac{x\mathbf{i} + y\mathbf{j} + z\mathbf{k}}{3}, \qquad \text{(Outer unit normal)}$$

$$d\sigma = \frac{3}{z}\,dA \qquad \text{(As in Section 14.4, Example 5, with } a = 3)$$

$$\nabla \times \mathbf{F} \cdot \mathbf{n}\,d\sigma = -\frac{2z}{3}\frac{3}{z}\,dA = -2\,dA,$$

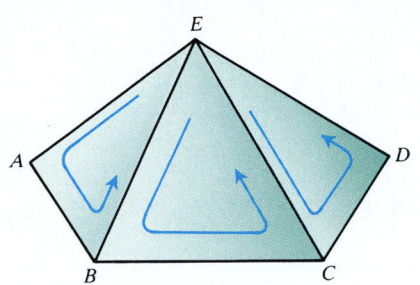

14.62 Part of a polyhedral surface.

and

$$\iint\limits_{S} \nabla \times \mathbf{F} \cdot \mathbf{n}\, d\sigma = \iint\limits_{x^2 + y^2 \leq 9} -2\, dA = -18\pi.$$

The circulation around the circle equals the integral of the curl over the hemisphere, as it should.

Proof of Stokes's Theorem for Polyhedral Surfaces

Let the surface S be a polyhedral surface consisting of a finite number of plane regions. (Think of one of Buckminster Fuller's geodesic domes.) We apply Green's theorem to each separate panel of S. There are two types of panels:

1. those that are surrounded on all sides by other panels,
2. those that have one or more edges that are not adjacent to other panels.

The boundary Δ of S consists of edges of type 2 panels that are not adjacent to other panels. In Fig. 14.62, the triangles EAB, BCE, and CDE represent a part of S, with $ABCD$ part of the boundary Δ. Applying Green's theorem to the three triangles in turn and adding the results, we get

$$\left(\oint_{EAB} + \oint_{BCE} + \oint_{CDE} \right) \mathbf{F} \cdot d\mathbf{r} = \left(\iint_{EAB} + \iint_{BCE} + \iint_{CDE} \right) \nabla \times \mathbf{F} \cdot \mathbf{n}\, d\sigma. \tag{6}$$

The three line integrals on the left-hand side of Eq. (6) combine into a single line integral taken around the periphery $ABCDEA$ because the integrals along interior segments cancel in pairs. For example, the integral along the segment BE in triangle ABE is opposite in sign to the integral along the same segment in triangle EBC. Similarly for the segment CE. Hence (6) reduces to

$$\oint_{ABCDEA} \mathbf{F} \cdot d\mathbf{r} = \iint_{ABCDE} \nabla \times \mathbf{F} \cdot \mathbf{n}\, d\sigma.$$

When we apply Green's theorem to all the panels and add the results, we get

$$\oint_{\Delta} \mathbf{F} \cdot d\mathbf{r} = \iint_{S} \nabla \times \mathbf{F} \cdot \mathbf{n}\, d\sigma. \tag{7}$$

This is Stokes's theorem for a polyhedral surface S.

A rigorous proof of Stokes's theorem for a more general oriented surface S and its bounding curve C is beyond the level of this course. However, the following intuitive argument shows why we would expect Stokes's equation to hold. Imagine a sequence of polyhedral surfaces $S_1, S_2, \ldots, S_n, \ldots$ approximating S. The surface S_n is constructed so that its boundary Δ_n is inscribed in or tangent to C and so that the length of Δ_n approaches the length of C as $n \to \infty$. The faces of S_n are polygonal regions approximating pieces of S, and the area of S_n approaches the area of S as $n \to \infty$. If the first partial derivatives of M, N, and P are continuous throughout an open region D containing S and C, it is plausible to expect that

$$\oint_{\Delta_n} \mathbf{F} \cdot d\mathbf{R} \quad \text{approaches} \quad \oint_{C} \mathbf{F} \cdot d\mathbf{R}$$

and that

$$\iint\limits_{S_n} \boldsymbol{\nabla} \times \mathbf{F} \cdot \mathbf{n} \, d\sigma_n \quad \text{approaches} \quad \iint\limits_{S} \boldsymbol{\nabla} \times \mathbf{F} \cdot \mathbf{n} \, d\sigma$$

as $n \to \infty$. But if

$$\oint_{\Delta_n} \mathbf{F} \cdot d\mathbf{R} \to \oint_{C} \mathbf{F} \cdot d\mathbf{R} \tag{8}$$

and

$$\iint\limits_{S_n} \boldsymbol{\nabla} \times \mathbf{F} \cdot \mathbf{n} \, d\sigma \to \iint\limits_{S} \boldsymbol{\nabla} \times \mathbf{F} \cdot \mathbf{n} \, d\sigma, \tag{9}$$

and if the left-hand sides of (8) and (9) are equal by Stokes's theorem for polyhedra, then their limits are equal.

Stokes's Theorem for Surfaces with Holes

Stokes's theorem can also be extended to a surface S that has one or more holes in it (like a curved slice of Swiss cheese), in a way analogous to the extension of Green's theorem: The surface integral over S of the normal component of $\boldsymbol{\nabla} \times \mathbf{F}$ is equal to the sum of the line integrals around all the boundaries of S (including boundaries of the holes) of the tangential component of $\mathbf{F}$, where the boundary curves are to be traced in the positive direction induced by the positive orientation of S.

Circulation of a Fluid

Stokes's theorem provides the following vector interpretation for $\boldsymbol{\nabla} \times \mathbf{F}$. As in the discussion of divergence, let $\mathbf{v}$ be the velocity field of a moving fluid, δ the density, and $\mathbf{F} = \delta \mathbf{v}$. Then

$$\oint_{C} \mathbf{F} \cdot d\mathbf{r}$$

is the circulation of the fluid around the closed curve C. By Stokes's theorem, the circulation is equal to the flux of $\boldsymbol{\nabla} \times \mathbf{F}$ through a surface S spanning C:

$$\oint_{C} \mathbf{F} \cdot d\mathbf{r} = \iint\limits_{S} \boldsymbol{\nabla} \times \mathbf{F} \cdot \mathbf{n} \, d\sigma.$$

Suppose we fix a point Q in the domain of $\mathbf{F}$ and a direction $\mathbf{u}$ at Q. Let C be a circle of radius ρ, with center at Q, whose plane is normal to $\mathbf{u}$. If $\boldsymbol{\nabla} \times \mathbf{F}$ is continuous at Q, then the average value of the $\mathbf{u}$-component of $\boldsymbol{\nabla} \times \mathbf{F}$ over the circular disk S bounded by C approaches the $\mathbf{u}$-component of $\boldsymbol{\nabla} \times \mathbf{F}$ at Q as $\rho \to 0$:

$$(\boldsymbol{\nabla} \times \mathbf{F} \cdot \mathbf{u})_Q = \lim_{\rho \to 0} \frac{1}{\pi \rho^2} \iint\limits_{S} \boldsymbol{\nabla} \times \mathbf{F} \cdot \mathbf{u} \, d\sigma. \tag{10}$$

If we replace the double integral on the right-hand side of Eq. (10) by the circulation, we get

$$(\boldsymbol{\nabla} \times \mathbf{F} \cdot \mathbf{u})_Q = \lim_{\rho \to 0} \frac{1}{\pi \rho^2} \oint_{C} \mathbf{F} \cdot d\mathbf{r}. \tag{11}$$

The left-hand side of Eq. (11) has its maximum value when $\mathbf{u}$ is the direction of $\boldsymbol{\nabla} \times \mathbf{F}$. When ρ is small, the limit on the right-hand side of Eq. (11) is

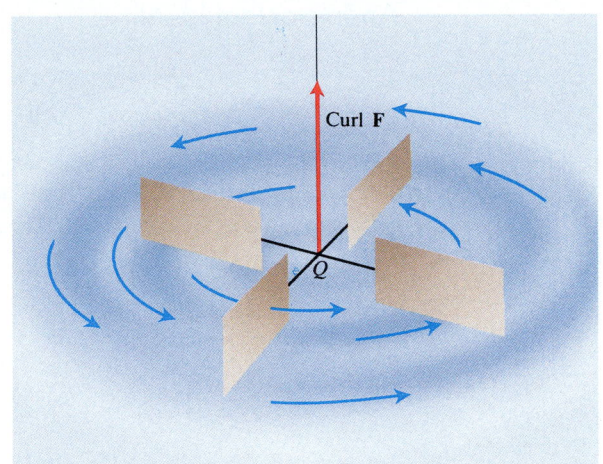

14.63 The paddle wheel interpretation of curl **F**.

approximately

$$\frac{1}{\pi\rho^2} \oint_C \mathbf{F} \cdot d\mathbf{r},$$

which is the circulation around C divided by the area of the disk (circulation density). Suppose that a small paddle wheel of radius ρ is introduced into the fluid at Q, with its axle directed along **u**. The circulation of the fluid around C will affect the rate of spin of the paddle wheel. The wheel will spin fastest when the circulation integral is maximized; therefore it will spin fastest when the axle of the paddle wheel points in the direction of $\nabla \times \mathbf{F}$ (Fig. 14.63).

Example 2   A fluid of constant density δ rotates around the z-axis with velocity $\mathbf{v} = \omega(-y\mathbf{i} + x\mathbf{j})$, where ω is a positive constant called the *angular velocity* of the rotation (Fig. 14.64). If $\mathbf{F} = \delta\mathbf{v}$, find $\nabla \times \mathbf{F}$ and show its relation to the circulation density.

Solution   With $\mathbf{F} = \delta\mathbf{v} = -\delta\omega y\mathbf{i} + \delta\omega x\mathbf{j}$,

$$\nabla \times \mathbf{F} = \left(\frac{\partial P}{\partial y} - \frac{\partial N}{\partial z}\right)\mathbf{i} + \left(\frac{\partial M}{\partial z} - \frac{\partial P}{\partial x}\right)\mathbf{j} + \left(\frac{\partial N}{\partial x} - \frac{\partial M}{\partial y}\right)\mathbf{k}$$

$$= (0 - 0)\mathbf{i} + (0 - 0)\mathbf{j} + (\delta\omega - (-\delta\omega))\mathbf{k}$$

$$= 2\,\delta\omega\mathbf{k}.$$

By Stokes's theorem, the circulation of **F** around a circle C of radius ρ bounding a disk S in a plane normal to $\nabla \times \mathbf{F}$, say the xy-plane, is

$$\oint_C \mathbf{F} \cdot d\mathbf{r} = \iint_S \nabla \times \mathbf{F} \cdot \mathbf{n}\, d\sigma = \iint_S 2\,\delta\omega\mathbf{k} \cdot \mathbf{k}\, dx\, dy = (2\,\delta\omega)(\pi\rho^2).$$

Thus,

$$(\nabla \times \mathbf{F}) \cdot \mathbf{k} = 2\,\delta\omega = \frac{1}{\pi\rho^2} \oint_C \mathbf{F} \cdot d\mathbf{r},$$

in agreement with Eq. (11) with $\mathbf{u} = \mathbf{k}$.

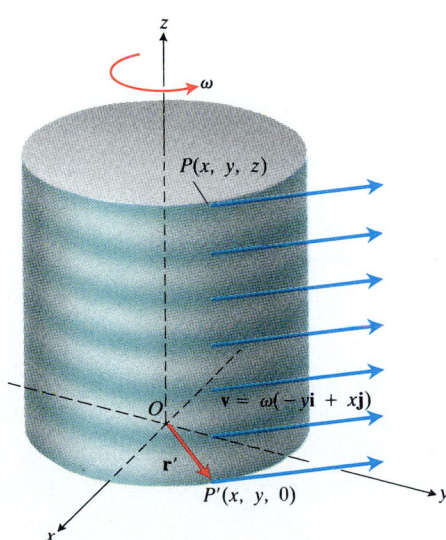

14.64 A steady rotational flow parallel to the xy-plane, with constant angular velocity ω in the positive (counter-clockwise) direction.

An Important Identity

The identity

$$\text{curl grad} f = \mathbf{0} \qquad \text{or} \qquad \nabla \times \nabla f = \mathbf{0} \tag{12}$$

arises frequently in mathematics and the physical sciences. (We shall use it ourselves in Section 14.7.) It holds for any function $f(x, y, z)$ whose second partial derivatives are continuous. The proof goes like this:

$$\nabla \times \nabla f = \begin{vmatrix} \mathbf{i} & \mathbf{j} & \mathbf{k} \\ \dfrac{\partial}{\partial x} & \dfrac{\partial}{\partial y} & \dfrac{\partial}{\partial z} \\ \dfrac{\partial f}{\partial x} & \dfrac{\partial f}{\partial y} & \dfrac{\partial f}{\partial z} \end{vmatrix} \tag{13}$$

$$= (f_{zy} - f_{yz})\,\mathbf{i} - (f_{zx} - f_{xz})\,\mathbf{j} + (f_{yx} - f_{xy})\,\mathbf{k}.$$

When the second partial derivatives of f are continuous, the mixed second derivatives in parentheses are equal and the vector is zero.

✳ Ampère's Law

In electromagnetic theory, Stokes's theorem gives Ampère's law:

$$\oint_C \mathbf{B} \cdot d\mathbf{r} = \iint_S \nabla \times \mathbf{B} \cdot \mathbf{n}\, d\sigma$$

$$= \frac{4\pi}{c} \iint_S \mathbf{J} \cdot \mathbf{n}\, d\sigma \tag{14}$$

$$= \frac{4\pi}{c} I.$$

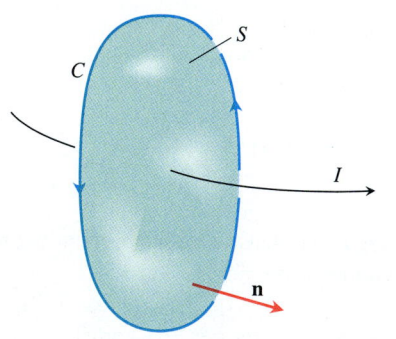

14.65   The directions of C, $\mathbf{n}$, and the current flow in Ampère's law.

In this equation, $\mathbf{B}$ is a magnetic field in space, I is a current passing through a loop C in the domain of $\mathbf{B}$, S is a surface spanning C, $\mathbf{J}$ is the current density on S (current per unit area), and c is the speed of light. The law tells us that the integral of $\mathbf{B}$ around a closed loop is proportional to the current through any surface bounded by the loop. The constant of proportionality is $4\pi/c$. The direction of $\mathbf{n}$ and the direction of I are assumed to have the same right-hand relationship to the direction of integration around C (Fig. 14.65).

Ampère's law is named after its discoverer, André-Marie Ampère (1775–1836), French physicist, mathematician, and natural philosopher known for his contributions to electrodynamics. The ampere, the SI unit of electrical current, is named for him.

Summary of Integral Theorems in del Notation

We close with Table 14.3, which lists the equations from Green's theorem and its generalizations to three dimensions.

TABLE 14.3
Green's theorem and its generalizations

Green's theorem:	$\oint_C \mathbf{F} \cdot \mathbf{n}\, ds = \int\int_R \nabla \cdot \mathbf{F}\, dA$
	$\oint_C \mathbf{F} \cdot d\mathbf{r} = \int\int_R \nabla \times \mathbf{F} \cdot \mathbf{k}\, dA$
The Divergence Theorem:	$\int\int_S \mathbf{F} \cdot \mathbf{n}\, d\sigma = \int\int\int_D \nabla \cdot \mathbf{F}\, dV$
Stokes's theorem:	$\oint_C \mathbf{F} \cdot d\mathbf{r} = \int\int_S \nabla \times \mathbf{F} \cdot \mathbf{n}\, d\sigma$

EXERCISES 14.6

In Exercises 1–6, use the surface integral in Stokes's theorem to calculate the circulation of the field $\mathbf{F}$ around the curve C in the indicated direction.

1. $\mathbf{F} = x^2\mathbf{i} + 2x\mathbf{j} + z^2\mathbf{k}$
 C: The ellipse $4x^2 + y^2 = 4$, counterclockwise as viewed from above

2. $\mathbf{F} = 2y\mathbf{i} + 3x\mathbf{j} - z^2\mathbf{k}$
 C: The circle $x^2 + y^2 = 9$ in the xy-plane, counterclockwise when viewed from above

3. $\mathbf{F} = y\mathbf{i} + xz\mathbf{j} + x^2\mathbf{k}$
 C: The boundary of the triangle cut from the plane $x + y + z = 1$ by the first octant, counterclockwise when viewed from above

4. $\mathbf{F} = (y^2 + z^2)\mathbf{i} + (x^2 + z^2)\mathbf{j} + (x^2 + y^2)\mathbf{k}$
 C: The boundary of the triangle cut from the plane $x + y + z = 1$ by the first octant, counterclockwise when viewed from above

5. $\mathbf{F} = (y^2 + z^2)\mathbf{i} + (x^2 + y^2)\mathbf{j} + (x^2 + y^2)\mathbf{k}$
 C: The square bounded by the lines $x = \pm 1$ and $y = \pm 1$ in the xy-plane, counterclockwise when viewed from above

6. $\mathbf{F} = x^2y^3\mathbf{i} + \mathbf{j} + z\mathbf{k}$
 C: The boundary of the circle $x^2 + y^2 = a^2$ in the xy-plane, counterclockwise when viewed from above

7. Let $\mathbf{n}$ be the outer unit normal of the elliptical shell
$$S: \quad 4x^2 + 9y^2 + 36z^2 = 36, \quad z \geq 0,$$
and let
$$\mathbf{F} = y\mathbf{i} + x^2\mathbf{j} + (x^2 + y^4)^{3/2} \sin e^{\sqrt{xyz}}\, \mathbf{k}.$$
Find the value of
$$\int\int_S \nabla \times \mathbf{F} \cdot \mathbf{n}\, d\sigma.$$

(*Hint:* One parametrization of the ellipse at the base of the shell is $x = 3 \cos t$, $y = 2 \sin t$, $0 \leq t \leq 2\pi$.)

8. Let $\mathbf{n}$ be the outer unit normal of the parabolic shell
$$S: 4x^2 + y + z^2 = 4, \quad y \geq 0$$
and let
$$\mathbf{F} = \left(-z + \frac{1}{2 + x}\right)\mathbf{i} + (\tan^{-1}y)\mathbf{j} + \left(x + \frac{1}{4 + z}\right)\mathbf{k}.$$
Find the value of
$$\int\int_S \nabla \times \mathbf{F} \cdot \mathbf{n}\, d\sigma.$$

9. Let S be the cylinder $x^2 + y^2 = a^2$, $0 \leq z \leq h$, together with its top, $x^2 + y^2 \leq a^2$, $z = h$. Let $\mathbf{F} = -y\mathbf{i} + x\mathbf{j} + x^2\mathbf{k}$. Use Stokes's theorem to calculate the flux of $\nabla \times \mathbf{F}$ outward through S.

10. Evaluate
$$\int\int_S \nabla \times (y\mathbf{i}) \cdot \mathbf{n}\, d\sigma$$
where S is the hemisphere $x^2 + y^2 + z^2 = 1$, $z \geq 0$.

11. Show that
$$\int\int_S \nabla \times \mathbf{F} \cdot \mathbf{n}\, d\sigma$$
has the same value for all oriented surfaces S that span C and that induce the same positive direction on C.

12. Show that if $\mathbf{F} = x\mathbf{i} + y\mathbf{j} + z\mathbf{k}$, then $\nabla \cdot \mathbf{F} = 3$ and $\nabla \times \mathbf{F} = \mathbf{0}$.

13. *div(curl F) = 0.* a) Show that if the necessary partial derivatives of the components of $\mathbf{F} = M\mathbf{i} + N\mathbf{j} + P\mathbf{k}$ are continuous, then $\nabla \cdot \nabla \times \mathbf{F} = 0$.
 b) Use the result in (a) to show that
$$\int\int_S \nabla \times \mathbf{F} \cdot \mathbf{n}\, d\sigma = 0.$$

for any surface to which the Divergence Theorem applies.

14. Use the identity $\mathbf{\nabla} \times \mathbf{\nabla} f = 0$ and Stokes's theorem to show that the circulations of the following fields around the boundary of any smooth orientable surface in space are zero.

a) $\mathbf{F} = 2x\mathbf{i} + 2y\mathbf{j} + 2z\mathbf{k}$
b) $\mathbf{F} = \mathbf{\nabla}(xy^2z^3)$
c) $\mathbf{F} = \mathbf{\nabla} \times (x\mathbf{i} + y\mathbf{j} + z\mathbf{k})$
d) $\mathbf{F} = \mathbf{\nabla} f$

15. Let $f(x, y, z) = (x^2 + y^2 + z^2)^{-1/2}$. Show that the clockwise circulation of the field $\mathbf{F} = \mathbf{\nabla} f$ around the circle $x^2 + y^2 = a^2$ in the xy-plane is zero

a) by taking $\mathbf{r} = (a \cos t)\mathbf{i} + (a \sin t)\mathbf{j}$, $0 \leq t \leq 2\pi$, and integrating $\mathbf{F} \cdot d\mathbf{r}$ over the circle, and
b) by applying Stokes's theorem.

16. Let C be a simple closed smooth curve in the plane $2x + 2y + z = 2$, oriented as in Fig. 14.66. Show that

$$\oint_C 2y \, dx + 3z \, dy - x \, dz$$

depends only on the area of the region enclosed by C and not on the position or shape of C.

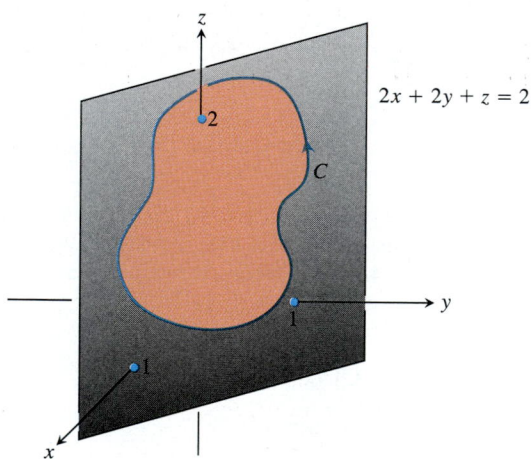

$2x + 2y + z = 2$

14.66 The curve in Exercise 16.

17. Suppose that a smooth closed oriented surface S in space is the union of two surfaces S_1 and S_2 joined to each other along a simple closed smooth curve C and let $\mathbf{n}$ be the unit normal vector field on S. Show that

$$\iint_S \mathbf{\nabla} \times \mathbf{F} \cdot \mathbf{n} \, d\sigma = 0$$

for any differentiable vector field defined on a region containing S and its interior.

18. Find a vector field with twice-differentiable components whose curl is $x\mathbf{i} + y\mathbf{j} + z\mathbf{k}$ or prove that no such field exists.

19. Let R be a region in the xy-plane that is bounded by a piecewise smooth simple closed curve C and suppose that the moments of inertia of R about the x- and y-axes are known to be I_x and I_y. Evaluate the line integral

$$\oint_C \mathbf{\nabla}(r^4) \cdot \mathbf{n} \, ds,$$

where $r = \sqrt{x^2 + y^2}$, in terms of I_x and I_y.

20. *Basic properties of curl and divergence.* Let $g(x, y, z)$ be a differentiable scalar function, let $\mathbf{F}$, $\mathbf{F}_1$, and $\mathbf{F}_2$ be differentiable vector fields, and let a and b be arbitrary real constants. Verify the following identities.

a) $\mathbf{\nabla} \cdot (a\mathbf{F}_1 + b\mathbf{F}_2) = a\mathbf{\nabla} \cdot \mathbf{F}_1 + b\mathbf{\nabla} \cdot \mathbf{F}_2$
b) $\mathbf{\nabla} \times (a\mathbf{F}_1 + b\mathbf{F}_2) = a\mathbf{\nabla} \times \mathbf{F}_1 + b\mathbf{\nabla} \times \mathbf{F}_2$
c) $\mathbf{\nabla} \cdot (g\mathbf{F}) = g\mathbf{\nabla} \cdot \mathbf{F} + \mathbf{\nabla}g \cdot \mathbf{F}$
d) $\mathbf{\nabla} \times (g\mathbf{F}) = g\mathbf{\nabla} \times \mathbf{F} + \mathbf{\nabla}g \times \mathbf{F}$
e) $\mathbf{\nabla} \cdot (\mathbf{F}_1 \times \mathbf{F}_2) = \mathbf{F}_2 \cdot \mathbf{\nabla} \times \mathbf{F}_1 - \mathbf{F}_1 \cdot \mathbf{\nabla} \times \mathbf{F}_2$

■ **EXPLORER PROGRAM**

Vector Fields — Displays two-dimensional vector fields and integrates the curl over rectangular regions. Also calculates flux across rectangular boundaries

14.7 Path Independence, Potential Functions, and Conservative Fields

This section introduces the notion of path independence of work integrals and discusses the remarkable properties of conservative fields, fields in which all work integrals are path independent. In an electric field, for example, the amount of work it takes to move a charge from one point to another depends only on the locations of the points and not on the path taken to get from one to the other. Electric fields are conservative. Gravitational fields are conservative, too: The amount of work it takes to move a mass from one point to another against gravity depends only on the locations of the points and not on the path taken in between. As we shall see, we can evaluate work integrals in such fields without resorting to parametrizations. We use the fields' so-called potential functions instead.

Path Independence

If A and B are two points in an open region D in space, the work $\int \mathbf{F} \cdot d\mathbf{r}$ done in moving a particle from A to B by a field $\mathbf{F}$ defined on D usually depends on the path taken. For some special fields, however, the integral's value depends only on the points A and B and is the same for all paths from A to B. If this is true for all points A and B in D, we say that the integral $\int \mathbf{F} \cdot d\mathbf{r}$ is path independent in D and that $\mathbf{F}$ is conservative on D.

DEFINITIONS

Let $\mathbf{F}$ be a field defined on an open region D in space, and suppose that for any two points A and B in D the work $\int_A^B \mathbf{F} \cdot d\mathbf{r}$ done in moving from A to B is the same over all paths in D from A to B. Then the integral $\int \mathbf{F} \cdot d\mathbf{r}$ is **path independent in D** and the field $\mathbf{F}$ is **conservative on D**.

The word *conservative* comes from physics, where it refers to fields in which the principle of conservation of energy holds (it does, in conservative fields).

Under conditions normally met in practice, a field $\mathbf{F}$ is conservative if and only if it is the gradient field of a scalar function f; that is, if and only if $\mathbf{F} = \nabla f$ for some f. The function f is then called a potential function for $\mathbf{F}$.

DEFINITIONS

If a scalar function f is differentiable throughout a domain D, the field of vectors ∇f is called the **gradient field** of f on D. If $\mathbf{F}$ is a field defined on D and $\mathbf{F} = \nabla f$ for some scalar function f on D, then f is called a **potential function for $\mathbf{F}$**.

An electric potential is a scalar function whose gradient field is an electric field. A gravitational potential is a scalar function whose gradient field is a gravitational field, and so on. As we shall see, once we have found a potential function f for a field $\mathbf{F}$, we can evaluate all work integrals in the domain of $\mathbf{F}$ by the rule

$$\int_A^B \mathbf{F} \cdot d\mathbf{r} = f(B) - f(A). \tag{1}$$

This is like the Fundamental Theorem of Calculus, and we call it the Fundamental Theorem of Line Integrals. The difference $f(B) - f(A)$ in the values of the potential function in Eq. (1) is called the **potential difference** of B with respect to A or, less precisely, the potential difference *between B and A*.

Conservative fields have other remarkable properties. Saying that a field $\mathbf{F}$ is conservative on D is equivalent to saying that the integral of $\mathbf{F}$ around every closed loop in D is zero. This, in turn, is equivalent in a great many domains to saying that $\nabla \times \mathbf{F} = \mathbf{0}$. We explain these relationships by discussing the implications in the following diagram.

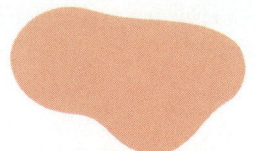

Connected and simply connected.

Connected but not simply connected.

Connected and simply connected.

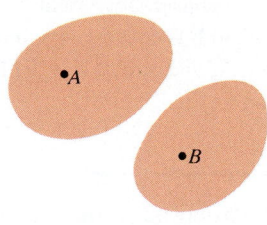

Simply connected but not connected.
No path from A to B lies entirely in the region.

14.67 Connectivity and simple connectivity are not the same thing. In fact, neither implies the other, as these pictures of plane regions illustrate. To make three-dimensional regions with these properties, thicken the plane regions into cylinders.

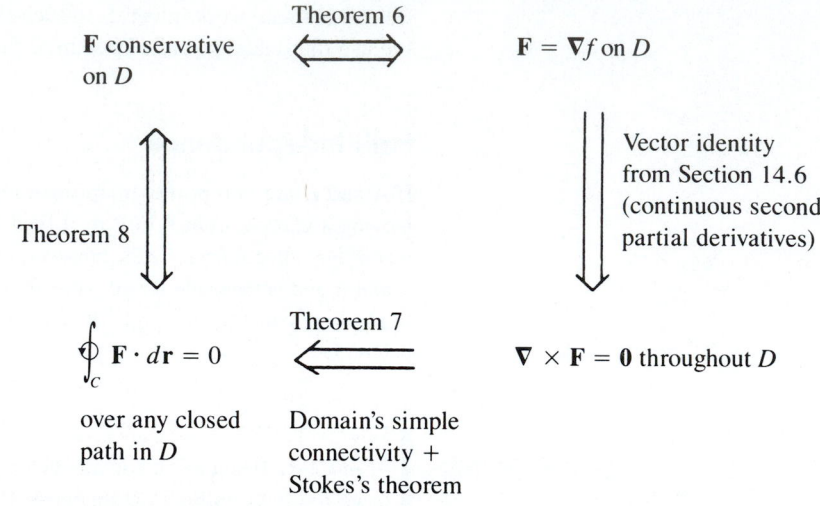

Naturally, we need to impose conditions on the curves, fields, and domains to make these implications hold.

We assume that all of the curves are piecewise smooth.

We also assume that the components of $\mathbf{F}$ have continuous second partial derivatives. Without this assumption, $\mathbf{F} = \nabla f$ need not imply $\nabla \times \mathbf{F} = \mathbf{0}$. As we saw in the previous section, the implication requires the mixed second derivatives of f to be equal. We assume their continuity to ensure that.

We assume D to be an *open* region in space. This means that every point in D is the center of a solid sphere that lies entirely in D. We need an added restriction on D if $\nabla \times \mathbf{F} = \mathbf{0}$ is to imply that the integral of $\mathbf{F}$ around every closed path in D is zero. For this, we assume D to be **simply connected.** This means that every closed path in D can be contracted to a point in D without ever leaving D. If D consisted of space with one of the axes removed, for example, D would not be simply connected. There would be no way to contract a loop around the axis to a point without leaving D. (See Exercise 33.) On the other hand, space itself *is* simply connected.

Finally, we assume D to be **connected,** which in an open region means that every point can be connected to every other point by a smooth curve that lies in the region. See Fig. 14.67.

The Implications in the Diagram

We now examine the implications in the preceding diagram.

THEOREM 6

The Fundamental Theorem of Line Integrals

1. Let $\mathbf{F} = M\mathbf{i} + N\mathbf{j} + P\mathbf{k}$ be a vector field whose components are continuous throughout an open connected region D in space. Then there exists a differentiable function f such that

$$\mathbf{F} = \nabla f = \frac{\partial f}{\partial x}\mathbf{i} + \frac{\partial f}{\partial y}\mathbf{j} + \frac{\partial f}{\partial z}\mathbf{k}$$

if and only if for all points A and B in D the value of $\displaystyle\int_A^B \mathbf{F} \cdot d\mathbf{r}$ is independent of the path joining A to B in D.

2. If the integral is independent of the path from A to B, its value is

$$\int_A^B \mathbf{F} \cdot d\mathbf{r} = f(B) - f(A).$$

Proof That $\mathbf{F} = \nabla f$ Implies Path Independence of $\int \mathbf{F} \cdot d\mathbf{r}$ Suppose that A and B are two points in D and that

$$C: \quad \mathbf{r} = g(t)\,\mathbf{i} + h(t)\,\mathbf{j} + k(t)\,\mathbf{k}, \quad a \le t \le b,$$

is a smooth curve in D joining A and B. Along the curve, f is a differentiable function of t and

$$\frac{df}{dt} = \frac{\partial f}{\partial x}\frac{dx}{dt} + \frac{\partial f}{\partial y}\frac{dy}{dt} + \frac{\partial f}{\partial z}\frac{dz}{dt} \qquad \text{(Chain Rule)}$$

$$= \nabla f \cdot \left[\frac{dx}{dt}\mathbf{i} + \frac{dy}{dt}\mathbf{j} + \frac{dz}{dt}\mathbf{k}\right] = \nabla f \cdot \frac{d\mathbf{r}}{dt} = \mathbf{F} \cdot \frac{d\mathbf{r}}{dt}. \qquad \left(\begin{matrix}\text{Because}\\ \mathbf{F} = \nabla f\end{matrix}\right)$$

(2)

Therefore,

$$\int_C \mathbf{F} \cdot d\mathbf{r} = \int_a^b \mathbf{F} \cdot \frac{d\mathbf{r}}{dt}\, dt = \int_a^b \frac{df}{dt}\, dt = f(g(t), h(t), k(t))\Big]_a^b = f(B) - f(A). \quad \text{(Eq. (2))}$$

Thus, the value of the work integral depends only on the values of f at A and B and not on the path in between. This proves Part 2 as well as the forward implication in Part 1. ∎

Proof That Path Independence of $\int \mathbf{F} \cdot d\mathbf{r}$ Implies $\mathbf{F} = \nabla f$

We need to show that if $\mathbf{F} = M\mathbf{i} + N\mathbf{j} + P\mathbf{k}$ is conservative on D then there exists a scalar function f defined on D for which

$$\frac{\partial f}{\partial x} = M, \qquad \frac{\partial f}{\partial y} = N, \qquad \text{and} \qquad \frac{\partial f}{\partial z} = P. \qquad (3)$$

To define f, we choose an arbitrary basepoint A in D. We define the value of f at A to be zero and the value of f at every other point B of D to be the value of the integral of $\mathbf{F} \cdot d\mathbf{r}$ along some smooth path in D from A to B. Such paths exist because D is a connected open region, and all such paths assign the same value to $f(B)$ because the integral of $\mathbf{F} \cdot d\mathbf{r}$ is path independent in D. Since the integral of $\mathbf{F} \cdot d\mathbf{r}$ from A to A is zero, the values of f are all defined by the single equation

$$f(B) = \int_A^B \mathbf{F} \cdot d\mathbf{r}. \qquad (4)$$

One consequence of this definition is that for any points B and B_0 in D,

$$f(B) - f(B_0) = \int_A^B \mathbf{F} \cdot d\mathbf{r} - \int_A^{B_0} \mathbf{F} \cdot d\mathbf{r}$$

$$= \int_{B_0}^A \mathbf{F} \cdot d\mathbf{r} + \int_A^B \mathbf{F} \cdot d\mathbf{r} = \int_{B_0}^B \mathbf{F} \cdot d\mathbf{r}.$$

(5)

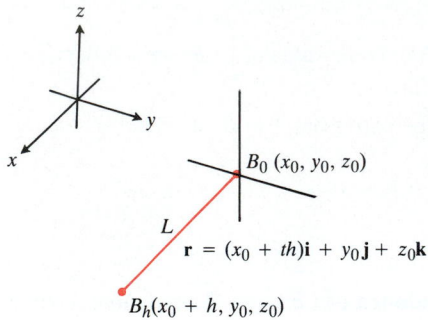

14.68   The line segment L from B_0 to B_h lies parallel to the x-axis. The partial derivative with respect to x of the function

$$f(x, y, z) = \int_A^{(x, y, z)} \mathbf{F} \cdot d\mathbf{r}$$

at B_0 is the limit of $(1/h)\int_L \mathbf{F} \cdot d\mathbf{r}$ as $h \to 0$.

Let us now show that f has partial derivatives that give the components of $\mathbf{F}$ at every point $B_0(x_0, y_0, z_0)$ in D.

Since D is open, B_0 is the center of a solid sphere that lies entirely in D. We can therefore find a positive number h (it might be small) for which the point $B_h(x_0 + h, y_0, z_0)$ and the line segment L joining B_0 to B_h both lie in D (Fig. 14.68). When we calculate the value of the integral of $\mathbf{F} \cdot d\mathbf{r}$ along L with the parametrization

$$\mathbf{r} = (x_0 + th)\,\mathbf{i} + y_0\,\mathbf{j} + z_0\,\mathbf{k}, \qquad 0 \le t \le 1, \tag{6}$$

we find that

$$\int_{B_0}^{B_h} \mathbf{F} \cdot d\mathbf{r} = \int_0^1 (M\mathbf{i} + N\mathbf{j} + P\mathbf{k}) \cdot (h\,dt\,\mathbf{i})$$

$$= h \int_0^1 M(x_0 + th, y_0, z_0)\,dt. \tag{7}$$

The difference quotient for defining $\partial f/\partial x$ at B_0 is therefore

$$\frac{f(x_0 + h, y_0, z_0) - f(x_0, y_0, z_0)}{h} = \frac{f(B_h) - f(B_0)}{h}$$

$$= \frac{1}{h} \int_{B_0}^{B_h} \mathbf{F} \cdot d\mathbf{r} \qquad \text{(Eq. (5))}$$

$$= \int_0^1 M(x_0 + th, y_0, z_0)\,dt. \qquad \text{(Eq. (7))}$$

Since M is continuous, given any $\epsilon > 0$ there exists a $\delta > 0$ such that

$$|th| < \delta \quad \Rightarrow \quad |M(x_0 + th, y_0, z_0) - M(x_0, y_0, z_0)| < \epsilon.$$

This implies that whenever $|h| < \delta$

$$\left| \int_0^1 M(x_0 + h, y_0, z_0)\,dt - M(x_0, y_0, z_0) \right|$$

$$= \left| \int_0^1 (M(x_0 + h, y_0, z_0) - M(x_0, y_0, z_0))\,dt \right|$$

$$\le \int_0^1 \left| M(x_0 + h, y_0, z_0) - M(x_0, y_0, z_0) \right| dt < \int_0^1 \epsilon\,dt = \epsilon.$$

We can find such a positive δ for every positive ϵ, so

$$\left. \frac{\partial f}{\partial x} \right|_{(x_0, y_0, z_0)} = \lim_{h \to 0} \int_0^1 M(x_0 + th, y_0, z_0)\,dt \tag{8}$$

$$= M(x_0, y_0, z_0).$$

In other words, the partial derivative of f with respect to x exists at B_0 and equals the value of M at B_0. Since B_0 was chosen arbitrarily, we can conclude that $\partial f/\partial x$ exists and equals M at every point of D.

The equations $\partial f/\partial y = N$ and $\partial f/\partial z = P$ are derived in a similar way. This concludes the proof.

THEOREM 7

If $\nabla \times \mathbf{F} = \mathbf{0}$ at every point of a simply connected open region D in space, then

$$\oint_C \mathbf{F} \cdot d\mathbf{r} = 0$$

on any piecewise smooth closed path C in D.

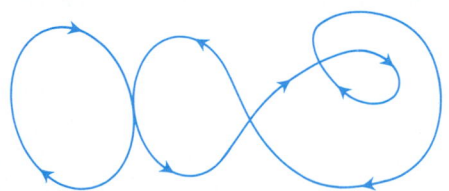

14.69 In a simply connected open region in space, differentiable curves that cross themselves can be divided into loops to which Stokes's theorem applies.

Sketch of a Proof The theorem is usually proved in two steps. The first step is for simple closed curves. A theorem from topology, a branch of advanced mathematics, states that every differentiable simple closed curve C in a simply connected open region D in space is the boundary of a smooth two-sided surface S that also lies in D. Hence, by Stokes's theorem,

$$\oint_C \mathbf{F} \cdot d\mathbf{r} = \iint_S \nabla \times \mathbf{F} \cdot \mathbf{n} \, d\sigma = 0.$$

The second step is for curves that cross themselves, like the one in Fig. 14.69. The idea is to break these into simple loops spanned by orientable surfaces, apply Stokes's theorem one loop at a time, and add the results. ∎

THEOREM 8

The following statements are equivalent:

1. $\int \mathbf{F} \cdot d\mathbf{r} = 0$ around every closed loop in D.
2. The field $\mathbf{F}$ is conservative on D.

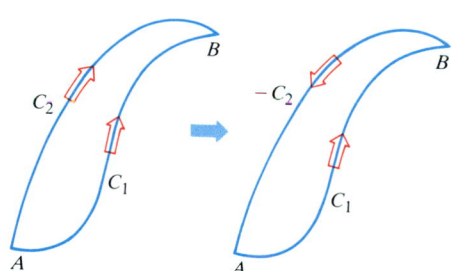

14.70 If we have two paths from A to B, one of them can be reversed to make a loop.

Proof That (1) $\Rightarrow$ (2) We want to show that for any two points A and B in D the integral of $\mathbf{F} \cdot d\mathbf{r}$ has the same value over any two paths C_1 and C_2 from A to B. We reverse the direction on C_2 to make a path $-C_2$ from B to A (Fig. 14.70). Together, C_1 and $-C_2$ make a closed loop C, and

$$\int_{C_1} \mathbf{F} \cdot d\mathbf{r} - \int_{C_2} \mathbf{F} \cdot d\mathbf{r} = \int_{C_1} \mathbf{F} \cdot d\mathbf{r} + \int_{-C_2} \mathbf{F} \cdot d\mathbf{r} = \int_C \mathbf{F} \cdot d\mathbf{r} = 0.$$

Thus the integrals over C_1 and C_2 give the same value.

Proof That (2) $\Rightarrow$ (1) We want to show that the integral of $\mathbf{F} \cdot d\mathbf{r}$ is zero over any closed loop C. We pick two points A and B on C and use them to break C into two pieces: C_1 from A to B followed by C_2 from B back to A (Fig. 14.71). Then

$$\oint_C \mathbf{F} \cdot d\mathbf{r} = \int_{C_1} \mathbf{F} \cdot d\mathbf{r} + \int_{C_2} \mathbf{F} \cdot d\mathbf{r} = \int_A^B \mathbf{F} \cdot d\mathbf{r} - \int_A^B \mathbf{F} \cdot d\mathbf{r} = 0.$$ ∎

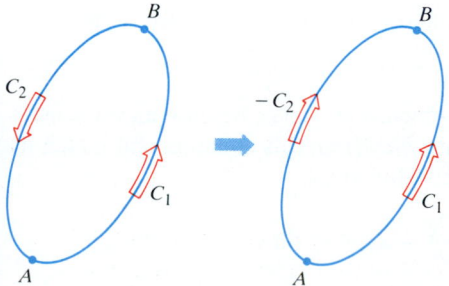

14.71 If A and B lie on a loop, we can reverse part of the loop to make two paths from A to B.

Finding Potentials for Conservative Fields

As we said earlier, conservative fields get their name from the fact that the principle of conservation of energy holds in them. The test for being conservative, as we have just seen, is that $\nabla \times \mathbf{F} = \mathbf{0}$. In terms of components, this gives us the following test.

Component Test for Conservative Fields

The field $\mathbf{F} = M(x, y, z)\,\mathbf{i} + N(x, y, z)\,\mathbf{j} + P(x, y, z)\,\mathbf{k}$ is conservative if and only if

$$\frac{\partial P}{\partial y} = \frac{\partial N}{\partial z}, \qquad \frac{\partial M}{\partial z} = \frac{\partial P}{\partial x}, \qquad \text{and} \qquad \frac{\partial N}{\partial x} = \frac{\partial M}{\partial y}. \tag{9}$$

When we know that $\mathbf{F}$ is conservative, we usually want to find a potential function for $\mathbf{F}$. This requires solving the equation

$$\nabla f = \frac{\partial f}{\partial x}\,\mathbf{i} + \frac{\partial f}{\partial y}\,\mathbf{j} + \frac{\partial f}{\partial z}\,\mathbf{k} = M\mathbf{i} + N\mathbf{j} + P\mathbf{k} \tag{10}$$

for f. We accomplish this by integrating the three equations

$$\frac{\partial f}{\partial x} = M, \qquad \frac{\partial f}{\partial y} = N, \qquad \frac{\partial f}{\partial z} = P, \tag{11}$$

as in the following example.

Example 1   Show that

$$\mathbf{F} = (e^x \cos y + yz)\,\mathbf{i} + (xz - e^x \sin y)\,\mathbf{j} + (xy + z)\,\mathbf{k}$$

is conservative and find a potential function for it. Then use the potential function to find the value of the integral of $\mathbf{F}$ along paths from $(0, 0, 0)$ to $(-1, \pi/2, 2)$.

Solution   To show that $\mathbf{F}$ is conservative, we apply the test in Eq. (9) to

$$M = e^x \cos y + yz, \qquad N = xz - e^x \sin y, \qquad P = xy + z$$

and calculate

$$\frac{\partial M}{\partial z} = y = \frac{\partial P}{\partial x}, \qquad \frac{\partial N}{\partial z} = x = \frac{\partial P}{\partial y}, \qquad \frac{\partial M}{\partial y} = -e^x \sin y + z = \frac{\partial N}{\partial x}.$$

Together, these equalities tell us that $\mathbf{F}$ is conservative and that there is a function f with $\nabla f = \mathbf{F}$.

We find a potential function f by integrating the equations

$$\frac{\partial f}{\partial x} = e^x \cos y + yz, \qquad \frac{\partial f}{\partial y} = xz - e^x \sin y, \qquad \frac{\partial f}{\partial z} = xy + z. \tag{12}$$

We integrate the first equation with respect to x, holding y and z fixed, to get

$$f(x, y, z) = e^x \cos y + xyz + g(y, z).$$

We write the constant of integration as a function of y and z because its value may change if y and z change. We then calculate $\partial f / \partial y$ from this equation and match it with the expression for $\partial f / \partial y$ in Eq. (12). This gives

$$-e^x \sin y + xz + \frac{\partial g}{\partial y} = xz - e^x \sin y,$$

so

$$\frac{\partial g}{\partial y} = 0.$$

Therefore, g is a function of z alone, and

$$f(x, y, z) = e^x \cos y + xyz + h(z).$$

We now calculate $\partial f/\partial z$ from this equation and match it to the formula for $\partial f/\partial z$ in Eq. (12). This gives

$$xy + \frac{dh}{dz} = xy + z, \qquad \text{or} \qquad \frac{dh}{dz} = z,$$

so

$$h(z) = \frac{z^2}{2} + C.$$

Hence,

$$f(x, y, z) = e^x \cos y + xyz + \frac{z^2}{2} + C.$$

We have found infinitely many potential functions for $\mathbf{F}$, one for each value of C.

By the Fundamental Theorem of Line Integrals, the value of the integral of $\mathbf{F}$ over every path from $(0, 0, 0)$ to $(-1, \pi/2, 2)$ is

$$\int_{(0,0,0)}^{(-1,\,\pi/2,\,2)} \mathbf{F} \cdot d\mathbf{r} = f(-1, \pi/2, 2) - f(0, 0, 0)$$

$$= \left(0 + (-1)\left(\frac{\pi}{2}\right)(2) + \frac{(2)^2}{2} + C\right) - (1 + 0 + 0 + C)$$

$$= -\pi + 2 - 1 = -\pi + 1.$$

Example 2   Show that $\mathbf{F} = (2x - 3)\,\mathbf{i} - z\mathbf{j} + (\cos z)\,\mathbf{k}$ is not conservative.

Solution   We apply the component test in Eq. (9) and find right away that

$$\frac{\partial P}{\partial y} = \frac{\partial}{\partial y}(\cos z) = 0, \qquad \frac{\partial N}{\partial z} = \frac{\partial}{\partial z}(-z) = -1.$$

The two are unequal, so $\mathbf{F}$ is not conservative. No further testing is required.

Exact Differential Forms

The forms $M\,dx + N\,dy + P\,dz$ that have been appearing in our line integrals are called differential forms. The most useful differential forms are differentials of scalar functions.

DEFINITIONS

The form

$$M(x, y, z)\,dx + N(x, y, z)\,dy + P(x, y, z)\,dz \tag{13}$$

is called a **differential form**. A differential form is **exact** on a domain D in space if

$$M\,dx + N\,dy + P\,dz = \frac{\partial f}{\partial x}\,dx + \frac{\partial f}{\partial y}\,dy + \frac{\partial f}{\partial z}\,dz = df \tag{14}$$

for some scalar function f throughout D.

Notice that if $M\,dx + N\,dy + P\,dz = df$ on D, then the field $\mathbf{F} = M\mathbf{i} + N\mathbf{j} + P\mathbf{k}$ is the gradient field of f on D. Conversely, if $\mathbf{F} = \nabla f$, then the form $M\,dx + N\,dy + P\,dz$ is exact. The test for the form's being exact is therefore the same as the test for $\mathbf{F}$'s being conservative.

The Test for Exactness of $M\,dx + N\,dy + P\,dz$

The differential form $M\,dx + N\,dy + P\,dz$ is exact if and only if

$$\frac{\partial P}{\partial y} = \frac{\partial N}{\partial z}, \qquad \frac{\partial M}{\partial z} = \frac{\partial P}{\partial x}, \qquad \text{and} \qquad \frac{\partial N}{\partial x} = \frac{\partial M}{\partial y}. \tag{15}$$

This is equivalent to saying that the field $\mathbf{F} = M\mathbf{i} + N\mathbf{j} + P\mathbf{k}$ is conservative or that

$$\nabla \times \mathbf{F} = \mathbf{0}. \tag{16}$$

Example 3   Show that $y\,dx + x\,dy + 4\,dz$ is exact and evaluate the integral

$$\int_{(1,1,1)}^{(2,3,-1)} y\,dx + x\,dy + 4\,dz$$

over the line segment from $(1, 1, 1)$ to $(2, 3, -1)$.

Solution   We let

$$M = y, \qquad N = x, \qquad P = 4$$

and apply the test of Eq. (15):

$$\frac{\partial M}{\partial z} = 0 = \frac{\partial P}{\partial x}, \qquad \frac{\partial N}{\partial z} = 0 = \frac{\partial P}{\partial y}, \qquad \frac{\partial M}{\partial y} = 1 = \frac{\partial N}{\partial x}.$$

These equalities tell us that $y\,dx + x\,dy + 4\,dz$ is exact, so

$$y\,dx + x\,dy + 4\,dz = df$$

for some function f, and the integral's value is $f(2, 3, -1) - f(1, 1, 1)$.

We find f up to a constant by integrating the equations

$$\frac{\partial f}{\partial x} = y, \qquad \frac{\partial f}{\partial y} = x, \qquad \frac{\partial f}{\partial z} = 4. \tag{17}$$

From the first equation we get

$$f(x, y, z) = xy + g(y, z).$$

The second equation tells us that

$$\frac{\partial f}{\partial y} = x + \frac{\partial g}{\partial y} = x, \qquad \text{or} \qquad \frac{\partial g}{\partial y} = 0.$$

Hence, g is a function of z alone, and

$$f(x, y, z) = xy + h(z).$$

The third of Eqs. (17) tells us that

$$\frac{\partial f}{\partial z} = 0 + \frac{dh}{dz} = 4, \qquad \text{or} \qquad h(z) = 4z + C.$$

Therefore,

$$f(x, y, z) = xy + 4z + C,$$

and the value of the integral is

$$f(2, 3, -1) - f(1, 1, 1) = 2 + C - (5 + C) = -3.$$

In the next chapter, we shall see how exact differential forms sometimes arise in differential equations.

EXERCISES 14.7

Which fields in Exercises 1–6 are conservative and which are not? Explain.

1. $\mathbf{F} = yz\mathbf{i} + xz\mathbf{j} + xy\mathbf{k}$

2. $\mathbf{F} = (y \sin z)\mathbf{i} + (x \sin z)\mathbf{j} + (xy \cos z)\mathbf{k}$

3. $\mathbf{F} = y\mathbf{i} + (x + z)\mathbf{j} - y\mathbf{k}$

4. $\mathbf{F} = -y\mathbf{i} + x\mathbf{j}$

5. $\mathbf{F} = (z + y)\mathbf{i} + z\mathbf{j} + (y + x)\mathbf{k}$

6. $\mathbf{F} = (e^x \cos y)\mathbf{i} - (e^x \sin y)\mathbf{j} + z\mathbf{k}$

In Exercises 7–12, find a potential function f for the field $\mathbf{F}$.

7. $\mathbf{F} = 2x\mathbf{i} + 3y\mathbf{j} + 4z\mathbf{k}$

8. $\mathbf{F} = (y + z)\mathbf{i} + (x + z)\mathbf{j} + (x + y)\mathbf{k}$

9. $\mathbf{F} = e^{y + 2z}(\mathbf{i} + x\mathbf{j} + 2x\mathbf{k})$

10. $\mathbf{F} = (y \sin z)\mathbf{i} + (x \sin z)\mathbf{j} + (xy \cos z)\mathbf{k}$

11. $\mathbf{F} = (\ln x + \sec^2(x + y))\mathbf{i} +$
$$\left(\sec^2(x + y) + \frac{y}{y^2 + z^2}\right)\mathbf{j} + \frac{z}{y^2 + z^2}\mathbf{k}$$

12. $\mathbf{F} = \dfrac{y}{1 + x^2 y^2}\mathbf{i} + \left(\dfrac{x}{1 + x^2 y^2} + \dfrac{z}{\sqrt{1 - y^2 z^2}}\right)\mathbf{j} +$
$$\left(\frac{y}{\sqrt{1 - y^2 z^2}} + \frac{1}{z}\right)\mathbf{k}$$

Evaluate the integrals in Exercises 13–22.

13. $\displaystyle\int_{(0, 0, 0)}^{(2, 3, -6)} 2x \, dx + 2y \, dy + 2z \, dz$

14. $\displaystyle\int_{(1, 1, 2)}^{(3, 5, 0)} yz \, dx + xz \, dy + xy \, dz$

15. $\displaystyle\int_{(0, 0, 0)}^{(1, 2, 3)} 2xy \, dx + (x^2 - z^2) \, dy - 2yz \, dz$

16. $\displaystyle\int_{(1, 1, 1)}^{(1, 2, 3)} 3x^2 \, dx + \frac{z^2}{y} \, dy + 2z \ln y \, dz$

17. $\displaystyle\int_{(1, 0, 0)}^{(0, 1, 1)} \sin y \cos x \, dx + \cos y \sin x \, dy + dz$

18. $\displaystyle\int_{(0, 0, 0)}^{(3, 3, 1)} 2x \, dx - y^2 \, dy - \frac{4}{1 + z^2} \, dz$

19. $\displaystyle\int_{(0, 2, 1)}^{(1, \pi/2, 2)} 2 \cos y \, dx + \left(\frac{1}{y} - 2x \sin y\right) dy + \frac{1}{z} \, dz$

20. $\displaystyle\int_{(1, 2, 1)}^{(2, 1, 1)} (2x \ln y - yz) \, dx + \left(\frac{x^2}{y} - xz\right) dy - xy \, dz$

21. $\displaystyle\int_{(1, 1, 1)}^{(2, 2, 2)} \frac{1}{y} \, dx + \left(\frac{1}{z} - \frac{x}{y^2}\right) dy - \frac{y}{z^2} \, dz$

22. $\displaystyle\int_{(-1, -1, -1)}^{(2, 2, 2)} \frac{2x \, dx + 2y \, dy + 2z \, dz}{x^2 + y^2 + z^2}$

23. Evaluate the integral
$$\int_{(1, 1, 1)}^{(2, 3, -1)} y \, dx + x \, dy + 4 \, dz$$
from Example 3 by finding parametric equations for the line segment from $(1, 1, 1)$ to $(2, 3, -1)$ and evaluating the line integral of $\mathbf{F} = y\mathbf{i} + x\mathbf{j} + 4\mathbf{k}$ along the segment. This is permissible in this case because $\mathbf{F}$ is conservative and the integral is path independent.

24. Evaluate
$$\int_C x^2 \, dx + yz \, dy + (y^2/2) \, dz$$
along the line segment C joining $(0, 0, 0)$ to $(0, 3, 4)$.

Show that the values of the integrals in Exercises 25 and 26 do not depend on the path taken from A to B.

25. $\displaystyle\int_A^B z^2 \, dx + 2y \, dy + 2xz \, dz$

26. $\displaystyle\int_A^B \frac{x\,dx + y\,dy + z\,dz}{\sqrt{x^2 + y^2 + z^2}}$

27. Express $\mathbf{F}$ in the form ∇f if

a) $\mathbf{F} = \dfrac{2x}{y}\,\mathbf{i} + \left(\dfrac{1 - x^2}{y^2}\right)\mathbf{j}$

b) $\mathbf{F} = (e^x \ln y)\,\mathbf{i} + \left(\dfrac{e^x}{y} + \sin z\right)\mathbf{j} + (y \cos z)\,\mathbf{k}$

28. Let $\mathbf{F} = \nabla(x^3 y^2)$ and let C be the path in the xy-plane from $(-1, 1)$ to $(1, 1)$ that consists of the line segment from $(-1, 1)$ to $(0, 0)$ followed by the line segment from $(0, 0)$ to $(1, 1)$. Evaluate $\int_C \mathbf{F} \cdot d\mathbf{r}$ in two ways:

a) by finding parametrizations for the segments that make up C and evaluating the integral;

b) by using the fact that $f(x, y) = x^3 y^2$ is a potential function for $\mathbf{F}$.

29. Find the work done by $\mathbf{F} = (x^2 + y)\,\mathbf{i} + (y^2 + x)\,\mathbf{j} + ze^z\,\mathbf{k}$ over the following paths from $(1, 0, 0)$ to $(1, 0, 1)$.

a) The line segment $x = 1$, $y = 0$, $0 \le z \le 1$

b) The helix $\mathbf{r} = (\cos t)\,\mathbf{i} + (\sin t)\,\mathbf{j} + (t/2\pi)\,\mathbf{k}$, $0 \le t \le 2\pi$

c) The x-axis from $(1, 0, 0)$ to $(0, 0, 0)$ followed by the parabola $z = x^2$, $y = 0$ from $(0, 0, 0)$ to $(1, 0, 1)$

30. Evaluate $\int_C 2x \cos y\,dx - x^2 \sin y\,dy$ along the following paths C in the xy-plane.

a) The parabola $y = (x - 1)^2$ from $(1, 0)$ to $(0, 1)$

b) The line segment from $(-1, \pi)$ to $(1, 0)$

c) The x-axis from $(-1, 0)$ to $(1, 0)$

d) The astroid $\mathbf{r} = (\cos^3 t)\,\mathbf{i} + (\sin^3 t)\,\mathbf{j}$, $0 \le t \le 2\pi$, counterclockwise from $(1, 0)$ back to $(1, 0)$

31. a) How are the constants a, b, and c related if the differential form

$(ay^2 + 2czx)\,dx + y(bx + cz)\,dy + (ay^2 + cx^2)\,dz$

is exact?

b) For what values of b and c will

$\mathbf{F} = (y^2 + 2czx)\,\mathbf{i} + y(bx + cz)\,\mathbf{j} + (y^2 + cx^2)\,\mathbf{k}$

be a gradient field?

32. Find a potential function for the gravitational field

$$\mathbf{F} = -GmM\,\frac{x\mathbf{i} + y\mathbf{j} + z\mathbf{k}}{(x^2 + y^2 + z^2)^{3/2}}$$

$(G, m,$ and M are constants).

33. Show that the curl of

$$\mathbf{F} = \frac{-y}{x^2 + y^2}\,\mathbf{i} + \frac{x}{x^2 + y^2}\,\mathbf{j} + z\mathbf{k}$$

is zero but that

$$\int_C \mathbf{F} \cdot d\mathbf{r}$$

is not zero if C is the circle $x^2 + y^2 = 1$ in the xy-plane. (Theorem 7 does not apply here because the domain of $\mathbf{F}$ is not simply connected. The field $\mathbf{F}$ is not defined along the z-axis so there is no way to contract C to a point without leaving the domain of $\mathbf{F}$.)

EXPLORER PROGRAM

Scalar Fields — Plots two-dimensional scalar fields and their associated gradient fields. Integrates $\nabla f \cdot d\mathbf{r}$ along line segments parallel to the coordinate axes

REVIEW QUESTIONS

1. What is a line integral? What are line integrals used for? Give examples.

2. What is a vector field? Give examples.

3. How do you calculate the work done by a force in moving a particle along a curve?

4. What are the flux density and circulation density of a vector field? How are they related to the divergence and curl of the field?

5. What theorem relates flux to the divergence for a vector field defined in the plane? Give an example.

6. What theorem relates circulation to the curl for a vector field defined in the plane? Give an example.

7. How is surface area calculated? Give an example.

8. What is a surface integral? How are surface integrals used to calculate masses and moments? Give examples.

9. What theorem relates flux to the divergence of a vector field in space? Give an example.

10. What theorem relates circulation to the curl of a vector field in space? Give an example.

11. What does it mean for a vector field to be conservative on a region D in space?

12. What is a potential function for a conservative field? Give an example.

13. How can you evaluate $\int_A^B \mathbf{F} \cdot d\mathbf{r}$ over a path in the domain of a conservative field? Give an example.

14. What special properties do conservative fields have?

15. How do you find a potential function for a conservative field?

16. What is an exact differential form? Give an example.

MISCELLANEOUS EXERCISES

1. Figure 14.72 shows two polygonal paths in space from the origin to the point $(1, 1, 1)$. Integrate $f(x, y, z) = 2x - 3y^2 - 2z + 3$ over each path.

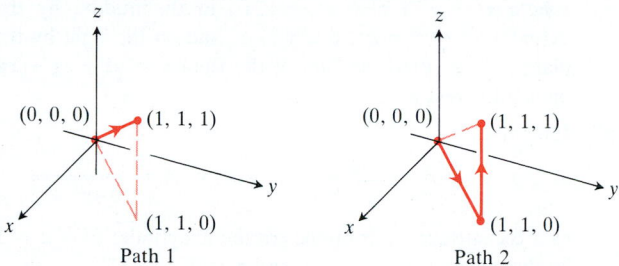

14.72   The paths in Exercise 1.

2. Figure 14.73 shows three polygonal paths from the origin to the point $(1, 1, 1)$. Integrate $f(x, y, z) = x^2 + y - z$ over each path.

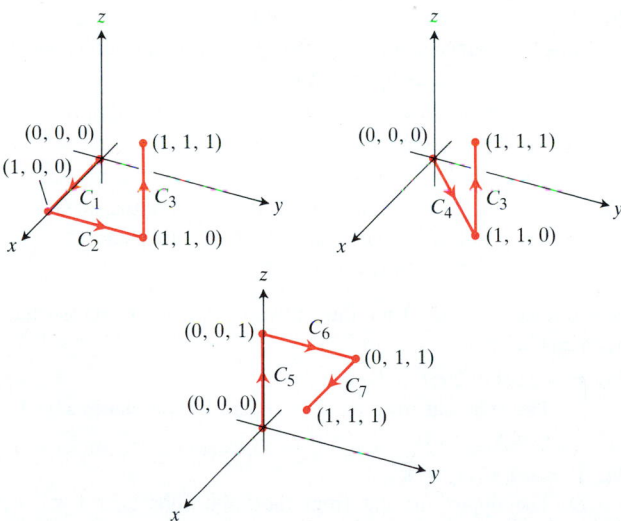

14.73   The paths in Exercise 2.

3. Integrate $f(x, y, z) = \sqrt{x^2 + z^2}$ around the circle $\mathbf{r} = (a \cos t) \mathbf{j} + (a \sin t) \mathbf{k}, \ 0 \le t \le 2\pi$, in the direction of increasing t.

4. Integrate $f(x, y, z) = \sqrt{x^2 + y^2}$ over the involute curve $\mathbf{r} = (\cos t + t \sin t) \mathbf{i} + (\sin t - t \cos t) \mathbf{j}, \ 0 \le t \le \sqrt{3}$, in the direction of increasing t. (Figure 11.21 shows the curve.)

5. CALCULATOR   Find the value of the integral of $f(x, y, z) = y$ over the curve $y = 2\sqrt{x}, \ z = 0$, from $(1, 2, 0)$ to $(4, 4, 0)$ to two decimal places.

6. A slender metal arch lies along the semicircle $y = \sqrt{a^2 - x^2}$ in the xy-plane. The density at the point (x, y) on the arch is $\delta(x, y) = 2a - y$. Find the center of mass.

7. A wire of constant density $\delta = 1$ lies along the curve $\mathbf{r} = (e^t \cos t) \mathbf{i} + (e^t \sin t) \mathbf{j} + e^t \mathbf{k}, \ 0 \le t \le \ln 2$. Find $\bar{z}$, I_z, and R_z.

8. A wire of constant density $\delta = 1$ lies along the curve $\mathbf{r} = (t \cos t) \mathbf{i} + (t \sin t) \mathbf{j} + (2\sqrt{2}/3)t^{3/2} \mathbf{k}, \ 0 \le t \le 1$. Find $\bar{z}$, I_z, and R_z.

9. Suppose $\mathbf{F}(x, y) = (x + y) \mathbf{i} - (x^2 + y^2) \mathbf{j}$ is the velocity field of a fluid flowing across the xy-plane. Find the flow along each of the following paths from $(1, 0)$ to $(-1, 0)$.
a) The upper half of the circle $x^2 + y^2 = 1$
b) The line segment from $(1, 0)$ to $(-1, 0)$
c) The line segment from $(1, 0)$ to $(0, -1)$ followed by the line segment from $(0, -1)$ to $(-1, 0)$

10. Find the circulation of $\mathbf{F} = 2x \mathbf{i} + 2z \mathbf{j} + 2y \mathbf{k}$ along the closed path consisting of the helix $\mathbf{r}_1 = (\cos t) \mathbf{i} + (\sin t) \mathbf{j} + t \mathbf{k}, \ 0 \le t \le \pi/2$, followed by the line segments $\mathbf{r}_2 = \mathbf{j} + (\pi/2)(1 - t) \mathbf{k}, \ 0 \le t \le 1$, and $\mathbf{r}_3 = t \mathbf{i} + (1 - t) \mathbf{j}, 0 \le t \le 1$.

Sketch the vector fields in Exercises 11 and 12 in the xy-plane.

11. $\mathbf{F} = (x - y) \mathbf{i} + (x + y) \mathbf{j}$

12. $\mathbf{F} = (x \mathbf{i} + y \mathbf{j})/(x^2 + y^2)$

Use Green's theorem to find the counterclockwise circulation and outward flux for the fields and curves in Exercises 13 and 14.

13. $\mathbf{F} = (2xy + x) \mathbf{i} + (xy - y) \mathbf{j}$
C: The square bounded by $x = 0, \ x = 1, \ y = 0, \ y = 1$

14. $\mathbf{F} = (y - 6x^2) \mathbf{i} + (x + y^2) \mathbf{j}$
C: The triangle made by the lines $y = 0, \ y = x$, and $x = 1$

15. Show that

$$\oint_C \ln x \sin y \, dy - \frac{\cos y}{x} \, dx = 0$$

for any closed curve C to which Green's theorem applies.

16. a) Show that the outward flux of the position vector field $\mathbf{F} = x \mathbf{i} + y \mathbf{j}$ across any closed curve to which Green's theorem applies is twice the area of the region enclosed by the curve.
b) Let $\mathbf{n}$ be the outward unit normal vector to a closed curve to which Green's theorem applies. Show that it is not possible for $\mathbf{F} = x \mathbf{i} + y \mathbf{j}$ to be orthogonal to $\mathbf{n}$ at every point of C.

Apply Green's theorem in one of its two forms to evaluate the line integrals in Exercises 17 and 18.

17. $\displaystyle\int_C 8x \sin y \, dx - 8y \cos x \, dy$

    C is the square cut from the first quadrant by the lines $x = \pi/2$ and $y = \pi/2$.

18. $\displaystyle\int_C y^2 \, dx + x^2 \, dy$

    C is the circle $x^2 + y^2 = 4$.

19. a) Find the area enclosed by the curve

$$\mathbf{r} = \left(\frac{2 \cos t - \sin t}{2}\right) \mathbf{i} + (\sin t) \mathbf{j}, \quad 0 \le t \le 2\pi.$$

   b) Find an equation for the curve in Cartesian coordinates and use the equation to identify the curve.

20. Among all rectangular regions $0 \le x \le a$, $0 \le y \le b$, find the one for which the total outward flux of $\mathbf{F} = (x^2 + 4xy)\,\mathbf{i} - 6y\mathbf{j}$ across the four sides is least.

21. Find the area of the elliptical region cut from the plane $x + y + z = 1$ by the cylinder $x^2 + y^2 = 1$.

22. Find the area of the cap cut from the paraboloid $y^2 + z^2 = 3x$ by the plane $x = 1$.

23. Find the area of the cap cut from the top of the sphere $x^2 + y^2 + z^2 = 1$ by the plane $z = \sqrt{2}/2$.

24. a) Find the area of the surface cut from the hemisphere $x^2 + y^2 + z^2 = 4$, $z \ge 0$, by the cylinder $x^2 + y^2 = 2x$ (Fig. 14.74).

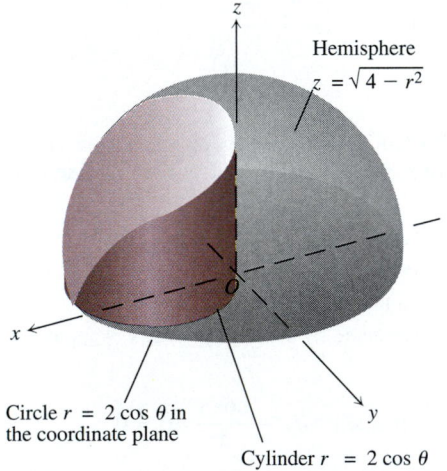

Circle $r = 2 \cos \theta$ in the coordinate plane

Cylinder $r = 2 \cos \theta$

14.74 The surface cut from the hemisphere $x^2 + y^2 + z^2 = 4$, $z \ge 0$, by the cylinder $x^2 + y^2 = 2x$ in Exercise 24. In cylindrical coordinates, the equations of the hemisphere and the cylinder are $z = \sqrt{4 - r^2}$ and $r = 2 \cos \theta$.

   b) Find the area of the portion of the cylinder that lies inside the hemisphere. (*Hint:* Project onto the xz-plane.

Or evaluate the integral $\int h \, ds$, where h is the altitude of the cylinder and ds is the element of arc length on the circle $x^2 + y^2 = 2x$ in the xy-plane.)

25. Find the area of the region in which the plane $(x/a) + (y/b) + (z/c) = 1$ (a, b, $c > 0$) intersects the first octant. Check your answer with an appropriate vector calculation.

26. Let S be the surface that is bounded on the left by the hemisphere $x^2 + y^2 + z^2 = a^2$, $y \le 0$, in the middle by the cylinder $x^2 + z^2 = a^2$, $0 \le y \le a$, and on the right by the plane $y = a$. Find the flux of the field $\mathbf{F} = y\mathbf{i} + z\mathbf{j} + x\mathbf{k}$ outward across S.

27. Integrate

   a) $g(x, y, z) = \dfrac{yz}{\sqrt{4y^2 + 1}}$ b) $g(x, y, z) = \dfrac{z}{\sqrt{4y^2 + 1}}$

   over the surface cut from the parabolic cylinder $y^2 - z = 1$ by the planes $x = 0$, $x = 3$, and $z = 0$.

28. Integrate $g(x, y, z) = x^4 y (y^2 + z^2)$ over the surface in the first quadrant cut from the cylinder $y^2 + z^2 = 25$ by the coordinate planes and the planes $x = 1$, $y = 4$, and $z = 3$.

29. Find the outward flux of the field $\mathbf{F} = 2xy\mathbf{i} + 2yz\mathbf{j} + 2xz\mathbf{k}$ across the surface of the cube cut from the first octant by the planes $x = 1$, $y = 1$, $z = 1$.

30. Find the outward flux of the field $\mathbf{F} = xz\mathbf{i} + yz\mathbf{j} + \mathbf{k}$ across the entire surface of the upper cap cut from the solid sphere $x^2 + y^2 + z^2 \le 25$ by the plane $z = 3$.

31. Find I_z, R_z, and the center of mass of a thin shell of density $\delta(x, y, z) = z$ cut from the upper portion of the sphere $x^2 + y^2 + z^2 = 25$ by the plane $z = 3$.

32. Find the moment of inertia about the z-axis of the surface of the cube cut from the first octant by the planes $x = 1$, $y = 1$, and $z = 1$ if the density is $\delta = 1$.

In Exercises 33–36, find the outward flux of $\mathbf{F}$ across the boundary of D.

33. $\mathbf{F} = 2xy\mathbf{i} + 2yz\mathbf{j} + 2xz\mathbf{k}$
    D: The cube cut from the first octant by the planes $x = 1$, $y = 1$, $z = 1$

34. $\mathbf{F} = xz\mathbf{i} + yz\mathbf{j} + \mathbf{k}$
    D: The upper cap cut from the solid sphere $x^2 + y^2 + z^2 \le 25$ by the plane $z = 3$

35. $\mathbf{F} = -2x\mathbf{i} - 3y\mathbf{j} + z\mathbf{k}$
    D: The region cut from the solid sphere $x^2 + y^2 + z^2 \le 2$ by the paraboloid $z = x^2 + y^2$

36. $\mathbf{F} = (6x + y)\mathbf{i} - (x + z)\mathbf{j} + 4yz\mathbf{k}$
    D: The region in the first octant bounded by the cone $z = \sqrt{x^2 + y^2}$, the cylinder $x^2 + y^2 = 1$, and the coordinate planes

37. Let S be the surface defined by the relations

$$y - \ln x = 0, \quad 1 \le x \le e, \quad 0 \le z \le 1,$$

   and let $\mathbf{n}$ be the unit normal field on S that has $\mathbf{n} \cdot \mathbf{j} > 0$. Find the flux of the field $\mathbf{F} = 2y\mathbf{j} + z\mathbf{k}$ across S in the direction of $\mathbf{n}$.

38. CALCULATOR   A lead radiation shield 2 in. thick is to be constructed in the shape of a circular cylinder surmounted by a parabolic dome. The equations for the dome and cylinder (measurements in feet) are

   Dome: $z = 45 - 0.09x^2 - 0.09y^2$, $9 \le z \le 45$,

   Cylinder: $x^2 + y^2 = 400$, $0 \le z \le 9$.

Lead weighs about 707 lb/ft³. About how many pounds of lead will the shield contain?

39. Find the outward flux of the field $\mathbf{F} = 3xz^2\mathbf{i} + yj - z^3\mathbf{k}$ across the surface of the solid in the first octant that is bounded by the cylinder $x^2 + 4y^2 = 16$ and the planes $y = 2z$, $x = 0$, and $z = 0$.

40. Find the flux of $\mathbf{F} = (3z + 1)\mathbf{k}$ upward across the hemisphere $x^2 + y^2 + z^2 = a^2$, $z \ge 0$ (a) with the Divergence Theorem, (b) by evaluating the flux integral directly.

In Exercises 41 and 42, use the surface integral in Stokes's theorem to find the circulation of the field $\mathbf{F}$ around the curve C in the indicated direction.

41. $\mathbf{F} = y^2\mathbf{i} - y\mathbf{j} + 3z^2\mathbf{k}$

   C: The ellipse in which the plane $2x + 6y - 3z = 6$ meets the cylinder $x^2 + y^2 = 1$, counterclockwise as viewed from above

42. $\mathbf{F} = (x^2 + y)\mathbf{i} + (x + y)\mathbf{j} + (4y^2 - z)\mathbf{k}$

   C: The circle in which the plane $z = -y$ meets the sphere $x^2 + y^2 + z^2 = 4$, counterclockwise as viewed from above

Which of the fields in Exercises 43–46 are conservative and which are not?

43. $\mathbf{F} = x\mathbf{i} + y\mathbf{j} + z\mathbf{k}$

44. $\mathbf{F} = (x\mathbf{i} + y\mathbf{j} + z\mathbf{k})/(x^2 + y^2 + z^2)^{3/2}$

45. $\mathbf{F} = x\,e^y\mathbf{i} + y\,e^z\mathbf{j} + z\,e^x\mathbf{k}$

46. $\mathbf{F} = (\mathbf{i} + z\mathbf{j} + y\mathbf{k})/(x + yz)$

Find potential functions for the fields in Exercises 47 and 48.

47. $\mathbf{F} = 2\mathbf{i} + (2y + z)\mathbf{j} + (y + 1)\mathbf{k}$

48. $\mathbf{F} = (z \cos xz)\mathbf{i} + e^y\mathbf{j} + (x \cos xz)\mathbf{k}$

Evaluate the integrals in Exercises 49 and 50.

49. $\displaystyle\int_{(-1,1,1)}^{(4,-3,0)} \frac{dx + dy + dz}{\sqrt{x + y + z}}$

50. $\displaystyle\int_{(1,1,1)}^{(10,3,3)} dx - \sqrt{\frac{z}{y}}\,dy - \sqrt{\frac{y}{z}}\,dz$

In Exercises 51 and 52, find the work done by each field along the paths from $(0, 0, 0)$ to $(1, 1, 1)$ in Fig. 14.72.

51. $\mathbf{F} = 2xy\mathbf{i} + \mathbf{j} + x^2\mathbf{k}$

52. $\mathbf{F} = 2xy\mathbf{i} + x^2\mathbf{j} + \mathbf{k}$

53. Let C be the ellipse in which the plane $2x + 3y - z = 0$ meets the cylinder $x^2 + y^2 = 12$. Show, without evaluating either line integral directly, that the circulation of the field $\mathbf{F} = x\mathbf{i} + y\mathbf{j} + z\mathbf{k}$ around C in either direction is zero.

54. Find the flow of the field $\mathbf{F} = \nabla(xy^2z^3)$
   a) once around the curve C in Exercise 53, clockwise as viewed from above,
   b) along the line segment from $(1, 1, 1)$ to $(2, 1, -1)$.

55. Find the work done by

$$\mathbf{F} = \frac{x\mathbf{i} + y\mathbf{j}}{(x^2 + y^2)^{3/2}}$$

over the plane curve $\mathbf{r} = (e^t\cos t)\mathbf{i} + (e^t \sin t)\mathbf{j}$ from the point $(1, 0)$ to the point $(e^{2\pi}, 0)$ in two ways:
   a) by using the parametrization of the curve to evaluate the work integral,
   b) by evaluating a potential function for f.

56. Integrate $\mathbf{F} = -(y \sin z)\mathbf{i} + (x \sin z)\mathbf{j} + (xy \cos z)\mathbf{k}$ around the circle cut from the sphere $x^2 + y^2 + z^2 = 5$ by the plane $z = -1$, clockwise as viewed from above.

Finding Areas with Green's Theorem

Use Green's theorem area formula, Eq. (29) in Exercises 14.3, to find the areas of the regions enclosed by the curves in Exercises 57–60.

57. The limaçon $x = 2 \cos t - \cos 2t$, $y = 2 \sin t - \sin 2t$, $0 \le t \le 2\pi$ (Fig. 14.75)

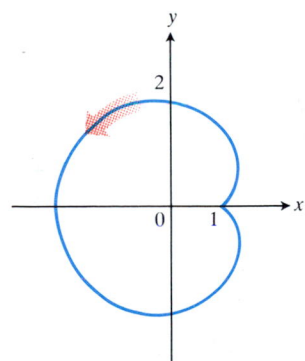

14.75   The limaçon in Exercise 57.

58. The deltoid $x = 2 \cos t + \cos 2t$, $y = 2 \sin t - \sin 2t$, $0 \le t \le 2\pi$ (Fig. 14.76)

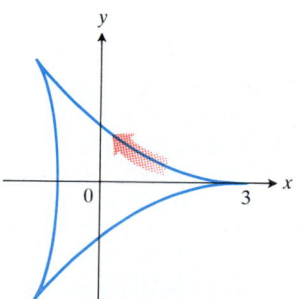

14.76   The deltoid in Exercise 58.

59. The eight curve $x = (1/2)\sin 2t$, $y = \sin t$, $0 \le t \le \pi$ (one loop) (Fig. 14.77)

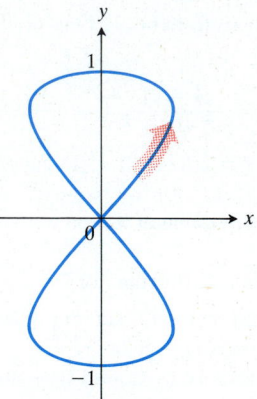

14.77   The eight curve in Exercise 59.

60. The teardrop $x = 2a \cos t - a \sin 2t$, $y = b \sin t$, $0 \le t \le 2\pi$ (Fig. 14.78)

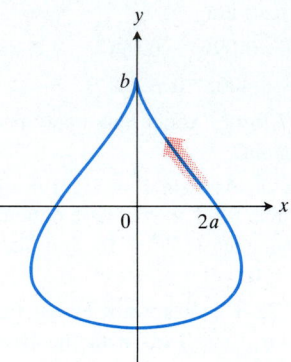

14.78   The teardrop in Exercise 60.

CHAPTER

15

DIFFERENTIAL EQUATIONS

Overview   Differential equations arise when we model the effects of change, motion, and growth. They arise when we study moving particles, changing business conditions, oscillating voltages in neural networks, changing concentrations in chemical reactions, and flowing resources in a market economy. Our space program more or less runs on differential equations (implemented by computers), and mathematical models of our environment involve huge systems of differential equations. The basic tool for solving differential equations is calculus. This chapter previews common types of differential equations and the methods we use to solve them.

15.1 Separable First Order Equations

We study differential equations for their mathematical importance and for what they tell us about reality. When we study unchecked bacterial growth, for example, and assume that the rate of increase in the population at any given time t is proportional to the number of bacteria then present, we are assuming that the population size y obeys the differential equation $dy/dt = ky$. If we also know that the size of the population is y_0 when $t = 0$, we can find a formula for the size of the population at any time $t > 0$ by solving the initial value problem

Differential equation: $\dfrac{dy}{dt} = ky$,

Initial condition: $y = y_0$ when $t = 0$.

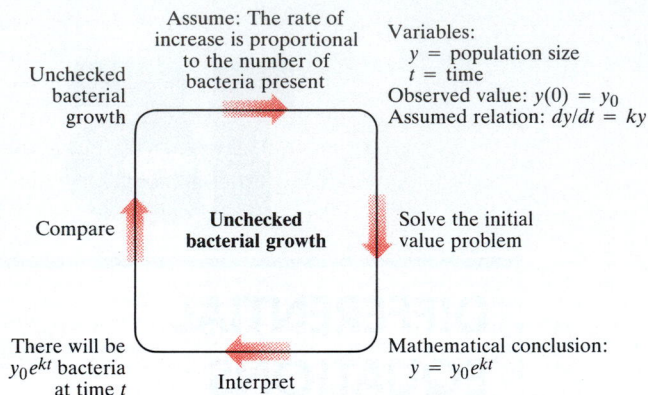

As we saw in Section 6.5, the general solution of the differential equation (the formula that gives all possible solutions) is

$$y = Ae^{kt} \qquad (A \text{ an arbitrary constant})$$

and the particular solution that satisfies the initial condition is $y = y_0 e^{kt}$.

We interpret this as saying that unchecked bacterial growth is exponential. There should be $y_0 e^{kt}$ bacteria present at time t. We know from experience that there are limits to all growth, and we do not expect this prediction to be accurate for large values of t. But for the early stages of growth the formula is a good predictor and gives valuable information about the population's size.

Definitions

A **differential equation** is an equation that contains one or more derivatives of a differentiable function. An equation with partial derivatives is called a **partial differential equation.** The wave equation and the Laplace equations in Exercises 12.3 are partial differential equations. An equation with ordinary derivatives, that is, derivatives of a function of a single variable, is called an **ordinary differential equation.** The equations in this chapter are ordinary differential equations.

The **order** of a differential equation is the order of the equation's highest order derivative. A differential equation is **linear** if it can be put in the form

$$a_n(x) \frac{d^n y}{dx^n} + a_{n-1}(x) \frac{d^{n-1} y}{dx^{n-1}} + \cdots + a_1(x) \frac{dy}{dx} + a_0(x)\, y = F(x), \qquad (1)$$

where the a's are functions of x.

Example 1

First order, linear: $\qquad \dfrac{dy}{dx} = 5y; \qquad 3\dfrac{dy}{dx} - \sin x = 0$

Third order, nonlinear: $\qquad \left(\dfrac{d^3 y}{dx^3}\right)^2 + \left(\dfrac{d^2 y}{dx^2}\right)^5 - \dfrac{dy}{dx} = e^x$

Almost all the differential equations we have solved so far in this book have been linear equations of first or second order. This will continue to be true, but the equations will now have more variety and model a broader range of applications.

We call a function $y = f(x)$ a **solution** of a differential equation if y and its derivatives satisfy the equation. To test whether a given function solves a particular equation, we substitute the function and its derivatives into the equation. If the equation then reduces to an identity, the function solves the equation; otherwise, it does not.

Example 2   Show that for any values of the arbitrary constants C_1 and C_2 the function $y = C_1 \cos x + C_2 \sin x$ is a solution of the differential equation

$$\frac{d^2y}{dx^2} + y = 0.$$

Solution   We differentiate the function twice to find d^2y/dx^2:

$$y = \quad C_1 \cos x + C_2 \sin x$$

$$\frac{dy}{dx} = -C_1 \sin x + C_2 \cos x$$

$$\frac{d^2y}{dx^2} = -C_1 \cos x - C_2 \sin x.$$

We then substitute the expressions for y and d^2y/dx^2 into the differential equation to see whether the left-hand side reduces to zero. It does because

$$\frac{d^2y}{dx^2} + y = (-C_1 \cos x - C_2 \sin x) + (C_1 \cos x + C_2 \sin x) = 0.$$

So the function is a solution of the differential equation.

It can be shown that the formula $y = C_1 \cos x + C_2 \sin x$ gives all possible solutions of the equation $d^2y/dx^2 + y = 0$. A formula that gives all the solutions of a differential equation is called the **general solution** of the equation. In this sense, $y = C_1 \cos x + C_2 \sin x$ is the general solution of $d^2y/dx^2 + y = 0$. To **solve** a differential equation means to find its general solution.

Notice that the equation $d^2y/dx^2 + y = 0$ has order two and that its general solution has two arbitrary constants. The general solution of an nth order ordinary differential equation can be expected to contain n arbitrary constants.

Separable Equations

A method that sometimes works for solving a first order differential equation involves treating the derivative dy/dx as a quotient of differentials and rearranging the equation to group the y-terms alone with dy and the x-terms alone with dx. We cannot always accomplish this, but when we can, the solution may then be found by separate integrations with respect to x and y.

DEFINITIONS

A first order differential equation is **separable** if it can be put in the form

$$M(x) + N(y)\frac{dy}{dx} = 0 \tag{2}$$

or in the equivalent differential form

$$M(x)\,dx + N(y)\,dy = 0. \tag{3}$$

When we write the equation this way we say that we have **separated the variables.**

We can solve Eq. (2) by integrating both sides with respect to x to get

$$\int M(x)\,dx + \int N(y)\frac{dy}{dx}\,dx = C. \tag{4}$$

However, the second integral in this equation is equivalent to

$$\int N(y)\,dy$$

and it usually saves time to integrate the equivalent equation

$$\int M(x)\,dx + \int N(y)\,dy = C \tag{5}$$

instead. The result of the integration will express y either explicitly or implicitly as a function of x that solves Eq. (2).

Example 3   Solve the differential equation

$$\frac{dy}{dx} = (1 + y^2)\,e^x.$$

Steps for Solving a Separable First Order Differential Equation

1. Write the equation in the form $M(x)\,dx + N(y)\,dy = 0$.

2. Integrate M with respect to x and N with respect to y to obtain an equation that relates y and x.

Solution   We use algebra to separate the variables and write the equation in the form $M(x)\,dx + N(y)\,dy = 0$, obtaining

$$e^x\,dx - \frac{1}{1 + y^2}\,dy = 0.$$

We then integrate to get

$$\int e^x\,dx - \int \frac{1}{1 + y^2}\,dy = C$$

$$e^x - \tan^{-1}y = C \qquad \text{(Constants of integration combined)}$$

$$\tan^{-1}y = e^x - C.$$

In this case, we can solve the resulting equation explicitly for y by taking the tangent of each side:

$$y = \tan(e^x - C).$$

Homogeneous First Order Equations

We can sometimes use a change of variable to transform a differential equation whose variables cannot be separated into one whose variables can be separated. This is the case with homogeneous first order equations.

A first order differential equation is **homogeneous** if it can be put into the form

$$\frac{dy}{dx} = F\left(\frac{y}{x}\right). \tag{6}$$

We can change Eq. (6) into a separable equation with the substitutions

$$y = vx, \qquad \frac{dy}{dx} = v + x\frac{dv}{dx}. \tag{7}$$

Then Eq. (6) becomes

$$v + x\frac{dv}{dx} = F(v), \tag{8}$$

which can be rearranged algebraically to give

$$\frac{dx}{x} + \frac{dv}{v - F(v)} = 0. \tag{9}$$

With the variables separated, we can now solve the equation by integrating with respect to x and v. We can then return to x and y by substituting $v = y/x$.

Example 4   Show that the equation

$$\frac{dy}{dx} = -\frac{x^2 + y^2}{2xy}$$

is homogeneous and find the solution that satisfies the condition $y(1) = 1$.

Solution   Dividing the numerator and denominator of the right-hand side by x^2 gives

$$\frac{dy}{dx} = -\frac{1 + (y/x)^2}{2(y/x)}.$$

This has the form of Eq. (6) with

$$F(v) = -\frac{1 + v^2}{2v}, \qquad \text{where} \quad v = \frac{y}{x},$$

so the original equation is homogeneous. Equation (9) becomes

$$\frac{dx}{x} + \frac{dv}{v + \dfrac{1 + v^2}{2v}} = 0, \qquad \text{or} \qquad \frac{dx}{x} + \frac{2v\,dv}{1 + 3v^2} = 0.$$

The solution of this equation can be written as

$$\ln|x| + \frac{1}{3}\ln(1 + 3v^2) = C \qquad \text{or} \qquad x^3(1 + 3v^2) = \pm e^{3C} = C'.$$

We substitute $v = y/x$ to find the corresponding xy-equation, obtaining

$$x^3\left(1 + 3\frac{y^2}{x^2}\right) = C' \qquad \text{or} \qquad x^3 + 3xy^2 = C'.$$

The value of C' that gives $y = 1$ when $x = 1$ is

$$(1)^3 + 3(1)(1)^2 = C' \qquad \text{or} \qquad C' = 4.$$

The corresponding solution (Fig. 15.1) is $x^3 + 3xy^2 = 4$.

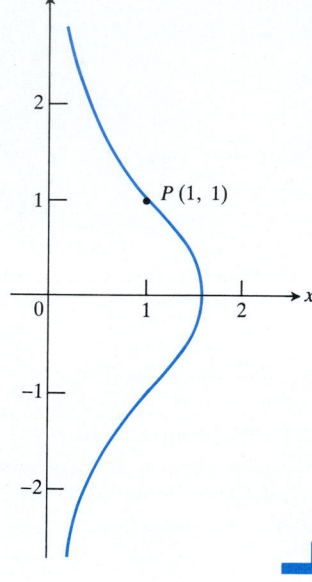

15.1   The solution of $(x^2 + y^2)\, dx + 2xy\, dy = 0$ with $y = 1$ when $x = 1$ is $x^3 + 3xy^2 = 4$ (Example 4).

EXERCISES 15.1

Find the orders of the differential equations in Exercises 1–4. Which equations are linear and which are nonlinear?

1. $y' = x^2 + y^2$

2. $(y''')^2 - 3x^2y'' + y^5 = 10$

3. $y^{(4)} + y = 1$

4. $y'' - \frac{1}{x}y' + \frac{2}{x^2}y = e^x$

In Exercises 5–8, show that each function $y = f(x)$ is a solution of the accompanying differential equation.

5. $xy'' - y' = 0$
   a) $y = x^2$ b) $y = 1$ c) $y = C_1 x^2 + C_2$

6. $y' + \frac{1}{x}y = 1$

   a) $y = \frac{x}{2}$ b) $y = \frac{1}{x} + \frac{x}{2}$ c) $y = \frac{C}{x} + \frac{x}{2}$

7. $2y' + 3y = e^{-x}$
   a) $y = e^{-x}$
   b) $y = e^{-x} + e^{-(3/2)x}$
   c) $y = e^{-x} + Ce^{-(3/2)x}$

8. $yy'' = 2(y')^2 - 2y'$
   a) $y = 1$
   b) $y = \tan x$

In Exercises 9 and 10, show that the function $y = f(x)$ is a solution of the given differential equation.

9. $y = \frac{1}{x}\int_1^x \frac{e^t}{t}\, dt, \quad x^2y' + xy = e^x$

10. $y = \frac{1}{\sqrt{1 + x^4}}\int_1^x \sqrt{1 + t^4}\, dt, \quad y' + \frac{2x^3}{1 + x^4}y = 1$

In Exercises 11–14, show that each function $y = f(x)$ is a solution of the accompanying initial value problem.

Differential equation	Initial condition(s)	Solution candidate
11. $y'' = -32$	$y(5) = 400,$ $y'(5) = 0$	$y = 160x - 16x^2$
12. $2\frac{dy}{dx} + 3y = 6$	$y(0) = 0$	$y = 2\left(1 - e^{-3x/2}\right)$
13. $y'' + 4y = 0$	$y(0) = 3,$ $y'(0) = -2$	$y = 3\cos 2x - \sin 2x$
14. $y'' - (y')^2 = 1$	$y(1) = 2,$ $y'(1) = 0$	$y = 2 - \ln\cos(x - 1)$

Solve the initial value problems in Exercises 15–20.

15. $\frac{dy}{dx} = \frac{2y^2 + 2}{x^2 - 1}, \quad y(0) = 1$

16. $\frac{dy}{dx} = \frac{e^{(y^2 + \sin x)}}{y \sec x}, \quad y(0) = \sqrt{\ln 2}$

17. $\frac{dy}{dx} = \frac{e^{(y + \sqrt{x})}}{\sqrt{x}\, y^3}, \quad y(\ln^2 2) = 0$

18. $\frac{dy}{dx} = (y^2 + 2y + 2)(x + 1)\ln x, \quad y(1) = -1$

19. $\dfrac{dy}{dx} = \sqrt{x^2 - x^2 y^2 + 1 - y^2}$, $y(0) = -1/2$

20. $\dfrac{dy}{dx} = \dfrac{\cot y}{x\sqrt{4x^2 - 1}}$, $y(1) = 0$

21. CALCULATOR   *Newton's law of cooling.*   Newton's law of cooling assumes that the temperature T of a small hot object placed in a surrounding cooling medium of constant temperature T_s decreases at a rate proportional to $T - T_s$. An object cooled from 100°C to 40°C in 20 minutes when the surrounding temperature was 20°C. How long did it take the temperature to reach 60°C on the way down?

22. *The snow plow problem.*   One morning it began to snow and kept on snowing at a constant rate. A snow plow began to plow at noon. It plowed clean and at a constant rate (volume per unit time). From one o'clock until two o'clock it went only half as far as it had gone between noon and one o'clock. When did the snow begin falling?

Homogeneous Equations

Show that the equations in Exercises 23–26 are homogeneous and find their general solutions.

23. $(x^2 + y^2)\,dx + xy\,dy = 0$

24. $(y^2 - xy)\,dx + x^2\,dy = 0$

25. $(xe^{y/x} + y)\,dx - x\,dy = 0$

26. $(x - y)\,dx + (x + y)\,dy = 0$

Solve the equations in Exercises 27 and 28 subject to the given initial conditions.

27. $\dfrac{dy}{dx} = \dfrac{y}{x} + \cos\dfrac{y - x}{x}$, $y(2) = 2$

28. $\left(x \sin\dfrac{y}{x} - y\cos\dfrac{y}{x}\right)dx + x\cos\dfrac{y}{x}\,dy = 0$, $y(2) = \pi$

Orthogonal Trajectories

If every member of one family of curves is a solution of the differential equation

$$M(x, y)\,dx + N(x, y)\,dy = 0$$

and every member of a second family of curves is a solution of the related equation

$$N(x, y)\,dx - M(x, y)\,dy = 0,$$

then each curve in each family is orthogonal to all the curves of the other family. Under these circumstances each family is said to be a family of **orthogonal trajectories** of the other family. If the curves in one family were electric field lines, the curves in the other family would be paths of constant electric potential.

Solve the differential equations in Exercises 29 and 30 to find equations for the families of orthogonal trajectories illustrated in Figs. 15.2 and 15.3.

29. $x\,dx + 2y\,dy = 0$ and $2y\,dx - x\,dy = 0$ (Fig. 15.2)

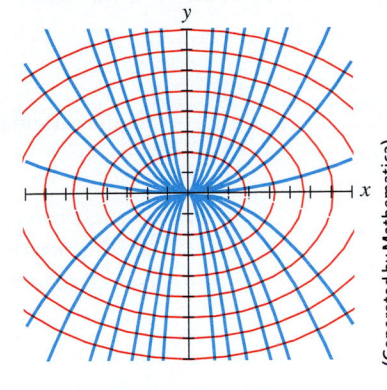

15.2   The curves in Exercise 29.

(Generated by Mathematica)

30. $3x\,dx + 2y\,dy = 0$ and $2y\,dx - 3x\,dy = 0$ (Fig. 15.3)

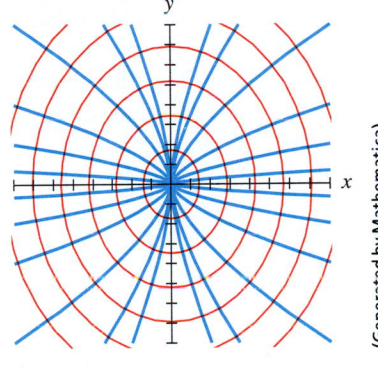

15.3   The curves in Exercise 30.

(Generated by Mathematica)

31. Identify the curves that satisfy the differential equation $y\,dx + x\,dy = 0$ and find equations for their orthogonal trajectories (Fig. 15.4).

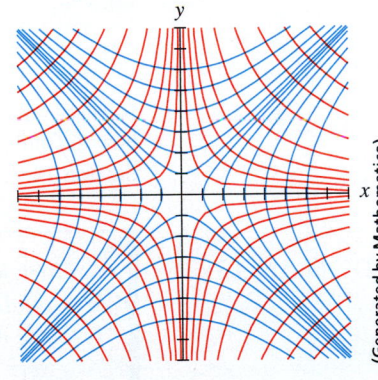

15.4   The curves in Exercise 31.

(Generated by Mathematica)

EXPLORER PROGRAM

Slope Fields

Graphs solutions of the initial value problem $y' = f(x, y)$, $y = y_0$ when $x = x_0$, for your choice of f, x_0, and y_0

15.2 Exact Differential Equations

If we write a (not necessarily separable) first order differential equation

$$M(x, y) + N(x, y) \frac{dy}{dx} = 0 \tag{1}$$

in the differential form

$$M(x, y)\, dx + N(x, y)\, dy = 0, \tag{2}$$

the left-hand side has the same form as the differential of a function of x and y. If the left-hand side *is* the differential df of a function $f(x, y)$, then Eq. (2) can be solved in terms of f:

$$M(x, y)\, dx + N(x, y)\, dy = 0$$
$$df(x, y) = 0 \tag{3}$$
$$f(x, y) = C.$$

The equation $f(x, y) = C$ defines y implicitly as one or more differentiable functions of x. Each of these functions, it can be shown, is a solution of the original differential equation. In writing $f(x, y) = C$ we therefore say that we have solved the differential equation.

Example 1

Since

$$d(\sin xy) = y \cos xy\, dx + x \cos xy\, dy = 0,$$

the equation

$$y \cos xy\, dx + x \cos xy\, dy = 0 \tag{4}$$

is equivalent to

$$d(\sin xy) = 0, \qquad \text{or} \qquad \sin xy = C.$$

The equation $\sin xy = C$ defines y implicitly as one or more functions of x that solve Eq. (4). ⌐

Not all differential forms are differentials of functions. Before we attempt to solve an equation this way we must know whether such a function f exists and then how to find it when it does. The test for the existence of f is the test for exactness in Section 14.7, repeated here for functions of two variables.

DEFINITION AND TEST

A differential form $M(x, y)\, dx + N(x, y)\, dy$ and the associated equation $M(x, y)\, dx + N(x, y)\, dy = 0$ are both said to be **exact** on a region R in the xy-plane if for some function $f(x, y)$ defined on R,

$$M(x, y)\, dx + N(x, y)\, dy = \frac{\partial f}{\partial x} dx + \frac{\partial f}{\partial y} dy = df. \tag{5}$$

The form and equation are exact on R if and only if, throughout R,

$$\frac{\partial M}{\partial y} = \frac{\partial N}{\partial x}. \tag{6}$$

Example 2

a) The equation $(x^2 + y^2)\, dx + (2xy + \cos y)\, dy = 0$ is exact because the partial derivatives

$$\frac{\partial M}{\partial y} = \frac{\partial}{\partial y}(x^2 + y^2) = 2y \qquad \text{and} \qquad \frac{\partial N}{\partial x} = \frac{\partial}{\partial x}(2xy + \cos y) = 2y$$

are equal.

b) The equation $(x + 3y)\, dx + (x^2 + \cos y)\, dy = 0$ is not exact because the partial derivatives

$$\frac{\partial M}{\partial y} = \frac{\partial}{\partial y}(x + 3y) = 3 \qquad \text{and} \qquad \frac{\partial N}{\partial x} = \frac{\partial}{\partial x}(x^2 + \cos y) = 2x$$

are not equal.

After we have tested an equation and found it to be exact, we can find the solution $f(x, y) = C$ by taking the steps described in the following example.

Steps for Solving an Equation You Know to Be Exact

1. Match the equation to the form $df = (\partial f/\partial x)dx + (\partial f/\partial y)dy$ to identify $\partial f/\partial x$ and $\partial f/\partial y$.

2. Integrate $\partial f/\partial x$ with respect to x, writing the constant of integration as $k(y)$.

3. Differentiate with respect to y and set the result equal to $\partial f/\partial y$ to find $k'(y)$.

4. Integrate to find $k(y)$ and determine f.

5. Write the solution of the exact equation as $f(x, y) = C$.

Example 3

Solve the differential equation

$$(x^2 + y^2)\, dx + (2xy + \cos y)\, dy = 0.$$

Solution   We already know that the equation is exact (Example 2), so it is safe to set about finding a function $f(x, y)$ whose differential is the equation's left-hand side. For the equality

$$\frac{\partial f}{\partial x}\, dx + \frac{\partial f}{\partial y}\, dy = (x^2 + y^2)\, dx + (2xy + \cos y)\, dy$$

to hold, we must have

$$\frac{\partial f}{\partial x} = x^2 + y^2 \qquad \text{and} \qquad \frac{\partial f}{\partial y} = 2xy + \cos y. \tag{7}$$

The partial derivative $\partial f/\partial x = x^2 + y^2$ was calculated by holding y fixed and differentiating f with respect to x. We may therefore find f by holding y at a constant value and integrating with respect to x. When we do so we get

$$f(x, y) = \int_{y\,\text{const.}} (x^2 + y^2)\, dx = \frac{x^3}{3} + y^2x + k(y). \tag{8}$$

The constant of integration is written as a function of y because its value may change with each new y.

To find $k(y)$, we calculate $\partial f/\partial y$ from Eq. (8) and set the result equal to the known partial derivative $\partial f/\partial y = 2xy + \cos y$:

$$\frac{\partial}{\partial y}\left(\frac{x^3}{3} + y^2x + k(y)\right) = 2xy + \cos y$$

$$2xy + k'(y) = 2xy + \cos y$$

$$k'(y) = \cos y.$$

A single integration with respect to y then shows that $k(y) = \sin y$ plus a constant, so

$$f(x, y) = \frac{x^3}{3} + y^2 x + \sin y + \text{a constant}.$$

If we wanted only to find a function whose differential is $(x^2 + y^2)\, dx + (2xy + \cos y)\, dy$, we'd be done now. We have found infinitely many such functions, one for each possible constant. Our goal, however, is to solve the original differential equation, and for that we must take one more step. We must write down the equation $f(x, y) = C$ because it is this equation, and not the formula for f alone, that defines the differential equation's solution functions. The solution of the differential equation is

$$\frac{x^3}{3} + y^2 x + \sin y = C. \qquad \text{(Constants combined)}$$

Integrating Factors

It can be shown that every nonexact differential equation $M(x, y)\, dx + N(x, y)\, dy = 0$ can be made exact by multiplying both sides by a suitable **integrating factor** $\rho(x, y)$. In other words, the equation

$$\rho(x, y)\, M(x, y)\, dx + \rho(x, y)\, N(x, y)\, dy = 0$$

is always exact for an appropriate choice of ρ.

Example 4   The equation

$$2y\, dx + x\, dy = 0 \qquad\qquad (9)$$

is not exact because $\partial(2y)/\partial y = 2$ while $\partial(x)/\partial x = 1$. The equation

$$2xy\, dx + x^2\, dy = 0,$$

obtained by multiplying both sides of Eq. (9) by x, *is* exact because $\partial(2xy)/\partial y = 2x$ and $\partial(x^2)/\partial x = 2x$.

Unfortunately, there is no general technique for finding an integrating factor when you need one, and the search for one can be a frustrating experience. Practice does help, however, and Exercises 19–22 supply integrating factors for you to work with. As you will see there, integrating factors need not be unique.

EXERCISES 15.2

Test the equations in Exercises 1–8 for exactness.

1. $x\, dx - y\, dy = 0$

2. $y\, dx - x\, dy = 0$

3. $\dfrac{1}{y}\, dx - \dfrac{x}{y^2}\, dy = 0$

4. $(x + y^2)\, dx + (2xy + 1)\, dy = 0$

5. $(x + e^y)\, dx + (y + xe^y)\, dy = 0$

6. $y \cos xy\, dx + x \cos xy\, dy = 0$

7. $(x + 2y)\, dx - (2x - y)\, dy = 0$

8. $(x + 2y)\, dx - (2x + y)\, dy = 0$

Solve the equations in Exercises 9–18.

9. $(x + y)\, dx + (x + y^2)\, dy = 0$

10. $(2x e^y + e^x) \, dx + (x^2 + 1)e^y \, dy = 0$

11. $\dfrac{dy}{dx} = \dfrac{2xy + y^2}{y - 2xy - x^2}$ **12.** $\dfrac{dy}{dx} = \dfrac{x^3 - y}{x + \sqrt{1 + \sin 2y}}$

13. $\left(e^x + \ln y + \dfrac{y}{x}\right) dx + \left(\dfrac{x}{y} + \ln x + \sin y\right) dy = 0$

14. $\dfrac{dy}{dx} = \dfrac{y \cos xy + 1}{3 - 2y - x \cos xy}$

15. $\left(\cos x \displaystyle\int_0^y \sin^2 t \, dt\right) dx + \sin x \sin^2 y \, dy = 0$

16. $\dfrac{2x - y}{x^2 + y^2} \, dx + \dfrac{x + 2y}{x^2 + y^2} \, dy = 0$

17. $\left(\dfrac{1}{x} + \displaystyle\sum_{n=0}^{\infty} \dfrac{x^{n-1} y^n}{n!}\right) dx + \left(\dfrac{1}{y} + \displaystyle\sum_{n=0}^{\infty} \dfrac{x^n y^{n-1}}{n!}\right) dy = 0$

18. $\left(\displaystyle\sum_{n=0}^{\infty} \dfrac{(-1)^{n+1}}{x^2 y^n}\right) dx + \left(\displaystyle\sum_{n=0}^{\infty} \dfrac{(-1)^{n+1} n}{x y^{n+1}}\right) dy = 0$

Show that the integrating factors ρ in Exercises 19–22 make the differential equations exact. Then solve the equations using those integrating factors.

19. $(xy^2 + y) \, dx - x \, dy = 0; \quad \rho = 1/y^2$

20. $(x + 2y) \, dx - x \, dy = 0; \quad \rho = 1/x^3$

21. $y \, dx + x \, dy = 0,$

    a) $\rho = \dfrac{1}{xy}$ b) $\rho = \dfrac{1}{(xy)^2}$

22. $y \, dx - x \, dy = 0,$

    a) $\rho = \dfrac{1}{y^2}$ b) $\rho = \dfrac{1}{x^2}$

    c) $\rho = \dfrac{1}{xy}$ d) $\rho = \dfrac{1}{x^2 + y^2}$

Solve the initial value problems in Exercises 23–26 for y as a function of x.

23. $\dfrac{dy}{dx} = -\dfrac{x + y^2}{2xy + 1}, \quad y(0) = 2$

24. $\dfrac{dy}{dx} = -\dfrac{x + e^y}{y + xe^y}, \quad y(2) = 0$

25. $\left(\dfrac{1}{x} - y\right) dx + \left(\dfrac{1}{y} - x\right) dy = 0, \quad y(1) = 1$

26. $(2x + y + 1) \, dx + (2y + x + 1) \, dy = 0, \quad y(1) = 5$

Solve the initial value problems in Exercises 27 and 28.

27. $\left(2x + \dfrac{1}{x} \displaystyle\int_1^y \dfrac{\ln t}{t} \, dt\right) dx + \dfrac{\ln x \ln y}{y} \, dy = 0, \quad y(2) = 1$

28. $\left(\sin^{-1} y + \dfrac{y}{\sqrt{1 - x^2}}\right) dx +$

           $\left(\sin^{-1} x + \dfrac{x}{\sqrt{1 - y^2}}\right) dy = 0, \quad y(1) = 1$

29. Find the value of a that makes the equation
$$(3x^2 + y^2) \, dx + a \, xy \, dy = 0$$
exact and solve the equation for this value of a.

30. The equation $(x^2 + by^2) \, dx + 6 \, xy \, dy = 0$ is not exact but is known to have $\rho = 1/x^2$ as an integrating factor. Use this information to find the value of b and solve the equation.

EXPLORER PROGRAM

Slope Fields   Graphs solutions of the initial value problem $y' = f(x, y)$, $y = y_0$ when $x = x_0$, for your choice of f, x_0, and y_0.

15.3 Linear First Order Equations

The equation $dy/dx = ky$, which we use to model bacterial growth, radioactive decay, and temperature change, is a special case of a more general equation called a linear first order equation. This section shows how to solve the more general equation and uses it to predict current flow in an electric circuit.

DEFINITIONS

A differential equation that can be written in the form

$$\dfrac{dy}{dx} + P(x)y = Q(x), \tag{1}$$

where P and Q are functions of x, is called a **linear first order equation**. Equation (1) is the equation's **standard form**.

Example 1   The standard form of the equation

$$x\frac{dy}{dx} + 3y = x^2$$

is

$$\frac{dy}{dx} + \frac{3}{x}y = x.$$

This is Eq. (1) with $P(x) = 3/x$ and $Q(x) = x$.

The way we solve the equation

$$\frac{dy}{dx} + P(x)\,y = Q(x)$$

is to multiply both sides by an integrating factor $\rho(x)$ that turns the left-hand side of the multiplied equation,

$$\rho\frac{dy}{dx} + P\rho y = \rho Q, \qquad (2)$$

into the derivative of the product ρy.

Once we have chosen ρ (we shall do so in a moment) and carried out the multiplication, we can solve Eq. (2) by integrating both sides with respect to x:

$$\rho\frac{dy}{dx} + P\rho y = \rho Q \qquad \text{(Eq. (2))}$$

$$\frac{d}{dx}(\rho y) = \rho Q \qquad \text{(Choice of } \rho\text{)}$$

$$\rho y = \int \rho Q\, dx \qquad \text{(Integration with respect to } x\text{)}$$

$$y = \frac{1}{\rho}\int \rho Q\, dx. \qquad \text{(Solved for } y\text{)}$$

To find ρ, we find the function of x that satisfies the equations

$$\frac{d}{dx}(\rho y) = \rho\frac{dy}{dx} + P\rho y,$$

$$\rho\frac{dy}{dx} + y\frac{d\rho}{dx} = \rho\frac{dy}{dx} + P\rho y, \qquad \text{(Product Rule for derivatives)}$$

$$y\frac{d\rho}{dx} = P\rho y, \qquad \text{(The terms } \rho\frac{dy}{dx} \text{ cancel.)}$$

$$\frac{d\rho}{dx} = P\rho, \qquad \text{(The } y\text{'s cancel.)}$$

$$\frac{d\rho}{\rho} = P\,dx, \qquad \text{(Variables separated)}$$

$$\rho = Ce^{\int P\,dx}. \qquad \text{(Equation integrated and solved for } \rho\text{)}$$

Since we do not need the most general function ρ, we may take C to be 1. We summarize our conclusions with the following theorem.

THEOREM 1

The solution of the equation

$$\frac{dy}{dx} + P(x)y = Q(x) \tag{3}$$

is

$$y = \frac{1}{\rho(x)} \int \rho(x)\, Q(x)\, dx, \tag{4}$$

where

$$\rho(x) = e^{\int P(x)\, dx}. \tag{5}$$

In the formula for ρ we do not need the most general antiderivative of $P(x)$. Any antiderivative will do.

Example 2   Solve the equation $x\dfrac{dy}{dx} - 3y = x^2, \quad x > 0.$

Solution   We solve the equation in four steps.

STEP 1: *Put the equation in standard form and identify the functions P and Q.* To do so, we divide both sides of the equation by the coefficient of dy/dx, in this case x, obtaining

$$\frac{dy}{dx} - \frac{3}{x}y = x, \qquad P(x) = -\frac{3}{x}, \qquad Q(x) = x.$$

STEP 2: *Find an antiderivative of $P(x)$ (any one will do):*

$$\int P(x)\, dx = \int -\frac{3}{x}\, dx = -3\int \frac{1}{x}\, dx = -3\ln|x| = -3\ln x \qquad (x > 0).$$

STEP 3: *Use Eq. (5) to find ρ:*

$$\rho(x) = e^{\int P(x)\, dx} = e^{-3\ln x} = e^{\ln x^{-3}} = \frac{1}{x^3}.$$

STEP 4: *Use Eq. (4) to find the solution:*

$$y = \frac{1}{\rho(x)} \int \rho(x)\, Q(x)\, dx = \frac{1}{(1/x^3)} \int \left(\frac{1}{x^3}\right)(x)\, dx \qquad \text{(Values from steps 1–3)}$$

$$= x^3 \int \frac{1}{x^2}\, dx = x^3 \left(-\frac{1}{x} + C\right) = Cx^3 - x^2.$$

The solution is the function $y = Cx^3 - x^2$.

Steps for Solving a Linear First Order Equation

1. Put it in standard form.
2. Find an antiderivative of $P(x)$.
3. Find $\rho = e^{\int P(x)\, dx}$.
4. Use Eq. (4) to find y.

RL Circuits

The diagram in Fig. 15.5 represents an electrical circuit whose total resistance is R ohms and whose self-inductance, shown schematically as a coil, is L henries (hence the name "*RL* circuit"). There is also a switch whose terminals at a and b can be closed to connect a constant electrical source of V volts.

Ohm's law, $V = RI$, has to be modified for such a circuit. The modified

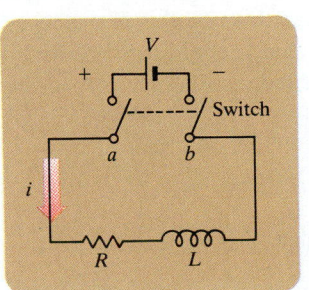

15.5   The *RL* circuit in Example 3.

form is

$$L\frac{di}{dt} + Ri = V, \tag{6}$$

where i is the intensity of the current in amperes and t is the time in seconds. By solving this equation, we can predict how the current will flow after the switch is closed.

Example 3   The switch in the RL circuit in Fig. 15.5 is closed at time $t = 0$. How will the current flow as a function of time?

Solution   Equation (6) is a linear first order differential equation for i as a function of t. When we put it in the standard form

$$\frac{di}{dt} + \frac{R}{L}i = \frac{V}{L}$$

and carry out the solution steps with $P = R/L$ and $Q = V/L$, we find that

$$i = \frac{1}{e^{(R/L)t}}\int e^{(R/L)t}\left(\frac{V}{L}\right)dt = \frac{V}{R} + C\frac{V}{L}e^{-(R/L)t}. \tag{7}$$

Imposing the initial condition that $i(0)$ be 0 determines the value of C to be $-L/R$, so

$$i = \frac{V}{R} + \left(-\frac{L}{R}\right)\frac{V}{L}e^{-(R/L)t} = \frac{V}{R}\left(1 - e^{-(R/L)t}\right). \tag{8}$$

We see from this that the current i is always less than V/R but that it approaches V/R as a **steady-state value:**

$$\lim_{t\to\infty}\frac{V}{R}\left(1 - e^{-Rt/L}\right) = \frac{V}{R}(1 - 0) = \frac{V}{R}. \tag{9}$$

The current $I = V/R$ is the current that will flow in the circuit if either $L = 0$ (no inductance) or $di/dt = 0$ (steady current, $i = $ constant) (Fig. 15.6).

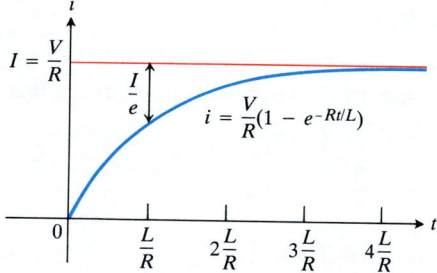

15.6   The growth of the current in the RL circuit in Example 3. I is the current's steady-state value. The number $t = L/R$ is the time constant of the circuit. The current gets to within 5% of its steady-state value in 3 time constants (Exercise 17).

Resistance Proportional to Velocity

In some cases it makes sense to assume that, other forces being absent, the resistance encountered by a moving object, like a car coasting to a stop, is proportional to the object's velocity. The slower the object moves, the less its forward progress is resisted by the air through which it passes. We can describe this in mathematical terms if we picture the object as a mass m moving along a coordinate line with position s and velocity v at time t. The magnitude of the resisting force opposing the motion is then $m\,(dv/dt)$ and we can write

$$m\frac{dv}{dt} = -kv \qquad (k > 0) \tag{10}$$

to say that the force decreases in proportion to velocity. If we rewrite (10) as

$$\frac{dv}{dt} = -\frac{k}{m}v,$$

we can see that the solution is $v = v_0\,e^{-(k/m)t}$.

Resistance Proportional to Velocity

$$v = v_0\, e^{-(k/m)t} \tag{11}$$

What can we learn from this equation? For one thing, we can see that if m is something large, like the mass of a 20,000-ton ore boat in Lake Erie, it will take a long time for the velocity to get near zero. For another, we can integrate the equation to find s as a function of t.

Suppose a body is coasting to a stop and the only force acting on it is a resistance proportional to its speed. How far will it coast? To find out, we start with Eq. (11) and solve the initial value problem

Differential equation: $\qquad \dfrac{ds}{dt} = v_0\, e^{-(k/m)t},$

Initial condition: $\qquad s = 0 \quad \text{when} \quad t = 0.$

Integrating with respect to t gives

$$s = -\frac{v_0 m}{k}\, e^{-(k/m)t} + C. \tag{12}$$

Substituting $s = 0$ when $t = 0$ gives

$$0 = -\frac{v_0 m}{k} + C \qquad \text{and} \qquad C = \frac{v_0 m}{k}.$$

The body's position at time t is therefore

$$s(t) = -\frac{v_0 m}{k} e^{-(k/m)t} + \frac{v_0 m}{k} = \frac{v_0 m}{k}\left(1 - e^{-(k/m)t}\right). \tag{13}$$

To find how far the body will coast, we find the limit of $s(t)$ as $t \to \infty$. Since $\lim_{t\to\infty} e^{-(k/m)t} = 0$, the limit of $s(t)$ is $v_0\, m/k$.

$$\text{Distance coasted} = \frac{v_0 m}{k} \tag{14}$$

This is an ideal figure, of course. Only in mathematics can time stretch to infinity. The number $v_0\, m/k$ is only an upper bound (albeit a useful one). It is true to life in one respect, at least—if m is large, it will take a lot of energy to stop the body. That is why ocean liners have to be docked by tugboats. Any liner of conventional design entering a slip with enough speed to steer would smash into the pier before it could stop.

Example 4 For a 192-lb ice skater, the k in Eq. (11) is about 1/3 slug/sec and $m = 192/32 = 6$ slugs. How long will it take the skater to coast from 11 ft/sec (7.5 mph) to 1 ft/sec? How far will the skater coast before coming to a complete stop?

Solution We answer the first question by solving Eq. (11) for t:

$11e^{-t/18} = 1$ $\qquad \left(\begin{array}{l}\text{Eq. (11) with } k = 1/3,\ m = 6,\\ v_0 = 11,\ v = 1\end{array}\right)$

$e^{-t/18} = 1/11$

$-t/18 = \ln(1/11) = -\ln 11$

$t = 18 \ln 11 = 43 \text{ sec.} \qquad \text{(Calculator, rounded)}$

We answer the second question with Eq. (14):

$$\text{Distance coasted} = \frac{v_0 m}{k} = \frac{11 \cdot 6}{1/3} = 198 \text{ ft}.$$

EXERCISES 15.3

Solve the differential equations in Exercises 1–8.

1. $e^x \dfrac{dy}{dx} + 2e^x y = 1$

2. $2 \dfrac{dy}{dx} - y = e^{x/2}$

3. $x \dfrac{dy}{dx} + 3y = \dfrac{\sin x}{x^2}$

4. $x \dfrac{dy}{dx} + y = x \cos x$

5. $(x - 1)^3 \dfrac{dy}{dx} + 4(x - 1)^2 y = x + 1$

6. $e^{2x} \dfrac{dy}{dx} + 2e^{2x} y = 2x$

7. $\sin x \dfrac{dy}{dx} + (\cos x)\, y = \tan x$

8. $\cosh x \dfrac{dy}{dx} + (\sinh x)\, y = e^{-x}$

Solve the initial value problems in Exercises 9–12 for y as a function of x.

Differential equation	Initial condition
9. $\dfrac{dy}{dx} + 2y = x$	$y(0) = 1$
10. $x \dfrac{dy}{dx} + 2y = x^3$	$y(2) = 1$
11. $x \dfrac{dy}{dx} + y = \sin x$	$y(\pi/2) = 1$
12. $x \dfrac{dy}{dx} - 2y = x^3 \sec x \tan x$	$y(\pi/3) = 2$

13. Solve the initial value problem $dy/dx = ky$, $y(0) = y_0$, with Eq. (4) and see what you get.

14. *Blood sugar.* If glucose is fed intravenously at a constant rate, the change in the overall concentration $c(t)$ of glucose in the blood with respect to time may be described by the differential equation

$$\frac{dc}{dt} = \frac{G}{100V} - kc.$$

In this equation, G, V, and k are positive constants, G being the rate at which glucose is admitted, in milligrams per minute, and V the volume of blood in the body, in liters (around 5 liters for an adult). The concentration $c(t)$ is measured in milligrams per centiliter. The term $-kc$ is included because the glucose is assumed to be changing continually into other molecules at a rate proportional to its concentration.

a) Solve the equation for $c(t)$, using c_0 to denote $c(0)$.

b) Find the steady-state concentration, $\lim_{t \to \infty} c(t)$.

RL Circuits

15. *Current in a closed RL circuit.* How many seconds after the switch in an RL circuit is closed will it take the current i to reach half of its steady-state value? Notice that the time depends only on R and L and not on how much voltage is applied.

16. *Current in an open RL circuit.* If the switch is thrown open after the current in an RL circuit has built up to its steady-state value, the decaying current (Fig. 15.7) obeys the equation

$$L \frac{di}{dt} + Ri = 0, \tag{15}$$

which is Eq. (6) with $V = 0$.

a) Solve Eq. (15) to express i as a function of t.

b) How long after the switch is thrown will it take the current to fall to half its original value?

c) What is the value of the current when $t = L/R$? (The significance of this time is explained in the next exercise.)

17. CALCULATOR *Time constants.* Engineers call the number L/R the *time constant* of the RL circuit in Fig. 15.5. The significance of the time constant is that the current will reach 95% of its final value within 3 time constants of the time the switch is closed (Fig. 15.6). Thus, the time constant gives a

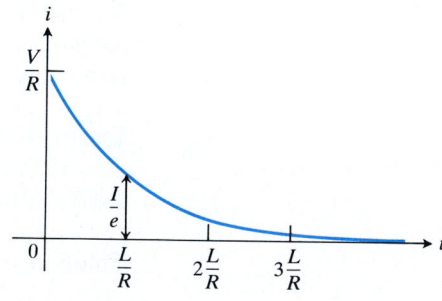

15.7 The current in an RL circuit decays exponentially when the power is turned off (Exercise 16).

built-in measure of how rapidly an individual circuit will reach equilibrium. Find the value of i in Eq. (8) that corresponds to $t = 3L/R$ and show that it is about 95% of the steady-state value $I = V/R$.

Additional Exercises

18. *Continuous compounding.*   You have $1000 with which to open an account and plan to add $1000 per year. All funds in the account will earn 10% interest per year compounded continuously. If the added deposits are also credited to your account continuously, the number of dollars x in your account at time t (years) will satisfy the initial value problem

Differential equation: $\dfrac{dx}{dt} = 1000 + 0.10x$,

Initial condition: $x(0) = 1000$.

a) Solve the initial value problem for x as a function of t.
b) CALCULATOR   About how many years will it take for the amount in your account to reach $100,000?

19. For a 145-lb cyclist on a 15-lb bicycle on level ground, the k

in Eq. (11) is about 1/5 slug/sec and $m = 160/32 = 5$ slugs. The cyclist starts coasting at 22 ft/sec (15 mph).
a) About how far will the cyclist coast before reaching a complete stop?
b) How long will it take the cyclist's speed to drop to 1 ft/sec?

20. For a 56,000-ton Iowa class battleship, $m = 1,750,000$ slugs and the k in Eq. (11) might be 3000 slugs/sec. Suppose the battleship loses power when it is moving at a speed of 22 ft/sec (13.2 knots).
a) About how far will the ship coast before it stops?
b) About how long will it take the ship's speed to drop to 1 ft/sec?

EXPLORER PROGRAM

Slope Fields

Graphs solutions of the initial value problem $y' = f(x, y)$, $y = y_0$ when $x = x_0$, for your choice of f, x_0, and y_0

15.4 Linear Homogeneous Second Order Equations

If $F(x) = 0$, the linear equation

$$a_n(x)\frac{d^n y}{dx^n} + a_{n-1}(x)\frac{d^{n-1}y}{dx^{n-1}} + \cdots + a_1(x)\frac{dy}{dx} + a_0(x)y = F(x) \qquad (1)$$

is called **homogeneous;** otherwise it is called **nonhomogeneous.** (This is different from the way we used the term *homogeneous* when we applied it to first order equations in Section 15.1.) This section shows how to solve the homogeneous equation

$$\frac{d^2 y}{dx^2} + a\frac{dy}{dx} + by = 0 \qquad (a \text{ and } b \text{ constant}).$$

Section 15.5 shows how to solve the nonhomogeneous equation

$$\frac{d^2 y}{dx^2} + a\frac{dy}{dx} + by = F(x).$$

Section 15.6 shows how to use these equations to model oscillation.

Linear Differential Operators

At this point, it is convenient to introduce the symbol D to represent the operation of differentiation with respect to x. That is, we write $Df(x)$ to mean $(d/dx)f(x)$. We define powers of D to mean taking successive derivatives:

$$D^2f(x) = D\{Df(x)\} = \frac{d^2 f(x)}{dx^2}, \qquad D^3f(x) = D\{D^2f(x)\} = \frac{d^3 f(x)}{dx^3},$$

and so on. A polynomial in D is to be interpreted as an operator that, when applied to $f(x)$, produces a linear combination of f and its successive derivatives. For example,

$$(D^2 + D - 2)f(x) = D^2f(x) + Df(x) - 2f(x) = \frac{d^2f(x)}{dx^2} + \frac{df(x)}{dx} - 2f(x).$$

DEFINITION

A polynomial in D is a **linear differential operator.**

We often denote linear differential operators by single letters. If L_1 and L_2 are two such linear operators, we define their sum and product by the equations

$$(L_1 + L_2)f(x) = L_1 f(x) + L_2 f(x), \qquad L_1 L_2 f(x) = L_1(L_2 f(x)).$$

Linear differential operators that are polynomials in D with constant coefficients satisfy basic algebraic laws that make it possible for us to treat them like ordinary polynomials so far as addition, multiplication, and factoring are concerned. Thus,

$$(D^2 + D - 2)f(x) = (D + 2)(D - 1)f(x) = (D - 1)(D + 2)f(x).$$

Since this equation holds for any twice-differentiable function f, we can also write the equality between operators:

$$D^2 + D - 2 = (D + 2)(D - 1) = (D - 1)(D + 2).$$

The Characteristic Equation

In the remainder of this section, we consider only linear *second order* equations with constant real-number coefficients. Because the solutions of nonhomogeneous equations depend on the solutions of the corresponding homogeneous equations, we focus on equations of the form

$$\frac{d^2y}{dx^2} + a\frac{dy}{dx} + by = 0 \tag{2}$$

or, in operator notation,

$$(D^2 + aD + b)y = 0. \tag{3}$$

The usual method for solving Eq. (3) is to begin by factoring the operator:

$$D^2 + aD + b = (D - r_1)(D - r_2).$$

We do this by finding the two roots r_1 and r_2 of the equation $r^2 + 2ar + b = 0$.

DEFINITION

The equation

$$r^2 + ar + b = 0 \tag{4}$$

is the **characteristic equation** of

$$\frac{d^2y}{dx^2} + a\frac{dy}{dx} + by = 0. \tag{5}$$

Equation (5) is equivalent to

$$(D - r_1)(D - r_2)y = 0. \tag{6}$$

If we now let

$$(D - r_2)y = u, \tag{7}$$

we get

$$(D - r_1)u = 0. \tag{8}$$

From here, we can solve Eq. (6) in two steps. From Eq. (8), which is first order separable in u and x, we find

$$u = C_1 e^{r_1 x}.$$

We substitute this into (7), which becomes

$$(D - r_2)y = C_1 e^{r_1 x} \qquad \text{or} \qquad \frac{dy}{dx} - r_2 y = C_1 e^{r_1 x}.$$

This equation is linear in y. Its integrating factor is $\rho = e^{-r_2 x}$ (Section 15.3), and its solution is

$$e^{-r_2 x} y = C_1 \int e^{(r_1 - r_2)x} \, dx + C_2. \tag{9}$$

How we proceed at this point depends on whether r_1 and r_2 are real and unequal, real and equal, or complex.

Real, Unequal Roots

If r_1 and r_2 are real and $r_1 \neq r_2$, the evaluation of the integral in Eq. (9) leads to

$$e^{-r_2 x} y = \frac{C_1}{r_1 - r_2} e^{(r_1 - r_2)x} + C_2 \qquad \text{or} \qquad y = \frac{C_1}{r_1 - r_2} e^{r_1 x} + C_2 e^{r_2 x}. \tag{10}$$

Since C_1 is an arbitrary constant, so is $C_1/(r_1 - r_2)$, and we can write the solution in Eq. (10) as

$$y = C_1 e^{r_1 x} + C_2 e^{r_2 x}. \tag{11}$$

Example 1   Solve the initial value problem

Differential equation: $\dfrac{d^2y}{dx^2} + \dfrac{dy}{dx} - 2y = 0,$

Initial conditions: $y(0) = 1, \quad y'(0) = -4.$

Solution   The characteristic equation $r^2 + r - 2 = 0$ has roots $r_1 = 1, r_2 = -2$. Hence, by Eq. (11), the solution of the differential equation is

$$y = C_1 e^x + C_2 e^{-2x}.$$

The initial conditions tell us that

$$C_1 + C_2 = 1, \qquad (y(0) = 1)$$

$$C_1 - 2C_2 = -4. \qquad (y'(0) = -4)$$

We solve these equations simultaneously and find that

$$C_1 = -\frac{2}{3}, \qquad C_2 = \frac{5}{3}.$$

The solution of the initial value problem is

$$y = -\frac{2}{3}e^x + \frac{5}{3}e^{-2x}.$$

Real, Equal Roots

If $r_1 = r_2$, then $e^{(r_1 - r_2)x} = e^0 = 1$ and Eq. (9) reduces to $e^{-r_2 x}y = C_1 x + C_2$ or

$$y = (C_1 x + C_2)e^{r_2 x}. \tag{12}$$

Example 2   Solve the equation $\dfrac{d^2y}{dx^2} + 4\dfrac{dy}{dx} + 4y = 0.$

Solution
$$r^2 + 4r + 4 = (r + 2)^2,$$
$$r_1 = r_2 = -2,$$
$$y = (C_1 x + C_2)e^{-2x}.$$

Complex Roots

From the theory of equations, we know that if a and b are real and the roots of the characteristic equation $r^2 + ar + b = 0$ are not real, then the roots must be a pair of complex conjugate numbers

$$r_1 = \alpha + \beta i, \qquad r_2 = \alpha - \beta i. \tag{13}$$

If $\beta \neq 0$ then Eq. (11) applies once again to give

$$y = c_1 e^{(\alpha + i\beta)x} + c_2 e^{(\alpha - i\beta)x} = e^{\alpha x}(c_1 e^{i\beta x} + c_2 e^{-i\beta x}), \tag{14}$$

where c_1 and c_2 are complex constants. By Euler's formula (Section 8.7),

$$e^{i\beta x} = \cos \beta x + i \sin \beta x \qquad \text{and} \qquad e^{-i\beta x} = \cos \beta x - i \sin \beta x.$$

Hence, we may replace Eq. (14) by

$$y = e^{\alpha x}((c_1 + c_2) \cos \beta x + i(c_1 - c_2) \sin \beta x). \tag{15}$$

Finally, we introduce new arbitrary constants, $C_1 = c_1 + c_2$ and $C_2 = i(c_1 - c_2)$, to give the solution in Eq. (15) a shorter form:

$$y = e^{\alpha x}(C_1 \cos \beta x + C_2 \sin \beta x). \tag{16}$$

(The constants C_1 and C_2 are real because c_1 and c_2 are complex conjugates.)

Example 3   Solve the initial value problem

Differential equation: $\quad \dfrac{d^2y}{dx^2} + 4\dfrac{dy}{dx} + 6y = 0,$

Initial conditions: $\quad y(0) = -1, \quad y'(0) = 0.$

Solution   We solve the characteristic equation $r^2 + 4r + 6 = 0$ with the quadratic formula and find its roots to be $r_1 = -2 + \sqrt{2}i$ and $r_2 = -2 - \sqrt{2}i$. With $\alpha = -2$ and $\beta = \sqrt{2}$, Eq. (16) gives

$$y = e^{-2x}(C_1 \cos \sqrt{2}x + C_2 \sin \sqrt{2}x). \tag{17}$$

A routine differentiation shows that

$$y' = e^{-2x}\left(\left(\sqrt{2}C_2 - 2C_1\right) \cos \sqrt{2}x - \left(\sqrt{2}C_1 + 2C_2\right) \sin \sqrt{2}x\right). \tag{18}$$

We substitute the initial values into Eqs. (17) and (18) and solve the resulting equations simultaneously for C_1 and C_2, finding that

$$C_1 = -1 \quad \text{and} \quad C_2 = -\sqrt{2}.$$

The solution of the initial value problem is

$$y = e^{-2x}(-\cos \sqrt{2}x - \sqrt{2} \sin \sqrt{2}x).$$

Example 4   *Imaginary Roots.*   Solve the equation

$$\frac{d^2y}{dx^2} + 4y = 0.$$

Solution   The roots of the characteristic equation $r^2 + 4 = 0$ are $r = \pm 2i$. With $\alpha = 0$ and $\beta = 2$, Eq. (16) gives

$$y = C_1 \cos 2x + C_2 \sin 2x.$$

Solutions of $\dfrac{d^2y}{dx^2} + a\dfrac{dy}{dx} + by = 0$

Roots of $r^2 + ar + b = 0$	Solution
r_1, r_2 real and unequal	$y = C_1 e^{r_1 x} + C_2 e^{r_2 x}$
r_1, r_2 real and equal	$y = (C_1 x + C_2)e^{r_2 x}$
r_1, r_2 complex conjugates, $\alpha \pm \beta i$	$y = e^{\alpha x}(C_1 \cos \beta x + C_2 \sin \beta x)$

EXERCISES 15.4

Solve the equations in Exercises 1–12.

1. $\dfrac{d^2y}{dx^2} + 2\dfrac{dy}{dx} = 0$

2. $\dfrac{d^2y}{dx^2} + 5\dfrac{dy}{dx} + 6y = 0$

3. $\dfrac{d^2y}{dx^2} + 6\dfrac{dy}{dx} + 5y = 0$

4. $\dfrac{d^2y}{dx^2} - 2\dfrac{dy}{dx} - 3y = 0$

5. $\dfrac{d^2y}{dx^2} - 4\dfrac{dy}{dx} + 4y = 0$

6. $\dfrac{d^2y}{dx^2} + 6\dfrac{dy}{dx} + 9y = 0$

7. $\dfrac{d^2y}{dx^2} - 10\dfrac{dy}{dx} + 25y = 0$

8. $\dfrac{d^2y}{dx^2} - 2\sqrt{2}\dfrac{dy}{dx} + 2y = 0$

9. $\dfrac{d^2y}{dx^2} + \dfrac{dy}{dx} + y = 0$

10. $\dfrac{d^2y}{dx^2} - 6\dfrac{dy}{dx} + 10y = 0$

11. $\dfrac{d^2y}{dx^2} - 2\dfrac{dy}{dx} + 4y = 0$

12. $\dfrac{d^2y}{dx^2} + 8\dfrac{dy}{dx} + 25y = 0$

Solve the initial value problems in Exercises 13–24 ($y'' = d^2y/dx^2$ and $y' = dy/dx$).

13. $y'' - y = 0$, $y(0) = 1$, $y'(0) = -2$

14. $2y'' - y' - y = 0$, $y(0) = -1$, $y'(0) = 0$

15. $y'' - 4y = 0$, $y(0) = 0$, $y'(0) = 3$

16. $y'' - 9y = 0$, $y(\ln 2) = 1$, $y'(\ln 2) = -3$

17. $y'' + 2y' + y = 0$, $y(0) = 0$, $y'(0) = 1$

18. $4y'' - 2y' + y = 0$, $y(0) = 4$, $y'(0) = 2$

19. $4y'' + 12y' + 9y = 0$, $y(0) = 0$, $y'(0) = -1$

20. $y'' = 0$, $y(0) = -3$, $y'(0) = 5$

21. $y'' + 4y = 0$, $y(0) = 0$, $y'(0) = 2$

22. $y'' + 9y = 0$, $y(0) = 0$, $y'(0) = \sqrt{3}$

23. $y'' - 2y' + 3y = 0$, $y(0) = 2$, $y'(0) = 1$

24. $y'' - 6y' + 10y = 0$, $y(0) = 7$, $y'(0) = 1$

25. *Still another way to solve* $(d^2y/dx^2) + y = 0$. Let $y' = $

dy/dx and define the functions c_1 and c_2 by the equations

$$c_1 = y \cos x - y' \sin x,$$
$$c_2 = y \sin x + y' \cos x. \tag{19}$$

Show that the condition $y'' + y = 0$ implies that $c_1' = c_2' = 0$ so that c_1 and c_2 are constants. Then eliminate y' from Eqs. (19) to find the general solution for y.

26. *Superposition.* Show that if $y = f_1(x)$ and $y = f_2(x)$ are both solutions of the homogeneous linear differential equation

$$a_n(x) \frac{d^ny}{dx^n} + a_{n-1}(x) \frac{d^{n-1}y}{dx^{n-1}} + \cdots + a_1(x) \frac{dy}{dx} + a_0(x)y = 0 \tag{20}$$

on an interval I then $y = C_1 f_1(x) + C_2 f_2(x)$ is also a solution of (20) on I. This fact is called the **principle of superposition.** It allows us to construct new solutions of homogeneous equations by superimposing one solution on another. If $y = \cos x$ is one solution of a homogeneous equation and $y = \sin x$ is another, then every linear combination $y = C_1 \cos x + C_2 \sin x$ is also a solution.

15.5 Second Order Equations; Reduction of Order

This section shows how to solve linear nonhomogeneous equations of the form

$$\frac{d^2y}{dx^2} + a \frac{dy}{dx} + by = F(x) \qquad (a \text{ and } b \text{ real constants}). \tag{1}$$

It also shows how to use an unrelated technique called reduction of order that enables us to solve some nonlinear second order equations by changing them into first order equations to which our earlier techniques may apply.

Linear Nonhomogeneous Second Order Equations

The procedure for solving Eq. (1) has three basic steps. First, we use the techniques of Section 15.4 to find the general solution y_h (h stands for homogeneous) of the **reduced equation**

$$\frac{d^2y}{dx^2} + a \frac{dy}{dx} + by = 0. \tag{2}$$

Then we find a particular solution y_p of the **complete equation,** Eq. (1). Then we add y_p to y_h to form the general solution of the complete equation. Why the procedure works and how we find y_p are the main topics of this section.

The main idea is that we can obtain all solutions of the complete equation (1) by adding a particular solution of Eq. (1) to the general solution of the reduced equation (2). The crux of the matter is that if L is any linear differential operator (not necessarily one of second order or having constant coefficients), then

$$L(k_1 y_1 + k_2 y_2) = k_1 L(y_1) + k_2 L(y_2)$$

for any constants k_1 and k_2 and functions y_1 and y_2 to which the operator L can be applied. In the present context, we can take

$$L = D^2 + aD + b$$

and write Eqs. (1) and (2) as $L(y) = F(x)$ and $L(y) = 0$, respectively. The following theorem states the relation between solutions of complete and reduced linear equations.

THEOREM 2

Let $L(y) = F(x)$ be a linear nonhomogeneous differential equation and $L(y) = 0$ be the reduced equation with the same operator L. If y_h is the general solution of the reduced equation and y_p is a particular solution of the complete equation, then the general solution of the complete equation is

$$y = y_h(x) + y_p(x).$$

Proof Let $L(y_h(x)) = 0$ and $L(y_p(x)) = F(x)$. Then

$$L(y_h(x) + y_p(x)) = L(y_h(x)) + L(y_p(x)) = 0 + F(x) = F(x),$$

which establishes that $y = y_h + y_p$ is a solution of the nonhomogeneous equation. To show that all solutions have this form, let y be any solution of the nonhomogeneous equation $L(y) = F(x)$. Then

$$L(y - y_p) = L(y) - L(y_p) = F(x) - F(x) = 0,$$

so that $y - y_p$ satisfies the reduced equation. Therefore, if $y_h = C_1 u_1(x) + C_2 u_2(x)$ is the general solution of $(D^2 + aD + b)y = 0$, there is some choice of the constants C_1 and C_2 such that

$$y - y_p = C_1 u_1(x) + C_2 u_2(x),$$

which means that

$$y = y_p + C_1 u_1(x) + C_2 u_2(x).$$

We now discuss three ways to find a particular solution of the complete equation (Eq. 1):

1. Inspired guessing (Try this first—it is a real time-saver when it works. Experience will help.)

2. The method of variation of parameters

3. The method of undetermined coefficients

Inspired Guessing

In the following example, we guess that there exists a particular solution of the form $y_p(x) = C$. We find the right value of C by substituting y_p and its derivatives into the differential equation.

Example 1   Find a particular solution of $\dfrac{d^2y}{dx^2} + 2\dfrac{dy}{dx} - 3y = 6.$

Solution   We guess that there is a solution of the form $y_p = C$ and substitute y_p and its derivatives $y_p' = 0$, $y_p'' = 0$ into the equation. This leads to

$$\frac{d^2y_p}{dx^2} + 2\frac{dy_p}{dx} - 3y_p = 6, \qquad 0 + 2(0) - 3(C) = 6, \qquad C = -2.$$

The equation has $y_p = -2$ as a solution.

Example 2   Find the general solution of $\dfrac{d^2y}{dx^2} + 2\dfrac{dy}{dx} - 3y = 6.$

Solution   We find the solution in three steps.

We always use these steps:
1. Find y_h.
2. Find y_p.
3. Add y_p to y_h.

STEP 1:   *Find the general solution y_h of the reduced equation*

$$\frac{d^2y}{dx^2} + 2\frac{dy}{dx} - 3y = 0.$$

The roots of the characteristic equation $r^2 + 2r - 3 = 0$ are $r_1 = 1$ and $r_2 = -3$, so

$$y_h = C_1\,e^x + C_2\,e^{-3x}.$$

STEP 2:   *Find a particular solution y_p of the complete equation:*

$y_p = -2.$ (From Example 1)

STEP 3:   *Add y_p to y_h to form the general solution y of the complete equation:*

$$y = y_h + y_p = C_1\,e^x + C_2\,e^{-3x} - 2.$$

Variation of Parameters

When guessing doesn't work (it usually doesn't), we find a particular solution of the complete equation (Eq. 1) by a method called **variation of parameters.** The method assumes we already know the general solution

$$y_h = C_1\,u_1(x) + C_2\,u_2(x) \tag{3}$$

of the reduced equation (Eq. 2), which is why we make finding y_h our first solution step.

The method of variation of parameters consists of replacing the constants C_1 and C_2 on the right side of Eq. (3) by functions $v_1(x)$ and $v_2(x)$ and then requiring that the new expression

$$y = v_1 u_1 + v_2 u_2$$

be a solution of the complete equation (1). There are two functions v_1 and v_2 to be determined, and requiring that Eq. (1) be satisfied is only one condition. It turns out to simplify things if we also require that

$$v_1' u_1 + v_2' u_2 = 0. \tag{4}$$

Then we have

$$y = v_1 u_1 + v_2 u_2,$$

$$\frac{dy}{dx} = v_1 u_1' + v_2 u_2' + \underbrace{v_1' u_1 + v_2' u_2}_{0} = v_1 u_1' + v_2 u_2',$$

$$\frac{d^2y}{dx^2} = v_1 u_1'' + v_2 u_2'' + v_1' u_1' + v_2' u_2'.$$

If we substitute these expressions into the left-hand side of Eq. (1) and rearrange terms, we obtain

$$v_1\left[\frac{d^2u_1}{dx^2} + a\frac{du_1}{dx} + bu_1\right] + v_2\left[\frac{d^2u_2}{dx^2} + a\frac{du_2}{dx} + bu_2\right] + v_1' u_1' + v_2' u_2' = F(x).$$

The two bracketed terms are zero, since u_1 and u_2 are solutions of the reduced equation (2). Hence Eq. (1) is satisfied if, in addition to Eq. (4), we require that

$$v_1' u_1' + v_2' u_2' = F(x). \tag{5}$$

Equations (4) and (5) may be solved together as a system,

$$v_1' u_1 + v_2' u_2 = 0, \qquad v_1' u_1' + v_2' u_2' = F(x),$$

for the unknown functions v_1' and v_2'. Cramer's rule (Appendix 10) gives

$$v_1' = \frac{\begin{vmatrix} 0 & u_2 \\ F(x) & u_2' \end{vmatrix}}{\begin{vmatrix} u_1 & u_2 \\ u_1' & u_2' \end{vmatrix}} = \frac{-u_2 F(x)}{D}, \qquad v_2' = \frac{\begin{vmatrix} u_1 & 0 \\ u_1' & F(x) \end{vmatrix}}{\begin{vmatrix} u_1 & u_2 \\ u_1' & u_2' \end{vmatrix}} = \frac{u_1 F(x)}{D}, \tag{6}$$

where

$$D = \begin{vmatrix} u_1 & u_2 \\ u_1' & u_2' \end{vmatrix}.$$

Now v_1 and v_2 can be found by integration.

How to Use the Method of Variation of Parameters

In applying the method of variation of parameters to solve the equation

$$\frac{d^2 y}{dx^2} + a\frac{dy}{dx} + by = F(x), \tag{1}$$

we can work directly with the equations in Eq. (6). It is not necessary to rederive them. The following are the steps to take.

STEP 1: Solve the reduced equation,

$$\frac{d^2 y}{dx^2} + a\frac{dy}{dx} + by = 0,$$

to find the functions u_1 and u_2.

STEP 2: Solve the equations

$$v_1' u_1 + v_2' u_2 = 0$$
$$v_1' u_1' + v_2' u_2' = F(x)$$

together for v_1' and v_2' (as in Eqs. (6)).

STEP 3: Integrate v_1' and v_2' to find v_1 and v_2.

STEP 4: Write down the general solution of Eq. (1) as $y = v_1 u_1 + v_2 u_2$.

Example 3   Solve the equation $\dfrac{d^2 y}{dx^2} + y = \tan x$, $-\dfrac{\pi}{2} < x < \dfrac{\pi}{2}$.

Solution   We first solve the reduced equation

$$\frac{d^2 y}{dx^2} + y = 0.$$

The solutions of the characteristic equation $r^2 + 1 = 0$ are $r = \pm i$, so

$$y = C_1 \cos x + C_2 \sin x.$$

With $u_1 = \cos x$ and $u_2 = \sin x$, we have

$$D = \begin{vmatrix} \cos x & \sin x \\ -\sin x & \cos x \end{vmatrix} = \cos^2 x + \sin^2 x = 1.$$

With $F(x) = \tan x$, Eqs. (6) then give

$$v_1' = \frac{-\sin x \tan x}{1} = -\frac{\sin^2 x}{\cos x} = \frac{\cos^2 x - 1}{\cos x} = \cos x - \sec x,$$

$$v_2' = \frac{\cos x \tan x}{1} = \sin x.$$

$$(7)$$

Hence

$$v_1 = \int (\cos x - \sec x)\, dx = \sin x - \ln|\sec x + \tan x| + C_1,$$

$$v_2 = \int \sin x\, dx = -\cos x + C_2,$$

and

$$y = v_1 u_1 + v_2 u_2$$

$$= (\sin x - \ln|\sec x + \tan x| + C_1) \cos x + (-\cos x + C_2) \sin x$$

$$= C_1 \cos x + C_2 \sin x - \cos x \ln|\sec x + \tan x|.$$

Undetermined Coefficients

The method of variation of parameters is completely general, for it produces a particular solution of Eq. (1) for any continuous function $F(x)$. But the calculations involved can be complicated, and in special cases there may be easier methods to use. For instance, we do not need to use variation of parameters to find a particular solution of

$$\frac{d^2y}{dx^2} - \frac{dy}{dx} + 5y = 3,$$

$$(8)$$

if we can find the particular solution $y_p = 3/5$ by inspection. And even for an equation like

$$\frac{d^2y}{dx^2} + 3y = e^x,$$

$$(9)$$

we might guess that there is a solution of the form

$$y_p = Ae^x$$

and substitute $y_p = Ae^x$ and its second derivative into Eq. (9) to find A. If we do so, we find the solution $y_p = (1/4)e^x$.

Again, we might guess that the equation

$$\frac{d^2y}{dx^2} + y = 3x^2 + 4$$

$$(10)$$

has a particular solution of the form

$$y_p = Cx^2 + Dx + E.$$

If we substitute this polynomial and its second derivative into Eq. (10) to look for appropriate values for the constants C, D, and E, we get

$$2C + (Cx^2 + Dx + E) = 3x^2 + 4,$$

$$Cx^2 + Dx + 2C + E = 3x^2 + 4. \tag{11}$$

This latter equation will hold for all values of x if its two sides are identical as polynomials in x; that is, if

$$C = 3, \quad D = 0, \quad \text{and} \quad 2C + E = 4, \tag{12}$$

or

$$C = 3, \quad D = 0, \quad E = -2.$$

We conclude that

$$y_p = 3x^2 + 0x - 2 = 3x^2 - 2 \tag{13}$$

is a solution of Eq. (10).

In each of the foregoing examples, the particular solution we found resembled the function $F(x)$ on the right side of the given differential equation. This was no accident, for we guessed the form of the particular solution by looking at $F(x)$ first. The method of first guessing the form of the solution up to certain undetermined constants and then determining the values of these constants by using the differential equation is known as the **method of undetermined coefficients.** It depends for its success on our ability to recognize the form of a particular solution and for this reason, among others, it lacks the generality of the method of variation of parameters. Nevertheless, its simplicity makes it the method of choice in a number of special cases.

We shall limit our discussion of the method of undetermined coefficients to selected equations in which the function $F(x)$ in Eq. (1) is the sum of one or more terms like

$$e^{rx}, \quad \cos kx, \quad \sin kx, \quad ax^2 + bx + c.$$

Even so, we will find that the particular solutions of Eq. (1) do not always resemble $F(x)$ as closely as the ones we have seen.

Example 4   Find a particular solution of

$$\frac{d^2y}{dx^2} - \frac{dy}{dx} = 2 \sin x. \tag{14}$$

Solution   If we try to find a particular solution of the form

$$y_p = A \sin x$$

and substitute the derivatives of y_p in the given equation, we find that A must satisfy the equation

$$-A \sin x - A \cos x = 2 \sin x \tag{15}$$

for all values of x. Since this requires A to be equal to -2 and 0 at the same time, we conclude that Eq. (14) has no solution of the form $A \sin x$.

It turns out that the required form is the sum

$$y_p = A \sin x + B \cos x. \tag{16}$$

The result of substituting the derivatives of this new candidate into Eq. (14) is

$$-A \sin x - B \cos x - (A \cos x - B \sin x) = 2 \sin x,$$

$$(B - A) \sin x - (A + B) \cos x = 2 \sin x. \tag{17}$$

Equation (17) will be an identity if

$$B - A = 2 \quad \text{and} \quad A + B = 0, \quad \text{or} \quad A = -1, \quad B = 1.$$

Our particular solution is $y_p = \cos x - \sin x$.

Example 5   Find a particular solution of

$$\frac{d^2y}{dx^2} - 3\frac{dy}{dx} + 2y = 5e^x. \tag{18}$$

Solution   If we substitute

$$y_p = Ae^x$$

and its derivatives in Eq. (18), we find that

$$Ae^x - 3Ae^x + 2Ae^x = 5e^x,$$

$$0 = 5e^x.$$

The trouble can be traced to the fact that $y = e^x$ is already a solution of the reduced equation

$$\frac{d^2y}{dx^2} - 3\frac{dy}{dx} + 2y = 0. \tag{19}$$

The characteristic equation of Eq. (19) is

$$r^2 - 3r + 2 = (r - 1)(r - 2) = 0,$$

which has $r = 1$ as a simple root. We may therefore expect Ae^x to "vanish" when substituted into the left-hand side of Eq. (18).

The appropriate way to modify the trial solution in this case is to replace Ae^x by Axe^x. When we substitute

$$y_p = Axe^x$$

and its derivatives into Eq. (18), we get

$$(Axe^x + 2Ae^x) - 3(Axe^x + Ae^x) + 2Axe^x = 5e^x$$

$$-Ae^x = 5e^x$$

$$A = -5.$$

Our solution is $y_p = -5xe^x$.

Example 6   Find a particular solution of

$$\frac{d^2y}{dx^2} - 6\frac{dy}{dx} + 9y = e^{3x}. \tag{20}$$

Solution   The characteristic equation,

$$r^2 - 6r + 9 = (r - 3)^2 = 0,$$

has $r = 3$ as a *double* root. The appropriate choice for y_p in this case is neither Ae^{3x} nor Axe^{3x}, but Ax^2e^{3x}. When we substitute

$$y_p = Ax^2e^{3x}$$

and its derivatives in the given differential equation, we get

$$(9Ax^2e^{3x} + 12Axe^{3x} + 2Ae^{3x}) - 6(3Ax^2e^{3x} + 2Axe^{3x}) + 9Ax^2e^{3x} = e^{3x}$$

$$2Ae^{3x} = e^{3x}$$

$$A = \frac{1}{2}.$$

Our solution is $y_p = (1/2)x^2e^{3x}$.

When we wish to find a particular solution of Eq. (1), and $F(x)$ has two or more terms, we include a trial function for each term in $F(x)$.

Example 7   Solve the equation

$$\frac{d^2y}{dx^2} - \frac{dy}{dx} = 5e^x - \sin 2x. \tag{21}$$

Solution   The roots of the characteristic equation $r^2 - r = 0$ are

$$r_1 = 1, \qquad r_2 = 0,$$

so the general solution of the reduced equation is

$$y_h = C_1e^x + C_2.$$

We now seek a particular solution y_p. That is, we seek a function that will produce $5e^x - \sin 2x$ when substituted into the left side of Eq. (21). One part of y_p is to produce $5e^x$, the other $-\sin 2x$.

Since any function of the form C_1e^x is a solution of the reduced equation, we choose our trial y_p to be the sum

$$y_p = Axe^x + B \cos 2x + C \sin 2x,$$

including xe^x where we might otherwise have included e^x. Substituting the derivatives of y_p in Eq. (21) gives

$$(Axe^x + 2Ae^x - 4B \cos 2x - 4C \sin 2x)$$

$$- (Axe^x + Ae^x - 2B \sin 2x + 2C \cos 2x) = 5e^x - \sin 2x,$$

$$Ae^x - (4B + 2C) \cos 2x + (2B - 4C) \sin 2x = 5e^x - \sin 2x.$$

These equations will hold if

$$A = 5, \qquad (4B + 2C) = 0, \qquad (2B - 4C) = -1,$$

or

$$A = 5, \qquad B = -\frac{1}{10}, \qquad C = \frac{1}{5}.$$

Our particular solution is

$$y_p = 5xe^x - \frac{1}{10} \cos 2x + \frac{1}{5} \sin 2x.$$

The complete solution of Eq. (21) is

$$y = y_h + y_p = C_1 e^x + C_2 + 5xe^x - \frac{1}{10} \cos 2x + \frac{1}{5} \sin 2x.$$

Reduction of Order

We can solve some (not necessarily linear) second order equations by changing them into first order equations with a substitution and solving the resulting first order equations with first order techniques. We did this in Section 6.10 when we solved the nonlinear hanging-cable equation

$$\frac{d^2y}{dx^2} = a\sqrt{1 + \left(\frac{dy}{dx}\right)^2}.$$

The equation is first order in dy/dx and the substitution $p = dy/dx$ changed it into

$$\frac{dp}{dx} = a\sqrt{1 + p^2},$$

which we solved by separation of variables. Having found the solution $p = \sinh ax$ of the associated initial value problem, we replaced p by the original dy/dx and went on to solve the first order equation

$$\frac{dy}{dx} = \sinh ax$$

for y as a function of x. We solved the second order equation by solving two first order equations. Because we reduced the equation's order to solve it, the technique is called **reduction of order.** The general form of a second order differential equation is

$$F\left(x, y, \frac{dy}{dx}, \frac{d^2y}{dx^2}\right) = 0. \tag{22}$$

The equations that can be reduced to first order by substitution normally arise when either the x or the y variable is missing.

TYPE 1: *Equations with dependent variable missing.* When Eq. (22) has the special form

$$F\left(x, \frac{dy}{dx}, \frac{d^2y}{dx^2}\right) = 0, \tag{23}$$

we can reduce it to a first order equation by substituting

$$p = \frac{dy}{dx}, \qquad \frac{d^2y}{dx^2} = \frac{dp}{dx}.$$

Then Eq. (23) takes the form

$$F\left(x, p, \frac{dp}{dx}\right) = 0,$$

which is of the first order in p. If this equation can be solved for p as a function of x, say $p = \phi(x, C_1)$, then we can find y by one additional integration:

$$y = \int \frac{dy}{dx} \, dx = \int p \, dx = \int \phi(x, C_1) \, dx + C_2.$$

Example 8   Solve the equation

$$x\frac{d^2y}{dx^2} + \frac{dy}{dx} = x^2, \qquad x > 0.$$

Solution   The dependent variable y does not occur (except in the derivatives) and the substitutions

$$\frac{dy}{dx} = p, \qquad \frac{d^2y}{dx^2} = \frac{dp}{dx}$$

reduce the equation to

$$x\frac{dp}{dx} + p = x^2,$$

a linear first order equation. Dividing by x puts the equation in standard form:

$$\frac{dp}{dx} + \frac{1}{x}p = x. \tag{24}$$

We use Eq. (4) of Section 15.3 with

$$\rho(x) = e^{\int (1/x)dx} = e^{\ln x} = x$$

and $Q(x) = x$ to write the solution of Eq. (24) as

$$p = \frac{1}{x}\int x \cdot x \, dx = \frac{1}{x}\left(\frac{x^3}{3} + C_1\right)$$

$$= \frac{x^2}{3} + \frac{C_1}{x}.$$

We then replace p by dy/dx,

$$\frac{dy}{dx} = \frac{x^2}{3} + \frac{C_1}{x},$$

and integrate again to get

$$y = \frac{x^3}{9} + C_1 \ln x + C_2.$$

(We can omit the usual absolute value bars in the logarithm because $x > 0$.)

TYPE 2: *Equations with independent variable missing.* When Eq. (22) does not contain x explicitly but has the form

$$F\left(y, \frac{dy}{dx}, \frac{d^2y}{dx^2}\right) = 0, \tag{25}$$

the substitutions to use are

$$p = \frac{dy}{dx} \quad \text{and} \quad \frac{d^2y}{dx^2} = \frac{dp}{dx} = \frac{dp}{dy}\frac{dy}{dx} = \frac{dp}{dy}p.$$

Then Eq. (25) takes the form

$$F\left(y, p, p\frac{dp}{dy}\right) = 0,$$

which is of the first order in p. Its solution gives p in terms of y, and then a further integration gives the solution of Eq. (25).

Example 9   Solve the initial value problem

$$y\frac{d^2y}{dx^2} + \left(\frac{dy}{dx}\right)^2 = 0, \quad y = 2 \quad \text{and} \quad \frac{dy}{dx} = \frac{1}{2} \text{ when } x = 1.$$

Solution   We let

$$\frac{dy}{dx} = p, \quad \frac{d^2y}{dx^2} = \frac{dp}{dy}p.$$

Then we have, in turn,

$$y\left(\frac{dp}{dy}p\right) + p^2 = 0$$

$$y\frac{dp}{dy} + p = 0$$

$$\frac{d}{dy}(yp) = 0$$

$$yp = C_1$$

$$y\frac{dy}{dx} = C_1.$$

The condition that $y = 2$ when $dy/dx = 1/2$ determines C_1:

$$2 \cdot \frac{1}{2} = C_1, \qquad C_1 = 1.$$

Proceeding with $C_1 = 1$, we have

$$y\frac{dy}{dx} = 1,$$

$$\frac{1}{2}y^2 = x + C_2. \qquad \text{(Integrated with respect to } x\text{)}$$

The condition that $y = 2$ when $x = 1$ determines C_2:

$$\frac{1}{2}(2)^2 = 1 + C_2, \qquad C_2 = 1.$$

The solution of the initial value problem is

$$\frac{1}{2}y^2 = x + 1 \qquad \text{or} \qquad y^2 = 2x + 2.$$

Finding the General Solution When One Solution Is Known

When we are fortunate enough to know one solution $u(x)$ of a second order equation (we might have guessed it or identified it from special information), we can sometimes find the equation's general solution by substituting

$$y = uv \tag{26}$$

and solving for the function $v(x)$. The next example shows how this is done.

Example 10   Show that $y = e^x$ is a solution of the equation

$$y'' - 2y' + y = 0,$$

and use the result to find the general solution.

Solution   If $y = e^x$, then $y' = y'' = e^x$ and

$$y'' - 2y' + y = e^x - 2e^x + e^x = 0.$$

This shows that $y = e^x$ is one solution. We now substitute $u = e^x$, $y = uv = e^x v$ and calculate the first and second derivatives to get

$$y' = e^x v' + e^x v, \qquad y'' = e^x v'' + 2e^x v' + e^x v.$$

Next we substitute into the equation $y'' - 2y' + y = 0$ and get

$$e^x(v'' + 2v' + v - 2v' - 2v + v) = 0,$$

which simplifies to $v'' = 0$. The solution of this equation is simply $v = C_1 x + C_2$, and this gives

$$y = e^x v = e^x(C_1 x + C_2)$$

as the general solution of $y'' - 2y' + y = 0$.

TABLE 15.1

The method of undetermined coefficients for selected equations of the form $\dfrac{d^2y}{dx^2} + a\,\dfrac{dy}{dx} + by = F(x)$

If $F(x)$ has a term that is a constant multiple of	And if	Then include this expression in the trial function for y_p
e^{rx}	r is not a root of the characteristic equation	Ae^{rx}
	r is a single root of the characteristic equation	Axe^{rx}
	r is a double root of the characteristic equation	Ax^2e^{rx}
$\sin kx$, $\cos kx$	ki is not a root of the characteristic equation	$B \cos kx + C \sin kx$
	ki is a root of the characteristic equation	$Bx \cos kx + Cx \sin kx$
$ax^2 + bx + c$	0 is not a root of the characteristic equation	$Dx^2 + Ex + F$ (chosen to match the degree of $ax^2 + bx + c$)
	0 is a single root of the characteristic equation	$Dx^3 + Ex^2 + Fx$ (degree one higher than the degree of $ax^2 + bx + c$)
	0 is a double root of the characteristic equation	$Dx^4 + Ex^3 + Fx^2$ (degree two higher than the degree of $ax^2 + bx + c$)

As we saw in the preceding example, $y = e^x$ is a solution of $y'' - 2y' + y = 0$. As you will see if you do Exercise 73, $y = e^x$ is a solution of any linear homogeneous nth order differential equation whose coefficients add to zero. It does not matter whether the coefficients are constant or variable. For instance, $y = e^x$ is a solution of $xy'' + (2 - x)y' - 2y = 0$.

EXERCISES 15.5

Solve the equations in Exercises 1–12 by variation of parameters.

1. $\dfrac{d^2y}{dx^2} + \dfrac{dy}{dx} = x$

2. $\dfrac{d^2y}{dx^2} + y = \tan^2 x, \quad -\dfrac{\pi}{2} < x < \dfrac{\pi}{2}$

3. $\dfrac{d^2y}{dx^2} + y = \sin x$

4. $\dfrac{d^2y}{dx^2} + 2\dfrac{dy}{dx} + y = e^x$

5. $\dfrac{d^2y}{dx^2} + 2\dfrac{dy}{dx} + y = e^{-x}$

6. $\dfrac{d^2y}{dx^2} - y = x$

7. $\dfrac{d^2y}{dx^2} - y = e^x$

8. $\dfrac{d^2y}{dx^2} - y = \sin x$

9. $\dfrac{d^2y}{dx^2} + 4\dfrac{dy}{dx} + 5y = 10$

10. $\dfrac{d^2y}{dx^2} - \dfrac{dy}{dx} = 2^x$

11. $\dfrac{d^2y}{dx^2} + y = \sec x, \quad -\dfrac{\pi}{2} < x < \dfrac{\pi}{2}$

12. $\dfrac{d^2y}{dx^2} - \dfrac{dy}{dx} = e^x \cos x, \quad x > 0$

Solve the equations in Exercises 13–28 by the method of undetermined coefficients.

13. $\dfrac{d^2y}{dx^2} - 3\dfrac{dy}{dx} - 10y = -3$

14. $\dfrac{d^2y}{dx^2} - 3\dfrac{dy}{dx} - 10y = 2x - 3$

15. $\dfrac{d^2y}{dx^2} - \dfrac{dy}{dx} = \sin x$

16. $\dfrac{d^2y}{dx^2} + 2\dfrac{dy}{dx} + y = x^2$

17. $\dfrac{d^2y}{dx^2} + y = \cos 3x$

18. $\dfrac{d^2y}{dx^2} + y = e^{2x}$

19. $\dfrac{d^2y}{dx^2} - \dfrac{dy}{dx} - 2y = 20 \cos x$

20. $\dfrac{d^2y}{dx^2} + y = 2x + 3e^x$

21. $\dfrac{d^2y}{dx^2} - y = e^x + x^2$

22. $\dfrac{d^2y}{dx^2} + 2\dfrac{dy}{dx} + y = 6 \sin 2x$

23. $\dfrac{d^2y}{dx^2} - \dfrac{dy}{dx} - 6y = e^{-x} - 7 \cos x$

24. $\dfrac{d^2y}{dx^2} + 3\dfrac{dy}{dx} + 2y = e^{-x} + e^{-2x} - x$

25. $\dfrac{d^2y}{dx^2} + 5\dfrac{dy}{dx} = 15x^2$

26. $\dfrac{d^2y}{dx^2} - \dfrac{dy}{dx} = -8x + 3$

27. $\dfrac{d^2y}{dx^2} - 3\dfrac{dy}{dx} = e^{3x} - 12x$

28. $\dfrac{d^2y}{dx^2} + 7\dfrac{dy}{dx} = 42x^2 + 5x + 1$

In each of Exercises 29–31, the given differential equation has a particular solution y_p of the form given. Determine the coefficients in y_p. Then solve the differential equation.

29. $\dfrac{d^2y}{dx^2} - 5\dfrac{dy}{dx} = xe^{5x}, \quad y_p = Ax^2e^{5x} + Bxe^{5x}$

30. $\dfrac{d^2y}{dx^2} - \dfrac{dy}{dx} = \cos x + \sin x, \quad y_p = A \cos x + B \sin x$

31. $\dfrac{d^2y}{dx^2} + y = 2 \cos x + \sin x, \quad y_p = Ax \cos x + Bx \sin x$

In Exercises 32–35, solve the given differential equations (a) by variation of parameters and (b) by the method of undetermined coefficients.

32. $\dfrac{d^2y}{dx^2} - 4\dfrac{dy}{dx} + 4y = 2e^{2x}$

33. $\dfrac{d^2y}{dx^2} - \dfrac{dy}{dx} = e^x + e^{-x}$

34. $\dfrac{d^2y}{dx^2} - 9\dfrac{dy}{dx} = 9e^{9x}$

35. $\dfrac{d^2y}{dx^2} - 4\dfrac{dy}{dx} - 5y = e^x + 4$

Solve the differential equations in Exercises 36–45. Some of the equations can be solved by the method of undetermined coefficients, but others cannot.

36. $\dfrac{d^2y}{dx^2} + y = \csc x, \quad 0 < x < \pi$

37. $\dfrac{d^2y}{dx^2} + y = \cot x, \quad 0 < x < \pi$

38. $\dfrac{d^2y}{dx^2} + 4y = \sin x$

39. $\dfrac{d^2y}{dx^2} - 8\dfrac{dy}{dx} = e^{8x}$

40. $\dfrac{d^2y}{dx^2} + 4\dfrac{dy}{dx} + 5y = x + 2$

41. $\dfrac{d^2y}{dx^2} - \dfrac{dy}{dx} = x^3$

42. $\dfrac{d^2y}{dx^2} + 9y = 9x - \cos x$

43. $\dfrac{d^2y}{dx^2} + 2\dfrac{dy}{dx} = x^2 - e^x$

44. $\dfrac{d^2y}{dx^2} - 3\dfrac{dy}{dx} + 2y = e^x - e^{2x}$

45. $\dfrac{d^2y}{dx^2} + y = \sec x \tan x, \quad -\dfrac{\pi}{2} < x < \dfrac{\pi}{2}$

The method of undetermined coefficients can sometimes be used to solve first order ordinary differential equations. Use the method to solve the equations in Exercises 46–49.

46. $\dfrac{dy}{dx} + 4y = x$

47. $\dfrac{dy}{dx} - 3y = e^x$

48. $\dfrac{dy}{dx} + y = \sin x$

49. $\dfrac{dy}{dx} - 3y = 5e^{3x}$

Solve the initial value problems in Exercises 50 and 51.

50. $\dfrac{d^2y}{dx^2} + y = e^{2x}; \quad y(0) = 0, \quad y'(0) = \dfrac{2}{5}$

51. $\dfrac{d^2y}{dx^2} + y = \sec^2 x, \quad -\dfrac{\pi}{2} < x < \dfrac{\pi}{2}; \quad y(0) = y'(0) = 1$

52. *Bernoulli's equation of order 2.*   Solve the equation

$$\dfrac{dy}{dx} + y = (xy)^2$$

by carrying out the following steps: (1) divide both sides of the equation by y^2; (2) make the change of variable $u = y^{-1}$; (3) solve the resulting equation for u in terms of x; (4) let $y = u^{-1}$.

Reduction of Order

Solve the equations in Exercises 53–56.

53. $\dfrac{d^2y}{dx^2} + \dfrac{dy}{dx} = 0$

54. $\dfrac{d^2y}{dx^2} + y\dfrac{dy}{dx} = 0$

55. $y'' = (y')^2$

56. $y'' = 2yy'$

Solve the initial value problems in Exercises 57–61.

57. $x\dfrac{d^2y}{dx^2} + \dfrac{dy}{dx} = 0, \quad y = -3 \text{ and } y' = 2 \quad \text{when } x = 1$

58. $x\dfrac{d^3y}{dx^3} - 2\dfrac{d^2y}{dx^2} = 0; \quad y = -5, y' = 2, \text{ and } y'' = 3 \quad \text{when } x = 1$

59. $x\dfrac{d^2y}{dx^2} + 2\dfrac{dy}{dx} = 1, \quad y = 2 \text{ and } y' = 1 \quad \text{when } x = 2$

60. $2\dfrac{d^2y}{dx^2} + \left(\dfrac{dy}{dx}\right)^2 = -1, \quad y = 2 \text{ and } y' = 0 \quad \text{when } x = -1$

61. $x\dfrac{d^2y}{dx^2} + \dfrac{dy}{dx} = x^2, \quad y = 0 \text{ and } y' = 1 \quad \text{when } x = 1$

62. *An integral equation.*   Solve the integral equation

$$y(x) + \int_0^x y(t)\, dt = x.$$

(*Hint:* Differentiate first.)

63. *Hooke's law.*   A body of mass m hangs at rest from one end of a vertical spring whose other end is attached to a rigid support. The body is pulled down an amount s_0 and then released. Find the body's subsequent position as a function of time t. (*Hint:* Assume Hooke's law and Newton's second law of motion and let s denote the body's displacement from its rest position. Then $m(d^2s/dt^2) = -ks$, where k is the spring constant.)

64. *Air resistance proportional to velocity.*   A car suspended from a parachute for a television ad falls through the air under the pull of gravity. If air resistance produces a retarding force proportional to the car's downward velocity and the car starts from rest at time $t = 0$, find the distance fallen as a function of the elapsed time t.

Equations with One Known Solution

In Exercises 65–72, use the given solution $y = u(x)$ to find the equation's general solution.

65. $u = \cos x, \quad y'' + y = 0$

66. $u = x^2, \quad x^2y'' - 2y = 0$

67. $u = e^{-x}, \quad y'' - y' - 2y = 0$

68. $u = \sqrt{x}, \quad 4x^2y'' + y = 0, \quad x > 0$

69. $u = x, \quad x^2y'' + xy' - y = 0$

70. $u = x^3, \quad x^2y'' - 5xy' + 9y = 0$

71. $u = x, \quad (1 - x^2)y'' - 2xy' + 2y = 0$

72. $u = (\sin x)/\sqrt{x}, \quad x^2y'' + xy' + (x^2 - (1/4))y = 0$

73. Show that if $\sum_{k=0}^{n} a_k(x) = 0$, then $y = e^x$ is a solution of $a_n(x)y^{(n)} + a_{n-1}(x)y^{(n-1)} + \cdots + a_1(x)y' + a_0(x)y = 0$.

74. Find the general solution of
$$xy'' - (2x + 1)y' + (x + 1)y = 0.$$

Higher Order Linear Equations with Constant Coefficients

We solve third and higher order linear equations with constant coefficients the same way we solve second order equations. We find the roots of the characteristic equation, use them to construct the general solution y_h of the reduced equation, and add a particular solution y_p of the complete equation to form the general solution $y = y_h + y_p$ of the complete equation.

Find the general solutions of the equations in Exercises 75–78.

75. $y''' - 3y'' + 2y = 3\sin x + \cos x$

76. $y''' - 7y' + 6y = 6x + 1$

77. $y^{(4)} - 8y'' + 16y = 8x - 16$

78. $y^{(4)} - 4y''' + 6y'' - 4y' + y = 7$

15.6 Oscillation

Linear second order equations with constant coefficients are important because they model oscillation. Oscillation occurs, for example, in atoms and molecules, in machinery and electrical circuits, and in the chemical and electrical systems of our bodies. The classical example of mechanical oscillation is the motion of a weighted spring, and that is the example we shall study in this section. But all oscillations are modeled with the same basic equations, and the mathematics we present here comes up in all the other applications as well.

Undamped Oscillation (Simple Harmonic Motion)

Suppose we have a spring of natural length L and spring constant k, with its upper end fastened to a rigid support (Fig. 15.8). We hang a mass m from the spring. The weight of the mass stretches the spring to a length $L + s$ when allowed to come to rest in a new equilibrium position. By Hooke's law, the tension in the spring is ks. The force of gravity pulling down on the mass is mg. Equilibrium requires

$$ks = mg. \tag{1}$$

How will the mass behave if we pull it down an additional amount x_0 beyond the equilibrium position and release it? To find out, let x, positive direction downward, denote the displacement of the mass from equilibrium t seconds after the motion has started. Then the forces acting on the mass are

$+ mg$ \qquad (weight due to gravity),

$- k(s + x)$ \qquad (spring tension).

By Newton's second law, the sum of these forces is $m(d^2x/dt^2)$, so

$$m \frac{d^2x}{dt^2} = mg - ks - kx. \tag{2}$$

Since $mg = ks$ from Eq. (1), Eq. (2) simplifies to

$$m \frac{d^2x}{dt^2} + kx = 0. \tag{3}$$

In addition to satisfying this differential equation, the position of the mass satisfies the initial conditions

$$x = x_0 \qquad \text{and} \qquad \frac{dx}{dt} = 0 \qquad \text{when } t = 0. \tag{4}$$

If we divide both sides of Eq. (3) by m and write ω for $\sqrt{k/m}$, the equation becomes

$$\frac{d^2x}{dt^2} + \omega^2 x = 0. \tag{5}$$

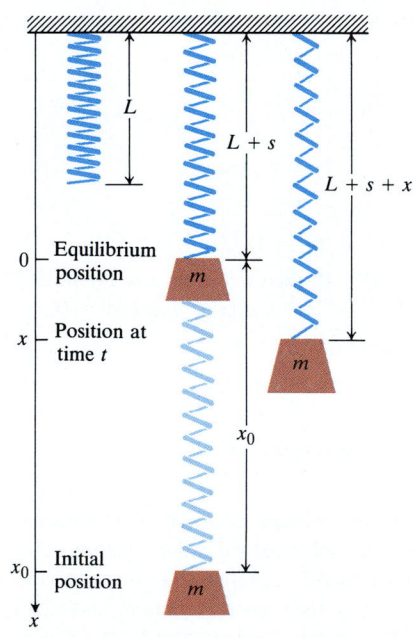

15.8 A spring of natural length L is stretched a distance s by the weight of a mass m. It is then stretched an additional distance x_0 and released. The position of the mass at any subsequent time is described by the solution of an initial value problem.

The roots of the characteristic equation $r^2 + \omega^2 = 0$ are $r = \pm \omega i$, so the general solution of Eq. (5) is

$$x = C_1 \cos \omega t + C_2 \sin \omega t. \tag{6}$$

Applying the initial conditions in Eq. (4) determines the constants to be

$$C_1 = x_0 \qquad \text{and} \qquad C_2 = 0.$$

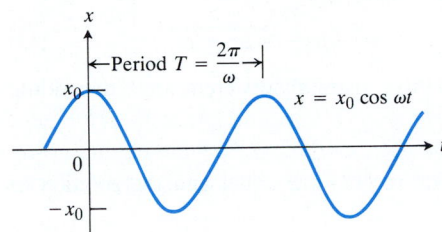

15.9   Simple harmonic motion.

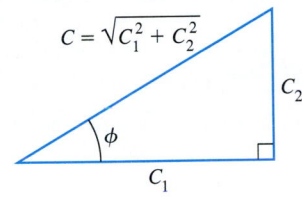

15.10 $C_1 = C \cos \phi$ and $C_2 = C \sin \phi$.

The mass's displacement from equilibrium t seconds into the motion is

$$x = x_0 \cos \omega t. \tag{7}$$

This equation represents a **simple harmonic motion** of amplitude x_0 and period $T = 2\pi/\omega$ (Fig. 15.9).

We normally combine the two terms in the general solution in Eq. (6) into a single term, using the trigonometric identity

$$C \cos(\omega t - \phi) = C \cos \phi \cos \omega t + C \sin \phi \sin \omega t.$$

To apply the identity, we take

$$C_1 = C \cos \phi, \qquad C_2 = C \sin \phi, \tag{8}$$

where

$$C = \sqrt{C_1^2 + C_2^2}, \qquad \tan \phi = \frac{C_2}{C_1}, \tag{9}$$

as in Fig. 15.10. With these substitutions, Eq. (6) becomes

$$x = C \cos(\omega t - \phi). \tag{10}$$

We treat C and ϕ as two new arbitrary constants, replacing the arbitrary constants C_1 and C_2 of Eq. (6). Equation (10) represents a simple harmonic motion of amplitude C and period $T = 2\pi/\omega$ (Fig. 15.11). The angle ϕ is the **phase angle** of the motion.

Damped Oscillation

Suppose that the mass is slowed by a frictional force $c\,(dx/dt)$ that is proportional to velocity, where c is a positive constant. Then the equation that replaces Eq. (3) is

$$m\frac{d^2x}{dt^2} + c\frac{dx}{dt} + kx = 0 \tag{11}$$

or

$$\frac{d^2x}{dt^2} + 2b\frac{dx}{dt} + \omega^2 x = 0, \tag{12}$$

where

$$2b = \frac{c}{m} \qquad \text{and} \qquad \omega = \sqrt{\frac{k}{m}}.$$

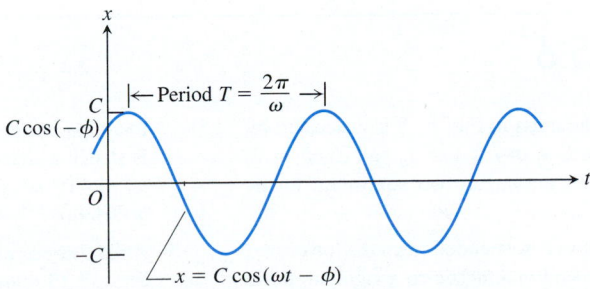

15.11   Graph of displacement vs. time for a simple harmonic motion.

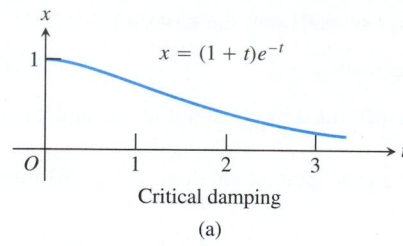

$x = (1 + t)e^{-t}$

Critical damping

(a)

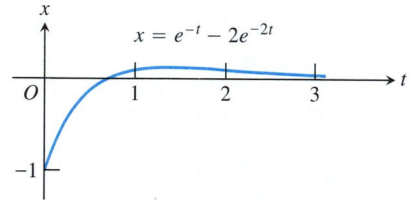

$x = e^{-t} - 2e^{-2t}$

Overcritical damping

(b)

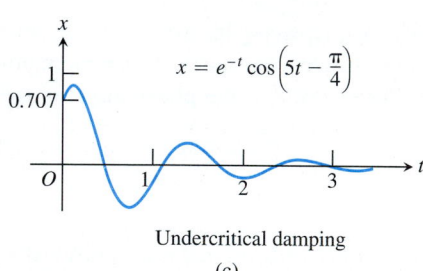

$x = e^{-t} \cos\left(5t - \dfrac{\pi}{4}\right)$

Undercritical damping

(c)

15.12   Examples of damping; x is displacement and t is time.

The roots of the characteristic equation $r^2 + 2br + \omega^2 = 0$ are

$$r_1 = -b + \sqrt{b^2 - \omega^2}, \qquad r_2 = -b - \sqrt{b^2 - \omega^2}. \tag{13}$$

As we shall now see, the mass behaves in three distinctly different ways, depending on the relative sizes of b and ω.

Critical Damping   If $b = \omega$, the roots in (13) are equal and the solution of (12) is

$$x = (C_1 t + C_2)e^{-\omega t}. \tag{14}$$

As time passes, x approaches zero. The mass does not oscillate (Fig. 15.12a).

Overcritical Damping   If $b > \omega$, the roots in (13) are real and unequal and the solution of (12) is

$$x = C_1 e^{r_1 t} + C_2 e^{r_2 t}. \tag{15}$$

Here again, the mass does not oscillate. Both r_1 and r_2 are negative and x approaches zero as time passes (Fig. 15.12b).

Undercritical Damping   If $0 < b < \omega$, let $\omega^2 - b^2 = \alpha^2$. Then

$$r_1 = -b + \alpha i, \qquad r_2 = -b - \alpha i$$

and

$$x = e^{-bt}(C_1 \cos \alpha t + C_2 \sin \alpha t). \tag{16}$$

If we introduce the substitutions (8), we may also write Eq. (16) in the equivalent form

$$x = Ce^{-bt} \cos (\alpha t - \phi). \tag{17}$$

This equation represents damped oscillation. It is analogous to simple harmonic motion, of period $T = 2\pi/\alpha$, except that the amplitude is not constant but is given by Ce^{-bt}. Since this tends to zero as t increases, the oscillations die out as time goes on (Fig. 15.12c). Observe, however, that Eq. (17) reduces to Eq. (10) in the absence of friction. The effect of friction is twofold:

1. $b = c/(2m)$ appears in the exponential **damping factor** e^{-bt}. The larger b is, the more quickly the oscillations tend to become unnoticeable.

2. The period $T = 2\pi/\alpha = 2\pi/\sqrt{\omega^2 - b^2}$ is longer than the period $T_0 = 2\pi/\omega$ in the friction-free system. The motion is slower.

EXERCISES 15.6

1. Suppose the motion of the mass in Fig. 15.8 is described by Eq. (3). Find x as a function of t if $x = x_0$ and $dx/dt = v_0$ when $t = 0$. Express your answer in two equivalent forms (Eqs. 6 and 10).

2. CALCULATOR   A 5-lb mass is suspended from the lower end of a spring whose upper end is attached to a rigid support. The weight of the mass extends the spring by 6 in. If, after the mass has come to rest in its new equilibrium position, it is struck a sharp blow that starts it downward with a velocity of 4 ft/sec, find its subsequent motion, assuming there is no friction.

3. *An RLC series circuit.*   The electrical series circuit shown in Fig. 15.13 contains a capacitor of capacitance C farads, a coil of inductance L henries, a resistance of R ohms, and a

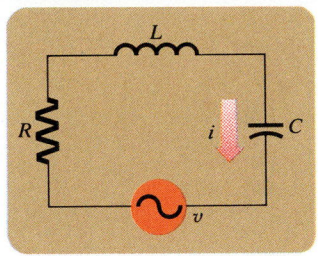

15.13   An *RLC* series circuit (Exercise 3).

generator that produces an electromotive force of v volts. If the current intensity in the circuit at time t is i amperes, the differential equation describing the current i is

$$L\frac{d^2i}{dt^2} + R\frac{di}{dt} + \frac{1}{C}i = \frac{dv}{dt}.$$

Find i as a function of t if
a) $R = 0$, $1/(LC) = \omega^2$, $v = $ constant;
b) $R = 0$, $1/(LC) = \omega^2$, $v = V\sin\alpha t$, $V = $ constant, $\alpha = $ constant $\neq \omega$;
c) $R = 0$, $1/(LC) = \omega^2$, $v = V\sin\omega t$, $V = $ constant;
d) $R = 50$, $L = 5$, $C = 9 \times 10^{-6}$, $v = $ constant.

4. A simple pendulum of length L makes an angle θ with the vertical. As it swings back and forth, its motion, neglecting friction, is described by the differential equation

$$\frac{d^2\theta}{dt^2} = -\frac{g}{L}\sin\theta,$$

where g is the (constant) acceleration of gravity (Fig. 15.14). Solve the differential equation of motion, under the assumption that θ is so small that $\sin\theta$ may be replaced by θ without appreciable error. Assume that $\theta = \theta_0$ and $d\theta/dt = 0$ when $t = 0$.

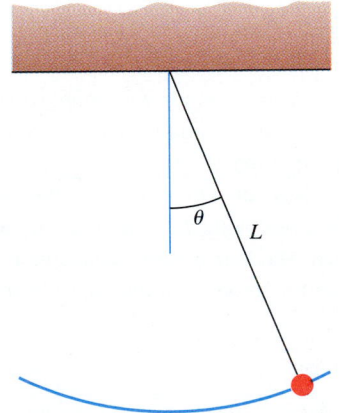

15.14   The pendulum in Exercise 4.

5. A circular disk of mass m and radius r is suspended by a thin wire attached to the center of one of its flat faces (Fig. 15.15). If the disk is twisted through an angle θ, torsion in the wire tends to turn the disk back in the opposite direction. The differential equation for the motion is

$$\frac{1}{2}mr^2\frac{d^2\theta}{dt^2} = -k\theta,$$

where k is the coefficient of torsion of the wire. Find θ as a function of t if $\theta = \theta_0$ and $d\theta/dt = v_0$ when $t = 0$.

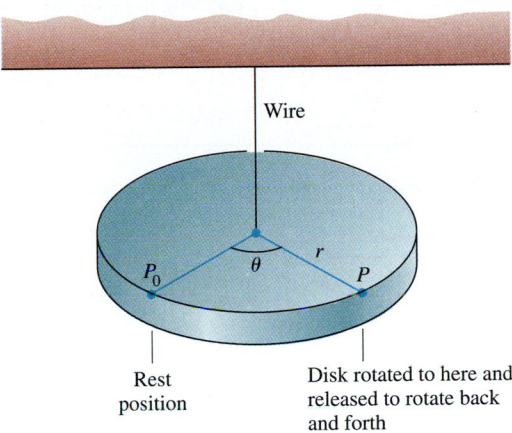

15.15   The suspended disk in Exercise 5.

6. CALCULATOR   A cylindrical spar buoy, diameter 1 ft, weight 100 lb, floats partially submerged in an upright position. When it is depressed slightly from its equilibrium position and released, it bobs up and down according to the differential equation

$$\frac{100}{g}\frac{d^2x}{dt^2} = -16\pi x - c\frac{dx}{dt}.$$

Here x (ft) is the vertical displacement of the buoy from its equilibrium position and $c(dx/dt)$ is the frictional resistance of the water. Find c if the period of oscillation is observed to be 1.6 sec. (Take $g = 32$ ft/sec².)

7. Suppose the upper end of the spring in Fig. 15.8 is attached not to a rigid support but to a member that itself undergoes up and down motion given by a function of time t, say $y = f(t)$. If the positive direction of y is downward, the differential equation of motion is

$$m\frac{d^2x}{dt^2} + kx = kf(t).$$

Let $x = x_0$ and $dx/dt = 0$ when $t = 0$, and solve for x
a) if $f(t) = A\sin\alpha t$ and $\alpha \neq \sqrt{k/m}$,
b) if $f(t) = A\sin\alpha t$ and $\alpha = \sqrt{k/m}$.

EXPLORER PROGRAMS

Oscillator

Enables you to explore solutions of the equation $x'' + 2bx' + \omega^2 x = 0$. The program can help you to learn more about amplitude, frequency, angular frequency (ω), period, and conservation of energy.

Forced Oscillator

Enables you to explore solutions of the equation $x'' + 2bx' + \omega^2 x = F(t)$, which models the behavior of an oscillator driven by an external force $F(t)$

15.7 Power Series Solutions

When we cannot find a nice compact expression for the solution of a differential equation, we try to get our information in other ways (which may be preferable anyway). One way is to try to find a power series representation for the solution. If we can do so, we immediately have a source of convenient and accurate polynomial approximations of the solution, which may be all we really need. Another way, which works for first order equations, is to graph the solutions directly from the equation. How can we do this without solving the equation first? We use the equation's slope field, as we shall see in Section 15.8. Still another method is to solve the equation numerically to generate tables of values for the solutions of particular initial value problems. The tables are what we usually want anyway and we can get them directly from the equation without the intermediate step of finding the general solution. (Having a computer does help.)

The present section introduces power series solutions. The next section covers graphical solutions, and Section 15.9 deals with numerical methods. Our discussions are too brief to be rigorous but you will get a sense of the main ideas and a chance to practice important techniques on a variety of equations. There is much more known about differential equations than we can describe here—enough to make the field one of the most valuable and far-reaching fields of applied mathematics and current mathematical research.

Equations with Analytic Coefficients

Our first example deals with the equation $y'' + x^2 y = 0$, which we cannot solve by any of our previous methods. But we can solve it by assuming that the solution has a power series representation $\Sigma a_n x^n$. The equation $y'' + x^2 y = 0$ is a linear second order equation that belongs to a class of equations of the form $y'' + a_1(x)y' + a_0(x)y = 0$, where the coefficient functions $a_1(x)$ and $a_0(x)$ are **analytic** in the neighborhood of $x = 0$ (that is, their Maclaurin series converge to the functions in some interval $|x| < h$ about the origin). In our example, $a_1(x) = 0$ and $a_0(x) = x^2$.

Example 1   Find a power series solution for

$$y'' + x^2 y = 0. \tag{1}$$

Solution   We assume that there is a solution of the form

$$y = a_0 + a_1 x + a_2 x^2 + \cdots + a_n x^n + \cdots, \tag{2}$$

and find what the coefficients a_k have to be to make the series and its second derivative

$$y'' = 2a_2 + 3 \cdot 2a_3 x + \cdots + n(n-1)a_n x^{n-2} + \cdots \tag{3}$$

satisfy Eq. (1). The series for $x^2 y$ is x^2 times the right-hand side of Eq. (2):

$$x^2 y = a_0 x^2 + a_1 x^3 + a_2 x^4 + \cdots + a_n x^{n+2} + \cdots. \tag{4}$$

The series for $y'' + x^2 y$ is the sum of the series in Eqs. (3) and (4):

$$y'' + x^2 y = 2a_2 + 6a_3 x + (12a_4 + a_0)x^2 + (20a_5 + a_1)x^3$$
$$+ \cdots + (n(n-1)a_n + a_{n-4})x^{n-2} + \cdots. \tag{5}$$

Notice that the coefficient of x^{n-2} in Eq. (4) is a_{n-4}. If y and its second derivative y'' are to satisfy Eq. (1), the coefficients of the individual powers of x on the right-hand side of Eq. (5) must all be zero (power series representations are unique, as you saw if you did Exercise 45 in Section 8.6):

$$2a_2 = 0, \qquad 6a_3 = 0, \qquad 12a_4 + a_0 = 0, \qquad 20a_5 + a_1 = 0, \tag{6}$$

and for all $n \geq 4$,

$$n(n-1)a_n + a_{n-4} = 0. \tag{7}$$

We can see from Eq. (2) that

$$a_0 = y(0), \qquad a_1 = y'(0). \tag{8}$$

In other words, the first two coefficients of the series are the values of y and y' at $x = 0$. The equations in (6) and the recursion formula in (7) enable us to evaluate all the other coefficients in terms of a_0 and a_1.

The first two of Eqs. (6) give

$$a_2 = 0, \qquad a_3 = 0.$$

Equation (7) shows that if $a_{n-4} = 0$, then $a_n = 0$; so we conclude that

$$a_6 = 0, \qquad a_7 = 0, \qquad a_{10} = 0, \qquad a_{11} = 0,$$

and whenever $n = 4k + 2$ or $4k + 3$, a_n is zero. For the other coefficients we have

$$a_n = \frac{-a_{n-4}}{n(n-1)}$$

so that

$$a_4 = \frac{-a_0}{4 \cdot 3}, \qquad a_8 = \frac{-a_4}{8 \cdot 7} = \frac{a_0}{3 \cdot 4 \cdot 7 \cdot 8},$$

$$a_{12} = \frac{-a_8}{11 \cdot 12} = \frac{-a_0}{3 \cdot 4 \cdot 7 \cdot 8 \cdot 11 \cdot 12},$$

and

$$a_5 = \frac{-a_1}{5 \cdot 4}, \qquad a_9 = \frac{-a_5}{9 \cdot 8} = \frac{a_1}{4 \cdot 5 \cdot 8 \cdot 9},$$

$$a_{13} = \frac{-a_9}{12 \cdot 13} = \frac{-a_1}{4 \cdot 5 \cdot 8 \cdot 9 \cdot 12 \cdot 13}.$$

The answer is best expressed as the sum of two separate series—one multiplied by a_0, the other by a_1:

$$y = a_0\left(1 - \frac{x^4}{3\cdot 4} + \frac{x^8}{3\cdot 4\cdot 7\cdot 8} - \frac{x^{12}}{3\cdot 4\cdot 7\cdot 8\cdot 11\cdot 12} + \cdots\right)$$

$$+ a_1\left(x - \frac{x^5}{4\cdot 5} + \frac{x^9}{4\cdot 5\cdot 8\cdot 9} - \frac{x^{13}}{4\cdot 5\cdot 8\cdot 9\cdot 12\cdot 13} + \cdots\right). \tag{9}$$

Both series converge absolutely for all values of x, as is readily seen by the ratio test.

Next we show how to find a series solution for an initial value problem of the form

Differential equation: $y' = f(x, y)$ (10)

Initial condition: $y = y_0$ when $x = x_0$.   (11)

If there is a Taylor series in powers of $x - x_0$ that converges to a solution of the problem, it must have the form

$$y = y_0 + y'(x_0)(x - x_0) + \frac{1}{2!} y''(x_0)(x - x_0)^2 + \cdots.$$

The idea is to use the differential equation itself to calculate $y'(x_0) = f(x_0, y_0)$ and higher order derivatives by differentiating both sides of Eq. (10) repeatedly with respect to x and then substituting previously calculated values, as shown in the next example. In the example, $x_0 = 1$, $y_0 = 0$, and $f(x, y) = x + y$.

Example 2   Solve the differential equation

$$y' = x + y \tag{12}$$

subject to the initial condition

$$y = 0 \qquad \text{when} \qquad x = 1. \tag{13}$$

Solution   The Taylor series for $y(x)$, in powers of $x - 1$, is

$$y(x) = y(1) + y'(1)(x - 1) + \frac{1}{2!} y''(1)(x - 1)^2 + \cdots.$$

Substituting $x = 1$ and $y = 0$ in Eq. (12), we get

$$y'(1) = 1 + 0 = 1. \tag{14}$$

If we differentiate both sides of Eq. (12) with respect to x, we get

$$y'' = 1 + y'. \tag{15}$$

We know that $y'(1) = 1$ and now we can calculate $y''(1)$:

$$y''(1) = 1 + y'(1) = 2. \tag{16}$$

By differentiating Eq. (15) with respect to x, we get

$$y''' = 0 + y'' = y''. \tag{17}$$

Since we already know that $y''(1) = 2$, we now find

$$y'''(1) = y''(1) = 2.$$

The pattern continues: all higher order derivatives satisfy

$$y^{(n+1)}(x) = y^{(n)}(x) \qquad \text{for } n \geq 2. \tag{18}$$

At $x = 1$, all of these derivatives have the value 2:

$$y^{(n)}(1) = 2, \qquad \text{for } n \geq 2. \tag{19}$$

When the values of $y(1)$, $y'(1)$, $y''(1)$, and so on are substituted into the Taylor series for $y(x)$, we get

$$y(x) = 0 + 1(x - 1) + 2\frac{1}{2!}(x-1)^2 + 2\frac{1}{3!}(x-1)^3 + \cdots$$
$$= x - 1 + 2 \sum_{n=2}^{\infty} \frac{(x-1)^n}{n!}. \tag{20}$$

This series brings to mind a comparable series, namely the series for $2e^{x-1}$, which is

$$2e^{x-1} = 2 + 2(x-1) + 2\sum_{n=2}^{\infty} \frac{(x-1)^n}{n!}. \tag{21}$$

Comparing Eqs. (20) and (21), we see that our series representation for $y(x)$, Eq. (20), is the same as

$$y(x) = 2e^{x-1} - 2 - (x - 1). \tag{22}$$

Exercise 1 asks you to verify that Eq. (22) does satisfy the differential equation $y' = x + y$ with $y = 0$ when $x = 1$.

The technique used in Example 1 could also have been used in Example 2. To apply the method of Example 1 in solving $y' = x + y$, we could use a series

$$y(x) = b_0 + b_1(x-1) + b_2(x-1)^2 + b_3(x-1)^3 + \cdots$$

together with the series for x as $x = 1 + (x - 1)$. Then

$$y' = b_1 + 2b_2(x-1) + 3b_3(x-1)^2 + \cdots$$

and the equation $y' = x + y$ becomes

$$b_1 + 2b_2(x-1) + 3b_3(x-1)^2 + \cdots + nb_n(x-1)^{n-1}$$
$$+ (n+1)b_{n+1}(x-1)^n + \cdots$$
$$= 1 + (x-1) + b_0 + b_1(x-1) + b_2(x-1)^2 + \cdots + b_n(x-1)^n + \cdots$$
$$= (1+b_0) + (1+b_1)(x-1) + b_2(x-1)^2 + \cdots + b_n(x-1)^n + \cdots.$$

The coefficients of corresponding powers of $x - 1$ must be the same on both sides of this equation. In other words,

$$b_1 = 1 + b_0,$$
$$2b_2 = 1 + b_1,$$
$$3b_3 = b_2, \quad \ldots, \quad (n+1)b_{n+1} = b_n \quad \text{for } n \geq 2.$$

These equations can be solved recursively to give all the coefficients in terms of b_0, where $b_0 = y(1)$. In particular, if $y(1) = 0$, then $b_0 = 0$, $b_1 = 1$, $2b_2 = 1 + b_1 = 2$, so $b_2 = 1$; $3b_3 = b_2 = 1$, so $b_3 = 1/3$; and so on.

EXERCISES 15.7

1. Show that $y = 2e^{x-1} - 2 - (x - 1)$ solves the initial value problem

$$y' = x + y, \quad y(1) = 0.$$

2. a) Show that

$$y_1 = \frac{a_0}{x}\left(1 - \frac{x^2}{2!} + \frac{x^4}{4!} - \cdots\right),$$

$$y_2 = \frac{a_1}{x}\left(x - \frac{x^3}{3!} + \frac{x^5}{5!} - \cdots\right)$$

are both solutions of the differential equation

$$xy'' + 2y' + xy = 0.$$

Hence, by the principle of superposition (Section 15.4, Exercise 26), $y = y_1 + y_2$ is a solution of the equation.

b) Express y_1 and y_2 in terms of elementary functions.

c) Solve the *boundary value problem*

Differential equation: $xy'' + 2y' + xy = 0$

Boundary values: $y(\pi/2) = 2$ and $y(\pi) = -1$

on the interval $\pi/2 \leq x \leq \pi$.

Find series solutions (Taylor or Maclaurin) for the initial value problems in Exercises 3–14.

3. $y' = y, \quad y = 1$ when $x = 0$

4. $y' + y = 0, \quad y = 1$ when $x = 0$

5. $y' = 2y, \quad y = 2$ when $x = 0$

6. $y' + 2y = 0, \quad y = -1$ when $x = 1$

7. $y'' = y, \quad y = 0$ and $y' = 1$ when $x = 0$

8. $y'' + y = 0, \quad y = 1$ and $y' = 0$ when $x = 0$

9. $y'' + y = x, \quad y = 2$ and $y' = 1$ when $x = 0$

10. $y'' - y = x, \quad y = -1$ and $y' = 2$ when $x = 0$

11. $y'' - y = -x, \quad y = 0$ and $y' = -2$ when $x = 2$

12. $y'' - x^2y = 0, \quad y = a$ and $y' = b$ when $x = 0$

13. $y'' + x^2y = x, \quad y = a$ and $y' = b$ when $x = 0$

14. $y'' - 2y' + y = 0, \quad y = 0$ and $y' = 1$ when $x = 0$

15.8 Slope Fields and Picard's Theorem

This section shows how to sketch the solutions of a differential equation $y' = f(x, y)$ without having to solve the equation first. It also describes conditions under which we can be sure that solutions exist and shows how we can sometimes generate polynomial approximations of the solutions by integration.

A **solution curve** is a curve whose equation satisfies the differential equation. For example, any curve whose equation is $y = Ce^x$ is a solution curve for the equation $y' = y$.

A **slope field** assigns to each point $P(x, y)$ in the domain of f the number $f(x, y)$, which is the slope of any solution curve through P.

An **isocline** of a differential equation $y' = f(x, y)$ is a curve with equation $f(x, y) = $ constant. For example, the isoclines of the equation $y' = y$ are the lines $y = C$, where C is a constant.

We could represent a slope field for the equation $y' = f(x, y)$ by the surface $z = f(x, y)$, but here we prefer to use a different approach that helps us visualize what a solution curve might look like without actually solving the differential equation. Figure 15.16 is a graphical representation for a portion of the slope field of the equation

$$y' = x + y. \tag{1}$$

(To simplify the language, we shall say simply that it *is* the slope field.) This figure was done by a computer, but it can be visualized as a pattern of iron filings that have been sprinkled onto a piece of paper and have arranged themselves so that the bit at the point $P(x, y)$ has slope $x + y$. Since iron filings won't actually do that for us, we resort to other approaches.

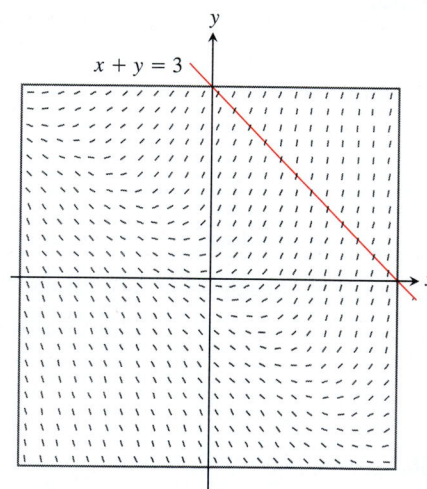

15.16   The slope field for $y' = x + y$, $-3 \leq x \leq 3$, $-3 \leq y \leq 3$.

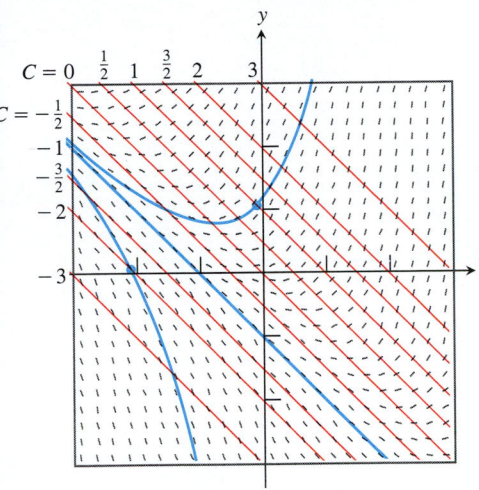

15.17   Slope field, isoclines $x + y = C$, and selected solution curves for the equation $y' = x + y$, $-3 \leq x \leq 3$, $-3 \leq y \leq 3$. The line $x + y = -1$ is both an isocline and a solution curve.

One simple approach is first to construct a set of isoclines. For Eq. (1), each isocline is a straight line satisfying the equation

$$x + y = C,$$

where C is a constant. Through some point on the line $x + y = 3$, for example, we draw a short line segment of slope 3. Having drawn one such segment, we then draw others parallel to it. Figure 15.16 shows 12 such segments on that line. We repeat this process on each of the isoclines, starting with a short segment of slope C at a point on the line $x + y = C$. The resulting pattern is a portion of the slope field for the differential equation $y' = x + y$.

Notice one special line, the line $x + y = -1$, which is both a solution curve and an isocline. Above this line the solution curves are concave up because $y'' = 1 + y' = 1 + x + y$ is positive if $x + y > -1$. Below the line, y'' is negative and the solution curves are concave down.

Figure 15.17 shows selected solution curves and isoclines in the slope field of $y' = x + y$ in the region $|x| \leq 3$, $|y| \leq 3$. The upper curve is for the solution $y = -1 - x + 2e^x$ of the initial value problem $y' = x + y$, $y(0) = 1$. The lower curve is for the solution $y = -1 - x - e^{x+2}$ of the initial value problem $y' = x + y$, $y(-2) = 0$. The curve in between is the special isocline $x + y = -1$. Exercise 21 asks you to show that the solution of the initial value problem $y' = x + y$, $y(x_0) = y_0$ is

$$y = -1 - x + (1 + x_0 + y_0) e^{x - x_0}.$$

Picard's Existence and Uniqueness Theorem and Iteration Scheme

We now turn our attention to a theorem of Charles Émile Picard's. The proof hinges on an iterative process that produces a sequence of functions that converge to a solution of the problem $y' = f(x, y)$ with $y(x_0) = y_0$, provided the hypotheses of the theorem are satisfied. The iteration process, known as **Picard's iteration scheme,** is an ingenious method for finding analytic approximations to a solution. We shall illustrate the scheme after we look at the theorem.

Suppose we are given an initial value problem

$$y' = f(x, y), \qquad y(x_0) = y_0, \tag{2}$$

where f is defined and continuous inside a rectangle R in the xy-plane. Does the initial value problem always have at least one solution if (x_0, y_0) is inside R? Might it have more than one solution? The next example shows that the answer to the second question is yes, unless something more is required of f.

Example 1   The initial value problem

$$y' = y^{4/5}, \qquad y(0) = 0, \tag{3}$$

has the obvious solution $y = 0$. Another solution is found by separating the variables and integrating:

$$y = \left(\frac{x}{5}\right)^5.$$

15.18 Part of the graph of $y = (x/5)^5$, one of the solutions of $y' = y^{4/5}$, $y(0) = 0$. Another solution is $y = 0$ (Example 1).

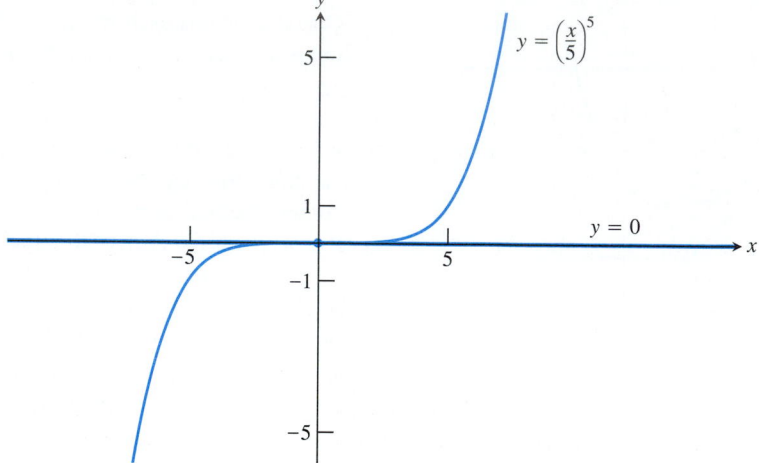

There are many more solutions (Fig. 15.18), among them

$$y = \begin{cases} 0, & x \leq 0, \\ \left(\dfrac{x}{5}\right)^5, & x > 0, \end{cases} \quad \text{and} \quad y = \begin{cases} \left(\dfrac{x}{5}\right)^5, & x \leq 0, \\ 0, & x > 0. \end{cases}$$

In this example, $f(x, y) = y^{4/5}$ is continuous in the entire xy-plane but its partial derivative $f_y = (4/5)y^{-1/5}$ is not continuous where $y = 0$.

The following theorem gives sufficient conditions for the existence and uniqueness of a solution of the initial value problem in (2).

THEOREM 3

Picard's Existence and Uniqueness Theorem

If (x_0, y_0) is a point in the interior of a rectangle R on which $f(x, y)$ is continuous, then the initial value problem

$$\frac{dy}{dx} = f(x, y), \qquad y(x_0) = y_0 \tag{4}$$

has at least one solution on some open interval of x-values containing x_0. If, in addition, the partial derivative f_y is continuous on R, then the solution is unique on some (perhaps smaller) open interval containing x_0 (Fig. 15.19).

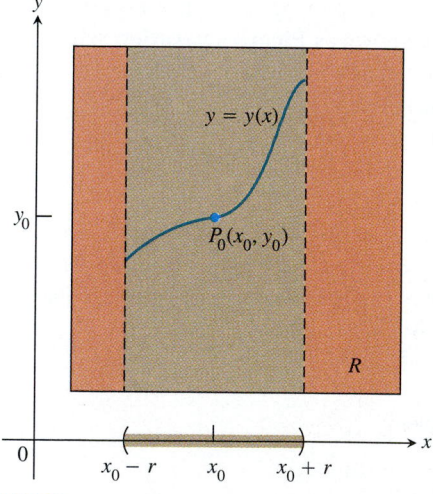

15.19 If f and f_y are continuous in R, and (x_0, y_0) is any point in R, then the initial value problem $y' = f(x, y)$, $y(x_0) = y_0$, has a unique solution on some open interval containing x_0.

We cannot take the time to prove the theorem but we can describe Picard's iteration scheme, which plays a key role in the proof and is important in its own right. We begin by noticing that any solution y of the initial value problem of Eqs. (4) must also satisfy the integral equation

$$y(x) = y_0 + \int_{x_0}^{x} f(t, y(t))\, dt \tag{5}$$

because

$$\int_{x_0}^{x} \frac{dy}{dt}\, dt = y(x) - y(x_0).$$

The converse is also true: If $y(x)$ satisfies Eq. (5), then $y' = f(x, y(x))$ and $y(x_0) = y_0$. So Eqs. (4) may be replaced by Eq. (5). This sets the stage for Picard's iteration method: In the integrand in Eq. (5), replace $y(t)$ by the constant y_0, then integrate and call the resulting right-hand side of Eq. (5) $y_1(x)$:

$$y_1(x) = y_0 + \int_{x_0}^{x} f(t, y_0) \, dt. \tag{6}$$

This starts the process. To keep it going, we use the iterative formulas

$$y_{n+1}(x) = y_0 + \int_{x_0}^{x} f(t, y_n(t)) \, dt. \tag{7}$$

The proof of Picard's theorem consists of showing that this process produces a sequence of functions $\{y_n(x)\}$ that converge to a function $y(x)$ that satisfies Eqs. (4) and (5), for values of x sufficiently near x_0. (The proof also shows that the solution is unique: that is, no other method will lead to a different solution.)

The following examples illustrate the Picard iteration scheme, but in most practical cases the computations soon become too burdensome to continue.

Example 2   Illustrate the Picard iteration scheme for the initial value problem

$$y' = x - y, \qquad y(0) = 1. \tag{8}$$

Solution   For the problem at hand, Eq. (6) becomes

$$y_1(x) = 1 + \int_{0}^{x} (t - 1) \, dt$$

$$= 1 + \frac{x^2}{2} - x. \tag{9a}$$

If we now use Eq. (7) with $n = 1$, we get

$$y_2(x) = 1 + \int_{0}^{x} \left(t - 1 - \frac{t^2}{2} + t \right) dt$$

$$= 1 - x + x^2 - \frac{x^3}{6}. \tag{9b}$$

The next iteration, with $n = 2$, gives

$$y_3(x) = 1 + \int_{0}^{x} \left(t - 1 + t - t^2 + \frac{t^3}{6} \right) dt$$

$$= 1 - x + x^2 - \frac{x^3}{3} + \frac{x^4}{4!}. \tag{9c}$$

In this example, it is possible to find the exact solution, because

$$\frac{dy}{dx} + y = x$$

is a first order equation that is linear in y. It has an integrating factor e^x and the general solution is

$$y = x - 1 + Ce^{-x}.$$

The solution of the initial value problem is

$$y = x - 1 + 2e^{-x}. \tag{10}$$

If we substitute the Maclaurin series for e^{-x} in Eq. (10), we get

$$y = x - 1 + 2\left(1 - x + \frac{x^2}{2!} - \frac{x^3}{3!} + \frac{x^4}{4!} - \cdots\right)$$

$$= 1 - x + x^2 - \frac{x^3}{3} + 2\left(\frac{x^4}{4!} - \frac{x^5}{5!} + \cdots\right),$$

and we see that the Picard method has given us the first four terms of this expansion in Eq. (9c) in $y_3(x)$.

In the next example, we cannot find a solution in terms of elementary functions. The Picard scheme is one way we could get an idea of how the solution behaves near the initial point.

Example 3   Find $y_n(x)$ for $n = 0, 1, 2,$ and 3 for the initial value problem

$$y' = x^2 + y^2, \qquad y(0) = 0.$$

Solution   By definition, $y_0(x) = y(0) = 0$. The other functions $y_n(x)$ are generated by the integral representation

$$y_{n+1}(x) = 0 + \int_0^x [t^2 + (y_n(t))^2]\, dt$$

$$= \frac{x^3}{3} + \int_0^x (y_n(t))^2\, dt.$$

We successively calculate

$$y_1(x) = \frac{x^3}{3},$$

$$y_2(x) = \frac{x^3}{3} + \frac{x^7}{63},$$

$$y_3(x) = \frac{x^3}{3} + \frac{x^7}{63} + \frac{2x^{11}}{2079} + \frac{x^{15}}{59535}.$$

In the next section we introduce numerical methods for solving initial value problems like those in Examples 2 and 3. When we program such a numerical solution for a calculator or computer, it is helpful to have an independent check on the program. For x near zero, we would expect $y_2(x)$ or $y_3(x)$ to provide such a check.

EXERCISES 15.8

In Exercises 1–6, sketch some of the isoclines and part of the slope field. Using the slope field, sketch the solution curve that passes through the point $P(1, -1)$.

1. $y' = x$

2. $y' = y$

3. $y' = 1/x$

4. $y' = 1/y$

5. $y' = xy$

6. $y' = x^2 + y^2$

7. Sketch part of the slope field of the equation $y' = (x + y)^2$. Show isoclines where the slope is 0, 1/4, 1, 4. Sketch (roughly) the solution curves through the points (a) $P(0, 0)$, (b) $Q(-1, 1)$, and (c) $R(1, 0)$. Now make the substitution $u = x + y$ and find the general solution of $y' = (x + y)^2$. Find an equation for the solution curve that passes through the origin. Sketch that solution more accurately.

8. Show that every solution of the equation $y' = (x + y)^2$ has a graph that has a point of inflection but no maximum or minimum. (The graph also has vertical asymptotes, as you may discover by letting $x + y = u$.)

9. Use isoclines to sketch part of the slope field for the equation $y' = x - y$. Include some part of all four quadrants.
   a) Which isocline is also a solution curve?
   b) Use the differential equation to determine where the solution curves are concave up and concave down.

In Exercises 10–13, write an equivalent first order differential equation and initial condition for y.

10. $y = 1 + \int_0^x y(t) \, dt$

11. $y = -1 + \int_1^x (t - y(t)) \, dt$

12. $y = \int_1^x \frac{1}{t} \, dt$

13. $y = 2 - \int_0^x (1 + y(t)) \sin t \, dt$

14. What integral equation is equivalent to the initial value problem $y' = f(x)$, $y(x_0) = y_0$?

Use Picard's iteration scheme to find $y_n(x)$ for $n = 0, 1, 2, 3$ in Exercises 15–20.

15. $y' = x$, $y(1) = 2$

16. $y' = y$, $y(0) = 1$

17. $y' = xy$, $y(1) = 1$

18. $y' = x + y$, $y(0) = 0$

19. $y' = x + y$, $y(0) = 1$

20. $y' = 2x - y$, $y(-1) = 1$

21. Show that the solution of the initial value problem
$$y' = x + y, \quad y(x_0) = y_0$$
is
$$y = -1 - x + (1 + x_0 + y_0) \, e^{x - x_0}.$$

22. Verify the formula for $y_3(x)$ in Example 3.

EXPLORER PROGRAM

Slope Fields

Enables you to study solutions of $y' = f(x, y)$ geometrically by constructing their graphs in the slope field defined by the equation

15.9 Numerical Methods

We complete our preview of differential equations by describing three numerical methods for solving the initial value problem $y' = f(x, y)$, $y(a) = y_0$ over an interval $a \le x \le b$. These methods do not produce a general solution of $y' = f(x, y)$; they produce tables of values of y for preselected values of x. Instead of being a drawback, however, this is a definite advantage, especially if we want to solve a differential equation like $y' = x^2 + y^2$ whose solution has no closed-form algebraic expression. Also, in solving equations of motion in real situations like launching a rocket or intercepting one in orbit, we find that numerical answers are much more useful than algebraic expressions would be. We shall keep our examples simple but the ideas can be extended to far more complicated equations and systems of equations. Naturally we would not attempt to solve such complicated systems with pencil and paper; instead, we would turn to well-designed computer programs that had been carefully checked for accuracy.

The first method we consider dates back to Euler, the second is an improved Euler method, and the third and most accurate method of the three is a so-called Runge–Kutta method. You will find BASIC computer programs for the three methods following the section exercises.

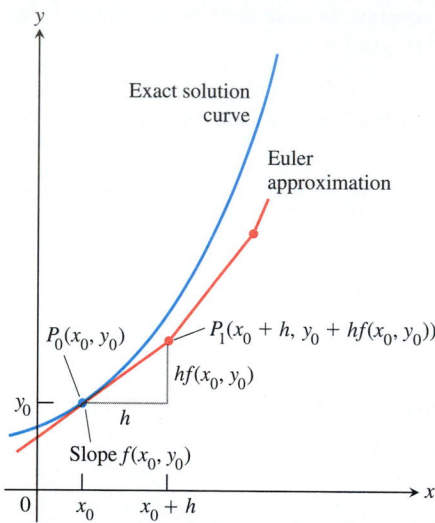

15.20 The Euler approximation to the solution of the initial value problem $y' = f(x, y)$, $y = y_0$ when $x = x_0$. The errors involved usually accumulate as we take more steps.

The Euler Method

The Euler method is a numerical process for generating a table of approximate values of the function that solves the initial value problem

$$y' = f(x, y), \qquad y(x_0) = y_0. \qquad (1)$$

The problem furnishes a starting point, $P_0(x_0, y_0)$, and a slope $f(x_0, y_0)$. We know that the graph of the solution must be a curve through P_0 with that slope. If we use the tangent through P_0 to approximate the actual solution curve, the approximation may be fairly good from $x_0 - h$ to $x_0 + h$, for small values of h. Thus, we might choose $h = 0.1$, say, and move along the tangent line from P_0 to P_1 (x_1, y_1), where $x_1 = x_0 + h$ and $y_1 = y_0 + hf(x_0, y_0)$. If we think of P_1 as a new starting point, we can move from P_1 to P_2 (x_2, y_2), where $x_2 = x_1 + h$ and $y_2 = y_1 + hf(x_1, y_1)$. If we replace h by $-h$, we move to the left from P_0 instead of to the right. The process can be continued, but the errors are likely to accumulate as we take more steps (Fig. 15.20).

Example 1   Take $h = 0.1$ and investigate the accuracy of the Euler approximation method for the initial value problem

$$y' = 1 + y, \qquad y(0) = 1, \qquad (2)$$

over the interval $0 \leq x \leq 1$ by letting

$$x_{n+1} = x_n + h, \qquad y_{n+1} = y_n + h(1 + y_n). \qquad (3)$$

Solution   The exact solution of Eqs. (2) is $y = 2e^x - 1$. Table 15.2 shows the results using Eqs. (3) and the exact results rounded to four decimals for comparison. By the time we get to $x = 1$, the error is about 5.6%.

TABLE 15.2
Euler solution of $y' = 1 + y$, $y(0) = 1$, step size $h = 0.1$

x	y (approx)	y (exact)	Error $= y$ (exact) $- y$ (approx)
0	1	1	0
0.1	1.2	1.2103	0.0103
0.2	1.42	1.4428	0.0228
0.3	1.662	1.6997	0.0377
0.4	1.9282	1.9836	0.0554
0.5	2.2210	2.2974	0.0764
0.6	2.5431	2.6442	0.1011
0.7	2.8974	3.0275	0.1301
0.8	3.2872	3.4511	0.1639
0.9	3.7159	3.9192	0.2033
1.0	4.1875	4.4366	0.2491

Carl Runge

German scientist Carl Runge (1856–1927) was a mathematical physicist of Max Planck's caliber who developed numerical methods for solving the differential equations that arose in his studies of atomic spectra. He used so much mathematics in his research that physicists thought he was a mathematician, and did so much physics that mathematicians thought he was a physicist. Neither group claimed him as their own and it was years before anyone could find him a professorship. In 1904, Felix Klein finally persuaded his Göttingen colleagues to create for Runge Germany's only full professorship in applied mathematics. Runge was the professorship's first and only occupant, there being no one of his accomplishments to assume the post when he died.

The Improved Euler Method

With this method we first get an estimate of y_{n+1}, as in the original Euler method, but call the result z_{n+1}. We then take the average of $f(x_n, y_n)$ and $f(x_{n+1}, z_{n+1})$ in place of $f(x_n, y_n)$ in the next step. Thus

$$z_{n+1} = y_n + hf(x_n, y_n), \tag{4}$$

$$y_{n+1} = y_n + \frac{h}{2}[f(x_n, y_n) + f(x_{n+1}, z_{n+1})]. \tag{5}$$

If we apply this improved method to Example 1, again with $h = 0.1$, we get the following results at $x = 1$:

$$y\ (\text{approx}) = 4.4281\ 61693,$$

$$y\ (\text{exact}) = 4.4365\ 63656,$$

$$\text{Error} = y\ (\text{exact}) - y\ (\text{approx}) = 0.0084\ 01963,$$

and the error is less than 2/10 of 1%.

A Runge–Kutta Method[*]

The Runge–Kutta method we use requires four intermediate calculations, as given in the following equations:

$$k_1 = hf(x_n, y_n) \qquad k_2 = hf\left(x_n + \frac{h}{2}, y_n + \frac{k_1}{2}\right)$$

$$k_3 = hf\left(x_n + \frac{h}{2}, y_n + \frac{k_2}{2}\right) \qquad k_4 = hf\left(x_n + h, y_n + k_3\right). \tag{6}$$

We then calculate y_{n+1} from y_n with the formula

$$y_{n+1} = y_n + \frac{1}{6}(k_1 + 2k_2 + 2k_3 + k_4). \tag{7}$$

When we apply this method to the problem of estimating $y(1)$ for the problem $y' = 1 + y$, $y(0) = 1$, still using $h = 0.1$, we get

$$y(1) = 4.4365\ 59490$$

with an error $0.0000\ 04166$, which is less than 1/10,000 of 1%. This is clearly the most accurate of the three methods.

The next example shows that the error in the Runge–Kutta approximation need not continue to increase as the process is continued. In fact, with $h = 0.1$, the difference between the exact solutions and the approximations remain less than 10^{-6} for the two initial value problems:

(a) $y' = x - y$, $y(0) = 1$,   (b) $y' = x - y$, $y(0) = -2$.

The fact that the differential equation is linear in y is significant in discussing the accuracy of the Runge–Kutta approximation. Such accuracy is not attained for the initial value problem

$$y' = x^2 + y^2, \qquad y(0) = 0.$$

---

[*]The method described here is one of many Runge–Kutta methods.

TABLE 15.3

	x	y (Runge–Kutta)	y (true value)	Difference
a) $y' = x - y$, $y(0) = 1$	0	1	1	0
	0.5	0.7130 61869	0.7130 61319	5.50×10^{-7}
	1.0	0.7357 59549	0.7357 58882	6.67×10^{-7}
	1.5	0.9462 60927	0.9462 60320	6.07×10^{-7}
	2.0	1.2706 71057	1.2706 70566	4.91×10^{-7}
	2.5	1.6641 70370	1.6641 69997	3.73×10^{-7}
	3.0	2.0995 74407	2.0995 74137	2.70×10^{-7}
b) $y' = x - y$, $y(0) = -2$	0	-2	-2	0
	0.5	$-1.1065\ 30935$	$-1.1065\ 30660$	-2.75×10^{-7}
	1.0	$-0.3678\ 79775$	$-0.3678\ 79441$	-3.34×10^{-7}
	1.5	$+0.2768\ 69537$	$+0.2768\ 69840$	-3.03×10^{-7}
	2.0	0.8646 64472	0.8646 64717	-2.46×10^{-7}
	2.5	1.4179 14816	1.4179 15001	-1.85×10^{-7}
	3.0	1.9502 12796	1.9502 12932	-1.36×10^{-7}

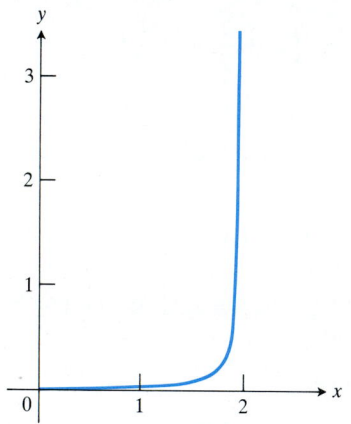

15.21   The graph of a Runge–Kutta solution of the initial value problem $y' = x^2 + y^2$, $y(0) = 0$, for $x > 0$. Data from Table 15.4.

The Runge–Kutta approximation to $y(2.1)$, using $h = 0.1$, is 1.47×10^{11}. The solution curve for this problem has a vertical asymptote at just beyond $x = 2$ (Fig. 15.21). No matter how small we take h, we cannot assert any accuracy for our approximations as the curve approaches this asymptote.

Example 2   Table 15.3 shows the comparison of $y(x)$ as estimated by the Runge–Kutta method with $h = 0.1$ and the true value, for solutions of $y' = x - y$ (a) with $y(0) = 1$ and (b) with $y(0) = -2$.

More points were actually computed and plotted to give the graphs in Fig. 15.22. The upper curve, $y = x - 1 + 2e^{-x}$, is concave up and has a minimum

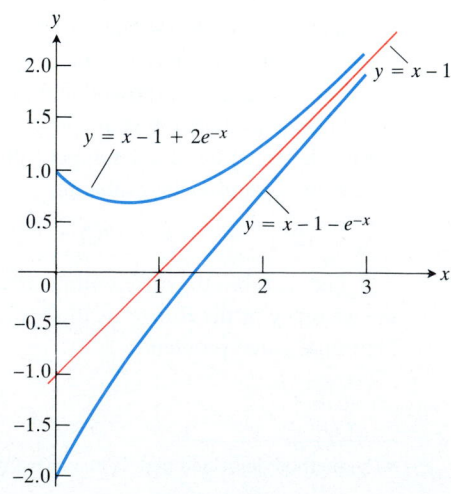

15.22   Two solutions of $y' = x - y$:
a) $y(0) = 1$, $y = x - 1 + 2e^{-x}$,
b) $y(0) = -2$, $y = x - 1 - e^{-x}$

TABLE 15.4
$y' = x^2 + y^2, \quad y(0) = 0$

x	y (Runge–Kutta)	y (actual)
0	0	0
0.5	0.0417 91288	0.0417 91146
1.0	0.3502 33742	0.3502 31844
1.5	1.5174 73414	1.5174 47544
2.0	71.5789 9545	317.2244 00
2.1	1.4700 1E + 11	
2.2	1.6666 7E + **	(meaning, "you broke the bank!")

when $x = y = \ln 2$. The lower curve is concave down, is always rising as x increases, and crosses the x-axis at a value of x near 1.3. Both curves approach the line $y = x - 1$ as $x \to \infty$.

Example 3 Table 15.4 lists Runge–Kutta approximations for the initial value problem

$$y' = x^2 + y^2, \qquad y(0) = 0.$$

We obtained the approximations with a step size of $h = 0.1$. Figure 15.21 shows the graph of y as a function of x.

EXERCISES 15.9

1. Use the Euler method with $h = 1/5$ to estimate $y(1)$ if $y' = y$ and $y(0) = 1$. What is the exact value of $y(1)$?

2. Show that the Euler method leads to the estimate $(1 + (1/n))^n$ for $y(1)$ if $h = 1/n$, $y' = y$, and $y(0) = 1$. What is the limit as $n \to \infty$?

3. Use the improved Euler method with $h = 1/5$ to estimate $y(1)$ if $y' = y$ and $y(0) = 1$.

4. Use the Runge–Kutta method with $h = 1/5$ to estimate $y(1)$ if $y' = y$ and $y(0) = 1$.

5. Show that the solution of the initial value problem $y' = x^2 + y^2$, $y(0) = 1$, increases faster on the interval $0 \le x < 1$ than does the solution of the initial value problem $y' = y^2$, $y(0) = 1$. Solve the latter problem by separation of variables and thus show that the solution of the original problem becomes infinite at a value of x not greater than 1. (The value is about 0.9698 10654.)

6. Solve the initial value problem $y' = 1 + y^2$, $y(0) = 0$, (a) by separation of variables, and (b) by using the substitution $y = -u'/u$ and solving the equivalent problem for u. (Notice the similarity with the initial value problem $y' = x^2 + y^2$, $y(0) = 0$.)

Computer Exercises (or Programmable Calculator)

Find numerical solutions of the initial value problems in Exercises 7–10. Each exercise gives a differential equation in the form $y' = f(x, y)$, a solution interval $a \le x \le b$, the initial value $y(a)$, the step size, and the number of steps. Your answer should be a table showing y vs. x.

7. $y' = x/7$, $a = 0$, $b = 4$, $y(0) = 1$, $h = 0.1$, $n = 20$

8. $y' = -y^2/x$, $a = 2$, $b = 4$, $y(2) = 2$, $h = 0.1$, $n = 20$

9. $y' = (x^2 + y^2)/2y$, $a = 0$, $b = 2$, $y(0) = 0.1$, $h = 0.2$, $n = 10$

10. Repeat Exercise 9, but with $h = 0.1$ and $n = 20$.

11. *Evaluating a nonelementary integral.* The value of the nonelementary integral

$$\int_0^1 \sin(t^2)\, dt$$

is the value of the function

$$y(x) = \int_0^x \sin(t^2)\, dt$$

at $x = 1$. The function, in turn, is the solution of the initial value problem

$$y' = \sin(x^2), \quad y(0) = 0.$$

Thus, by solving the initial value problem numerically on the interval $0 \le x \le 1$ we can find the value of the integral as the value of y that corresponds to $x = 1$. Estimate the integral's value by solving the initial value problem with (a) $n = 20$ steps, and (b) $n = 40$ steps.

12. *Continuation of Exercise 11.*   Estimate the value of

$$\int_0^1 x^2 e^{-x}\, dx$$

by solving an appropriate initial value problem on the interval $[0, 1]$ with (a) $n = 20$ steps, and (b) $n = 40$ steps.

EXPLORER PROGRAM

First Order Equations   Provides numerical solutions of $y' = f(x, y)$, $y(a) = y_0$ over an interval $a \le x \le b$ with a Runge–Kutta method with up to 100 steps. Also prints tables and displays graphs of y and y'.

✳ COMPUTER PROGRAMS: SOLVING FIRST ORDER INITIAL VALUE PROBLEMS

Here are three BASIC computer programs for approximating the solutions of the initial value problem $y' = f(x, y)$, $y = y_0$ when $x = x_0$, over a finite interval starting at x_0.

The Euler Method

PROGRAM		COMMENT
10	INPUT "ENTER INITIAL X-VALUE ", X	Initial value of x
20	INPUT "ENTER INITIAL Y-VALUE ", Y	Initial value of y
30	INPUT "ENTER STEP SIZE ", H	Step size
40	INPUT "ENTER NUMBER OF STEPS ", N	Number of steps
50	DEF FND(X, Y) = 1 + Y	Key in formula for the derivative, $f(x, y)$
60	FOR J = 0 TO N	Each value of J gives one Euler approximation
70	PRINT X;Y	Prints the table
80	Y = Y + H*FND(X, Y)	Next y
90	X = X + H	Next x
100	NEXT J	Returns to line 60 with new value for J
110	END	

An Improved Euler Method

```
10      INPUT "ENTER INITIAL X-VALUE    ", X
20      INPUT "ENTER INITIAL Y-VALUE    ", Y
30      INPUT "ENTER STEP SIZE     ", H
40      INPUT "ENTER NUMBER OF STEPS    ", N
50      DEF FND(X, Y) = 1 + Y
60      FOR J = 0 TO N
70      PRINT X;Y
80      Z = Y + H*FND(X, Y)
90      XN = X + H
100     Y = Y + H*(FND(X, Y) + FND(XN, Z))/2
110     X = XN
120     NEXT J
130     END
```

Runge–Kutta Method

```
10        INPUT "ENTER INITIAL X-VALUE     ", X
20        INPUT "ENTER INITIAL Y-VALUE     ", Y
30        INPUT "ENTER STEP SIZE      ", H
40        INPUT "ENTER NUMBER OF STEPS     ", N
50        DEF FND(X, Y) = X − Y
60        FOR J = 0 TO N
70        PRINT X;Y
80        K1 = H*FND(X, Y)
90        X = X + H/2
100       K2 = H*FND(X, Y + K1/2)
110       K3 = H*FND(X, Y + K2/2)
120       X = X + H/2
130       K4 = H*FND(X, Y + K3)
140       Y = Y + (K1 + 2*K2 + 2*K3 + K4)/6
150       NEXT J
160       END
```

REVIEW QUESTIONS

1. What is a differential equation?

2. What is a solution of a differential equation?

3. Describe methods for solving first order equations that are (a) separable, (b) homogeneous, (c) exact, (d) linear. Give examples.

4. Describe methods for solving linear second order equations with constant coefficients if the equations are (a) homogeneous, (b) nonhomogeneous. Give examples.

5. How do we describe oscillation with second order differential equations? Give an example.

6. How can we sometimes solve differential equations with power series? Give an example.

7. How can we graph the solutions of a differential equation $y' = f(x, y)$ without first solving the equation? Give an example.

8. Describe three numerical methods for solving the initial value problem $y' = f(x, y)$, $y = y_0$ when $x = x_0$.

MISCELLANEOUS EXERCISES

Solve the initial value problems in Exercises 1–20.

1. $e^{y-2}\, dx - e^{x+2y}\, dy = 0$, $y(0) = -2$

2. $y \ln y\, dx + (1 + x^2)\, dy = 0$, $y(0) = e$

3. $\dfrac{dy}{dx} = \dfrac{x^2 + y^2}{2xy}$, $y(5) = 0$

4. $\dfrac{dy}{dx} = \dfrac{y(1 + \ln y - \ln x)}{x(\ln y - \ln x)}$, $y(1) = 1$

5. $(x^2 + y)\, dx + (e^y + x)\, dy = 0$, $y(3) = 0$

6. $(e^x + \ln y)\, dx + \left(\dfrac{x+y}{y}\right) dy = 0$, $y(\ln 2) = 1$

7. $(x + 1)\dfrac{dy}{dx} + 2y = x$, $y(0) = 1$

8. $x\dfrac{dy}{dx} + 2y = x^2 + 1$, $y(1) = 1$

9. $\dfrac{d^2y}{dx^2} - \left(\dfrac{dy}{dx}\right)^2 = 1$, $y(\pi/3) = 0$, $y'(\pi/3) = \sqrt{3}$

10. $x^2\dfrac{d^2y}{dx^2} + x\dfrac{dy}{dx} = 1$, $y(1) = 1$, $y'(1) = 1$

11. $\dfrac{d^2y}{dx^2} - 4\dfrac{dy}{dx} + 3y = 0$, $y(0) = 2$, $y'(0) = -2$

12. $\dfrac{d^2y}{dx^2} + 5\dfrac{dy}{dx} + 6y = 0$, $y(0) = 5/6$, $y'(0) = -2$

13. $\dfrac{d^2y}{dx^2} + 4\dfrac{dy}{dx} + 4y = 0$, $y(0) = 0$, $y'(0) = 7$

14. $\dfrac{d^2y}{dx^2} - 8\dfrac{dy}{dx} + 16y = 0, \quad y(0) = 4, \quad y'(0) = -4$

15. $\dfrac{d^2y}{dx^2} + 2\dfrac{dy}{dx} + 2y = 0, \quad y(0) = 1, \quad y'(0) = 0$

16. $\dfrac{d^2y}{dx^2} - 2\dfrac{dy}{dx} - 4y = 0, \quad y(0) = 1, \quad y'(0) = -1$

17. $\dfrac{d^2y}{dx^2} + 2\dfrac{dy}{dx} = 4x, \quad y(0) = 1, \quad y'(0) = -3$

18. $\dfrac{d^2y}{dx^2} + y = \csc x, \quad y(\pi/2) = 1, \quad y'(\pi/2) = \pi/2$

19. $\dfrac{d^2y}{dx^2} - \dfrac{dy}{dx} - 2y = 3e^{2x}, \quad y(0) = -2, \quad y'(0) = 0$

20. $\dfrac{d^2y}{dx^2} - 2\dfrac{dy}{dx} + 5y = 4e^{-x}, \quad y(0) = 1, \quad y'(0) = -1/2$

21. Find the orthogonal trajectories of the family of curves $x^2 = Cy^3$. (*Caution:* The differential equation for the family should not contain C.)

22. Find the orthogonal trajectories of the circles $(x - C)^2 + y^2 = C^2$.

23. Find the orthogonal trajectories of the parabolas $y^2 = 4C(C - x)$.

24. Which of the following is not an exact first order differential equation?
a) $y^2\,dx + 2xy\,dy = 0$
b) $(2x \sin y + y^3 e^x)\,dx + (x^2 \cos y + 3y^2 e^x)\,dy = 0$
c) $(2x \cos y + 3x^2 y)\,dx + (x^3 - x^2 \sin y - y)\,dy = 0$
d) $(y \ln y - e^{-xy})\,dx + \left(\dfrac{1}{y} + x \ln y\right)dy = 0$

25. The differential equation $(x^2 - y^3)y' = 2xy$ has an integrating factor of the form y^n. Find n and solve the equation.

26. Solve the equation
$$(x + y + 1)\,dx + (y - x - 3)\,dy = 0$$
by substituting $x = r + a$, $y = b - s$, and choosing values for the constants a and b that allow the resulting equation to assume the form
$$(r + s)\,dr + (r + s)\,ds = 0.$$
Then solve this equation and express its solution in terms of x and y.

27. What values must the constants a, b, and c have to make
$$(ax^2 e^y + by^2 + cy)\,dx + (cxy + 2x + x^3 e^y)\,dy = 0$$
exact? What is the equation's solution when a, b, and c have these values?

28. The equation
$$(x^2 + y^2)\,dx + cxy\,dy = 0$$
has the integrating factor $1/x^2$. What is the value of the constant c? What is the equation's solution when c has this value?

29. Suppose that P and Q are functions of x, that the function $y = u(x)$ satisfies the equation $y'' + Py' + Qy = 0$, and that u is never zero. Show that the substitution $y = uv$ transforms the equation $y'' + Py' + Qy = F(x)$ into the equation $v'' + v'(P + (2u'/u)) = F(x)/u$. The further substitution $w = v'$ changes this last equation into a first order equation for w as a function of x. Show that this procedure can be applied to the equation $y'' - 2y' + y = e^x$ with $u(x) = e^x$ to lead successively to
$$w' = 1, \quad w = x + C_1, \quad v = \frac{x^2}{2} + C_1 x + C_2,$$
and
$$y = uv = e^x\left(\frac{x^2}{2} + C_1 x + C_2\right).$$

30. *A rocket with variable mass.* If an external force F acts on a system whose mass varies with time, Newton's law of motion is
$$\frac{d(mv)}{dt} = F + (v + u)\frac{dm}{dt}.$$
In this equation, m is the mass of the system at time t, v is its velocity, and $v + u$ is the velocity of the mass that is entering (or leaving) the system at the rate dm/dt. Suppose that a rocket of initial mass m_0 starts from rest but is driven upward by firing some of its mass directly backward at the constant rate of $dm/dt = -b$ units per second and at constant speed relative to the rocket $u = -c$. The only external force acting on the rocket is $F = -mg$ due to gravity. Under these assumptions, show that the height of the rocket above the ground at the end of t seconds (t small compared with m_0/b) is
$$y = c\left(t + \frac{m_0 - bt}{b}\ln\frac{m_0 - bt}{m_0}\right) - \frac{1}{2}gt^2.$$

31. If y is a solution of $(D^2 + 4)y = e^x$, show that y satisfies the equation $(D - 1)(D^2 + 4)y = 0$. Find the general solution of $(D - 1)(D^2 + 4)y = 0$ and use the result to solve the equation $(D^2 + 4)y = e^x$.

32. The equation $d^2y/dt^2 + 100y = 0$ describes a simple harmonic motion. Find the solution that satisfies the initial conditions that $y = 10$ and $dy/dt = 50$ when $t = 0$. Find the period and amplitude of the motion.

33. Find the first five nonzero terms in the Maclaurin series for the solution of the initial value problem $y' = x^2 + y^2$, $y(0) = 1$. (*Hint:* If
$$y = a_0 + a_1 x + a_2 x^2 + \cdots + a_n x^n + \cdots$$
and
$$y^2 = c_0 + c_1 x + c_2 x^2 + \cdots + c_n x^n + \cdots,$$
then
$$c_n = \sum_{k=0}^{n} a_k a_{n-k}$$
and
$$y' = a_1 + 2a_2 x + 3a_3 x^2 + \cdots + na_n x^{n-1} + \cdots.$$

To satisfy the differential equation, you must have $a_1 = c_0$, $2a_2 = c_1$, $3a_3 = 1 + c_2$, and $na_n = c_{n-1}$ for $n \geq 4$. You can now determine a_0 (from the initial value), c_0, a_1, c_1, a_2, c_2, a_3, and so on.)

34. If you were to solve the initial value problem $y' = y^2$, $y = 1$, when $x = 0$, by finding the Maclaurin series for y as a function of x, for what values of x would you expect the series to converge? (*Hint:* Solve the initial value problem without series and then expand your answer in a series.)

Exercises 35–37 refer to the differential equation $y' = x + \sin y$, whose slope field appears in Figure 15.23.

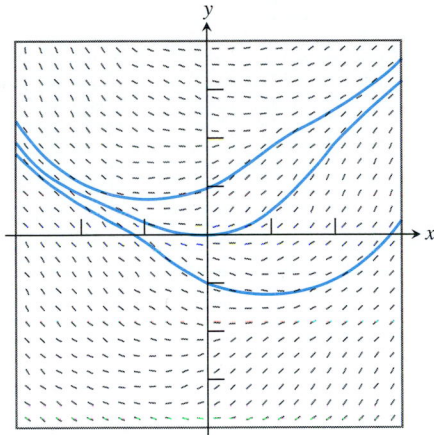

15.23 Solutions to $dy/dx = x + \sin y$ passing through $(0, 0)$, $(0, \pi/2)$, $(0, -\pi/2)$ (Exercises 35–37).

35. For solutions near $y = 0$, we might use the approximation $\sin y \approx y$ and replace the original problem with $dy/dx = x + y$. Solve this equation with $y(0) = 0$. What is the Maclaurin series for your answer?

36. *Series.* Starting with the initial value problem $y' = x + \sin y$, $y(0) = 0$, we can calculate successive derivatives by implicit differentiation. For example, $y'' = 1 + (\cos y)y'$, $y''' = -(\sin y)(y')^2 + (\cos y)y''$. These can in turn be evaluated at $x = 0$ by using the given initial value $y(0) = 0$. Using this procedure, find the terms of the Maclaurin series for $y(x)$ through x^4. (Compare the answer with that of Exercise 35.)

37. Using the method of the preceding exercise, find the terms through x^4 in the Maclaurin series for $y(x)$ if y satisfies $y' = x + \sin y$, and (a) $y(0) = \pi/2$. (b) Repeat for $y(0) = -\pi/2$.

38. Suppose you used a computer to graph solutions of

$$y' = (1 + y^2) \cos x,$$

(a) for $y(0) = 0$ and (b) for $y(0) = 1$. One solution was continuous and very well behaved. The other blew up! By solving the equation for an arbitrary initial value $y(0) = y_0$, find the values of y_0 for which the solution remains bounded. If a solution does not remain bounded, locate the asymptotes of its graph. See Fig. 15.24.

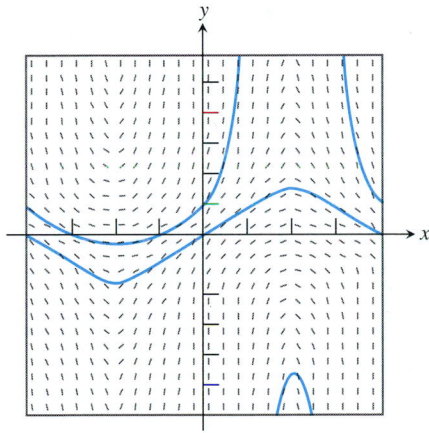

15.24 Solutions: $y = \tan(\sin x + C)$ for $C = 0$ and $\pi/4$ (Exercise 38).

Appendices

Formulas from Precalculus Mathematics

Algebra

1. Laws of Exponents

$$a^m a^n = a^{m+n}, \qquad (ab)^m = a^m b^m, \qquad (a^m)^n = a^{mn}, \qquad a^{m/n} = \sqrt[n]{a^m}$$

If $a \neq 0$,

$$\frac{a^m}{a^n} = a^{m-n}, \quad a^0 = 1, \quad a^{-m} = \frac{1}{a^m}.$$

2. Zero   Division by zero is not defined.

If $a \neq 0$: $\quad \dfrac{0}{a} = 0, \quad a^0 = 1, \quad 0^a = 0$

For any number a: $\quad a \cdot 0 = 0 \cdot a = 0$

3. Fractions

$$\frac{a}{b} + \frac{c}{d} = \frac{ad + bc}{bd}, \qquad \frac{a}{b} \cdot \frac{c}{d} = \frac{ac}{bd}, \qquad \frac{a/b}{c/d} = \frac{a}{b} \cdot \frac{d}{c}, \qquad \frac{-a}{b} = -\frac{a}{b} = \frac{a}{-b},$$

$$\frac{(a/b) + (c/d)}{(e/f) + (g/h)} = \frac{(a/b) + (c/d)}{(e/f) + (g/h)} \cdot \frac{bdfh}{bdfh} = \frac{(ad + bc)fh}{(eh + fg)bd}$$

4. The Binomial Theorem

For any positive integer n,

$$(a + b)^n = a^n + na^{n-1}b + \frac{n(n-1)}{1 \cdot 2} a^{n-2}b^2$$

$$+ \frac{n(n-1)(n-2)}{1 \cdot 2 \cdot 3} a^{n-3}b^3 + \cdots + nab^{n-1} + b^n.$$

For instance,

$$(a + b)^1 = a + b,$$

$$(a + b)^2 = a^2 + 2ab + b^2,$$

$$(a + b)^3 = a^3 + 3a^2b + 3ab^2 + b^3,$$

$$(a + b)^4 = a^4 + 4a^3b + 6a^2b^2 + 4ab^3 + b^4.$$

5. Difference of Like Integer Powers, $n > 1$

$$a^n - b^n = (a - b)(a^{n-1} + a^{n-2}b + a^{n-3}b^2 + \cdots + ab^{n-2} + b^{n-1})$$

For instance,

$$a^2 - b^2 = (a - b)(a + b),$$

$$a^3 - b^3 = (a - b)(a^2 + ab + b^2),$$

$$a^4 - b^4 = (a - b)(a^3 + a^2b + ab^2 + b^3).$$

6. Completing the Square

If $a \neq 0$, we can rewrite the quadratic $ax^2 + bx + c$ in the form $au^2 + C$ by a process called completing the square:

$$ax^2 + bx + c = a\left(x^2 + \frac{b}{a}x\right) + c \qquad \left(\begin{matrix}\text{Factor } a \text{ from the}\\ \text{first two terms.}\end{matrix}\right)$$

$$= a\left(x^2 + \frac{b}{a}x + \frac{b^2}{4a^2} - \frac{b^2}{4a^2}\right) + c \qquad \left(\begin{matrix}\text{Add and subtract}\\ \text{the square of half}\\ \text{the coefficient of } x.\end{matrix}\right)$$

$$= a\left(x^2 + \frac{b}{a}x + \frac{b^2}{4a^2}\right) + a\left(-\frac{b^2}{4a^2}\right) + c \qquad \left(\begin{matrix}\text{Bring out}\\ \text{the } -b^2/4a^2.\end{matrix}\right)$$

$$= a\underbrace{\left(x^2 + \frac{b}{a}x + \frac{b^2}{4a^2}\right)}_{\text{This is } \left(x + \frac{b}{2a}\right)^2.} + \underbrace{c - \frac{b^2}{4a}}_{\text{Call this part } C.}$$

$$= au^2 + C \qquad (u = x + b/2a)$$

7. The Quadratic Formula

By completing the square on the first two terms of the equation

$$ax^2 + bx + c = 0$$

and solving the resulting equation for x (details omitted), we obtain the formula

$$x = \frac{-b \pm \sqrt{b^2 - 4ac}}{2a}.$$

This equation is called the **quadratic formula.**

The solutions of the equation $2x^2 + 3x - 1 = 0$ are

$$x = \frac{-3 \pm \sqrt{(3)^2 - 4(2)(-1)}}{2(2)} = \frac{-3 \pm \sqrt{9 + 8}}{4},$$

or

$$x = \frac{-3 + \sqrt{17}}{4} \qquad \text{and} \qquad x = \frac{-3 - \sqrt{17}}{4}.$$

The solutions of the equation $x^2 + 4x + 6 = 0$ are

$$x = \frac{-4 \pm \sqrt{(4)^2 - 4 \cdot 1 \cdot 6}}{2} = \frac{-4 \pm \sqrt{16 - 24}}{2}$$

$$= \frac{-4 \pm \sqrt{-8}}{2} = \frac{-4 \pm 2\sqrt{2}\sqrt{-1}}{2} = -2 \pm \sqrt{2}\,i.$$

The solutions are the complex numbers $-2 + \sqrt{2}\,i$ and $-2 - \sqrt{2}\,i$. Appendix A.6 has more on complex numbers.

Geometry

(A = area, B = area of base, C = circumference, S = lateral area or surface area, V = volume)

1. Triangle

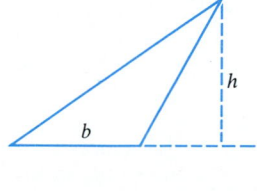

$$A = \frac{1}{2}bh$$

2. Similar Triangles

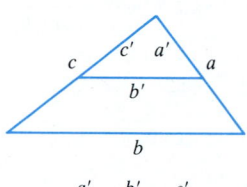

$$\frac{a'}{a} = \frac{b'}{b} = \frac{c'}{c}$$

3. Pythagorean Theorem

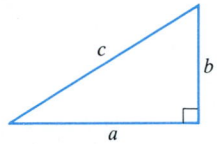

$$a^2 + b^2 = c^2$$

4. Parallelogram

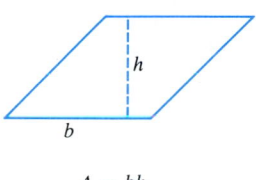

$$A = bh$$

5. Trapezoid

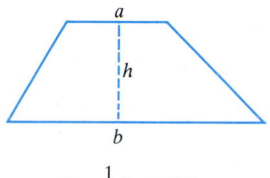

$$A = \frac{1}{2}(a + b)h$$

6. Circle

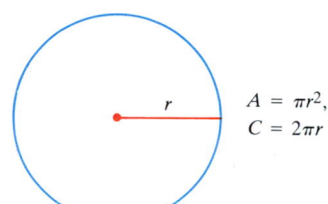

$$A = \pi r^2, \quad C = 2\pi r$$

7. Any Cylinder or Prism with Parallel Bases

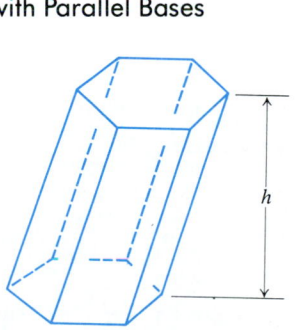

$$V = Bh$$

8. Right Circular Cylinder

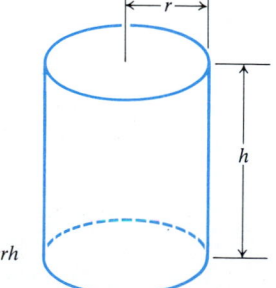

$$V = \pi r^2 h, \; S = 2\pi rh$$

9. Any Cone or Pyramid

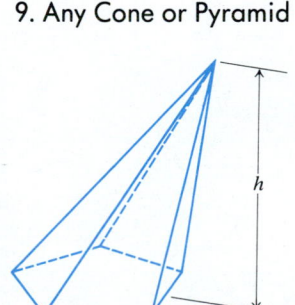

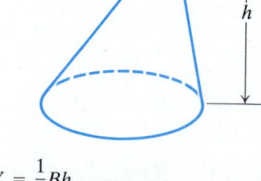

$$V = \frac{1}{3}Bh$$

10. Right Circular Cone

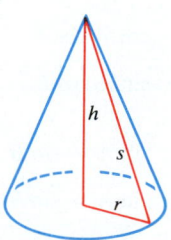

$$V = \frac{1}{3}\pi r^2 h, \; S = \pi rs$$

11. Sphere

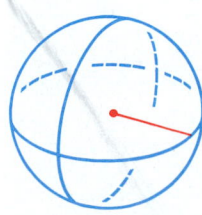

$$V = \frac{4}{3}\pi r^3, \; S = 4\pi r^2$$

Trigonometry

1. Definitions and Fundamental Identities

Sine: $\quad \sin \theta = \dfrac{y}{r} = \dfrac{1}{\csc \theta}$

Cosine: $\quad \cos \theta = \dfrac{x}{r} = \dfrac{1}{\sec \theta}$

Tangent: $\quad \tan \theta = \dfrac{y}{x} = \dfrac{1}{\cot \theta}$

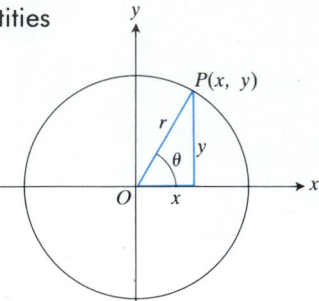

2. Identities

$\sin(-\theta) = -\sin \theta, \qquad \cos(-\theta) = \cos \theta$

$\sin^2\theta + \cos^2\theta = 1, \qquad \sec^2\theta = 1 + \tan^2\theta, \qquad \csc^2\theta = 1 + \cot^2\theta$

$\sin 2\theta = 2 \sin \theta \cos \theta, \qquad \cos 2\theta = \cos^2\theta - \sin^2\theta$

$\cos^2\theta = \dfrac{1 + \cos 2\theta}{2}, \qquad \sin^2\theta = \dfrac{1 - \cos 2\theta}{2}$

$\sin(A + B) = \sin A \cos B + \cos A \sin B$

$\sin(A - B) = \sin A \cos B - \cos A \sin B$

$\cos(A + B) = \cos A \cos B - \sin A \sin B$

$\cos(A - B) = \cos A \cos B + \sin A \sin B$

$\tan(A + B) = \dfrac{\tan A + \tan B}{1 - \tan A \tan B}$

$\tan(A - B) = \dfrac{\tan A - \tan B}{1 + \tan A \tan B}$

$\sin\left(A - \dfrac{\pi}{2}\right) = -\cos A, \qquad \cos\left(A - \dfrac{\pi}{2}\right) = \sin A$

$\sin\left(A + \dfrac{\pi}{2}\right) = \cos A, \qquad \cos\left(A + \dfrac{\pi}{2}\right) = -\sin A$

$\sin A \sin B = \tfrac{1}{2}\cos(A - B) - \tfrac{1}{2}\cos(A + B)$

$\cos A \cos B = \tfrac{1}{2}\cos(A - B) + \tfrac{1}{2}\cos(A + B)$

$\sin A \cos B = \tfrac{1}{2}\sin(A - B) + \tfrac{1}{2}\sin(A + B)$

$\sin A + \sin B = 2 \sin \tfrac{1}{2}(A + B) \cos \tfrac{1}{2}(A - B)$

$\sin A - \sin B = 2 \cos \tfrac{1}{2}(A + B) \sin \tfrac{1}{2}(A - B)$

$\cos A + \cos B = 2 \cos \tfrac{1}{2}(A + B) \cos \tfrac{1}{2}(A - B)$

$\cos A - \cos B = -2 \sin \tfrac{1}{2}(A + B) \sin \tfrac{1}{2}(A - B)$

3. Common Reference Triangles

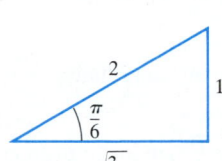 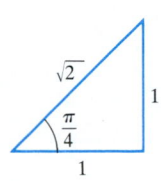 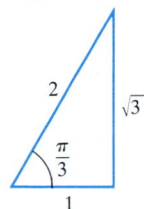

4. Angles and Sides of a Triangle

Law of cosines: $\quad c^2 = a^2 + b^2 - 2ab \cos C$

Law of sines: $\quad \dfrac{\sin A}{a} = \dfrac{\sin B}{b} = \dfrac{\sin C}{c}$

Area $= \dfrac{1}{2} bc \sin A = \dfrac{1}{2} ac \sin B = \dfrac{1}{2} ab \sin C$

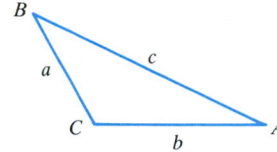

A.2
A Brief Review of Trigonometric Functions

Radian Measure

The **radian measure** of the angle ACB at the center of the unit circle (Fig. A.1) equals the length of the arc that the angle cuts from the unit circle.

If angle ACB cuts an arc $A'B'$ from a second circle centered at C, then circular sector $A'CB'$ will be similar to circular sector ACB. In particular,

$$\frac{\text{Length of arc } A'B'}{\text{Radius of second circle}} = \frac{\text{Length of arc } AB}{\text{Radius of first circle}}. \tag{1}$$

In the notation of Fig. A.1, Eq. (1) says that

$$\frac{s}{r} = \frac{\theta}{1} = \theta \qquad \text{or} \qquad \theta = \frac{s}{r}. \tag{2}$$

When you know r and s, you can calculate the angle's radian measure θ from this equation. Notice that the units of length for r and s cancel out and that radian measure, like degree measure, is a dimensionless number.

We find the relation between degree measure and radian measure by observing that a semicircle of radius r, which we know has length $s = \pi r$, subtends a central angle of 180°. Therefore,

$$180° = \pi \text{ radians.} \tag{3}$$

We can restate this relation in several useful ways.

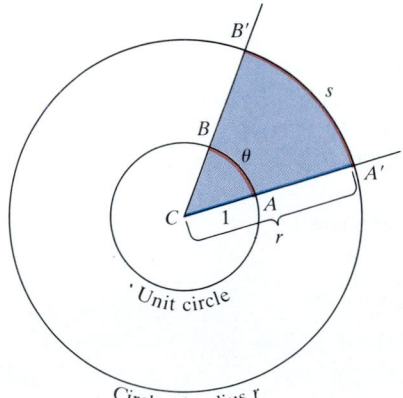

A.1 The radian measure of angle ACB is the length θ of arc AB on the unit circle centered at C. The value of θ can be found from any other circle, however, as the ratio of s to r.

Degrees to radians:

1 degree makes $\dfrac{\pi}{180}$ radians (about 0.02 rad).

To change degrees to radians, multiply degrees by $\dfrac{\pi}{180}$.

Radians to degrees:

1 radian makes $\dfrac{180}{\pi}$ degrees (about 57°).

To change radians to degrees, multiply radians by $\dfrac{180}{\pi}$.

Degrees	Radians

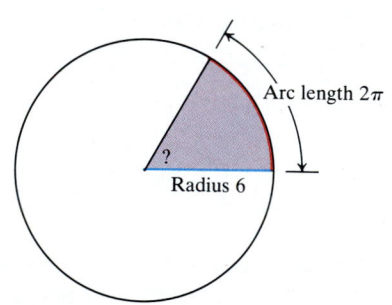

A.2   The angles of two common triangles, in degrees and radians.

Example 1   *Conversions*   (Fig. A.2)

Change 45° to radians: $45 \cdot \dfrac{\pi}{180} = \dfrac{\pi}{4}$ rad

Change 90° to radians: $90 \cdot \dfrac{\pi}{180} = \dfrac{\pi}{2}$ rad

Change $\dfrac{\pi}{6}$ radians to degrees: $\dfrac{\pi}{6} \cdot \dfrac{180}{\pi} = 30°$

Change $\dfrac{\pi}{3}$ radians to degrees: $\dfrac{\pi}{3} \cdot \dfrac{180}{\pi} = 60°$

Example 2   An acute angle whose vertex lies at the center of a circle of radius 6 subtends an arc of length 2π (Fig. A.3). The angle's radian measure is

$$\theta = \frac{s}{r} = \frac{2\pi}{6} = \frac{\pi}{3}. \qquad \text{(Eq. (2) with } s = 2\pi, r = 6\text{)}$$

The equation $\theta = s/r$ is sometimes written

$$s = r\theta. \tag{4}$$

This equation gives a handy way to find s when you know r and θ.

A.3   What is the radian measure of this angle? See Example 2.

Arc length 2π

? Radius 6

Example 3   An angle of $3\pi/4$ radians at the center of a circle of radius 8 subtends an arc

$$s = r\theta = 8 \cdot \frac{3\pi}{4} = 6\pi \qquad \left(\begin{array}{l} \text{Eq. (4) with } r = 8 \\ \text{and } \theta = 3\pi/4 \end{array} \right)$$

units long.

Example 4   How long is the arc subtended by a central angle of 120° in a circle of radius 4?

Solution   The equation $s = r\theta$ holds only when the angle is measured in radians, so we must find the angle's radian measure before finding s:

$$\theta = 120 \cdot \frac{\pi}{180} = \frac{2\pi}{3} \text{ rad} \qquad \text{(Convert to radians.)}$$

$$s = r\theta = 4 \cdot \frac{2\pi}{3} = \frac{8\pi}{3}. \qquad \text{(Then find } s = r\theta.\text{)}$$

The arc is $8\pi/3$ units long.

When angles are used to describe counterclockwise rotations, our measurements can go arbitrarily far beyond 2π radians, or 360°. Similarly, angles that describe clockwise rotations can have negative measures of all sizes (Fig. A.4).

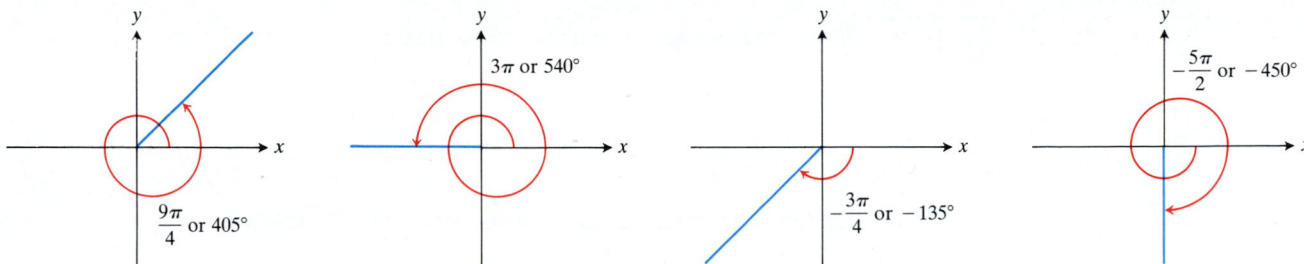

A.4   Angles can have any measure.

The Six Basic Trigonometric Functions

When an angle of measure θ is placed in standard position at the center of a circle of radius r (Fig. A.5), the six basic trigonometric functions of θ are defined in the following way:

REVIEW

$$\text{Sine:} \qquad \sin\theta = \frac{y}{r} \qquad\qquad \text{Cosecant:} \qquad \csc\theta = \frac{r}{y}$$

$$\text{Cosine:} \qquad \cos\theta = \frac{x}{r} \qquad\qquad \text{Secant:} \qquad \sec\theta = \frac{r}{x} \qquad (5)$$

$$\text{Tangent:} \qquad \tan\theta = \frac{y}{x} \qquad\qquad \text{Cotangent:} \qquad \cot\theta = \frac{x}{y}$$

As you can see, $\tan\theta$ and $\sec\theta$ are not defined if $x = 0$. In terms of radian measure, this means they are not defined when θ is $\pm\pi/2$, $\pm 3\pi/2$, ... Similarly, $\cot\theta$ and $\csc\theta$ are not defined for values of θ for which $y = 0$, namely $\theta = 0$, $\pm\pi$, $\pm 2\pi$, ... Notice also that

$$\tan\theta = \frac{\sin\theta}{\cos\theta}, \qquad \csc\theta = \frac{1}{\sin\theta}$$

$$\sec\theta = \frac{1}{\cos\theta}, \qquad \cot\theta = \frac{1}{\tan\theta} \qquad (6)$$

whenever the quotients on the right-hand sides are defined.

Because $x^2 + y^2 = r^2$ (Pythagorean theorem),

$$\cos^2\theta + \sin^2\theta = \frac{x^2}{r^2} + \frac{y^2}{r^2} = \frac{x^2 + y^2}{r^2} = 1. \qquad (7)$$

The equation $\cos^2\theta + \sin^2\theta = 1$, true for all values of θ, is probably the most frequently used identity in trigonometry.

The coordinates of the point $P(x, y)$ in Fig. A.5 can be expressed in terms of r and θ as

$$x = r\cos\theta \qquad (\text{Because } x/r = \cos\theta)$$

$$y = r\sin\theta \qquad (\text{Because } y/r = \sin\theta) \qquad (8)$$

We use these equations when we study circular motion and when we work with polar coordinates. Notice that if $\theta = 0$ in Fig. A.5, then $x = r$ and $y = 0$, so

$$\cos 0 = 1 \qquad \text{and} \qquad \sin 0 = 0.$$

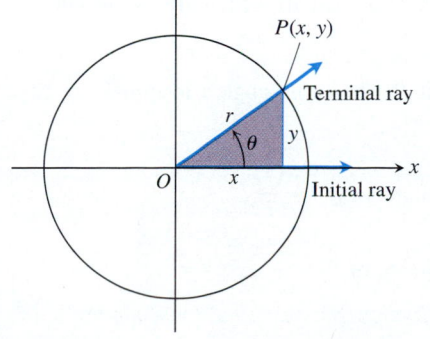

A.5   An angle θ in standard position.

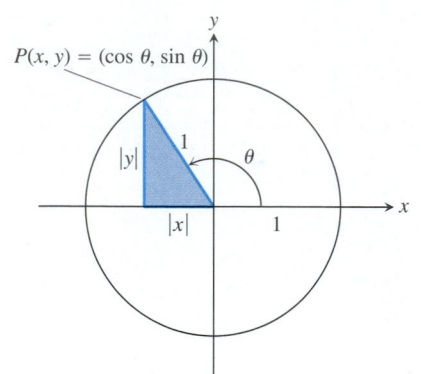

A.6 The acute reference triangle for an angle θ.

If $\theta = \pi/2$, we have $x = 0$ and $y = r$. Hence,

$$\cos\frac{\pi}{2} = 0 \qquad \text{and} \qquad \sin\frac{\pi}{2} = 1.$$

Using Triangles to Calculate Sines and Cosines

If the circle in Fig. A.5 has radius $r = 1$ unit, Eqs. (8) simplify to

$$x = \cos\theta, \qquad y = \sin\theta.$$

We can therefore calculate the values of the cosine and sine from the acute reference triangle made by dropping a perpendicular from the point $P(x, y)$ to the x-axis (Fig. A.6). The numerical values of x and y are read from the triangle's sides. The signs of x and y are determined by the quadrant in which the triangle lies.

Example 5 Find the sine and cosine of $2\pi/3$ radians.

Solution

STEP 1: Draw the angle in standard position and write in the lengths of the sides of the reference triangle (Fig. A.7).

STEP 2: Find the coordinates of the point P where the angle's terminal ray cuts the circle:

$$\cos\frac{2\pi}{3} = x\text{-coordinate of } P = -\frac{1}{2}$$

$$\sin\frac{2\pi}{3} = y\text{-coordinate of } P = \frac{\sqrt{3}}{2}.$$

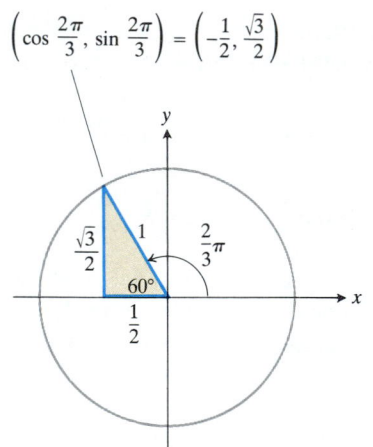

A.7 The triangle for calculating the sine and cosine of $2\pi/3$ radians (Example 5).

Example 6 Find the sine and cosine of $-\pi/4$ radians.

Solution

STEP 1: Draw the angle in standard position and write in the lengths of the sides of the reference triangle (Fig. A.8).

STEP 2: Find the coordinates of the point P where the angle's terminal ray cuts the circle:

$$\cos-\frac{\pi}{4} = x\text{-coordinate of } P = \frac{\sqrt{2}}{2},$$

$$\sin-\frac{\pi}{4} = y\text{-coordinate of } P = -\frac{\sqrt{2}}{2}.$$

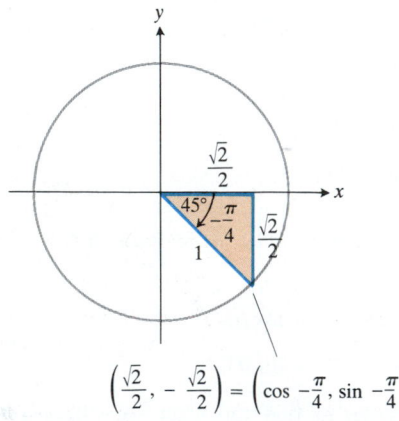

A.8 The triangle for calculating the sine and cosine of $-\pi/4$ radians (Example 6).

Table A.1 gives the values of the sine, cosine, and tangent for selected values of θ.

TABLE A.1
Values of sin θ, cos θ, and tan θ for selected values of θ

Degrees	-180	-135	-90	-45	0	45	90	135	180
θ (radians)	$-\pi$	$-3\pi/4$	$-\pi/2$	$-\pi/4$	0	$\pi/4$	$\pi/2$	$3\pi/4$	π
sin θ	0	$-\sqrt{2}/2$	-1	$-\sqrt{2}/2$	0	$\sqrt{2}/2$	1	$\sqrt{2}/2$	0
cos θ	-1	$-\sqrt{2}/2$	0	$\sqrt{2}/2$	1	$\sqrt{2}/2$	0	$-\sqrt{2}/2$	-1
tan θ	0	1		-1	0	1		-1	0

Periodicity

When an angle of measure θ and an angle of measure $\theta + 2\pi$ are in standard position, their terminal rays coincide. The two angles therefore have the same trigonometric function values:

$$\cos(\theta + 2\pi) = \cos\theta$$
$$\sin(\theta + 2\pi) = \sin\theta$$
$$\tan(\theta + 2\pi) = \tan\theta$$
$$\cot(\theta + 2\pi) = \cot\theta \tag{9}$$
$$\sec(\theta + 2\pi) = \sec\theta$$
$$\csc(\theta + 2\pi) = \csc\theta$$

Similarly, $\cos(\theta - 2\pi) = \cos\theta$, $\sin(\theta - 2\pi) = \sin\theta$, and so on.

From another point of view, Eqs. (9) tell us that if we start at any particular value $\theta = \theta_0$ and let θ increase or decrease steadily, we see the values of the trigonometric functions start to repeat after any interval of length 2π. We describe this behavior by saying that the six basic trigonometric functions are **periodic** and that they repeat after a fixed **period** of θ-values.

DEFINITION

A function $f(x)$ is **periodic** with **period** $p > 0$ if $f(x + p) = f(x)$ for every value of x.

Equations (9) tell us that the six basic trigonometric functions are periodic with period 2π. Other periods include 4π, 6π, and so on (positive integer multiples of 2π).

Notice that the tangent and cotangent functions (Fig. A.9) also have period π. Other periods include 2π, 3π, and so on (positive integer multiples of π).

In naming the period of a function it is conventional to name the *smallest* positive value of p for which $f(x + p) = f(x)$ for all x. The longer periods can all be

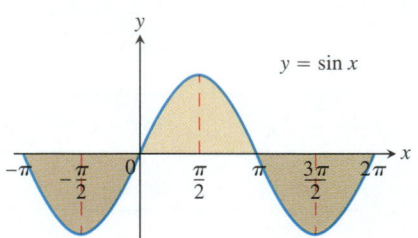

Domain: $-\infty < x < \infty$
Range: $-1 \le y \le 1$

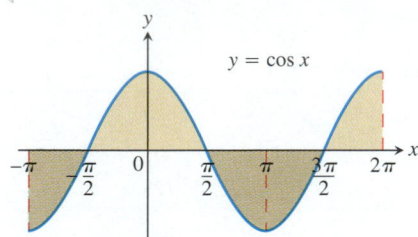

Domain: $-\infty < x < \infty$
Range: $-1 \le y \le 1$

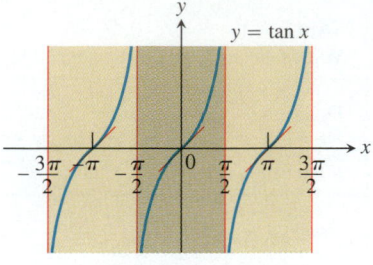

Domain: All real numbers except odd
integer multiples of $\pi/2$
Range: $-\infty < y < \infty$

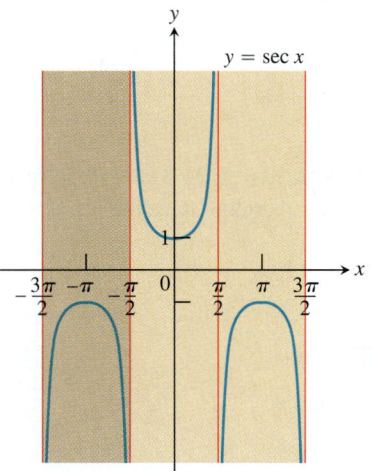

Domain: $x \ne \pm\dfrac{\pi}{2}, \pm\dfrac{3\pi}{2}, \ldots$
Range: $y \le -1$ and $y \ge 1$

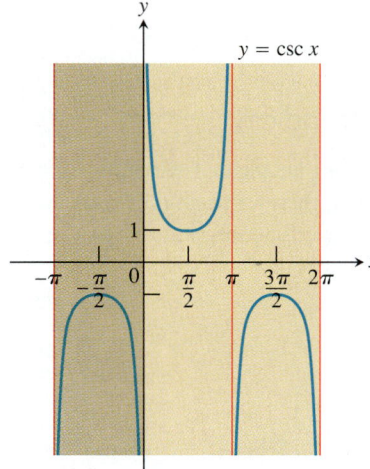

Domain: $x \ne 0, \pm\pi, \pm 2\pi, \ldots$
Range: $y \le -1$ and $y \ge 1$

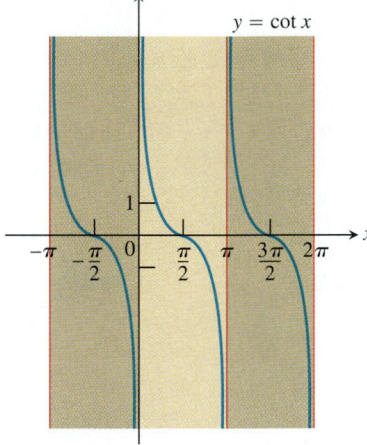

Domain: $x \ne 0, \pm\pi, \pm 2\pi, \ldots$
Range: $-\infty < y < \infty$

A.9   The graphs of the six basic trigonometric functions as functions of radian measure. Each function's periodicity shows clearly in its graph.

constructed from p. Thus, we say that the period of $\tan x$ is π while the period of $\sin x$ is 2π.

The importance of periodic functions stems from the fact that much of the behavior we study in science is periodic. Brain waves and heartbeats are periodic, as are household voltage and electric current. The electromagnetic field that heats food in a microwave oven is periodic, as are cash flows in seasonal businesses and the behavior of rotational machinery. The seasons are periodic—so is the weather. The phases of the moon are periodic, as are the motions of the planets. There is strong evidence that the ice ages are periodic, with a period of 90,000–100,000 years.

If so many things are periodic, why limit our discussion to trigonometric functions? The answer lies in a surprising and beautiful theorem from advanced calculus that says that every periodic function we want to use in mathematical modeling can be written as an algebraic combination of sines and cosines. Thus, once we learn the calculus of sines and cosines, we will know everything we need to know to model the mathematical behavior of periodic phenomena.

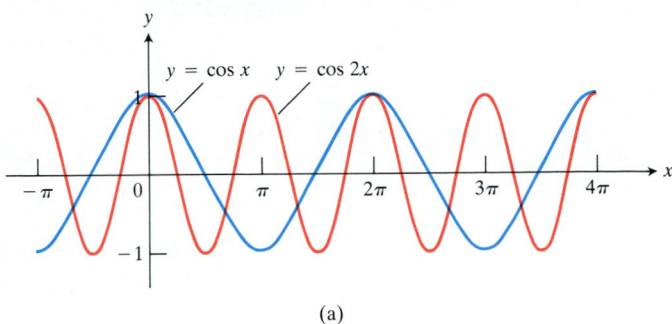

(a)

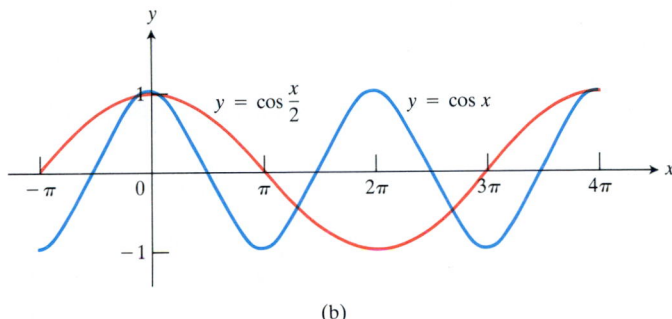

(b)

A.10 (a) Multiplying x by a number greater than 1 speeds the cosine up. (b) Multiplying x by a number less than 1 slows the cosine down.

Graphs

When we graph trigonometric functions in the coordinate plane, we usually denote the independent variable by x instead of θ. Figure A.9 shows how graphs of the six basic trigonometric functions appear as graphs in the xy-plane. Notice that the graph of $\tan x = (\sin x)/(\cos x)$ "blows up" as x approaches odd-integer multiples of $\pi/2$. Notice, too, the similar behavior of $\cot x = (\cos x)/(\sin x)$ as x approaches integer multiples of π.

Figure A.10 shows the graphs of $y = \cos 2x$ and $y = \cos(x/2)$ plotted against the graph of $y = \cos x$. Multiplying x by 2 speeds the cosine up and shortens the period from 2π to π. Multiplying x by 1/2 slows the cosine down and lengthens its period from 2π to 4π.

Odd vs. Even

The two angles in Fig. A.11 have the same magnitude but opposite signs. The points where the terminal rays cross the circle have the same x-coordinate, and their y-coordinates differ only in sign. Hence,

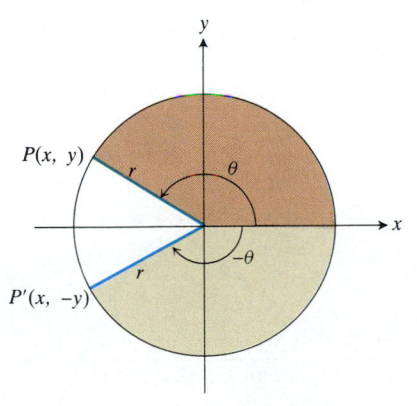

A.11 Angles of opposite sign.

$$\cos(-\theta) = \frac{x}{r} = \cos\theta, \qquad \text{(The cosine is an even function.)} \qquad (10a)$$

$$\sin(-\theta) = \frac{-y}{r} = -\sin\theta. \qquad \text{(The sine is an odd function.)} \qquad (10b)$$

Example 7

$$\cos\left(-\frac{\pi}{3}\right) = \cos\frac{\pi}{3} = \frac{1}{2},$$

$$\sin\left(-\frac{\pi}{3}\right) = -\sin\frac{\pi}{3} = -\frac{\sqrt{3}}{2}$$

As for the other basic trigonometric functions, the secant is even and the cosecant, tangent, and cotangent are odd. For the secant and tangent,

$$\sec(-\theta) = \frac{1}{\cos(-\theta)} = \frac{1}{\cos\theta} = \sec\theta, \tag{11}$$

$$\tan(-\theta) = \frac{\sin(-\theta)}{\cos(-\theta)} = \frac{-\sin\theta}{\cos\theta} = -\tan\theta. \tag{12}$$

Similar calculations show that the contangent and cosecant are odd.

Shift Formulas

If you look once again at Fig. A.9, you will see that the cosine curve is the same as the sine curve shifted $\pi/2$ units to the left. Also, the sine curve is the same as the cosine curve shifted $\pi/2$ units to the right. In symbols,

$$\sin\left(x + \frac{\pi}{2}\right) = \cos x, \qquad \cos\left(x - \frac{\pi}{2}\right) = \sin x. \tag{13}$$

Figure A.12(a) shows the cosine shifted to the left $\pi/2$ units to become the reflection of the sine curve across the x-axis. Next to it, Fig. A.12(b) shows the sine curve shifted $\pi/2$ units to the right to become the reflection of the cosine curve across the x-axis. In symbols,

$$\cos\left(x + \frac{\pi}{2}\right) = -\sin x, \qquad \sin\left(x - \frac{\pi}{2}\right) = -\cos x. \tag{14}$$

Example 8
The builders of the Trans-Alaska Pipeline used insulated pads to keep the heat from the hot oil in the pipeline from melting the permanently frozen

A.12 (a) The reflection of the sine as a shifted cosine. (b) The reflection of the cosine as a shifted sine.

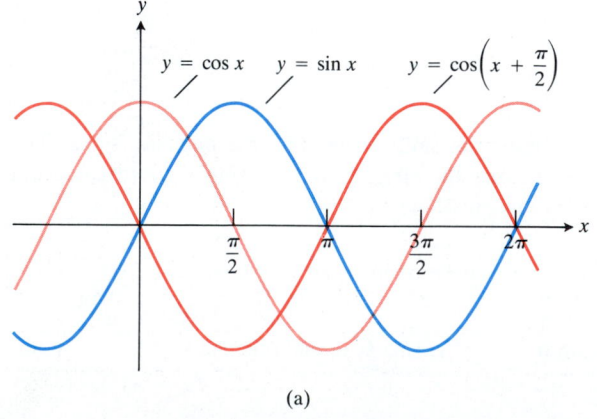

(a)

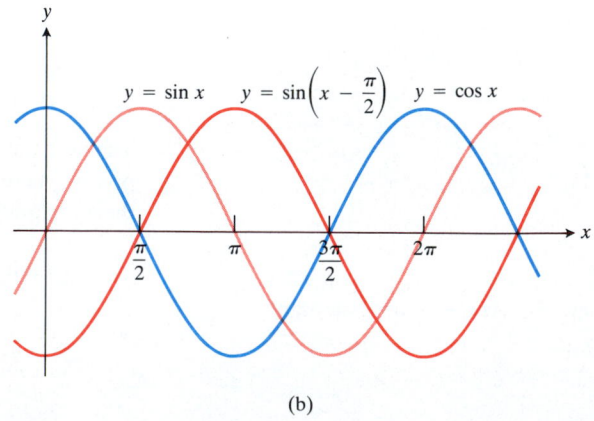

(b)

soil beneath. To design the pads, it was necessary to take into account the variation in air temperature throughout the year. The variation was represented in the calculations by a *general sine function*

$$f(x) = A \sin\left[\frac{2\pi}{B}(x - C)\right] + D.$$

In this formula, $|A|$ is the amplitude, $|B|$ is the period, C is the horizontal shift, and D is the vertical shift (Fig. A.13).

Figure A.14 shows how we can use such a function to model temperature data. The data points in the figure are plots of the mean air temperature for Fairbanks, Alaska, based on records of the National Weather Service from 1941 to 1970. The sine function used to fit the data is

$$f(x) = 37 \sin\left[\frac{2\pi}{365}(x - 101)\right] + 25,$$

where f is temperature in degrees Fahrenheit and x is the number of the day counting from the beginning of the year. The fit is remarkably good.

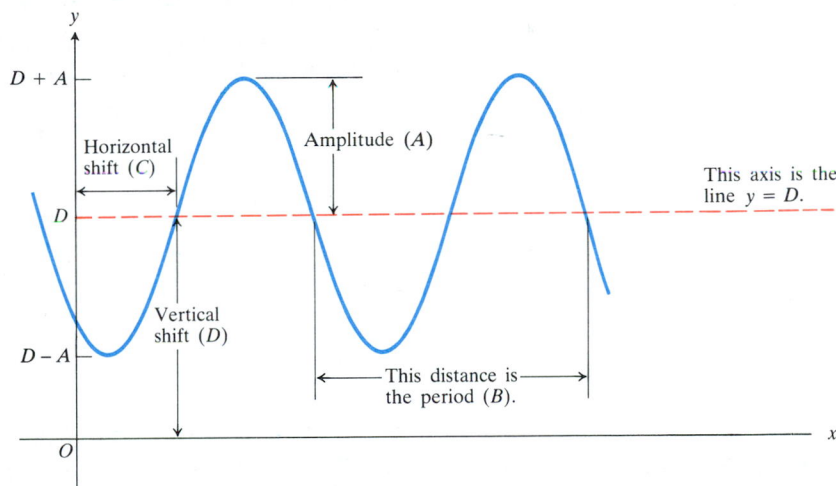

A.13   The general sine curve

$$y = A \sin[(2\pi/B)(x - C)] + D,$$

shown for A, B, C, and D positive.

A.14   Normal mean air temperature at Fairbanks, Alaska, plotted as data points. The approximating sine function is

$$f(x) = 37 \sin\left[\frac{2\pi}{365}(x - 101)\right] + 25.$$

(*Source:* "Is the Curve of Temperature Variation a Sine Curve?" by B. M. Lando and C. A. Lando, *The Mathematics Teacher*, 7:6, Fig. 2, p. 535 (September 1977).)

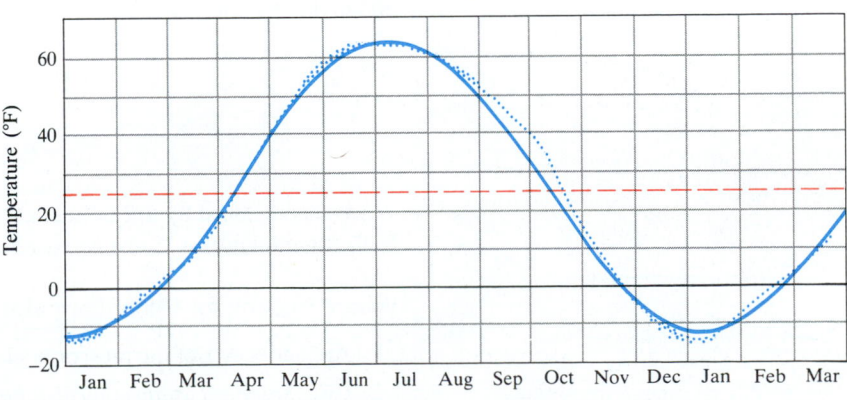

Angle Sum and Difference Formulas

As you may recall from an earlier course,

$$\cos (A + B) = \cos A \cos B - \sin A \sin B, \tag{15}$$

$$\sin (A + B) = \sin A \cos B + \cos A \sin B. \tag{16}$$

These formulas hold for all angles A and B.

If we replace B by $-B$ in Eqs. (15) and (16), we get

$$\cos (A - B) = \cos A \cos (-B) - \sin A \sin (-B)$$
$$= \cos A \cos B - \sin A (-\sin B) \tag{17}$$
$$= \cos A \cos B + \sin A \sin B,$$

$$\sin (A - B) = \sin A \cos (-B) + \cos A \sin (-B)$$
$$= \sin A \cos B + \cos A (-\sin B) \tag{18}$$
$$= \sin A \cos B - \cos A \sin B$$

Double-angle (Half-angle) Formulas

It is sometimes possible to simplify a calculation by changing trigonometric functions of θ into trigonometric functions of 2θ. There are four basic formulas for doing this, called **double-angle formulas.** The first two come from setting A and B equal to θ in Eqs. (15) and (16):

$$\cos 2\theta = \cos^2\theta - \sin^2\theta, \qquad \text{(Eq. (15) with } A = B = \theta) \tag{19}$$

$$\sin 2\theta = 2 \sin \theta \cos \theta. \qquad \text{(Eq. (16) with } A = B = \theta) \tag{20}$$

The other two double-angle formulas come from the equations

$$\cos^2\theta + \sin^2\theta = 1, \qquad \cos^2\theta - \sin^2\theta = \cos 2\theta.$$

We add to get

$$2 \cos^2\theta = 1 + \cos 2\theta,$$

subtract to get

$$2 \sin^2\theta = 1 - \cos 2\theta,$$

and divide by 2 to get

$$\cos^2\theta = \frac{1 + \cos 2\theta}{2} \tag{21}$$

$$\sin^2\theta = \frac{1 - \cos 2\theta}{2}. \tag{22}$$

When θ is replaced by $\theta/2$ in Eqs. (21) and (22), the resulting formulas are called **half-angle formulas.** Some books refer to Eqs. (21) and (22) by this name as well.

Where to Look for Other Formulas   Additional information is available in

1. Appendix A.1 of the present book.

2. *CRC Standard Mathematical Tables* (any recent edition), CRC Press, Inc.

Tips for Graphing sines and cosines: Curve first, scaled axes later

1. The one basic sine and cosine curve:

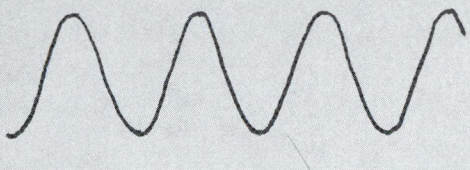

2. The basic curve with approximately scaled axes in different positions:

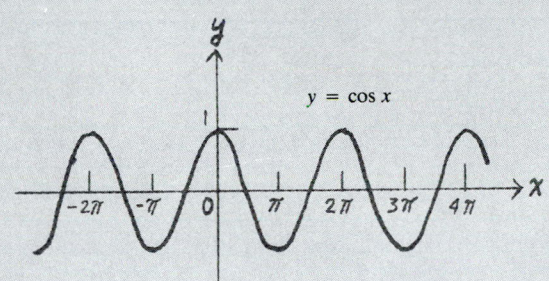

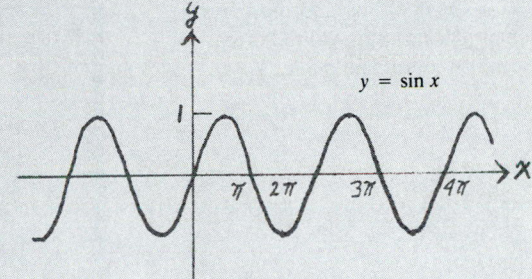

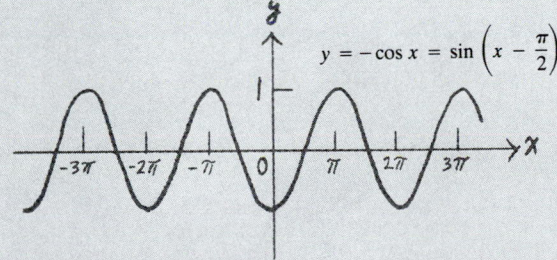

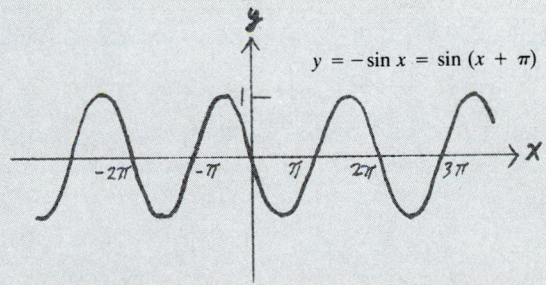

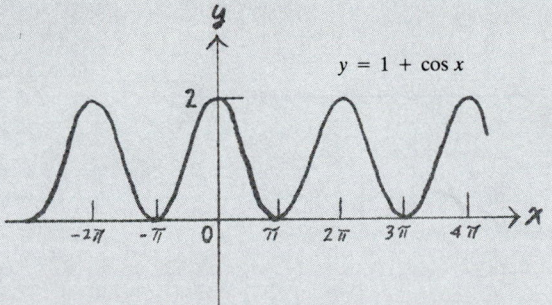

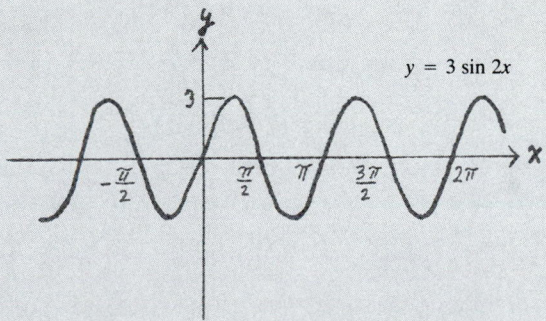

A.3
Symmetry in the xy-Plane

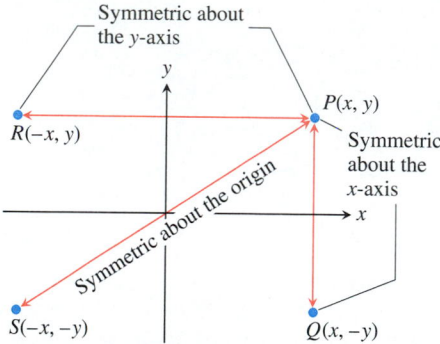

A.15   The coordinate formulas for symmetry with respect to the origin and axes in the coordinate plane.

We can use coordinate formulas to describe important symmetries in the coordinate plane. Figure A.15 shows how this is done.

Example 1   Symmetric points:

$P(5, 2)$ and $Q(5, -2)$ symmetric about the x-axis
$P(5, 2)$ and $R(-5, 2)$ symmetric about the y-axis
$P(5, 2)$ and $S(-5, -2)$ symmetric about the origin

The coordinate relations in Fig. A.15 provide the following symmetry tests for graphs.

> ### Symmetry Tests for Graphs
>
> 1. *Symmetry about the x-axis:*
>   If the point (x, y) lies on the graph, then the point $(x, -y)$ lies on the graph (Fig. A.16a).
>
> 2. *Symmetry about the y-axis:*
>   If the point (x, y) lies on the graph, the point $(-x, y)$ lies on the graph (Fig. A.16b).
>
> 3. *Symmetry about the origin:*
>   If the point (x, y) lies on the graph, the point $(-x, -y)$ lies on the graph (Fig. A.16c).

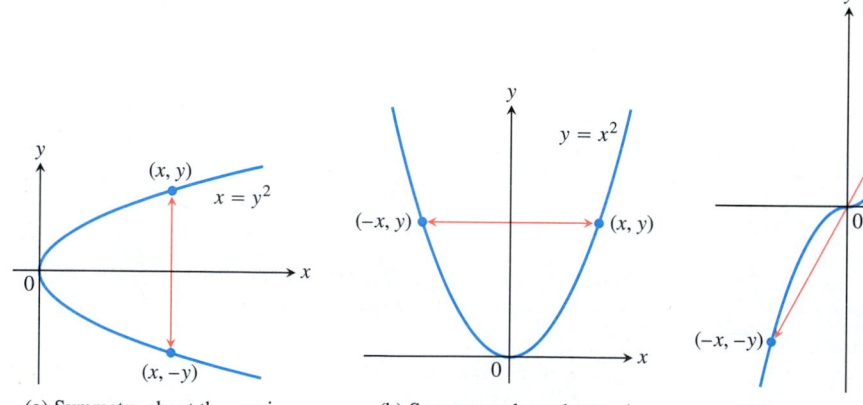

(a) Symmetry about the x-axis   (b) Symmetry about the y-axis   (c) Symmetry about the origin

A.16   Symmetry tests for graphs in the xy-plane.

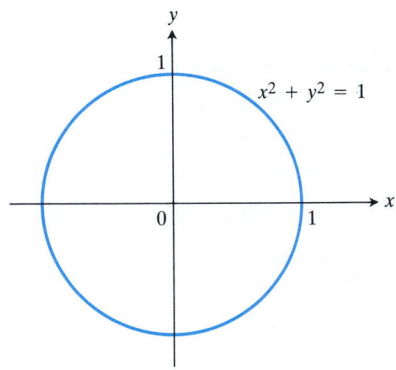

A.17 The graph of the equation $x^2 + y^2 = 1$ is the circle of radius 1, centered at the origin. It is symmetric about both axes and about the origin. Example 2 shows how to predict these symmetries before you graph the equation.

Example 2   The graph of $x^2 + y^2 = 1$ has all three of the symmetries listed above (Fig. A.17).

1. Symmetry about the x-axis:
(x, y) on the graph $\Rightarrow$ $x^2 + y^2 = 1$ ($\Rightarrow$ means "implies.")

$\Rightarrow$ $x^2 + (-y)^2 = 1$ ($y^2 = (-y)^2$)

$\Rightarrow$ $(x, -y)$ on the graph

2. Symmetry about the y-axis:
(x, y) on the graph $\Rightarrow$ $x^2 + y^2 = 1$

$\Rightarrow$ $(-x)^2 + y^2 = 1$ ($x^2 = (-x)^2$)

$\Rightarrow$ $(-x, y)$ on the graph

3. Symmetry about the origin:
(x, y) on the graph $\Rightarrow$ $x^2 + y^2 = 1$

$\Rightarrow$ $(-x)^2 + (-y)^2 = 1$

$\Rightarrow$ $(-x, -y)$ on the graph

A.4
Proofs of the Limit Theorems in Chapter 1

This appendix furnishes the ε-δ proofs for Theorems 1 and 6 in Chapter 1. Theorem 1 tells us that the limit of an algebraic combination of functions is the corresponding combination of the functions' limits, when the latter exists.

THEOREM 1 (SECTION 1.7)

The following rules hold if $\lim_{x \to c} f_1(x) = L_1$ and $\lim_{x \to c} f_2(x) = L_2$.

1. *Sum Rule:* $\lim (f_1(x) + f_2(x)) = L_1 + L_2$

2. *Difference Rule:* $\lim (f_1(x) - f_2(x)) = L_1 - L_2$

3. *Product Rule:* $\lim f_1(x) \cdot f_2(x) = L_1 \cdot L_2$

4. *Constant Multiple Rule:* $\lim k \cdot f_2(x) = k \cdot L_2$ (Any number k)

5. *Quotient Rule:* $\lim \dfrac{f_1(x)}{f_2(x)} = \dfrac{L_1}{L_2}$ if $L_2 \neq 0$

The limits are all taken as $x \to c$, and L_1 and L_2 are real numbers.

We proved the Sum Rule in Section 1.7. We obtain the Difference Rule by replacing $f_2(x)$ by $-f_2(x)$ and L_2 by $-L_2$ in the Sum Rule. The Constant Multiple Rule is the special case $f_1(x) = k$ of the Product Rule. This leaves only the Product and Quotient Rules to prove.

Proof of the Limit Product Rule   We need to show that for any $\varepsilon > 0$ there exists a $\delta > 0$ such that for all x in the functions' common domain,

$$0 < |x - c| < \delta \quad \Rightarrow \quad |f_1(x) f_2(x) - L_1 L_2| < \varepsilon. \tag{1}$$

Suppose then that ε is a positive number, and write $f_1(x)$ and $f_2(x)$ as

$$f_1(x) = L_1 + (f_1(x) - L_1) \quad \text{and} \quad f_2(x) = L_2 + (f_2(x) - L_2).$$

Multiply these expressions and subtract L_1L_2:

$$f_1(x) \cdot f_2(x) - L_1L_2 = (L_1 + (f_1(x) - L_1))(L_2 + (f_2(x) - L_2)) - L_1L_2$$

$$= L_1L_2 + L_1(f_2(x) - L_2) + L_2(f_1(x) - L_1)$$

$$+ (f_1(x) - L_1)(f_2(x) - L_2) - L_1L_2 \tag{2}$$

$$= L_1(f_2(x) - L_2) + L_2(f_1(x) - L_1) + (f_1(x) - L_1)(f_2(x) - L_2).$$

Since f_1 and f_2 have limits L_1 and L_2 as $x \to c$, there exist positive numbers δ_1, δ_2, δ_3, and δ_4 such that for all x,

$$0 < |x - c| < \delta_1 \quad \Rightarrow \quad |f_1(x) - L_1| < \sqrt{\varepsilon/3},$$

$$0 < |x - c| < \delta_2 \quad \Rightarrow \quad |f_2(x) - L_2| < \sqrt{\varepsilon/3},$$

$$0 < |x - c| < \delta_3 \quad \Rightarrow \quad |f_1(x) - L_1| < \varepsilon/(3(1 + |L_2|)), \tag{3}$$

$$0 < |x - c| < \delta_4 \quad \Rightarrow \quad |f_2(x) - L_2| < \varepsilon/(3(1 + |L_1|)).$$

All four of the inequalities on the right-hand side of (3) will hold for $0 < |x - c| < \delta$ if we take δ to be the smallest of the numbers δ_1 through δ_4. Therefore for all x, $0 < |x - c| < \delta$ implies

$$|f_1(x) \cdot f_2(x) - L_1L_2|$$

$$\leq |L_1||f_2(x) - L_2| + |L_2||f_1(x) - L_1| + |f_1(x) - L_1||f_2(x) - L_2|$$

(Triangle inequality applied to Eq. (2))

$$\leq (1 + |L_1|)|f_2(x) - L_2| + (1 + |L_2|)|f_1(x) - L_1| + |f_1(x) - L_1||f_2(x) - L_2|$$

$$\leq \frac{\varepsilon}{3} + \frac{\varepsilon}{3} + \sqrt{\frac{\varepsilon}{3}}\sqrt{\frac{\varepsilon}{3}} = \varepsilon. \qquad \text{(Values from (3))}$$

This completes the proof of the Limit Product Rule. ∎

Proof of the Limit Quotient Rule   We show that

$$\lim_{x \to c} \frac{1}{f_2(x)} = \frac{1}{L_2}.$$

Then we can apply the Limit Product Rule to show that

$$\lim_{x \to c} \frac{f_1(x)}{f_2(x)} = \lim_{x \to c} f_1(x) \cdot \frac{1}{f_2(x)} = \lim_{x \to c} f_1(x) \cdot \lim_{x \to c} \frac{1}{f_2(x)} = L_1 \cdot \frac{1}{L_2} = \frac{L_1}{L_2}.$$

To show that $\lim_{x \to c} (1/f_2(x)) = 1/L_2$, we need to show that for any $\varepsilon > 0$ there exists a $\delta > 0$ such that for all x,

$$0 < |x - c| < \delta \quad \Rightarrow \quad \left| \frac{1}{f_2(x)} - \frac{1}{L_2} \right| < \varepsilon.$$

Since $|L_2| > 0$, there exists a positive number δ_1 such that for all x,

$$0 < |x - c| < \delta_1 \quad \Rightarrow \quad |f_2(x) - L_2| < \frac{|L_2|}{2}. \tag{4}$$

For any numbers A and B it can be shown that $|A| - |B| \leq |A - B|$ and $|B| - |A| \leq |A - B|$, from which it follows that

$$||A| - |B|| \leq |A - B|. \tag{5}$$

With $A = f_2(x)$ and $B = L_2$, this gives

$$\big|\,|f_2(x)| - |L_2|\,\big| \le |f_2(x) - L_2|,$$

which we can combine with the right-hand inequality in (4) to get, in turn,

$$\big|\,|f_2(x)| - |L_2|\,\big| < \frac{|L_2|}{2},$$

$$-\frac{|L_2|}{2} < |f_2(x)| - |L_2| < \frac{|L_2|}{2},$$

$$\frac{|L_2|}{2} < |f_2(x)| < \frac{3|L_2|}{2}, \tag{6}$$

$$|L_2| < 2|f_2(x)| < 3|L_2|,$$

$$\frac{1}{|f_2(x)|} < \frac{2}{|L_2|} < \frac{3}{|f_2(x)|}.$$

Therefore $0 < |x - c| < \delta_1$ implies that

$$\left| \frac{1}{f_2(x)} - \frac{1}{L_2} \right| = \left| \frac{L_2 - f_2(x)}{L_2 f_2(x)} \right| \le \frac{1}{|L_2|} \cdot \frac{1}{|f_2(x)|} \cdot |L_2 - f_2(x)|$$

$$< \frac{1}{|L_2|} \cdot \frac{2}{|L_2|} \cdot |L_2 - f_2(x)|. \qquad \text{(Eq. (6))}$$

Suppose now that ε is an arbitrary positive number. Then $\frac{1}{2}|L_2|^2\,\varepsilon > 0$, so there exists a number $\delta_2 > 0$ such that for all x,

$$0 < |x - c| < \delta_2 \quad \Rightarrow \quad |L_2 - f_2(x)| < \frac{\varepsilon}{2}|L_2|^2. \tag{7}$$

The conclusions in (6) and (7) both hold for all x such that $0 < |x - c| < \delta$ if we take δ to be the smaller of the positive values δ_1 and δ_2. Combining (6) and (7) then gives

$$0 < |x - c| < \delta \quad \Rightarrow \quad \left| \frac{1}{f_2(x)} - \frac{1}{L_2} \right| < \varepsilon. \tag{8}$$

This concludes the proof of the Limit Quotient Rule.

THEOREM 6 (SECTION 1.9)

The Sandwich Theorem

Suppose that $g(x) \le f(x) \le h(x)$ for all $x \ne c$ in some interval about c and that $\lim_{x \to c} g(x) = \lim_{x \to c} h(x) = L$. Then $\lim_{x \to c} f(x) = L$.

Proof for Right-hand Limits   Suppose $\lim_{x \to c^+} g(x) = \lim_{x \to c^+} h(x) = L$. Then for any $\varepsilon > 0$ there exists a $\delta > 0$ such that for all x the inequality $c < x < c + \delta$ implies

$$L - \varepsilon < g(x) < L + \varepsilon \qquad \text{and} \qquad L - \varepsilon < h(x) < L + \varepsilon. \tag{9}$$

These inequalities combine with the inequality $g(x) \le f(x) \le h(x)$ to give

$$L - \varepsilon < g(x) \le f(x) \le h(x) < L + \varepsilon,$$

$$L - \varepsilon < f(x) < L + \varepsilon, \tag{10}$$

$$-\varepsilon < f(x) - L < \varepsilon.$$

Therefore, for all x, the inequality $c < x < c + \delta$ implies $|f(x) - L| < \varepsilon$.

Proof for Left-hand Limits   Suppose $\lim_{x \to c^-} g(x) = \lim_{x \to c^-} h(x) = L$. Then for any $\varepsilon > 0$ there exists a $\delta > 0$ such that for all x the inequality $c - \delta < x < c$ implies

$$L - \varepsilon < g(x) < L + \varepsilon \qquad \text{and} \qquad L - \varepsilon < h(x) < L + \varepsilon. \tag{11}$$

We conclude as before that for all x the inequality $c - \delta < x < c$ implies $|f(x) - L| < \varepsilon$.

Proof for Two-sided Limits   If $\lim_{x \to c} g(x) = \lim_{x \to c} h(x) = L$, then $g(x)$ and $h(x)$ both approach L as $x \to c^+$ and as $x \to c^-$; so $\lim_{x \to c^+} f(x) = L$ and $\lim_{x \to c^-} f(x) = L$. Hence $\lim_{x \to c} f(x)$ exists and equals L.

EXERCISES A.4

1. Suppose that functions $f_1(x), f_2(x)$, and $f_3(x)$ have limits L_1, L_2, and L_3, respectively, as $x \to c$. Show that their sum has limit $L_1 + L_2 + L_3$. Use mathematical induction (Appendix A.5) to generalize this result to the sum of any finite number of functions.

2. Use mathematical induction and the Limit Product Rule in Theorem 1 to show that if functions $f_1(x), f_2(x), \ldots, f_n(x)$ have limits $L_1, L_2, \ldots, L_n$ as $x \to c$, then $\lim_{x \to c} f_1(x) f_2(x) \cdot \cdots \cdot f_n(x) = L_1 \cdot L_2 \cdot \cdots \cdot L_n$.

3. Use the fact that $\lim_{x \to c} x = c$ and the result of Exercise 2 to show that $\lim_{x \to c} x^n = c^n$ for any integer $n > 1$.

4. *Limits of polynomials.*   Use the fact that $\lim_{x \to c}(k) = k$ for any number k together with the results of Exercises 1 and 3 to show that $\lim_{x \to c} f(x) = f(c)$ for any polynomial function

$$f(x) = a_0 x^n + a_1 x^{n-1} + \cdots + a_{n-1} x + a_n.$$

5. *Limits of rational functions.*   Use Theorem 1 and the result of Exercise 4 to show that if $f(x)$ and $g(x)$ are polynomial functions and $g(c) \ne 0$, then

$$\lim_{x \to c} \frac{f(x)}{g(x)} = \frac{f(c)}{g(c)}.$$

6. *Composites of continuous functions.*   Figure A.18 gives the diagram for a proof that the composite of two continuous functions is continuous. Reconstruct the proof from the diagram. The statement to be proved is this: If g is continuous at $x = c$ and f is continuous at $g(c)$, then $f \circ g$ is continuous at c.

Assume that c is an interior point of the domain of g and that $g(c)$ is an interior point of the domain of f. This will make the limits involved two-sided. (The arguments for the cases that involve one-sided limits are similar.)

A.18   The diagram for a proof that the composite of two continuous functions is continuous. The continuity of composites holds for any finite number of functions. The only requirement is that each function be continuous where it is applied. In the figure, g is to be continuous at c, and f is to be continuous at $g(c)$.

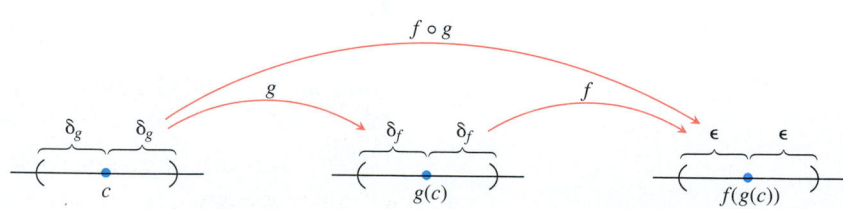

A.5
Mathematical Induction

Many formulas, like

$$1 + 2 + \cdots + n = \frac{n(n + 1)}{2},$$

can be shown to hold for every positive integer n by applying an axiom called the *mathematical induction principle*. A proof that uses this axiom is called a *proof by mathematical induction* or a *proof by induction*.

The steps in proving a formula by induction are

STEP 1: Check that it holds for $n = 1$.

STEP 2: Prove that if it holds for any positive integer $n = k$, then it also holds for $n = k + 1$.

Once these steps are completed (the axiom says), we know that the formula holds for all positive integers n. By step 1 it holds for $n = 1$. By step 2 it holds for $n = 2$, and therefore by step 2 also for $n = 3$, and by step 2 again for $n = 4$, and so on. If the first domino falls, and the kth domino always knocks over the $(k + 1)$st when it falls, all the dominoes fall.

From another point of view, suppose we have a sequence of statements

$$S_1, S_2, \ldots, S_n, \ldots,$$

one for each positive integer. Suppose we can show that assuming any one of the statements to be true implies that the next statement in line is true. Suppose that we can also show that S_1 is true. Then we may conclude that the statements are true from S_1 on.

Example 1   Show that for every positive integer n,

$$1 + 2 + \cdots + n = \frac{n(n + 1)}{2}.$$

Solution   We accomplish the proof by carrying out the two steps of mathematical induction.

STEP 1: The formula holds for $n = 1$ because

$$1 = \frac{1(1 + 1)}{2}.$$

STEP 2: If the formula holds for $n = k$, does it also hold for $n = k + 1$? The answer is yes, and here's why: If

$$1 + 2 + \cdots + k = \frac{k(k + 1)}{2},$$

then

$$1 + 2 + \cdots + k + (k + 1) = \frac{k(k + 1)}{2} + (k + 1) = \frac{k^2 + k + 2k + 2}{2}$$

$$= \frac{(k + 1)(k + 2)}{2} = \frac{(k + 1)((k + 1) + 1)}{2}.$$

The last expression in this string of equalities is the expression $n(n + 1)/2$ for $n = (k + 1)$.

The mathematical induction principle now guarantees the original formula for all positive integers n.

Notice that all *we* have to do is carry out steps 1 and 2. The mathematical induction principle does the rest.

Example 2   Show that for all positive integers n,

$$\frac{1}{2^1} + \frac{1}{2^2} + \cdots + \frac{1}{2^n} = 1 - \frac{1}{2^n}.$$

Solution   We accomplish the proof by carrying out the two steps of mathematical induction.

STEP 1:   The formula holds for $n = 1$ because

$$\frac{1}{2^1} = 1 - \frac{1}{2^1}.$$

STEP 2:   If

$$\frac{1}{2^1} + \frac{1}{2^2} + \cdots + \frac{1}{2^k} = 1 - \frac{1}{2^k},$$

then

$$\frac{1}{2^1} + \frac{1}{2^2} + \cdots + \frac{1}{2^k} + \frac{1}{2^{k+1}} = 1 - \frac{1}{2^k} + \frac{1}{2^{k+1}} = 1 - \frac{1 \cdot 2}{2^k \cdot 2} + \frac{1}{2^{k+1}}$$

$$= 1 - \frac{2}{2^{k+1}} + \frac{1}{2^{k+1}} = 1 - \frac{1}{2^{k+1}}.$$

Thus, the original formula holds for $n = k + 1$ whenever it holds for $n = k$.

With these two steps verified, the mathematical induction principle now guarantees the formula for every positive integer n.

Other Starting Integers

Instead of starting at $n = 1$, some induction arguments start at another integer. The steps for such an argument are

STEP 1:   Check that the formula holds for $n = n_1$ (whatever the appropriate first integer is).

STEP 2:   Prove that if the formula holds for any integer $n = k \geq n_1$, then it also holds for $n = k + 1$.

Once these steps are completed, the mathematical induction principle will guarantee the formula for all $n \geq n_1$.

Example 3   Show that $n! > 3^n$ if n is large enough.

Solution   How large is large enough? We experiment:

n	1	2	3	4	5	6	7
$n!$	1	2	6	24	120	720	5040
3^n	3	9	27	81	243	729	2187

It looks as if $n! > 3^n$ for $n \geq 7$. To be sure, we apply mathematical induction. We take $n_1 = 7$ in step 1 and try for step 2.

Suppose $k! > 3^k$ for some $k \geq 7$. Then

$$(k + 1)! = (k + 1)(k!) > (k + 1)3^k > 7 \cdot 3^k > 3^{k+1}.$$

Thus, for $k \geq 7$,

$$k! > 3^k \quad \Rightarrow \quad (k + 1)! > 3^{k+1}.$$

The mathematical induction principle now guarantees $n! \geq 3^n$ for all $n \geq 7$.

EXERCISES A.5

1. Assuming that the triangle inequality $|a + b| \leq |a| + |b|$ holds for any two numbers a and b, show that

$$|x_1 + x_2 + \cdots + x_n| \leq |x_1| + |x_2| + \cdots + |x_n|$$

for any n numbers.

2. Show that if $r \neq 1$, then

$$1 + r + r^2 + \cdots + r^n = \frac{1 - r^{n+1}}{1 - r}$$

for all positive integers n.

3. Use the Product Rule,

$$\frac{d}{dx}(uv) = u\frac{dv}{dx} + v\frac{du}{dx},$$

and the fact that

$$\frac{d}{dx}(x) = 1$$

to show that

$$\frac{d}{dx}(x^n) = nx^{n-1}$$

for all positive integers n.

4. Suppose that a function $f(x)$ has the property that $f(x_1 x_2) = f(x_1) + f(x_2)$ for any two positive numbers x_1 and x_2. Show that

$$f(x_1 x_2 \cdots x_n) = f(x_1) + f(x_2) + \cdots + f(x_n)$$

for the product of any n positive numbers $x_1, x_2, \ldots, x_n$.

5. Show that

$$\frac{2}{3^1} + \frac{2}{3^2} + \cdots + \frac{2}{3^n} = 1 - \frac{1}{3^n}$$

for all positive integers n.

6. Show that $n! > n^3$ if n is large enough.

7. Show that $2^n > n^2$ if n is large enough.

8. Show that $2^n \geq 1/8$ for $n \geq -3$.

9. *Sums of squares.*   Show that the sum of the squares of the first n positive integers is

$$\frac{n\left(n + \frac{1}{2}\right)(n + 1)}{3}.$$

10. *Sums of cubes.*   Show that the sum of the cubes of the first n positive integers is $(n(n + 1)/2)^2$.

11. Show that the following finite-sum rules hold for every positive integer n.

a) $\displaystyle\sum_{k=1}^{n} (a_k + b_k) = \sum_{k=1}^{n} a_k + \sum_{k=1}^{n} b_k$

b) $\displaystyle\sum_{k=1}^{n} (a_k - b_k) = \sum_{k=1}^{n} a_k - \sum_{k=1}^{n} b_k$

c) $\displaystyle\sum_{k=1}^{n} ca_k = c \cdot \sum_{k=1}^{n} a_k$ (Any number c)

d) $\displaystyle\sum_{k=1}^{n} a_k = n \cdot c$ if a_k has the constant value c

12. *Sums of products of consecutive positive integers.*   You are probably already familiar with the sums

$$\sum_{k=1}^{n} k^0 = \sum_{k=1}^{n} 1 = \frac{n}{1} \quad \text{and} \quad \sum_{k=1}^{n} k = \frac{n(n + 1)}{2}.$$

Show that

a) $\displaystyle\sum_{k=1}^{n} k(k + 1) = \frac{n(n + 1)(n + 2)}{3}$

b) $\displaystyle\sum_{k=1}^{n} k(k + 1)(k + 2) = \frac{n(n + 1)(n + 2)(n + 3)}{4}$

c) $\displaystyle\sum_{k=1}^{n} k(k + 1)(k + 2)(k + 3) =$

$$\frac{n(n + 1)(n + 2)(n + 3)(n + 4)}{5}$$

Then show, in general, that if m and n are integers with $m \geq 0$ and $n \geq 1$, then

d) $\displaystyle\sum_{k=1}^{n} \left(\prod_{j=0}^{m} (k + j)\right) = \frac{1}{m + 2} \prod_{j=0}^{m+1} (n + j)$

A.6
Complex Numbers

Complex numbers are numbers of the form $a + bi$, where a and b are real numbers and $i = \sqrt{-1}$. The number a is the **real part** of $a + bi$, and b is the **imaginary part**.

Complex numbers $a + bi$ and $c + di$ are equal if and only if $a = c$ and $b = d$. In particular, $a + bi = 0$ if and only if $a = 0$ and $b = 0$ or, equivalently, if and only if $a^2 + b^2 = 0$.

We add and subtract complex numbers by adding and subtracting their real and imaginary parts:

$$(a + bi) + (c + di) = (a + c) + (b + d)i,$$

$$(a + bi) - (c + di) = (a - c) + (b - d)i.$$

We multiply complex numbers the way we multiply other binomials, using the fact that $i^2 = -1$ to simplify the final result:

$$(a + bi)(c + di) = ac + adi + bci + bdi^2$$

$$= ac + adi + bci - bd \qquad (i^2 = -1)$$

$$= (ac - bd) + (ad + bc)i. \qquad \left(\begin{array}{c}\text{Real and imaginary}\\\text{parts combined}\end{array}\right)$$

To divide a complex number $c + di$ by a nonzero complex number $a + bi$, multiply the numerator and denominator of the quotient by $a - bi$ (the number $a - bi$ is the **complex conjugate** of $a + bi$):

$$\frac{c + di}{a + bi} = \frac{c + di}{a + bi} \frac{a - bi}{a - bi} \qquad \left(\begin{array}{c}\text{Multiply numerator and}\\\text{denominator by the}\\\text{complex conjugate of}\\a + bi.\end{array}\right)$$

$$= \frac{ac - bci + adi - bdi^2}{a^2 - abi + abi - b^2i^2}$$

$$= \frac{(ac + bd) + (ad - bc)i}{a^2 + b^2}$$

$$= \frac{ac + bd}{a^2 + b^2} + \frac{ad - bc}{a^2 + b^2} i.$$

Example

a) $(2 + 3i) + (6 - 2i) = (2 + 6) + (3 - 2)i = 8 + i$

b) $(2 + 3i) - (6 - 2i) = (2 - 6) + (3 - (-2))i = -4 + 5i$

c) $(2 + 3i)(6 - 2i) = (2)(6) + (2)(-2i) + (3i)(6) + (3i)(-2i)$

$$= 12 - 4i + 18i - 6i^2 = 12 + 14i + 6 = 18 + 14i$$

d)
$$\frac{2 + 3i}{6 - 2i} = \frac{2 + 3i}{6 - 2i} \frac{6 + 2i}{6 + 2i}$$

$$= \frac{12 + 4i + 18i + 6i^2}{36 + 12i - 12i - 4i^2}$$

$$= \frac{6 + 22i}{40} = \frac{3}{20} + \frac{11}{20} i$$

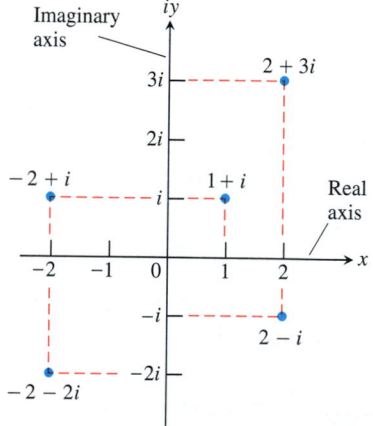

A.19 The complex plane.

We plot complex numbers as points in a plane by identifying the number $x + iy$ with the point (x, y) in the Cartesian plane. We then call the x-axis the **real**

axis, the y-axis the **imaginary axis** (sometimes, *iy-axis*), and the relabeled plane the **complex plane** (Fig. A.19).

A.7
Cauchy's Mean Value Theorem and the Stronger Form of L'Hôpital's Rule

In this appendix, we prove the stronger form of l'Hôpital's rule (Theorem 6, Section 3.6). When the limit of a quotient of differentiable functions leads to an indeterminate form, l'Hôpital's rule is the one to apply.

THEOREM 6

L'Hôpital's Rule (Stronger Form)

Suppose that

$$f(x_0) = g(x_0) = 0$$

and that the functions f and g are both differentiable on an open interval (a, b) that contains the point x_0. Suppose also that $g' \neq 0$ at every point in (a, b) except possibly x_0. Then

$$\lim_{x \to x_0} \frac{f(x)}{g(x)} = \lim_{x \to x_0} \frac{f'(x)}{g'(x)}, \tag{1}$$

provided the limit on the right exists.

The proof of the stronger form of l'Hôpital's rule is based on Cauchy's Mean Value Theorem, a mean value theorem that involves two functions instead of one. We prove Cauchy's theorem first and then show how it leads to l'Hôpital's rule.

THEOREM

Cauchy's Mean Value Theorem

Suppose that functions f and g are continuous on $[a, b]$ and differentiable throughout (a, b) and suppose also that $g' \neq 0$ throughout (a, b). Then there exists a number c in (a, b) at which

$$\frac{f'(c)}{g'(c)} = \frac{f(b) - f(a)}{g(b) - g(a)}. \tag{2}$$

Notice that the ordinary Mean Value Theorem (Theorem 3, Section 3.2) is the case $g(x) = x$.

Proof of Cauchy's Mean Value Theorem   We apply the Mean Value Theorem of Section 3.2 twice. First we use it to show that $g(a) \neq g(b)$. For if $g(b)$ did equal $g(a)$, then the Mean Value Theorem would give

$$g'(c) = \frac{g(b) - g(a)}{b - a} = 0$$

for some c between a and b. This cannot happen because $g'(x) \neq 0$ in (a, b).

We next apply the Mean Value Theorem to the function

$$F(x) = f(x) - f(a) - \frac{f(b) - f(a)}{g(b) - g(a)} [g(x) - g(a)]. \tag{3}$$

This function is continuous and differentiable where f and g are, and $F(b) = F(a) = 0$. Therefore there is a number c between a and b for which $F'(c) = 0$. In terms of f and g this says

$$F'(c) = f'(c) - \frac{f(b) - f(a)}{g(b) - g(a)}[g'(c)] = 0, \tag{4}$$

or

$$\frac{f'(c)}{g'(c)} = \frac{f(b) - f(a)}{g(b) - g(a)},$$

which is Eq. (2).

Proof of the Stronger Form of L'Hôpital's Rule   We first establish Eq. (1) for the case $x \to x_0^+$. The method needs almost no change to apply to $x \to x_0^-$, and the combination of these two cases establishes the result.

Suppose that x lies to the right of x_0. Then $g'(x) \neq 0$ and we can apply Cauchy's Mean Value Theorem to the closed interval from x_0 to x. This produces a number c between x_0 and x such that

$$\frac{f'(c)}{g'(c)} = \frac{f(x) - f(x_0)}{g(x) - g(x_0)}. \tag{5}$$

But $f(x_0) = g(x_0) = 0$, so

$$\frac{f'(c)}{g'(c)} = \frac{f(x)}{g(x)}. \tag{6}$$

As x approaches x_0, c approaches x_0 because it lies between x and x_0. Therefore,

$$\lim_{x \to x_0^+} \frac{f(x)}{g(x)} = \lim_{c \to x_0^+} \frac{f'(c)}{g'(c)} = \lim_{x \to x_0^+} \frac{f'(x)}{g'(x)}.$$

This establishes l'Hôpital's rule for the case where x approaches x_0 from above. The case where x approaches x_0 from below is proved by applying Cauchy's Mean Value Theorem to the closed interval $[x, x_0]$, $x < x_0$.

EXERCISES A.7

Although the importance of Cauchy's Mean Value Theorem lies elsewhere, we can sometimes satisfy our curiosity about the identity of the number c in Eq. (2), as in Exercises 1–4.

In Exercises 1–4, find all values of c that satisfy Eq. (2) in the conclusion of Cauchy's Mean Value Theorem.

1. $f(x) = x$, $\quad g(x) = x^2$, $\quad [a, b] = [-2, 0]$
2. $f(x) = x$, $\quad g(x) = x^2$, $\quad [a, b]$ arbitrary
3. $f(x) = \dfrac{x^3}{3} - 4x$, $\quad g(x) = x^2$, $\quad [a, b] = [0, 3]$
4. $f(x) = \sin x$, $\quad g(x) = \cos x$, $\quad [a, b] = [0, \pi/2]$

A.8
Limits That Arise Frequently

This appendix verifies the limits in Table 8.1 of Section 8.1.

1. $\displaystyle \lim_{n \to \infty} \frac{\ln n}{n} = 0$

2. $\displaystyle \lim_{n \to \infty} \sqrt[n]{n} = 1$

3. $\displaystyle \lim_{n \to \infty} x^{1/n} = 1 \quad (x > 0)$

4. $\displaystyle \lim_{n \to \infty} x^n = 0 \quad (|x| < 1)$

5. $\displaystyle \lim_{n \to \infty} \left(1 + \frac{x}{n}\right)^n = e^x \quad (\text{Any } x)$

6. $\displaystyle \lim_{n \to \infty} \frac{x^n}{n!} = 0 \quad (\text{Any } x)$

In Formulas 3–6, x remains fixed while $n \to \infty$.

1. $\lim\limits_{n\to\infty} \dfrac{\ln n}{n} = 0$ We proved this in Section 8.1, Example 9.

2. $\lim\limits_{n\to\infty} \sqrt[n]{n} = 1$ Let $a_n = n^{1/n}$. Then

$$\ln a_n = \ln n^{1/n} = \frac{1}{n}\ln n \to 0. \tag{1}$$

Applying Theorem 3, Section 8.1, with $f(x) = e^x$ gives

$$n^{1/n} = a_n = e^{\ln a_n} = f(\ln a_n) \to f(0) = e^0 = 1. \tag{2}$$

3. If $x > 0$, $\lim\limits_{n\to\infty} x^{1/n} = 1$ Let $a_n = x^{1/n}$. Then

$$\ln a_n = \ln x^{1/n} = \frac{1}{n}\ln x \to 0 \tag{3}$$

because x remains fixed as $n\to\infty$. Applying Theorem 3, Section 8.1, with $f(x) = e^x$ gives

$$x^{1/n} = a_n = e^{\ln a_n} \to e^0 = 1. \tag{4}$$

4. If $|x| < 1$, $\lim\limits_{n\to\infty} x^n = 0$ We need to show that to each $\varepsilon > 0$ there corresponds an integer N so large that $|x^n| < \varepsilon$ for all n greater than N. Since $\varepsilon^{1/n} \to 1$, while $|x| < 1$, there exists an integer N for which

$$\varepsilon^{1/N} > |x|. \tag{5}$$

In other words,

$$|x^N| = |x|^N < \varepsilon. \tag{6}$$

This is the integer we seek because, if $|x| < 1$, then

$$|x^n| < |x^N| \qquad \text{for all } n > N. \tag{7}$$

Combining (6) and (7) produces

$$|x^n| < \varepsilon \qquad \text{for all } n > N, \tag{8}$$

and we're done.

5. For any number x, $\lim\limits_{n\to\infty}\left(1 + \dfrac{x}{n}\right)^n = e^x$ Let

$$a_n = \left(1 + \frac{x}{n}\right)^n.$$

Then

$$\ln a_n = \ln\left(1 + \frac{x}{n}\right)^n = n\ln\left(1 + \frac{x}{n}\right) \to x,$$

as we can see by the following application of l'Hôpital's rule, in which we differentiate with respect to n:

$$\lim_{n\to\infty} n\ln\left(1 + \frac{x}{n}\right) = \lim_{n\to\infty} \frac{\ln(1 + x/n)}{1/n}$$

$$= \lim_{n\to\infty} \frac{\left(\dfrac{1}{1 + x/n}\right)\cdot\left(-\dfrac{x}{n^2}\right)}{-1/n^2} = \lim_{n\to\infty} \frac{x}{1 + x/n} = x.$$

Apply Theorem 3, Section 8.1, with $f(x) = e^x$ to conclude that

$$\left(1 + \frac{x}{n}\right)^n = a_n = e^{\ln a_n} \rightarrow e^x.$$

6. For any number x, $\lim\limits_{n \to \infty} \dfrac{x^n}{n!} = 0$ Since

$$-\frac{|x|^n}{n!} \leq \frac{x^n}{n!} \leq \frac{|x|^n}{n!},$$

all we need to show is that $|x|^n/n! \rightarrow 0$. We can then apply the Sandwich Theorem for Sequences (Section 8.1, Theorem 2) to conclude that $x^n/n! \rightarrow 0$.

The first step in showing that $|x|^n/n! \rightarrow 0$ is to choose an integer $M > |x|$, so that

$$\frac{|x|}{M} < 1 \qquad \text{and} \qquad \left(\frac{|x|}{M}\right)^n \rightarrow 0.$$

We then restrict our attention to values of $n > M$. For these values of n, we can write

$$\frac{|x|^n}{n!} = \frac{|x|^n}{1 \cdot 2 \cdot \cdots \cdot M \cdot \underbrace{(M + 1)(M + 2) \cdot \cdots \cdot n}_{(n - M) \text{ factors}}}$$

$$\leq \frac{|x|^n}{M! M^{n-M}} = \frac{|x|^n M^M}{M! M^n} = \frac{M^M}{M!}\left(\frac{|x|}{M}\right)^n.$$

Thus,

$$0 \leq \frac{|x|^n}{n!} \leq \frac{M^M}{M!}\left(\frac{|x|}{M}\right)^n.$$

Now, the constant $M^M/M!$ does not change as n increases. Thus the Sandwich Theorem tells us that

$$\frac{|x|^n}{n!} \rightarrow 0 \qquad \text{because} \left(\frac{|x|}{M}\right)^n \rightarrow 0.$$

A.9
The Distributive Law for Vector Cross Products

In this appendix we prove the distributive law

$$\mathbf{A} \times (\mathbf{B} + \mathbf{C}) = \mathbf{A} \times \mathbf{B} + \mathbf{A} \times \mathbf{C} \tag{1}$$

from Eq. (6) in Section 10.4.

Proof   To derive Eq. (1), we construct $\mathbf{A} \times \mathbf{B}$ a new way. We draw $\mathbf{A}$ and $\mathbf{B}$ from the common point O and construct a plane M perpendicular to $\mathbf{A}$ at O (Fig. A.20). We then project $\mathbf{B}$ orthogonally onto M, yielding a vector $\mathbf{B}'$ with length $|\mathbf{B}|\sin \theta$. We rotate $\mathbf{B}'$ 90° about $\mathbf{A}$ in the positive sense to produce a vector $\mathbf{B}''$. Finally, we multiply $\mathbf{B}''$ by the length of $\mathbf{A}$. The resulting vector $|\mathbf{A}|\mathbf{B}''$ is equal to $\mathbf{A} \times \mathbf{B}$ since $\mathbf{B}''$ has the same direction as $\mathbf{A} \times \mathbf{B}$ by its construction (Fig. A.20) and

$$|\mathbf{A}||\mathbf{B}''| = |\mathbf{A}||\mathbf{B}'| = |\mathbf{A}||\mathbf{B}|\sin \theta = |\mathbf{A} \times \mathbf{B}|.$$

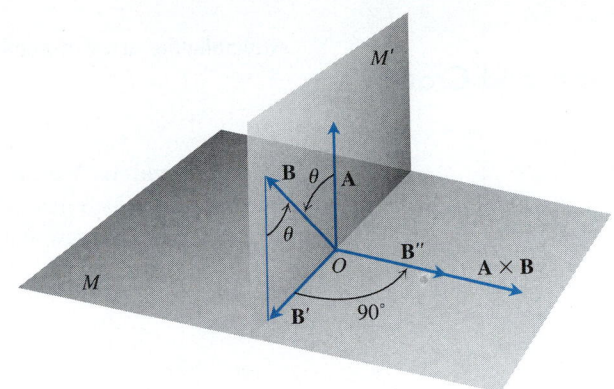

A.20   As explained in the text,
$\mathbf{A} \times \mathbf{B} = |\mathbf{A}|\mathbf{B}''$.

Now each of these three operations, namely,

1. projection onto M,

2. rotation about $\mathbf{A}$ through $90°$,

3. multiplication by the scalar $|\mathbf{A}|$,

when applied to a triangle whose plane is not parallel to $\mathbf{A}$, will produce another triangle. If we start with the triangle whose sides are $\mathbf{B}$, $\mathbf{C}$, and $\mathbf{B} + \mathbf{C}$ (Fig. A.21) and apply these three steps, we successively obtain

1. a triangle whose sides are $\mathbf{B}'$, $\mathbf{C}'$, and $(\mathbf{B} + \mathbf{C})'$ satisfying the vector equation

$$\mathbf{B}' + \mathbf{C}' = (\mathbf{B} + \mathbf{C})';$$

2. a triangle whose sides are $\mathbf{B}''$, $\mathbf{C}''$, and $(\mathbf{B} + \mathbf{C})''$ satisfying the vector equation

$$\mathbf{B}'' + \mathbf{C}'' = (\mathbf{B} + \mathbf{C})''$$

(the double prime on each vector has the same meaning as in Fig. A.20); and, finally,

3. a triangle whose sides are $|\mathbf{A}|\mathbf{B}''$, $|\mathbf{A}|\mathbf{C}''$, and $|\mathbf{A}|(\mathbf{B} + \mathbf{C})''$ satisfying the vector equation

$$|\mathbf{A}|\mathbf{B}'' + |\mathbf{A}|\mathbf{C}'' = |\mathbf{A}|(\mathbf{B} + \mathbf{C})''. \tag{2}$$

When we use the equations $|\mathbf{A}|\mathbf{B}'' = \mathbf{A} \times \mathbf{B}$, $|\mathbf{A}|\mathbf{C}'' = \mathbf{A} \times \mathbf{C}$, and $|\mathbf{A}|(\mathbf{B} + \mathbf{C})'' = \mathbf{A} \times (\mathbf{B} + \mathbf{C})$, which result from our discussion above, Eq. (2) becomes

$$\mathbf{A} \times \mathbf{B} + \mathbf{A} \times \mathbf{C} = \mathbf{A} \times (\mathbf{B} + \mathbf{C}),$$

which is the law we wanted to establish.

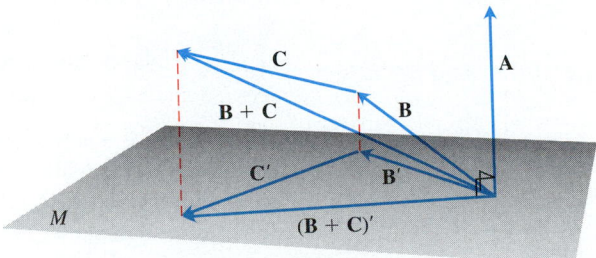

A.21   The vectors, $\mathbf{B}$, $\mathbf{C}$, $\mathbf{B} + \mathbf{C}$, and their projections onto a plane perpendicular to $\mathbf{A}$.

A.10
Determinants and Cramer's Rule

A rectangular array of numbers like

$$A = \begin{bmatrix} 2 & 1 & 3 \\ 1 & 0 & -2 \end{bmatrix}$$

is called a **matrix**. We call A a 2 by 3 matrix because it has two rows and three columns. An m by n matrix has m rows and n columns, and the **entry** or **element** (number) in the ith row and jth column is often denoted by a_{ij}:

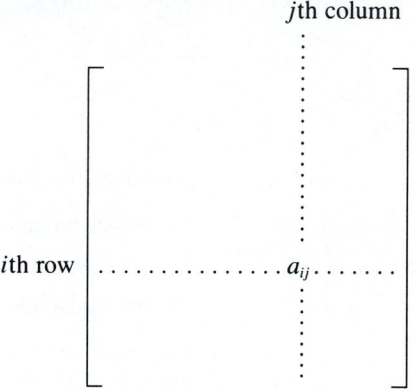

The matrix

$$A = \begin{bmatrix} 2 & 1 & 3 \\ 1 & 0 & -2 \end{bmatrix}$$

has

$$a_{11} = 2, \qquad a_{12} = 1, \qquad a_{13} = 3,$$
$$a_{21} = 1, \qquad a_{22} = 0, \qquad a_{23} = -2.$$

A matrix with the same number of rows as columns is a **square matrix.** It is a **matrix of order n** if the number of rows and columns is n.

With each square matrix A we associate a number det A or $|a_{ij}|$, called the **determinant** of A, calculated from the entries of A in the following way. (The vertical bars in the notation $|a_{ij}|$ do not mean absolute value.) For $n = 1$ and $n = 2$, we define

$$\det [a] = a, \tag{1}$$

$$\det \begin{bmatrix} a_{11} & a_{12} \\ a_{21} & a_{22} \end{bmatrix} = a_{11}a_{22} - a_{21}a_{12}. \tag{2}$$

For a matrix of order 3, we define

$$\det A = \det \begin{bmatrix} a_{11} & a_{12} & a_{13} \\ a_{21} & a_{22} & a_{23} \\ a_{31} & a_{32} & a_{33} \end{bmatrix} = \begin{array}{l} \text{Sum of all signed products} \\ \text{of the form } \pm\, a_{1i}a_{2j}a_{3k}, \end{array} \tag{3}$$

where i, j, k is a permutation of 1, 2, 3 in some order. There are $3! = 6$ such permutations, so there are six terms in the sum. Half of these have plus signs and the other half have minus signs, according to the index of the permutation, where the index is the number we define next. The sign is positive when the index is even and negative when the index is odd.

DEFINITION

Index of a Permutation

Given any permutation of the numbers $1, 2, 3, \ldots, n$, denote the permutation by $i_1, i_2, i_3, \ldots, i_n$. In this arrangement, some of the numbers following i_1 may be less than i_1, and the number of these is called the **number of inversions** in the arrangement pertaining to i_1. Likewise, there is a number of inversions pertaining to each of the other i's; it is the number of indices that come after that particular i in the arrangement and are less than it. The **index** of the permutation is the sum of all of the numbers of inversions pertaining to the separate indices.

Example 1   For $n = 5$, the permutation

$$5 \quad 3 \quad 1 \quad 2 \quad 4$$

has

4 inversions pertaining to the first element, 5,

2 inversions pertaining to the second element, 3,

and no further inversions, so the index is $4 + 2 = 6$.

The following table shows the permutations of 1, 2, 3, the index of each permutation, and the signed product in the determinant of Eq. (3).

Permutation	Index	Signed product
1 2 3	0	$+a_{11}a_{22}a_{33}$
1 3 2	1	$-a_{11}a_{23}a_{32}$
2 1 3	1	$-a_{12}a_{21}a_{33}$
2 3 1	2	$+a_{12}a_{23}a_{31}$
3 1 2	2	$+a_{13}a_{21}a_{32}$
3 2 1	3	$-a_{13}a_{22}a_{31}$

(4)

The sum of the six signed products is

$$a_{11}(a_{22}a_{33} - a_{23}a_{32}) - a_{12}(a_{21}a_{33} - a_{23}a_{31}) + a_{13}(a_{21}a_{32} - a_{22}a_{31})$$

$$= a_{11}\begin{vmatrix} a_{22} & a_{23} \\ a_{32} & a_{33} \end{vmatrix} - a_{12}\begin{vmatrix} a_{21} & a_{23} \\ a_{31} & a_{33} \end{vmatrix} + a_{13}\begin{vmatrix} a_{21} & a_{22} \\ a_{31} & a_{32} \end{vmatrix} = \begin{vmatrix} a_{11} & a_{12} & a_{13} \\ a_{21} & a_{22} & a_{23} \\ a_{31} & a_{32} & a_{33} \end{vmatrix}. \quad (5)$$

The formula

$$\begin{vmatrix} a_{11} & a_{12} & a_{13} \\ a_{21} & a_{22} & a_{23} \\ a_{31} & a_{32} & a_{33} \end{vmatrix} = a_{11}\begin{vmatrix} a_{22} & a_{23} \\ a_{32} & a_{33} \end{vmatrix} - a_{12}\begin{vmatrix} a_{21} & a_{23} \\ a_{31} & a_{33} \end{vmatrix} + a_{13}\begin{vmatrix} a_{21} & a_{22} \\ a_{31} & a_{32} \end{vmatrix} \quad (6)$$

reduces the calculation of a 3 by 3 determinant to the calculation of three 2 by 2 determinants.

Many people prefer to remember the following scheme for calculating the six signed products in the determinant of a 3 by 3 matrix:

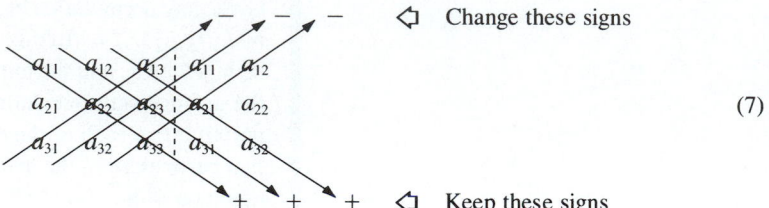

(7)

Minors and Cofactors

The second order determinants on the right-hand side of Eq. (6) are called the **minors** (short for "minor determinants") of the entries they multiply. Thus,

$$\begin{vmatrix} a_{22} & a_{23} \\ a_{32} & a_{33} \end{vmatrix} \text{ is the minor of } a_{11},$$

$$\begin{vmatrix} a_{21} & a_{23} \\ a_{31} & a_{33} \end{vmatrix} \text{ is the minor of } a_{12},$$

and so on. The minor of the element a_{ij} in a matrix A is the determinant of the matrix that remains after we delete the row and column containing a_{ij}:

$$\begin{vmatrix} a_{11} & a_{12} & a_{13} \\ a_{21} & a_{22} & a_{23} \\ a_{31} & a_{32} & a_{33} \end{vmatrix}. \qquad \text{The minor of } a_{22} \text{ is } \begin{vmatrix} a_{11} & a_{13} \\ a_{31} & a_{33} \end{vmatrix}.$$

$$\begin{vmatrix} a_{11} & a_{12} & a_{13} \\ a_{21} & a_{22} & a_{23} \\ a_{31} & a_{32} & a_{33} \end{vmatrix}. \qquad \text{The minor of } a_{23} \text{ is } \begin{vmatrix} a_{11} & a_{12} \\ a_{31} & a_{32} \end{vmatrix}.$$

The **cofactor** A_{ij} of a_{ij} is $(-1)^{i+j}$ times the minor of a_{ij}. Thus,

$$A_{22} = (-1)^{2+2} \begin{vmatrix} a_{11} & a_{13} \\ a_{31} & a_{33} \end{vmatrix} = \begin{vmatrix} a_{11} & a_{13} \\ a_{31} & a_{33} \end{vmatrix},$$

$$A_{23} = (-1)^{2+3} \begin{vmatrix} a_{11} & a_{12} \\ a_{31} & a_{32} \end{vmatrix} = - \begin{vmatrix} a_{11} & a_{12} \\ a_{31} & a_{32} \end{vmatrix}.$$

The effect of the factor $(-1)^{i+j}$ is to change the sign of the minor when the sum $i + j$ is odd. There is a checkerboard pattern for remembering these sign changes.

$$+ \quad - \quad +$$
$$- \quad + \quad -$$
$$+ \quad - \quad +$$

In the upper left corner, $i = 1, j = 1$ and $(-1)^{1+1} = +1$. In going from any cell to an adjacent cell in the same row or column, we change i by 1 or j by 1, but not both, so we change the exponent from even to odd or from odd to even, which changes the sign from + to − or from − to + .

When we rewrite Eq. (6) in terms of cofactors we get

$$\det A = a_{11}A_{11} + a_{12}A_{12} + a_{13}A_{13}. \tag{8}$$

Example 2   Find the determinant of the matrix

$$A = \begin{bmatrix} 2 & 1 & 3 \\ 3 & -1 & -2 \\ 2 & 3 & 1 \end{bmatrix}.$$

Solution 1   The cofactors are

$$A_{11} = (-1)^{1+1} \begin{vmatrix} -1 & -2 \\ 3 & 1 \end{vmatrix}, \qquad A_{12} = (-1)^{1+2} \begin{vmatrix} 3 & -2 \\ 2 & 1 \end{vmatrix},$$

$$A_{13} = (-1)^{1+3} \begin{vmatrix} 3 & -1 \\ 2 & 3 \end{vmatrix}.$$

To find $\det A$, we multiply each element of the first row of A by its cofactor and add:

$$\det A = 2 \begin{vmatrix} -1 & -2 \\ 3 & 1 \end{vmatrix} + (-1) \begin{vmatrix} 3 & -2 \\ 2 & 1 \end{vmatrix} + 3 \begin{vmatrix} 3 & -1 \\ 2 & 3 \end{vmatrix}$$

$$= 2(-1 + 6) - 1(3 + 4) + 3(9 + 2) = 10 - 7 + 33 = 36.$$

Solution 2   From (7) we find

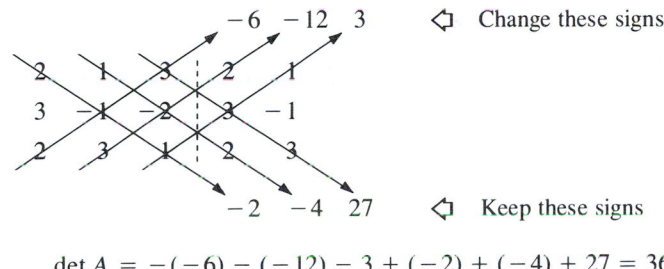

$$\det A = -(-6) - (-12) - 3 + (-2) + (-4) + 27 = 36$$

Expanding by Columns or by Other Rows

The determinant of a square matrix can be calculated from the cofactors of any row or any column.

If we were to expand the determinant in Example 2 by cofactors according to elements of its third column, say, we would get

$$+ 3 \begin{vmatrix} 3 & -1 \\ 2 & 3 \end{vmatrix} - (-2) \begin{vmatrix} 2 & 1 \\ 2 & 3 \end{vmatrix} + 1 \begin{vmatrix} 2 & 1 \\ 3 & -1 \end{vmatrix}$$

$$= 3(9 + 2) + 2(6 - 2) + 1(-2 - 3) = 33 + 8 - 5 = 36.$$

Useful Facts About Determinants

FACT 1: If two rows (or columns) of a matrix are identical, the determinant is zero.

FACT 2: Interchanging two rows (or columns) of a matrix changes the sign of its determinant.

FACT 3: The determinant of a matrix is the sum of the products of the elements of the ith row (or column) by their cofactors, for any i.

FACT 4: The determinant of the transpose of a matrix is the same as the determinant of the original matrix. (The "**transpose**" of a matrix is obtained by writing the rows as columns.)

FACT 5: Multiplying each element of some row (or column) of a matrix by a constant c multiplies the determinant by c.

FACT 6: If all elements of a matrix above the main diagonal (or all below it) are zero, the determinant of the matrix is the product of the elements on the main diagonal. (The **main diagonal** is the diagonal from upper left to lower right.)

Example 3

$$\begin{vmatrix} 3 & 4 & 7 \\ 0 & -2 & 5 \\ 0 & 0 & 5 \end{vmatrix} = (3)(-2)(5) = -30.$$

FACT 7: If the elements of any row of a matrix are multiplied by the cofactors of the corresponding elements of a different row and these products are summed, the sum is zero.

Example 4   If A_{11}, A_{12}, A_{13} are the cofactors of the elements of the first row of $A = (a_{ij})$, then the sums

$$a_{21}A_{11} + a_{22}A_{12} + a_{23}A_{13}$$

(elements of second row times cofactors of elements of first row) and

$$a_{31}A_{11} + a_{32}A_{12} + a_{33}A_{13}$$

are both zero.

FACT 8: If the elements of any column of a matrix are multiplied by the cofactors of the corresponding elements of a different column and these products are summed, the sum is zero.

FACT 9: If each element of a row of a matrix is multiplied by a constant c and the results added to a different row, the determinant is not changed. A similar result holds for columns.

Example 5   If we start with

$$A = \begin{bmatrix} 2 & 1 & 3 \\ 3 & -1 & -2 \\ 2 & 3 & 1 \end{bmatrix}$$

and add -2 times row 1 to row 2 (subtract 2 times row 1 from row 2) we get

$$B = \begin{bmatrix} 2 & 1 & 3 \\ -1 & -3 & -8 \\ 2 & 3 & 1 \end{bmatrix}.$$

Since det $A = 36$ (Example 2), we should find that det $B = 36$ as well. Indeed we do, as the following calculation shows:

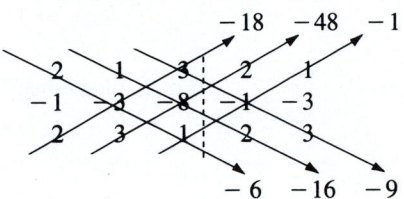

$$\det B = -(-18) - (-48) - (-1) + (-6) + (-16) + (-9)$$

$$= 18 + 48 + 1 - 6 - 16 - 9 = 67 - 31 = 36$$

Example 6   Evaluate the fourth order determinant

$$D = \begin{vmatrix} 1 & -2 & 3 & 1 \\ 2 & 1 & 0 & 2 \\ -1 & 2 & 1 & -2 \\ 0 & 1 & 2 & 1 \end{vmatrix}.$$

Solution   We subtract 2 times row 1 from row 2 and add row 1 to row 3 to get

$$D = \begin{vmatrix} 1 & -2 & 3 & 1 \\ 0 & 5 & -6 & 0 \\ 0 & 0 & 4 & -1 \\ 0 & 1 & 2 & 1 \end{vmatrix}.$$

We then multiply the elements of the first column by their cofactors to get

$$D = \begin{vmatrix} 5 & -6 & 0 \\ 0 & 4 & -1 \\ 1 & 2 & 1 \end{vmatrix} = 5(4 + 2) - (-6)(0 + 1) + 0 = 36.$$

Cramer's Rule

If the determinant

$$D = \det A = \begin{vmatrix} a_{11} & a_{12} \\ a_{21} & a_{22} \end{vmatrix}$$

of the coefficient matrix of the system

$$a_{11}x + a_{12}y = b_1,$$
$$a_{21}x + a_{22}y = b_2 \tag{9}$$

of linear equations is 0, the system has either infinitely many solutions or no solution at all. The system

$$x + y = 0,$$
$$2x + 2y = 0$$

whose determinant is

$$D = \begin{vmatrix} 1 & 1 \\ 2 & 2 \end{vmatrix} = 2 - 2 = 0$$

has infinitely many solutions. We can find an x to match any given y. The system

$$x + y = 0,$$
$$2x + 2y = 2$$

has no solution. If $x + y = 0$, then $2x + 2y = 2(x + y)$ cannot be 2.

If $D \neq 0$, the system (9) has a unique solution, and Cramer's rule states that it may be found from the formulas

$$x = \frac{\begin{vmatrix} b_1 & a_{12} \\ b_2 & a_{22} \end{vmatrix}}{D}, \qquad y = \frac{\begin{vmatrix} a_{11} & b_1 \\ a_{21} & b_2 \end{vmatrix}}{D}. \tag{10}$$

The numerator in the formula for x comes from replacing the first column in A (the x-column) by the column of constants b_1 and b_2 (the b-column). Replacing the y-column by the b-column gives the numerator of the y-solution.

Example 7   Solve the system

$$3x - y = 9,$$
$$x + 2y = -4.$$

Solution   We use Eqs. (10). The determinant of the coefficient matrix is

$$D = \begin{vmatrix} 3 & -1 \\ 1 & 2 \end{vmatrix} = 6 + 1 = 7.$$

Hence,

$$x = \frac{\begin{vmatrix} 9 & -1 \\ -4 & 2 \end{vmatrix}}{D} = \frac{18 - 4}{7} = \frac{14}{7} = 2,$$

$$y = \frac{\begin{vmatrix} 3 & 9 \\ 1 & -4 \end{vmatrix}}{D} = \frac{-12 - 9}{7} = \frac{-21}{7} = -3.$$

Systems of three equations in three unknowns work the same way. If the determinant

$$D = \det A = \begin{vmatrix} a_{11} & a_{12} & a_{13} \\ a_{21} & a_{22} & a_{23} \\ a_{31} & a_{32} & a_{33} \end{vmatrix}$$

of the system

$$a_{11}x + a_{12}y + a_{13}z = b_1,$$
$$a_{21}x + a_{22}y + a_{23}z = b_2, \tag{11}$$
$$a_{31}x + a_{32}y + a_{33}z = b_3$$

is zero, the system has either infinitely many solutions or no solution at all. If $D \neq 0$, the system has a unique solution, given by Cramer's rule:

$$x = \frac{1}{D}\begin{vmatrix} b_1 & a_{12} & a_{13} \\ b_2 & a_{22} & a_{23} \\ b_3 & a_{32} & a_{33} \end{vmatrix}, \qquad y = \frac{1}{D}\begin{vmatrix} a_{11} & b_1 & a_{13} \\ a_{21} & b_2 & a_{23} \\ a_{31} & b_3 & a_{33} \end{vmatrix}, \qquad z = \frac{1}{D}\begin{vmatrix} a_{11} & a_{12} & b_1 \\ a_{21} & a_{22} & b_2 \\ a_{31} & a_{32} & b_3 \end{vmatrix}.$$

The pattern continues in higher dimensions.

EXERCISES A.10

Evaluate the following determinants.

1. $\begin{vmatrix} 2 & 3 & 1 \\ 4 & 5 & 2 \\ 1 & 2 & 3 \end{vmatrix}$
2. $\begin{vmatrix} 2 & -1 & -2 \\ -1 & 2 & 1 \\ 3 & 0 & -3 \end{vmatrix}$

3. $\begin{vmatrix} 1 & 2 & 3 & 4 \\ 0 & 1 & 2 & 3 \\ 0 & 0 & 2 & 1 \\ 0 & 0 & 3 & 2 \end{vmatrix}$
4. $\begin{vmatrix} 1 & -1 & 2 & 3 \\ 2 & 1 & 2 & 6 \\ 1 & 0 & 2 & 3 \\ -2 & 2 & 0 & -5 \end{vmatrix}$

Evaluate the following determinants by expanding according to the cofactors of (a) the third row and (b) the second column.

5. $\begin{vmatrix} 2 & -1 & 2 \\ 1 & 0 & 3 \\ 0 & 2 & 1 \end{vmatrix}$
6. $\begin{vmatrix} 1 & 0 & -1 \\ 0 & 2 & -2 \\ 2 & 0 & 1 \end{vmatrix}$

7. $\begin{vmatrix} 1 & 1 & 0 & 0 \\ 0 & 0 & -2 & 1 \\ 0 & -1 & 0 & 7 \\ 3 & 0 & 2 & 1 \end{vmatrix}$
8. $\begin{vmatrix} 0 & 1 & 0 & 0 \\ 0 & 1 & 1 & 0 \\ 1 & 1 & 1 & 1 \\ 1 & 1 & 0 & 0 \end{vmatrix}$

Solve the following systems of equations by Cramer's rule.

9. $x + 8y = 4$
   $3x - y = -13$

10. $2x + 3y = 5$
    $3x - y = 2$

11. $4x - 3y = 6$
    $3x - 2y = 5$

12. $x + y + z = 2$
    $2x - y + z = 0$
    $x + 2y - z = 4$

13. $2x + y - z = 2$
    $x - y + z = 7$
    $2x + 2y + z = 4$

14. $2x - 4y = 6$
    $x + y + z = 1$
    $5y + 7z = 10$

15. $x - z = 3$
    $2y - 2z = 2$
    $2x + z = 3$

16. $x_1 + x_2 - x_3 + x_4 = 2$
    $x_1 - x_2 + x_3 + x_4 = -1$
    $x_1 + x_2 + x_3 - x_4 = 2$
    $x_1 + x_3 + x_4 = -1$

17. Find values of h and k for which the system
    $$2x + hy = 8,$$
    $$x + 3y = k$$
    has (a) infinitely many solutions, (b) no solution at all.

18. For what value of x will
    $$\begin{vmatrix} x & x & 1 \\ 2 & 0 & 5 \\ 6 & 7 & 1 \end{vmatrix} = 0?$$

19. Suppose u, v, and w are twice-differentiable functions of x that satisfy the relation $au + bv + cw = 0$, where a, b, and c are constants, not all zero. Show that
    $$\begin{vmatrix} u & v & w \\ u' & v' & w' \\ u'' & v'' & w'' \end{vmatrix} = 0.$$

20. *Partial fractions.* Expanding the quotient
    $$\frac{ax + b}{(x - r_1)(x - r_2)}$$
    by partial fractions calls for finding the values of C and D that make the equation
    $$\frac{ax + b}{(x - r_1)(x - r_2)} = \frac{C}{x - r_1} + \frac{D}{x - r_2}$$
    hold for all x.
    a) Find a system of linear equations that determines C and D.
    b) Under what circumstances does the system of equations in part (a) have a unique solution? That is, when is the determinant of the coefficient matrix of the system different from zero?

A.11
Euler's Theorem and the Increment Theorem

This appendix derives Euler's Mixed Derivative Theorem (Theorem 2, Section 12.3) and the Increment Theorem for Functions of Two Variables (Theorem 3, Section 12.4). Euler first published the Mixed Derivative Theorem in 1734, in one of a series of papers he wrote on hydrodynamics.

THEOREM 2

Euler's Mixed-Derivative Theorem

If $f(x, y)$ and its partial derivatives f_x, f_y, f_{xy}, and f_{yx} are defined throughout an open region containing a point (a, b) and are all continuous at (a, b), then

$$f_{xy}(a, b) = f_{yx}(a, b). \tag{1}$$

Proof The equality of $f_{xy}(a, b)$ and $f_{yx}(a, b)$ can be established by four applications of the Mean Value Theorem (Theorem 3, Section 3.2). By hypothesis, the point (a, b) lies in the interior of a rectangle R in the xy-plane on which f, f_x, f_y, f_{xy}, and f_{yx} are all defined. We let h and k be numbers such that the point $(a + h, b + k)$ also lies in the rectangle R, and we consider the difference

$$\Delta = F(a + h) - F(a), \tag{2}$$

where

$$F(x) = f(x, b + k) - f(x, b). \tag{3}$$

We apply the Mean Value Theorem to F (which is continuous because it is differentiable), and Eq. (2) becomes

$$\Delta = hF'(c_1), \tag{4}$$

where c_1 lies between a and $a + h$. From Eq. (3),

$$F'(x) = f_x(x, b + k) - f_x(x, b),$$

so Eq. (4) becomes

$$\Delta = h[f_x(c_1, b + k) - f_x(c_1, b)]. \tag{5}$$

Now we apply the Mean Value Theorem to the function $g(y) = f_x(c_1, y)$ and have

$$g(b + k) - g(b) = kg'(d_1), \tag{6}$$

or

$$f_x(c_1, b + k) - f_x(c_1, b) = kf_{xy}(c_1, d_1), \tag{7}$$

for some d_1 between b and $b + k$. By substituting this into Eq. (5), we get

$$\Delta = hkf_{xy}(c_1, d_1), \tag{8}$$

for some point (c_1, d_1) in the rectangle R' whose vertices are the four points (a, b), $(a + h, b)$, $(a + h, b + k)$, and $(a, b + k)$. (See Fig. A.22.)

By substituting from Eq. (3) into Eq. (2), we may also write

$$\begin{aligned}\Delta &= f(a + h, b + k) - f(a + h, b) - f(a, b + k) + f(a, b) \\ &= [f(a + h, b + k) - f(a, b + k)] - [f(a + h, b) - f(a, b)] \tag{9} \\ &= \phi(b + k) - \phi(b),\end{aligned}$$

where

$$\phi(y) = f(a + h, y) - f(a, y). \tag{10}$$

The Mean Value Theorem applied to Eq. (9) now gives

$$\Delta = k\phi'(d_2), \tag{11}$$

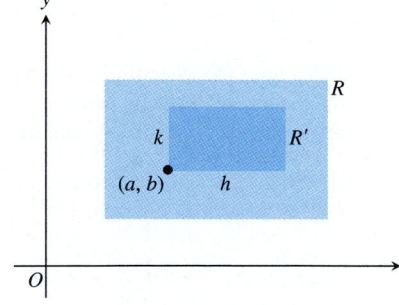

A.22 The key to proving $f_{xy}(a, b) = f_{yx}(a, b)$ is the fact that no matter how small R' is, f_{xy} and f_{yx} take on equal values somewhere inside R' (although not necessarily at the same point).

for some d_2 between b and $b + k$. By Eq. (10),

$$\phi'(y) = f_y(a + h, y) - f_y(a, y). \tag{12}$$

Substituting from Eq. (12) into Eq. (11) gives

$$\Delta = k[f_y(a + h, d_2) - f_y(a, d_2)]. \tag{13}$$

Finally, we apply the Mean Value Theorem to the expression in brackets and get

$$\Delta = khf_{yx}(c_2, d_2), \tag{14}$$

for some c_2 between a and $a + h$.

Together, Eqs. (8) and (14) show that

$$f_{xy}(c_1, d_1) = f_{yx}(c_2, d_2), \tag{15}$$

where (c_1, d_1) and (c_2, d_2) both lie in the rectangle R' (Fig. A.22). Equation (15) is not quite the result we want, since it says only that f_{xy} has the same value at (c_1, d_1) that f_{yx} has at (c_2, d_2). But the numbers h and k in our discussion may be made as small as we wish. The hypothesis that f_{xy} and f_{yx} are both continuous at (a, b) means that $f_{xy}(c_1, d_1) = f_{xy}(a, b) + \varepsilon_1$ and $f_{yx}(c_2, d_2) = f_{yx}(a, b) + \varepsilon_2$, where ε_1, $\varepsilon_2 \to 0$ as h, $k \to 0$. Hence, if we let h and $k \to 0$, we have $f_{xy}(a, b) = f_{yx}(a, b)$.   ∎

The equality of $f_{xy}(a, b)$ and $f_{yx}(a, b)$ can be proved with weaker hypotheses than the ones we assumed. For example, it is enough for f, f_x, and f_y to exist in R and for f_{xy} to be continuous at (a, b). Then f_{yx} will exist at (a, b) and will equal f_{xy} at that point.

THEOREM 3 (SECTION 12.4)

> **The Increment Theorem for Functions of Two Variables**
>
> Suppose that the first partial derivatives of $z = f(x, y)$ are defined throughout an open region R containing the point (x_0, y_0) and that f_x and f_y are continuous at (x_0, y_0). Then the change $\Delta z = f(x_0 + \Delta x, y_0 + \Delta y) - f(x_0, y_0)$ in the value of f that results from moving from (x_0, y_0) to another point $(x_0 + \Delta x, y_0 + \Delta y)$ in R satisfies an equation of the form
>
> $$\Delta z = f_x(x_0, y_0)\Delta x + f_y(x_0, y_0)\Delta y + \varepsilon_1\Delta x + \varepsilon_2\Delta y, \tag{16}$$
>
> in which ε_1, $\varepsilon_2 \to 0$ as Δx, $\Delta y \to 0$.

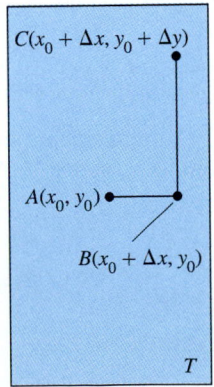

A.23 The rectangular region T in the proof of the Increment Theorem. The figure is drawn for Δx and Δy positive, but either increment might be zero or negative.

Proof   We work within a rectangle T centered at $A(x_0, y_0)$ and lying within R, and we assume that Δx and Δy are already so small that the line segment joining A to $B(x_0 + \Delta x, y_0)$ and the line segment joining B to $C(x_0 + \Delta x, y_0 + \Delta y)$ lie in the interior of T (Fig. A.23).

We may think of Δz as the sum $\Delta z = \Delta z_1 + \Delta z_2$ of two increments, where

$$\Delta z_1 = f(x_0 + \Delta x, y_0) - f(x_0, y_0) \tag{17}$$

is the change in the value of f from A to B and

$$\Delta z_2 = f(x_0 + \Delta x, y_0 + \Delta y) - f(x_0 + \Delta x, y_0) \tag{18}$$

is the change in the value of f from B to C (Fig. A.24).

A.24   Part of the surface $z = f(x, y)$ near $P_0(x_0, y_0, f(x_0, y_0))$. The points P_0, P', and P'' have the same height $z_0 = f(x_0, y_0)$ above the xy-plane. The change in z is $\Delta z = P'S$. The change

$$\Delta z_1 = f(x_0 + \Delta x, y_0) - f(x_0, y_0),$$

shown as $P''Q = P'Q'$, is caused by changing x from x_0 to $x_0 + \Delta x$ while holding y equal to y_0. Then, with x held equal to $x_0 + \Delta x$,

$$\Delta z_2 = f(x_0 + \Delta x, y_0 + \Delta y) \\ - f(x_0 + \Delta x, y_0)$$

is the change in z caused by changing y from y_0 to $y_0 + \Delta y$. This is represented by $Q'S$. The total change in z is the sum of Δz_1 and Δz_2.

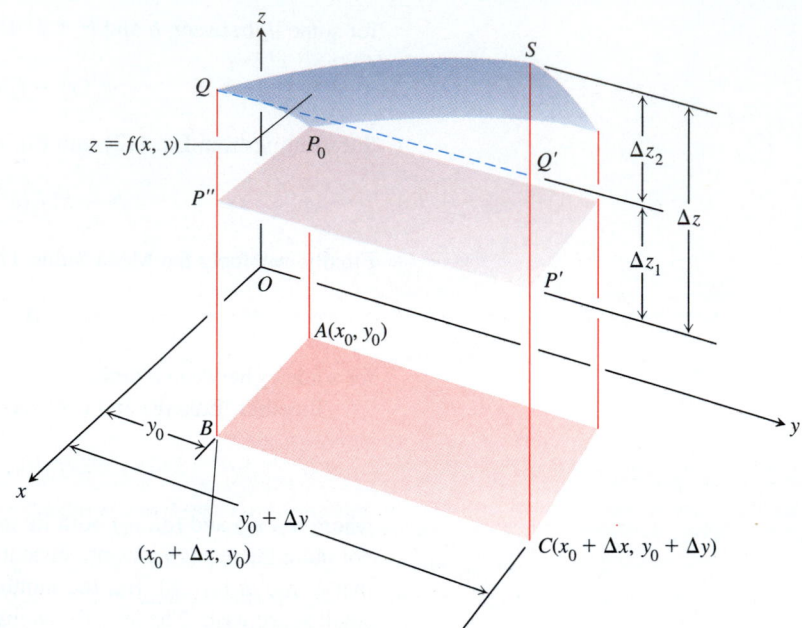

On the closed interval of x-values joining x_0 to $x_0 + \Delta x$, the function $F(x) = f(x, x_0)$ is a differentiable (and hence continuous) function of x, with derivative

$$F'(x) = f_x(x, y_0).$$

By the Mean Value Theorem (Theorem 3, Section 3.2), there is an x-value c between x_0 and $x_0 + \Delta x$ at which

$$F(x_0 + \Delta x) - F(x_0) = F'(c)\Delta x$$

or

$$f(x_0 + \Delta x, y_0) - f(x_0, y_0) = f_x(c, y_0)\Delta x$$

or

$$\Delta z_1 = f_x(c, y_0)\Delta x. \tag{19}$$

Similarly, the function $G(y) = f(x_0 + \Delta x, y)$ is a differentiable (and hence continuous) function of y on the closed y-interval joining y_0 and $y_0 + \Delta y$, with derivative

$$G'(y) = f_y(x_0 + \Delta x, y).$$

Hence there is a y-value d between y_0 and $y_0 + \Delta y$ at which

$$G(y_0 + \Delta y) - G(y_0) = G'(d)\Delta y$$

or

$$f(x_0 + \Delta x, y_0 + \Delta y) - f(x_0 + \Delta x, y) = f_y(x_0 + \Delta x, d)\Delta y$$

or

$$\Delta z_2 = f_y(x_0 + \Delta x, d)\Delta y. \tag{20}$$

Now, as Δx and $\Delta y \to 0$, we know $c \to x_0$ and $d \to y_0$. Therefore, since f_x and f_y are continuous at (x_0, y_0), the quantities

$$\varepsilon_1 = f_x(c, y_0) - f_x(x_0, y_0),$$

$$\varepsilon_2 = f_y(x_0 + \Delta x, d) - f_y(x_0, y_0) \tag{21}$$

both approach zero as Δx and $\Delta y \to 0$.

Finally,

$$\Delta z = \Delta z_1 + \Delta z_2$$

$$= f_x(c, y_0)\, \Delta x + f_y(x_0 + \Delta x, d)\, \Delta y \qquad \text{(From (19) and (20))}$$

$$= [f_x(x_0, y_0) + \varepsilon_1]\, \Delta x + [f_y(x_0, y_0) + \varepsilon_2]\, \Delta y \qquad \text{(From (21))}$$

$$= f_x(x_0, y_0)\, \Delta x + f_y(x_0, y_0)\, \Delta y + \varepsilon_1\, \Delta x + \varepsilon_2\, \Delta y,$$

where ε_1 and $\varepsilon_2 \to 0$ as Δx and $\Delta y \to 0$. This is what we set out to prove.   ◼

Analogous results hold for functions of any finite number of independent variables. Suppose that the first partial derivatives of

$$w = f(x, y, z)$$

are defined throughout an open region containing the point (x_0, y_0, z_0) and that f_x, f_y, and f_z are continuous at (x_0, y_0). Then

$$\Delta w = f(x_0 + \Delta x, y_0 + \Delta y, z_0 + \Delta z) - f(x_0, y_0, z_0)$$

$$= f_x\, \Delta x + f_y\, \Delta y + f_z\, \Delta z + \varepsilon_1\, \Delta x + \varepsilon_2\, \Delta y + \varepsilon_3\, \Delta z, \tag{22}$$

where

$$\varepsilon_1, \varepsilon_2, \varepsilon_3 \to 0 \qquad \text{when} \qquad \Delta x, \Delta y, \text{ and } \Delta z \to 0.$$

The partial derivatives f_x, f_y, f_z in this formula are to be evaluated at the point (x_0, y_0, z_0).

The result (22) can be proved by treating Δw as the sum of three increments,

$$\Delta w_1 = f(x_0 + \Delta x, y_0, z_0) - f(x_0, y_0, z_0), \tag{23}$$

$$\Delta w_2 = f(x_0 + \Delta x, y_0 + \Delta y, z_0) - f(x_0 + \Delta x, y_0, z_0), \tag{24}$$

$$\Delta w_3 = f(x_0 + \Delta x, y_0 + \Delta y, z_0 + \Delta z) - f(x_0 + \Delta x, y_0 + \Delta y, z_0), \tag{25}$$

and applying the Mean Value Theorem to each of these separately. Two coordinates remain constant and only one varies in each of these partial increments $\Delta w_1, \Delta w_2, \Delta w_3$. In (24), for example, only y varies, since x is held equal to $x_0 + \Delta x$ and z is held equal to z_0. Since the function $f(x_0 + \Delta x, y, z_0)$ is a continuous function of y with a derivative f_y, it is subject to the Mean Value Theorem, and we have

$$\Delta w_2 = f_y(x_0 + \Delta x, y_1, z_0)\, \Delta y$$

for some y_1 between y_0 and $y_0 + \Delta y$.

A.12
Numerical Tables for sin x, cos x, tan x, e^x, e^{-x}, and ln x

TABLE A.2
Natural trigonometric functions

Angle Degree	Angle Radian	Sine	Cosine	Tangent	Angle Degree	Angle Radian	Sine	Cosine	Tangent
0°	0.000	0.000	1.000	0.000					
1°	0.017	0.017	1.000	0.017	46°	0.803	0.719	0.695	1.036
2°	0.035	0.035	0.999	0.035	47°	0.820	0.731	0.682	1.072
3°	0.052	0.052	0.999	0.052	48°	0.838	0.743	0.669	1.111
4°	0.070	0.070	0.998	0.070	49°	0.855	0.755	0.656	1.150
5°	0.087	0.087	0.996	0.087	50°	0.873	0.766	0.643	1.192
6°	0.105	0.105	0.995	0.105	51°	0.890	0.777	0.629	1.235
7°	0.122	0.122	0.993	0.123	52°	0.908	0.788	0.616	1.280
8°	0.140	0.139	0.990	0.141	53°	0.925	0.799	0.602	1.327
9°	0.157	0.156	0.988	0.158	54°	0.942	0.809	0.588	1.376
10°	0.175	0.174	0.985	0.176	55°	0.960	0.819	0.574	1.428
11°	0.192	0.191	0.982	0.194	56°	0.977	0.829	0.559	1.483
12°	0.209	0.208	0.978	0.213	57°	0.995	0.839	0.545	1.540
13°	0.227	0.225	0.974	0.231	58°	1.012	0.848	0.530	1.600
14°	0.244	0.242	0.970	0.249	59°	1.030	0.857	0.515	1.664
15°	0.262	0.259	0.966	0.268	60°	1.047	0.866	0.500	1.732
16°	0.279	0.276	0.961	0.287	61°	1.065	0.875	0.485	1.804
17°	0.297	0.292	0.956	0.306	62°	1.082	0.883	0.469	1.881
18°	0.314	0.309	0.951	0.325	63°	1.100	0.891	0.454	1.963
19°	0.332	0.326	0.946	0.344	64°	1.117	0.899	0.438	2.050
20°	0.349	0.342	0.940	0.364	65°	1.134	0.906	0.423	2.145
21°	0.367	0.358	0.934	0.384	66°	1.152	0.914	0.407	2.246
22°	0.384	0.375	0.927	0.404	67°	1.169	0.921	0.391	2.356
23°	0.401	0.391	0.921	0.424	68°	1.187	0.927	0.375	2.475
24°	0.419	0.407	0.914	0.445	69°	1.204	0.934	0.358	2.605
25°	0.436	0.423	0.906	0.466	70°	1.222	0.940	0.342	2.748
26°	0.454	0.438	0.899	0.488	71°	1.239	0.946	0.326	2.904
27°	0.471	0.454	0.891	0.510	72°	1.257	0.951	0.309	3.078
28°	0.489	0.469	0.883	0.532	73°	1.274	0.956	0.292	3.271
29°	0.506	0.485	0.875	0.554	74°	1.292	0.961	0.276	3.487
30°	0.524	0.500	0.866	0.577	75°	1.309	0.966	0.259	3.732
31°	0.541	0.515	0.857	0.601	76°	1.326	0.970	0.242	4.011
32°	0.559	0.530	0.848	0.625	77°	1.344	0.974	0.225	4.332
33°	0.576	0.545	0.839	0.649	78°	1.361	0.978	0.208	4.705
34°	0.593	0.559	0.829	0.675	79°	1.379	0.982	0.191	5.145
35°	0.611	0.574	0.819	0.700	80°	1.396	0.985	0.174	5.671
36°	0.628	0.588	0.809	0.727	81°	1.414	0.988	0.156	6.314
37°	0.646	0.602	0.799	0.754	82°	1.431	0.990	0.139	7.115
38°	0.663	0.616	0.788	0.781	83°	1.449	0.993	0.122	8.144
39°	0.681	0.629	0.777	0.810	84°	1.466	0.995	0.105	9.514
40°	0.698	0.643	0.766	0.839	85°	1.484	0.996	0.087	11.43
41°	0.716	0.656	0.755	0.869	86°	1.501	0.998	0.070	14.30
42°	0.733	0.669	0.743	0.900	87°	1.518	0.999	0.052	19.08
43°	0.750	0.682	0.731	0.933	88°	1.536	0.999	0.035	28.64
44°	0.768	0.695	0.719	0.966	89°	1.553	1.000	0.017	57.29
45°	0.785	0.707	0.707	1.000	90°	1.571	1.000	0.000	

TABLE A.3
Exponential functions

x	e^x	e^{-x}	x	e^x	e^{-x}
0.00	1.0000	1.0000	2.5	12.182	0.0821
0.05	1.0513	0.9512	2.6	13.464	0.0743
0.10	1.1052	0.9048	2.7	14.880	0.0672
0.15	1.1618	0.8607	2.8	16.445	0.0608
0.20	1.2214	0.8187	2.9	18.174	0.0550
0.25	1.2840	0.7788	3.0	20.086	0.0498
0.30	1.3499	0.7408	3.1	22.198	0.0450
0.35	1.4191	0.7047	3.2	24.533	0.0408
0.40	1.4918	0.6703	3.3	27.113	0.0369
0.45	1.5683	0.6376	3.4	29.964	0.0334
0.50	1.6487	0.6065	3.5	33.115	0.0302
0.55	1.7333	0.5769	3.6	36.598	0.0273
0.60	1.8221	0.5488	3.7	40.447	0.0247
0.65	1.9155	0.5220	3.8	44.701	0.0224
0.70	2.0138	0.4966	3.9	49.402	0.0202
0.75	2.1170	0.4724	4.0	54.598	0.0183
0.80	2.2255	0.4493	4.1	60.340	0.0166
0.85	2.3396	0.4274	4.2	66.686	0.0150
0.90	2.4596	0.4066	4.3	73.700	0.0136
0.95	2.5857	0.3867	4.4	81.451	0.0123
1.0	2.7183	0.3679	4.5	90.017	0.0111
1.1	3.0042	0.3329	4.6	99.484	0.0101
1.2	3.3201	0.3012	4.7	109.95	0.0091
1.3	3.6693	0.2725	4.8	121.51	0.0082
1.4	4.0552	0.2466	4.9	134.29	0.0074
1.5	4.4817	0.2231	5	148.41	0.0067
1.6	4.9530	0.2019	6	403.43	0.0025
1.7	5.4739	0.1827	7	1096.6	0.0009
1.8	6.0496	0.1653	8	2981.0	0.0003
1.9	6.6859	0.1496	9	8103.1	0.0001
2.0	7.3891	0.1353	10	22026	0.00005
2.1	8.1662	0.1225			
2.2	9.0250	0.1108			
2.3	9.9742	0.1003			
2.4	11.023	0.0907			

TABLE A.4
Natural logarithms

x	$\log_e x$	x	$\log_e x$	x	$\log_e x$
0.0	*	4.5	1.5041	9.0	2.1972
0.1	7.6974	4.6	1.5261	9.1	2.2083
0.2	8.3906	4.7	1.5476	9.2	2.2192
0.3	8.7960	4.8	1.5686	9.3	2.2300
0.4	9.0837	4.9	1.5892	9.4	2.2407
0.5	9.3069	5.0	1.6094	9.5	2.2513
0.6	9.4892	5.1	1.6292	9.6	2.2618
0.7	9.6433	5.2	1.6487	9.7	2.2721
0.8	9.7769	5.3	1.6677	9.8	2.2824
0.9	9.8946	5.4	1.6864	9.9	2.2925
1.0	0.0000	5.5	1.7047	10	2.3026
1.1	0.0953	5.6	1.7228	11	2.3979
1.2	0.1823	5.7	1.7405	12	2.4849
1.3	0.2624	5.8	1.7579	13	2.5649
1.4	0.3365	5.9	1.7750	14	2.6391
1.5	0.4055	6.0	1.7918	15	2.7081
1.6	0.4700	6.1	1.8083	16	2.7726
1.7	0.5306	6.2	1.8245	17	2.8332
1.8	0.5878	6.3	1.8405	18	2.8904
1.9	0.6419	6.4	1.8563	19	2.9444
2.0	0.6931	6.5	1.8718	20	2.9957
2.1	0.7419	6.6	1.8871	25	3.2189
2.2	0.7885	6.7	1.9021	30	3.4012
2.3	0.8329	6.8	1.9169	35	3.5553
2.4	0.8755	6.9	1.9315	40	3.6889
2.5	0.9163	7.0	1.9459	45	3.8067
2.6	0.9555	7.1	1.9601	50	3.9120
2.7	0.9933	7.2	1.9741	55	4.0073
2.8	1.0296	7.3	1.9879	60	4.0943
2.9	1.0647	7.4	2.0015	65	4.1744
3.0	1.0986	7.5	2.0149	70	4.2485
3.1	1.1314	7.6	2.0281	75	4.3175
3.2	1.1632	7.7	2.0412	80	4.3820
3.3	1.1939	7.8	2.0541	85	4.4427
3.4	1.2238	7.9	2.0669	90	4.4998
3.5	1.2528	8.0	2.0794	95	4.5539
3.6	1.2809	8.1	2.0919	100	4.6052
3.7	1.3083	8.2	2.1041		
3.8	1.3350	8.3	2.1163		
3.9	1.3610	8.4	2.1282		
4.0	1.3863	8.5	2.1401		
4.1	1.4110	8.6	2.1518		
4.2	1.4351	8.7	2.1633		
4.3	1.4586	8.8	2.1748		
4.4	1.4816	8.9	2.1861		

*Subtract 10 from $\log_e x$ entries for $x < 1.0$.

A.13
Vector Operator Formulas in Cartesian, Cylindrical, and Spherical Coordinates; Vector Identities

Formulas for Grad, Div, Curl, and the Laplacian

	Cartesian (x, y, z)	Cylindrical (r, θ, z)	Spherical (ρ, ϕ, θ)
	$\mathbf{i}$, $\mathbf{j}$, and $\mathbf{k}$ are unit vectors in the directions of increasing x, y, and z.	$\mathbf{u}_r$, $\mathbf{u}_\theta$, and $\mathbf{k}$ are unit vectors in the directions of increasing r, θ, and z.	$\mathbf{u}_\rho$, and $\mathbf{u}_\phi$, and $\mathbf{u}_\theta$ are unit vectors in the directions of increasing ρ, ϕ, and θ.
	F_x, F_y, and F_z are the scalar components of $\mathbf{F}(x, y, z)$ in these directions.	F_r, F_θ, and F_z are the scalar components of $\mathbf{F}(r, \theta, z)$ in these directions.	F_ρ, F_ϕ, and F_θ are the scalar components of $\mathbf{F}(\rho, \phi, \theta)$ in these directions.
Gradient	$\nabla f = \dfrac{\partial f}{\partial x}\mathbf{i} + \dfrac{\partial f}{\partial y}\mathbf{j} + \dfrac{\partial f}{\partial z}\mathbf{k}$	$\nabla f = \dfrac{\partial f}{\partial r}\mathbf{u}_r + \dfrac{1}{r}\dfrac{\partial f}{\partial \theta}\mathbf{u}_\theta + \dfrac{\partial f}{\partial z}\mathbf{k}$	$\nabla f = \dfrac{\partial f}{\partial \rho}\mathbf{u}_\rho + \dfrac{1}{\rho}\dfrac{\partial f}{\partial \phi}\mathbf{u}_\phi + \dfrac{1}{\rho \sin \phi}\dfrac{\partial f}{\partial \theta}\mathbf{u}_\theta$
Divergence	$\nabla \cdot \mathbf{F} = \dfrac{\partial F_x}{\partial x} + \dfrac{\partial F_y}{\partial y} + \dfrac{\partial F_z}{\partial z}$	$\nabla \cdot \mathbf{F} = \dfrac{1}{r}\dfrac{\partial}{\partial r}(rF_r) + \dfrac{1}{r}\dfrac{\partial F_\theta}{\partial \theta} + \dfrac{\partial F_z}{\partial z}$	$\nabla \cdot \mathbf{F} = \dfrac{1}{\rho^2}\dfrac{\partial}{\partial \rho}(\rho^2 F_\rho)$ $+ \dfrac{1}{\rho \sin \phi}\dfrac{\partial}{\partial \phi}(F_\phi \sin \phi) + \dfrac{1}{\rho \sin \phi}\dfrac{\partial F_\theta}{\partial \theta}$
Curl	$\nabla \times \mathbf{F} = \begin{vmatrix} \mathbf{i} & \mathbf{j} & \mathbf{k} \\ \dfrac{\partial}{\partial x} & \dfrac{\partial}{\partial y} & \dfrac{\partial}{\partial z} \\ F_x & F_y & F_z \end{vmatrix}$	$\nabla \times \mathbf{F} = \begin{vmatrix} \dfrac{1}{r}\mathbf{u}_r & \mathbf{u}_\theta & \dfrac{1}{r}\mathbf{k} \\ \dfrac{\partial}{\partial r} & \dfrac{\partial}{\partial \theta} & \dfrac{\partial}{\partial z} \\ F_r & F_\theta & F_z \end{vmatrix}$	$\nabla \times \mathbf{F} = \begin{vmatrix} \dfrac{\mathbf{u}_\rho}{\rho^2 \sin \phi} & \dfrac{\mathbf{u}_\phi}{\rho \sin \phi} & \dfrac{\mathbf{u}_\theta}{\rho} \\ \dfrac{\partial}{\partial \rho} & \dfrac{\partial}{\partial \phi} & \dfrac{\partial}{\partial \theta} \\ F_\rho & \rho F_\phi & \rho \sin \phi\, F_\theta \end{vmatrix}$
Laplacian	$\nabla^2 f = \dfrac{\partial^2 f}{\partial x^2} + \dfrac{\partial^2 f}{\partial y^2} + \dfrac{\partial^2 f}{\partial z^2}$	$\nabla^2 f = \dfrac{1}{r}\dfrac{\partial}{\partial r}\left(r\dfrac{\partial f}{\partial r}\right) + \dfrac{1}{r^2}\dfrac{\partial^2 f}{\partial \theta^2} + \dfrac{\partial^2 f}{\partial z^2}$	$\nabla^2 f = \dfrac{1}{\rho^2}\dfrac{\partial}{\partial \rho}\left(\rho^2\dfrac{\partial f}{\partial \rho}\right)$ $+ \dfrac{1}{\rho^2 \sin \phi}\dfrac{\partial}{\partial \phi}\left(\sin \phi\dfrac{\partial f}{\partial \phi}\right) + \dfrac{1}{\rho^2 \sin^2\phi}\dfrac{\partial^2 f}{\partial \theta^2}$

Vector Triple Products

$$(\mathbf{A} \times \mathbf{B}) \cdot \mathbf{C} = (\mathbf{B} \times \mathbf{C}) \cdot \mathbf{A} = (\mathbf{C} \times \mathbf{A}) \cdot \mathbf{B}$$

$$\mathbf{A} \times (\mathbf{B} \times \mathbf{C}) = (\mathbf{A} \cdot \mathbf{C})\mathbf{B} - (\mathbf{A} \cdot \mathbf{B})\mathbf{C}$$

Vector Identities for the Cartesian Form of the Operator ∇

In the identities listed here, $f(x, y, z)$ and $g(x, y, z)$ are differentiable scalar functions and $\mathbf{u}(x, y, z)$ and $\mathbf{v}(x, y, z)$ are differentiable vector functions.

$$\nabla \cdot f\mathbf{v} = f\nabla \cdot \mathbf{v} + \mathbf{v} \cdot \nabla f = f\nabla \cdot \mathbf{v} + (\mathbf{v} \cdot \nabla)f$$

$$\nabla \times f\mathbf{v} = f\nabla \times \mathbf{v} + \nabla f \times \mathbf{v}$$

$$\nabla \cdot (\nabla \times \mathbf{v}) = 0$$

$$\nabla \times (\nabla f) = \mathbf{0}$$

$$\nabla (fg) = f\nabla g + g\nabla f$$

$$\nabla (\mathbf{u} \cdot \mathbf{v}) = (\mathbf{u} \cdot \nabla)\mathbf{v} + (\mathbf{v} \cdot \nabla)\mathbf{u} + \mathbf{u} \times (\nabla \times \mathbf{v}) + \mathbf{v} \times (\nabla \times \mathbf{u})$$

$$\nabla \cdot (\mathbf{u} \times \mathbf{v}) = \mathbf{v} \cdot (\nabla \times \mathbf{u}) - \mathbf{u} \cdot (\nabla \times \mathbf{v})$$

$$\nabla \times (\mathbf{u} \times \mathbf{v}) = (\mathbf{v} \cdot \nabla)\mathbf{u} - (\mathbf{u} \cdot \nabla)\mathbf{v} + \mathbf{u}(\nabla \cdot \mathbf{v}) - \mathbf{v}(\nabla \cdot \mathbf{u})$$

$$\nabla \times (\nabla \times \mathbf{v}) = \nabla(\nabla \cdot \mathbf{v}) - (\nabla \cdot \nabla)\mathbf{v} = \nabla(\nabla \cdot \mathbf{v}) - \nabla^2 \mathbf{v}$$

$$(\nabla \times \mathbf{v}) \times \mathbf{v} = (\mathbf{v} \cdot \nabla)\mathbf{v} - \frac{1}{2}\nabla(\mathbf{v} \cdot \mathbf{v})$$

ANSWERS

Chapter 1

Section 1.1, pp. 10–12

1. $2, -4$ **3.** $-4.9, 0$ **5.** $m = 3, m_\perp = -\dfrac{1}{3}$ **7.** $m = 0, m_\perp$ does

not exist **9.** a) -1 b) $\dfrac{4}{3}$ **11.** a) 0 b) $-\sqrt{2}$ **13.** $y = x + 2$

15. $y = 2x + b$ **17.** $y = 1$ **19.** $y = \dfrac{3}{2}x$ **21.** $y = x + \sqrt{2}$

23. $y = -5x + 2.5$

25. **27.**

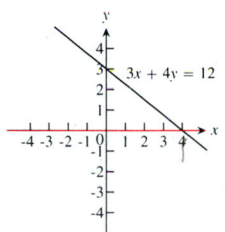

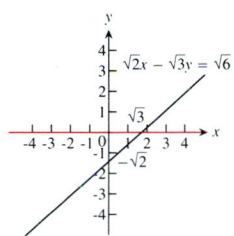

29. a) $y = x - 1$ b) $y = -x + 3$ **31.** a) $y = -\dfrac{1}{2}x + \dfrac{5}{2}$

b) $y = 2x$ **33.** a) $x = -2$ b) $y = 4$ **35.** $(3, -3)$

37. $(-2, -9)$ **39.** a) $-2.5°/\text{in.}$ b) $-16.1°/\text{in.}$ c) $-8.3°/\text{in.}$

41. 5.97 atm **43.** a) b) $F = \dfrac{9}{5}C + 32$

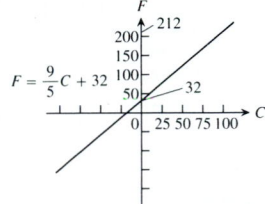

c) $-40°$ **45.** a) $(-4, -2)$ and $(3, -7)$ b) 35 **47.** $b = 2$

49. $D_1 = (-1, 4), D_2 = (-1, -2), D_3 = (5, 2)$ **51.** $k = -2$

Section 1.2, pp. 22–24

1. D: All reals; $R: y \geq 1$ **3.** $D: -x \geq 0 \Rightarrow x \leq 0; R: y \geq 0$

5. D: All reals except $x = \dfrac{(2n + 1)\pi}{2}$, n an integer; $R: y \geq 0$

7. D: All reals; $R: y = -1, 1, 0$ **9.** Odd **11.** Odd

13. Neither **15.** Even **17.** Even **19.** Odd

21. $D: -\infty < x < \infty; R: y \geq 0$; symmetric to y-axis

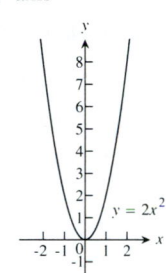

23. $D: -\infty < x < \infty; R: y \geq -9$; symmetric to y-axis

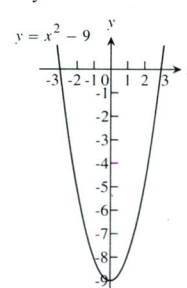

25. D: All reals; R: All reals; symmetric to origin

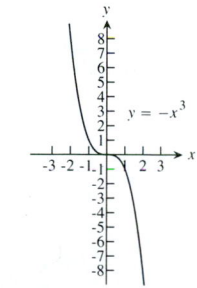

27. $D: x \neq 0; R: y \neq 0$; symmetric to origin

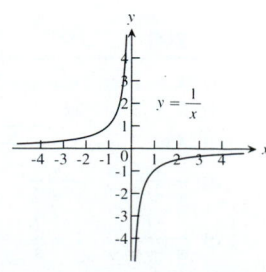

29. Symmetric to origin

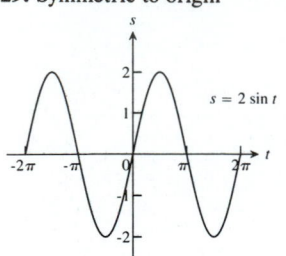

31. Symmetric to s-axis

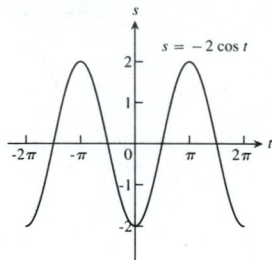

33. a) No b) No c) $D: s > 0$ **35.** a) i b) iv   **37.** a) $0 \le x < 1$
b) $-1 < x \le 0$ **39.** a)

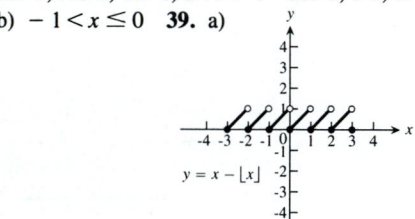

b)

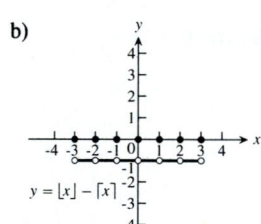

41.

x	0	1	2
y	0	1	0

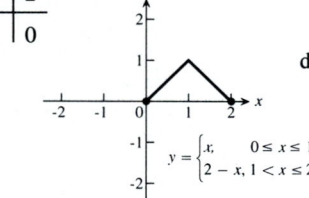

43.

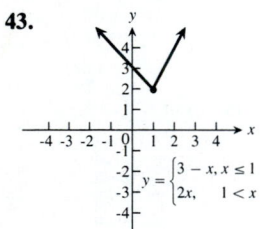

45.

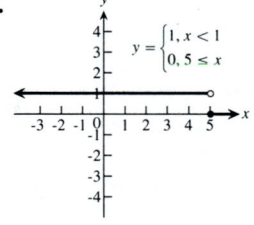

47. a) $y = \begin{cases} x, & 0 \le x \le 1 \\ 2 - x, & 1 < x \le 2 \end{cases}$ b) $y = \begin{cases} 2, & 0 \le x < 1 \\ 0, & 1 \le x < 2 \\ 2, & 2 \le x < 3 \\ 0, & 3 \le x \le 4 \end{cases}$

49. $D_{f+g} = D_{f-g} = D_{fg} = D_{g/f}: x \ge 1; D_{f/g}: x > 1$ **51.** a) 2 b) 22
c) $x^2 + 2$ d) $x^2 + 10x + 22$ e) 5 f) -2 g) $x + 10$
h) $x^4 - 6x^2 + 6$
53.

$g(x)$	$f(x)$	$(f \circ g)(x)$
a) $x - 7$	$\sqrt{x}$	$\sqrt{x - 7}$
b) $x + 2$	$3x$	$3x + 6$
c) x^2	$\sqrt{x - 5}$	$\sqrt{x^2 - 5}$
d) $\dfrac{x}{x - 1}$	$\dfrac{x}{x - 1}$	x
e) $\dfrac{1}{x - 1}$	$1 + \dfrac{1}{x}$	x
f) $\dfrac{1}{x}$	$\dfrac{1}{x}$	x

57.

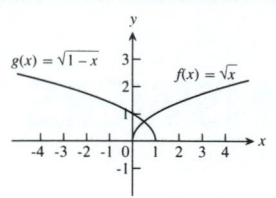

a)

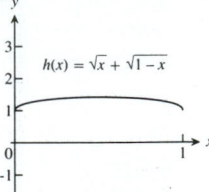

b)

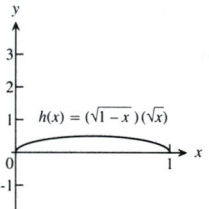

c)

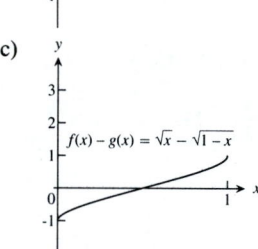   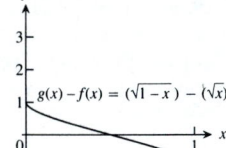

d)

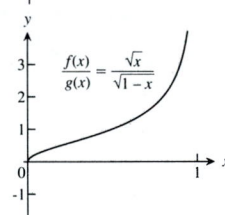   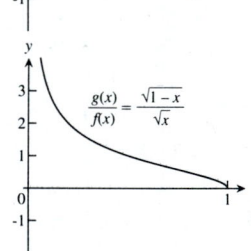

59. Graphs intersect at $x = \dfrac{(2n + 1)\pi}{4}$, n an integer; $\tan x = 0$
when $\cot x$ is undefined; $\cot x = 0$ when $\tan x$ is undefined.

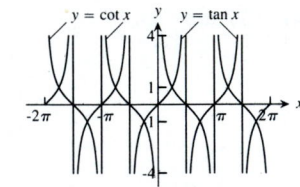

Section 1.3, pp. 29–31

1. a) Answers vary b) $e^{-1} = 0.367879441$, $e^{-10} = 0.000045399$,
$e^{-100} = 3.7200759 \times 10^{-44}$, $e^{-1000} = 0$ (Answers vary)
3. Answers vary   **5.** $y = x$ **7.** $c = 1$ if $x \ne -1$ **9.** 2 for every
$x \ne \dfrac{n\pi}{2}$, n an odd integer   **11.** Constant for $x > 0$

13. a) $\sqrt{3} = 1.732050808$, $\sqrt{1.732050808} = 1.316074013$,
$\sqrt{1.316074013} = 1.14720269$, etc. b) $\sqrt{5} = 2.236067978$,
$\sqrt{2.236067978} = 1.495348781$, $\sqrt{1.495348781} = 1.222844545$,
etc.   **15.** $\sqrt[10]{2} = 1.071773463$; $\sqrt[10]{1.071773463} = 1.00695555$,
$\sqrt[10]{1.00695555} = 1.000693387$, $\sqrt[10]{1.000693387} = 1.000069317$,
etc.   **17.** $x_0 = 1$, $x_1 = 1.540302306$, $x_2 = 1.570791601$,
$x_3 = 1.570796327$, $x_4 = 1.570796327$, etc.

19.

x	x^x
0.1	0.7943282
0.01	0.9549925
0.001	0.993116
0.0001	0.9990793
0.00001	0.9998848
0.000001	0.9999861

21. $m_{sec} = \dfrac{\ln(1 + \Delta x)}{\Delta x}$ Slope = 1 (See table)

Δx	m_{sec}
0.1	0.9531018
0.01	0.995033
0.001	0.9995003
0.0001	0.99995
0.00001	0.999995
0.000001	0.9999995
0.0000001	0.9999999

Section 1.4, pp. 37–38

1. $\sqrt{2}$ **3.** 6 **5.** $\sqrt{a^2 + b^2}$ **7.** 3 **9.** 5 **11.** a) False b) True c) True d) True e) True f) True g) True h) True **13.** ± 2

15. $-\dfrac{1}{2}, -\dfrac{9}{2}$ **17.** $-\dfrac{1}{3}, \dfrac{17}{3}$ **19.** e **21.** b **23.** b

25. $-2 < y < 2$ **27.** $-1 \le r \le 3$ **29.** $\dfrac{5}{3} < s < 3$

31. $0 \le t \le 4$ **33.** $|x - 6| < 3$ **35.** $|x + 1| < 4$
37. $\sqrt{99.9} < x < \sqrt{100.1}$ **39.** $22.21 < x < 23.81$
41. $30 > x > 20$ **43.** $|x - 3| < 0.5$ **45.** $|x - 6| < 1$
47. $|x - 3.385| < 0.002$ **49.** a) a any negative real b) $a \ge 0$
51. $D: -\infty < x < \infty$ and $R: y \ge 0$ for $y = \sqrt{x^2}$; $D: x \ge 0$ and $R:$ $y \ge 0$ for $y = (\sqrt{x})^2$ **53.** $g(x) = \sqrt{x}$

Section 1.5, pp. 46–48

1. a) $y = (x + 4)^2$ b) $y = (x - 7)^2$ **3.** a) Position 4 b) Position 1 c) Position 2 d) Position 3
5.   **7.**

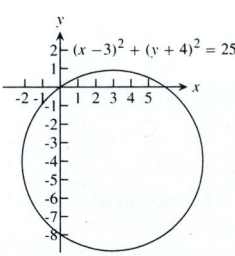

9. $x^2 + y^2 = 4$ **11.** $(x - 3)^2 + (y - 3)^2 = 9$ **13.** a) Exterior of circle with center (0, 0) and radius = 1 b) Interior of circle with center (0, 0) and radius = 2 c) Interior of concentric ring centered at (0, 0) with interior radius = 1 and exterior radius = 2

15. $(x + 2)^2 + (y + 1)^2 < 6$ **17.** $y = \dfrac{1}{16}x^2$ **19.** $y = -\dfrac{1}{12}x^2$

21.

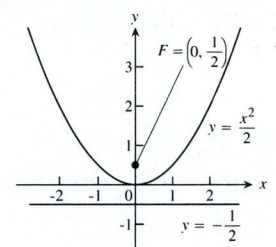

23.

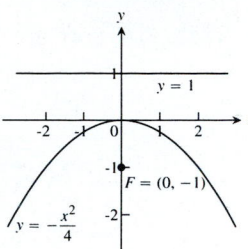

25. $y + 3 = (x + 2)^2$ **27.** $y = x^2$ **29.** $y = \sqrt{x + 4}$
31. $y = \dfrac{1}{2}x$ **33.** $y = \sqrt{-(x - 9)}$

35.

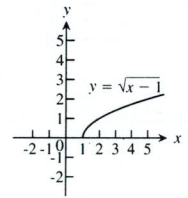

37.

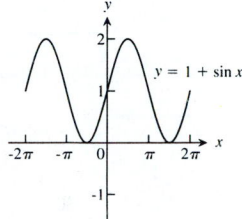

39.

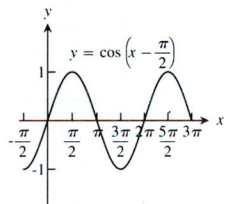

41. a) iii b) ii c) i d) iv

43. $y - y_0 = m(x - x_0)$

Section 1.6, pp. 56–58

1.

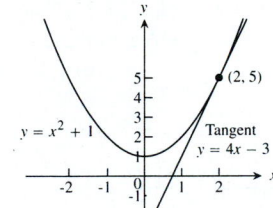

3.

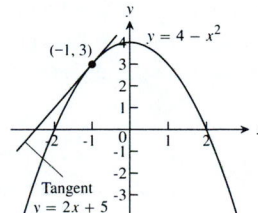

5.

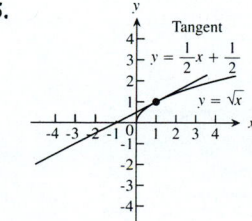

7. $f'(x) = 2x$, $m = 2$ **9.** $f'(x) = \dfrac{-2}{x^2}$, $m = -\dfrac{1}{2}$

11. $f'(x) = 1 - \dfrac{9}{x^2}$, $m = 0$ **13.** $f'(x) = \dfrac{1}{\sqrt{2x}}$, $m = 1$

15. $f'(x) = \dfrac{-1}{(2x + 3)\sqrt{2x + 3}}$, $m = -1$ **17.** $\sqrt{2}$ **19.** $-\dfrac{1}{9}$

21. $\dfrac{1}{4}$ **23.** a) 0, 1, 4   b)

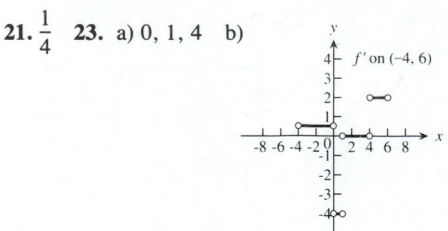

25. a) They are the same.   b)

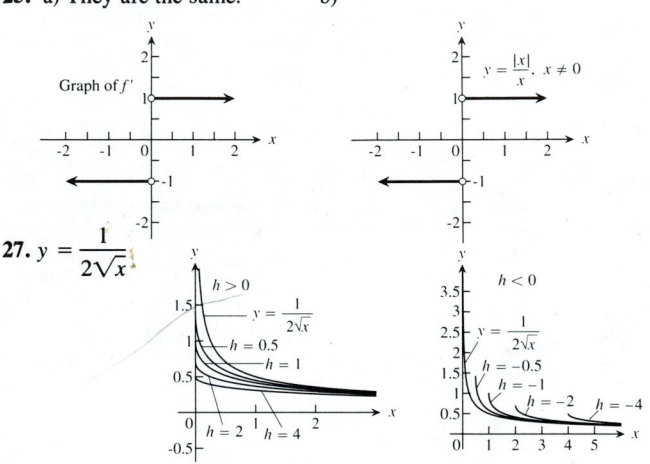

27. $y = \dfrac{1}{2\sqrt{x}}$

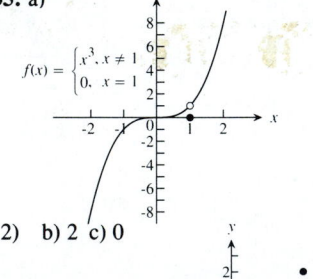

Section 1.7, pp. 69–72

1. 2   **3.** 25   **5.** -5 **7.** 45   **9.** -15 **11.** $\dfrac{5}{8}$ **13.** -2 **15.** 5

17. 0   **19.** $\dfrac{11}{4}$ **21.** -7 **23.** $\dfrac{1}{2}$ **25.** $-\dfrac{1}{2}$ **27.** $\dfrac{1}{10}$ **29.** $\dfrac{1}{2}$

31. a) $\lim_{x\to 2^+} f(x) = 2$; $\lim_{x\to 2^-} f(x) = 1$ b) No   **33.** a) True
b) True c) False d) True e) True f) True g) False h) False
i) False j) False   **35.** a)   b) 1,1 c) 1

$$f(x) = \begin{cases} x^3, & x \neq 1 \\ 0, & x = 1 \end{cases}$$

37. a) $(0, 1) \cup (1, 2)$ b) 2 c) 0

$$f(x) = \begin{cases} \sqrt{1 - x^2}, & 0 \leq x < 1 \\ 1, & 1 \leq x < 2 \\ 2, & x = 2 \end{cases}$$

39. a) Does not exist b) 0 c) Does not exist   **41.** a) 0 b) 0 c) 1
43. a) 0 b) 0 c) 9 d) 3   **45.** Right-hand derivative: 2; Left-hand
derivative: $\dfrac{1}{2}$; derivative doesn't exist   **47.** $\delta = 2$

49. $\delta = \dfrac{1}{2}$ **51.** $\delta = 2$

53. $\delta = \dfrac{1}{18}$

55. 0.1   **57.** $\sqrt{5} - 2$ **59.** $\dfrac{7}{16}$ **61.** $2.99 < x < 3.01$

63. $\sqrt{8.95} < x < \sqrt{9.05}$ **65.** $3.2391 > x > 2.7591$

67. $\dfrac{10}{39} > x > \dfrac{10}{41}$ **69.** 5; $\delta \leq 0.005$ **71.** 0; $\delta \leq 0.005$ **73.** 4;

$\delta \leq 0.05$ **75.** 2; $\delta \leq 0.0399$ **77.** 2; $\delta \leq 1/3$ **79.** For

$\epsilon = 0.01$, $\delta = 0.01$; for $\epsilon = 0.001$, $\delta = 0.001$; for $\epsilon = 0.0001$,

$\delta = 0.0001$; for arbitrary ϵ, $\delta = \epsilon$ **81.** $\delta = \epsilon^2$, limit is 0

83. a) 2 b) 2   **85.** 2   **87.** $\dfrac{\sqrt{9 - h^2} - 3}{h}$; the value of the

derivative at $x = 0$ appears to be 0   **89.** $\dfrac{1}{2}$

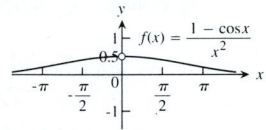

Section 1.8, pp. 78–79

1. a) $\dfrac{2}{5}$ b) $\dfrac{2}{5}$ **3.** a) 0 b) 0   **5.** a) -3 b) -3 **7.** a) $+\infty$

b) $-\infty$ **9.** a) 0 b) 0   **11.** a) 7 b) 7   **13.** a) $-\dfrac{2}{3}$ b) $-\dfrac{2}{3}$

15. a) -1 b) -1 **17.** a) -1 b) -1 **19.** $+\infty$

21. $+\infty$ **23.** $+\infty$ **25.** $+\infty$ **27.** $-\infty$ **29.** $\dfrac{1}{2}$ **31.** 1

33. $+\infty$ **35.** $\dfrac{1}{3} - \dfrac{5}{6} = -\dfrac{1}{2}$ **37.** a) ∞ b) $-\infty$ c) $-\infty$

d) ∞ **39.** a) ∞ b) $-\infty$ c) ∞ d) $-\infty$ **41.** 0   **43.** 1
45. $\lim_{x\to -\infty} f(x) = 0$; $\lim_{x\to 0^+} f(x) = -1$; $\lim_{x\to 0^-} f(x) = -\infty$;

$\lim_{x\to\infty} f(x) = -1$ **49.** 0.4   **51.** Appears to be 0

Section 1.9, pp. 83–84

1. $\dfrac{1}{2}$ **3.** -1 **5.** 1   **7.** 2   **9.** 1   **11.** 0   **13.** 2   **15.** 1   **17.** 1

19. Does not exist   **21.** a) 0.5 b) $\dfrac{1}{2}$ **23.** 2

Section 1.10, pp. 93–94

1. a) Yes b) Yes c) Yes d) Yes   **3.** a) No b) No   **5.** a) 0 b) 0
7. All except $x = 2$ **9.** $[-1, 0) \cup (0, 1) \cup (1, 2]$ **11.** 2   **13.** -1
15. Continuous everywhere   **17.** ± 1 **19.** 0   **21.** Yes   **23.** 7

25. $\dfrac{8}{5}$ **27.** $\dfrac{4}{3}$ **29.** a) $\dfrac{\sqrt{2}}{2}$ b) 1   **31.** $x = 2$ and 3   **33.** No

maximum   **41.** $x \approx 1.87939, -1.53209, -0.347296$
43. $x \approx 0.73908513$ **45.** $x = 3.515625$

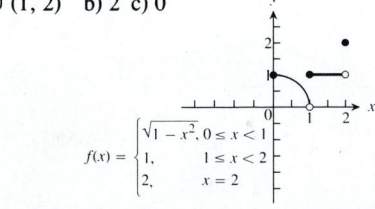

Section 1 Miscellaneous Exercises, pp. 95–99

1. $(0, 11)$ **3. a)**

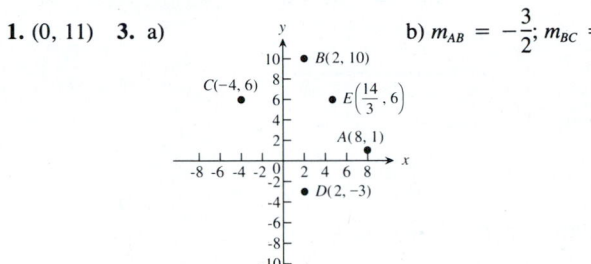

b) $m_{AB} = -\dfrac{3}{2}$; $m_{BC} = \dfrac{2}{3}$;

$m_{CD} = -\dfrac{3}{2}$; $m_{DA} = \dfrac{2}{3}$; $m_{CE} = 0$; m_{BD} undefined **c)** Yes **d)** Yes

e) CD **7.** $A = \dfrac{C^2}{4\pi}$ **9.** $f(x) = \begin{cases} -x + 1, & 0 \le x < 1 \\ -x + 2, & 1 \le x \le 2 \end{cases}$

11. No; no **13. a)**   **b)**

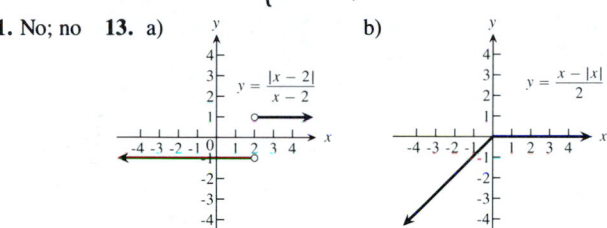

c)   **15. a)**

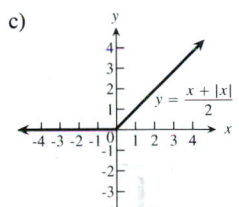   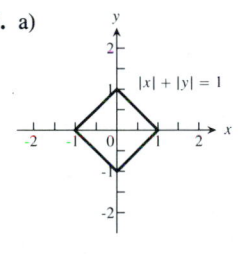

b)   **17.** $2.56 < x < 5.76$; $3.24 < x < 4.84$

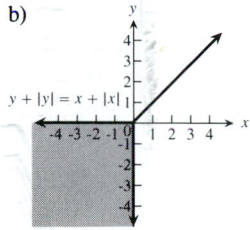

21. a) -21 **b)** 49 **c)** 0 **d)** 1 **e)** 1 **f)** 7 **23.**

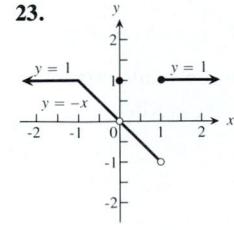

a) $\lim_{x \to -1^+} f(x) = 1$; $\lim_{x \to -1^-} f(x) = 1$; $\lim_{x \to 0^+} f(x) = 0$;
$\lim_{x \to 1^+} f(x) = 1$; $\lim_{x \to 1^-} f(x) = -1$; $\lim_{x \to 0^-} f(x) = 0$
b) $\lim_{x \to -1} f(x) = 1$; $\lim_{x \to 0} f(x) = 0$; no **c)** f is continuous at
$x = -1$ **25.** 0

27. a)

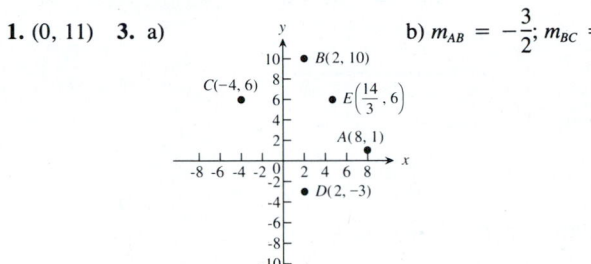

$f(x) = \begin{cases} x \sin \dfrac{1}{x}, & x \ne 0 \\ 0, & x = 0 \end{cases}$

29. a) $\lim_{x \to 0} x^2 = 0$.

But as $x \to 0$, x^2 gets closer to -1 (yes it does—think about it).
But $L \ne -1$. (There are other answers to this exercise.)
b) Let $f(x) = \begin{cases} 0, & x \ge 0 \\ x + 1, & x < 0 \end{cases}$ and let $L = 0$.
Then $|f(x) - 0| = |0 - 0| = 0 < \epsilon$ (where $\epsilon > 0$ is given) for every
$x \ge 0$. But $\lim_{x \to 0} f(x)$ does not exist because $\lim_{x \to 0^+} f(x) = 0 \ne$
$1 = \lim_{x \to 0^-} f(x)$. **31.** For $\epsilon = 4$, true for all x; for $\epsilon = 2$, true
for all $x > 5$; for $\epsilon = \dfrac{1}{2}$, true for all $x > 5$; for $\epsilon = 1$, true for all
$x > 5$ **33.** 5; $\delta \le 0.01$ **35.** -5; $\delta \le 0.01$ **37.** 2; $\delta \le 3$ **39.** 1;
$\delta \le \dfrac{2}{11}$ **41.** $\dfrac{1}{2}$ **43.** Does not exist **45.** $2a$ **47.** $2x$ **51.** No
53. True

Chapter 2

Section 2.1, pp. 111–112

1. $\dfrac{dy}{dx} = -20x$, $\dfrac{d^2y}{dx^2} = -20$ **3.** $\dfrac{dy}{dx} = x^3 - x^2 + x - 1$,

$\dfrac{d^2y}{dx^2} = 3x^2 - 2x + 1$ **5.** $y' = 2x$, $y'' = 2$ **7.** $y' =$
$2x^3 - 3x - 1$, $y'' = 6x^2 - 3$, $y''' = 12x$, $y^{(4)} = 12$ **9.** $y' =$
$-3x^2 - 2x + 3$ **11.** $y' = 3x^2$ **13.** $\dfrac{dy}{dx} = \dfrac{-19}{(3x - 2)^2}$

15. $\dfrac{dy}{dx} = \dfrac{x^2 + x + 4}{(x + 0.5)^2}$ **17.** $\dfrac{dy}{dx} = \dfrac{x^2 - 2x - 1}{(1 + x^2)^2}$

19. $\dfrac{dy}{dx} = \dfrac{1 - 3x^2 - 4x^3}{(x^2 - 1)^2(x^2 + x + 1)^2}$ **21.** $\dfrac{dy}{dx} = \dfrac{-6}{x^3}$, $\dfrac{d^2y}{dx^2} = \dfrac{18}{x^4}$

23. $\dfrac{dy}{dx} = \dfrac{-5}{3x^2}$, $\dfrac{d^2y}{dx^2} = \dfrac{10}{3x^3}$ **25.** $\dfrac{dy}{dx} = 1 - \dfrac{1}{x^2}$, $\dfrac{d^2y}{dx^2} = \dfrac{2}{x^3}$

27. $\dfrac{dy}{dx} = 2x - \dfrac{7}{x^2}$, $\dfrac{d^2y}{dx^2} = 2 + \dfrac{14}{x^3}$ **29.** $\dfrac{dy}{dx} = \dfrac{3}{x^4}$, $\dfrac{d^2y}{dx^2} = -\dfrac{12}{x^5}$

31. $\dfrac{dy}{dx} = 3x^2 - 1$ **33.** $\dfrac{dy}{dx} = 4x^3 - 8x$ **35.** $\dfrac{dy}{dx} = 4x^3 - 10x$

37. a) $\dfrac{ds}{dt} = \dfrac{1 - t^2}{(1 + t^2)^2}$ **b)** $\dfrac{ds}{dt} = 2t + 3t^2 - 4t^3 - 5t^4$

39. a) $\dfrac{dw}{dz} = \dfrac{2(z^2 - 1)}{(z + 1)^4}$ **b)** $\dfrac{dw}{dz} = -z^{-2} + 1$ **41. a)** 13
b) -7 **c)** $\dfrac{7}{25}$ **d)** 20 **43.** $y = \dfrac{-1}{9}x + \dfrac{29}{9}$ **45.** $(2, 0)$
and $(-1, 27)$ **47.** $y = 2$ **49.** $y = x^2 + x$
51. $\dfrac{dP}{dV} = -\dfrac{nRT}{(V - nb)^2} + \dfrac{2an^2}{V^3}$, $\dfrac{d^2P}{dV^2} = \dfrac{2nRT}{(V - nb)^3} - \dfrac{6an^2}{V^4}$

Section 2.2, pp. 121–125

1. a) $v = 0.0$ m/sec and $a = 1.6$ m/sec² at $t = 0$; $v = 16$ m/sec
and $a = 1.6$ m/sec² at $t = 10$ b) $\Delta s = s(10) - s(0) = 80 - 0 =$
80 m; $v_{av} = \dfrac{\Delta s}{\Delta t} = 8.0$ m/sec **3.** a) $v = 2$ m/sec and $a =$
-6 m/sec² at $t = 0$; $v = 2.0$ m/sec and $a = 6$ m/sec² at $t = 2$
b) $\Delta s = s(2) - s(0) = 0 - 0 = 0.0$ m; $v_{av} = \dfrac{\Delta s}{\Delta t} = 0.0$ m/sec
5. a) $v = 1750$ m/sec and $a = -7250$ m/sec² at $t = 1$; $v =$
-2 m/sec and $a = -0.4$ m/sec² at $t = 5$ b) $\Delta s = s(5) - s(1) =$
$25 - 5 = 20$ m; $v_{av} = \dfrac{\Delta s}{\Delta t} = 5$ m/sec **7.** 4.46 sec on Mars;
0.726 sec on Jupiter **9.** 24/9.8 sec is the time it takes to reach its
maximum height; the maximum is about 29.39 m **11.** a) 10^4
bacteria/h b) 0 bacteria/h c) -10^4 bacteria/h **13.** a) $110 per
machine b) The marginal cost of producing 100 machines is
$80. c) The cost of producing the 101st machine is $79.90.
15. -6 m/sec² and 6 m/sec² **17.** a) 190 ft/sec b) 2 sec
c) at 8 sec, 0 ft/sec **19.** a) $\dfrac{4}{7}$ sec; 280 cm/sec b) 560 cm/sec;
980 cm/sec² c) 29.75 flashes per second **21.** a) 0, 0 b) Largest
1700, smallest about 1400 **23.** b **25.** d
27. a)   b) 0, 2, 4, and 5

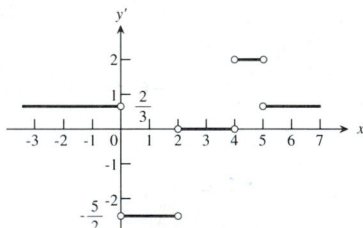

29. a)

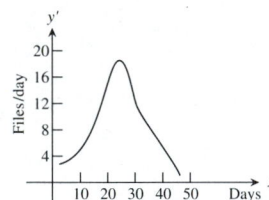

b) The fastest is between the 20th and 30th days; slowest is
between the 40th and 50th days. **31.**

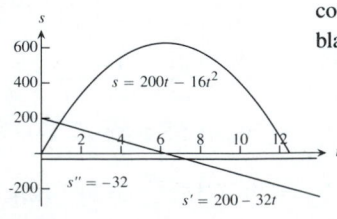

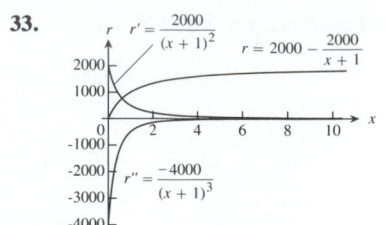

33.

Section 2.3, pp. 130–132

1. $\dfrac{dy}{dx} = 1 + \sin x$ **3.** $\dfrac{dy}{dx} = -\csc x \cot x - 5$ **5.** $\dfrac{dy}{dx} =$
$2x \cot x - x^2 \csc^2 x$, product rule **7.** $\dfrac{dy}{dx} = \sec^2 x$ **9.** $\dfrac{dy}{dx} =$
$4 \sec x \tan x$ **11.** $\dfrac{dy}{dx} = \dfrac{1 + \cos x + x \sin x}{(1 + \cos x)^2}$, quotient rule
13. $\dfrac{ds}{dt} = 2 \cos t - \sec^2 t$ **15.** $\dfrac{ds}{dt} = 2 - \csc^2 t$ **17.** $\dfrac{dr}{d\theta} =$
$-\theta[2 \sin \theta + \theta \cos \theta]$, product rule **19.** $\dfrac{dr}{d\theta} = \sec^2\theta - \csc^2\theta$,
product rule **21.** $\dfrac{dp}{dq} = -\csc^2 q$ **23.** $\dfrac{dp}{dq} = -\dfrac{\csc^2 q}{(1 + \cot q)^2}$,
quotient rule **25.** $y' = 2 \csc^3 x - \csc x$ **27.** Tangent, $y = x$;
normal, $y = -x$ **29.** Tangent, $y = -1$; normal, $x = \pi$
31. Horizontal tangents when $x = 0$ **33.** A horizontal tangent at
$x = \pi$ **35.** A horizontal tangent at $x = \pi/2$ **37.** 0 **39.** 1
41. $b = 1$ **43.** Tangent, $y = -x + \pi/4 + 1$; normal, $y =$
$x + 1 - \pi/4$ **45.** a) The horizontal tangent, $y = 4 - \sqrt{3}$
b) Tangent, $y = -x + \dfrac{\pi + 4}{2}$

47.

h	1	0.1	0.01	0.001
$\dfrac{1 - \cos h}{h^2}$	0.459697694	0.499583474	0.499958	0.5

The limit appears to be 1/2.
51.

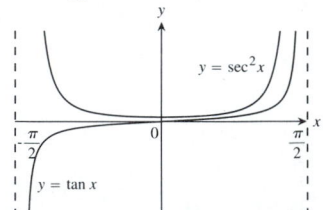

53. a) As h takes on the values of 1, 0.5, 0.3, and 0.1, the
corresponding curves of $y = \dfrac{\sin(x + h) - \sin x}{h}$ move from gray to
black, which is $y = \cos x$.

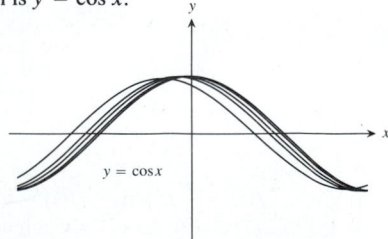

b) As h takes on the values of -1, -0.5, -0.3, and -0.1, the corresponding curves of $y = \dfrac{\sin(x + h) - \sin x}{h}$ move from gray to black, which is $y = \cos x$.

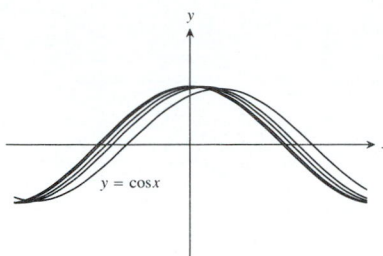

If $h \to 0^+$ or $h \to 0^-$, then $y = \dfrac{\sin(x + h) - \sin x}{h} \to y = \cos x$.

Section 2.4, pp. 139–141

1. $\dfrac{dy}{dx} = \cos(x + 1)$ **3.** $\dfrac{dy}{dx} = \cos(5x) - 5x \sin(5x)$

5. $\dfrac{dy}{dx} = 3(\sin^2 x)(\cos x)$ **7.** $\dfrac{dy}{dx} = 2x \tan\!\left(\dfrac{1}{x}\right) - \sec^2\!\left(\dfrac{1}{x}\right)$ **9.** $\dfrac{dy}{dx} =$

$1 + \sec\!\left(x^2 + 1\right)\tan\!\left(x^2 + 1\right)(2x)$ **11.** $\dfrac{dy}{dx} = -[\sin(\sin x)](\cos x)$

13. $\dfrac{dy}{dx} = 4(x - 2)^3$ **15.** $\dfrac{dy}{dx} = (-5)(2 \sin x + 5)^{-6}(2 \cos x)$

17. $\dfrac{dy}{dx} = -\left(1 - \dfrac{x}{7}\right)^6$ **19.** $\dfrac{dy}{dx} = (-4)\left(1 + x - \dfrac{1}{x}\right)^{-5}\left(1 + \dfrac{1}{x^2}\right)$

21. $\dfrac{dy}{dx} = 8(2x - 5)^3(x + 1)^7(3x - 4)$ **23.** $\dfrac{dy}{dx} =$

$36\,(x^2 + x + 1)(2x^3 + 3x^2 + 6x + 6)^5$ **25.** $\dfrac{dy}{dx} =$

$(\sec x)(\sec x + \tan x)^{-1}$ **27.** $\dfrac{dy}{dx} = \cos\!\left(\dfrac{x - 2}{x + 3}\right)\!\left[\dfrac{5}{(x + 3)^2}\right]$

29. $\dfrac{ds}{dt} = -8(2t + 1)^{-5}$ **31.** $\dfrac{ds}{dt} = \dfrac{4}{\pi}\cos 3t + \dfrac{4}{\pi}\cos 5t$

33. $\dfrac{dr}{d\theta} = -\sec^2(2 - \theta)$ **35.** $\dfrac{dr}{d\theta} = \dfrac{-6 \csc^2 3\theta}{(2 + \cot 3\theta)^2}$ **37.** $\dfrac{dy}{dx} =$

$6[\sin(3x - 2)]\cos(3x - 2)$ **39.** $\dfrac{dy}{dx} = -4(1 + \cos 2x)(\sin 2x)$

41. $\dfrac{dy}{dx} = -2[\cos(\cos(2x - 5))]\,[\sin(2x - 5)]$

43. $y'' = 2(\sec^2 x)\tan x$ **45.** $y'' = 2(\csc^2 x)(\cot x)$ **47.** $5/2$

49. $-\dfrac{\pi}{4}$ **51.** 0 **53.** a) $-6 \sin(6x + 2)$ b) $-6 \sin(6x + 2)$

55. a) 1 b) 1   **57.** 5   **59.** 1/2   **61.** The tangent line is

$y = \pi x + 2 - \pi$; the normal line is $y = -\dfrac{1}{\pi}x + 2 + \dfrac{1}{\pi}$.

63. a) 2/3 b) $2\pi + 5$ c) $-8\pi + 15$ d) 37/6 e) -1 f) $\sqrt{2}/24$ g) 5/32 h) $-5/3\sqrt{17}$ **65.** The range of the velocity doubles while the range of the acceleration quadruples.

69. a)

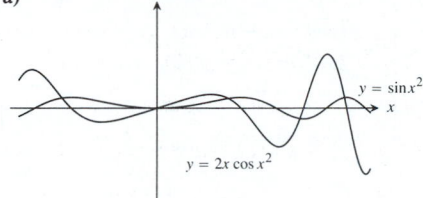

b)

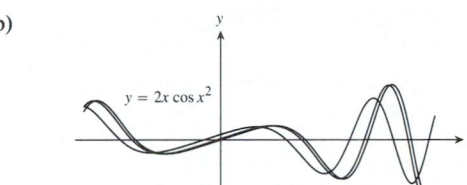

As h takes on the values of 0.5 and 0.1, the corresponding curves of $y = \dfrac{\sin(x + h)^2 - \sin x^2}{h}$ in the graph of (b) move from gray to black, which is $y = 2x \cos x^2$. If $h \to 0^+$ or $h \to 0^-$, then $y = \dfrac{\sin(x + h)^2 - \sin x^2}{h} \to y = 2x \cos x^2$.

Section 2.5, pp. 146–148

1. $\dfrac{dy}{dx} = (9/4)x^{5/4}$ **3.** $\dfrac{dy}{dx} = \dfrac{1}{3\sqrt[3]{x^2}}$ **5.** $y' = -(2x + 5)^{-3/2}$

7. $y' = \dfrac{2x^2 + 1}{\sqrt{x^2 + 1}}$ **9.** $\dfrac{dy}{dx} = \dfrac{-2xy - y^2}{x^2 + 2xy}$

11. $\dfrac{dy}{dx} = \dfrac{1 - 2y}{2x + 2y - 1}$ **13.** $\dfrac{dy}{dx} = \dfrac{x(1 - y^2)}{y(x^2 - 1)}$ **15.** $\dfrac{dy}{dx} = -1$

17. $\dfrac{dy}{dx} = \dfrac{-1}{4\sqrt{x}\sqrt{1 - \sqrt{x}}}$ **19.** $\dfrac{dy}{dx} = \dfrac{-\sin 2x}{\sqrt{1 + \cos 2x}}$

21. $\dfrac{dy}{dx} = (-9/2)(\csc^{3/2} x)(\cot x)$ **23.** $\dfrac{dy}{dx} = \cos^2 y$

25. $\dfrac{dy}{dx} = \dfrac{-\cos^2(xy) - y}{x}$ **27.** $\dfrac{dy}{dx} = -\dfrac{x}{y},\ \dfrac{d^2y}{dx^2} = \dfrac{-y^2 - x^2}{y^3}$

29. $\dfrac{dy}{dx} = \dfrac{x + 1}{y},\ \dfrac{d^2y}{dx^2} = \dfrac{y^2 - (x + 1)^2}{y^3}$ **31.** -2 **33.** The

tangent line is $y = \dfrac{7}{4}x - \dfrac{1}{2}$; the normal line is $y = -\dfrac{4}{7}x + \dfrac{29}{7}$.

35. The tangent line is $y = 3x + 6$; the normal line is

$y = -\dfrac{1}{3}x + \dfrac{8}{3}$.   **37.** $(-\sqrt{7}, 0)$ and $(\sqrt{7}, 0)$; $m = -2$

39. $-\dfrac{\pi}{2}$ **41.** $-1, \sqrt{3}$ **43.** a) False b) True c) True d) True

45. $a = \dfrac{3}{4}$ **47.** $y + 2x = -3$ **51.** $\dfrac{kT}{2}$

Section 2.6, pp. 158–160

1. $L(x) = 4x - 3$ **3.** $L(x) = 2x - 2$ **5.** $L(x) = \dfrac{1}{4}x + 1$

7. $L(x) = 2x$ **9.** $L(x) = -5$ **11.** $L(x) = \dfrac{1}{12}x + \dfrac{4}{3}$

13. $L(x) = x$

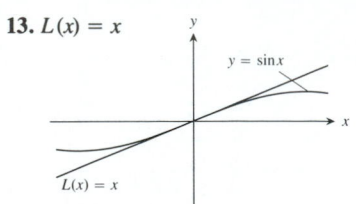

15. $L(x) = -x + \pi$

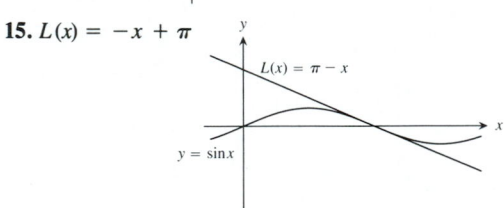

17. $L(x) = 2x + 1 - \pi/2$ **19.** a) $1 + 2x$

b) $1 + (-5)x = 1 - 5x$ c) $2[1 + (-1)(-x)] = 2 + 2x$
d) $1 + (6)(-x) = 1 - 6x$ e) $3[1 + (1/3)(x)] = 3 + x$

f) $1 + (-1/2)(x) = 1 - \dfrac{x}{2}$

21. $\dfrac{3}{2}x + 1$; it is equal to the sum of the linearizations.

23. 1.414213562, 1.189207115, 1.090507733, 1.044273782,
1.021897149, 1.010889286, 1.005429901, . . .   **25.** a) 0.21
b) 0.2 c) 0.01   **27.** a) 0.231 b) 0.2 c) 0.031   **29.** a) $-1/3$
b) $-2/5$ c) 1/15   **31.** $dV = 4\pi r_0^2\, dr$ **33.** $dV = 3x_0^2\, dx$
35. $dV = 2\pi r_0 h\, dr$ **37.** a) $0.08\pi\,\text{m}^2$ b) 2%   **39.** 3%

41. 3%   **43.** $\dfrac{1}{3}$%   **45.** 0.05%   **47.** 40% increase

49. Volume $= (x + \Delta x)^3 = x^3 + 3x^2(\Delta x) + 3x(\Delta x)^2 + (\Delta x)^3$

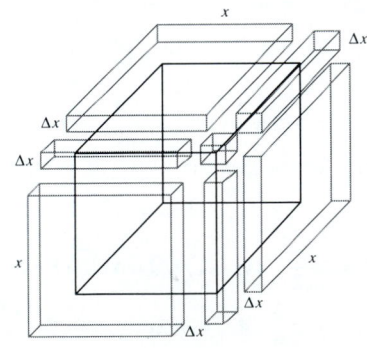

51. $dy = (3x^2 - 3)\, dx$ **53.** $dy = \dfrac{2}{(1 + x)^2}\, dx$

55. $dy = \dfrac{(1 - y)}{1 + x}\, dx$ **57.** $dy = 4\tan\left(\dfrac{x}{2}\right)\sec^2\left(\dfrac{x}{2}\right) dx$

59. $dy = 2x(1 + \sin x^2)\, dx$ **61.** a) $x = 1$

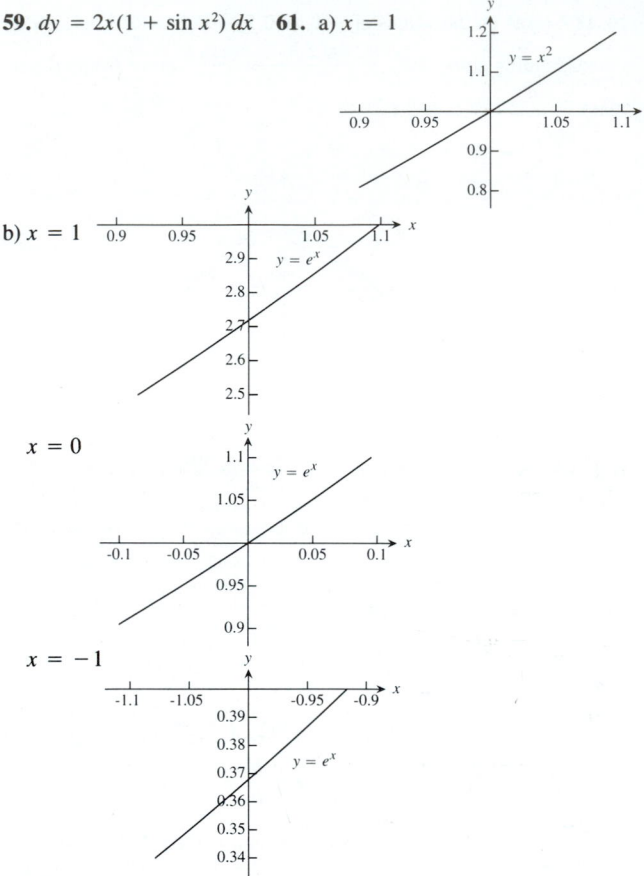

Section 2.7, pp. 166–168

1. 0.618033989, -1.61803399 **3.** 1.16403514, -1.45262688
5. 1.18920711   **7.** If $f'(x_0) \neq 0$ then $x_0 = x_1 = x_2 = x_3 = \cdots$
11. 1.17951   **13.** 3.14159   **15.** 1.49870   **17.** 1.16556119
19. 0.5411961, 1.30656296   **21.** 0.11051
23.

x_0	Approximation of corresponding root
-1.0	-0.976823589
0.1	0.100363332
0.6	0.64274667
2.0	1.98371359

Section 2 Miscellaneous Exercises, pp. 171–175

1. $\dfrac{dy}{dx} = 2(x + 1)^3 (3x^2 + 6x + 1)$ **3.** $\dfrac{dy}{dx} =$

$3(x^2 + \sin x + 1)^2 (2x + \cos x)$ **5.** $\dfrac{dy}{dx} = 5(\sec x)(\sec x + \tan x)^5$

7. $\dfrac{dy}{dx} = \dfrac{5 - 2x - y}{x + 2y}$ **9.** $\dfrac{ds}{dt} = 3(4 - t)^{-2}$

11. $\dfrac{ds}{dt} = \left(1 + \dfrac{t}{2} + \dfrac{t^2}{4}\right)(1 + t)$ **13.** $\dfrac{ds}{dt} = (4 - t)(t^2 - 8t)^{-3/2}$

15. $\dfrac{dp}{dq} = \dfrac{6q - 4p}{3p^2 + 4q}$ **17.** $\dfrac{dr}{ds} = (2r - 1)\tan(2s)$ **19.** $\dfrac{1}{4}$

23. a) 2 sec, 64 ft/sec b) 12.31 sec, 393.85 ft   **25.** a) 32 ft/sec b) -16 ft/sec c) 0 ft/sec

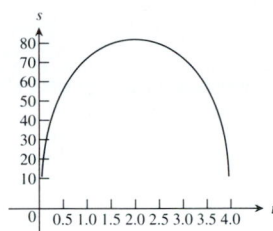

27. 40 people, \$4.00   **31.** a) $m = -\dfrac{b}{\pi}$ b) $m = -1, b = \pi$

33. The curve has an infinite number of horizontal tangents.

37. 9/2   **39.** (0, 1), $(-4, 0)$ **41.** $y = -\dfrac{x}{27} + \dfrac{28}{27}$

43. a) $\dfrac{d^3y}{dx^3} = 3(2x - 1)^{-5/2}$ b) $\dfrac{d^3y}{dx^3} = -162(3x + 2)^{-4}$

c) $\dfrac{d^3y}{dx^3} = 6a$ **47.** a) Tangent, $y = -\dfrac{1}{4}x + \dfrac{9}{4}$; normal,

$y = 4x - 2$ b) Tangent, $y = -\dfrac{3}{2}x + \dfrac{5}{2}$; normal, $y = \dfrac{2}{3}x + \dfrac{1}{3}$

c) Tangent, $y = 2x - 4$; normal, $y = -\dfrac{1}{2}x + \dfrac{7}{2}$ **49.** 3 ft

51. $-1/2$ **57.** $L(x) = 2.5x - 0.1$ **59.** a) The measurement must have an error less than 1%. b) 3%   **61.** $\pm\dfrac{20\pi}{27}$ ft

63. a) 0.8156 ft b) 0.006 sec c) It will gain about 8.64 min/day.
65. 2.19582

Chapter 3

Section 3.1, pp. 181–183

1. $\dfrac{dA}{dt} = 2\pi r \dfrac{dr}{dt}$ **3.** $\dfrac{dV}{dt} = 3x^2 \dfrac{dx}{dt}$ **5.** $\dfrac{dV}{dt} = (1/3)\pi r^2 \dfrac{dh}{dt}$

7. π cm^2/min   **9.** a) 14 cm^2/sec, increasing b) 0 cm/sec,

constant c) $-\dfrac{14}{13}$ cm/sec, decreasing   **11.** -12 ft/sec,

$-\dfrac{119}{2}$ ft^2/sec   **13.** 0.5093 ft/min   **15.** 40π ft^2/min

17. 11 ft/sec   **19.** $\dfrac{466}{1681}$ L/min   **21.** 0 rad/sec   **23.** a) $\dfrac{dr}{dt} = 0.9$,

$\dfrac{dc}{dt} = 0.3$, $\dfrac{dp}{dt} = 0.6$ b) $\dfrac{dr}{dt} = 3.5$, $\dfrac{dc}{dt} = -1.5625$, $\dfrac{dp}{dt} =$

5.0625   **25.** 1500 ft/sec   **27.** 80 mph   **29.** $\dfrac{3\pi}{16}$ ft/min

Section 3.2, pp. 191–192

5. 1/2   **7.** 1   **11.** a)

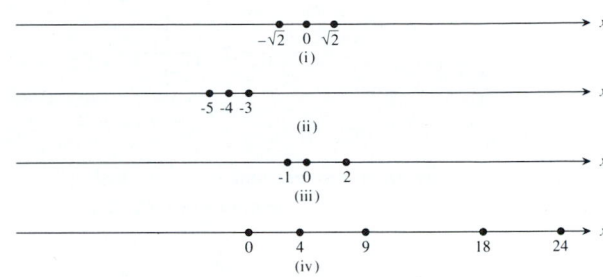

13. $a = 3, m = 1, b = 4$ **15.** $f(x)$ is not continuous
27. $f(0.1) \approx 1.099995025$

Section 3.3, pp. 199–201

1. The graph is rising on $(-\infty, -1) \cup (2, \infty)$, falling on $(-1, 2)$, concave upward on $(1/2, \infty)$, and concave downward on $(-\infty, 1/2)$. The local maximum is 3/2 at $x = -1$, the local minimum is -3 at $x = 2$, and $(1/2, -3/4)$ is a point of inflection.   **3.** The graph is rising on $(-3\pi/2, -\pi/2) \cup (0, \pi/2) \cup (3\pi/2, 2\pi)$, falling on $(-2\pi, -3\pi/2) \cup (-\pi/2, 0) \cup (\pi/2, 3\pi/2)$, concave upward on $(-2\pi, -\pi) \cup (\pi, 2\pi)$, and concave downward on $(-\pi, 0) \cup (0, \pi)$. The maximum is 1 at $x = \pm\pi/2$, the minimum is -1 at $x = \pm 3\pi/2$, and the points of inflection are $(-\pi, 0)$ and $(\pi, 0)$.   **5.** At $x = 2$ there is a minimum.

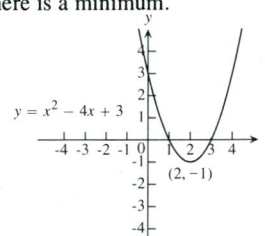

7. At $x = 0$ there is a local maximum and at $x = 1$ a local minimum. At $x = 1/2$ there is a point of inflection.

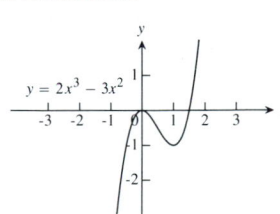

9. At $x = -1$ there is a local minimum and at $x = 1$ a local maximum. At $x = 0$ there is a point of inflection.

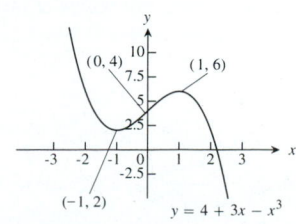

11. At $x = 2$ there is a point of inflection.

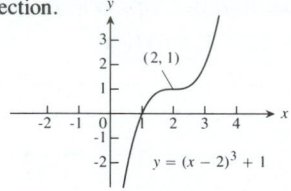

13. At $x = \pm 1$ there are minimums and at $x = 0$ a local maximum. At $x = \pm \sqrt{3}/3$ there are points of inflection.

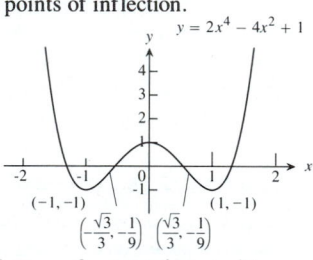

15. At $x = 0$ there is a minimum and at $x = 2\pi$ a maximum. At $x = \pi$ there is a point of inflection.

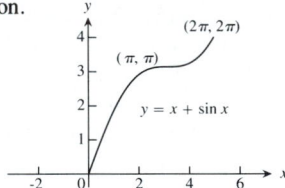

17. The curve rises on $(-1, \infty)$ and never falls. There are no points of inflection.

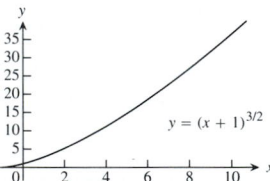

19. At $x = \pm 1$ there are minimums and at $x = 0$ a local maximum. At $x = \pm 1$ there are points of inflection.

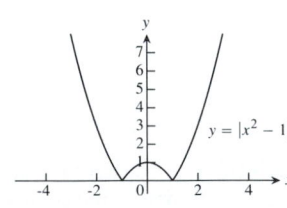

21. No extrema nor points of inflection

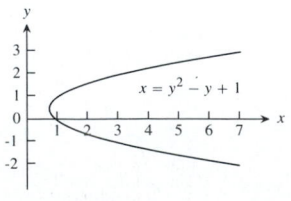

23. At $y = 0$ there is a point of inflection.

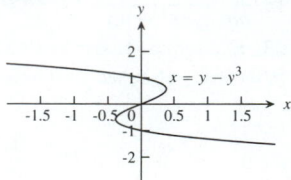

25. There is a vertical tangent at $x = 0$.

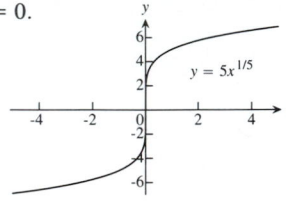

27. There is a vertical tangent at $x = -1$.

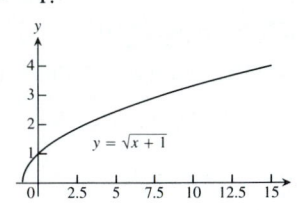

29. There is a cusp at $x = 8$.

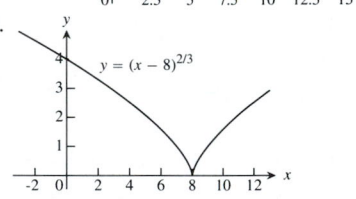

31. There are cusps at $x = \pm 1$.

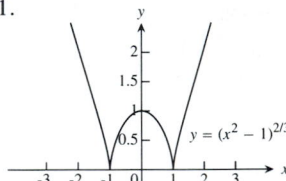

33. There is a cusp at $x = 0$.

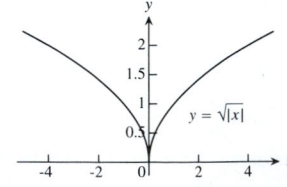

35. The velocity is zero when t is approximately 2, 6, or 9.5 sec. The acceleration is zero when t is approximately 4, 8, or 12.5 sec.

37.

Point	y'	y''
P	$-$	$+$
Q	$+$	0
R	$+$	$-$
S	0	$-$
T	$-$	$-$

39.

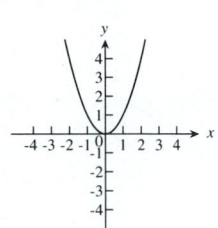

41.

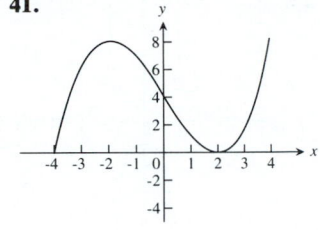

43. a)

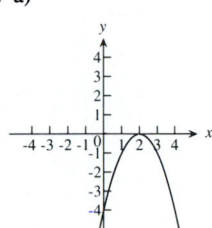

b)

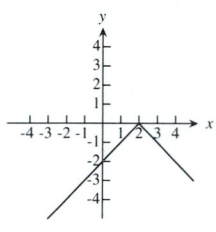

c)

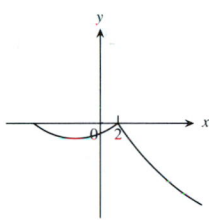

45. At $x = 2$ there is a local minimum. At $x = 1$ or $x = 5/3$ there are inflection points.   **47.** No. When $f(x) = x^3$, then $f'(x) = 3x^2$ and $f'(0) = 0$, but $x = 0$ is neither a local minimum nor maximum.
49. a) True **b)** True   **51.** If $c \to -\infty$, then $y \to mx$ where $m \to \infty$. If $c \to \infty$, then $y \to mx$ where $m \to -\infty$.

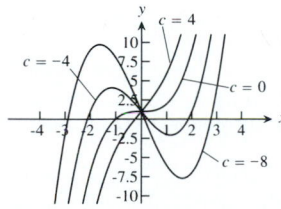

53. The zeros of $y' = 0$ and $y'' = 0$ are extrema and points of inflection, respectively.

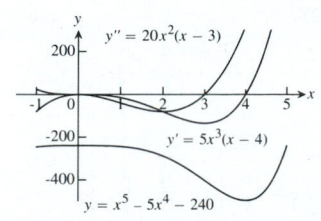

55. The zeros of $y' = 0$ and $y'' = 0$ are extrema and points of inflection, respectively.

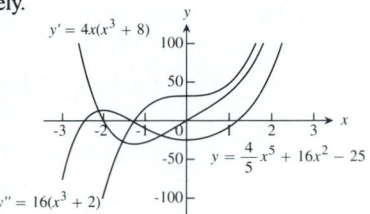

57. b) A cusp since $\lim_{x \to 0^-} y' = \infty$ and $\lim_{x \to 0^+} y' = -\infty$

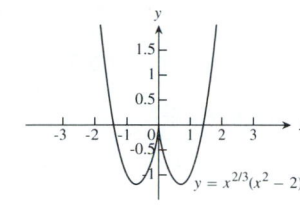

59. a) Computer-generated

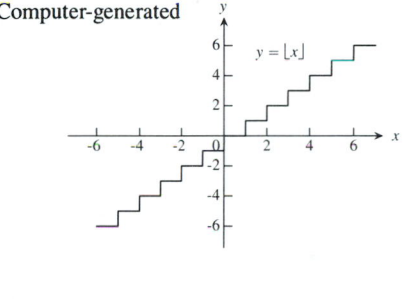

b) The correct graph

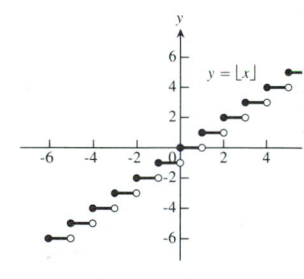

61. a) Computer-generated   **b)** The correct graph

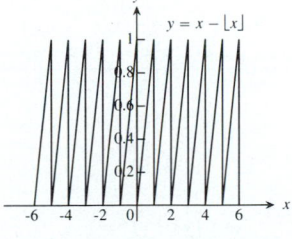

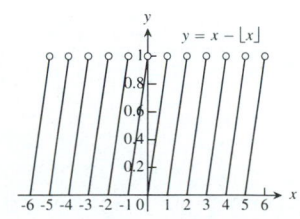

63. a) Computer-generated

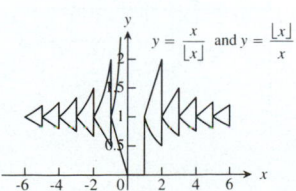

b) The correct graphs

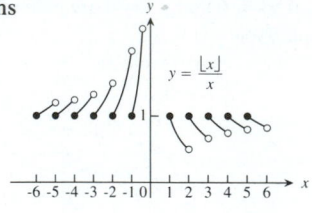

$y = \dfrac{\lfloor x \rfloor}{x}$

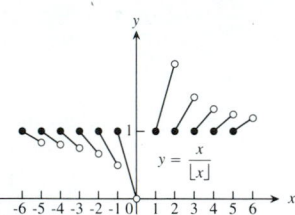

$y = \dfrac{x}{\lfloor x \rfloor}$

65. No difference

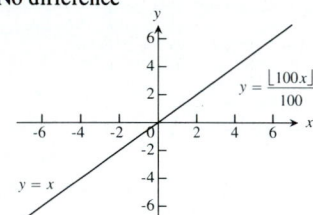

$y = \dfrac{\lfloor 100x \rfloor}{100}$

$y = x$

Section 3.4, pp. 206–207

1.

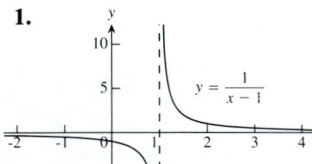

$y = \dfrac{1}{x - 1}$

3.

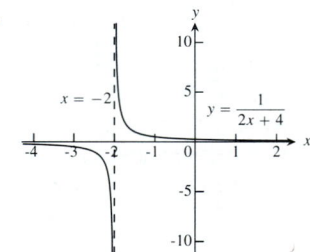

$x = -2$

$y = \dfrac{1}{2x + 4}$

5.

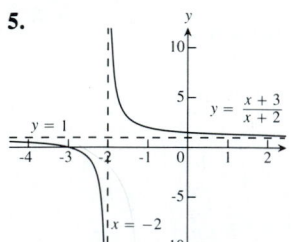

$y = 1$

$y = \dfrac{x + 3}{x + 2}$

$x = -2$

7.

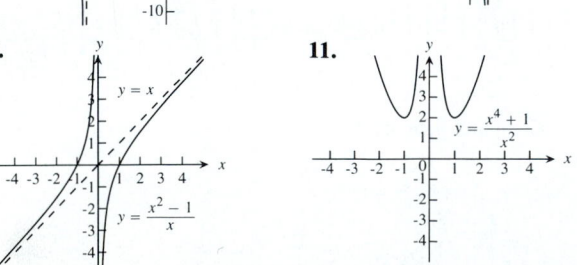

$y = \dfrac{3}{2}$

$y = \dfrac{3x + 1}{2x - 1}$

9.

$y = x$

$y = \dfrac{x^2 - 1}{x}$

11.

$y = \dfrac{x^4 + 1}{x^2}$

13.

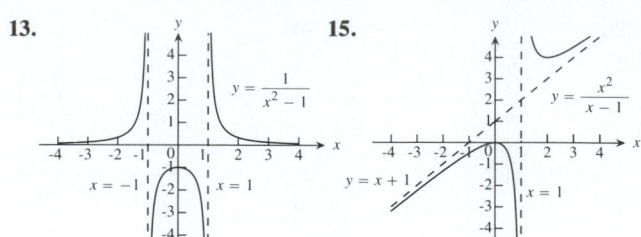

$y = \dfrac{1}{x^2 - 1}$

$x = -1$ $x = 1$

15.

$y = \dfrac{x^2}{x - 1}$

$y = x + 1$ $x = 1$

17.

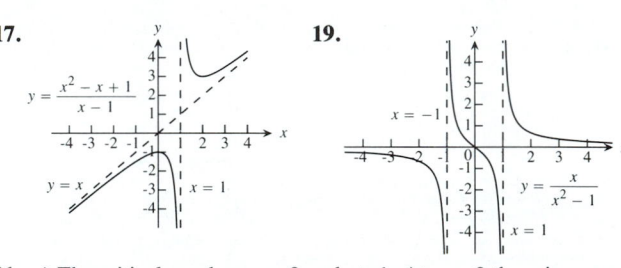

$y = \dfrac{x^2 - x + 1}{x - 1}$

$y = x$ $x = 1$

19.

$x = -1$

$y = \dfrac{x}{x^2 - 1}$

$x = 1$

21. a) The critical numbers are 0 and ± 1. At $x = 0$ there is a local minimum. b) The window is $-5 \le x \le 5$.

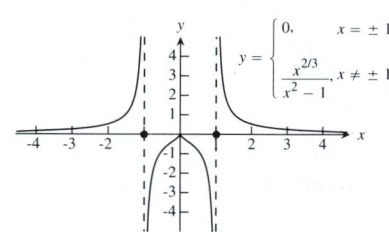

$y = \begin{cases} 0, & x = \pm 1 \\ \dfrac{x^{2/3}}{x^2 - 1}, & x \ne \pm 1 \end{cases}$

The window is $-0.25, x \le 0.25$.

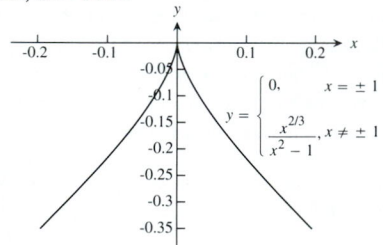

$y = \begin{cases} 0, & x = \pm 1 \\ \dfrac{x^{2/3}}{x^2 - 1}, & x \ne \pm 1 \end{cases}$

23. a) Increasing b) Decreasing **25.**

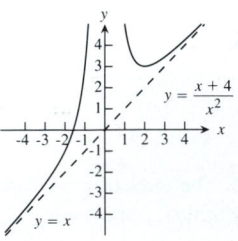

$y = \dfrac{x + 4}{x^2}$

$y = x$

27.

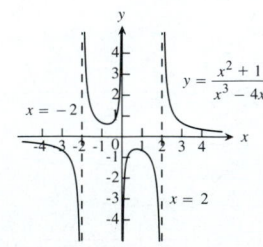

$x = -2$

$y = \dfrac{x^2 + 1}{x^3 - 4x}$

$x = 2$

29.

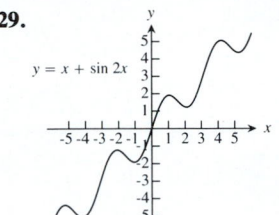

$y = x + \sin 2x$

31.

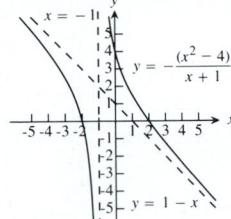

$x = -1$

$y = -\dfrac{(x^2 - 4)}{x + 1}$

$y = 1 - x$

33. a)

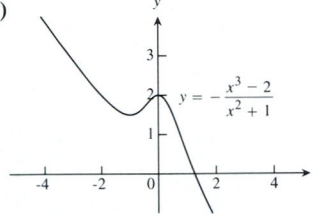

$y = -\dfrac{x^3 - 2}{x^2 + 1}$

b)

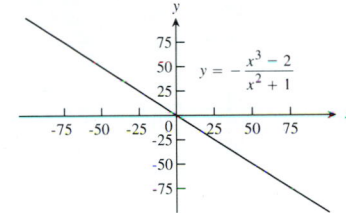

$y = -\dfrac{x^3 - 2}{x^2 + 1}$

c) The distance in part c is so great that small movements are not visible.

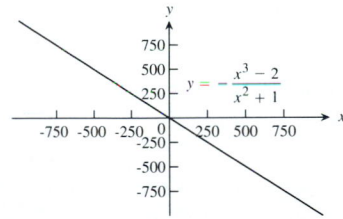

$y = -\dfrac{x^3 - 2}{x^2 + 1}$

35. See Exercise 21. The reason that the cusp did not appear in the first graph is that the size of the interval was too large to show such detail.

Section 3.5, pp. 215–220

1. $r = 25$ ft, $s = 50$ ft   **3.** 16 in.   **5. a)** $y = 1 - x$
b) $A = 2x(1 - x)$, where $0 \le x \le 1$ **c)** 1/2 sq. units
7. $14/3 \times 35/3 \times 5/3$ inches   **9.** 80,000 m² **11.** 10 ft on the base edge and 5 ft for the height   **13.** 18×9 in.   **15.** $\pi/2$
17. $r = h = 10/\sqrt[3]{\pi}$ **19. a)** $18 \times 18 \times 36$ in. **b)** The graph indicates that the maximum volume occurs near $I = 36$, which is compatible with the result of part (a).

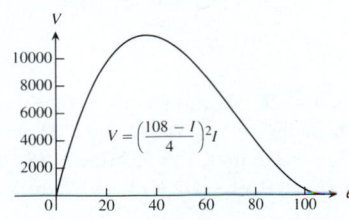

$V = \left(\dfrac{108 - I}{4}\right)^2 I$

21. a) $x = 12, y = 6$ **b)** $x = 12, y = 6$ **23. a)** 16 **b)** -1
25. a) $a = -3, b = -9$ **b)** $a = -3, b = -24$
27. a) $4\sqrt{3} \times 4\sqrt{6}$ in. **b)**

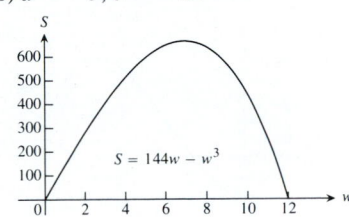

$S = 144w - w^3$

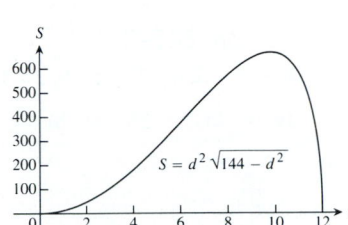

$S = d^2 \sqrt{144 - d^2}$

Both graphs indicate the same maximum and are compatible with each other. The changing of k has no effect.
29. a) -1 **b)**

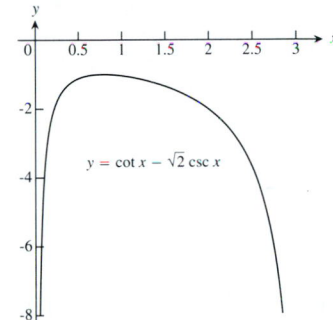

$y = \cot x - \sqrt{2}\csc x$

31. $\sqrt{5}/2$ **33.** $\pi/6$ **37.** $a/2$, $ka^2/4$ **39. a)** The circumference of the circle is 4 m. **b)** This graph indicates the same maximum and is compatible with part (a).

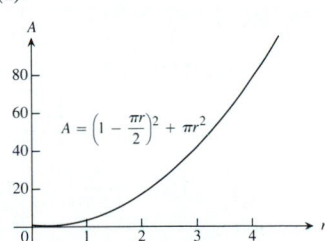

$A = \left(1 - \dfrac{\pi r}{2}\right)^2 + \pi r^2$

c) This graph indicates the same maximum and is compatible with part (a).

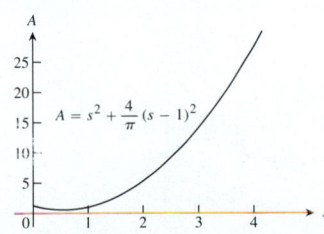

$A = s^2 + \dfrac{4}{\pi}(s - 1)^2$

41. a) $\left(c - \frac{1}{2}, \sqrt{c - \frac{1}{2}}\right)$ b) $(0, 0)$

45. b)

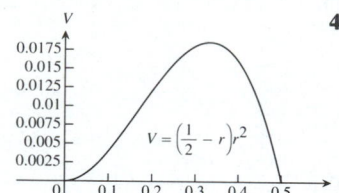

$V = \left(\frac{1}{2} - r\right)r^2$

47. $\frac{c}{2} + 50$ **49.** $\sqrt{\frac{2km}{h}}$

Section 3.6, pp. 225–226

1. 1/4   **3.** 3/11   **5.** 0   **7.** 1/2   **9.** 2   **11.** 0   **13.** 0   **15.** $-5/3$

17. ∞ **19.** 0   **21.** 3   **23.** a) 0 b) 1 c) ∞ **27.** $\frac{4\sqrt{2} - \pi}{4}$

29. 6

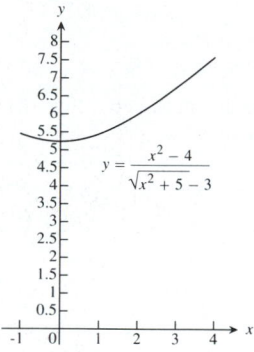

$y = \dfrac{x^2 - 4}{\sqrt{x^2 + 5} - 3}$

Section 3.7, pp. 233–234

1. $Q(x) = x^2$ **3.** $Q(x) = \frac{1}{2} - \frac{x}{2}$ **5.** $Q(x) = 1 + x + \frac{3x^2}{2}$

7. 0.00146   **9.** a) $Q(x) = 1 + \frac{x}{2} - \frac{x^2}{8}$ b) 0.0000813

11. a) $Q(x) = 1 + \frac{x^2}{2}$ b) 0.00078   **13.** a) $Q(x) =$
$1 - \frac{1}{2}(x - \pi/2)^2$ b) 0.000167   **15.** a) $Q(x) = \pi/2 - x$

b) 0.000167   **17.** $|e_2(x)| \leq \max\left\{\frac{f'''(x)}{6}\right\}x^3 = \frac{x^3}{6}$;

a) $|x| \leq \left(\frac{6}{100}\right)^{1/3}$ b) $|x| \leq \left(\frac{6}{100}\right)^{1/2}$ **19.** a) $Q(x) = 3 + 2x$
b) $Q(x) = b + mx$

Section 3 Miscellaneous Exercises, pp. 234–238

1. $\dfrac{r}{10\sqrt{3}}$ m/sec   **3.** 4%   **5.** b) 0.8555996772

11.

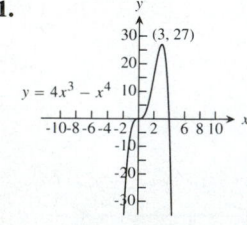

$y = 4x^3 - x^4$

13.

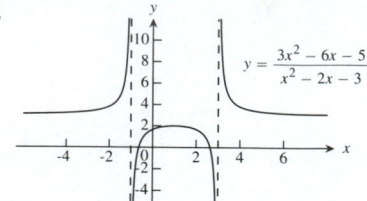

$y = \dfrac{3x^2 - 6x - 5}{x^2 - 2x - 3}$

15.

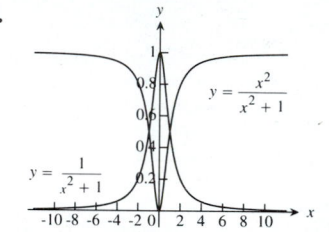

$y = \dfrac{x^2}{x^2 + 1}$

$y = \dfrac{1}{x^2 + 1}$

17. A local maximum at $x = 1$ and a local minimum at $x = 3$

19. $\lim_{x \to -1^-} f'(x) = 4$, $\lim_{x \to -1^+} f'(x) = -4$,
$\lim_{x \to 1^-} f'(x) = 0$, $\lim_{x \to 1^+} f'(x) = 0$, $\lim_{x \to 3^-} f'(x) = 4$,
$\lim_{x \to 3^+} f'(x) = -4$

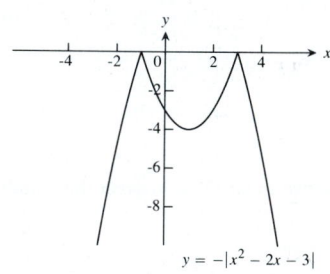

$y = -|x^2 - 2x - 3|$

21. a) The graph does not indicate any local extrema.

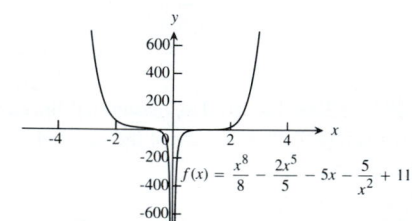

$f(x) = \dfrac{x^8}{8} - \dfrac{2x^5}{5} - 5x - \dfrac{5}{x^2} + 11$

c)

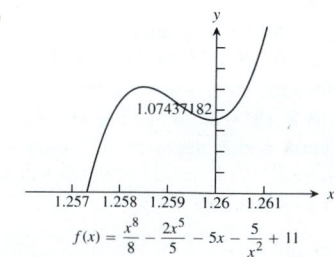

$f(x) = \dfrac{x^8}{8} - \dfrac{2x^5}{5} - 5x - \dfrac{5}{x^2} + 11$

25. $h = 24$ **27.** a) 10, 10 b) 0, 20 c) 79/4, 1/4
29. Radius $= \sqrt{2}$, height $= 2$ **31.** $x = 15$ mi, $y = 9$ mi
33. $x \approx 276$ tires, $y \approx 553$ tires   **35.** 20 ft   **41.** a) 10/3 b) 5/3
c) 1/2 d) 0 e) $-1/2$ f) 1 g) 1/2 h) 3 i) 1/4 j) $-\sqrt{3}/2$

Chapter 4

Section 4.1, pp. 245–246

1. a) $3x^2 + C$ b) $-2x + C$ c) $3x^2 - 2x + C$

3. a) $x^{-3} + C$ b) $-\frac{1}{3}x^{-3} + C$ c) $-\frac{1}{3}x^{-3} + x^2 + 3x + C$

5. a) $x^{3/2} + C$ b) $\frac{8}{3}x^{3/2} + C$ c) $\frac{x^3}{3} - \frac{8}{3}x^{3/2} + C$ **7.** a) $x^{2/3} + C$

b) $x^{1/3} + C$ c) $x^{-1/3} + C$ **9.** a) $\frac{\cos 3x}{3} + C$ b) $-3\cos(x) + C$

c) $-3\cos x + \frac{\cos 3x}{3} + C$ **11.** a) $\tan(x) + C$ b) $\tan(5x) + C$

c) $\frac{\tan 5x}{5} + C$ **13.** a) $\sec(x) + C$ b) $\sec(2x) + C$

c) $2\sec(2x) + C$ **15.** $x + \frac{\cos 2x}{2} + C$ **17.** $7x + C$ **19.** $2x^{3/2} + C$

21. $\frac{3}{2}t^{2/3} + C$ **23.** $\frac{5v^3}{3} + v^2 + C$ **25.** $\frac{1}{2}s^4 - \frac{5}{2}s^2 + 7s + C$

27. $x^4 - 4x^3 - \frac{7x^2}{2} + 21x + C$ **29.** $\frac{t^3}{3} + \frac{t^2}{2} + t + C$

31. $\frac{x^2}{2} + 2\sin x + C$ **33.** $-3\cos\frac{t}{3} + C$ **35.** $-3\cot x + C$

37. $-\frac{1}{2}\csc x + C$ **39.** $\tan\theta - \cos\theta + C$ **41.** $-\frac{1}{2}\cos 2y +$

$\cot y + C$ **43.** $2v - \sin 2v + C$ **45.** $-\frac{1}{4}\cos 2x + C$

47. $\tan\theta + C$ **49.** $-\cos\theta + \theta + C$ **55.** a) Wrong
b) Wrong c) Right   **57.** a) $-\sqrt{x} + C$ b) $x + C$ c) $\sqrt{x} + C$
d) $-x + C$ e) $x - \sqrt{x} + C$ f) $-x - \sqrt{x} + C$

g) $\frac{x^2}{2} - \sqrt{x} + C$ h) $-3x + C$ **59.**

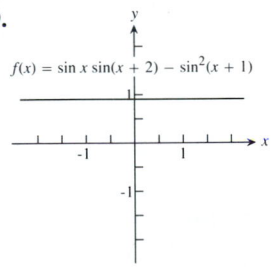

$f(x) = \sin x \sin(x + 2) - \sin^2(x + 1)$

Section 4.2, pp. 253–255

1. $y = x^2 - 7x + 10$ **3.** $y = -\frac{1}{x} + \frac{x^2}{2} - \frac{1}{2}$ **5.** $y = 2x^{3/2} - 50$

7. $s = t + \sin t + 4$ **9.** $v = \cos(\pi t) - 1$

11. $y = x^2 - x^3 + 4x + 1$ **13.** $r = \frac{1}{t} + 2t - 2$

15. $y = x^3 - 4x^2 + 5$ **17.** $s = 4.9t^2 + 10$ **19.** $s = 16t^2 + 20t$
21. $y = 2\sqrt{x^3} - 50$ **23.** 48 m/sec   **25.** 14 m/sec

27. $y = \frac{(C - kt)^2}{4}$ **29.** $k = 16$ **31.** 1.24 sec

Section 4.3, pp. 267–269

1. $\frac{6(1)}{1 + 1} + \frac{6(2)}{2 + 1} = 7$ **3.** $-\cos\pi + \cos 2\pi - \cos 3\pi +$

$\cos 4\pi = 4$ **5.** $\sum_{k=1}^{6} k$ **7.** $\sum_{k=1}^{4} \frac{1}{2^k}$ **9.** $\sum_{k=1}^{5} (-1)^{k+1}\frac{1}{k}$ **11.** All of
them represent $1 + 2 + 4 + 8 + 16 + 32$.   **13.** a) -15 b) 1 c) 1
d) -11 e) 16   **15.**

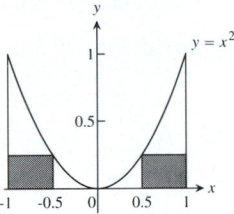

Rectangles associated with $\sum_{k=1}^{4} \min_k \Delta x$, $\sum_{k=1}^{4} \min_k \Delta x = \frac{1}{4}$

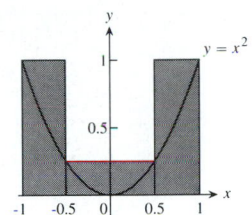

Rectangles associated with $\sum_{k=1}^{4} \max_k \Delta x$,

$\sum_{k=1}^{4} \max_k \Delta x = \frac{5}{4}$ **17.** $\int_0^2 x^2\, dx$ **19.** $\int_{-7}^5 (x^2 - 3x)\, dx$

21. $\int_2^3 \frac{1}{1 - x}\, dx$ **23.** $\int_0^4 \cos x\, dx$ **25.** $\int_{-\pi/4}^0 \sec x\, dx$

27. $\int_0^5 \sqrt{25 - x^2}\, dx$ **29.** $\int_0^{-2} (x^2 - 4)\, dx$

31. $\int_1^{\sqrt{2}} (x^2 - 1)\, dx - \int_0^1 (x^2 - 1)\, dx$ **33.** $\frac{1}{2}\pi$ **35.** 1
37. $x = 0$ and 1 maximize the integral.

Section 4.4, pp. 276–278

1. a) 0 b) -8 c) -12 d) 10 e) -2 f) 16   **3.** a) 5 b) 5 c) -5
5. Upper bound is 1, lower bound is 1/2.   **7.** 17.6   **9.** -14

11. 10   **13.** 2   **15.** $\frac{1}{2}$ **17.** $\frac{7}{3}$ **19.** $\frac{1}{3}$ **21.** 12   **23.** 1   **25.** $\frac{\pi}{4}$

27. $2\sqrt{2} + 1$ **33.** 3   **35.** $-\frac{23}{8}$ **37.** $\frac{b^2}{2} - \frac{a^2}{2}$ **39.** $\frac{1}{2}$ **41.** b) 1

Section 4.5, pp. 285–288

1. 8   **3.** $-33/4$ **5.** 1   **7.** 5/2   **9.** 2   **11.** $2\sqrt{3}$ **13.** 0

15. $-\pi$ **17.** $\frac{2\pi^3}{3}$ **19.** 8/3   **21.** 0   **23.** $\sqrt{2} - \sqrt[4]{8} + 1$

25. 16   **27.** 5/2   **29.** 1/2   **31.** 5/6   **33.** π **37.** $\frac{dy}{dx} = \sqrt{1 + x^2}$

39. $\dfrac{dy}{dx} = \dfrac{1}{2}x^{-1/2}\sin x$ **41.** $\dfrac{dy}{dx} = -\sqrt{x^4 + 7}$ **43.** $\dfrac{dy}{dx} = -1$

45. d   **47.** b   **49.** $y = \displaystyle\int_{2}^{x}\sec t\,dt + 3$ **55.** $2x - 2$ **57.** $1/3$

59. a) True b) True c) True d) False e) True f) False g) True

61.

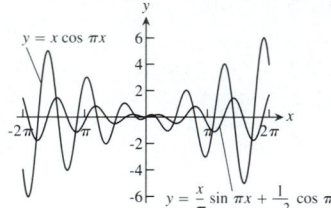

63.

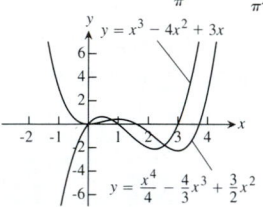

65.

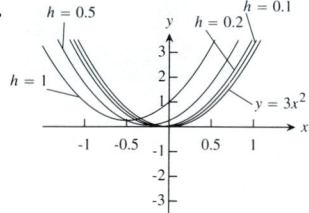

Section 4.6, pp. 295–297

1. $-\dfrac{1}{4}\cos 2x^2 + C$ **3.** $\dfrac{2}{3}\left(1 - \cos\dfrac{t}{2}\right)^3 + C$

5. $-6\left(1 - r^3\right)^{1/2} + C$ **7.** a) $-\dfrac{\cot^2 2\theta}{4} + C$

b) $-\dfrac{\csc^2 2\theta}{4} + C$ **9.** $1/2$ **11.** $2/3$ **13.** $\dfrac{1}{1 - x} + C$

15. $\tan(x + 2) + C$ **17.** $3(r^2 - 1)^{4/3} + C$ **19.** $\sec\left(\theta + \dfrac{\pi}{2}\right) + C$

21. $2(1 + x^4)^{3/4} + C$ **23.** a) $14/3$ b) $2/3$ **25.** a) $1/2$ b) $-1/2$

27. a) $\dfrac{\sqrt{10} - 3}{2}$ b) $\dfrac{3 - \sqrt{10}}{2}$ **29.** a) $45/8$ b) $-45/8$ **31.** a) $1/6$

b) $1/2$ **33.** a) 0 b) 0 **35.** $2\sqrt{3}$ **37.** 0 **39.** 8 **41.** $38/3$

43. $16/3$ **47.** $s = (3t^2 - 1)^4 - 1$ **49.** $s = -6\cos(t + \pi) - 6$

51. $F(6) - F(2)$ **53.** a) -3 b) 3 **55.** a) 1 b) 1 c) 1

57. $-\dfrac{\sin^2(1/x)}{2} + C$ **59.** $-\dfrac{2}{\sin\sqrt{\theta}} + C$

Section 4.7, pp. 305–309

1. a) 2 b) 2 c) 2   **3.** a) 17/4 b) 4 c) 4   **5.** a) 5.1463
b) 5.252 c) 16/3   **7.** 0.001667   **9.** a) $n = 1$ b) $n = 2$
11. a) $n = 283$ b) $n = 2$ **13.** a) $n = 75$ b) $n = 12$ **15.** 1013
17. 466.7 in.^{2} **19.** No. 22 flashbulb, 40.25 lumens; no. 31
flashbulb, 55.5 lumens   **21.** 4, 4   **23.** 3.13791233
25. 1.37074081

Section 4 Miscellaneous Exercises, pp. 311–316

1. $y = x - \dfrac{1}{x} - 1$ **3.** $s = -\dfrac{2}{1 + t^2} + 2$ **5.** $y = 4x^{5/2} +$
$4x^{3/2} - 8x$ **7.** $F(1) - F(0)$ **9.** $y = x$ **11.** DUCK!!!
13. $k \approx 21.5$ ft/sec^2 **15.** a) 6144 ft b) 4296 ft c) B hits first.

17. $\displaystyle\int_{1}^{2}\dfrac{1}{x}\,dx$ **19.** a) True b) True c) False   **21.** 17/3   **23.** 6/5

25. $-\dfrac{3}{5}\sqrt[3]{2} + \dfrac{3}{5}\sqrt[3]{7}$ **27.** 4   **29.** $6\sqrt{3} - 2\pi$ **31.** 2   **33.** -2

35. $\sqrt{3}$ **37.** $-1 + \sqrt{2}$ **39.** 25°F   **41.** $f(x) = \dfrac{x}{\sqrt{x^2 + 1}}$

43. 4   **45.** a) 1/4 b) $\sqrt[3]{12}$ **47.** $h \leq 1/8$ **49.** $T = \pi$, $S = \pi$
51. The job cannot be done for \$11,000.   **53.** $2/x$

55. $\dfrac{\sin 4y}{\sqrt{y}} - \dfrac{\sin y}{2\sqrt{y}}$ **57.** 36/5

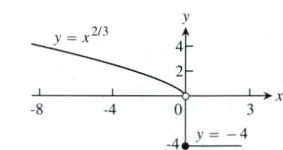

59. $\dfrac{1}{2} - \dfrac{2}{\pi}$

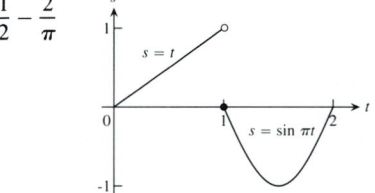

61. 13/3   **63.** 1/2   **65.** 1/6   **67.** $\displaystyle\int_{0}^{1}f(x)\,dx$

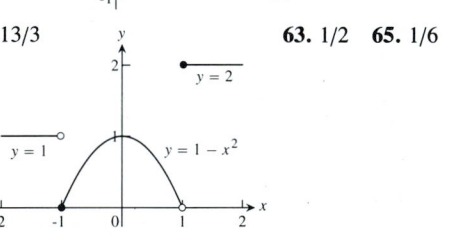

Chapter 5

Section 5.1, pp. 323–324

1. $\pi/2$ **3.** 1/6   **5.** 1/12   **7.** 32/3   **9.** 48/5   **11.** 8/3
13. 104/15   **15.** 34/15   **17.** 8   **19.** 18   **21.** 243/8   **23.** 8/3

25. 2   **27.** 4   **29.** $1 + \tan^2 x$ **31.** $\dfrac{4\pi - 12}{3\pi}$ **33.** 2   **35.** 1/6

37. 9/2   **39.** 1   **41.** $\sqrt{2} - 1$ **43.** a) $4^{2/3}$ b) $4^{2/3}$ **45.** 4

Section 5.2, pp. 335–338

1. $\dfrac{32\pi}{5}$ **3.** $\pi/30$ **5.** π **7.** 8π **9.** $\dfrac{16\pi}{15}$ **11.** 2π **13.** $\dfrac{2\pi}{3}$

15. $\dfrac{128\pi}{5}$ **17.** $\dfrac{117\pi}{5}$ **19.** $\pi(\pi - 2)$ **21.** $\dfrac{4\pi}{3}$ **23.** 8π

25. $\dfrac{500\pi}{3}$ **27.** $\pi^2 - 2\pi$ **29.** a) 8π b) $\dfrac{32\pi}{5}$ c) $\dfrac{8\pi}{3}$ d) $\dfrac{224\pi}{15}$

31. a) $\dfrac{16\pi}{15}$ b) $\dfrac{56\pi}{15}$ c) $\dfrac{64\pi}{15}$ **33.** $2/\pi$ **35.** 1053π cm^3

37. $2a^2b\pi^2$ **39.** a) $\dfrac{4a^3\pi}{3}$ b) $\dfrac{\pi r^2 h}{3}$ **43.** 16   **45.** 16/3

47. $\dfrac{24\sqrt{3} - \pi}{24}$ **49.** 8/3   **51.** $s^2h,\ s^2h$

Section 5.3, pp. 343–344

1. 8π **3.** $\dfrac{3\pi}{2}$ **5.** 3π **7.** $\dfrac{4\pi}{3}$ **9.** $\dfrac{16\pi}{3}$ **11.** $\dfrac{8\pi}{3}$ **13.** $\dfrac{14\pi}{3}$

15. a) $\dfrac{6\pi}{5}$ b) $\dfrac{4\pi}{5}$ c) 2π d) 2π **17.** a) $\dfrac{5\pi}{3}$ b) $\dfrac{4\pi}{3}$ c) 2π d) $\dfrac{2\pi}{9}$

19. a) $\dfrac{11\pi}{15}$ b) $\dfrac{97\pi}{105}$ c) $\dfrac{121\pi}{210}$ d) $\dfrac{23\pi}{30}$ **21.** a) $\dfrac{512\pi}{21}$ b) $\dfrac{832\pi}{21}$

23. a) $\pi/6$ b) $\pi/6$ **25.** $f(x) = 3x$

Section 5.4, pp. 350–352

1. $\displaystyle\int_{-1}^{2} \sqrt{1 + 4x^2}\, dx$ **3.** $\displaystyle\int_{-1}^{3} \sqrt{1 + (y + 1)^2}\, dy$

5. $\displaystyle\int_{0}^{\pi} \sqrt{1 + \cos^2 y}\, dy$ **7.** $\displaystyle\int_{0}^{\pi/6} \sec x\, dx$ **9.** 12   **11.** 14/3

13. 53/6   **15.** 123/32   **17.** 99/8   **19.** 2   **21.** 6   **23.** 1
25. $\sqrt{a^2 + b^2}$ **27.** \$38,419.50

Section 5.5, pp. 357–358

1. $2\pi \displaystyle\int_{-\pi/4}^{\pi/4} \tan x \sqrt{1 + \sec^4 x}\, dx$ **3.** $2\pi \displaystyle\int_{1}^{2} \dfrac{1}{y}\sqrt{1 + y^{-4}}\, dy$

5. $2\pi \displaystyle\int_{1}^{4} (3 - \sqrt{x})^2 \sqrt{1 + (1 - 3x^{-1/2})^2}\, dx$ **7.** $2\pi \displaystyle\int_{0}^{\pi/3}$

$\left(\displaystyle\int_{0}^{y} \tan t\, dt\right) \sec y\, dy$ **9.** $4\pi\sqrt{5}$ **11.** $3\pi\sqrt{5}$ **13.** $\dfrac{98\pi}{81}$

15. 4π **17.** $\dfrac{\pi(\sqrt{8} - 1)}{9}$ **19.** $\dfrac{35\pi\sqrt{5}}{3}$ **21.** $\dfrac{253\pi}{20}$ **23.** $4\pi a^2$

25. a) $2\pi \displaystyle\int_{-\pi/2}^{\pi/2} (\cos x) \sqrt{1 + \sin^2 x}\, dx$ b) 14.4236

27. Order 1446.39 liters of each color.   **29.** $5\sqrt{2}\pi$ **31.** 14.4
33. 54.9

Section 5.6, pp. 367–369

1. 4 ft   **3.** 1   **5.** 17/7   **7.** $\bar{x} = 0, \bar{y} = 12/5$ **9.** $\bar{x} = 1, \bar{y} = -\dfrac{3}{5}$

11. $\bar{x} = 16/105, \bar{y} = 8/15$ **13.** $\bar{x} = 0, \bar{y} = \pi/8$ **15.** $\bar{x} = 1,$

$\bar{y} = -2/5$ **17.** $\bar{x} = \bar{y} = \dfrac{2}{4 - \pi}$ **19.** $\dfrac{13\delta}{6}$ **21.** $\bar{x} = 0, \bar{y} = \dfrac{a\pi}{4}$

23. $\bar{x} = 0, \bar{y} = 1$ **25.** $\bar{x} = \bar{y} = a/3$

Section 5.7, pp. 376–378

1. 400 ft · lb   **3.** 925 N · m   **5.** 72900 ft · lb   **7.** 8 N · m
9. a) 7238 lb/in. b) 904.75 in. · lb, 2714.25 in. · lb

11. a) 1500000 ft · lb b) 6000 sec   **13.** 245436.93 ft · lb
15. 7238229.48 ft · lb   **17.** 137258.28 N · m
19. 967610.54 ft · lb; yes   **21.** 85.1 ft · lb   **23.** 64.6 ft · lb
25. 110.6 ft · lb   **27.** 5.144×10^{10} N · m

Section 5.8, pp. 382–384

1. 2812.5 lb   **3.** a) 375 lb against each end; no b) About
7.43 in.   **5.** 229.17 lb   **7.** a) Each side, 420.14 lb; each end
210.07 lb b) 840.28 lb   **9.** 4.2 lb   **11.** a) 93.33 lb b) 3 ft

13. a) $333\dfrac{1}{3}$ lb b) 9 ft   **15.** The tank will overflow.

Section 5.9, pp. 391–393

1. a) 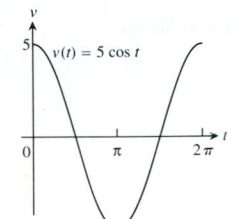 b) 20 m c) 0 m

3. a) 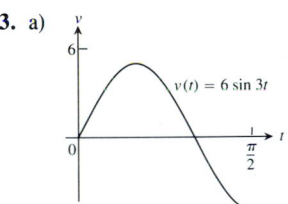 b) 6 m c) 2 m

5. a) 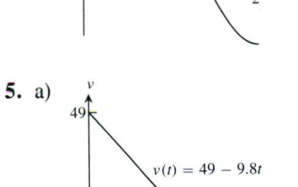 b) 245 m c) 0 m

7. a) 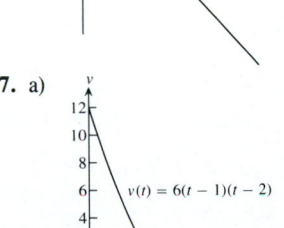 b) 6 m c) 4 m

9. a) $v(0) = 8$ b) $2 < t < 4$ c) 6m d) 22/3 m   **11.** 70%
13. $\sqrt{3}\pi$ **15.** V $= 32\pi$, S $= 32\sqrt{2}\pi$ **17.** $4\pi^2$ **19.** $\bar{x} = 0,$
$\bar{y} = \dfrac{2a}{\pi}$ **21.** $\bar{x} = 0, \bar{y} = \dfrac{4b}{3\pi}$ **23.** $\dfrac{\sqrt{2}\pi a^3(4 + 3\pi)}{6}$ **25.** $\dfrac{2a^3}{3}$

Section 5 Miscellaneous Exercises, pp. 393–396

1. 9/2   **3.** 0.0155   **5.** $\dfrac{8\sqrt{2} - 7}{6}$ **7.** min − 4; max 0; area 27/4

9. 6/5   **11.** 0.10097   **13.** π^2 **15.** $\dfrac{72\pi}{35}$ **17.** a) 2π b) π

c) $\dfrac{12\pi}{5}$ d) $\dfrac{26\pi}{5}$ **19.** a) 8π b) 227.87 c) 107.233

21. $\dfrac{\pi(3\sqrt{3} - \pi)}{3}$ **23.** a) $\dfrac{16\pi}{15}$ b) $\dfrac{8\pi}{5}$ c) $\dfrac{8\pi}{3}$ d) $\dfrac{32\pi}{5}$ **25.** $\dfrac{28\pi}{3}$

27. $\sqrt{\dfrac{2x - a}{\pi}}$ **29.** $2\sqrt{3}$ **31.** 285/8   **33.** 271.739   **35.** $\dfrac{425\pi}{9}$

37. 28/3   **39.** $\bar{x} = 0, \bar{y} = \dfrac{6}{5}$ **41.** $\bar{x} = 8/5, \bar{y} = 1$ **43.** $\bar{x} = 0,$

$\bar{y} = \dfrac{n}{2n + 1}; \left(0, \dfrac{1}{2}\right)$ **45.** 4560 N · m   **47.** 10 ft · lb, 30 ft · lb

49. $\dfrac{k(b - a)}{ab}$ **51.** 335153.25 ft · lb   **53.** $\dfrac{4h\sqrt{3mh}}{3}$

55. 333.33 lb   **57.** 2200 lb   **59.** $216\omega_1 + 360\omega_2$

61. a)   b) 29/2 m c) 27/2 m

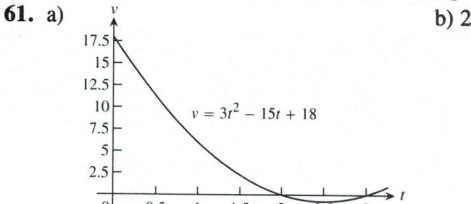

63. a)   b) 15 ft c) -5 ft

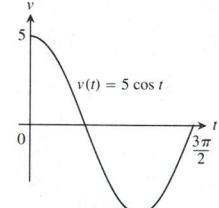

Chapter 6

Section 6.1, pp. 401–402

1. a) $f^{-1}(x) = \dfrac{x - 3}{2}$ b)

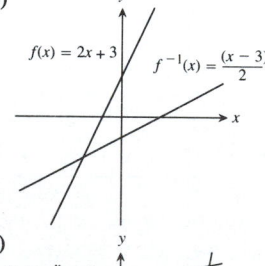

3. a) $f^{-1}(x) = 5x - 35$ b)

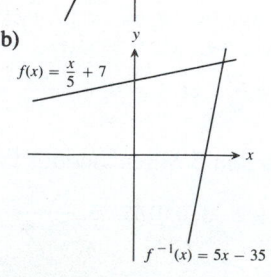

5. $f^{-1}(x) = \sqrt{x - 1}$ **7.** $f^{-1}(x) = \sqrt[3]{x + 1}$ **9.** $f^{-1}(x) = \sqrt[5]{x}$

11. $f^{-1}(x) = \sqrt[3]{x - 1}$ **13.** $f^{-1}(x) = \dfrac{1}{\sqrt{x}}$ **15.** $f^{-1}(x) = \sqrt{x} - 1$

17. a)   b) $y = 0, x = 0$ **19.** $\dfrac{1}{4}$

Section 6.2, pp. 410–412

1. a) $\ln 3 - \ln 4$ b) $2(\ln 2 - \ln 3)$ c) $-\ln 2$ d) $\dfrac{1}{3}\ln 3$

e) $\ln 3 + \dfrac{1}{2}\ln 2$ f) $\dfrac{1}{2}(3\ln 3 - \ln 2)$ **3.** $\dfrac{1}{x}$ **5.** $\dfrac{3}{x}$ **7.** $\dfrac{3}{2x}$

9. $\dfrac{1}{x + 2}$ **11.** $\dfrac{\sin x}{2 - \cos x}$ **13.** $\dfrac{1}{x \ln x}$ **15.** $\sin(\ln x) + \cos(\ln x) -$

$\dfrac{\sin(\ln x)}{x}$ **17.** $-\dfrac{3x + 2}{2x(x + 1)}$ **19.** $\dfrac{2}{x(1 - \ln x)^2}$ **21.** $\dfrac{\tan(\ln x)}{x}$

23. $\dfrac{10x}{x^2 + 1} + \dfrac{1}{2(1 - x)}$ **25.** $2x \ln|x| - x \ln\dfrac{|x|}{\sqrt{2}}$

27. $\left(\dfrac{1}{2}\right)\sqrt{x(x + 1)}\left(\dfrac{1}{x} + \dfrac{1}{x + 1}\right)$

29. $\sqrt{x + 3}\,(\sin x)\left(\dfrac{1}{2(x + 3)} + \cot x\right)$

31. $x(x + 1)(x + 2)\left[\dfrac{1}{x} + \dfrac{1}{x + 1} + \dfrac{1}{x + 2}\right]$

33. $\dfrac{x + 5}{x \cos x}\left[\dfrac{1}{x + 5} - \dfrac{1}{x} + \tan x\right]$

35. $\dfrac{x\sqrt{x^2 + 1}}{(x + 1)^{2/3}}\left[\dfrac{1}{x} + \dfrac{x}{x^2 + 1} - \dfrac{2}{3(x + 1)}\right]$

37. $\dfrac{1}{3}\sqrt[3]{\dfrac{x(x - 2)}{x^2 + 1}}\left[\dfrac{1}{x} + \dfrac{1}{x - 2} - \dfrac{2x}{x^2 + 1}\right]$ **39.** $-\infty$ **41.** $\ln 2$

43. -1 **45.** $\ln\left(\dfrac{2}{3}\right)$ **47.** $\ln|y^2 - 25| + C$ **49.** $\ln 3$ **51.** $(\ln 2)^2$

53. $\dfrac{1}{\ln 4}$ **55.** $\ln|6 + 3\tan t| + C$ **57.** $\ln 2$ **59.** $\ln 27$

61. $\ln|1 + \sqrt{x}| + C$ **63.** b) 0.005, 0.00033   **65.** $\ln 16$

67. $4\pi \ln 4$ **69.** $\pi \ln 16$ **71.** a) $6 + \ln 2$ b) $\dfrac{\sqrt{3}}{2} +$

$\dfrac{1}{4}\ln\left(2 + \sqrt{3}\right)$ **73.** a) $\bar{x} \approx 1.44$ and $\bar{y} \approx 0.36$

b)

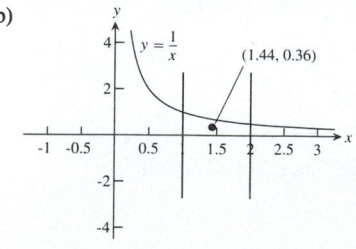

75. $y = x + \ln(x) + 2$ **77.** $\left(4 \ln \left(\dfrac{3}{2}\right) + 1\right)m$

79. $x = 0, x = 0.5, 0.09453, 0.2195$

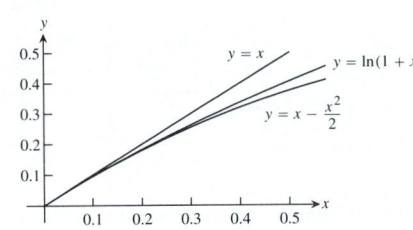

81. The graph of $y = -\ln|\sin x|$ would have inverted arches.

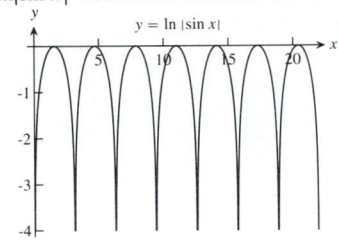

83. a)

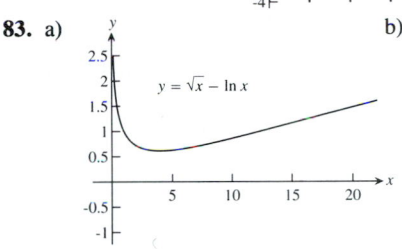

b) $x = 16$

Section 6.3, pp. 418–420

1. a) 7 b) 1 c) 1 **3.** a) $\dfrac{1}{x^2}$ b) xy c) $\dfrac{1}{x}$ **5.** a) ln 2 b) $-\dfrac{\ln 4}{5}$

c) $1000 \ln a$ **7.** a) 0 b) $-10 \ln 5$ c) $-10 \ln 3$ **9.** $e^{(2t + 4)}$

11. $e^{5t} + 40$ **13.** $4(\ln x)^2$ **15.** $x\,e^x + 1$ **17.** $-5e^{-5x}$

19. $-7e^{(5 - 7x)}$ **21.** $x\,e^x$ **23.** $x^2 e^x$ **25.** $2e^x \cos x$

27. $2xe^{-x^2} \sin e^{-x^2}$ **29.** $e^{\cos x}(1 - x \sin x)$ **31.** $(x^3 + 6)e^x - 6$

33. $\dfrac{\sin x}{x}$ **35.** $(\sin x + x \cos x)e^{x \sin x}$ **37.** $\dfrac{2e^{2x} - \cos(x + 3y)}{3 \cos(x + 3y)}$

39. 2 **41.** 1 **43.** 8 **45.** $e^2 - e$ **47.** 1 **49.** 1 **51.** $\tan \theta + e^{\tan \theta}$

$+ C$ **53.** $\lim_{\theta \to 0} \dfrac{\cos \theta - 1}{e^\theta - \theta - 1} = \lim_{\theta \to 0} \dfrac{-\sin \theta}{e^\theta - 1} = \lim_{\theta \to 0} \dfrac{-\cos \theta}{e^\theta} =$

-1 **55.** 1 **57.** b) 0.015, 0.0005 **59.** 1.09861 **61.** A local

minimum **63.** 2 **65.** a) 1 b) $\dfrac{\pi}{2}$ c) π **67.** $y = e^{x/2} - 1$

69. $y = -\cos\left(e^x - 2\right) - 1$ **71.** 2.71828183 **73.** a) $Q(x)$

overestimates e^x when $x < 0$ and underestimates e^x when $x > 0$.

$L(x)$ always underestimates e^x.

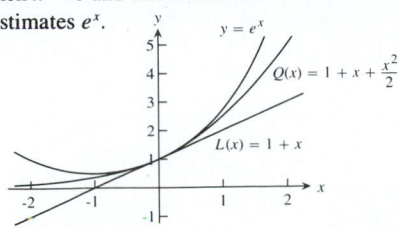

b) $x = 1; 0.71828, 0.21828$

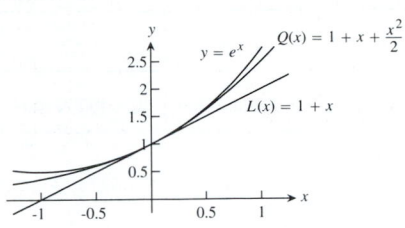

Section 6.4, pp. 428–431

1. $\pi x^{(\pi - 1)}$ **3.** $-\sqrt{2}r^{-\sqrt{2} - 1}$ **5.** $2^x \ln 2$ **7.** $\left(\dfrac{\ln 5}{2\sqrt{s}}\right)5^{\sqrt{s}}$

9. $2^{\sec x}(\ln 2)(\sec x)(\tan x)$ **11.** $(x^{\ln x})\left(\dfrac{\ln x^2}{x}\right)$ **13.** $(\sqrt{t})^t \left(\dfrac{\ln t}{2} + \dfrac{1}{2}\right)$

15. $((\cos x)(\ln(\sin x)) + \cos x)(\sin x)^{\sin x}$ **17.** $\left(\dfrac{\ln(\ln x) + 1}{x}\right)(\ln x)^{\ln x}$

19. $\dfrac{3}{\sqrt{3} + 1}$ **21.** $3^{\sqrt{2} + 1}$ **23.** $\dfrac{4}{\ln 5}$ **25.** $\dfrac{1}{\ln 4}$ **27.** $\dfrac{3}{2}$ **29.** $\dfrac{1}{\ln 2}$

31. 32760 **33.** a) 2 b) $\dfrac{3}{5}$ c) $\dfrac{1}{4}$ **35.** a) $\sqrt{x}$ b) x^2 c) $\sin x$

37. a) $\dfrac{\ln 3}{\ln 2}$ b) 3 c) 2 **39.** 12 **41.** $\dfrac{3}{x \ln 4}$ **43.** $\dfrac{2(\ln r)}{r(\ln 2)(\ln 3)}$

45. $\dfrac{1}{(\ln 100)(s + 1)}$ **47.** $\dfrac{-\ln 6}{t(\ln t)^2}$ **49.** $\dfrac{\ln 2}{x \ln(\ln 2)}$ **51.** $\dfrac{1}{\ln 5}$

53. $\dfrac{3 + 2x}{2x \ln 2}$ **55.** $\dfrac{1}{t(\ln 2)(\ln t)}$ **57.** $\dfrac{1}{\ln 10}\left[\dfrac{(\ln x)^2}{2}\right] + C$ **59.** ln 4

61. $\dfrac{3 \ln 2}{2}$ **63.** ln 10 **65.** $\displaystyle\int \dfrac{1}{x \log_{10} x}\,dx = (\ln 10)\displaystyle\int \dfrac{1/x}{\ln x}\,dx =$

$(\ln 10)\ln(\ln x) + C$ **67.** ln 3 **69.** $\dfrac{1}{e}$ **71.** 1 **73.** 1

75. a) $0.30341x + 0.08976, 1 + 0.30341x - 0.05057(x - 3)^2$

b) $0.00059, 0.00001$ **77.** $(-0.766664696, 0.587774756)$

79. a) 10^{-7} b) 7 c) 1 **81.** 10

83. a) 1

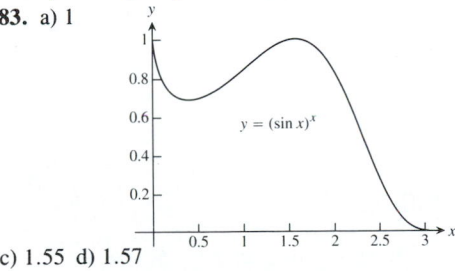

c) 1.55 d) 1.57

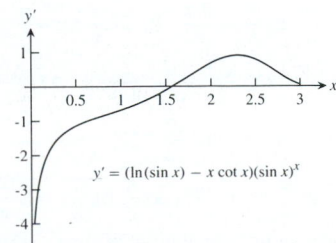

e) 1.570796327

f)

x	1.55	1.57	1.57070796327
$(\sin x)^x$	0.999664854	0.999999502	1

85. a)

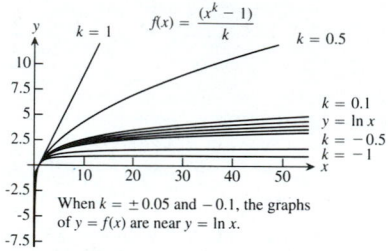

When $k = \pm 0.05$ and -0.1, the graphs of $y = f(x)$ are near $y = \ln x$.

Section 6.5, pp. 437–440

1. a) -0.00001 b) 10536 yr c) 82% **3.** 54.88 grams **5.** 59.8 ft
7. 2.8147497×10^{14} **9.** a) 8.00 (to the nearest hundredth of a year) b) 32.02 (to the nearest hundredth of a year) **11.** 16.09 yr
13. a) $A(t) = A_o e^t$ b) 1.099 years c) More than 2.7 times the original amount, A_o **15.** 4.875% **17.** 0.585 days
21. a) 17.5 min b) 13.26 min **23.** $-3°C$ **25.** 6658.3 yr
27. 41.22 yr

Section 6.6, pp. 444–446

1. a) Yes b) Yes c) Yes d) No e) Yes f) Yes g) Yes h) Yes
i) No j) No **3.** a) Yes b) Yes c) Yes d) No e) No f) No
g) No h) Yes i) No j) No **7.** a) False b) False c) True d) True
e) True f) True g) False h) True **11.** b) 24154952.75 c) $x \approx$
$3.430631121 \times 10^{15}$ d) In the interval $[3.41 \times 10^{15}, 3.45 \times 10^{15}]$
we have $\ln x = 10 \ln(\ln x)$.

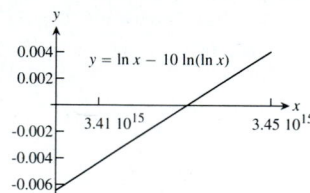

13. n

Section 6.7, pp. 450–451

1. a) $\dfrac{\pi}{4}$ b) $-\dfrac{\pi}{3}$ c) $\dfrac{\pi}{6}$ **3.** a) $-\dfrac{\pi}{6}$ b) $\dfrac{\pi}{4}$ c) $-\dfrac{\pi}{3}$ **5.** a) $\dfrac{\pi}{3}$
b) $\dfrac{3\pi}{4}$ c) $\dfrac{\pi}{6}$ **7.** a) $\dfrac{3\pi}{4}$ b) $\dfrac{\pi}{6}$ c) $\dfrac{2\pi}{3}$ **9.** a) $\dfrac{\pi}{4}$ b) $-\dfrac{\pi}{3}$ c) $\dfrac{\pi}{6}$
11. a) $\dfrac{3\pi}{4}$ b) $\dfrac{\pi}{6}$ c) $\dfrac{2\pi}{3}$ **13.** $\dfrac{\sqrt{3}}{2}, \dfrac{1}{\sqrt{3}}, \dfrac{2}{\sqrt{3}}, 2$ **15.** $\dfrac{\sqrt{3}}{2}$ **17.** $\dfrac{\pi}{2}$
19. $\dfrac{4 + \sqrt{3}}{2\sqrt{3}}$ **21.** 0 **23.** $\dfrac{\pi}{6}$ **25.** $\dfrac{12}{13}$ **27.** $\dfrac{x}{\sqrt{x^2 + 4}}$ **29.** $\dfrac{2|x|}{3}$
31. x **33.** π **35.** a) $\dfrac{\pi}{2}$ b) π **37.** a) $\dfrac{\pi}{2}$ b) $\dfrac{\pi}{2}$ **41.** $\dfrac{1}{\pi}$ sec; no,
because it would take an infinite amount of time **43.** 42.23°

47. The graphs are identical.

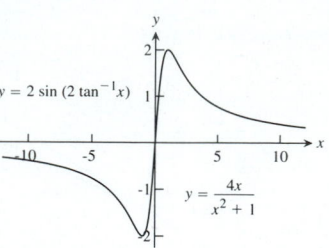

49. a) Domain, all reals except those having the form $\dfrac{\pi}{2} + k\pi$,
where k is an integer; range, $(-\pi/2, \pi/2)$; no, many computer graphs will not be correct.

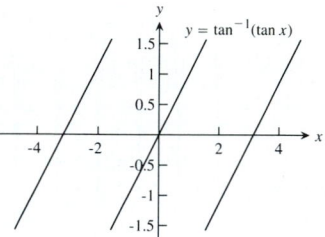

b) Domain, all reals; range, all reals; yes, many computer graphs will be correct.

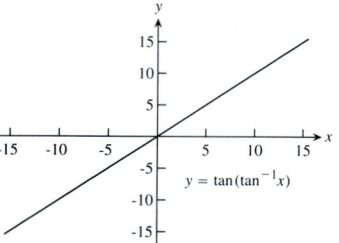

51. a) Domain, all real; range, $[0, \pi]$; yes, many computer graphs will be correct.

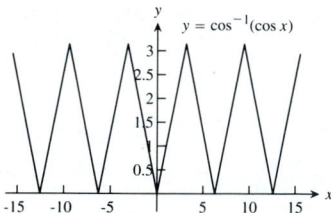

b) Domain, $[-1, 1]$; range, $[-1, 1]$; no, many computer graphs will not be correct.

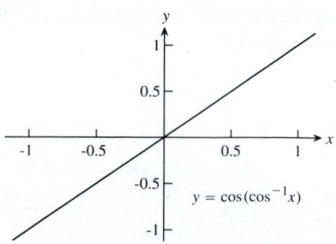

Section 6.8, pp. 456–457

1. $\dfrac{-2x}{\sqrt{1-x^4}}$ **3.** $\dfrac{15}{1+9x^2}$ **5.** $\dfrac{1}{\sqrt{4-x^2}}$ **7.** $\dfrac{1}{|x|\sqrt{25x^2-1}}$

9. $\dfrac{-2x}{(x^2+1)\sqrt{x^4+2x^2}}$ **11.** 0 **13.** $\dfrac{-1}{2x\sqrt{x-1}}$ **15.** $\dfrac{x|x|-1}{|x|\sqrt{x^2-1}}$

17. $\sin^{-1}x$ **23.** $\dfrac{\pi}{6}$ **25.** $\dfrac{\pi}{12}$ **27.** $\dfrac{\pi}{12}$ **29.** $\dfrac{\pi}{2}$ **31.** π **33.** $\dfrac{\pi}{6}$

35. $\dfrac{\pi}{6}$ **37.** $\dfrac{1}{2}\sec^{-1}(x^2)+C$ **39.** $\dfrac{\pi}{12}$ **41.** $-\dfrac{\pi}{12}$ **43.** 1

45. $\dfrac{\pi^2}{2}$ **47.** $3\sqrt{5}$ ft **49.** $1,\dfrac{\pi}{2}$ **51.** $y=\sec^{-1}x+\dfrac{2\pi}{3}$

53. $y=\cos^{-1}(x)-\dfrac{\pi}{4}$ **55.** 0.643517104 **57.** Yes

63.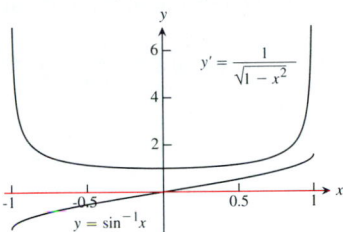

65.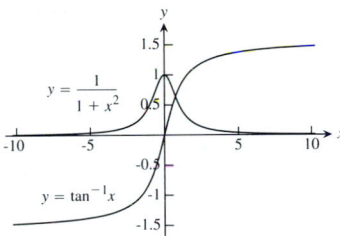

Section 6.9, pp. 464–468

1. $\cosh x=\dfrac{5}{4}$, $\tanh x=-\dfrac{3}{5}$, $\coth x=-\dfrac{5}{3}$, $\operatorname{sech} x=\dfrac{4}{5}$,
$\operatorname{csch}=-\dfrac{4}{3}$ **3.** $\sinh x=\dfrac{8}{15}$, $\tanh x=\dfrac{8}{17}$, $\coth x=\dfrac{17}{8}$,
$\operatorname{sech} x=\dfrac{15}{17}$, $\operatorname{csch}=\dfrac{15}{8}$ **5.** $x+\dfrac{1}{x}$ **7.** e^{5x} **9.** e^{-3x}

13. $3\cosh(3x)$ **15.** $\operatorname{sech}^2\left(\dfrac{x}{2}\right)$ **17.** $-\tanh(x)$ **19.** $-\operatorname{csch}(x)$

21. $\operatorname{csch}(2x)$ **23.** a) $\sinh(2x)$ b) $\sinh(2x)$ c) $\sinh(2x)$

25. $\dfrac{2}{\sqrt{1+4x^2}}$ **27.** $\dfrac{1}{1+x}-\tanh^{-1}(x)$ **29.** $\operatorname{sech}^{-1}(x)-\dfrac{1}{\sqrt{1-x^2}}$

31. $\sec x$ **33.** $\sec x$ **35.** $-\csc x$ **45.** $\left(\dfrac{2}{5}\right)\sinh 5$ **47.** 0

49. e **51.** $\sinh(\ln x)+C$ **53.** $\dfrac{4}{5}$ **55.** $2(\sinh(2)-\sinh(1))$

57. $\dfrac{\sinh(\ln 64)}{12}+\ln\sqrt{2}$ **59.** $\dfrac{(\operatorname{sech} 5x)^3}{3}+C$ **61.** $\ln\left(\dfrac{2}{3}\right)$

63. $\ln 3$ **65.** $-\dfrac{\ln 3}{2}$ **67.** $\left(\dfrac{1}{2}\right)\ln 9=\ln 3$ **69.** $\ln 3$

71. $\ln(-\sqrt{3}+2)$ **73.** a) $\sinh^{-1}(1)$ b) $\ln(1+\sqrt{2})$

75. a) $\cosh^{-1}\left(\dfrac{5}{3}\right)-\cosh^{-1}\left(\dfrac{5}{4}\right)$ b) $\ln\left(\dfrac{3}{2}\right)$

77. a) $\coth^{-1}(2)-\coth^{-1}\left(\dfrac{5}{4}\right)$ b) $\left(\dfrac{1}{2}\right)\ln\left(\dfrac{1}{3}\right)$

79. a) $\left(\dfrac{1}{2}\right)\left(\operatorname{csch}^{-1}\left(\dfrac{1}{2}\right)-\operatorname{csch}^{-1}(1)\right)$ b) $\left(\dfrac{1}{2}\right)\ln\left(\dfrac{2+\sqrt{5}}{1+\sqrt{2}}\right)$

81. 2π **83.** $\bar{x}=0,\bar{y}=\dfrac{3}{2\pi}$ **85.** a) $\bar{x}=0,\bar{y}=\dfrac{5}{8}+\dfrac{\ln 4}{3}$

b)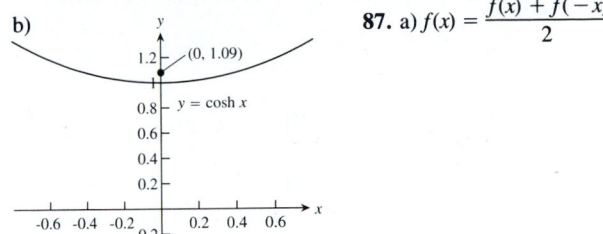

87. a) $f(x)=\dfrac{f(x)+f(-x)}{2}$

b) $f(x)=\dfrac{f(x)-f(-x)}{2}$; $v=\sqrt{\dfrac{mg}{k}}\tanh\left(\sqrt{\dfrac{gk}{m}}\,t\right)+C$ and

$v(0)=0\Rightarrow C=0\therefore v=\sqrt{\dfrac{mg}{k}}\tanh\left(\sqrt{\dfrac{gk}{m}}\,t\right)$ b) $\lim_{t\to\infty}v=$

$\lim_{t\to\infty}\sqrt{\dfrac{mg}{k}}\tanh\left(\sqrt{\dfrac{gk}{m}}\,t\right)=\sqrt{\dfrac{mg}{k}}\lim_{t\to\infty}\tanh\left(\sqrt{\dfrac{gk}{m}}\,t\right)=$

$\sqrt{\dfrac{mg}{k}}\,(1)=\sqrt{\dfrac{mg}{k}}$ **89.** $y=\operatorname{sech}^{-1}(x)-\sqrt{1-x^2}$

91. c) $A(u)=\dfrac{u}{2},\,A(0)=0$

Section 6.10, p. 474

1. a) $\dfrac{6}{5}$ b) $\dfrac{\sinh ab}{a}$

Section 6 Miscellaneous Exercises, pp. 474–478

1. $\dfrac{-1}{\sqrt{1-x^2}\cos^{-1}x}$ **3.** $\dfrac{2}{(\ln 2)x}$ **5.** $2xe^{\tan^{-1}x}+e^{\tan^{-1}x}$

7. $(\ln(x+2)+1)(x+2)^{x+2}$ **9.** $\dfrac{-1}{2\sqrt{x-x^2}}$

11. $\tan^{-1}(x)+\dfrac{x}{1+x^2}-\dfrac{1}{2x}$ **13.** $\dfrac{\sec^{-1}\sqrt{x}}{\sqrt{x-1}}+\dfrac{1}{x}$ **17.** $-\dfrac{y\ln y}{x\ln x}$

19. $\coth^2 s$ **21.** $-s\operatorname{csch}^2 s$ **23.** $\operatorname{sech} s$ **25.** $\dfrac{v\sinh^{-1}v}{\sqrt{1+v^2}}+1$

27. $\dfrac{1}{v^2-1}$ **29.** $2\sec 2v$ **31.** $\ln 4$ **33.** $\ln 2$ **35.** $\ln\left(\dfrac{32}{31}\right)$

37. 0 **39.** 4 **41.** $\dfrac{1}{\ln 2}$ **43.** $3\sin^{-1}\left(\ln(x)-\dfrac{1}{2}\right)+C$ **45.** $\dfrac{\pi}{4}$

47. $\ln|\sec^{-1}x|+C$, where $x>1$ **49.** $3+\ln 4$ **51.** $8\sqrt{2}$

53. $-5\operatorname{csch}^2 s+C$ **55.** a) $\sinh^{-1}(1)$ b) $\ln(1+\sqrt{2})$

57. a) $2\left[\left(\tanh^{-1}(1/2)\right)^2-\left(\tanh^{-1}(1/5)\right)^2\right]$ b) $\ln\left(\dfrac{9}{2}\right)\ln\sqrt{2}$

59. a) $\left(\operatorname{sech}^{-1}\dfrac{3}{5}\right)^2-\left(\operatorname{sech}^{-1}\dfrac{4}{5}\right)^2$ b) $(\ln 6)\left(\ln\dfrac{3}{2}\right)$

61. $\ln|y+1|=\dfrac{1}{x}-1$ **63.** a) $\ln x$ b) $\dfrac{\ln 2}{\ln\left(\dfrac{3}{2}\right)}$ **65.** $\ln 2$ **67.** $2\pi^2$

69. e^3 **71.** $\ln 2$ **73.** $\dfrac{1}{3}$ **75.** 1   **79.** $8 + \ln 9$ **81.** $\dfrac{1}{\ln (6/5)}$

83. a) $\dfrac{\pi \ln 4}{2}$ b) $\bar{x} = \dfrac{7}{8}, \bar{y} = \dfrac{\ln 4}{12}$ **85.** x **87.** $\dfrac{2}{17}$ **89.** a) No

b) Yes   **91.** a) 0.698970004   b) 1.584962501   c) 0.356207187

93. $\dfrac{1}{\ln 2}, \dfrac{1}{2 \ln 2}$; 2:1   **97.** a) About $x = 11$ there is a minimum.

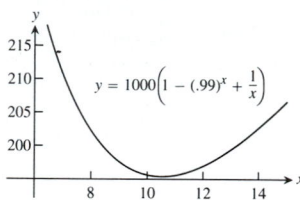

$$y = 1000\left(1 - (.99)^x + \dfrac{1}{x}\right)$$

b) There is no maximum; however, the curve is asymptotic to $y = 1000$. The curve is near 1000 when $x \geq 643$.

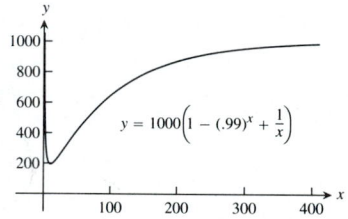

$$y = 1000\left(1 - (.99)^x + \dfrac{1}{x}\right)$$

99. 18935 yr   **101.** a) $c - (c - y_o)e^{-\frac{kA}{V}t}$ b) c

103. a) All the growth rates are the same.   b) $x^{\log 2^x}$ grows faster than $x^{\ln x}$.   c) $\left(\dfrac{1}{2}\right)^x$ grows faster than $\left(\dfrac{1}{3}\right)^x$.   **105.** a) True   b) True

c) False   **107.** $\dfrac{\pi}{2}$ **109.** $20(5 - \sqrt{17})m$ **113.** $y = \dfrac{x^2}{4} - \ln\sqrt{x} - \dfrac{1}{4}$

115. 3.809730

Chapter 7

Section 7.1, pp. 484–485

1. $2\sqrt{8x^2 + 1} + C$ **3.** 2   **5.** $\ln 5$ **7.** $2 \ln(\sqrt{x} - 1) + C$
9. $\ln(2 + \sqrt{3})^6$ **11.** $\ln(\sqrt{2} + 1)$ **13.** 1   **15.** $e^{\sqrt{3}} - 1$
17. $\dfrac{2}{\ln 3}$ **19.** $\dfrac{2^{\sqrt{w}}}{\ln 2} + C$ **21.** π **23.** $12 \tan^{-1}(\sqrt{y}) + C$
25. $\dfrac{\pi}{18}$ **27.** $\dfrac{\pi}{6}$ **29.** $6 \sec^{-1}|5x| + C$ **31.** $\sec^{-1}(e^y) + C$
33. $\ln(2 + \sqrt{3})$ **35.** $\dfrac{1}{8} \ln(1 + 4 \ln^2 x) + C$ **37.** $[\sin^{-1}(x - 2)] + C$
39. 2π **41.** $[\sec^{-1}|x + 1|] + C$, when $|x + 1| > 1$
43. $4 - \dfrac{\pi}{2}$ **45.** 0   **47.** $x - \ln|x + 1| + C$ **49.** $7 + \ln 8$
51. $\dfrac{2\pi - 3}{6}$ **53.** $\sqrt{2}$ **55.** $\ln(2 + \sqrt{3})$ **57.** $\bar{x} = 0$,

$\bar{y} = \dfrac{1}{\ln(2\sqrt{2} + 3)}$ **59.** $\bar{x} = \dfrac{\pi}{2}, \bar{y} = \dfrac{\dfrac{4\pi}{3} - \sqrt{3}}{\dfrac{4\pi}{3} + \ln(7 - 4\sqrt{3})}$

63. 8.30 cm

Section 7.2, pp. 491–492

1. $-x \cos x + \sin x + C$ **3.** $t^2 \sin t + 2t \cos t - 2 \sin t + C$
5. $\ln(4) - \dfrac{3}{4}$ **7.** $x \tan^{-1}(x) - \ln\sqrt{1 + x^2} + C$ **9.** $x \tan x +$
$\ln|\cos x| + C$ **11.** $(x^3 - 3x^2 + 6x - 6)e^x + C$
13. $(x^2 - 7x + 7)e^x + C$ **15.** $(x^5 - 5x^4 + 20x^3 - 60x^2 +$
$120x - 120)e^x + C$ **17.** $\dfrac{\pi^2 - 4}{8}$ **19.** $\dfrac{5\pi - 3\sqrt{3}}{9}$
21. $\dfrac{1}{2}[-e^\theta \cos \theta + e^\theta \sin \theta] + C$ **23.** $\dfrac{e^{2x}}{13}(3 \sin 3x + 2 \cos 3x)$
$+ C$ **25.** $\dfrac{2}{3}(\sqrt{3s + 9}\,e^{\sqrt{3s + 9}} - e^{\sqrt{3s + 9}}) + C$
27. $\dfrac{\pi\sqrt{3}}{3} - \ln(2) - \dfrac{\pi^2}{18}$ **29.** $\dfrac{1}{2}[-x \cos(\ln x) + x \sin(\ln x)] + C$
31. a) π b) 3π **33.** $2\pi - \dfrac{4\pi}{e}$ **35.** a) $\bar{x} = \dfrac{e^2 + 1}{4}, \bar{y} = \dfrac{e - 2}{2}$
b) $\bar{x} \approx 2.1$ and $\bar{y} \approx 0.36$

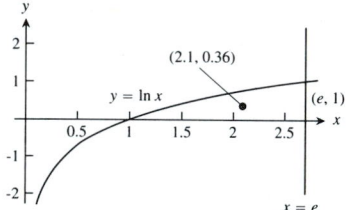

37. $\pi^2 + \pi - 4$ **39.** $x \sin^{-1}x + \cos(\sin^{-1}x) + C$
41. $x \sec^{-1}x - \ln|x + \sqrt{x^2 - 1}| + C$ **43.** Yes   **45.** a) $x \sinh^{-1}x - \cosh(\sinh^{-1}x) + C$ b) $x \sinh^{-1}x + (1 + x^2)^{1/2} + C$

Section 7.3, pp. 499–501

1. $\dfrac{8}{15}$ **3.** $\dfrac{4}{3}$ **5.** $\cos y + \cos^3 y - \dfrac{3 \cos^5 y}{5} + \dfrac{\cos^7 y}{7} + C$ **7.** $3x -$
$2 \sin 2x + \dfrac{\sin 4x}{4} + C$ **9.** $10 - 3\pi$ **11.** $\dfrac{13\sqrt{2} - 16}{2}$ **13.** 0
15. $\dfrac{3\pi}{2}$ **17.** $35\left[\dfrac{\sin^5 x}{5} - \dfrac{\sin^7}{7}\right] + C$ **19.** $\dfrac{1}{2}\left[\ln x - \dfrac{\sin(\ln x^4)}{4}\right] + C$
21. 4   **23.** 2   **25.** $\ln(3 + 2\sqrt{2})$ **27.** $\sqrt{2}[-\theta \cos \theta + \sin \theta] + C$
29. 4   **31.** $2(\ln(\sin x) - 1)(\sin x) + C$ **33.** $2\sqrt{3} -$
$\ln(2 - \sqrt{3})$ **35.** $\sqrt{2} + \ln(\sqrt{2} - 1)$ **37.** $\dfrac{\sqrt{2} - \ln(\sqrt{2} + 1)}{2}$
39. $\dfrac{4}{3}$ **41.** $-\cot \theta + \dfrac{\cot^3\theta}{3} + C$ **43.** $2 - \ln 4$ **45.** $\dfrac{6\pi - 16}{3}$
47. $(\tan t) \ln(\cos t) + \tan t - t + C$ **49.** $\dfrac{2}{7} \sec^{7/2} - \dfrac{2}{3} \sec^{3/2} + C$
51. $-\dfrac{6}{5}$ **53.** $\dfrac{3\sqrt{3} - 1}{3}$ **55.** $\dfrac{24\sqrt{3} - 18}{7}$ **57.** a, b, c, e, g, and i
are zero; d, f, and h are not zero.   **59.** d, f, h; 2, $\dfrac{4}{3}, \dfrac{4}{3}$

Section 7.4, pp. 506–507

1. $\dfrac{\pi}{4}$ **3.** $\dfrac{\pi}{6}$ **5.** $\ln|x + \sqrt{x^2 - 4}| + C$ **7.** $2 \tan^{-1}2\sqrt{t} + C$
9. $\dfrac{\pi}{8}$ **11.** $\dfrac{\pi}{3}$ **13.** $\sinh^{-1}\left(\dfrac{\ln y}{2}\right) + C$ **15.** π **17.** $\dfrac{\pi}{8}$

19. $\dfrac{2 - \sqrt{3}}{2}$ **21.** $\ln|(x - 1) + \sqrt{x^2 - 2x}| + C$ **23.** 2

25. $\sinh^{-1}\left(\dfrac{s - 1}{2}\right) + C$ **27.** $\dfrac{1}{2}\tan^{-1}\left(\dfrac{3x - 1}{2}\right) + C$

29. $\cosh^{-1}\left(\dfrac{r - 1}{2}\right) + C$ **31.** $\ln(26) - 4\tan^{-1}(5)$

33. $\dfrac{1}{3}\sinh^{-1}\left(\dfrac{3x - 1}{2}\right) + C$ **35.** $\sinh^{-1}(1)$ **37.** $\cosh^{-1}\left(\dfrac{5}{3}\right) -$

$\cosh^{-1}\left(\dfrac{4}{3}\right)$ **39.** $\sqrt{r^2 + 4r + 5} - 2\sinh^{-1}(r + 2) + C$

41. $\dfrac{4x}{\sqrt{1 - x^2}} - 4\sin^{-1}x + C$ **43.** $\dfrac{3\pi}{4}$ **45.** 1.195 **47.** 576.102

49. $\bar{x} \approx 2.82,\ \bar{y} \approx 0.28$ **51.** $\dfrac{1}{2}\operatorname{arcsec} 2$

Section 7.5, pp. 511–513

1. $\dfrac{A}{x - 3} + \dfrac{B}{x - 2} = \dfrac{2}{x - 3} + \dfrac{3}{x - 2}$ **3.** $\dfrac{1}{x + 1} + \dfrac{3}{(x + 1)^2}$

5. $\dfrac{-2}{x} + \dfrac{-1}{x^2} + \dfrac{2}{x - 1}$ **7.** $1 + \dfrac{17}{x - 3} + \dfrac{-12}{x - 2}$ **9.** $\ln\sqrt{3}$

11. $\dfrac{1}{7}\ln|(x + 6)^2(x - 1)^5| + C$ **13.** $\ln\sqrt{15}$ **15.** $-\dfrac{1}{2}\ln|t| +$

$\dfrac{1}{6}\ln|t + 2| + \dfrac{1}{3}\ln|t - 1| + C$ **17.** $\dfrac{x^2}{2} - 2x + 3\ln|x + 1| +$

$\dfrac{1}{x + 1} + C$ **19.** $\dfrac{1}{4}\ln\left|\dfrac{x + 1}{x - 1}\right| - \dfrac{x}{2(x^2 - 1)} + C$

21. $\ln\left(\dfrac{9}{8}\right)$ **23.** $\dfrac{1}{5}\ln\left|\dfrac{x - 2}{x + 3}\right| + C$ **25.** $-x + 4\sqrt{x} -$

$4\ln(\sqrt{x} + 1) + C$ **27.** $\dfrac{t^2\ln(t + 5)}{2} + \dfrac{5x}{2} - \dfrac{x^2}{4} - \dfrac{25}{2}\ln|x + 5| + C$

29. $4 - \ln 3$ **31.** $\ln\left(\dfrac{2\sqrt{2}}{\sqrt{5}}\right)$ **33.** $5\sqrt{3} - \dfrac{5\pi}{3}$

35. $\ln\left(\dfrac{x^4}{(x^2 + 1)^2}\right) - \dfrac{4}{x} - 4\tan^{-1}x + C$ **37.** $\dfrac{2 + \pi - \ln 4}{4}$

39. $\dfrac{\pi + 2}{4}$ **41.** $2 + \ln 3$ **43.** $\dfrac{\pi - 5}{4}$ **45.** $\dfrac{1}{8}\ln|4w^2 - 4w - 1| +$

$\dfrac{\sqrt{2}}{8}\ln\left|\dfrac{1 + \sqrt{2}w - \sqrt{2}}{1 - \sqrt{2}w + \sqrt{2}}\right| + C$ **47.** $\ln\left|\dfrac{\theta - 1}{\theta^2 + \theta + 1}\right| -$

$\dfrac{2}{\sqrt{3}}\tan^{-1}\left(\dfrac{2\theta + 1}{\sqrt{3}}\right) + C$ **49.** $\dfrac{-1}{x^2 + 2x + 2} + \ln|x^2 + 2x + 2| -$

$\tan^{-1}(x + 1) + C$ **51.** $3\pi\ln 25$ **53.** $\bar{x} \approx 1.10$

55. a) $x = \dfrac{1000e^{4t}}{499 + e^{4t}}$ b) 1.55 days **57.** $\dfrac{22}{7} - \pi$

Section 7.6, pp. 518–519

1. $\dfrac{x^2}{2}\cos^{-1}(x) + \dfrac{1}{4}\sin^{-1}(x) - \dfrac{1}{4}x\sqrt{1 - x^2} + C$ **3.** $\dfrac{\pi - 2}{16}$

5. $\dfrac{2}{3}\ln\left(\dfrac{8}{5}\right) + \dfrac{11}{20}$ **7.** $-\dfrac{\sqrt{7 - x^2}}{7x} + C$ **9.** $\dfrac{\pi}{18}$ **11.** $\dfrac{26}{3}$

13. $2\sqrt{3x - 4} - 4\tan^{-1}\sqrt{\dfrac{3x - 4}{4}} + C$ **15.** $\dfrac{\pi}{4}$ **17.** $\dfrac{\pi - 2}{4}$

19. $\bar{x} = \dfrac{4}{3},\ \bar{y} = \dfrac{\ln 2}{2} = \ln\sqrt{2}$ **21.** 7.62 **27.** $\dfrac{3\pi}{4}$ **29.** $\dfrac{1 - \ln 2}{4}$

33. $\dfrac{3\pi - 8}{6}$ **35.** $4\sqrt{3}$ **37.** $\dfrac{4}{3}$ **39.** $\dfrac{5e^4 - 1}{2}$ **41.** $\dfrac{1}{\sqrt{2}}$

43. $\dfrac{735 + 384\ln 2}{1024}$

Section 7.7, pp. 528–530

1. $\dfrac{\pi}{2}$ **3.** 6 **5.** 4 **7.** 1000 **9.** Diverges **13.** $\sqrt{3}$ **15.** -1

17. 24 **19.** Converges **21.** Converges **23.** Diverges

25. Converges **27.** Diverges **29.** Diverges **31.** Diverges

33. Converges **35.** Converges **37.** Diverges **39.** Diverges

41. Converges **43.** Diverges **45.** Converges **47.** Converges

49. Converges **51.** $\ln 2$ **53.** $\dfrac{\pi}{4}$ **55.** $\dfrac{-\ln\left(\dfrac{5}{36}\right)}{16}$ **57.** $\cosh^{-1}2$

59. $-\sqrt{8}$ **61.** $0.000041136;\ 0.88617257$ **65.** 1 **67.** 2π

69. $\ln 2$ **73.** Diverges **75.** Converges **77.** 2 **79.** Diverges

83. a) 0 b) 1

Section 7 Miscellaneous Exercises, pp. 531–535

1. $2\sqrt{1 + \sin x} + C$ **3.** $\dfrac{\tan^2 x}{2} + C$ **5.** 18

7. $\ln|\sqrt{x^2 + 2x + 2} + x + 1| + C$ **9.** $e^x(x^2 - 2x + 2) + C$

11. $2\sqrt{r} - 2\ln|1 + \sqrt{r}| + C$ **13.** $\ln|x^4 - 10x^2 + 9| + C$

15. $\dfrac{9}{25}(t^{5/3} + 1)^{5/3} + C$ **17.** $\ln\left|\dfrac{u - 1}{u + 1}\right| + C$ **19.** $\dfrac{\pi}{4}$ **21.** $\dfrac{\pi}{4}$

23. $\ln\sqrt{3}$ **25.** 2 **27.** $\dfrac{3}{5}(1 + e^x)^{5/3} - \dfrac{3}{2}(1 + e^x)^{2/3} + C$ **29.** $\dfrac{5}{12}$

31. $\ln|2 + \ln x| + C$ **33.** $\dfrac{4}{21}(\sqrt[4]{8} + 2\sqrt[4]{27})$ **35.** $\dfrac{3}{80}$

37. $-2\sqrt{x + 1}\cos\sqrt{x + 1} + 2\sin\sqrt{x + 1} + C$ **39.** $\dfrac{\ln 3}{4}$

41. $\ln|y| - \dfrac{1}{3}\ln|2y^3 + 1| + \dfrac{1}{3}(2y^3 + 1)^{-1} + C$ **43.** $\ln\left|\dfrac{x}{\sqrt{x^2 + 1}}\right| +$

$\dfrac{1}{2}(x^2 + 1)^{-1} + C$ **45.** $\ln|1 - e^{-x}| + C$ **47.** $\dfrac{1}{2}\ln\left|\dfrac{(x + 3)^3}{x + 1}\right| + C$

49. $\ln\left|\dfrac{x}{\sqrt{x^2 + 4}}\right| + C$ **51.** $\sqrt{x^2 - 1} - \sec^{-1}|x| + C$

53. $\ln\left(\dfrac{x}{(3\sqrt{x} + 1)^2}\right) + C$ **55.** $\ln\sqrt{\dfrac{5}{2}}$ **57.** $\dfrac{3(e^{2t} + 1)^{4/3}}{8} -$

$\dfrac{3(e^{2t} + 1)^{1/3}}{2} + C$ **59.** $\dfrac{x^2}{2} + \dfrac{1}{3}\ln((x + 2)^4(x - 1)^2) + C$

61. $\ln|x - 1| - \dfrac{1}{x - 1} + C$ **63.** $\sec^{-1}(2y + 1) + C$

65. $\dfrac{3\sin^{-1}x}{8} + \dfrac{x\sqrt{1 - x^2}}{2} + \dfrac{x\sqrt{1 - x^2}(1 - 2x^2)}{8} + C$

67. $p\tan p + \ln|\cos p| - \dfrac{p^2}{2} + C$ **71.** $\dfrac{(9 - 2\sqrt{3})\pi}{72}$ **73.** $\dfrac{1}{2} + \ln 2$

75. $q\sec^{-1}q - \ln|q + \sqrt{q^2 - 1}| + C$ **77.** $\dfrac{1}{16}\ln\left|\dfrac{x^2 - 4}{x^2 + 4}\right| + C$

79. $\ln|\csc 2x + \cot 2x| + C$ **81.** $\dfrac{2\pi\sqrt{3} - 9}{9}$

83. $\sqrt{2}\tan^{-1}(\sqrt{2}\tan t) - t + C$ **85.** $e^t\sin(e^t) + \cos(e^t) + C$

87. $\dfrac{x^2\ln(x^3 + x)}{2} - \dfrac{3x^2}{4} + \dfrac{\ln(x^2 + 1)}{2} + C$

89. $\ln|\sqrt{3 + \sin^2 x} + \sin x| + C$ **91.** $x^2\cos(1 - x) +$
$2x \sin(1 - x) - 2 \cos(1 - x) + C$ **93.** $\dfrac{\tan^2 x}{2} + \ln|\cos x| + C$
95. $\ln(\sqrt{x^2 + 1}) + \dfrac{1}{2(x^2 + 1)} + C$ **97.** $\dfrac{298}{15}$
99. $\ln(x - \sqrt{x^2 - 1}) + \sqrt{x^2 - 1} + C$
101. $x \ln(x + \sqrt{x}) - x + \sqrt{x} - \ln|\sqrt{x} + 1| + C$ **103.** $\dfrac{10\pi}{3} -$
$4\sqrt{3} + 4$ **105.** 2 **107.** $\dfrac{(\sin^{-1}x)^2}{4} + \dfrac{x \sin^{-1}(x)\sqrt{1 - x^2}}{2} - \dfrac{x^2}{4} + C$
109. $x - \tan x + \sec x + C$ **111.** $\dfrac{-1}{3(\sin x - 2)} -$
$\dfrac{4}{9} \ln|\sin x - 2| + \dfrac{1}{2} \ln|\sin x - 1| - \dfrac{1}{18} \ln|\sin x + 1| + C$
113. $-2 \sin^{-1}\left(\dfrac{\cos\left(\dfrac{x}{2}\right)}{\cos\left(\dfrac{\alpha}{2}\right)}\right) + C$ **115.** $\ln 6 - 2 + 2\sqrt{2} \tan^{-1}\left(\dfrac{1}{\sqrt{2}}\right)$
117. $\ln\left|\dfrac{1 + \sqrt{1 - e^{-t}}}{1 - \sqrt{1 - e^{-t}}}\right| + C$ **119.** $\dfrac{\sqrt{8} - 1}{6}$ **121.** $\ln\left(\dfrac{729}{5}\right)$
123. $\dfrac{1}{2}$ **125.** $x(\sin^{-1}x)^2 + 2(\sin^{-1}x)\sqrt{1 - x^2} - 2x + C$
127. $\dfrac{x^2\sin^{-1}x}{2} + \dfrac{x\sqrt{1 - x^2} - \sin^{-1}x}{4} + C$
129. $\dfrac{\ln|\sec 2\theta + \tan 2\theta| + 2\theta}{4} + C$
131. $\dfrac{1}{2} \left(\ln(t - \sqrt{1 - t^2}) - \sin^{-1}t\right) + C$
133. $\dfrac{1}{16} \ln\left|\dfrac{x^2 + 2x + 2}{x^2 - 2x + 2}\right| + \dfrac{1}{8} (\tan^{-1}(x + 1) + \tan^{-1}(x - 1)) + C$
135. $\ln(4) - 1$ **137.** a) $\pi(2e - 5)$ b) $\dfrac{\pi(e^2 - 3)}{2}$
c) $\dfrac{\pi(5 + e^2 - 4e)}{2}$ d) $\pi(2e - 5)$ **139.** $\bar{x} = \dfrac{2}{\pi}, \bar{y} = 0$
141. $\pi\left(\dfrac{8(\ln 2)^2}{3} - \dfrac{16(\ln 2)}{9} + \dfrac{16}{27}\right)$
143. $2\pi\left[e\sqrt{1 + e^2} + \ln\left(\dfrac{e + \sqrt{1 + e^2}}{1 + \sqrt{2}}\right) - \sqrt{2}\right]$
147. Diverges   **149.** Diverges
151. a)

153. 1 **155.** $\dfrac{2}{1 - \tan\left(\dfrac{x}{2}\right)} + C$ **157.** $-\cot\left(\dfrac{x}{2}\right) - x + C$
159. $\dfrac{1}{\sqrt{2}} \ln\left|\dfrac{\tan\left(\dfrac{x}{2}\right) + 1 - \sqrt{2}}{\tan\left(\dfrac{x}{2}\right) + 1 + \sqrt{2}}\right| + C$

Chapter 8

Section 8.1, pp. 549–552

1. $a_1 = 0, a_2 = -\dfrac{1}{4}, a_3 = -\dfrac{2}{9}, a_4 = -\dfrac{3}{16}$ **3.** $a_1 = 1, a_2 = -\dfrac{1}{3},$
$a_3 = \dfrac{1}{5}, a_4 = -\dfrac{1}{7}$ **5.** $1, \dfrac{3}{2}, \dfrac{7}{4}, \dfrac{15}{8}, \dfrac{31}{16}, \dfrac{63}{32}, \dfrac{127}{64}, \dfrac{255}{128}, \dfrac{511}{256}, \dfrac{1023}{512}$
7. $2, 1, -\dfrac{1}{2}, -\dfrac{1}{4}, \dfrac{1}{8}, \dfrac{1}{16}, -\dfrac{1}{32}, -\dfrac{1}{64}, \dfrac{1}{128}, \dfrac{1}{256}$ **9.** $1, 1, 2, 3, 5,$
$8, 13, 21, 34, 55$ **11.** Converges, 2 **13.** Converges, -1
15. Converges, -5 **17.** Diverges   **19.** Diverges   **21.** Converges, $\dfrac{1}{2}$
23. Converges, 0 **25.** Converges, 0 **27.** Converges, $\sqrt{2}$
29. Converges, $\dfrac{\pi}{2}$ **31.** Converges, 0 **33.** Converges, 0
35. Converges, 1 **37.** Converges, e^7 **39.** Diverges
41. Converges, 1 **43.** Converges, 1 **45.** Diverges
47. Converges, 0 **49.** Converges, 0 **51.** Converges, e^{-1}
53. Diverges   **55.** Converges, 1 **57.** Converges, 1
59. Converges, $e^{2/3}$ **61.** Converges, x **63.** Converges, 0
65. Converges, 1 **67.** Converges, $\dfrac{1}{2}$ **69.** Converges, $\dfrac{1}{2}$ **71.** 693
73. 66 **75.** $x_1 = 1, x_2 \approx 1.540302305, x_3 \approx 1.570791601,$
$x_4 \approx 1.570796327$ **79.** a) $f(x) = x^2 - 2; 1.414213562 \approx \sqrt{2}$
b) $f(x) = \tan(x) - 1; 0.7853981635 \approx \dfrac{\pi}{4}$ c) $f(x) = e^x;$
the sequence $1, 0, -1, -2, -3, -4, -5, \ldots$, diverges.
81. b) 1 **85.** 1 **91.** 0.999998902 **93.** 0.73908456
95. 0.853748068 **99.** a) -3 b) 0.2

Section 8.2, pp. 561–562

1. $s_n = \dfrac{2(1 - (1/3)^n)}{1 - 1/3}, 3$ **3.** $s_n = \dfrac{1 - (-1/2)^n}{1 - (-1/2)}, \dfrac{2}{3}$
5. $s_n = \dfrac{1}{2} - \dfrac{1}{n + 2}, \dfrac{1}{2}$ **7.** $1 - \dfrac{1}{4} + \dfrac{1}{16} - \dfrac{1}{64} + \cdots, \dfrac{4}{5}$
9. $\dfrac{7}{4} + \dfrac{7}{16} + \dfrac{7}{64} + \cdots, \dfrac{7}{3}$ **11.** $(5 + 1) + \left(\dfrac{5}{2} + \dfrac{1}{3}\right) +$
$\left(\dfrac{5}{4} + \dfrac{1}{9}\right) + \left(\dfrac{5}{8} + \dfrac{1}{27}\right) + \cdots, \dfrac{23}{2}$ **13.** $(1 + 1) + \left(\dfrac{1}{2} - \dfrac{1}{5}\right) +$
$\left(\dfrac{1}{4} + \dfrac{1}{25}\right) + \left(\dfrac{1}{8} - \dfrac{1}{125}\right) + \cdots, \dfrac{17}{6}$ **15.** 1 **17.** $\dfrac{1}{9}$ **19.** $2 + \sqrt{2}$
21. 1 **23.** Diverges   **25.** $\dfrac{e^2}{e^2 - 1}$ **27.** Diverges   **29.** $\dfrac{3}{2}$
31. Diverges   **33.** $\dfrac{\pi}{\pi - e}$ **35.** Diverges   **37.** 1 **39.** $a = 1,$
$r = -x$, where $|x| < 1$.   **41.** $a = 2, r = \dfrac{x}{3}$, where $|x| < 3$.
43. $|x| < \dfrac{1}{2}, \dfrac{1}{1 - 2x}$ **45.** $-2 < x < 0, \dfrac{1}{2 + x}$ **47.** $x \neq k(\pi),$
where k is an integer, $\dfrac{1}{1 - \sin x}$.   **49.** $\dfrac{23}{99}$ **51.** $\dfrac{7}{9}$ **53.** $\dfrac{1}{15}$
55. $\dfrac{41333}{33300}$ **57.** 28 m   **59.** a) $\displaystyle\sum_{n = -2}^{\infty} \dfrac{1}{(n + 4)(n + 5)}$

b) $\sum_{n=0}^{\infty} \dfrac{1}{(n+2)(n+3)}$ c) $\sum_{n=5}^{\infty} \dfrac{1}{(n-3)(n-2)}$

61. $S_{n} = \dfrac{1+(-1)^{n+1}}{2}$ **63.** 8 m^2 **65.** a) $3\left(\dfrac{4}{3}\right)^n$

b) $\dfrac{\sqrt{3}}{4} + \dfrac{27\sqrt{3}}{64}\left(\dfrac{4}{9}\right)^2 + \dfrac{27\sqrt{3}}{64}\left(\dfrac{4}{9}\right)^3 + \cdots, \dfrac{2\sqrt{3}}{5}$

Section 8.3, pp. 570–572

1. Converges **3.** Converges **5.** Converges **7.** Diverges
9. Converges **11.** Diverges **13.** Diverges **15.** Converges
17. Diverges **19.** Diverges **21.** Converges **23.** Converges
25. Converges **27.** Converges **29.** Converges **31.** Converges
33. Diverges **35.** About 41.55

Section 8.4, p. 577

1. Converges **3.** Converges **5.** Diverges **7.** Diverges
9. Diverges **11.** Converges **13.** Diverges **15.** Converges
17. Converges **19.** Converges **21.** Converges **23.** Converges
25. Diverges **27.** Converges **29.** Converges **31.** Converges
33. Diverges **35.** Converges **37.** Converges **39.** Converges
41. Diverges

Section 8.5, pp. 584–585

1. Converges **3.** Diverges **5.** Converges **7.** Converges
9. Diverges **11.** Diverges **13.** Converges absolutely
15. Converges absolutely **17.** Converges conditionally
19. Diverges **21.** Converges conditionally **23.** Converges
absolutely **25.** Converges absolutely **27.** Diverges
29. Diverges **31.** Converges absolutely **33.** Converges
absolutely **35.** Converges absolutely **37.** Converges
conditionally **39.** Diverges **41.** Converges conditionally
43. Converges absolutely **45.** $|\text{error}| < 0.2$ **47.** $|\text{error}|$
$< 2 \times 10^{-11}$ **49.** 0.54030 **51.** a) $a_n \geq a_{n+1}$ b) $-\dfrac{1}{2}$

59. $S_1 = -\dfrac{1}{2}$

$S_2 = -\dfrac{1}{2} + 1$

$S_3 = -\dfrac{1}{2} + 1 - \dfrac{1}{4} - \dfrac{1}{4} - \dfrac{1}{6} - \dfrac{1}{8} - \dfrac{1}{10} - \dfrac{1}{12} - \dfrac{1}{14} - \dfrac{1}{16} - \dfrac{1}{18}$

$- \dfrac{1}{20} - \dfrac{1}{22} \approx -0.5099$

$S_4 = -\dfrac{1}{2} + 1 - \dfrac{1}{4} - \dfrac{1}{4} - \dfrac{1}{6} - \dfrac{1}{8} - \dfrac{1}{10} - \dfrac{1}{12} - \dfrac{1}{14} - \dfrac{1}{16} - \dfrac{1}{18}$

$- \dfrac{1}{20} - \dfrac{1}{22} + \dfrac{1}{3} \approx -0.1766$

$S_5 = -\dfrac{1}{2} + 1 - \dfrac{1}{4} - \dfrac{1}{4} - \dfrac{1}{6} - \dfrac{1}{8} - \dfrac{1}{10} - \dfrac{1}{12} - \dfrac{1}{14} - \dfrac{1}{16} - \dfrac{1}{18}$

$- \dfrac{1}{20} - \dfrac{1}{22} + \dfrac{1}{3} - \dfrac{1}{24} - \dfrac{1}{26} - \dfrac{1}{28} - \dfrac{1}{30} - \dfrac{1}{32} - \dfrac{1}{34} - \dfrac{1}{36}$

$- \dfrac{1}{38} - \dfrac{1}{40} - \dfrac{1}{42} - \dfrac{1}{44} \approx -0.512$

$S_6 = -\dfrac{1}{2} + 1 - \dfrac{1}{4} - \dfrac{1}{4} - \dfrac{1}{6} - \dfrac{1}{8} - \dfrac{1}{10} - \dfrac{1}{12} - \dfrac{1}{14} - \dfrac{1}{16} - \dfrac{1}{18}$

$- \dfrac{1}{20} - \dfrac{1}{22} + \dfrac{1}{3} - \dfrac{1}{24} - \dfrac{1}{26} - \dfrac{1}{28} - \dfrac{1}{30} - \dfrac{1}{32} - \dfrac{1}{34} - \dfrac{1}{36}$

$- \dfrac{1}{38} - \dfrac{1}{40} - \dfrac{1}{42} - \dfrac{1}{44} - \dfrac{1}{46} - \dfrac{1}{48} - \dfrac{1}{50} - \dfrac{1}{52} - \dfrac{1}{54} - \dfrac{1}{56}$

$- \dfrac{1}{58} - \dfrac{1}{60} - \dfrac{1}{62} - \dfrac{1}{64} - \dfrac{1}{66} \approx -0.511065$

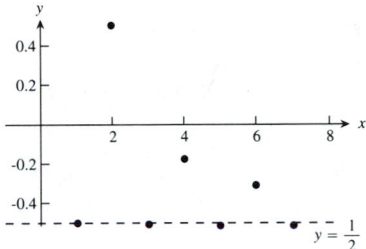

Section 8.6; pp. 592–593

1. a) 1, $-1 < x < 1$ b) $-1 < x < 1$ c) $\{\}$ **3.** a) 1,
$-2 < x < 0$ b) $-2 < x < 0$ c) $\{\}$ **5.** a) 10, $-8 < x < 12$
b) $-8 < x < 12$ c) $\{\}$ **7.** a) 1, $-1 < x < 1$ b) $-1 < x < 1$
c) $\{\}$ **9.** a) 3, $[-3, 3]$ b) $[-3, 3]$ c) $\{\}$ **11.** a) ∞, for all x
b) for all x c) $\{\}$ **13.** a) ∞, for all x b) for all x c) $\{\}$
15. a) 1, $-1 \leq x < 1$ b) $-1 < x < 1$ c) $\{-1\}$ **17.** a) 4, $[-4, 4)$
b) $(-4, 4)$ c) $\{-4\}$ **19.** a) 3, $-3 < x < 3$ b) $-3 < x < 3$
c) $\{\}$ **21.** a) 1, $-1 < x < 1$ b) $-1 < x < 1$ c) $\{\}$ **23.** a) 0,
$x = 0$ b) $x = 0$ c) $\{\}$ **25.** a) 2, $(0, 4]$ b) $0 < x < 4$ c) $\{4\}$
27. a) 1, $[-1, 1)$ b) $(-1, 1)$ c) $\{-1\}$ **29.** a) ∞,
$(-\infty, -1) \cup (1, \infty)$ b) $x < -1$ or $x > 1$ c) $\{\}$ **31.** a) $\dfrac{1}{e}$,
$\left(-\dfrac{1}{e}, \dfrac{1}{e}\right)$ b) $-\dfrac{1}{e} < x < \dfrac{1}{e}$ c) $\{\}$ **33.** a) 1, $(-1, 1)$
b) $-1 < x < 1$ c) $\{\}$ **35.** $(1 - \sqrt{2}, 1 + \sqrt{2})$, $\dfrac{2}{1 + 2x - x^2}$
37. $(-\sqrt{3}, \sqrt{3})$, $\dfrac{2}{3 - x^2}$ **39.** $(-\infty, -\sqrt{5}) \cup (\sqrt{5}, \infty)$,
$\dfrac{x^2 - 2}{x^2 + 1}$ **41.** $1 < x < 5$, $\dfrac{2}{x - 1}$; $1 < x < 5$, $\dfrac{-2}{(x - 1)^2}$
43. a) $\dfrac{x^2}{2} + \dfrac{x^4}{12} + \dfrac{x^6}{45} + \dfrac{17x^8}{2520} + \dfrac{31x^{10}}{14175}$, when $-\dfrac{\pi}{2} < x < \dfrac{\pi}{2}$
b) $1 + x^2 + \dfrac{2x^4}{3}$, when $-\dfrac{\pi}{2} < x < \dfrac{\pi}{2}$

Section 8.7, pp. 606–608

1. $P_0(x) = 0$, $P_1(x) = x - 1$, $P_2(x) = (x - 1) - \dfrac{1}{2}(x - 1)^2$,
$P_3(x) = (x - 1) - \dfrac{1}{2}(x - 1)^2 + \dfrac{1}{3}(x - 1)^3$ **3.** $P_0(x) = \dfrac{1}{2}$,
$P_1(x) = \dfrac{1}{2} - \dfrac{1}{4}(x - 2)$, $P_2(x) = \dfrac{1}{2} - \dfrac{1}{4}(x - 2) + \dfrac{1}{8}(x - 2)^2$,
$P_3(x) = \dfrac{1}{2} - \dfrac{1}{4}(x - 2) + \dfrac{1}{8}(x - 2)^2 - \dfrac{1}{16}(x - 2)^3$

5. $P_1(x) = \dfrac{\sqrt{2}}{2} + \dfrac{\sqrt{2}}{2}\left(x - \dfrac{\pi}{4}\right)$, $P_2(x) = \dfrac{\sqrt{2}}{2} +$

$\dfrac{\sqrt{2}}{2}\left(x - \dfrac{\pi}{4}\right) - \dfrac{\sqrt{2}}{4}\left(x - \dfrac{\pi}{4}\right)^2$, $P_3(x) = \dfrac{\sqrt{2}}{2} + \dfrac{\sqrt{2}}{2}\left(x - \dfrac{\pi}{4}\right) -$

$\dfrac{\sqrt{2}}{4}\left(x - \dfrac{\pi}{4}\right)^2 - \dfrac{\sqrt{2}}{12}\left(x - \dfrac{\pi}{4}\right)^3$ **7.** $P_0(x) = 2$, $P_1(x) = 2 +$

$\dfrac{1}{4}(x - 4)$, $P_2(x) = 2 + \dfrac{1}{4}(x - 4) - \dfrac{1}{64}(x - 4)^2$, $P_3(x) = 2 +$

$\dfrac{1}{4}(x - 4) - \dfrac{1}{64}(x - 4)^2 + \dfrac{1}{512}(x - 4)^3$ **9.** $1 - x + \dfrac{x^2}{2!} -$

$\dfrac{x^3}{3!} + \dfrac{x^4}{4!} - \cdots$ **11.** $3x - \dfrac{(3x)^3}{3!} + \dfrac{(3x)^5}{5!} - \cdots$

13. $1 - \dfrac{x^2}{2!} + \dfrac{x^4}{4!} - \dfrac{x^6}{6!} + \cdots$ **15.** $1 + \dfrac{x^2}{2!} + \dfrac{x^4}{4!} + \dfrac{x^6}{6!} + \cdots$

17. $\dfrac{x^4}{4!} - \dfrac{x^6}{6!} + \dfrac{x^8}{8!} - \dfrac{x^{10}}{10!} + \cdots$ **19.** $1 + x + \dfrac{x^2}{2} + \dfrac{f'''(c)}{3!}x^3$, where

c is between 0 and x.   **21.** $-x^2 + \dfrac{f'''(c)}{3!}x^3$, where c is between 0

and x.   **23.** $x + \dfrac{f'''(c)}{3!}x^3$, where c is between 0 and x.

27. $-0.569679052 < x < 0.569679052$ **29.** $|\text{error}| < 1.67 \times$
10^{-10}, $-10^{-3} < x < 0$ **31.** 0.00018602 **33.** 0.000260416
35. $\sin x$, when $x = 0.1$; 0.099833416

39. $2x - \dfrac{8x^3}{3!} + \dfrac{32x^5}{5!} - \dfrac{128x^7}{7!} + \dfrac{512x^9}{9!}$ **45.** a) -1

b) $\left(\dfrac{1}{\sqrt{2}}\right)(1 + i)$ c) $-i$ **49.** $x + x^2 + \dfrac{1}{3}x^3 - \dfrac{1}{30}x^5$, for all x

Section 8.8, p. 614

1. $\cos x = \dfrac{1}{2} \cdot \dfrac{\sqrt{3}}{2}(x - \pi/3) - \dfrac{1}{4}(x - \pi/3)^2 + \dfrac{\sqrt{3}}{12}(x - \pi/3)^3$

3. $1 + x + \dfrac{x^2}{2!} + \dfrac{x^3}{3!}$ **5.** $1 - \dfrac{1}{2}(x - 22\pi)^2 + \dfrac{1}{4!}(x - 22\pi)^4 -$

$\dfrac{1}{6!}(x - 22\pi)^6$ **7.** 0.0027; 0.0000003 **9.** 0.00033; 0.000002
11. 0.96356; 0.000266 **13.** 0.1; 0.000001 **15.** 0.099944461
17. 0.100001 **21.** 500 **23.** 3 **25.** a) $x + \dfrac{x^3}{6} + \dfrac{3x^5}{40} + \dfrac{5x^7}{112}$; 1

b) $\dfrac{\pi}{2} - x - \dfrac{x^3}{6} - \dfrac{3x^5}{40} - \dfrac{5x^7}{112}$ **27.** $1 - 2x + 3x^2 - 4x^3 + \cdots$

29. $\dfrac{x^2}{2} + \dfrac{x^4}{12} + \dfrac{x^6}{45}$

Section 8 Miscellaneous Exercises, pp. 618–621

1. Converges, 1 **3.** Diverges   **5.** Diverges   **7.** Converges, $\sqrt{3}$
9. Converges, 0 **11.** Converges, $\ln 2$ **13.** Converges, $\dfrac{1}{e}$

15. Diverges   **17.** $\dfrac{e}{e - 1}$ **19.** Diverges   **21.** Converges
conditionally   **23.** Converges conditionally   **25.** Converges
27. Converges absolutely   **29.** Converges absolutely
31. Converges   **33.** Converges   **35.** Diverges   **37.** Converges
absolutely   **39.** Converges

41. a) 3, $[-5, 1)$ b) $(-5, 1)$ c) $\{-5\}$ **43.** a) ∞, for all x
b) for all x c) $\{\}$ **45.** a) 1, $[0, 2]$ b) $[0, 2]$ c) $\{\}$ **47.** a) e,
$(-e, e)$ b) $(-e, e)$ c) $\{\}$ **49.** a) $\sqrt{3}$, $(-\sqrt{3}, \sqrt{3})$
b) $(-\sqrt{3}, \sqrt{3})$ c) $\{\}$ **51.** $\dfrac{2}{3}$ **53.** $\dfrac{1}{1 + x}$, $\dfrac{1}{4}$, $\dfrac{4}{5}$ **55.** $\sin x$, π, 0

57. e^x, $\ln 2$, 2 **59.** $2 - \dfrac{(x + 1)}{2 \cdot 1!} + \dfrac{3(x + 1)^2}{2^3 \cdot 2!} + \dfrac{9(x + 1)^3}{2^5 \cdot 3!}$

61. Yes, $L^2 = \sin L$ **63.** $n\left(\dfrac{n + 1}{2n}\right)$, $\ln\left(\dfrac{1}{2}\right)$ **71.** Diverges

75. b) Yes   **77.** a) ∞ **79.** $-\dfrac{x^2}{2} - \dfrac{x^4}{12} - \dfrac{x^6}{45}$ **81.** $(1, 3)$, 1

83. ∞, for all x **87.** 0.185330149 **93.** a) $\displaystyle\sum_{n=1}^{\infty} nx^{n-1}$ b) 6 c) $\dfrac{1}{q}$

97. $\dfrac{1}{2}$ **99.** 1 **101.** 1 **103.** 2 **105.** $\dfrac{1}{2}$ **107.** 2 **109.** 1
111. 0 **113.** ∞

Chapter 9

Section 9.1, pp. 637–640

1. $y^2 = 8x$, $(2, 0)$, $x = -2$ **3.** $x^2 = -6y$, $\left(0, -\dfrac{3}{2}\right)$, $y = \dfrac{3}{2}$

5. $\dfrac{x^2}{4} - \dfrac{y^2}{9} = 1$, $(\pm\sqrt{13}, 0)$, $e = \sqrt{13}/2$; Directrices:

$x = \pm 4/\sqrt{13}$; Asymptotes: $y = \pm(3/2)x$ **7.** $\dfrac{x^2}{2} + y^2 = 1$,

$(\pm 1, 0)$, $e = 1/\sqrt{2}$; Directrices: $x = \pm 2$

9. $\dfrac{x^2}{25} + \dfrac{y^2}{16} = 1$, $e = \dfrac{3}{5}$ **11.** $x^2 + \dfrac{y^2}{2} = 1$, $e = \dfrac{1}{\sqrt{2}}$

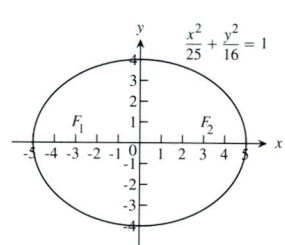 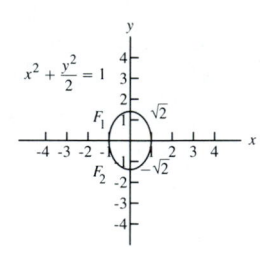

13. $\dfrac{x^2}{2} + \dfrac{y^2}{3} = 1$, $e = \dfrac{1}{\sqrt{3}}$ **15.** $\dfrac{x^2}{9} + \dfrac{y^2}{6} = 1$, $e = \dfrac{\sqrt{3}}{3}$

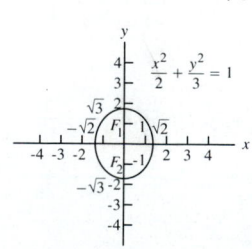 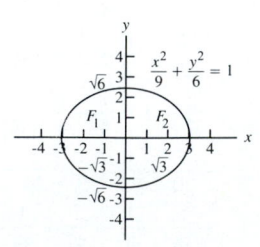

17. $x^2 - y^2 = 1$, $e = \sqrt{2}$, $y = \pm x$

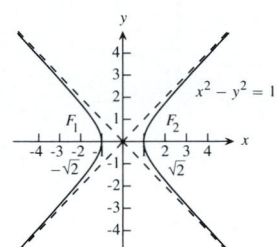

19. $\dfrac{y^2}{8} - \dfrac{x^2}{8} = 1$, $e = \sqrt{2}$, $y = \pm x$

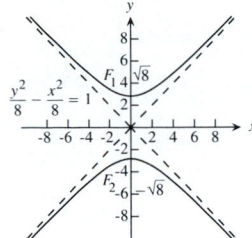

21. $\dfrac{x^2}{2} - \dfrac{y^2}{8} = 1$, $e = \sqrt{5}$, $y = \pm 2x$

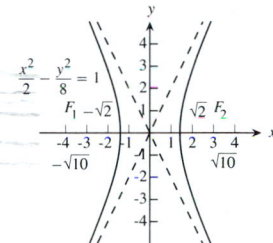

23. $\dfrac{y^2}{2} - \dfrac{x^2}{8} = 1$, $e = \sqrt{5}$, $y = \pm \dfrac{x}{2}$

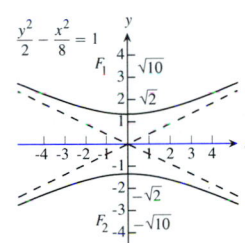

25. $\dfrac{x^2}{4} + \dfrac{y^2}{2} = 1$ **27.** $\dfrac{x^2}{27} + \dfrac{y^2}{36} = 1$ **29.** $\dfrac{x^2}{4851} + \dfrac{y^2}{4900} = 1$

31. $e = \dfrac{\sqrt{5}}{3}$, $\dfrac{x^2}{9} + \dfrac{y^2}{4} = 1$ **33.** $e = \dfrac{1}{2}$, $\dfrac{x^2}{64} + \dfrac{y^2}{48} = 1$

35. $y^2 - x^2 = 1$ **37.** $\dfrac{x^2}{9} - \dfrac{y^2}{16} = 1$ **39.** $y^2 - \dfrac{x^2}{8} = 1$

41. $x^2 - \dfrac{y^2}{8} = 1$ **43.** $e = \sqrt{2}$, $\dfrac{x^2}{8} - \dfrac{y^2}{8} = 1$

45. $e = 2$, $x^2 - \dfrac{y^2}{3} = 1$ **47.**

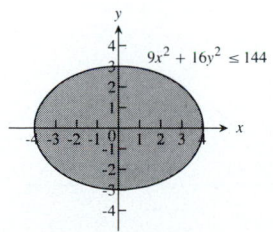

49.

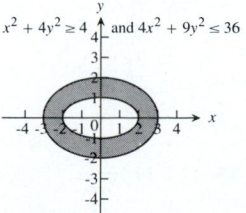

51.

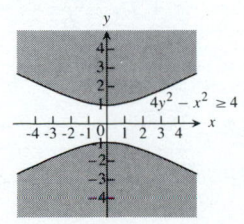

55. At $(4, 0)$, $y = 2x - 8$; at $(0, 0)$, $y = -2x$; at $(0, 2)$, $y = 2x + 2$ **57.** b) 1:1   **59.** Length, $2\sqrt{2}$; height, $\sqrt{2}$
61. a) 24π b) 16π **63.** 24π

Section 9.2, pp. 648–649

1. Hyperbola   **3.** Parabola   **5.** Ellipse   **7.** Parabola
9. Hyperbola   **11.** Hyperbola   **13.** Ellipse   **15.** Ellipse
17. $x'^2 - y'^2 = 4$, hyperbola   **19.** $4x'^2 + 16y' = 0$, parabola
21. $y'^2 = 1$, parallel horizontal lines   **23.** $2\sqrt{2}x'^2 + 8\sqrt{2}y' = 0$,
parabola   **25.** $4x'^2 + 2y'^2 = 19$, ellipse   **27.** $A' = 0.88$, $B' = 0$,
$C' = 3.12$, $D' = 0.74$, $E' = -1.20$, $F' = -3$, ellipse
29. $A' = 0.00$, $B' = 0$, $C' = 5.00$, $D' = 0$, $E' = 0$, $F' = -5$,
parallel lines   **31.** $A' = 5.05$, $B' = 0$, $C' = -0.05$,
$D' = -5.07$, $E' = -6.19$, $F' = -1$, hyperbola

33. a) $\dfrac{x'^2}{b^2} + \dfrac{y'^2}{a^2} = 1$ b) $\dfrac{y'^2}{a^2} - \dfrac{x'^2}{b^2} = 1$ c) $x'^2 + y'^2 = a^2$

d) $y' = -\dfrac{1}{m}x'$ e) $y' = -\dfrac{1}{m}x' + \dfrac{b}{m}$ **35.** a) $x'^2 - y'^2 = 2$

b) $x'^2 - y'^2 = 2a$ **39.** a) Hyperbola b)

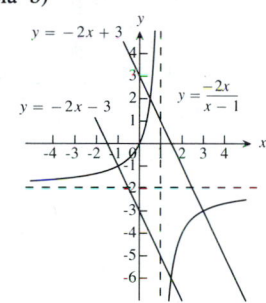

c) $y = -2x - 3$, $y = -2x + 3$ **41.** $x^2 + 4xy + 5y^2 - 1 = 0$,
ellipse   **49.** There is no unique line L, a directrix.

Section 9.3, pp. 656–657

1. $x^2 + y^2 = 1$

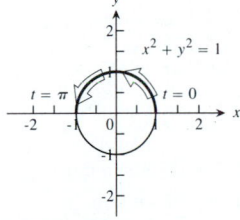

3. $x^2 + y^2 = 1$

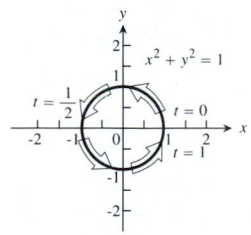

5. $\dfrac{x^2}{16} + \dfrac{y^2}{4} = 1$

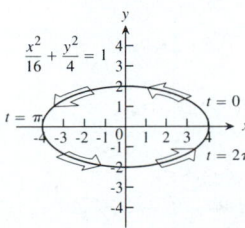

7. $\dfrac{x^2}{16} + \dfrac{y^2}{25} = 1$

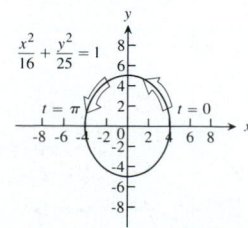
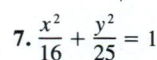

9. $y = x^2$

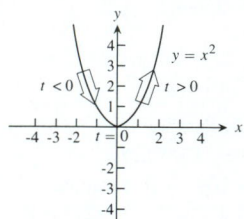

11. $x = y^2$

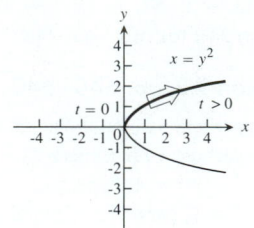

c)

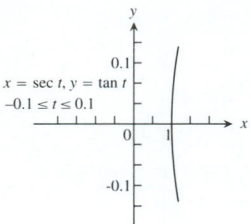

35. a)

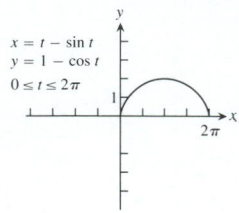

13. $x^2 - y^2 = 1$

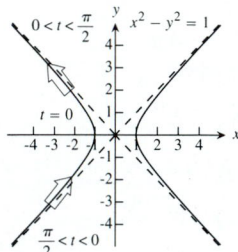

15. $y = 2x + 3$

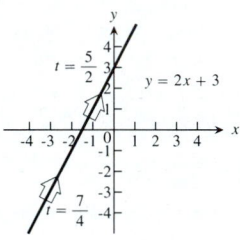

b)

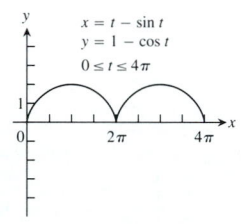

c)

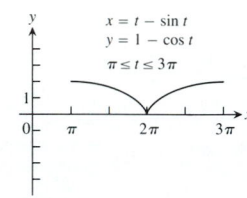

17. $y = 1 - x$

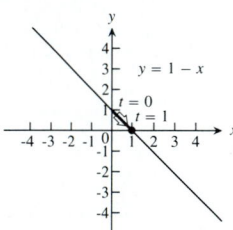

19. $y = \sqrt{1 - x^2}$

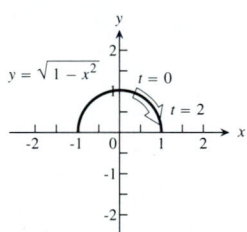

37. a)

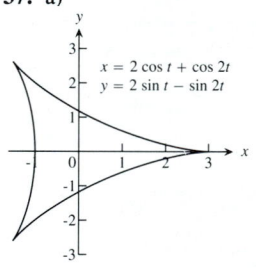

b)

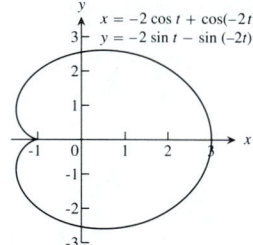

21. $y = \sqrt{x^2 + 1}, \; x \geq 0$

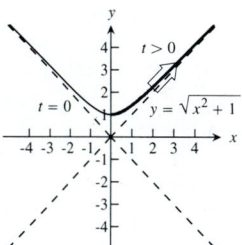

23. $x^2 - y^2 = 1$

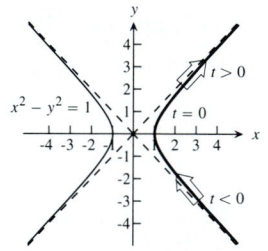

39. a)

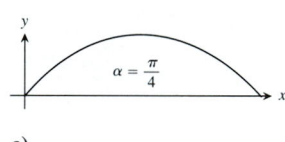

b)

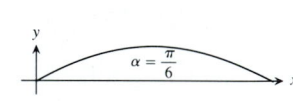

c)

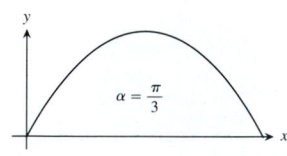

d)

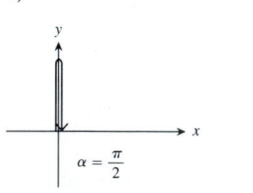

25. a) $x = a \cos t, \; y = -a \sin t, \; 0 \leq t \leq 2\pi$ b) $x = a \cos t,$
$y = a \sin t, \; 0 \leq t \leq 2\pi$ c) $x = a \cos t, \; y = -a \sin t,$
$0 \leq t \leq 4\pi$ d) $x = a \cos t, \; y = a \sin t, \; 0 \leq t \leq 4\pi$ **29.** (1, 1)
31. b) $x = x_1 t, \; y = y_1 t$ c) $x = -1 + t, \; y = t,$ or $x = -t,$
$y = 1 - t$

33. a)

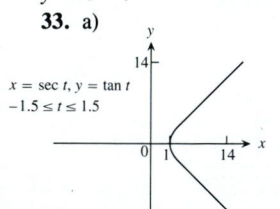

b)

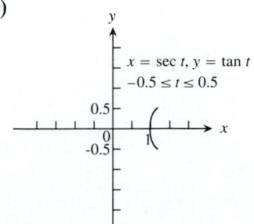

Section 9.4, pp. 662–664

1. $y = -x + 2\sqrt{2}, \dfrac{d^2y}{dx^2} = -\sqrt{2}$ **3.** $y = -\dfrac{1}{2}x + 2\sqrt{2},$
$\dfrac{d^2y}{dx^2} = -\dfrac{\sqrt{2}}{4}$ **5.** $y = x + \dfrac{1}{4}, \dfrac{d^2y}{dx^2} = -2$ **7.** $y = 2x - \sqrt{3},$
$\dfrac{d^2y}{dx^2} = -3\sqrt{3}$ **9.** $y = x - 4, \dfrac{d^2y}{dx^2} = \dfrac{1}{2}$ **11.** $y =$
$\sqrt{3}x - \dfrac{\pi\sqrt{3}}{3} + 2, \dfrac{d^2y}{dx^2} = -4$ **13.** 4   **15.** 12   **17.** π^2

19. $8\pi^2$ **21.** $\dfrac{52\pi}{3}$ **23.** $3\pi\sqrt{5}$ **25.** a) $(\bar{x}, \bar{y}) =$

$\left(\dfrac{12}{\pi} - \dfrac{24}{\pi^2}, \dfrac{24}{\pi^2} - 2\right)$ b)

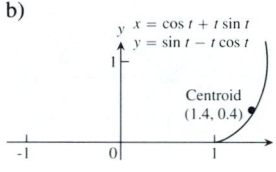

27. a) $(\bar{x}, \bar{y}) = \left(\dfrac{1}{3}, \pi - \dfrac{4}{3}\right)$ b)

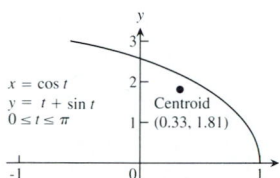

29. a) π b) π **31.** $\left(\dfrac{\sqrt{2}}{2}, 1\right)$, $y = 2x$, $y = -2x$

33. 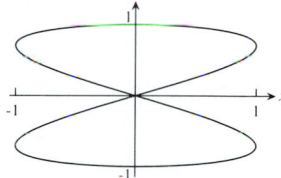   **35.**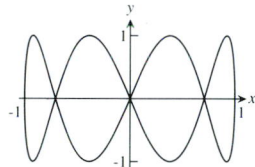

37. 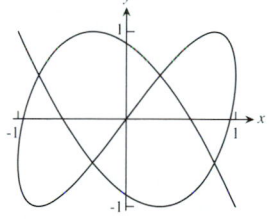   **39.**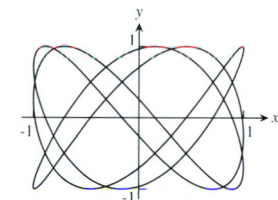

Section 9.5, pp. 669–670

1. a, c; b, d; e, k; g, j; h, f; i, l; m, o; n, p   **3.** a) $\left(2, \dfrac{\pi}{2} + 2n\pi\right)$

and $\left(-2, \dfrac{\pi}{2} + (2n+1)\pi\right)$, n an integer b) $(2, 2n\pi)$ and

$(-2, (2n+1)\pi)$, n an integer c) $\left(2, \dfrac{3\pi}{2} + 2n\pi\right)$ and

$\left(-2, \dfrac{3\pi}{2} + (2n+1)\pi\right)$, n an integer d) $(2, (2n+1)\pi)$ and

$(-2, 2n\pi)$, n an integer

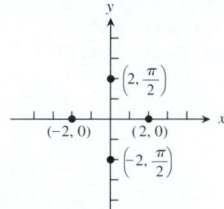

5. a) $(1, 1)$ b) $(1, 0)$ c) $(0, 0)$ d) $(-1, -1)$ e) $\left(\dfrac{3\sqrt{3}}{2}, -\dfrac{3}{2}\right)$

f) $(3, 4)$ g) $(1, 0)$ h) $(-\sqrt{3}, 3)$ **7.**

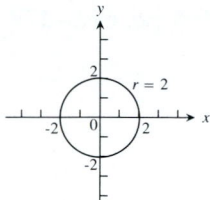

9.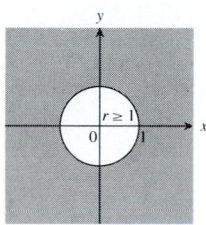

11.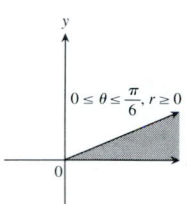

13.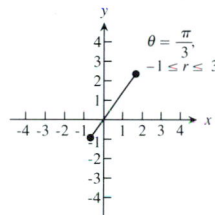

15.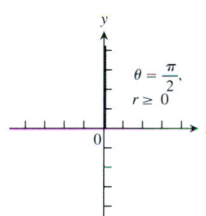

17.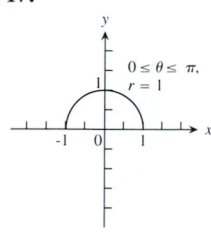

19.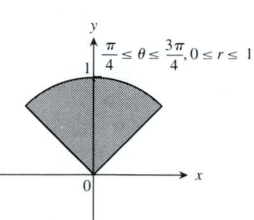

21.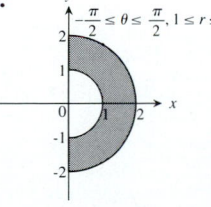

23. $x = 2$, vertical line through $(2, 0)$ **25.** $y = 4$, horizontal line
through $(0, 4)$ **27.** $y = 0$, the x-axis   **29.** $x + y = 1$,
line, $m = -1$, $b = 1$ **31.** $x^2 + y^2 = 1$, circle, $C(0, 0)$,
$r = 1$ **33.** $y - 2x = 5$, line, $m = 2$, $b = 5$ **35.** $y^2 = x$,
parabola, $V(0, 0)$, opens right   **37.** $y = e^x$, natural exponential
function   **39.** $x^2 + 2xy + y^2 = 1$, parabola, rotated $45°$
41. $(x - 1)^2 + (y - 1)^2 = 2$, circle, $C(1, 1)$, $r = \sqrt{2}$

43. $r \cos \theta = 7$ **45.** $\theta = \dfrac{\pi}{4}$ **47.** $r = 2$ or $r = -2$

49. $4r^2\cos^2\theta + 9r^2\sin^2\theta = 36$ **51.** $r^2\sin^2\theta = 4r \cos \theta$

Section 9.6, pp. 676–677

1.

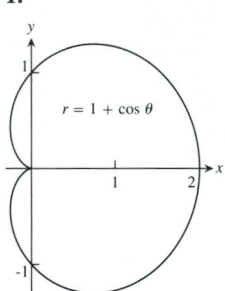

$r = 1 + \cos \theta$

3.

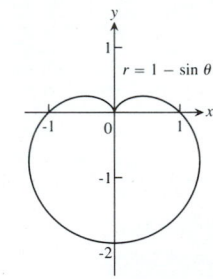

$r = 1 - \sin \theta$

5.

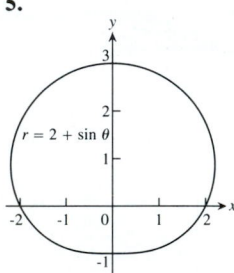

$r = 2 + \sin \theta$

7.

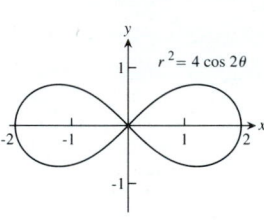

$r^2 = 4 \cos 2\theta$

9.

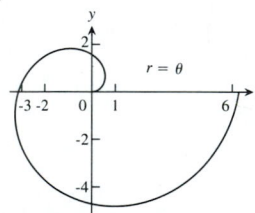

$r = \theta$

11. Slope at $(-1, \pi/2) = -1$; slope at $(-1, -\pi/2) = 1$

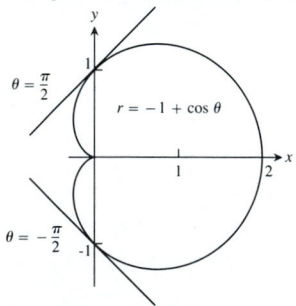

$\theta = \dfrac{\pi}{2}$

$r = -1 + \cos \theta$

$\theta = -\dfrac{\pi}{2}$

13. Slope at $(1, \pi/4) = -1$;
slope at $(-1, -\pi/4) = 1$;
slope at $(-1, 3\pi/4) = 1$;
slope at $(1, -3\pi/4) = -1$;
slope at $(0, 0)$ and $(0, \pi) = 0$;
slope at $(0, \pi/2)$ and $(0, 3\pi/2)$
undefined

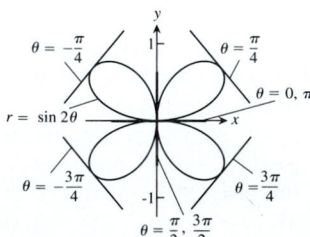

$\theta = -\dfrac{\pi}{4}$ $\theta = \dfrac{\pi}{4}$

$\theta = 0, \pi$

$r = \sin 2\theta$

$\theta = -\dfrac{3\pi}{4}$ $\theta = \dfrac{3\pi}{4}$

$\theta = \dfrac{\pi}{2}, \dfrac{3\pi}{2}$

15.

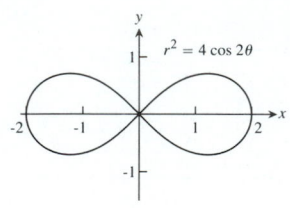

$r^2 = 4 \cos 2\theta$

17. a)

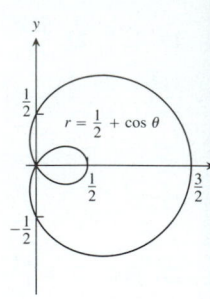

$r = \dfrac{1}{2} + \cos \theta$

b)

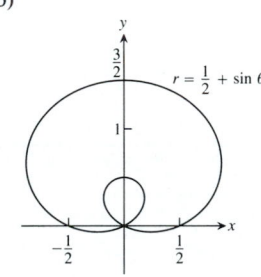

$r = \dfrac{1}{2} + \sin \theta$

b)

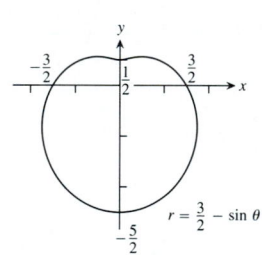

$r = \dfrac{3}{2} - \sin \theta$

19. a)

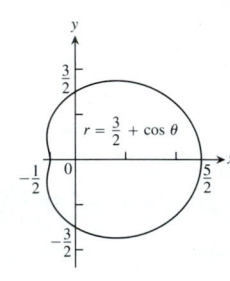

$r = \dfrac{3}{2} + \cos \theta$

21.

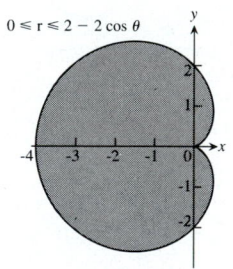

$0 \le r \le 2 - 2 \cos \theta$

25. $(1, \pi/2), (1, 3\pi/2), (0, 0)$

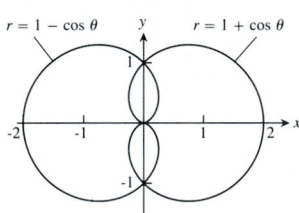

$r = 1 - \cos \theta$ $r = 1 + \cos \theta$

27. $(2\sqrt{2} - 1, \sin^{-1}(3 - 2\sqrt{2}))$,
$(0, 0), (2, 3\pi/2), (-2(\sqrt{2} - 1), -\sin^{-1}(3 - 2\sqrt{2}))$

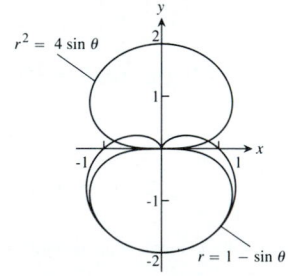

$r^2 = 4 \sin \theta$

$r = 1 - \sin \theta$

29. $(1, \pi/12), (1, 5\pi/12),$
$(1, 7\pi/12), (1, 11\pi/12),$
$(1, 13\pi/12), (1, 17\pi/12),$
$(1, 19\pi/12), (1, 23\pi/12)$

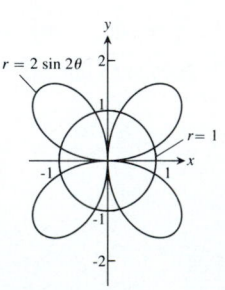

$r = 2 \sin 2\theta$

$r = 1$

31. $(1, \pi/8), (1, 9\pi/8), (-1, 5\pi/8), (-1, 13\pi/8), (1, 3\pi/8),$
$(1, 7\pi/8), (1, 11\pi/8), (1, 15\pi/8)$

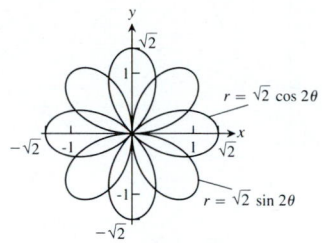

35. $(\pm 1/\sqrt[4]{2}, \pi/8), (0, 0)$

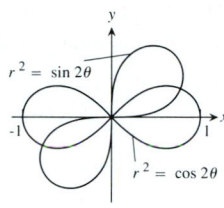

37. $(1, \pi/12), (1, 5\pi/12), (1, 13\pi/12), (1, 17\pi/12), (1, 7\pi/12),$
$(1, 11\pi/12), (1, 19\pi/12), (1, 23\pi/12)$

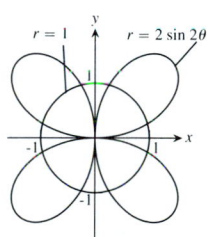

39.

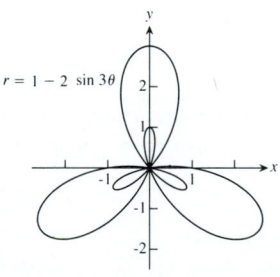

41. a)

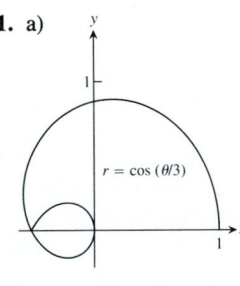

b)

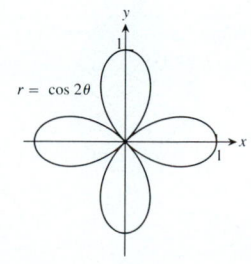

c)

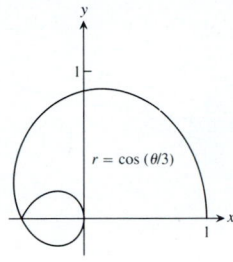

d)

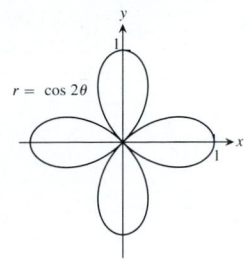

Section 9.7, pp. 682–683

1. $r \cos(\theta - \pi/6) = 5; \sqrt{3}x + y = 10$
3. $r \cos(\theta - 4\pi/3) = 3; x + \sqrt{3}y = -6$
5. $x + y = 2$ **7.** $y = -1$

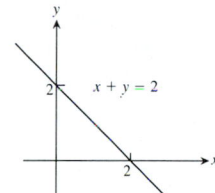

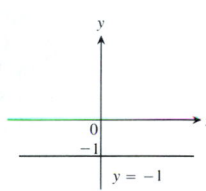

9. $r = 8 \cos\theta$ **11.** $r = 2\sqrt{2} \sin\theta$
13.   **15.**

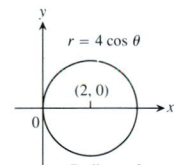

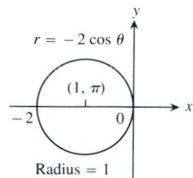

17. $r = \dfrac{2}{1 + \cos\theta}$ **19.** $r = \dfrac{8}{1 + 2\cos\theta}$ **21.** $r = \dfrac{1}{2 + \cos\theta}$

23. $r = \dfrac{10}{5 - \sin\theta}$ **25.**

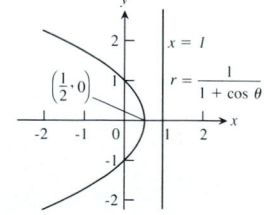

27.   **29.**

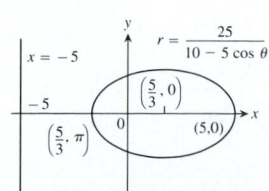

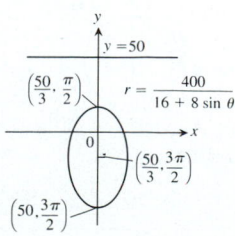

31.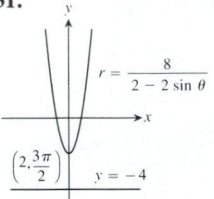

33.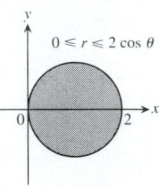

35. b)

Planet	Perihelion	Aphelion
Mercury	0.3075 AU	0.4667 AU
Venus	0.7184 AU	0.7282 AU
Earth	0.9833 AU	1.0167 AU
Mars	1.3817 AU	1.6663 AU
Jupiter	4.9512 AU	5.4548 AU
Saturn	9.0210 AU	10.0570 AU
Uranus	18.2977 AU	20.0623 AU
Neptune	29.8135 AU	30.3065 AU
Pluto	29.6549 AU	49.2251 AU

37. a) $x^2 + (y - 1)^2 = 1$; $y = 1$ b)

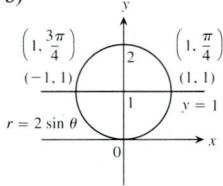

39. $r = \dfrac{4}{1 + \cos \theta}$ **43.** 2 in. apart

45.

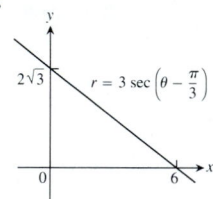

47.

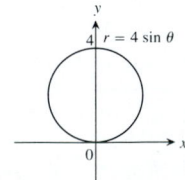

49.

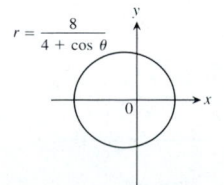

51.

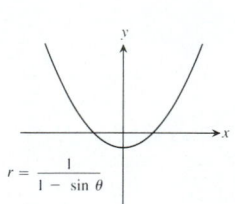

53.

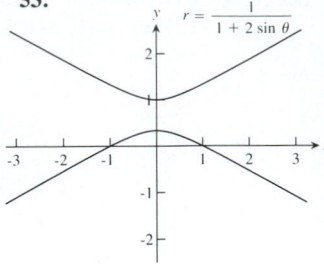

Section 9.8, pp. 689–690

1. 18π **3.** $\pi/8$ **5.** 2   **7.** $\pi/2 - 1$ **9.** $5\pi - 8$ **11.** $3\sqrt{3} - \pi$
13. $\pi/3 + \sqrt{3}/2$ **15.** $9\pi - 18$ **17.** a) 1.12 b) Yes   **19.** 19/3
21. 8   **23.** $\pi/4 + 3/8$ **25.** $1/2 \ln(2 + \sqrt{3})$ **27.** $\pi\sqrt{2}$
29. $\pi(2 - \sqrt{2})$ **33.** $\left(\dfrac{5}{6}a, 0\right)$

Section 9 Miscellaneous Exercises, pp. 691–698

1. 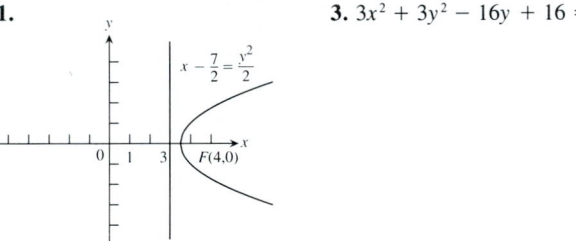   **3.** $3x^2 + 3y^2 - 16y + 16 = 0$

5. $(0, \pm 1)$ **7.** a) $\dfrac{(y - 1)^2}{16} - \dfrac{x^2}{48} = 1$ b) $\dfrac{16\left(y + \dfrac{3}{4}\right)^2}{25} - \dfrac{2x^2}{75} = 1$

13. 6.5×10^6 mi   **15.** a) $\displaystyle\int_0^a \sqrt{a^2 - x^2}\, dx$

b) $\displaystyle\int_0^a b/a\sqrt{a^2 - x^2}\, dx$ c) $b/a(\pi a^2)$

17. $(5, 10\sqrt{10}/3)$, where $(0, 0)$ is midway between the two
stations   **19.** The listener is on a branch of a hyperbola with foci at
the rifle and the target.
23.   **25.**

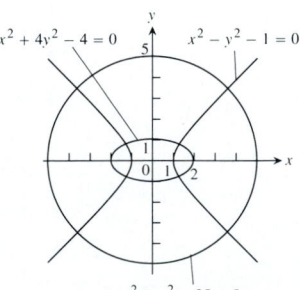   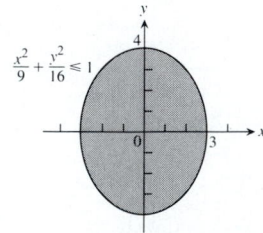

27.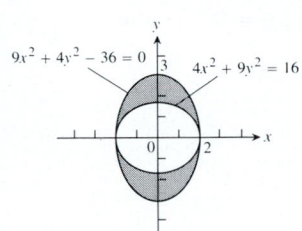

29.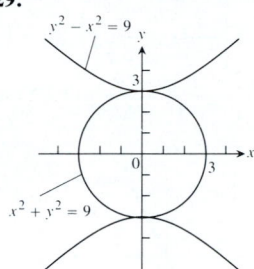

31. Ellipse   **33.** Parabola   **35.** Ellipse; $5x'^2 + 3y'^2 - 30 = 0$
37. $(0, 3), (\pm 2, 1)$ **39.** 0.82 **41.** $\sqrt{2}$ **43.**

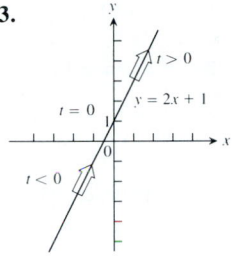

45. 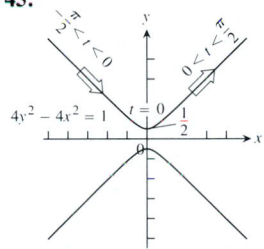   **47.**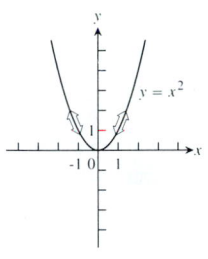

49. $x = 3 \cos t, y = 4 \sin t, 0 \le t \le 2\pi$

51. $x = (a + b)\cos \theta - b \cos\left(\dfrac{a + b}{b}\theta\right),$

$y = (a + b)\sin \theta - b \sin\left(\dfrac{a + b}{b}\theta\right)$

53. P traces the diameter of the circle.   **55.** $y = \dfrac{\sqrt{3}}{2}x + \dfrac{1}{4},$

$\dfrac{d^2y}{dx^2} = \dfrac{1}{4}$ **57.** $3 + \ln 2/8$ **59.** $\dfrac{76\pi}{3}$ **61.** $\left(\pi a, \dfrac{5}{6}a\right)$

63. a) $x = \ln(t + 2), y = t^2 + 1$ b) $y = e^{2x} - 4e^x + 5$

c) $x = \ln(\sqrt{y - 1} + 2)$ d) $\dfrac{4}{\ln 2}$ e) 6   **65.** $M_x = \dfrac{36}{5}\sqrt{3},$

$M_y = \dfrac{72}{7}\sqrt{3}$ **67.** d   **69.** l   **71.** k   **73.** i

75.

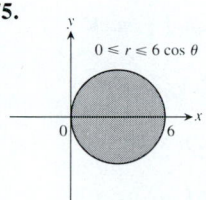

77. $(0, 0)$ **79.** All points of $r = 1 + \sin \theta$
81. $(a, 0), (a, \pi)$ **83.** $(0, 0)$ **85.** $y = x, y = -x$

87. $r \cos(\theta - \pi/4) = 1; r \cos(\theta - 3\pi/4) = 1;$
$r \cos(\theta - 5\pi/4) = 1; r \cos(\theta - 7\pi/4) = 1$

89. 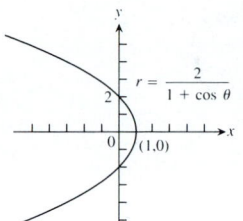   **91.**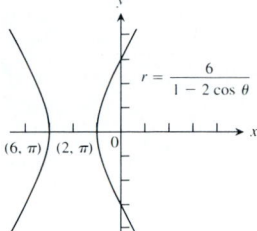

93. $r = \dfrac{4}{1 + 2 \cos \theta}$ **95.** $r = \dfrac{2}{2 + \sin \theta}$

97. a) $r = \dfrac{2a}{1 + \cos\left(\theta - \dfrac{\pi}{4}\right)}$ b) $r = \dfrac{8}{3 - \cos \theta}$ c) $r = \dfrac{3}{1 + 2 \sin \theta}$

99. $\dfrac{9}{2}\pi$ **101.** $2 + \dfrac{\pi}{4}$ **103.** $2\pi\left(1 - \dfrac{\sqrt{2}}{2}\right)$

105. a) $r = e^{2\theta}$ b)

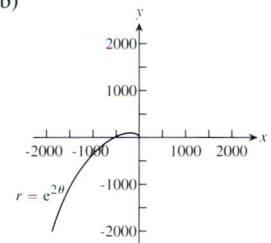

c) $\dfrac{\sqrt{5}e^{4\pi}}{2} - \dfrac{\sqrt{5}}{2}$ **107.** $\dfrac{32\pi - 4\pi\sqrt{2}}{5}$ **111.** 0

115. $\left(0, \pm\dfrac{\pi}{3}\right), \dfrac{\pi}{2}$ **119.** $\dfrac{\pi}{2}$ **121.** $\dfrac{\pi}{4}$

Chapter 10

Section 10.1, pp. 707–708

1. a)   b)   c)

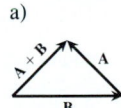   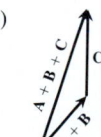   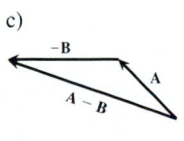

d)   **3.** $3\mathbf{i} - \mathbf{j}$ **5.** $-\mathbf{i} - 9\mathbf{j}$

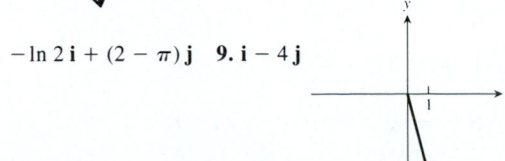

7. $-\ln 2\,\mathbf{i} + (2 - \pi)\mathbf{j}$ **9.** $\mathbf{i} - 4\mathbf{j}$

11. $-2\,\mathbf{i} - 3\,\mathbf{j}$

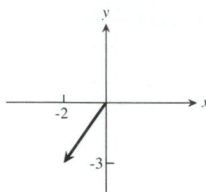

13. $\mathbf{u} = \dfrac{\sqrt{3}}{2}\,\mathbf{i} + \dfrac{1}{2}\,\mathbf{j}$, when $\theta = \dfrac{\pi}{6}$; $\mathbf{u} = -\dfrac{1}{2}\,\mathbf{i} + \dfrac{\sqrt{3}}{2}\,\mathbf{j}$,

when $\theta = \dfrac{2\pi}{3}$

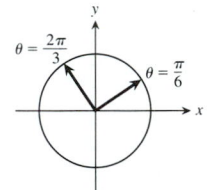

15. $\mathbf{u} = \dfrac{\sqrt{3}}{2}\,\mathbf{i} - \dfrac{1}{2}\,\mathbf{j}$

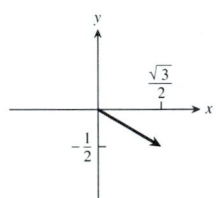

17. $\mathbf{u} = \dfrac{1}{\sqrt{17}}\,\mathbf{i} + \dfrac{4}{\sqrt{17}}\,\mathbf{j}$; $-\mathbf{u} = -\dfrac{1}{\sqrt{17}}\,\mathbf{i} - \dfrac{4}{\sqrt{17}}\,\mathbf{j}$; $\mathbf{n} = \dfrac{4}{\sqrt{17}}\,\mathbf{i} - \dfrac{1}{\sqrt{17}}\,\mathbf{j}$; $-\mathbf{n} = -\dfrac{4}{\sqrt{17}}\,\mathbf{i} + \dfrac{1}{\sqrt{17}}\,\mathbf{j}$

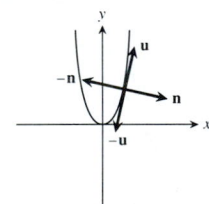

19. $\mathbf{u} = \dfrac{1}{\sqrt{5}}\,(2\,\mathbf{i} + \mathbf{j})$; $-\mathbf{u} = \dfrac{1}{\sqrt{5}}\,(-2\,\mathbf{i} - \mathbf{j})$; $\mathbf{n} = \dfrac{1}{\sqrt{5}}\,(-\mathbf{i} + 2\,\mathbf{j})$;

$-\mathbf{n} = \dfrac{1}{\sqrt{5}}\,(\mathbf{i} - 2\,\mathbf{j})$

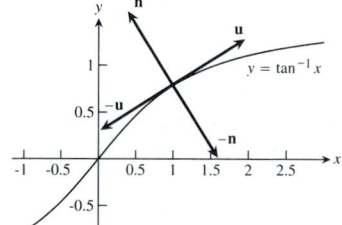

21. $\mathbf{u} = \dfrac{\pm 1}{5}\,(-4\,\mathbf{i} + 3\,\mathbf{j})$, $\mathbf{v} = \dfrac{\pm 1}{5}\,(3\,\mathbf{i} + 4\,\mathbf{j})$

23. $\mathbf{u} = \dfrac{\pm 1}{2}\,(\mathbf{i} + \sqrt{3}\,\mathbf{j})$, $\mathbf{v} = \dfrac{\pm 1}{2}\,(-\sqrt{3}\,\mathbf{i} + \mathbf{j})$

25. $\sqrt{2}\left[\dfrac{1}{\sqrt{2}}\,\mathbf{i} + \dfrac{1}{\sqrt{2}}\,\mathbf{j}\right]$ **27.** $2\left[\dfrac{\sqrt{3}}{2}\,\mathbf{i} + \dfrac{1}{2}\,\mathbf{j}\right]$ **29.** $13\left[\dfrac{5}{13}\,\mathbf{i} + \dfrac{12}{13}\,\mathbf{j}\right]$

31. $|\mathbf{A}| = |3\,\mathbf{i} + 6\,\mathbf{j}| = \sqrt{3^2 + 6^2} = 3\sqrt{5} \Rightarrow \mathbf{A} =$
$3\sqrt{5}\left[\dfrac{1}{\sqrt{5}}\,\mathbf{i} + \dfrac{2}{\sqrt{5}}\,\mathbf{j}\right]$; $|\mathbf{B}| = |-\mathbf{i} - 2\,\mathbf{j}| = \sqrt{5} \Rightarrow \mathbf{B} =$
$\sqrt{5}\left[-\dfrac{1}{\sqrt{5}}\,\mathbf{i} - \dfrac{2}{\sqrt{5}}\,\mathbf{j}\right]$ KEY: AIT 11.1.31

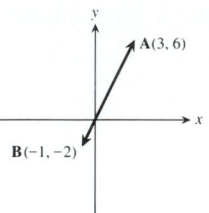

33. $\dfrac{2}{\sqrt{2}}\,(-\mathbf{i} - \mathbf{j})$, one **35.** Same

Section 10.2, pp. 716–718

1. A line through the point $(2, 3, 0)$ parallel to the z-axis **3.** The x-axis **5.** The circle, $x^2 + y^2 = 4$ in the xy-plane **7.** The circle, $x^2 + z^2 = 4$ in the xz-plane **9.** The circle, $y^2 + z^2 = 1$ in the yz-plane **11.** The circle, $x^2 + y^2 = 16$ in the xy-plane
13. a) The first quadrant of the xy-plane b) The fourth quadrant of the xy-plane **15.** a) A solid sphere of radius 1 centered at the origin b) All points that are greater than 1 unit from the origin
17. a) The upper hemisphere of radius 1 centered at the origin b) The solid upper hemisphere of radius 1 centered at the origin
19. a) $x = 3$ b) $y = -1$ c) $z = -2$ **21.** a) $z = 1$ b) $x = 3$
c) $y = -1$ **23.** a) $x^2 + (y - 2)^2 = 4$ b) $(y - 2)^2 + z^2 = 4$
c) $x^2 + z^2 = 4$ **25.** a) $y = 3, z = -1$ b) $x = 1, z = -1$
c) $x = 1, y = 3$ **27.** $x^2 + y^2 + z^2 = 25, z = 3$ **29.** $0 \le z \le 1$
31. $z \le 0$ **33.** a) $(x - 1)^2 + (y - 1)^2 + (z - 1)^2 < 1$
b) $(x - 1)^2 + (y - 1)^2 + (z - 1)^2 > 1$ **35.** $3\left[\dfrac{2}{3}\,\mathbf{i} + \dfrac{1}{3}\,\mathbf{j} - \dfrac{2}{3}\,\mathbf{k}\right]$
37. $9\left[\dfrac{1}{9}\,\mathbf{i} + \dfrac{4}{9}\,\mathbf{j} - \dfrac{8}{9}\,\mathbf{k}\right]$ **39.** $5[\mathbf{k}]$ **41.** $4[-\mathbf{j}]$ **43.** $\dfrac{5}{12}\left[-\dfrac{4}{5}\,\mathbf{j} + \dfrac{3}{5}\,\mathbf{k}\right]$
45. $\sqrt{\dfrac{1}{2}}\left[\dfrac{1}{\sqrt{3}}\,\mathbf{i} - \dfrac{1}{\sqrt{3}}\,\mathbf{j} - \dfrac{1}{\sqrt{3}}\,\mathbf{k}\right]$ **47.** $3, \dfrac{2}{3}\,\mathbf{i} + \dfrac{2}{3}\,\mathbf{j} - \dfrac{1}{3}\,\mathbf{k}, (2, 2, 1/2)$
49. $7, \dfrac{3}{7}\,\mathbf{i} - \dfrac{6}{7}\,\mathbf{j} + \dfrac{2}{7}\,\mathbf{k}, (5/2, 1, 6)$ **51.** $2\sqrt{3}, \dfrac{1}{\sqrt{3}}\,\mathbf{i} - \dfrac{1}{\sqrt{3}}\,\mathbf{j} -$
$\dfrac{1}{\sqrt{3}}\,\mathbf{k}, (1, -1, -1)$ **53.** a) $2\,\mathbf{i}$ b) $4\,\mathbf{j}$ c) $\sqrt{3}\,\mathbf{k}$
d) $\dfrac{3}{10}\,\mathbf{j} + \dfrac{2}{5}\,\mathbf{k}$ e) $6\,\mathbf{i} + 2\,\mathbf{j} + 3\,\mathbf{k}$ f) $au_1\,\mathbf{i} + au_2\,\mathbf{j} + au_3\,\mathbf{k}$
55. $7\left[\dfrac{1}{\sqrt{3}}\,\mathbf{i} + \dfrac{1}{\sqrt{3}}\,\mathbf{j} + \dfrac{1}{\sqrt{3}}\,\mathbf{k}\right]$ **57.** a) $(-2, 0, 2), 2\sqrt{2}$
b) $(-\tfrac{1}{2}, -\tfrac{1}{2}, -\tfrac{1}{2}), \dfrac{\sqrt{21}}{2}$ c) $(\sqrt{2}, \sqrt{2}, -\sqrt{2}), \sqrt{2}$
d) $(0, -\tfrac{1}{3}, \tfrac{1}{3}), \dfrac{\sqrt{29}}{3}$ **59.** $(-2, 0, 2), \sqrt{8}$ **61.** $(0, 0, a), a$
63. a) $\sqrt{y^2 + z^2}$ b) $\sqrt{x^2 + z^2}$ c) $\sqrt{x^2 + y^2}$ **65.** a) $\dfrac{3}{2}\,\mathbf{i} + \dfrac{3}{2}\,\mathbf{j} - 3\,\mathbf{k}$
b) $\mathbf{i} + \mathbf{j} - 2\,\mathbf{k}$ c) $(2, 2, 1)$

Section 10.3, pp. 724–725

	$\mathbf{A} \cdot \mathbf{B}$	$|\mathbf{A}|$	$|\mathbf{B}|$	$\cos \theta$	$|\mathbf{B}| \cos \theta$	$\text{Proj}_{\mathbf{A}}\mathbf{B}$
1.	10	$\sqrt{13}$	$\sqrt{26}$	$\dfrac{10}{13\sqrt{2}}$	$\dfrac{10}{\sqrt{13}}$	$\dfrac{10}{13}[3\mathbf{i}+2\mathbf{j}]$
3.	4	$\sqrt{14}$	2	$\dfrac{2}{\sqrt{14}}$	$\dfrac{4}{\sqrt{14}}$	$\dfrac{2}{7}[3\mathbf{i}-2\mathbf{j}-\mathbf{k}]$
5.	2	$\sqrt{34}$	$\sqrt{3}$	$\dfrac{2}{\sqrt{3}\sqrt{34}}$	$\dfrac{2}{\sqrt{34}}$	$\dfrac{1}{17}[5\mathbf{j}-3\mathbf{k}]$
7.	$\sqrt{3}-\sqrt{2}$	$\sqrt{2}$	3	$\dfrac{\sqrt{3}-\sqrt{2}}{3\sqrt{2}}$	$\dfrac{\sqrt{3}-\sqrt{2}}{\sqrt{2}}$	$\dfrac{\sqrt{3}-\sqrt{2}}{2}[-\mathbf{i}+\mathbf{j}]$
9.	-25	5	5	-1	-5	$-2\mathbf{i}+4\mathbf{j}-\sqrt{5}\,\mathbf{k}$
11.	25	15	5	$\dfrac{1}{3}$	$\dfrac{5}{3}$	$\dfrac{1}{9}[10\mathbf{i}+11\mathbf{j}-2\mathbf{k}]$

13. $\left[\dfrac{3}{2}\mathbf{i}+\dfrac{3}{2}\mathbf{j}\right]+\left[-\dfrac{3}{2}\mathbf{i}+\dfrac{3}{2}\mathbf{j}+4\mathbf{k}\right]$

15. $\left[\dfrac{14}{3}\mathbf{i}+\dfrac{28}{3}\mathbf{j}-\dfrac{14}{3}\mathbf{k}\right]+\left[\dfrac{10}{3}\mathbf{i}-\dfrac{16}{3}\mathbf{j}-\dfrac{22}{3}\mathbf{k}\right]$

17. $x+2y=4$ **19.** $(-2\mathbf{i}-\mathbf{j})\cdot((x+1)\mathbf{i}+(y-2)\mathbf{j})=$
$0 \Rightarrow -2x-y=0$

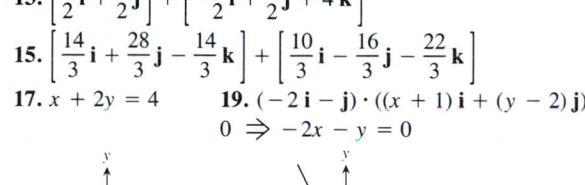

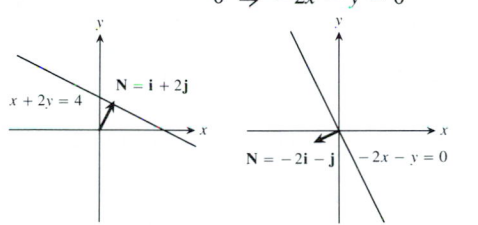

21. $2\sqrt{10}$ **23.** $\sqrt{2}$ **31.** $71.1°,\ 37.9°,\ 71.1°$ **33.** $35.3°$
35. $-5\,\text{N}\cdot\text{m}$ **37.** $3464.10\,\text{N}\cdot\text{m}$ **39.** $45°$ or $135°$
41. $\dfrac{\pi}{3}$ or $\dfrac{2\pi}{3}$ **43.** $45°$ or $135°$

11. $\mathbf{A}\times\mathbf{B}=\mathbf{i}-\mathbf{j}+\mathbf{k}$

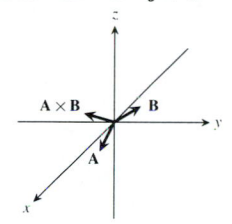

13. $\mathbf{A}\times\mathbf{B}=3\mathbf{i}-\mathbf{j}$

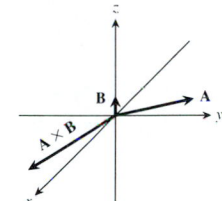

15. a) $\pm(8\mathbf{i}+4\mathbf{j}+4\mathbf{k})$ b) $2\sqrt{6}$ c) $\pm\left[\dfrac{2}{\sqrt{6}}\mathbf{i}+\dfrac{1}{\sqrt{6}}\mathbf{j}+\dfrac{1}{\sqrt{6}}\mathbf{k}\right]$

17. a) $\pm(-\mathbf{i}+\mathbf{j})$ b) $\dfrac{\sqrt{2}}{2}$ c) $\pm\left[\dfrac{-1}{\sqrt{2}}\mathbf{i}+\dfrac{1}{\sqrt{2}}\mathbf{j}\right]$

19. a) None b) $\mathbf{A}$ and $\mathbf{C}$ **21.** $10\sqrt{3}$ ft · lb

23. $0, 0$ **25.** a) $\dfrac{\mathbf{A}\cdot\mathbf{B}}{\mathbf{B}\cdot\mathbf{B}}\mathbf{B}$ b) $(\pm)(\mathbf{A}\times\mathbf{B})$ c) $\sqrt{\mathbf{A}\cdot\mathbf{A}}\,\dfrac{\mathbf{B}}{\sqrt{\mathbf{B}\cdot\mathbf{B}}}$
d) $(\pm)(\mathbf{A}\times\mathbf{B})\times\mathbf{C}$ e) $(\pm)(\mathbf{B}\times\mathbf{C})\times\mathbf{A}$ **27.** 3

Section 10.4, p. 730

1. $\mathbf{A}\times\mathbf{B},\ 3,\ \dfrac{2}{3}\mathbf{i}+\dfrac{1}{3}\mathbf{j}+\dfrac{2}{3}\mathbf{k};\ \mathbf{B}\times\mathbf{A},\ 3,\ -\dfrac{2}{3}\mathbf{i}-\dfrac{1}{3}\mathbf{j}-\dfrac{2}{3}\mathbf{k}$

3. $\mathbf{A}\times\mathbf{B},\ 0$, no direction; $\mathbf{B}\times\mathbf{A},\ 0$, no direction **5.** $\mathbf{A}\times\mathbf{B},\ 6,$
$-\mathbf{k};\ \mathbf{B}\times\mathbf{A},\ 6,\ \mathbf{k}$ **7.** $\mathbf{A}\times\mathbf{B},\ 6\sqrt{5},\ \dfrac{1}{\sqrt{5}}\mathbf{i}-\dfrac{2}{\sqrt{5}}\mathbf{k};\ \mathbf{B}\times\mathbf{A},$
$6\sqrt{5},\ -\dfrac{1}{\sqrt{5}}\mathbf{i}+\dfrac{2}{\sqrt{5}}\mathbf{k}$ **9.** $\mathbf{A}\times\mathbf{B}=\mathbf{k}$

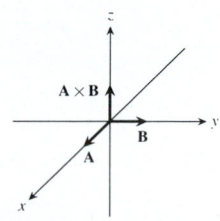

Section 10.5, pp. 736–738

1. $x=3+t,\ y=-4+t,\ z=-1+t$ **3.** $x=-2+5t,$
$y=5t,\ z=3-5t$ **5.** $x=0,\ y=2t,\ z=t$ **7.** $x=1,\ y=1,$
$z=1+t$ **9.** $x=t,\ y=-7+2t,\ z=2t$ **11.** $x=t,\ y=0,$
$z=0$

13. The direction $\overrightarrow{PQ}=\mathbf{i}+\mathbf{j}+\mathbf{k}$
and $P(0,0,0)\Rightarrow x=t,\ y=t,$
$z=t$, where $0\le t\le 1$.

15. The direction $\overrightarrow{PQ}=\mathbf{j}$ and
$P(1,0,0)\Rightarrow x=1,$
$y=1+t,\ z=0,$
where $-1\le t\le 0$.

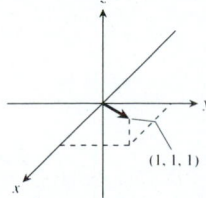

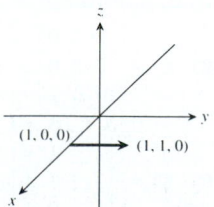

17. The direction $\overrightarrow{PQ} = 2\,\mathbf{j}$ and P$(0, -1, 1) \Rightarrow x = 0$, $y = -1 + 2t$, $z = 1$, where $0 \le t \le 1$.

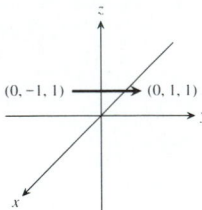

19. The direction $\overrightarrow{PQ} = -\mathbf{i} - 2\,\mathbf{k}$ and $P(2, 2, 0) \Rightarrow x = 2 - t$, $y = 2$, $z = -2t$, where $0 \le t \le 1$.

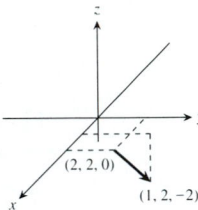

21. $3x - 2y - z = -3$ **23.** $7x - 5y - 4z = 6$
25. $x + 3y + 4z = 34$ **27.** $x = 2 + 2t$, $y = -4 - t$, $z = 7 + 3t$; $x = -2 - t$, $y = -2 + \frac{1}{2}t$, $z = 1 - \frac{3}{2}t$ **29.** $2\sqrt{30}$

31. 0   **33.** $\dfrac{9\sqrt{42}}{7}$ **35.** 3   **37.** $\dfrac{19}{5}$ **39.** $\dfrac{5}{3}$ **41.** $(3/2, -3/2, 1/2)$

43. $(1, 1, 0)$ **45.** $\dfrac{\pi}{4}$ **47.** $101.1°$ **49.** $47.1°$ **51.** $x = 1 - t$, $y = 1 + t$, $z = -1$ **53.** $x = 4$, $y = 3 + 6t$, $z = 1 + 3t$
55. $(1, -1, 0)$ **59.** $|\overrightarrow{OA}| = 1.41421356$; $|\overrightarrow{OB}| = 2$; $|\overrightarrow{AB}| = 2.44948974$; the midpoint of AB is $(.5, 1, -.5)$; $\mathbf{A} \cdot \mathbf{B} = 0$; the angle between $\overrightarrow{OA}$ and $\overrightarrow{OB}$ is 1.5707965 radians $\approx 90°$; $\mathbf{A} \times \mathbf{B} = 2\,\mathbf{i} + 0\,\mathbf{j} + 2\,\mathbf{k}$; the line through A and B: $x = 1 - t$, $y = 2t$, $z = -1 + t$; the line through C parallel to AB: $x = -1 - t$, $y = 2t$, $z = 1 + t$; the distance from C to AB is 2.30940108; the equation of the ABC plane is $-4x - 4z = 0$; the distance from S to the ABC plane is 2.12132034.   **61.** $|\overrightarrow{OA}| = 4.47213595$; $|\overrightarrow{OB}| = 8.24621126$; $|\overrightarrow{AB}| = 9.79795897$; the midpoint of AB is $(0, 4, 2)$; $\mathbf{A} \cdot \mathbf{B} = -4$; the angle between $\overrightarrow{OA}$ and $\overrightarrow{OB}$ is 1.62032412 radians $\approx 92.8377235°$; $\mathbf{A} \times \mathbf{B} = -32\,\mathbf{i} + 8\,\mathbf{j} - 16\,\mathbf{k}$; the line through A and B: $x = -2 + 4t$, $y = 8t$, $z = 4 - 4t$; the line through C parallel to AB: $x = 2 + 4t$, $y = 4 + 8t$, $z = -2 - 4t$; the distance from C to AB is 3.74165739; the equation of the ABC plane is $32x - 8y + 16z = 0$; the distance from S to the ABC plane is 0.

Section 10.6, p. 746

1. $|(\mathbf{A} \times \mathbf{B}) \cdot \mathbf{C}| = \begin{vmatrix} 2 & 0 & 0 \\ 0 & 2 & 0 \\ 0 & 0 & 2 \end{vmatrix} = 8$,
$(\mathbf{A} \times \mathbf{B}) \times \mathbf{C} = 0$, $\mathbf{A} \times (\mathbf{B} \times \mathbf{C}) = 0$

3. $|(\mathbf{A} \times \mathbf{B}) \cdot \mathbf{C}| = \begin{vmatrix} 2 & 1 & 0 \\ 2 & -1 & 1 \\ 1 & 0 & 2 \end{vmatrix} = 7$, $(\mathbf{A} \times \mathbf{B}) \times \mathbf{C} = -4\,\mathbf{i} - 6\,\mathbf{j} + 2\,\mathbf{k}$, $\mathbf{A} \times (\mathbf{B} \times \mathbf{C}) = \mathbf{i} - 2\,\mathbf{j} - 4\,\mathbf{k}$

5. a) True b) False c) True d) True e) False f) True g) True h) True i) True   **7.** a) 0 b) 2 c) 4   **11.** a) $(0, 9, -3)$ b) $\sqrt{963}$ c) xy, 11; yz, 29; xz, 1

Section 10.7, pp. 756–758

1. $x^2 + y^2 = 4$

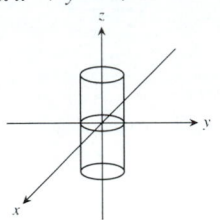

3. $z = y^2 - 1$

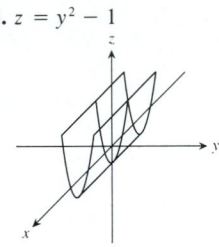

5. $x^2 + 4z^2 = 16$

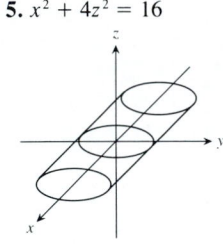

7. $z^2 - y^2 = 1$

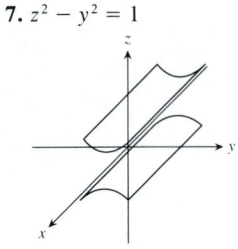

9. $9x^2 + y^2 + z^2 = 9$

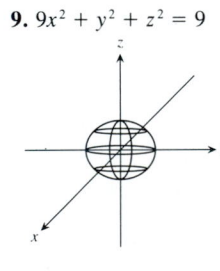

11. $4x^2 + 9y^2 + 4z^2 = 36$

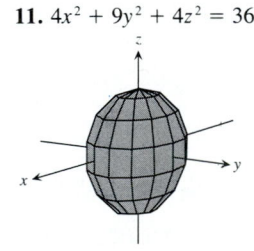

13. $x^2 + 4y^2 = z$

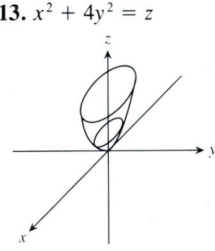

15. $z = 8 - x^2 - y^2$

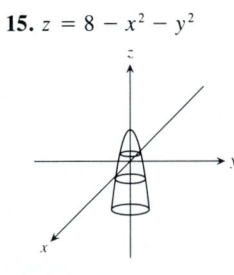

17. $x = 4 - 4y^2 - z^2$

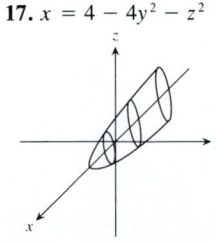

19. $x^2 + y^2 = z^2$

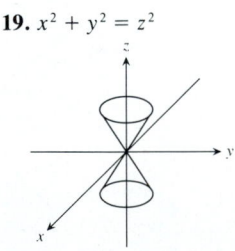

21. $4x^2 + 9z^2 = 9y^2$

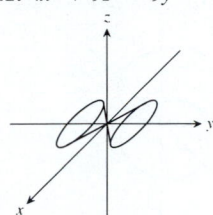

23. $x^2 + y^2 - z^2 = 1$

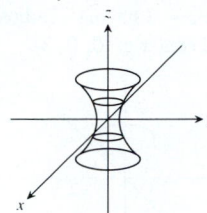

45. $x^2 + z^2 = 1$

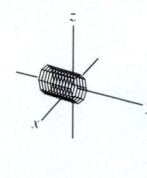

47. $16y^2 + 9z^2 = 4x^2$

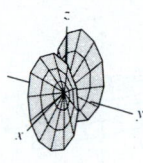

49. $9x^2 + 4y^2 + z^2 = 36$

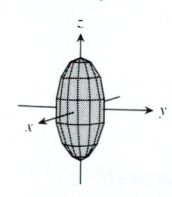

51. $x^2 + y^2 - 16z^2 = 16$

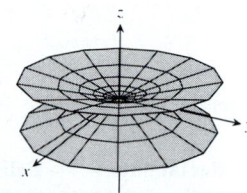

25. $(y^2/4) + (z^2/9) - (x^2/4) = 1$

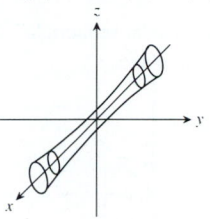

27. $z^2 - x^2 - y^2 = 1$

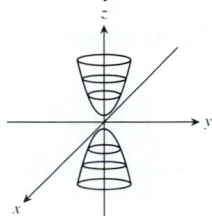

53. $z = -(x^2 + y^2)$

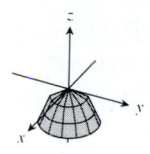

55. $x^2 - 4y^2 = 1$

29. $x^2 - y^2 - (z^2/4) = 1$

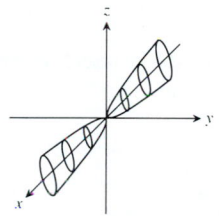

31. $y^2 - x^2 = z$

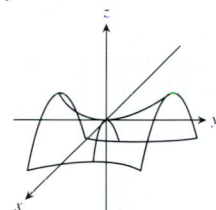

57. $4y^2 + z^2 - 4x^2 = 4$

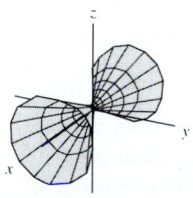

59. $x^2 + y^2 = z$

33. $x^2 + y^2 + z^2 = 4$

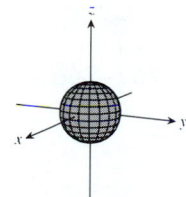

35. $z = 1 + y^2 - x^2$

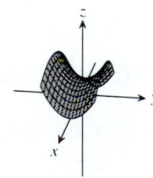

61. $yz = 1$

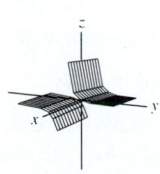

63. $9x^2 + 16y^2 = 4z^2$

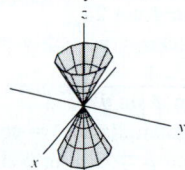

37. $y = -(x^2 + z^2)$

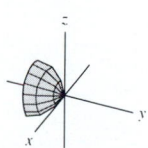

39. $16x^2 + 4y^2 = 1$

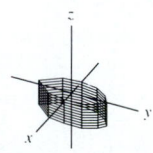

65. a) $\dfrac{2\pi(9 - c^2)}{9}$ b) 8π c) $\dfrac{4\pi\, abc}{3}$, yes

71. $z = y^2$

73. $z = x^2 + y^2$

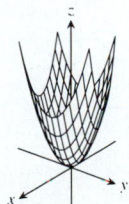

41. $x^2 + y^2 - z^2 = 4$

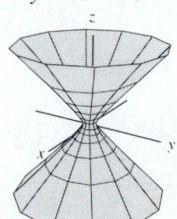

43. $x^2 + z^2 = y$

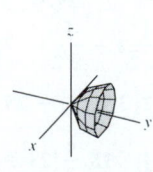

75. $z = \sqrt{1 - x^2}$

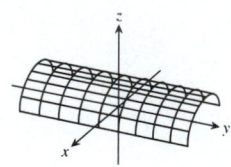

77. $z = \sqrt{x^2 + 2y^2 + 4}$

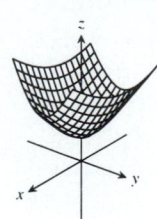

Section 10.8, p. 761

Rectangular	Cylindrical	Spherical
1. $(0, 0, 0)$	$(0, 0, 0)$	$(0, 0, 0)$
3. $(0, 1, 0)$	$(1, \pi/2, 0)$	$(1, \pi/2, \pi/2)$
5. $(1, 0, 0)$	$(1, 0, 0)$	$(1, \pi/2, 0)$
7. $(0, 1, 1)$	$(1, \pi/2, 1)$	$(\sqrt{2}, \pi/4, \pi/2)$
9. $(0, -2\sqrt{2}, 0)$	$(2\sqrt{2}, 3\pi/2, 0)$	$(2\sqrt{2}, \pi/2, 3\pi/2)$

11. $r = 0 \Rightarrow$ rectangular, $x^2 + y^2 = 0$; spherical, $\phi = 0$ and $\phi = \pi$; the z-axis   **13.** $z = 0 \Rightarrow$ cylindrical, $z = 0$; spherical, $\phi = \dfrac{\pi}{2}$; the xy-plane   **15.** $\rho \cos \phi = 3 \Rightarrow$ rectangular, $z = 3$; cylindrical, $z = 3$; the plane $z = 3$ **17.** $\rho \sin \phi \cos \phi = 0 \Rightarrow$ rectangular, $x = 0$; cylindrical $\theta = \dfrac{\pi}{2}$; the yz-plane

19. $x^2 + y^2 + z^2 = 4 \Rightarrow$ cylindrical, $r^2 + z^2 = 4$; spherical, $\rho = 2$; a sphere centered at the origin with a radius of 2
21. $z = r^2 \cos 2\theta \Rightarrow$ rectangular, $z = r^2(\cos^2\theta - \sin^2\theta) \Rightarrow$ $z = x^2 - y^2$; spherical, $\rho \cos^2\phi = \rho^2 \sin^2\phi \cos 2\theta \Rightarrow$ $\rho = \dfrac{\cos \phi}{\sin^2\phi \cos 2\theta}$; hyperbolic paraboloid   **23.** $r = \csc \theta \Rightarrow$ rectangular, $r = \dfrac{r}{y} \Rightarrow y = 1$; spherical, $\rho \sin \phi = \csc \theta \Rightarrow$ $\rho = \dfrac{1}{\sin \phi \sin \theta}$; the plane $y = 1$ **25.** $3 \tan^2\phi = 1 \Rightarrow$ rectangular, $3(\sin^2\phi) = \cos^2\phi \Rightarrow 3(\rho^2 \sin^2\phi)(\sin^2\theta + \cos^2\theta)$ $= \rho^2\cos^2\phi \Rightarrow 3(\rho \sin \phi \sin \theta)^2 + 3(\rho \sin \phi \cos \theta)^2 = (\rho \cos \phi)^2$ $\Rightarrow 3x^2 + 3y^2 = z^2$; cylindrical, $3r^2 = z^2$, a cone
27. A right circular cylinder whose generating curve is a circle of radius 4 in the $r\theta$-plane   **29.** A cylinder whose generations curve is a cardioid in the $r\theta$-plane

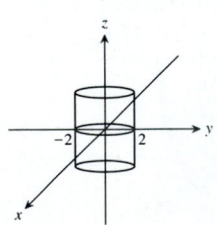

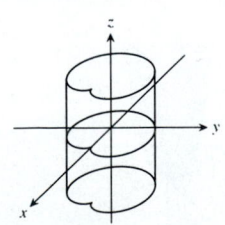

31. A circle contained in the plane $z = 3$ having a radius of 2 and center at $(0, 0, 3)$

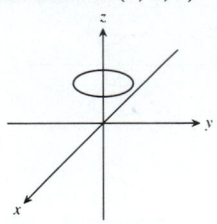

33. A space curve called a helix

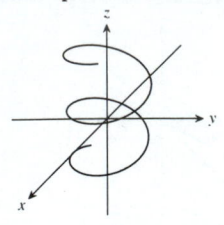

35. $(2, 3, 1)$

37. The upper nappe of a cone

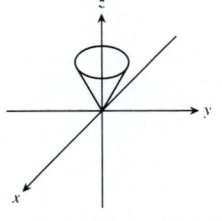

39. A "vertical semicircle" in the $y = x$ plane

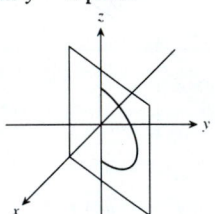

41. A sphere whose center is $(0, 0, 1/2)$ with a radius of $1/2$

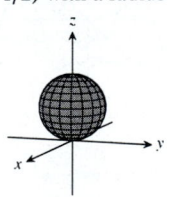

43. A torus-like object centered at the origin

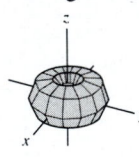

Section 10 Miscellaneous Exercises, pp. 762–766

1. $\theta = 0 \Rightarrow \mathbf{u} = \mathbf{i}$; $\theta = \dfrac{\pi}{2} \Rightarrow \mathbf{u} = \mathbf{j}$; $\theta = \dfrac{2\pi}{3} \Rightarrow$ $\mathbf{u} = -\dfrac{1}{2}\mathbf{i} + \dfrac{\sqrt{3}}{2}\mathbf{j}$; $\theta = \dfrac{5\pi}{4} \Rightarrow \mathbf{u} = -\dfrac{1}{\sqrt{2}}\mathbf{i} - \dfrac{1}{\sqrt{2}}\mathbf{j}$; $\theta = \dfrac{5\pi}{3} \Rightarrow$ $\mathbf{u} = \dfrac{1}{2}\mathbf{i} - \dfrac{\sqrt{3}}{2}\mathbf{j}$

3. Tangents $\pm \left(\dfrac{1}{\sqrt{5}}\mathbf{i} + \dfrac{2}{\sqrt{5}}\mathbf{j} \right)$; normals are $\pm \left(-\dfrac{2}{\sqrt{5}}\mathbf{i} + \dfrac{1}{\sqrt{5}}\mathbf{j} \right)$
5. $2, \dfrac{1}{\sqrt{2}}\mathbf{i} + \dfrac{1}{\sqrt{2}}\mathbf{j}$ **7.** $17, \dfrac{2}{7}\mathbf{i} - \dfrac{3}{7}\mathbf{j} + \dfrac{6}{7}\mathbf{k}$ **11.** 0
13. $\sqrt{2}, 3, 3, 3, -2\mathbf{i} + 2\mathbf{j} - \mathbf{k}, 2\mathbf{i} - 2\mathbf{j} + \mathbf{k}, 3, \dfrac{\pi}{4}, \dfrac{3}{\sqrt{2}},$
$\dfrac{3}{2}[\mathbf{i} + \mathbf{j}]$ **15.** $\dfrac{4}{3}[2\mathbf{i} + \mathbf{j} - \mathbf{k}] - \dfrac{1}{3}[5\mathbf{i} + \mathbf{j} + 11\mathbf{k}]$

17. $\mathbf{A} \times \mathbf{B} = \mathbf{k}$

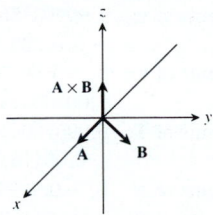

19. $28000 \sqrt{3}$ ft · lb   **27.** 3   **29.** $\dfrac{\sqrt{78}}{3}$ **31.** $x = 1 - 3t$, $y = 2$,

$z = 3 + 7t$ **33.** $\sqrt{2}$ **35.** $2x + y - z = 3$ **39.** $\dfrac{\pi}{6}$ **41.** $\dfrac{25}{\sqrt{38}}$

43. $\dfrac{1}{\sqrt{14}} (-2\mathbf{i} - 3\mathbf{j} + \mathbf{k})$ **45.** $\left(\dfrac{4}{3}, -\dfrac{2}{3}, -\dfrac{2}{3}\right)$ **49.** $59.5°$

53. $\left(26, 23, -\dfrac{1}{3}\right)$ **55.** $\dfrac{11}{\sqrt{107}}$ **57.** a) $(2, 1, 8)$ b) $\dfrac{3}{\sqrt{15}}$

c) $\dfrac{9}{5}(\mathbf{i} + 2\mathbf{k})$ d) $6\sqrt{6}$ e) $7x + 2y - z = 8$ f) yz, 14; xz, 4; xy, 2

63. b) $\dfrac{6}{\sqrt{14}}$ c) $2x - y + 2x = 8$ d) $x - 2y + z = 3 - 5\sqrt{6}$

67. The y-axis in the xy-plane and the yz-plane in three-dimensional space   **69.** A circle centered at $(0, 0)$ with a radius of 2 in the xy-plane; a cylinder parallel with the z-axis in three-dimensional space with the circle as its generating curve   **71.** A horizontal parabola opening to the right with its vertex at $(0, 0)$ in the xy-plane; a cylinder parallel with the z-axis in three-dimensional space with the parabola as the generating curve   **73.** A horizontal cardioid in the $r\theta$-plane; a cylinder parallel with the z-axis in three-dimensional space with the cardioid as the generating curve   **75.** A horizontal lemniscate of length $2\sqrt{2}$ in the $r\theta$-plane; a cylinder parallel with the z-axis in three-dimensional space with the lemniscate as the generating curve.   **77.** A sphere with a radius of 2 centered at the origin   **79.** The upper nappe of a cone whose surface makes a $\pi/6$ angle with the z-axis   **81.** The upper hemisphere of a sphere with a radius 1 centered at the origin   **83.** $z = 2 \Rightarrow$ cylindrical, $z = 2$; spherical, $\rho \cos \phi = 2$; a plane parallel to the xy-plane   **85.** $z = r^2 \cos 2\theta \Rightarrow$ rectangular, $z = r^2(\cos^2\theta - \sin^2\theta) \Rightarrow$ $z = x^2 - y^2$; spherical, $\rho \cos \phi = \rho^2 \sin^2\phi \cos 2\theta \Rightarrow \rho =$ $\dfrac{\cos \phi}{\sin^2\phi \cos 2\theta}$; hyperbolic paraboloid   **87.** $\rho = 4 \sec \phi \Rightarrow$ rectangular, $\rho \cos \phi = 4 \Rightarrow z = 4$; cylindrical, $z = 4$, the plane $z = 4$ **89.** 2   **91.** 13   **93.** $\dfrac{11}{2}$ **95.** $\dfrac{25}{2}$

Chapter 11

Section 11.1, pp. 777–779

1. $\mathbf{v} = (-2 \sin t)\mathbf{i} + (3 \cos t)\mathbf{j} + 4\mathbf{k}$, $\mathbf{a} = (-2 \cos t)\mathbf{i} - (3 \sin t)\mathbf{j}$,

speed $= 2\sqrt{5}$, direction $= -\dfrac{1}{\sqrt{5}}\mathbf{i} + \dfrac{2}{\sqrt{5}}\mathbf{k}$,

$\mathbf{v}\left(\dfrac{\pi}{2}\right) = 2\sqrt{5}\left[-\dfrac{1}{\sqrt{5}}\mathbf{i} + \dfrac{2}{\sqrt{5}}\mathbf{k}\right]$ **3.** $\mathbf{v} = (-2 \sin 2t)\mathbf{j} +$

$(2 \cos t)\mathbf{k}$, $\mathbf{a} = (-4 \cos 2t)\mathbf{j} - (2 \sin t)\mathbf{k}$, speed $= 2$,

direction $= \mathbf{k}$, $\mathbf{v}(0) = 2\mathbf{k}$ **5.** $\mathbf{v} = (\sec t \tan t)\mathbf{i} + (\sec^2 t)\mathbf{j} + \dfrac{4}{3}\mathbf{k}$,

$\mathbf{a} = (\sec t \tan^2 t + \sec^3 t)\mathbf{i} + (2 \sec^2 t \tan t)\mathbf{j}$, speed $= 2$,

direction $= \dfrac{1}{3}\mathbf{i} + \dfrac{2}{3}\mathbf{j} + \dfrac{2}{3}\mathbf{k}$, $\mathbf{v}\left(\dfrac{\pi}{6}\right) = 2\left[\dfrac{1}{3}\mathbf{i} + \dfrac{2}{3}\mathbf{j} + \dfrac{2}{3}\mathbf{k}\right]$

7. $\theta = \dfrac{\pi}{2}$ **9.** $\theta = \dfrac{\pi}{2}$ **11.** $t = 0$, π, 2π **13.** $\dfrac{1}{4}\mathbf{i} + 7\mathbf{j} + \dfrac{3}{2}\mathbf{k}$

15. $\left(\dfrac{\pi + 2\sqrt{2}}{2}\right)\mathbf{j} + 2\mathbf{k}$ **17.** $\mathbf{v}\left(\dfrac{\pi}{4}\right) = \dfrac{\sqrt{2}}{2}\mathbf{i} - \dfrac{\sqrt{2}}{2}\mathbf{j}$,

$\mathbf{a}\left(\dfrac{\pi}{4}\right) = -\dfrac{\sqrt{2}}{2}\mathbf{i} - \dfrac{\sqrt{2}}{2}\mathbf{j}$, $\mathbf{v}\left(\dfrac{\pi}{2}\right) = -\mathbf{j}$, $\mathbf{a}\left(\dfrac{\pi}{2}\right) = -\mathbf{i}$

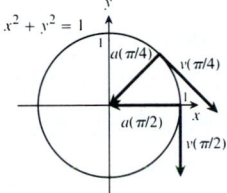

19. $\mathbf{v}(\pi) = 2\mathbf{i}$, $\mathbf{a}(\pi) = -\mathbf{j}$, $\mathbf{v}\left(\dfrac{3\pi}{2}\right) = \mathbf{i} - \mathbf{j}$, $\mathbf{a}\left(\dfrac{3\pi}{2}\right) = -\mathbf{i}$

21. $\mathbf{r} = \left(-\dfrac{t^2}{2} + 1\right)\mathbf{i} + \left(-\dfrac{t^2}{2} + 2\right)\mathbf{j} + \left(-\dfrac{t^2}{2} + 3\right)\mathbf{k}$

23. $\mathbf{r} = ((t + 1)^{3/2} - 1)\mathbf{i} + (1 - e^{-t})\mathbf{j} + (1 + \ln(t + 1))\mathbf{k}$

25. $\mathbf{r} = 8t\mathbf{i} + 8t\mathbf{j} + (100 - 16t^2)\mathbf{k}$ **27.** $\max|\mathbf{a}| = \min|\mathbf{a}| = 1$

Section 11.2, pp. 784–786

1. 50 sec   **3.** a) 72.2 sec, 25,510.2 m b) 4020.3 m c) 6377.6 m
7. $v_0 = 9.9$ m/s, $\alpha = 18.44°$ or $71.56°$ **9.** a) 189.6 mph
b) 120.8 ft · lb   **13.** The golf ball will clip the leaves at the top.
15. 3.72 in.   **17.** 2.25 sec, 148.98 ft/sec   **21.** $\mathbf{v}(t) = -gt\mathbf{k} + \mathbf{v}_0$,

$\mathbf{r}(t) = -\dfrac{1}{2} gt^2 \mathbf{k} + \mathbf{v}_0 t$ **23.** b) $\mathbf{v}_0$ would bisect $\angle AOR$.

Section 11.3, pp. 790–791

1. $\mathbf{T} = \left(-\dfrac{2}{3} \sin t\right)\mathbf{i} + \left(\dfrac{2}{3} \cos t\right)\mathbf{j} + \dfrac{\sqrt{5}}{3}\mathbf{k}$, 3π

3. $\mathbf{T} = \dfrac{1}{\sqrt{1 + t}}\mathbf{i} + \dfrac{\sqrt{t}}{\sqrt{1 + t}}\mathbf{k}$, $\dfrac{52}{3}$ **5.** $\mathbf{T} = \dfrac{1}{\sqrt{3}}\mathbf{i} - \dfrac{1}{\sqrt{3}}\mathbf{j} +$

$\dfrac{1}{\sqrt{3}}\mathbf{k}$, $3\sqrt{3}$ **7.** $\mathbf{T} = \left(\dfrac{\cos t - t \sin t}{t + 1}\right)\mathbf{i} + \left(\dfrac{\sin t + t \cos t}{t + 1}\right)\mathbf{j} +$

$\left(\dfrac{\sqrt{2}\, t^{1/2}}{t + 1}\right)\mathbf{k}$, $\dfrac{\pi^2}{2} + \pi$ **9.** $s(t) = 5t$, $L = \dfrac{5\pi}{2}$

11. $s(t) = \sqrt{3}\, e^t - \sqrt{3}$, $L = -\dfrac{3\sqrt{3}}{4}$ **13.** $\sqrt{2} + \ln(1 + \sqrt{2})$

15. a) Cylinder is $x^2 + y^2 = 1$.

b)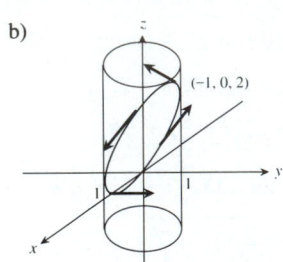

c)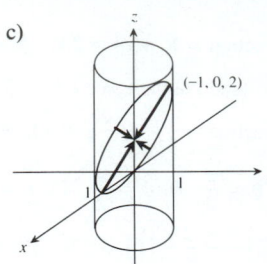

d) $L = \int_0^{2\pi} \sqrt{1 + \sin^2 t}\, dt$ e) $L \approx 7.64$

Section 11.4, pp. 799–800

1. $\mathbf{T} = (\cos t)\,\mathbf{i} - (\sin t)\,\mathbf{j}$, $\mathbf{N} = (-\sin t)\,\mathbf{i} - (\cos t)\,\mathbf{j}$, $\kappa = \cos t$

3. $\mathbf{T} = \dfrac{1}{\sqrt{1 + t^2}}\,\mathbf{i} - \dfrac{1}{\sqrt{1 + t^2}}\,\mathbf{j}$, $\mathbf{N} = \dfrac{-t}{\sqrt{1 + t^2}}\,\mathbf{i} - \dfrac{1}{\sqrt{1 + t^2}}\,\mathbf{j}$,
$\kappa = \dfrac{1}{2(\sqrt{1 + t^2})^3}$ **5.** $\mathbf{T} = \dfrac{3 \cos t}{5}\,\mathbf{i} - \dfrac{3 \sin t}{5}\,\mathbf{j} + \dfrac{4}{5}\,\mathbf{k}$, $\mathbf{N} =$
$(-\sin t)\,\mathbf{i} - (\cos t)\,\mathbf{j}$, $\mathbf{B} = \left(\dfrac{4}{5}\cos t\right)\mathbf{i} - \left(\dfrac{4}{5}\sin t\right)\mathbf{j} - \dfrac{3}{5}\,\mathbf{k}$, $\kappa = \dfrac{3}{25}$,
$\tau = -\dfrac{4}{25}$ **7.** $\mathbf{T} = \left(\dfrac{e^t\cos t - e^t\sin t}{e^t\sqrt{2}}\right)\mathbf{i} + \left(\dfrac{e^t\cos t + e^t\sin t}{e^t\sqrt{2}}\right)\mathbf{j}$,
$\mathbf{N} = \left(\dfrac{-\cos t - \sin t}{\sqrt{2}}\right)\mathbf{i} + \left(\dfrac{-\sin t + \cos t}{\sqrt{2}}\right)\mathbf{j}$, $\mathbf{B} = \mathbf{k}$, $\kappa = \dfrac{1}{e^t\sqrt{2}}$,
$\tau = 0$ **9.** $\mathbf{T} = \dfrac{t}{\sqrt{t^2 + 1}}\,\mathbf{i} + \dfrac{1}{\sqrt{t^2 + 1}}\,\mathbf{j}$, $\mathbf{N} = \dfrac{\mathbf{i}}{\sqrt{t^2 + 1}} - \dfrac{t\mathbf{j}}{\sqrt{t^2 + 1}}$,
$\mathbf{B} = -\mathbf{k}$, $\kappa = \dfrac{1}{t(t^2 + 1)^{3/2}}$, $\tau = 0$
11. $\mathbf{T} = \left(\operatorname{sech}\dfrac{t}{a}\right)\mathbf{i} + \left(\tanh\dfrac{t}{a}\right)\mathbf{j}$, $\mathbf{N} = \left(-\tanh\dfrac{t}{a}\right)\mathbf{i} + \left(\operatorname{sech}\dfrac{t}{a}\right)\mathbf{j}$,
$\mathbf{B} = \mathbf{k}$, $\kappa = \dfrac{1}{a\cosh^2\dfrac{t}{a}}$, $\tau = 0$ **13.** $\mathbf{a} = \dfrac{2t}{\sqrt{1 + t^2}}\,\mathbf{T} + \dfrac{2}{\sqrt{1 + t^2}}\,\mathbf{N}$
15. $\mathbf{a} = |a|\mathbf{N}$ **17.** $\mathbf{a}(1) = \dfrac{4}{3}\,\mathbf{T} + \dfrac{2\sqrt{5}}{3}\,\mathbf{N}$ **19.** $\mathbf{a}(0) = 2\mathbf{N}$
21. $\mathbf{r}\left(\dfrac{\pi}{4}\right) = \dfrac{\sqrt{2}}{2}\,\mathbf{i} + \dfrac{\sqrt{2}}{2}\,\mathbf{j} - \mathbf{k}$, $\mathbf{T}\left(\dfrac{\pi}{4}\right) = -\dfrac{\sqrt{2}}{2}\,\mathbf{i} + \dfrac{\sqrt{2}}{2}\,\mathbf{j}$,
$\mathbf{N}\left(\dfrac{\pi}{4}\right) = -\dfrac{\sqrt{2}}{2}\,\mathbf{i} - \dfrac{\sqrt{2}}{2}\,\mathbf{j}$, $\mathbf{B}\left(\dfrac{\pi}{4}\right) = \mathbf{k}$; osculating plane: $z = -1$;
rectifying plane: $x + y = \sqrt{2}$; normal plane: $-x + y = 0$
23. Yes, if the car is moving around a circle at a constant speed.
27. $\kappa = \dfrac{1}{t}$, $\rho = t$ **31.** $\dfrac{1}{2b}$ **35.** $\left(x - \dfrac{\pi}{2}\right)^2 + y^2 = 1$ **37.** a) $\kappa =$
$\sin t$ b) $\kappa = \operatorname{sech} t$ **39.** a) $\delta = \dfrac{1}{2\sqrt{2}}$, $(x + 2)^2 + (y - 3)^2 = 8$
b) $f'(0) = 1$, $f''(0) = 1$
41. Components of **v**: $-1.87001408, 0.708992989, 0.999999977$
    Components of **a**: $-1.69606646, -2.03053933, 0$
    Speed: 2.23598383
    Components of **T**: $-0.836327193, 0.317083237,$
        0.447230417
    Components of **N**: $-0.641065166, -0.76748645, 0$
    Components of **B**: $0.343243285, -0.28670381, 0.845140807$
    Curvature: 0.505990677

43. Components of **v**: $1.99998356, -796280802 \times 10^{-6},$
        -0.162867596
    Components of **a**: $0, -1.00000296, -8.65198672 \times 10^{-3}$
    Speed: 2.00660413
    Components of **T**: $0.996700614, -3.96830043 \times 10^{-6},$
        -0.0811657837
    Components of **N**: $0, -0.999962572, -8.65163731 \times 10^{-3}$
    Components of **B**: $-0.0811627115, 8.62309222 \times 10^{-3},$
        -0.99666331
    Curvature: 0.248367078

Section 11.5, pp. 809–810

1. 93.17 min   **3.** 6765 km   **5.** 1655 min   **7.** 20,430 km
9. $1.9966 \times 10^7 r^{-1/2}$ m/s   **11.** Circle: $v_0 = \sqrt{\dfrac{GM}{r_0}}$; Ellipse:
$\sqrt{\dfrac{GM}{r_0}} < v_0 < \sqrt{\dfrac{2GM}{r_0}}$; Parabola: $v_0 = \sqrt{\dfrac{2GM}{r_0}}$; Hyperbola:
$v_0 > \sqrt{\dfrac{2GM}{r_0}}$ **15.** $\mathbf{v} = (\cos t - t \sin t)\,\mathbf{i} + (\sin t + t \cos t)\,\mathbf{j}$;
$\mathbf{a} = (-2 \sin t - t \cos t)\,\mathbf{i} + (2 \cos t - t \sin t)\,\mathbf{j}$;
$\kappa = (t^2 + 1)\sqrt{t^2 + 1}$

Section 11 Miscellaneous Exercises, pp. 811–814

1.

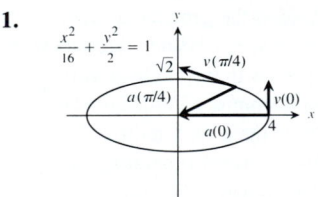

3. $6\,\mathbf{i} + 8\,\mathbf{j}$ **5.** $\mathbf{r} = ((\cos t) - 1)\,\mathbf{i} + ((\sin t) + 1)\,\mathbf{j} + t\mathbf{k}$
7. $\mathbf{r} = \mathbf{i} + t^2\mathbf{j} + t\mathbf{k}$ **9.** $L = \dfrac{\pi}{4}\sqrt{1 + \dfrac{\pi^2}{16}} + \ln\left(\dfrac{\pi}{4} + \sqrt{1 + \dfrac{\pi^2}{16}}\right)$
11. $\mathbf{T}(0) = \dfrac{2}{3}\,\mathbf{i} - \dfrac{2}{3}\,\mathbf{j} + \dfrac{1}{3}\,\mathbf{k}$; $\mathbf{N}(0) = \dfrac{1}{\sqrt{2}}\,\mathbf{i} + \dfrac{1}{\sqrt{2}}\,\mathbf{j}$; $\mathbf{B}(0) =$
$-\dfrac{1}{3\sqrt{2}}\,\mathbf{i} + \dfrac{1}{3\sqrt{2}}\,\mathbf{j} + \dfrac{4}{3\sqrt{2}}\,\mathbf{k}$; $\kappa = \dfrac{\sqrt{2}}{3}$, $\tau = \dfrac{1}{6}$; osculating plane:
$x - y - 4z = 0$; rectifying plane: $x + y = \dfrac{8}{9}$; normal plane:
$2x - 2y + z = 0$ **13.** $\mathbf{T}(\ln 2) = \dfrac{1}{\sqrt{17}}\,\mathbf{i} + \dfrac{4}{\sqrt{17}}\,\mathbf{j}$; $\mathbf{N}(\ln 2) =$
$-\dfrac{4}{\sqrt{17}}\,\mathbf{i} + \dfrac{1}{\sqrt{17}}\,\mathbf{j}$; $\mathbf{B}(\ln 2) = \mathbf{k}$; $\kappa = \dfrac{8}{17\sqrt{17}}$, $\tau = 0$; osculating
plane: $z = 0$; rectifying plane: $4x - y = 4 \ln 2 - 2$;
normal plane: $x + 4y = \ln 2 + 8$ **15.** $\mathbf{a}(0) = 10\mathbf{T} + 6\mathbf{N}$
17. $\mathbf{T} = \left(\dfrac{1}{\sqrt{2}}\cos t\right)\mathbf{i} - (\sin t)\,\mathbf{j} + \left(\dfrac{1}{\sqrt{2}}\cos t\right)\mathbf{k}$; $\mathbf{N} =$
$\left(-\dfrac{1}{\sqrt{2}}\sin t\right)\mathbf{i} - (\cos t)\,\mathbf{j} - \left(\dfrac{1}{\sqrt{2}}\sin t\right)\mathbf{k}$; $\mathbf{B} = \dfrac{1}{\sqrt{2}}\,\mathbf{i} - \dfrac{1}{\sqrt{2}}\,\mathbf{k}$;
$\kappa = \dfrac{1}{\sqrt{2}}$, $\tau = 0$ **19.** $\dfrac{\pi}{2}$ for all t **21.** $\dfrac{\pi}{3}$ **23.** 26

25. a)

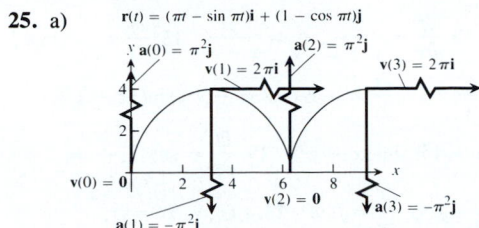

$\mathbf{r}(t) = (\pi t - \sin \pi t)\mathbf{i} + (1 - \cos \pi t)\mathbf{j}$

c) 3.14 ft/sec   **27.** $\mathbf{r}(t) = \dfrac{a \cos \theta}{\sqrt{1 + \sin^2\theta}} \mathbf{i} + \dfrac{a \sin \theta}{\sqrt{1 + \sin^2\theta}} \mathbf{j} -$

$\dfrac{a \sin \theta}{\sqrt{1 + \sin^2\theta}} \mathbf{k}; L = 2\pi a$ **29. a)** 90.91 ft/sec **b)** 60.30 ft

c) 170.46 ft · lb   **35.** $\kappa = \pi s$ **37.** $\kappa = \dfrac{1}{a}$ if $a > 0$

39. a) $\dfrac{d\theta}{dt} = \dfrac{\sqrt{2gb\theta}}{\sqrt{a^2 + a^2\theta^2 + b^2}}$ **b)** $s =$

$\dfrac{a}{2}\left(\theta\sqrt{c^2 + \theta^2} + c^2\ln|\theta + \sqrt{c^2 + \theta^2}| - c^2\ln c\right)$, where $c =$

$\dfrac{a^2 + b^2}{a^2}$ **43.** $\kappa = \dfrac{f^2 - ff'' + 2(f')^2}{((f')^2 + f^2)^{3/2}}$ **45. a)** $\dfrac{dx}{dt} = \dot{r}\cos\theta -$

$r\dot{\theta}\sin\theta, \dfrac{dy}{dt} = \dot{r}\sin\theta + r\dot{\theta}\cos\theta$ **b)** $\dfrac{dr}{dt} = \dot{x}\cos\theta + \dot{y}\sin\theta$,

$r\dfrac{d\theta}{dt} = -\dot{x}\sin\theta + \dot{y}\cos\theta$ **47. a)** $\mathbf{u}_\rho = \sin\phi\cos\theta\mathbf{i} +$

$\sin\phi\sin\theta\mathbf{j} + \cos\phi\mathbf{k}, \mathbf{u}_\phi = \cos\phi\cos\theta\mathbf{i} + \cos\phi\sin\theta\mathbf{j} - \sin\phi\mathbf{k}$,

$\mathbf{u}_\theta = -\sin\theta\mathbf{i} + \cos\theta\mathbf{j}$ **49. b)**

c) $7\sqrt{3}$

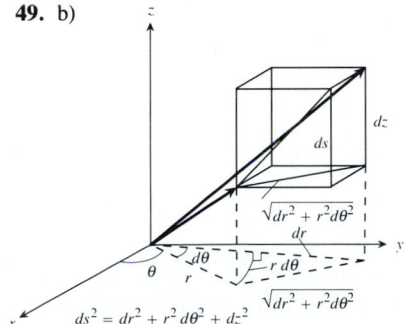

Chapter 12

Section 12.1, pp. 824–825

1. Domain: set of all (x, y) so that $y - x \geq 0 \Rightarrow y \geq x$; range: $z \geq 0$. Level curves are straight lines of the form $y - x = c$, where $c \geq 0$.   **3.** Domain: all $(x, y) \neq (0, y)$; range: all real numbers. Level curves are parabolas with vertex $(0, 0)$ and the y-axis as axis.   **5.** Domain: all points in the xy-plane; range: all positive real numbers. Level curves are hyperbolas with the x- and y-axes as asymptotes.   **7.** Domain: set of all (x, y) so that

$-1 \leq y - x \leq 1$; range: $-\dfrac{\pi}{2} \leq z \leq \dfrac{\pi}{2}$. Level curves are straight

lines of the form $y - x = c$, where $-1 \leq c \leq 1$.   **9.** Domain: set of all (x, y) so that $x > 0$ and $y > 0$; range: all real numbers. Level curves are straight lines of the form $y - cx$, where $c > 0$, $x > 0$, and $y > 0$.

11. a)

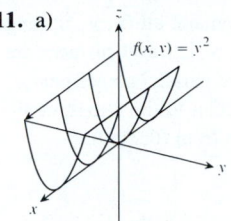

b)

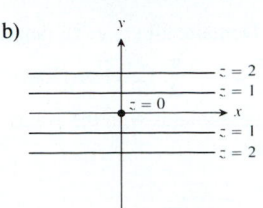

13. a)

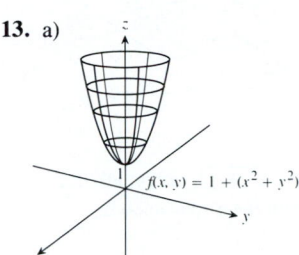

b)

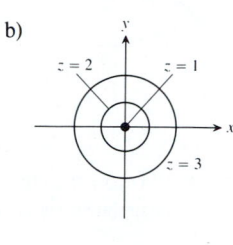

15. a)

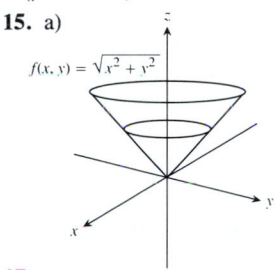

b)

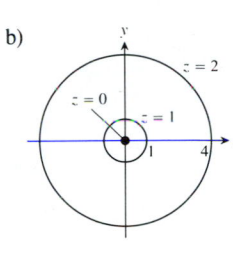

17. a)

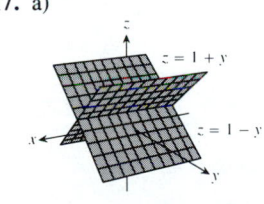

b)

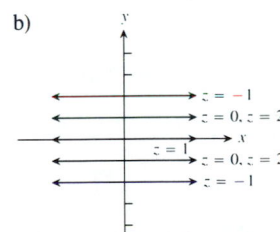

19. a)

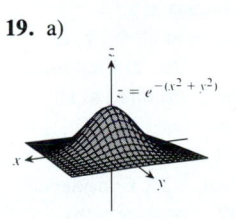

b)

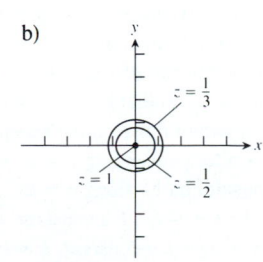

21. f   **23.** a   **25.** d   **27.** Domain: all (x, y, z); range: all real numbers. Level surfaces are spheres with center $(0, 0, 0)$.

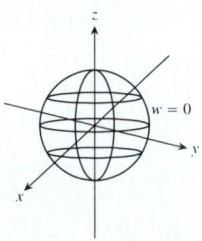

29. Domain: all (x, y, z); range: $-\frac{\pi}{2} < w < \frac{\pi}{2}$. Level surfaces are paraboloids with the z-axis as axis.

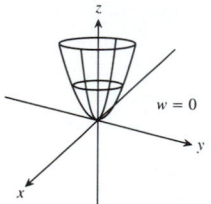

31. Domain: all (x, y, z); range: $0 < w \leq 1$. Level surfaces are pairs of parallel planes perpendicular to the z-axis, equidistant from $(0, 0, 0)$.

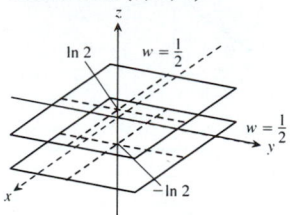

33. Domain: all (x, y, z) so that $z > \sqrt{x^2 + y^2}$; range: all real numbers. Level surfaces are cones with the z-axis as axis.

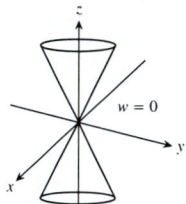

35. $x^2 + y^2 = 10$ **37.** $-2\sqrt{2} - 2\sqrt{2}xy = y - x$
39. $z = e^{\sqrt{x-y}-2}$ **41.** $z \ln 2 = x + y$ **43.** 62.43 km south of Nantucket

Section 12.2, pp. 831–833

1. $\frac{5}{2}$ **3.** $\frac{1}{2}$ **5.** 5   **7.** 1   **9.** 1   **11.** 1   **13.** 0   **15.** -1 **17.** 7

19. 2   **21.** $\tan^{-1}\left(-\frac{\pi}{4}\right)$ **23.** a) Continuous at all (x, y)
b) Continuous at all (x, y) except $(0, 0)$ **25.** a) Continuous at all (x, y) except where $x = 0$ or $y = 0$ b) Continuous at all (x, y)
27. a) Continuous at all (x, y, z) b) Continuous at all (x, y, z) except the interior of the cylinder $x^2 + y^2 = 1$ **29.** a) Continuous at all (x, y, z) so that $(x, y, z) \neq (x, y, 0)$ b) Continuous at all (x, y, z) except those on the sphere $x^2 + y^2 + z^2 = 1$
31. Consider paths along $y = x$, where $x > 0$ or $x < 0$.
33. Consider paths along $y = kx^2$, k a constant.   **35.** Consider paths along $y = kx$, k a constant, $k \neq 1$.   **37.** Consider paths along $y = kx^2$, k a constant, $k \neq 0$.   **41.** 1   **43.** 0   **45.** Does not exist   **47.** $\frac{\pi}{2}$ **49.** $f(0, 0) = \ln 3$ **51.** $\delta = 0.1$ **53.** $\delta = 0.005$
55. $\delta = \sqrt{0.015}$ **57.** $\delta = 0.005$

Section 12.3, pp. 840–842

1. $\frac{\partial f}{\partial x} = 4x$, $\frac{\partial f}{\partial y} = -3$ **3.** $\frac{\partial f}{\partial x} = 2x(y - 2)$, $\frac{\partial f}{\partial y} = x^2 - 1$
5. $\frac{\partial f}{\partial x} = 2y(xy - 1)$, $\frac{\partial f}{\partial y} = 2x(xy - 1)$ **7.** $\frac{\partial f}{\partial x} = \frac{x}{\sqrt{x^2 + y^2}}$, $\frac{\partial f}{\partial y} = \frac{y}{\sqrt{x^2 + y^2}}$ **9.** $\frac{\partial f}{\partial x} = \frac{-1}{(x + y)^2}$, $\frac{\partial f}{\partial y} = \frac{-1}{(x + y)^2}$ **11.** $\frac{\partial f}{\partial x} = \frac{-y^2 - 1}{(xy - 1)^2}$,

$\frac{\partial f}{\partial y} = \frac{-x^2 - 1}{(xy - 1)^2}$ **13.** $\frac{\partial f}{\partial x} = \frac{1}{x + y}$, $\frac{\partial f}{\partial y} = \frac{1}{x + y}$ **15.** $\frac{\partial f}{\partial x} = ye^{xy}\ln y$,
$\frac{\partial f}{\partial y} = xe^{xy}\ln y + \frac{e^{xy}}{y}$ **17.** $\frac{\partial f}{\partial x} = -6\cos(3x - y^2)\sin(3x - y^2)$;
$\frac{\partial f}{\partial y} = -4y\cos(3x - y^2)\sin(3x - y^2)$ **19.** $\frac{\partial f}{\partial x} = yx^{y-1}$, $\frac{\partial f}{\partial y} = x^y\ln y$ **21.** $\frac{\partial f}{\partial x} = -f(x)$, $\frac{\partial f}{\partial y} = f(y)$ **23.** $f_x(x, y, z) = y^2$,
$f_y(x, y, z) = 2xy$, $f_z(x, y, z) = -4z$ **25.** $f_x(x, y, z) = 1$,
$f_y(x, y, z) = -y(y^2 + z^2)^{-1/2}$, $f_z(x, y, z) = -z(y^2 + z^2)^{-1/2}$
27. $f_x(x, y, z) = \frac{yz}{\sqrt{1 - x^2y^2z^2}}$, $f_y(x, y, z) = \frac{xz}{\sqrt{1 - x^2y^2z^2}}$,
$f_z(x, y, z) = \frac{xy}{\sqrt{1 - x^2y^2z^2}}$
29. $f_x(x, y, z) = \frac{1}{x + 2y + 3z}$, $f_y(x, y, z) = \frac{2}{x + 2y + 3z}$,
$f_z(x, y, z) = \frac{3}{x + 2y + 3z}$ **31.** $f_x(x, y, z) = -2xe^{-(x^2 + y^2 + z^2)}$,
$f_y(x, y, z) = -2ye^{-(x^2 + y^2 + z^2)}$, $f_z(x, y, z) = -2ze^{-(x^2 + y^2 + z^2)}$
33. $f_x(x, y, z) = \text{sech}^2(x + 2y + 3z)$, $f_y(x, y, z) = 2\,\text{sech}^2(x + 2y + 3z)$, $f_z(x, y, z) = 3\,\text{sech}^2(x + 2y + 3z)$
35. $\frac{\partial f}{\partial t} = -2\pi\sin(2\pi t - \alpha)$, $\frac{\partial f}{\partial a} = \sin(2\pi t - \alpha)$
37. $\frac{\partial h}{\partial \rho} = \sin\phi\cos\theta$, $\frac{\partial h}{\partial \phi} = \rho\cos\phi\cos\theta$,
$\frac{\partial h}{\partial \theta} = -\rho\sin\phi\sin\theta$ **39.** $W_P(P, V, v, g) = \frac{V}{3}$, $W_V(P, V, v, g) = \frac{P}{3} + \frac{v^2}{2g}$, $W_v(P, V, v, g) = \frac{Vv}{g}$, $W_g(P, V, v, g) = -\frac{Vv^2}{2g^2}$
41. $\frac{\partial^2 f}{\partial x^2} = 0$, $\frac{\partial^2 f}{\partial y^2} = 0$, $\frac{\partial^2 f}{\partial y \partial x} = \frac{\partial^2 f}{\partial x \partial y} = 1$ **43.** $\frac{\partial^2 g}{\partial x^2} = 2y - y\sin x$,
$\frac{\partial^2 g}{\partial y^2} = -\cos y$, $\frac{\partial^2 g}{\partial y \partial x} = \frac{\partial^2 g}{\partial x \partial y} = 2x + \cos x$ **45.** $\frac{\partial^2 r}{\partial x^2} = \frac{-1}{(x + y)^2}$,
$\frac{\partial^2 r}{\partial y^2} = \frac{-1}{(x + y)^2}$, $\frac{\partial^2 r}{\partial y \partial x} = \frac{\partial^2 r}{\partial x \partial y} = \frac{-1}{(x + y)^2}$ **51.** a) x first
b) y first c) x first d) x first e) y first f) y first   **53.** -2
55. $\frac{\partial A}{\partial b} = \frac{c\cos A - b}{bc\sin A}$ **57.** $v_x = \frac{u^2\ln v - vu}{u^2\ln u\ln v - v^2}$ **65.** $n = 0$
or $n = \frac{1}{2}$

Section 12.4, pp. 852–854

1. a) $L(x, y) = 1$ b) $L(x, y) = 2x + 2y - 1$ **3.** a) $L(x, y) = 1 + x$ b) $L(x, y) = -y + \frac{\pi}{2}$ **5.** a) $L(x, y) = 5 + 3x - 4y$
b) $L(x, y) = 3x - 4y + 5$ **7.** $L(x, y) = 7 + x - 6y$; 0.06
9. $L(x, y) = x + y + 1$; 0.08   **11.** $L(x, y) = 1 + x$; 0.0222
13. Pay more attention to the width.   **15.** 0.31   **17.** $\pm 4.83\%$
19. Let $|x - 1| \leq 0.014$, $|y - 1| \leq 0.014$ **21.** 0.099%
25. a) $L(x, y, z) = 2x + 2y + 2z - 3$ b) $L(x, y, z) = y + z$
c) $L(x, y, z) = 0$ **27.** a) $L(x, y, z) = x$ b) $L(x, y, z) = \frac{1}{\sqrt{2}}x + \frac{1}{\sqrt{2}}y$ c) $L(x, y, z) = \frac{1}{3}x + \frac{2}{3}y + \frac{2}{3}z$

29. a) $L(x, y, z) = 2 + x$ **b)** $L(x, y, z) = x - y - z + \dfrac{\pi}{2} + 1$

c) $L(x, y, z) = x - y - z + \dfrac{\pi}{2} + 1$ **31.** $L(x, y, z) = 2x - 6y -$

$2z + 6; 0.0024$ **33.** $L(x, y, z) = x + y - z - 1; 0.00135$

35. a) $S_0\left(\dfrac{1}{100} dp + dx - 5 dw - 30 dh\right)$ **b)** More sensitive to a

change in height **37.** 6% **39.** ± 319.23 ft^2 **41.** Q is most

sensitive to changes in h.

Section 12.5, pp. 860–862

1. $\dfrac{dw}{dt} = 0, \dfrac{dw}{dt}(\pi) = 0$ **3.** $\dfrac{dw}{dt} = 1, \dfrac{dw}{dt}(3) = 1$ **5.** $\dfrac{dw}{dt} =$

$4t \tan^{-1}t + 1, \dfrac{dw}{dt}(1) = \pi + 1$ **7. a)** $\dfrac{\partial z}{\partial r} = 4 \cos \theta \ln(r \sin \theta) +$

$4 \cos \theta; \dfrac{\partial z}{\partial \theta} = -4r \sin \theta \ln(r \sin \theta) + \dfrac{4r \cos^2\theta}{\sin \theta}$ **b)** $\dfrac{\partial z}{\partial r} =$

$\sqrt{2}(\ln 2 + 2); \dfrac{\partial z}{\partial \theta} = -2\sqrt{2} \ln 2 + 2\sqrt{2}$ **9. a)** $\dfrac{\partial w}{\partial u} = 2u + 4uv;$

$\dfrac{\partial w}{\partial v} = -2v + 2u^2$ **b)** $\dfrac{\partial w}{\partial u} = 3; \dfrac{\partial w}{\partial v} = -\dfrac{3}{2}$ **11. a)** $\dfrac{\partial u}{\partial y} = \dfrac{z}{(z - y)^2};$

$\dfrac{\partial u}{\partial z} = \dfrac{-y}{(z - y)^2}$ **b)** $\dfrac{\partial u}{\partial y} = 0; \dfrac{\partial u}{\partial z} = -2$ **13.** $\dfrac{dz}{dt} = \dfrac{\partial z}{\partial x}\dfrac{dx}{dt} + \dfrac{\partial z}{\partial y}\dfrac{dy}{dt}$

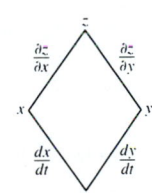

15. $\dfrac{\partial w}{\partial u} = \dfrac{\partial w}{\partial x}\dfrac{\partial x}{\partial u} + \dfrac{\partial w}{\partial y}\dfrac{\partial y}{\partial u} + \dfrac{\partial w}{\partial z}\dfrac{\partial z}{\partial u}$ $\dfrac{\partial w}{\partial v} = \dfrac{\partial w}{\partial x}\dfrac{\partial x}{\partial v} + \dfrac{\partial w}{\partial y}\dfrac{\partial y}{\partial v} + \dfrac{\partial w}{\partial z}\dfrac{\partial z}{\partial v}$

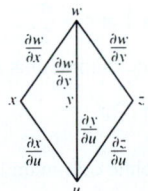 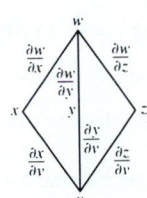

17. $\dfrac{\partial w}{\partial u} = \dfrac{\partial w}{\partial x}\dfrac{\partial x}{\partial u} + \dfrac{\partial w}{\partial y}\dfrac{\partial y}{\partial u}$ $\dfrac{\partial w}{\partial v} = \dfrac{\partial w}{\partial x}\dfrac{\partial x}{\partial v} + \dfrac{\partial w}{\partial y}\dfrac{\partial y}{\partial v}$

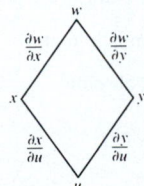 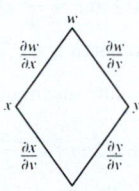

19. $\dfrac{\partial z}{\partial t} = \dfrac{\partial z}{\partial x}\dfrac{\partial x}{\partial t} + \dfrac{\partial z}{\partial y}\dfrac{\partial y}{\partial t}$ $\dfrac{\partial z}{\partial s} = \dfrac{\partial z}{\partial x}\dfrac{\partial x}{\partial s} + \dfrac{\partial z}{\partial y}\dfrac{\partial y}{\partial s}$

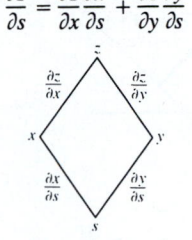

21. $\dfrac{\partial w}{\partial s} = \dfrac{dw}{du}\dfrac{\partial u}{\partial s}$ $\dfrac{\partial w}{\partial t} = \dfrac{dw}{du}\dfrac{\partial u}{\partial t}$

23. $\dfrac{\partial w}{\partial r} = \dfrac{\partial w}{\partial x}\dfrac{dx}{dr} + \dfrac{\partial w}{\partial y}\dfrac{dy}{dr} = \dfrac{\partial w}{\partial x}\dfrac{dx}{dr}$ since $\dfrac{dy}{dr} = 0$

$\dfrac{\partial w}{\partial s} = \dfrac{\partial w}{\partial x}\dfrac{dx}{ds} + \dfrac{\partial w}{\partial y}\dfrac{dy}{ds} = \dfrac{\partial w}{\partial y}\dfrac{dy}{ds}$ since $\dfrac{dx}{ds} = 0$

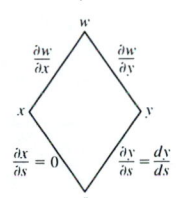

25. $\dfrac{4}{3}$ **27.** $-\dfrac{4}{5}$ **29.** $-\dfrac{3}{4}$ **31.** -1 **33.** 12 **35.** -7

37. $\dfrac{\partial z}{\partial u} = 2; \dfrac{\partial z}{\partial v} = 5$ **39.** 0 **41.** -0.00005 amps/sec

45. $x = \cos 1, y = \sin 1, z = 1$ **47. a)** T has a minimum

at $\dfrac{\pi}{4}, \dfrac{5\pi}{4}$; T has a maximum at $\dfrac{3\pi}{4}, \dfrac{7\pi}{4}$ **b)** $T_{max} = 6; T_{min} = 2$

Section 12.6, pp. 868–869

1. a) 0 **b)** $1 + 2z$ **c)** $1 + 2z$ **3. a)** $\dfrac{\partial U}{\partial p} + \dfrac{\partial U}{\partial T}\left(\dfrac{v}{nR}\right)$

b) $\dfrac{\partial U}{\partial p}\left(\dfrac{nR}{v}\right) + \dfrac{\partial U}{\partial T}$ **5. a)** 5 **b)** 5 **7. a)** $\cos \theta$ **b)** $\dfrac{x}{\sqrt{x^2 + y^2}}$

Section 12.7, pp. 877–879

1. $\nabla f = 3\,\mathbf{i} + 2\,\mathbf{j} - 4\,\mathbf{k}$ **3.** $\nabla f = -\dfrac{26}{27}\mathbf{i} + \dfrac{23}{54}\mathbf{j} - \dfrac{23}{54}\mathbf{k}$

5.

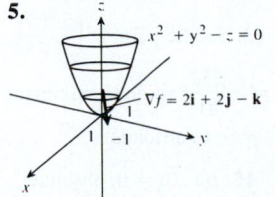

7.

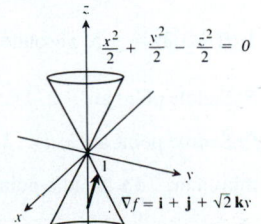

9.

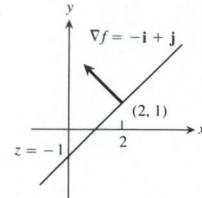

$\nabla f = -\mathbf{i} + \mathbf{j}$

$(2, 1)$

$z = -1$

11. $\nabla f = 2\mathbf{i} + \mathbf{j}$

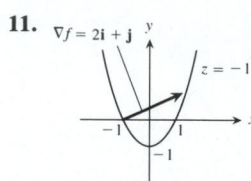

$z = -1$

13. 3   **15.** -2 **17.** $\dfrac{31}{13}$ **19.** $\mathbf{u} = \mathbf{j}, D_{\mathbf{u}}f = 2 \ -\mathbf{u} = -\mathbf{j},$

$D_{-\mathbf{u}}f = -2$ **21.** $\mathbf{u} = \dfrac{1}{\sqrt{5 + 4\pi^2}}(\mathbf{i} + \mathbf{j} + 2\pi\mathbf{k}),$

$D_{\mathbf{u}}f = \sqrt{5 + 4\pi^2}; \ -\mathbf{u} = -\dfrac{1}{\sqrt{5 + 4\pi^2}}(\mathbf{i} + \mathbf{j} + 2\pi\mathbf{k}),$

$D_{-\mathbf{u}}f = -\sqrt{5 + 4\pi^2}$ **23.** $\mathbf{u} = \dfrac{1}{\sqrt{3}}(\mathbf{i} + \mathbf{j} + \mathbf{k}), D_{\mathbf{u}}f = 2\sqrt{3};$

$-\mathbf{u} = -\dfrac{1}{\sqrt{3}}(\mathbf{i} + \mathbf{j} + \mathbf{k}), D_{-\mathbf{u}}f = -2\sqrt{3}$ **25.** Tangent:

$-4x + 2z = 4 = 0$; normal line: $x = 2 - 4t, y = 0, z = t$
27. Tangent: $2y + 3z = 7$; normal line: $x = 1, y = -1 + 4t,$
$z = 3 + 6t$ **29.** Tangent: $2x + 2y + z - 4 = 0$; normal line:
$x = 2t, y = 1 + 2t, z = 2 + t$ **31.** $x = 1, y = 1 + 2t,$
$z = 1 - 2t$ **33.** $x = 1 + 90t, y = 1 - 90t, z = 3$

35.

$\nabla f = 2\sqrt{2}\,\mathbf{i} + 2\sqrt{2}\,\mathbf{j}$

$x^2 + y^2 = 4$

$y = -x + 2\sqrt{2}$

37.

$xy = -4$

$y = x - 4$

$\nabla f = -2\mathbf{i} + 2\mathbf{j}$

39. $\dfrac{3\sqrt{2}}{20}$ **41.** $\dfrac{1}{15}$ **43.** $\mathbf{u} = \dfrac{7}{\sqrt{53}}\mathbf{i} - \dfrac{2}{\sqrt{53}}\mathbf{j}; \ -\mathbf{u} =$

$-\dfrac{7}{\sqrt{53}}\mathbf{i} + \dfrac{2}{\sqrt{53}}\mathbf{j}$ **45.** At $-\dfrac{\pi}{4}, \dfrac{-\pi}{2\sqrt{2}}$; at 0, 0; at $\dfrac{\pi}{4}, \dfrac{\pi}{2\sqrt{2}}$

47. $-\dfrac{7}{\sqrt{5}}$ **53.** ∇f: 0, 4, 0; Directional derivative: 2.309401;
Plane: $4y = 0$; Normal line: $x = 1, y = 4t, z = 1$ **55.** ∇f:
$-1.818396, -1.818396, -0.4161477$; Directional derivative:
-1.85944; Plane: $-1.818396x - 1.818396y -$
$0.416477z = -4.05294$; Normal line: $x = 1 - 1.818396\,t,$
$y = 1 - 1.818396\,t, z = 1 - 0.4161477t$

Section 12.8, pp. 887–891

1. $f(-3, 3) = -5$, absolute minimum   **3.** Saddle point at $\left(\dfrac{6}{5}, \dfrac{69}{25}\right)$

5. Saddle point at $(-2, 1)$ **7.** $f\left(\dfrac{4}{9}, \dfrac{2}{9}\right) = -\dfrac{252}{81}$, absolute maximum

9. Saddle point at $(2, 1)$ **11.** $f(2, -1) = -6$, absolute
minimum   **13.** Saddle point at $\left(\dfrac{1}{6}, \dfrac{4}{3}\right)$ **15.** $f(1, 0) = 0$, absolute

minimum   **17.** $f(0, 1) = 4$, absolute maximum   **19.** $f\left(-\dfrac{2}{3}, \dfrac{2}{3}\right) =$
$\dfrac{170}{27}$, local maximum   **21.** $f(0, 0) = 0$, local minimum; saddle
point at $(1, -1)$ **23.** Saddle point at $(0, 0); f(-1, -1) = 1,$
local maximum   **25.** $f(0, 0) = -1$, local maximum   **27.** $f(0, 0) =$
-4, absolute minimum   **29.** Saddle point at $(n\pi, 0), f(n\pi, 0) = 0$
for every integer n **31.** Absolute maximum is 1 at $(0, 0)$; absolute
minimum is -5 at $(1, 2)$ **33.** Absolute maximum is 4 at $(0, 2)$;
absolute minimum is 0 at $(0, 0)$.   **35.** Absolute maximum is 11 at
$(0, -3)$; absolute minimum is -10 at $(4, -2)$.   **37.** Absolute
maximum is 4 at $(2, 0)$; absolute minimum is $\dfrac{3\sqrt{2}}{2}$ at $\left(3, -\dfrac{\pi}{4}\right),$
$\left(3, \dfrac{\pi}{4}\right), \left(1, -\dfrac{\pi}{4}\right),$ and $\left(1, \dfrac{\pi}{4}\right)$.   **39.** Hottest is 2.25° at $\left(-\dfrac{1}{2}, \dfrac{\sqrt{3}}{2}\right)$
and $\left(-\dfrac{1}{2}, -\dfrac{\sqrt{3}}{2}\right)$; coldest is $-0.25°$ at $\left(\dfrac{1}{2}, 0\right)$.   **41.** a) Saddle
point at $(0, 0)$ b) Local minimum at $(1, 2)$ c) Local minimum at
$(1, -2)$; saddle point at $(-1, -2)$

43. a) i) Absolute minimum is 0 at $t = 0, \dfrac{\pi}{2}$; absolute maximum is
2 at $t = \dfrac{\pi}{4}$.   ii) Absolute minimum is -2 at $t = \dfrac{3\pi}{4}$; absolute
maximum is 2 at $t = \dfrac{\pi}{4}$.   iii) Absolute minimum is -2 at $t = \dfrac{3\pi}{4},$
$\dfrac{7\pi}{4}$; absolute maximum is 2 at $t = \dfrac{\pi}{4}, \dfrac{5\pi}{4}$.   b) i) Absolute minimum
is 2 at $t = 0, \dfrac{\pi}{2}$; absolute maximum is $2\sqrt{2}$ at $t = \dfrac{\pi}{4}$.   ii) Absolute
minimum is -2 at $t = \pi$; absolute maximum is $2\sqrt{2}$ at $t = \dfrac{\pi}{4}$.
iii) Absolute minimum is $-2\sqrt{2}$ at $t = \dfrac{5\pi}{4}$; absolute maximum is
$2\sqrt{2}$ at $t = \dfrac{\pi}{4}$.   c) i) Absolute minimum is 1 at $t = \dfrac{\pi}{2}$; absolute
maximum is 8 at $t = 0$.   ii) Absolute minimum is 1 at $t = \dfrac{\pi}{2}$;
absolute maximum is 8 at $t = 0, \pi$.   iii) Absolute minimum is 1
at $t = \dfrac{\pi}{2}, \dfrac{3\pi}{2}$; absolute maximum is 8 at $t = 0, \pi, 2\pi$.

45. a) i) Absolute maximum is 12 at $t = \dfrac{\pi}{2}$; absolute minimum is
9 at $t = 0$. ii) Absolute minimum is 9 at $t = 0, \pi$; absolute
maximum is 12 at $t = \dfrac{\pi}{2}$. iii) Absolute maximum is 12 at $t = \dfrac{\pi}{2},$
$\dfrac{3\pi}{2}$; absolute minimum is 9 at $t = 0, \pi, 2\pi$. b) i) Absolute
maximum is $6\sqrt{2}$ at $t = \dfrac{\pi}{4}$; absolute minimum is 6 at $t = 0, \dfrac{\pi}{2}$.
ii) Absolute maximum is $6\sqrt{2}$; absolute minimum is -6 at
$t = \pi$. iii) Absolute maximum is $6\sqrt{2}$ at $t = \dfrac{\pi}{4}$; absolute
minimum is $-6\sqrt{2}$ at $t = \dfrac{5\pi}{4}$.   **47.** $y = 1.5x + 0.7$;
$y|_{x = 4} = -5.3$ **49.** $y = 1.5x + 0.2; y|_{x = 4} = 6.2$

51. $y = 0.122x + 3.59$ **53.** a)

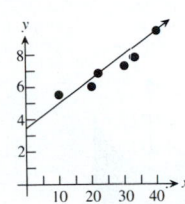

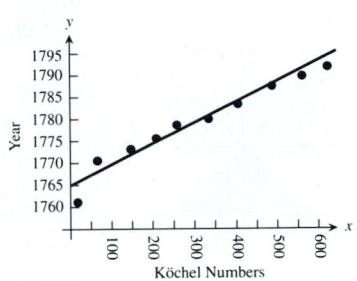

b) $y = 0.0427K + 1764.8$ c) 1780

Section 12.9, pp. 900–902

1. $\left(\pm\sqrt{2}, \frac{1}{2}\right), \left(\pm\sqrt{2}, \frac{1}{2}\right)$ **3.** 39 **5.** $f(0, 3) = 0, f(2, 1) = 4$

7. a) 8 b) 64 **9.** $r = 2$ cm, $h = 4$ cm **11.** Minimum $= 0°$, maximum $= 125°$ **13.** $f(0, 0) = 0$ is minimum, $f(2, 4) = 20$ is maximum. **15.** $f(1, -2, 5) = 30$ is maximum, $f(-1, 2, -5) = -30$ is minimum. **17.** $(0, 0, 1)$ **19.** $f\left(\frac{5}{\sqrt{14}}, \frac{10}{\sqrt{14}}, \frac{15}{\sqrt{14}}\right) = 5\sqrt{14}$ is maximum, $f\left(-\frac{5}{\sqrt{14}}, -\frac{10}{\sqrt{14}}, -\frac{15}{\sqrt{14}}\right) = -5\sqrt{14}$ is minimum. **21.** $\frac{4096}{25\sqrt{5}}$ **23.** $U(8, 14) = \$128$ **27.** $(2, 4, 4)$ **29.** Maximum is $1 + 6\sqrt{3}$ at $(\pm\sqrt{6}, \sqrt{3}, 1)$, minimum is $1 - 6\sqrt{3}$ at $(\pm\sqrt{60}, -\sqrt{3}, 1)$. **31.** Maximum is 4 at $(0, 0, \pm2)$, minimum is 2 at $(\pm\sqrt{2}, \pm\sqrt{2}, 0)$.

Section 12.10, p. 908

1. Quadratic: $1 + x + \frac{1}{2}(x^2 - y^2)$; cubic: $1 + x + \frac{1}{2}(x^2 - y^2) + \frac{1}{6}(x^3 - 3xy^2)$ **3.** Quadratic: $\frac{1}{2}(2x^2) = x^2$; cubic: x^2 **5.** Quadratic: $1 + x + y + x^2 + 2xy + y^2$; cubic: $1 + x + y + x^2 + 2xy + y^2 + x^3 + 3x^2y + 3xy^2 + y^3$ **7.** Quadratic: $1 - \frac{1}{2}x^2 - \frac{1}{2}y^2$; $E(x, y) \le 0.0013$

Section 12 Miscellaneous Exercises, pp. 909–914

1.

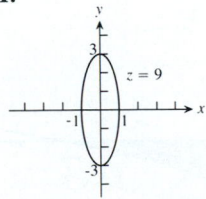

3.

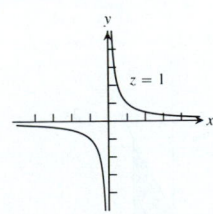

Domain: all points in the xy-plane; range: $f(x, y) \ge 0$. Level curves are ellipses with major axis along the y-axis and minor axis along the x-axis.

Domain: all (x, y) such that $x \ne 0$ or $y \ne 0$; range: $f(x, y) \ne 0$. Level curves are hyperbolas rotated 45° or 135°.

5.

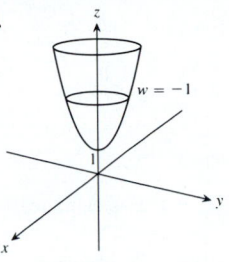

7.

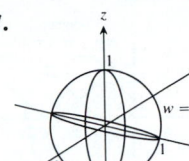

Domain: all (x, y, z) such that $(x, y, z) \ne (0, 0, 0)$; range: all real numbers. Level surfaces are paraboloids of revolution with the z-axis as axis.

Domain: all (x, y, z) such that $(x, y, z) \ne (0, 0, 0)$; range: $f(x, y, z) > 0$. Level surfaces are spheres with center $(0, 0, 0)$ and radius $r > 0$.

9. -2 **11.** 2 **13.** 0 **17.** a) Does not exist b) Not continuous at $(0, 0)$ **19.** $\frac{\partial g}{\partial r} = \cos\theta + \sin\theta$, $\frac{\partial g}{\partial\theta} = -r\sin\theta + r\cos\theta$

21. $\frac{\partial f}{\partial R_1} = -\frac{1}{R_1^2}$ **23.** $\frac{\partial P}{\partial n} = \frac{RT}{V}, \frac{\partial P}{\partial R} = \frac{nT}{V}, \frac{\partial P}{\partial T} = \frac{nR}{V}, \frac{\partial P}{\partial V} = -\frac{nRT}{V^2}$ **25.** $\frac{\partial^2 f}{\partial x^2} = 0, \frac{\partial^2 f}{\partial y^2} = \frac{2x}{y^3}, \frac{\partial^2 f}{\partial y \partial x} = \frac{\partial^2 f}{\partial x \partial y} = -\frac{1}{y^2}$

27. $(\partial^2 f, \partial x^2) = -30x + \frac{2 - 2x^2}{(x^2 + 1)^2}, \frac{\partial^2 f}{\partial y^2} = 0, \frac{\partial^2 f}{\partial y \partial x} = \frac{\partial^2 f}{\partial x \partial y} = 1$

29. $f_{xy}(0, 0) = -1, f_{yx}(0, 0) = 1$ **31.** $L(x, y) = \frac{1}{2} + \frac{1}{2}x - \frac{1}{2}y$, $|E(x, y)| \le 0.02$ **33.** a) $L(x, y, z) = y - 3z$ b) $L(x, y, z) = x + y + 2z - 1$ **35.** Be more careful with the diameter. **37.** $dl = 0.038$, % change in $V = -4.17\%$, % change in $R = -20\%$, % change in $l = 15.83\%$ **39.** a) 5% **41.** -1

43. $\frac{\partial w}{\partial r} = 2, \frac{\partial w}{\partial s} = 2 - \pi$ **45.** -1 **47.** $-(\sin 1 + \cos 2)\sin 1 + (\cos 1 + \cos 2)\cos 1 - 2(\sin 1 + \cos 1)\sin 2$

49. $\frac{\partial w}{\partial x} = \cos\theta \frac{\partial w}{\partial r} - \frac{\sin\theta}{r}\frac{\partial w}{\partial\theta}; \frac{\partial w}{\partial y} = \sin\theta \frac{\partial w}{\partial r} + \frac{\cos\theta}{r}\frac{\partial w}{\partial\theta}$

59. $\mathbf{u} = -\frac{\sqrt{2}}{2}\mathbf{i} - \frac{\sqrt{2}}{2}\mathbf{j}, -\mathbf{u} = \frac{\sqrt{2}}{2}\mathbf{i} + \frac{\sqrt{2}}{2}\mathbf{j}, (D_{\mathbf{u}}f)P_0 = \frac{\sqrt{2}}{2}$, $(D_{-\mathbf{u}}f)P_0 = -\frac{\sqrt{2}}{2}, (D_{\mathbf{u}}f)P_0 = -\frac{7}{10}$ **61.** $\mathbf{u} = \frac{2}{7}\mathbf{i} + \frac{3}{7}\mathbf{j} + \frac{6}{7}\mathbf{k}$, $-\mathbf{u} = -\frac{2}{7}\mathbf{i} - \frac{3}{7}\mathbf{j} - \frac{6}{7}\mathbf{k}, (D_{\mathbf{u}}f)P_0 = 7, (D_{-\mathbf{u}}f)P_0 = -7$, $(D_{\mathbf{u}_1}f)P_0 = 7$ **63.**

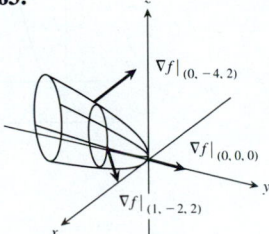

65. Tangent: $4x - y - 5z = 4$; normal line: $x = 2 + 4t$, $y = -1 - t, z = 1 - 5t$

67.

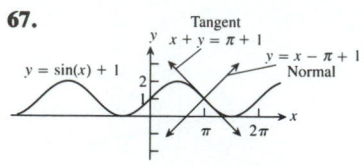

69. $x = 1 - 2t$, $y = 1$, $z = \frac{1}{2} + 2t$ **71.** $(t, -t \pm 4, t)$, t a real

number   **73.** $\dfrac{\pi}{\sqrt{2}}$ **75.** $\dfrac{14}{5}$ **87.** $f(-2, -2) = -8$, absolute

minimum   **89.** $f(0, 0) = 0$, saddle point; $f(6, 18) = 108$, local

maximum   **91.** $f(0, 0) = 15$, saddle point; $f(1, 1) = 14$, local

minimum   **93.** Absolute maximum is 28 at $(0, 4)$; absolute

minimum is $-\dfrac{9}{4}$ at $(3/2, 0)$.   **95.** Absolute maximum is 18 at

$(2, -2)$; absolute minimum is $-\dfrac{17}{4}$ at $\left(-2, \dfrac{1}{2}\right)$.   **97.** Absolute

maximum is 8 at $(-2, 0)$; absolute minimum is -1 at $(1, 0)$.

99. Absolute maximum is 4 at $(1, 0)$; absolute minimum is -4 at

$(0, -1)$.   **101.** Maximum   **103.** $(3, \pm 3\sqrt{2})$ **105.** 50 is

maximum; -50 is minimum.   **107.** $(1, 1, 1)$, $(1, -1, -1)$,

$(-1, -1, -1)$, $(-1, 1, -1)$ **109.** $V = \dfrac{\sqrt{3}\,abc}{2}$ **111.** $w =$

$e^{-c^2\pi^2 t} \sin \pi x$ **113.** 0.213%

Chapter 13

Section 13.1, pp. 924–926

1. 16

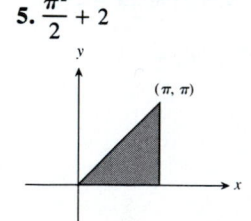

3. 1

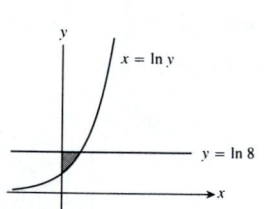

5. $\dfrac{\pi^2}{2} + 2$

7. $8 \ln 8 - 16 + e$

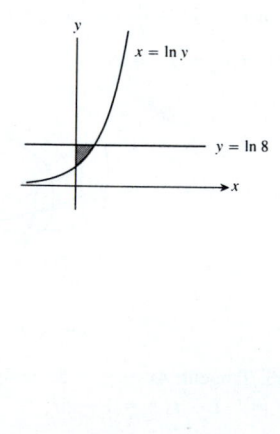

9. $9 - 9e$

11. $\dfrac{3}{2} \ln 2$ **13.** $(\ln 2)^2$ **15.** $-\dfrac{1}{10}$

17. 8

19. 2π

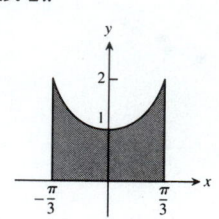

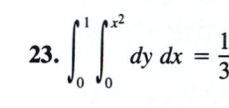

21. $\displaystyle\int_2^4 \int_0^{\frac{4-y}{2}} dx\, dy = 1$

23. $\displaystyle\int_0^1 \int_0^{x^2} dy\, dx = \dfrac{1}{3}$

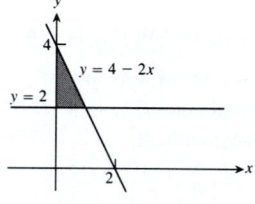

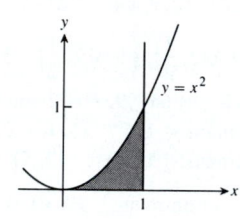

25. $\displaystyle\int_0^9 \int_0^{\frac{1}{2}\sqrt{9-y}} 16x\, dx\, dy = 81$ **27.** 1   **29.** π^2

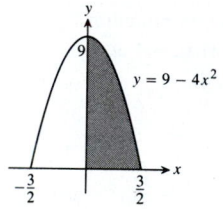

31. 2

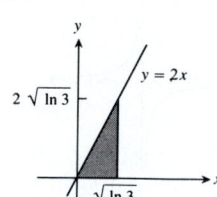

33. $4 - \sin 4$

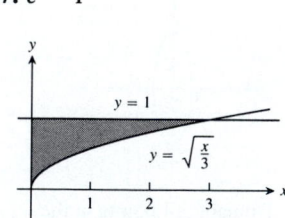

35. 2

37. $e - 1$

39. $\dfrac{4}{3}$ **41.** $\dfrac{625}{12}$ **43.** 16   **45.** 20   **47.** $2(1 + \ln 2)$

49. $\dfrac{20\sqrt{3}}{9}$

51. $\int_0^1 \int_x^{2-x} (x^2 + y^2)\, dy\, dx = \dfrac{4}{3}$

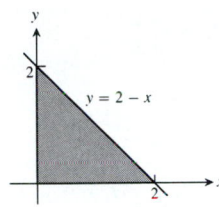

53. Interior of the ellipse $x^2 + 2y^2 = 4$ **55.** 0.603   **57.** 0.233

Section 13.2, pp. 934–937

1. 2

3. 4

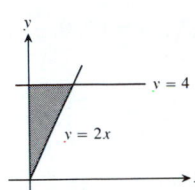

5. $\dfrac{1}{3}$

7. $\pi + 2$

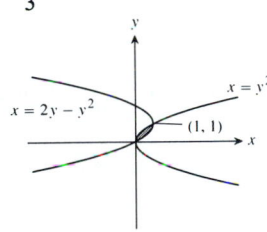

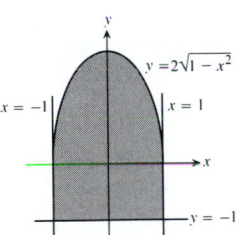

9. 12

11. $\sqrt{2} - 1$

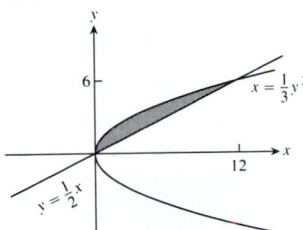

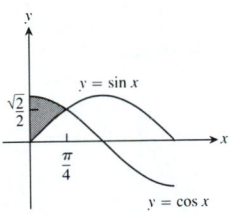

13. $\dfrac{3}{2}$

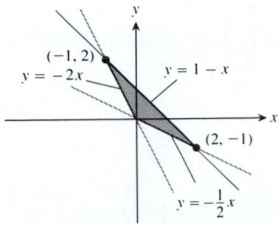

15. $(\bar{x}, \bar{y}) = \left(\dfrac{5}{14}, \dfrac{38}{35}\right)$ **17.** $(\bar{x}, \bar{y}) = \left(\dfrac{64}{35}, \dfrac{5}{7}\right)$ **19.** $(\bar{x}, \bar{y}) =$

$\left(0, \dfrac{4}{3\pi}\right)$ **21.** $(\bar{x}, \bar{y}) = \left(\dfrac{4a}{3\pi}, \dfrac{4a}{3\pi}\right)$ **23.** $(\bar{x}, \bar{y}) = \left(\dfrac{\pi}{2}, \dfrac{\pi}{8}\right)$

25. $\dfrac{4}{3} ab(a^2 + b^2)$ **27.** $(\bar{x}, \bar{y}) = \left(-1, \dfrac{1}{4}\right)$ **29.** $I_x = \dfrac{64}{105}$,

$R_x = \sqrt{\dfrac{8}{7}}$ **31.** $(\bar{x}, \bar{y}) = \left(\dfrac{3}{8}, \dfrac{17}{16}\right)$ **33.** $\bar{x} = \dfrac{11}{3}, \bar{y} = \dfrac{14}{27}$,

$I_y = 432, R_y = 4$ **35.** $\bar{x} = 0, \bar{y} = \dfrac{13}{31}, I_y = \dfrac{7}{5}, R_y = \sqrt{\dfrac{21}{31}}$

37. $\bar{x} = 0, \bar{y} = \dfrac{7}{10}, I_x = \dfrac{9}{10}, I_y = \dfrac{3}{10}, I_0 = \dfrac{6}{5}, R_x = \dfrac{3\sqrt{6}}{10}, R_y =$

$\dfrac{3\sqrt{2}}{10}, R_0 = \dfrac{3\sqrt{2}}{5}$ **39.** a) 0 b) $\dfrac{4}{\pi^2}$ **41.** $\dfrac{8}{3}$ **43.** If $0 < a \le \dfrac{5}{2}$, then

the appliance will have to be tipped more than 45° to fall over.

45. $(\bar{x}, \bar{y}) = \left(\dfrac{2}{\pi}, 0\right)$ **47.** a) By the hint, $M_{\text{c.m.}} = M_y = 0$.

b) $I_L = I_y - 2M_y h + mh^2$, but $M_y = 0$ and $I_{\text{c.m.}} = I_y$.
49. Use the domain additivity property for $M, M_x,$ and M_y.

51. a) $\left(\dfrac{7}{5}, \dfrac{31}{10}\right)$ b) $\left(\dfrac{19}{7}, \dfrac{18}{7}\right)$ c) $\left(\dfrac{9}{2}, \dfrac{19}{8}\right)$ d) $\left(\dfrac{11}{4}, \dfrac{43}{16}\right)$ **53.** In order

for c.m. to be on the common boundary, $h = a\sqrt{2}$. In order
for c.m. to be inside T, $h > a\sqrt{2}$.

Section 13.3, pp. 942–943

1. $\dfrac{\pi}{2}$ **3.** $\dfrac{\pi}{8}$ **5.** πa^2 **7.** $\dfrac{2}{3}$ **9.** $3 \ln (2 + \sqrt{3})$ **11.** $\dfrac{5\pi}{6}$

13. $\dfrac{\pi}{2} + 1$ **15.** π **17.** $2(\pi - 1)$ **19.** 12π **21.** $\dfrac{3\pi}{8} + 1$

23. $\dfrac{8}{3}$ **25.** $\pi \ln 4$ **27.** 4   **29.** $\bar{x} = \dfrac{5}{6}, \bar{y} = 0$ **31.** $\dfrac{2}{3} a(3\sqrt{3} - \pi)$,

assuming $\delta(r) = \dfrac{1}{r}$ **33.** $\dfrac{4}{3} + \dfrac{5\pi}{8}$ **35.** $\dfrac{2}{3}$ **37.** $\dfrac{1}{2}(a^2 + 2h^2)$

Section 13.4, pp. 948–950

1. 1   **3.** $\displaystyle\int_0^1 \int_0^{2-2x} \int_0^{\frac{6-6x-3y}{2}} dz\, dy\, dx$; $\displaystyle\int_0^2 \int_0^{1-\frac{y}{2}} \int_0^{\frac{6-6x-3y}{2}} dz\, dx\, dy$;

$\displaystyle\int_0^1 \int_0^{3-3x} \int_0^{\frac{6-6x-2z}{3}} dy\, dz\, dx$; $\displaystyle\int_0^3 \int_0^{1-\frac{z}{3}} \int_0^{\frac{6-6x-2z}{3}} dy\, dx\, dz$;

$\displaystyle\int_0^2 \int_0^{3-\frac{3y}{2}} \int_0^{\frac{6-3y-2z}{6}} dx\, dz\, dy$; $\displaystyle\int_0^3 \int_0^{2-\frac{2z}{3}} \int_0^{\frac{6-3y-2z}{6}} dx\, dy\, dz$; the

value of all six integrals is 1.   **5.** 1   **7.** 1   **9.** $\dfrac{\pi^3}{2}(1 - \cos 1)$

11. 18   **13.** $\dfrac{7}{6}$ **15.** 0   **17.** $\dfrac{1}{2} - \dfrac{\pi}{8}$

19. a) $\displaystyle\int_{-1}^1 \int_0^{1-x^2} \int_{x^2}^{1-z} dy\, dz\, dx$ b) $\displaystyle\int_0^1 \int_{-\sqrt{1-z}}^{\sqrt{1-z}} \int_{x^2}^{1-z} dy\, dx\, dz$

c) $\displaystyle\int_0^1 \int_0^{1-z} \int_{-\sqrt{y}}^{\sqrt{y}} dx\, dy\, dz$ d) $\displaystyle\int_0^1 \int_0^{1-y} \int_{-\sqrt{y}}^{\sqrt{y}} dx\, dz\, dy$

e) $\int_0^1 \int_{-\sqrt{y}}^{\sqrt{y}} \int_0^{1-y} dz\, dx\, dy$ **21.** 2 sin 4 **23.** 4 **25.** $\dfrac{31}{3}$ **27.** 1

29. $\dfrac{2}{3}$ **31.** $\dfrac{20}{3}$ **33.** $\dfrac{32}{15}$ **35.** $\dfrac{16}{3}$ **37.** 2 **39.** 12π **41.** $\dfrac{4}{7}$

43. $a = 3$ or $a = \dfrac{13}{3}$ **45.** Interior of the ellipsoid $x^2 + y^2 + 2z^2 = 18$

Section 13.5, pp. 952–955

1. $R_x = \sqrt{\dfrac{b^2 + c^2}{12}}, R_y = \sqrt{\dfrac{a^2 + c^2}{12}}, R_z = \sqrt{\dfrac{a^2 + b^2}{12}}$

3. $I_x = \dfrac{M}{3}(b^2 + c^2), I_y = \dfrac{M}{3}(a^2 + c^2), I_z = \dfrac{M}{3}(a^2 + b^2)$

5. $\bar{x} = \bar{y} = 0, \bar{z} = \dfrac{12}{5}, I_x = \dfrac{7904}{105}, I_y = \dfrac{4832}{63}, I_z = \dfrac{256}{45}$

7. a) $(\bar{x}, \bar{y}, \bar{z}) = \left(0, 0, \dfrac{8}{3}\right)$ b) $c = 2\sqrt{2}$ **9.** $I_L = 1386, R_L = \sqrt{\dfrac{77}{2}}$

11. $I_L = \dfrac{40}{3}, R_L = \sqrt{\dfrac{5}{3}}$ **13.** $(\bar{x}, \bar{y}, \bar{z}) = \left(\dfrac{4}{5}, \dfrac{2}{5}, \dfrac{2}{5}\right)$ **15.** $(\bar{x}, \bar{y}, \bar{z}) =$

$\left(\dfrac{8}{15}, \dfrac{8}{15}, \dfrac{8}{15}\right), I_x = I_y = I_z = \dfrac{11}{6}, R_x = R_y = R_z = \sqrt{\dfrac{11}{15}}$

17. a) By the hint, $M_{\text{c.m.}} = M_{yz} = 0$. b) $I_L = I_{yz} - 2M_{yz}h + mh^2$,

but $M_{yz} = 0$ and $I_{\text{c.m.}} = I_{yz}$ **19.** a) $I_{\text{c.m.}} = \dfrac{abc(a^2 + b^2)}{12}$,

$R_{\text{c.m.}} = \sqrt{\dfrac{a^2 + b^2}{12}}$ b) $I_L = \dfrac{abc(a^2 + 7b^2)}{3}, R_L = \sqrt{\dfrac{a^2 + 7b^2}{3}}$

21. Use the domain additivity property for M, M_{yz}, M_{xz}, and M_{xy}. **23.** a) $h = a\sqrt{3}$ b) $h = a\sqrt{2}$

Section 13.6, pp. 960–962

1. $\dfrac{4\pi(\sqrt{2} - 1)}{3}$ **3.** $\dfrac{17\pi}{5}$ **5.** $\pi(6\sqrt{2} - 8)$ **7.** π^2 **9.** $\dfrac{\pi}{3}$

11. 5π **13.** $\dfrac{3\pi}{10}$ **15.** $\dfrac{\pi}{3}$ **17.** 2π **19.** $\left(\dfrac{8 - 5\sqrt{2}}{2}\right)\pi$

21. a) $8\int_0^{\frac{\pi}{2}} \int_0^{\frac{\pi}{2}} \int_0^2 \rho^2 \sin\phi\, d\rho\, d\phi\, d\theta$ b) $8\int_0^{\frac{\pi}{2}} \int_0^2 \int_0^{\sqrt{4-r^2}} r\, dz\, dr\, d\theta$

c) $8\int_0^2 \int_0^{\sqrt{4-x^2}} \int_0^{\sqrt{4-x^2-y^2}} dz\, dy\, dx$

23. $\int_{-\frac{\pi}{2}}^{\frac{\pi}{2}} \int_0^{\cos\theta} \int_0^{3r^2} f(r, \theta, z)\, r\, dz\, dr\, d\theta$

25. a) $\int_0^{2\pi} \int_0^{\frac{\pi}{3}} \int_{\sec\phi}^2 \rho^2 \sin\phi\, d\rho\, d\phi\, d\theta$ b) $\int_0^{2\pi} \int_0^{\sqrt{3}} \int_1^{\sqrt{4-r^2}} r\, dz\, dr\, d\theta$

c) $\int_{-\sqrt{3}}^{\sqrt{3}} \int_{-\sqrt{3-x^2}}^{\sqrt{3-x^2}} \int_1^{\sqrt{4-x^2-y^2}} dz\, dy\, dx$ d) $\dfrac{5\pi}{3}$ **27.** $\dfrac{\pi}{2}$ **29.** 8π

31. $\dfrac{5\pi}{2}$ **33.** $\dfrac{4\pi(8 - 3\sqrt{3})}{3}$ **35.** $\dfrac{2}{3}$ **37.** $(\bar{x}, \bar{y}, \bar{z}) = \left(0, 0, \dfrac{3}{8}\right)$

39. $I_z = 30\pi$, $R_z = \sqrt{\dfrac{5}{2}}$ **41.** a) π b) $\dfrac{7\pi}{6}$ **43.** $\dfrac{8\pi a^5}{15}$

45. $M = \int_0^{2\pi} \int_0^a \int_0^{h\sqrt{1 - \frac{r^2}{a^2}}} r\, dz\, dr\, d\theta = \dfrac{2\pi a^2 h}{3}$ and

$M_{xy} = \int_0^{2\pi} \int_0^a \int_0^{h\sqrt{1 - \frac{r^2}{a^2}}} rz\, dz\, dr\, d\theta = \dfrac{\pi a^2 h^2}{4}$

47. $M = \int_0^{2\pi} \int_0^a \int_0^{h\sec\phi} \rho^2 \sin\phi\, d\rho\, d\phi\, d\theta = \dfrac{\pi h^3}{6}(\sec^4 a - 1)$ and

$M_{xy} = \int_0^{2\pi} \int_0^a \int_0^{h\sec\phi} \rho^3 \sin\phi\cos\phi\, d\rho\, d\phi\, d\theta = \dfrac{3h}{4}M$

49. a) $(\bar{x}, \bar{y}, \bar{z}) = \left(0, 0, \dfrac{4}{5}\right), I_z = \dfrac{\pi}{12}, R_z = \sqrt{\dfrac{1}{3}}$

b) $(\bar{x}, \bar{y}, \bar{z}) = \left(0, 0, \dfrac{5}{6}\right), I_z = \dfrac{\pi}{14}, R_z = \sqrt{\dfrac{5}{14}}$

51. $\dfrac{2\pi a^3}{3}$ **53.** $\dfrac{5\pi}{3}$ **55.** $\dfrac{4(2\sqrt{2} - 1)\pi}{3}$ **57.** $\dfrac{3}{4}$ **59.** $\bar{x} = \bar{y} = 0$,

$\bar{z} = \dfrac{3(2 + \sqrt{2})a}{16}$

Section 13.7, pp. 968–969

1. 2 **3.** a) $x = \dfrac{u + v}{3}, y = \dfrac{v - 2u}{3}, J = \dfrac{1}{3}$ b) $\dfrac{33}{4}$ **5.** $8 + \dfrac{52 \ln 2}{3}$

7. $\dfrac{\pi ab(a^2 + b^2)}{4}$ **9.** $\rho^2 \sin\phi$ **11.** $\dfrac{4\pi abc}{3}$ **13.** $2 + 3\ln 2$

Section 13 Miscellaneous Exercises, pp. 969–972

1. a) $\int_{-2}^0 \int_{2x+4}^{4-x^2} dy\, dx = \dfrac{4}{3}$ b) $\int_0^1 \int_{-\sqrt{v}}^{\sqrt{v}} du\, dv = \dfrac{4}{3}$

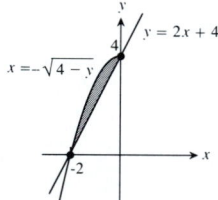

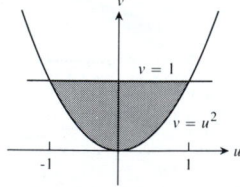

c) $\int_{-3}^3 \int_0^{\frac{\sqrt{9-s^2}}{2}} t\, dt\, ds = \dfrac{9}{2}$ d) $\int_0^4 \int_0^{\sqrt{4-z}} 2w\, dw\, dz = 8$

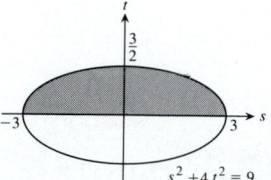

$s^2 + 4t^2 = 9$

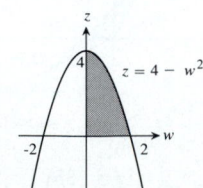

3. π **5.** $\ln\left(\dfrac{b}{a}\right)$ **7.** $\dfrac{e^{a^2 b^2} - 1}{ab}$ **11.** a) $\displaystyle\int_{-3}^{2}\int_{x}^{6-x^2} x^2 \, dy \, dx$

b) $\displaystyle\int_{-3}^{2}\int_{x}^{6-x^2}\int_{0}^{x^2} dz \, dy \, dx$ c) $\dfrac{125}{4}$ **13.** $\bar{x} = \bar{y} = \dfrac{1}{2 - \ln 4}$

15. $\left(-\dfrac{12}{5}, 2\right)$ **17.** $M = 4$, $M_x = M_y = 0$ **19.** $I_x = \dfrac{\delta}{12} bh^3$,

$R_x = \dfrac{h}{\sqrt{6}}$ **21.** $\bar{x} = \dfrac{15\pi + 32}{6\pi + 48}$, $\bar{y} = 0$

23. $\bar{x} = \dfrac{13}{3\pi}$, $\bar{y} = \dfrac{13}{3\pi}$ **25.** Area $= a^2 \cos^{-1}\left(\dfrac{b}{a}\right) - b\sqrt{a^2 - b^2}$,

$I_0 = \dfrac{a^4}{2}\cos^{-1}\left(\dfrac{b}{a}\right) - \dfrac{b^3}{2}\sqrt{a^2 - b^2} - \dfrac{b^3}{6}(a^2 - b^2)^{\frac{3}{2}}$

27. $a^2\beta\left(\ln a - \dfrac{1}{2}\right)$ **29.** $\dfrac{2\pi k R^3}{3}$ coulomb **31.** 4π

33. a) $4\displaystyle\int_{0}^{1}\int_{0}^{\sqrt{1-x^2}}\int_{0}^{\sqrt{4-x^2-y^2}} \sqrt{x^2 + y^2} \, dz \, dy \, dx$

b) $4\displaystyle\int_{0}^{\frac{\pi}{2}}\int_{0}^{\frac{\pi}{6}}\int_{0}^{2} \rho^3\sin^2\phi \, d\rho \, d\phi \, d\theta + 4\int_{0}^{\frac{\pi}{2}}\int_{\frac{\pi}{6}}^{\frac{\pi}{2}}\int_{0}^{\csc\phi} \rho^3\sin^2\phi \, d\rho \, d\phi \, d\theta$

35. $\displaystyle\int_{0}^{1}\int_{\sqrt{1-x^2}}^{\sqrt{3-x^2}}\int_{1}^{\sqrt{4-x^2-y^2}} z^2xy \, dz \, dy \, dx + \int_{1}^{\sqrt{3}}\int_{0}^{\sqrt{3-x^2}}\int_{1}^{\sqrt{4-x^2-y^2}}$

$z^2xy \, dz \, dy \, dx$ **37.** 2π **39.** $\dfrac{15}{8}$ **41.** $\dfrac{3\pi}{2}$ **43.** a) Sphere

radius $= 2$; hole radius $= 1$. b) $4\sqrt{3}\pi$ **45.** $\dfrac{\pi}{4}$

47. $\dfrac{8\pi}{3}$ **49.** $\dfrac{64\pi}{35}$

Chapter 14

Section 14.1, pp. 978–979

1. c **3.** g **5.** d **7.** f **9.** $\sqrt{2}$ **11.** $\dfrac{13}{2}$ **13.** 0

15. $\dfrac{1}{6}(5\sqrt{5} + 9)$ **17.** $+\infty$ **19.** $I_x = 2\pi - 2$, $R_x = 1$

21. $I_z = 2\pi\delta a^3$, $R_z = a$ **23.** a) $I_z = 2\pi\sqrt{2}\delta$, $R_z = 1$

b) $I_z = 4\pi\sqrt{2}\delta$, $R_z = 1$ **25.** a) $4\sqrt{2} - 2$ b) $\sqrt{2} +$

$\dfrac{1}{2}\ln(1 + \sqrt{2})$ **27.** $(\bar{x}, \bar{y}, \bar{z}) = \left(1, \dfrac{16}{15}, \dfrac{2}{3}\right)$, $I_x = \dfrac{29}{5}$, $I_y = \dfrac{64}{15}$,

$I_z = \dfrac{56}{9}$, $R_x = \sqrt{\dfrac{29}{10}}$, $R_y = \sqrt{\dfrac{32}{15}}$, $R_z = \sqrt{\dfrac{28}{9}}$

Section 14.2, pp. 987–988

1. $F(x, y) = -\dfrac{kx}{x^2 + y^2}i - \dfrac{ky}{x^2 + y^2}j$, any $k > 0$

3. a)

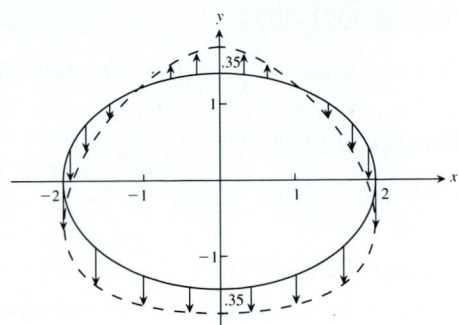

b)

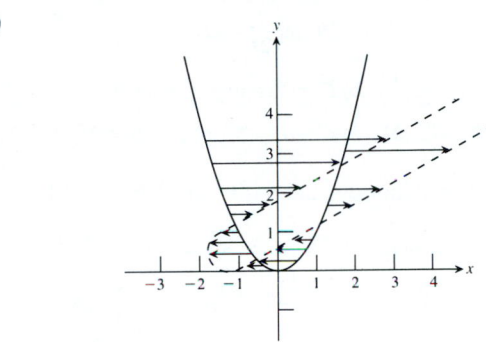

5. a) $\dfrac{9}{2}$ b) $\dfrac{13}{3}$ c) $\dfrac{9}{2}$ **7.** a) $\dfrac{1}{3}$ b) $-\dfrac{1}{5}$ c) -1 **9.** a) 2 b) $\dfrac{3}{2}$

c) $-\dfrac{1}{2}$ **11.** $\dfrac{1}{2}$ **13.** $-\pi$ **15.** 48 **17.** π **19.** a) F_1: circ. $= 0$,

flux $= 2\pi$; F_2: circ. $= 2\pi$, flux $= 0$ b) F_1: circ. $= 0$, flux $= 8\pi$;

F_2; circ. $= 8\pi$, flux $= 0$ **21.** Circ. $= 0$, flux $= \pi a^2$

23. Circ. $= \pi a^2$, flux $= 0$ **25.** a) $-\dfrac{\pi}{2}$ b) 0 c) 1 **27.** $\dfrac{9}{2}$

Section 14.3, pp. 999–1001

1. Flux $= 0$, circ. $= 2\pi a^2$ **3.** Flux $= -\pi a^2$, circ. $= 0$

5. Flux $= 2$, circ. $= 0$ **7.** -9 **9.** Flux $= \dfrac{1}{2}$, circ. $= \dfrac{1}{2}$

11. Flux $= \dfrac{1}{5}$, circ. $= -\dfrac{1}{12}$ **13.** $\dfrac{3}{2}\pi a^2$ **15.** 0 **17.** -16π

19. $\dfrac{2}{33}$ **35.** πa^2 **37.** $\dfrac{3}{8}\pi$

Section 14.4, pp. 1010–1012

1. $\dfrac{13}{3}\pi$ **3.** 4 **5.** $6\sqrt{6} - 2\sqrt{2}$ **7.** $2\pi(2 - \sqrt{2})$ **9.** $\dfrac{2}{3}\pi$

11. $\dfrac{\pi}{6}(5\sqrt{5} - 1)$ **13.** $9a^3$ **15.** $\dfrac{abc}{4}(ab + ac + bc)$ **17.** 2

19. $\dfrac{\pi a^3}{6}$ **21.** $\dfrac{\pi a^2}{4}$ **23.** $\dfrac{\pi a^3}{2}$ **25.** -32 **27.** -4 **29.** $3a^4$

31. $\left(\dfrac{a}{2}, \dfrac{a}{2}, \dfrac{a}{2}\right)$ **33.** $(\bar{x}, \bar{y}, \bar{z}) = \left(0, 0, \dfrac{14}{9}\right)$, $I_z = \dfrac{15\pi\sqrt{2}\delta}{2}$,

$R_z = \sqrt{\dfrac{5}{2}}$ **35.** a) $2\pi a^4\delta$ b) $2\pi a^4\delta + 4\pi a^2\delta$

37. $\dfrac{\pi}{6}\left(4\sqrt{3} + 1)^{\frac{3}{2}} - 1\right)$ **39.** $\dfrac{\pi}{6}(5\sqrt{5} - 1)$ **41.** $2\sqrt{5}$

Section 14.5, pp. 1021–1022

1. -16 **3.** -8π **5.** 3π **7.** $\dfrac{76}{3}$ **9.** 12π **11.** $12\pi(4\sqrt{2}-1)$

Section 14.6, pp. 1029–1030

1. 4π **3.** $-\dfrac{5}{6}$ **5.** 0 **7.** -6π **9.** $2\pi a^2$

Section 14.7, pp. 1039–1040

1. Conservative   **3.** Not conservative   **5.** Not conservative
7. Conservative   **9.** Conservative   **11.** Not conservative

13. 49 **15.** -16 **17.** 1 **19.** $\ln\left(\dfrac{\pi}{2}\right)$ **21.** 0

27. a) $F = \nabla\left(\dfrac{x^2-1}{y}\right)$ b) $F = \nabla(e^x \ln y + y \sin z)$

29. a) 0 b) 0 c) 0 **31.** a) $c = b = 2a$ b) $c = b = 2$

Section 14 Miscellaneous Exercises, pp. 1041–1044

1. Path 1: $2\sqrt{3}$; path 2: $1 + 3\sqrt{2}$ **3.** $4a$ **5.** 11.08 **7.** $\bar{z} = \dfrac{2}{3}$,

$I_z = \dfrac{14}{3}, R_z = \sqrt{\dfrac{7}{3}}$ **9.** a) $-\dfrac{a^2\pi}{2}$ b) 0 c) 1 **13.** $\dfrac{3}{2}$ **17.** 0

19. a) π b) An ellipse   **21.** $\pi\sqrt{3}$ **23.** $2\pi\left(1 - \dfrac{1}{\sqrt{2}}\right)$

25. $\dfrac{abc}{2}\sqrt{\dfrac{1}{a^2}+\dfrac{1}{b^2}+\dfrac{1}{c^2}}$ **27.** a) $\dfrac{3}{4}$ b) 2 **29.** 3

31. $(\bar{x}, \bar{y}, \bar{z}) = \left(0, 0, \dfrac{98}{24}\right), I_z = 640\pi, R_z = 2\sqrt{2}$ **33.** 3

35. $\dfrac{4\pi}{3}(2 - 2\sqrt{2})$ **37.** 3 **39.** $\dfrac{8}{3}$ **41.** 0 **43.** Conservative

45. Not conservative   **47.** $f(x, y, z) = y^2 + yz + 2x + z$ **49.** 0

51. Path 1: $\dfrac{3}{2}$; path 2: $\dfrac{5}{3}$ **57.** 6π **59.** $\dfrac{2}{3}$

Chapter 15

Section 15.1, pp. 1050–1051

1. First order, nonlinear   **3.** Fourth order, nonlinear
5. a) $xy'' - y' = x(2) - 2x = 0$ b) $xy'' - y' = x(0) - 0 = 0$
c) $xy'' - y' = x(2C_1) - 2C_1 x = 0$ **7.** a) $2y' + 3y =$

$2(-e^{-x}) + 3(e^{-x}) = e^{-x}$ b) $2y' + 3y = 2\left(-e^{-x} - \dfrac{3}{2} e^{-\left(\frac{3}{2}\right)x}\right) +$

$3\left(e^{-x} + e^{-\left(\frac{3}{2}\right)x}\right) = e^{-x}$ c) $2y' + 3y = 2\left(-e^{-x} - \dfrac{3}{2} Ce^{-\left(\frac{3}{2}\right)x}\right) +$

$3\left(e^{-x} + Ce^{-\left(\frac{3}{2}\right)x}\right) = e^{-x}$ **9.** $x^2 y' + xy = -\displaystyle\int_1^x \dfrac{e^t}{t}\,dt + e^x +$

$\displaystyle\int_1^x \dfrac{e^t}{t}\,dt = e^x$ **11.** $y' = 160 - 32x, y'' = -32, y(5) = 400,$
$y'(5) = 0$ **13.** $y' = -6 \sin 2t - 2 \cos 2t, y'' = -12 \cos 2t +$
$4 \sin 2t, y(0) = 3, y'(0) = -2, y'' + 4y = 0$ **15.** $y =$
$\tan\left(\ln\left|\dfrac{x-1}{x+1}\right| + \dfrac{\pi}{4}\right)$ **17.** $e^{-y}(y^3 + 3y^2 + 6y + 6) = 2e^{\sqrt{x}} +$

$38e^{-2} - 2$ **19.** $y = \sin\left[\dfrac{x}{2}\sqrt{x^2+1} + \dfrac{1}{2}\sinh^{-1}x - \dfrac{\pi}{6}\right]$ **21.** 10 min

23. $\ln|x| = -\dfrac{1}{4}\ln\left(2\left(\dfrac{y}{x}\right)^2 + 1\right) + C$ **25.** $\ln|x| + e^{-\frac{y}{x}} = C$

27. $\ln|x| - \ln\left|\sec\left(\dfrac{y}{x} - 1\right) + \tan\left(\dfrac{y}{x} - 1\right)\right| = \ln 2$ **29.** $y = Cx^2$, a

family of parabolas   **31.** $\dfrac{x^2}{2} - \dfrac{y^2}{2} = C$, a family of hyperbolas

Section 15.2, pp. 1054–1055

1. Exact   **3.** Exact   **5.** Exact   **7.** Not exact   **9.** $\dfrac{1}{2}x^2 + xy +$

$\dfrac{1}{3}y^3 + C = 0$ **11.** $x^2 y + xy^2 - \dfrac{1}{2}y^2 + C = 0$ **13.** $e^x + x \ln y +$

$y \ln x - \cos y + C = 0$ **15.** $\left(\dfrac{y}{2} - \dfrac{1}{4}\sin 2y\right)\sin x + C = 0$

17. $\ln x^2 y^2 + \displaystyle\sum_{n=1}^{\infty} \dfrac{x^n y^n}{n \cdot n!} + C = 0$ **19.** $\dfrac{1}{2}x^2 + \dfrac{x}{y} + C = 0$

21. a) $\ln|xy| + C = 0$ b) $y = \dfrac{C}{x}$ **23.** $\dfrac{1}{2}x^2 + xy^2 + y - 2 = 0$

25. $\ln|xy| - xy + 1 = 0$ **27.** $x^2 + \dfrac{1}{2}(\ln x)(\ln y)^2 - 4 = 0$

29. $x^3 + xy^2 + C = 0$

Section 15.3, pp. 1060–1061

1. $y = c^{-x} + Ce^{-2x}$ **3.** $y = \dfrac{-\cos x + C}{x^3}$ **5.** $y = \dfrac{\frac{1}{3}x^3 - x + C}{(x-1)^4}$

7. $y = (\ln|\sec x| + C)\csc x$ **9.** $y = \dfrac{x}{2} - \dfrac{1}{4} + \dfrac{3}{4}e^{-2x}$

11. $y = \dfrac{-\cos x + \dfrac{\pi}{2}}{x}$ **13.** $y = y_0 e^{kx}$ **15.** $t = \dfrac{L}{R}\ln 2$

17. $.95\left(\dfrac{V}{R}\right)$ (calculator, rounded)   **19.** a) 550 ft b) 77.28 sec
(calculator, rounded)

Section 15.4, pp. 1065–1066

1. $y = C_1 + C_2 e^{-2x}$ **3.** $y = C_1 e^{-x} + C_2 e^{-5x}$ **5.** $y =$
$(C_1 x + C_2)e^{2x}$ **7.** $y = (C_1 x + C_2)e^{5x}$ **9.** $y =$
$e^{-\frac{1}{2}x}\left(C_1 \cos\left(\dfrac{\sqrt{3}}{2}x\right) + C_2 \sin\left(\dfrac{\sqrt{3}}{2}x\right)\right)$ **11.** $y =$
$c^x(C_1 \cos(\sqrt{3}x) + C_2 \sin(\sqrt{3}x))$ **13.** $y = -\dfrac{1}{2}e^x + \dfrac{3}{2}e^{-x}$

15. $y = \dfrac{3}{4}e^{2x} - \dfrac{3}{4}c^{-2x}$ **17.** $y = xe^{-x}$ **19.** $y = 2xe^{-\frac{3}{2}x}$

21. $y = \sin 2x$ **23.** $y = e^x\left(2 \cos \sqrt{2}x - \dfrac{1}{\sqrt{2}}\sin \sqrt{2}x\right)$

25. Multiply $C_1 = y \cos x - y' \sin x$ by $\cos x$ and multiply $C_2 =$
$y \sin x + y' \cos x$ by $\sin x$. Add the equations together to obtain
$C_1 \cos x + C_2 \sin x = y(\cos^2 x + \sin^2 x) = y$.

Section 15.5, pp. 1078–1079

1. $y = 1 - x + \frac{1}{2}x^2 + C_2 e^{-x} + C_1$ **3.** $y = C_1\cos x +$

$C_3 \sin x - \frac{1}{2}x \cos x$ **5.** $y = C_1 e^{-x} + C_2 x e^{-x} + \frac{1}{2}x^2 e^{-x}$

7. $y = C_1 e^x + C_2 e^{-x} + \frac{1}{2}xe^x$ **9.** $y =$

$e^{-2x}(C_1 \cos x + C_2 \sin x) + 2$ **11.** $y = C_1 \cos x + C_2 \sin x +$

$\cos x \ln|\cos x| + x \sin x$ **13.** $y = C_1 e^{5x} + C_2 e^{-2x} + \frac{3}{10}$

15. $y = C_1 + C_2 e^x + \frac{1}{2}(\cos x - \sin x)$ **17.** $y = C_1 \cos x +$

$C_2 \sin x - \frac{1}{8}\cos 3x$ **19.** $y = C_1 e^{2x} + C_2 e^{-x} - 6\cos x - 2\sin x$

21. $y = C_1 e^x + C_2 e^{-x} + \frac{1}{2}xe^x - x^2 - 2$ **23.** $y =$

$C_1 e^{-2x} + C_2 e^{3x} - \frac{1}{4}e^{-x} + \frac{49}{50}\cos x + \frac{7}{50}\sin x$ **25.** $y =$

$C_1 + C_2 e^{-5x} + x^3 - \frac{3}{5}x^2 + \frac{6}{25}x$ **27.** $y = C_1 + C_2 e^{3x} +$

$\frac{1}{3}xe^{3x} + 2x^2 + \frac{4}{3}x$ **29.** $y = C_1 + C_2 e^{5x} + \left(\frac{1}{10}x^2 - \frac{1}{25}x\right)e^{5x}$

31. $y = C_1\cos x + C_2\sin x + x\left(\sin x - \frac{1}{2}\cos x\right)$

33. a) $y_p = (x - 1)e^x + \frac{1}{2}e^{-x}$ **b)** $y = (C_1 + C_2 e^x) + xe^x + \frac{1}{2}e^{-x}$

35. a) $y_p = -\frac{1}{8}e^x - \frac{4}{5}$ **b)** $y = C_1 e^{5x} + C_2 e^{-x} - \frac{1}{8}e^x - \frac{4}{5}$

37. $y = C_1 \cos x + C_2 \sin x - \sin x \ln|\csc x + \cot x|$

39. $y = C_1 + C_2 e^{8x} + \frac{1}{8}xe^{8x}$ **41.** $y = C_1 + C_2 e^x - \frac{1}{4}x^4 - x^3 -$

$3x^2 - 6x$ **43.** $y = C_1 + C_2 e^{-2x} + \frac{1}{6}x^3 - \frac{1}{4}x^2 + \frac{1}{4}x - \frac{1}{3}e^x$

45. $y = C_1 \cos x + C_2 \sin x + x \cos x - \sin x + \sin x \ln|\sec x|$

47. $y = Ce^{3x} - \frac{1}{2}e^x$ **49.** $y = Ce^{3x} + 5xe^{3x}$ **51.** $y = 2\cos x +$

$\sin x - 1 + \sin x \ln|\sec x + \tan x|$ **53.** $y = -C_1 e^{-x} + C_2$

55. $y = -\ln|x + C_1| + C_2$ **57.** $y = \ln x^2 - 3$ **59.** $y = \frac{1}{2}x -$

$\frac{2}{x} + 2$ **61.** $y = \frac{1}{9}x^3 + \frac{2}{3}\ln|x| - \frac{1}{9}$ **63.** $s = s_0 \cos\left(\sqrt{\frac{k}{m}}\,t\right)$

65. $y = C_1 \sin x + C_2 \cos x$ **67.** $y = Ae^{2x} + Be^{-x}$

69. $y = C_1 x^{-1} + C_2 x$ **71.** $y = C_1\left(\frac{x}{2}\ln\left|\frac{x+1}{x-1}\right| - 1\right) + C_2 x$

73. For all k, $y^{(k)} = y = e^x$, so $\displaystyle\sum_{k=0}^{n} a_n(x)y^{(k)} = \sum_{k=0}^{n} a_n(x)e^x =$

$e^x\left(\displaystyle\sum_{k=0}^{n} a_n(x)\right) = 0$.   **75.** $y = C_1 e^x + C_2 e^{(1 + \sqrt{3})x} +$

$C_3 e^{(1-\sqrt{3})x} + \frac{7}{13}\sin x + \frac{4}{13}\cos x$ **77.** $y = C_1 e^{-2x} + C_2 xe^{-2x} +$

$C_3 e^{2x} + C_4 xe^{2x} + \frac{1}{2}x - 1$

Section 15.6, pp. 1082–1083

1. $x = C \sin(\omega t + \phi)$, where $C = \sqrt{x_0^2 + \left(\frac{v_0}{\omega}\right)^2}$ and $\phi =$

$\tan^{-1}\left(\frac{v_0}{\omega x_0}\right)$ **3. a)** $i = C_1 \cos \omega t + C_2 \sin \omega t$ **b)** $i = C_1 \cos \omega t +$

$C_2 \sin \omega t + \frac{\alpha V}{L(\omega^2 - \alpha^2)}\cos \alpha t$ **c)** $i = C_1 \cos \omega t + C_2 \sin \omega t +$

$\frac{V}{2L}t \cos \alpha t$ **d)** $i = e^{-5t}(C_1 \cos 149t + C_2 \sin 149t)$ (rounded)

5. $0 = \theta_0 \cos \omega t + \frac{v_0}{\omega}\sin \omega t$, $\omega = \sqrt{\frac{2k}{mr^2}}$ **7. a)** $x =$

$C_1 \cos \omega t + C_2 \sin \omega t + \frac{A\omega^2}{\omega^2 - \alpha^2}\sin \alpha t$, where $C_1 = x_0$ and

$C_2 = -\frac{\alpha\omega A}{\omega^2 - \alpha^2}$ **b)** $x = x_0 \cos \omega t + \frac{A}{2}\sin \omega t - \frac{\omega A}{2}t \cos \omega t$

Section 15.7, p. 1088

1. $y' = 2e^{x-1} - 1$, and $x + y = x + 2e^{x-1} - 2 - (x - 1) = 2e^{x-1} - 1$. Also, $y(1) = 0$.   **3.** $y = 1 + x + \frac{1}{2}x^2 + \cdots =$

$\displaystyle\sum_{n=0}^{\infty} \frac{x^n}{n!}$ **5.** $y = 2 + 2^2x + \frac{2^3}{2!}x^2 + \frac{2^4}{3!}x^3 + \cdots = 2\sum_{n=0}^{\infty} \frac{(2x)^n}{n!}$

7. $y = x + \frac{x^3}{3!} + \frac{x^5}{5!} + \cdots = \displaystyle\sum_{n=0}^{\infty} \frac{x^{2n+1}}{(2n+1)!}$ **9.** $y = 2 + x -$

$\frac{2x^2}{2!} + \frac{2x^4}{4!} + \cdots = 2 + x + 2\displaystyle\sum_{n=1}^{\infty} \frac{(-1)^n x^{2n}}{n!}$ **11.** $y = -(x - 2) -$

$2\displaystyle\sum_{n=1}^{\infty} \frac{(x-2)^{2n}}{(2n)!} - 3\sum_{n=1}^{\infty} \frac{(x-2)^{2n+1}}{(2n+1)!}$ **13.** $y = \displaystyle\sum_{n=0}^{\infty} a_n \frac{x^n}{n!}$, where

$a_0 = a$, $a_1 = b$, $a_2 = 0$, $a_3 = 1$, $a_4 = -2a$, $a_5 = -6b$, and

$a_n = (n - 2)(n - 3)a_{n-4}$ for $n \geq 6$.

Section 15.8, pp. 1092–1093

1. The isoclines are the vertical lines $x = $ constant.

3. The isoclines are the vertical lines $x = $ constant.

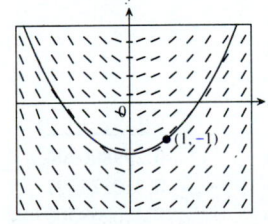

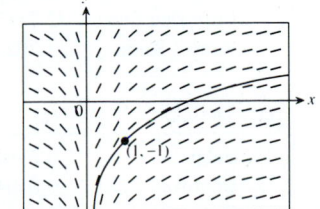

5. The isoclines are the hyperbolas $xy - $ constant.

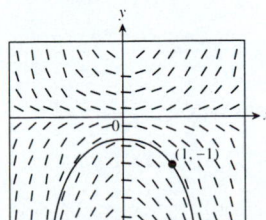

7. The isoclines are the diagonal lines $x + y = $ constant; $y = \tan(x + C) - x$; passes through $(0, 0)$ when $C = 0$.

9. a) $y = x - 1$ b) Concave up for $y > x - 1$; concave down for $y < x - 1$

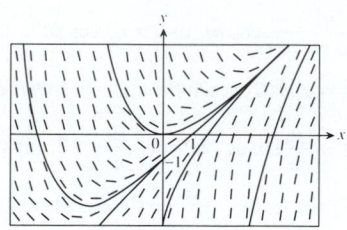

11. $y' = x - y$; $y(1) = -1$ **13.** $y' = -(1 + y)\sin x$; $y(0) = 2$

15. $y_0 = 2$; $y_1 = (1/2)x^2 + 3/2$; $y_2 = (1/2)x^2 + 3/2$; $y_3 = (1/2)x^2 + 3/2$

17. $y_0 = 1$; $y_1 = (1/2)x^2 + 1/2$; $y_2 = 5/8 + (1/4)x^2 + (1/8)x^4$; $y_3 = 29/48 + (5/16)x^2 + (1/16)x^4 + (1/48)x^6$ **19.** $y_0 = 1$; $y_1 - 1 + x + (1/2)x^2$; $y_2 = 1 + x + x^2 + (1/6)x^3$; $y_3 = 1 + x + x^2 + (1/3)x^3 + (1/24)x^4$ **21.** $y = (x_0 + y_0 + 1)e^{x - x_0} - (x + 1)$

Section 15.9, pp. 1097–1098

1. $y \approx 2.48832$. The exact value is e.   **3.** $y \approx 2.702708163$

5. $y = \dfrac{1}{x - 1}$. Since $x^2 + y^2 \geq y^2$, the solution to $y' = x^2 + y^2$ grows faster than that of $y' = y^2$, which grows infinite at $x = 1$.

7.

Method	Approximation
Euler	1.271428571
Improved Euler	1.285714286
Runge–Kutta	1.285714286

9.

Method	Approximation
Euler	2.073343406
Improved Euler	2.211571264
Runge–Kutta	2.202834625

11.

Method	(a)	(b)
Euler	0.289456778	0.299806204
Improved Euler	0.310493553	0.310324591
Runge–Kutta	0.310268270	0.310268300

Section 15 Miscellaneous Exercises, pp. 1099–1101

1. $y = \ln|2 - e^{-x}| - 2$ **3.** $\ln\left|x^2\left(\dfrac{x^2 - y^2}{y^2}\right)\right| + C = 0$

5. $\dfrac{1}{3}x^3 + xy + e^y - 9 = 0$ **7.** $y = (x + 1)^{-2}\left(\dfrac{1}{3}x^3 + \dfrac{1}{2}x^2 + 1\right)$

9. $y = \ln\left|\dfrac{\sec x}{2}\right|$ **11.** $y = 2e^{-x}$ **13.** $y = 7xe^{-2x}$

15. $y = e^{-x}(\cos x + \sin x)$ **17.** $y = -\dfrac{1}{2} + \dfrac{3}{2}e^{-2x} + x^2 - x$

19. $y = -\dfrac{7}{6}e^{2x} - \dfrac{5}{6}e^{-x} + \dfrac{3}{2}xe^{2x}$ **21.** $y^2 = -\dfrac{3}{2}x^2 + C$

23. $y^2 = 4B(B - x)$ **25.** $-x^2y^{-1} - \dfrac{1}{2}y^2 + C = 0$

27. $x^3e^y + xy^2 + 2xy + C = 0$ **29.** $y = e^x\left(\dfrac{1}{2}x^2 + C_1 + xC_2\right)$

31. $y = \dfrac{1}{5}e^x + C_2\cos 2x + C_3\sin 2x$ **33.** $a_0 = 1$, $a_1 = 2$, $a_3 = \dfrac{13}{3}$, $a_4 = \dfrac{19}{4}$ **35.** $y = \sum_{k=2}^{\infty} \dfrac{x^n}{n!}$ **37.** $a_0 = 0$, $a_1 = 0$, $a_2 = \dfrac{1}{2}$, $a_3 = \dfrac{1}{12}$, $a_4 = \dfrac{1}{288}$

Appendix

Section A.4, pp. A-20

1. If $F_1(t) \to L_1$, $F_2(t) \to L_2$, ..., $F_n(t) \to L_n$ as $t \to c$, then $F_1(t) + F_2(t) + \cdots + F_n(t) \to L_1 + L_2 + \cdots + L_n$ as $t \to c$, for any positive integer n.

Section A.5, pp. A-23

7. Statement true for $n \geq 5$.

Section A.7, pp. A-26

1. $c = -1$ **3.** $c = \dfrac{-1 + \sqrt{37}}{3}$

Section A.10, pp. A-37

1. -5 **3.** 1 **5.** a) -7 b) -7 **7.** a) 38 b) 38 **9.** $x = -4$, $y = 1$ **11.** $x = 3$, $y = 2$ **13.** $x = 3$, $y = -2$, $z = 2$ **15.** $x = 2$, $y = 0$, $z = -1$ **17.** a) $h = 6$, $k = 4$ b) $h = 6$, $k \neq 4$

Chapter 1

Section 1.1, pp. 10–12

2. $-2, 4$ **4.** $-\sqrt{2}, -2.5$ **6.** $m = -\frac{1}{2}, m_\perp = 2$ **8.** m does not

exist $m_\perp = 0$ **10.** a) $\sqrt{2}$ b) -1.3 **12.** a) $-\pi$ b) -1

14. $y = \frac{1}{2}x + 3$ **16.** $y = -2x + 2a$ **18.** $x = -2$

20. $y = -\frac{3}{4}x - \frac{1}{2}$ **22.** $y = -\frac{1}{2}x - 3$ **24.** $y = \frac{1}{3}x - 1$

26.

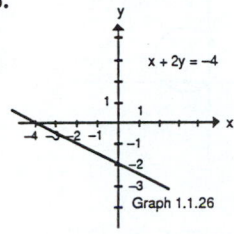

Graph 1.1.26

28.

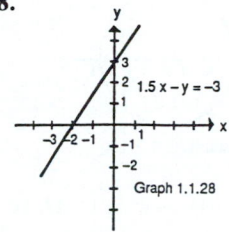

Graph 1.1.28

30. a) $y = 3x$ b) $y = -\frac{1}{3}x$ **32.** a) $y = -2x - 2$

b) $y = \frac{1}{2}x + 3$ **34.** a) $y = -2$ b) $x = -3$ **36.** $(0,0)$ **38.** $0, 0$

40. Fiberglass, because the temperature differential is $-16.1°/\text{in.}$

42. $y = x - 1$ **44.** 40.2 ft **46.** $A = (-12, 2), B = (-12, -5),$

$C = (9, -5)$ **48.** a) $(-1, 4)$ b) $(3, -2)$ c) $(5, 2)$ d) $(0, x)$

e) $(-y, 0)$ f) $(-y, x)$ g) $(3, -10)$ **50.** Answers vary with

different calculators **52.** $x = 1$

Section 1.2, pp. 22–24

2. $D: x \geq 0; R: y \leq 1$ **4.** $D: x = 0; R: y = 0$ **6.** $D:$ All Reals

except $x = \frac{(2n+1)\pi}{2}$, n an integer; $R: y \geq 1$ **8.** $D:$ All Reals;

$R:$ All even integers **10.** Even **12.** Even **14.** Neither **16.** Odd

18. Neither **20.** Even **24.** $D: -\infty < x < \infty;$

22. $D: -\infty < x < \infty; R: y \leq 0;$ $R: y \leq 4;$ Symmetric to

Symmetric to y axis y axis

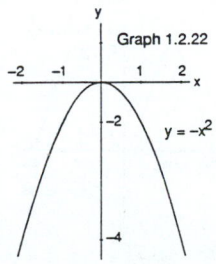

Graph 1.2.22

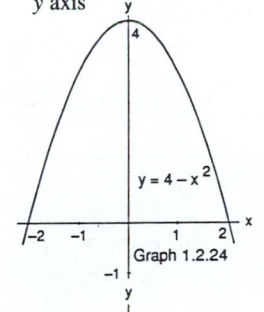

Graph 1.2.24

26. $D:$ All Reals; $R:$ All Reals;
Symmetric to origin

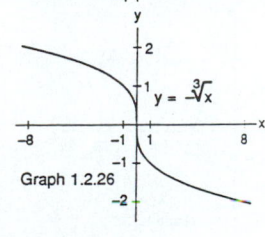

Graph 1.2.26

28. $D: x \neq 0; R: y \neq 0;$
Symmetric to y axis

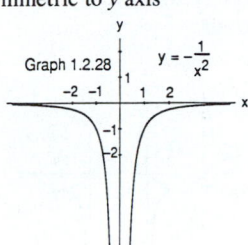

Graph 1.2.28

30. Symmetric to origin

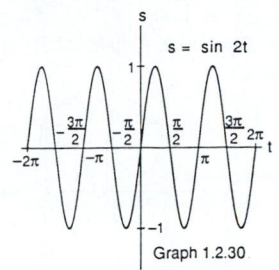

Graph 1.2.30

32. Symmetric to s-axis

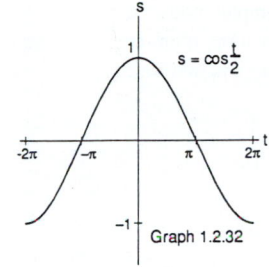

Graph 1.2.32

34. a) No. b) No. c) No. d) $0 < r \leq 1$ **36.** i, iii, and iv
38. Yes, when x is an integer. **40.** $\boxed{x}$ is the integer part of the
decimal representation of x

42.

x	0	1	2
y	1	0	0

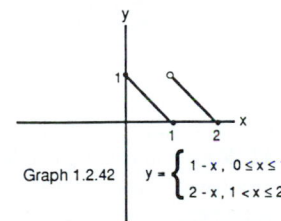

Graph 1.2.42 $y = \begin{cases} 1 - x, & 0 \leq x \leq 1 \\ 2 - x, & 1 < x \leq 2 \end{cases}$

44.

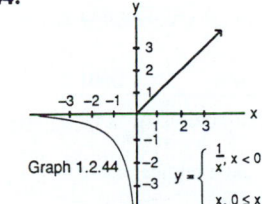

Graph 1.2.44 $y = \begin{cases} \frac{1}{x}, & x < 0 \\ x, & 0 \leq x \end{cases}$

46.

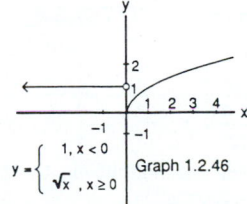

$y = \begin{cases} 1, & x < 0 \\ \sqrt{x}, & x \geq 0 \end{cases}$ Graph 1.2.46

48. a) $y = \begin{cases} 0, & 0 \leq x \leq \frac{T}{2} \\ \frac{2}{T}x - 1, & \frac{T}{2} < x \leq T \end{cases}$

b) $y = \begin{cases} A, & 0 \leq x < \frac{T}{2} \\ -A, & \frac{T}{2} \leq x < T \\ A, & T \leq x < \frac{3T}{2} \\ -A, & \frac{3T}{2} \leq x \leq 2T \end{cases}$

50. $D_f: x \geq -1, D_g: x \geq 1 \therefore D_{f+g} = D_{f-g} = D_{fg} =$
$D_{gf}: x \geq 1; D_{f/g}: x > 1$ **52.** a) 0 b) 0 c) 1 d) 1 e) x f) x
54. Don't look it up here! Try it for yourself.
58. a)   b)

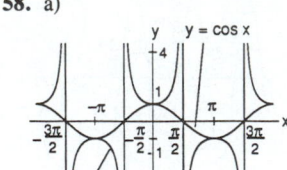
Graph 1.2.58 a

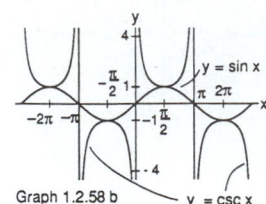
Graph 1.2.58 b

a) $x = n\pi$, n an integer is where the graphs intersect; $\cos x = 0$
when $\sec x$ is undefined. b) $x = (2n + 1)\pi$, n an integer is where
the graphs intersect; $\sin x = 0$ when $\csc x$ is undefined.

Section 1.3, pp. 29–31

2. a) Answers vary b) Answers vary
4.

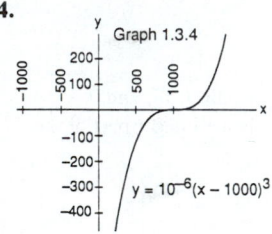
Graph 1.3.4
$y = 10^{-6}(x - 1000)^3$

6. $y = x$ **8.** $c = 1$ for all x **10.** Constant for all x.
12. Not constant if $x < 0$ **14.** With a positive number less than
one, the difference between 1 and the square is approximately
halved each time. **16.** $\sqrt[10]{0.5} = 0.9330329 \Rightarrow \sqrt[10]{0.9330329} =$
$0.9930924 \Rightarrow \sqrt[10]{0.9930924} = 0.999307 \Rightarrow \sqrt[10]{0.999307} =$
$0.9999306 \Rightarrow \sqrt[10]{0.9999306} = 0.999993 \Rightarrow \sqrt[10]{0.999993} =$
$0.9999993 \Rightarrow \sqrt[10]{0.9999993} = 0.9999999 \Rightarrow \sqrt[10]{0.9999999} = 1$

18.

Δx	.1	.01	.001	.0001
$\dfrac{\sin \Delta x}{\Delta x}$	.017453283	.017453292	.017453292	.017453292 etc.

When you multiply m_{sec} by $180/\pi$, you get a number close to 1
(0.999999472).

20.

x	10	100	1000	10000
$\left[\dfrac{1}{x}\right]^{1/\ln x}$	0.367879441	0.367879441	0.367879441	0.367879441 etc.

Section 1.4, pp. 37–38

2. 5 **4.** $\dfrac{5}{3}$ **6.** $\sqrt{x^2 + y^2}$ **8.** 5 **10.** 4.1 **12.** a) True b) True
c) True d) True e) True f) True g) True h) False **14.** 10, -4
16. 0, 2 **18.** 4, 0 **20.** g **22.** f **24.** a **26.** $-2 \leq x \leq 2$
28. $-3 < r < -1$ **30.** $-3 < s < -2$ **32.** 5 $> t > 3$
34. $|x - 3| < 6$ **36.** $|x + 4| < 3$ **38.** $\sqrt{15} < x < \sqrt{17}$

40. $15 > x > 3$ **42.** $\dfrac{1}{2} > x > \dfrac{1}{6}$ **44.** $|x - (-2)| < \dfrac{1}{2}$
46. $|x - (-3)| < 0.1$ **48.** $24.5 \geq R \geq 23.5$ **50.** $x > 1$
52. $f(x) = x^2$ **54.** Let $g(x) = \sqrt{x}$ and $f(x) = \sin^2 x$

Section 1.5, pp. 46–48

2. a) $y = x^2 + 3$ b) $y = x^2 - 5$ **4.** a) $y = (x + 2)^2$
b) $y = (x - 4)^2 + 1$ c) $y = (x + 1)^2 - 4$ d) $y = (x - 2)^2 - 3$
6.   **8.**

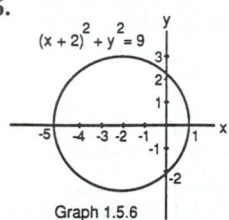
$(x + 2)^2 + y^2 = 9$
Graph 1.5.6

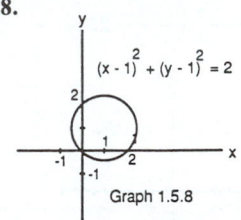
$(x - 1)^2 + (y - 1)^2 = 2$
Graph 1.5.8

10. $(x - 1)^2 + y^2 = 1$ **12.** $(x + 1)^2 + (y + 1)^2 = 1$
14. a) Circle with $C(0,0)$, $a = 1$, and its exterior b) Circle with
$C(0,0)$, $a = 2$, and its interior c) Region between the two
circles in a) and b) along with the two circles
16. $(x + 4)^2 + (y - 2)^2 > 16$ **18.** $y = x^2$
20. $y = -\dfrac{1}{2}x^2$

22.   **24.**

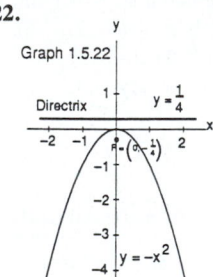
Graph 1.5.22

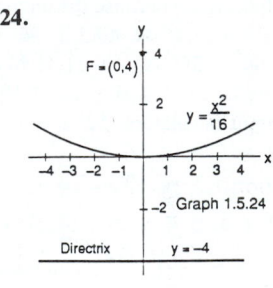
Graph 1.5.24
$y = \dfrac{x^2}{16}$

26. $(x + 4)^2 + (y - 3)^2 = 5$ **28.** $y = (x - 1)^3$ **30.** $y = 2x$
32. $x + 1 = y^2$ **34.** $y = \dfrac{x}{x - 1}$
36.   **38.**

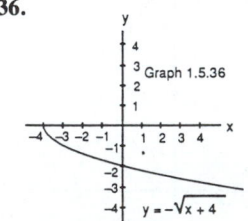
Graph 1.5.36
$y = -\sqrt{x + 4}$

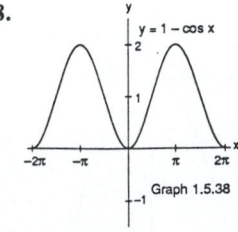
$y = 1 - \cos x$
Graph 1.5.38

40.   **42.** a)

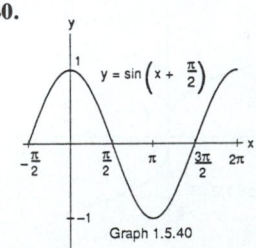
$y = \sin\left(x + \dfrac{\pi}{2}\right)$
Graph 1.5.40

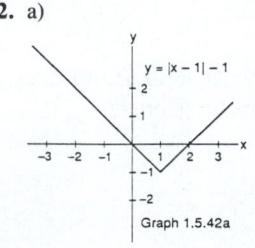
$y = |x - 1| - 1$
Graph 1.5.42a

b)

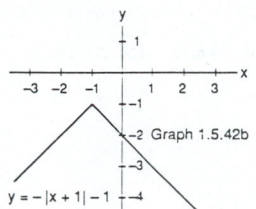

Graph 1.5.42b

$y = -|x + 1| - 1$

Section 1.6, pp. 56–58

2.

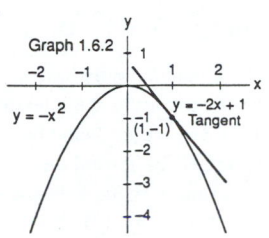

Graph 1.6.2

$y = -x^2$

$y = -2x + 1$ Tangent

4.

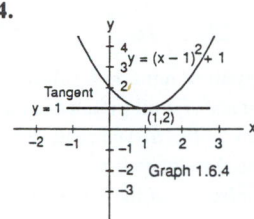

Tangent $y = 1$

$y = (x - 1)^2 + 1$

$(1,2)$

Graph 1.6.4

6.

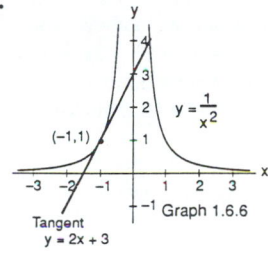

$(-1,1)$

$y = \dfrac{1}{x^2}$

Tangent $y = 2x + 3$

Graph 1.6.6

8. $f'(x) = 4x, m = 8$ **10.** $f'(x) = \dfrac{1}{(x + 1)^2}, m = 1$

12. $f'(x) = \dfrac{1}{2\sqrt{x}}, m = 1$ **14.** $f'(x) = \dfrac{-1}{2x\sqrt{x}}, m = -\dfrac{1}{16}$

16. $f'(x) = 3x^2, m = 12$ **18.** -2 **20.** 2 **22.** $\dfrac{1}{2}$

24. a)

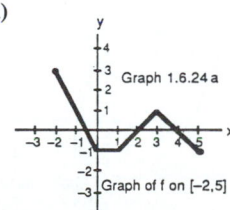

Graph 1.6.24 a

Graph of f on [-2,5]

b)

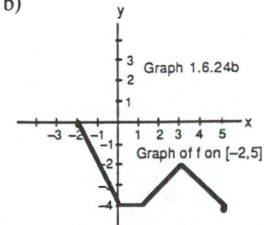

Graph 1.6.24b

Graph of f on [-2,5]

26. The difference quotient approaches $y = 2x$.

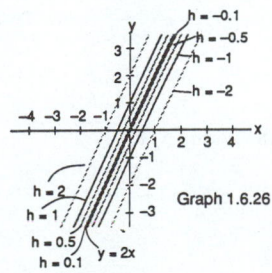

$h = -0.1$
$h = -0.5$
$h = -1$
$h = -2$

$h = 2$
$h = 1$
$h = 0.5$
$h = 0.1$ $y = 2x$

Graph 1.6.26

28.

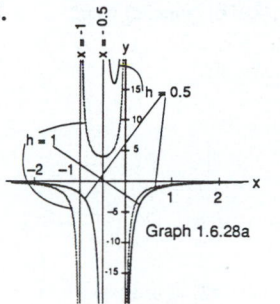

$x = -1$ $x = -0.5$ $h = 0.5$

$h = 1$

Graph 1.6.28a

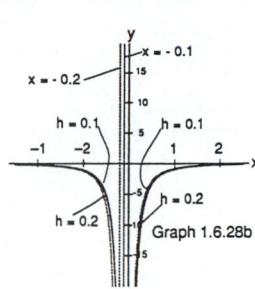

$x = -0.1$
$x = -0.2$
$h = 0.1$ $h = 0.1$
$h = 0.2$ $h = 0.2$

Graph 1.6.28b

Graphs 1.6.28a and 1.6.28b are the graphs of the difference quotient when h is positive. Graphs 1.6.28d and 1.6.28e are the graphs of the difference quotient when h is negative.

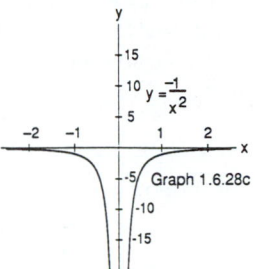

$y = \dfrac{-1}{x^2}$

Graph 1.6.28c

As h approaches 0, the difference quotient approaches $y = -1/x^2$

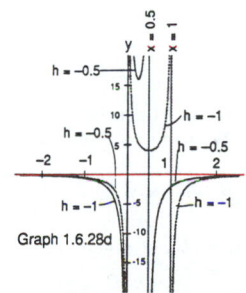

$x = 0.5$ $x = 1$
$h = -0.5$
$h = -1$
$h = -0.5$
$h = -0.5$
$h = -1$ $h = -1$

Graph 1.6.28d

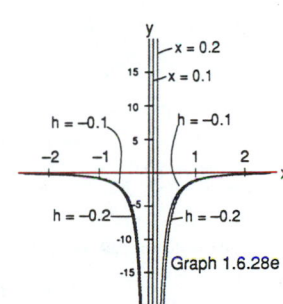

$x = 0.2$
$x = 0.1$
$h = -0.1$ $h = -0.1$
$h = -0.2$ $h = -0.2$

Graph 1.6.28e

Section 1.7, pp. 69–72

2. 0 **4.** 0 **6.** -9 **8.** -8 **10.** -16 **12.** -2 **14.** $\dfrac{5}{2}$ **16.** $\dfrac{4}{3}$

18. $\dfrac{1}{3}$ **20.** $\dfrac{1}{2}$ **22.** -3 **24.** $\dfrac{3}{2}$ **26.** $\dfrac{3}{4}$ **28.** $-\dfrac{1}{10}$ **30.** $-\dfrac{1}{2}$

32. a) $\lim_{x \to 2} + f(x) = 1$; $\lim_{x \to 2} - f(x) = 1$ b) Yes

34. a) True b) False c) False d) True e) True f) True g) True
h) True i) True

36. a)   b) 0,0 c) 0

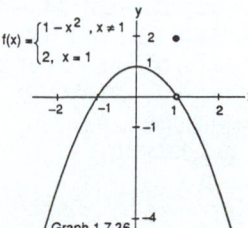

$f(x) = \begin{cases} 1 - x^2, & x \neq 1 \\ 2, & x = 1 \end{cases}$

Graph 1.7.36

38. a) $(-\infty, -1) \cup (-1,1) \cup (1,\infty)$ b) None c) None

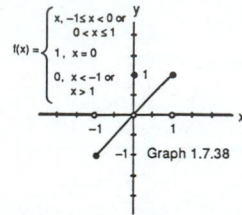

$f(x) = \begin{cases} x, -1 \le x < 0 \text{ or} \\ \quad 0 < x \le 1 \\ 1, \; x = 0 \\ 0, \; x < -1 \text{ or} \\ \quad x > 1 \end{cases}$

Graph 1.7.38

40. a) 0 b) 0 c) 0   **41.** a) 0 b) 0 c) 1   **42.** a) 1 b) −1
44. a) 4 b) −21 c) −12 d) 3
46. Right Hand Derivative: −1; Left Hand Derivative: 1;
Derivative doesn't exist.
48. $\delta = 1$ **50.** $\delta = 1$

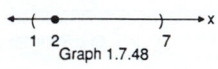

Graph 1.7.48

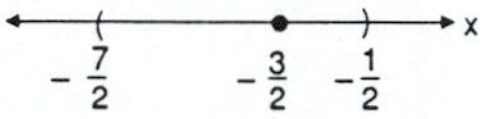

Graph 1.7.50

52. $\delta = 3$ **54.** $\delta = 0.2391$

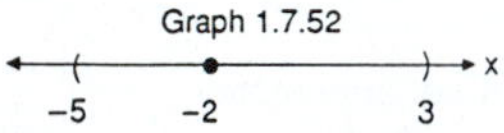

Graph 1.7.52

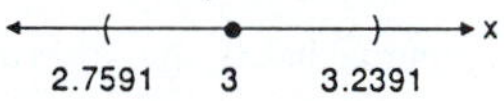

Graph 21.7.54

56. 0.1   **58.** $\dfrac{-2 + \sqrt{5}}{2}$ **60.** 0.39   **62.** $2.99 < x < 3.01$
64. $-0.2 < x < 0.2$ **66.** $14.9201 < x < 15.81$
68. $\dfrac{16}{7} > x > \dfrac{20}{9}$ **70.** $-3; \delta \le 0.01$ **72.** $1; \delta \le 0.01$
74. $-4; \delta \le 0.05$ **76.** $4; \delta \le 0.75$ **78.** $8; \delta \le 0.0024$
80. For $\epsilon = 0.003$, $\delta = 0.001$; For $\epsilon = 0.0003$, $\delta = 0.0001$; For
arbitrary ϵ, $\delta = \dfrac{\epsilon}{3}$ **82.** $\delta = \epsilon^2$, limit is 0   **84.** a) 2 b) 2
86. a) $\dfrac{1}{4}$ b) Finding the derivative of $f(x) = \sqrt{x}$ at $x = 4$.
88. 3   **90.** 0

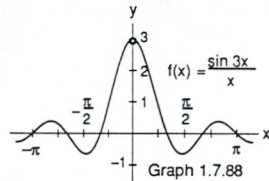

$f(x) = \dfrac{\sin 3x}{x}$

Graph 1.7.88

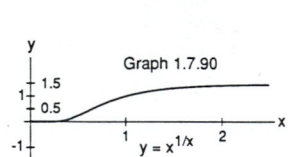

Graph 1.7.90

$y = x^{1/x}$

Section 1.8, pp. 78–79

2. a) 2 b) 2   **4.** a) 0 b) 0   **6.** a) −2 b) −2   **8.** a) $+\infty$
b) $-\infty$ **10.** a) 0 b) 0   **12.** a) $\dfrac{9}{2}$ b) $\dfrac{9}{2}$ **14.** a) −1 b) −1
16. a) 5 b) 5   **18.** a) 0 b) 0   **20.** $+\infty$ **22.** $-\infty$ **24.** $-\infty$
26. $-\infty$ **28.** $+\infty$ **30.** 0   **32.** 9   **34.** 4   **36.** $1 - 1 = 0$
38. a) ∞ b) $-\infty$ c) ∞ d) $-\infty$ **40.** a) ∞ b) $-\infty$ c) ∞
d) ∞ **42.** ∞ **44.** All equal -1 **46.** $\lim_{x \to -\infty} f(x) = 1$;
$\lim_{x \to 0} + f(x) = +\infty$; $\lim_{x \to 0} - f(x) = 2$; $\lim_{x \to \infty} f(x) = 0$
48. a) 1 b) 1 c) 2 (Approximately) d) 2.302585093
50. 2.72 (Approximately)

Section 1.9, pp. 83–84

2. 1   **4.** 2   **6.** 1   **8.** $\dfrac{1}{3}$ **10.** 2   **12.** 0   **14.** 1   **16.** 2   **18.** 1
20. 1   **22.** $A\dfrac{1}{2} r^2 \left(\dfrac{\pi}{180°}\right)\alpha$ **24.** 1

Section 1.10, pp. 93–94

2. a) Yes b) Yes c) No d) No   **4.** $[-1,0) \cup (0,1) \cup (1,2) \cup$
$(2,3]$ **6.** 2   **8.** All except $x = 2$ **10.** $[-1,1) \cup (1,2) \cup (2,3]$
12. -2 **14.** 3,1   **16.** 5, -2 **18.** Continuous everywhere.
20. 0   **22.** 6   **24.** $\dfrac{3}{2}$ **26.** 1   **28.** $\dfrac{1}{2}$ **30.** a) $\sqrt{3}$ b) 1   **32.** No
maximum or minimum values.   **34.** No, a function **may** have a
maximum or a minimum even if it is discontinuous or the interval
is open.   **40.** a) The roots of $f(x) = x^3 - 3x - 1$ are found by
solving the equation $x^3 - 3x - 1 = 0$ (Part e). b) To find the
x-coordinates of the intersection of $y = x^3$ and $y = 3x + 1$, you
solve the system $y = x^3$ and $y = 3x + 1 \Rightarrow x^3 = 3x + 1 \Rightarrow$
$x^3 - 3x - 1 = 0$ (Part e). c) $x^3 - 3x = 1$ is equivalent to
$x^3 - 3x - 1 = 0$ (Part e). d) In the same way as b) above, you
must solve the system $y = x^3 - 3x$ and $y = 1 \Rightarrow x^3 - 3x = 1 \Rightarrow$
$x^3 - 3x - 1 = 0$. e) All of the above are equivalent to solving
$x^3 - 3x - 1 = 0$.   **42.** $x \approx 1.45161, -0.854638, 0.403032$
44. $x \approx 1.5596105$

Section 1 Miscellaneous Exercises, pp. 95–99

4. a) $y = -\dfrac{2}{3}x - \dfrac{7}{3}$ b) $(-2, -1)$ c) $\sqrt{13}$

10. $y = \begin{cases} \dfrac{5}{2}x, \; 0 \le x \le 2 \\ -\dfrac{5}{2}x + 10, \; 2 < x \le 4 \end{cases}$ **12.** $\left(\dfrac{17}{18}, \dfrac{23}{6}\right)$

14. a)   b)

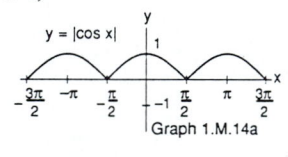

$y = |\cos x|$

Graph 1.M.14a

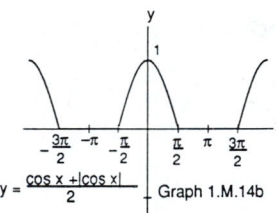

$y = \dfrac{\cos x + |\cos x|}{2}$ Graph 1.M.14b

c)

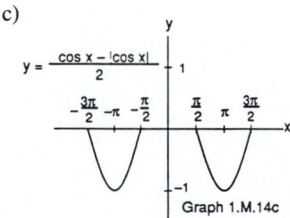

$y = \dfrac{\cos x - |\cos x|}{2}$

Graph 1.M.14c

16. $\min(a,b) = \dfrac{a + b}{2} - \dfrac{|a - b|}{2}$ **18.** Within 5° F.

22. a) $-\sqrt{2}$ b) $\dfrac{1}{2}\sqrt{2}$ c) $\dfrac{1}{2} + \sqrt{2}$ d) 2 e) $\dfrac{1}{2}$ f) $\dfrac{1}{2}$

24. a) $\lim_{x \to -1} + f(x) = 2$; $\lim_{x \to 1} + f(x) = 1$; $\lim_{x \to -1} - f(x) = 0$; $\lim_{x \to 1} - f(x) = 2$; $\lim_{x \to 0} + f(x) = 0$; $\lim_{x \to 0} - f(x) = 0$; b) No limit exists at -1 and 1 c) f is continuous at $x = 0$

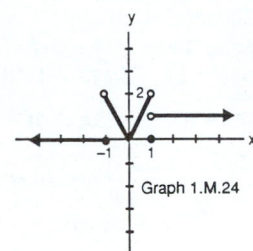

Graph 1.M.24

26. 0 **30.** $\dfrac{\epsilon}{6}$

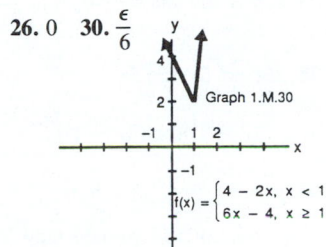

Graph 1.M.30

$$f(x) = \begin{cases} 4 - 2x, & x < 1 \\ 6x - 4, & x \geq 1 \end{cases}$$

32. Let $\delta = \sqrt{4 + \epsilon} - 2$

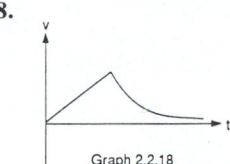

$\delta = \sqrt{4 + \epsilon} - 2$

Graph 1.M.32

34. 0; $\delta \leq 0.01$ **36.** -10; $\delta \leq 0.01$ **38.** 1; $\delta \leq \dfrac{3}{8}$

40. 1; $\delta \leq \dfrac{1}{84}$ **42.** $\dfrac{1}{3}$ **44.** $\dfrac{3}{7}$ **46.** $\dfrac{1}{2}$ **48.** $\dfrac{1}{2\sqrt{x}}$

52. Maximum value is 1 at $x = -1$; minimum value is 0 at $x = 0$. This is not consistent with the Max-Min Theorem because the function must be continuous over a **closed** interval to **guarantee** a maximum and minimum value on the interval.

Chapter 2

Section 2.1, pp. 111–112

2. $-1 + 2x - 3x^2$, $2 - 6x$ **4.** $8x^3 - 8x$, $24x^2 - 8$

6. $x^2 + x$, $2x + 1$, 2 **8.** $\dfrac{x^4}{24}, \dfrac{x^3}{6}, \dfrac{x^2}{2}$, x, 1 **10.** $3x^2 + 10x + 1$

12. $2x + \dfrac{2}{x^3}$ **14.** $\dfrac{-2(x^2 + x + 1)}{(x^2 - 1)^2}$ **16.** $\dfrac{1}{(x + 2)^2}$

18. $\dfrac{-17}{(2x - 7)^2}$ **20.** $\dfrac{12 - 6x^2}{(x^2 - 3x + 2)^2}$ **22.** $\dfrac{1}{x^2}, \dfrac{-2}{x^3}$

24. $\dfrac{21}{x^8}, \dfrac{-168}{x^9}$ **26.** $\dfrac{-12}{x^2} + \dfrac{12}{x^4} - \dfrac{4}{x^5}, \dfrac{24}{x^3} - \dfrac{48}{x^5}$

28. $\dfrac{-5}{x^2} + \dfrac{2}{x^3}, \dfrac{10}{x^3} - \dfrac{6}{x^4}$ **30.** $\dfrac{-3}{x^4}, \dfrac{12}{x^5}$ **32.** $4x^3$ **34.** $5x^4 - 2x$

36. $4x^3 + 6x^2 - 26x - 14$ **38.** a) $\dfrac{-6\theta^5}{(2 - \theta^3)^2}$ b) $-\dfrac{4}{3}\theta^{-3} - 3\theta$

40. a) $\dfrac{-q}{(q^2 + 1)^2}$ b) $-4q^{-5} + 6q^{-7}$ **42.** a) 0 b) 0 c) 0 d) 0

44. The smallest value of $m = 3x^2 + 1$ is 1, when $x = 0$.

46. $(-4/3, 0)$, $(0, 16)$ **48.** $y = -\dfrac{1}{2}x + 2$ **50.** $a = -3$,

$b = 2$ **52.** $\dfrac{dR}{dM} = CM - M^2$

Section 2.2, pp. 121–125

2. a) 0.0 m/sec, 3.72 m/sec²; 1.86 m/sec, 3.72 m/sec² b) 0.465 m, 0.93 m/sec **4.** a) 0 m/sec, -6 m/sec²; 0.0 m/sec, $a = 6$ m/sec² b) -4 m, -2 m/sec **6.** a) 1750 m/sec,

$a = -7250$ m/sec²; -1 m/sec, $\dfrac{2}{5}$ m/sec² b) -20 m, -5 m/sec

8. a) $24 - 1.6t$, -1.6 b) 15 sec c) 180 m d) 4.39 sec e) 30 sec

10. 320 sec on the moon; 52 sec on the earth. **12.** -8000 gallons/min; -10000 gallons/min **14.** a) $\dfrac{2000}{(x + 1)^2}$ b) \$55.56

c) 0 **16.** 0 m/sec, 1 m/sec

18.

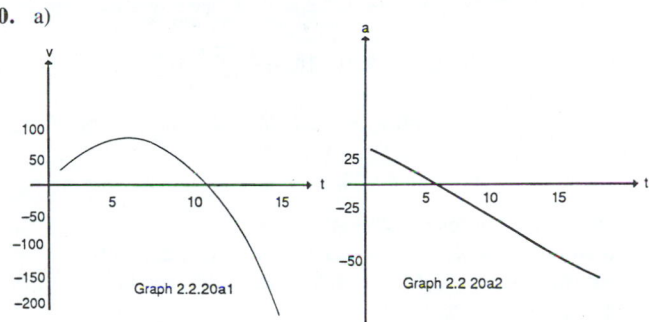

Graph 2.2.18

20. a)

Graph 2.2.20a1

Graph 2.2 20a2

b)

$v = 30t - 3t^2$

Graph 2.2 20b1

$a = 30 - 6t$

Graph 2.2 20b2

22. rabbits/day and foxes/day **24.** a **26.** c

28.

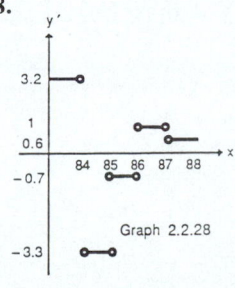

Graph 2.2.28

30. 348.7 ft/sec (350 ft/sec two significant digits), 237.75 MPH (240 MPH two significant digits)

32.

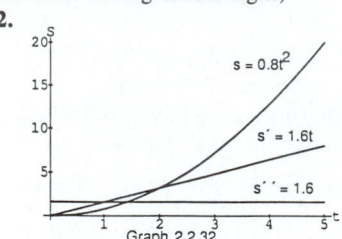

Graph 2.2.32

34.

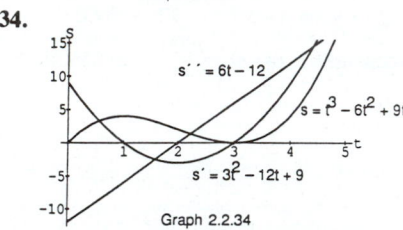

Graph 2.2.34

Section 2.3, pp. 130–132

2. $-x^{-2} + 5 \cos x = -\dfrac{1}{x^2} + 5 \cos x$ **4.** $\sec x + x \sec x \tan x$

6. $3 + \tan x + x \sec^2 x$ **8.** 0 **10.** $-\dfrac{x \sin x + \cos x}{x^2}$

12. $-\dfrac{\csc^2 x}{(1 + \cot x)^2}$ **14.** $2t - \sec t \tan t$ **16.** $\csc t - t \csc t \cot t$

18. $\theta \cos \theta$ **20.** $-\sin \theta$ **22.** $2 \cos q + \sin q$ **24.** $\dfrac{-1}{(1 + \sin q)}$

26. a) $\sin x$ b) $\cos x$ **28.** $y = x, y = -x$

30. $y = -x + \pi/2 + 1, y = x - \pi/2 + 1$ **32.** Proof

34. No   **36.** $x = \pi/6, x = 5\pi/6$ **38.** 1   **40.** $\cos(1)$

42. a) $-\sqrt{2}, \sqrt{2}, \sqrt{2}$ b) $0, 0, -\sqrt{2}$

44. $(-\pi/4, -1), (\pi/4, 1)$ **46.** a) $y = 2$

b) $y = -4x + \pi + 4$.

48.

h	$\dfrac{\sin h}{h}$	$\left[\dfrac{\sin h}{h}\right]\left[\dfrac{180}{\pi}\right]$
1	.017453283	.999999492
.01	.017453292	1
.001	.017453292	1
.0001	.017453292	1

The derivative of $\sin x$ would be $\dfrac{\pi}{180} \cos x \approx .017453292 \cos x$

52.

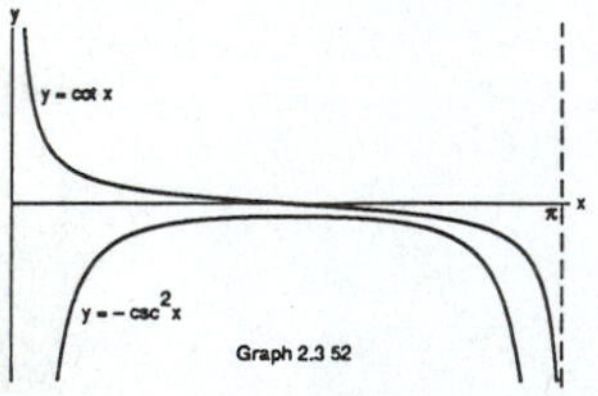

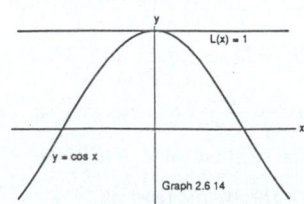

Graph 2.3 52

Section 2.4, pp. 139–141

2. $\dfrac{\pi}{2} \sec\left(\dfrac{\pi x}{2}\right) \tan\left(\dfrac{\pi x}{2}\right)$ **4.** $\dfrac{-2}{1 + \sin 2x}$ **6.** $2x \cos x^2$

8. $2 \cos 2x \cos 3x - 3 \sin 2x \sin 3x$

10. $-(2x + 7) \csc(x^2 + 7x) \cot(x^2 + 7x)$

12. $\sec(\tan x) \tan(\tan x) \sec^2 x$ **14.** $-27(4 - 3x)^8$

16. $6 \csc x \cot x (1 + \cot x)^{-7}$ **18.** $-5(x/2 - 1)^{-11}$

20. $\left[\dfrac{x}{5} - \dfrac{1}{5x}\right]^4 \left[1 - \dfrac{1}{x^2}\right]$ **22.** $-3(x + 1)^{-3}(x - 5)^{-2}$

$(x - 3)$ **24.** $-4x \sin x (\sin x - x \cos x)^{-5}$

26. $(\csc x)(\cot x + \csc x)^{-1}$ **28.** $\dfrac{2 \sin x}{(1 + \cos x)^2}$

30. $\dfrac{6t^5}{(t + 1)^7}$ **32.** $-4 \cos(\pi - 4t) + 4(\pi - 4t) \sin(\pi - 4t)$

34. $4 \sec^3 2\theta - 2 \sec 2\theta$

36. $[\cos(\theta + \sin \theta)](1 + \cos \theta)$ **38.** $10(\sec^2 5x) \tan 5x$

42. $-42(1 + \cos^2 7x)^2 (\cos 7x)(\sin 7x)$

44. $2(\sec^2(x/3)) (\tan(x/3))$ **46.** $18(\csc^2(3x - 1))(\cot(3x - 1))$

48. 1   **50.** 5π **52.** -8 **54.** $2x \cos(x^2 + 1)$

56. $2[\cos(\sin 2x)] \cos 2x$ **58.** 3   **60.** $y = mx$ **64.** a) 1 b) 6

c) 1 d) $-\dfrac{1}{9}$ e) $-\dfrac{40}{3}$ f) $-1/3$ g) $-\dfrac{4}{9}$ **66.** a) 101st b) $0.64°$/day

68. No, $g(x)$ is not differentiable at $x = 0$ and thus does not satisfy the hypothesis of the Chain Rule.

Section 2.5, pp. 146–148

2. $-3/5x^{-8/5}$ **4.** $\dfrac{1}{4\sqrt[4]{x^3}}$ **6.** $-4(1 - 6x)^{-1/3}$ **8.** $\dfrac{1}{\sqrt{(x^2 + 1)^3}}$

10. $\dfrac{6y - x^2}{y^2 - 6x}$ **12.** $\dfrac{y - 3x^2}{3y^2 - x}$ **14.** $\dfrac{3x + 7}{y^2}$ **16.** $\dfrac{1 - 3x^2 - 2xy}{x^2 + 1}$

18. $(2x^{-1/2} + 1)^{-4/3}x^{-3/2}$ **20.** $(\tan 2x)\sqrt{\sec 2x}$

22. $5/4[\sin(x + 5)]^{1/4}[\cos(x + 5)]$ **24.** $\sec y$ **30.** $\dfrac{1}{y + 1}$,

$\dfrac{-1}{(y + 1)^3}$ **32.** $-\dfrac{1}{4}$ **34.** $y = \dfrac{3}{4}x - \dfrac{25}{4}, y = -\dfrac{4}{3}x$

36. $y = -x - 1, y = x + 3$ **38.** a) $x = \pm\sqrt{7/3}$

b) $x = \pm 2\sqrt{7/3}$ **40.** $y = 2x$ **42.** $y = 2x - 1, y = -\dfrac{1}{2}x + \dfrac{3}{2}$

44. $(3, -1)$ **45.** $a = \dfrac{3}{4}$ **48.** $\dfrac{2}{5}$ m/sec, $-\dfrac{4}{125}$ m/sec^2

Section 2.6, pp. 158–160

2. $-\dfrac{1}{4}x + 1$ **4.** $10x - 13$ **6.** $-\dfrac{4}{5}x + \dfrac{9}{5}$ **8.** $-x + 2$

10. $x + 1$ **12.** $\dfrac{1}{4}x + \dfrac{1}{4}$ **14.** $L(x) = 1$

Graph 2.6 14

16. $L(x) = x + \pi/2$

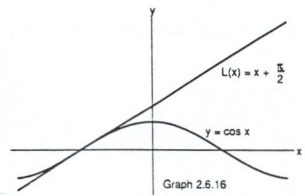

Graph 2.6.16

18. $L(x) = \sqrt{2}x + \sqrt{2}(4 - \pi)/4$

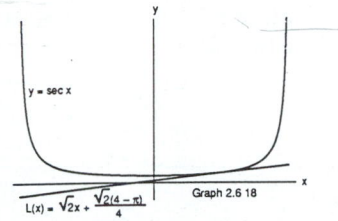

Graph 2.6 18

20. a) 1.2 b) 1.003   **24.** section 1.7 exercise 14: .5, .707106781, .840896415, .95760328, .978572062, .989228013, .994599423, . . .; section 1.7 exercise 15: 2, 1.071773463, 1.00695555, 1.000693387, . . .; The decimal part of each display has an additional zero.   **28.** a) $4x^3$ b) 0.4641 c) 0.4
30. a) 1.061 b) 1 c) 0.061   **32.** $8\pi r_0 dr$ **34.** $12x_0 dx$

36. $2\pi rdh$ **38.** $\dfrac{2}{\pi}$ in, 10 in²   **40.** 1% of the true value

42. 1%   **44.** the measurement must have an error less than 1/2%
46. Proof   **48.** 37.87   **50.** a) $-\pi\sqrt{L}g^{-3/2}dg$
b) the speed of both increases c) -0.976 cm/sec², 979 cm/sec²
52. $\dfrac{(1 - 2x^2)dx}{\sqrt{1 - x^2}}$ **54.** $\dfrac{dx}{1 + \cos x}$ **56.** $-\dfrac{y^2 + 2xy}{2xy + x^2}dx$
58. $\sin^2\left(\dfrac{x}{3}\right)\cos\left(\dfrac{x}{3}\right)dx$ **60.** $-\left[\dfrac{\csc^2(\sqrt{x})}{\sqrt{x}}\right]dx$

Section 2.7, pp. 166–168

2. $-0.321321, -0.322185$ **4.** $-0.416666667, 2.416666667$
6. $-1.1935, -1.18920711$ **8.** Yes, you must let x_0 be near $\pi/2$.
If you let $x_0 = 4.7$, then Newton's method would approach $3\pi/2$
not $\pi/2$.   **10.** $|x_n| = 2x_{n-1}, |x_n| \to \infty$ **14.** 1.313811607
16. 0.000245   **18.** 0.630115396, 2.57327196   **20.** 0.68233
22. If $x_0 = -0.8$, then $x_6 = -1$. If $x_0 = -0.7$, then $x_{11} = 1$.
24. Newton's method yields the following:

the initial value	2	i	$\sqrt{3} + i$
the approached value	1	$-5.55931i$	$-29.5815 - 17.0789i$

Section 2 Miscellaneous Exercises, pp. 171–175

2. $\left(\dfrac{1}{1 - 2x}\right)^2$ **4.** $\dfrac{-\sin x - 2\tan x}{(1 + \sec x)^2}$ **6.** $-2x^{-3}\sin x^3 + 3\cos x^3$
8. $-\dfrac{1}{3(xy)^{1/5}}$ **10.** $\dfrac{t + 2}{2\sqrt{(t + 1)^3}}$ **12.** $\sec^2(1 + t/2)\tan(1 + t/2)$
14. $t\csc 5t[2 - 5t\cot 5t]$

16. $-\dfrac{(5p^2 + 2p)^{5/2}}{2(5p + 1)}$ **18.** $\dfrac{1 - 2s - 2r}{2s - 1}$ **22.** $\dfrac{10}{\sqrt{2}}$ **24.** 1 or
2.5 sec   **26.** a) iii b) i c) ii   **28.** 0.09y, increasing at 1% per
year   **30.** $\dfrac{a^2k}{(akt + 10)^2}$, at $t = 0$, a^2k **32.** yes   **34.** a) $\tan\theta = x$
b) $-\dfrac{3\sec^2\theta}{5} = \dfrac{dx}{dt}$ c) $\dfrac{3}{5}$ m/sec d) $\dfrac{18}{\pi}$ RPM   **36.** 0   **38.** $\sqrt{3}$

40. $y = 1/2$ **42.** $y = -x + 3$ **44.** 4   **46.** $y = \dfrac{3}{4}x - \dfrac{5}{2}$,
$y = -\dfrac{4}{3}x + 10$ **48.** d
50. $\dfrac{-1}{2r^2l}\sqrt{\dfrac{T}{\pi d}}, \dfrac{-1}{2rl^2}\sqrt{\dfrac{T}{\pi d}}, \dfrac{-1}{4rl}\sqrt{\dfrac{T}{\pi d^3}}, \dfrac{1}{4rl}\sqrt{\dfrac{1}{\pi Td}}$,
52. $\dfrac{1}{6}$ **54.** $k = \dfrac{9}{2}, h = -4, a = \dfrac{5\sqrt{5}}{2}$
56. $1.5x + 0.5$ **57.** $L(x) = 2.5x - 0.1$ **58.** $\dfrac{2}{3}\pi r_0 hdr$,
$\dfrac{\pi r h_0}{\sqrt{r^2 + h_0^2}}\dfrac{dh}{dt}$ **60.** a) 4% b) 8% c) 12%
62. ± 0.53 in   **66.** 0.82836

Chapter 3

Section 3.1, pp. 181–183

2. $\dfrac{dS}{dt} = 8\pi r\dfrac{dr}{dt}$ **4.** $\dfrac{dV}{dt} = (2/3)\pi hr\dfrac{dr}{dt}$ **6.** $\dfrac{ds}{dt} = \dfrac{x}{\sqrt{x^2 + y^2}}\dfrac{dx}{dt}$ and
$\dfrac{ds}{dt} = \dfrac{y}{\sqrt{x^2 + y^2}}\dfrac{dy}{dt}$ **8.** a) 1 volt/sec b) $-\dfrac{1}{3}$ amp/sec
c) $\dfrac{dR}{dt} = \dfrac{1}{l}\left[\dfrac{dV}{dt} - \dfrac{V}{l}\dfrac{dl}{dt}\right]$ d) $\dfrac{3}{2}$ ohms/sec, R is increasing
10. -680 MPH   **12.** 0.0239 in³/min   **16.** 2.5 ft/sec
18. $\dfrac{10}{9\pi}$ in/min, $-\dfrac{8}{5\pi}$ in/min   **20.** -0.318 ft/min   **22.** -5 m/sec
24. 8 ft/sec, -5 ft/sec   **26.** $\dfrac{5}{72\pi}$ in/min, 10/3 in²/min
28. a) -2 rad/sec b) 1 rad/sec   **30.** 29.5 knots apart

Section 3.2, pp. 191–192

2. With $f(-\pi) = 1 - 2\pi < 0, f(\pi) = 2\pi + 1 > 0$, and the
Intermediate Value Theorem, we may conclude that $f(x) = 0$ has at
least one root between $-\pi$ and π when $f(x) = 2x - \cos x$.
$-1 \le \sin x \le 1 \Rightarrow 1 \le 2 + \sin x \le 3 \Rightarrow 1 \le f'(x) \le 3 \Rightarrow$
$f'(x) > 0$ for $-\pi < x < \pi \Rightarrow f(x)$ is increasing for all
$-\pi < x < \pi \Rightarrow f(x) = 0$ has exactly one solution when
$-\pi < x < \pi$.   **4.** With $f(0) = 1 > 0, f(1) = -3 < 0$, and the
Intermediate Value Theorem, we may conclude that $f(x) = 0$ has at
least one root between 0 and 1 when $f(x) = -x^3 - 3x + 1$. When
$0 < x < 1 \Rightarrow 0 < x^2 < 1 \Rightarrow 1 < x^2 + 1 < 2 \Rightarrow$
$-3 > -3x^2 - 3 \ge -6 \Rightarrow f'(x) < 0$ for $0 < x < 1 \Rightarrow f(x)$ is
decreasing for all $0 < x < 1 \Rightarrow f(x) = 0$ has exactly one solution
when $0 < x < 1$.   **6.** $c = 8/27$ **8.** $c = 3/2$ **12.** a) No
b) Yes c) Yes d) No   **22.** b) Yes,

because $f'(x) < 0$ when $x < 1$, $f'(x) > 0$ when $x > 1$ and
$\lim_{x \to 1^-} f'(x) = \lim_{x \to 1^+} f'(x) \Rightarrow f'(1) = 0$.   **28.** 2.1000050001

Section 3.3, pp. 199–201

2. The graph is rising on $(-2, 0) \cup (2, \infty)$, falling on
$(-\infty, -2) \cup (0, 2)$, concave upward on $(-\infty, -2/\sqrt{3}) \cup$
$(2/\sqrt{3}, \infty)$, and concave downward on $(-2/\sqrt{3}, 2/\sqrt{3})$.
Consequently, the local maximum is 4 at $x = 0$, the minimum is 0
at $x = \pm 2$, and $(-2/\sqrt{3}, 16/9)$ and $(2/\sqrt{3}, 16/9)$ are points of
inflection.
4. The graph is rising on $(-\pi/3, \pi/3)$, falling on $(-2\pi/3,$
$-\pi/3) \cup (\pi/3, 2\pi/3)$, concave upward on
$(-\pi/2, 0) \cup (\pi/2, 2\pi/3)$, and concave downward on
$(-2\pi/3, -\pi/2) \cup (0, \pi/2)$. Consequently, the maximum is
$\pi/3 + \sqrt{3}/2$ at $x = \pi/3$, the minimum is $-\pi/3 - \sqrt{3}/2$ at
$x = -\pi/3$, and the points of inflection are $(-\pi/2, -\pi/2)$, $(0, 0)$
and $(\pi/2, \pi/2)$.   **6.** At $x = 1$ there is a maximum.

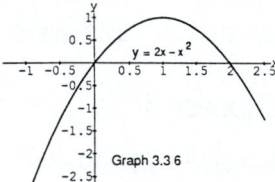

Graph 3.3 6

8.

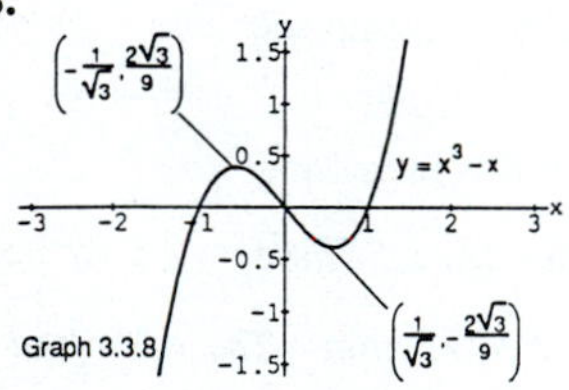

Graph 3.3.8

10.

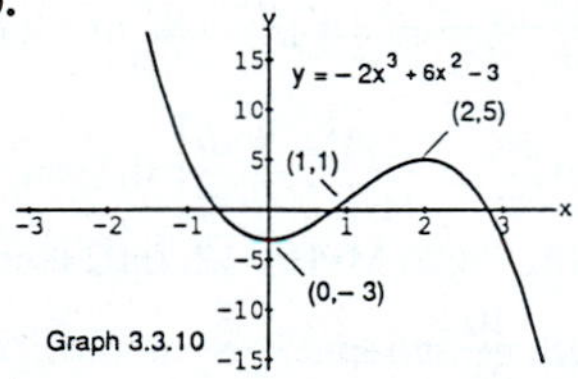

Graph 3.3.10

12.

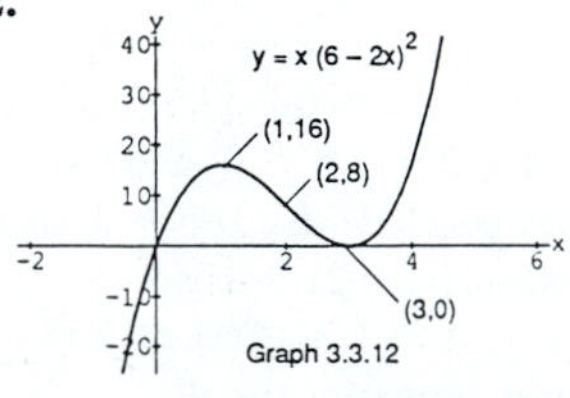

Graph 3.3.12

14.

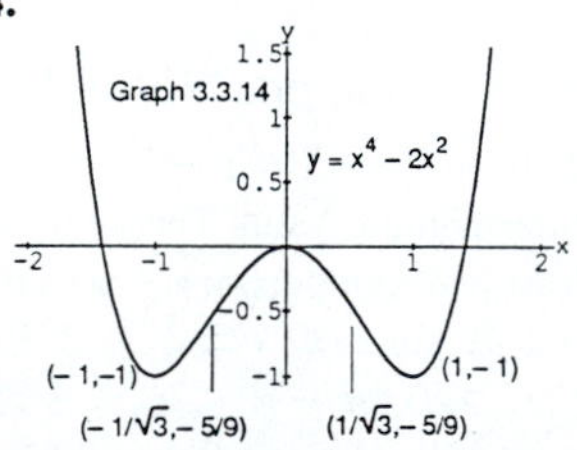

Graph 3.3.14

16.

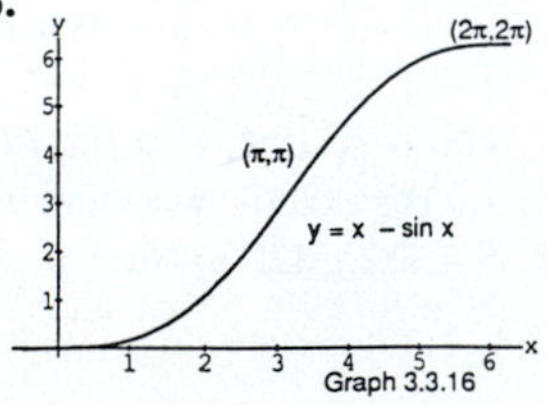

Graph 3.3.16

18.

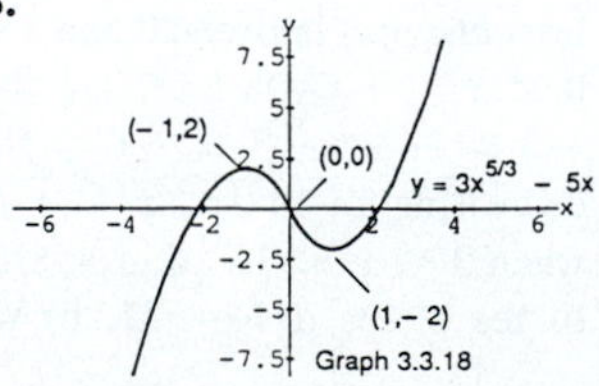

Graph 3.3.18

20. There are local maximums at $x = \pm\sqrt{\dfrac{1}{3}}$ and absolute
minimums at $x = \pm 1$. At $x = \pm 1$ there are points of inflection.

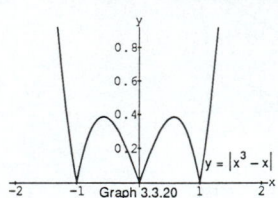

Graph 3.3.20

22.

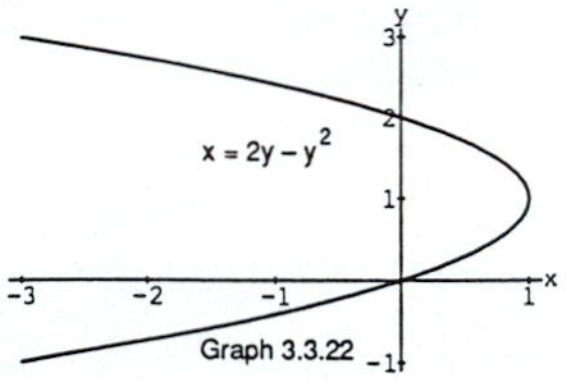

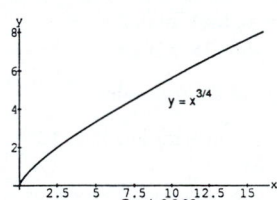

Graph 3.3.22

24. At $y = 2$ there is a point of
inflection.

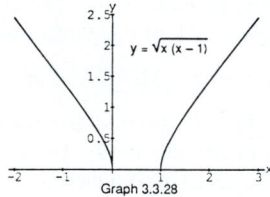

Graph 3.3.24

26. There is a vertical tangent
at $x = 0$.

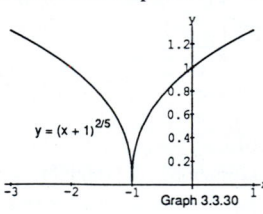

Graph 3.3.26

28. There are vertical tangents
at $x = 0$ and $x = 1$.

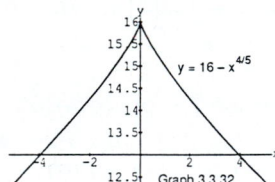

Graph 3.3.28

30. There is a cusp at $x = -1$.

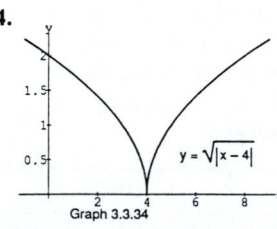

Graph 3.3.30

32. There is a cusp at $x = 0$.

Graph 3.3.32

34.

Graph 3.3.34

36. 0, 4, 12 or 16 sec; 2, 6, 9, 10.5, or 14 sec   **38.** $b = -3$
40.

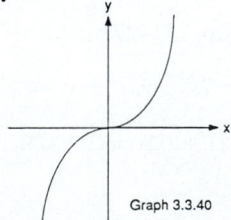

Graph 3.3.40

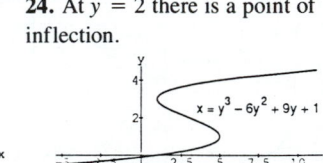

42.

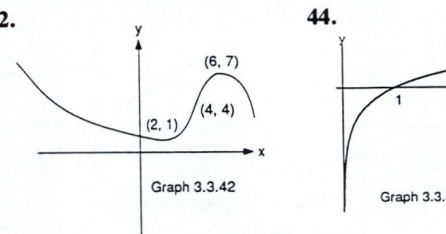

Graph 3.3.42

44.

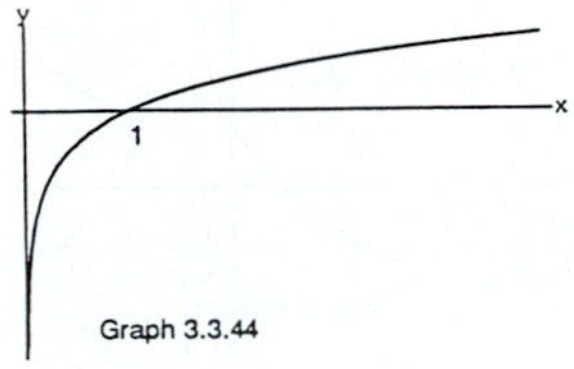

Graph 3.3.44

46. At $x = 2$ there is a local maximum and at $x = 4$ a local minimum. At $x = 1$, $[5 - \sqrt{3}]/2$ and $[5 + \sqrt{3}]/2$ there are inflection points.   **48.** No   **50.** a) True b) False

52. a)   b)

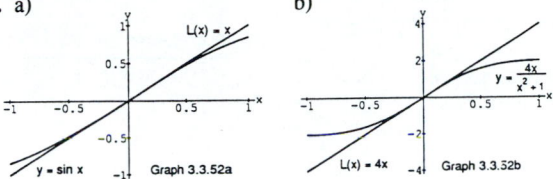

Graph 3.3.52a   Graph 3.3.52b

c)

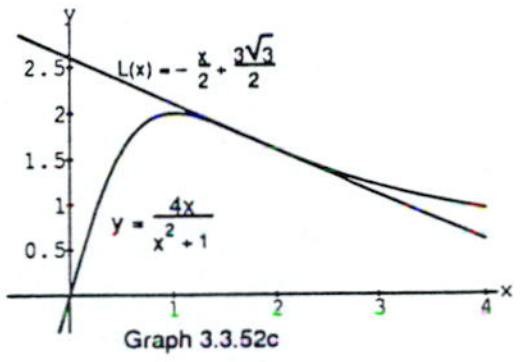

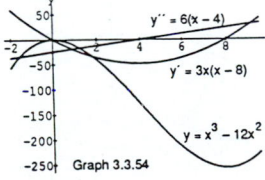

Graph 3.3.52c

54. The zeros of $y' = 0$ and $y'' = 0$ are extrema and points of inflection respectively.

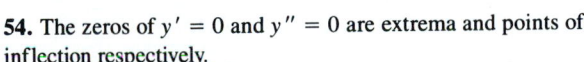

Graph 3.3.54

56. The zeros of $y' = 0$ and $y'' = 0$ are extrema and points of inflection respectively.

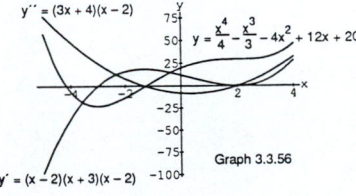

Graph 3.3.56

58. A cusp

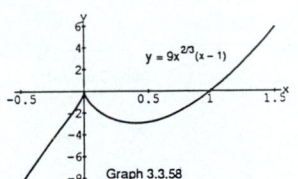

Graph 3.3.58

60. a) computer generated

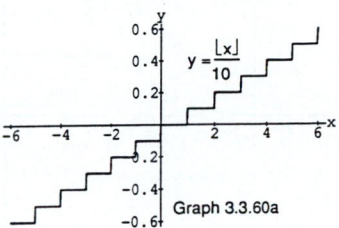

Graph 3.3.60a

b) the correct graph

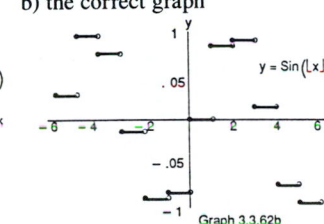

Graph 3.3.60b

62. a) computer generated   b) the correct graph

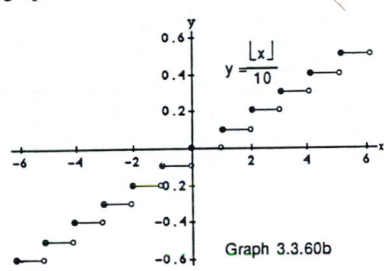

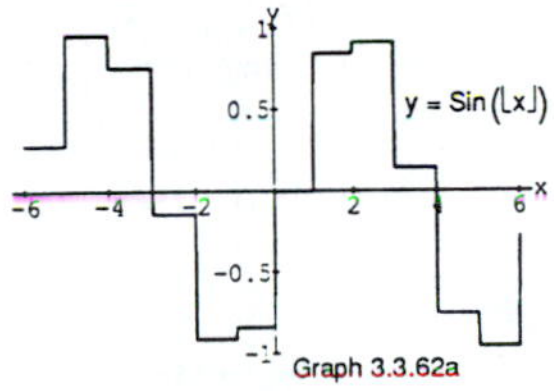

Graph 3.3.62a   Graph 3.3.62b

64. As x approaches zero, then $\dfrac{1}{x}$ approaches infinity which implies there will be an infinite number of values where the tangent is not defined. This is demonstrated by the following graphs.

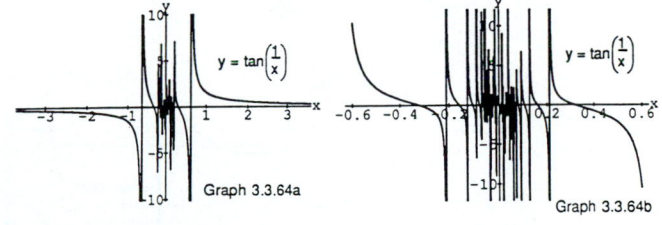

Graph 3.3.64a   Graph 3.3.64b

Section 3.4, pp. 206–207

2.

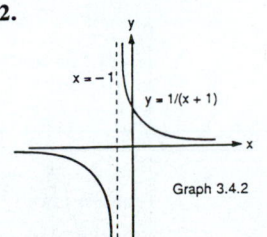

Graph 3.4.2

4.

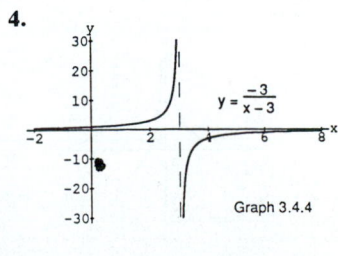

Graph 3.4.4

6.

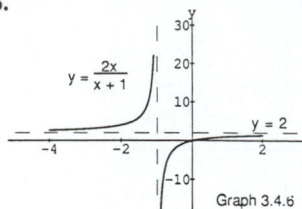

Graph 3.4.6

8.

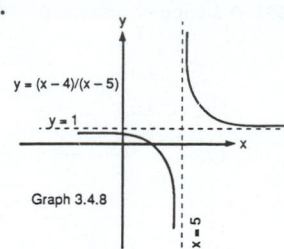

Graph 3.4.8

10.

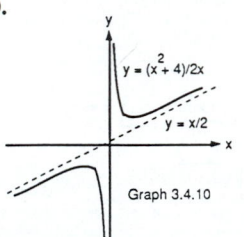

Graph 3.4.10

12.

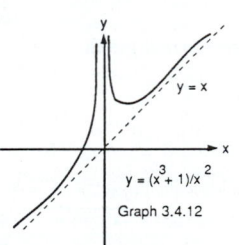

Graph 3.4.12

14.

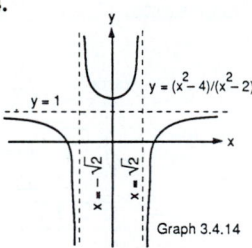

Graph 3.4.14

16.

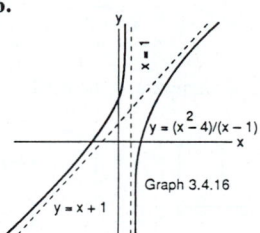

Graph 3.4.16

18.

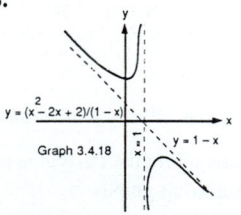

Graph 3.4.18

20.

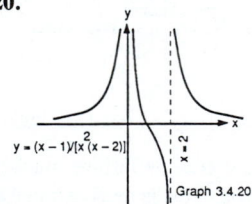

Graph 3.4.20

22.

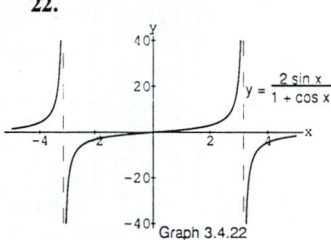

Graph 3.4.22

26.

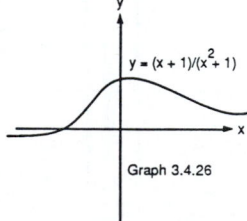

Graph 3.4.26

28.

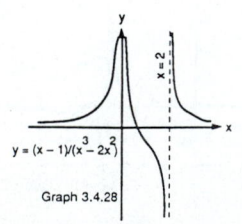

Graph 3.4.28

30.

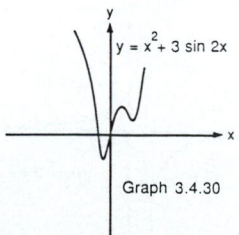

Graph 3.4.30

32.

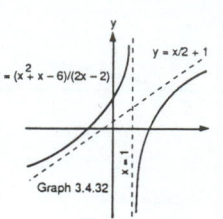

Graph 3.4.32

34. a)

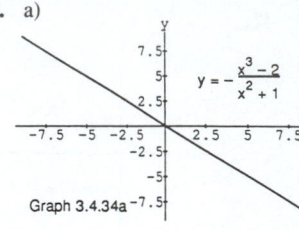

Graph 3.4.34a

b)

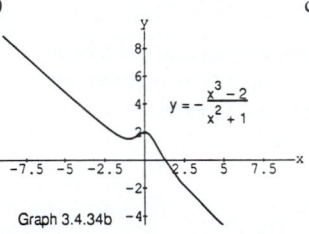

Graph 3.4.34b

c)

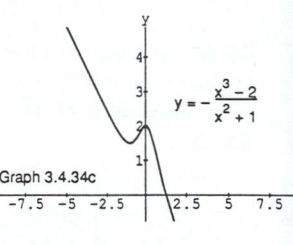

Graph 3.4.34c

d)

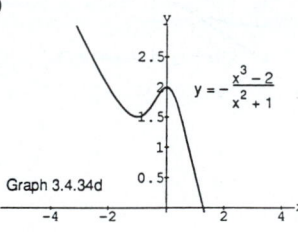

Graph 3.4.34d

Section 3.5, pp. 215–220

2. $\frac{25}{4}$ cm^2 **6.** 32 **10.** 18 m by 12 m; 72 m **12.** $x = 15$ ft, $y = 5$ ft **14.** a) 96 ft/sec b) 256 ft c) -128 ft/sec **16.** 5

18. $r = 8$: π.

20. b)

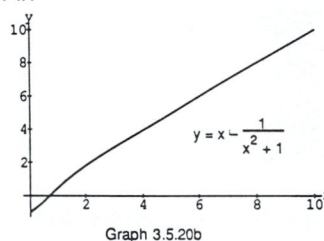

Graph 3.5.20b

22. $r = \sqrt{2}$, $h = 1$, $V = \frac{2\pi}{3}$ **24.** Proof **26.** $\frac{32\pi}{3}$

28. a) 6 by $6\sqrt{3}$ inches

b)

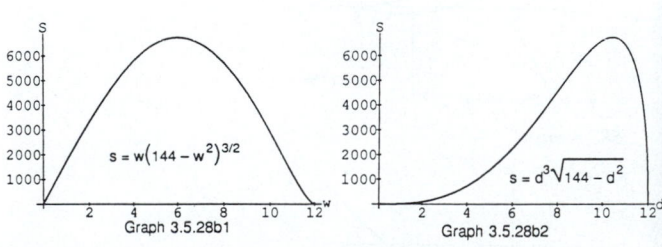

Graph 3.5.28b1 Graph 3.5.28b2

30. a) 2.632

b)

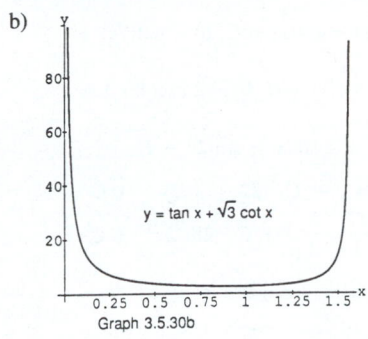

Graph 3.5.30b

32. 2   **34.** c) L^2 is minimized when $x = 5\ 1/8$ d) 11 in

e)

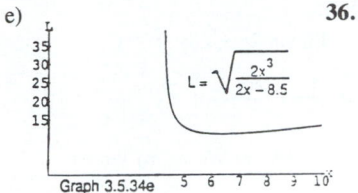

Graph 3.5.34e

b) $8/5 < t < 4$ c) $t = 11/5$, $2187/125$ units per time

d)

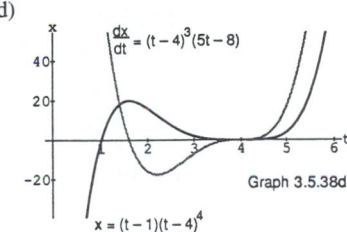

Graph 3.5.38d

40. a) $s = \sqrt{(12 - 12t)^2 + (8t)^2}$ b) -12 knots c) No

d)

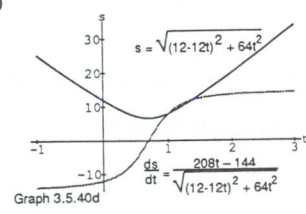

Graph 3.5.40d

e) $4\sqrt{13}$; the limit is the square root of the sums of the individual speeds.   **42.** 46.87 ft   **44.** a) 2 : 1 b) If $x = 0$, there is no ratio because the cube does not exist and we have only the sphere.

46. $M = C/2$ **48.** 67   **50.** $q = \sqrt{\dfrac{2km}{h}}$

Section 3.6, pp. 225–226

2. 2   **4.** -2 **6.** 5   **8.** $\dfrac{5}{7}$ **10.** $-\dfrac{1}{2}$ **12.** 1   **14.** $-\dfrac{2}{3}$ **16.** 3

18. $-\dfrac{3}{2}$ **20.** 1   **22.** 1   **24.** b **26.** $\dfrac{27}{10}$

28.

h	.1	.01	.001
$\dfrac{1 - \cos h}{h^2}$	.4996	.499996	.5

30. -1

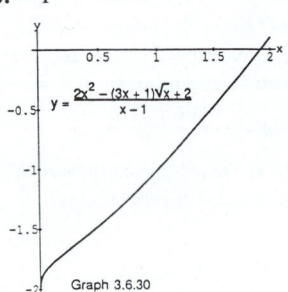

Graph 3.6.30

Section 3.7, pp. 233–234

2. $1 + \dfrac{x}{2} - \dfrac{x^2}{8}$ **4.** $\dfrac{3}{8} + \dfrac{3x}{4} - \dfrac{x^2}{8}$ **6.** $1 + \dfrac{x^2}{2}$ **8.** 0.005

10. a) $1 - \dfrac{x^2}{2}$ b) 0.000167   **12.** a) x b) 0.00089

14. a) $\pi - x$ b) 0.000167   **16.** a) $-1 + \dfrac{1}{2}(x - \pi)^2$

b) 0.000167   **18.** a) $|x| \le \left(\dfrac{6}{100}\right)^{1/3}$ b) $|x| \le \left(\dfrac{6}{100}\right)^{1/2}$

20. Proof   **22.** 0.9612

24. a)   b)

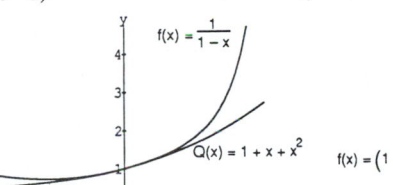

Graph 3.7.24a

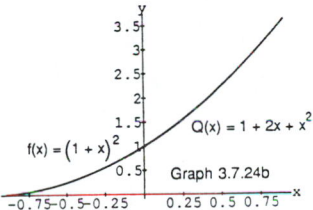

Graph 3.7.24b

f(x) and Q(x) are the same

c)

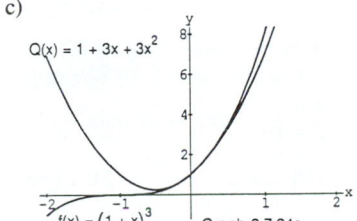

Graph 3.7.24c

Section 3 Miscellaneous Exercises, pp. 234–238

2. 1 cm/min   **4.** a) $-\dfrac{3}{400}$ ft/min, $-\dfrac{6\pi}{5}$ ft²/min

b) $-\dfrac{3}{4000}$ ft; $-\dfrac{3\pi}{25}$ ft²   **6.** 12

12.   **14.**

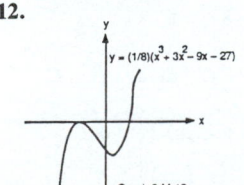

Graph 3.M.12

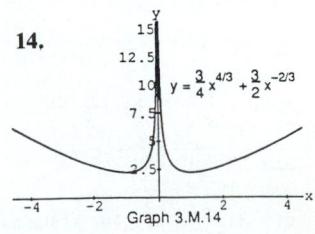

Graph 3.M.14

16. a) f has a local minimum at $x = -1$. f has points of inflection at $x = 0$ and 2. b) If $y' = 6x(x + 1)(x - 2)$, then $y' < 0$ for $x < -1$ and $0 < x < 2$, $y' > 0$ for $-1 < x < 0$ and $x > 2$. f has a local maximum at $x = 0$ with local minimums at $x = -1$ and $x = 2$. f has points of inflection at $x = \dfrac{1 \pm \sqrt{7}}{3}$.

18. $a = 1$, $b = 0$ and $c = 1$ **20.** a) There appears to be local minimums at $x = -1.75$ and 1.8. A point of inflection is indicated at $x = 0$.

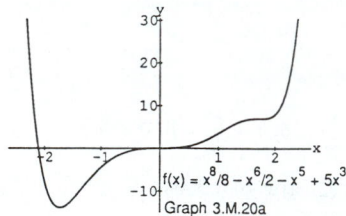

Graph 3.M.20a

c)

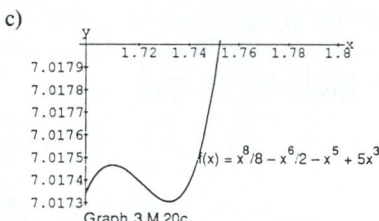

Graph 3.M.20c

22. Proof   **24.** a) $x = \dfrac{1}{3}$

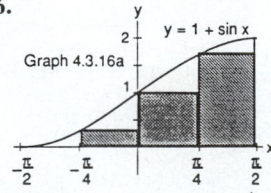

Graph 3.M.24a

26. If $H \epsilon (0, R]$ or $H = 2R$, then the maximum is at $r = R$. If $H \epsilon (R, 2R)$, then $r > R$ which is not possible. If $H \epsilon (2R, \infty)$, then the maximum is at $r = \dfrac{RH}{2(H - R)}$.   **28.** 54 sq units

30. $r = 4$ and $h = 4$ **32.** $x = 100$ m and $r = \dfrac{100}{\pi}$ **34.** 1 unit

36. $y = \dfrac{h}{2}$ **38.** $\dfrac{1}{4}$ **40.** a) Yes, at $x = a$ b) No, let $f(x) = g(x) = x^3$ **42.** a) 1 b) 2

Chapter 4

Section 4.1, pp. 245–246

2. a) $x^8 + C$ b) $\dfrac{x^8}{8} + C$ c) $\dfrac{x^8}{8} - 3x^2 + 8x + C$

4. a) $-\dfrac{1}{x} + C$ b) $-\dfrac{5}{x} + C$ c) $2x + \dfrac{5}{x} + C$ **6.** a) $x^{1/2} + C$

b) $x^{-1/2} + C$ c) $x^{-3/2} + C$ **8.** a) $\dfrac{3}{2}x^{2/3} + C$ b) $3x^{1/3} + C$

c) $-3x^{-1/3} + C$ **10.** a) $\sin(\pi x) + C$ b) $\sin\left(\dfrac{\pi x}{2}\right) + C$

c) $\left(\dfrac{2}{\pi}\right) \sin\left(\dfrac{\pi x}{2}\right) + C$ **12.** a) $-\cot(x) + C$ b) $-\cot(7x) + C$

c) $-\dfrac{\cot 7x}{7} + C$ **14.** a) $-\csc(x) + C$ b) $-2\csc(4x) + C$

c) $-\dfrac{1}{4}\csc(4x) + C$ **16.** $3x + 4\sin x + \sin 2x + C$

18. $6x - 3x^2 + C$ **20.** $-4x^{-1} + C$ **22.** $y + 2y^{-2} + C$

24. $\dfrac{z^3}{6} + \dfrac{z^4}{12} + C$ **26.** $x - \dfrac{x^3}{3} - \dfrac{1}{2}x^6 + C$ **28.** $2x^{5/2} + C$

30. $2r^{1/2} + \dfrac{2}{3}r^{3/2} + C$ **32.** $2\theta^{1/2} - \cos \theta + C$

34. $\dfrac{3}{5}\sin 5y + C$ **36.** $\dfrac{\tan x}{3} + C$ **38.** $\dfrac{2}{5}\sec t + C$

40. $-\dfrac{1}{2}\cot x + \dfrac{1}{2}\csc x + C$ **42.** $\sin 2t + \cos 3t + C$

44. $\dfrac{1}{14}x + \dfrac{1}{28}\sin 2x + C$ **46.** $\dfrac{1}{2}t - \dfrac{\sin 2t}{4} + C$

48. $-\dfrac{\cot x}{2} + C$ **50.** $\tan \theta + C$ **56.** a) Wrong b) Wrong

c) Right   **58.** a) $e^x + C$ b) $x \sin x + C$ c) $-e^x + C$

d) $-x \sin x + C$ e) $e^x + x \sin x + C$ f) $e^x - x \sin x + C$

g) $\dfrac{x^2}{2} + e^x + C$ h) $x \sin x - 4x + C$

Section 4.2, pp. 253–255

2. $y = 10x - \dfrac{x^2}{2} - 1$ **4.** $y = 3x^3 - 2x^2 + 5x + 10$

6. $y = x^{1/2} - 2$ **8.** $s = \sin t - \cos t$ **10.** $v = \dfrac{1}{2}\sec t + \dfrac{1}{2}$

12. $y = 2x$ **14.** $s = \dfrac{t^3}{16}$ **16.** $y = \cos x + x^2 + 2x + 2$

18. $s = -\cos t + 1$ **20.** $s = -\sin t + 1$ **22.** a) $y = x^3 + 1$

b) One   **24.** 1200 m/s   **26.** 1162.5 m   **28.** a) $y = \dfrac{(6 - 0.1\,t)^2}{4}$

b) 1 hr   **30.** a) $v = 10t^{3/2} - 6t^{1/2}$ b) $s = 4t^{5/2} - 4t^{3/2}$

Section 4.3, pp. 267–269

2. $\dfrac{1 - 1}{1} + \dfrac{2 - 1}{2} + \dfrac{3 - 1}{3} = \dfrac{7}{6}$ **4.** $\sin \pi - \sin \dfrac{\pi}{2} +$

$\sin \dfrac{\pi}{3} = \dfrac{\sqrt{3} - 2}{2}$ **6.** $\displaystyle\sum_{k=1}^{4} k^2$ **8.** $\displaystyle\sum_{k=1}^{5} 2k$ **10.** $\displaystyle\sum_{k=1}^{5} (-1)^k \dfrac{k}{5}$

12. b is not equivalent to the others   **14.** a) 0 b) 250 c) n

d) $1 - n$

16.

Graph 4.3.16a

Rectangles associated with $\displaystyle\sum_{k=1}^{4} \min_k \Delta x$

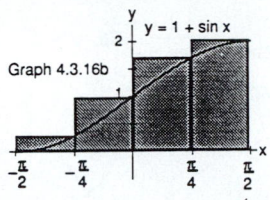

Graph 4.3.16b

Rectangles associated with $\displaystyle\sum_{k=1}^{4} \max_k \Delta x$

$$\sum_{k=1}^{4} \min_k \Delta x = \frac{3\pi}{4} \quad \sum_{k=1}^{4} \max_k \Delta x = \left(\frac{2-\sqrt{2}}{2}\right)\left(\frac{\pi}{4}\right) + (1)$$

$\left(\dfrac{\pi}{4}\right) = \dfrac{5\pi}{4}$ **18.** $\displaystyle\int_{-1}^{0} 2x^3 dx$ **20.** $\displaystyle\int_{1}^{4} \frac{1}{x} dx$ **22.** $\displaystyle\int_{0}^{1} \sqrt{4-x^2}\, dx$

24. $\displaystyle\int_{0}^{\pi/4} \tan x\, dx$ **26.** $\displaystyle\int_{-\pi}^{\pi} \sin^3 x\, dx$ **28.** $\displaystyle\int_{-2}^{0} \sqrt{-x}\, dx$

30. $\displaystyle\int_{3\pi/2}^{\pi/2} 3\cos x\, dx$ $\displaystyle\int_{0}^{1} (x^2 - 1)\, dx$ **32.** $\displaystyle\int_{-\pi/2}^{\pi/2} 2\cos x\, dx - \int_{-\pi/2}^{\pi}$

$2\cos x\, dx$ **34.** π **36.** $2 + \dfrac{1}{2}\pi$ **38.** $x = \pm\sqrt{2}$ will minimize

the integral.
42.

n	upper endpoint sum	lower endpoint sum
4	1.46875	1.21875
10	1.385	1.285
20	1.35875	1.30875
50	1.3434	1.3234

44.

n	upper endpoint sum	lower endpoint sum
4	1.1108675	0.914517956
10	1.04132188	0.962782062
20	1.02014873	0.980878825
50	1.00793622	0.99222826

46.

n	upper endpoint sum	lower endpoint sum
4	0.75952381	0.63452381
10	0.718771403	0.668771403
20	0.705803382	0.680803383
50	0.698172179	0.688172179

Section 4.4, pp. 276–278

2. a) 2 b) 9 c) -2 d) 1 e) -6 f) 1   **6.** Upper bound is $\dfrac{1}{2}$

Lower bound is $\dfrac{2}{5}$ **8.** $\dfrac{\pi}{2}$ **10.** -2 **12.** 1   **14.** -2 **16.** $-\dfrac{1}{4}$

18. 0   **20.** $-4\sqrt{2} - 3$ **22.** 12   **24.** 0   **26.** $\dfrac{4}{3}$ **28.** 4

34. -4 **36.** 0   **38.** $\dfrac{b^3}{3}$ **40.** $\dfrac{1}{3}$

Section 4.5, pp. 285–288

2. $\dfrac{4}{3}$ **4.** $\dfrac{7}{3}$ **6.** $10\sqrt{5}$ **8.** 1   **10.** π **12.** $2\sqrt{3}$ **14.** 4

16. $\dfrac{2\pi}{3} + \dfrac{\sqrt{3}}{2}$ **18.** $8 - \pi$ **20.** 3   **22.** $10\sqrt{3}$ **24.** $-\dfrac{5}{6}$

26. 1   **28.** 12   **30.** 8   **32.** 3   **34.** $\sqrt{3} - \dfrac{\pi}{3}$ **36.** $\cos x$

38. $\dfrac{dy}{dx} = \dfrac{1}{x}$, $x > 0$ **40.** $\dfrac{dy}{dx} = 1$ **42.** $\dfrac{dy}{dx} = -2x\cos|x|$

44. $\dfrac{dy}{dx} = 1$ **46.** c **48.** a **50.** $s = \displaystyle\int_{t_0}^{t} f(x)\, dx + s_0$ **52.** a) $\dfrac{125}{6}$

b) $\dfrac{25}{4}$ c) $\dfrac{125}{6}$ d) $\dfrac{2}{3} bh$

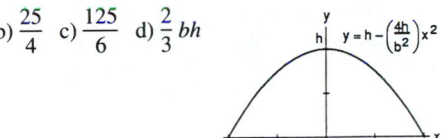

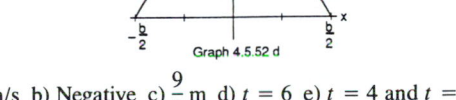

Graph 4.5.52 d

54. a) 2 m/s b) Negative c) $\dfrac{9}{2}$ m d) $t = 6$ e) $t = 4$ and $t = 7$

f) Toward the origin between $t = 6$ and $t = 9$; away from the
origin between $t = 0$ and $t = 6$.   **56.** 1   **58.** a) $L(x) = 20x$
b) $Q(x) = 20x$

60.

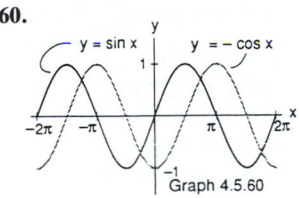

Graph 4.5.60

62.

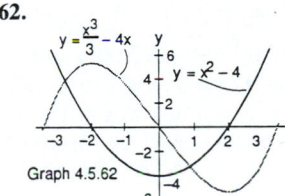

Graph 4.5.62

64.

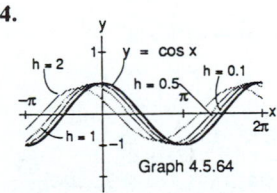

Graph 4.5.64

Section 4.6, pp. 295–297

2. $\dfrac{1}{2}\sec 2x + C$ **4.** $(7x - 2)^4 + C$ **6.** $(y^4 + 4y^2 + 1)^3 + C$

8. a) $\dfrac{2}{5}\sqrt{5t} + C$ b) $\dfrac{2}{5}\sqrt{5t} + C$ **10.** $\dfrac{7}{24}$ **12.** 2

14. $2(2y^2 + 1)^{1/2} + C$ **16.** $4\tan\left(\dfrac{x}{4}\right) + C$

18. $-\dfrac{1}{20}(7 - x^5)^4 + C$ **20.** $\dfrac{2}{3}(\tan x)^{3/2} + C$

22. $\dfrac{(s^3 + 2s^2 - 5s + 6)^3}{3} + C$ **24.** a) $\dfrac{1}{3}$ b) 0 **26.** a) $\dfrac{15}{16}$

b) 0 **28.** a) 0 b) $\dfrac{1}{4}$ **30.** a) 2 b) 2 **32.** a) 4 b) 0

34. a) $-\dfrac{1}{12}$ b) $\dfrac{1}{12}$ **36.** $\dfrac{1}{6}$ **38.** $\dfrac{2}{3}$ **40.** $\dfrac{1}{5}$ **42.** $-\dfrac{25}{6}$ **44.** 2

48. $y = 3(x^2 + 8)^{2/3} - 12$ **50.** $s = \sin\left(2t - \dfrac{\pi}{2}\right) + 100t + 1$

56. a) $\dfrac{1}{3}\left(1 + \sin^2(x - 1)\right)^{3/2} + C$ b) $\dfrac{1}{3}\left(1 + \sin^2(x - 1)\right)^{3/2} + C$

c) $\dfrac{1}{3}\left(1 + \sin^2(x - 1)\right)^{3/2} + C$ **58.** $\dfrac{1}{6}\sin^4\left(\dfrac{3t + 1}{2}\right) + C$

60. $\dfrac{1}{6}\sin\sqrt{3(2r - 1)^2 + 6} + C$

62. a)

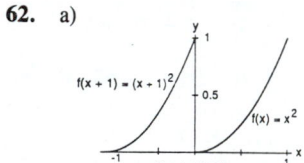

Graph 4.6.62 a

b)

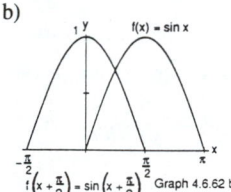

Graph 4.6.62 b

c)

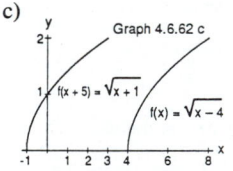

Graph 4.6.62 c

Section 4.7, pp. 305–309

2. a) $\dfrac{11}{4}$ b) $\dfrac{8}{3}$ c) $\dfrac{8}{3}$ **4.** a) 0.508994 b) 0.5004 c) $\dfrac{1}{2}$

6. a) 1.8961 b) 2.0046 c) 2 **8.** $|E_s| \le 0.000013$

10. a) $n = 116$ b) $n = 2$ **12.** a) $n = 71$ b) $n = 10$

14. a) $n = 161$ b) $n = 12$ **16.** 10.6 ft   **18.** a) 541.5 mg/L

b) 0.620 L/min   **20.** a) 0.0002075 b) 1.370792988

22. 3.1415926550 **24.** 1.08942941 **26.** 0.828116331

Section 4 Miscellaneous Exercises, pp. 311–316

2. $y = \dfrac{x^3}{3} + 2x - \dfrac{1}{x} - \dfrac{1}{3}$ **4.** $r = \dfrac{1}{3}(s^2 - 4)^{3/2} + 3$

6. $u = \sin r - r - 1$ **10.** $y = x^3 + 2x - 4$

12. $v = 2t + 3t^2 + 4$, $s = 6$ m   **14.** b   **18.** $\displaystyle\int_0^1 e^x dx$

20. a) π b) $-\pi$ c) -3π **22.** 7 **24.** $4\sqrt{3} - \dfrac{8}{3}\sqrt{2}$ **26.** $\dfrac{63}{512}$

28. $\dfrac{1}{3}$ **30.** $3\sqrt{3} - \pi$ **32.** 0 **34.** 1 **36.** $1 - \dfrac{\pi}{4}$

38. $4 - 2\sqrt{2}$ **40.** Proof   **42.** a) $\dfrac{1}{x + \sqrt{x^2 + 1}}$ b) $2f(2)$

44. Proof   **46.** $\dfrac{1}{2}$ **48.** $n = 26$ **50.** $n = 8$

52. a) 0.844 b) 6.7×10^{-6} **54.** $\dfrac{1}{\cos x} + \dfrac{1}{\sin x}$ **56.** $x = 1$

58. $\dfrac{7}{3}$ **60.** $\dfrac{55}{42}$

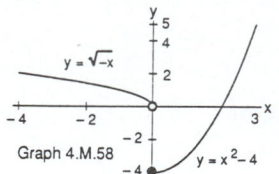

Graph 4.M.58

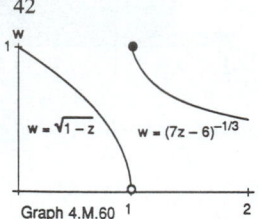

Graph 4.M.60

62. $\dfrac{7}{6}$

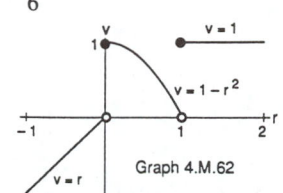

Graph 4.M.62

64. $\dfrac{2}{3}$ **66.** $\dfrac{1}{4}$ **68.** a) 1 b) $\dfrac{1}{16}$ c) $\dfrac{2}{\pi}$ d) 0 e) ∞

Chapter 5

Section 5.1, pp. 323–324

2. $2\sqrt{2}$ **4.** $\dfrac{128}{15}$ **6.** 1 **8.** $\dfrac{32}{3}$ **10.** $\dfrac{9}{2}$ **12.** 4 **14.** $\dfrac{27}{4}$

16. $\dfrac{2a^3}{3}$ **18.** $\dfrac{64}{3}$ **20.** $\dfrac{9}{2}$ **22.** 4 **24.** $\dfrac{12}{5}$ **26.** $\dfrac{37}{12}$ **28.** $6\sqrt{3}$

30. $\dfrac{6\sqrt{3}}{\pi}$ **32.** 4π **34.** $4 - \pi$ **36.** $\dfrac{9}{2}$ **38.** $\dfrac{4}{3}$ **40.** $\dfrac{11}{3}$ **42.** $\dfrac{32}{3}$

44. $\dfrac{3}{4}$

Section 5.2, pp. 335–338

2. 36π **4.** $\dfrac{128\pi}{7}$ **6.** 2π **8.** 2π **10.** 4π **12.** 2π **14.** π

16. $\dfrac{48\pi}{5}$ **18.** $\dfrac{108\pi}{5}$ **20.** π **22.** $\dfrac{20\pi}{3}$ **24.** $\dfrac{2\pi}{15}$ **26.** 36π

28. $\dfrac{\pi}{3}$ **30.** a) $2\pi/3$ b) $14\pi/3$ **32.** a) $\dfrac{\pi h^2 b}{3}$ b) $\dfrac{\pi b^2 h}{3}$

34. 192.27 gm   **36.** a) 13 ft^3 b) 195 mi   **38.** a) $\dfrac{\pi h^2(3a - h)}{3}$

b) 0.00265 m/sec   **44.** $\dfrac{8}{3}$ **46.** $\dfrac{16\pi}{15}$ **48.** 8 **50.** $2\sqrt{3}$ **52.** 8

Section 5.3, pp. 343–344

2. $\dfrac{128\pi}{5}$ **4.** $\dfrac{7\pi}{15}$ **6.** 14π **8.** $\dfrac{4\pi}{3}$ **10.** $\dfrac{16\pi(3\sqrt{2} + 5)}{15}$ **12.** $\dfrac{\pi}{6}$

14. 36π **16.** a) $\dfrac{8\pi}{3}$ b) $\dfrac{8\pi}{5}$ c) 8π d) $\dfrac{283\pi}{48}$ **18.** a) $\dfrac{4\pi}{15}$ b) $\dfrac{7\pi}{30}$

20. a) 16π b) $\dfrac{32\pi}{3}$ c) $\dfrac{64\pi}{3}$ d) 48π **24.** a) 8π b) $\dfrac{32\pi}{5}$

c) $224\pi/15$ d) $\dfrac{8\pi}{3}$

Section 5.4, pp. 350–352

2. $L = \displaystyle\int_{-\pi/3}^{0} \sqrt{1 + \sec^4 x}\, dx$ **4.** $L = \displaystyle\int_{0}^{\pi} \sqrt{1 + x^2 \sin^2 x}\, dx$

6. $L = \displaystyle\int_{-1/2}^{1/2} \sqrt{\dfrac{1}{1 - y^2}}\, dy$ **8.** $L = \displaystyle\int_{-\pi/3}^{\pi/4} \sec y\, dy$

10. $\dfrac{8}{27}[10\sqrt{10} - 1]$ **12.** 4.807 **14.** $\dfrac{32}{3}$ **16.** $\dfrac{13}{4}$ **18.** $\dfrac{28}{3}$

20. $\dfrac{7\sqrt{3}}{3}$ **22.** $y = x^{1/2}$ **24.** $f(x) = \pm x + C$ where C is any real number.   **26.** 21.07

Section 5.5, pp. 367–369

2. $S = 2\pi \displaystyle\int_{0}^{2} x^2 \sqrt{1 + 4x^2}\, dx$ **4.** $S = 2\pi \displaystyle\int_{0}^{\pi} (\sin y)$ $\sqrt{1 + \cos^2 y}\, dy$ **6.** $S = 2\pi \displaystyle\int_{1}^{2} \left(y + 2\sqrt{y}\right) \sqrt{1 + (1 + y^{-1/2})^2}\, dy$

8. $S = 2\pi \displaystyle\int_{1}^{\sqrt{5}} \left[\displaystyle\int_{1}^{x} \sqrt{t^2 - 1}\, dt\right] x\, dx$ **10.** $8\pi\sqrt{5}$

12. $4\pi\sqrt{5}$ **14.** $\dfrac{28\pi}{3}$ **16.** $\dfrac{49\pi}{3}$ **18.** 3π

20. $\dfrac{2\pi(2\sqrt{2} - 1)}{3}$ **22.** 4π **26.** $\dfrac{12\pi}{5}$

30. $\dfrac{2\pi}{3}\left(\sqrt{8} - 1\right)$ **32.** 5.28 **34.** 41.8

Section 5.6, pp. 367–369

2. $(I/4,\ I/4)$ **4.** $\dfrac{15}{2}, \dfrac{5}{3}$ **6.** $\dfrac{45}{2}, \dfrac{5}{2}$ **8.** $\bar{y} = 10, \bar{x} = 0$ **10.** $\bar{x} = 0,$ $\bar{y} = -\dfrac{8}{5}$ **12.** $\bar{x} = \dfrac{3}{5}, \bar{y} = 1$ **14.** $\bar{y} = \dfrac{1}{2\ln(\sqrt{2} + 1)}, \bar{x} = 0$

16. a) $\bar{x} = \bar{y} = \dfrac{4}{\pi}$ b) $\bar{y} = \dfrac{4}{\pi}, \bar{x} = 0$ **18.** $\bar{y} = 0, \bar{x} = 2$

20. $\dfrac{\delta}{54}\left[10^{3/2} - 1\right]$ **24.** $\bar{x} = \bar{y} = \dfrac{1}{3}$ **26.** $\bar{x} = \dfrac{a}{3}$ and $\bar{y} = \dfrac{b}{3}$

30. b)

α	0.2	0.4	0.6	0.8	1.0
$\dfrac{\sin\alpha - \alpha\cos\alpha}{\alpha - \alpha\cos\alpha}$	0.666222	0.664879	0.662615	0.659389	0.655145

Section 5.7, pp. 376–378

2. 240 ft $\cdot$ lb   **4.** $\dfrac{400}{7}$ N/m   **6.** 1944 ft $\cdot$ lb   **8.** 1125 N $\cdot$ m

10. 300 lb, $\dfrac{25}{16}$ ft $\cdot$ lb   **12.** a) 2250000 ft $\cdot$ lb b) 2 hr and 16.4 min c) 1 hr and 9.4 min   **14.** 282252.5 ft $\cdot$ lb   **16.** Through the

valve   **18.** 96129.46 ft $\cdot$ lb   **20.** 30 hr and 50.1 min

22. 122.5 ft $\cdot$ lb   **24.** 109.7 ft $\cdot$ lb   **26.** $\dfrac{9}{2}$ ft

28. a) 11.5×10^{-29} N $\cdot$ m b) 7.67×10^{-29} N $\cdot$ m

Section 5.8, pp. 382–384

2. 114511052 lb, 85883289 lb   **4.** 1166.67 lb   **6.** 41.67 lb
8. 2261952 lb   **10.** 1161 lb   **12.** 1034 ft^3 **14.** Proof

Section 5.9, pp. 391–393

2. a)
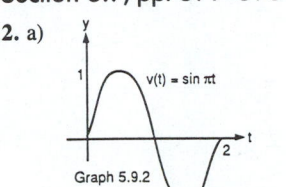
Graph 5.9.2

b) $\dfrac{4}{\pi}$ m c) 0 m

4. a)
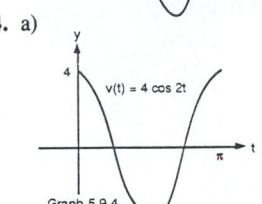
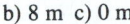
Graph 5.9.4

b) 8 m c) 0 m

6. a)
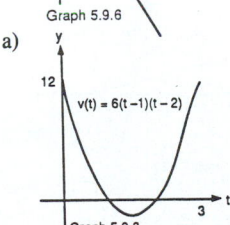
Graph 5.9.6

b) 40 m c) 0 m

8. a)
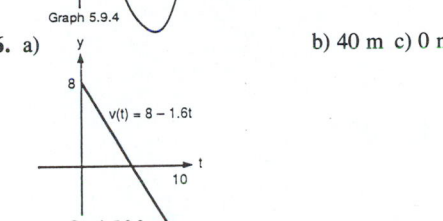
Graph 5.9.8

b) 11 m c) 9 m   **10.** $\dfrac{52}{3}, \dfrac{268}{3}$ **12.** 74%   **14.** Proof   **16.** 72π

18. $\pi r \sqrt{r^2 + h^2}, \dfrac{1}{3}\pi r^2 h$ **20.** $2\pi a^2(\pi - 2)$ **22.** $\dfrac{\pi a^3(3\pi + 4)}{3}$

24. $\sqrt{2}\pi a^2(2 + \pi)$

Section 5 Miscellaneous Exercises, pp. 393–396

2. $\dfrac{9}{8}$ **4.** $\pi + 2$ **6.** $\dfrac{13}{6}$ **8.** $\dfrac{a^2}{6}$ **10.** $4\sqrt{2} - 2$ **12.** $\dfrac{8\sqrt{3}}{15}$

14. $\dfrac{72}{5}$ **16.** $2\sqrt{3}a^2$ **18.** a) $\dfrac{57\pi}{20}$ b) $\dfrac{5\pi}{2}$ c) $\dfrac{3\pi}{2}$ d) $\dfrac{103\pi}{20}$

20. a) $\dfrac{32\pi}{3}$ b) $\dfrac{128\pi}{15}$ c) $\dfrac{64\pi}{5}$ d) $\dfrac{32\pi}{3}$ **22.** $\dfrac{\pi(16 - \pi)}{2}$

24. $2\pi(4 - \pi)$ **26.** 276 in^3 **28.** $f(x) = \sqrt{\dfrac{2x + 1}{\pi}}$ **30.** 7.634

32. a) $\sqrt{C^2 - 1}\, x + a$, where $C \geq 1$ **34.** 0.359 **36.** $\dfrac{49\pi}{3}$

38. $2\pi^2 a^2 k$　**40.** $\bar{x} = \dfrac{3}{2}, \bar{y} = \dfrac{12}{5}$　**42.** a) $\bar{x} = 3, \bar{y} = \dfrac{5}{9}$

b) $\bar{x} = \dfrac{13}{3}, \bar{y} = \dfrac{1}{3}$　**44.** a) 22 ft from the top　b) 61.86 lb/ft^3

46. 633333333.3 ft · lb　**48.** 4.28 min　**50.** $\dfrac{w(2Rr - r^2)}{2R}$

52. a) Proof　b) -37968.75 in · lb　**54.** 30 ft　**56.** 118.63 lb
58. 6.638 lb, 5.773 lb, 9.484 lb　**60.** 18750 lb

62. a)　　　　　　　　　　　　b) $\dfrac{1}{2}$ m　c) 0 m

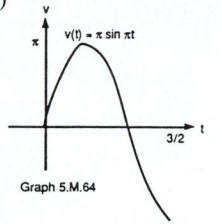
Graph 5.M.62

64. a)　　　　　　　　　　　　b) 3 ft　c) 1 ft

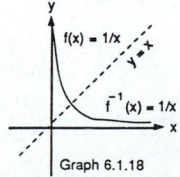
Graph 5.M.64

Chapter 6

Section 6.1, pp. 401–402

2. a) $\dfrac{5 - x}{4}$　　　　　b)

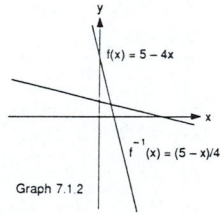
Graph 7.1.2

4. a) $\sqrt{\dfrac{x}{2}}, x \geq 0$　　　b)

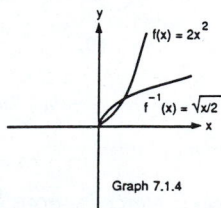
Graph 7.1.4

6. $-\sqrt{x}$　**8.** $1 + \sqrt{x}$　**10.** $\sqrt[4]{x}$　**12.** $2x + 7$　**14.** $\sqrt[3]{\dfrac{1}{x}}$　**16.** $x^{3/2}$

18. a)　　　　　　　b) $\dfrac{1}{x}, x > 0$

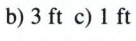

Graph 6.1.18

Section 6.2, pp. 410–412

2. a) $l - 3\ln 5$　b) $2\ln 7 - \ln 5$　c) $\dfrac{3}{2}\ln 7$　d) $2\ln 5 + 2\ln 7$

e) $\ln 7 - 3\ln 5$　f) $\dfrac{1}{2}$　**4.** $\dfrac{1}{x}$　**6.** $\dfrac{1}{x}$　**8.** $-\dfrac{1}{x}$　**10.** $\dfrac{2x + 2}{x^2 + 2x}$

12. $\ln x$　**14.** $\dfrac{1}{x(\ln x)\ln(\ln x)}$　**16.** $\sec x$　**18.** $\dfrac{1}{1 - x^2}$

20. $\dfrac{1}{4x\sqrt{\ln\sqrt{x}}}$　**22.** $\dfrac{1}{2}\left[\cot x - \tan x - \dfrac{1}{x(1 + 2\ln x)}\right]$

24. $-\dfrac{5}{2}\left[\dfrac{3x + 2}{(x + 1)(x + 2)}\right]$　**26.** $\dfrac{\ln\sqrt[3]{x}}{3\sqrt[3]{x^2}} - \dfrac{\ln\sqrt{x}}{2\sqrt{x}}$

28. $\left(\dfrac{1}{2}\right)\sqrt{\dfrac{x}{x + 1}}\left[\dfrac{1}{x} - \dfrac{1}{x + 1}\right]$　**30.** $\dfrac{\tan x}{\sqrt{2x + 1}}\left[\dfrac{\sec^2 x}{\tan x} - \dfrac{1}{2x + 1}\right]$

32. $\dfrac{1}{x(x + 1)(x + 2)}\left[-\dfrac{1}{x} - \dfrac{1}{x + 1} - \dfrac{1}{x + 2}\right]$

34. $\dfrac{x\sin x}{\sqrt{\sec x}}\left[\dfrac{1}{x} + \cot x - \dfrac{1}{2}\tan x\right]$

36. $\sqrt{\dfrac{(x + 1)^{10}}{(2x + 1)^5}}\left[\dfrac{5}{x + 1} - \dfrac{5}{2x + 1}\right]$

38. $\left(\dfrac{1}{3}\right)\sqrt[3]{\dfrac{x(x + 1)(x - 2)}{(x^2 + 1)(2x + 3)}}\left[\dfrac{1}{x} + \dfrac{1}{x + 1} + \dfrac{1}{x - 2} - \dfrac{2x}{x^2 + 1}\right.$

$\left. - \dfrac{2}{2x + 3}\right]$　**40.** 0　**42.** 1　**44.** -2　**46.** $\ln\left(\dfrac{2}{5}\right)$

48. $\ln|4r^2 - 5| + C$　**50.** $\ln\left(\dfrac{1}{3}\right)$　**52.** $\ln 2$　**54.** $\sqrt{\ln 2}$

56. $\ln|2 + \sec y| + C$　**58.** $\ln\sqrt{2}$　**60.** $\ln 2$

62. $2\sqrt{\ln(\sec x + \tan x)} + C$　b) 0.00033　**64.** a) $x = \dfrac{\pi}{3}$ there is

a minimum and at $x = 0$ a maximum　b) at $x = \dfrac{1}{2}$ and 2 a

minimum and at $x = 1$ a maximum　**66.** $\dfrac{3}{2}\ln 2$　**68.** $\pi\ln 2$

70. $27\pi\ln\left(\dfrac{17}{9}\right)$　**72.** $y = \ln x$

74. a) $\bar{x} = 7, \bar{y} = \dfrac{\ln 4}{6}$　b) $\bar{x} = \dfrac{15}{\ln 16}, \bar{y} = \dfrac{3}{4\ln 16}$

76. $y = \ln|\sec x| + x$　**78.** 0.182323232, -0.223148148
80. Each graph $y = \ln kx = \ln x + \ln k$ has a vertical shift of $\ln k$
when compared with $y = \ln x$.

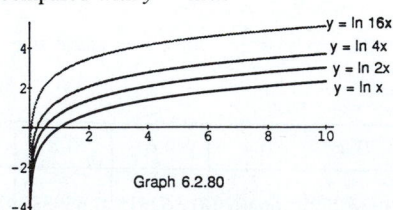
Graph 6.2.80

82. a)

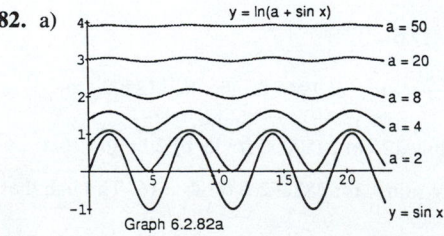
Graph 6.2.82a

b) The slope approaches zero as a increases without bound.

Section 6.3, pp. 418–420

2. a) $\frac{1}{8}$ b) $\frac{1}{32}$ c) 6 **4.** a) $-x$ b) $-x^2$ c) $\frac{x}{y^2}$ **6.** a) $\frac{\ln 2}{10}$

b) $-\ln 80$ c) $\frac{\ln 0.1}{\ln 0.8}$ **8.** a) $-300 \ln 10$ b) $-\frac{\ln 2}{k}$ c) $\frac{\ln 0.4}{\ln 0.2}$

10. $y = e^{(5-t)}$ **12.** $y = \frac{1-e^t}{2}$ **14.** $y = x^2 + 2x + 1$

16. $y = e^{\sin x} + 1$ **18.** $\frac{2}{3} e^{2x/3}$ **20.** $e(4\sqrt{x} + x^2)\left(\frac{2}{\sqrt{x}} + 2x\right)$

22. $-4x e^{-2x}$ **24.** $27x^2 e^{3x}$ **26.** $\frac{1}{x} - 1$ **28.** $x^2 e^{-2x}(3 \cos 5x - 2x$

$\cos 5x - 5x \sin 5x)$ **30.** $\frac{1}{1 + e^x}$ **32.** $x e^{ax}$ **34.** $4xe^{2x} - 8$

36. $\frac{y(xe^{x+y} - 1)}{x(1 - ye^{x+y})}$ **38.** $\frac{(xe^x + 1) \cos^2 y}{x}$ **40.** 1

42. $e^2 - 1$ **44.** 0 **46.** $\ln\left(\frac{e + e^2}{e + 1}\right)$ **48.** $\ln 3$

50. $-\ln|\cos \ln v| + C$ **52.** $-\ln|e^{-x} + 1| + C$ **54.** $\frac{1}{2}$ **56.** 0

58. a) ay b) $y = 3e^{-2x}$ **60.** 1, 0.613706 **62.** $e, \frac{1}{e}$ **64.** 1

68. $\pi\left(\frac{15}{16} + \ln 2\right)$ **70.** $y = 3\sqrt{x} - 1$ **72.** Calculator

Section 6.4, pp. 428–431

2. $\left(1 + \sqrt{2}\right)x^{\sqrt{2}}$ **4.** $(1 - e)r^{-e}$ **6.** $-9^{-x}(\ln 9)$ **8.** $(\ln 4)s\, 2^{s^2}$

10. $3^{\tan x} (\ln 3)^2 (\sec x)^2$ **12.** $x^{(x+1)}\left(\ln(x) + 1 + \frac{1}{x}\right)$

14. $\left[\ln(1 - t) - \frac{t}{1 - t}\right](1 - t)^t$

16. $\left(\frac{\ln(\cos v)}{2\sqrt{v}} - \sqrt{v} \tan v\right)(\cos v)^{\sqrt{v}}$ **18.** $\left(\frac{2}{x} \log_2 x\right) x^{\log_2 x}$

20. $2^{(\sqrt{2} - 1/2)}$ **22.** $\frac{1}{\ln 2}$ **24.** $\frac{1}{\ln 4}$ **26.** $\frac{3}{\ln 5}$ **28.** $\frac{24}{\ln 5}$ **30.** $\frac{1}{\ln 2}$

32. $\frac{2^{\ln 2} - 1}{\ln 2}$ **34.** a) $\frac{\ln 0.04}{\ln 5} = \frac{-2 \ln 5}{\ln 5} = -2$

b) $\frac{\ln 4}{\ln 0.5} = \frac{2 \ln 2}{-\ln 2} = -2$ c) $\frac{\ln 5}{-2 \ln 5} = -\frac{1}{2}$ **36.** a) $9x^4$ b) x

c) $\frac{e^x \sin x}{2}$ **38.** a) $\frac{1}{2}$ b) $\frac{\ln 2}{\ln 10}$ c) $\left(\frac{\ln b}{\ln a}\right)^2$ **40.** 1, 2 **42.** $\frac{x - 1}{2x \ln 5}$

44. $-\left(\frac{\ln r + 1}{\ln 4}\right)$ **46.** $\frac{6s}{(\ln 3)(3s^2 + 1)}$ **48.** $\frac{1}{t}$

50. $\sin (\log_7 x) + \frac{1}{\ln 7} \cos(\log_7 x)$ **52.** $\left(\frac{\ln 3}{\ln 2}\right) 3^{\log_2 \theta}$

54. $(\cot x - \tan x - 1 - \ln 2)\frac{1}{\ln 7}$ **56.** $\sin t + t \cos t$

58. $\ln 4$ **60.** 1 **62.** $2 \ln 10$ **64.** $\ln 2$ **66.** $-\frac{(\ln 8)^2}{\ln x} + C$

68. $-\ln 2$ **70.** e^2 **72.** e **74.** $\sqrt{e}$ **76.** a) $0.69315x + 1$,
$1 + 0.69315x + 0.24023x^2$ b) 0.01, 0.00033

78. $[10^{-7.44}, 10^{-7.37}]$ **80.** a) $e^{1/e}$ b) $e^{1/2e}$ c) $e^{1/ne}$

82.

n	10	100	1000
$\left(1 + \dfrac{1}{n}\right)^n$	2.59374246	2,704813829	2.716923932

10000	10000
2.718145927	2.718268237

84. a) $\sin x < 0$, π b) 1

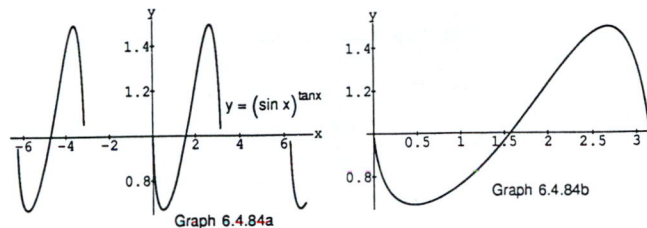

Graph 6.4.84a

Graph 6.4.84b

c) minimum of about 0.665 at $x \approx 0.47$, maximum is about 1.491 at $x \approx 2.66$

Section 6.5, pp. 437–440

2. a) -0.121 b) 2.389 millibars c) 0.977 km **4.** 585.35 kg
6. 92.1 sec **8.** 1250 **10.** a) 0.0059 b) 266123121
12. a) $\$54.61e^{-0.01x}$ b) \$22.20
d)

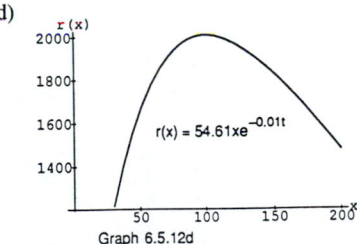

$r(x) = 54.61xe^{-0.01t}$

Graph 6.5.12d

14. $A(100) = 90000 \Rightarrow 90000 = 1000e^{r(100)} \Rightarrow 90 = e^{100r} \Rightarrow \ln 90 = 100r \Rightarrow r = \frac{\ln 90}{100} \approx 0.0450$ or 4.50% **18.** 605 days
20. a) 0.262 b) 3.816 yr c) 11.431 yr **22.** 5° **24.** 53.45°,
232.47 min **26.** a) 12571 BC b) 12101 BC c) 13070 BC

Section 6.6, pp. 444–446

2. a) Same b) Faster c) Faster d) Slower e) Same f) Same
g) Slower h) Slower i) Slower j) Faster **4.** x^x, $(\ln x)^x$, e^x, and
$e^{x/2}$ **8.** a) True b) True c) False d) True e) True f) False
g) True h) False
12. Proof **14.** 11 million, 20 **16.** a) Polynomials of a greater
degree grow at a greater rate than polynomials of a lesser degree
and polynomials of the same degree grow at the same rate.
b) When it has a degree smaller than g. When it has a degree less
than or equal to the degree of g.

Section 6.7, pp. 450–451

2. a) $-\dfrac{\pi}{4}$ b) $\dfrac{\pi}{3}$ c) $-\dfrac{\pi}{6}$ **4.** a) $\dfrac{\pi}{6}$ b) $-\dfrac{\pi}{4}$ c) $\dfrac{\pi}{3}$ **6.** a) $\dfrac{2\pi}{3}$

b) $\dfrac{\pi}{4}$ c) $\dfrac{5\pi}{6}$ **8.** a) $\dfrac{\pi}{4}$ b) $\dfrac{5\pi}{6}$ c) $\dfrac{\pi}{3}$ **10.** a) $-\dfrac{\pi}{4}$ b) $\dfrac{\pi}{3}$ c) $-\dfrac{\pi}{6}$

12. a) $\dfrac{\pi}{4}$ b) $\dfrac{5\pi}{6}$ c) $\dfrac{\pi}{3}$ **14.** $\sin\alpha = \dfrac{\sqrt{3}}{2}$, $\tan\alpha = -\sqrt{3}$, $\sec$

$\alpha = -2$, $\csc\alpha = \dfrac{2}{\sqrt{3}}$ **16.** $-\dfrac{1}{\sqrt{3}}$ **18.** 0 **20.** 1 **22.** $\dfrac{3\pi}{4}$

24. $\dfrac{3\pi}{4}$ **26.** $\dfrac{3}{5}$ **28.** $\dfrac{|x-1|}{x+1}$ **30.** $\dfrac{-1}{|x|}$ **32.** x **34.** $\dfrac{2\pi}{3}$

36. a) $\dfrac{\pi}{2}$ b) π **38.** a) $\dfrac{\pi}{2}$ b) 0 **40.** $\pi\left(\dfrac{\pi}{3} - \sqrt{3}\right)$

42. 54.7° **48.** The graphs are identical

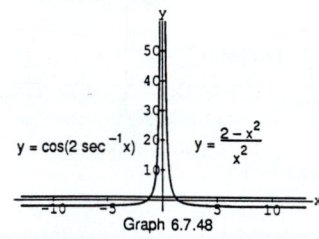
Graph 6.7.48

50. a) Domain, all reals; Range, $[-\pi/2, \pi/2]$; Yes, many computer graphs will be correct

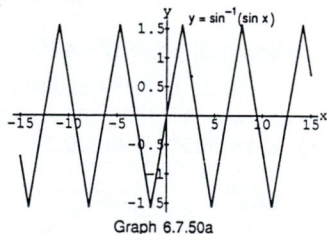
Graph 6.7.50a

b) Domain, $[-1, 1]$; Range, $[-1, 1]$; No, many computer graphs will not be correct

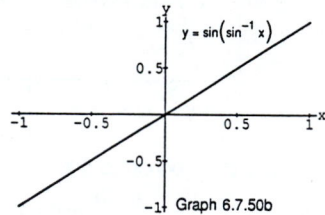
Graph 6.7.50b

Section 6.8, pp. 456–457

2. $\dfrac{-1}{|x|\sqrt{x^2-1}}$ **4.** $\dfrac{-1}{2\sqrt{x}(1+x)}$ **6.** $\dfrac{-1}{\sqrt{2x-x^2}}$ **8.** $\dfrac{1}{9+x^2}$

10. $\dfrac{-2}{\sqrt{1-4x^2}}$ **12.** $\dfrac{1}{\sqrt{1-x^2}}$ **14.** $2\sqrt{1-x^2}$ **16.** 0

18. $-\tan^{-1}\left(\dfrac{x}{2}\right)$ **24.** $\dfrac{\pi}{2}$ **26.** $-\dfrac{\pi}{12}$ **28.** $\dfrac{\pi}{3}$ **30.** $\dfrac{\pi}{12}$

32. $4\tan^{-1}(\ln t) + C$ **34.** $\dfrac{\pi}{12}$ **36.** $6\sin^{-1}(\tan\theta) + C$

38. $2\ln\left(\dfrac{4}{3}\right)$ **40.** $\dfrac{\pi}{6}$ **42.** $12\sec^{-1}(e^x) + C$ **44.** 1 **46.** $\dfrac{\pi}{3}$

48. $\bar{x} = \dfrac{\ln 4}{\pi}$, $\bar{y} = 0$ **50.** x, $-x + \dfrac{\pi}{2}$, x, $-x + \dfrac{\pi}{2}$

52. $y = \sin^{-1}(x) + 1$ **54.** $y = \dfrac{\pi}{2} - \tan^{-1}(x)$ **56.** T3.1415925

58. b) $C = -\dfrac{\pi}{2}$

64.

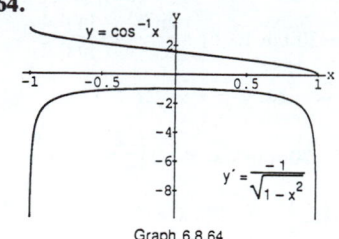
Graph 6.8.64

66.

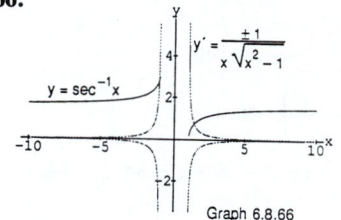
Graph 6.8.66

Section 6.9, pp. 464–468

2. $\cos hx = \dfrac{5}{3}$, $\tan hx = \dfrac{4}{5}$, $\cot hx = \dfrac{5}{4}$, $\sec hx = \dfrac{3}{5}$, $\csc$

$h = \dfrac{3}{4}$ **4.** $\sin hx = \dfrac{12}{5}$, $\tan hx = \dfrac{12}{13}$, $\cot hx = \dfrac{13}{12}$, $\sec hx = \dfrac{5}{13}$,

$\csc h = \dfrac{5}{12}$ **6.** $\dfrac{x^4-1}{2x^2}$ **8.** e^{4x} **10.** 0

14. $\cosh(2x + 1)$ **16.** $1 - \text{sech}^2(x)$ **18.** $-\coth(x)$

20. $x\sinh(x)$ **22.** $\text{sech}(x)$ **24.** $\dfrac{2x^4 + 8x^2 - 2}{(x^2 + 1)^2}$ **26.** $\dfrac{1}{\sqrt{x^2-x}}$

28. $1 - 2x\coth^{-1}(x)$ **30.** $2x\,\text{csch}^{-1}(x^2) - \dfrac{2x}{\sqrt{1+x^4}}$ **32.** $\sec x$

34. $-\csc x$ **36.** $-\csc x$ **46.** $\sinh 1$ **48.** 0 **50.** $\dfrac{1}{e}$

52. $\ln|\cosh(x)| + C$ **54.** $\ln(2) - \dfrac{3}{5}$ **56.** $\dfrac{5}{12}$ **58.** 0

60. $\ln(\cosh x) - \tan^2 x + C$ **62.** $-\ln 3$ **64.** $\ln\sqrt{3}$ **66.** $-\ln 2$

68. $\left(-\dfrac{1}{2}\right)\ln 3$ **70.** $\ln 2$ **72.** $\ln 5$

74. a) $\text{sech}^{-1}\left(\dfrac{4}{5}\right) - \text{sech}^{-1}\left(\dfrac{12}{13}\right)$ b) $\ln\left(\dfrac{4}{3}\right)$ **76.** a) $\tanh^{-1}\left(\dfrac{1}{2}\right)$

b) $\left(\dfrac{1}{2}\right)\ln 3$ **78.** a) $\sinh^{-1}(\sqrt{3})$ b) $\ln(\sqrt{3} + 2)$ **80.** a) 0 b) 0

82. π **84.** $2\pi\ln\left(\dfrac{199}{100}\right) - \dfrac{99\pi}{100}$

Section 6.10, pp. 474

4. $\dfrac{455}{18} + 4\ln 36$

6. b) (0.042, 0.672)

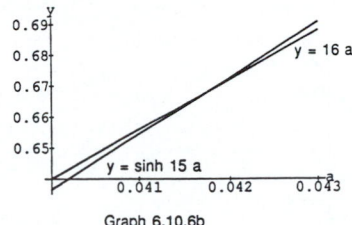

Graph 6.10.6b

c) 0.0417525 d) 47.90 lb
e) 4.8 ft

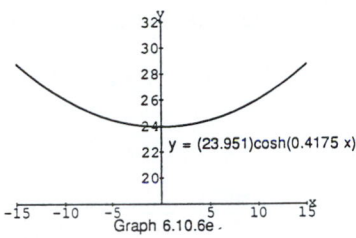

Graph 6.10.6e

Section 6 Miscellaneous Exercises, pp. 474–478

2. $\dfrac{1}{\sqrt{1-x^2}\sin^{-1}x}$ **4.** $\dfrac{1}{(\ln 5)(x-7)}$

6. $\begin{cases} 2x\sec^{-1}x, \; x > 1 \\ 2x\sec^{-1}x - \dfrac{2x}{\sqrt{x^2-1}}, \, x < -1 \end{cases}$ **8.** $\dfrac{8^{\sin^{-1}x}}{\sqrt{1-x^2}}$

10. $2x\cot^{-1}(2x) + (1 + x^2)\left(\dfrac{-2}{1+4x^2}\right)$ **12.** $\cos^{-1}(x)$ **14.** -1

18. $\dfrac{y(xe^{x+y}-1)}{x(1-ye^{x+y})}$ **20.** $s\cosh s$ **22.** $s\,\text{sech}^2\,s$ **24.** $\text{sech}\,s$

26. $\dfrac{v\cosh^{-1}v}{\sqrt{v^2-1}} + 1$ **28.** 1   **30.** $\sec v$ **32.** $\dfrac{2}{3}$ **34.** $\dfrac{\pi}{6}$ **36.** $\dfrac{\pi}{4}$

38. $\dfrac{9\ln 2}{4}$ **40.** $\dfrac{1}{\ln 3}$ **42.** $\dfrac{e^2-1}{e^2}$ **44.** $\dfrac{\pi}{4}$

46. $\ln\left|\dfrac{\pi}{4} + \sin^{-1}x\right| + C$ **48.** 1   **50.** $\ln\left(\dfrac{9}{8}\right)$

52. $5[\sin^{-1}(\tanh(\ln 2))]$ **54.** $4\left(\dfrac{\tanh^2 x}{2} - \dfrac{\tanh^4 x}{4}\right) + C$

56. a) $\cosh^{-1}(17) - \cosh^{-1}(2)$ b) $\ln\left(\dfrac{17+12\sqrt2}{2+\sqrt3}\right)$

58. a) $\text{csch}^{-1}\left(\dfrac{1}{2}\right) - \text{csch}^{-1}\left(\dfrac{1}{\sqrt2}\right)$ b) $\ln\left(\dfrac{2+\sqrt5}{\sqrt{2+\sqrt3}}\right)$

60. a) $e^{\coth^{-1}\sqrt3} - e^{\coth^{-1}\sqrt8}$ b) $\sqrt{2+\sqrt3} - \sqrt{\dfrac{9+4\sqrt2}{7}}$

62. $y = e^{-x} - x - 1$ **64.** a) $x = \dfrac{1}{1-e^y}$ b) $-\dfrac{\ln 9}{\ln 12}$ **66.** $-\dfrac{1}{4}$

68. $\dfrac{\pi}{2}$ **70.** $\dfrac{\pi}{2}$ **72.** $e - 1$ **74.** $\dfrac{1}{x-1}$ **76.** $\ln\left(\dfrac{5}{3}\right)$

78. $4\ln\left(\dfrac{2}{3}\right) + 3$ **80.** $f(1+x)e^x, (2+x)e^x, (3+x)e^x,$

$f^{(n)}(x) = (n+x)e^x$ **82.** a) x b) $-\dfrac{3}{2} + 2x - \dfrac{1}{2}x^2$

86. a) At $x = e^2$ there is a maximum and the curve is always concave downward.

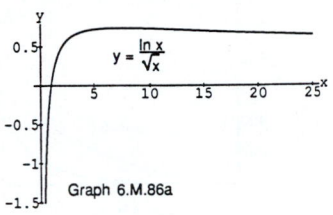

Graph 6.M.86a

b) At $x = 0$ there is a maximum and points of inflection at $\left(-\dfrac{1}{\sqrt2}, \dfrac{1}{\sqrt2}\right)$ and $\left(\dfrac{1}{\sqrt2}, \dfrac{1}{\sqrt2}\right)$.

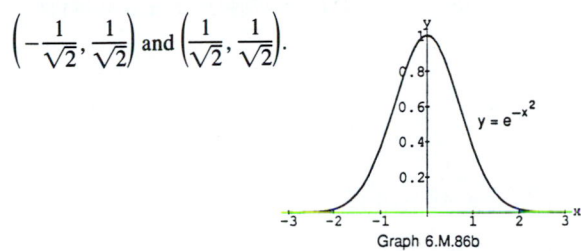

Graph 6.M.86b

c) At $x = 0$ there is a maximum and a point of inflection at $\left(1, \dfrac{2}{e}\right)$.

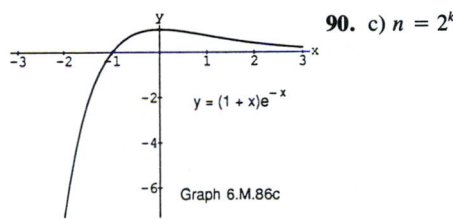

Graph 6.M.86c

90. c) $n = 2^k$

k	2	4	5
$n\left(\sqrt[n]{x}-1\right)$	0.75682846001	0.70838051884	0.70070875693

10	15
0.69338182970	0.69315451174

92. a) $0,\ -\infty,\ \infty,\ 0$ b)

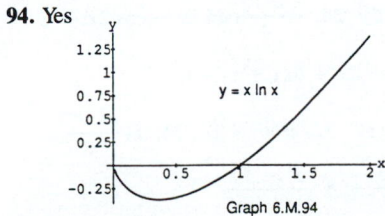

Graph 6.M.92b

94. Yes

Graph 6.M.94

96. 2 **98.** The 14th century model of free fall was exponential.

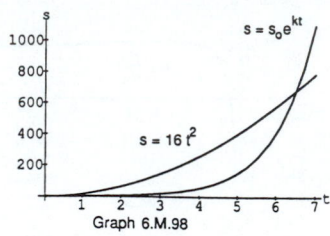

Graph 6.M.98

b) $\bar{x} \approx 0.78$ and $\bar{y} \approx 0.76$

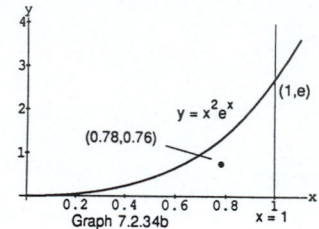

Graph 7.2.34b

100. 2 min **102.** a) same b) same c) same **104.** a) true b) false c) true **110.** b) 61° **114.** 1.478943 **116.** 0.3814522

36. a) $\dfrac{\pi^4}{6} - \dfrac{\pi^2}{4}$ b) 8π **38.** $\dfrac{1}{2}$ **40.** $x \tan^{-1} x + \ln|\cos(\tan^{-1} x)|$ $+ C$ **42.** $x \log_2 x - \dfrac{x}{\ln 2} + C$ **44.** Yes **46.** a) $x \tanh^{-1} x - \ln$ $|\cosh(\tanh^{-1} x)| + C$ b) $x \tanh^{-1} x + \dfrac{1}{2}\ln|1 - x^2| + C$

Chapter 7

Section 7.1, pp. 484–485

2. 2 **4.** $\dfrac{9}{4}$ **6.** $\ln 4$ **8.** $\ln\left(\dfrac{1}{2}\right)$ **10.** $\ln(\sqrt{\sec 1 + \tan 1})$

12. $-\ln\sqrt{3}$ **14.** $\dfrac{e-1}{e}$ **16.** 2 **18.** $\dfrac{2^{\ln 2} - 1}{\ln 2}$ **20.** $\dfrac{99}{\ln 100}$

22. π **24.** $\tan^{-1} e^x + C$ **26.** $\dfrac{\pi}{6}$ **28.** $\sin^{-1}(2 \ln x) + C$

30. $-\dfrac{\pi}{36}$ **32.** $\sec^{-1}(2x) + C$ **34.** $\dfrac{1}{\pi}\ln\left(\dfrac{2\sqrt{3}}{3} + 1\right)$

36. $-\ln(2 + \cos\theta) + C$ **38.** $[\sin^{-1}(x - 1)] + C$ **40.** π

42. $\left[\sec^{-1}|x - 2|\right] + C$, when $|x - 2| > 1$ **44.** $5 + 4\pi$

46. $-\cos x + C$ **48.** $\dfrac{4 - \pi}{4}$ **50.** $\ln(9) - 4$

52. $(x - 1)^{1/2} + \ln|x| + C$ **54.** $\dfrac{\pi}{4} - \ln 2$ **56.** $\ln(\sqrt{2} + 1)$

58. $2\sqrt{2} - \ln(3 + 2\sqrt{2})$ **62.** $\dfrac{\pi}{4}$ **64.** 10.89 cm

Section 7.2, pp. 491–492

2. $\dfrac{x \sin(2x)}{2} + \dfrac{\cos(2x)}{4} + C$ **4.** $-x^2 \cos x + 2x \sin x + 2 \cos x$ $+ C$ **6.** $\dfrac{x^4 \ln x}{4} - \dfrac{x^4}{16} + C$ **8.** $x \sin^{-1}(x) + \sqrt{1 - x^2} + C$

10. $2x \tan 2x - \ln|\sec 2x| + C$

12. $(-p^4 - 4p^3 - 12p^2 - 24p - 24)e^{-p} + C$

14. $[(r^2 + r + 1) - (2r + 1) + 2]e^r + C$

16. $\left(\dfrac{t^2}{4} - \dfrac{t}{8} + \dfrac{1}{32}\right)e^{4t} + C$ **18.** $\dfrac{3(4 - \pi^2)}{16}$ **20.** $\dfrac{\pi + 6\sqrt{3} - 12}{12}$

22. $\dfrac{e^{-y}}{2}(\sin y - \cos y) + C$ **24.** $-\dfrac{e^{-2x}}{4}(\sin 2x + \cos 2x) + C$

26. $\dfrac{4}{15}$ **28.** $x \ln(x + x^2) - 2x + \ln|1 + x| + C$

30. $\left(\dfrac{\ln z^2}{2} - \dfrac{\ln z}{2} + \dfrac{1}{4}\right)z^2 + C$ **32.** $\pi(\pi - 2)$ **34.** a) $\dfrac{e^2 - 3}{8}$;

$6 - 2e; e - 2; \bar{x} = \dfrac{6 - 2e}{e - 2}, \bar{y} = \dfrac{e^2 - 3}{4(e - 2)}$

Section 7.3, pp. 499–501

2. $\dfrac{16}{15}$ **4.** $\dfrac{8}{15}$ **6.** $7\left(\sin t - \sin^3 t + \dfrac{3 \sin^5 t}{5} - \dfrac{\sin^7 t}{7}\right) + C$

8. $2\left[\dfrac{3}{2}x + \dfrac{\sin(4\pi x)}{2\pi} + \dfrac{\sin(8\pi x)}{16\pi}\right] + C$ **10.** $-\cot\theta - \dfrac{15\theta}{8} -$

$\dfrac{\sin 2\theta}{2} - \dfrac{\sin 4\theta}{32} + C$ **12.** 0 **14.** $\dfrac{8}{77}$

16. $2\left[\dfrac{3y}{2} - 2 \sin y + \dfrac{\sin 4y}{8}\right] + \left[\dfrac{5}{2}y - 2 \sin 2y +\right.$

$\left.\dfrac{3 \sin 4y}{8} + \dfrac{\sin^3 2y}{6}\right] + C$ **18.** $\dfrac{1}{16}\left[\dfrac{3t}{2} - \dfrac{\sin 4t}{2} + \dfrac{\sin 8t}{16}\right] + C$

20. $12\left[-\cos(\tan^{-1}y) + \dfrac{\cos^3(\tan^{-1} y)}{3}\right] + C$ **22.** $2\sqrt{2}$ **24.** 2

26. $\ln 2$ **28.** $\dfrac{8}{3}$ **30.** $4\sqrt{2} - 4$ **32.** $\dfrac{1}{\sqrt{2}}\ln(\sin x) + C$ if $\sin x$

> 0 **34.** $\dfrac{1}{2}\left[\sec(e^x - 1)\tan(e^x - 1) + \ln|\sec(e^x - 1) + \tan\right.$

$\left.(e^x - 1)|\right] + C$ **36.** $-\csc\sqrt{\theta}\cot\sqrt{\theta} - \ln|\csc\sqrt{\theta} + \cot$

$\sqrt{\theta}| + C$ **38.** $\sqrt{3} - \ln\sqrt{2 + \sqrt{3}}$ **40.** $\dfrac{4}{3}$ **42.** 8

44. $\dfrac{4}{3} - \ln\sqrt{3}$ **46.** $6\left[\dfrac{\tan^3 x}{3} - \tan x + x\right] + C$ **48.** $2\sqrt{3}$

50. $-\dfrac{2}{3}\csc^{3/2} + 2 \csc^{-1/2} x + C$ **52.** $-\dfrac{2}{5}$ **54.** $\dfrac{1}{2}$ **56.** $-\dfrac{14\sqrt{2}}{195}$

58. a, b, c, d, e, h, and i **60.** f, 0; g, $\dfrac{2}{3}$

Section 7.4, pp. 506–507

2. $\dfrac{\pi}{16}$ **4.** $\dfrac{\pi}{4}$ **6.** $\ln|3x + \sqrt{9x^2 - 1}| + C$ **8.** $\ln(\sqrt{2} + 1)$

10. $\dfrac{25\pi}{4}$ **12.** π **14.** $\dfrac{2 - \sqrt{2}}{3}$ **16.** $-\dfrac{\sqrt{1 - w^2}}{w}$ **18.** $\dfrac{\pi}{12}$

20. $-\sqrt{5 + 4x - x^2} + C$ **22.** $\cosh^{-1}(1 + \sin t) + C$

24. $3 - 2\sqrt{2}$ **26.** $\dfrac{\pi}{4} + \ln 2$ **28.** π **30.** $\dfrac{\pi}{12}$

32. $\dfrac{1}{2} + \sinh^{-1}\left(\dfrac{3}{4}\right)$ **34.** $\dfrac{1}{6}\ln(z^2 + 4) + \dfrac{1}{6}\tan^{-1}\left(\dfrac{z}{2}\right) + C$

36. $\sqrt{z^2 - 4z + 3} + \cosh^{-1}(z - 2) + C$ **38.** $\frac{2}{9}(z^2 + 1)^{1/2} +$

$\frac{1}{9}\sinh^{-1}z + C$ **40.** $\frac{1}{4}\ln(4x^2 + 4x + 5) - \frac{1}{2}\tan^{-1}(x + 1/2) + C$

42. $\frac{1}{\sqrt{3}}$ **44.** $\frac{\pi}{4}$ **46.** 0.644   **48.** $2\pi + \sqrt{2}\pi\ln(\sqrt{2} + 1)$

52. $\frac{\pi}{2}$

Section 7.5, pp. 511–513

2. $\frac{3}{x - 2} + \frac{2}{x - 1}$ **4.** $\frac{2}{x - 1} + \frac{4}{(x - 1)^2}$ **6.** $\frac{1/5}{z - 3} + \frac{-1/5}{z + 2}$

8. $1 + \frac{1}{x^2} - \frac{10}{x^2 + 9}$ **10.** $\frac{1}{2}\ln\left(\frac{3}{2}\right)$ **12.** $\ln\left|\frac{(x - 4)^9}{(x - 3)^7}\right| + C$

14. $y + 2\ln(y^2 + 1) - \tan^{-1}y + C$

16. $\frac{1}{16}\ln\left|\frac{(x - 2)^5(x + 2)}{x^6}\right| + C$ **18.** $\frac{124}{5} + \ln(125)$

20. $\frac{\ln|(x - 1)(x + 1)^3|}{4} + \frac{1}{2(x + 1)} + C$

22. $\frac{(\tan^{-1}x)^2}{2} + \ln|x + 1| + \frac{1}{x + 1} + C$ **24.** $\frac{1}{3}\ln\left(\frac{2}{5}\right)$

26. $6\sqrt[6]{\theta} - 3(\sqrt[6]{\theta})^2 + 2(\sqrt[6]{\theta})^3 - 6\ln|\sqrt[6]{\theta} + 1| + C$

28. $\ln 2 - 2 + \frac{\pi}{2}$ **30.** $\frac{3\pi - 8}{12} + \ln 2$ **32.** $\frac{\pi + \ln 4}{8}$

34. $\ln\left(\frac{t^4}{\sqrt{t^2 + 1}}\right) + \tan^{-1}t + C$ **36.** $\frac{\pi}{4} - \ln 2$

38. $-\frac{2}{3} + \frac{\pi}{4} + \ln 2$ **40.** $\frac{\pi}{2}$ **42.** $\frac{\pi + 2}{8}$

44. $-\frac{5}{8} + \ln\left(\frac{8}{\sqrt{125}}\right) - \frac{\sqrt{5}}{10}\ln\left|\frac{3 - \sqrt{5}}{2}\right|$

46. $\frac{1}{4}\ln|4x^2 + 8x + 1| + \frac{\sqrt{3}}{4}\ln\left|\frac{\sqrt{3} + 2x + 2}{\sqrt{3} - 2x - 2}\right| + C$

48. $\frac{1}{2}\ln|y^2 + 2y - 1| - \frac{1}{2\sqrt{2}}\ln\left|\frac{\sqrt{2} + y + 1}{\sqrt{2} - y - 1}\right| + C$

50. $-\frac{1}{4}(x^2 + 1)^{-2} + 2(x^2 + 1)^{-1} + \tan^{-1}x + C$ **52.** $\frac{\ln(9) - 1}{2}$

54. 3.90   **56.** a) $a = b \Rightarrow x = \dfrac{a^2kt}{akt + 1}$ b) $a \neq b \Rightarrow x =$

$\dfrac{ab[1 - e^{(a - b)kt}]}{b - a\,e^{(a - b)kt}}$ **58.** $x = \dfrac{a}{1 - \left(\dfrac{x_0 - a}{x_0}\right)e^{akt}}$

Section 7.6, pp. 518–519

2. $\frac{2}{\sqrt{3}}\left(\tan^{-1}\sqrt{2} - \frac{\pi}{4}\right)$ **4.** $\frac{x}{18(9 - x^2)} + \frac{1}{108}\ln\left|\frac{x + 3}{x - 3}\right| + C$

6. $\frac{44 - 24\sqrt{2}}{231}$ **8.** $\sqrt{2}\left(\frac{\pi}{4} - 1\right)$

10. $-\frac{1}{6}\left[\ln\left|\frac{5 + 4\sin 2x + 3\cos 2x}{4 + 5\sin 2x}\right|\right] + C$

12. $\frac{(2x - 3)^{3/2}(x + 1)}{5} + C$ **14.** $\frac{\pi}{12} - \frac{1}{6} + \frac{\ln 2}{6}$ **16.** $\frac{6 - \pi\sqrt{3}}{6}$

18. $-\frac{1}{\sqrt{2}} + \ln(\sqrt{2} + 1)$ **20.** $54 - 27\ln 3$ **22.** a) Proof

---

b) $2L\left[\left(\frac{d - r}{2}\right)\sqrt{2rd - d^2} + \left(\frac{r^2}{2}\right)\left(\sin^{-1}\left(\frac{d - r}{r}\right) + \frac{\pi}{2}\right)\right]$

24. Proof   **26.** Proof   **28.** $\frac{5\pi}{96}$ **30.** $\frac{4}{15}$ **32.** $\frac{4\sqrt{2} - 8}{3} + \frac{\pi}{4}$

34. $\frac{3\pi - 8}{12}$ **36.** $\frac{7\sqrt{2} + 3\ln(\sqrt{2} + 1)}{8}$

38. $\frac{7\sqrt{2} + 3\ln(\sqrt{2} + 1)}{8}$ **40.** $6 - 2e$

42. $\frac{7\sqrt{2} + 3\ln(\sqrt{2} + 1)}{8}$ **43.** $\frac{735 + 384\ln 2}{1024}$ **44.** $\frac{\pi}{3}$

48. b) $\displaystyle\int x^n \cos ax\,dx = \frac{x^n \sin ax}{a} - \frac{n}{a}\int x^{n - 1}\sin ax\,dx$

Section 7.7, pp. 528–530

2. 2   **4.** 1000   **6.** $\frac{\pi}{2}$ **8.** $\ln 3$ **10.** π **12.** π **14.** $\frac{4 + \pi}{2}$

16. $-\frac{1}{4}$ **18.** 1   **20.** diverges   **22.** converges   **24.** diverges

26. diverges   **28.** diverges   **30.** converges   **32.** converges
34. converges   **36.** converges   **38.** diverges   **40.** diverges
42. diverges   **44.** converges   **46.** converges   **48.** converges

50. diverges   **52.** $\frac{\pi}{4}$ **54.** $\frac{3\pi - 2}{4}$ **56.** ∞ **58.** $\frac{\pi}{2}$ **60.** $\frac{\pi}{2}$

62. π **64.** a) $p < 1$ b) $p > 1$ **66.** $\bar{x} = 1, \bar{y} = \frac{1}{4}$ **68.** $\frac{\pi}{2}$

74. converges   **76.** diverges   **78.** diverges   **80.** converges

82. $\displaystyle\int_3^\infty \left(\frac{1}{x - 2} - \frac{1}{x}\right)dx \neq \int_3^\infty \frac{1}{x - 2}\,dx - \int_3^\infty \frac{1}{x}\,dx$ **84.** $\frac{1}{2}$

Section 7 Miscellaneous Exercises, pp. 531–535

2. $\frac{\pi^2}{32}$ **4.** $\frac{\tan^{-1}y^2}{2} + C$ **6.** $2\sin\sqrt{x} + C$ **8.** $\ln\left(\frac{27}{16}\right)$

10. $\frac{1}{2}\left(z\sqrt{z^2 + 1} + \ln|z + \sqrt{z^2 + 1}|\right) + C$

12. $25\tan^{-1}(2) - 10$ **14.** $\frac{80}{3}$ **16.** $-\ln(1 - \ln 2)$

18. $e^{\sec x} + C$ **20.** $\sin^{-1}\left(\frac{\sqrt{2} - 1}{\sqrt{2}}\right) + \frac{\pi}{3}$ **22.** $\ln\sqrt{2}$

24. $-2\sqrt{1 - \sin x} + C$ **26.** $\frac{x}{a^2\sqrt{x^2 + a^2}} + C$

28. $\frac{\ln(4) - 4 + \pi}{4}$ **30.** $\frac{14}{375}$ **32.** $x - \csc 2x + \cot 2x + C$

34. $-\cos 2\sqrt{x} + C$ **36.** $2\sqrt{(e - 1)^2 + 1} - 2$

38. $-2u\sin u - 2\cos u + C$ **40.** $\frac{\ln 2}{3} + \frac{\pi}{3\sqrt{3}}$

42. $\frac{2x^{3/2}}{3} - x + 2\sqrt{x} - 2\ln(\sqrt{x} + 1) + C$

44. $\frac{1}{2}\left(x\ln|x - 1| - x - \ln|x - 1|\right) + C$

46. $2\ln\left|\frac{x - 1}{x}\right| + \frac{1}{x} + C$ **48.** $\frac{25}{16}$ **50.** $\frac{1}{3}\tan^{-1}\left(\frac{5a + 4}{3}\right) + C$

52. $\frac{2e^x\sin 2x}{5} + \frac{e^x\cos 2x}{5} + C$ **54.** $3\ln\left|\frac{\sqrt[3]{x}}{1 + \sqrt[3]{x}}\right| + C$

56. $\dfrac{(z^2+1)^{5/2}}{5} - \dfrac{(z^2+1)^3}{3} + (z^2+1)^{1/2} + C$

58. $\dfrac{5\sqrt{1+x^{4/5}}}{2} + C$ **60.** $1 + \ln\left(\dfrac{4}{3}\right)$

62. $\dfrac{3}{2}(x^2+2x-3)^{1/3} + C$

64. $-\dfrac{\sqrt{a^2-s^2}}{a^2 s} + C$ **66.** $x\ln\left(x+\sqrt{1+x^2}\right) - \sqrt{1+x^2} + C$

68. $\dfrac{x^2+x\sin 2x}{4} + \dfrac{\cos 2x}{8} + C$ **70.** $\dfrac{x^2-x\sin 2x}{4} - \dfrac{\cos 2x}{8} + C$

72. $\dfrac{u}{3} + \dfrac{\ln|e^{2u}+3|}{12} - \dfrac{\ln|e^{2u}+1|}{4} + C$

74. $(x+1)^2 e^x - 2(x+1)e^x + 2e^x + C$

76. $\ln\left|\dfrac{x}{x+2}\right| + \dfrac{2}{x} - \dfrac{2}{x^2} + C$ **78.** $2 - \sqrt{2}\ln(\sqrt{2}+1)$ **80.** $\dfrac{3}{20}$

82. $\ln\left|\dfrac{\tan t - 2}{\tan t - 1}\right| + C$ **84.** $\dfrac{\pi\sqrt{2}}{8}$ **86.** $\ln\sqrt{2} + \dfrac{\pi}{4} - 1$ **88.** $\dfrac{1}{2}$

90. $\sin^{-1}\left(\dfrac{1}{\sqrt{3}}\right)$ **92.** $\ln\left(\dfrac{\pi+8}{8}\right)$ **94.** $\dfrac{1}{108}$ **96.** $\dfrac{5}{12}$

98. $x\ln\left(x+\sqrt{x^2-1}\right) - \sqrt{x^2-1} + C$

100. $\dfrac{-\pi}{3\sqrt{3}} + \dfrac{\pi}{4} + \ln\sqrt{\dfrac{3}{2}}$ **102.** $x\tan^{-1}\sqrt{x} - \sqrt{x} + \tan^{-1}$

$\sqrt{x} + C$ **104.** -4 **106.** $\dfrac{5\pi}{6} + 1 - \sqrt{3}$ **108.** $\dfrac{\pi^2}{4}$

110. $\dfrac{1}{2}\ln\left|\dfrac{\sqrt{e^{2t}+1}-1}{\sqrt{e^{2t}+1}+1}\right| + C$ **112.** $-\dfrac{1}{ac}\ln|ae^{-ct}+b| + C,$

where $a, b, c \neq 0$ **114.** $-\dfrac{1}{x} - 3\tan^{-1}(3x) + C$

116. $\dfrac{1}{3}\ln\left|\dfrac{\cos x - 1}{\cos x - 4}\right| + C$

118. $-\dfrac{\tan^{-1}x}{x} + \ln|x| - \ln\sqrt{1+x^2} + C$ **120.** $\ln|t + t\ln t| + C$

122. $\dfrac{\pi}{2} + \ln(5) + \dfrac{1}{2} - \tan^{-1}2$ **124.** 1

126. $\displaystyle\sum_{k=0}^{m}\left(\dfrac{(-1)^k}{(k!)(m-k)!}\ln|x+k|\right) + C$

128. $x\sin^{-1}\sqrt{x} - \dfrac{\sin^{-1}\sqrt{x}}{2} + \dfrac{\sqrt{x-x^2}}{2} + C$

130. $x\ln\left(\sqrt{x}+\sqrt{1+x}\right) -$

$\dfrac{2\sqrt{x^2+x} - \ln|2x+1+2\sqrt{x^2+x}|}{4} + C$

132. $\dfrac{1}{\sqrt{3}}\left(2\sqrt{e^{2x}-2e^x-1/3} + \ln|e^x - 1 + \right.$

$\left.\sqrt{e^{2x}-2e^x-1/3}|\right) + C$

134. $\dfrac{1}{6}\ln\left|\dfrac{x-1}{x+1}\right| + \dfrac{1}{12}\left[\ln\left|\dfrac{x^2-x+1}{x^2+x+1}\right| - 2\sqrt{3}\right.$

$\left.\tan^{-1}\left(\dfrac{2x-1}{\sqrt{3}}\right) - 2\sqrt{3}\tan^{-1}\left(\dfrac{2x+1}{\sqrt{3}}\right)\right] + C$ **136.** $\dfrac{\pi}{2}$ **138.** $\dfrac{32\pi}{35}$

140. 2π

142. $\sqrt{1+e^2} - \ln\left|\dfrac{\sqrt{1+e^2}}{e} + \dfrac{1}{e}\right| - \sqrt{2} + \ln(1+\sqrt{2})$

144. $2\pi\left(\sqrt{2} + \ln(1+\sqrt{2})\right)$ **148.** converges **150.** diverges

152. $\dfrac{1}{2}$ **154.** $\sqrt{3} - 1$ **156.** $\dfrac{\sqrt{3}\pi}{9}$ **158.** $\ln\left|\tan\left(\dfrac{x}{2}\right) + 1\right| + C$

160. $\dfrac{\ln(\sqrt{3})-1}{2}$ **162.** a) Proof

b)

n	$\left(\dfrac{n}{e}\right)^n \sqrt{2n\pi}$	calculator
10	3598695.619	3628800
20	2.4227868×10^{18}	2.432902×10^{18}
30	2.6451709×10^{32}	2.652528×10^{32}
40	8.1421726×10^{47}	8.1591528×10^{47}
50	3.0363445×10^{64}	3.0414093×10^{64}
60	8.3094383×10^{81}	8.3209871×10^{81}

c)

n	$\left(\dfrac{n}{e}\right)^n \sqrt{2n\pi}$	$\left(\dfrac{n}{e}\right)^n \sqrt{2n\pi}\,e^{1/12n}$	calculator
10	3598695.619	3628810.051	3628800

Chapter 8

Section 8.1, pp. 549–552

2. $a_1 = 1, a_2 = \dfrac{1}{2}, a_3 = \dfrac{1}{6}, a_4 = \dfrac{1}{24}$ **4.** $a_1 = 1, a_2 = 3, a_3 = 1,$

$a_4 = 3$ **6.** $1, \dfrac{1}{2}, \dfrac{1}{6}, \dfrac{1}{24}, \dfrac{1}{120}, \dfrac{1}{720}, \dfrac{1}{5040}, \dfrac{1}{40320}, \dfrac{1}{362880},$

$\dfrac{1}{3628800}$ **8.** $-2, -1, -\dfrac{2}{3}, -\dfrac{1}{2}, -\dfrac{2}{5}, -\dfrac{1}{3}, -\dfrac{2}{7}, -\dfrac{1}{4}, -\dfrac{2}{9},$

$-\dfrac{1}{5}$ **10.** $2, -1, -\dfrac{1}{2}, \dfrac{1}{2}, -1, -2, 2, -1, -\dfrac{1}{2}, \dfrac{1}{2}$ **12.** $1,$

converges **14.** diverges **16.** 0, converges **18.** diverges

20. diverges **22.** 6, converges **24.** 0 **26.** 0 **28.** 1 **30.** 0

32. diverges **34.** 1 **36.** 1 **38.** e^{-1} **40.** diverges **42.** 1

44. 1 **46.** 4 **48.** 9 **50.** 0 **52.** diverges **54.** 0 **56.** diverges

58. $\dfrac{1}{e}$ **60.** 1 **62.** 0 **64.** $\dfrac{1}{p-1}$, converges when $p > 1$

66. 0 **68.** 0 **70.** -2 **72.** 9124 **74.** 15 **76.** $x_n = 2^{n-2}$ for

$n \geq 2$

78.

n	a_n	b_n
1	1	1
2	2	3
3	5	8
4	13	21
5	34	55
6	89	144

The first 12 terms of the Fibonacci sequence are 1, 1, 2, 3, 5, 8, 13, 21, 34, 55, 89, 144.
80. 1.73205081   **82.** a) Proof

b)

n	$\sqrt[n]{n!}$	$\dfrac{n}{e}$
40	15.76852702	14.71517765
50	19.48325423	18.39397206
60	23.19189561	22.07276647

84. 1   **86.** 2   **92.** 0   **94.** 0   **96.** 3.51562548   **98.** At $x = \pm 1$ this iterative method does not converge nor diverge
100. a) $g(x) = \sqrt{2x + 3}$ we have a root at 3; with $g(x) = -\sqrt{2x + 3}$ the computer program will not work;
b) Picard's method does not converge for most starting values;
c) -1

Section 8.2, pp. 561–562

2. $\dfrac{(9/100)(1 - (1/100)^n)}{(1 - 1/100)}, \dfrac{1}{11}$ **4.** diverges   **6.** $\dfrac{5}{n} - \dfrac{5}{n + 1}, 5$

8. $\dfrac{1}{16} + \dfrac{1}{64} + \dfrac{1}{256} + \cdots, \dfrac{1}{12}$ **10.** $5 - \dfrac{5}{4} + \dfrac{5}{16} - \dfrac{5}{64} + \cdots, 4$

12. $(5 - 1) + \left(\dfrac{5}{2} - \dfrac{1}{3}\right) + \left(\dfrac{5}{4} - \dfrac{1}{9}\right) + \left(\dfrac{5}{8} - \dfrac{1}{27}\right) + \cdots, \dfrac{17}{2}$

14. $2 + \dfrac{4}{5} + \dfrac{8}{25} + \dfrac{16}{125} + \cdots, \dfrac{10}{3}$ **16.** $\dfrac{1}{4}$ **18.** 1   **20.** diverges

22. diverges   **24.** $\dfrac{5}{6}$ **26.** diverges   **28.** $\dfrac{2}{9}$ **30.** diverges

32. $\dfrac{x}{x - 1}$ **34.** $\dfrac{\ln(2)}{\ln(4/3)}$ **36.** diverges   **38.** diverges   **40.** $a = 1$, $r = -x^2$ where $|x| < 1$ **42.** $a = \dfrac{2 + \sin x}{8}, r = -\dfrac{\sin x}{4}$ where x is real   **44.** $|x| < 1, \dfrac{1}{1 + x^2}$ **46.** $1 < x < 5, \dfrac{2}{x - 1}$

48. $e^{-1} < x < e, \dfrac{1}{1 - \ln x}$ **50.** $\dfrac{234}{999}$ **52.** $\dfrac{d}{9}$ **54.** $\dfrac{1413}{999}$

56. $\dfrac{349206}{111111}$ **58.** 12.58 sec   **60.** a) $\displaystyle\sum_{n = -1}^{\infty} \dfrac{5}{(n + 2)(n + 3)}$

b) $\displaystyle\sum_{n = 3}^{\infty} \dfrac{5}{(n - 2)(n - 1)}$ c) $\displaystyle\sum_{n = 20}^{\infty} \dfrac{5}{(n - 19)(n - 18)}$

62. $-\ln(n + 1)$ **64.** $\dfrac{\pi}{2}$ **66.** $\displaystyle\sum_{n = 1}^{\infty} n$ and $\displaystyle\sum_{n = 1}^{\infty} (-n)$

Section 8.3, pp. 570–572

2. diverges   **4.** diverges   **6.** converges   **8.** converges
10. diverges   **12.** diverges   **14.** converges   **16.** diverges
18. diverges   **20.** converges   **22.** diverges   **24.** converges
26. converges   **28.** converges   **30.** converges   **32.** converges
34. 1   **42.** b) $p = 1 \Rightarrow$ the series diverges c) $p = 1.01 \Rightarrow$ the series converges d) diverges e) diverges f) diverges

Section 8.4, pp. 577

2. diverges   **4.** converges   **6.** converges   **8.** converges
10. diverges   **12.** diverges   **14.** converges   **16.** converges
18. converges   **20.** converges   **22.** diverges   **24.** diverges
26. converges   **28.** diverges   **30.** diverges   **32.** diverges
34. converges   **36.** diverges   **38.** diverges   **40.** converges
42. converges

Section 8.5, pp. 584–585

2. converges   **4.** diverges   **6.** converges   **8.** diverges
10. converges   **12.** converges   **14.** converges conditionally
16. diverges   **18.** converges absolutely   **20.** converges conditionally   **22.** converges absolutely   **24.** diverges
26. converges conditionally   **28.** converges absolutely
30. converges conditionally   **32.** converges   **34.** converges absolutely   **36.** converges conditionally   **38.** converges absolutely   **40.** converges absolutely   **42.** diverges
44. converges absolutely   **46.** 0.00001   **48.** t^4
50. 0.367881944   **52.** 0.692580927   **58.** The terms in this conditionally convergent series were not added in the order given.

Section 8.6, pp. 592–593

2. a) 1, $-6 < x < -4$ b) $-6 < x < -4$ c) $\{\ \}$
4. a) 1, $1 \le x < 3$ b) $1 < x < 3$ c) $\{1\}$
6. a) $\dfrac{1}{2}, -\dfrac{1}{2} < x < \dfrac{1}{2}$ b) $-\dfrac{1}{2} < x < \dfrac{1}{2}$ c) $\{\ \}$
8. a) 1, $-3 < x \le -1$ b) $-3 < x < -1$ c) $\{-1\}$
10. a) 1, $0 \le x < 2$ b) $0 < x < 2$ c) $\{0\}$
12. a) ∞, For all x b) For all x c) $\{\ \}$
14. a) ∞, For all x b) For all x c) $\{\ \}$
16. a) 1, $-1 < x \le 1$ b) $-1 < x < 1$ c) $\{1\}$
18. a) 5, $(-2, 8)$ b) $(-2, 8)$ c) $\{\ \}$
20. a) 1, $0 < x < 2$ b) $0 < x < 2$ c) $\{\ \}$
22. a) 1, $-1 < x < 1$ b) $-1 < x < 1$ c) $\{\ \}$
24. a) 0, $x = 4$ b) $x = 4$ c) $\{\ \}$ **26.** a) $\dfrac{1}{2}, \left(\dfrac{1}{2}, \dfrac{3}{2}\right)$
b) $\dfrac{1}{2} < x < \dfrac{3}{2}$ c) $\{\ \}$ **28.** a) 1, $[0, 2]$ b) $0 \le x \le 2$
c) $\{\ \}$ **30.** a) $\sqrt{2}, \left(-\sqrt{2}, \sqrt{2}\right)$ b) $-\sqrt{2} < x < \sqrt{2}$
c) $\{\ \}$ **32.** a) $\dfrac{1}{e}, \left(-\dfrac{1}{e}, \dfrac{1}{e}\right)$ b) $-\dfrac{1}{e} < x < \dfrac{1}{e}$ c) $\{\ \}$
34. a) $e, [-e, e]$ b) $[-e, e]$ c) $\{\ \}$ **36.** $(-e, e), \dfrac{\ln x}{1 - \ln x}$
38. $(0, 2], -\dfrac{2}{\sqrt{x}}$ **40.** $(-\infty, 0) \cup (2, \infty), \dfrac{x - 1}{x - 2}$
42. $x - \dfrac{(x - 3)^2}{4} + \dfrac{(x - 3)^3}{12} + \cdots + \left(-\dfrac{1}{2}\right)^n \dfrac{(x - 3)^{n + 1}}{n + 1} + \cdots$
$1 < x \le 5, 2 \ln|x - 1| + 3 - \ln 4$
44. a) $x + \dfrac{x^3}{6} + \dfrac{x^5}{24} + \dfrac{61x^7}{5040} + \cdots$, when $-\dfrac{\pi}{2} < x < \dfrac{\pi}{2}$
b) $x + \dfrac{5x^3}{6} + \dfrac{61x^5}{120} + \dfrac{277x^7}{1008} + \cdots$, when $-\dfrac{\pi}{2} < x < \dfrac{\pi}{2}$
46. 10

Section 8.7, pp. 606–608

2. $P_0(x) = 0$, $P_1(x) = x$, $P_2(x) = x - \dfrac{x^2}{2}$, $P_3(x) = x - \dfrac{x^2}{2} + \dfrac{x^3}{3}$

4. $P_0(x) = \dfrac{1}{2}$, $P_1(x) = \dfrac{1}{2} - \dfrac{x}{4}$, $P_2(x) = \dfrac{1}{2} - \dfrac{x}{4} + \dfrac{x^2}{8}$, $P_3(x) = \dfrac{1}{2} -$

$\dfrac{x}{4} + \dfrac{x^2}{8} - \dfrac{x^3}{16}$ **6.** $P_1(x) = \dfrac{1}{\sqrt{2}} - \dfrac{1}{\sqrt{2}}\left(x - \dfrac{\pi}{4}\right)$, $P_2(x) = \dfrac{1}{\sqrt{2}} -$

$\dfrac{1}{\sqrt{2}}\left(x - \dfrac{\pi}{4}\right) - \dfrac{1}{2\sqrt{2}}\left(x - \dfrac{\pi}{4}\right)^2$, $P_3(x) = \dfrac{1}{\sqrt{2}} -$

$\dfrac{1}{\sqrt{2}}\left(x - \dfrac{\pi}{4}\right) - \dfrac{1}{2\sqrt{2}}\left(x - \dfrac{\pi}{4}\right)^2 + \dfrac{1}{6\sqrt{2}}\left(x - \dfrac{\pi}{4}\right)^3$

8. $P_0(x) = 2$, $P_1(x) = 2 + \dfrac{1}{4}(x)$, $P_2(x) = 2 + \dfrac{1}{4}x - \dfrac{1}{64}x^2$,

$P_3(x) = 2 + \dfrac{1}{4}x - \dfrac{1}{64}x^2 + \dfrac{1}{512}x^3$ **10.** $1 + \dfrac{x}{2} + \dfrac{x^2}{4 \cdot 2!} +$

$\dfrac{x^3}{2^3 \cdot 3!} + \cdots$

12. $5\left(1 - \dfrac{(\pi x)^2}{2!} + \dfrac{(\pi x)^4}{4!} - \dfrac{(\pi x)^6}{6!} + \cdots\right)$

14. $\displaystyle\sum_{n=0}^{\infty} \dfrac{(-1)^n x^{2n+2}}{(2n+1)!}$ **16.** $x + \dfrac{x^3}{3!} + \dfrac{x^5}{5!} + \cdots$

18. $1 - \dfrac{(2x)^2}{2 \cdot 2!} + \dfrac{(2x)^4}{2 \cdot 4!} - \dfrac{(2x)^6}{2 \cdot 6!} + \dfrac{(2x)^8}{2 \cdot 8!} - \cdots$

20. $1 + x + \dfrac{x^2}{2} + \dfrac{f'''(c)}{3!}x^3$ where c is between 0 and x

22. $1 + \dfrac{x^2}{2} + \dfrac{f'''(c)}{3!}x^3$ where c is between 0 and x

24. $1 + \dfrac{x^2}{2} + \dfrac{f'''(c)}{3!}x^3$ where c is between 0 and x

26. $e\left[1 + (x - 1) + \dfrac{(x-1)^2}{2!} + \cdots\right]$ **28.** the approximation,

$1 - \dfrac{x^2}{2}$, is low   **30.** 0.0000125   **32.** 0.0001, $\overline{6}$ **36.** $\cos x$, $\dfrac{\pi}{4}$,

0.707106781   **40.** $1 + x - \dfrac{x^3}{3} - \dfrac{x^4}{6} - \dfrac{x^5}{30} + \cdots$

44. a) the Maclaurin series is $0 + 0 + \cdots$, since $f^{(n)}(0) = 0$ for all n; this series converges for all x; $f(x)$ is zero only at $x = 0$

b) $f(x) = 0 + 0 + \cdots + 0 + R_n(x) \Rightarrow R_n(x) = \exp\left(-\dfrac{1}{x^2}\right)$

52. $\displaystyle\int e^{ax}\cos bx\, dx = \dfrac{e^{ax}[a\cos bx + b\sin bx]}{a^2 + b^2} + C_1$ and $\displaystyle\int e^{ax}\sin$

$bx\, dx = \dfrac{e^{ax}[a\sin bx - b\cos bx]}{a^2 + b^2} + iC_2$

56. a)

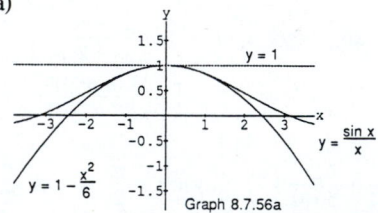

Graph 8.7.56a

Section 8.8, pp. 614

2. $(x - 2\pi) - \dfrac{(x - 2\pi)^3}{3!} + \dfrac{(x - 2\pi)^5}{5!} - \dfrac{(x - 2\pi)^7}{7!} + \cdots$

4. $(x - 1) - \dfrac{(x-1)^2}{2} + \dfrac{(x-1)^3}{3} - \dfrac{(x-1)^4}{4} + \cdots$

6. $\dfrac{\pi}{4} + \dfrac{(x-1)}{2} - \dfrac{(x-1)^2}{4} + \dfrac{(x-1)^3}{12} + \cdots$ **8.** 0.005

10. 0.1   **11.** 0.96356   **12.** 0.0975   **14.** 0.2517

15. 0.099944461   **16.** 0.099667666   **18.** 0.486385345   **20.** 8

23. 3   **26.** a) $x - \dfrac{x^3}{6} + \dfrac{3x^5}{40} - \dfrac{5x^7}{112}$ b) 0.24746635,

0.000000173   **28.** $2x + 4x^3 + 6x^5 + \cdots =$

Section 8 Miscellaneous Exercises, pp. 618–721

2. converges to -1 **4.** converges to 1   **6.** converges to e^5

10. converges to 0   **12.** 1, converges   **14.** diverges   **16.** 1

18. $-\dfrac{3}{5}$ **20.** diverges   **22.** converges absolutely

24. converges   **26.** converges conditionally   **28.** diverges

30. converges absolutely   **32.** converges   **34.** converges

36. conditionally convergent   **38.** converges   **40.** converges

42. a) ∞, for all x b) for all x c) $\{\ \}$ **44.** a) 2, $(-1, 3)$

b) $(-1, 3)$ c) $\{\ \}$ **46.** a) 1, $[-1, 1)$ b) $(-1, 1)$ c) $\{-1\}$

48. a) 1, $(-1, 1)$ b) $(-1, 1)$ c) $\{\ \}$ **50.** a) $\dfrac{\pi}{2}$, $(0, \pi]$ b) $(0, \pi)$

c) $\{\pi\}$ **52.** b) too small   **54.** $\ln(1 + x)$, $\dfrac{2}{3}$, 0.510825623.

55. $\sin x$, π, 0   **56.** $\cos x$, $\dfrac{\pi}{3}$, $\dfrac{1}{2}$ **58.** $\tan^{-1}x$, $\dfrac{1}{\sqrt{3}}$, $\dfrac{\pi}{6}$

60. $-1 + (x - 2) - (x - 2)^2 + (x - 2)^3 - \cdots$ **64.** $\dfrac{3}{4}$

78. a) $x^2 - x^3 + x^4 - x^5 + \cdots$ b) diverges

80. $\sqrt{2} + \dfrac{1}{\sqrt{2}}(x - 1) + \dfrac{\sqrt{2}}{8}(x - 1)^2$ **82.** 4, $(-1, 7)$

84. a) $(0, 2a)$ b) $(2a, 0)$ **86.** Yes   **88.** 0.001302037

90. a) $lx^2 - \dfrac{x^3}{2} - \dfrac{x^5}{12} + \cdots$ b) $x + x^2 + \dfrac{2x^3}{3} + \dfrac{2x^4}{3} + \dfrac{13x^5}{15} +$

$\cdots$ **92.** a) Proof   b) Proof   c) $\dfrac{3\pi}{4}$ **94.** a) 2   b) 6   c) diverges

96. a)

$1 + t \overline{\sqrt{1}} \, \dfrac{1 - t + t^2 - t^3 + \cdots + (-1)^n t^n}{}$

$\dfrac{1 + t}{}$

$-t$

$\dfrac{-t - t^2}{t^2}$

$\dfrac{t^2 + t^3}{-t^3}$

$\cdot$

$\cdot$

$(-1)^{n+1}t^{n+1}$

98. $-\dfrac{1}{24}$ **100.** 0   **102.** $\dfrac{1}{120}$ **104.** $\dfrac{1}{3}$ **106.** -1 **108.** 3

110. 2   **112.** $\dfrac{1}{12}$ **114.** $r = -3$ and $s = \dfrac{9}{2}$

Chapter 9

Section 9.1, pp. 637–640

2. $y^2 = -4x$, $(-1, 0)$, $x = 1$ **4.** $x^2 = 2y$, $(0, \frac{1}{2})$, $y = -\frac{1}{2}$

6. $\frac{x^2}{4} + \frac{y^2}{9} = 1$, $(0, \pm\sqrt{5})$, $e = \frac{\sqrt{5}}{3}$, Directrices:

$y = \pm\frac{9}{\sqrt{5}}$ **8.** $\frac{y^2}{4} - x^2 = 1$, $(0, \pm\sqrt{5})$, $e = \frac{\sqrt{5}}{2}$,

Directrices: $y = \pm 4/\sqrt{5}$ Asymptotes: $y = \pm 2x$

10. $\frac{x^2}{16} + \frac{y^2}{7} = 1$, $e = \frac{3}{4}$ **12.** $\frac{x^2}{2} + \frac{y^2}{4} = 1$, $e = \frac{\sqrt{2}}{2}$

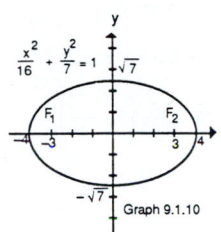

Graph 9.1.10

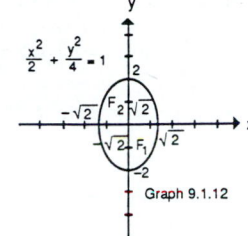

Graph 9.1.12

14. $\frac{x^2}{10} + \frac{y^2}{9} = 1$, $e = \frac{1}{\sqrt{10}}$ **16.** $\frac{x^2}{25} + \frac{y^2}{169} = 1$, $e = \frac{12}{13}$

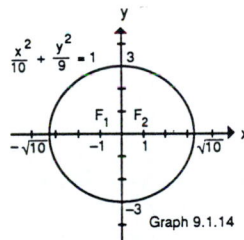

Graph 9.1.14

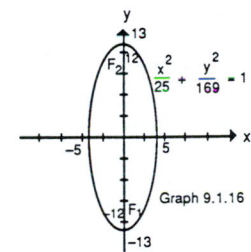

Graph 9.1.16

18. $\frac{x^2}{16} - \frac{y^2}{9} = 1$,

$e = \frac{5}{4}$, $y = \pm\frac{3}{4}x$

20. $\frac{y^2}{4} - \frac{x^2}{4} = 1$,

$e = \sqrt{2}$, $y = \pm x$

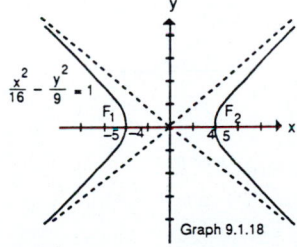

Graph 9.1.18

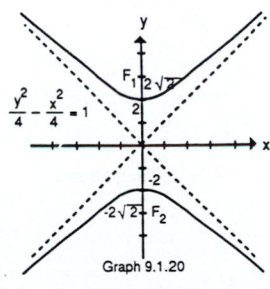

Graph 9.1.20

22. $\frac{y^2}{3} - x^2 = 1$, $e = \frac{2}{\sqrt{3}}$,

$y = \pm\sqrt{3}x$

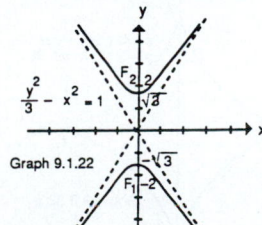

Graph 9.1.22

24. $\frac{x^2}{36} - \frac{y^2}{64} = 1$, $e = \frac{5}{3}$ $y = \pm\frac{4}{3}x$

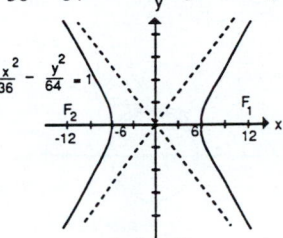

Graph 9.1.24

26. $\frac{x^2}{9} + \frac{y^2}{25} = 1$ **28.** $\frac{x^2}{1600} + \frac{y^2}{1536} = 1$ **30.** $\frac{x^2}{100} + \frac{y^2}{94.24} = 1$

32. $e = \frac{\sqrt{3}}{2}$, $\frac{x^2}{64/3} + \frac{y^2}{16/3} = 1$ **34.** $e = \frac{1}{\sqrt{2}}$, $\frac{x^2}{4} + \frac{y^2}{2} = 1$

36. $\frac{x^2}{3} - y^2 = 1$ **38.** $\frac{y^2}{4} - \frac{x^2}{16} = 1$ **40.** $\frac{x^2}{4} - \frac{y^2}{12} = 1$

42. $\frac{y^2}{16} - \frac{x^2}{9} = 1$ **44.** $e = \sqrt[4]{5}$, $\frac{x^2}{2\sqrt{5}} - \frac{y^2}{10 - 2\sqrt{5}} = 1$

46. $e = \sqrt{3}$, $\frac{x^2}{12} - \frac{y^2}{24} = 1$

48.

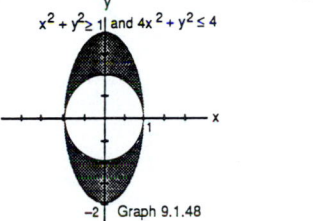

Graph 9.1.48

50.

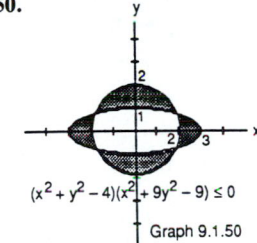

Graph 9.1.50

52.

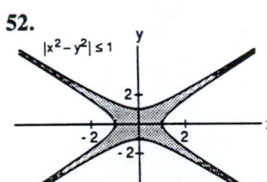

Graph 9.1.52

56. $(x + 2)^2 + (y - 1)^2 = 13$;

inside   **60.** $(0, \frac{16}{3\pi})$ **62.** 24π

64. $\frac{\pi}{\sqrt{2}}(2\sqrt{5} + \ln(2 + \sqrt{5}))$

Section 9.2, pp. 648–649

2. Parabola   **4.** Ellipse   **6.** Hyperbola   **8.** Ellipse (Circle)
10. Parabola   **12.** Hyperbola   **14.** Hyperbola   **16.** Parabola
18. $3x'^2 + y'^2 = 2$, Ellipse   **20.** $x'^2 + 5y'^2 = 2$, Ellipse
22. $5y'^2 - x'^2 = 10$, Hyperbola
24. $x'^2 - y'^2 - 2\sqrt{2}x' + 2 = 0$, Hyperbola
26. $5x'^2 - 3y'^2 = 7$, Hyperbola   **28.** $A' = 2.05$, $B' = 0$,
$C' = -3.05$, $D' = 2.99$, $E' = -0.30$, $F' = -7$, Hyperbola
30. $A' = 0.00$, $B' = 0$, $C' = 20.00$, $D' = 0$, $E' = 0$, $F' = -49$,
Parallel Lines   **32.** $A' = 0.55$, $B' = 0$, $C' = 10.45$, $D' = 18.48$,

$E' = 7.65$, $F' = -86$, Ellipse   **34.** a) $\frac{x'^2}{a^2} + \frac{y'^2}{b^2} = 1$

b) $\frac{x'^2}{a^2} - \frac{y'^2}{b^2} = 1$ c) $x'^2 + y'^2 = a^2$ d) $y' = mx'$

e) $y' = mx' - b$ **36.** $e = \sqrt{2}$ **40.** Any a, b so that

$a - 2b = 8$ **42.** $x' = \pm\frac{\sqrt{2}}{2}$, two vertical lines.

Section 9.3, pp. 656–657

2. $x^2 + y^2 = 1$

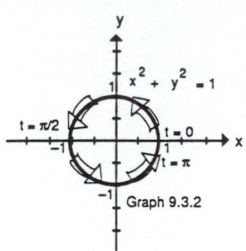

Graph 9.3.2

4. $x^2 + y^2 = 1$

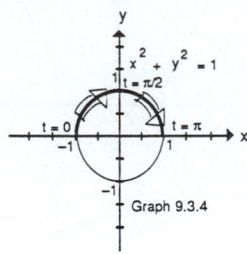

Graph 9.3.4

6. $\dfrac{x^2}{16} + \dfrac{y^2}{4} = 1$

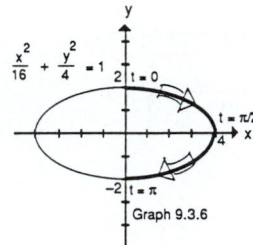

Graph 9.3.6

8. $\dfrac{x^2}{16} + \dfrac{y^2}{25} = 1$

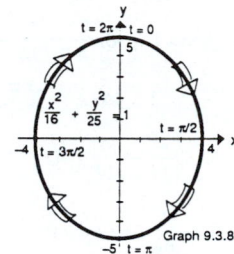

Graph 9.3.8

10. $x = -\sqrt{y}$

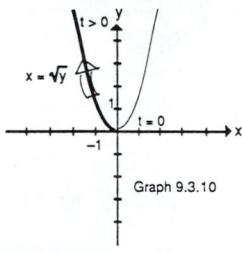

Graph 9.3.10

12. $x = y^2$

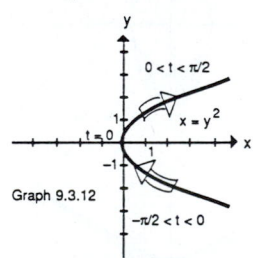

Graph 9.3.12

14. $x^2 - y^2 = 1$

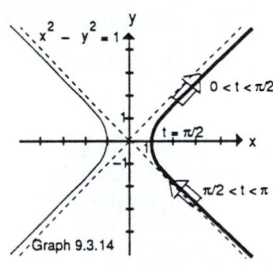

Graph 9.3.14

16. $y = -x + 2$

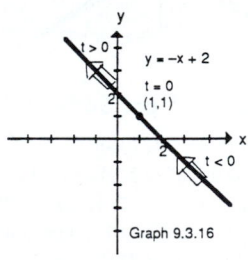

Graph 9.3.16

18. $y = 2 - \dfrac{2}{3}x$

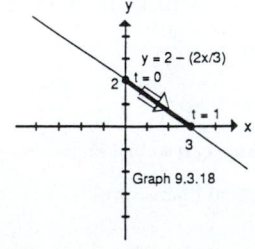

Graph 9.3.18

20. $y = \sqrt{4 - x^2}$

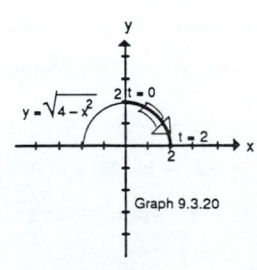

Graph 9.3.20

22. $x = \sqrt{y^2 + 1},\ y \geq 0$

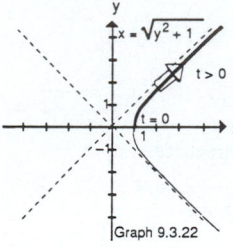

Graph 9.3.22

24. $y^2 - x^2 = 4$

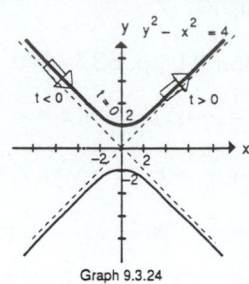

Graph 9.3.24

26. a) $x = a \sin t,\ y = b \cos t,\ \dfrac{\pi}{2} \leq t \leq \dfrac{5\pi}{2}$ b) $x = a \cos t,$
$y = b \sin t,\ 0 \leq t \leq 2\pi$ c) $x = a \sin t,\ y = b \cos t,$
$\dfrac{\pi}{2} \leq t \leq \dfrac{9\pi}{2}$ d) $x = a \cos t,\ y = b \sin t,\ 0 \leq t \leq 4\pi$

28. $x = 2 \cot t,\ y = 2 \sin^2 t,\ 0 < t < \pi$ **30.** $\left(1, \dfrac{\sqrt{3}}{2}\right)$

32. a)

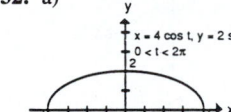

Graph 9.3.32a

b)

Graph 9.3.32b

c)

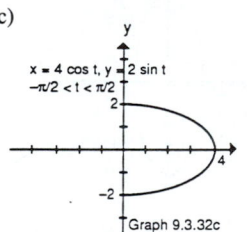

Graph 9.3.32c

34.

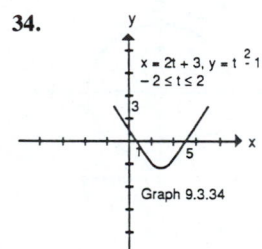

Graph 9.3.34

36. a)

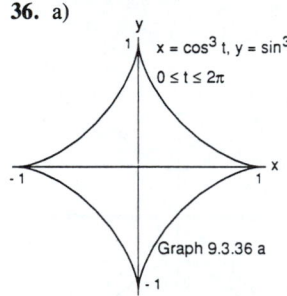

Graph 9.3.36 a

b)

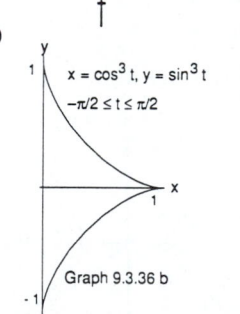

Graph 9.3.36 b

38. a)

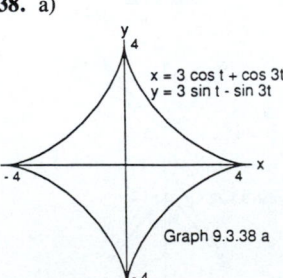

Graph 9.3.38 a

b)

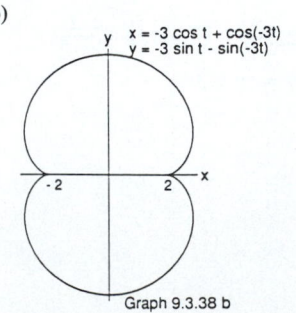

Graph 9.3.38 b

40. a)

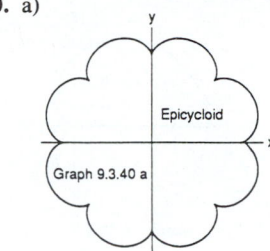

Graph 9.3.40 a

Epicycloid

b)

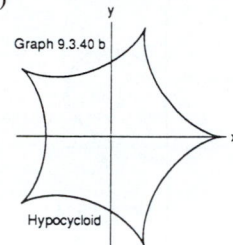

Graph 9.3.40 b

Hypocycloid

c)

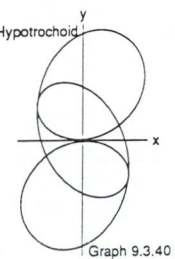

Hypotrochoid

Graph 9.3.40 c

Section 9.4, pp. 662–664

2. $y = \sqrt{3}\,x + 2, \dfrac{d^2y}{dx^2} = -8$ **4.** $y = \sqrt{3}\,x, \dfrac{d^2y}{dx^2} = 0$

6. $y = -\dfrac{1}{2}x - \dfrac{1}{2}, \dfrac{d^2y}{dx^2} = \dfrac{1}{4}$ **8.** $y = -2x - 1, \dfrac{d^2y}{dx^2} = -\dfrac{1}{3}$

10. $y = -x - 1, \dfrac{d^2y}{dx^2} = 1$ **12.** $y = 2, \dfrac{d^2y}{dx^2} = -1$ **14.** 7   **16.** $\dfrac{21}{2}$

18. $-\ln\dfrac{1}{2}$ or $\ln 2$ **20.** $\dfrac{28\pi}{9}$ **22.** π **24.** $\pi r \sqrt{h^2 + r^2}$

26. a) $(\bar{x}, \bar{y}) = \left(-\dfrac{e^{2\pi} + 2}{5(e^{\pi} - 1)}, \dfrac{e^{2\pi} + 1}{5(e^{\pi} - 1)}\right)$

b)

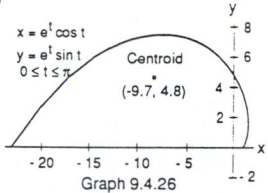

$x = e^t \cos t$
$y = e^t \sin t$
$0 \le t \le \pi$

Centroid
(-9.7, 4.8)

Graph 9.4.26

28. $(\bar{x}, \bar{y}) = (2.35, 2.49)$
30. a) 5.869848837 b) 0.00032
32. $\left(\dfrac{\sqrt{3}}{2}, 1\right) y = \dfrac{3}{2}x, y = -\dfrac{3}{2}x$

34.

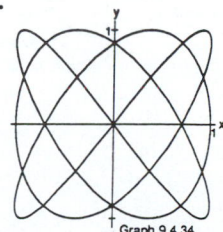

Graph 9.4.34

36.

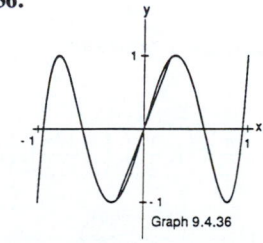

Graph 9.4.36

38.

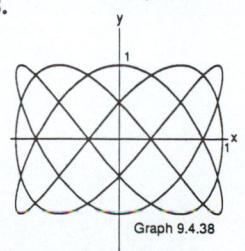

Graph 9.4.38

Section 9.5, pp. 669–670

2. a) $(3, 0)$ b) $(-3, 0)$ c) $(3, 0)$ d) $(-3, 0)$ e) $(-1, \sqrt{3})$
f) $(-1, -\sqrt{3})$ g) $(1, \sqrt{3})$ h) $(-1, -\sqrt{3})$ i) $(1, -\sqrt{3})$
j) $(1, \sqrt{3})$ k) $(-1, \sqrt{3})$ l) $(1, -\sqrt{3})$ **4.** a) $\left(3, \dfrac{\pi}{4} + 2n\pi\right)$ and

$\left(-3, \dfrac{5\pi}{4} + 2n\pi\right)$, n an integer b) $\left(-3, \dfrac{\pi}{4} + 2n\pi\right)$ and

$\left(3, \dfrac{5\pi}{4} + 2n\pi\right)$, n an integer c) $\left(3, -\dfrac{\pi}{4} + 2n\pi\right)$ and

$\left(-3, \dfrac{3\pi}{4} + 2n\pi\right)$, n an integer d) $\left(-3, -\dfrac{\pi}{4} + 2n\pi\right)$ and

$\left(3, \dfrac{3\pi}{4} + 2n\pi\right)$, n an integer

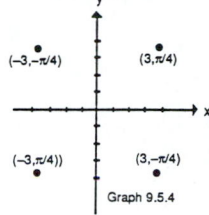

$(-3,-\pi/4)$ $(3,\pi/4)$

$(-3,\pi/4))$ $(3,-\pi/4)$

Graph 9.5.4

6. $(0, \theta)$ where θ is any angle.

8.

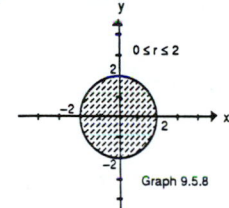

$0 \le r \le 2$

Graph 9.5.8

10.

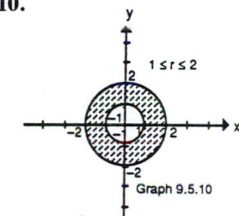

$1 \le r \le 2$

Graph 9.5.10

12.

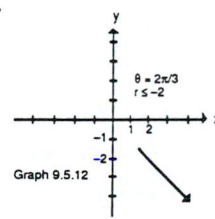

$\theta = 2\pi/3$
$r \le -2$

Graph 9.5.12

14.

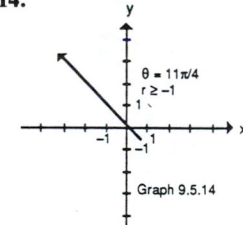

$\theta = 11\pi/4$
$r \ge -1$

Graph 9.5.14

16.

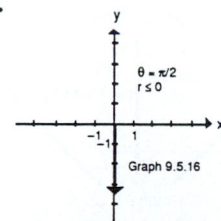

$\theta = \pi/2$
$r \le 0$

Graph 9.5.16

18.

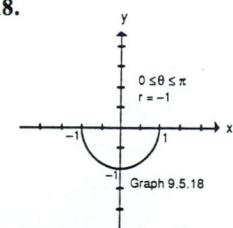

$0 \le \theta \le \pi$
$r = -1$

Graph 9.5.18

20.

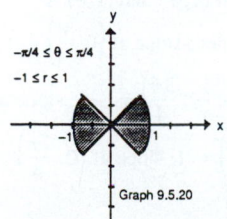

$-\pi/4 \le \theta \le \pi/4$
$-1 \le r \le 1$

Graph 9.5.20

22.

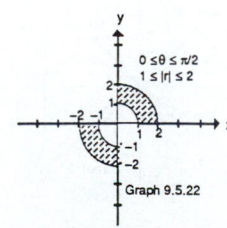

$0 \le \theta \le \pi/2$
$1 \le |r| \le 2$

Graph 9.5.22

24. $y = -1$, horizontal line through $(0, -1)$.   **26.** $x = 0$, the
y-axis.   **28.** $x = -3$, vertical line through $(-3, 0)$.   **30.** $y = x$,
line, $m = 1$, $b = 0$ **32.** $x^2 + y^2 = 4y$, circle $C(0, 2)$, $r = 2$
34. $xy = 1$, hyperbola, rotated $45°$ **36.** $x^2 = 4y$, parabola,
$V(0, 0)$, opens up   **38.** $y = \ln x$, natural logarithmic function
40. $x - y = 0$ or $x + y = 0$, two lines through origin
42. $(x - 1)^2 + \left(y + \frac{1}{2}\right)^2 = \frac{5}{4}$, circle, $C\left(1, -\frac{1}{2}\right)$, $r = \frac{\sqrt{5}}{2}$
44. $r \sin \theta = 1$ **46.** $r \cos \theta - r \sin \theta = 3$ **48.** $r^2 \cos 2\theta = 1$
50. $r^2 \sin 2\theta = 4$ **52.** $r^2 \cos 2\theta = 25r$

Section 9.6, pp. 676–677

2.
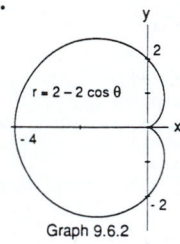
Graph 9.6.2

4.
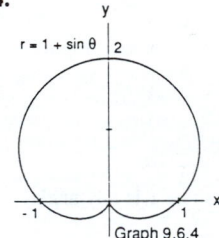
Graph 9.6.4

6.
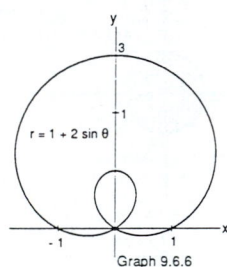
Graph 9.6.6

8.
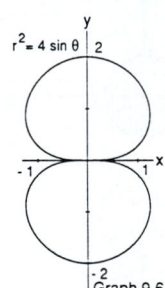
Graph 9.6.8

10.
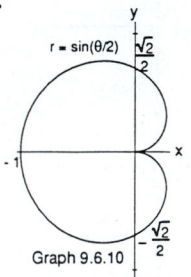
Graph 9.6.10

12. Slope at $(-1, 0) = -1$;
slope at $\left(0, \frac{\pi}{2}\right)$ is undefined;
slope at $(-1, \pi) = 1$

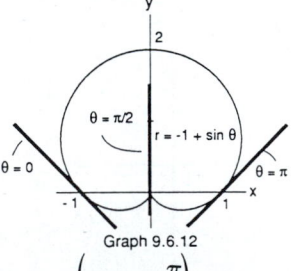
Graph 9.6.12

Slope at $\left(-1, \pm\frac{\pi}{2}\right) = 0$;
slope at $(1, \pi)$ and $(1, 0)$ is
undefined; slope at $\left(0, \frac{\pi}{4}\right) = 1$;
slope at $\left(0, \frac{3\pi}{4}\right) = -1$; slope at
$\left(0, \frac{5\pi}{4}\right) = 1$; slope at $\left(0, \frac{7\pi}{4}\right)$
$= -1$

14.
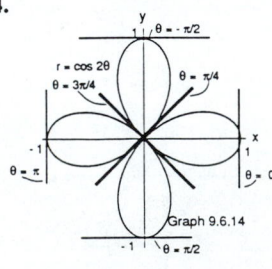
Graph 9.6.14

16.
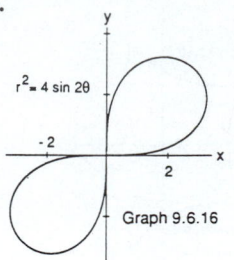
Graph 9.6.16

b)
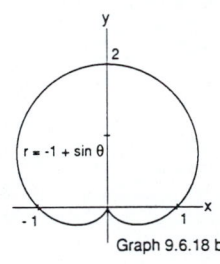
Graph 9.6.18 b

b)
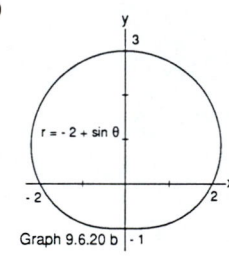
Graph 9.6.20 b

26. $(1, 0), (1, \pi), (0, 0)$

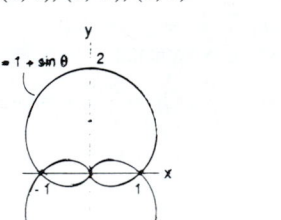
Graph 9.6.26

30. $\left(1, \frac{\pi}{12}\right), \left(1, \frac{5\pi}{12}\right), \left(1, \frac{13\pi}{12}\right), \left(1, \frac{17\pi}{12}\right)$

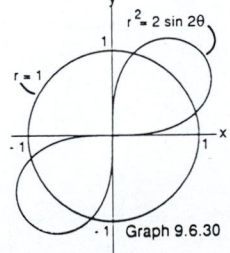
Graph 9.6.30

18. a)
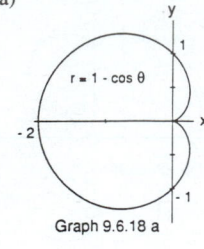
Graph 9.6.18 a

20. a)
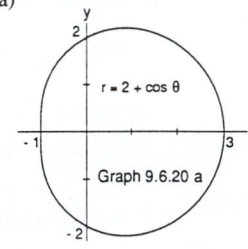
Graph 9.6.20 a

22.
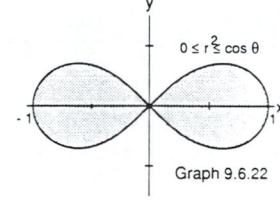
Graph 9.6.22

28. $\left(\pm 1, \frac{\pi}{4}\right), \left(\pm 1, \frac{3\pi}{4}\right), (0, 0)$

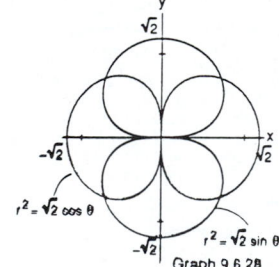
Graph 9.6.28

32. $\left(1, \frac{\pi}{8}\right), \left(1, \frac{9\pi}{8}\right)$, $(0, 0)$ **36.** $\left(1 - \frac{\sqrt{2}}{2}, \frac{3\pi}{2}\right)$, $(0, 0)$

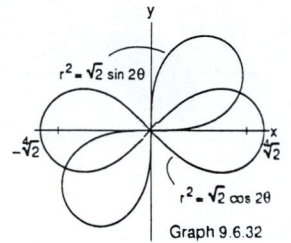

Graph 9.6.32

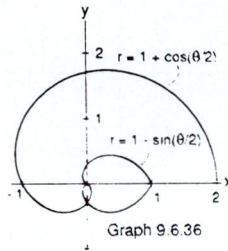

Graph 9.6.36

38. $\left(\pm 1, \frac{\pi}{12}\right), \left(\pm 1, \frac{5\pi}{12}\right)$ **40.**

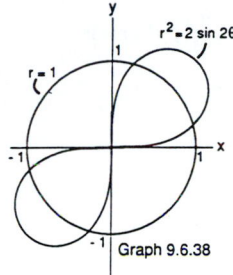

Graph 9.6.38

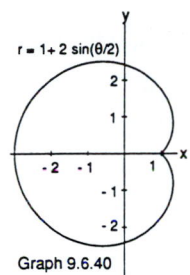

Graph 9.6.40

42. a)

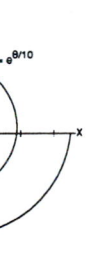

Graph 9.6.42 a

b)

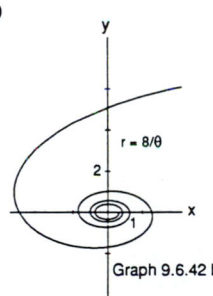

Graph 9.6.42 b

c)

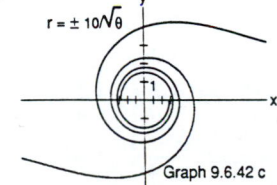

Graph 9.6.42 c

Section 9.7, pp. 682–683

2. $r \cos\left(\theta - \frac{3\pi}{4}\right) = 2$; $-\sqrt{2}\,x + \sqrt{2}\,y = 4$ **4.** $r \cos\left(\theta + \frac{\pi}{4}\right) = 4$; $\sqrt{2}\,x - \sqrt{2}\,y = 8$ **6.** $-x + \sqrt{3}\,y = 6$

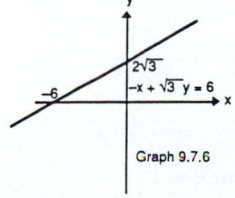

Graph 9.7.6

8. $x - \sqrt{3}\,y = 4$ **10.** $r = -2 \sin\theta$

12. $r = -\cos\theta$

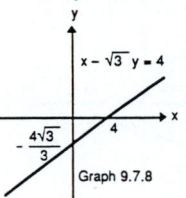

Graph 9.7.8

14.

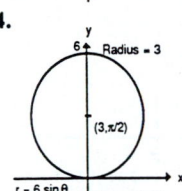

Graph 9.7.14

16.

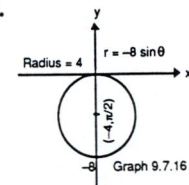

Graph 9.7.16

18. $r = \dfrac{2}{1 + \sin\theta}$ **20.** $r = \dfrac{30}{1 - 5 \sin\theta}$ **22.** $r = \dfrac{2}{4 - \cos\theta}$

24. $r = \dfrac{6}{3 + \sin\theta}$ **26.**

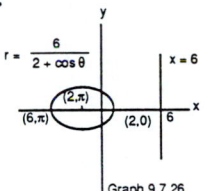

Graph 9.7.26

28.

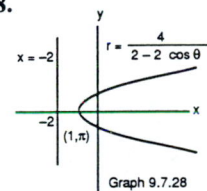

Graph 9.7.28

30.

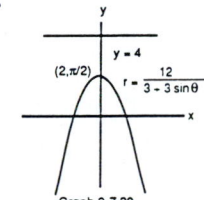

Graph 9.7.30

32.

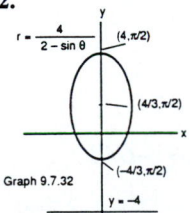

Graph 9.7.32

34.

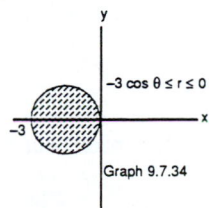

Graph 9.7.34

36. Mercury: $r = \dfrac{0.3707}{1 + 0.2056 \cos\theta}$

Venus: $r = \dfrac{0.7233}{1 + 0.0068 \cos\theta}$ Earth: $r = \dfrac{0.9997}{1 + 0.0617 \cos\theta}$

Mars: $r = \dfrac{1.511}{1 + 0.0934 \cos\theta}$ Jupiter: $r = \dfrac{5.191}{1 + 0.0484 \cos\theta}$

Saturn: $r = \dfrac{9.511}{1 + 0.0543 \cos\theta}$ Uranus: $r = \dfrac{19.14}{1 + 0.0460 \cos\theta}$

Neptune: $r = \dfrac{30.06}{1 + 0.0082 \cos\theta}$ **38.** $(x - 1)^2 + y^2 = 1$; $x = 1$

40. $r = \dfrac{2}{1 + \sin\theta}$ **42.** a) $r = \dfrac{2.14}{1 + 0.97 \cos\theta}$

b) 1.09 AU or 100 500 000 miles c) 71.3 AU or 6 600 000 000 miles

46.

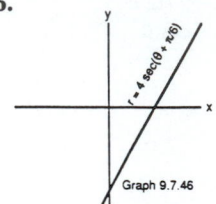

Graph 9.7.46

48.

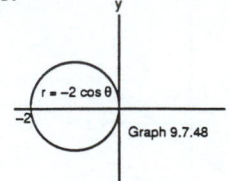

Graph 9.7.48

50.

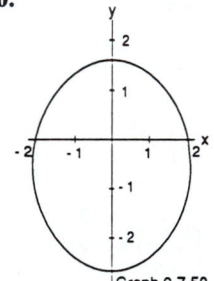

Graph 9.7.50

52.

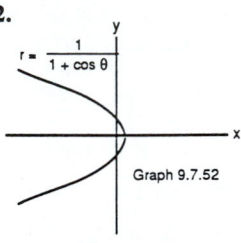

Graph 9.7.52

54.

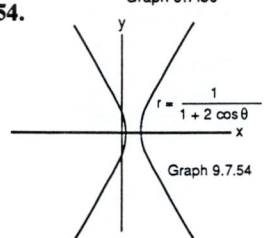

Graph 9.7.54

Section 9.8, pp. 689–690

2. $\frac{3}{2}\pi a^2$ **4.** $2a^2$ **6.** 4   **8.** $\frac{4\pi - 3\sqrt{3}}{6}$ **10.** $6\pi - 16$

12. πa^2 **14.** a) $2\pi + \frac{3\sqrt{3}}{2}$ b) $\pi + 3\sqrt{3}$ **16.** $\frac{3\sqrt{3}}{2}$ **18.** $\frac{5\pi}{4}$

20. $e^\pi - 1$ **22.** $2a$ **24.** $2\pi\sqrt{2}$ **26.** $2\pi a, a \geq 0$

28. $\pi\sqrt{5}(e^{\pi/2} + 1)$ **30.** $4a^2\pi^2$ **32.** a) a b) a c) $\frac{2a}{\pi}$

34. $\left(0, \frac{4a}{3\pi}\right)$

Section 9 Miscellaneous Exercises, pp. 691–698

2. (3, 3), $y = -3$ **6.** $\frac{25\left(x - \frac{28}{5}\right)^2}{144} + \frac{25y^2}{64} = 1$ **8.** $y^2 - \frac{x^2}{3} = 1$

14. The reflective property of an ellipse states that all the ripples must intersect at the other focus.   **16.** a) 120°   **18.** The ship is 259.8 miles closer to A than B.

24.

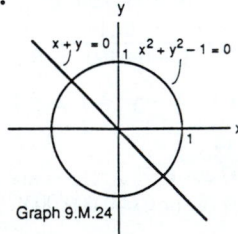

Graph 9.M.24

26.

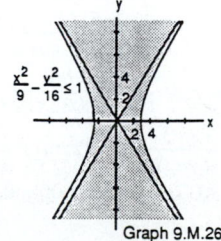

Graph 9.M.26

28.

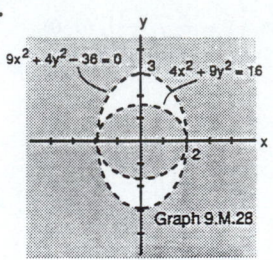

Graph 9.M.28

30.

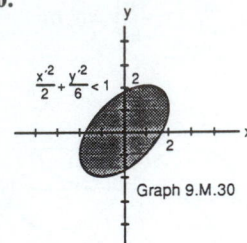

Graph 9.M.30

32. Hyperbola   **34.** Hyperbola   **36.** Hyperbola; $x'^2 - y'^2 = 2$

38. Parabola   **42.** Center: $\left(\frac{\sqrt{2}}{2}, 0\right)$; vertices: $\left(\frac{\sqrt{2}}{2} - 1, 1\right)$,

$\left(\frac{\sqrt{2}}{2} + 1, -1\right)$; foci: $\left(-\frac{\sqrt{2}}{2}, \sqrt{2}\right)$, $\left(\frac{3\sqrt{2}}{2}, -\sqrt{2}\right)$; asymptotes:

$x = \frac{1}{\sqrt{2}}, y = 0$; axes: x'-axis is $y = x - \frac{\sqrt{2}}{2}$, y'-axis is

$y = -x + \frac{\sqrt{2}}{2}$

44.

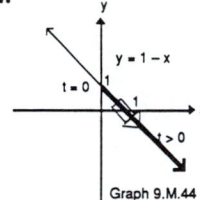

Graph 9.M.44

46.

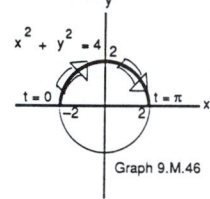

Graph 9.M.46

48.

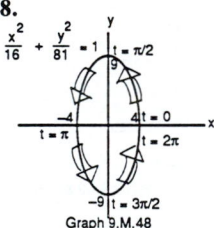

Graph 9.M.48

50. $x = -2\cos t, y = 2\sin t, 0 \leq t \leq 6\pi$

56. $y = -3x + \frac{13}{4}, \frac{d^2y}{dx^2} = 3$ **58.** $4\sqrt{3}$ **60.** $2\pi\left(2 - \frac{3\sqrt{2}}{4}\right)$

62. $5\pi^2 a^3$ **64.** $x = 2\cos t + 1, y = \sin t, \frac{(x - 1)^2}{4} + y^2 = 1$

68. e   **70.** f   **72.** h   **74.** j

76.

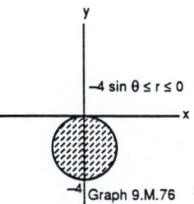

Graph 9.M.76

78. $\left(\frac{1}{2}, \pm\frac{\pi}{3}\right)$, (0, 0)   **80.** (0, π), $\left(1, \frac{\pi}{2}\right)$, $\left(1, \frac{3\pi}{2}\right)$

82. $\left(a\sqrt{2}, \frac{\pi}{4}\right)$ **84.** (0, 0)   **86.** $y = \sqrt{3}\,x, y = -\sqrt{3}\,x$

88. $r\sin\theta = r\cos\theta - 1, r\sin\theta = -r\cos\theta - 1$

90.

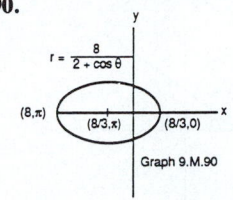

$r = \dfrac{8}{2 + \cos \theta}$

$(8,\pi)$ $(8/3,\pi)$ $(8/3,0)$

Graph 9.M.90

92.

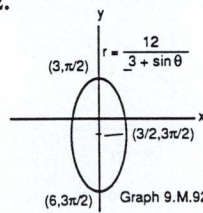

$r = \dfrac{12}{3 + \sin \theta}$

$(3,\pi/2)$ $(3/2,3\pi/2)$ $(6,3\pi/2)$

Graph 9.M.92

94. $r = \dfrac{4}{1 - \cos \theta}$ **96.** $r = \dfrac{6}{3 - \sin \theta}$ **98.** a) $\dfrac{7}{93}$

b) $\dfrac{430\,000}{93 + 7 \cos \theta}$ **100.** $\dfrac{\pi}{12}$ **102.** 5π **104.** 4π **106.** 3π

108. $3a\pi$ **110.** $\tan \dfrac{\theta}{4}$ **114.** $\dfrac{\pi}{2}$

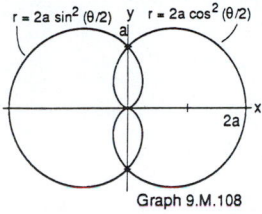

$r = 2a \sin^2 (\theta/2)$ $r = 2a \cos^2 (\theta/2)$

$2a$

Graph 9.M.108

116. a) $\left(\dfrac{3a}{2}, \pm \dfrac{\pi}{3} \right)$, $(0, \pi)$ b) $\left(\dfrac{a}{2}, \pm \dfrac{2\pi}{3} \right)$, $(2a, 0)$

118. $\dfrac{\pi}{4}$ **120.** -1

122. a) b) $xy = 1$ c) $\dfrac{\pi}{2}$

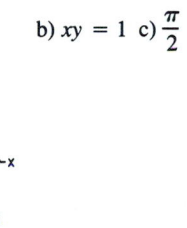

$r^2 = 2 \csc 2\theta$

Graph 9.M.122

Chapter 10

Section 10.1, pp. 707–708

2. a)

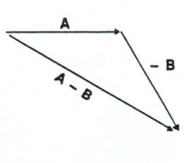

A $-B$ $A - B$

b)

A $-B$ $A - B + C$ C

c)

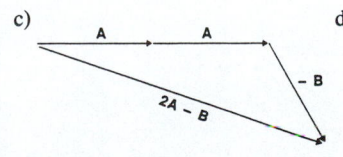

A   A $-B$ $2A - B$

d)

C   B   A $A + B + C = 0$

4. $\left(\sqrt{3} + 6 \right) \mathbf{i} - 3 \mathbf{j}$ **6.** $-\mathbf{i}$ **8.** $\left(-6 - 2\sqrt{2} \right) \mathbf{i} + 7 \mathbf{j}$
10. $-\mathbf{i} + \mathbf{j}$ **12.** $\vec{0}$

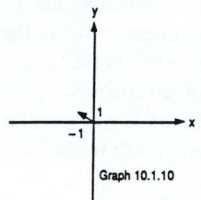

Graph 10.1.10

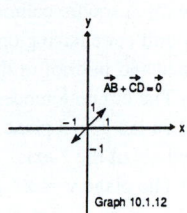

$\vec{AB} + \vec{CD} = \vec{0}$

Graph 10.1.12

14. $\mathbf{u} = \dfrac{1}{\sqrt{2}} \mathbf{i} - \dfrac{1}{\sqrt{2}} \mathbf{j}$, when $\theta = -\dfrac{\pi}{4}$

$\mathbf{u} = -\dfrac{1}{\sqrt{2}} \mathbf{i} - \dfrac{1}{\sqrt{2}} \mathbf{j}$, when $\theta = -\dfrac{3\pi}{4}$

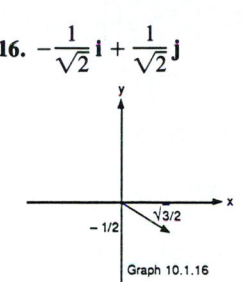

Graph 10.1.14

$\theta = -3\pi/4$ $\theta = -\pi/4$

16. $-\dfrac{1}{\sqrt{2}} \mathbf{i} + \dfrac{1}{\sqrt{2}} \mathbf{j}$

$\sqrt{3}/2$ $-1/2$

Graph 10.1.16

18. $\mathbf{u} = \dfrac{1}{\sqrt{2}} (\mathbf{i} - \mathbf{j})$, $-\mathbf{u} = \dfrac{1}{\sqrt{2}} (-\mathbf{i} + \mathbf{j})$

$\mathbf{n} = \dfrac{1}{\sqrt{2}} (\mathbf{i} + \mathbf{j})$, $-\mathbf{n} = \dfrac{1}{\sqrt{2}} (-\mathbf{i} - \mathbf{j})$

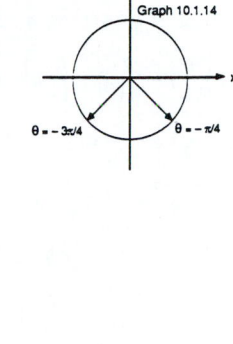

$-\mathbf{u}$ $\mathbf{n}$ $-\mathbf{n}$ $\mathbf{u}$

Graph 10.1.18

20. $\mathbf{u} = \dfrac{1}{\sqrt{2}} (\mathbf{i} + \mathbf{j})$, $-\mathbf{u} = \dfrac{1}{\sqrt{2}} (-\mathbf{i} - \mathbf{j})$, $\mathbf{v} = \dfrac{1}{\sqrt{2}} (-\mathbf{i} + \mathbf{j})$,

$-\mathbf{v} = \dfrac{1}{\sqrt{2}} (\mathbf{i} - \mathbf{j})$

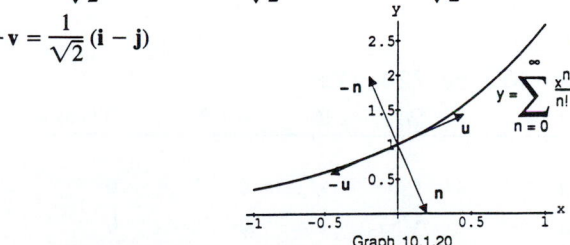

$-\mathbf{n}$ $-\mathbf{u}$ $\mathbf{u}$ $\mathbf{n}$ $y = \displaystyle\sum_{n=0}^{\infty} \dfrac{x^n}{n!}$

Graph 10.1.20

22. $\mathbf{u} = \dfrac{\pm 1}{\sqrt{34}} (5\mathbf{i} + 3\mathbf{j})$, $\mathbf{v} = \dfrac{\pm 1}{\sqrt{34}} (-3\mathbf{i} + 5\mathbf{j})$ **24.** $\mathbf{u} = \mathbf{v} = 0$

26. $\sqrt{13} \left[\dfrac{2}{\sqrt{13}} \mathbf{i} - \dfrac{3}{\sqrt{13}} \mathbf{j} \right]$ **28.** $\sqrt{13} \left[-\dfrac{2}{\sqrt{13}} \mathbf{i} + \dfrac{3}{\sqrt{13}} \mathbf{j} \right]$

30. $13 \left[-\dfrac{5}{13} \mathbf{i} - \dfrac{12}{13} \mathbf{j} \right]$ **34.** $-5 \left(\dfrac{3}{5} \mathbf{i} + \dfrac{4}{5} \mathbf{j} \right)$, one

Section 10.2, pp. 716–718

2. A line through the point $(-1,0,0)$ parallel to the y-axis   **4.** A line through the point $(1,0,0)$ parallel to the z-axis   **6.** The circle, $x^2 + y^2 = 4$ in the $z = -2$ plane   **8.** The circle, $y^2 + z^2 = 1$ in the yz-plane

10. The circle, $x^2 + z^2 = 9$ in the $y = -4$ plane   **12.** The circle, $x^2 + z^2 = 3$ in the xz-plane   **14.** a) A slab 1 unit thick containing the yz-plane b) A square column one unit wide containing the z-axis c) A unit cube having one vertex at the origin   **16.** a) The circumference and interior of the circle $x^2 + y^2 = 1$ in the xy-plane b) The circumference and interior of the circle $x^2 + y^2 = 1$ in the $z = 3$ plane c) A solid column having a radius of 1 centered about the z-axis   **18.** a) The line $y = x$ in the xy-plane b) The plane $y = x$ **20.** a) $x = 3$ b) $y = -1$ c) $z = 2$ **22.** a) $x^2 + y^2 = 4$, $z = 0$ b) $y^2 + z^2 = 4$, $x = 0$ c) $x^2 + z^2 = 4$, $y = 0$ **24.** a) $(x + 3)^2 + (y - 4)^2 = 1$, $z = 1$ b) $(y - 4)^2 + (z - 1)^2 = 1$, $x = -3$ c) $(x + 3)^2 + (z - 1)^2 = 1$, $y = 4$ **26.** $y = 1$ **28.** $z = 0$, $x^2 + y^2 = 3$ **30.** $0 \le x \le 2$, $0 \le y \le 2$, $0 \le z \le 2$ **32.** $z = \sqrt{1 - x^2 - y^2}$

34. $1 \le x^2 + y^2 + z^2 \le 4$ **36.** $17\left[\dfrac{3}{7}\mathbf{i} - \dfrac{6}{7}\mathbf{j} + \dfrac{2}{7}\mathbf{k}\right]$

38. $11\left[\dfrac{9}{11}\mathbf{i} - \dfrac{2}{11}\mathbf{j} + \dfrac{6}{11}\mathbf{k}\right]$ **40.** $6[\mathbf{i}]$ **42.** $1\left[\dfrac{3}{5}\mathbf{i} + \dfrac{4}{5}\mathbf{k}\right]$

44. $1\left[\dfrac{1}{\sqrt{2}}\mathbf{i} - \dfrac{1}{\sqrt{2}}\mathbf{k}\right]$ **46.** $1\left[\dfrac{1}{\sqrt{3}}\mathbf{i} + \dfrac{1}{\sqrt{3}}\mathbf{j} + \dfrac{1}{\sqrt{3}}\mathbf{k}\right]$

48. $5\sqrt{2}$, $\dfrac{3}{5\sqrt{2}}\mathbf{i} + \dfrac{4}{5\sqrt{2}}\mathbf{j} - \dfrac{1}{\sqrt{2}}\mathbf{k}$, $(1/2, 3, 5/2)$ **50.** $\sqrt{3}$,

$-\dfrac{1}{\sqrt{3}}\mathbf{i} - \dfrac{1}{\sqrt{3}}\mathbf{j} - \dfrac{1}{\sqrt{3}}\mathbf{k}$, $(5/2, 7/2, 9/2)$ **52.** $\sqrt{38}$,

$-\dfrac{5}{\sqrt{38}}\mathbf{i} - \dfrac{3}{\sqrt{38}}\mathbf{j} + \dfrac{2}{\sqrt{38}}\mathbf{k}$, $(5/2, 3/2, -1)$ **54.** $\dfrac{7}{13}(12\,\mathbf{i} - 5\,\mathbf{k})$

56. $\dfrac{1}{2}\left[\dfrac{1}{\sqrt{2}}\mathbf{i} - \dfrac{1}{\sqrt{2}}\mathbf{j}\right]$

58. a) $(x - 1)^2 + (y - 2)^2 + (z - 3)^2 = 14$ b) $x^2 + (y + 1)^2 + (z - 5)^2 = 4$ c) $(x + 2)^2 + y^2 + z^2 = 3$ d) $x^2 + (y + 7)^2 + z^2 = 49$ **60.** $(-1/4, -1/4, -1/4)$, $\dfrac{\sqrt{147}}{4}$

62. $(0, -1/3, 1/3)$, $\dfrac{\sqrt{29}}{3}$ **64.** a) z b) x c) y

18. $x - y = -3$

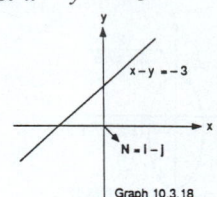

Graph 10.3.18

20. $2x - 3y = -8$

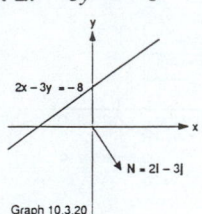

Graph 10.3.20

22. $\dfrac{6}{\sqrt{10}}$ **24.** $\sqrt{5}$ **26.** $\dfrac{1}{3}(\mathbf{i} - \mathbf{j} + \mathbf{k})$, $(\mathbf{j} + \mathbf{k})$, $\dfrac{2}{3}\mathbf{i} + \dfrac{1}{3}\mathbf{j} - \dfrac{1}{3}\mathbf{k}$

28. $\mathbf{V}_1$ and $\mathbf{V}_2$ must have the same length or both must be zero vectors   **32.** 73.2°   **34.** 54.7°   **36.** $6 \times 10^7\,\text{N} \cdot \text{m}$

38. 2640000 ft · lb   **40.** $\dfrac{\pi}{3}$ or $\dfrac{2\pi}{3}$ **42.** 45° or 135°

Section 10.4, p. 730

2. 5, $\mathbf{k}$, $-\mathbf{k}$ **4.** 0 and has no direction

6. $\sqrt{3}$, $\dfrac{1}{\sqrt{3}}(\mathbf{i} - \mathbf{j} + \mathbf{k})$, $\dfrac{-1}{\sqrt{3}}(\mathbf{i} - \mathbf{j} + \mathbf{k})$

8. $2\sqrt{3}$, $-\dfrac{1}{\sqrt{3}}\mathbf{i} - \dfrac{1}{\sqrt{3}}\mathbf{j} + \dfrac{1}{\sqrt{3}}\mathbf{k}$, $\dfrac{1}{\sqrt{3}}\mathbf{i} + \dfrac{1}{\sqrt{3}}\mathbf{j} - \dfrac{1}{\sqrt{3}}\mathbf{k}$

10.

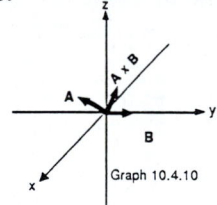

Graph 10.4.10

12.

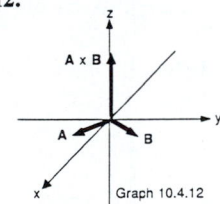

Graph 10.4.12

14.

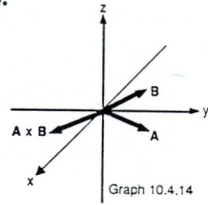

Graph 10.4.14

Section 10.3, pp. 724–725

	$\mathbf{A} \cdot \mathbf{B}$	$|\mathbf{A}|$	$|\mathbf{B}|$	$\cos \theta$	$|\mathbf{B}| \cos \theta$	$\text{Proj}_{\mathbf{A}}\mathbf{B}$
2.	0	1	$\sqrt{34}$	0	0	0
4.	0	$\sqrt{53}$	1	0	0	0
6.	0	1	$\sqrt{\dfrac{3}{2}}$	0	0	0
8.	2	$\sqrt{2}$	$\sqrt{3}$	$\sqrt{\dfrac{2}{3}}$	$\sqrt{2}$	$\mathbf{i} + \mathbf{k}$
10.	$-10 + \sqrt{17}$	$\sqrt{26}$	11	$\dfrac{\sqrt{17} - 10}{11\sqrt{26}}$	$\dfrac{\sqrt{17} - 10}{\sqrt{26}}$	$\dfrac{\sqrt{17} - 10}{26}[-5\mathbf{i} + \mathbf{j}]$
12.	13	15	3	$\dfrac{13}{45}$	$\dfrac{13}{15}$	$\dfrac{13}{225}[2\,\mathbf{i} + 10\,\mathbf{j} - 11\,\mathbf{k}]$

14. $\left[\dfrac{1}{2}\mathbf{i} + \dfrac{1}{2}\mathbf{j}\right] + \left[-\dfrac{1}{2}\mathbf{i} + \dfrac{1}{2}\mathbf{j} + \mathbf{k}\right]$ **16.** Yes

16. a) $\pm(4\mathbf{i} + 4\mathbf{j} - 2\mathbf{k})$ b) 3 c) $\pm\left[\frac{2}{3}\mathbf{i} + \frac{2}{3}\mathbf{j} - \frac{1}{3}\mathbf{k}\right]$

18. a) $\pm(2\mathbf{i} + 3\mathbf{j} + \mathbf{k})$ b) $\frac{\sqrt{14}}{2}$

c) $\pm\left[\frac{2}{\sqrt{14}}\mathbf{i} + \frac{3}{\sqrt{14}}\mathbf{j} + \frac{1}{\sqrt{14}}\mathbf{k}\right]$ **20.** a) $\mathbf{A}\perp\mathbf{B}, \mathbf{A}\perp\mathbf{C}, \mathbf{B}\perp\mathbf{C},$

$\mathbf{B}\perp\mathbf{D}$ and $\mathbf{C}\perp\mathbf{D}$ b) none **22.** $10\sqrt{2}$ ft · lb **24.** $\mathbf{A} \times \mathbf{B}$ is
perpendicular to the plane containing A and B
d) $(\pm)(\mathbf{A} \times \mathbf{B}) \times \mathbf{C}$ e) $(\pm)(\mathbf{B} \times \mathbf{C}) \times \mathbf{A}$ **26.** $\mathbf{i} \times \mathbf{j} = \mathbf{k}$ and
$\mathbf{i} \times (\mathbf{i} + \mathbf{j}) = \mathbf{k}$

Section 10.5, pp. 736–738

2. $x = 1 - 2t, y = 2 - 2t, z = -1 + 2t$
4. $x = 1, y = 2 - t, z = -t$ **6.** $x = 3 + 2t, y = -2 - t,$
$z = 1 + 3t$ **8.** $x = 2 + 3t, y = 4 + 7t, z = 5 - 5t$
10. $x = 2 - 2t, y = 3 + 4t, z = -2t$ **12.** $x = 0, y = 0, z = t$
14. $x = t, y = 0, z = 0,$ **16.** $x = 1, y = 1, z = t,$
where $0 \le t \le 1$ where $0 \le t \le 1$

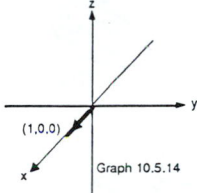

Graph 10.5.14

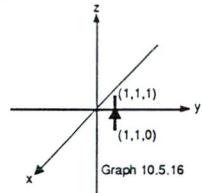

Graph 10.5.16

18. $x = 3 - 3t, y = 2t,$ **20.** $x = 1 - t,$
$z = 0, 0 \le t \le 1$ $y = -1 + 3t, z = -2 + 3t,$
where $0 \le t \le 1$

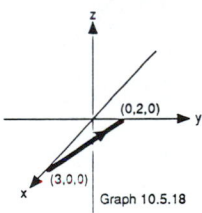

Graph 10.5.18

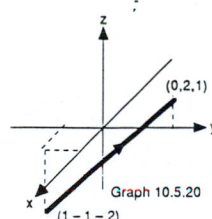

Graph 10.5.20

22. $3x + y + z = 5$ **24.** $x + 3y - z = 9$
26. $x - 2y + z = 6$ **28.** $-\sqrt{2}x + 2\sqrt{2}y - \sqrt{2}z = -7\sqrt{2}$
30. 3 **32.** $\sqrt{\frac{14}{3}}$ **34.** $7\sqrt{3}$ **36.** $\frac{6}{7}$ **38.** $\frac{8}{3}$ **40.** $\frac{3\sqrt{2}}{2}$
42. $(2, -20/7, 27/7)$ **44.** $(-4, -2, -5)$ **46.** $\frac{\pi}{2}$ **48.** $54.7°$
50. $42°$ **52.** $x = 1 + 14t, y = 2t, z = 15t$ **54.** $x = 1 + 10t,$
$y = -3 + 25t, z = 1 + 20t$ **56.** $x = \sqrt{3}t, y = t, z = 3.$
58. $\frac{2}{3}$ **60.** $|\overrightarrow{OA}| = 1.41421356; |\overrightarrow{OB}| = 1.41421356;$
$|\overrightarrow{AB}| = 1.41421356$; the midpoint of AB is (.5, .5, 1); $\mathbf{A} \cdot \mathbf{B} = 1$;
the angle between $\overrightarrow{OA}$ and $\overrightarrow{OB}$ is 1.04719773 radians $\approx 60°$;
$\mathbf{A} \times \mathbf{B} = -1\mathbf{i} - 1\mathbf{j} + 1\mathbf{k}$; the line through A and B: $x = 1 - t,$
$y = t, z = 1$; the line through C parallel to AB: $x = 1 - t,$
$y = 1 + t, z = 0$; the distance from C to AB is 1.22474487; the
equation of the ABC plane is $x + y + z = 2$; the distance from S
to the ABC plane is 2.30940108 **62.** $|\overrightarrow{OA}| = 3.74165739;$

$|\overrightarrow{OB}| = 4.5825757; |\overrightarrow{AB}| = 4.5825757$; the midpoint of AB is (1.5,
1, 3); $\mathbf{A} \cdot \mathbf{B} = 7$; the angle between $\overrightarrow{OA}$ and $\overrightarrow{OB}$ is 1.23095959
radians $\approx 70.5287815°$; $\mathbf{A} \times \mathbf{B} = 14\mathbf{i} - 7\mathbf{k}$; the line through A
and B: $x = 1 + t, y = 3 - 4t, z = 2 + 2t$; the line through C
parallel to AB: $x = 3 + t, y = -4t, z = -1 + 2t$; the distance
from C to AB is 4.35343324; the equation of the ABC plane is
$-18x - 7y - 5z = -49$; the distance from S to the ABC plane is
2.45614806

Section 10.6, p. 746

2. 4, $(\mathbf{A} \times \mathbf{B}) \times \mathbf{C} = -10\mathbf{i} - 2\mathbf{j} + 6\mathbf{k},$
$\mathbf{A} \times (\mathbf{B} \times \mathbf{C}) = -9\mathbf{i} - 2\mathbf{j} + 7\mathbf{k}$
4. 8, $(\mathbf{A} \times \mathbf{B}) \times \mathbf{C} = -10\mathbf{i} - 10\mathbf{k},$
$\mathbf{A} \times (\mathbf{B} \times \mathbf{C}) = -12\mathbf{i} - 4\mathbf{j} - 8\mathbf{k}$ **6.** All parts are true.

Section 10.7, pp. 756–758

2.

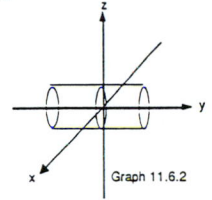

Graph 11.6.2

4.

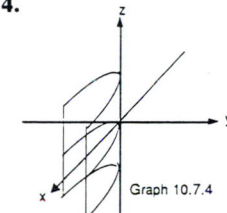

Graph 10.7.4

6.

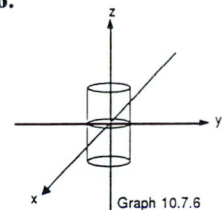

Graph 10.7.6

8.

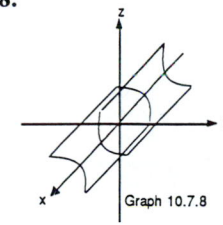

Graph 10.7.8

10.

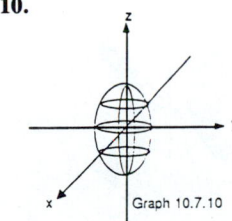

Graph 10.7.10

12.

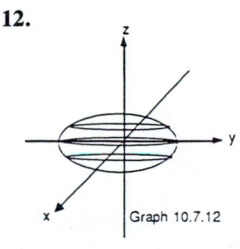

Graph 10.7.12

14.

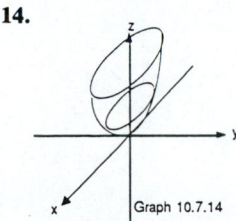

Graph 10.7.14

16.

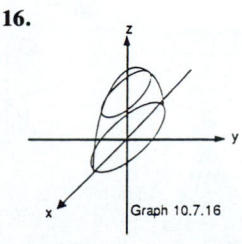

Graph 10.7.16

18.

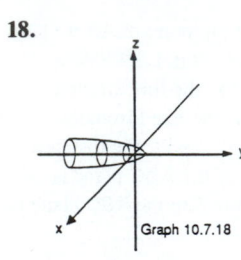

Graph 10.7.18

20.

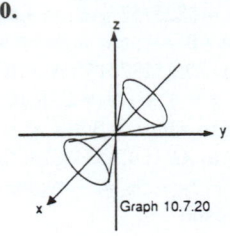

Graph 10.7.20

22.

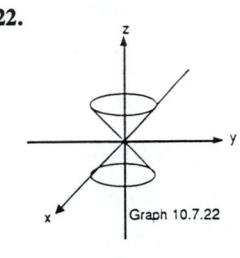

Graph 10.7.22

24.

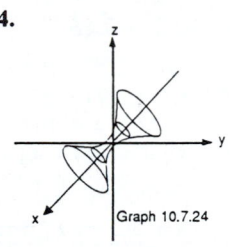

Graph 10.7.24

26.

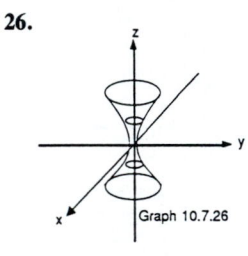

Graph 10.7.26

28.

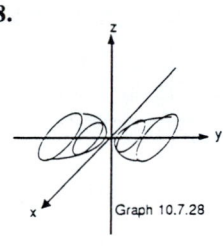

Graph 10.7.28

30.

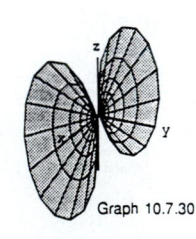

Graph 10.7.30

32.

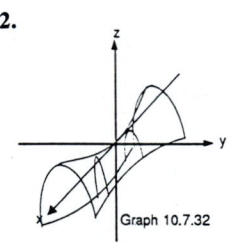

Graph 10.7.32

34.

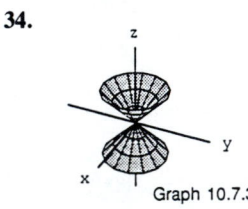

Graph 10.7.34

36.

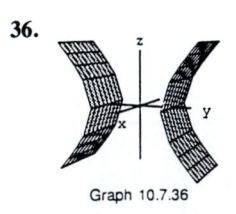

Graph 10.7.36

38.   **40.**   **42.**

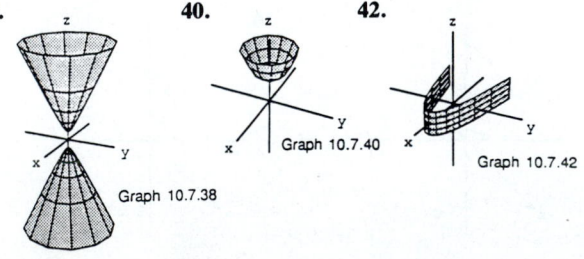

Graph 10.7.38   Graph 10.7.40   Graph 10.7.42

44.   **46.**   **48.**

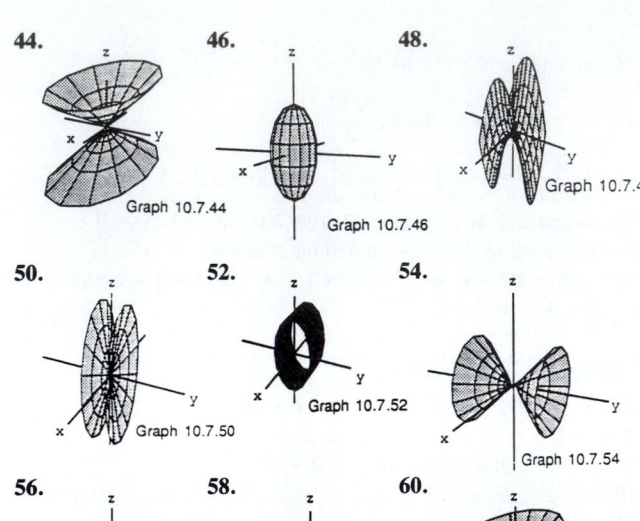

Graph 10.7.44   Graph 10.7.46   Graph 10.7.48

50.   **52.**   **54.**

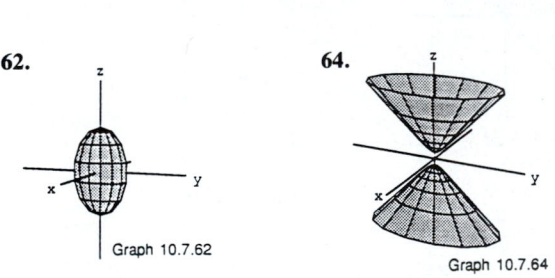

Graph 10.7.50   Graph 10.7.52   Graph 10.7.54

56.   **58.**   **60.**

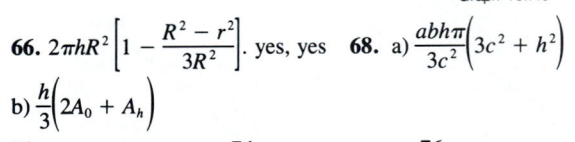

Graph 10.7.56   Graph 10.7.58   Graph 10.7.60

62.   **64.**

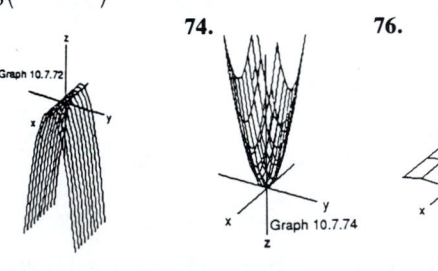

Graph 10.7.62   Graph 10.7.64

66. $2\pi hR^2\left[1 - \dfrac{R^2 - r^2}{3R^2}\right]$. yes, yes   **68.** a) $\dfrac{abh\pi}{3c^2}\left(3c^2 + h^2\right)$

b) $\dfrac{h}{3}\left(2A_0 + A_h\right)$

72.   **74.**   **76.**

Graph 10.7.72   Graph 10.7.74   Graph 10.7.76

Section 10.8, p. 761

	Rectangular	Cylindrical	Spherical
2.	$(1, 0, 0)$	$(1, 0, 0)$	$(1, \pi/2, 0)$
4.	$(0, 0, 1)$	$0, *, 1$	$(1, 0, *)$
6.	$(\sqrt{2}, 0, 1)$	$(\sqrt{2}, 0, 1)$	$(\sqrt{3}, \cos^{-1}(1/\sqrt{3}), 0)$
8.	$(0, -3/2, \sqrt{3}/2)$	$(3/2, 3\pi/2, \sqrt{3}/2)$	$(\sqrt{3}, \pi/3, -\pi/2)$
10.	$(0, 0, -\sqrt{2})$	$(0, *, \sqrt{2})$	$(\sqrt{2}, \pi, 3\pi/2)$

*any angle may be used

12. $x^2 + y^2 = 5 \Rightarrow$ cylindrical, $r = \sqrt{5}$; spherical, $\rho \sin \phi = \sqrt{5}$; a cylinder   **14.** $z = -2 \Rightarrow$ cylindrical, $z = -2$; spherical, $\rho \cos \phi = -2$; the plane $z = -2$ **16.** $\sqrt{x^2 + y^2} = z \Rightarrow$ cylindrical, $z = r$; spherical, $\phi = \dfrac{\pi}{4}$; a cone   **18.** $\tan^2 \phi = 1 \Rightarrow$ rectangular $x^2 + y^2 = z^2$; cylindrical, $r^2 = z^2$; a circular cone symmetric to the z-axis   **20.** $\rho = 6 \cos \phi \Rightarrow$ rectangular, $x^2 + y^2 + (z - 3)^2 = 9$; cylindrical, $r^2 + z^2 = 6z$; a sphere   **22.** $z^2 - r^2 = 1 \Rightarrow$ rectangular, $z^2 - (x^2 + y^2) = 1$; spherical, $\rho^2\cos^2\phi = \rho^2\sin^2\phi = 1 \Rightarrow \rho^2 = \dfrac{1}{\cos 2\phi}$; hyperboloid of 2 sheets   **24.** $z = x^2 + y^2 \Rightarrow$ cylindrical, $z = r^2$; spherical, $\rho \cos \phi = \rho^2[\sin^2\phi \cos^2\theta + \sin^2\phi \sin^2\theta] \Rightarrow$

$\rho[\cos \phi - \rho \sin^2\phi] = 0 \Rightarrow \rho = \dfrac{\cos \phi}{\sin^2\phi} = \cot \phi \csc \phi$;

paraboloid
26. $\rho^2 \cos 2\phi = -1 \Rightarrow$ rectangular, $\rho^2(\sin^2\phi - \cos^2\phi) = 1 \Rightarrow \rho^2 \sin^2\phi(\sin^2\theta + \cos^2\theta) = \rho^2 \cos^2\phi + 1 \Rightarrow x^2 + y^2 = z^2 + 1$; cylindrical $r^2 = z^2 + 1$, a hyperboloid of 1 sheet
28. A sphere with center $(0, 0, 0)$ and radius 1

30. A cylinder whose generating curve is the circle $(x - 1)^2 + y^2 = 1$ in the xy-plane

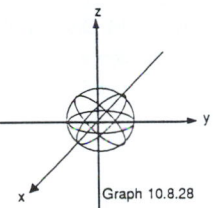

Graph 10.8.28

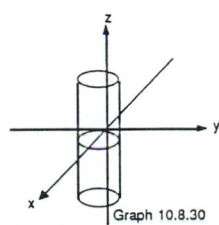

Graph 10.8.30

32. The intersection of the plane $\theta = \dfrac{\pi}{6}$ and the upper nappe of the cone $z = \sqrt{x^2 + y^2}$

34. A cylinder whose generating curve is a lemniscate in the $r\theta$-plane; the generating curve a lemniscate similar to exercise 15 in section 10.6.

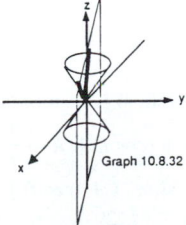

Graph 10.8.32

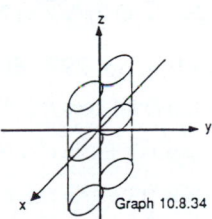

Graph 10.8.34

36. Symmetry to the z-axis.
38. The intersection of the sphere $\rho = 6$ and cone $\phi = \dfrac{\pi}{6}$

40. The intersection of the cone $\phi = \dfrac{\pi}{4}$ and the plane $\theta = \dfrac{\pi}{4}$

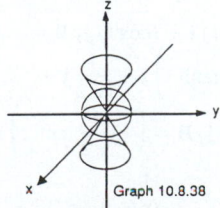

Graph 10.8.38

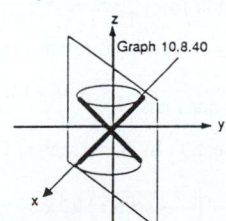

Graph 10.8.40

42. A "3D cardioid" symmetric to the z-axis

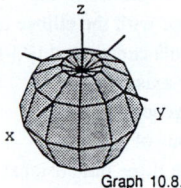

Graph 10.8.42

44. The intersection of the sphere, $x^2 + y^2 + (z - 2)^2 = 4$, and the yz-plane when $y \geq 0$

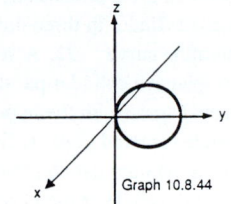

Graph 10.8.44

Section 10 Miscellaneous Exercises, pp. 762–766

2. a) $\dfrac{1}{\sqrt{2}}\mathbf{i} - \dfrac{1}{\sqrt{2}}\mathbf{j}$ b) $-\dfrac{\sqrt{3}}{2}\mathbf{i} - \dfrac{1}{2}\mathbf{j}$ **4.** $\pm\dfrac{1}{5}(4\mathbf{i} - 3\mathbf{j})$,

$\pm\dfrac{1}{5}(3\mathbf{i} + 4\mathbf{j})$ **6.** $\sqrt{2}\left[-\dfrac{1}{\sqrt{2}}\mathbf{i} - \dfrac{1}{\sqrt{2}}\mathbf{j}\right]$

8. $\sqrt{6}\left[\dfrac{1}{\sqrt{6}}\mathbf{i} + \dfrac{2}{\sqrt{6}}\mathbf{j} - \dfrac{1}{\sqrt{6}}\mathbf{k}\right]$

14. $|\mathbf{A}| = \sqrt{27}$, $|\mathbf{B}| = \sqrt{14}$, $\mathbf{A} \cdot \mathbf{B} = 6$,

$\mathbf{B} \cdot \mathbf{A} = 6$, $\mathbf{A} \times \mathbf{B} = \begin{vmatrix} \mathbf{i} & \mathbf{j} & \mathbf{k} \\ 5 & 1 & 1 \\ 1 & -2 & 3 \end{vmatrix} = 5\mathbf{i} - 14\mathbf{j} - 11\mathbf{k}$,

$\mathbf{B} \times \mathbf{A} = \begin{vmatrix} \mathbf{i} & \mathbf{j} & \mathbf{k} \\ 1 & -2 & 3 \\ 5 & 1 & 1 \end{vmatrix} = -5\mathbf{i} + 14\mathbf{j} + 11\mathbf{k}$,

$|\mathbf{A} \times \mathbf{B}| = \sqrt{25 + 196 + 121} = \sqrt{342}$,

$\theta = \cos^{-1}\left(\dfrac{\mathbf{A} \cdot \mathbf{B}}{|\mathbf{A}| |\mathbf{B}|}\right) = \cos^{-1}\left(\dfrac{6}{\sqrt{27} \sqrt{14}}\right) = 72.02°$,

$\text{comp}_\mathbf{A} \mathbf{B} = \dfrac{\mathbf{A} \cdot \mathbf{B}}{|\mathbf{A}|} = \dfrac{6}{\sqrt{27}} = \dfrac{2}{\sqrt{3}}$, $\text{proj}_\mathbf{A} \mathbf{B} = \dfrac{\mathbf{A} \cdot \mathbf{B}}{\mathbf{A} \cdot \mathbf{A}}\mathbf{A} =$

$\dfrac{2}{\sqrt{3}}[5\mathbf{i} + \mathbf{j} + \mathbf{k}]$ **16.** $-\dfrac{1}{5}[\mathbf{i} - 2\mathbf{j}] + \left[\dfrac{6}{5}\mathbf{i} + \dfrac{3}{5}\mathbf{j} + \mathbf{k}\right]$

18.

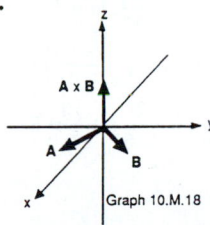

Graph 10.M.18

20. 20 lb   **22.** $\mathbf{C} = \text{proj}_\mathbf{B} \mathbf{A} = \left(\dfrac{\mathbf{A} \cdot \mathbf{B}}{\mathbf{B} \cdot \mathbf{B}}\right)\mathbf{B}$ and

$\mathbf{D} = \mathbf{A} - \mathbf{C} = \mathbf{A} - \left(\dfrac{\mathbf{A} \cdot \mathbf{B}}{\mathbf{B} \cdot \mathbf{B}}\right)\mathbf{B}$ **28.** 2   **29.** $\dfrac{\sqrt{78}}{3}$

30. $\dfrac{\sqrt{78}}{3}$ **32.** $x = 1$, $y = 2 + t$, $z = -t$, $0 \leq t \leq 1$

34. $\sqrt{14}$ **36.** $x - 2y + 3z = -13$ **38.** $85.9°$ **40.** 3

42. $-3\mathbf{j} + 3\mathbf{k}$ **44.** $\dfrac{2}{\sqrt{35}}(5\mathbf{i} - \mathbf{j} - 3\mathbf{k})$.   **46.** $x = -5 + 5t$,

$y = 3 - t$, $z = -3t$ **48.** $\left(\dfrac{11}{9}, \dfrac{26}{9}, -\dfrac{7}{9}\right)$ **50.** $(1, -2, -1)$, $x = 1 - 5t$, $y = -2 + 3t$, $z = -1 + 4t$

52. $2x + 7y + 2z + 10 = 0$.   **54.** 1.37 hours   **56.** $\dfrac{12}{\sqrt{62}}$.

58. a) No b) No c) No d) No e) Yes   **68.** A line in the xy-plane and a plane in three dimensional space   **70.** An ellipse in the xy-plane; a cylinder in three dimensional space with the ellipse as the generating curve   **72.** A vertical hyperbola centered at $(0,0)$ in the xy-plane; a cylinder parallel with the z-axis in three dimensional space with the hyperbola as the generating curve   **74.** A circle centered at $(0, 1/2)$ having a radius of $1/2$ in the xy-plane; a cylinder parallel with the z-axis in three dimensional space with the circle as the generating curve   **76.** A rose in the $r\theta$-plane; a cylinder in three dimensional space with the rose as the generating curve   **78.** A plane which intersects the xy-plane at a right angle on the line $y = x$ **80.** The circle $x^2 + y^2 = 1$ in the xy-plane   **82.** Two concentric spheres centered at the origin having radii of 1 and 2; all points contained on and between the spheres   **84.** $y^2 = x^2 + z^2 \Rightarrow$ spherical, $\rho^2 \sin^2 \phi \sin^2\theta = \rho^2 \sin^2\phi \cos^2\theta + \rho^2 \cos^2\phi \Rightarrow \sin^2\phi(\sin^2\theta - \cos^2\theta) = \cos^2\phi \Rightarrow \cos2\theta + \cot^2\phi = 0$; cylindrical, $r^2 \sin\theta = r^2 \cos^2\theta + z^2 \Rightarrow z^2 = r^2(\sin^2\theta - \cos^2\theta) \Rightarrow z^2 = -r^2 \cos2\theta$; a cone   **86.** $r = \cos\theta \Rightarrow$ rectangular, $\left(x - \dfrac{1}{2}\right)^2 + y^2 = \dfrac{1}{4}$; spherical, $\rho \sin\phi = \cos\theta$; a cylinder whose generating curve is the circle   **88.** $\rho \tan\phi \sin\phi = -1 \Rightarrow$ rectangular, $\rho \sin^2\phi = -\cos\phi \Rightarrow \dfrac{r^2}{\rho} = -\dfrac{z}{\rho} \Rightarrow -(x^2 + y^2) = z$; cylindrical, $z = -r^2$, paraboloid   **90.** 29   **92.** 43   **94.** 2   **96.** 42

Chapter 11

Section 11.1, pp. 777–779

2. $\mathbf{v} = \mathbf{i} + 2t\,\mathbf{j} + 2\,\mathbf{k}$, $\mathbf{a} = 2\,\mathbf{j}$, Speed $= 3$, Direction $= \dfrac{1}{3}\mathbf{i} + \dfrac{2}{3}\mathbf{j} + \dfrac{2}{3}\mathbf{k}$, $\mathbf{v}(1) = 3\left[\dfrac{1}{3}\mathbf{i} + \dfrac{2}{3}\mathbf{j} + \dfrac{2}{3}\mathbf{k}\right]$ **4.** $\mathbf{v} = e^t\mathbf{i} + \dfrac{4}{9}e^{2t}\mathbf{j}$, $\mathbf{a} = e^t\mathbf{i} + \dfrac{8}{9}e^{2t}\mathbf{j}$, Speed $= 5$, Direction $= \dfrac{3}{5}\mathbf{i} + \dfrac{4}{5}\mathbf{j}$, $\mathbf{v}(\ln 3) = 5\left[\dfrac{3}{5}\mathbf{i} + \dfrac{4}{5}\mathbf{j}\right]$ **6.** $\mathbf{v} = \left(\dfrac{2}{t+1}\right)\mathbf{i} + 2t\,\mathbf{j} + t\,\mathbf{k}$, $\mathbf{a} = \left(\dfrac{-2}{(t+1)^2}\right)\mathbf{i} + 2\,\mathbf{j} + \mathbf{k}$, Speed $= \sqrt{6}$, Direction $= \dfrac{1}{\sqrt{6}}\mathbf{i} + \dfrac{2}{\sqrt{6}}\mathbf{j} + \dfrac{1}{\sqrt{6}}\mathbf{k}$, $\mathbf{v}(1) = \sqrt{6}\left[\dfrac{1}{\sqrt{6}}\mathbf{i} + \dfrac{2}{\sqrt{6}}\mathbf{j} + \dfrac{1}{\sqrt{6}}\mathbf{k}\right]$

8. $\theta = \dfrac{3\pi}{4}$ **10.** $\theta = \dfrac{\pi}{2}$ **12.** All $t \geq 0$ **14.** $(\ln 4)\,\mathbf{i} + (\ln 4)\,\mathbf{j} + (\ln 2)\,\mathbf{k}$ **16.** $\mathbf{i} + (\ln 2)\,\mathbf{j} + \dfrac{3}{4}\mathbf{k}$

18. $\mathbf{v}(\pi) = -2\,\mathbf{i}$, $\mathbf{a}(\pi) = -\mathbf{j}$, $\mathbf{v}\left(\dfrac{3\pi}{2}\right) = -\sqrt{2}\,\mathbf{i} - \sqrt{2}\,\mathbf{j}$, $\mathbf{a}\left(\dfrac{3\pi}{2}\right) = \dfrac{\sqrt{2}}{2}\mathbf{i} - \dfrac{\sqrt{2}}{2}\mathbf{j}$,

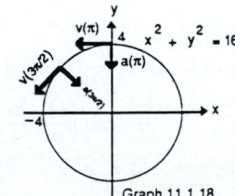

Graph 11.1.18

20. $\mathbf{v}(-1) = \mathbf{i} - 2\,\mathbf{j}$, $\mathbf{a}(-1) = 2\,\mathbf{j}$, $\mathbf{v}(0) = \mathbf{i}$, $\mathbf{a}(0) = 2\,\mathbf{j}$, $\mathbf{v}(1) = \mathbf{i} + 2\,\mathbf{j}$, $\mathbf{a}(1) = 2\,\mathbf{j}$

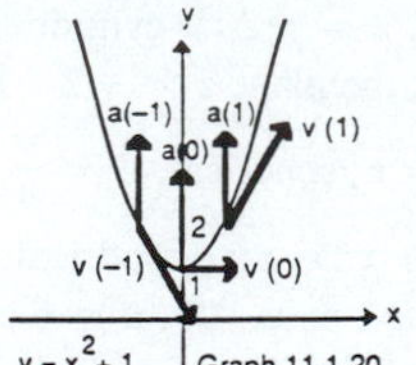

Graph 11.1.20

22. $\mathbf{r} = 90t^2\,\mathbf{i} + \left(90t^2 - \dfrac{16}{3}t^3 + 100\right)\mathbf{j}$

24. $\mathbf{r} = \left(\dfrac{t^4}{4} + 2t^2 + 1\right)\mathbf{i} + \left(\dfrac{t^2}{2} + 1\right)\mathbf{j} + \dfrac{2t^3}{3}\mathbf{k}$

26. $\mathbf{r} = \left(-\dfrac{t^2}{2} + 10\right)\mathbf{i} + \left(-\dfrac{t^2}{2} + 10\right)\mathbf{j} + \left(-\dfrac{t^2}{2} + 10\right)\mathbf{k}$

28. $\max|\mathbf{v}|$ is 3 when $t = \dfrac{\pi}{2}, \dfrac{3\pi}{2}$; $\min|\mathbf{v}| = 2$ when $t = 0, \pi$; $\max|\mathbf{a}| = 3$ when $t = 0, \pi$; $\min|\mathbf{a}| = 2$ when $t = \dfrac{\pi}{2}, \dfrac{3\pi}{2}$.

Section 11.2, pp. 784–786

2. 490 m/s   **4.** $t = 2$, 55.4 ft   **6.** $\alpha = 39.25°$ or $50.75°$ **8.** Drops 3.136×10^{-12} cm   **10.** 141%   **12.** 32.1°, 31.4 ft   **14.** It should reach the green.   **16.** a) 45.6 ft/sec b) 286.3 ft-lbs   **18.** 7.80 ft, 7.80 ft/sec   **24.** See Exercise 39, Section 9.3

Section 11.3, pp. 790–791

2. $\mathbf{T} = \left(\dfrac{12}{13}\cos 2t\right)\mathbf{i} - \left(\dfrac{12}{13}\sin 2t\right)\mathbf{j} + \dfrac{5}{13}\mathbf{k}$, 13π

4. $\mathbf{T} = (-\cos t)\,\mathbf{j} + (\sin t)\,\mathbf{k}$ if $0 \leq t \leq \dfrac{\pi}{2}$, $L = \dfrac{3}{2}$

6. $\mathbf{T} = \dfrac{6}{7}\mathbf{i} - \dfrac{2}{7}\mathbf{j} - \dfrac{3}{7}\mathbf{k}$, 14   **8.** $\mathbf{T} = (\cos t)\,\mathbf{i} - (\sin t)\,\mathbf{j}$, 1

10. $s(t) = \dfrac{t^2}{2}$, $L = \dfrac{3\pi^2}{8}$ **12.** $s(t) = 7t$, $L = -7$

16. a) $2\pi\sqrt{2}$ b) $2\pi\sqrt{2}$ c) $2\pi\sqrt{2}$

Section 11.4, pp. 799–800

2. $\mathbf{T} = (\sin t)\,\mathbf{i} + (\cos t)\,\mathbf{j}$, $\mathbf{N} = (\cos t)\,\mathbf{i} - (\sin t)\,\mathbf{j}$, $\kappa = \cos t$

4. $\mathbf{T} = (\cos t)\,\mathbf{i} + (\sin t)\,\mathbf{j}$, $\mathbf{N} = (-\sin t)\,\mathbf{i} + (\cos t)\,\mathbf{j}$, $\kappa = \dfrac{1}{t}$

6. $\mathbf{T} = (\cos t)\,\mathbf{i} + (\sin t)\,\mathbf{j}$, $t > 0$, $\mathbf{N} = (-\sin t)\,\mathbf{i} + (\cos t)\,\mathbf{j}$, $\mathbf{B} = \mathbf{k}$, $\kappa = \dfrac{1}{t}$, $\tau = 0$ **8.** $\mathbf{T} = \left(\dfrac{12}{13}\cos 2t\right)\mathbf{i} - \left(\dfrac{12}{13}\sin 2t\right)\mathbf{j} + \dfrac{5}{13}\mathbf{k}$, $\mathbf{N} = (-\sin 2t)\,\mathbf{i} - (\cos 2t)\,\mathbf{j}$, $\mathbf{B} = \left(\dfrac{5}{13}\cos 2t\right)\mathbf{i} - \left(\dfrac{5}{13}\sin 2t\right)\mathbf{j} - \dfrac{12}{13}\mathbf{k}$, $\kappa = \dfrac{124}{169}$, $\tau = -0.0592$

10. $\mathbf{T} = (-\cos t)\,\mathbf{i} + (\sin t)\,\mathbf{j}$, $\mathbf{N} = (\sin t)\,\mathbf{i} + (\cos t)\,\mathbf{j}$, $\mathbf{B} = -\mathbf{k}$, $\kappa = \dfrac{1}{3\cos t \sin t}$, $\tau = 0$ **12.** $\mathbf{T} = \left(\dfrac{1}{\sqrt{2}}\tanh t\right)\mathbf{i} + \dfrac{1}{\sqrt{2}}\mathbf{j} + \left(\dfrac{1}{\sqrt{2}}\text{sech}\,t\right)\mathbf{k}$, $\mathbf{N} = (\text{sech}\,t)\,\mathbf{i} - (\tanh t)\,\mathbf{j}$, $\mathbf{B} = \left(-\dfrac{1}{\sqrt{2}}\tanh t\right)\mathbf{i} + \dfrac{1}{\sqrt{2}}\mathbf{j} - \left(\dfrac{1}{\sqrt{2}}\text{sech}\,t\right)\mathbf{k}$, $\kappa = \dfrac{1}{2\cosh^2 t}$, $\tau = \dfrac{1}{2\cosh^2 t}$

14. $\mathbf{a} = \dfrac{2}{(t^2 + 1)}\mathbf{N}$ **16.** $\mathbf{a} = 0$ **18.** $\mathbf{a}(0) = 2\sqrt{2}\,\mathbf{N}$

20. $\mathbf{a}(0) = 2\,\mathbf{T} + \sqrt{2}\,\mathbf{N}$ **22.** $\mathbf{r}(0) = \mathbf{i}$; $\mathbf{T}(0) = \dfrac{1}{\sqrt{2}}\mathbf{j} + \dfrac{1}{\sqrt{2}}\mathbf{k}$;

$\mathbf{N}(0) = -\mathbf{i}$; $\mathbf{B}(0) = -\dfrac{1}{\sqrt{2}}\mathbf{j} + \dfrac{1}{\sqrt{2}}\mathbf{k}$

Osculating Plane: $y - z = 0$; Rectifying Plane: $x = 1$; Normal

Plane: $y + z = 0$ **32.** $\tau = \dfrac{b}{a^2 + b^2}$, ^{r}max $= \dfrac{1}{2a}$ **34.** $\tau = \dfrac{1}{2}$

36. a)

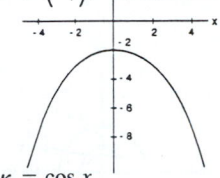

$r = (2 \ln t)\mathbf{i} - \left(t + \dfrac{1}{t}\right)\mathbf{j}$ Graph 11.4.36 a

b) $x^2 + (y + 4)^2 = 4$

38. b) $\kappa = \cos x$

42. Components of **v:**   0.260460144, 2.81640687, 2.00003104
Components of **a:** -2.55589373, 3.07690352, 2.000001057
Speed: 3.46411768
Components of **T:**   0.0751880183, 0.813022861,
   0.577356551
Components of **N:** -0.571515637, 0.688017053,
   0.447216291
Components of **B:** -0.0336340846, -0.3635933603,
   0.516385917
Curvature: 0.235695767

44. Components of **v:**   1.0000005, 5.99999982, 6.00009984
Components of **a:** -2.00001057, 6.00001775, 5.99999912
Speed: 8.54407374
Components of **T:**   0.117040194, 0.702241109, 0.702252816
Components of **N:** -0.229416577, 0.688248129,
   0.688245992
Components of **B:** $-9.55761189 \times 10^{-6}$, -0.241660882,
   0.241658446
Curvature: 0.0408128276

Section 11.5, pp. 809–810

2. 3.00×10^4 m/s   **4.** a) 21900 km b) 6501 km   **6.** 42166 km
8. 376 821 km   **10.** Solar System: 2.97×10^{-19} s²/m³; Earth:
9.903×10^{-14} s²/m³; Moon: 8.046×10^{-12} s²/m³
16. $\mathbf{a} = (2\omega^2 - 2)\,\mathbf{i}$

Section 11 Miscellaneous Exercises, pp. 811–814

2.

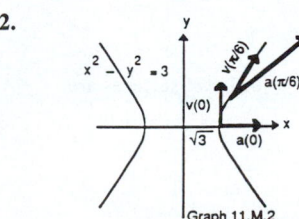

Graph 11.M.2

4. $3\,\mathbf{i} + (\ln 2)\,\mathbf{j} + \mathbf{k}$ **6.** $\mathbf{r} = (\tan^{-1}t)\,\mathbf{i} - ((\sin^{-1}t) - 1)\,\mathbf{j} + (\sinh^{-1}(t) + 1)\,\mathbf{k}$ **8.** $\mathbf{r} = (6t - t^2 - 2)\mathbf{i} + (4t - 2t^2 + 1)\,\mathbf{j}$

10. $L = 14$ **12.** $\mathbf{T}(0) = \dfrac{2}{3}\mathbf{i} + \dfrac{1}{3}\mathbf{j} + \dfrac{2}{3}\mathbf{k}$; $\mathbf{N}(0) = \dfrac{2}{5}\mathbf{i} - \dfrac{4}{5}\mathbf{j}$;

$\mathbf{B}(0) = \dfrac{8}{15}\mathbf{i} + f4, 15)\,\mathbf{j} - \dfrac{2}{3}\mathbf{k}$; $\kappa = \dfrac{2\sqrt{5}}{9}$, $\tau = -\dfrac{4}{9}$; Osculating

Plane: $4x + 2y - 5z = -8$; Rectifying Plane: $x - 2y = -2$;
Normal Plane: $2x + y + 2z = 5$ **14.** $\mathbf{T}(\ln 2) = \dfrac{15}{17\sqrt{2}}\mathbf{i} +$

$\dfrac{1}{\sqrt{2}}\mathbf{j} + \dfrac{8}{17\sqrt{2}}\mathbf{k}$; $\mathbf{N}(\ln 2) = \dfrac{8}{17}\mathbf{i} - \dfrac{15}{17}\mathbf{k}$; $\mathbf{B}(\ln 2) = -\dfrac{15}{17\sqrt{2}}$

$\mathbf{i} + \dfrac{1}{\sqrt{2}}\mathbf{j} - \dfrac{8}{17\sqrt{2}}\mathbf{k}$; $\kappa = \dfrac{32}{867}$, $\tau = \dfrac{32}{867}$; Osculating Plane:

$15x - 17y + z = 48 \ln 2$; Rectifying Plane: $8x - 15z = 3 - 90$

$\ln 2$; Normal Plane: $15x + 17y + 8z = \dfrac{765}{4} + 48 \ln 2$

16. $\mathbf{a}(0) = 2\sqrt{2}\,\mathbf{T} + 2\sqrt{3}\,\mathbf{N}$ **18.** 1   **20.** $0, \dfrac{\pi}{2}, \pi$

22. $\dfrac{dy}{dt} = -x$, clockwise   **26.** $\dfrac{x^2}{4} + \dfrac{y^2}{9} + z^2 = 1$, an ellipsoid,

upper half.   **28.** $\mathbf{v} = 2\sqrt{3}\,\mathbf{i} + 2\sqrt{3}\,\mathbf{j} - 4\sqrt{3}\,\mathbf{k}$ **32.** $\kappa = \dfrac{1}{5}$

34. $\left(-\ln\sqrt{2}, \dfrac{1}{\sqrt{2}}\right)$ **36.** Normal Plane: $x + 2y + 3z = 6$;

Rectifying Plane: $11x + 8y - 9z = 10$; Osculating Plane:

$3x - 3y + z = 1$ **38.** a) $2\sqrt{\dfrac{\pi g b}{a^2 + b^2}}$ b) $\theta = \dfrac{gbt^2}{2(a^2 + b^2)}$,

$z = \dfrac{gb^2t^2}{2(a^2 + b^2)}$ c) $\mathbf{v}(t) = \dfrac{gbt}{\sqrt{a^2 + b^2}}\,\mathbf{T}$, $\dfrac{d^2r}{dt^2} = \dfrac{bg}{\sqrt{a^2 + b^2}}$

$\mathbf{T} + a\left(\dfrac{bgt}{a^2 + b^2}\right)\mathbf{N}$. No component in the direction of **B.**

40. 6.63%   **42.** b) 1.49870113   **44.** a) $\mathbf{a}(1) = -9\,\mathbf{u}_r - 6\,\mathbf{u}_\theta$
b) 6.50 m   **48.** $\mathbf{r}(t) = \rho[(\sin\phi\cos\theta)\,\mathbf{i} + (\sin\phi\sin\theta)$

$\mathbf{j} + (\cos\phi)\,\mathbf{k}]$. $\mathbf{v}(t) = \dfrac{d\rho}{dt}\,\mathbf{u}_\rho + \rho\dfrac{d\phi}{dt}\,\mathbf{u}_\phi + \rho\sin\phi\dfrac{d\theta}{dt}\,\mathbf{u}_\theta$

50. b) Concerning the graph:
This relationship is very difficult to see because several arcs of
circles are used to approximate line segments.

Arc AB in the xy-plane is congruent to arc CD in the bottom face
of the box. Both are equal to $\rho\sin\phi\,d\phi$. Arc CD approximates
segment CD. Arc CE approximates segment CE (not drawn) and is
equal to the square root.

$\therefore$ right $\triangle CDE$ in the left face of the box is approximated by arc
CE (hypotenuse), arc CD (leg), and segment DE (leg).

A second right triangle is formed by segments CE, EF, and CF,
and we again use arc CE to approximate segment CE.

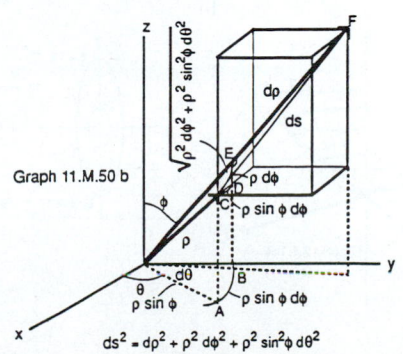

Graph 11.M.50 b

Chapter 12

Section 12.1, pp. 824–825

2. Domain: All points in the xy-plane; Range: $z \geq 0$. Level curves are ellipses with center $(0, 0)$ and major and minor axes along the x and y axes. **4.** Domain: All points on the xy-plane; Range: All Real Numbers. Level curves are hyperbolas centered at $(0, 0)$ and transverse and conjugate axes along the x and y axes. **6.** Domain: Set of all (x, y) so that $(x, y) \neq (0, 0)$; Range: All Real Numbers. Level curves are circles with center $(0, 0)$ and radii ≥ 0.

8. Domain: Set of all (x, y) so that $x \neq 0$; Range: $-\dfrac{\pi}{2} < z < \dfrac{\pi}{2}$.

Level curves are straight lines of the form $y = cx$ where c is any Real Number and $x \neq 0$. **10.** Domain: Set of all (x, y) so that $x \geq y$ and $y > -x$ or $x \leq y$ and $y < -x$; Range: $z \geq 0$. Level curves are straight lines of the form $y = -\left(\dfrac{c-1}{c+1}\right) x$ where $c > 0$.

12. a)   b)

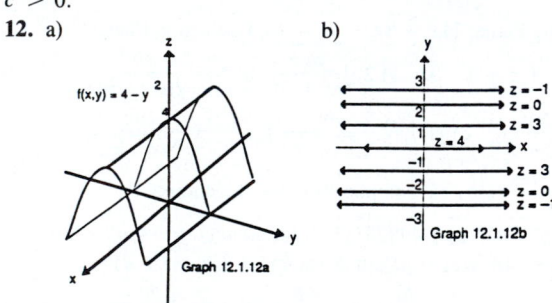

Graph 12.1.12a   Graph 12.1.12b

14. a)   b)

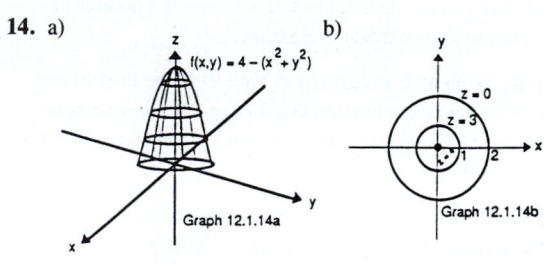

Graph 12.1.14a   Graph 12.1.14b

16. a)   b)

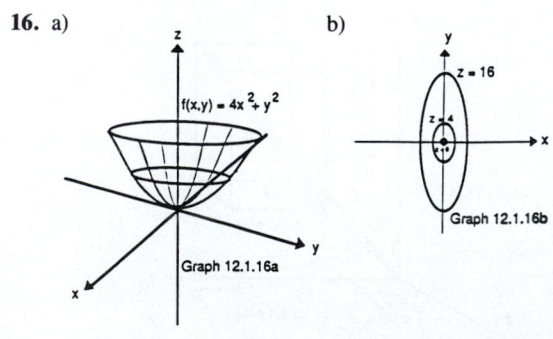

Graph 12.1.16a   Graph 12.1.16b

18. a)   b)

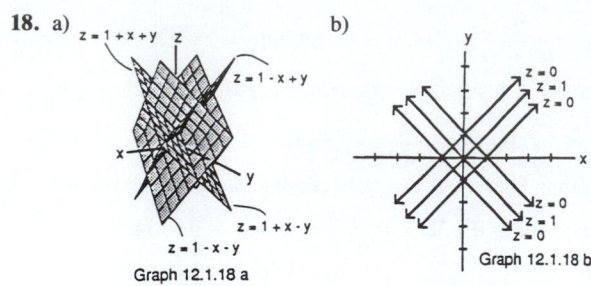

Graph 12.1.18 a   Graph 12.1.18 b

20. a)   b)

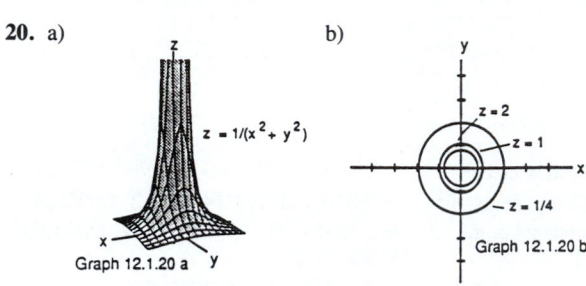

Graph 12.1.20 a   Graph 12.1.20 b

22. e   **24.** c   **26.** b   **28.** Domain: All (x, y, z); Range: $w \geq 1$. Level surfaces are right circular cylinders with the y-axis as axis.

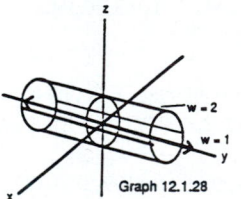

Graph 12.1.28

30. Domain: All (x, y, z) so that $-1 \leq z \leq 1$ Range: $-\dfrac{\pi}{2} \leq w \leq \dfrac{\pi}{2}$. Level surfaces are planes perpendicular to the z-axis so that $-1 \leq z \leq 1$.

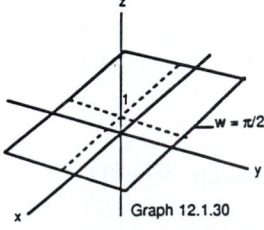

Graph 12.1.30

32. Domain: All (x, y, z); Range: $w \geq 0$. Level surfaces are ellipsoids with center $(0, 0, 0)$.

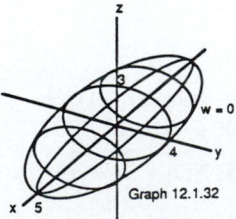

Graph 12.1.32

34. Domain: All (x, y, z); Range: $w \geq 0$. Level surfaces are planes.

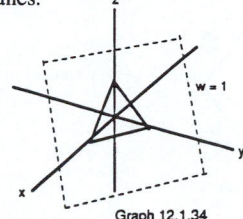

Graph 12.1.34

36. $x = 1$ or $x = -1$ **38.** $1 = \lim\limits_{n \to \infty} \dfrac{y(1 - (x/y)^n)}{y - x}$

40. $y = x^2 + z^2$ **42.** $\dfrac{\pi}{2} = \sin^{-1} y - \sin^{-1} x + \sec^{-1}|z|$ **44.** 2000

Section 12.2, pp. 831–833

2. $\ln 2$ **4.** 0   **6.** $\dfrac{5}{3}$ **8.** 1   **10.** 0   **12.** $\dfrac{\pi}{2}$ **14.** 2   **16.** $\dfrac{1}{2}$

18. $\ln 2$ **20.** 3   **22.** $-\dfrac{1}{2}$ **24.** a) Continuous at all (x, y) so that $x \neq y$ b) Continuous at all (x, y) **26.** a) Continuous at all (x, y) so that $x^2 - 3x + 2 \neq 0 \Rightarrow (x - 2)(x - 1) \neq 0 \Rightarrow x \neq 2$ and $x \neq 1$ b) Continuous at all (x, y) so that $y \neq x^2$
28. a) Continuous at all (x, y, z) so that $xyz > 0$ b) Continuous at all (x, y, z) **30.** a) Continuous at all (x, y, z) except $(0, 0, 0)$ b) Continuous at all (x, y, z) except those of the form $(0, y, 0)$ or $(x, 0, 0)$, that is all (x, y, z) except those on the x and y axes
32. Consider paths along $y = 0$ and $y = x^2$ **34.** Consider paths along $y = kx$, $k \neq 0$, k a constant.   **36.** Consider paths along $y = kx$, k a constant, $k \neq 1$.   **38.** Consider paths along $y = kx^2$, k a constant, $k \neq 0$.   **40.** $f(0, 0) = 0$ **42.** 2   **44.** 1   **46.** Does not exist   **48.** Does not exist   **50.** $f(0, 0) = -1$ **52.** $\delta = 0.05$ **54.** $\delta = 0.01$ **56.** $\delta = 0.2$ **58.** $\delta = \tan^{-1} 0.1$

Section 12.3, pp. 840–842

2. $\dfrac{\partial f}{\partial x} = 2x - y, \dfrac{\partial f}{\partial y} = -x + 2y$ **4.** $\dfrac{\partial f}{\partial x} = 5y - 14x + 3,$
$\dfrac{\partial f}{\partial y} = 5x - 2y - 6$ **6.** $\dfrac{\partial f}{\partial x} = 6(2x - 3y)^2, \dfrac{\partial f}{\partial y} = -9(2x - 3y)^2$
8. $\dfrac{\partial f}{\partial x} = \dfrac{2x^2}{\sqrt[3]{x^3 + (y/2)}}, \dfrac{\partial f}{\partial y} = \dfrac{1}{3\sqrt[3]{x^3 + (y/2)}}$ **10.** $\dfrac{\partial f}{\partial x} = \dfrac{y^2 - x^2}{(x^2 + y^2)^2},$
$\dfrac{\partial f}{\partial y} = \dfrac{-2xy}{(x^2 + y^2)^2}$ **12.** $\dfrac{\partial f}{\partial x} = \dfrac{-y}{y^2 + x^2}, \dfrac{\partial f}{\partial y} = \dfrac{x}{y^2 + x^2}$
14. $\dfrac{\partial f}{\partial x} = -e^{-x}\sin(x + y) + e^{-x}\cos(x + y),$
$\dfrac{\partial f}{\partial y} = e^{-x}\cos(x + y)$ **16.** $\dfrac{\partial f}{\partial x} = e^{x + y + 1}, \dfrac{\partial f}{\partial y} = e^{x + y + 1}$
18. $\dfrac{\partial f}{\partial x} = \dfrac{y}{2\sqrt{xy - x^2 y^2}}, \dfrac{\partial f}{\partial y} = \dfrac{x}{2\sqrt{xy - x^2 y^2}}$
20. $\dfrac{\partial f}{\partial x} = \dfrac{1}{x \ln y}, \dfrac{\partial f}{\partial y} = \dfrac{-\ln x}{y(\ln y)^2}$ **22.** $\dfrac{\partial f}{\partial x} = \dfrac{y}{(1 - xy)^2},$
$\dfrac{\partial f}{\partial y} = \dfrac{x}{(1 - xy)^2}$
24. $f_x(x, y, z) = y + z, f_y(x, y, z) = x + z, f_z(x, y, z) = y + x$
26. $f_x(x, y, z) = -x(x^2 + y^2 + z^2)^{-3/2}, f_y(x, y, z) = -y(x^2 + y^2 + z^2)^{-3/2}, f_z(x, y, z) = -z(x^2 + y^2 + z^2)^{-3/2}$

28. $f_x(x, y, z) = \dfrac{1}{|x + yz|\sqrt{(x + yz)^2 - 1}},$
$f_y(x, y, z) = \dfrac{z}{|x + yz|\sqrt{(x + yz)^2 - 1}},$
$f_z(x, y, z) = \dfrac{y}{|x + yz|\sqrt{(x + yz)^2 - 1}},$ all with $|x + yz| > 1$.
30. $f_x(x, y, z) = \dfrac{yz}{x}, f_y(x, y, z) = z \ln(xy) + z, f_z(x, y, z) = y \ln(xy)$ **32.** $f_x(x, y, z) = -yz\, e^{-xyz}, f_y(x, y, z) = -xz\, e^{-xyz}, f_z(x, y, z) = -xy\, e^{-xyz}$ **34.** $f_x(x, y, z) = y \cosh(xy - z^2), f_y(x, y, z) = x \cosh(xy - z^2), f_z(x, y, z) = -2z \cosh(xy - z^2)$
36. $\dfrac{\partial g}{\partial u} = 2ve^{2u/v}, \dfrac{\partial g}{\partial v} = 2ve^{2u/v} - 2ue^{2u/v}$ **38.** $\dfrac{\partial g}{\partial r} = 1 - \cos \theta,$
$\dfrac{\partial g}{\partial \theta} = r \sin \theta, \dfrac{\partial g}{\partial z} = -1$ **40.** $\dfrac{\partial A}{\partial c} = m, \dfrac{\partial A}{\partial h} = \dfrac{q}{2},$
$\dfrac{\partial A}{\partial k} = \dfrac{m}{q}, \dfrac{\partial A}{\partial m} = \dfrac{k}{q} + c, \dfrac{\partial A}{\partial q} = -\dfrac{km}{q^2} + \dfrac{h}{2}$ **42.** $\dfrac{\partial^2 f}{\partial x^2} = -y^2 \sin xy,$
$\dfrac{\partial^2 f}{\partial y^2} = -x^2 \sin xy, \dfrac{\partial^2 f}{\partial y \partial x} = \dfrac{\partial^2 f}{\partial x \partial y} = \cos xy - xy \sin xy$
44. $\dfrac{\partial^2 h}{\partial x^2} = 0, \dfrac{\partial^2 h}{\partial y^2} = xe^y, \dfrac{\partial^2 h}{\partial y \partial x} = \dfrac{\partial^2 h}{\partial x \partial y} = e^y$ **46.** $\dfrac{\partial^2 s}{\partial x^2} = \dfrac{2xy}{(y^2 + x^2)^2},$
$\dfrac{\partial^2 s}{\partial y^2} = \dfrac{-2xy}{(y^2 + x^2)^2}, \dfrac{\partial^2 s}{\partial y \partial x} = \dfrac{\partial^2 s}{\partial x \partial y} = \dfrac{y^2 - x^2}{(y^2 + x^2)^2}$ **52.** a) y first three times b) y first three times c) y first twice d) x first twice
54. $\dfrac{1}{6}$ **56.** $\dfrac{\partial a}{\partial A} = \dfrac{a \cos A}{\sin A}, \dfrac{\partial a}{\partial B} = -b \cos B \cot B \sin A$
58. $x_u = \dfrac{1}{2x - 4xy}, y_u = \dfrac{1}{1 - 2y}, s_u = \dfrac{1 + 2y}{1 - 2y}$
66. a) $a = -c$ b) $a = -\dfrac{1}{3}c$ and $b = -3d$

Section 12.4, pp. 852–854

2. a) $L(x, y) = 3x + 4y - 6$ b) $L(x, y) = 0$
4. a) $4 + 4x + 4y$ b) $L(x, y) = 10x + 10y - 5$
6. a) $L(x, y) = 1 - x + 2y$ b) $L(x, y) = -e^3x + 2e^3y - 2e^3$
8. $L(x, y) = 7x - 3; 0.02$ **10.** $L(x, y) = x + y - 2; 0.0832$
12. $L(x, y) = 4x + 5y - 8; 0.0860$ **14.** a) More sensitive to x
b) $-\dfrac{1}{2}$ **16.** 3%   **18.** $E \leq 0.002\, m$ **20.** b) More sensitive to a change in R_1
22. a) 0.28%, 0.30% b) r to y; θ to x
26. a) $L(x, y, z) = 2x + 2y + 2z - 3$
b) $L(x, y, z) = 2y - 1$ c) $L(x, y, z) = 2x - 1$
28. a) $L(x, y, z) = 2 - z$ b) $L(x, y, z) = 2y$
30. a) $L(x, y, z) = 0$ b) $L(x, y, z) = z$ c) $L(x, y, z) = \dfrac{1}{2}x + \dfrac{1}{2}y + \dfrac{1}{2}z + \dfrac{\pi}{4} - \dfrac{3}{2}$ **32.** $L(x, y, z) = 3x + 3y + 2z - 5; 0.01$
34. $L(x, y, z) = y + z - \dfrac{\pi}{4} + 1; 0.000636$ **36.** Most sensitive to a change in d **38.** $\dfrac{47}{24}$ ft^3 **40.** 4.8

Section 12.5, pp. 860–862

2. $\dfrac{dw}{dt} = 0, \dfrac{dw}{dt}(0) = 0$ **4.** $\dfrac{dw}{dt} = \dfrac{16}{1 + 16t}, \dfrac{dw}{dt}(3) = \dfrac{16}{49}$

6. $\dfrac{dw}{dt} = -\ln(\cos(t \ln t)) - \cos(t \ln t) + e^{t-1};$ $\dfrac{dw}{dt}(1) = 0$

8. a) $\dfrac{\partial z}{\partial r} = 0, \dfrac{\partial z}{\partial \theta} = -1$ b) $\dfrac{\partial z}{\partial r} = 0, \dfrac{\partial z}{\partial \theta} \approx -1$

10. a) $\dfrac{\partial w}{\partial u} = \dfrac{2}{u};$ $\dfrac{\partial w}{\partial v} = 2$ b) $\dfrac{\partial w}{\partial u} = -1; \dfrac{\partial w}{\partial v} = 2$

12. a) $\dfrac{\partial u}{\partial x} = y^z$ if $-\dfrac{\pi}{2} < x < \dfrac{\pi}{2}; \dfrac{\partial u}{\partial y} = xzy^{z-1}; \dfrac{\partial u}{\partial z} = xy^z \ln y$

b) $\dfrac{\partial u}{\partial x} = \sqrt{2}; \dfrac{\partial u}{\partial y} = -\dfrac{\pi\sqrt{2}}{4}; \dfrac{\partial u}{\partial z} = -\dfrac{\pi\sqrt{2}\ln 2}{4}$

14. $\dfrac{dz}{dt} = \dfrac{\partial z}{\partial u}\dfrac{du}{dt} + \dfrac{\partial z}{\partial v}\dfrac{dv}{dt} + \dfrac{\partial z}{\partial w}\dfrac{dw}{dt}$

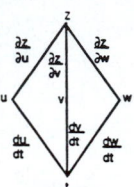
Diagram 12.5.14

16. a) $\dfrac{\partial w}{\partial x} = \dfrac{\partial w}{\partial r}\dfrac{\partial r}{\partial x} + \dfrac{\partial w}{\partial s}\dfrac{\partial s}{\partial x} + \dfrac{\partial w}{\partial t}\dfrac{\partial t}{\partial x}$

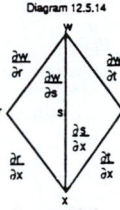
Diagram 12.5.16 a

b) $\dfrac{\partial w}{\partial y} = \dfrac{\partial w}{\partial r}\dfrac{\partial r}{\partial y} + \dfrac{\partial w}{\partial s}\dfrac{\partial s}{\partial y} + \dfrac{\partial w}{\partial t}\dfrac{\partial t}{\partial y}$

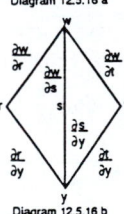
Diagram 12.5.16 b

18. a) $\dfrac{\partial w}{\partial x} = \dfrac{\partial w}{\partial u}\dfrac{\partial u}{\partial x} + \dfrac{\partial w}{\partial v}\dfrac{\partial v}{\partial x}$

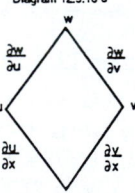
Diagram 12.5.18 a

b) $\dfrac{\partial w}{\partial y} = \dfrac{\partial w}{\partial u}\dfrac{\partial u}{\partial y} + \dfrac{\partial w}{\partial v}\dfrac{\partial v}{\partial y}$

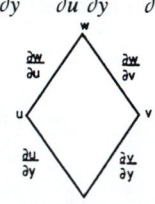
Diagram 12.5.18 b

20. $\dfrac{\partial y}{\partial r} = \dfrac{dy}{du}\dfrac{\partial u}{\partial r}$

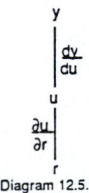
Diagram 12.5.20

22. $\dfrac{\partial w}{\partial p} = \dfrac{\partial w}{\partial x}\dfrac{\partial x}{\partial p} + \dfrac{\partial w}{\partial y}\dfrac{\partial y}{\partial p} + \dfrac{\partial w}{\partial z}\dfrac{\partial z}{\partial p} + \dfrac{\partial w}{\partial v}\dfrac{\partial v}{\partial p}$

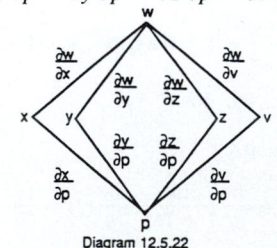
Diagram 12.5.22

24. $\dfrac{\partial w}{\partial s} = \dfrac{\partial w}{\partial x}\dfrac{\partial x}{\partial s} + \dfrac{\partial w}{\partial y}\dfrac{\partial y}{\partial s}$

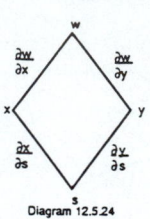
Diagram 12.5.24

26. 2 **28.** $-(2 + \ln 2)$ **30.** -4 **32.** $-\dfrac{5}{3\ln 2}$ **34.** -8

36. 2 **38.** $3\pi + 6$ **42.** a) $-1 \text{ cm}^3/\text{sec}$, decreasing; $130 \text{ cm}^2/\text{sec}$, increasing b) $\dfrac{-9}{\sqrt{275}}$ cm/sec, decreasing **46.** 4

Section 12.6, pp. 868–869

2. a) $1 + \cos(x + y)$ b) $1 - 2x$ c) -1 d) -1 e) $1 + \cos t$ f) $\cos t + 2(t - y)$ **4.** a) $2\pi^2$ b) 2π **6.** -1

12.

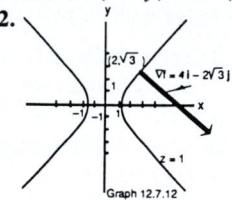
Graph 12.7.12

14. 0 **16.** $\dfrac{5}{4} - \sqrt{2}$ **18.** $-\dfrac{3}{2\sqrt{13}}$ **20.** $\mathbf{u} = \dfrac{1}{\sqrt{2}}(\mathbf{i} - \mathbf{j})$,

$D_{\mathbf{u}}f = \dfrac{\sqrt{2}}{3} - \mathbf{u} = -\dfrac{1}{\sqrt{2}}(\mathbf{i} - \mathbf{j}), D_{-\mathbf{u}}f = -\dfrac{\sqrt{2}}{3}$ **22.** $\mathbf{u} = \dfrac{1}{\sqrt{106}}$

$(-9\mathbf{j} - 5\mathbf{k}), D_{\mathbf{u}}f = \sqrt{106} - \mathbf{u} = \dfrac{1}{\sqrt{106}}(9\mathbf{j} + 5\mathbf{k})$,

$D_{-\mathbf{u}}f = -\sqrt{106}$ **24.** $\mathbf{u} = \dfrac{1}{3}(2\mathbf{i} + 2\mathbf{j} + \mathbf{k}), D_{\mathbf{u}}f = 3 - \mathbf{u} = -\dfrac{1}{3}$

$(2\mathbf{i} + 2\mathbf{j} + \mathbf{k}), D_{-\mathbf{u}}f = -3$ **26.** Tangent: $-2x + z = -2$; normal line: $x = 1 - 2t, y = 0, z = t$ **28.** Tangent: $x - 11y - 6z - 12 = 0$; normal line: $x = -4 + t, y = -2 - 11t, z = 1 - 6t$ **30.** Tangent: $9x - 7y - z = 21$; normal line: $x = 2 + 9t, y = -3 - 7t, z = 18 - t$

32. $x = 1 + 2t, y = 1 - 4t, z = 1 + 2t$ **34.** $x = \sqrt{2} - 2\sqrt{2}\, t, y = \sqrt{2} + 2\sqrt{2}\, t, z = 4$

36.

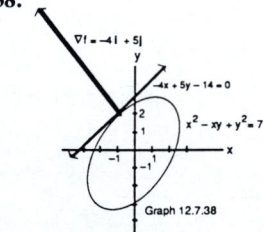
Graph 12.7.36 ... Graph 12.7.38

38.

40. 0.00076 **42.** 0 **44.** $\mathbf{u} = \dfrac{1}{\sqrt{2}}\mathbf{i} + \dfrac{1}{\sqrt{2}}\mathbf{j} - \mathbf{u} = -\dfrac{1}{\sqrt{2}}\mathbf{i} - \dfrac{1}{\sqrt{2}}\mathbf{j}$ **46.** 2 **48.** a) $2\mathbf{i} + 2\mathbf{j} - 2\mathbf{k}$ b) $2\sqrt{2}$

54. ∇f: 5.999995, 3.000009, 0.9999991; Directional Derivative: 3.666659; Plane: $5.999995\,x + 3.000009\,y + 0.9999991\,z = 11.00001$; Normal Line: $x = 1 + 5.999995\,t, y = 2 + 3.000009\,t$,

$z = -1 + 0.9999991\,t$ **56.** ∇f: -3.000096, -0.9999871, 0.9999991; Directional Derivative: -0.5071441; Plane: $-3.000096\,x - 0.9999871\,y + 0.9999991\,z = -4.000095$; Normal Line: $x = 1 - 3.000096\,t$, $y = 0 - 0.9999871\,t$, $z = -1 + 0.9999991\,t$

Section 12.7, pp. 877–879

2. $\nabla f = -\dfrac{11}{2}\mathbf{i} - 6\mathbf{j} - \dfrac{23}{2}\mathbf{k}$

4. $\nabla f = \left(\dfrac{\sqrt{3}+2}{2}\right)\mathbf{i} + \dfrac{\sqrt{3}}{2}\mathbf{j} - \dfrac{1}{2}\mathbf{k}$

6.

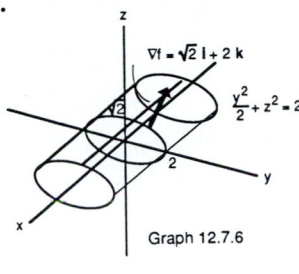

Graph 12.7.6

8.

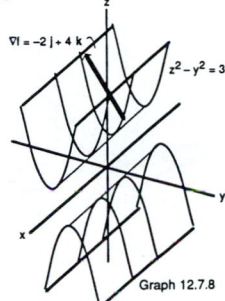

Graph 12.7.8

10.

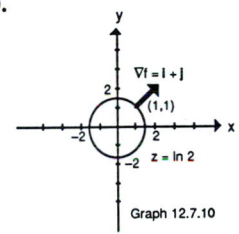

Graph 12.7.10

Section 12.8, pp. 887–891

2. $f(15, -8) = -63$, absolute minimum   **4.** $f\left(\dfrac{2}{3}, \dfrac{4}{3}\right) = 0$, absolute maximum   **6.** Saddle point at $(-2, 2)$ **8.** $f\left(3, \dfrac{3}{2}\right) = \dfrac{17}{2}$, absolute maximum   **10.** $f\left(\dfrac{13}{12}, -\dfrac{3}{4}\right) = -\dfrac{31}{12}$, absolute minimum

12. $f(1, -2) = -36$, absolute minimum   **14.** Saddle point at $(1, 2)$ **16.** Saddle point at $(0, 0)$ **18.** $f(-2, 3) = -2$, absolute minimum   **20.** Saddle points at $(0, 0)$; and $(-2, -2)$ $f(0, 2)$ -12, local minimum; $f(-2, 0) = -4$, local maximum

22. Saddle point at $(0, 0)$; $f\left(\dfrac{4}{9}, \dfrac{4}{3}\right) = -\dfrac{64}{81}$, local minimum

24. Saddle point at $(0, 0)$; $f(1, 1) = f(-1, -1) = 2$, absolute maximum   **26.** $f(1, 1) = 3$, local minimum   **28.** Saddle point at $(0, 0)$, $f(0, 0) = \dfrac{14}{3}$ **30.** No extrema and no saddle points

32. Absolute maximum is 17 at $(0, 4)$ and $(4, 4)$; Absolute minimum is 1 at $(0, 0)$ **34.** Absolute maximum is 19 at $(5, 3)$; absolute minimum is -12 at $(4, -2)$ **36.** Absolute maximum is 2 at $\left(\dfrac{1}{2}, \dfrac{1}{2}\right)$; absolute minimum is -32 at $(1, 0)$ **38.** Absolute maximum is 5 at $(1, 0)$; absolute minimum is 1 at $(0, 0)$

40. $x = \dfrac{1}{2}$, $y = 2$ **42.** a) Minimum at $(0, 0)$ since $f(x, y) > 0$

for all other (x, y). b) Maximum of 1 at $(0, 0)$ since $f(x, y) < 1$ for all other (x, y). c) Neither since $f(x, y) < 0$ for $x < 0$ and $f(x, y) > 0$ for $x > 0$. d) Neither since $f(x, y) < 0$ for $x < 0$, $y > 0$ and $f(x, y) > 0$ for $x > 0$, $y > 0$. e) Neither since $f(x, y) < 0$ for $x < 0$, $y > 0$ and $f(x, y) > 0$ for $x > 0$, $y > 0$. f) Minimum since $f(x, y) > 0$ for all other (x, y).

44. a) $f\left(-\dfrac{1}{2}\right)$ is an absolute minimum. No absolute maximum.

b) Absolute minimum is $-\dfrac{1}{2}$ at $t = -\dfrac{1}{2}$; absolute maximum is 0 at $t = 0, -1$. c) Absolute minimum is 0 at $t = 0$; absolute maximum is 4 at $t = 1$.

46. Absolute maximum is $\dfrac{1}{2}\,t$ at $= \dfrac{\pi}{4}, \dfrac{5\pi}{4}$; absolute minimum is $-\dfrac{1}{2}$ at $t = \dfrac{3\pi}{4}, \dfrac{7\pi}{4}$.   **48.** $y = \dfrac{3}{4}x + \dfrac{5}{3}$; $y\big|_{x=4} = \dfrac{14}{3}$

50. $y = \dfrac{5}{14}x + \dfrac{15}{14}$; $y\big|_{x=4} = \dfrac{35}{14}$ **52.** $F = 57000\dfrac{1}{D^2} - 3.72$.

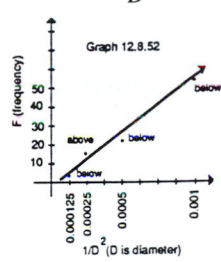

Graph 12.8.52

54. $y = 1.10x + 0.294$

Section 12.9, pp. 900–902

2. $5, -5$ **4.** $\dfrac{3\sqrt{2}}{8}$ **6.** $\left(\pm\sqrt{2}, 1\right)$ **8.** $(1, -1)$, $(-1, 1)$ are farthest, $\left(\dfrac{1}{\sqrt{3}}, \dfrac{1}{\sqrt{3}}\right) \left(-\dfrac{1}{\sqrt{3}}, -\dfrac{1}{\sqrt{3}}\right)$ are closest.

10. Length $= 4\sqrt{2}$, width $= 3\sqrt{2}$ **12.** The tank is a sphere, there is no cylindrical part.   **14.** $\left(\dfrac{3}{2}, 2, \dfrac{5}{2}\right)$ is closest.   **16.** 1

18. $(0, 0, 2)$ and $(0, 0, -2)$ **20.** Each of the numbers is 3.

22. $\left(\pm\dfrac{4}{3}, -\dfrac{4}{3}, -\dfrac{4}{3}\right)$ **24.** $(-3\sqrt{3}, 0, 3)$

26. $z = -\dfrac{1}{2}x + \dfrac{3}{2}y - \dfrac{1}{4}$ **28.** $\dfrac{4}{3}$ **30.** a) 2000

Section 12.10, p. 908

2. Quadratic: $y + \dfrac{1}{2}(2xy - y^2)$; cubic: $y + \dfrac{1}{2}(2xy - y^2) + \dfrac{1}{6}(3x^2y - 3xy^2 + 2y^3)$ **4.** Quadratic: 1; cubic: 1
6. Quadratic: $1 + x + y + x^2 + xy + y^2$; cubic: $1 + x + y + x^2 + xy + y^2 + x^3 + x^2y + xy^2 + y^3$
8. Quadratic: $y + xy$; $E(x, y) \le 0.00081$

Section 12 Miscellaneous Exercises, pp. 909–914

2. Domain: All points in the xy-plane; Range: $-1 \le f(x, y) \le 1$.

Level curves are lines with $m = 1$.

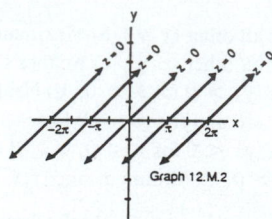

Graph 12.M.2

4. Domain: All (x, y) so that $x^2 - y \geq 0$; Range: $f(x, y) \geq 0$. Level curves are parabolas with vertex on y-axis.

6. Domain: All points (x, y, z) in space; Range: $f(x, y, z) \geq 0$. Level surfaces are ellipsoids with center $(0, 0, 0)$.

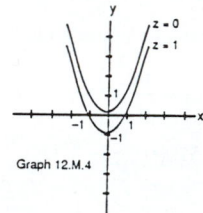

Graph 12.M.4

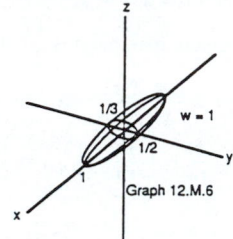

Graph 12.M.6

8. Domain: All points (x, y, z) in space; Range: $0 < f(x, y, z) \leq 1$. Level surfaces are spheres with center $(0, 0, 0)$ and radius $r > 0$.

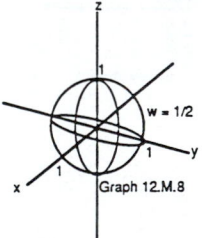

Graph 12.M.8

10. 2 **12.** 1 **14.** 3 **18.** a) 1 b) 0 c) $0, r \neq 0$
20. $f_x(x, y) = \dfrac{x - y}{x^2 + y^2}, f_y(x, y) = \dfrac{x + y}{x^2 + y^2}$ **22.** $h_x(x, y, z) =$

$- 3 \cos(2\pi x + y - 3z)$ **24.** $f_r(r, l, T, d) = -\dfrac{1}{2r^2 l}\sqrt{\dfrac{T}{\pi d}}$,

$f_l(r, l, T, d) = -\dfrac{1}{2rl^2}\sqrt{\dfrac{T}{\pi d}}, f_T(r, l, T, d) = \dfrac{1}{4\pi rld}\sqrt{\dfrac{\pi d}{T}}$,

$f_d(r, l, T, d) = \dfrac{-T}{2\pi rld^2}\sqrt{\dfrac{\pi d}{T}}$ **26.** $f_{xx}(x, y) = e^x - y \sin x$,

$f_{yy}(x, y) = 0, f_{xy}(x, y) = f_{yx}(x, y) = \cos x$ **28.** $f_{xx}(x, y) = 0$,
$f_{yy}(x, y) = 2 - \cos y + 7e^y, f_{xy}(x, y) = f_{yx}(x, y) = -3$
30. $w = x + e^x \cos y + y^2 - 2$ **32.** $L(x, y) = x - 5y + 4$,

$|E(x, y)| \leq 0.27$ **34.** a) $L(x, y, z) = 1 - \dfrac{\pi}{4} + y + z$

b) $L(x, y, z) = -\dfrac{\sqrt{2}}{2}x + \dfrac{\sqrt{2}}{2}y + \dfrac{\sqrt{2}}{2}z + \dfrac{\sqrt{2}}{2}$ **36.** More

sensitive to changes in y.   **38.** 0.8125% **40.** 1 cm change in

height has more effect.   **42.** 5 **44.** $\dfrac{\partial w}{\partial u} = \dfrac{2}{5}, \dfrac{\partial w}{\partial v} = 0$

44. $\dfrac{\partial w}{\partial u} = \dfrac{2}{5}, \dfrac{\partial w}{\partial v} = 0$ **46.** $-(\ln(2) + 1)$

50. $\dfrac{\partial z}{\partial x} = f_u\dfrac{\partial u}{\partial x} + f_v\dfrac{\partial v}{\partial x} = af_u + af_v, \dfrac{\partial z}{\partial y} = f_u\dfrac{\partial u}{\partial y} + f_v\dfrac{\partial v}{\partial y} =$

$bf_u - bf_v$ **52.** $\dfrac{\partial w}{\partial r} = \dfrac{2}{r + s}, \dfrac{\partial w}{\partial s} = \dfrac{2}{r + s}$

54. $\left(2, \dfrac{\pi}{2}\right)$ **58.** $\dfrac{\partial \mathbf{r}}{\partial \rho} = \mathbf{u}_\rho, \dfrac{\partial \mathbf{r}}{\partial \phi} = \rho\, \mathbf{u}_\phi, \dfrac{\partial \mathbf{r}}{\partial \theta} = \rho \sin \phi\, \mathbf{u}_\theta$

60. $\mathbf{u} = \dfrac{1}{\sqrt{2}}\mathbf{i} - \dfrac{1}{\sqrt{2}}\mathbf{j}, -\mathbf{u} = -\dfrac{1}{\sqrt{2}}\mathbf{i} + \dfrac{1}{\sqrt{2}}\mathbf{j}, (D_\mathbf{u}f)_{P_0} = 2\sqrt{2}$,

$(D_{-\mathbf{u}}f)_{P_0} = -2\sqrt{2}, (D_{\mathbf{u}_1}f)_{P_0} = 0$ **62.** $\mathbf{u} = \dfrac{2}{\sqrt{5}}\mathbf{i} + \dfrac{1}{\sqrt{5}}\mathbf{k}$,

$-\mathbf{u} = -\dfrac{2}{\sqrt{5}}\mathbf{i} + -\dfrac{1}{\sqrt{5}}\mathbf{k}, (D_\mathbf{u}f)_{P_0} = \sqrt{5}$,

$(D_{-\mathbf{u}}f)_{P_0} = -\sqrt{5}, (D_{\mathbf{u}_1}f)_{P_0} = \sqrt{3}$
64.

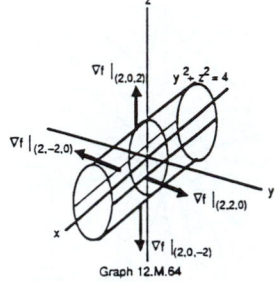

Graph 12.M.64

66. Tangent: $2x + 2y + z - 6 = 0$; normal line: $x = 1 + 2t$, $y = 1 + 2t, z = 2 + t$
68.

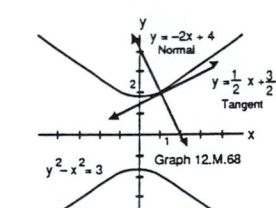

Graph 12.M.68

70. $x = \dfrac{1}{2} - t, y = 1\ z = \dfrac{1}{2} + t$ **72.** $(0, 1, 0)$ **74.** $\sqrt{3}$
80.

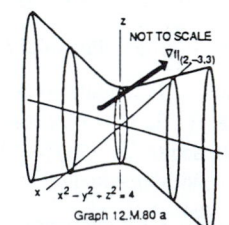

Graph 12.M.80 a

b) $\nabla f = 4\mathbf{i} + 6\mathbf{j} + 6\mathbf{k}$ c) Tangent: $2x + 3y + 3z = 4$; normal
line: $x = 2 + 4t, y = -3 + 6t, z = 3 + 6t$ **88.** $f(0, -1) = 2$,

saddle point   **90.** $f(0, 0) = 0$, saddle point $f\left(-\dfrac{1}{\sqrt[3]{4}}, -\dfrac{1}{\sqrt[3]{2}}\right) = \dfrac{1}{2}$,

local maximum   **92.** $f(0, 0) = 0$, saddle point;
$f(0, 2) = -4, f(-2, 0) = 4$, local maximum; $f(-2, 2) = 0$,
saddle point   **94.** Absolute maximum is 13 at $(4, 2)$; absolute
minimum is 0 at $(1, 0)$ **96.** Absolute maximum is 2 at $(1, 1)$;
absolute minimum is 0 at $(0, 0), (0, 2), (2, 2), (2, 0)$
98. Absolute maximum is 18 at $(1, 1)$ and $(-1, -1)$; absolute
minimum is -32 at $(2, -2)$

100. Absolute maximum is 6 at $(1, 1)$; absolute minimum is -2 at $(1, -1), (-1, 1)$ **102.** Maximum of $\dfrac{1}{e^2}$ at $\left(\dfrac{1}{2}, \dfrac{1}{3}\right)$ **104.** Absolute maximum is 1 at $(0, \pm 1)$ and $(1, 0)$; absolute minimum is -1 at $(-1, 0)$ **106.** Absolute maximum is 5 at $(0, 1)$; absolute minimum is $-\dfrac{1}{3}$ at $\left(0, -\dfrac{1}{3}\right)$ **108.** $x = \left(\dfrac{c^2 V}{ab}\right)^{1/3}, y = \left(\dfrac{b^2 V}{ac}\right)^{1/3}, z = \left(\dfrac{a^2 V}{bc}\right)^{1/3}$

112. $w = e^{-c^2 n^2 \pi^2 t / L^2} \sin\left(\dfrac{n\pi}{L} x\right)$; as $t \to \infty$, $w \to 0$

Chapter 13

Section 13.1, pp. 924–926

2. 0

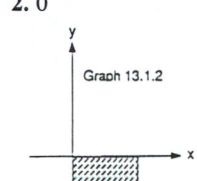

Graph 13.1.2

4. 2π

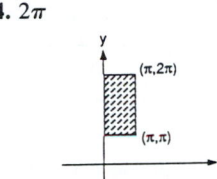

Graph 13.1.4

6. $\dfrac{\pi}{4}$

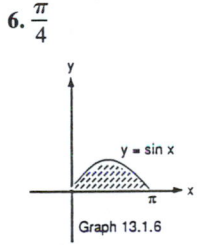

Graph 13.1.6

8. $\dfrac{5}{6}$

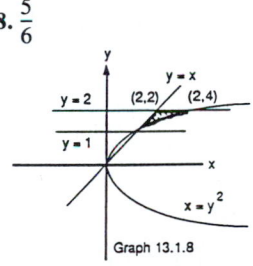

Graph 13.1.8

10. $\dfrac{e}{2} - 1$

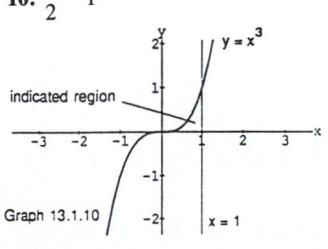

Graph 13.1.10

12. $\dfrac{1}{6}$ **14.** $\dfrac{2}{\pi}$ **16.** $\dfrac{1}{4}$

18. $\dfrac{8}{3}$

20. $-\dfrac{9}{2}$

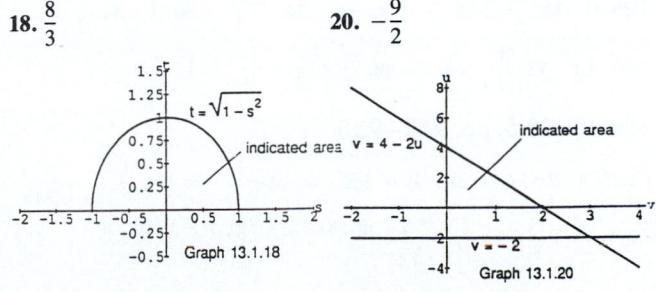

Graph 13.1.18   Graph 13.1.20

22. $\displaystyle\int_0^1 \int_{x^2}^x dy\, dx = \dfrac{1}{6}$

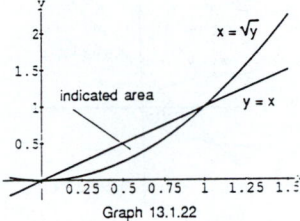

Graph 13.1.22

24. $\displaystyle\int_1^e \int_{\ln y}^1 dx\, dy = e - 2$

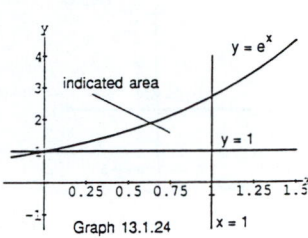

Graph 13.1.24

26. $\displaystyle\int_{-1}^1 \int_0^{\sqrt{1-x^2}} 3y\, dy\, dx = 2$ **28.** 2π **30.** $\dfrac{1}{2}$

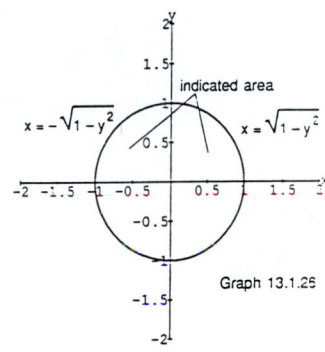

Graph 13.1.26

32. $\dfrac{e - 2}{2}$

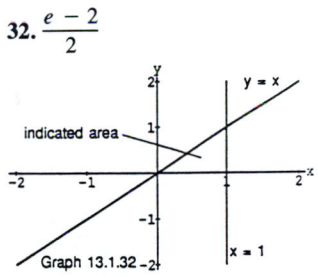

Graph 13.1.32

34. $\dfrac{e^8 - 1}{4}$

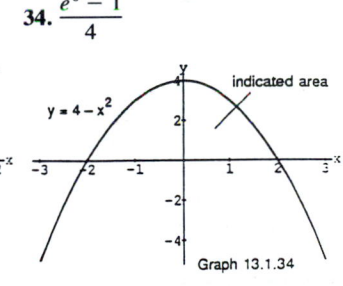

Graph 13.1.34

36. $\dfrac{1}{80\pi}$

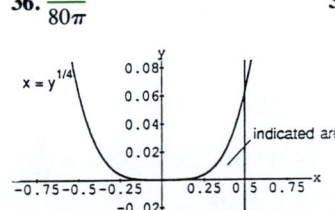

Graph 13.1.36

38. $\dfrac{\ln 17}{4}$

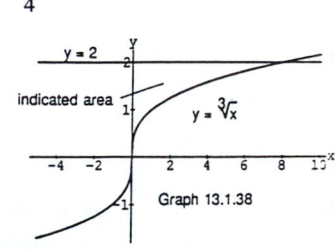

Graph 13.1.38

40. $\dfrac{187\sqrt{17}}{60}$ **42.** $\dfrac{9\pi - 8}{3}$ **44.** $\dfrac{128}{15}$ **46.** 6

48. $\dfrac{2}{3}\left[7\ln\left(2 + \sqrt{3}\right) + 2\sqrt{3}\right]$ **50.** $6\ln 2$ **52.** $2\arctan(2\pi) - 2\arctan(2) - \dfrac{\ln(1 + 4\pi^2)}{2\pi} + \dfrac{\ln 5}{2}$ **54.** Between the concentric circles, $\{(x, y) \mid 1 < x^2 + y^2 < 4\}$ **56.** 0.557   **58.** 3.142

Section 13.2, pp. 934–937

2. 1

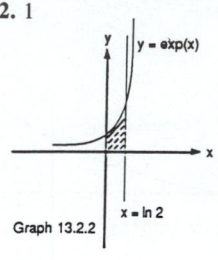

Graph 13.2.2

4. $\dfrac{9}{2}$

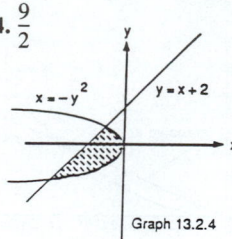

Graph 13.2.4

6. $\dfrac{4}{3}$

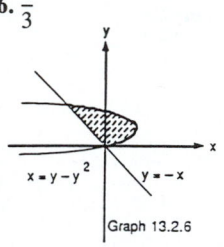

Graph 13.2.6

8. 5

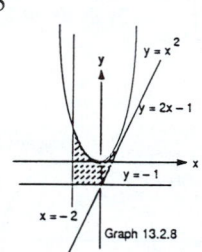

Graph 13.2.8

10. $\dfrac{9}{2}$

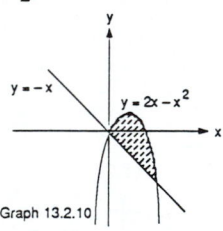

Graph 13.2.10

12. $\dfrac{9}{2}$

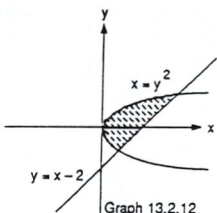

Graph 13.2.12

14. $\dfrac{32}{3}$

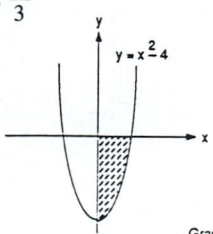

Graph 13.2.14

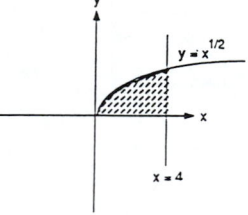

16. $I_x = 27, R_x = \sqrt{3}, I_y = 27, R_y = \sqrt{3}$ **18.** $\bar{x} = 1, \bar{y} = 1$

20. $\bar{x} = \dfrac{5}{2}, \bar{y} = 5$ **22.** $I_x = 4\pi, I_y = 4\pi, I_0 = 8\pi$ **24.** $\dfrac{\pi}{2}$

26. $I_x = \dfrac{ab^3\pi}{4}, R_x = \sqrt{\dfrac{I_x}{M}} = \dfrac{b}{2}$ **28.** 0   **30.** $23\sqrt{3}$ **32.** $\bar{x} = \dfrac{8}{15},$

$\bar{y} = \dfrac{8}{15}; I_x = \dfrac{1}{6}$ **34.** $\bar{x} = 0, \bar{y} = \dfrac{9}{14}, I_y = \dfrac{16}{35}, R_y = \sqrt{\dfrac{3}{14}}$

36. $\bar{x} = \dfrac{100}{9}, \bar{y} = 0, I_x = 20, R_x = \sqrt{\dfrac{1}{3}}$ **38.** $\bar{x} = 0, \bar{y} = \dfrac{32}{45};$

$I_x = \dfrac{5}{6}, R_x = \dfrac{\sqrt{5}}{3}; I_y = \dfrac{11}{30}, R_y = \sqrt{\dfrac{11}{45}}; I_0 = \dfrac{6}{5}, R_0 = \dfrac{2}{\sqrt{5}}$

40. Average value over the square $= \dfrac{1}{4}$, average value over the

circle $= \dfrac{1}{2\pi}$ **42.** 1   **44.** 1   **46.** a) $\dfrac{L}{2\sqrt{3}}$ b) $\dfrac{L}{\sqrt{3}}$

48. a) $I_{x = 5/7} = \dfrac{48}{35}; I_{y = 11/14} = \dfrac{47}{14}$ b) $I_{x = 1} = \dfrac{88}{35}; I_{y = 2} = 24$

52. $\bar{x} = 9.32, \bar{y} = 2.48$ **54.** $h = 3s$, then the centroid is on the base of the triangle.

Section 13.3, pp. 942–943

2. π **4.** $\dfrac{\pi}{2}$ **6.** 2π **8.** $\dfrac{4}{3}$ **10.** $\dfrac{4}{3}$ **12.** $\dfrac{\pi(e-1)}{4e}$

14. $\pi(\ln(4) - 1)$ **16.** a) Proof b) $\sqrt{\pi}$ **18.** $\dfrac{8+\pi}{4}$ **20.** $\dfrac{64\pi^3}{27}$

22. $\dfrac{3\pi - 8}{2}$ **24.** $\dfrac{9\pi + 16}{3}$ **26.** 2π **28.** $I_x = \dfrac{ka^6\pi}{6}; I_0 = \dfrac{ka^6\pi}{3}$

30. $\dfrac{35\pi}{16}$ **32.** $\dfrac{27\pi}{32} + \dfrac{10}{3}$ **34.** $\dfrac{6\pi\sqrt{2} + 40\sqrt{2} - 64}{9}$ **36.** $\dfrac{3}{2}$

38. a) $\dfrac{\pi}{2}$

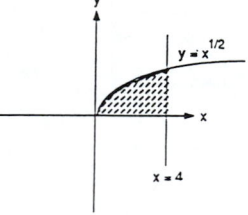

indicated area   Graph 13.3.38a

b) $\dfrac{2\pi(3\pi + 4)}{3}$

Section 13.4, pp. 948–950

2. $\displaystyle\int_0^1\int_0^2\int_0^3 dz\,dy\,dx = 6, \int_0^2\int_0^1\int_0^3 dz\,dx\,dy, \int_0^3\int_0^2\int_0^1 dx\,dy\,dz, \cdots$

$\displaystyle\int_0^2\int_0^3\int_0^1 dx\,dz\,dy, \int_0^3\int_0^1\int_0^2 dy\,dx\,dz, \int_0^1\int_0^3\int_0^2 dy\,dz\,dx$

4. $\displaystyle\int_0^2\int_0^3\int_0^{\sqrt{4-x^2}} dz\,dy\,dx = 3\pi, \int_0^3\int_0^2\int_0^{\sqrt{4-x^2}} dz\,dx\,dy, \int_0^2\int_0^{\sqrt{4-x^2}}\int_0^3$

$dy\,dz\,dx, \displaystyle\int_0^2\int_0^{\sqrt{4-z^2}}\int_0^3 dy\,dx\,dz, \int_0^3\int_0^2\int_0^{\sqrt{4-z^2}} dx\,dy\,dz, \int_0^3\int_0^2\int_0^{\sqrt{4-z^2}}$

$dx\,dz\,dy$ **6.** -6 **8.** $\dfrac{3}{2}$ **10.** 0   **12.** $\dfrac{16}{3}$ **14.** $\dfrac{1}{12}$ **16.** 1

18. $\dfrac{8\ln 8}{3}$ **20.** a) $\displaystyle\int_0^1\int_0^1\int_{-1}^{-\sqrt{z}} dy\,dz\,dx$

b) $\displaystyle\int_0^1\int_0^1\int_{-1}^{-\sqrt{z}} dy\,dx\,dz$ c) $\displaystyle\int_0^1\int_{-1}^{-\sqrt{z}}\int_0^1 dx\,dy\,dz$

d) $\displaystyle\int_{-1}^0\int_0^{y^2}\int_0^1 dx\,dz\,dy$ e) $\displaystyle\int_{-1}^0\int_0^{y^2}\int_0^1 dz\,dx\,dy$ **22.** $3e - 6$ **24.** $\dfrac{\sin^2 4}{2}$

26. 0   **28.** $\dfrac{a^3}{8}$ **30.** $\dfrac{2}{3}$ **32.** $\dfrac{4}{\pi^2}$ **34.** $-\dfrac{32}{3}$ **36.** 1   **38.** $\dfrac{2}{3}$

40. 4π **42.** $\dfrac{\pi}{2}$ **44.** 3   **46.** $\dfrac{x^2}{4} + \dfrac{y^2}{4} + \dfrac{z^2}{9} > 1$

Section 13.5, pp. 952–955

2. $I_x = 208, I_y = 280, I_z = 360$ **4.** a) $\bar{x} = \bar{y} = \bar{z} = \dfrac{1}{4}; I_z = \dfrac{1}{30},$

$I_y = I_x = \dfrac{1}{30}$ b) $R_x = \sqrt{\dfrac{1}{5}}$ **6.** a) $\bar{x} = -\dfrac{1}{2}, \bar{y} = 0$ b) 5π

8. $\bar{x} = 0, \bar{y} = 4, \bar{z} = 0, I_x = \dfrac{400}{3}, I_y = \dfrac{16}{3}, I_z = \dfrac{400}{3},$

$R_x = R_z = \sqrt{\dfrac{50}{3}}, R_y = \sqrt{\dfrac{2}{3}}$ **10.** $I_L = 696, R_L = \sqrt{\dfrac{58}{3}}$

12. $I_L = \dfrac{160}{3}, R_L = \sqrt{\dfrac{20}{3}}$ **14.** $\bar{x} = \dfrac{4}{3}, \bar{y} = \bar{z} = 0, I_x = 90,$

$I_y = 66, I_z = 120, R_x = \sqrt{\dfrac{5}{2}}, R_y = \sqrt{\dfrac{11}{6}}, R_z = \sqrt{\dfrac{10}{3}}$ **16.** a) $\dfrac{32k}{15}$

b) 1.25 c) 0.734656 d) 1.142857   **18.** $\dfrac{7}{5} ma^2$

20. $I_4 = 208 + 8abc, I_{-4/3} = 16\left(\dfrac{39 + 2abc}{3}\right)$ **22.** a) $\bar{x} = \dfrac{13}{14},$

$\bar{y} = \dfrac{13}{7}, \bar{z} = \dfrac{13}{14}$ b) $\bar{x} = \dfrac{3}{2}, \bar{y} = \dfrac{7}{2}, \bar{z} = \dfrac{1}{4}$ c) $\bar{x} = \dfrac{25}{14}, \bar{y} = \dfrac{37}{7},$

$\bar{z} = -\dfrac{5}{14}$ d) $\bar{x} = \dfrac{37}{36}, \bar{y} = \dfrac{46}{13}, \bar{z} = \dfrac{7}{26}$ **24.** $h < \sqrt{6}\, s$

Section 13.6, pp. 960–962

2. $\dfrac{9\pi\left(8\sqrt{2} - 7\right)}{2}$ **4.** $\dfrac{37\pi}{15}$ **6.** $\dfrac{\pi}{3}$ **8.** 2π **10.** $\dfrac{5\pi}{2}$

12. $\dfrac{\pi(2\sqrt{2} - 1)}{9}$ **14.** 12π **16.** 8π **18.** $3\pi \ln\left(\dfrac{2 + \sqrt{3}}{2 - \sqrt{3}}\right)$

20. $11\pi\sqrt{3}$ **22.** a) $\displaystyle\int_0^{\pi/2}\int_0^{3/\sqrt{2}}\int_r^{\sqrt{9 - r^2}} r\, dz\, dr\, d\theta$ b) $\displaystyle\int_0^{\pi/2}\int_0^{\pi/4}\int_0^3 \rho^2$

$\sin\phi\, d\rho\, d\phi\, d\theta$ c) $\dfrac{9\pi\left(2 - \sqrt{2}\right)}{4}$ **24.** $\displaystyle\int_{-\pi/2}^{\pi/2}\int_0^1\int_0^{r\cos\theta} r^3\, dz\, dr$

$d\theta = \dfrac{2}{5}$ **26.** a) $I_z = \displaystyle\int_0^{2\pi}\int_0^1\int_0^{\sqrt{1 - r^2}} r^3\, dz\, dr\, d\theta$

b) $I_z = \displaystyle\int_0^{2\pi}\int_0^{\pi/2}\int_0^1 (\rho^2 \sin^2\phi)\, \rho^2 \sin\phi\, d\rho\, d\phi\, d\theta$ c) $\dfrac{4\pi}{15}$

28. π **30.** 16π **32.** 32π **34.** $\dfrac{\pi\left(8\sqrt{3} - 7\right)}{6}$ **36.** $\dfrac{3\pi}{16}$

38. $\bar{z} = \dfrac{10}{3\left(5 - 2\sqrt{2}\right)}, \bar{x} = \bar{y} = 0$ **40.** $\bar{x} = \bar{y} = \dfrac{3}{\pi}, \bar{z} = \dfrac{3}{4}$

42. $\dfrac{\pi}{4}$ **44.** $\bar{x} = \dfrac{9\sqrt{3}}{32}, \bar{y} = \bar{z} = 0$

46. $I_z = \dfrac{a^4 h\pi}{10}$, where a is the base radius   **48.** a) $\bar{z} = \dfrac{1}{2},$

$\bar{x} = \bar{y} = 0, I_z = \dfrac{\pi}{8}, R_z = \dfrac{\sqrt{3}}{2}$ b) $\bar{z} = \dfrac{5}{14}, \bar{x} = \bar{y} = 0, I_z = \dfrac{2\pi}{7},$

$R_z = \sqrt{\dfrac{5}{7}}$ **50.** $\bar{z} = \dfrac{2h^2 + 3h}{3h + 6}, \bar{x} = \bar{y} = 0, I_z = \dfrac{\pi a^4\, (h^2 + 2h)}{4},$

$R_z = \dfrac{a}{\sqrt{2}}$ **52.** $\dfrac{a^3\pi}{18}$ **54.** $\dfrac{8\pi}{3}$ **56.** $\dfrac{4\pi}{3}$ **58.** $\bar{z} = \dfrac{3}{8}, \bar{x} = \bar{y} = 0$

60. a) $I_z = \dfrac{8a^7\pi}{21}, R_z = \sqrt{\dfrac{10}{21}}\, a$ b) $I_z = \dfrac{a^6\pi^2}{8}, R_z = \dfrac{a}{\sqrt{2}}$

Section 13.7, pp. 968–969

2. a) u b) $-u$

4. a) u b) $\dfrac{3 \ln 2}{2}$

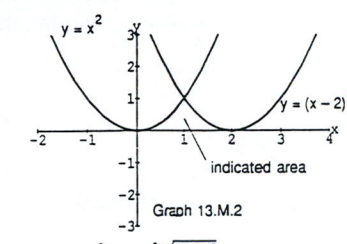

Graph 13.7.4

6. $ab\pi$ **8.** $e^{16} - 1$
10. 12   **12.** $a^2 b^2 c^2$

Section 13 Miscellaneous Exercises, pp. 969–972

2. $\dfrac{1}{5}$ **4.** $\dfrac{1 - e + e\sqrt{\pi}\, erf(1)}{2e}$

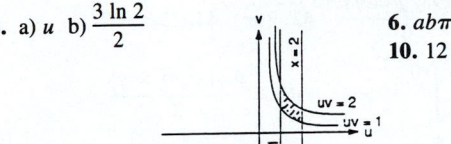

Graph 13.M.2

8. b) $\displaystyle\int_0^{a\cos\beta}\int_0^{(\tan\beta)x} \ln\,(x^2 + y^2)\, dy\, dx + \int_{a\cos\beta}^{a}\int_0^{\sqrt{a^2 - x^2}}$

$\ln\,(x^2 + y^2)\, dy\, dx$ **12.** a) 1 b) 0   **14.** 104   **16.** $\bar{x} = \dfrac{7}{6}, \bar{y} = 0$

18. $\bar{x} = \dfrac{8}{15}, \bar{y} = \dfrac{13}{30}, I_x = \dfrac{17}{280}, R_x = \sqrt{\dfrac{17}{70}},$

$I_y = \dfrac{1}{12}, R_y = \sqrt{\dfrac{1}{3}}$ **20.** a) $\dfrac{u_1 + u_2}{3}$ b) Proof

22. a) $\bar{x} = \dfrac{2a \sin\alpha}{3\alpha}, \bar{y} = 0; \underset{\alpha \to \pi}{\text{Lim}}\, \bar{x} = 0$ b) $\bar{x} = \dfrac{2a}{5\pi}, \bar{y} = 0$

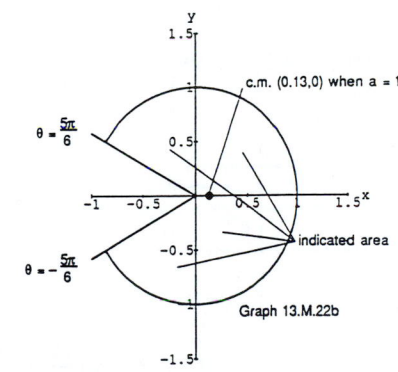

c.m. (0.13,0) when $a = 1$

Graph 13.M.22b

26. $R_x = \dfrac{a\sqrt{3\pi - 8}}{2\sqrt{6}}, R_y = \dfrac{a\sqrt{3\pi + 8}}{2\sqrt{6}}$ **28.** $\dfrac{\pi - 2}{4}$

32. $\dfrac{2(31 - 3^{5/2})}{3}$ **34.** a) $\displaystyle\int_0^1\int_0^{\sqrt{1 - x^2}}\int_0^{\sqrt{x^2 + y^2}} 6 + 4y\, dz\, dy\, dx =$

b) $\displaystyle\int_0^{\pi/2}\int_0^1\int_0^r (6 + 4r \sin\theta)\, r\, dz\, dr\, d\theta =$ c) $\displaystyle\int_0^{\pi/2}\int_{\pi/4}^{\pi/2}\int_0^{\csc\phi}$

$(6 + 4\rho \sin\phi \sin\theta)\, \rho^2 \sin\phi\, d\rho\, d\phi\, d\theta$ d) $\pi + 1$ **36.** a) The
sphere $x^2 + y^2 + z^2 = 4$ forms the top and bottom of the solid.
The sides of the solid are the right circular cylinder

$(x - 1)^2 + y^2 = 1$. b) $V = \displaystyle\int_0^{\pi/2}\int_0^{2\cos\theta}\int_{-\sqrt{4 - r^2}}^{\sqrt{4 - r^2}} r\, dz\, dr\, d\theta$

38. $\dfrac{52\pi}{3}$ **40.** $\dfrac{\pi\left(8\sqrt{2}-7\right)}{6}$ **42.** $2\pi^2$ **44.** $18\pi-24$

46. a) $V=\displaystyle\int_0^{2\pi}\int_0^2\int_2^{\sqrt{8-r^2}} r\,dz\,dr\,d\theta=\dfrac{8\pi\left(4\sqrt{2}-5\right)}{3}$

b) $V=\displaystyle\int_0^{2\pi}\int_0^{\pi/4}\int_{2\sec\phi}^{\sqrt{8}} \rho^2\sin\phi\,d\rho\,d\phi\,d\theta$ **48.** $\dfrac{8\pi(b^5-a^5)}{15}$

50. $\dfrac{4\pi a^3}{3}+\dfrac{16}{3}\left[\dfrac{b}{2}\,(3a^2-b^2)\sin^{-1}\dfrac{b}{\sqrt{a^2-b^2}}+\dfrac{b^2\sqrt{a^2-2b^2}}{2}\right.$

$\left.-\,a^3\tan^{-1}\dfrac{a}{\sqrt{a^2-2b^2}}\right]$, this integration was done by a
computer.   **52.** $ac-b^2=\pi^2$.

Chapter 14

Section 14.1, pp. 978–979

2. e   **4.** a   **6.** b   **8.** h   **10.** $-\sqrt{2}$ **12.** 80π **14.** 1   **16.** $-\dfrac{1}{6}$

18. $-4a^2$ **20.** $2\sqrt{2}-1$ **22.** $I_x=\dfrac{5}{3}a\sqrt{5},\ R_x=\sqrt{\dfrac{5}{3}},\ I_y=\dfrac{4}{3}$

$a\sqrt{5},\ R_y=\dfrac{2}{\sqrt{3}},\ I_z=\dfrac{1}{3}a\sqrt{5},\ R_z=\dfrac{1}{\sqrt{3}}$

24. $\left(0,-\dfrac{3}{5},0\right)$ **26.** $\left(\dfrac{19}{9},\dfrac{23}{18},\dfrac{4}{7}\sqrt{2}\right)$

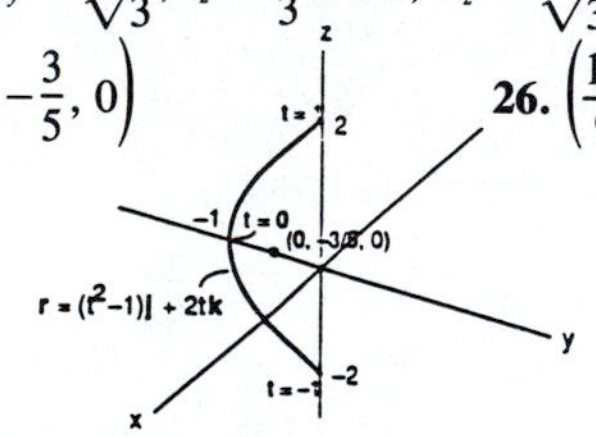

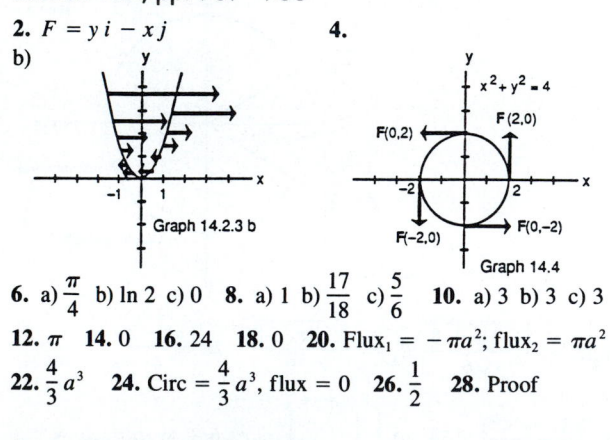

Section 14.2, pp. 987–988

2. $F=y\,i-x\,j$ **4.**
b)

Graph 14.2.3 b

$x^2+y^2=4$

$F(0,2)$ $F(2,0)$ $F(-2,0)$ $F(0,-2)$

Graph 14.4

6. a) $\dfrac{\pi}{4}$ b) $\ln 2$ c) 0   **8.** a) 1   b) $\dfrac{17}{18}$ c) $\dfrac{5}{6}$ **10.** a) 3   b) 3   c) 3

12. π **14.** 0   **16.** 24   **18.** 0   **20.** $\text{Flux}_1=-\pi a^2$; $\text{flux}_2=\pi a^2$

22. $\dfrac{4}{3}a^3$ **24.** Circ $=\dfrac{4}{3}a^3$, flux $=0$ **26.** $\dfrac{1}{2}$ **28.** Proof

Section 14.3, pp. 999–1001

2. $0,\ -a^2\pi$ **4.** $0,\ \dfrac{a^4\pi}{2}$ **6.** Flux $=2$, circ $=-3$ **8.** Flux $=\dfrac{1}{6}$,

circ $=-\dfrac{7}{6}$ **10.** Flux $=2$, circ $=0$ **12.** Flux $=-\dfrac{\pi^2}{8}$,

circ $=\pi$ **14.** $\dfrac{14\sqrt{2}}{15}$ **16.** -2 **18.** 0   **20.** 16π

26. Region enclosed by $1=\dfrac{1}{4}x^2+y^2$ **36.** πab **38.** $\dfrac{8}{5}\sqrt{3}$

Section 14.4, pp. 1010–1012

2. $\dfrac{49}{3}\pi$ **4.** $\dfrac{7}{3}$ **6.** $3+2\ln 2$ **8.** $\pi\sqrt{c^2+1}$ **10.** $\dfrac{\pi\sqrt{3}}{2}-\dfrac{\pi}{6}$

12. $2\pi ah$ **14.** $\dfrac{8}{3}+2\sqrt{2}$ **16.** 0   **18.** $\dfrac{56}{3}$ **20.** 0   **22.** $\dfrac{1}{4}\pi a^4$

24. $\dfrac{a^2\pi}{2}$ **26.** 2π **28.** 2   **30.** 128π **32.** $\left(\dfrac{3}{2},0,\dfrac{6}{\pi}\right)$

36. a) $\left(0,0,\dfrac{2h}{3}\right)$ b) $\left(0,0,\dfrac{2h\sqrt{h^2+a^2}+3ah}{3(\sqrt{h^2+a^2}+a)}\right)$ c) $\dfrac{\sqrt{2\sqrt{37}-2}}{2}\,a$

38. $5\pi\sqrt{2}$ **40.** $\dfrac{7}{2}$ **42.** $\dfrac{16\sqrt{2}}{3}$

Section 14.5, pp. 1021–1022

2. a) 3   b) 0   c) 4π **4.** 0   **6.** $42\pi+16$ **8.** $4\pi a^5$ **10.** 60π

12. $3\sqrt{2}\pi-3\pi$ **14.** Area of S **16.** $a=3,\ b=\dfrac{3}{2},\dfrac{27}{2}$ **20.** $\dfrac{\pi a}{6}$

Section 14.6, pp. 1029–1030

2. 9π **4.** 0   **6.** $-\dfrac{\pi a^6}{8}$ **8.** $-2\sqrt{2}\pi$ **10.** $-\pi$

18. No such field exists.

Section 14.7, pp. 1039–1040

2. Conservative   **4.** Not conservative   **6.** Conservative
8. $f(x,y,z)=(y+z)x+zy+C$ **10.** $f(x,y,z)=xy\sin z+C$
14. -2 **16.** $9\ln 2$ **18.** $-\pi$ **20.** $-\ln 2$ **22.** $\ln 4$ **24.** 18
30. a) -1 b) 2   c) 0   d) 0   **32.** $f(x,y,z)=\dfrac{-GmM}{(x^2+y^2+z^2)^{1/2}}$

Section 14 Miscellaneous Exercises, pp. 1041–1044

2. Path 1: $\dfrac{1}{3}$; path 2: $\dfrac{5}{6}\sqrt{2}-\dfrac{1}{2}$; path 3: $\dfrac{1}{3}$ **4.** $\dfrac{7}{3}$

6. $\left(0,\dfrac{8a-a\pi}{4\pi-4a},0\right)$ **8.** $\bar{z}=\dfrac{32\sqrt{2}}{105}$; $I_z=\dfrac{7}{12}$; $R_z=\sqrt{\dfrac{7}{18}}$ **10.** 0

12.

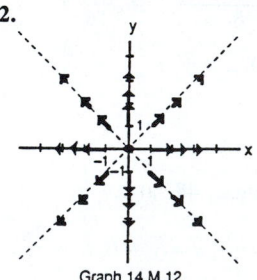

Graph 14.M.12

14. Flux $=-\dfrac{11}{3}$; circ $=0$ **18.** 0   **20.** $a=2,\ b=1$

22. $\dfrac{1}{18}\left(21\sqrt{21}-27\right)$ **24.** a) $4\pi-8$ b) 8   **28.** 50   **30.** 128π

32. $\dfrac{14}{3}$ **34.** 256π **36.** $\dfrac{3\pi}{2}+\dfrac{4}{3}$ **38.** 861.1 tons   **40.** $2\pi a^3$

42. 0   **44.** Conservative   **46.** Conservative   **48.** $f(x,y,z)=$
$\sin xz+e^y+C$ **50.** 5   **52.** 2   **54.** a) 0   b) -3 **56.** 0
58. 2π **60.** $2ab\pi$

Chapter 15

Section 15.1, pp. 1050–1051

2. Third Order, Non-linear   **4.** Second Order, Linear

16. $-\frac{1}{2}e^{-y} = e^{\sin x} - \frac{5}{4}$ **18.** $\tan^{-1}(y+1) = \frac{x^2}{2}\ln x - \frac{x^2}{4} +$

$x \ln x - x + \frac{5}{4}$ **20.** $-\ln|\cos y| = \sec^{-1}(2x) - \frac{\pi}{3}$

22. $\approx 11{:}23$ a.m.   **28.** $x \sin \frac{y}{x} = 2$ **30.** $\sqrt[3]{3x} = C_1 y$

Section 15.2, pp. 1054–1055

2. Not exact   **4.** Exact   **6.** Exact   **8.** Not exact

10. $x^2 e^y + e^x + e^y = C$ **12.** $xy - \frac{x^4}{4} + \sin\left(y - \frac{\pi}{4}\right) = C$

14. $\sin xy + x + y^2 - 3y = C$ **16.** $\ln(x^2+y^2) - \tan^{-1}\left(\frac{x}{y}\right) = C$

18. $\sum_{n=0}^{\infty} \frac{(-1)^n}{xy^n} = C$ **20.** $-\frac{1}{x} - \frac{y}{x^2} = C$ **22. a)** $\frac{x}{y} = C$

b) $\frac{x}{y} = C$ **c)** $\frac{x}{y} = C$ **d)** $\frac{x}{y} = C$ **24.** $x^2 + y^2 + 2x e^y = 8$

26. $x^2 + y^2 + xy + x + y = 37$ **28.** $x \sin^{-1} y + y \sin^{-1} x = \pi$

30. $b = -3, x + \frac{3y^2}{x} = C$

Section 15.3, pp. 1060–1061

2. $y = \frac{1}{2}x e^{x/2} + C e^{x/2}$ **4.** $y = \sin x + \frac{\cos x}{x} + \frac{C}{x}$

6. $y = x^2 e^{-2x} + C e^{-2x}$ **8.** $y = -e^{-x} \operatorname{sech} x + C \operatorname{sech} x$

10. $y = \frac{x^3}{5} - \frac{12}{5x^2}$ **12.** $y = x^2 \sec x + \left(\frac{18}{\pi^2} - 2\right)x^2$

14. a) $c = \frac{G}{100Vk} + \left(c_0 - \frac{G}{100Vk}\right)e^{-kt}$ **b)** $\frac{G}{100Vk}$

16. a) $i = i_0 e^{-Rt/L}$ **b)** $t = \frac{L}{R}\ln 2$ **c)** $i = i_0 e^{-1}$

18. a) $x = 11000 e^{0.10t} - 10000$ **b)** ≈ 23.03

20. a) 12833.3 ft or 2.43 miles **b)** 1807.6 sec or 30.1 minutes

Section 15.4, pp. 1065–1066

2. $y = C_1 e^{-2x} + C_2 e^{-3x}$ **4.** $y = C_1 e^{3x} + C_2 e^{-x}$
6. $y = (C_1 x + C_2) e^{-3x}$ **8.** $y = (C_1 x + C_2) e^{\sqrt{2}x}$
10. $y = e^{3x}(C_1 \cos x + C_2 \sin x)$ **12.** $y = e^{-4x}(C_1 \cos 3x +$

$C_2 \sin 3x)$ **14.** $y = -\frac{2}{3}e^{-x/2} - \frac{1}{3}e^x$ **16.** $y = 8 e^{-3x}$

18. $y = e^{x/4}\left(4 \cos \frac{\sqrt{3}}{4}x + \frac{4}{\sqrt{3}} \sin \frac{\sqrt{3}}{4}x\right)$ **20.** $y = 5x - 3$

22. $y = \frac{\sqrt{3}}{3} \sin 3x$ **24.** $y = e^{3x}(7 \cos x - 20 \sin x)$

Section 15.5, pp. 1078–1079

2. $y = -2 + \sin x \ln|\sec x + \tan x| + C_1 \cos x + C_2 \sin x$

4. $y = C_1 x e^{-x} + \frac{e^x}{4} + C_2 e^{-x}$ **6.** $y = C_1 e^x + C_2 e^{-x} - x$

8. $y = C_1 e^x + C_2 e^{-x} - \frac{1}{2}\sin x$ **10.** $y = -\frac{2^x}{\ln 2} + C_1 + \frac{2^x}{\ln 2 - 1}$

$+ C_2 e^x$ **12.** $y = \frac{e^x \sin x - e^x \cos x}{2} + C_1 + C_2 e^x$

14. $y = C_1 e^{5x} + C_2 e^{-2x} - \frac{1}{5}x + \frac{9}{25}$ **16.** $y = C_1 x e^x +$

$C_2 e^{-x} + x^2 - 4x + 6$ **18.** $y = C_1 \cos x + C_2 \sin x + \frac{1}{5}e^{2x}$

20. $y = C_1 \cos x + C_2 \sin x + 2x + \frac{3}{2}e^x$ **22.** $y = C_1 x e^{-x} +$

$C_2 e^{-x} - \frac{24}{25}\cos 2x - \frac{18}{25}\sin 2x$ **24.** $y = -x e^{-2x} +$

$x e^{-x} - \frac{1}{2} + \frac{3}{4} + C_1 e^{-2x} + C_2 e^{-x}$ **26.** $y = C_1 + C_2 e^x +$

$4x^2 + 5x$ **28.** $y = 2x^3 - \frac{1}{2}x^2 + \frac{2}{7}x + C_1 + C_2 e^{-7x}$

30. $A = 0, B = -1; y = C_1 + C_2 e^x - \sin x$ **32.** $y = C_1 x e^{2x} + C_2 e^{2x} + x^2 e^{2x}$ for both a and b.   **34.** $y = C_1 + C_2 e^{9x} + x e^{9x}$ for both a and b.   **36.** $y = -x \cos x - \sin x \ln(\sin x) + C_1 \cos x + C_2 \sin x$ **38.** $y = C_1 \cos 2x + C_2 \sin 2x + \frac{1}{3}\sin x$

40. $y = e^{-2x}(C_1 \cos x + C_2 \sin x) + \frac{1}{5}x + \frac{6}{25}$ **42.** $y = C_1 \cos 3x$

$+ C_2 \sin 3x - \frac{1}{8}\cos x + x$ **44.** $y = -x e^x - x e^{2x} + C_1 e^x + C_2 e^{2x}$

46. $y = C_1 e^{-4x} + \frac{1}{4}x - \frac{1}{16}$ **48.** $y = C_1 e^{-x} - \frac{1}{2}\cos x + \frac{1}{2}\sin x$

50. $y = -\frac{1}{5}\cos x + \frac{1}{5}e^{2x}$ **52.** $y = (C_1 e^x + x^2 + 2x + 2)^{-1}$

54. $C_1 \ln\left(\frac{C_2 + y}{C_2 - y}\right) = \frac{1}{2}x + C_3$

56. $C_1 \tan^{-1}\left(\frac{y}{C_1}\right) = x + C_2$ **58.** $y = \frac{x^4}{4} + x - \frac{25}{4}$ **60.** $y = 2 \ln$

$\left|\cos\left(\frac{1}{2}(x+1)\right)\right| + 2$ **62.** $y = -e^{-x} + 1$ **64.** $s = \frac{mg}{k}t + \frac{m^2 g}{k^2}$

$e^{-(k/m)t} - \frac{m^2 g}{k^2}$ **66.** $y = \frac{C_1}{x} + C_2 x^2$ **68.** $y = C_1 x^{1/2} \ln x +$

$C_2 x^{1/2}$ **70.** $y = C_1 x^3 \ln x + C_2 x^3$ **72.** $y = \frac{C_1 \cos x}{\sqrt{x}} + \frac{C_2 \sin x}{\sqrt{x}}$

74. $y = C_1 x^2 e^x + C_2 e^x$ **76.** $y = x + \frac{4}{3} + C_1 e^x + C_2 e^{-3x} +$

$C_3 e^{2x}$ **78.** $y = 7 + (C_1 + C_2 x + C_3 x^2 + C_4 x^3) e^x$

Section 15.6, pp. 1082–1083

2. $x = 0.5 \sin 8t$ **4.** $\theta = \theta_0 \cos \sqrt{\frac{g}{L}}\, t$ **6.** $c \approx 5.09$

Section 15.7, p. 1088

2. b) $y_1 = \frac{a_0}{x}\cos x, y_2 = \frac{a_1}{x}\sin x$ **c)** $y = -\frac{\pi}{x}\cos x + \frac{\pi}{x}\sin x$.

4. $y(x) = 1 - x + \frac{x^2}{2!} - \frac{x^3}{3!} + \cdots = \sum_{n=0}^{\infty} \frac{(-1)^n x^n}{n!}$

6. $y(x) = -1 + 2(x-1) - \frac{4(x-1)^2}{2!} + \frac{8(x-1)^3}{3!} - \cdots = \sum_{n=0}^{\infty}$

$\frac{(-1)^{n+1}(2^n)(x-1)^n}{n!}$ **8.** $y(x) = 1 + 0 - \frac{x^2}{2!} + 0 + \frac{x^4}{4!} +$

$\cdots = \sum_{n=0}^{\infty} \frac{(-1)^n x^{2n}}{(2n)!}$

10. $y(x) = -1 + 2x - \dfrac{x^2}{2!} + \dfrac{3x^3}{3!} - \dfrac{x^4}{4!} + \dfrac{3x^5}{5!} - \cdots =$

$-1 + 2x - \displaystyle\sum_{n=1}^{\infty} \dfrac{x^{2n}}{(2n)!} + 3 \sum_{n=1}^{\infty} \dfrac{x^{2n+1}}{(2n+1)!}$

14. $y(x) = 0 + x + \dfrac{2x^2}{2!} + \dfrac{3x^3}{3!} + \dfrac{4x^4}{4!} + \dfrac{5x^5}{5!} + \cdots = \displaystyle\sum_{n=0}^{\infty} \dfrac{x^n}{(n-1)!}$

Section 15.8, pp. 1092–1093

2.

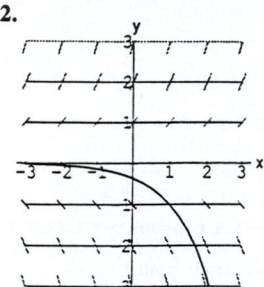

Graph 15.8.2

4.

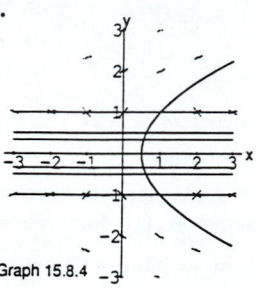

Graph 15.8.4

6.

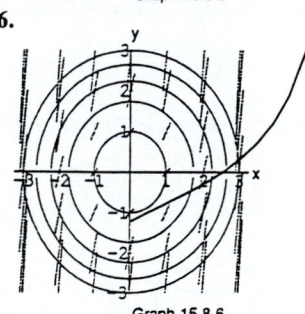

Graph 15.8.6

10. $y' = y,\ y(0) = 1$ **12.** $y' = \dfrac{1}{x},\ y(1) = 0$ **14.** $y = y_0 + \displaystyle\int_{x_0}^{x} f(t)\,dt$ **16.** $y_0 = 1,\ y_1 = 1 + x,\ y_2 = 1 + x + \dfrac{x^2}{2},\ y_3 = +x + \dfrac{1}{2}x^2 + \dfrac{1}{6}x^3$ **18.** $y_0 = 1,\ y_1 = \dfrac{1}{2}x^2,\ y_2 = {,}6)\,x^3 + \dfrac{1}{2}x^2,$

$y_3 = x^2 + \dfrac{1}{6}x^3 + \dfrac{1}{24}x^4$ **20.** $y_0 = 1,\ y_1 = -1 - x + x^2,$

$y_2 = \dfrac{1}{6} + x + \dfrac{3}{2}x^2 - \dfrac{1}{3}x^3,\ y_3 = -\dfrac{1}{4} - \dfrac{1}{6}x + \dfrac{1}{2}x^2 - \dfrac{1}{2}x^3 + \dfrac{1}{12}x^4$

22. $y_3 = \dfrac{1}{3}x^3 + \dfrac{x^7}{63} + \dfrac{2x^{11}}{2079} + \dfrac{x^{15}}{59535}$

Section 15.9, pp. 1097–1098

2. e **4.** $y(1) \approx 2.7182507$ **6.** $y = \tan x$

8. $y = -0.175611085$ **10.** $y = 0.810317873$

12. a) $y = 0.367879689$ b) $y = 0.367879688$

Section 15 Miscellaneous Exercises, pp. 1099–1101

2. $\tan^{-1}x + \ln(\ln y) = 0$ **4.** $\ln x - \dfrac{1}{2}\left(\ln\left(\dfrac{y}{x}\right)\right)^2 = 0$ **6.** $e^x + x \ln y + y - 3 = 0$ **8.** $y = \dfrac{x^2}{4} + \dfrac{1}{2} + \dfrac{1}{4x^2}$ **10.** $y = \dfrac{(\ln x)^2}{2} + \ln x + 1$ **12.** $y = \dfrac{1}{2}e^{-2x} + \dfrac{1}{3}e^{-3x}$ **14.** $y = (-20x + 4)e^{4x}$

16. $y = \left(\dfrac{-2 + \sqrt{5}}{2\sqrt{5}}\right)e^{(1+\sqrt{5})x} + \left(\dfrac{2 + \sqrt{5}}{2\sqrt{5}}\right)e^{(1-\sqrt{5})x}$

18. $y = -x \cos x + \sin x \ln(\sin x) + \sin x$

20. $y = e^x\left(\dfrac{1}{2}\cos 2x - \dfrac{1}{4}\sin 2x\right) + \dfrac{1}{2}e^{-x}$ **22.** $x^2 + y^2 = Cy$

24. d   **26.** $\ln|x + 2| + \dfrac{1}{2}\ln\left(\dfrac{(x + 2)^2 + (y - 1)^2}{(y - 1)^2}\right) + \tan^{-1}\left(\dfrac{x + 2}{y - 1}\right) = C$ **28.** $c = -2;\ x - \dfrac{y^2}{x} + C = 0$

32. $y = 10 \cos 10t + 5 \sin 10t$; amplitude $= 5\sqrt{5}$;

period $= \dfrac{2\pi}{10} \approx 6.28$

34. $|x| < 1$ **36.** $y = 0 + 0 + \dfrac{x^2}{2!} + \dfrac{x^3}{3!} + \dfrac{x^4}{4!} + \cdots$

38. $x = \sin^{-1}\left(\pm\dfrac{n\pi}{2} - \tan^{-1}y_0\right)$, n odd.

Appendices

Appendix A.7, p. A–26

2. $\dfrac{g(b) - g(a)}{2(f(b) - f(a))}$ **4.** $\dfrac{\pi}{4}$

Appendix A.10, p. A–37

2. 0   **4.** 2   **6.** a) 6 b) 6   **8.** a) -1 b) -1 **10.** $\left(1, \dfrac{19}{11}\right)$

12. $\left(\dfrac{6}{7}, \dfrac{10}{7}, -\dfrac{2}{7}\right)$ **14.** $\left(0, -\dfrac{3}{2}, \dfrac{5}{2}\right)$ **16.** $\left(2, 0, -\dfrac{3}{2}, -\dfrac{3}{2}\right)$ **18.** 2

20. a) $C + D = a$, $-Cr_2 - Dr_1 = b$ b) Not 0 when $r_1 \neq r_2$

Index

A Brief Table of Integrals

1. $\int u\,dv = uv - \int v\,du$

2. $\int a^u\,du = \dfrac{a^u}{\ln a} + C, \qquad a \neq 1, \qquad a > 0$

3. $\int \cos u\,du = \sin u + C$

4. $\int \sin u\,du = -\cos u + C$

5. $\int (ax+b)^n\,dx = \dfrac{(ax+b)^{n+1}}{a(n+1)} + C, \qquad n \neq -1$

6. $\int (ax+b)^{-1}\,dx = \dfrac{1}{a}\ln|ax+b| + C$

7. $\int x(ax+b)^n\,dx = \dfrac{(ax+b)^{n+1}}{a^2}\left[\dfrac{ax+b}{n+2} - \dfrac{b}{n+1}\right] + C, \qquad n \neq -1, -2$

8. $\int x(ax+b)^{-1}\,dx = \dfrac{x}{a} - \dfrac{b}{a^2}\ln|ax+b| + C$

9. $\int x(ax+b)^{-2}\,dx = \dfrac{1}{a^2}\left[\ln|ax+b| + \dfrac{b}{ax+b}\right] + C$

10. $\int \dfrac{dx}{x(ax+b)} = \dfrac{1}{b}\ln\left|\dfrac{x}{ax+b}\right| + C$

11. $\int (\sqrt{ax+b})^n\,dx = \dfrac{2}{a}\dfrac{(\sqrt{ax+b})^{n+2}}{n+2} + C, \qquad n \neq -2$

12. $\int \dfrac{\sqrt{ax+b}}{x}\,dx = 2\sqrt{ax+b} + b\int \dfrac{dx}{x\sqrt{ax+b}}$

13. (a) $\int \dfrac{dx}{x\sqrt{ax+b}} = \dfrac{2}{\sqrt{-b}}\tan^{-1}\sqrt{\dfrac{ax+b}{-b}} + C, \qquad \text{if } b < 0$

    (b) $\int \dfrac{dx}{x\sqrt{ax+b}} = \dfrac{1}{\sqrt{b}}\ln\left|\dfrac{\sqrt{ax+b} - \sqrt{b}}{\sqrt{ax+b} + \sqrt{b}}\right| + C, \qquad \text{if } b > 0$

14. $\int \dfrac{\sqrt{ax+b}}{x^2}\,dx = -\dfrac{\sqrt{ax+b}}{x} + \dfrac{a}{2}\int \dfrac{dx}{x\sqrt{ax+b}} + C$

15. $\int \dfrac{dx}{x^2\sqrt{ax+b}} = -\dfrac{\sqrt{ax+b}}{bx} - \dfrac{a}{2b}\int \dfrac{dx}{x\sqrt{ax+b}} + C$

16. $\int \dfrac{dx}{a^2+x^2} = \dfrac{1}{a}\tan^{-1}\dfrac{x}{a} + C$

17. $\int \dfrac{dx}{(a^2+x^2)^2} = \dfrac{x}{2a^2(a^2+x^2)} + \dfrac{1}{2a^3}\tan^{-1}\dfrac{x}{a} + C$

18. $\int \dfrac{dx}{a^2-x^2} = \dfrac{1}{2a}\ln\left|\dfrac{x+a}{x-a}\right| + C$

19. $\int \dfrac{dx}{(a^2-x^2)^2} = \dfrac{x}{2a^2(a^2-x^2)} + \dfrac{1}{2a^2}\int \dfrac{dx}{a^2-x^2}$

20. $\int \dfrac{dx}{\sqrt{a^2+x^2}} = \sinh^{-1}\dfrac{x}{a} + C = \ln|x + \sqrt{a^2+x^2}| + C$

Continued

21. $\displaystyle\int \sqrt{a^2 + x^2}\, dx = \frac{x}{2}\sqrt{a^2 + x^2} + \frac{a^2}{2}\sinh^{-1}\frac{x}{a} + C$

22. $\displaystyle\int x^2\sqrt{a^2 + x^2}\, dx = \frac{x(a^2 + 2x^2)\sqrt{a^2 + x^2}}{8} - \frac{a^4}{8}\sinh^{-1}\frac{x}{a} + C$

23. $\displaystyle\int \frac{\sqrt{a^2 + x^2}}{x}\, dx = \sqrt{a^2 + x^2} - a\sinh^{-1}\left|\frac{a}{x}\right| + C$

24. $\displaystyle\int \frac{\sqrt{a^2 + x^2}}{x^2}\, dx = \sinh^{-1}\frac{x}{a} - \frac{\sqrt{a^2 + x^2}}{x} + C$

25. $\displaystyle\int \frac{x^2}{\sqrt{a^2 + x^2}}\, dx = -\frac{a^2}{2}\sinh^{-1}\frac{x}{a} + \frac{x\sqrt{a^2 + x^2}}{2} + C$

26. $\displaystyle\int \frac{dx}{x\sqrt{a^2 + x^2}} = -\frac{1}{a}\ln\left|\frac{a + \sqrt{a^2 + x^2}}{x}\right| + C$

27. $\displaystyle\int \frac{dx}{x^2\sqrt{a^2 + x^2}} = -\frac{\sqrt{a^2 + x^2}}{a^2 x} + C$ 28. $\displaystyle\int \frac{dx}{\sqrt{a^2 - x^2}} = \sin^{-1}\frac{x}{a} + C$

29. $\displaystyle\int \sqrt{a^2 - x^2}\, dx = \frac{x}{2}\sqrt{a^2 - x^2} + \frac{a^2}{2}\sin^{-1}\frac{x}{a} + C$

30. $\displaystyle\int x^2\sqrt{a^2 - x^2}\, dx = \frac{a^4}{8}\sin^{-1}\frac{x}{a} - \frac{1}{8}x\sqrt{a^2 - x^2}\,(a^2 - 2x^2) + C$

31. $\displaystyle\int \frac{\sqrt{a^2 - x^2}}{x}\, dx = \sqrt{a^2 - x^2} - a\ln\left|\frac{a + \sqrt{a^2 - x^2}}{x}\right| + C$

32. $\displaystyle\int \frac{\sqrt{a^2 - x^2}}{x^2}\, dx = -\sin^{-1}\frac{x}{a} - \frac{\sqrt{a^2 - x^2}}{x} + C$

33. $\displaystyle\int \frac{x^2}{\sqrt{a^2 - x^2}}\, dx = \frac{a^2}{2}\sin^{-1}\frac{x}{a} - \frac{1}{2}x\sqrt{a^2 - x^2} + C$

34. $\displaystyle\int \frac{dx}{x\sqrt{a^2 - x^2}} = -\frac{1}{a}\ln\left|\frac{a + \sqrt{a^2 - x^2}}{x}\right| + C$ 35. $\displaystyle\int \frac{dx}{x^2\sqrt{a^2 - x^2}} = -\frac{\sqrt{a^2 - x^2}}{a^2 x} + C$

36. $\displaystyle\int \frac{dx}{\sqrt{x^2 - a^2}} = \cosh^{-1}\frac{x}{a} + C = \ln\left|x + \sqrt{x^2 - a^2}\right| + C$

37. $\displaystyle\int \sqrt{x^2 - a^2}\, dx = \frac{x}{2}\sqrt{x^2 - a^2} - \frac{a^2}{2}\cosh^{-1}\frac{x}{a} + C$

38. $\displaystyle\int \left(\sqrt{x^2 - a^2}\right)^n dx = \frac{x\left(\sqrt{x^2 - a^2}\right)^n}{n + 1} - \frac{na^2}{n + 1}\int \left(\sqrt{x^2 - a^2}\right)^{n-2} dx, \quad n \neq -1$

39. $\displaystyle\int \frac{dx}{\left(\sqrt{x^2 - a^2}\right)^n} = \frac{x\left(\sqrt{x^2 - a^2}\right)^{2-n}}{(2 - n)a^2} - \frac{n - 3}{(n - 2)a^2}\int \frac{dx}{\left(\sqrt{x^2 - a^2}\right)^{n-2}}, \quad n \neq 2$

40. $\displaystyle\int x\left(\sqrt{x^2 - a^2}\right)^n dx = \frac{\left(\sqrt{x^2 - a^2}\right)^{n+2}}{n + 2} + C, \quad n \neq -2$

41. $\displaystyle\int x^2\sqrt{x^2 - a^2}\, dx = \frac{x}{8}(2x^2 - a^2)\sqrt{x^2 - a^2} - \frac{a^4}{8}\cosh^{-1}\frac{x}{a} + C$

42. $\displaystyle\int \frac{\sqrt{x^2 - a^2}}{x}\, dx = \sqrt{x^2 - a^2} - a\sec^{-1}\left|\frac{x}{a}\right| + C$

43. $\displaystyle\int \frac{\sqrt{x^2-a^2}}{x^2}\,dx = \cosh^{-1}\frac{x}{a} - \frac{\sqrt{x^2-a^2}}{x} + C$

44. $\displaystyle\int \frac{x^2}{\sqrt{x^2-a^2}}\,dx = \frac{a^2}{2}\cosh^{-1}\frac{x}{a} + \frac{x}{2}\sqrt{x^2-a^2} + C$

45. $\displaystyle\int \frac{dx}{x\sqrt{x^2-a^2}} = \frac{1}{a}\sec^{-1}\left|\frac{x}{a}\right| + C = \frac{1}{a}\cos^{-1}\left|\frac{a}{x}\right| + C$

46. $\displaystyle\int \frac{dx}{x^2\sqrt{x^2-a^2}} = \frac{\sqrt{x^2-a^2}}{a^2 x} + C$

47. $\displaystyle\int \frac{dx}{\sqrt{2ax-x^2}} = \sin^{-1}\left(\frac{x-a}{a}\right) + C$

48. $\displaystyle\int \sqrt{2ax-x^2}\,dx = \frac{x-a}{2}\sqrt{2ax-x^2} + \frac{a^2}{2}\sin^{-1}\left(\frac{x-a}{a}\right) + C$

49. $\displaystyle\int (\sqrt{2ax-x^2})^n\,dx = \frac{(x-a)(\sqrt{2ax-x^2})^n}{n+1} + \frac{na^2}{n+1}\int (\sqrt{2ax-x^2})^{n-2}\,dx,$

50. $\displaystyle\int \frac{dx}{(\sqrt{2ax-x^2})^n} = \frac{(x-a)(\sqrt{2ax-x^2})^{2-n}}{(n-2)a^2} + \frac{(n-3)}{(n-2)a^2}\int \frac{dx}{(\sqrt{2ax-x^2})^{n-2}}$

51. $\displaystyle\int x\sqrt{2ax-x^2}\,dx = \frac{(x+a)(2x-3a)\sqrt{2ax-x^2}}{6} + \frac{a^3}{2}\sin^{-1}\frac{x-a}{a} + C$

52. $\displaystyle\int \frac{\sqrt{2ax-x^2}}{x}\,dx = \sqrt{2ax-x^2} + a\sin^{-1}\frac{x-a}{a} + C$

53. $\displaystyle\int \frac{\sqrt{2ax-x^2}}{x^2}\,dx = -2\sqrt{\frac{2a-x}{x}} - \sin^{-1}\left(\frac{x-a}{a}\right) + C$

54. $\displaystyle\int \frac{x\,dx}{\sqrt{2ax-x^2}} = a\sin^{-1}\frac{x-a}{a} - \sqrt{2ax-x^2} + C$

55. $\displaystyle\int \frac{dx}{x\sqrt{2ax-x^2}} = -\frac{1}{a}\sqrt{\frac{2a-x}{x}} + C$

56. $\displaystyle\int \sin ax\,dx = -\frac{1}{a}\cos ax + C$

57. $\displaystyle\int \cos ax\,dx = \frac{1}{a}\sin ax + C$

58. $\displaystyle\int \sin^2 ax\,dx = \frac{x}{2} - \frac{\sin 2ax}{4a} + C$

59. $\displaystyle\int \cos^2 ax\,dx = \frac{x}{2} + \frac{\sin 2ax}{4a} + C$

60. $\displaystyle\int \sin^n ax\,dx = \frac{-\sin^{n-1} ax\cos ax}{na} + \frac{n-1}{n}\int \sin^{n-2} ax\,dx$

61. $\displaystyle\int \cos^n ax\,dx = \frac{\cos^{n-1} ax\sin ax}{na} + \frac{n-1}{n}\int \cos^{n-2} ax\,dx$

62. (a) $\displaystyle\int \sin ax\cos bx\,dx = -\frac{\cos(a+b)x}{2(a+b)} - \frac{\cos(a-b)x}{2(a-b)} + C, \quad a^2 \neq b^2$

(b) $\displaystyle\int \sin ax\sin bx\,dx = \frac{\sin(a-b)x}{2(a-b)} - \frac{\sin(a+b)x}{2(a+b)}, \quad a^2 \neq b^2$

(c) $\displaystyle\int \cos ax\cos bx\,dx = \frac{\sin(a-b)x}{2(a-b)} + \frac{\sin(a+b)x}{2(a+b)}, \quad a^2 \neq b^2$

Continued

63. $\int \sin ax \cos ax \, dx = -\dfrac{\cos 2ax}{4a} + C$

64. $\int \sin^n ax \cos ax \, dx = \dfrac{\sin^{n+1} ax}{(n+1)a} + C, \qquad n \neq -1$

65. $\int \dfrac{\cos ax}{\sin ax} \, dx = \dfrac{1}{a} \ln |\sin ax| + C$

66. $\int \cos^n ax \sin ax \, dx = -\dfrac{\cos^{n+1} ax}{(n+1)a} + C, \qquad n \neq -1$

67. $\int \dfrac{\sin ax}{\cos ax} \, dx = -\dfrac{1}{a} \ln |\cos ax| + C$

68. $\int \sin^n ax \cos^m ax \, dx = -\dfrac{\sin^{n-1} ax \cos^{m+1} ax}{a(m+n)} + \dfrac{n-1}{m+n} \int \sin^{n-2} ax \cos^m ax \, dx,$

$$n \neq -m \qquad \text{(If } n = -m, \text{ use No. 86.)}$$

69. $\int \sin^n ax \cos^m ax \, dx = \dfrac{\sin^{n+1} ax \cos^{m-1} ax}{a(m+n)} + \dfrac{m-1}{m+n} \int \sin^n ax \cos^{m-2} ax \, dx,$

$$m \neq -n \qquad \text{(If } m = -n, \text{ use No. 87.)}$$

70. $\int \dfrac{dx}{b + c \sin ax} = \dfrac{-2}{a\sqrt{b^2 - c^2}} \tan^{-1}\left[\sqrt{\dfrac{b-c}{b+c}} \tan\left(\dfrac{\pi}{4} - \dfrac{ax}{2} \right) \right] + C, \qquad b^2 > c^2$

71. $\int \dfrac{dx}{b + c \sin ax} = \dfrac{-1}{a\sqrt{c^2 - b^2}} \ln\left| \dfrac{c + b \sin ax + \sqrt{c^2 - b^2} \cos ax}{b + c \sin ax} \right| + C, \qquad b^2 < c^2$

72. $\int \dfrac{dx}{1 + \sin ax} = -\dfrac{1}{a} \tan\left(\dfrac{\pi}{4} - \dfrac{ax}{2} \right) + C$

73. $\int \dfrac{dx}{1 - \sin ax} = \dfrac{1}{a} \tan\left(\dfrac{\pi}{4} + \dfrac{ax}{2} \right) + C$

74. $\int \dfrac{dx}{b + c \cos ax} = \dfrac{2}{a\sqrt{b^2 - c^2}} \tan^{-1}\left[\sqrt{\dfrac{b-c}{b+c}} \tan \dfrac{ax}{2} \right] + C, \qquad b^2 > c^2$

75. $\int \dfrac{dx}{b + c \cos ax} = \dfrac{1}{a\sqrt{c^2 - b^2}} \ln\left| \dfrac{c + b \cos ax + \sqrt{c^2 - b^2} \sin ax}{b + c \cos ax} \right| + C, \qquad b^2 < c^2$

76. $\int \dfrac{dx}{1 + \cos ax} = \dfrac{1}{a} \tan \dfrac{ax}{2} + C$

77. $\int \dfrac{dx}{1 - \cos ax} = -\dfrac{1}{a} \cot \dfrac{ax}{2} + C$

78. $\int x \sin ax \, dx = \dfrac{1}{a^2} \sin ax - \dfrac{x}{a} \cos ax + C$

79. $\int x \cos ax \, dx = \dfrac{1}{a^2} \cos ax + \dfrac{x}{a} \sin ax + C$

80. $\int x^n \sin ax \, dx = -\dfrac{x^n}{a} \cos ax + \dfrac{n}{a} \int x^{n-1} \cos ax \, dx$

81. $\int x^n \cos ax \, dx = \dfrac{x^n}{a} \sin ax - \dfrac{n}{a} \int x^{n-1} \sin ax \, dx$

82. $\int \tan ax \, dx = \dfrac{1}{a} \ln |\sec ax| + C$

83. $\int \cot ax \, dx = \dfrac{1}{a} \ln |\sin ax| + C$

84. $\int \tan^2 ax \, dx = \dfrac{1}{a} \tan ax - x + C$

85. $\int \cot^2 ax \, dx = -\dfrac{1}{a} \cot ax - x + C$

86. $\int \tan^n ax \, dx = \dfrac{\tan^{n-1} ax}{a(n-1)} - \int \tan^{n-2} ax \, dx, \qquad n \neq 1$

87. $\int \cot^n ax \, dx = -\dfrac{\cot^{n-1} ax}{a(n-1)} - \int \cot^{n-2} ax \, dx, \qquad n \neq 1$

88. $\int \sec ax \, dx = \dfrac{1}{a} \ln |\sec ax + \tan ax| + C$

89. $\int \csc ax \, dx = -\dfrac{1}{a} \ln |\csc ax + \cot ax| + C$

90. $\int \sec^2 ax \, dx = \dfrac{1}{a} \tan ax + C$

91. $\int \csc^2 ax \, dx = -\dfrac{1}{a} \cot ax + C$

92. $\int \sec^n ax \, dx = \dfrac{\sec^{n-2} ax \tan ax}{a(n-1)} + \dfrac{n-2}{n-1} \int \sec^{n-2} ax \, dx, \quad n \neq 1$

93. $\int \csc^n ax \, dx = -\dfrac{\csc^{n-2} ax \cot ax}{a(n-1)} + \dfrac{n-2}{n-1} \int \csc^{n-2} ax \, dx, \quad n \neq 1$

94. $\int \sec^n ax \tan ax \, dx = \dfrac{\sec^n ax}{na} + C, \quad n \neq 0$

95. $\int \csc^n ax \cot ax \, dx = -\dfrac{\csc^n ax}{na} + C, \quad n \neq 0$

96. $\int \sin^{-1} ax \, dx = x \sin^{-1} ax + \dfrac{1}{a} \sqrt{1 - a^2 x^2} + C$

97. $\int \cos^{-1} ax \, dx = x \cos^{-1} ax - \dfrac{1}{a} \sqrt{1 - a^2 x^2} + C$

98. $\int \tan^{-1} ax \, dx = x \tan^{-1} ax - \dfrac{1}{2a} \ln(1 + a^2 x^2) + C$

99. $\int x^n \sin^{-1} ax \, dx = \dfrac{x^{n+1}}{n+1} \sin^{-1} ax - \dfrac{a}{n+1} \int \dfrac{x^{n+1} \, dx}{\sqrt{1 - a^2 x^2}}, \quad n \neq -1$

100. $\int x^n \cos^{-1} ax \, dx = \dfrac{x^{n+1}}{n+1} \cos^{-1} ax + \dfrac{a}{n+1} \int \dfrac{x^{n+1} \, dx}{\sqrt{1 - a^2 x^2}}, \quad n \neq -1$

101. $\int x^n \tan^{-1} ax \, dx = \dfrac{x^{n+1}}{n+1} \tan^{-1} ax - \dfrac{a}{n+1} \int \dfrac{x^{n+1} \, dx}{1 + a^2 x^2}, \quad n \neq -1$

102. $\int e^{ax} \, dx = \dfrac{1}{a} e^{ax} + C$

103. $\int b^{ax} \, dx = \dfrac{1}{a} \dfrac{b^{ax}}{\ln b} + C, \quad b > 0, \ b \neq 1$

104. $\int x e^{ax} \, dx = \dfrac{e^{ax}}{a^2} (ax - 1) + C$

105. $\int x^n e^{ax} \, dx = \dfrac{1}{a} x^n e^{ax} - \dfrac{n}{a} \int x^{n-1} e^{ax} \, dx$

106. $\int x^n b^{ax} \, dx = \dfrac{x^n b^{ax}}{a \ln b} - \dfrac{n}{a \ln b} \int x^{n-1} b^{ax} \, dx, \quad b > 0, \ b \neq 1$

107. $\int e^{ax} \sin bx \, dx = \dfrac{e^{ax}}{a^2 + b^2} (a \sin bx - b \cos bx) + C$

108. $\int e^{ax} \cos bx \, dx = \dfrac{e^{ax}}{a^2 + b^2} (a \cos bx + b \sin bx) + C$

109. $\int \ln ax \, dx = x \ln ax - x + C$

110. $\int x^n (\ln ax)^m \, dx = \dfrac{x^{n+1} (\ln ax)^m}{n+1} - \dfrac{m}{n+1} \int x^n (\ln ax)^{m-1} \, dx, \quad n \neq -1$

111. $\int x^{-1} (\ln ax)^m = \dfrac{(\ln ax)^{m+1}}{m+1} + C, \quad m \neq -1$

112. $\int \dfrac{dx}{x \ln ax} = \ln |\ln ax| + C$

113. $\int \sinh ax \, dx = \dfrac{1}{a} \cosh ax + C$

114. $\int \cosh ax \, dx = \dfrac{1}{a} \sinh ax + C$

115. $\int \sinh^2 ax \, dx = \dfrac{\sinh 2ax}{4a} - \dfrac{x}{2} + C$

116. $\int \cosh^2 ax \, dx = \dfrac{\sinh 2ax}{4a} + \dfrac{x}{2} + C$

117. $\int \sinh^n ax \, dx = \dfrac{\sinh^{n-1} ax \cosh ax}{na} - \dfrac{n-1}{n} \int \sinh^{n-2} ax \, dx, \quad n \neq 0$

Continued

118. $\displaystyle\int \cosh^n ax\, dx = \frac{\cosh^{n-1} ax \sinh ax}{na} + \frac{n-1}{n}\int \cosh^{n-2} ax\, dx, \qquad n \neq 0$

119. $\displaystyle\int x \sinh ax\, dx = \frac{x}{a}\cosh ax - \frac{1}{a^2}\sinh ax + C$

120. $\displaystyle\int x \cosh ax\, dx = \frac{x}{a}\sinh ax - \frac{1}{a^2}\cosh ax + C$

121. $\displaystyle\int x^n \sinh ax\, dx = \frac{x^n}{a}\cosh ax - \frac{n}{a}\int x^{n-1}\cosh ax\, dx$

122. $\displaystyle\int x^n \cosh ax\, dx = \frac{x^n}{a}\sinh ax - \frac{n}{a}\int x^{n-1}\sinh ax\, dx$

123. $\displaystyle\int \tanh ax\, dx = \frac{1}{a}\ln(\cosh ax) + C$
124. $\displaystyle\int \coth ax\, dx = \frac{1}{a}\ln|\sinh ax| + C$

125. $\displaystyle\int \tanh^2 ax\, dx = x - \frac{1}{a}\tanh ax + C$
126. $\displaystyle\int \coth^2 ax\, dx = x - \frac{1}{a}\coth ax + C$

127. $\displaystyle\int \tanh^n ax\, dx = -\frac{\tanh^{n-1} ax}{(n-1)a} + \int \tanh^{n-2} ax\, dx, \qquad n \neq 1$

128. $\displaystyle\int \coth^n ax\, dx = -\frac{\coth^{n-1} ax}{(n-1)a} + \int \coth^{n-2} ax\, dx, \qquad n \neq 1$

129. $\displaystyle\int \operatorname{sech} ax\, dx = \frac{1}{a}\sin^{-1}(\tanh ax) + C$
130. $\displaystyle\int \operatorname{csch} ax\, dx = \frac{1}{a}\ln\left|\tanh \frac{ax}{2}\right| + C$

131. $\displaystyle\int \operatorname{sech}^2 ax\, dx = \frac{1}{a}\tanh ax + C$
132. $\displaystyle\int \operatorname{csch}^2 ax\, dx = -\frac{1}{a}\coth ax + C$

133. $\displaystyle\int \operatorname{sech}^n ax\, dx = \frac{\operatorname{sech}^{n-2} ax \tanh ax}{(n-1)a} + \frac{n-2}{n-1}\int \operatorname{sech}^{n-2} ax\, dx, \qquad n \neq 1$

134. $\displaystyle\int \operatorname{csch}^n ax\, dx = -\frac{\operatorname{csch}^{n-2} ax \coth ax}{(n-1)a} - \frac{n-2}{n-1}\int \operatorname{csch}^{n-2} ax\, dx, \qquad n \neq 1$

135. $\displaystyle\int \operatorname{sech}^n ax \tanh ax\, dx = -\frac{\operatorname{sech}^n ax}{na} + C, \qquad n \neq 0$

136. $\displaystyle\int \operatorname{csch}^n ax \coth ax\, dx = -\frac{\operatorname{csch}^n ax}{na} + C, \qquad n \neq 0$

137. $\displaystyle\int e^{ax} \sinh bx\, dx = \frac{e^{ax}}{2}\left[\frac{e^{bx}}{a+b} - \frac{e^{-bx}}{a-b}\right] + C, \qquad a^2 \neq b^2$

138. $\displaystyle\int e^{ax} \cosh bx\, dx = \frac{e^{ax}}{2}\left[\frac{e^{bx}}{a+b} + \frac{e^{-bx}}{a-b}\right] + C, \qquad a^2 \neq b^2$

139. $\displaystyle\int_0^\infty x^{n-1}e^{-x}\, dx = \Gamma(n) = (n-1)!, \qquad n > 0.$
140. $\displaystyle\int_0^\infty e^{-ax^2}\, dx = \frac{1}{2}\sqrt{\frac{\pi}{a}}, \qquad a > 0$

141. $\displaystyle\int_0^{\pi/2} \sin^n x\, dx = \int_0^{\pi/2} \cos^n x\, dx = \begin{cases} \dfrac{1\cdot 3\cdot 5\cdots(n-1)}{2\cdot 4\cdot 6\cdots n}\cdot\dfrac{\pi}{2}, & \text{if } n \text{ is an even integer} \geq 2, \\[2mm] \dfrac{2\cdot 4\cdot 6\cdots(n-1)}{3\cdot 5\cdot 7\cdots n}, & \text{if } n \text{ is an odd integer} \geq 3 \end{cases}$